Student Study Guide and Solutions Manual, 4e

for

Organic Chemistry, 4e

David Klein
Johns Hopkins University

WILEY

SVP, PUBLISHING STRATEGY AND QUALITY: Elizabeth Widdicombe
ASSOCIATE PUBLISHER: Sladjana Bruno
MARKETING MANAGER: Michael Olsen
SENIOR MANAGING EDITOR: Mary Donovan
EXECUTIVE MANAGING EDITOR: Valerie Zaborski
EDITORIAL ASSISTANT: Samantha Hart
COURSE CONTENT DEVELOPER: Andrew Moore
SENIOR COURSE PROUCTION OPERATIONS SPECIALIST: Patricia Gutierrez
SENIOR MANAGER, COURSE DEVELOPMENT AND PRODUCTION: Svetlana Barskaya
ART DIRECTOR/COVER DESIGNER: Wendy Lai
Cover: Wiley
Cover Images: Abstract geometric texture ©bgblue/Getty Images, Flask – Norm Christiansen, Colorful Paintbrushes © Maartje van Caspel/Getty Images, Honeycomb slice © eli_asenova/Getty Images, Color diffuse © Korolkoff/Getty Images, Rosemary © Tetiana Rostopira/Getty Images

ISBN: 978-1-119-65958-7

Printed in the United States of America

SKY10036900_101722

The inside back cover will contain printing identification and country of origin if omitted from this page. In addition, if the ISBN on the back cover differs from the ISBN on this page, the one on the back cover is correct.

CONTENTS

HOW TO USE THIS BOOK

Organic chemistry is much like bicycle riding. You cannot learn how to ride a bike by watching other people ride bikes. Some people might fool themselves into believing that it's possible to become an expert bike rider without ever getting on a bike. But you know that to be incorrect (and very naïve). In order to learn how to ride a bike, you must be willing to get on the bike, and you must be willing to fall. With time (and dedication), you can quickly train yourself to avoid falling, and to ride the bike with ease and confidence. The same is true of organic chemistry. In order to become proficient at solving problems, you must "ride the bike". You must try to solve the problems yourself (*without* the solutions manual open in front of you). Once you have solved the problems, this book will allow you to check your solutions. If, however, you don't attempt to solve each problem on your own, and instead, you read the problem statement and then immediately read the solution, you are only hurting yourself. You are not learning how to avoid falling. Many students make this mistake every year. They use the solutions manual as a crutch, and then they never really attempt to solve the problems on their own. It really is like believing that you can become an expert bike rider by watching hundreds of people riding bikes. The world doesn't work that way!

The textbook has thousands of problems to solve. Each of these problems should be viewed as an opportunity to develop your problem-solving skills. By reading a problem statement and then reading the solution immediately (without trying to solve the problem yourself), you are robbing yourself of the opportunity provided by the problem. If you repeat that poor study habit too many times, you will not learn how to solve problems on your own, and you will not get the grade that you want.

Why do so many students adopt this bad habit (of using the solutions manual too liberally)? The answer is simple. Students often wait until a day or two before the exam, and then they spend all night cramming. Sound familiar? Unfortunately, organic chemistry is the type of course where cramming is insufficient, because you need time in order to ride the bike yourself. You need time to think about each problem until you have developed a solution *on your own*. For some problems, it might take days before you think of a solution. This process is critical for learning this subject. Make sure to allot time every day for studying organic chemistry, and use this book to check your solutions. This book has also been designed to serve as a study guide, as described below.

WHAT'S IN THIS BOOK

This book contains more than just solutions to all of the problems in the textbook. Each chapter of this book also contains a series of exercises that will help you review the concepts, skills and reactions presented in the corresponding chapter of the textbook. These exercises are designed to serve as study tools that can help you identify your weak areas. Each chapter of this solutions manual/study guide has the following parts:

- ***Review of Concepts***. These exercises are designed to help you identify which concepts are the least familiar to you. Each section contains sentences with missing words (blanks). Your job is to fill in the blanks, demonstrating mastery of the concepts. To verify that your answers are correct, you can open your textbook to the end of the corresponding chapter, where you will find a section entitled *Review of Concepts and Vocabulary*. In that section, you will find each of the sentences, verbatim.
- ***Review of Skills***. These exercises are designed to help you identify which skills are the least familiar to you. Each section contains exercises in which you must demonstrate mastery of the skills developed in the *SkillBuilders* of the corresponding textbook chapter. To verify that your answers are correct, you can open your textbook to the end of the corresponding chapter, where you will find a section entitled *SkillBuilder Review*. In that section, you will find the answers to each of these exercises.
- ***Review of Reactions***. These exercises are designed to help you identify which reagents are not at your fingertips. Each section contains exercises in which you must demonstrate familiarity with the reactions covered in the textbook. Your job is to fill in the reagents necessary to achieve each reaction. To verify that your answers are correct, you can open your textbook to the end of the corresponding chapter, where you will find a section entitled *Review of Reactions*. In that section, you will find the answers to each of these exercises.
- ***Review of Mechanisms***. These exercises are designed to help you practice drawing the mechanisms. To verify that you have drawn the mechanism correctly, you can open your textbook to the corresponding chapter, where you will find the mechanisms appearing in numbered boxes throughout the chapter. In those numbered boxes, you will find the answers to each of these exercises.
- ***Common Mistakes to Avoid***. This is a new feature to this edition. The most common student mistakes are described, so that you can avoid them when solving problems.
- ***A List of Useful Reagents***. This is a new feature to this edition. This list provides a review of the reagents that appear in each chapter, as well as a description of how each reagent is used.
- ***Solutions***. At the end of each chapter, you'll find detailed solutions to all problems in the textbook, including all SkillBuilders, conceptual checkpoints, additional problems, integrated problems, and challenge problems.

The sections described above have been designed to serve as useful tools as you study and learn organic chemistry. Good luck!

David Klein
Johns Hopkins University

Chapter 1
A Review of General Chemistry: Electrons, Bonds and Molecular Properties

Review of Concepts

Fill in the blanks below. To verify that your answers are correct, look in your textbook at the end of Chapter 1. Each of the sentences below appears verbatim in the section entitled *Review of Concepts and Vocabulary*.

- ____________ **isomers** share the same molecular formula but have different connectivity of atoms and different physical properties.
- Second-row elements generally obey the _______ **rule**, bonding to achieve noble gas electron configuration.
- A pair of unshared electrons is called a ____________.
- A **formal charge** occurs when atoms do not exhibit the appropriate number of ________________.
- An **atomic orbital** is a region of space associated with ________________, while a **molecular orbital** is associated with ________________.
- Methane's tetrahedral geometry can be explained using four degenerate _____**-hybridized orbitals** to achieve its four single bonds.
- Ethylene's planar geometry can be explained using three degenerate _____**-hybridized orbitals.**
- Acetylene's linear geometry is achieved via _____**-hybridized** carbon atoms.
- The geometry of small compounds can be predicted using valence shell electron pair repulsion (**VSEPR**) theory, which focuses on the number of _______ bonds and ____________ exhibited by each atom.
- The physical properties of compounds are determined by ______________ forces, the attractive forces between molecules.
- **London dispersion forces** result from the interaction between transient ____________ and are stronger for larger alkanes due to their larger surface area and ability to accommodate more interactions.

Review of Skills

Fill in the blanks and empty boxes below. To verify that your answers are correct, look in your textbook at the end of Chapter 1. The answers appear in the section entitled *SkillBuilder Review*.

SkillBuilder 1.1 Drawing Constitutional Isomers of Small Molecules

Step 1 Determine the valency of each atom in C_3H_8O.	**Step 2** Connect the atoms of highest valency, and place the monovalent atoms at the periphery.	**Step 3** Consider other ways to connect the atoms.
—C— (four bonds); —O—; —H	—C—C—C—O— ⟹ H—C—C—C—O—H (each C bearing two H)	[ ] [ ]

SkillBuilder 1.2 Drawing the Lewis Structure of a Small Molecule

Step 1 Draw the Lewis dot structure of each atom in CH_2O.	**Step 2** First connect atoms that form more than one bond.	**Step 3** Connect the hydrogen atoms.	**Step 4** Pair any unpaired electrons, so that each atom achieves an octet.
[ ] [ ] [ ] [ ]	[ ]—[ ]	[ ]	[ ]

SkillBuilder 1.3 Calculating Formal Charge

Step 1 Determine the appropriate number of valence electrons.	**Step 2** Determine the number of valence electrons in this case.	**Step 3** Assign a formal charge to the nitrogen atom.
H–N(–H)(–H) Nitrogen is in Group ____ of the periodic table, and is expected to have _____ valence electrons.	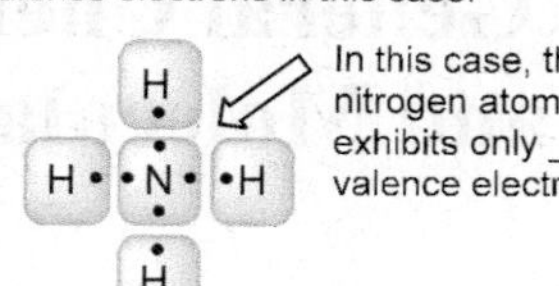 In this case, the nitrogen atom exhibits only _____ valence electrons.	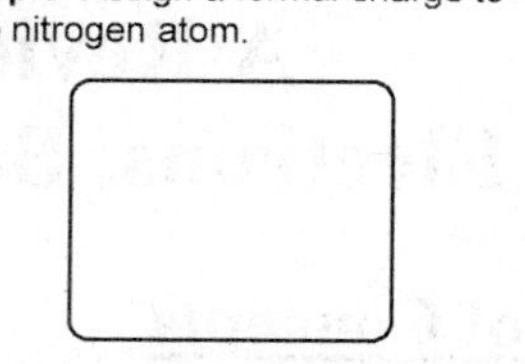

SkillBuilder 1.4 Locating Partial Charges Resulting from Induction

Step 1 Circle the bonds below that are polar covalent.	**Step 2** For each polar covalent bond, draw an arrow that shows the direction of the dipole moment.	**Step 3** Indicate the location of all partial charges (δ+ and δ–).
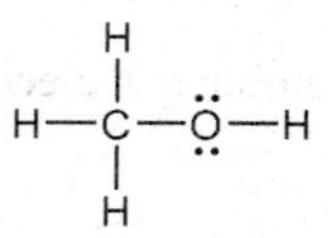	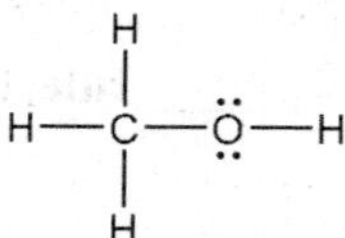	H–C(H)(H)–Ö–H

SkillBuilder 1.5 Reading Bond-Line Structures

Circle all carbon atoms in the compound below	Draw all hydrogen atoms in the compound below
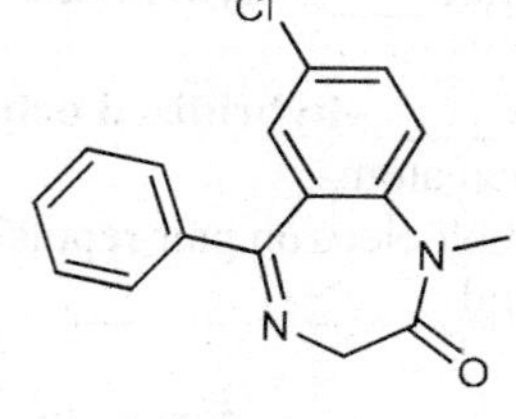	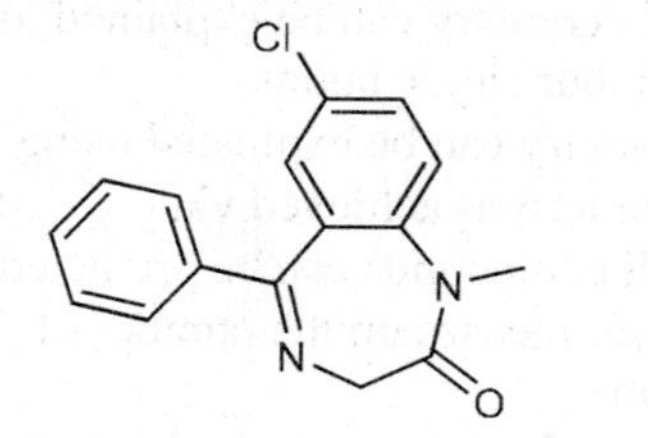

SkillBuilder 1.6 Identifying Electron Configurations

Step 1 In the energy diagram shown here, draw the electron configuration of nitrogen (using arrows to represent electrons).	**Step 2** Fill in the boxes below with the numbers that correctly describe the electron configuration of nitrogen.
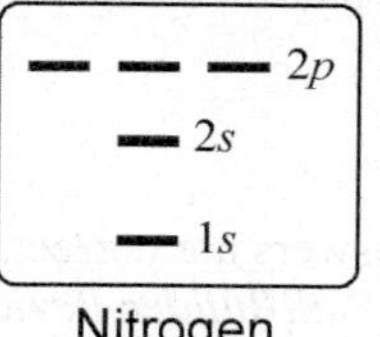	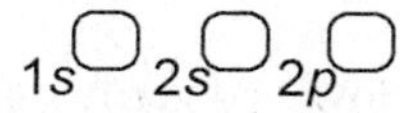

SkillBuilder 1.7 Identifying Hybridization States

A carbon atom with four single bonds will be _______ hybridized.	A carbon atom with one double bond will be _______ hybridized.	A carbon atom with a triple bond will be _______ hybridized.
–C– (four single bonds)	=C<	–C≡

SkillBuilder 1.8 Predicting Geometry

Step 1 Determine the steric number of the nitrogen atom here by adding the number of single bonds and lone pairs.

H—N̈—H
|
H

\# of single bonds = ☐

\# of lone pairs = ☐

Steric Number = ☐

Step 1 The steric number determines the arrangement of electron pairs. Fill in the chart below:

Steric #	Arrangement of Electron Pairs
4	
3	
2	

Step 3 Identify the geometry in each case below:

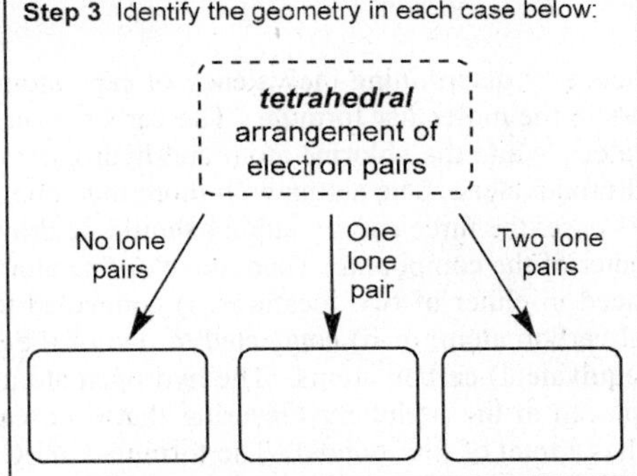

SkillBuilder 1.9 Identifying the Presence of Molecular Dipole Moments

Step 1 Identify the geometry of the oxygen atom below:

H_3C—CH(H)—Ö—CH(H)—CH_3

Geometry = ☐

Step 2 Redraw the compound. Draw an arrow for each dipole moment (showing its direction):

☐

Step 3 Redraw the compound, and draw the net dipole moment:

☐

SkillBuilder 1.10 Predicting Physical Properties

Dipole-Dipole Interactions
Circle the compound below that is expected to have the higher boiling point.

H_3C—C(=CH_2)—CH_3 H_3C—C(=Ö)—CH_3

Hydrogen-Bonding Interactions
Circle the compound below that is expected to have the higher boiling point.

H—C(H)(H)—Ö—C(H)(H)—H   H—C(H)(H)—C(H)(H)—Ö—H

Carbon Skeleton
Circle the compound below that is expected to have the higher boiling point.

H—C(H)(H)—C(H)(H)—C(H)(H)—H   H—C(H)(H)—C(H)(H)—C(H)(H)—C(H)(H)—C(H)(H)—H

A Common Mistake to Avoid

When drawing a structure, don't forget to draw formal charges, as forgetting to do so is a common error. If a formal charge is present, it MUST be drawn. For example, in the following case, the nitrogen atom bears a positive charge, so the charge must be drawn:

INCORRECT

H—N(H)(H)—H

CORRECT

H—N⊕(H)(H)—H

As we progress though the course, we will see structures of increasing complexity. If formal charges are present, failure to draw them constitutes an error, and must be scrupulously avoided. If you have trouble drawing formal charges, go back and master that skill. You can't go on without it. Don't make the mistake of underestimating the importance of being able to draw formal charges with confidence.

Solutions

1.1.

(a) Begin by determining the valency of each atom that appears in the molecular formula. The carbon atoms are tetravalent, while the chlorine atom and hydrogen atoms are all monovalent. The atoms with more than one bond (in this case, the three carbon atoms) should be drawn in the center of the compound. Then, the chlorine atom can be placed in either of two locations: i) connected to the central carbon atom, or ii) connected to one of the other two (equivalent) carbon atoms. The hydrogen atoms are then placed at the periphery (ensuring that each carbon atom has a total of four bonds). The formula C_3H_7Cl has two constitutional isomers.

```
   H  Cl H              H  H  H
   |  |  |              |  |  |
H—C—C—C—H          H—C—C—C—Cl
   |  |  |              |  |  |
   H  H  H              H  H  H
```

(b) Begin by determining the valency of each atom that appears in the molecular formula. The carbon atoms are tetravalent, while the hydrogen atoms are all monovalent. The atoms with more than one bond (in this case, the four carbon atoms) should be drawn in the center of the compound. There are two different ways to connect four carbon atoms. They can either be arranged in a linear fashion or in a branched fashion:

```
                        C
                        |
C—C—C—C              C—C—C
1 2 3 4              1 2 3
 Linear             Branched
```

We then place the hydrogen atoms at the periphery (ensuring that each carbon atom has a total of four bonds). The formula C_4H_{10} has two constitutional isomers:

```
                          H
                          |
                        H—C—H
  H  H  H  H          H   |   H
  |  |  |  |          |   |   |
H—C—C—C—C—H        H—C———C———C—H
  |  |  |  |          |   |   |
  H  H  H  H          H   H   H
```

(c) Begin by determining the valency of each atom that appears in the molecular formula. The carbon atoms are tetravalent, while the hydrogen atoms are all monovalent. The atoms with more than one bond (in this case, the five carbon atoms) should be drawn in the center of the compound. So we must explore all of the different ways to connect five carbon atoms. First, we can connect all five carbon atoms in a linear fashion:

```
C—C—C—C—C
1 2 3 4 5
```

Alternatively, we can draw four carbon atoms in a linear fashion, and then draw the fifth carbon atom on a branch. There are many ways to draw this possibility:

```
  C                    C               1 C
  |                    |                 |
C—C—C—C    ≡    C—C—C—C    ≡    C—C—C
1 2 3 4         4 3 2 1          |2 3 4
                                 C
```

Finally, we can draw three carbon atoms in a linear fashion, and then draw the remaining two carbon atoms on separate branches.

```
  C
  |
C—C—C
1 |2 3
  C
```

Note that we cannot draw a unique carbon skeleton (a unique arrangement of carbon atoms) simply by placing the last two carbon atoms together as one branch, because that possibility has already been drawn earlier (a linear chain of four carbon atoms with a single branch):

```
  C 4
  |
  C 3             C
  |        ≡      |
C—C—C           C—C—C—C
1 2             1 2 3 4
```

In summary, there are three different ways to connect five carbon atoms:

```
                    C           C
                    |           |
C—C—C—C—C        C—C—C—C       C—C—C
1 2 3 4 5        1 2 3 4       1 |2 3
                                 C
```

We then place the hydrogen atoms at the periphery (ensuring that each carbon atom has a total of four bonds). The formula C_5H_{12} has three constitutional isomers:

```
  H  H  H  H  H
  |  |  |  |  |
H—C—C—C—C—C—H
  |  |  |  |  |
  H  H  H  H  H

         H                            H
         |                            |
       H—C—H                        H—C—H
  H      |     H  H            H      |      H
  |      |     |  |            |      |      |
H—C——————C—————C—C—H         H—C——————C——————C—H
  |      |     |  |            |      |      |
  H      H     H  H            H      |      H
                                    H—C—H
                                      |
                                      H
```

(d) Begin by determining the valency of each atom that appears in the molecular formula. The carbon atoms are tetravalent, the oxygen atom is divalent, and the hydrogen atoms are all monovalent. Any atoms with more than one bond (in this case, the four carbon atoms and the one oxygen atom) should be drawn in the center of the compound, with the hydrogen atoms at the periphery. There are several different ways to connect four carbon atoms and one oxygen atom. Let's begin with the four carbon atoms. There are two different ways to connect four carbon atoms. They can either be arranged in a linear fashion or in a branched fashion.

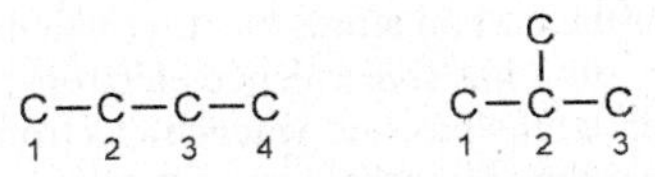

C–C–C–C (1 2 3 4) Linear; C–C(C)–C (1 2 3) Branched

Next, the oxygen atom must be inserted. For each of the two skeletons above (linear or branched), there are several different locations to insert the oxygen atom. The linear skeleton has four possibilities, shown here:

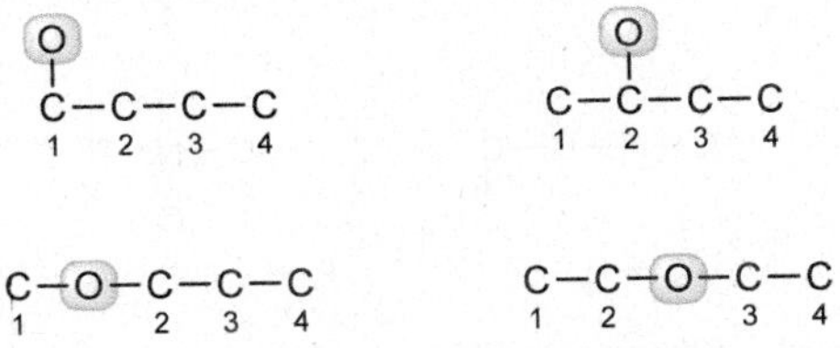

and the branched skeleton has three possibilities shown here:

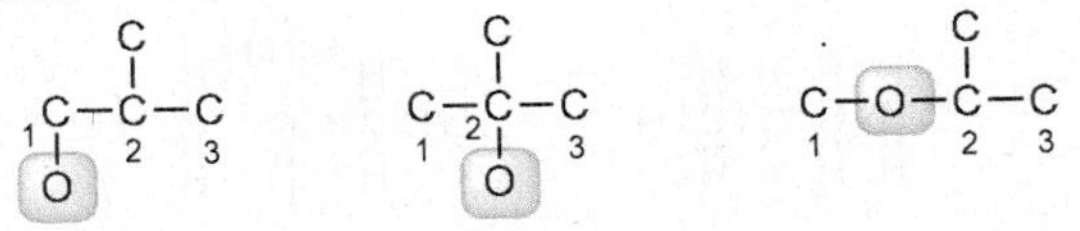

Finally, we complete all of the structures by drawing the bonds to hydrogen atoms (ensuring that each carbon atom has four bonds, and each oxygen atoms has two bonds). The formula $C_4H_{10}O$ has seven constitutional isomers:

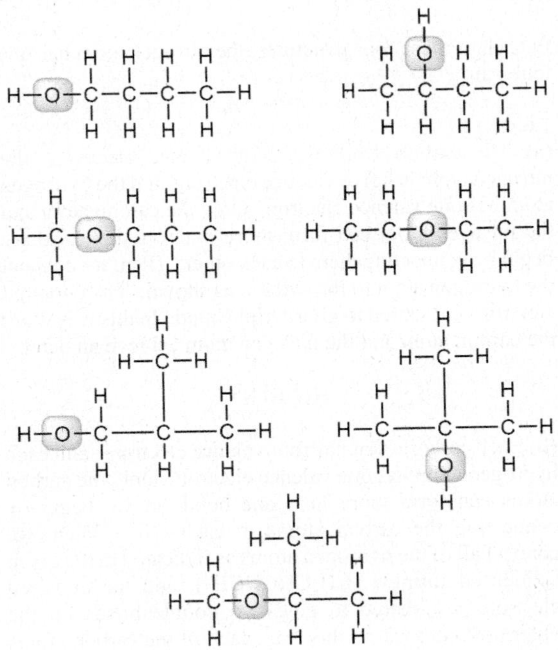

(e) Begin by determining the valency of each atom that appears in the molecular formula. The carbon atoms are tetravalent, while the chlorine atom and hydrogen atoms are all monovalent. The atoms with more than one bond (in this case, the three carbon atoms) should be drawn in the center of the compound. There is only way to connect three carbon atoms:

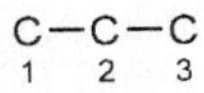

C–C–C (1 2 3)

Next, we must determine all of the different possible ways of connecting two chlorine atoms to the chain of three carbon atoms. If we place one chlorine atom at C1, then the second chlorine atom can be placed at C1, at C2 or at C3:

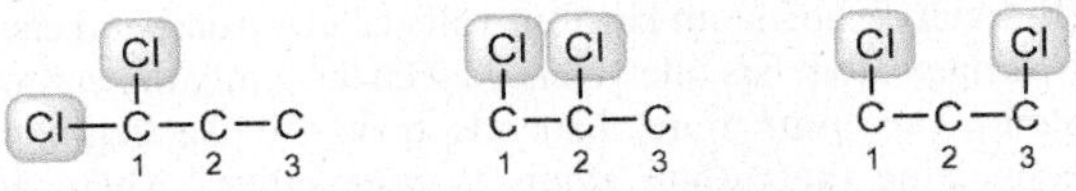

Furthermore, we can place both chlorine atoms at C2, giving a new possibility not shown above:

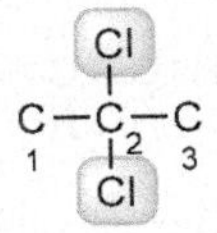

There are no other possibilities. For example, placing the two chlorine atoms at C2 and C3 is equivalent to placing them at C1 and C2:

Finally, the hydrogen atoms are placed at the periphery (ensuring that each carbon atom has a total of four bonds). The formula $C_3H_6Cl_2$ has four constitutional isomers:

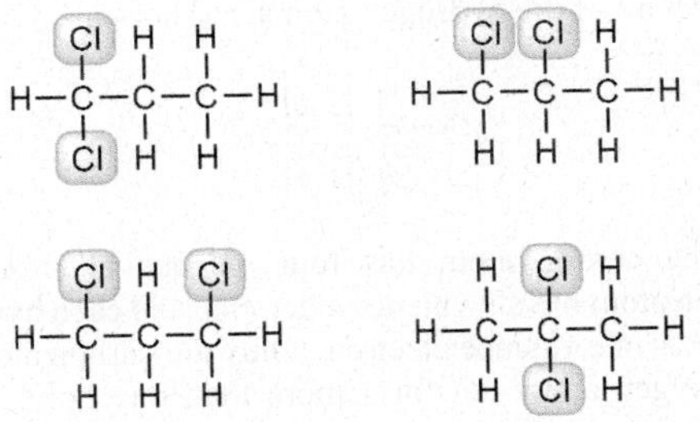

1.2. The carbon atoms are tetravalent, while the chlorine atoms and fluorine atoms are all monovalent. The atoms with more than one bond (in this case, the two carbon atoms) should be drawn in the center of the compound. The chlorine atoms and fluorine atoms are then placed at the periphery, as shown. There are only two possible constitutional isomers: one with the three chlorine atoms all connected to the same carbon, and one in which they are distributed over both carbon atoms. Any other representations that one may draw must be one of these structures drawn in a different orientation.

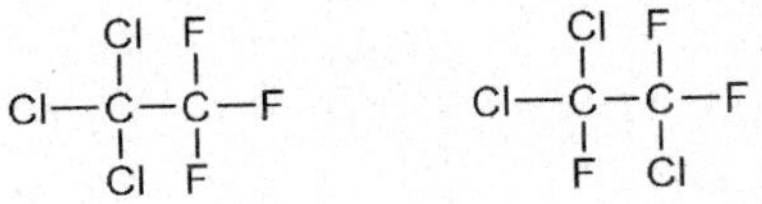

1.3.

(a) Each carbon atom has four valence electrons, and each hydrogen atom has one valence electron. Only the carbon atoms can form more than one bond, so we begin by connecting the carbon atoms to each other. Then, we connect all of the hydrogen atoms, as shown.

```
   H H
H:C:C:H
   H H
```

(b) Each carbon atom has four valence electrons, and each hydrogen atom has one valence electron. Only the carbon atoms can form more than one bond, so we begin by connecting the carbon atoms to each other. Then, we connect all of the hydrogen atoms, and the unpaired electrons are shared to give a double bond. In this way, each of the carbon atoms achieves an octet.

```
H:C::C:H
  H  H
```

(c) Each carbon atom has four valence electrons, and each hydrogen atom has one valence electron. Only the carbon atoms can form more than one bond, so we begin by connecting the carbon atoms to each other. Then, we connect all of the hydrogen atoms, and the unpaired electrons are shared to give a triple bond. In this way, each of the carbon atoms achieves an octet.

```
H:C:::C:H
```

(d) Each carbon atom has four valence electrons, and each hydrogen atom has one valence electron. Only the carbon atoms can form more than one bond, so we begin by connecting the carbon atoms to each other. Then, we connect all of the hydrogen atoms, as shown.

```
  H H H
H:C:C:C:H
  H H H
```

(e) The carbon atom has four valence electrons, the oxygen atom has six valence electrons, and each hydrogen atom has one valence electron. Only the carbon atom and the oxygen atom can form more than one bond, so we begin by connecting them to each other. Then, we connect all of the hydrogen atoms, as shown.

```
  H
H:C:Ö:H
  H
```

1.4. Boron is in column 3A of the periodic table, so it has three valence electrons. Each of these valence electrons is shared with a hydrogen atom, shown below. The central boron atom lacks an octet of electrons, and it is therefore very unstable and reactive.

```
  H
H:B
  H
```

1.5. Each of the carbon atoms has four valence electrons; the nitrogen atom has five valence electrons; and each of the hydrogen atoms has one valence electron. We begin by connecting the atoms that have more than one bond (in this case, the three carbon atoms and the nitrogen atom). There are four different ways that these four atoms can be connected to each other, shown here.

```
N—C—C—C          N
  1 2 3          |
               C—C—C
               1 2 3

C—N—C—C        C—N—C
1   2 3        1 | 3
                 C
                 2
```

For each of these possible arrangements, we connect the hydrogen atoms, giving the following four constitutional isomers.

```
  H H H H              H:N:H
:N:C:C:C:H           H   ..   H
  H H H H          H:C : C : C:H
                     H   H    H

  H   H H          H   ..   H
H:C:N:C:C:H      H:C : N : C:H
  H H H H          H  H:C:H  H
                         H
```

In each of these four structures, the nitrogen atom has one lone pair.

1.6.

(a) The carbon atom has four valence electrons, the nitrogen atom has five valence electrons and the hydrogen atom has one valence electron. Only the carbon atom and the nitrogen atom can form more than one bond, so we begin by connecting them to each other. Then, we connect the hydrogen atom to the carbon, as shown. The unpaired electrons are shared to give a triple bond. In this way, both the carbon atom and the nitrogen atom achieve an octet.

```
H:C:::N:
```

(b) Each carbon atom has four valence electrons, and each hydrogen atom has one valence electron. Only the carbon atoms can form more than one bond, so we begin by connecting the carbon atoms to each other. Then, we connect all of the hydrogen atoms as indicated in the given condensed formula ($CH_2CHCHCH_2$), and the unpaired electrons are shared to give two double bonds on the outermost carbons. In this way, each of the carbon atoms achieves an octet.

```
H  H H  H
C::C:C::C
H       H
```

1.7.

(a) Aluminum is in group 3A of the periodic table, and it should therefore have three valence electrons. In this case, the aluminum atom exhibits four valence electrons (one for each bond). With one extra electron, this aluminum atom will bear a negative charge.

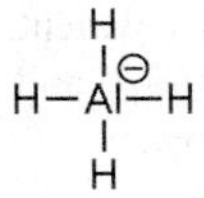

(b) Oxygen is in group 6A of the periodic table, and it should therefore have six valence electrons. In this case, the oxygen atom exhibits only five valence electrons (one for each bond, and two for the lone pair). This oxygen atom is missing an electron, and it therefore bears a positive charge.

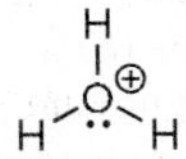

(c) Nitrogen is in group 5A of the periodic table, and it should therefore have five valence electrons. In this case, the nitrogen atom exhibits six valence electrons (one for each bond and two for each lone pair). With one extra electron, this nitrogen atom will bear a negative charge.

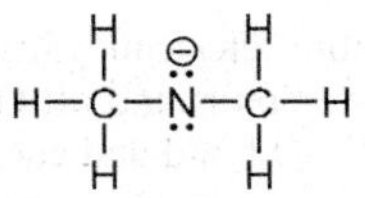

(d) Oxygen is in group 6A of the periodic table, and it should therefore have six valence electrons. In this case, the oxygen atom exhibits only five valence electrons (one for each bond, and two for the lone pair). This oxygen atom is missing an electron, and it therefore bears a positive charge.

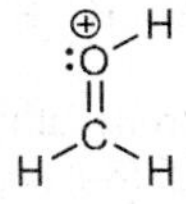

(e) Carbon is in group 4A of the periodic table, and it should therefore have four valence electrons. In this case, the carbon atom exhibits five valence electrons (one for each bond and two for the lone pair). With one extra electron, this carbon atom will bear a negative charge.

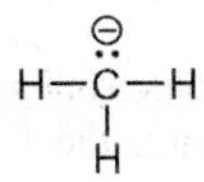

(f) Carbon is in group 4A of the periodic table, and it should therefore have four valence electrons. In this case, the carbon atom exhibits only three valence electrons (one for each bond). This carbon atom is missing an electron, and it therefore bears a positive charge.

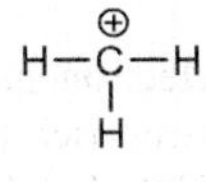

(g) Oxygen is in group 6A of the periodic table, and it should therefore have six valence electrons. In this case, the oxygen atom exhibits only five valence electrons (one for each bond, and two for the lone pair). This oxygen atom is missing an electron, and it therefore bears a positive charge.

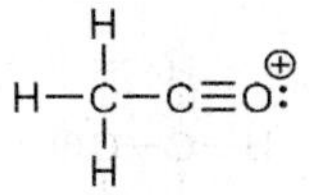

(h) Two of the atoms in this structure exhibit a formal charge because each of these atoms does not exhibit the appropriate number of valence electrons. The aluminum atom (group 3A) should have three valence electrons, but it exhibits four (one for each bond). With one extra electron, this aluminum atom will bear a negative charge. The neighboring chlorine atom (to the right) should have seven valence electrons, but it exhibits only six (one for each bond and two for each lone pair). It is missing one electron, so this chlorine atom will bear a positive charge.

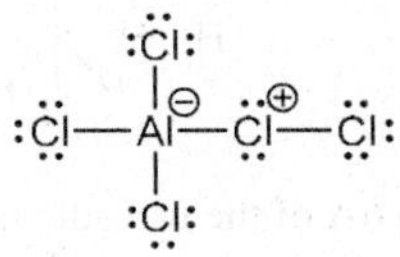

(i) Two of the atoms in this structure exhibit a formal charge because each of these atoms does not exhibit the appropriate number of valence electrons. The nitrogen atom (group 5A) should have five valence electrons, but it exhibits four (one for each bond). It is missing one electron, so this nitrogen atom will bear a positive charge. One of the two oxygen atoms (the one on the right) exhibits seven valence electrons (one for the bond, and two for each lone pair), although it should have only six. With one extra electron, this oxygen atom will bear a negative charge.

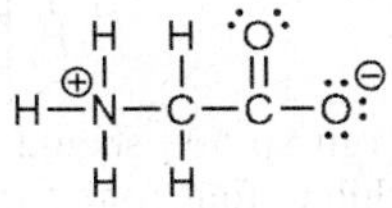

1.8.

(a) The boron atom in this case exhibits four valence electrons (one for each bond), although boron (group 3A) should only have three valence electrons. With one extra electron, this boron atom bears a negative charge.

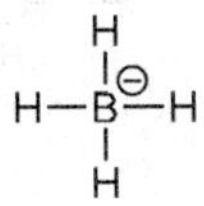

(b) Nitrogen is in group 5A of the periodic table, so a nitrogen atom should have five valence electrons. A negative charge indicates one extra electron, so this nitrogen atom must exhibit six valence electrons (one for each bond and two for each lone pair).

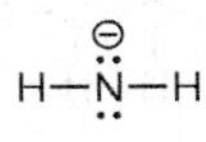

(c) One of the carbon atoms (below right) exhibits three valence electrons (one for each bond), but carbon (group 4A) is supposed to have four valence electrons. It is missing one electron, so this carbon atom therefore bears a positive charge.

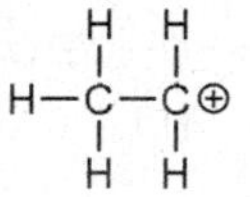

1.9. Carbon is in group 4A of the periodic table, and it should therefore have four valence electrons. Every carbon atom in acetylcholine has four bonds, thus exhibiting the correct number of valence electrons (four) and having no formal charge.

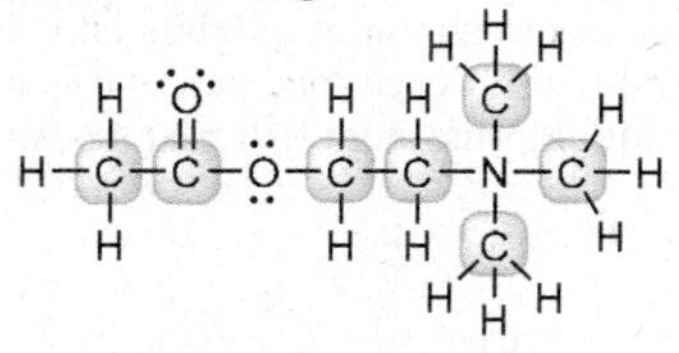

Oxygen is in group 6A of the periodic table, and it should therefore have six valence electrons. Each oxygen atom in acetylcholine has two bonds and two lone pairs of electrons, so each oxygen atom exhibits six valence electrons (one for each bond, and two for each lone pair). With the correct number of valence electrons, each oxygen atom will lack a formal charge.

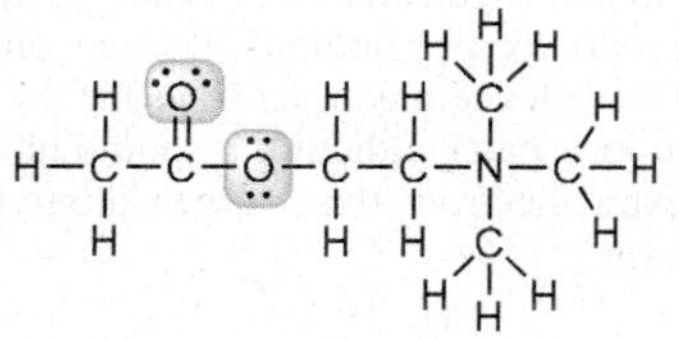

The nitrogen atom (group 5A) should have five valence electrons, but it exhibits four (one for each bond). It is missing one electron, so this nitrogen atom will bear a positive charge.

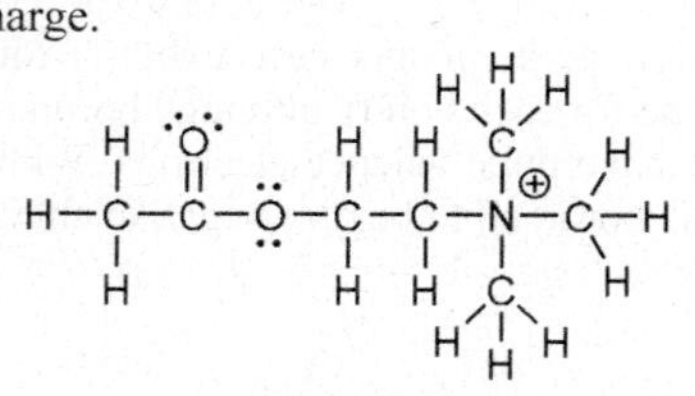

1.10.

(a) Oxygen is more electronegative than carbon, and a C–O bond is polar covalent. For each C–O bond, the O will be electron-rich (δ–), and the C will be electron-poor (δ+), as shown below.

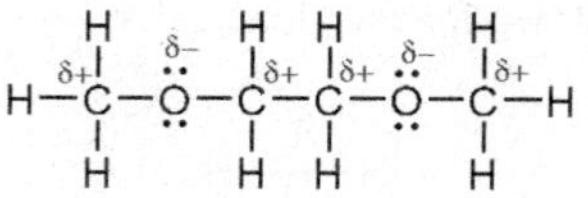

(b) Fluorine is more electronegative than carbon, and a C–F bond is polar covalent. For a C–F bond, the F will be electron-rich (δ–), and the C will be electron-poor (δ+). Chlorine is also more electronegative than carbon, so a C–Cl bond is also polar covalent. For a C–Cl bond, the Cl will be electron-rich (δ–), and the C will be electron-poor (δ+), as shown below.

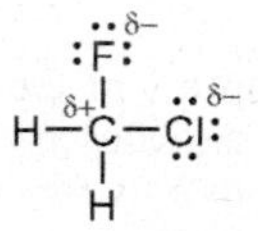

(c) Carbon is more electronegative than magnesium, so the C will be electron-rich (δ–) in a C–Mg bond, and the Mg will be electron-poor (δ+). Also, bromine is more electronegative than magnesium. So in a Mg–Br bond, the Br will be electron-rich (δ–), and the Mg will be electron-poor (δ+), as shown below.

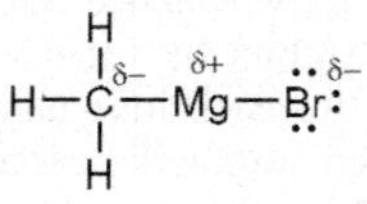

(d) Oxygen is more electronegative than carbon or hydrogen, so all C–O bonds and all O–H bond are polar covalent. For each C–O bond and each O–H bond, the O will be electron-rich (δ–), and the C or H will be electron-poor (δ+), as shown below.

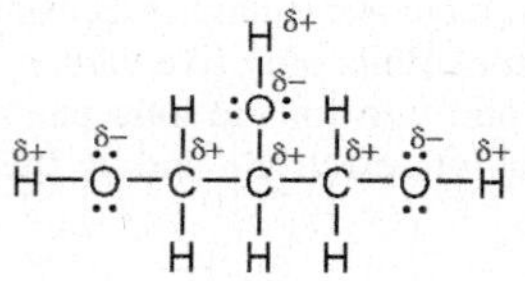

(e) Oxygen is more electronegative than carbon. As such, the O will be electron-rich (δ–) and the C will be electron-poor (δ+) in a C=O bond, as shown below.

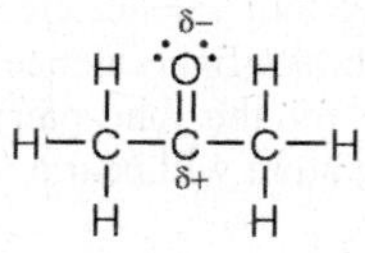

(f) Chlorine is more electronegative than carbon. As such, for each C–Cl bond, the Cl will be electron-rich (δ–) and the C will be electron-poor (δ+), as shown below.

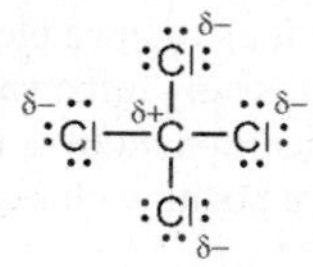

1.11. Oxygen is more electronegative than carbon. As such, the O will be electron-rich (δ–) and the C will be electron-poor (δ+) in a C=O bond. In addition, chlorine is more electronegative than carbon. So for a C–Cl bond, the Cl will be electron-rich (δ–) and the C will be electron-poor (δ+), as shown below.

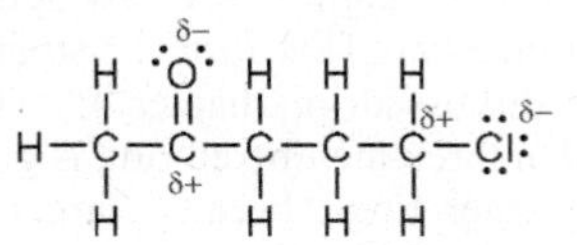

Notice that two carbon atoms are electron-poor (δ+). These are the positions that are most likely to be attacked by an electron-rich anion, such as hydroxide.

1.12. Oxygen is more electronegative than carbon. As such, the O will be electron-rich (δ–) and the C will be electron-poor (δ+) in a C—O bond. In addition, chlorine is more electronegative than carbon. So for a C—Cl bond, the Cl will be electron-rich (δ–) and the C will be electron-poor (δ+), as shown below. As you might imagine, epichlorohydrin is a very reactive molecule!

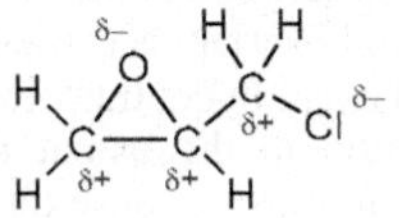

1.13.

(a) Each corner and each endpoint represents a carbon atom (highlighted below), so this compound has six carbon atoms. Each carbon atom has enough attached hydrogen atoms to have exactly four bonds, as shown:

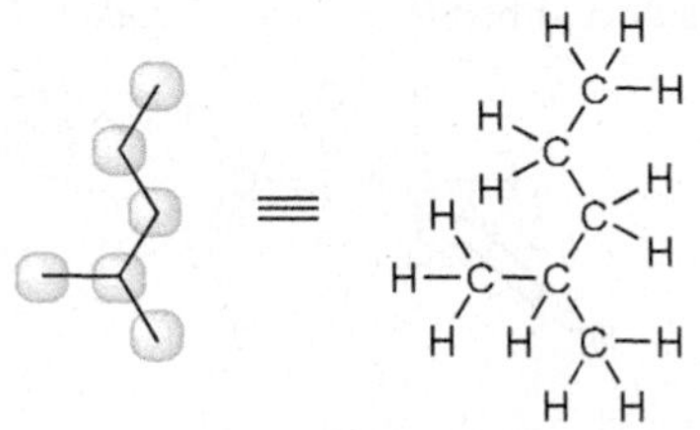

(b) Each corner and each endpoint represents a carbon atom (highlighted below), so this compound has twelve carbon atoms. Each carbon atom has enough attached hydrogen atoms to have exactly four bonds, as shown:

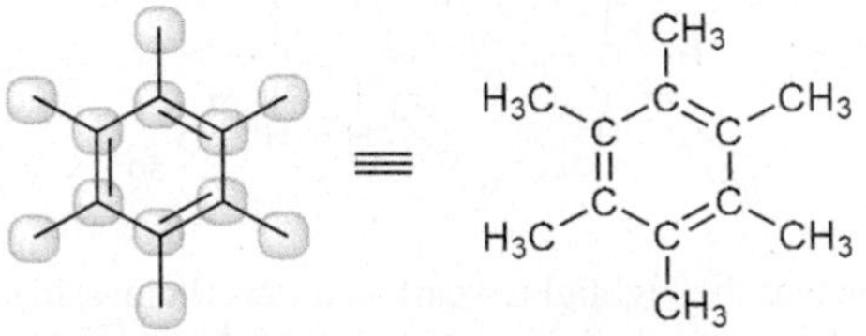

(c) Each corner represents a carbon atom (highlighted below), so this compound has six carbon atoms. Each carbon atom has enough attached hydrogen atoms to have exactly four bonds, as shown:

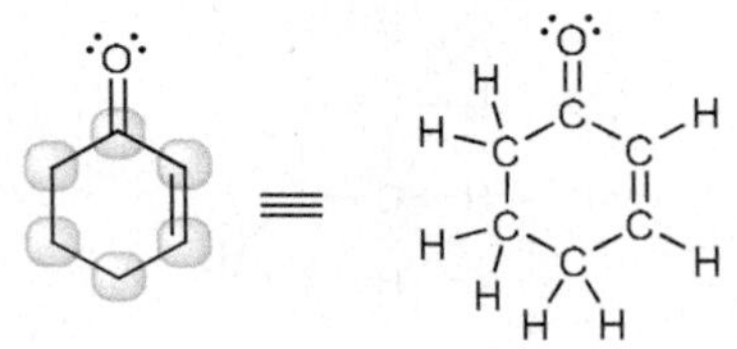

(d) Each corner represents a carbon atom (highlighted below), so this compound has seven carbon atoms. Each carbon atom has enough attached hydrogen atoms to have exactly four bonds, as shown:

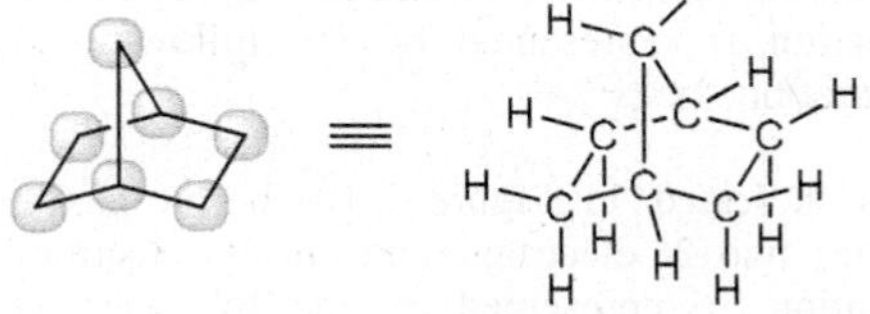

(e) Each corner and each endpoint represents a carbon atom (highlighted below), so this compound has seven carbon atoms. Each carbon atom has enough attached hydrogen atoms to have exactly four bonds, as shown:

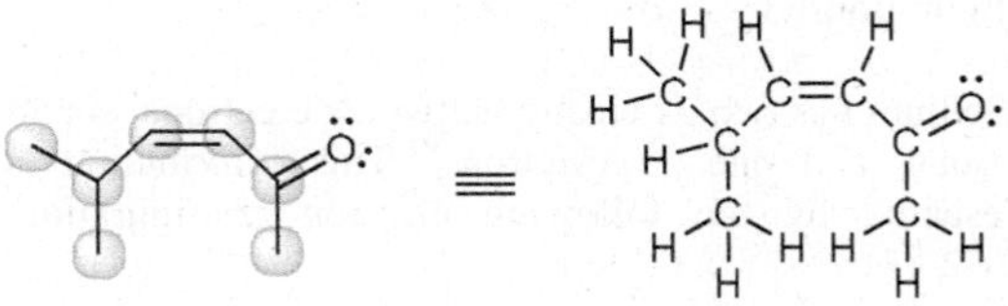

(f) Each corner represents a carbon atom (highlighted below), so this compound has seven carbon atoms. Each carbon atom has enough attached hydrogen atoms to have exactly four bonds, as shown:

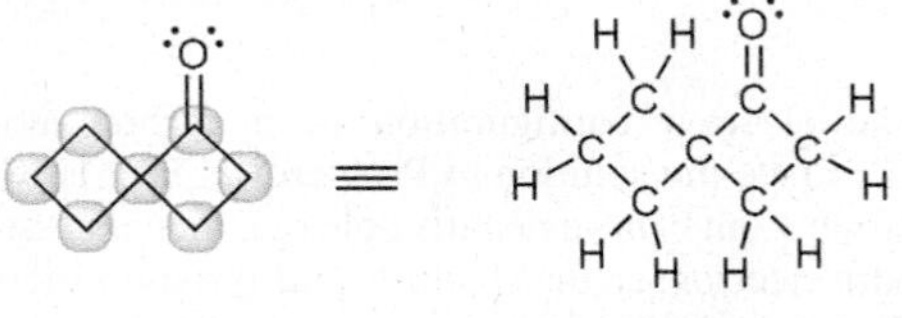

1.14. Remember that each corner and each endpoint represents a carbon atom. This compound therefore has 16 carbon atoms, highlighted below:

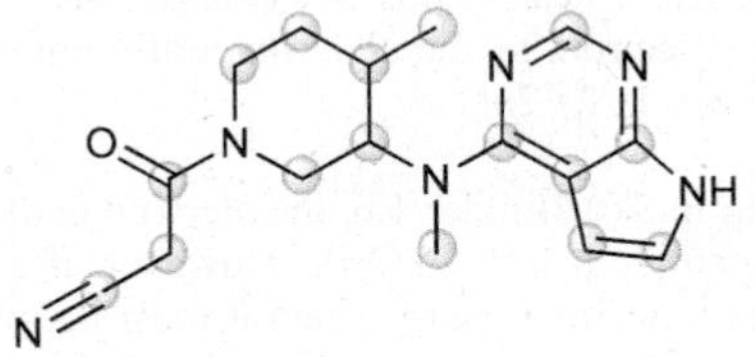

Each carbon atom should have four bonds. We therefore draw enough hydrogen atoms in order to give each carbon atom a total of four bonds. Any carbon atoms that already have four bonds will not have any hydrogen atoms:

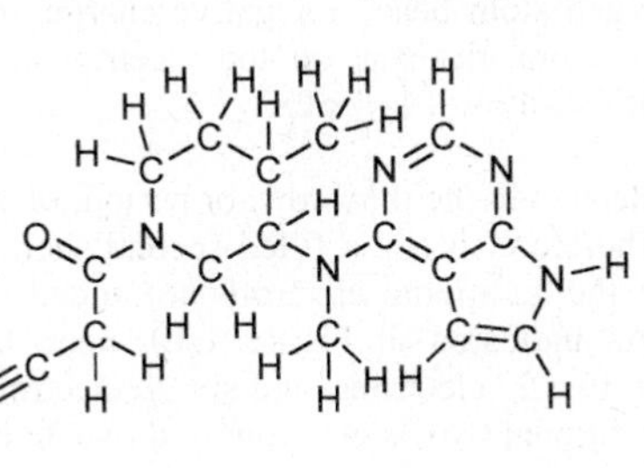

1.15.

(a) As indicated in Figure 1.10, carbon has two *1s* electrons, two *2s* electrons, and two *2p* electrons. This information is represented by the following electron configuration: $1s^22s^22p^2$

(b) As indicated in Figure 1.10, oxygen has two *1s* electrons, two *2s* electrons, and four *2p* electrons. This information is represented by the following electron configuration: $1s^22s^22p^4$

(c) As indicated in Figure 1.10, boron has two *1s* electrons, two *2s* electrons, and one *2p* electron. This information is represented by the following electron configuration: $1s^22s^22p^1$

(d) As indicated in Figure 1.10, fluorine has two *1s* electrons, two *2s* electrons, and five *2p* electrons. This information is represented by the following electron configuration: $1s^22s^22p^5$

(e) Sodium has two *1s* electrons, two *2s* electrons, six *2p* electrons, and one *3s* electron. This information is represented by the following electron configuration: $1s^22s^22p^63s^1$

(f) Aluminum has two *1s* electrons, two *2s* electrons, six *2p* electrons, two *3s* electrons, and one *3p* electron. This information is represented by the following electron configuration: $1s^22s^22p^63s^23p^1$

1.16.

(a) The electron configuration of a carbon atom is $1s^22s^22p^2$ (see the solution to Problem 1.15a). However, if a carbon atom bears a negative charge, then it must have one extra electron, so the electron configuration should be as follows: $1s^22s^22p^3$

(b) The electron configuration of a carbon atom is $1s^22s^22p^2$ (see the solution to Problem 1.15a). However, if a carbon atom bears a positive charge, then it must be missing an electron, so the electron configuration should be as follows: $1s^22s^22p^1$

(c) As seen in SkillBuilder 1.6, the electron configuration of a nitrogen atom is $1s^22s^22p^3$. However, if a nitrogen atom bears a positive charge, then it must be missing an electron, so the electron configuration should be as follows: $1s^22s^22p^2$

(d) The electron configuration of an oxygen atom is $1s^22s^22p^4$ (see the solution to Problem 1.15b). However, if an oxygen atom bears a negative charge, then it must have one extra electron, so the electron configuration should be as follows: $1s^22s^22p^5$

1.17. Silicon is in the third row, or period, of the periodic table. Therefore, it has a filled second shell, like neon, and then the additional electrons are added to the third shell. As indicated in Figure 1.10, neon has two 1*s* electrons, two 2*s* electrons, and six 2*p* electrons. Silicon has an additional two 3*s* electrons and two 3*p* electrons to give a total of 14 electrons and an electron configuration of $1s^22s^22p^63s^23p^2$.

1.18. The angles of an equilateral triangle are 60°, but each bond angle of cyclopropane is supposed to be 109.5°. Therefore, each bond angle is severely strained, causing an increase in energy. This form of strain, called ring strain, will be discussed in Chapter 4. The ring strain associated with a three-membered ring is greater than the ring strain of larger rings, because larger rings do not require bond angles of 60°.

1.19.

(a) The C=O bond of formaldehyde is comprised of one σ bond and one π bond.

(b) Each C–H bond is formed from the interaction between an sp^2-hybridized orbital from carbon and an *s* orbital from hydrogen.

(c) The oxygen atom is sp^2 hybridized, so the lone pairs occupy sp^2-hybridized orbitals.

1.20. Rotation of a single bond does not cause a reduction in the extent of orbital overlap, because the orbital overlap occurs on the bond axis. In contrast, rotation of a π bond results in a reduction in the extent of orbital overlap between the two *p* orbitals, because the orbital overlap is NOT on the bond axis.

1.21.

(a) The highlighted carbon atom (below) has four σ bonds, and is therefore sp^3 hybridized. The other carbon atoms in this structure are all sp^2 hybridized, because each of them has three σ bonds and one π bond.

*sp*³

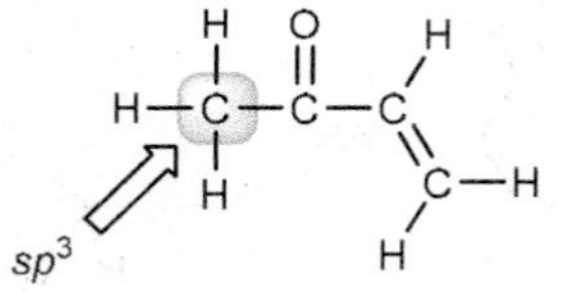

(b) Each of the highlighted carbon atoms has four σ bonds, and is therefore sp^3 hybridized. The other two carbon atoms in this structure are *sp* hybridized, because each has two σ bonds and two π bonds.

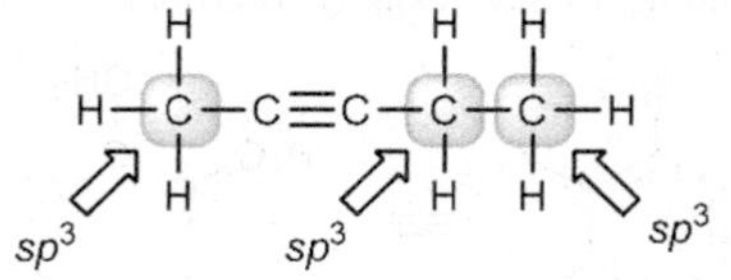

*sp*³ *sp*³ *sp*³

(c) Each of the highlighted carbon atoms (below) has four σ bonds, and is therefore sp^3 hybridized. The other two carbon atoms in this structure are sp^2 hybridized, because each has three σ bonds and one π bond.

*sp*³ *sp*³

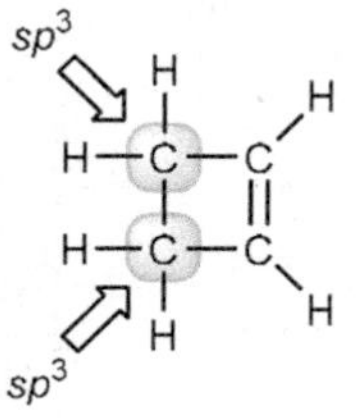

(d) Each of the two central carbon atoms has two σ bonds and two π bonds, and as such, each of these carbon atoms is *sp* hybridized. The other two carbon atoms (the outer ones) are sp^2 hybridized because each has three σ bonds and one π bond.

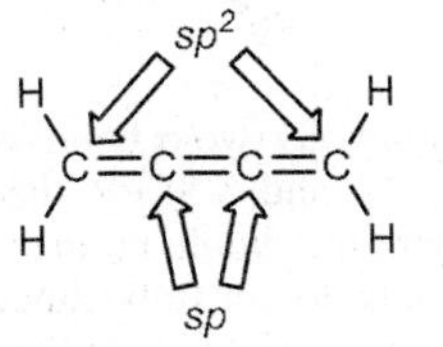

(e) One of the carbon atoms (the one connected to oxygen) has two σ bonds and two π bonds, and as such, it is *sp* hybridized. The other carbon atom is sp^2 hybridized because it has three σ bonds and one π bond.

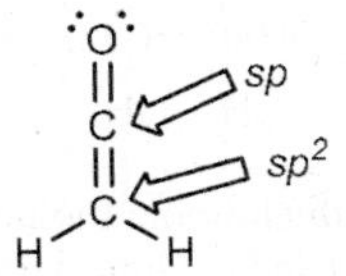

1.22. Each of the following three highlighted carbon atoms has four σ bonds, and is therefore sp^3 hybridized:

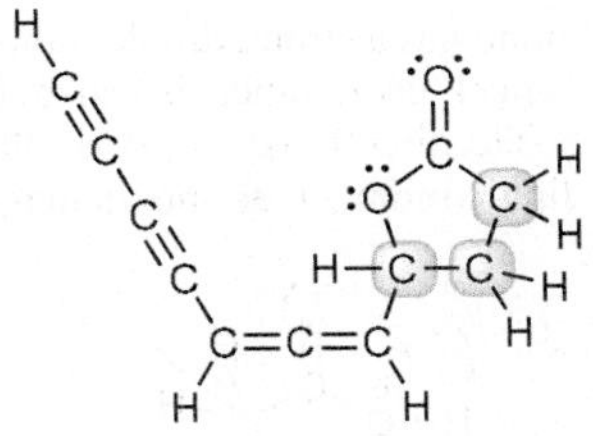

And each of the following three highlighted carbon atoms has three σ bonds and one π bond, and is therefore sp^2 hybridized:

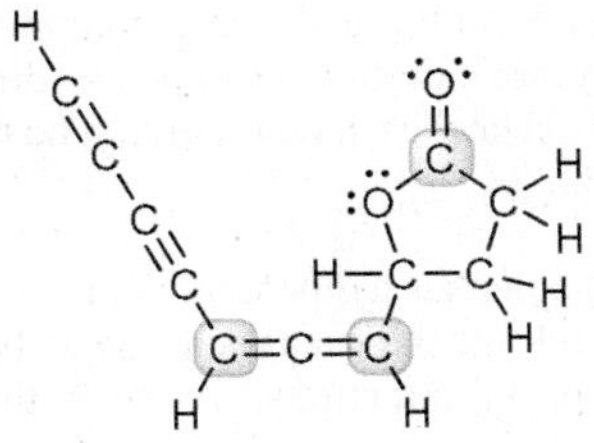

Finally, each of the following five highlighted carbon atoms has two σ bonds and two π bonds, and is therefore *sp* hybridized.

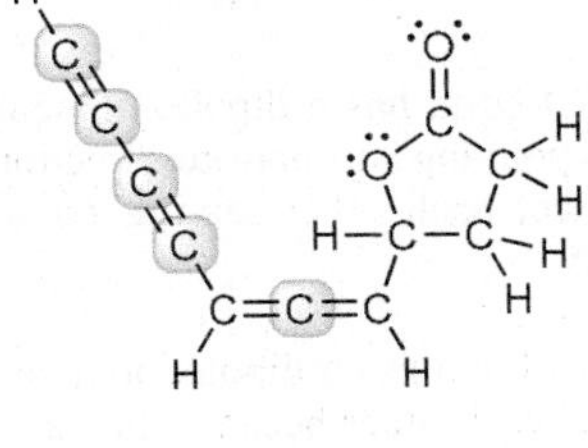

1.23. Carbon-carbon triple bonds generally have a shorter bond length than carbon-carbon double bonds, which are generally shorter than carbon-carbon single bonds (see Table 1.2).

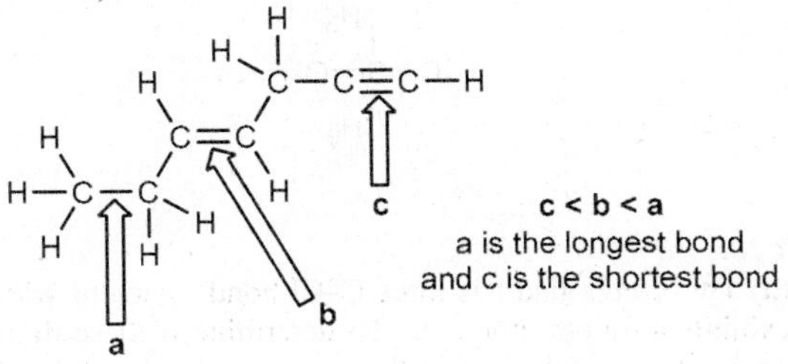

c < b < a
a is the longest bond and c is the shortest bond

1.24.

(a) In this structure, the boron atom has four σ bonds and no lone pairs, giving a total of four electron pairs (steric number = 4). VSEPR theory therefore predicts a tetrahedral arrangement of electron pairs. Since all of the electron pairs are bonds, the structure is expected to have tetrahedral geometry.

(b) In this structure, the boron atom has three σ bonds and no lone pairs, giving a total of three electron pairs (steric number = 3). VSEPR theory therefore predicts a trigonal planar geometry.

(c) In this structure, the nitrogen atom has four σ sigma bonds and no lone pairs, giving a total of four electron pairs (steric number = 4). VSEPR theory therefore predicts a tetrahedral arrangement of electron pairs. Since all of the electron pairs are bonds, the structure is expected to have tetrahedral geometry.

(d) The carbon atom has four σ bonds and no lone pairs, giving a total of four electron pairs (steric number = 4). VSEPR theory therefore predicts a tetrahedral arrangement of electron pairs. Since all of the electron pairs are bonds, the structure is expected to have tetrahedral geometry.

1.25. In the carbocation, the carbon atom has three bonds and no lone pairs. Since there are a total of three electron pairs (steric number = 3), and all three are bonds, VSEPR theory predicts trigonal planar geometry, with bond angles of 120°. In contrast, the carbon atom of the carbanion has three bonds and one lone pair, giving a total of four electron pairs (steric number = 4). For this ion, VSEPR theory predicts a tetrahedral arrangement of electron pairs, with a lone pair positioned at one corner of the tetrahedron, giving rise to trigonal pyramidal geometry with bond angles approximately 107°.

1.26. In ammonia, the nitrogen atom has three bonds and one lone pair. Therefore, VSEPR theory predicts trigonal pyramidal geometry, with bond angles of approximately 107°. In the ammonium ion, the nitrogen atom has four bonds and no lone pairs, so VSEPR theory predicts tetrahedral geometry, with bond angles of 109.5°. Therefore, we predict that the bond angles will increase (by approximately 2.5°) as a result of the reaction.

1.27. The silicon atom has four σ bonds and no lone pairs, so the steric number is 4 (sp^3 hybridization), which means that the arrangement of electron pairs will be tetrahedral. With no lone pairs, the arrangement of the atoms (geometry) is the same as the electronic arrangement. It is tetrahedral.

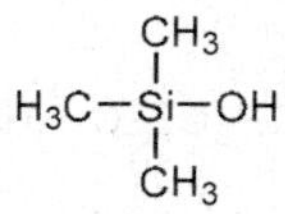

1.28.

(a) This compound has three C–Cl bonds, each of which exhibits a dipole moment. To determine if these dipole moments cancel each other, we must identify the molecular geometry. The central carbon atom has four σ bonds so we expect tetrahedral geometry. As such, the three polar C–Cl bonds do not lie in the same plane, and they do not completely cancel each other out. There is a net molecular dipole moment, as shown:

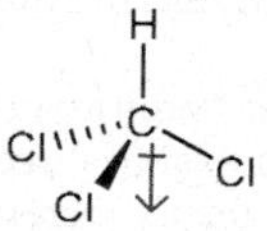

(b) The oxygen atom has two σ bonds and two lone pairs (steric number = 4), and VSEPR theory predicts bent geometry. As such, the dipole moments associated with the polar C–O bonds do not fully cancel each other, and the dipole moments associated with the lone pairs also do not fully cancel each other. As a result, there is a net molecular dipole moment, as shown:

(c) The nitrogen atom has three σ bonds and one lone pair (steric number = 4), and VSEPR theory predicts trigonal pyramidal geometry (because one corner of the tetrahedron is occupied by a lone pair). As such, the dipole moments associated with the polar N–H bonds do not fully cancel each other, and there is also a dipole moment associated with the lone pair (pointing up). As a result, there is a net molecular dipole moment, as shown:

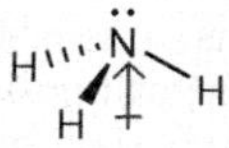

(d) The central carbon atom has four σ bonds (steric number = 4), and VSEPR theory predicts tetrahedral geometry. There are individual dipole moments associated with each of the C–Cl bonds and each of the C–Br bonds. If all four dipole moments had the same magnitude, then we would expect them to completely cancel each other to give no molecular dipole moment (as in the case of CCl_4). However, because Cl is more electronegative than Br, each C–Cl bond is more polar than each C–Br bond. Therefore, the dipole moments for the C–Cl bonds are larger than the dipole moments of the C–Br bonds, and as such, there is a net molecular dipole moment, shown here:

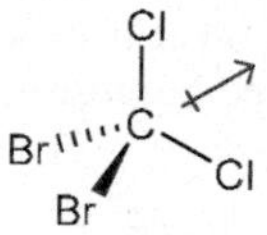

(e) The oxygen atom has two σ bonds and two lone pairs (steric number = 4), and VSEPR theory predicts bent geometry. As such, the dipole moments associated with the polar C–O bonds do not fully cancel each other, and the dipole moments associated with the lone pairs also do not fully cancel each other. As a result, there is a net molecular dipole moment, as shown:

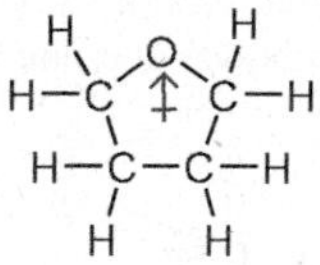

(f) There are individual dipole moments associated with each polar C–O bond and the lone pairs (as in the previous solution), but due to the symmetrical shape of the molecule in this case, they fully cancel each other to give no net molecular dipole moment.

(g) Each C=O bond has a strong dipole moment, and they do not fully cancel each other because they are not pointing in opposite directions. As such, there will be a net molecular dipole moment, as shown here:

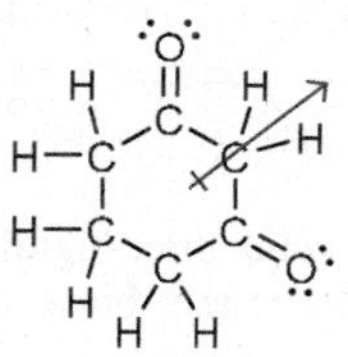

(h) Each C=O bond has a strong dipole moment, and in this case, they are pointing in opposite directions. As such, they fully cancel each other, giving no net molecular dipole moment.

(i) Each C–Cl bond has a dipole moment, and they do not fully cancel each other because the polar bonds are not pointing in opposite directions. As such, there will be a net molecular dipole moment, as shown here:

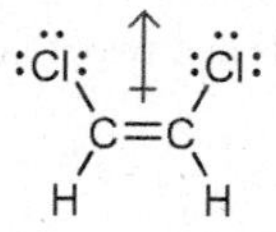

(j) Each C–Cl bond has a dipole moment, and in this case, they are pointing in opposite directions. As such, they fully cancel each other, giving no net molecular dipole moment.

(k) Each C–Cl bond has a dipole moment, and they do not fully cancel each other because they are not pointing in opposite directions. As such, there will be a net molecular dipole moment, as shown here:

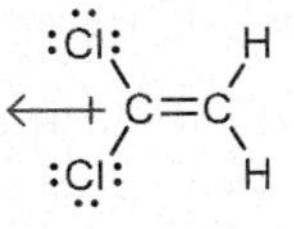

(l) Each C–Cl bond has a dipole moment, but in this case, they fully cancel each other to give no net molecular dipole moment.

1.29. Each of the C–O bonds has an individual dipole moment, shown here:

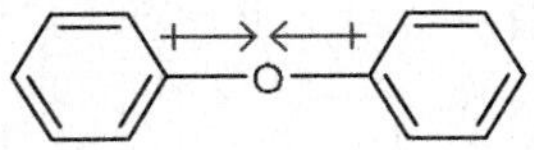

To determine if these individual dipole moments fully cancel each other, we must determine the geometry around the oxygen atom. The oxygen atom has two σ bonds and two lone pairs, giving rise to a bent geometry. As such, the dipole moments associated with the polar C–O bonds do NOT fully cancel each other,

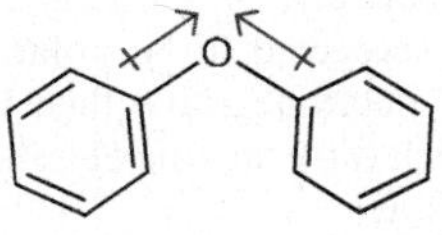

and the dipole moments associated with the lone pairs also do not fully cancel each other. As a result, there is a net molecular dipole moment, as shown:

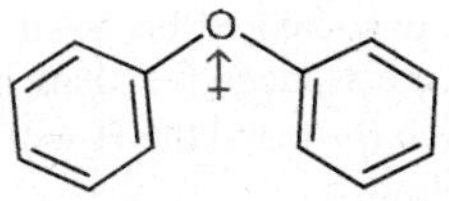

1.30.

(a) Both compounds have the same molecular weight because they are isomers. The second compound is expected to have a higher boiling point, because it is less branched. The greater surface area in the second compound results in greater London dispersion forces and a higher boiling point (b.p.):

higher b.p.

(b) The second compound is expected to have a higher boiling point, because more carbon atoms results in a higher molecular weight. Larger compounds have greater London dispersion forces and a higher boiling point (b.p.):

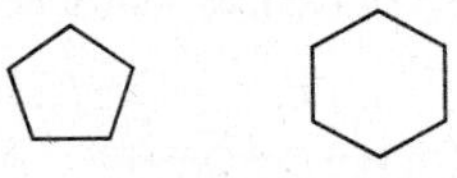

higher b.p.

(c) The second compound is expected to have a higher boiling point, because it has an O–H bond, which will lead to hydrogen-bonding interactions between molecules.

higher b.p.

(d) Both compounds have the same molecular weight because they are isomers, and they are both capable of forming hydrogen bonds. The first compound is expected to have a higher boiling point, however, because it is less branched. The greater surface area in the first compound results in greater London dispersion forces and a higher boiling point (b.p.):

higher b.p.

1.31. Compound **3** is expected to have a higher boiling point than compound **4**, because only compound **3** has an O-H group. Compound **4** does not form hydrogen-bonds, so it will have a lower boiling point. When this mixture is heated, the lower boiling compound (**4**) can be collected first, leaving behind compound **3**.

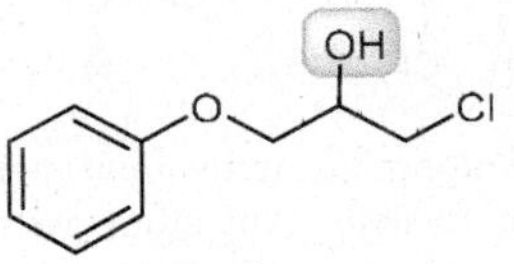

3

1.32.

(a) The carbon atoms are tetravalent, while the chlorine atom and hydrogen atoms are all monovalent. The atoms with more than one bond (in this case, the two carbon atoms) should be drawn in the center of the compound. The chlorine atom and hydrogen atoms are then placed at the periphery (ensuring that each carbon atom has a total of four bonds), as shown:

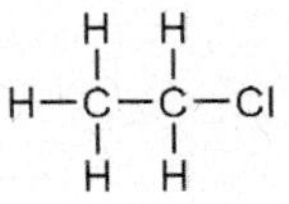

The chlorine atom can be placed in any one of the six available positions. The following six drawings all represent the same compound, in which the two carbon atoms are connected to each other, and the chlorine atom is connected to one of the carbon atoms.

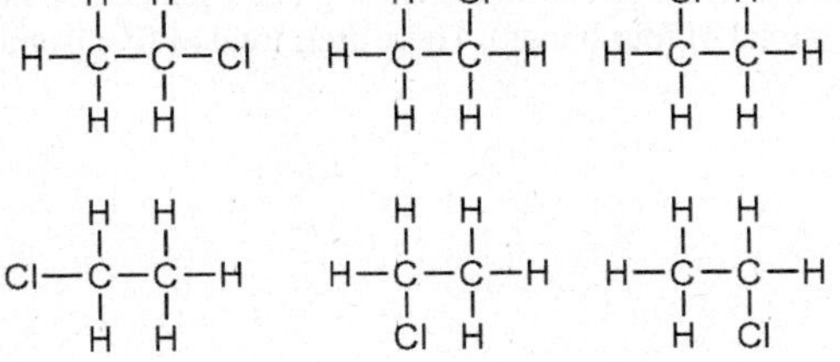

(b) The carbon atoms are tetravalent, while the chlorine atoms and hydrogen atoms are all monovalent. The atoms with more than one bond (in this case, the two carbon atoms) should be drawn in the center of the compound. The chlorine atoms and hydrogen atoms are then placed at the periphery, and there are two different ways to do this. The two chlorine atoms can either be connected to the same carbon atom or to different carbon atoms, as shown.

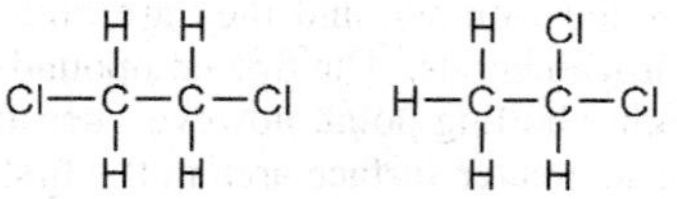

```
   H  H          H  Cl
   |  |          |  |
Cl—C—C—Cl     H—C—C—Cl
   |  |          |  |
   H  H          H  H
```

(c) The carbon atoms are tetravalent, while the chlorine atoms and hydrogen atoms are all monovalent. The atoms with more than one bond (in this case, the two carbon atoms) should be drawn in the center of the compound. The chlorine atoms and hydrogen atoms are then placed at the periphery, and there are two different ways to do this. One way is to connect all three chlorine atoms to the same carbon atom. Alternatively, we can connect two chlorine atoms to one carbon atom, and then connect the third chlorine atom to the other carbon atom, as shown here:

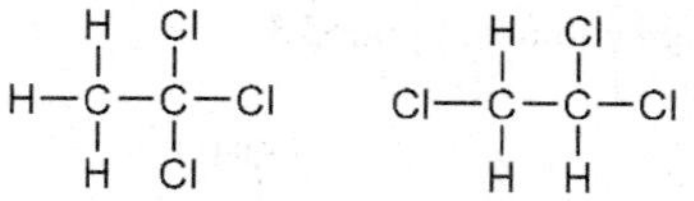

```
   H  Cl          H  Cl
   |  |           |  |
H—C—C—Cl     Cl—C—C—Cl
   |  |           |  |
   H  Cl          H  H
```

(d) The carbon atoms are tetravalent, and the hydrogen atoms are all monovalent. Any atoms with more than one bond (in this case, the six carbon atoms) should be drawn in the center of the compound, with the hydrogen atoms at the periphery. There are five different ways to connect six carbon atoms, which we will organize based on the length of the longest chain.

In a 6-carbon chain:

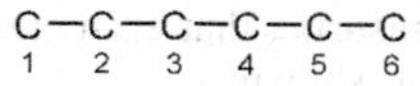

```
C—C—C—C—C—C
1  2  3  4  5  6
```

In a 5-carbon chain:

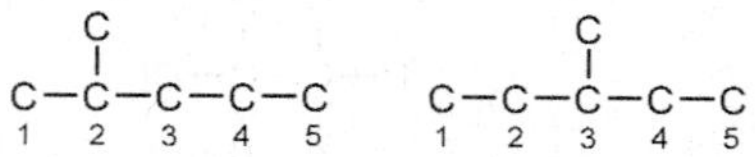

```
   C                  C
   |                  |
C—C—C—C—C       C—C—C—C—C
1  2  3  4  5     1  2  3  4  5
```

In a 4-carbon chain:

```
   C              C  C
   |              |  |
C—C—C—C       C—C—C—C
1 2| 3  4      1  2  3  4
   C
```

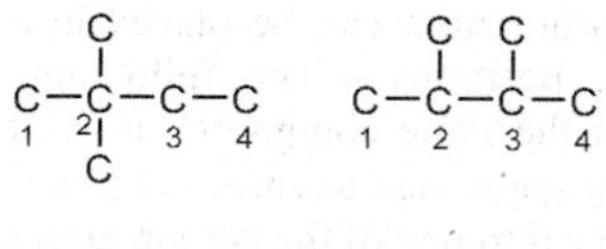

Finally, we complete all of the structures by drawing the bonds to hydrogen atoms (ensuring that each carbon atom has a total of four bonds). There are a total of five isomers:

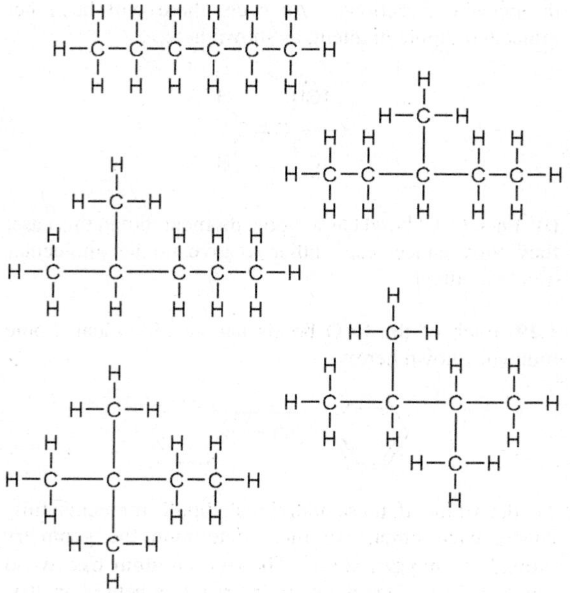

1.33.

(a) According to Table 1.1, the difference in electronegativity between Br and H is 2.8 – 2.1 = 0.7, so an H–Br bond is expected to be polar covalent. Since bromine is more electronegative than hydrogen, the Br will be electron-rich (δ–), and the H will be electron-poor (δ+), as shown below:

```
δ+   δ–
H——Br
```

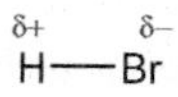

(b) According to Table 1.1, the difference in electronegativity between Cl and H is 3.0 – 2.1 = 0.9, so an H–Cl bond is expected to be polar covalent. Since chlorine is more electronegative than hydrogen, the Cl will be electron-rich (δ–), and the H will be electron-poor (δ+), as shown below:

```
δ+   δ–
H——Cl
```

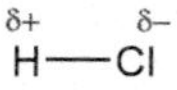

(c) According to Table 1.1, the difference in electronegativity between O and H is 3.5 – 2.1 = 1.4, so an O–H bond is expected to be polar covalent. Oxygen is more electronegative than hydrogen, so for each O–H bond, the O will be electron-rich (δ–) and the H will be electron-poor (δ+), as shown below:

```
      δ–
δ+   O   δ+
   H     H
```

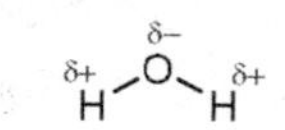

(d) According to Table 1.1, oxygen (3.5) is more electronegative than carbon (2.5) or hydrogen (2.1), and a C–O or H–O bond is polar covalent. For each C–O or H–O bond, the O will be electron-rich (δ–), and the C or H will be electron-poor (δ+), as shown below:

```
    H
  δ+|  δ–  δ+
H—C—Ö—H
    |
    H
```

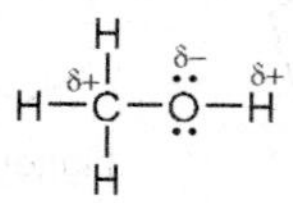

1.34.

(a) The difference in electronegativity between Na (0.9) and Br (2.8) is greater than the difference in electronegativity between H (2.1) and Br (2.8). Therefore, NaBr is expected to have more ionic character than HBr.

(b) The difference in electronegativity between F (4.0) and Cl (3.0) is greater than the difference in electronegativity between Br (2.8) and Cl (3.0). Therefore, FCl is expected to have more ionic character than BrCl.

1.35.

(a) Each carbon atom has four valence electrons, the oxygen atom has six valence electrons, and each hydrogen atom has one valence electron. In this case, the information provided in the problem statement (CH_3CH_2OH) indicates how the atoms are connected to each other:

H· ·C· ·C· ·O· ·H ⟶ H:C:C:O:H

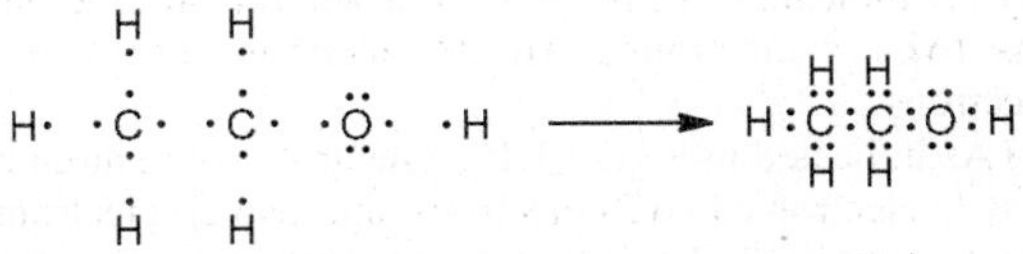

(b) Each carbon atom has four valence electrons, the nitrogen atom has five valence electrons, and each hydrogen atom has one valence electron. In this case, the information provided in the problem statement (CH_3CN) indicates how the atoms are connected to each other:

H· ·C· ·C· ·N: ⟶ H:C:C:N:

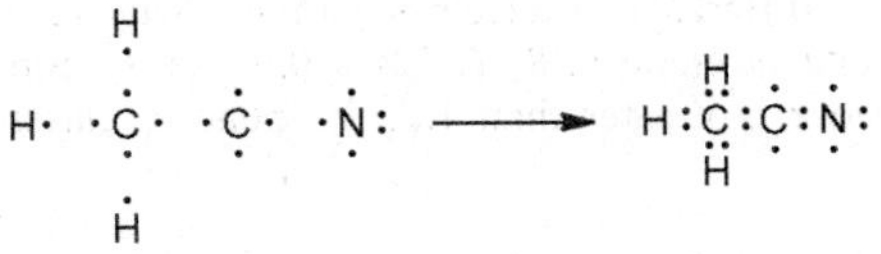

The unpaired electrons are then paired up to give a triple bond. In this way, each of the atoms achieves an octet.

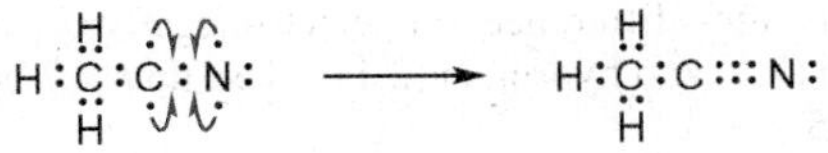

H:C:C:N: ⟶ H:C:C:::N:

1.36. Each of the carbon atoms has four valence electrons; the nitrogen atom has five valence electrons, and each of the hydrogen atoms has one valence electron. We begin by connecting the atoms that have more than one bond (in this case, the four carbon atoms and the nitrogen atom). The problem statement indicates how we should connect them:

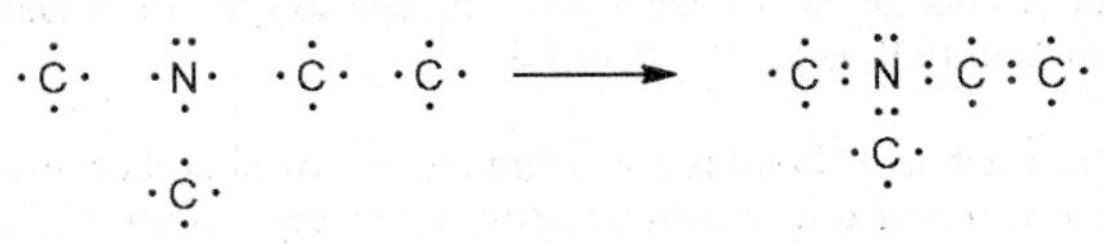

·C· ·N· ·C· ·C· ·C· ⟶ ·C : N : C : C·

Then, we connect all of the hydrogen atoms (ensuring that each carbon atom has four bonds), as shown.

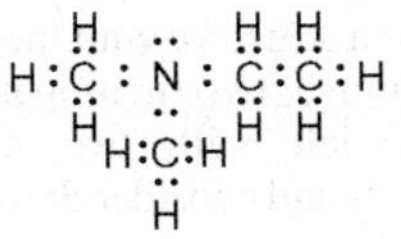

H:C : N : C:C:H

The nitrogen atom has three σ bonds and one lone pair, so the steric number is 4, which means that the arrangement of electron pairs is expected to be tetrahedral. One corner of the tetrahedron is occupied by a lone pair, so the geometry of the nitrogen atom (the arrangement of atoms around that nitrogen atom) is trigonal pyramidal. As such, the individual dipole moments associated with the C–N bonds do not fully cancel each other, and there is also a dipole moment associated with the lone pair (pointing up). As a result, there is a net molecular dipole moment, as shown:

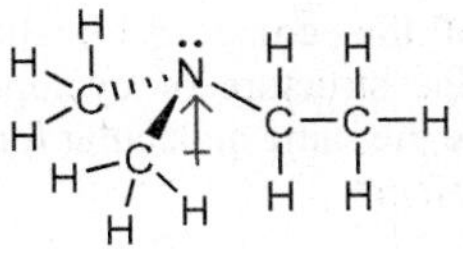

1.37. Bromine is in group 7A of the periodic table, so each bromine atom has seven valence electrons. Aluminum is in group 3A of the periodic table, so aluminum is supposed to have three valence electrons, but the structure bears a negative charge, which means that there is one extra electron. That is, the aluminum atom has four valence electrons, rather than three, which is why it has a formal negative charge. This gives the following Lewis structure:

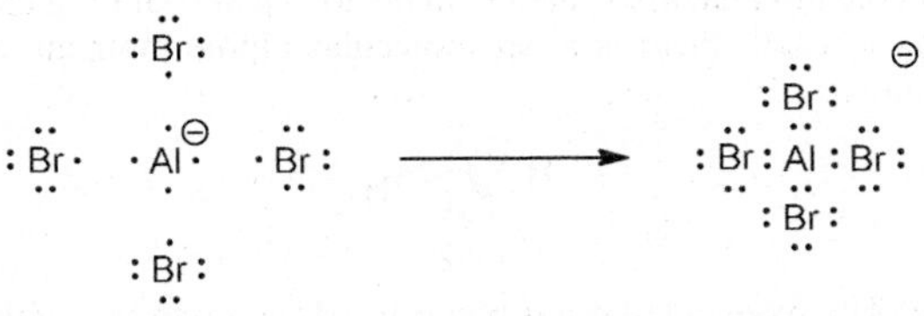

:Br· ·Al⊖· ·Br: ⟶ :Br : Al : Br: ⊖

The aluminum atom has four bonds and no lone pairs, so the steric number is 4, which means that this aluminum atom will have tetrahedral geometry.

1.38. The molecular formula of cyclopropane is C_3H_6, so we are looking for a different compound that has the same molecular formula, C_3H_6. That is, we need to find another way to connect the carbon atoms, other than in a ring (there is only one way to connect three carbon atoms in a ring, so we must be looking for something other than a ring). If we connect the three carbon atoms in a linear fashion and then complete the drawing by placing hydrogen atoms at the periphery, we notice that the molecular formula (C_3H_8) is not correct:

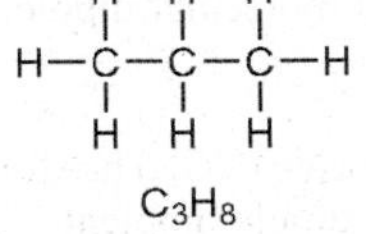

H–C–C–C–H

C_3H_8

We are looking for a structure with the molecular formula C_3H_6. If we remove two hydrogen atoms from our drawing, we are left with two unpaired electrons, indicating that we should consider drawing a double bond:

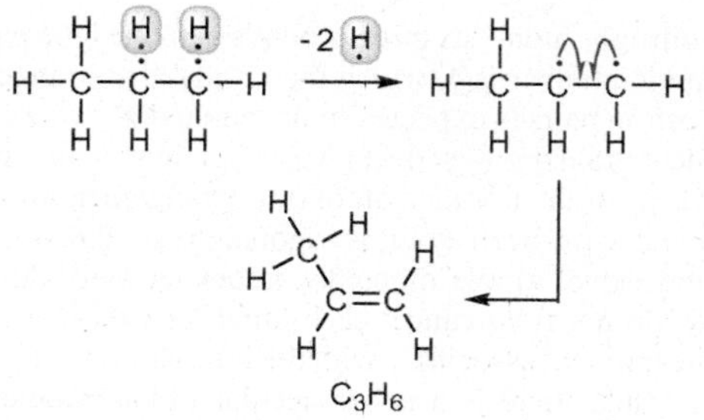

C_3H_6

The structure of this compound (called propylene) is different from the structure of cyclopropane, but both compounds share the same molecular formula, so they are constitutional isomers.

1.39.

(a) C–H bonds are considered to be nonpolar, although they do have a very small dipole moment, because there is a small difference in electronegativity between carbon (2.5) and hydrogen (2.1). With no polar bonds present, the molecule does not have a molecular dipole moment.

(b) The nitrogen atom has trigonal pyramidal geometry. As such, the dipole moments associated with the polar N–H bonds do not fully cancel each other, and there is also a dipole moment associated with the lone pair (pointing up). As a result, there is a net molecular dipole moment, as shown:

(c) The oxygen atom has two σ bonds and two lone pairs (steric number = 4), and VSEPR predicts bent geometry. As such, the dipole moments associated with the polar O–H bonds do not cancel each other, and the dipole moments associated with the lone pairs also do not fully cancel each other. As a result, there is a net molecular dipole moment, as shown:

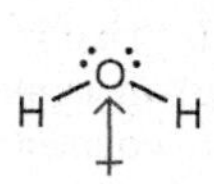

(d) The central carbon atom of carbon dioxide (CO_2) has two σ bonds and no lone pairs, so it is *sp* hybridized and is expected to have linear geometry. Each C=O bond has a strong dipole moment, but in this case, they are pointing in opposite directions. As such, they fully cancel each other, giving no net molecular dipole moment.

(e) Carbon tetrachloride (CCl_4) has four C–Cl bonds, each of which exhibits a dipole moment. However, the central carbon atom has four σ bonds so it is expected to have tetrahedral geometry. As such, the four dipole moments completely cancel each other out, and there is no net molecular dipole moment.

(f) This compound has two C–Br bonds, each of which exhibits a dipole moment. To determine if these dipole moments cancel each other, we must identify the molecular geometry. The central carbon atom has four σ bonds so it is expected to have tetrahedral geometry. As such, the polar C–Br bonds do not completely cancel each other out. There is a net molecular dipole moment, as shown:

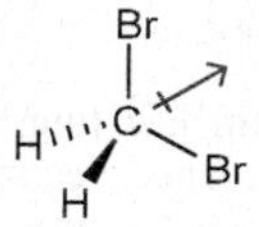

1.40.

(a) As indicated in Figure 1.10, a neutral oxygen atom has two *1s* electrons, two *2s* electrons, and four *2p* electrons.

(b) As indicated in Figure 1.10, a neutral fluorine atom has two *1s* electrons, two *2s* electrons, and five *2p* electrons.

(c) As indicated in Figure 1.10, a neutral carbon atom has two *1s* electrons, two *2s* electrons, and two *2p* electrons.

(d) As seen in SkillBuilder 1.6, the electron configuration of a neutral nitrogen atom is $1s^22s^22p^3$

(e) This is the electron configuration of a neutral chlorine atom.

1.41.

(a) The difference in electronegativity between sodium (0.9) and bromine (2.8) is 2.8 – 0.9 = 1.9. Since this difference is greater than 1.7, the bond is classified as ionic.

(b) The difference in electronegativity between sodium (0.9) and oxygen (3.5) is 3.5 – 0.9 = 2.6. Since this difference is greater than 1.7, the Na–O bond is classified as ionic. In contrast, the O–H bond is polar covalent, because the difference in electronegativity between oxygen (3.5) and hydrogen (2.1) is less than 1.7 but more than 0.5.

(c) Each C–H bond is considered to be covalent, because the difference in electronegativity between carbon (2.5) and hydrogen (2.1) is less than 0.5.
The C–O bond is polar covalent, because the difference in electronegativity between oxygen (3.5) and carbon (2.5) is less than 1.7 but more than 0.5.
The Na–O bond is classified as ionic, because the difference in electronegativity between oxygen (3.5) and sodium (0.9) is greater than 1.7.

(d) Each C–H bond is considered to be covalent, because the difference in electronegativity between carbon (2.5) and hydrogen (2.1) is less than 0.5.
The C–O bond is polar covalent, because the difference in electronegativity between oxygen (3.5) and carbon (2.5) is less than 1.7 but more than 0.5.

The O–H bond is polar covalent, because the difference in electronegativity between oxygen (3.5) and hydrogen (2.1) is less than 1.7 but more than 0.5.

(e) Each C–H bond is considered to be covalent, because the difference in electronegativity between carbon (2.5) and hydrogen (2.1) is less than 0.5.
The C=O bond is polar covalent, because the difference in electronegativity between oxygen (3.5) and carbon (2.5) is less than 1.7 but more than 0.5.

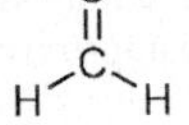

1.42.
(a) Begin by determining the valency of each atom in the compound. The carbon atoms are tetravalent, the oxygen atom is divalent, and the hydrogen atoms are all monovalent. Any atoms with more than one bond (in this case, the two carbon atoms and the oxygen atom) should be drawn in the center of the compound, with the hydrogen atoms at the periphery. There are two different ways to connect two carbon atoms and an oxygen atom, shown here:

C–C–O   C–O–C

We then complete both structures by drawing the remaining bonds to hydrogen atoms (ensuring that each carbon atom has four bonds, and each oxygen atom has two bonds):

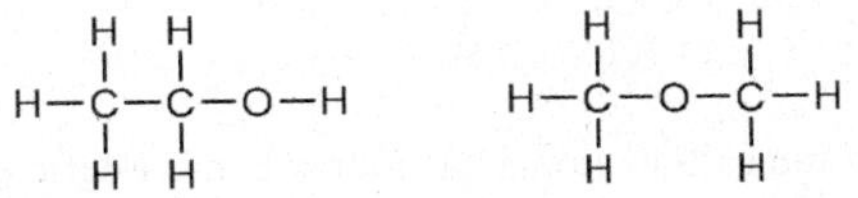

(b) Begin by determining the valency of each atom in the compound. The carbon atoms are tetravalent, the oxygen atoms are divalent, and the hydrogen atoms are all monovalent. Any atoms with more than one bond (in this case, the two carbon atoms and the two oxygen atoms) should be drawn in the center of the compound, with the hydrogen atoms at the periphery. There are several different ways to connect two carbon atoms and two oxygen atoms (highlighted, for clarity of comparison), shown here:

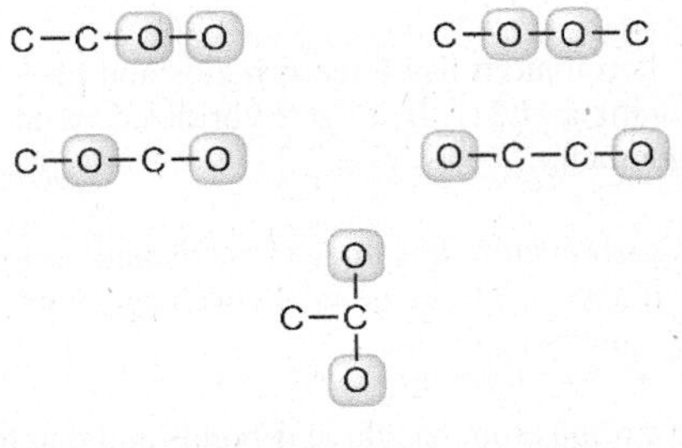

C–C–O–O   C–O–O–C

C–O–C–O   O–C–C–O

We then complete all of these structures by drawing the remaining bonds to hydrogen atoms:

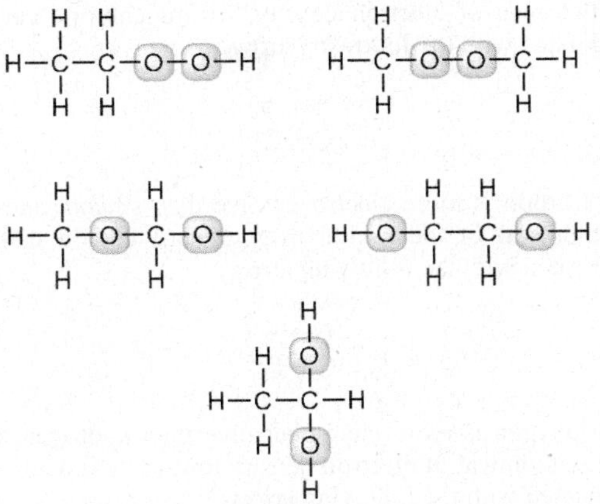

(c) The carbon atoms are tetravalent, while the bromine atoms and hydrogen atoms are all monovalent. The atoms with more than one bond (in this case, the two carbon atoms) should be drawn in the center of the compound. The bromine atoms and hydrogen atoms are then placed at the periphery, and there are two different ways to do this. The two bromine atoms can either be connected to the same carbon atom or to different carbon atoms, as shown.

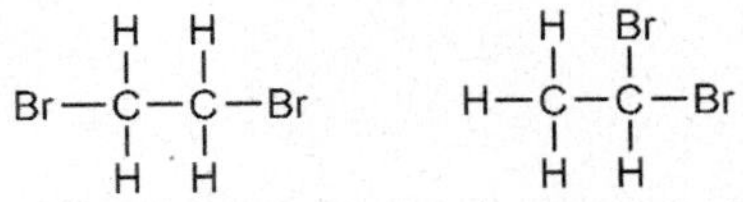

1.43.
(a) Oxygen is more electronegative than carbon, and the withdrawal of electron density toward oxygen can be indicated with the following arrow:

C–O

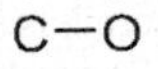

(b) Carbon is more electronegative than magnesium, and the withdrawal of electron density toward carbon can be indicated with the following arrow:

C–Mg

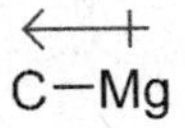

(c) Nitrogen is more electronegative than carbon, and the withdrawal of electron density toward nitrogen can be indicated with the following arrow:

C–N

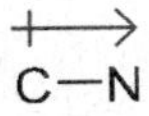

(d) Carbon is more electronegative than lithium, and the withdrawal of electron density toward carbon can be indicated with the following arrow:

C–Li

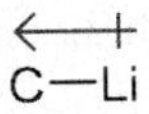

(e) Chlorine is more electronegative than carbon, and the withdrawal of electron density toward chlorine can be indicated with the following arrow:

```
+⟶
C—Cl
```

(f) Carbon is more electronegative than silicon, and the withdrawal of electron density toward carbon can be indicated with the following arrow:

```
⟵+
C—Si
```

(g) Oxygen is more electronegative than hydrogen, and the withdrawal of electron density toward oxygen can be indicated with the following arrow:

```
⟵+
O—H
```

(h) Nitrogen is more electronegative than hydrogen, and the withdrawal of electron density toward nitrogen can be indicated with the following arrow:

```
⟵+
N—H
```

1.44.

(a) The oxygen atom has two σ bonds and two lone pairs (steric number = 4), and VSEPR theory predicts bent geometry. The C-O-H bond angle is expected to be approximately 105°, and all other bonds angles are expected to be approximately 109.5° (because each carbon atom has four σ bonds and tetrahedral geometry).

(b) The central carbon atom has three σ bonds and no lone pairs (steric number = 3), and VSEPR theory predicts trigonal planar geometry. As such, all bond angles are approximately 120°.

```
   O
   ||
   C
  / \
 H   H
```

(c) Each of the carbon atoms has three σ bonds and no lone pairs (steric number = 3), and VSEPR theory predicts trigonal planar geometry. As such, all bond angles are approximately 120°.

```
H     H
 \   /
  C=C
 /   \
H     H
```

(d) Each of the carbon atoms has two σ bonds and no lone pairs (steric number = 2), and VSEPR theory predicts linear geometry. As such, all bond angles are 180°.

```
H—C≡C—H
```

(e) The oxygen atom has two σ bonds and two lone pairs (steric number = 4), and VSEPR theory predicts bent geometry. Therefore, the C-O-C bond angle is expected to be around 105°. The remaining bond angles are all expected to be approximately 109.5° (because each carbon atom has four σ bonds and tetrahedral geometry).

(f) The nitrogen atom has three σ bonds and one lone pair (steric number = 4), and VSEPR theory predicts trigonal pyramidal geometry, with bond angles of 107°. The carbon atom is also tetrahedral (because it has four σ bonds), although the bond angles around the carbon atom are expected to be approximately 109.5°.

```
    H   H
    |  /
H—C—N:
    |  \
    H   H
```

(g) Each of the carbon atoms has four σ bonds (steric number = 4), so each of these carbon atoms has tetrahedral geometry. Therefore, all bond angles are expected to be approximately 109.5°.

```
    H  H  H
    |  |  |
H—C—C—C—H
    |  |  |
    H  H  H
```

(h) The structure of acetonitrile (CH_3CN) is shown below (see the solution to Problem **1.35b**).

```
    H
    |
H—C—C≡N:
    |
    H
```

One of the carbon atoms has four σ bonds (steric number = 4), and is expected to have tetrahedral geometry. The other carbon atom (connected to nitrogen) has two σ bonds and no lone pairs (steric number = 2), so we expect linear geometry.
As such, the C–C≡N bond angle is 180°, and all other bond angles are approximately 109.5°.

1.45.

(a) The nitrogen atom has three σ bonds and one lone pair (steric number = 4). It is sp^3 hybridized (electronically tetrahedral), with trigonal pyramidal geometry (because one corner of the tetrahedron is occupied by a lone pair).

(b) The boron atom has three σ bonds and no lone pairs (steric number = 3). It is sp^2 hybridized, with trigonal planar geometry.

(c) This carbon atom has three σ bonds and no lone pairs (steric number = 3). It is sp^2 hybridized, with trigonal planar geometry.

(d) This carbon atom has three σ bonds and one lone pair (steric number = 4). It is sp^3 hybridized (electronically tetrahedral), with trigonal pyramidal geometry (because one corner of the tetrahedron is occupied by a lone pair).

1.46.

(a) Each corner and each endpoint represents a carbon atom (highlighted), so this compound has nine carbon atoms. Each carbon atom will have enough hydrogen atoms to have exactly four bonds, as shown.

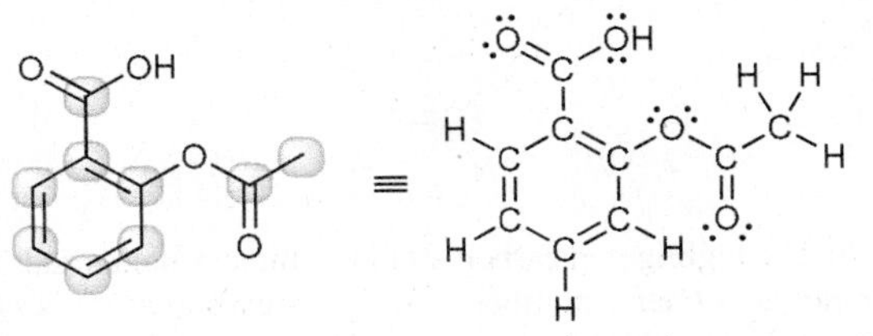

(b) Each corner and each endpoint represents a carbon atom (highlighted below), so this compound has eight carbon atoms. Each carbon atom will have enough hydrogen atoms to have exactly four bonds, as shown.

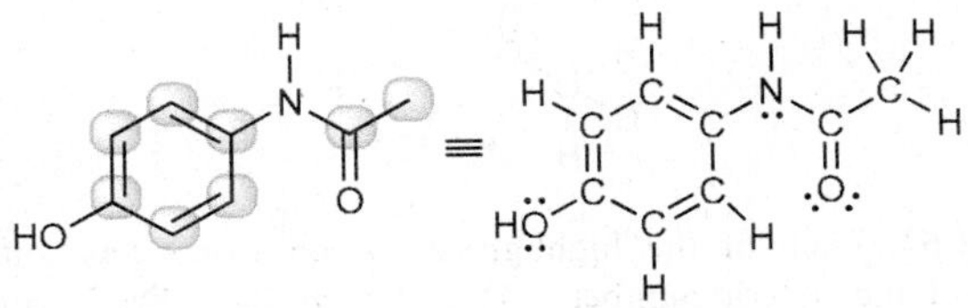

(c) Each corner and each endpoint represents a carbon atom (highlighted below), so this compound has eight carbon atoms. Each carbon atom will have enough hydrogen atoms to have exactly four bonds, as shown.

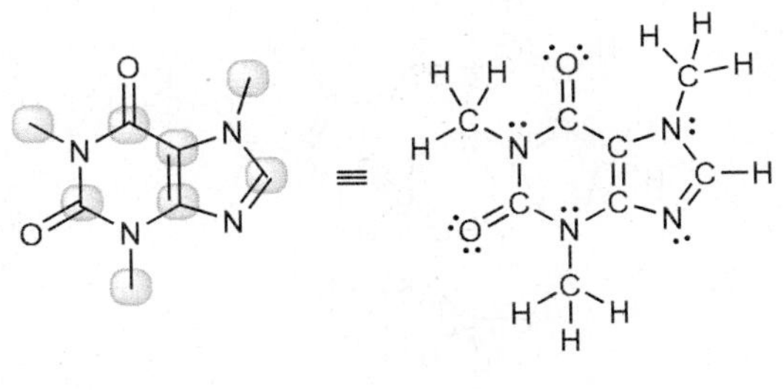

1.47. Each corner and each endpoint represents a carbon atom, so this compound has fifteen carbon atoms. Each carbon atom will have enough hydrogen atoms to have exactly four bonds, giving a total of eighteen hydrogen atoms, as shown here:

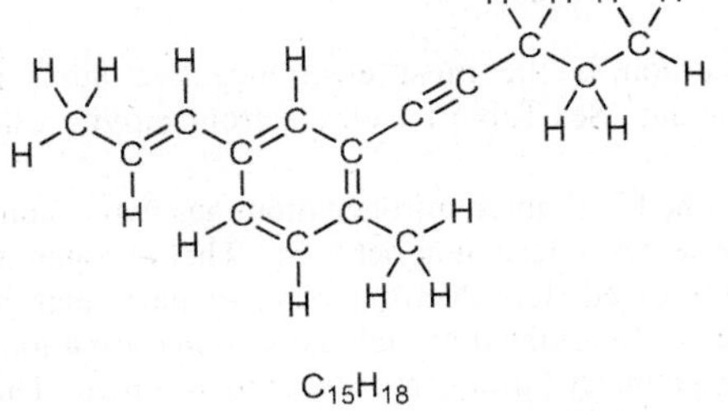

$C_{15}H_{18}$

1.48. The double bond represents one σ bond and one π bond, while the triple bond represents one σ bond and two π bonds. All single bonds are σ bonds. Therefore, this compound has sixteen σ bonds and three π bonds.

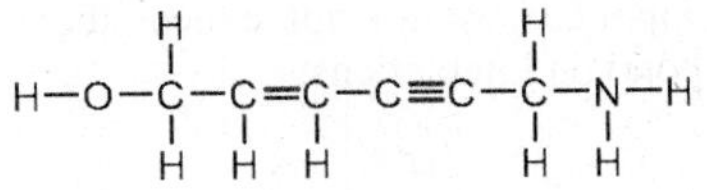

1.49.

(a) The second compound is expected to have a higher boiling point, because it has an O–H bond, which will lead to hydrogen bonding interactions.

(b) The second compound is expected to have a higher boiling point, because it has more carbon atoms, and thus a higher molecular weight and more opportunity for London dispersion forces.

(c) The first compound has a C=O bond, which has a strong dipole moment, while the second compound is nonpolar. The first compound is therefore expected to exhibit strong dipole-dipole interactions and to have a higher boiling point than the second compound.

1.50.

(a) This compound possesses an O–H bond, so it is expected to exhibit hydrogen bonding interactions.

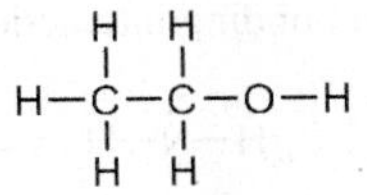

(b) This compound lacks a hydrogen atom that is connected to an oxygen or nitrogen atom. Therefore, this compound cannot serve as a hydrogen-bond donor (although the lone pairs can serve as hydrogen-bond acceptors, in the presence of a hydrogen-bond donor). As a pure compound, we do not expect there to be any hydrogen bonding interactions.

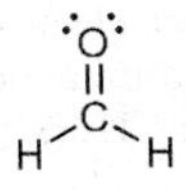

(c) This compound lacks a hydrogen atom that is connected to an oxygen or nitrogen atom. Therefore, this compound will not exhibit hydrogen bonding interactions.

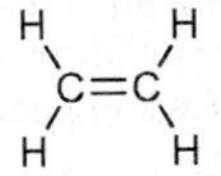

(d) This compound lacks a hydrogen atom that is connected to an oxygen or nitrogen atom . Therefore, this compound will not exhibit hydrogen bonding interactions.

H–C≡C–H

(e) This compound lacks a hydrogen atom that is connected to an oxygen or nitrogen atom. Therefore, this compound cannot serve as a hydrogen-bond donor (although the lone pairs can serve as hydrogen-bond acceptors, in the presence of a hydrogen-bond donor). As a pure compound, we do not expect there to be any hydrogen bonding interactions.

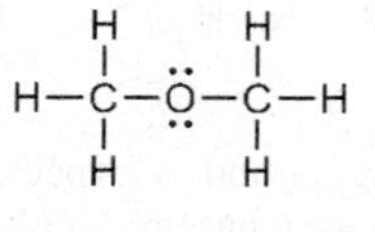

(f) This compound possesses two N–H bonds, so it is expected to exhibit hydrogen bonding interactions.

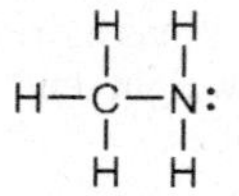

(g) This compound lacks a hydrogen atom that is connected to an oxygen or nitrogen atom. Therefore, this compound will not exhibit hydrogen bonding interactions.

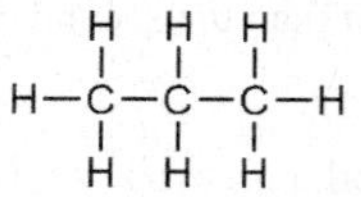

(h) This compound possesses N–H bonds, so it is expected to exhibit hydrogen bonding interactions.

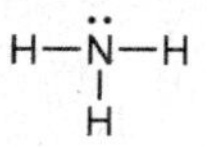

1.51.

(a) Boron is in group 3A of the periodic table, and therefore has three valence electrons. It can use each of its valence electrons to form a bond, so we expect the molecular formula to be BH_3 (x=3).

(b) Carbon is in group 4A of the periodic table, and therefore has four valence electrons. It can use each of its valence electrons to form a bond, so we expect the molecular formula to be CH_4 (x=4).

(c) Nitrogen is in group 5A of the periodic table, and therefore has five valence electrons. But it cannot form five bonds, because it only has four orbitals with which to form bonds. One of those orbitals must be occupied by a lone pair (two electrons), and each of the remaining three electrons is available to form a bond. Nitrogen is therefore trivalent, and we expect the molecular formula to be NH_3 (x=3).

(d) Carbon is in group 4A of the periodic table, and therefore has four valence electrons. It can use each of its valence electrons to form a bond, and indeed, we expect the carbon atom to have four bonds. Two of the bonds are with hydrogen atoms, so the other two bonds must be with chlorine atoms. The molecular formula is CH_2Cl_2 (x=2).

1.52.

(a) Each of the highlighted carbon atoms has three σ bonds and no lone pairs (steric number = 3). Each of these carbon atoms is sp^2 hybridized, with trigonal planar geometry. Each of the other four carbon atoms has two σ bonds and no lone pairs (steric number = 2). Those four carbon atoms are all *sp* hybridized, with linear geometry.

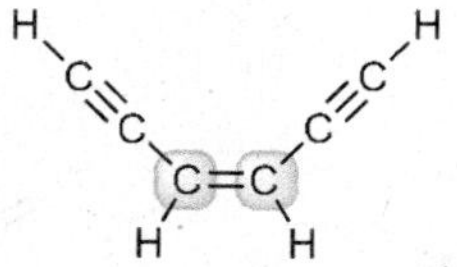

(b) The highlighted carbon atom has three σ bonds and no lone pairs (steric number = 3). This carbon atom is sp^2 hybridized, with trigonal planar geometry. Each of the other three carbon atoms has four σ bonds (steric number = 4). Those three carbon atoms are all sp^3 hybridized, with tetrahedral geometry.

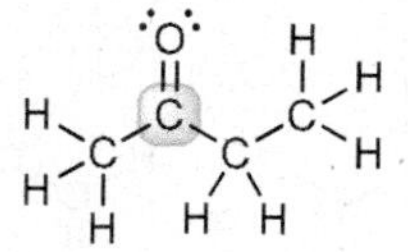

1.53. Each of the highlighted carbon atoms has four σ bonds (steric number = 4), and is sp^3 hybridized, with tetrahedral geometry. Each of the other fourteen carbon atoms in this structure has three σ bonds and no lone pairs (steric number = 3). Each of these fourteen carbon atoms is sp^2 hybridized, with trigonal planar geometry.

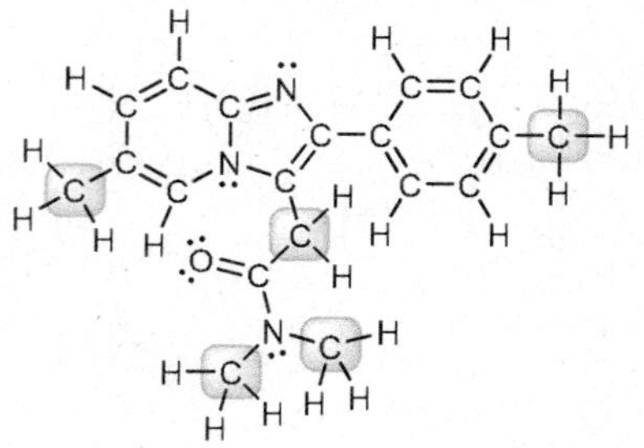

1.54.

(a) Oxygen is the most electronegative atom in this compound. See Table 1.1 for electronegativity values.

(b) Fluorine is the most electronegative atom. See Table 1.1 for electronegativity values.

(c) Carbon is the most electronegative atom in this compound. See Table 1.1 for electronegativity values.

1.55. The highlighted nitrogen atom has two σ bonds and one lone pair (steric number = 3). This nitrogen atom is sp^2 hybridized. It is electronically trigonal planar, but one of the sp^2 hybridized orbitals is occupied by a lone pair, so the geometry (arrangement of atoms) is bent. The other nitrogen atom (not highlighted) has three σ bonds and a lone pair (steric number = 4). That nitrogen atom is sp^3

hybridized and electronically tetrahedral. One corner of the tetrahedron is occupied by a lone pair, so the geometry (arrangement of atoms) is trigonal pyramidal.

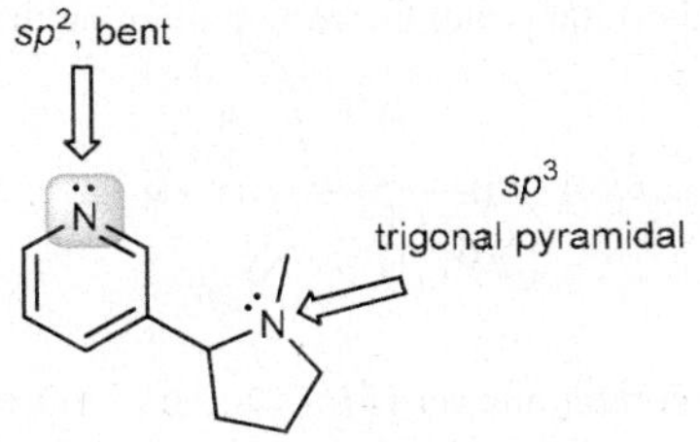

sp², bent

sp³
trigonal pyramidal

1.56. Each of the nitrogen atoms in caffeine achieves an octet with three bonds and one lone pair, while each oxygen atom in this structure achieves an octet with two bonds and two lone pairs, as shown:

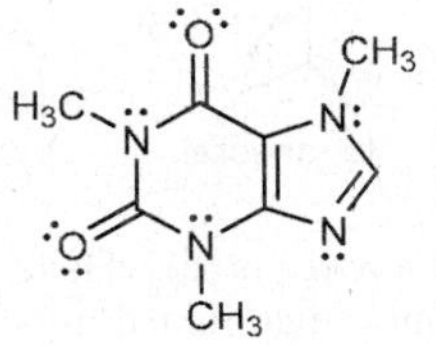

1.57. As seen in Section 1.2, the following two compounds have the molecular formula C_2H_6O.

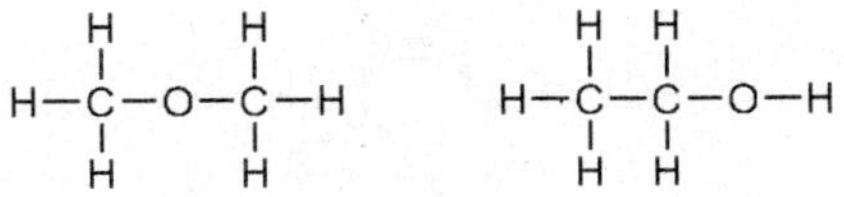

The second compound will have a higher boiling point because it possesses an OH group which can exhibit hydrogen bonding interactions.

1.58.

(a) Each C–Cl bond has a dipole moment, and the two dipole moments do not fully cancel each other because the polar bonds are not pointing in opposite directions. As such, there will be a net molecular dipole moment, as shown here:

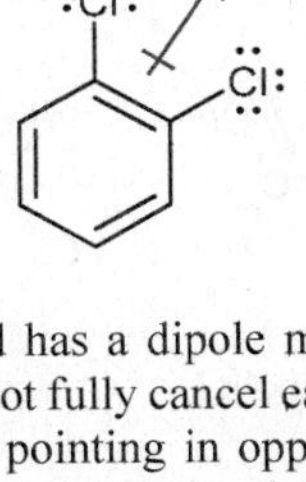

(b) Each C–Cl bond has a dipole moment, and the two dipole moments do not fully cancel each other because the polar bonds are not pointing in opposite directions. As such, there will be a net molecular dipole moment, as shown here:

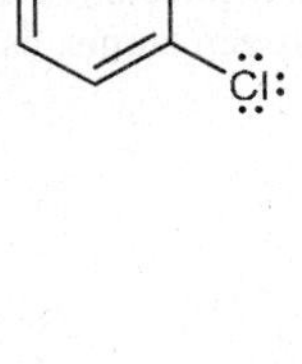

(c) Each C–Cl bond has a dipole moment, and in this case, the two dipole moments are pointing in opposite directions. As such, they fully cancel each other, giving no net molecular dipole moment.

(d) The C–Cl bond has a dipole moment, and the C–Br bond also has a dipole moment. These two dipole moments are in opposite directions, but they do not have the same magnitude. The C–Cl bond has a larger dipole moment than the C–Br bond, because chlorine is more electronegative than bromine. Therefore, there will be a net molecular dipole moment, as shown here:

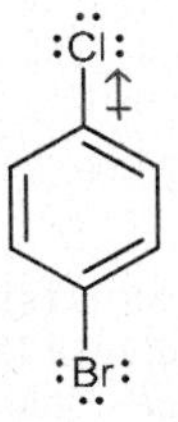

1.59. The C–Cl bond in chloroform partially cancels the dipole moments of the other two C–Cl bonds, thereby reducing the molecular dipole moment relative to methylene chloride.

1.60. $CHCl_3$ is expected to have a larger molecular dipole moment than $CBrCl_3$, because the bromine atom in $CBrCl_3$ serves to partially cancel out the dipole moments of the three C–Cl bonds (as is the case for CCl_4).

1.61. The carbon atom of O=C=O has two σ bonds and no lone pairs (steric number = 2) and VSEPR theory predicts linear geometry. As a result, the individual dipole moments of each C=O bond cancel each other completely to give no overall molecular dipole moment. In contrast, the sulfur atom in SO_2 has a steric number of three (because it also has a lone pair, in addition to the two S=O bonds), which means that it has bent geometry. As a result, the individual dipole moments of each S=O bond do NOT cancel each other completely, and the molecule does have a molecular dipole moment.

1.62. Two compounds possess OH groups. These two compounds will have the highest boiling points, because they can form hydrogen bonds. Among these two compounds, the one with more carbon atoms (six) will be higher boiling than the one with fewer carbon atoms (four), because the higher molecular weight results in greater London dispersion forces. The remaining three compounds all have equal molecular weights (five carbon atoms) and lack an OH group. The difference between these three compounds is the extent of branching. Among these three compounds, the compound with the greatest extent of branching has smallest surface area and therefore the lowest boiling point, and the one with the least branching has the highest boiling point.

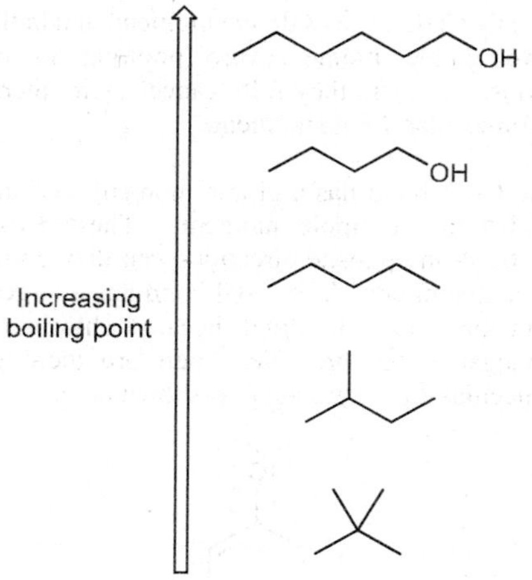

Increasing boiling point

OH

OH

1.63. The correct answer is (a). We must first draw the structure of HCN. To draw a Lewis structure, we begin by counting the valence electrons (hydrogen has one valence electron, carbon has four valence electrons, and nitrogen has five valence electrons, for a total of ten valence electrons). The structure must have ten valence electrons (no more and no less). Carbon should have four bonds, and it can only form a single bond with the hydrogen atom, so there must be a triple bond between carbon and nitrogen:

H—C≡N:

The single bond accounts for two electrons, and the triple bonds accounts for another six electrons. The remaining two electrons must be a lone pair on nitrogen. This accounts for all ten valence electrons, and it gives all atoms an octet.
Since the carbon atom has a triple bond, it must be *sp* hybridized, with linear geometry.

1.64. The molecular formula of cyclobutane is C_4H_8. Of the four structures shown, only structure (c) has the same molecular formula (C_4H_8).

1.65. The correct answer is (b). Each of the structures has two carbon atoms and one oxygen atom. However, only the second structure has an OH group. This compound will have an elevated boiling point, relative to the other three structures, because of hydrogen bonding.

1.66. The first statement (a) is the correct answer, because an oxygen atom has a negative charge, and the nitrogen atom has a positive charge, as shown here:

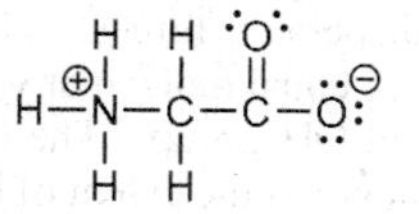

1.67. The correct answer is (d). Carbon-carbon σ bonds are formed by overlapping hybridized orbitals, so we must determine the hybridization of both carbon atoms (highlighted below). The carbon on the left has four σ bonds, so it is sp^3-hybridized, and the carbon on the right has three σ bonds, so it is sp^2-hybridized. The indicated σ bond results from the overlap of one sp^3 hybridized orbital (from the carbon atom on the left), and one sp^2 hybridized orbital (from the carbon atom on the right).

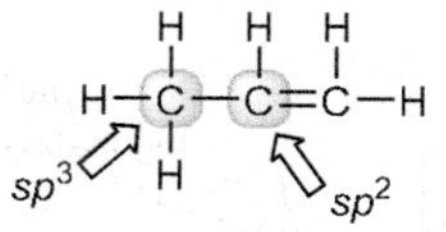

sp^3 sp^2

1.68. The correct answer is (c). An atom with a trigonal planar arrangement of bonds has a bond angle of 120°, the largest bond angle among the choices listed. A tetrahedral arrangement of bonds corresponds to an angle of 109.5°, while trigonal pyramidal and bent geometries typically have bond angles that are slightly smaller than 109.5°.

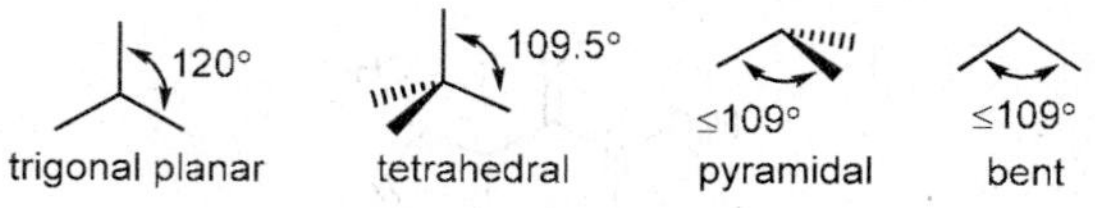

120° trigonal planar; 109.5° tetrahedral; ≤109° pyramidal; ≤109° bent

1.69. The correct answer is (c). There are 13 σ bonds in the given compound (highlighted in bold below). All of the single bonds are σ bonds, and the double bond is comprised of one σ bond and one π bond.

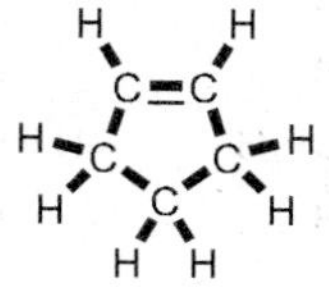

1.70. The correct answer is (b). Each C–Cl bond in **Y** has a dipole moment, and compound **Y** has a molecular dipole moment. Each C–Cl bond in **X** also has a dipole moment, but in this case, the two polar bonds are pointing in opposite directions. As such, the dipole moments fully cancel each other, giving no net molecular dipole moment for compound **X**. The polar isomer (**Y**) has more significant dipole-dipole attractions between molecules and, therefore, has a higher boiling point than the nonpolar isomer (**X**).

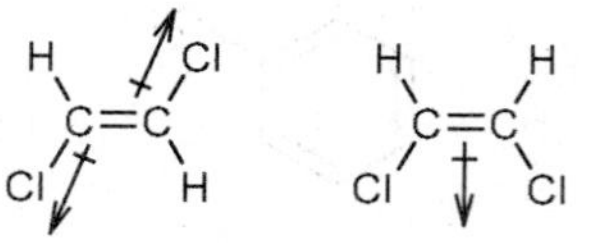

X has no net dipole

Y has a net dipole

1.71. The correct answer is (b). In the bond-line drawing, each corner and each endpoint represents a carbon atom, so this compound has three carbon atoms (highlighted below). Each carbon atom will have enough hydrogen atoms to have exactly four bonds. Together with the hydrogen atom that is connected to the oxygen atom, there are a total of six hydrogen atoms, as shown below:

1.72. The correct answer is (d). In the bond-line drawing, each corner represents a carbon atom (highlighted below), so this compound has nine carbon atoms. Each double bond is comprised of one σ bond and one π bond, so there are five π bonds (in bold below).

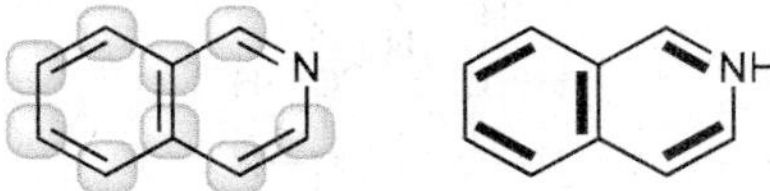

1.73.

(a) Compounds A and B share the same molecular formula (C_4H_9N) but differ in their constitution (connectivity of atoms), and they are therefore constitutional isomers.

(b) The nitrogen atom in compound B has three σ bonds and one lone pair (steric number = 4). It is sp^3 hybridized (electronically tetrahedral), with trigonal pyramidal geometry (because one corner of the tetrahedron is occupied by a lone pair).

(c) A double bond represents one σ bond and one π bond, while a triple bond represents one σ bond and two π bonds. A single bond represents a σ bond. With this in mind, compound B has 14 σ bonds, as compared with compounds A and C, which have 13 and 11 σ bonds, respectively.

(d) As explained in the solution to part (c), compound C has the fewest σ bonds.

(e) A double bond represents one σ bond and one π bond, while a triple bond represents one σ bond and two π bonds. As such, compound C exhibits two π bonds.

(f) Compound A has a C=N bond, in which the carbon atom has three σ bonds and no lone pairs (steric number = 3). It is sp^2 hybridized.

(g) Each of the carbon atoms in compound B is sp^3 hybridized with four σ bonds (steric number = 4). Similarly, the nitrogen atom in compound B has three σ bonds and one lone pair (steric number = 4). This nitrogen atom is also sp^3 hybridized.

(h) Compound A has an N–H bond, and is therefore expected to form hydrogen bonding interactions. Compounds B and C do not contain an N–H bond, so compound A is expected to have the highest boiling point.

1.74.

(a) In each of the following two compounds, all of the carbon atoms are sp^2 hybridized (each carbon atom has three σ bonds and one π bond). There are certainly many other possible compounds for which all of the carbon atoms are sp^2 hybridized.

(b) In each of the following two compounds, all of the carbon atoms are sp^3 hybridized (because each carbon atom has four σ bonds) with the exception of the carbon atom connected to the nitrogen atom. That carbon atom has two σ bonds and is therefore *sp* hybridized. There are certainly many other acceptable answers.

(c) In each of the following two compounds, there is a ring, and all of the carbon atoms are sp^3 hybridized (because each carbon atom has four σ bonds). There are certainly many other acceptable answers.

(d) In each of the following two compounds, all of the carbon atoms are *sp* hybridized (because each carbon atom has two σ bonds). There are certainly many other acceptable answers.

N≡C−C≡C−C≡C−C≡N

F−C≡C−C≡C−C≡C−F

1.75. In each of the following two compounds, the molecular formula is $C_4H_{10}N_2$, there is a ring (as suggested in the hint given in the problem statement), there are no π bonds, there is no net dipole moment, and there is an N-H bond, which enables hydrogen bonding interactions.

1.76. If we try to draw a linear skeleton with five carbon atoms and one nitrogen atom, we find that the number of hydrogen atoms is not correct (there are thirteen, rather than eleven):

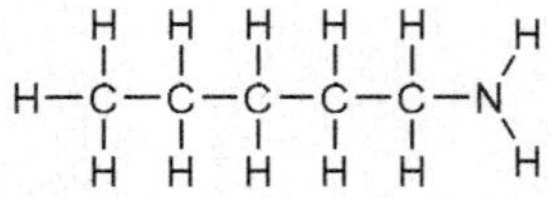

$C_5H_{13}N$

This will be the case even if try to draw a branched skeleton:

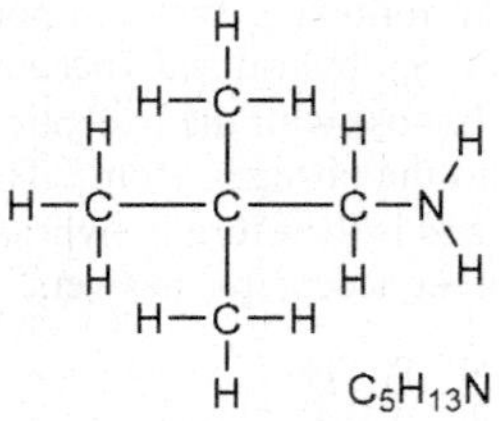

$C_5H_{13}N$

In fact, regardless of how the skeleton is branched, it will still have thirteen hydrogen atoms. But we need to draw a structure with only eleven hydrogen atoms ($C_5H_{11}N$). So we must remove two hydrogen atoms, which gives two unpaired electrons:

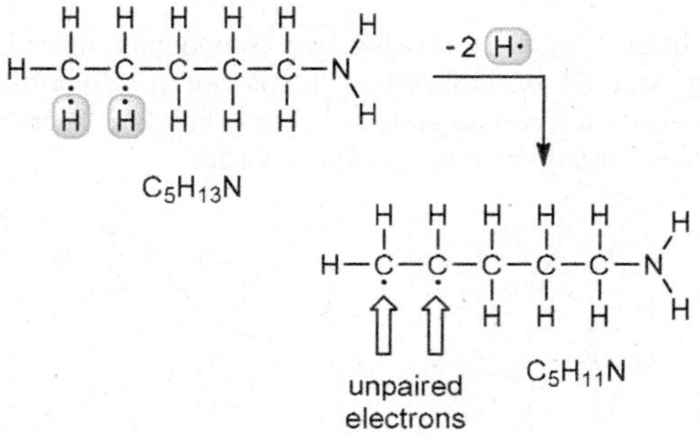

$C_5H_{13}N$ → -2 H· → $C_5H_{11}N$

unpaired electrons

This indicates that we should consider pairing these electrons as a double bond. However, the problem statement specifically indicates that the structure cannot contain a double bond. So, we must find another way to pair the unpaired electrons. Let's consider forming a ring instead of a double bond:

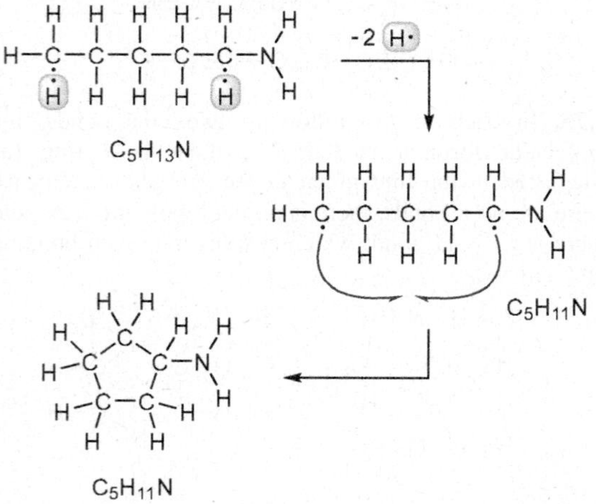

$C_5H_{13}N$ → -2 H· → $C_5H_{11}N$ → $C_5H_{11}N$

Now we have the correct number of hydrogen atoms (eleven), which means that our structure must indeed contain a ring. But this particular cyclic structure (cyclic = containing a ring) does not meet all of the criteria described in the problem statement. Specifically, each carbon atom must be connected to exactly two hydrogen atoms. This is not the case in the structure above. This issue can be remedied in the following structure, which has a ring, and each of the carbon atoms is connected to exactly two hydrogen atoms, as required by the problem statement.

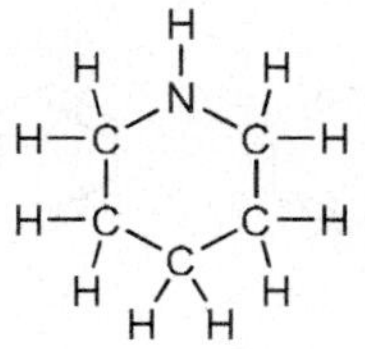

1.77.

(a) In compound **A**, the nitrogen atom has two σ bonds and no lone pairs (steric number = 2). It is *sp* hybridized. The highlighted carbon atom has one σ bond and one lone pair (steric number = 2), so that carbon atom is also *sp* hybridized.

(b) The highlighted carbon atom is *sp* hybridized, so the lone pair occupies an *sp*-hybridized orbital.

(c) The nitrogen atom is *sp* hybridized and therefore has linear geometry. As such, the C-N-C bond angle in **A** is expected to be 180°.

(d) The nitrogen atom in **B** has two σ bonds and one lone pair (steric number = 3). It is sp^2 hybridized. The highlighted carbon atom has three σ bonds and no lone pairs (steric number = 3), and that carbon atom is sp^2 hybridized. Each of the chlorine atoms has three lone pairs and one bond (steric number = 4), and the chlorine atoms are sp^3 hybridized.

(e) The nitrogen atom is sp^2 hybridized, so the lone pair occupies an sp^2-hybridized orbital.

(f) The nitrogen atom is sp^2 hybridized so the C-N-C bond angle in **B** is expected to be approximately 120°.

1.78. By analyzing the data, we can see that C(sp^2)–Cl must be shorter than 1.79Å [compare with C(sp^3)–Cl], while C(*sp*)–I must be longer than 1.79Å [compare with C(*sp*)–Br]. Therefore, C(*sp*)–I must be longer than C(sp^2)–Cl.

1.79.

(a) In the first compound, the fluorine isotope (^{18}F) has no formal charge. Therefore, it must have three lone pairs (see Section 1.4 for a review of how formal charges are calculated). Since it has one σ bond and three lone pairs, it must have a steric number of 4, and is sp^3 hybridized. The bromine atom also has no formal charge. So, it too, like the fluorine isotope, must have three lone pairs. Once again, one σ bond and three lone pairs give a steric number of 4, so the bromine atom is sp^3 hybridized.

In the second compound, the nitrogen atom has no formal charge. Therefore, it must have one lone pair. Since the nitrogen atom has three σ bonds and one lone pair, it must have a steric number of 4, and is sp^3 hybridized.
In the product, the fluorine isotope (^{18}F) has no formal charge. Therefore, it must have three lone pairs. Since it has one σ bond and three lone pairs, it must have a steric number of 4, and is sp^3 hybridized. The nitrogen atom does have a positive formal charge. Therefore, it must have no lone pairs. Since it has four σ bonds and no lone pairs, it must have a steric number of 4, and is sp^3 hybridized. Finally, the bromine atom has a negative charge and no bonds. So it must have four lone pairs. With four lone pairs and no bonds, it will have a steric number of 4, and is expected to be sp^3 hybridized.
In summary, all of the atoms that we analyzed are sp^3 hybridized.

(b) The nitrogen atom is sp^3 hybridized. With four bonds, we expect the geometry around the nitrogen atom to be tetrahedral. So, the bond angle for each C-N-C bond is expected to be approximately 109.5°.

1.80.
(a) Boron is in group 3A of the periodic table and is therefore expected to be trivalent. That is, it has three valence electrons, and it uses each one of those valence electrons to form a bond, giving rise to three bonds. It does not have any electrons left over for a lone pair (as in the case of nitrogen). With three σ bonds and no lone pairs, the boron atom has a steric number of three, and is sp^2 hybridized.

(b) Since the boron atom is sp^2 hybridized with three σ bonds, we expect the geometry to be trigonal planar and the bond angles to be approximately 120°. However, in this case, the O-B-O system is part of a five-membered ring. That is, there are five different bond angles (of which the O-B-O angle is one of them) that together must form a closed loop. That requirement could conceivably force some of the bond angles (including the O-B-O bond angle) to deviate from the predicted value. In fact, we will explore this very phenomenon, called ring strain, in Chapter 4, and we will see that five-membered rings actually possess very little ring strain compared with smaller rings.

(c) Each of the oxygen atoms has no formal charge, and must therefore have two bonds and two lone pairs. The boron atom has no lone pairs, as explained in the solution to part (a) of this problem.

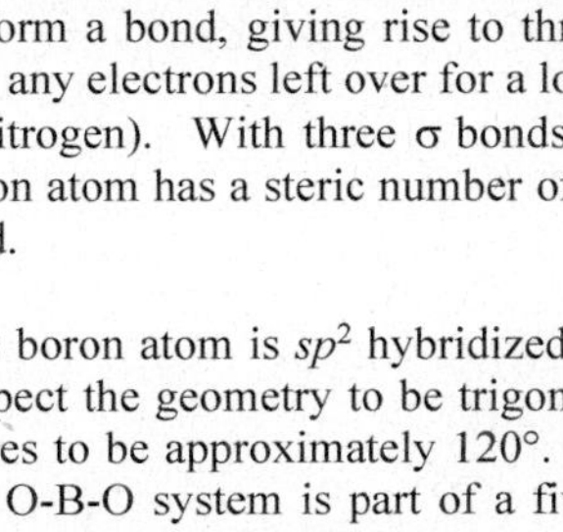

1.81.
(a) If we analyze each atom (in both **1** and **2**) using the procedure outlined in Section 1.4, we find that none of the atoms in compound **1** have a formal charge, while compound **2** possesses two formal charges:

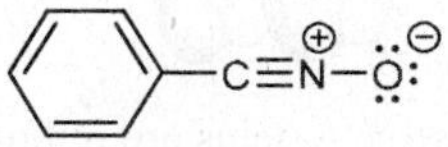

The nitrogen atom has a positive charge (it should have five valence electrons, but it is exhibiting only four), and the oxygen atom has a negative charge (it should have six valence electrons, but it is exhibiting seven).

(b) Compound **1** possesses polar bonds, as a result of the presence of partial charges (δ+ and δ-). The associated dipole moments can form favorable interactions with the dipole moments present in the polar solvent molecules (dipole-dipole interactions). However, compound **2** has formal charges (negative on O and positive on N), so the dipole moment of the N-O bond is expected to be much more significant than the dipole moments in compound **1**. The dipole moment of the N-O bond in compound **2** is the result of full charges, rather than partial charges. As such, compound **2** is expected to experience much stronger interactions with the solvent molecules, and therefore, **2** should be more soluble than **1** in a polar solvent.

(c) In compound **1**, the carbon atom (attached to nitrogen) has three σ bonds and no lone pairs (steric number = 3). That carbon atom is sp^2 hybridized, with trigonal planar geometry. As such, the C-C-N bond angle in compound **1** is expected to be approximately 120°. However, in compound **2**, the same carbon atom has two σ bonds and no lone pairs (steric number = 2). This carbon atom is *sp* hybridized, with linear geometry. As such, the C-C-N bond angle in **2** is expected to be 180°. The conversion of **1** to **2** therefore involves an increase in the C-C-N bond angle of approximately 60°.

1.82.
(a) C_a has three σ bonds and no lone pairs, so it has a steric number of 3, and is sp^2 hybridized. The same is true for C_c. In contrast, C_b has two σ bonds and no lone pairs, so it has a steric number of 2, and is therefore *sp* hybridized.

(b) Since C_a is sp^2 hybridized, we expect its geometry to be trigonal planar, so the bond angle should be approximately 120°.

(c) Since C_b is *sp* hybridized, we expect its geometry to be linear, so the bond angle should be approximately 180°.

(d) The central carbon atom (C_b) is *sp* hybridized, so it is using two *sp* hybridized orbitals to form its two σ bonds, which will be arranged in a linear fashion. The remaining two *p* orbitals of C_b used for π bonding will be 90° apart from one another (just as we saw for the carbon atoms of a triple bond; see Figure 1.33).

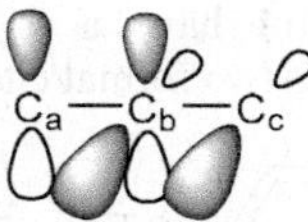

As a result, the two π systems are orthogonal (or 90°) to each other. Therefore, the p orbitals on C_a and C_c are orthogonal. The following is another drawing from a different perspective (looking down the axis of the linear C_a-C_b-C_c system.

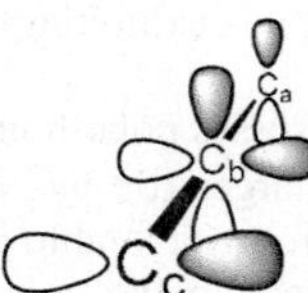

1.83.

(a) The N-C-N unit highlighted below exhibits a central carbon atom that is sp^3 hybridized and is therefore expected to have tetrahedral geometry. Accordingly, the bond angles about that carbon atom are expected to be approximately 109.5°.

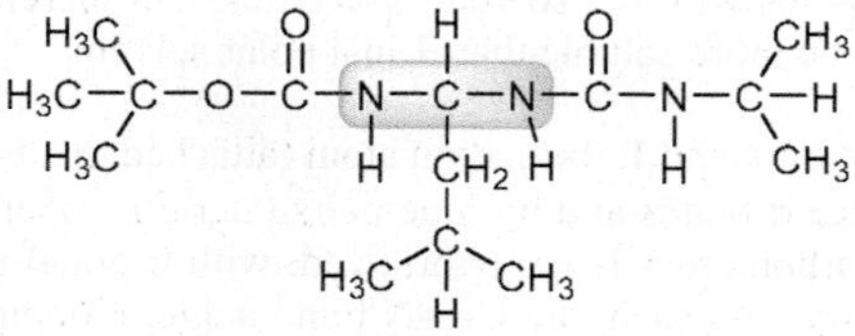

The other N-C-N unit (highlighted below) exhibits a central carbon atom that is sp^2 hybridized and is therefore expected to have trigonal planar geometry. Accordingly, the bond angles about that carbon atom are expected to be approximately 120°.

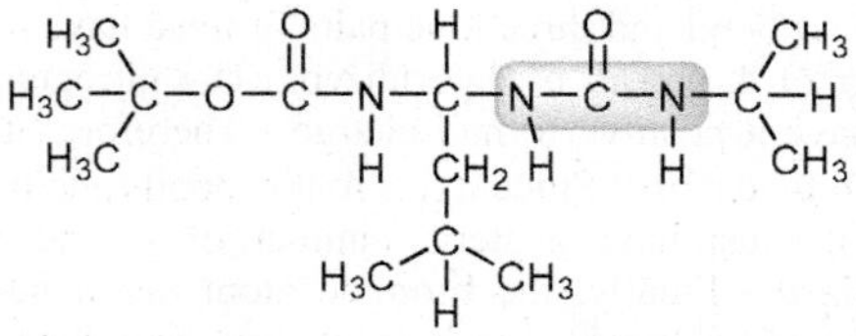

(b) The interaction is an intramolecular hydrogen bond that forms between the δ+ H (connected to the highlighted nitrogen atom) and the lone pair of the δ– oxygen atom:

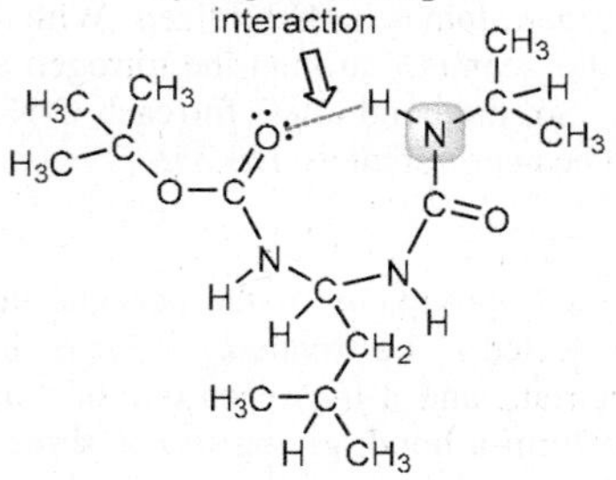

Chapter 2
Molecular Representations

Review of Concepts

Fill in the blanks below. To verify that your answers are correct, look in your textbook at the end of Chapter 2. Each of the sentences below appears verbatim in the section entitled *Review of Concepts and Vocabulary.*

- In **bond-line structures,** ________atoms and most ________ atoms are not drawn.
- A ________________ is a characteristic group of atoms/bonds that show a predictable behavior.
- When a carbon atom bears either a positive charge or a negative charge, it will have ____________, rather than four, bonds.
- In bond-line structures, a **wedge** represents a group coming ______ the page, while a **dash** represents a group __________ the page.
- ____________ **arrows** are tools for drawing resonance structures.
- When drawing curved arrows for resonance structures, avoid breaking a ________ bond and never exceed ______________ for second-row elements.
- The following rules can be used to identify the significance of resonance structures:
    1. The most significant resonance forms have the greatest number of filled ___________.
    2. The structure with fewer _________________ is more significant.
    3. Other things being equal, a structure with a negative charge on the more _____________ element will be more significant. Similarly, a positive charge will be more stable on the less ______________ element.
    4. Resonance forms that have equally good Lewis structures are described as ____________ and contribute equally to the resonance hybrid.
- A ______________ lone pair participates in resonance and is said to occupy a ____ orbital.
- A ______________ lone pair does not participate in resonance.

Review of Skills

Fill in the blanks and empty boxes below. To verify that your answers are correct, look in your textbook at the end of Chapter 2. The answers appear in the section entitled *SkillBuilder Review.*

SkillBuilder 2.1 Converting a Condensed Structure into a Lewis Structure

Draw a Lewis structure of the following compound

$(CH_3)_2CC(CH_3)OCH_2CH(CH_3)CHO$ ⟶

Lewis Structure

SkillBuilder 2.2 Drawing Bond-Line Structures

Draw a bond-line drawing of the following compound:

SkillBuilder 2.3 Identifying Lone Pairs on Oxygen Atoms

An oxygen atom with a negative charge will have ______ lone pair(s).	An oxygen atom with no formal charge will have ______ lone pair(s).	An oxygen atom with a positive charge will have ______ lone pair(s).

SkillBuilder 2.4 Identifying Lone Pairs on Nitrogen Atoms

A nitrogen atom with a negative charge will have ______ lone pair(s).	A nitrogen atom with no formal charge will have ______ lone pair(s).	A nitrogen atom with a positive charge will have ______ lone pair(s).

SkillBuilder 2.5 Identifying Valid Resonance Arrows

Rule 1 The tail of a curved arrow cannot be placed on a ______.	**Rule 2** The head of a curved arrow cannot result in ______.
TAIL	HEAD

SkillBuilder 2.6 Assigning Formal Charges in Resonance Structures

Indicate the location of the negative charge in the second resonance structure below:

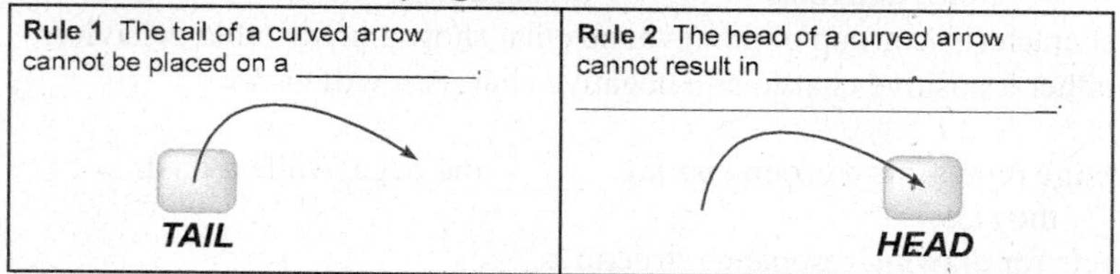

SkillBuilder 2.7 Ranking the Significance of Resonance Structures

Step 1 The most significant resonance forms have the greatest number of filled octets. Identify which form below is the most significant.	**Step 2** The structure with the fewer formal charges is more significant. Based on steps one and two, identify which resonance form below is the most significant and which is the least significant.	**Step 3** Other things being equal, the most significant resonance structure has the negative charge on the more electronegative atom. Consider the location of the formal charge(s), and identify the more significant contributor below.
		Step 4 Rank resonance forms. As a general rule, if one or more of the resonance forms has all filled octets, then any resonance form missing an octet will be a minor contributor.

SkillBuilder 2.8 Drawing a Resonance Hybrid

Steps 1 and 2 After drawing all resonance structures, identify which one is more significant.	**Step 3** Redraw the structure, showing partial bonds and partial charges.	**Step 4** Revise the size of the partial charges to indicate distribution of electron density.

SkillBuilder 2.9 Identifying Localized and Delocalized Lone Pairs

Identify whether the lone pair on the nitrogen atom below is delocalized.	Identify the hybridization state of the nitrogen atom.

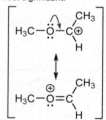

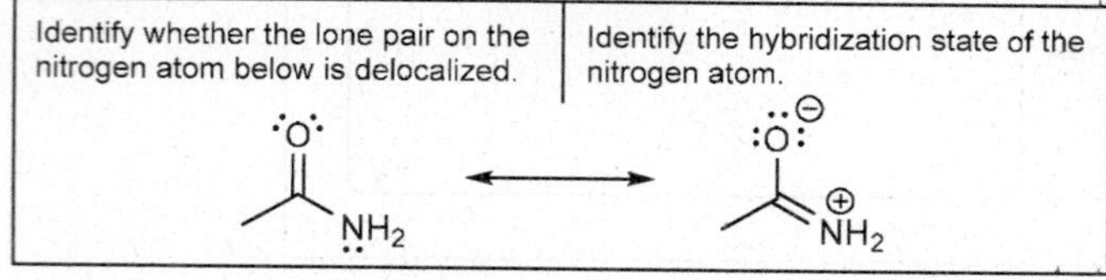

Common Mistakes to Avoid

When drawing a structure, make sure to avoid drawing a pentavalent, hexavalent, or heptavalent carbon atom:

INCORRECT	INCORRECT	INCORRECT
	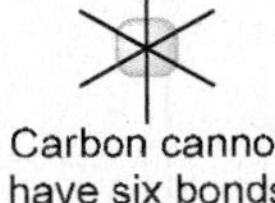	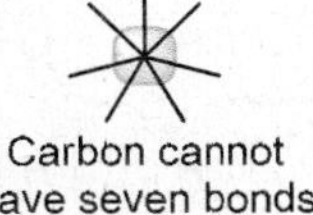
Carbon cannot have five bonds	Carbon cannot have six bonds	Carbon cannot have seven bonds

Carbon cannot have more than four bonds. *Never draw a carbon atom with more than four bonds!* While this type of mistake is not uncommon among new students of organic chemistry, it is important to correct it right away, so that it does not lead to further confusion.
Also, when drawing a structure, either draw all carbon atom labels (C) and all hydrogen atom labels (H), like this Lewis structure:

```
  H   H   H   H
  |   |   |   |
H-C---C---C---C-O-H
  |   |   |   |
  H   H   H   H
```

or don't draw any labels (except H attached to a heteroatom), like this bond-line structure:

OH

That is, if you draw all C labels, then you should draw all H labels also. Avoid drawings in which the C labels are drawn and the H labels are not, as shown here:

INCORRECT	INCORRECT
C–C–C–C–OH	C C C C OH

These types of drawings (where C labels are shown and H labels are not shown) should only be used when you are working on a scratch piece of paper and trying to draw constitutional isomers. For example, if you are considering all constitutional isomers with the molecular formula C_3H_8O, you might find it helpful to use drawings like these as a form of "short-hand" so that you can identify all of the different ways of connecting three carbon atoms and one oxygen atom:

O–C–C–C   C–C(O)–C   C–O–C–C

But your final structures should either show all C and H labels, or no labels at all. The latter is the more commonly used method:

HO   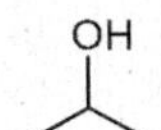

O

Solutions

2.1.

(a) We begin by drawing the carbon chain and any atoms attached to the carbon chain, and then we look for atoms that do not have the correct number of bonds (highlighted):

$H_2CCHOCH_2CH(CH_3)_2$ ⟶

To resolve this issue, we draw a pi bond between the two carbon atoms. Finally, we draw the lone pairs on the oxygen atom, giving the following Lewis structure, in which all atoms have filled octets:

(b) We begin by drawing the carbon chain and any atoms attached to the carbon chain:

$(CH_3CH_2)_2CHCH_2CH_2OH$ ⟶

In the structure drawn above, all atoms have the correct number of bonds. So, we complete the Lewis structure by drawing the lone pairs on the oxygen atom, so that the oxygen atom has a filled octet of electrons:

(c) We begin by drawing the carbon chain and any atoms attached to the carbon chain:

$(CH_3CH_2)_3COH$ ⟶

In the structure drawn above, all atoms have the correct number of bonds. So, we complete the Lewis structure by drawing the lone pairs on the oxygen atom, so that the oxygen atom has a filled octet of electrons:

(d) We begin by drawing the carbon chain and any atoms attached to the carbon chain, and then we look for atoms that do not have the correct number of bonds (highlighted):

$(CH_3)_2CC(CH_3)OCH_3$ ⟶

To resolve this issue, we draw a pi bond between the two carbon atoms. Finally, we draw the lone pairs on the oxygen atom, giving the following Lewis structure, in which all atoms have filled octets:

(e) We begin by drawing the carbon chain and any atoms attached to the carbon chain:

$CH_3CH_2C(CH_3)_2CH_2CH_3$ ⟶

In the structure drawn above, all atoms have the correct number of bonds, and there are no atoms with lone pairs, so this is the correct Lewis structure.

(f) We begin by drawing the carbon chain and any atoms attached to the carbon chain:

$CH_3CHBrCH_2CH_3$ ⟶

In the structure drawn above, all atoms have the correct number of bonds. So, we complete the Lewis structure by drawing the lone pairs on the bromine atom, so that the bromine atom has a filled octet of electrons:

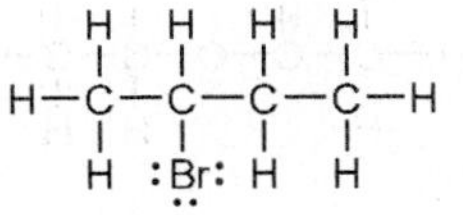

2.2.

(a) We begin by drawing the carbon chain and any atoms attached to the carbon chain:

$HOCH_2CH_2CH_2CHO$ →

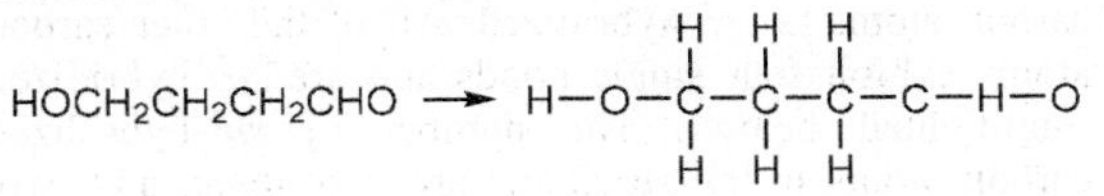

Before we can deal with missing octets, we must correct the sigma bond framework shown, because it cannot be correct. Notice that there is a hydrogen atom on the right side of the structure that appears to have two bonds:

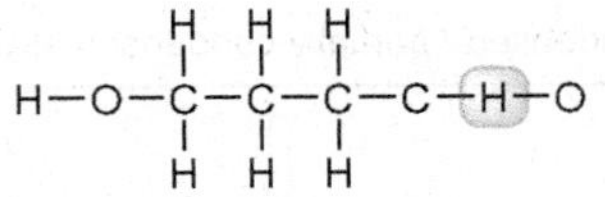

This is not possible. Hydrogen is monovalent – it can only form one bond, so let's look more closely at the condensed structure, and see where we went wrong. The condensed structure ends with -CHO. In this grouping of atoms, notice that H is listed immediately after C, indicating that the H is connected directly to the C. Then, O is listed next, but it cannot be connected to the H (because hydrogen cannot have more than one bond), so it is understood that the O must also be connected directly to the C, like this:

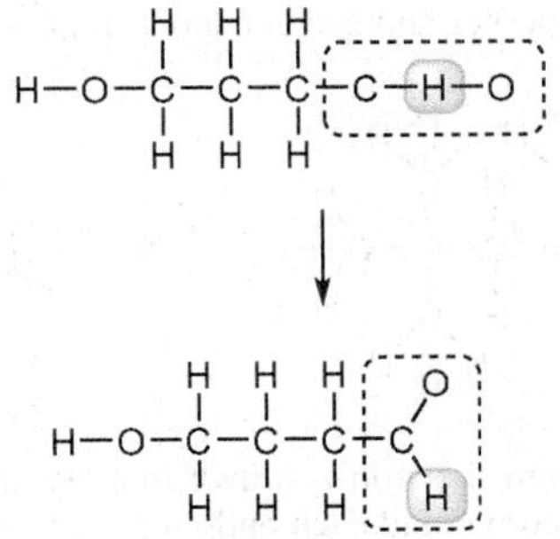

Now that we have drawn a framework of sigma bonds, the next step is to look for any atoms that do not have the correct number of bonds:

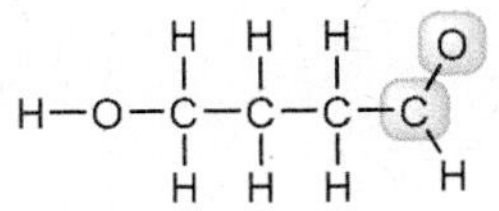

The highlighted carbon has only three bonds (it should have four bonds), and the highlighted oxygen atom has only one bond (it should have two bonds). This issue can be resolved by drawing a pi bond between the carbon and oxygen atoms:

π bond

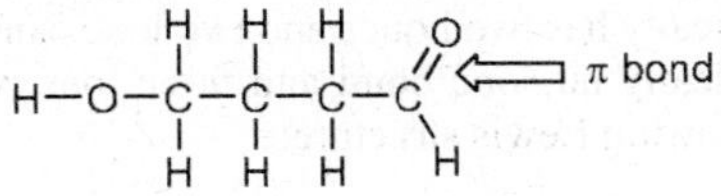

Finally, we complete the octets on each oxygen atom by drawing the lone pairs (an uncharged oxygen atom has two bonds and two lone pairs), giving the following Lewis structure:

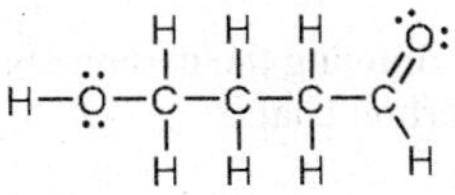

(b) We begin by drawing the carbon chain and any atoms attached to the carbon chain:

Cl_3CCO_2H →

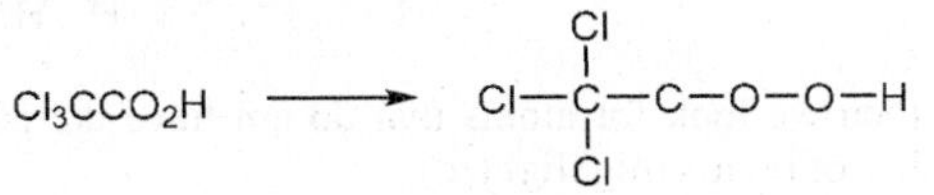

and then we look for atoms that do not have the correct number of bonds (highlighted):

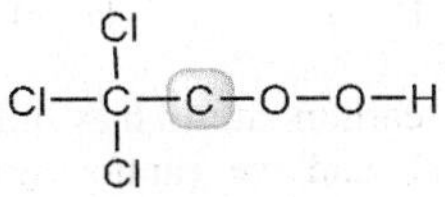

The highlighted carbon atom has only two bonds (it should have four), and we cannot introduce pi bonds, because both neighboring atoms already have the correct number of bonds. So we must consider whether we have incorrectly drawn the framework of sigma bonds. Let's look more closely at the condensed structure, which ends with $-CO_2H$. In this grouping of atoms, there are two oxygen atoms listed. Certainly the first one is connected to C, but let's consider the possibility that the second oxygen atom is also connected to C, like this:

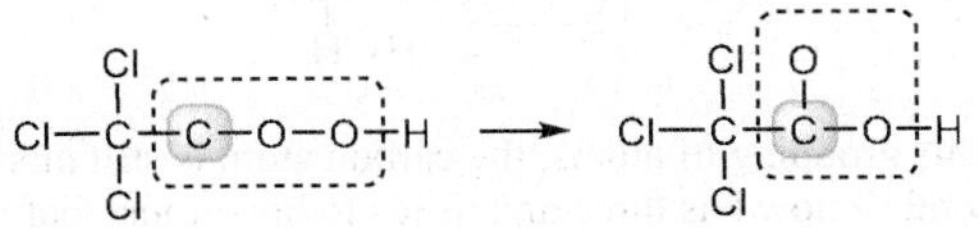

In this grouping of atoms, the carbon atom is still missing a bond (it now has three and still needs to have four), but now the neighboring oxygen atom is also missing a bond,

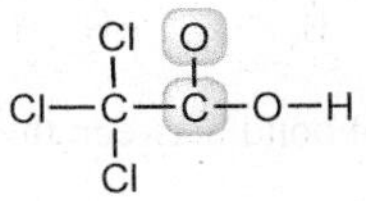

so we can draw a pi bond between the carbon and oxygen atoms:

π bond

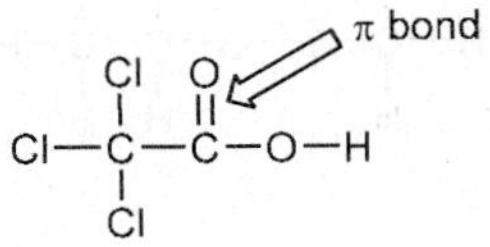

Finally, we complete the octets on each oxygen atom and each chlorine atom by drawing the lone pairs (oxygen typically has two bonds and two lone pairs, while chlorine typically has one bond and three lone pairs), giving the following Lewis structure:

(c) We begin by drawing the carbon chain and any atoms attached to the carbon chain:

$CH_3CO_2CH_2CH_3$ ⟶

and then we look for atoms that do not have the correct number of bonds (highlighted):

The highlighted carbon atom has only two bonds (it should have four), and we cannot introduce pi bonds, because both neighboring atoms already have the correct number of bonds. So we must consider whether we have incorrectly drawn the framework of sigma bonds. Let's look more closely at the condensed structure, which has a $-CO_2-$ group. In this grouping of atoms, there are two oxygen atoms listed. Certainly, the first one is connected to C, but let's consider the possibility that the second oxygen atom is also connected to C, like this:

In this grouping of atoms, the carbon atom is still missing a bond (it now has three and needs to have four), but now the neighboring oxygen atom is also missing a bond,

so we can draw a pi bond between the carbon and oxygen atoms:

π bond

Finally, we complete the octets on each oxygen atom by drawing the lone pairs (oxygen typically has two bonds and two lone pairs), giving the following Lewis structure:

2.3. Begin by drawing a Lewis structure for each isomer, so that the bonding of the carbon atoms is shown more clearly. Notice that in two of the isomers, a carbon atom is sharing a double bond with oxygen. Each of these carbon atoms is sp^2 hybridized. All of the other carbon atoms exhibit four single bonds and are sp^3 hybridized (highlighted below). The number of sp^3-hybridized carbon atoms in the structures are two, three, and two, respectively:

$(CH_3)_2CO$ $\quad$ H_2C, H_2C CHOH $\quad$ CH_3CH_2CHO

condensed / partially condensed structures

⇩

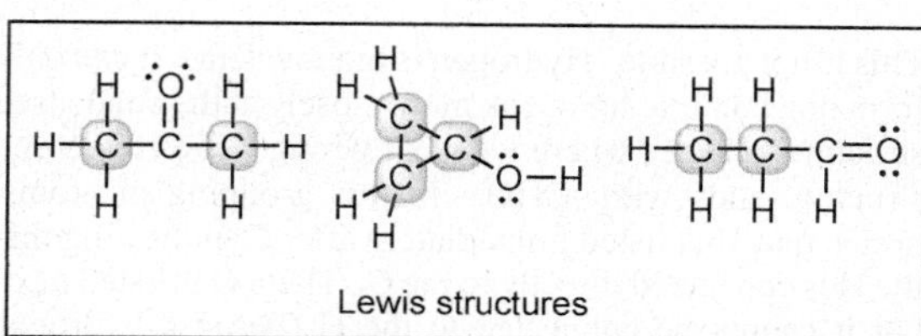

2.4.

(a) The carbon skeleton is drawn in a zig-zag format, in which each corner and each endpoint represents a carbon atom:

(b) The carbon skeleton is drawn in a zig-zag format, in which each corner and each endpoint represents a carbon atom:

(c) The carbon skeleton is drawn in a zig-zag format, in which each corner and each endpoint represents a carbon atom. Hydrogen atoms generally don't have to be drawn, but make sure to draw any hydrogen atoms that are connected to a heteroatom (such as oxygen). In this structure, there are three such hydrogen atoms, and all three must be drawn:

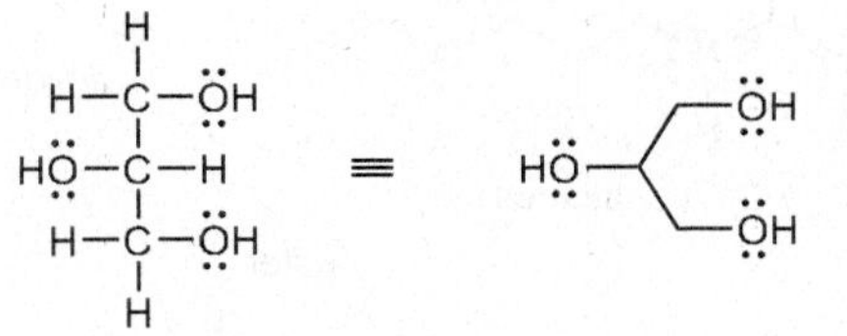

2.5.

(a) The carbon skeleton is drawn in a zig-zag format, in which each corner and each endpoint represents a carbon atom:

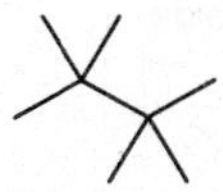

$(CH_3)_3CC(CH_3)_3$

(b) The carbon skeleton is drawn in a zig-zag format, in which each corner and each endpoint represents a carbon atom:

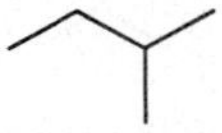

$CH_3CH_2CH(CH_3)_2$

(c) The carbon skeleton is drawn in a zig-zag format, in which each corner and each endpoint represents a carbon atom. The hydrogen atom connected to oxygen must be drawn:

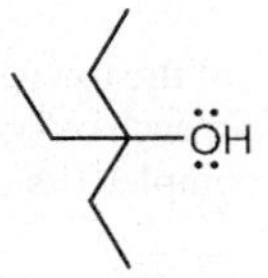

$(CH_3CH_2)_3COH$

(d) The carbon skeleton is drawn in a zig-zag format, in which each corner and each endpoint represents a carbon atom. The hydrogen atom connected to oxygen must be drawn:

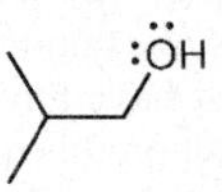

$(CH_3)_2CHCH_2OH$

(e) The carbon skeleton is drawn in a zig-zag format, in which each corner and each endpoint represents a carbon atom:

$CH_3CH_2CH_2OCH_3$

(f) The carbon skeleton is drawn in a zig-zag format, in which each corner and each endpoint represents a carbon atom. In this case, a pi bond must be drawn between two carbon atoms, so that all atoms have filled octets:

$(CH_3CH_2)_2CCH_2$

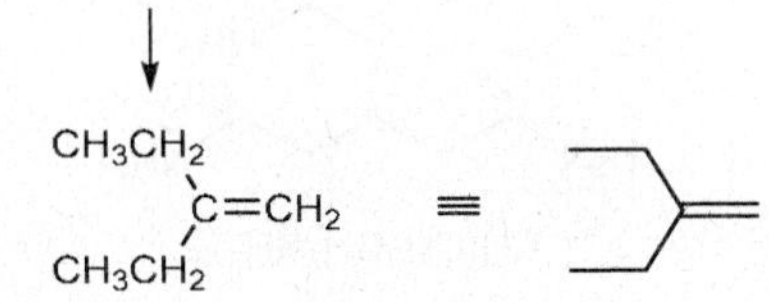

(g) The carbon skeleton is drawn in a zig-zag format, in which each corner and each endpoint represents a carbon atom. In this case, a pi bond must be drawn between two carbon atoms, so that all atoms have filled octets:

$H_2CCHOCH_2CH(CH_3)_2$

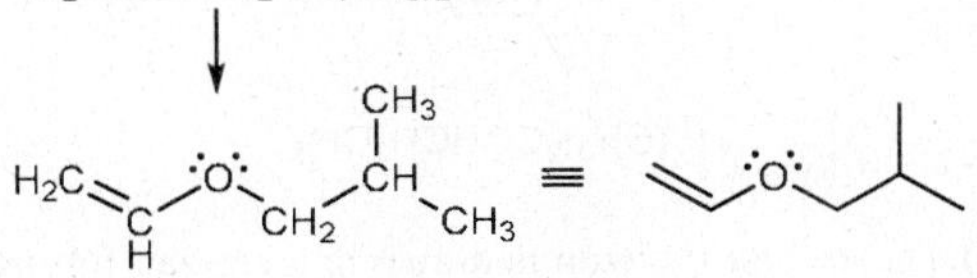

(h) The carbon skeleton is drawn in a zig-zag format, in which each corner and each endpoint represents a carbon atom. The hydrogen atom connected to oxygen must be drawn:

$CH_3(CH_2)_6OH$

(i) The carbon skeleton is drawn in a zig-zag format, in which each corner and each endpoint represents a carbon atom. In this case, a pi bond must be drawn between two carbon atoms, so that all atoms have filled octets:

$H_2CCHCH_2OCH_2CH(CH_3)_2$

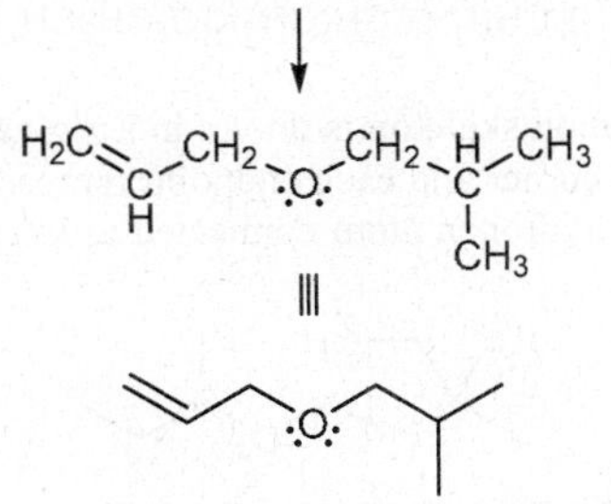

(j) The carbon skeleton is drawn in a zig-zag format, in which each corner and each endpoint represents a carbon atom:

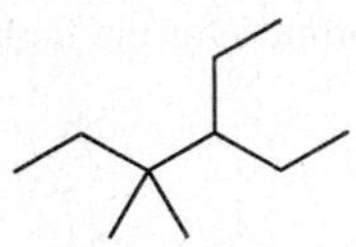

$CH_3CH_2C(CH_3)_2CH(CH_2CH_3)_2$

(k) The carbon skeleton is drawn in a zig-zag format, in which each corner and each endpoint represents a carbon atom. Hydrogen atoms generally don't have to be drawn, but make sure to draw any hydrogen atoms that are connected to a heteroatom (such as nitrogen). In this structure, there are many hydrogen atoms that must be drawn:

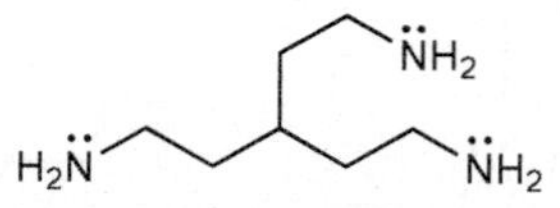

$CH(CH_2CH_2NH_2)_3$

(l) The carbon skeleton is drawn in a zig-zag format, in which each corner and each endpoint represents a carbon atom. In this case, a pi bond must be drawn between two carbon atoms, so that all atoms have filled octets:

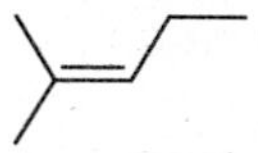

$(CH_3)_2CCHCH_2CH_3$

(m) The carbon skeleton is drawn in a zig-zag format, in which each corner and each endpoint represents a carbon atom. Notice that there is a $-CO_2-$ group in the condensed structure. If you try to draw this group in a linear fashion (-C-O-O-), you will find that the carbon atom does not have the correct number of bonds. This is resolved by drawing both oxygen atoms being connected to the same carbon atom, and by also drawing a pi bond between the carbon atom and one of the oxygen atoms, as shown below. In this way, all atoms have filled octets:

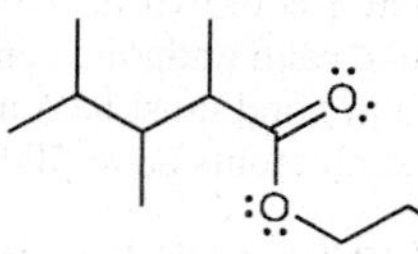

$(CH_3)_2CH(CH_3)CH(CH_3)CO_2CH_2CH_2CH_3$

(n) The carbon skeleton is drawn in a zig-zag format, in which each corner and each endpoint represents a carbon atom. The hydrogen atom connected to oxygen must be drawn:

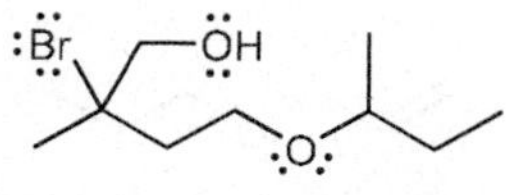

$CH_3CBr(CH_2OH)CH_2CH_2OCH(CH_3)CH_2CH_3$

2.6. The carbon skeleton is drawn in a zig-zag format, in which each corner and endpoint represents a carbon atom. The condensed formula for the highlighted substituent is provided below:

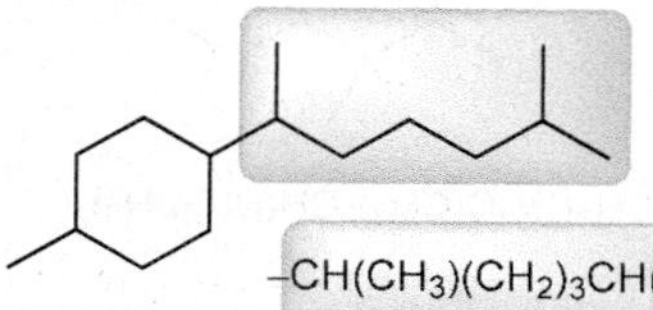

$-CH(CH_3)(CH_2)_3CH(CH_3)_2$

2.7. The functional groups in the following compounds are highlighted and identified, using the terminology found in Table 2.1.

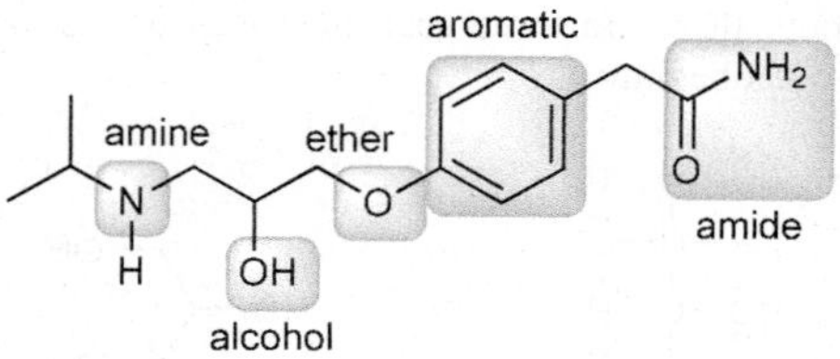

aromatic
amine
ether
amide
alcohol

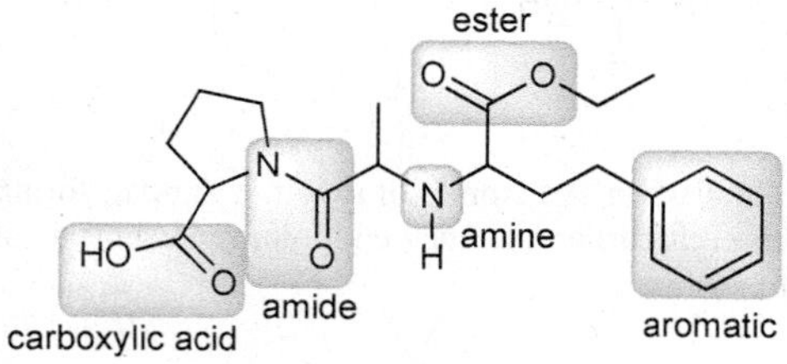

ester
amine
amide
carboxylic acid
aromatic

2.8.

(a) In this case, the oxygen atom has two bonds and no formal charge, so it must have two lone pairs in order to complete its octet (see Table 2.2).

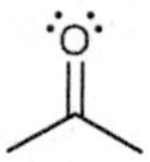

(b) In this case, each of the oxygen atoms has two bonds and no formal charge, so each oxygen atom must have two lone pairs in order to complete its octet (see Table 2.2).

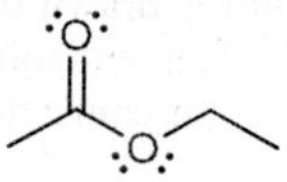

(c) In this case, each of the oxygen atoms has two bonds and no formal charge, so each oxygen atom must have two lone pairs in order to complete its octet (see Table 2.2).

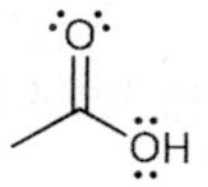

(d) One of the oxygen atoms has two bonds and no formal charge, so that oxygen atom must have two lone pairs in order to complete its octet (see Table 2.2). The other oxygen atom has one bond and a negative charge, so that oxygen atom must have three lone pairs. Three lone pairs complete the octet on this oxygen atom, and it has a negative charge because it exhibits seven electrons.

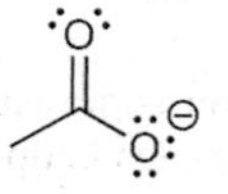

(e) In this case, the oxygen atom has one bond and a negative charge, so it must have three lone pairs (see Table 2.2). Three lone pairs complete the octet on the oxygen atom, and it has a negative charge because it exhibits seven electrons.

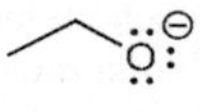

(f) In this case, the oxygen atom has two bonds and no formal charge, so it must have two lone pairs in order to fill its octet (see Table 2.2).

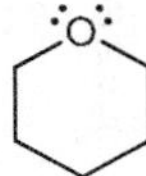

(g) In this case, the oxygen atom has three bonds and a positive charge, so it must have one lone pair (see Table 2.2). One lone pair completes the octet on the oxygen atom, and it has a positive charge because it exhibits only five electrons.

(h) In this case, the oxygen atom has three bonds and a positive charge, so it must have one lone pair (see Table 2.2). One lone pair completes the octet on the oxygen atom, and it has a positive charge because it exhibits only five electrons.

(i) From left to right, the first oxygen atom has two bonds and no formal charge, so it must have two lone pairs to complete its octet (see Table 2.2). The second oxygen atom has three bonds and a positive charge, so it must have one lone pair. One lone pair completes the octet on the second oxygen atom, and it has a positive charge because it exhibits only five electrons. Finally, the third oxygen atom has one bond and a negative charge, so it has three lone pairs. Three lone pairs complete the octet on the third oxygen atom, and it has a negative charge because it exhibits seven electrons.

(j) One of the oxygen atoms has two bonds and no formal charge, so that oxygen atom must have two lone pairs to complete its octet (see Table 2.2). The other oxygen atom has three bonds and a positive charge, so that oxygen atom must have one lone pair. One lone pair completes the octet on the oxygen atom, and it has a positive charge because it exhibits only five electrons.

2.9. Each oxygen atom in hydroxymethylfurfural lacks a charge and has two bonds, so each oxygen atom must have two lone pairs to complete its octet.

2.10.

(a) In this case, the nitrogen atom has three bonds and no formal charge, so it must have one lone pair to complete its octet (see Table 2.3).

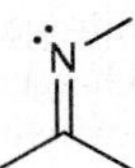

(b) In this case, the nitrogen atom has three bonds and no formal charge, so it must have one lone pair to complete its octet (see Table 2.3).

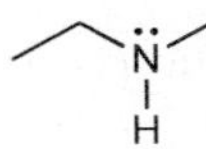

(c) In this case, the nitrogen atom has three bonds and no formal charge, so it must have one lone pair to complete its octet (see Table 2.3).

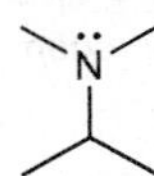

(d) In this case, the nitrogen atom has four bonds and a positive charge, so it must have no lone pairs (see Table 2.3). Four bonds complete the octet on the nitrogen atom, and it has a positive charge because it exhibits only four electrons.

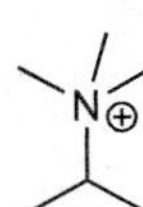

(e) In this case, the nitrogen atom has two bonds and a negative charge, so it must have two lone pairs (see Table 2.3). Two lone pairs complete the octet on the nitrogen atom, and it has a negative charge because it exhibits six electrons.

(f) In this case, the nitrogen atom has three bonds and no formal charge, so it must have one lone pair in order to complete its octet (see Table 2.3).

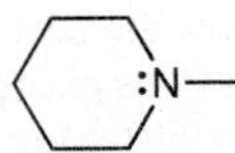

(g) In this case, the nitrogen atom has four bonds and a positive charge, so it must have no lone pairs (see Table 2.3). Four bonds complete the octet on the nitrogen atom, and it has a positive charge because it exhibits only four electrons.

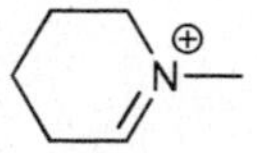

(h) One of the nitrogen atoms has four bonds and a positive charge, so it must have no lone pairs (see Table 2.3). Four bonds complete the octet on the nitrogen atom, and it has a positive charge because it exhibits only four electrons. The other nitrogen atom has three bonds and no formal charge, so it must have one lone pair to complete its octet.

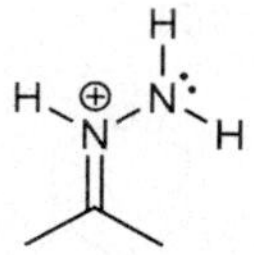

2.11. Every uncharged nitrogen atom in this compound has three bonds and needs one lone pair of electrons to fill its octet. Every positively charged nitrogen atom has four bonds and no lone pairs. Four bonds complete the octet on a nitrogen atom, and it is positively charged because it exhibits only four electrons. The missing lone pairs that have been added to nitrogen atoms are highlighted below:

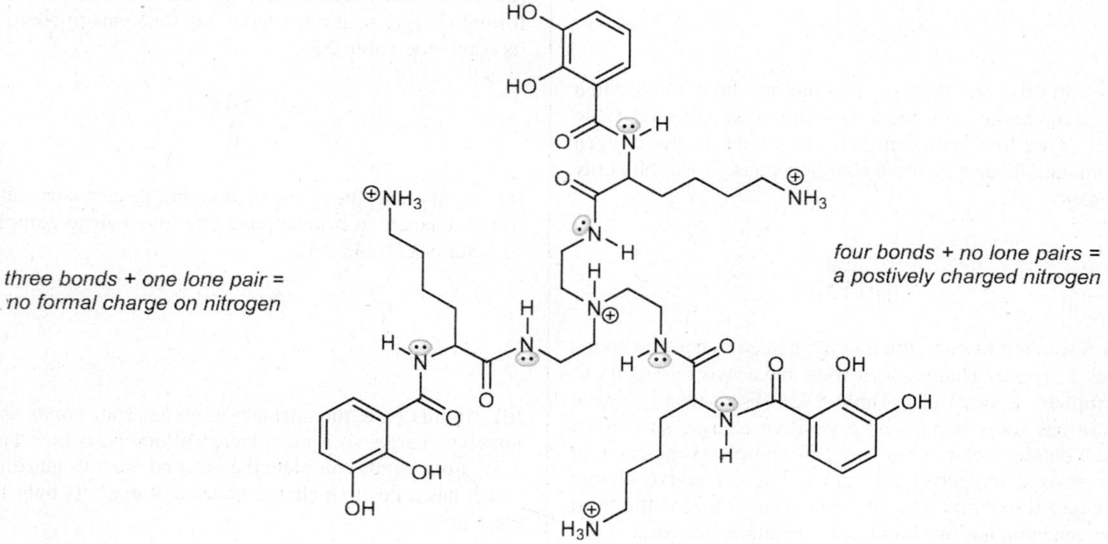

three bonds + one lone pair = no formal charge on nitrogen

four bonds + no lone pairs = a postively charged nitrogen

2.12.
(a) This curved arrow violates the second rule by giving a fifth bond to a nitrogen atom (thus exceeding its octet).
(b) This curved arrow does not violate either rule.
(c) This curved arrow violates the second rule by giving five bonds to a carbon atom (thus exceeding its octet).
(d) This curved arrow violates the second rule by giving three bonds and two lone pairs to an oxygen atom (thus exceeding its octet).
(e) This curved arrow violates the second rule by giving five bonds to a carbon atom (thus exceeding its octet).
(f) This curved arrow violates the second rule by giving five bonds to a carbon atom (thus exceeding its octet).
(g) This curved arrow violates the first rule by breaking a single bond, and violates the second rule by giving five bonds to a carbon atom (thus exceeding its octet).
(h) This curved arrow violates the first rule by breaking a single bond, and violates the second rule by giving five bonds to a carbon atom (thus exceeding its octet).
(i) This curved arrow does not violate either rule.
(j) This curved arrow does not violate either rule.
(k) This curved arrow violates the second rule by giving five bonds to a carbon atom (thus exceeding its octet).
(l) This curved arrow violates the second rule by giving five bonds to a carbon atom (thus exceeding its octet).

2.13. The tail of the curved arrow must be placed on the double bond in order to avoid violating the first rule (avoid breaking a single bond).

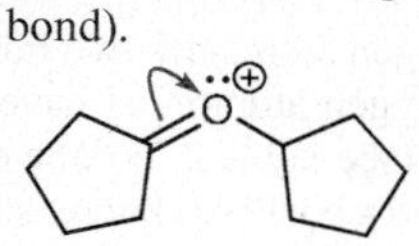

2.14.
(a) This curved arrow violates the first rule (avoid breaking a single bond).
(b) This curved arrow does not violate either rule.
(c) This curved arrow violates the second rule (never exceed an octet for second-row elements) by giving five bonds to a carbon atom.
(d) This curved arrow violates the second rule by giving five bonds to a carbon atom (thus exceeding its octet).

2.15.
(a) The curved arrow indicates that we should draw a resonance structure in which the π bond has been pushed

over. We then complete the resonance structure by assigning any formal charges. Notice that both resonance structures show a positive charge, but in different locations:

(b) The curved arrows indicate that we should draw a resonance structure in which the lone pair has been pushed to become a π bond, and the π bond has been pushed to become a lone pair. We then complete the resonance structure by assigning any formal charges. Notice that both resonance structures show a negative charge, but in different locations:

(c) The curved arrows indicate that we should draw a resonance structure in which a lone pair has been pushed to become a π bond, and the π bond has been pushed to become a lone pair. We then complete the resonance structure by assigning any formal charges. Notice that both resonance structures have zero net charge:

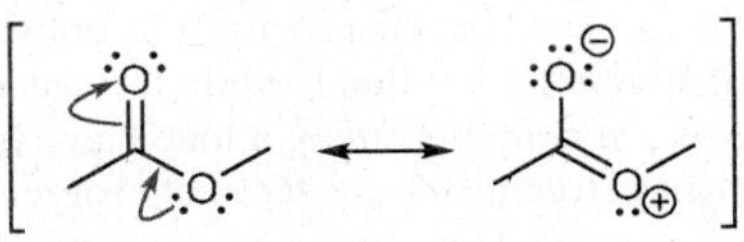

(d) The curved arrows indicate that we should draw a resonance structure in which a lone pair has been pushed to become a π bond, and the π bond has been pushed to become a lone pair. We then complete the resonance structure by assigning any formal charges. Notice that both resonance structures show a negative charge, but in different locations:

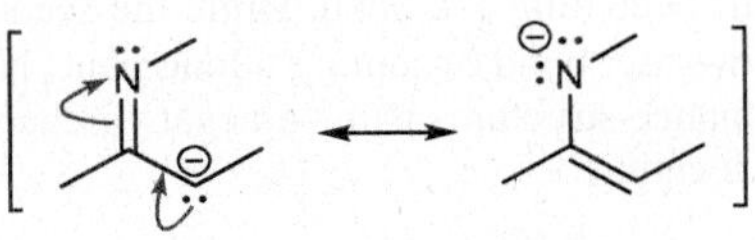

(e) The curved arrows indicate that we should draw the following resonance structure. Notice that both resonance structures have zero net charge:

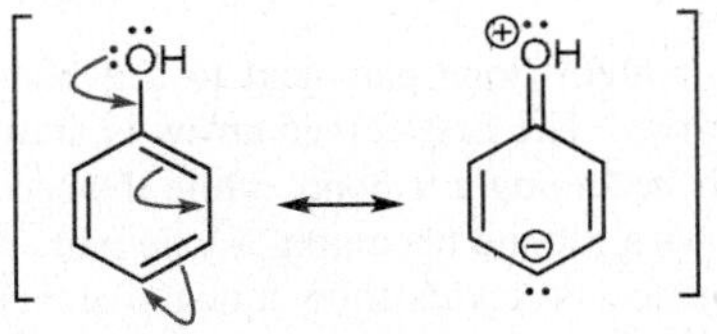

(f) The curved arrows indicate that we should draw the following resonance structure. Notice that both resonance structures have zero net charge:

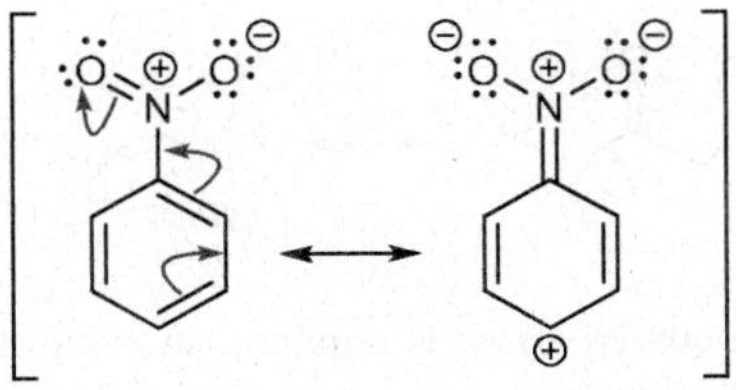

(g) The curved arrows indicate that we should draw a resonance structure in which a lone pair has been pushed to become a π bond, and a π bond has been pushed to become a lone pair. We then complete the resonance structure by assigning any formal charges. Notice that both resonance structures have zero net charge, but they differ in the location of the negative charge:

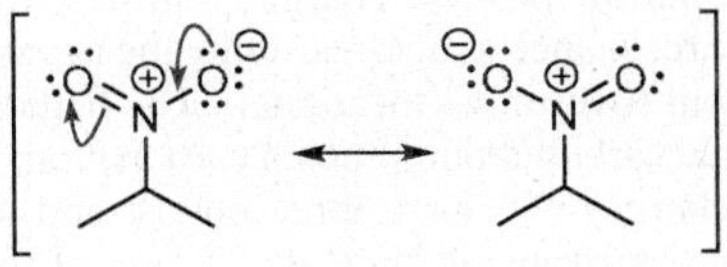

(h) The curved arrows indicate that we should draw the following resonance structure. Notice that both resonance structures have zero net charge:

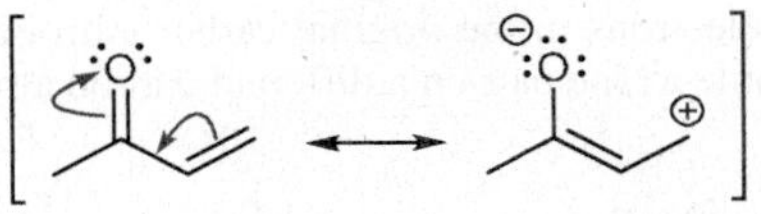

2.16.

(a) One curved arrow is required, showing the π bond being pushed to become a lone pair:

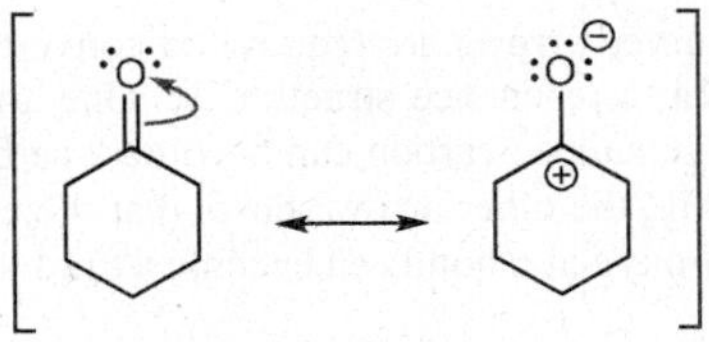

(b) Two curved arrows are required. One curved arrow shows the carbon-carbon π bond being pushed up, and the other curved arrow shows the carbon-oxygen π bond becoming a lone pair:

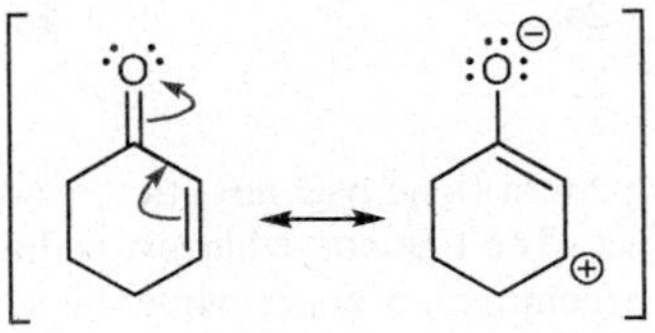

(c) Two curved arrows are required. One curved arrow shows a lone pair from the nitrogen atom becoming a π bond, and the other curved arrow shows the carbon-oxygen π bond becoming a lone pair:

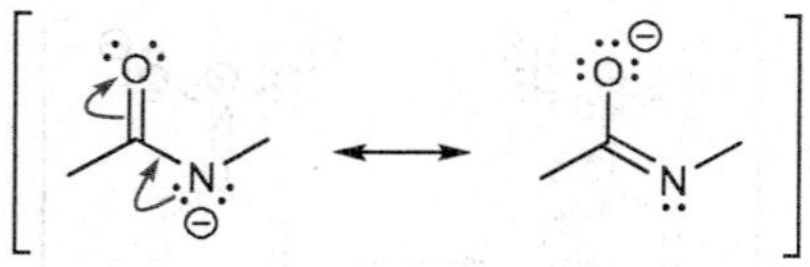

(d) One curved arrow is required, showing the π bond being pushed over:

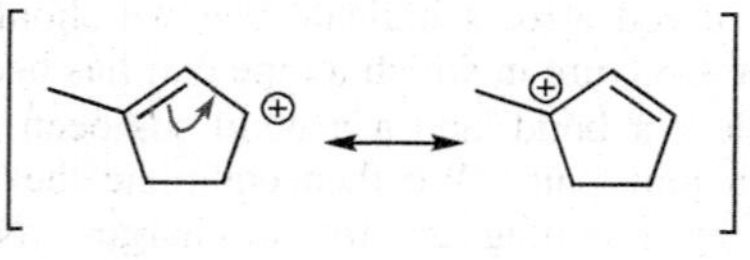

2.17.

(a) Notice that there are two formal charges (positive and negative), but the positive charge is in the same location in all three resonance structures. Only the negative charge is spread out (over three locations). The negative charge in **2a** is on a carbon atom. Such a carbon atom (bearing a negative charge) will have three bonds, and in order to exhibit five electrons, it must also have one lone pair of electrons. Two curved arrows are required to convert from resonance structure **2a** to resonance structure **2b**. One arrow shows that the lone pair on carbon can become a new carbon-nitrogen π bond while the other arrow shows that the electrons in the original carbon-nitrogen π bond can become a lone pair on a different carbon atom.

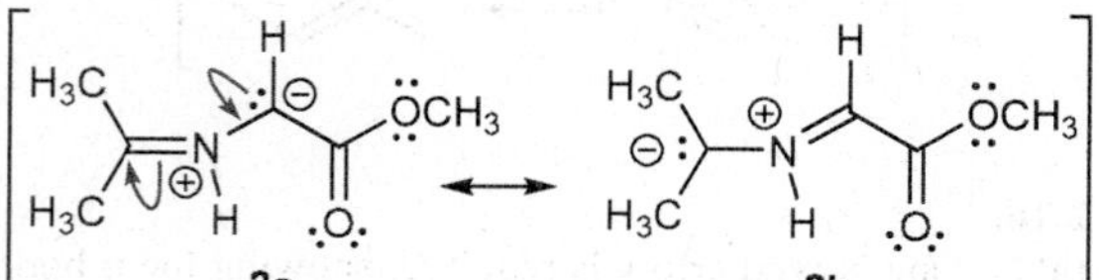

(b) Two curved arrows are required to convert resonance structure **2a** to resonance structure **2c**. One arrow shows that the lone pair on carbon can become a carbon-carbon π bond while the other arrow shows that the electrons in the carbon-oxygen π bond can become a third lone pair on oxygen.

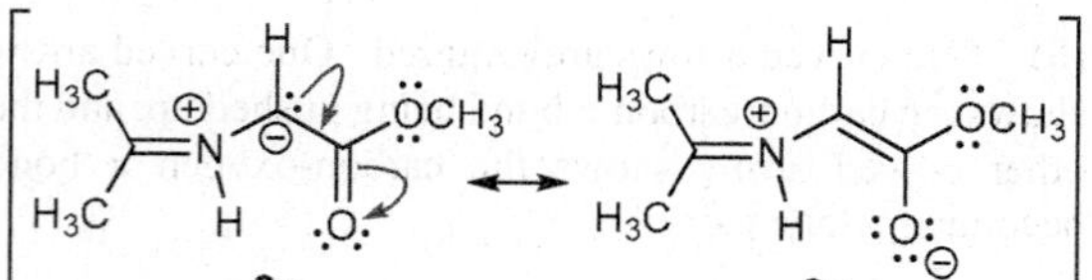

2.18.

(a) This pattern (lone pair next to a π bond) has two curved arrows. The first curved arrow is drawn showing a lone pair becoming a π bond, while the second curved arrow shows a π bond becoming a lone pair. Notice that both resonance structures have zero net charge:

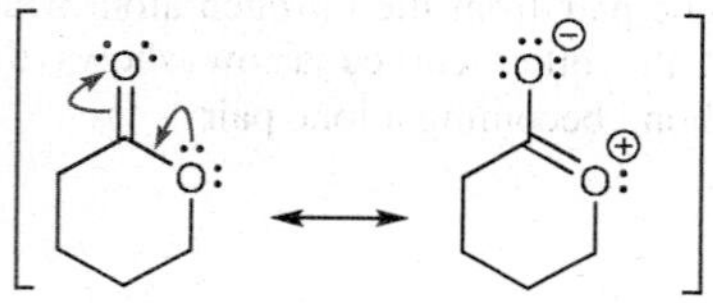

(b) This pattern (lone pair next to a π bond) has two curved arrows. The first curved arrow is drawn showing a lone pair becoming a π bond, while the second curved arrow shows a π bond becoming a lone pair. Notice that both resonance structures show a negative charge, but in different locations:

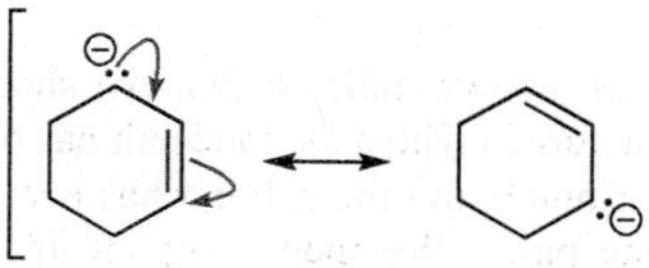

(c) This pattern (lone pair next to a π bond) has two curved arrows. The first curved arrow is drawn showing a lone pair becoming a π bond, while the second curved arrow shows a π bond becoming a lone pair. Notice that both resonance structures have zero net charge:

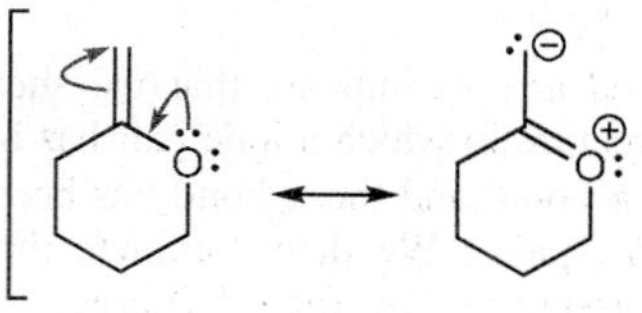

(d) This pattern (lone pair next to a π bond) has two curved arrows. The first curved arrow is drawn showing a lone pair becoming a π bond, while the second curved arrow shows a π bond becoming a lone pair. Notice that both resonance structures have zero net charge:

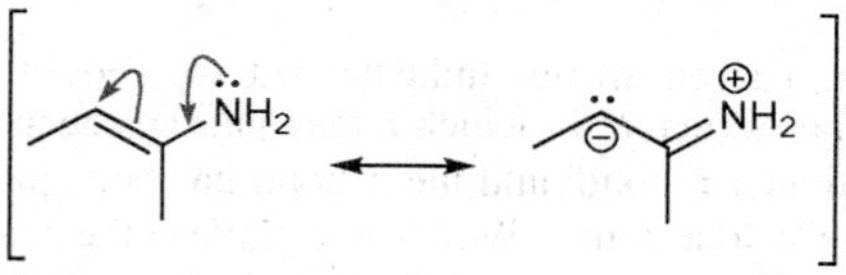

(e) This pattern (lone pair next to a π bond) has two curved arrows. The first curved arrow is drawn showing a lone pair becoming a π bond, while the second curved arrow shows a π bond becoming a lone pair. Notice that both resonance structures show a negative charge, but in different locations:

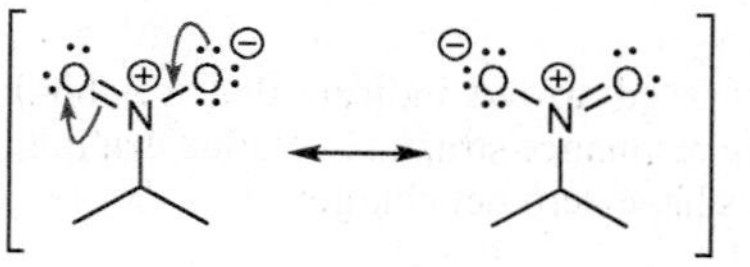

(f) This pattern (lone pair next to a π bond) has two curved arrows. The first curved arrow is drawn showing a lone pair becoming a π bond, while the second curved arrow shows a π bond becoming a lone pair. Notice that both resonance structures show a negative charge, but in different locations:

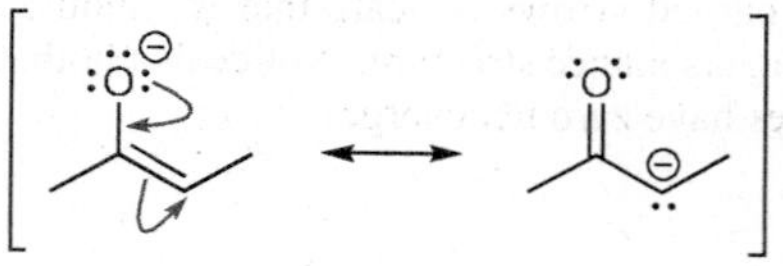

(g) This pattern (lone pair next to a π bond) has two curved arrows. The first curved arrow is drawn showing a lone pair becoming a π bond, while the second curved arrow shows a π bond becoming a lone pair. Notice that both resonance structures have an overall +1 net charge:

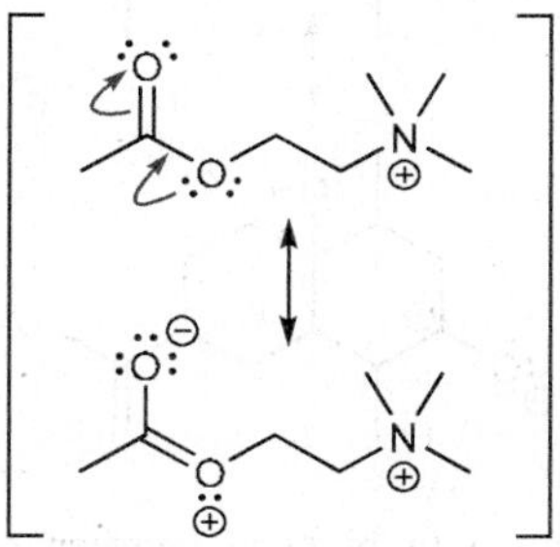

(h) This pattern (lone pair next to a π bond) has two curved arrows. The first curved arrow is drawn showing a lone pair becoming a π bond, while the second curved arrow shows a π bond becoming a lone pair. Notice that both resonance structures have zero net charge:

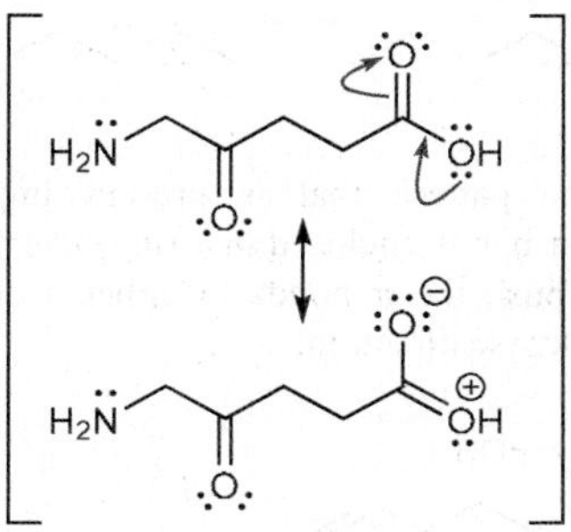

2.19.

(a) This pattern (an allylic C+) has just one curved arrow, showing the π bond being pushed over. Notice that both resonance structures have an overall +1 net charge, but they show the positive charge in different locations:

(b) This pattern (an allylic C+) has just one curved arrow, showing the π bond being pushed over. Notice that both resonance structures have an overall +1 net charge, but they show the positive charge in different locations:

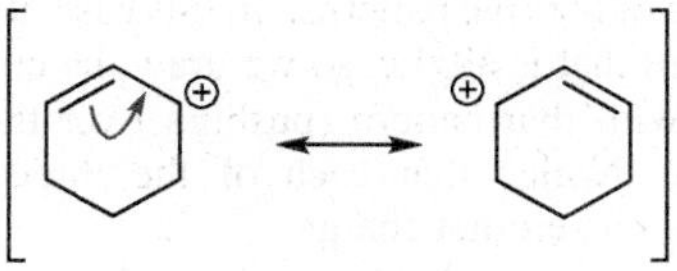

(c) This pattern (an allylic C+) has just one curved arrow, showing the π bond being pushed over. But when we draw the resulting resonance structure, we find that the same pattern can be applied again, giving another resonance structure, as shown:

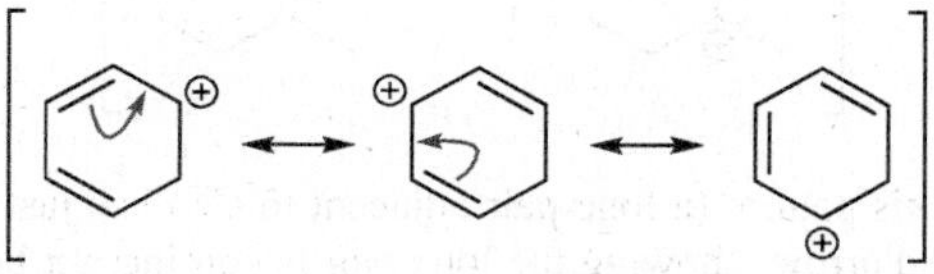

Notice that each resonance structure has an overall +1 net charge, and that each structure shows the positive charge in a different location.

(d) This pattern (an allylic C+) has just one curved arrow, showing the π bond being pushed over. But when we draw the resulting resonance structure, we find that the same pattern can be applied again, giving another resonance structure. This process continues several more times:

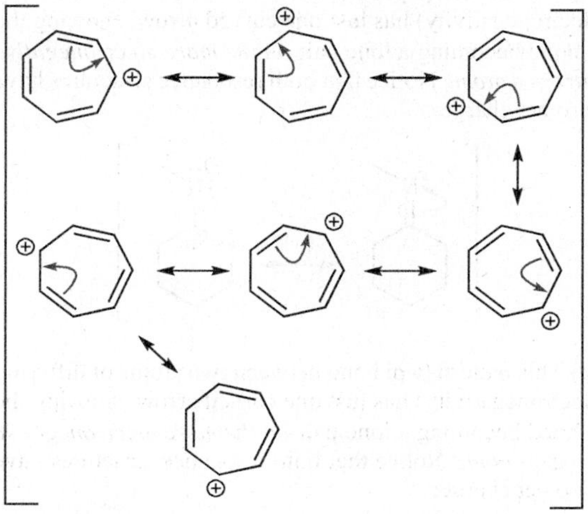

Notice that each resonance structure has an overall +1 net charge, and that each structure shows the positive charge in a different location. We can see that the positive charge is spread (via resonance) over all seven carbon atoms of the ring

2.20.

(a) This pattern (a lone pair adjacent to C+) has just one curved arrow, showing the lone pair becoming a π bond. Notice that both resonance structures have an overall +1 net charge, but they show the positive charge in different locations:

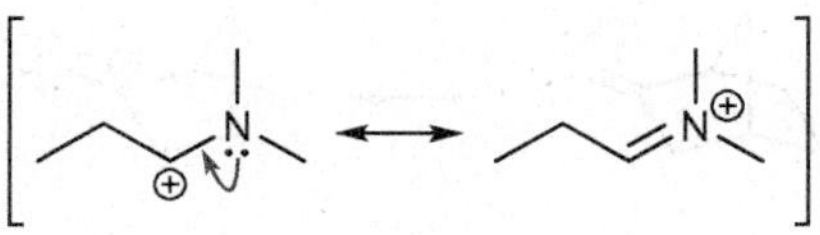

(b) This pattern (a lone pair adjacent to C+) has just one curved arrow, showing the lone pair becoming a π bond. Notice that both resonance structures have zero net charge:

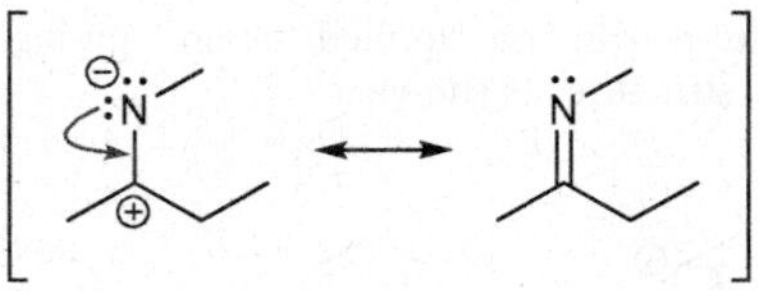

(c) This pattern (a lone pair adjacent to C+) has just one curved arrow, showing the lone pair becoming a π bond. Notice that both resonance structures have an overall +1 net charge, but they show the positive charge in different locations:

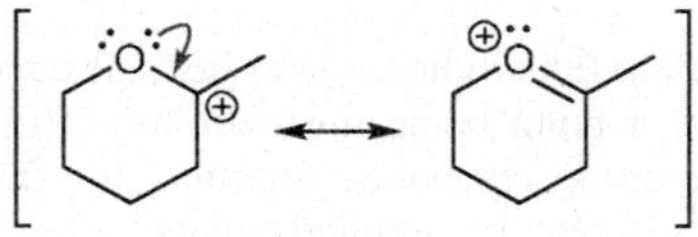

2.21.

(a) This pattern (a pi bond between two atoms of differing electronegativity) has just one curved arrow, showing the π bond becoming a lone pair *on the more electronegative nitrogen atom*. Notice that both resonance structures have zero net charge:

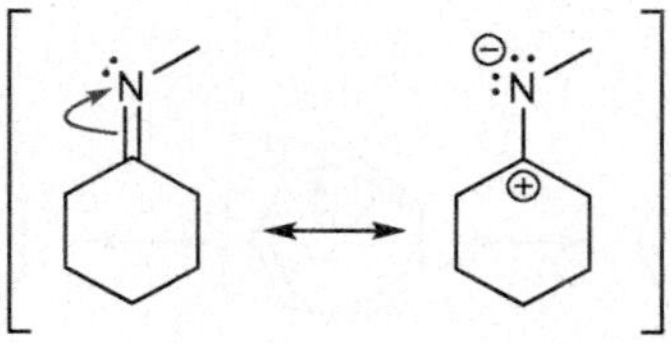

(b) This pattern (a pi bond between two atoms of differing electronegativity) has just one curved arrow, showing the π bond becoming a lone pair *on the more electronegative oxygen atom*. Notice that both resonance structures have zero net charge:

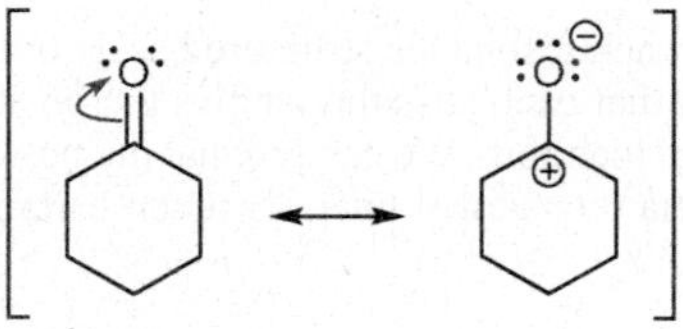

(c) This pattern (a pi bond between two atoms of differing electronegativity) has just one curved arrow, showing the π bond becoming a lone pair *on the more electronegative oxygen atom*. Notice that both resonance structures have zero net charge:

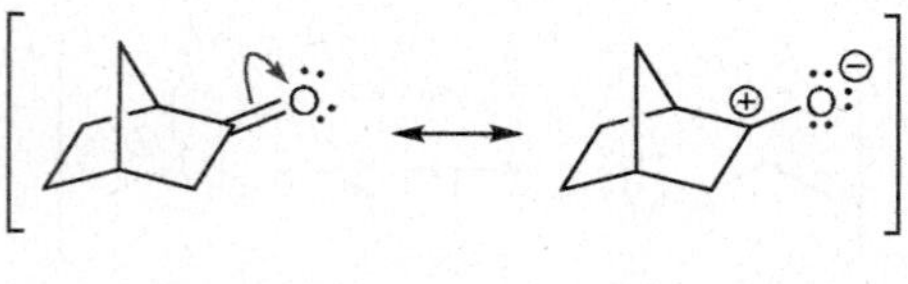

2.22. This pattern (a pi bond between two atoms of differing electronegativity) has just one curved arrow, showing the π bond becoming a lone pair *on the more electronegative oxygen atom*. Notice that both resonance structures have zero net charge:

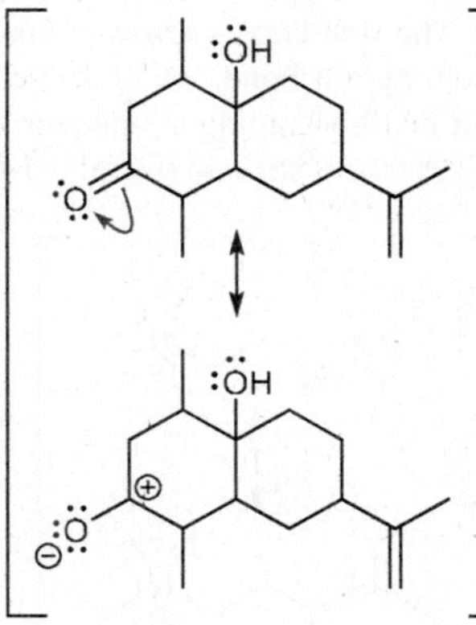

2.23. This pattern (a pi bond between two atoms of differing electronegativity) has just one curved arrow, showing the π bond becoming a lone pair *on the more electronegative oxygen atom*. Notice that both resonance structures have zero net charge:

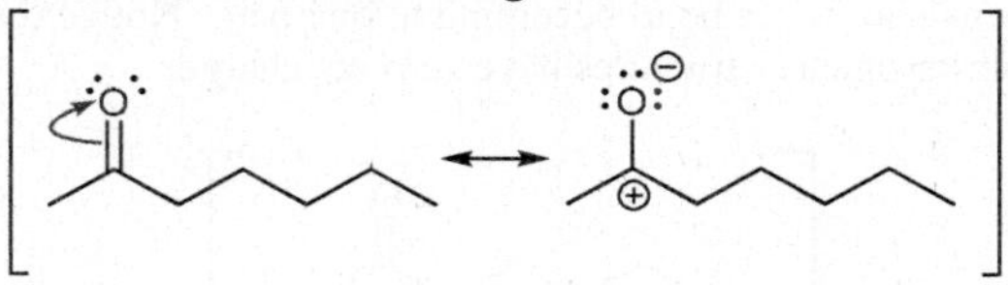

2.24. The pattern that is present in fingolimod (conjugated π bonds enclosed in a ring) has three curved arrows that push the π bonds in either a clockwise or counter-clockwise direction:

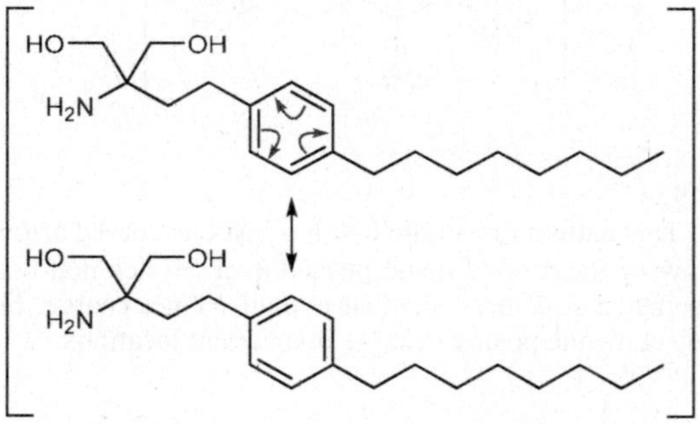

2.25.

(a) We begin by looking for the five patterns. In this case, there is a C=O bond (a π bond between two atoms of differing electronegativity), so we draw one curved arrow showing the π bond becoming a lone pair *on the more electronegative oxygen atom*. We then draw the resulting resonance structure and assess whether it exhibits one of the five patterns. In this case, there is a C+ (carbocation) that is allylic, so we draw the curved arrow associated with that pattern (pushing over the π bond), shown here. Notice that each of the three resonance structures has a zero net charge:

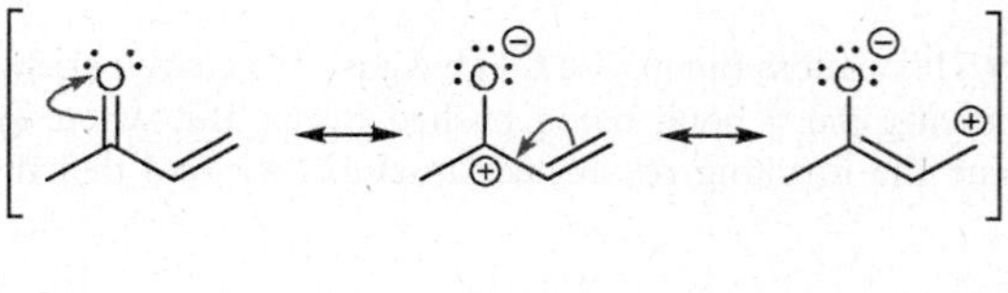

(b) The C+ (a carbocation) occupies an allylic position, so we draw the one curved arrow associated with that pattern (pushing over the π bond). The C+ in the resulting resonance structure is again next to another π bond, so we draw one curved arrow and another resonance structure, as shown here:

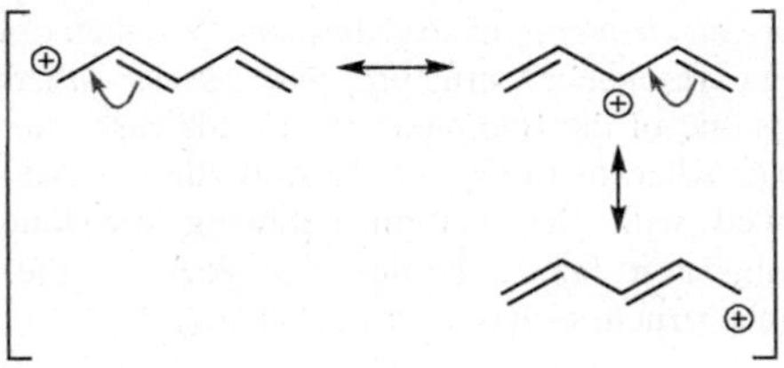

Notice that each resonance structure has an overall +1 net charge, and that each structure shows the positive charge in a different location.

(c) The lone pair (associated with the negative charge) occupies an allylic position, so we draw the two curved arrows associated with that pattern. The first curved arrow is drawn showing a lone pair becoming a π bond, while the second curved arrow shows a π bond becoming a lone pair. Notice that each resonance structure has an overall −1 net charge, and that each structure shows the negative charge in a different location:

(d) We begin by looking for the five patterns, focusing first on any patterns that employ just one curved arrow (in this case, there is another pattern that requires two curved arrows, but we will start with the pattern using just one curved arrow). There is a C=O bond (a π bond between two atoms of differing electronegativity), so we draw one curved arrow showing the π bond becoming a lone pair *on the more electronegative oxygen atom.* We then draw the resulting resonance structure and assess whether it exhibits one of the five patterns. In this case, there is a lone pair adjacent to a C+ (a carbocation), so we draw the curved arrow associated with that pattern (showing the lone pair becoming a π bond), shown here. Notice that each of the three resonance structures has a zero net charge:

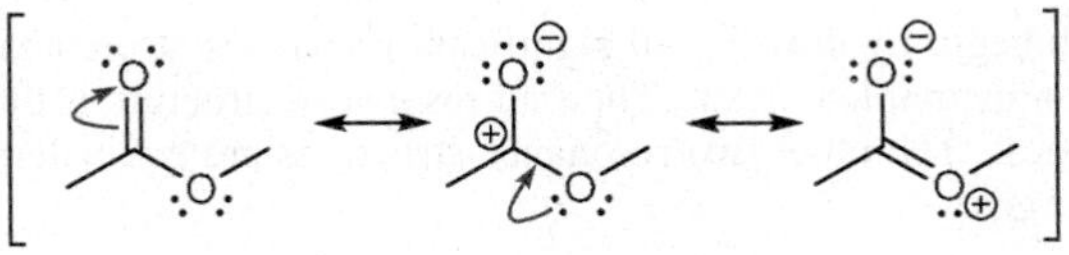

(e) This structure exhibits a lone pair that is adjacent to a C+ (a carbocation), so we draw one curved arrow, showing a lone pair becoming a π bond. Notice that both resonance structures have an overall +1 net charge, but they show the positive charge in different locations:

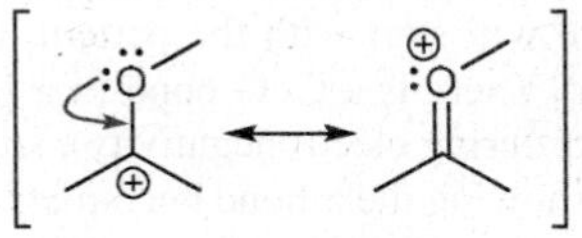

(f) This compound exhibits a C=O bond (a π bond between two atoms of differing electronegativity), so we draw one curved arrow showing the π bond becoming a lone pair *on the more electronegative oxygen atom.* We then draw the resulting resonance structure and assess whether it exhibits one of the five patterns. In this case, there is a C+ (carbocation) that is allylic, so we draw the curved arrow associated with that pattern (pushing over the π bond). The C+ in the resulting resonance structure is next to another π bond, so we draw one more resonance structure. Notice that all four resonance structures have zero net charge:

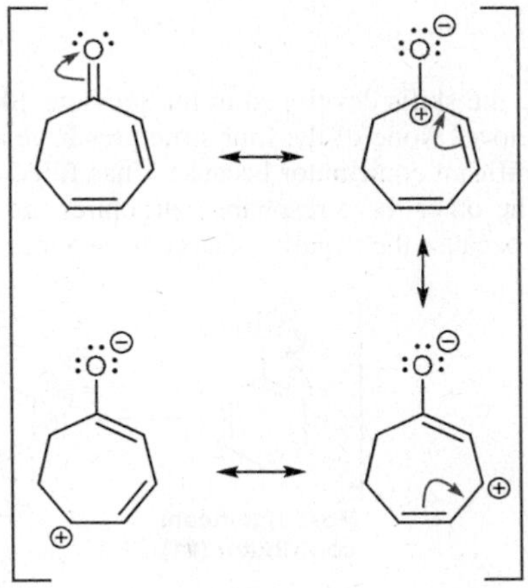

(g) This structure exhibits a lone pair that is adjacent to C+, so we draw one curved arrow, showing a lone pair becoming a π bond. Notice that both resonance structures have an overall +1 net charge, but they show the positive charge in different locations:

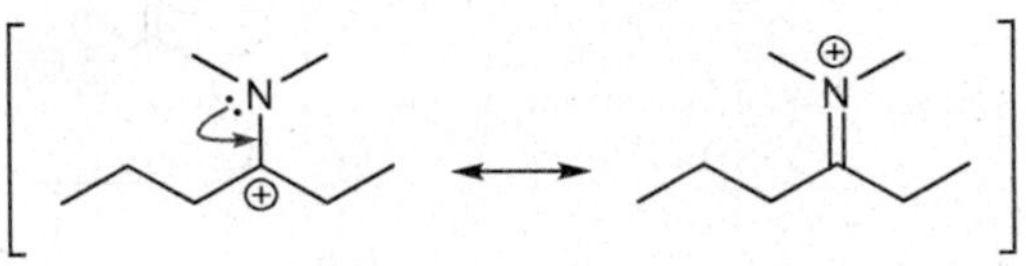

(h) This compound exhibits a C=N bond (a π bond between two atoms of differing electronegativity), so we draw one curved arrow showing the π bond becoming a lone pair *on the more electronegative nitrogen atom.* Notice that both resonance structures have zero net charge:

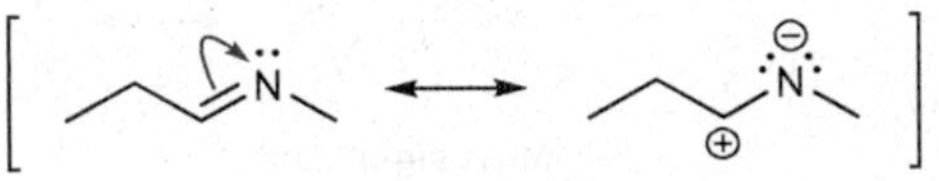

(i) We begin by looking for the five patterns, focusing first on any patterns that employ just one curved arrow (in this case, there is another pattern that requires two curved arrows, but we will start with the pattern using just one curved arrow). There is a C=O bond (a π bond between two atoms of differing electronegativity), so we draw one curved arrow showing the π bond becoming a lone pair *on the more electronegative oxygen atom.* We then draw the resulting resonance structure and assess whether it exhibits one of the five patterns. In this case, there is a lone pair adjacent to C+, so we draw the curved arrow associated with that pattern (showing the lone pair becoming a π bond). Notice that each of the three resonance structures has a zero net charge:

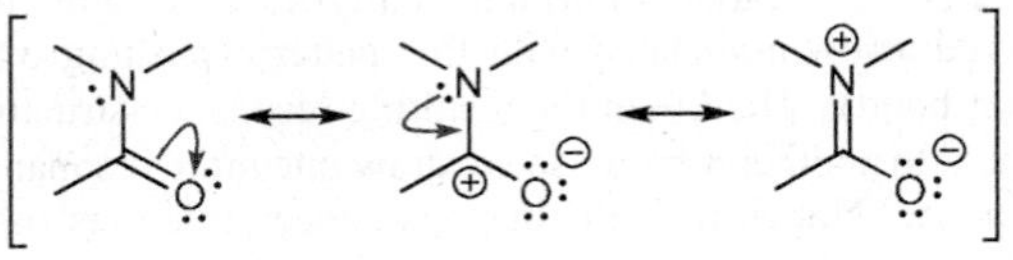

(j) We begin by looking for the five patterns, focusing first on any patterns that employ just one curved arrow (in this case, there is another pattern that requires two curved arrows, but we will start with the pattern using just one curved arrow). There is a C=O bond (a π bond between two atoms of differing electronegativity), so we draw one curved arrow showing the π bond becoming a lone pair *on the more electronegative oxygen atom.* We then draw the resulting resonance structure and assess whether it exhibits one of the five patterns. In this case, there is a lone pair adjacent to C+, so we draw the curved arrow associated with that pattern (showing the lone pair becoming a π bond). Notice that each of the three resonance structures has a zero net charge:

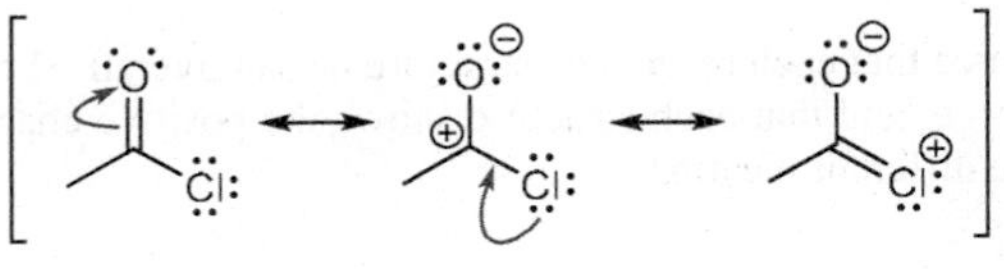

2.26.

(a) Using the skills developed in the previous SkillBuilders, we begin by drawing all significant resonance structures, shown below. None of the four structures have an atom with an incomplete octet. The first resonance structure is the most significant contributor because it has filled octets and the negative charge is on the more electronegative nitrogen atom. The other three resonance structures are approximately equivalent and they are less significant than the first structure because the negative charge is on a less electronegative carbon atom.

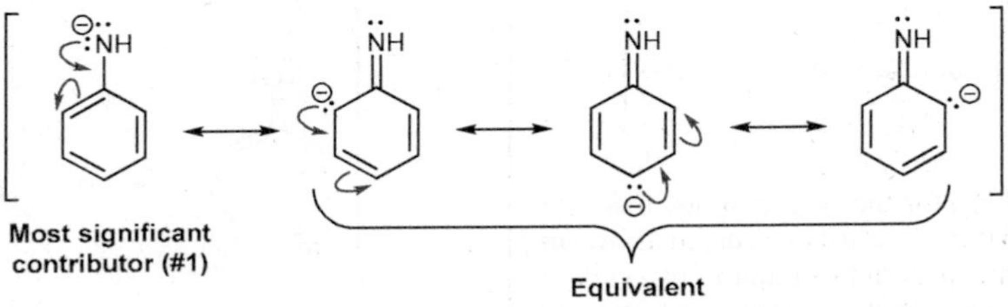

Most significant contributor (#1)

Equivalent

(b) Using the skills developed in the previous SkillBuilders, we begin by drawing all significant resonance structures, shown below. Both resonance structures have filled octets and a negative charge on an oxygen atom. Both resonance structures are equivalent and thus equally significant.

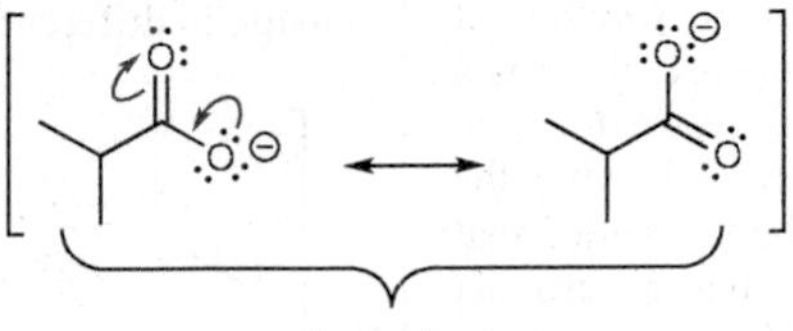

Equivalent

(c) Using the skills developed in the previous SkillBuilders, we begin by drawing all significant resonance structures, shown below. None of the three structures have an atom with an incomplete octet. The first resonance structure is the most significant because it has filled octets and no formal charges. The other two resonance structures are equivalent and less significant contributors because they contain formal charges.

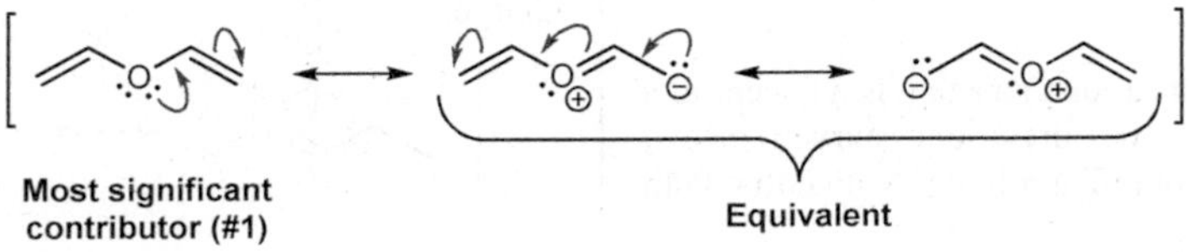

Most significant contributor (#1)

Equivalent

(d) Using the skills developed in the previous SkillBuilders, we begin by drawing all significant resonance structures, shown below. The first resonance structure is the most significant because it is the only one with every second-row element having a filled octet. Recall that a structure with filled octets and no formal charges is an ideal Lewis structure. The other three resonance structures are approximately equivalent and they are less significant than the first structure because each one is missing an octet, highlighted below (they also have formal charges, but that is a less significant feature to consider when ranking resonance forms):

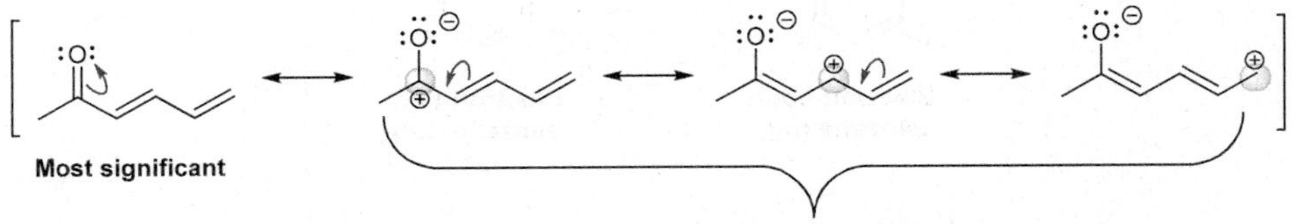

Most significant

Approximately equivalent contributors (each is missing an octet)

(e) Using the skills developed in the previous SkillBuilders, we begin by drawing all significant resonance structures, shown below. Neither resonance structure has an incomplete octet. The second resonance structure is the more significant contributor because it has the negative charge on the more electronegative nitrogen atom. The first resonance structure is less significant because the negative charge is on the less electronegative carbon atom.

$[CH_3-\overset{\ominus}{\ddot{C}}H-C\equiv N: \longleftrightarrow CH_3-CH=C=\ddot{N}:^{\ominus}]$

Less significant — More significant

(f) This cation has two different resonance patterns that can be employed, using the lone pair or the π bond to fill the vacancy on carbon, giving a total of three resonance structures. The middle resonance structure is the most significant contributor because it is the only one with filled octets. The other two structures are approximately equivalent, each having one missing octet (highlighted below) and a positive charge on a carbon atom. They are less significant and contribute equally to the hybrid. Note that the location of the charge (C+ vs. N+) is not relevant in this problem, because filled octets are more important.

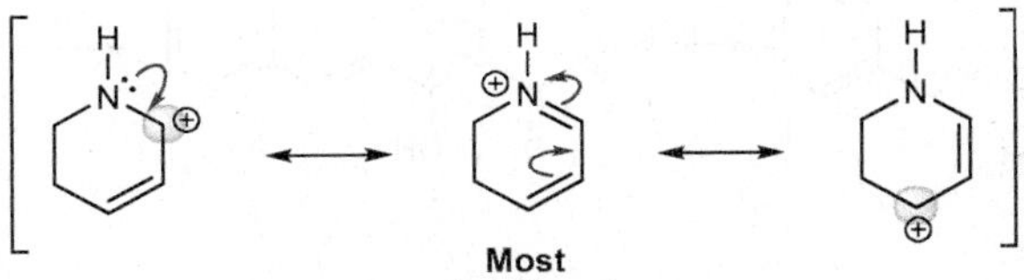

Most significant

2.27. When looking for resonance in the first structure, we can begin with the carbonyl (C=O) group by relocating the π bond electrons to the more electronegative oxygen atom. This provides a resonance structure with a C+ (carbocation) that is allylic. Allylic resonance throughout the ring provides for three more resonance structures. Be careful to use just one π bond at a time so you don't accidentally "jump" over a possible resonance structure. Notice that each of the five resonance structures has a zero net charge:

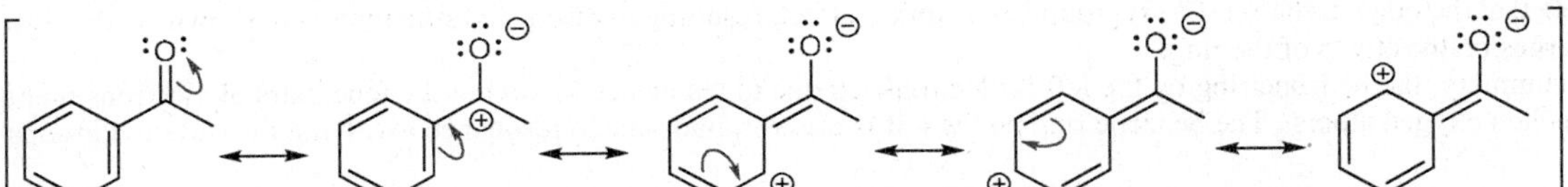

For the second compound, the allylic lone pair can be delocalized using one of the π bonds in the benzene ring, and this pattern can continue to use the remaining π bonds in the ring (again, one at a time!). Notice that each of the four resonance structures has a zero net charge:

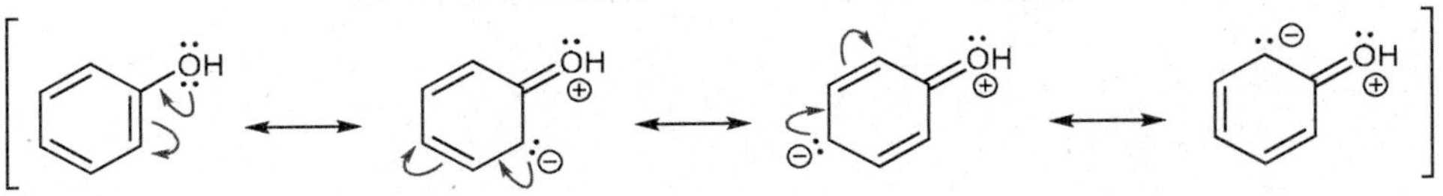

Overall, when we consider the contributions made by all resonance structures, we find that the ring in the first compound is electron-poor, with several electron-deficient sites on the ring, and the ring in the second compound is electron-rich, with several δ– sites on the ring.

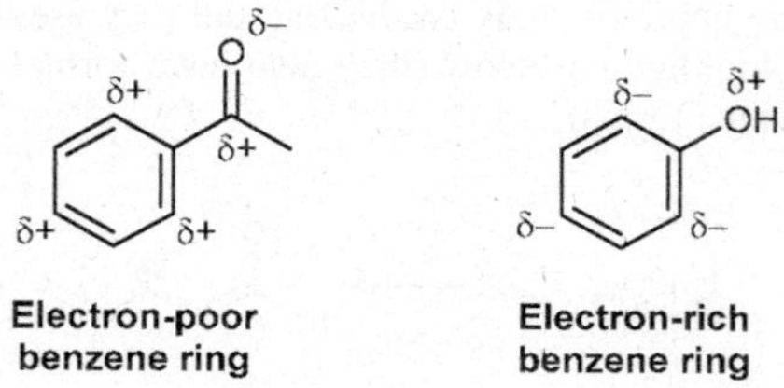

Electron-poor benzene ring

Electron-rich benzene ring

2.28. The benzene ring on the left has two oxygen atoms attached. Both oxygen atoms can *donate* electron density via allylic lone pair resonance, making this benzene ring electron-rich:

Electron-rich ring

Electron-rich ring

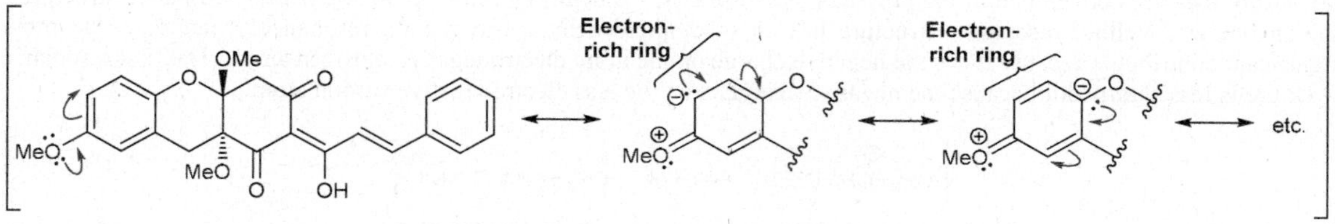

Note that the other oxygen atom directly attached to the benzene ring has a similar effect, resulting in additional resonance structures (not shown) with negative charges on the atoms of the ring.

The benzene ring on the right is "connected" to the two carbonyl (C=O) groups by a series of conjugated π bonds, so the resonance of each carbonyl group extends into the benzene ring. The carbonyl groups *withdraw* electron density via allylic carbocation (C+) resonance, making this ring electron-deficient:

Electron-deficient ring

Electron-deficient ring

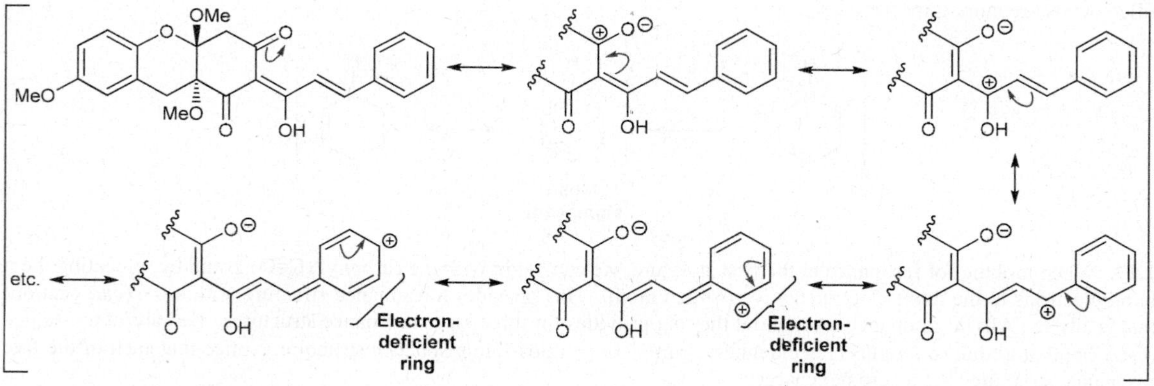

Note that the other carbonyl (C=O) group has a similar effect, resulting in resonance structures (not shown) with positive charges on the atoms of the ring.

In summary, the benzene ring on the left is electron-rich due to resonance involving the lone pairs of electrons on both attached oxygen atoms. The benzene ring on the left is electron-poor due to resonance involving the carbonyl groups.

Electron-rich ring

Electron-deficient ring

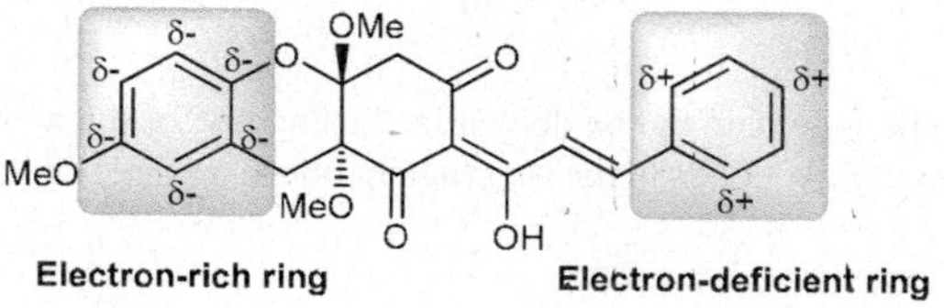

2.29.

(a) Begin by drawing all significant resonance structures. In this case, there are two:

Both resonance structures are equally significant, so the resonance hybrid is the simple average of these two resonance structures. There are no formal charges, so only partial bonds need to be drawn.

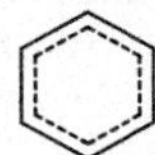

Resonance hybrid

(b) Begin by drawing all significant resonance structures. In this case, there are two:

Both are equally significant, so the resonance hybrid is the simple average of these two resonance structures. Both partial bonds and partial charges are required, as shown:

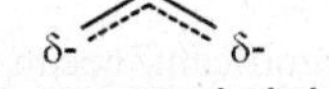

Resonance hybrid

(c) Begin by drawing all significant resonance structures. In this case, there are two:

The left-hand structure is more significant because every atom has an octet. The resonance hybrid is a weighted average of these two resonance structures in which the oxygen atom has more of the charge than the carbon atom.

Resonance hybrid

(d) Begin by drawing all significant resonance structures. In this case, there are two:

Both are equally significant, so the resonance hybrid is the simple average of these two resonance structures. Only the negative charge is delocalized, so partial charges are used for the negative charge but not for the positive charge.

Resonance hybrid

(e) Begin by drawing all significant resonance structures. In this case, there are two:

The structure on the right is more significant because every atom has an octet. The resonance hybrid is a weighted average of these two resonance structures in which the nitrogen atom has more of the charge than the carbon atom.

Resonance hybrid

(f) Begin by drawing all significant resonance structures. In this case, there are two:

The structure on the left is more significant because every atom has an octet and it has no formal charges. The resonance hybrid is a weighted average of these two resonance structures, although we do not denote that by making the partial charges different sizes. In this example, the charges are opposite in sign, but they must be equal in magnitude so that the overall charge will be zero.

Resonance hybrid

(g) Begin by drawing all significant resonance structures. In this case, there are three:

All three resonance structures are approximately equally significant, so the resonance hybrid is approximately the average of these three resonance structures, illustrating that the positive charge is delocalized over three carbon atoms.

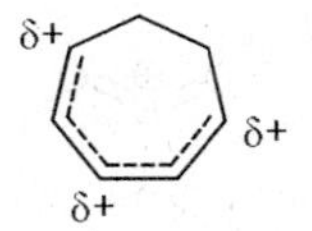

Resonance hybrid

(h) Begin by drawing all significant resonance structures. In this case, there are three:

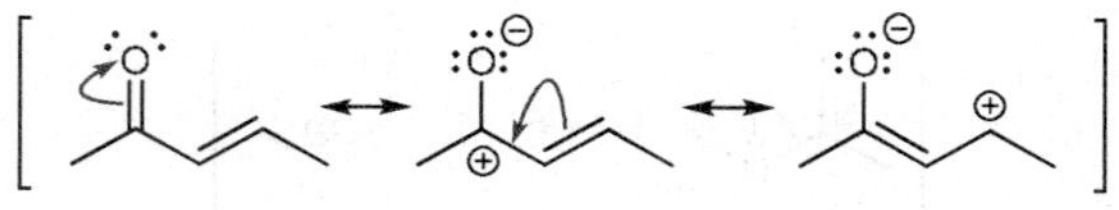

The structure on the left is most significant because every atom has an octet and it has no formal charges. The resonance hybrid is a weighted average of these three resonance structures. Since the partial positive charge is delocalized over two carbon atoms and the partial negative charge is localized on only one oxygen atom, the partial negative charge is drawn larger than each of the individual partial positive charges.

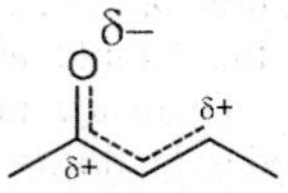

Resonance hybrid

2.30. Begin by drawing all significant resonance structures that show delocalization of the positive charge. In this case, there are four:

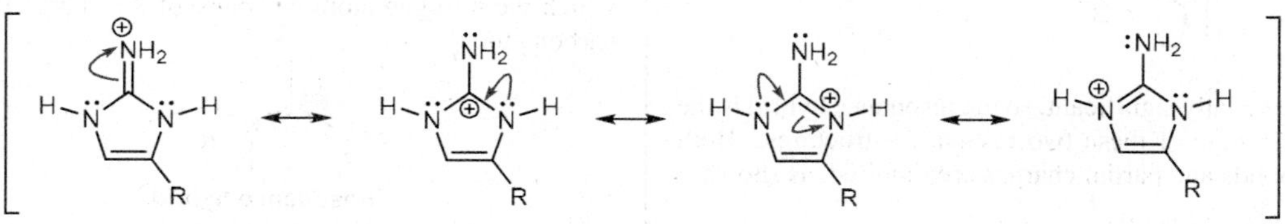

The first resonance structure and the last two resonance structures are the most significant, because all atoms have an octet in each of these three resonance structures. The second resonance structure (in which the carbon atom bears the positive charge) is the least significant because a carbon atom lacks an octet. If we compare the three most significant resonance structures, each has a positive charge on a nitrogen atom, so we expect these three resonance structures to contribute roughly equally to the resonance hybrid. To show this, we indicate δ+ at all three positions, with a smaller δ+ symbol at the central carbon atom (indicating the lower contribution of the second resonance structure). Also, if we compare the three significant resonance structures, we find that the π bond is spread over three locations, and these locations are indicated with dashed lines in the resonance hybrid:

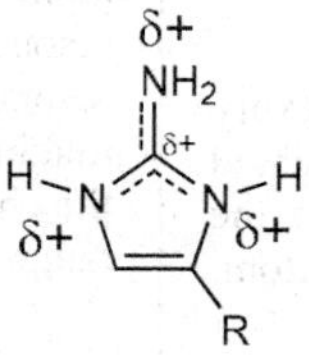

2.31.

(a) Let's begin with the nitrogen atom on the left side of the structure. The lone pair on this nitrogen atom is delocalized by resonance (because it is next to a π bond). Therefore, this lone pair occupies a *p* orbital, which means that the nitrogen atom is sp^2 hybridized. As a result, the geometry is trigonal planar.

On the right side of the structure, there is a nitrogen atom with a localized lone pair (it does not participate in resonance). This nitrogen atom is therefore sp^3 hybridized, with trigonal pyramidal geometry, just as expected for a nitrogen atom with σ sigma bonds and a localized lone pair.

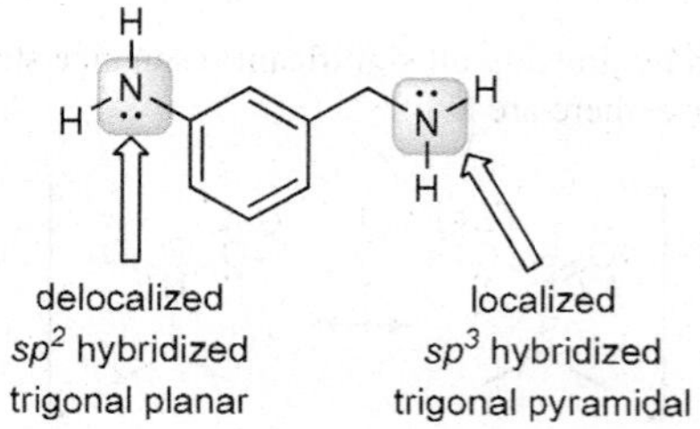

delocalized
sp^2 hybridized
trigonal planar

localized
sp^3 hybridized
trigonal pyramidal

(b) As we saw with pyridine, the lone pair on this nitrogen atom is not participating in resonance, because the nitrogen atom is already using a *p* orbital for the π bond. As a result, the lone pair cannot join in the conduit of overlapping *p* orbitals, and therefore, it cannot participate in resonance. In this case, the lone pair occupies an sp^2-hybridized orbital, which is in the plane of the ring. Since this lone pair is not participating in resonance, it is localized. The nitrogen atom is sp^2 hybridized, and the geometry is bent.

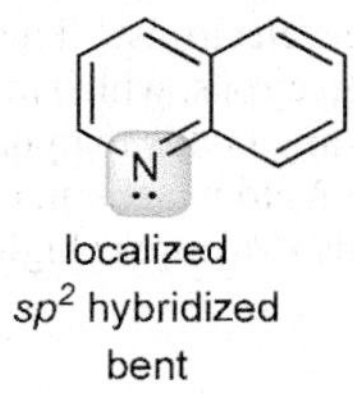

localized
sp^2 hybridized
bent

(c) The lone pair on this nitrogen atom is participating in resonance (it is next to a π bond), so it is delocalized via resonance. As such, the nitrogen atom is sp^2 hybridized, with trigonal planar geometry.

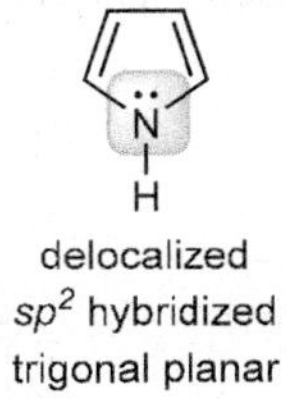

delocalized
sp^2 hybridized
trigonal planar

2.32. As we saw with pyridine, the lone pair on the nitrogen atom in the 6-membered aromatic ring is not participating in resonance, because the nitrogen atom is already using a *p* orbital for the π bond. As a result, the lone pair cannot join in the conduit of overlapping *p* orbitals, and therefore, it cannot participate in resonance. In this case, the lone pair occupies an sp^2-hybridized orbital, which is in the plane of the ring. Since this lone pair is not participating in resonance, it is localized. On the right side of the structure, the nitrogen atom in the 5-membered ring also has a localized lone pair (it does not participate in resonance). Each of these lone pairs is localized, and, therefore, both lone pairs are expected to be reactive.

2.33. Lone pairs that participate in resonance are delocalized, while those that do not participate in resonance are localized:

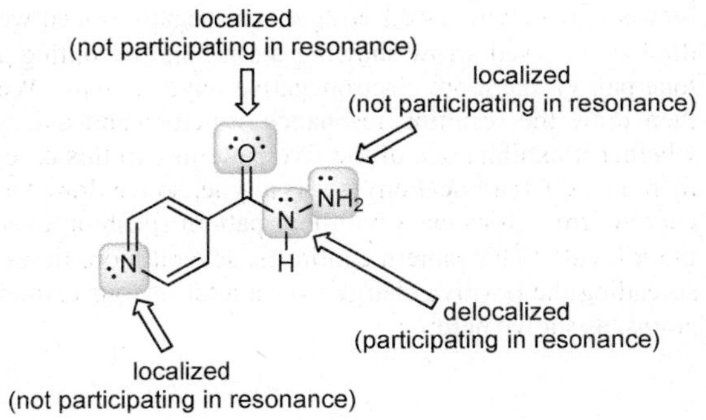

localized
(not participating in resonance)

localized
(not participating in resonance)

delocalized
(participating in resonance)

localized
(not participating in resonance)

2.34.

(a) Each of the oxygen atoms has two bonds and no formal charge, so each oxygen atom will have two lone pairs to complete its octet.

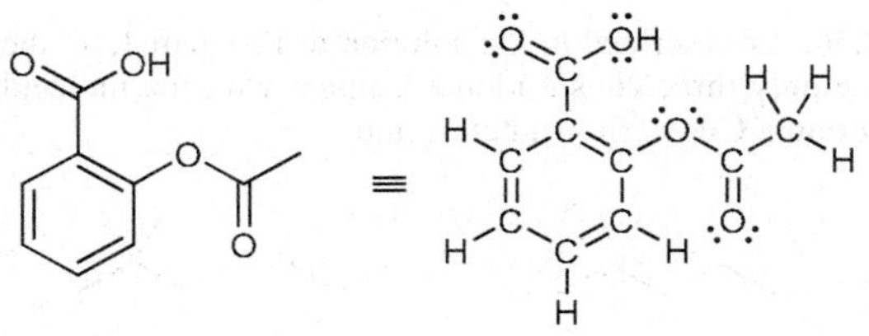

(b) Each of the oxygen atoms has two bonds and no formal charge, so each oxygen atom will have two lone pairs to complete its octet. The nitrogen atom has three bonds and no formal charge, so it must have one lone pair of electrons to complete its octet.

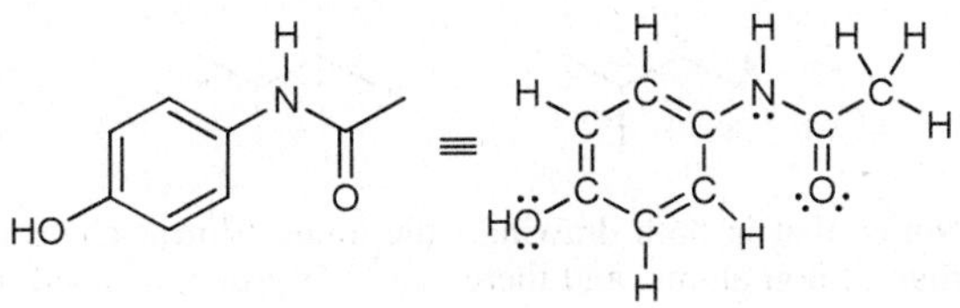

(c) Each of the oxygen atoms has two bonds and no formal charge, so each oxygen atom will have two lone pairs to complete its octet. Each nitrogen atom has three bonds and no formal charge, so each nitrogen atom must have one lone pair of electrons to complete its octet.

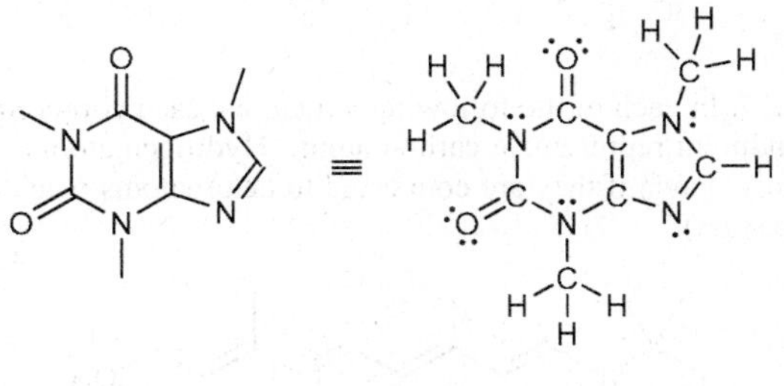

2.35. The molecular formula indicates that there are four carbon atoms. Recall that constitutional isomers are compounds that share the same molecular formula, but differ in constitution (the connectivity of atoms). So we are looking for different ways that four carbon atoms can be connected together. As described in the solution to Problem 1.1b, the carbon atoms can be connected in a linear fashion (below left), or they can be connected with a branch (below right).

C_4H_{10} C_4H_{10}

These two compounds are the only constitutional isomers that have the molecular formula C_4H_{10}, because there are no other ways to connect four carbon atoms without changing the number of hydrogen atoms. For example, if we try to connect the carbon atoms into a ring, we find that the number of hydrogen atoms is reduced:

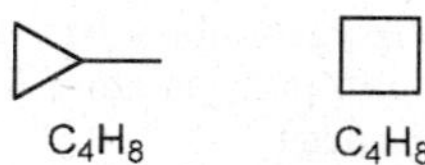

2.36. As described in the solution to Problem 1.1c, there are only three constitutional isomers with the molecular formula C_5H_{12}, shown here again.

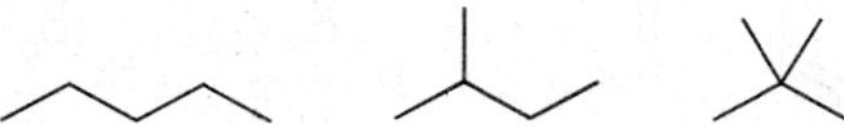

There are no other constitutional isomers with the molecular formula C_5H_{12}. The following two structures do NOT represent constitutional isomers, but are in fact two drawings of the same compound, as can be seen when the carbon skeletons are numbered, as shown:

Notice that in both drawings, the longest linear chain is four carbon atoms, and there is a CH_3 group attached to the second carbon atom of the chain. As such, these two drawings represent the same compound. In contrast, we can see that all three constitutional isomers with the molecular formula C_5H_{12} exhibit different connectivity of the carbon atoms:

2.37. In each of the following structures, each corner and endpoint represents a carbon atom. Hydrogen atoms are only drawn if they are connected to heteroatoms (such as oxygen).

Vitamin A

Vitamin C

2.38. Each oxygen atom has two bonds and no formal charge. Therefore, each oxygen atom has two lone pairs, for a total of twelve lone pairs.

2.39. Carbon is in group **4A** of the periodic table, and it therefore has four valence electrons. We are told that, in this case, the central carbon atom does not bear a formal charge.

Therefore, it must exhibit the appropriate number of valence electrons (four). This carbon atom already has two bonds (each of which requires one valence electron) and a lone pair (which represents two electrons), for a total of 1+1+2=4 valence electrons. This is the appropriate number of valence electrons, which means that this carbon atom does not have any bonds to hydrogen.

Notice that the carbon atom lacks an octet, so it should not be surprising that this structure is highly reactive and very short-lived.

2.40. An oxygen atom will bear a negative charge if it has one bond and three lone pairs, and it will bear a positive charge if it has three bonds and one lone pair (see Table 2.2). A nitrogen atom will bear a negative charge if it has two bonds and two lone pairs, and it will bear a positive charge if it has four bonds and no lone pairs (see Table 2.3).

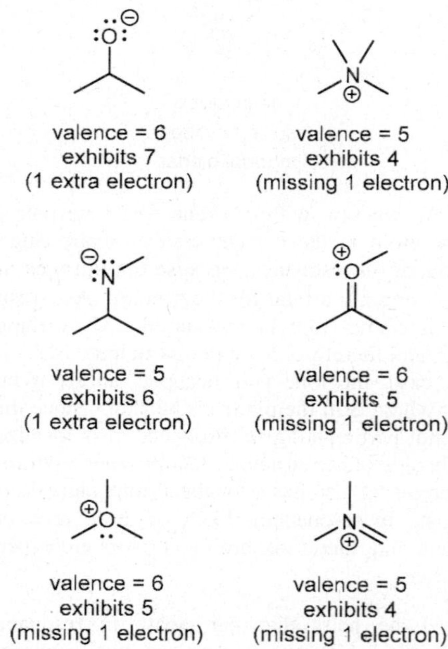

2.41. This compound exhibits a C=O bond (a π bond between two atoms of differing electronegativity), so we draw one curved arrow showing the π bond becoming a lone pair on the more electronegative oxygen atom. We then draw the resulting resonance structure and assess whether it exhibits one of the five patterns. In this case, there is a C+ (carbocation) that is allylic, so we draw the curved arrow associated with that pattern (pushing over the π bond). This pattern continues, several more times, spreading the positive charge over a total of four carbon atoms, as shown here:

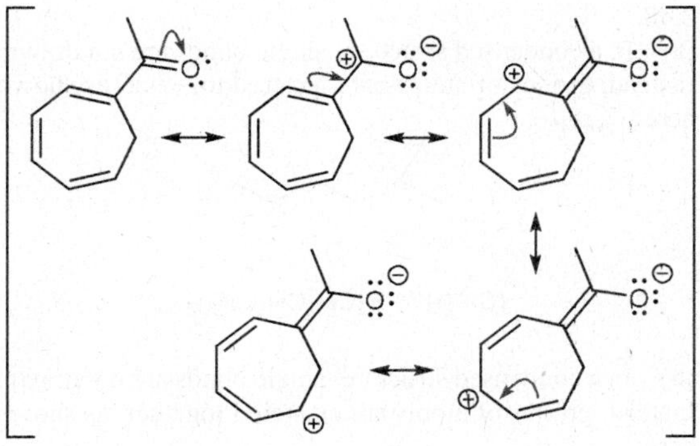

2.42. Recall that constitutional isomers are compounds that share the same molecular formula, but differ in constitution (the connectivity of atoms). The problem statement shows a compound with the molecular formula C_5H_{12} and the following structure:

$CH_3CH_2CH(CH_3)_2$ ≡

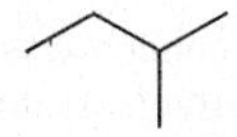

So we are looking for other compounds that also have the molecular formula C_5H_{12} but show a different connectivity of atoms. As seen in the solution to Problem 2.36, there are only two such compounds:

2.43. The following two compounds are constitutional isomers because they share the same molecular formula ($C_5H_{12}O$). The third compound (not shown here) has a different molecular formula ($C_4H_{10}O$).

$(CH_3)_3COCH_3$ $(CH_3)_2CHOCH_2CH_3$

2.44. Begin by drawing a Lewis structure, so that the bonding of each carbon atom is shown more clearly:

$(CH_3)_2C{=}CHC(CH_3)_3$
Condensed structure

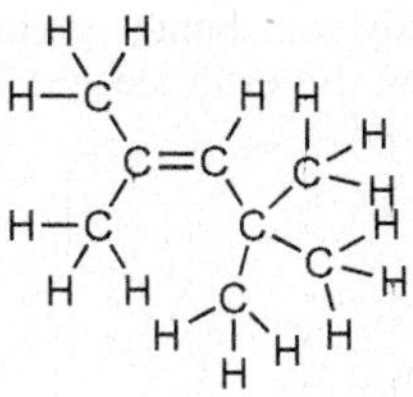

Lewis structure

Notice that two of the carbon atoms are sharing a double bond. These two atoms are sp^2 hybridized. Each of the other six carbon atoms exhibits four single bonds, and as such, each of these six carbon atoms (highlighted) is sp^3 hybridized.

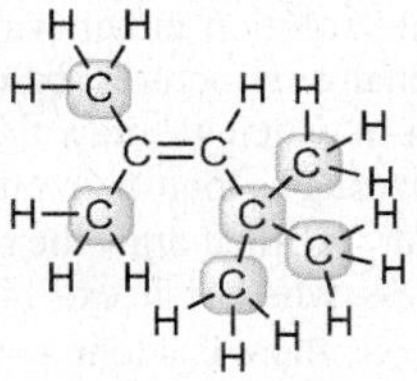

2.45. One of the oxygen atoms has two bonds and no formal charge, so that oxygen atom must have two lone pairs (see Table 2.2). The other oxygen atom has one bond and a negative charge, so that oxygen atom must have three lone pairs. The nitrogen atom has four bonds and a positive charge, so it does not have any lone pairs (see Table 2.3). Therefore, there are a total of five lone pairs in this structure.

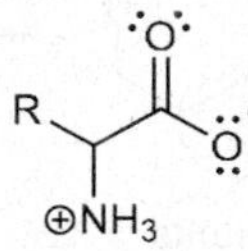

2.46.

(a) The lone pair on this nitrogen atom is not participating in resonance, because the nitrogen atom is already using a *p* orbital for the π bond. As a result, the lone pair cannot join in the conduit of overlapping *p* orbitals, and therefore, it cannot participate in resonance. In this case, the lone pair occupies an sp^2-hybridized orbital, which is in the plane of the ring.

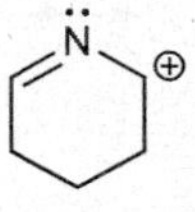

(b) There is a lone pair associated with the negative charge, and this lone pair is delocalized via resonance (the lone pair is allylic):

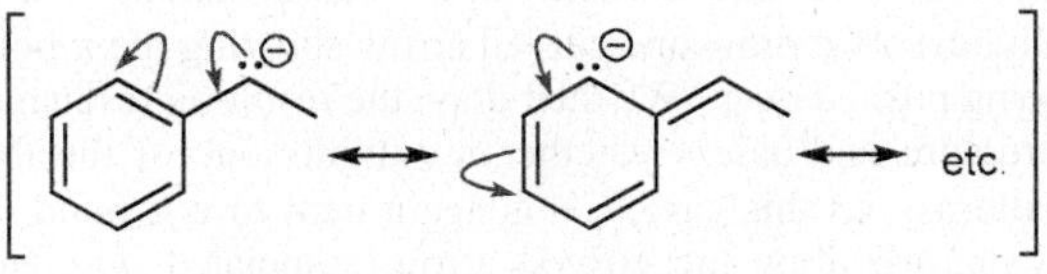

As such, the lone pair must occupy a *p* orbital.

(c) The nitrogen atom has a lone pair, which is delocalized via resonance (there is an adjacent positive charge):

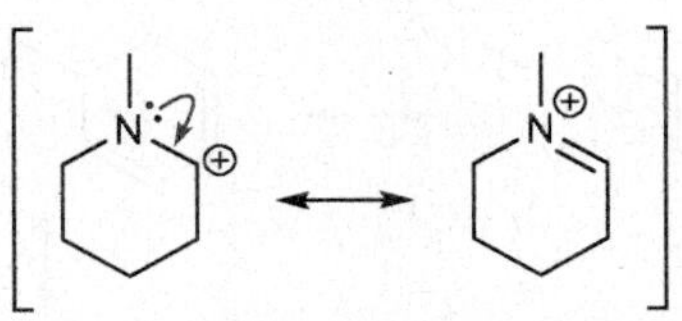

As such, the lone pair must occupy a *p* orbital.

2.47.

(a) This structure exhibits a lone pair that is adjacent to a C+ (carbocation). In fact, there are two such lone pairs (on the nitrogen and oxygen atoms). We will begin with a lone pair on the oxygen atom, although we would have arrived at the same solution either way (we will draw a total of three resonance structures, below, and it is just a matter of the order in which we draw them). We draw one curved arrow, showing a lone pair on the oxygen atom becoming a π bond. We then draw the resulting resonance structure and assess whether it exhibits one of the five patterns. In this case, there is a lone pair next to a π bond, so we draw the two curved arrows associated with that pattern. The first curved arrow is drawn showing a lone pair on the nitrogen atom becoming a π bond, while the second curved arrow shows a π bond becoming a lone pair on the oxygen atom:

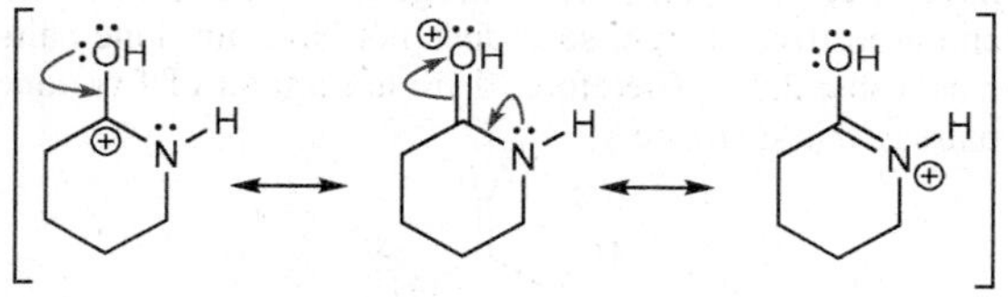

(b) This structure exhibits a C+ (carbocation) that is allylic, so we draw one curved arrow showing the π bond being pushed over. We then draw the resulting resonance structure and assess whether it exhibits one of the five patterns. In this case, C+ is adjacent to a lone pair, so we draw the curved arrow associated with that pattern (the lone pair is shown becoming a π bond):

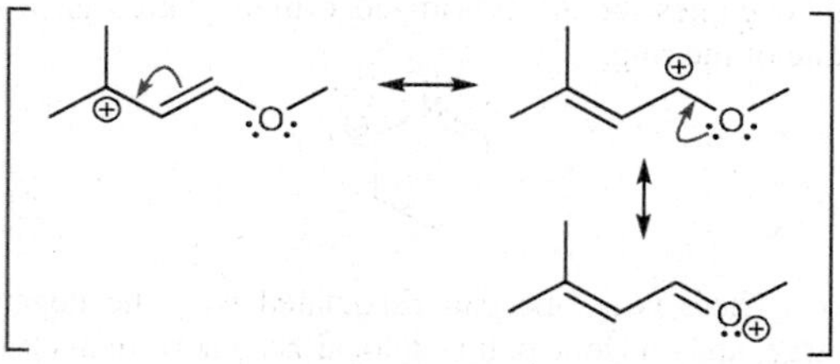

(c) This structure exhibits a C+ (carbocation) that is allylic, so we draw one curved arrow showing the π bond being pushed over. We then draw the resulting resonance structure and assess whether it exhibits one of the five patterns. In this case, C+ is again next to a π bond, so again, we draw the curved arrow associated with that pattern (pushing over the π bond again). The resulting resonance structure has C+ next to yet another π bond, so we draw a curved arrow showing the π bond being pushed over one more time to give our final resonance structure:

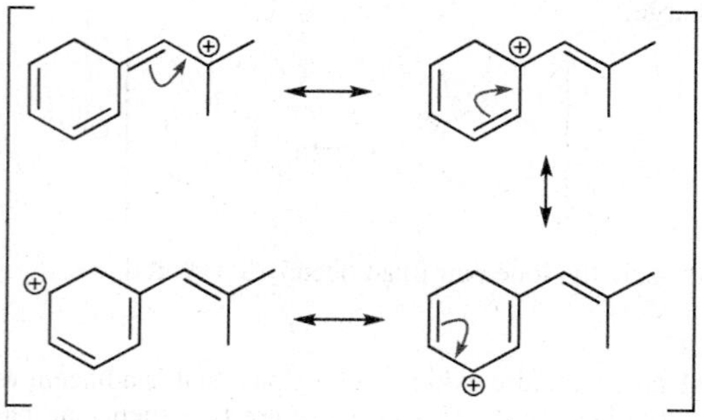

2.48.

(a) In a condensed structure, single bonds are not drawn. Instead, groups of atoms are clustered together, as shown here:

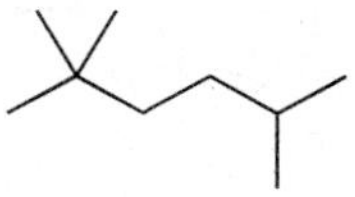

$(CH_3)_3CCH_2CH_2CH(CH_3)_2$

(b) In a condensed structure, single bonds are not drawn. Instead, groups of atoms are clustered together, as shown here:

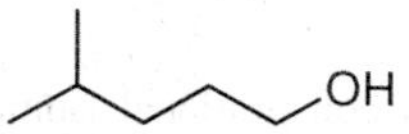

$(CH_3)_2CHCH_2CH_2CH_2OH$

or

$(CH_3)_2CH(CH_2)_3OH$

(c) In a condensed structure, single bonds are not drawn. Instead, groups of atoms are clustered together, as shown here:

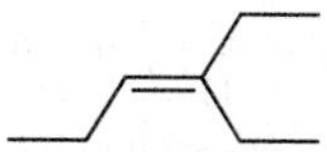

$CH_3CH_2CH{=}C(CH_2CH_3)_2$

or

$CH_3CH_2CHC(CH_2CH_3)_2$

2.49.

(a) Each corner and each endpoint represents a carbon atom, so this compound has nine carbon atoms. Each carbon atom will have enough hydrogen atoms to have exactly four bonds, giving a total of twenty hydrogen atoms. So the molecular formula is C_9H_{20}.

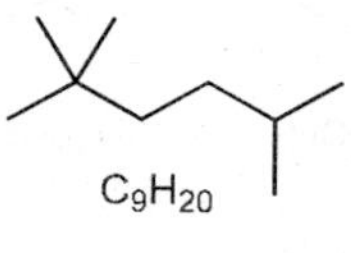

C_9H_{20}

(b) Each corner and each endpoint represents a carbon atom, so this compound has six carbon atoms. Each carbon atom will have enough hydrogen atoms to have exactly four bonds, giving a total of fourteen hydrogen atoms. So the molecular formula is $C_6H_{14}O$.

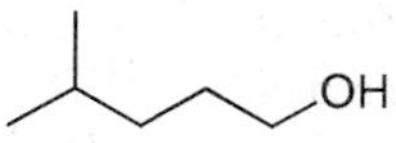

$C_6H_{14}O$

(c) Each corner and each endpoint represents a carbon atom, so this compound has eight carbon atoms. Each carbon atom will have enough hydrogen atoms to have exactly four bonds, giving a total of sixteen hydrogen atoms. So the molecular formula is C_8H_{16}.

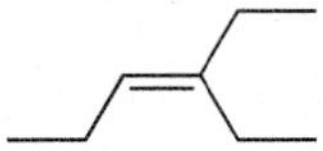

C_8H_{16}

2.50. As seen in the solution to Problem 2.35, there are only two ways to connect four carbon atoms in a compound with the molecular formula C_4H_{10}:

linear branched

In our case, the molecular formula is C_4H_9Cl, which is similar to C_4H_{10}, but one H has been replaced with a chlorine atom. So, we must explore all of the different locations where a chlorine atom can be placed on each of the carbon skeletons above (the linear skeleton and the branched skeleton).

Let's begin with the linear skeleton. There are two distinctly different locations where a chlorine atom can be placed on this skeleton: either at position 1 or position 2, shown here:

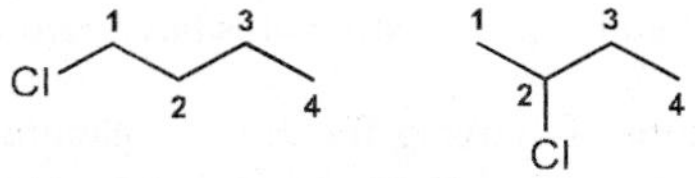

Placing the chlorine atom at position 3 would be the same as placing it at position 2; and placing the chlorine atom at position 4 would be the same as it as position 1:

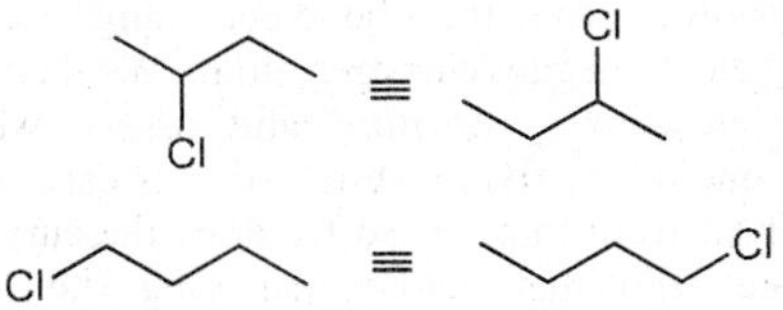

Next, we move on to the other carbon skeleton, containing a branch. Once again, there are two distinctly different locations where a chlorine atom can be placed: either at position 1 or position 2, shown here:

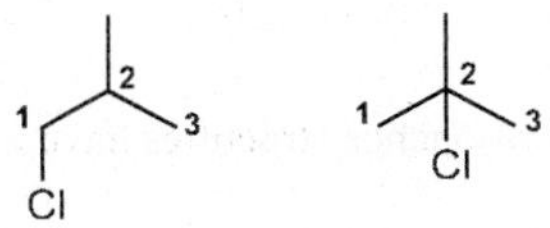

Placing the chlorine atom on any of the peripheral carbon atoms will lead to the same compound:

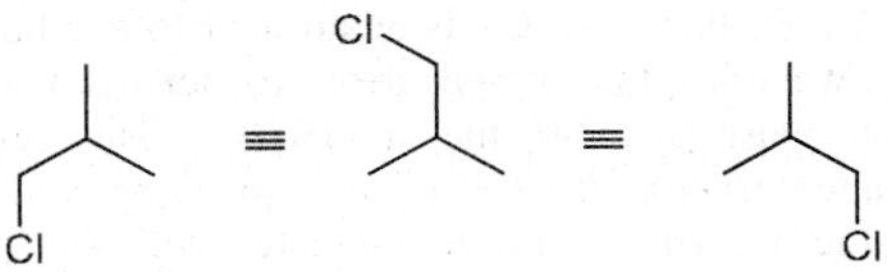

In summary, there are a total of four constitutional isomers with the molecular formula C_4H_9Cl:

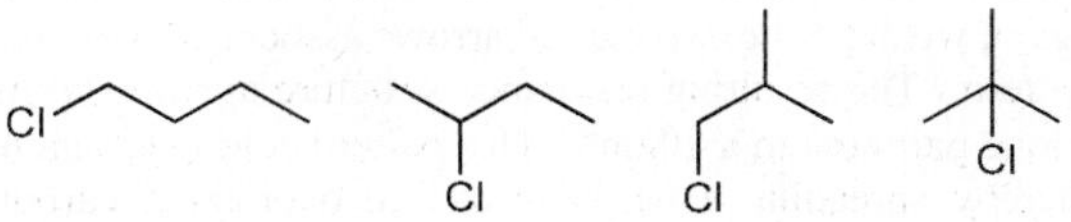

2.51.

(a) This compound exhibits a lone pair next to a π bond, so we draw the two curved arrows associated with that pattern. The first curved arrow is drawn showing a lone pair becoming a π bond, while the second curved arrow shows a π bond becoming a lone pair. Notice that both resonance structures have zero net charge:

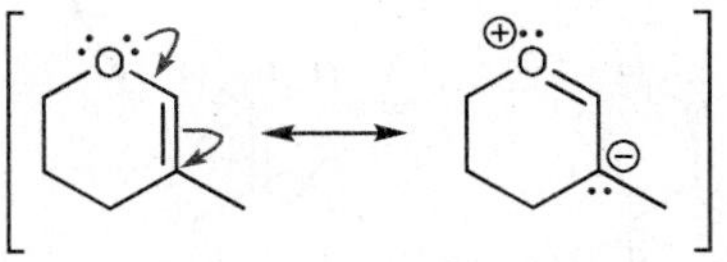

(b) This structure exhibits a lone pair that is adjacent to C+ so we draw one curved arrow, showing a lone pair becoming a π bond. Notice that both resonance structures have an overall +1 net charge, but they show the positive charge in different locations:

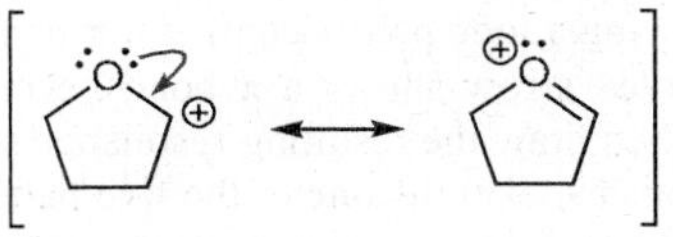

(c) This structure exhibits a C+ (carbocation) that is allylic, so we draw one curved arrow showing the π bond being pushed over. Notice that both resonance structures have an overall +1 net charge, but they show the positive charge in different locations:

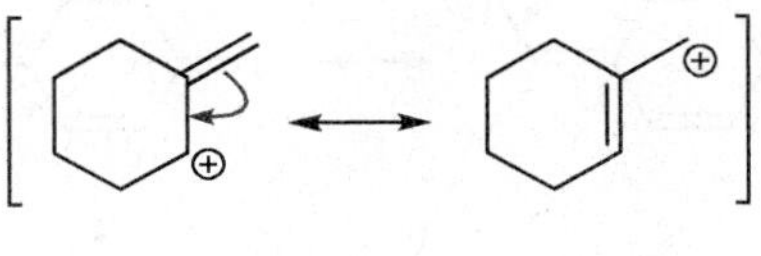

(d) This compound exhibits a C=N bond (a π bond between two atoms of differing electronegativity), so we draw one curved arrow showing the π bond becoming a lone pair on the more electronegative nitrogen atom. Notice that both resonance structures have zero net charge:

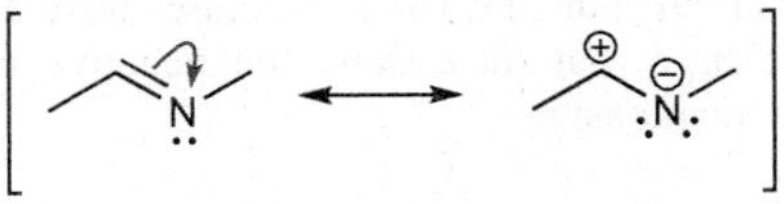

2.52.

(a) This compound exhibits a lone pair next to a π bond, so we draw two curved arrows. The first curved arrow is drawn showing a lone pair becoming a π bond, while the second curved arrow shows a π bond becoming a lone pair. We then draw the resulting resonance structure and assess whether it exhibits one of the five patterns. In this case, the lone pair is now next to another π bond, so once again, we draw the two curved arrows associated with that pattern. The resulting resonance structure again exhibits a lone pair next to a π bond. This pattern continues again, thereby spreading a negative charge over three carbon atoms in the ring, as shown here:

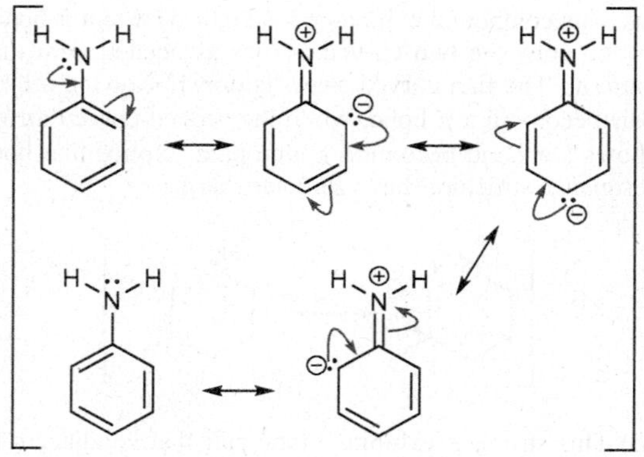

Notice that all resonance structures have zero net charge.

(b) This structure exhibits a lone pair next to a π bond, so we draw two curved arrows. The first curved arrow is drawn showing a lone pair becoming a π bond, while the second curved arrow shows a π bond becoming a lone pair. We then draw the resulting resonance structure and assess whether it exhibits one of the five patterns. In this case, the lone pair is now next to another π bond, so once again, we draw the two curved arrows associated with that pattern. The resulting resonance structure again exhibits a lone pair next to a π bond, so we draw one more resonance structure, as shown here:

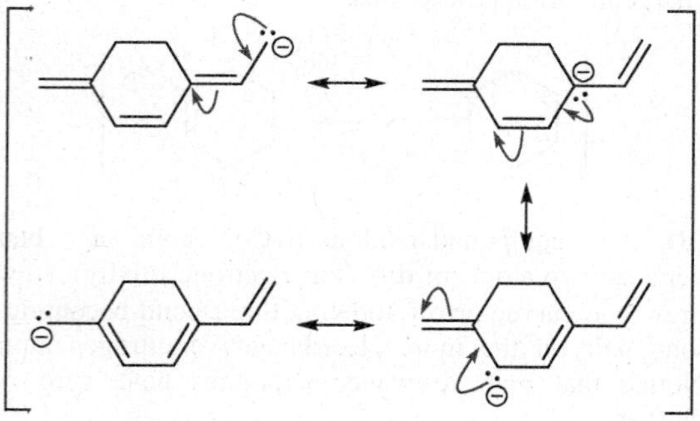

Notice that all four resonance structures have an overall –1 net charge, but they show the negative charge in different locations:

(c) This compound exhibits a C=O bond (a π bond between two atoms of differing electronegativity), so we draw one curved arrow showing the π bond becoming a lone pair on the more electronegative oxygen atom. We then draw the resulting resonance structure and assess whether it exhibits one of the five patterns. In this case, there is a C+ (carbocation) that is allylic, so we draw the curved arrow associated with that pattern (pushing over the π bond). This pattern continues, several more times, spreading the positive charge over four carbon atoms, as shown here:

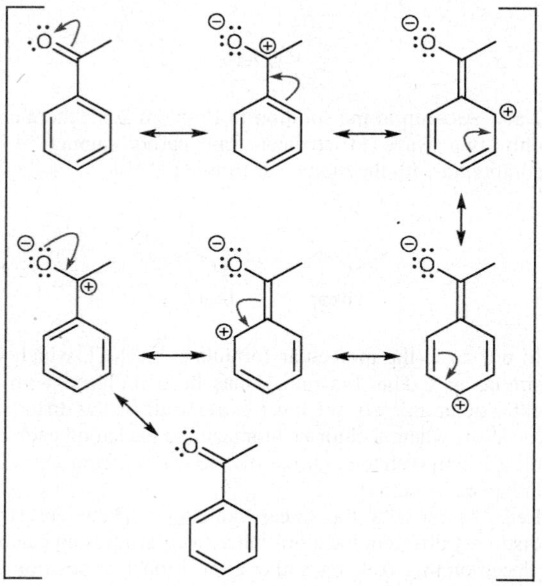

Notice that all resonance structures have zero net charge.

(d) We begin by looking for the five patterns, focusing first on any patterns that employ just one curved arrow (in this case, there is another pattern that requires two curved arrows, but we will start with the pattern using just one curved arrow). There is a C=O bond (a π bond between two atoms of differing electronegativity), so we draw one curved arrow showing the π bond becoming a lone pair on the more electronegative oxygen atom. We then draw the resulting resonance structure and assess whether it exhibits one of the five patterns. In this case, there is a lone pair adjacent to a C+, so we draw the curved arrow associated with that pattern (showing the lone pair becoming a π bond), shown here:

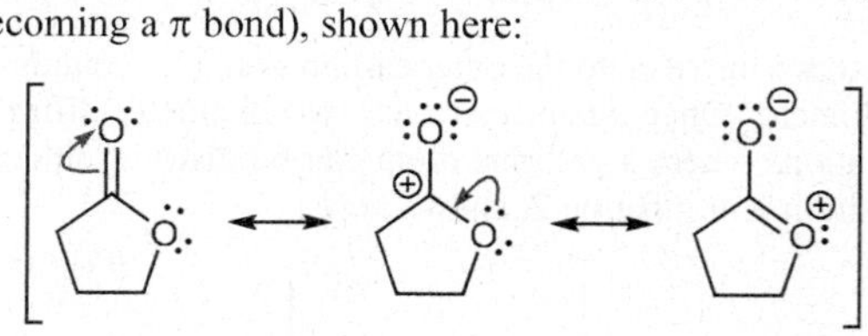

Notice that all resonance structures have zero net charge.

(e) This structure exhibits a C+ (carbocation) that is allylic, so we draw one curved arrow showing the π bond being pushed over. We then draw the resulting resonance structure and assess whether it exhibits one of the five patterns. In this case, C+ is again next to a π bond, so again, we draw the curved arrow associated with that pattern (pushing over the π bond). The resulting resonance structure has C+ adjacent to a lone pair, so we draw the one curved arrow associated with that pattern (showing the lone pair becoming a π bond), shown here:

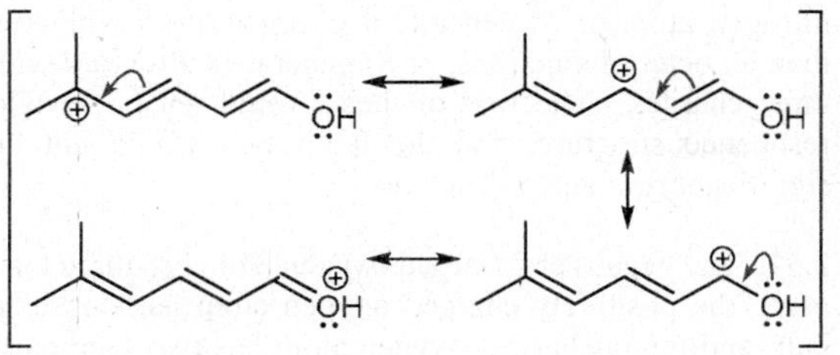

Notice that all resonance structures have an overall +1 net charge, but they show the positive charge in different locations.

(f) This structure exhibits a C+ (carbocation) that is allylic, so we draw one curved arrow showing the π bond being pushed over. We then draw the resulting resonance structure and assess whether it exhibits one of the five patterns. In this case, C+ is again next to a π bond, so again, we draw the curved arrow associated with that pattern (pushing over the π bond). The resulting resonance structure again has C+ next to a π bond, so again, we draw the curved arrow associated with that pattern (pushing over the π bond). The resulting resonance structure has C+ adjacent to a lone pair, so we draw the one curved arrow associated with that pattern (showing the lone pair becoming a π bond), shown here:

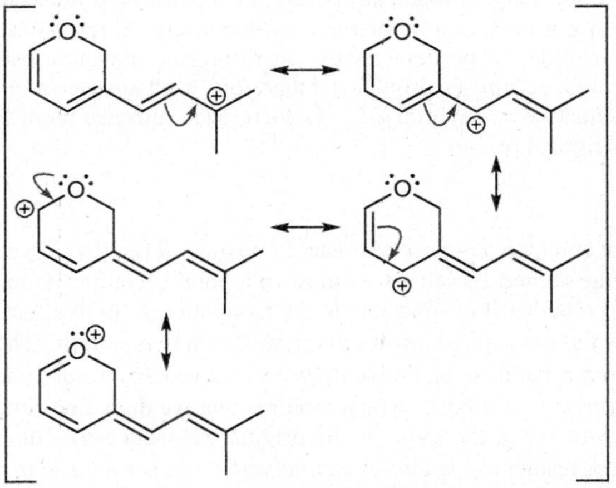

Notice that all resonance structures have an overall +1 net charge, but they show the positive charge in different locations.

2.53. These structures do not differ in their connectivity of atoms. They differ only in the placement of electrons that are in pi bonds and lone pairs. Therefore, these structures are resonance structures, as shown here:

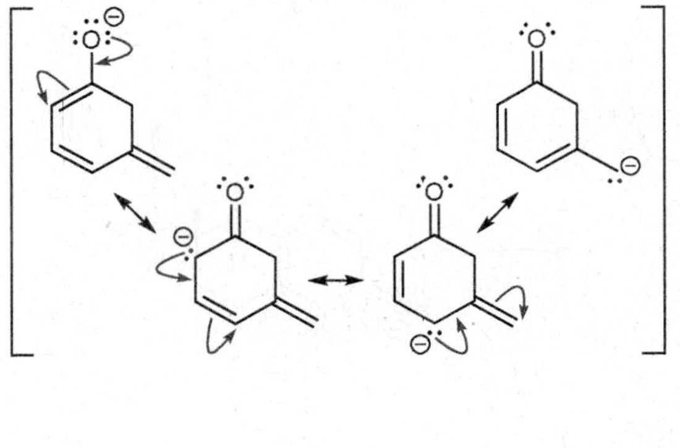

2.54.
(a) These compounds both have the same molecular formula (C_7H_{12}), but they differ in their connectivity of atoms, or constitution. Therefore, they are constitutional isomers.
(b) These structures have the same molecular formula (C_7H_{16}), AND they have the same constitution (connectivity of atoms), so they represent the same compound.
(c) The first compound has the molecular formula C_5H_{10}, while the second compound has the molecular formula C_5H_8. As such, they are different compounds that are not isomeric.
(d) These compounds both have the same molecular formula (C_5H_8), but they differ in their connectivity of atoms, or constitution. Therefore, they are constitutional isomers.

2.55.
(a) The condensed structure (shown in the problem statement) indicates the constitution (how the atoms are connected to each other). In the bond-line structure, hydrogen atoms are not drawn (they are implied). Each corner and each endpoint represents a carbon atom, so the carbon skeleton is shown more clearly.

$CH_2{=}CHCH_2C(CH_3)_3$

Condensed structure — Bond-line structure

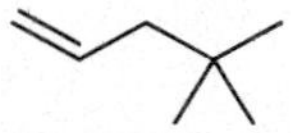

(b) The condensed structure indicates how the atoms are connected to each other. In the bond-line structure, hydrogen atoms are not drawn (they are implied), except for the hydrogen atom attached to the oxygen atom (hydrogen atoms must be drawn if they are connected to a heteroatom, such as oxygen). Each corner and each endpoint represents a carbon atom, so the carbon skeleton is shown more clearly.

$(CH_3CH_2)_2CHCH_2CH_2OH$

Condensed structure — Bond-line structure

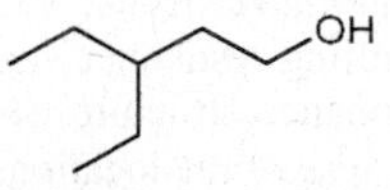

(c) The condensed structure indicates how the atoms are connected to each other. In the bond-line structure, hydrogen atoms are not drawn (they are implied). Each corner and each endpoint represents a carbon atom, so the carbon skeleton is shown more clearly.

$CH{\equiv}COCH_2CH(CH_3)_2$

Condensed structure — Bond-line structure

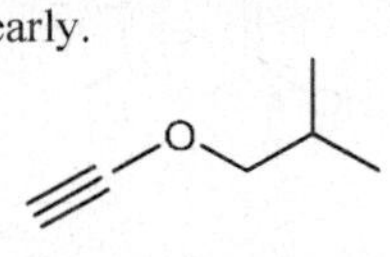

(d) The condensed structure indicates how the atoms are connected to each other. In the bond-line structure, hydrogen atoms are not drawn (they are implied). Each corner and each endpoint represents a carbon atom, so the carbon skeleton is shown more clearly.

$CH_3CH_2OCH_2CH_2OCH_2CH_3$

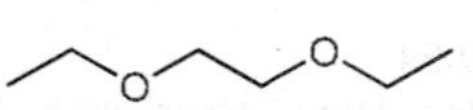

Condensed structure　Bond-line structure

(e) The condensed structure indicates how the atoms are connected to each other. In the bond-line structure, hydrogen atoms are not drawn (they are implied). Each corner and each endpoint represents a carbon atom, so the carbon skeleton is shown more clearly.

$(CH_3CH_2)_3CBr$

Condensed structure　Bond-line structure

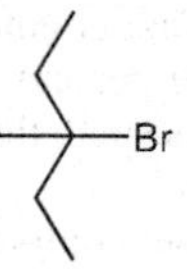

(f) The condensed structure indicates how the atoms are connected to each other. In the bond-line structure, hydrogen atoms are not drawn (they are implied). Each corner and each endpoint represents a carbon atom, so the carbon skeleton is shown more clearly.

$(CH_3)_2C{=}CHCH_3$

Condensed structure　Bond-line structure

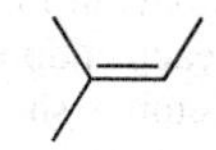

2.56. The nitronium ion does *not* have any significant resonance structures because any attempts to draw a resonance structure will either 1) exceed an octet for the nitrogen atom or 2) generate a nitrogen atom with less than an octet of electrons, or 3) generate a structure with three charges. The first of these would not be a valid resonance structure, and the latter two would not be significant resonance structures.

2.57. The negatively charged oxygen atom has three lone pairs, the positively charged oxygen atom has one lone pair, and the uncharged oxygen atom has two lone pairs (see Table 2.2). Notice that this compound exhibits a lone pair that is next to a π bond, so we must draw two curved arrows associated with that pattern. The first curved arrow is drawn showing a lone pair becoming a π bond, while the second curved arrow shows a π bond becoming a lone pair:

These two resonance forms have completely filled octets and are equivalent (they contribute equally to the resonance hybrid). There are no other valid resonance structures that are significant.

2.58. Each nitrogen atom has a lone pair that is adjacent to a π bond and is, therefore, delocalized via resonance. In order to be delocalized via resonance, the lone pair must occupy a *p* orbital, and therefore, each nitrogen atom must be sp^2 hybridized. As such, each nitrogen atom is trigonal planar.

2.59.

(a) This compound (estradiol) exhibits a lone pair next to a π bond, so we draw two curved arrows. The first curved arrow is drawn showing a lone pair becoming a π bond, while the second curved arrow shows a π bond becoming a lone pair. We then draw the resulting resonance structure and assess whether it exhibits one of the five patterns. In this case, the lone pair is next to another π bond, so once again, we draw the two curved arrows associated with that pattern. The resulting resonance structure again exhibits a lone pair next to a π bond, so again we draw two curved arrows and the resulting resonance structure. Once again, there is a lone pair next to a π bond, which requires that we draw one final resonance structure, shown below. This last resonance structure is not the same as the original resonance structure, because of the locations in which the π bonds are drawn (*i.e.*, the resonance pattern of *conjugated π bonds enclosed in a ring*). Notice that all resonance structures have zero net charge:

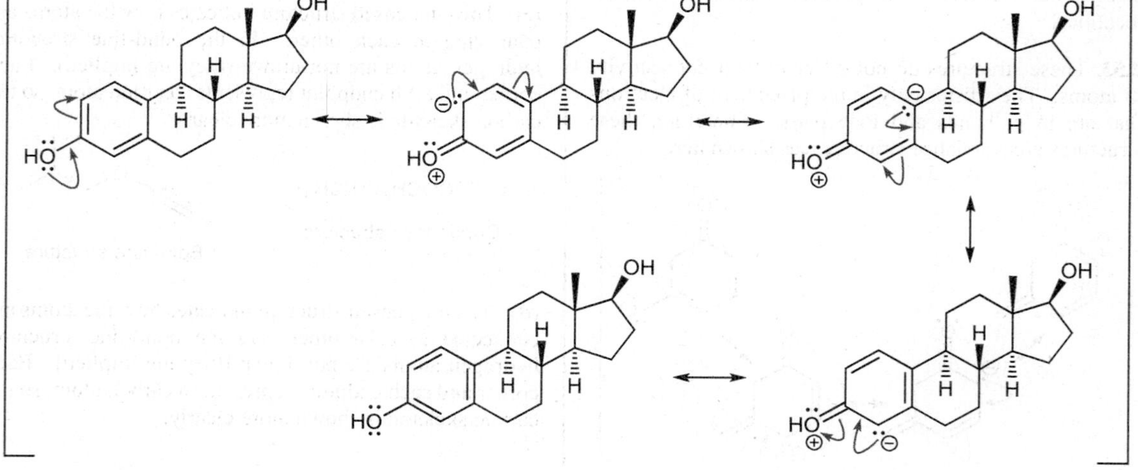

(b) The following compound (testosterone) exhibits a C=O bond (a π bond between two atoms of differing electronegativity), so we draw one curved arrow showing the π bond becoming a lone pair on the more electronegative oxygen atom. We then draw the resulting resonance structure and assess whether it exhibits one of the five patterns. In this case, there is a C+ (carbocation) that is allylic, so we draw the curved arrow associated with that pattern (pushing over the π bond). Notice that all resonance structures have zero net charge:

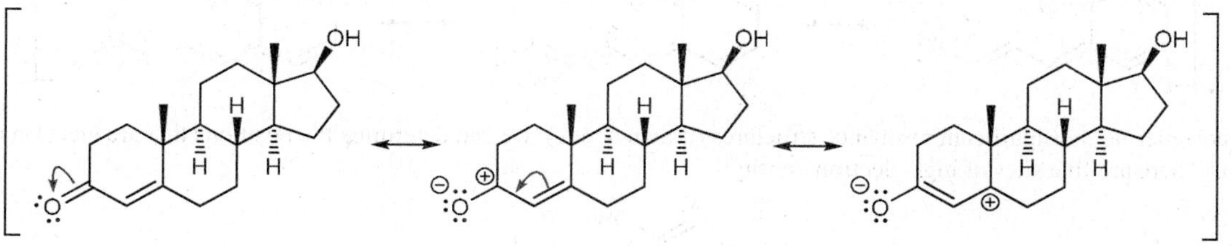

2.60. This compound exhibits a C=O bond (a π bond between two atoms of differing electronegativity), so we draw one curved arrow showing the π bond becoming a lone pair on the more electronegative oxygen atom. We then draw the resulting resonance structure and assess whether it exhibits one of the five patterns. In this case, there is an allylic C+ (carbocation), so we draw the curved arrow associated with that pattern (pushing over the π bond), shown here. This pattern continues, several more times, spreading the positive charge over a total of six carbon atoms. Notice that all resonance structures have zero net charge:

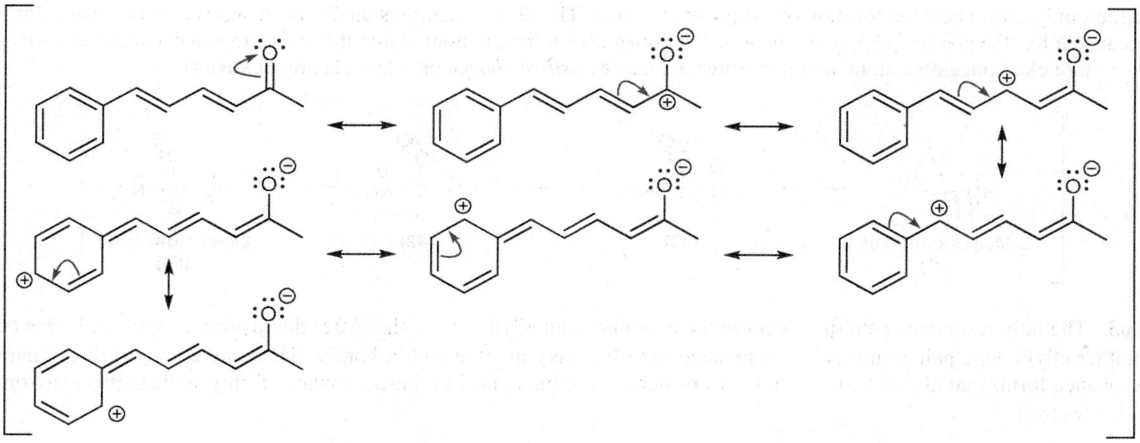

By considering the significant resonance structures (drawn above), we can determine the positions that are electron deficient (δ+). There are six sites of low electron density:

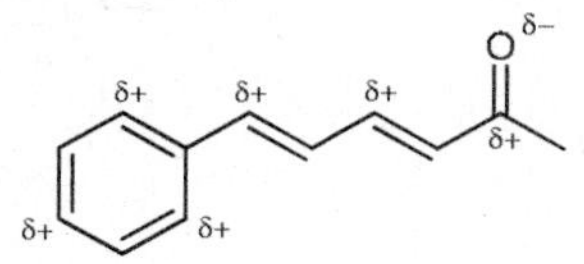

2.61. This compound exhibits a lone pair next to a π bond, so we draw two curved arrows. The first curved arrow is drawn showing a lone pair becoming a π bond, while the second curved arrow shows a π bond becoming a lone pair. We then draw the resulting resonance structure and assess whether it exhibits one of the five patterns. In this case, there is a lone pair next to a π bond, so once again, we draw the two curved arrows associated with that pattern. The resulting resonance structure again exhibits a lone pair next to a π bond. This pattern continues, several more times, spreading a negative charge over a total of five carbon atoms. Notice that all resonance structures have zero net charge:

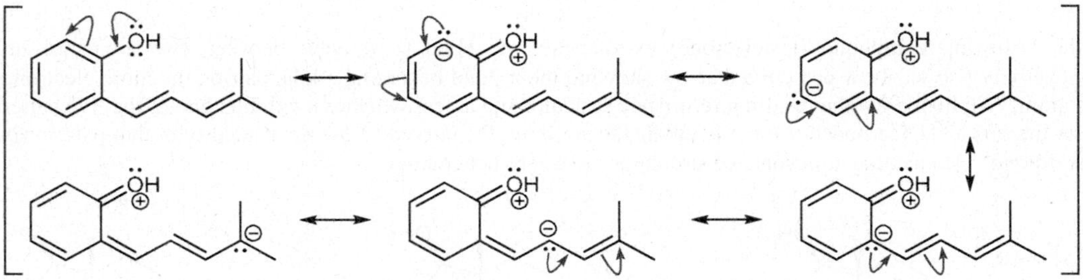

By considering the significant resonance structures (drawn above), we can determine the positions that are electron rich (δ–). There are five sites of high electron density:

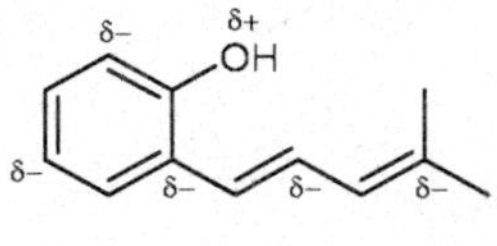

2.62. Two patterns of resonance can be identified on the given structure: carbonyl resonance and allylic lone pair resonance (involving either the oxygen or nitrogen lone pairs). All three of these options will be used. Since it is possible to start with any one of the three, you may have developed the resonance forms in a different order than presented here, but you still should have found four reasonable resonance forms. Only one of the four structures has an atom with an incomplete octet (the last resonance structure shown), so that is identified as the least significant contributor to the hybrid. The first resonance form is the most significant contributor because it has filled octets and no formal charges. The middle two resonance structures both have filled octets and a negative charge on an oxygen atom, so they are ranked according to their only difference: the location of the positive charge. The third structure is the 2nd most significant resonance form because it has the positive charge on the less electronegative nitrogen atom. Note it is better to place a *negative charge* on a more electronegative atom, and it is better to place a *positive charge* on a less electronegative atom.

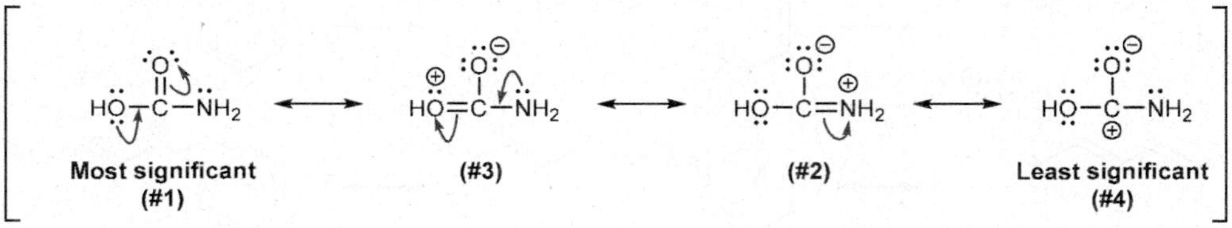

Most significant (#1) (#3) (#2) **Least significant (#4)**

2.63. The only resonance pattern evident in the enamine is an allylic lone pair. After that pattern is applied, however, another allylic lone pair results so the resonance can ultimately involve both π bonds. There are a total of three major resonance forms that all have filled octets. Consideration of the hybrid of these resonance forms predicts two electron-rich sites (δ–).

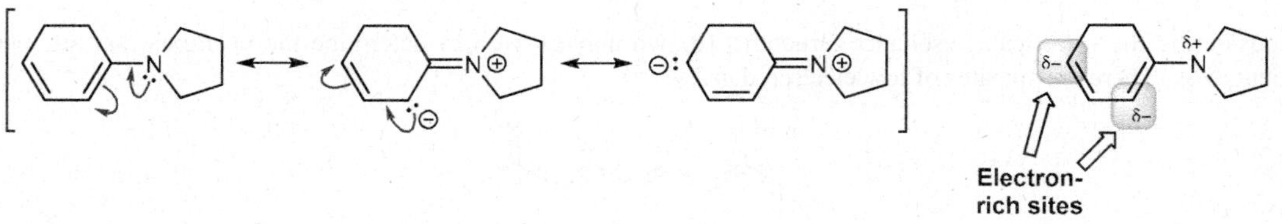

Electron-rich sites

Did you draw the following additional structure (or something similar, with C+ and C–) and wonder why it was not shown in this solution?

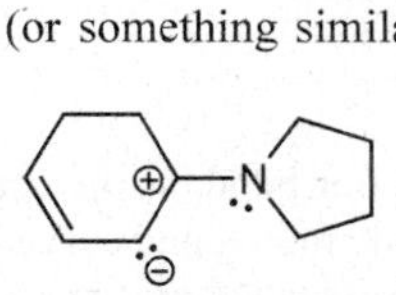

Insignificant

This resonance form suffers from two major deficiencies: 1) it does not have filled octets, while the other resonance forms shown above all have filled octets, and 2) it has a negative charge on a carbon atom (which is not an electronegative atom). Either of these deficiencies alone would render the resonance form a minor contributor. But with both deficiencies together (C+ and C–), this resonance form is insignificant. The same is true for any resonance form that has both C+

and C–. Such a resonance form will generally be insignificant (there are very few exceptions, one of which will be seen in Chapter 17).

Also, note that the π bonds cannot be moved to other parts of the six-membered ring since the CH_2 groups are sp^3 hybridized. These carbon atoms cannot accommodate an additional bond without violating the octet rule.

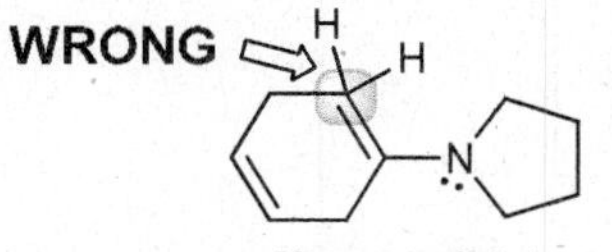

WRONG

Not possible!
(pentavalent carbon)

2.64. The correct answer is (c). Structures (a), (b) and (d) are all significant resonance structures, as shown:

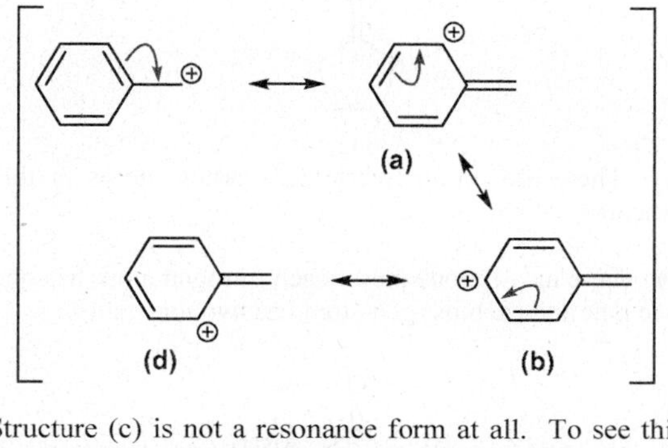

(a) (d) (b)

Structure (c) is not a resonance form at all. To see this more clearly, notice that the benzyl carbocation does not have any CH_2 groups in the ring, but structure (c) does have a CH_2 group in the ring:

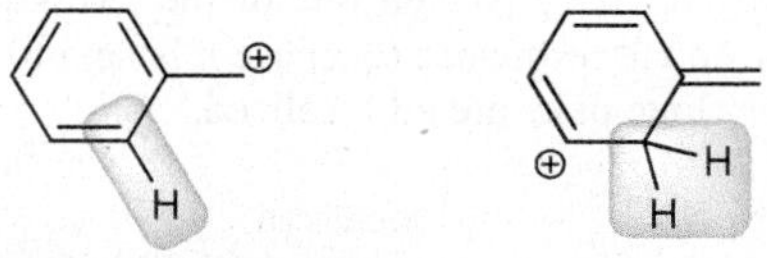

Benzyl carbocation (c)

Resonance structures differ only in the placement of electrons. Since structure (c) differs in the connectivity of atoms, it cannot be considered a resonance structure of the benzyl carbocation.

2.65. The correct answer is (a). The atoms in all four structures have complete octets. So we must consider the location of the negative charge. Structure (a) has a negative charge on an electronegative atom (oxygen). A negative charge is more stable on the more electronegative atom (oxygen) than it is on a nitrogen atom or a carbon atom. Therefore, structure (a) will contribute the most character to the resonance hybrid:

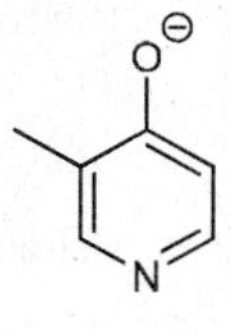

It turns out that this resonance structure is also significant for another reason that we will explore in Chapter 17.

Specifically, this resonance structure has alternating single and double bonds in the ring and is considered to be aromatic (a stabilizing effect).

2.66. The correct answer is (d). The nitrogen atom in structure (a) is delocalized by resonance, and is therefore sp^2 hybridized:

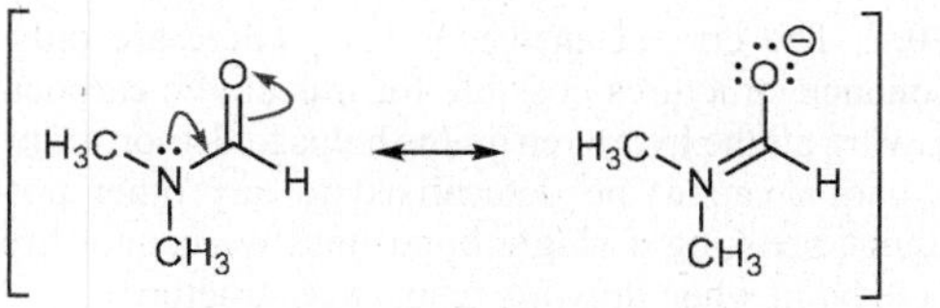

The nitrogen atom in structure (b) is also delocalized, so it is sp^2 hybridized as well:

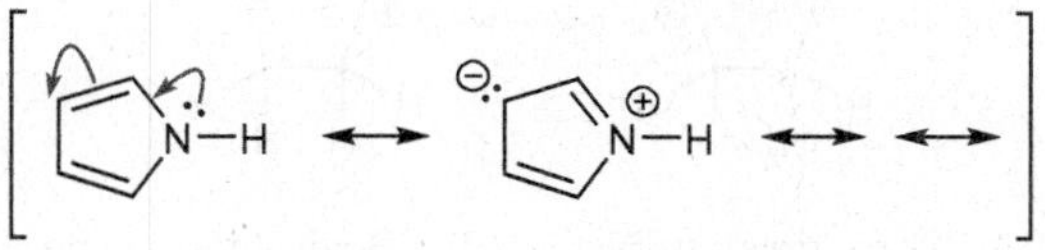

The nitrogen atom in structure (c) is also sp^2 hybridized, because this nitrogen atom must be using a *p* orbital to participate in π bonding (C=N). The nitrogen atom in structure (d) has three σ bonds, and its lone pair is localized. Therefore, this nitrogen atom is sp^3 hybridized:

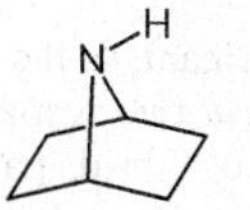

2.67. The correct answer is (c). We begin by considering all bonds in the compound, which is easier to do if we redraw the compound as shown:

$(CH_3)_2CHCH_2OC(CH_3)_3$ ≡

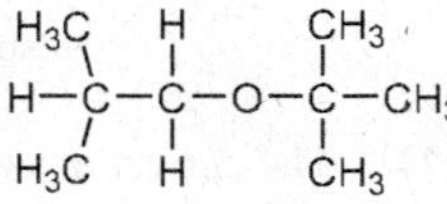

This compound corresponds structure (c):

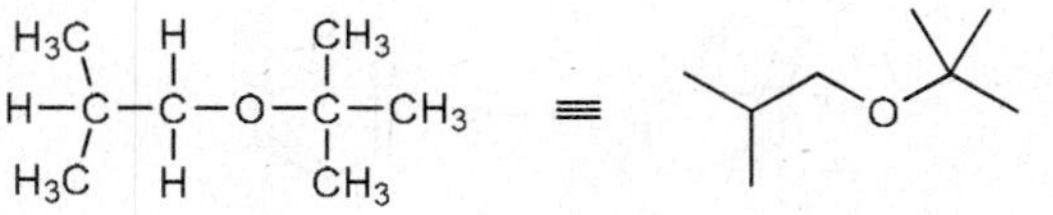

2.68. The correct answer is (c). For C+ that is allylic, only one curved arrow will be required to draw the resonance structures (pushing over the π bond):

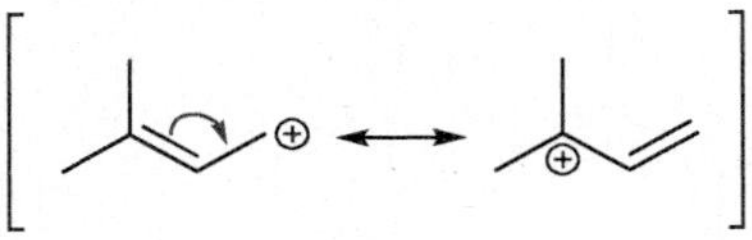

2.69. The correct answer is (a). An anion will be resonance delocalized if the atom bearing the negative charge has a lone pair that is next to a π bond. Only compound **I** fits this pattern:

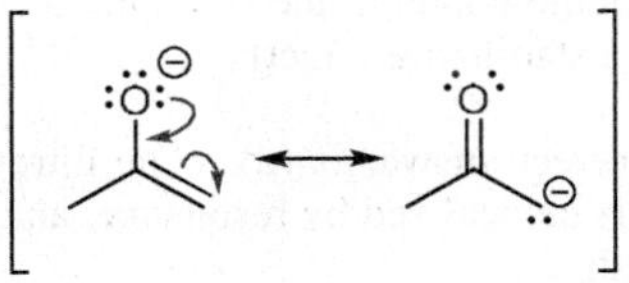

2.70. The correct answer is (a). There are only two resonance structures possible for this allylic carbocation. Drawing all the hydrogen atoms helps to demonstrate why the cation cannot be delocalized to any other position without breaking a single bond (and we cannot break a single bond when drawing resonance structures).

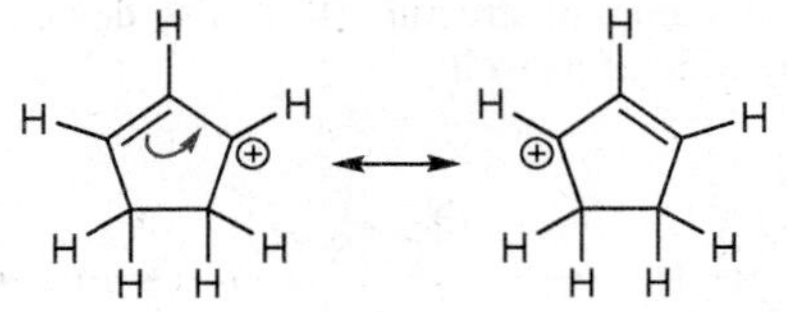

2.71. The correct answer is (d). Begin by drawing all significant resonance structures. In this case, there are two:

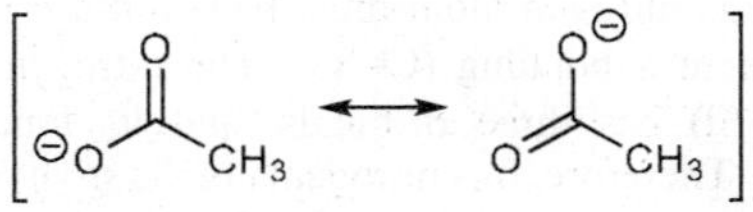

Both are equally significant, so the resonance hybrid is the simple average of these two resonance structures. When drawing the resonance hybrid, partial bonds and partial charges are required, as shown below:

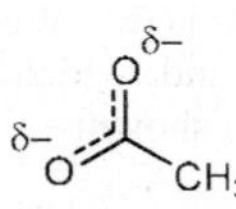

Resonance hybrid

2.72. The correct answer is (b). The following do not represent a pair of resonance structures, because the first rule of resonance has been violated (by breaking a single bond):

2.73.

(a) The molecular formula is $C_3H_6N_2O_2$.

(b) Each of the highlighted carbon atoms (below) has four sigma bonds (the bonds to hydrogen are not shown). As such, these two carbon atoms are sp^3 hybridized.

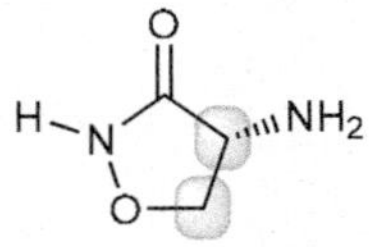

(c) There is one carbon atom that is using a *p* orbital to form a π bond. As such, this carbon atom (highlighted) is sp^2 hybridized.

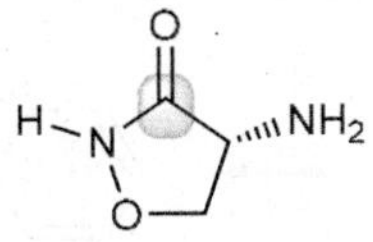

(d) There are no *sp* hybridized carbon atoms in this structure.

(e) There are six lone pairs (each nitrogen atom has one lone pair and each oxygen atom has two lone pairs):

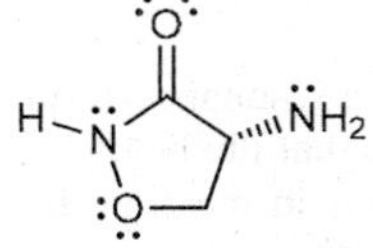

(f) Only the lone pair on one of the nitrogen atoms is delocalized via resonance (because it is next to a π bond). The other lone pairs are all localized.

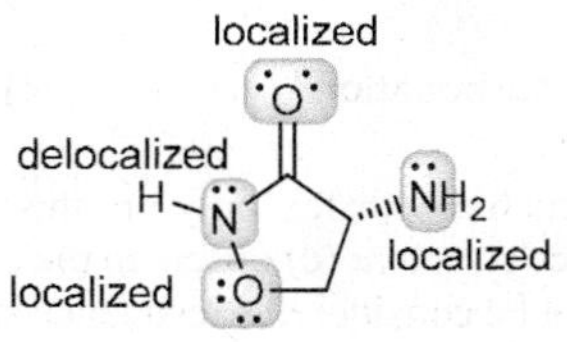

localized
delocalized
localized
localized

(g) The geometry of each atom is shown below (see SkillBuilder 1.8):

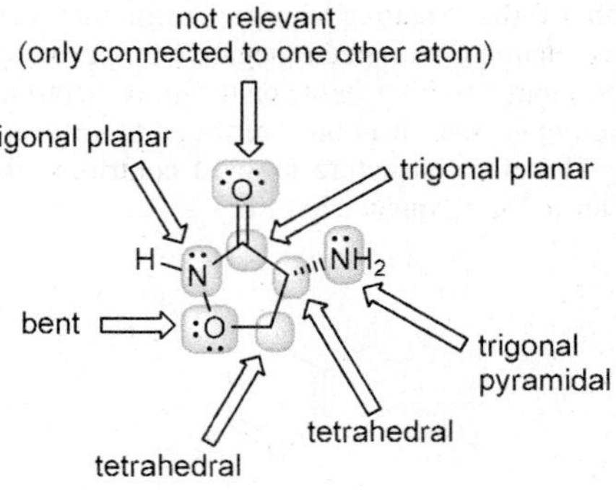

not relevant (only connected to one other atom)
trigonal planar
trigonal planar
bent
trigonal pyramidal
tetrahedral
tetrahedral

(h) We begin by looking for the five patterns, focusing first on any patterns that employ just one curved arrow (in this case, there is another pattern that requires two curved arrows, but we will start with the pattern using just one curved arrow). There is a C=O bond (a π bond between two atoms of differing electronegativity), so we draw one curved arrow showing the π bond becoming a lone pair on the more electronegative oxygen atom. We then draw the resulting resonance structure and assess whether it exhibits one of the five patterns. In this case, there is a lone pair adjacent to a C+, so we draw the curved arrow associated with that pattern (showing the lone pair becoming a π bond). Notice that all three resonance structures have zero net charge:

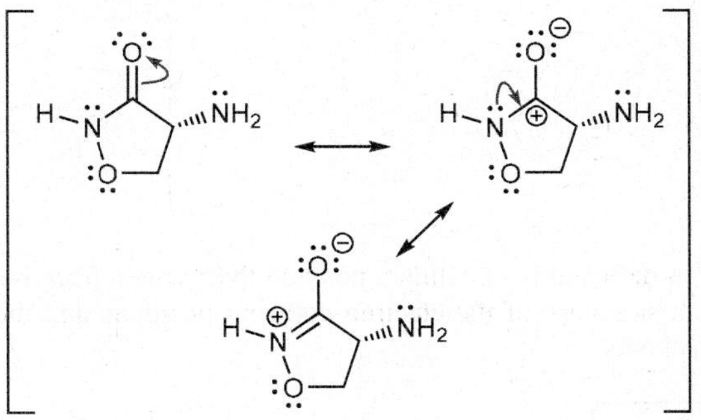

2.74.

(a) The molecular formula is $C_{16}H_{21}NO_2$.

(b) Each of the highlighted carbon atoms (below) has four sigma bonds (the bonds to hydrogen are not shown). As such, these nine carbon atoms are sp^3 hybridized.

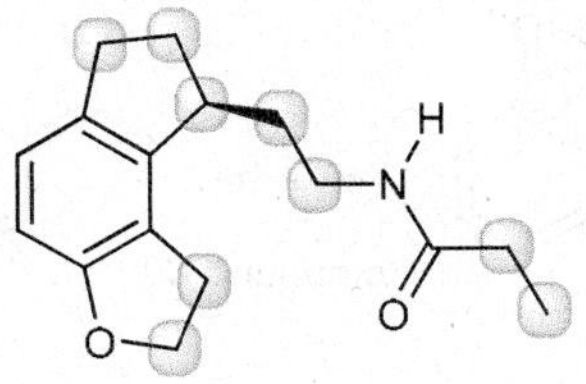

(c) There are seven carbon atoms that are each using a *p* orbital to form a π bond. As such, these seven carbon atoms (highlighted) are sp^2 hybridized.

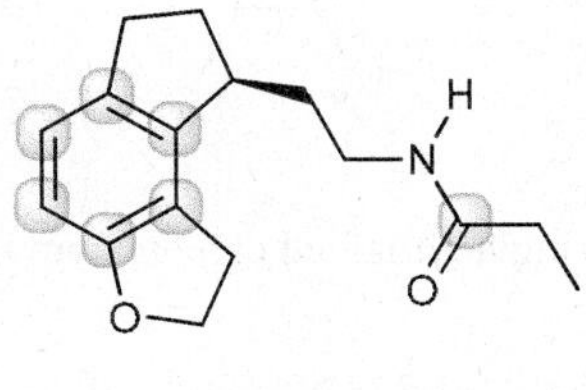

(d) There are no *sp*-hybridized carbon atoms in this structure.

(e) There are five lone pairs (the nitrogen atom has one lone pair and each oxygen atom has two lone pairs):

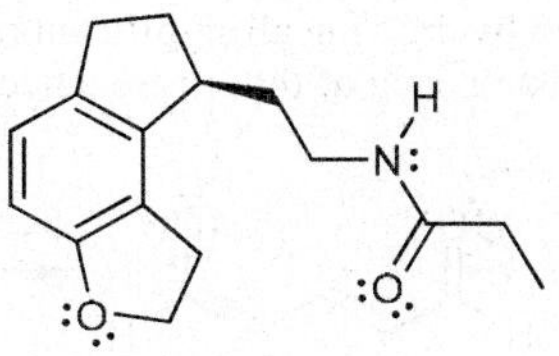

(f) Recall that a lone pair next to a π bond will be delocalized by resonance. The lone pairs on the oxygen of the C=O bond are localized. One of the lone pairs on the other oxygen atom (attached to the aromatic ring) is delocalized via resonance. The lone pair on the nitrogen atom is delocalized via resonance.

(g) All sp^2-hybridized carbon atoms are trigonal planar. All sp^3 hybridized carbon atoms are tetrahedral. The nitrogen atom is trigonal planar. The oxygen atom of the C=O bond does not have a geometry because it is connected to only one other atom, and the other oxygen atom has bent geometry (see SkillBuilder 1.8).

2.75.

(a) In Section 1.5, we discussed inductive effects and we learned how to identify polar covalent bonds. In this case, there are two carbon atoms that participate in polar covalent bonds (the C–Br bond and the C–O bond). Each of these carbon atoms will be poor in electron density (δ+) because oxygen and bromine are each more electronegative than carbon:

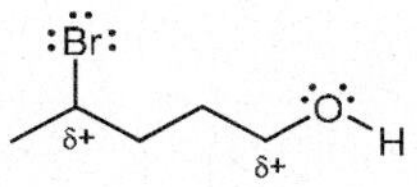

(b) There are two carbon atoms that are adjacent to one or both oxygen atoms. These carbon atoms will be poor in electron density (δ+), because oxygen is more electronegative than carbon:

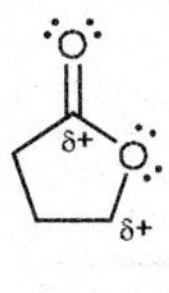

The carbon atom of the carbonyl (C=O) group is especially electron deficient, as a result of resonance.

(c) There are two carbon atoms that are adjacent to electronegative atoms. These carbon atoms will be poor in electron density (δ+), because oxygen and chlorine are each more electronegative than carbon:

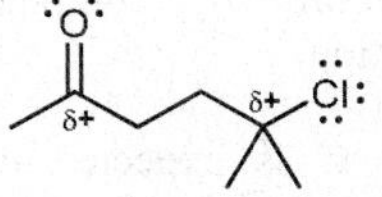

The carbon atom of the carbonyl (C=O) group is especially electron deficient, as a result of resonance.

2.76. We begin by drawing all significant resonance structures, and then considering the placement of the formal charges on carbon atoms in each of those resonance structures (highlighted below)

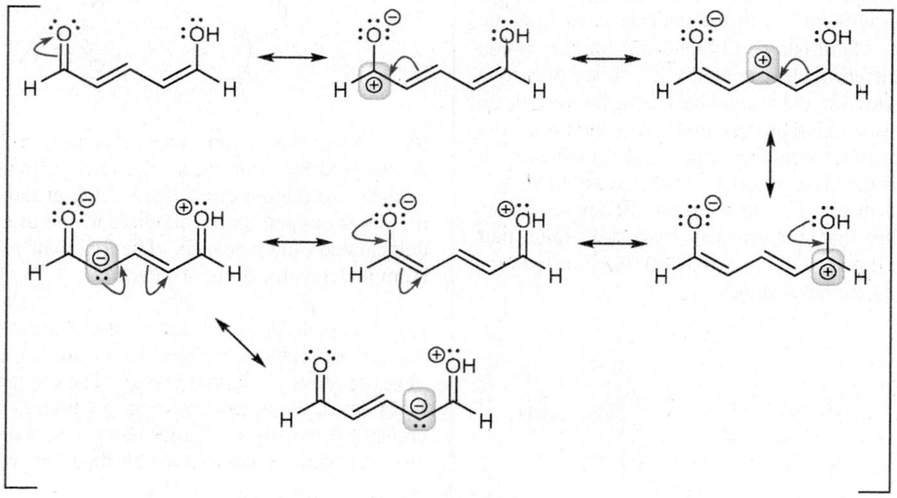

A position that bears a positive charge is expected to be electron deficient (δ+), while a position that bears a negative charge is expected to be electron rich (δ–). The following is a summary of the electron-deficient positions and the electron-poor positions, as indicated by the resonance structures above.

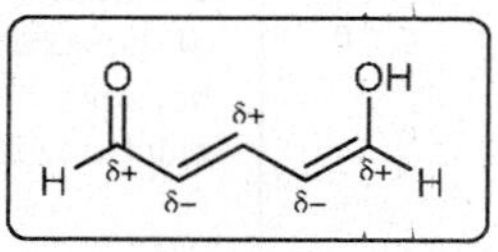

2.77.

(a) Compound **B** has one additional resonance structure that compound **A** lacks, because of the relative positions of the two groups on the aromatic ring. Specifically, compound **B** has a resonance structure in which one oxygen atom has a negative charge and the other oxygen atom has a positive charge:

Compound **B**

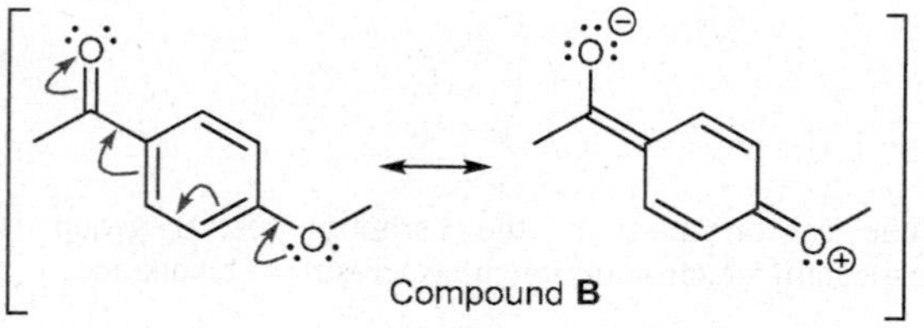

Compound **A** does *not* have a significant resonance structure in which one oxygen atom has a negative charge and the other oxygen atom has a positive charge. That is, compound **A** has fewer resonance structures than compound **B**. Accordingly, compound **B** has greater resonance stabilization.

(b) Compound **C** is expected to have resonance stabilization similar to that of compound **B**, because compound **C** also has a resonance structure in which one oxygen atom has a negative charge and the other oxygen atom has a positive charge:

Compound **C**

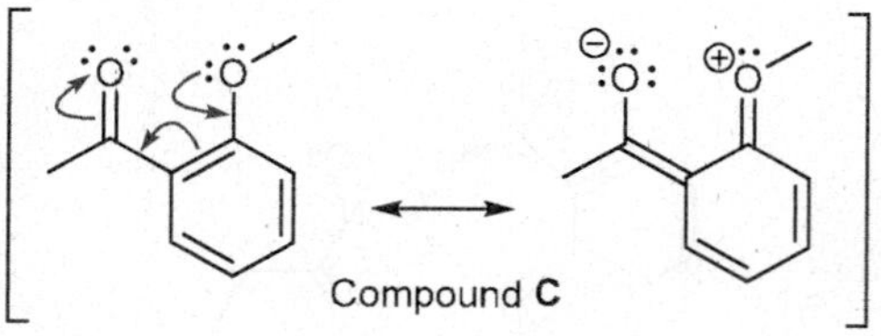

2.78.

(a) The following group is introduced, and it contains five carbon atoms:

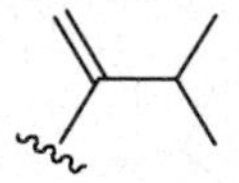

(b) The following highlighted carbon atom is involved in the reaction:

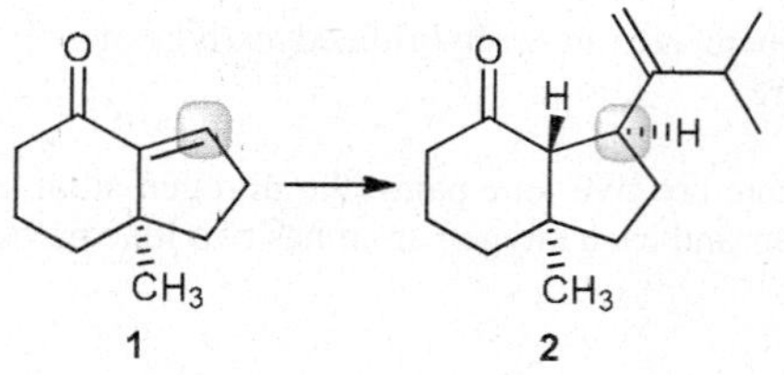

1

2

(c) Compound **1** has three significant resonance structures, shown here:

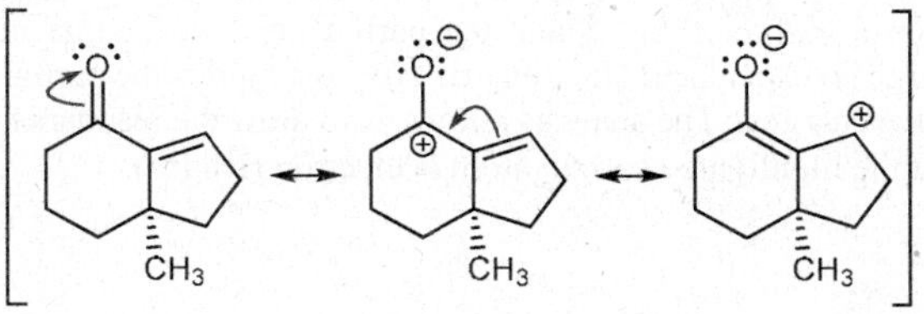

The structure on the left is the most significant, because every atom has an octet and it has no formal charges. The resonance hybrid is a weighted average of these three resonance structures. Since the partial positive charge is delocalized onto two carbon atoms and the partial negative charge is localized on only one oxygen atom, the partial negative charge is drawn larger than each of the individual partial positive charges.

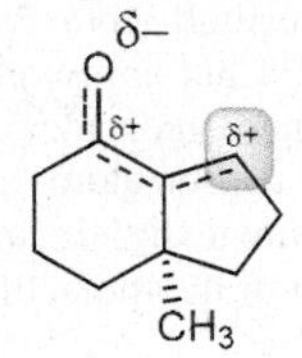

(d) The reactive site (highlighted above) has partial positive character, which means that it is electron deficient. This is what makes it reactive. In the actual synthesis, this compound is treated with a carbanion (a structure containing a carbon atom with a negative charge). The reaction causes formation of a bond between the electron-deficient carbon atom and the electron-rich carbon atom. We will learn about such reactions in Chapter 21.

2.79. We will need to draw two resonance hybrids, one for each of the highlighted carbon atoms. One highlighted position is part of an aromatic ring, and we will begin by focusing on that position. In doing so, we can save time by redrawing only the relevant portion of the molecule, like this:

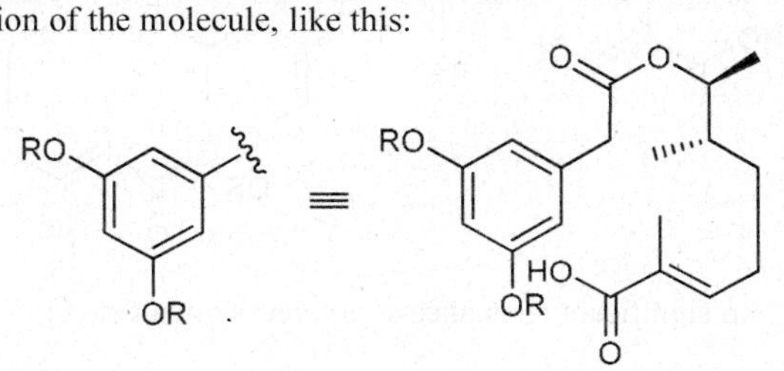

This aromatic ring has eight significant resonance structures, shown here:

resonance structure 1

resonance structure 2

resonance structure 1

resonance structure 3

resonance structure 4

resonance structure 5

resonance structure 1

resonance structure 6

resonance structure 7

resonance structure 8

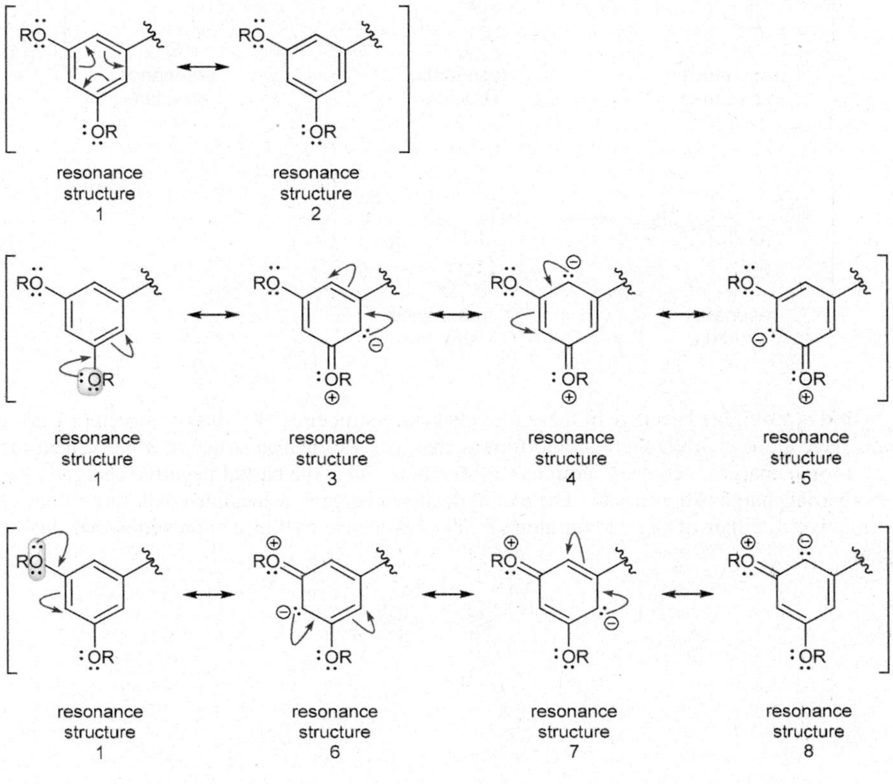

The resonance hybrid is a weighted average of these eight resonance structures. Resonance structures **1** and **2** are equally most significant because all atoms have an octet AND there are no formal charges. Resonance structures **3-8** are less significant than **1** and **2** but approximately equally significant to each other because every atom has a full octet with a positive charge on an oxygen atom and a negative charge on a carbon atom. Since the partial negative charge is delocalized onto three carbon atoms and the partial positive charge is delocalized onto only two oxygen atoms, the partial positive charges are drawn slightly larger than the partial negative charges. The analysis allows us to draw the resonance hybrid for this portion of the molecule, and it demonstrates that the highlighted carbon atom is electron rich (δ-).

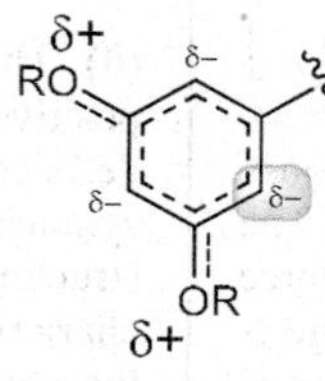

Now let's focus on our attention on the other highlighted carbon atom (the one that is part of a carboxylic acid group). Just as we did before, we will draw only the portion of the molecule that is of interest:

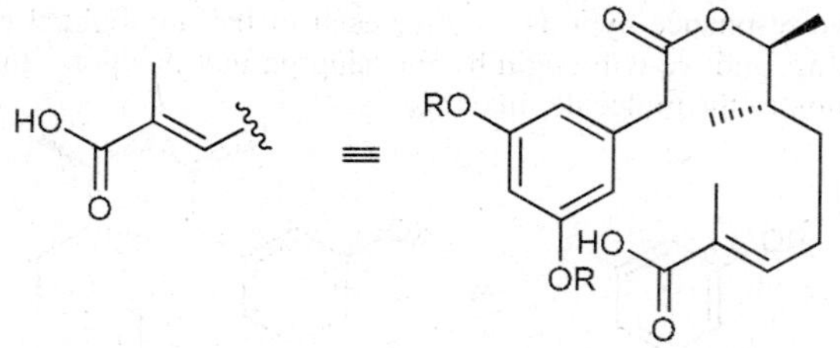

This portion of the molecule has four significant resonance structures, shown below:

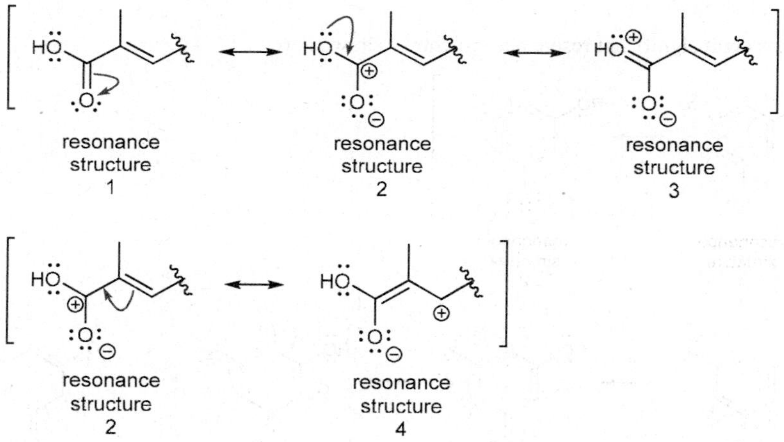

resonance structure 1

resonance structure 2

resonance structure 3

resonance structure 2

resonance structure 4

The resonance hybrid is a weighted average of these four resonance structures. Resonance structure **1** is most significant because all atoms have an octet AND there are no formal charges. Resonance structure **3** is the next most significant because there are formal charges, yet every atom has a full octet. Since the partial negative charge is localized on one oxygen atom, this partial charge is the largest. The partial positive charge is delocalized over three atoms, but it is larger on the oxygen (relative to either of the carbon atoms). The resonance hybrid demonstrates that this carbon atom is electron poor (δ+).

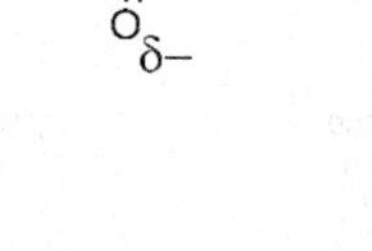

2.80.

(a) The molecular formula for CL-20 is $C_6H_6N_{12}O_{12}$. The molecular formula for HMX is $C_4H_8N_8O_8$.

(b) The lone pair is next to a π bond, and therefore delocalized (see resonance structures below).

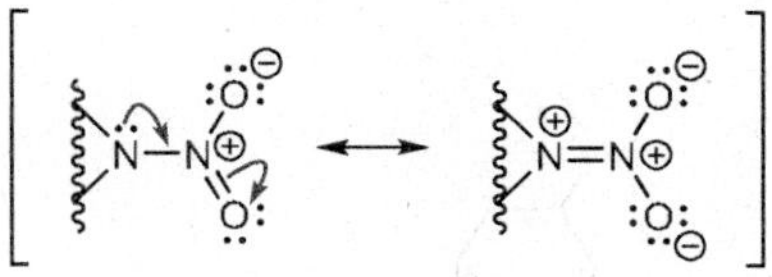

2.81.

This intermediate is highly stabilized by resonance. The positive charge is spread over one carbon atom and three oxygen atoms.

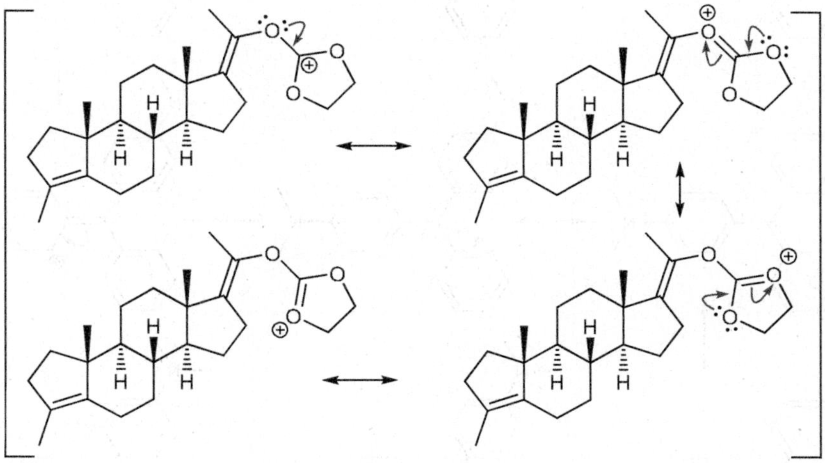

2.82.

(a) Both molecules have identical functional groups (alcohol + alkene), and both compounds have the same molecular formula ($C_{27}H_{46}O$):

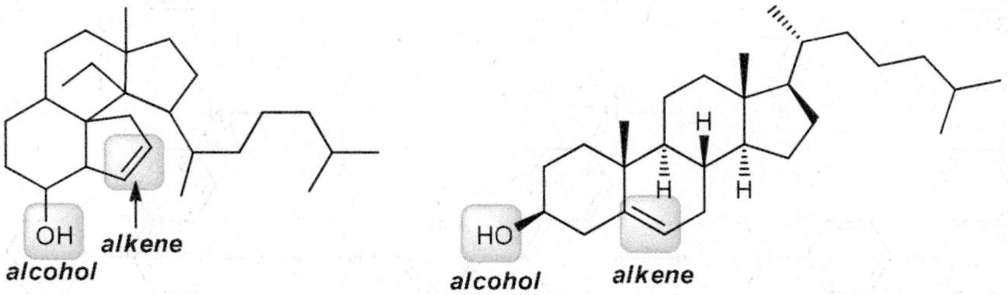

The structure on the left exhibits two six-membered rings and two five-membered rings, while the structure on the right has three six-membered rings and only one five-membered ring.

The long alkane group is apparently located in the wrong position on the five-membered ring of the incorrect structure.

(b) Both structures contain an alkene group, an aromatic ring, an amide group, and two ether functional groups. But the incorrect structure has a third ether functional group (in the eight-membered ring), while the correct structure has an alcohol functional group. The incorrect structure has an eight-membered ring, while the correct structure has a five-membered ring. The two carbon atoms and oxygen atom in the ring of the incorrect structure are not part of the ring for the correct structure.

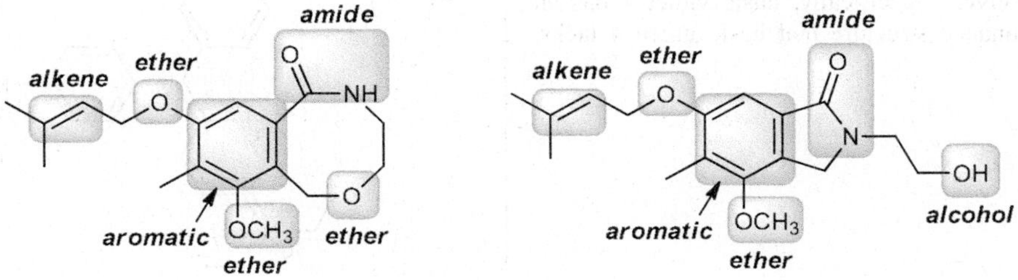

2.83.

(a) The positive charge in basic green 4 is resonance-stabilized (delocalized) over twelve positions (two nitrogen atoms and ten carbon atoms), as seen in the following resonance structures.

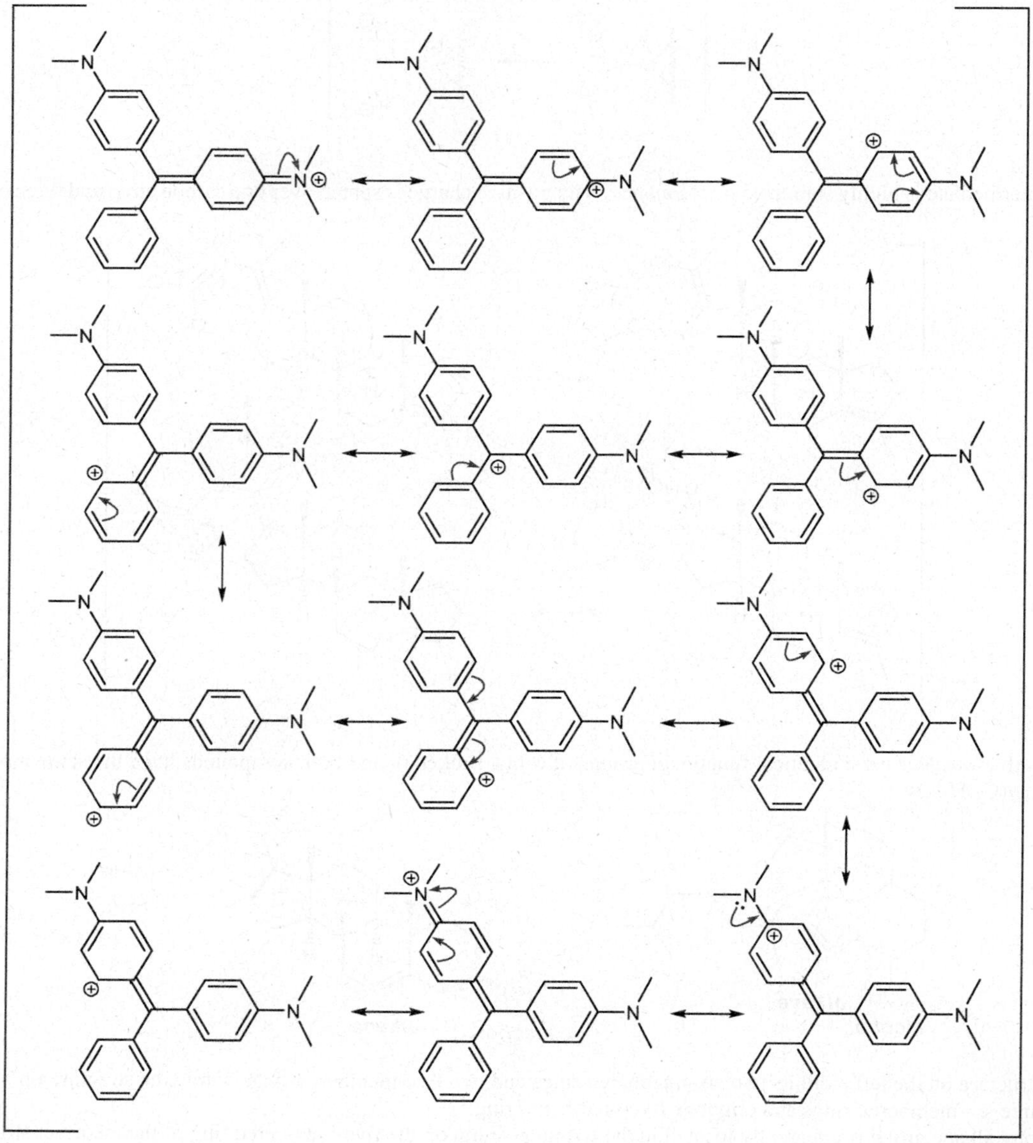

(b) The positive charge in basic violet 4 is expected to be more stabilized than the positive charge in basic green 4, because the former is delocalized over thirteen positions, rather than twelve. Specifically, basic violet 4 has an additional resonance structure that basic green 4 lacks, shown here:

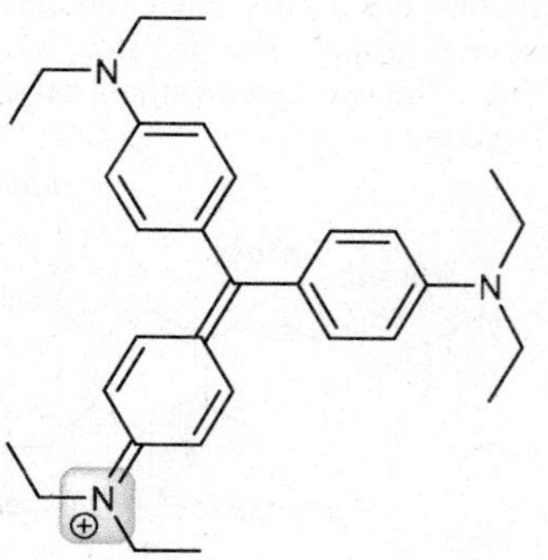

In basic violet 4, the positive charge is spread over ***three*** nitrogen atoms and ten carbon atoms.

2.84. Polymer 2 contains only ester groups, so the IR spectrum of polymer **2** is expected to exhibit a signal near 1740 cm^{-1} (typical for esters), associated with vibrational excitation (stretching) of the C=O bond. Polymer **4** lacks any ester groups, so the signal near 1740 cm^{-1} is expected to be absent in the IR spectrum of polymer **4**. Instead, polymer **4** has OH groups, which are expected to produce a broad signal in the range 3200-3600 cm^{-1}. Polymer **3** has both functional groups (alcohol group and ester group), so an IR spectrum of polymer **3** is expected to exhibit both characteristic signals. When polymer **3** is converted to polymer **4**, the signal near 1740 cm^{-1} is expected to vanish, which would indicate complete hydrolysis of polymer **3**.

In practice, the signal for the C=O stretch in polymer **2** appears at 1733 cm^{-1}, which is very close to our estimated value of 1740 cm^{-1}.

2.85. Compound **1** has an OH group, which is absent in compound **2**. Therefore, the IR spectrum of **1** should exhibit a broad signal in the range 3200-3600 cm^{-1} (associated with O-H stretching), while the IR spectrum of **2** would be expected to lack such a signal. The conversion of **1** to **2** could therefore be confirmed with the disappearance of the signal corresponding with excitation of the O-H bond.

Another way to monitor the conversion of **1** to **2** is to focus on the C-H bond of the aldehyde group in compound **1**, which is expected to produce a signal in the range 2750-2850 cm^{-1}. Since the aldehyde group is not present in compound **2**, we expect this signal to vanish when **1** is converted to **2**.

There is yet another way to monitor this reaction with IR spectroscopy. Compound **1** possesses only one C=O bond, while compound **2** has two C=O bonds. As such, the latter should exhibit two C=O signals. One signal is expected to be near 1680 cm^{-1} (for the conjugated ketone), and the other signal should be near 1700 cm^{-1} (corresponding to the conjugated ester). In contrast compound **1** has only one C=O bond, which is expected to produce a signal near 1680 cm^{-1} (for the conjugated aldehyde). Therefore, the conversion of **1** to **2** can be monitored by the appearance of a signal near 1700 cm^{-1}.

2.86.

(a) This compound contains the following functional groups:

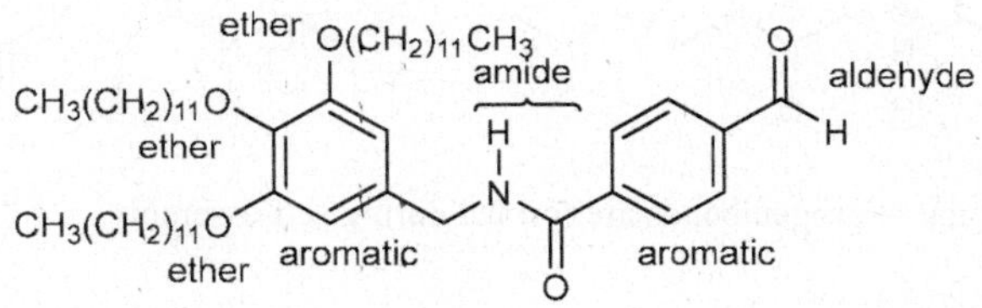

(b) The nitrogen atom has a lone pair that is delocalized via resonance:

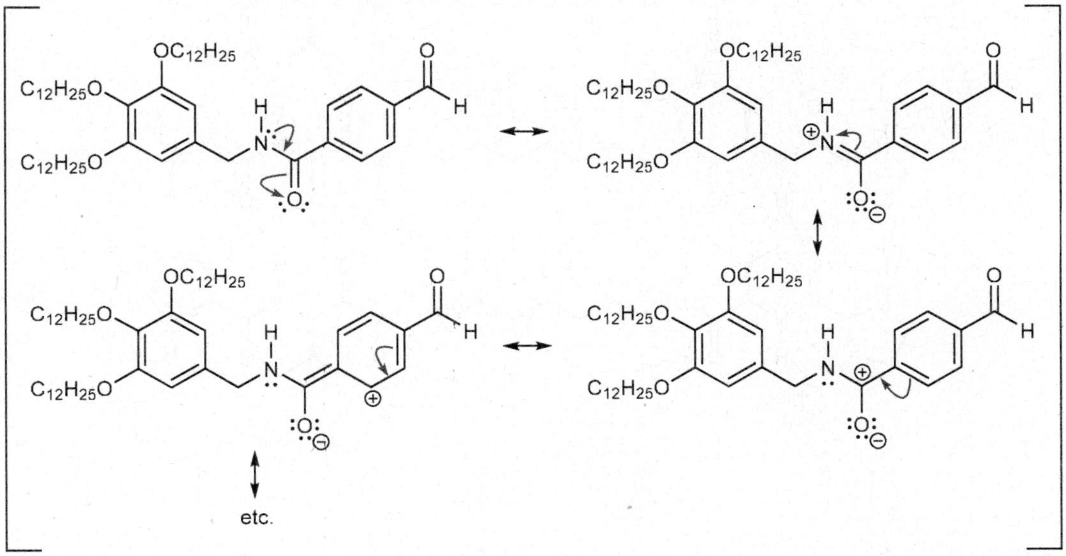

In the second resonance structure shown, the C-N bond is drawn as a double bond, indicating partial double-bond character. This bond is thus a hybrid between a single and double bond; the partial double bond character results in partially restricted rotation around this bond. In contrast, the C-N bond on the left experiences free rotation because that bond has only single-bond character.

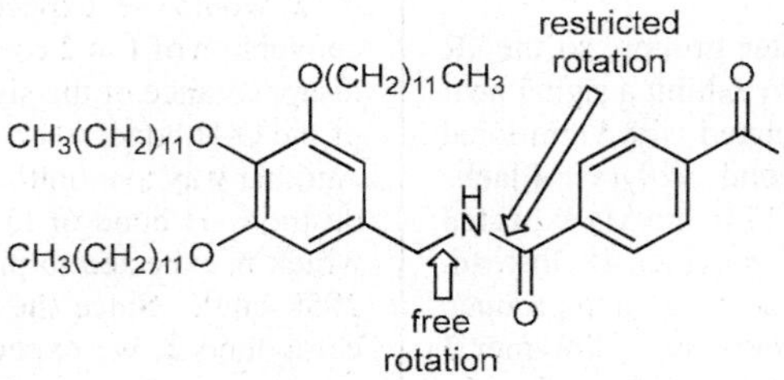

restricted rotation

free rotation

2.87.

(a) Each of the four amides can be represented as a resonance hybrid (one example shown below). The charge-separated resonance structure indicates that there is a δ+ on the amide nitrogen, which thus pulls the electrons in the N-H bond closer to the nitrogen atom, leaving the hydrogen atom with a greater δ+. This resonance effect is not present in the N-H bond of the amines. Thus, the δ+ on an amide H is greater than that on an amine H, leading to a stronger hydrogen bond.

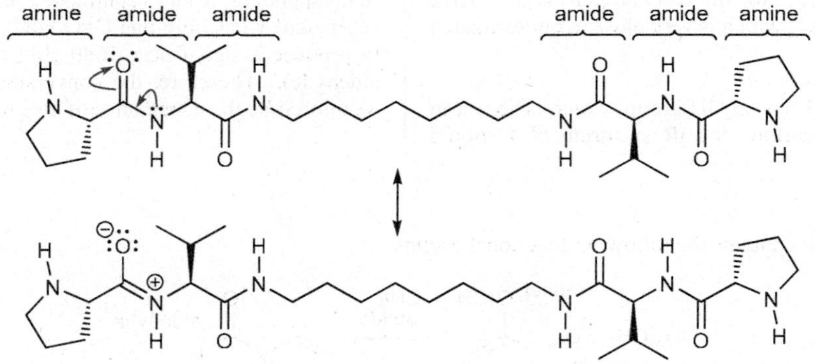

amine amide amide amide amide amine

(b) The following intermolecular hydrogen bonds are formed during self-assembly:

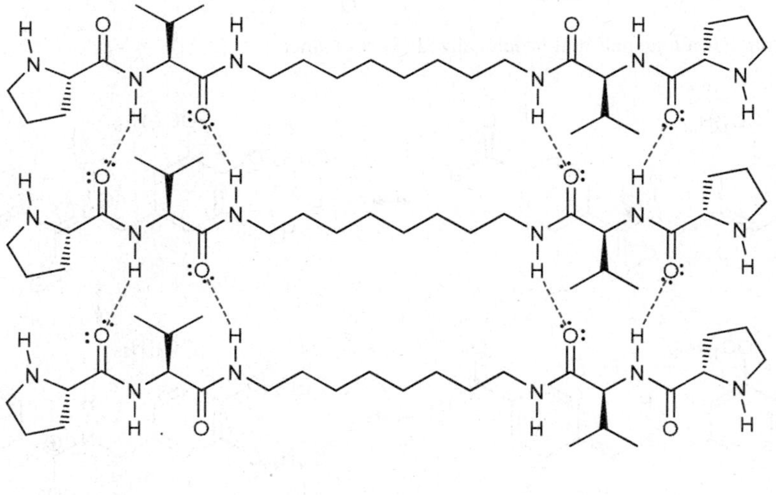

2.88.

(a) Anion **2** is highly stabilized by resonance (the negative charge is delocalized over two oxygen atoms and three carbon atoms). The resonance structures for **2** are as follows:

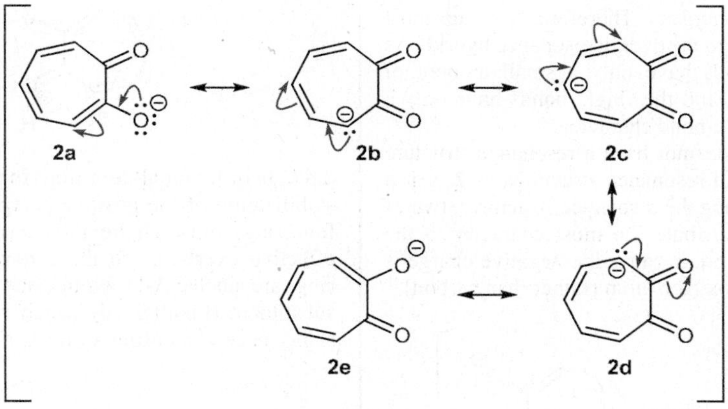

2a 2b 2c 2d 2e

Cation **3** is highly stabilized by resonance (the positive charge is delocalized over two oxygen atoms and four carbon atoms). The resonance structures of **3** are as follows:

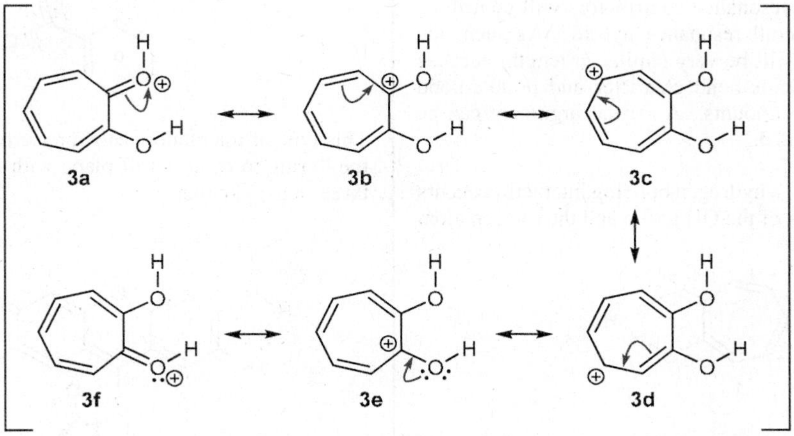

3a 3b 3c 3d 3e 3f

(b) Double bonds are shorter in length than single bonds (see Table 1.2). As such, the C-C bonds in compound **1** will alternate in length (double, single, double, etc.):

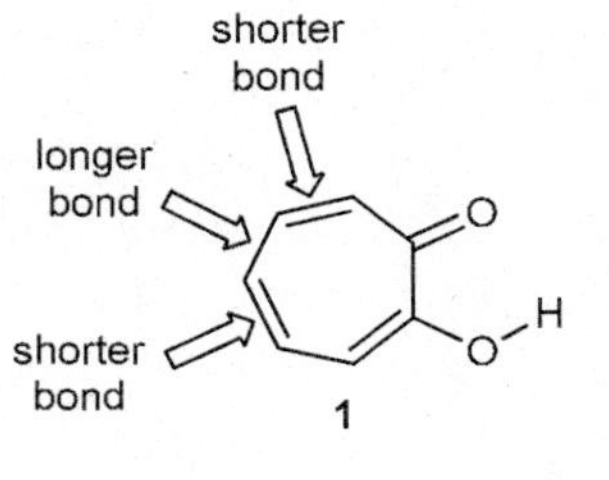

shorter bond
longer bond
shorter bond
1

The double bonds do have some single-bond character as a result of resonance, as can be seen in resonance structure **1c**:

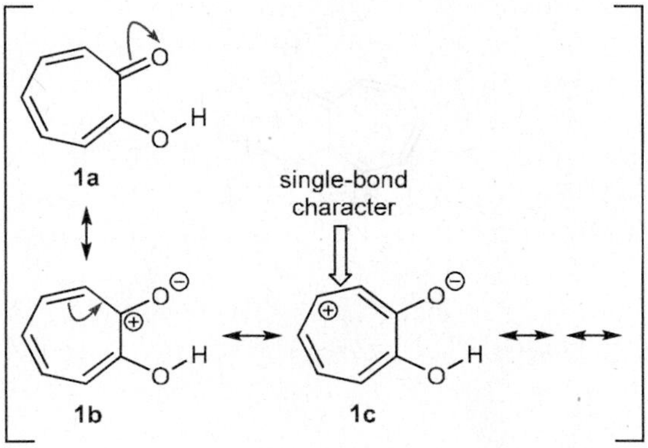

1a
single-bond character
1b 1c

Similarly, the single bonds have some double-bond character, also because of resonance. However, this effect is relatively small, because there is only one resonance structure (**1a** above) in which all atoms have an octet AND there are no formal charges. Therefore, it is the most significant contributor to the overall resonance hybrid. As such, the double bonds have only a small amount of single-bond character, and the single bonds have only a small amount of double-bond character.

In contrast, anion **2** does not have a resonance structure that lacks charges. All resonance structures of **2** bear a negative charge. Among the resonance structures, two of them (**2a** and **2e**) contribute the most character to the overall resonance hybrid, because the negative charge is on an electronegative oxygen atom (rather than carbon).

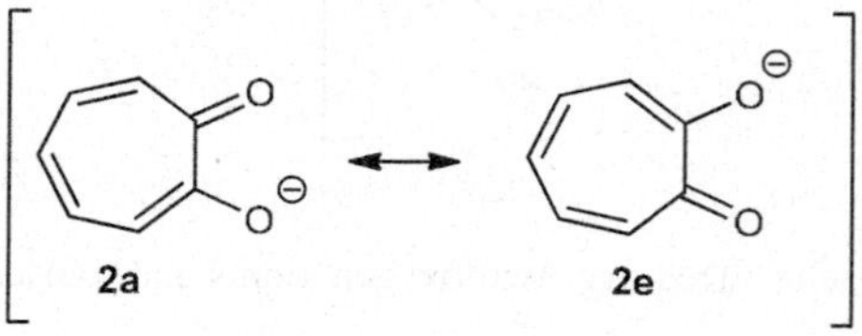

2a 2e

In fact, these two resonance contributors will contribute equally to the overall resonance hybrid. As such, the bonds of the ring will be very similar in length, because they have both single-bond character and double-bond character in equal amounts. A similar argument can be made for compound **3**.

(c) In compound **1**, a hydrogen bonding interaction occurs between the proton of the OH group and the oxygen atom of the C=O bond:

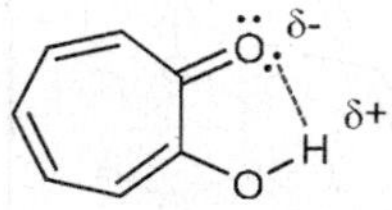

1

This interaction is the result of the attraction between partial charges (δ+ and δ-). However, in cation **3**, a similar type of interaction is less effective because the O of the C=O bond is now poor in electron density, and therefore less capable of forming a hydrogen bonding interaction, as can be seen in resonance structure **3a**.

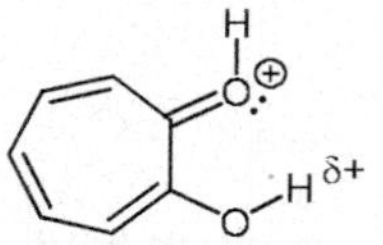

3a

The other oxygen atom is also ineffective at forming an intramolecular hydrogen bond because it too is poor in electron density, as can be seen in resonance structure **3f**:

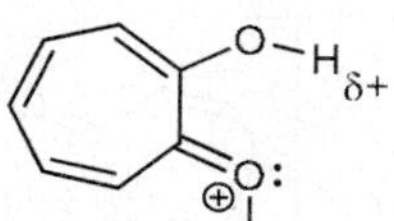

3f

2.89. In order for all four rings to participate in resonance stabilization of the positive charge, the *p* orbitals in the four rings must all lie in the same plane (to achieve effective overlap). In the following drawing, the four rings are labeled A-D. Notice that the D ring bears a large substituent (highlighted) which is trying to occupy the same space as a portion of the C ring:

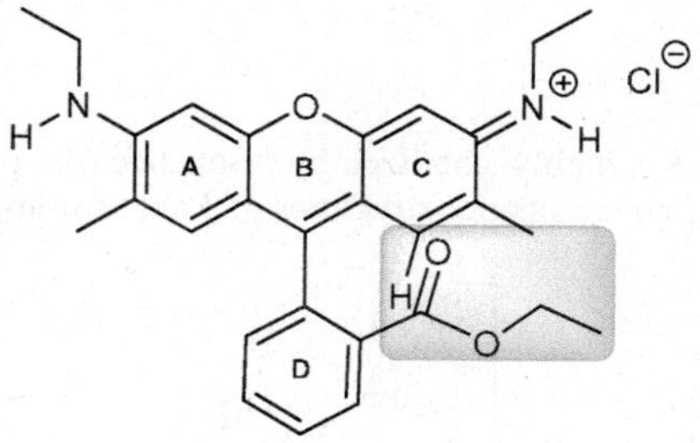

This type of interaction, called a steric interaction, forces the D ring to twist out of plane with respect to the other three rings, like this:

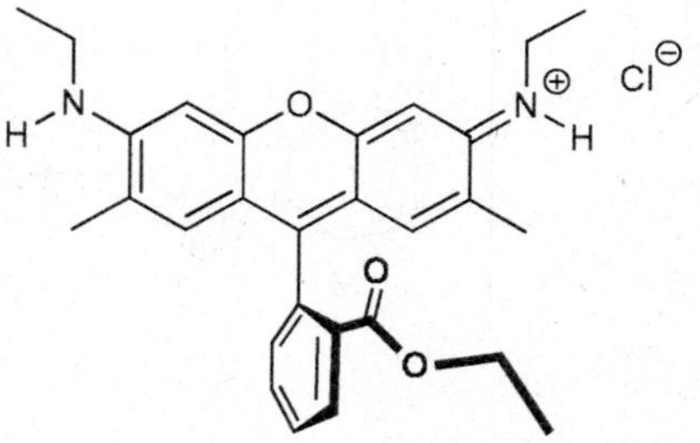

In this way, the overlap between the *p* orbitals of the D ring and the *p* orbitals of the other three rings is expected to be less effective. As such, participation of the D ring in resonance stabilization is expected to be diminished with respect to the participation of the other three rings.

Chapter 3
Acids and Bases

Review of Concepts

Fill in the blanks below. To verify that your answers are correct, look in your textbook at the end of Chapter 3. Each of the sentences below appears verbatim in the section entitled *Review of Concepts and Vocabulary*.

- A **Brønsted-Lowry acid** is a proton ________, while a **Brønsted-Lowry base** is a proton ____________.
- A **reaction mechanism** utilizes curved arrows to show the flow of ____________ that account for a chemical reaction.
- The mechanism of proton transfer always involves at least _____ curved arrows.
- A strong acid has a ________ pK_a, while a weak acid has a _________ pK_a.
- There are four factors to consider when comparing the __________ of anionic conjugate bases.
- The equilibrium of an acid-base reaction always favors the more __________ negative charge.
- A **Lewis acid** is an electron-pair __________, while a **Lewis base** is an electron-pair ________.

Review of Skills

Fill in the blanks and empty boxes below. To verify that your answers are correct, look in your textbook at the end of Chapter 3. The answers appear in the section entitled *SkillBuilder Review*.

SkillBuilder 3.1 Drawing the Mechanism of a Proton Transfer

Draw the curved arrows for the following acid-base reaction:

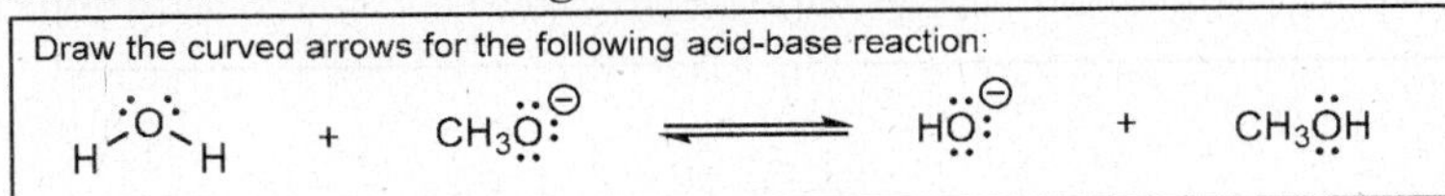

SkillBuilder 3.2 Using pK_a Values to Compare Acids

Circle the compound below that is more acidic:

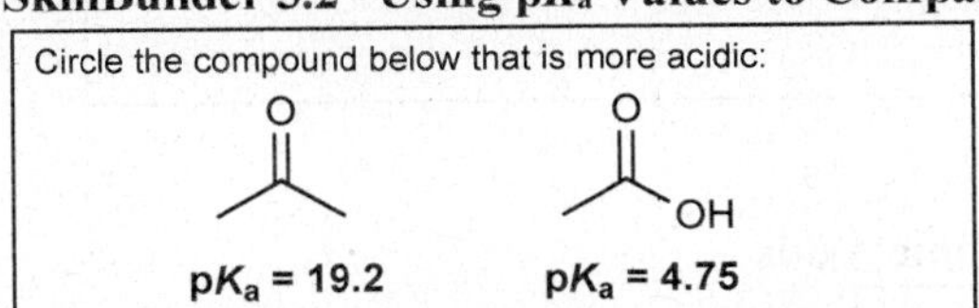

SkillBuilder 3.3 Using pK_a Values to Compare Basicity

Compare the following pK_a values, and circle the stronger base below:

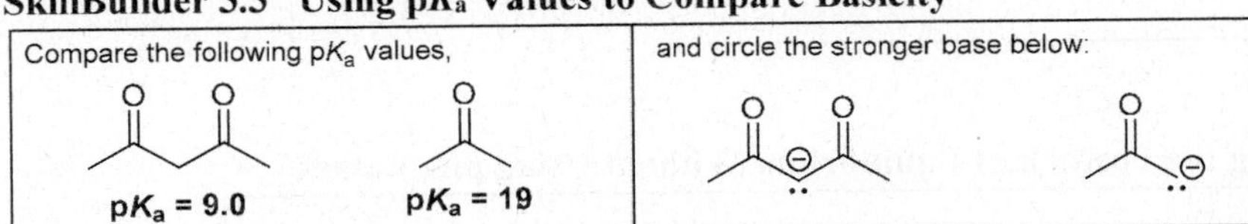

SkillBuilder 3.4 Using pK_a Values to Predict the Position of Equilibrium

Circle the side of the equilibrium that is favored:

pK_a = 9.0 pK_a = 15.7

SkillBuilder 3.5 Assessing Relative Stability. Factor 1 – Atom

Compare the two highlighted protons, and circle the one that is more acidic. Use the extra space to draw the two possible conjugate bases.

SkillBuilder 3.6 Assessing Relative Stability. Factor 2 – Resonance

Compare the two highlighted protons, and circle the one that is more acidic. Use the extra space to draw the two possible conjugate bases.

SkillBuilder 3.7 Assessing Relative Stability. Factor 3 – Induction

Compare the two highlighted protons, and circle the one that is more acidic. Use the extra space to draw the two possible conjugate bases.

CCl_3

SkillBuilder 3.8 Assessing Relative Stability. Factor 4 – Orbital

Compare the two highlighted protons, and circle the one that is more acidic. Use the extra space to draw the two possible conjugate bases.

SkillBuilder 3.9 Assessing Relative Stability. Using All Four Factors

Compare the two highlighted protons, and circle the one that is more acidic. Use the extra space to draw the two possible conjugate bases.

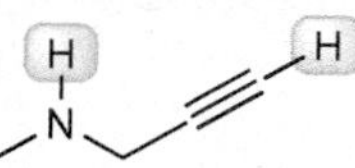

SkillBuilder 3.10 Assessing the Relative Acidity of Cationic Acids

Compare the two highlighted protons, and circle the one that is more acidic.

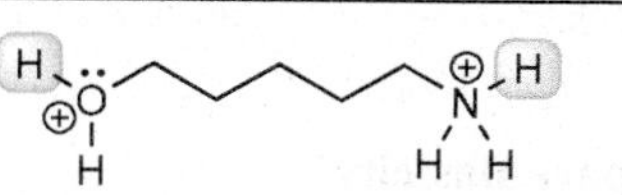

SkillBuilder 3.11 Predicting the Position of Equilibrium Without Using pK_a Values

Which side of the equilibrium is favored?

SkillBuilder 3.12 Choosing the Appropriate Reagent for a Proton Transfer Reaction

Determine whether water is a suitable proton source to protonate the acetate ion, as shown below:

SkillBuilder 3.13 Identifying Lewis Acids and Lewis Bases

Identify the Lewis acid and the Lewis base in the following reaction:

Common Mistakes to Avoid

When drawing the mechanism of a proton transfer, two curved arrows are required. The first curved arrow shows the base attacking the proton, and the second curved arrow shows the bond to H being broken.

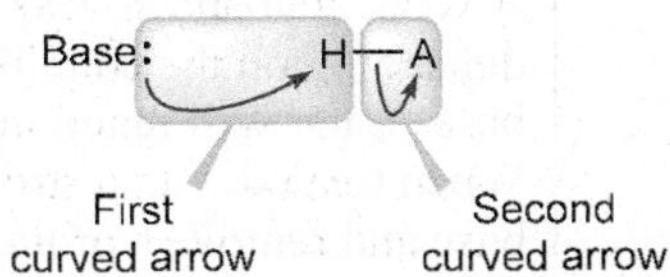

It is a common mistake to draw only the first curved arrow and not the second, so make sure to draw both curved arrows. When drawing the second curved arrow, make sure that the tail is placed on the middle of the bond to the H, as shown:

Correct

Base: H—A

NOT correct

Base: HA

When the acid is H_3O^+, you must draw at least one of the O–H bonds in order to draw the second curved arrow properly.

Correct

Base: H—O⊕(H)(H)

NOT correct

Base: H_3O^+

This is also the case for other acids:

Correct

Base: H—O—S(=O)(=O)—OH

NOT correct

Base: H_2SO_4

Useful reagents

In Chapter 3, we explored the behavior of acids and bases. Throughout the remainder of the textbook, many acids and bases will be frequently encountered. It would be wise to become familiar with the following reagents (and their uses), as they will appear many times in the upcoming chapters:

Structure	Name	Use
H–O–S(=O)(=O)–OH	Sulfuric acid	A very strong acid. Commonly used as a source of protons. Concentrated sulfuric acid is a mixture of H_2SO_4 and H_2O, and the acid present in solution is actually H_3O^+, because of the leveling effect. That is, H_3O^+ is a weaker acid than H_2SO_4, so the protons are transferred from H_2SO_4 to water, giving a high concentration of H_3O^+.
H–Cl	Hydrochloric acid	A very strong acid. Similar in function to H_2SO_4. In an aqueous solution of HCl, the acid that is present is H_3O^+, because of the leveling effect, as described above for H_2SO_4.
$CH_3C(=O)OH$	Acetic acid	A weak acid. Mild source of protons.
H–O–H	Water	A weak acid and a weak base. It can function as either, depending on the conditions. When treated with a strong base, water will function as an acid (a source of protons). When treated with a strong acid, water will function as a base and remove a proton from the strong acid.
R–O–H	An alcohol	R represents the rest of the compound. Alcohols (the subject of Chapter 12) are compounds that possess an OH group connected to an sp^3-hybridized carbon atom. Alcohols can function very much like water (either as weak acids or as weak bases).
NH_3 (H–N(–H)–H)	Ammonia	A fairly strong base, despite the absence of a negative charge. It is a strong base, because its conjugate acid (NH_4^+), called an ammonium ion, is a weak acid ($pK_a = 9.2$).
CH_3CH_2ONa	Sodium ethoxide	The ethoxide ion ($CH_3CH_2O^-$) is a strong base, and Na^+ is the counterion. Other alkoxide ions (RO^-, where R is an alkyl group), such as methoxide (CH_3O^-), are also strong bases.
$NaNH_2$	Sodium amide	H_2N^- is a very strong base, and Na^+ is the counterion.
$CH_3CH_2CH_2CH_2^{\ominus}\ Li^{\oplus}$	*n*-Butyllithium	An extremely strong base. This is one of the strongest bases that you will encounter. It is often abbreviated as *n*-BuLi

Solutions

3.1.

(a) Phenol (C_6H_5OH) loses a proton and is therefore functioning as an acid. Hydroxide (HO^-) functions as the base that removes the proton. Two curved arrows must be drawn. The first curved arrow shows a lone pair of the base attacking the proton, and the second curved arrow comes from the O–H bond (being broken) and goes to the oxygen atom, as shown:

Acid Base Conjugate Base Conjugate Acid

(b) H_3O^+ loses a proton and is therefore functioning as an acid. The ketone functions as the base that removes the proton. Two curved arrows must be drawn. The first curved arrow shows a lone pair of the base attacking the proton; the second curved arrow comes from the O–H bond (being broken) and goes to the oxygen atom, as shown:

Base Acid Conjugate Acid Conjugate Base

(c) The ketone loses a proton and is therefore functioning as an acid. The other reagent functions as the base that removes the proton. Two curved arrows must be drawn. The first curved arrow shows a lone pair of the base attacking the proton, and the second curved arrow comes from the C–H bond (being broken) and goes to the carbon atom, as shown:

Acid Base

Conjugate Base Conjugate Acid

(d) Benzoic acid ($C_6H_5CO_2H$) loses a proton and is therefore functioning as an acid. Hydroxide (HO^-) functions as the base that removes the proton. Two curved arrows must be drawn. The first curved arrow shows a lone pair of the base attacking the proton, and the second curved arrow comes from the O–H bond (being broken) and goes to the oxygen atom, as shown:

Acid Base

Conjugate Base Conjugate Acid

3.2. Two curved arrows must be drawn. The first curved arrow shows a lone pair of the base attacking the proton, and the second curved arrow comes from the O–H bond (being broken) and goes to the oxygen atom, as shown:

3.3. The nitrogen atom is accepting a proton, so dextromethorphan is the base in this proton-transfer reaction, and HBr is the acid (proton donor). As usual, the mechanism involves two curved arrows. The salt product contains both the conjugate acid (protonated dextromethorphan) and the conjugate base (Br^-).

Acid Base

Conjugate base Conjugate acid

3.4.

(a) According to Table 3.1, phenol (C_6H_5OH) has a pK_a of 9.9, while water has a pK_a of 15.7. Phenol is more acidic because it has a lower pK_a value.

(b) According to Table 3.1, $(CH_3)_3COH$ has a pK_a of 18.0, while water has a pK_a of 15.7. Water is more acidic because it has a lower pK_a value.

(c) According to Table 3.1, ammonia (NH_3) has a pK_a of 38, while acetylene (H–C≡C–H) has a pK_a of 25. As such, acetylene is more acidic because it has a lower pK_a value.

(d) According to Table 3.1, H_3O^+ has a pK_a of -1.74, while HCl has a pK_a of -7. As such, HCl is more acidic because it has a lower pK_a value.

(e) According to Table 3.1, ethane (C_2H_6) has a pK_a of 50, while acetylene (H–C≡C–H) has a pK_a of 25. As such, acetylene is more acidic because it has a lower pK_a value.

(f) According to Table 3.1, a protonated ketone has a pK_a of -7.3, while sulfuric acid (H_2SO_4) has a pK_a of -9. As such, sulfuric acid is more acidic because it has a lower pK_a value.

3.5. According to Table 3.1, the proton connected to the oxygen atom is expected to be the most acidic proton in the compound. That proton is expected to have a pK_a value near 16 (similar to CH_3CH_2OH), though it may be somewhat different from this value, based on the effect of adjacent groups. The proton connected to the nitrogen atom is expected to have a pK_a value near 38 (similar to NH_3), so it is not the most acidic proton.

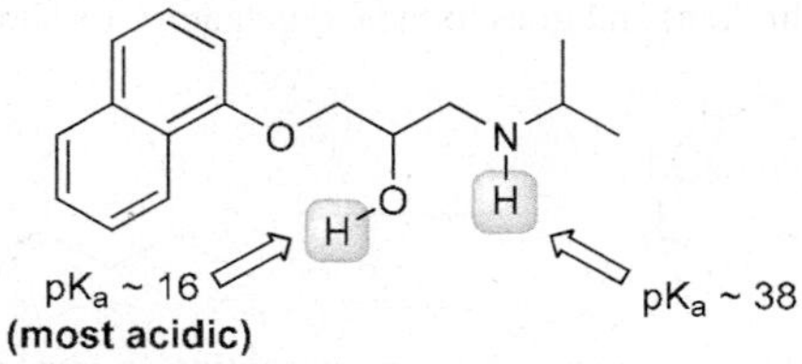

pKa ~ 16 (most acidic)

pKa ~ 38

3.6. According to Table 3.1, the proton of the carboxylic acid group (H_a below) is expected to be the most acidic (pK_a ~ 5). The two protons labeled H_b and H_c are expected to have a pK_a near 10 (like C_6H_5OH), and the protons labeled H_d are expected to have a pK_a near 38 (like NH_3). The order of acidity is shown below:

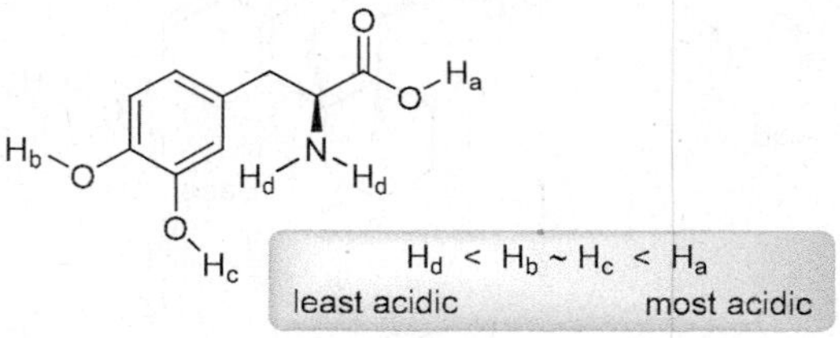

H_d < H_b ~ H_c < H_a

least acidic — most acidic

3.7.

(a) We first imagine protonating each base, and then we compare the pK_a values of the resulting compounds (using Table 3.1):

H–C≡C–H	NH_3
pK_a = 25	pK_a = 38
stronger acid	weaker acid

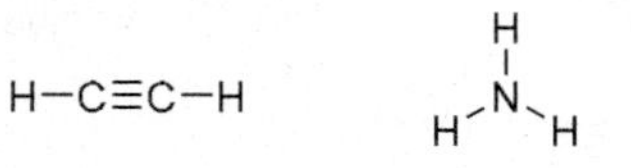

The first compound is more acidic because it has a lower pK_a value. We know that a stronger acid has a weaker conjugate base, so the conjugate base of the first compound will be a weaker base than the conjugate base of the second compound:

$H-C\equiv C^{\ominus}$	$NH_2^{\ominus}$
weaker base	stronger base

(b) We first imagine protonating each base, and then we compare the pK_a values of the resulting compounds (using Table 3.1):

$(CH_3)_3COH$	CH_3CH_2OH
pK_a = 18.0	pK_a = 16.0
weaker acid	stronger acid

The second compound (CH_3CH_2OH) is more acidic because it has a lower pK_a value. We know that a stronger acid has a weaker conjugate base, so the conjugate base of the second compound will be a weaker base than the conjugate base of the first compound:

$(CH_3)_3CO^{\ominus}$	$CH_3CH_2O^{\ominus}$
stronger base	weaker base

(c) We first imagine protonating each base, and then we compare the pK_a values of the resulting ions (using Table 3.1):

protonated acetone	H_3O^+
pK_a = -7.3	pK_a = -1.74
stronger acid	weaker acid

The first ion is more acidic because it has a lower pK_a value. We know that a stronger acid has a weaker conjugate base, so the conjugate base of the first ion will be a weaker base than the conjugate base of the second ion:

acetone	H_2O
weaker base	stronger base

(d) We first imagine protonating each base, and then we compare the pK_a values of the resulting compounds (using Table 3.1):

H_2O	CH_3CH_2OH
pK_a = 15.7	pK_a = 16.0
stronger acid	weaker acid

The first compound (water) is more acidic because it has a lower pK_a value. We know that a stronger acid has a weaker conjugate base, so the conjugate base of the second compound will be a stronger base than the conjugate base of water:

H–O⊖ weaker base; ethoxide (O⊖) stronger base

(e) We first imagine protonating each base, and then we compare the pK_a values of the resulting compounds (using Table 3.1):

$H_3C–CH_3$: pK_a = 50, weaker acid; H–C≡C–H: pK_a = 25, stronger acid

The second compound is more acidic because it has a lower pK_a value. We know that a stronger acid has a weaker conjugate base, so the conjugate base of the second compound will be a weaker base than the conjugate base of the first compound:

$H_3C–CH_2^{\ominus}$ stronger base; H–C≡C⊖ weaker base

(f) We first imagine protonating each base, and then we compare the pK_a values of the resulting compounds (using Table 3.1). HCl has a pK_a of -7, while H_2O has a pK_a of 15.7. Since HCl has a lower pK_a value, it is more acidic than water. We know that a stronger acid has a weaker conjugate base, so the conjugate base of HCl will be a weaker base than the conjugate base of H_2O.

Cl⊖ weaker base (HCl pK_a = -7); HO⊖ stronger base (H_2O pK_a = 15.7)

3.8. The pK_a values indicate that a proton is more acidic when it is connected to a positively charged oxygen atom (pK_a ~ -2.2) than when it is connected to a positively charged nitrogen atom (pK_a ~ 10.6). The weaker acid, $(CH_3)_2NH_2^+$, gives the stronger base upon deprotonation:

$(CH_3)_2NH_2^+$ weaker acid ⟹ $(CH_3)_2NH$ stronger base

$(CH_3)_2OH^+$ stronger acid ⟹ $H_3C–O–H$ weaker base

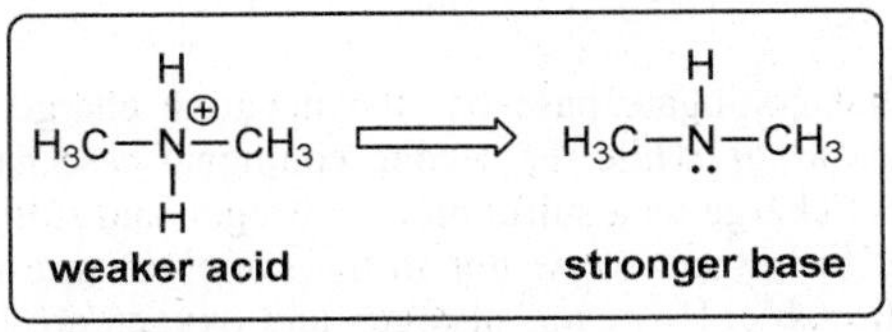

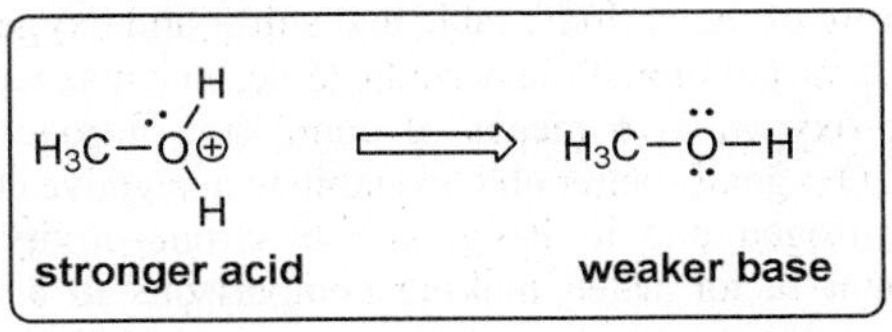

Therefore, in the structure of compound **1**, we expect that the lone pair on the nitrogen atom will be more basic than the lone pairs on the oxygen atom. When compound **1** is protonated, we expect the nitrogen atom to be protonated (rather than the oxygen atom) to give the following conjugate acid:

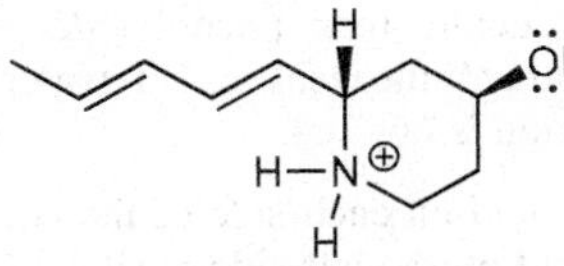

conjugate acid of 1

3.9. Inspection of structures **1** and **3** shows that the nitrogen atoms, and more specifically their lone pairs, serve as bases in the protonation of nicotine. In order to determine which nitrogen atom in structure **1** is more basic, we can compare the pK_a values for each of the acidic protons in structure **3**, using similar acids found in Table 3.1 for reference:

pyridinium pK_a = 5.3; ammonium pK_a = 9.3

The second has a higher pK_a value, and is therefore a weaker acid. The weaker acid gives the stronger base upon deprotonation, so we expect the lone pair on the nitrogen atom on the right in structure **1** to be more basic:

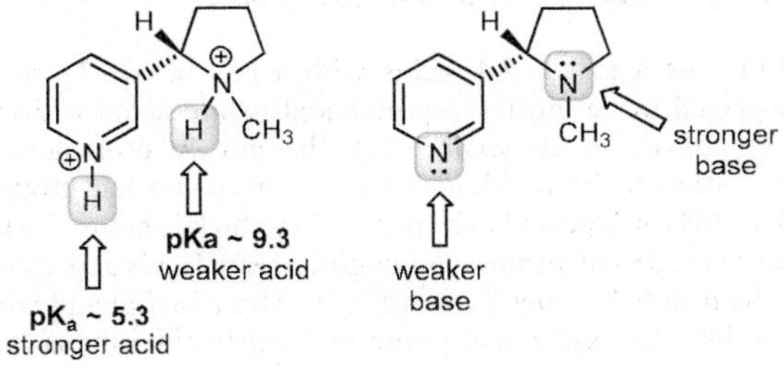

pK_a ~ 5.3 stronger acid; pKa ~ 9.3 weaker acid; weaker base; stronger base

Therefore, when **1** is mono-protonated, the sp^3 hybridized nitrogen atom is protonated, as shown:

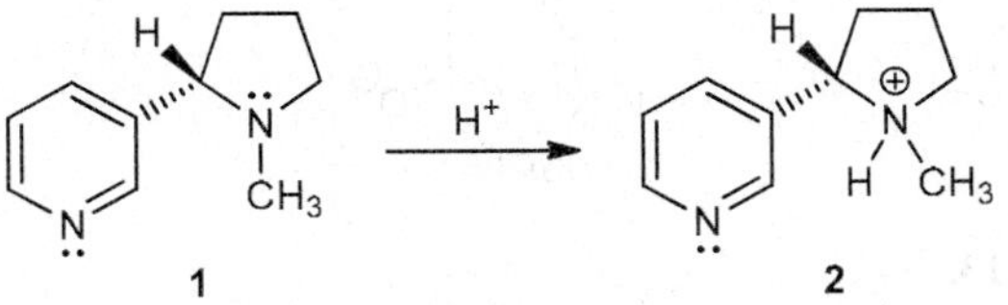

1 —H^+→ **2**

3.10.

(a) We begin by identifying the acid on each side of the equilibrium. In this case, the acid on the left side is ethanol (CH_3CH_2OH) and the acid on the right side is water (H_2O). We compare their pK_a values (Table 3.1), and we find that ethanol (pK_a = 16.0) is less acidic than water (pK_a = 15.7). The equilibrium will favor the weaker acid (ethanol), so the left side of the reaction is favored.

(b) Identify the acid on each side of the equilibrium. In this case, the acid on the left side is phenol (C_6H_5OH) and the acid on the right side is water (H_2O). We compare their pK_a values (Table 3.1), and we find that water (pK_a = 15.7) is less acidic than phenol (pK_a = 9.9). The equilibrium will favor the weaker acid (water), so the right side of the reaction is favored.

(c) Identify the acid on each side of the equilibrium. In this case, the acid on the left side is HCl and the acid on the right side is H_3O^+. We compare their pK_a values (Table 3.1), and we find that H_3O^+ (pK_a = -1.74) is less acidic than HCl (pK_a = -7). The equilibrium will favor the weaker acid (H_3O^+), so the right side of the reaction is favored.

(d) Identify the acid on each side of the equilibrium. In this case, the acid on the left side of the equilibrium is acetylene (H–C≡C–H), and the acid on the right side is ammonia (NH_3). We compare their pK_a values (Table 3.1), and we find that ammonia (pK_a = 38) is less acidic than acetylene (pK_a = 25). The equilibrium will favor the weaker acid (ammonia), so the right side of the reaction is favored.

3.11. One way to answer this question is to look up (or at least estimate) the pK_a of the acid on each side of the equilibrium. Because phenolic protons (pK_a ~ 10) are more acidic than water (pK_a = 15.7), we can determine that the equilibrium should favor deprotonation of hydroquinone by hydroxide. The equilibrium will favor the weaker acid (water), so the forward reaction is favored and hydroxide is a sufficiently strong base.

3.12. At a pH of 7.4, acids with a pK_a below 7.4 are expected to be mostly deprotonated, while acids with a pK_a above 7.4 are expected to be mostly protonated. According to Table 3.1, the pK_a of a carboxylic acid group (RCO_2H) is expected to be near 4.75 (which is below 7.4), while the pK_a of an ammonium group (RNH_3^+) is expected to be near 9.3 (which is above 7.4) Therefore, at a pH of 7.4, the carboxylic acid group will exist primarily as its conjugate base (RCO_2^-), called a carboxylate ion, while the ammonium group (RNH_3^+) will retain its proton, and will primarily exist in the charged form, as shown here:

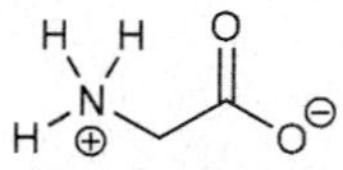

glycine

3.13.

(a) Carbon and oxygen are in the same row of the periodic table, so we must compare their electronegativity values. Oxygen is more electronegative than carbon and can better stabilize the negative charge that will be generated upon deprotonation. A more stable (weaker) conjugate base indicates a stronger parent acid. Therefore, a proton connected to an oxygen atom is expected to be more acidic than a proton connected to a carbon atom:

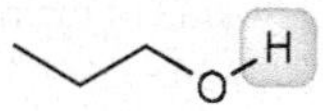

(b) Carbon and nitrogen are in the same row of the periodic table, so we must compare their electronegativity values. Nitrogen is more electronegative than carbon and can better stabilize the negative charge that will be generated upon deprotonation. A more stable (weaker) conjugate base indicates a stronger parent acid. Therefore, a proton connected to a nitrogen atom is expected to be more acidic than a proton connected to a carbon atom:

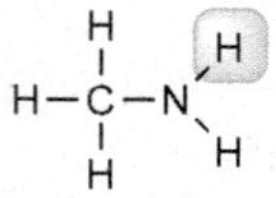

(c) Sulfur and oxygen are in the same column of the periodic table, so we must compare their size. Sulfur is larger than oxygen and can better stabilize the negative charge that will be generated upon deprotonation. A more stable (weaker) conjugate base indicates a stronger parent acid. Therefore, a proton connected to a sulfur atom is expected to be more acidic than a proton connected to an oxygen atom:

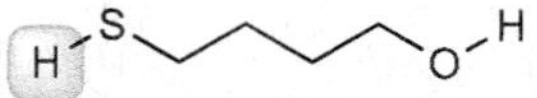

(d) Nitrogen and oxygen are in the same row of the periodic table, so we must compare their electronegativity values. Oxygen is more electronegative than nitrogen and can better stabilize the negative charge that will be generated upon deprotonation. A more stable (weaker) conjugate base indicates a stronger parent acid. Therefore, a proton connected to an oxygen atom is expected to be more acidic than a proton connected to a nitrogen atom:

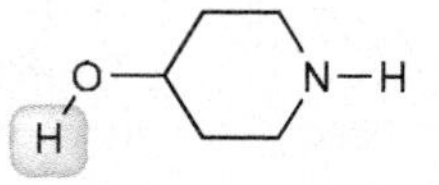

3.14. We start by drawing the two possible conjugate bases:

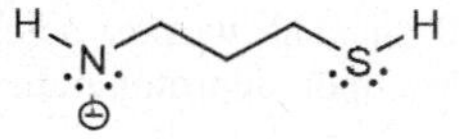

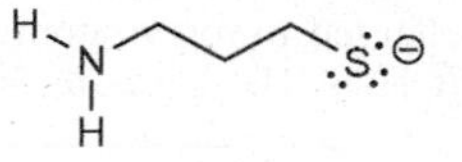

The first conjugate base has the negative charge on a nitrogen atom, while the second conjugate base has the negative charge on a sulfur atom. Nitrogen and sulfur are neither in the same row nor in the same column of the periodic table. However, nitrogen and oxygen are in the same row of the periodic table and sulfur and oxygen are in the same column of the periodic table, so it makes sense to use oxygen as a means of comparing nitrogen and sulfur. Oxygen is better able to stabilize a negative charge than nitrogen due to its greater electronegativity (the dominant factor when making comparisons in a row). Sulfur is better able to stabilize a negative charge than oxygen due to its larger size (the dominant factor when making comparisons in a column). So, we can deduce that sulfur must be more capable of stabilizing the negative charge than nitrogen:

less stable

more stable

A more stable (weaker) conjugate base indicates a stronger parent acid. Therefore, a proton connected to a sulfur atom is expected to be more acidic than a proton connected to a nitrogen atom:

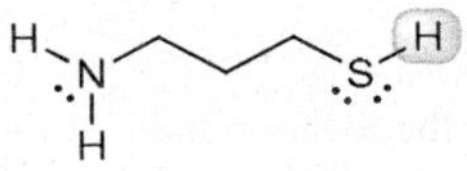

3.15.

(a) The proton marked in blue is expected to be more acidic than the proton marked in red, because removal of the blue proton leads to a resonance-stabilized anion, shown here:

A more stable (weaker) conjugate base indicates a stronger parent acid.
Removal of the red proton would result in a carbanion that is less stable, because it is *not* resonance stabilized.

(b) The proton marked in blue is expected to be more acidic than the proton marked in red, because removal of the blue proton leads to a resonance-stabilized anion, shown here:

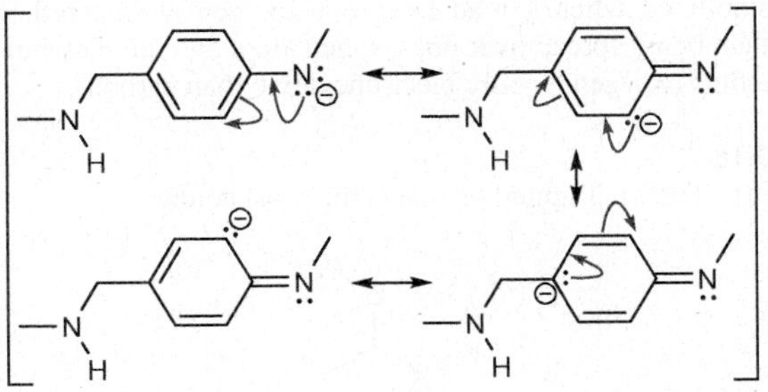

A more stable (weaker) conjugate base indicates a stronger parent acid.
Removal of the red proton would result in a nitrogen anion that is less stable, because it is *not* resonance stabilized.

(c) The proton marked in red is expected to be more acidic than the proton marked in blue, because removal of the red proton leads to a resonance-stabilized anion, shown here:

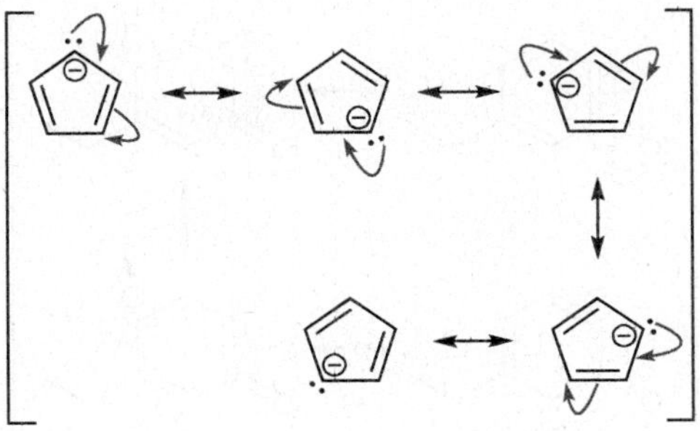

A more stable (weaker) conjugate base indicates a stronger parent acid.
Removal of the blue proton would result in a carbanion that is less stable, because it is *not* resonance stabilized.

(d) The proton marked in red is expected to be more acidic than the proton marked in blue, because removal of the red proton leads to a resonance-stabilized anion, shown here:

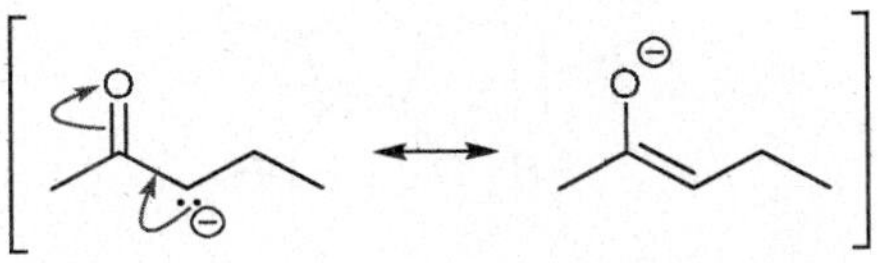

A more stable (weaker) conjugate base indicates a stronger parent acid.
Removal of the blue proton would result in a carbanion that is less stable, because it is *not* resonance stabilized

(e) The proton marked in blue is expected to be more acidic than the proton marked in red, because removal of the blue proton leads to a resonance-stabilized anion in which the negative charge is spread over two oxygen atoms.

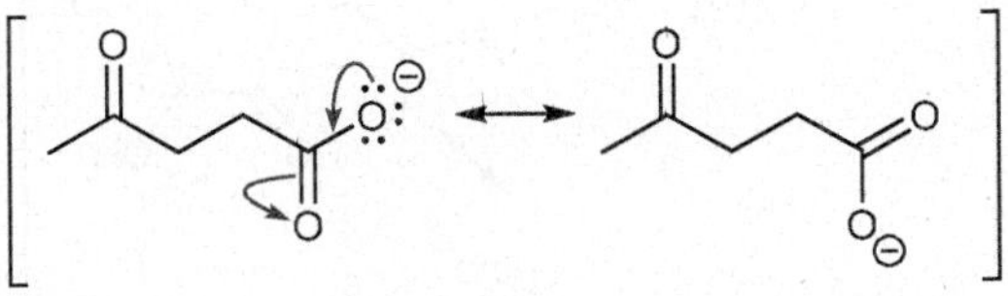

A more stable (weaker) conjugate base indicates a stronger parent acid. Removal of the red proton also leads to a resonance-stabilized anion, but the negative charge is spread over an oxygen atom and a carbon atom, which is less stable than spreading the charge over two oxygen atoms.

(f) The proton marked in red is expected to be more acidic than the proton marked in blue, because removal of the red proton leads to a resonance-stabilized anion in which the negative charge is spread over one oxygen atom and three carbon atoms.

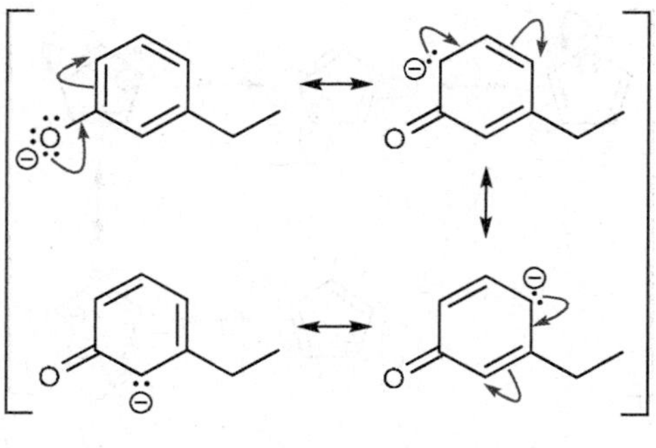

A more stable (weaker) conjugate base indicates a stronger parent acid. Deprotonation at the location of the blue proton also leads to a resonance-stabilized anion, but the negative charge is spread over four carbon atoms, which is less stable than spreading the charge over one oxygen atom and three carbon atoms.

3.16. The proton highlighted is the most acidic proton in the structure:

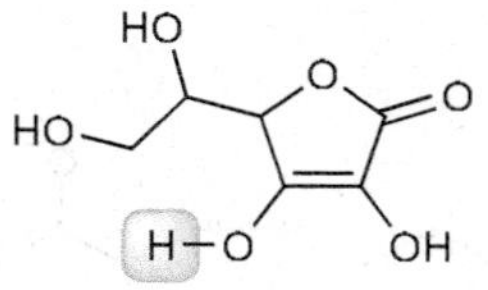

This proton is the most acidic, because deprotonation at that location generates a resonance-stabilized anion, in which the negative charge is spread over two oxygen atoms and one carbon atom:

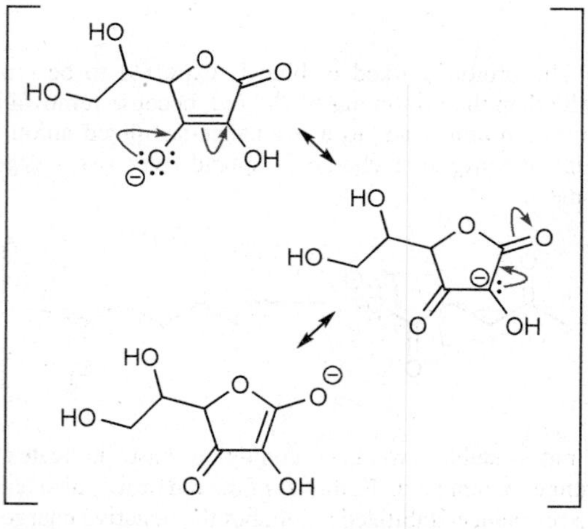

A more stable (weaker) conjugate base indicates a stronger parent acid.
Deprotonation of any of the other oxygen atoms would result in an oxyanion (or oxygen anion) that is less stable, because it is either *not* resonance stabilized (for the two OH groups that are not part of the ring), or less resonance stabilized (for the other OH group attached to the ring).

3.17. Deprotonation at the location marked by the red proton leads to a resonance-stabilized anion in which the negative charge is spread over one oxygen atom and three carbon atoms:

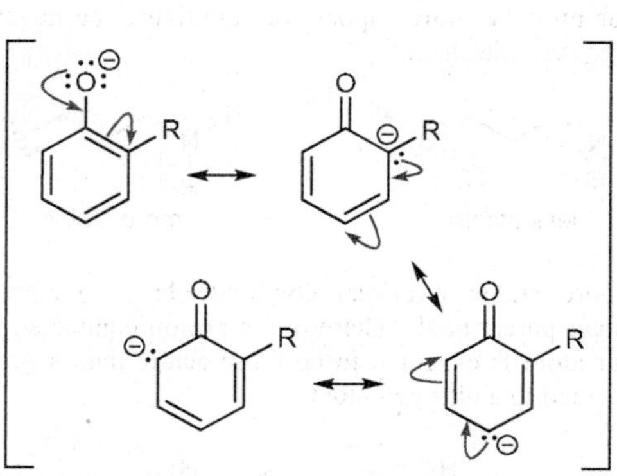

Deprotonation at the location marked by the blue proton leads to a resonance-stabilized anion in which the negative charge is spread over two oxygen atoms:

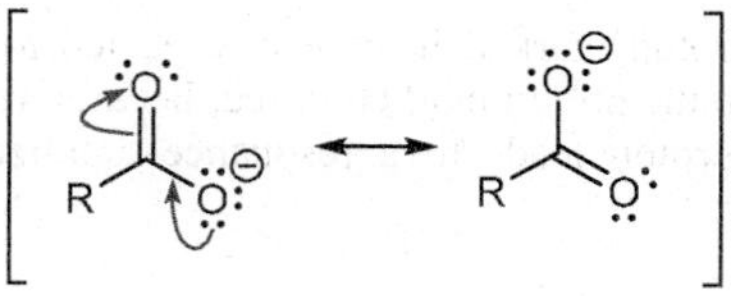

By comparing pK_a values from Table 3.1, we conclude that the carboxylic acid proton is more acidic (lower pK_a value):

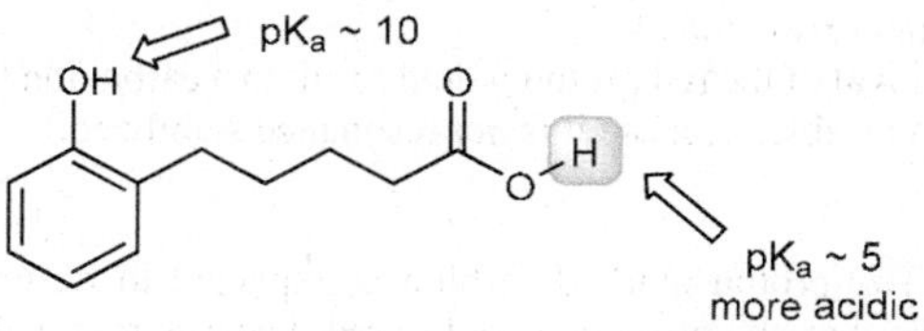

pK_a ~ 10

pK_a ~ 5
more acidic

The stronger acid has a more stable (weaker) conjugate base. This suggests that a negative charge will be more stabilized when spread over two oxygen atoms, rather than being spread over one oxygen atom and three carbon atoms (oxygen is more electronegative than carbon).

3.18.

(a) The highlighted proton is the most acidic:

more acidic

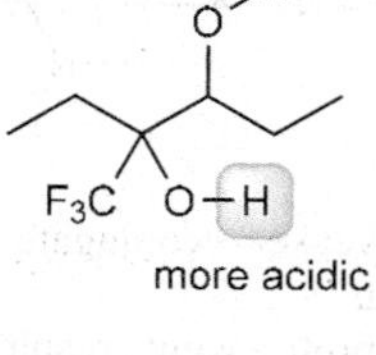

When this proton is removed, the resulting anionic conjugate base is stabilized by the electron-withdrawing effects of the nearby electronegative fluorine atoms. A more stable (weaker) conjugate base indicates a stronger parent acid.

(b) The highlighted proton is more acidic:

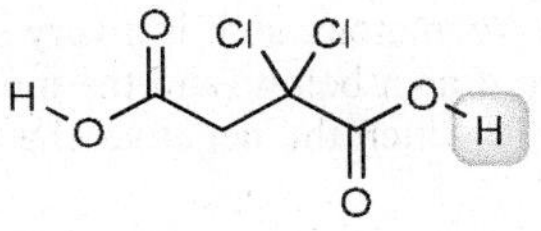

When this proton is removed, the resulting anionic conjugate base is stabilized by the electron-withdrawing effects of the electronegative chlorine atoms, which are closer to this proton than the other acidic proton (far left).

3.19.

(a) The compound with two chlorine atoms is more acidic:

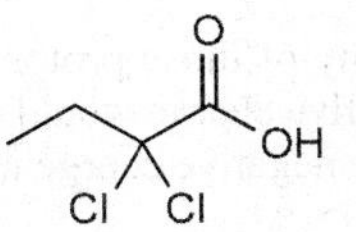

The electron-withdrawing effects of the additional chlorine atom help stabilize the conjugate base that is formed when the proton is removed. A more stable (weaker) conjugate base indicates a stronger parent acid.

(b) The more acidic compound is the one in which the bromine atom is closer to the acidic proton:

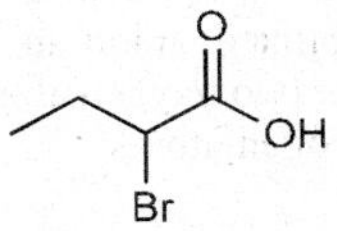

The electron-withdrawing effects of the bromine atom stabilize the conjugate base that is formed when the proton is removed. A more stable (weaker) conjugate base indicates a stronger parent acid.

3.20.

(a) In the following compound, one of the chlorine atoms has been moved closer to the acidic proton of the carboxylic acid group, which further stabilizes the conjugate base that is formed when the proton is removed.

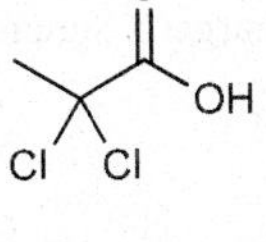

(b) In the following compound, one of the chlorine atoms has been moved farther away from the acidic proton of the carboxylic acid group, and the distant chlorine atom is less capable of stabilizing the conjugate base that is formed when the proton is removed.

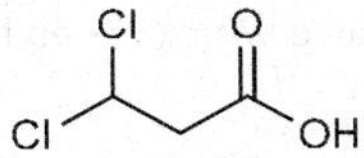

(c) There are many acceptable answers to this question, since there are many constitutional isomers that lack the carboxylic acid functional group. One example is shown below. This compound is not a carboxylic acid, so its conjugate base is not resonance-stabilized:

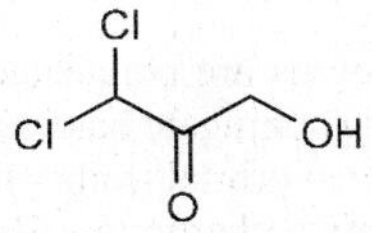

Without resonance stabilization, the conjugate base of this isomer is a fairly strong base, indicating a weaker parent acid.

3.21. The most acidic proton is highlighted in each of the following compounds. For each of the first two compounds, deprotonation leads to a conjugate base in which the negative charge is associated with an *sp*-hybridized orbital (which is more stable than being associated with an sp^2- or sp^3-hybridized orbital). In the final compound, deprotonation leads to a conjugate base in which the negative charge is associated with an sp^2-hybridized orbital (which is more stable than being associated with an sp^3-hybridized orbital).

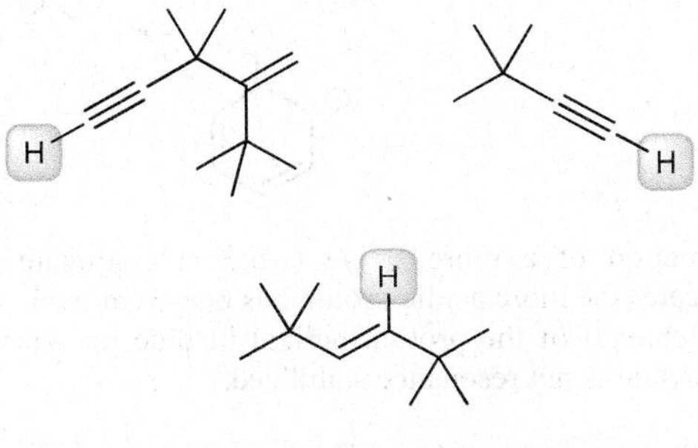

Note that in the final compound, there are two equivalent sp^2-hybridized protons (removal of either one of them would produce the same conjugate base).

3.22. *n*-Butyllithium (*n*-BuLi) has a negative charge on an sp^3-hybridized carbon atom, so it is a very strong base and will remove the most acidic proton in the compound. All protons have been drawn below, and the most acidic one is highlighted. Removal of this proton gives a conjugate base (shown below) in which the negative charge is stabilized because it is associated with an *sp*-hybridized orbital.

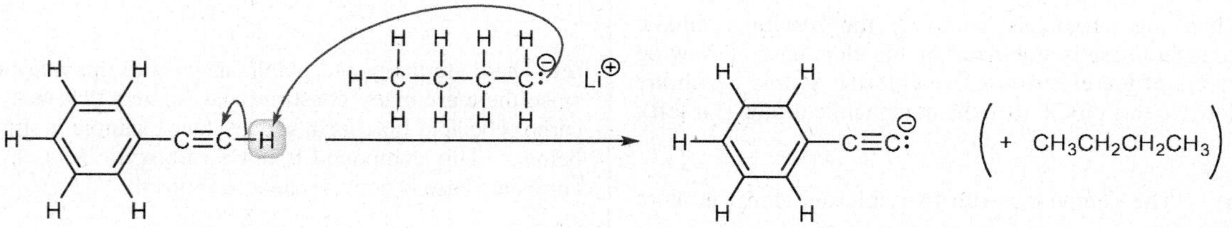

All of the other protons are connected to sp^2-hybridized carbon atoms. So if any of those protons had been removed instead, the resulting conjugate base would have been less stable because the negative charge would have been associated with an sp^2-hybridized orbital (rather than an *sp*-hybridized orbital). Note that the negative charge in the product is more stable than the negative charge in *n*-BuLi, so the forward reaction is favored.

3.23.

(a) The proton marked in red is expected to be more acidic than the proton marked in blue, because removal of the red proton leads to a resonance-stabilized anion in which the negative charge is spread over one oxygen atom and three carbon atoms:

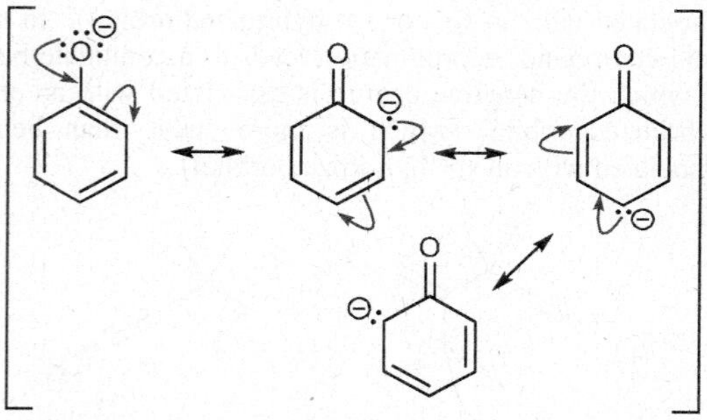

Formation of a more stable (weaker) conjugate base indicates the more acidic proton has been removed.

Removal of the proton marked in blue results in an anion that is not resonance stabilized.

(b) The proton marked in red is expected to be more acidic than the proton marked in blue. Both conjugate bases are resonance-stabilized, but removal of the red proton leads to a resonance-stabilized anion in which the negative charge is spread over one carbon atom and *two* oxygen atoms, rather than just being spread over one carbon atom and *one* oxygen atom:

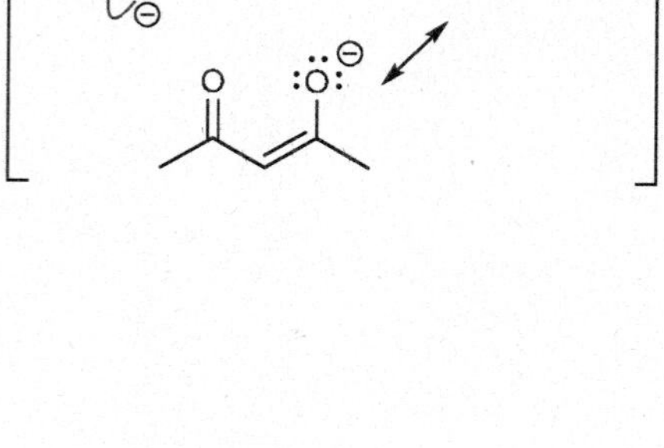

Formation of a more stable (weaker) conjugate base indicates the more acidic proton has been removed.

(c) The proton marked in red is expected to be more acidic than the proton marked in blue. Both conjugate bases are resonance stabilized, but removal of the red proton leads to a resonance-stabilized anion in which the negative charge is spread over two oxygen atoms, rather than being spread over two nitrogen atoms:

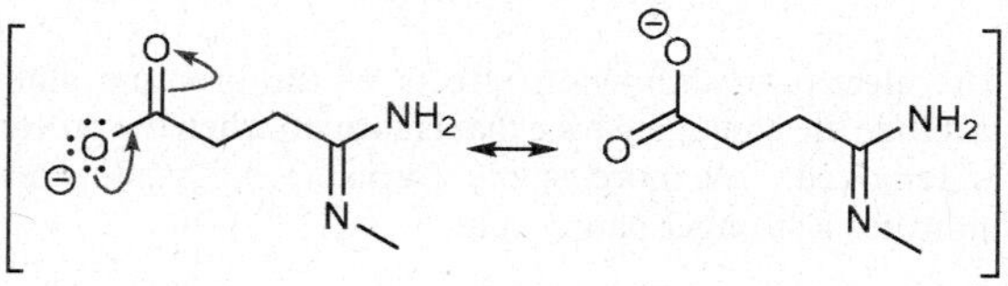

Formation of a more stable (weaker) conjugate base indicates the more acidic proton has been removed.

(d) The proton marked in blue is expected to be more acidic than the proton marked in red, because removal of the blue proton leads to a resonance-stabilized anion in which the negative charge is spread over a nitrogen atom and a carbon atom:

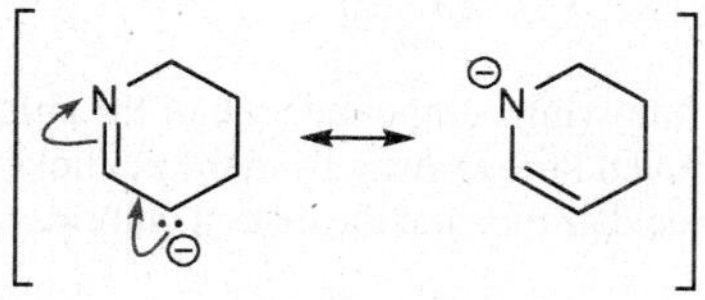

Formation of a more stable (weaker) conjugate base indicates the more acidic proton has been removed. Removal of the red proton would result in a less stable carbanion, because it is *not* resonance stabilized.

(e) The proton marked in red is expected to be more acidic than the proton marked in blue, because removal of the red proton leads to a resonance-stabilized anion in which the negative charge is spread over two nitrogen atoms, rather than being spread over two carbon atoms:

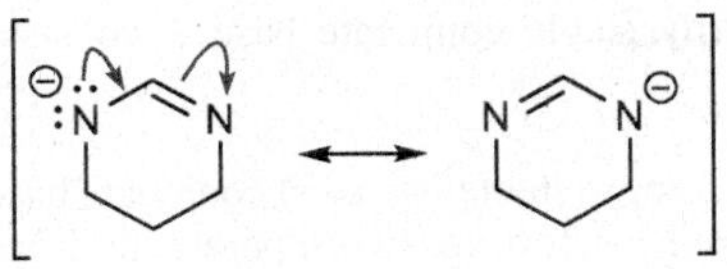

Formation of a more stable (weaker) conjugate base indicates the more acidic proton has been removed. Removal of the blue proton would result in a less stable anion because the negative charge would be spread over two carbon atoms (rather than two nitrogen atoms).

(f) The proton marked in blue is expected to be more acidic than the proton marked in red. Both conjugate bases are resonance-stabilized, but removal of the blue proton leads to a resonance-stabilized anion in which the negative charge is spread over two sulfur atoms, rather than two oxygen atoms:

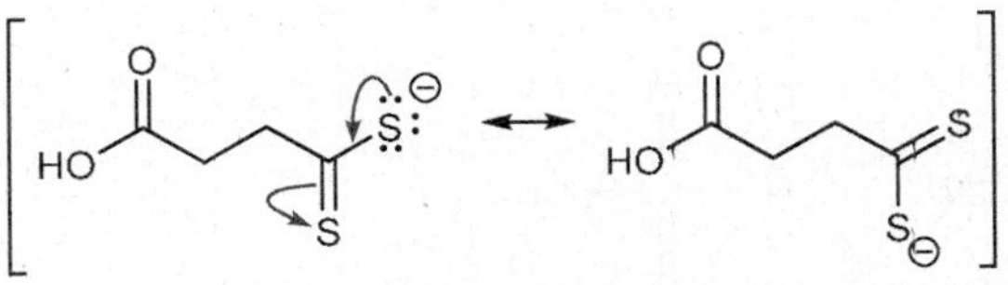

A negative charge on sulfur is more stable than a negative charge on oxygen, because sulfur is larger, and therefore more polarizable than oxygen. Formation of a more stable (weaker) conjugate base indicates the more acidic proton has been removed.

(g) The proton marked in red is expected to be more acidic than the proton marked in blue, because removal of the red proton leads to a more stable conjugate base in which the negative charge is associated with an *sp* hybridized orbital. This case represents an exception to the ARIO priority scheme: factor 4 (orbital) trumps factor 1 (atom), and this can be confirmed by looking at Table 3.1 (the pK_a of acetylene is 25, and the pK_a of a proton on nitrogen is ~38).

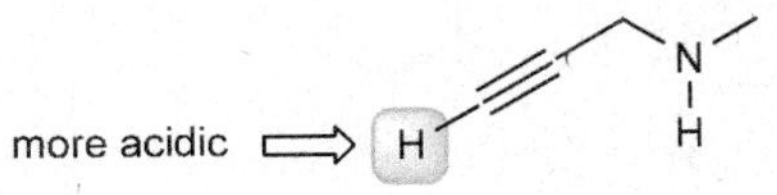

more acidic ⇨ H

(h) The proton marked in blue is expected to be more acidic than the proton marked in red. Both conjugate bases are resonance-stabilized, but removal of the blue proton leads to a conjugate base in which the negative charge is spread over two oxygen atoms, rather than being spread over one oxygen atom and three carbon atoms:

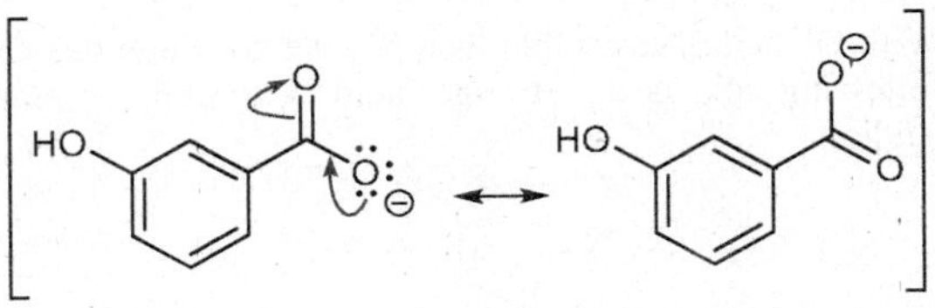

Formation of a more stable (weaker) conjugate base indicates the more acidic proton has been removed.

(i) The proton marked in blue is expected to be more acidic than the proton marked in red. Both conjugate bases are resonance-stabilized, but removal of the blue proton leads to a conjugate base in which the negative charge is spread over three oxygen atoms, rather than being spread over one nitrogen atom and one oxygen atom:

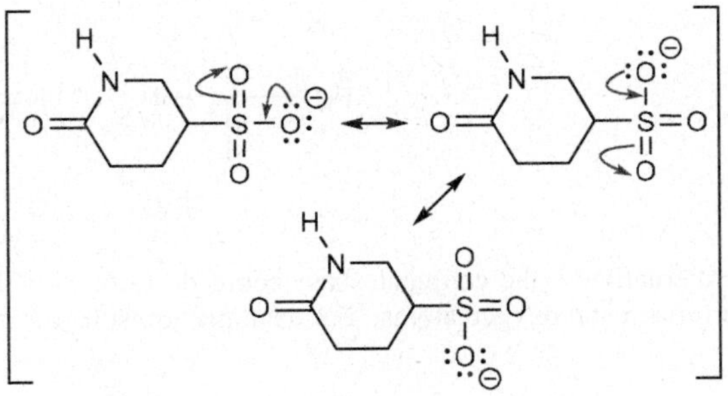

Formation of a more stable (weaker) conjugate base indicates the more acidic proton has been removed.

3.24.

(a) Bromine and chlorine are in the same column of the periodic table (group 7A), so we must compare their size. Bromine is larger than chlorine and can better stabilize the negative charge that will be generated upon deprotonation. A more stable (weaker) conjugate base indicates a stronger parent acid, and therefore, HBr is expected to be more acidic than HCl.

(b) Sulfur and oxygen are in the same column of the periodic table (group 6A), so we must compare their size. Sulfur is larger than oxygen and can better stabilize the negative charge that will be generated upon deprotonation. A more stable (weaker) conjugate base indicates a stronger parent acid, and therefore, H_2S is expected to be more acidic than H_2O.

(c) Carbon and nitrogen are in the same row of the periodic table, so we must compare their electronegativity values. Nitrogen is more electronegative than carbon and can better stabilize the negative charge that will be generated upon deprotonation. A more stable (weaker) conjugate base indicates a stronger parent acid, and therefore, NH_3 is expected to be more acidic than CH_4.

(d) Acetylene (H–C≡C–H) is more acidic. The conjugate base of acetylene has a negative charge associated with a lone pair in an *sp*-hybridized orbital, which is more stable than a negative charge associated with a lone pair in an *sp*^{2}-hybridized orbital. A more stable (weaker) conjugate base indicates a stronger parent acid.

3.25.

(a) When the proton is removed, the resulting conjugate base is highly resonance-stabilized because the negative charge is spread over four nitrogen atoms and seven oxygen atoms. In addition, the inductive effects of the trifluoromethyl groups (-CF_3) further stabilize the negative charge. This extraordinarily stable conjugate base is consistent with a particularly strong parent acid.

(b) There are certainly many, many acceptable answers to this problem. A more stable (weaker) conjugate base indicates a stronger parent acid, so the goal is to further stabilize the conjugate base. Below are two separate modifications that would render the compound even more acidic:
The OH group can be replaced with an SH group. Sulfur is larger than oxygen and more capable of stabilizing a negative charge:

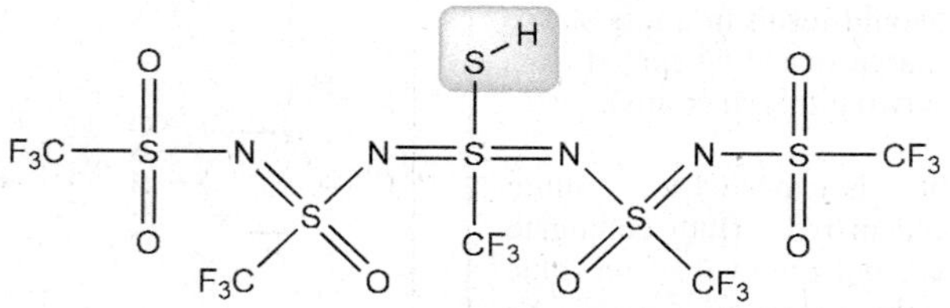

Alternatively, the conjugate base could be further stabilized by spreading the charge over an even larger number of nitrogen and oxygen atoms. For example, consider the structural changes, highlighted here:

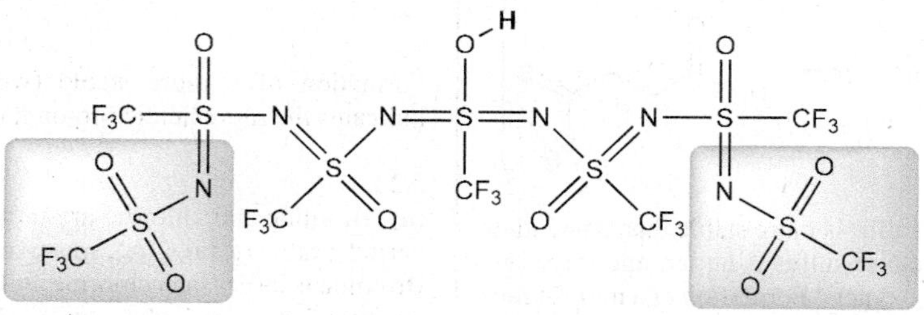

These additional structural units would enable the conjugate base to spread its negative charge over six nitrogen atoms and nine oxygen atoms, which should be even more stable than being spread over four nitrogen atoms and seven oxygen atoms.

3.26. The most acidic proton belongs to the carboxylic acid group (COOH). Deprotonation of this functional group gives a resonance-stabilized anion in which the negative charge is spread over two oxygen atoms. Formation of a more stable (weaker) conjugate base indicates the more acidic proton has been removed. Removal of any of the other protons would lead to a less stable conjugate base.

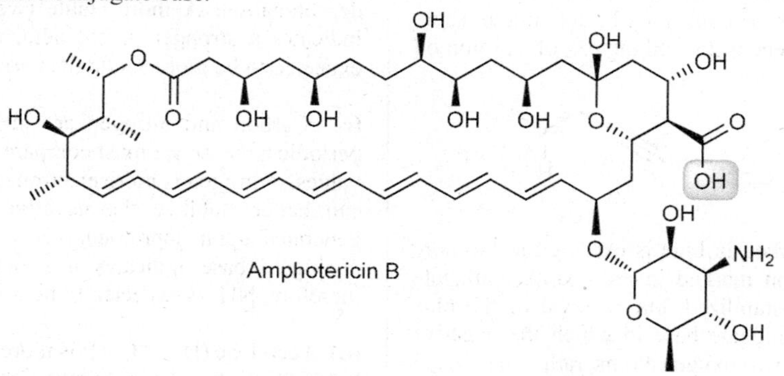

Amphotericin B

3.27.

(a) To assess the relative acidity of cationic acids, we compare the stability of the positive charges. We are comparing a positive charge on an oxygen atom with a positive charge on a nitrogen atom. Nitrogen is better at stabilizing a positive charge (compared with oxygen), because nitrogen is less electronegative than oxygen. With O^+ being less stable than N^+, we conclude that the following acid is the stronger acid (*less* stable = *more* acidic):

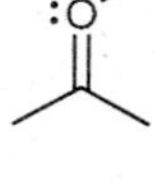

(b) To assess the relative acidity of cationic acids, we compare the stability of the positive charges:

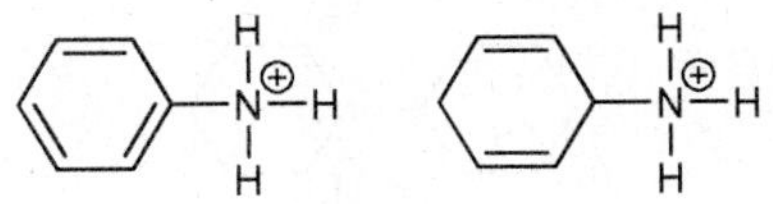

In this case, we cannot directly assess the relative stability of the two acids because the positive charge is on the same element (a nitrogen atom) in each structure. Therefore, we must draw the conjugate bases and compare the availability of the lone pairs in the conjugate bases:

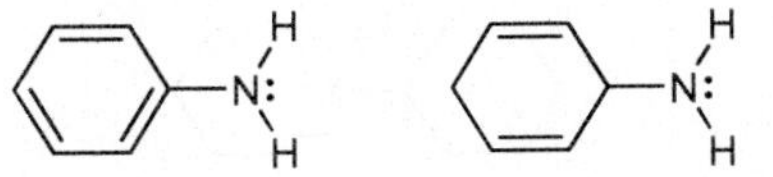

The first conjugate base above has a lone pair that is delocalized by resonance:

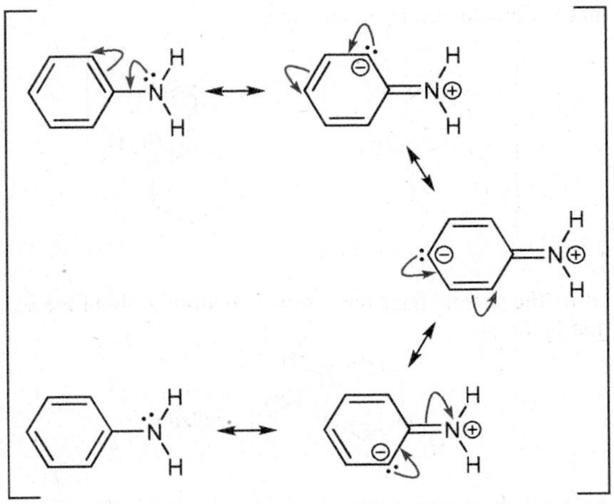

In the other conjugate base, the nitrogen atom is connected to an sp^3-hybridized atom, so the lone pair is localized:

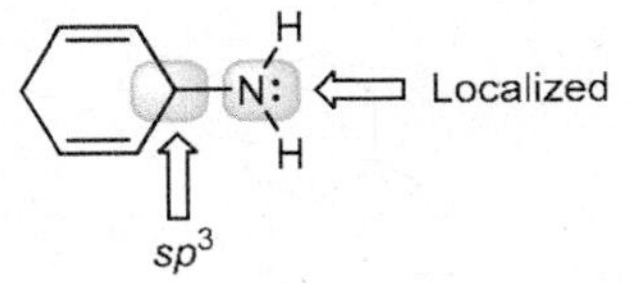

Localized

sp^3

As a result, the first conjugate base is a weaker base (its lone pair is less available to function as a base). Therefore, we conclude that the first acid is the stronger acid (because the stronger acid has a weaker conjugate base):

weaker base (delocalized lone pair) | stronger base (localized lone pair)

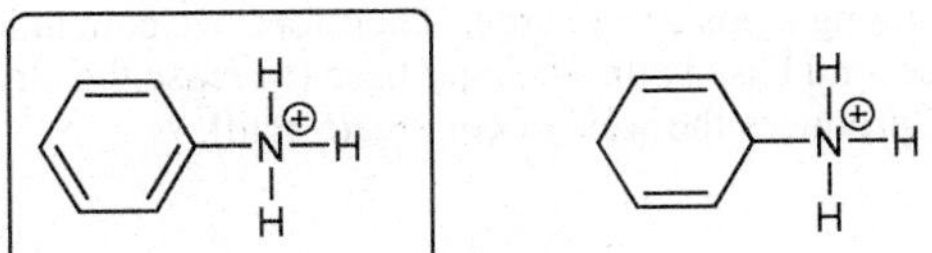

stronger acid | weaker acid

(c) To assess the relative acidity of cationic acids, we compare the stability of the positive charges:

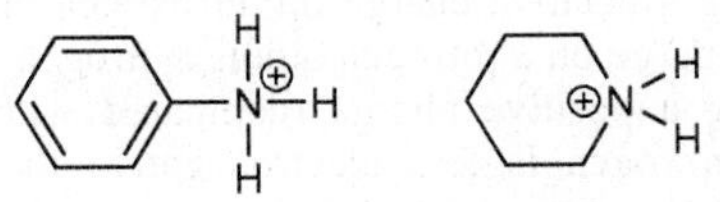

In this case, we cannot directly assess the relative stability of the two acids because the positive charge is on the same element (a nitrogen atom) in each structure. Therefore, we must draw the conjugate bases and compare the availability of the lone pairs in the conjugate bases:

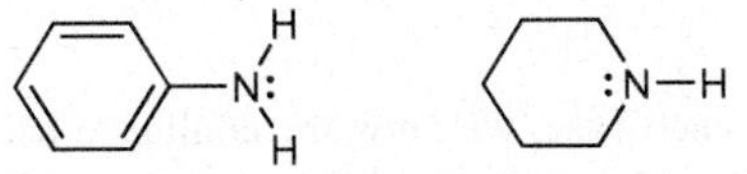

The first conjugate base above has a lone pair that is delocalized by resonance:

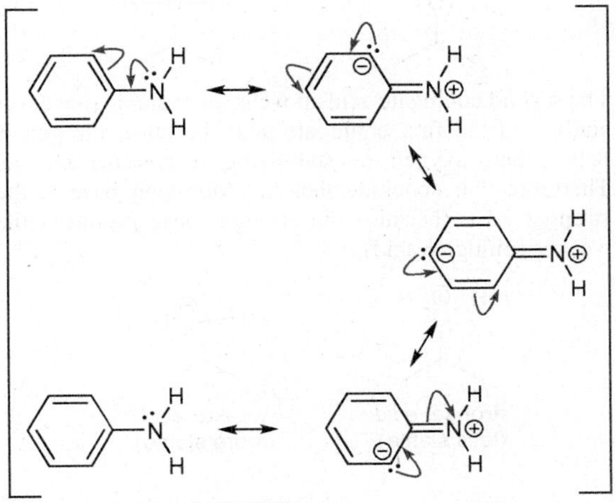

In the other conjugate base, the lone pair is localized:

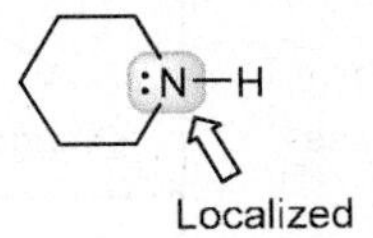

Localized

As a result, the first conjugate base is a weaker base (its lone pair is less available to function as a base). Therefore, we conclude that the first acid is the stronger acid (because the stronger acid has a weaker conjugate base).

weaker base (delocalized lone pair) | stronger base (localized lone pair)

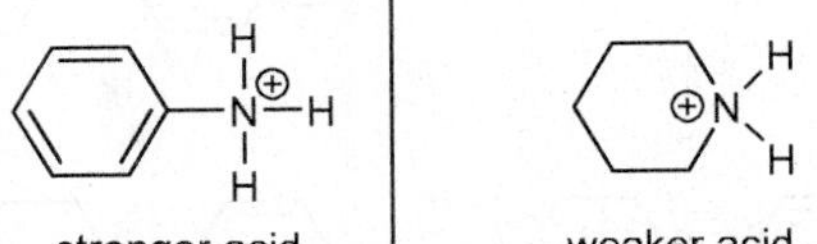

stronger acid | weaker acid

(d) To assess the relative acidity of cationic acids, we compare the stability of the positive charges. We are comparing a positive charge on an oxygen atom with a positive charge on a nitrogen atom. Nitrogen is better at stabilizing a positive charge (compared with oxygen), because nitrogen is less electronegative than oxygen. With O^+ being less stable than N^+, we conclude that the following acid is the stronger acid (*less* stable = *more* acidic):

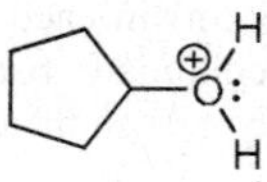

3.28.

(a) For each base, we draw the conjugate acid, and we then compare the stability of the conjugate acids:

The second conjugate acid above is more stable (a weaker acid) than the first conjugate acid, because nitrogen is better than oxygen at stabilizing a positive charge. Therefore, we conclude that the following base is the stronger base (because the stronger base produces the weaker conjugate acid).

stronger acid (less stable) — weaker acid (more stable)

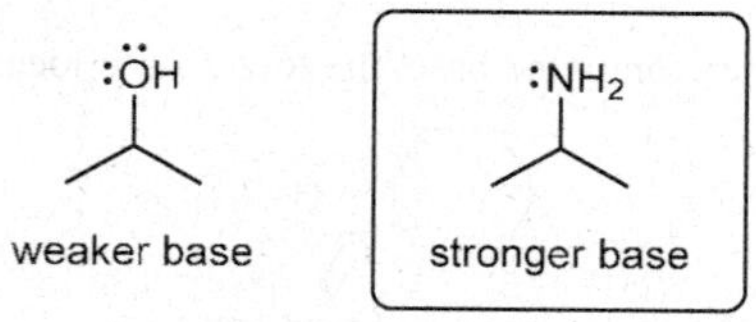

weaker base — stronger base

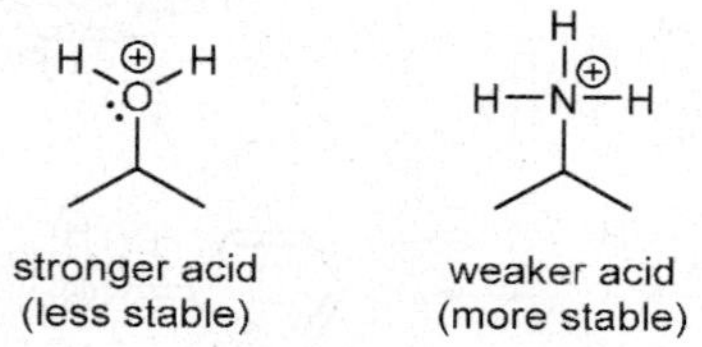

Indeed, this follows the general trend that amines (RNH_2) are stronger bases than alcohols (ROH).

(b) The first base has a lone pair that is delocalized by resonance,

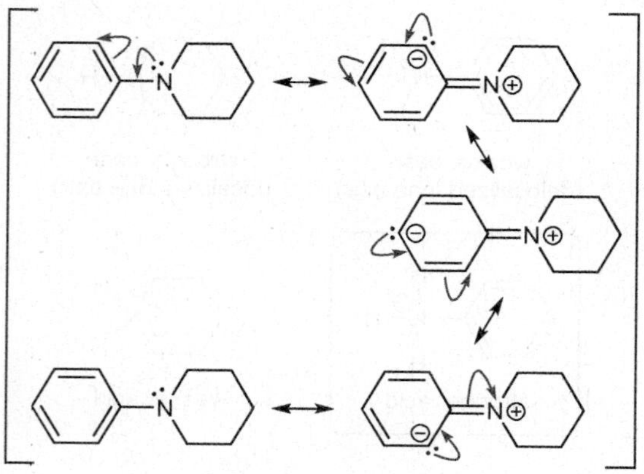

while the second base has a lone pair that is localized:

Localized

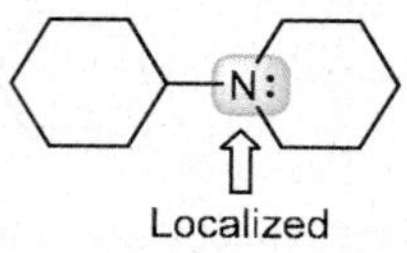

As a result, the first base is weaker base (its lone pair is less available to function as a base). Therefore, the following base is the stronger base:

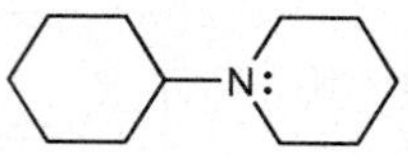

(c) The first base has a nitrogen atom with a lone pair that is delocalized by resonance,

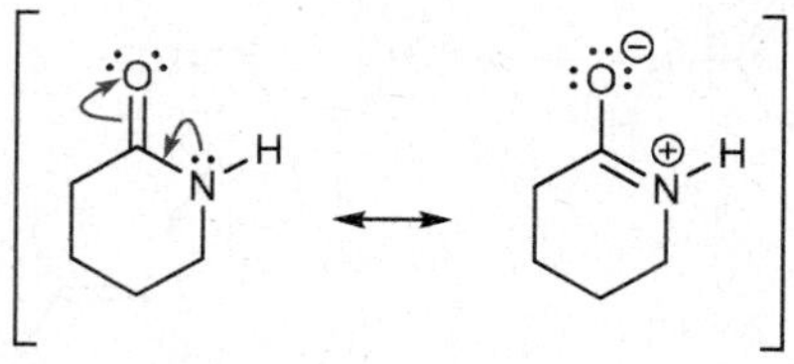

while the second base has a nitrogen atom with a lone pair that is localized:

Localized

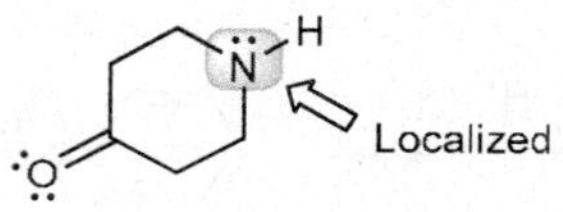

As a result, the first base is weaker base (its lone pair is less available to function as a base). Therefore, the following base is the stronger base:

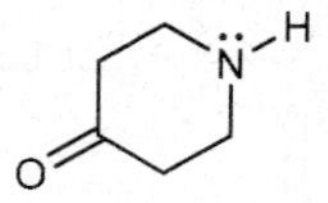

(d) For each base, we draw the conjugate acid, and we then compare the stability of the conjugate acids:

The second conjugate acid above is more stable (a weaker acid) than the first conjugate acid, because the less electronegative nitrogen atom is better than oxygen at stabilizing a positive charge. Therefore, we conclude that the second base is the stronger base (because the stronger base produces the weaker conjugate acid).

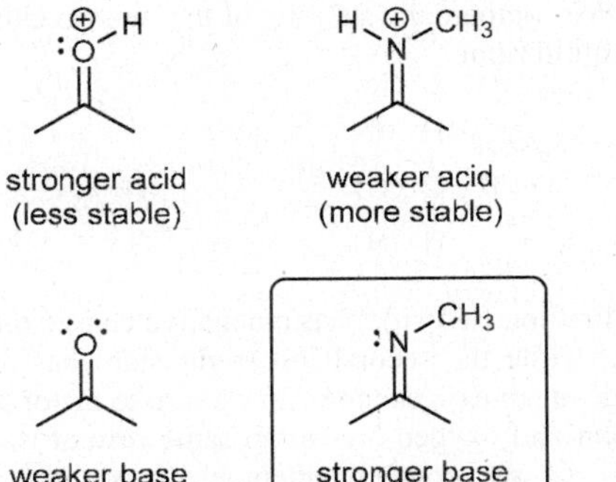

3.29. There are three nitrogen atoms in guanidine, although two of them (highlighted below) are identical because of symmetry:

So we only need to compare the availability of the lone pairs in the following two locations:

One of these lone pairs is localized,

Localized lone pair

while the other lone pair is delocalized by resonance:

Delocalized lone pair

The localized lone pair is more available to function as a base, so the following position is the most basic nitrogen atom in guanidine:

The same conclusion can be reached if we compare the structures of the two possible conjugate acids:

The charge in the first conjugate acid is localized (not stabilized by resonance), while the charge in the second conjugate acid is delocalized via resonance, as shown here:

Because of resonance stabilization, this conjugate acid is more stable and a weaker acid, so it must correspond with the most strongly basic nitrogen atom:

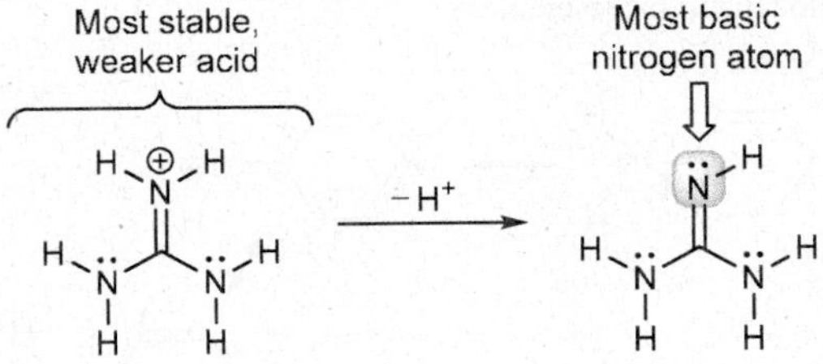

In summary, the same conclusion can be reached by comparing the availability of the lone pairs in the bases or by comparing the stability of the positive charges in the conjugate acids.

3.30.

(a) We compare the stability of the ions on either side of the equilibrium:

The first ion (left side) has a negative charge on a carbon atom, while the second ion (right side) has a negative charge on a nitrogen atom, so we turn to factor #1 (atom). Carbon and nitrogen are in the same row of the periodic table, so we compare their electronegativity values. Nitrogen is more electronegative than carbon, so a negative charge will be more stable on a nitrogen atom. The base on the product side is the more stable (weaker) base. As such, the reaction favors the products.

(b) We compare the stability of the ions on either side of the equilibrium:

The first ion (left side) has a negative charge on an oxygen atom, while the second ion (right side) has a negative charge on a nitrogen atom, so we turn to factor #1 (atom). Oxygen and nitrogen are in the same row of the periodic table, so we compare their electronegativity values. Oxygen is more electronegative than nitrogen, so a negative charge will be more stable on an oxygen atom. The base on the reactant side is the more stable (weaker) base. As such, the reaction favors the reactants, and the equilibrium lies to the left.

(c) We begin by comparing the stability of the ions on either side of the equilibrium. Both ions have a positive charge on nitrogen, so we must look at their conjugate bases (shown in the reaction) and compare the availability of the lone pairs:

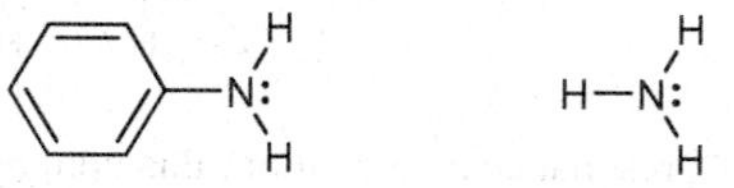

The second conjugate base above has a localized lone pair, while the first conjugate base above has a lone pair that is delocalized by resonance:

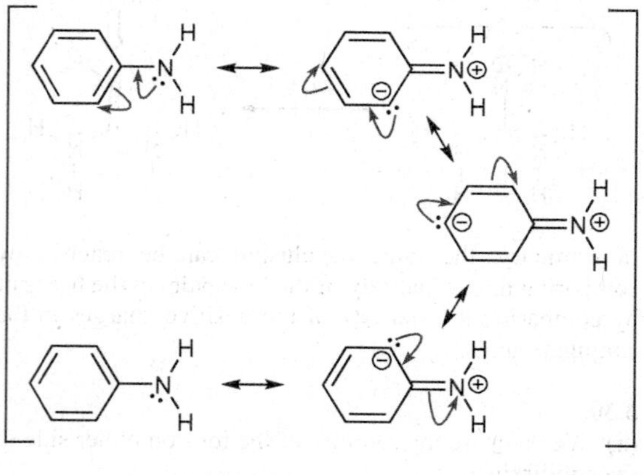

As a result, this conjugate base is a weaker base (its lone pair is less available to function as a base). The base on the reactant side is the more stable (weaker) base. Therefore, the equilibrium favors the reactants, not the products, and the equilibrium lies to the left.

(d) We compare the stability of the ions on either side of the equilibrium:

The first ion (left side) has a negative charge on a carbon atom, while the second ion (right side) has a negative charge on an oxygen atom, so we turn to factor #1 (atom). Carbon and oxygen are in the same row of the periodic table, so we compare their electronegativity values. Oxygen is more electronegative than carbon, so a negative charge will be more stable on an oxygen atom. The base on the product side is the more stable (weaker) base. As such, the reaction favors the products, and the equilibrium lies to the right.

(e) We compare the stability of the ions on either side of the equilibrium:

In this case, we are comparing a positive charge on a nitrogen atom with a positive charge on an oxygen atom. Nitrogen is better at stabilizing a positive charge, because nitrogen is less electronegative than oxygen. The acid on the reactant side is the more stable (weaker) acid. Therefore, the equilibrium favors the reactants, not the products, and the equilibrium lies to the left.

(f) We compare the stability of the ions on either side of the equilibrium:

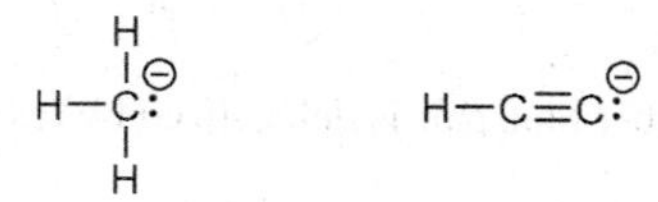

Both ions have a negative charge on a carbon atom, and neither ion is stabilized by resonance or induction. The difference between these ions is the type of orbital that is accommodating the lone pair (and its associated negative charge). The first ion above has the negative charge associated with an sp^3-hybridized orbital, while the second ion has the negative charge associated with an *sp*-hybridized orbital, and the latter is more stable. The base on the product side is the more stable (weaker) base. Therefore, the reaction favors the products, and the equilibrium lies to the right.

3.31.

(a) First identify the acid and the base in the conversion of **2** into **3**. Anion **2** has its negative charge on an oxygen atom, and this oxygen atom functions as a base and accepts a proton during the conversion of **2** to **3**. The nearby hydroxyl group loses its proton during the reaction, so it is functioning as the acid.

To draw a mechanism, remember to use two curved arrows. The tail of the first curved arrow is placed on a lone pair from the base. The head of the arrow is placed on the proton of the acid. The tail of the second curved arrow is placed on the bond between the acidic proton and the attached oxygen atom. The head of that arrow is placed on the oxygen atom that receives the negative charge as a result of the process.

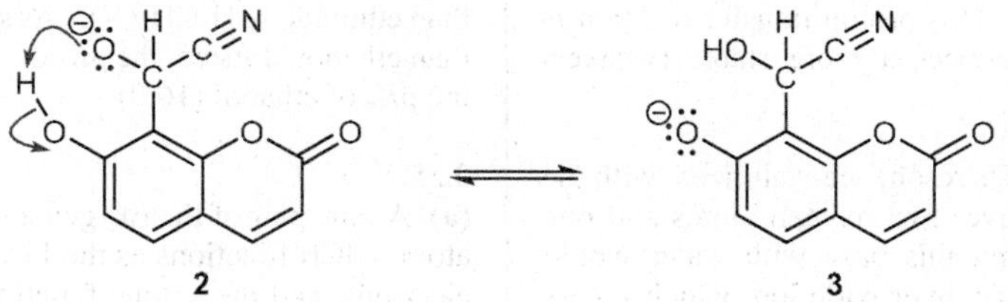

(b) To determine the position of the equilibrium, we compare the stability of the negative charge in **2** with the negative charge in **3** using the four factors (ARIO).

1. *Atom.* In each case, the negative charge is on an oxygen atom. Thus, the atom effect is not relevant.

2. *Resonance.* The anionic oxygen in **2** has no resonance stabilization, because it is attached to an sp^3-hybridized carbon atom. In contrast, anion **3** is highly stabilized by resonance, as shown here. Notice that the negative charge is spread over many locations, including two oxygen atoms.

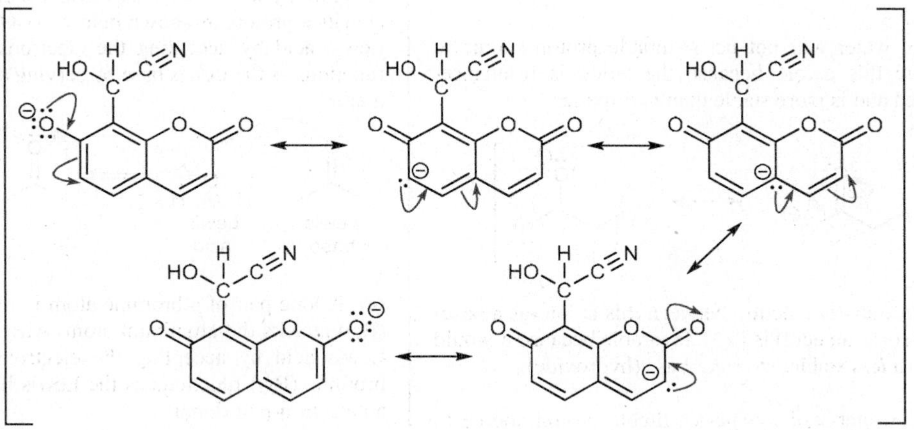

3. *Induction.* The anionic oxygen in **2** is stabilized by the inductive effect of the nitrile nitrogen, the hydroxyl oxygen, and the ester oxygen atoms. However, each of these electronegative atoms are fairly far removed from the negative charge and thus probably play a minimal role in stabilizing **2**.
Anion **3** is also stabilized by the nitrile nitrogen and the other oxygen atoms in the molecule, but here the electronegative atoms are even more distant, thus the effects are even less than in **2**.

4. *Orbitals.* Orbitals are not a relevant factor in this case.

Recall that factors 1-4 are listed in order of priority. Thus, based on factor 2 (resonance), we would expect that anion **3** should be significantly more stable than anion **2**. The inductive effects that stabilize **2** are significantly less important than the resonance effects that stabilize **3**. Therefore, the equilibrium should favor **3** over **2**.

3.32.

(a) Yes, because a negative charge on an oxygen atom will be more stable than a negative charge on a nitrogen atom. This proton transfer reaction is favorable because it produces a more stable (weaker) base.

(b) Yes, because a negative charge on a nitrogen atom will be more stable than a negative charge on an sp^3-hybridized carbon atom. This proton transfer reaction is favorable because it produces a more stable (weaker) base.

(c) No, because this base is resonance-stabilized, with the negative charge spread over two oxygen atoms and one carbon atom. Protonating this base with water would result in the formation of a hydroxide ion, which is less stable because the negative charge is localized on one oxygen atom. This proton transfer reaction is NOT favorable because it would produce a *less* stable, *stronger* base.

(d) Yes, because a negative charge on an oxygen atom will be more stable than a negative charge on a carbon atom. This proton transfer reaction is favorable because it produces a more stable (weaker) base.

3.33.

(a) No, water will not be a suitable proton source to protonate this anion, because the anion is resonance-stabilized and is more stable than hydroxide:

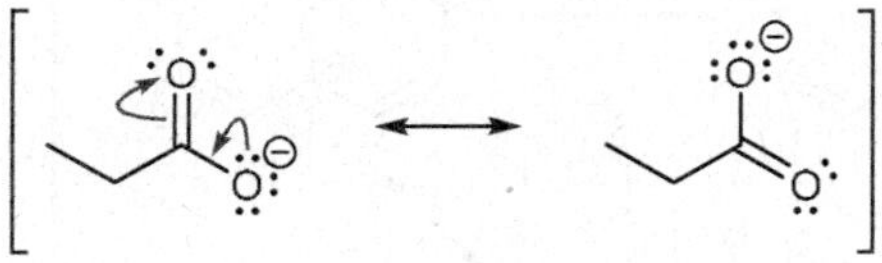

A proton transfer reaction between this anion (as a base) and water (as an acid) is NOT favorable because it would produce a *less* stable, *stronger* base (hydroxide).

(b) No, water will not be a suitable proton source to protonate this anion, because the anion is resonance-stabilized (with the negative charge being spread over two oxygen atoms) and is therefore more stable than hydroxide:

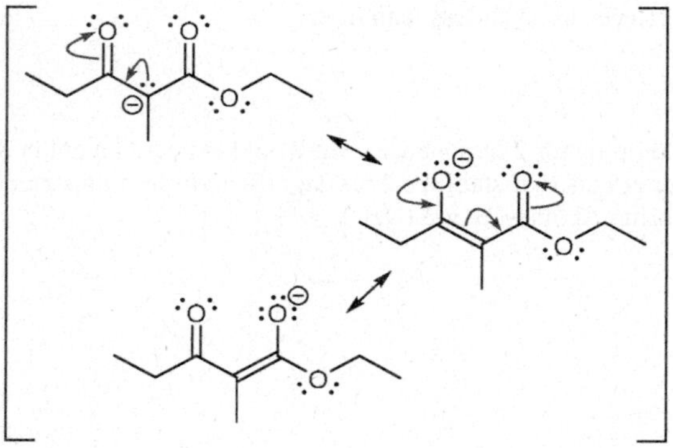

A proton transfer reaction between this anion (as a base) and water (as an acid) is NOT favorable because it would produce a *less* stable, *stronger* base (hydroxide).

3.34. The CH_3CH_2 group in ethanol provides a small amount of steric bulk that is absent in water. As a result, hydroxide is better solvated and is therefore more stable. This improved solvation makes hydroxide a weaker base than ethoxide ($CH_3CH_2O^-$). As such, water is more acidic than ethanol. Indeed, the pK_a of water (15.7) is lower than the pK_a of ethanol (16.0).

3.35.

(a) A lone pair of the oxygen atom attacks the aluminum atom. $AlCl_3$ functions as the Lewis acid by accepting the electrons, and the ketone functions as the Lewis base by serving as an electron-pair donor.

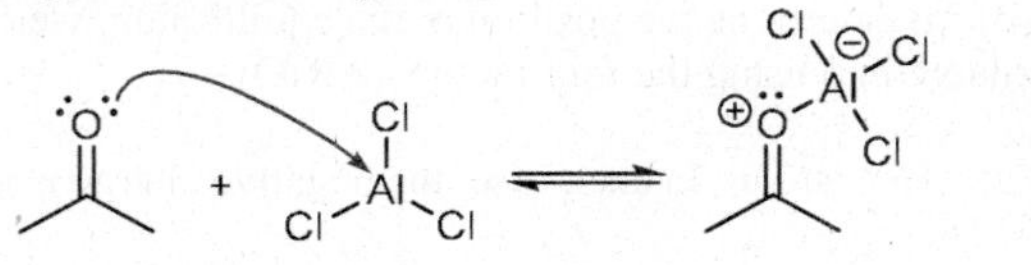

(b) A lone pair of the oxygen atom (in the starting ketone) attacks a proton, as shown below. H_3O^+ functions as the Lewis acid by accepting the electrons, and the ketone functions as the Lewis base by serving as an electron-pair donor.

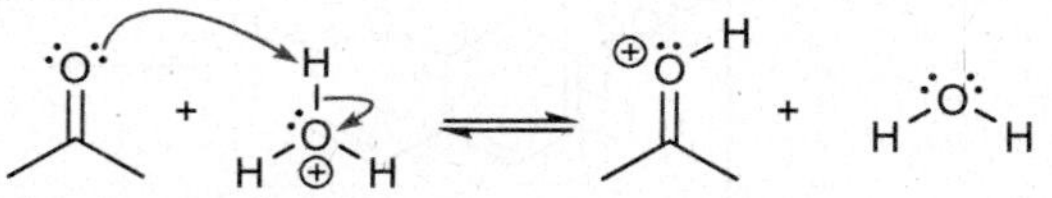

(c) A lone pair of a bromine atom in molecular bromine (Br_2) attacks the aluminum atom. $AlBr_3$ functions as the Lewis acid by accepting the electrons, and molecular bromine (Br_2) functions as the Lewis base by serving as an electron-pair donor.

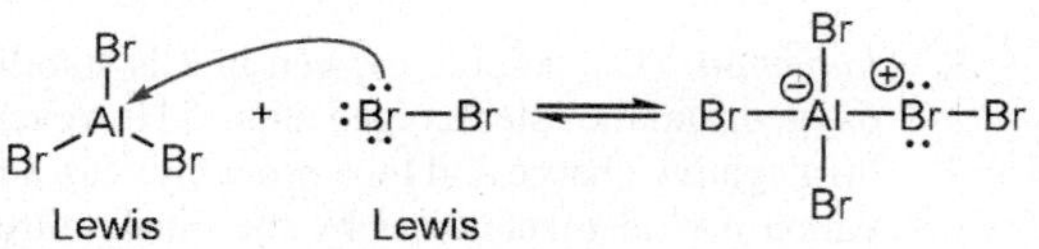

(d) A lone pair on one of the oxygen atoms of the ester attacks a proton on H_3O^+, as shown below. H_3O^+ functions as the Lewis acid by accepting the electrons, and the ester functions as the Lewis base by serving as an electron-pair donor.

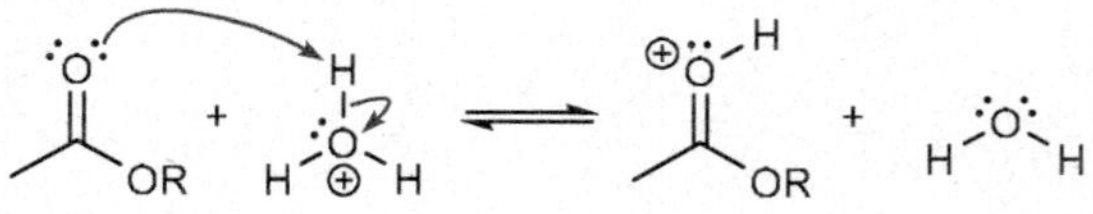

(e) A lone pair of the oxygen atom attacks the boron atom. BF_3 functions as the Lewis acid by accepting the electrons, and the ketone functions as the Lewis base by serving as an electron-pair donor.

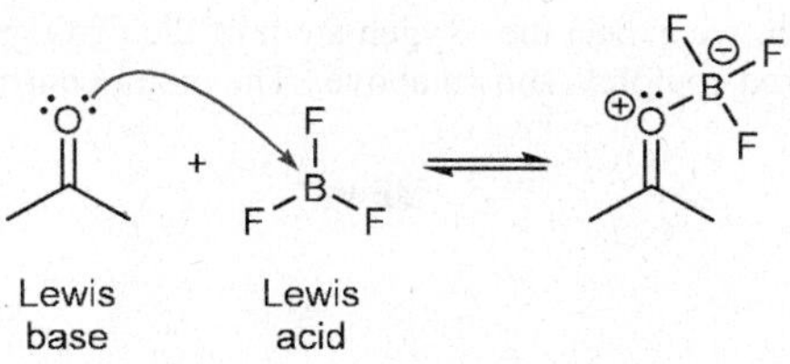

Lewis base

Lewis acid

3.36. In compound **1**, there are two oxygen atoms that bear lone pairs, and either location could certainly function as a Lewis base to donate an electron pair. But the problem statement indicates that the carbonyl (C=O) group interacts with $AlCl_3$. Let's explore why the carbonyl groups is the stronger Lewis base, by considering the structure of the complex that is obtained in each scenario. If the oxygen atom of the ether group (left) interacts with $AlCl_3$, the following complex is obtained:

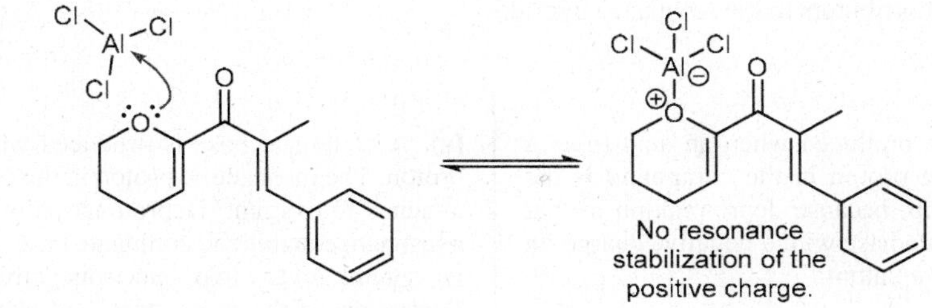

No resonance stabilization of the positive charge.

Notice that the positive charge in this complex is not resonance stabilized. In contrast, if the oxygen atom of the C=O bond interacts with $AlCl_3$, then the following resonance-stabilized complex is obtained:

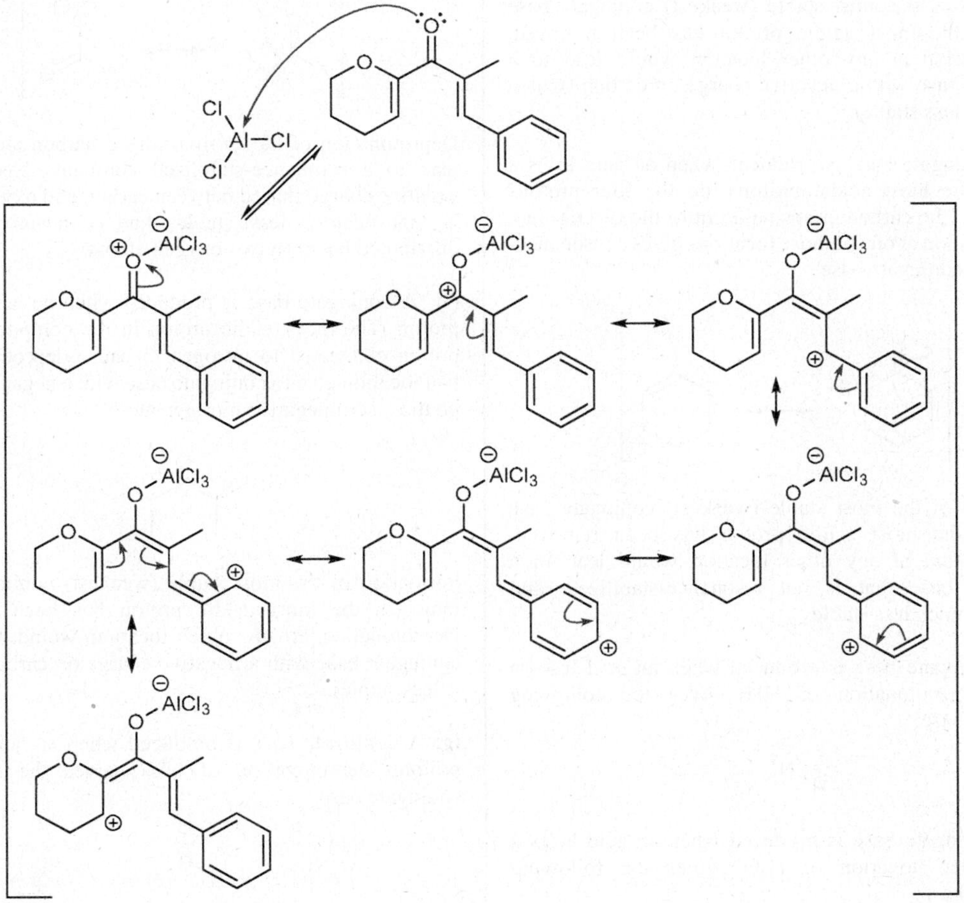

As such, we expect the oxygen atom of the C=O group to interact with the $AlCl_3$ to form the lower-energy, resonance-stabilized complex, shown above. The major contributor is the one that exhibits filled octets:

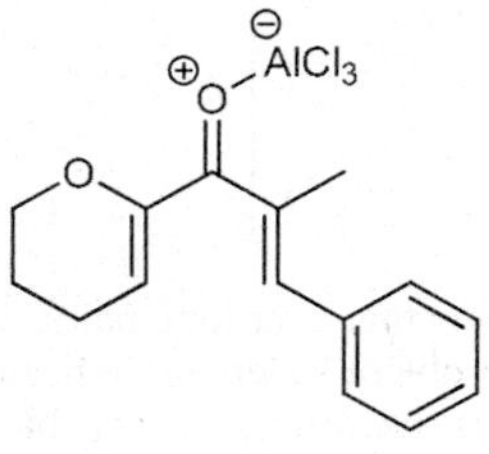

All other resonance structures exhibit a C+, which lacks an octet of electrons. Those resonance structures are less significant, and minor contributors to the resonance hybrid.

3.37.

(a) A conjugate base is produced when an acid loses a proton. The most acidic proton in the compound is the proton of the O-H group, because deprotonation at that location gives a conjugate base with a negative charge on the electronegative oxygen atom:

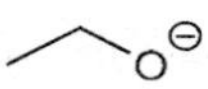

Formation of the most stable (weakest) conjugate base indicates the most acidic proton has been removed. Deprotonation at any other location would lead to a conjugate base with a negative charge on carbon (which is MUCH less stable).

(b) A conjugate base is produced when an acid loses a proton. The most acidic protons are the four protons attached to the carbon atoms adjacent to the C=O group. Deprotonation at one of those locations gives a resonance-stabilized conjugate base:

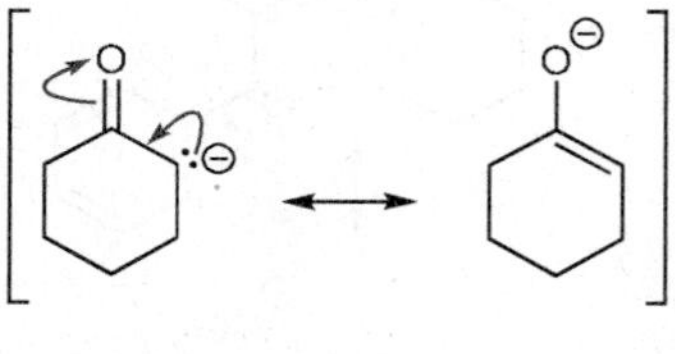

Formation of the most stable (weakest) conjugate base indicates the most acidic proton has been removed. Deprotonation at any other location would lead to a conjugate base that is not resonance-stabilized, and therefore much less stable.

(c) A conjugate base is produced when an acid loses a proton. Deprotonation of NH_3 gives the following conjugate base:

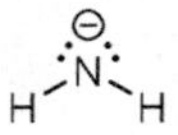

(d) A conjugate base is produced when an acid loses a proton. Deprotonation of H_3O^+ gives the following conjugate base:

(e) A conjugate base is produced when an acid loses a proton. The most acidic proton in the compound is the one attached to oxygen. Deprotonation at that location gives a resonance-stabilized conjugate base, in which the charge is spread over two electronegative oxygen atoms. Formation of the most stable (weakest) conjugate base indicates the most acidic proton has been removed.

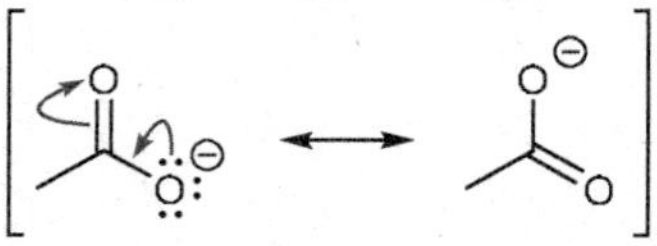

Deprotonation of the sp^3-hybridized carbon atom would lead to a resonance-stabilized conjugate base with a negative charge shared between carbon and oxygen. This is significantly less stable than a negative charge distributed between two oxygen atoms.

(f) A conjugate base is produced when an acid loses a proton. The most acidic proton in the compound is the proton connected to nitrogen, because deprotonation at that location gives a conjugate base with a negative charge on the electronegative nitrogen atom:

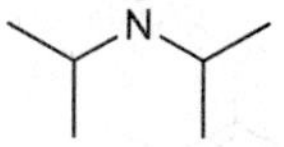

Formation of the most stable (weakest) conjugate base indicates the most acidic proton has been removed. Deprotonation at any other location would lead to a conjugate base with a negative charge on carbon (which is less stable).

(g) A conjugate base is produced when an acid loses a proton. Deprotonation of NH_4^+ gives the following conjugate base:

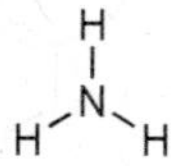

3.38.

(a) A conjugate acid is produced when a base accepts a proton. Protonation occurs at the site bearing the negative charge, giving the following compound:

(b) A conjugate acid is produced when a base accepts a proton. Protonation occurs at the site bearing the negative charge, giving the following ketone:

(c) A conjugate acid is produced when a base accepts a proton. In this case, the base is H_2N^- (the sodium cation is simply a counterion, also called a spectator ion, and does not participate). Protonation of H_2N^- gives NH_3, as shown below:

(d) A conjugate acid is produced when a base accepts a proton. Protonation of H_2O gives H_3O^+, as shown below:

(e) A conjugate acid is produced when a base accepts a proton. One of the lone pairs of the oxygen atom can serve as a base. Protonation gives the following oxonium ion (a cation in which the positive charge is located on an oxygen atom):

(f) A conjugate acid is produced when a base accepts a proton. The lone pair of the nitrogen atom can serve as a base. Protonation gives the following ammonium ion (a cation in which the positive charge is located on a nitrogen atom):

(g) A conjugate acid is produced when a base accepts a proton. One of the lone pairs of the oxygen atom can serve as a base. Protonation gives the following resonance-stabilized cation:

(h) A conjugate acid is produced when a base accepts a proton. In this case, the base is HO^- (the sodium cation is simply a counterion, also called a spectator ion, and does not participate). Protonation of HO^- gives H_2O, as show below:

3.39. The difference in acidity between compounds A and B is 10 – 7 = 3 pK_a units. Each pK_a unit represents one order of magnitude (a power of 10), so compound A is 10^3, or 1000 times more acidic than compound B.

3.40.

(a) A lone pair of the oxygen atom attacks the carbocation (C+), as shown below. The carbocation functions as the Lewis acid by accepting the electrons, and ethanol (CH_3CH_2OH) functions as the Lewis base by serving as an electron-pair donor.

Lewis base | Lewis acid

(b) A lone pair of the oxygen atom attacks the boron atom. BF_3 functions as the Lewis acid by accepting the electrons, and ethanol (CH_3CH_2OH) functions as the Lewis base by serving as an electron-pair donor.

Lewis base | Lewis acid

(c) A lone pair of a chlorine atom attacks the aluminum atom. $AlCl_3$ functions as the Lewis acid by accepting the electrons, and ethyl chloride (CH_3CH_2Cl) functions as the Lewis base by serving as an electron-pair donor.

Lewis base | Lewis acid

3.41. The amide ion (H_2N^-) is a strong base, and in the presence of H_2O, a proton transfer reaction will occur, generating hydroxide, which is a more stable (weaker) base than the amide ion (Factor #1: oxygen is more electronegative than nitrogen). The reaction will greatly favor products (ammonia and hydroxide).

3.42. No, the reaction cannot be performed in the presence of ethanol, because the leveling effect would cause deprotonation of ethanol to form more-stable ethoxide ions ($CH_3CH_2O^-$) (Factor #1: oxygen is more electronegative than nitrogen and carbon). The desired anion would not be formed under these conditions.

3.43. No, water would not be a suitable proton source in this case. Note that the sodium cation (Na^+) serves as a counterion. The anion is the conjugate base of a carboxylic acid, and the negative charge is resonance stabilized. A proton transfer reaction between this anion (as a base) and water (as an acid) is NOT favorable because it would produce a *less* stable, *stronger* base (hydroxide). Hydroxide is less stable (and thus more strongly basic) than the compound shown, because the negative charge in hydroxide is not resonance stabilized.

3.44.

(a) Water functions as a base and deprotonates HBr. Two curved arrows are required. The first curved arrow shows a lone pair of the base attacking the proton, and the second curved arrow comes from the H–Br bond (being broken) and goes to the bromine atom, as shown:

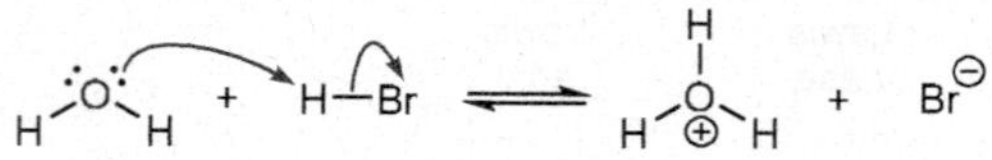

(b) Water functions as a base and deprotonates sulfuric acid (H_2SO_4). Two curved arrows are required. The first curved arrow shows a lone pair of the base attacking the proton. The second curved arrow comes from the O–H bond (being broken) and goes to the oxygen atom, as shown:

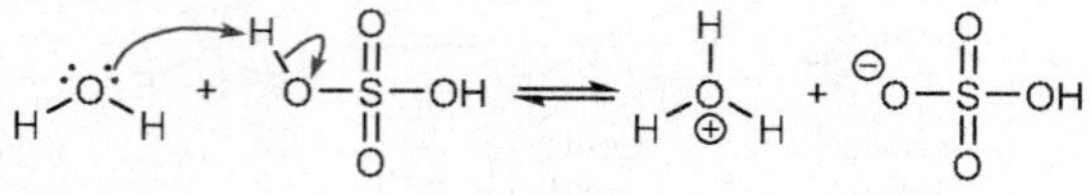

(c) Water functions as a base and deprotonates this strong acid (See Table 3.1). Two curved arrows are required. The first curved arrow shows a lone pair of the base attacking the proton, and the second curved arrow comes from the O–H bond (being broken) and goes to the oxygen atom, as shown:

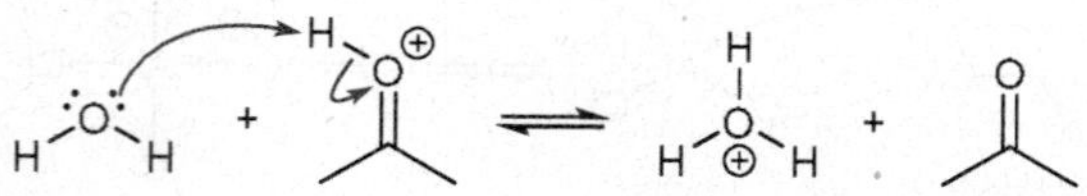

3.45.

(a) Water functions as an acid in this case, by giving a proton to the strong base, as shown below. Two curved arrows are required. The first curved arrow shows a lone pair of the base attacking the proton, and the second curved arrow comes from the O–H bond (being broken) and goes to the oxygen atom, as shown:

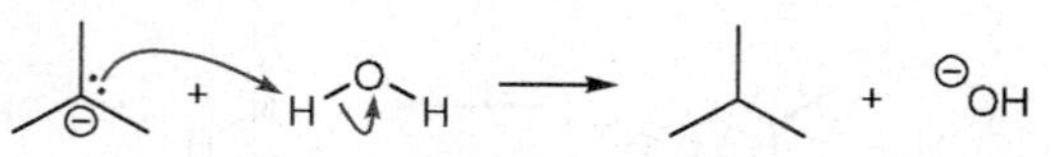

(b) Water functions as an acid in this case, by giving a proton to the base, as shown below. Two curved arrows are required. The first curved arrow shows a lone pair of the base attacking the proton, and the second curved arrow comes from the O–H bond (being broken) and goes to the oxygen atom, as shown:

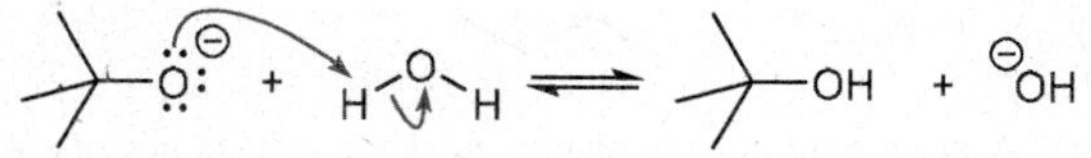

(c) Water functions as an acid in this case, by giving a proton to the strong base, as shown below. Two curved arrows are required. The first curved arrow shows a lone pair of the base attacking the proton, and the second curved arrow comes from the O–H bond (being broken) and goes to the oxygen atom, as shown:

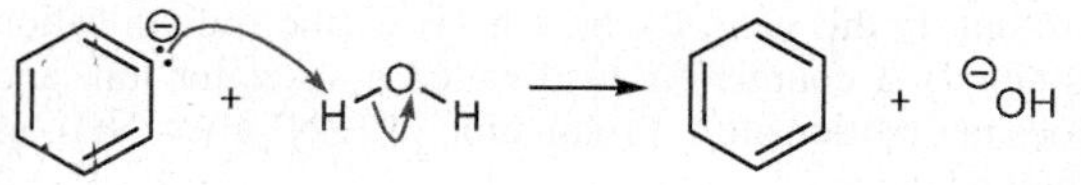

(d) Water functions as an acid in this case, by giving a proton to the strong base, as shown below. Two curved arrows are required. The first curved arrow shows a lone pair of the base attacking the proton, and the second curved arrow comes from the O–H bond (being broken) and goes to the oxygen atom, as shown:

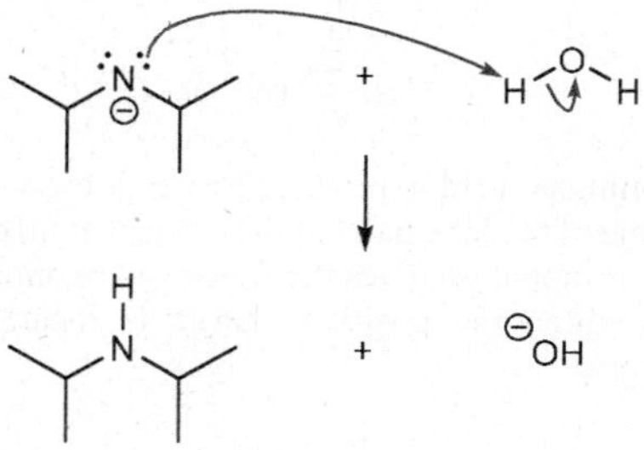

3.46.

(a) The second anion is more stable because it has an allylic lone pair, and the negative charge is delocalized by resonance.

(b) The second anion is more stable because the negative charge is on a more electronegative nitrogen atom (factor #1 of ARIO), rather than an sp^3 hybridized carbon atom.

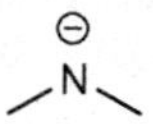

(c) The second anion is more stable because the negative charge is on an *sp*-hybridized carbon atom, rather than an sp^3-hybridized carbon atom.

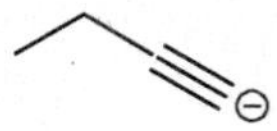

3.47.

(a) The second compound is more acidic, because its conjugate base (shown below) has a negative charge on a sulfur atom, which is more stable than a negative charge on an oxygen atom (factor #1 of ARIO), A more stable (weaker) conjugate base indicates a stronger parent acid.

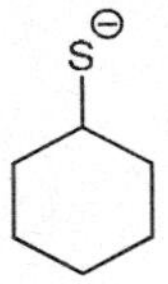

(b) The first compound (called phenol) is more acidic, because its conjugate base (shown below) is resonance stabilized (factor #2 of ARIO). A more stable (weaker) conjugate base indicates a stronger parent acid.

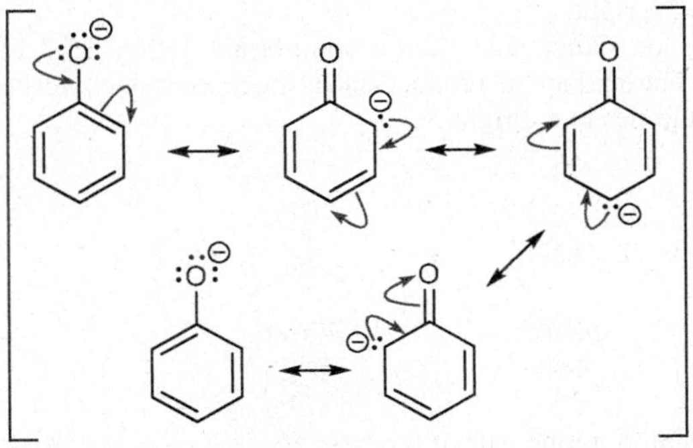

(c) The conjugate base for each of these compounds is resonance-stabilized (as in part b of this problem). The difference between these compounds is the presence of electron-withdrawing chlorine atoms. The first compound is more acidic as a result of the combined inductive effects of the five chlorine atoms which stabilize the conjugate base shown below (factor #3 of ARIO). A more stable (weaker) conjugate base indicates a stronger parent acid.

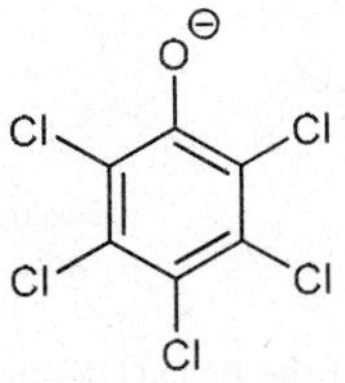

(d) The second compound is more acidic, because its conjugate base (shown below) has a negative charge associated with an *sp*-hybridized orbital, while the first compound is less acidic because its conjugate base has a negative charge on a nitrogen atom. We learned that this example constitutes an exception to the order of priorities, ARIO. In this case, factor #4 (orbital) trumps factor #1 (atom). This exception applies whenever we compare a negative charge on an *sp*-hybridized carbon atom and a negative charge on an sp^3-hybridized nitrogen atom.

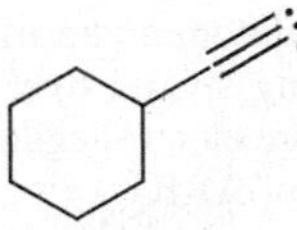

(e) The first compound is more acidic because its conjugate base (shown below) is resonance stabilized.

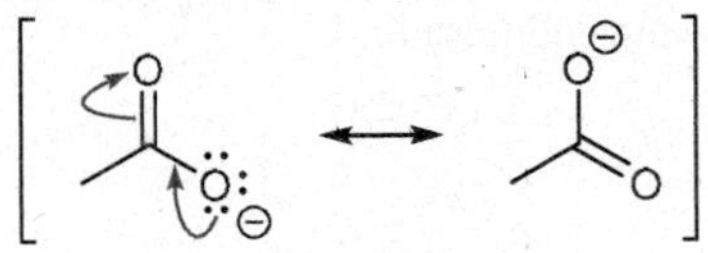

A more stable (weaker) conjugate base indicates a stronger parent acid. The conjugate base of the second compound is not resonance stabilized.

(f) The first compound is more acidic because its conjugate base (shown below) is resonance stabilized, and one of the resonance structures has the negative charge on an electronegative oxygen atom.

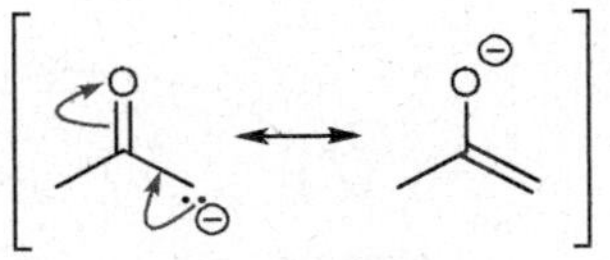

A more stable (weaker) conjugate base indicates a stronger parent acid. The conjugate base of the second compound is not resonance stabilized (the negative charge would be localized on a carbon atom).

(g) The first compound is more acidic because its conjugate base (shown below) is resonance stabilized, and one of the resonance structures has the negative charge on an oxygen atom.

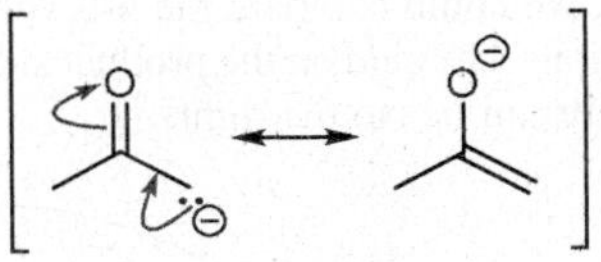

A more stable (weaker) conjugate base indicates a stronger parent acid. The conjugate base of the second compound is not resonance stabilized (the negative charge would be localized on a carbon atom).

(h) The second compound is more acidic because its conjugate base is resonance stabilized, with the negative charge being spread over two oxygen atoms (shown below). A more stable (weaker) conjugate base indicates a stronger parent acid.

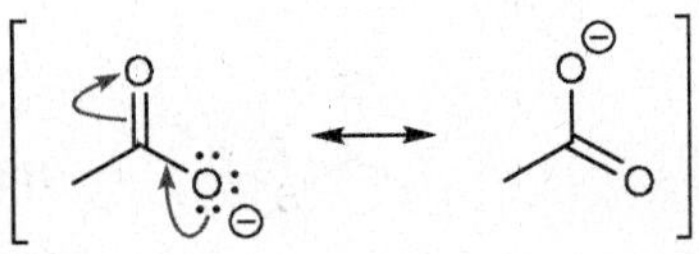

The conjugate base of the first compound is also resonance stabilized, but the negative charge would be spread over an oxygen atom and a nitrogen atom, which is less stable than being spread over two oxygen atoms (because oxygen is more electronegative than nitrogen, as described in factor #1 of ARIO).

3.48. NaA represents an ionic compound, comprised of cations (Na^+) and anions (A^-). When H–B is treated with A^-, a proton can be transferred from H–B to A^-, as shown in the following equilibrium:

$$\underset{pK_a = 5}{H{-}B} + Na^{\oplus}\,:A^{\ominus} \rightleftharpoons \underset{pK_a = 15}{H{-}A} + Na^{\oplus}\,:B^{\ominus}$$

The equilibrium will favor the weaker acid (the acid with the higher pK_a value). In this case, the equilibrium favors formation of HA.

3.49.

(a) Water (H_2O) loses a proton and is therefore functioning as an acid. Two curved arrows must be drawn. The first curved arrow shows a lone pair of the base attacking the proton of water, and the second curved arrow comes from the O–H bond (being broken) and goes to the oxygen atom of water, as shown. To determine the position of the equilibrium, we compare the stability of the ions on either side of the equilibrium (the competing bases, in this case). Because of better solvation, hydroxide is the more stable anion. The base on the product side is the more stable (weaker) base. As such, the reaction favors the products, and the equilibrium lies to the right.
Alternatively, we could compare the pK_a values of the acids on either side of the equilibrium [H_2O, 15.7 and $(CH_3)_2CHOH$, ~16] and we would arrive at the same conclusion. The acid on the product side is the more stable (weaker) acid. As such, the reaction favors the products, and the equilibrium lies to the right.

stronger acid | stronger base | weaker base | weaker acid

(b) Two curved arrows must be drawn. The first curved arrow shows a lone pair of the base attacking the proton, and the second curved arrow comes from the S–H bond (being broken) and goes to the sulfur atom, as shown below. To determine the position of the equilibrium, we compare the stability of the ions on either side of the equilibrium (the competing bases, in this case). A negative charge on a larger sulfur atom is expected to be more stable than the negative charge on an oxygen atom (factor #1 of ARIO). The base on the product side is the more stable (weaker) base. As such, the reaction favors the products, and the equilibrium lies to the right.

Alternatively, we could compare the pK_a values of the acids on either side of the equilibrium and we would arrive at the same conclusion. The acid on the product side is the more stable (weaker) acid. As such, the reaction favors the products, and the equilibrium lies to the right.

stronger acid | stronger base | weaker base | weaker acid

(c) Two curved arrows must be drawn. The first curved arrow shows a lone pair of the base (HS^- in this case) attacking the proton, and the second curved arrow comes from the S–H bond (being broken) and goes to the sulfur atom, as shown. To determine the position of the equilibrium, we compare the stability of the ions on either side of the equilibrium (the competing bases, in this case). The base on the right side of the equilibrium is resonance stabilized, with the negative charge being spread over two sulfur atoms. The base on the left side of the equilibrium is not resonance stabilized, and the negative charge is localized on one sulfur atom. The equilibrium favors the more stable (weaker) base, and the product side is favored.

stronger base | weaker base

(d) Two curved arrows must be drawn. The first curved arrow shows a lone pair of the base (the nitrogen atom) attacking the proton, and the second curved arrow comes from the O–H bond (being broken) and goes to the oxygen atom, as shown. The equilibrium favors the product, which can be determined by comparing the stability of the ions on either side of the equilibrium (the competing bases, in this case). Factor #1 of ARIO indicates that the product is favored in this case, because a negative charge is more stable on an oxygen atom than a nitrogen atom. The equilibrium favors the more stable (weaker) base, and the product side is favored.

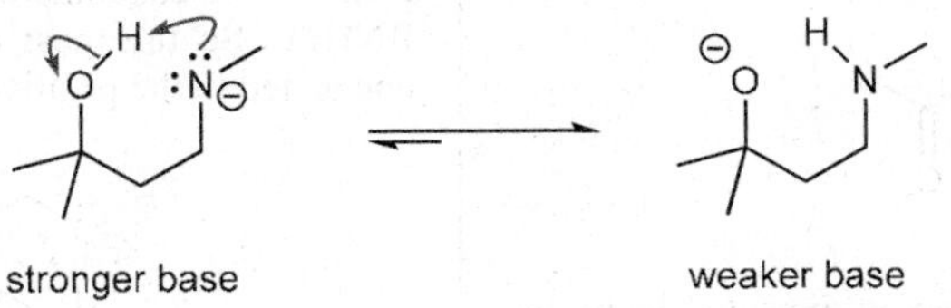

stronger base

weaker base

3.50. One of the anions is resonance stabilized, with the negative charge spread over two oxygen atoms. That anion is the weakest (most stable) base. Among the remaining three anions, they do not differ from each other in any of the four factors (ARIO), but they are expected to differ from each other in terms of solvent effects. That is, an anion will be less stable (stronger base) if it has steric bulk in close proximity with the negative charge. The steric bulk reduces the stability of the anion by limiting its ability to interact with solvent molecules, as described in Section 3.8.

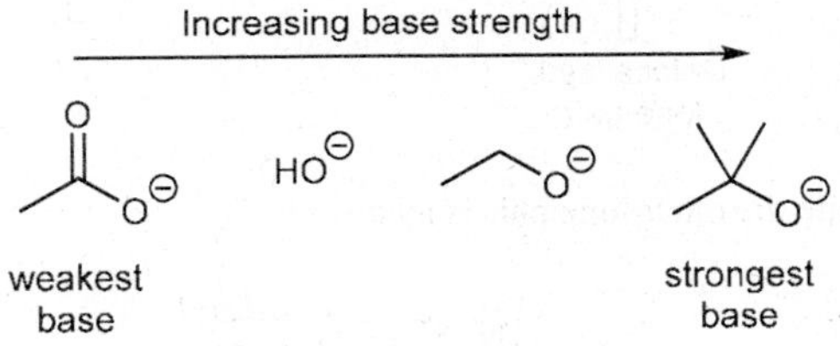

Increasing base strength

weakest base

strongest base

3.51.

(a) The proton highlighted below is expected to be the most acidic, because deprotonation leads to a conjugate base in which the negative charge is associated with an *sp*-hybridized orbital. This case represents an exception to the ARIO priority scheme, because factor #4 (orbital) trumps factor #1 (atom). This exception applies whenever we compare a negative charge on an *sp*-hybridized carbon atom and a negative charge on an sp^3-hybridized nitrogen atom.

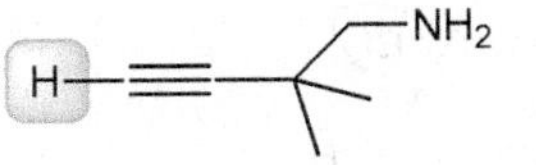

(b) The proton highlighted below is expected to be the most acidic, because deprotonation leads to a conjugate base in which the negative charge is on a sulfur atom (more stable than being on an oxygen, nitrogen, or carbon atom, according to factor #1 of ARIO). Formation of the most stable (weakest) conjugate base indicates that the most acidic proton has been removed.

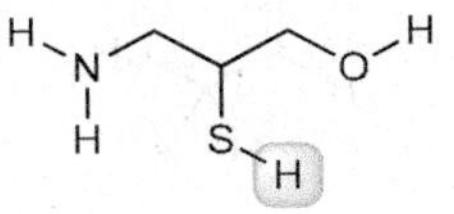

(c) The proton highlighted below is expected to be the most acidic, because deprotonation leads to a resonance-stabilized conjugate base. Formation of the most stable (weakest) conjugate base indicates that the most acidic proton has been removed. Removal of any of the other protons would lead to an anion that is not resonance stabilized.

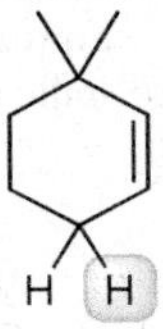

(d) The proton highlighted below is expected to be the most acidic, because deprotonation leads to a resonance-stabilized conjugate base in which the negative charge is spread over two carbon atoms and one oxygen atom. Formation of the most stable (weakest) conjugate base indicates that the most acidic proton has been removed. Removal of any of the other protons would lead to an anion that is either not resonance stabilized or less resonance stabilized.

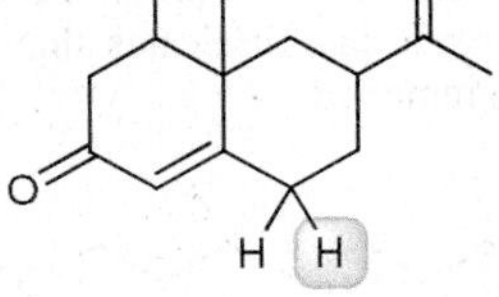

(e) The proton highlighted below is expected to be the most acidic, because deprotonation leads to a resonance-stabilized conjugate base in which the negative charge is spread over two oxygen atoms (factor #2 of ARIO). Formation of the most stable (weakest) conjugate base indicates that the most acidic proton has been removed. Removal of any other proton would lead to a less stable anion.

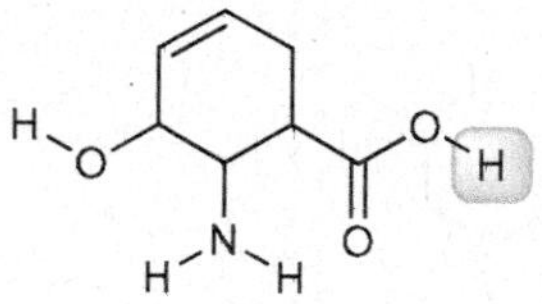

(f) The proton highlighted below is expected to be the most acidic, because deprotonation leads to a resonance-stabilized conjugate base in which the negative charge is spread over one carbon atom and one oxygen atom (factor #2 of ARIO). Formation of the most stable (weakest) conjugate base indicates that the most acidic proton has been removed. Deprotonation of either of the other carbon atoms would lead to an anion that is not resonance stabilized, and therefore less stable.

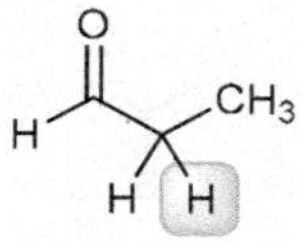

(g) There are three carboxylic acid (CO_2H) groups, each of which bears an acidic proton. Removing any one of these protons will result in a resonance-stabilized conjugate base. Among these three protons, the highlighted proton is the most acidic because of the electron-withdrawing effects of the nearby chlorine atoms. When this proton is removed, the conjugate base is stabilized not only by resonance, but also by induction (factor #3 of ARIO). Formation of the most stable (weakest) conjugate base indicates that the most acidic proton has been removed.

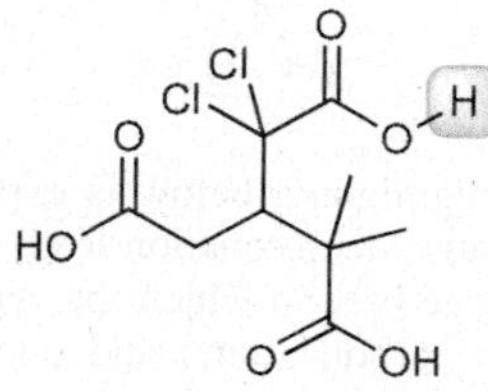

(h) The highlighted proton is expected to be the most acidic, because deprotonation leads to a conjugate base in which the negative charge is on a large sulfur atom, which is more stable than being on an oxygen or carbon atom (factor #1 of ARIO). Formation of the most stable (weakest) conjugate base indicates that the most acidic proton has been removed.

3.52. In this case, we are assessing the relative acidity of two cationic acids: ROH_2^+ and RNH_3^+. To compare the acidity of cationic acids, we compare the stability of the positive charges. We are comparing a positive charge on an oxygen atom with a positive charge on a nitrogen atom. As we have seen, nitrogen is better at stabilizing a positive charge (compared with oxygen), because nitrogen is less electronegative than oxygen. With O^+ being less stable than N^+, we conclude that ROH_2^+ is a stronger acid than RNH_3^+. So the most acidic proton in this structure is connected to the positively charged oxygen atom.

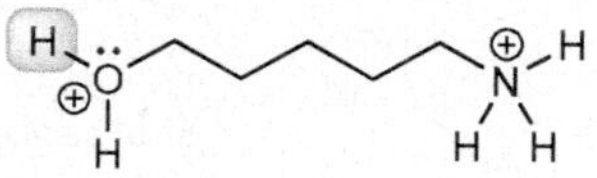

3.53.

(a) Ammonia is protonated by HCl to give an ammonium ion, as shown:

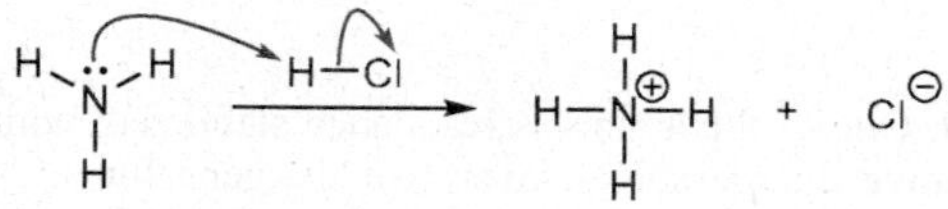

(b) There are two nitrogen atoms in DBN. One of the lone pairs is delocalized via resonance:

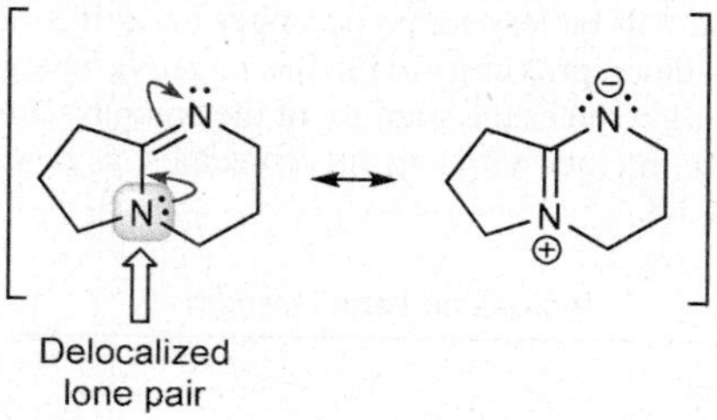

Delocalized lone pair

while the other lone pair is localized:

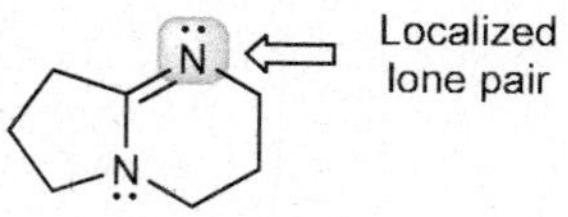

Localized lone pair

The localized lone pair is more available to function as a base, and protonation at that location leads to the following resonance-stabilized cation:

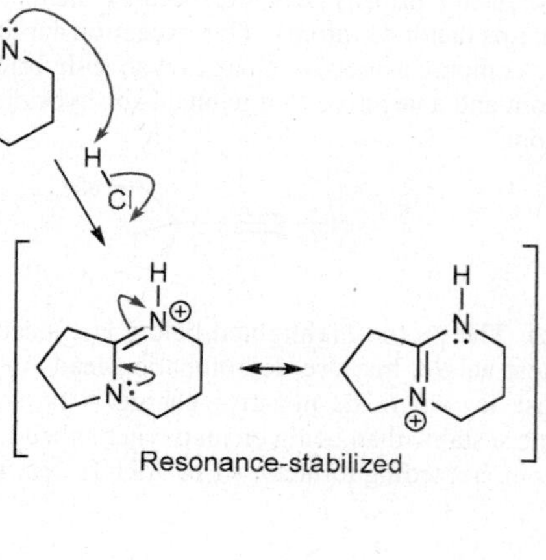

Resonance-stabilized

Protonation of the other nitrogen atom would lead to a conjugate acid that is not resonance stabilized.

(c) When we compare the two conjugate acids to each other, we find that the conjugate acid of DBN is more stable (because the charge is delocalized by resonance). A more stable (weaker) conjugate acid indicates a stronger parent base, and therefore, DBN is expected to be a stronger base than NH_3.

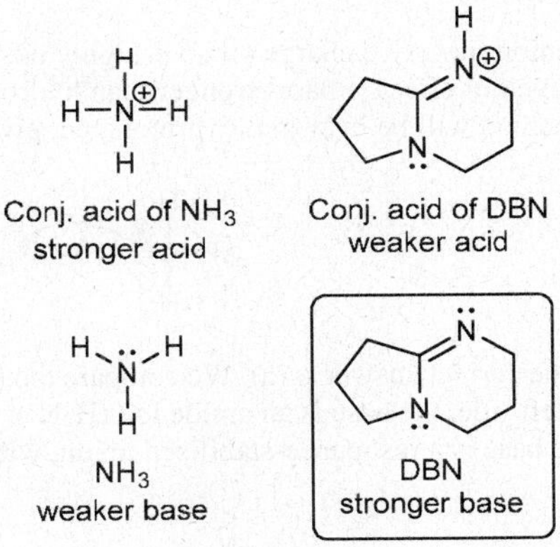

Conj. acid of NH_3
stronger acid

Conj. acid of DBN
weaker acid

NH_3
weaker base

DBN
stronger base

3.54. In each case, the lone pair on the nitrogen atom is delocalized via resonance throughout the ring.

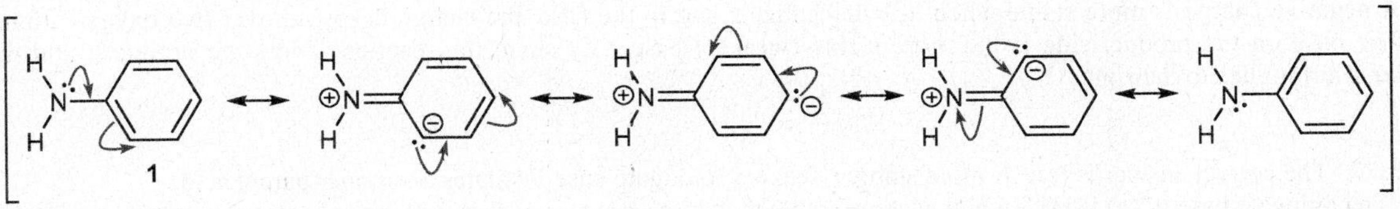

1

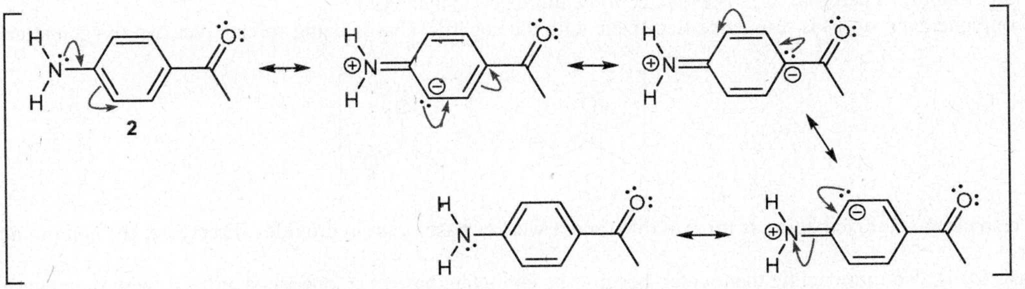

2

But compound **2** has additional delocalization of electrons that compound **1** lacks, because of the additional resonance structure shown below:

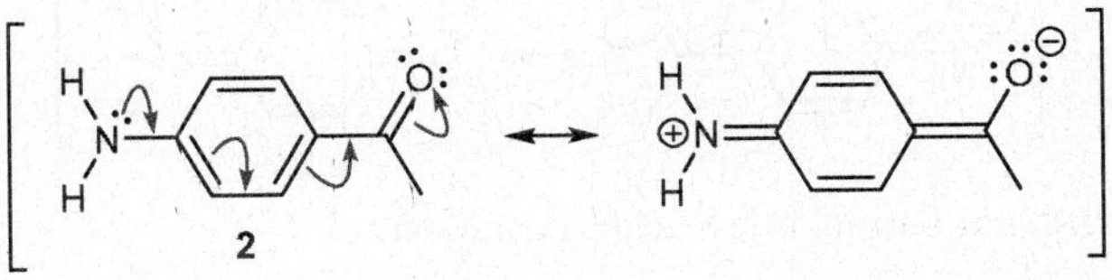

2

Therefore, the lone pair on the nitrogen atom in compound **2** is less available to function as a base, rendering compound **2** a weaker base than compound **1**.

3.55. The correct answer is (c). We begin by drawing methyl magnesium bromide (CH_3MgBr):

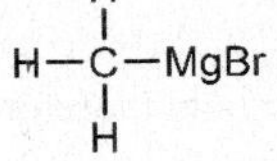

The difference in electronegativity between C and Mg is so great, that the C-Mg bond is sometimes drawn as ionic:

A carbanion (negative charge on C) is highly unstable. That is, the carbanion is a very strong base. However, as a result of the leveling effect, a base stronger than hydroxide is not possible in the presence of water. When treated with water, the carbanion will be irreversibly protonated, giving the more stable (weaker) hydroxide base (highlighted below).

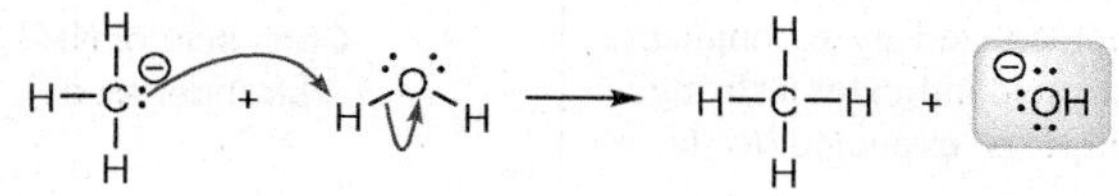

3.56. The correct answer is (a). We compare the ions on either side of the equilibrium (the competing bases, in this case). On the left side, the base is an amide ion (H_2N^-), which has a negative charge localized on a nitrogen atom. On the right side, the base is a resonance-stabilized anion, with the negative charge being delocalized:

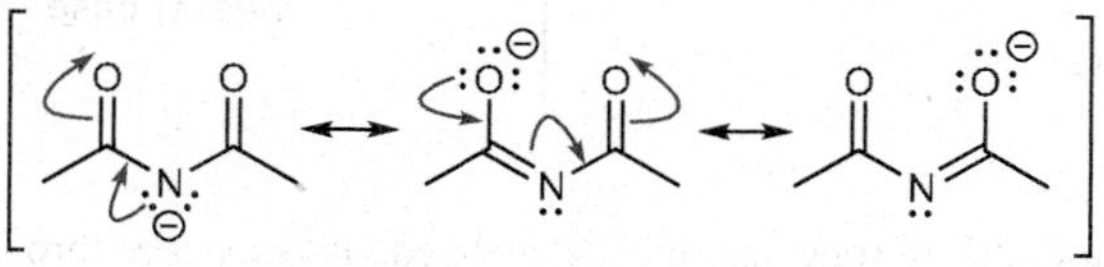

A negative charge is more stable when it is delocalized, and in the case, the charge is spread over two oxygen atoms. The base on the product side is the more stable (weaker) base. As such, the reaction favors the products, and the equilibrium lies to the right.

3.57. The correct answer is (c). A more stable (weaker) conjugate base indicates a stronger parent acid.
The conjugate base of (a) is HS^-, which is more stable than HO^-, because of factor #1 (the negative charge is more stable on a larger atom). Therefore, H_2S is expected to be more acidic than H_2O.
The conjugate base of (b) is resonance stabilized, with the negative charge being spread over two oxygen atoms:

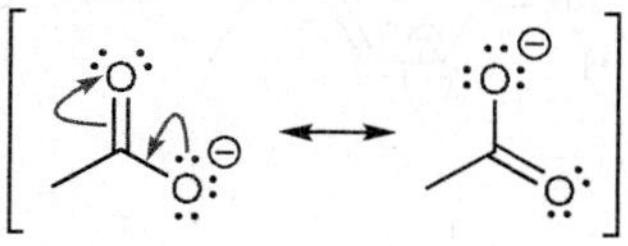

This resonance-stabilized anion is more stable (and a weaker base) than hydroxide. Therefore, (b) is more acidic than water.
Answer (d) is also more acidic than water, because its conjugate base is resonance-stabilized, with the negative charge delocalized over two oxygen atoms:

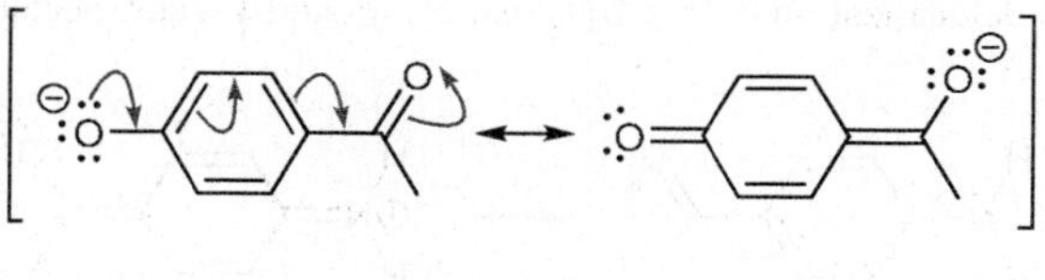

As a result of solvent effects, (c) *tert*-butanol is less acidic than water.

3.58. The correct answer is (d). Each of the first three structures has at least one lone pair and can therefore function as a Lewis base:

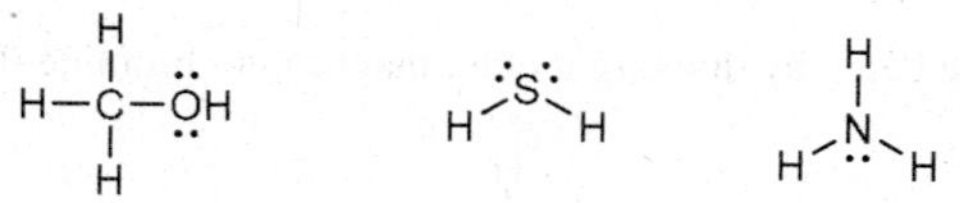

The final structure (CH_4) does not possess a lone pair, and therefore cannot function as a Lewis base.

3.59. The correct answer is (b). The highlighted proton in compound **I** (below) is the most acidic proton. When this proton is removed, the resulting conjugate base is stabilized by the electron-withdrawing effects of the electronegative fluorine atom. The conjugate base of acid **III** is also stabilized by an electron-withdrawing group, but chlorine is less electronegative than fluorine, so the effects are smaller. The least stable conjugate base comes from deprotonating acid **II**. In the conjugate base of acid **II**, the chlorine atom is farther away from the negative charge, so the electron-withdrawing effects are the smallest. The least stable conjugate base (the strongest conjugate base) has the weakest parent acid. When arranged from the weakest acid to the strongest acid, the answer is: II < III < I.

$ClCH_2CH_2CO_2H$ < $ClCH_2CO_2H$ < FCH_2CO_2H

II (weakest acid) III I (strongest acid)

3.60. The correct answer is (d). In this example, we are comparing the stability of the conjugate bases, Cl^- and Br^-, which appear in the same column of the periodic table. In such a case, electronegativity is not the dominant effect. Instead, the dominant effect is size. Bromine is larger than chlorine and can therefore better stabilize a negative charge by spreading the charge over a larger volume of space. As such, Br^- is the more stable (weaker) conjugate base, and it has the stronger parent acid. Therefore, HBr is a stronger acid than HCl.

3.61. The correct answer is (d). Proton **D** (highlighted below) is the most acidic proton. Deprotonation at this position leads to a conjugate base in which the negative charge is associated with an *sp*-hybridized orbital (which is more stable than being associated with an sp^2- or sp^3-hybridized orbital). Formation of the most stable (weakest) conjugate base indicates that the most acidic proton has been removed.

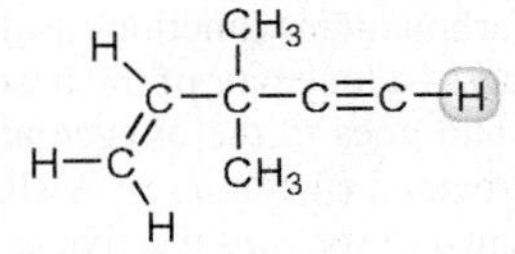

$CH_2=C(H)–C(CH_3)_2–C\equiv C–H$

3.62. The correct answer is (c). Removing a proton from compound **X** results in an anion that is not resonance-stabilized, while removing a proton from compound **Y** results in an anion that is resonance-stabilized. The more stable conjugate base (the weaker conjugate base) has the stronger parent acid, and therefore compound **Y** is the stronger acid.

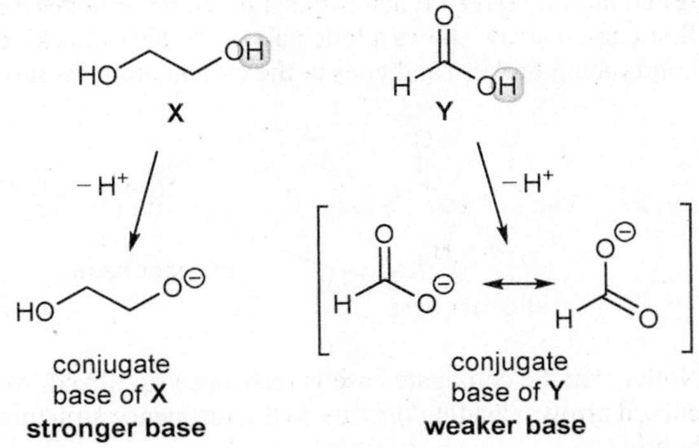

HO–CH₂CH₂–OH (X) → $-H^+$ → conjugate base of X, **stronger base**

HCO₂H (Y) → $-H^+$ → conjugate base of Y (resonance forms), **weaker base**

3.63. The correct answer is (a). Carbon, nitrogen and oxygen are in the same row of the periodic table, so we must compare their electronegativity values. Oxygen is the most electronegative and can best stabilize the negative charge. Carbon is the least electronegative and is the least effective at stabilizing the negative charge. Nitrogen has an electronegativity value between that of carbon and oxygen, so it has an intermediate ability to stabilize the negative charge. As such, the carbanion is the *least* stable, *strongest* base.

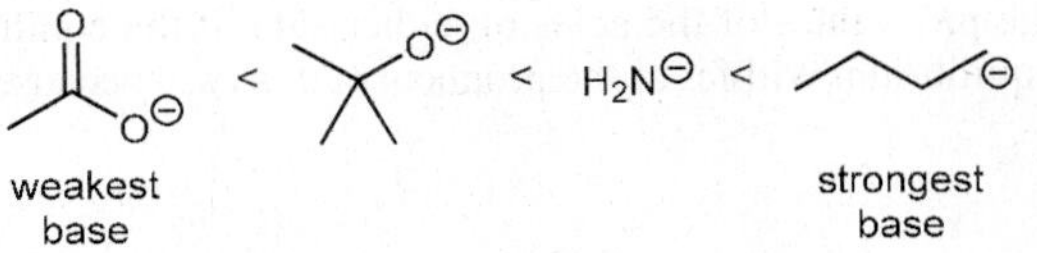

weakest base (acetate) < tert-butoxide < $H_2N^{\ominus}$ < carbanion (strongest base)

3.64.

(a) Acetic acid (CH_3CO_2H) loses a proton and is therefore functioning as an acid. Hydroxide (HO^-) functions as the base that removes the proton. Two curved arrows must be drawn. The first curved arrow shows a lone pair of the base attacking the acidic proton, and the second curved arrow comes from the O–H bond (being broken) and goes to the oxygen atom, as shown. The equilibrium favors the products, because the base on the right side of the equilibrium is resonance stabilized and therefore more stable than a hydroxide ion. The equilibrium favors the more stable (weaker) base. Alternatively, we could compare the pK_a values of the acids on either side of the equilibrium and we would arrive at the same conclusion (the equilibrium will favor the products in this case because the equilibrium favors the weaker acid).

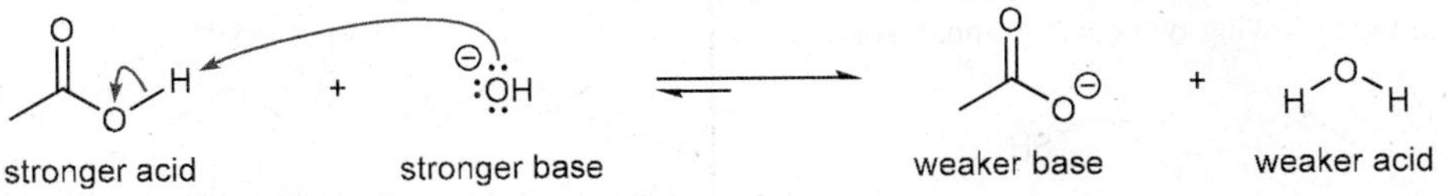

stronger acid + stronger base ⇌ weaker base + weaker acid

(b) Water (H_2O) loses a proton and is therefore functioning as an acid. The carbanion (an anion in which the negative charge is on a carbon atom) functions as the base that removes the proton. Two curved arrows must be drawn. The first curved arrow shows a lone pair of the base attacking the proton, and the second curved arrow comes from the O–H bond (being broken) and goes to the oxygen atom, as shown. The reaction favors the products, because hydroxide is more stable than a carbanion (factor #1 of ARIO). Alternatively, we could compare the pK_a values of the acids on either side of the equilibrium and we would arrive at the same conclusion (the reaction favors the products). In fact, the pK_a values of H_2O (15.7) and C_4H_{10} (~50) are so vastly different that the reaction is essentially irreversible.

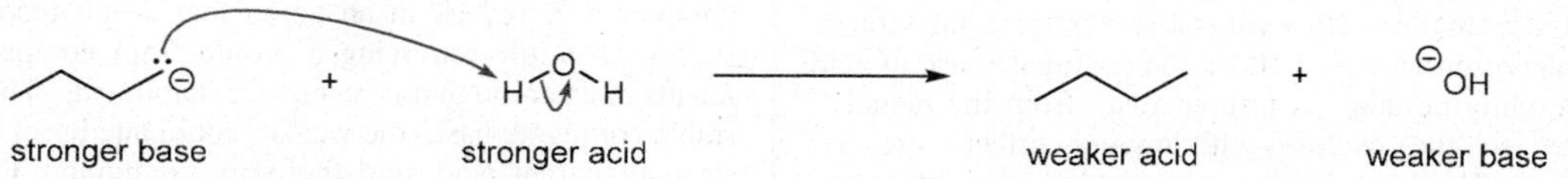

stronger base + stronger acid → weaker acid + weaker base

(c) Hydroxide (HO^-) functions as a base and removes a proton from the acid. Two curved arrows must be drawn. The first curved arrow shows a lone pair of the base attacking the proton, and the second curved arrow comes from the C–H bond (being broken) and goes to the carbon atom, as shown.

stronger acid + stronger base ⇌ weaker base + weaker acid

Notice that the conjugate base is resonance stabilized, so its formation can alternatively be shown with more than two curved arrows (leading directly to the resonance structure that contributes the most character to the overall resonance hybrid).

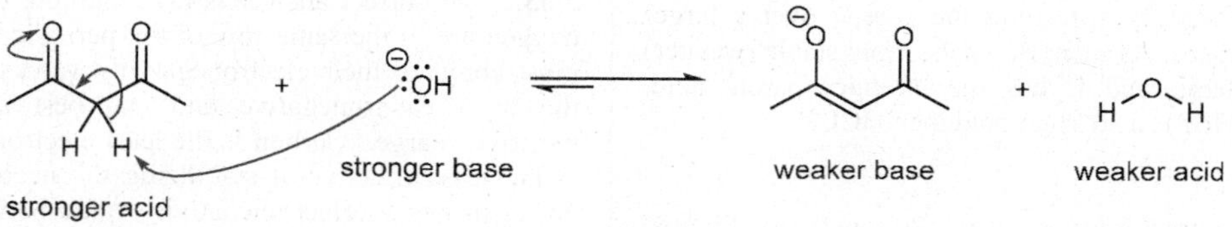

stronger acid + stronger base ⇌ weaker base + weaker acid

The equilibrium favors the products, because the base on the right side of the equilibrium is resonance stabilized, with the negative charge spread over two oxygen atoms. This is more stable than the negative charge being localized on one oxygen atom, as in hydroxide. The equilibrium favors the more stable (weaker) base. Alternatively, we could compare the pK_a values of the acids on either side of the equilibrium and we would arrive at the same conclusion. That is, the equilibrium will favor the products in this case because the equilibrium favors formation of the weaker acid (H_2O).

3.65. Each of the carbon atoms is tetravalent; the sulfur atom is divalent; and each of the hydrogen atoms is monovalent. We begin by connecting the atoms that have more than one bond (in this case, the two carbon atoms and the sulfur atom). There are only two different ways that these three atoms can be connected to each other, shown below:

C–S–C C–C–S

For each of these arrangements, we connect the hydrogen atoms, giving the following two constitutional isomers:

The second isomer is more acidic because deprotonation of that isomer gives a conjugate base with a negative charge on a sulfur atom. A more stable (weaker) conjugate base indicates a stronger parent acid. The first compound above is less acidic, because its conjugate base would have a negative charge on a carbon atom, which is much less stable (factor #1 of ARIO).

3.66. There are three constitutional isomers with the molecular formula C_3H_8O, shown here:

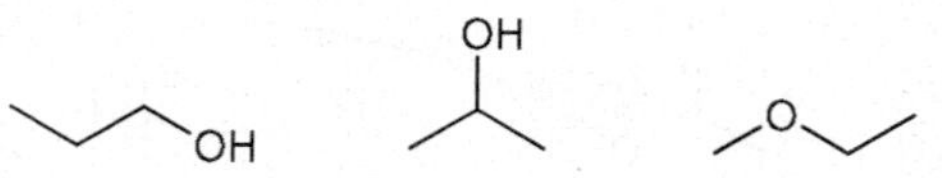

One of these compounds lacks an O-H group, so that compound will be the least acidic (its conjugate base will have an unstable negative charge on a carbon atom). Of the two remaining compounds, the compound with the least branching will be the most acidic, because its conjugate base is the most stable (due to better solvating effects, discussed in Section 3.8). A more stable (weaker) conjugate base indicates a stronger parent acid.

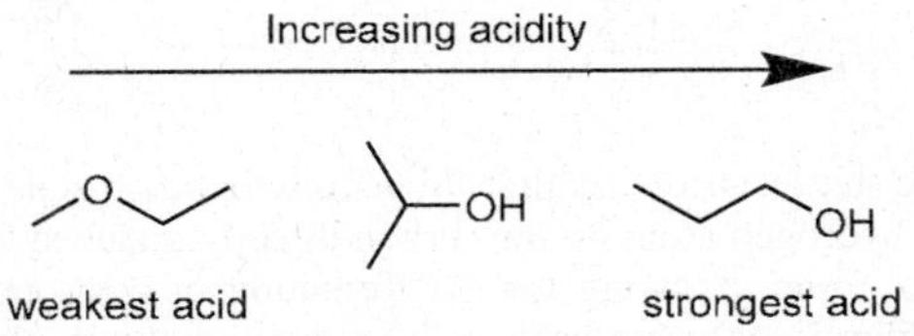

Increasing acidity

weakest acid strongest acid

3.67.

(a) A carbon atom must have four sigma bonds in order to be sp^3 hybridized. There is only one such carbon atom in cyclopentadiene, highlighted below.

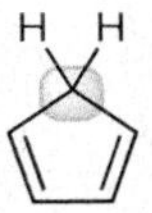

(b) The most acidic proton in cyclopentadiene is highlighted below:

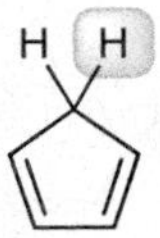

The corresponding conjugate base is highly resonance stabilized, as shown in the solution to part (c), below. In addition, the conjugate base is further stabilized by yet another factor that we will discuss in Chapter 18. Formation of the most stable (weakest) conjugate base indicates that the most acidic proton has been removed.

(c) Deprotonation of cyclopentadiene gives a conjugate base that is highly stabilized by resonance, as shown here:

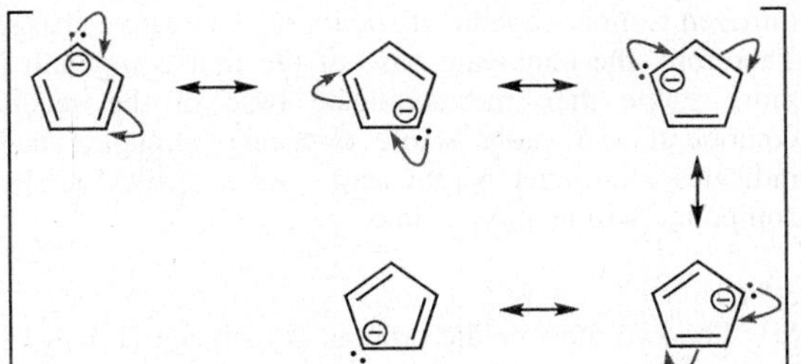

(d) There are no sp^3-hybridized carbon atoms in the conjugate base. All five carbon atoms are sp^2 hybridized.

(e) In order to be delocalized via resonance, the lone pair must occupy a *p* orbital, and therefore, all carbon atoms must be sp^2 hybridized and trigonal planar. Therefore, the entire compound has planar geometry.

(f) There are five hydrogen atoms in the conjugate base.

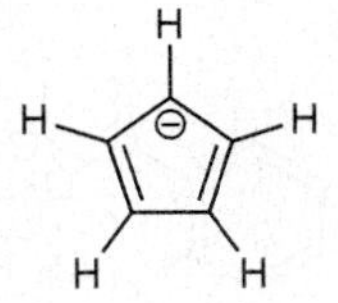

(g) As seen in the resonance structures (see the solution to part c of this problem), there is one lone pair in the conjugate base, and it is highly delocalized.

3.68. We begin by drawing the conjugate base of each compound and comparing them:

The first conjugate base is stabilized by resonance, while the second conjugate base is not. However, the second conjugate base exhibits a negative charge on a sulfur atom, which is larger than an oxygen atom. Therefore, there is a competition between two factors. Using the ARIO order of priority that is generally applied ("atom" is more important than "resonance"), we would expect that the second conjugate base should be more stable than the first. Yet, when we compare the pK_a values, we find that our prediction is not correct. Therefore, this is an exception, in which "resonance" is more important than "atom."

3.69. In an intramolecular acid-base reaction, a proton is transferred from one region of the molecule to another region within the same molecule, because the acid and base are tethered together, as in this case.

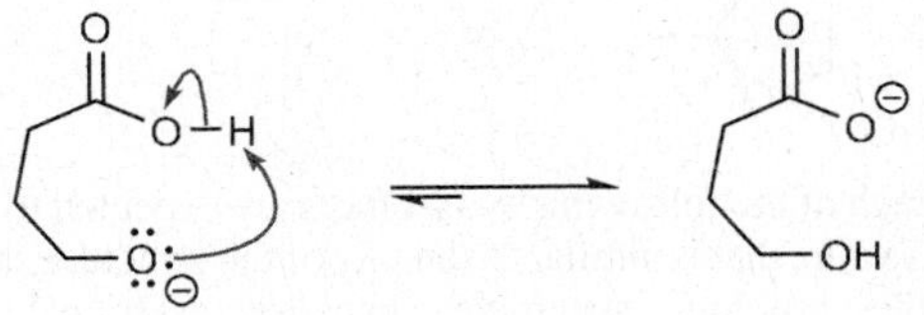

The equilibrium favors the product shown, because the negative charge in the product is resonance-stabilized (factor 2), while the negative charge on the reactant is not. The reaction will therefore favor the formation of a weaker base from a stronger one.

3.70. The stronger acid has a more stable, (weaker) conjugate base. Two possible explanations for the increased acidity of salicylic acid can be given:

1) In each compound, the carboxylic acid proton (CO_2H) is the most acidic proton. In salicylic acid, the inductive effect (electron-withdrawal) of the OH group is expected to be more pronounced because of its proximity to the negative charge in the conjugate base.

2) When salicylic acid is deprotonated, the resulting conjugate base can be significantly stabilized by

intramolecular hydrogen bonding (this explanation is likely more significant than the first explanation):

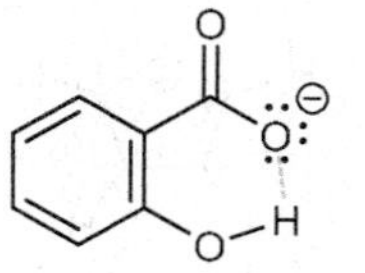

3.71.

(a) We are looking for a constitutional isomer with a highly acidic proton. The following two compounds are both carboxylic acids, and in each case, the conjugate base is resonance stabilized, with the negative charge being spread over two oxygen atoms. Carboxylic acids typically have a pK_a in the range of 4-5, while the alcohol shown in the problem statement is expected to have a pK_a in the range of 16-18. Therefore, the structures shown here satisfy the criteria described in the problem statement (they are both expected to be approximately one trillion times more acidic that the structure shown in the problem statement).

or

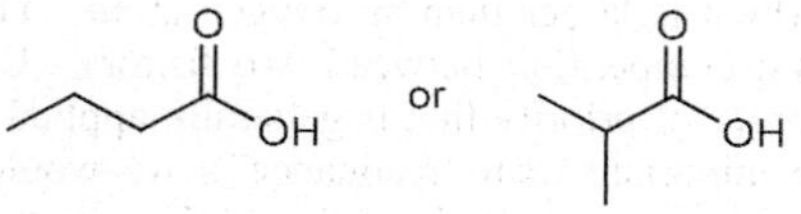

(b) The structure shown in the problem statement has an O-H group, and that proton is the most acidic proton in the compound. If we draw a constitutional isomer that lacks an O-H group, we would expect the resulting isomer to be significantly less acidic. The following two compounds fit this criterion. There are certainly many other constitutional isomers that also lack an O-H group, so there are many correct answers to this problem.

or

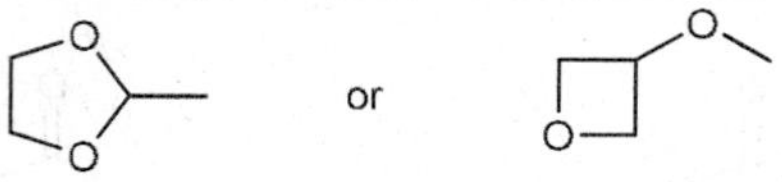

(c) Each of the following two isomers are expected to have a pK_a value that is similar to the pK_a of the structure shown in the problem statement, because each of these compounds possesses an alcohol O-H group. There are certainly many other constitutional isomers that also contain an O-H group, so there are many correct answers to this problem.

or

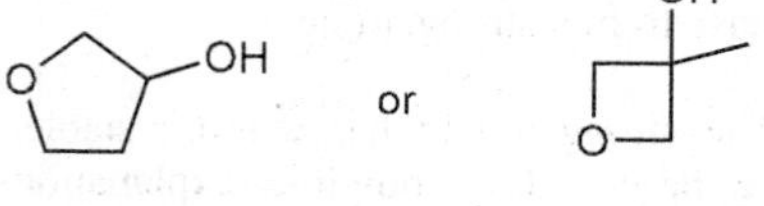

3.72. The four constitutional isomers are shown below.

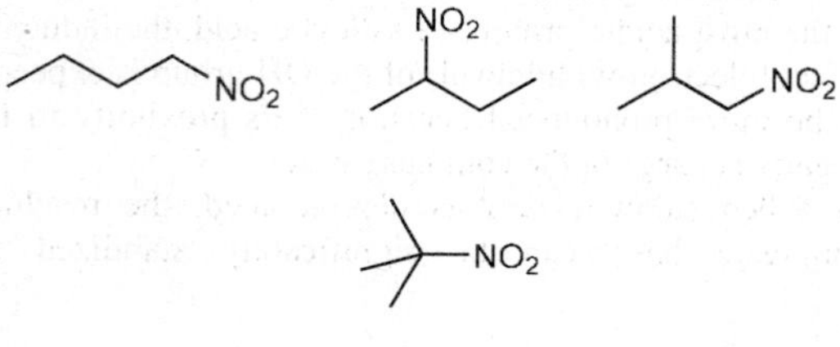

The last compound is expected to have the highest pK_a (it is the weakest acid) because its conjugate base is not resonance stabilized. The other three compounds have resonance-stabilized conjugate bases, for example:

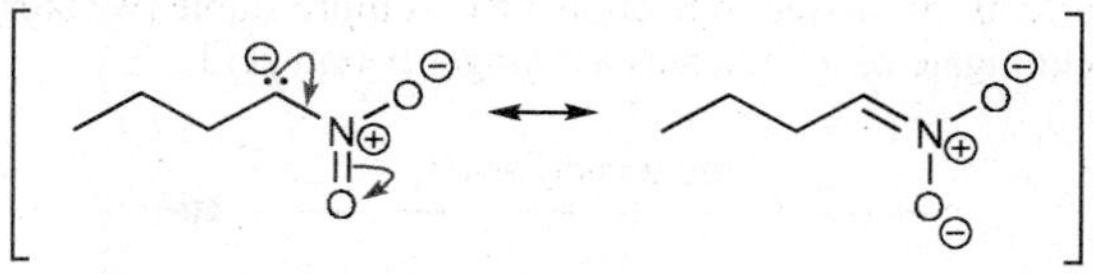

Note that each of the first three compounds has at least one hydrogen atom on the carbon directly attached to the nitro group, allowing for the formation of a resonance-stabilized conjugate base such as the one above. In the last compound, the carbon attached to the nitro group has no attached hydrogen atoms. A less stable (stronger) conjugate base indicates a weaker parent acid (higher pK_a).

3.73. Compare the conjugate bases:

Both are resonance stabilized. But the conjugate base of the first compound has a negative charge spread over two nitrogen atoms, as well as two carbon atoms:

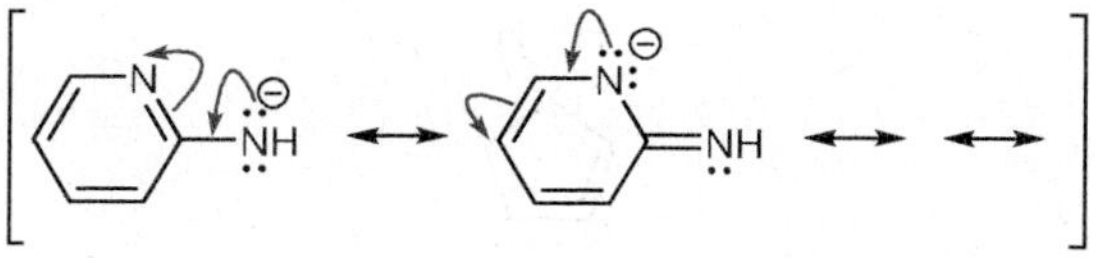

The conjugate base of the second compound has a negative charge spread over one nitrogen atom and three carbon atoms:

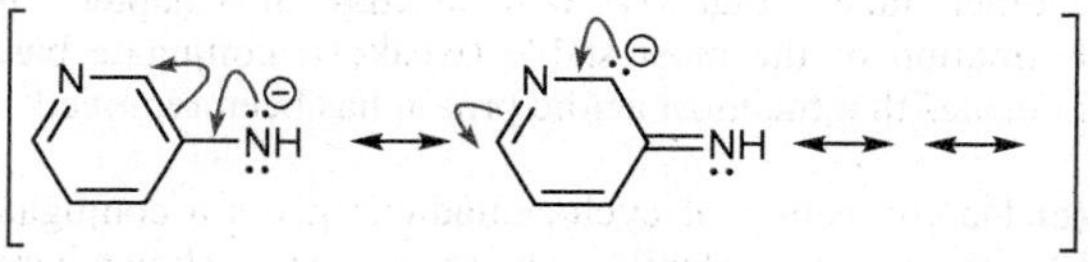

Since nitrogen is more electronegative than carbon, nitrogen is more capable of stabilizing a negative charge. Therefore, the conjugate base of the first compound is more stable than the conjugate base of the second compound. A more stable (weaker) conjugate base indicates a stronger parent acid. As a result, the first compound will be more acidic.

3.74.

(a) The two most acidic protons are labeled H_a and H_b. Deprotonation at either site will lead to a resonance-stabilized anion in which the negative charge is highly delocalized (spread over many nitrogen atoms).

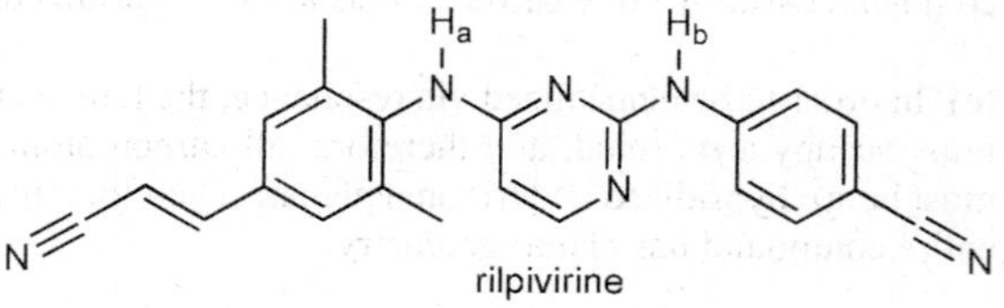

rilpivirine

(b) H_a is expected to be slightly more acidic than H_b, because removal of H_a produces a conjugate base that has one more resonance structure than the conjugate base formed from removal of H_b. The conjugate base from the removal of H_a has the negative charge spread over four nitrogen atoms and *five* carbon atoms, while the conjugate base from the removal of H_b has the negative charge spread over four nitrogen atoms and *four* carbon atoms. Formation of a more stable (weaker) conjugate base indicates the more acidic proton has been removed.

3.75.

(a) When R is a cyano group, the conjugate base is resonance stabilized:

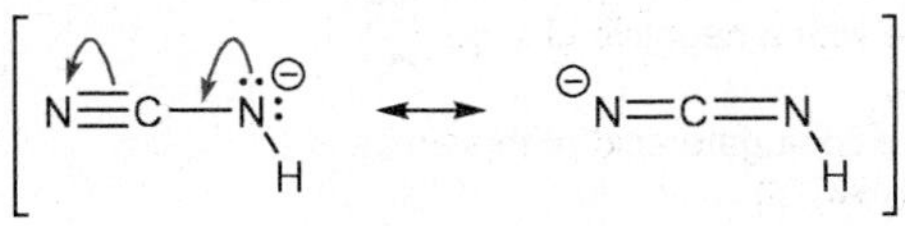

This resonance stabilization is not seen when R is an alkyl group. A more stable (weaker) conjugate base indicates a stronger parent acid (lower pK_a).

(b) There are many possible answers. Here is one example. It is expected to be more acidic, because its conjugate base is more stable (the negative charge would be spread over three nitrogen atoms, rather than just two nitrogen atoms):

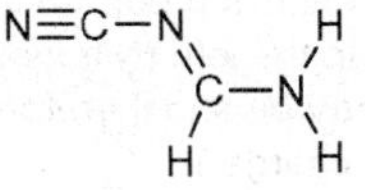

3.76.

(a) The positively charged structure serves as the acid, while the lone pair on the nitrogen atom serves as the base. Two curved arrows are required. The first curved arrow shows a lone pair of the base attacking the proton, and the second curved arrow comes from the O–H bond (being broken) and goes to the oxygen atom, as shown:

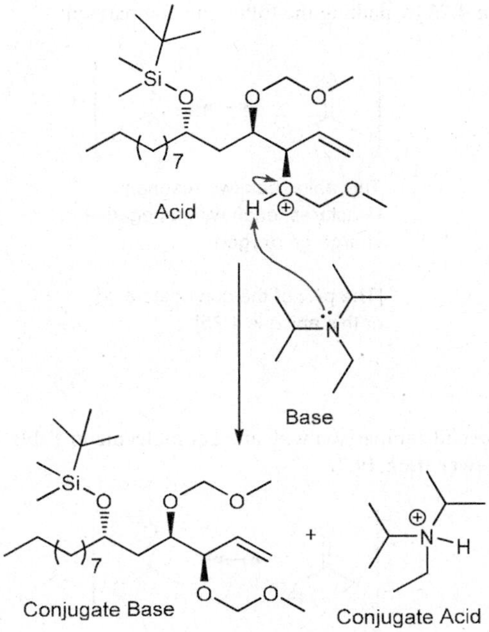

Acid

Base

Conjugate Base

Conjugate Acid

(b) Using Table 3.1, we see that oxonium ions have a pK_a of approximately -2, while ammonium ions have a pK_a of approximately 9.3. Therefore, we expect the oxonium ion labeled "Acid" above to be far more acidic than the ammonium ion labeled "Conjugate Acid" above. The acid on the product side is the more stable (weaker) acid. As such, the reaction favors the products, and the equilibrium lies in the forward direction.

3.77. Let's estimate the relative acidity of the protons in the compound by comparing them to similar acids in Table 3.1. We focus on the protons directly bound to electronegative heteroatoms (O, N, etc.) and protons bound to *sp*-hybridized carbon atoms. Three of the four most acidic protons are directly comparable to acids in Table 3.1. Note that the pK_a values shown are <u>not</u> meant to be precise (due to other structural factors that can affect these values), but rather, they should be considered as rough approximations.

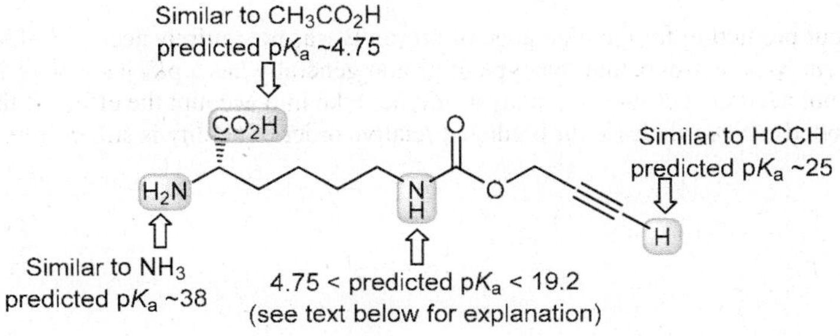

Similar to CH_3CO_2H
predicted pK_a ~4.75

Similar to HCCH
predicted pK_a ~25

Similar to NH_3
predicted pK_a ~38

4.75 < predicted pK_a < 19.2
(see text below for explanation)

The proton on the nitrogen atom adjacent to the C=O group is also a good candidate for one of our four most acidic protons, because it is bound to an electronegative atom (nitrogen). In this case there is no obvious acid to be used for comparison in Table 3.1. However, we can still make a reasonable prediction of pK_a range by using our knowledge of the relationship between structure and acidity. First of all, we can predict that the pK_a of the indicated proton is **greater than 4.75** by making the following comparison:

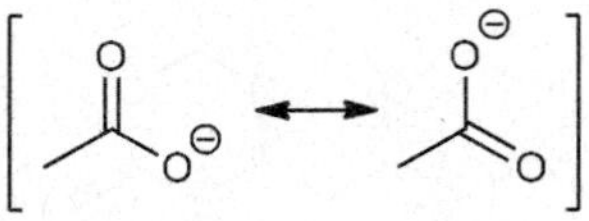

This anion has two resonance structures, each with a negative charge on oxygen.

[The pK_a of the conjugate acid of this anion is 4.75]

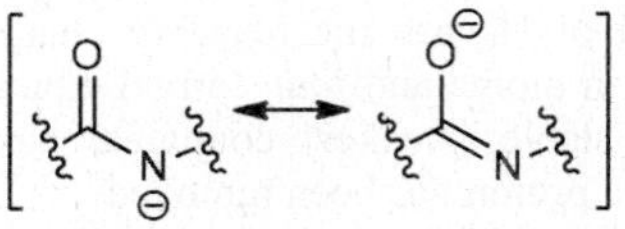

In this case, the negative charge is shared between nitrogen and oxygen. Therefore, this anion is predicted to be *less stable* than the anion to the left. (Nitrogen is less electronegative than oxygen, so it is not as stable with a negative charge.)

[The pK_a of the conjugate acid of this anion is predicted to be >4.75]

A second comparison with another molecule in Table 3.1 allows us to predict that the pK_a of the proton of interest might be **lower than 19.2.**

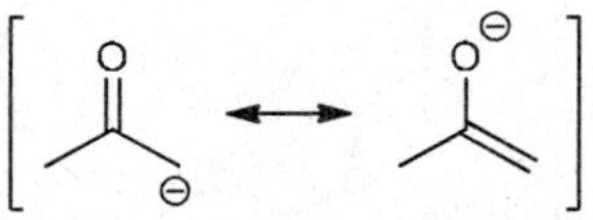

This anion has two resonance structures, with the negative charge spread over carbon and oxygen.

[The pK_a of the conjugate acid of this anion is 19.2]

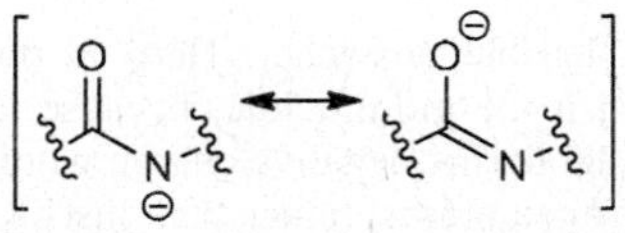

In this case, the negative charge is shared between nitrogen and oxygen. This anion is predicted to be *more stable* than the anion to the left. (Nitrogen is more electronegative than carbon and therefore more stable with a negative charge.)

[The pK_a of the conjugate acid of this anion is predicted to be <19.2]

After determining the approximate pK_a values for each of the four most acidic protons, we can now rank them in order of increasing acidity. Recall that the lowest pK_a value is associated with the most acidic proton:

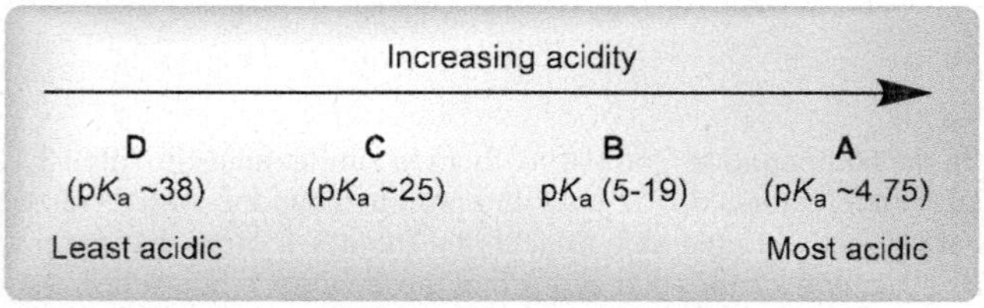

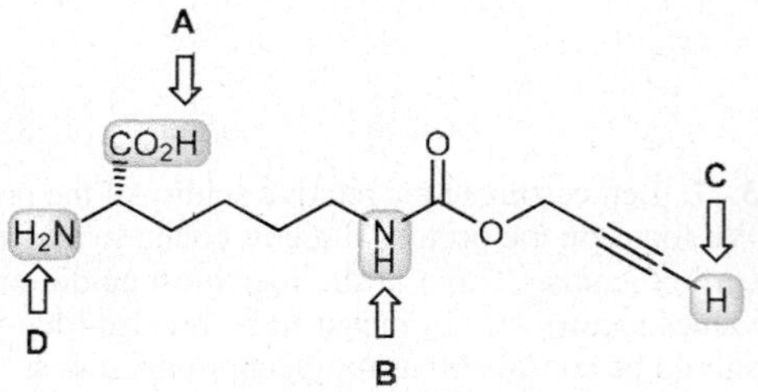

Increasing acidity →

D	C	B	A
(pK_a ~38)	(pK_a ~25)	pK_a (5-19)	(pK_a ~4.75)
Least acidic			Most acidic

It should be noted that our prediction for the pK_a range of proton B was not entirely accurate. This type of proton is part of a group called a *carbamate* group, and this type of proton generally has a pK_a just above 20. Our estimate (pK_a between 5 and 19) was not accurate because our analysis did not take into account the effect of the additional oxygen atom (next to the C=O bond). Nevertheless, our predicted relative order of acidity is still correct.

3.78. All significant resonance structures for anion **1** are shown here:

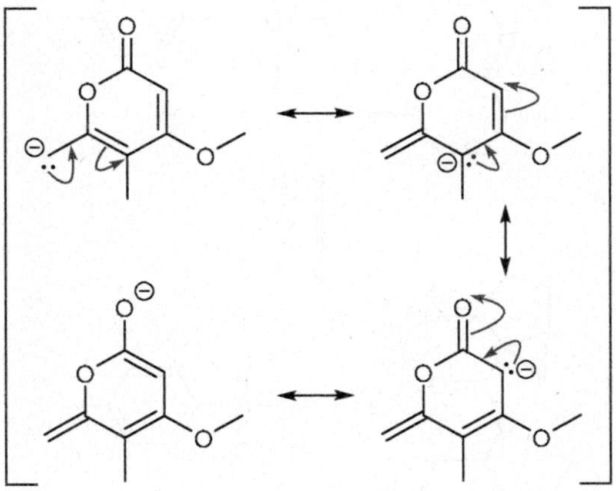

Only one of these structures has the negative charge on an oxygen atom. The other resonance structures have a negative charge on a carbon atom. Since oxygen is more electronegative than carbon, oxygen is more capable of stabilizing the charge. Therefore, the following resonance structure is the most significant and will contribute the most character to the resonance hybrid. The other resonance structures are all less significant.

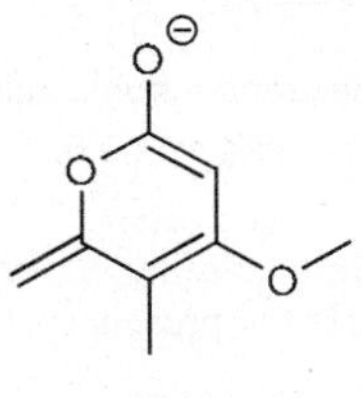

major contributor

3.79.

(a) As seen in Table 1.1, the electronegativity value for carbon is 2.5, while the electronegativity value for magnesium is only 1.2. The difference (1.3) is significant, and as explained in Section 1.5, this bond can be drawn as either covalent or ionic:

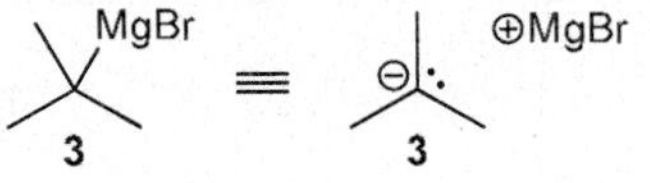

When drawn with an ionic bond, the anion exhibits a negative charge on a carbon atom (called a carbanion), which is a very strong base (because it is the conjugate base of a very, very weak acid). If compound **3** were treated with H_2O, we would expect the following proton-transfer reaction:

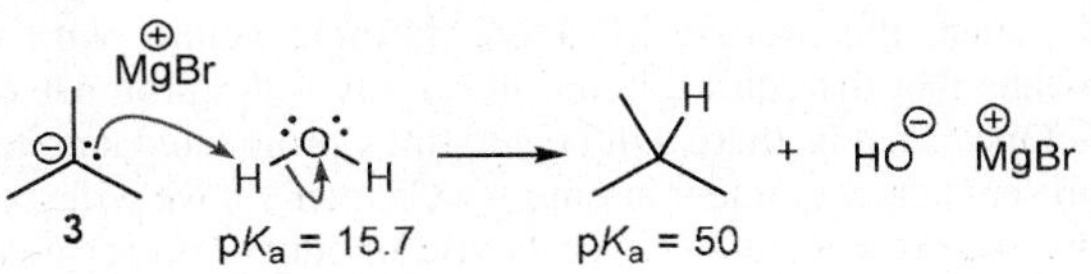

$pK_a = 15.7$ $pK_a = 50$

This reaction would be essentially irreversible as a result of the enormous difference in pK_a values (see Table 3.1). In a similar way, if compound **3** is treated with D_2O, rather than H_2O, the following reaction occurs, in which a deuteron (rather than a proton) is transferred:

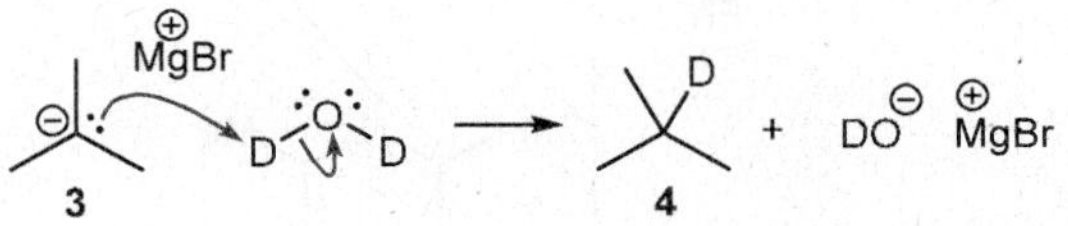

(b) As explained in Section 14.3, signals associated with the stretching of single bonds generally appear in the fingerprint region of an IR spectrum (400 – 1500 cm^{-1}), while signals associated with the stretching of double bonds and triple bonds appear in the diagnostic region (1500 – 4000 cm^{-1}). Notable exceptions are C-H bonds, which produce high-energy signals in the diagnostic region (2800 – 3000 cm^{-1}). This is explained in Section 14.3, by exploring the following equation, derived from Hooke's law:

$$\tilde{\nu} = \left(\frac{1}{2\pi c}\right)\left(\frac{f}{m_{red}}\right)^{\frac{1}{2}}$$

$$\text{reduced mass} = \left(\frac{m_1 m_2}{m_1 + m_2}\right)$$

Specifically, mass is found in the denominator, rather than the numerator, and as such, atoms with lower mass produce higher energy signals. Hydrogen has the smallest mass of all atoms, which explains why C-H bonds appear in the range 2800 – 3000 cm^{-1}. Extending this logic, a C-D bond is also expected to produce a higher energy signal, although not quite as high as C-H, because D has greater mass than H. So, we expect the C-D signal to appear somewhere below 2800 cm^{-1}, but still within the diagnostic region (greater than 1500 cm^{-1}). There is only one signal in the IR spectrum of compound **4** that fits this description, and that is the signal at 2180 cm^{-1}.
Also, 2180 cm^{-1} is in the region of the IR spectrum where signals for triple bonds are expected to appear (2100 – 2300 cm^{-1}). But we know that compound **4** lacks a triple bond. So this signal must be attributed to something else, and the C-D bond is the only candidate, because the C-H bonds appear in the range 2800 – 3000 cm^{-1} and the C-C bonds appear in the range between 1250 and 1500 cm^{-1}.

3.80.

(a) The lone pair in compound **4** functions as a base and deprotonates intermediate **3**. This requires two curved arrows, as shown:

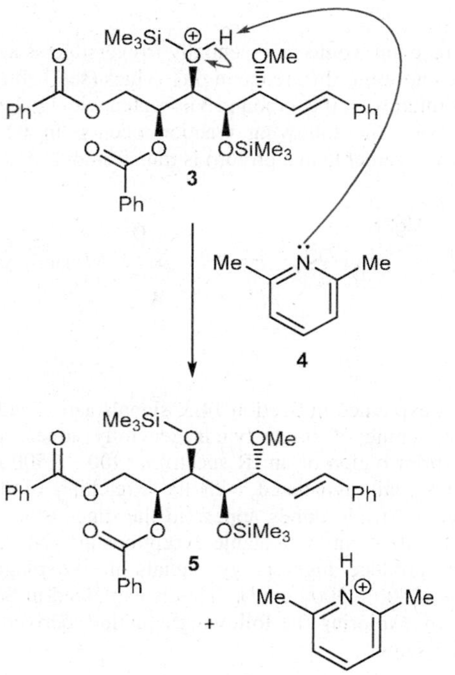

(b) If we consult Table 3.1, we find the following entries to be used for reference:

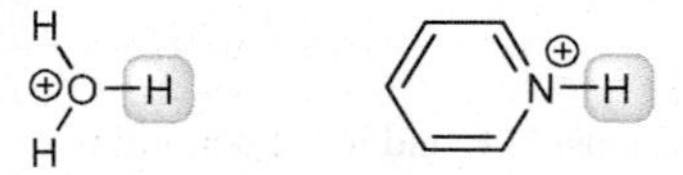

pK_a −1.74 pK_a 5.3

Notice that the first cation (an oxonium ion) has a pK_a value of -1.74, while the second cation (a pyridinium ion) has a pK_a value of 5.3. The difference between them is seven pK_a units. In other words, an oxonium ion is approximately 10^7 (or 10,000,000) times more acidic than a pyridinium ion. Since a proton transfer step will proceed in the direction that favors the weaker acid, we expect compound **4** to be successful in deprotonating intermediate **3** to give compound **5**.

(c) The conversion of **1** to **5** involves the loss of an alcohol O-H bond. That is, the IR spectrum of compound **1** should exhibit a broad signal between 3200 and 3600 cm^{-1}, due to the O-H stretching vibration. This signal is expected to be absent in the IR spectrum of compound **5**, which lacks an O-H bond. This can be used to verify that the desired reaction has occurred. Specifically, the disappearance of the O-H signal indicates the conversion of **1** to **5**.

3.81.

(a) One of the lone pairs on the ketone group functions as a base and abstracts a proton from *p*-TsOH. Two curved arrows are required, as shown:

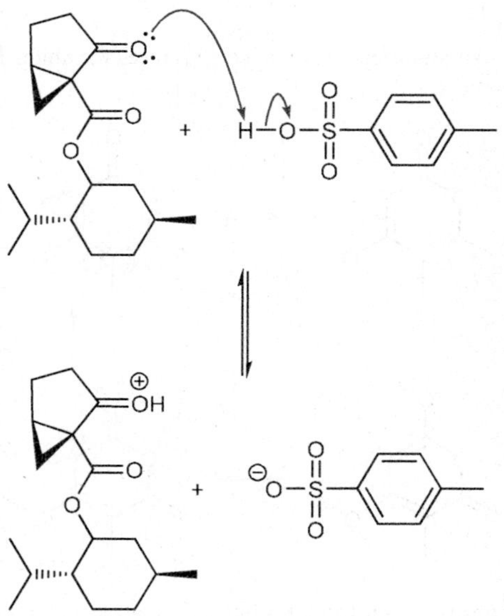

(b) Consult the pK_a table in the front endsheet of the textbook. Benzenesulfonic acid has a pK_a of −0.6:

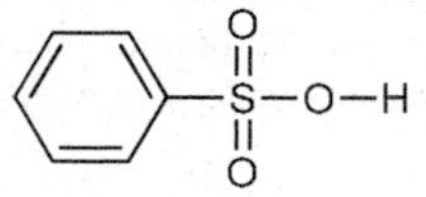

benzenesulfonic acid
pK_a = −0.6

p-TsOH is structurally similar to benzenesulfonic acid (the only difference is the presence of a methyl group),

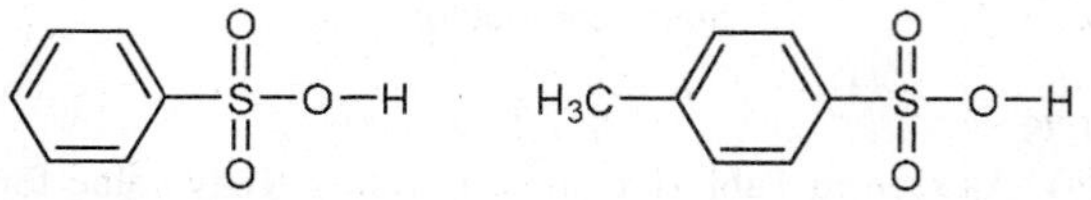

benzenesulfonic acid   *p*-toluenesulfonic acid

so we expect the pK_a of *p*-TsOH to be somewhere near −0.6. In contrast, the protonated ketone will have a pK_a that is approximately −7.3 (as seen in Table 3.1).

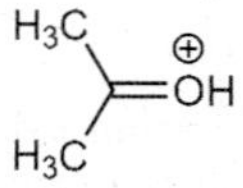

pK_a = −7.3

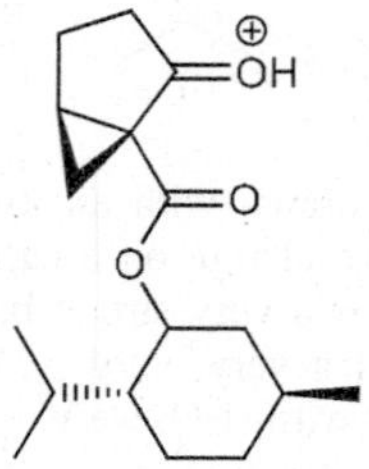

As such, the protonated ketone is more acidic, which means that the equilibrium will not favor the protonated ketone. That is, there will be very little protonated ketone present at any moment in time. In Chapter 19, we will see that the presence of even a catalytic amount of protonated ketone is sufficient to achieve the transformation shown in the problem statement.

3.82.

(a) If we take the structure of **1**, as drawn, and rotate it 180 degrees, the same image is obtained. As such, there are only four different locations (rather than eight) where deprotonation can occur.

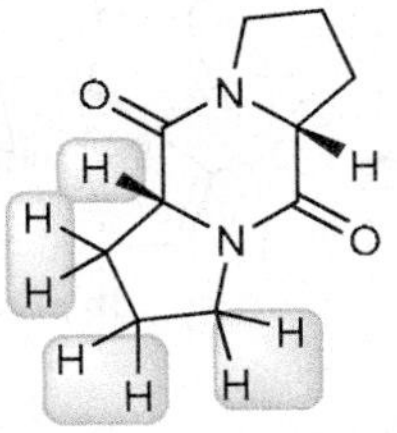

The most acidic proton is the one whose removal generates a resonance-stabilized conjugate base.

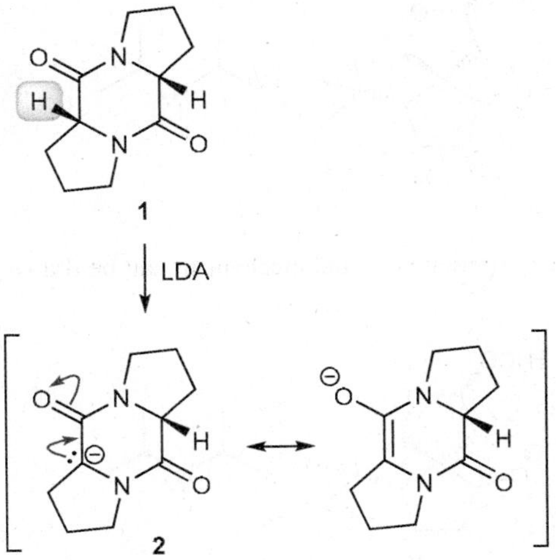

Formation of a more stable (weaker) conjugate base indicates the most acidic proton has been removed.

(b) pK_a values: Comparing the pK_a values at the front of the textbook, we see that amines (pK_a ~ 38–40) are *significantly* less acidic than ketones (16–19), or esters (24–25), both of which bear an acidic proton on the carbon atom connected to the C=O group, allowing for a resonance stabilized conjugate base.

pK_a ~ 16-19 pK_a ~ 24-25

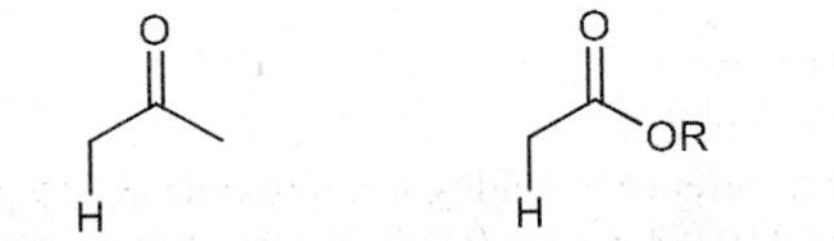

With this in mind, it is reasonable to expect the following proton of an amide to exhibit a lower pK_a than an amine:

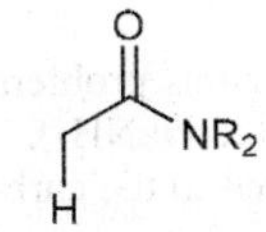

As such, the conjugate base of this compound should be more stable than the conjugate base of an amine. Therefore, LDA should be strong enough to remove a proton from compound **1**.

Structural comparison: Compare the structures of the anions. LDA exhibits a negative charge that is localized on a nitrogen atom. In contrast, the conjugate base of **1** is resonance-stabilized, with a major resonance contributor that places the negative charge on the more electronegative oxygen atom. As such, the conjugate base of **1** is more stable than LDA, so LDA should be a suitable base.

(c) LDA functions as a base and deprotonates compound **1**. This proton transfer step can either be drawn with two curved arrows, like this,

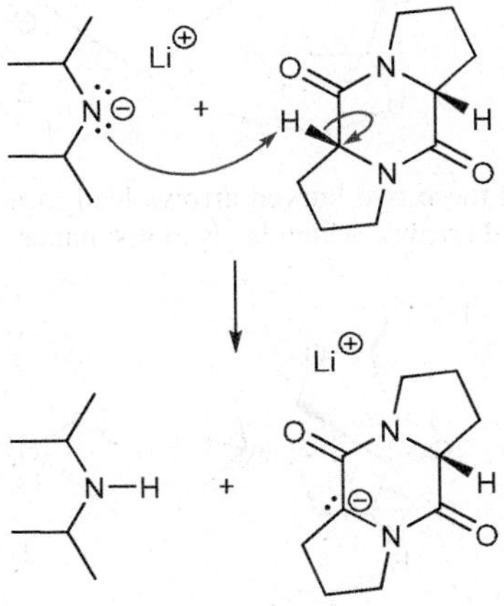

or with three curved arrows, like this:

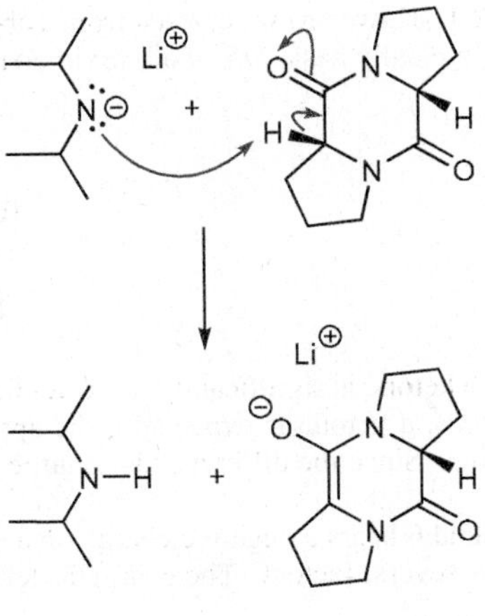

3.83.

(a) As seen in the previous problem, LDA (compound **2**) is a very strong base because it has a negative charge on a nitrogen atom (much like $NaNH_2$). So it will deprotonate compound **1** to give an anion. The most acidic proton in compound **1** is connected to the carbon atom adjacent to the C=O bond, since deprotonation at that location generates a resonance-stabilized, weak conjugate base:

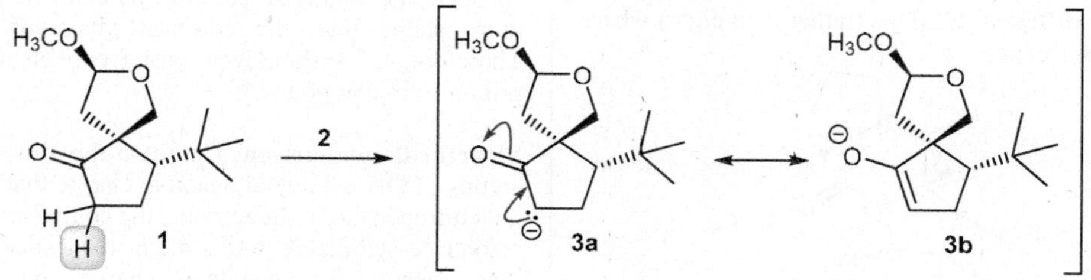

H3CO O O H H 1 2 H3CO O O 3a H3CO O O 3b

(b) Anion **3** is produced via a proton transfer step, in which the base removes a proton from compound **1**. The mechanism for this step requires at least two curved arrows:

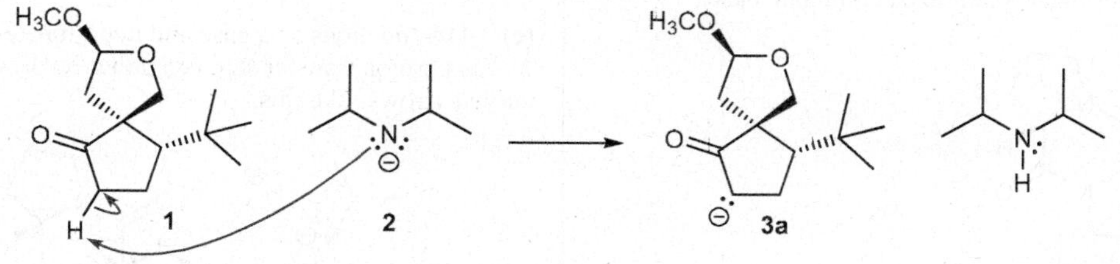

H3CO O O H 1 N 2 H3CO O O 3a N H

Notice that these two curved arrows lead to resonance structure **3a**. Alternatively, the mechanism can be drawn with three curved arrows, which leads to resonance structure **3b**:

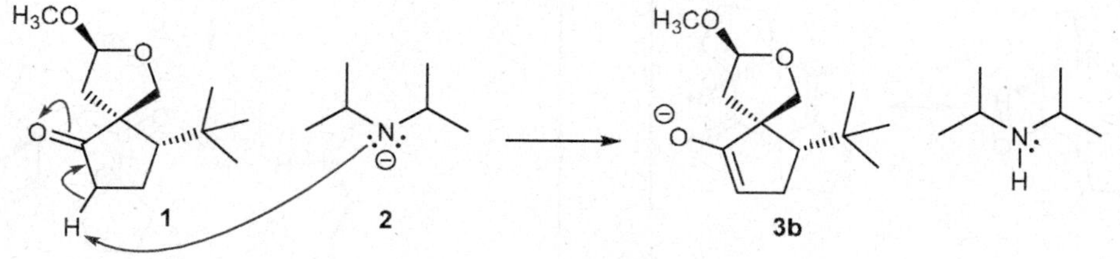

H3CO O O H 1 N 2 H3CO O O 3b N H

To evaluate the position of equilibrium for this process, we compare the pK_a values for the acid on either side of the equilibrium (compound **1** on the left, and R_2NH on the right). Clearly, we are not going to find the structure of compound **1** in Table 3.1. However, we can see from Table 3.1 that a proton connected to a carbon atom adjacent to the C=O bond of a ketone typically has a pK_a of approximately 19. In contrast, an amine is expected to have a pK_a near 38:

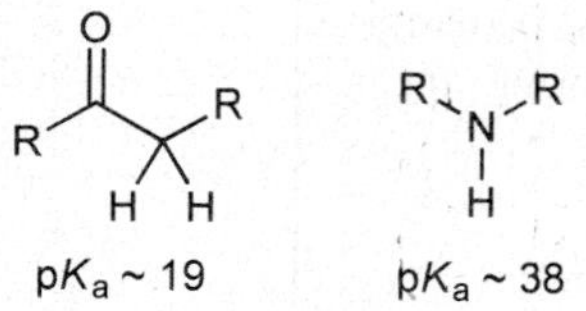

O R R H H pK_a ~ 19 R N R H pK_a ~ 38

Notice that a ketone is significantly more acidic than an amine. The difference in acidity is approximately 19 pK_a units. In other words, a ketone is expected to be approximately 10^{19} times (10,000,000,000,000,000,000 times) more acidic than an amine. Since the difference is so large, we can treat this proton transfer step as essentially irreversible.

(c) Compound **6** bears a negative charge on a nitrogen atom, which is generally fairly unstable. However, this anion is stabilized by several factors. The charge is delocalized into the neighboring triflate (Tf) group via resonance:

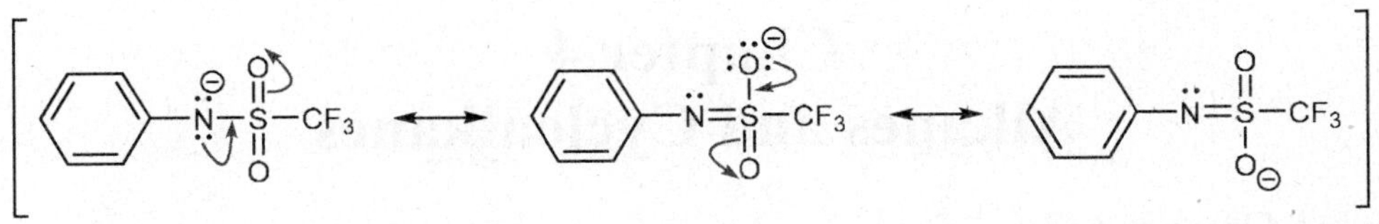

And the charge is further delocalized into the ring via resonance:

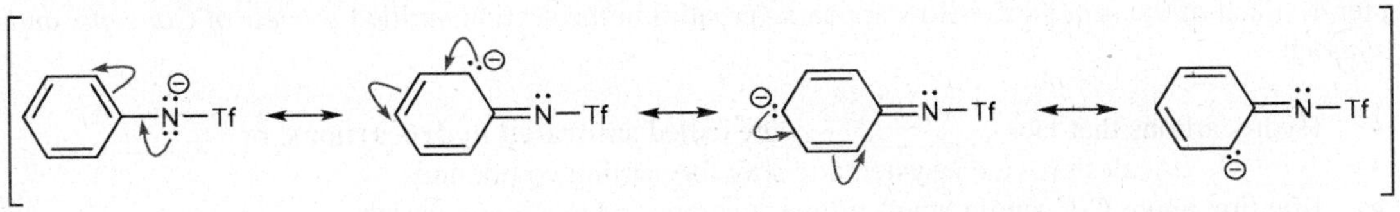

In total, the negative charge is spread over three carbon atoms, one nitrogen atom, and two oxygen atoms. The charge is therefore highly delocalized.
In addition, the electronegative fluorine atoms in the triflate group withdraw electron density via induction, thereby stabilizing the negative charge even further. As such the negative charge in this case is stabilized by resonance as well as induction, and is therefore highly stabilized.

3.84. The conjugate base of compound **5** possesses many resonance contributors. The negative charge can be delocalized into both aromatic rings and so the conjugate base is highly stabilized. Notice, however, that the placement of the negative charge is exclusively on carbon atoms in the ring and never on an electronegative atom such as an oxygen atom that could better stabilize the charge.

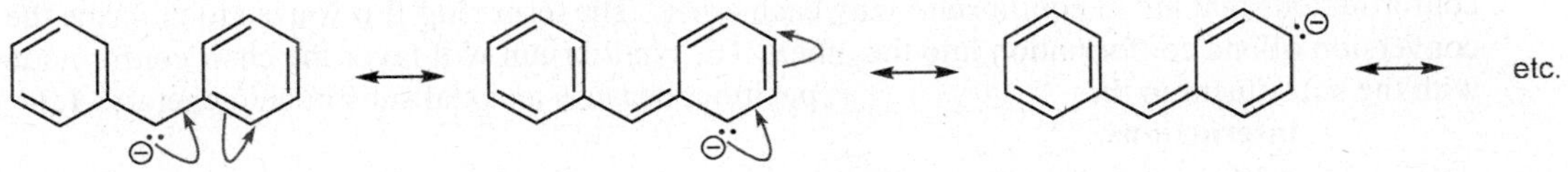

etc.

In contrast, in the conjugate base of **6**, the negative charge is delocalized onto the carbonyl oxygen atom:

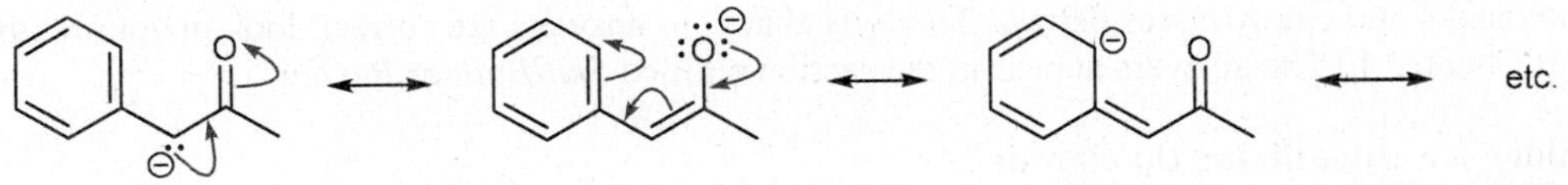

etc.

We now have to contend with two competing arguments: on the one hand, the conjugate base of **6**, with a resonance structure in which the negative charge is on an oxygen atom, should be more stable than the conjugate base of **5** (which lacks such a resonance structure). On the other hand, the conjugate base of **5** has more resonance structures (seven) than the conjugate base of **6** (which has only five resonance structures). In other words, is it better to spread a negative charge over seven carbon atoms or to spread the negative charge over four carbon atoms and an oxygen atom? The given table of pK_a values offers guidance on this question. It is apparent that compounds **1** and **3** represent the comparison we are asked to make in this problem. Recall that the stronger acid has a more stable (weaker) conjugate base. The conjugate base of compound **1** involves delocalization of the negative charge into the C=O group, while that of compound **3** will involve the aromatic ring. From the given pK_a values, we see that compound **1** is the stronger acid. So, compared to the phenyl group in **3**, the C=O group in **1** better stabilizes the negative charge in the conjugate base. It is reasonable to conclude that the same trend must hold when comparing the acidities of **5** and **6**. Therefore, the C=O group of the conjugate base of **6** is better able to stabilize the negative charge than the phenyl group of the conjugate base of **5**, despite the fact that there are overall more resonance contributors in the conjugate base of **5**.

Chapter 4
Alkanes and Cycloalkanes

Review of Concepts

Fill in the blanks below. To verify that your answers are correct, look in your textbook at the end of Chapter 4. Each of the sentences below appears verbatim in the section entitled *Review of Concepts and Vocabulary*.

- Hydrocarbons that lack ____________ are called **saturated hydrocarbons**, or ____________.
- ___________rules provide a systematic way for naming compounds.
- Rotation about C-C single bonds allows a compound to adopt a variety of _________________.
- ___________ **projections** are often used to draw the various conformations of a compound.
- ____________ **conformations** are lower in energy, while ____________ **conformations** are higher in energy.
- The difference in energy between staggered and eclipsed conformations of ethane is referred to as ____________ **strain**.
- _________ **strain** occurs in cycloalkanes when bond angles are less than the preferred _____°.
- The _______ conformation of cyclohexane has no torsional strain and very little angle strain.
- For a monosubstituted cyclohexane in a chair conformation, the substituent can occupy either an axial or an ______________ position. These two possibilities represent two different conformations that are in equilibrium with each other. The term **ring flip** is used to describe the conversion of one conformation into the other. The equilibrium will favor the chair conformation with the substituent in the _____________ position, because an axial substituent generates **1,3-** _________ **interactions.**

Review of Skills

Fill in the blanks and empty boxes below. To verify that your answers are correct, look in your textbook at the end of Chapter 4. The answers appear in the section entitled *SkillBuilder Review*.

SkillBuilder 4.1 Identifying the Parent

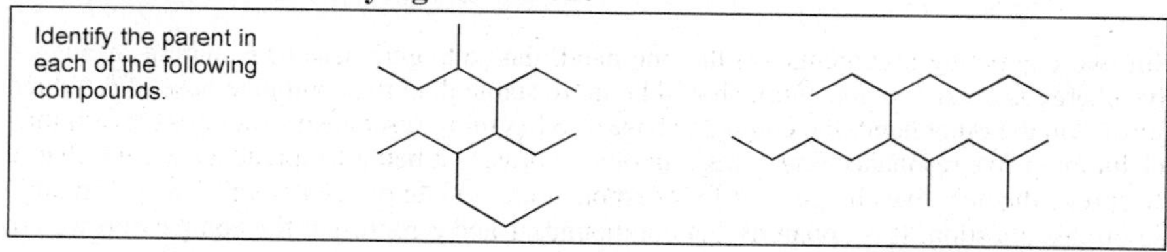

Identify the parent in each of the following compounds.

SkillBuilder 4.2 Identifying and Naming Substituents

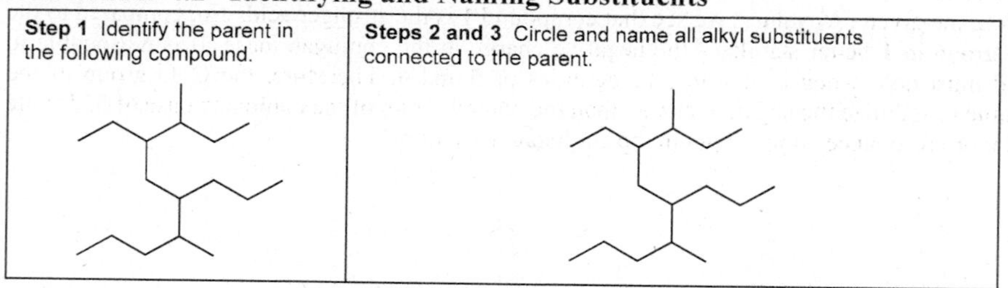

Step 1 Identify the parent in the following compound.
Steps 2 and 3 Circle and name all alkyl substituents connected to the parent.

SkillBuilder 4.3 Identifying and Naming Complex Substituents

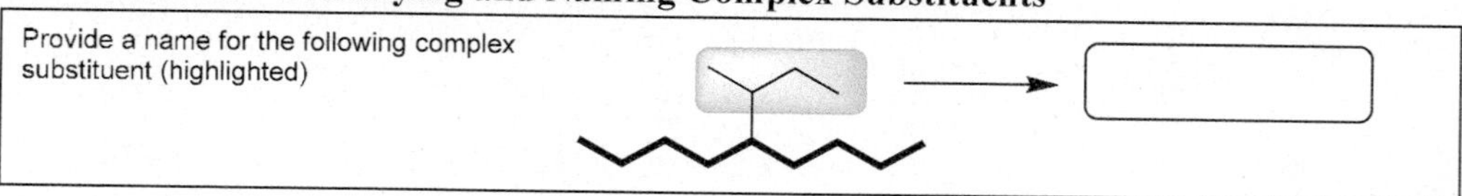

Provide a name for the following complex substituent (highlighted)

SkillBuilder 4.4 Assembling the Systematic Name of an Alkane

Provide a systematic name for the following compound:

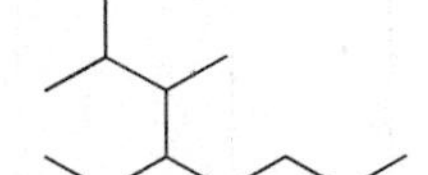

1) Identify the parent
2) Identify and name substituents
3) Assign locants to each substituent
4) Alphabetize

SkillBuilder 4.5 Assembling the Name of a Bicyclic Compound

Provide a systematic name for the following compound:

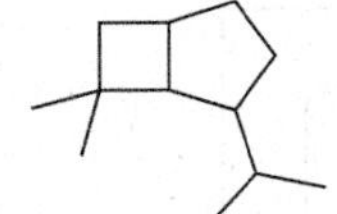

1) Identify the parent
2) Identify and name substituents
3) Assign locants to each substituent
4) Alphabetize

SkillBuilder 4.6 Identifying Constitutional Isomers

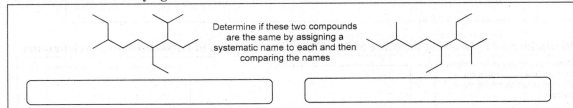

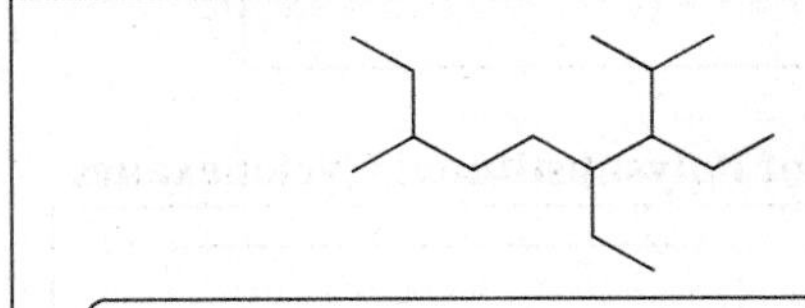

Determine if these two compounds are the same by assigning a systematic name to each and then comparing the names

SkillBuilder 4.7 Drawing Newman Projections

Step 1 Identify the three groups connected to the front carbon atom

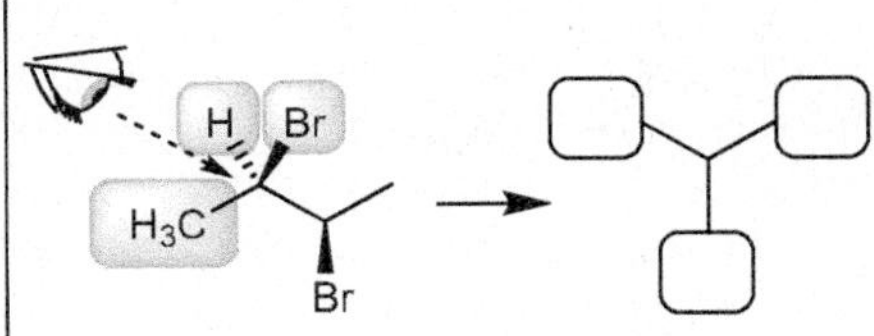

Step 2 Identify the three groups connected to the back carbon atom

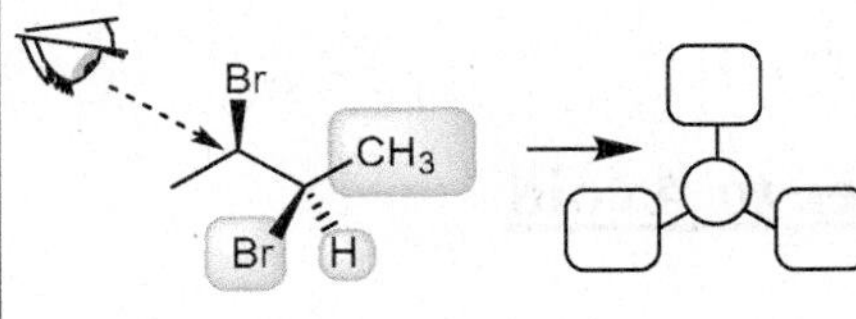

Step 3 Assemble the Newman projection from the two pieces obtained in the previous steps.

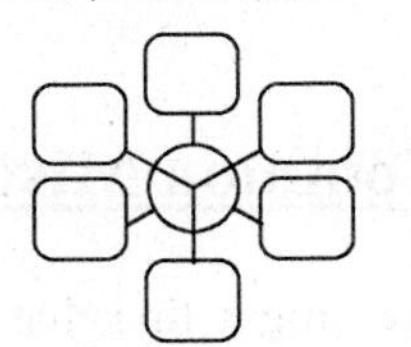

SkillBuilder 4.8 Identifying Relative Energy of Conformations

Step 1 Draw a Newman projection looking down the bond indicated.

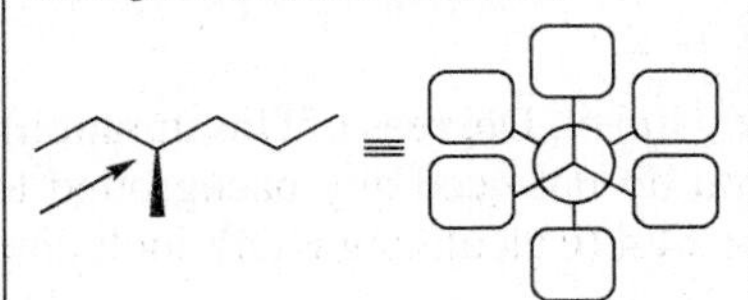

Step 2 Draw all three staggered conformations and determine which one has the fewest or least severe gauche interactions.

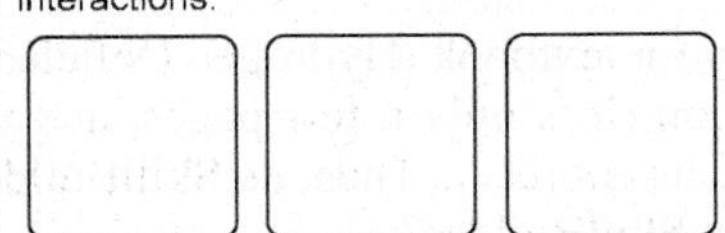

Step 3 Draw all three eclipsed conformations and determine which one has the highest energy interactions.

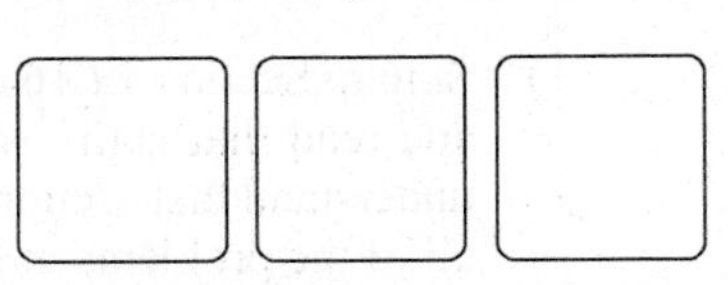

SkillBuilder 4.9 Drawing a Chair Conformation

Draw a chair conformation:

SkillBuilder 4.10 Drawing Axial and Equatorial Positions

Draw a chair conformation showing all six axial positions and all six equatorial positions.

SkillBuilder 4.11 Drawing Both Chair Conformations of a Monosubstituted Cyclohexane

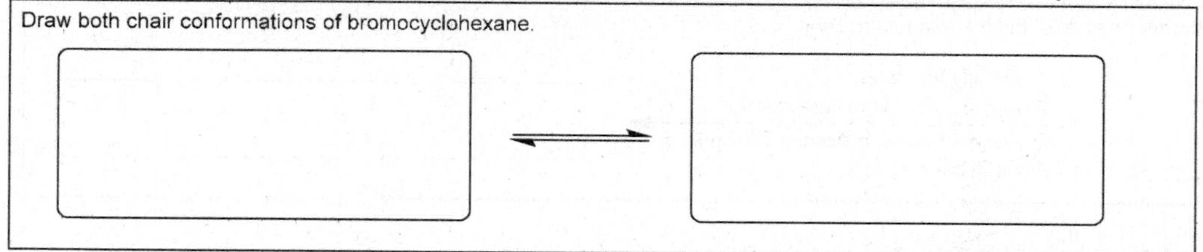

SkillBuilder 4.12 Drawing Both Chair Conformations of Disubstituted Cyclohexanes

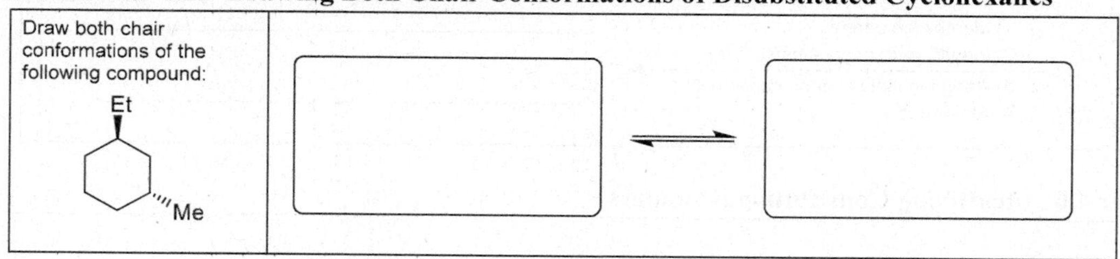

Draw both chair conformations of the following compound:
Et
Me

SkillBuilder 4.13 Drawing the More Stable Chair Conformation of Polysubstituted Cyclohexanes

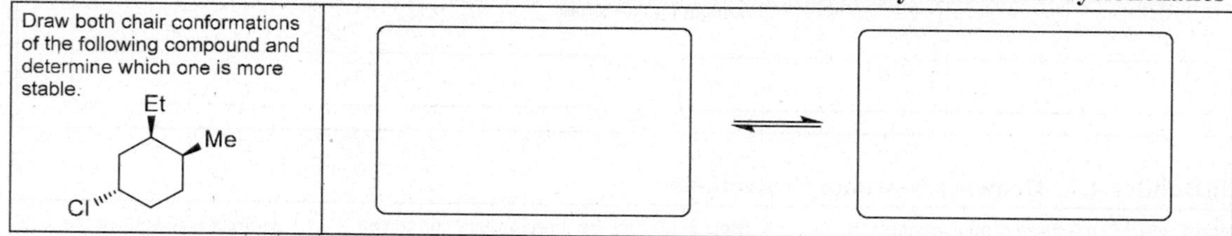

Draw both chair conformations of the following compound and determine which one is more stable.
Et
Me
Cl

Common Mistakes to Avoid

You might find that you struggle with problems that ask you to draw constitutional isomers. Don't be discouraged. Many students struggle with drawing constitutional isomers. In particular, it is sometimes difficult to find ALL of the constitutional isomers with a particular molecular formula, and it is also difficult to avoid drawing the same compound more than once. However, these skills will be critical as we progress through the upcoming chapters. So, in order to get more proficient with constitutional isomers, do the following:

1) Skip to Section 14.16 of your textbook (Hydrogen Deficiency Index: Degrees of Unsaturation), and read that entire section (it is only a few pages, and you do not need any background to understand that section in its entirety). Then, do SkillBuilder 14.4 (Calculating HDI), including all of the problems in that SkillBuilder.
2) Review the methodical approach for drawing constitutional isomers that is presented in the solution to problem 4.3.
3) Try to do problem 4.15, using the approach outlined in the solution to problem 4.3, and then check the solution to problem 4.15 to make sure that you applied the approach correctly.

If you complete the three tasks above, you should gain confidence in your ability to draw constitutional isomers and to avoid drawing the same isomer twice.

Solutions

4.1.

(a) The longest chain is six carbon atoms (shown below), so the parent is hexane:

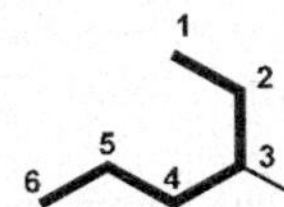

Parent = hexane

(b) The longest chain is seven carbon atoms (shown below), so the parent is heptane:

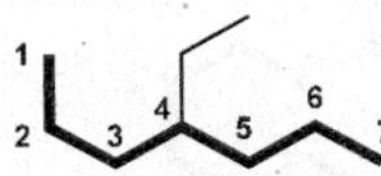

Parent = heptane

(c) The longest chain is seven carbon atoms (shown below), so the parent is heptane:

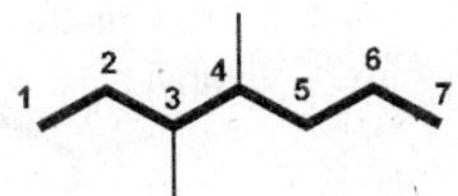

Parent = heptane

(d) The longest chain is nine carbon atoms (shown below), so the parent is nonane:

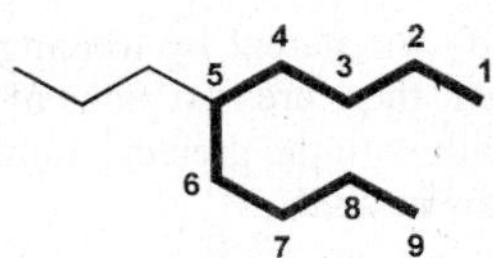

Parent = nonane

(e) The longest chain is eight carbon atoms (shown below), so the parent is octane:

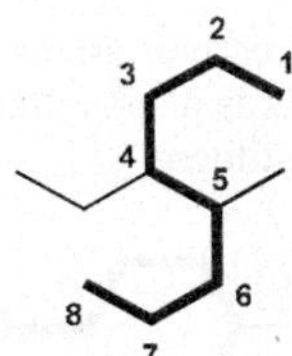

Parent = octane

(f) The longest chain is seven carbon atoms, so the parent is heptane. In this case, there is more than one seven-carbon chain. While the parent will be heptane regardless of which chain we choose, the correct parent chain is the one with the most substituents, shown below (this will be important later when we must use the numbering scheme to identify the locations of the substituents connected to the chain):

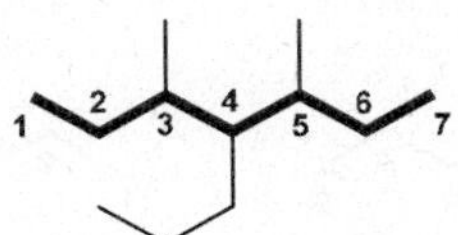

Parent = heptane

(g) This compound has a five-membered ring (shown below), so the parent is cyclopentane:

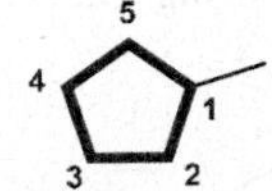

Parent = cyclopentane

(h) This compound has a seven-membered ring (shown below), so the parent is cycloheptane:

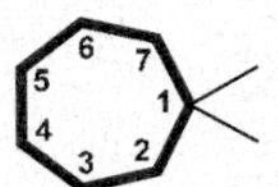

Parent = cycloheptane

(i) This compound has a three-membered ring (shown below), so the parent is cyclopropane:

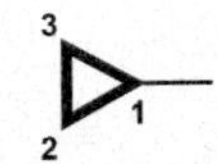

Parent = cyclopropane

4.2. Compounds B and D (shown below) each have a parent chain of eight carbon atoms (octane). Compounds A and C (not shown here) have parent chains of seven and nine carbon atoms, respectively.

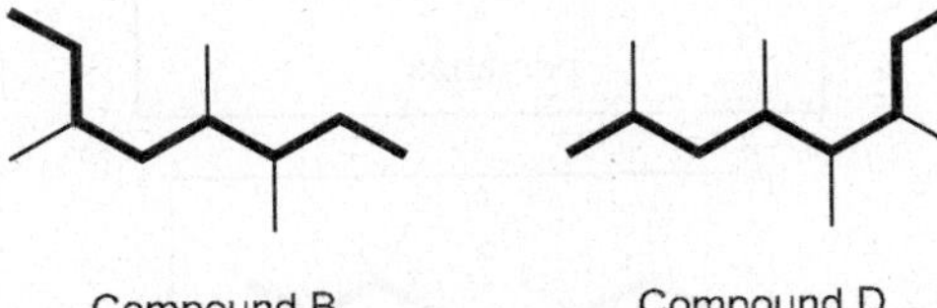

Compound B
Parent = octane

Compound D
Parent = octane

4.3. Recall that constitutional isomers are compounds that have the same molecular formula but differ in their constitution (connectivity of atoms). We must draw all constitutional isomers of hexane. Hexane is a linear chain of six carbon atoms:

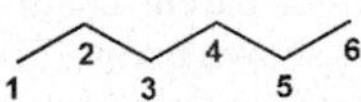

The first step is to look for any constitutional isomers where the parent is pentane (five carbon atoms). There are only two such isomers. Specifically, we can either connect the extra CH_3 group to positions C2 or C3 of the pentane chain:

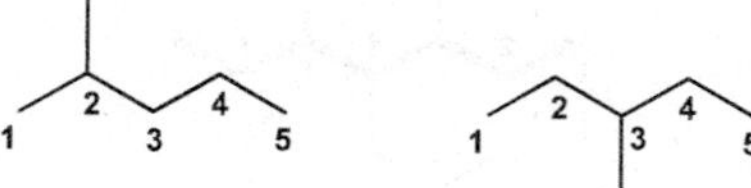

We cannot connect the CH_3 group to positions C1 or C5, as that would simply give us the linear chain (hexane), which we already drew (above). We also cannot connect the CH_3 group to position C4 as that would generate the same structure as placing the CH_3 group at the C2 position:

Next, we look for any constitutional isomers where the parent is butane (four carbon atoms). There are only two such isomers. Specifically, we can either connect two CH_3 groups to adjacent positions (C2 and C3) or the same position:

If we try to connect a CH_3CH_2 group to a butane chain, we end up with a pentane chain (an isomer already drawn above):

In summary, we have found the four isomers of hexane (two pentanes and two butanes):

pentanes

butanes

4.4.

(a) First, we identify the parent. In this case, there is more than one possible parent chain of nine carbon atoms. While the parent will be nonane regardless of which chain we choose, the correct parent chain is the one with the most substituents, as shown below. The alkyl substituents (highlighted) connected to the parent are identified below.

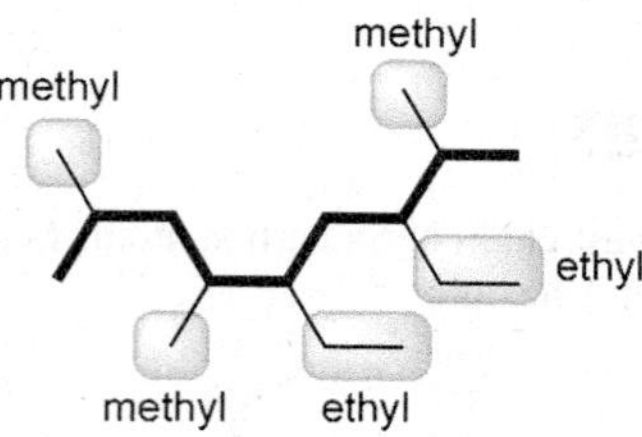

(b) First, we identify the parent (nonane), and then we identify any alkyl substituents (highlighted) that are connected to the parent.

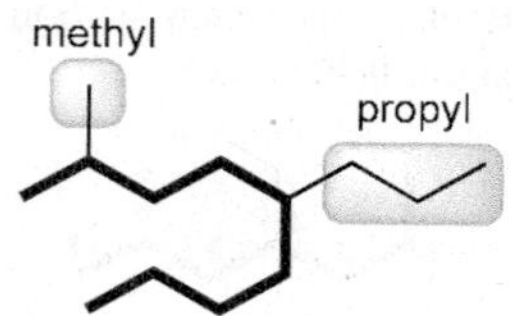

(c) First, we identify the parent (undecane), and then we identify any alkyl substituents (highlighted) that are connected to the parent.

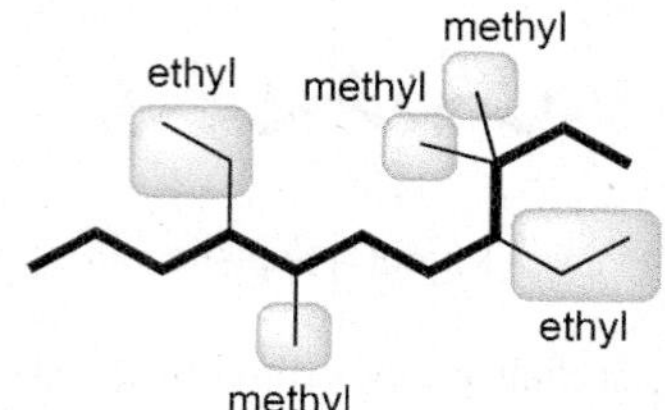

4.5. First identify the parent by looking for the longest chain. In this case, there are two paths of equal length, so we choose the path with the greatest number of branches, as indicated below in bold.

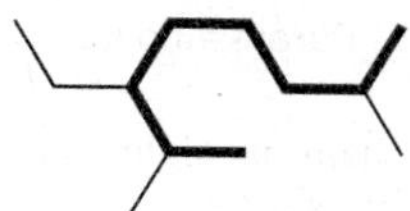

The parent has eight carbon atoms (octane). Everything connected to the chain is a substituent, and we use Table 4.2 to name each substituent:

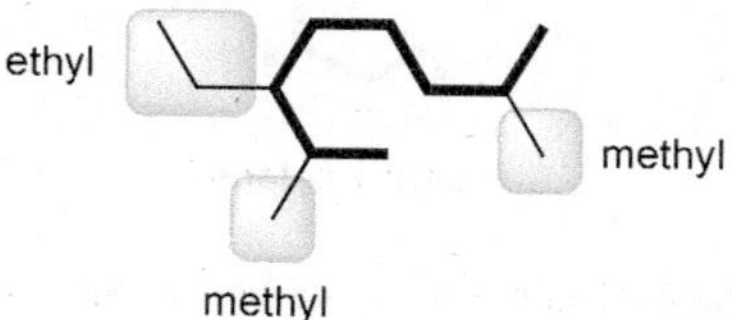

4.6.

(a) First we identify the parent chain (heptane), and then we identify any alkyl substituents connected to the parent. In this case, the substituent (highlighted) is complex, so we treat it as a "substituent on a substituent," and we assign a name based on numbers going away from the parent:

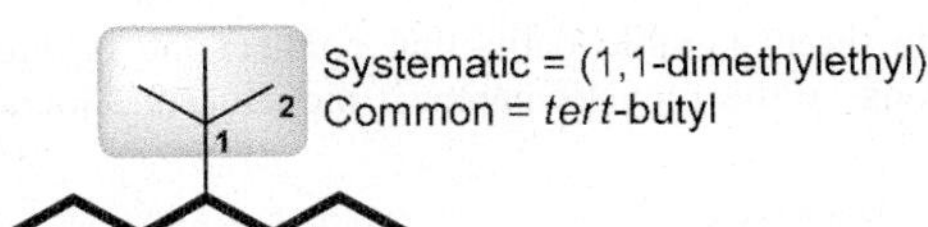

(b) First we identify the parent chain (nonane), and then we identify any alkyl substituents connected to the parent. In this case, one of the highlighted substituents is complex, so we treat it as a "substituent on a substituent," and we assign a name based on numbers going away from the parent:

Systematic = (1-methylethyl)
Common = isopropyl

Systematic = methyl
Common = methyl

(c) First we identify the parent chain (nonane), and then we identify any alkyl substituents connected to the parent. In this case, the substituent (highlighted) is complex, so we treat it as a "substituent on a substituent," and we assign a name based on numbers going away from the parent:

Systematic = (2,2-dimethylpropyl)
Common = neopentyl

(d) First we identify the parent chain (cyclohexane), and then we identify any alkyl substituents connected to the parent. In this case, all four substituents (highlighted) are complex, so we treat each of them as a "substituent on a substituent," and we assign a name based on numbers going away from the parent:

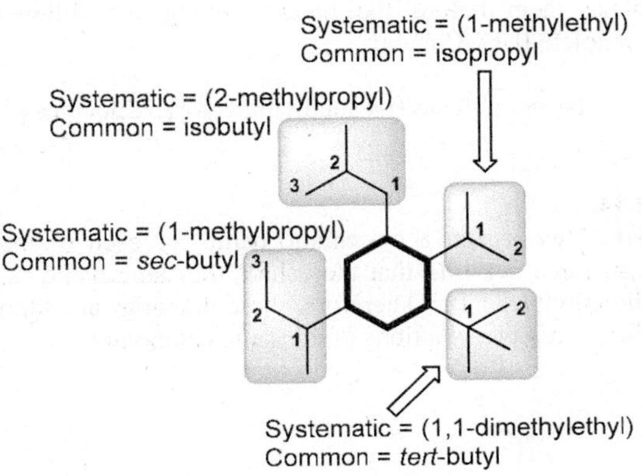

4.7. We treat this complex substituent as a "substituent on a substituent." The longest chain that starts at the attachment point has six carbon atoms, so this is a hexyl group. Then we assign numbers such that the attachment point is numbered C1. With this numbering scheme, shown below, there are two methyl substituents at positions 1 and 5, so the name of the side chain is (1,5-dimethylhexyl).

methyl
methyl

systematic name = (1,5-dimethylhexyl)

4.8. For each of the following compounds, we assign its name via a four-step process: First identify the parent, then the substituents, then assign locants, and finally, arrange the substituents alphabetically. In each case, use commas to separate numbers from each other, and use hyphens to separate letters from numbers.

(a) 3,4,6-trimethyloctane
(b) *sec*-butylcyclohexane *or* (1-methylpropyl)cyclohexane
(c) 3-ethyl-2-methylheptane
(d) 3-isopropyl-2,4-dimethylpentane *or* 2,4-dimethyl-3-(1-methylethyl)pentane
(e) 3-ethyl-2,2-dimethylhexane
(f) 2-cyclohexyl-4-ethyl-5,6-dimethyloctane
(g) 3-ethyl-2,5-dimethyl-4-propylheptane
(h) 2,2,6,6,7,7-hexamethylnonane
(i) 4-*tert*-butylheptane *or* 4-(1,1-dimethylethyl)heptane
(j) 1,3-diisopropylcyclopentane *or* 1,3-bis(1-methylethyl)cyclopentane
(note: for complex substituents, the prefix "bis" is used, rather than the prefix "di")
(k) 3-ethyl-2,5-dimethylheptane

4.9.

(a) The name 3-isopropyl-2,4-dimethylpentane indicates that the parent is a five-carbon chain and there are three substituents (a methyl group at C2, an isopropyl group at C3 and another methyl group at C4):

(b) The name 4-ethyl-2-methylhexane indicates that the parent is a six-carbon chain and there are two substituents (a methyl group at C2 and an ethyl group at C4):

(c) The name 1,1,2,2-tetramethylcyclopropane indicates that the parent is a three-membered ring and there are four substituents (two methyl groups at C1, and two methyl groups at C2), as shown:

4.10. For each compound, we assign its name in the following way: First identify the parent, then the substituents, and then assign locants (the final step, arranging the substituents alphabetically, is not needed for any of these examples). In each case, use commas to separate numbers from each other, and use hyphens to separate letters from numbers.

(a) 2,2-dimethylundecane

(b) 2-methyldodecane

(c) 2,2-dimethyloctane

(d) butylcyclohexane

4.11. For each of the following compounds, we assign its name via a four-step process: First identify the parent, then the substituents, then assign locants, and finally, arrange the substituents alphabetically. When assigning locants, make sure to start at a bridgehead and continue numbering along the longest path to the second bridgehead. Then continue assigning locants along the second longest path, and then finally, along the shortest path that connects the two bridgehead positions.

(a) 4-ethyl-1-methylbicyclo[3.2.1]octane

(b) 2,2,5,7-tetramethylbicyclo[4.2.0]octane

(c) 2,7,7-trimethylbicyclo[4.2.2]decane

(d) 3-*sec*-butyl-2-methylbicyclo[3.1.0]hexane *or* 2-methyl-3-(1-methylpropyl)bicyclo[3.1.0]hexane

(e) 2,2-dimethylbicyclo[2.2.2]octane

4.12.

(a) The name 2,2,3,3-tetramethylbicyclo[2.2.1]heptane indicates a bicyclic parent with two methyl groups at C2 and two methyl groups at C3, as shown:

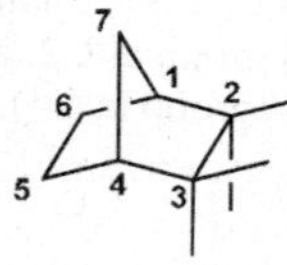

(b) The name 8,8-diethylbicyclo[3.2.1]octane indicates a bicyclic parent with two ethyl groups at C8, as shown:

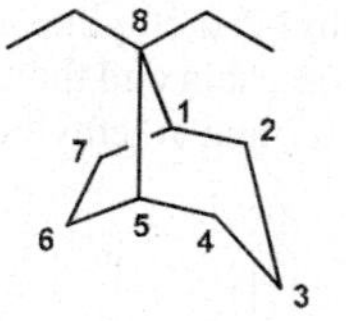

(c) The name 3-isopropylbicyclo[3.2.0]heptane indicates a bicyclic parent with an isopropyl group at C3, as shown:

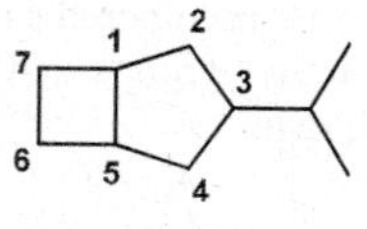

4.13. The bicyclic parent contains eight carbon atoms (highlighted below). There are two carbon atoms in each of the three possible paths that connect the bridgehead carbons, so this bicyclic parent is bicyclo[2.2.2.]octane.

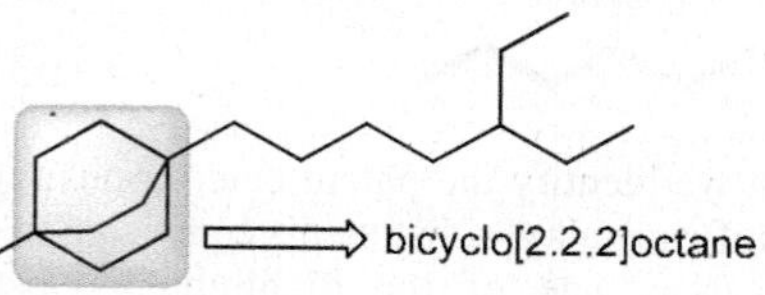

bicyclo[2.2.2]octane

There are two substituents on this bicyclic system. One is a methyl substituent and the other one is a complex substituent that is called 5-ethylheptyl, as shown below:

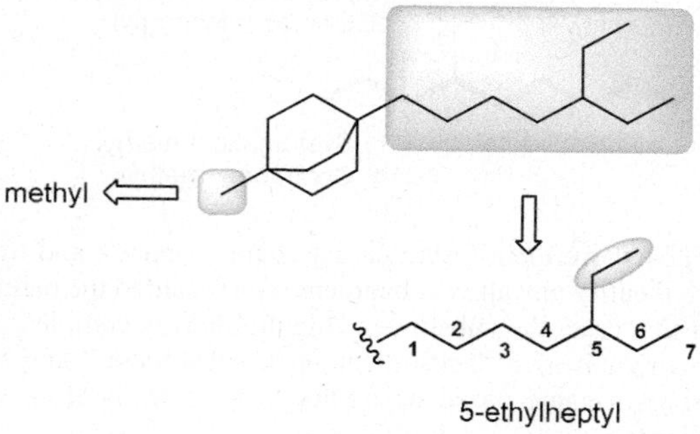

methyl

5-ethylheptyl

The locants for the two substituents are positions 1 and 4 on the bicyclic parent, and the lower number should be assigned according to which substituent comes first alphabetically.

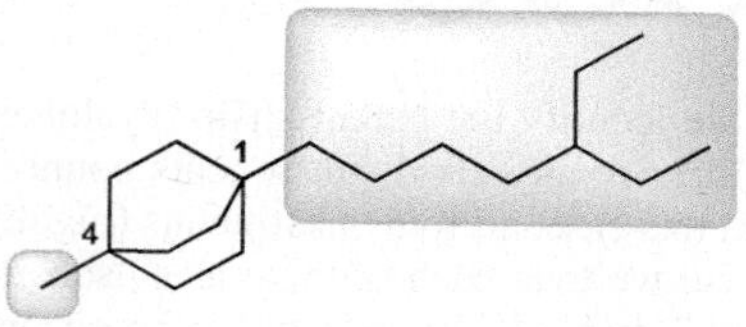

Finally, we arrange the substituents alphabetically and place them before the parent, giving the following complete IUPAC name:

1-(5-ethylheptyl)-4-methylbicyclo[2.2.2]octane

4.14.

(a) If we assign a systematic name for each of these structures, we find that they share the same name (2,3-dimethylpentane). Therefore, these drawings are simply different representations of the same compound.

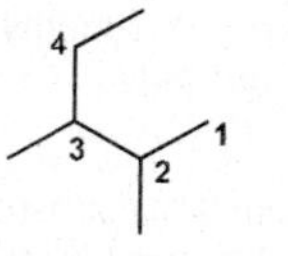

2,3-dimethylpentane 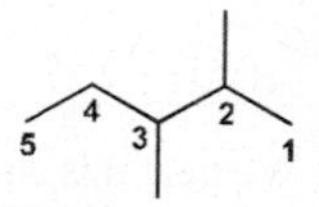 2,3-dimethylpentane

(b) If we assign a systematic name for each of these structures, we find that they share the same name (3-ethyl-2,4-dimethylpentane). Therefore, these drawings are simply different representations of the same compound.

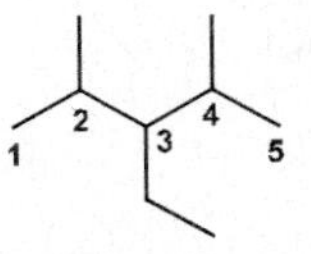

3-ethyl-2,4-dimethyl pentane

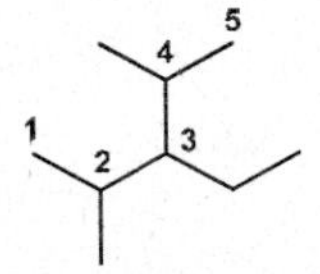

3-ethyl-2,4-dimethyl pentane

(c) If we assign a systematic name for each of these structures, we find that they share the same name (4-isobutyl-2,8-dimethylnonane). Therefore, these drawings are simply different representations of the same compound.

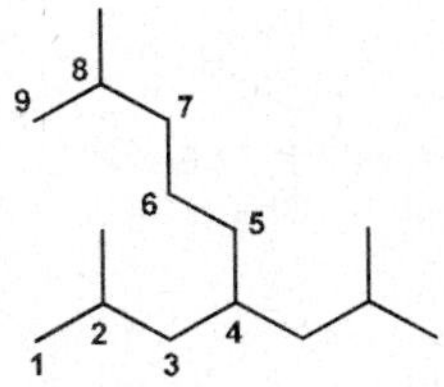

4-isobutyl-2,8-dimethyl nonane

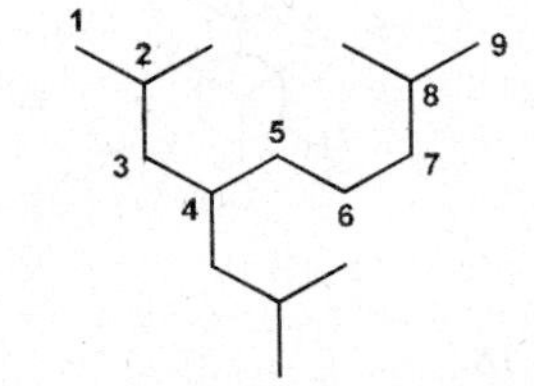

4-isobutyl-2,8-dimethyl nonane

(d) We assign a systematic name to each structure, and find that they have different names (shown below). Therefore, they must differ in their constitution (connectivity of atoms). These compounds are therefore different from each other, but they share the same molecular formula ($C_{11}H_{24}$), so they are constitutional isomers.

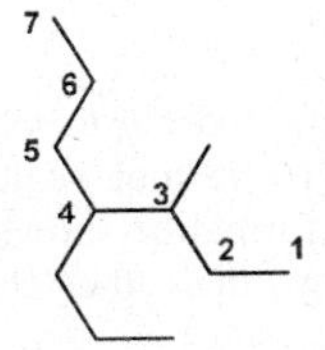

3-methyl-4-propyl heptane

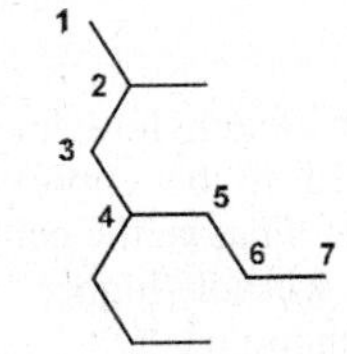

2-methyl-4-propyl heptane

4.15. We must draw all of the constitutional isomers of heptane:

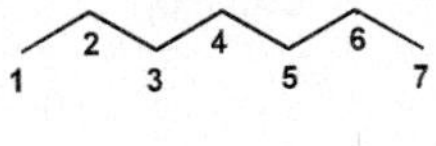

To accomplish the goal, we will follow the same methodical approach that we used in Problem **4.3**. We begin by drawing all possible substituted hexanes with the molecular formula C_7H_{16}. There are only two possibilities - the methyl group can be placed at either C2 or C3. Then, we move on to the pentanes, and finally any possible butanes. This methodical analysis gives the following constitutional isomers:

Parent = hexane

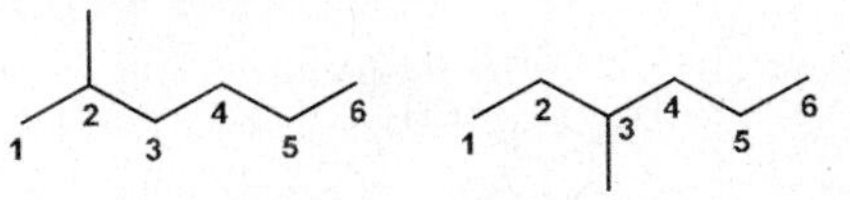

Parent = pentane

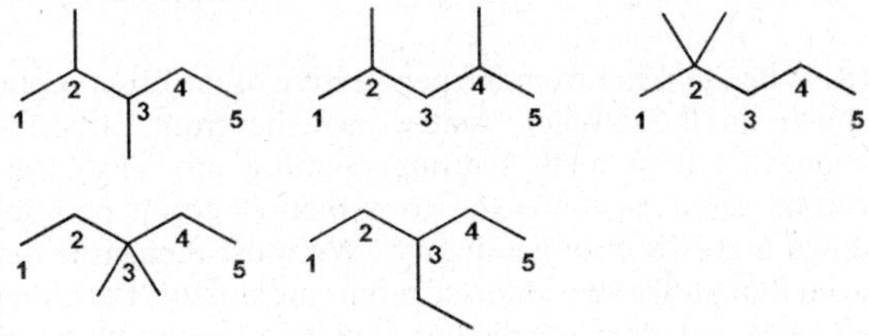

Parent = butane

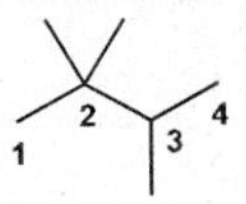

4.16.

(a) When looking from the perspective of the observer (as shown in the problem statement), the front carbon is connected to three methyl groups, and the back carbon is connected to one methyl group. Even though the bond-line structure is shown "flat" (i.e., without indicating the 3D structure), we must recognize that each carbon atom is tetrahedral:

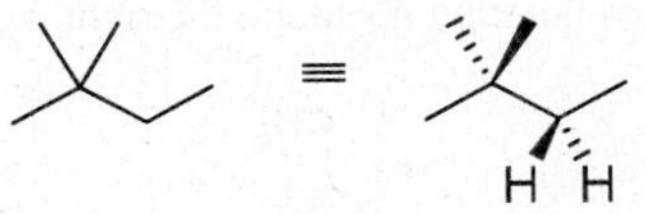

H H

In the Newman projection, two methyl groups on C2 are drawn pointing "up and to the right" and "up and to the left", while the methyl group on C3 is pointing up.

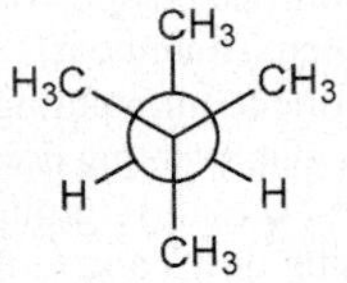

CH_3 H_3C CH_3 H H CH_3

(b) When looking from the perspective of the observer (as shown in the problem statement), the front carbon is connected to a chlorine atom pointing up and to the right, as well as a methyl group pointing down. The back carbon atom is connected to a chlorine atom pointing down and to the left, as well as a methyl group pointing up.

CH_3 H Cl Cl H CH_3

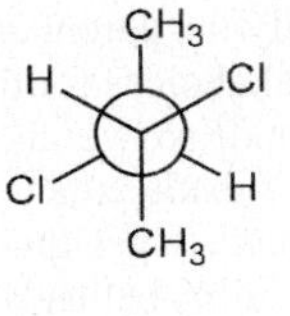

(c) When looking from the perspective of the observer (as shown in the problem statement), the front carbon is connected to a methyl group pointing up and to the right, as well as an ethyl group pointing down. The back carbon atom is connected to an ethyl group pointing up.

(d) When looking from the perspective of the observer (as shown in the problem statement), the front carbon is connected to a methyl group pointing up. The back carbon atom is connected to a methyl group pointing down and two chlorine atoms. We must recognize that even though the two chlorine atoms are drawn "flat", they are attached to a tetrahedral carbon atom, so they are accordingly drawn "up and to the left" and "up and to the right" on the Newman projection, as shown.

(e) When looking from the perspective of the observer (as shown in the problem statement), the front carbon is connected to a methyl group pointing up and a chlorine atom pointing down and to the right. The back carbon atom is connected to a methyl group pointing down and a chlorine atom pointing up and to the right, as shown.

(f) When looking from the perspective of the observer (as shown in the problem statement), the front carbon is connected to a chlorine atom pointing up and to the right, as well as a methyl group pointing down. The back carbon atom is connected to a methyl group pointing up and a bromine atom pointing down and to the left.

4.17. First we identify the parent chain (pentane). In this case, there are several choices for the parent (all of which are five carbon atoms), so we choose the one with the greatest number of substituents. This parent has five substituents: four methyl groups and one ethyl group, which are arranged alphabetically, together with their appropriate locants, to give the following IUPAC name:

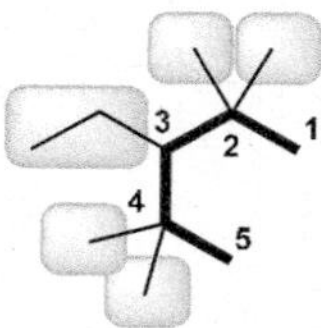

3-ethyl-2,2,4,4-tetramethylpentane

For the Newman projection, when looking from the perspective of the observer (as shown in the problem statement), the front carbon is connected to one methyl group pointing down, and the back carbon is connected to two *tert*-butyl groups: one pointing up and the other pointing down and to the right:

4.18.

(a) For 2,2-dimethylpropane, the energy barrier to rotation is expected to be approximately 18 kJ/mol (calculation below). That is, the eclipsed conformation shown below is 18 kJ/mol higher in energy than the staggered conformation:

(b) For 2-methylpropane, the energy barrier to rotation is expected to be approximately 16 kJ/mol (calculation below). That is, the eclipsed conformation shown below is 16 kJ/mol higher in energy than the staggered conformation:

4.19.

(a) In the Newman projection, the front carbon atom has three methyl groups, and the back carbon has one methyl group. Since the front carbon atom has three identical groups, we expect all staggered conformations to be degenerate. Similarly, we expect all eclipsed conformations to be degenerate as well. The lowest energy conformation is the staggered conformation, and the highest energy conformation is the eclipsed conformation.

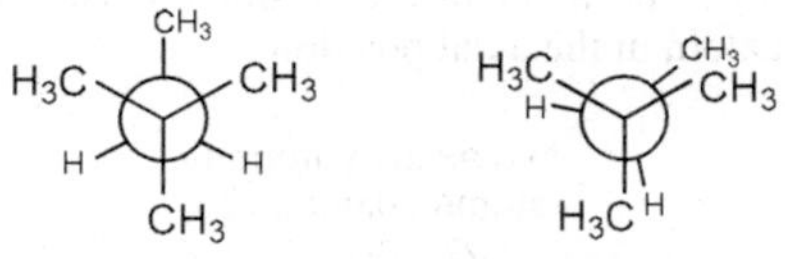

Lowest energy Highest energy

(b) We begin by converting the bond-line drawing into a Newman projection.

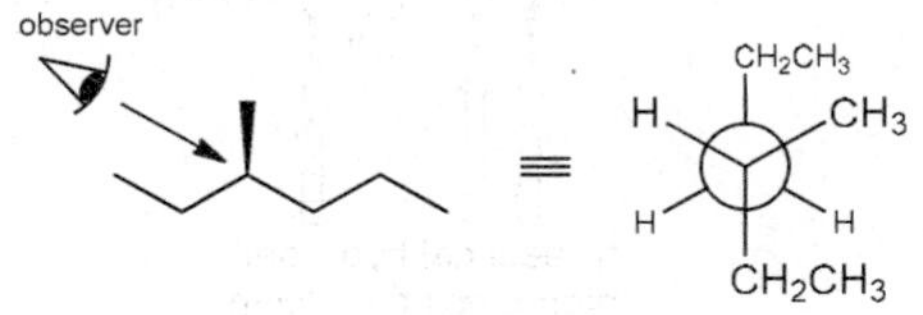

We expect the lowest energy conformation to be staggered and the highest energy conformation to be eclipsed. The lowest energy conformation is the staggered conformation in which the two largest groups (the ethyl groups) are *anti* to each other, while the highest energy conformation is the eclipsed conformation in which the two ethyl groups are eclipsing each other:

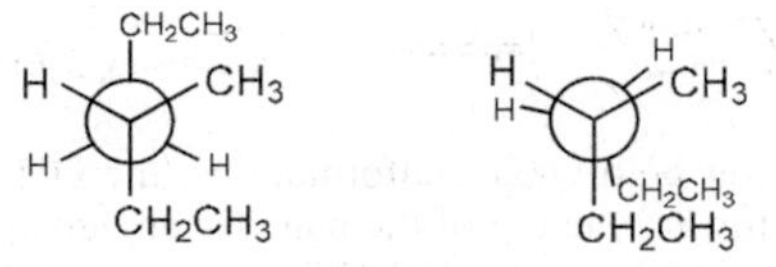

Lowest energy Highest energy

(c) We begin by converting the bond-line drawing into a Newman projection.

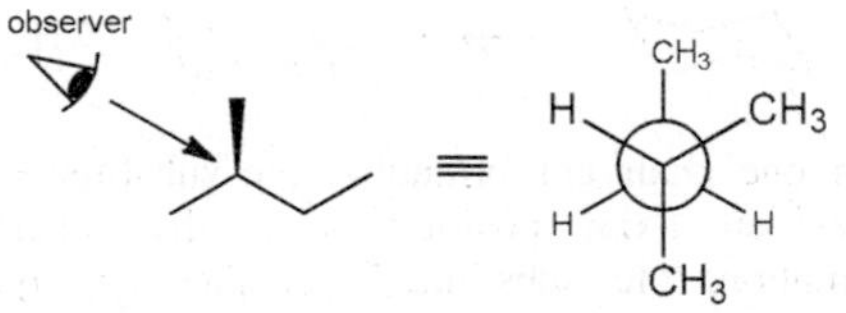

We expect the lowest energy conformation to be staggered and the highest energy conformation to be eclipsed. Let's first explore the three staggered conformations, shown below. One of these conformations exhibits two gauche interactions, while the other two staggered conformations are degenerate, with each having only one gauche interaction:

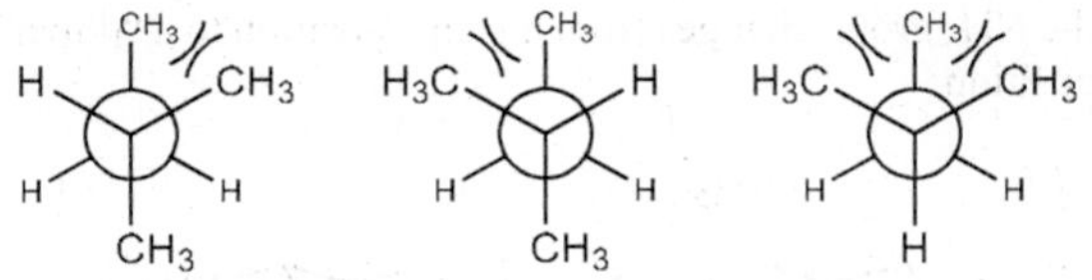

One gauche interaction One gauche interaction Two gauche interactions

degenerate, and lowest energy

The two degenerate conformations are lowest in energy among all available conformations.
In order to determine which conformation is highest in energy, we must examine the eclipsed conformations. There are three eclipsed conformations, shown below:

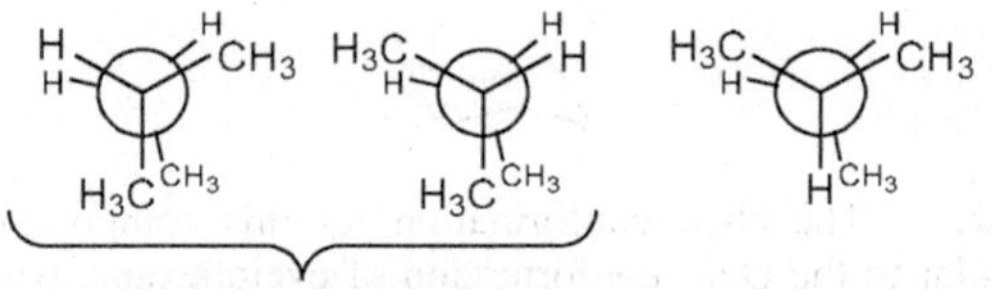

degenerate, and highest energy

Two of these conformations (the first two above) are degenerate, because they each exhibit one Me/Me eclipsing interaction and one Me/H eclipsing interaction. These two degenerate conformations are highest in energy among all available conformations, because these conformations are less stable than the eclipsed conformation with two Me/H eclipsing interactions.

(d) We begin by converting the bond-line drawing into a Newman projection.

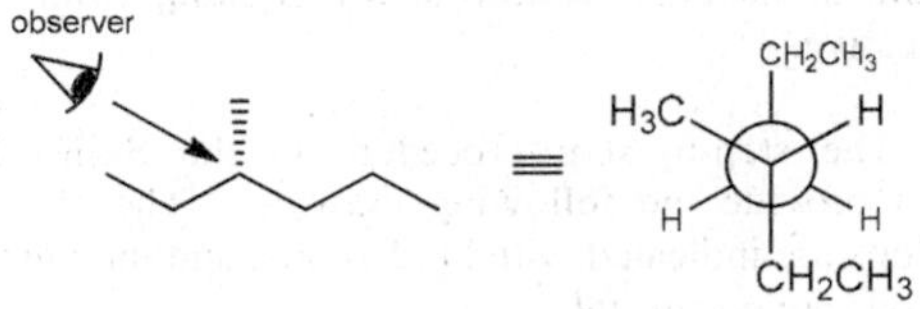

We expect the lowest energy conformation to be staggered and the highest energy conformation to be eclipsed. The lowest energy conformation is the staggered conformation in which the two largest groups (the ethyl groups) are *anti* to each other, while the highest energy conformation is the eclipsed conformation in which the two ethyl groups are eclipsing each other:

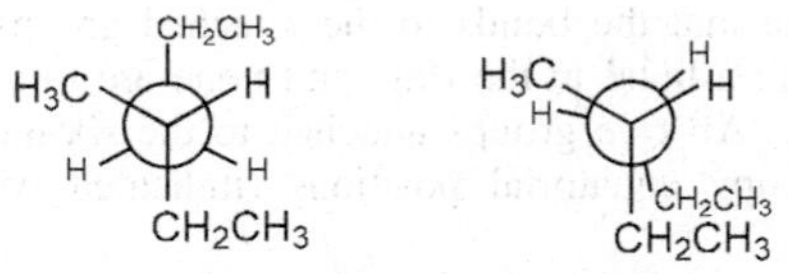

Lowest energy Highest energy

4.20.

(a) The three possible staggered conformations of **1**, viewed along the C_a-C_b bond, are as follows:

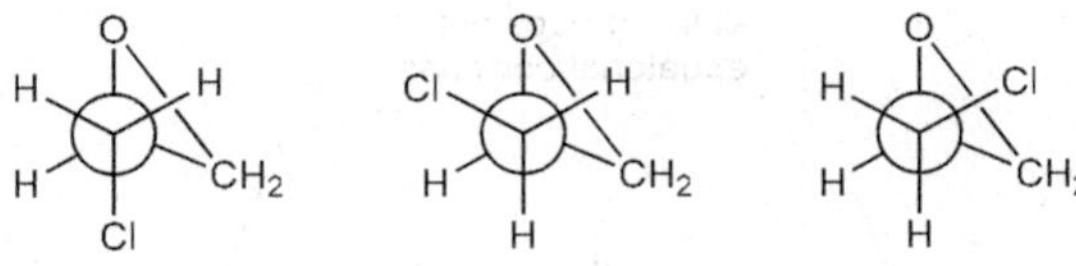

A B C

(b) In conformation **C**, there are two gauche interactions: one between the chlorine atom and the CH_2 group, and another between the chlorine atom and the oxygen atom. This conformation is the highest in energy (least stable) due to these two interactions. Conformations **A** and **B** each exhibit only one of these two gauche interactions.

4.21. The step-by-step procedure in the SkillBuilder should provide the following drawing:

4.22. The chair conformation for this compound is similar to the chair conformation of cyclohexane, where two of the CH_2 groups (C1 and C4) have been replaced with oxygen atoms. Each of the following drawings represents dioxane, with oxygen atoms in the 1 and 4 positions:

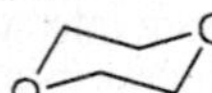

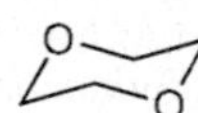

All three drawings represent the same compound (viewed from different angles). This can be seen clearly if you build a molecular model of dioxane (in a chair conformation) and view it from different angles. Indeed, there are still more representations that could be drawn for dioxane in its chair conformation, looking from other perspectives.

4.23. The step-by-step procedure in the SkillBuilder should provide the following drawing. The six axial positions are indicated with bold bonds, and the other six positions are equatorial:

4.24.

(a) Note that the bonds to the attached groups are all parallel to a bond in the ring, and none are in a vertical position. All five groups attached to the six-membered ring occupy equatorial positions (indicated with bold bonds).

(−)-Cassioside

all five groups are in equatorial positions

(b) Each carbon atom in the six-membered ring has a hydrogen atom in the axial position:

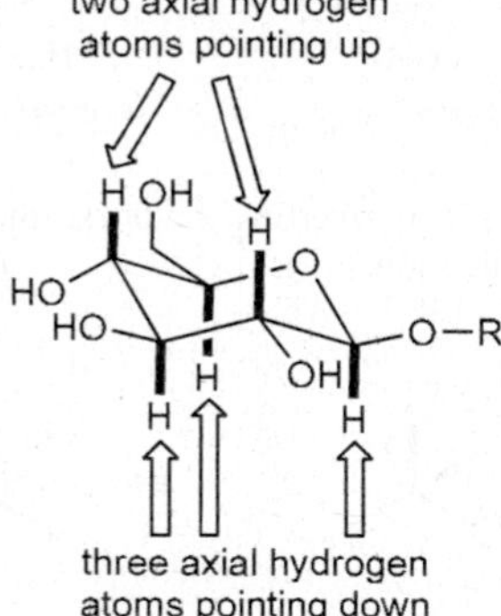

4.25.

(a) In one chair conformation, the substituent (OH) occupies an axial position. In the other chair conformation, the substituent occupies an equatorial position.

Note that in both chair conformations, the OH group is pointing toward the top of the page (in an "up" position). The following drawings do NOT represent a proper chair flip, because the two chairs have the same orientation, and the OH group changes from an "up" position to a "down" position:

(b) In one chair conformation, the substituent (NH_2) occupies an axial position. In the other chair conformation, the substituent occupies an equatorial position:

Note that in both chair conformations, the NH_2 group is pointing toward the top of the page (in an "up" position). The following drawings do NOT represent a proper chair flip, because the two chairs have the same orientation, and the NH_2 group changes from an "up" position to a "down" position:

(c) In one chair conformation, the substituent (chlorine) occupies an axial position. In the other chair conformation, the substituent occupies an equatorial position:

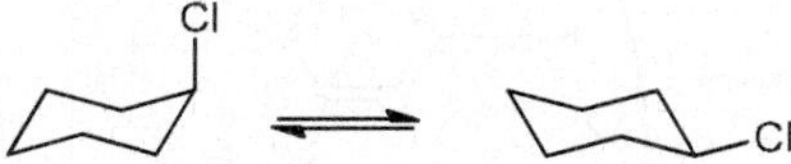

Note that in both chair conformations, the chlorine atom is pointing toward the top of the page (in an "up" position). The following drawings do NOT represent a proper chair flip, because the two chairs have the same orientation, and the chorine atom changes from an "up" position to a "down" position:

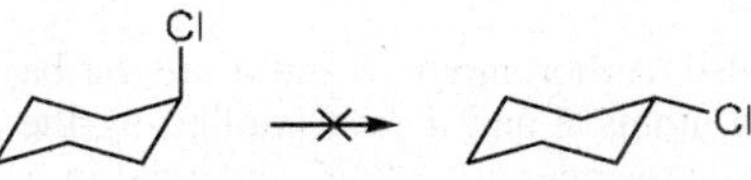

(d) In one chair conformation, the methyl group occupies an axial position. In the other chair conformation, the methyl group occupies an equatorial position:

Note that in both chair conformations, the methyl group is pointing toward the top of the page (in an "up" position). The following drawings do NOT represent a proper chair flip, because the two chairs have the same orientation, and the methyl group changes from an "up" position to a "down" position:

(e) In one chair conformation, the *tert*-butyl group occupies an axial position. In the other chair conformation, the *tert*-butyl group occupies an equatorial position:

Note that in both chair conformations, the *tert*-butyl group is pointing toward the top of the page (in an "up" position). The following drawings do NOT represent a proper chair flip, because the two chairs have the same orientation, and the *tert*-butyl group changes from an "up" position to a "down" position:

4.26.

(a) In a Newman projection of cyclohexane, the four hydrogen atoms that point straight up or straight down occupy axial positions, while the four hydrogen atoms that point out to either side are equatorial:

Notice that four of the twelve hydrogens are not drawn explicitly; rather they are implied by the bond line intersections (highlighted):

In the structure shown in the problem statement, the adenine group occupies an axial position and the CH_2OH group occupies an equatorial position.

(b) The carbon atoms of the cyclohexane ring are numbered in the Newman projection and in the chair conformation (the bond-line structure is also included for reference). Note that these numbers do not necessarily have to be in accordance with IUPAC rules, but that is OK, because we are not assigning a name. Rather, we are using the numbers just to redraw the compound. In all depictions, the CH_2OH group is attached to carbon atom **1** and the adenine group is attached to carbon atom **2** (a clockwise relationship going from **1** to **2**).

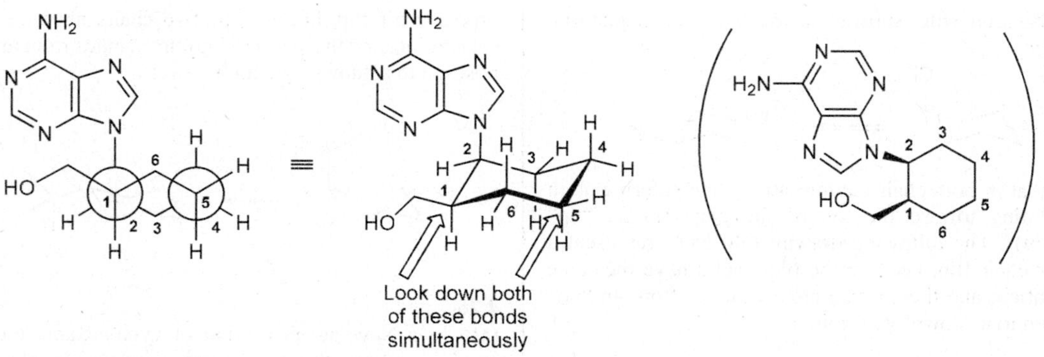

Look down both of these bonds simultaneously

Carbon atoms **1** and **5** are the front carbon atoms in both forms; likewise, carbon atoms **2** and **4** are the back carbon atoms. The bond between carbon atoms **1** and **2** (and between carbon atoms **5** and **4**) are implied by the Newman projections. The CH_2OH group is "up" and equatorial in both forms, while the adenine is "up" and axial in both forms. The clockwise location of the adenine group compared to the CH_2OH is an important feature which must be maintained in all depictions of the structure; this will be explored in more detail in Chapter 5.

4.27. Although the OH group is in an axial position, this conformation is capable of intramolecular hydrogen bonding, which is a stabilizing effect:

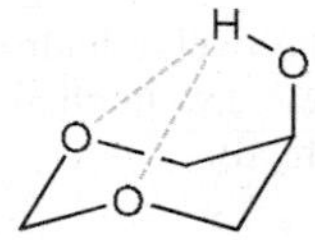

4.28.

(a) Begin by assigning a numbering system (which does not need to adhere to IUPAC rules), and then determine the location and three-dimensional orientation of each substituent. In this case, there is a methyl group at C1, which is UP, and there is an ethyl group at C2, which is DOWN:

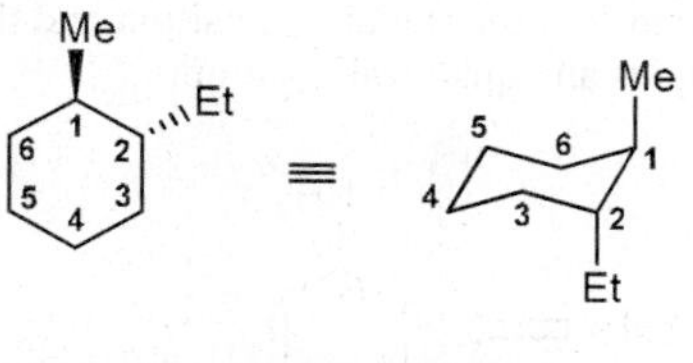

When assigning the numbers to the chair drawing, the first number can be placed anywhere on the ring. For example, it would be perfectly acceptable to draw the chair like this:

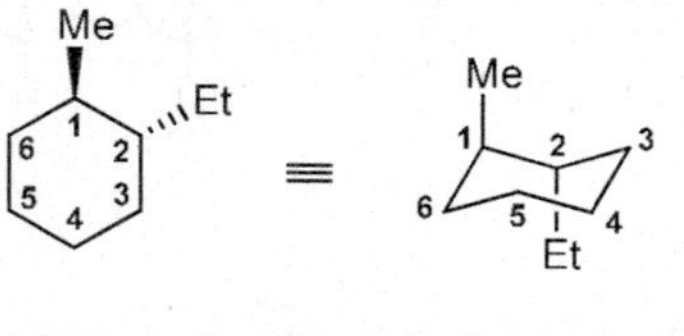

At first it might look like this is a different chair than the one previously drawn. But let's compare them:

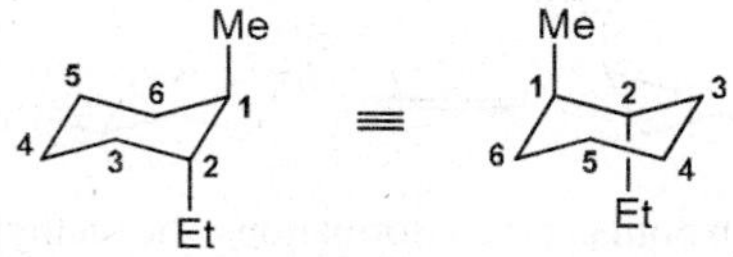

In both drawings, if we travel clockwise around the ring, we will encounter the methyl group first and the ethyl group second. Also, in both chair drawings, the methyl group is up and the ethyl group is down. Therefore, either of these drawings is an acceptable chair representation. Neither one is "more correct" than the other (to see this more clearly, you may find it helpful to build a molecular model and view it from different angles).

Once we have drawn the first chair conformation, we then draw the second chair conformation, once again using a numbering system. After doing a chair flip, the methyl group is still up and the ethyl group is still down. Notice that a ring flip causes both axial groups to become equatorial.

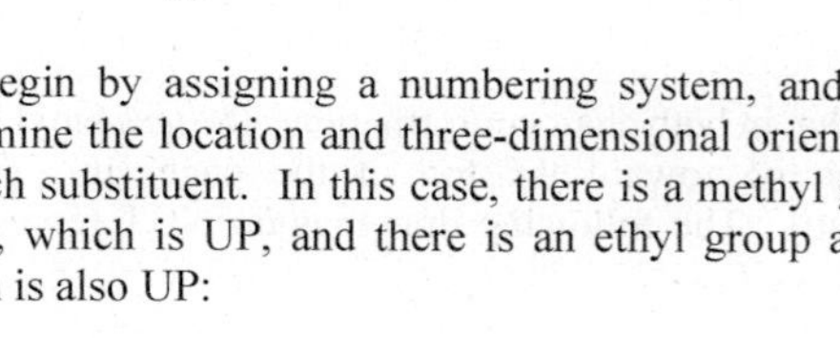

(b) Begin by assigning a numbering system, and then determine the location and three-dimensional orientation of each substituent. In this case, there is a methyl group at C1, which is UP, and there is an ethyl group at C2, which is also UP:

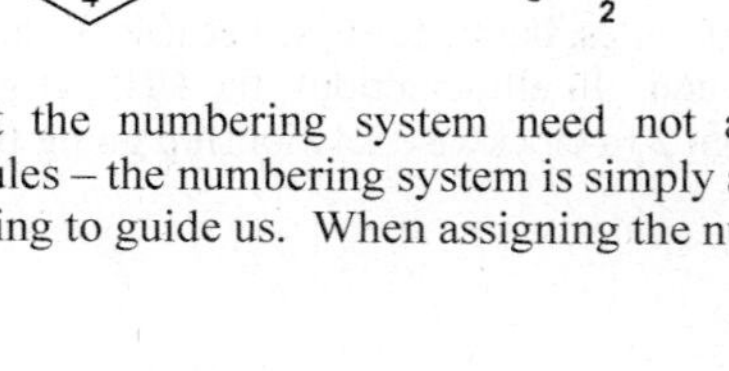

Note that the numbering system need not adhere to IUPAC rules – the numbering system is simply a tool that we are using to guide us. When assigning the numbers to the chair drawing, the first number can be placed anywhere on the ring (as long as the numbers go clockwise), as seen in the solution to part (a) of this problem.

Finally, we draw the second chair conformation, once again using a numbering system. After doing a chair flip, the methyl group is still up and the ethyl group is still up. Notice that a ring flip causes all equatorial groups to become axial groups, and vice versa.

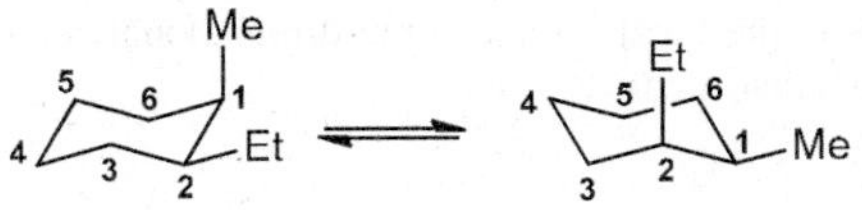

(c) Begin by assigning a numbering system, and then determine the location and three-dimensional orientation of each substituent. In this case, there is a methyl group at C1, which is UP, and there is a bromine atom at C3, which is also UP:

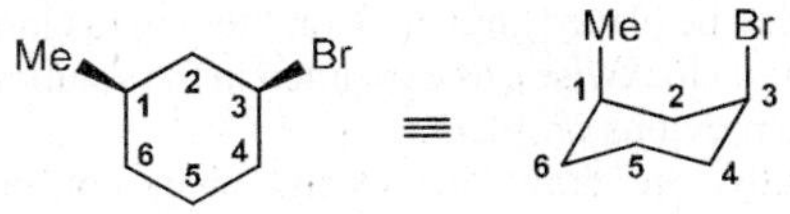

Note that the numbering system need not adhere to IUPAC rules – the numbering system is simply a tool that we are using to guide us. When assigning the numbers to the chair drawing, the first number can be placed anywhere on the ring (as long as the numbers go clockwise), as seen in the solution to part (a) of this problem.

Finally, we draw the second chair conformation, once again using a numbering system. After doing a chair flip, the methyl group is still up and the bromine atom is still up. Notice that a ring flip causes both axial groups to become equatorial.

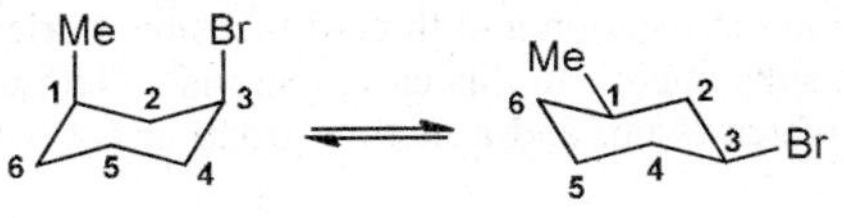

(d) Begin by assigning a numbering system, and then determine the location and three-dimensional orientation of each substituent. In this case, there is a bromine atom at C1, which is UP, and there is a methyl group at C3, which is also UP:

Note that the numbering system need not adhere to IUPAC rules – the numbering system is simply a tool that we are using to guide us. When assigning the numbers to the chair drawing, the first number can be placed anywhere on the ring (as long as the numbers go clockwise), as seen in the solution to part (a) of this problem.

Finally, we draw the second chair conformation, once again using a numbering system. After doing a chair flip, the bromine atom is still up and the methyl group is still up. Notice that a ring flip causes both axial groups to become equatorial.

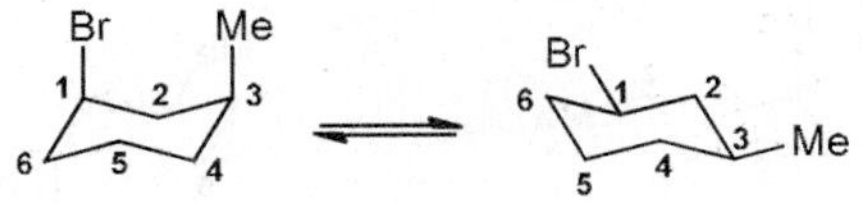

(e) Begin by assigning a numbering system, and then determine the location and three-dimensional orientation of each substituent. In this case, there is a methyl group at C1, which is UP, and a *tert*-butyl group at C4, which is DOWN:

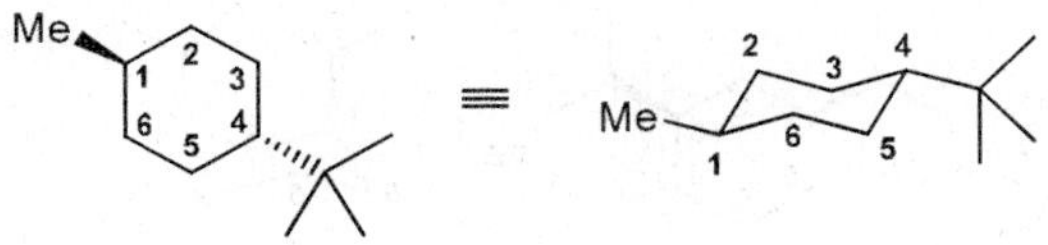

Notice that the numbering system need not adhere to IUPAC rules – the numbering system is simply a tool that we are using to guide us. When assigning the numbers to the chair drawing, the first number can be placed anywhere on the ring (as long as the numbers go clockwise), as seen in the solution to part (a) of this problem.

Finally, we draw the second chair conformation, once again using a numbering system. After doing a chair flip, the methyl group is still up and the *tert*-butyl group is still down. Notice that a ring flip causes all equatorial groups to become axial groups, and vice versa.

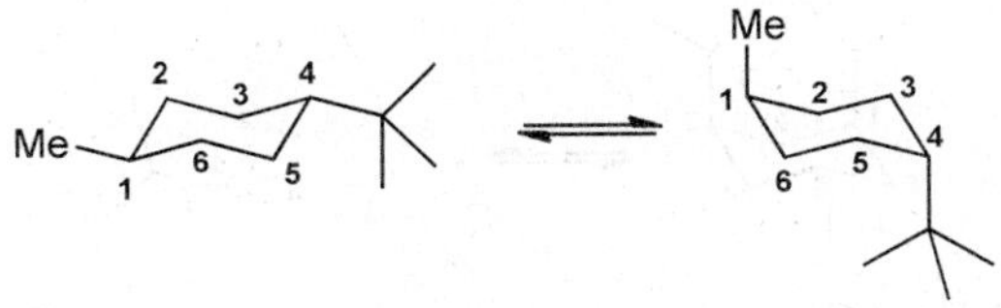

(f) Begin by assigning a numbering system, and then determine the location and three-dimensional orientation of each substituent. In this case, there is a methyl group at C1, which is UP, and another methyl group at C3, which is DOWN:

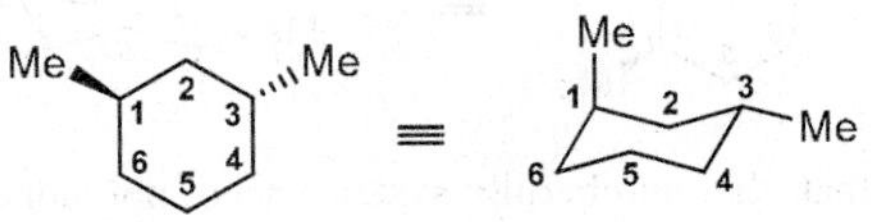

Note that the numbering system need not adhere to IUPAC rules – the numbering system is simply a tool that we are using to guide us. When assigning the numbers to the chair drawing, the first number can be placed anywhere on the ring (as long as the numbers go clockwise), as seen in the solution to part (a) of this problem.

Finally, we draw the second chair conformation, once again using a numbering system. After doing a chair flip, the C1 methyl group is still up and the C3 methyl group is still down. Notice that a ring flip causes all equatorial groups to become axial groups, and vice versa.

(g) Begin by assigning a numbering system, and then determine the location and three-dimensional orientation of each substituent. In this case, there is an isopropyl group at C1, which is UP, and another isopropyl group at C3, which is also UP:

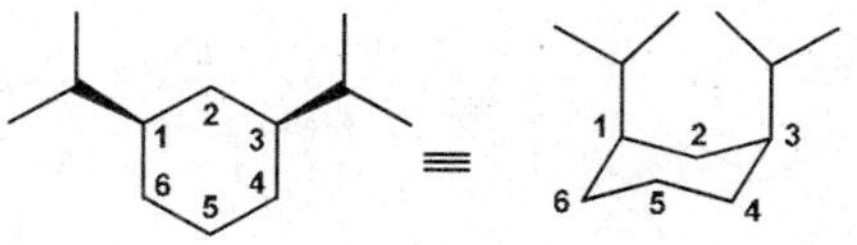

Note that the numbering system need not adhere to IUPAC rules – the numbering system is simply a tool that we are using to guide us. When assigning the numbers to the chair drawing, the first number can be placed anywhere on the ring (as long as the numbers go clockwise), as seen in the solution to part (a) of this problem.

Finally, we draw the second chair conformation, once again using a numbering system. After doing a chair flip, both isopropyl groups are still pointing up. Notice that a ring flip causes both axial groups to become equatorial.

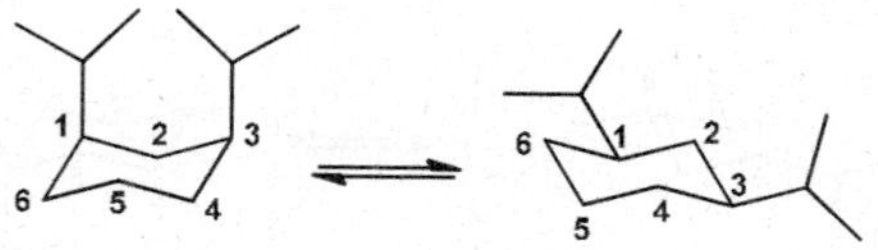

(h) Begin by assigning a numbering system, and then determine the location and three-dimensional orientation of each substituent. In this case, there is a methyl group at C1, which is UP, and another methyl group at C4, which is also UP:

Note that the numbering system need not adhere to IUPAC rules – the numbering system is simply a tool that we are using to guide us. When assigning the numbers to the chair drawing, the first number can be placed anywhere on the ring (as long as the numbers go clockwise), as seen in the solution to part (a) of this problem.

Finally, we draw the second chair conformation, once again using a numbering system. After doing a chair flip, both methyl groups are still pointing up. Notice that a ring flip causes all equatorial groups to become axial groups, and vice versa.

4.29. Begin by assigning a numbering system, and then determine the location and three-dimensional orientation of each substituent:

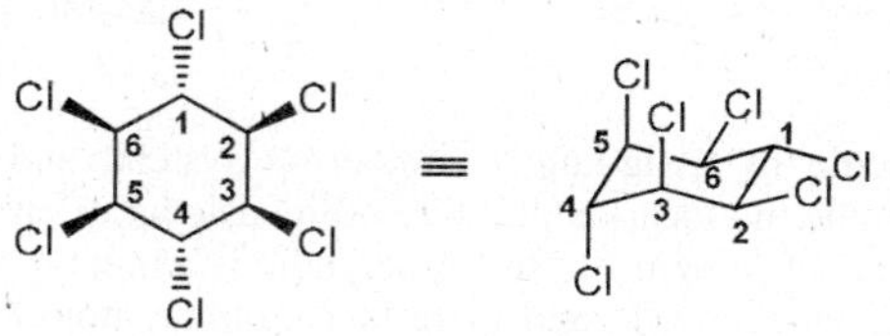

When assigning the numbers to the chair drawing, the first number can be placed anywhere on the ring (as long as the numbers go clockwise), as explained in the solution to part (a) of the previous problem.

Finally, we draw the second chair conformation, once again using a numbering system. After doing a chair flip, the chlorine atoms on C1 and C4 are still pointing down, and the chlorine atoms on C2, C3, C5 and C6 are all still pointing up. Notice that a ring flip causes all equatorial groups to become axial groups, and vice versa.

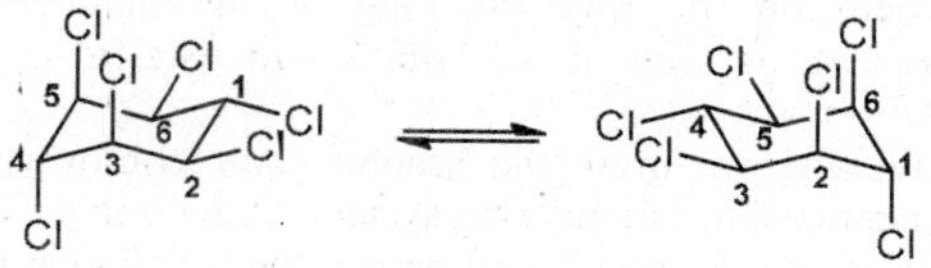

4.30.

(a) Begin by assigning a numbering system, and then determine the location and three-dimensional orientation of each substituent. In this case, there is a methyl group at C1, which is up, and a methyl group at C2, which is down:

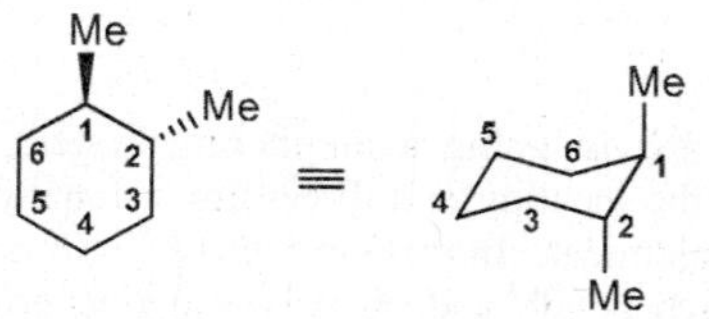

When assigning the numbers to the chair drawing, the first number can be placed anywhere on the ring (as long as the numbers go clockwise), as explained in the solution to Problem **4.28a**.

Once we have drawn the first chair conformation, we then draw the second chair conformation, again using the same numbering system. After doing a chair flip, the C1 methyl group is still up and the C2 methyl group is still down. Notice that a ring flip causes both axial groups to become equatorial.

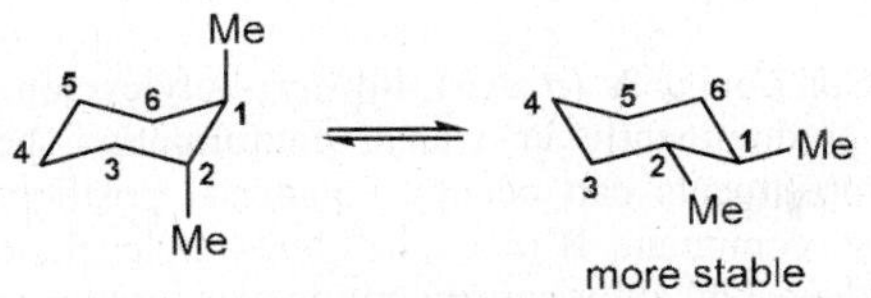

Me 5 6 1 4 3 2 Me ⇌ 4 5 6 1 Me 3 2 Me more stable

Finally, we compare both chair drawings and determine which one has fewer or less severe 1,3-diaxial interactions. In the first drawing, both substituents occupy axial positions, but in the second drawing, both substituents occupy equatorial positions. As such, the second chair conformation is more stable, since it lacks 1,3-diaxial interactions.

(b) Begin by assigning a numbering system, and then determine the location and three-dimensional orientation of each substituent. In this case, there is a methyl group at C1, which is up, and an isopropyl group at C2, which is also up:

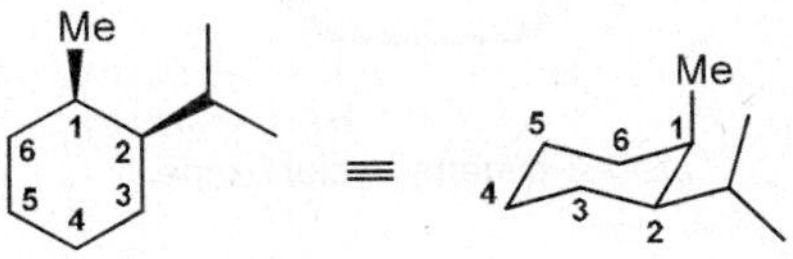

When assigning the numbers to the chair drawing, the first number can be placed anywhere on the ring (as long as the numbers go clockwise), as explained in the solution to Problem **4.28a**.

Once we have drawn the first chair conformation, we then draw the second chair conformation, again using the same numbering system. After doing a chair flip, the methyl group is still up and the isopropyl group is still up. Notice that a ring flip causes all equatorial groups to become axial groups, and vice versa.

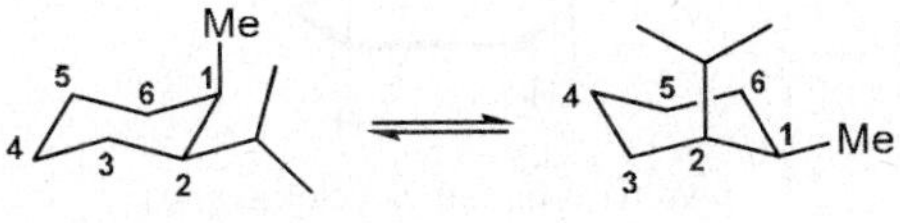

more stable

Finally, we compare both chair drawings and determine which one has fewer or less severe 1,3-diaxial interactions. In the first drawing, the methyl group occupies an axial position, but in the second drawing, the larger isopropyl group occupies an axial position. As such, the first chair conformation is more stable, since it is expected to have less severe 1,3-diaxial interactions.

(c) Begin by assigning a numbering system, and then determine the location and three-dimensional orientation of each substituent:

When assigning the numbers to the chair drawing, the first number can be placed anywhere on the ring (as long as the numbers go clockwise), as explained in the solution to Problem **4.28a**.

Once we have drawn the first chair conformation, we then draw the second chair conformation, again using the same numbering system. After doing a chair flip, the methyl group is still down, the chlorine atom at C2 is still down, and the chlorine atom at C3 is still up. Notice that a ring flip causes all equatorial groups to become axial groups, and vice versa.

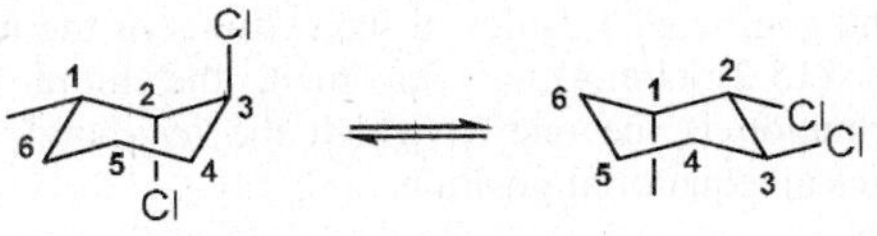

more stable

Finally, we compare both chair drawings and determine which one has fewer or less severe 1,3-diaxial interactions. In the first drawing, the methyl group occupies an equatorial position, while both chlorine atoms occupy axial positions. In the second drawing, the methyl group occupies an axial position, while both chlorine atoms occupy equatorial positions. According to the data presented in Table 4.8, the methyl group experiences 1,3-diaxial interactions of 7.6 kJ/mol, while each chlorine atom experiences 1,3-diaxial interactions of 2.0 kJ/mol. As such, the 1,3-diaxial interactions from the methyl group are more severe than the combined 1,3-diaxial interactions of the two chlorine atoms (4.0 kJ/mol). As such, the more stable conformation is the one in which the methyl group occupies an equatorial position.

(d) Begin by assigning a numbering system, and then determine the location and three-dimensional orientation of each substituent:

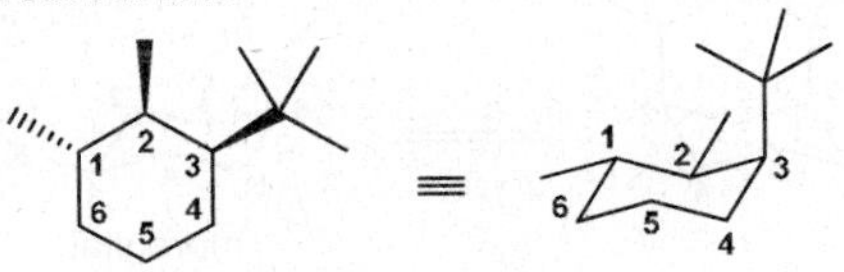

When assigning the numbers to the chair drawing, the first number can be placed anywhere on the ring (as long as the numbers go clockwise), as explained in the solution to Problem **4.28a**.

Once we have drawn the first chair conformation, we then draw the second chair conformation, again using the same numbering system. After doing a chair flip, the C1 methyl group is still down, the C2 methyl group is still up, and the *tert*-butyl group is still up. Notice that a ring flip causes all equatorial groups to become axial groups, and vice versa.

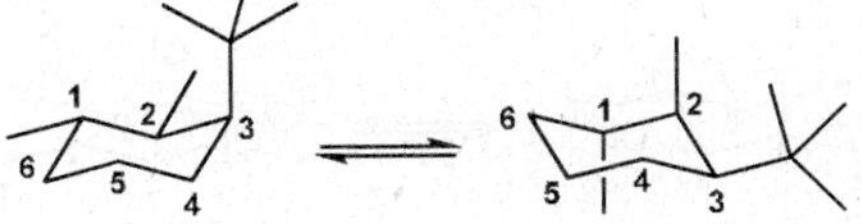

more stable

Finally, we compare both chair drawings and determine which one has fewer or less severe 1,3-diaxial interactions. In the first drawing, the *tert*-butyl group occupies an axial position, while both methyl groups occupy equatorial positions. In the second drawing, the *tert*-butyl group occupies an equatorial position, while both methyl groups occupy equatorial positions. According to the data presented in Table 4.8, each methyl group experiences 1,3-diaxial interactions of 7.6 kJ/mol, while a *tert*-butyl group experiences 1,3-diaxial interactions of 22.8 kJ/mol. As such, the 1,3-diaxial interactions from the one *tert*-butyl group are more severe than the combined 1,3-diaxial interactions of the methyl groups (15.2 kJ/mol). As such, the more stable conformation is the one in which the *tert*-butyl group occupies an equatorial position.

(e) Begin by assigning a numbering system, and then determine the location and three-dimensional orientation of each substituent:

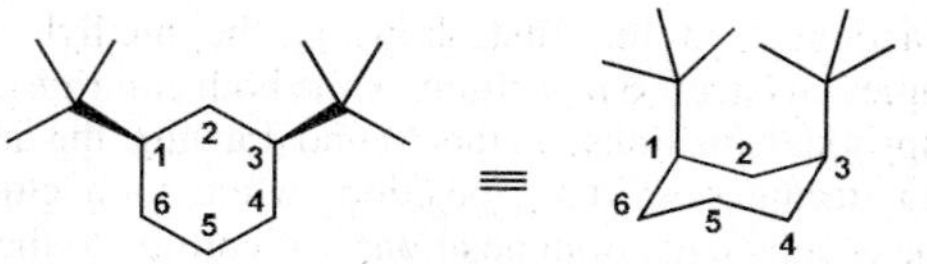

When assigning the numbers to the chair drawing, the first number can be placed anywhere on the ring (as long as the numbers go clockwise), as explained in the solution to Problem **4.28a**.

Once we have drawn the first chair conformation, we then draw the second chair conformation, again using the same numbering system. After doing a chair flip, both *tert*-butyl groups are still pointing up. Notice that a ring flip causes both axial groups to become equatorial.

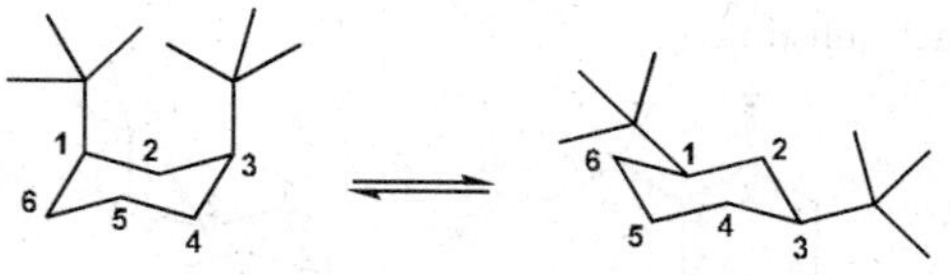

more stable

Finally, we compare both chair drawings and determine which one has fewer or less severe 1,3-diaxial interactions. In the first drawing, both *tert*-butyl groups occupy axial positions, but in the second drawing, both occupy equatorial positions. As such, the second chair conformation is more stable, since it lacks 1,3-diaxial interactions.

4.31. Each chair conformation has three substituents occupying equatorial positions and three substituents occupying axial positions:

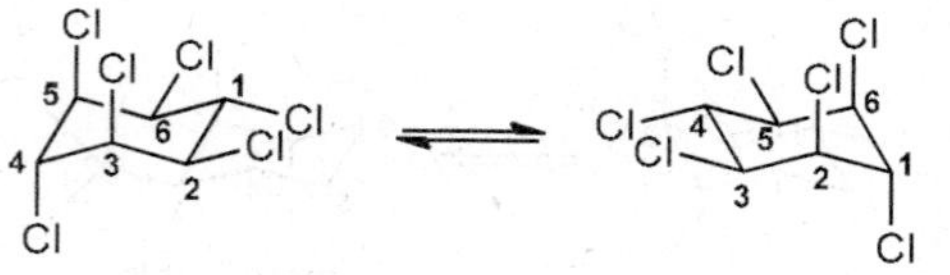

As such, the two chair conformations of lindane are degenerate. There is no difference in energy between them.

4.32. Compound **A** (*trans*-1,4-di-*tert*-butylcyclohexane) exists predominantly in a chair conformation, because both substituents can occupy equatorial positions. In contrast, compound **B** (*cis*-1,4-di-*tert*-butylcyclohexane) cannot have both of its substituents in equatorial positions. Each chair conformation has one of the substituents in an axial position, which is high in energy due to severe 1,3-diaxial interactions. Compound **B** can achieve a lower energy state by adopting a twist boat conformation.

4.33. *cis*-1,3-Dimethylcyclohexane is expected to be more stable than *trans*-1,3-dimethylcyclohexane because the *cis* isomer can adopt a chair conformation in which both substituents are in equatorial positions (highlighted below):

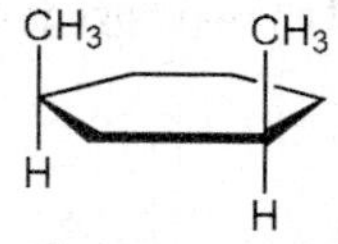

cis-1,3-dimethylcyclohexane

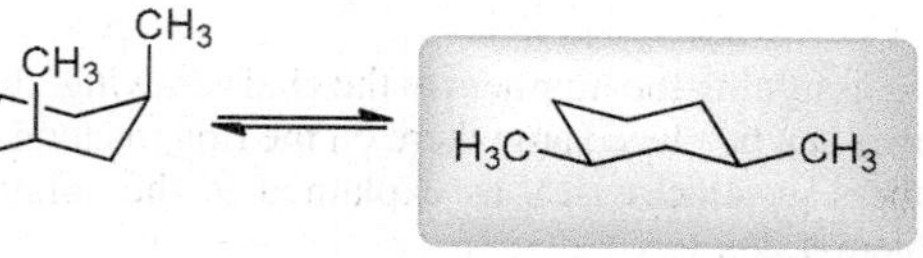

In contrast, *trans*-1,3-dimethylcyclohexane cannot adopt a chair conformation in which both substituents are in equatorial positions. Each chair conformation has one methyl group in an axial position:

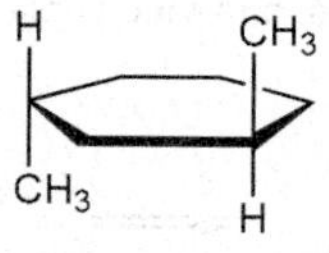

trans-1,3-dimethylcyclohexane

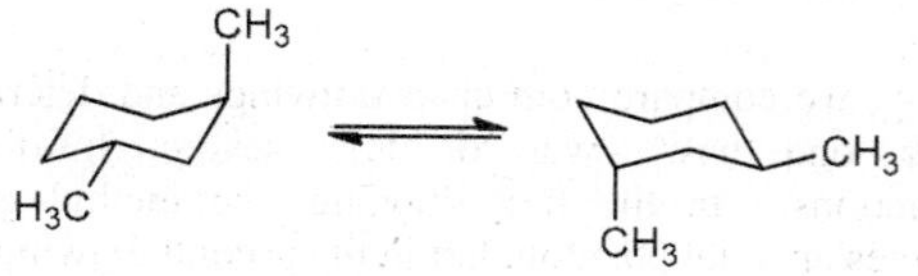

4.34. *trans*-1,4-Dimethylcyclohexane is expected to be more stable than *cis*-1,4-dimethylcyclohexane because the *trans* isomer can adopt a chair conformation in which both substituents are in equatorial positions (highlighted below):

trans-1,4-dimethylcyclohexane

In contrast, *cis*-1,4-dimethylcyclohexane cannot adopt a chair conformation in which both substituents are in equatorial positions. Each chair conformation has one methyl group in an axial position:

cis-1,4-dimethylcyclohexane

4.35. *cis*-1,3-Di-*tert*-butylcyclohexane can adopt a chair conformation in which both *tert*-butyl groups occupy equatorial positions (highlighted below), and as a result, it is expected to exist primarily in that conformation.

cis-1,3-di-*tert*-butylcyclohexane

(R = *tert*-butyl group)

In contrast, *trans*-1,3-di-*tert*-butylcyclohexane cannot adopt a chair conformation in which both *tert*-butyl groups occupy equatorial positions. In either chair conformation, one of the *tert*-butyl groups occupies an axial position.

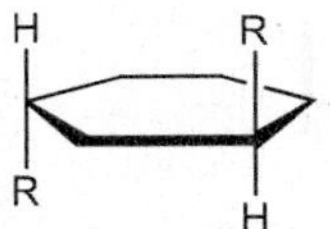

trans-1,3-di-*tert*-butylcyclohexane

(R = *tert*-butyl group)

This compound can achieve a lower energy state by adopting a twist-boat conformation, as explained in the solution to Problem 4.32 for the 1,4-isomer.

4.36. For each of the following compounds, we assign its name via a four-step process: First identify the parent, then the substituents, then assign locants, and finally, arrange the substituents alphabetically. In each case, use commas to separate numbers from each other, and use hyphens to separate letters from numbers.

(a) 4-ethyl-3-methyloctane
(b) 5-isopropylnonane *or*
5-(1-methylethyl)nonane
(c) 2-methyl-4-propyloctane
(d) 4-*tert*-butylheptane *or*
4-(1,1-dimethylethyl)heptane
(e) 5-*sec*-butyl-4-ethyl-2-methyldecane *or*
4-ethyl-2-methyl-5-(1-methylpropyl)decane
(f) 3-ethyl-6-isopropyl-2,4-dimethyldecane *or*
3-ethyl-2,4-dimethyl-6-(1-methylethyl)decane
(g) 3,5-diethyl-2-methyloctane
(h) 2,3,5-trimethyl-4-propylheptane
(i) 1,2,4,5-tetramethyl-3-propylcyclohexane
(j) 2,3,5,9-tetramethylbicyclo[4.4.0]decane
(k) 1,4-dimethylbicyclo[2.2.2]octane

4.37.
(a) If we assign a systematic name for each of these structures, we find that they share the same name (2-methylpentane). Therefore, these drawings are simply different representations of the same compound.

(b) We assign a systematic name to each structure, and find that they have different names (shown below). Therefore, they must differ in their constitution (connectivity of atoms). These compounds are therefore different from each other, but they share the same molecular formula (C_7H_{16}), so they are constitutional isomers.

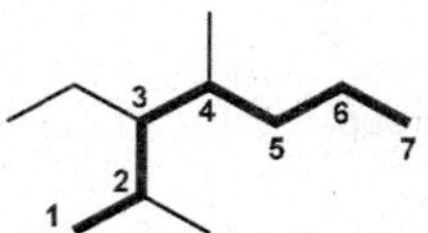

2,2-dimethylpentane

3,3-dimethylpentane

(c) If we assign a systematic name for each of these structures, we find that they share the same name (3-ethyl-2,4-dimethylheptane). Therefore, these drawings are simply different representations of the same compound.

3-ethyl-2,4-dimethyl heptane

3-ethyl-2,4-dimethyl heptane

4.38. When looking down the C2-C3 bond, the front carbon atom of 3-methylpentane has one methyl group and two hydrogen atoms, while the back carbon atom has an ethyl group, a methyl group, and a hydrogen atom. The lowest energy conformation is the staggered conformation with the fewest and least severe gauche interactions, shown here. In this conformation, there is only one Me/Me gauche interaction.

4.39. Both compounds share the same molecular formula (C_6H_{14}). That is, they are constitutional isomers, and the unbranched isomer is expected to have the larger heat of combustion:

4.40.

(a) The name 2,2,4-trimethylpentane indicates that the parent is a five-carbon chain and there are three substituents (two methyl groups at C2, and one methyl group at C4):

(b) The name 1,2,3,4-tetramethylcycloheptane indicates that the parent is a seven-membered ring and there are four substituents (all methyl groups), as shown:

(c) The name 2,2,4,4-tetraethylbicyclo[1.1.0]butane indicates a bicyclic parent with two ethyl groups at C2 and two ethyl groups at C4, as shown:

4.41. We begin by drawing a Newman projection of 2,2-dimethylpropane:

Notice that the front carbon has three identical groups (all hydrogen atoms), and the back carbon atom also has three identical groups (all methyl groups). As such, we expect all staggered conformations to be degenerate, and we expect all eclipsed conformations to be degenerate as well. Therefore, the energy diagram will more closely resemble the shape of the energy diagram for the conformational analysis of ethane.

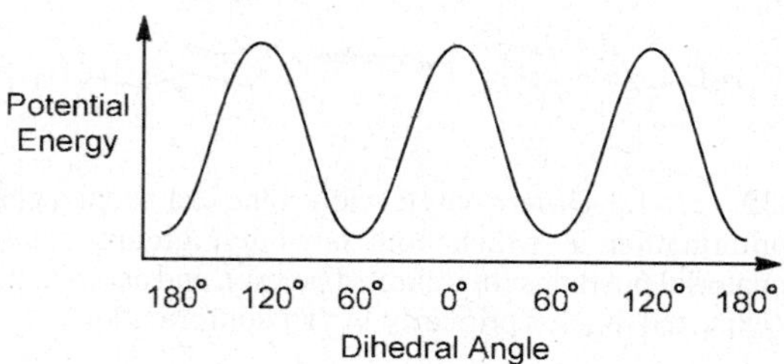

4.42. Two of the staggered conformations of 2,3-dimethylbutane are degenerate (each of these two conformations has three Me/Me gauche interactions). The remaining staggered conformation is lower in energy than the other two, because it exhibits only two Me/Me gauche interactions, as shown:

4.43. For each of the following cases, we draw the second chair conformation using a numbering system to ensure the substituents are placed correctly. The numbering system does NOT need to adhere to IUPAC rules, as it is just a tool that we are using to draw both chair conformations correctly. After doing a chair flip, all UP groups are still pointing up, and all DOWN groups are still pointing down. Notice that a ring flip causes all equatorial groups to become axial groups, and vice versa.

(a)

(b)

(c)

4.44.

(a) The second compound is expected to have a higher heat of combustion because it has more carbon atoms, and thus more C–C and C–H bonds that can break (and release heat) upon combustion.

(b) The first compound is expected to have a higher heat of combustion because it cannot adopt a chair conformation in which both methyl groups occupy equatorial positions.

(c) The second compound is expected to have a higher heat of combustion because it cannot adopt a chair conformation in which both methyl groups occupy equatorial positions.

(d) The first compound is expected to have a higher heat of combustion because it cannot adopt a chair conformation in which both methyl groups occupy equatorial positions.

4.45. The energy diagram of 1,2-dichloroethane is similar to the energy diagram of butane (see Figure 4.11). The CH_3 groups have simply been replaced with chlorine atoms.

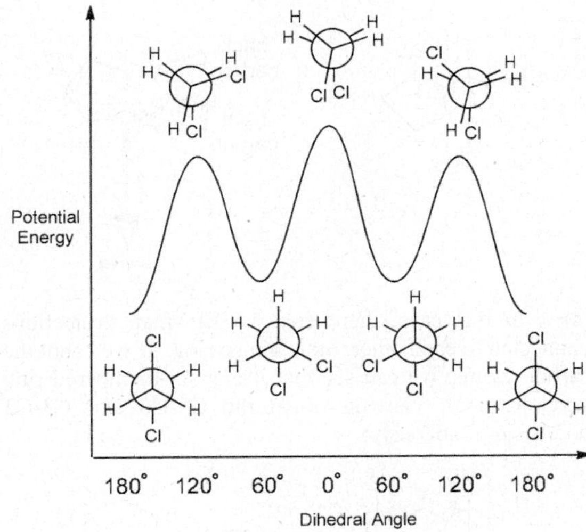

4.46. There are eight hydrogen atoms in axial positions (in bold) and seven hydrogen atoms in equatorial positions.

4.47.

(a) The Newman projection indicates that the front carbon atom is connected to two methyl groups, while the back carbon atom is connected to two ethyl groups and a methyl group. This corresponds with the following bond-line drawing:

(b) The Newman projection indicates that the front carbon atom and the back carbon atom are part of a five-membered ring:

The front carbon atom is connected to a methyl group pointing above the ring, while the back carbon atom is connected to a methyl group pointing below the ring. This corresponds with the following bond-line drawing, with *trans* methyl groups on adjacent carbon atoms:

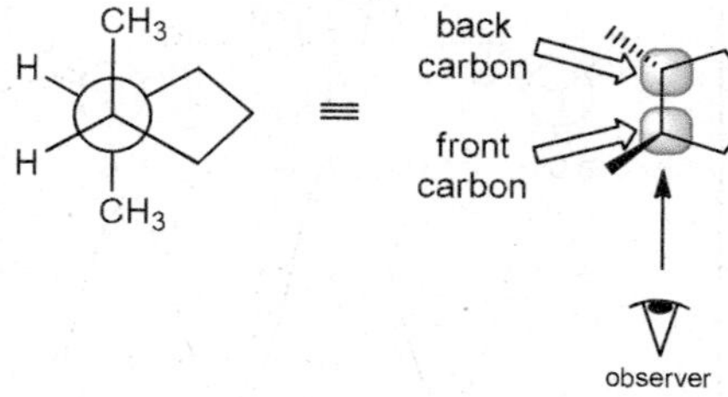

(c) In this case, there are two Newman projections connected to each other, indicating a ring. If we count the carbon atoms, we can see that it is a six-membered ring (cyclohexane), viewing down the C5-C6 and C3-C2 bonds simultaneously:

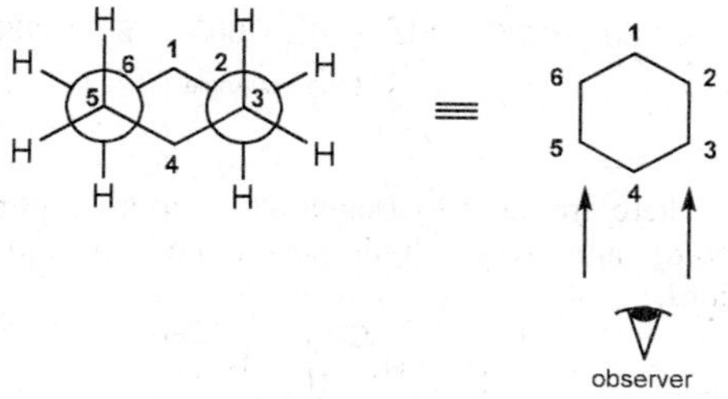

4.48.

(a) We first convert the Newman projection into a bond-line drawing, because it is easier to assign a systematic name to a bond-line drawing. In this case, the compound is hexane:

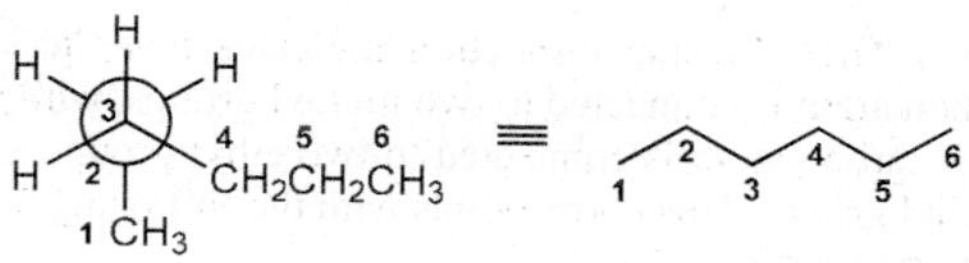

(b) We first convert the Newman projection into a bond-line drawing, because it is easier to assign a systematic name to a bond-line drawing. In this case, the compound is methylcyclohexane:

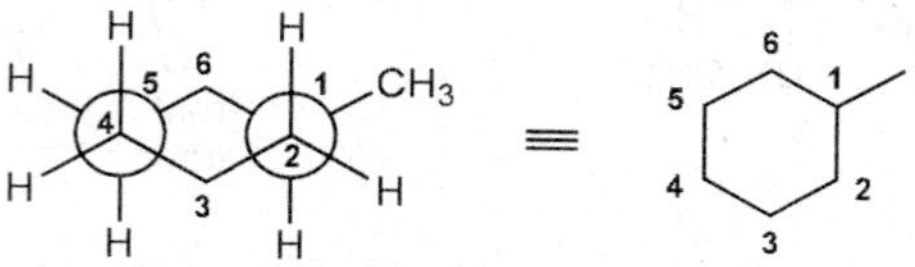

(c) We first convert the Newman projection into a bond-line drawing, because it is easier to assign a systematic name to a bond-line drawing. In this case, the compound is methylcyclopentane:

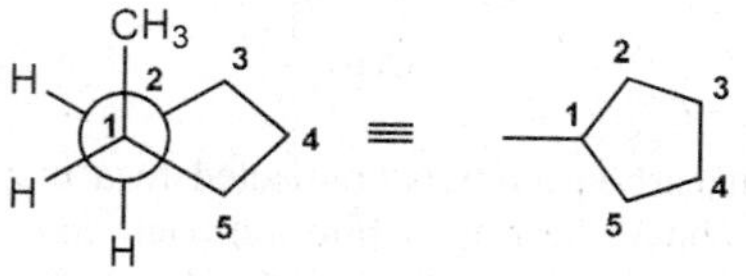

(d) As seen in the solution to Problem **4.47b**, this compound is *trans*-1,2-dimethylcyclopentane:

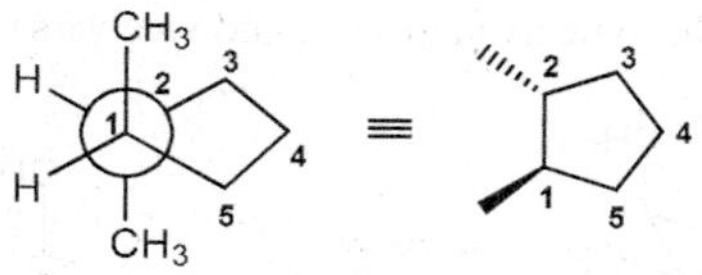

4.49. In the eclipsed conformation of bromoethane, each H/H eclipsing interaction is 4 kJ/mol, and there are two of them (for a total of 8 kJ/mol). The remaining energy cost is associated with the Br/H eclipsing interaction: 15 – 8 = 7 kJ/mol.

4.50. In order to draw the first chair conformation, begin by assigning a numbering system, and then determine the location and three-dimensional orientation of each substituent:

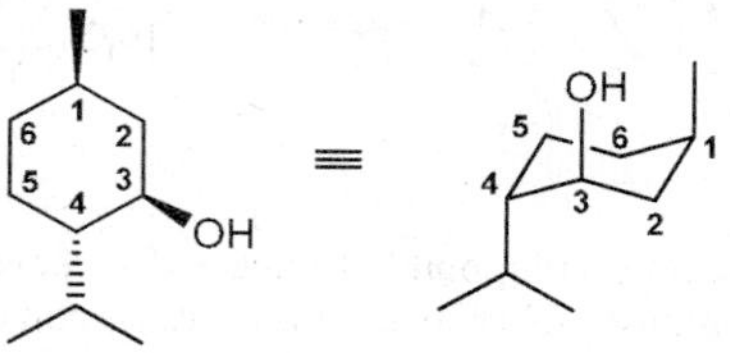

Note that the numbering system need not adhere to IUPAC rules – the numbering system is simply a tool that we are using to guide us. When assigning the numbers to the chair drawing, the first number can be placed anywhere on the ring (as long as the numbers go clockwise), as seen in the solution to Problem **4.28a**.

Once we have drawn the first chair conformation, we then draw the second chair conformation, again using the same numbering system. After doing a chair flip, the methyl group is still up, the OH group is still up, and the isopropyl group is still down. Notice that a ring flip causes the three axial groups to become equatorial.

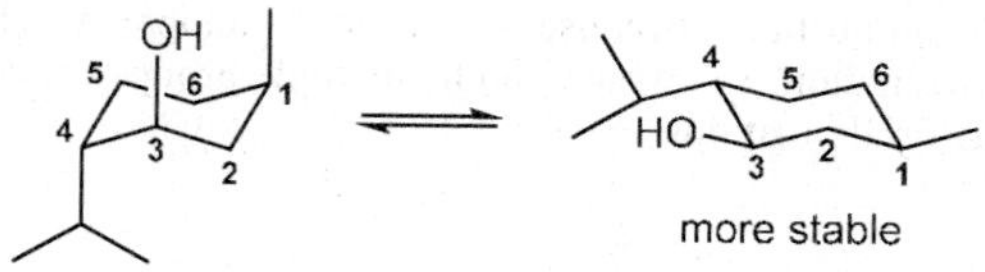

Finally, we compare both chair drawings and determine which one has fewer or less severe 1,3-diaxial interactions. In the first drawing, all three substituents occupy axial positions, but in the second drawing, they all occupy equatorial positions. As such, the second chair conformation is more stable.

4.51.

(a) The methyl group occupies an axial position in one chair conformation, and occupies an equatorial position in the other chair conformation. The second chair conformation is more stable because it lacks 1,3-diaxial interactions.

(b) Both isopropyl groups occupy axial positions in one chair conformation, and both occupy equatorial positions in the other chair conformation. The second chair conformation is more stable because it lacks 1,3-diaxial interactions.

more stable

(c) Both isopropyl groups occupy axial positions in one chair conformation, and both occupy equatorial positions in the other chair conformation. The second chair conformation is more stable because it lacks 1,3-diaxial interactions.

more stable

(d) Both isopropyl groups occupy axial positions in one chair conformation, and both occupy equatorial positions in the other chair conformation. The second chair conformation is more stable because it lacks 1,3-diaxial interactions.

more stable

4.52.

(a) The compound below can adopt a chair conformation in which all three substituents occupy equatorial positions, as shown. Therefore, this compound is expected to be more stable.

(b) The compound below can adopt a chair conformation in which all three substituents occupy equatorial positions, as shown. Therefore, this compound is expected to be more stable.

(c) The compound below can adopt a chair conformation in which both substituents occupy equatorial positions, as shown. Therefore, this compound is expected to be more stable.

(d) The compound below can adopt a chair conformation in which all four substituents occupy equatorial positions, as shown. Therefore, this compound is expected to be more stable.

4.53. When looking from the perspective of the observer (as shown in the problem statement), the front carbon has a chlorine atom pointing up and to the right, a bromine atom pointing up and the left, and a methyl group pointing down. The back carbon atom has a chlorine atom pointing down and to the right, a bromine atom pointing down and to the left, as well as a methyl group pointing up.

observer Cl Br Cl Br ≡ CH_3 Br Cl Br Cl CH_3

4.54. Two chair conformations can be drawn. In one of these conformations, all substituents occupy equatorial positions. In the other conformation, all substituents occupy axial positions. The all-equatorial conformation lacks 1,3-diaxial interactions, and is therefore the most stable conformation of glucose:

HO HO HO O OH OH

4.55. Begin by drawing a Newman projection:

CH_3 H_3C CH_3 H_3C CH_3 CH_3

2,2,3,3-tetramethylbutane

All staggered conformations are degenerate, and the same is true for all eclipsed conformations. As such, the energy diagram has a shape that is similar to the energy diagram for the conformational analysis of ethane:

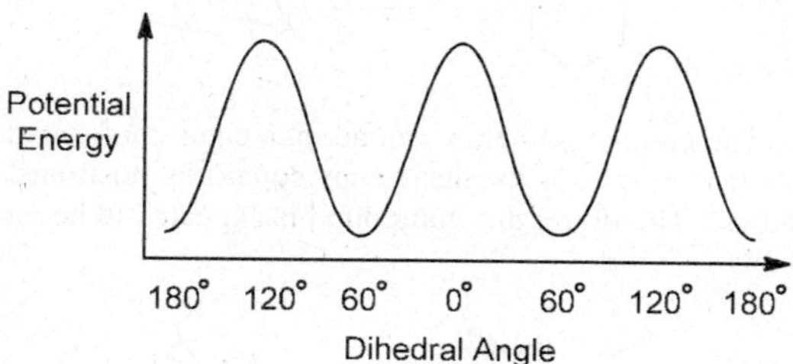

The staggered conformations have six gauche interactions, each of which has an energy cost of 3.8 kJ/mol. Therefore, each staggered conformation has an energy cost of 22.8 kJ/mol. The eclipsed conformations have three methyl-methyl eclipsing interactions, each of which has an energy cost of 11 kJ/mol. Therefore, each eclipsed conformation has an energy cost of 33 kJ/mol. The difference in energy between staggered and eclipsed conformations is therefore expected to be approximately 10.2 kJ/mol.

4.56. The two staggered conformations are lower in energy than the two eclipsed conformations. Among the staggered conformations, the *anti* conformation is the lowest in energy. Among the eclipsed conformations, the highest energy conformation is the one in which the bromine atoms are eclipsing each other. This information is summarized here:

Increasing energy

most stable — least stable

4.57.
(a) This conformation has three Me/Me gauche interactions, each of which has an energy cost of 3.8 kJ/mol. Therefore, this conformation has a total energy cost of 11.4 kJ/mol associated with steric strain.

(b) This conformation has two Me/H eclipsing interactions, each of which has an energy cost of 6 kJ/mol. In addition, it also has one Me/Me eclipsing interaction, which has an energy cost of 11 kJ/mol. Therefore, this conformation has a total energy cost of 23 kJ/mol associated with torsional strain and steric strain.

4.58. There are two chair conformations that can be drawn. In one chair conformation, all groups are equatorial except for one. In the other chair conformation, all groups are axial except for one. The conformation with only one axial group is more stable because it has fewer 1,3-diaxial interactions.

4.59.
(a) A group at C2, pointing up, will occupy an equatorial position, as seen here:

(b) A group at C3, pointing down, will occupy an equatorial position, as seen here:

(c) A group at C4, pointing down, will occupy an axial position, as seen here:

(d) A group at C7, pointing down, will occupy an equatorial position, as seen here:

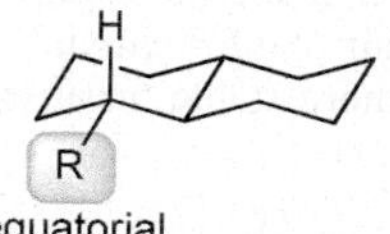

(e) A group at C8, pointing up, will occupy an equatorial position, as seen here:

(f) A group at C9, pointing up, will occupy an axial position, as seen here:

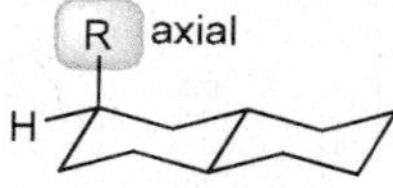

4.60. We are looking for constitutional isomers of C_8H_{18} that have a parent chain of seven carbon atoms (heptane). The extra CH_3 group can be placed in any of three positions (C2, C3, or C4) giving the following three constitutional isomers:

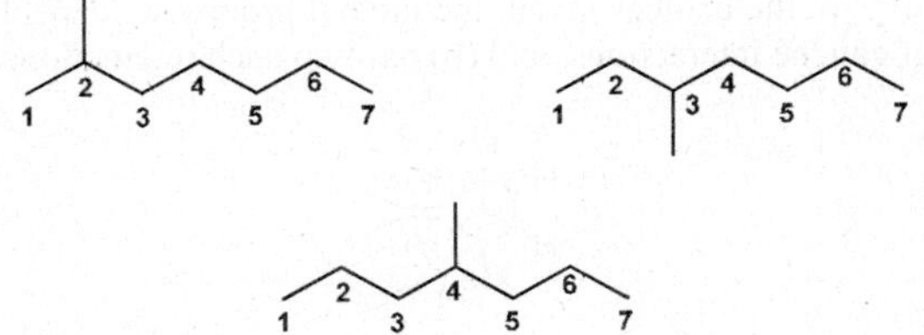

We cannot connect the CH_3 group to positions C1 or C7, as that would give a parent chain of eight carbon atoms (octane). We also cannot connect the CH_3 group to position C5 as that would generate the same structure as placing the CH_3 group at the C3 position:

Similarly, we cannot connect the CH_3 group to position C6 as that would generate the same structure as placing the CH_3 group at the C2 position. In summary, there are only three constitutional isomers of C_8H_{18} that have a parent name of heptane (shown above).

4.61. The correct answer is (d). Compound (a) is *cis*-1,2-dimethylcyclohexane, which can be seen more clearly when the hydrogen atoms are drawn, and it is clear that both methyl groups are DOWN:

Compound (b) is a Haworth projection of *cis*-1,2-dimethylcyclohexane.
Compound (c) is also *cis*-1,2-dimethylcyclohexane:

Compound (d) is the correct answer, because this compound is NOT *cis*-1,2-dimethylcyclohexane. Rather, it is *trans*-1,2-dimethylcyclohexane, which can be seen more clearly when the hydrogen atoms are drawn:

4.62. The correct answer is (b). As shown below, compound (a) has a five-carbon parent chain, with a methyl group (highlighted) connected to C2. Therefore, this compound is 2-methylpentane:

2-methylpentane

Compound (c) also has a five-carbon parent chain, with a methyl group connected to C2, so this compound is also 2-methylpentane:

2-methylpentane

Similarly, compound (d) also has a five-carbon parent chain, with a methyl group connected to C2. Therefore, this compound is also 2-methylpentane:

2-methylpentane

The correct answer is compound (b). If we assign numbers to the carbon atoms in compound (b), we find that compound (b) is not 2-methylpentane. Indeed, it is 2,3-dimethylbutane:

2,3-dimethylbutane

4.63. The correct answer is (c). The compound with the largest heat of combustion will be the compound that is highest in energy. All of the compounds are cycloalkanes with the molecular formula C_5H_{10}, but they differ in the size of the rings. Three-membered rings have more ring strain (higher energy), so compounds (a) and (b) are not the correct answers. The answer must be compound (c) or (d), both of which have a highly strained, cyclopropane ring. The difference between them is the relative orientation of the methyl groups. The *cis* isomer is expected to have more steric strain than the *trans* isomer, because the methyl groups are necessarily eclipsed (or close to being so), so compound (c) is the highest energy isomer. Compound (c) is therefore the correct answer.

4.64. The correct answer is (d). Heptane has the molecular formula C_7H_{16}, so we are looking for another compound with the same molecular formula. Compound (a) does not have the correct molecular formula (it is C_7H_{14}, rather than C_7H_{16}) so it is not a constitutional isomer of heptane. Similarly, compound (b) also has the molecular formula C_7H_{14}, so it is also not a constitutional isomer of heptane. Compound (c) has eight carbon atoms, so it is not a constitutional isomer of heptane. Compound (d) is the correct answer, since it has the molecular formula C_7H_{16}.

4.65. The correct answer is (d). The parent chain has six carbon atoms, so the parent will be hexane. This rules out two of the choices (with a parent of pentane). When numbering the parent chain, we number in the direction that gives the lower number to the first substituent. So the systematic name of the given compound is (d) 2,3-dimethylhexane.

2,3-dimethylhexane

4.66. The correct answer is (c). The correct drawing of *cis*-1,4-dimethylcyclohexane has two methyl groups in the proper positions (1,4), and with a *cis* relationship (both up or both down). Only one drawing (c) represents *cis*-1,4-dimethylcyclohexane [note that drawing (a) is actually *trans*-1,4-dimethylcyclohexane].
The conformation given is the most stable, with one substituent occupying an axial position and the other occupying an equatorial position:

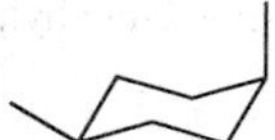

The other chair would have the exact same 1,3-diaxial interactions (with one substituent occupying an axial position and the other occupying an equatorial position):

4.67. The correct answer is (b). A gauche interaction is a steric interaction that results when two groups in a Newman projection are separated by a dihedral angle of 60°. In the choices given, the methyl groups are capable of gauche interactions, and (b) has two such interactions:

4.68. The correct answer is (c). Three of the answer choices are staggered conformations, and we must look for the lowest energy (most stable) among them:

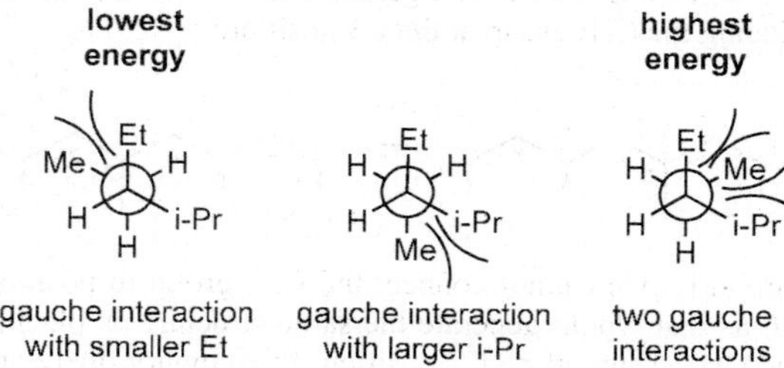

The most stable conformation is the staggered conformation with one gauche interaction between the methyl and ethyl groups. The correct answer is the first drawing above, corresponding with choice (c). This conformation is the lowest in energy. The conformation with one gauche interaction between the methyl group and the larger isopropyl groups has larger steric interactions and is higher in energy, and the conformation with two gauche interactions is the highest energy staggered conformation given.
The remaining (fourth) conformation is eclipsed, and is therefore the highest energy (least stable) of all given choices.

4.69. The correct answer is (a). The *tert*-butyl group is a complex substituent, and it can be viewed as an ethyl group with two methyl groups (highlighted below) connected to C1 of the ethyl group. The systematic name for the *tert*-butyl group is (1,1-dimethylethyl):

(1,1-dimethylethyl)

4.70. There are five highlighted groups: two methyl groups, one cyclohexyl group and two complex substituents. For each of the complex substituents, we treat it as a "substituent on a substituent," and we assign a name based on numbers going away from the parent, as shown:

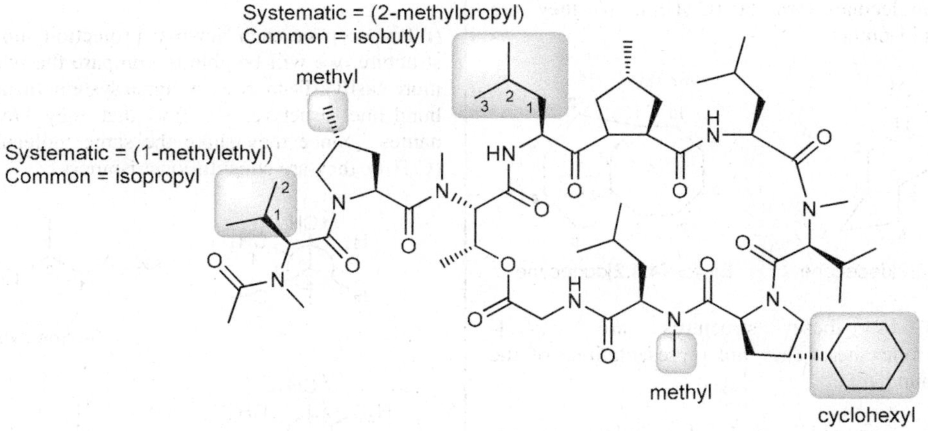

4.71. As mentioned in Section 4.9, cyclobutane adopts a slightly puckered conformation in order to alleviate some of the torsional strain associated with the eclipsing hydrogen atoms:

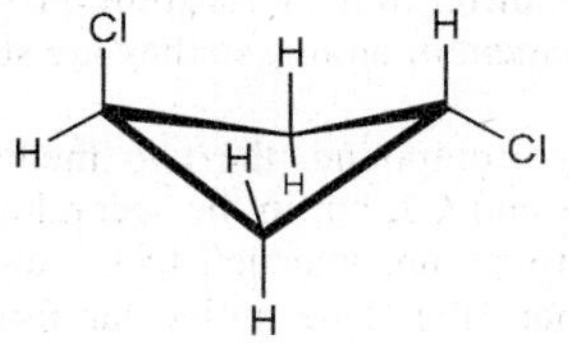

In this non-planar conformation, the individual dipole moments of the C-Cl bonds in *trans*-1,3-dichlorocyclobutane do not fully cancel each other, giving rise to a small molecular dipole moment.

4.72.

(a) If we convert each Newman projection into a bond-line structure, we will be able to compare the two structures more easily.

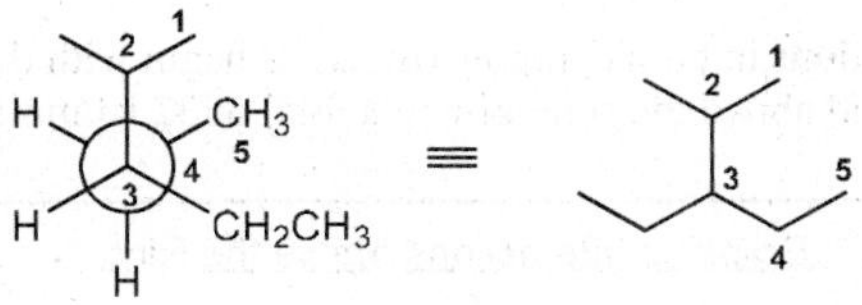

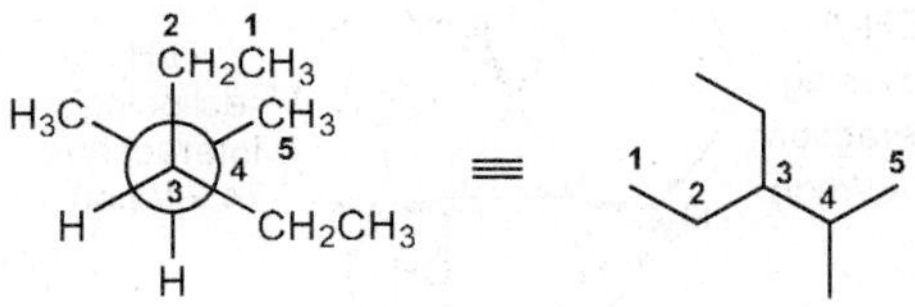

Now let's compare these bond-line drawings by naming them (using IUPAC rules). Note that in the second bond-line drawing above, the numbers shown do *not* adhere to IUPAC rules. This was OK when we were using the numbers as a tool for redrawing the structures in bond-line format, but it is not OK to use the wrong numbers when naming structures. To name these structures, we must use IUPAC rules, and we find that these two structures have the same name:

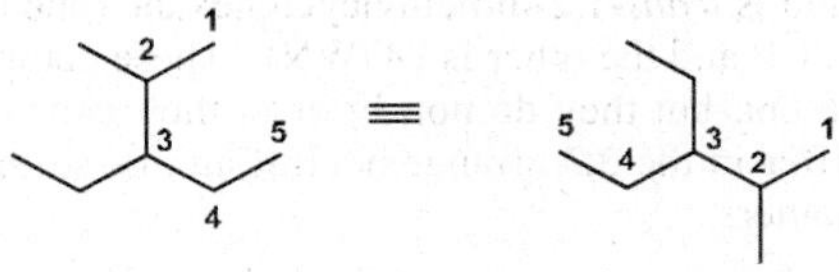

3-ethyl-2-methyl-pentane   3-ethyl-2-methyl-pentane

Therefore, these two drawings represent the same compound.

(b) In the first compound, the two methyl groups are attached to C1 and C2, but in the second compound, the two methyl groups are attached to C1 and C3. These compounds share the same molecular formula (C_8H_{16}), but they have different constitution (connectivity of atoms). Therefore, these two compounds are constitutional isomers.

(c) These two structures are both representations of bicyclo[2.2.1]heptane, as can be seen when we apply the numbering system below. Therefore, these two drawings represent the same compound.

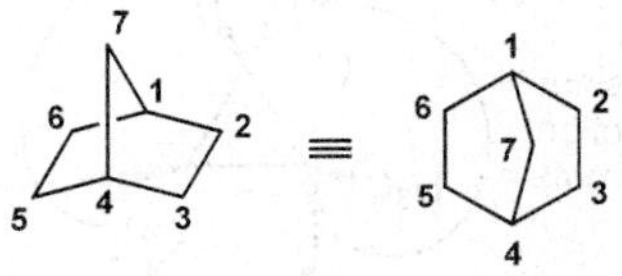

(d) We assign a systematic name to each structure, and find that they have different names (because they have different ring fusions, shown below). Therefore, they differ in their constitution (connectivity of atoms). These compounds are different from each other, but they share the same molecular formula ($C_{12}H_{22}$), so they are constitutional isomers.

Bicyclo[5.3.2]dodecane   Bicyclo[4.4.2]dodecane

(e) Both of these structures are *cis*-1,4-dimethylcyclohexane. They are representations of the same compound.

(f) The first compound is *cis*-1,4-dimethylcyclohexane, and the second compound is *trans*-1,4-dimethylcyclohexane. These compounds are different, but they do not differ in their constitution. They differ in the 3D arrangement of atoms, so they are stereoisomers.

(g) The first compound is *cis*-1,2-dimethylcyclohexane (both methyl groups are in UP positions), and the second compound is *trans*-1,2-dimethylcyclohexane (one methyl group is UP and the other is DOWN). These compounds are different, but they do not differ in their constitution. They differ in the 3D arrangement of atoms, so they are stereoisomers.

(h) The first compound is *cis*-1,2-dimethylcyclohexane (both methyl groups are in UP positions), and the second compound is *trans*-1,2-dimethylcyclohexane (one methyl group is UP and the other is DOWN). These compounds are different, but they do not differ in their constitution. They differ in the 3D arrangement of atoms, so they are stereoisomers.

(i) If we convert each Newman projection into a bond-line structure, we will be able to compare the two structures more easily. Then, if we assign a systematic name to each bond-line structure, we find that they have different names. Since they share the same molecular formula (C_6H_{14}), they are constitutional isomers.

2,3-dimethylbutane

2,2-dimethylbutane

(j) The first compound is *cis*-1,3-dimethylcyclohexane, and the second compound is *trans*-1,3-dimethylcyclohexane. These compounds are different, but they do not differ in their constitution. They differ in the 3D arrangement of atoms, so they are stereoisomers.

(k) In the first compound, the two methyl groups are attached to C1 and C3, but in the second compound, the two methyl groups are attached to C1 and C2. These compounds share the same molecular formula (C_8H_{16}), but they have different constitution (connectivity of atoms). Therefore, these two compounds are constitutional isomers.

4.73.

(a) The *trans* isomer is expected to be more stable, because the *cis* isomer has a very high-energy methyl-methyl eclipsing interaction (11 kJ/mol). See calculation below.

(b) We calculate the energy cost associated with all eclipsing interactions in both compounds. Let's begin with the *trans* isomer. It has the following eclipsing interactions, below the ring and above the ring, giving a total of 32 kJ/mol:

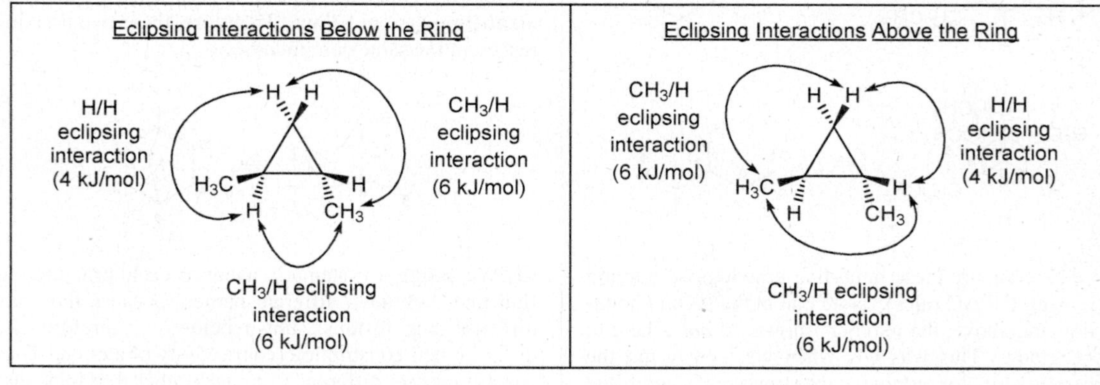

Now let's focus on the *cis* isomer. It has the following eclipsing interactions, below the ring and above the ring, giving a total of 35 kJ/mol:

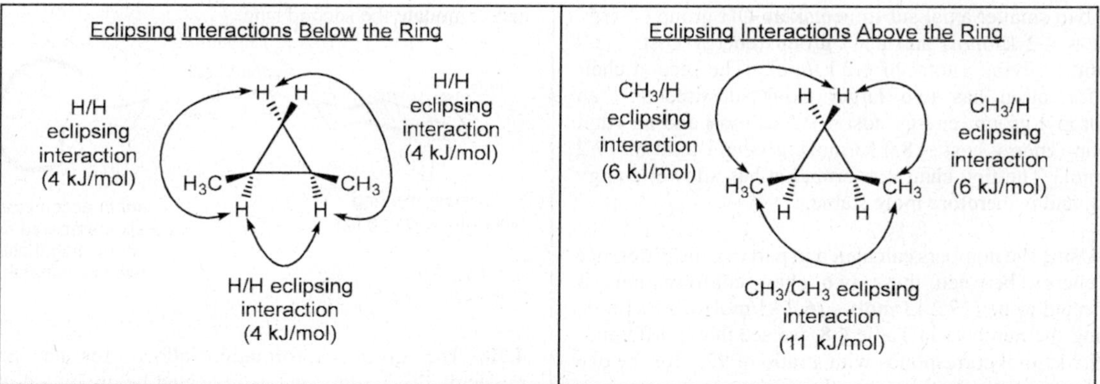

The difference between these two isomers is therefore predicted to be (35 kJ/mol) – (32 kJ/mol) = 3 kJ/mol.

4.74. The gauche conformations are capable of intramolecular hydrogen bonding, as shown below. The *anti* conformation lacks this stabilizing effect, because the two OH groups are too far apart to form an intramolecular hydrogen bond.

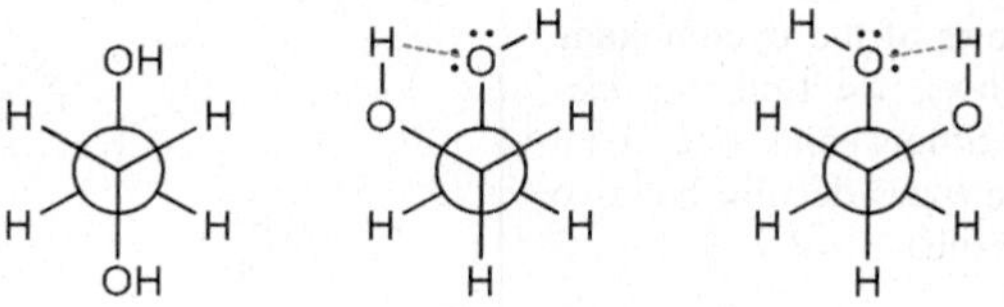

Anti Gauche Gauche

4.75. If we convert each Newman projection into a bond-line structure, we see that these structures are representations of the same compound (2,3-dimethylbutane). They are not constitutional isomers.

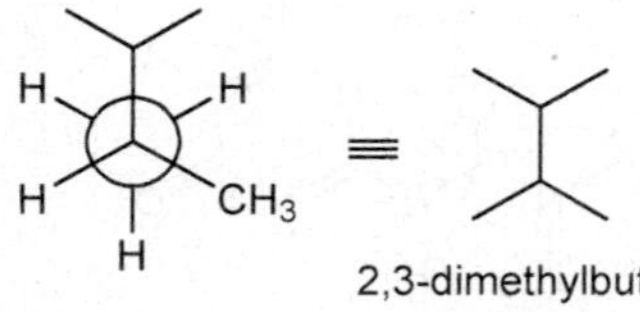

2,3-dimethylbutane

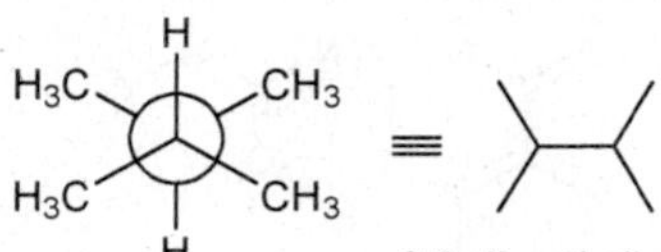

2,3-dimethylbutane

4.76.

(a) Begin by assigning a numbering system, and then determine the location and three-dimensional orientation of each substituent:

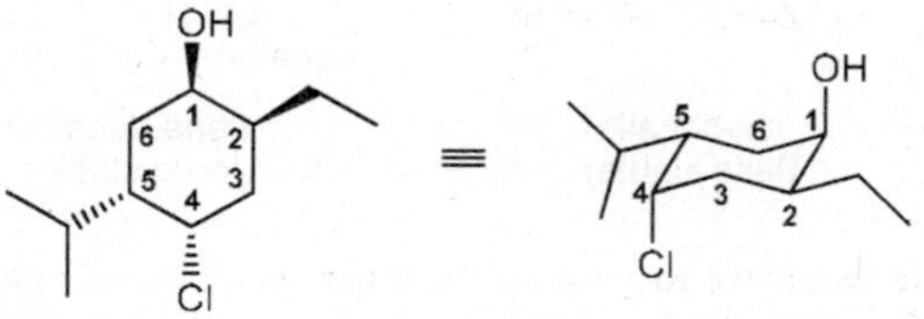

When assigning the numbers to the chair drawing, the first number can be placed anywhere on the ring (as long as the numbers go clockwise), as explained in the solution to Problem **4.28a**.

Once we have drawn the first chair conformation, we then draw the second chair conformation, again using a numbering system. After doing a chair flip, the OH and ethyl groups are still up, and the chloro and isopropyl groups are still down. Notice that a ring flip causes all equatorial groups to become axial groups, and vice versa.

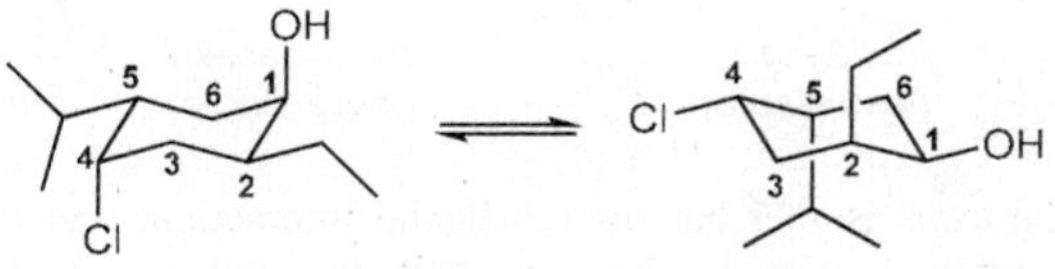

more stable

(b) Comparison of these chair conformations requires a comparison of the energy costs associated with all axial substituents (see Table 4.8). The first chair conformation has two smaller axial substituents: an OH group (energy cost = 4.2 kJ/mol) and a Cl group (energy cost = 2.0 kJ/mol), giving a total of 6.2 kJ/mol. The second chair conformation has two larger axial substituents: an isopropyl group (energy cost = 9.2 kJ/mol) and an ethyl group (energy cost = 8.0 kJ/mol), giving a total of 17.2 kJ/mol. The first chair conformation has a lower energy cost, and is therefore more stable.

(c) Using the numbers calculated in part (b), the difference in energy between these two chair conformations is expected to be (17.2 kJ/mol) – (6.2 kJ/mol) = 11 kJ/mol. Using the numbers in Table 4.8, we see that a difference of 9.2 kJ/mol corresponds with a ratio of 97:3 for the two conformations. In this case, the difference in energy is more than 9.2 kJ/mol, so the ratio should be even higher (more than 97%). Therefore, we do expect the compound to spend more than 95% of its time in the more stable chair conformation.

4.77.

(a) If we focus our attention on one of the cyclohexane rings in decalin (highlighted below), we find that *cis*-decalin has one equatorial substituent and one axial substituent (shown in bold), while *trans*-decalin has two equatorial substituents (shown in bold).

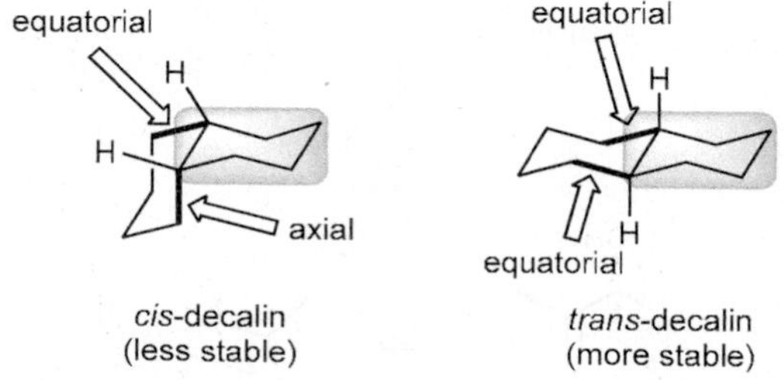

cis-decalin (less stable)   *trans*-decalin (more stable)

Note that if we focused on the other cyclohexane ring in each isomer (highlighted below), we would get the same result. Specifically, we see that *cis*-decalin has one equatorial substituent and one axial substituent (shown in bold), while *trans*-decalin has two equatorial substituents (shown in bold).

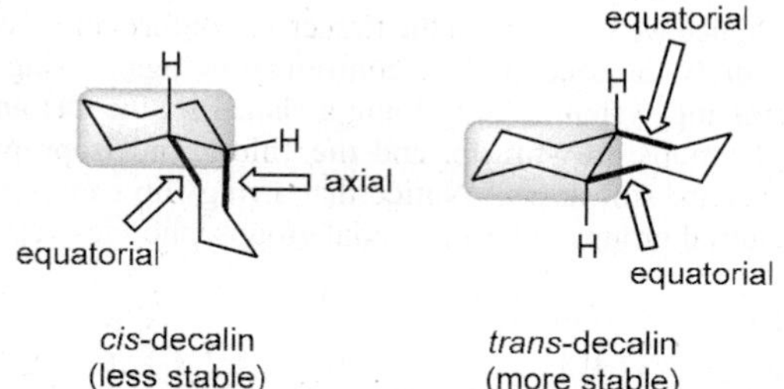

cis-decalin (less stable)   *trans*-decalin (more stable)

The *trans* isomer has no 1,3-diaxial interactions and is, therefore, expected to be more stable than the *cis* isomer.

(b) *trans*-Decalin is incapable of ring flipping, because a ring flip of one ring would cause its two alkyl substituents (which comprise the second ring) to be too far apart to accommodate the second ring.

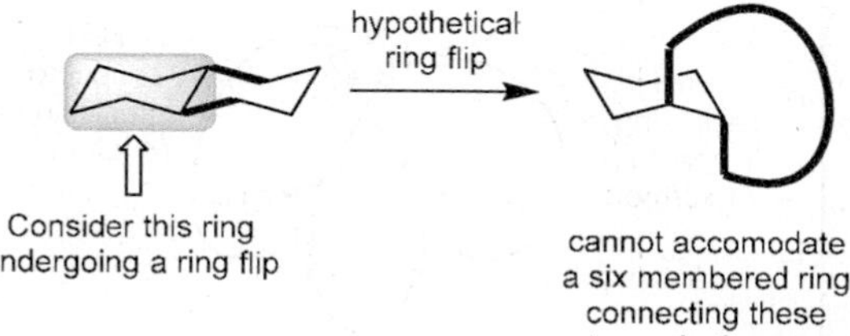

4.78. The given conformation allows for an intramolecular hydrogen bond as indicated below by a dotted line. In this conformation, there are two hydrogen-bond acceptors ("A") and two hydrogen bond donors ("D") along the bottom edge of the molecule, as drawn.

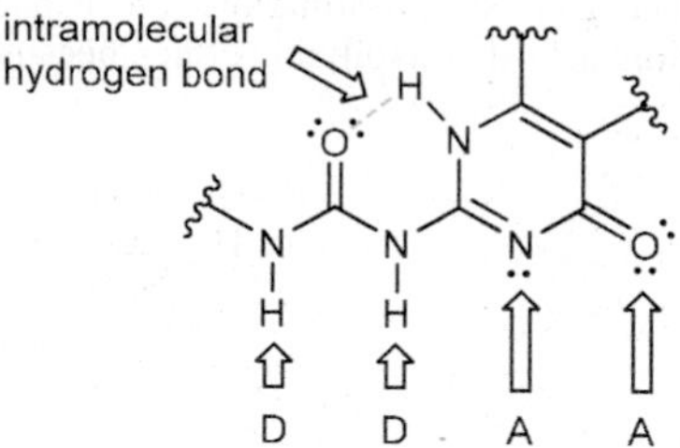

Next, to show the four intermolecular hydrogen bonds, we draw a second molecule rotated by 180° relative to the first. In this orientation the donor-donor-acceptor-acceptor pattern of the top molecule is complementary to the acceptor-acceptor-donor-donor motif on the bottom molecule, resulting in a bimolecular complex with the four intermolecular hydrogen bonds shown below.

4.79.

(a) The two substituents at the bridgehead carbons are *cis* to each other, analogous to the two bridgehead hydrogen atoms on *cis*-decalin.

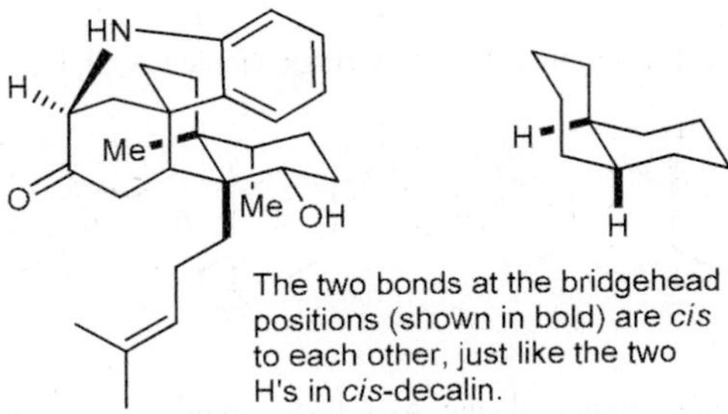
HN
H
Me
O
Me
OH
H
H
The two bonds at the bridgehead positions (shown in bold) are *cis* to each other, just like the two H's in *cis*-decalin.

(b) In order to clearly identify the position of the aromatic ring, we rotate the molecule counter-clockwise so that the 6-membered ring shown in bold is oriented as shown below. In this orientation, it is clear that the aromatic ring (highlighted) is in an axial position:

(c) The branch is *equatorial* to the left chair,

equatorial relative to the left chair

but *axial* to the right chair (note, this structure has been rotated to see the other chair more clearly):

axial relative to the right chair

(d) The dihedral angle between the two methyl groups should be approximately 60° (From the perspective of the chair on the right, one is equatorial down, and the other is axial down. They are gauche to each other.)

equatorial down (relative to the right chair)

axial down

4.80. The three staggered conformations are as follows:

A B C

Several types of gauche interactions are present in these conformers (Me/Me, Me/OH, and/or OH/OH). The strain of a methyl-methyl gauche interaction is 3.8 kJ/mol (see Table 4.6). The destabilization due to a Me/OH gauche interaction can be estimated as being roughly half of the value for the 1,3-diaxial interaction associated with an OH group (an OH group in an axial position experiences two gauche interactions, each of which might be expected to be somewhat similar to a Me/OH gauche interaction). Therefore, a Me/OH gauche interaction is expected to be approximately 4.2 / 2 = 2.1 kJ/mol (see Table 4.8).

Before exploring the OH/OH gauche interaction, our analysis thus far gives the following calculations for each conformer

Conformer **A**: 2 x 2.1 kJ/mol = 4.2 kJ/mol
Conformer **B**: 3.8 kJ/mol
Conformer **C**: 2 x 2.1 kJ/mol + 3.8 kJ/mol = 8.0 kJ/mol

This calculation must be modified when we take into account the effect of two OH groups that are gauche to each other, as seen in conformers **A** and **B**. We should expect an OH/OH gauche interaction to be less than a Me/Me gauche interaction (less than 3.8 kJ/mol), because an OH group appears to be less sterically encumbering than a methyl group (compare CH_3 and OH in Table 4.8). Therefore, the destabilizing effect associated with an OH/OH gauche interaction (less than 3.8kJ/mol) should be overshadowed by the *stabilizing* effect that results from the hydrogen bonding interactions between the two OH groups, which is expected to be approximately 20 kJ/mol (see section 1.12). As a result, we expect extra stabilization to be associated with any conformer in which two OH groups are gauche to each other. This occurs in conformers **A** and **B**, but the hydroxyl groups in **C** are too far to form this type of interaction. If we assume that the stabilization achieved through hydrogen bonding is the same for conformers **A** and **B**, then conformer **B** should be the lowest-energy staggered conformation for this compound.

4.81.

(a) The all-equatorial chair conformation of compound **2** experiences severe steric repulsion due to the nearby isopropyl groups as evident in the following drawing, in which we focus on the interactions between a pair of neighboring isopropyl groups. These interactions, highlighted below, occur for each pair of neighboring isopropyl groups.

A double Newman projection of **2**, shows how crowded the equatorial groups are as the methyl groups are in constant contact with each other. These interactions, highlighted below, occur for each pair of neighboring isopropyl groups.

(b) In the all-axial chair conformation for **2**, the C–H bond of each isopropyl group can all point towards each other (highlighted below) such that steric repulsion can be minimized. For clarity, this is shown only for the top face of the cyclohexane ring in the following structure.

In this conformation, the methyl groups shown above avoid steric strain. In the all-equatorial chair conformation, it is not possible for the methyl groups to avoid steric strain entirely. In the all-equatorial chair conformation, each isopropyl group is in close proximity with two neighboring isopropyl groups (one on either adjacent carbon atom), and the hydrogen atom of each isopropyl group can only point to only one of its isopropyl neighbors, not both. So, in the all-equatorial chair conformation, the methyl groups of each isopropyl group will experience steric strain that cannot be avoided. In contrast, in the all-axial chair conformation, the isopropyl groups can achieve less steric strain (because neighboring isopropyl groups are pointing in opposite directions, relative to the ring, and are not close to one another).

4.82.

(a) Each compound has three rings, labeled A, B, and C.

In each compound, the A-B fusion represents a *trans*-decalin system:

The *trans* fusion imposes a severe restriction on the conformational flexibility of the system. Specifically, in a *trans*-decalin system, neither ring can undergo a ring-flip to give a different chair conformation. For each ring, the chair conformation shown above is the only chair conformation that is achievable. However, both rings are still free to adopt a higher energy boat conformation. For example, the B ring of *trans*-decalin can adopt the following boat conformation:

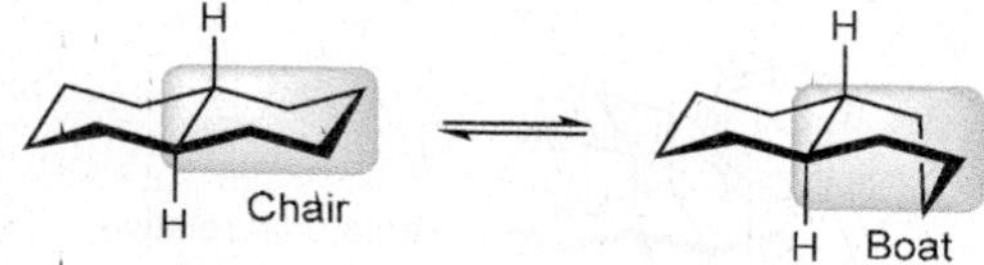

In compound **1**, the C ring imposes a further conformational restriction, by locking the B ring into a chair conformation:

In contrast, the C ring in compound **2** imposes the restriction of locking the B ring in a boat conformation:

(b) The boat conformation of compound **2** is expected to be higher in energy than the chair conformation of compound **1**. Therefore, compound **2** is expected to have the higher heat of combustion. In fact, the investigators prepared compounds **1** and **2** for the purpose of measuring the difference in energy between chair and boat conformations.

Chapter 5
Stereoisomerism

Review of Concepts

Fill in the blanks below. To verify that your answers are correct, look in your textbook at the end of Chapter 5. Each of the sentences below appears verbatim in the section entitled *Review of Concepts and Vocabulary.*

- _______ **isomers** have the same connectivity of atoms but differ in their spatial arrangement.
- **Chiral** objects are not **superimposable** on their ____________________. The most common source of molecular chirality is the presence of a ________________, a carbon atom bearing _______ different groups.
- A compound with one chiral center will have one nonsuperimposable mirror image, called its ________________.
- The Cahn-Ingold-Prelog system is used to assign the ______________ of a chiral center.
- A **polarimeter** is a device used to measure the ability of chiral organic compounds to rotate the plane of _____________________ light. Such compounds are said to be ____________ **active**.
- A solution containing equal amounts of both enantiomers is called a __________ **mixture**. A solution containing a pair of enantiomers in unequal amounts is described in terms of **enantiomeric** __________.
- For a compound with multiple chiral centers, a family of stereoisomers exists. Each stereoisomer will have at most one enantiomer, with the remaining members of the family being ______________.
- A ______ **compound** contains multiple chiral centers but is nevertheless achiral, because it possesses reflectional symmetry.
- __________ **projections** are drawings that convey the configuration of chiral centers, without the use of wedges and dashes.
- Compounds that contain two adjacent C=C bonds are called ____________, and they are another common class of compounds that can be chiral despite the absence of a chiral center.
- The stereodescriptors *cis* and *trans* are generally reserved for alkenes that are disubstituted. For trisubstituted and tetrasubstituted alkenes, the stereodescriptors ____ and ____ must be used. ____ indicates priority groups on the same side, while ____ indicates priority groups on opposite sides.

Review of Skills

Fill in the blanks and empty boxes below. To verify that your answers are correct, look in your textbook at the end of Chapter 5. The answers appear in the section entitled *SkillBuilder Review.*

SkillBuilder 5.1 Locating Chiral Centers

Circle the chiral center in the following compound

SkillBuilder 5.2 Drawing an Enantiomer

Show three ways to draw the enantiomer of the following compound. Place your answers in the boxes shown.

NH_2

SkillBuilder 5.3 Assigning the Configuration of a Chiral Center

Assign the configuration of the chiral center (*R* or *S*) in the following compound.

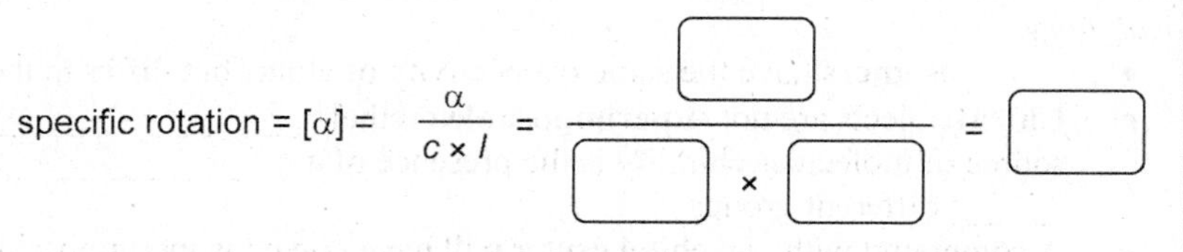

SkillBuilder 5.4 Calculating Specific Rotation

Calculate the specific rotation given the following information:

0.300 grams sucrose dissolved in 10.0 mL of water
sample cell = 10.0 cm
observed rotation = +1.99°

specific rotation = $[\alpha] = \frac{\alpha}{c \times l} = \frac{\square}{\square \times \square} = \square$

SkillBuilder 5.5 Calculating % *ee*

Calculate the enantiomeric excess given the following information:

The specific rotation of optically pure adrenaline is −53. A mixture of (*R*)- and (*S*)-adrenaline was found to have a specific rotation of −45. Calculate the % *ee* of the mixture.

$\% \, ee = \frac{[\alpha] \text{ of mixture}}{[\alpha] \text{ of pure enantiomer}} \times 100\,\%$

$= \frac{\square}{\square} \times 100\,\% = \square$

SkillBuilder 5.6 Determining Stereoisomeric Relationship

Identify whether the given compounds are enantiomers or diastereomers:

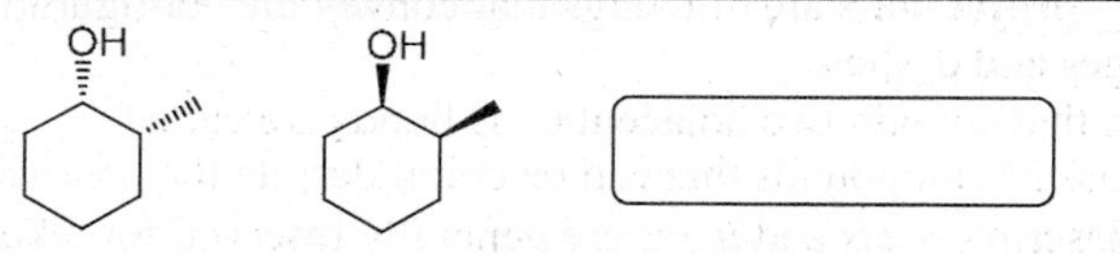

SkillBuilder 5.7 Identifying *Meso* Compounds

Draw all possible stereoisomers of 1,2-cyclohexanediol (shown left) and then look for a plane of symmetry in any of the drawings. The presence of a plane of symmetry indicates a *meso* compound.

□ + □ enantiomers

□ = □ **a *meso* compound**

SkillBuilder 5.8 Assigning Configuration from a Fischer projection

Assign the configuration of the chiral center in the following compound.

SkillBuilder 5.9 Assigning the Configuration of a Double Bond

Assign the configuration of the central double bond in the following compound.

Common Mistakes to Avoid

When drawing a chiral center, the four groups connected to the chiral center are typically drawn so that one group is on a wedge (which indicates that it is coming out of the page), and one group is on a dash (which indicates that it is going behind the page), and two groups are on straight lines (which indicate that these two groups are in the plane of the page), as seen in each of the following drawings:

Notice that in all of these cases, the two straight lines form a V, and neither the dash nor the wedge is placed inside that V. This is very important. If either the dash or the wedge is placed inside the V, the drawing becomes ambiguous and inaccurate. Avoid making this common mistake:

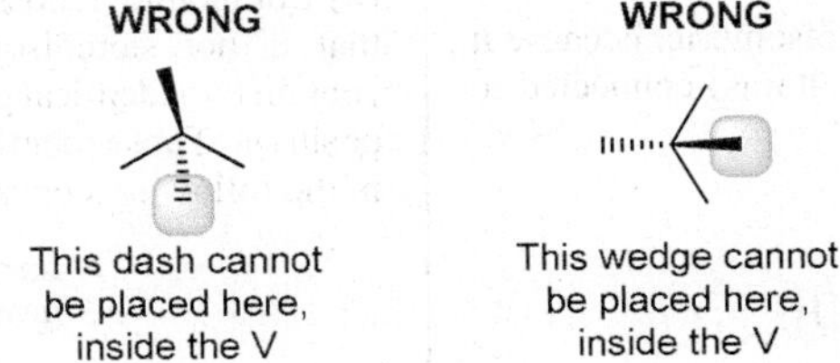

The drawings above do not make any sense, and if a chiral center is drawn like either of the drawings above, it would be impossible to assign a configuration to the chiral center. Never draw a chiral center that way. For the same reason, never draw a chiral center like this:

These two drawings imply square planar geometry, which is not the case for an sp^3 hybridized carbon atom (the geometry is tetrahedral). In some rare cases, you might find a chiral center for which three of the lines are drawn as straight lines, as in the following example:

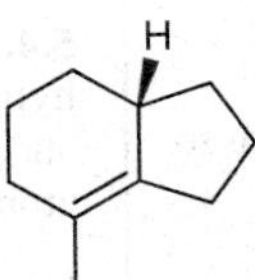

This compound has one chiral center, and its configuration is unambiguous (and therefore acceptable), although you will not encounter this convention often. In most cases that you will encounter in this course, a chiral center will be drawn as two lines (making a V), and one wedge and one dash that are both *outside* of the V:

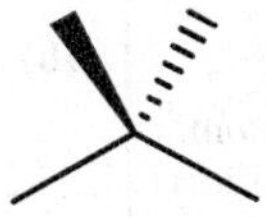

Solutions

5.1.
There are two C=C units in the ring (highlighted), and their configurations are shown below.

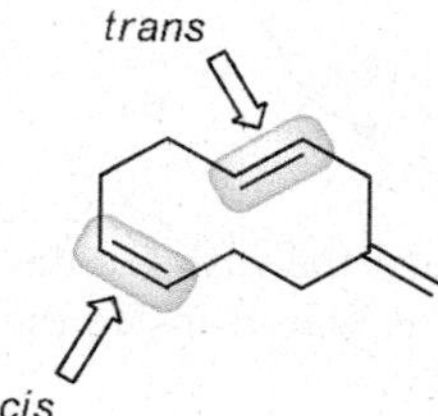

The remaining C=C unit is not stereoisomeric, because it has two identical groups (hydrogen atoms) connected to the same position.

5.2. We first draw a bond-line structure, which makes it easier to see the groups that are connected to each of the double bonds.

Each of the double bonds has two identical groups (hydrogen atoms) connected to the same position.

As such, neither double bond exhibits stereoisomerism, so this compound does not have any stereoisomers.

5.3.
(a) Compound X must contain a carbon-carbon double bond in the *trans* configuration, which accounts for four of the five carbon atoms:

Now we must decide where to place the fifth carbon atom. We cannot attach this carbon atom to a vinylic position (C2 or C3), as that would give a double bond that is not stereoisomeric, and compound X is supposed to have the *trans* configuration.

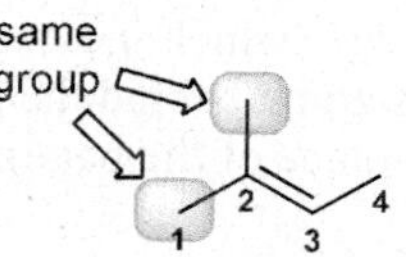

Therefore, we must attach the fifth carbon atom to an allylic position, giving the following compound:

(b) Compound Y possesses a carbon-carbon double bond that is not stereoisomeric, which means that it must contain two identical groups connected to the same vinylic position. Those identical groups can be methyl groups, as in the following compound,

or the identical groups can be hydrogen atoms, as in the following three compounds:

5.4. In each of the following cases, we ignore all sp^2-hybridized carbon atoms, all *sp*-hybridized carbon atoms, and all CH_2 and CH_3 groups. We identify those carbon atoms bearing four different groups (highlighted below):

(a) This compound has two chiral centers:

(b) This compound has five chiral centers:

(c) This compound has five chiral centers:

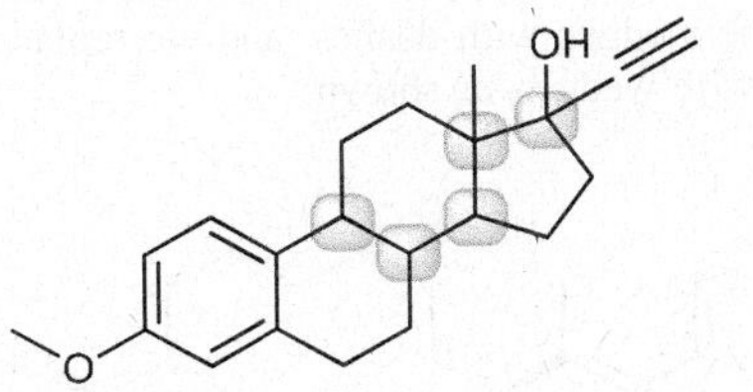

(d) This compound has only one chiral center:

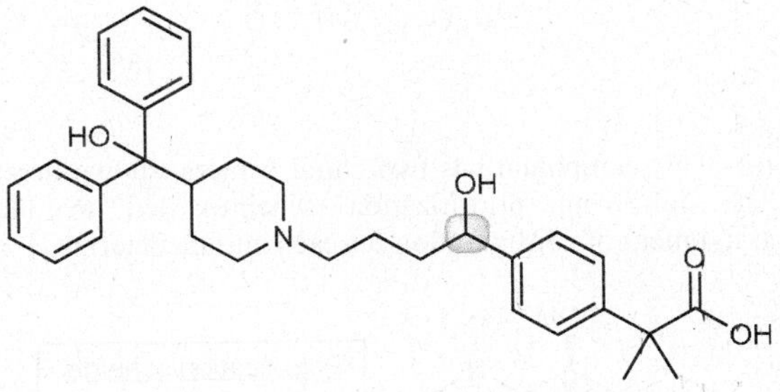

5.5. Recall that constitutional isomers are compounds that share the same molecular formula, but differ in constitution (the connectivity of atoms). There are two different ways that four carbon atoms can be connected together. They can be connected in a linear fashion (below left), or they can be connected with a branch (below right).

linear branched

For each of these skeletons, we must consider the different locations where a bromine atom can be placed. In the linear skeleton, the bromine atom can either be placed at C1 or at C2.

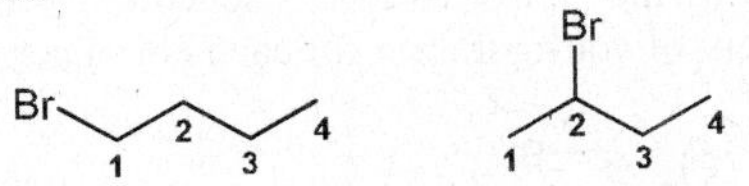

Placing the bromine atom at C3 is the same as placing it at C2:

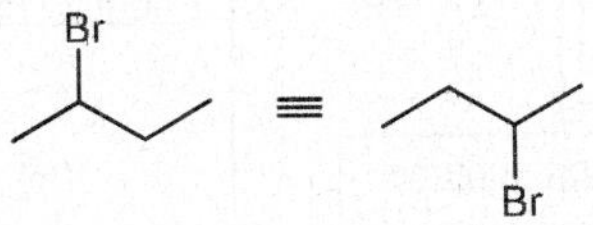

Similarly, placing the bromine atom at C4 is the same as placing it at C1.

Now let's consider the branched skeleton. There are two unique locations where the bromine atom can be placed (either at C1 or at C2).

Placing the bromine atom at C3 or C4 is the same as placing it at C1:

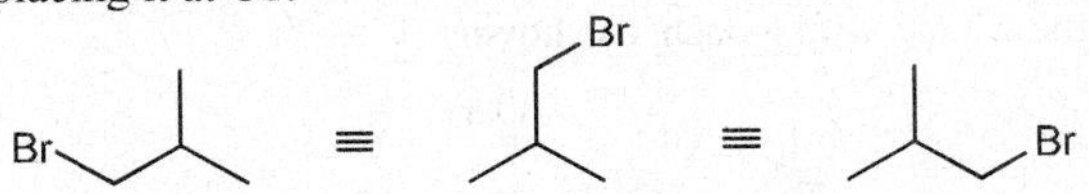

In summary, we have found four constitutional isomers with the molecular formula C_4H_9Br, shown again here:

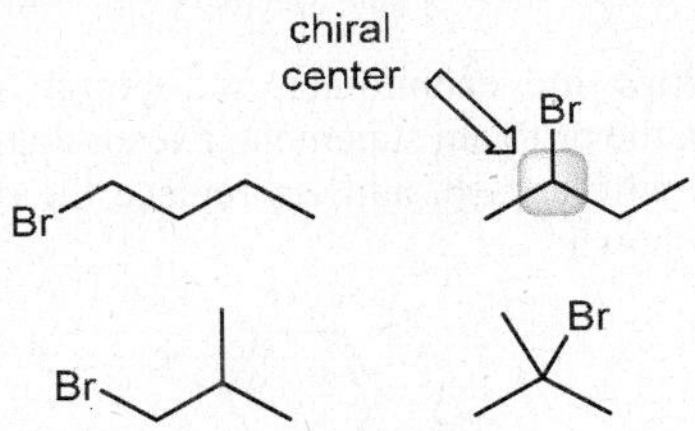

chiral center

Note that only one of these isomers exhibits a carbon atom that is connected to four different groups, which makes it a chiral center.

5.6. The phosphorus atom has four different groups attached to it (a methyl group, an ethyl group, a phenyl group, and a lone pair). This phosphorous atom therefore represents a chiral center. This compound is not superimposable on its mirror image (this can be seen more clearly by building and comparing molecular models).

5.7.

(a) To draw the enantiomer, we simply redraw the structure in the problem statement, except that we replace the wedge with a dash, as shown:

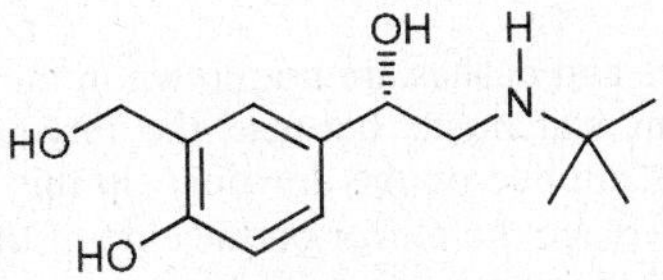

(b) To draw the enantiomer, we simply redraw the structure in the problem statement, except that we replace the wedge with a dash, as shown:

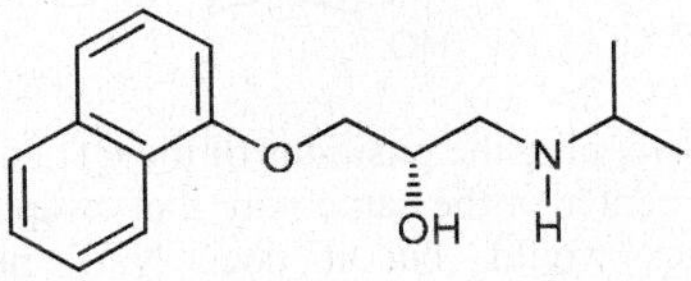

(c) To draw the enantiomer, we simply redraw the structure in the problem statement, except that we replace the wedge with a dash, and we replace the dash with a wedge, as shown:

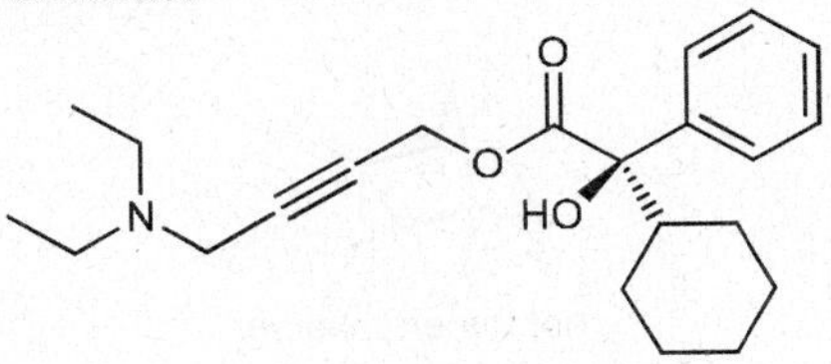

(d) To draw the enantiomer, we simply redraw the structure in the problem statement, except that we replace the wedge with a dash, as shown:

(e) To draw the enantiomer, we simply redraw the structure in the problem statement, except that we replace the wedge with a dash, and we replace the dash with a wedge, as shown:

(f) To draw the enantiomer, we simply redraw the structure in the problem statement, except that we replace the wedge with a dash, and we replace the dash with a wedge, as shown:

(g) Wedges and dashes are not drawn in the structure in the problem statement, because the three-dimensional geometry is implied by the drawing. In this case, it will be easier to place the mirror on the side of the molecule, giving the following structure for its enantiomer:

Note that swapping the positions of the OH and H inverts that chiral center in the same way that swapping dashes and wedges would, but it does NOT produce the enantiomer because the resulting drawing is not a mirror image of the given compound. This compound has three chiral centers (marked with * below), and all three would have to be modified to draw the enantiomer. Drawing the mirror image is the easiest method in this case.

not the enantiomer

5.8. To draw the enantiomer, we simply redraw the structure in the problem statement, except that we replace all of the wedges with dashes, and we replace all of the dashes with wedges, as shown:

5.9.

(a) This compound has two chiral centers, shown here. The following prioritization schemes led to the assignment of configuration for each chiral center.

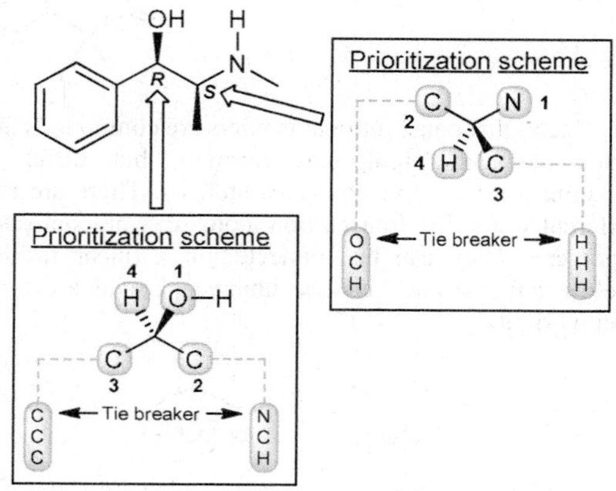

(b) This compound has two chiral centers, shown below. The following prioritization schemes led to the assignment of configuration for each chiral center.

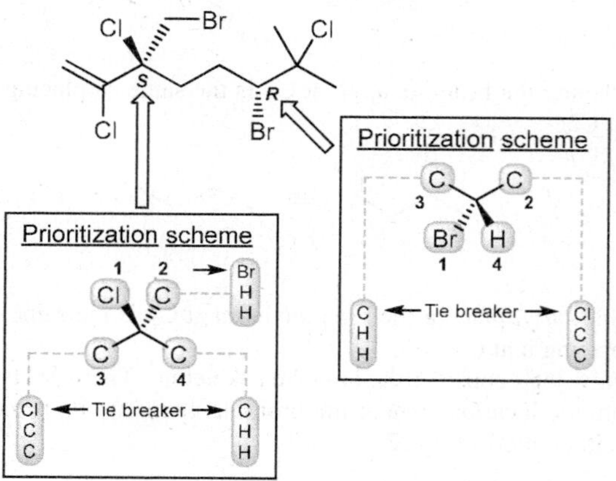

Note that in both cases the lowest priority group (4) is not on a dash, so your point of view must be adjusted before assigning the configuration (e.g., by switching groups).

(c) This compound has two chiral centers, shown below. The following prioritization schemes led to the assignment of configuration for each chiral center.

For the chiral center on the right, the lowest priority group (H) is not on a dash, so your point of view must be adjusted before assigning the configuration (e.g., by switching groups).

(d) This compound has three chiral centers. The following prioritization schemes led to the assignment of configuration for each chiral center.

For all three chiral centers, the lowest priority group (H) is not on a dash, so your point of view must be adjusted before assigning the configuration (e.g., by switching groups).

(e) This compound has four chiral centers, shown below. The following prioritization schemes led to the assignment of configuration for each chiral center.

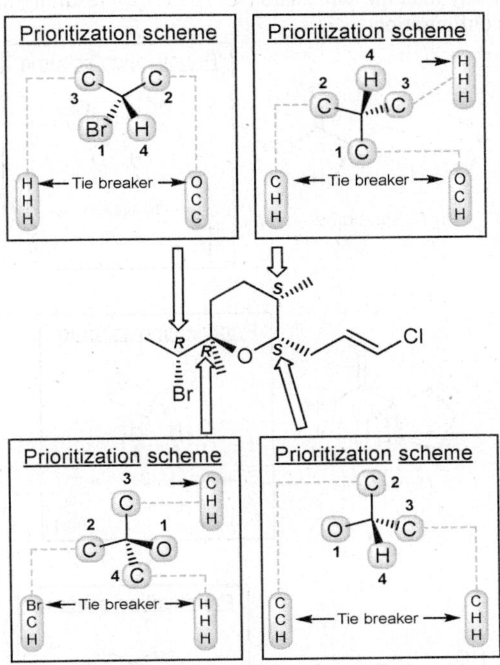

For three out of the four chiral centers, the lowest priority group (H) is not on a dash, so your point of view must be adjusted before assigning the configuration (e.g., by switching groups).

(f) This compound has two chiral centers, shown here. The following prioritization schemes led to the assignment of configuration for each chiral center.

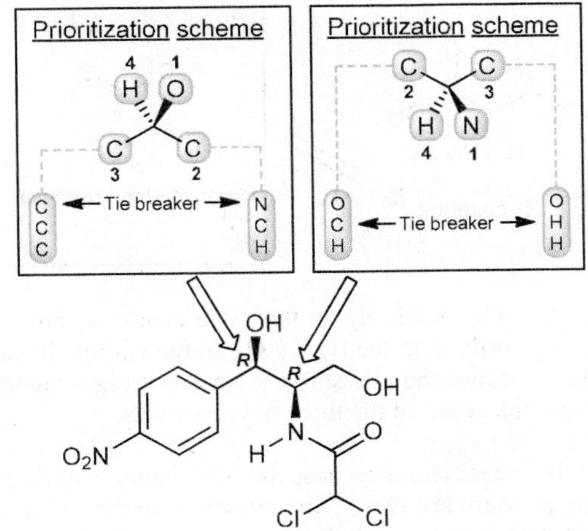

5.10.

(a) At the alpha carbon of every naturally occurring chiral amino acid, hydrogen is the lowest priority (#4) and nitrogen is the highest priority (#1). The other two atoms at the chiral center are both carbon (the carboxylic acid group and the so-called "side chain," R, of the amino acid). In almost every amino acid, including histidine, proline and aspartic acid, the oxygen atoms of the carboxylic acid functional group give that carbon a higher priority than the side chain (-CO_2H is #2), resulting in an *S* configuration.

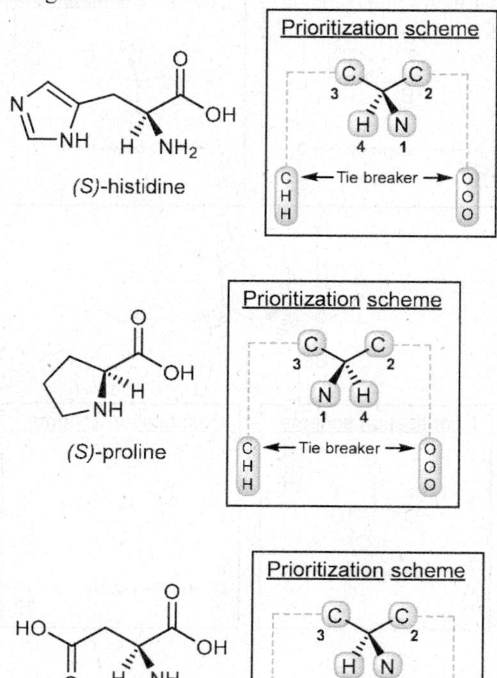

The only exception is cysteine; because the attached sulfur atom has a higher atomic number than oxygen, cysteine has the *R* configuration.

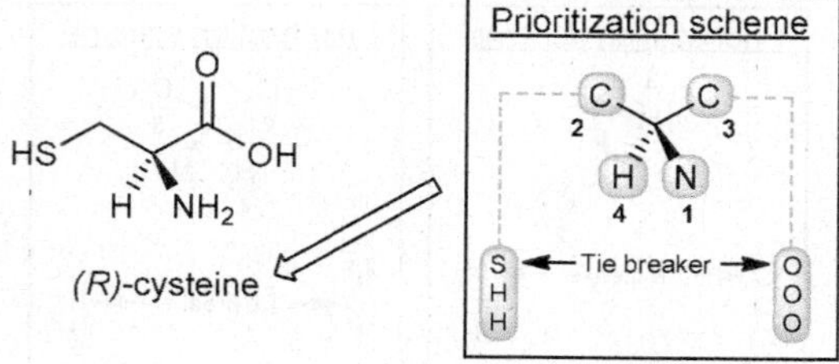

Note that the (S, H, H) on the "side chain" carbon is a higher priority than the (O, O, O) on the carboxylic acid carbon because the one sulfur atom has a higher atomic number than each of the three oxygen atoms.

(b) Glycine is achiral because the side chain is a hydrogen atom, so there are two hydrogen atoms connected to the alpha carbon. As a result, it cannot be a chiral center, because it doesn't have four unique groups.

two groups on α-carbon are identical

glycine (achiral)

5.11. The problem statement indicates that 0.575 grams are dissolved in 10.0 mL, so the concentration is 0.575g/10.0 mL = 0.0575g/mL. The path length is 10.0 cm, which is equivalent to 1.00 dm, and the observed rotation is +1.47°. Plugging these values into the equation, we get the following:

$$\text{specific rotation} = [\alpha] = \frac{\alpha}{c \times l}$$

$$= \frac{(+1.47^\circ)}{(0.0575\text{ g/mL}) \times (1.00\text{ dm})} = \mathbf{+25.6}$$

5.12. The problem statement indicates that 0.095 grams are dissolved in 1.00 mL, so the concentration is 0.095g/1.00 mL = 0.095g/mL. The path length is 10.0 cm, which is equivalent to 1.00 dm, and the observed rotation is -2.99°. Plugging these values into the equation, we get the following:

$$\text{specific rotation} = [\alpha] = \frac{\alpha}{c \times l}$$

$$= \frac{(-2.99^\circ)}{(0.095\text{ g/mL}) \times (1.00\text{ dm})} = \mathbf{-31.5}$$

5.13. The problem statement indicates that 1.30 grams are dissolved in 5.00 mL, so the concentration is 1.30g/5.00 mL = 0.260g/mL. The path length is 10.0 cm, which is equivalent to 1.00 dm, and the observed rotation is +0.57°. We then plug this value into the equation, as shown:

$$\text{specific rotation} = [\alpha] = \frac{\alpha}{c \times l}$$

$$= \frac{(+0.57^\circ)}{(0.260\text{ g/mL}) \times (1.00\text{ dm})} = \mathbf{+2.2}$$

5.14.

(a) In this problem we are solving for the observed rotation, α, with a known specific rotation, [α]. The concentration (*c*) must be calculated to obtain the proper units of grams per milliliter. A 500 mg tablet is equivalent to 0.500 grams, and with 10.0 mL of solvent, the concentration (*c*) is calculated to be 0.0500 g/mL. The pathlength (*l*) is given as 10.0 cm, which is equivalent to 1.00 dm.

$[\alpha]$ = –93.4

α = ?

c = (0.500 g) / (10.0 mL) = 0.0500 g/mL

l = 10.0 cm = 1.00 dm

Plugging these values into the equation and solving for α gives an expected observed rotation of –4.67°.

$$[\alpha] = \frac{\alpha}{c \times l}$$

$$-93.4 = \frac{\alpha}{0.0500 \text{ g/mL} \times 1.00 \text{ dm}}$$

$$\alpha = (-93.4)(0.0500)(1.00) = -4.67^{\circ}$$

(b) Because this enantiomer has a negative $[\alpha]^{20}$ value, it is described as levorotatory. A clever combination of levorotatory and etiracetam leads to the drug's name, levetiracetam. It is good to point out the sometimes confused, but completely separate, properties of having an *S* configuration (a result of applying IUPAC rules) and being levorotatory (as measured by a polarimeter). You will recall that a compound with an *S* configuration might be either dextrorotatory (optical rotation $\alpha > 0$) or levorotatory (optical rotation $\alpha < 0$), which can only be determined experimentally.

5.15. The % ee is calculated in the following way:

$$\% \, ee = \frac{[\alpha] \text{ of mixture}}{[\alpha] \text{ of pure enantiomer}} \times 100\,\%$$

$$= \frac{-37}{-39.5} \times 100\,\%$$

$$= 94\,\%$$

5.16. The % ee is calculated in the following way:

$$\% \, ee = \frac{[\alpha] \text{ of mixture}}{[\alpha] \text{ of pure enantiomer}} \times 100\,\%$$

$$= \frac{-6.0}{-6.3} \times 100\,\%$$

$$= 95\,\%$$

5.17. The % ee is calculated in the following way:

$$\% \, ee = \frac{[\alpha] \text{ of mixture}}{[\alpha] \text{ of pure enantiomer}} \times 100\,\%$$

$$= \frac{+85}{+92} \times 100\,\%$$

$$= 92\,\%$$

5.18. The magnitude of the specific rotation for **1** is calculated using the reported *ee*; the calculation below indicates a value of 74.4.

$$\% \, ee = \frac{[\alpha] \text{ of mixture}}{[\alpha] \text{ of pure enantiomer}} \times 100\%$$

$$[\alpha] \text{ of mixture} = \frac{\% \, ee \times [\alpha] \text{ of pure enantiomer}}{100\%}$$

$$= \frac{84\% \times 88.6}{100\%}$$

$$= 74$$

Next, we must determine the sign (+ or –) of the specific rotation. The problem statement indicates that the pure (*S*) enantiomer is dextrorotatory. Since **1** has the (*R*) configuration (see below), we conclude that it must be levorotatory. Therefore, the specific rotation [α] should be –74.

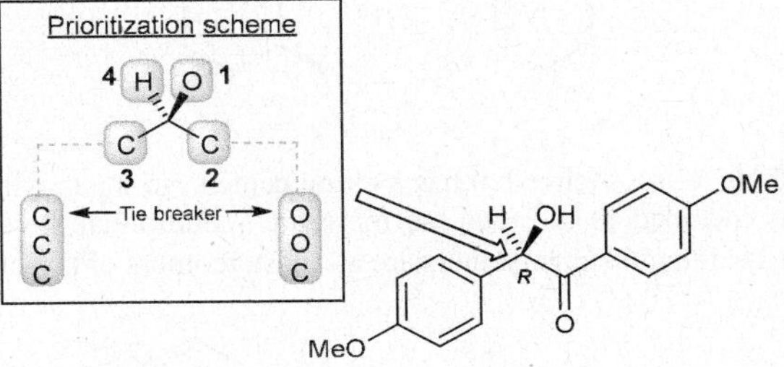

5.19.
(a) We assign a configuration to each chiral center in each compound, and we compare:

In the first compound, both chiral centers have the *R* configuration, while in the second compound, both chiral centers have the *S* configuration. These compounds are mirror images of each other, but they are nonsuperimposable. That is, if you try to rotate the first compound 180 degrees about a horizontal axis, you will not generate the second compound (if you have trouble seeing this, you may find it helpful to build a molecular model). These compounds are therefore enantiomers.

(b) We assign a configuration to each chiral center in each compound, and we compare:

In the first compound, the configurations of the chiral centers are *R* and *S*, while in the second compound, they are *S* and *S*. These compounds are stereoisomers, but they are not mirror images of each other. Therefore, they are diastereomers.

(c) We assign a configuration to each chiral center in each compound, and we compare:

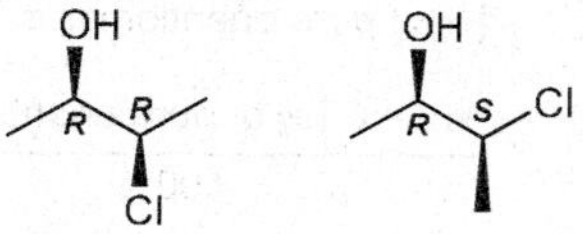

In the first compound, the configurations of the chiral centers are *R* and *R*, while in the second compound, they are *R* and *S*. These compounds are stereoisomers, but they are not mirror images of each other. Therefore, they are diastereomers.

(d) We assign a configuration to each chiral center in each compound, and we compare:

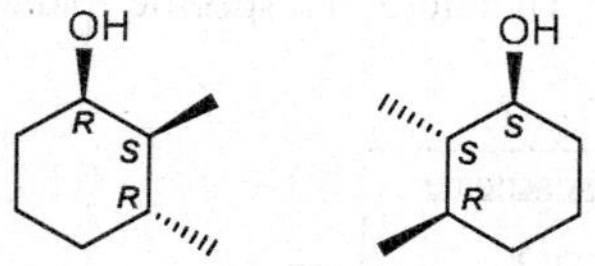

In the first compound, the configurations of the chiral centers are *R*, *S*, and *R*, respectively. In the second compound, they are *S*, *S*, and *R*. These compounds are stereoisomers, but they are not mirror images of each other. Therefore, they are diastereomers.

(e) We assign a configuration to each chiral center in each compound, and we compare:

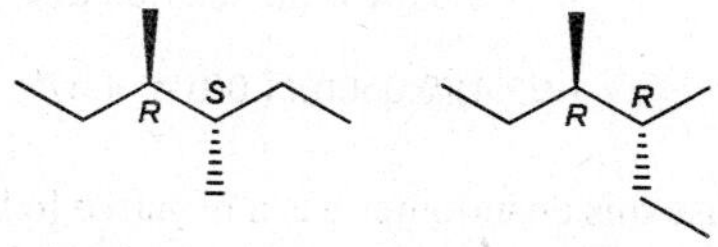

In the first compound, the configurations of the chiral centers are *R* and *S*, while in the second compound, they are *R* and *R*. These compounds are stereoisomers, but they are not mirror images of each other. Therefore, they are diastereomers.

5.20. (–)-Lariciresinol has 3 chiral centers, so the maximum number of stereoisomers = $2^3 = 8$. The original compound is included in the total, so there are 7 additional stereoisomers. A molecule can only have one enantiomer, so the remaining 6 stereoisomers are all diastereomers of (–)-lariciresinol.

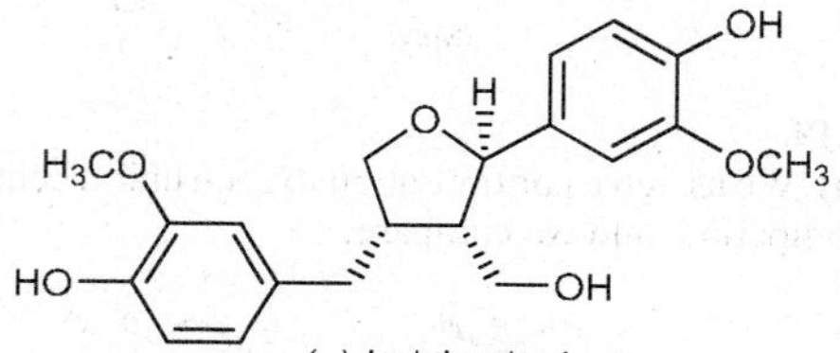

(–)-lariciresinol

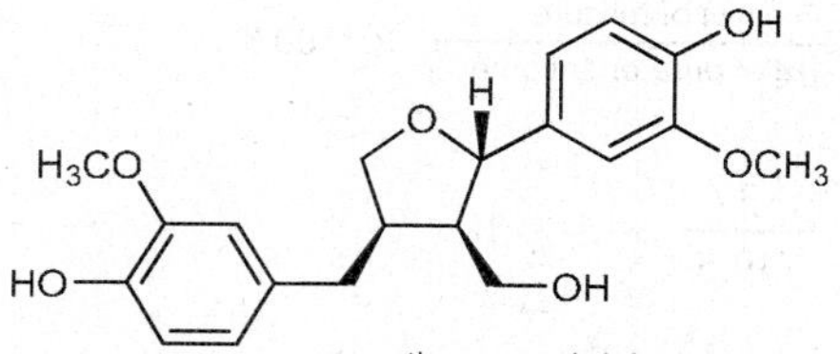

enantiomer = (+)-lariciresinol

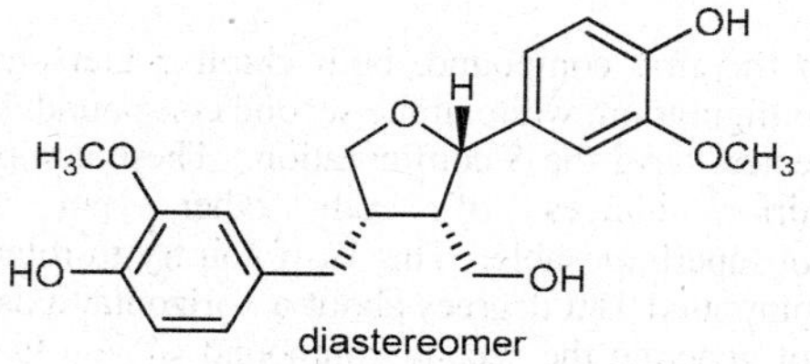

diastereomer

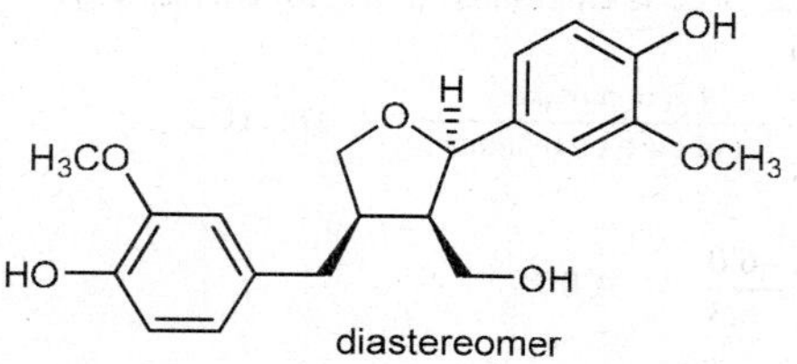

diastereomer

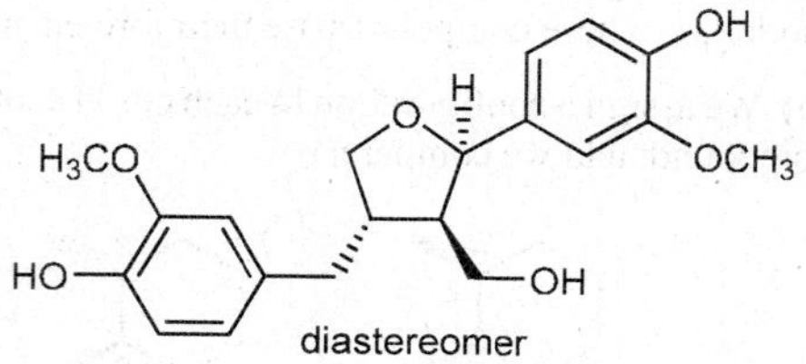

diastereomer

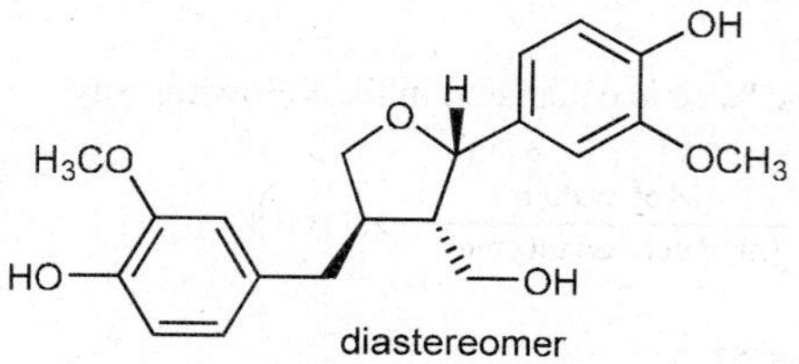

diastereomer

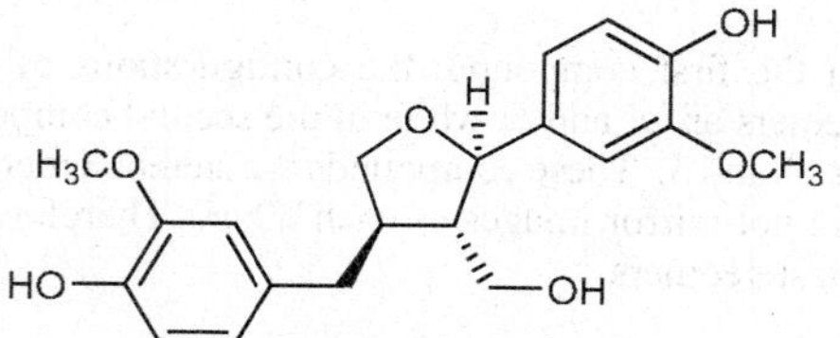

diastereomer

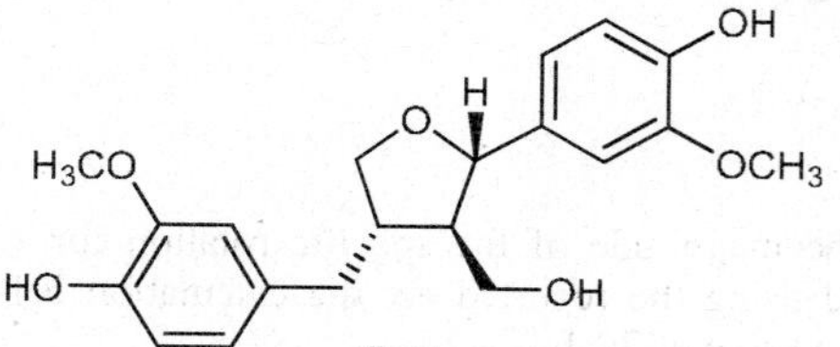

diastereomer

5.21.
(a) Yes, there is a plane of symmetry that chops the goggles in half, with the right side reflecting the left side.
(b) Yes, there is a plane of symmetry that goes through the handle of the cup.
(c) No, there is no plane of symmetry. At first glance, the pliers look fairly symmetrical, but if you look more closely you will notice that one handle attaches on the top face of the other handle (much like the difference between left-handed scissors and right-handed scissors, neither of which has a plane of symmetry).
(d) Yes, there is a plane of symmetry that chops the whistle in half.
(e) Yes, there are three planes of symmetry in this cinder block.
(f) No, there are no planes of symmetry in a hand.

5.22. The cinder block (**5.21e**) has three planes of symmetry, each of which chops the block in half.

5.23. Each of the following compounds has a plane of symmetry, as shown:

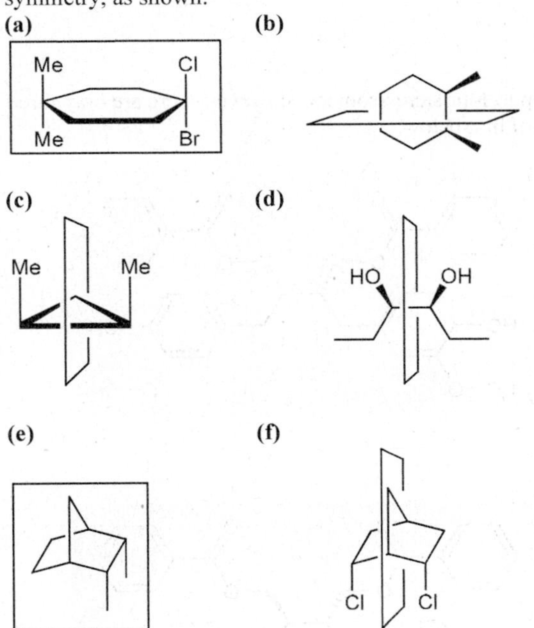

5.24.
(a) With two chiral centers, we would expect a maximum of four stereoisomers. However, there are only three stereoisomers in this case, because the first one shown below is a *meso* compound.

meso

The *meso* compound has a plane of symmetry and is, therefore, achiral. It does not have an enantiomer (replacing the wedges with dashes results in the same compound).

A *meso* compound has no enantiomer.

(b) With two chiral centers, we would expect a maximum of four stereoisomers. However, there are only three stereoisomers in this case, because the first one shown below is a *meso* compound.

meso

The *meso* compound has a plane of symmetry and is, therefore, achiral. It does not have an enantiomer (replacing the wedges with dashes results in the same compound).

A *meso* compound has no enantiomer.

(c) With two chiral centers, we would expect a maximum of four stereoisomers. However, there are only three stereoisomers in this case, because the first one shown below is a *meso* compound.

meso

The *meso* compound has a plane of symmetry and is, therefore, achiral. It does not have an enantiomer (replacing the wedges with dashes results in the same compound).

is the same as:

(d) With two chiral centers, there are four stereoisomers. In this case, there are no *meso* compounds, due to a lack of symmetry in the molecule.

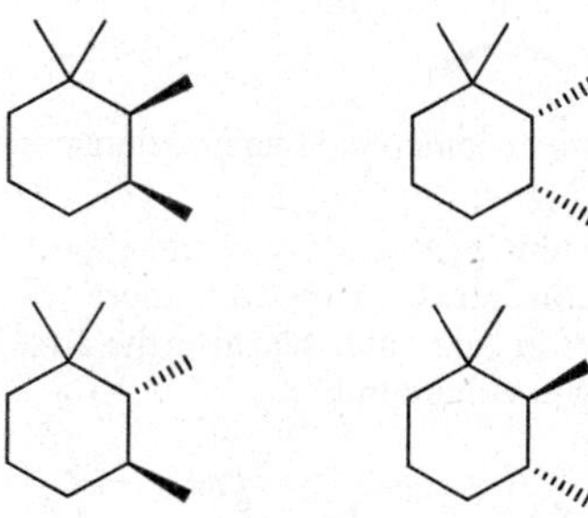

(e) With two chiral centers, we would expect four stereoisomers. However, there are only three stereoisomers in this case, because the first one shown below is a *meso* compound.

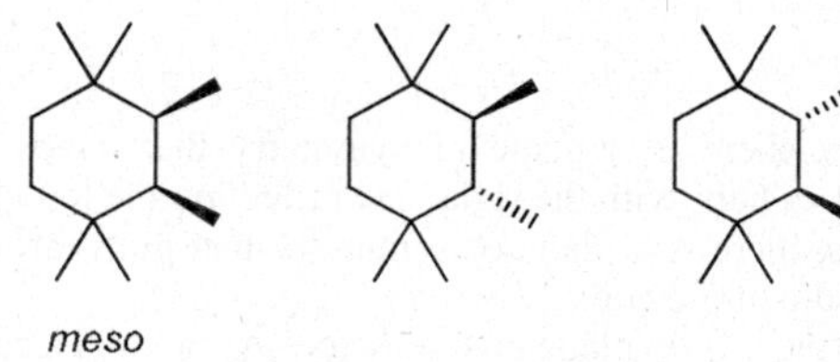

meso

The *meso* compound has a plane of symmetry and is, therefore, achiral. It does not have an enantiomer (replacing the wedges with dashes results in the same compound).

A *meso* compound has no enantiomer.

5.25. For a molecule with two chiral centers, we would expect up to four stereoisomers. However, there are only three stereoisomers in this case, because one is a *meso* compound, shown first below.

meso
(achiral)

chiral — enantiomers — chiral

The *meso* compound is achiral. The other two stereoisomers are chiral, and they are enantiomers of each other. The problem statement indicates that a racemic mixture (a pair of enantiomers) was isolated from the plant. Therefore, the two chiral stereoisomers must have been isolated from the plant.

5.26.

(a) This chiral center has the *R* configuration, as shown below:

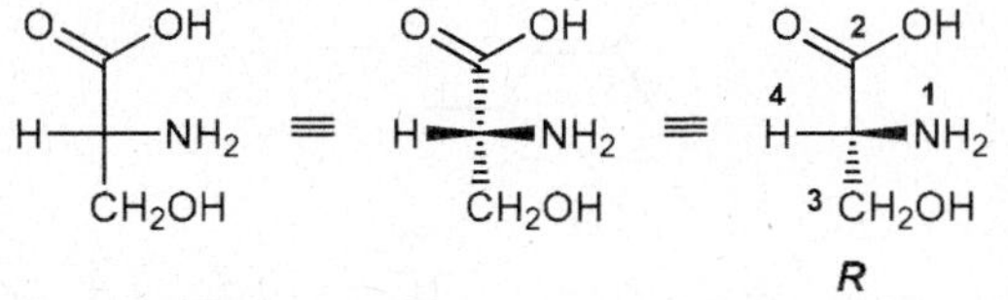

Note that the lowest priority group (H) is not on a dash, so your point of view must be adjusted before assigning the configuration (e.g., by switching groups).

(b) This chiral center has the *S* configuration, as shown below:

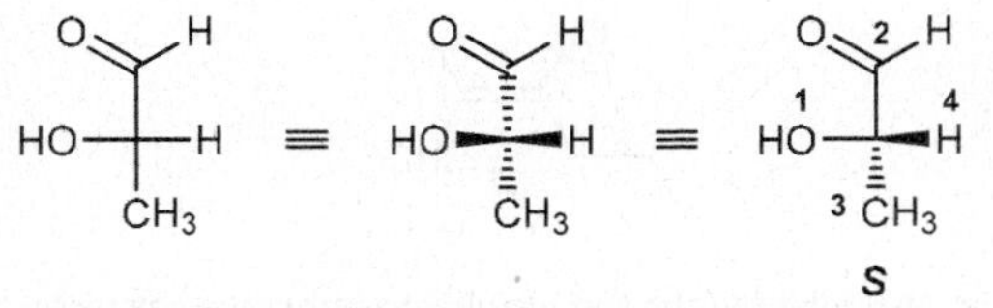

Note that the lowest priority group (H) is not on a dash, so your point of view must be adjusted before assigning the configuration (e.g., by switching groups).

(c) This chiral center has the *S* configuration, as shown below:

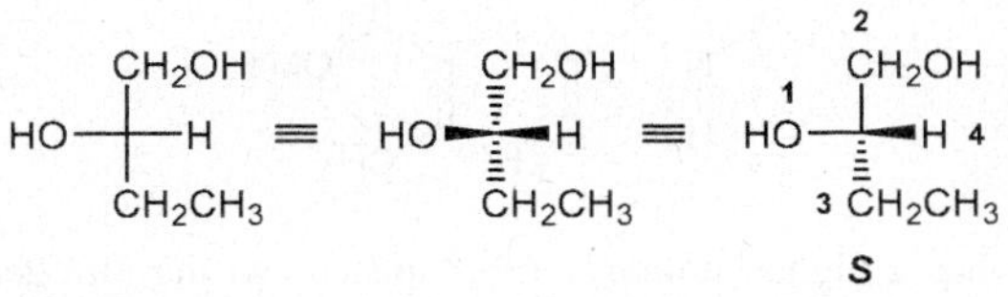

Note that the lowest priority group (H) is not on a dash, so your point of view must be adjusted before assigning the configuration (e.g., by switching groups).

(d) This chiral center has the *S* configuration, as shown below:

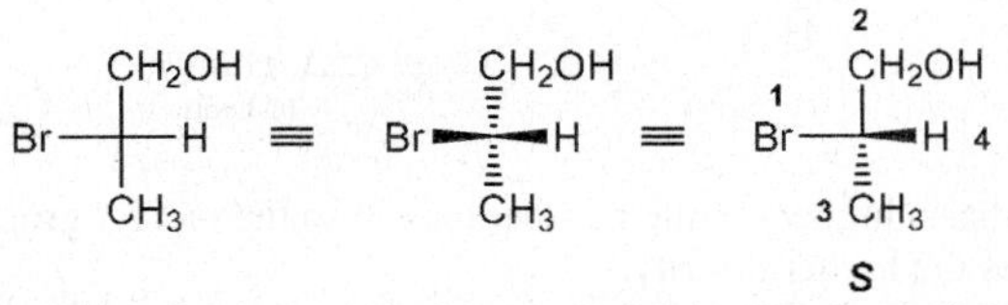

Note that the lowest priority group (H) is not on a dash, so your point of view must be adjusted before assigning the configuration (e.g., by switching groups).

5.27.

(a) For each chiral center, we follow the same procedure that we used in the previous problem. We first redraw the chiral center so that there are two lines, one wedge, and one dash:

Then, we assign priorities, and determine the configuration. If we repeat this process for each chiral center, we find the following configurations:

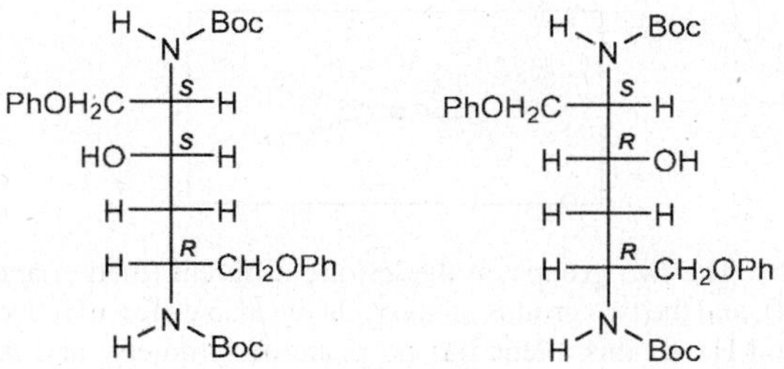

(b) These compounds differ in the configuration of only one chiral center, so they are diastereomers. That is, they are stereoisomers that are not mirror images.

(c) Fischer projections show the eclipsed conformations of neighboring bonds, whereas bond-line structures show the staggered conformations of neighboring bonds. Visualizing the conversion between Fischer projections and bond-line structures therefore involves bond rotations. Instead, let's use R/S assignments to compare bond-line structures and Fischer projections, without the need to rotate bonds. For convenience, we start by drawing the bond-line structure using wedges for all of the chiral centers so that the low-priority hydrogen atoms will all be in the back (on dashes). Then, we can assign each configuration and determine which chiral centers were drawn correctly (and which need to be changed). In this case, if we use this method and assign configurations, we find that the R/S assignments all match that of compound **2**. So this is compound **2**.

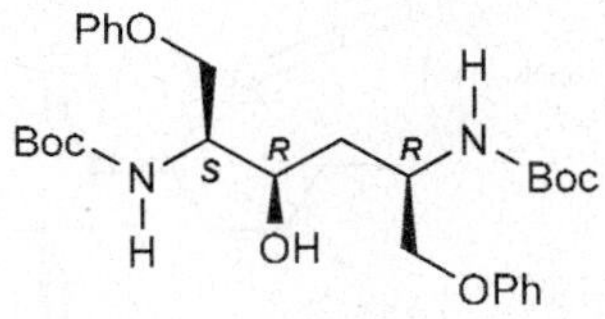

Compound **2**

We can easily draw compound **1** from compound **2**, because these two compounds differ in the configuration of only one chiral center. Changing the wedge (on the OH group) to a dash will necessarily change the configuration from *R* to *S*.

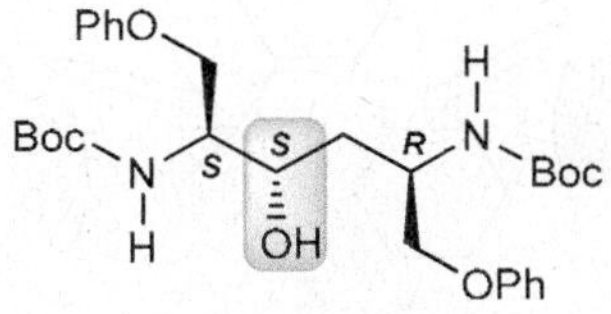

Compound **1**

5.28.

(a) On the left side of the structure, the two groups are different (methyl and H), but the two groups on the right side are the same (both are methyl groups). In order to be chiral, the two groups on the right would have to be different also. This compound has a plane of symmetry and, therefore, is not chiral.

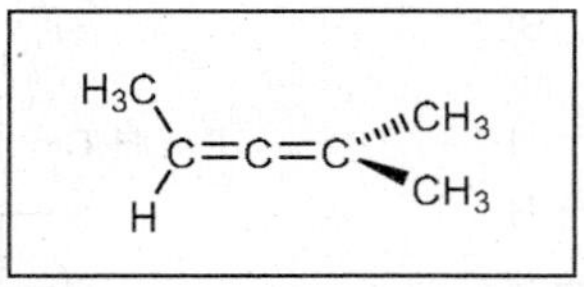

(b) The two groups on the left are different (methyl and H), and the two groups on the right are also different (ethyl and H), so this allene has no plane of symmetry and is, therefore, chiral.

(c) The two groups on the left are the same (going either way around the cyclopentane ring). In order to be chiral, the two groups would have to be different. This compound has a plane of symmetry and, therefore is not chiral.

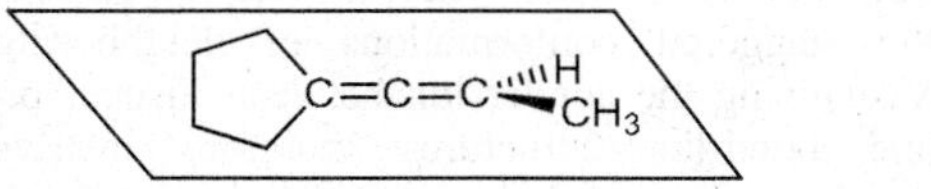

(d) The two groups on the left are different (because of asymmetry imparted by the position of the double bond in the ring), and the two groups on the right are also different, so this allene has no plane of symmetry and is, therefore, is chiral.

5.29.

(a) The priorities (as shown below) are on opposite sides of the double bond, so this alkene has the *E* configuration.

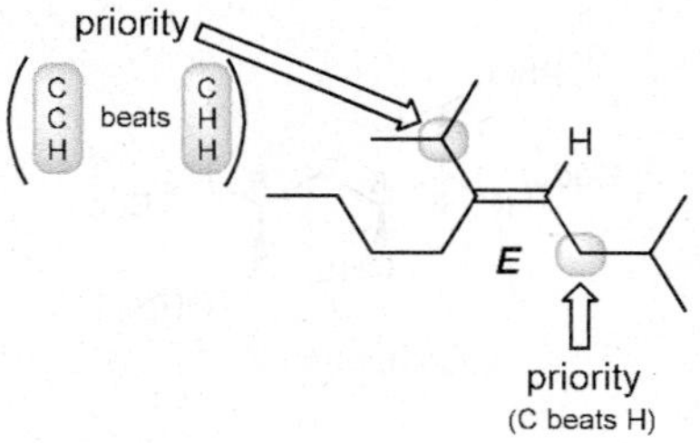

(b) The priorities (as shown below) are on the same sides of the double bond, so this alkene has the *Z* configuration.

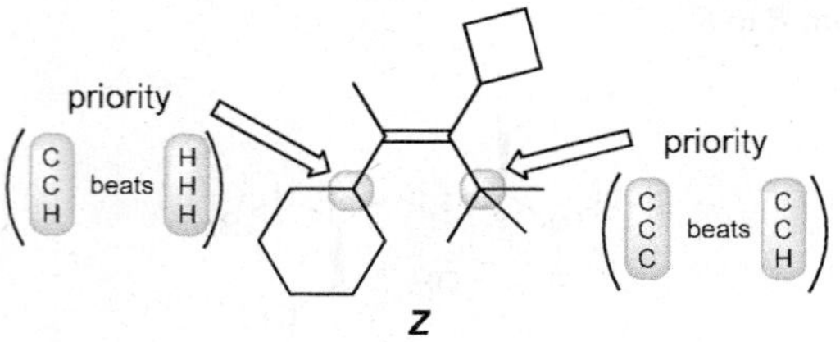

(c) The priorities (as shown below) are on the same sides of the double bond, so this alkene has the *Z* configuration.

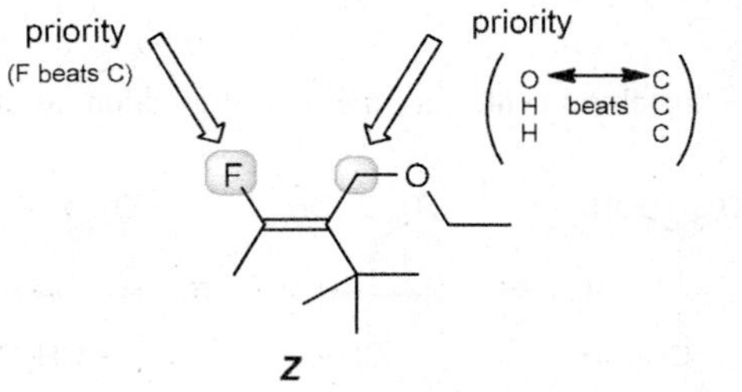

(d) The priorities (as shown below) are on the same sides of the double bond, so this alkene has the *Z* configuration.

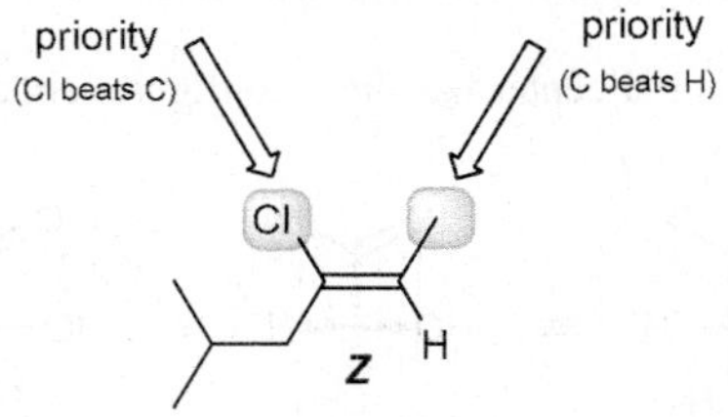

5.30. Attached to the top vinylic position, we see nitrogen and carbon (highlighted):

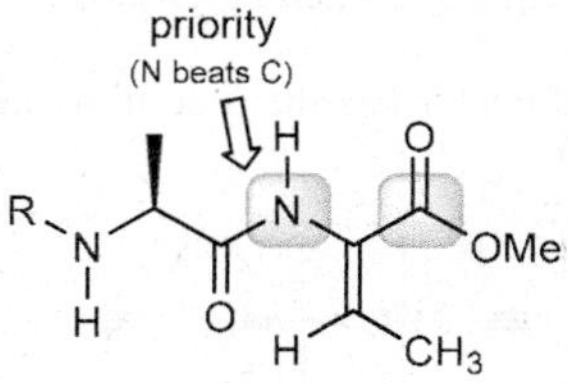

N has a higher atomic number than C, so the nitrogen group has the higher priority. Attached to the bottom vinylic position, we see hydrogen and carbon (highlighted):

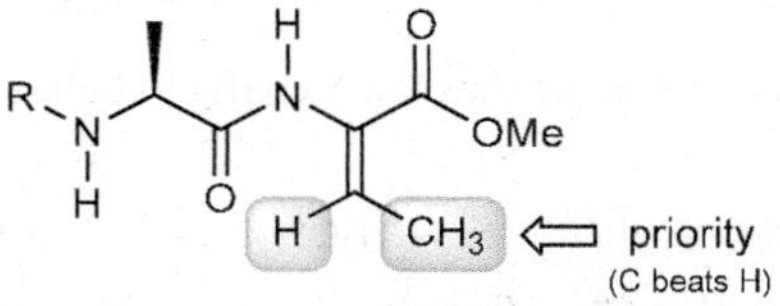

C has a higher atomic number than H so the methyl group has the higher priority.

Since the higher priority groups are on opposite sides of the double bond, the configuration is *E*:

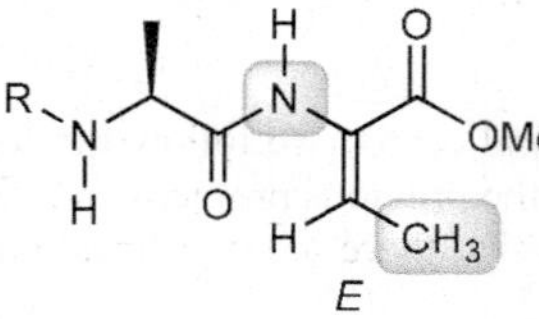

5.31. This compound has two chiral centers, shown below. The following prioritization schemes led to the assignment of configuration for each chiral center.

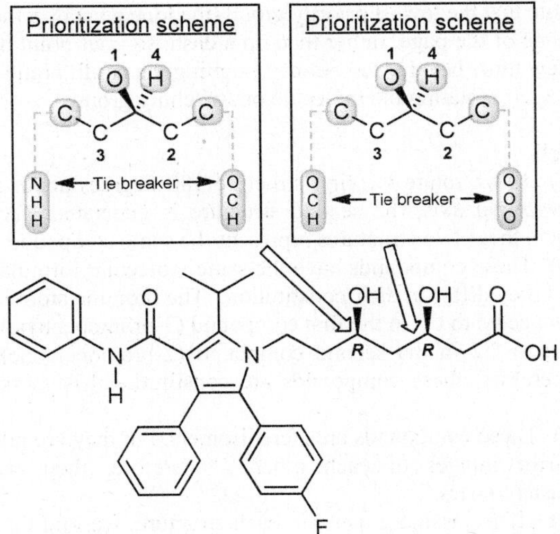

5.32. In this case, it will be easiest to draw a mirror image. The mirror can be placed *behind* the molecule, giving the structure on the left, or the mirror can be placed *next to* the molecule, giving the structure on the right. Both drawings represent the enantiomer of atropine (can you see that they are superimposable?):

5.33.
(a) In order to draw the enantiomer of paclitaxel, we convert every wedge into a dash, and we convert every dash into a wedge, as shown:

(b) Paclitaxel has eleven chiral centers (highlighted below):

5.34. To assign the configuration, we must first assign a prioritization scheme to the four atoms connected to the chiral center:

One of these atoms is H, so that atom is immediately assigned the fourth priority. The remaining three atoms are all carbon atoms, so for each of them, we prepare a list of the atoms attached to them. Let's begin with the double bond. Recall that a double bond is comprised of one σ bond and one π bond. For purposes of assigning configurations, we treat the π bond as if it were a σ bond to another carbon atom, like this:

is treated as

We treat a triple bond similarly. Recall that triple bonds are comprised of one σ bond and two π bonds. As such, each of the two π bonds is treated as if it were a σ bond to another carbon atom:

is treated as

This gives the following competition:

Tie breaker

Among these groups, the double bond is assigned the third priority in our prioritization scheme, and we must continue to assign the first and second priorities.

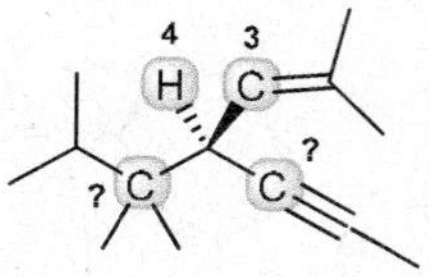

We move one carbon atom away from the chiral center, and we compare the following two positions:

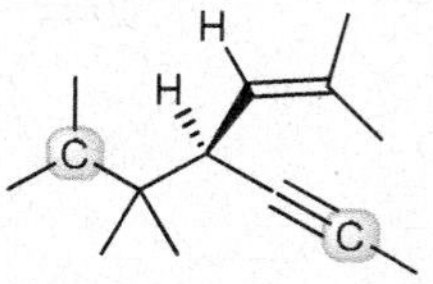

In each case, we must construct a list of the three atoms that are connected to each position. This is easy to do for the left side, which is connected to two carbon atoms and one hydrogen atom:

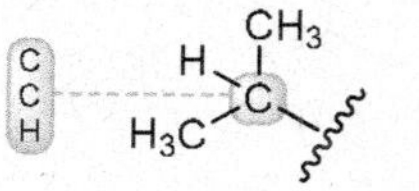

For the right side, the highlighted carbon atom is part of a triple bond, so we treat each π bond as if it were a sigma bond to another carbon atom:

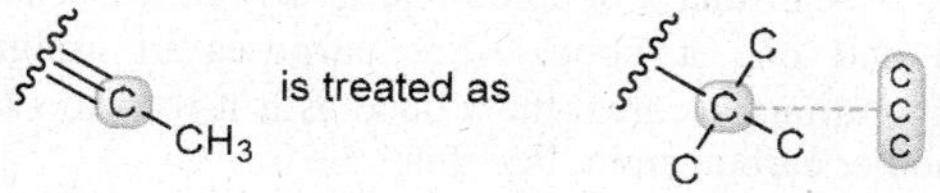

is treated as

As such, the right side wins the tie-breaker:

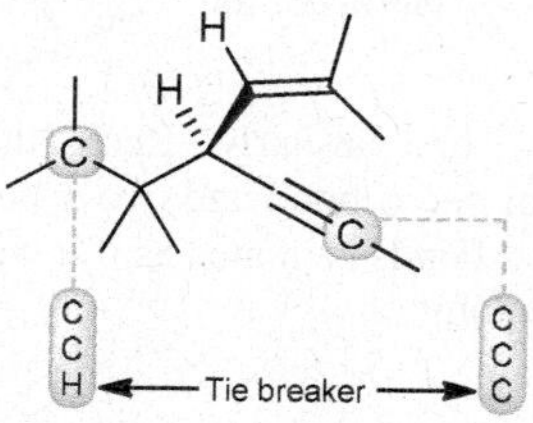

Tie breaker

Our prioritization scheme indicates that the chiral center has the *R* configuration:

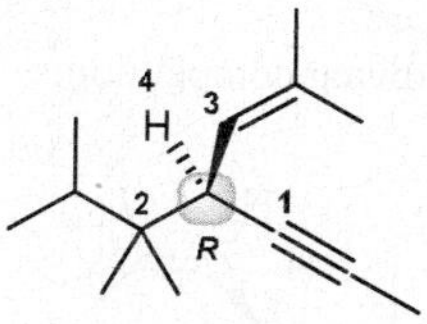

5.35. The chiral center has the *S* configuration, determined by the prioritization scheme shown here.

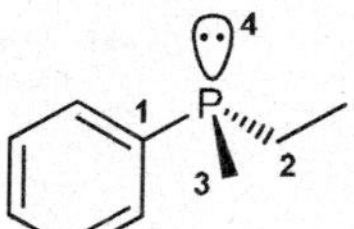

(S)

Note that the lowest priority group (the lone pair) is in the plane of the page, rather than on a dash, so your point of view must be adjusted before assigning the configuration (e.g., by rotating the molecule or switching groups).

5.36.

(a) If we rotate the first structure 180 degrees about a horizontal axis, the second structure is generated. As such, these two structures represent the same compound.

(b) These compounds have the same molecular formula, but they differ in their constitution. The bromine atom is connected to C3 in the first compound (3-bromopentane), and to C2 in the second compound (2-bromopentane). Therefore, these compounds are constitutional isomers.

(c) These compounds are stereoisomers, but they are not mirror images of each other. Therefore, they are diastereomers.

(d) If we assign a name to each structure, we find that both have the same name:

(*S*)-3-methylhexane

Therefore, these structures represent the same compound.

(e) These compounds are nonsuperimposable mirror images of each other. Therefore, they are enantiomers.

(f) These structures represent the same compound (rotating the first compound 180 degrees about a vertical axis generates the second compound).

(g) These structures represent the same compound, which does not contain a chiral center, because there are two ethyl groups connected to the central carbon atom.

(h) These structures represent the same compound (rotating the first compound 180 degrees about a vertical axis generates the second compound). This is a *meso* compound, so it does not have an enantiomer (it is superimposable on its mirror image).

5.37.

(a) There are three chiral centers (n=3), highlighted below, so we expect $2^n = 2^3 = 8$ stereoisomers.

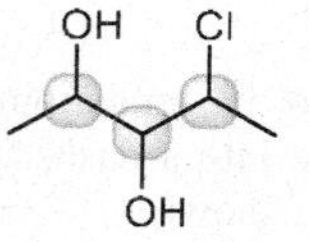

(b) There are two chiral centers (n=2), so we might initially expect $2^n = 2^2 = 4$ stereoisomers. However, one of the stereoisomers is a *meso* compound, so there will only be 3 stereoisomers:

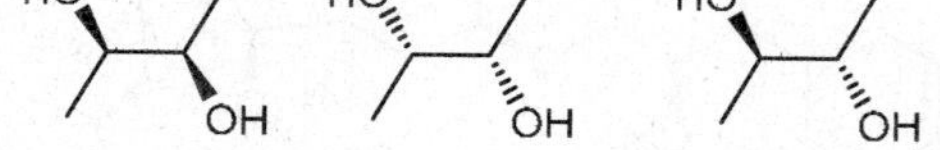

meso

To see the plane of symmetry in the *meso* isomer, rotate about the middle C–C bond. A 180° rotation results in an eclipsed conformation with both methyl groups aligned and both OH groups aligned.

(c) There are four chiral centers (n=4), highlighted below, so we expect $2^n = 2^4 = 16$ stereoisomers.

(d) There are two chiral centers (n=2), so we might initially expect $2^n = 2^2 = 4$ stereoisomers. However, one of the stereoisomers is a *meso* compound, so there will only be 3 stereoisomers:

meso

(e) There are two chiral centers (n=2), so we might initially expect $2^n = 2^2 = 4$ stereoisomers. However, one of the stereoisomers is a *meso* compound, so there will only be 3 stereoisomers:

meso

(f) There are five chiral centers (n=5), highlighted below, so we expect $2^n = 2^5 = 32$ stereoisomers.

5.38. In each case, we draw the enantiomer by replacing all wedges with dashes, and all dashes with wedges. For Fischer projections, we simply change the configuration at every chiral center by switching the groups on the left with the groups on the right:

(a) **(b)**

(c) **(d)**

(e)

5.39. The configuration of each chiral center is shown below:

(a) **(b)**

(c)

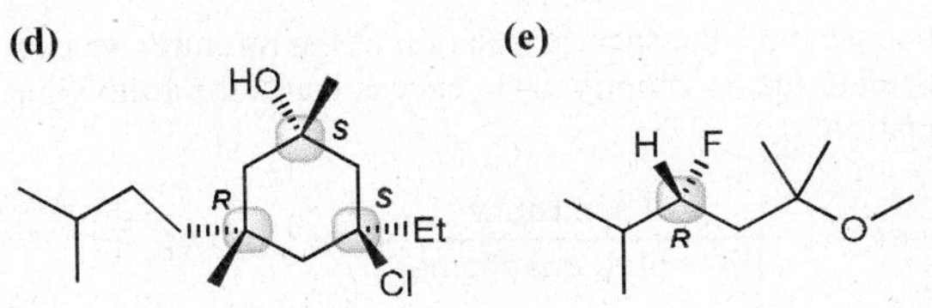

(d) **(e)**

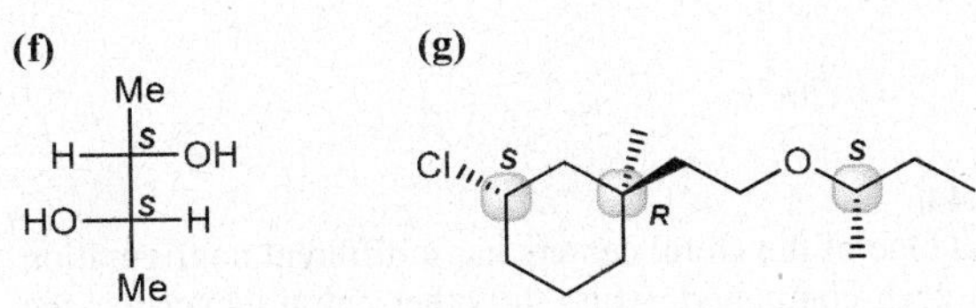

(f) **(g)**

5.40. In this case, there is a 96% *excess* of A (98 – 2 = 96). The remainder of the solution is a racemic mixture of both enantiomers (2% A and 2% B). Therefore, the enantiomeric excess (*ee*) is 96%.

5.41. This compound does not have a chiral center, because two of the groups are identical:

Accordingly, the compound is achiral and is not optically active. We thus predict a specific rotation of zero, $[\alpha] = 0$.

5.42. The specific rotation of (*S*)-2-butanol is +13.5, so its enantiomer has a specific rotation of –13.5. Dissolving

1.00 g of sample in 10.0 mL of solvent gives a concentration (c) of 0.100 g/mL:

$$[\alpha] = \frac{\alpha}{c \times l} \quad \text{and therefore, } \alpha = [\alpha] \times c \times l$$

$$= (-13.5)(0.100 \text{ g/mL})(1.00 \text{ dm}) = -1.35°$$

5.43. We begin by determining the specific rotation of the mixture, using the following equation:

$$[\alpha] \text{ of mixture} = \frac{\alpha}{c \times l}$$

The concentration (c) is calculated by dividing 3.50 g by 10.0 mL, giving 0.350 g/mL.

$$[\alpha] = \frac{(+0.78°)}{(0.350 \text{ g/mL}) \times (1.00 \text{ dm})}$$

$$[\alpha] = +2.2$$

Once we have the specific rotation of the mixture, we can calculate the % enantiomeric excess using the following equation:

$$\%\ ee = \frac{[\alpha] \text{ of mixture}}{[\alpha] \text{ of pure enantiomer}} \times 100\ \%$$

$$= \frac{+2.2}{+2.8} \times 100\ \%$$

$$= 79\ \%$$

5.44.

(a) One of the chiral centers has a different configuration in each compound, while the other chiral center has the same configuration in each compound. As such, these compounds are stereoisomers, but they are not mirror images of each other. They are diastereomers.

(b) Each of the chiral centers has a different configuration when comparing these compounds. As such, these compounds are nonsuperimposable mirror images. They are enantiomers.

(c) These structures represent the same compound (rotating the first compound 180 degrees about a horizontal axis generates the second compound). This is a *meso* compound, so it does not have an enantiomer (it is superimposable on its mirror image).

(d) These compounds are nonsuperimposable mirror images. Therefore, they are enantiomers.

5.45.

(a) One of the chiral centers has a different configuration in each compound, while the other chiral center has the same configuration in each compound. As such, these compounds are stereoisomers, but they are not mirror images of each other. They are diastereomers.

(b) Each of the chiral centers has a different configuration when comparing these compounds. As such, these compounds are nonsuperimposable mirror images. They are enantiomers.

(c) Each of the chiral centers has a different configuration when comparing these compounds. As such, these compounds are nonsuperimposable mirror images. They are enantiomers.

(d) One compound has *cis* methyl groups, and the other compound has *trans* methyl groups. As such, these compounds are stereoisomers that are not mirror images of each other. Therefore, they are diastereomers.

5.46.

(a) True. Each enantiomer has a specific rotation that is equal in magnitude, but opposite in direction (one is levorotatory and the other is dextrorotatory). As a result, the rotations exactly cancel each other ($\alpha = 0°$).

(b) False. A *meso* compound cannot have an enantiomer because it has a plane of symmetry and is therefore achiral. Its mirror image IS superimposable.

(c) True. In a Fischer projection, all horizontal lines represent wedges, and all vertical lines represent dashes. If we rotate the structure by 90 degrees, we are changing all wedges into dashes and all dashes into wedges. Therefore, the new structure will be an enantiomer of the original structure.

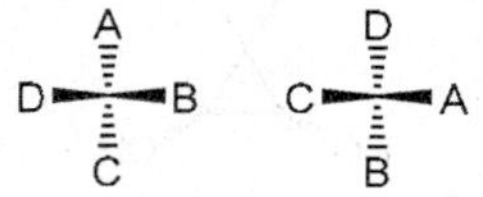

enantiomers

Be careful though – you can only rotate a Fischer projection by 90 degrees (to draw the enantiomer) if there is one chiral center. If there is more than one chiral center, then you cannot rotate the Fischer projection by 90 degrees in order to draw its enantiomer. Instead, you must change the configuration at every chiral center by switching the groups on the left with the groups on the right.

5.47. For each chiral center, we first redraw the chiral center so that there are two lines, one wedge and one dash:

Then, we assign priorities, and determine the configuration. At each chiral center, note that the lowest priority group (H) is not on a dash, so your point of view must be adjusted before assigning the configuration (e.g., by switching groups). This process gives the following configurations:

(a) *R*, *S*, *R*

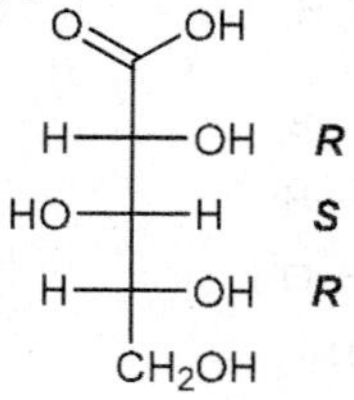

(b) *S*, *S*, *S*

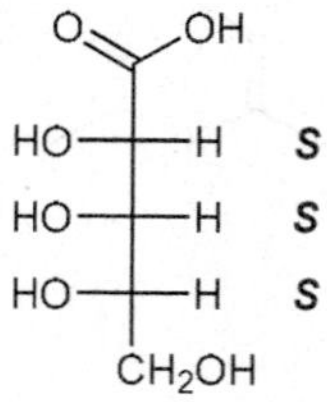

(c)

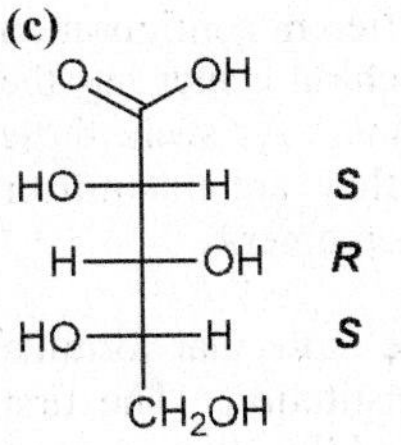

HO—C—H		*S*
H—C—OH		*R*
HO—C—H		*S*

CH_2OH

5.48. In order to draw the enantiomer for each compound, we simply change the configuration at every chiral center by switching the groups on the left with the groups on the right. Note that this process results in a drawing that is the mirror image of the original drawing.

(a) HO—H *S*; H—OH *R*; HO—H *S*; CH_2OH

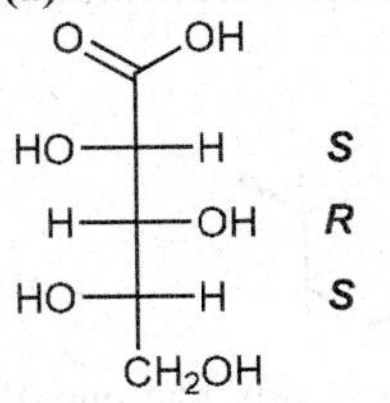

(b) H—OH *R*; H—OH *R*; H—OH *R*; CH_2OH

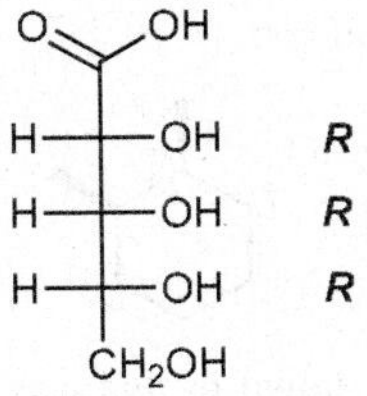

(c) H—OH *R*; HO—H *S*; H—OH *R*; CH_2OH

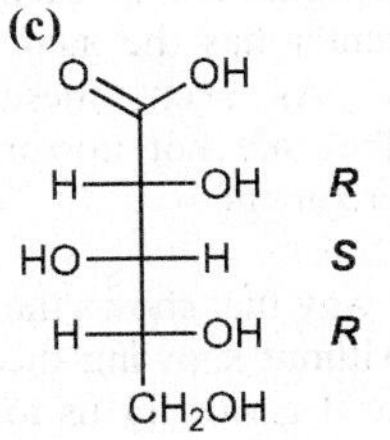

5.49. Dissolving 0.075 g of penicillamine in 10.0 mL of solvent gives a concentration (c) of 0.0075 g/mL:

$$\text{specific rotation} = [\alpha] = \frac{\alpha}{c \times l}$$

$$[\alpha] = \frac{(-0.47°)}{(0.0075\ \text{g}/\text{mL}) \times (1.00\ \text{dm})}$$

$$[\alpha] = -63$$

5.50.

(a) This compound is (*S*)-limonene, as determined by the following prioritization scheme:

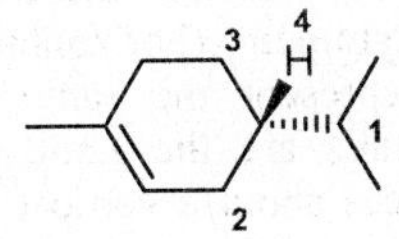

You might also recognize that this drawing is the mirror image of the given drawing of (*R*)-limonene, so it must represent the enantiomer of (*R*)-limonene.

(b) This compound is (*R*)-limonene, as determined by the following prioritization scheme:

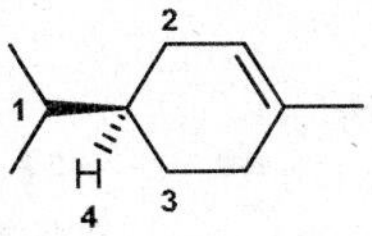

You might also recognize that this drawing is superimposable with the given drawing of (*R*)-limonene (rotating the given drawing 180 degrees about a horizontal axis generates this drawing).

(c) This compound is (*S*)-limonene, as determined by the following prioritization scheme:

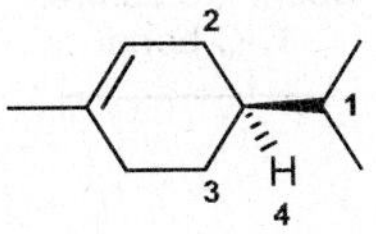

(d) This compound is (*R*)-limonene, as determined by the following prioritization scheme:

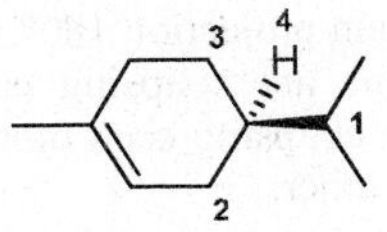

You might also recognize that this drawing is superimposable with the given drawing of (*R*)-limonene (rotating the given drawing 180 degrees about a vertical axis generates this drawing).

5.51.

(a) We must first rotate 180 degrees about the central carbon-carbon bond, in order to see more clearly that the compound possesses a plane of symmetry, shown below:

Rotate 180° about the central C-C bond

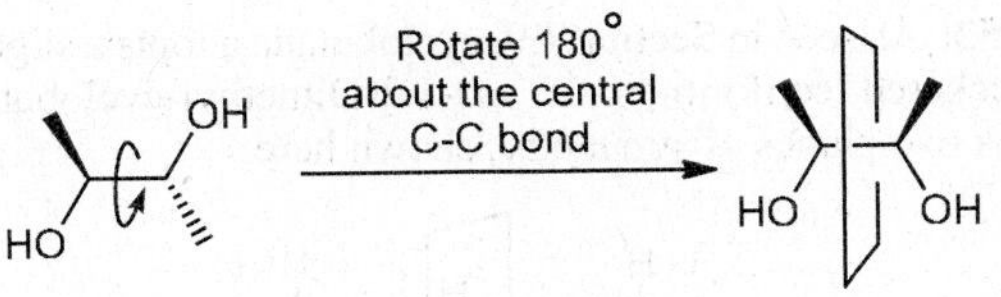

(b) We must first rotate 180 degrees about the central carbon-carbon bond, in order to see more clearly that the compound possesses a plane of symmetry, shown below:

Rotate 180° about the central C-C bond

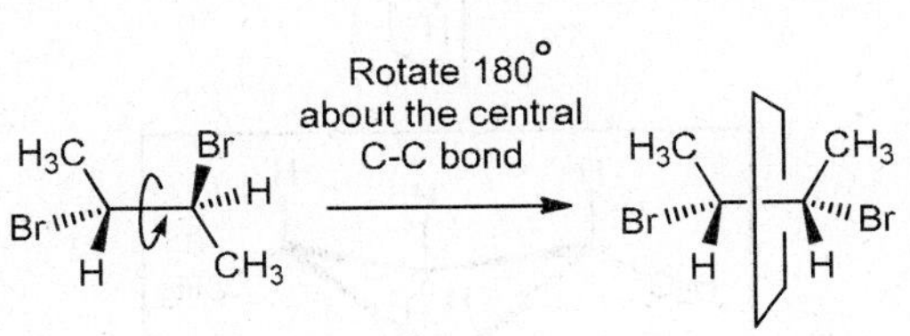

(c) We first convert the Newman projection into a bond-line drawing:

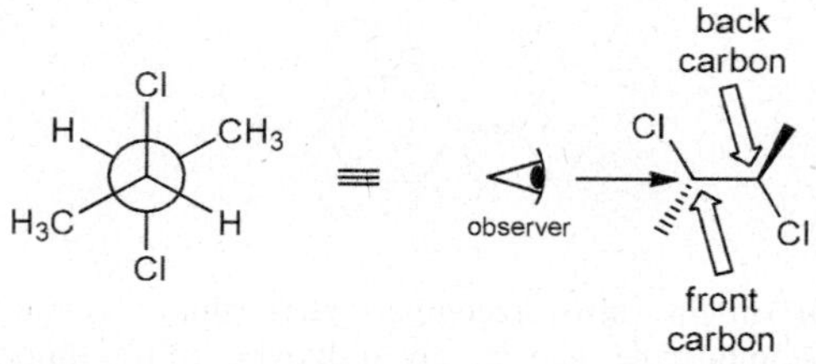

Then, we rotate 180 degrees about the central carbon-carbon bond, in order to see more clearly that the compound possesses a plane of symmetry, shown below:

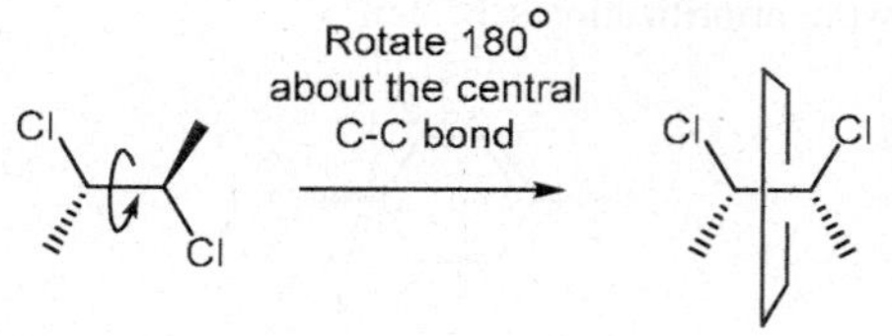

Alternatively, we can identify the plane of symmetry without converting the Newman projection into a bond-line drawing. Instead, we rotate the central carbon-carbon bond of the Newman projection 180°, and we see that the two chlorine atoms are eclipsing each other, the two methyl groups are eclipsing each other, and the two H's are eclipsing each other.

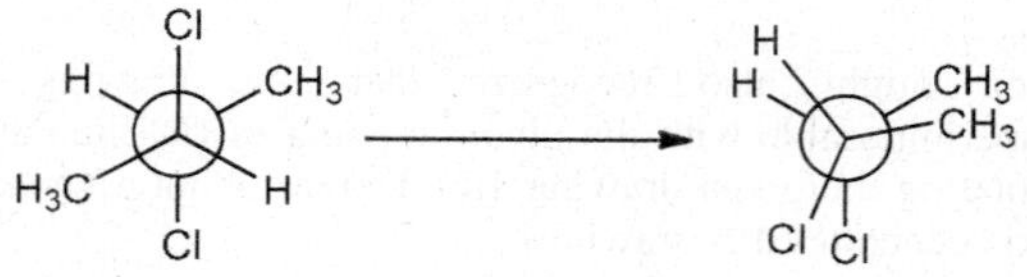

In this eclipsed conformation, we can see more clearly that the molecule has an internal plane of symmetry (it is a *meso* compound). The plane of symmetry is the plane that is parallel to the page, halfway between the front carbon atom and the back carbon atom of the Newman projection.

5.52. As seen in Section 4.9, cyclobutane adopts a slightly puckered conformation. *cis*-1,3-Dimethyl-cyclobutane has two planes of symmetry, shown here:

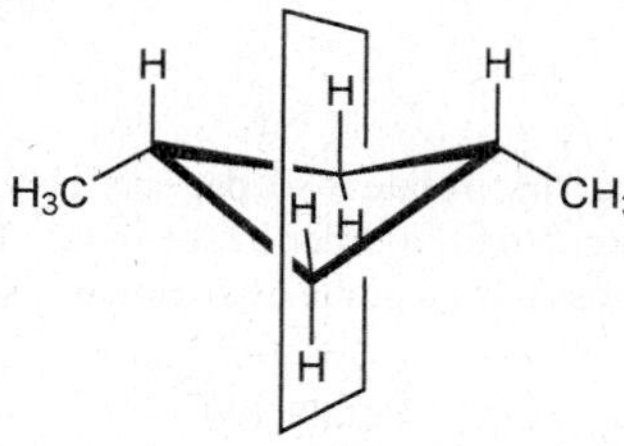

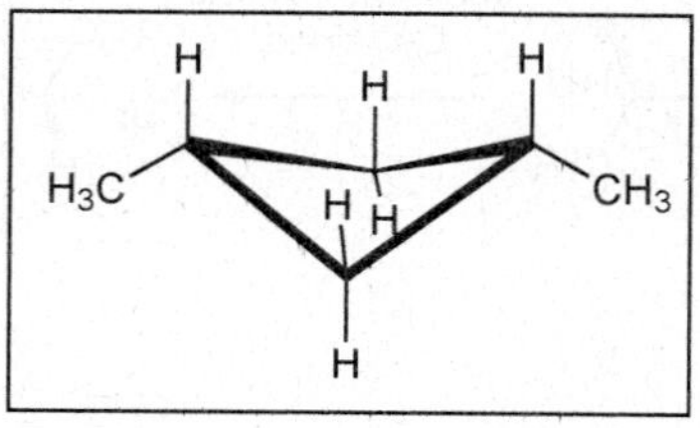

5.53.

(a) One of the chiral centers has a different configuration in each compound, while the other chiral center has the same configuration in each compound. As such, these compounds are stereoisomers, but they are not mirror images of each other. They are diastereomers.

(b) These compounds have the same molecular formula (C_8H_{16}), but they differ in their constitution. The first compound is 1,2-disubstituted, while the second compound is 1,3-disubstituted. Therefore, they are constitutional isomers.

(c) Let's redraw the compounds in a way that shows the configuration of each chiral center without showing the conformation (chair). This will make it easier for us to determine the stereoisomeric relationship between the two compounds:

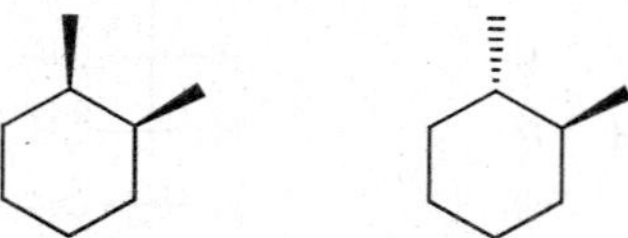

When drawn in this way, we can see clearly that one of the chiral centers has a different configuration in each compound, while the other chiral center has the same configuration in each compound. As such, these compounds are stereoisomers, but they are not mirror images of each other. They are diastereomers.

(d) Let's redraw the compounds in a way that shows the configuration of each chiral center without showing the conformation (chair). This will make it easier for us to determine the stereoisomeric relationship between the two compounds:

When drawn in this way, we can see clearly that one of the chiral centers has a different configuration in each compound, while the other chiral center has the same configuration in each compound. As such, these compounds are stereoisomers, but they are not mirror images of each other. They are diastereomers.

(e) It might appear as if each of the chiral centers has a different configuration when comparing these compounds. However, each of these compounds has an internal plane of symmetry (horizontal plane). As such, both structures represent the same *meso* compound. These two structures are the same (rotating the first drawing 180 degrees about a vertical axis generates the second drawing).

(f) Each of the chiral centers has a different configuration when comparing these compounds. As such, these compounds are nonsuperimposable mirror images. They are enantiomers.

5.54.

(a) The specific rotation of (*R*)-carvone should be the same magnitude but opposite sign as the specific rotation of (*S*)-carvone (assuming both are measured under the same conditions). Therefore, we expect the specific rotation of (*R*)-carvone at 20°C to be –61.

(b) The specific rotation of pure (*R*)-carvone is –61, while the specific rotation of this mixture of (*R*)-carvone and its enantiomer is –55. We use the following equation to calculate the enantiomeric excess:

$$\% \, ee = \frac{[\alpha] \text{ of mixture}}{[\alpha] \text{ of pure enantiomer}} \times 100\,\%$$

$$= \frac{-55}{-61} \times 100\,\%$$

$$= 90\,\%$$

(c) Since the *ee* is 90%, the mixture must be comprised of 95% (*R*)-carvone and 5% (*S*)-carvone (95 – 5 = 90).

5.55.

(a) This compound has a non-superimposable mirror image, and therefore it is chiral.

Enantiomers

(b) This compound has a non-superimposable mirror image, and therefore it is chiral.

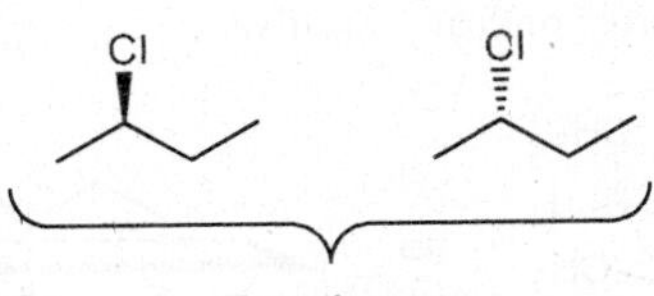

Enantiomers

(c) This compound lacks a chiral center and is therefore achiral. Notice that the carbon atom bearing a wedge and a dash is not a chiral center because that position is connected to two methyl groups.

(d) This compound has a non-superimposable mirror image, and therefore it is chiral.

Enantiomers

(e) This compound is a *meso* compound, which we can see more clearly if we rotate the central carbon-carbon bond by 180 degrees, shown below. Since the compound has a plane of symmetry, it must be achiral. This compound has a superimposable mirror image.

(f) This compound has chiral centers and an internal plane of symmetry and is therefore a *meso* compound. Since the compound has a plane of symmetry, it is achiral. This compound has a superimposable mirror image.

(g) This compound has an internal plane of symmetry (chopping the OH group in half and chopping the methyl group in half). As such, it must be achiral. This compound has a superimposable mirror image.

(h) This compound has a non-superimposable mirror image, and therefore it is chiral.

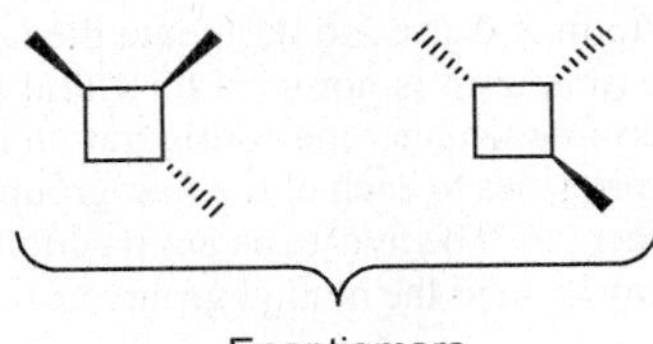

Enantiomers

(i) This compound has a non-superimposable mirror image, and therefore it is chiral.

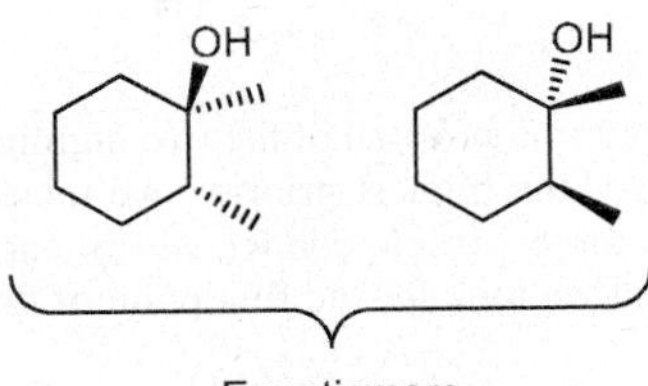

Enantiomers

(j) This compound has no chiral centers and is therefore achiral. This compound has a superimposable mirror image.

(k) This compound has chiral centers and an internal plane of symmetry and is therefore a *meso* compound. As such, it must be achiral. This compound has a superimposable mirror image.

(l) This compound has chiral centers and an internal plane of symmetry (shown below), and is therefore a *meso* compound. Since the compound has a plane of symmetry, it is achiral. This compound has a superimposable mirror image.

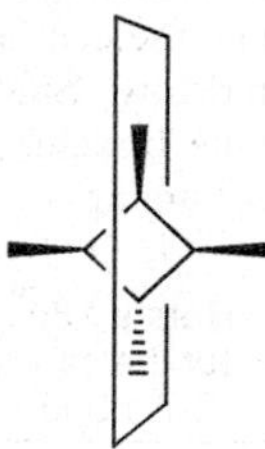

5.56. The concentration (c) is calculated by dividing 0.100 g by 10.0 mL, giving 0.0100 g/mL, which is then plugged into the following equation:

$$[\alpha] = \frac{\alpha}{c \times l}$$

$$\alpha = [\alpha] \times c \times l$$

$$\alpha = (+24)(0.0100 \text{ g/mL})(1.00 \text{ dm}) = +0.24\ ^\circ$$

5.57. The correct answer is (a). The configuration of a chiral center does not depend on temperature, so (d) is not the correct answer. The answer also cannot be (c), because the term Z is used to designate the configuration of an alkene (this term is not used for chiral centers). In order to determine whether the configuration is R or S, we must assign priorities to each of the four groups connected to the chiral center. The hydrogen atom certainly receives the fourth priority, and the methyl group receives the third priority:

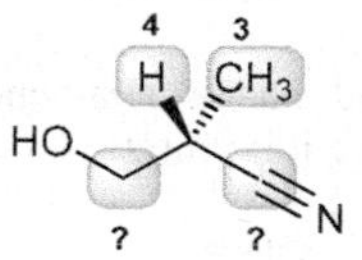

In order to determine which of the two highlighted carbon atoms receives the highest priority, we must construct a list of the three atoms connected to each of those positions, and we look for the first point of difference:

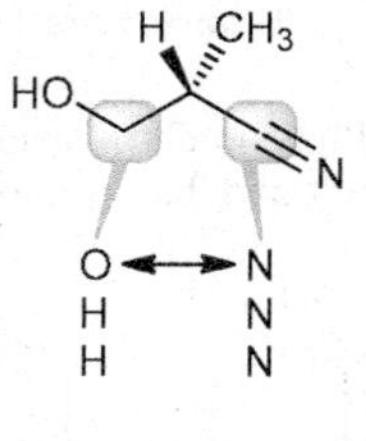

Since O has a higher atomic number than N, the priorities are as follows:

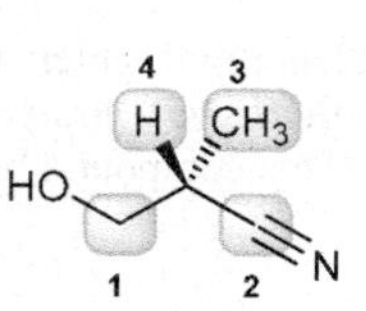

Since the fourth priority (H) is on a wedge (rather than a dash), these priorities correspond with an R configuration, so the correct answer is (a).

5.58. The correct answer is (b).
These compounds have the same connectivity – they differ only in the spatial arrangement of atoms. Therefore, they must be stereoisomers, so the answer is not (c) or (d). To determine whether these compounds are enantiomers or diastereomers, we must decide whether the compounds are mirror images or not. These compounds are NOT mirror images of each other, so they cannot be enantiomers. Since they are stereoisomers but not enantiomers, they must be diastereomers.
Note that there are three chiral centers (highlighted below) in each of these compounds. These compounds differ from each other only in the configuration of one of these chiral centers, thereby justifying their designation as diastereomers:

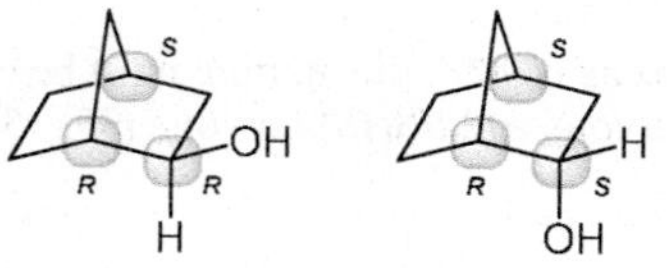

5.59. The correct answer is (d). Compound (a) has a plane of symmetry and is therefore a *meso* compound. This compound is achiral and will be optically inactive:

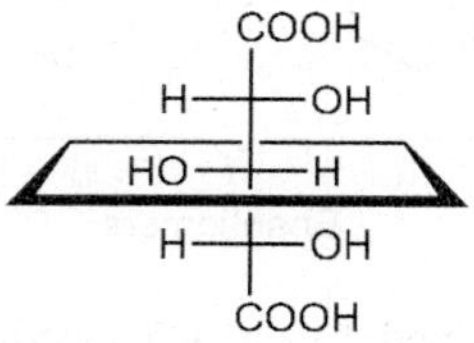

Compound (b) is also a *meso* compound (as shown below) and is therefore optically inactive.

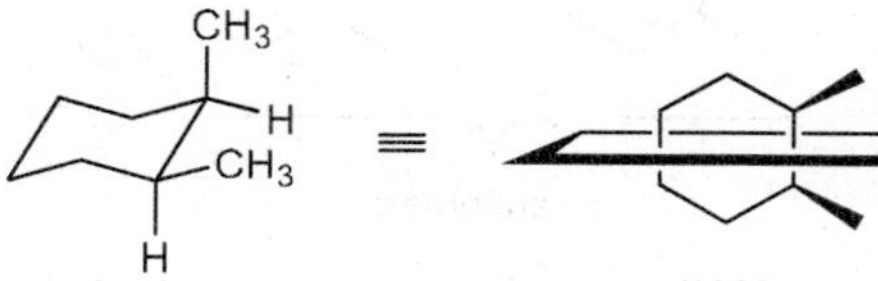

Compound (c) is also a *meso* compound, as shown below, so it is also achiral and optically inactive:

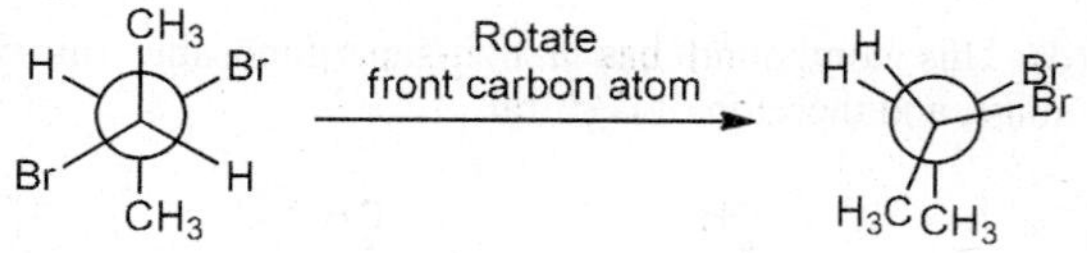

By process of elimination, the correct answer must be (d). Indeed, this compound is chiral (no plane of symmetry) and therefore optically active:

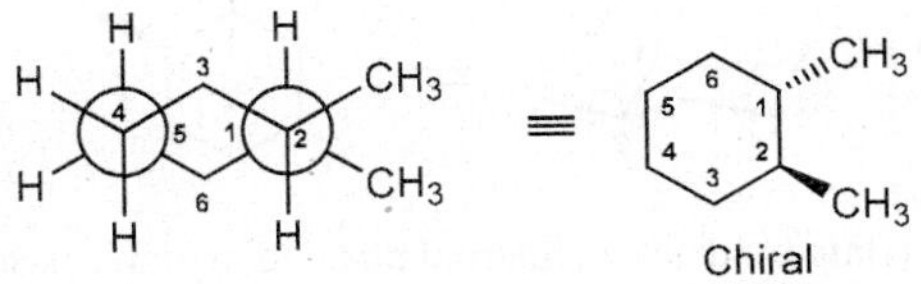

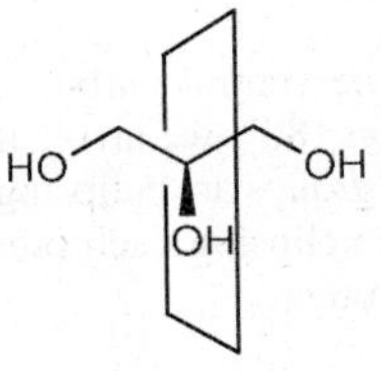

Chiral

5.60. The correct answer is (b). Only compound II is a *meso* compound. A *meso* compound is a molecule that has chiral centers but also has reflectional symmetry (making it achiral). At first glance, it might be tempting to suggest that compound I is a *meso* compound, because both OH groups are on wedges. But close inspection reveals that the compound does not have a plane of symmetry:

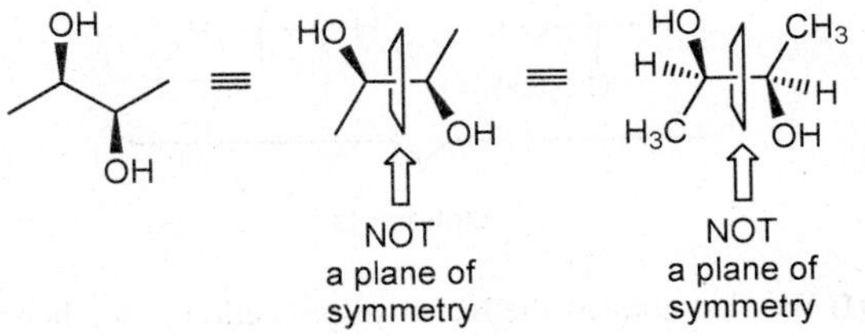

NOT a plane of symmetry

NOT a plane of symmetry

This is perhaps more apparent if we rotate the central C–C bond, and then we can clearly see that the groups are not reflecting each other.

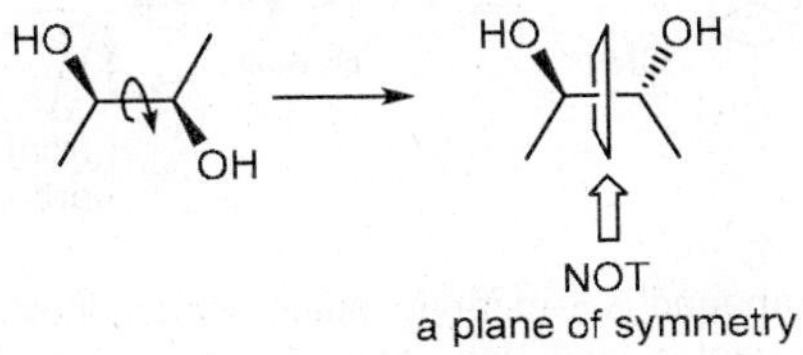

NOT a plane of symmetry

Compound I is not a *meso* compound because it lacks reflectional symmetry (it is chiral).
Only compound II is a *meso* compound, because it has chiral centers AND it has a plane of symmetry:

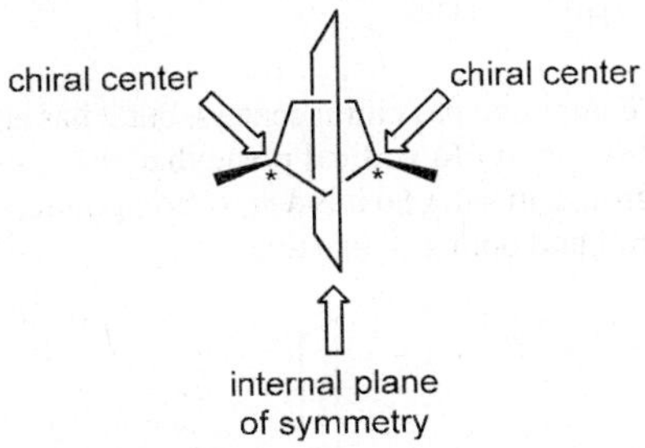

chiral center chiral center

internal plane of symmetry

Compound III is simply an achiral molecule with no chiral centers.

internal plane of symmetry, but NO chiral centers

5.61. Choices (a) and (b) are superimposable with compound **A**; in other words, they are identical to compound **A**. Choice (d) is a diastereomer of compound **A**. The correct answer is (c). Compound **A** is (2*R*,3*R*)-2-bromo-3-chloropentane, and choice (c) is its enantiomer, (2*S*,3*S*)-2-bromo-3-chloropentane. When viewed from the correct perspective, the mirror image relationship of the two enantiomers can be observed.

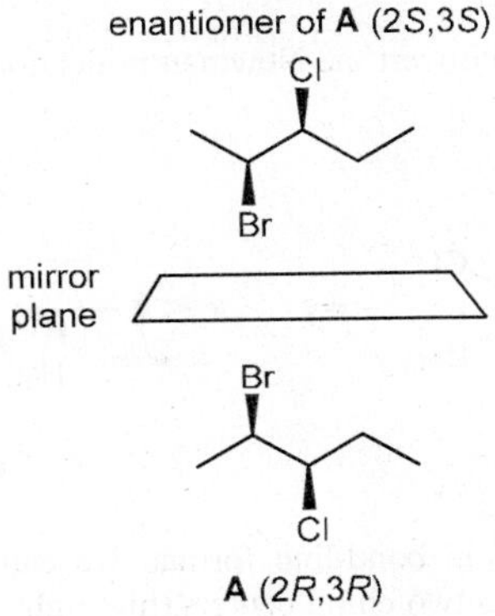

enantiomer of **A** (2*S*,3*S*)

mirror plane

A (2*R*,3*R*)

5.62. The correct answer is (a). If the given compound is converted to a bond-line drawing, we can clearly see that the molecule has an internal plane of symmetry (it is a *meso* compound). Therefore, the compound is achiral and does not have an enantiomer (it is superimposable with its mirror image).

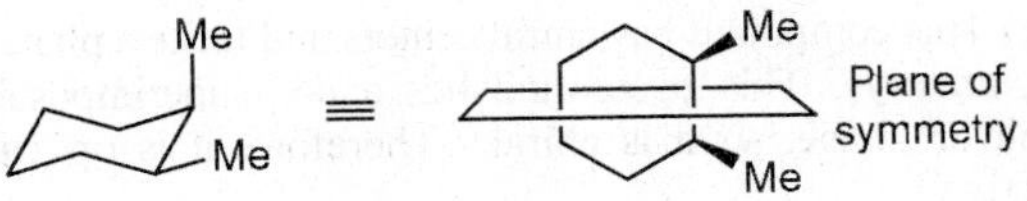

Plane of symmetry

5.63. The correct answer is (c). Prioritization begins by comparing the atoms attached to the chiral center (O, C, and S, highlighted below):

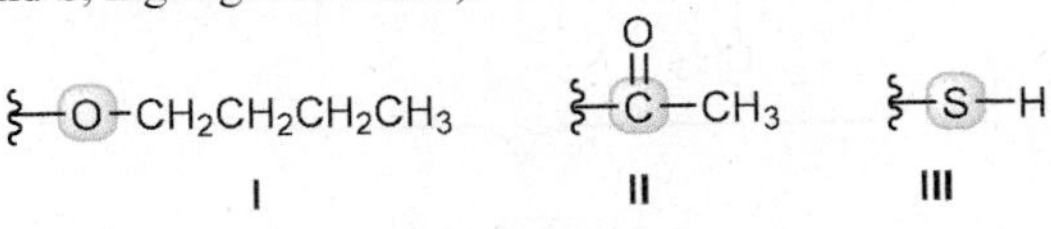

$-O-CH_2CH_2CH_2CH_3$ (I) $-C(=O)-CH_3$ (II) $-S-H$ (III)

Sulfur has the highest atomic number, so it is assigned priority 1. Oxygen has the second highest atomic number, so it is assigned priority 2. Carbon is assigned priority 3. When arranged from lowest to highest priority, the answer is:

II < I < III

5.64.

(a) If we rotate the central carbon-carbon bond of the Newman projection 180°, we arrive at a conformation in which the two OH groups are eclipsing each other, the two methyl groups are eclipsing each other, and the two H's are eclipsing each other:

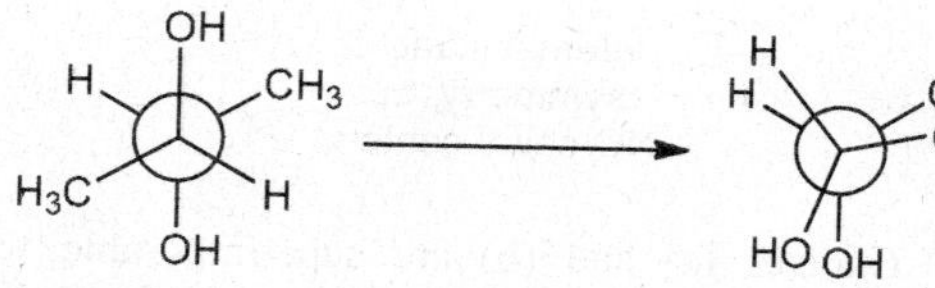

In this eclipsed conformation, we can see more clearly that the molecule has an internal plane of symmetry in the plane of the page, in between the front carbon atom and the back carbon atom of the Newman projection (this compound is a *meso* compound). Therefore, the compound is achiral and optically inactive.

(b) We first convert the Newman projection into a bond-line drawing:

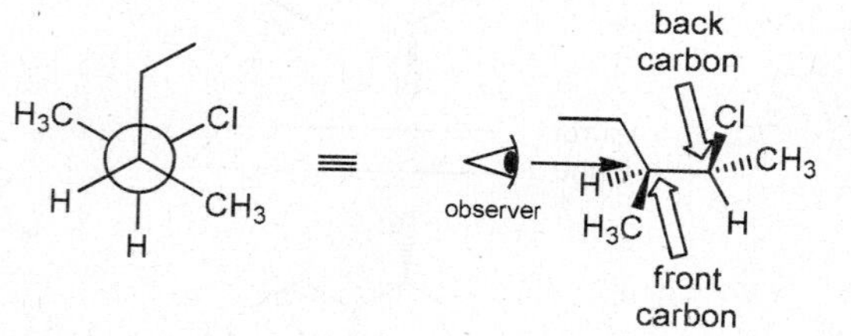

When drawn in bond-line format, we can see that the compound has two chiral centers (highlighted):

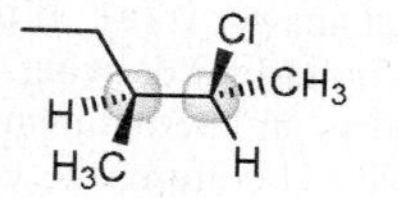

This compound is chiral and therefore optically active.

(c) This compound has chiral centers and lacks a plane of symmetry. This compound has a non-superimposable mirror image, so it is chiral. Therefore, it is optically active.

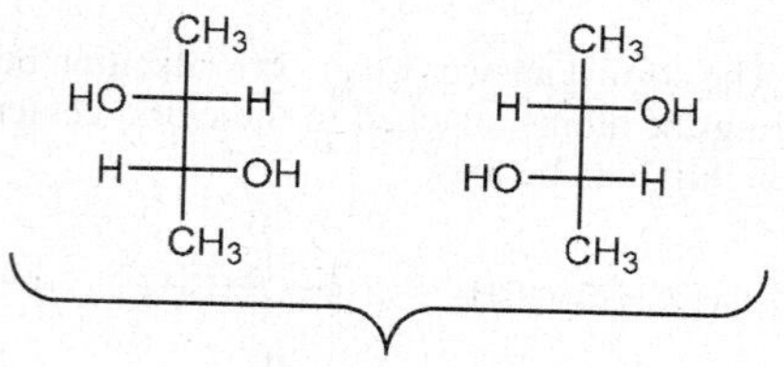

Enantiomers

(d) Let's redraw the compound in a way that shows the configuration of each chiral center without showing the chair conformation. This will make it easier for us to evaluate:

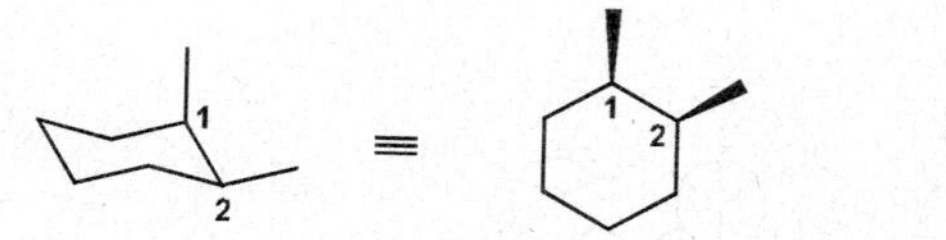

This compound has an internal plane of symmetry, so it is a *meso* compound. As such, it is achiral and optically inactive.

(e) This compound has chiral centers and lacks a plane of symmetry. This compound has a non-superimposable mirror image, so it is chiral. Therefore, it is optically active.

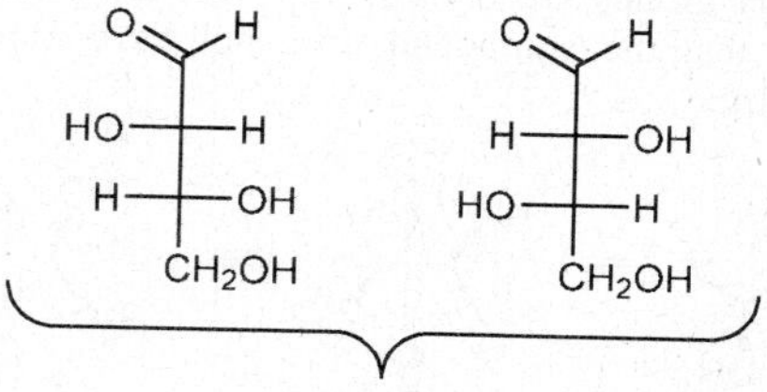

Enantiomers

(f) We first convert the Newman projection into a bond-line drawing:

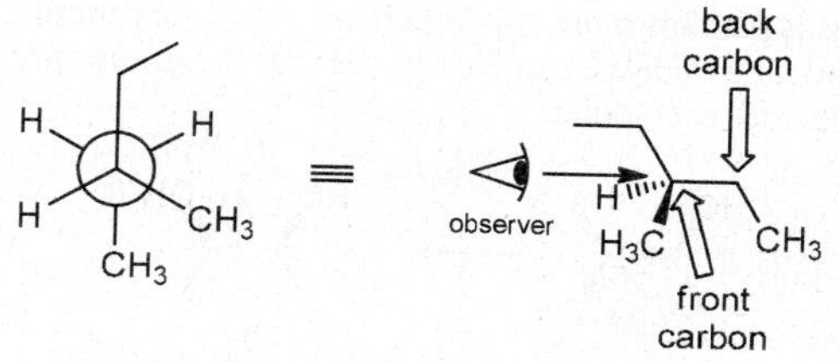

This compound is 3-methylpentane, which does not have a chiral center (and does have a plane of symmetry). Therefore, it is optically inactive.

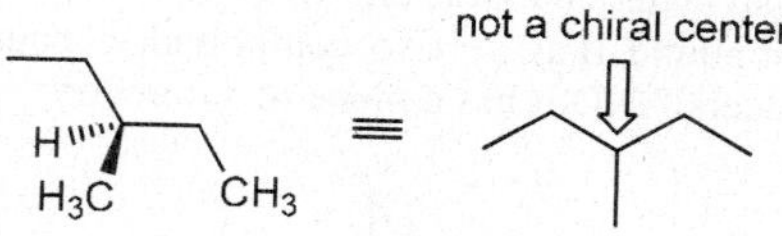

not a chiral center

(g) This compound has chiral centers, but it has an internal plane of symmetry (a vertical plane that chops one of the methyl groups in half), so it is a *meso* compound. As such, it is achiral and optically inactive.

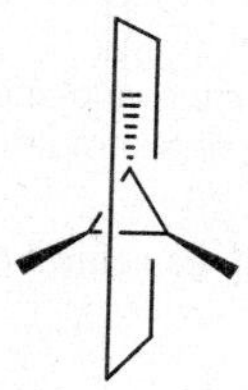

(h) This compound has an internal plane of symmetry, as shown below. As such, it is achiral and optically inactive.

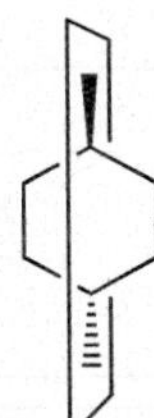

5.65. In each case, begin by numbering the carbon atoms in the Fischer projection (from top to bottom) and then draw the skeleton of a bond-line drawing with the same number of carbon atoms (note that this numbering system does not need to follow IUPAC rules). Then, place the substituents in their correct locations (by comparing the numbering system in the Fischer projection with the numbering system in the bond-line drawing). When drawing each substituent in the bond-line drawing, you must decide whether it is on a dash or a wedge. For each chiral center, make sure that the configuration is the same as the configuration in the Fischer projection. If necessary, assign the configuration of each chiral center in both the Fischer projection and the bond-line drawing to ensure that you drew the configuration correctly. With enough practice, you may begin to notice some trends (rules of thumb) that will allow you to draw the configurations more quickly.

(a)

(b)

(c)

(d)

(e)

5.66.

(a) The compound in part (a) of the previous problem has an internal plane of symmetry, and is therefore a *meso* compound:

(b) No, the structures shown in parts (b) and (c) of the previous problem are enantiomers. An equal mixture of these two compounds is a racemic mixture, which will be optically inactive.

(c) Yes, this mixture is expected to be optically active, because the structures shown in parts (d) and (e) of the previous problem are not enantiomers. They are diastereomers, which are not expected to exhibit equal and opposite rotations.

5.67. As we saw in problem **5.65**, it is helpful to use a numbering system when converting one type of drawing into another. When drawing each substituent in the Fischer projection, you must decide whether it is on the right or left side of the Fischer projection. For each chiral center, make sure that the configuration is the same as the configuration in the bond-line drawing. If necessary, assign the configuration of each chiral center in both the Fischer projection and the bond-line drawing to ensure that you drew the configuration correctly. With enough practice, you may begin to notice some trends (rules of thumb) that will allow you to draw the configurations more quickly.

(a)

(2*S*, 3*R*)

(b)

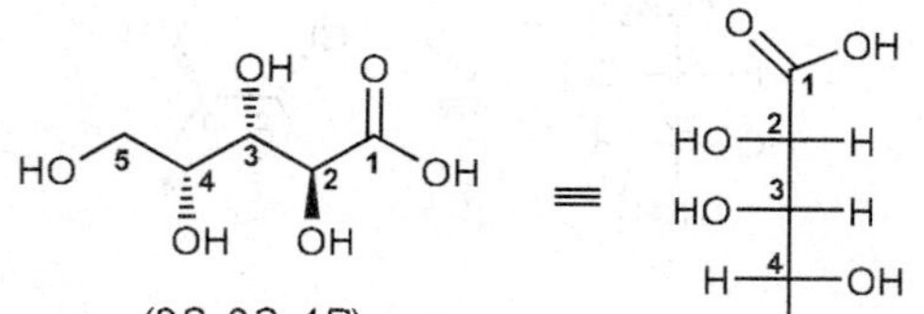

(2S, 3S, 4R)

(c)

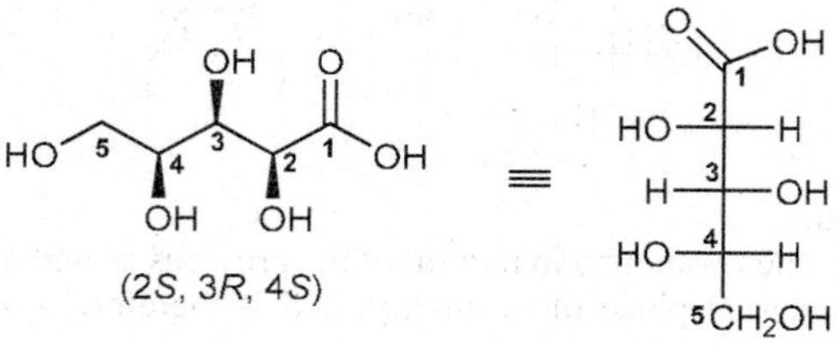

(2S, 3R, 4S)

5.68. As shown below, there are only two stereoisomers (the *cis* isomer and the *trans* isomer).

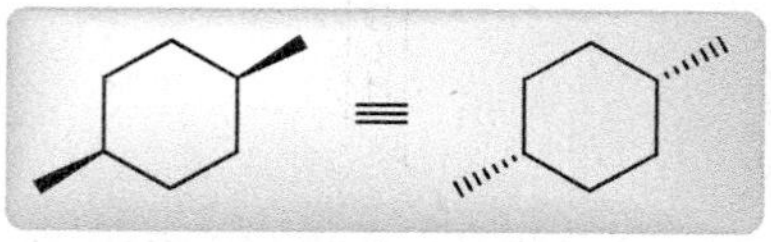

cis

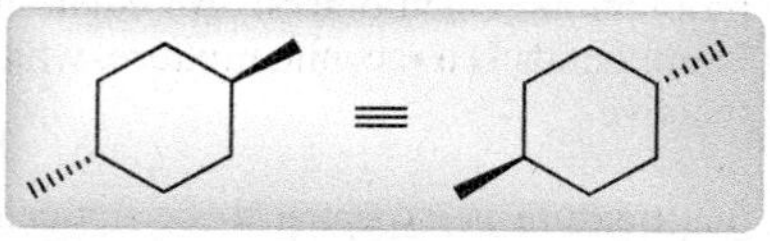

trans

While a structure with two chiral centers would be expected to have four possible stereoisomers, each of these compounds has no chiral centers and a plane of symmetry (and each, therefore, is superimposable on its mirror image).

5.69. With three chiral centers, we would expect a maximum of eight stereoisomers ($2^3 = 8$), labeled **1–8**. However, structures **1** and **2** represent one compound (an achiral *meso* compound), while structures **3** and **4** also represent one compound (an achiral *meso* compound). In addition, structure **5** is the same as structure **7**, while structure **6** is the same as structure **8** (for each pair, rotating one compound 180 degrees about a vertical axis generates the other compound). Structures **5-8** are chiral and are thus not *meso* structures. The reason for the equivalence of structures **5** and **7** (and also for the equivalence of **6** and **8**) is that the central carbon atom in each of these four structures is actually not a chiral center. For structures **5-8**, changing the "configuration" at the central carbon atom does not produce a stereoisomer, which proves that the central carbon atom is not a chiral center in these cases.
In summary, there are only four stereoisomers (structures **1/2**, **3/4**, **5/7**, and **6/8**).

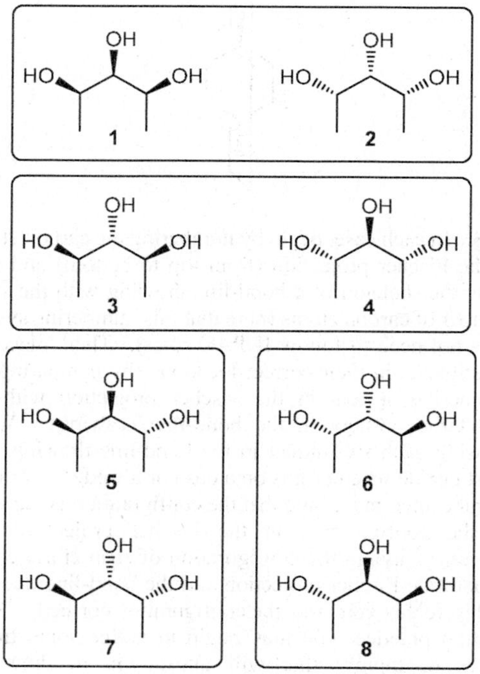

5.70.
(a) The second compound is 2-methylpentane. If we redraw the first compound as a bond-line drawing, rather than a Newman projection, we see that the first compound is 3-methylpentane.

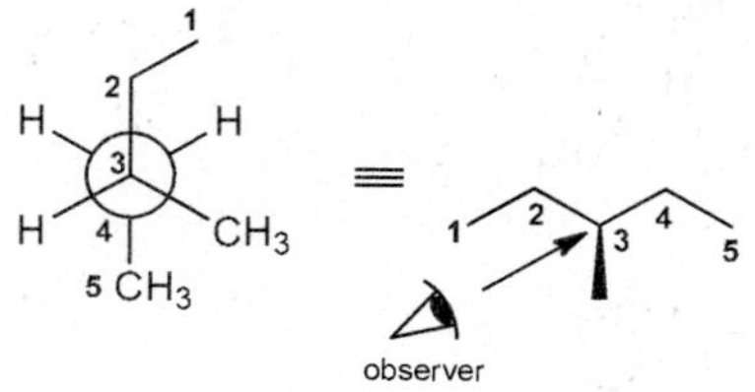

observer

These two compounds, 2-methylpentane and 3-methylpentane, have the same molecular formula (C_6H_{14}) but different constitution, so they are constitutional isomers.

(b) The first compound is *trans*-1,2-dimethyl-cyclohexane:

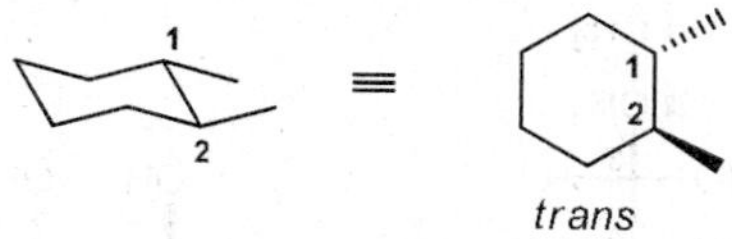

trans

In contrast, the second compound is *cis*-1,2-dimethyl-cyclohexane. These compounds are stereoisomers, but they are not mirror images of each other. Therefore, they are diastereomers.

5.71. This compound lacks a chiral center but is nevertheless chiral because of the relative orientations of the two methyl groups (just like chiral allenes). As an achiral compound, this compound is not superimposable on its mirror image, shown below.

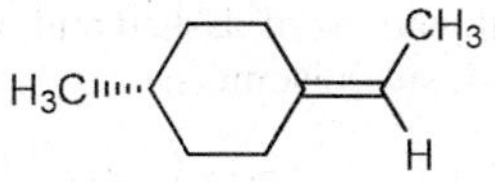

5.72. This compound has a center of inversion, which is a form of reflection symmetry. As a result, this compound is achiral (it is *superimposable* on its mirror image) and is therefore optically inactive.

5.73. The compound contains three chiral centers, with the following assignments:

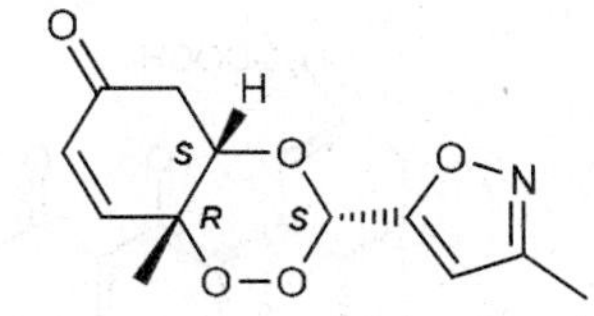

Note that in all three cases the lowest priority group (4) is not on a dash, so your point of view must be adjusted before assigning the configuration (e.g., by switching groups).

With three chiral centers, a total of $2^3 = 8$ stereoisomers are possible, shown below. Pairs of enantiomers are highlighted together. All other relationships are diastereomeric.

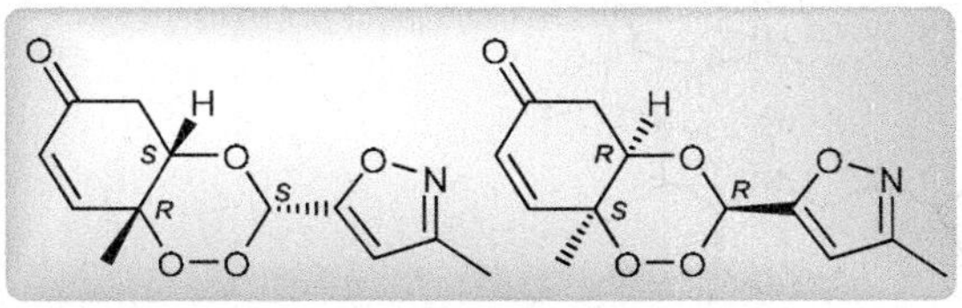

enantiomers

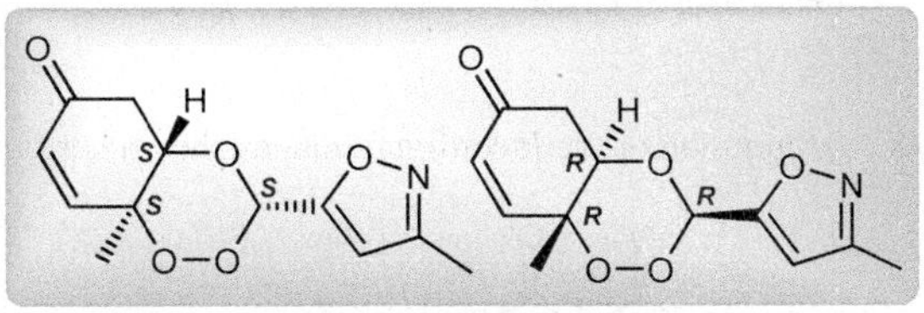

enantiomers

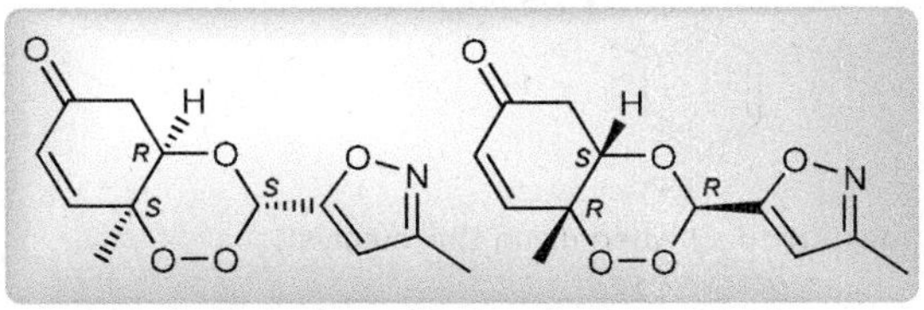

enantiomers

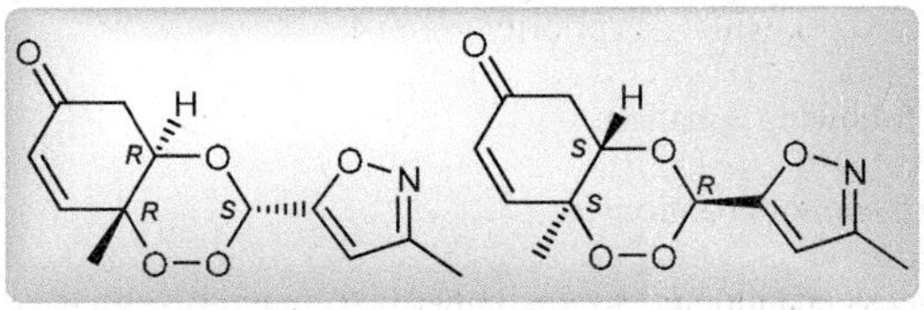

enantiomers

5.74.

(a) The following are the prioritization schemes that give rise to the correct assignment of configuration for each chiral center.

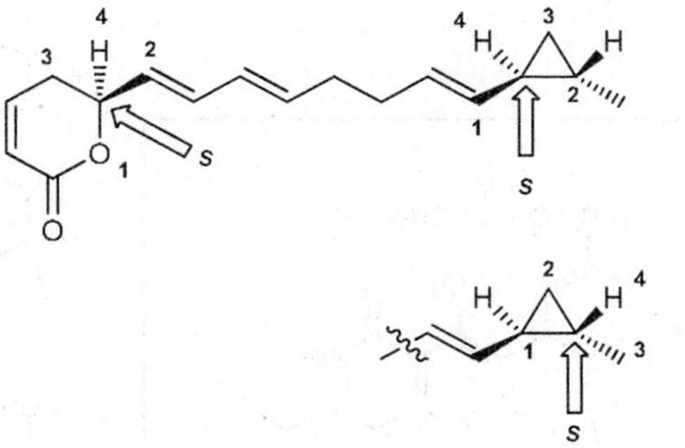

(b) The total number of possible stereoisomers is 2^n (where n = the number of chiral centers). With three chiral centers, we expect $2^3 = 8$ possible stereoisomers, one of which is the natural product coibacin B.

5.75. There are two different C=C π bonds in this compound. For clarity, some of the hydrogen atoms have been drawn explicitly in the drawing below. The C=C π bond on the left is not stereoisomeric, because there are two hydrogen atoms (highlighted) attached to the same position.

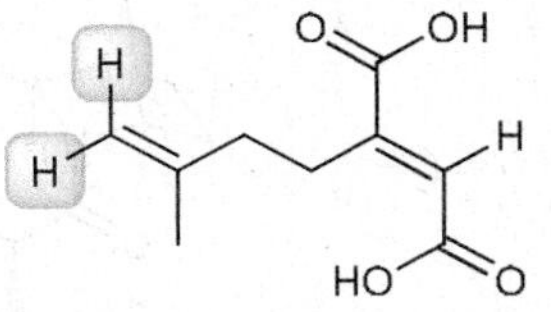

The C=C π bond on the right is stereoisomeric. The two high-priority carboxylic acid groups are on opposite sides of the π bond, giving the *E* configuration.

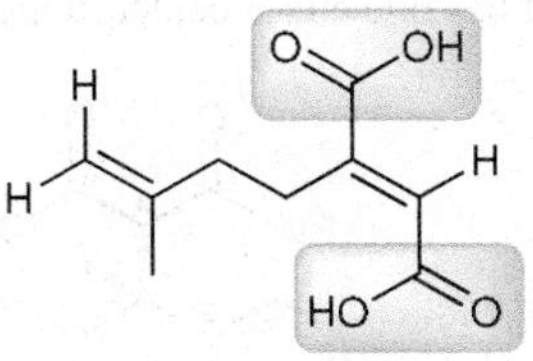

This compound has one stereoisomer, in which the carboxylic acid groups are on the same side of the π bond (the *Z* configuration), shown below:

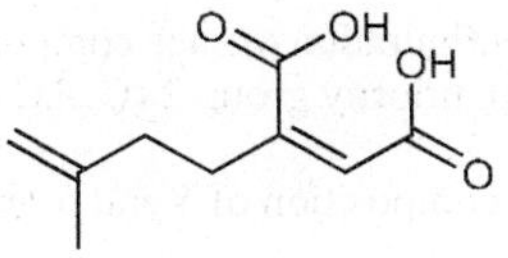

5.76. When you draw out the condensed group at the top, you will find that, as an ester, it contains no chiral center. There are only two carbon atoms, marked with asterisks below, that each bear four different groups (and are therefore chiral centers). To draw the four possible stereoisomers, draw the various combinations of dashed and wedged bonds at these two chiral centers. Since there are two chiral centers, there will be 2^2, or 4, stereoisomers.

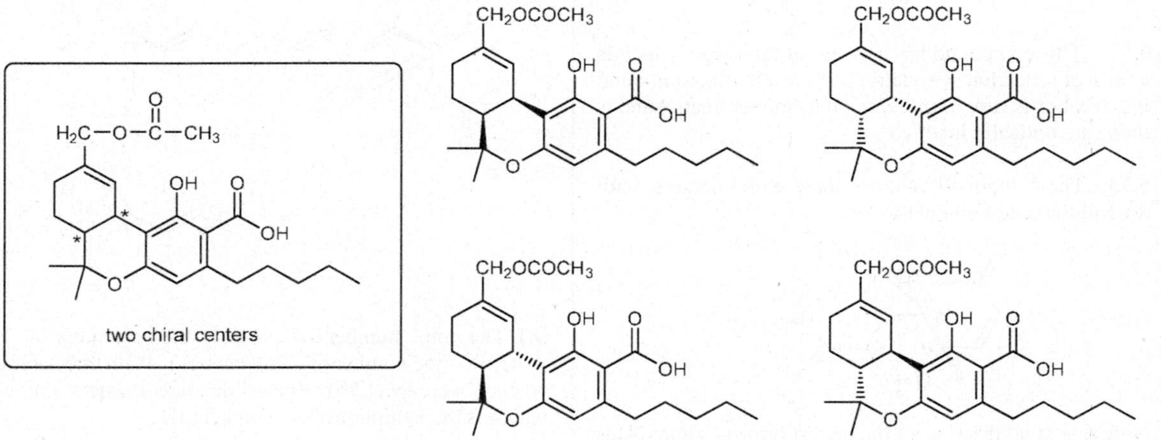

two chiral centers

5.77. The enantiomer of a chiral molecule is its mirror image. The mirror can be placed anywhere, so there are multiple ways of correctly representing the enantiomer depending on where you place the mirror. In this case, the stereochemistry of the bicyclic part of the compound is implied by the drawing, so the mirror is most easily placed on the side of the molecule. Notice that the wedge remains a wedge because a mirror placed on the side reverses left and right sides only; it does not exchange front and back.

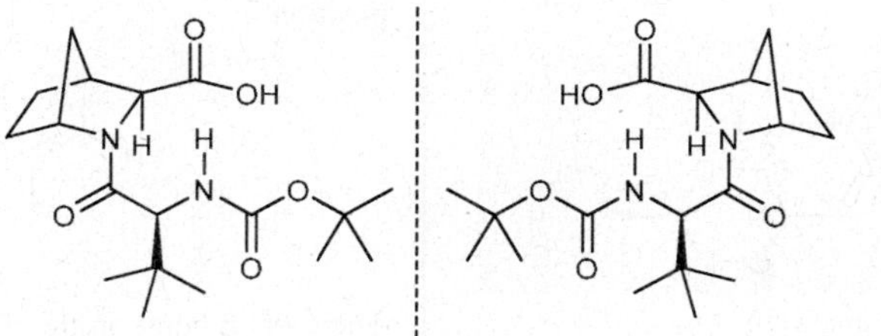

1 | Enantiomer of **1**

5.78.

(a) The product has one chiral center, which can either have the *S* configuration or the *R* configuration, as shown here.

(*S*)-product
(major product when using Cu[I] catalyst)

(*R*)-product
(major product when using Cu[II] catalyst)

To assign the configuration of each compound, the priorities of the three attached carbon groups are shown above: priority group 1 (N,H,H), priority group 2 (C,C,C) and priority group 3 (C,C,H), with H as priority group 4.

(b) The percent composition of *S* and *R* are calculated using the following equation:

$$\%\ ee = (\%\ \text{major enantiomer} - \%\ \text{minor enantiomer})$$

The problem statement indicates that the *S* enantiomer is the excess enantiomer, so the higher percentage is associated with *S*, as shown in the table below:

Solvent	% *ee*	% *S*	% *R*
toluene	24	62	38
tetrahydrofuran	48	74	26
CH_3CN	72	86	14
$CHCl_3$	30	65	35
CH_2Cl_2	46	73	27
hexane	51	75.5	24.5

Acetonitrile (CH_3CN) is the best choice of solvent because it results in the highest combination of *ee* (72%) and percent yield (55%). While toluene gives the same percent yield (55%), the *ee* in this solvent is significantly lower (only 24%).

5.79. There is one chiral center, which was incorrectly assigned. So, it must have the *R* configuration, rather than the *S* configuration (as originally thought). To draw the enantiomer, we replace the dash with a wedge, and we replace the wedge with a dash, as shown:

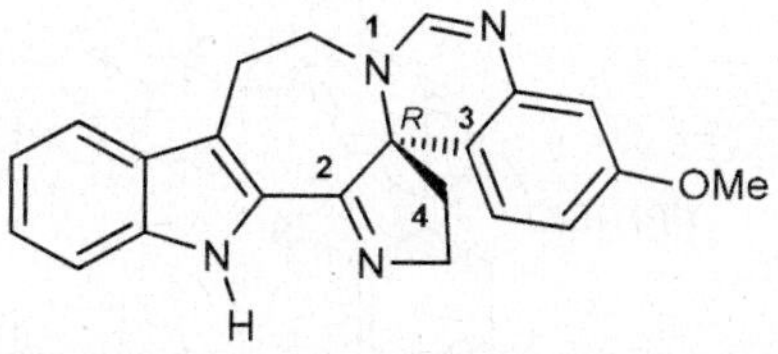

Correct structure of (+)-trigonoliimine A
(enantiomer of the structure reported in 2010).

5.80.
(a) To draw the enantiomer, we simply redraw the structure in the problem statement, except that we replace all dashes with wedges, and all wedges with dashes, as shown:

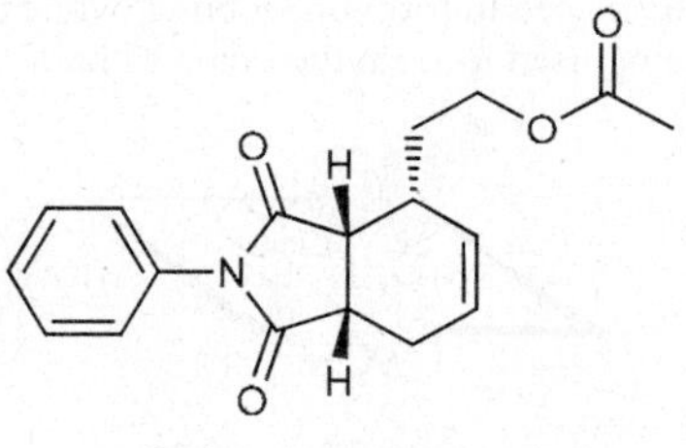

(2nd major product)

(b) The following compounds are the minor products, as described in the problem statement (*cis* connectivity at the bridgehead carbons).

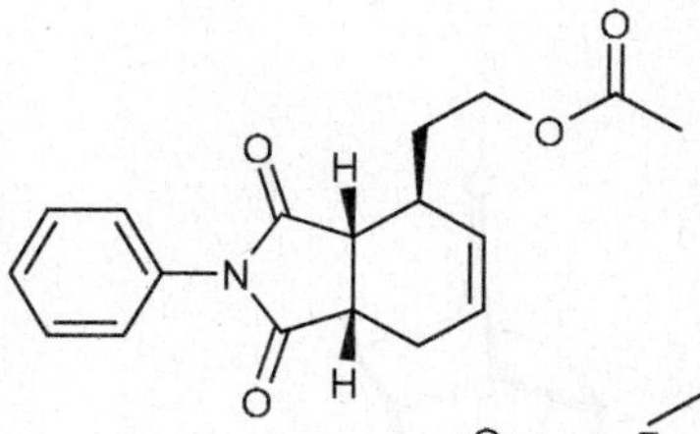

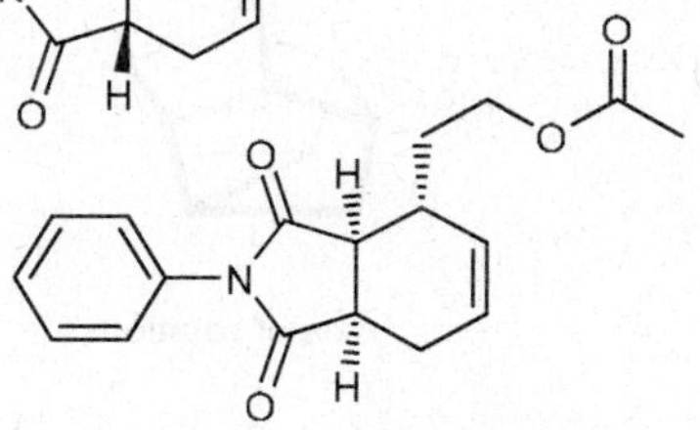

(c) The minor products are nonsuperimposable mirror images of each other. Therefore, they are enantiomers.

(d) They are stereoisomers, but they are not mirror images of each other. Therefore, they are diastereomers.

5.81.
The following are two examples of correct answers, where the molecule is viewed from different perspectives.

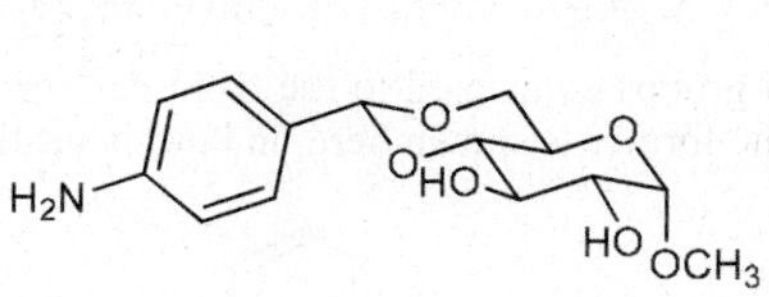

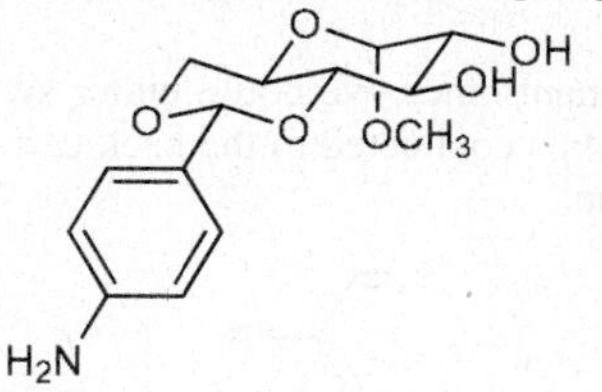

A suggested approach to this problem:

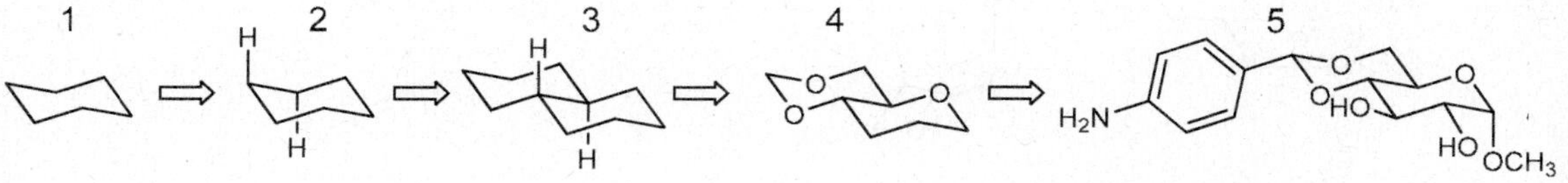

1) Draw a chair structure of the ring on the right side of the compound.
2) Now, to find an appropriate place to connect the second ring, find two axial positions on adjacent carbon atoms so that they are down and up when going counterclockwise around the ring, as they are in the wedge and dash drawing. These are the two bridgehead positions.
3) Draw the second ring (with connecting bonds equatorial to the first ring).
4) Replace appropriate methylene groups in the rings with oxygen atoms.
5) Draw all substituents with those on dashes pointing "down" in the chair form, and those on wedges pointing "up."

5.82.

(a) The compound exhibits rotational symmetry, because it possesses an axis of symmetry (consider rotating the molecule 180° about this vertical axis). You might find it helpful to construct a molecular model of this compound.

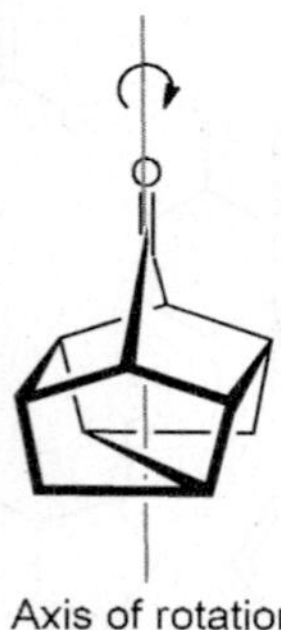

Axis of rotation

(b) The compound lacks reflectional symmetry; it does not have a plane of symmetry.

(c) Chirality is not dependent on the presence or absence of rotational symmetry. It is only dependent on the presence or absence of reflectional symmetry. This compound lacks reflectional symmetry and is therefore chiral. That is, it has a nonsuperimposable mirror image, drawn here:

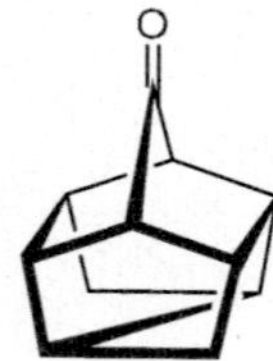

5.83.

(a) In the following Newman projection, the front carbon atom is connected to only two groups, and the back carbon atom is also connected to only two groups:

Notice that the two groups connected to the front carbon atom are twisted 90° with respect to the two groups connected to the back carbon atom. This is because the central carbon atom (in between the front carbon atom and the back carbon atom) is *sp* hybridized – it has two *p* orbitals, which are 90° apart from each other. One *p* orbital is being used to form one π bond, while the other *p* orbital is being used to form the other π bond.

(b) To draw the enantiomer, we could either switch the two groups connected to the front carbon atom, or we could switch the two groups connected to the back carbon atom. The former is shown here, in both bond-line format and in a Newman projection.

(c) To draw a diastereomer of the original compound, simply convert the *trans* configuration of the alkene to a *cis* configuration. We can make the same change on the enantiomer of the original compound (convert from *trans* to *cis*) to draw another diastereomer. The pair of enantiomers below are both diastereomers of the original compound.

5.84.

(a) Glucuronolactone **1L** is the enantiomer of **1D**, which is shown in the problem statement. To draw the enantiomer of **1D**, we simply redraw it, except that we replace all dashes with wedges, and all wedges with dashes, as shown:

(b) There are 5 chiral centers, so there are 32 (or 2^5) possible stereoisomers.

(c)

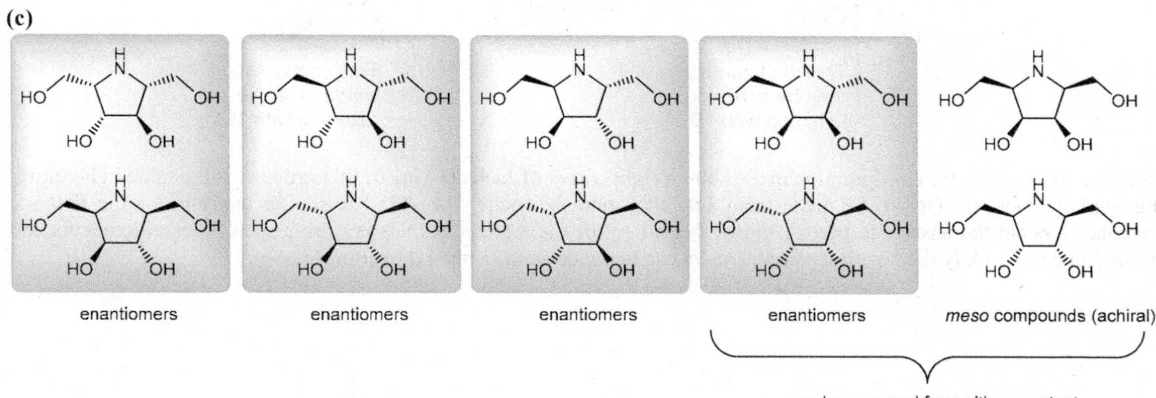

(d) The four products that are accessible from either of the reactants are the four products shown on the right in the solution to part c, as indicated above. Recall that the synthetic protocol allows for control of configurations at C2, C3 and C5, but not at C4. Therefore, in order for a specific stereoisomer to be accessible from either **1D** or from **1L**, that stereoisomer must display a specific feature. To understand this feature, we must draw one of the ten stereoisomers and then redraw it again after rotating it 180 degrees about a vertical axis. For example, let's do this for one of the *meso* compounds:

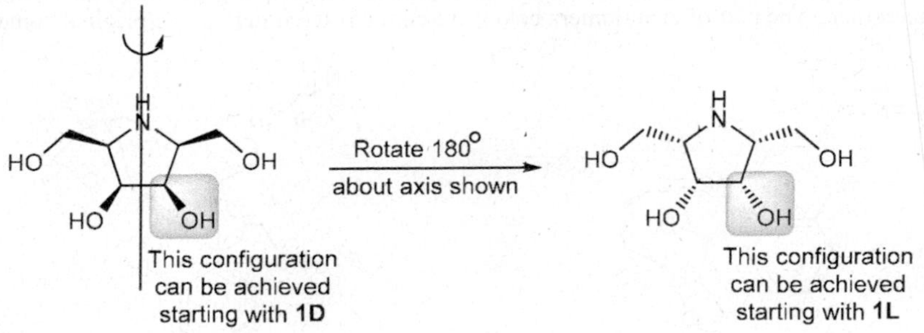

Rotate 180° about axis shown

This configuration can be achieved starting with **1D**

This configuration can be achieved starting with **1L**

Now we look at the configuration of the chiral center in the bottom right corner of each drawing above (highlighted in gray). Notice that they have opposite configuration. This is the necessary feature that enables this compound to be accessible from either **1D** or from **1L**. Here is another example:

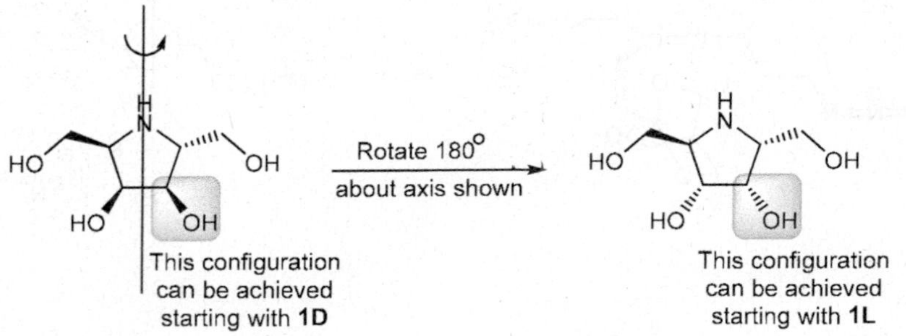

Rotate 180° about axis shown

This configuration can be achieved starting with **1D**

This configuration can be achieved starting with **1L**

Once again, this stereoisomer will be accessible from either **1D** or from **1L**. In contrast, the first six structures (in the answer to part c) do not have this feature. For example, consider the first structure: let's draw it, rotate it 180 degrees, and then inspect the configuration in the bottom right corner of each drawing:

Rotate 180° about axis shown

This configuration can be achieved starting with **1D**

This configuration *cannot* be achieved starting with **1L**

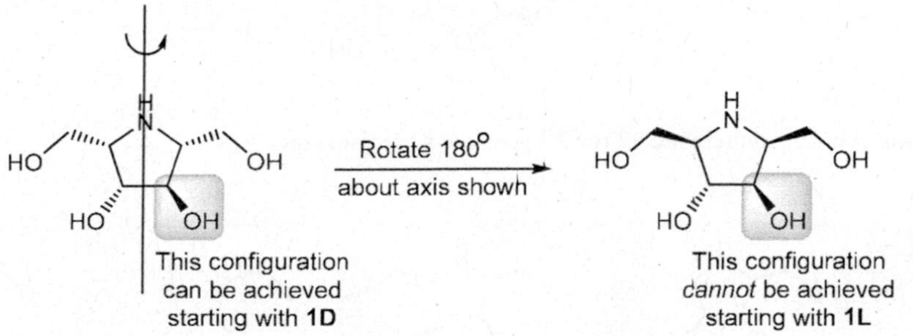

Note that in this case, the configuration in the bottom right corner of each drawing of this structure is the same. Therefore, this stereoisomer can ONLY be made from **1D**. It cannot be made from **1L**. A similar analysis for the first six stereoisomers (in the answer to part c) shows that all six of these stereoisomers require a specific enantiomer for the starting material. Only the last four stereoisomers can be made from either **1D** or from **1L**.

Chapter 6
Chemical Reactivity and Mechanisms

Review of Concepts

Fill in the blanks below. To verify that your answers are correct, look in your textbook at the end of Chapter 6. Each of the sentences below appears verbatim in the section entitled *Review of Concepts and Vocabulary*.

- _______________ reactions involve a transfer of energy from the system to the surroundings, while _____________ reactions involve a transfer of energy from the surroundings to the system.
- Each type of bond has a unique _______________________ **energy**, which is the amount of energy necessary to accomplish **homolytic bond cleavage**.
- **Entropy** is loosely defined as the __________ of a system.
- In order for a process to be spontaneous, the change in _________________ must be negative.
- The study of relative energy levels (ΔG) and equilibrium concentrations (K_{eq}) is called _________________.
- The study of reaction rates is called ____________.
- ____________ speed up the rate of a reaction by providing an alternate pathway with a lower **energy of activation**.
- On an energy diagram, each peak represents a _______________ while each valley represents _____________________.
- A ________________ is an electron-rich molecule or ion that is capable of donating a pair of electrons.
- An _________________ is an electron-deficient molecule or ion that is capable of accepting a pair of electrons.
- For **ionic reactions**, there are four characteristic arrow-pushing patterns: 1) _________________, 2) _________________, 3) _________________, and 4) _________________.
- As a result of **hyperconjugation**, __________ carbocations are more stable than secondary carbocations, which are more stable than __________ carbocations.

Review of Skills

Fill in the empty boxes below. To verify that your answers are correct, look in your textbook at the end of Chapter 6. The answers appear in the section entitled *SkillBuilder Review*.

SkillBuilder 6.1 Predicting $\Delta H°$ of a Reaction

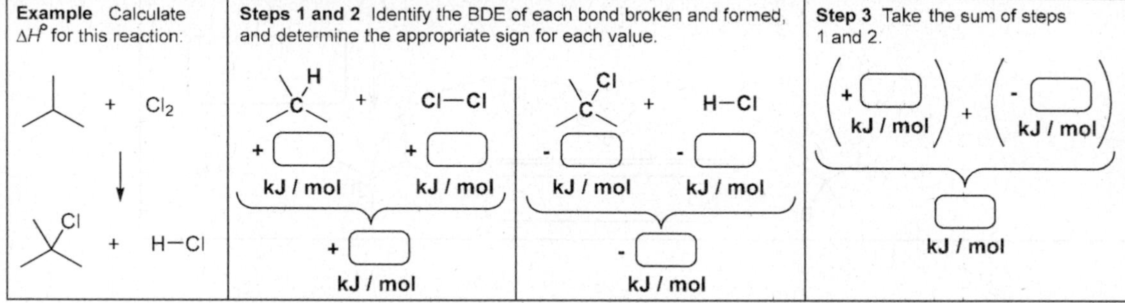

SkillBuilder 6.2 Identifying Nucleophilic and Electrophilic Centers

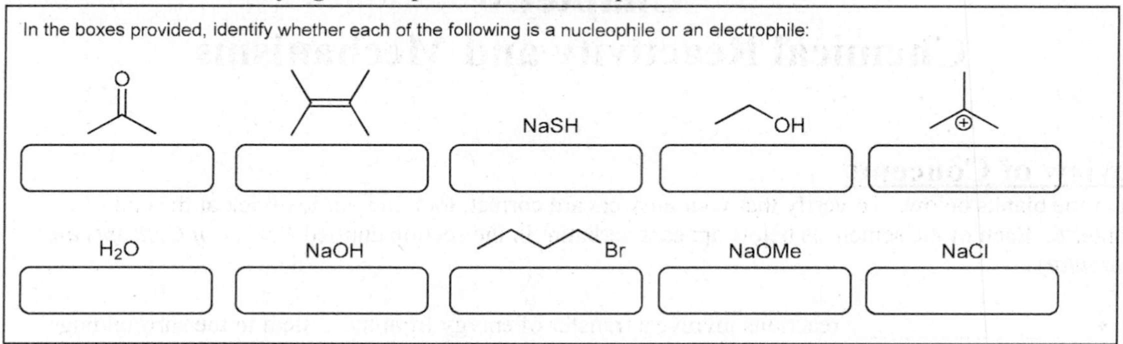

SkillBuilder 6.3 Identifying an Arrow-Pushing Pattern

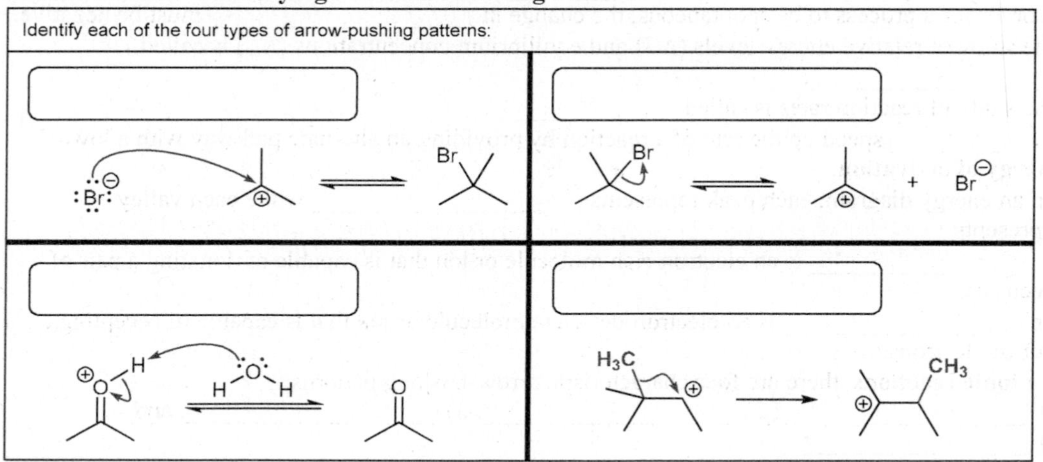

SkillBuilder 6.4 Identifying a Sequence of Arrow-Pushing Patterns

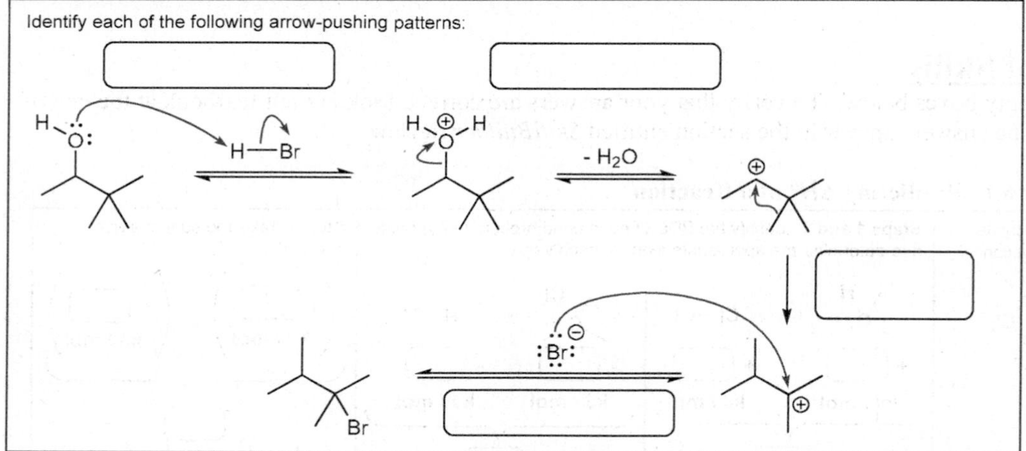

SkillBuilder 6.5 Drawing Curved Arrows

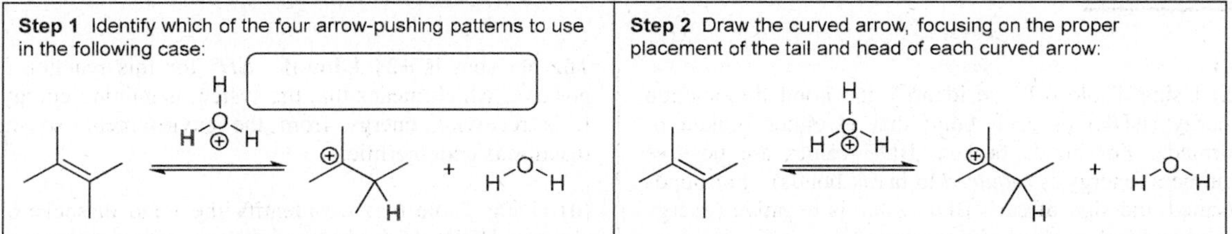

Step 1 Identify which of the four arrow-pushing patterns to use in the following case:
Step 2 Draw the curved arrow, focusing on the proper placement of the tail and head of each curved arrow:

SkillBuilder 6.6 Predicting Carbocation Rearrangements

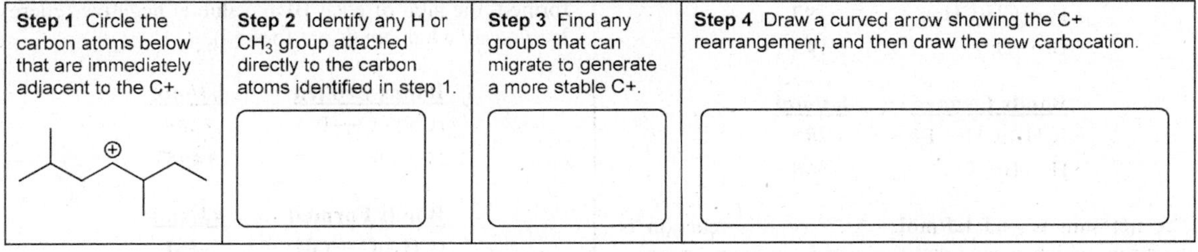

Step 1 Circle the carbon atoms below that are immediately adjacent to the C+.
Step 2 Identify any H or CH_3 group attached directly to the carbon atoms identified in step 1.
Step 3 Find any groups that can migrate to generate a more stable C+.
Step 4 Draw a curved arrow showing the C+ rearrangement, and then draw the new carbocation.

Common Mistakes to Avoid

In this chapter, we learned many skills that are necessary for drawing reaction mechanisms (identifying nucleophilic and electrophilic centers, drawing curved arrows, identifying arrow-pushing patterns, etc.). We will use these skills frequently in the upcoming chapters. In particular, it is important to become proficient with curved arrows, as they represent the language of reaction mechanisms, and you will have to become fluent in that language as we progress through the textbook. There are many common mistakes that students make when drawing curved arrows, and most of those mistakes can be avoided if you always remember that curved arrows represent the motion of electrons. That is, the tail of every curved arrow must identify which electrons are moving, and the head of every curved arrow must show where those electrons are going. Let's first focus on the tail. The tail must always be placed on electrons, which means that it must be placed either on a lone pair or on a bond. If we examine each of the four characteristic arrow-pushing patterns below, we see this clearly:

Nucleophilic attack	**Loss of a leaving group**
Proton transfer	**Rearrangement**

Notice that the tail of each curved arrow is placed on a lone pair or on a bond. Similarly, the head of every curved arrow must either show the formation of a lone pair or the formation of a bond. Look at all of the curved arrows above and convince yourself that this is correct. If you keep this in mind when drawing curved arrows, you can avoid many unnecessary errors.

Solutions

6.1.

(a) Using Table 6.1, we identify the bond dissociation energy (BDE) of each bond that is either broken or formed. For bonds broken, BDE values are positive numbers (energy is *required* to break bonds). For bonds formed, the sign of each BDE value is negative (energy is *released* when bonds are formed).

Bonds Broken	kJ/mol
H—CH$(CH_3)_2$	**+397**
Br—Br	**+193**

Bonds Formed	kJ/mol
$(CH_3)_2$CH—Br	**–285**
H—Br	**–368**

The net sum is **–63 kJ/mol**. $\Delta H°$ for this reaction is negative, which means that the system is losing energy. It is giving off energy to the environment, so the reaction is exothermic.

(b) Using Table 6.1, we identify the bond dissociation energy (BDE) of each bond that is either broken or formed. For bonds broken, BDE values are positive numbers (energy is *required* to break bonds). For bonds formed, the sign of each BDE value is negative (energy is *released* when bonds are formed).

Bonds Broken	kJ/mol
$(CH_3)_3$C—Cl	**+331**
H—OH	**+498**

Bonds Formed	kJ/mol
$(CH_3)_3$C—OH	**–381**
H—Cl	**–431**

The net sum is **+17 kJ/mol**. $\Delta H°$ for this reaction is positive, which means that the system is gaining energy. It is receiving energy from the environment, so the reaction is endothermic.

(c) Using Table 6.1, we identify the bond dissociation energy (BDE) of each bond that is either broken or formed. For bonds broken, BDE values are positive numbers (energy is *required* to break bonds). For bonds formed, the sign of each BDE value is negative (energy is *released* when bonds are formed).

Bonds Broken	kJ/mol
$(CH_3)_3$C—Br	**+272**
H—OH	**+498**

Bonds Formed	kJ/mol
$(CH_3)_3$C—OH	**–381**
H—Br	**–368**

The net sum is **+21 kJ/mol**. $\Delta H°$ for this reaction is positive, which means that the system is gaining energy. It is receiving energy from the environment, so the reaction is endothermic.

(d) Using Table 6.1, we identify the bond dissociation energy (BDE) of each bond that is either broken or formed. For bonds broken, BDE values are positive numbers (energy is *required* to break bonds). For bonds formed, the sign of each BDE value is negative (energy is *released* when bonds are formed).

Bonds Broken	kJ/mol
$(CH_3)_3$C—I	**+209**
H—OH	**+498**

Bonds Formed	kJ/mol
$(CH_3)_3$C—OH	**–381**
H—I	**–297**

The net sum is **+29 kJ/mol**. $\Delta H°$ for this reaction is positive, which means that the system is gaining energy. It is receiving energy from the environment, so the reaction is endothermic.

6.2. Begin by identifying all of the bonds that are being made or broken in the reaction. For simplicity, we will focus on the region of the molecule where the change is taking place, and we redraw the reaction as follows:

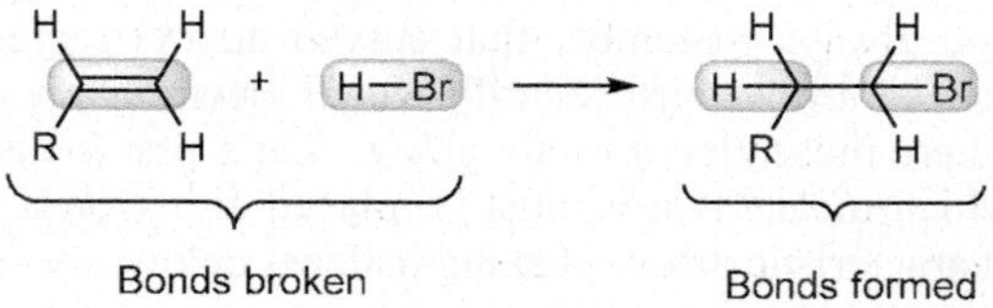

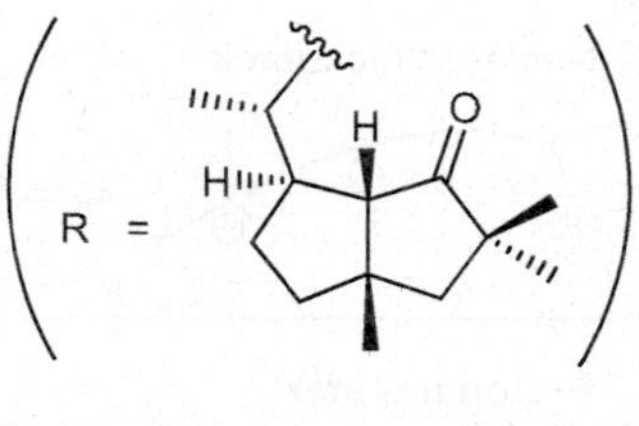

The highlighted bonds represent the bonds that are broken or formed. Notice that the C=C bond is not completely broken, but it is converted to a single bond. That is, only the π bond is broken, not the σ bond. Table 6.1 lists only σ bonds (it does not list the π component of a C=C bond), but we can determine the value of the π component, given the value of the entire double bond:

$$\text{BDE}(\sigma) + \text{BDE}(\pi) = \text{BDE (double)}$$
$$356\text{ kJ/mol} + \text{BDE}(\pi) = 607\text{ kJ/mol}$$
$$\text{BDE}(\pi) = 251\text{ kJ/mol}$$

We are using the BDE of $CH_3CH_2—CH_3$ as the best estimate for the σ bond of our double bond because the substitution pattern is the same. Notice that the π component is not as strong as the σ component of the C=C bond. Indeed, this is why the π component of the C=C bond is broken while the σ component remains intact.

The calculation for ΔH is as follows:

Bonds Broken	**kJ/mol**
C=C (just the π component)	+251
H—Br	+368

Bonds Formed	**kJ/mol**
primary C—Br (approximated by $CH_3CH_2—Br$)	−285
secondary C—H (approximated by $H—CH(CH_3)_2$)	−397

The net sum is −63 kJ/mol. Since ΔH^o is predicted to be negative, we can also predict that this reaction will be exothermic.

6.3.

(a) ΔS_{sys} is expected to be negative ($\Delta S_{sys} < 0$, a decrease in entropy) because two molecules are converted into one molecule. In addition, acyclic compounds are converted into a cyclic compound. Both of these features are consistent with a decrease in entropy

(b) ΔS_{sys} is expected to be negative ($\Delta S_{sys} < 0$, a decrease in entropy) because an acylic compound is converted into a cyclic compound.

(c) ΔS_{sys} is expected to be positive ($\Delta S_{sys} > 0$, an increase in entropy) because one molecule is converted into two molecules. In addition, a cyclic compound is converted into two acyclic compounds. Both of these features are consistent with an increase in entropy.

(d) ΔS_{sys} is expected to be positive ($\Delta S_{sys} > 0$, an increase in entropy) because one molecule is converted into two ions.

(e) ΔS_{sys} is expected to be negative ($\Delta S_{sys} < 0$, a decrease in entropy) because two chemical entities are converted into one.

(f) ΔS_{sys} is expected to be positive (an increase in entropy) because a cyclic compound is converted into an acyclic compound.

6.4.

(a) There is a competition between the two terms contributing to ΔG. In this case, the reaction is endothermic, which contributes to a positive value for ΔG, but the second term contributes to a negative value for ΔG:

$$\Delta G = \underset{\oplus}{\Delta H} + \underset{\ominus}{(-T\Delta S)}$$

The sign of ΔG will therefore depend on the competition between these two terms, which is affected by temperature. A high temperature will cause the second term to dominate, giving rise to a negative value of ΔG. A low temperature will render the second term insignificant, and the first term will dominate, giving rise to a positive value of ΔG. The process is expected to be spontaneous above a certain temperature.

(b) In this case, both terms contribute to a negative value for ΔG, so ΔG will always be negative (the process will be spontaneous at all temperatures).

(c) In this case, both terms contribute to a positive value for ΔG, so ΔG will always be positive (the process will not be spontaneous at any temperature).

(d) There is a competition between the two terms contributing to ΔG. In this case, the reaction is exothermic, which contributes to a negative value for ΔG, but the second term contributes to a positive value for ΔG:

$$\Delta G = \underset{\ominus}{\Delta H} + \underset{\oplus}{(-T\Delta S)}$$

The sign of ΔG will therefore depend on the competition between these two terms, which is affected by temperature. A high temperature will cause the second term to dominate, giving rise to a positive value of ΔG. A low temperature will render the second term insignificant, and the first term will dominate, giving rise to a negative value of ΔG. The process is expected to be spontaneous only below a certain temperature.

6.5. A system can only achieve a lower energy state by transferring energy to its surroundings (conservation of energy). This increases the entropy of the surroundings, which more than offsets the decrease in entropy of the system. As a result, ΔS_{tot} increases.

6.6.

(a) A positive value of ΔG favors reactants.

(b) A reaction for which $K_{eq} < 1$ will favor reactants.

(c) $\Delta G = \Delta H - T\Delta S = (33\ \text{kJ/mol}) - (298\ \text{K})(0.150\ \text{kJ/mol} \bullet \text{K}) = -11.7\ \text{kJ/mol}$

A negative value of ΔG favors products.

(d) Both terms contribute to a negative value of ΔG, which favors products (because the forward reaction is spontaneous).

(e) Both terms contribute to a positive value of ΔG, which favors reactants (because the *reverse* reaction is spontaneous).

6.7.

(a) Process D will occur more rapidly because it has a lower energy of activation than process A.

(b) Process A will more greatly favor products than process B, because process A is exergonic (the products are lower in energy than the reactants).

(c) None of these processes exhibits an intermediate, because none of the energy diagrams has a local minimum (a valley) in between the reactants and products. However, all of the processes proceed via a transition state, because all of the energy diagrams have a local maximum (a peak).

(d) In process A, the transition state resembles the reactants more than products because the transition state is closer in energy to the reactants than the products (the Hammond postulate).
(e) Process A will occur more rapidly because it has a lower energy of activation than process B.
(f) Process D will more greatly favor products at equilibrium than process B, because process D is exergonic (the products are lower in energy than the reactants).
(g) In process C, the transition state resembles the products more than reactants because the transition state is closer in energy to the products than the reactants (the Hammond postulate).

6.8.
(a) This compound is an alkyl halide. The carbon atom that is connected directly to the electronegative halogen (Br) is an electrophilic center, because this carbon atom bears a partial positive charge:

Br
δ+

(b) This compound is an alcohol, which is a weak nucleophile because of the lone pairs on the OH group:

O–H

(c) This compound has a carbonyl (C=O) group. The carbon atom of the carbonyl group is an electrophilic center, because of resonance and inductive effects making it partially positive:

δ+ =O

(d) This compound is an alkene. The electron-rich π bond (highlighted) is a nucleophilic center:

(e) This structure is a carbocation. Carbocations are electron-deficient and can serve as electrophiles:

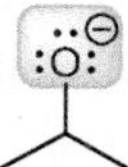

(f) This is an alkoxide ion, which is electron-rich and can serve as a nucleophile:

:O:⊖

(g) The sodium cation (Na^+) is the counter-ion, so the reagent in this case is the electron-rich Br^- anion, which is a nucleophile.

(h) The lithium cation (Li^+) is the counter-ion, so the reagent in this case is the electron-rich hydroxide (HO^-) anion, which is a nucleophile.

(i) This compound is an alkyl halide. The carbon atom that is connected directly to the electronegative halogen (I) is an electrophilic center, because the carbon atom is partially positive:

H–C(δ+)(H)(H)–I

(j) The potassium cation (K^+) is the counter-ion, so the reagent in this case is an electron-rich alkoxide ion, which is a nucleophile.

$KOC(CH_3)_3 \equiv H_3C–C(CH_3)_2–O^{\ominus}\ K^{\oplus}$

6.9.
This compound has two functional groups. On the left side of the structure, there is an OH group. This group is the characteristic functional group for an alcohol, which can function as a weak nucleophile, because of the lone pairs. And on the right side of the structure, there is a halogen (Cl) connected to an sp^3-hybridized carbon atom, so that elctron-deficient carbon atom is an electrophilic center:

H–O ... δ+ Br

A nucleophilic center

An electrophilic center

6.10. When we draw all significant resonance structures, we find that there are two positions (highlighted) that are deficient in electron density and, therefore, are electrophilic. Of the two carbon-carbon π bonds, the one that is conjugated with the carbon-oxygen π bond is the one that is electrophilic:

Electrophilic centers

6.11.
(a) The curved arrow indicates a hydride shift, which is a type of carbocation rearrangement.
(b) The curved arrow indicates a nucleophilic attack. In this case, water functions as a nucleophile and attacks the carbocation.
(c) The curved arrows indicate a proton transfer. In this case, water functions as the base that removes the proton.
(d) The curved arrows indicate a nucleophilic attack. In this case, one of the lone pairs on the oxygen atom functions as the nucleophilic center that attacks the carbocation.
(e) The curved arrow indicates loss of a leaving group (Cl^-).

6.12. The curved arrow indicates a proton transfer. In this case, the nitrogen atom functions as the base that removes the proton from the oxygen atom.

6.13.
(a) The sequence of arrow-pushing patterns is as follows:
(i) nucleophilic attack
(ii) proton transfer
(iii) proton transfer

(b) The sequence of arrow-pushing patterns is as follows:
(i) proton transfer
(ii) nucleophilic attack
(iii) loss of a leaving group

(c) The sequence of arrow-pushing patterns is as follows:
(i) proton transfer
(ii) loss of a leaving group
(iii) nucleophilic attack
(iv) proton transfer

(d) The sequence of arrow-pushing patterns is as follows:
(i) proton transfer
(ii) nucleophilic attack
(iii) proton transfer

6.14. The first step is a typical proton transfer, requiring two curved arrows. The second step is a nucleophilic attack (recall that a π bond can function as a nucleophile) combined with the loss of a leaving group, each of which requires one arrow, for a total of two curved arrows. The third step is another proton transfer, which requires two curved arrows. In this last step, the electrons from the breaking C–H bond are used to make a new π bond.

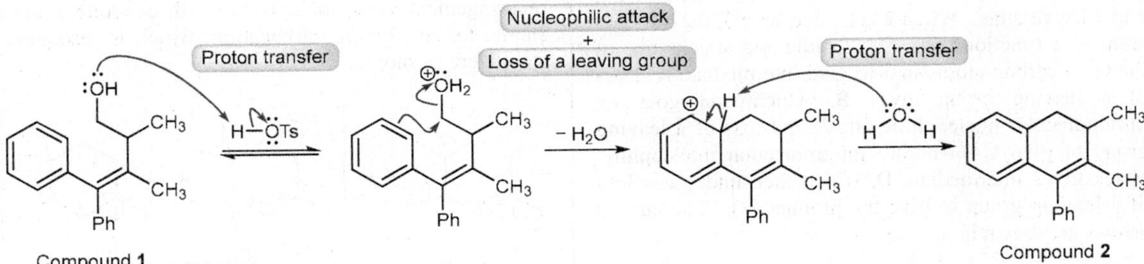

6.15.

(a) In this case, a C–O bond is formed, indicating a nucleophilic attack. Water (H_2O) functions as a nucleophile and attacks the carbocation. This is shown with one curved arrow. The tail of this curved arrow is placed on the lone pair of the oxygen atom, and the head is placed on the electrophilic center (the empty *p* orbital of the carbocation), as shown here:

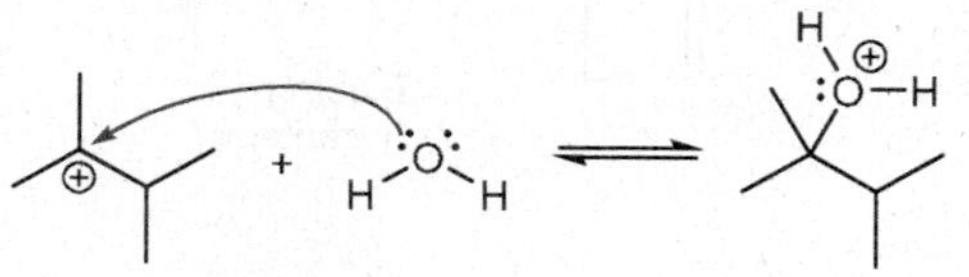

(b) This is a proton transfer step, in which water functions as a base and removes a proton, thereby generating H_3O^+. A proton transfer step requires two curved arrows. The tail of the first curved arrow is placed on a lone pair of H_2O, and the head is placed on the proton that is being transferred. Don't forget the second curved arrow. The tail is placed on the O–H bond (that is being broken) and the head is placed on the cationic oxygen atom, as shown:

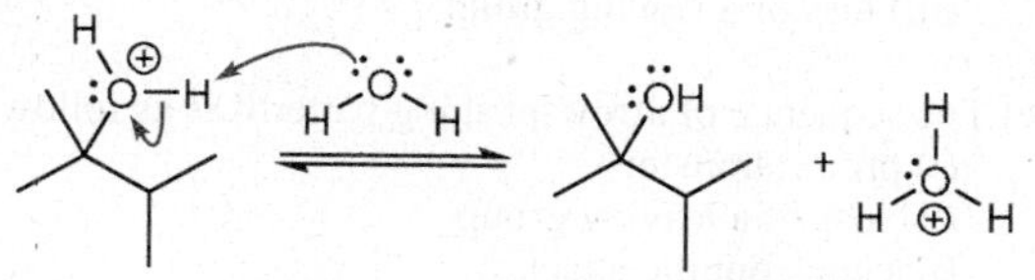

(c) This step represents the loss of a leaving group (where the leaving group is H_2O). One curved arrow is required. The tail is placed on the C–O bond that is broken, and the head is placed on the oxygen atom.

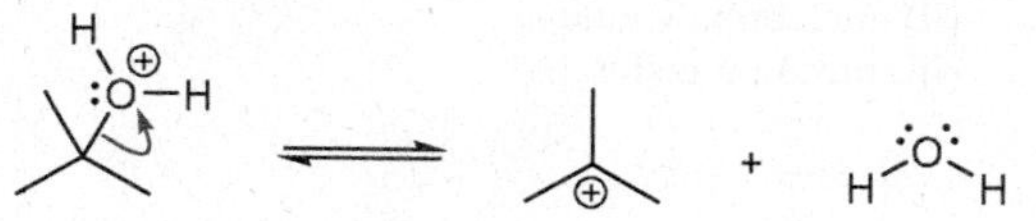

6.16. The conversion of **1** to **2** involves two steps: the lone pair on the phosphorus atom functions as a nucleophile and attacks compound **1**, followed by loss of a leaving group to give compound **2**. Note that nitrogen gas (N_2) is liberated, which renders the conversion from **1** to **2** irreversible. When **2** is treated with **3**, the nitrogen atom of **2** functions as a nucleophile and attacks one of the C=O carbon atoms in **3** to give intermediate **A**. Loss of a leaving group gives **B**, which undergoes an intramolecular nucleophilic attack and loss of a leaving group to give **C**. Another intramolecular nucleophilic attack gives intermediate **D**, which then undergoes loss of a leaving group to give the product (**4**). The curved arrows are drawn here:

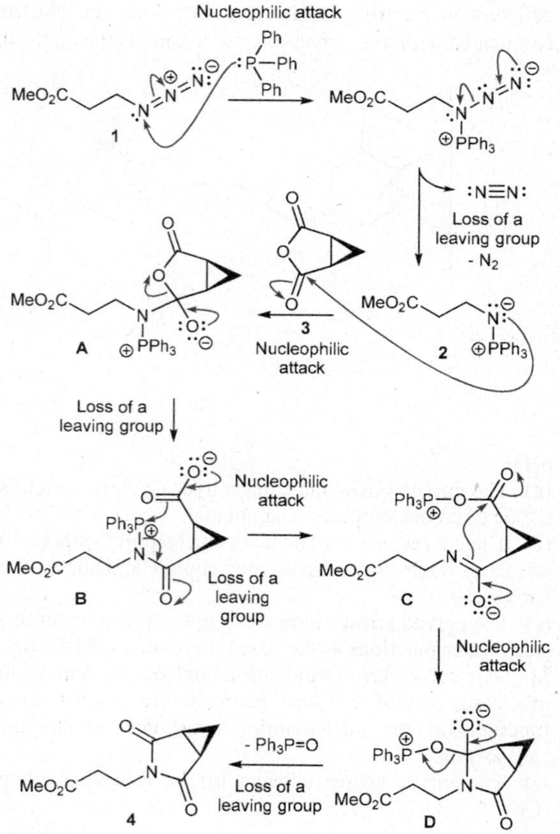

Nucleophilic attack

Loss of a leaving group - N_2

Nucleophilic attack

Loss of a leaving group

Nucleophilic attack

Loss of a leaving group

Nucleophilic attack

- $Ph_3P=O$

Loss of a leaving group

6.17.

(a) This carbocation is secondary, and it can rearrange via a hydride shift (shown below) to give a more stable, tertiary carbocation:

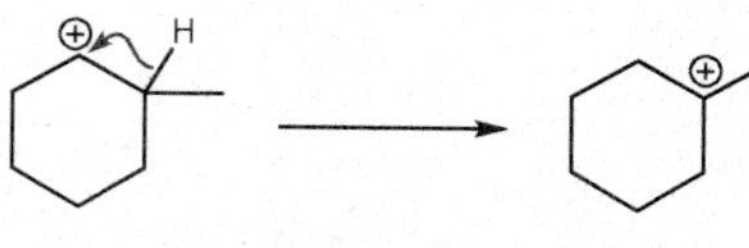

(b) This carbocation is tertiary, and it cannot become more stable via a rearrangement.

(c) This carbocation is tertiary. Yet, in this case, rearrangement via a methyl shift will generate a more stable, tertiary *allylic* carbocation, which is resonance-stabilized, as shown:

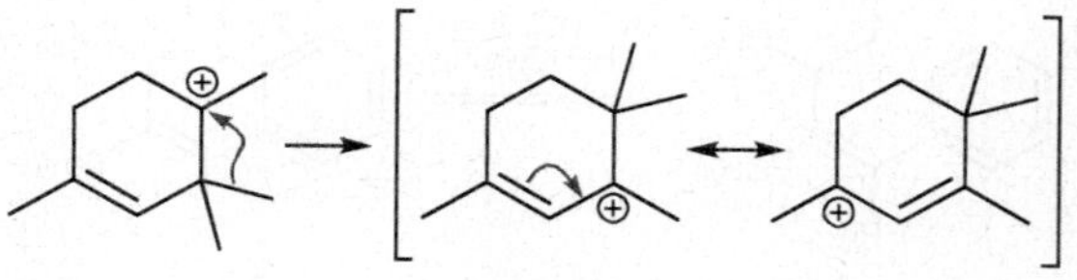

tertiary

tertiary allylic

(d) This carbocation is secondary, but there is no way for it to rearrange to form a tertiary carbocation.

(e) This carbocation is secondary, and it can rearrange via a hydride shift (shown below) to give a more stable, tertiary carbocation:

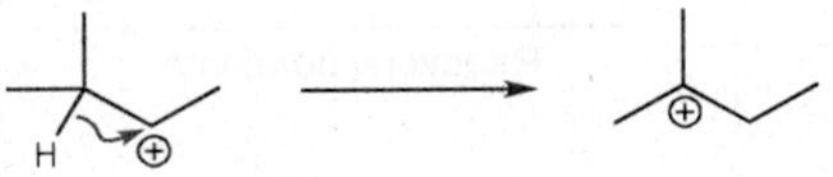

(f) This carbocation is secondary, and it can rearrange via a methyl shift (shown below) to give a more stable, tertiary carbocation:

(g) This carbocation is primary, and it can rearrange via a hydride shift (shown below) to give a secondary, resonance-stabilized carbocation (we will see in Chapter 7 that this carbocation is called a benzylic carbocation):

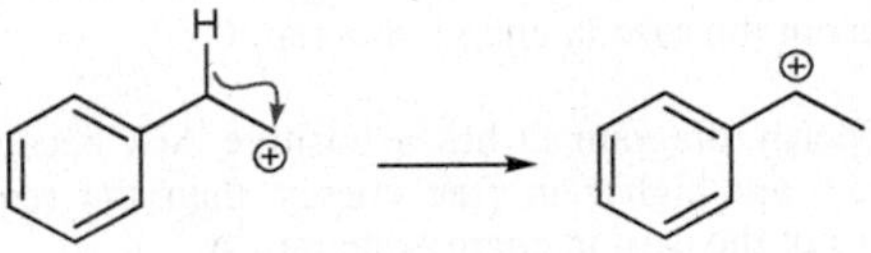

(h) This carbocation is tertiary and it is resonance-stabilized (we will see in Chapter 7 that this carbocation is called a benzylic carbocation). It will not rearrange.

6.18. In the first rearrangement, a three-membered ring is converted into a four-membered ring (a process called ring expansion). In order for this to occur, the migrating methylene group (highlighted) must be part of the three-membered ring:

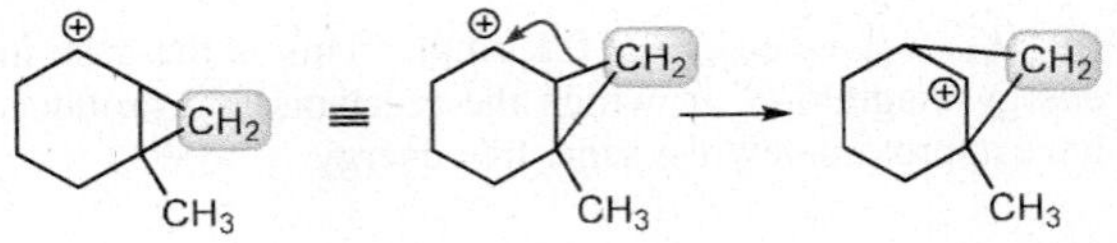

In the second rearrangement, a four-membered ring is converted into a three-membered ring (a process called ring contraction). In order for this to occur, the migrating carbon atom must be part of the four-membered ring. Once again, the same methylene group migrates (highlighted):

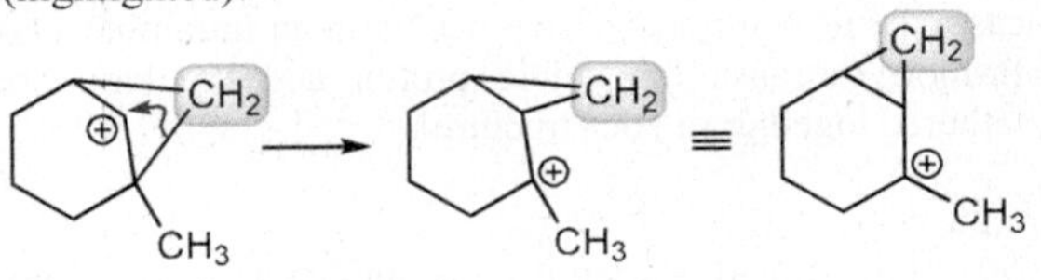

6.19.

(a) A carbon-carbon triple bond is comprised of one σ bond and two π bonds, and is therefore stronger than a carbon-carbon double bond (one σ and one π bond) or a carbon-carbon single bond (only one σ bond). Therefore, the carbon-carbon triple bond is expected to have the largest bond dissociation energy.

(b) In this molecule, the values for the C–X bonds (where X = halogen) can be estimated from the secondary alkyl halides [$(CH_3)_2CH$–X] in Table 6.1. These data indicate that the C–F bond will have the largest bond dissociation energy.

6.20.

(a) Using Table 6.1, we identify the bond dissociation energy (BDE) of each bond that is either broken or formed. For bonds broken, BDE values are positive numbers (energy is *required* to break bonds). For bonds formed, the sign of each BDE value is negative (energy is *released* when bonds are formed).

Bonds Broken	**kJ/mol**
RCH_2—Br	**+285**
RCH_2O—H	**+435**

Bonds Formed	**kJ/mol**
RCH_2—OR	**– 381**
H—Br	**– 368**

The net sum is **–29 kJ/mol**. $\Delta H°$ for this reaction is negative, which means that the system is losing energy. It is giving off energy to the environment, so the reaction is exothermic.

(b) One mole of reactant is converted into two moles of product (resulting in an increase in entropy), although an acyclic molecule is converted into a cyclic molecule (resulting in a decrease in energy). If ΔS for this reaction is positive, then we can conclude that the first factor (conversion of one mole into two moles) is more significant and is the dominating factor.

(c) Both terms (ΔH) and ($-T\Delta S$) contribute to a negative value of ΔG ($\Delta G < 0$).

(d) No, the sign of ΔG is not dependent on temperature. Since both terms (ΔH and $-T\Delta S$) have negative values, the value of ΔG will be negative at all temperatures.

(e) Yes, the magnitude of ΔG is dependent on temperature At high temperatures, the value of $-T\Delta S$ is large and negative, while at low temperatures, the value of $-T\Delta S$ is small and negative.
Since $\Delta G = \Delta H + (-T\Delta S)$, the magnitude of ΔG will be dependent on temperature.

6.21.

(a) A reaction for which $K_{eq} > 1$ will favor products.

(b) A reaction for which $K_{eq} < 1$ will favor reactants.

(c) A positive value of ΔG favors reactants.

(d) Both terms contribute to a negative value of ΔG, which favors products.

(e) Both terms contribute to a positive value of ΔG, which favors reactants.

6.22. Choice (b) is the correct answer. $K_{eq} = 1$ when $\Delta G = 0$ kJ/mol (See Table 6.2).

6.23. $K_{eq} < 1$ when ΔG has a positive value. The answer is therefore: (a) +1 kJ/mol.

6.24.

(a) ΔS_{sys} is expected to be negative ($\Delta S_{sys} < 0$, a decrease in entropy) because two moles of reactant are converted into one mole of product.

(b) ΔS_{sys} is expected to be positive ($\Delta S_{sys} > 0$, an increase in entropy) because one mole of reactant is converted into two moles of product.

(c) ΔS_{sys} is expected to be approximately zero, because two moles of reactant are converted into two moles of product.

(d) ΔS_{sys} is expected to be negative ($\Delta S_{sys} < 0$, a decrease in entropy) because an acylic compound is converted into a cyclic compound.

(e) ΔS_{sys} is expected to be approximately zero, because one mole of reactant is converted into one mole of product, and both the reactant and the product are acyclic.

6.25.

(a) If the reaction has only one step, then the energy diagram will have only one peak. Since ΔG for this reaction is negative, the product will be lower in free energy than the reactant, as shown here.

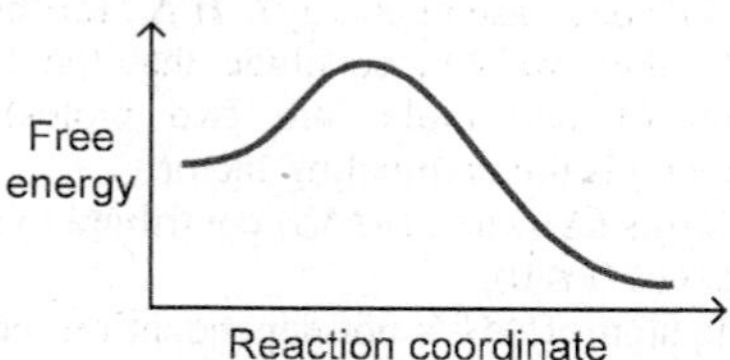

(b) If the reaction has only one step, then the energy diagram will have only one peak. Since ΔG for this reaction is positive, the product will be higher in free energy than the reactant, as shown here.

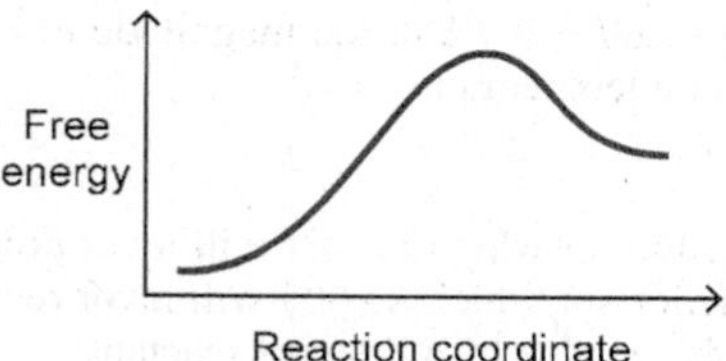

(c) If the reaction has two steps, then the energy diagram will have two peaks. Since ΔG for this reaction is negative, the product will be lower in free energy than the reactant. And the problem statement indicates that the transition state for the first step is higher in energy than the transition state for the second step, as shown here.

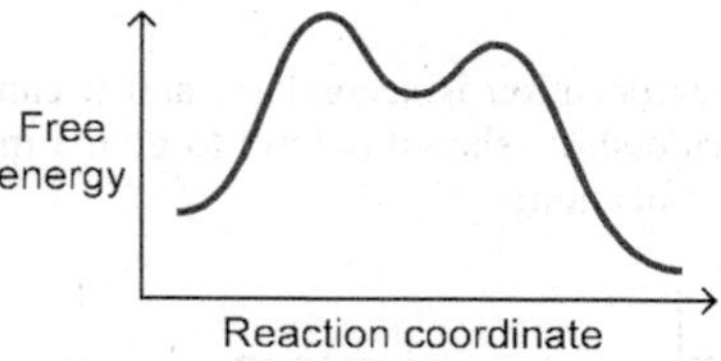

6.26.

(a) Energy diagrams B and D each exhibit two peaks, characteristic of a two-step process.

(b) Energy diagrams A and C each exhibit only one peak, characteristic of a one-step process.

(c) The energy of activation (E_a) is determined by the difference in energy between the reactants and the transition state (the top of the peak in the energy diagram). This energy difference is greater in C than it is in A, so C has a larger E_a.

(d) Energy diagram A has a negative ΔG, because the products are lower in free energy than the reactants. This is not the case in energy diagram C.

(e) Energy diagram D has a positive ΔG, because the products are higher in free energy than the reactants. This is not the case in energy diagram A.

(f) The energy of activation (E_a) is determined by the difference in energy between the reactants and the transition state (the top of the peak in the energy diagram). This energy difference is greatest in D.

(g) $K_{eq} > 1$ when ΔG has a negative value. This is the case in energy diagrams A and B, because in each of these energy diagrams, the products are lower in free energy than the reactants.

(h) $K_{eq} = 1$ when $\Delta G = 0$ kJ/mol. This is the case in energy diagram C, in which the reactants and products have approximately the same free energy.

6.27.

(a) The curved arrow indicates loss of a leaving group (Cl^-).
(b) The curved arrow indicates a methyl shift, which is a type of carbocation rearrangement.
(c) The curved arrows indicate a nucleophilic attack.
(d) The curved arrows indicate a proton transfer. In this case, the proton transfer step occurs in an intramolecular fashion (because the acidic proton and the base are tethered together in one structure).

6.28.

(a) A tertiary carbocation is more stable than a secondary carbocation, which is more stable than a primary carbocation.

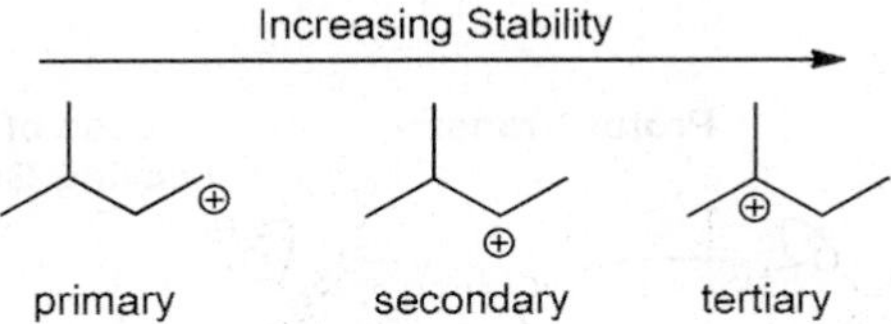

(b) The most stable carbocation is the one that is resonance-stabilized. Among the other two carbocations, the secondary carbocation is more stable than the primary carbocation, as shown here.

Increasing Stability

primary secondary secondary resonance-stabilized

6.29. CH_3I is an alkyl halide, and the electron-deficient carbon atom can function as an electrophile. In the other reagent, $NaOCH_2CH_3$, sodium (Na^+) is the counter-ion and can be ignored, so this reagent is ethoxide (an electron-rich alkoxide ion) which is a strong nucleophile.

6.30.

(a) When we draw all significant resonance structures, we find that there are two positions (highlighted) that are deficient in electron density:

Therefore, these two positions are electrophilic:

(b) When we draw all significant resonance structures, we find that there are two positions (highlighted) that are deficient in electron density:

Therefore, these two positions are electrophilic:

6.31. The sequence of arrow-pushing patterns is as follows:

Nucleophilic Attack

Loss of Leaving Group

6.32. The sequence of arrow-pushing patterns is as follows:

Proton Transfer

Proton Transfer

Loss of a Leaving Group

+ SO_3

6.33. Both reactions have the same sequence: (i) nucleophilic attack, followed by (ii) loss of a leaving group. In both cases, a hydroxide ion functions as a nucleophile and attacks a compound that can accept the negative charge and store it temporarily. The charge is then expelled as a chloride ion in both cases.

6.34. The sequence of arrow-pushing patterns is as follows:

Nucleophilic Attack

Loss of a Leaving Group

Proton Transfer

6.35. The sequence of arrow-pushing patterns is as follows:

Nucleophilic Attack

Loss of a Leaving Group

Nucleophilic Attack

Proton Transfer

6.36. The sequence of arrow-pushing patterns is as follows:

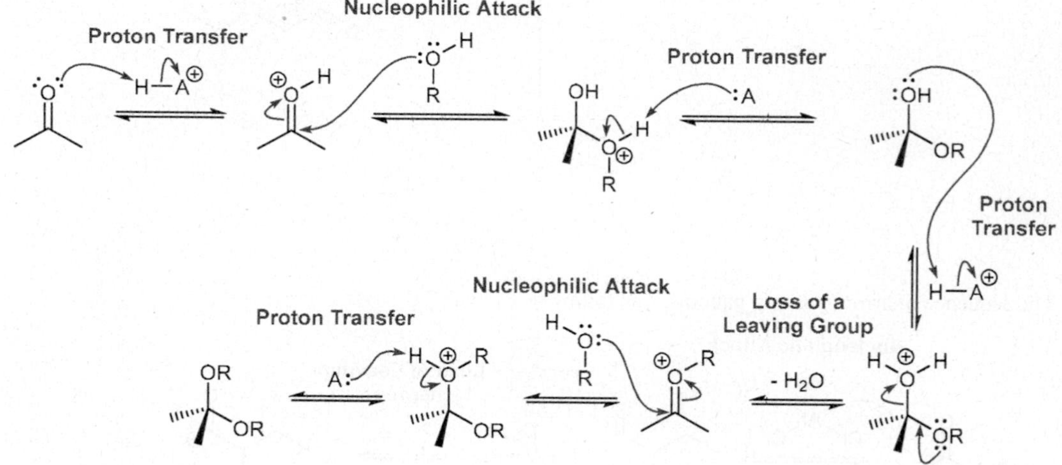

6.37. The first step is a nucleophilic attack, the second step is loss of a leaving group, and the final step is a proton transfer. In this case, each of the first two steps requires several curved arrows, and the final step requires two curved arrows, as shown:

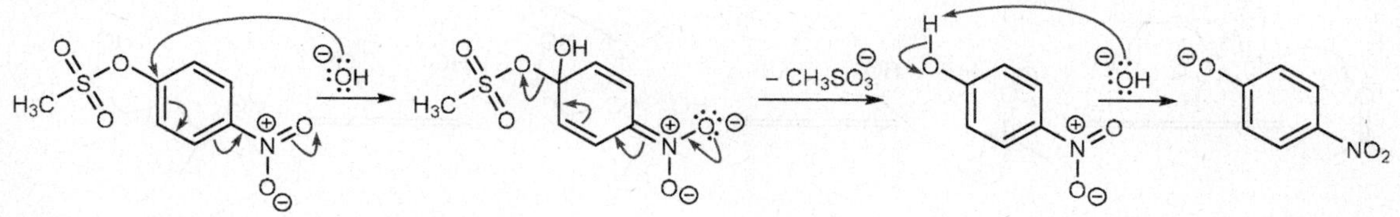

6.38. The following curved arrows show the flow of electrons that achieve the transformation as shown:

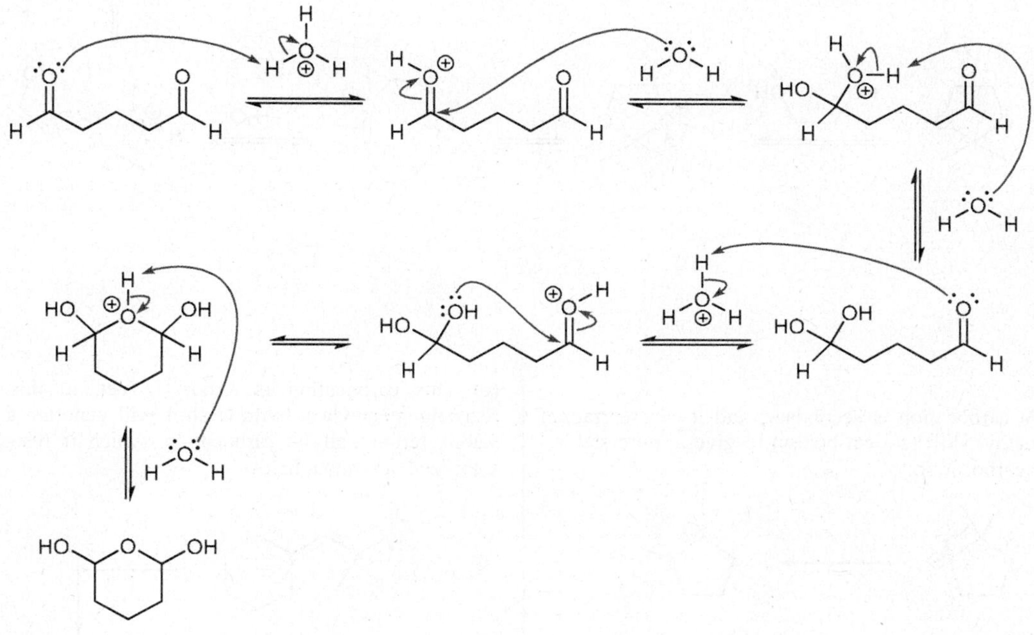

6.39. The following curved arrows show the flow of electrons that achieve the transformation as shown:

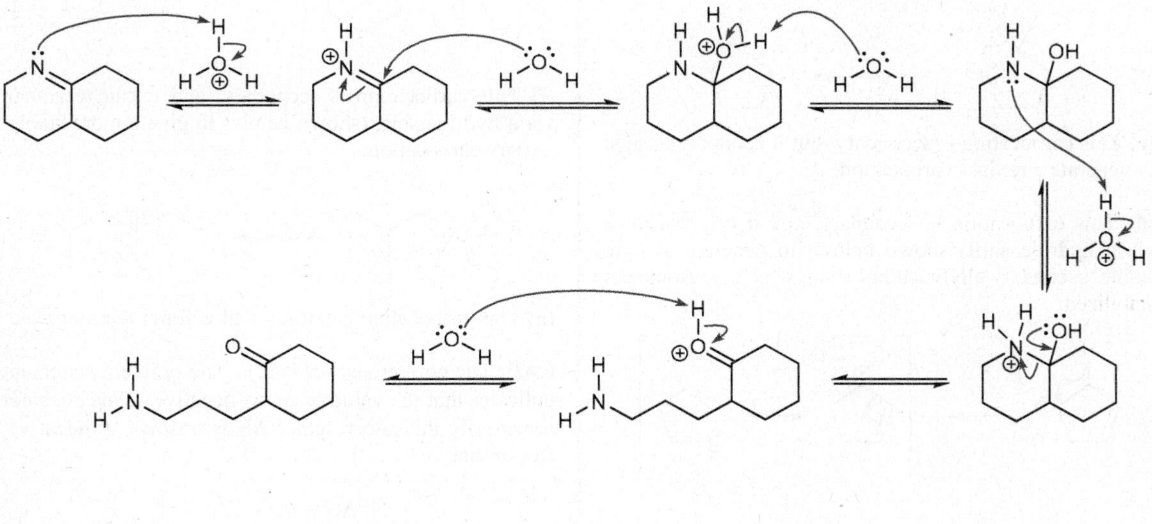

6.40. The following curved arrows show the flow of electrons that achieve the transformation as shown:

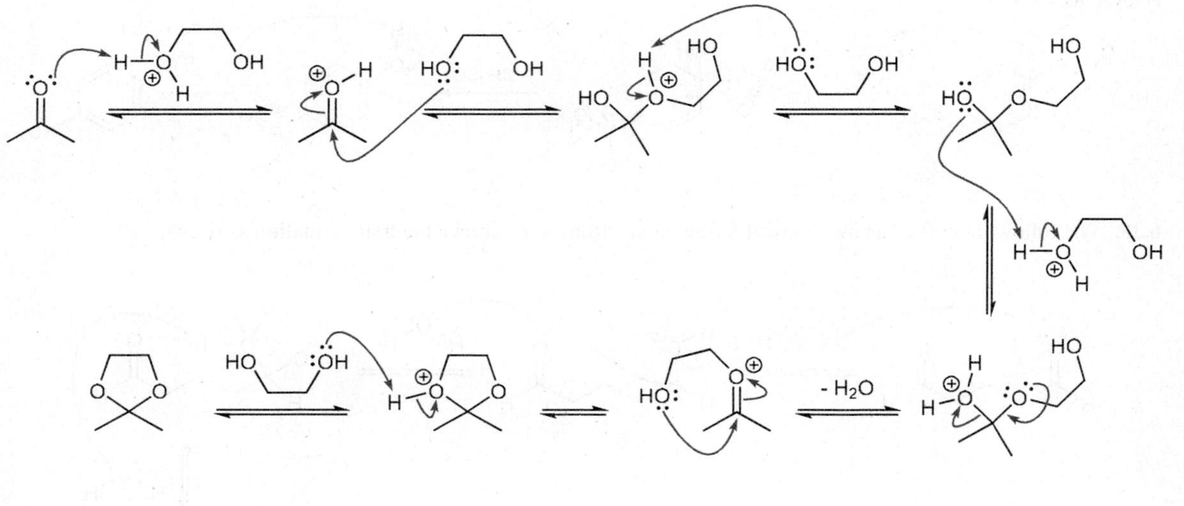

6.41.

(a) This carbocation is secondary, and it can rearrange via a methyl shift (shown below) to give a more stable, tertiary carbocation:

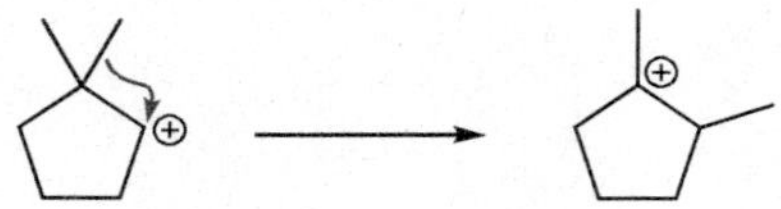

(b) This carbocation is secondary, and it can rearrange via a hydride shift (shown below) to give a more stable, tertiary carbocation:

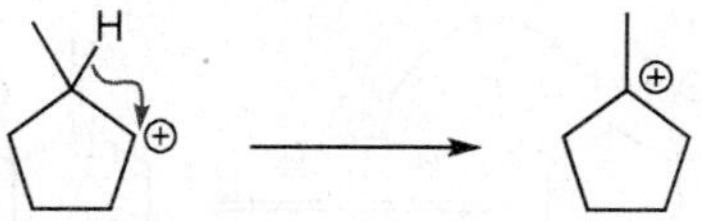

(c) This carbocation is secondary, but it cannot rearrange to generate a tertiary carbocation.

(d) This carbocation is secondary, and it can rearrange via a hydride shift (shown below) to generate a more stable, secondary allylic carbocation, which is resonance-stabilized:

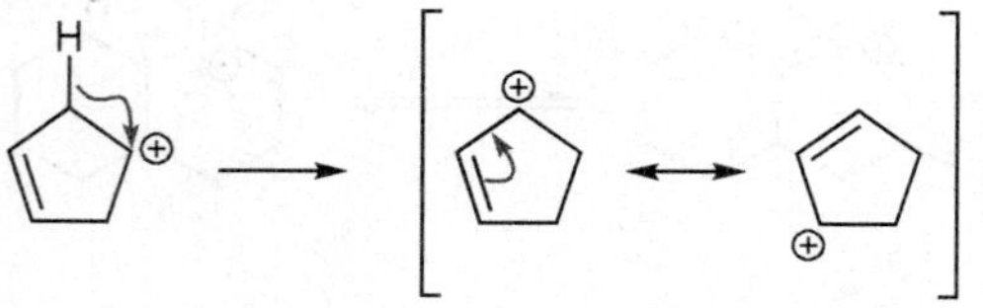

(e) This carbocation is tertiary. Yet, in this case, rearrangement via a hydride shift will generate a more stable, tertiary allylic carbocation, which is resonance-stabilized, as shown below.

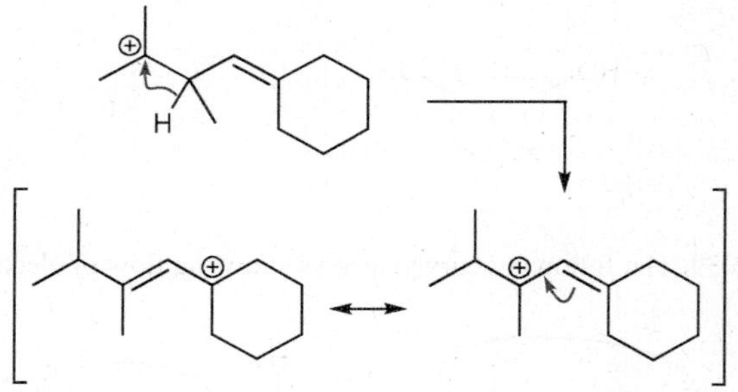

(f) This carbocation is secondary, and it can rearrange via a hydride shift (shown below) to give a more stable, tertiary carbocation:

(g) This carbocation is tertiary and will not rearrange.

6.42. The correct answer is (c). The problem statement indicates that the value of ΔG is negative. This does not necessarily indicate whether ΔH is positive or negative. Recall that:

$$\Delta G = \Delta H - T\Delta S$$

We would need to know ΔS and T, in order to know the sign of ΔH. So we don't know whether the reaction is exothermic or endothermic. Therefore, answers (a) and (b) are not correct.
Answer (c) is correct, because an exergonic process is thermodynamically favorable, which means that the products are favored over the reactants at equilibrium. That is, $K_{eq} > 1$.

6.43. The correct answer is (d). Compound (a) lacks a lone pair or a pi bond, so it cannot function as a nucleophile. The same is true for compounds (b) and (c). In contrast, the oxygen atom in compound (d) does have lone pairs, which are localized, so water can function as a nucleophile (albeit a weak one).

6.44. The correct answer is (b). Intermediates are represented by local minima (valleys) between the reactant(s) and product(s) on an energy diagram. On the energy diagram shown, there are two local minima, II and IV.

6.45. The correct answer is (c). ΔS_{sys} is expected to be negative ($\Delta S < 0$, a decrease in entropy) because two moles of reactant are converted into one mole of product.

6.46. The correct answer is (a). This reaction step represents a nucleophilic attack.

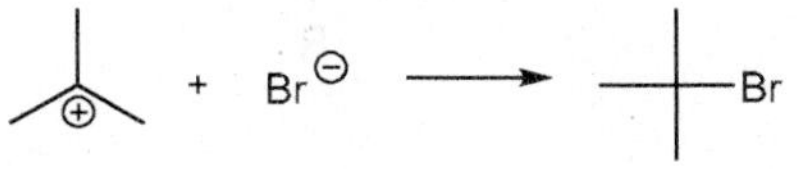

+ Br⊖ ⟶ Br

A new bond is forming (represented as a partial bond), the positive charge is dissipating (represented as a partial positive charge on the carbon atom), and the negative charge is also dissipating (represented as a partial negative charge on the bromine atom). All of these changes must be shown in the transition state:

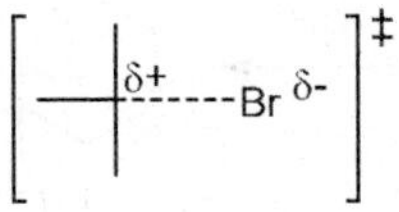

[δ+-----Br δ-]‡

6.47. The correct answer is (d). ΔH = (sum of BDE bonds broken) – (sum of BDE bonds formed). In this case, the ΔH for this reaction is negative, which means that the system is releasing energy. It is giving off energy to the environment, so the reaction is exothermic. The energy required to break the bonds must be less than the energy being released when bonds are formed. Therefore, to have a negative value for ΔH, the bonds broken must be weaker than the bonds formed.

6.48. The correct answer is (b). The rate expression has the following form:

$$\text{Rate} = k\,[\text{EtBr}][\text{NaSH}]$$

The sum of the exponents in this case is 2, and the reaction is said to be second order. This is consistent with the term "bimolecular", meaning two molecules are involved in the rate-determining step.

6.49. The correct answer is (b). In the first step, the oxygen atom of the alcohol functions as a base and removes a proton from HBr (in a proton transfer process). Then, in the second step, a single bond is broken to release H_2O (loss of a leaving group). Therefore, the order of steps is: proton transfer, followed by loss of a leaving group.

6.50.

(a) The C–Br bond is broken, indicating the loss of a leaving group (Br^-), while the C–O bond is formed, indicating a nucleophilic attack. This is, in fact, a concerted process in which nucleophilic attack and loss of the leaving group occur in a simultaneous fashion. One curved arrow is required to show the nucleophilic attack, and another curved arrow is required to show loss of the leaving group:

HO:⊖ + H₃C–C(H)(H)–Br ⟶ HO–C(CH₃)(H)(H) + Br⊖

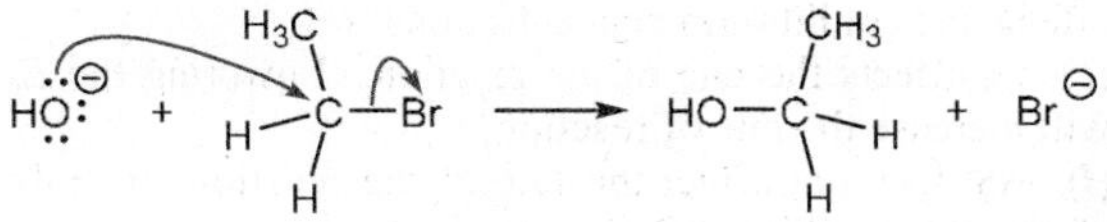

(b) We identify the bond broken (CH_3CH_2—Br), and the bond formed (CH_3CH_2—OH). Using the data in Table 6.1, ΔH for this reaction is expected to be approximately (285 kJ/mol) – (381 kJ/mol). The sign of ΔH is therefore predicted to be negative, which means that the reaction should be exothermic.
(c) Two chemical entities are converted into two chemical entities. Both the reactants and products are acyclic. Therefore, ΔS for this process is expected to be approximately zero.

(d) ΔG has two components: (ΔH) and ($-T\Delta S$). Based on the answers to the previous questions, the first term has a negative value and the second term is insignificant, because ΔS is approximately zero. Therefore, ΔG is expected to have a negative value. This is consistent with the energy diagram, which shows the products having lower free energy than the reactants.

(e) This transition state corresponds with the peak of the curve, and has the following structure:

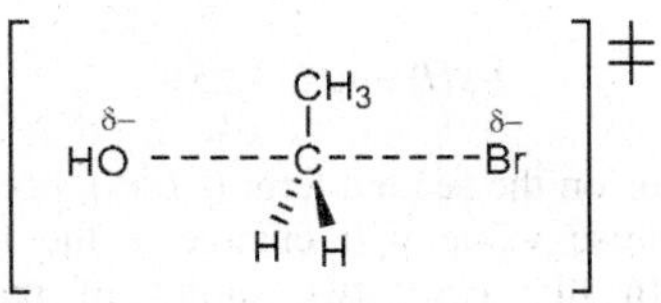

[δ– HO -----C(CH₃)(H)(H)------Br δ–]‡

(f) The transition state in this case is closer in energy to the reactants than the products, and therefore, it is closer in structure to the reactants than the products (the Hammond postulate).

(g) If we inspect the rate equation, we see that the sum of the exponents is two, so this reaction is second order.

(h) According to the rate equation, the rate is linearly dependent on the concentration of hydroxide. Therefore, the rate will be doubled if the concentration of hydroxide is doubled.

(i) At a higher temperature, more molecules will have the requisite energy to achieve the energy of activation necessary for the reaction to occur, so the rate will increase with increasing temperature.

6.51.
(a) K_{eq} does not affect the rate of the reaction. It only affects the equilibrium concentrations.
(b) ΔG does not affect the rate of the reaction. It only affects the equilibrium concentrations.
(c) Temperature affects the rate of the reaction. When more molecules move faster (*i.e.*, as temperature increases), there will be more collisions with sufficient energy to achieve the energy of activation.
(d) ΔH does not affect the rate of the reaction. It only affects the equilibrium concentrations.
(e) E_a affects the rate of the reaction. Lowering the E_a will increase the rate of reaction.
(f) ΔS does not affect the rate of the reaction. It only affects the equilibrium concentrations.

6.52. In order to determine if reactants or products are favored at high temperature, we must consider the effect of temperature on the sign of ΔG. Recall that ΔG has two components: (ΔH) and ($-T\Delta S$). The reaction is exothermic, so the first term (ΔH) has a negative value, which contributes to a negative value of ΔG. This favors products. At low temperatures, the second term will be insignificant, and the first term will be dominant. Therefore, the process will be thermodynamically favorable, and the reaction will favor the formation of products. However, above a certain temperature, the second term becomes dominant. In this case, two molecules of reactants are converted into one molecule of product. Therefore, ΔS for this process is negative, which means that ($-T\Delta S$) is positive. At a high enough temperature, the second term ($-T\Delta S$) should dominate over the first term (ΔH), generating a positive value for ΔG. Therefore, the reaction will favor reactants at high temperatures.

6.53. Recall that ΔG has two components:

$$(\Delta H) \text{ and } (-T\Delta S)$$

We will focus on the second term ($-T\Delta S$), because this is the term whose value will change if the temperature changes. In this case, two moles of reactants are converted into one mole of product, which represents a decrease in entropy. Therefore, ΔS for this process is negative, which means that ($-T\Delta S$) is positive. As the temperature increases, the value of ($-T\Delta S$) becomes a larger positive number, contributing to a positive value for ΔG. Therefore, the reaction will favor reactants at high temperatures.

6.54. The nitrogen atom of an ammonium ion is positively charged, but that does not render it electrophilic. The nitrogen atom in this case does not have an empty orbital (like a carbocation), because nitrogen is a second row element and therefore only has four orbitals with which to form bonds. All four orbitals are being used for bonding, leaving none of the orbitals vacant. As a result, the nitrogen atom is not electrophilic, despite the fact that it is positively charged. In contrast, the positive charge in an iminium ion is delocalized by resonance:

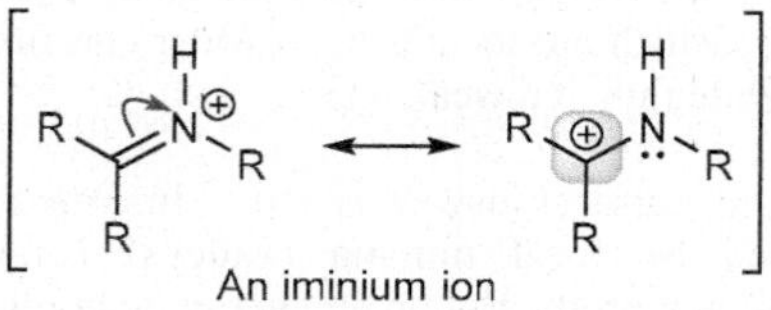

An iminium ion

The second resonance structure exhibits a positive charge on a carbon atom, which serves as an electrophilic center (because it has an empty *p* orbital that can accept a pair of electrons). Therefore, an iminium ion is an electrophile and is subject to attack by a nucleophile:

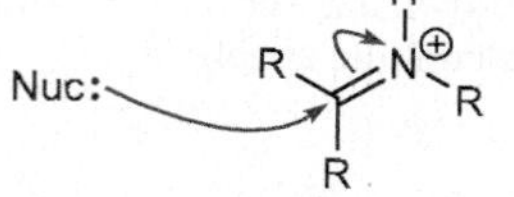

6.55. The following curved arrows show the flow of electrons that achieve the transformation as shown:

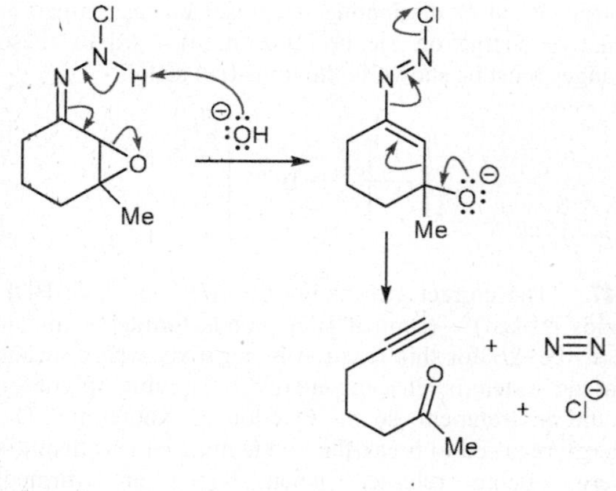

6.56. In this case, a six-membered ring is being converted into a five-membered ring (a process called ring-contraction). The migrating carbon atom (highlighted) is bonded to the position that is adjacent to the C+:

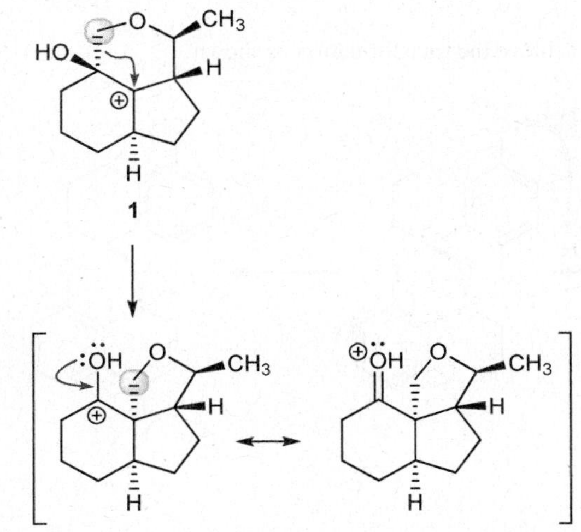

Carbocation **1** is not resonance-stabilized, while carbocation **2** is stabilized by resonance, as shown above. Indeed, the second resonance structure (**2b**) is the more significant contributor to the resonance hybrid of **2**, because all atoms have an octet of electrons. This renders carbocation **2** particularly stable. This additional stability is absent in carbocation **1**.

6.57.

(a) The resonance structure of **2** with an anionic oxygen (highlighted) is the greatest contributor to the resonance hybrid. Oxygen is more electronegative than carbon, so oxygen is more stable with the negative charge.

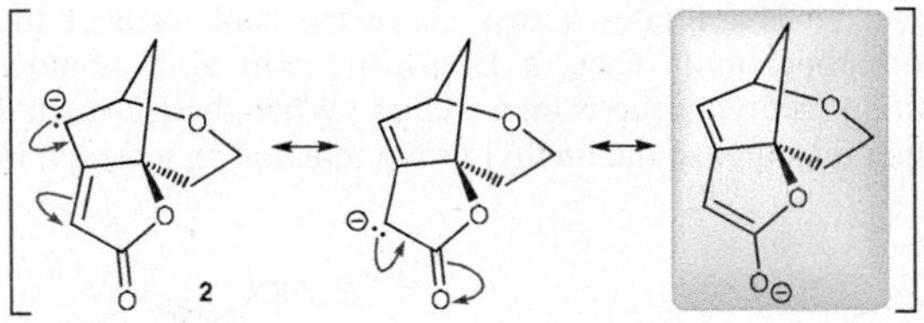

(b) The first step (**1** → **2**) is a proton transfer, in which MeO^- deprotonates **1** to produce **2**, the conjugate base of **1**. The transformation of **2** to **3** includes the formation of a C=O π bond, the relocation of two C=C π bonds, and the breaking of a C–O σ bond. The resulting anionic oxygen of **3** then serves as a nucleophile, attacking the indicated carbon, pushing electrons up to the oxygen of the C=O bond, as shown. In the final step (**4** → **5**), the C=O π bond is re-formed and a proton is transferred from MeOH to the cyclic structure, regenerating MeO^- (a proton transfer step).

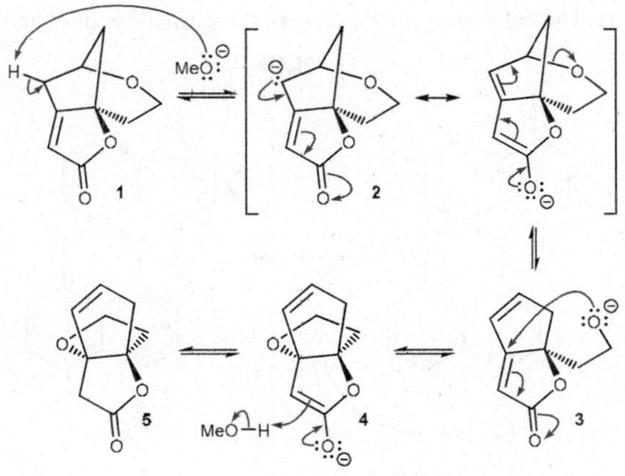

6.58.

In step 1, the hydroxide ion functions as a nucleophile and attacks the carbon atom of the ester group on the left, pushing the π electrons up to the oxygen atom. In step 2, the anionic oxygen atom serves as a base and removes a proton in an intramolecular fashion (proton transfer), as shown. The electrons in the C–H bond form a C=C π bond; the C–O σ bond breaks thus converting the adjacent C–O bond to a double bond, expelling HO^- as a leaving group. Step 3 is a proton transfer, with hydroxide (HO^-) serving as the base, and the carboxylic acid serving as the acid.

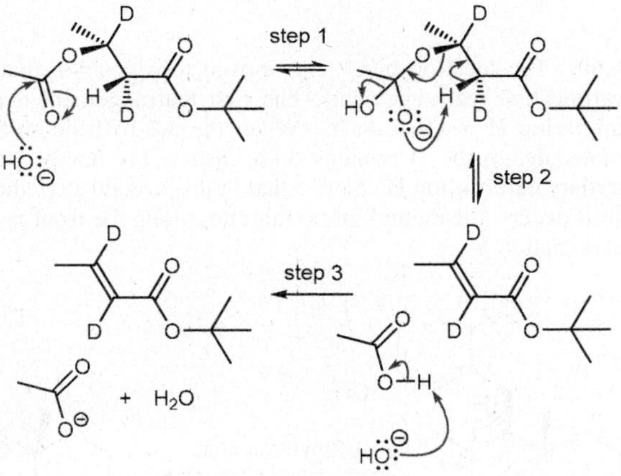

6.59.

(a) The following curved arrows show the flow of electrons that achieve the transformation as shown:

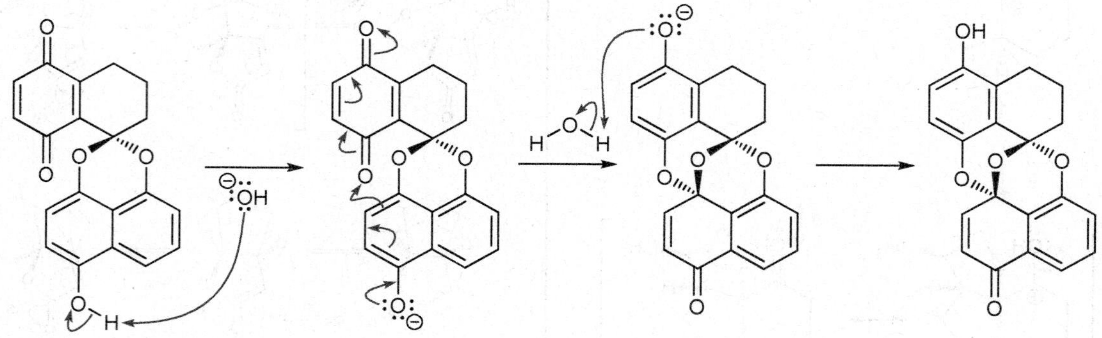

(b) When the first intermediate is redrawn from the following perspective, it becomes clear that the electrons in the carbon-carbon π-bond come from above the carbon-oxygen π bond, allowing for the chiral center to be generated as shown. If you have trouble seeing this, you might find it helpful to build a molecular model.

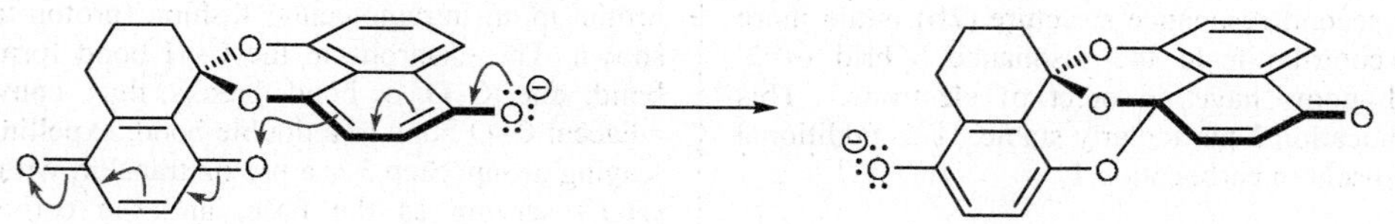

6.60. The location of C+ has moved to a position that is two carbon atoms away, which requires two consecutive carbocation rearrangements. The first rearrangement is a 1,2-hydride shift to give a 3° carbocation. Notice that the migrating H is on a dash. When the 1,2-hydride shift occurs, the H migrates across along the back face of the molecule, so the H remains on a dash in the newly generated carbocation. Then, a 1,2-methyl shift gives another tertiary carbocation **B**. Notice that in this second step, the migrating methyl group is on a wedge. When the 1,2-methyl shift occurs, the methyl group migrates along the front face of the molecule, so the methyl group remains on a wedge in carbocation **B**.

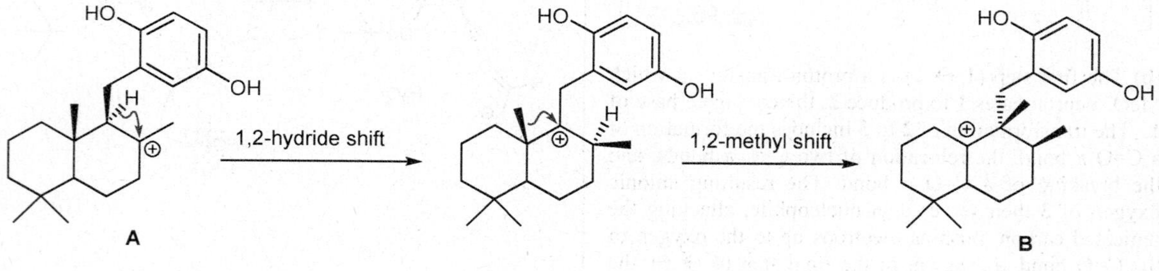

1,2-hydride shift

1,2-methyl shift

A

B

6.61.

(a) The sequence of arrow-pushing patterns is shown below, together with all curved arrows:

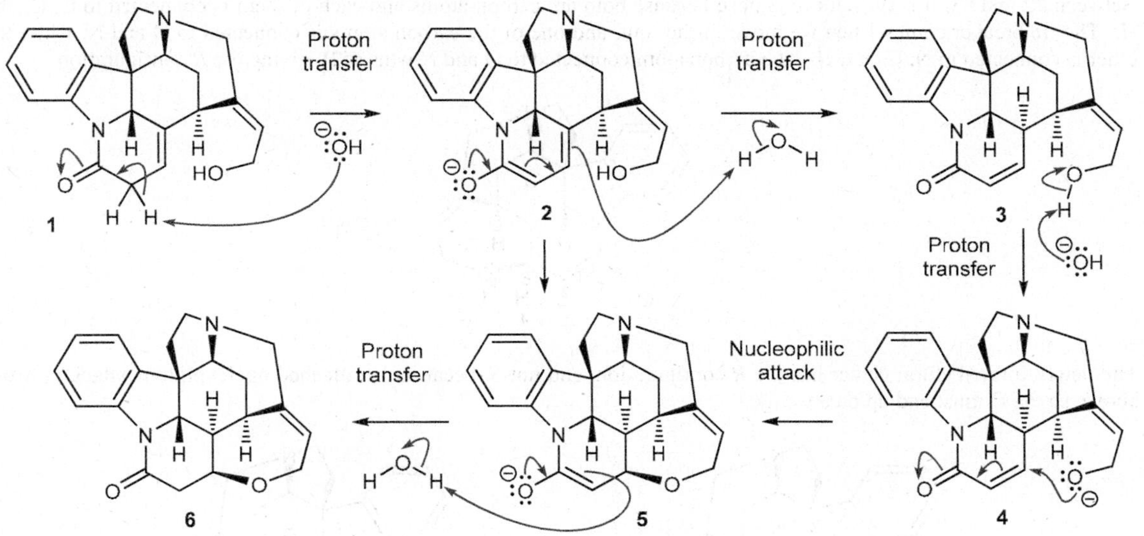

(b) Consider where the nucleophilic attack is occurring. The lone pairs on the oxygen atom (highlighted) are attacking the highlighted carbon atom.

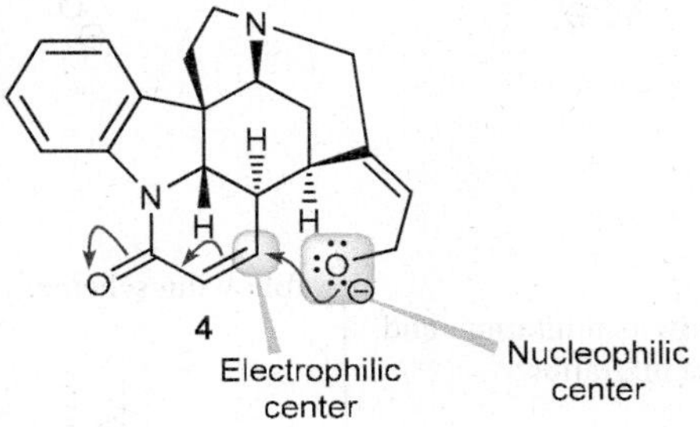

The oxygen atom is the nucleophilic center, and the carbon atom is the electrophilic center. In order to understand why this carbon atom is electron-poor, we draw resonance structures of **4**:

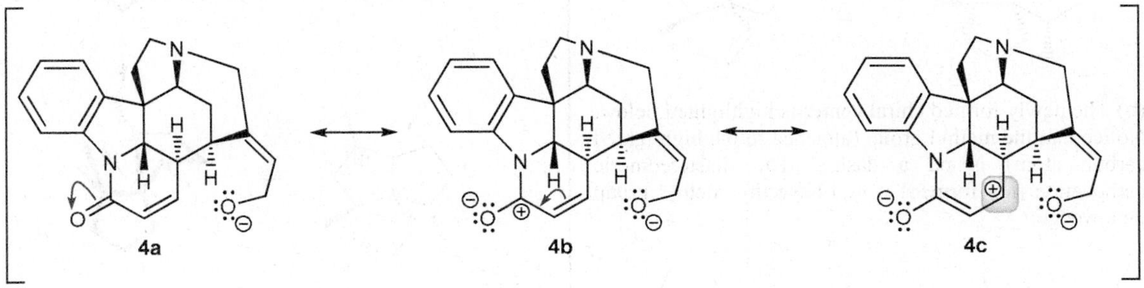

If we inspect resonance structure 4c, we see why the highlighted carbon atom is electrophilic. In fact, when we draw this resonance structure, we can see the nucleophilic attack more clearly:

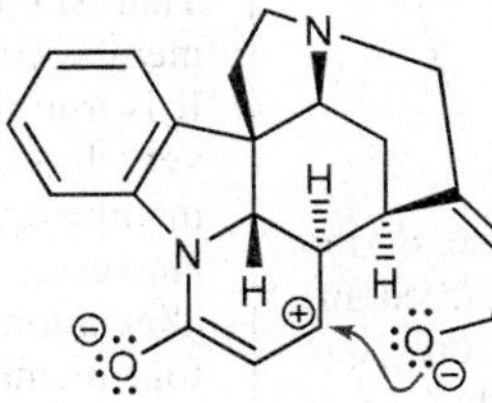

(c) The new chiral center is highlighted below. The oxygen atom takes priority #1, while the H has priority #4. Between #2 and #3, it is difficult to choose because both are carbon atoms and each of them is connected to C, C, and H. The tie breaker comes when we move farther out, and one of the carbon atoms is connected to O and N, while the other is connected to N, C, and H. The carbon atom connected to O and N wins (#2), giving the *R* configuration:

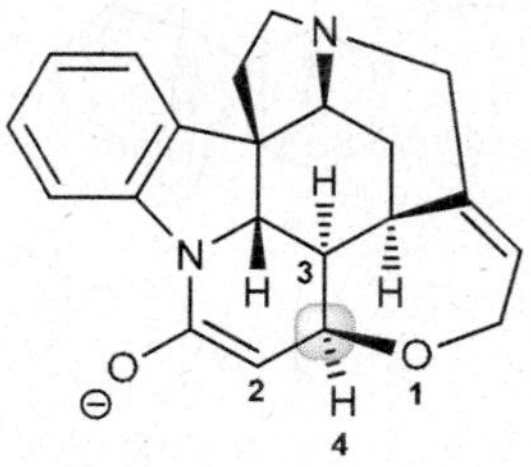

The newly formed chiral center has the *R* configuration, and not *S*, because the attached nucleophile is attacking from above, so the O must end up on a wedge:

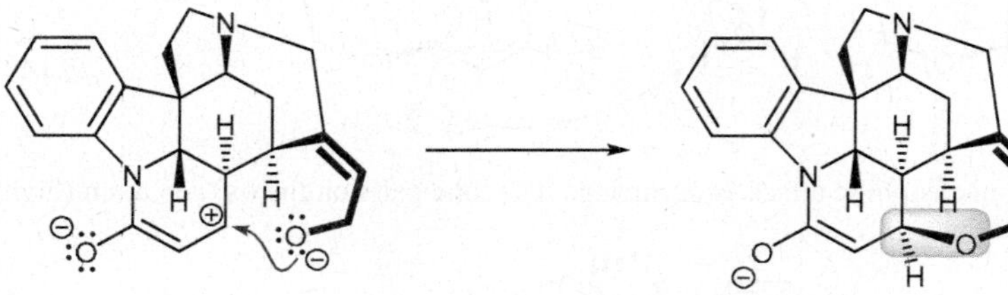

6.62.

(a) The carbon atom highlighted below is migrating, and the following curved arrow shows the migration:

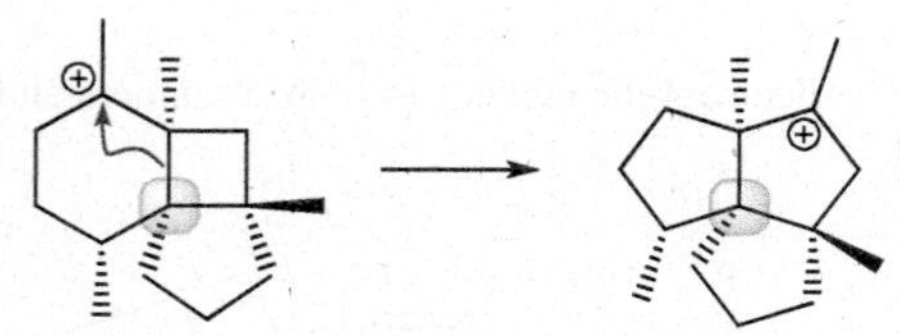

(b) The newly formed chiral center is highlighted below. Notice that the methyl group (attached to the highlighted carbon atom) is on a dash. The diastereomeric carbocation (not formed) would have this methyl group on a wedge:

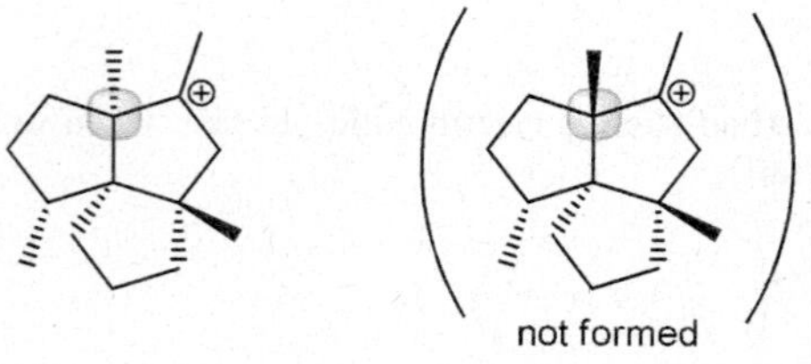

not formed

The other configuration is not formed because of the structural rigidity (and lack of conformational freedom) imposed by the tricyclic system. Specifically, only one face of the empty *p* orbital (associated with C+) is accessible to the migrating carbon atom, as seen in the following scheme:

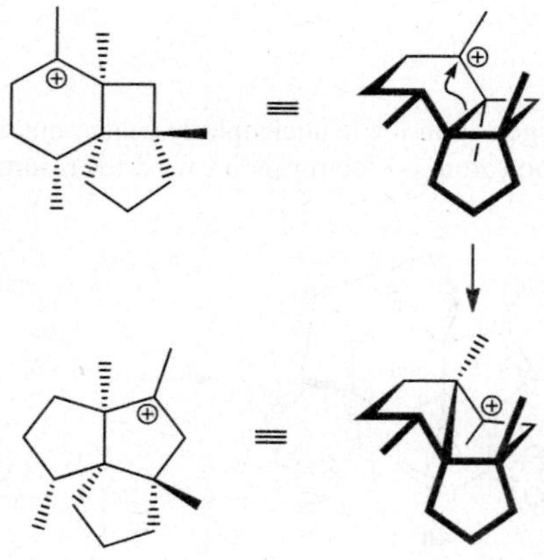

(c) The initial carbocation has three rings. One of them is a four-membered ring. The ring strain associated with this ring is alleviated as a result of the rearrangement. That is, the four-membered ring is converted into a five-membered ring, which has considerably less ring strain. It is true that a six-membered ring (which generally has very little, if any, ring strain) is converted into a five-membered ring, which does possess some ring strain. However, this energy cost is more than offset by the alleviation of ring strain resulting from enlarging the four-membered ring.

6.63.

(a) The following curved arrows show an intramolecular nucleophilic attack, with the simultaneous loss of a leaving group:

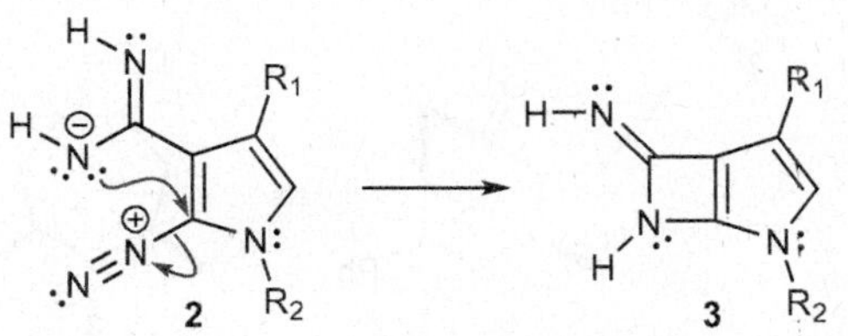

(b) Resonance structures (**2a** and **2b**) can be drawn for intermediate **2**:

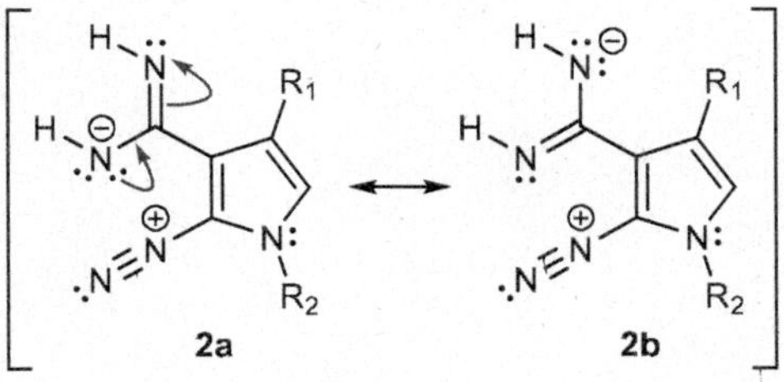

Notice that the negative charge is spread over two nitrogen atoms via resonance. As such, either of these nitrogen atoms can function as the nucleophilic center during the nucleophilic attack. Either the isotopically labeled nitrogen atom can attack, like this:

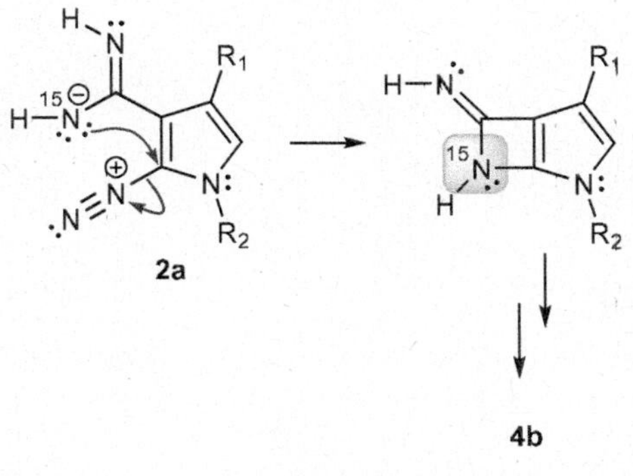

or the other nitrogen atom can attack, like this:

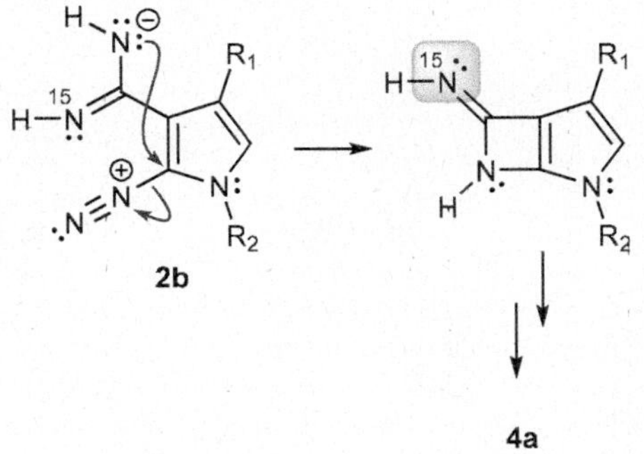

So, if the proposed mechanism were truly operating, we would expect that either nitrogen atom would have an equal probability of forming the four-membered ring, and so only 50% of the ^{15}N atom would be incorporated into the nitrile. That is, both **4a** and **4b** should both be formed. If that had been the case, we would have said that the ^{15}N atom was "scrambled" during the reaction. Since that did not occur, the proposed mechanism was refuted.

6.64.

(a) Compound **1** is converted to intermediate **2** upon loss of a leaving group.

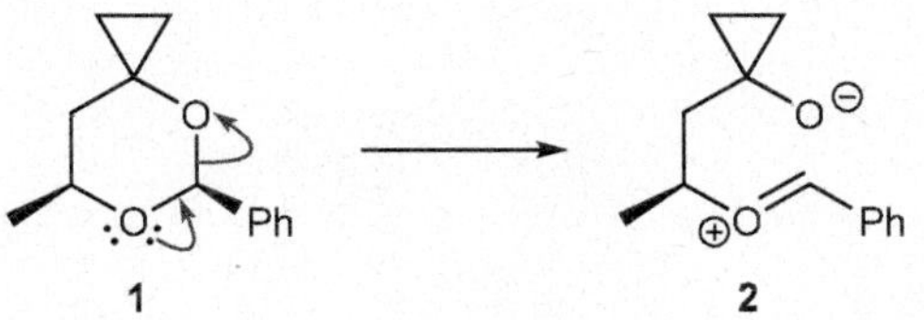

Notice that two curved arrows are employed in this case. The arrow with its tail on a C–O bond represents loss of the leaving group. The other curved arrow (with its tail on a lone pair) can be viewed in two ways: It can be viewed as the electrons coming from the O to push out the leaving group, as shown above, or it can be viewed as a resonance arrow that allows us to draw resonance structure **2**, thereby bypassing resonance structure **2a**:

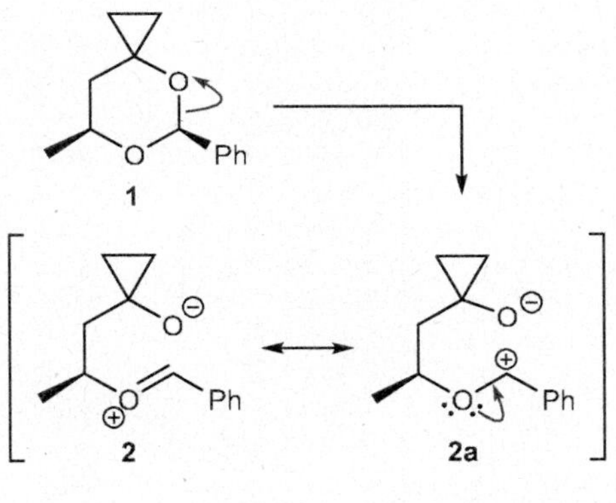

Structure **2** is a greater contributor than **2a** to the overall resonance hybrid because all atoms possess an octet of electrons. (Resonance structure **2a** has a C+ that is missing an octet.)

(b) In order to draw the curved arrows in this case, it will be helpful if we first rotate about the C–C bond indicated below.

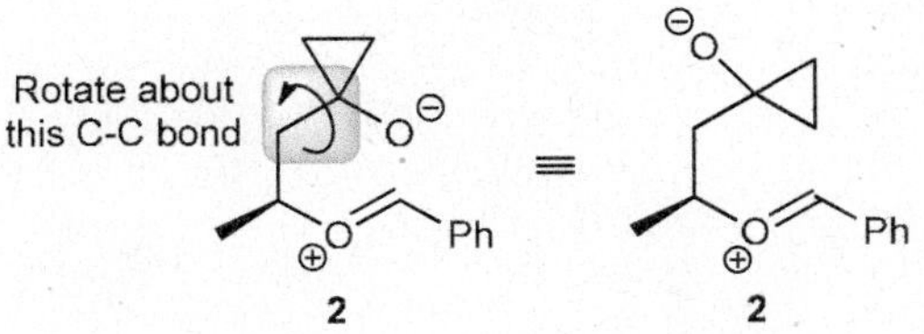

Rotate about this C-C bond

This gives a conformation in which the atoms are arranged in the proper orientation necessary to show the conversion of **2** into **3**. Three curved arrows are required to show the flow of electrons that correspond with the transformation of **2** into **3**:

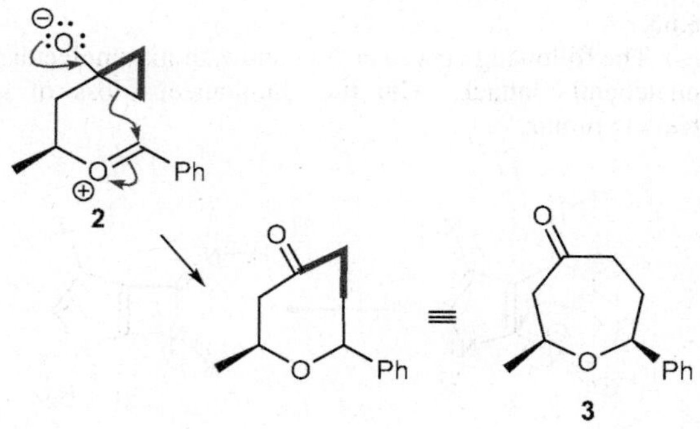

Chapter 7
Alkyl Halides: Nucleophilic Substitution and Elimination Reactions

Review of Concepts

Fill in the blanks below. To verify that your answers are correct, look in your textbook at the end of Chapter 7. Each of the sentences below appears verbatim in the section entitled *Review of Concepts and Vocabulary.*

- Good leaving groups are the conjugate bases of __________ acids.
- S_N2 reactions proceed via ______________ **of configuration**, because the nucleophile can only attack from the **back side.**
- S_N2 reactions cannot be performed with ___________ alkyl halides.
- A *trans* π bond cannot be incorporated into a small ring. When applied to bicyclic systems, this rule is called ___________ **rule**, which states that it is not possible for a ____________ carbon of a bicyclic system to possess a C=C double bond if it involves a *trans* π bond being incorporated in a small ring.
- E2 reactions are **regioselective** and generally favor the more substituted alkene, called the ___________ **product**.
- When both the substrate and the base are sterically hindered, an E2 reaction can favor the less substituted alkene, called the ______________ **product**.
- If the β position has two different protons, the resulting E2 reaction can be stereoselective, because the ________ isomer will be favored over the _________ isomer (when applicable).
- If the β position has only one proton, an E2 reaction is said to be ________________, because the proton and the leaving group must be ______________ to one another.
- **Unimolecular** nucleophilic substitution reactions are called ______ reactions. An S_N1 mechanism is comprised of two core steps: 1) ________________________ to give a carbocation intermediate, and 2) ____________________________.
- When a solvent molecule functions as the attacking nucleophile, the resulting S_N1 process is called a __________________.
- Unimolecular elimination reactions are called _____ reactions.
- S_N1 processes are favored by _________________ solvents.
- S_N1 and E1 processes are observed for tertiary alkyl halides, as well as allylic and ____________ halides.
- When the α position is a chiral center, an S_N1 reaction gives nearly a racemic mixture. In practice, there is generally a slight preference for ______________ over **retention of configuration**, as a result of the effect of ion pairs.
- Substitution and elimination reactions often compete with each other. To predict the products, three steps are required: 1) determine the function of the ____________; 2) analyze the ___________ and determine the expected mechanism(s); and 3) consider any relevant regiochemical and stereochemical requirements.
- Alcohols react with HBr to give alkyl halides, either via an _____ pathway (for primary and secondary substrates) or via an _____ pathway (for tertiary substrates).
- When treated with concentrated sulfuric acid, tertiary alcohols are converted into _________ via an E1 process.
- A _____________ **analysis** shows the product first, followed by reagents that can be used to make that product.
- _______________ **solvents** contain a hydrogen atom connected directly to an electronegative atom, while ________________ **solvents** lack such a hydrogen atom.
- A ______________ solvent will speed up the rate of an S_N2 process by many orders of magnitude.
- S_N1 processes are favored by ________________ solvents.

Review of Skills

Follow the instructions below. To verify that your answers are correct, look in your textbook at the end of Chapter 7. The answers appear in the section entitled *SkillBuilder Review*.

SkillBuilder 7.1 Drawing the Product of an SN2 Process

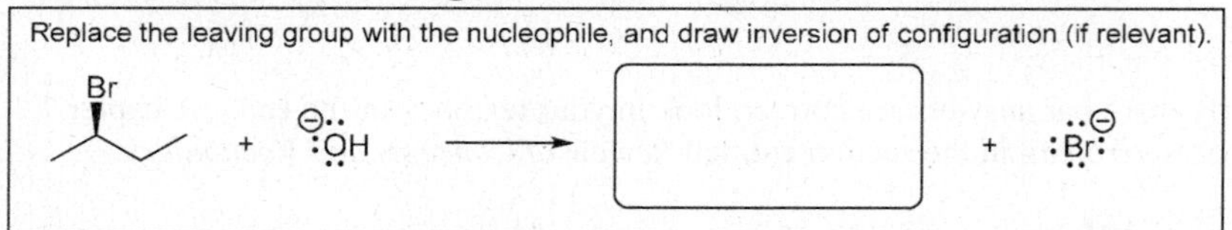

SkillBuilder 7.2 Drawing the Transition State of an SN2 Process

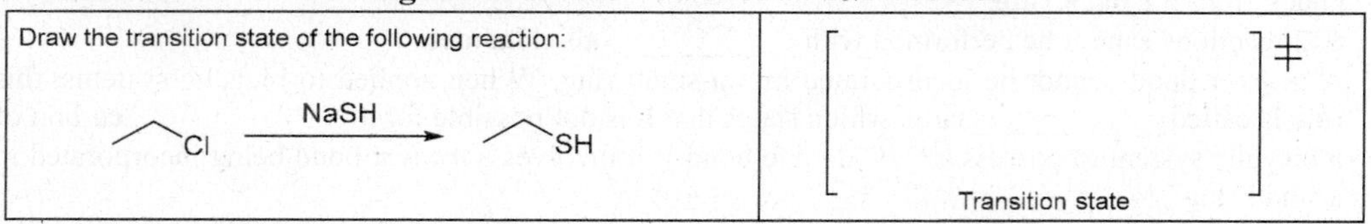

SkillBuilder 7.3 Predicting the Regiochemical Outcome of an E2 Reaction

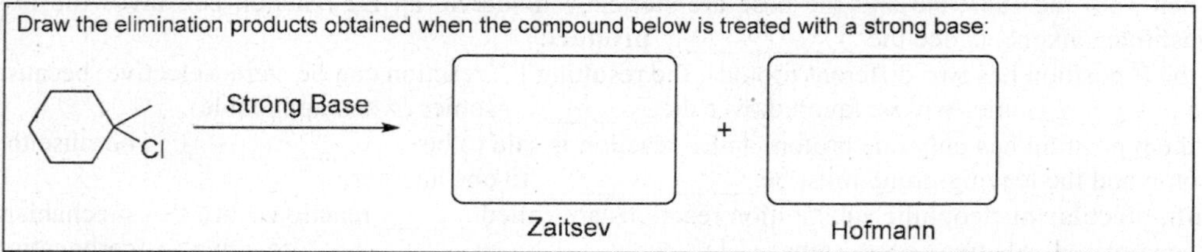

SkillBuilder 7.4 Predicting the Stereochemical Outcome of an E2 Reaction

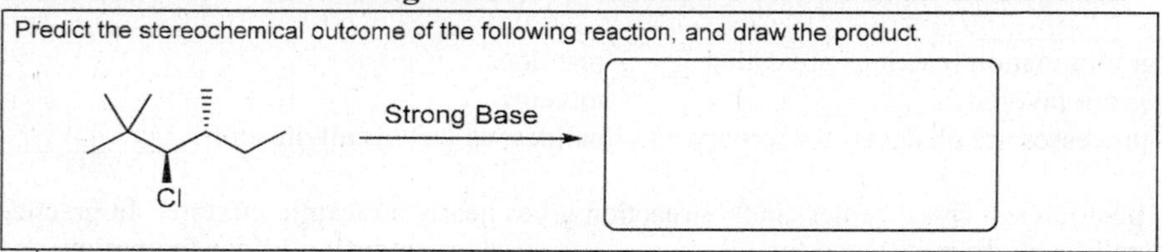

SkillBuilder 7.5 Drawing the Products of an E2 Reaction

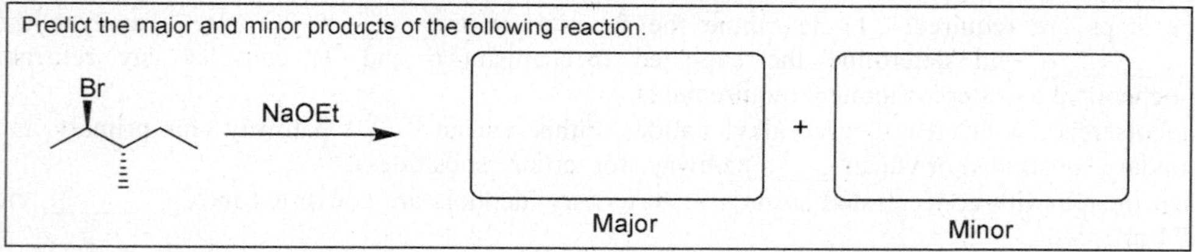

SkillBuilder 7.6 Drawing the Carbocation Intermediate of an SN1 or E1 Process

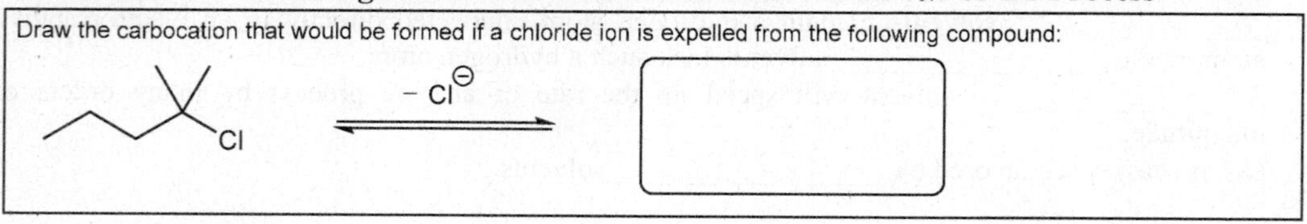

SkillBuilder 7.7 Predicting the Products of Substitution and Elimination Reactions of Alkyl Halides

Fill in the blanks below:		
Step 1 Determine the function of the _______________.	**Step 2** Analyze the ___________ and determine the expected mechanism(s).	**Step 3** Consider any relevant regiochemical and ___________ requirements.

SkillBuilder 7.8 Performing a Retrosynthesis and Providing a Synthesis of a Target Molecule

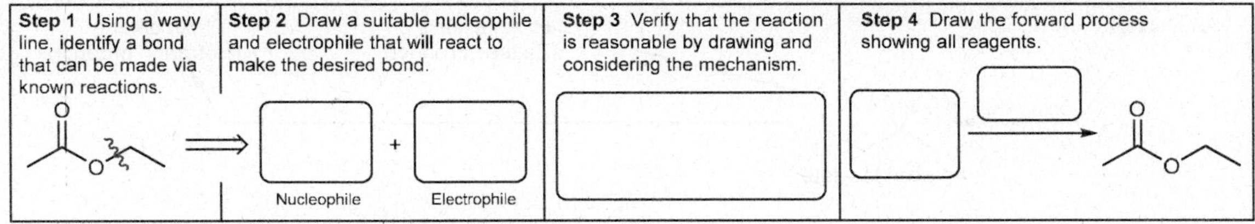

Step 1 Using a wavy line, identify a bond that can be made via known reactions.
Step 2 Draw a suitable nucleophile and electrophile that will react to make the desired bond.
Step 3 Verify that the reaction is reasonable by drawing and considering the mechanism.
Step 4 Draw the forward process showing all reagents.
O
O
+
Nucleophile
Electrophile
O
O

SkillBuilder 7.9 Retrosynthesis and Synthesis of an Alkene Target Molecule: Elimination Reactions

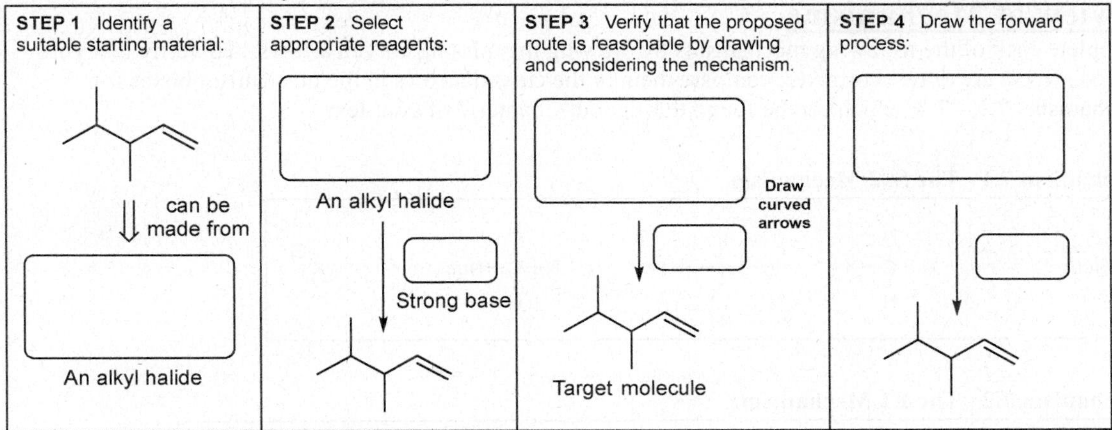

Review of Reactions

Identify reagents that can be used to achieve each of the following transformations. To verify that your answers are correct, look in your textbook at the end of Chapter 7. The answers appear in the section entitled *Review of Reactions*.

Primary Substrates

R Br
An alkyl halide

R OH
An alcohol

R CN
A nitrile

R OTs
An alkyl tosylate

R
An alkene

R OCH_3
An ether

R SH
A thiol

R O O
An ester

Tertiary Substrates

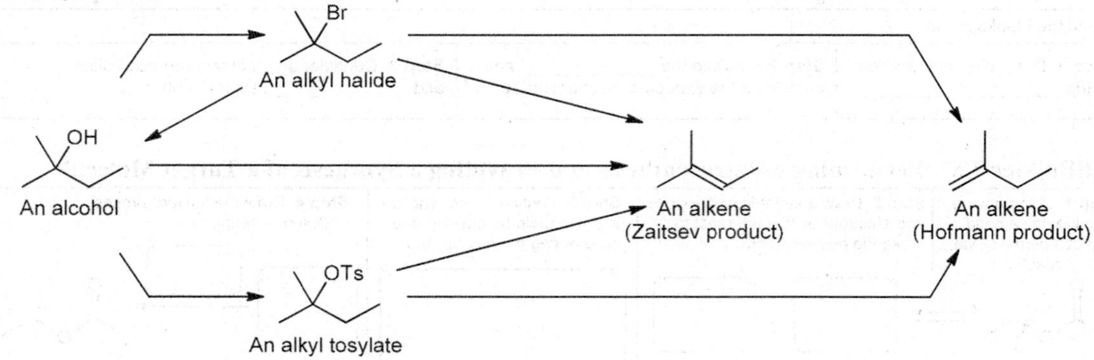

Review of Mechanisms

Complete each of the following mechanisms by drawing the missing curved arrows. To verify that your curved arrows are drawn correctly, compare them to the curved arrows in the mechanism boxes for Mechanisms 7.1 – 7.4, which can be found throughout Chapter 7 of your text.

Mechanism 7.1 The SN2 Mechanism

Nuc:⊖ + R—CH₂—X ⟶ R—CH₂—Nuc + :X:⊖

Mechanism 7.2 The E2 Mechanism

H—C—C—X + :Base⊖ ⟶ C=C + H—Base + :X:⊖

Mechanism 7.3 The SN1 Mechanism

X ⇌ (– X⊖) ⊕ ⟶ (:Nuc⊖) Nuc

Mechanism 7.4 The E1 Mechanism

X ⇌ (– X⊖) ⊕ H ⟶ (:Base) alkene + H—Base⊕

Useful reagents

The following is a list of commonly encountered reagents for substitution and elimination reactions:

Reagent	Name	Function
NaCl	Sodium chloride	An ionic salt consisting of Na^+ and Cl^- ions. The former (Na^+) can be ignored in most cases, while the latter (chloride) is a strong nucleophile. NaCl is a source of chloride ions.
NaBr	Sodium bromide	An ionic salt consisting of Na^+ and Br^- ions. The former (Na^+) can be ignored in most cases, while the latter (bromide) is a strong nucleophile. NaBr is a source of bromide ions.
NaI	Sodium iodide	An ionic salt consisting of Na^+ and I^- ions. The former (Na^+) can be ignored in most cases, while the latter (iodide) is a strong nucleophile. NaI is a source of iodide ions.
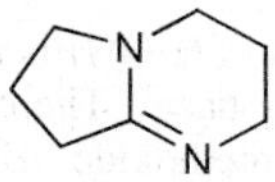	DBN	A strong base that can be used in E2 reactions.
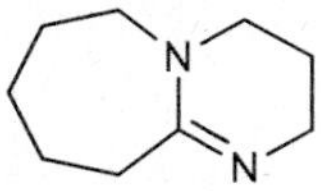	DBU	A strong base that can be used in E2 reactions.
NaOH	Sodium hydroxide	Hydroxide (HO^-) is both a strong nucleophile AND a strong base, and can therefore be used for either E2 or S_N2 reactions, depending on the substrate (S_N2 is favored for primary substrates, E2 is favored for secondary substrates, and E2 is the exclusive pathway for tertiary substrates).
NaOR	Sodium alkoxide	R is an alkyl group. Examples include sodium methoxide (NaOMe) and sodium ethoxide (NaOEt). Alkoxide ions are both strong nucleophiles and strong bases. They can therefore be used for either E2 or S_N2 reactions, depending on the substrate (S_N2 is favored for primary substrates, E2 is favored for secondary substrates, and E2 is the exclusive pathway for tertiary substrates).
t-BuOK	Potassium *tert*-butoxide	*tert*-Butoxide is both a strong nucleophile and a strong base. But it is sterically hindered, which favors E2 over S_N2 even for primary substrates. For E2 reactions, when more than one regiochemical outcome is possible, *tert*-butoxide will favor formation of the less substituted alkene.
NaSH	Sodium hydrosulfide	HS^- is a strong nucleophile, used in S_N2 reactions.
H_2O	Water	Water is a weak nucleophile and a weak base, used in S_N1 and E1 reactions. Heat will often favor E1 over S_N1.
ROH	An alcohol	Examples include methanol (CH_3OH) and ethanol (CH_3CH_2OH). Alcohols are weak nucleophiles and weak bases, used in S_N1 and E1 reactions. Heat will often favor E1 over S_N1. Alcohols can also serve as substrates for S_N1 or E1 reactions under acidic conditions.
HX (X = Cl, Br, or I)	Hydrogen halides	A strong acid that serves as both a source of H^+ and nucleophilic X^- where X = Cl, Br, or I.
conc. H_2SO_4	Concentrated sulfuric acid	A strong acid, used to convert alcohols into alkenes (acid-catalyzed dehydration of an alcohol).

Common Mistakes to Avoid

When drawing the mechanism of a reaction, you must always consider what reagents are being used, and your mechanism must be consistent with the conditions employed. As an example, consider the following S_N1 reaction:

The following proposed mechanism is unacceptable, because the reagent employed in the second step is not present:

WRONG

This is a common student error. To see what's wrong, let's look closely at the reagent. Methanol (CH_3OH) is not a strong acid. Rather, it is a weak acid, because its conjugate base, methoxide (CH_3O^-), is a strong base. Therefore, methoxide is not present in substantial quantities, so the mechanism in this case should not employ methoxide. Below is the correct mechanism:

CORRECT

Methanol (rather than methoxide) functions as the nucleophile in the second step, because methoxide was not indicated as a reagent, and it is not expected to be present. The result of the nucleophilic attack is an oxonium ion (an intermediate with a positive charge on an oxygen atom), which is then deprotonated by another molecule of methanol. Once again, in this final step of the mechanism, methanol functions as the base, rather than methoxide, because the latter is not present.

This example is just one illustration of the importance of analyzing the reagent and considering what entities can be used in your mechanism. This will become increasingly important in upcoming chapters.

Now let's consider another reaction, so that we can identity another common student error. We have seen that an OH group is a bad leaving group (because hydroxide is a strong base). Therefore, in order for an alcohol (ROH) to serve as a substrate in a substitution or elimination reaction, the OH group must first be converted into a better leaving group. We have seen two ways to do this. One method involves converting the alcohol into a tosylate:

$$\text{R–OH} \xrightarrow[\text{pyridine}]{\text{TsCl}} \text{R–OTs} \rightarrow \text{Subst. / Elim.}$$

Bad leaving group — Good leaving group

The other method involves protonation of the OH group, as seen in the following example:

$$\text{R–}\ddot{\text{O}}\text{–H} + H_3O^+ \rightleftharpoons \text{R–}\overset{+}{\text{O}}H_2$$

Bad leaving group — Good leaving group

This latter approach (protonation) has a serious limitation. Specifically, it cannot be used if the reagent is a strong base. For example, the following reaction sequence does not work:

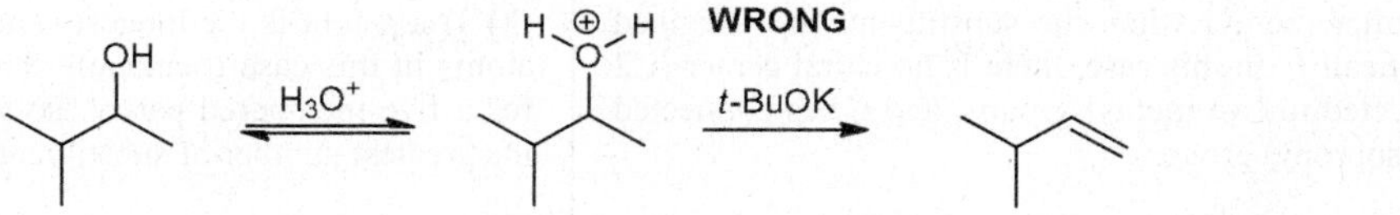

H⊕H WRONG
OH
H_3O^+
t-BuOK

It doesn't work because it is not possible to have a strong acid (H_3O^+) and a strong base (*t*-BuOK) present in the same reaction flask at the same time (they would simply neutralize each other). In order to achieve the desired transformation, the OH group must first be converted to a tosylate (rather than simply being protonated), and then the desired reaction can be performed, as shown here:

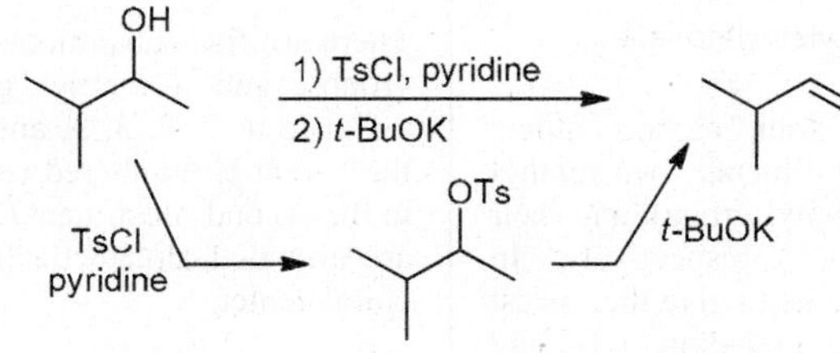

OH
1) TsCl, pyridine
2) t-BuOK
OTs
TsCl pyridine
t-BuOK

Solutions

7.1.

(a) The parent is the longest chain, which is five carbon atoms in this case (pentane). There are three substituents (bromo, bromo, and chloro), and their locants are assigned as 4, 4, and 1, respectively. In this case, the parent was numbered from right to left, so as to give the lowest number to the first substituent (1,4,4 rather than 2,2,5). Notice that two locants are necessary (rather than one) to indicate the locations of the two bromine atoms, even though they are connected to the same position (4,4-dibromo rather than 4-dibromo).

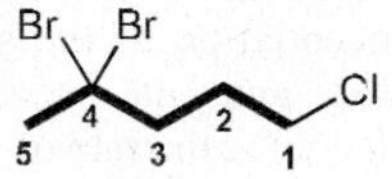

4,4-dibromo-1-chloropentane

(b) The parent is a six-membered ring (cyclohexane). There are two substituents (methyl and bromo), both of which are located at the C1 position. Substituents are alphabetized in the name (bromo precedes methyl).

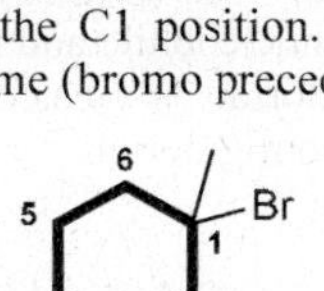

1-bromo-1-methylcyclohexane

(c) The parent is the longest chain, which is eight carbon atoms in this case (octane). There are two substituents (ethyl and chloro), both of which are located at the C4 position. These substituents are alphabetized (chloro precedes ethyl). In this case, there is also a chiral center, so we must assign the configuration (*R*), which must be indicated at the beginning of the name.

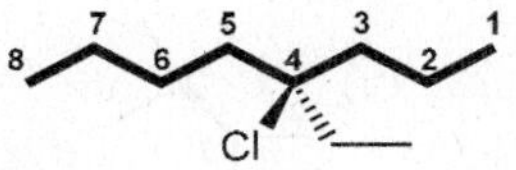

(*R*)-4-chloro-4-ethyloctane

(d) The parent is the longest chain, which is six carbon atoms in this case (hexane). There are three substituents (fluoro, methyl, and methyl), and their locants are assigned as 5, 2, and 2, respectively. In this case, the parent was numbered from right to left, so as to give the lowest number to the second substituent (2,2,5 rather than 2,5,5). The substituents are arranged alphabetically in the name, so fluoro precedes dimethyl (the former is "f" and the latter is "m"). In this case, there is also a chiral center, so we must assign the configuration (*R*), which must be indicated at the beginning of the name.

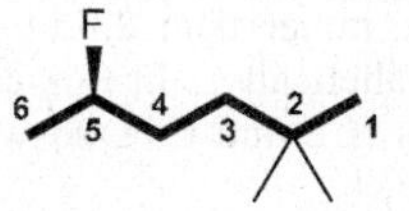

(*R*)-5-fluoro-2,2-dimethylhexane

(e) The parent is the longest chain, which is nine carbon atoms in this case (nonane). There are three substituents (methyl, bromo, and isopropyl), and their locants are assigned as 2, 3, and 3, respectively. The substituents are arranged alphabetically in the name (note that isopropyl is alphabetized as "i" rather than as "p", so it comes before methyl). If we choose to use the systematic name for isopropyl, (1-methylethyl), it is placed after methyl when the substituents are arranged alphabetically. In this case, there is no chiral center (C2 is connected to two methyl groups, and C3 is connected to two isopropyl groups).

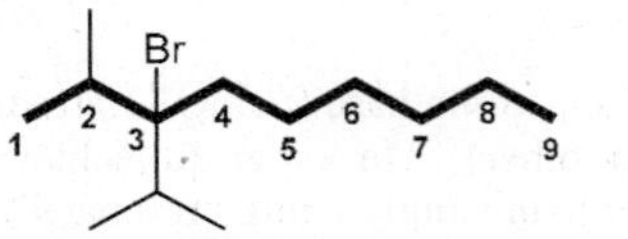

3-bromo-3-isopropyl-2-methylnonane
or
3-bromo-2-methyl-3-(1-methylethyl)nonane

(f) The parent is a ring of four carbon atoms (cyclobutane). There are four substituents (two methyl groups, a chloro group, and a *tert*-butyl group), and their locants are assigned as 1, 1, 2, and 3, respectively. In this case, the parent is numbered so as to give the lowest number to the second substituent (1,**1**-dimethyl), and then continuing clockwise. The substituents are arranged alphabetically (note that *tert*-butyl is alphabetized as "b"). If we choose to use the systematic name for *tert*-butyl, (1,1-dimethylethyl), it is alphabetized as "methylethyl," so it is placed after the "methyl" (in 1,2-dimethyl) when the substituents are arranged alphabetically. In this case, there are also two chiral centers (C2 and C3), so we must assign the configuration of each. Note that C1 is not a chiral center, because it bears two methyl groups.

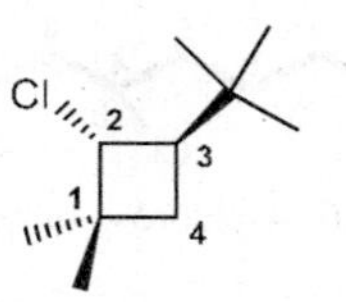

(2*R*,3*R*)-3-*tert*-butyl-2-chloro-1,1-dimethylcyclobutane
or
(2*R*,3*R*)-2-chloro-1,1-dimethyl-3-(1,1-dimethylethyl)cyclobutane

(g) The parent is the longest chain, which is nine carbon atoms in this case (nonane). There are three substituents (chloro, ethyl, and methyl), and their locants are assigned as 2, 4, and 8, respectively. In this case, the parent is numbered so as to give the lowest number to the second substituent (2,**4**,8, rather than 2,**6**,8). The substituents are arranged alphabetically. In this case, there are also two chiral centers (C2 and C4), so we must assign the configuration of each.

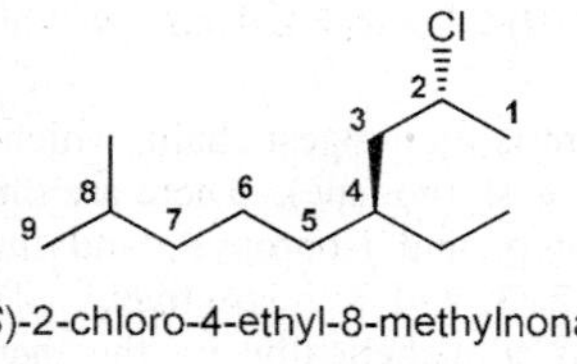

(2*R*,4*S*)-2-chloro-4-ethyl-8-methylnonane

(h) The parent is the longest chain, which is five carbon atoms in this case (pentane). There are several choices for a five-membered parent, so we choose the one with the greatest number of substituents.

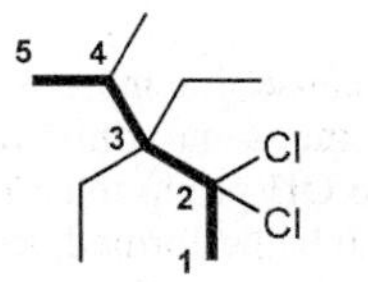

There are five substituents (two chloro groups, two ethyl groups, and a methyl group), and their locants are assigned as 2, 2, 3, 3, and 4, respectively. In this case, the parent is numbered so as to give the lowest number to the second substituent (2,2-dichloro). The substituents are arranged alphabetically. In this case, there are no chiral centers.

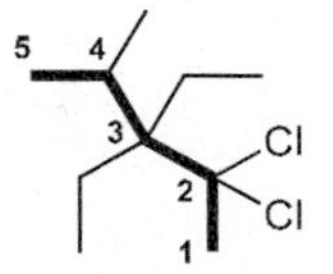

2,2-dichloro-3,3-diethyl-4-methylpentane

7.2.

(a) The reaction has a second-order rate equation, and the rate is linearly dependent on the concentrations of two compounds (the nucleophile AND the alkyl halide). If the concentration of the substrate (the alkyl halide) is tripled, the rate should also be tripled.

(b) As described above, the rate is linearly dependent on the concentrations of both the nucleophile and the alkyl halide. If the concentration of the nucleophile is doubled, the rate of the reaction is doubled.

(c) As described above, the rate is linearly dependent on the concentrations of both the nucleophile and the alkyl halide. If the concentration of the substrate (the alkyl halide) is doubled and the concentration of the nucleophile is tripled, then the rate of the reaction will be six times faster (×2×3).

7.3.

(a) The substrate is (*S*)-2-chloropentane, and the nucleophile is HS^- (note that Na^+ serves as the counterion for the nucleophile and plays no active role in the reaction). Chloride is ejected as a leaving group, with inversion of configuration.

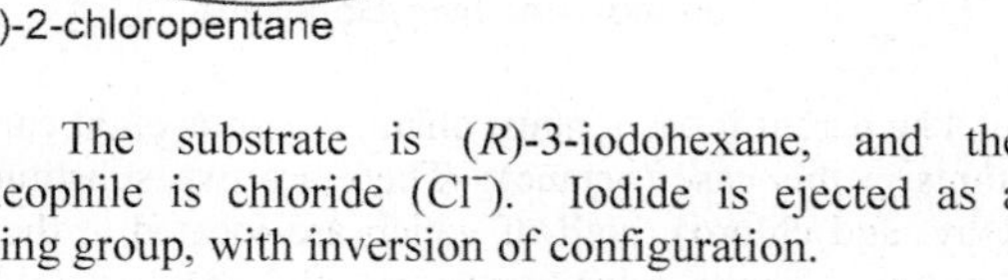

(*S*)-2-chloropentane

(b) The substrate is (*R*)-3-iodohexane, and the nucleophile is chloride (Cl^-). Iodide is ejected as a leaving group, with inversion of configuration.

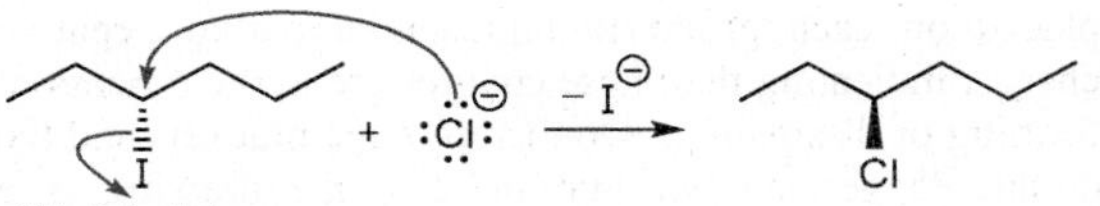

(R)-3-iodohexane

(c) The substrate is (R)-2-bromohexane, and the nucleophile is cyanide (N≡C⁻). Bromide is ejected as a leaving group, with inversion of configuration.

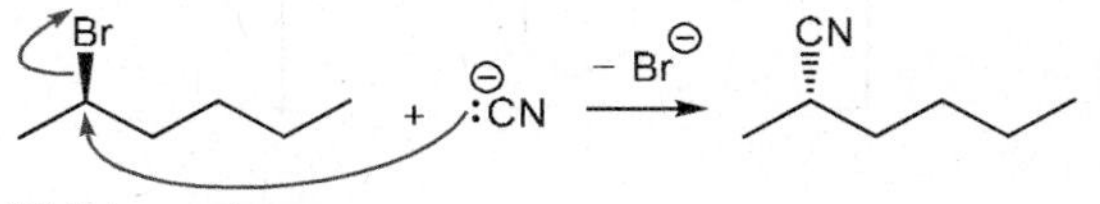

(R)-2-bromohexane

(d) The substrate is 1-bromoheptane, and the nucleophile is hydroxide (HO⁻). Bromide is ejected as a leaving group. There is no inversion of configuration in this case, because there is no chiral center:

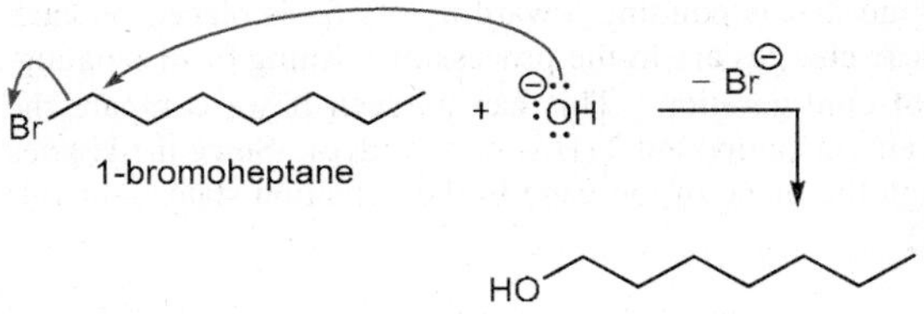

1-bromoheptane

7.4. The problem statement indicates that *tert*-butoxide functions as a base, removing a proton from compound **1** to give the carbanion intermediate. This is a proton transfer step, so it requires two curved arrows, as shown:

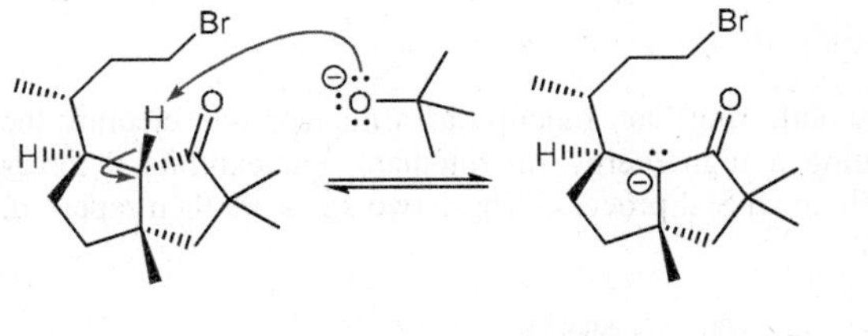

1 **carbanion intermediate**

This intermediate then undergoes an intramolecular SN2-type process, because it has both a nucleophilic center and an electrophilic center:

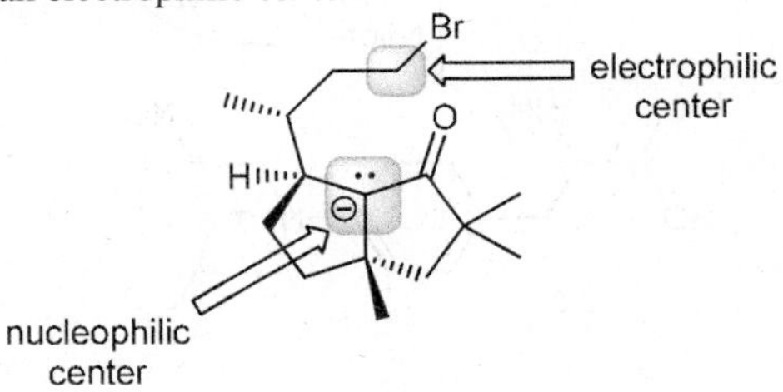

electrophilic center

nucleophilic center

That is, the nucleophile and electrophile are tethered to each other (rather than being separate compounds), and the reaction occurs in an intramolecular fashion, as shown:

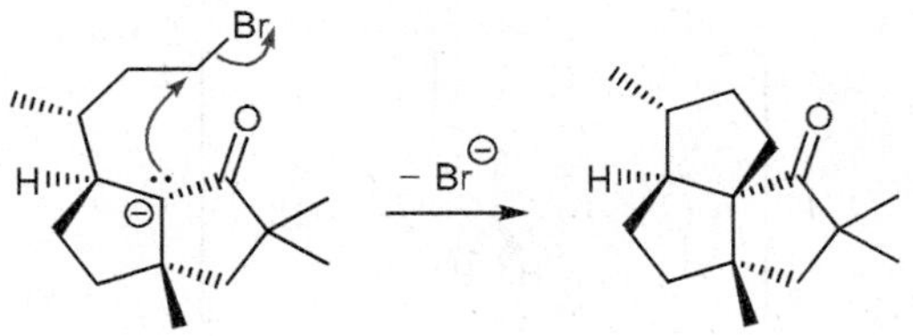

This process is not bimolecular, but it is similar to an SN2 mechanism (concerted, back-side attack).

7.5.
(a) The leaving group is a bromide ion (Br⁻) and the nucleophile is a hydroxide ion (HO⁻). In the transition state, each of these groups is drawn as being connected to the α position with a dotted line to represent a partial bond (a bond that is either in the process of forming or breaking), and at a bond angle of 180° (indicating back-side attack). A δ– is placed on each group to represent a partial negative charge, indicating that these charges are in the process of forming or dissipating. Don't forget the brackets and the double-dagger symbol that indicate the drawing is a transition state.

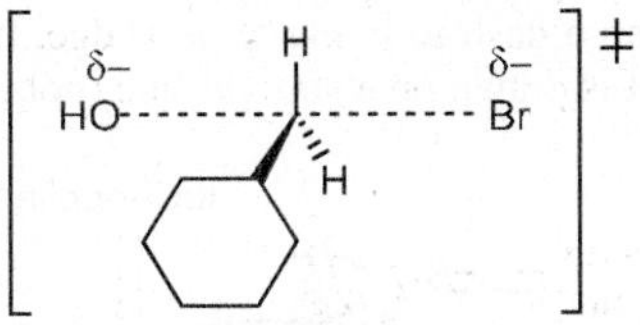

(b) The leaving group is an iodide ion (I⁻) and the nucleophile is an acetate ion ($CH_3CO_2^-$). In the transition state, each of these groups is drawn as being connected to the α position with a dotted line to represent a partial bond (a bond that is either in the process of forming or breaking), and at a bond angle of 180° (indicating back-side attack). A δ– is placed on each group to represent a partial negative charge, indicating that these charges are in the process of forming or dissipating. Don't forget the brackets and the double-dagger symbol that indicate the drawing is a transition state.

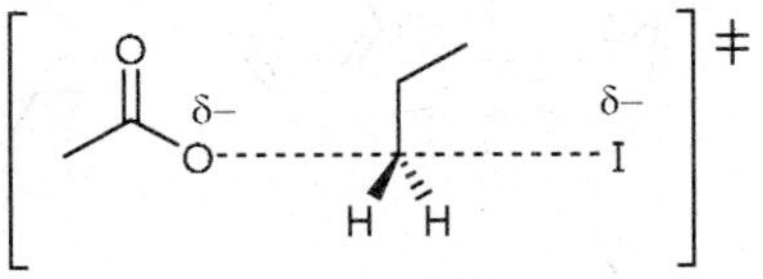

(c) The leaving group is a chloride ion (Cl⁻) and the nucleophile is a hydroxide ion (HO⁻). In the transition state, each of these groups is drawn as being connected to the α position with a dotted line to represent a partial bond (a bond that is either in the process of forming or breaking), and at a bond angle of 180° (indicating back-side attack). A δ– is placed on each group to represent a partial negative charge, indicating that these charges are in the process of forming or dissipating. Don't forget the brackets and the double-dagger symbol that indicate the drawing is a transition state.

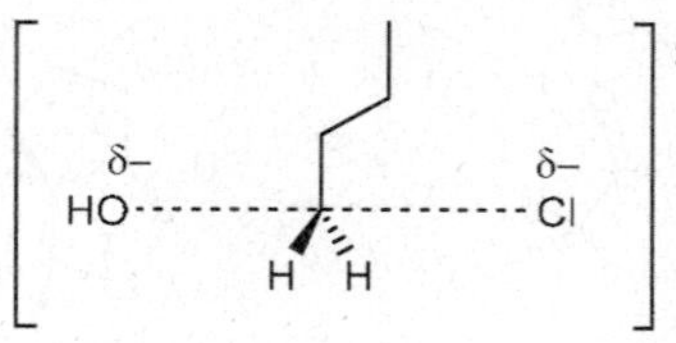

(d) The leaving group is a bromide ion (Br^-) and the nucleophile is HS^-. In the transition state, each of these groups is drawn as being connected to the α position with a dotted line to represent a partial bond (a bond that is either in the process of forming or breaking), and at a bond angle of 180° (indicating back-side attack). A δ– is placed on each group to represent a partial negative charge, indicating that these charges are in the process of forming or dissipating. Don't forget the brackets and the double-dagger symbol that indicate the drawing is a transition state.

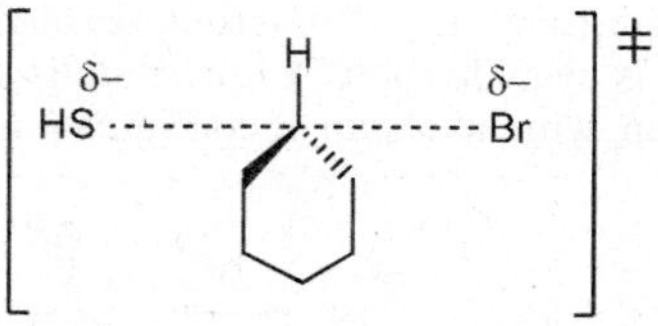

7.6. The leaving group is the oxygen atom of the three-membered ring, and the nucleophile is the oxygen atom bearing the negative charge. In the transition state, each of these groups is drawn as being connected to the α carbon (highlighted below) with a dotted line (indicating a bond that is either in the process of forming or breaking), and at a bond angle of 180° (indicating back-side attack). In this case, the nucleophilic oxygen atom is approaching from behind the page, because the leaving group is on a wedged bond that is pointing toward us. A δ– is placed on each group to represent a partial negative charge, indicating that these charges are in the process of forming or dissipating. Notice that the α carbon (highlighted) undergoes inversion of configuration. This can be seen if we compare the location of the hydrogen atom in compound **1** (H is on a dash) and in compound **2** (H is on a wedge). Since the H goes from being on a dash to being on a wedge, it must pass through the plane of the page in the transition state. For this reason, the H is drawn on a straight line (not a wedge or a dash):

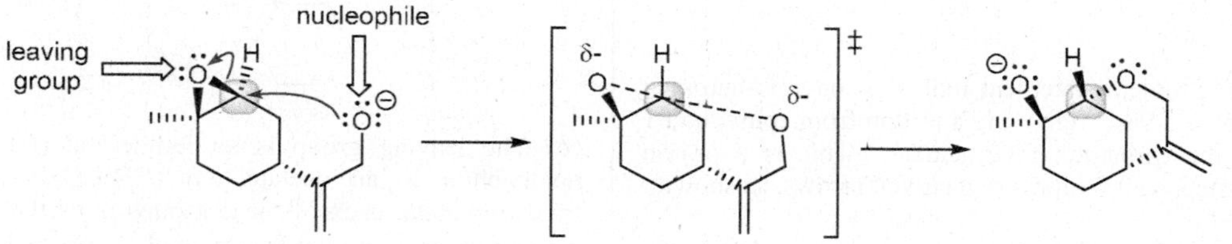

7.7. The lone pair of the nitrogen atom (connected to the aromatic ring) can function as a nucleophile, ejecting the chloride ion in an intramolecular S_N2-type reaction, generating a high-energy intermediate that exhibits a three-membered ring. The ring is opened upon attack of a nucleophile in an S_N2 process. These two steps are then repeated, as shown here.

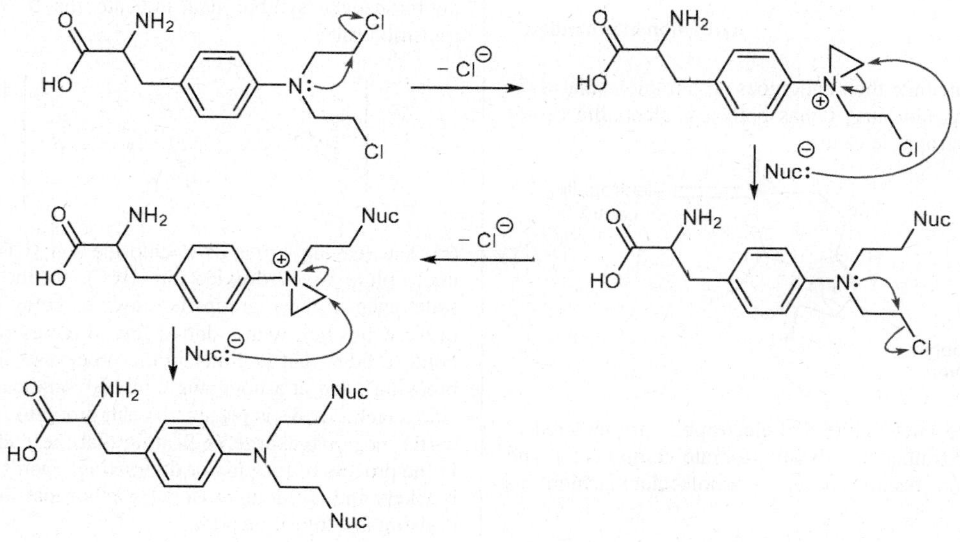

7.8. The nitrogen atom functions as a nucleophilic center and attacks the electrophilic methyl group in SAM, forming an ammonium ion.

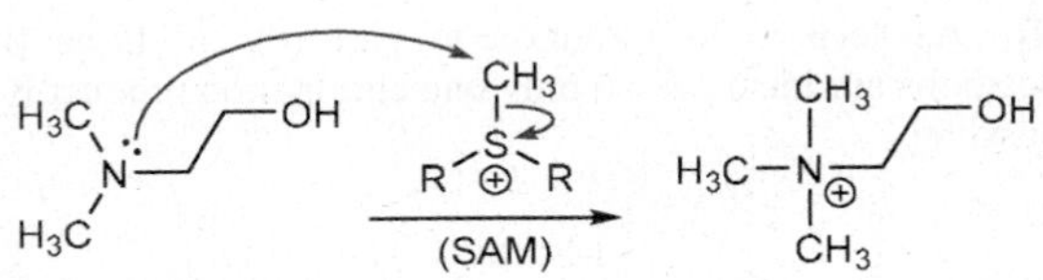

7.9.

(a) With a second-order rate equation, the rate is expected to be linearly dependent on the concentrations of the alkyl halide and the base. If the concentration of the alkyl halide is tripled, then the rate is expected to be three times faster.

(b) With a second-order rate equation, the rate is expected to be linearly dependent on the concentrations of the alkyl halide and the base. If the concentration of the base is doubled, then the rate is expected to be two times faster.

(c) With a second-order rate equation, the rate is expected to be linearly dependent on the concentrations of the alkyl halide and the base. If the concentration of the alkyl halide is doubled and the concentration of the base is tripled, then the rate is expected to be six times faster (×2×3).

7.10.

(a) Each alkene is classified according to its degree of substitution. Highlighted below are the alkyl substituents attached to each C=C double bond. The most highly substituted alkene will be the most stable, and therefore, the following order of stability is expected.

increasing stability

di-substituted tri-substituted tetra-substituted

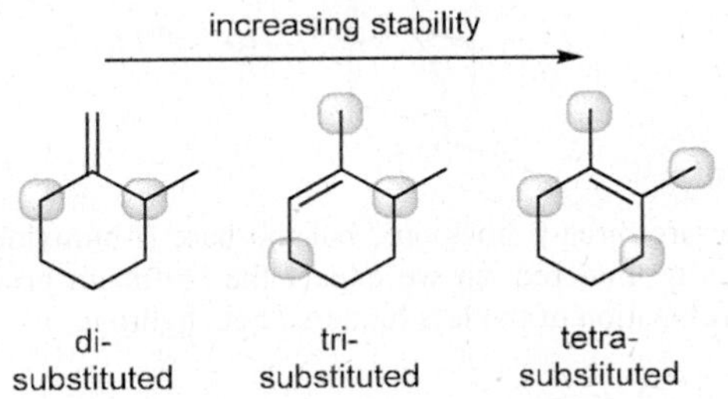

(b) Each alkene is classified according to its degree of substitution. Highlighted below are the alkyl substituents attached to each C=C double bond. The most highly substituted alkene will be the most stable, and therefore, the following order of stability is expected.

increasing stability

mono-substituted di-substituted tri-substituted

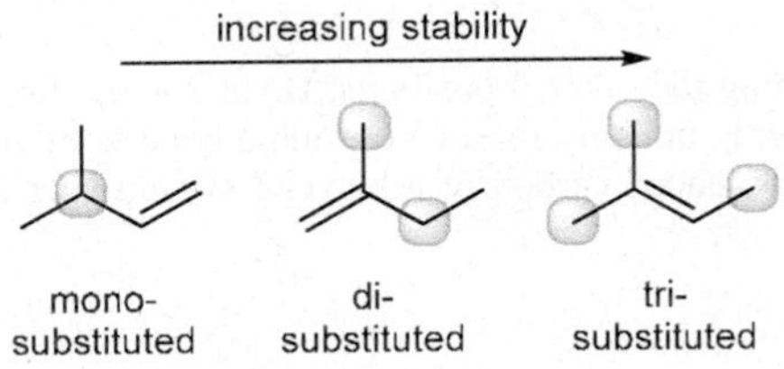

7.11. In the first compound, all of the carbon atoms of the cyclobutane ring are sp^3 hybridized and tetrahedral. As a result, they are supposed to have bond angles of approximately 109.5°, but their bond angles are compressed due to the ring (and are almost 90°). In other words, the compound exhibits angle strain characteristic of small rings. In the second compound, two of the carbon atoms in the ring (the two joined by a double bond) are sp^2 hybridized and trigonal planar. As a result, these two carbon atoms are each supposed to have bond angles of approximately 120°, but their bond angles are compressed due to the ring (and are almost 90°). The resulting angle strain (120° → 90°) is greater than the angle strain in the first compound (109.5° → 90°). Therefore, the second compound is less stable and higher in energy, despite the fact that it has a more highly substituted double bond.

7.12.

(a) This compound has three β positions, but one of them (highlighted) does not bear protons:

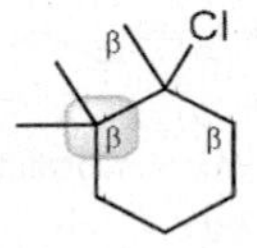

Since there are two β positions bearing protons, there are two possible elimination products. Since the base (ethoxide) is not sterically hindered, we expect that the major product will be the more-substituted alkene (the Zaitsev product), and the minor product will be the less-substituted alkene (the Hofmann product).

NaOEt

major (more substituted) minor (less substituted)

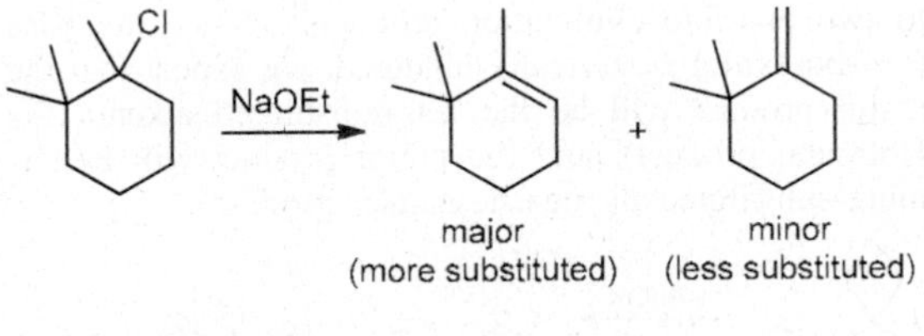

(b) This compound has three β positions that bear protons, but two of them (highlighted) are identical:

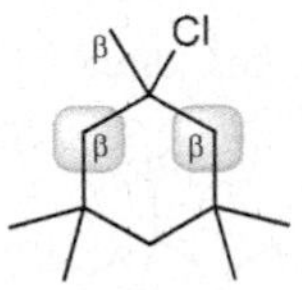

Thus, there are only two unique β positions, giving rise to two possible elimination products, shown below. Since the base (*tert*-butoxide) is sterically hindered, we expect that the major product will be the less-substituted alkene (the Hofmann product), and the minor product will be the more-substituted alkene (the Zaitsev product).

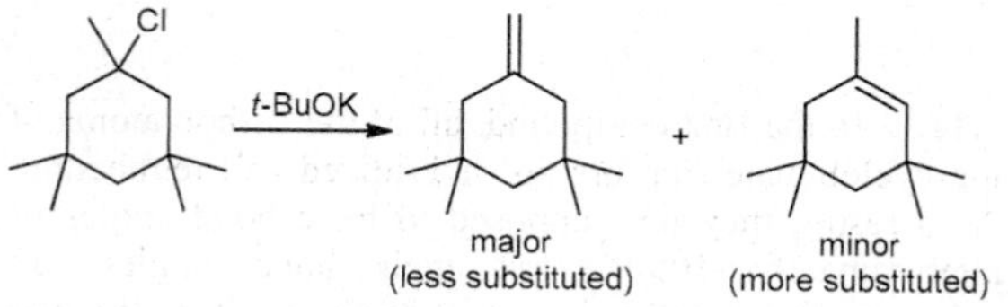

t-BuOK

major (less substituted)

minor (more substituted)

(c) This compound has three β positions that bear protons, but two of them (highlighted) are identical:

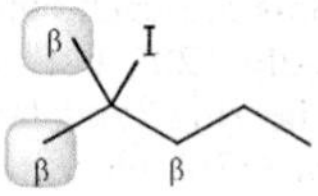

Thus, there are only two unique β positions, giving rise to two possible elimination products, shown below. Since the base (hydroxide) is not sterically hindered, we expect that the major product will be the more-substituted alkene (the Zaitsev product), and the minor product will be the less-substituted alkene (the Hofmann product).

NaOH

major (more substituted)

minor (less substituted)

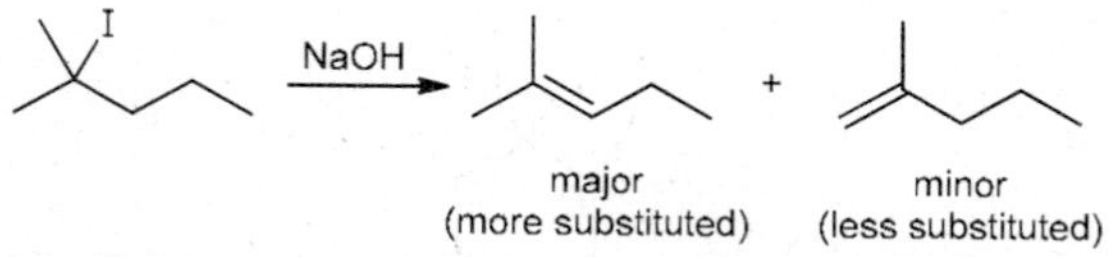

(d) This compound has three β positions that bear protons, but two of them (highlighted) are identical:

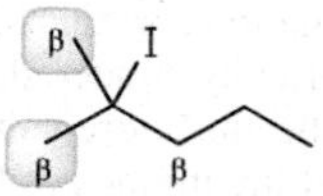

Thus, there are only two unique β positions, giving rise to two possible elimination products. Since the base (*tert*-butoxide) is sterically hindered, we expect that the major product will be the less-substituted alkene (the Hofmann product), and the minor product will be the more-substituted alkene (the Zaitsev product).

t-BuOK

major (less substituted)

minor (more substituted)

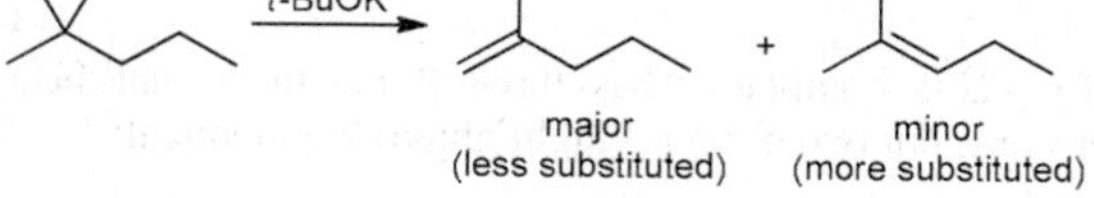

(e) This compound has three β positions that bear protons:

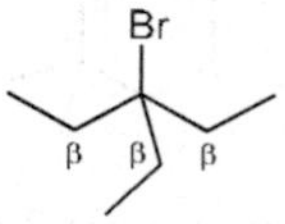

In this case, all three β positions are identical, because removing a proton from any one of these positions will lead to the same product. As such, there is only one possible elimination product:

NaOH

only product

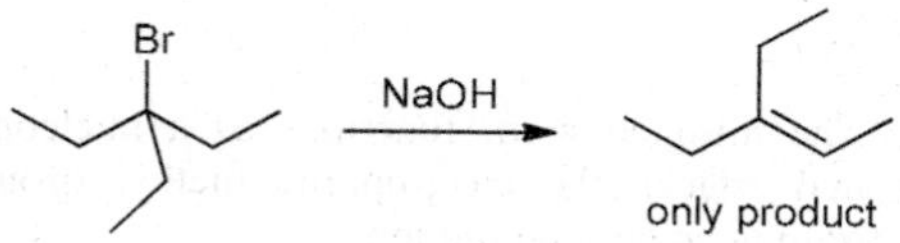

(f) As seen in the solution to part (e), all three β positions are identical, so only one elimination product is possible.

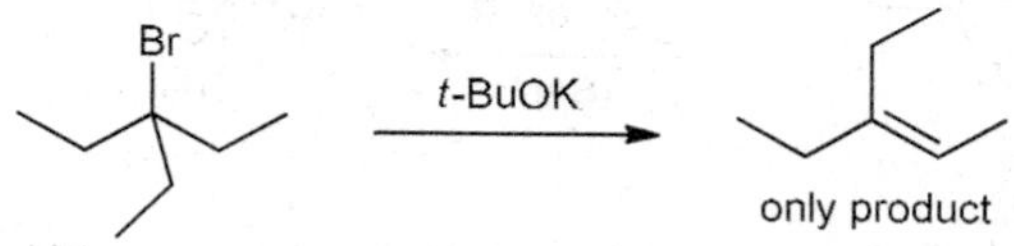

t-BuOK

only product

7.13.

(a) The more substituted alkene is desired (the Zaitsev product), so hydroxide (not sterically hindered) should be used.

(b) The less substituted alkene is desired (Hofmann product), so *tert*-butoxide (a sterically hindered base) should be used.

7.14. Both bromide leaving groups should be considered for elimination, but the primary bromide cannot be eliminated because it has no beta hydrogen.

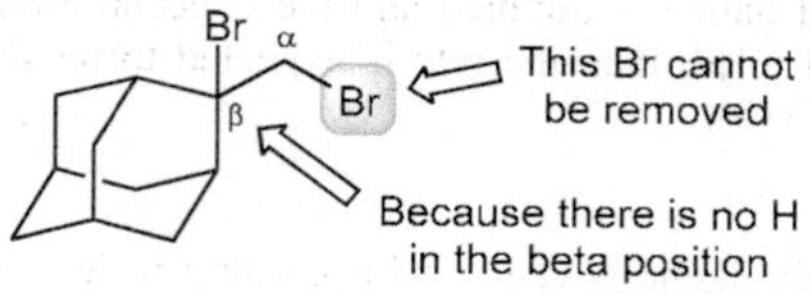

This Br cannot be removed

Because there is no H in the beta position

Therefore, the product must be formed via an E2 reaction involving the tertiary bromide.

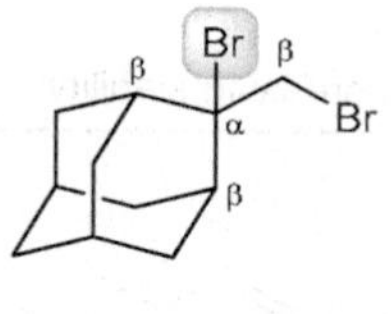

There are three β positions, but the base (*t*-butoxide) is sterically hindered, so we expect the Hofmann product (deprotonation of the less hindered beta hydrogen):

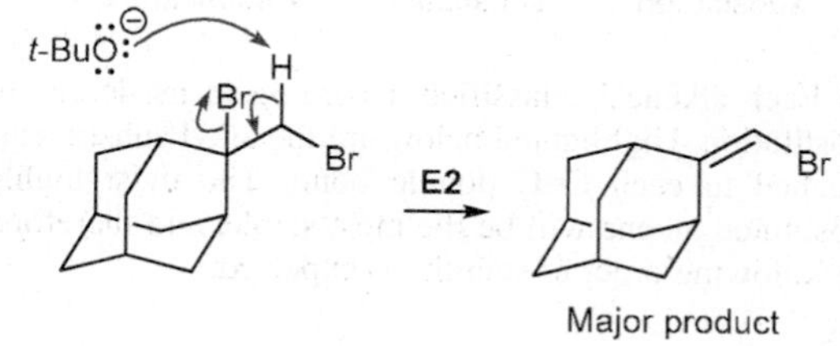

E2

Major product

Regarding the other β positions, Bredt's Rule states that it is highly unlikely for a C=C double bond to be formed at a bridgehead carbon of a bicyclic system such as the one shown.

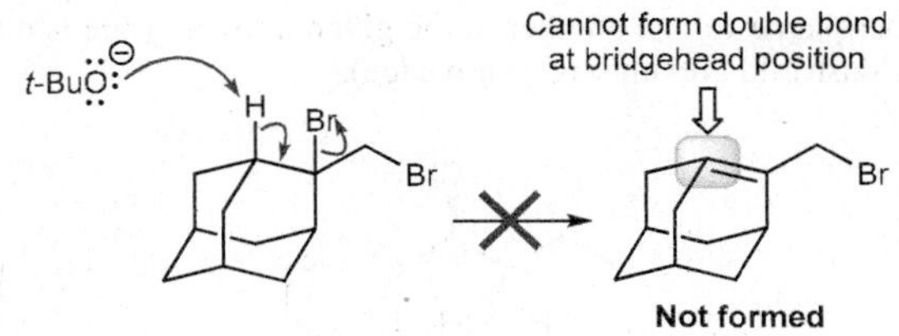

Cannot form double bond at bridgehead position

Not formed

The same is true regarding deprotonation at the other beta position:

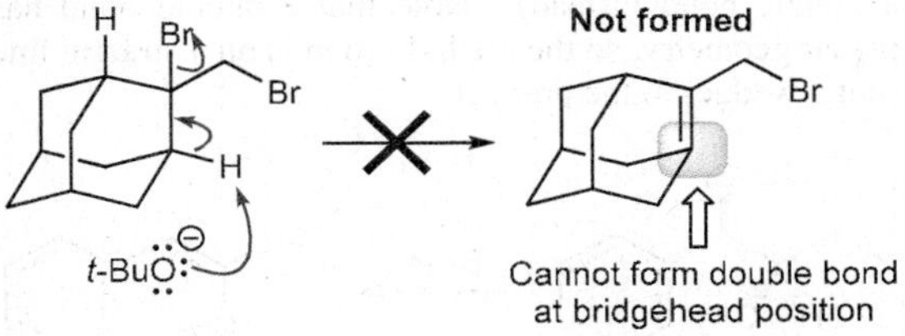

Not formed

Cannot form double bond at bridgehead position

As a result, this reaction affords only one product.

7.15.

(a) The substrate has two β positions, but only one of these positions (highlighted) bears protons.

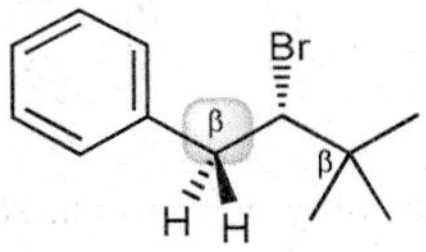

This β position has two protons, and both can achieve the required *anti*-periplanar conformation, so the reaction will be stereoselective. That is, we expect both *cis* and *trans* isomers, with a preference for the more stable *trans* isomer.

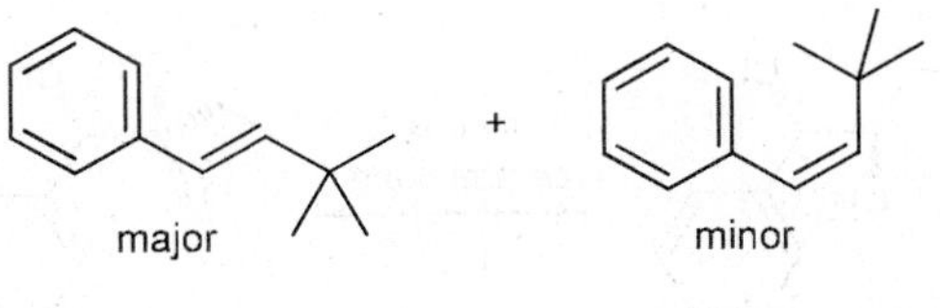

major + minor

(b) The substrate has two β positions, but only one of these positions (highlighted) bears a proton.

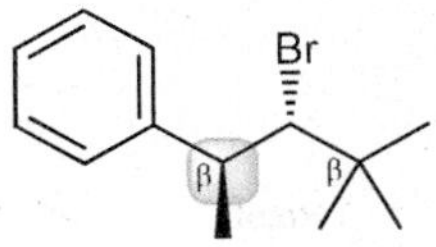

This β position has only one proton, so the reaction will be stereospecific. That is, only one particular stereoisomeric product will be obtained. To determine which product to expect, we must rotate the central C–C bond so that the β proton is *anti*-periplanar to the leaving group. We will do so in two stages. First, we rotate the central C–C bond in a manner that places the β proton in the plane of the page (rather than on a dash):

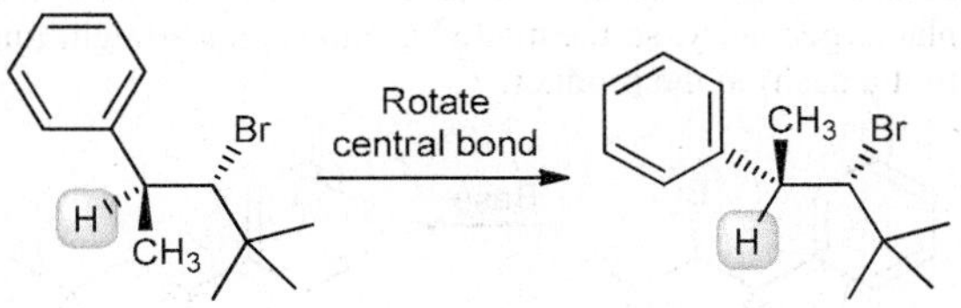

Rotate central bond

Then, we rotate the central C–C bond again, in a manner that places the leaving group (Br) in the plane of the page:

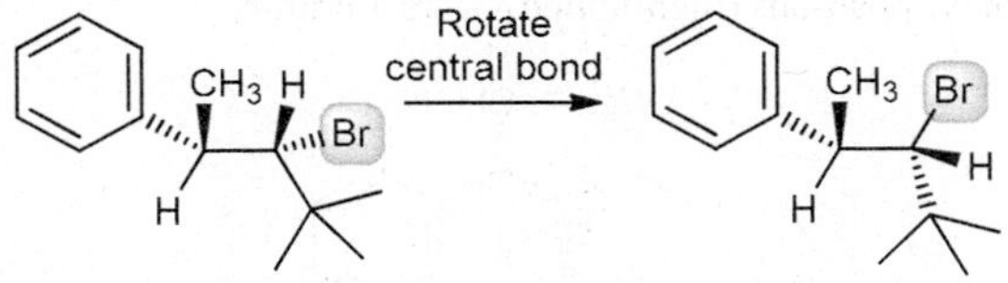

Rotate central bond

In this conformation, the proton and the leaving group are *anti*-periplanar. To draw the product, use the wedges and dashes as guides. In this case, the *tert*-butyl group and the phenyl group are both on dashes, so they will be *cis* to each other in the product:

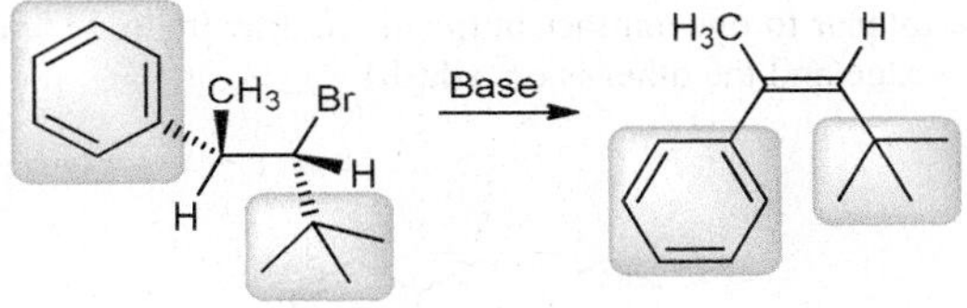

Base

(c) The substrate has two β positions, but only one of these positions (highlighted) bears a proton.

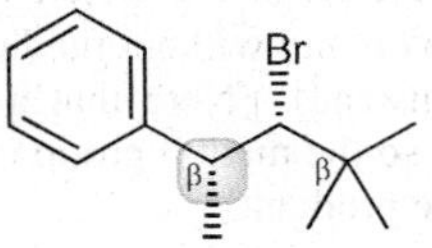

This β position has only one proton, so the reaction will be stereospecific. That is, only one particular stereoisomeric product will be obtained. To determine which product to expect, we can rotate the central C–C bond so as to place the β proton and the leaving group in the plane of the page. But in this case, that is not necessary, because the β proton and the leaving group are already *anti*-periplanar to one another in the given drawing (one is on a wedge and the other is on a dash):

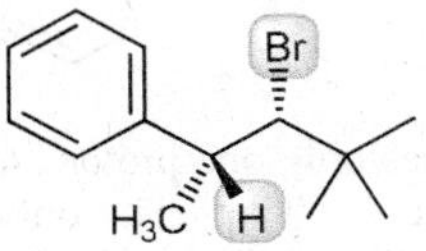

In such a case, it is relatively easy to draw the product, because the carbon skeleton is simply redrawn without the β proton and without the leaving group (with a double bond instead). Note that a double bond has planar geometry, so the methyl group is on a straight line (not a dash) in the product:

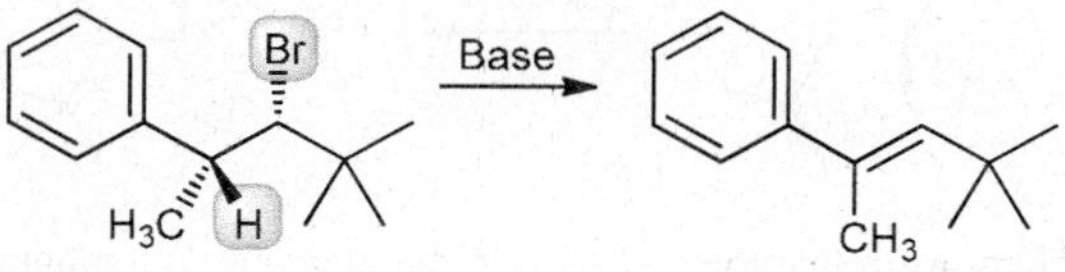

(d) The substrate has two β positions, but only one of these positions (highlighted) bears a proton.

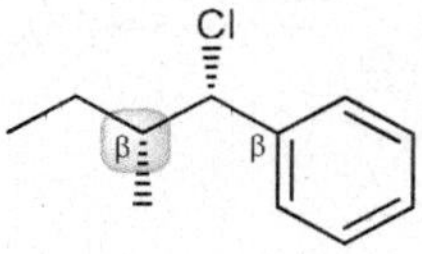

This β position has only one proton, so the reaction will be stereospecific. That is, only one particular stereoisomeric product will be obtained. To determine which product to expect, we will use the same method employed in the solution to part (c). In this case, the β proton and the leaving group are already *anti*-periplanar to one another in the given drawing (one is on a wedge and the other is on a dash).

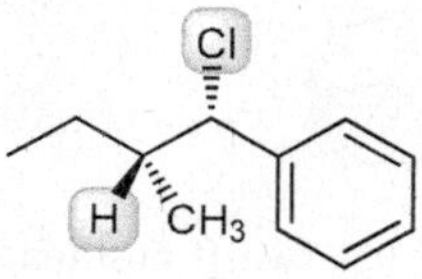

In such a case, the carbon skeleton is simply redrawn without the β proton and without the leaving group (with a double bond instead). Note that a double bond has planar geometry, so the methyl group is on a straight line (not a dash) in the product:

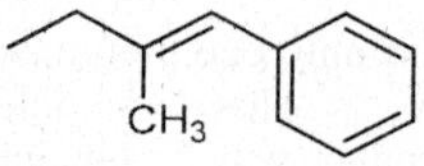

(e) The substrate has two β positions, but only one of these positions (highlighted) bears a proton.

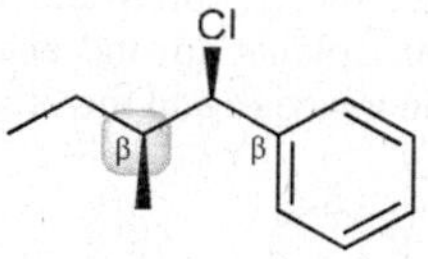

This β position has only one proton, so the reaction will be stereospecific. That is, only one particular stereoisomeric product will be obtained. To determine which product to expect, we will use the same method employed in the solution to part (d). In this case, the β proton and the leaving group are already *anti*-periplanar to one another in the given drawing (one is on a dash and the other is on a wedge).

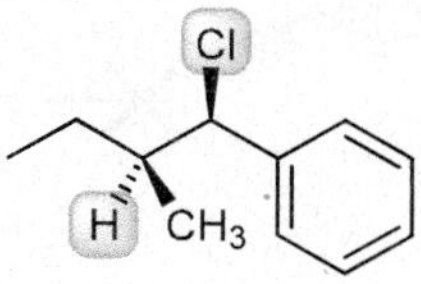

In such a case, the carbon skeleton is simply redrawn without the β proton and without the leaving group (with a double bond instead). Note that a double bond has planar geometry, so the methyl group is on a straight line (not a wedge) in the product:

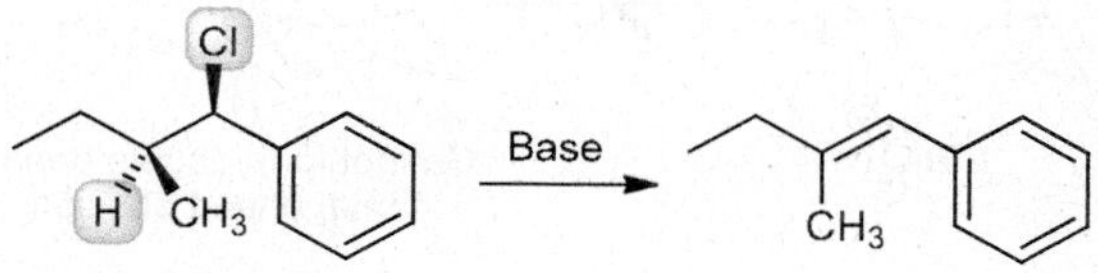

(f) The substrate has two β positions, but only one of these positions (highlighted) bears a proton.

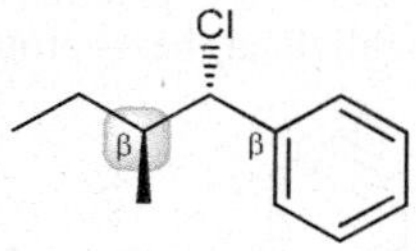

This β position has only one proton, so the reaction will be stereospecific. That is, only one particular stereoisomeric product will be obtained. To determine which product to expect, we must rotate the central C–C bond so that the β proton is *anti*-periplanar to the leaving group. We will do so in two stages. First, we rotate the central C–C bond in a manner that places the β proton in the plane of the page (rather than on a dash):

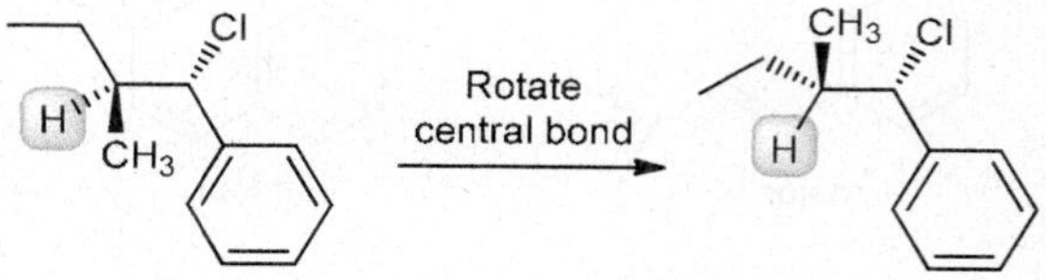

Then, we rotate the central C–C bond again, in a manner that places the leaving group (Cl) in the plane of the page:

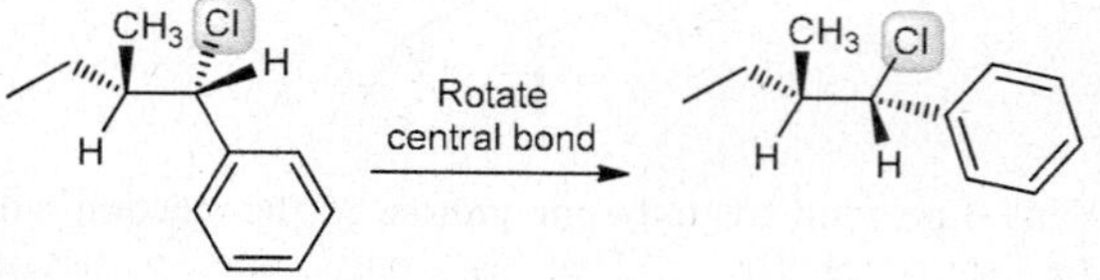

In this conformation, the proton and the leaving group are *anti*-periplanar. To draw the product, use the wedges and dashes as guides. In this case, the ethyl group and the phenyl group are both on dashes, so they will be *cis* to each other in the product:

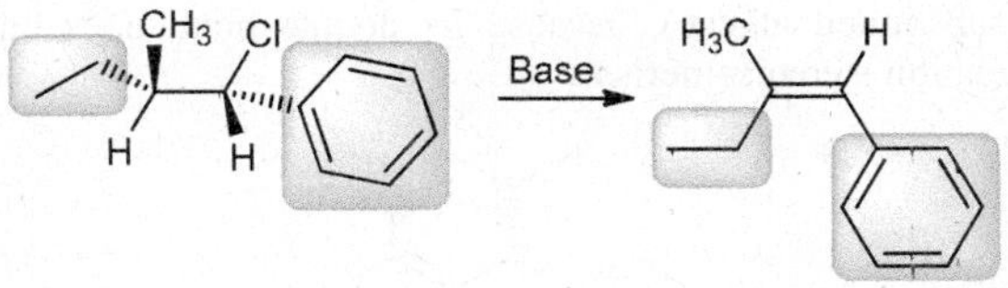

Base

(g) The substrate has two β positions, but only one of these positions (highlighted) bears protons.

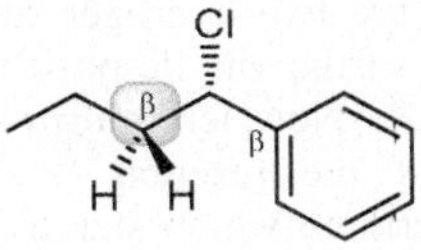

This β position has two protons, and both can achieve the required *anti*-periplanar conformation, so the reaction will be stereoselective. That is, we expect both *cis* and *trans* isomers, with a preference for the more stable *trans* isomer.

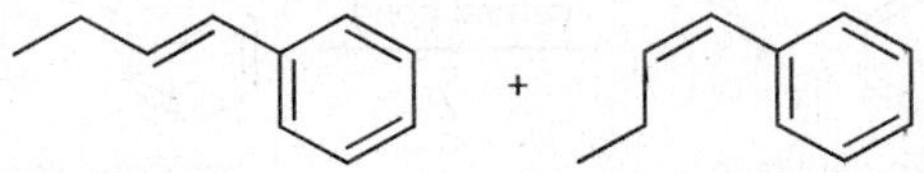

major minor

(h) The substrate has two β positions, but only one of these positions (highlighted) bears protons.

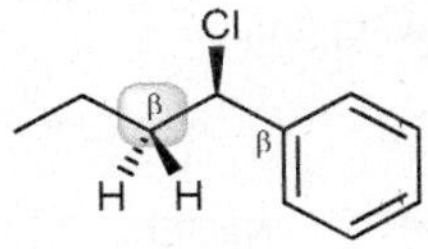

This β position has two protons, and both can achieve the required *anti*-periplanar conformation, so the reaction will be stereoselective. That is, we expect both *cis* and *trans* isomers, with a preference for the more stable *trans* isomer.

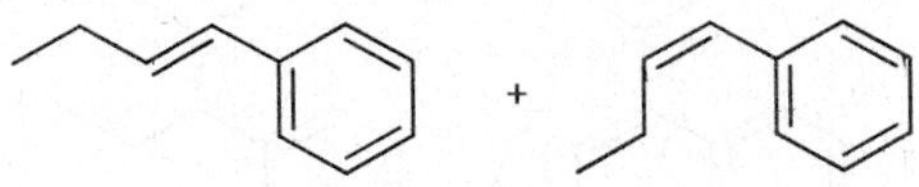

major minor

7.16. Since the two alkyl bromides are identical with the exception of the configuration at C6, the stereospecificity of the reaction will dictate the configuration (*E* or *Z*) of the newly formed C=C unit in each case. Let's begin with compound **1**, which can be easily drawn because the β proton and the leaving group are already *anti*-periplanar to one another (one is on a dash and the other is on a wedge):

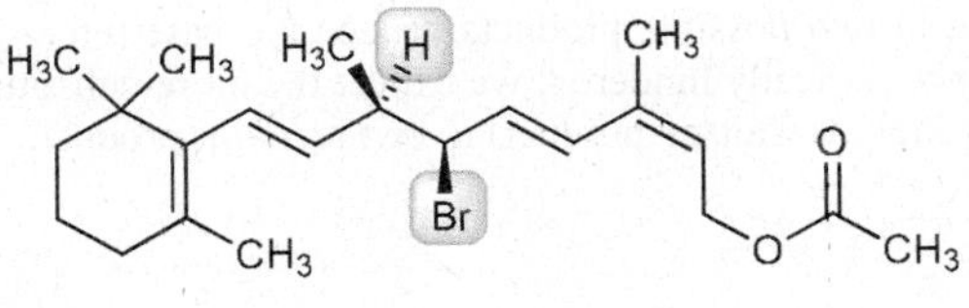

In such a case, the carbon skeleton is simply redrawn without the β proton and without the leaving group (with a double bond instead). Note that a double bond has planar geometry, so the methyl group is on a straight line (not a dash) in the product:

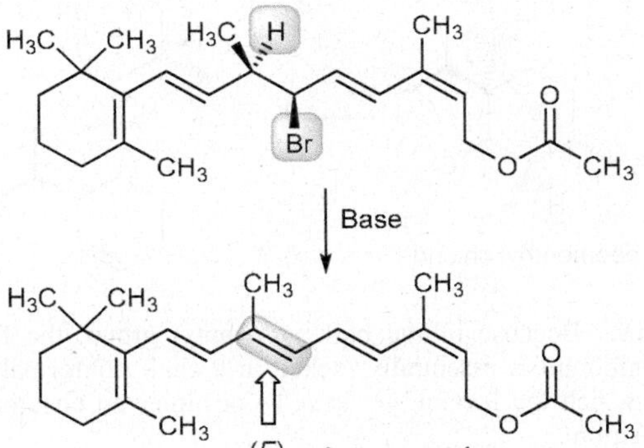

Base

(*E*) Compound **1**

In this case, the *E* product predominates. Since compound **1** has the *E* configuration for the double bond between C6 and C7, compound **2** must have the *Z* configuration for that double bond:

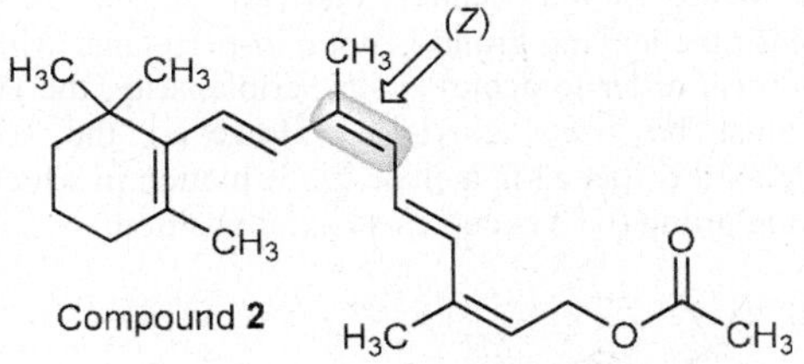

Compound **2**

Compound **1** is expected to be more stable than compound **2**, because the steric strain in the *Z* alkene (**2**) causes it to be higher in energy than the *E* alkene (**1**).

7.17. In the structure of menthyl chloride (shown below), the leaving group (Cl^-) is on a dash. Therefore, we are looking for a β proton that is on a wedge, in order that it should be *anti*-periplanar with the leaving group. In this case, there is only one β proton on a wedge (highlighted below). Therefore, only one elimination product is observed:

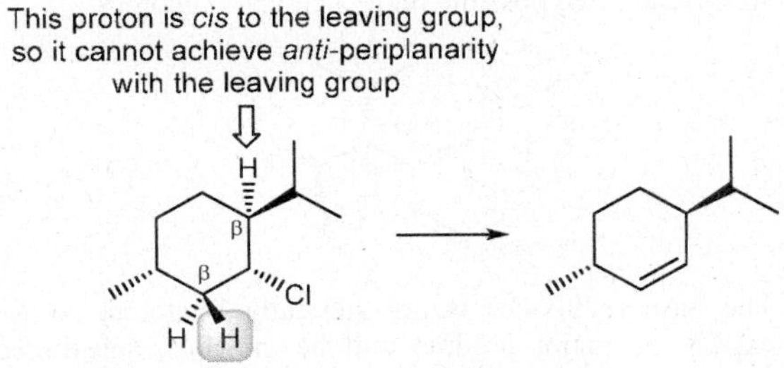

This proton is *cis* to the leaving group, so it cannot achieve *anti*-periplanarity with the leaving group

menthyl chloride

In contrast, the leaving group in neomenthyl chloride is *anti*-periplanar with two different β protons (highlighted), giving rise to two possible products:

neomenthyl chloride

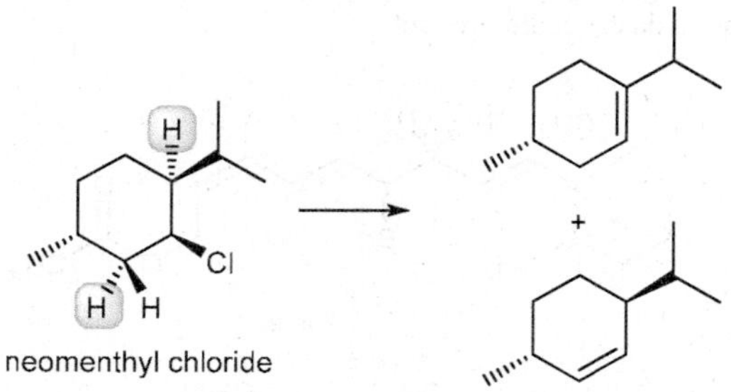

7.18. Because of the bulky *tert*-butyl group, the first compound is essentially locked in a chair conformation in which the leaving group (Cl^-) occupies an equatorial position.

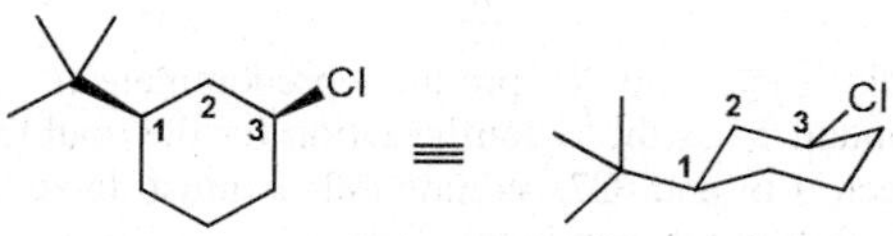

This conformation cannot undergo an E2 reaction because the leaving group is not *anti*-periplanar with a β proton (in order to achieve *anti*-periplanarity, the H and Cl must be *trans* diaxial). However, the second compound is locked in a chair conformation in which the leaving group (Cl^-) occupies an axial position:

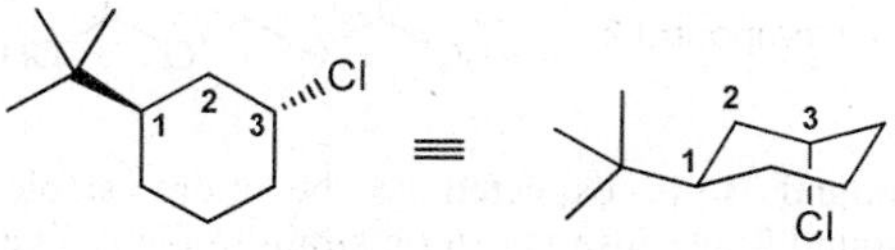

This conformation rapidly undergoes an E2 reaction, because it has *anti*-periplanar β protons readily available. Therefore, the second compound is expected to be more reactive towards an E2 process than the first compound.

7.19.

(a) We must determine both the regiochemical outcome and the stereochemical outcome. Let's begin with regiochemistry. There are two β positions in this case, so there are two possible regiochemical outcomes.

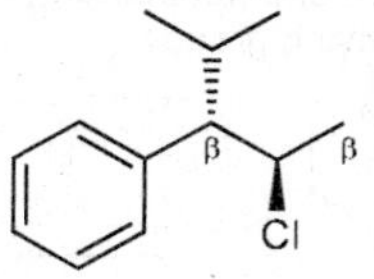

The base (ethoxide) is not sterically hindered, so we expect the major product will be the more-substituted alkene (the Zaitsev product), while the minor product will be the less-substituted alkene. Next, we must identify the stereochemistry of formation of each of the products. Let's begin with the minor product (the less-substituted alkene), because its double bond does not exhibit stereoisomerism:

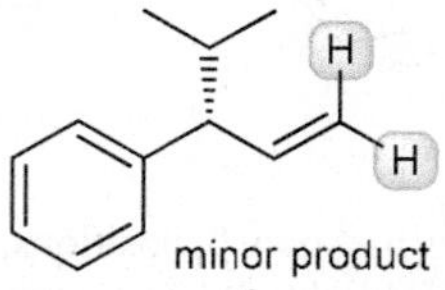

minor product

As shown, there are two hydrogen atoms (highlighted) connected to one of the vinylic positions, so this alkene is neither *E* nor *Z*. Now let's turn our attention to the major product of the reaction (the more-substituted alkene). To determine which stereoisomer is obtained, we must first redraw the starting material in a way that shows the β proton and the leaving group in an *anti*-periplanar arrangement:

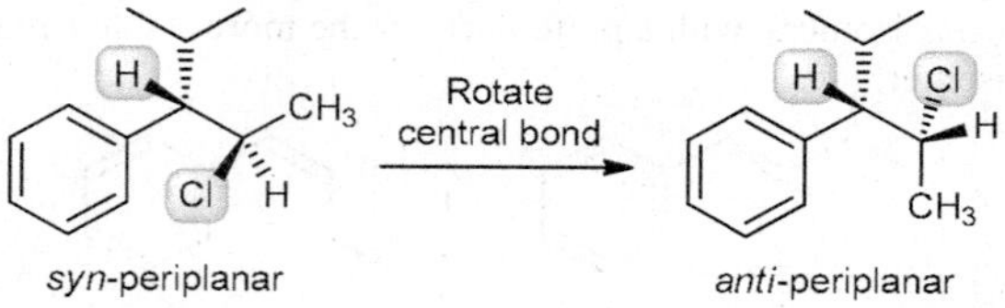

Rotate central bond

syn-periplanar

anti-periplanar

When drawn in an *anti*-periplanar conformation (with the proton on a wedge and the leaving group on a dash), the product can be easily drawn by redrawing the skeleton, with a double bond taking the place of the β proton and the leaving group:

NaOEt

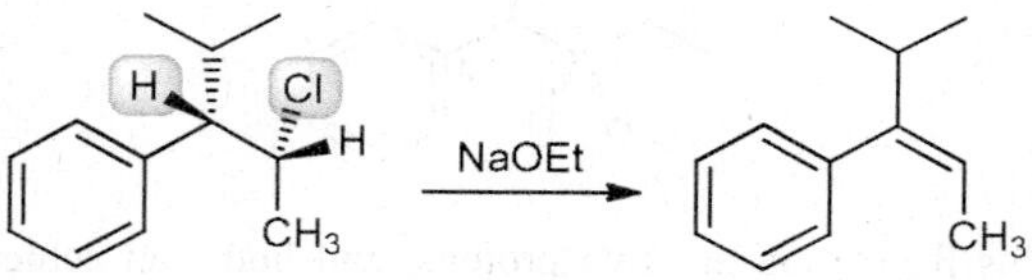

Notice that the isopropyl group is drawn on a straight line (double bonds have planar geometry).
In summary, we expect the following two products:

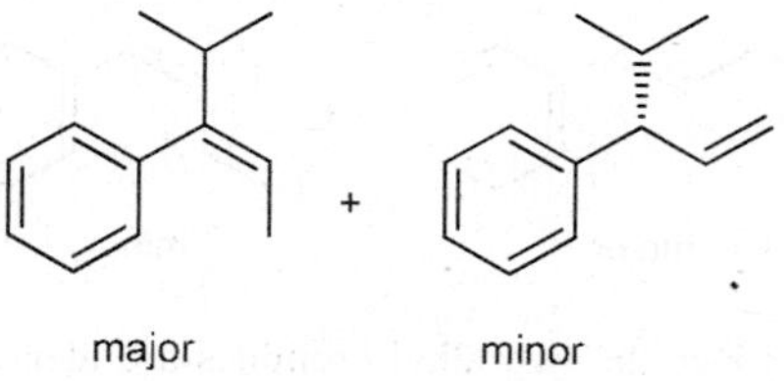

major minor

(b) For substituted cyclohexanes, an E2 reaction will occur if the leaving group and the β proton can achieve antiperiplanarity. In order to achieve this, one must be on a wedge and the other must be on a dash. The leaving group (Br^-) is on a wedge. Therefore, we are looking for a β proton that is on a dash. In this case, there are two different β protons on a dash (highlighted below), giving rise to two possible products. Since the base (ethoxide) is not sterically hindered, we expect the more-substituted alkene (the Zaitsev product) to be the major product:

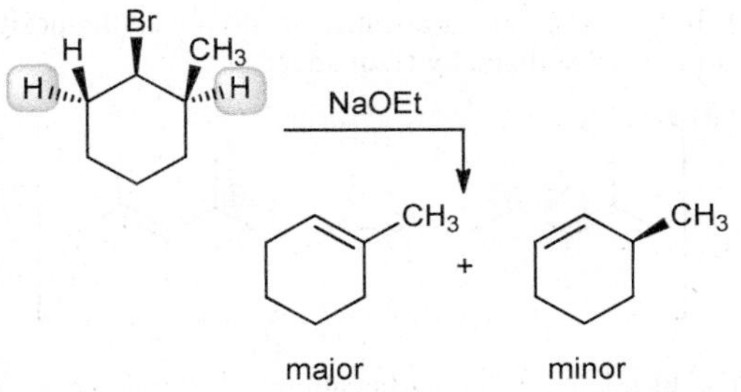

Br H CH3 H H NaOEt → CH3 (major) + CH3 (minor)

(c) For substituted cyclohexanes, an E2 reaction will occur if the leaving group and the β proton can achieve *anti*-periplanarity. In order to achieve this, one must be on a wedge and the other must be on a dash. The leaving group (Br^-) is on a wedge. Therefore, we are looking for a β proton that is on a dash. In this case, there is only one β proton on a dash (highlighted below), giving rise to only one possible product:

Br H NaOEt → only E2 product

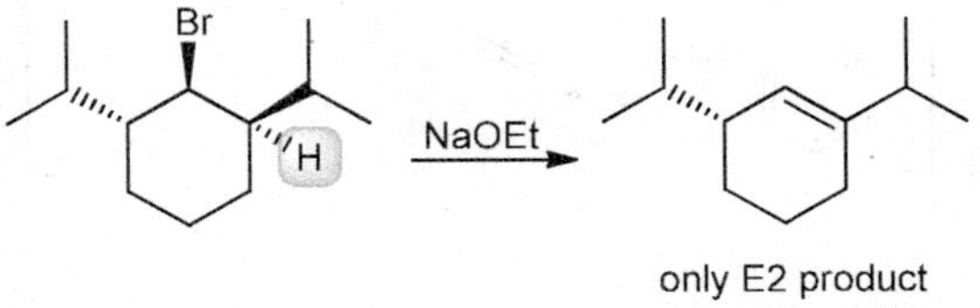

(d) We must determine both the regiochemical outcome and the stereochemical outcome. Let's begin with regiochemistry. This compound has three β positions, but two of them (highlighted) are identical because deprotonation at either of these locations will result in the same alkene:

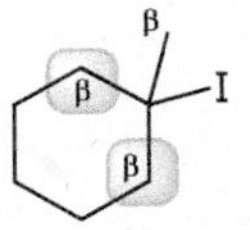

β β β I

As such, we expect two possible regiochemical outcomes. Since the base (ethoxide) is not sterically hindered, we expect the more-substituted alkene (the Zaitsev product) as the major product.

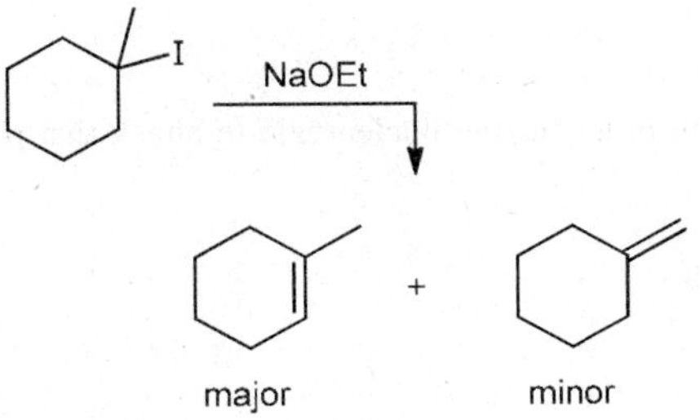

I NaOEt → major + minor

Stereochemisty is not a consideration for either product. The minor product is not stereoisomeric, and the major product cannot exist as an *E* isomer (because the ring makes that impossible).

(e) For substituted cyclohexanes, an E2 reaction will occur if the leaving group and the β proton can achieve antiperiplanarity. In order to achieve this, one must be on a wedge and the other must be on a dash. The leaving group (Br^-) is on a wedge. Therefore, we are looking for a β proton that is on a dash. In this case, there are two different β protons on a dash (highlighted below), giving rise to two products. Since the base (*tert*-butoxide) is sterically hindered, we expect the less-substituted alkene (the Hofmann product) to be the major product:

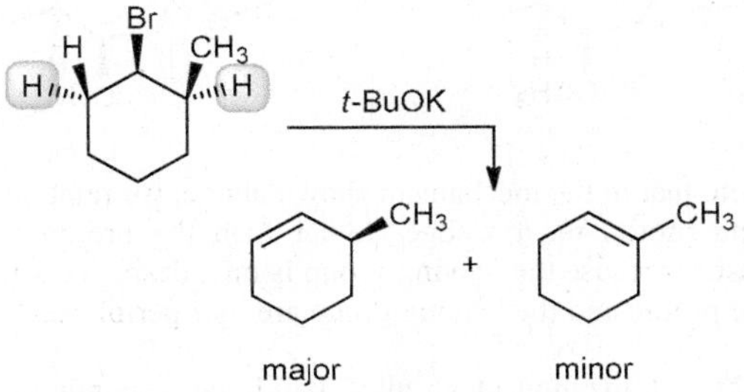

Br H CH3 H H *t*-BuOK → CH3 (major) + CH3 (minor)

(f) For substituted cyclohexanes, an E2 reaction will occur if the leaving group and the β proton can achieve *anti*-periplanarity. In order to achieve this, one must be on a wedge and the other must be on a dash. The leaving group (Br^-) is on a wedge. Therefore, we are looking for a β proton that is on a dash. In this case, there is only one β proton on a dash (highlighted below), giving rise to only one possible product:

Br H CH3 H H *t*-BuOK → CH3 only E2 product

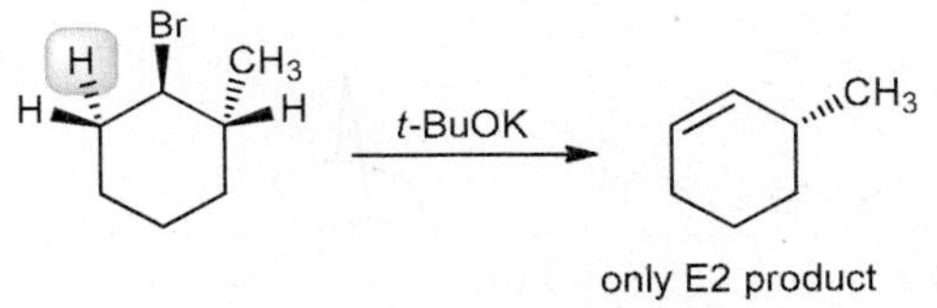

7.20. Bromide is the best (and only) leaving group in compound **1**, so the carbon atom connected to Br is (by definition) the α carbon. There are two β carbons, but only one of these positions (highlighted) has protons.

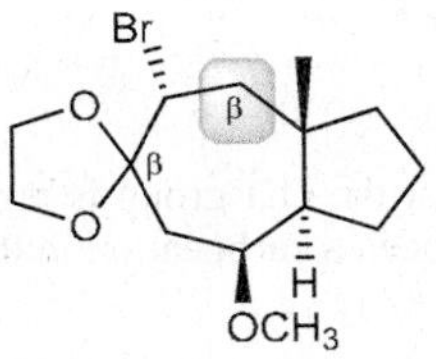

Br O O β β H OCH3

The base (*tert*-butoxide) is sterically hindered; however, there is only one regioisomer possible in this reaction because there is only one β carbon with protons. In this case, *tert*-butoxide was probably chosen as the base (rather than hydroxide, methoxide or some other unhindered base) to favor an E2 reaction over an S_N2 reaction. As we will discover in Section 7.11, substitution and elimination reactions often compete with each other, especially when the substrate is secondary.

Since there are two protons on the β position, we might predict that this reaction would be stereoselective, favoring the *trans* isomer. However, *trans* double bonds are not stable in rings with fewer than eight carbons. Thus, we expect formation of only the *cis* double bond due to the geometric constraints of the seven-membered ring.

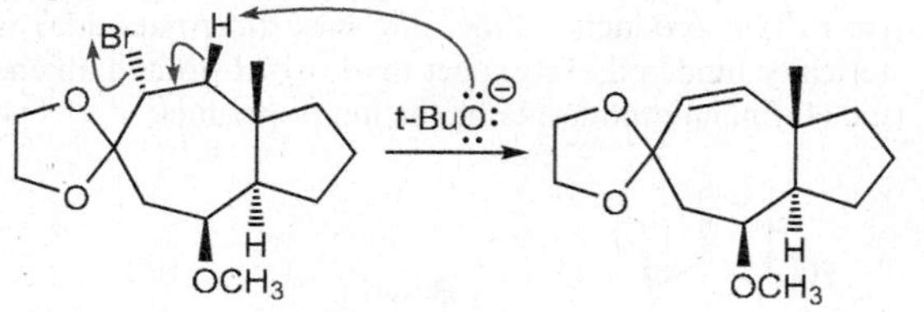

Br H t-BuO: H OCH3 H OCH3

Note that in the mechanism shown above, we remove the beta proton on a wedge (rather than the proton on a dash), because the leaving group is on a dash. This way, the proton and the leaving group are *anti*-periplanar.

7.21. Ionization of an alkyl halide corresponds to the loss of a halide leaving group to form a carbocation. In each case, the bond between the α position and the leaving group is broken, and the carbon atom obtains a positive charge. Resonance structures are drawn where applicable:

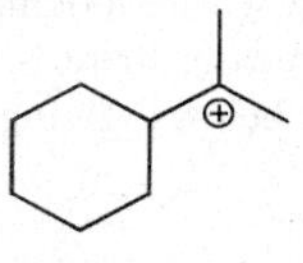

(a) 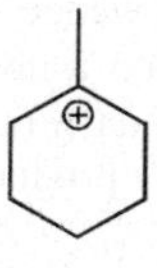 **(b)**

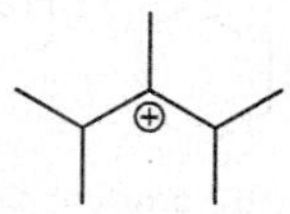

(c) 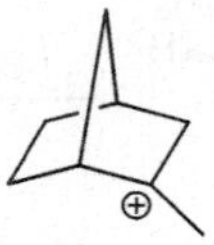 **(d)**

(e) In this case, the carbocation is allylic, so the positive charge is delocalized by resonance:

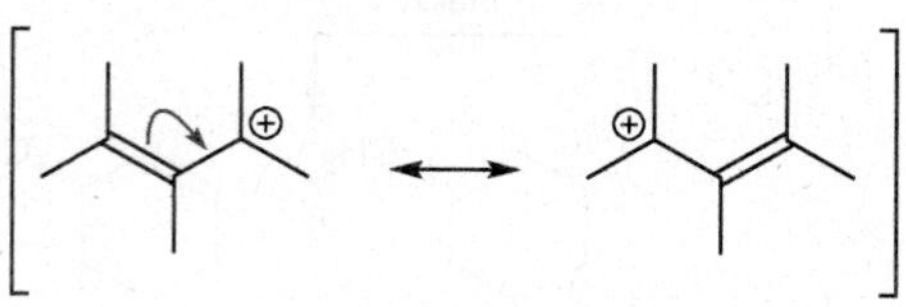

(f) In this case, the carbocation is benzylic (next to a benzene ring), so the positive charge is delocalized by resonance involving all three π bonds in the aromatic ring:

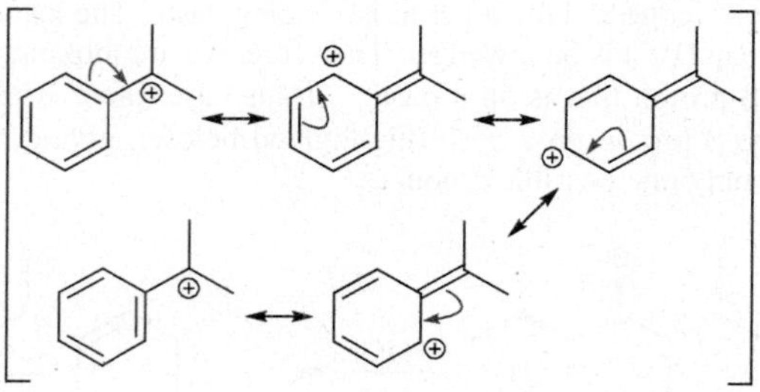

(g)

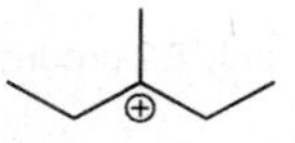

(h) In this case, the carbocation is doubly allylic, so the positive charge is delocalized by resonance involving both π bonds:

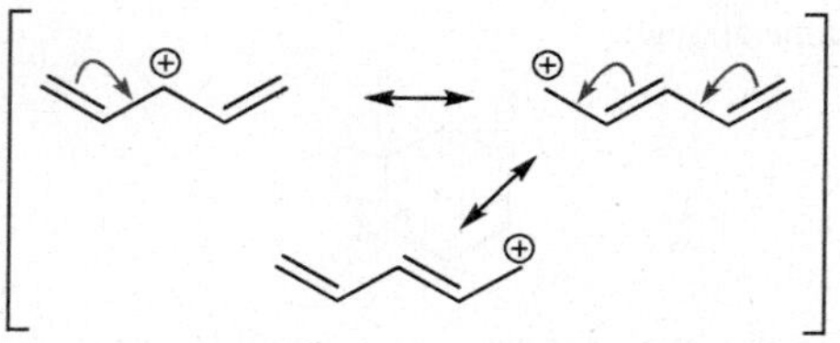

7.22. Notice that the OH group is replaced with a nucleophile. In order for the nucleophile to attack that position, there must have been a carbocation in that position:

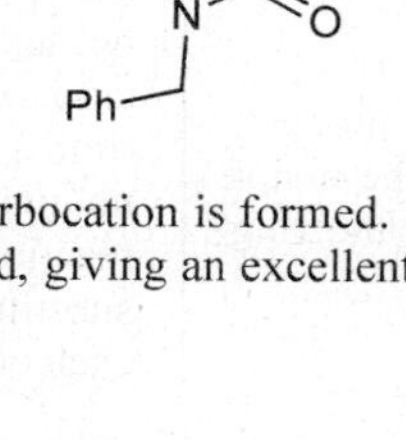

CH_3 N O Ph

Later in this chapter, we will explore how this carbocation is formed. Specifically, we will see that the OH group can be protonated in the presence of concentrated acid, giving an excellent leaving group. Then, loss of the leaving group (H_2O) gives the carbocation intermediate:

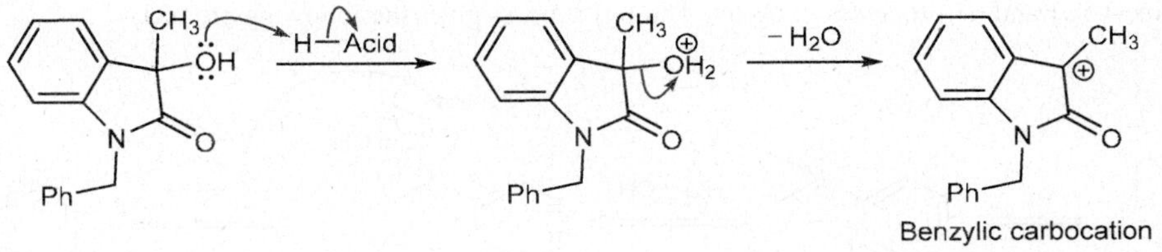

Benzylic carbocation

This carbocation is benzylic (next to a benzene ring), so it has extensive resonance stabilization, as shown here:

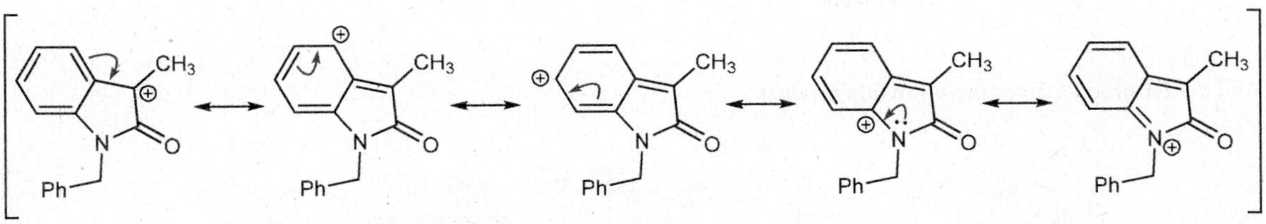

Note that the following resonance structure is not significant because the electronegative oxygen atom (highlighted) does not have an octet of electrons:

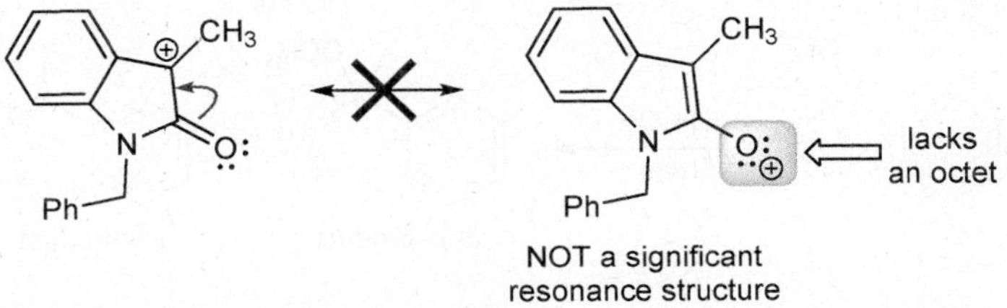

lacks an octet

NOT a significant resonance structure

7.23.

(a) The first compound is expected to undergo solvolysis more rapidly because it will proceed via a tertiary benzylic carbocation (stabilized by resonance with the adjacent benzene ring). In contrast, solvolysis of the second compound will proceed via a tertiary carbocation, which is not resonance-stabilized.

Tertiary benzylic

Tertiary

(b) The first compound is a primary alkyl halide and is therefore not expected to undergo solvolysis at an appreciable rate (because it would proceed via an unstable primary carbocation). In contrast, the second compound is an allylic bromide. Solvolysis of an allylic bromide will proceed via an allylic carbocation, which is resonance-stabilized.

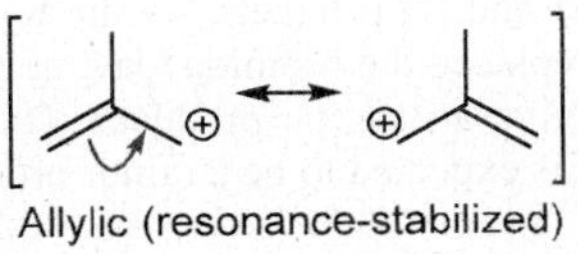

Allylic (resonance-stabilized)

(c) Both compounds have the same carbon skeleton, and therefore give the same allylic, resonance-stabilized carbocation (during solvolysis). Nevertheless, the second compound is expected to undergo solvolysis more rapidly, because it has a better leaving group. Bromide is a better leaving group than chloride because bromide is a larger, more stable anion (making it a weaker base).

7.24. We expect S_N1 and E1 processes to occur. The S_N1 process gives the following product:

Resonance-stabilized

Product

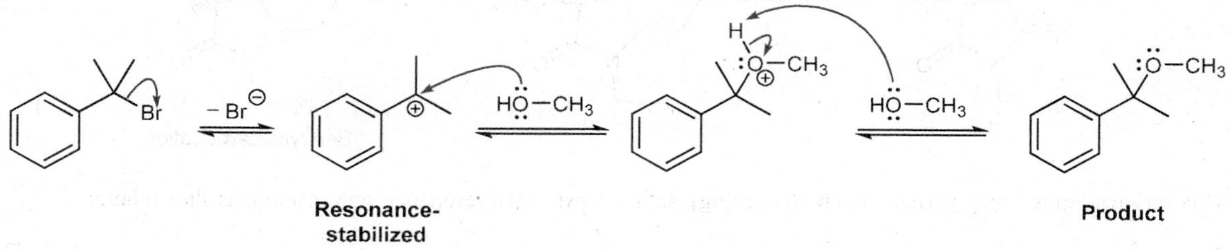

And the E1 process gives the following product:

Product

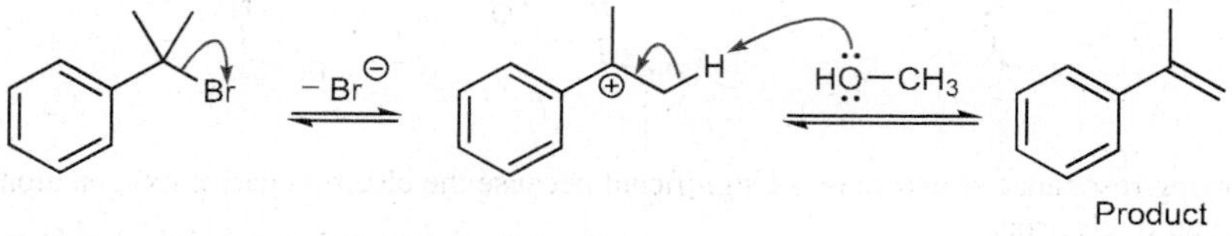

In summary, solvolysis in methanol should give the following products:

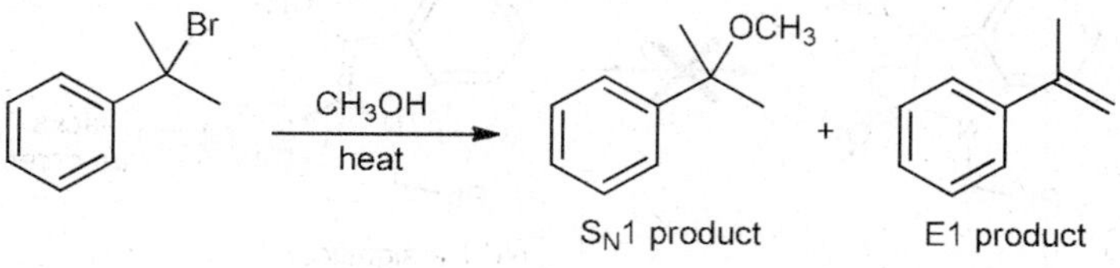

S_N1 product

E1 product

7.25. In each case, the leaving group X is lost in the reaction, so the bond between a carbon atom and the leaving group is broken, and a carbocation intermediate is formed. The rate of any S_N1 reaction is dependent on this step of the reaction. More stable carbocations are formed faster because the transition state that leads to them is lower in energy. Ionization of either compound **1** or **3** will result in a tertiary carbocation. Notice, however, that the carbocation that is formed from compound **1** is tertiary AND benzylic (adjacent to a benzene ring). Because compound **1** produces a resonance-stabilized carbocation (shown below), compound **1** should undergo a faster S_N1 reaction.

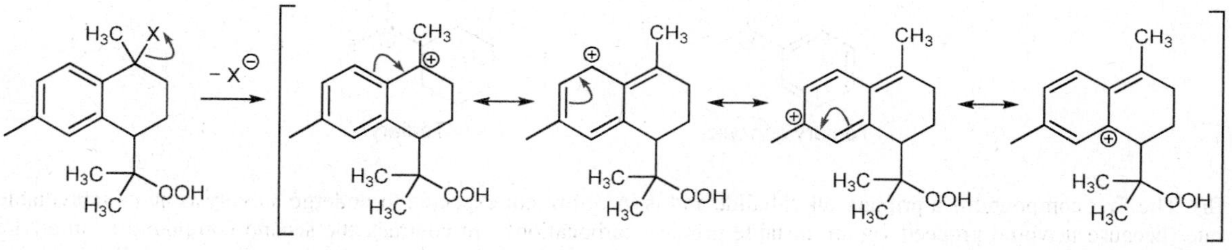

7.26.

(a) Solvolysis is expected to afford both S_N1 and E1 products. In the S_N1 product, a *substitution* has taken place, and the nucleophile, an ethoxy (OEt) group has replaced the bromide leaving group. The E1 reaction *eliminates* the leaving group and an adjacent proton (– HBr), to give an alkene product. There are two E1 products (two regiochemical outcomes), although the disubstituted alkene is expected to be a minor product.

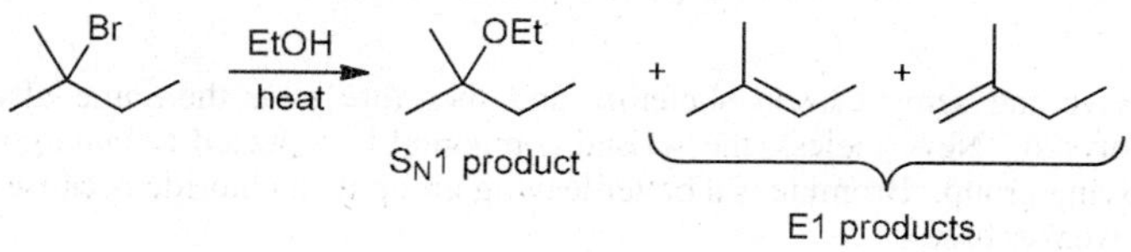

S_N1 product

E1 products

(b) Solvolysis is expected to afford both S_N1 and E1 products. In the S_N1 product, a *substitution* has taken place, and the nucleophile, a hydroxy (OH) group has replaced the chloride leaving group. The E1 reaction *eliminates* the leaving group and an adjacent proton (– HCl), to give an alkene product. There are two E1 products (two regiochemical outcomes), although the disubstituted alkene is expected to be a minor product.

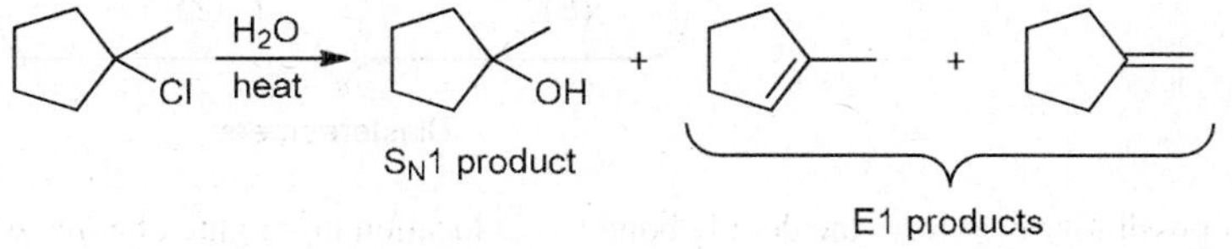

(c) Solvolysis is expected to afford both SN1 and E1 products. In the SN1 product, a *substitution* has taken place, and the nucleophile, a methoxy (OMe) group has replaced the bromide leaving group. The E1 reaction *eliminates* the leaving group and an adjacent proton (– HBr), to give an alkene product. In this case, there is only one E1 product (there is only one possible regiochemical outcome), although the disubstituted alkene is expected to be a minor product.

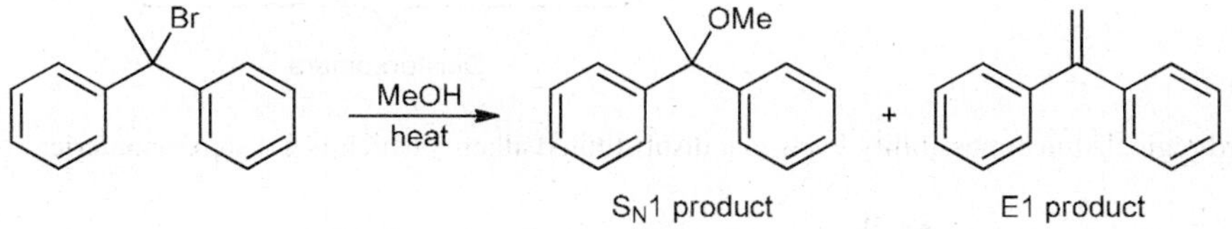

(d) Solvolysis is expected to afford both SN1 and E1 products. In the SN1 product, a *substitution* has taken place, and the nucleophile, a methoxy (OMe) group has replaced the chloride leaving group. In this case, the α position is a chiral center, so we expect a pair of enantiomers (with a small preference for the inverted product, as a result of ion pairs). Both SN1 products are shown below. The E1 reaction *eliminates* the leaving group and an adjacent proton (– HCl), to give an alkene product. There are three E1 products in this case, shown below as well.

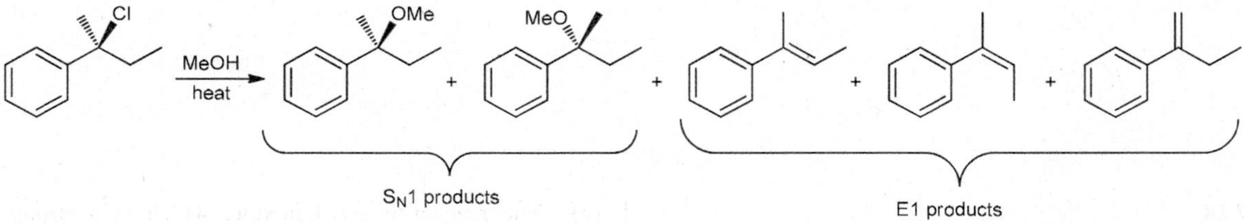

7.27.

(a) This is a solvolysis reaction involving a weak nucleophile, so the substitution products arise from an SN1 mechanism. To draw the SN1 product, we replace the bromide leaving group with the nucleophile, an ethoxy (OEt) group. In this case, the α position is a chiral center, so we expect a pair of enantiomers:

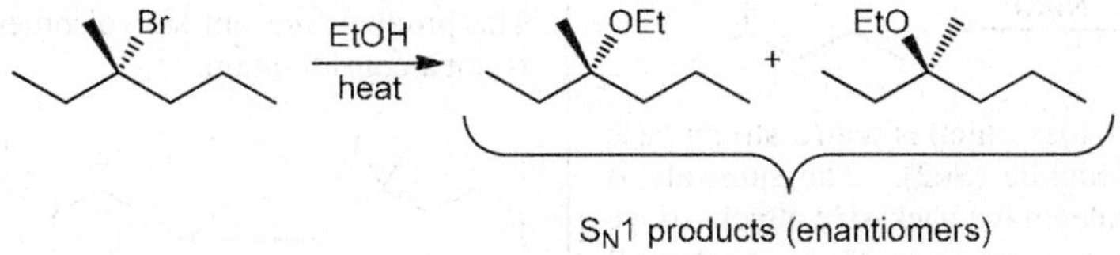

(b) The inverted product is expected to predominate slightly, as a result of ion pairs.

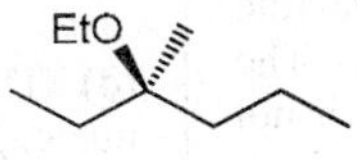

(c) This is a solvolysis reaction involving a weak base, so the elimination products arise from an E1 mechanism. To draw the E1 products, we remove the leaving group and an adjacent proton (– HBr), to give an alkene product. There are three β positions bearing protons, so there are three regiochemical possibilities for an E1 process. Let's explore each of the three possibilities. One regiochemical possibility is to form the double bond in the location highlighted below, which leads to two diastereomeric alkenes.

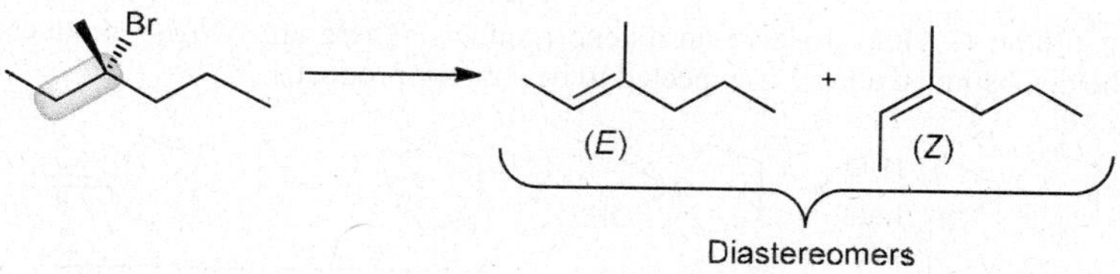

(E) + (Z)

Diastereomers

Another regiochemical possibility is to form the double bond in the location highlighted below, which also leads to two diastereomeric alkenes.

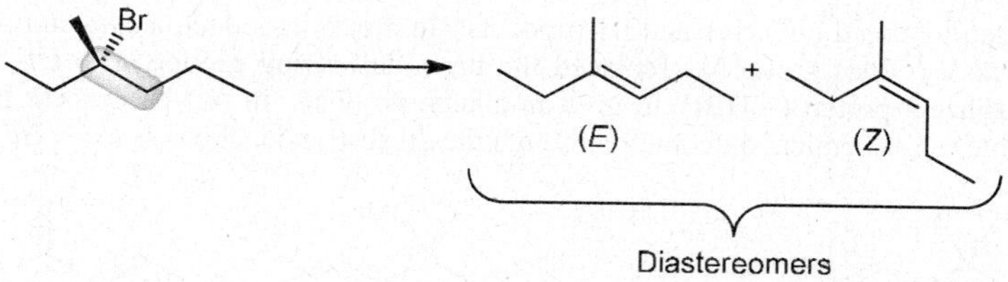

(E) + (Z)

Diastereomers

Finally, the third regiochemical possibility leads to a disubstituted alkene, which is not stereoisomeric:

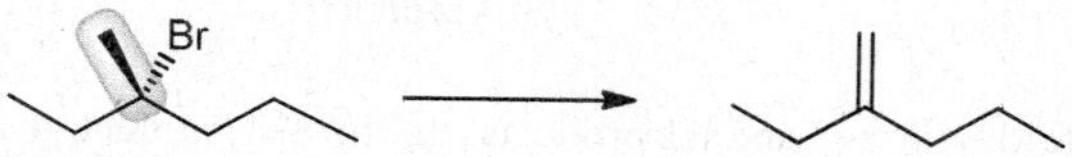

In total, there are five alkene products, all shown above.

(d) For each regiochemical possibility (shown above), the alkene with fewer or weaker steric interactions is favored. In each of these cases, the *E* isomer has weaker steric interactions, and is therefore the more stable, favored isomer.

7.28.

(a) The reagent is chloride, which functions as a nucleophile, so we expect a substitution reaction. The substrate is secondary, indicating an SN2 process. As such, we expect inversion of configuration, as shown:

NaCl

(b) The reagent is hydroxide, which is both a strong base (E2) and a strong nucleophile (SN2). The substrate is tertiary, which is too hindered for back-side attack, so we expect an E2 process to predominate. There are three β positions, but two of them are identical, so there are two possible regiochemical outcomes. The base is not sterically hindered so the more-substituted alkene (the Zaitsev product) is the major product, as shown. The products are not stereoisomeric, so stereochemistry is not a consideration.

NaOH

major (more substituted) + minor (less substituted)

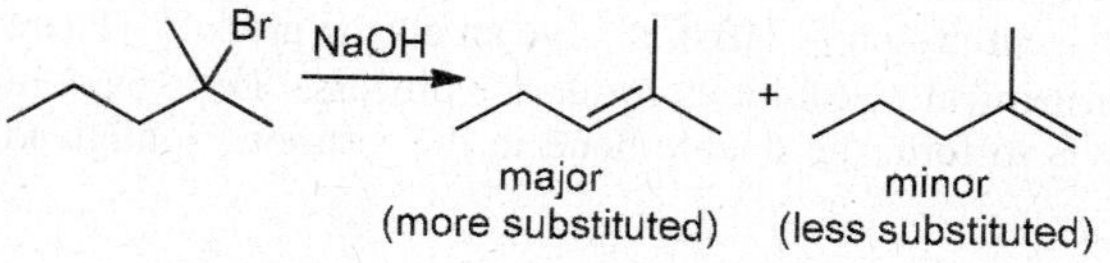

(c) The reagent is *tert*-butoxide, which is a strong, sterically hindered base. The substrate is tertiary, so we expect an E2 process. There are three β positions, but two of them are identical, so there are two possible regiochemical outcomes. Since the base is sterically hindered, we expect that the less-substituted alkene (the Hofmann product) will be the major product, as shown. The products are not stereoisomeric, so stereochemistry is not a consideration.

t-BuOK

major (less substituted) + minor (more substituted)

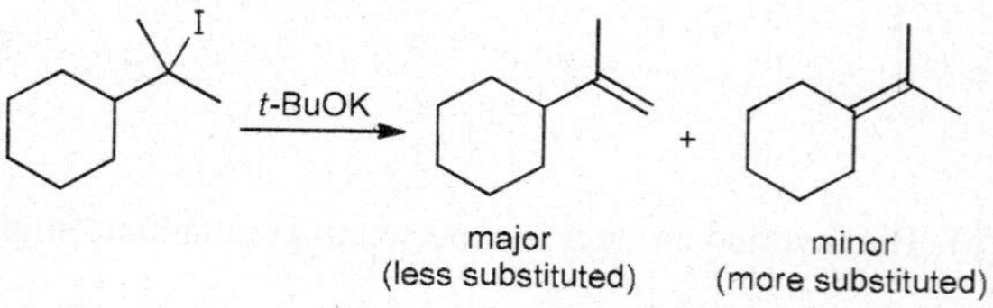

(d) The reagent is DBN, which is a strong base (not a nucleophile). The substrate is tertiary, so we expect an E2 process. There are two β positions bearing protons, and both of these positions are identical, so there is only one possible regiochemical outcome. The product is not stereoisomeric, so stereochemistry is not a consideration.

Br

DBN

(e) The reagent is DBU, which is a strong base (not a nucleophile). Treating a primary alkyl halide with DBU will give E2 as the exclusive pathway. The product is not stereoisomeric, so stereochemistry is not a consideration.

DBU

I

(f) The reagent is HS^-, which is a strong nucleophile and a weak base. The substrate is primary, so we expect an S_N2 process. There is no inversion of configuration in this case, because there is no chiral center:

NaSH

I

SH

(g) The reagent is hydroxide, which is both a strong base (E2) and a strong nucleophile (S_N2). The substrate is primary, so we expect the major product to result from an S_N2 process, and the minor product to result from an E2 process, as shown:

Br

NaOH

OH

+

major minor

In this case, there are no stereochemical considerations for either product. There is no observed inversion of configuration in the S_N2 product, because there is no chiral center, and the alkene product is not stereoisomeric.

(h) The reagent is ethoxide, which is both a strong base (E2) and a strong nucleophile (S_N2). The substrate is primary, so we expect the major product to result from an S_N2 process, and the minor product to result from an E2 process, as shown:

Br

NaOEt

OEt

+

major minor

In this case, there are no stereochemical considerations for either product. There is no observed inversion of configuration in the S_N2 product, because there is no chiral center, and the alkene product is not stereoisomeric.

(i) The reagent is ethanol, which is both a weak base and a weak nucleophile. The substrate is tertiary, so we expect E1 and S_N1 processes. One of the alkene products is trisubstituted, so we expect E1 to predominate over S_N1. For the E1 pathway, two regiochemical outcomes are possible. The more-substituted alkene (the Zaitsev product) is the major product, while the minor products are the less-substituted alkene and the S_N1 product, shown below:

I

EtOH

heat

major

+ +

OEt

minor minor

(j) The reagent is hydroxide, which is both a strong base (E2) and a strong nucleophile (S_N2). The substrate is secondary so we expect both E2 and S_N2 processes, although E2 will be responsible for the major product. Because the base is not sterically hindered, the major product is the more substituted alkene (the Zaitsev product), with the *trans* configuration (because the reaction is stereoselective, favoring the more stable *trans* isomer over the *cis* isomer). The minor products include the *cis* isomer, the less-substituted alkene, and the S_N2 product (with the expected inversion of configuration):

Br

NaOH

major

+

minor

+

OH

minor minor

(k) The reagent is methoxide, which is both a strong base (E2) and a strong nucleophile (S_N2). The substrate is secondary so we expect both E2 and S_N2 processes, although E2 will be responsible for the major product. Because the base is not sterically hindered, the major product is the more substituted alkene (the Zaitsev product), with the *trans* configuration (because the reaction is stereoselective, favoring the more stable *trans* isomer over the *cis* isomer). The minor products include the *cis* isomer, the less-substituted alkene, and the S_N2 product (with the expected inversion of configuration):

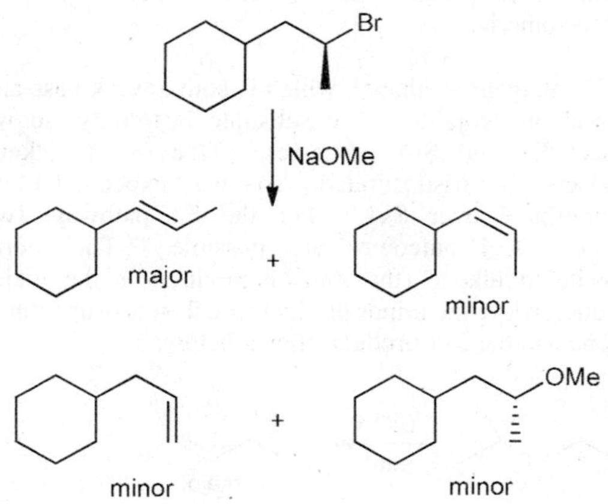

NaOMe

major

minor

minor

minor

(l) The reagent is methoxide, which is both a strong base (E2) and a strong nucleophile (S_N2). The substrate is secondary so we expect both E2 and S_N2 processes, although E2 will be responsible for the major product. Let's begin by drawing the major product, and then we will move on to the minor products.

To draw the major product (E2), we note that there are two β positions in this case, so there are two possible regiochemical outcomes.

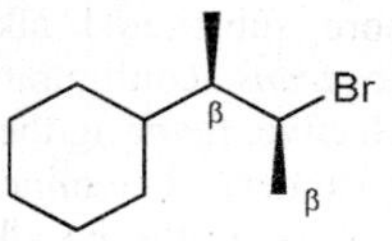

The base is not sterically hindered, so the Zaitsev product (the more substituted alkene) is expected to be the major product. To determine which stereoisomer is obtained, we must first redraw the starting material in a way that shows the β proton and the leaving group in an *anti*-periplanar arrangement:

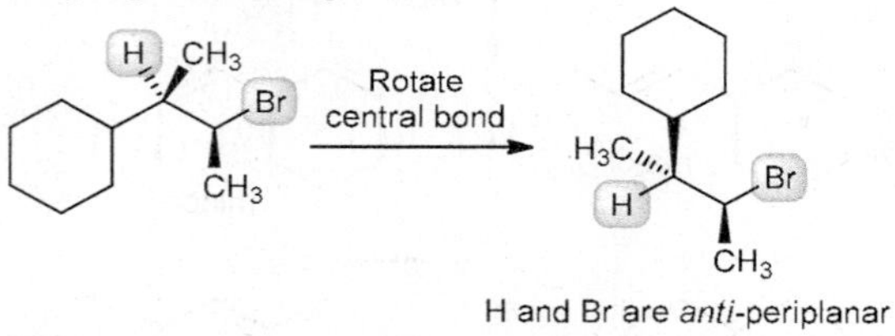

Rotate central bond

H and Br are *anti*-periplanar

When drawn in an *anti*-periplanar conformation, we can use the dashes and wedges as guides to draw the correct stereoisomer (methyl groups are *trans* to each other in the product):

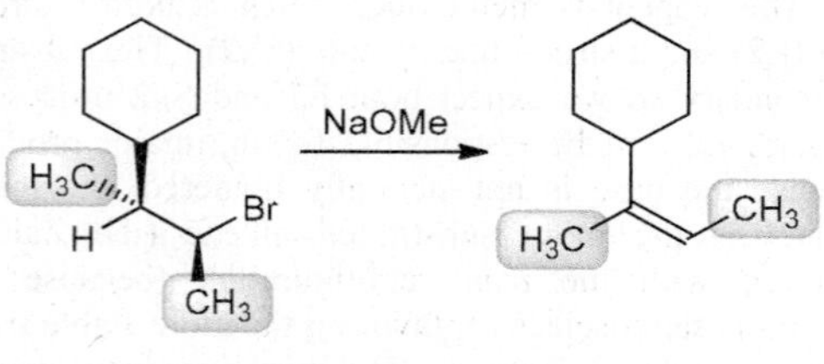

NaOMe

Now let's consider the minor products. We noted before that there are two possible regiochemical outcomes for an E2 process. The Hofmann product (the less substituted alkene) will be a minor product:

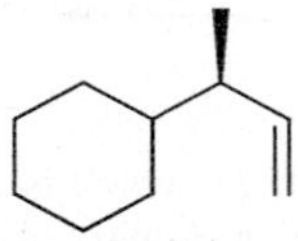

minor

Another minor product is obtained via an S_N2 process (with inversion of configuration), as shown:

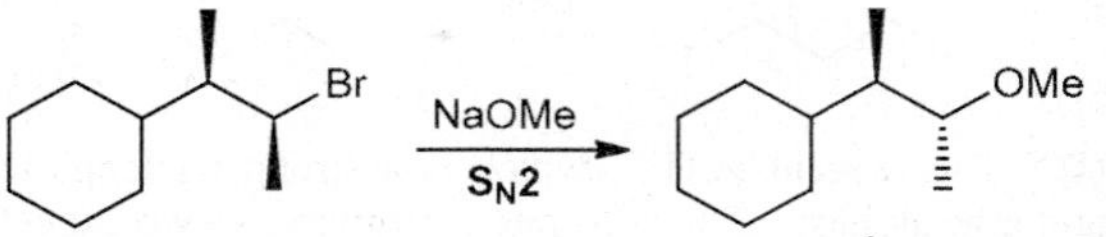

NaOMe
S_N2

minor

In summary, we expect the following products:

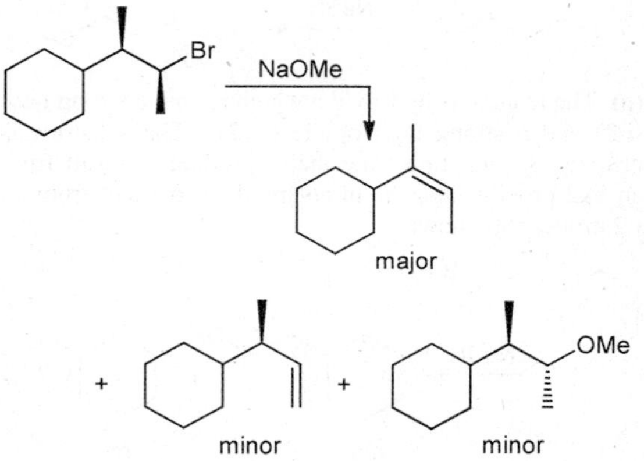

NaOMe

major

minor

minor

(m) The reagent is hydroxide, which is both a strong base (E2) and a strong nucleophile (S_N2). The substrate is secondary so we expect both E2 and S_N2 processes, although E2 will be responsible for the major product. For substituted cyclohexanes, an E2 reaction occurs via a conformation in which the leaving group and the β proton are *anti*-periplanar to one another (one must be on a wedge and the other must be on a dash). The leaving group (Br^-) is on a wedge. Therefore, we are looking for a β proton that is on a dash. In this case, there is only one β proton that is on a dash (highlighted below), giving rise to only one elimination product, as shown:

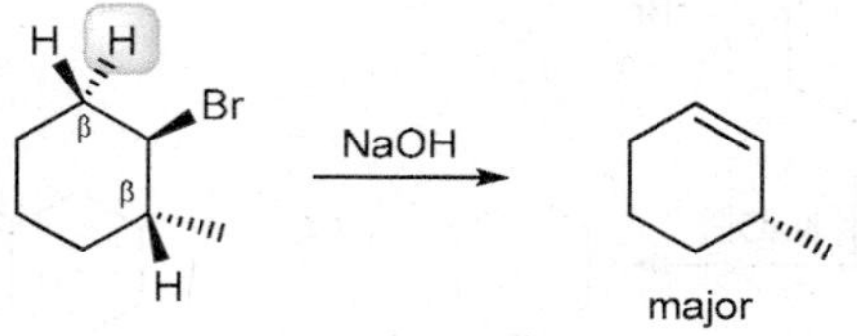

NaOH

major

The minor product is generated via an S_N2 pathway (with inversion of configuration, as expected):

Br NaOH OH

major minor

(n) The reagent is ethoxide, which is both a strong base (E2) and a strong nucleophile (S_N2). The substrate is secondary so we expect both E2 and S_N2 processes, although E2 will be responsible for the major product. For substituted cyclohexanes, an E2 reaction occurs via a conformation in which the leaving group and the β proton are *anti*-periplanar to one another (one must be on a wedge and the other must be on a dash). The leaving group (Br^-) is on a wedge. Therefore, we are looking for a β proton that is on a dash. In this case, there are two such protons (highlighted below), giving rise to two elimination products, as shown. The major product is the more-substituted alkene (we expect the Zaitsev product, because the base is not sterically hindered), while the minor products include the less-substituted alkene and the S_N2 product (with the expected inversion of configuration):

H H β Br β H NaOEt

major

OEt

minor minor

7.29. There are only two constitutional isomers with the molecular formula C_3H_7Cl:

Cl Cl

a primary alkyl halide a secondary alkyl halide

Sodium methoxide is both a strong nucleophile (S_N2) and a strong base (E2). When compound **A** is treated with sodium methoxide, a substitution reaction predominates. Therefore, compound **A** must be the primary alkyl chloride above. When compound **B** is treated with sodium methoxide, an elimination reaction predominates. Therefore, compound **B** must be the secondary alkyl chloride:

Cl Cl

Compound **A** Compound **B**

7.30.

(a) There are four constitutional isomers with the molecular formula C_4H_9Cl, shown below. Only one of them affords the desired product when treated with methoxide, so the compound highlighted below must be compound **A**:

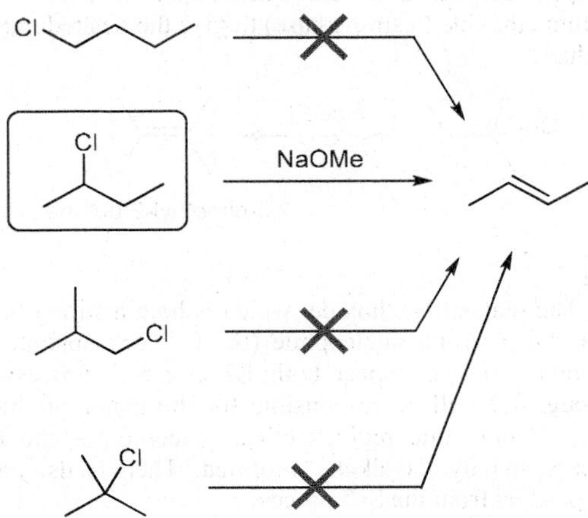

(b) We saw in the solution to part (a) that there are four constitutional isomers with the molecular formula C_4H_9Cl. Let's consider the **major product** that is expected when each of these isomers is treated with methoxide. The primary alkyl halides are expected to undergo substitution to form the major product (S_N2), while E2 elimination is expected to predominate for the secondary and tertiary halides. Notice that only one of these cases produces a disubstituted alkene that is different from *trans*-2-butene:

Cl NaOMe MeO

Cl NaOMe

Cl NaOMe OMe

Cl NaOMe

Therefore, the following structure must be compound **B**:

Cl

7.31. The starting alkyl halide must have the same carbon skeleton as the product (2,3-dimethyl-2-butene), and there are only two such isomers with the molecular formula $C_6H_{13}Cl$.

Cl Cl

Only one of these isomers (the one with a tertiary leaving group) undergoes an E2 elimination upon treatment with sodium ethoxide (a strong base) to give the desired major product:

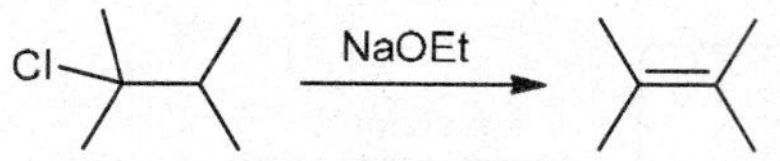

Cl NaOEt

2,3-dimethyl-2-butene

7.32.

(a) The reagent is ethoxide, which is both a strong base (E2) and a strong nucleophile (SN2). The substrate is secondary so we expect both E2 and SN2 processes, although E2 will be responsible for the major product. There is only one regiochemical outcome for the E2 process, so only one alkene is formed. There is also only one product from the SN2 process:

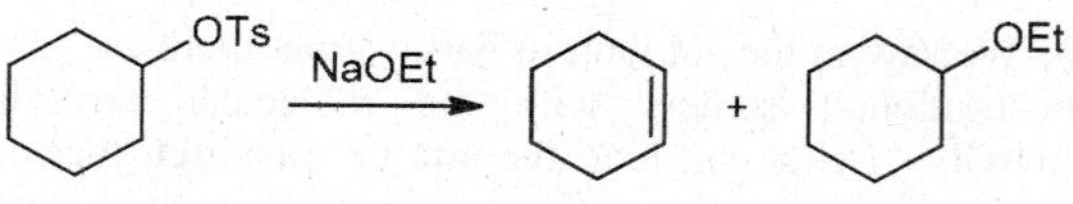

OTs NaOEt + OEt

Major Minor

In this case, there are no stereochemical considerations for either product. There is no observed inversion of configuration in the SN2 product, because there is no chiral center, and the alkene product is not stereoisomeric (only the *cis* alkene is possible in a six-membered ring).

(b) The reagent is hydroxide, which is both a strong base (E2) and a strong nucleophile (SN2). The substrate is secondary so we expect both E2 and SN2 processes, although the E2 process will be responsible for the major product. There are two regiochemical outcomes for the E2 process. One of these outcomes leads to a disubstituted alkene, which is stereoisomeric. Both stereoisomers are expected, although the more stable *trans* isomer will be favored over the *cis* isomer because the reaction is stereoselective. Indeed, the *trans* isomer is the major product (the Zaitsev product), because the base is not sterically hindered (a sterically hindered base would have favored the monosubstituted alkene). There is also one minor product that is formed via an SN2 process (with the expected inversion of configuration):

OTs NaOH

major + minor

\+ minor + OH minor

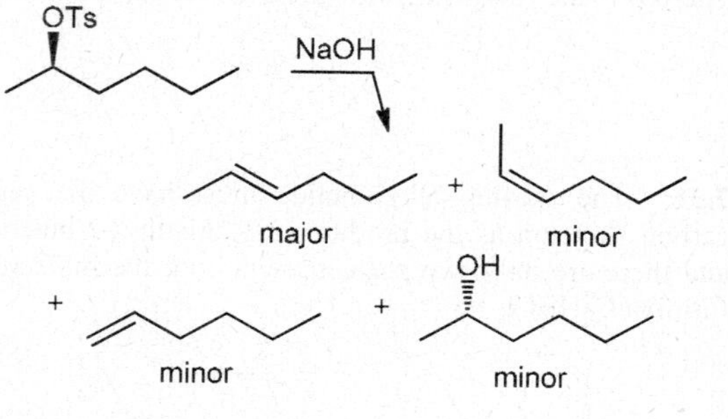

(c) The reagent (HS^-) is a strong nucleophile (not a strong base), and the substrate is secondary, so we expect an SN2 process. There is no observed inversion of configuration, because there is no chiral center.

OTs NaSH SH

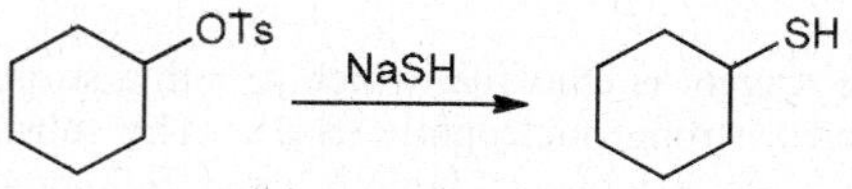

(d) The reagent is DBU, which is a strong base (not a nucleophile), so we expect an E2 process (no SN2). For substituted cyclohexanes, an E2 reaction occurs via a conformation in which the leaving group and the β proton are *anti*-periplanar to one another (one must be on a wedge and the other must be on a dash). The leaving group (OTs) is on a wedge. Therefore, we are looking for a β proton that is on a dash. In this case, there is only one β proton that is on a dash (highlighted below), giving rise to only one possible elimination product, as shown:

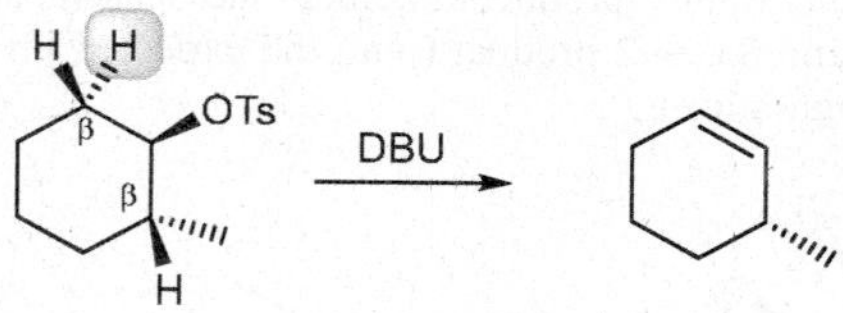

H H OTs β β H DBU

(e) In the first step, the alcohol is converted into a tosylate. Note that there is no change in configuration at the chiral center, because it is not involved in the formation of the tosylate (the C—O bond is not broken):

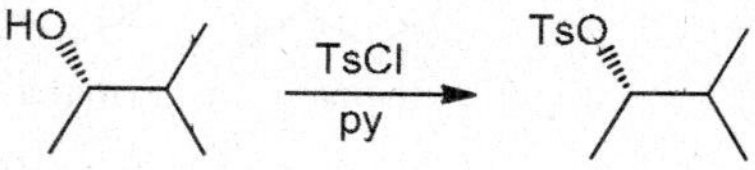

HO TsCl py TsO

This tosylate is a secondary substrate. When treated with a strong base, we expect E2 to be favored. The base (*tert*-butoxide) is sterically hindered, so the Hofmann product will be favored (the less-substituted alkene is the major product). The Zaitsev product is expected to be a minor product:

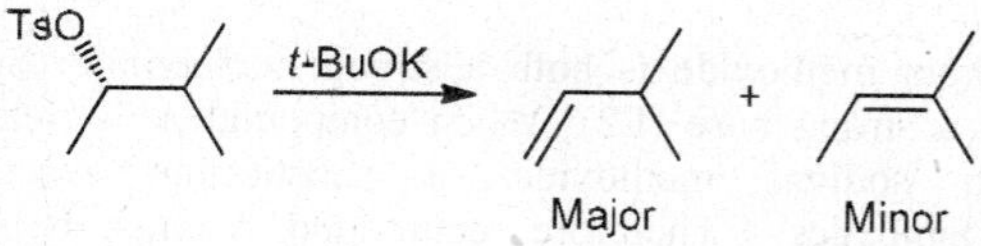

TsO *t*-BuOK +

Major Minor

In general, when a secondary substrate is treated with a reagent that is both a strong base and a strong nucleophile, we would expect a minor product from an SN2 process. However, in this case, the nucleophile (*tert*-butoxide) is sterically hindered, so we would expect there to be very little, if any, substitution product formed.

(f) In the first step, the alcohol is converted into a tosylate:

OH TsCl py OTs

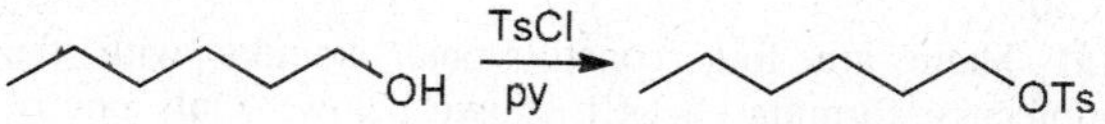

In the next step, the tosylate (a primary substrate) is treated with ethoxide, which is both a strong base (E2) and a strong nucleophile (SN2). Since the substrate is primary, SN2 is expected to predominate, giving the major product, shown below. An E2 process is responsible for the minor product, which is an alkene:

NaOEt

OTs

OEt +

Major Minor

In this case, there are no stereochemical considerations for either product. There is no observed inversion of configuration in the SN2 product, because there is no chiral center, and the alkene product is not stereoisomeric.

7.33.

(a) The reagent HBr is both a strong acid (can protonate the alcohol to make a good leaving group) and a source of bromide (a nucleophile). Therefore, we expect a *substitution* reaction to take place. A tertiary alcohol will react with HBr to give the corresponding tertiary alkyl bromide via an SN1 process:

OH HBr Br

(b) Concentrated sulfuric acid is a strong acid that can protonate the alcohol to make a good leaving group. In the conjugate base of sulfuric acid, the negative charge is highly stabilized by resonance, so the conjugate base is not a nucleophilic species. In the absence of a nucleophile, we expect an *elimination* reaction to take place (E1 process). A tertiary alcohol undergoes a dehydration reaction upon treatment with concentrated sulfuric acid, and is converted into an alkene. The more substituted alkene (the Zaitsev product) is favored:

OH conc. H_2SO_4 heat

7.34.

(a) The starting alcohol is tertiary, so we expect the reaction to proceed via an SN1 process (rather than SN2). The first step of the process is protonation of the OH group, thereby converting a bad leaving group into an excellent leaving group. Loss of the leaving group gives a carbocation, which is then attacked by the bromide nucleophile to give the product:

:ÖH HBr Br

H—Br :Br:⊖

H ⊕ :O—H − H_2O ⊕

(b) Upon treatment with concentrated sulfuric acid, a tertiary alcohol undergoes dehydration. The alcohol is converted into an alkene via an E1 process. The OH group is first protonated, thereby converting a bad leaving group into an excellent leaving group. Loss of the leaving group gives a carbocation, which is then deprotonated by a molecule of the solvent to give the product:

conc. H_2SO_4 heat

ÖH H—O⊕ H H H—O—H

− H_2O O⊕ H H ⊕ H H

(c) The OH group is first protonated, thereby converting a bad leaving group into an excellent leaving group. This substrate is primary, so an SN2 process is expected (rather than SN1) to give the product. Back-side attack by the bromide nucleophile displaces the leaving group:

HBr ÖH Br

H—Br :Br:⊖ SN2

O⊕ H H

(d) The carbon skeleton has rearranged, which indicates an E1 process (the expected mechanism for an alcohol dehydration). The OH group is first protonated, thereby converting a bad leaving group into an excellent leaving group. Loss of the leaving group gives a secondary carbocation, which can rearrange via a methyl shift to give a tertiary carbocation. The tertiary carbocation is then deprotonated by a molecule of water to give the product:

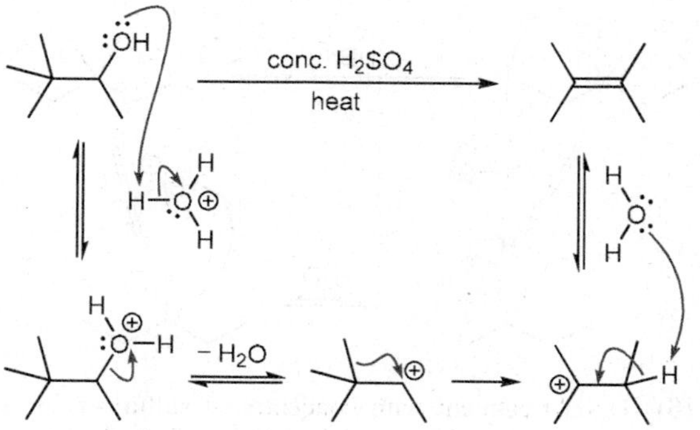

conc. H_2SO_4
heat

$- H_2O$

7.35.

(a) Our retrosynthesis begins with a disconnection at the cyano group, at the C–C bond:

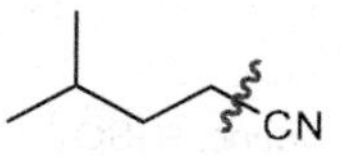

Cyanide ($N{\equiv}C^-$) is a stable and familiar anion, so we will use cyanide as the nucleophile. The other carbon (at the disconnected bond) must have started out as an electrophile, so we draw a leaving group (such as Cl, Br, or I) at that position.

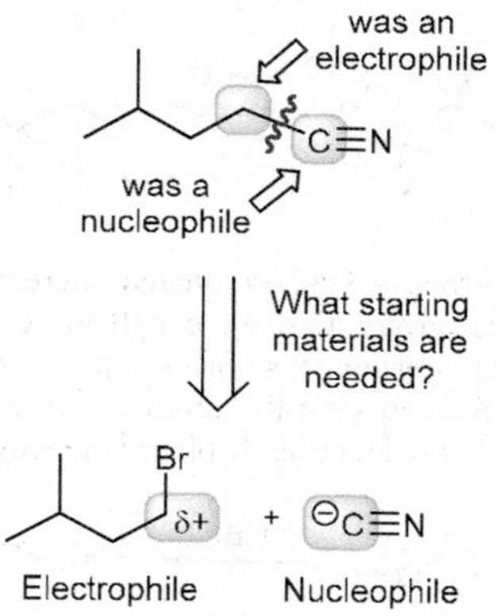

was an electrophile

was a nucleophile

What starting materials are needed?

Electrophile Nucleophile

The last step of the planning process is to confirm that the reaction mechanism is favorable. Cyanide is a strong nucleophile but not a strong base, so the S_N2 pathway does not have any significant competition from the E2 pathway. With a primary substrate, we expect an S_N2 process to proceed smoothly to give the desired target molecule.

Now we draw the forward process. As shown here, bromide was selected as the leaving group and NaCN can be used as a source of cyanide anion. Other variations are equally acceptable:

NaCN

Target molecule

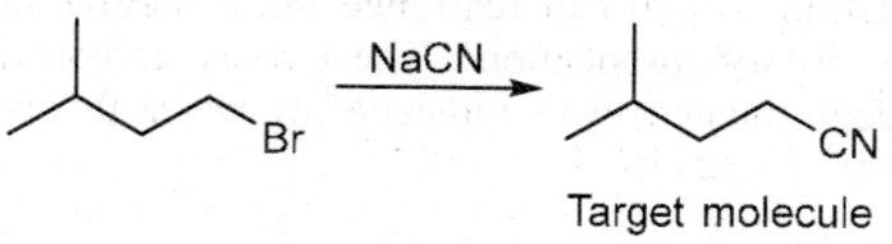

(b) Our retrosynthesis begins with a disconnection at the thiol group, at the C–S bond:

The more electronegative sulfur atom is more likely to have served as the nucleophile, so we draw a negative charge on the sulfur atom to give a suitable nucleophile (the hydrosulfide ion). The carbon atom (at the disconnected bond) must have started out as an electrophile, so we draw a leaving group (such as Cl, Br, or I) at that position.

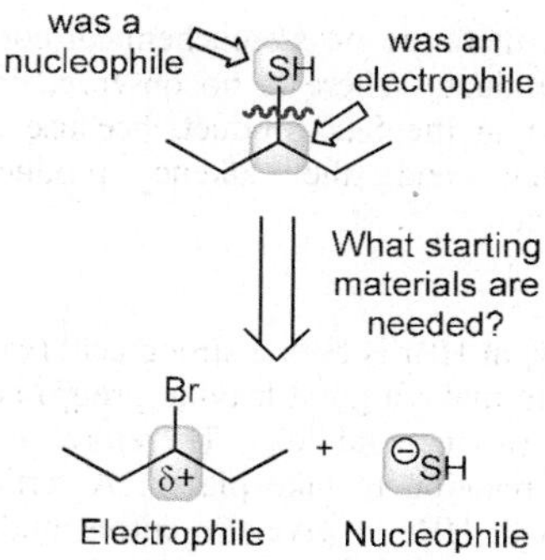

was a nucleophile

was an electrophile

What starting materials are needed?

Electrophile Nucleophile

The last step of the planning process is to confirm that the reaction mechanism is favorable. The large hydrosulfide anion is a strong nucleophile but not a strong base, so the S_N2 pathway does not have any significant competition from the E2 pathway. With a secondary substrate, we expect an S_N2 process to proceed smoothly to give the desired target molecule.

Now we draw the forward process. As shown here, bromide was selected as the leaving group and NaSH can be used as a source of hydrosulfide anion. Other variations are equally acceptable:

NaSH

Target molecule

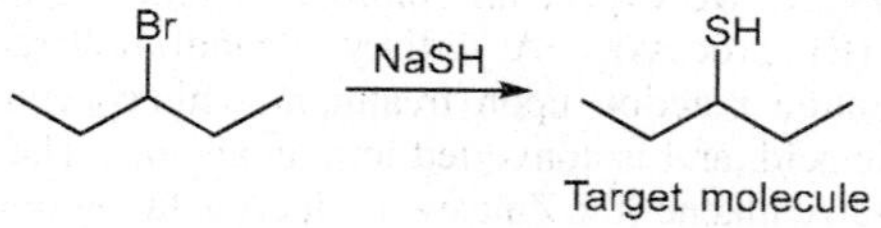

(c) Our retrosynthesis begins with a disconnection at the ether group, at either C–O bond (drawing out the methyl group reveals the second potential disconnection):

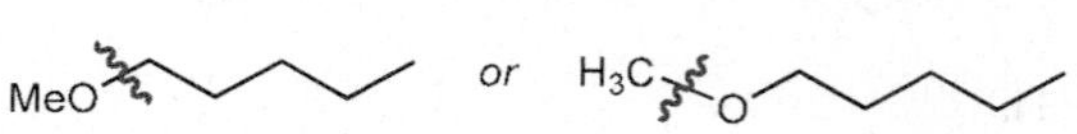

MeO *or* H_3C O

Both solutions will be presented, but let's take a look at the retrosynthesis using the first disconnection. The more electronegative oxygen atom is more likely to have served as the nucleophile, so we draw a negative charge on the oxygen atom to give a suitable nucleophile (methoxide ion). The carbon atom (at the disconnected bond) must have started out as an electrophile, so we draw a leaving group (such as Cl, Br, or I) at that position.

was an electrophile
MeO
was a nucleophile
What starting materials are needed?
Br
⊖OMe + δ+
Nucleophile Electrophile

The last step of the planning process is to confirm that the reaction mechanism is favorable. The methoxide anion is a strong nucleophile and a strong base, so we must consider both SN2 and E2 pathways. In this case, we have a primary substrate and a small nucleophile, so there is very little steric hindrance and we expect an SN2 process to give the desired target molecule as the major product.

Now we draw the forward process. As shown here, bromide was selected as the leaving group and NaOMe can be used as a source of the methoxide anion. Other variations are equally acceptable:

Br NaOMe → MeO
Target molecule

The other disconnection leads to an equally good SN2 process and represents an equally acceptable synthesis. Note that we select methyl iodide for the methyl halide, because methyl iodide is a liquid at room temperate and easier to handle than methyl chloride or methyl bromide, which are gases at room temperature):

NaO MeI → MeO
Target molecule

(d) Our retrosynthesis begins with a disconnection at the ester group. We must avoid making a disconnection at the sp^2 hybridized carbon atom, as such atoms do not participate in SN2 reactions, leaving the C–O bond the only reasonable option:

O

The electronegative oxygen atom is more likely to have served as the nucleophile, so we draw a negative charge on the oxygen atom to give a suitable nucleophile (a carboxylate ion). The carbon atom (at the disconnected bond) must have started out as an electrophile, so we draw a leaving group (such as Cl, Br, or I) at that position.

was an electrophile
O
was a nucleophile
What starting materials are needed?
Br O
δ+ + ⊖O
Electrophile Nucleophile

The last step of the planning process is to confirm that the reaction mechanism is favorable. The carboxylate anion is electron-rich, making it a strong nucleophile. However, the negative charge is delocalized (stabilized by resonance), making it a relatively weak base (second factor in ARIO). Because it is not a strong base, the SN2 pathway does not have significant competition from the E2 pathway. With a primary substrate, we expect an SN2 process to proceed smoothly to give the desired target molecule.

Now we draw the forward process. Notice that bromide was selected as the leaving group, and the carboxylate ion is shown as a sodium salt. Other variations are equally acceptable:

O O
NaO Br → O
Target molecule

(e) Our retrosynthesis begins with a disconnection at the alcohol group, at the C–O bond:

OH

The more electronegative oxygen atom is more likely to have served as the nucleophile, so we draw a negative charge on the oxygen atom to give a suitable nucleophile (hydroxide ion). The carbon atom (at the disconnected bond) must have started out as an electrophile, so we draw a leaving group (such as Cl, Br, or I) at that position.

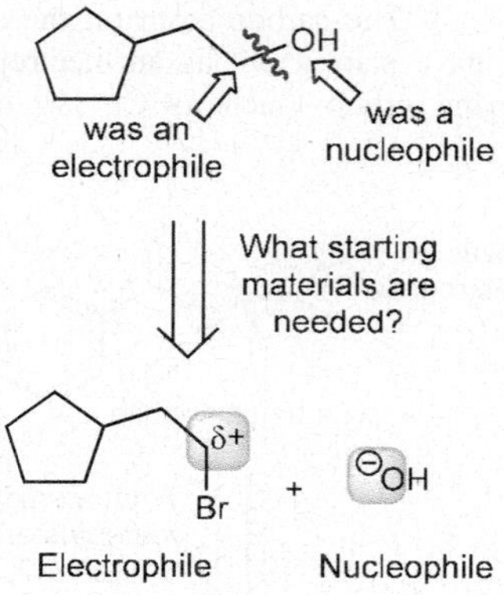

was an electrophile

was a nucleophile

What starting materials are needed?

Electrophile Nucleophile

The last step of the planning process is to confirm that the reaction mechanism is favorable. The hydroxide anion is a strong nucleophile and a strong base, so we must consider both S_N2 and E2 pathways. In this case, we have a primary substrate and a small nucleophile, so there is very little steric hindrance and we expect an S_N2 process to give the desired target molecule as the major product.

Now we draw the forward process. As shown here, bromide was selected as the leaving group and NaOH can be used as a source of the hydroxide anion. Other variations are equally acceptable:

NaOH

Target molecule

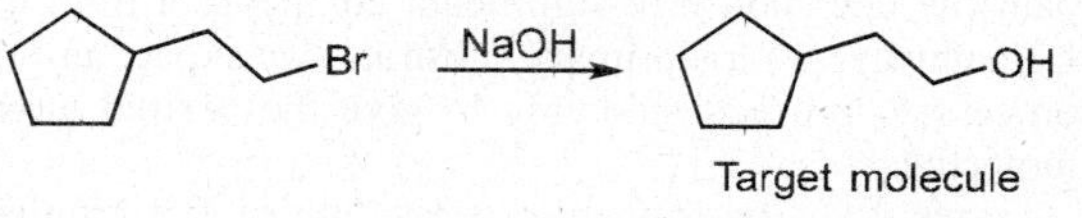

(f) Our retrosynthesis begins with a disconnection at the cyano group, at the C–C bond:

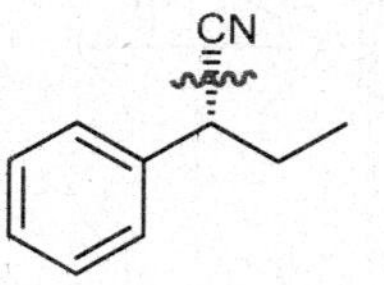

Cyanide ($N{\equiv}C^-$) is a stable and familiar anion, so we will use cyanide as the nucleophile. The other carbon (at the disconnected bond) must have started out as an electrophile, so we draw a leaving group (such as Cl, Br, or I) at that position. To account for the required stereochemistry in the product, we must consider the mechanism of the proposed reaction. An S_N2 process involves back-side attack, resulting in inversion of stereochemistry at the carbon atom bearing the leaving group. In order to furnish a product with the nucleophile positioned on a dash, the leaving group must be placed on a wedge:

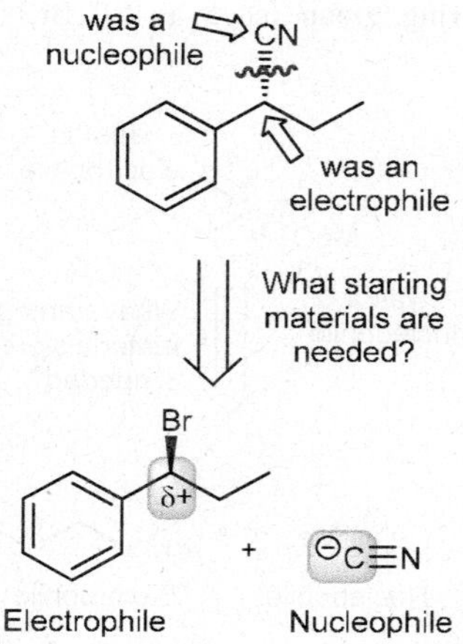

was a nucleophile

was an electrophile

What starting materials are needed?

Electrophile Nucleophile

The last step of the planning process is to confirm that the reaction mechanism is favorable. Cyanide is a strong nucleophile but not a strong base, so the S_N2 pathway does not have any significant competition from the E2 pathway. With a secondary substrate, we expect an S_N2 process to proceed smoothly to give the desired target molecule.

Now we draw the forward process. As shown here, bromide was selected as the leaving group and NaCN can be used as a source of cyanide anion. Other variations are equally acceptable:

NaCN

Target molecule

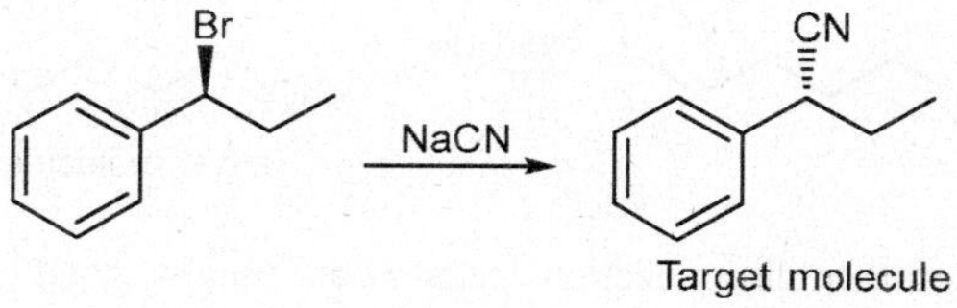

7.36.

(a) Even though this synthesis problem is presented as a transform problem in which a starting material is provided, it is still helpful to begin with a retrosynthetic analysis of the target molecule. Identifying the portion of the target molecule that came from the given starting material determines a suitable place to make a disconnection (at the new C–S bond):

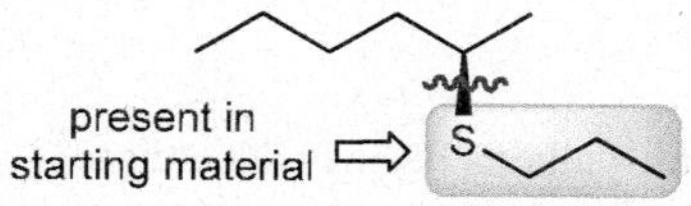

present in starting material

The more electronegative sulfur atom is more likely to have served as the nucleophile; this is confirmed by the presence of the negative charge on the sulfur atom in the given starting material (a thiolate ion). The carbon atom (at the disconnected bond) must have started out as an electrophile, so we draw a leaving group (such as Cl, Br, I or OTs) at that position. To account for the required stereochemistry in the product, we must consider the mechanism of the proposed reaction. An S_N2 process involves back-side attack, resulting in inversion of stereochemistry at the carbon atom bearing the leaving group. In order to furnish a product with the nucleophile positioned on a wedge, the leaving group must be placed on a dash:

S

What starting materials are needed?

δ+ + ⊖S

Electrophile Br Nucleophile

The last step of the planning process is to confirm that the reaction mechanism is favorable. The large thiolate ion is a strong nucleophile but not a strong base, so the S_N2 pathway does not have any significant competition from the E2 pathway. With a secondary substrate, we expect an S_N2 process to proceed smoothly to give the desired target molecule.

Now we draw the forward process. As shown here, bromide was selected as the leaving group, but other variations are equally acceptable:

Br

SNa → S

(b) Even though this synthesis problem is presented as a transform problem in which a starting material is provided, it is still helpful to begin with a retrosynthetic analysis of the target molecule. We can consider either C–O bond for a disconnection, but because we know how to convert OH into a good leaving group (and we have not yet learned how to turn the OH into a strong nucleophile), the disconnection shown is the better one:

O

The more electronegative oxygen atom is more likely to have served as the nucleophile, so we draw a negative charge on the oxygen atom to give a suitable nucleophile (butoxide ion). The carbon atom (at the disconnected bond) must have started out as an electrophile, so we draw a leaving group (such as Cl, Br, I or OTs) at that position:

was a nucleophile

O

was an electrophile

What starting materials are needed?

δ+ + ⊖O

LG

Electrophile Nucleophile

Because we are starting with an alcohol substrate, it is convenient to select the tosylate as a convenient substrate. Another option to install a good leaving group involves substitution of the OH group with Br, by treatment with HBr.

TsCl pyridine → OTs

OH

HBr → Br

The last step of the planning process is to confirm that the reaction mechanism is favorable. The alkoxide anion is a strong nucleophile and a strong base, so we must consider both S_N2 and E2 pathways. In this case, we have a primary substrate and a small nucleophile, so there is very little steric hindrance and we expect an S_N2 process to give the desired target molecule as the major product.

Now we draw the forward process. This transformation requires two steps: conversion of the alcohol into a good leaving group (tosylate or bromide), followed by reaction with *n*-butoxide:

HBr → Br → NaO

OH O

TsCl pyridine → OTs → NaO

7.37. Let's begin by exploring the retrosynthesis based on disconnection of bond **a**. This retrosynthesis involves a very poor electrophile, because S_N2 reactions generally do not occur at sp^2 hybridized centers:

a

S R' R

sp^2

S⊖ + Br R'

R

Good nucleophile

Not a good electrophile for S_N2 reactions

The better retrosynthesis is based on disconnection of bond **b**, because it involves both a good electrophile and a good nucleophile:

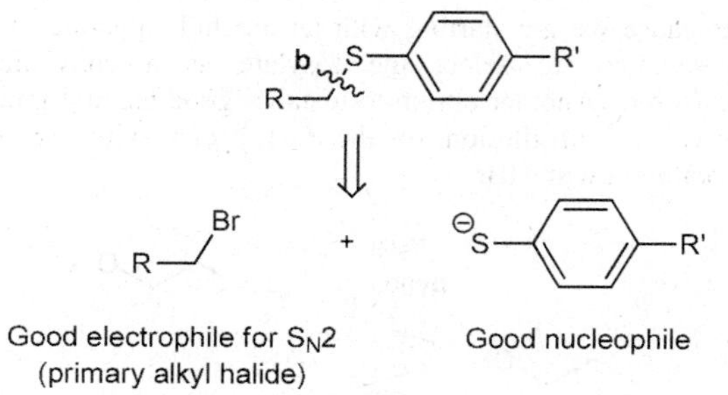

Good electrophile for S_N2 (primary alkyl halide)

Good nucleophile

The S_N2 process portrayed in retrosynthesis **b** is favorable, because it involves a strong nucleophile and a primary substrate, so this reaction is expected to give the desired target molecule.

Target molecule

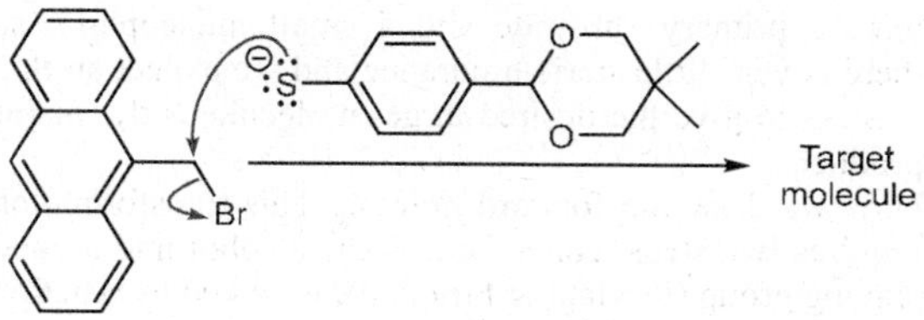

7.38.

(a) The π bond in the target molecule can be formed by an elimination reaction. In this case, the target molecule is a stable alkene, so the potential starting material can be either an alkyl halide or an alcohol. To draw these starting materials, we delete the C=C double bond, we place a H atom on one of the carbon atoms, and we place either a Br atom or an OH group on the other carbon atom:

What starting materials are needed?

+ strong base or + conc. H_2SO_4

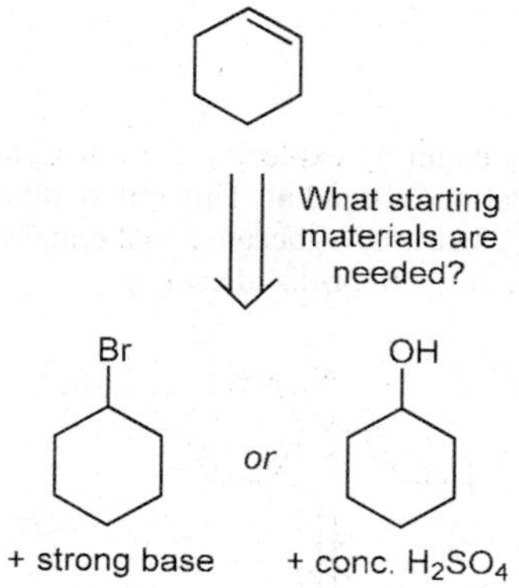

Next, we must select the appropriate reagents, and also confirm that the desired target molecule would be the major product formed. When starting with an alkyl halide, a strong base must be used for the E2 elimination. Because the leaving group is on a secondary carbon, elimination is favored over S_N2 substitution. Only one elimination product is possible, and using a sterically hindered strong base (such as *tert*-butoxide) would further inhibit the S_N2 process.

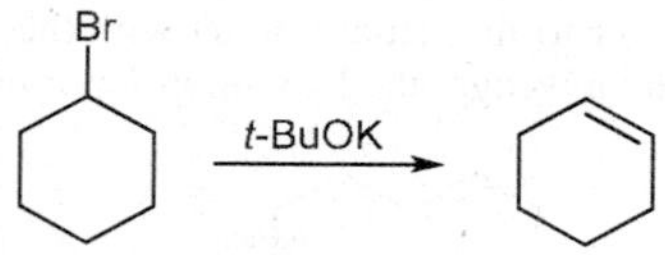

t-BuOK

Keep in mind that any good leaving group can be used for the E2 elimination (Cl, Br, I, OTs), and such variations are certainly acceptable. Notice that the *tert*-butoxide ion is shown with a potassium counterion, as this is the most commonly encountered salt.
Alternatively, synthesis of the target molecule by dehydration of an alcohol starting material can be achieved by treating the alcohol with a strong, concentrated acid (such as H_2SO_4) and heat.

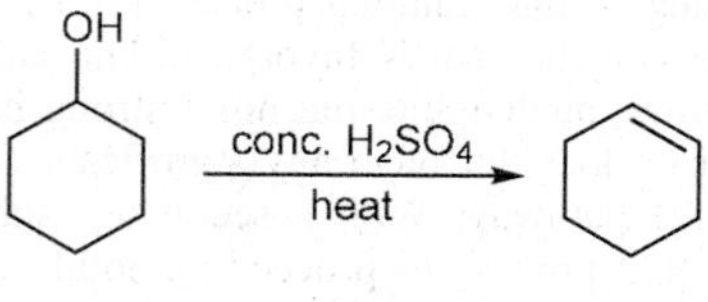

conc. H_2SO_4 heat

(b) The π bond in the target molecule can be formed by an elimination reaction, so the potential starting material can be either an alkyl halide or an alcohol. In this case, the target molecule is not a highly substituted alkene, so it cannot be synthesized in high yield via acid-catalyzed dehydration of an alcohol (such a process would yield a more-substituted alkene). The ideal starting material, therefore, is an alkyl halide. To draw the alkyl halide, we delete the C=C double bond, and we place a H atom on one of the carbon atoms and a Br atom on the other carbon atom. In this example, the retrosynthesis reveals two suitable alkyl halide starting materials for an E2 elimination:

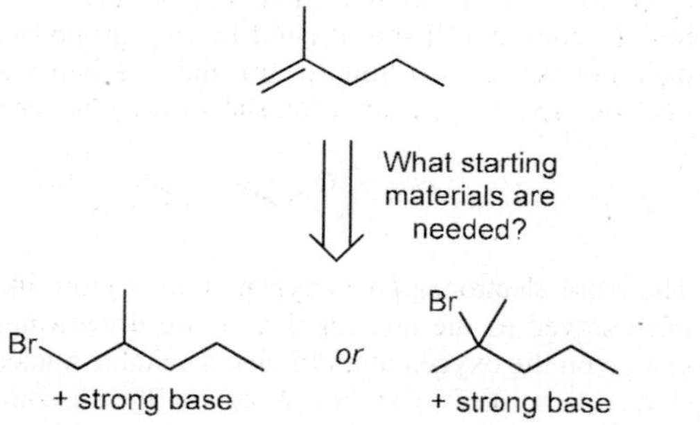

What starting materials are needed?

+ strong base or + strong base

Next, we must determine which strong base to use for the required E2 elimination, and also confirm that the desired target molecule would be the major product formed. For the primary alkyl halide, a strong, sterically hindered base (such as *tert*-butoxide) must be used to favor E2 elimination over S_N2 substitution.

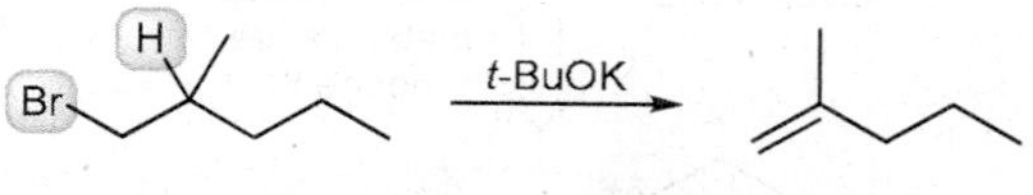

t-BuOK

E2 favored over S_N2 with bulky base

If we selected the secondary alkyl halide as our starting material, we clearly have the possibility of forming either the Hofmann or Zaitsev product, so *tert*-butoxide would be the appropriate choice of base in this case as well, to give the target molecule (the less-substituted alkene).

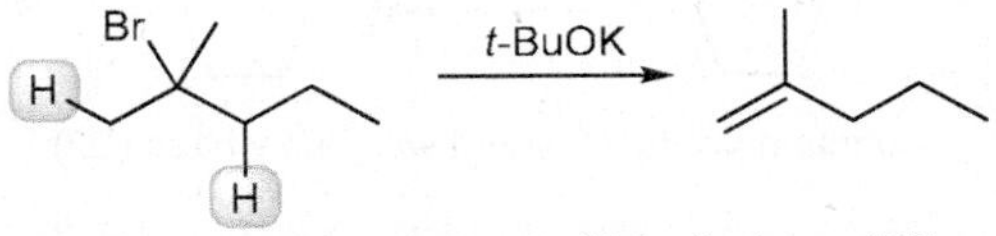

Br H H *t*-BuOK

Hofmann product favored with bulky base (E2)

Note that removal of a β hydrogen atom from either methyl group would lead to the desired target molecule. Keep in mind that any good leaving group can be used for the E2 elimination (Cl, Br, I, OTs), and such variations are certainly acceptable. Notice that the *tert*-butoxide ion is shown with a potassium counterion, as this is the most commonly encountered salt.

(c) The π bond in the target molecule can be formed by an elimination reaction, so the potential starting material can be either an alkyl halide or an alcohol. In this case, the target molecule is not a highly substituted alkene, so it cannot be synthesized in high yield via acid-catalyzed dehydration of an alcohol (such a process would yield a more-substituted alkene). The ideal starting material, therefore, is an alkyl halide. To draw the alkyl halide, we delete the C=C double bond, and we place a H atom on one of the carbon atoms and a Br atom on the other carbon atom. In this example, the retrosynthesis reveals two suitable alkyl halide starting materials for an E2 elimination:

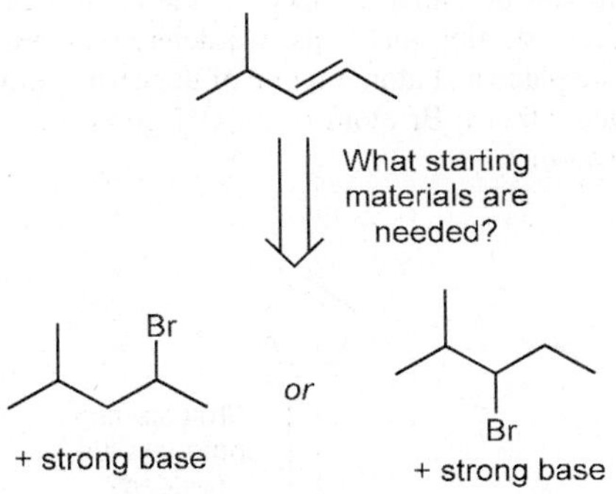

What starting materials are needed?

Br + strong base *or* Br + strong base

Next, we must determine which strong base to use for the required E2 elimination, and also confirm that the desired target molecule would be the major product formed. For both alkyl halides, we have the possibility of forming either the Hofmann or Zaitsev product. If we select the first alkyl halide as our starting material, the more-substituted Zaitsev product is desired, so we must a use a strong base that is not sterically hindered, such as hydroxide, methoxide or ethoxide.

Br H H NaOMe

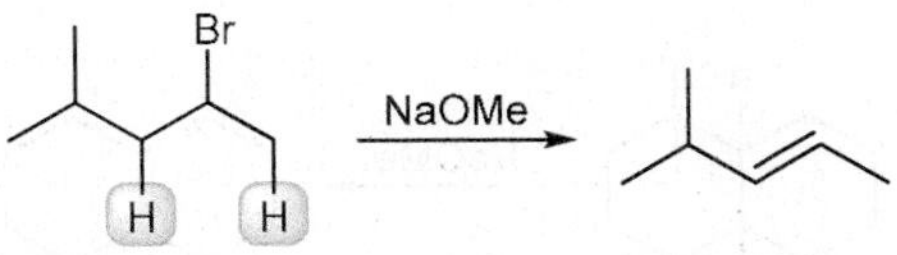

Zaitsev product favored with small base (E2)

Alternatively, if we start with the other alkyl halide, *tert*-butoxide would be the appropriate choice of base, to give the target molecule (the less-substituted Hofmann product).

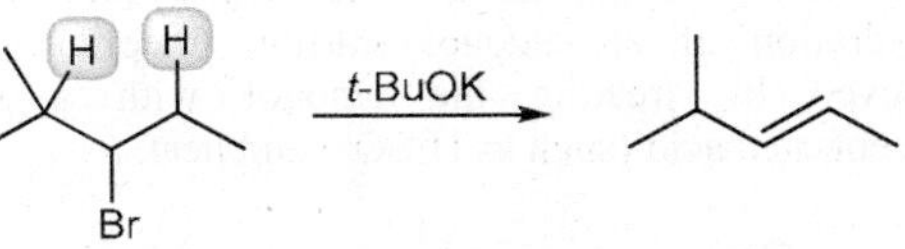

H H Br *t*-BuOK

Hofmann product favored with bulky base (E2)

Keep in mind that any good leaving group can be used for the E2 elimination (Cl, Br, I, OTs), and such variations are certainly acceptable. Notice that the *tert*-butoxide ion is shown with a potassium counterion, as this is the most commonly encountered salt.

(d) The π bond in the target molecule can be formed by an elimination reaction. In this case, the target molecule is a tetrasubstituted alkene, so the potential starting material can be either an alkyl halide or an alcohol. To draw these starting materials, we delete the C=C double bond, we place a H atom on one of the carbon atoms, and we place either a Br atom or an OH group on the other carbon atom:

What starting materials are needed?

Br + strong base *or* OH + conc. H_2SO_4

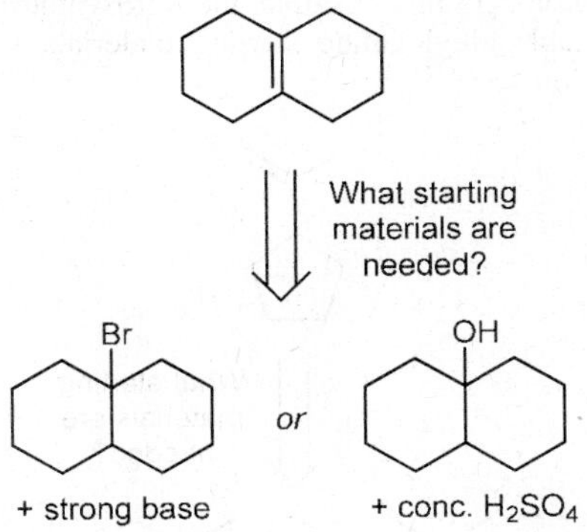

Next, we must select the appropriate reagents, and also confirm that the desired target molecule would be the major product formed. When starting with an alkyl halide, a strong base must be used for the E2 elimination. Because the leaving group is on a tertiary carbon, E2 elimination is guaranteed over S_N2 substitution. The more-substituted Zaitsev product is desired, so we must a use a strong base that is not sterically hindered, such as hydroxide, methoxide or ethoxide:

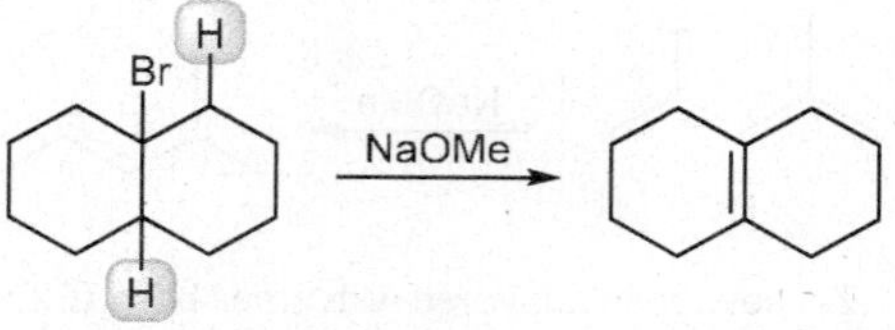

NaOMe

Zaitsev product favored with small base (E2)

Keep in mind that any good leaving group can be used for the E2 elimination (Cl, Br, I, OTs), and such variations are certainly acceptable.
Alternatively, synthesis of the target molecule by dehydration of an alcohol starting material can be achieved by treating the alcohol with a strong, concentrated acid (such as H_2SO_4) and heat.

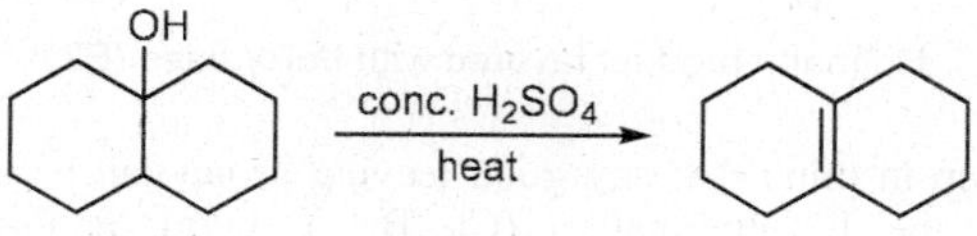

OH

conc. H_2SO_4
heat

(e) The π bond in the target molecule can be formed by an elimination reaction, so the potential starting material can be either an alkyl halide or an alcohol. In this case, the target molecule is not a highly substituted alkene, so it cannot be synthesized in high yield via acid-catalyzed dehydration of an alcohol (such a process would yield a more-substituted alkene). The ideal starting material, therefore, is an alkyl halide. To draw the alkyl halide, we delete the C=C double bond, and we place a H atom on one of the carbon atoms and a Br atom on the other carbon atom. In this example, the retrosynthesis reveals two suitable alkyl halide starting materials for an E2 elimination:

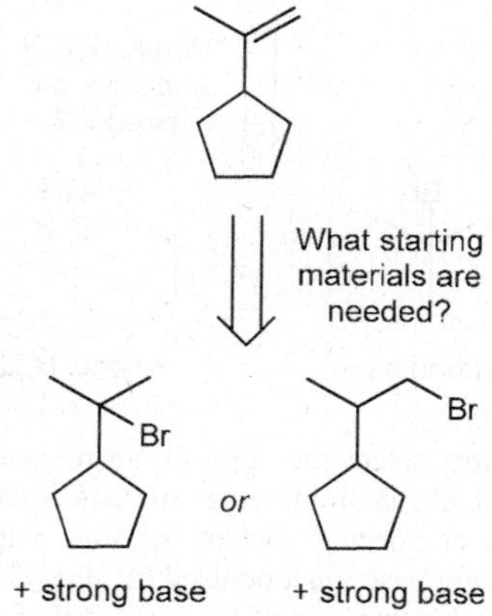

What starting materials are needed?

Br *or* Br

+ strong base + strong base

Next, we must determine which strong base to use for the required E2 elimination, and also confirm that the desired target molecule would be the major product formed. If we selected the tertiary alkyl halide as our starting material, we clearly have the possibility of forming either the Hofmann or Zaitsev product, so *tert*-butoxide would be the appropriate choice of base, to give the target molecule (the less-substituted alkene). Note that removal of a β hydrogen atom from either methyl group would lead to the desired target molecule.

t-BuOK

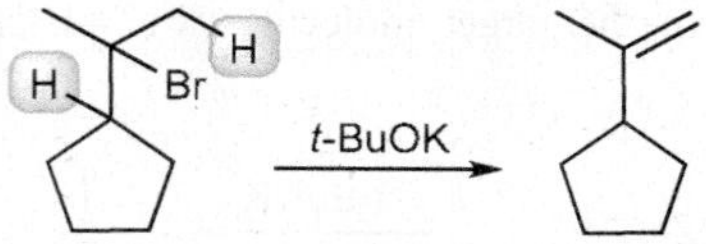

Hofmann product favored with bulky base (E2)

If we selected the primary alkyl halide as our starting material, a strong, sterically hindered base (such as *tert*-butoxide) must be used to favor E2 elimination over S_N2 substitution.

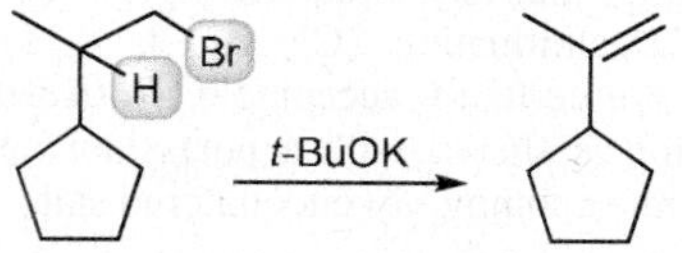

t-BuOK

E2 favored over S_N2 with bulky base

Keep in mind that any good leaving group can be used for the E2 elimination (Cl, Br, I, OTs), and such variations are certainly acceptable. Notice that the *tert*-butoxide ion is shown with a potassium counterion, as this is the most commonly encountered salt.

(f) The π bond in the target molecule can be formed by an elimination reaction. In this case, the target molecule is a highly substituted alkene, so the potential starting material can be either an alkyl halide or an alcohol. To draw these starting materials, we delete the C=C double bond, we place a H atom on one of the carbon atoms, and we place either a Br atom or an OH group on the other carbon atom:

What starting materials are needed?

Br *or* OH

+ strong base + conc. H_2SO_4

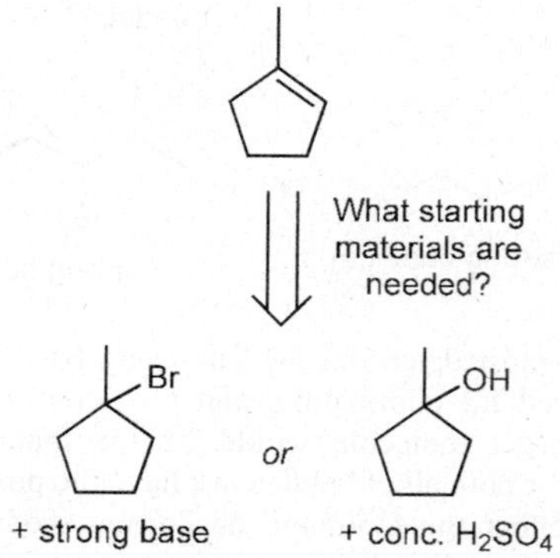

Another option for drawing starting materials places the Br atom or OH group on a different carbon, so there are four reasonable solutions to this synthesis problem:

What starting materials are needed?

+ strong base or + conc. H_2SO_4

Next, we must select the appropriate reagents, and also confirm that the desired target molecule would be the major product formed. When starting with one of the alkyl halides, a strong base must be used for the E2 elimination. Because the leaving group is on either a tertiary or secondary carbon, E2 elimination is favored over S_N2 substitution. The more-substituted Zaitsev product is desired, so we must a use a strong base that is not sterically hindered, such as hydroxide, methoxide or ethoxide:

NaOMe

Zaitsev product favored with small base (E2)

Keep in mind that any good leaving group can be used for the E2 elimination (Cl, Br, I, OTs), and such variations are certainly acceptable.
Alternatively, synthesis of the target molecule by acid-catalyzed dehydration of one of the alcohol starting materials can be achieved by treating the alcohol with a strong, concentrated acid (such as H_2SO_4) and heat.

conc. H_2SO_4, heat

Alcohol dehydration favors Zaitsev product (E1)

7.39.

(a) This transformation represents an elimination reaction (dehydrohalogenation), and a strong base is required. The Hofmann product is desired, so a strong, sterically hindered base should be used, such as potassium *tert*-butoxide.

t-BuOK

(b) This transformation represents an elimination reaction (dehydrohalogenation), and a strong base is required. The Zaitsev product is desired, so we must a use a strong base that is not sterically hindered, such as hydroxide, methoxide or ethoxide.

NaOMe

(c) This transformation represents an elimination reaction to give the Zaitsev product (dehydration), and a strong, concentrated acid is required, such as sulfuric acid.

conc. H_2SO_4, heat

(d) This transformation represents an elimination reaction to give the less-substituted, Hofmann product, so acid-catalyzed dehydration of the alcohol is not a suitable approach (that would afford the more-substituted, Zaitsev product). Instead, a good leaving group is required (*e.g.*, Cl, Br, I, OTs) to facilitate an E2 elimination reaction.

What starting materials are needed?

Therefore, the synthesis requires a two-step procedure: conversion of the OH group into a good leaving group, followed by treatment with a strong, sterically hindered base, such as potassium *tert*-butoxide, to generate the Hofmann product. Reaction of the alcohol with tosyl chloride produces a good leaving group (tosylate). Alternatively, the alcohol can be converted to an alkyl halide by reaction with HX, (*e.g.*, HBr).

HBr; *t*-BuOK

TsCl, pyridine; *t*-BuOK

Because acidic conditions favor the formation of carbocations that can rearrange, the tosylate approach is likely to be the better choice for this transformation. Reaction of the given alcohol substrate with HBr is likely to give a mixture of products, via a combination of S_N2 and S_N1 mechanisms, shown below:

S_N2 Mechanism

HBr / S_N2

S_N1 Mechanism

HBr / S_N1

– H_2O

7.40. Consider the mechanism involved in each possible transformation, and for each one, consider whether or not the target molecule is expected to be the major (or only) product. We begin with the first possibility, which employs an alkyl halide as the starting material. It is true that alkyl halides can be converted to alkenes upon treatment with a strong base (via an E2 process). However, in this case, there are two possible regiochemical outcomes, so we expect a mixture of two alkene products:

NaOEt / EtOH

Poor synthesis (gives mixture of products)

In contrast, treating the second halide with a strong base is expected to give only one product (the desired target molecule):

NaOEt / EtOH

In this case there is only one possible regiochemical outcome:

Therefore, the better synthesis begins with the second alkyl halide shown. Any strong base, such as NaOEt, can be used:

NaOEt / EtOH

7.41. The π bond in the target molecule can be formed by an elimination reaction, so the potential starting material can be either an alkyl halide or an alcohol. In this case, the reaction used a strong base (DBU) as the reagent, indicating an E2 elimination, so the starting material must have been an alkyl halide. To draw the alkyl halide, we delete the C=C double bond, and we place a H atom on one of the carbon atoms and a Br atom on the other carbon atom. In this example, the retrosynthesis reveals two suitable alkyl halide starting materials for an E2 elimination:

What starting materials are needed?

+ strong base *or* + strong base

To determine which is the better substrate, we need to confirm that the desired target molecule would be the major product formed. The first alkyl bromide has only one type of β hydrogen available, and only one elimination product is possible (the desired target molecule), while the second choice has two types of β hydrogens, leading to a mixture of alkenes upon treatment with strong base:

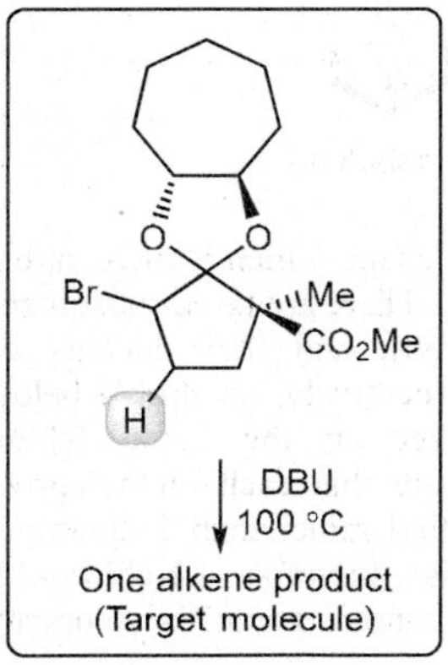

One alkene product (Target molecule)

Two alkene products (poor synthesis)

Therefore, the first alkyl halide is the more suitable starting material. As usual, any good leaving group can be used for the E2 elimination (Cl, Br, I, OTs).

7.42. This is a substitution reaction with a strong nucleophile (HO^-) and a primary leaving group, so an S_N2 process is expected. The second S_N2 reaction employs a polar aprotic solvent (DMSO) and is therefore expected to occur at a faster rate.

7.43. This is a substitution reaction with a strong nucleophile (X^-) and a primary leaving group, so an S_N2 process is expected. In a polar protic solvent (EtOH), iodide is a stronger nucleophile than chloride. With all other factors being the same (the alkyl halide is the same in both reactions, and the solvent is the same for both reactions), the first S_N2 reaction is expected to occur at a faster rate, because it involves the use of a stronger nucleophile.

7.44.

(a) Solvolysis is expected to afford both S_N1 and E1 products. The first step of solvolysis is loss of the leaving group to give a carbocation. This carbocation can be deprotonated by the solvent to give an alkene product (E1),

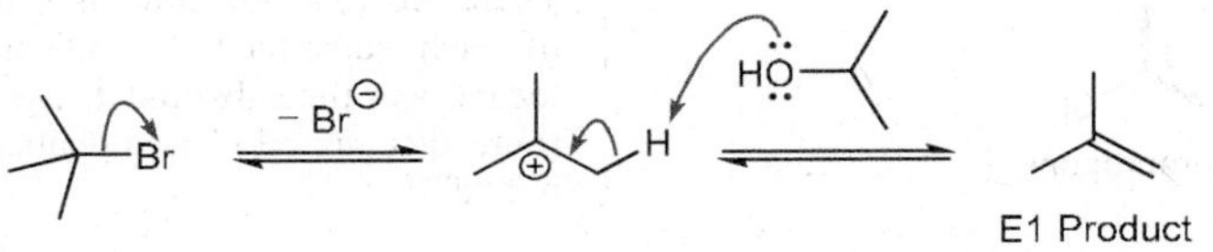

E1 Product

or the solvent can act as a nucleophile and attack the carbocation (S_N1). If the solvent is isopropanol, $(CH_3)_2CHOH$, then the carbocation can be captured by a molecule of isopropanol, giving an oxonium ion, which is then deprotonated to give the substitution product, shown below:

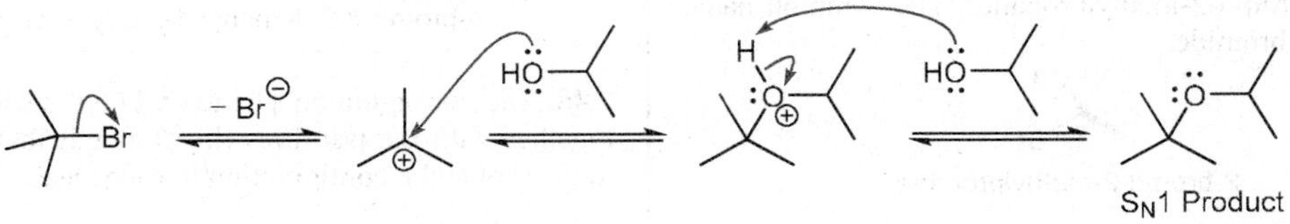

S_N1 Product

(b) The rate-determining step in a solvolysis reaction is the formation of the carbocation intermediate. A more polar solvent better stabilizes a carbocation, resulting in a faster solvolysis reaction. Therefore the rate of solvolysis is dependent on solvent polarity, measured by the dielectric constant of the solvent. A higher dielectric constant corresponds with a faster rate of solvolysis. The dielectric constant of ethanol is 24, while the dielectric constant of isopropanol is only 18. Therefore, solvolysis is expected to occur more rapidly in ethanol, which is a more polar solvent than isopropanol.

7.45.

(a) In the first compound, the β positions are deuterated. As such, an elimination reaction will involve loss of DBr, which occurs at a slower rate than loss of HBr. Therefore, the second compound is expected to undergo elimination more rapidly.

(b) In the first compound, the β positions are deuterated. In the second compound, the α position is deuterated. Since elimination involves removal of H (or D) from the β position, the rate of reaction for the first compound will be more affected by the presence of deuterium. That is, the first compound is expected to undergo elimination at a slower rate. Therefore, the second compound is expected to undergo elimination more rapidly.

(c) In the first compound, the β position is deuterated. In the second compound, the α position is deuterated. Since elimination involves removal of H (or D) from the β position, the rate of reaction for the first compound will be more affected by the presence of deuterium. That is, the first compound is expected to undergo elimination at a slower rate. Therefore, the second compound is expected to undergo elimination more rapidly.

7.46.

(a) Ethoxide is a strong base, so we expect an E2 reaction, which involves deprotonation at the β position. If all of the β protons are replaced with D (as indicated in the problem statement), then the reaction is expected to occur at a slower rate, as a result of a primary isotope effect.

(b) In this case, the reagent is ethanol, which is a weak base. These conditions favor an E1 process. In an E1 process, deprotonation of the β position occurs AFTER the rate determining step (loss of the leaving group). Therefore, we do not expect a primary isotope effect.

7.47.

(a) The parent is the longest chain, which is three carbon atoms in this case (propane). There is only one substituent (chloro), and its locant is assigned as 2 (as shown below), so the systematic name for this compound is 2-chloropropane. The common name is isopropyl chloride.

2-chloropropane

(b) The parent is the longest chain, which is three carbon atoms in this case (propane). There are two substituents (bromo and methyl), and their locants are assigned as 2 and 2, as shown below. Substituents are alphabetized in the name (bromo precedes methyl), so the systematic name is 2-bromo-2-methylpropane. The common name is *tert*-butyl bromide.

2-bromo-2-methylpropane

(c) The parent is the longest chain, which is three carbon atoms in this case (propane). There is only one substituent (iodo), and its locant is assigned as 1 (as shown), so the systematic name for this compound is 1-iodopropane. The common name is *n*-propyl iodide.

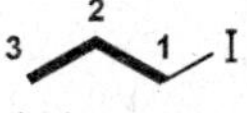

1-iodopropane

(d) The parent is the longest chain, which is four carbon atoms in this case (butane). There is only one substituent (bromo), and its locant is assigned as 2 (as shown below). The compound has a chiral center, so the configuration must be indicated at the beginning of the name: (*R*)-2-bromobutane. The common name is (*R*)-*sec*-butyl bromide.

(*R*)-2-bromobutane

(e) The parent is the longest chain, which is three carbon atoms in this case (propane). There are three substituents (chloro, methyl, and methyl), and their locants are assigned as 1, 2, and 2, respectively, as shown below. Substituents are alphabetized in the name (chloro precedes methyl). Make sure that each methyl group receives a locant (2,2-dimethyl rather than 2-dimethyl). The systematic name is therefore 1-chloro-2,2-dimethylpropane. The common name is neopentyl chloride.

1-chloro-2,2-dimethylpropane

(f) We begin by identifying the parent. The longest chain is seven carbon atoms, so the parent is heptane. In this case, there are choices for the parent, and we choose the path that gives the maximum number of substituents. There are four substituents (highlighted). The location of each substituent is indicated with the appropriate locant, and the substituents are alphabetized in the name (note that dimethyl is alphabetized as "m" for methyl, not "d").

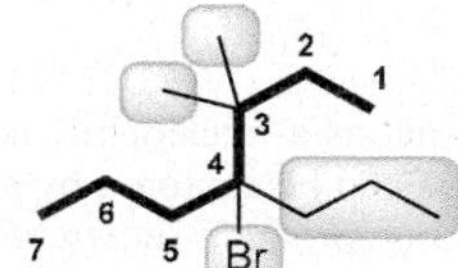

4-bromo-3,3-dimethyl-4-propylheptane

7.48. The configuration of each π bond is shown below, together with the priorities (highlighted) that were used to determine the configuration in each case.

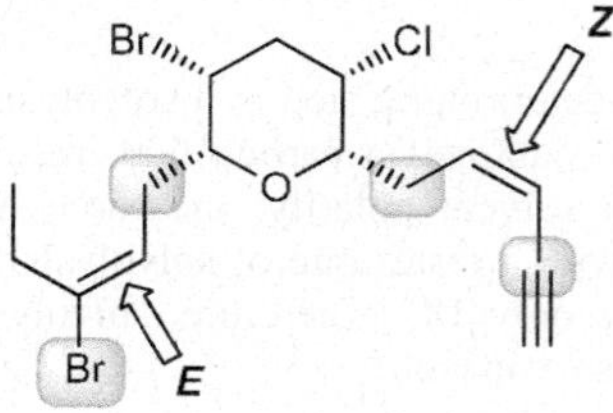

7.49. We begin by drawing all constitutional isomers with the molecular formula C_4H_9Br, shown below. For help, see the solution to Problem 2.51.

(a) The four constitutional isomers are arranged below in order of increasing reactivity toward S_N2. Notice that the tertiary substrate is the least reactive because it is the most hindered. Among the two primary substrates, 1-

bromobutane is the least sterically hindered (the β position is less substituted). Therefore, 1-bromobutane is the most reactive toward S_N2.

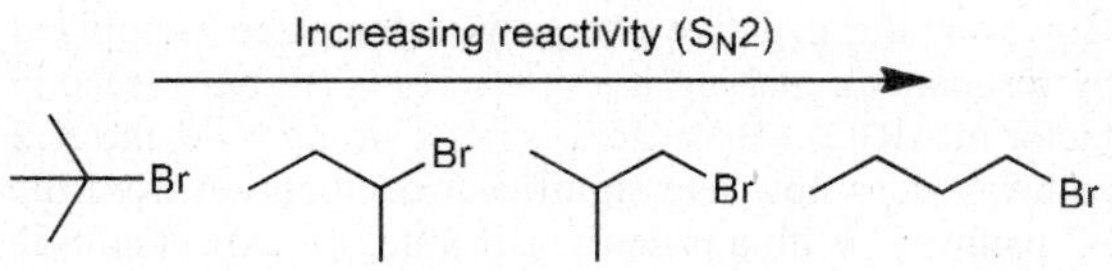

Increasing reactivity (S_N2)

(b) The four constitutional isomers are arranged below in order of increasing reactivity toward E2. Notice that the tertiary alkyl bromide is the most reactive towards E2, followed by the secondary alkyl bromide. Among the two primary alkyl bromides, the one leading to a disubstituted alkene will be more reactive towards E2 than the one leading to a monosubstituted alkene.

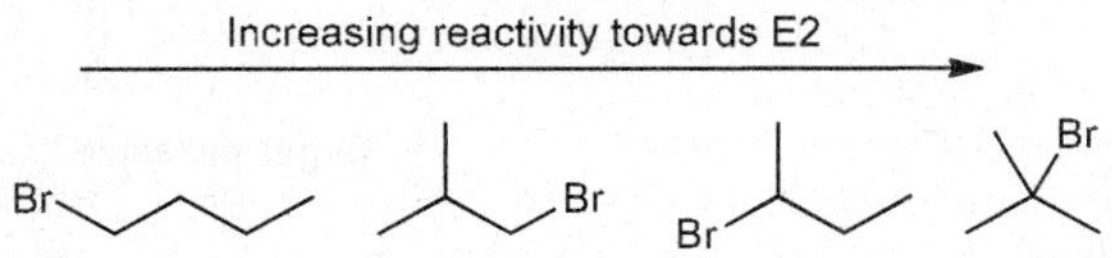

Increasing reactivity towards E2

7.50. The leaving group is an iodide ion (I^-) and the nucleophile is an acetate ion ($CH_3CO_2^-$). In the transition state, each of these groups is drawn as being connected to the α position with a dotted line to represent a partial bond (a bond that is either in the process of forming or breaking), and at a bond angle of 180° (indicating back-side attack). A δ– is placed on each group to represent a partial negative charge, indicating that these charges are in the process of forming or dissipating. Don't forget the brackets and the double-dagger symbol that indicate the drawing is a transition state.

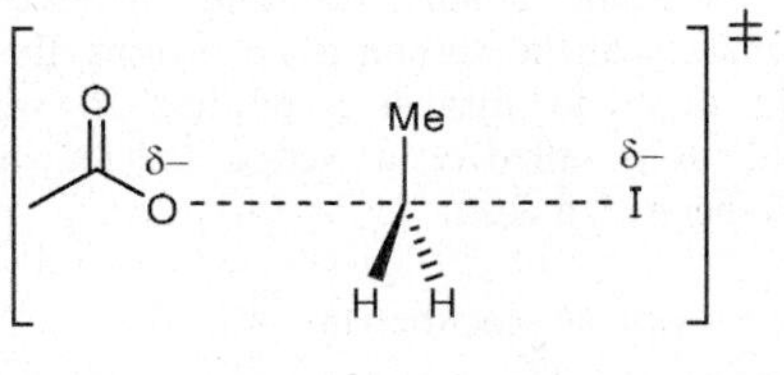

7.51. The substrate is secondary, the solvent is DMSO (a polar aprotic solvent), and the nucleophile (iodide) is a very strong nucleophile. All of these factors suggest an S_N2 process. Iodide functions as a nucleophile and attacks (*S*)-2-iodopentane, displacing iodide as a leaving group. Since the reaction is an S_N2 process, we expect inversion of configuration. The product is (*R*)-2-iodopentane, which is the enantiomer of the initial substrate. The (*S*) enantiomer continues to be converted to the (*R*) enantiomer, and vice versa, until a racemic mixture is eventually obtained.

7.52. No. Preparation of this compound via an acetylide ion would require the use of the following tertiary alkyl halide, which will not participate in an S_N2 reaction because of steric crowding.

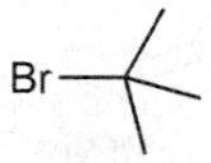

7.53.

(a) The chiral center in the substrate has the *R* configuration, as shown below.

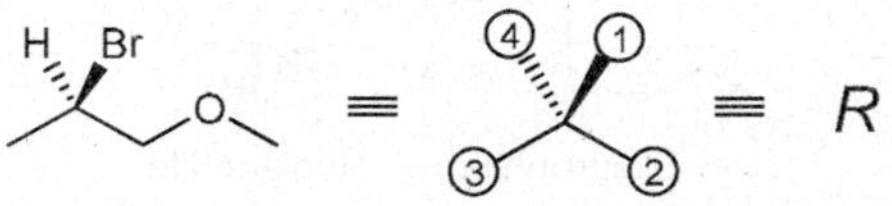

(b) The chiral center in the product has the *R* configuration, as shown below.

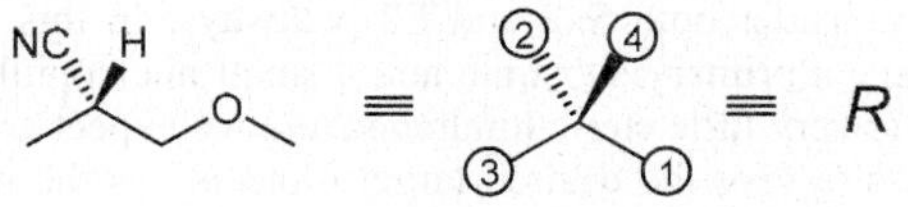

(c) The reaction is an S_N2 process, and it does proceed with inversion of configuration. However, the prioritization scheme changes when the bromo group (#1) is replaced with a cyano group (#2). As a result, the Cahn-Ingold-Prelog system assigns the same configuration to the reactant and the product, even though an inversion has indeed occurred.

7.54. The dianion has two nucleophilic centers, and the electrophile has two electrophilic centers. As such, these compounds can react with each other via two successive S_N2 reactions, as shown below, giving a six-membered ring with the molecular formula $C_4H_8O_2$.

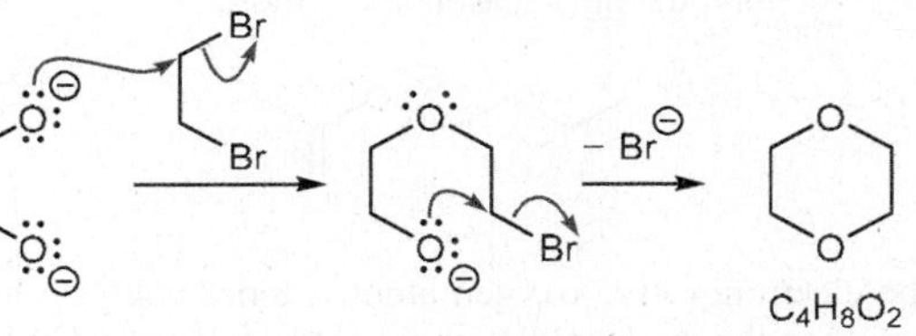

$C_4H_8O_2$

7.55.

(a) Our retrosynthesis begins with a disconnection at the alcohol group, at the C–O bond:

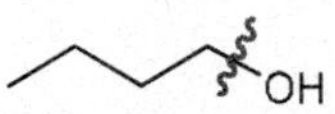

The more electronegative oxygen atom is more likely to have served as the nucleophile, so we draw a negative charge on the oxygen atom to give a suitable nucleophile (hydroxide ion). The carbon atom (at the disconnected bond) must have started out as the alkyl iodide electrophile, so we draw the leaving group (I) at that position.

was an electrophile
OH
was a nucleophile
What starting materials are needed?
I
δ+
+
⊖OH
Electrophile Nucleophile

The last step of the planning process is to confirm that the reaction mechanism is favorable. The hydroxide anion is a strong nucleophile and a strong base, so we must consider both SN2 and E2 pathways. In this case, we have a primary substrate and a small nucleophile, so there is very little steric hindrance and we expect an SN2 process to give the desired target molecule as the major product.

Now we draw the forward process. As shown here, NaOH can be used as the source of hydroxide. Other counterions (such as lithium or potassium) are equally acceptable:

I NaOH OH
Target molecule

(b) Our retrosynthesis begins with a disconnection at the ester group. We must avoid making a disconnection at the sp^2 hybridized carbon atom (because SN2 reactions generally do not occur at sp^2 hybridized centers), leaving the C–O bond the only reasonable option:

O
O

The electronegative oxygen atom is more likely to have served as the nucleophile, so we draw a negative charge on the oxygen atom to give a suitable nucleophile (a carboxylate ion). The carbon atom (at the disconnected bond) must have started out as the alkyl iodide electrophile, so we draw the leaving group (I) at that position.

was a nucleophile
O
was an electrophile
O
What starting materials are needed?
δ+
I
+
⊖O
O
Electrophile Nucleophile

The last step of the planning process is to confirm that the reaction mechanism is favorable. The carboxylate anion is electron-rich, making it a strong nucleophile. However, the negative charge is delocalized (stabilized by resonance), making it a relatively weak base (second factor in ARIO). Because it is not a strong base, the SN2 pathway does not have significant competition from the E2 pathway. With a primary substrate, we expect an SN2 process to proceed smoothly to give the desired target molecule.

Now we draw the forward process. Notice that the carboxylate ion is shown as a sodium salt. Other counterions (such as lithium or potassium) are equally acceptable:

I NaO (O) O O
Target molecule

(c) Our retrosynthesis begins with a disconnection at the cyano group, at the C–C bond:

C≡N

Cyanide (N≡C$^-$) is a stable and familiar anion, so we will use cyanide as the nucleophile. The other carbon atom (at the disconnected bond) must have started out as the alkyl iodide electrophile, so we draw the leaving group (I) at that position. To account for the required stereochemistry in the product, we must consider the mechanism of the proposed reaction. An SN2 process involves back-side attack, resulting in inversion of stereochemistry at the carbon atom bearing the leaving group. In order to furnish a product in which the nucleophile is installed on a wedge, the leaving group must have been on a dash:

was an electrophile
C≡N
was a nucleophile
What starting materials are needed?
I
δ+
+
⊖C≡N
Electrophile Nucleophile

The last step of the planning process is to confirm that the reaction mechanism is favorable. Cyanide is a strong nucleophile but not a strong base, so the SN2 pathway does not have any significant competition from the E2 pathway. With a secondary substrate, we expect an SN2

process to proceed smoothly to give the desired target molecule.
Now we draw the forward process. As shown here, NaCN can be used as the source of cyanide. Other counterions (such as lithium or potassium) are equally acceptable:

NaCN

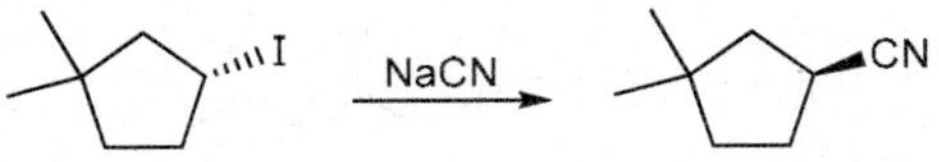

Target molecule

(d) Our retrosynthesis begins with a disconnection at the thiol group, at the C–S bond:

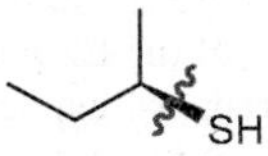

The more electronegative sulfur atom is more likely to have served as the nucleophile, so we draw a negative charge on the sulfur atom to give a suitable nucleophile (a thiolate ion). The carbon atom (at the disconnected bond) must have started out as the alkyl iodide electrophile, so we draw the leaving group (I) at that position. To account for the required stereochemistry in the product, we must consider the mechanism of the proposed reaction. An SN2 process involves back-side attack, resulting in inversion of stereochemistry at the carbon atom bearing the leaving group. In order to furnish a product in which the nucleophile is installed on a wedge, the leaving group must have been on a dash:

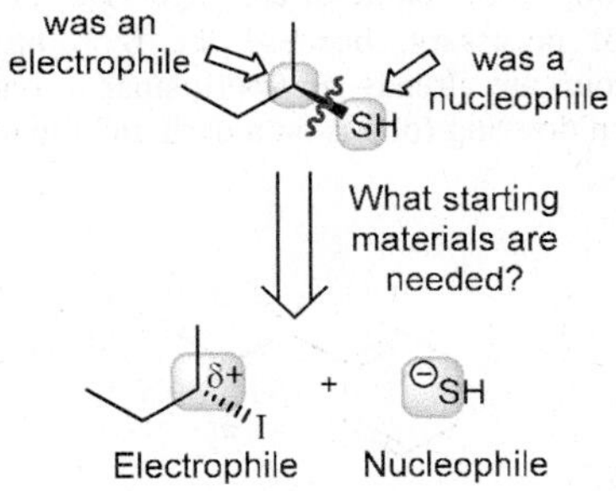

was an electrophile

was a nucleophile

What starting materials are needed?

Electrophile Nucleophile

The last step of the planning process is to confirm that the reaction mechanism is favorable. The large thiolate ion is a strong nucleophile (because it is very polarizable) but not a strong base, so the SN2 pathway does not have any significant competition from the E2 pathway. With a secondary substrate, we expect an SN2 process to proceed smoothly to give the desired target molecule.
Now we draw the forward process. As shown here, NaSH can be used as the source of thiolate (HS^-). Other counterions (such as lithium or potassium) are equally acceptable:

NaSH

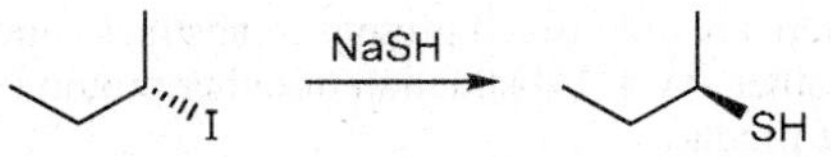

Target molecule

7.56. Each proposed method is a substitution process. In the first method, the nucleophile is a strong nucleophile (methoxide), which favors SN2, but the substrate is tertiary. SN2 reactions do not occur at tertiary substrates, so this method will not work (E2 is expected to be the major pathway under these conditions).
The second method should be efficient, because the substrate (methyl iodide) is not sterically hindered, the nucleophile is a strong nucleophile (*tert*-butoxide), and E2 is impossible with a one-carbon substrate. These conditions favor an SN2 process.

7.57. We begin by drawing the substrate and identifying the β positions (highlighted):

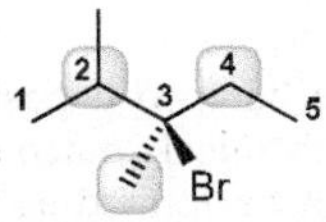

(*R*)-3-Bromo-2,3-dimethylpentane

There are three β positions, each of which contains protons, so there are three possible regiochemical outcomes. If the double bond is formed between C3 and C4, then both *E* and *Z* stereoisomers are possible, giving a total of four alkenes:

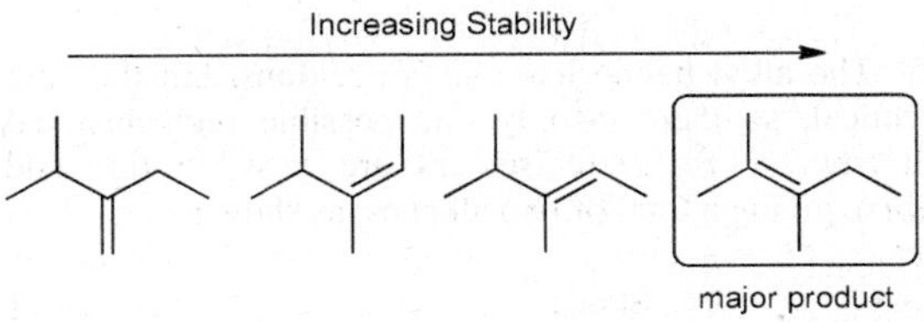

Increasing Stability

major product

The tetrasubstituted alkene is the most stable, while the disubstituted alkene is the least stable. Among the two trisubstituted alkenes, the *E* isomer is more stable, because it exhibits fewer steric interactions than the *Z* isomer.

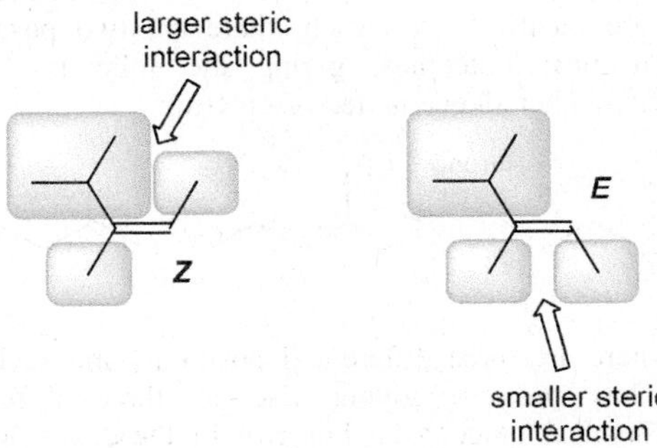

larger steric interaction

Z

E

smaller steric interaction

7.58. There are only two β protons to abstract: one at C2 and the other at C4. Abstraction of either proton leads to the same product.

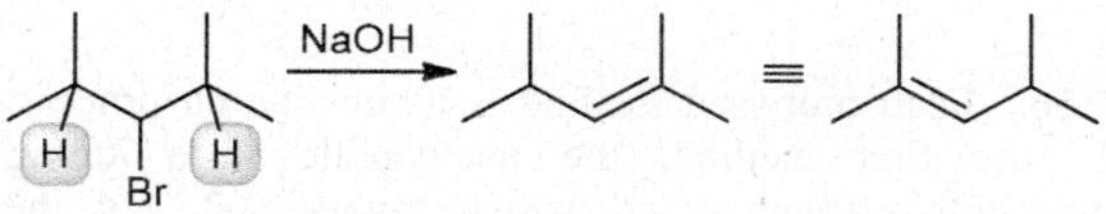

NaOH

7.59.

(a) There is only one β position, and the resulting alkene is not stereoisomeric, so only one alkene will be produced, as shown:

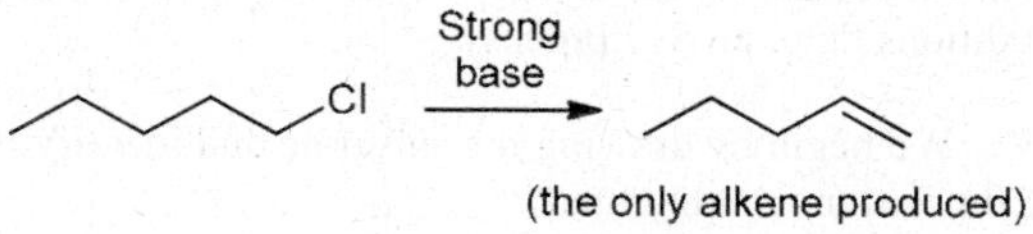

Strong base

(the only alkene produced)

(b) The alkyl halide has two β positions, so there are two possible regiochemical outcomes. The more-substituted alkene can be formed as the *E* or *Z* isomer, giving a total of three alkenes, as shown:

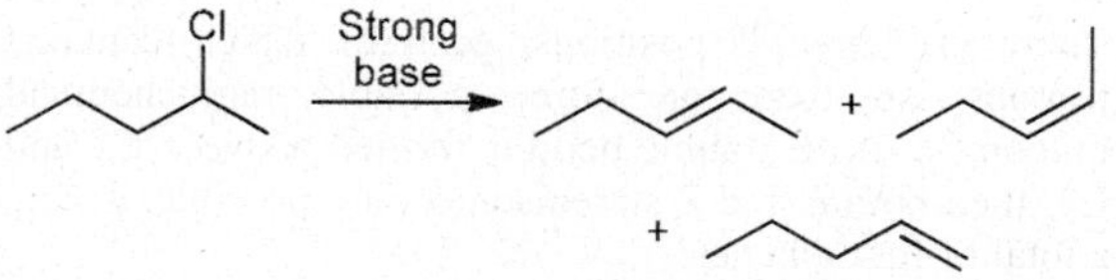

Strong base

(c) The alkyl halide has two β positions, but they are identical, so there is only one possible regiochemical outcome. Two stereoisomers are possible (*cis* and *trans*), giving a total of two alkenes, as shown:

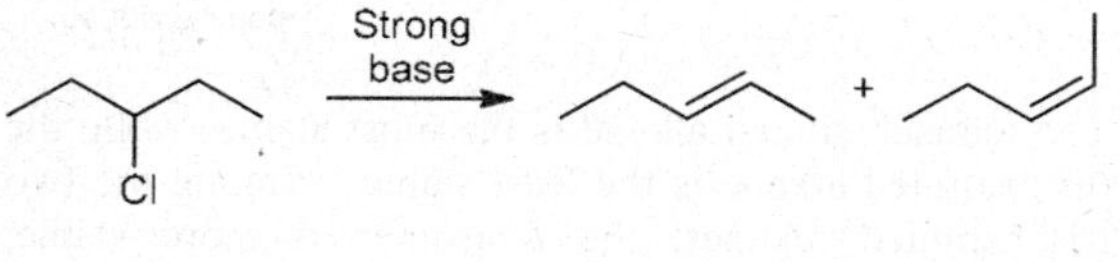

Strong base

(d) The alkyl halide has three β positions, but two of them are identical. As such, there are two possible regiochemical outcomes, giving the following two alkenes (neither alkene is stereoisomeric):

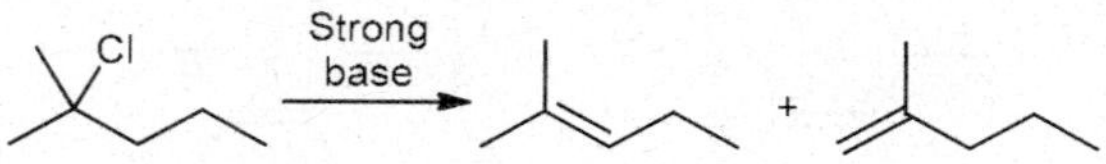

Strong base

(e) There are three different β positions, and each of them has protons, giving rise to three different regiochemical outcomes. For two of these outcomes, both *E* and *Z* isomers are possible, giving a total of five alkenes, as shown:

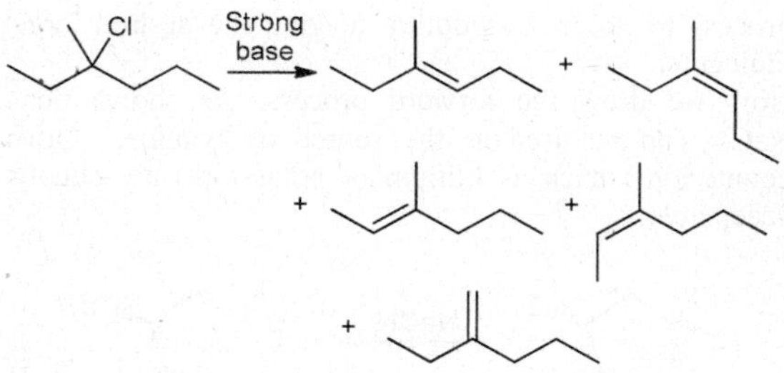

Strong base

7.60. The reagent (NaOEt) is both a strong base (E2) and a strong nucleophile (S_N2). The substrate is secondary, so we expect the major product to be obtained via an E2 process. The substrate has two β positions, both of which bear protons, so we must identify the regiochemical outcome of the E2 process. Since the base is not sterically hindered, we expect the Zaitsev product (more-substituted alkene) to be major. Formation of the Zaitsev product requires deprotonation at the following, highlighted β position:

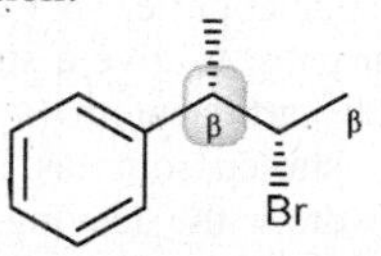

(2*S*,3*S*)-2-bromo-3-phenylbutane

This position bears only one proton, so the reaction is expected to be stereospecific. That is, only one particular stereoisomeric product will be obtained. To determine which product to expect, we can rotate the central C–C bond so as to place the β proton and the leaving group in the plane of the page. But in this case, that is not necessary, because the β proton and the leaving group are already *anti*-periplanar to one another in the given drawing (one is on a dash and the other is on a wedge):

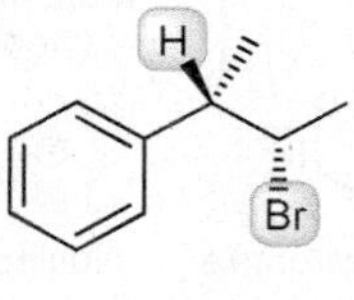

In such a case, it is relatively easy to draw the product, because the carbon skeleton is simply redrawn without the β proton and without the leaving group (with a double bond instead). Note that a double bond has planar geometry, so the methyl group is on a straight line (not a dash) in the product:

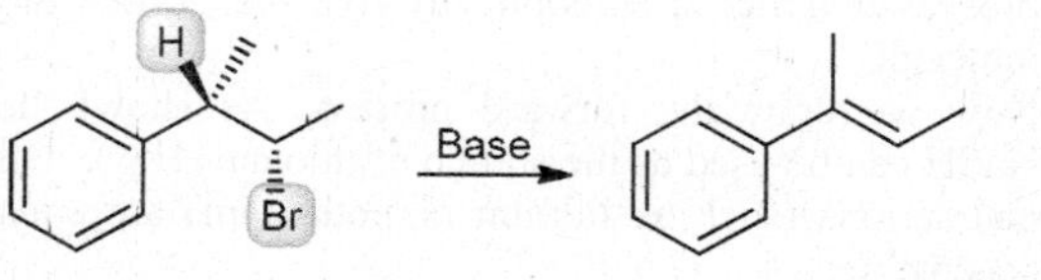

Base

7.61.

(a) The problem statement indicates that the major product is obtained via an E2 process. The substrate has two β positions, both of which bear protons, so we must identify the regiochemical outcome of the E2 process. Since the base is not sterically hindered, we expect the Zaitsev product (the more-substituted alkene) to be major. Formation of the Zaitsev product requires deprotonation at the following, highlighted β position:

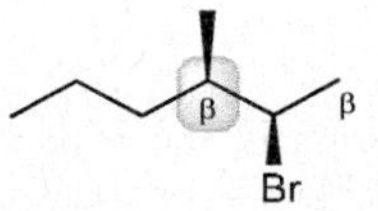

This position bears only one proton, so the reaction is expected to be stereospecific. That is, only one particular stereoisomeric product will be obtained. To determine which product to expect, we can rotate the central C–C bond so as to place the β proton and the leaving group in the plane of the page. But in this case, that is not necessary, because the β proton and the leaving group are already *anti*-periplanar to one another in the given drawing (one is on a dash and the other is on a wedge):

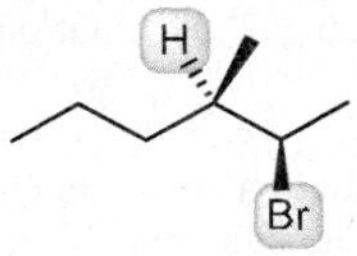

In such a case, it is relatively easy to draw the product, because the carbon skeleton is simply redrawn without the β proton and without the leaving group (with a double bond instead). Note that a double bond has planar geometry, so the methyl group is on a straight line (not a wedge) in the product:

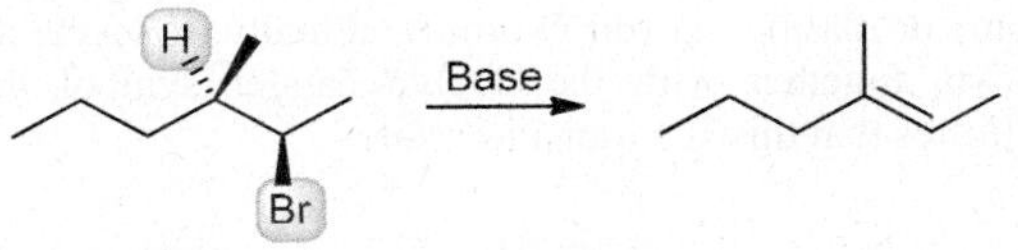

(b) The problem statement indicates that the major product is obtained via an E2 process. There are three β positions, but only two of them bear protons, so there are two possible regiochemical outcomes. Since the base is not sterically hindered, we expect the Zaitsev product (the more-substituted alkene) to be major. Formation of the Zaitsev product requires deprotonation at the following, highlighted β position:

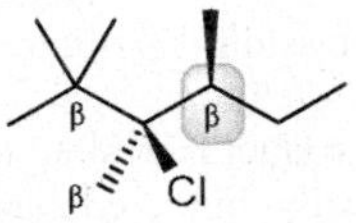

This position bears only one proton, so the reaction is expected to be stereospecific. That is, only one particular stereoisomeric product will be obtained. To determine which product to expect, we can rotate the central C–C bond so as to place the β proton and the leaving group in the plane of the page. But in this case, that is not necessary, because the β proton and the leaving group are already *anti*-periplanar to one another in the given drawing (one is on a dash and the other is on a wedge):

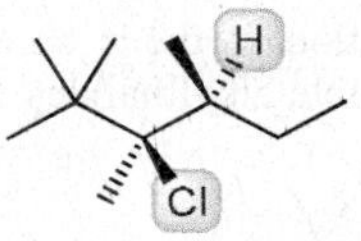

In such a case, it is relatively easy to draw the product, because the carbon skeleton is simply redrawn without the β proton and without the leaving group (with a double bond instead). Note that a double bond has planar geometry, so the methyl groups are drawn on straight lines in the product:

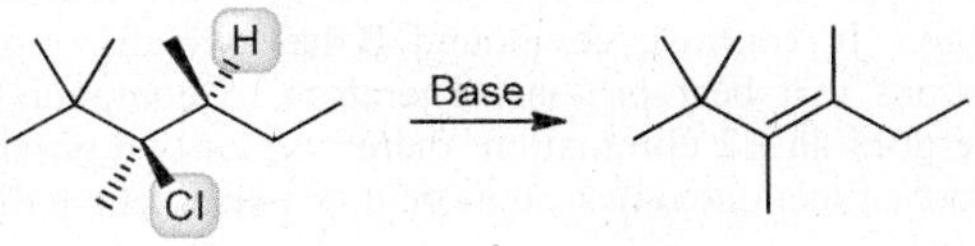

7.62. The reagent is a strong nucleophile (SN2) and a strong base (E2), so we expect a bimolecular reaction. The substrate is tertiary so only E2 can operate (SN2 is too sterically hindered to occur). There is only one possible regiochemical outcome for the E2 process, because the other β positions lack protons.

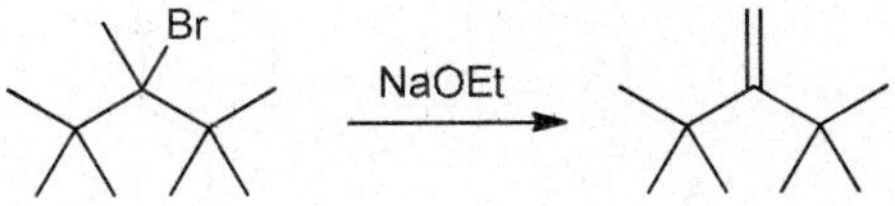

7.63.

(a) The reagent is hydroxide, which is both a strong nucleophile (SN2) and a strong base (E2). The substrate is secondary, so we expect the E2 pathway to predominate. There are two β positions that bear protons, so there are two possible regiochemical outcomes for an E2 process. The base is not sterically hindered, so the major product will be the more-substituted alkene (the Zaitsev product). Two stereoisomers are possible (*cis* and *trans*), and the more stable *trans* isomer is favored:

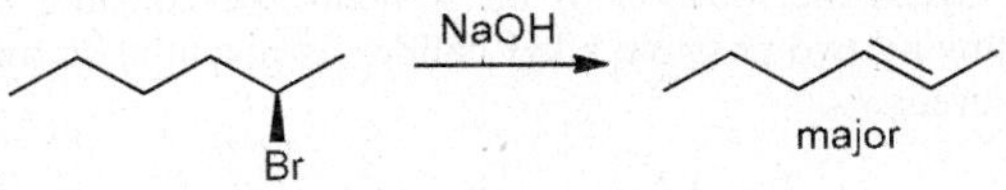

(b) The reagent is *tert*-butoxide, which is a strong, sterically hindered base. The substrate is secondary so we expect E2 processes to predominate (SN2 is highly disfavored because of steric interactions). There are two β positions that bear protons, so there are two possible regiochemical outcomes. The base is sterically hindered, so the major product will be the less-substituted alkene (the Hofmann product):

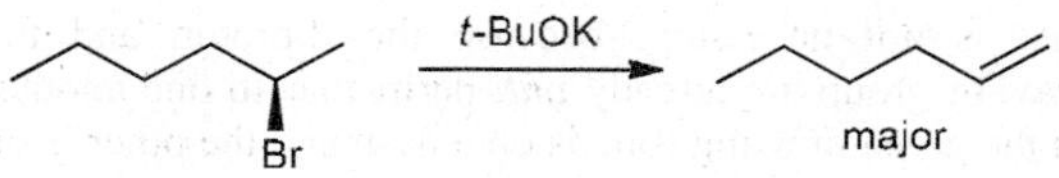

7.64.

(a) Given the location of the π bond, we consider the following two possible alkyl halides as potential starting materials.

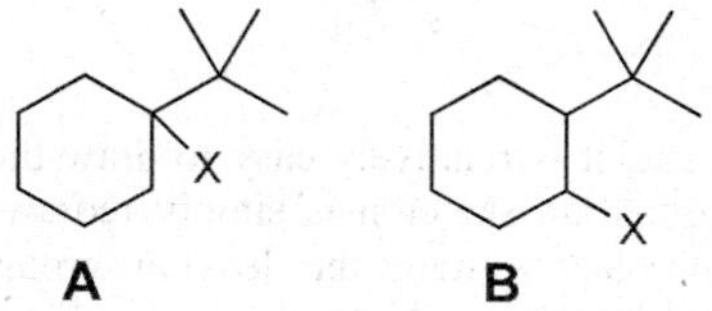

Compound **A** has three β positions, but only two of them bear protons, and those two positions are identical. Deprotonation at either location will result in the desired alkene. In contrast, compound **B** has two different β positions that bear protons. Therefore, if compound **B** undergoes an E2 elimination, there will be two possible regiochemical outcomes, so more than one alkene will be formed.

Compound **A** (where X = Cl, Br or I) is the alkyl halide that will undergo an E2 elimination to give only the desired alkene.

(b) Given the location of the π bond, we consider the following two possible alkyl halides as potential starting materials.

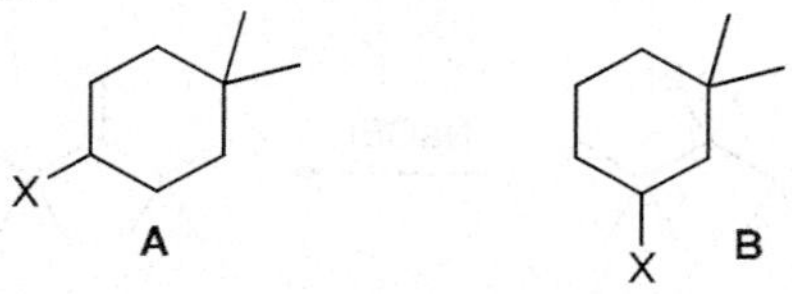

Compound **A** has two β positions, and those two positions are identical. Deprotonation at either location will result in the desired alkene. In contrast, compound **B** has two different β positions that bear protons. Therefore, if compound **B** undergoes an E2 elimination, there will be two possible regiochemical outcomes, so more than one alkene will be formed.

Compound **A** (where X = Cl, Br or I) is the alkyl halide that will undergo an E2 elimination to give only the desired alkene.

(c) Given the location of the π bond, we consider the following two possible alkyl halides as potential starting materials.

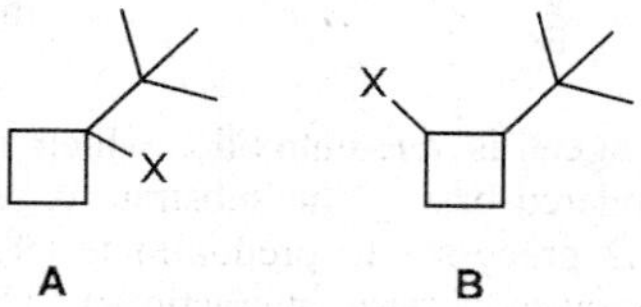

Compound **A** has three β positions, but only two of them bear protons, and those two positions are identical. Deprotonation at either location will result in the desired alkene. In contrast, compound **B** has two different β positions that bear protons. Therefore, if compound **B** undergoes an E2 elimination, there will be two possible regiochemical outcomes, so more than one alkene will be formed.

Compound **A** (where X = Cl, Br or I) is the alkyl halide that will undergo an E2 elimination to give only the desired alkene.

(d) Given the location of the π bond, we consider the following two possible alkyl halides as potential starting materials.

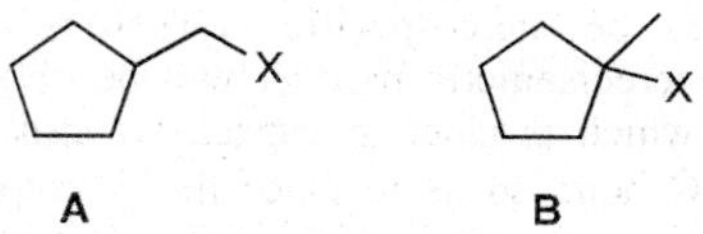

Compound **A** has only one β position, giving rise to only one alkene. In contrast, compound **B** has more than one β position, giving rise to more than one alkene.

Compound **A** (where X = Cl, Br or I) is the alkyl halide that will undergo an E2 elimination to give only the desired alkene.

7.65. Hydroxide is both a strong nucleophile (S_N2) and a strong base (E2). The substrate is tertiary, so we expect an E2 process only. In the transition state, the hydroxide ion is in the process of removing the proton, the double bond is in the process of forming, and the leaving group is in the process of leaving. We use dotted lines to represent partial bonds (those bonds that are either in the process of being formed or being broken), and we use δ– symbols to represent partial negative charges, indicating that these charges are in the process of forming (on Cl atom) or dissipating (on O atom). Finally, brackets are drawn, together with the double dagger symbol that indicates that this is a transition state:

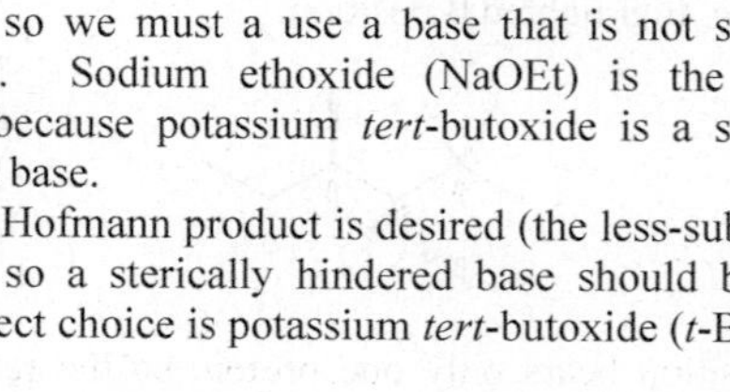

7.66.

(a) The Zaitsev product is desired (the more-substituted alkene), so we must a use a base that is not sterically hindered. Sodium ethoxide (NaOEt) is the correct choice, because potassium *tert*-butoxide is a sterically hindered base.

(b) The Hofmann product is desired (the less-substituted alkene), so a sterically hindered base should be used. The correct choice is potassium *tert*-butoxide (*t*-BuOK).

7.67. π bonds cannot be formed at the bridgehead of a bicyclic compound, unless one of the rings is large (at least eight carbon atoms). This rule is known as Bredt's rule. Thus, in this case, no reaction would be expected.

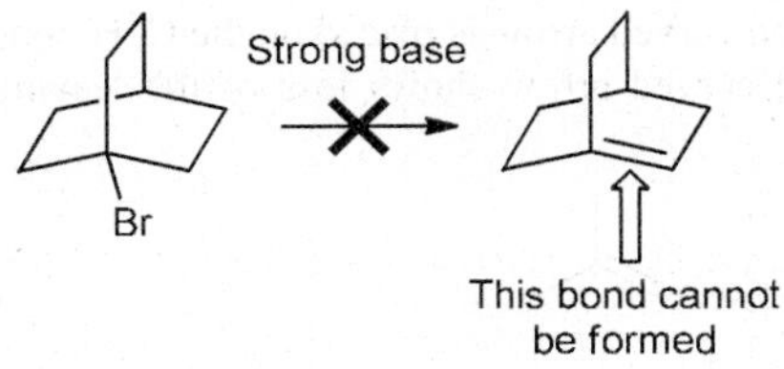

7.68. For substituted cyclohexanes, an E2 reaction will occur if the leaving group and the β proton can achieve *anti*-periplanarity. In order to achieve this, one must be on a wedge and the other must be on a dash. The leaving group (Br^-) is on a wedge. Therefore, we are looking for a β proton that is on a dash. In this case, there is only one such proton, highlighted below, so there is only one possible regiochemical outcome. In this case, the Hofmann product (the less-substituted alkene) is formed regardless of the choice of base.

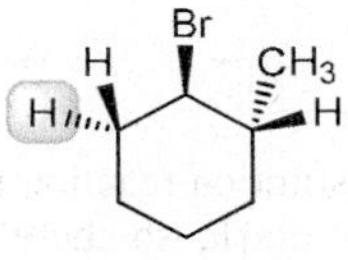

7.69.

(a) This is an acid-catalyzed dehydration reaction, which is an E1 process. In acidic conditions, the OH group is protonated, which converts it from a bad leaving group to a good leaving group. In aqueous sulfuric acid, the acid that is present in solution is H_3O^+ (because of the leveling effect, as explained in Section 3.6), so we use H_3O^+ as the proton source in the first step of the mechanism. The next two steps of the mechanism constitute the core steps of an E1 process: (i) loss of a leaving group (H_2O), which requires one curved arrow, and (ii) removal of the β proton, which requires two curved arrows, as shown:

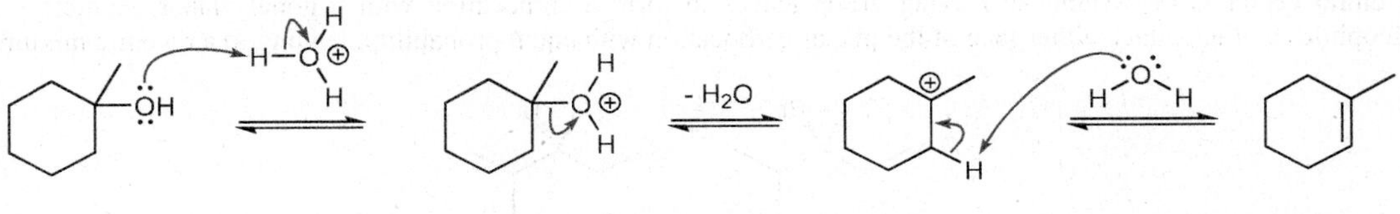

(b) This is an acid-catalyzed dehydration reaction, which is an E1 process. In acidic conditions, the OH group is protonated, which converts it from a bad leaving group to a good leaving group. In aqueous sulfuric acid, the acid that is present in solution is H_3O^+ (because of the leveling effect, as explained in Section 3.6), so we use H_3O^+ as the proton source in the first step of the mechanism. Loss of the leaving group gives a secondary carbocation that can undergo a rearrangement (a methyl shift), giving a more stable, tertiary carbocation. Finally, water serves as the base that deprotonates the carbocation to generate the product:

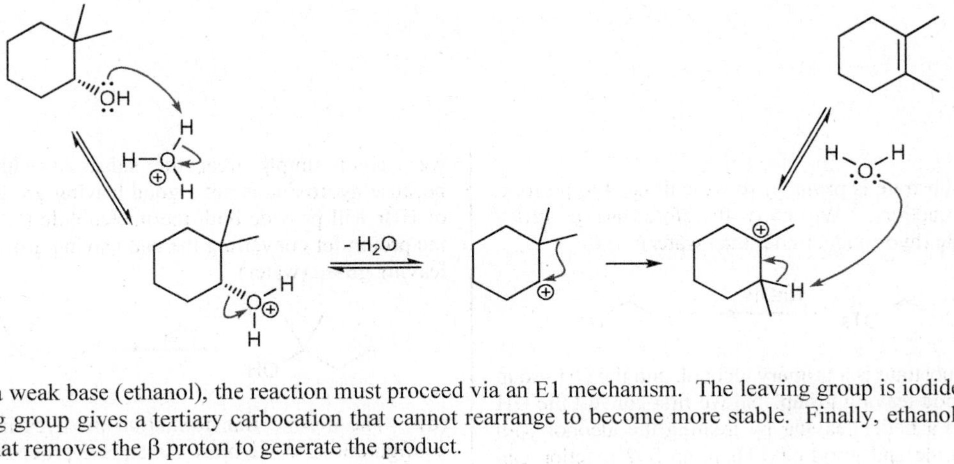

(c) With a weak base (ethanol), the reaction must proceed via an E1 mechanism. The leaving group is iodide. Loss of the leaving group gives a tertiary carbocation that cannot rearrange to become more stable. Finally, ethanol serves as the base that removes the β proton to generate the product.

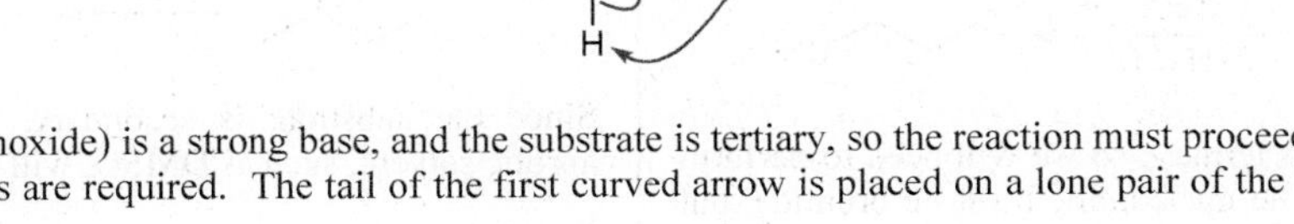

(d) The reagent (ethoxide) is a strong base, and the substrate is tertiary, so the reaction must proceed via an E2 process. Three curved arrows are required. The tail of the first curved arrow is placed on a lone pair of the base (ethoxide) and the head is placed on the proton that is removed. The tail of the second curved arrow is placed on the C–H bond that is breaking, and the head shows formation of the π bond. The third curved arrow shows loss of the leaving group (iodide), as shown here.

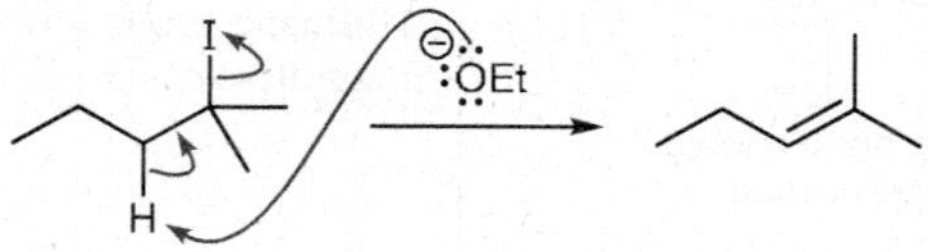

7.70. This is a substitution reaction that takes place at a tertiary center, so it must be an S_N1 process (steric hindrance prohibits back-side attack, so there is no S_N2 mechanism). The substrate is an alcohol, so acidic conditions are employed so that the OH group can be protonated, rendering it a better leaving group. Then, loss of a leaving group generates a carbocation, which is then attacked by the bromide nucleophile to give the substitution product. Notice that the mechanism is comprised of a proton transfer, followed by the two core steps of an S_N1 process (loss of a leaving group and nucleophilic attack).

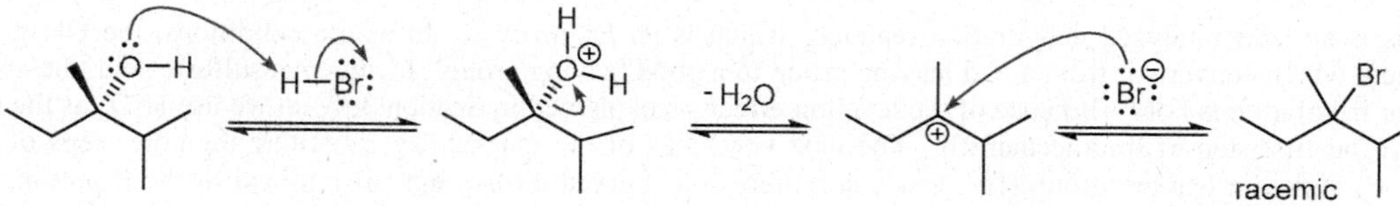

The chiral center is lost when the leaving group leaves to form a carbocation with trigonal planar geometry. The nucleophile can then attack either face of the planar carbocation with equal probability, leading to a racemic mixture:

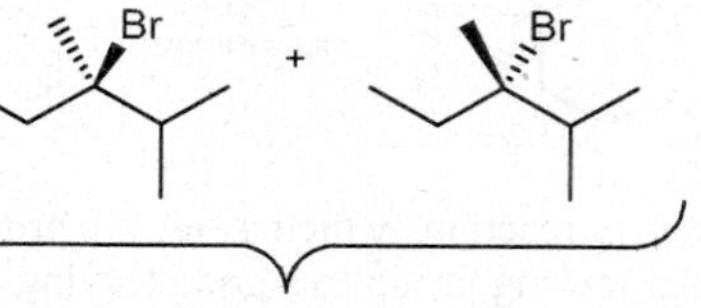

7.71.

(a) The substrate is primary, so we will need to perform an S_N2 reaction. We must therefore use a strong nucleophile (hydroxide, rather than water).

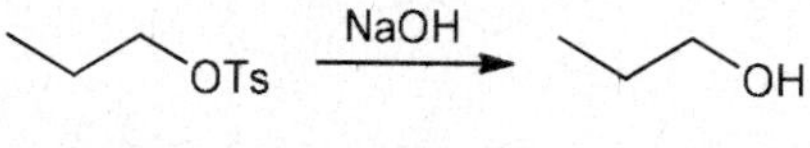

(b) The substrate is a primary alcohol, and the OH group is not a good leaving group. So we first convert the OH group into a better leaving by treating the alcohol with tosyl chloride and pyridine. Then, an S_N2 reaction can be performed (since the substrate is primary) with cyanide as the nucleophile, giving the desired product.

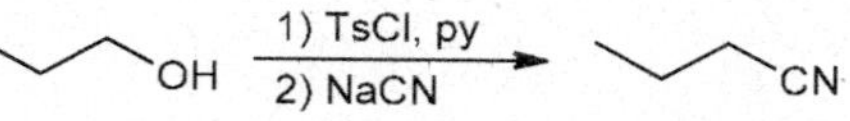

(c) The substrate is tertiary, so we will need to perform an S_N1 reaction. The nucleophile must be bromide, but we cannot simply treat the substrate with bromide, because hydroxide is not a good leaving group. The use of HBr will provide both the nucleophile (bromide) and the proton for converting the bad leaving group to a good leaving group (water).

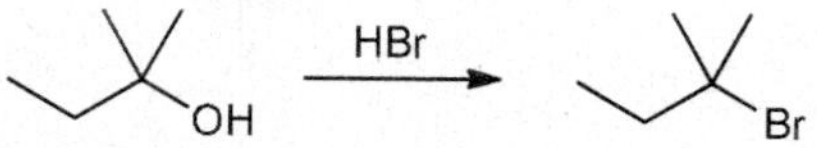

(d) The desired transformation involves inversion of configuration, so we must use an S_N2 process. As such, we want to use a strong nucleophile (HS^-).

Since the substrate is secondary, the use of a polar aprotic solvent, such as DMSO, will be helpful.

7.72. Solvolysis is expected to afford both S_N1 and E1 products, via a carbocation intermediate. The first compound is a tertiary substrate. The second compound is a tertiary allylic substrate (the leaving group is adjacent to a π bond). The allylic substrate will undergo solvolysis reactions more rapidly because a tertiary allylic carbocation is more highly stabilized than a tertiary carbocation (because of resonance, shown here).

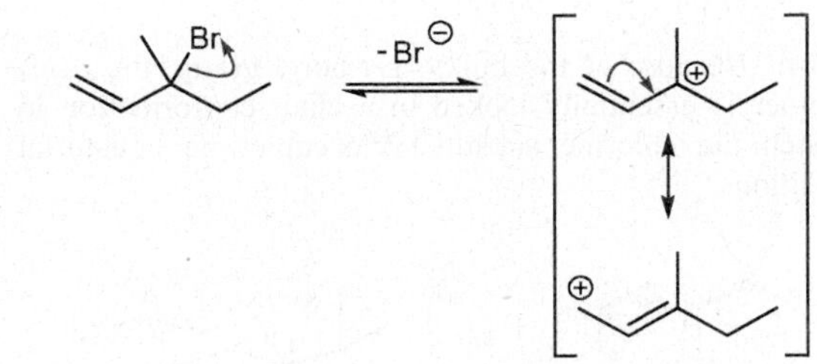

7.73.
(a) We begin by drawing a Newman projection, and we find that the front carbon atom bears the leaving group (bromide), while the back carbon atom bears two β protons, either of which can be removed. The following two Newman projections represent the two conformations in which a β proton is *anti*-periplanar to the leaving group. In the first Newman projection, the phenyl groups (highlighted) are *anti* to each other, so the transition state is not expected to exhibit a steric interaction between the phenyl groups. In contrast, the second Newman projection exhibits a gauche interaction between the two phenyl groups (highlighted), so the transition state is expected to exhibit a steric interaction. Therefore, *trans*-stilbene is formed at a faster rate than *cis*-stilbene (because formation of the less stable *cis*-stilbene involves a higher energy transition state).

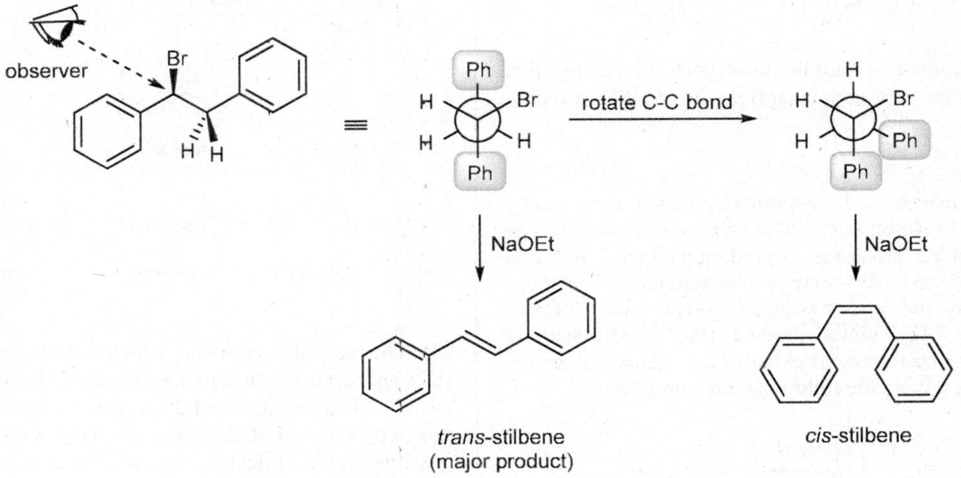

trans-stilbene
(major product)

cis-stilbene

(b) The same argument, as seen in the solution to part (a) of this problem, can be applied again in this case. That is, there are still two β protons that can be abstracted in a β elimination, so both products are still possible, as shown below.

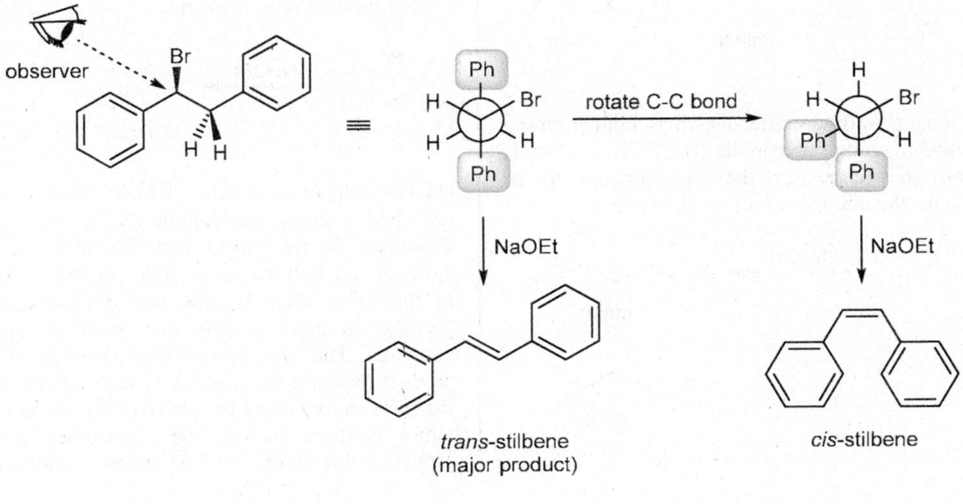

trans-stilbene
(major product)

cis-stilbene

7.74. Because of the bulky *tert*-butyl group, the *trans* isomer is essentially locked in a chair conformation in which the chlorine substituent occupies an equatorial position.

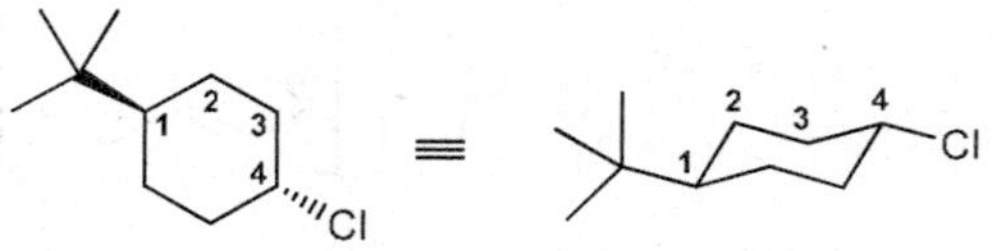

This conformation cannot readily undergo an E2 reaction because the leaving group is not *anti*-periplanar to a β proton (in order to achieve *anti*-periplanarity, the H and Cl must be *trans* diaxial). However, the *cis* isomer is locked in a chair conformation in which the chlorine occupies an axial position:

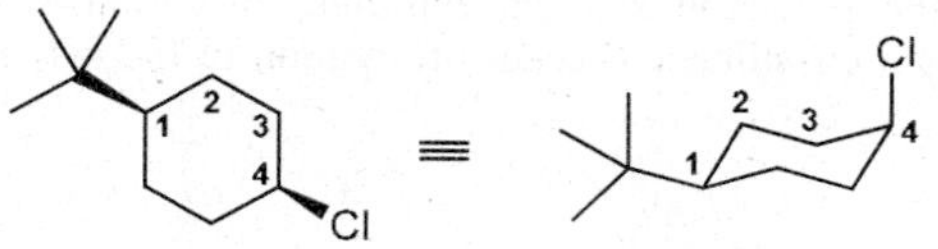

This conformation rapidly undergoes an E2 reaction, because it has *anti*-periplanar β protons readily available.

7.75.

(a) The reagent is *tert*-butoxide, which is a strong, sterically hindered base. The substrate is secondary so we expect E2 processes to predominate (SN2 is highly disfavored because of steric interactions). The major product is the less-substituted alkene (the Hofmann product). The minor product (the more-substituted alkene) can be formed as either of two stereoisomers (*cis* and *trans*), giving the following three products:

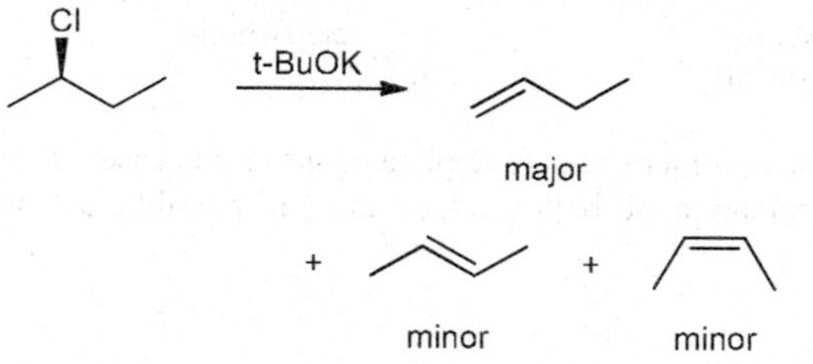

t-BuOK

major

minor minor

(b) The reagent is hydroxide, which is both a strong base (E2) and a strong nucleophile (SN2). The substrate is primary, so we expect the SN2 process to be responsible for the major product, as shown here.

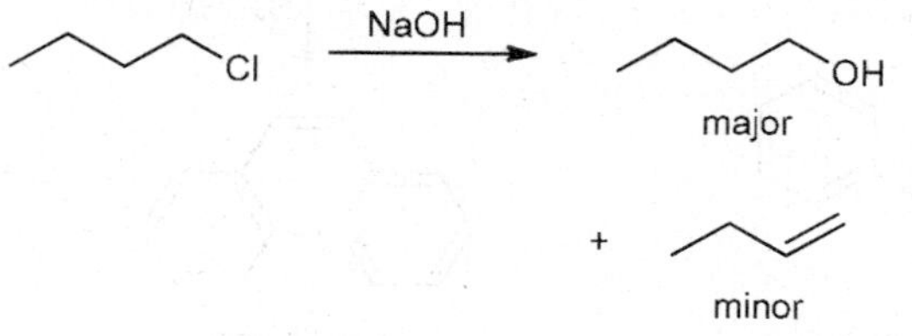

NaOH

major

minor

In this case, there are no stereochemical considerations for either product. There is no observed inversion of configuration in the SN2 product, because there is no chiral center, and the alkene product is not stereoisomeric.

(c) The reagent is ethoxide, which is both a strong base (E2) and a strong nucleophile (SN2). The substrate is secondary so we expect both E2 and SN2 processes, although E2 will be responsible for the major product. Because the base is not sterically hindered, the major product is the more-substituted alkene (the Zaitsev product), with the *trans* configuration (because the reaction is stereoselective, favoring the more stable *trans* isomer over the *cis* isomer). The minor products include the *cis* isomer, the less-substituted alkene, and the SN2 product (with the expected inversion of configuration):

NaOEt

major

minor minor minor

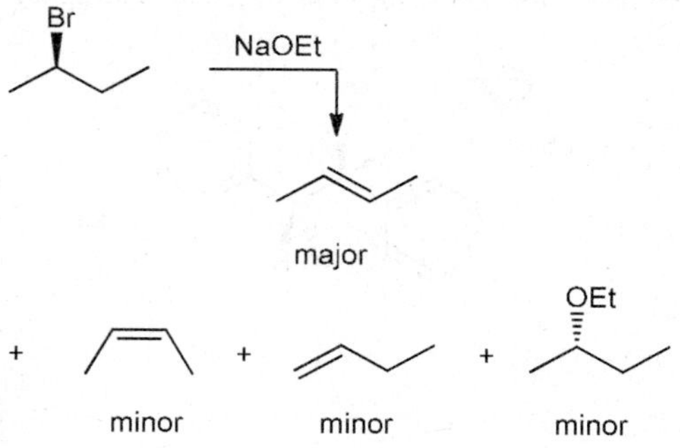

(d) The reagent is ethoxide, which is both a strong base (E2) and a strong nucleophile (SN2). The substrate is tertiary, so we expect an E2 process. There are three β positions, but two of them are identical, so there are two possible regiochemical outcomes. Since the base is not sterically hindered, we expect that the more-substituted alkene (the Zaitsev product) will be the major product, as shown. The less-substituted alkene is the minor product. There are no stereochemical considerations, because neither product is stereoisomeric.

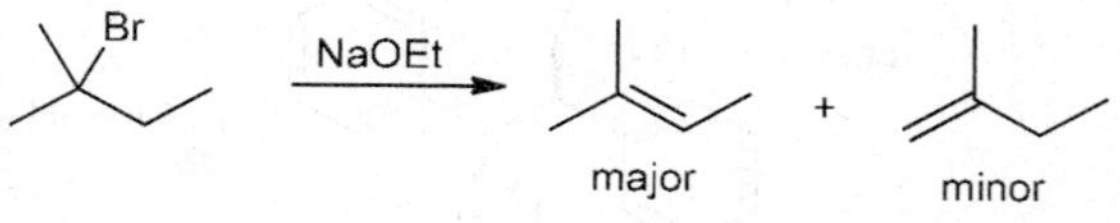

NaOEt

major minor

(e) The reagent is ethoxide, which is both a strong base (E2) and a strong nucleophile (SN2). The substrate is secondary so we expect both E2 and SN2 processes, although E2 will be responsible for the major product. In this case, there is only one β position that bears protons, so there is only one possible regiochemical outcome. The *trans* isomer is expected to be the major product (because the reaction is stereoselective, favoring the more stable *trans* isomer over the *cis* isomer). The minor products include the *cis* isomer and the SN2 product (with the expected inversion of configuration):

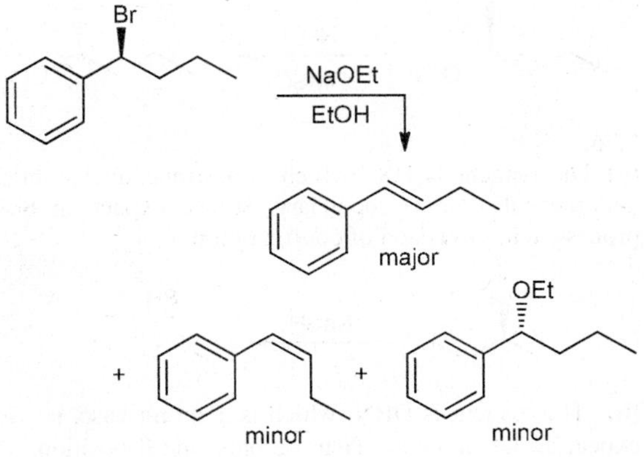

(f) The reagent is ethoxide, which is both a strong base (E2) and a strong nucleophile (S$_N$2). The substrate is secondary so we expect both E2 and S$_N$2 processes, although E2 will be responsible for the major product. In this case, there is only one β position that bears a proton,

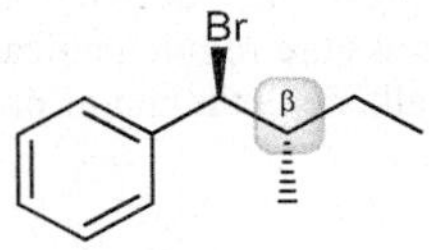

so there is only one possible regiochemical outcome for the E2 process.

Moreover, since there is only one β proton, the E2 reaction is expected to be stereospecific. That is, only one particular stereoisomeric product will be obtained. To determine which product to expect, we must rotate the central C–C bond so that the β proton is *anti*-periplanar to the leaving group. We will do so in two stages. First, we rotate the central C–C bond in a manner that places the β proton in the plane of the page (rather than on a wedge):

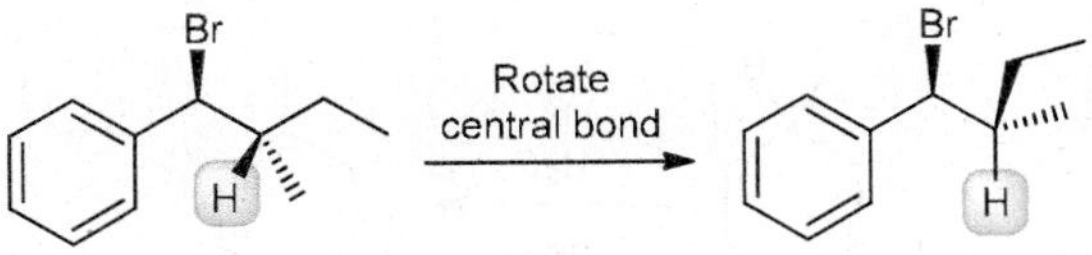

Then, we rotate the central C–C bond again, in a manner that places the leaving group (Br) in the plane of the page:

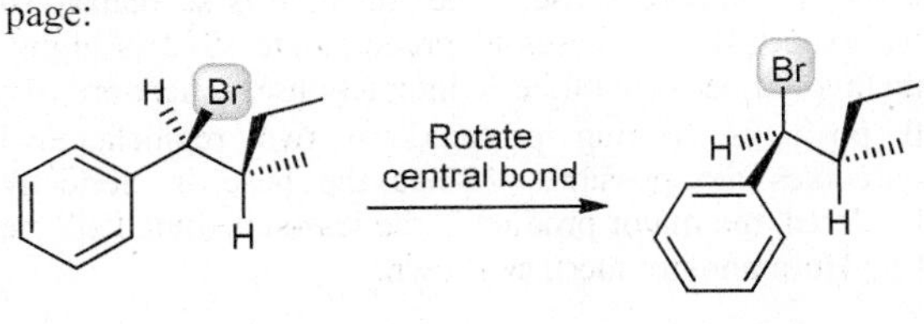

In this conformation, the proton and the leaving group are *anti*-periplanar. To draw the product, use the wedges and dashes as guides. In this case, the phenyl group and the ethyl group are both on wedges, so they will be *cis* to each other in the product:

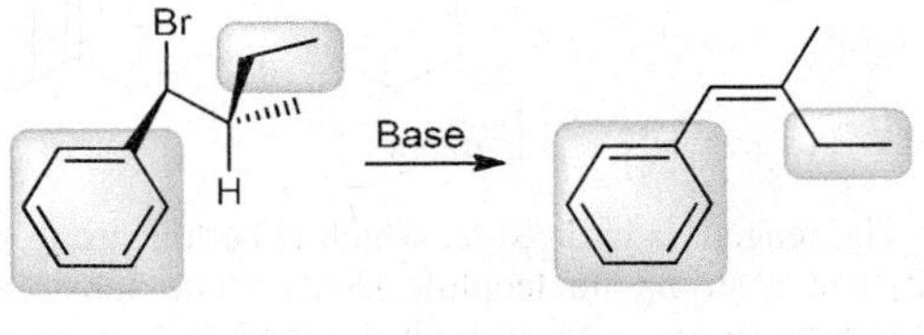

The minor product is formed via an S$_N$2 process (with inversion of configuration).
In summary, the following products are expected:

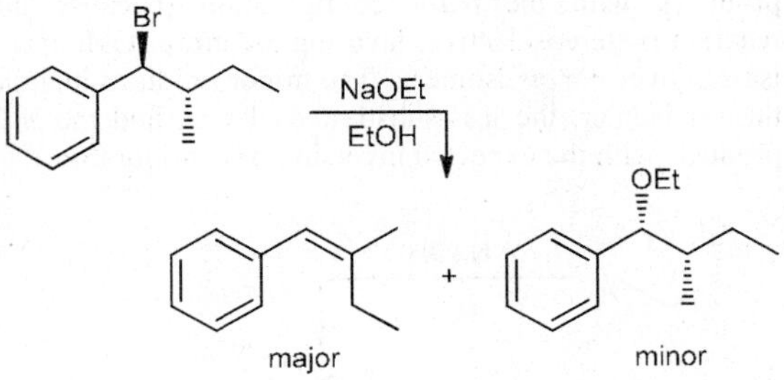

(g) The reagent is *tert*-butoxide, which is a strong base (E2). The substrate is tertiary, so we expect only E2 (no S$_N$2). The substrate has three β positions that bear protons, but two of them are identical, giving rise to two possible regiochemical outcomes for the E2 process. Since the base (*tert*-butoxide) is sterically hindered, we expect that the major product will be the less-substituted alkene (the Hofmann product), and the minor product will be the more-substituted alkene. The Zaitsev product is formed as a mixture of *cis* and *trans* stereoisomers, giving a total of three products, shown here:

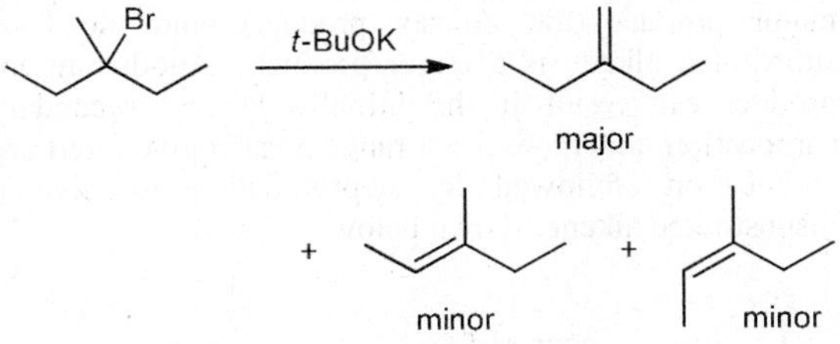

(h) The reagent is methoxide, which is both a strong base (E2) and a strong nucleophile (S$_N$2). The substrate is tertiary so we expect only E2 (no S$_N$2). The substrate has three β positions that bear protons, but two of them are identical, giving rise to two possible regiochemical outcomes. Since the base (methoxide) is not sterically hindered, we expect that the major product will be the more-substituted alkene (the Zaitsev product) as the more stable *E* isomer, because the process is stereoselective. The minor products include the *Z* isomer and the less-substituted alkene:

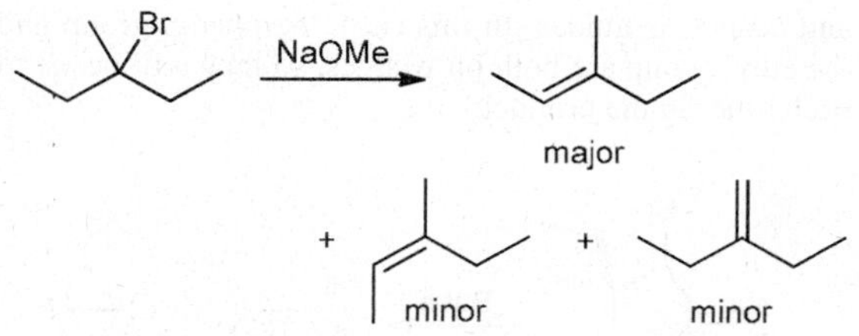

Br

NaOMe

major

+ minor + minor

(i) The reagent is hydroxide, which is both a strong base (E2) and a strong nucleophile (S$_N$2). The substrate is secondary so we expect both E2 and S$_N$2 processes, although E2 will be responsible for the major product. Because the base is not sterically hindered, the major product is the more-substituted alkene (the Zaitsev product), with the *trans* configuration (because the reaction is stereoselective, favoring the more stable *trans* isomer over the *cis* isomer). The minor products include the *cis* isomer, the less-substituted alkene, and the S$_N$2 product (with the expected inversion of configuration):

Cl

NaOH

major

+ minor + minor + OH minor

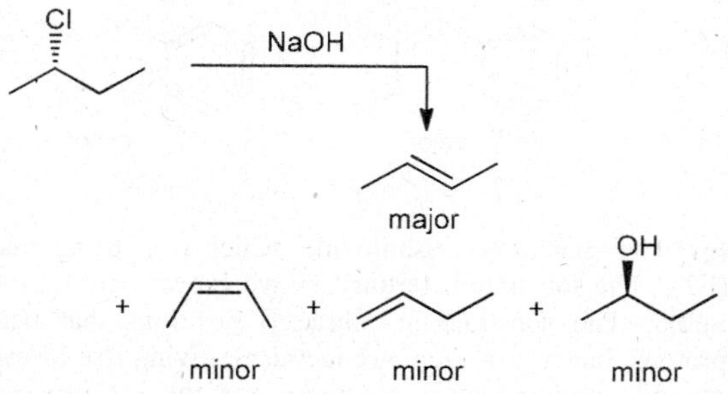

(j) Treatment of an alcohol with concentrated sulfuric acid results in a dehydration reaction (an E1 process) to afford an alkene (or a mixture of alkenes). In this case, there are two different β positions that bear protons, so there are two possible regiochemical outcomes. As with all E1 mechanisms, the more-substituted alkene is the major product (the Zaitsev product), and the less-substituted alkene is a minor product. Another minor product can result if the initially formed secondary carbocation undergoes a rearrangement to give a tertiary carbocation, followed by deprotonation to give a disubstituted alkene, shown below:

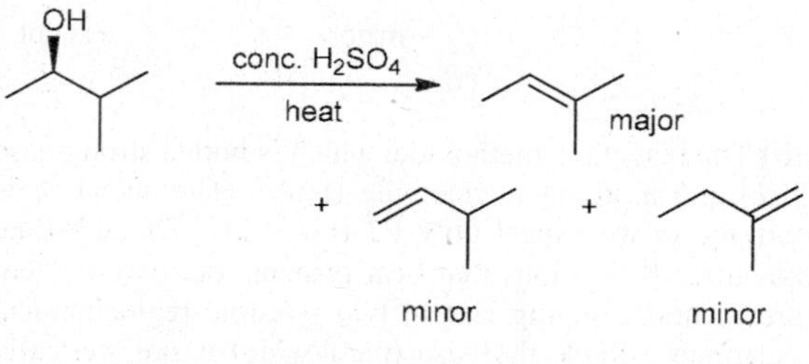

OH

conc. H_2SO_4

heat

major

+ minor + minor

(k) The reagent is chloride, which functions as a nucleophile, so we expect a substitution reaction (we don't expect elimination to occur, because chloride is a very weak base). The substrate is secondary and the solvent is polar aprotic, indicating an S$_N$2 process. As such, we expect inversion of configuration, as shown:

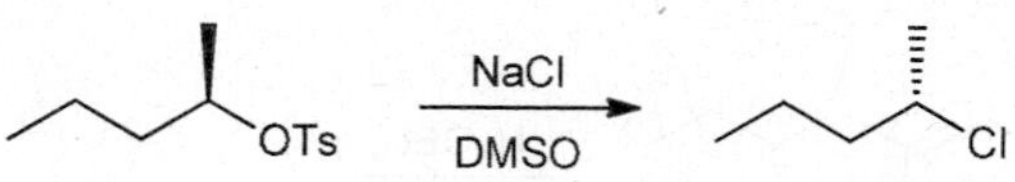

OTs

NaCl

DMSO

Cl

7.76.

(a) The reagent is HS$^-$, which is a strong nucleophile, and the substrate is secondary, so we expect an S$_N$2 process, with inversion of configuration:

Cl

NaSH

SH

(b) The reagent is DBN, which is a strong base, so we expect an E2 process. There is only one β position, so only one regiochemical outcome is possible:

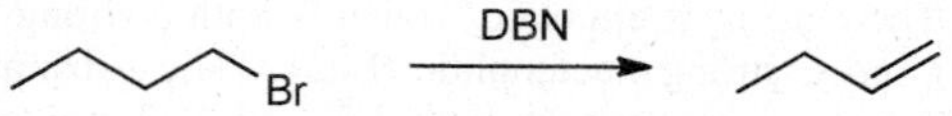

Br

DBN

(c) The reagent is hydroxide, which is both a strong base (E2) and a strong nucleophile (S$_N$2). The substrate is tertiary, so we expect only an E2 process (no S$_N$2). There are three β positions, but two of them are identical, so there are two possible regiochemical outcomes. The more-substituted alkene is the major product (the Zaitsev product):

I

NaOH

H_2O

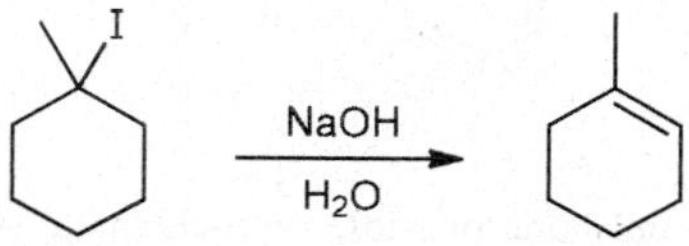

(d) The reagent is water, which is both a weak base and a weak nucleophile, so we expect E1 and S$_N$1 processes. Furthermore, the substrate is tertiary, so the expected carbocation is stable and the solvolysis reaction is reasonable. One of the alkene products is trisubstituted, so we expect the E1 pathway to predominate. For the E1 pathway, two regiochemical outcomes are possible. The base is not sterically hindered, so the more-substituted alkene is the major product (the Zaitsev product), as shown:

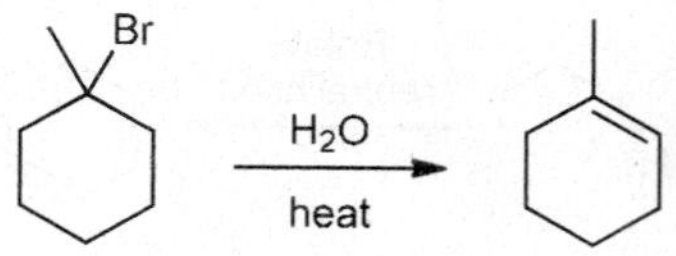

Br

H_2O

heat

(e) The reagent is *tert*-butoxide, which is a strong, sterically hindered base. The substrate is secondary so we expect E2 processes to predominate (S$_N$2 is highly disfavored because of steric interactions). There are two β positions bearing protons, so two regiochemical outcomes are possible. Since the base is sterically hindered, the major product is the less-substituted alkene (the Hofmann product), as shown:

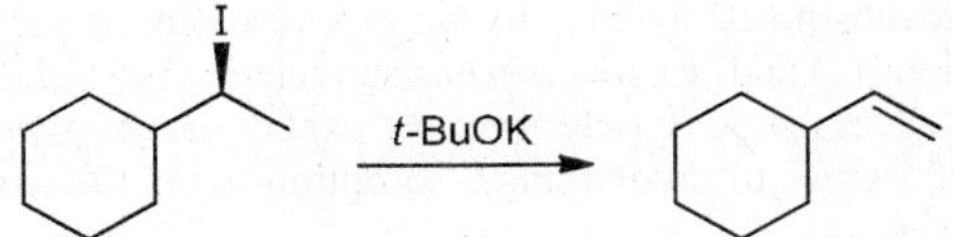

(f) The reagent is methoxide, which is both a strong base and a strong nucleophile. The substrate is tertiary, so we expect an E2 process. There are three β positions, but two of them are identical, so there are two possible regiochemical outcomes. The more-substituted alkene is the major product (the Zaitsev product), as shown:

Br NaOMe / MeOH

(g) The reagent is *tert*-butoxide, which is a strong, sterically hindered base. The substrate is tertiary, so we expect an E2 process. There are three β positions, but two of them are identical, so there are two possible regiochemical outcomes. Since the base is sterically hindered, we expect that the less-substituted alkene will be the major product (the Hofmann product), as shown:

Br *t*-BuOK

7.77.

(a) A polar aprotic solvent is used, and the reaction occurs with inversion of configuration. These factors indicate an SN2 process.

In an SN2 process, nucleophilic attack and loss of the leaving group occur in a concerted fashion (in one step), as shown below:

Br ⊖:C≡N CN

(b) Since the reaction is an SN2 process, we expect a second-order rate equation that is linearly dependent on both the concentration of the substrate and the concentration of the nucleophile.

Rate = *k* [Br] [NaCN]

(c) As given in the solution to part (b) of this problem, the rate of an SN2 reaction is linearly dependent on the concentration of the nucleophile. As such, if the concentration of the nucleophile (cyanide) is doubled, the reaction rate is expected to double.

(d) As seen in the solution to part (a) of this problem, the reaction occurs via an SN2 process, which is comprised of one concerted step (in which the nucleophile attacks with simultaneous loss of the leaving group). As such, there is only one transition state, and the energy diagram will have only one maximum (only one hump).

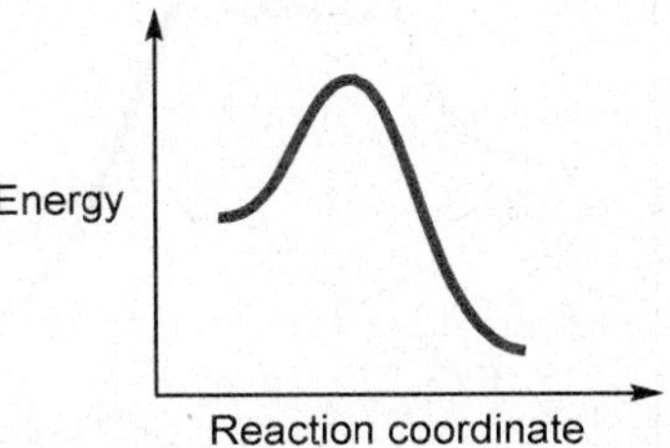

7.78.

(a) The reagent (ethoxide) is a strong base, and the substrate is tertiary, so the reaction must proceed via an E2 process. In this case, there are two possible regiochemical outcomes (Zaitsev or Hofmann), and we expect the Zaitsev product (the more-substituted alkene) to be the major product, because ethoxide is not a sterically hindered base:

Br NaOEt

2-Bromo-2-methylhexane — Zaitsev product (major)

The mechanism for this E2 process must involve three curved arrows. The tail of the first curved arrow is placed on a lone pair of the base (ethoxide) and the head is placed on the proton that is removed. The tail of the second curved arrow is placed on the C–H bond that is breaking, and the head shows formation of the π bond. The third curved arrow shows loss of the leaving group (bromide), as shown here.

Br H ⊖:ÖEt

(b) For an E2 process, the rate is dependent on the concentrations of the substrate and the base:

Rate = *k* [alkyl halide] [NaOEt]

(c) As given in the solution to part (b) of this problem, the rate of an E2 reaction is linearly dependent on the concentration of the base (NaOEt). As such, if the concentration of base is doubled, the rate will be doubled.

(d) The mechanism has one step and one transition state, so the energy diagram must have only one maximum (only one hump). The products are lower in energy than the reactants, in part because bromide is more stable than ethoxide.

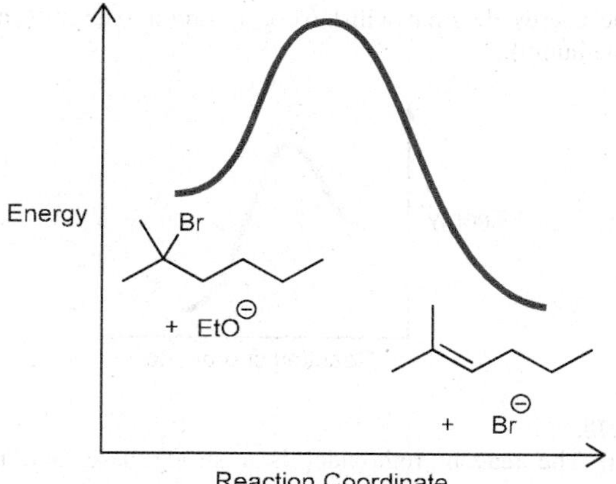

(e) In the transition state, the ethoxide ion is in the process of removing the proton, the double bond is in the process of forming, and the leaving group (bromide) is in the process of leaving. We use dotted lines to indicate the bonds that are either in the process of being formed or being broken, and we use δ– symbols to represent partial negative charges, indicating that these charges are in the process of forming (on the Br atom) or dissipating (on the O atom). Finally, brackets are drawn, together with the double dagger symbol that indicates that this is a transition state:

7.79. The correct answer is (d). The substrate is primary and the reagent is a strong nucleophile, so this substitution reaction must occur via an SN2 pathway. SN2 reactions are bimolecular processes, so statement (a) is true. In an SN2 process, the rate is linearly dependent on the concentration of the nucleophile, so statement (b) is also true. Polar aprotic solvents enhance the rate of SN2 processes, so statement (c) is also true. Therefore, statement (d) must be false, and indeed, it is false. SN2 reactions do not proceed via carbocation intermediates. Carbocation intermediates are involved in SN1 reactions, but the alkyl halide is primary in this case, and a primary carbocation is too unstable to form.

7.80. The correct answer is (d). The reagent is both a strong base (E2) and a strong nucleophile (SN2). With a secondary substrate, the E2 product is favored over the SN2 product. This rules out option (a). Since the base is not sterically hindered, we expect the Zaitsev product. That is, we expect a disubstituted alkene, rather than a monosubstituted alkene, so we can rule out option (b). Options (c) and (d) are *cis/trans* stereoisomers. Since the E2 process is stereoselective, we expect the more stable *trans* isomer to predominate, so option (d) is the correct answer.

7.81. The correct answer is (c). Option (a) involves a primary substrate being treated with hydroxide (which is both a strong nucleophile and a strong base). Under these conditions, the SN2 product is expected to predominate over the E2 product. The alcohol dehydration in option (b) is an E1 process that would give the more-substituted alkene as the major product (the Zaitsev product), not 1-butene. The strong base in option (d) is not sterically hindered so it would also give the more-substituted alkene as the major product (the Zaitsev product). Only option (c) will give 1-butene, because the sterically hindered base will produce the less-substituted alkene (the Hofmann product) as the major product, after the initial step of converting the alcohol (bad leaving group) into a tosylate (good leaving group).

7.82. The correct answer is (d), increased steric hindrance. An SN2 reaction is a one-step mechanism: back-side attack of a strong nucleophile on an unhindered carbon bearing a leaving group. The mechanism is facilitated by polar aprotic solvents, and is slowed by steric hindrance.

7.83. The correct answer is (c). The E1 mechanism involves at least two steps: formation of a carbocation by loss of a leaving group, and then attack by a weak nucleophile to afford the substitution product.

7.84. The correct answer is (b). This substitution reaction is taking place at a tertiary carbon atom, so an SN1 mechanism is expected. The mechanism has a total of three steps: protonation of the alcohol, loss of water as a leaving group, and nucleophilic attack on the resulting carbocation by the bromide ion.

The rate-determining step for an SN1 process is formation of the carbocation (loss of the leaving group). Therefore, the transition state for the rate-determining step for the given reaction has one partial bond (representing the breaking C–O bond), a partial positive charge on the oxygen atom, and a partial positive charge on the carbon atom.

7.85. The correct answer is (c). Sodium ethoxide is a strong nucleophile that can do a back-side attack to displace a leaving group (SN2). It is also a strong base that can initiate the *anti*-periplanar elimination of a β proton and a leaving group (E2). A *strong* nucleophile/base is associated with one-step, *bimolecular* reactions, while unimolecular reactions (SN1 and E1, with carbocation intermediates) require a weak nucleophile/base.

7.86. The correct answer is (a). The Hofmann product (the less-substituted alkene) is desired, so a strong, sterically hindered base should be used, such as potassium *tert*-butoxide.

7.87. The correct answer is (d). The solvolysis reaction of an alkyl halide with methanol describes a substitution reaction:

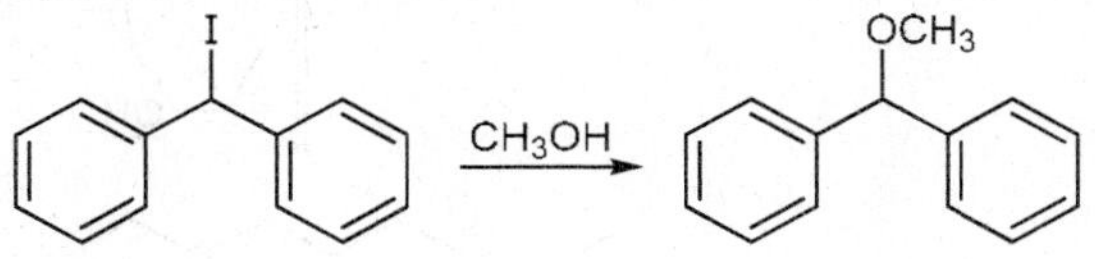

I

CH_3OH

OCH_3

Methanol is a weak nucleophile, so solvolysis typically follows an SN1 process. The fastest SN1 is expected with the alkyl halide that would generate the most stable carbocation. The correct answer has a leaving group on a secondary, benzylic carbon, and the carbocation resulting from loss of this iodide leaving group is highly stabilized by resonance that involves both benzene rings. Shown below are two of many possible resonance structures:

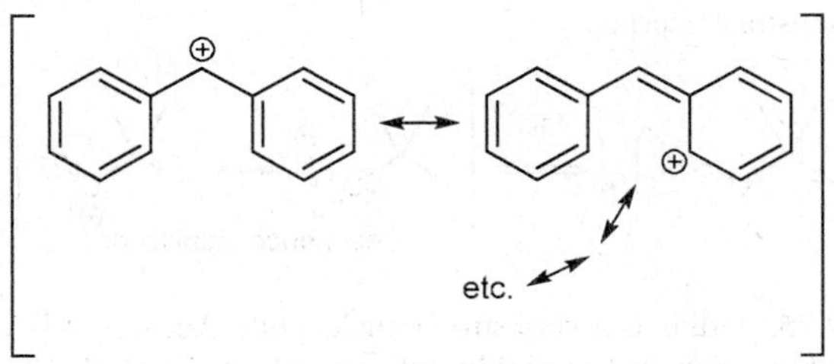

etc.

7.88. The correct answer is (b). Reaction I is faster because it has less steric hindrance. The HS^- anion is a strong nucleophile, so the reaction is expected to be bimolecular (SN2). In this case, we are comparing substrates with a leaving group on a primary or tertiary carbon atom. The one-step back-side attack is faster if there is less steric hindrance, so the primary alkyl bromide will have the faster reaction.

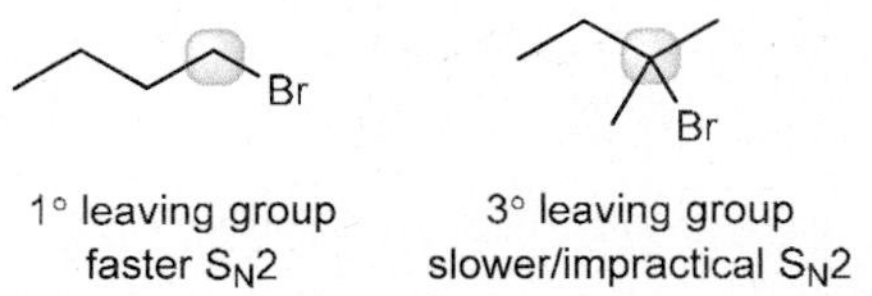

Br Br

1° leaving group faster S_N2

3° leaving group slower/impractical S_N2

7.89. The correct answer is (c). The most reasonable retrosynthesis begins with a disconnection at the cyano group, at the C–C bond:

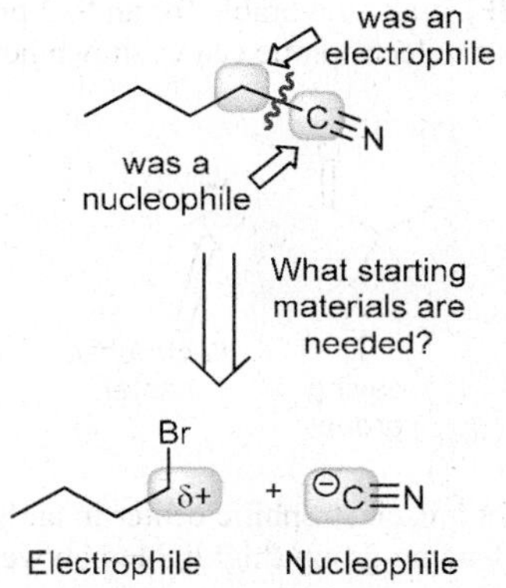

was an electrophile

C≡N

was a nucleophile

What starting materials are needed?

Br

δ+ + ⊖C≡N

Electrophile Nucleophile

Cyanide ($N{\equiv}C^-$) is a stable and familiar anion, so we will use cyanide as the nucleophile. The other carbon (at the disconnected bond) must have started out as an electrophile, so we draw a leaving group (such as Cl, Br, I or OTs) at that position.

7.90. The correct answer is (a). A reasonable retrosynthesis reveals a potential starting material from which the desired target molecule can be made. The π bond in the target molecule can be formed by an elimination reaction, either by dehydration of an alcohol or E2 elimination with a substrate bearing a good leaving group. The retrosyntheses leading to an alcohol, tosylate or bromide are all reasonable, as these substrates could all be used to prepare the given target molecule:

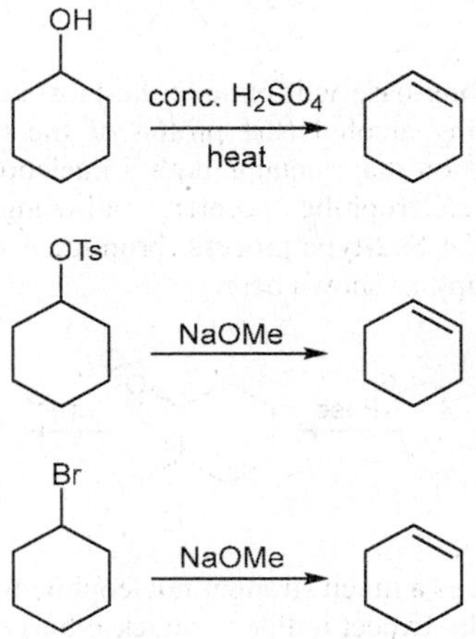

OH

conc. H_2SO_4, heat

OTs

NaOMe

Br

NaOMe

The methoxy group (OMe) is not a good leaving group, so an ether is not a suitable starting material for an elimination reaction:

OMe

7.91.

(a) Iodide is a strong nucleophile, and the substrate is primary, indicating an S_N2 reaction. The polar aprotic solvent (DMF) is also favorable for an S_N2 process. The substrate has an electrophilic center shown here.

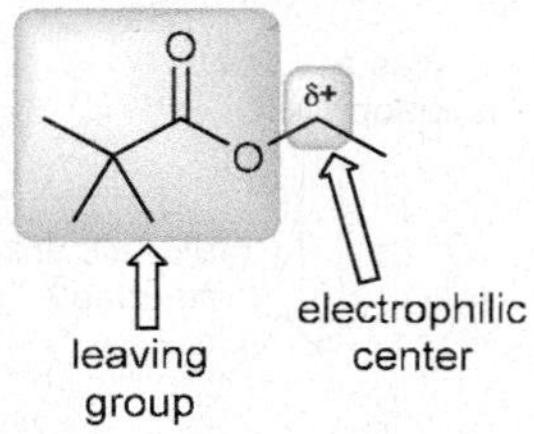

δ+

leaving group

electrophilic center

Iodide attacks this electrophilic center in an S_N2 process, ejecting the leaving group (highlighted above), as shown here:

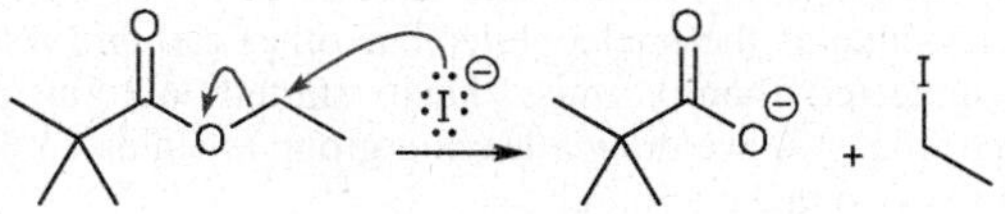

(b) This reaction occurs via an S_N2 process. As such, the rate of the reaction is highly sensitive to the nature of the substrate. The reaction will be faster in this case, because the methyl ester (a methyl substrate) is less sterically hindered than the ethyl ester (a primary substrate).

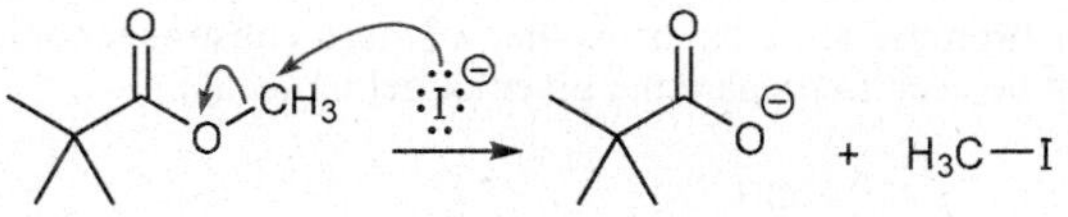

+ $H_3C—I$

7.92. A strong base will remove the most acidic proton in the starting alcohol (the proton of the OH group), giving an anion that contains both a nucleophilic center and an electrophilic center, allowing for an intramolecular S_N2-type process (bromide is ejected as a leaving group), as shown here.

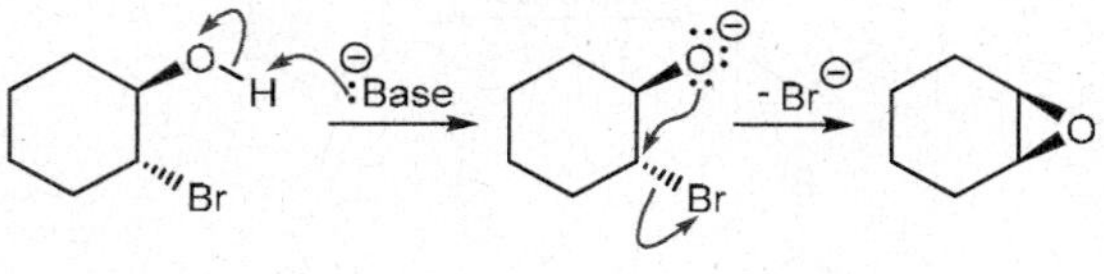

:Base

$- Br^{\ominus}$

7.93. Iodide is a much stronger nucleophile than ethanol, so we initially expect iodide to attack *n*-butyl bromide (a primary substrate) in an S_N2 reaction to give *n*-butyl iodide.

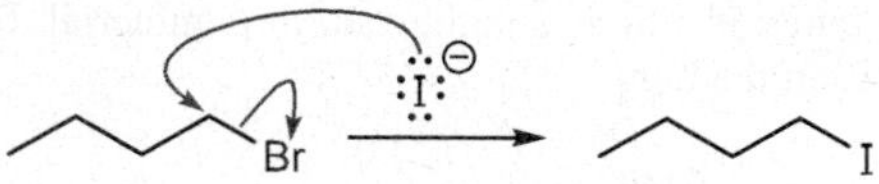

As a result of this rapid reaction, the concentration of iodide quickly decreases. Then, the slow rise in concentration of iodide indicates that another nucleophile is slowly ejecting the iodide ions. Indeed, there is a weak nucleophile present (ethanol). In the presence of ethanol, a slow S_N2 process can occur in which *n*-butyl iodide functions as the substrate (iodide is an excellent leaving group) and ethanol functions as the nucleophile. The resulting oxonium ion is then deprotonated (by ethanol, this time functioning as a base), giving the product shown below, which is an ether.

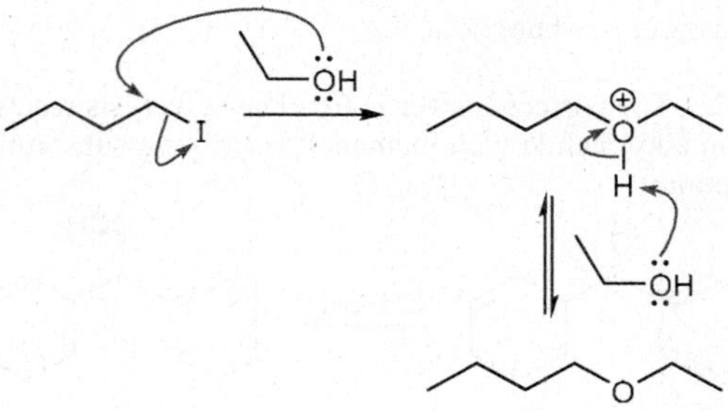

Product

7.94. Primary substrates generally do not undergo S_N1 reactions, because primary carbocations are too high in energy to form at an appreciable rate. However, in this case, loss of the leaving group generates a resonance-stabilized cation. Because this intermediate is stabilized, it can form at an appreciable rate, allowing an S_N1 process to proceed successfully, despite the fact that the substrate is primary.

$- TsO^{\ominus}$

Resonance stabilized

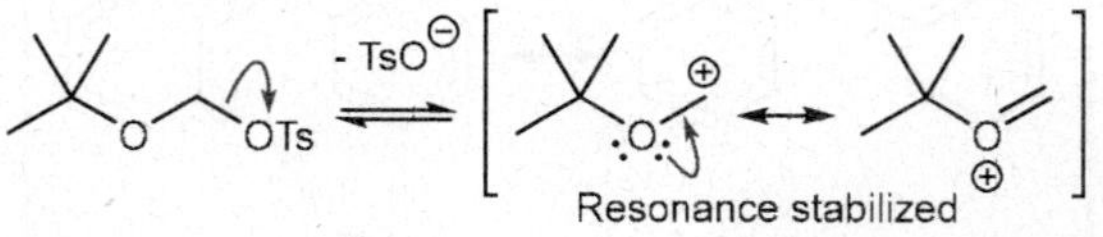

7.95. Iodide is a very strong nucleophile (because it is a large anion that is highly polarizable), and it is also an excellent leaving group (because it is a very weak base). As such, iodide will function as a nucleophile to displace the chloride ion. Once installed, the iodide group is a better leaving group than chloride, thereby increasing the rate of the reaction with cyanide.

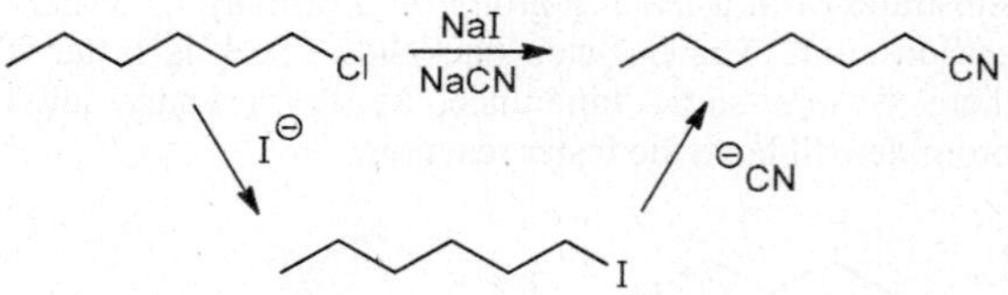

NaI

NaCN

7.96. The conditions (no strong nucleophile or strong base; polar protic solvent) favor unimolecular processes (E1 and S_N1), so we must explain the formation of the products with those mechanisms.
Formation of the first product can be explained with the following S_N1 mechanism, in which the first step is loss of the leaving group to generate a resonance-stabilized cation (resonance structures are not shown here). This carbocation is then captured by ethanol, which functions as a nucleophile. The resulting oxonium ion is then deprotonated by another molecule of ethanol, which functions as a base:

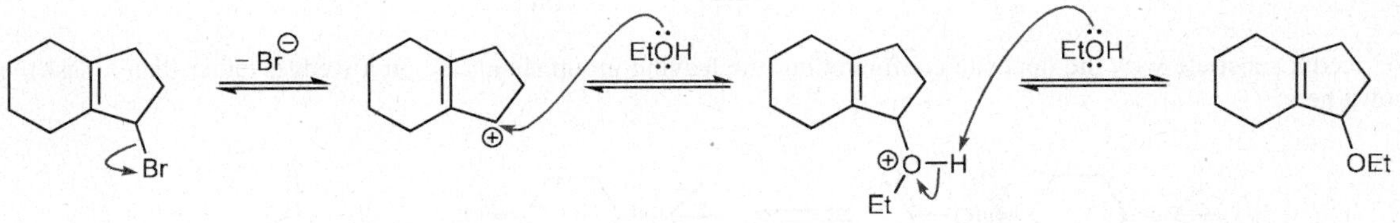

If we draw the resonance structure of the carbocation above, we can see that the positive charge is delocalized over two different carbon atoms. Attack by the ethanol nucleophile on the tertiary center (as shown below) leads to formation of the second product:

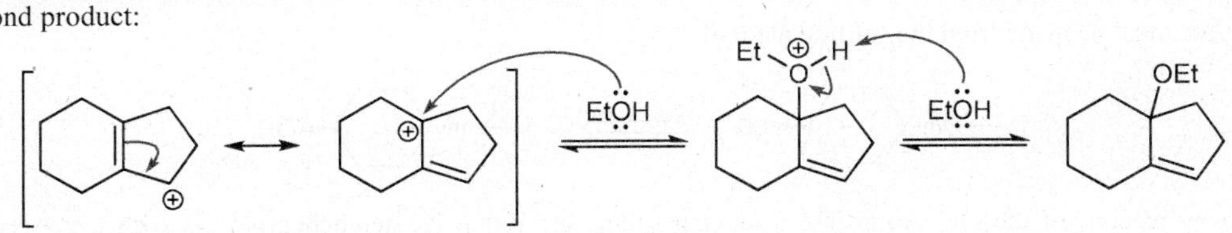

Finally, the third product is formed via an E1 process. In the first step of the mechanism, loss of the leaving group generates a resonance-stabilized cation. Then, in the second (and final) step, ethanol functions as a base and removes a proton, giving the observed product.

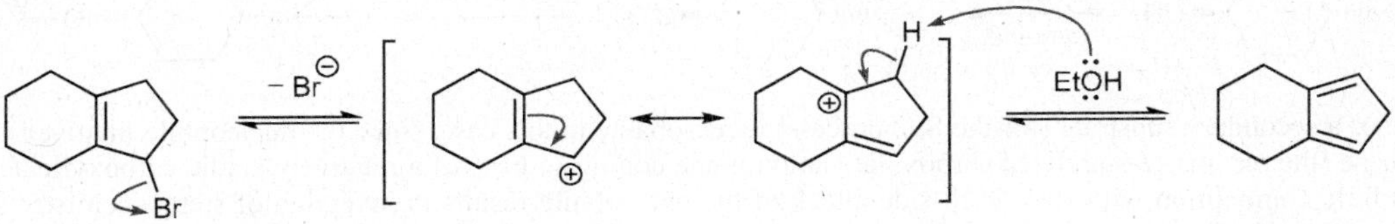

7.97. In this target molecule, the only bond we know how to make (based on reactions covered thus far) is the one next to the cyano group:

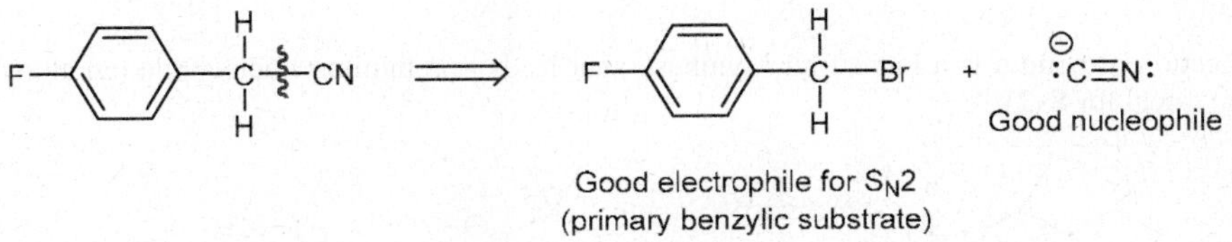

Good electrophile for S_N2 (primary benzylic substrate)

Good nucleophile

Cyanide is a good nucleophile, and a primary substrate is ideal for an S_N2 process, so these starting materials would lead to a good synthesis of 4-fluorophenylacetonitrile. As shown here, bromide was selected as the leaving group and NaCN can be used as a source of cyanide anion. Other variations are equally acceptable, including the use of different leaving groups (e.g., Cl, I, OTs):

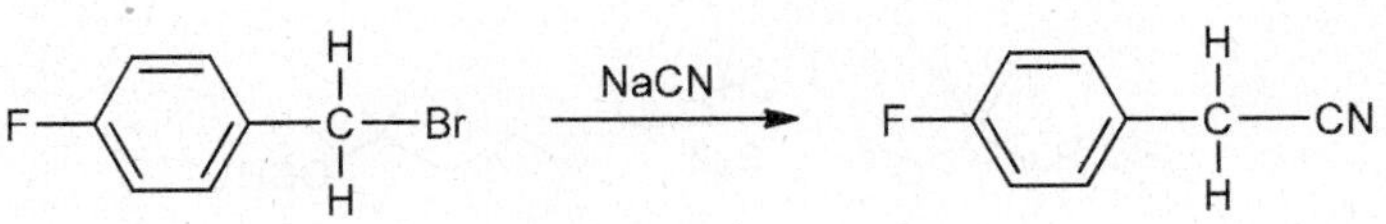

7.98. When comparing the starting compound to the target molecule, we note that the ester group is new, and the configuration of the adjacent chiral center has been inverted. The latter observation indicates that an S_N2 process must be involved. So our retrosynthesis should focus on disconnection of the following bond:

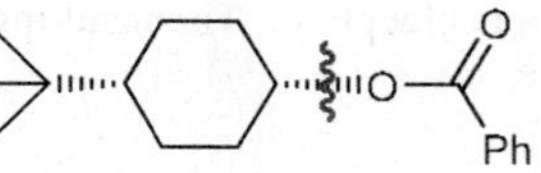

We need a substrate with the opposite configuration (the leaving group should be on a wedge, rather than a dash), as shown here:

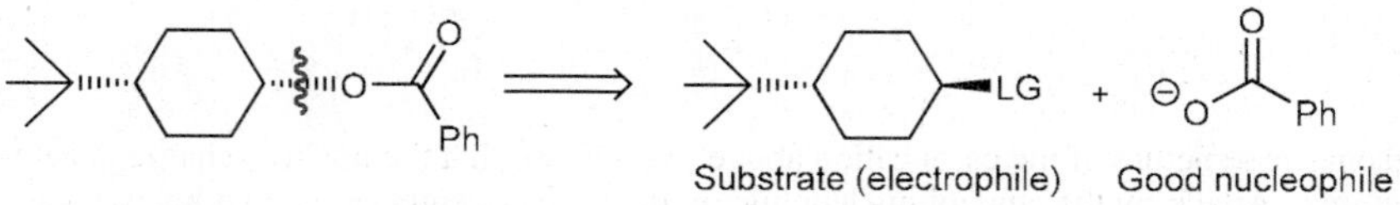

Substrate (electrophile) Good nucleophile

This substrate must be made from the starting alcohol:

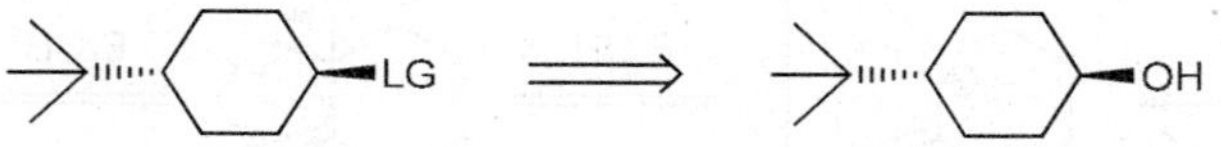

The best way to convert the OH group into a leaving group and retain its stereochemistry is with the use of tosyl chloride to make the tosylate. This tosylate can then function as the substrate in an S_N2 reaction. The forward synthesis is shown here:

TsCl pyridine

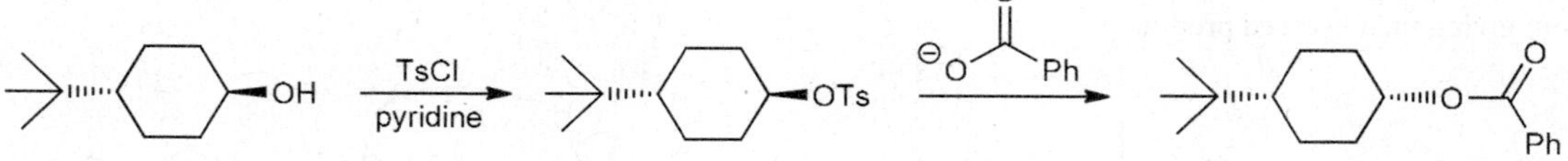

The use of a secondary substrate (for the S_N2 process) is reasonable in this case, since the nucleophile involved is not a strong base (the resonance-stabilized carboxylate anion is the conjugate base of a relatively acidic carboxylic acid), so there is little competition with E2. Back-side attack of the nucleophile results in inversion of stereochemistry to give the desired target molecule.

7.99. Disconnection of bond **a** is a logical retrosynthesis that leads to a familiar nucleophile (an alkoxide ion) and a methyl substrate (ideal for S_N2):

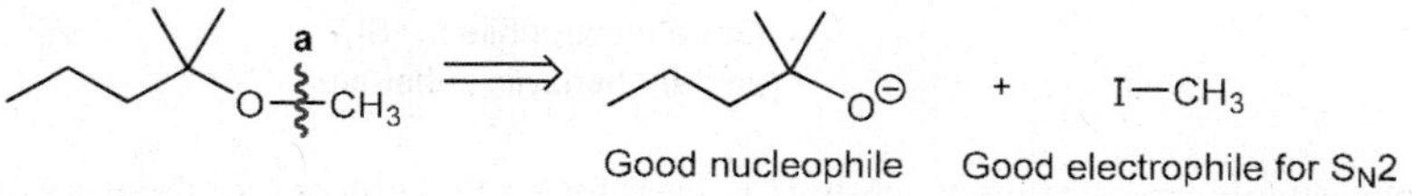

Good nucleophile Good electrophile for S_N2

Note that methyl iodide is chosen, rather than methyl bromide or methyl chloride, because methyl iodide is a liquid at room temperature (while methyl bromide and methyl chloride are gases at room temperature, making them more difficult to work with). The forward synthesis is shown here:

CH_3I S_N2

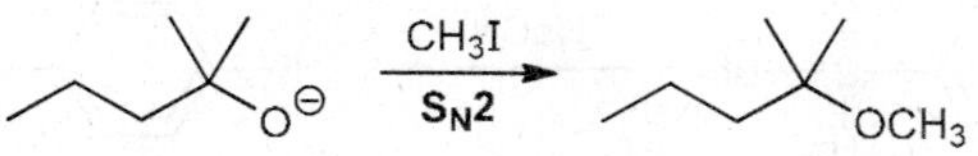

A similar disconnection of bond **b** reveals that a tertiary substrate would be required, and we have seen that S_N2 processes do not occur with tertiary substrates.

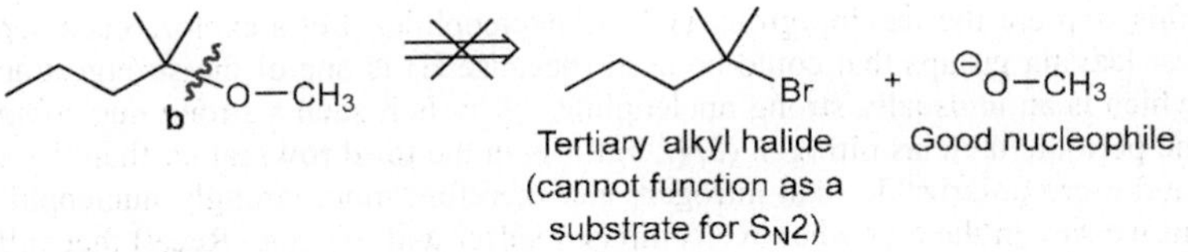

Tertiary alkyl halide (cannot function as a substrate for S_N2)

Good nucleophile

An S_N2 process won't work here. Instead, the alkoxide ion is expected to function as a base, giving E2 products. This issue can be overcome if we employ an S_N1 process instead, by using a weak nucleophile (an alcohol).

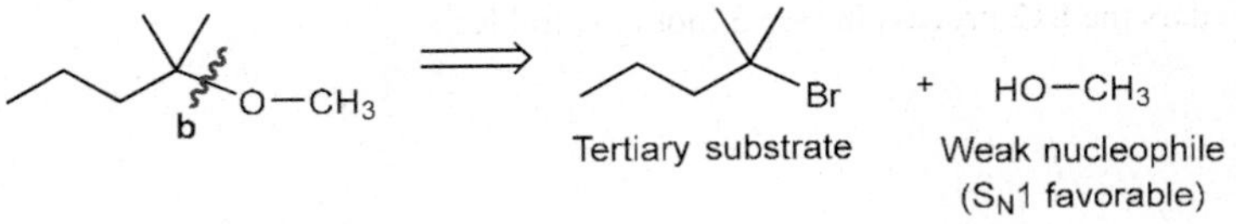

Tertiary substrate

Weak nucleophile (S_N1 favorable)

This S_N1 process would involve a tertiary carbocation (which would not rearrange), so it is expected to afford the target molecule, as shown:

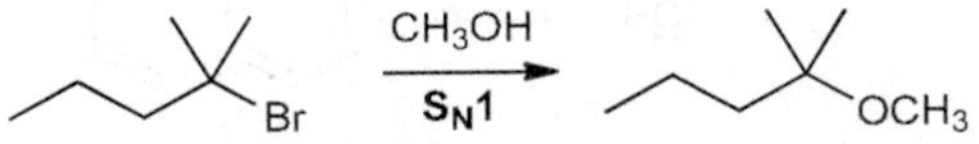

However, this process (solvolysis) is likely to produce competing E1 products, as well:

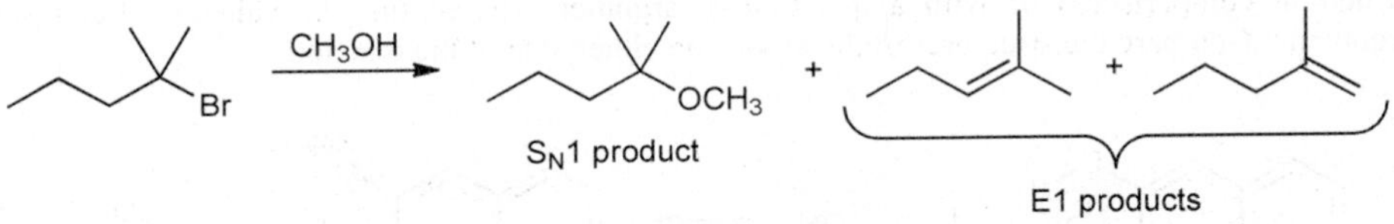

S_N1 product

E1 products

Since one of the E1 products is a trisubstituted alkene, we might even expect for that E1 product to predominate. As such, this synthesis would not be efficient, because it may not provide the desired S_N1 product as the major product.

In summary, our retrosynthetic analysis has uncovered two possible synthetic routes: an S_N2 process and an S_N1 process. The S_N2 process is expected to be more efficient:

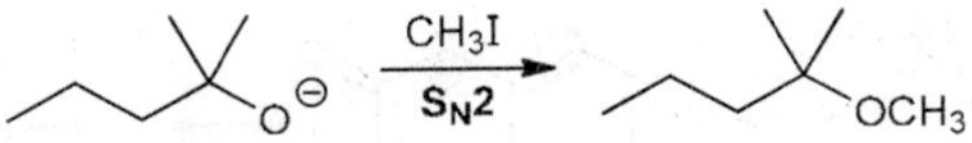

7.100.

(a) The first reaction (with TsCl and pyridine) transforms the OH group into a good leaving group (a tosylate) that undergoes an S_N2 reaction with sodium iodide. The net result of these two steps is the conversion of an alcohol to an alkyl iodide.

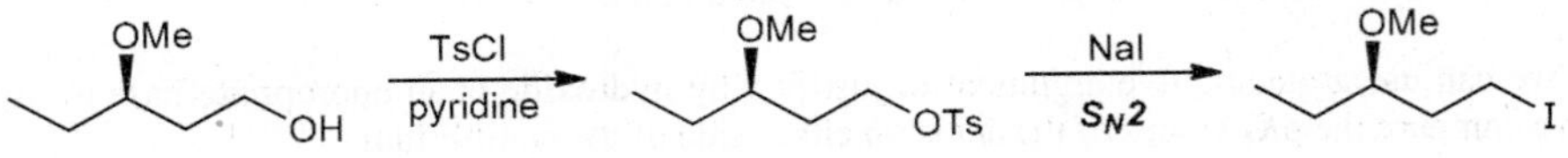

In the final step of the process, the primary alkyl iodide is then treated with triphenylphosphine (PPh_3), which functions as a nucleophile, giving another S_N2 reaction, to afford a phosphonium salt.

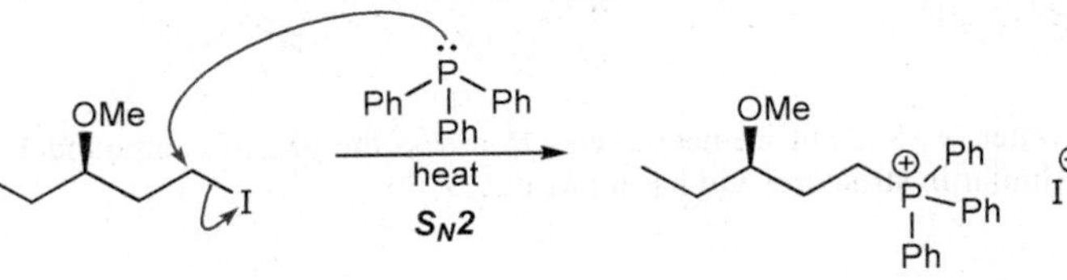

The factors favoring this step are the leaving group and the nucleophile. Let's explore each separately. The leaving group is one of the best leaving groups that could be used (because HI is one of the strongest acids, pK_a = -10). The nucleophile is PPh_3, which is an unusually strong nucleophile. Why is it such a strong nucleophile? Phosphorus is in the same column of the periodic table as nitrogen (5A), but it is in the third row, rather than the second row. As such, phosphorus is larger and more polarizable than nitrogen, and therefore more strongly nucleophilic. This argument is similar to the argument we saw in the text when we compared sulfur and oxygen. Recall that sulfur is larger and more polarizable than oxygen, and therefore, sulfur is very strongly nucleophilic (even if it lacks a negative charge). Similarly, PPh_3 is a powerful nucleophile, even though it lacks a negative charge.

(b) In the second step, one leaving group (tosylate) is replaced with another (iodide). Iodide is a better leaving group than tosylate, which renders the S_N2 process in step 3 more favorable.

7.101.

(a) The proton connected to the oxygen atom is the most acidic proton in compound **1**, so it is removed upon treatment with a strong base.

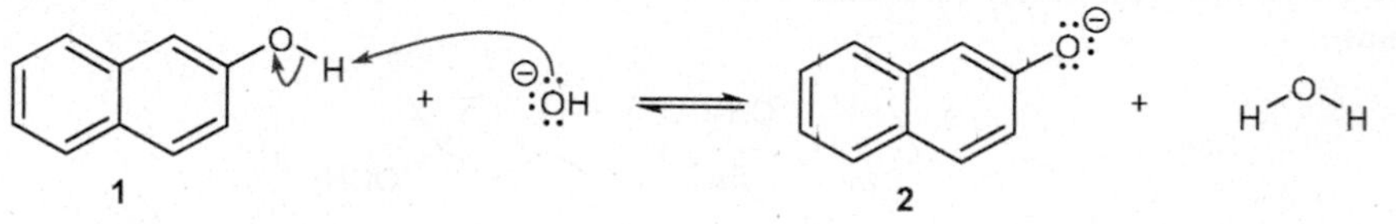

We can justify that hydroxide is a suitable base to achieve the conversion of **1** to **2**, with either a qualitative argument (based on structural comparisons) or with a quantitative argument (based on pK_a values). Let's start with the qualitative argument. Compare the structures of the anions on either side of the reaction.

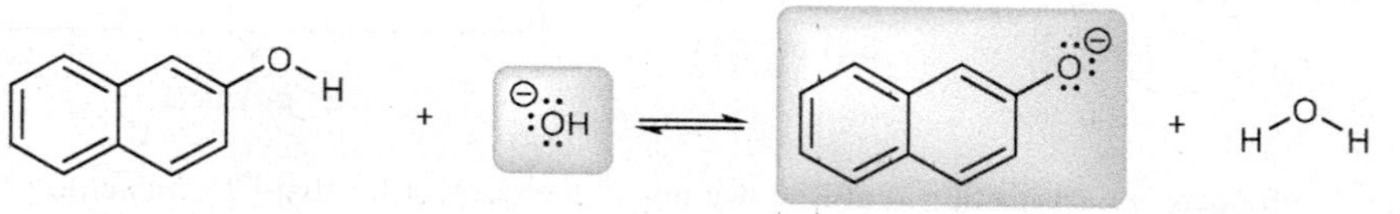

The negative charge in a hydroxide ion is localized on one oxygen atom, while the negative charge in the other anion is delocalized over one oxygen atom and five carbon atoms:

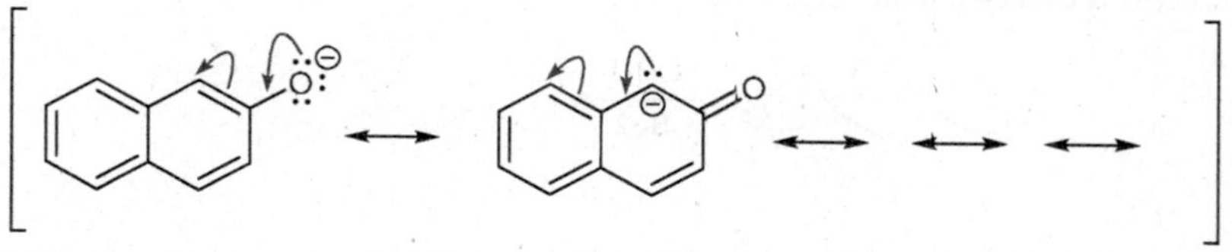

We expect the anion on the product side to be more stabilized (via resonance delocalization), and a weaker base than hydroxide. The equilibrium will favor formation of the weaker base (the more stable anion). That is, hydroxide is a sufficiently strong base, because it is stronger (less stable) than anion **2**.

Alternatively, we can use a quantitative argument to justify why hydroxide is an appropriate base to use in this case. Specifically, we compare the pK_a values of the acids on either side of the equilibrium.

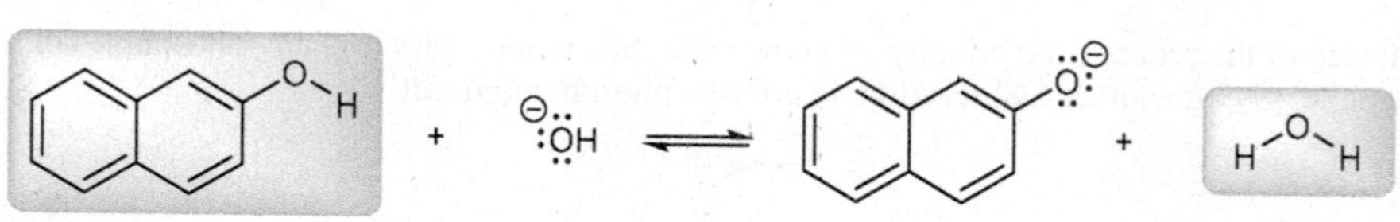

We know that the pK_a of water is 15.7, but we need a way to assess the pK_a of compound **1**. When we explore Table 3.1, we see that phenol is similar in structure, and has a pK_a of 9.9.

2-naphthol phenol

We expect the pK_a of compound **1** (2-naphthol) to be more similar to the pK_a of phenol than to the pK_a of water. Therefore, we expect the pK_a of compound **1** to be lower than the pK_a of water, and the equilibrium will favor formation of the weaker acid.

1 pK_a ~ 10 2 pK_a = 15.7

As such, hydroxide is a suitable base to favor deprotonation of compound **1**.

(b) Anion **2** is a strong nucleophile, and the substrate is primary, so an S_N2 process is expected. Anion **2** attacks *n*-butyl iodide and displaces the iodide leaving group (in a concerted, back-side attack), giving compound **3**, as shown:

2 3

7.102.

(a) The substrate is tertiary, so we expect the reaction to proceed exclusively through an S_N1 pathway (steric crowding prevents S_N2 from competing). The first step of the mechanism involves loss of a bromide to give a carbocation, which is then captured by the MeOH nucleophile to produce an oxonium ion. Deprotonation of the oxonium ion gives the product, **2b**.

1 - $Br^{\ominus}$ 2b

(b) When bromide leaves, the resulting carbocation is benzylic to three different aromatic rings. As such, it is highly stabilized because the positive charge is delocalized over 10 carbon atoms via resonance (as shown). Since this intermediate is so stabilized (low in energy), we can infer that the transition state for formation of the carbocation will also be very low in energy (because any developing charge in the transition state is stabilized by resonance, just as seen in the intermediate carbocation). Since the transition state is low in energy, formation of the carbocation (which is the rate-determining step) will occur very rapidly.

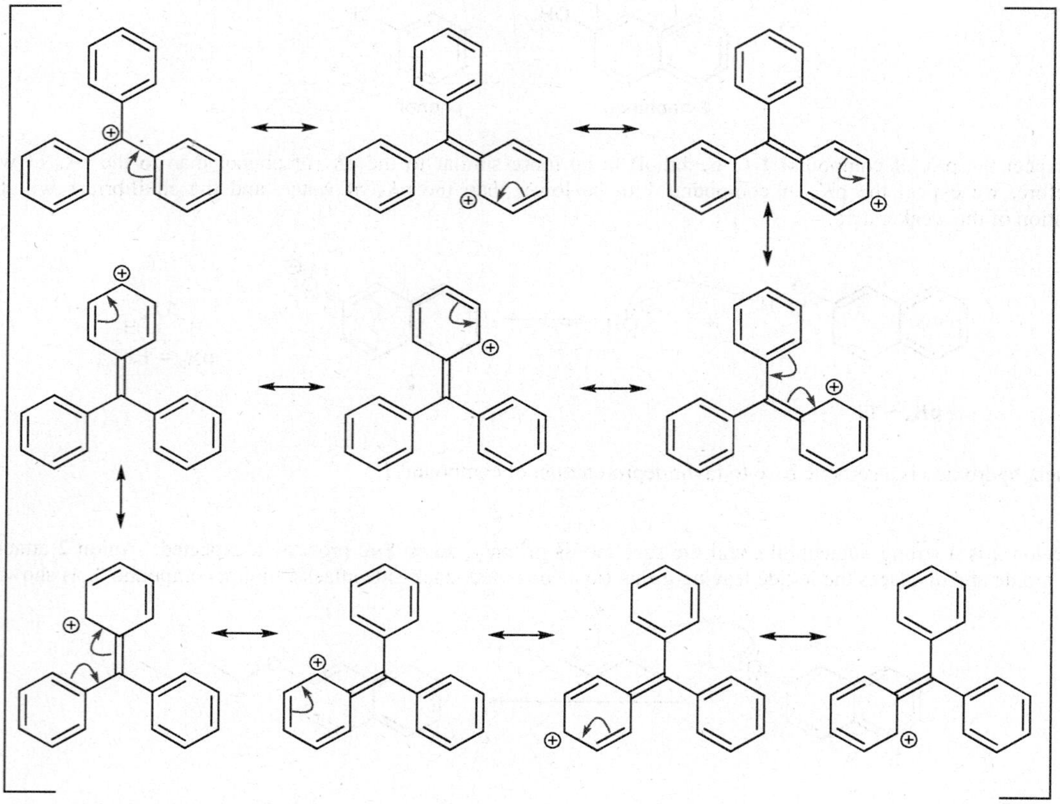

7.103.

(a) The trisubstituted π bond, indicated below, has the *E* configuration. The higher priority groups (highlighted below) are on opposite sides (*E*).

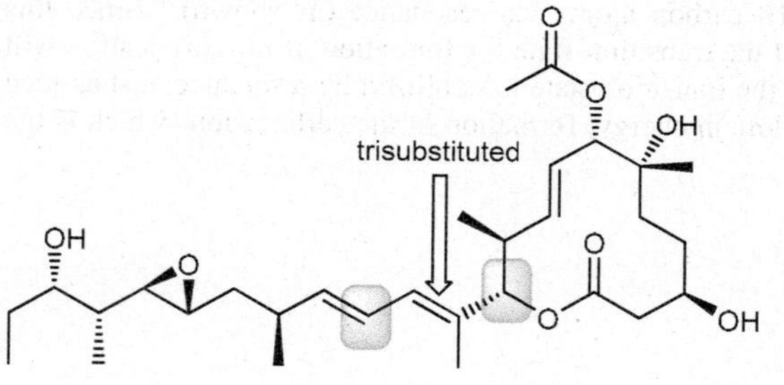

trisubstituted

(b) There are 10 chiral centers (indicated below, each of which can be *R* or *S*), and there are three C=C π bonds (each of which can be *E* or *Z*). Thus, the number of possible stereoisomers is $2^{13} = 8192$.

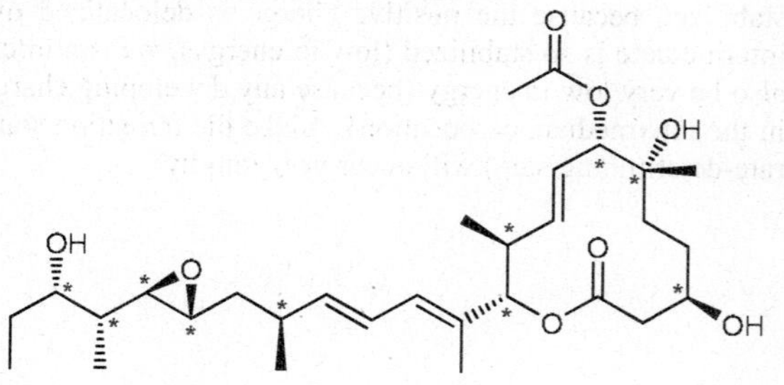

(c) The enantiomer of pladienolide B has the opposite configuration at each and every chiral center, but the configuration of each C=C π bond remains the same as in pladienolide B (in order to retain the mirror-image relationship with pladienolide B).

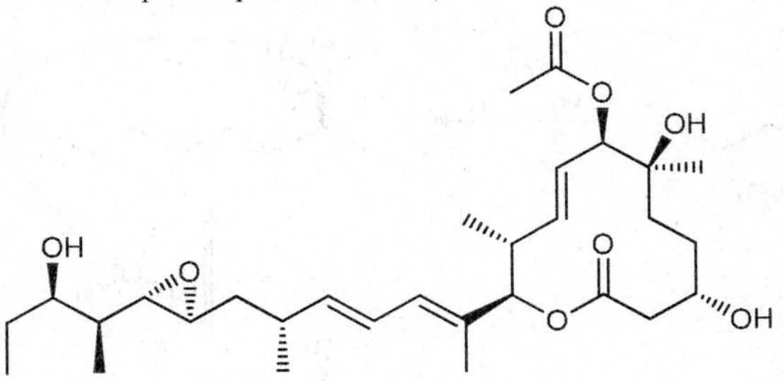

(d) There are two ester groups (highlighted below) in pladienolide B. The following diastereomer is the result of inverting the configuration of the two chiral centers adjacent to the ester groups (indicated below with *).

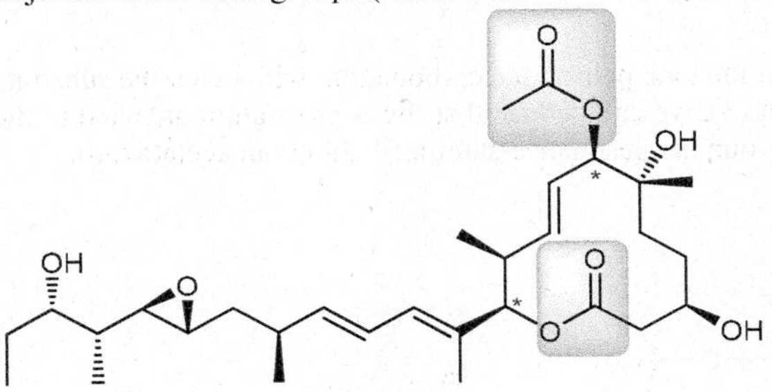

(e) In pladienolide B, the disubstituted π bond outside of the ring has the *E* configuration. The following is a diastereomer in which this disubstituted π bond (highlighted below) has the *Z* configuration:

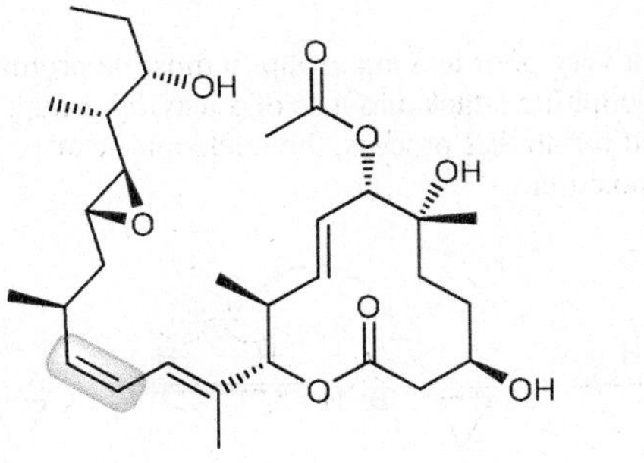

7.104

(a) If compound **3** was properly dried, then it could be distinguished from compound **1** with IR spectroscopy. Specifically, compound **1** has an O-H bond, so we expect a broad signal in the range 3200-3600 cm^{-1}, while compound **3** lacks such a bond, so its IR spectrum should lack a signal in the same range.

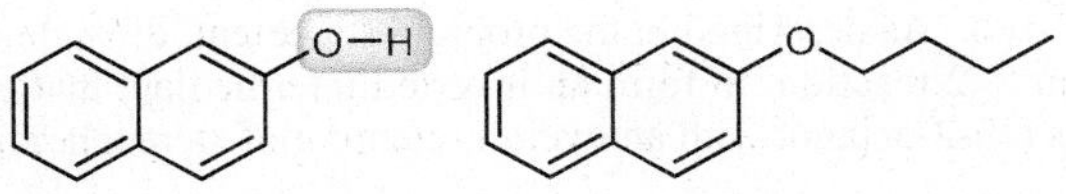

In this way, we can verify whether the reaction has gone to completion, by looking for a signal in the range 3200-3600 cm^{-1} in the IR spectrum of the product. The absence of this signal verifies completion of the reaction. Since we are looking for the absence of an OH stretching signal, it is essential that the product is dried. Otherwise, the water molecules would give a signal exactly in the region of interest (because water has O-H bonds). This would prevent from us from being able to determine whether the reaction had gone to completion.

(b) The conversion of **1** to **2a** involves introduction of an OH group, which should produce a broad, easily detectable signal in the range 3200-3600 cm^{-1}. Therefore, by taking an IR spectrum of the product, we can verify formation of **2a** by looking for a broad signal in the range 3200-3600 cm^{-1}. In contrast, **2b** and **2c** do not have an OH group. As such, it will be difficult to distinguish the diagnostic regions of the IR spectra of compounds **1**, **2b**, and **2c**. The utility of IR in spectroscopy in these cases must rely on analysis of the fingerprint region (C-Br stretch vs. C-O stretch), although fingerprint regions are often more difficult to interpret (except in the hands of a trained expert). Therefore, IR spectroscopy is not the best tool for verifying the conversion of **1** to either **2b** or **2c**. NMR spectroscopy would be a better tool for confirming completion of those reactions.

7.105. Compound **2** is the nucleophile in this S_N2 reaction. To see why, recall from Chapter 1 that a C-Li bond can be viewed as an ionic bond, in which the carbon atom has a lone pair and negative charge.

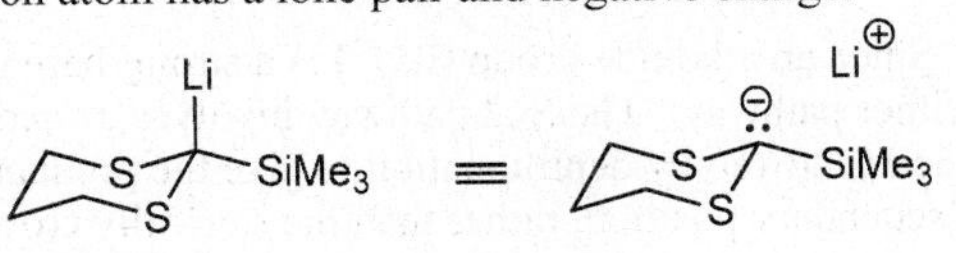

This compound is indeed a very strong nucleophile, and it attacks the alkyl halide in an S_N2 process, as shown here:

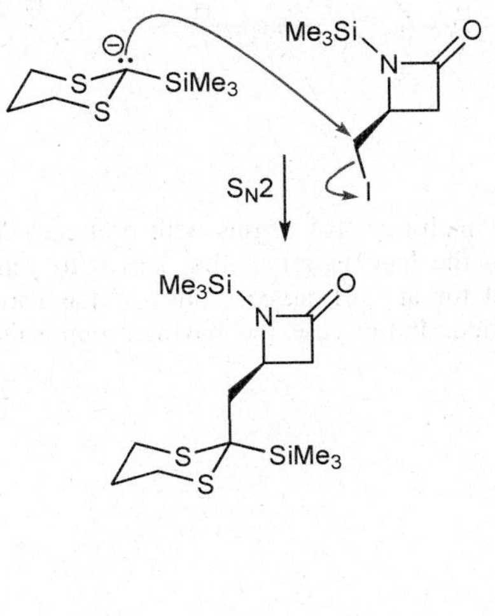

7.106. As described in the problem statement, dioxane can function as a nucleophile and attack (*R*)-2-octyl sulfonate in an SN2 reaction, to form an inverted intermediate that can then undergo another SN2 reaction with water. The product is (*R*)-2-octanol with an overall retention of stereochemistry, due to two successive SN2 steps taking place.

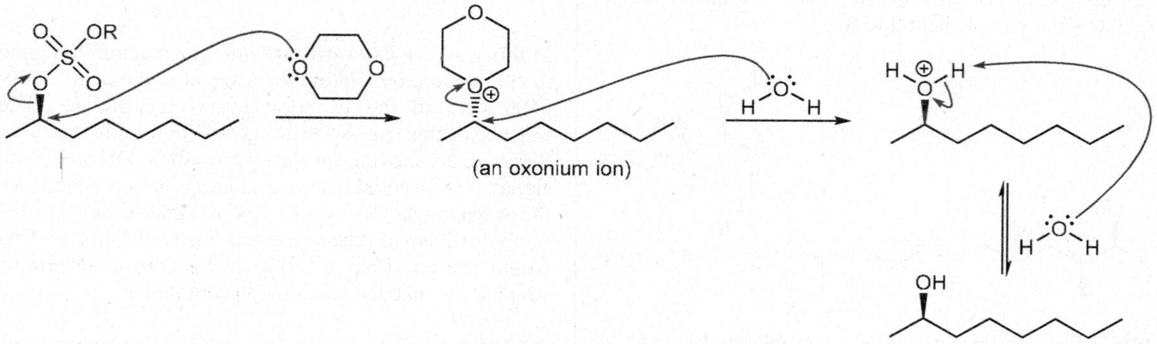

(an oxonium ion)

Since this process increases in frequency as the concentration of dioxane increases, the optical purity of (*S*)-2-octanol decreases as dioxane's concentration is increased.

7.107.

(a) Compound **2** functions as a nucleophile, which means that the lone pair on the carbon atom will attack the substrate in an SN2-type process. Based on the structure of the product (**3**), we can deduce that the oxygen atom attached to the *tert*-butyl group is attacked by the nucleophile. The leaving group is a resonance-stabilized anion (an acetate ion).

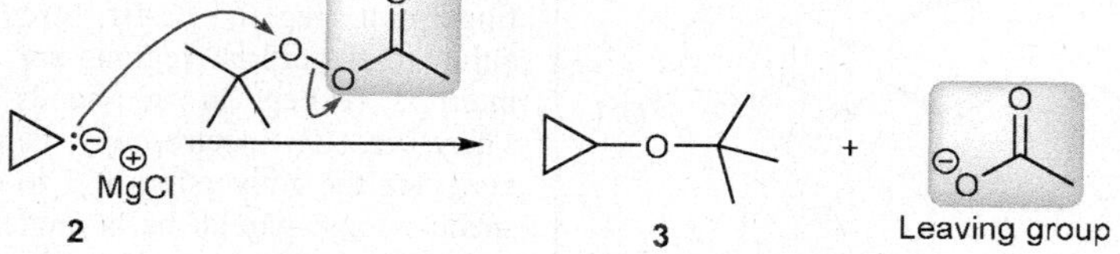

MgCl
2
3
Leaving group

(b) The reverse process would involve an acetate ion functioning as a nucleophile and the expulsion of **2** as a leaving group. That is extremely unlikely to occur, because **2** is not a good leaving group. It is a very strong base, because its conjugate acid is an alkane, which is an extremely weak acid (compare pK_a values of alkanes with other organic acids). Since **2** is not a stable, weak base, it cannot function as a leaving group. And as a result, the reaction is irreversible.

(c) Since an alkoxide group (RO^-) is a strong base and therefore a very poor leaving group, it must be protonated first, in either pathway. The SN2 pathway involves a *simultaneous* nucleophilic attack and loss of a leaving group. This step is then followed by deprotonation to give the product. As expected for an SN2 process, the nucleophilic attack occurs at the secondary position, rather than the sterically crowded tertiary position.

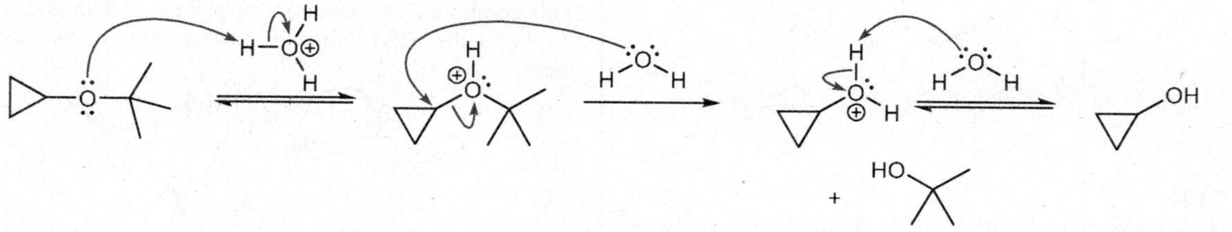

The SN1 pathway also begins with protonation of the oxygen atom to produce a better leaving group. But in this pathway, the leaving group first leaves to generate a carbocation, and only then does the nucleophile attack. As expected for an SN1 process, loss of the leaving group generates a tertiary carbocation, rather than a secondary carbocation. In this case, the leaving group is the product (cyclopropanol).

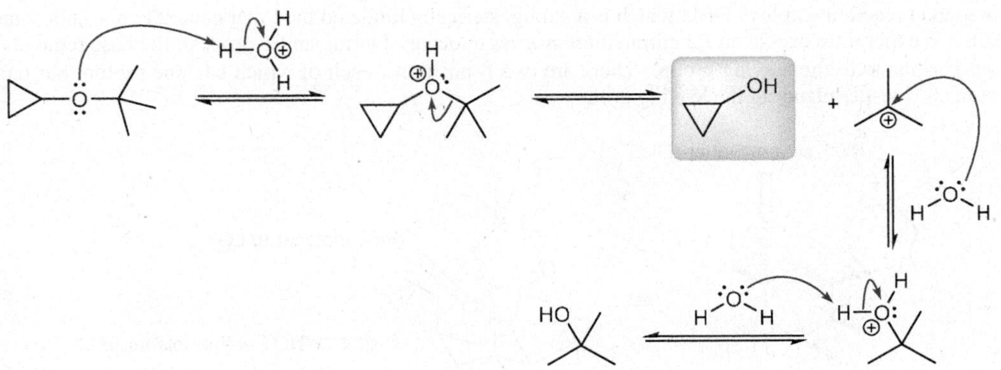

(d) As seen in both the SN1 and SN2 pathways, the by-product is *tert*-butanol.

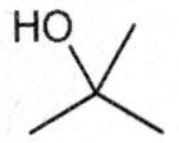

7.108.

(a) The proposed mechanism is a concerted process that very closely resembles an E2 process. As such, we would expect this process to occur when the two bromine atoms are *anti*-periplanar to one another, as shown in the problem statement. If we perform the same reaction with a diastereomer of **1**, we would expect a *Z*-alkene:

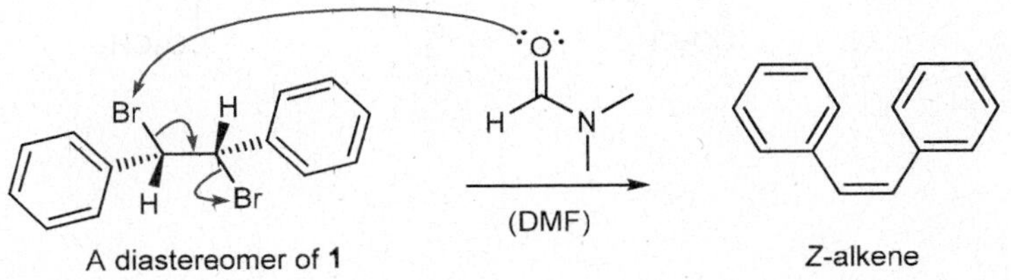

A diastereomer of **1**

(DMF)

Z-alkene

But none of the *Z*-alkene is formed. The *E* isomer is obtained exclusively, which means that the preference for the *E*-isomer is not dependent on the configuration of the starting dibromide. The *E*-alkene is obtained either from **1** or from a diastereomer of **1**. As such, the reaction does not appear to have a requirement for *anti*-periplanarity. This is evidence against a concerted mechanism.

(b) The reaction is not considered to be stereospecific, because the preference for the *E* isomer is not dependent on the configuration of the starting dibromide. The reaction is stereoselective, because one configuration of the alkene product (the *E* isomer) is favored over the other.

7.109.

In the first reaction, the OH group is converted into a better leaving group. Notice that the configuration of the chiral center bearing the OH group does NOT change in the process (the substituent remains on a dash):

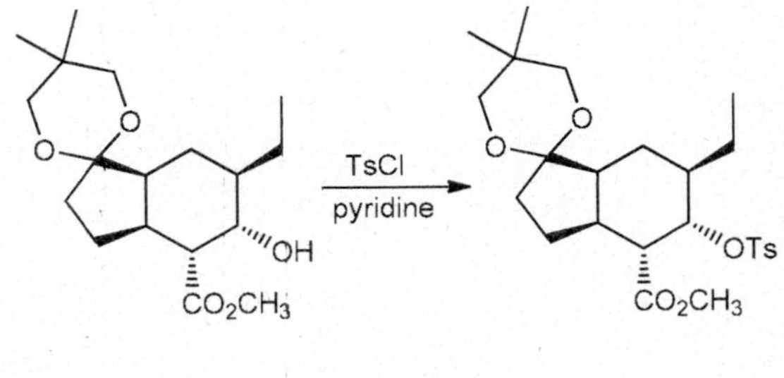

TsCl
pyridine

Then, the second reaction employs DBU which is a strong, sterically hindered base that generally does not function as a nucleophile. We therefore expect an E2 elimination process to occur. During an E2 process, the base removes a proton that is *anti*-periplanar to the leaving group. There are two β positions, each of which has one proton, but only one of these protons is *anti*-periplanar to the leaving group:

NOT antiperiplanar to LG

Antiperiplanar to LG

Antiperiplanar to LG

NOT antiperiplanar to LG

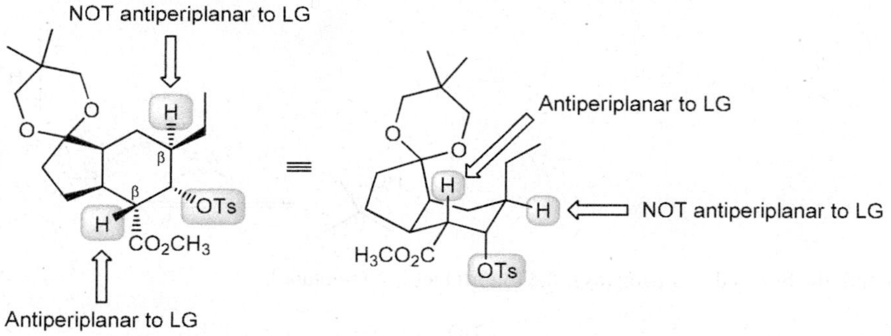

Therefore, we expect the following E2 product:

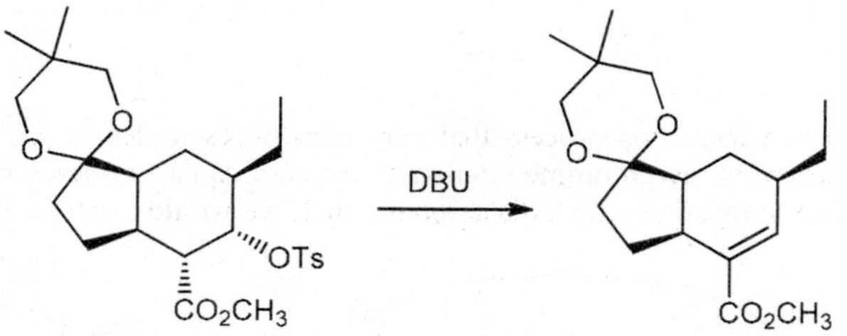

DBU

7.110. Compound **1** exhibits a mesylate group (OSO_2CH_3) as a good leaving group (described in Section 7.10). Upon its formation, compound **1** undergoes an intramolecular S_N2-type reaction, in which the nitrogen atom functions as the nucleophile. This results in the formation of a three-membered nitrogen-containing ring fused to a six-membered ring (**3**). Notice the stereochemistry: the nitrogen atom displaced the mesylate from the back face of the molecule, resulting in inversion at the original chiral center. Next, an intermolecular S_N2 reaction occurs: the chloride ion attacks the three-membered ring at the least hindered site, resulting in the formation of the product (**2**). Notice that the configuration of the chiral center (that underwent inversion in the first S_N2-type reaction) does not change during this second S_N2 reaction (in this step, chloride attacks a carbon atom that is not a chiral center).

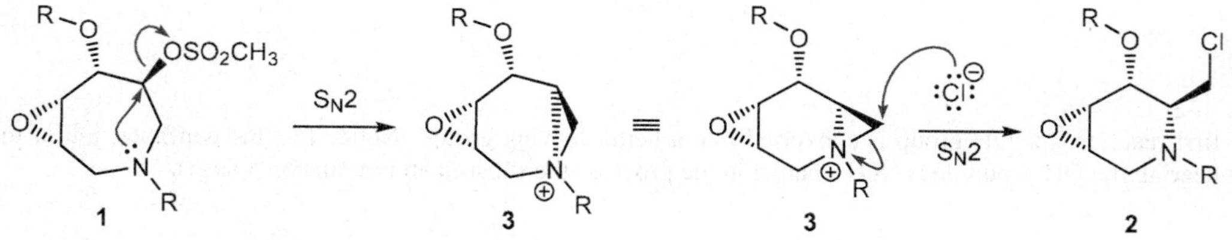

S_N2 S_N2

1 3 3 2

7.111.

(a) Table 7.1 indicates that bromide is expected to be a better leaving group than chloride, because bromide is a more stable base (HBr is a stronger acid than HCl). This is supported by the hydrolysis data. When we compare the rate of hydrolysis for $PhCHCl_2$ (2.21×10^4 /min) and PhCHBrCl (31.1×10^4 /min), we observe that the hydrolysis is faster for the substrate with the better leaving group.

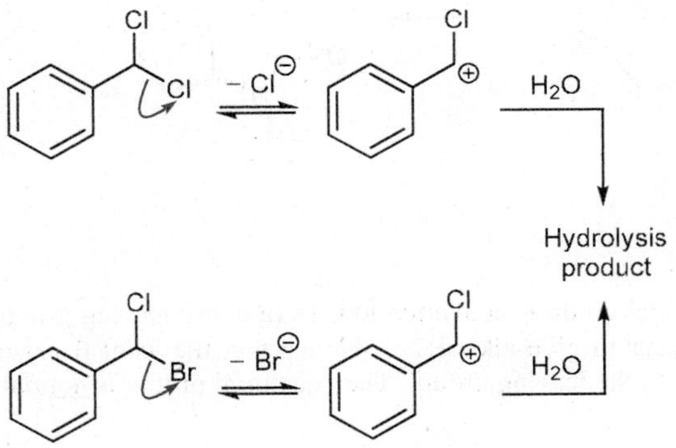

In both cases, the first step involves loss of a leaving group, and in both cases, the intermediate carbocation is the same. The only difference between these two reactions is the identity of the leaving group. According to the data provided, hydrolysis occurs more rapidly when the leaving group is bromide (31.1×10^4 /min) rather than chloride (2.21×10^4 /min). This is consistent with the expectation that bromide is a better leaving group than chloride.

(b) If we compare the rates of hydrolysis for $PhCH_2Cl$ (0.22×10^4 /min) and $PhCHCl_2$ (2.21×10^4 /min) we find that the presence of a chloro group (attached to C+) causes an increased rate of hydrolysis. A similar trend is observed if we compare $PhCHCl_2$ (2.21×10^4 /min) and $PhCCl_3$ (110.5×10^4 /min). Therefore, we can conclude that a chloro group will stabilize a carbocation (if the chloro group is attached directly to C+ of the carbocation). This stabilizing effect is unlikely to be caused by induction, because we expect the chloro group to be electron-withdrawing via induction, which would *destabilize* the carbocation (rather than stabilize it). Instead, the effect must be explained with resonance, which overwhelms the inductive effect. Specifically, the presence of a chloro group stabilizes the carbocation intermediate by delocalizing the charge via resonance.

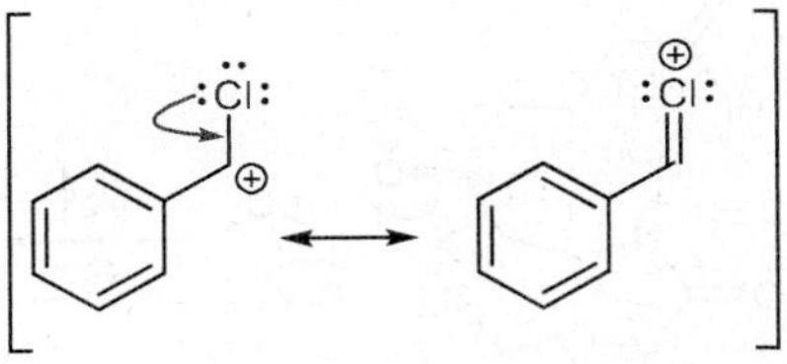

(c) If we compare the rates of hydrolysis for $PhCHBr_2$ (6.85×10^4 /min) and $PhCBr_3$ (1131×10^4 /min), we find that the presence of a bromo group (attached to C+) causes an increased rate of hydrolysis. This is likely explained by resonance stabilization of the carbocation, just as we saw in part (b).

(d) Compare hydrolysis of PhCHBrCl (31.1×10^4 /min) with hydrolysis of $PhCHBr_2$ (6.85×10^4 /min). In both cases, the identity of the leaving group is the same (bromide). But the resulting carbocations are different.

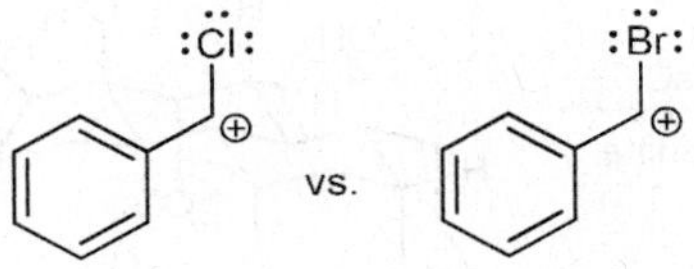

Comparing the rates of hydrolysis indicates that an adjacent chloro group more effectively stabilizes a carbocation than an adjacent bromo group.

(e) Comparison of $PhCHCl_2$ (2.21×10^4 /min) versus $PhCHBr_2$ (6.85×10^4 /min) suggests that the better leaving group ability of the bromide ion is more important than the greater carbocation stability afforded by the chlorine atom via resonance.

7.112. The sulfur atom provides anchimeric assistance via an intramolecular S_N2-type reaction. That is, a lone pair on the sulfur atom functions as a nucleophile, ejecting the leaving group (causing the liberation of SO_2 gas, as described in the problem statement) to form an intermediate with a positively charged sulfur atom in a three-membered ring. This intermediate is then attacked by a chloride ion (an intermolecular S_N2) to give **3**. Each of these two steps proceeds with inversion of configuration, as expected, which gives a net overall retention of configuration.

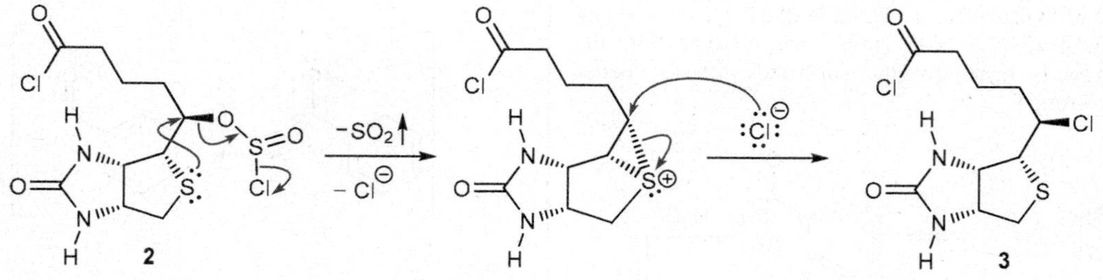

7.113.

(a) When compound **1a** is treated with TsCl and pyridine, the OH group is converted to OTs (a better leaving group). If sodium acetate functions as a base, an E2 process must occur to give alkene **2a**. Notice that the axial β proton (highlighted) is removed, since that proton is *anti*-periplanar with the leaving group. The equatorial proton is <u>not</u> *anti*-periplanar with the leaving group.

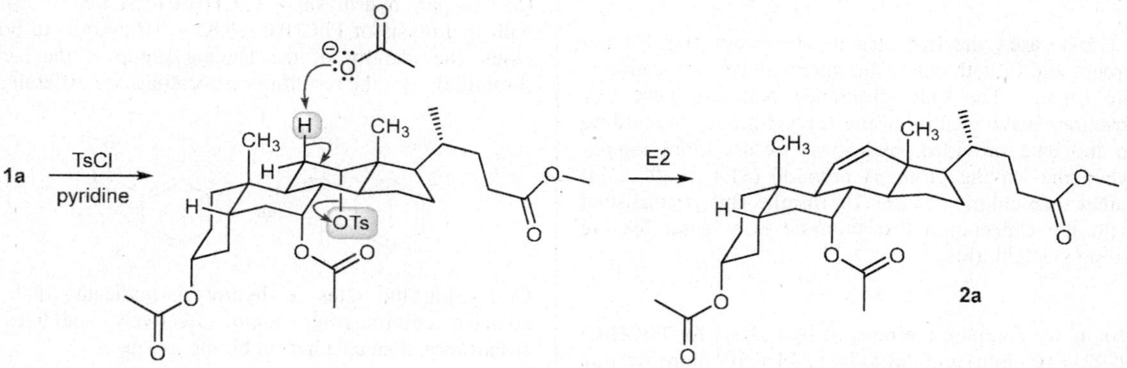

(b) We know that the axial proton (or deuteron) is removed in the elimination step. In compound **1b**, the axial position is occupied by a deuteron, so the deuteron is removed in the E2 elimination, and the product (**2b**) will not contain deuterium (and thus, **2b** is the same as **2a**). In compound **1c**, the axial position is occupied by a proton, which is removed during the elimination step. The deuteron in **1c** is in an equatorial position, so it survives the reaction. Compound **2c** will be deuterated, as seen below:

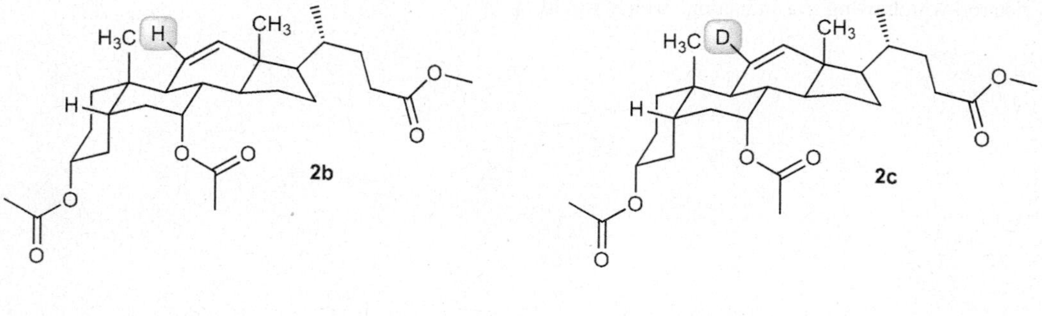

7.114.

(a) In order to compare the strength of these four bases, we can compare the pK_a values of their conjugate acids (see the pK_a table at the beginning of the textbook). Benzoic acid (pK_a = 4.8) is more acidic than phenol (pK_a = 9.9), which is in turn more acidic than trifluoroethanol (pK_a = 12.5), which is more acidic than ethanol (pK_a = 16.0):

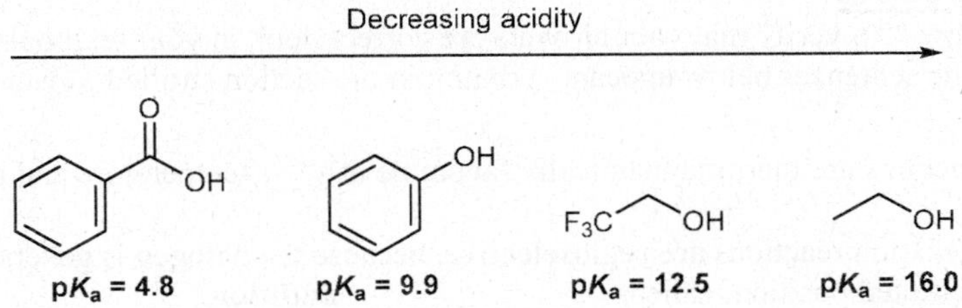

Decreasing acidity

pK_a = 4.8 pK_a = 9.9 pK_a = 12.5 pK_a = 16.0

Therefore, basicity is expected to increase in the order that the bases were presented. That is, potassium benzoate is the weakest base (among the four bases listed), because it is the conjugate base of the strongest acid. Ethoxide is the strongest base in the group, because it is the conjugate base of the weakest acid.

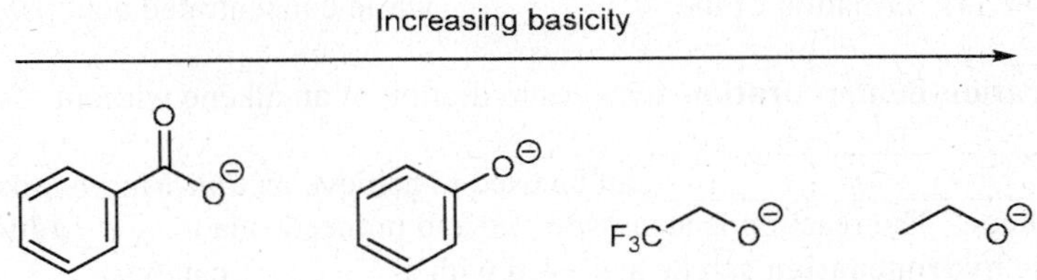

Increasing basicity

The data indicates that the percentage of 1-butene increases as the basicity of the base increases. This observation can also be stated in the following way: the preference for formation of the more-substituted alkene (the Zaitsev product) decreases as the base strength increases. A stronger base is a more reactive base. So we see that there is an inverse relationship between reactivity and selectivity. Specifically, a more-reactive reagent results in lower selectivity, while a less-reactive reagent results in higher selectivity. This is a trend that we will encounter several times throughout the remaining chapters of the textbook, so it would be wise to remember this trend.

(b) Based on the pK_a value of 4-nitrophenol, we can conclude that it is more acidic than phenol, but not quite as acidic as benzoic acid. Based on our answer for part (a), we would expect that the conjugate base of 4-nitrophenol will be a stronger base than potassium benzoate, but a weaker base than potassium phenoxide. As such, we expect the selectivity to be somewhere in between the selectivity of potassium benzoate and potassium phenoxide. So we would expect that the percentage of 1-butene should be somewhere between 7.2% and 11.4%.

Chapter 8
Addition Reactions of Alkenes

Review of Concepts

Fill in the blanks below. To verify that your answers are correct, look in your textbook at the end of Chapter 8. Each of the sentences below appears verbatim in the section entitled *Review of Concepts and Vocabulary*.

- Addition reactions are thermodynamically favorable at ____ temperature and disfavored at _____ temperature.
- Hydrohalogenation reactions are **regioselective**, because the halogen is generally installed at the ______ substituted position, called ______________ **addition.**
- In the presence of _____________, addition of HBr proceeds via an ***anti*-Markovnikov addition.**
- The regioselectivity of an ionic addition reaction is determined by the preference for the reaction to proceed through ______________________________.
- Acid-catalyzed hydration is inefficient when ___________________________ are a concern. Dilute acid favors formation of the ___________, while concentrated acid favors the ___________.
- **Oxymercuration-demercuration** achieves hydration of an alkene without ______________________________.
- _____________-_______________ can be used to achieve an *anti*-Markovnikov addition of water across an alkene. The reaction is stereospecific and proceeds via a _____ **addition.**
- **Asymmetric hydrogenation** can be achieved with a ________ catalyst.
- Bromination proceeds through a bridged intermediate, called a _______________ _______, which is opened by an S_N2 process that produces an _______ **addition.**
- A two-step procedure for *anti* dihydroxylation involves conversion of an alkene to an _________, followed by acid-catalyzed ring opening.
- **Ozonolysis** can be used to cleave a double bond and produce two ______ groups.
- The position of a halogen or OH group can be changed via _____________ followed by ___________.
- The position of a π bond can be changed via _____________ followed by ______________.

Review of Skills

Fill in the blanks and empty boxes below. To verify that your answers are correct, look in your textbook at the end of Chapter 8. The answers appear in the section entitled *SkillBuilder Review*.

8.1 Drawing a Mechanism for Hydrohalogenation

Step 1 - Draw two curved arrows showing protonation of the alkene, and draw the carbocation that is preferentially formed.	**Step 2** - Draw one curved arrow that shows the halide ion attacking the carbocation, and draw the product.

H—X

+ $X^{\ominus}$

8.2 Drawing a Mechanism for Hydrohalogenation with a Carbocation Rearrangement

Step 1 - Draw two curved arrows showing protonation of the alkene, and draw the carbocation that is initially formed.	**Step 2** - Draw one curved arrow showing a carbocation rearrangement, and draw the resulting, more stable carbocation.	**Step 3** - Draw one curved arrow showing the halide ion attacking the carbocation, and draw the product.

H—Cl

+ $Cl^{\ominus}$

:Cl:$^{\ominus}$

8.3 Drawing a Mechanism for an Acid-Catalyzed Hydration

Step 1 - Draw two curved arrows showing protonation of the alkene, and draw the resulting carbocation.	**Step 2** - Draw one curved arrow showing water attacking the carbocation, and draw the resulting oxonium ion.	**Step 3** - Draw two curved arrows showing deprotonation of the oxonium ion, and draw the resulting product.

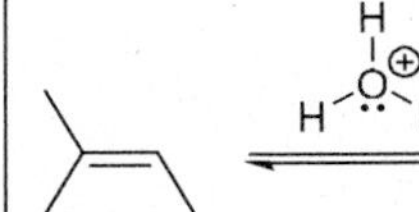
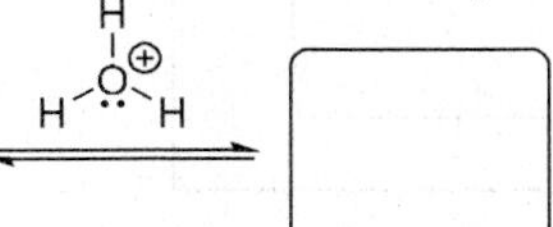
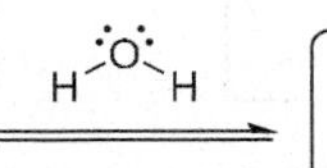
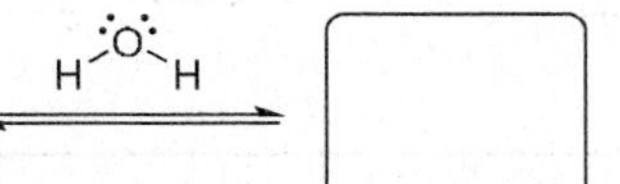

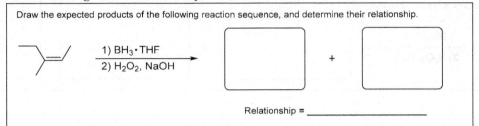

8.4 Predicting the Products of Hydroboration-Oxidation

Draw the expected products of the following reaction sequence, and determine their relationship.

1) $BH_3 \cdot THF$
2) H_2O_2, NaOH

+

Relationship = ____________________

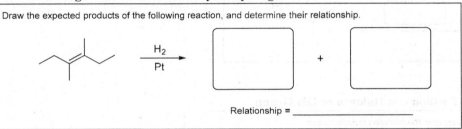

8.5 Predicting the Products of Catalytic Hydrogenation

Draw the expected products of the following reaction, and determine their relationship.

H_2
Pt

+

Relationship = ____________________

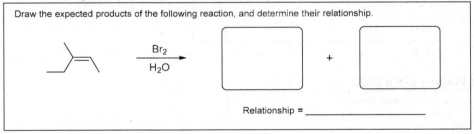

8.6 Predicting the Products of Halohydrin Formation

Draw the expected products of the following reaction, and determine their relationship.

Br_2
H_2O

+

Relationship = ____________________

8.7 Drawing the Products of *Anti* Dihydroxylation

Draw the expected products of the following reaction, and determine their relationship.

1) RCO_3H
2) H_3O^+

+

Relationship = ____________________

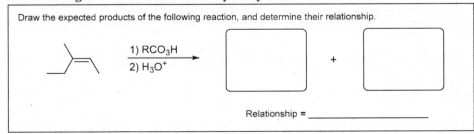

8.8 Predicting the Products of Ozonolysis

Draw the expected products of the following reaction:

1) O_3

2) DMS

+

8.9 Predicting the Products of an Addition Reaction

Draw the expected products of the following reaction:

1) $BH_3 \cdot THF$

2) H_2O_2, NaOH

+

8.10 Proposing a One-Step Synthesis

Identify reagents that will achieve the following transformation:

OH

8.11 Changing the Position of a Halogen or OH Group

Identify reagents that will achieve the following transformation:

Br

1)

2)

Br

8.12 Changing the Position of a π Bond

Identify reagents that will achieve the following transformation:

1)

2)

Review of Reactions

Identify the reagents necessary to achieve each of the following transformations. To verify that your answers are correct, look in your textbook at the end of Chapter 8. The answers appear in the section entitled *Review of Reactions*.

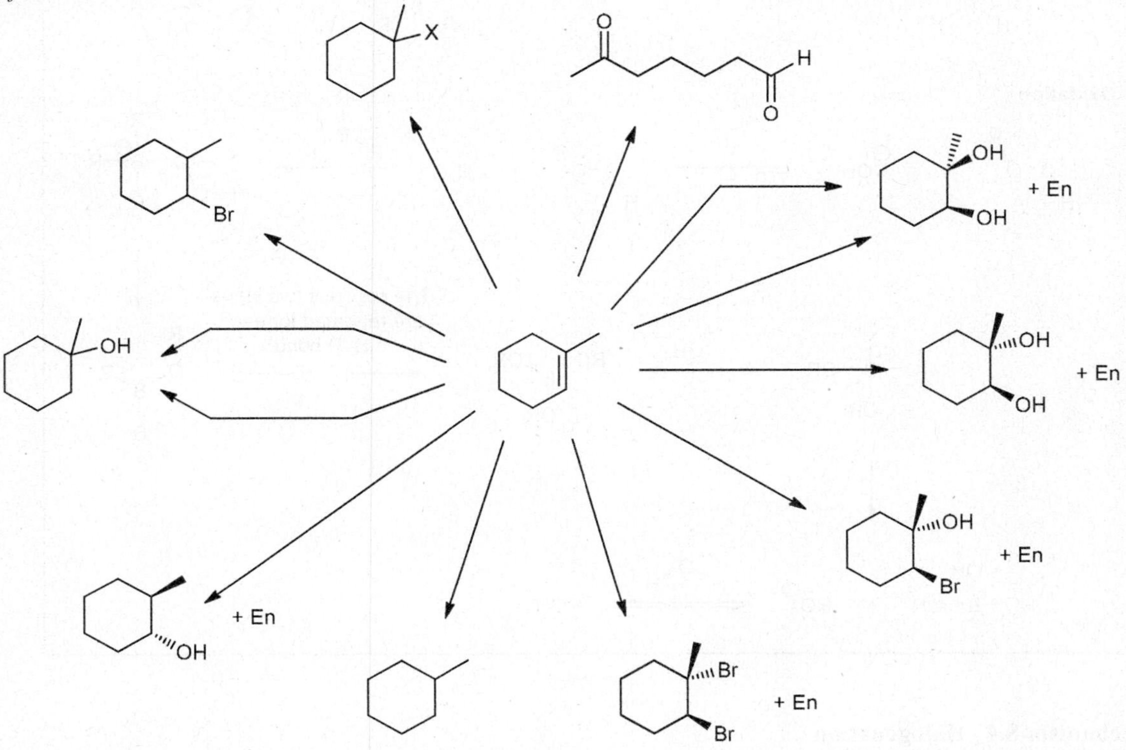

X

O

H

O

OH

+ En

OH

Br

OH

OH

+ En

OH

OH

+ En

Br

+ En

OH

Br

+ En

Br

Review of Mechanisms

Complete each of the following mechanisms by drawing the missing curved arrows. To verify that your curved arrows are drawn correctly, compare them to the curved arrows in the mechanism boxes for Mechanisms 8.1 – 8.6, which can be found throughout Chapter 8 of your text.

Mechanism 8.1 Hydrohalogenation

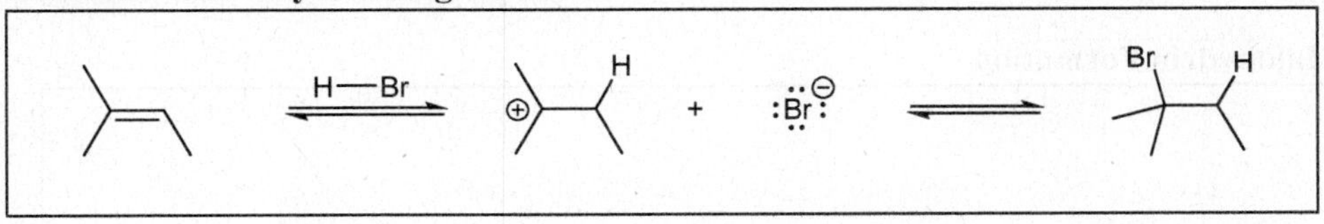

H—Br H + :Br: Br H

Mechanism 8.2 Acid-Catalyzed Hydration

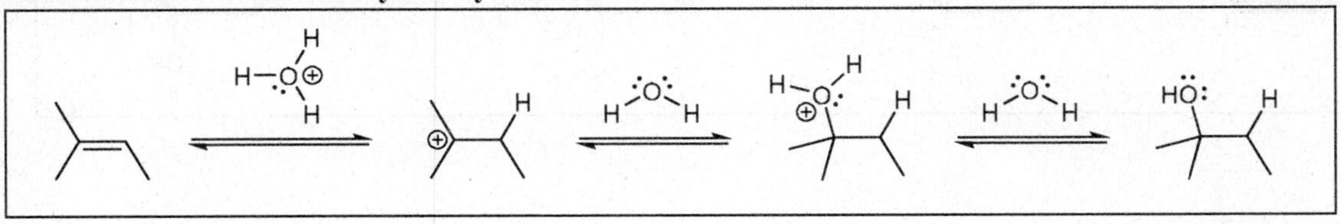

H H—O⊕ H H H—O—H H—O—H H H—O—H H HO: H

Mechanism 8.3 Hydroboration-Oxidation

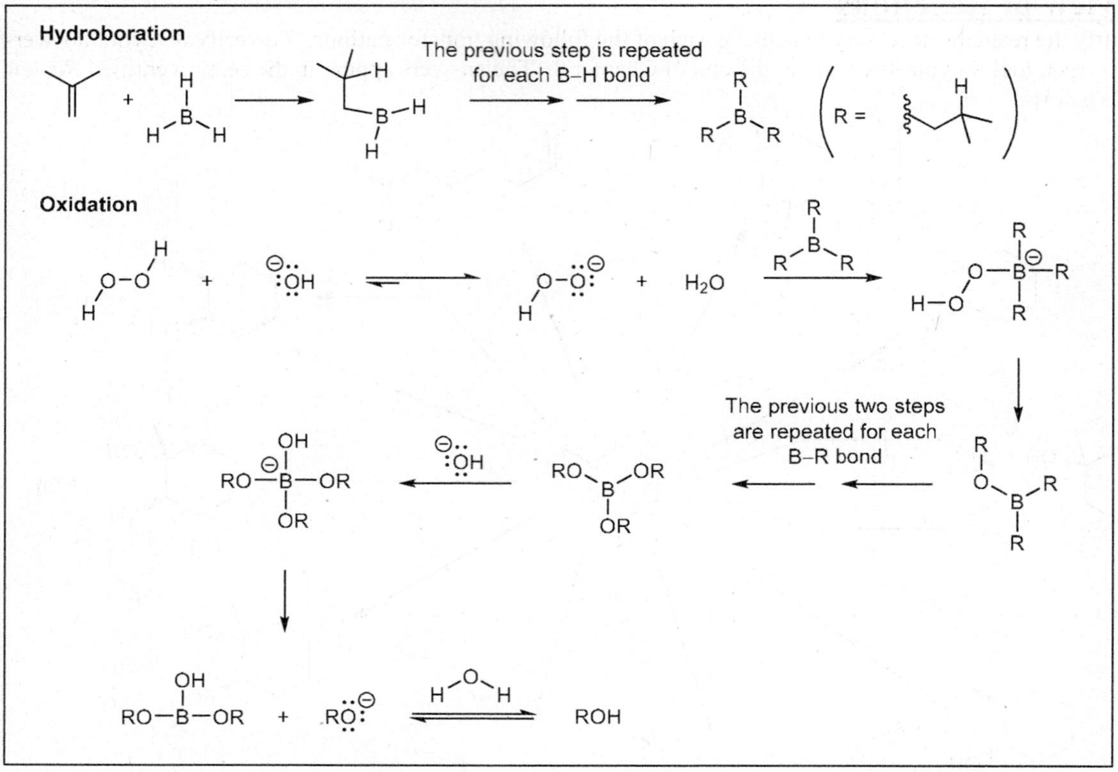

Hydroboration

The previous step is repeated for each B–H bond

R =

Oxidation

The previous two steps are repeated for each B–R bond

Mechanism 8.4 Halogenation

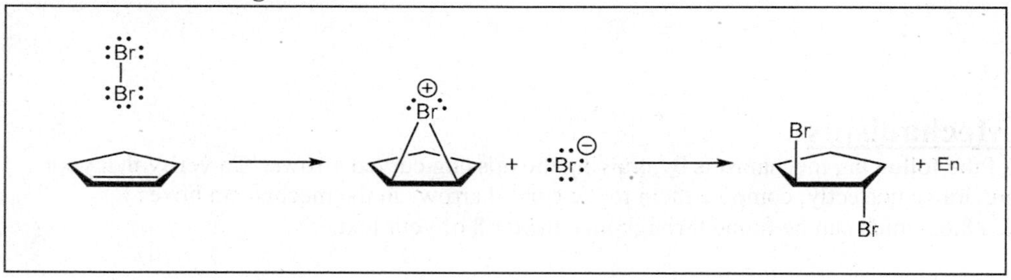

Mechanism 8.5 Halohydrin Formation

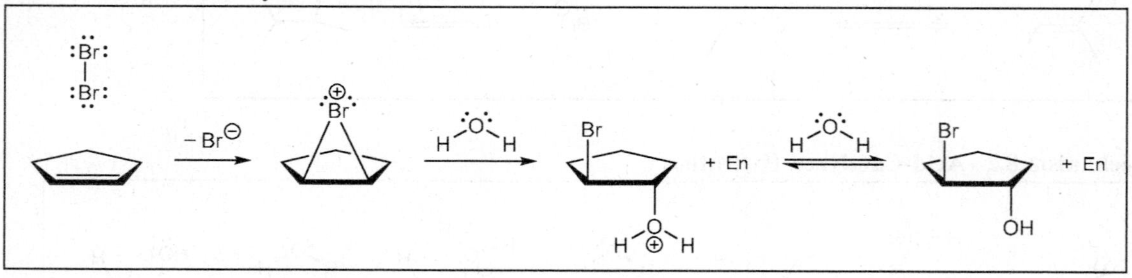

Mechanism 8.6 ***Anti*** **Dihydroxylation**

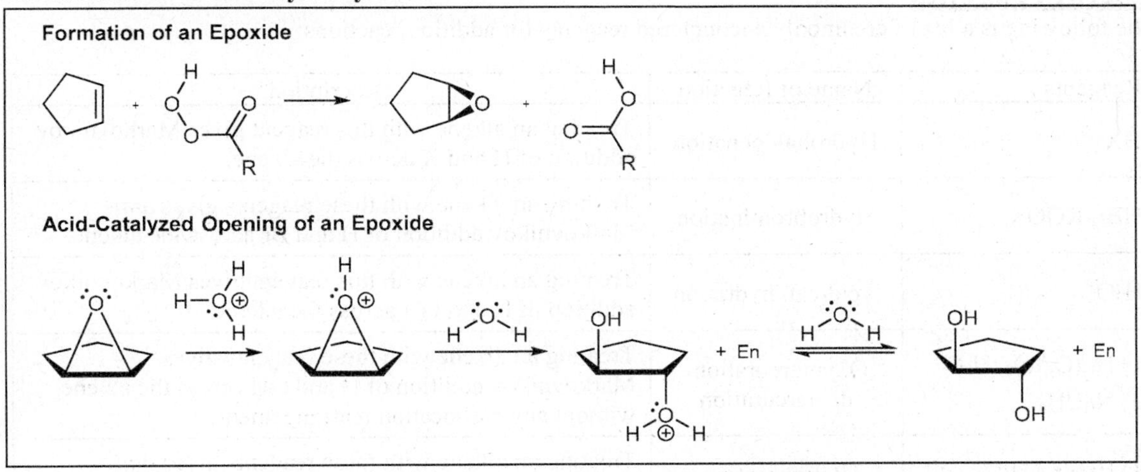

Formation of an Epoxide

+ → +

Acid-Catalyzed Opening of an Epoxide

→ + En ⇌ + En

Common Mistakes to Avoid

This chapter introduces several stereospecific addition reactions. Some of them occur exclusively via a *syn* addition (such as hydrogenation or hydroboration-oxidation), while others occur exclusively via an *anti* addition (such as bromination or halohydrin formation). When drawing the products of a stereospecific addition reaction, be careful to avoid drawing a wedge or a dash on a location that is not a chiral center. For example, consider the following *syn* dihydroxylation. In such a case, it is tempting for students to draw the products as if they have two chiral centers, like this:

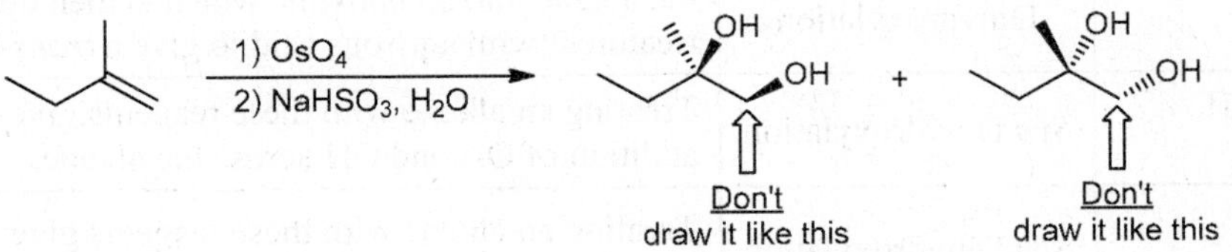

1) OsO_4
2) $NaHSO_3$, H_2O

Don't draw it like this   Don't draw it like this

This mistake is understandable – after all, the two OH groups are indeed added in a *syn* fashion. But the product does not contain two chiral centers. It has only one chiral center. As such, the products should be drawn like this:

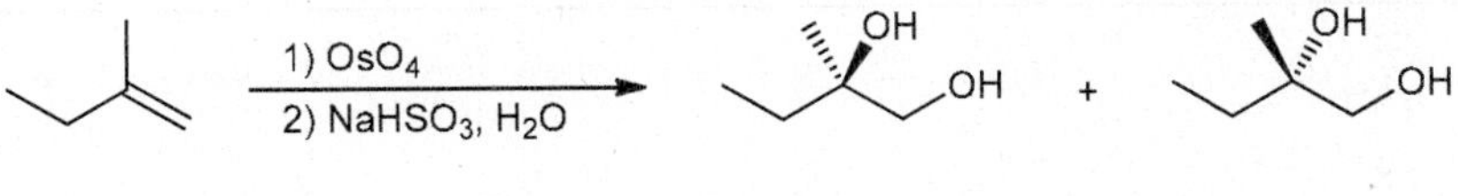

1) OsO_4
2) $NaHSO_3$, H_2O

Notice that the stereochemical requirement for *syn* addition is not relevant in this case, because only one chiral center is formed. As such, both enantiomers are produced, because *syn* addition can occur on either face of the alkene to give either enantiomer.

Useful reagents

The following is a list of commonly encountered reagents for addition reactions:

Reagents	Name of Reaction	Description
HX	Hydrohalogenation	Treating an alkene with this reagent gives Markovnikov addition of H and X across the alkene.
HBr, ROOR	Hydrobromination	Treating an alkene with these reagents gives *anti*-Markovnikov addition of H and Br across the alkene.
H_3O^+	Acid-cat. hydration	Treating an alkene with this reagent gives Markovnikov addition of H and OH across the alkene.
1) $Hg(OAc)_2$, H_2O 2) $NaBH_4$	Oxymercuration-demercuration	Treating an alkene with these reagents gives Markovnikov addition of H and OH across the alkene, without any carbocation rearrangements.
1) $BH_3 \bullet THF$ 2) H_2O_2, NaOH	Hydroboration-oxidation	Treating an alkene with these reagents gives *anti*-Markovnikov addition of H and OH across the alkene. The reaction proceeds exclusively via a *syn* addition.
H_2, Pt	Hydrogenation	Treating an alkene with these reagents gives *syn* addition of H and H across the alkene.
Br_2	Bromination	Treating an alkene with this reagent gives *anti* addition of Br and Br across the alkene.
Br_2, H_2O	Halohydrin formation	Treating an alkene with these reagents gives *anti* addition of Br and OH across the alkene, with the OH group being installed at the more substituted position.
1) RCO_3H 2) H_3O^+	*Anti* Dihydroxylation	Treating an alkene with a peroxy acid (RCO_3H) converts the alkene into an epoxide, which is then opened upon treatment with aqueous acid to give a *trans*-diol.
$KMnO_4$, NaOH, cold	*Syn* Dihydroxylation	Treating an alkene with these reagents gives *syn* addition of OH and OH across the alkene.
1) OsO_4 2) $NaHSO_3$, H_2O	*Syn* Dihydroxylation	Treating an alkene with these reagents gives *syn* addition of OH and OH across the alkene.
1) O_3 2) DMS	Ozonolysis	Ozonolysis of an alkene causes cleavage of the C=C bond, giving two compounds, each of which possesses a C=O bond.

Solutions

8.1.

(a) We begin by identifying the parent. The longest chain is six carbon atoms, so the parent is hexene. There is one substituent (highlighted), which is a methyl group.

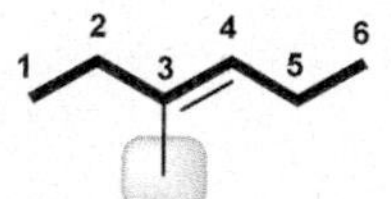

With respect to the π bond, the parent could be numbered from either direction (either way, the double bond will be between C3 and C4), but in this case, we must assign numbers from left to right to give the substituent the lower possible number (C3 rather than C4). We include a locant that identifies the position of the double bond ("3" indicates that the double bond is located between C3 and C4), as well as a locant to identify the position of the substituent. Furthermore, we must include the configuration of the double bond (*E*):

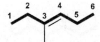

(*E*)-3-methyl-3-hexene

(b) We begin by identifying the parent, which must include the two carbon atoms bearing the double bond. The longest possible chain has five carbon atoms, so the parent is pentene. There is more than one choice for the parent, and we choose the parent with the greater number of substituents:

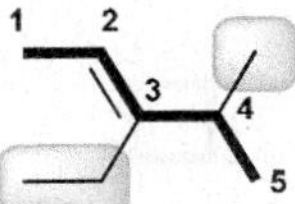

There are two substituents (highlighted): a methyl group and an ethyl group. The parent is numbered to give the double bond the lowest possible number (C2). Therefore, the ethyl group is located at C3, and the methyl group is located at C4. These groups are arranged alphabetically, together with their locants, in the name. Finally, we must include the configuration of the double bond (*E*) at the beginning of the name:

(*E*)-3-ethyl-4-methyl-2-pentene

(c) We begin by identifying the parent. The longest chain is seven carbon atoms, so the parent is heptene. There are three substituents (highlighted), all of which are methyl groups. Notice that the parent chain is numbered starting from the side that is closest to the π bond. According to this numbering scheme, the methyl groups are located at C2, C3, and C5. Finally, we use the prefix "tri" to indicate the presence of three methyl groups, and we include a locant that identifies the position of the double bond ("2" indicates that the double bond is located between C2 and C3):

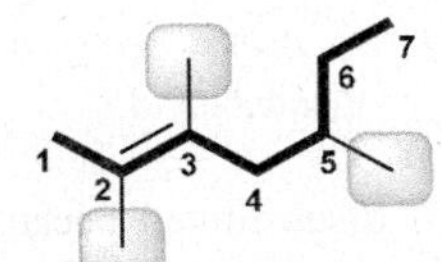

2,3,5-trimethyl-2-heptene

Note that the C2 position is connected to two methyl groups, so the double bond is not stereoisomeric (neither *E* nor *Z*).

(d) We begin by identifying the parent. The longest chain is seven carbon atoms, so the parent is heptene. There are two substituents – a methyl group and an ethyl group (highlighted). Notice that the parent chain is numbered starting from the side that is closest to the π bond. According to this numbering scheme, the methyl group is located at C2, and the ethyl group is located at C3. Finally, we arrange the substituents alphabetically, and we include a locant that identifies the position of the double bond:

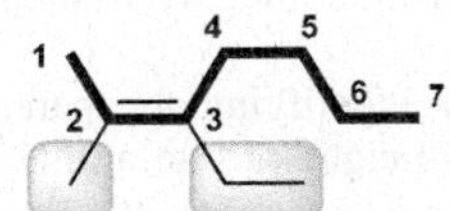

3-ethyl-2-methyl-2-heptene

Note that the C2 position is connected to two methyl groups, so the double bond is not stereoisomeric (neither *E* nor *Z*).

(e) We begin by identifying the parent. The longest chain (containing the double bond) is five carbon atoms, so the parent is pentene. There are three substituents – an isopropyl group and two methyl groups (highlighted). Notice that the parent chain is numbered starting from the side that is closest to the π bond. According to this numbering scheme, the isopropyl group is located at C3, and the methyl groups are located at C2 and C4. Finally, we arrange the substituents alphabetically, and we include a locant that identifies the position of the double bond:

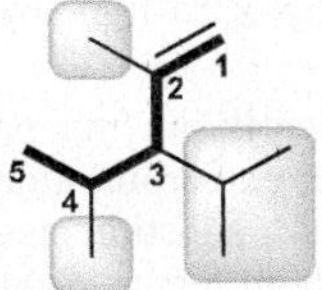

3-isopropyl-2,4-dimethyl-1-pentene

Note that the C1 position is connected to two hydrogen atoms, so the double bond is not stereoisomeric (not *E* or *Z*).
Recall that the systematic name for an isopropyl group is "1-methylethyl." In this case, renaming the substituent also changes the alphabetical order of the substituents. The following is an alternative name:

2,4-dimethyl-3-(1-methylethyl)-1-pentene.

(f) We begin by identifying the parent. The longest chain is seven carbon atoms, so the parent is heptene. There is one substituent – a *tert*-butyl group (highlighted). Notice that the parent chain is numbered starting from the side that is closest to the π bond. According to this numbering scheme, the *tert*-butyl group is located at C4. Finally, we include a locant that identifies the position of the double bond:

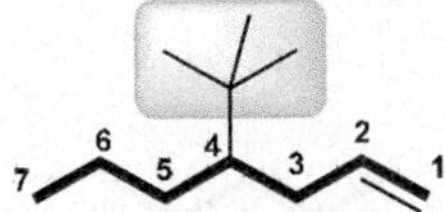

4-*tert*-butyl-1-heptene

Note that the C1 position is connected to two hydrogen atoms, so the double bond is not stereoisomeric (not *E* or *Z*).

Recall that the systematic name for a *tert*-butyl group is "1,1-dimethylethyl," so the following is an alternative name:

4-(1,1-dimethylethyl)-1-heptene

(g) We begin by identifying the parent. The longest possible chain has eight carbon atoms, so the parent is octene. There are five substituents (highlighted). With respect to the π bond, the parent could be numbered from either direction (either way, the double bond will be between C4 and C5), but in this case, we must assign numbers from left to right to give the second substituent the lower possible number (2,<u>2</u>,7-trimethyl, rather than 2,<u>7</u>,7-trimethyl). We include a locant to identify the position of the double bond. Substituents are arranged alphabetically, together with their locants, in the name. Furthermore, we must include the configuration of the double bond (*E*) at the beginning of the name:

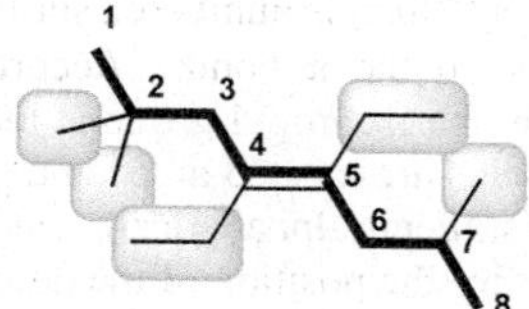

(*E*)-4,5-diethyl-2,2,7-trimethyl-4-octene

(h) We begin by identifying the parent. The longest possible chain has seven carbon atoms, so the parent is heptene. There are six substituents (highlighted). The parent is numbered from left to right to give the double bond the lower possible number. Substituents are arranged alphabetically (ethyl before methyl), together with their locants, in the name. Furthermore, we must include the configuration of the double bond (*E*) at the beginning of the name:

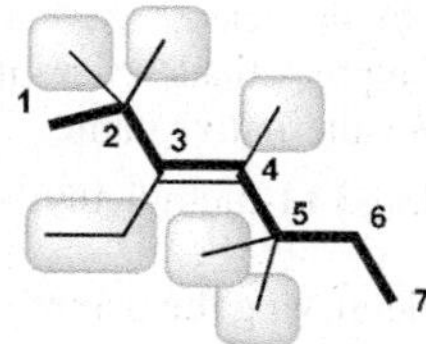

(*E*)-3-ethyl-2,2,4,5,5-pentamethyl-3-heptene

8.2.

(a) The parent is five carbon atoms (pentene), with the double bond between C2 and C3. There are three substituents – an isopropyl group at C3, and two methyl groups at C2 and C4:

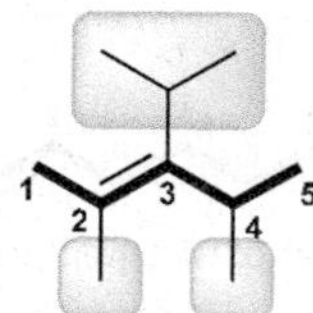

3-isopropyl-2,4-dimethyl-2-pentene

(b) The parent is six carbon atoms (hexene), with the double bond between C2 and C3. There are two substituents – an ethyl group at C4, and a methyl group at C2:

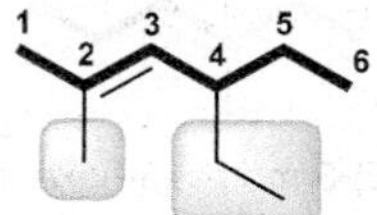

4-ethyl-2-methyl-2-hexene

(c) The parent is a four-membered ring (cyclobutene). There are two substituents located at C1 and C2, which are (by definition) the positions bearing the double bond:

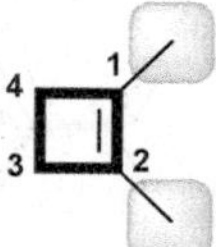

1,2-dimethylcyclobutene

8.3. We begin by identifying the parent, which is bicyclic in this case. The parent is bicyclo[2.2.1]heptene. There are two substituents (highlighted), both of which are methyl groups. Notice that the parent chain is numbered starting from one of the bridgeheads, as seen in Section 4.2, which places the double bond between C2 and C3. According to this numbering scheme, the methyl groups are also located at C2 and C3. Finally, we include a locant that identifies the position of the double bond (C2).

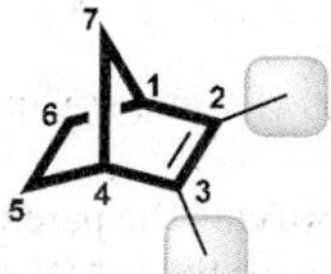

2,3-dimethylbicyclo[2.2.1]hept-2-ene

8.4.

(a) This alkene is trisubstituted because there are three carbon atoms (highlighted) connected to the double bond:

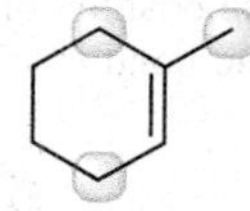

trisubstituted

(b) This alkene is disubstituted because there are two carbon atoms (highlighted) connected to the double bond:

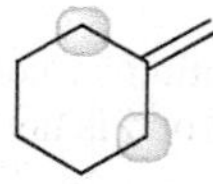

disubstituted

(c) This alkene is trisubstituted because there are three carbon atoms (highlighted) connected to the double bond.

Notice that one of the groups counts twice because it is connected to both vinylic positions:

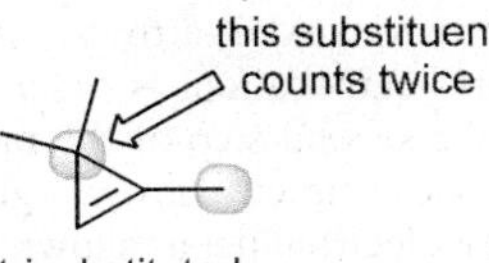

(d) This alkene is trisubstituted because there are three carbon atoms (highlighted) connected to the double bond:

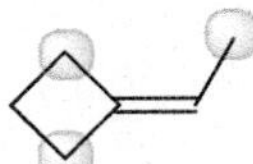

trisubstituted

(e) This alkene is monosubstituted because there is only one carbon atom (highlighted) connected to the double bond:

monosubstituted

8.5.

(a) An alkene is treated with HBr (in the absence of peroxides), so we expect a Markovnikov addition of H and Br across the π bond. That is, Br is installed at the more substituted position:

HBr → Br

(b) An alkene is treated with HBr in the presence of peroxides, so we expect an *anti*-Markovnikov addition of H and Br across the π bond. That is, Br is installed at the less substituted position:

HBr / ROOR → Br

(c) An alkene is treated with HBr (in the absence of peroxides), so we expect a Markovnikov addition of H and Br across the π bond. That is, Br is installed at the more substituted position:

HBr → Br

(d) An alkene is treated with HCl, so we expect a Markovnikov addition of H and Cl across the π bond. That is, Cl is installed at the more substituted position:

HCl → Cl

(e) An alkene is treated with HI, so we expect a Markovnikov addition of H and I across the π bond. That is, iodine is installed at the more substituted position:

HI → I

(f) An alkene is treated with HBr in the presence of peroxides, so we expect an *anti*-Markovnikov addition of H and Br across the π bond. That is, Br is installed at the less substituted position:

HBr / ROOR → Br

8.6.

(a) The desired transformation is a Markovnikov addition of H and Br across the π bond. This can be achieved by treating the alkene with HBr (in the absence of peroxides).

HBr → Br

(b) The desired transformation is an *anti*-Markovnikov addition of H and Br across the π bond. This can be achieved by treating the alkene with HBr in the presence of peroxides.

HBr / ROOR → Br

8.7.

(a) In this reaction, H and Br are added across the alkene in a Markovnikov addition, which indicates an ionic process. There are two mechanistic steps in the ionic addition of HBr across an alkene: 1) proton transfer, followed by 2) nucleophilic attack. In the first step, a proton is transferred from HBr to the alkene, which requires two curved arrows, as shown below. The resulting, tertiary carbocation is then captured by a bromide ion in the second step of the mechanism. This step requires one curved arrow, going from the nucleophile to the electrophile, as shown:

H—Br ⇌ ⊕ :Br:⊖ ⇌ Br

(b) In this reaction, H and Cl are added across the alkene in a Markovnikov addition. There are two mechanistic steps in the ionic addition of HCl across an alkene: 1) proton transfer, followed by 2) nucleophilic attack. In the first step, a proton is transferred from HCl to the alkene, which requires two curved arrows, as shown below. The resulting, tertiary carbocation is then captured by a chloride ion in the second step of the mechanism. This step requires one curved arrow, going from the nucleophile to the electrophile, as shown:

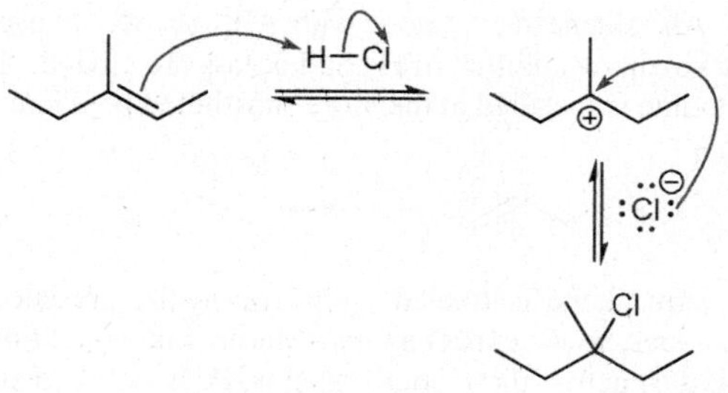

(c) In this reaction, H and Cl are added across the alkene in a Markovnikov addition. There are two mechanistic steps in the ionic addition of HCl across an alkene: 1) proton transfer, followed by 2) nucleophilic attack. In the first step, a proton is transferred from HCl to the alkene, which requires two curved arrows, as shown. The resulting, tertiary carbocation is then captured by a chloride ion in the second step of the mechanism. This step requires one curved arrow, going from the nucleophile to the electrophile, as shown:

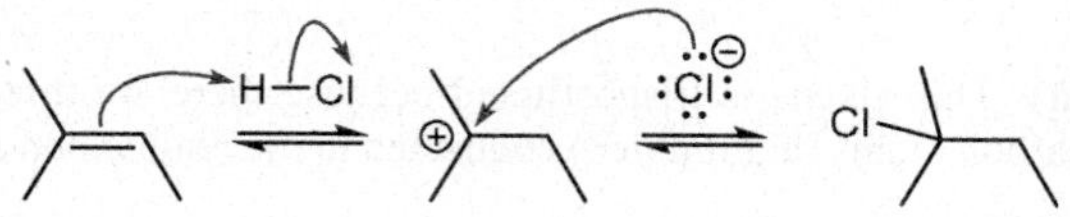

8.8. Compound **1** has two C=C double bonds and each of them can react with HCl in an ionic reaction (Markovnikov addition). Indeed, the molecular formula of compound **2** indicates that two addition reactions must have occurred, because compound **2** has two Cl atoms. We will consider each addition process separately, and in each case, we will focus on the regiochemical outcome. Let's begin by drawing a mechanism for the addition of HCl to the double bond that is adjacent to the methoxy (OCH_3) group). In the first step, a proton is transferred from HCl to the double bond, giving the more stable carbocation. This step requires two curved arrows:

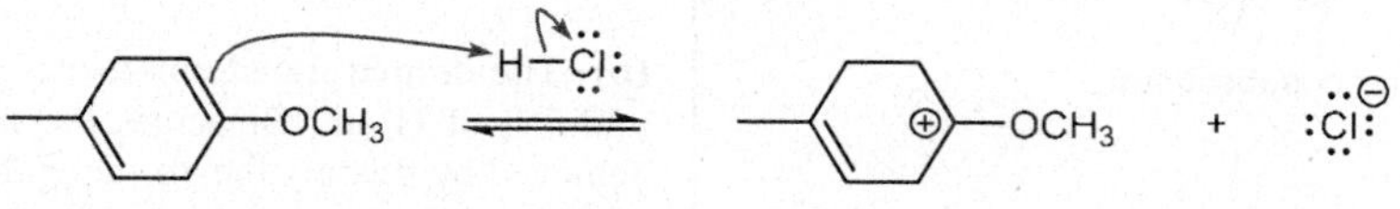

This carbocation is the more stable carbocation because it is stabilized by resonance, as shown:

Indeed, the second resonance structure has completely filled octets, rendering this cationic intermediate particularly stable. Then, in the second step of the process, the cation is captured by a chloride ion to generate an alkyl chloride. This step requires one curved arrow:

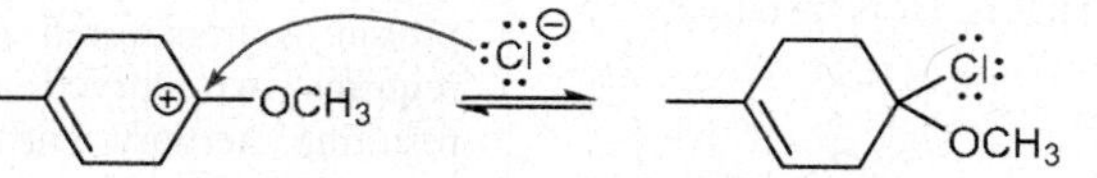

The other C=C bond also reacts with HCl via a similar two-step process:

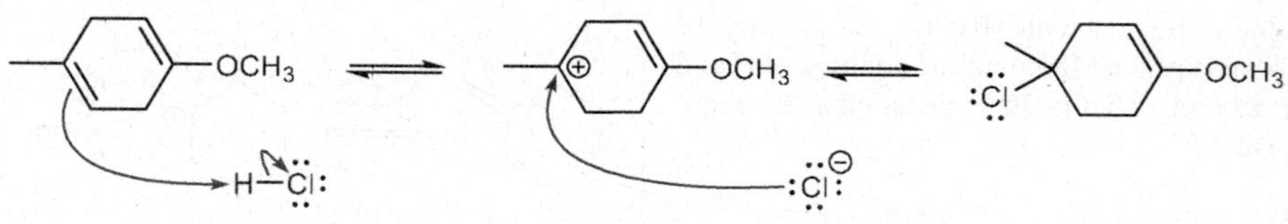

First, a proton is transferred from HCl to the double bond, giving the more stable tertiary carbocation (rather than a secondary carbocation). This carbocation is then captured by a chloride ion to generate a tertiary alkyl chloride.

In summary, there are two addition processes, and each of them occurs, one after the other, to produce a dichloride with predictable regiochemistry. When drawing a complete mechanism, make sure to draw each step separately. That is, draw one process followed by the other. Either process can be drawn first. Both possibilities are shown here:

Possibility #1

Possibility #2

Compound 2

Compound 2

While both possibilities are certainly acceptable, the order of events in possibility #1 is more likely, due to the initial formation of the more stable (resonance-stabilized) intermediate.

8.9.

(a) In this case, Markovnikov addition of HBr involves the formation of a new chiral center. As such, we expect both possible stereochemical outcomes. That is, we expect a pair of enantiomers, as shown:

HBr

enantiomers

(b) In this case, Markovnikov addition of HCl does not involve the formation of a new chiral center (the α carbon of the resulting alkyl halide bears two propyl groups):

HCl

(c) In this case, Markovnikov addition of HBr involves the formation of a new chiral center. As such, we expect both possible stereochemical outcomes. That is, we expect a pair of enantiomers, as shown:

HBr

enantiomers

(d) In this case, Markovnikov addition of HI involves the formation of a new chiral center. As such, we expect both possible stereochemical outcomes. That is, we expect a pair of enantiomers, as shown:

HI

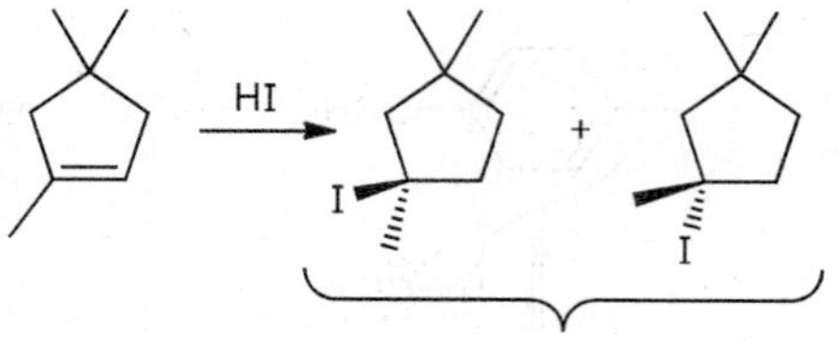

enantiomers

(e) In this case, Markovnikov addition of HCl does not involve the formation of a new chiral center (the α carbon of the resulting alkyl halide bears two methyl groups):

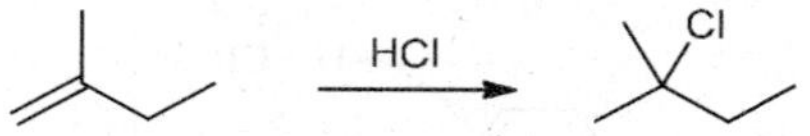

HCl

(f) In this case, addition of HCl involves the formation of a new chiral center. As such, we expect both possible stereochemical outcomes. That is, we expect a pair of enantiomers, as shown:

HCl

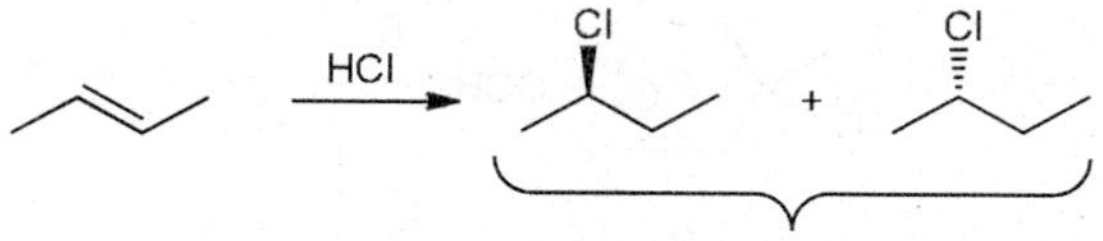

enantiomers

Note that the starting alkene is symmetrical, so there is only one regiochemical outcome.

8.10.

(a) Protonation of the alkene requires two curved arrows, as shown, and leads to the more stable, secondary carbocation (rather than a primary carbocation). This secondary carbocation then undergoes a hydride shift, shown with one curved arrow, generating a more stable, tertiary carbocation. In the final step (nucleophilic attack), the carbocation is captured by a bromide ion. This step requires one curved arrow, going from the nucleophile (bromide) to the electrophile (the carbocation), as shown:

Hydride Shift

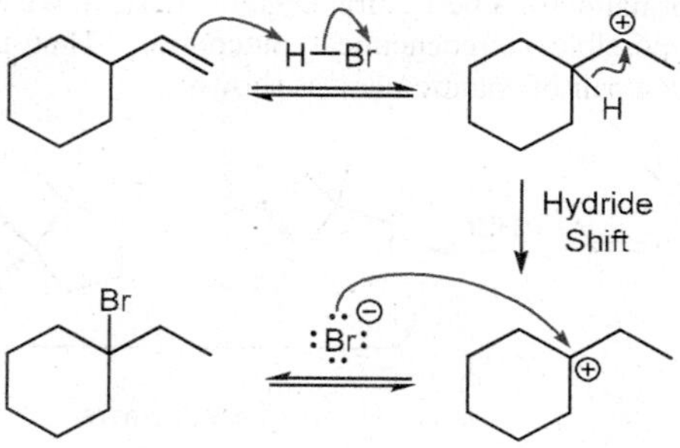

(b) Protonation of the alkene requires two curved arrows, as shown, and leads to the more stable, secondary carbocation (rather than a primary carbocation). This secondary carbocation then undergoes a hydride shift, shown with one curved arrow, generating a more stable, tertiary carbocation. In the final step (nucleophilic attack), the carbocation is captured by a bromide ion. This step requires one curved arrow, going from the nucleophile (bromide) to the electrophile (the carbocation), as shown:

Hydride Shift

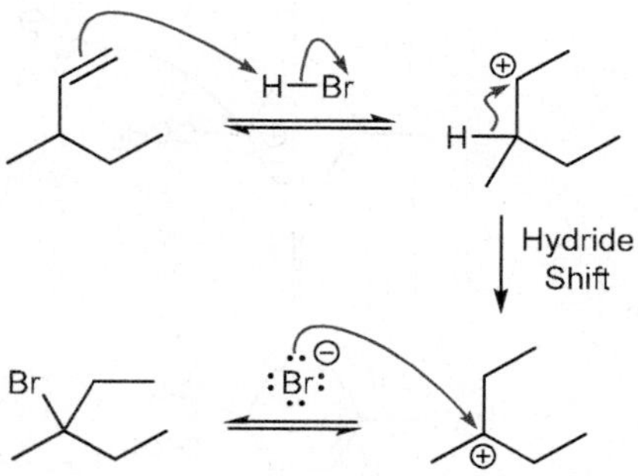

(c) Protonation of the alkene requires two curved arrows, generating a secondary carbocation. This secondary carbocation then undergoes a methyl shift, shown with one curved arrow, generating a more stable, tertiary carbocation. In the final step of the mechanism (nucleophilic attack), the carbocation is captured by a chloride ion. This step requires one curved arrow, going from the nucleophile (chloride) to the electrophile (the carbocation), as shown:

Methyl Shift

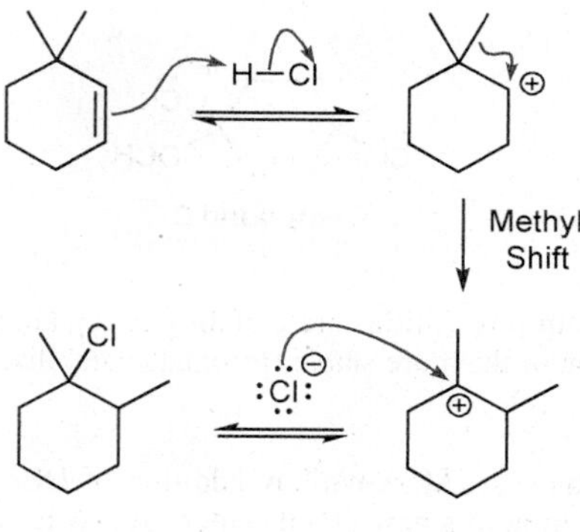

8.11.

(a) The mechanism begins with protonation of the carbon-carbon double bond to form a tertiary carbocation. This step requires two curved arrows:

1

2

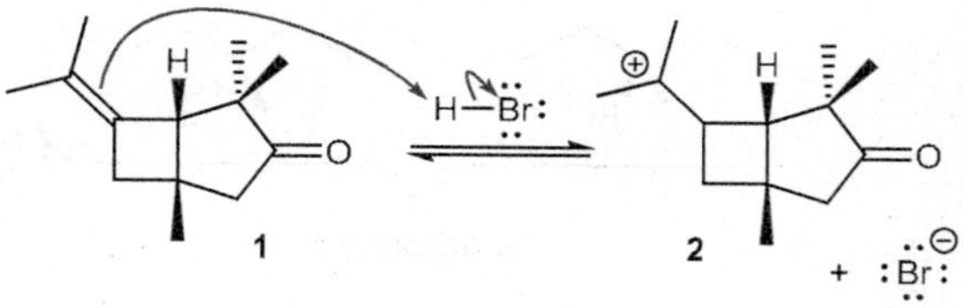

Although both ends of the double bond are doubly substituted, and therefore would yield a tertiary carbocation, only structure **2** can enable the subsequent rearrangement.

Next, the highlighted carbon atom shifts, giving the rearrangement shown. This step requires one curved arrow:

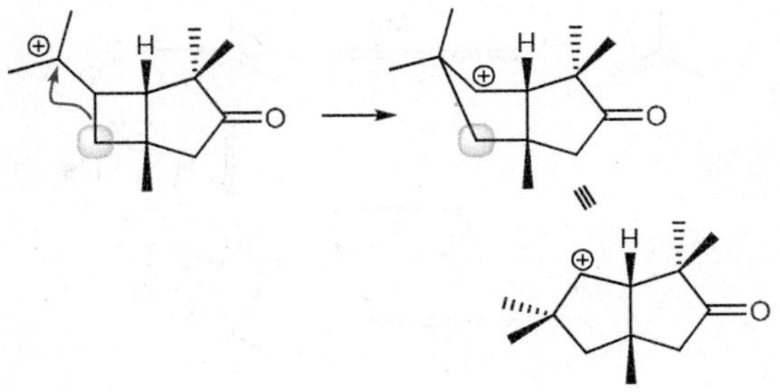

In the final step of the mechanism, a bromide ion (produced in the first step) captures the carbocation to form an alkyl bromide. This step requires one curved arrow:

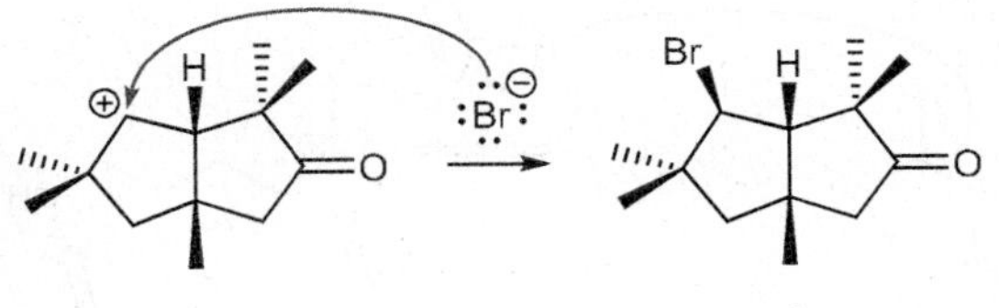

3 4

(b) This rearrangement involves a ring expansion, and it is favorable because ring strain is released when a strained four-membered ring becomes a more stable five-membered ring, even though the resulting carbocation is secondary rather than tertiary.

8.12.

(a) The second compound is expected to be more reactive toward acid-catalyzed hydration than the first compound, because the reaction proceeds via a tertiary carbocation, rather than via a secondary carbocation, as shown.

Secondary Tertiary (more stable)

(b) Begin by drawing the compounds:

2-Methyl-2-butene 3-Methyl-1-butene

The first compound (2-methyl-2-butene) is expected to be more reactive toward acid-catalyzed hydration than the second compound, because the reaction proceeds via a tertiary carbocation, rather than a secondary carbocation.

Tertiary (more stable) Secondary

8.13.

(a) To favor the alcohol, dilute sulfuric acid (mostly water) is used. The presence of a lot of water favors the alcohol, according to Le Châtelier's principle.

(b) To favor the alkene, concentrated sulfuric acid (which has less water than dilute acid) is used. With less water present, the alkene is favored, according to Le Châtelier's principle.

8.14.

(a) Water (H and OH) is added across the alkene in a Markovnikov fashion. The mechanism is expected to have three steps: 1) proton transfer, 2) nucleophilic attack, and 3) proton transfer. In the first step, a proton is transferred from H_3O^+ to the alkene, which requires two curved arrows, as shown. The resulting tertiary carbocation is then captured by a water molecule in the second step of the mechanism. This step requires one curved arrow, going from the nucleophile (water) to the electrophile (the carbocation). Then, in the final step of the mechanism, a molecule of water functions as a base and removes a proton, thereby generating the product. This final step is a proton transfer step, and therefore requires two curved arrows, as shown:

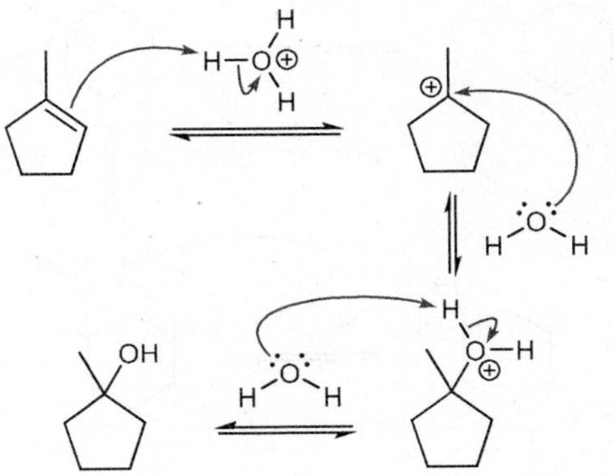

(b) Water (H and OH) is added across the alkene in a Markovnikov fashion. The mechanism is expected to have three steps: 1) proton transfer, 2) nucleophilic attack, and 3) proton transfer. In the first step, a proton is transferred from H_3O^+ to the alkene, which requires two curved arrows, as shown below. The resulting tertiary carbocation is then captured by a water molecule in the second step of the mechanism. This step requires one curved arrow, going from the nucleophile (water) to the electrophile (the carbocation). Then, in the final step of the mechanism, a molecule of water functions as a base and removes a proton, thereby generating the product.

This final step is a proton transfer step, and therefore requires two curved arrows, as shown:

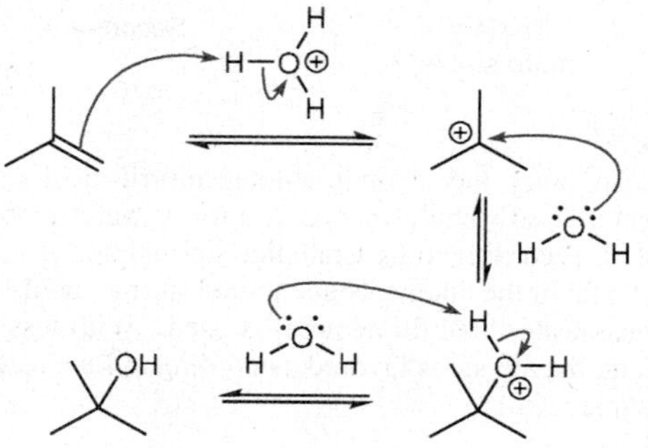

(c) Water (H and OH) is added across the alkene in a Markovnikov fashion. The mechanism is expected to have three steps: 1) proton transfer, 2) nucleophilic attack, and 3) proton transfer. In the first step, a proton is transferred from H_3O^+ to the alkene, which requires two curved arrows, as shown. The resulting tertiary carbocation is then captured by a water molecule in the second step of the mechanism. This step requires one curved arrow, going from the nucleophile (water) to the electrophile (the carbocation). Then, in the final step of the mechanism, a molecule of water functions as a base and removes a proton, thereby generating the product. This final step is a proton transfer step, and therefore requires two curved arrows, as shown:

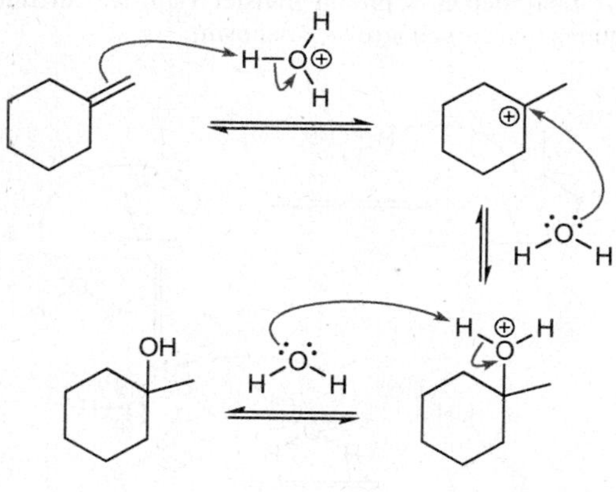

8.15.

(a) Methanol (H and OCH_3) is added across the alkene in a Markovnikov fashion. The reaction is extremely similar to the addition of water across an alkene under acid-catalyzed conditions, so we expect the mechanism to have three steps: 1) proton transfer, 2) nucleophilic attack, and 3) proton transfer. In the first step, a proton is transferred from $CH_3OH_2^+$ to the alkene, which requires two curved arrows, as shown below. The resulting tertiary carbocation is then captured by a molecule of methanol in the second step of the mechanism. This step requires one curved arrow, going from the nucleophile (methanol) to the electrophile (the carbocation). Then, in the final step of the mechanism, a molecule of methanol functions as a base and removes a proton, thereby generating the product. This final step is a proton transfer step, and therefore requires two curved arrows, as shown:

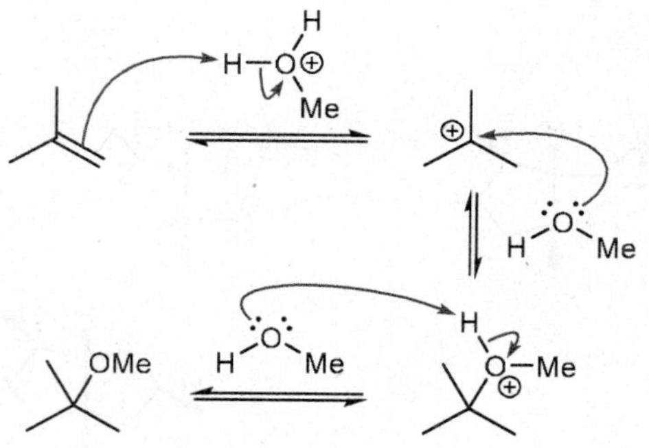

(b) The reactant is acyclic (it does not have a ring), and the product is cyclic, indicating an intramolecular reaction. We can justify an intramolecular reaction if we inspect the cation that is obtained upon protonation of the alkene:

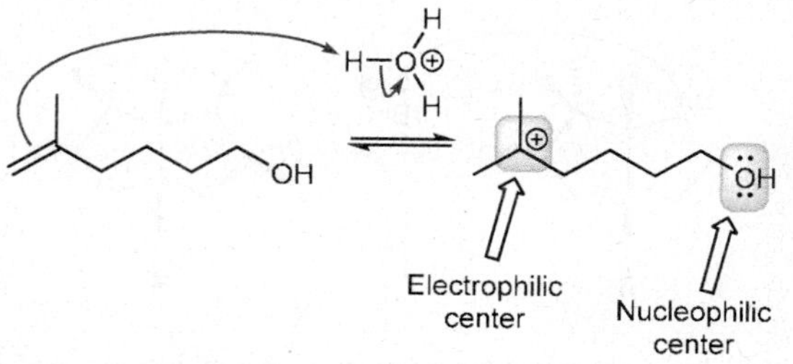

Electrophilic center

Nucleophilic center

Notice that this intermediate exhibits both an electrophilic center and a nucleophilic center. That is, the reactive centers are tethered together, via a chain of methylene (CH_2) groups. As such, a ring is formed in the following intramolecular nucleophilic attack, which is shown with one curved arrow:

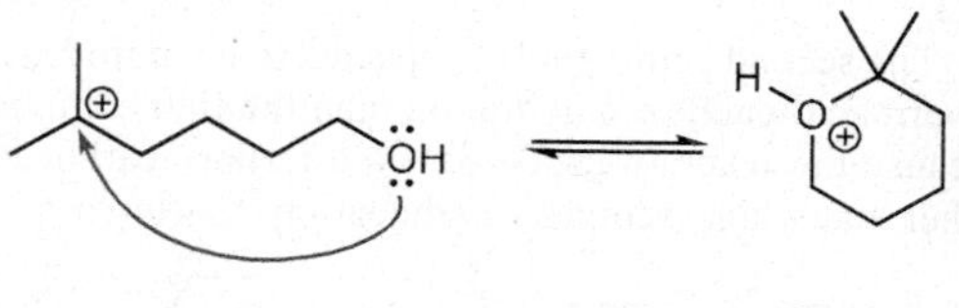

Finally, water functions as a base and removes a proton, thereby generating the product. This final step is a proton transfer step, and therefore requires two curved arrows, as shown:

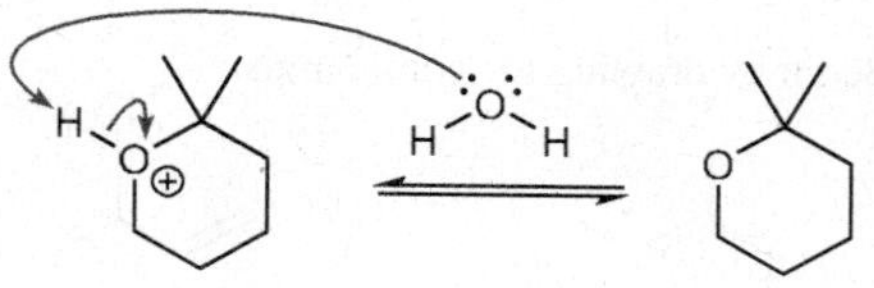

8.16.

(a) Oxymercuration-demercuration gives Markovnikov addition of water (H and OH) without carbocation rearrangements. That is, the OH group ends up at the more substituted (secondary) position, and the proton ends up at the less substituted (primary) position:

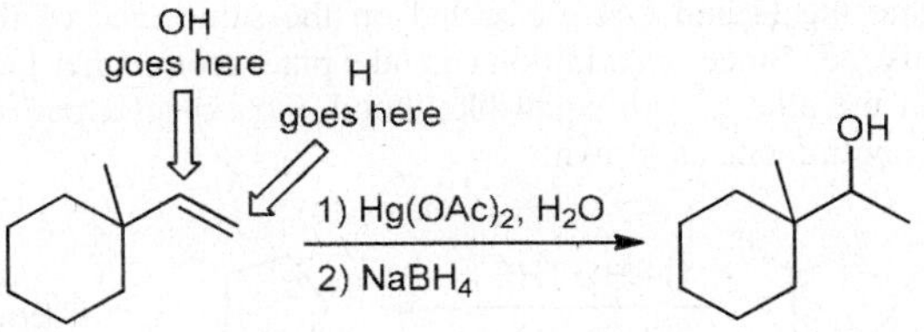

OH goes here H goes here

1) $Hg(OAc)_2$, H_2O
2) $NaBH_4$

If the same alkene were treated with aqueous acid, the resulting acid-catalyzed hydration would involve a carbocation rearrangement:

Methyl shift

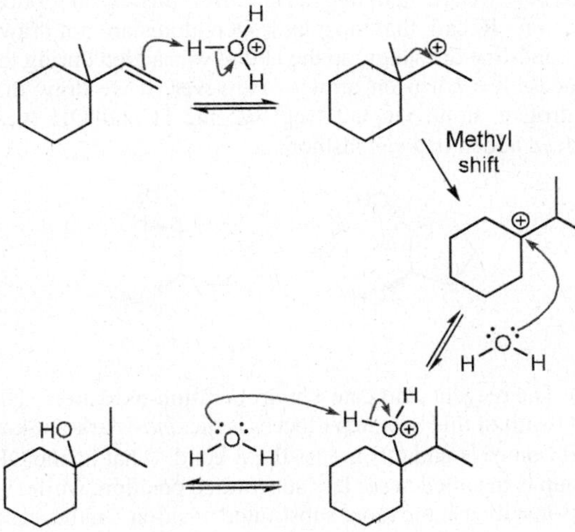

(b) Oxymercuration-demercuration gives Markovnikov addition of water (H and OH) without carbocation rearrangements. That is, the OH group ends up at the more substituted (secondary) position, and the proton ends up at the less substituted (primary) position:

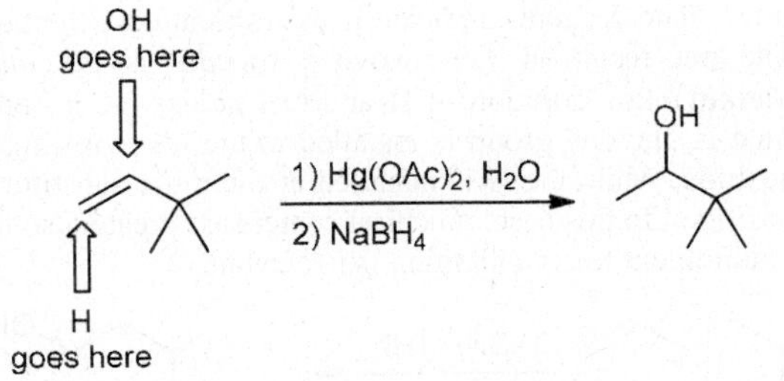

OH goes here

1) $Hg(OAc)_2$, H_2O
2) $NaBH_4$

H goes here

If the same alkene were treated with aqueous acid, the resulting acid-catalyzed hydration would involve a carbocation rearrangement:

Methyl shift

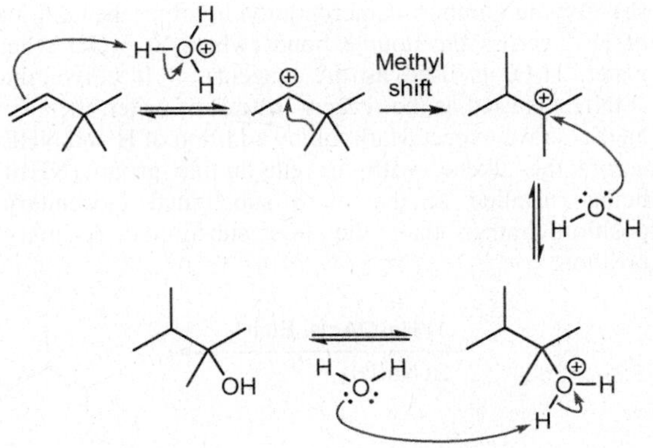

(c) Oxymercuration-demercuration gives Markovnikov addition of water (H and OH) without carbocation rearrangements. That is, the OH group ends up at the more substituted (tertiary) position, and the proton ends up at the less substituted (primary) position.

H goes here

1) $Hg(OAc)_2$, H_2O
2) $NaBH_4$

OH goes here

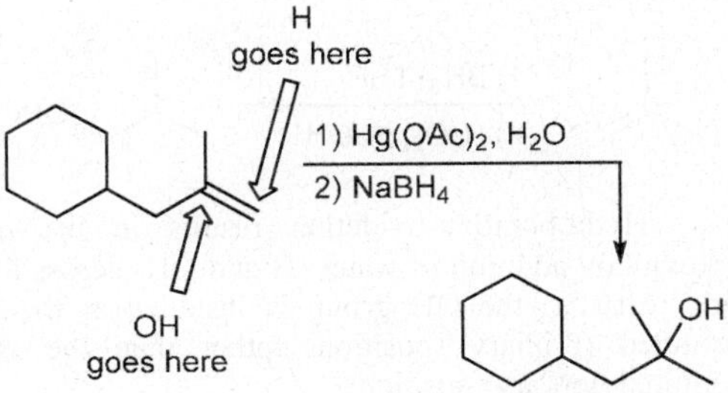

In this case, acid-catalyzed hydration gives the same product, because the intermediate tertiary carbocation does not undergo rearrangement:

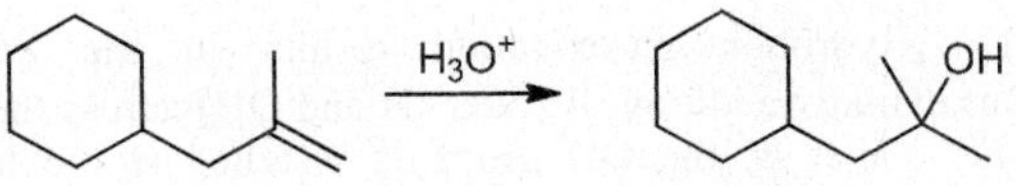

H_3O^+

8.17.

(a) Oxymercuration-demercuration involves the addition of H-Z across the double bond (where Z = OH when water, H_2O, is used as the reagent). If ethanol (EtOH) is used as the reagent instead of water, then Z = OEt, so we expect Markovnikov addition of ethanol (H and OEt) across the alkene, with the ethoxy (OEt) group being installed at the more substituted (secondary) position, rather than the less substituted (primary) position.

H goes here

1) $Hg(OAc)_2$, EtOH
2) $NaBH_4$

OEt goes here

OEt

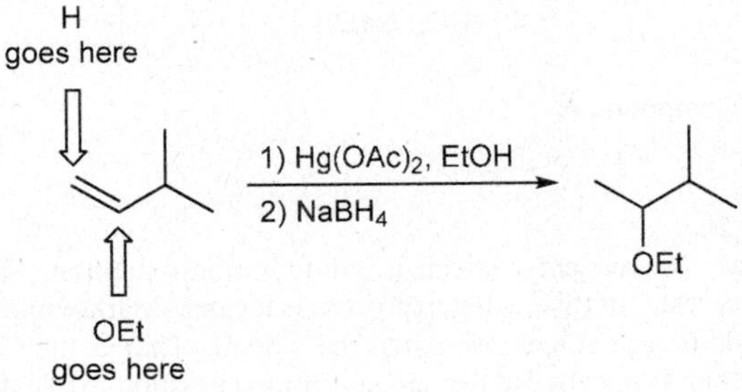

(b) Oxymercuration-demercuration involves the addition of H-Z across the double bond (where Z = OH when water, H_2O, is used as the reagent). If ethylamine ($EtNH_2$) is used as the reagent instead of water, then Z = NHEt, so we expect Markvnikov addition of H and NHEt across the alkene, with the ethylamino group (NHEt) being installed at the more substituted (secondary) position, rather than the less substituted (primary) position.

1) $Hg(OAc)_2$, $EtNH_2$
2) $NaBH_4$

H–N–Et

8.18.

(a) Hydroboration-oxidation results in the *anti*-Markovnikov addition of water (H and OH) across the π bond. That is, the OH group is installed at the less substituted (primary) position, rather than the more substituted (tertiary) position:

1) $BH_3 \cdot THF$
2) H_2O_2, NaOH

OH

(b) Hydroboration-oxidation results in the *anti*-Markovnikov addition of water (H and OH) across the π bond. That is, the OH group is installed at the less substituted (primary) position, rather than the more substituted (tertiary) position:

1) $BH_3 \cdot THF$
2) H_2O_2, NaOH

OH

(c) Hydroboration-oxidation results in the *anti*-Markovnikov addition of water (H and OH) across the π bond. That is, the OH group is installed at the less substituted (primary) position, rather than the more substituted (secondary) position:

1) $BH_3 \cdot THF$
2) H_2O_2, NaOH

OH

8.19. There is only one alkene that will afford the desired product upon hydroboration-oxidation:

1) $BH_3 \cdot THF$
2) H_2O_2, NaOH

OH

Compound **A**

8.20.

(a) The reagents indicate a hydroboration-oxidation. The net result of this two-step process is the *anti*-Markovnikov addition of H and OH across the π bond. That is, the OH group is installed at the less substituted position, while the H is installed at the more substituted position. In this case, two chiral centers are created. Therefore, the stereochemical requirement for *syn* addition determines that the H and OH are added on the same face of the alkene. Since *syn* addition can take place from either face of the alkene with equal likelihood, we expect a pair of enantiomers, as shown:

1) $BH_3 \cdot THF$
2) H_2O_2, NaOH

OH + En

In this case, it might seem as if there was an *anti* addition, rather than a *syn* addition, because we see that the product has one wedge and one dash. But this is an optical illusion. Recall, that most hydrogen atoms are not drawn in bond-line drawings, so the H that was added during the process has not been drawn. However, if we draw that hydrogen atom, we will see that the H and OH were indeed added in a *syn* fashion:

CH_3 OH ≡ CH_3 H OH

(b) The reagents indicate a hydroboration-oxidation. The net result of this two-step process is the *anti*-Markovnikov addition of H and OH across the π bond. That is, the OH group is installed at the less substituted position, while the H is installed at the more substituted position. In this case, only one chiral center is created. Since *syn* addition can take place from either face of the alkene with equal likelihood, we expect a pair of enantiomers, as shown:

1) $BH_3 \cdot THF$
2) H_2O_2, NaOH

OH + En

(c) The reagents indicate a hydroboration-oxidation. The net result of this two-step process is the *anti*-Markovnikov addition of H and OH across the π bond. That is, the OH group is installed at the less substituted position, while the H is installed at the more substituted position. In this case, no chiral centers are created, so the requirement for *syn* addition is irrelevant.

1) $BH_3 \cdot THF$
2) H_2O_2, NaOH

OH

(d) The reagents indicate a hydroboration-oxidation. The net result of this two-step process is the *anti*-Markovnikov addition of H and OH across the π bond. That is, the OH group is installed at the less substituted position, while the H is installed at the more substituted position. In this case, only one chiral center is created. Since *syn* addition can take place from either face of the alkene with equal likelihood, we expect a pair of enantiomers, as shown:

1) $BH_3 \cdot THF$
2) H_2O_2, NaOH
HO
+ En

(e) The reagents indicate a hydroboration-oxidation. The net result of this two-step process is the *anti*-Markovnikov addition of H and OH across the π bond. That is, the OH group is installed at the less substituted position, while the H is installed at the more substituted position. In this case, no chiral centers are created, so the requirement for *syn* addition is irrelevant.

1) $BH_3 \cdot THF$
2) H_2O_2, NaOH
HO

(f) The reagents indicate a hydroboration-oxidation. The net result of this two-step process is the *anti*-Markovnikov addition of H and OH across the π bond. That is, the OH group is installed at the less substituted position, while the H is installed at the more substituted position. In this case, two chiral centers are created. Therefore, the stereochemical requirement for *syn* addition determines that the H and OH are added on the same face of the alkene. Since *syn* addition can take place from either face of the alkene with equal likelihood, we expect a pair of enantiomers, as shown:

1) $BH_3 \cdot THF$
2) H_2O_2, NaOH
H
OH
+ En

8.21. The problem statement indicates that this is a hydroboration-oxidation sequence, the net result of which is an *anti*-Markovnikov addition of H and OH across the π bond. That is, the OH group is installed at the less substituted position, while the H is installed at the more substituted position. In this case, two chiral centers are created. Therefore, the stereochemical requirement for *syn* addition determines that the H and OH are added on the same face of the alkene, giving the products shown. Note that the highlighted methyl group creates steric hindrance on the top face of the ring, so the major product arises from hydroboration from the bottom face:

H H C H H_3C CH_3
(–)-α-Pinene
1) B_2H_6
2) H_2O_2, NaOH
H_3C CH_3 H CH_3 OH H
Major
+
H_3C CH_3 OH H H CH_3
Minor

8.22.

(a) The reagents indicate a catalytic hydrogenation process, so we expect the addition of H and H across the alkene. In this case, the product does not have a chiral center, so stereochemistry is not a relevant consideration.

H_2
Ni

(b) The reagents indicate a catalytic hydrogenation process, so we expect the addition of H and H across the alkene. In this case, the product does not have a chiral center, so stereochemistry is not a relevant consideration.

H_2
Pd

(c) The reagents indicate a catalytic hydrogenation process, so we expect the addition of H and H across the alkene. In this case, the product has one chiral center, so we expect both possible enantiomers (*syn* addition can occur from either face of the π bond).

H_2
Ni
+ En

(d) The reagents indicate a catalytic hydrogenation process, so we expect the addition of H and H across the alkene. In this case, the product does not have a chiral center, so stereochemistry is not a relevant consideration.

H_2
Pt

(e) The reagents indicate a catalytic hydrogenation process, so we expect the addition of H and H across the alkene. In this case, the reaction generates two chiral centers. The requirement for *syn* addition results in the formation of a *meso* compound, so there is only one product.

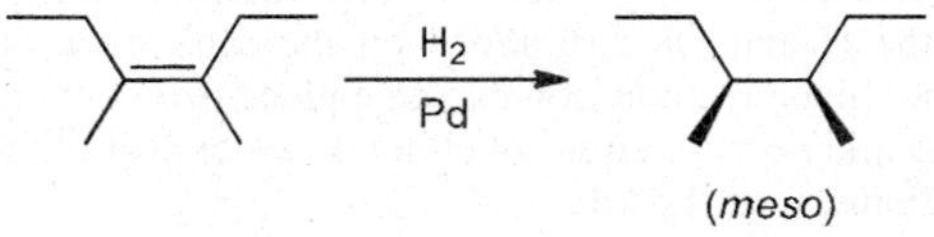

H_2
Pd
(*meso*)

8.23. The starting material has two carbon-carbon π bonds, each of which will undergo hydrogenation when treated with H_2 and a catalyst. This creates two new chiral centers in the product, highlighted below:

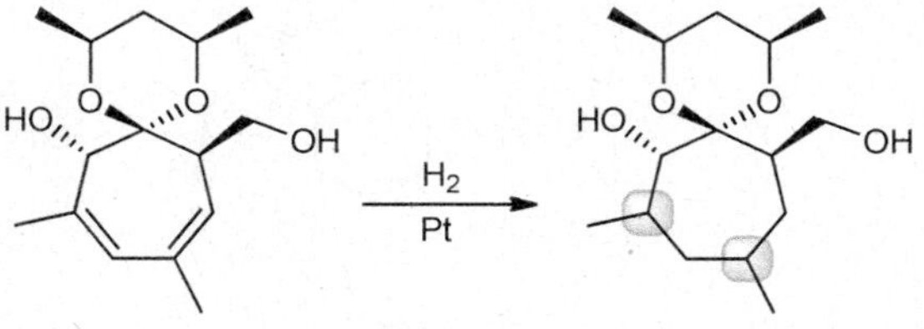

HO OH H_2 Pt HO OH

Each individual π bond undergoes a *syn* addition, which can occur either from the top face or from the bottom face of the π bond. Therefore, each of the two new chiral centers can have either the *R* configuration or the *S* configuration, giving four possible products (shown below). These products are all diastereomers of each other (because the starting compound already has other chiral centers).

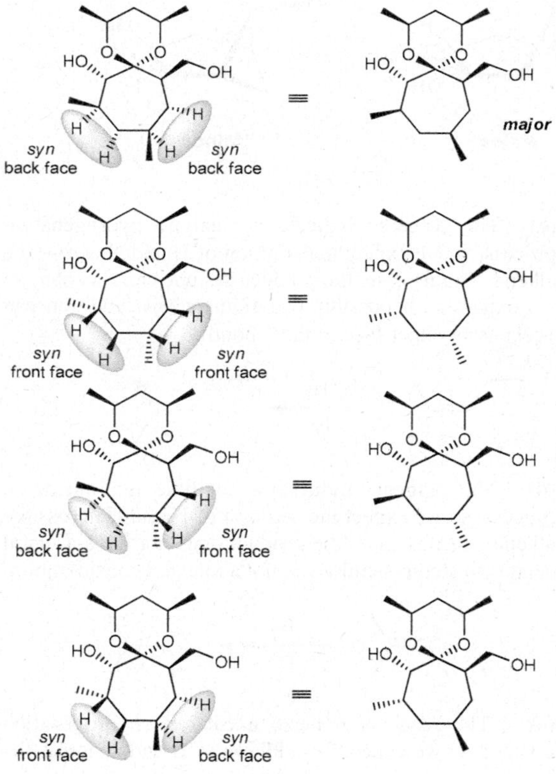

syn back face *syn* back face = *major*

syn front face *syn* front face =

syn back face *syn* front face =

syn front face *syn* back face =

Interestingly, only two stereoisomers are observed in this reaction and the major product results from *syn* addition from the back face of both π bonds. This is an example of an asymmetric catalytic hydrogenation since the chirality of the starting material clearly influences which face undergoes hydrogenation preferentially. Asymmetric catalytic hydrogenation is also possible using a chiral catalyst, as will be discussed later in Section 8.9.

8.24.

(a) When an alkene is treated with molecular bromine (Br_2), we expect an *anti* addition of Br and Br across the alkene, giving the following pair of enantiomers:

Br_2 Br Br + En

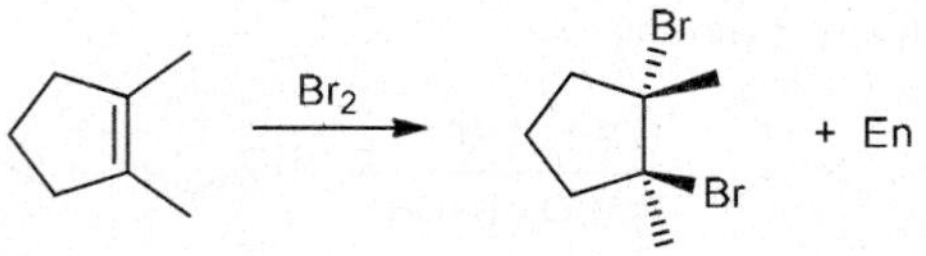

(b) When an alkene is treated with molecular bromine (Br_2), we expect an *anti* addition of Br and Br across the alkene, giving the following pair of enantiomers:

Br_2 Br Br + En

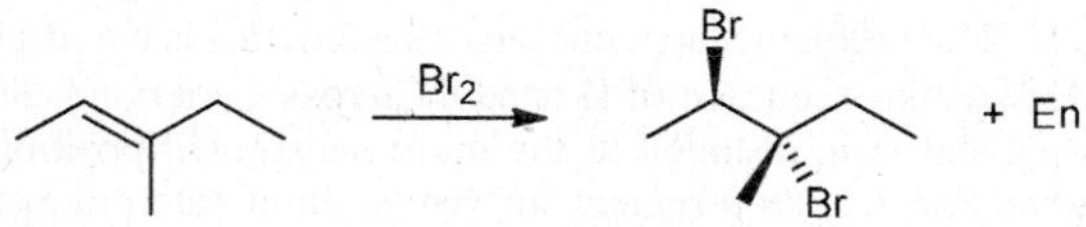

(c) When an alkene is treated with molecular bromine (Br_2), we expect an *anti* addition of Br and Br across the alkene. In this case, only one chiral center is created, so we expect both possible enantiomers (formation of the initial bromonium ion can occur on either face of the π bond with equal likelihood):

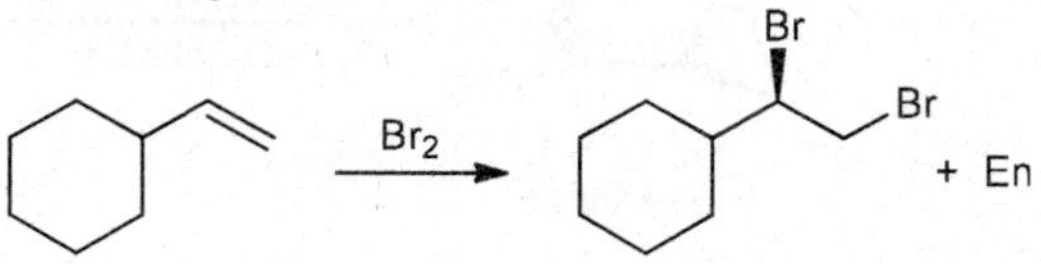

Br_2 Br Br + En

(d) When an alkene is treated with molecular bromine (Br_2), we expect an *anti* addition of Br and Br across the alkene, giving the following pair of enantiomers:

Br_2 Br Br + En

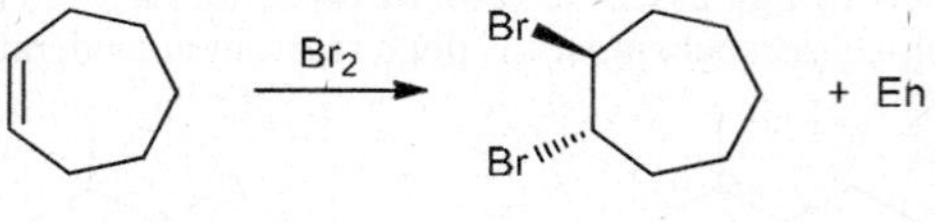

8.25.

(a) Treating an alkene with molecular bromine (Br_2) and water results in the addition of OH and Br across the alkene (halohydrin formation). The OH group is expected to be installed at the more substituted position, while Br is installed at the less substituted position. In this case, two new chiral centers are generated, so we expect only the pair of enantiomers that would result from *anti* addition.

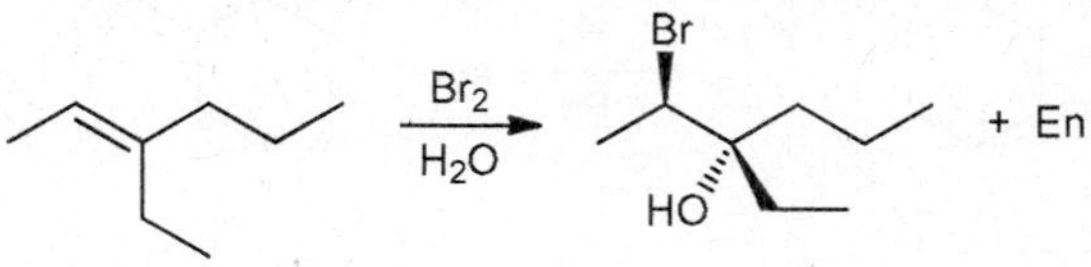

Br_2 H_2O Br HO + En

(b) Treating an alkene with molecular bromine (Br_2) and water results in the addition of OH and Br across the alkene (halohydrin formation). The OH group is expected to be installed at the more substituted position, while Br is installed at the less substituted position. In this case, two new chiral centers are generated, so we expect only the pair of enantiomers that would result from *anti* addition.

Br_2 / H_2O → OH, Br + En

(c) Treating an alkene with molecular bromine (Br_2) and water results in the addition of OH and Br across the alkene (halohydrin formation). The OH group is expected to be installed at the more substituted position, while Br is installed at the less substituted position. In this case, only one new chiral center is generated, so we expect both possible enantiomers (formation of the initial bromonium ion can occur on either face of the π bond with equal likelihood).

Br_2 / H_2O → OH, Br + En

(d) Treating an alkene with molecular bromine (Br_2) and water results in the addition of OH and Br across the alkene (halohydrin formation). In this case, two new chiral centers are generated, so we expect only the pair of enantiomers that would result from *anti* addition.

Br_2 / H_2O → Br, OH + En

8.26.

(a) The alkene reacts with molecular bromine to give a bromonium ion, which is then attacked by a molecule of solvent (EtOH, in this case, rather than H_2O). The result is the addition of Br and OEt (rather than the addition of Br and OH). The OEt group is expected to be installed at the more substituted position, while Br is installed at the less substituted position. In this case, two new chiral centers are generated, so we expect only the pair of enantiomers that would result from *anti* addition:

Br_2 / OH → OEt, Br + En

(b) The alkene reacts with molecular bromine to give a bromonium ion, which is then captured by a molecule of solvent ($EtNH_2$, in this case, rather than H_2O). The result is the addition of Br and NHEt (rather than the addition of Br and OH). The ethylamino group (NHEt) is expected to be installed at the more substituted position, while Br is installed at the less substituted position. In this case, two new chiral centers are generated, so we expect only the pair of enantiomers that would result from *anti* addition:

Br_2 / $EtNH_2$ → Et, N–H, Br + En

8.27. The bromonium ion can open (before a bromide ion attacks), forming a resonance-stabilized carbocation. This carbocation is trigonal planar and can be attacked from either side:

Br ⊕ ⇌ ⊕ Br

resonance-stabilized

8.28.

(a) Treating an alkene with a peroxy acid followed by aqueous acid results in the addition of OH and OH across the alkene. In this case, two new chiral centers are generated, so we expect only the pair of enantiomers that would result from *anti* addition:

1) RCO_3H 2) H_3O^+ → HO, OH + En

(b) Treating an alkene with a peroxy acid followed by aqueous acid results in the addition of OH and OH across the alkene. In this case, two new chiral centers are generated, so we expect only the pair of enantiomers that would result from *anti* addition:

1) RCO_3H 2) H_3O^+ → HO, OH + En

(c) Treating an alkene with a peroxy acid followed by aqueous acid results in the addition of OH and OH across the alkene. In this case, two new chiral centers are generated, so we expect only the pair of enantiomers that would result from *anti* addition:

1) RCO_3H 2) H_3O^+ → OH, OH + En

(d) Treating an alkene with a peroxy acid followed by aqueous acid results in the addition of OH and OH across the alkene. In this case, the product has no chiral centers, so stereochemistry is not a relevant consideration.

1) RCO_3H 2) H_3O^+ → OH, OH

(e) Treating an alkene with a peroxy acid followed by aqueous acid results in the addition of OH and OH across the alkene. In this case, two new chiral centers are generated. However, the *anti* addition results in the formation of a *meso* compound:

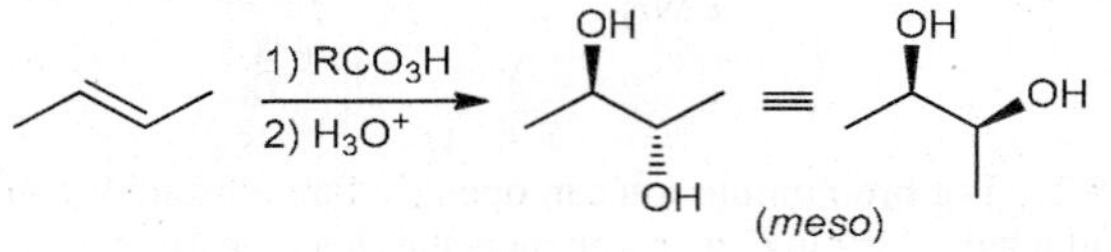

1) RCO_3H
2) H_3O^+
(*meso*)

(f) Treating an alkene with a peroxy acid followed by aqueous acid results in the addition of OH and OH across the alkene. In this case, two new chiral centers are generated, so we expect only the pair of enantiomers that would result from *anti* addition:

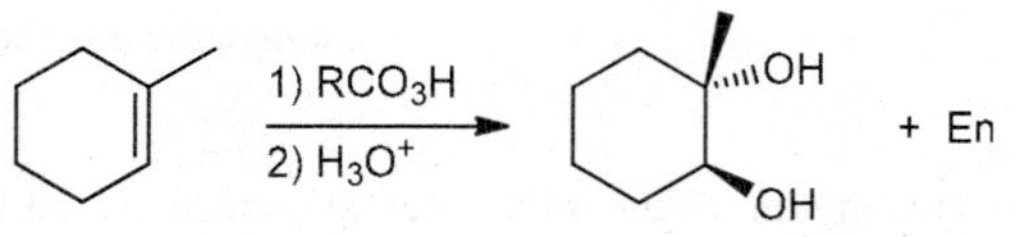

1) RCO_3H
2) H_3O^+
+ En

8.29.

(a) Treating an alkene with a peroxy acid results in an epoxide. Further treatment of the epoxide with ethanol under acidic conditions results in a ring opening reaction in which ethanol serves as the nucleophile. Nucleophilic attack occurs at the more substituted (tertiary) position, so the net result is the addition of OH and OEt across the alkene, with the latter being installed at the more substituted position, as shown:

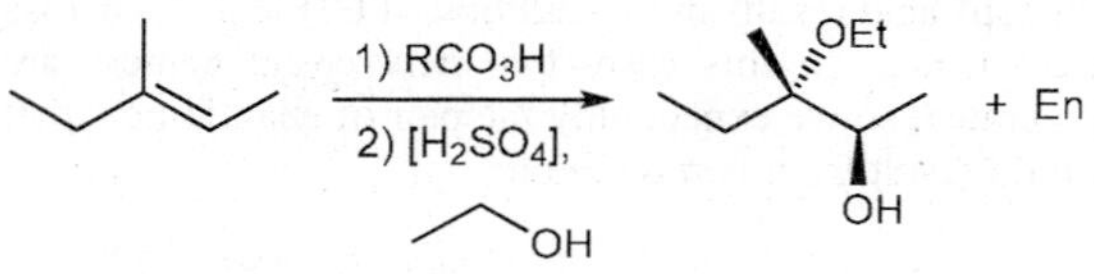

1) RCO_3H
2) $[H_2SO_4]$,
+ En

(b) Treatment of the epoxide with phenol (C_6H_5OH) under acidic conditions results in a ring opening reaction in which the oxygen atom of phenol serves as the nucleophilic center. Nucleophilic attack occurs at the more substituted (tertiary) position, so the net result is the addition of OH and OR (where R is C_6H_5) across the alkene, with the latter being installed at the more substituted position. Since the starting epoxide is enantiomerically pure (we are starting only with the enantiomer shown), we expect an enantiomerically pure product (not a mixture of enantiomers), as shown.

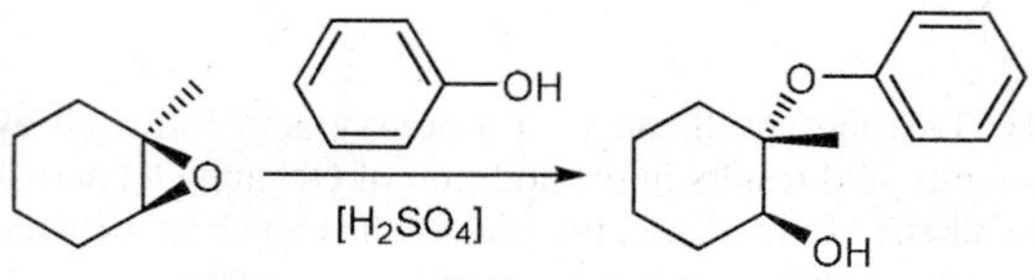

$[H_2SO_4]$

8.30.

(a) Compound **A** is converted to an epoxide upon treatment with a peroxy acid, so compound **A** must be an alkene. There are many alkenes with the molecular formula C_6H_{12}, and it would be time-consuming to try to draw them all. Instead, we notice the following: in order for the product to have no chiral centers, each of the vinylic positions must already contain two identical groups, like this:

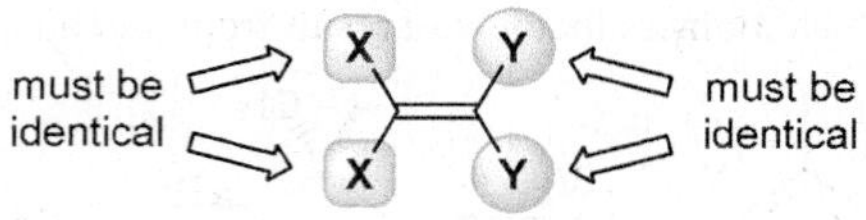

must be identical
must be identical

There are only two alkenes with the molecular formula C_6H_{12} that fit this criterion:

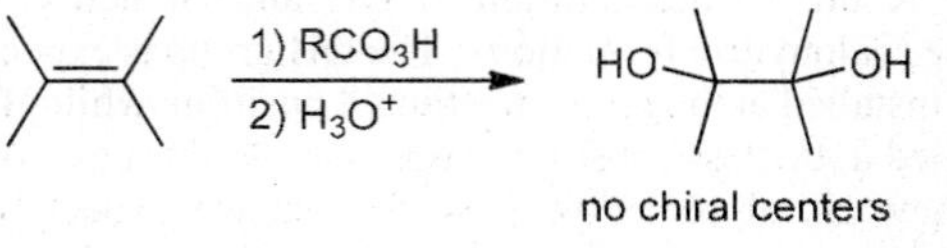

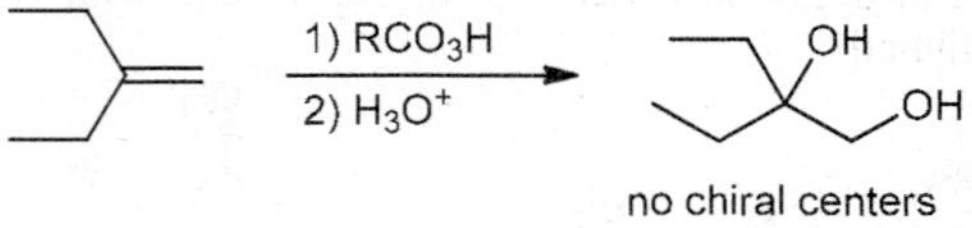

1) RCO_3H
2) H_3O^+
no chiral centers

1) RCO_3H
2) H_3O^+
no chiral centers

(b) In order to be a *meso* compound, the resulting diol must contain two chiral centers, as well as reflectional symmetry (such as an internal plane of symmetry). In order to achieve this result, the starting alkene must have the following structural features:

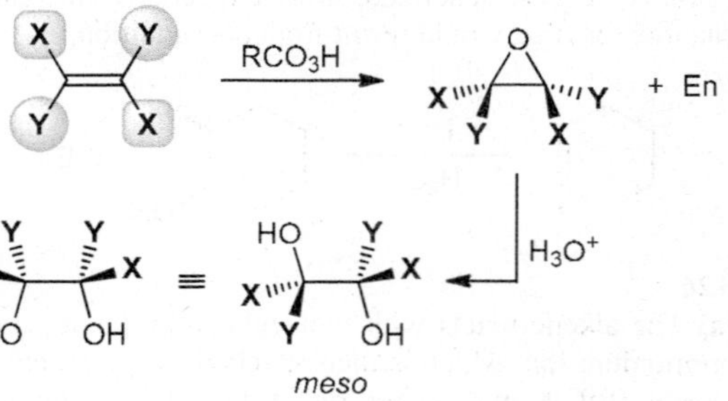

RCO_3H
+ En
H_3O^+
meso

The identity of X and Y must be different, or the resulting diol would have no chiral centers. There is only one alkene with the molecular formula C_6H_{12} that fits this criterion:

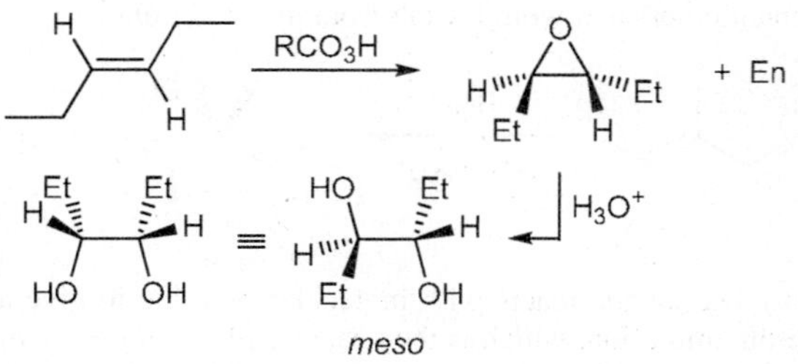

RCO_3H
+ En
H_3O^+
meso

8.31.

(a) Treating an alkene with catalytic osmium tetroxide and NMO results in the addition of OH and OH across the alkene. In this case, two new chiral centers are generated, so we expect only the pair of enantiomers that would result from *syn* addition:

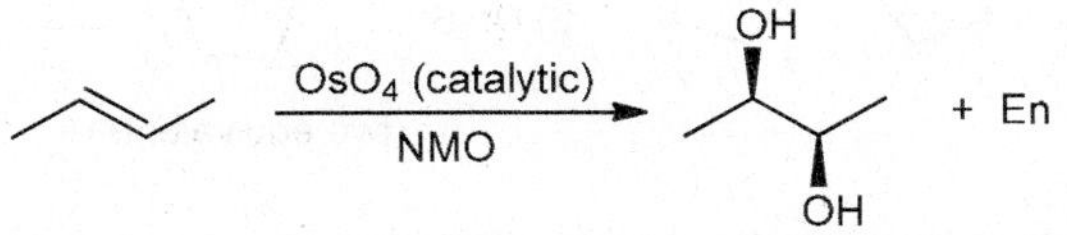
OsO4 (catalytic)
NMO
OH
OH
+ En

(b) Treating an alkene with osmium tetroxide followed by aqueous sodium bisulfite results in the addition of OH and OH across the alkene. In this case, only one chiral center is created, so we expect both possible enantiomers (formation of the initial cyclic osmate ester can occur on either face of the π bond with equal likelihood):

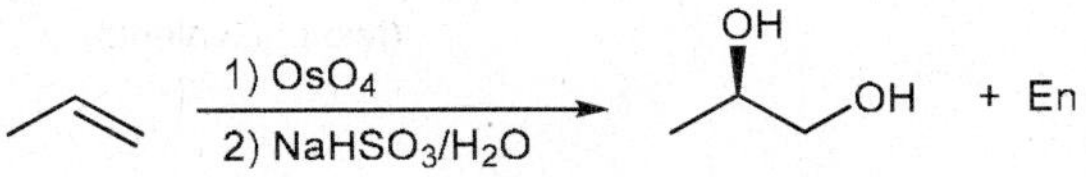
1) OsO4
2) NaHSO3/H2O
OH
OH
+ En

(c) Treating an alkene with cold potassium permanganate and sodium hydroxide results in the addition of OH and OH across the alkene. In this case, two new chiral centers are generated, and we expect a *syn* addition, giving the following *meso* compound:

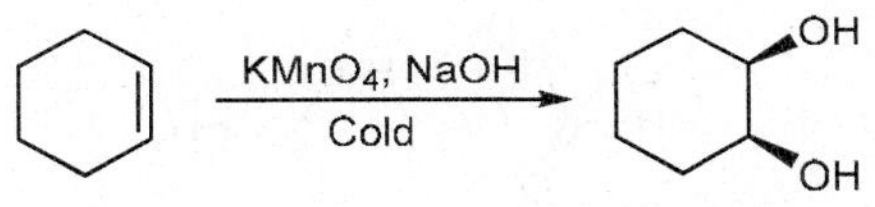
KMnO4, NaOH
Cold
OH
OH

(meso)

(d) Treating an alkene with cold potassium permanganate and sodium hydroxide results in the addition of OH and OH across the alkene. In this case, the product has no chiral centers, so stereochemistry is not a relevant consideration.

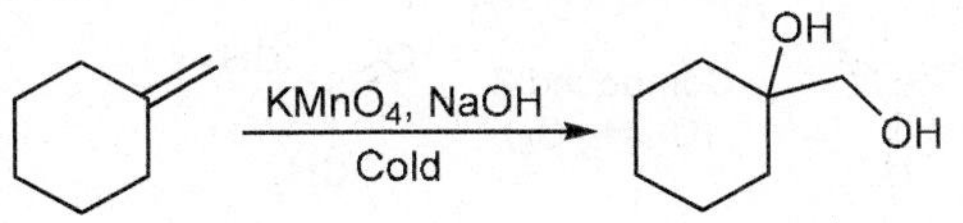
KMnO4, NaOH
Cold
OH
OH

(e) Treating an alkene with catalytic osmium tetroxide and a suitable co-oxidant (*tert*-butyl hydroperoxide) results in the addition of OH and OH across the alkene. In this case, only one chiral center is created, so we expect both possible enantiomers (formation of the initial cyclic osmate ester can occur on either face of the π bond with equal likelihood):

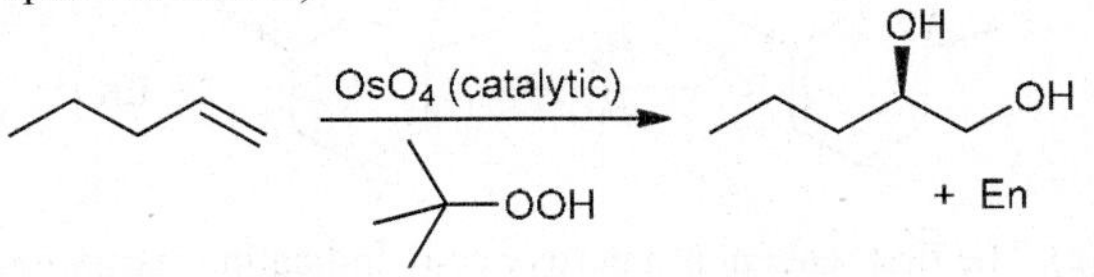
OsO4 (catalytic)
OOH
OH
OH
+ En

(f) Treating an alkene with catalytic osmium tetroxide and NMO results in the addition of OH and OH across the alkene. In this case, two new chiral centers are generated, so we expect only the pair of enantiomers that would result from *syn* addition:

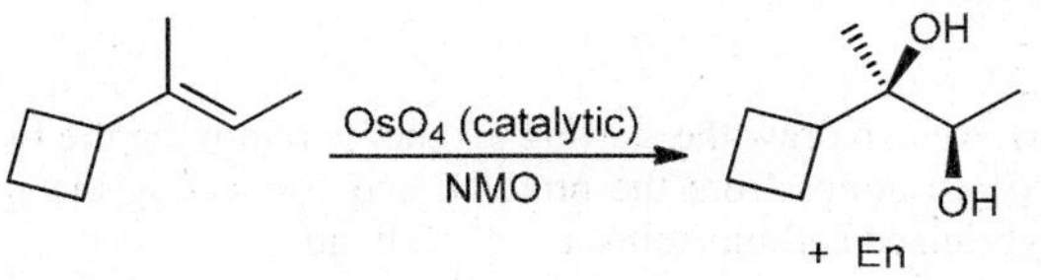
OsO4 (catalytic)
NMO
OH
OH
+ En

8.32.

(a) Each C=C bond is split apart and redrawn as two C=O bonds, giving the following products:

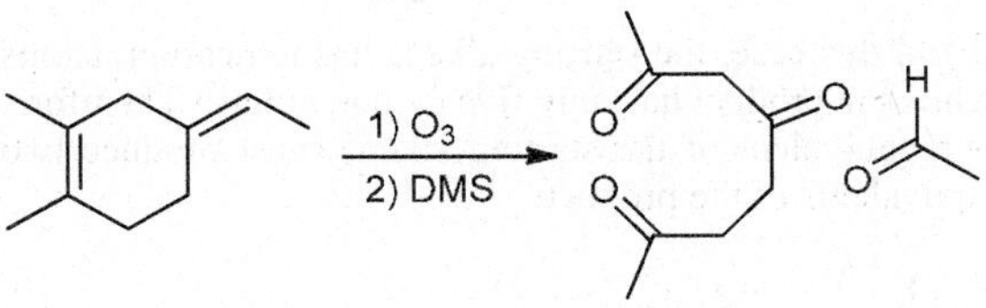
1) O3
2) DMS
O
O
H
O
O

(b) Each C=C bond is split apart and redrawn as two C=O bonds, giving two equivalents of the same product:

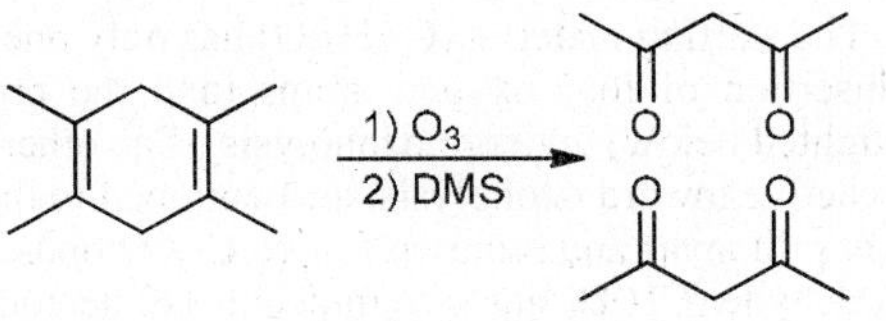
1) O3
2) DMS
O
O
O
O

(c) The C=C bond is split apart and redrawn as two C=O bonds, giving two equivalents of the same product:

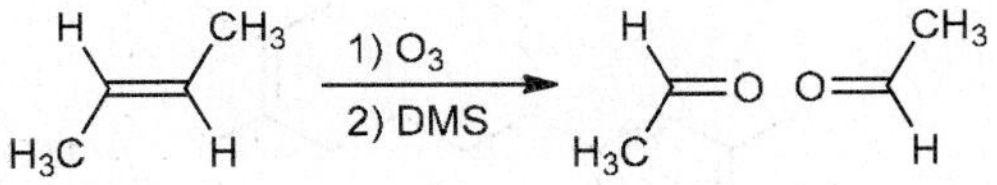
H
CH3
H3C
H
1) O3
2) DMS
H
O
H3C
O
CH3
H

(d) The C=C bond is split apart and redrawn as two C=O bonds, giving the following product:

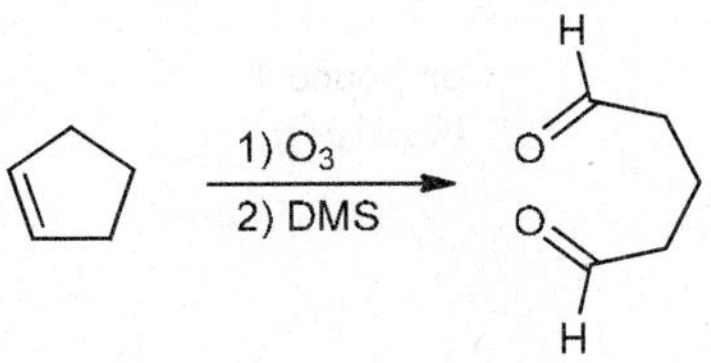
1) O3
2) DMS
H
O
O
H

(e) The C=C bond is split apart and redrawn as two C=O bonds, giving the following *meso* compound:

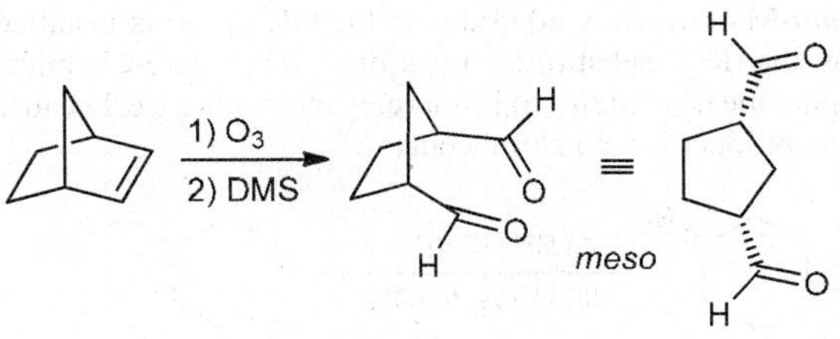
1) O3
2) DMS
H
O
O
H
≡
H
O
O
H
meso

(f) The C=C bond is split apart and redrawn as two C=O bonds, giving two equivalents of the same product:

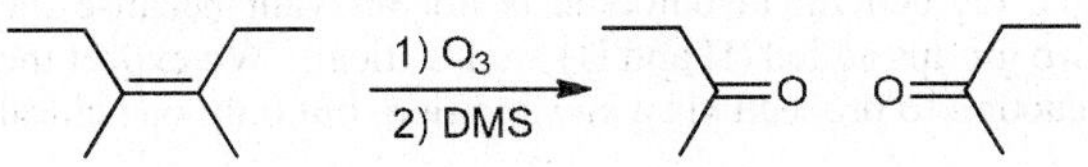
1) O3
2) DMS
O
O

8.33.

(a) We can draw the starting alkene by removing the two oxygen atoms from the product, and connecting the sp^2 hybridized carbon atoms as a C=C bond:

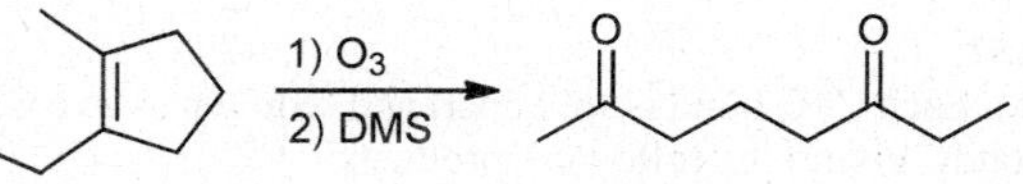

1) O_3
2) DMS

(b) In this case, the starting alkene has ten carbon atoms while the product has only five carbon atoms. Therefore, one equivalent of the starting alkene must produce two equivalents of the product:

1) O_3
2) DMS

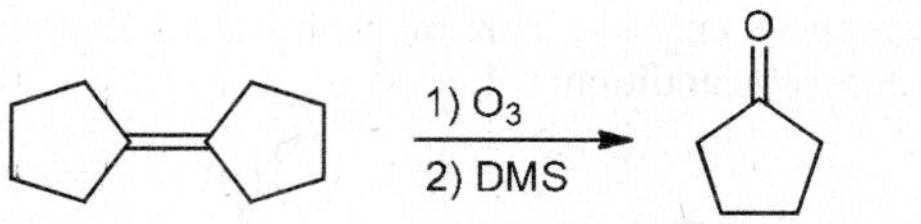

(two equivalents)

(c) In this case, the starting alkene has ten carbon atoms while the product has only five carbon atoms. Therefore, one equivalent of the starting alkene must produce two equivalents of the product:

1) O_3
2) DMS

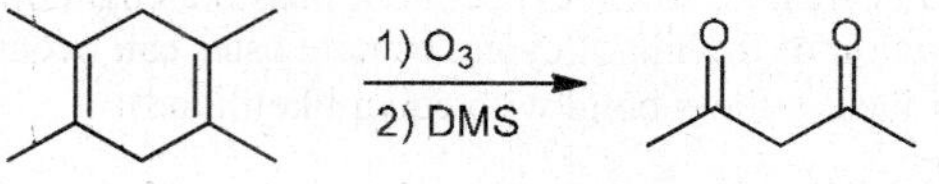

(two equivalents)

8.34. The starting material ($C_{27}H_{38}O$) has only one oxygen atom, while compound **2** ($C_{19}H_{22}O_5$) has five oxygen atoms. The insertion of four oxygen atoms (and the production of four small molecules) indicates that four C=C bonds (highlighted below) undergo ozonolysis. The other three C=C bonds (in the ring) are part of the aromatic system, which is unreactive toward ozonolysis, as mentioned in the problem statement. To draw the products of ozonolysis, each C=C bond is split apart and redrawn as two C=O bonds. This gives compound **2**, shown below, along with two molecules of formaldehyde (CH_2O) and two molecules of acetone ($(CH_3)_2CO$).

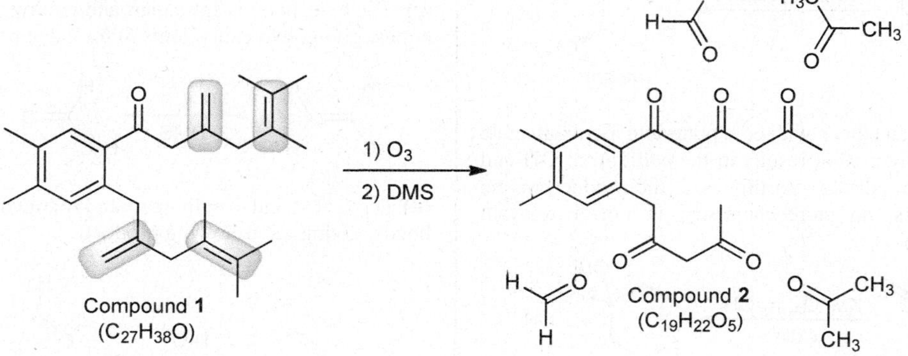

1) O_3
2) DMS

Compound **1**
($C_{27}H_{38}O$)

Compound **2**
($C_{19}H_{22}O_5$)

8.35.

(a) The reagents indicate a hydroboration-oxidation, so the net result will be the addition of H and OH across the alkene. For the regiochemical outcome, we expect an *anti*-Markovnikov addition, so the OH group is installed at the less substituted position. The stereochemical outcome (*syn* addition) is not relevant in this case, because the product has no chiral centers:

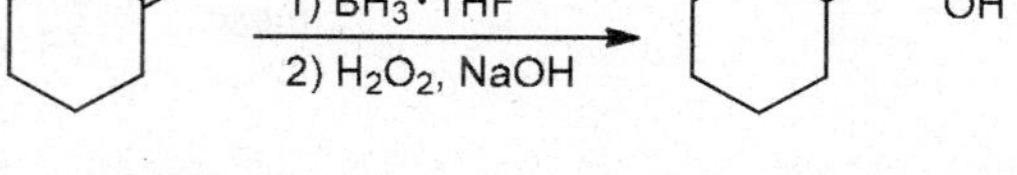

1) $BH_3 \cdot THF$
2) H_2O_2, NaOH

(b) The reagents indicate a hydrogenation reaction, so the net result will be the addition of H and H across the alkene. The regiochemical outcome is not relevant because the two groups added (H and H) are identical. We expect the reaction to proceed via a *syn* addition, but only one chiral center is formed. Therefore, both enantiomers are obtained because *syn* addition can occur from either face of the starting alkene:

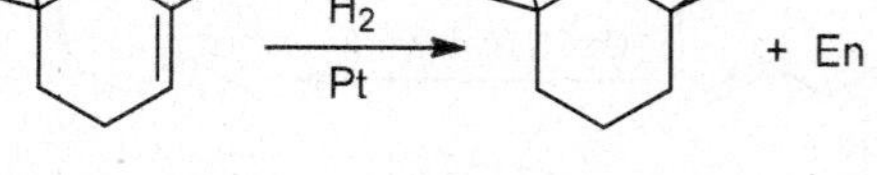

H_2
Pt

+ En

(c) The first reagent is a peroxy acid, indicating formation of an epoxide, which is then opened under aqueous acidic conditions. The net result is expected to be the addition of OH and OH across the alkene. The regiochemical outcome is not relevant because the two groups added (OH and OH) are identical. For the stereochemical outcome, we notice that two chiral centers are formed, and we expect only the pair of enantiomers resulting from an *anti* addition:

1) CH_3CO_3H
2) H_3O^+
+ En

(d) The reagents indicate a dihydroxylation reaction, so the net result will be the addition of OH and OH across the alkene. The regiochemical outcome is not relevant because the two groups added (OH and OH) are identical. We expect the reaction to proceed via a *syn* addition. In this case, two chiral centers are formed, so we expect only the pair of enantiomers resulting from a *syn* addition:

1) OsO_4
2) $NaHSO_3/H_2O$
+ En

(e) The reagent indicates an acid-catalyzed hydration, so the net result will be the addition of H and OH across the alkene. We expect a Markovnikov addition, so the OH group will be installed at the more substituted position. Only one chiral center is formed, so we expect the following pair of enantiomers:

H_3O^+
+ En

(f) The reagent indicates a hydrobromination reaction, so the net result will be the addition of H and Br across the alkene. We expect a Markovnikov addition, so the Br group will be installed at the more substituted position. No chiral centers are formed in this case, so stereochemistry is irrelevant:

HBr

(g) The reagents indicate a dihydroxylation process (via an epoxide), so the net result will be the addition of OH and OH across the alkene. The regiochemical outcome is not relevant because the two groups added (OH and OH) are identical. We expect the reaction to proceed via an *anti* addition. In this case, two chiral centers are formed, so we expect only the pair of enantiomers resulting from an *anti* addition:

1) RCO_3H
2) H_3O^+
+ En

(h) The reagents indicate a hydroboration-oxidation, so the net result will be the addition of H and OH across the alkene. For the regiochemical outcome, we expect an *anti*-Markovnikov addition, so the OH group is installed at the less substituted position. We expect the reaction to proceed via a *syn* addition, but only one chiral center is formed, so we expect both enantiomers (*syn* addition can occur on either face of the starting alkene):

1) $BH_3 \cdot THF$
2) H_2O_2, NaOH
+ En

(i) The reagents indicate a dihydroxylation reaction, so the net result will be the addition of OH and OH across the alkene. The regiochemical outcome is not relevant because the two groups added (OH and OH) are identical. We expect the reaction to proceed via a *syn* addition. In this case, two chiral centers are formed, so we expect only the pair of enantiomers resulting from a *syn* addition:

OsO_4 (catalytic)
NMO
+ En

8.36. The net result will be the addition of OH and OH across the alkene. The regiochemical outcome is not relevant because the two groups added (OH and OH) are identical. We expect the reaction to proceed via a *syn* addition. In this case, two chiral centers are formed, so we expect the two products shown below. Because of the presence of a third chiral center that has the same configuration in both products, these two products are diastereomers, rather than enantiomers.

$KMnO_4$, NaOH
Cold
Diastereomers

8.37. *syn*-Dihydroxylation of *trans*-2-butene results in the same products as *anti*-dihydroxylation of *cis*-2-butene, as shown below. Rotation about the middle C–C bond reveals that the two compounds are superimposable, and the configuration of each chiral center has been assigned to demonstrate that the products are indeed the same for these two reaction sequences:

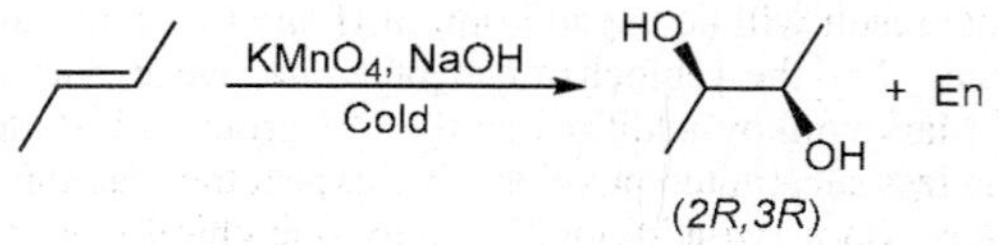

KMnO$_4$, NaOH / Cold → HO...OH (2R,3R) + En

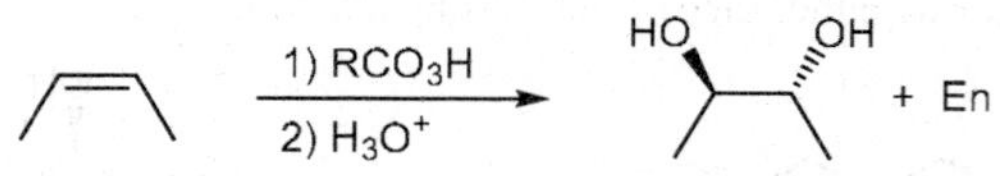

1) RCO$_3$H / 2) H$_3$O$^+$ → HO...OH (2R,3R) + En

8.38. Compound **A** must be an alkene (because it undergoes reactions that are typically observed for alkenes, such as hydroboration-oxidation, hydrobromination and ozonolysis). So, we begin by drawing all possible alkenes with the molecular formula C_5H_{10} (using a methodical approach similar to the one described in the solution to Problem 4.3):

Five-carbon chain

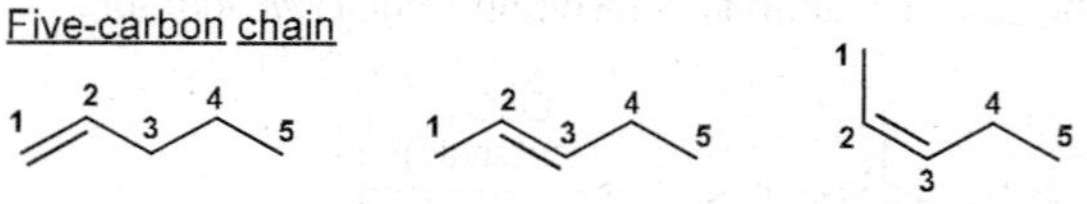

Four-carbon chain

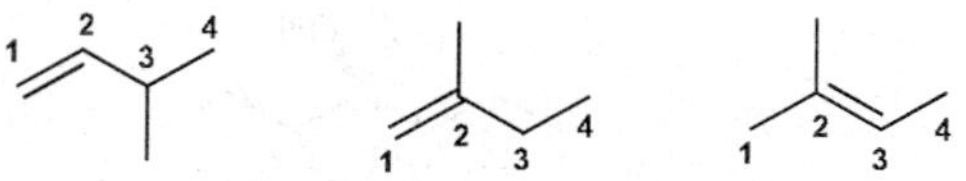

Among these isomers, only the last two will cleanly afford a tertiary alkyl halide upon treatment with HBr. And among these two alkenes, only the latter will undergo ozonolysis to produce a compound with three carbon atoms and another compound with two carbon atoms. Now that we have identified the starting alkene, we can draw the products **B–F**, as shown here:

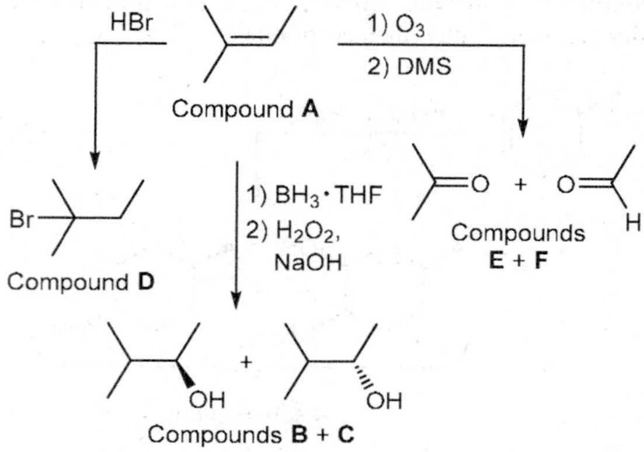

HBr; Compound **A**; 1) O$_3$ 2) DMS; Compound **D**; 1) BH$_3$·THF 2) H$_2$O$_2$, NaOH; Compounds **E** + **F**; Compounds **B** + **C**

8.39.

(a) The two groups being added across the alkene are H and OH. The OH group is installed at the more substituted carbon atom, so we must use conditions that give a Markovnikov addition of H and OH. This can be accomplished via oxymercuration-reduction. This is not a stereospecific reaction, so a racemic mixture of the product is expected:

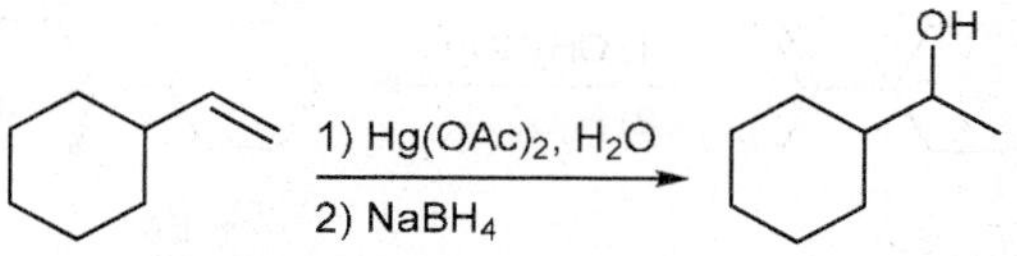

1) Hg(OAc)$_2$, H$_2$O / 2) NaBH$_4$

Oxymercuration-demercuration has no carbocation in mechanism

It should be noted that in this case, acid-catalyzed hydration with dilute H_2SO_4 is not a viable solution, because the carbocation intermediate is likely to rearrange to the tertiary position:

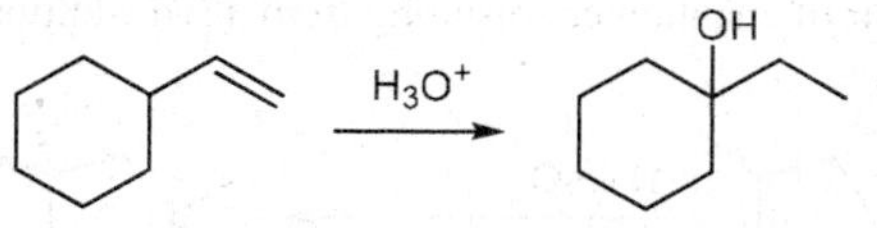

H$_3$O$^+$

Acid-catalyzed hydration leads to carbocation rearrangement

(b) This reaction involves elimination of H and Br to give the less substituted alkene, so a sterically hindered base (such as *tert*-butoxide) is required:

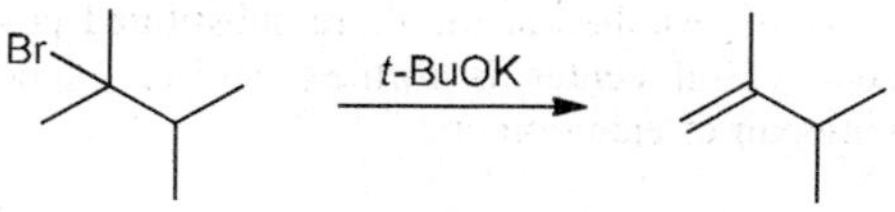

t-BuOK

(c) The two groups being added across the alkene are H and Br. The Br group is installed at the less substituted carbon atom, so we must use conditions that give an *anti*-Markovnikov addition of H and Br. This can be accomplished by treating the alkene with HBr in the presence of peroxides:

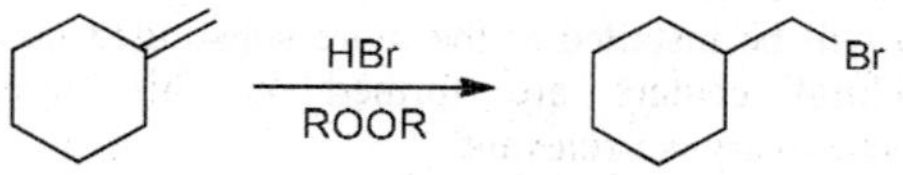

HBr / ROOR

(d) This is a substitution reaction. Our retrosynthesis begins with a disconnection at a C–O bond of the ether group. The carbon atom (at the disconnected bond below) started out as an electrophile in the given starting material, with a Cl leaving group at that position. Therefore, the oxygen atom served as the nucleophile, so we draw a negative charge on the oxygen atom to give a suitable nucleophile (ethoxide ion):

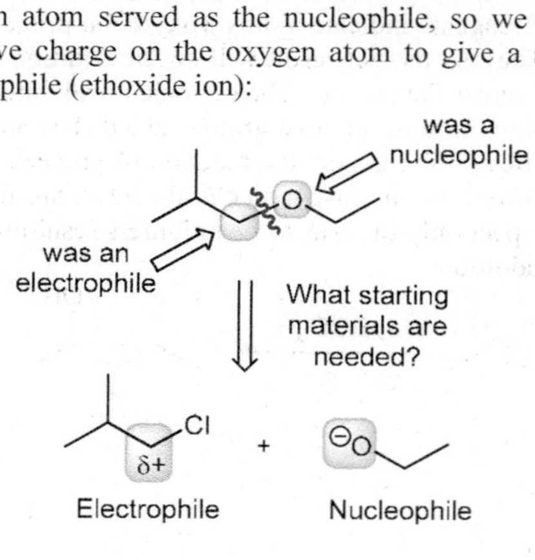

was a nucleophile; was an electrophile; What starting materials are needed?

δ+ Electrophile + Nucleophile

The last step of the planning process is to confirm that the reaction mechanism is favorable. The ethoxide anion is a strong nucleophile and a strong base, so we must consider both S_N2 and E2 pathways. In this case, we have a primary substrate and a small nucleophile, so there is very little steric hindrance and we expect an S_N2 process to give the desired target molecule as the major product:

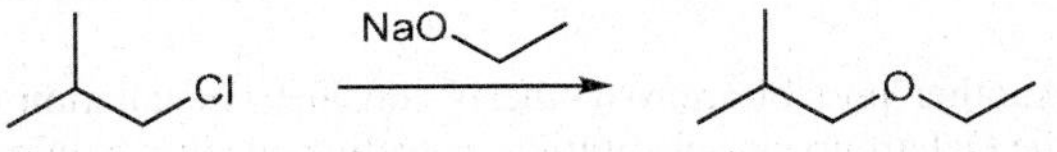

NaO

(e) The two groups being added across the alkene are H and Cl. The latter is installed at the more substituted carbon atom, so we must use conditions that give a Markovnikov addition of H and Cl. This can be accomplished by treating the alkene with HCl:

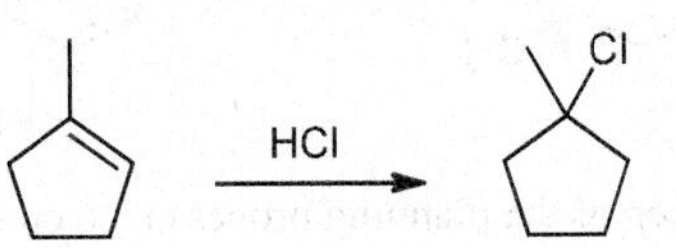

HCl

(f) The two groups being added across the alkene are H and OH. The OH group is installed at the less substituted position, so we must use conditions that give an *anti*-Markovnikov addition of H and OH. Also, the H and OH are added in a *syn* fashion (this can be seen more clearly if you draw the H that was installed, as shown below). This can be accomplished via hydroboration-oxidation:

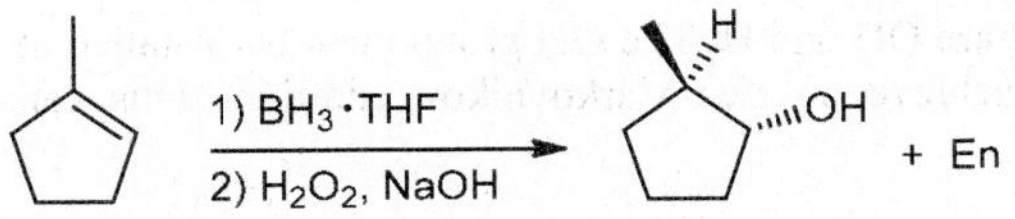

1) $BH_3 \cdot THF$
2) H_2O_2, NaOH
+ En

(g) This reaction involves elimination of H and Br to give the more substituted alkene, so we must use a strong base that is not sterically hindered. We can use hydroxide, methoxide or ethoxide as the base. All of these bases are suitable, as the substrate is tertiary so S_N2 reactions will not compete:

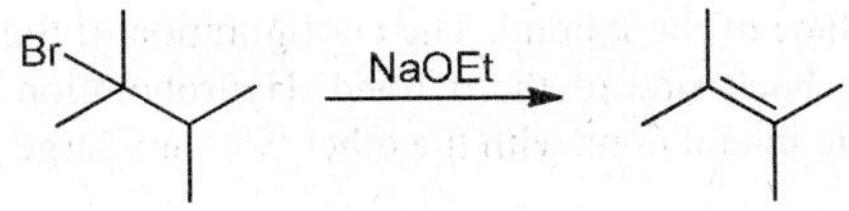

NaOEt

(h) The two groups being added across the alkene are H and Br. The latter is installed at the more substituted carbon atom, so we must use conditions that give a Markovnikov addition of H and Br. This can be accomplished by treating the alkene with HBr:

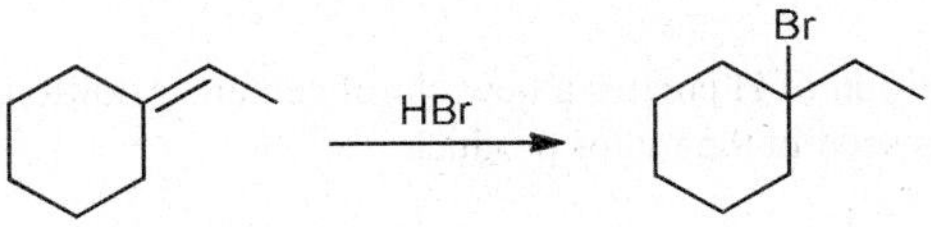

HBr

(i) This is a substitution reaction. The substrate is tertiary, so we will need to perform an S_N1 reaction. The nucleophile must be bromide, but we cannot simply treat the substrate with bromide, because hydroxide is not a good leaving group. The use of HBr will provide both the nucleophile (bromide) and the proton for converting the bad leaving group to a good leaving group (water):

HBr

(j) The two groups being added across the alkene are H and H, which can be accomplished by treating the alkene with molecular hydrogen (H_2) in the presence of a suitable catalyst:

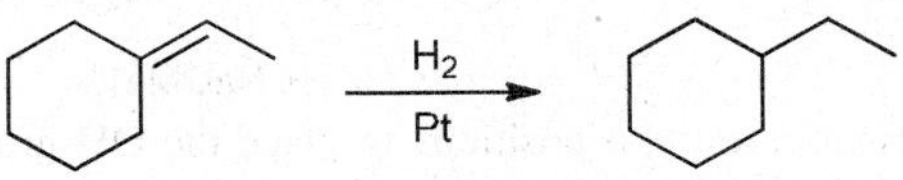

H_2
Pt

(k) The two groups being added across the alkene are OH and OH, and the alkene has undergone *anti* addition (one OH group has added to the top face, and the other OH has added to the bottom face). *Anti* dihydroxylation can be accomplished with a two-step procedure: formation of an epoxide by treatment with a peroxy acid (RCO_3H), followed by a ring-opening reaction with aqueous acid:

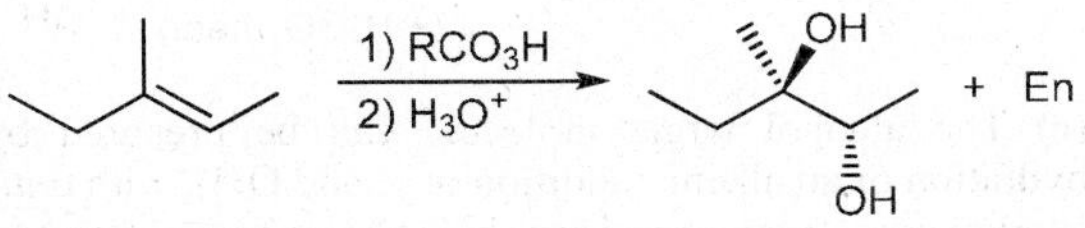

1) RCO_3H
2) H_3O^+
+ En

(l) The two groups being added across the alkene are OH and OH, and the alkene has undergone *syn* addition (both OH groups have been added to the same face, resulting in an achiral *meso* product). *Syn* dihydroxylation can be accomplished by treating the alkene with $KMnO_4$, or OsO_4:

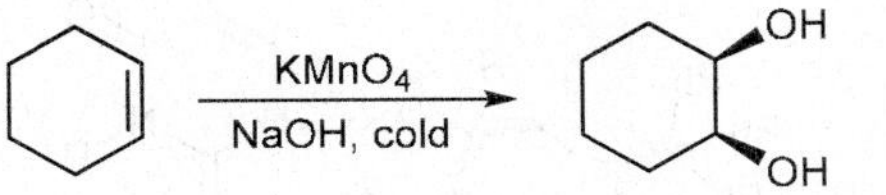

$KMnO_4$
NaOH, cold

8.40.

(a) The alkyl bromide target molecule can be prepared by addition of HBr to an alkene, and there are two suitable alkene starting materials. Either alkene will produce the target molecule upon treatment with HBr:

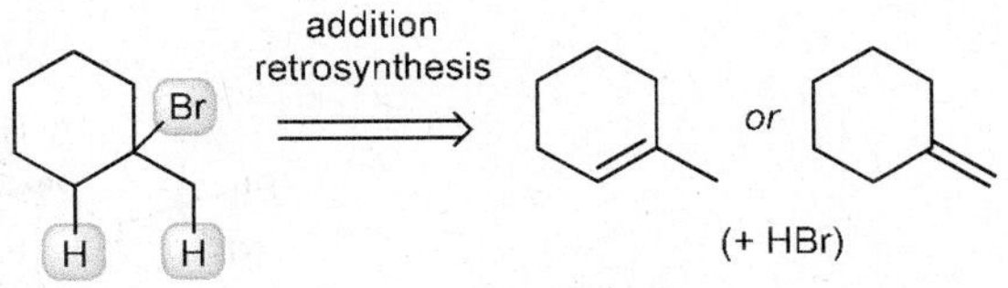

addition retrosynthesis
or
(+ HBr)

Another possible retrosynthesis considers installation of the Br atom via substitution, revealing an alcohol as a third viable starting material:

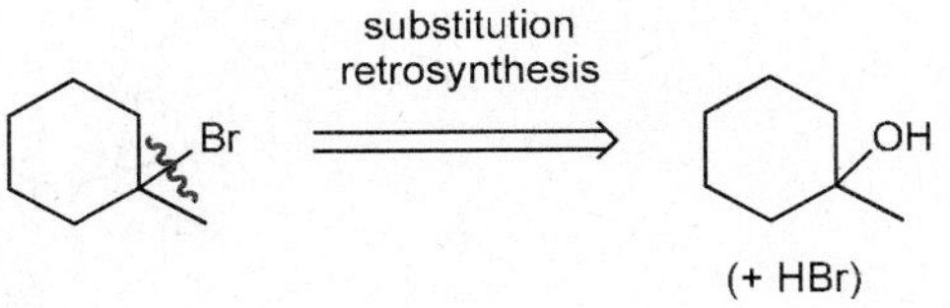

substitution retrosynthesis
(+ HBr)

(b) The pi bond in the target molecule can be formed by an elimination reaction. In this case, the target molecule is a stable alkene, so the potential starting material can be either an alkyl halide or an alcohol. There are two positions to place the halide leaving group to furnish the desired target molecule upon treatment with a strong base (E2 elimination gives the Zaitsev product):

E2 ⟹ Br *or* Br (+ NaOMe)

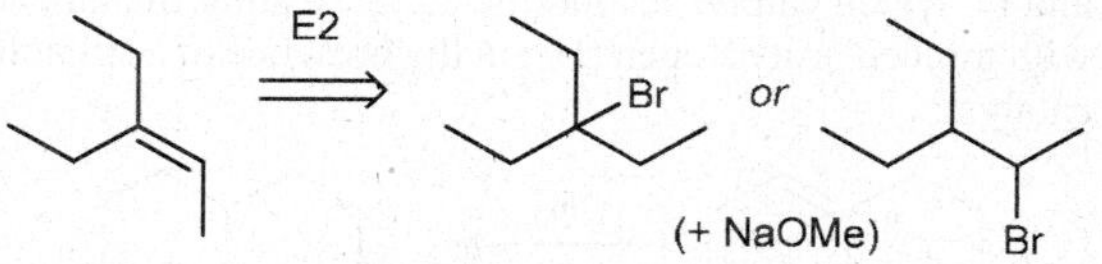

In addition, are two positions to place the OH group to furnish the desired target molecule upon treatment with a strong, concentrated acid (acid-catalyzed dehydration of an alcohol follows an E1 mechanism and gives the Zaitsev product):

E1 ⟹ OH *or* OH (+ H_2SO_4/heat)

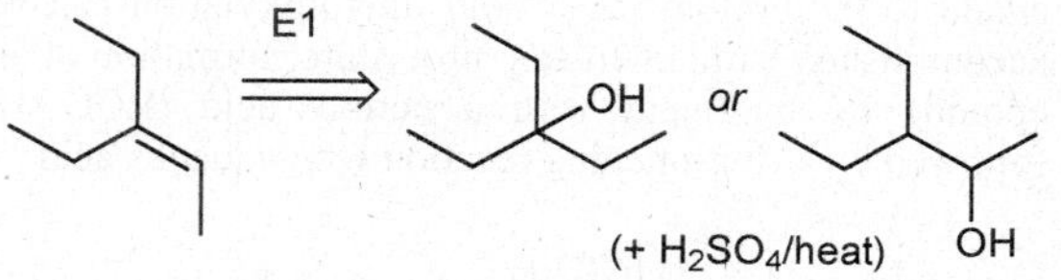

(c) The alcohol target molecule can be prepared by hydration of an alkene (addition of H and OH), with *anti*-Markovnikov regiochemistry (hydroboration-oxidation):

addition retrosynthesis ⟹ (+ hydroboration-oxidation)

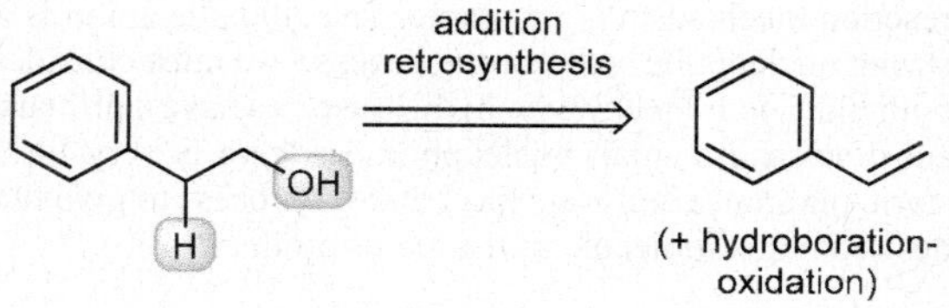

Another possible retrosynthesis considers installation of the OH group via substitution, revealing an alkyl halide as a second viable starting material:

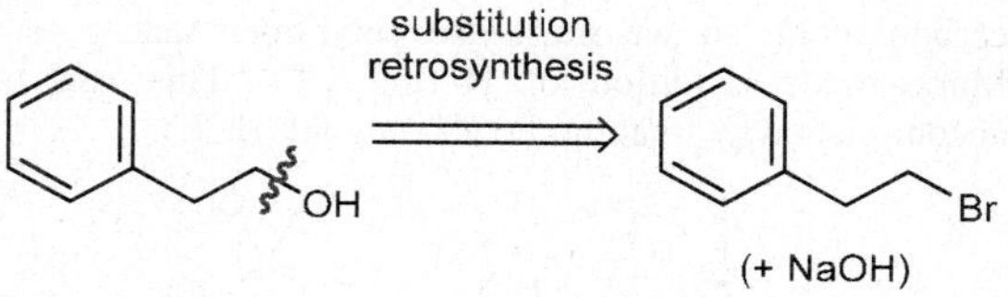

substitution retrosynthesis ⟹ Br (+ NaOH)

The last step of the planning process is to confirm that the reaction mechanism is favorable. The hydroxide anion is a strong nucleophile and a strong base, so we must consider both S_N2 and E2 pathways. In this case, we have a primary substrate and a small nucleophile, so there is very little steric hindrance and we expect an S_N2 process to give the desired target molecule as the major product.

8.41.

(a) The two groups that are being added across the double bond are OH and H. The OH group must be installed at the less substituted position, so we must choose reagents that achieve an *anti*-Markovnikov addition. This can be accomplished via hydroboration-oxidation.

1) $BH_3 \cdot THF$
2) H_2O_2, NaOH

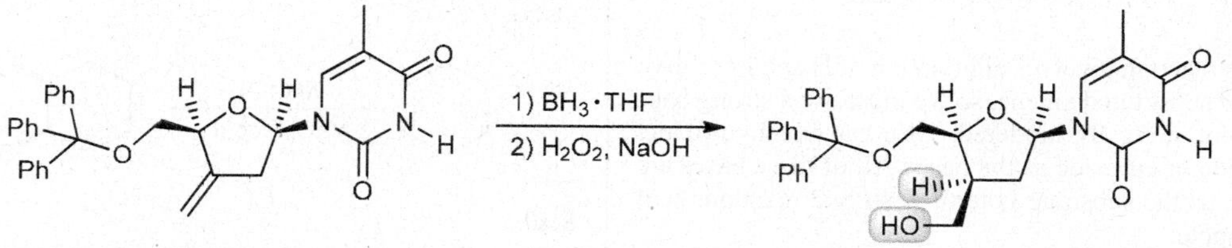

(b) The addition of OH and H can occur via a *syn* addition to either face of the π bond. The configuration of the newly formed chiral center (as shown above) results from addition to the back side of the π bond. Hydroboration occurs preferentially on the back face of the π bond in order to minimize steric interactions with the other two very large groups (both on wedges) which are both on the front face.

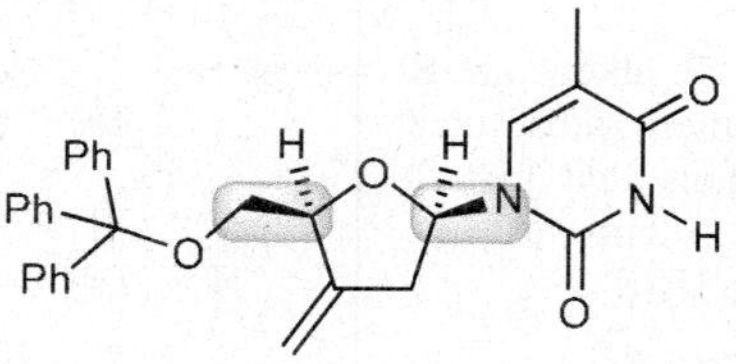

Although both H and OH are added across the π bond, only the addition of H creates a new chiral center. Addition of H to the back side of the π bond pushes the CH_2OH group forward, as seen in the major product.

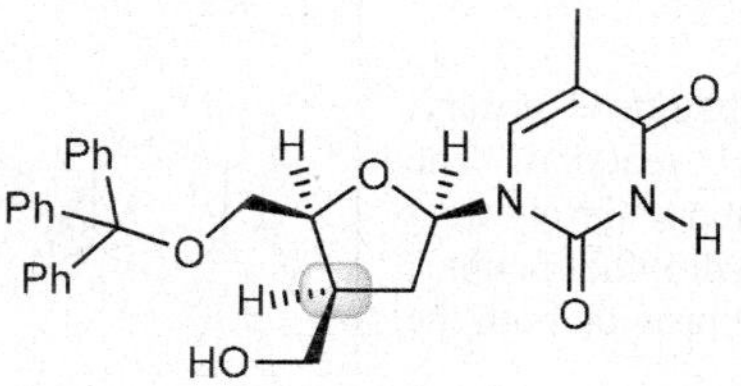

8.42.

(a) The desired transformation can be achieved via a two-step process (elimination, followed by addition). We must be careful to control the regiochemical outcome of each of these processes. During the elimination process, we want to form the more substituted alkene (Zaitsev product), so we must use a strong base that is not sterically hindered (such as hydroxide, methoxide, or ethoxide). Then, during the addition process, we want to add HCl in a Markovnikov fashion (with the Cl being installed at the more substituted position). This can be accomplished by treating the alkene with HCl, as shown here:

Cl NaOMe HCl Cl

(b) The desired transformation can be achieved via elimination, followed by addition. We must be careful to control the regiochemical outcome of each of these processes. During the elimination process, we want to form the less substituted alkene (Hofmann product), so we must use a strong, sterically hindered base, such as potassium *tert*-butoxide. Notice that the substrate is an alcohol, so we must first convert the OH group (bad leaving group) into a tosylate group (good leaving group) before performing the elimination process. Then, during the addition process, we want to add H and OH in an *anti*-Markovnikov fashion (with the OH being installed at the less substituted position). This can be accomplished via hydroboration-oxidation, as shown here:

OH 1) TsCl, py 2) *t*-BuOK HO 1) $BH_3 \cdot THF$ 2) H_2O_2, NaOH

(c) The desired transformation can be achieved via elimination, followed by addition. We must be careful to control the regiochemical outcome of each of these processes. During the elimination process, we want to form the less substituted alkene (Hofmann product), so we must use a strong, sterically hindered base, such as potassium *tert*-butoxide. Then, during the addition process, we want to add H and OH in an *anti*-Markovnikov fashion (with the OH being installed at the less substituted position). This can be accomplished via hydroboration-oxidation, as shown here:

Br *t*-BuOK HO 1) $BH_3 \cdot THF$ 2) H_2O_2, NaOH

(d) The desired transformation can be achieved via elimination, followed by addition. We must be careful to control the regiochemical outcome of each of these processes. During the elimination process, we want to form the more substituted alkene (Zaitsev product), so we will need a strong base that is not sterically hindered (such as hydroxide, methoxide, or ethoxide). Notice that the substrate is an alcohol, so we must first convert the OH group (bad leaving group) into a tosylate group (good leaving group) before performing the elimination process. Alternatively, we can simply perform the elimination process in one step by treating the alcohol with concentrated aqueous sulfuric acid (via an E1 process, as seen in Section 7.12). Then, during the addition process, we want to add H and OH in an *anti*-Markovnikov fashion (with the OH being installed at the less substituted position) via a *syn* addition (this can be seen more clearly if you draw the H that is installed, as shown). This can be accomplished via hydroboration-oxidation:

OH 1) TsCl, py 2) NaOEt conc. H_2SO_4 Heat H OH + En 1) $BH_3 \cdot THF$ 2) H_2O_2, NaOH

8.43.

(a) The retrosynthetic analysis of the target molecule asks, "What starting material can be used to make an epoxide?" The epoxide in the target molecule can be produced from an alkene intermediate, and the alkene can be produced from the given alkyl bromide:

O ⟹ ⟹ Br

The synthesis requires a two-step process: elimination followed by formation of an epoxide (called epoxidation). Treatment of the benzylic alkyl halide with a strong, hindered base (such as *t*-BuOK) favors E2 over S_N2, and reaction of the resulting alkene with a peroxy acid (RCO_3H) achieves the desired transformation. Formation of the epoxide can occur at either face of the alkene, resulting in the formation of a racemic mixture:

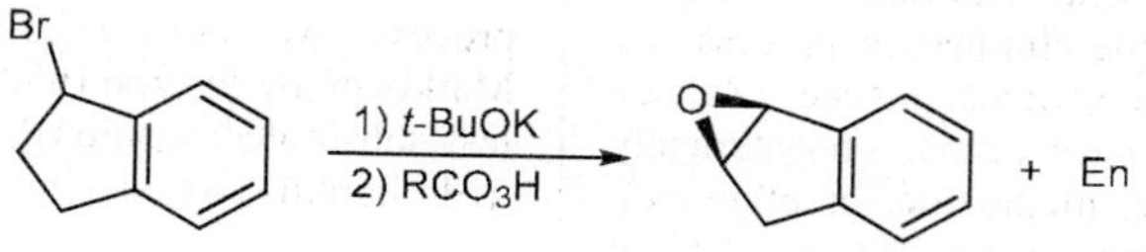

1) *t*-BuOK
2) RCO_3H
+ En

(b) The retrosynthetic analysis of the target molecule asks, "What starting material can be used to make a 1,2-dibromide?" The two Br atoms in the target molecule can be produced from an alkene intermediate, and the alkene can be produced from the given alkyl bromide:

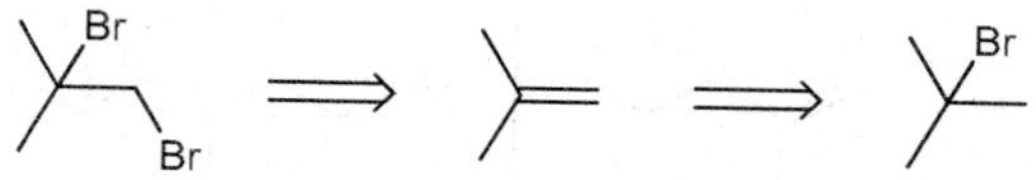

The synthesis requires a two-step process: elimination followed by bromination. The tertiary alkyl halide will undergo E2 elimination when treated with any strong base (such as hydroxide, methoxide or ethoxide), and reaction of the resulting alkene with molecular bromine (Br_2) achieves the desired transformation. In this case, no new chiral centers are generated, so the stereochemistry of the two Br atoms (*anti* addition) is not relevant:

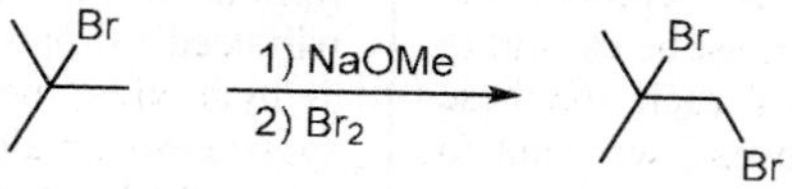

1) NaOMe
2) Br_2

(c) The retrosynthetic analysis of the target molecule asks, "What starting material can be used to make a 1,2-diol?" The two OH groups in the target molecule can be produced from an alkene intermediate, and the alkene can be produced from the given alcohol:

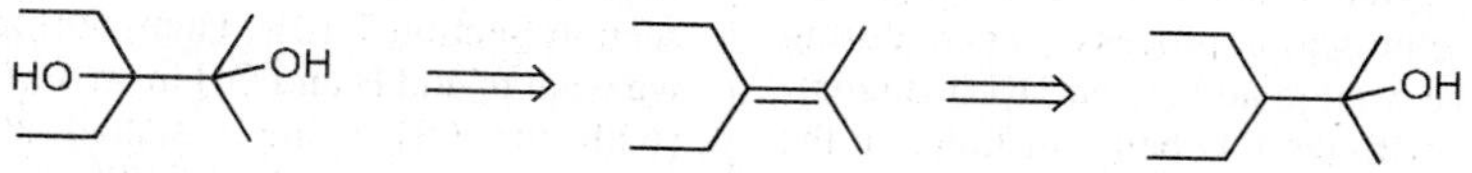

The synthesis requires a two-step process: elimination followed by dihydroxylation. The tertiary alcohol will undergo acid-catalyzed dehydration when treated with a strong, concentrated acid (such as H_2SO_4). Alternatively, we can first convert the OH group (bad leaving group) into a tosylate group (good leaving group) before performing an elimination process (strong, non-hindered base gives Zaitsev product). Treating the resulting alkene with cold potassium permanganate and sodium hydroxide results in the addition of two OH groups across the alkene, providing the target molecule. Alternatively, the dihydroxylation can be accomplished in a two-step process by epoxide ring-opening. Since there are no chiral centers in the product, *syn*- or *anti*-dihydroxylation of the alkene both give rise to the same diol.

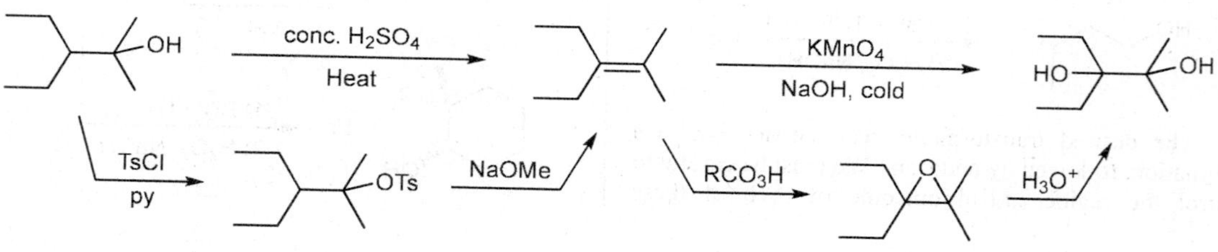

conc. H_2SO_4, Heat; $KMnO_4$, NaOH, cold; TsCl, py; NaOMe; RCO_3H; H_3O^+

(d) The retrosynthetic analysis of the target molecule asks, "What starting material can be used to make a product with a *trans* Br atom and OH group (a *trans*-bromohydrin)?" The Br atom and OH group in the target molecule can be produced from an alkene intermediate, and the alkene can be produced from the given alkyl iodide:

The synthesis requires a two-step process: elimination followed by halohydrin formation. Treatment of the secondary alkyl halide with a strong, hindered base (such as *t*-BuOK) further favors E2 over S_N2 (though any strong base would also give E2 as the major product), and reaction of the resulting alkene with molecular bromine and water (Br_2, H_2O) achieves the desired transformation. In this case, two chiral centers are formed, so we expect the pair of enantiomers resulting from an *anti* addition:

1) *t*-BuOK
2) Br_2, H_2O
Br OH + En

(e) The retrosynthetic analysis of the target molecule asks, "What starting material can be used to make a product with two aldehydes?" Oxidative cleavage of an alkene generates C=O bonds, so we can draw the starting alkene by removing the two oxygen atoms from the product, and connecting the sp^2 hybridized carbon atoms as a C=C bond. The required alkene can be produced from the given alkyl chloride:

O O ⟹ ⟹ Cl

The synthesis requires a two-step process: elimination followed by ozonolysis. Treatment of the secondary alkyl halide with a strong, hindered base (such as *t*-BuOK) favors E2 over S_N2, and reaction of the resulting alkene with ozone (O_3) followed by dimethylsulfide (DMS) achieves the desired transformation:

Cl
1) *t*-BuOK
2) O_3
3) DMS
O O

(f) The retrosynthetic analysis of the target molecule asks, "What starting material can be used to make an alkane?" The only reaction we've seen that generates a product *with no functional groups* is catalytic hydrogenation of an alkene. The required alkene can be produced from the given alcohol:

⟹ ⟹ OH

The synthesis requires a two-step process: elimination followed by hydrogenation. The tertiary alcohol will undergo acid-catalyzed dehydration when treated with a strong, concentrated acid (such as H_2SO_4). Alternatively, we can first convert the OH group (bad leaving group) into a tosylate group (good leaving group) before performing an elimination process with a non-hindered, strong base to give the Zaitsev product. In this case, no new chiral centers are generated, so the stereochemistry of the two added H atoms (*syn* addition) is not relevant:

OH
conc. H_2SO_4
Heat
H_2, Pt
TsCl
py
OTs
NaOMe

8.44. The retrosynthetic analysis of the target molecule asks, "What starting material can be used to make a 1,2-diol?" The two OH groups in the target molecule can be produced from an alkene intermediate, and the alkene can be produced from the given alcohol:

HO OH ⟹ $H_2C{=}CH_2$ ⟹ OH

The synthesis requires a two-step process: elimination followed by dihydroxylation. Notice that the substrate is an alcohol, so we can simply perform the elimination process by treating the alcohol with concentrated aqueous sulfuric acid. Alternatively, we can first convert the OH group (bad leaving group) into a tosylate group (good leaving group) before performing an elimination process. Treating the resulting ethylene with cold potassium permanganate and sodium hydroxide results in the addition of two OH groups across the alkene, providing the target structure ethylene glycol. Alternatively, the dihydroxylation can be accomplished in a two-step process by epoxide ring-opening. Since there are no chiral centers in the product, *syn*- or *anti*-dihydroxylation of the alkene both give rise to the same diol.

conc. H_2SO_4, Heat; $H_2C{=}CH_2$; $KMnO_4$, NaOH, cold; HO, OH; TsCl, py; OTs; DBU; RCO_3H; O; H_3O^+

8.45.

(a) The desired transformation can be achieved via a two-step process (addition, followed by elimination). We must be careful to control the regiochemical outcome of each step of the process. During the addition reaction, we want to install Br at the more substituted position, so we treat the alkene with HBr (without peroxides). Then, the elimination process must be performed in a way that gives the more substituted alkene (Zaitsev product), so we must use a strong base that is not sterically hindered, such as methoxide (hydroxide or ethoxide can also be used).

1) HBr 2) NaOMe

(b) These two alkenes can be interconverted via a two-step process (addition, followed by elimination). We must be careful to control the regiochemical outcome of each step of the process. In one case, a sterically hindered base is required, while in the other case, we must use a base that is not sterically hindered, as shown.

1) HBr 2) *t*-BuOK; 1) HBr 2) NaOMe

(c) The two-step process (addition followed by elimination) must be used twice in this case. First, we perform a Markovnikov addition of HBr to install Br at the more substituted position, followed by elimination with a base that is not sterically hindered, thereby giving the more substituted alkene (Zaitsev product). Then, we perform the two-step process again. But this time, we begin with an *anti*-Markovnikov addition of HBr (in the presence of peroxides) to install Br at the less substituted position, followed by elimination with a sterically hindered base to give the less substituted alkene (Hofmann product):

1) HBr 2) NaOMe; 1) HBr, ROOR 2) *t*-BuOK

(d) The two-step process (addition followed by elimination) must be used twice in this case. First, we perform a Markovnikov addition of HBr to install Br at the more substituted position, followed by elimination with a base that is not sterically hindered, thereby giving the more substituted alkene (Zaitsev product). Then, we perform the two-step process again. But this time, we begin with an *anti*-Markovnikov addition of HBr (in the presence of peroxides) to install Br at the less substituted position, followed by elimination with a sterically hindered base to give the less substituted alkene (Hofmann product).

1) HBr 2) NaOMe; 1) HBr, ROOR 2) *t*-BuOK

8.46. The retrosynthetic analysis of target molecule **2** asks, "What starting material can be used to make a diene?" Each of the two pi bonds in compound **2** can be formed by an elimination reaction, starting with a suitable alkyl halide. To draw the starting materials, we delete each C=C double bond, we place a H atom on one of the carbon atoms, and we place a Br atom on the other carbon atom (two E2 eliminations require two Br leaving groups). If we place the Br atoms on adjacent carbon atoms, then we have an intermediate structure that can be prepared from compound **1**.

O, OTBDMS, CH_3 — **2** ⟹ Br, Br, O, OTBDMS, CH_3 ⟹ O, OTBDMS, CH_3 — **1**

When treated with a strong base, the dibromide will undergo a double elimination to give the desired diene target molecule (compound **2**):

Strong base

Both elimination reactions are favored if we use a bulky base, such as *tert*-butoxide. The necessary dibromide can be made directly from compound **1**, via bromination of the π bond. This gives the following two-step synthesis:

Br_2 ; *t*-BuOK (2 equiv.)

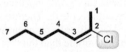

8.47.

(a) We begin by identifying the parent. The longest chain is seven carbon atoms, so the parent is heptene. There are four substituents (highlighted), all of which are methyl groups. Notice that the parent chain is numbered starting from the side that is closest to the π bond. According to this numbering scheme, the methyl groups are located at C3, C4, C5 and C5. Finally, we use the prefix "tetra" to indicate the presence of four methyl groups, and we include a locant that identifies the position of the double bond ("3" indicates that the double bond is located between C3 and C4). In addition, the configuration of the double bond (*E*) must be indicated at the beginning of the name:

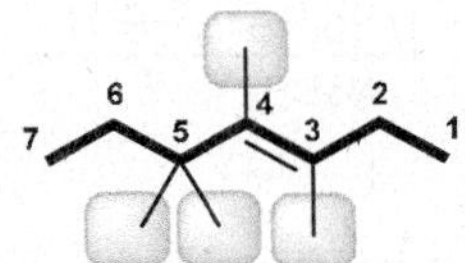

(*E*)-3,4,5,5-tetramethyl-3-heptene

(b) The parent is cyclohexene, which does not require a locant to identify the position of the double bond, because it is assumed to be between C1 and C2. There is only one substituent (an ethyl group, located at C1):

1-ethylcyclohexene

(c) We begin by identifying the parent, which is bicyclic in this case. The parent is bicyclo[2.2.2]octene. There is only one substituent (a methyl group). Notice that the parent chain is numbered starting from one of the bridgeheads, as seen in Section 4.2, which places the double bond between C2 and C3. According to this numbering scheme, the methyl group is located at C2. Finally, we include a locant that identifies the position of the double bond (C2).

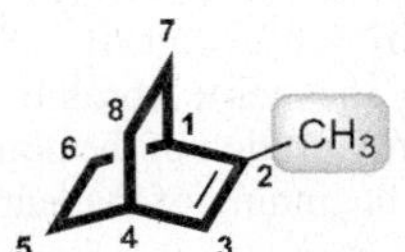

2-methylbicyclo[2.2.2]oct-2-ene

(d) We begin by identifying the parent. The longest chain is seven carbon atoms, so the parent is heptene. There is one substituent (a chloro group). The parent chain is numbered starting from the side that is closest to the π bond. According to this numbering scheme, the chloro group is located at C2. We include a locant that identifies the position of the double bond ("2" indicates that the double bond is located between C2 and C3). In addition, the configuration of the double bond (*E*) must be indicated at the beginning of the name:

(*E*)-2-chloro-2-heptene

(e) The parent is cyclohexene, which does not require a locant to identify the position of the double bond, because it is assumed to be between C1 and C2. Locants are assigned counter-clockwise, in order to give the substituent (Br) the lowest possible locant (C3 rather than C6). The configuration of the chiral center (*R*) must also be included at the beginning of the name.

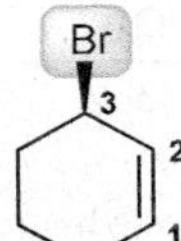

(*R*)-3-bromocyclohexene

(f) The parent is cyclopentene, which does not require a locant to identify the position of the double bond, because it is assumed to be between C1 and C2. Substituents are arranged alphabetically in the name. The configuration of the chiral center (*R*) must also be included at the beginning of the name.

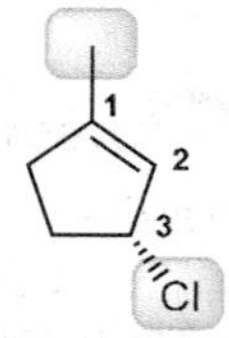

(*R*)-3-chloro-1-methylcyclopentene

(g) We begin by identifying the parent. The longest chain is eight carbon atoms, so the parent is octene. There are five substituents (highlighted). The location of each substituent is indicated with the appropriate locant. In this case, locants are assigned from right-to-left, to give the lowest possible numbers to the first two substituents (when numbering from right-to-left, the first two substituents are both at C2, which is not the case when numbering from left-to-right)). The substituents are alphabetized in the name (bromo before methyl). In addition, the configuration of the double bond (*E*) must be indicated at the beginning of the name:

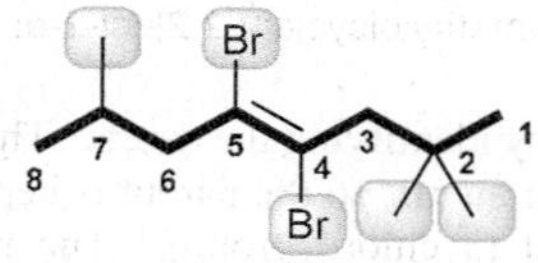

(*E*)-4,5-dibromo-2,2,7-trimethyl-4-octene

8.48. A reaction is only favorable if ΔG is negative. Recall that ΔG has two components: (ΔH) and ($-T\Delta S$). The first term (ΔH) is positive for this reaction (two strong σ bonds are broken, and a combination of weaker bonds are formed: one σ bond and one π bond). The second term ($-T\Delta S$) is negative because ΔS is positive (one molecule is converted into two molecules). Therefore, the reaction is only favorable (ΔG will only be negative) if the second term is greater in magnitude than the first term. This only occurs above a certain temperature.

8.49.

(a) Treating an alkene with cold potassium permanganate and sodium hydroxide results in the addition of OH and OH across the alkene. In this case, two new chiral centers are generated, so we expect only the pair of enantiomers that would result from *syn* addition:

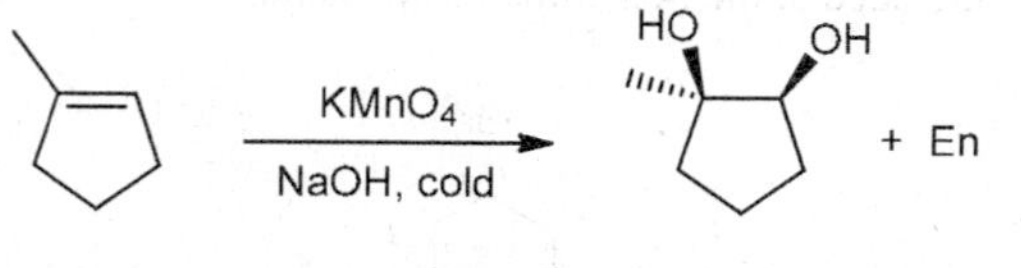

$KMnO_4$ / NaOH, cold → + En

(b) The reagent indicates a hydrochlorination reaction, so the net result will be the addition of H and Cl across the alkene. We expect a Markovnikov addition, so the Cl group will be installed at the more substituted position. No chiral centers are formed in this case, so stereochemistry is irrelevant:

HCl →

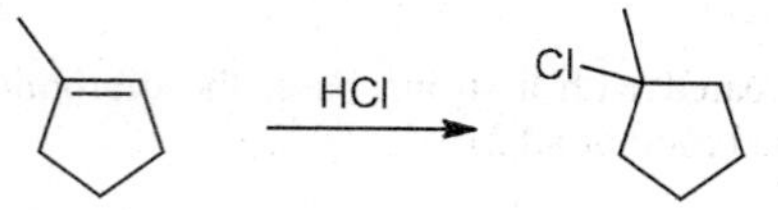

(c) The reagents indicate a hydrogenation reaction, so the net result will be the addition of H and H across the alkene. The regiochemical outcome is not relevant because the two groups added (H and H) are identical. The stereochemical requirement for the reaction (*syn* addition) is not relevant in this case, as no chiral centers are formed:

H_2 / Pt →

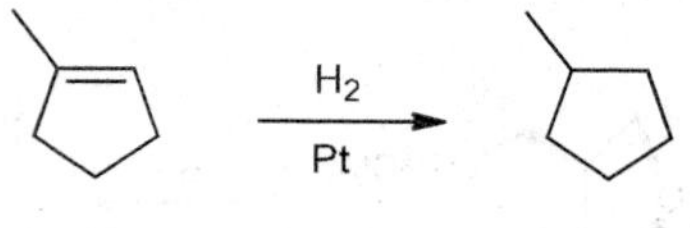

(d) The reagents indicate bromohydrin formation, so the net result will be the addition of Br and OH across the alkene. The OH group is expected to be installed at the more substituted position. The reaction proceeds via an *anti* addition, giving the following pair of enantiomers:

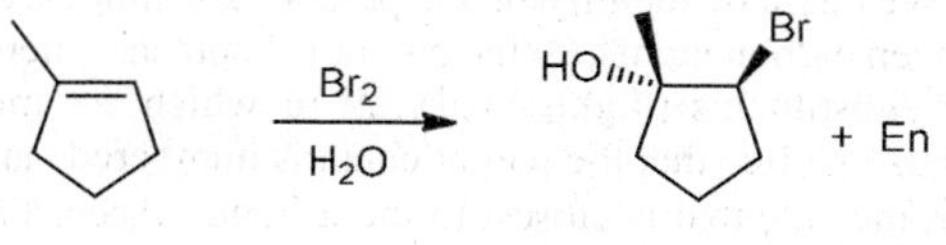

Br_2 / H_2O → + En

(e) The reagents indicate hydration of the alkene via oxymercuration-demercuration. The net result will be the addition of H and OH across the alkene, with the OH group being installed at the more substituted position. The product has no chiral centers, so stereochemistry is not a consideration:

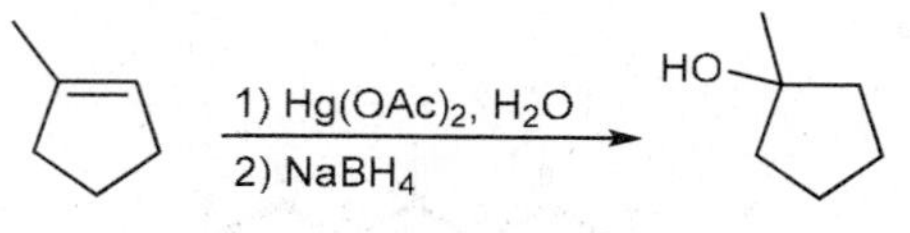

1) $Hg(OAc)_2$, H_2O 2) $NaBH_4$ →

8.50.

(a) The reagents indicate a dihydroxylation process, so the net result will be the addition of OH and OH across the alkene. The regiochemical outcome is not relevant because the two groups added (OH and OH) are identical. We expect the reaction to proceed via an *anti* addition, but only one chiral center is formed, so we expect both enantiomers:

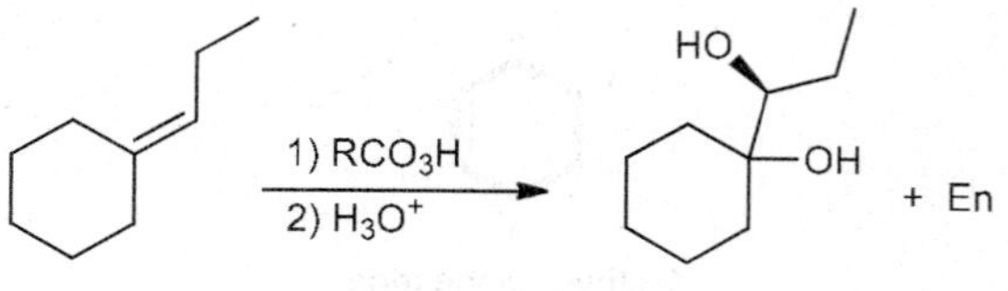

1) RCO_3H 2) H_3O^+ → + En

(b) The reagent indicates a hydrobromination reaction, so the net result will be the addition of H and Br across the alkene. We expect a Markovnikov addition, so Br will be installed at the more substituted position. No chiral centers are formed in this case, so stereochemistry is irrelevant:

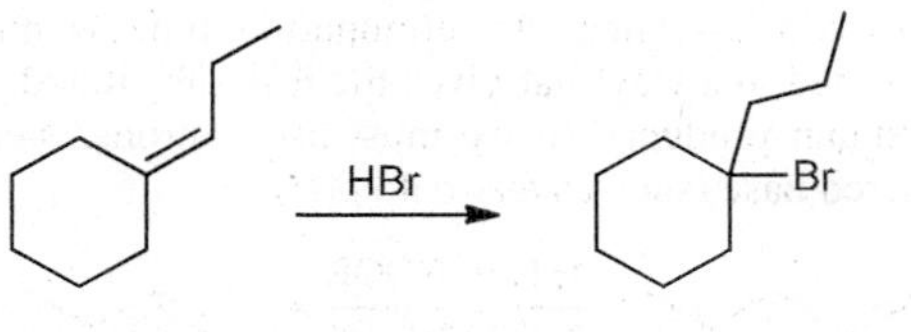

HBr

Br

(c) The reagents indicate a hydrogenation reaction, so the net result will be the addition of H and H across the alkene. The regiochemical outcome is not relevant because the two groups added (H and H) are identical. No chiral centers are formed in this case, so stereochemistry is also irrelevant.

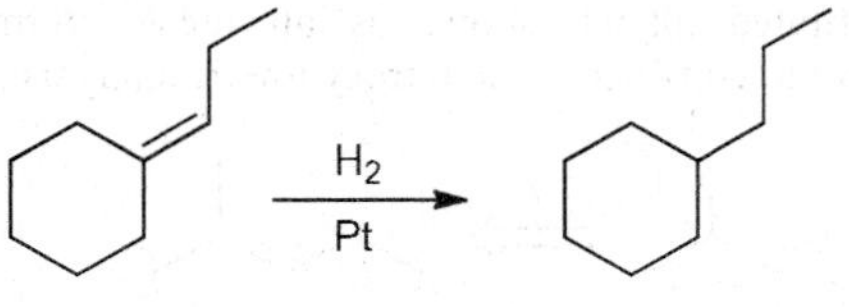

H_2

Pt

(d) The reagent indicates a bromination reaction, so the net result will be the addition of Br and Br across the alkene. The regiochemical outcome is not relevant because the two groups added (Br and Br) are identical. We expect the reaction to proceed via an *anti* addition, but only one chiral center is formed, so we expect both enantiomers:

Br_2

Br

Br

+ En

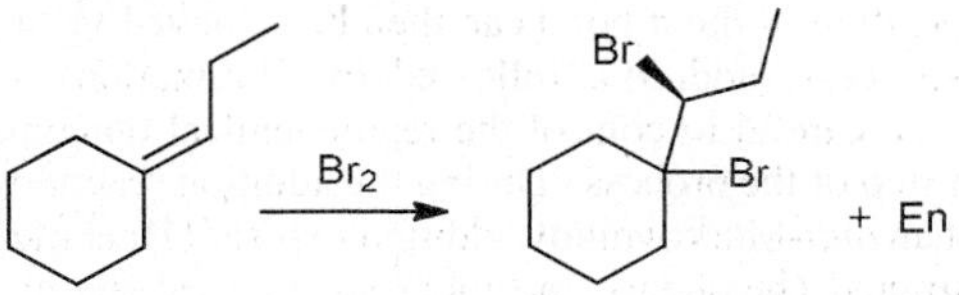

(e) The reagents indicate a hydroboration-oxidation, so the net result will be the addition of H and OH across the alkene. For the regiochemical outcome, we expect an *anti*-Markovnikov addition, so the OH group is installed at the less substituted (secondary) position. We expect the reaction to proceed via a *syn* addition, but only one chiral center is formed, so we expect both enantiomers (*syn* addition can occur on either face of the alkene):

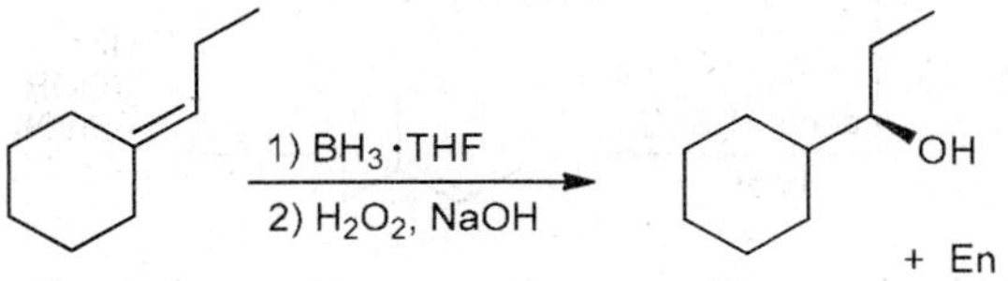

1) $BH_3 \cdot THF$

2) H_2O_2, NaOH

OH

+ En

8.51.

(a) Water (H and OH) is added across the alkene in a Markovnikov fashion. The mechanism is expected to have three steps: 1) proton transfer, 2) nucleophilic attack, and 3) proton transfer. In the first step, a proton is transferred from H_3O^+ to the alkene, which requires two curved arrows, as shown below. The resulting tertiary carbocation is then captured by a water molecule in the second step of the mechanism. This step requires one curved arrow, going from the nucleophile (water) to the electrophile (the carbocation). Then, in the final step of the mechanism, a molecule of water functions as a base and removes a proton, thereby generating the product. This final step is a proton transfer step, and therefore requires two curved arrows, as shown:

OH

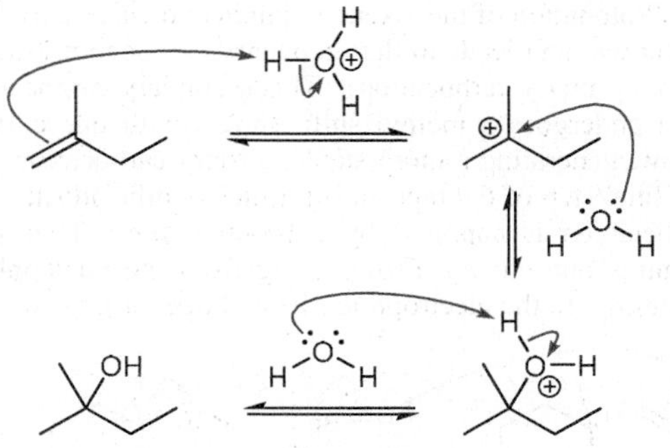

(b) In the first step of the mechanism, a proton is transferred from H_3O^+ to the alkene, which requires two curved arrows, as shown below. The resulting secondary carbocation then rearranges via a hydride shift, giving a more stable, tertiary carbocation. That step is shown with one curved arrow. The tertiary carbocation is then captured by a water molecule, which is shown with one curved arrow, going from the nucleophile (water) to the electrophile (the carbocation). Then, in the final step of the mechanism, a molecule of water functions as a base and removes a proton, thereby generating the product. This final step is a proton transfer step, and therefore requires two curved arrows, as shown:

Hydride shift

OH

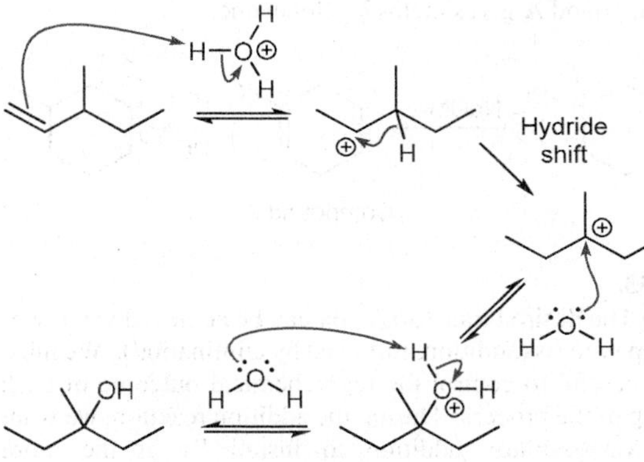

(c) In this reaction, H and Br are added across the alkene in the absence of peroxides (Markovnikov addition), which indicates an ionic process. There are two mechanistic steps in the ionic addition of HBr across an alkene: 1) proton transfer, followed by 2) nucleophilic attack. In the first step, a proton is transferred from HBr to the alkene, which requires two curved arrows, as shown below. The resulting tertiary carbocation is then captured by a bromide ion in the second step of the mechanism. This step requires one curved arrow, going from the nucleophile to the electrophile, as shown:

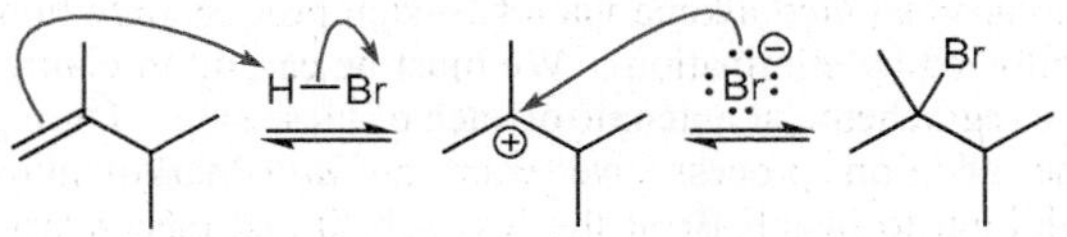

H—Br

Br

(d) Protonation of the alkene requires two curved arrows, as shown, and leads to the secondary carbocation (rather than a primary carbocation). This secondary carbocation then undergoes a methyl shift, shown with one curved arrow, generating a more stable, tertiary carbocation. In the final step of the mechanism (nucleophilic attack), the carbocation is captured by a bromide ion. This step requires one curved arrow, going from the nucleophile (bromide) to the electrophile (the carbocation), as shown:

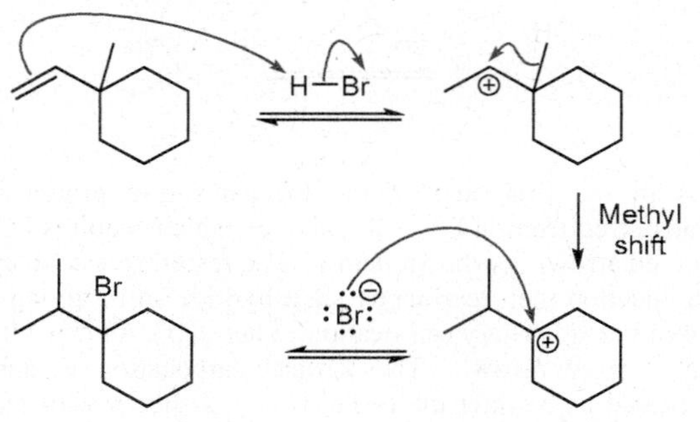

H—Br

Methyl shift

:Br:⊖

Br

8.52. The starting material (1-bromo-1-methylcyclohexane) is a tertiary alkyl halide, and will undergo an E2 reaction when treated with a strong base such as methoxide, to give the more substituted alkene (Zaitsev product, compound **A**). Hydrogenation of compound **A** gives methylcyclohexane:

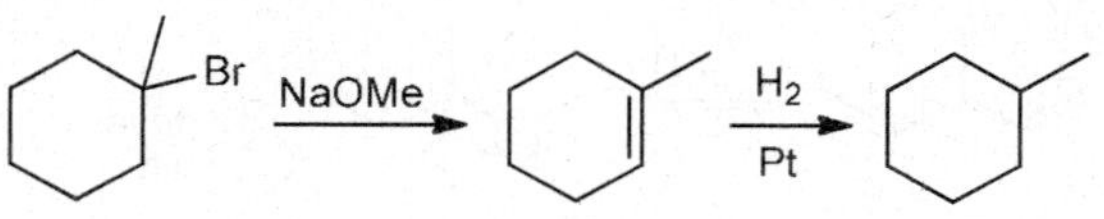

Br — NaOMe → Compound **A** — H_2 / Pt →

8.53.

(a) The desired transformation can be achieved via a two-step process (addition, followed by elimination). We must be careful to control the regiochemical outcome of each step of the process. During the addition reaction, we want a Markovnikov addition to install Br at the more substituted (tertiary) position, so we treat the alkene with HBr (without peroxides present). Then, the elimination process must be performed in a way that gives the more substituted alkene (Zaitsev product), so we must use a strong base that is not sterically hindered, such as methoxide (hydroxide or ethoxide can also be used).

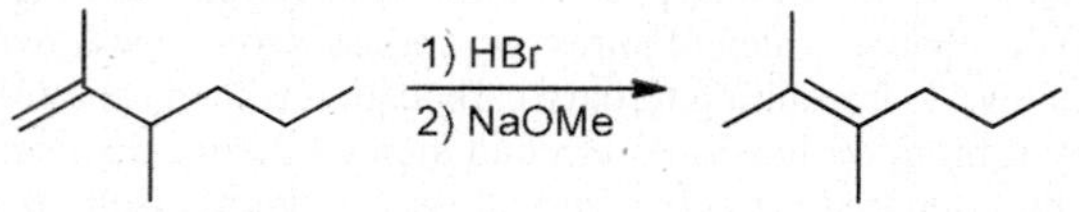

1) HBr
2) NaOMe

(b) This trisubstituted alkene can be converted into the monosubstituted alkene via a two-step process (addition, followed by elimination). We must be careful to control the regiochemical outcome of each of these steps. During the addition process, we want an *anti*-Markovnikov addition to install Br at the less substituted (secondary) position, so we treat the alkene with HBr in the presence of peroxides. Then, the elimination process must be performed in a way that gives the less substituted alkene (Hofmann product), so we must use a strong, sterically hindered base (such as *tert*-butoxide).

1) HBr, ROOR
2) *t*-BuOK

8.54. In this example, we are forming an alkene (so we will need to perform an elimination), and because the pi bond is not on the same carbon atom as the original OH group, we are also moving the position of the pi bond. The trisubstituted alkene shown is an ideal intermediate structure to consider in the retrosynthetic analysis:

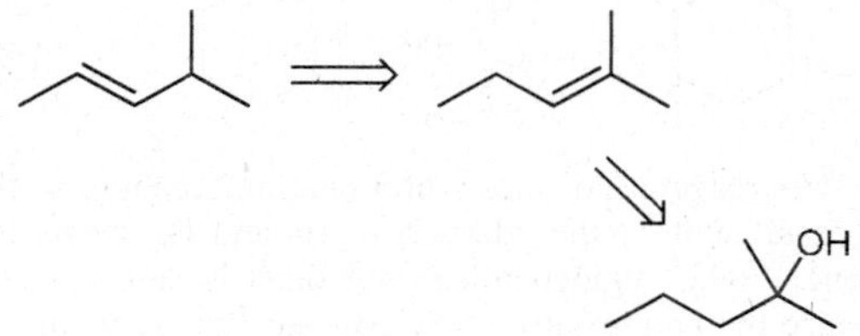

OH

Treatment of the starting alcohol with concentrated sulfuric acid affords the more substituted alkene. Moving the position of the π bond can then be achieved via a two step-process (addition, followed by elimination). We must be careful to control the regiochemical outcome of each step of the process. During the addition reaction, we want an *anti*-Markovnikov addition to install Br at the less substituted (secondary) position, so we treat the alkene with HBr in the presence of peroxides. Then, the elimination reaction must be performed in a way that gives the less substituted alkene (Hofmann product), so we must use a strong, sterically hindered base (such as *tert*-butoxide).

OH

1) conc. H_2SO_4, heat
2) HBr, ROOR
3) *t*-BuOK

conc. H_2SO_4
heat

1) HBr, ROOR
2) *t*-BuOK

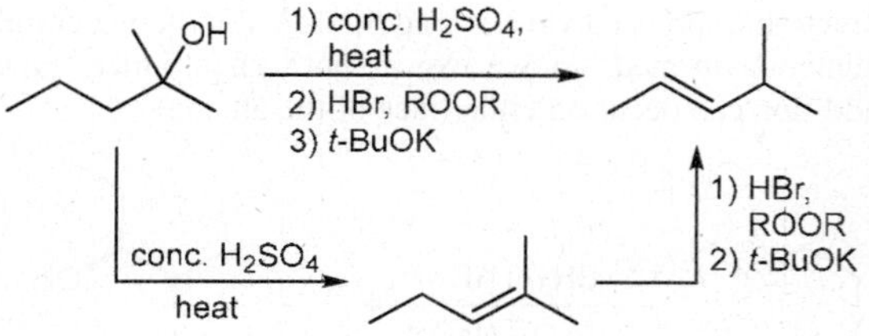

8.55. Two different alkenes will produce 2,4-dimethylpentane upon hydrogenation:

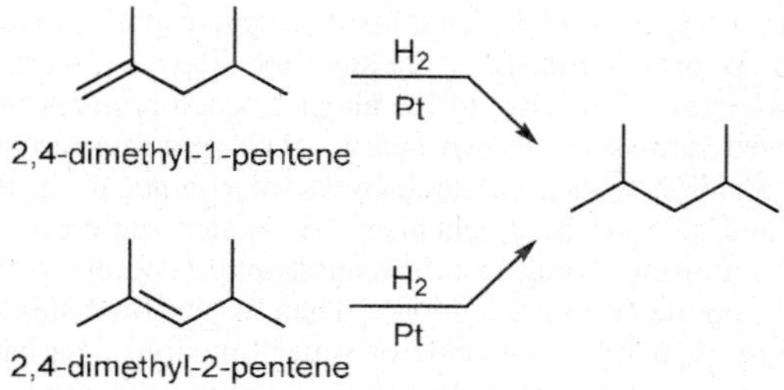

2,4-dimethyl-1-pentene — H_2 / Pt

2,4-dimethyl-2-pentene — H_2 / Pt

Note that the following four drawings all represent the same compound (2,4-dimethyl-1-pentene):

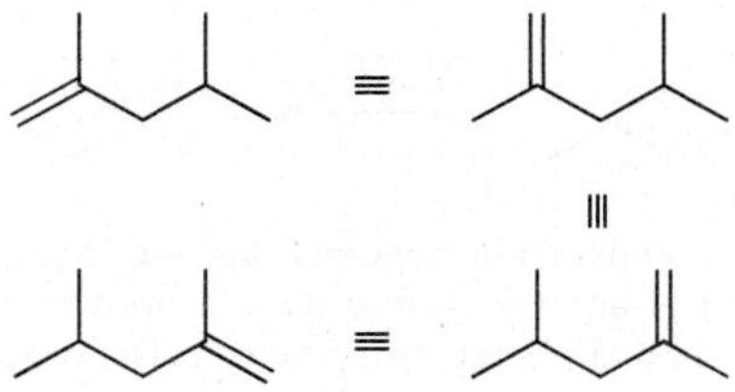

8.56. We must first determine the structure of compound **A**. The necessary information has been provided. Specifically, ozonolysis of compound **A** gives only one product, which has only one C=O bond. Therefore, the starting alkene must be symmetrical, leading to two equivalents of the ketone product:

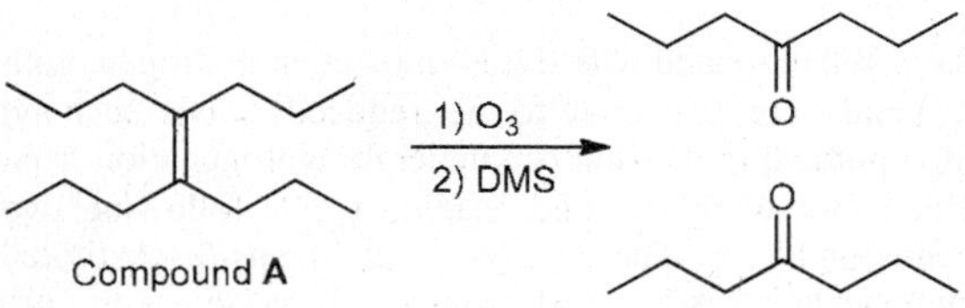

1) O_3
2) DMS

Compound **A**

Treatment of compound **A** with a peroxy acid, followed by aqueous acid, affords a diol. No chiral centers are formed, so stereochemistry is not a relevant consideration.

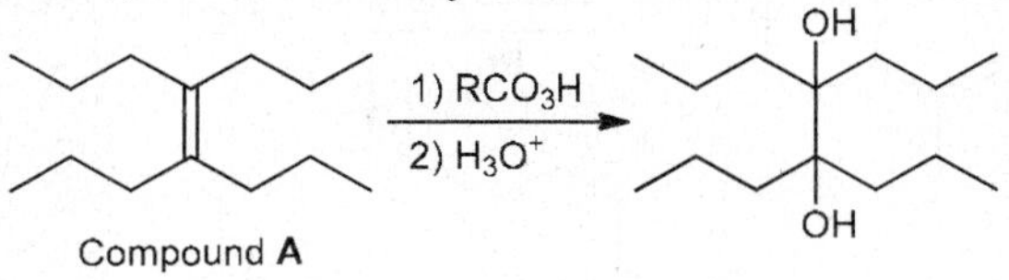

1) RCO_3H
2) H_3O^+

Compound **A**

8.57.

(a) Interconversion between the two alcohols requires moving the position of the OH group. In each case, this can be accomplished via a two step-process (elimination followed by addition). In each case, the elimination step can be achieved by treating the alcohol with concentrated sulfuric acid to give an alkene. For the addition step (adding H and OH to the alkene), the regiochemical outcome must be carefully considered. In the first case below, dilute aqueous acid is used to give a Markovnikov addition, while in the second case below, hydroboration-oxidation is employed to give an *anti*-Markovnikov addition.

1) conc. H_2SO_4, heat
2) dilute H_2SO_4

1) conc. H_2SO_4, heat
2) $BH_3 \cdot THF$
3) H_2O_2, NaOH

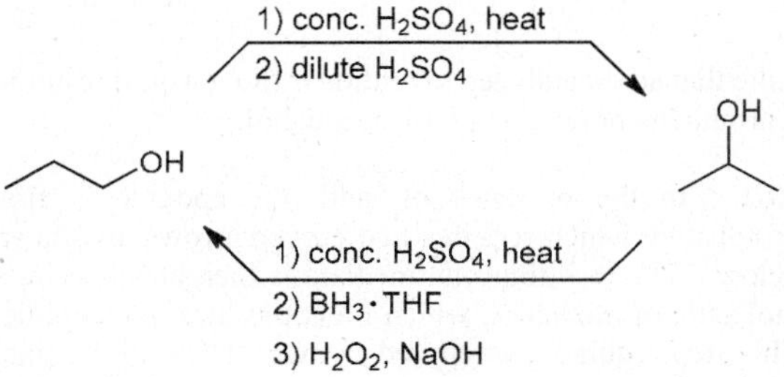

(b) Interconversion between the two alkyl halides requires moving the position of Br. In each case, this can be accomplished via a two step-process (elimination followed by addition). In each case, the elimination step can be achieved via an E2 reaction, using a strong base (such as hydroxide, or methoxide or ethoxide) to give the more substituted alkene (Zaitsev product). For the addition step (adding H and Br to the alkene), the regiochemical outcome must be carefully considered. In the first case below, HBr and peroxides are used to give an *anti*-Markovnikov addition, while in the second case below, HBr is used to give a Markovnikov addition.

1) NaOMe
2) HBr, ROOR

1) NaOMe
2) HBr

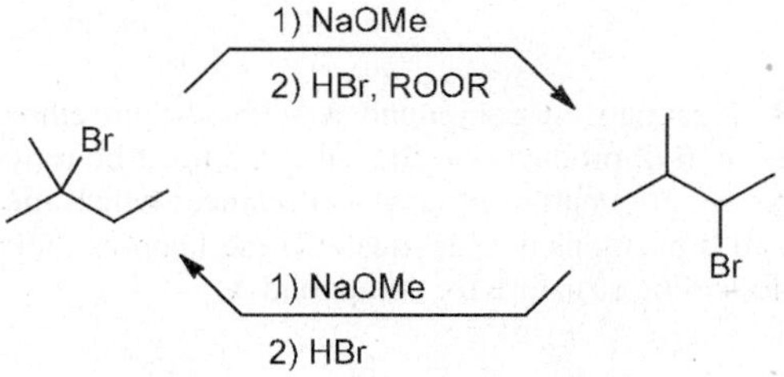

(c) The retrosynthetic analysis of the target molecule asks, "What starting material can be used to make an alkane?" The only reaction we've seen that generates a product *with no functional groups* is catalytic hydrogenation of an alkene. The required alkene can be produced from the given alkyl chloride:

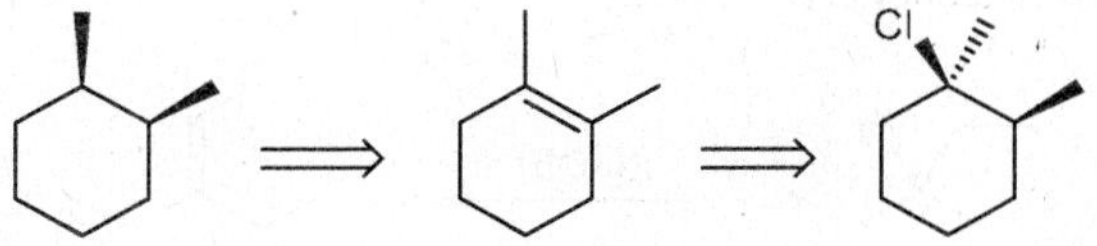

The synthesis requires a two-step process: elimination followed by hydrogenation. Treating the starting material with a strong base (such as hydroxide, methoxide or ethoxide) gives the more substituted alkene (Zaitsev product), which can then be converted to the desired *meso* compound upon hydrogenation.

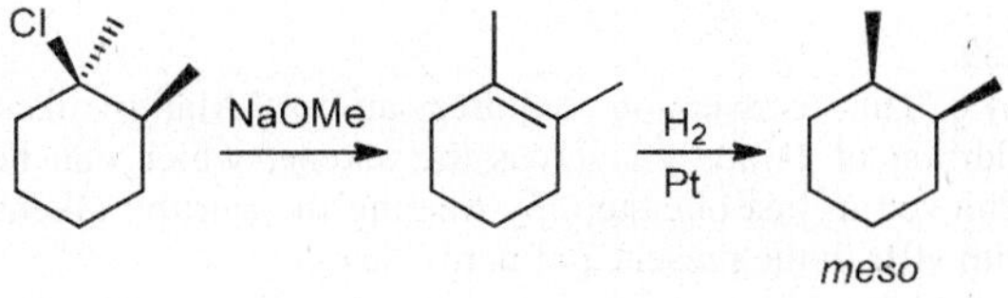

NaOMe

H_2
Pt

meso

(d) The retrosynthetic analysis of the target molecule asks, "What starting material can be used to make a *cis*-1,2-diol?" The two OH groups in the target molecule can be produced from an alkene intermediate, and the alkene can be produced from the given alcohol:

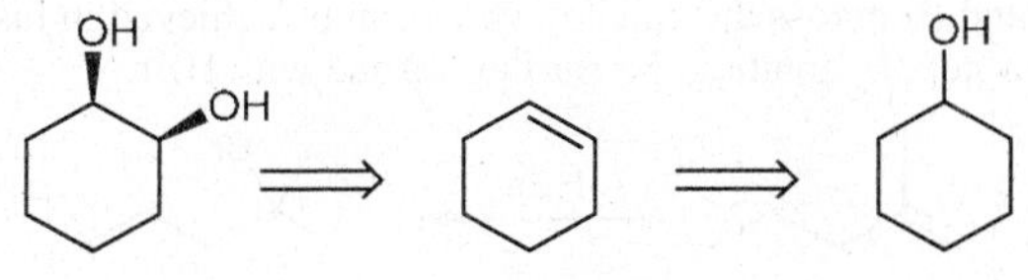

The synthesis requires a two-step process: elimination followed by *syn*-dihydroxylation. The necessary alkene (cyclohexene) can be made in one step from the starting alcohol, upon treatment with concentrated sulfuric acid and heat. Treating the alkene with catalytic osmium tetroxide and NMO results in the *syn* addition of OH and OH:

conc. H_2SO_4, Heat → OsO_4, NMO → *meso*

8.58. Treatment of compound **A** with sodium ethoxide gives no S_N2 products, so the substrate must be tertiary. Only one elimination product is obtained, which means that all β positions are identical. These features indicate the following structure for compound **A**:

Compound **A**

Treatment of compound **A** with ethoxide gives alkene **B** (Zaitsev product, shown below), which undergoes acid-catalyzed hydration (Markovnikov addition of water) to give alcohol **C** ($C_7H_{16}O$):

Compound **A** —NaOEt→ Compound **B** —dilute H_2SO_4→ Compound **C**

8.59.

(a) This conversion requires an *anti*-Markovnikov addition of H and Br across the alkene, which can be achieved in just one step, by treating the starting alkene with HBr in the presence of peroxides:

HBr, ROOR → + En

(b) This conversion requires a Markovnikov addition of H and Br across the alkene, which can be achieved in just one step, by treating the starting alkene with HBr:

HBr →

(c) This conversion requires a Markovnikov addition of H and OH across the alkene, which can be achieved via acid-catalyzed hydration:

H_3O^+ →

(d) This conversion requires an *anti*-Markovnikov addition of H and OH across the alkene, which can be achieved via hydroboration-oxidation. The process does proceed via *syn* addition, but only one chiral center is formed, and the *syn* addition can take place on either face of the alkene, giving a pair of enantiomers.

1) $BH_3 \cdot THF$ 2) H_2O_2, NaOH → + En

8.60. When treated with excess molecular hydrogen, both π bonds are expected to be reduced. The π bond incorporated in the ring can undergo hydrogenation from either face of the π bond, leading to the following two compounds. These *cis*- and *trans*-disubstituted cyclohexanes are diastereomers because they are stereoisomers that are not mirror images of each other.

excess H_2, Pt → *trans* + *cis* — Diastereomers

8.61. This conversion requires the Markovnikov addition of water *without* carbocation rearrangement. This can be achieved via oxymercuration-demercuration:

1) $Hg(OAc)_2$, H_2O 2) $NaBH_4$ → racemic

Note that acid-catalyzed hydration (H_3O^+) would result in rearrangement to give a tertiary alcohol.

8.62. In the presence of acid, the epoxide is first protonated, which requires two curved arrows, as shown below. The resulting intermediate is then attacked by a molecule of methanol, which functions as a nucleophile. This step requires two curved arrows. Then, in the final step of the mechanism, a molecule of methanol functions as a base and removes a proton, thereby generating the product. This final step is a proton transfer step, and therefore requires two curved arrows:

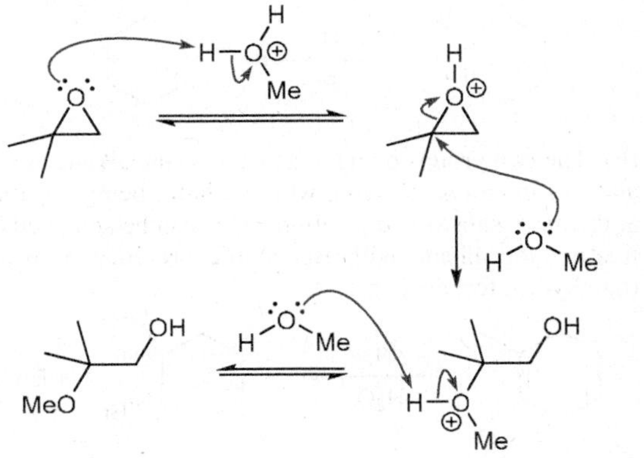

8.63. In the first step of the mechanism, a proton is transferred from H_3O^+ to the alkene, which requires two curved arrows, as shown. The resulting secondary carbocation then rearranges via a methyl shift, giving a more stable, tertiary carbocation. That step is shown with one curved arrow. The tertiary carbocation is then captured by a water molecule, which is shown with one curved arrow, going from the nucleophile (water) to the electrophile (the carbocation). Then, in the final step of the mechanism, a molecule of water functions as a base and removes a proton, thereby generating the product. This final step is a proton transfer step, and therefore requires two curved arrows, as shown:

Methyl shift

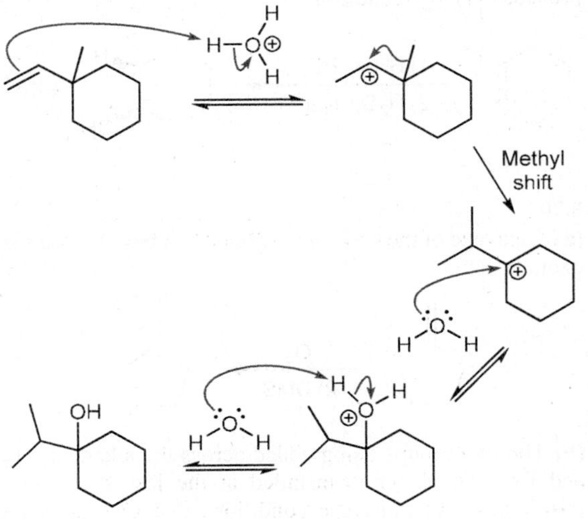

8.64.
(a) The reagents indicate a hydrogenation reaction, so the net result will be the addition of H and H across the alkene. The regiochemical outcome is not relevant because the two groups added (H and H) are identical. We expect the reaction to proceed via a *syn* addition, giving the following *meso* compound:

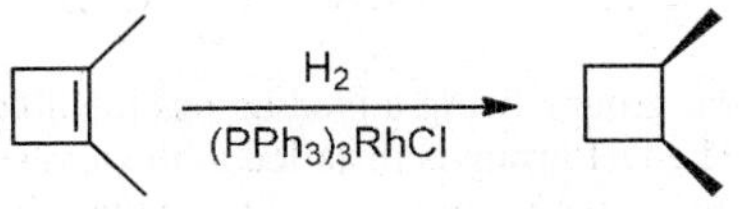

H_2, $(PPh_3)_3RhCl$

meso

(b) The reagents indicate an acid-catalyzed hydration, so the net result will be the addition of H and OH across the alkene. We expect a Markovnikov addition, so the OH group will be installed at the more substituted position. No chiral centers are formed in the process, so stereochemistry is not a relevant consideration:

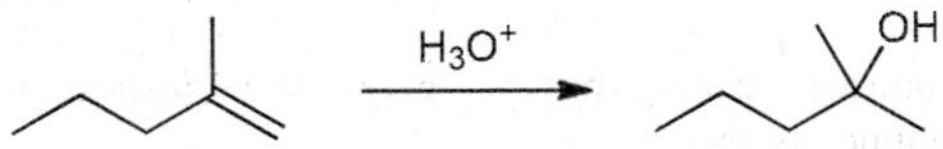

H_3O^+

(c) The reagents indicate a hydroboration-oxidation, so the net result will be the addition of H and OH across the alkene. For the regiochemical outcome, we expect an *anti*-Markovnikov addition, so the OH group is installed at the less substituted position. The stereochemical outcome (*syn* addition) is not relevant in this case, because the product has no chiral centers:

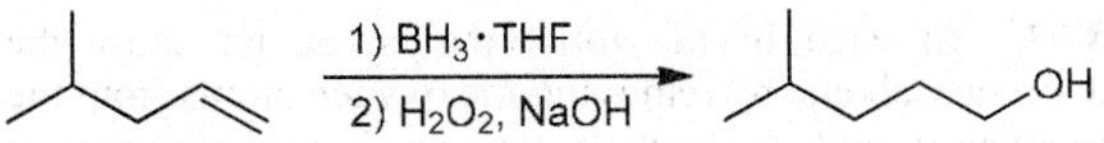

1) $BH_3 \cdot THF$
2) H_2O_2, NaOH

(d) The reagents indicate a dihydroxylation process (via an epoxide), so the net result will be the addition of OH and OH across the alkene. The regiochemical outcome is not relevant because the two groups added (OH and OH) are identical. We expect the reaction to proceed via an *anti* addition. In this case, two chiral centers are formed, so we expect the pair of enantiomers resulting from an *anti* addition:

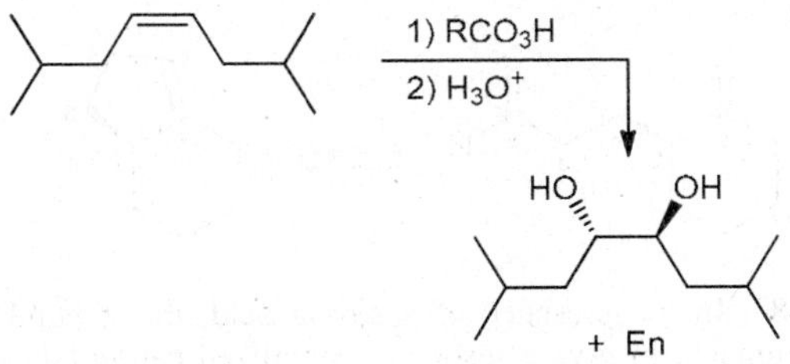

1) RCO_3H
2) H_3O^+

+ En

8.65.
(a) Hydroboration-oxidation gives an *anti*-Markovnikov addition. If 1-propene is the starting material, the OH group will not be installed in the correct location. Acid-catalyzed hydration of 1-propene would give the desired product.

(b) Hydroboration-oxidation gives a *syn* addition of H and OH across a double bond. This compound does not have an adjacent hydrogen atom that is *cis* to the OH group, and therefore, hydroboration-oxidation cannot be used to make this compound.

(c) Hydroboration-oxidation gives an *anti*-Markovnikov addition, but the OH group is on a tertiary position. There is no starting alkene that would yield the desired product via an *anti*-Markovnikov addition.

8.66. The bromination product is a four-carbon chain with Br atoms on C2 and C3, so compound **X** must be 2-butene.

Bromination of *cis*-2-butene does NOT give the desired *meso* compound:

NOT *meso*

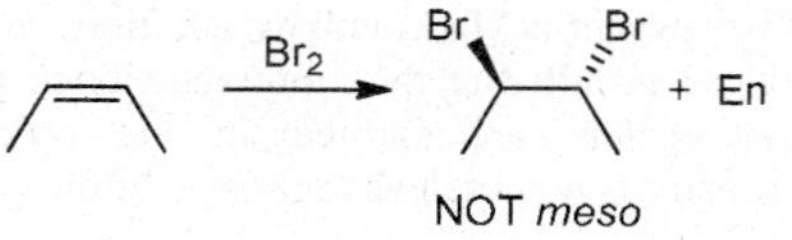

In contrast, *trans*-2-butene gives the desired *meso* compound, as shown:

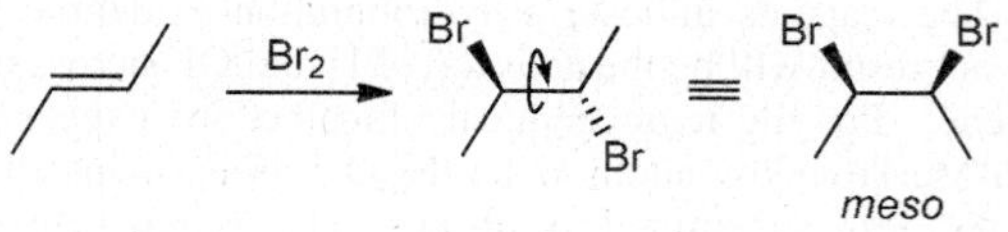

meso

Therefore, compound **X** is *trans*-2-butene.

8.67. In each of the following cases, we draw the necessary alkene by removing the oxygen atoms from the product(s) and connecting the sp^2 hybridized carbon atoms to form a C=C bond:

(a) **(b)** **(c)**

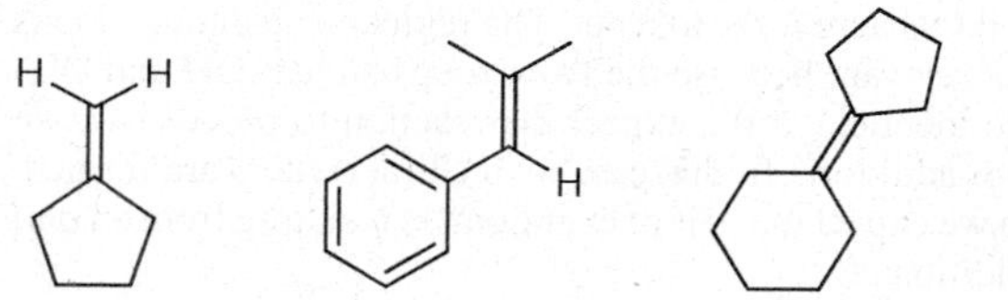

(d)

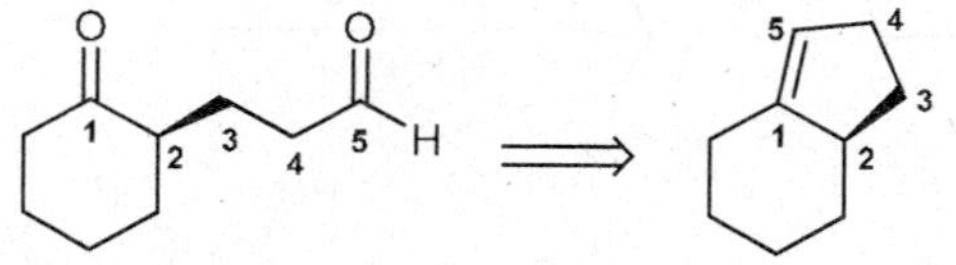

8.68. In the presence of a strong acid, the π bond is protonated to give a resonance-stabilized cation (shown below), which is even lower in energy than a tertiary carbocation. This protonation step determines the regiochemical outcome of the reaction, because the resonance-stabilized cation is captured by a bromide ion to give the product, as shown.

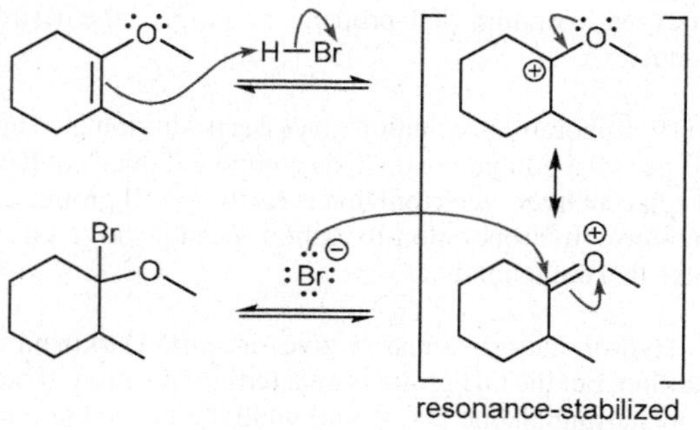

resonance-stabilized

8.69.

(a) The two groups being added across the alkene are H and H, which can be accomplished by treating the alkene with molecular hydrogen (H_2) in the presence of a suitable catalyst.

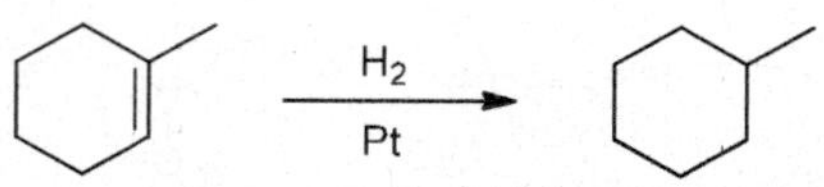

(b) The two groups being added across the alkene are Br and OH in an *anti* fashion, with the latter being installed at the more substituted position. This can be achieved by treating the alkene with Br_2 in the presence of water (halohydrin formation):

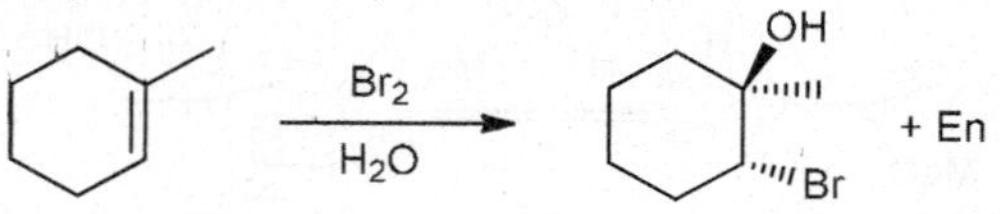

(c) Cleavage of the C=C double bond can be achieved via ozonolysis:

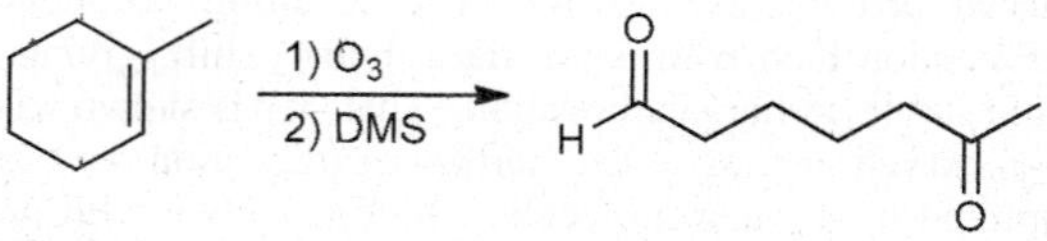

(d) The two groups being added across the alkene are H and OH. The OH group must be installed at the less substituted carbon atom, so we must use conditions that give an *anti*-Markovnikov addition of H and OH. This can be accomplished via hydroboration-oxidation, which proceeds via a *syn* addition:

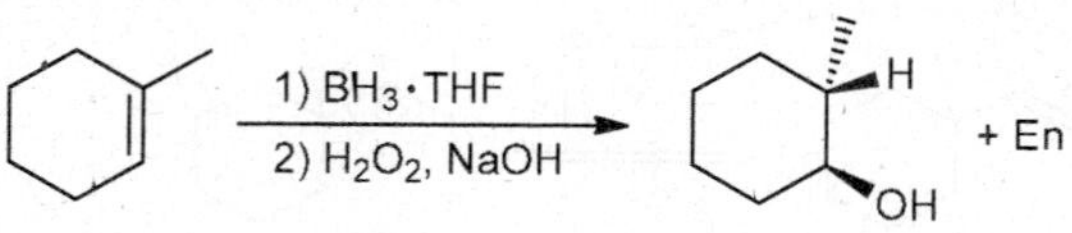

8.70.

(a) Cleavage of the C=C double bond can be achieved via ozonolysis:

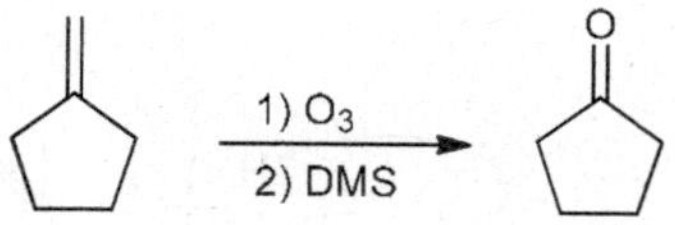

(b) The two groups being added across the alkene are H and Br. The latter is installed at the less substituted position, so we must use conditions that give an *anti*-Markovnikov addition of H and Br. This can be accomplished by treating the alkene with HBr in the presence of peroxides.

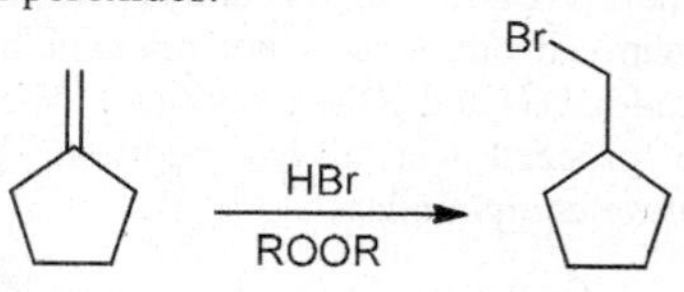

(c) The two groups being added across the alkene are H and OH. The OH group is installed at the less substituted carbon atom, so we must use conditions that give an *anti*-

Markovnikov addition of H and OH. This can be accomplished via hydroboration-oxidation.

1) $BH_3 \cdot THF$
2) H_2O_2, NaOH

(d) The two groups being added across the alkene are OH and OH. No chiral centers are formed, so stereochemistry is irrelevant. We have learned more than one way to achieve a dihydroxylation. For example, we can convert the alkene to an epoxide and then open the epoxide under aqueous acidic conditions.

RCO_3H → H_3O^+

Alternatively, we can treat the alkene with catalytic osmium tetroxide and a suitable co-oxidant, or even with cold potassium permanganate and NaOH.

OsO_4 (catalytic), NMO
$KMnO_4$, NaOH, Cold

(e) The two groups being added across the alkene are Br and OH, with the latter being installed at the more substituted position. This can be achieved by treating the alkene with Br_2 in the presence of water (halohydrin formation):

Br_2, H_2O

(f) The two groups being added across the alkene are H and OH, with the latter being installed at the more substituted position (Markovnikov addition). This can be achieved via acid-catalyzed hydration:

H_3O^+

(g) The two groups being added across the alkene are H and Br. The Br group must be installed at the more substituted, tertiary position, so we must use conditions that give a Markovnikov addition of H and Br. This can be accomplished by treating the alkene with HBr.

HBr

(h) This transformation requires the elimination of H and Br to give the more substituted alkene (Zaitsev product), so a strong base is required (such as hydroxide, methoxide, or ethoxide):

NaOMe

(i) The two groups being added across the alkene are H and H, which can be accomplished by treating the alkene with molecular hydrogen (H_2) in the presence of a suitable catalyst.

H_2, Pt

(j) The two groups being added across the alkene are OH and OH, and they must be installed via a *syn* addition. This can be achieved by treating the alkene with catalytic osmium tetroxide and a suitable co-oxidant, or with cold potassium permanganate:

OsO_4 (catalytic), NMO
$KMnO_4$, NaOH, Cold
+ En

(k) The two groups being added across the alkene are OH and OH, and they must be installed via an *anti* addition. This can be achieved by treating the alkene with a peroxy acid, followed by aqueous acid:

1) RCO_3H
2) H_3O^+
+ En

8.71. Let's begin by drawing the structures of the alkenes under comparison:

2-methyl-2-pentene   4-methyl-1-pentene

Addition of HBr to 2-methyl-2-pentene is expected to be more rapid because the reaction can proceed via a tertiary carbocation. In contrast, addition of HBr to 4-methyl-1-pentene proceeds via a less stable, secondary carbocation.

8.72. When treated with molecular bromine (Br_2), the alkene is converted to an intermediate bromonium ion, which is then subject to attack by a nucleophile. We have seen that the nucleophile can be water when the reaction is performed in the presence of water, so it is reasonable that the nucleophile can be H_2S in this case, as shown below:

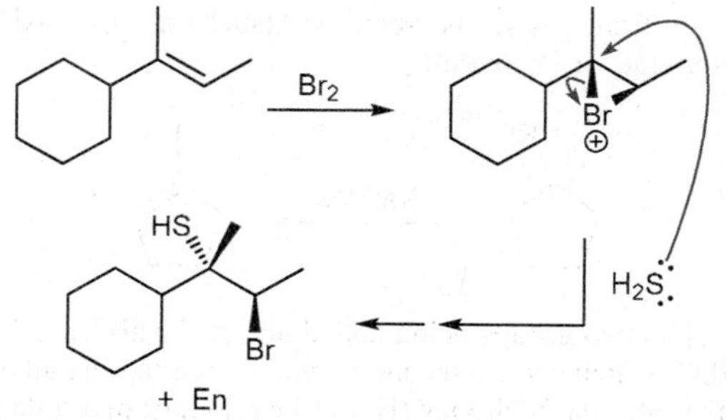

Br2

Br ⊕

H2S:

HS

Br

+ En

This should give the installation of an SH group (rather than an OH group) at the more substituted position.

8.73. We begin by drawing all possible alkenes with the molecular formula C_5H_{10}. We can use a methodical approach by first considering all isomers with an unbranched parent chain, and then branched isomers with a four-carbon chain:

Five-carbon chain

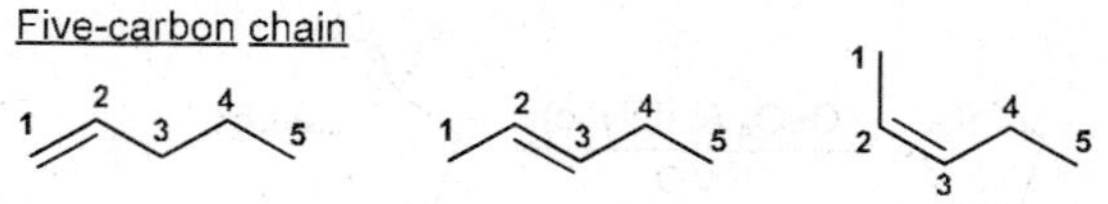

Four-carbon chain

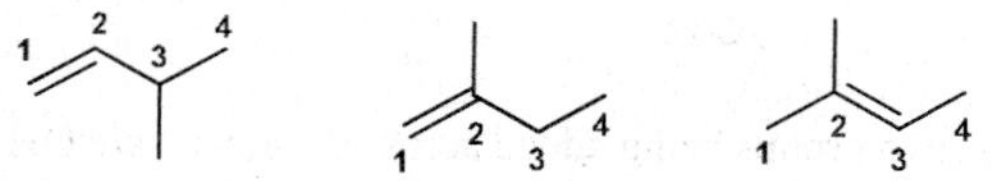

Among these isomers, only two of them will undergo hydroboration-oxidation to afford an alcohol with no chiral centers, shown here:

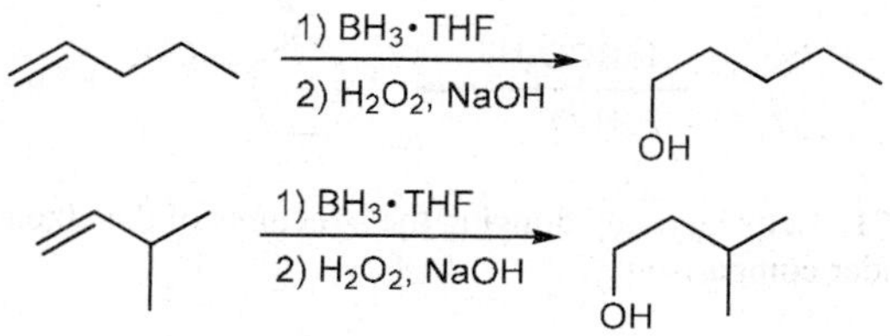

1) BH3•THF
2) H2O2, NaOH

OH

1) BH3•THF
2) H2O2, NaOH

OH

The remaining four isomers will undergo hydroboration-oxidation to produce alcohols that do possess a chiral center (marked with * below):

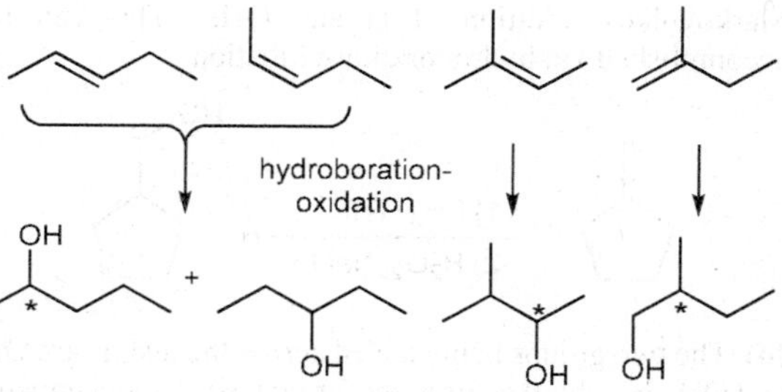

hydroboration-oxidation

OH

OH OH OH

8.74. We expect a *syn* addition of D and D across the alkene, giving a pair of enantiomers:

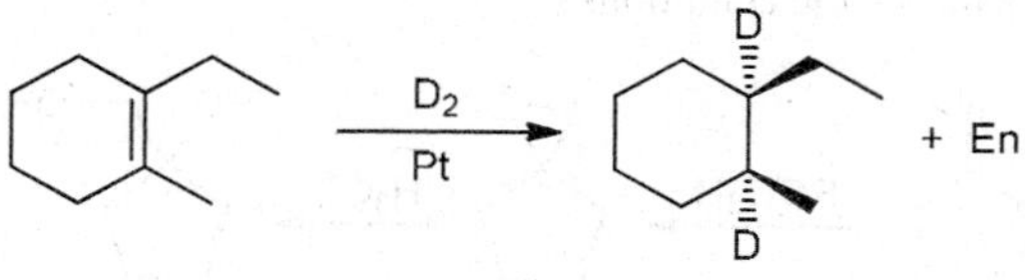

D2
Pt

D

D

+ En

8.75.

(a) First draw the starting alkene. Treating this alkene with HBr will result in a tertiary alkyl halide. But if peroxides are present, an *anti*-Markovnikov addition via a radical process will occur, resulting in the formation of a secondary alkyl halide, as shown:

HBr
ROOR

Br

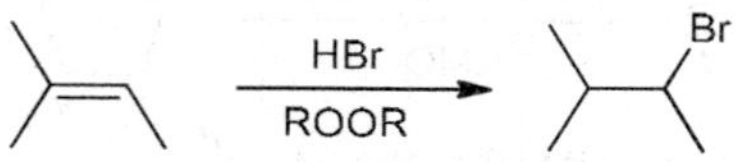

(b) First draw the starting alkene. Treating this alkene with HBr results in a Markovnikov addition to give a tertiary alkyl halide, as shown:

HBr

Br

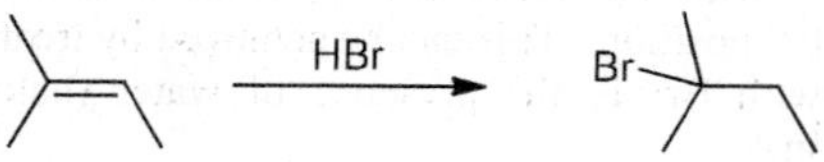

(c) First draw the starting alkene. In this case, a *syn* dihydroxylation is required in order to produce a *meso* diol. We have seen several reagents that can be used to accomplish a *syn* dihydroxylation, such as cold potassium permanganate:

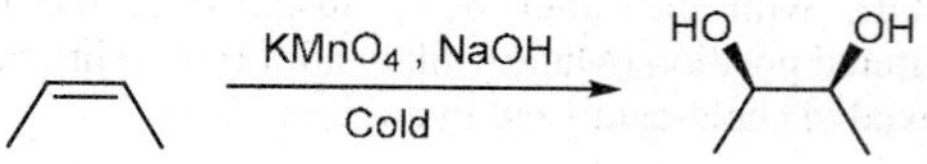

KMnO4 , NaOH
Cold

HO OH

Alternatively, we could achieve the same result with catalytic osmium tetroxide and a suitable co-oxidant.

(d) First draw the starting alkene. In this case, an *anti* dihydroxylation is required in order to prepare enantiomeric diols. This can be accomplished by converting the alkene into an epoxide, followed by acid-catalyzed ring opening of the epoxide, as shown.

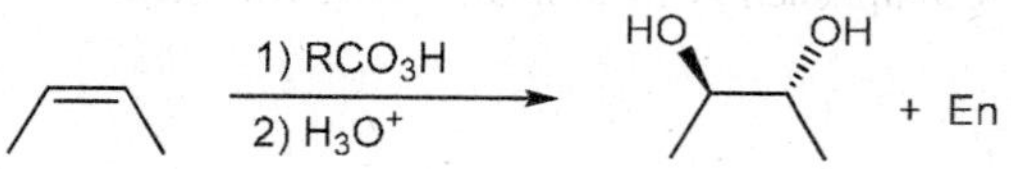

1) RCO3H
2) H3O+

HO OH

+ En

8.76.

(a) Begin by drawing the starting alkyl halide. This tertiary alkyl halide can be converted into a primary alkyl halide via a two-step process (elimination followed by addition). In each case, we must carefully consider the regiochemical outcome. During the elimination process, there is only one regiochemical outcome, so any strong base will work (even if it is sterically hindered, although that is not necessary). In the addition process, we want to install a halide at the less substituted position. This can be achieved *only if the halide is Br*, employing an *anti*-Markovnikov addition of HBr (with peroxides):

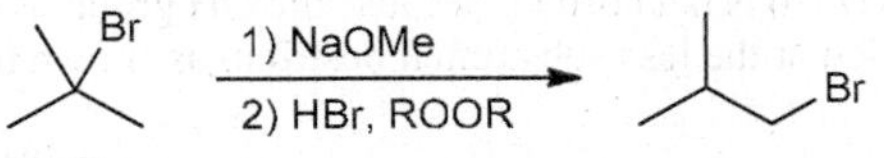

Br 1) NaOMe 2) HBr, ROOR Br

(b) Begin by drawing the starting alkyl halide. This secondary alkyl halide can be converted into a primary alkyl halide via a two-step process (elimination followed by addition). In each case, we must carefully consider the regiochemical outcome. During the elimination process, there is only one regiochemical outcome, so any strong base will work (even if it is sterically hindered). In fact, in this case, there is a distinct advantage to using a sterically hindered base. Specifically, it will suppress the competing S_N2 process (the substrate is secondary, so S_N2 should be a minor product, unless a sterically hindered base is used). During the addition process, we want to install Br at the less substituted position, so we will need an *anti*-Markovnikov addition of HBr (using peroxides):

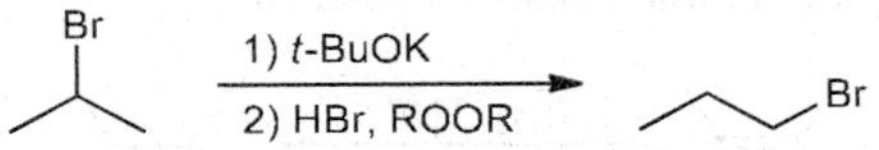

Br 1) *t*-BuOK 2) HBr, ROOR Br

8.77.

(a) Begin by drawing the starting alkene. This trisubstituted alkene can be converted into a monosubstituted alkene via a two-step process (addition, followed by elimination). We must be careful to control the regiochemical outcome of each step of the process. During the addition reaction, we want to install Br at the less substituted position (*anti*-Markovnikov), so we treat the alkene with HBr in the presence of peroxides. Then, the elimination reaction must be performed in a way that gives the less substituted alkene (Hofmann product), so we must use a strong, sterically hindered base (such as *tert*-butoxide).

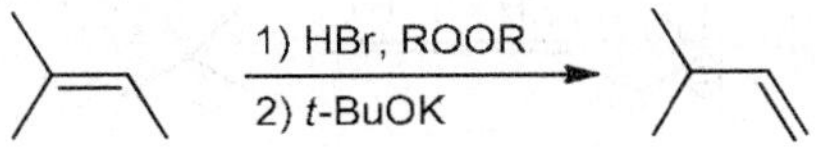

1) HBr, ROOR 2) *t*-BuOK

(b) Begin by drawing the starting alkene. This disubstituted alkene can be converted into a tetrasubstituted alkene via a two-step process (addition, followed by elimination). We must be careful to control the regiochemical outcome of each step of the process. During the addition reaction, we want to install Br at the more substituted position (Markovnikov), so we treat the alkene with HBr (without peroxides). Then, the elimination process must be performed in a way that gives the more substituted alkene (Zaitsev product), so we must use a strong base that is not sterically hindered, such as methoxide (hydroxide or ethoxide can also be used).

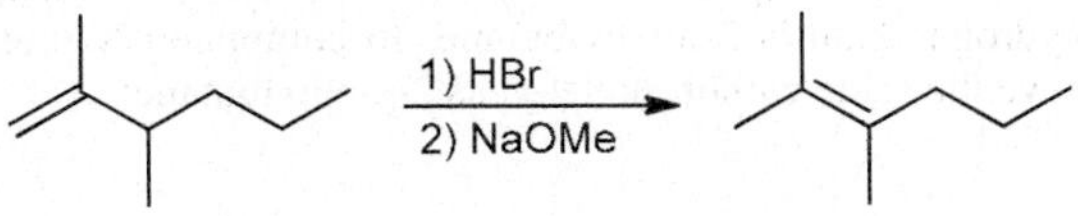

1) HBr 2) NaOMe

8.78. Since this reaction proceeds through an ionic mechanism, we expect the mechanism to be comprised of two steps: 1) proton transfer, followed by 2) nucleophilic attack. In the first step, a proton is transferred from HCl to the alkene, which requires two curved arrows, as shown below. There are two possible regiochemical outcomes for the protonation step, and we might have expected formation of a tertiary carbocation. However, in this particular case, the other regiochemical outcome is favored because it involves formation of a resonance-stabilized cation. As a result of resonance stabilization, this cation is even more stable than a tertiary carbocation, and the reaction proceeds via the more stable intermediate. This cation is then captured by a chloride ion in the second step of the mechanism, which requires two curved arrows, as shown:

H—Cl :Cl: Cl

resonance-stabilized

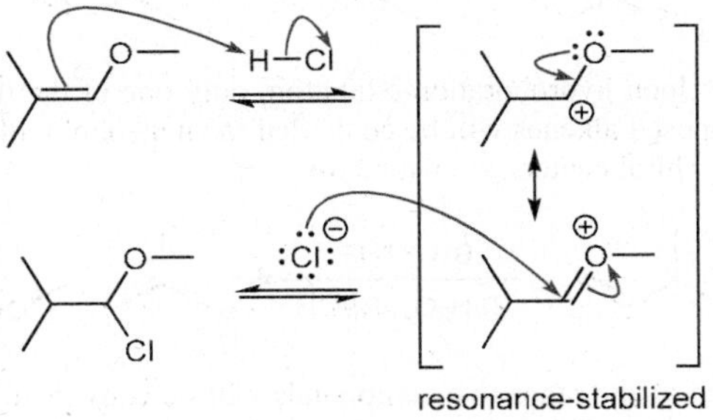

8.79. Protonation of the alkene requires two curved arrows, as shown in the first step of the following mechanism. This leads to the more stable, secondary carbocation (rather than a primary carbocation). This secondary carbocation then undergoes a rearrangement, in which one of the carbon atoms of the ring migrates (as described in the problem statement). This is represented with one curved arrow that shows the formation of a more stable, tertiary carbocation. In the final step of the mechanism (nucleophilic attack), the carbocation is captured by a bromide ion. This step requires one curved arrow, going from the nucleophile (bromide) to the electrophile (the carbocation), as shown:

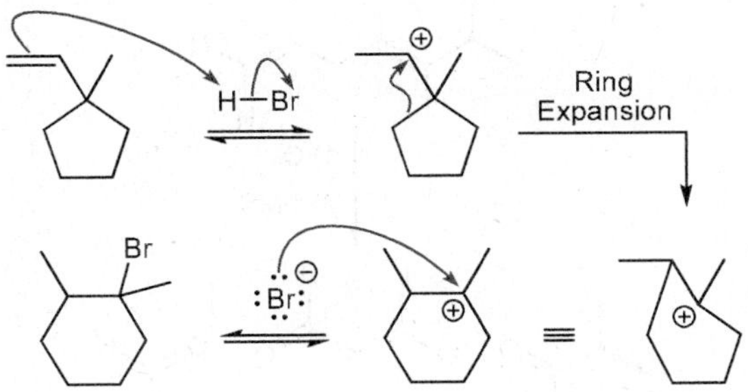

H—Br Ring Expansion Br :Br:

8.80.
(a) Compound **X** reacts with H_2 in the presence of a catalyst, so compound **X** is an alkene. The product of hydrogenation is 2-methylbutane, so compound **X** must have the same carbon skeleton as 2-methylbutane:

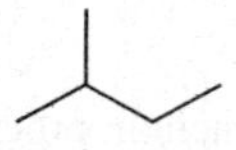

2-methylbutane

We just have to decide where to place the double bond in compound **X**. Keep in mind that the following two structures represent the same compound:

2-methyl-1-butene ≡ 2-methyl-1-butene

So, there are only three possible locations where we can place the double bond:

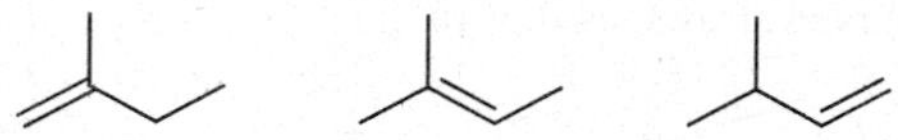

(b) Upon hydroboration-oxidation, only one of the three proposed alkenes will be converted to an alcohol without any chiral centers, shown below:

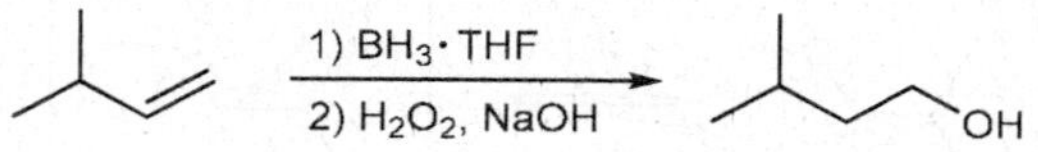

1) $BH_3 \cdot THF$
2) H_2O_2, NaOH

Each of the other two compounds will be converted into an alcohol with a chiral center (marked with * below):

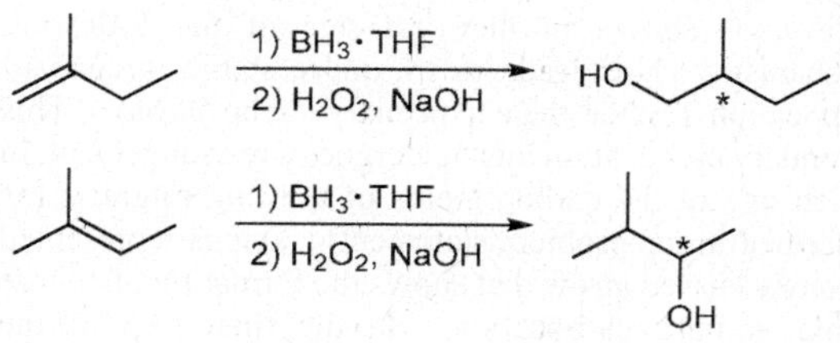

1) $BH_3 \cdot THF$
2) H_2O_2, NaOH

1) $BH_3 \cdot THF$
2) H_2O_2, NaOH

8.81. The correct answer is (c). Each of the double bonds is cleaved upon ozonolysis, giving the following products:

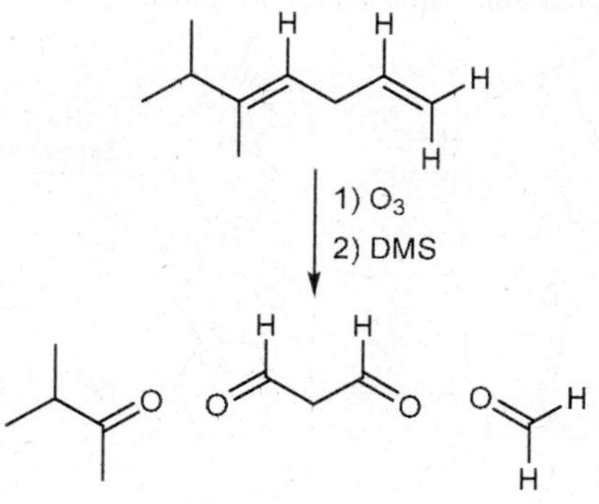

1) O_3
2) DMS

Answer (c) is the only structure that is not among these products.

8.82. The correct answer is (d). Answer (a) is not correct, because the OH group would be installed at the more substituted position, as shown here:

H_3O^+

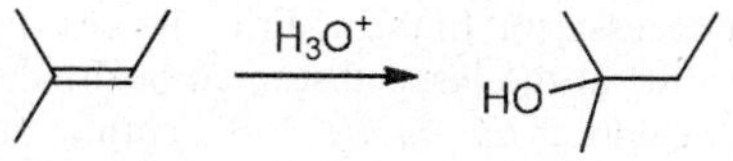

Answer (b) is not correct, because the OH group would be installed at the less substituted position, as shown here:

1) $BH_3 \cdot THF$
2) H_2O_2, NaOH

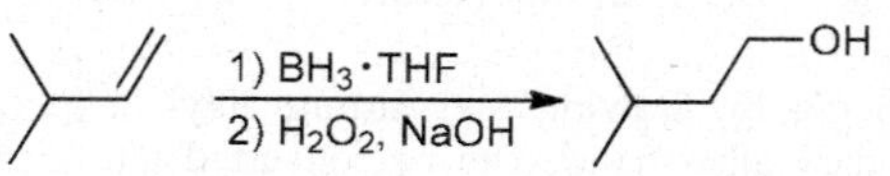

Answer (c) is not correct, because a rearrangement is possible, giving a mixture of products:

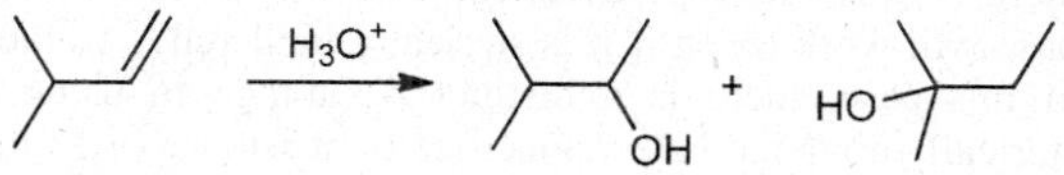

H_3O^+

Only one of these products is the desired product, so this method is not efficient.
Answer (d) is the correct answer, because hydroboration-oxidation involves installation of an OH group at the less substituted position (*anti*-Markovnikov):

1) $BH_3 \cdot THF$
2) H_2O_2, NaOH

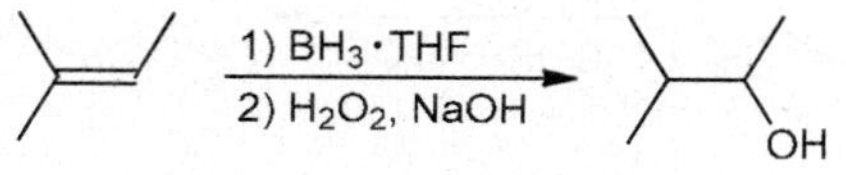

8.83. The correct answer is (b). Acid-catalyzed hydration is believed to occur via the following mechanism:

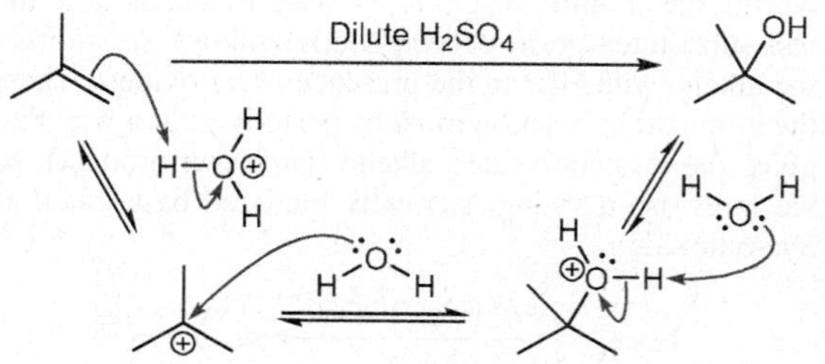

Dilute H_2SO_4

As shown, this mechanism has two intermediates, which correspond with answer (b), structures I and II.

8.84. The correct answer is (d). During the addition process (Step 1), we want to install Br at the less substituted position (*anti*-Markovnikov), so we treat the alkene with HBr in the presence of peroxides. Then, the elimination process (Step 2) must be performed in a way that gives the less substituted alkene (Hofmann product), so we must use a strong, sterically hindered base (such as *tert*-butoxide).

HBr
ROOR
Br
t-BuOK

8.85.
The correct answer is (b). Treating an alkene with molecular bromine (Br_2) and water results in the addition of OH and Br across the alkene (halohydrin formation). The OH group is expected to be installed at the more substituted position, while Br is installed at the less substituted position. In this case, two new chiral centers are generated, so we expect only the pair of enantiomers that would result from *anti* addition.

8.86.
The correct answer is (a). Treating an alkene with catalytic osmium tetroxide and NMO results in the addition of OH and OH across the alkene. In this case, two new chiral centers are generated, so we expect only the pair of enantiomers that would result from *syn* addition.

8.87.
The correct answer is (c). The mechanism begins with protonation of the alkene, which leads to the more stable, secondary carbocation (rather than a primary carbocation). This secondary carbocation then undergoes a hydride shift, shown with one curved arrow, generating a more stable, tertiary carbocation. In the final step (nucleophilic attack), the carbocation is captured by a bromide ion.

HBr
Br
H—Br
:Br:
H

8.88.
The correct answer is (d). Each alkene is classified according to its degree of substitution. The most highly substituted (trisubstituted) alkene will be the most stable. The four isomeric alkenes can be arranged in the following order of stability:

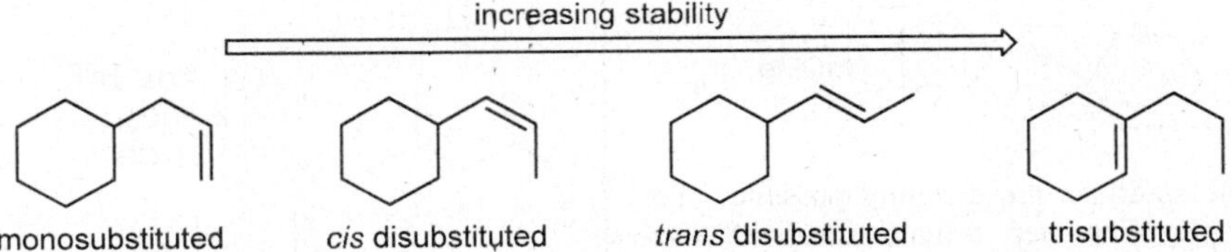

8.89.
The correct answer is (c). We begin by identifying the parent. The longest possible chain (including both carbon atoms of the C=C bond) has seven carbon atoms, so the parent is heptene. There is only one substituent (highlighted), an isopropyl group, or (1-methylethyl) group. The parent is numbered from right to left to give the double bond the lower possible number, so the parent is 2-heptene. Finally, we must determine the configuration of the double bond. Of the two groups attached to C3, the isopropyl group has a higher priority than the butyl group, and the isopropyl group is on the opposite side of the methyl group on C1, so the alkene has the (*E*) configuration.

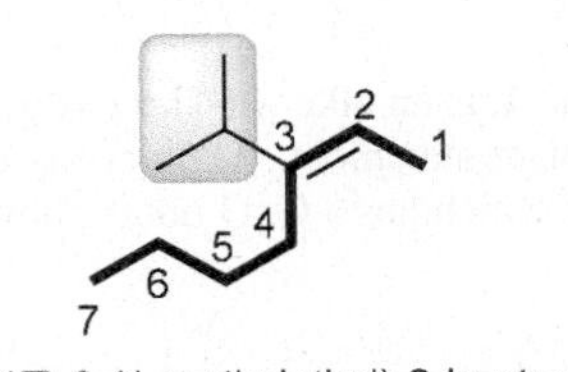

(*E*)-3-(1-methylethyl)-2-heptene

8.90. We begin our retrosynthetic analysis of the target molecule by asking the following question, "What starting material can be used to make an alcohol with the desired stereochemistry?" In this case, the target alcohol can be produced from an alkene intermediate, via hydroboration-oxidation:

OH
(+ hydroboration-oxidation)

Continuing our retrosynthetic analysis, we must ask how we can make this alkene from the given alkyl bromide?

?
Br

Notice the pi bond of the alkene is in a different location than the halide in the starting alkyl halide, so we will need to change the identity AND location of the functional group. The *geminal* disubstituted alkene shown below is an ideal intermediate structure to consider in the continued retrosynthetic analysis:

Br

The above plan suggests one possible route for the desired transformation

OH + En

1) *t*-BuOK
2) HBr
3) NaOMe
4) $BH_3 \cdot THF$
5) H_2O_2, NaOH

Br

t-BuOK

1) $BH_3 \cdot THF$
2) H_2O_2, NaOH

HBr

Br

NaOMe

Other acceptable solutions are certainly possible. For example, after the first step (elimination with *tert*-butoxide), the next two steps (addition of HBr, followed by elimination) could be replaced with acid-catalyzed hydration, followed by elimination with conc. H_2SO_4.

OH + En

1) *t*-BuOK
2) H_3O^+
3) conc. H_2SO_4, heat
4) $BH_3 \cdot THF$
5) H_2O_2, NaOH

Br

t-BuOK

1) $BH_3 \cdot THF$
2) H_2O_2, NaOH

conc. H_2SO_4 heat

H_3O^+

OH

8.91. Compound **X** reacts with H_2 in the presence of a catalyst, so compound **X** is an alkene. There is only one alkene (shown below) that can be converted to 2,4-dimethyl-1-pentanol via hydroboration-oxidation. Treatment of that alkene with aqueous acid affords the tertiary alcohol shown below (via Markovnikov addition):

H_2 Pt

Compound **X**

H_3O^+

OH

1) $BH_3 \cdot THF$
2) H_2O_2, NaOH

OH

2,4-dimethyl-1-pentanol (racemic)

8.92. The substrate is a secondary alkyl halide, and treatment with *tert*-butoxide gives the less substituted alkene (Hofmann product) as the major product. When that alkene is treated with HBr, the π bond is protonated to give a secondary carbocation (rather than a primary carbocation). This carbocation can either be captured by a bromide ion, giving enantiomeric products **A** and **B** below, or the carbocation can undergo a rearrangement (hydride shift) to give a tertiary carbocation, which is then captured by a bromide ion, affording product **C**.

Cl

t-BuOK

HBr

Br + Br + Br

A B C

8.93. Compound **Y** reacts with H_2 in the presence of a catalyst, so compound **Y** is an alkene. There is only one alkene (shown below) that is consistent with the information provided in the problem statement. Ozonolysis of that alkene results in cleavage of the C=C bond to give two separate compounds, each of which has a C=O bond, shown here:

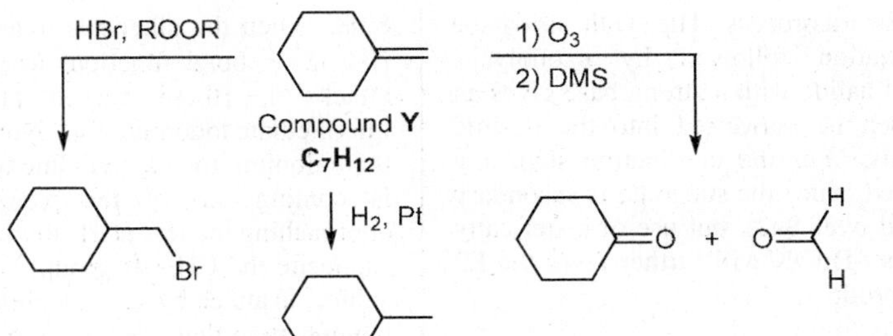

HBr, ROOR

Compound **Y**
C_7H_{12}

1) O_3
2) DMS

H_2, Pt

Br

8.94. Each of the products is an aldehyde, and their sp^2 hybridized carbon atoms were once connected to each other as a C=C bond in the original alkene. That gives the following two possibilities (*cis* and *trans* stereoisomers) for the structure of the original alkene:

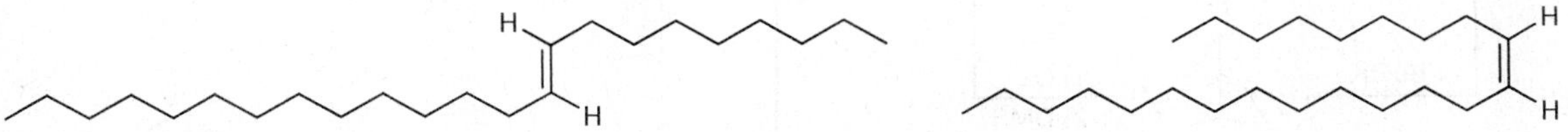

8.95.

(a) In the presence of aqueous acid, the epoxide is first protonated (two curved arrows), as shown below. The resulting intermediate can then undergo an S_N2-like, intramolecular attack (two curved arrows), in which the OH group functions as the nucleophilic center. Then, in the final step of the mechanism, a molecule of water functions as a base and removes a proton, thereby generating the product. This final step is a proton transfer step, and therefore requires two curved arrows:

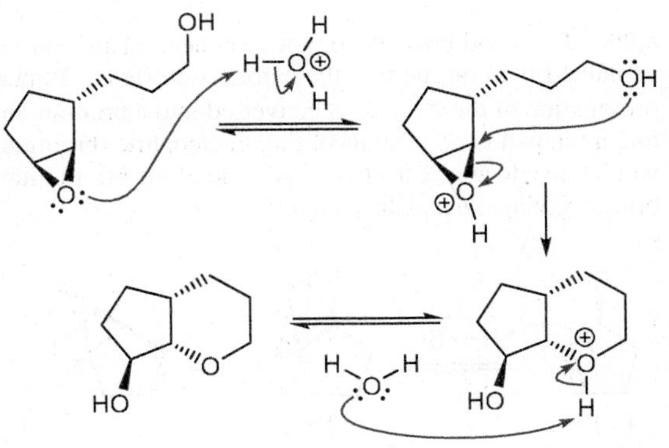

(b) In the presence of aqueous acid, the epoxide is first protonated (two curved arrows), as shown. The resulting intermediate can then undergo an S_N2-like, intramolecular attack (two curved arrows), in which the π bond functions as the nucleophilic center. Then, in the final step of the mechanism, a molecule of water functions as a base and removes a proton, thereby generating the product. This final step is a proton transfer step, and therefore requires two curved arrows:

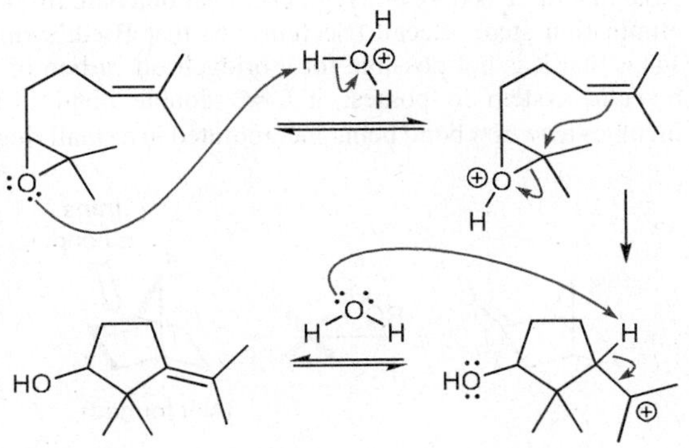

8.96. The retrosynthetic analysis of the target molecule begins with the following question: "What starting material can be used to make a product with two aldehyde groups?" We know that C=O bonds can be made from oxidative cleavage of an alkene, so we draw an alkene by removing the two oxygen atoms from the product, and connecting the sp^2 hybridized carbon atoms (highlighted below) as a C=C bond, like this:

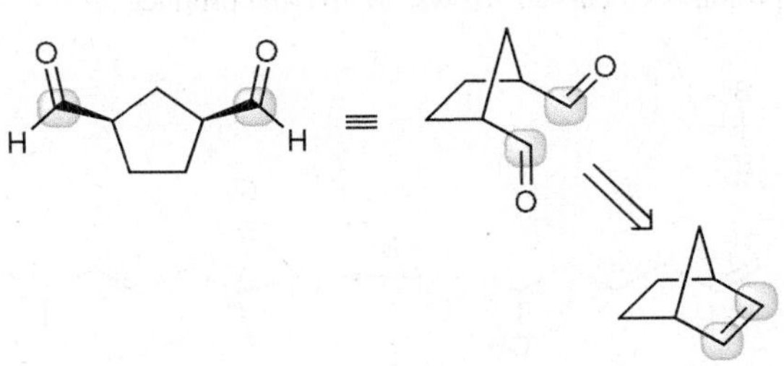

And this alkene can be made from the given alkyl chloride:

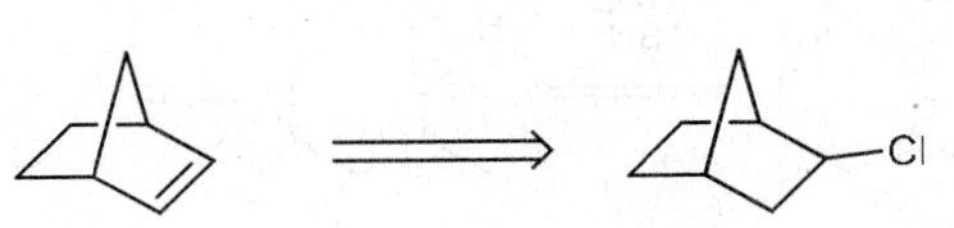

Now let's draw the forward process. The synthesis can be achieved via elimination followed by ozonolysis. Treatment of the alkyl halide with a strong base gives an alkene which can then be converted into the desired product via ozonolysis. For the elimination step, any strong base can be used, since the substrate is secondary (so E2 will be favored over S_N2), but use of a sterically hindered base, such as *t*-BuOK will further favor the E2 product over the S_N2 product.

1) *t*-BuOK
2) O_3
3) DMS

t-BuOK

1) O_3
2) DMS

Note that there is only one regiochemical outcome for the elimination step. Recall (Section 7.6) that Bredt's rule states that it is not possible for a bridgehead carbon of a bicyclic system to possess a C=C double bond if it involves a *trans* π bond being incorporated in a small ring:

trans π bond

(Not formed)

8.97. When the alkene is treated with molecular bromine (Br_2), the π bond functions as a nucleophilic center and attacks Br_2 (three curved arrows), resulting in an intermediate bromonium ion. The bromonium ion is then subject to attack by a nucleophilic center, such as the OH group that is tethered to the bromonium group. The resulting intramolecular nucleophilic attack (two curved arrows) generates an oxonium ion, which then loses a proton (two curved arrows) to give the product.

8.98. When the alkene is treated with molecular iodine (I_2), the π bond functions as a nucleophilic center and attacks I_2 (three curved arrows), resulting in an intermediate iodonium ion. Note that the configuration of the iodonium ion (with iodine on wedges, meaning that it is coming out of the page) is consistent with I_2 approaching the less sterically hindered face of the alkene (opposite the CH_2OH group). The iodonium ion is then subject to attack by a nucleophile, such as the nucleophilic oxygen atom that is tethered to the iodonium group. The resulting intramolecular nucleophilic attack (two curved arrows) generates an intermediate which then loses a proton (two curved arrows) to give the product.

8.99. The *cis*-dibromide is not obtained. This can be explained if we argue that the carbocation (formed upon protonation of the π bond) is converted into a bromonium ion, as shown here. The incoming nucleophile (bromide) would have to attack from the backside of the bromonium bridge, giving a *trans* dibromide:

8.100. Inspection of the molecular formula reveals that compound **A** contains the same number of carbon atoms as the product of ozonolysis, compound **B** (21 carbon atoms). Thus, we can predict the structure of compound **A** by choosing any two carbonyl (C=O) groups in the product, removing the oxygen atoms, and connecting the sp^2 hybridized carbon atoms with a double bond. There are three carbonyl groups, so there are three possible alkenes that should be considered.

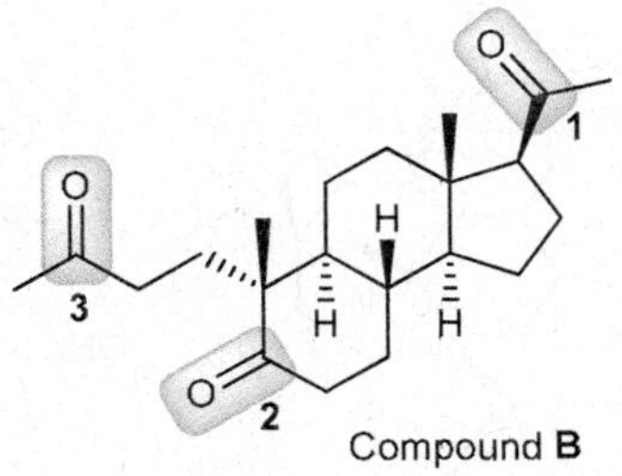

Compound B

Attachment of carbons **1** and **2** would result in a complex ring system with far too much strain to possibly exist. Carbons **1** and **2** are remotely located on opposite ends of the molecule, pointing in opposite directions. Indeed, we will later learn that this polycyclic skeletal structure is very rigid, and the hypothetical connection of carbons **1** and **2** is impossible.

Similarly, if carbons **1** and **3** were attached it would also result in significant strain (below). Though these carbons are closer than **1** and **2**, and they are linked by a slightly more flexible chain, the *trans* relationship of the two connecting groups creates too much angle strain:

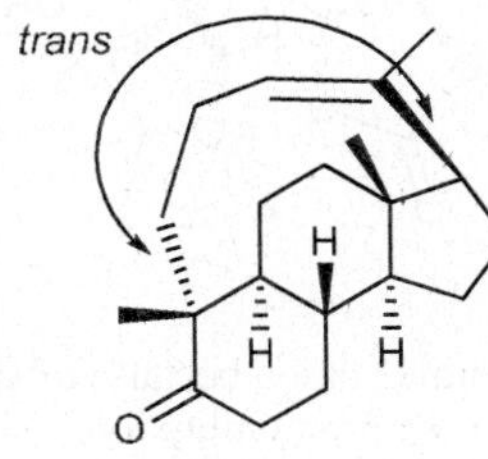

trans

The only logical reactant is one in which carbons **2** and **3** are connected in a fused cyclopentene precursor. This is the structure of compound **A**.

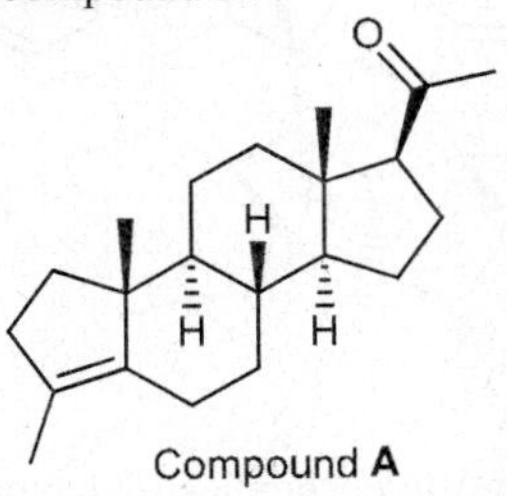

Compound **A**

Ozonolysis of compound **A** generates a single product, compound **B**.

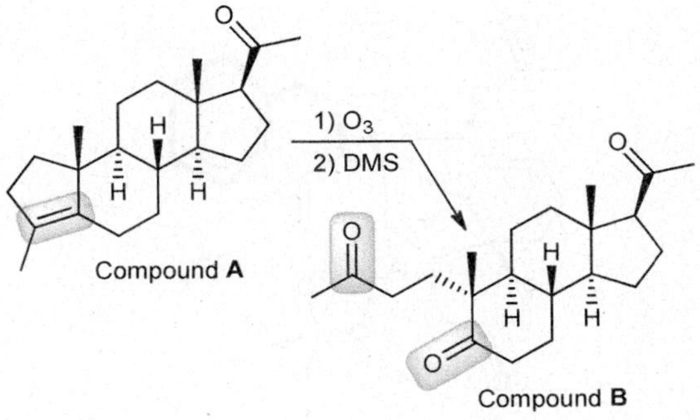

1) O_3
2) DMS

Compound **A**

Compound **B**

8.101.

(a) The reagents indicate a hydroboration-oxidation, in which an alkene is converted to an alcohol. In compound **1**, there are two alkene groups, so we must choose which one is more likely to react with BH_3. One alkene group is disubstituted, while the other is tetrasubstituted:

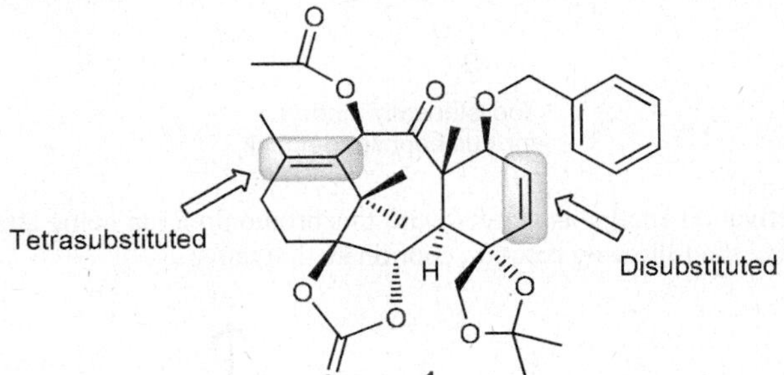

Tetrasubstituted

Disubstituted

1

Since the rate of hydroboration is particularly sensitive to steric factors, we expect the disubstituted alkene group to undergo hydroboration more readily.

(b) As mentioned in part (a), hydroboration is sensitive to steric considerations. When we inspect both vinylic positions, we find that both are equally substituted. The tie-breaker will likely be the nearby presence of a six-membered, aromatic ring, which provides significant steric crowding that favors the following regiochemical outcome:

In predicting the stereochemical outcome, we once again invoke the steric bulk of the six-membered, aromatic ring. Specifically, the front face of the alkene group is blocked by the large six-membered, aromatic ring. As a result, the back face of the alkene is more accessible, so the reaction occurs more readily on the back face, giving the following expected product:

8.102.

The bromonium ion is unusually resistant towards nucleophilic attack by the bromide anion because of significant steric hindrance involved when the anion approaches the electrophilic carbon atoms of the bromonium ion.

Too sterically hindered for nucleophile to attack

Thus, there is a very large activation energy associated with this bromonium ion being attacked by a nucleophile. This can be illustrated by comparing the following reaction coordinate diagrams:

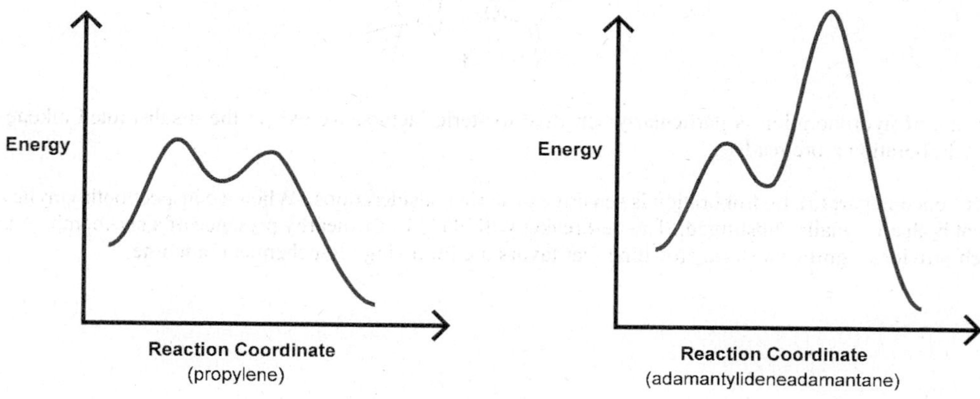

The first figure (left) is the expected energy diagram for bromination of propylene, while the second figure (right) is a proposed energy diagram for bromination of adamantylideneadamantane. The first step in each figure is similar, but compare the second step in each figure (nucleophilic attack of the bromonium ion). For bromination of adamantylideneadamantane, the magnitude of the activation energy for the second step is so large that this step does not take place at an appreciable rate.

8.103.

Hydroboration of an alkene typically occurs with equal probability from both faces of the π-bond. However, in the case of α-pinene, the top face of the π-bond is blocked by one of the methyl substituents, so the approach of borane on this face of the π-bond is severely hindered. Therefore, hydroboration cannot occur at an appreciable rate on the top face of the π-bond. However, the bottom face of the π-bond is unhindered; it is much more accessible to hydroboration. As a result, boron and hydrogen add in a *syn*-fashion to the bottom face of α-pinene resulting in the $IpcBH_2$ diastereomer shown; as always, boron is added to the less hindered position on the alkene. A racemic mixture is not formed, because the synthesis begins with a chiral alkene (single enantiomer).

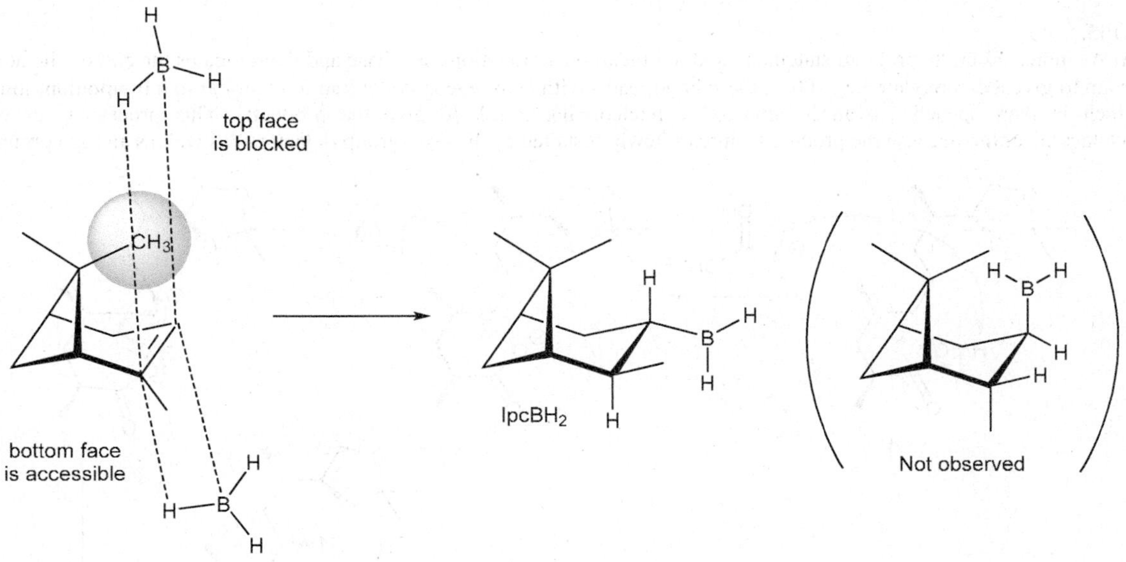

top face is blocked

CH_3

bottom face is accessible

$IpcBH_2$

Not observed

Notice that the methyl group of the alkene is now occupying an equatorial position in the product.

Next, the π-bond of a second molecule of α-pinene will react with $IpcBH_2$, also from the same face, to produce the observed diastereomer of Ipc_2BH, as a single enantiomer.

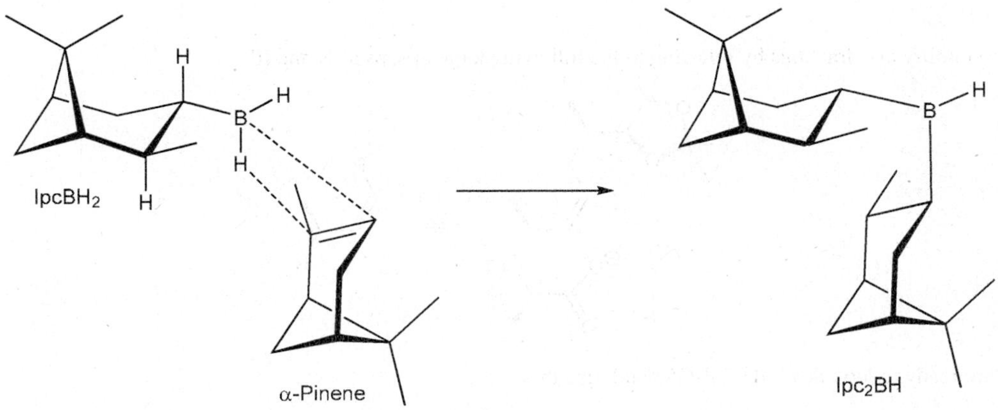

$IpcBH_2$

α-Pinene

Ipc_2BH

8.104.

Building a molecular model is perhaps the best way to see that the bottom face of the π bond is more hindered than the top face. Alternatively, this can be seen if we redraw the compound in a Haworth projection:

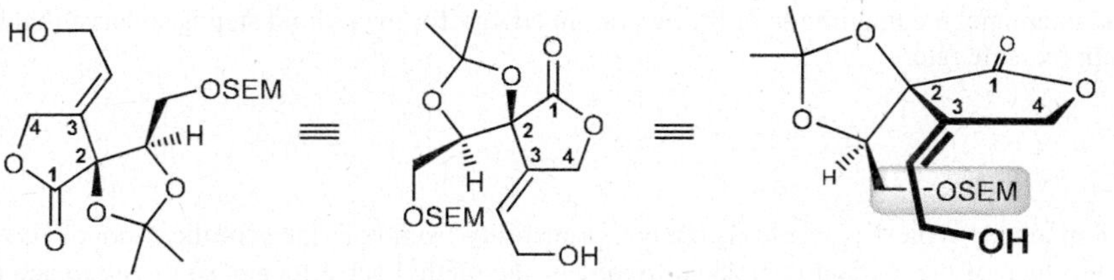

Notice that the π bond is in the plane of the ring, and the large and bulky OSEM group is positioned below the plane of the ring, directly underneath the π bond. As such, the bottom face of the π bond is sterically encumbered, so approach of the oxidizing agent (OsO_4) from that face is blocked (it would involve a transition state that is too high in energy). Attack on the top face is unencumbered, so it involves a lower energy transition state, and as a result, the reaction occurs more readily on this face.

8.105.

(a) As indicated in the problem statement, sodium bicarbonate functions as a base and deprotonates the carboxylic acid group to give a carboxylate ion. Then, the π bond reacts with I_2 to give an iodonium ion (similar to a bromonium ion), which is then opened via an intramolecular nucleophilic attack to give the product: This process is called iodolactonization, because the product features a newly installed cyclic ester group (a lactone) as well as an iodo group:

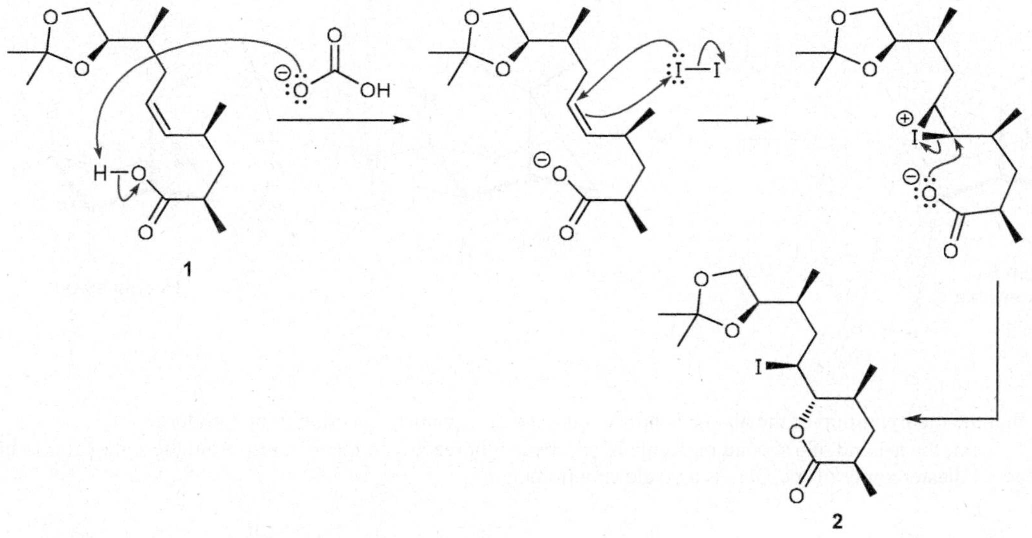

(b) Let's simplify our drawings by referring to the following large groups as R and R':

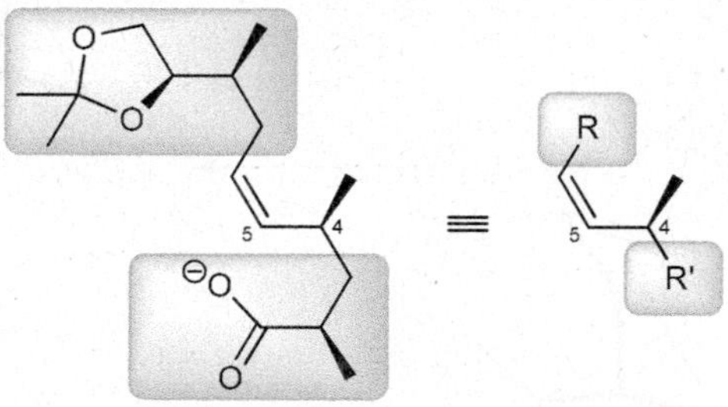

Now we are ready to look down the C4-C5 bond, like this:

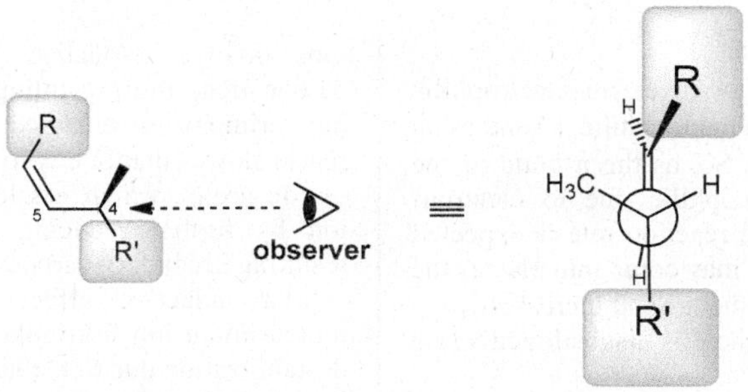

Let's rotate this entire Newman projection by 90° (which does not change the conformation at all) so that we can clearly see the top face and bottom face of the π bond:

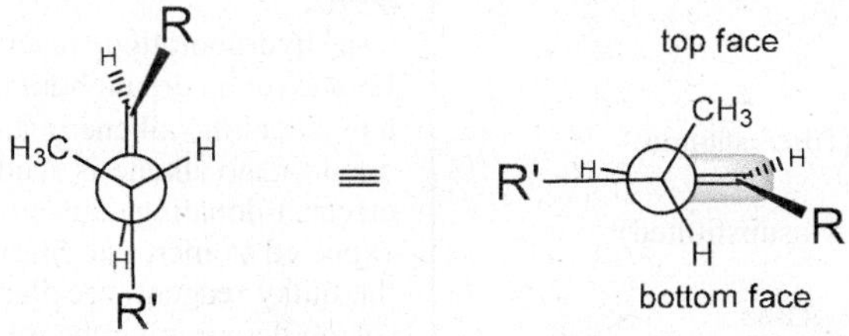

Notice that the two largest groups (R and R') are farthest away from each other. In our search for the lowest energy conformation, this conformation should be the first one that we examine, because significant steric interactions will be present if R and R' are near each other in space. When we analyze this conformation, we see two additional factors contributing to its overall energy: 1) gauche interactions between the methyl group and the R group, and 2) an eclipsing interaction between R' and a hydrogen atom. The former can be avoided by rotating the front carbon atom counterclockwise, like this:

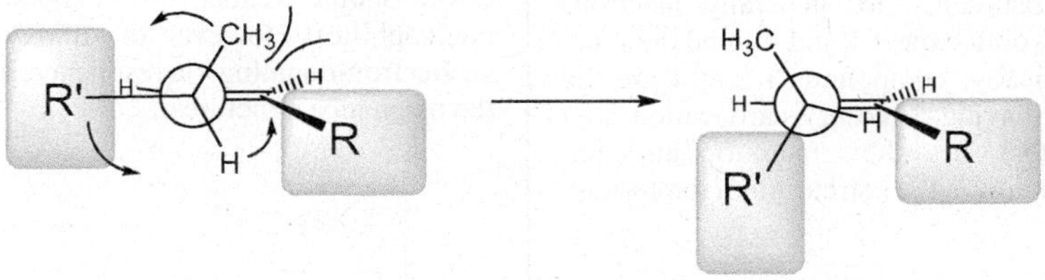

We have traded one eclipsing interaction for another (presumably similar), but notice that we have lost the gauche interaction. Accordingly, we expect this conformation to be the most stable conformation, looking down the C4-C5 bond. This means that the molecule will spend most of its time in this conformation. Notice that, in this lowest energy conformation, the bottom face of the π bond is sterically hindered, while the top face is relatively unhindered:

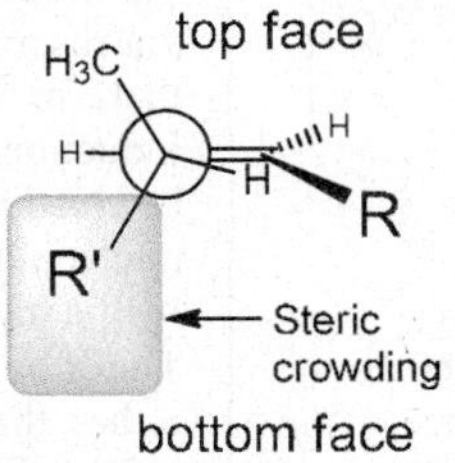

That is, the top face is more accessible (most of the time). As such, if I_2 approaches from the top face of the π bond, the transition state will be lower in energy then if attack occurs from the bottom face. So attack of the top face occurs more readily.

8.106.
The oxymercuration reaction involves an electrophilic mercuric cation reacting with a nucleophilic π bond of an alkene in an addition reaction. So, as the π bond of the alkene is rendered less nucleophilic due to electron-withdrawing substituent(s), the reaction rate is expected to decrease. Also, steric effects may come into play as the number of substituents around the π bond increases.
Among the alkenes listed, alkene **1** is disubstituted while **4** is trisubstituted:

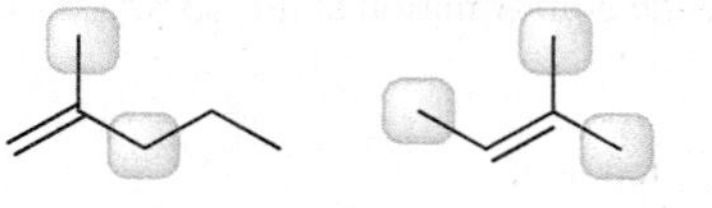

1 (Disubstituted) 4 (Trisubstituted)

The remaining alkenes are all monosubstituted:

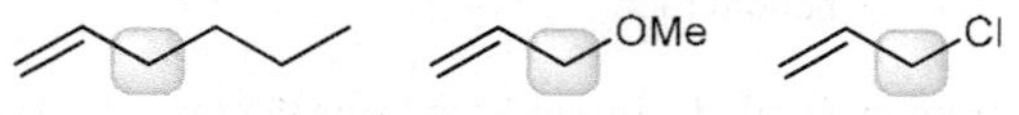

OMe Cl
2 3 5
(All are monosubstituted)

Given that alkyl substituents are generally electron-donating groups, we would expect **1** and **4** to be the most reactive. More specifically, compounds **1** and **4** are the only ones capable of having a tertiary carbocation as a resonance contributor in the mercurinium ion intermediate (shown below is the mercurinium ion formed from alkene **4**):

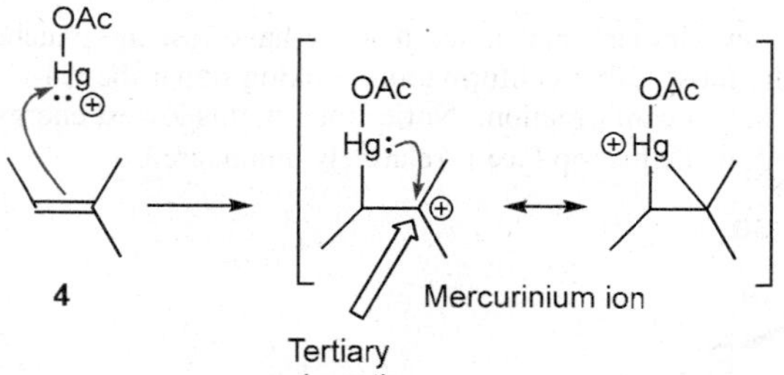

OAc Hg ⊕ 4 OAc Hg: ⊕ OAc ⊕Hg
Mercurinium ion
Tertiary carbocation

Therefore, these mercurinium ions are expected to be among the most stable ones, and hence, the oxymercuration reactions of alkenes **1** and **4** are expected to proceed the fastest. But this expectation does not bear itself out for alkene **4** in the relative reactivity data (alkene **4** is among the slower reacting compounds). This anomaly must be due to the steric repulsion associated when the mercuric cation tries to approach the π bond, or a destabilizing steric effect present in the resulting mercurinium ion intermediate between these substituents and the bound mercury ion.
The monosubstituted alkenes **2**, **3** and **5** are all less reactive than **1** because their corresponding mercurinium ions involve resonance structures with a secondary carbocation, thus resulting in higher energy than the mercurinium ion obtained from compound **1**. Alkene **3** reacts slower than **2** due to electron-withdrawal from the –OMe group, which would destabilize the mercurinium ion by further reducing the electron density of the resulting secondary carbocation resonance contributor. A similar inductive effect would also destabilize the mercurinium ion from alkene **5**, but an additional steric destabilization due to the large chlorine atom may also be in effect to make this alkene the slowest reacting compound among the series.

8.107.
The hydroboration reaction involves an electrophilic borane (or an organoborane such as 9-BBN) reacting with a nucleophilic alkene in an addition reaction. So, as the π bond of the alkene is rendered more nucleophilic due to electron-donating substituent(s), the reaction rate is expected to increase. Steric effects (that arise because of the bulky reagent) are also expected to play an important role in determining the relative rates of reactivity.

(a) Alkene **1** possesses an alkoxy substituent (OR) connected to the double bond. An alkoxy group is expected to be inductively electron-withdrawing, because oxygen is an electronegative atom and will therefore withdraw electron density away from the π bond. This effect should render the π bond less reactive (less nucleophilic). However, the alkoxy group is expected to be electron-donating via resonance, as seen when we draw the resonance structures:

R R ⊖ ⊕
1

So there are two effects in competition with each other. The alkoxy group is expected to be electron-withdrawing via induction, but it is expected to be electron-donating via resonance. Which effect is stronger? We have seen that, in general, resonance is a stronger effect than induction. As such, we would expect the alkoxy group to be electron-donating, which would render the alkene more nucleophilic (more reactive). This prediction is verified by the high rate of reactivity of compound **1**.
The π bond in compound **2** has an attached alkyl group, rather than an alkoxy group, so there is no resonance effect. The only effect is induction (we have seen that alkyl groups are generally electron donating). As such, the nucleophilicity of the π bond in compound **2** is expected to be enhanced by the presence of the alkyl group, but it is not expected to be quite as nucleophilic as the π bond in compound **1**.
Compounds **3** and **5** both exhibit a CH_2 group in between the π bond and the substituent. As such, the substituents in these compounds do not affect the π bond via resonance effects, only via inductive effects. Both substituents (OMe and OAc) are expected to be inductively electron-

withdrawing. Therefore, the π bonds in these cases are less electron-rich than in compounds **1** or **2**. The acetoxy group of alkene **5** is expected to be a stronger electron-withdrawing group than the methoxy group of **3** since the carbonyl group in the former compound will enhance the electron-withdrawing ability of the oxygen atom via resonance. Compound **4** is interesting in that the acetoxy oxygen atom is expected to be an electron-donating group with respect to the alkene π bond (via resonance), much like the oxygen atom in compound **1**. From the relative reactivity data, however, the lone pairs on the acetoxy oxygen are apparently less efficient at donating to the carbon-carbon π bond. This can be explained by recognizing that the lone pairs on the acetoxy oxygen are already partially delocalized into the carbonyl group:

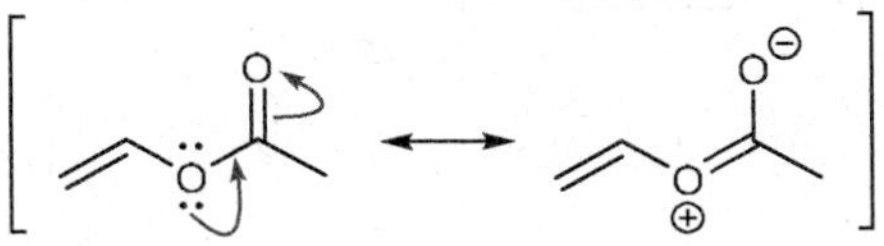

4

That is, the lone pairs on the acetoxy oxygen are less available to donate electron density to the π bond. Therefore, the acetoxy group of **4** influences the reactivity primarily through its inductive and steric effects, which will be much like that of the substituents in compounds **3** and **5**.

(b) Compound **6** is apparently able to impose the electron-withdrawing effect due to induction of the –CN group on the alkene π bond via the shorter σ bond (due to the sp^3-sp orbital overlap) between the CH_2 and the CN groups, and thus the closer proximity of the partial positive charge to the π electrons of the alkene. This is apparent when comparing the relative rates between compounds **5** and **6**. Here, though the oxygen atom in compound **5** is more electronegative than carbon, the carbon atom of the cyano group apparently possesses a very large partial positive change that effectively renders this carbon atom more electronegative than that of the acetoxy oxygen atom of **5**. The low reactivity associated with compound **7** may be due to both induction and steric effects, since the inductive withdrawal of electron density by the chlorine atom is not expected to exceed the substituent in compound **5**.

(c) Compounds **2**, **8** and **9** illustrate the consequences of steric effects on the hydroboration reaction. Compounds **8** and **9** are the only disubstituted alkenes among the series, and it is not surprising that they have the lowest relative reactivity, due to increased steric effects. They are more than 100 times lower in reactivity than compound **2**, the only monosubstituted alkene with no significant electronic effects (i.e. resonance and induction). In compound **9**, the electron-withdrawing effect of the chlorine atom, which is superimposed upon the increased steric effect, further lowers the reactivity.

8.108. When approaching this problem, it would be advisable to first label the carbon side chain coming off the benzene ring so that you can determine what new connections have been made. Your numbering system does not need to conform to IUPAC rules for assigning locants. Rather, it is OK to use an arbitrary numbering system, because the goal of the numbering system is to track the fate of all atoms during the transformation:

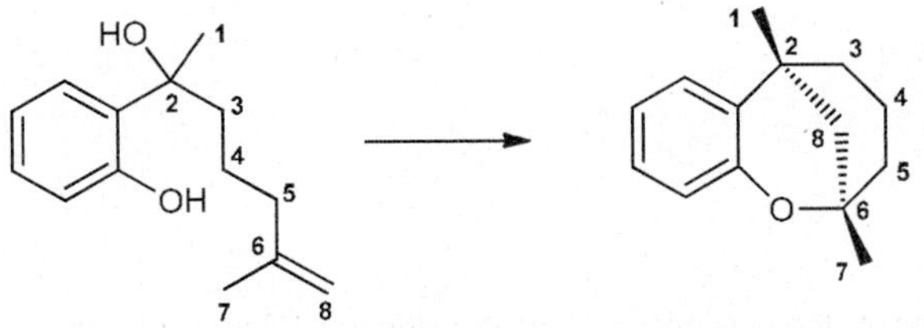

With this numbering system, the benzene ring is attached to C2 of the chain and the phenolic oxygen is attached to C6 of the chain in the product.

Based on this, a possible mechanism is illustrated below. Protonation of the tertiary alcohol, followed by loss of water, gives the stable tertiary benzylic carbocation (at C2). Markovnikov attack of the terminal alkene (C6 and C8) at the tertiary benzylic carbocation affords a tertiary carbocation at C6. This carbocation is then attacked by the phenolic oxygen to afford the final product after removal of the acidic proton.

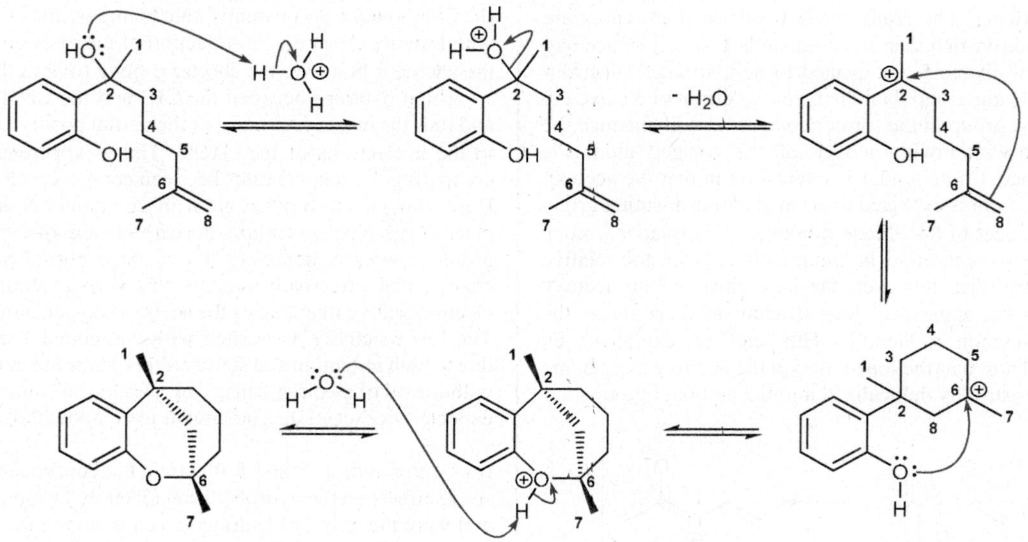

8.109.
The first step involves formation of a bromonium ion, which requires three curved arrows, followed by an intramolecular nucleophilic attack in which the OH group functions as a nucleophilic center and attacks the bromonium ion. Deprotonation then affords a cyclic ether. In the last step, a methoxide ion functions as a base and removes a proton, which leads to expulsion of bromide in an E2 process.

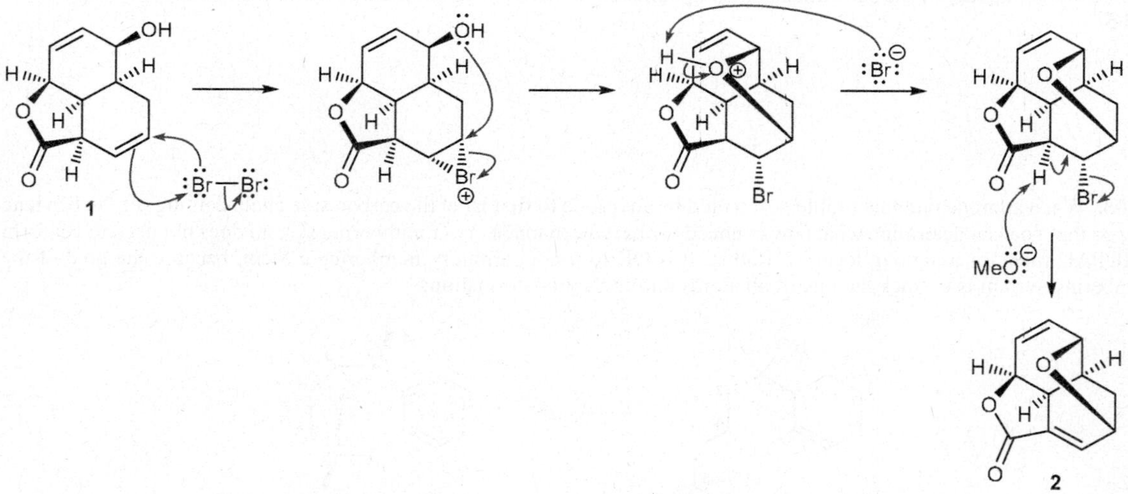

You might be wondering about the stereochemistry of the last step (the E2 process). In general, E2 processes occur more rapidly when the H and the leaving group are *anti*-periplanar in the transition state. However, it is possible for an E2 process to occur via a transition state in which the H and leaving group are *syn*-periplanar. In general, this is not favored (the transition state is high in energy because all groups are eclipsed rather than staggered), but in this case, the rigid geometry of the polycyclic structure essentially locks the H and the leaving group into a *syn*-periplanar arrangement, where the H and the leaving group are eclipsing each other. As such, the reaction can occur, because the minimum requirement of periplanarity is still met.

Chapter 9
Alkynes

Review of Concepts

Fill in the blanks below. To verify that your answers are correct, look in your textbook at the end of Chapter 9. Each of the sentences below appears verbatim in the section entitled *Review of Concepts and Vocabulary*.

- A triple bond is comprised of three separate bonds: one ____ bond and two ____ bonds.
- Alkynes exhibit ___________ geometry and can function as bases or as _______________.
- Monosubstituted acetylenes are **terminal alkynes**, while disubstituted acetylenes are _________ **alkynes**.
- Catalytic hydrogenation of an alkyne yields an _______.
- A **dissolving metal reduction** will convert an internal alkyne into a ________ alkene.
- Acid-catalyzed hydration of alkynes is catalyzed by mercuric sulfate ($HgSO_4$) to produce an ________ that cannot be isolated because it is rapidly converted into a ketone.
- Enols and ketones are ___________, which are constitutional isomers that rapidly interconvert via the migration of a proton.
- When treated with ozone followed by water, internal alkynes undergo oxidative cleavage to produce _______________.
- Alkynide ions undergo _____________ when treated with an alkyl halide (methyl or primary).

Review of Skills

Fill in the blanks and empty boxes below. To verify that your answers are correct, look in your textbook at the end of Chapter 9. The answers appear in the section entitled *SkillBuilder Review*.

9.1 Assembling the Systematic Name of an Alkyne

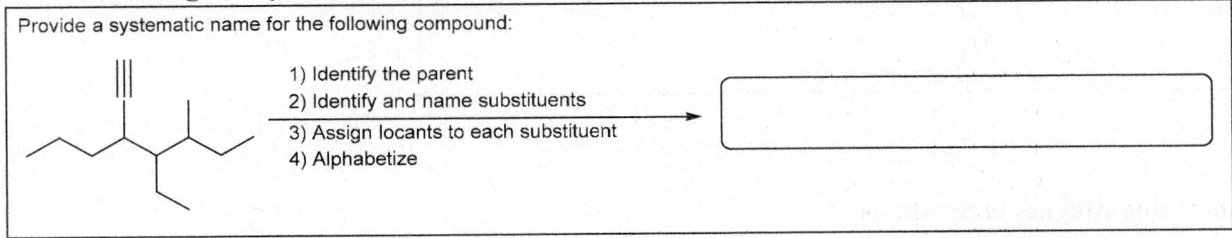

Provide a systematic name for the following compound:
1) Identify the parent
2) Identify and name substituents
3) Assign locants to each substituent
4) Alphabetize

9.2 Selecting a Base for Deprotonating a Terminal Alkyne

Circle the side of the equilibrium that is favored in the following acid-base reaction.

$$\mathrm{R{-}C{\equiv}C{-}H} \quad + \quad {:}\ddot{\mathrm{O}}\mathrm{H}^{\ominus} \quad \rightleftharpoons \quad \mathrm{R{-}C{\equiv}C{:}^{\ominus}} \quad + \quad \mathrm{H_2O}$$

9.3 Drawing a Mechanism for Acid-Catalyzed Keto-Enol Tautomerization

Draw two curved arrows showing protonation of the π bond.	Draw the resonance structures of the intermediate.	Draw two curved arrows showing deprotonation to form the ketone.

9.4 Choosing the Appropriate Reagents for the Hydration of an Alkyne

Identify reagents that can achieve each of the following transformations.

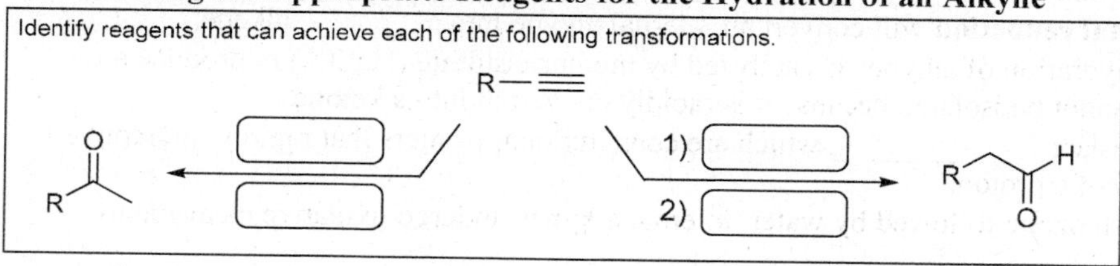

9.5 Alkylating Terminal Alkynes

Identify reagents that will achieve the following transformation:

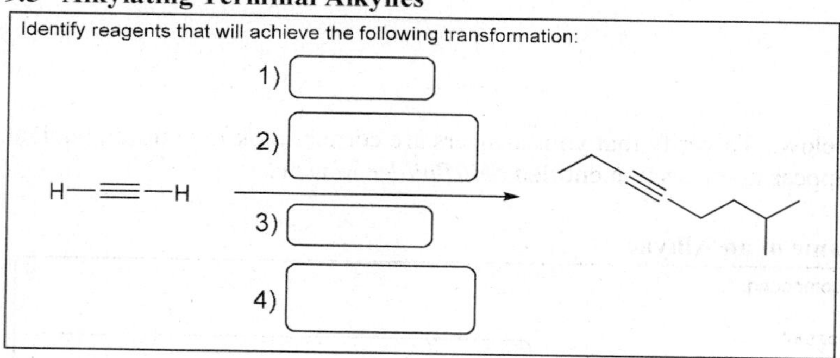

9.6 Using Alkynes in Synthesis

Identify reagents that will achieve the following transformation:

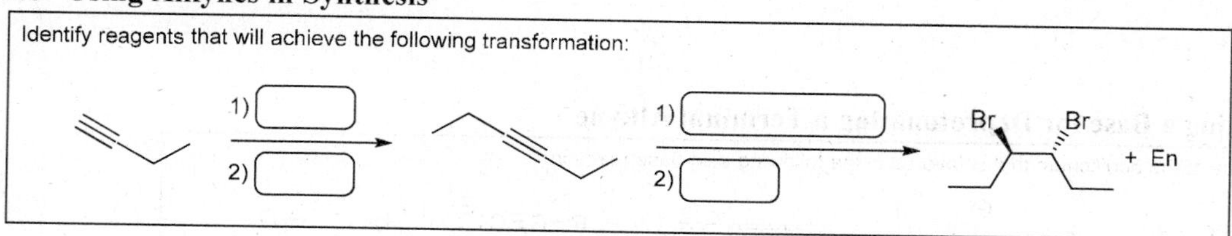

Review of Reactions

Identify the reagents necessary to achieve each of the following transformations. To verify that your answers are correct, look in your textbook at the end of Chapter 9. The answers appear in the section entitled *Review of Reactions*.

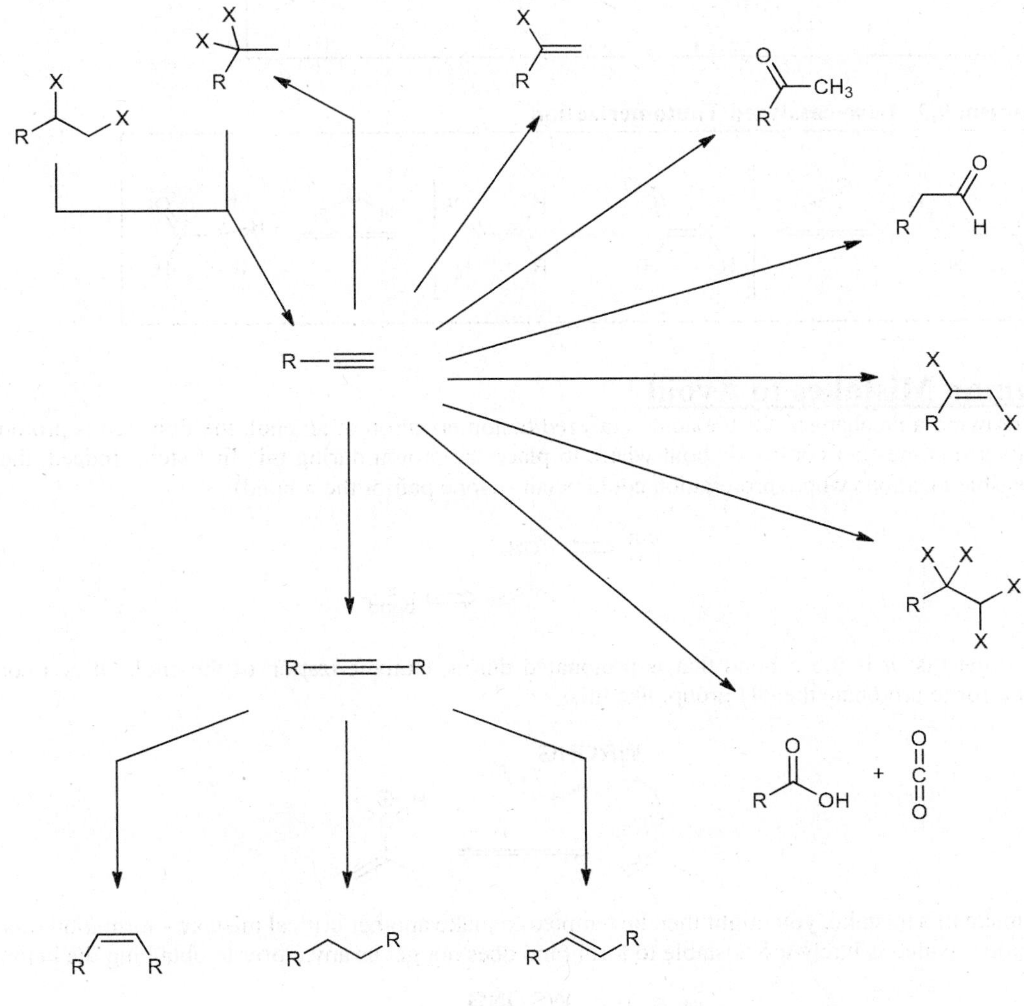

Review of Mechanisms

Complete each of the following mechanisms by drawing the missing curved arrows. To verify that your curved arrows are drawn correctly, compare them to the curved arrows in the mechanism boxes for Mechanisms 9.1 – 9.3, which can be found throughout Chapter 9 of your text.

Mechanism 9.1 Dissolving Metal Reductions

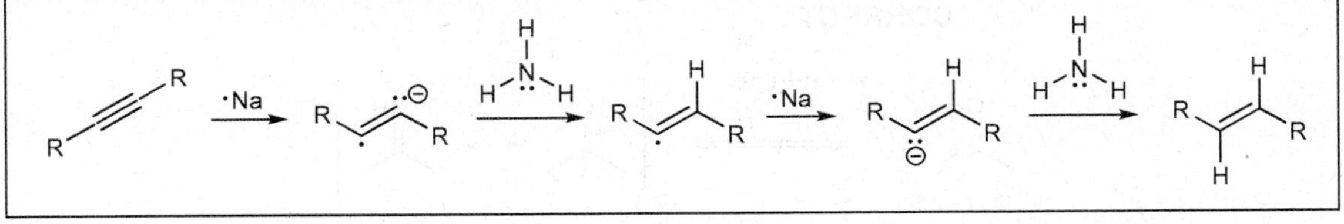

Mechanism 9.2 Acid-catalyzed Tautomerization

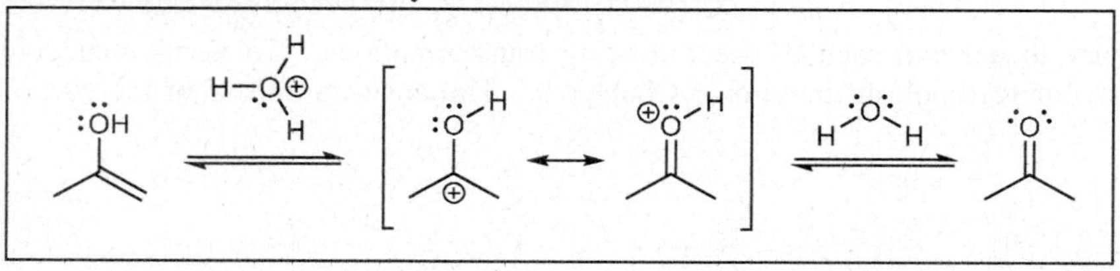

Mechanism 9.3 Base-catalyzed Tautomerization

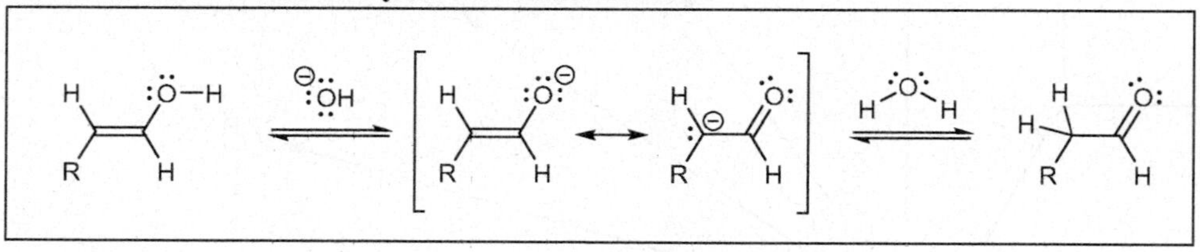

Common Mistakes to Avoid

When drawing a mechanism for the acid-catalyzed tautomerization of an enol, the first step is protonation. Students sometimes get confused about where to place the proton during this first step. Indeed, there are two possible locations where protonation could occur (a lone pair or the π bond):

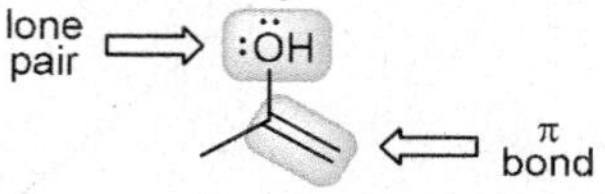

lone pair

π bond

Don't forget that it is the π bond that is protonated during tautomerization of the enol. It is a common student error to protonate the OH group, like this:

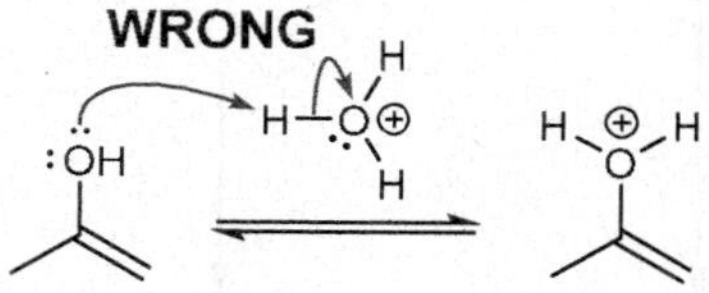

WRONG

If you make this mistake, you might then be tempted to make another critical mistake - formation of a vinyl carbocation, which is likely too unstable to form (and does not get us any closer to obtaining the ketone).

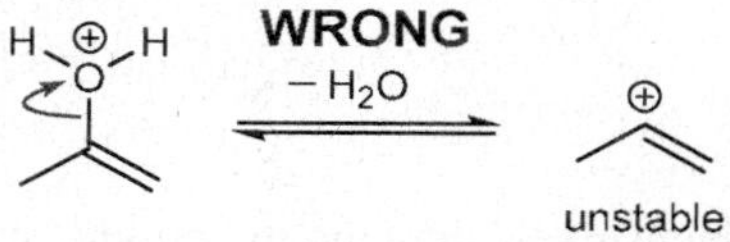

WRONG

$-H_2O$

unstable

Whenever possible, avoid the formation of high-energy intermediates that are unlikely to form. This is a general rule that should be followed whenever you are drawing a mechanism (exceptions are rare). The correct first step for acid-catalyzed tautomerization of an enol is protonation of the π bond to generate a resonance-stabilized cation:

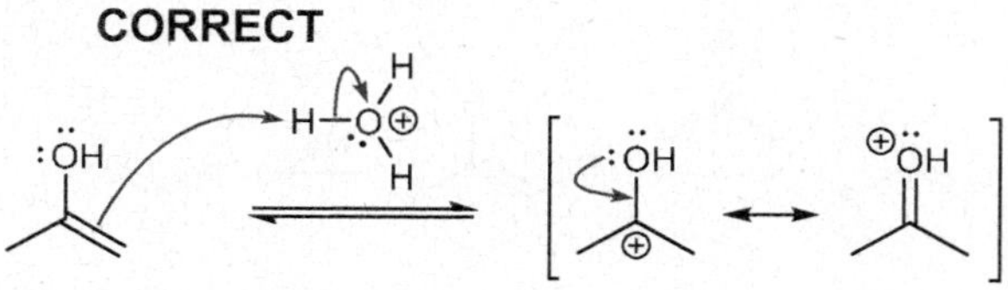

CORRECT

There is one other (unrelated) common error that should be avoided. When designing a synthesis in which acetylene is alkylated twice, make sure to alkylate each side separately, even if both alkyl groups are the same:

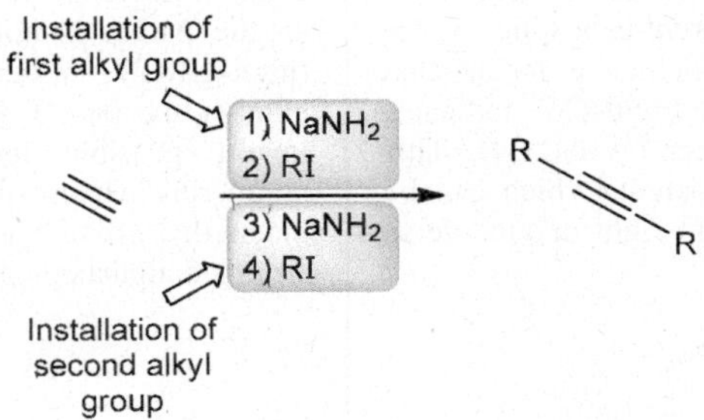

It is a common student error to show the reagents just one time, assuming that alkylation will occur on both sides of acetylene. If the reagents for alkylation are only shown once, then alkylation will only occur once:

1) $NaNH_2$
2) MeI
Me

Useful Reagents

The following is a list of commonly encountered reagents for reactions of alkynes:

Reagents	Name of Reaction	Description of Reaction
1) excess $NaNH_2$ 2) H_2O	Elimination	When treated with these reagents, a vicinal or geminal dibromide is converted to an alkyne.
HX	Hydrohalogenation	When treated with HX, an alkyne undergoes Markovnikov addition (excess HX gives two addition reactions to afford a geminal dihalide).
H_2SO_4, H_2O, $HgSO_4$	Acid-cat. hydration	When treated with these reagents, a terminal alkyne undergoes Markovnikov addition of H and OH to give an enol, which quickly tautomerizes to give a ketone.
1) R_2BH 2) H_2O_2, NaOH	Hydroboration-oxidation	When treated with these reagents, a terminal alkyne undergoes *anti*-Markovnikov addition of H and OH to give an enol, which quickly tautomerizes to give an aldehyde.
X_2	Halogenation	When treated with this reagent, an alkyne undergoes addition of X and X (excess X_2 gives a tetrahalide).
1) O_3 2) H_2O	Ozonolysis	When treated with these reagents, an alkyne undergoes oxidative cleavage of the C≡C bond. Internal alkynes are converted into two carboxylic acids, while terminal alkynes are converted into a carboxylic acid and carbon dioxide.
H_2, Lindlar's catalyst	Hydrogenation	When treated with these reagents, an alkyne is converted to a *cis* alkene.
H_2, Pt	Hydrogenation	When treated with these reagents, an alkyne is converted to an alkane.
Na, NH_3 (*l*)	Dissolving metal reduction	When treated with these reagents, an internal alkyne is converted to a *trans* alkene.

Solutions

9.1.

(a) We begin by identifying the parent. The longest chain is six carbon atoms, so the parent is hexyne. There are no substituents. We must include a locant that identifies the position of the triple bond ("3" indicates that the triple bond is located between C3 and C4). This is determined by numbering the parent, which can be done in this case either from left to right or vice versa (either way gives the same result).

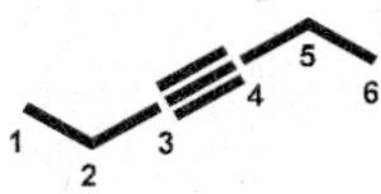

3-Hexyne

(b) We begin by identifying the parent. The longest chain is six carbon atoms, so the parent is hexyne. There is one substituent – a methyl group (highlighted). In this case, the triple bond is at C3 regardless of which way we number the parent, so the parent chain is numbered starting from the side that gives the substituent the lowest possible number. According to this numbering scheme, the methyl group is located at C2:

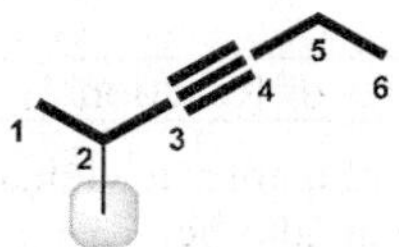

2-Methyl-3-hexyne

(c) We begin by identifying the parent. The longest chain is eight carbon atoms, so the parent is octyne. There are no substituents. We must include a locant that identifies the position of the triple bond. The parent chain is numbered so that the triple bond is assigned the lowest possible locant ("3" indicates that the triple bond is located between C3 and C4).

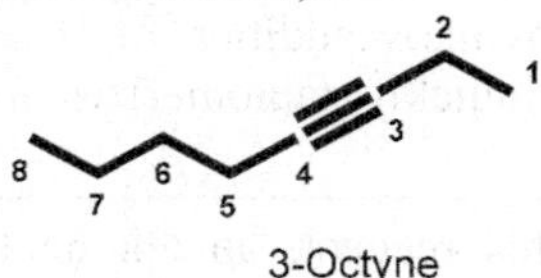

3-Octyne

(d) We begin by identifying the parent. The longest chain is four carbon atoms, so the parent is butyne. There are two substituents – both methyl groups (highlighted). We number the parent so that the triple bond receives the lowest possible locant (C1). According to this numbering scheme, the methyl groups are both located at C3:

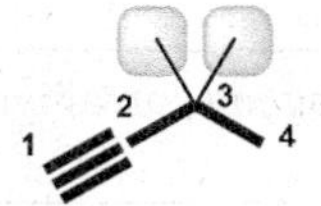

3,3-Dimethyl-1-butyne

(e) We begin by identifying the parent. The longest chain containing the triple bond has nine carbon atoms, so the parent is nonyne. There are two substituents (highlighted) – a methyl group and a propyl group. We number the parent so that the triple bond receives the lowest possible locant (C1). According to this numbering scheme, the propyl group is located at C5 and the methyl group is located at C7. These substituents are arranged alphabetically in the name:

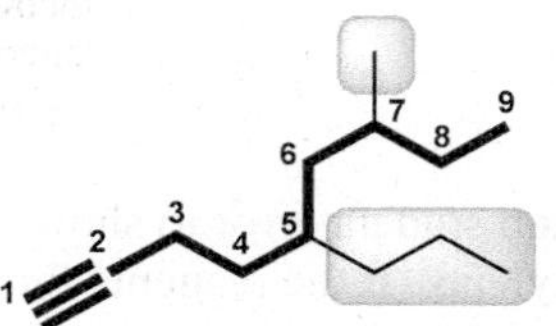

7-methyl-5-propyl-1-nonyne

(f) We begin by identifying the parent. The longest chain containing the triple bond has ten carbon atoms, so the parent is decyne. There are three substituents (highlighted) – two methyl groups and an ethyl group. We number the parent so that the triple bond receives the lowest possible locant (C3). According to this numbering scheme, the ethyl group is located at C5 and the methyl groups are both located at C7. These substituents are arranged alphabetically in the name (ethyl before methyl):

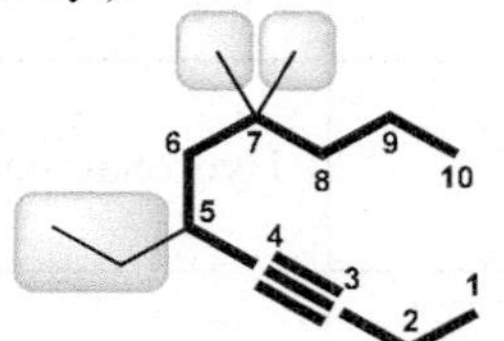

5-ethyl-7,7-dimethyl-3-decyne

(g) We begin by identifying the parent. The longest chain containing the triple bond has ten carbon atoms, so the parent is decyne. There are many substituents (highlighted). We number the parent so that the triple bond receives the lowest possible locant (C2), and the substituents are arranged alphabetically in the name (chloro, then ethyl, and then methyl):

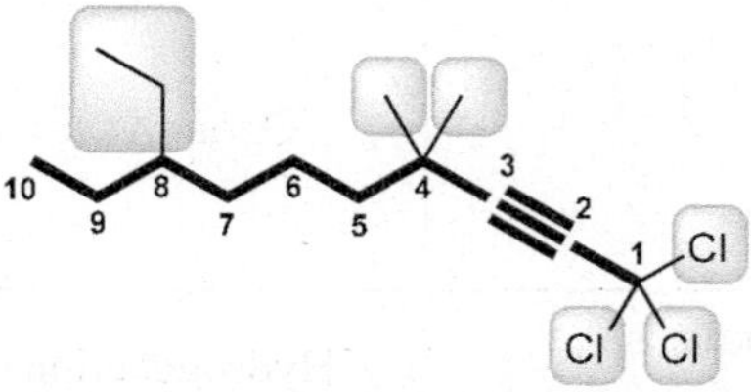

1,1,1-trichloro-8-ethyl-4,4-dimethyl-2-decyne

9.2.

(a) The parent (pentyne) indicates a chain of five carbon atoms. The triple bond is between C2 and C3, and there are two methyl groups (highlighted), both located at C4.

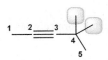

4,4-Dimethyl-2-pentyne

(b) The parent (heptyne) indicates a chain of seven carbon atoms. The triple bond is between C3 and C4, and there are three substituents (highlighted) – two methyl groups (at C2 and C5), as well as an ethyl group at C5.

5-Ethyl-2,5-dimethyl-3-heptyne

9.3. The parent is cyclononyne, so we draw a nine-membered ring that incorporates a triple bond. The triple bond is (by definition) between C1 and C2, and a methyl group is located at C3. This position (C3) is a chiral center, with the *R* configuration, shown here:

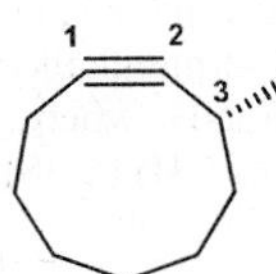

9.4. The parent chain contains eight carbon atoms with two methyl substituents. Numbering so that the triple bond receives the lowest possible locant, the two methyl groups are located at C3 and C7:

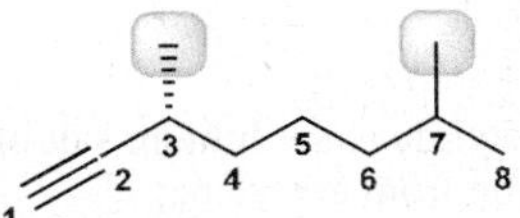

Notice that there is a chiral center at C3. The complete IUPAC name must include the configuration of this chiral center using the Cahn-Ingold-Prelog system. Priorities are assigned in the following way:

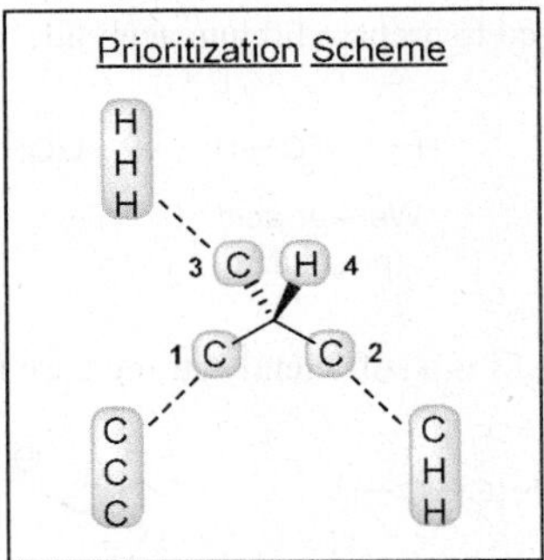

Notice that the low-priority group is in the front rather than the back. So, we rotate the molecule (thereby placing the fourth priority on a dash), and we assign the configuration (*R*):

Putting it all together, the IUPAC name of this alkyne is (*R*)-3,7-dimethyl-1-octyne (using the old IUPAC rules) or (*R*)-3,7-dimethyloct-1-yne (using the new IUPAC rules). Both names are acceptable.

9.5.

(a) Yes, as seen in Table 9.2, $NaNH_2$ is a sufficiently strong base to deprotonate a terminal alkyne.

t-Bu—C≡C—H + $NaNH_2$ ⟶ *t*-Bu—C≡C:$^{\ominus}$ Na$^{\oplus}$ + NH_3

Stronger acid (pK_a ~ 25) Weaker acid (pK_a = 38)

(b) Yes, as seen in Table 9.2, NaH is a sufficiently strong base to deprotonate a terminal alkyne.

Stronger acid (pK_a ~ 25) + NaH ⟶ Na$^{\oplus}$ + H_2 Weaker acid (pK_a = 35)

(c) No, as seen in Table 9.2, *t*-BuOK is not a sufficiently strong base to deprotonate a terminal alkyne.

—C≡C—H + *t*-BuOK ⇌ —C≡C:$^{\ominus}$ Li$^{\oplus}$ + *t*-BuOH

Weaker acid (pK_a ~ 25) Stronger acid (pK_a = 18)

9.6.

(a) Lithium acetylide is the lithium salt of the conjugate base of acetylene, which is formed when a suitably strong base removes a proton from acetylene:

H−C≡C−H + :Base⊖ Li⊕ ⟶ H−C≡C:⊖ Li⊕ + H−Base

Lithium acetylide

(b) As seen in Table 9.2, hydroxide is not a sufficiently strong base to deprotonate a terminal alkyne. Therefore, LiOH cannot be used to prepare lithium acetylide, since the reverse reaction would be favored:

H−C≡C−H + LiOH ⇌ H−C≡C:⊖ Li⊕ + H_2O

Weaker acid (pK_a = 25) — Stronger acid (pK_a = 15.7)

However, BuLi is a sufficiently strong base to deprotonate a terminal alkyne and can be used to prepare lithium acetylide:

H−C≡C−H + BuLi ⟶ H−C≡C:⊖ Li⊕ + butane

Stronger acid (pK_a = 25) — Weaker acid (pK_a = 50)

LDA is a base with a negative charge on a nitrogen atom, so we expect that it will be similar in base strength to sodium amide ($NaNH_2$). Therefore, LDA will be a sufficiently strong base to deprotonate a terminal alkyne and can be used to prepare lithium acetylide, as shown below. The pK_a value of diisopropyl amine (R_2NH, where R = isopropyl) is not given in Table 9.2, but we can assume that its pK_a is similar in magnitude to the pK_a of NH_3 (~ 38).

H−C≡C−H + LDA ⟶ H−C≡C:⊖ Li⊕ + diisopropylamine (N−H)

Stronger acid (pK_a = 25) — Weaker acid (pK_a = ~38)

9.7.

(a) The starting material is a geminal dibromide. When treated with excess sodium amide ($NaNH_2$), two successive E2 reactions occur (each of which requires three curved arrows, as shown below). The resulting terminal alkyne is then deprotonated to give an alkynide ion:

⊖$\ddot{N}H_2$; $H_2\ddot{N}$⊖ ; ⊖$\ddot{N}H_2$

After the reaction is complete, water is introduced into the reaction flask to protonate the alkynide ion, thereby giving the neutral terminal alkyne as the final product:

H−O−H

(b) The starting material is a vicinal dichloride. When treated with excess sodium amide ($NaNH_2$), two successive E2 reactions occur (each of which requires three curved arrows, as shown below). The resulting terminal alkyne is then deprotonated to give an alkynide ion:

⊖$\ddot{N}H_2$; $H_2\ddot{N}$⊖ ; ⊖$\ddot{N}H_2$

After the reaction is complete, water is introduced into the reaction flask to protonate the alkynide ion, thereby giving the neutral terminal alkyne as the final product:

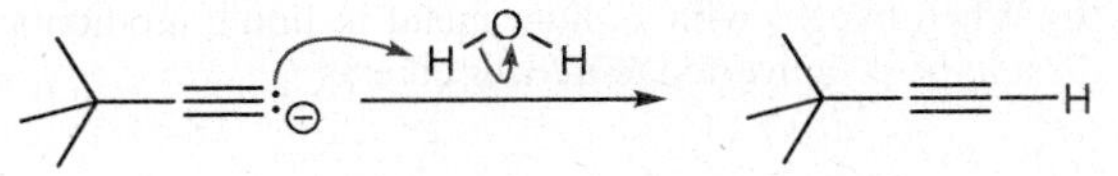

9.8. Deprotonation of 2-pentyne generates a resonance-stabilized anion, which is then protonated by NH_3 to give an allene (a compound featuring a C=C=C unit). The allene is then deprotonated to give a resonance-stabilized anion, which is then protonated by NH_3 to give 1-pentyne. Deprotonation of this terminal alkyne gives an alkynide ion. Formation of this alkynide ion pushes the equilibrium to favor this isomerization process.

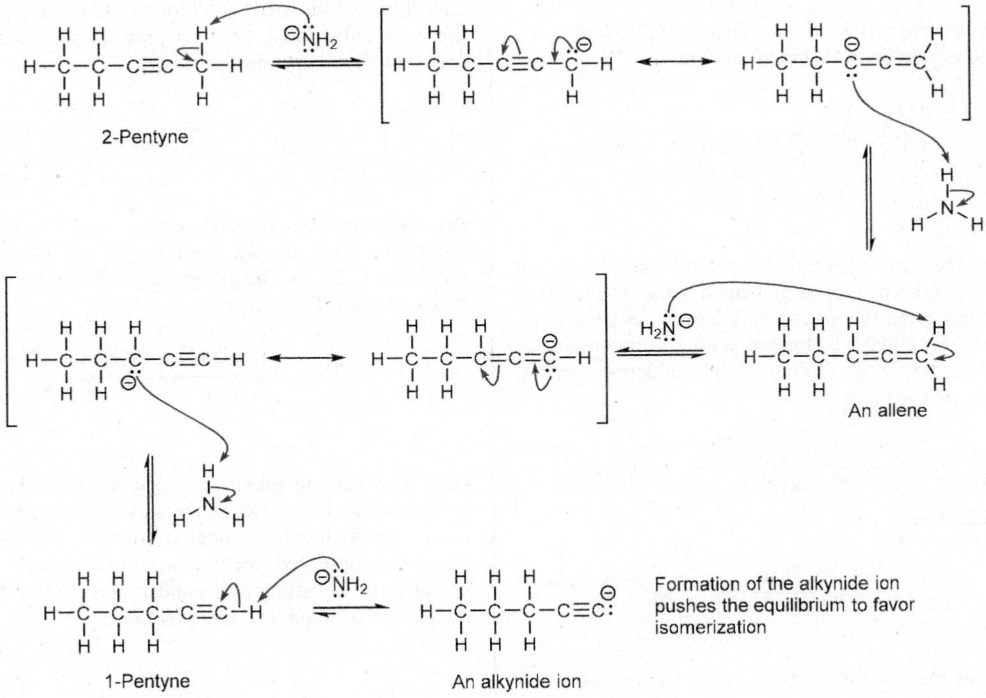

2-Pentyne

An allene

1-Pentyne

An alkynide ion

Formation of the alkynide ion pushes the equilibrium to favor isomerization

9.9.

(a) When hydrogenation is performed in the presence of a poisoned catalyst (such as Lindlar's catalyst), the alkyne is reduced to a *cis* alkene. When Pt is used as the catalyst, the alkyne is reduced all the way to an alkane, as shown here:

H_2 Lindlar's catalyst

H_2 Pt

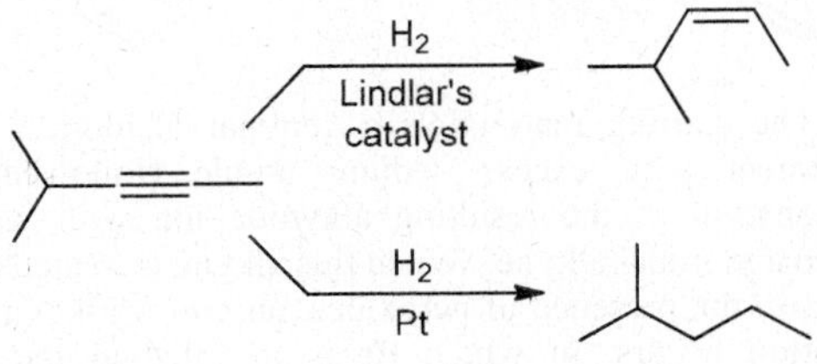

(b) When hydrogenation is performed in the presence of a poisoned catalyst (such as Ni_2B), the alkyne is reduced to a *cis* alkene. When nickel is used as the catalyst, the alkyne is reduced all the way to an alkane, as shown here:

H_2 Ni_2B

H_2 Ni

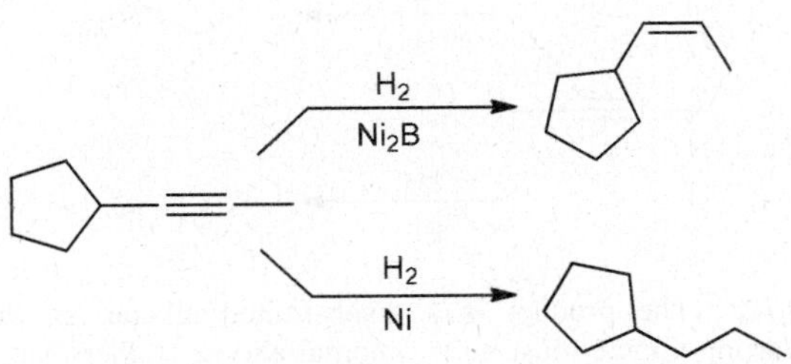

9.10.

(a) When treated with sodium metal in liquid ammonia, the alkyne is converted to a *trans* alkene:

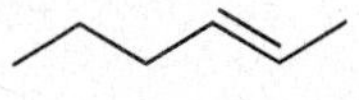

(b) When treated with sodium metal in liquid ammonia, the alkyne is converted to a *trans* alkene:

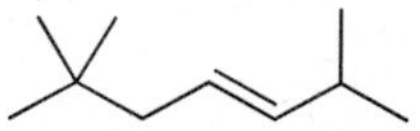

(c) When treated with sodium metal in liquid ammonia, the alkyne is converted to a *trans* alkene:

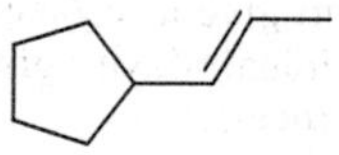

(d) When treated with sodium metal in liquid ammonia, the alkyne is converted to a *trans* alkene:

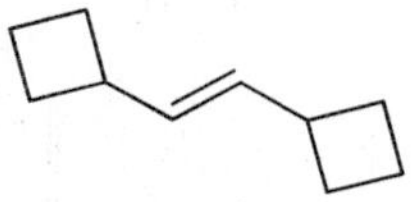

9.11.

(a) When the alkyne is treated with molecular hydrogen (H_2) in the presence of a poisoned catalyst (such as Lindlar's catalyst), the alkyne is reduced to a *cis* alkene. If instead, the alkyne is treated with sodium metal in liquid ammonia, a dissolving metal reduction occurs, giving a *trans* alkene, as shown:

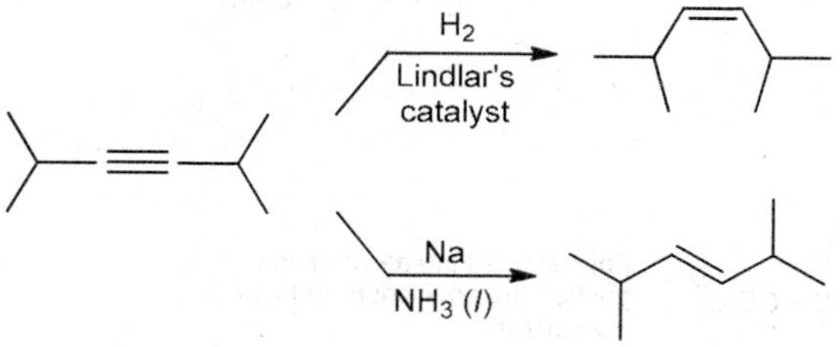

H_2 / Lindlar's catalyst

Na / NH_3 (*l*)

(b) When the alkyne is treated with sodium metal in liquid ammonia, a dissolving metal reduction occurs, giving a *trans* alkene. If instead, the alkyne is treated with molecular hydrogen (H_2) in the presence of a catalyst such as platinum (NOT a poisoned catalyst), the alkyne is reduced to an alkane, as shown:

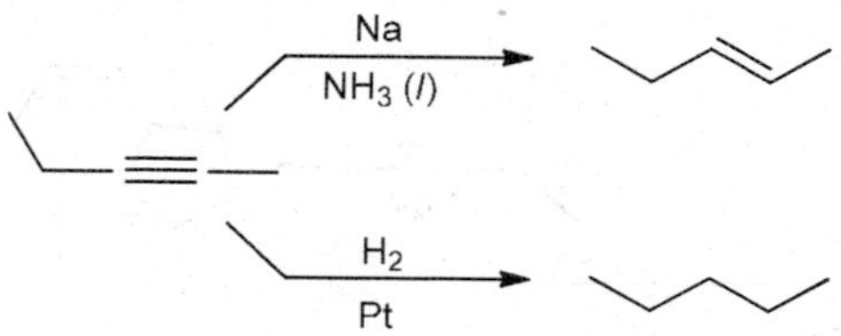

Na / NH_3 (*l*)

H_2 / Pt

9.12. The product is a disubstituted alkene, so the starting alkyne must be an internal alkyne (rather than a terminal alkyne):

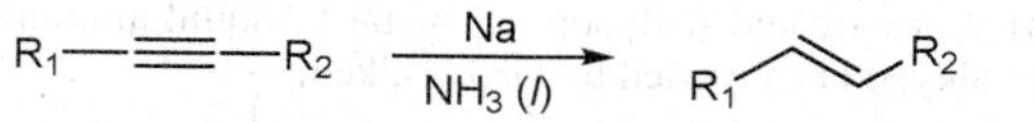

R_1—≡—R_2 → Na / NH_3 (*l*) → R_1, R_2

Internal alkyne

Disubstituted alkene

The molecular formula of the alkyne indicates five carbon atoms. Two of those atoms are the *sp*-hybridized carbon atoms of the triple bond. The remaining three carbon atoms must be in the R groups. So, one R group must be a methyl group, and the other R group must be an ethyl group:

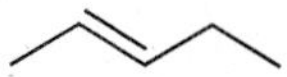

9.13.

(a) The starting alkyne is terminal, and when treated with excess HCl, two successive addition reactions occur, producing a geminal dihalide. Markovnikov addition is observed (two chlorine atoms are installed at the more substituted, secondary position, rather than the less substituted, primary position):

xs HCl

Cl Cl

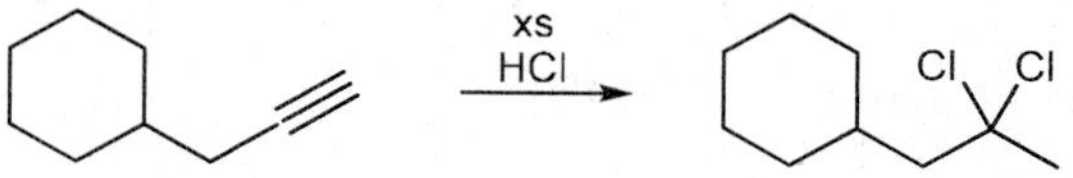

(b) The starting material is a geminal dichloride, and treatment with excess sodium amide (followed by protonation of the resulting alkynide ion with water) gives a terminal alkyne.

Cl Cl

1) xs $NaNH_2$ / NH_3

2) H_2O

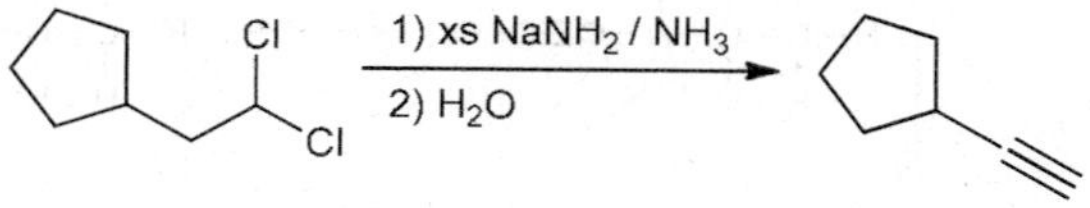

(c) The starting alkyne is terminal, and when treated with excess HBr, two successive addition reactions occur, producing a geminal dibromide. Markovnikov addition is observed (the two bromine atoms are installed at the more substituted, secondary position, rather than the less substituted, primary position):

xs HBr

Br Br

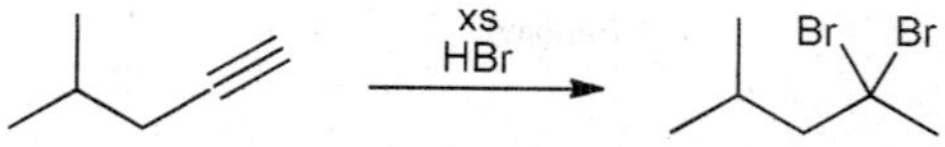

(d) The starting material is a geminal dibromide, and treatment with excess sodium amide (followed by protonation of the resulting alkynide ion with water) gives a terminal alkyne.

Br Br

1) xs $NaNH_2$ / NH_3

2) H_2O

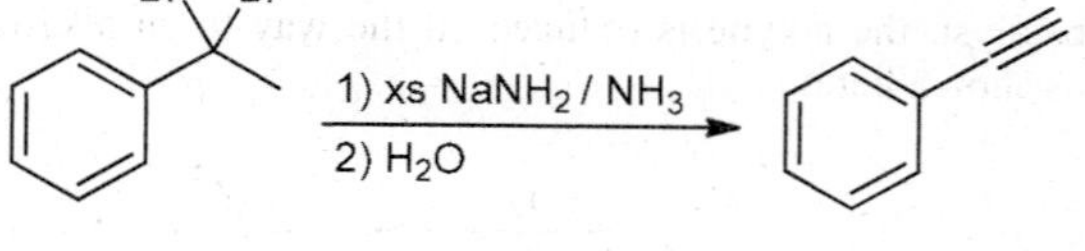

(e) The starting material is a geminal dichloride, and treatment with excess sodium amide (followed by protonation of the resulting alkynide ion with water) gives a terminal alkyne. When this alkyne is treated with HBr, in the presence of peroxides, an *anti*-Markovnikov addition occurs, in which Br is installed at the less substituted position. This gives rise to two stereoisomers, as shown:

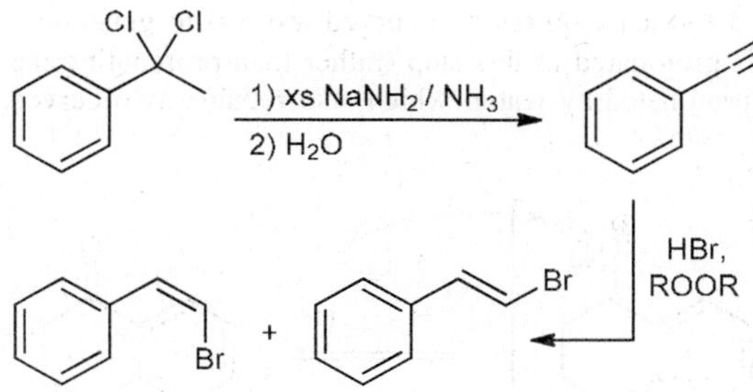

1) xs $NaNH_2$ / NH_3
2) H_2O

HBr, ROOR

(f) The starting material is a geminal dichloride, and treatment with excess sodium amide (followed by protonation of the resulting alkynide ion with water) gives a terminal alkyne. When this alkyne is treated with excess HBr, two addition reactions occur, installing two bromine atoms at the more substituted position (Markovnikov addition), as shown:

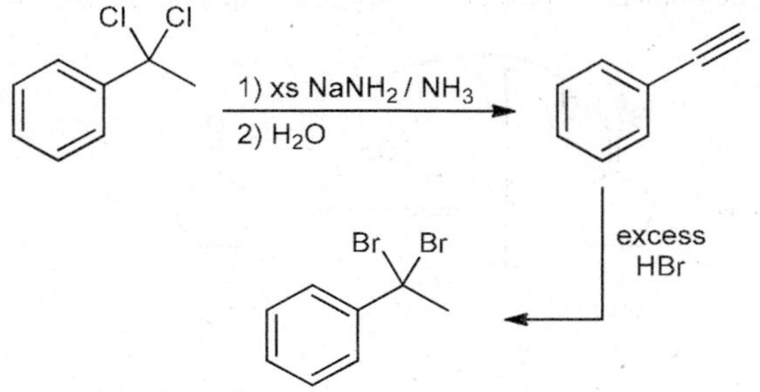

1) xs $NaNH_2$ / NH_3
2) H_2O

excess HBr

9.14. The starting material is a geminal dichloride, and the product is also a geminal dichloride. The difference between these compounds is the placement of the chlorine atoms. We did not learn a single reaction that will change the locations of the chlorine atoms. But the desired transformation can be achieved via an alkyne. Specifically, the starting material is treated with excess sodium amide (followed by protonation of the resulting alkynide ion with water) to give the terminal alkyne, which is then treated with excess HCl to give the desired compound (via Markovnikov addition):

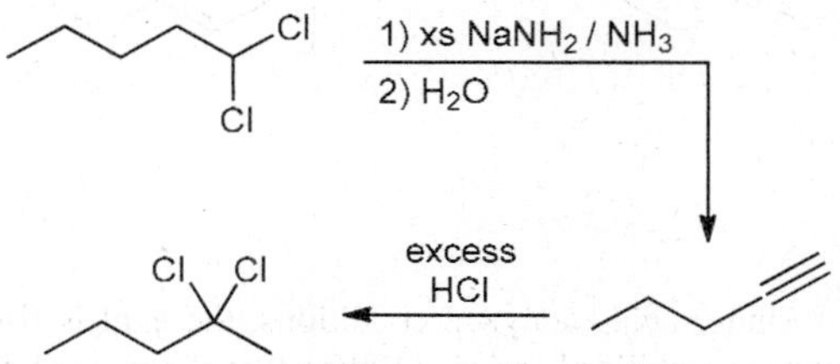

1) xs $NaNH_2$ / NH_3
2) H_2O

excess HCl

9.15. If two products are obtained, then the alkyne must be internal and unsymmetrical. There is only one such alkyne with the molecular formula C_5H_8 (shown below). The products are also shown:

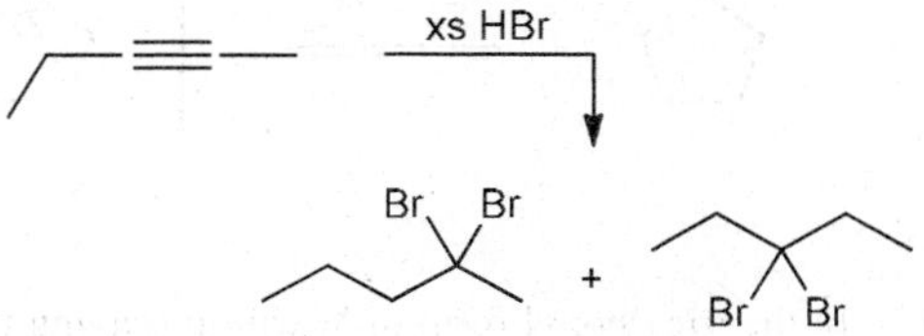

xs HBr

9.16.

(a) Under acid-catalyzed conditions, the enol is first protonated (which requires two curved arrows) to generate a resonance-stabilized cation. Notice that the π bond of the enol is protonated in this step (rather than protonating the OH group). The resulting resonance-stabilized cation is then deprotonated by water, which also requires two curved arrows, as shown:

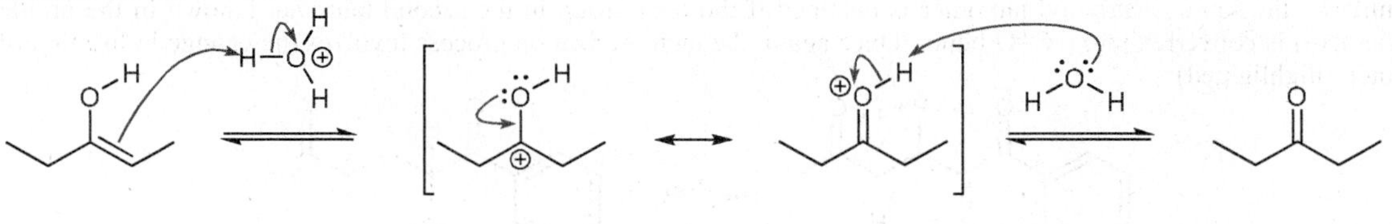

(b) Under acid-catalyzed conditions, the enol is first protonated (which requires two curved arrows) to generate a resonance-stabilized cation. Notice that the π bond of the enol is protonated in this step (rather than protonating the OH group). The resulting resonance-stabilized cation is then deprotonated by water, which also requires two curved arrows, as shown:

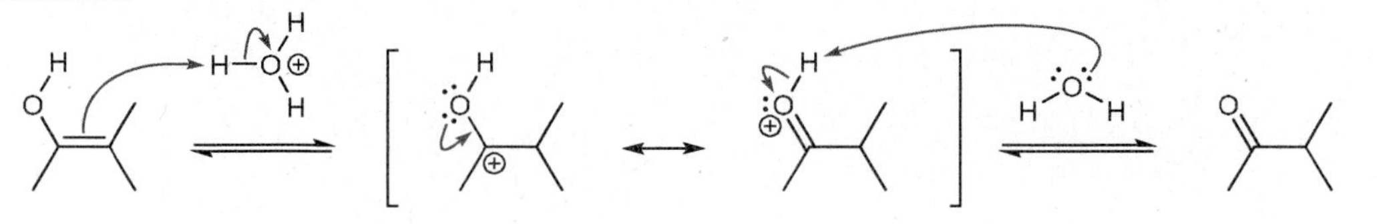

(c) Under acid-catalyzed conditions, the enol is first protonated (which requires two curved arrows) to generate a resonance-stabilized cation. Notice that the π bond of the enol is protonated in this step (rather than protonating the OH group). The resulting resonance-stabilized cation is then deprotonated by water, which also requires two curved arrows, as shown:

other resonance structures

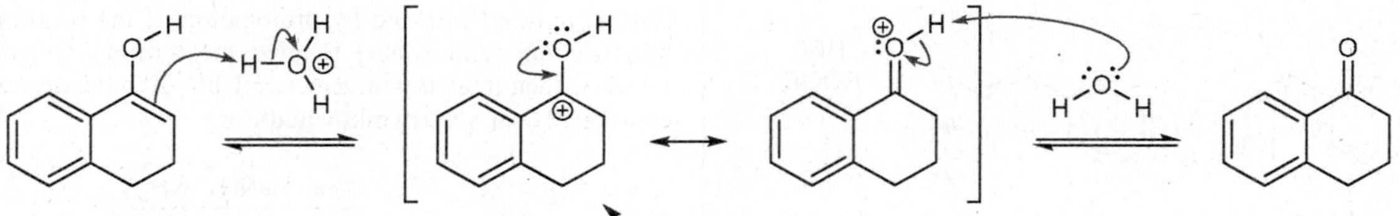

(d) Under acid-catalyzed conditions, the enol is first protonated (which requires two curved arrows) to generate a resonance-stabilized cation. Notice that the π bond of the enol is protonated in this step (rather than protonating the OH group). The resulting resonance-stabilized cation is then deprotonated by water, which also requires two curved arrows, as shown:

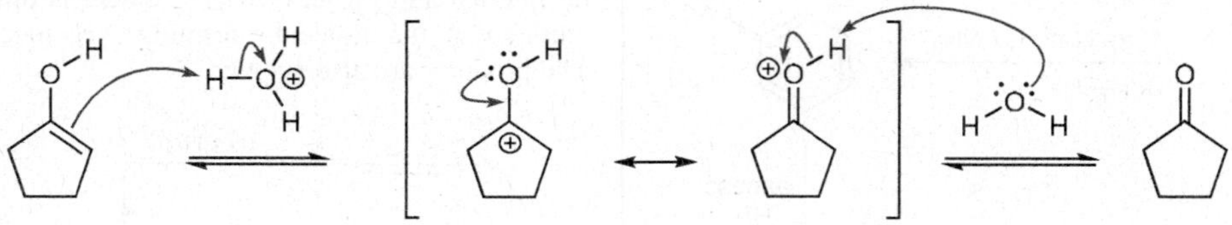

9.17. In the tricarbonyl form of Warfarin (having three C=O bonds), any enol groups are converted into C=O bonds. Let's begin by considering the structure of the first tautomer shown in the problem statement. This compound contains one enol group. If this enol group is converted into a ketone, the resulting tautomer would have three carbonyl groups. This tautomerization process involves the change in location of a proton (highlighted):

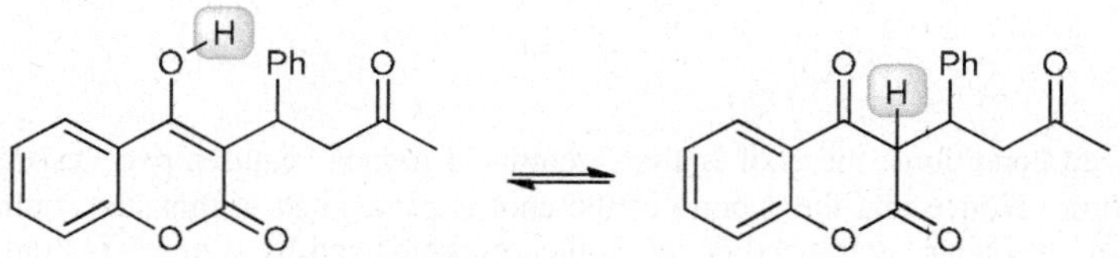

Similarly, the same tricarbonyl tautomer is obtained if the enol group in the second tautomer (shown in the problem statement) is converted into a C=O bond. Once again, the tautomerization process involves the change in location of a proton (highlighted):

Under acidic conditions, the tautomerization of either enol tautomer (to give the tricarbonyl form) occurs via the following steps. First, the C=C π bond is protonated (not the OH group) to give a resonance-stabilized carbocation intermediate. After drawing the resonance structures of the intermediate, deprotonation gives the tricarbonyl tautomer:

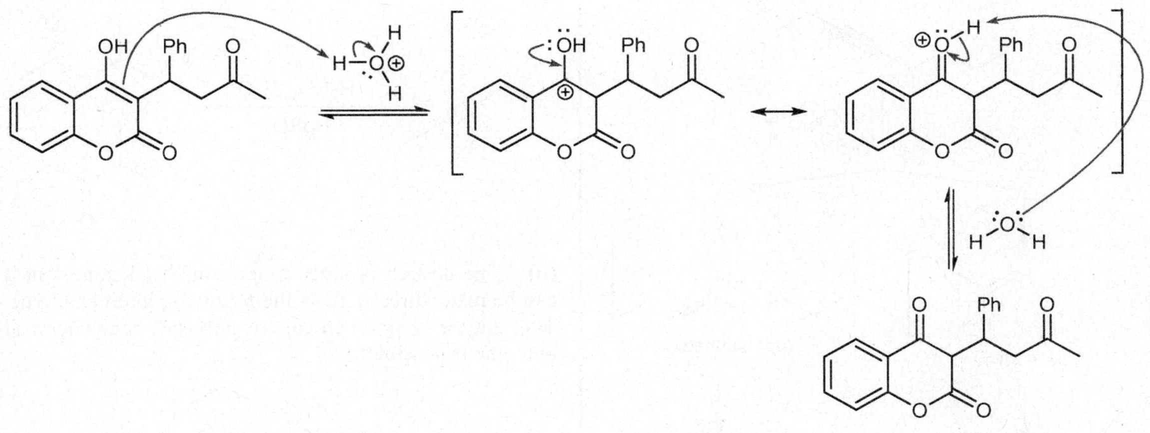

Starting with the other tautomer, we follow the same steps again. First, the π bond is protonated to give a resonance-stabilized carbocation intermediate. After drawing the resonance structures of the intermediate, deprotonation gives the tricarbonyl tautomer:

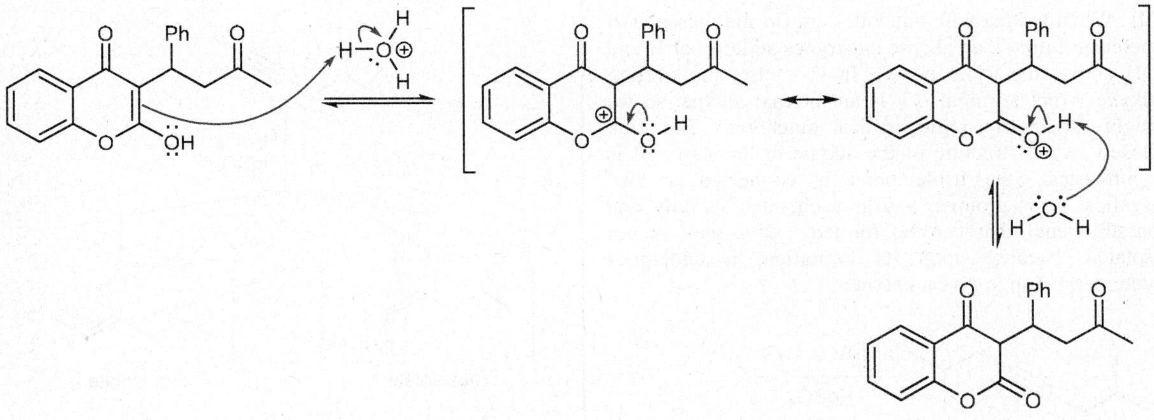

9.18.

(a) When treated with aqueous acid (in the presence of mercuric sulfate), the terminal alkyne undergoes Markovnikov addition of H and OH across the triple bond, giving an enol. The enol is not isolated, because upon its formation, it undergoes tautomerization to give a methyl ketone:

H_2SO_4, H_2O

$HgSO_4$

(not isolated)

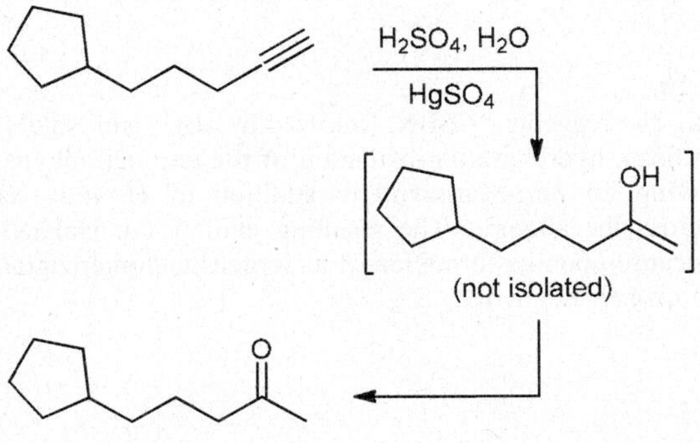

(b) When treated with aqueous acid (in the presence of mercuric sulfate), the terminal alkyne undergoes Markovnikov addition of H and OH across the triple bond, giving an enol. The enol is not isolated, because upon its formation, it undergoes tautomerization to give a methyl ketone:

H_2SO_4

H_2O

$HgSO_4$

(not isolated)

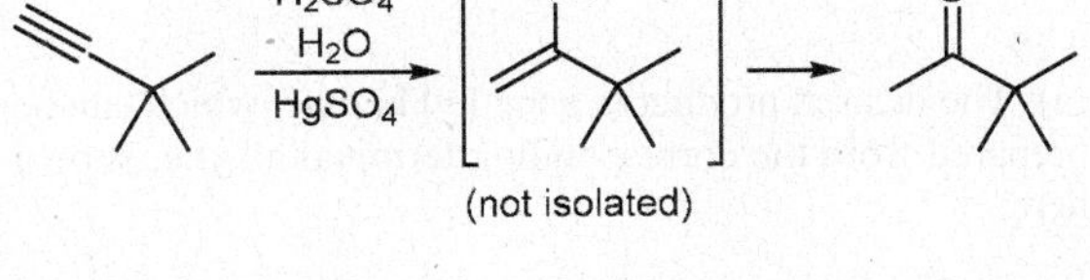

(c) When treated with aqueous acid (in the presence of mercuric sulfate), an alkyne undergoes addition of H and OH across the triple bond. In this case, the starting alkyne is not terminal – it is an internal alkyne. As such, there are two possible regiochemical outcomes, giving rise to two possible enols. Neither of these enols is isolated, because upon their formation, they each undergo tautomerization to give a ketone, as shown here:

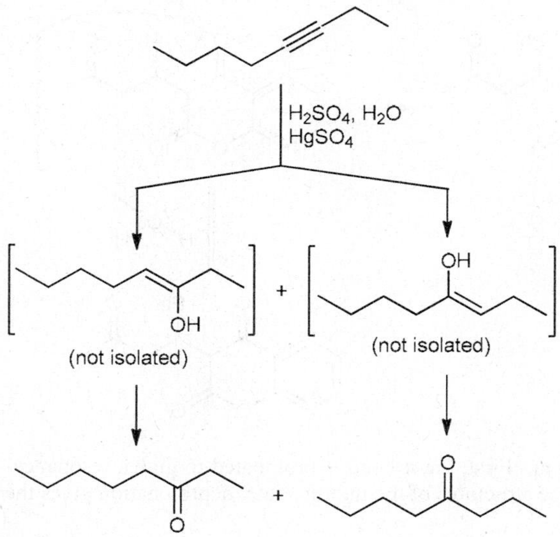

H_2SO_4, H_2O
$HgSO_4$

(not isolated) + OH (not isolated)

O + O

(d) When treated with aqueous acid (in the presence of mercuric sulfate), an alkyne undergoes addition of H and OH across the triple bond. In this case, the starting alkyne is not terminal – it is an internal alkyne, so we might expect two regiochemical outcomes. But look closely at the structure of the alkyne in this case. It is symmetrical (the triple bond is connected to two identical alkyl groups), and as such, there is only one possible enol that can be formed. This enol is not isolated, because upon its formation, it undergoes tautomerization to give a ketone:

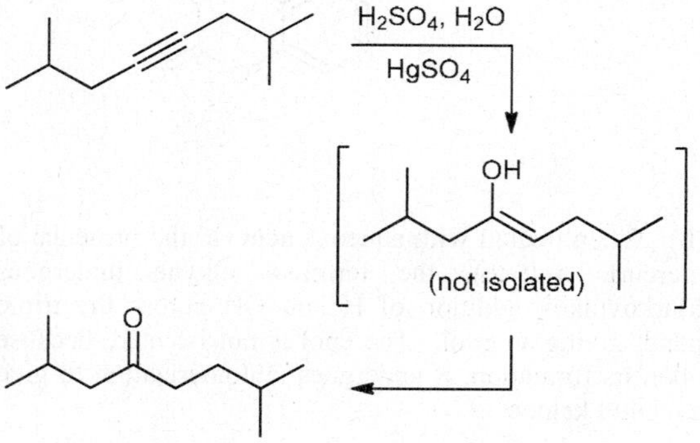

H_2SO_4, H_2O
$HgSO_4$

OH (not isolated)

O

9.19.

(a) The desired product is a methyl ketone, which can be prepared from the corresponding terminal alkyne, shown here:

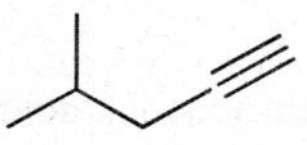

The following alkyne cannot be used because it would give a mixture of isomers:

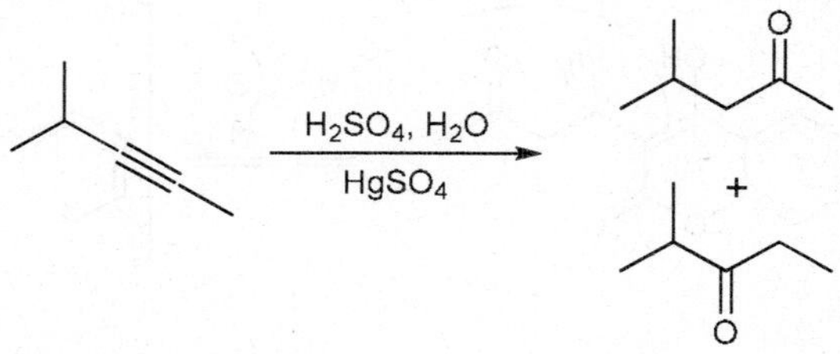

H_2SO_4, H_2O
$HgSO_4$

O +

O

(b) The desired product is not a methyl ketone, but it can be made directly from the following internal alkyne. This alkyne is symmetrical, so only one regiochemical outcome is possible:

The following alkyne would NOT be a suitable starting material, because hydration of this alkyne would produce a mixture of two ketones:

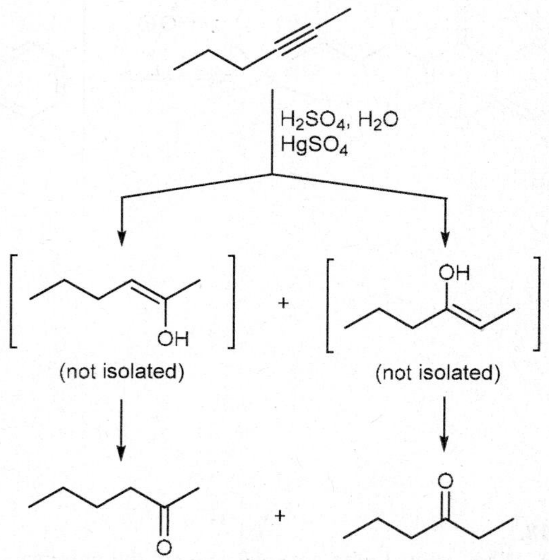

H_2SO_4, H_2O
$HgSO_4$

OH (not isolated) + OH (not isolated)

O + O

(c) The desired product is a methyl ketone, which can be prepared from the corresponding terminal alkyne, shown here:

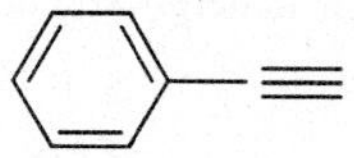

9.20.

(a) The reagents (9-BBN, followed by H_2O_2 and NaOH) indicate hydroboration-oxidation of the terminal alkyne, giving an *anti*-Markovnikov addition of H and OH across the alkyne. The resulting enol is not isolated, because upon its formation, it undergoes tautomerization to give an aldehyde:

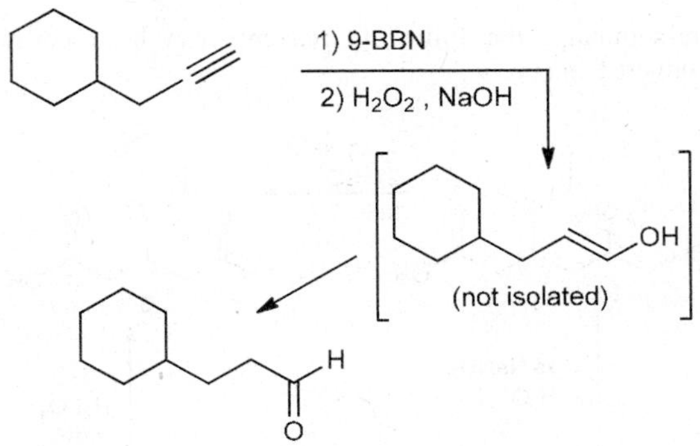

1) 9-BBN

2) H_2O_2 , NaOH

OH

(not isolated)

H

O

(b) The reagents (disiamylborane, followed by H_2O_2 and NaOH) indicate hydroboration-oxidation of the terminal alkyne, giving an *anti*-Markovnikov addition of H and OH across the alkyne. The resulting enol is not isolated, because upon its formation, it undergoes tautomerization to give an aldehyde:

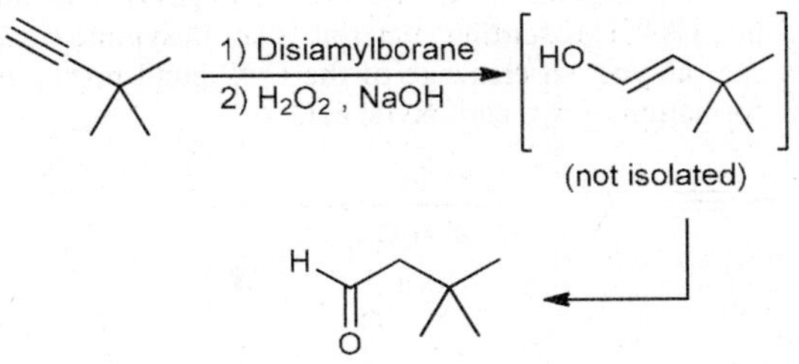

1) Disiamylborane

2) H_2O_2 , NaOH

HO

(not isolated)

H

O

(c) The reagents (R_2BH, followed by H_2O_2 and NaOH) indicate hydroboration-oxidation of the terminal alkyne, giving addition of H and OH. Since the alkyne is symmetrical, regiochemistry is not relevant in this case. The resulting enol is not isolated, because upon its formation, it undergoes tautomerization to give a ketone:

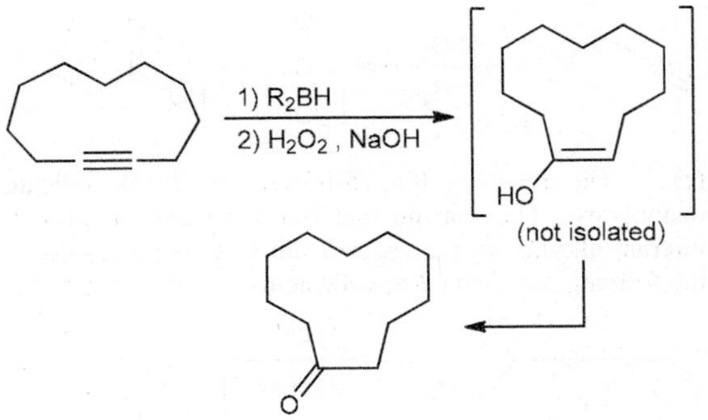

1) R_2BH

2) H_2O_2 , NaOH

HO

(not isolated)

O

9.21.

(a) The desired product is an aldehyde, which can be prepared from the corresponding terminal alkyne, shown here:

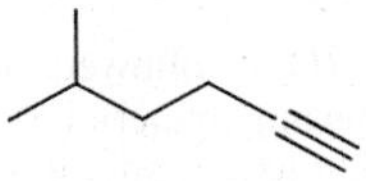

(b) The desired product is not an aldehyde, but it can be made directly from the following internal alkyne. This alkyne is symmetrical, so only one regiochemical outcome is possible:

The following alkyne would NOT be a suitable starting material, because hydration of this alkyne would produce a mixture of two ketones:

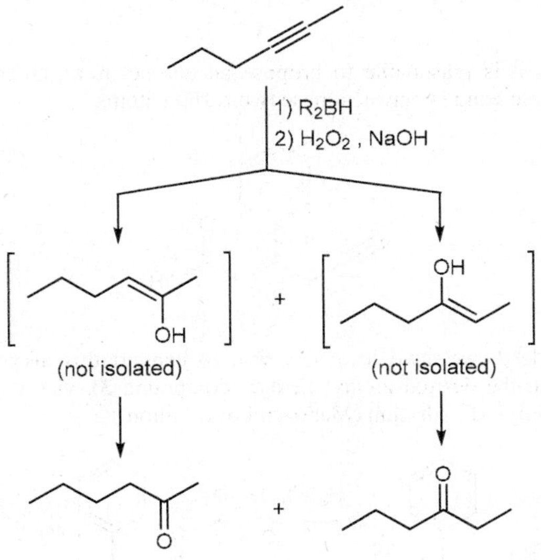

1) R_2BH

2) H_2O_2 , NaOH

OH

OH

(not isolated)

(not isolated)

O

O

(c) The desired product is an aldehyde, which can be prepared from the corresponding terminal alkyne, shown here:

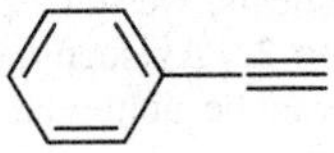

9.22.

(a) The starting material is a terminal alkyne and the product is a methyl ketone. This transformation requires a Markovnikov addition, which can be achieved via an acid-catalyzed hydration in the presence of mercuric sulfate.

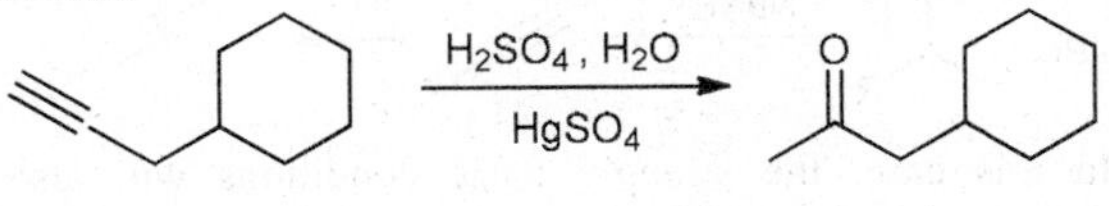

H_2SO_4 , H_2O

$HgSO_4$

O

(b) The starting material is a terminal alkyne and the product is an aldehyde. This transformation requires an *anti*-Markovnikov addition, which can be achieved via hydroboration-oxidation.

1) R_2BH

(9-BBN or disiamylborane)

2) H_2O_2 , NaOH

O

H

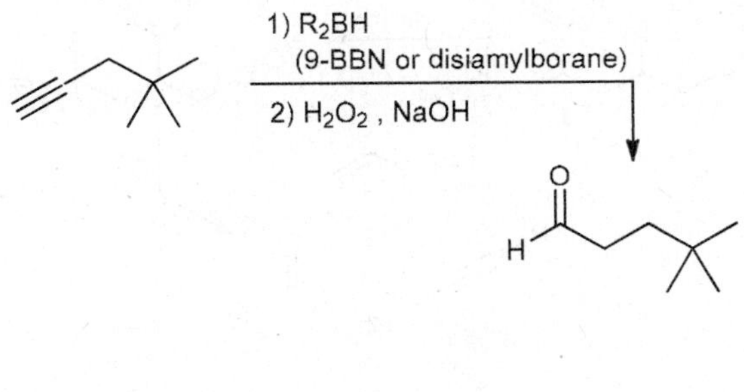

9.23. Compounds **1** and **3** have the same carbon skeleton. Only the identity and location of the functional groups have been changed. In compound **1**, the following two highlighted positions are functionalized (each is attached to Br):

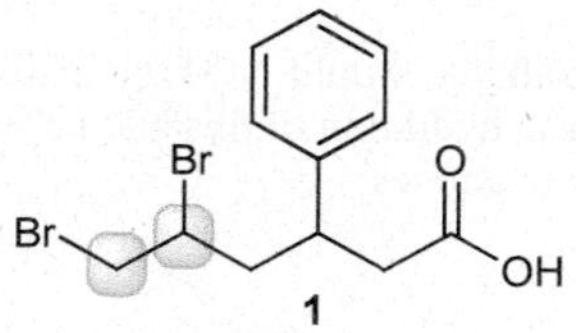

So it is reasonable to propose an alkyne in which the triple bond is between these two carbon atoms:

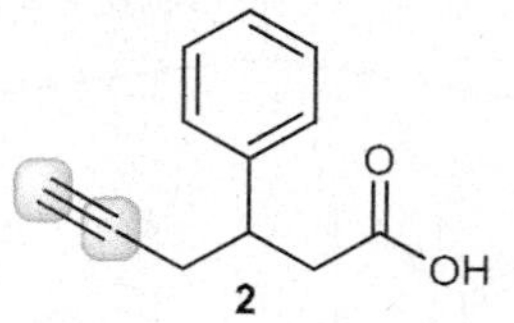

Indeed, we have learned a way to convert this alkyne into the desired methyl ketone (compound **3**), via acid-catalyzed hydration (Markovnikov addition):

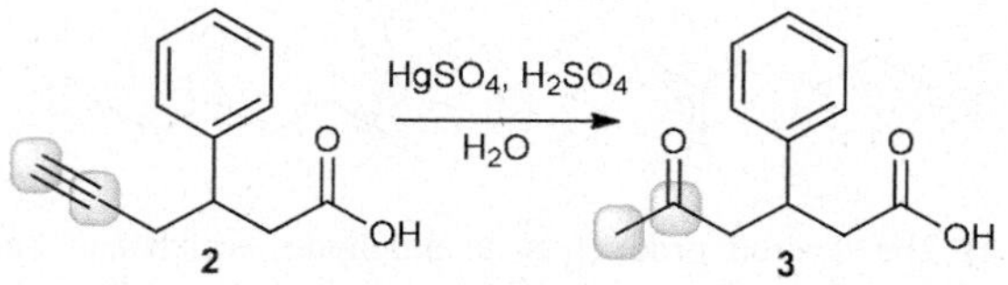

To complete the synthesis, we must propose reagents for the conversion of **1** to **2**. As seen earlier in the chapter, this transformation can be achieved via two successive E2 reactions, requiring a strong base such as $NaNH_2$. For this type of process, excess base is generally required. As shown below, the initial product of this general process is an alkynide ion, which is protonated upon treatment with a mild acid to give a terminal alkyne:

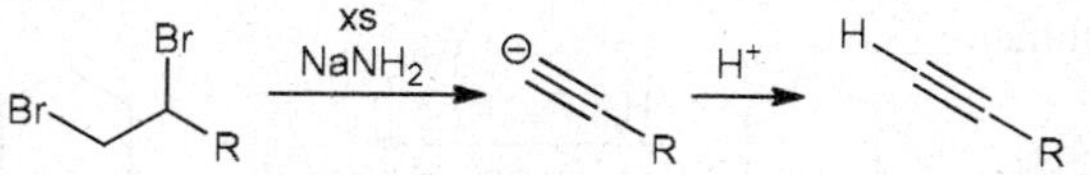

In this case, the strongly basic conditions will also deprotonate the carboxylic acid to give a carboxylate group. This carboxylate group is protonated (together with the alkynide ion) upon treatment with a mild acid:

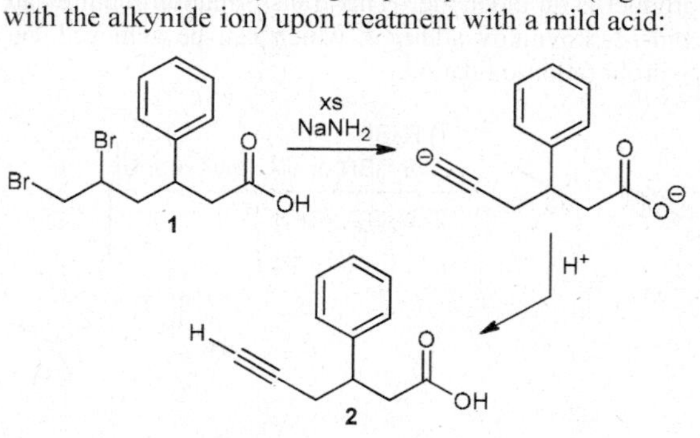

In summary, the following reagents can be used to convert **1** into **3**:

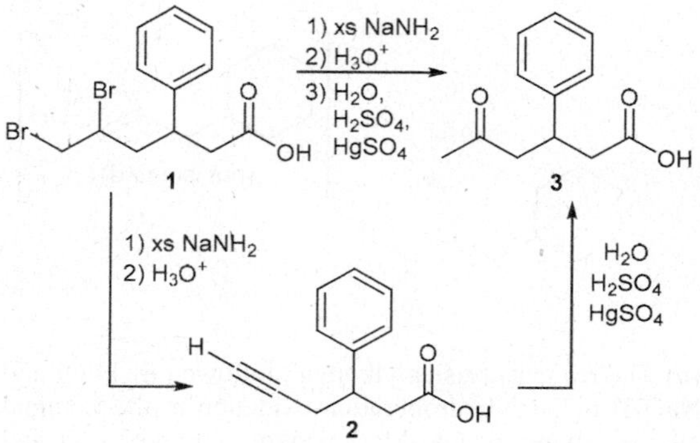

9.24.

(a) The reagents (O_3, followed by H_2O) indicate ozonolysis. The starting material is an unsymmetrical, internal alkyne, so cleavage of the C≡C bond results in the formation of two carboxylic acids:

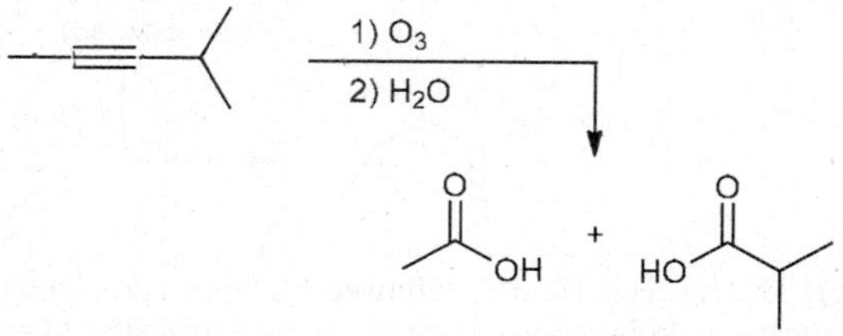

(b) The reagents (O_3, followed by H_2O) indicate ozonolysis. The starting material is a terminal alkyne, so cleavage of the C≡C bond results in the formation of a carboxylic acid and carbon dioxide (CO_2):

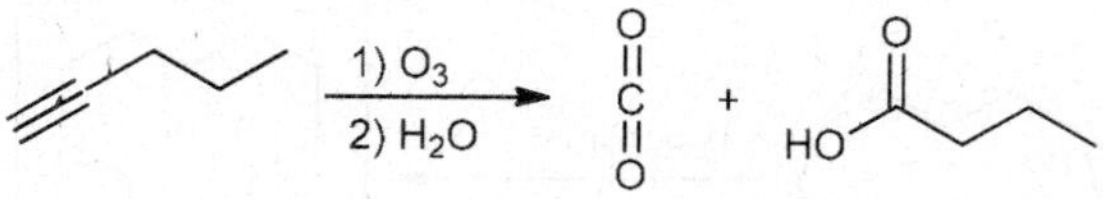

(c) The reagents (O_3, followed by H_2O) indicate ozonolysis. The starting material is an unsymmetrical, internal alkyne, so cleavage of the C≡C bond results in the formation of two carboxylic acids:

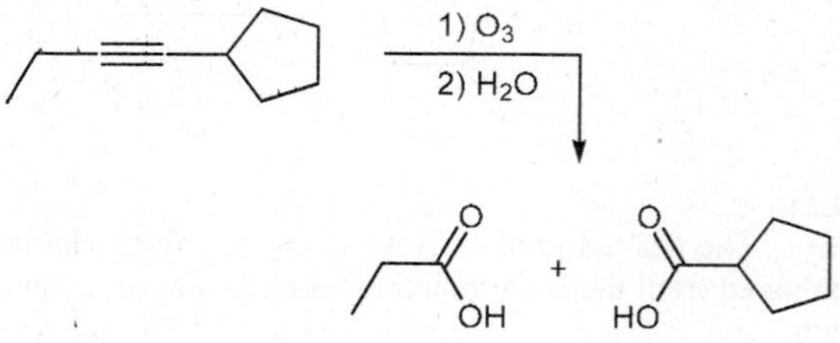

(d) The reagents (O_3, followed by H_2O) indicate ozonolysis. The starting material is a cycloalkyne, so cleavage of the C≡C bond results in the formation of an acyclic compound with two carboxylic acid groups:

1) O_3
2) H_2O

HO OH

9.25. If ozonolysis produces only one product, then the starting alkyne must be symmetrical. There is only one symmetrical alkyne with the molecular formula C_6H_{10}:

1) O_3
2) H_2O
2 OH

9.26. If ozonolysis produces a carboxylic acid and carbon dioxide, then the starting alkyne must be terminal. There is only one terminal alkyne with the molecular formula C_4H_6:

C_4H_6 1) O_3 2) H_2O OH + O=C=O

When this alkyne is treated with aqueous acid in the presence of mercuric sulfate, the alkyne is expected to undergo a Markovnikov addition of H and OH, generating an enol. This enol is not isolated, because upon its formation, it undergoes tautomerization to give a methyl ketone:

H_2O
H_2SO_4
$HgSO_4$
OH
(not isolated)

9.27.

(a) We begin by drawing the starting material (acetylene) and the product (1-hexyne):

H—≡—H ? → Acetylene 1-Hexyne

The desired transformation involves a single alkylation process, and a logical retrosynthesis reveals that a four-carbon alkyl halide is needed:

Br δ+ + Electrophile Nucleophile

Now let's draw the forward process. The alkylation is achieved by treating acetylene with sodium amide, followed by 1-bromobutane, as shown here:

1) $NaNH_2$
2) Br

(b) We begin by drawing the starting material (acetylene) and the product (2-hexyne):

H—≡—H ? → Acetylene 2-Hexyne

The transformation of acetylene to an internal alkyne involves two, successive alkylation processes. We can do a retrosynthesis to determine which alkyl halides are needed. The two groups can be installed in either order, and therefore can be disconnected in either order:

+ I—CH_3 δ+
Nucleophile Electrophile

Br δ+ +
Electrophile Nucleophile

Now let's draw the forward process. We must install a methyl group and a propyl group. The propyl group is installed by treating acetylene with sodium amide, followed by 1-bromopropane. And the methyl group is installed in a similar way (upon treatment with sodium amide, followed by methyl iodide). The following shows installation of the propyl group followed by installation of the methyl group:

1) $NaNH_2$
2) Br
3) $NaNH_2$
4) MeI

(c) We begin by drawing the starting material (acetylene) and the product (2,7-dimethyl-4-octyne):

H—≡—H ? → Acetylene

1 2 3 4 5 6 7 8
2,7-Dimethyl-4-octyne

The transformation of acetylene to an internal alkyne involves two, successive alkylation processes. We can do a retrosynthesis to determine which alkyl halides are needed:

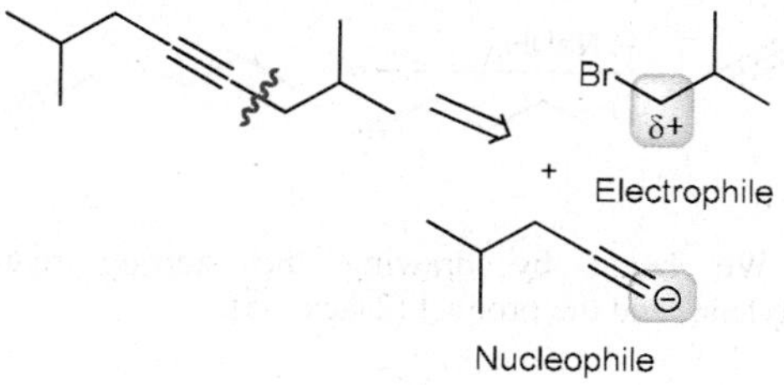

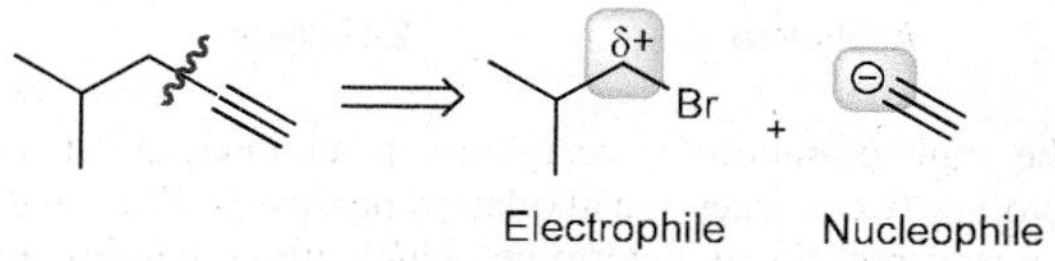

Br
δ+
Electrophile
Nucleophile

δ+
Br
Electrophile Nucleophile

Now let's draw the forward process. Two isobutyl groups must be installed, and each alkylation process must be performed separately. Each alkylation is achieved by treating the alkyne with sodium amide, followed by isobutyl bromide:

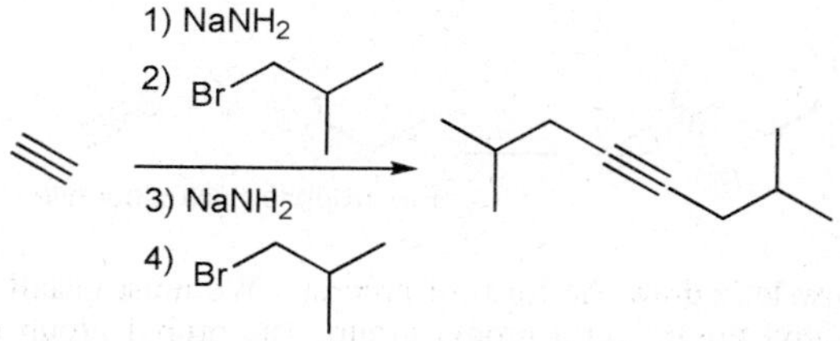

1) $NaNH_2$
2) Br
3) $NaNH_2$
4) Br

(d) We begin by drawing the starting material (acetylene) and the product (5-ethyl-1-heptyne):

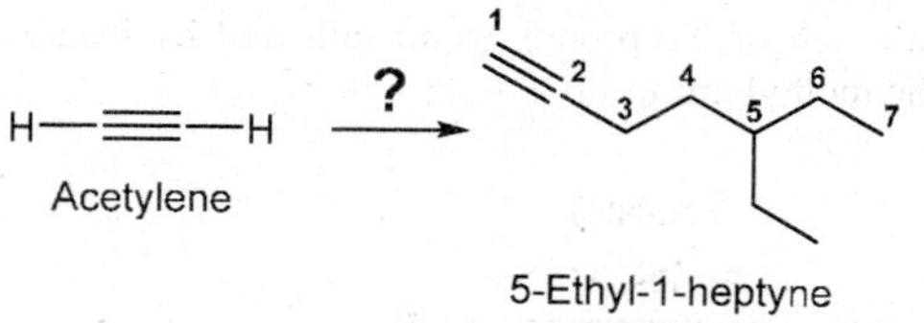

H—≡—H ? → 1 2 3 4 5 6 7

Acetylene

5-Ethyl-1-heptyne

The desired transformation involves a single alkylation process, and we can do a retrosynthesis to determine which alkyl halide is needed:

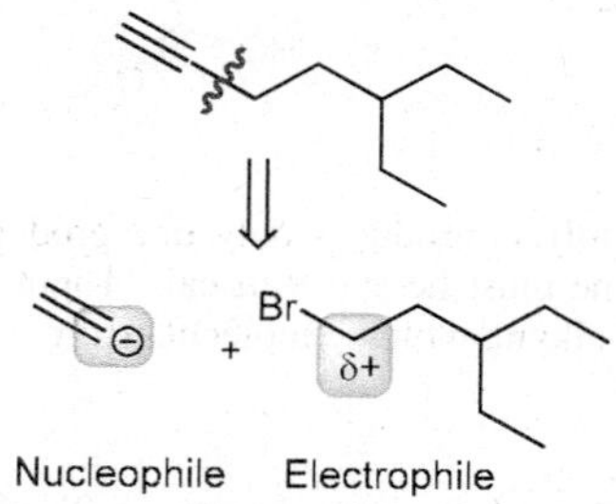

Br
δ+
Nucleophile Electrophile

Now let's draw the forward process. The alkylation is achieved by treating acetylene with sodium amide, followed by the alkyl bromide shown:

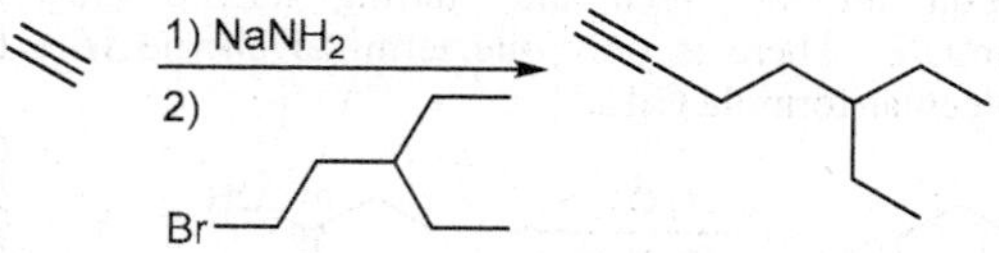

1) $NaNH_2$
2)
Br

(e) The desired transformation involves two, successive alkylation processes. Each alkylation is achieved by treating the alkyne with sodium amide, followed by benzyl iodide. Note that each alkylation process must be performed separately.

1) $NaNH_2$
2) I
3) $NaNH_2$
4) I

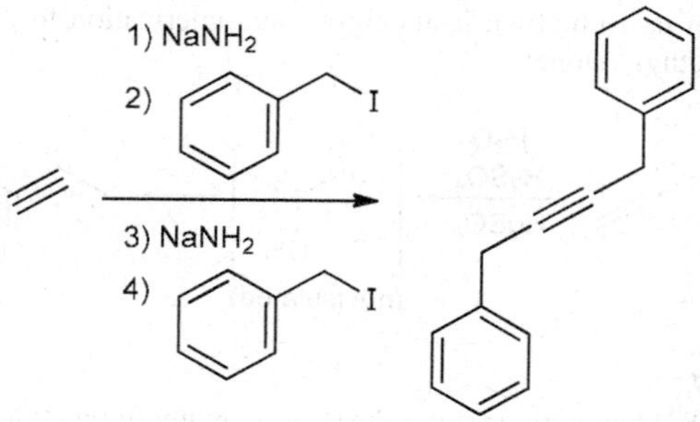

9.28. The carbon atoms present in compounds **1** and **2** completely account for the carbon skeleton of compound **3**, if we can connect the two highlighted carbon atoms:

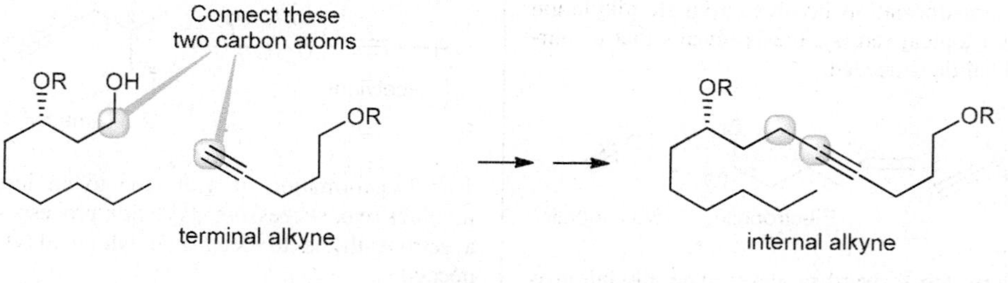

Connect these two carbon atoms

OR OH OR OR OR

terminal alkyne

internal alkyne

This process involves the conversion of a terminal alkyne (compound **2**) into an internal alkyne (compound **3**), which can be accomplished via an SN2 process. In order to perform the desired SN2 reaction, the starting materials must first be modified. The terminal alkyne (compound **2**) must be deprotonated with a strong base to convert it into a good nucleophile (using, for example, $NaNH_2$ to produce an alkynide ion):

$NaNH_2$

An alkynide ion

Also, the alcohol (compound **1**) must be converted into a substrate that contains a good leaving group, because hydroxide is not a good leaving group. For example, the alcohol can be treated with TsCl and pyridine to give a tosylate (see Section 7.10):

Bad leaving group

TsCl / py

Good leaving group

Treating this tosylate with the alkynide ion will result in an S_N2 reaction, giving compound **3**:

9.29.

(a) The desired transformation must lengthen the carbon chain by one carbon atom, and we need to make a *trans* alkene. To begin the retrosynthesis we ask, "What starting material is needed to make a *trans* alkene?" The *trans* alkene can be prepared by partial reduction of the corresponding alkyne, and the alkyne intermediate is suitable for making the required C–C bond disconnection:

$I-CH_3$ ($\delta+$) + (nucleophile)

Electrophile Nucleophile

Now let's draw the forward process. The first part of the synthesis (alkylation of the alkyne) can be accomplished by treating the starting alkyne with $NaNH_2$ followed by methyl iodide. Next, the alkyne is treated with sodium metal in liquid ammonia to give the *trans* alkene (dissolving metal reduction).

1) $NaNH_2$
2) MeI
3) Na, NH_3 (*l*)

1) $NaNH_2$
2) MeI

Na, NH_3 (*l*)

(b) The desired transformation must lengthen the carbon chain by one carbon atom, and we need to make a triple bond. The alkyne target molecule is suitable for making the required C–C bond disconnection, so that is how the retrosynthesis begins:

$I-CH_3$ ($\delta+$) + (nucleophile)

Electrophile Nucleophile

The required alkyne intermediate (1-propyne) can be prepared from a dibromide that leads back to the given starting material (1-propene):

Br Br

Now let's draw the forward process. The synthesis begins by converting the starting alkene into an alkyne upon treatment with Br_2, giving a dibromide, followed by treatment with excess sodium amide (and then water to protonate the resulting alkynide ion). At this point, the alkylation step can be accomplished via treatment with $NaNH_2$ followed by methyl iodide:

1) Br_2
2) xs $NaNH_2$
3) H_2O
4) $NaNH_2$
5) MeI

Br_2

1) $NaNH_2$
2) MeI

Br

Br

1) xs $NaNH_2$
2) H_2O

(racemic)

Note: The alkyne produced after step 3 does not need to be isolated and purified, and therefore, steps 3 and 4 can be omitted. That is, the synthesis can be presented like this:

1) Br_2
2) xs $NaNH_2$
3) MeI

Br_2

MeI

Br

Br

xs $NaNH_2$

(racemic)

(c) The desired transformation is a functional group interconversion (there is no change to the carbon skeleton). To begin the retrosynthesis we ask, "What starting material is needed to make an alcohol?" Selecting an alkene is a logical choice, because the alkene, in turn, can be prepared from the given alkyne starting material:

OH

Now let's draw the forward process. The synthesis begins by first reducing the alkyne with hydrogen gas in the presence of a poisoned catalyst, giving an alkene. This alkene can then be treated with dilute aqueous acid to give Markovnikov addition of H and OH, resulting in the desired product:

1) H_2, Lindlar's catalyst
2) dilute H_2SO_4

OH

H_2,
Lindlar's cat.

dilute H_2SO_4

(d) Much as we saw in the solution to the previous problem, the alcohol target molecule can be prepared from an alkene, which can be prepared from the starting alkyne:

HO

Now let's draw the forward process. The desired product can be made by first reducing the alkyne with hydrogen gas in the presence of a poisoned catalyst (to give an alkene), followed by addition of H and OH across the alkene. In this case, we need an *anti*-Markovnikov addition, so we employ a hydroboration-oxidation procedure:

1) H_2,
Lindlar's catalyst
2) $BH_3 \cdot THF$
3) H_2O_2, NaOH

HO

H_2,
Lindlar's cat.

1) $BH_3 \cdot THF$
2) H_2O_2, NaOH

(e) The desired transformation is a functional group interconversion (there is no change to the carbon skeleton). To begin the retrosynthesis we ask, "What starting material is needed to make an aldehyde?" Recall that aldehydes can be prepared by hydration of an alkyne:

H

O

The required alkyne intermediate (1-butyne) can be prepared from a dibromide that leads back to the given starting material (1-butene):

Br

Br

Now let's draw the forward process. The transformation begins by brominating the alkene to give a dibromide, followed by elimination with excess sodium amide (and then water to protonate the resulting alkynide ion) to give an alkyne. Hydroboration-oxidation of this alkyne affords the aldehyde target molecule:

1) Br_2
2) xs $NaNH_2$
3) H_2O
4) R_2BH
5) H_2O_2, NaOH

H

O

Br_2

1) R_2BH
2) H_2O_2,
NaOH

Br

1) xs $NaNH_2$
2) H_2O

Br (racemic)

(f) When we compare the starting and final compounds, we observe two differences: the triple bond is gone, and the carbon chain in the product is longer by two carbon atoms. The retrosynthetic analysis of the target molecule asks, "What starting material is required to make an alkane?" One reaction we've seen that generates a product *with no functional groups* is catalytic hydrogenation of an alkyne.

The alkyne intermediate is suitable for making the required C–C bond disconnection, so that is how the retrosynthesis continues:

Br δ+ + ⊖

Electrophile Nucleophile

Now let's draw the forward process. The first part of the synthesis (alkylation of the alkyne) can be accomplished by treating the starting alkyne with $NaNH_2$ followed by ethyl bromide. Finally, catalytic hydrogenation of the alkyne produces the desired alkane target molecule:

1) $NaNH_2$
2) EtBr
3) H_2, Pt

H_2, Pt

1) $NaNH_2$
2) EtBr

(g) When we compare the starting alkyne and the target compound, we observe three differences: the triple bond is gone, there is a carboxylic acid functional group, and the carbon chain in the product is *shorter* by one carbon atom. All these changes can be accomplished by an ozonolysis process (numbering the carbon atoms is a helpful way to keep track of the original carbon skeleton):

1) O_3
2) H_2O

OH

(h) The desired transformation must lengthen the carbon chain by two carbon atoms, and there are two Br atoms in the product (with a specific stereochemical relationship). To begin the retrosynthesis we ask, "What starting material is needed to make a dibromide with the required stereochemistry?" The Br atoms in the target molecule can be produced from the *trans* alkene shown *trans*-3-hexene), and the alkene can be prepared from the corresponding alkyne:

Br

Br

The alkyne intermediate is suitable for making the required C–C bond disconnection, so that is how the retrosynthesis continues:

δ+ I + ⊖

Electrophile Nucleophile

Now let's draw the forward process. The first part of the synthesis (alkylation of the alkyne) can be accomplished by treating the starting alkyne with $NaNH_2$ followed by ethyl iodide. After the alkylation, the resulting, symmetrical alkyne can be reduced to an alkene, followed by bromination. Since the last step (bromination) proceeds via an *anti* addition, the desired stereoisomer can only be obtained if the previous step (reduction of the alkyne) is performed to produce the *trans* alkene. That is, we must use a dissolving metal reduction, rather than hydrogenation with a poisoned catalyst.

1) $NaNH_2$
2) EtI
3) Na, NH_3 (*l*)
4) Br_2

Br Br

1) $NaNH_2$
2) EtI

Br_2

Na, NH_3 (*l*)

Notice that the product is a *meso* compound, which can be seen more clearly if we rotate about the central C–C bond, like this:

Br Br ≡ Br Br

(*meso*)

Internal plane of symmetry

9.30. The starting material has two carbon atoms, and the desired product has six carbon atoms:

O

Two carbon atoms Six carbon atoms

So, we must form new C–C bonds. This can be achieved with acetylene and two equivalents of ethyl bromide:

Br Ethyl bromide — Acetylene — Br Ethyl bromide

And the resulting alkyne can be converted into the product in just one step:

H_2O, H_2SO_4, $HgSO_4$

So, our goal is to convert ethylene into both acetylene and ethyl bromide:

ethyl bromide

acetylene

The synthesis of ethyl bromide can be achieved in just one step, by treating ethylene with HBr:

HBr

Acetylene can be made from ethylene via the following two step process:

Br_2 ; 1) xs $NaNH_2$ 2) H_2O

In summary, the following synthetic route converts ethylene into 3-hexanone:

Br_2 ; HBr ; 1) xs $NaNH_2$ 2) H_2O ; 1) $NaNH_2$ 2) ethyl bromide 3) $NaNH_2$ 4) ethyl bromide ; H_2O, H_2SO_4, $HgSO_4$

9.31. In this overall transformation, we need to add two hydrogen atoms (as the deuterium isotope, ^2H) and two hydroxyl groups to the two π bonds of the alkyne. We have also been told that an alkene is made, and then used, in each synthesis. It is reasonable, therefore, to begin by reducing the triple bond to a double bond (using deuterium instead of protium). Use of a dissolving metal reduction will generate a *trans* alkene, while hydrogenation with a poisoned catalyst will generate a *cis* alkene.

COOH — Na, ND_3 (*l*) → D, D, COOH ; D_2, Lindlar's catalyst → D, D, COOH

Dihydroxylation of these alkenes will provide diols; however, care must be taken to produce the desired stereochemical outcome. An *anti* addition of two OH groups to the *trans* alkene will produce the desired product (as a mixture of enantiomers). *Anti*-dihydroxylation can be accomplished upon treatment with a peroxyacid, to generate an epoxide intermediate, followed by aqueous acid to generate the diol.

D, D, COOH — 1) RCO_3H 2) H_3O^+ → D, OH, HO, D, COOH + En

correct relative stereochemistry

However, the same reagents will not provide the proper stereochemical outcome when used with the *cis* alkene. This is more apparent after rotating the newly formed single bond in the product.

incorrect relative stereochemistry

Alternatively, a *syn* addition of two OH groups to the *cis* alkene DOES provide the appropriate stereochemical outcome (again a single bond rotation in the product makes this easier to see). *Syn*-dihydroxylation can be accomplished using NMO and catalytic OsO_4.

correct relative stereochemistry

In summary, there are two ways to accomplish this overall transformation with the appropriate stereochemical outcome. One reduces the alkyne to the *trans* alkene, followed by an *anti* dihydroxylation. The other reduces the alkyne to the *cis* alkene, followed by a *syn* dihydroxylation.

9.32.

(a) We begin by identifying the parent. The longest chain is six carbon atoms, so the parent is hexyne. There are three substituents – all methyl groups (highlighted). Numbering the parent chain from either direction will place the triple bond at C3 (between C3 and C4), so we number in the direction that gives the lower number to the second substituent (2,2,5 rather than 2,5,5):

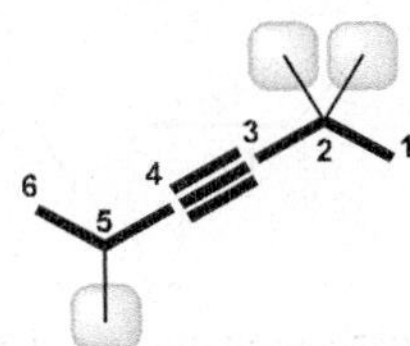

2,2,5-Trimethyl-3-hexyne

(b) We begin by identifying the parent. The chain is six carbon atoms, so the parent is hexyne. There are two substituents – both chloro groups. The parent chain is numbered to give the triple bond the lower number, C2 (because it is between C2 and C3). According to this numbering scheme, the chlorine atoms are both at C4:

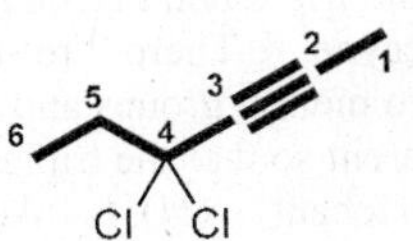

4,4-Dichloro-2-hexyne

(c) We begin by identifying the parent. The chain is six carbon atoms, so the parent is hexyne. There are no substituents. We must include a locant that identifies the position of the triple bond. The parent chain is numbered so that the triple bond is assigned the lowest possible locant ("1" indicates that the triple bond is located between C1 and C2).

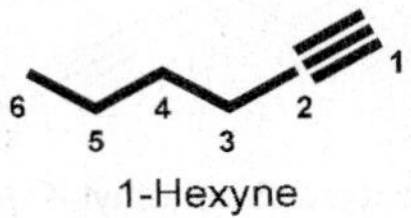

1-Hexyne

(d) We begin by identifying the parent. The longest chain is four carbon atoms, so the parent is butyne.

There are two substituents (highlighted) – a methyl group and a bromo group. The parent chain is numbered so that the triple bond is assigned the lowest possible locant ("1" indicates that the triple bond is located between C1 and C2). According to this numbering scheme, both substituents are located at C3. They are alphabetized in the name, so "bromo" precedes "methyl."

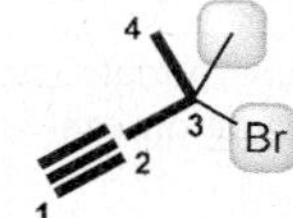

3-Bromo-3-methyl-1-butyne

(e) We begin by identifying the parent. The longest chain containing the triple bond has ten carbon atoms, so the parent is decyne. There are two substituents (highlighted) – a methyl group and a *sec*-butyl group. We number the parent so that the triple bond receives the lowest possible locant (C3). According to this numbering scheme, the *sec*-butyl group is located at C5 and the methyl group is located at C2. These substituents are arranged alphabetically in the name (butyl before methyl). If we use the systematic name for the *sec*-butyl group, (1-methylpropyl), then the methyl group comes first alphabetically:

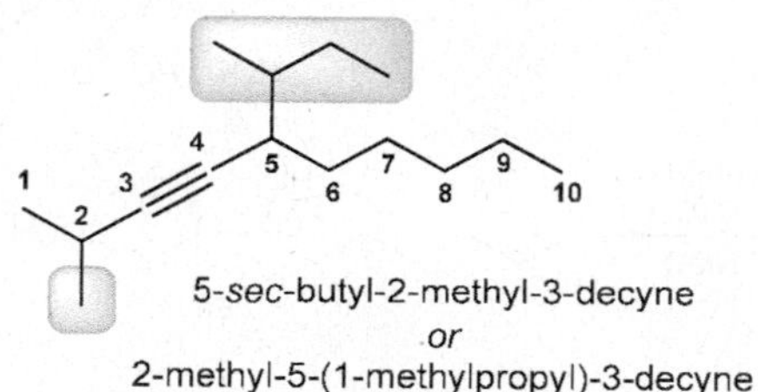

5-*sec*-butyl-2-methyl-3-decyne
or
2-methyl-5-(1-methylpropyl)-3-decyne

(f) We begin by identifying the parent. The longest chain containing the triple bond has ten carbon atoms, so the parent is decyne. There are three substituents (highlighted) – two methyl groups and an isobutyl group. We number the parent so that the triple bond receives the lowest possible locant (C4). According to this numbering scheme, the isobutyl group is located at C6 and the two methyl groups are located at C2. These substituents are arranged alphabetically in the name (isobutyl before methyl). If we use the systematic name for the isobutyl group, (2-methylpropyl), then the methyl group comes first alphabetically:

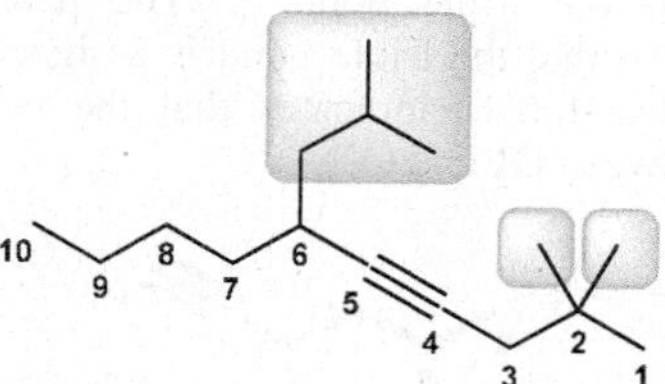

6-isobutyl-2,2-dimethyl-4-decyne
or
2,2-dimethyl-6-(2-methylpropyl)-4-decyne

9.33.

(a) The parent (heptyne) indicates a chain of seven carbon atoms. The triple bond is between C2 and C3, and there are no substituents.

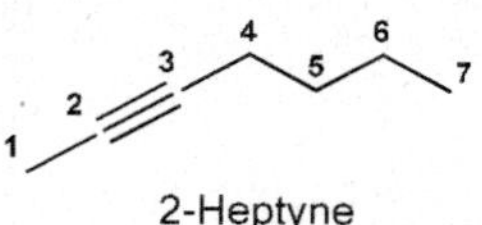

2-Heptyne

(b) The parent (octyne) indicates a chain of eight carbon atoms, with the triple bond located between C4 and C5. There are two methyl groups (highlighted), both located at C2.

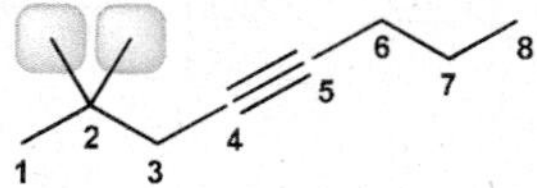

2,2-Dimethyl-4-octyne

(c) The parent (cyclodecyne) indicates a ring of ten carbon atoms. The numbering system is assigned such that the triple bond is between C1 and C2. There are two substituents (ethyl groups), each of which is located at C3:

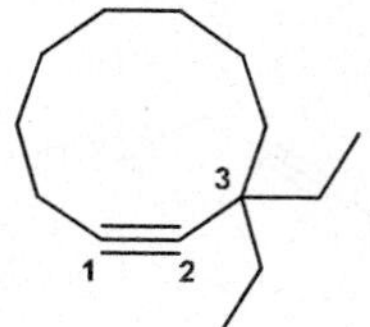

3,3-Diethylcyclodecyne

9.34. The starting material has both a double bond and a triple bond. When hydrogenation is performed in the presence of a poisoned catalyst (such as Lindlar's catalyst), only the triple bond is reduced (via *syn* addition to give a *cis* alkene). However, when Pt is used as the catalyst, both the double bond and the triple bond are reduced, giving an alkane:

H_2
Lindlar's cat.

H_2
Pt

9.35.

(a) The starting carbanion (C^-) is a strong base, and acetylene has an acidic proton. The resulting proton transfer step gives an acetylide ion and butane. When we compare the acids (highlighted below), we find a massive difference in pK_a values (See Table 9.2). The difference between acetylene (pK_a = 25) and butane (pK_a = 50) is 25 units, which represents a difference of 25 orders of magnitude. That is, acetylene is 10^{25} times more acidic than butane. With such a large difference in pK_a values, the reaction is considered to be irreversible.

stronger acid (pK_a = 25)

weaker acid (pK_a = 50)

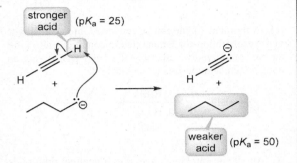

(b) Hydride (H^-) is a strong base, and the terminal alkyne has an acidic proton. The resulting proton transfer step gives an alkynide ion and hydrogen gas (H_2). We compare the pK_a values of the acids (highlighted below) using the values given in Table 9.2. Acetylene (pK_a = 25) is more acidic than H_2 (pK_a = 35). As such, the equilibrium favors the weaker acid (H_2).

stronger acid (pK_a ~ 25)

weaker acid (pK_a = 35)

In practice, H_2 bubbles out of solution as a gas, and as a result, the reaction is irreversible and proceeds to completion (Le Chatelier's principle).

9.36.

(a) When treated with aqueous acid (in the presence of mercuric sulfate), the terminal alkyne undergoes Markovnikov addition of H and OH across the alkyne, giving an enol. The enol is not isolated, because upon its formation, it undergoes tautomerization to give a methyl ketone:

H_2SO_4, H_2O / $HgSO_4$ (not isolated)

(b) The reagents (R_2BH, followed by H_2O_2 and NaOH) indicate hydroboration-oxidation of the terminal alkyne, giving an *anti*-Markovnikov addition of H and OH across the alkyne. The resulting enol is not isolated, because upon its formation, it undergoes tautomerization to give an aldehyde:

1) R_2BH 2) H_2O_2, NaOH (not isolated)

(c) When treated with an excess of HBr, a terminal alkyne undergoes two successive addition reactions, each of which proceeds in a Markovnikov fashion, giving the following geminal dibromide:

HBr (xs)

(d) When treated with one equivalent of HCl, a terminal alkyne undergoes an addition reaction that proceeds in a Markovnikov fashion, giving the following vinyl chloride:

HCl (1 eq)

(e) When treated with an excess of Br_2, the alkyne undergoes two successive addition reactions, giving the following tetrabromide:

Br_2 (xs) / CCl_4

(f) These reagents indicate an alkylation process. In step 1, the alkyne is deprotonated by the strong base (H_2N^-) to give an alkynide ion. In step 2, this alkynide ion is used as a nucleophile to attack methyl iodide (in an S_N2 reaction), thereby installing a methyl group:

1) $NaNH_2$, NH_3 2) MeI

(g) Since platinum is used as the catalyst (rather than a poisoned catalyst), hydrogenation of the alkyne gives an alkane:

H_2 / Pt

9.37.

(a) Markovnikov addition of addition of H and Cl can be achieved by treating the alkyne with one equivalent of HCl:

HCl (1 eq)

(b) Reduction of a terminal alkyne cannot be achieved with a dissolving metal reduction. Instead, the desired transformation can be achieved by performing a hydrogenation reaction in the presence of a poisoned catalyst, such as Lindlar's catalyst:

H_2 Lindlar's catalyst

(c) Ozonolysis of the alkyne achieves cleavage of the C≡C bond to give a carboxylic acid (and carbon dioxide as a by-product):

1) O_3 2) H_2O (+CO_2)

(d) A terminal alkyne can be converted into a methyl ketone upon treatment with aqueous acid in the presence of mercuric sulfate. These conditions allow for Markovnikov addition of H and OH, giving an enol, which tautomerizes to the methyl ketone:

H_2SO_4, H_2O $HgSO_4$

(e) A terminal alkyne can be converted into an aldehyde via hydroboration-oxidation. These conditions allow for an *anti*-Markovnikov addition of H and OH, giving an enol, which tautomerizes to the aldehyde:

1) R_2BH 2) H_2O_2, NaOH

(f) An alkyne can be converted to a geminal dibromide via two successive addition reactions with HBr. Markovnikov addition is required, so we use excess HBr without peroxides.

HBr (xs)

(g) Reduction of the alkyne to an alkane can be achieved by performing hydrogenation in the presence of a catalyst, such as platinum (not a poisoned catalyst):

H_2 Pt

9.38.

(a) As seen in Table 9.2, methoxide is not a sufficiently strong base to deprotonate a terminal alkyne, because the conjugate acid of methoxide (methanol, pK_a = 16) is a stronger acid than a terminal alkyne (pK_a ~ 25).

$$R{-}C{\equiv}C{-}H + NaOCH_3 \rightleftharpoons R{-}C{\equiv}C{:}^{\ominus}\ Na^{\oplus} + CH_3OH$$

Weaker acid (pK_a ~ 25) — Stronger acid (pK_a = 16)

(b) As seen in Table 9.2, hydride (H^-) is a sufficiently strong base to deprotonate a terminal alkyne, because the conjugate acid of hydride (H_2, pK_a = 35) is a weaker acid than a terminal alkyne (pK_a ~ 25).

$$R{-}C{\equiv}C{-}H + NaH \longrightarrow R{-}C{\equiv}C{:}^{\ominus}\ Na^{\oplus} + H_2$$

Stronger acid (pK_a ~ 25) — Stronger acid (pK_a = 35)

(c) As seen in Table 9.2, butyllithium is a sufficiently strong base to deprotonate a terminal alkyne, because the conjugate acid (butane, pK_a = 50) is a weaker acid than a terminal alkyne (pK_a ~ 25).

$$R{-}C{\equiv}C{-}H + Bu^{\ominus}\ Li^{\oplus} \longrightarrow R{-}C{\equiv}C{:}^{\ominus}\ Li^{\oplus} + \text{butane}$$

Stronger acid (pK_a ~ 25) — Weaker acid (pK_a = 50)

(d) As seen in Table 9.2, hydroxide is not a sufficiently strong base to deprotonate a terminal alkyne, because the conjugate acid of hydroxide (water, pK_a = 15.7) is a stronger acid than a terminal alkyne (pK_a ~ 25).

$$\underset{\substack{\text{Weaker acid} \\ (\text{p}K_a\ \sim 25)}}{\text{R}-\text{C}\equiv\text{C}-\text{H}} + \text{NaOH} \rightleftharpoons \text{R}-\text{C}\equiv\text{C}:^{\ominus}\ \text{Na}^{\oplus} + \underset{\substack{\text{Stronger acid} \\ (\text{p}K_a\ 15.7)}}{\text{H}_2\text{O}}$$

(e) As seen in Table 9.2, the amide ion (H_2N^-) is a sufficiently strong base to deprotonate a terminal alkyne, because the conjugate acid (NH_3, pK_a = 38) is a weaker acid than a terminal alkyne (pK_a ~ 25).

$$\underset{\substack{\text{Stronger acid} \\ (\text{p}K_a \sim 25)}}{\text{R}-\text{C}\equiv\text{C}-\text{H}} + \text{NaNH}_2 \longrightarrow \text{R}-\text{C}\equiv\text{C}:^{\ominus}\ \text{Na}^{\oplus} + \underset{\substack{\text{Weaker acid} \\ (\text{p}K_a = 38)}}{\text{NH}_3}$$

9.39.

(a) Yes, these compounds represent a pair of keto-enol tautomers. The only difference in the two structures is the location of a proton and the location of the π bond (C=C vs. C=O, and note that the molecule has also been rotated!). The two compounds can be interconverted by a two-step mechanism that is typically catalyzed by acid: protonation of the π bond (to give a resonance-stabilized cation), followed by deprotonation of the oxygen atom.

(b) Yes, these compounds represent a pair of keto-enol tautomers. The only difference in the two structures is the location of a proton and the location of the π bond (C=C vs. C=O). The two compounds can be interconverted by a two-step mechanism that is typically catalyzed by acid: protonation of the π bond (to give a resonance-stabilized cation), followed by deprotonation of the oxygen atom.

9.40. The difference between oleic acid and elaidic acid is the configuration of the C=C double bond. Oleic acid has the *cis* configuration, while elaidic acid has the *trans* configuration. Each of these stereoisomers can be obtained via reduction of the corresponding alkyne. Hydrogenation in the presence of a poisoned catalyst affords the *cis* alkene,

Stearolic Acid —(H_2, Lindlar's catalyst)→ Oleic Acid

while a dissolving metal reduction gives the *trans* alkene:

Stearolic Acid —(Na, NH_3 (*l*))→ Elaidic Acid

Note that, under these conditions, the carboxylic acid group will be deprotonated, so a workup will be required to protonate the carboxylate ion and regenerate the carboxylic acid group.

9.41.

(a) Upon treatment with excess sodium amide, the geminal dibromide undergoes elimination (twice) followed by deprotonation, to give an alkynide ion. This alkynide ion then undergoes alkylation when treated with ethyl chloride (S_N2 process). Finally, hydrogenation in the presence of a poisoned catalyst affords the *cis* alkene:

Br Br
1) excess $NaNH_2$
2) EtCl
3) H_2, Lindlar's catalyst

excess $NaNH_2$

Cl

H_2, Lindlar's catalyst

(b) Acetylene undergoes alkylation upon deprotonation with sodium amide, followed by treatment with methyl iodide (S_N2 process). The resulting alkyne (1-propyne) then undergoes hydroboration-oxidation when treated with a dialkylborane (R_2BH) followed by H_2O_2 and NaOH, giving an *anti*-Markovnikov addition of H and OH across the alkyne. The resulting enol is not isolated, because upon its formation, it undergoes tautomerization to give an aldehyde:

1) $NaNH_2$
2) MeI
3) R_2BH
4) H_2O_2 , NaOH

1) $NaNH_2$
2) MeI

1) R_2BH
2) H_2O_2 , NaOH

(c) Acetylene undergoes alkylation upon deprotonation with sodium amide, followed by treatment with ethyl iodide (S_N2 process). The resulting alkyne (1-butyne) then undergoes acid-catalyzed hydration when treated with aqueous acid in the presence of mercuric sulfate, giving a Markovnikov addition of H and OH across the alkyne. The resulting enol is not isolated, because upon its formation, it undergoes tautomerization to give a methyl ketone:

1) $NaNH_2$
2) EtI
3) H_2SO_4 , H_2O, $HgSO_4$

1) $NaNH_2$
2) EtI

$HgSO_4$
H_2SO_4 , H_2O

(d) Acetylene undergoes alkylation upon deprotonation with sodium amide, followed by treatment with methyl iodide (S_N2 process). The resulting alkyne (1-propyne) then undergoes alkylation, once again, upon treatment with sodium amide, followed by ethyl iodide. The first alkylation process installs a methyl group, while the second process installs an ethyl group. Finally, a dissolving metal reduction converts the alkyne to a *trans* alkene:

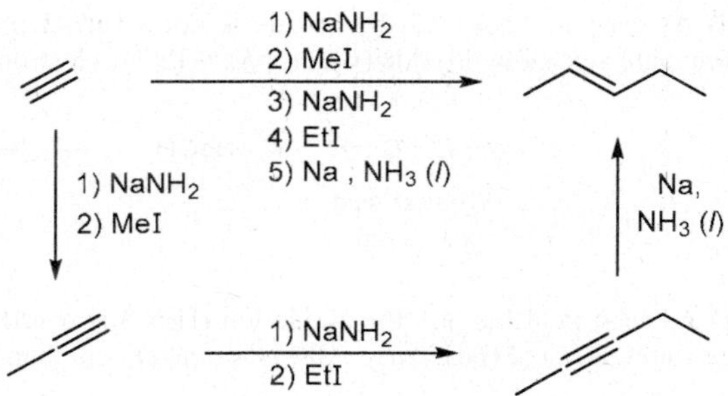

9.42. When *(R)*-4-bromohept-2-yne is treated with H_2 in the presence of Pt, the asymmetry is destroyed and C4 is no longer a chiral center:

Br — H_2 / Pt — Br

not a chiral center

The product, therefore, is optically inactive. This is not the case for *(R)*-4-bromohex-2-yne:

Br — H_2 / Pt — Br

a chiral center

In this case, the chiral center is retained throughout the reaction, so the product is optically active.

9.43. We are looking for an alkyne that will undergo hydrogenation to give the following alkane:

3-Ethylpentane

In order to produce this alkane, the starting alkyne must have the same carbon skeleton as this alkane. There is only one such alkyne (3-ethyl-1-pentyne, shown below), because the triple bond cannot be placed between C2 and C3 of the skeleton (as that would make the C3 position pentavalent, and carbon cannot accommodate more than four bonds):

H_2 / Pt

3-Ethyl-1-pentyne — 3-Ethylpentane

9.44.

(a) This process is a dissolving metal reduction, so it follows the four steps shown in Mechanism 9.1. In the first step, a single electron is transferred from the sodium atom to the alkyne, generating a radical anion intermediate. This intermediate is then protonated (ammonia is the proton source), generating a radical intermediate. A single electron is transferred once again from a sodium atom to the radical intermediate, generating an anion, which is then protonated in the final step of the mechanism. Be sure to use the correct arrows for each step of the mechanism. An ordinary arrow shows the movement of two electrons (e.g., proton transfer steps), while a fishhook arrow shows the movement of a single electron:

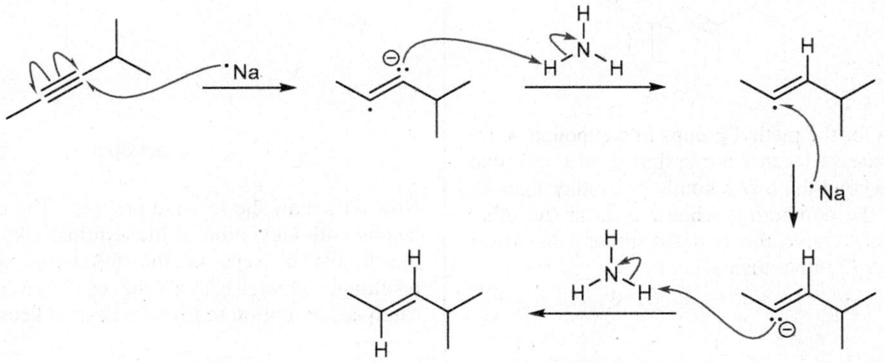

(b) The starting material has two enol groups, and the product has two ketone groups. As such, this transformation represents two tautomerization processes. Each tautomerization process must be shown separately, with two steps. In the first step, a double bond of one of the enol groups is protonated, generating a resonance-stabilized cation, which is then deprotonated (with water serving as the base) to generate the ketone. These two steps are then repeated for the other enol group, as shown:

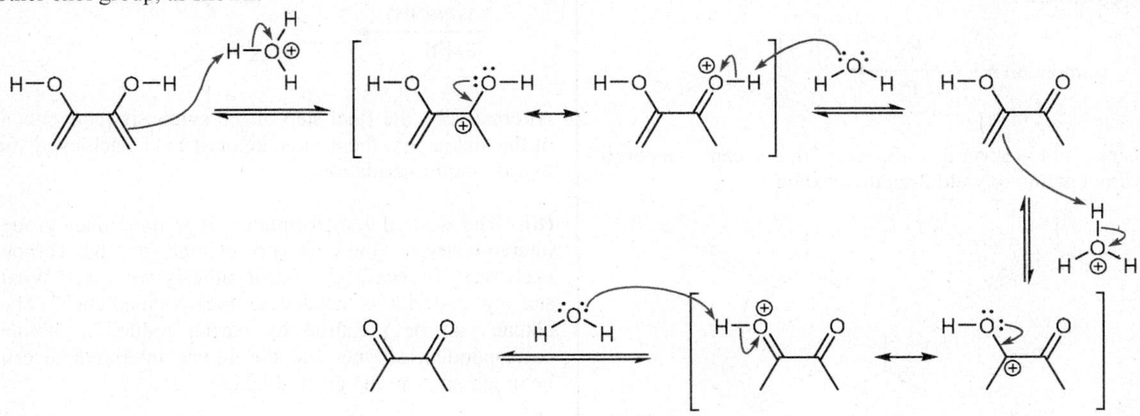

9.45. Treatment of the terminal alkyne with sodium amide results in the formation of alkynide ion, which then functions as a nucleophile in an S_N2 reaction. The stereochemical requirement for inversion determines the configuration of the chiral center in the product.

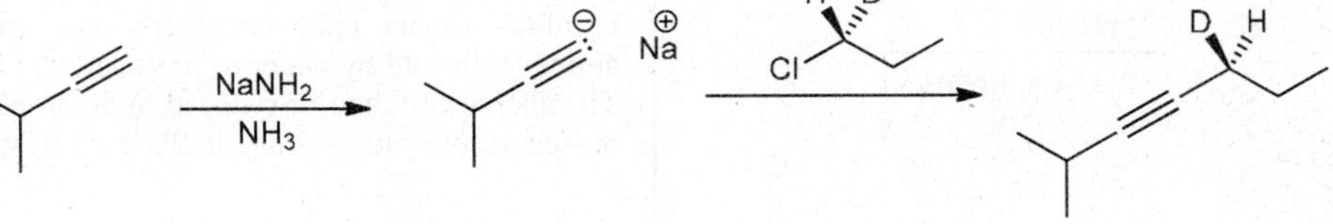

9.46.

(a) In order to produce 2,4,6-trimethyloctane, the starting alkyne must have the same carbon skeleton as this alkane. There is only one such alkyne (shown below), because the triple bond cannot be placed in any other location (as that would give a pentavalent carbon atom, which is not possible):

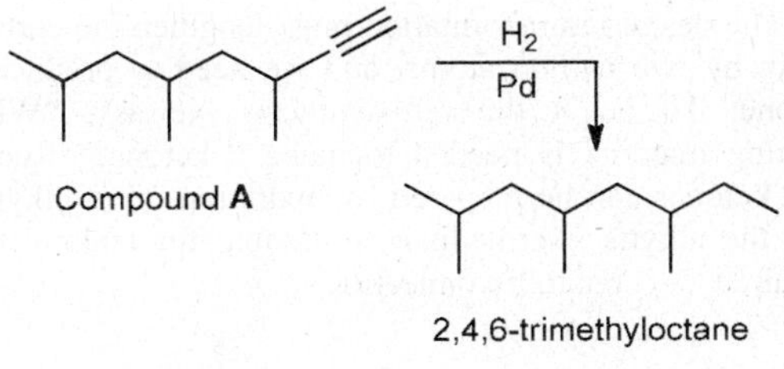

Compound A

2,4,6-trimethyloctane

(b) Compound **A** has two chiral centers, highlighted below:

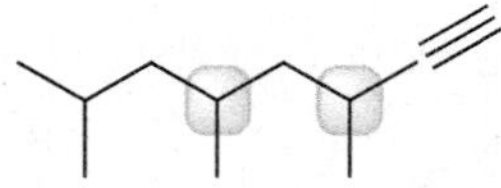

(c) The locants for the methyl groups in compound **A** are 3, 5, and 7, because locants are assigned in a way that gives the triple bond the lower number (1 rather than 7). In the alkane, the numbering scheme goes in the other direction, so as to give the first substituent the lower possible number (2 rather than 3)

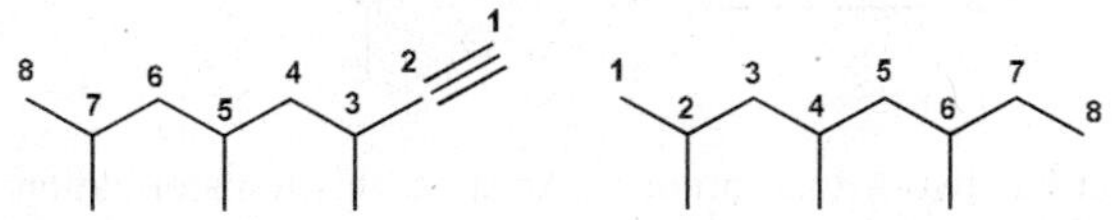

9.47. Hydrogenation of compound **A** produces 2-methylhexane:

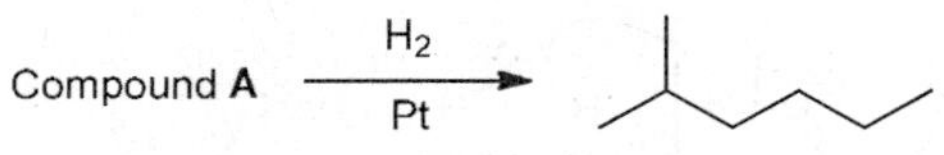

There are several alkynes that can undergo hydrogenation to yield 2-methylhexane.

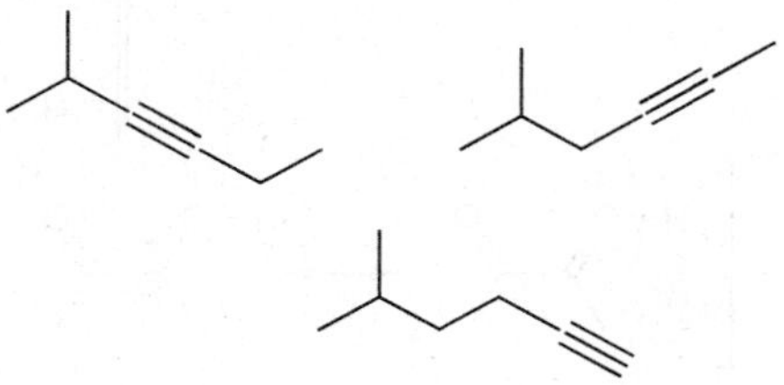

However, only one of these three possibilities will undergo hydroboration-oxidation to give an aldehyde. Specifically, the alkyne must be terminal in order to generate an aldehyde upon hydroboration-oxidation:

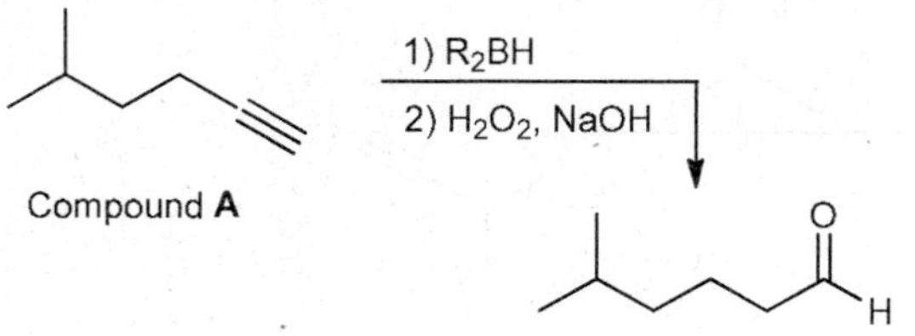

9.48.

(a) The desired transformation must lengthen the carbon chain by two carbon atoms, and we need to produce a ketone. To begin the retrosynthesis we ask, "What starting material is needed to make a ketone?" Recall that ketones can be prepared by hydration of an alkyne, and the alkyne intermediate is suitable for making the required C–C bond disconnection:

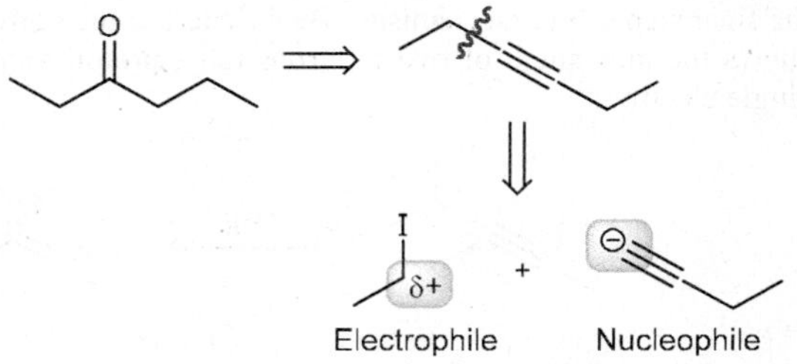

Now let's draw the forward process. The transformation begins with alkylation of the terminal alkyne, as seen in the first two steps of the following synthesis. The resulting, symmetrical alkyne can then undergo acid-catalyzed hydration to give the desired ketone.

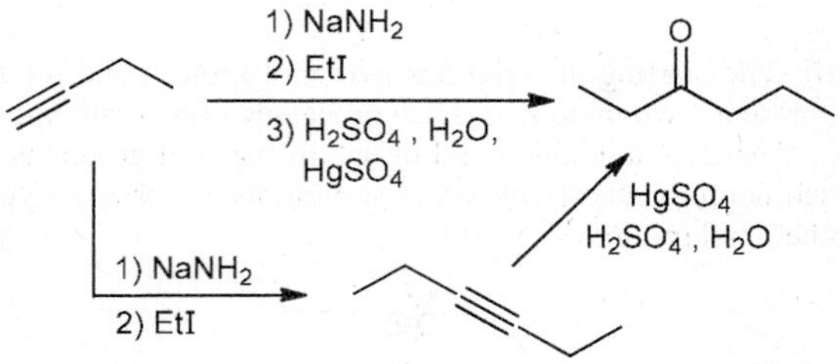

Alternatively, the final step of the synthesis (conversion of the alkyne into the desired ketone) can be achieved via hydroboration-oxidation.

(b) The desired transformation is a functional group interconversion (there is no change to the carbon skeleton). To begin the retrosynthesis we ask, "What starting material is needed to make an alkene?" The alkene can be prepared by partial reduction of the corresponding alkyne, and the alkyne intermediate can be prepared from the given dihalide:

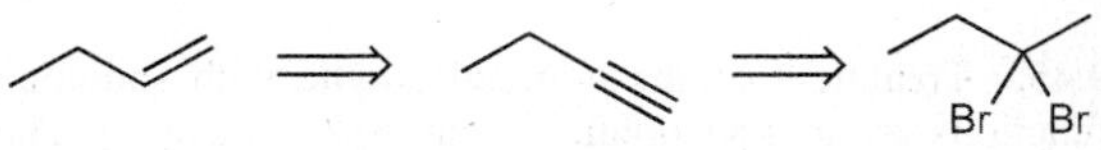

Now let's draw the forward process. The geminal dihalide starting material can be converted into a terminal alkyne upon treatment with excess sodium amide (followed by water to protonate the alkynide ion). The alkyne can then be reduced by hydrogenation with a poisoned catalyst, such as Lindlar's catalyst.

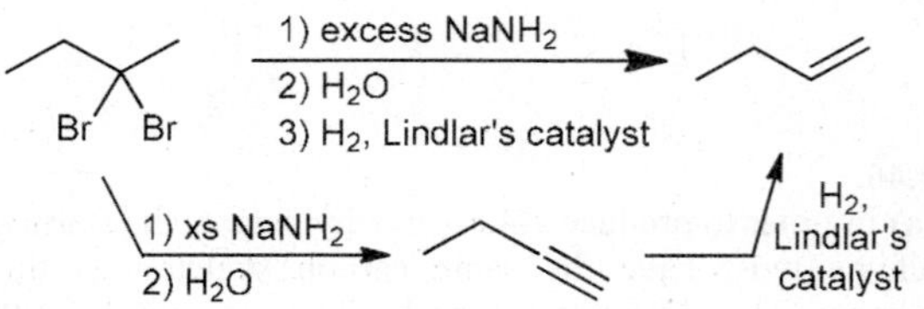

(c) The desired transformation must lengthen the carbon chain by one carbon atom, and we need to make a *trans* alkene. To begin the retrosynthesis we ask, "What starting material is needed to make a *trans* alkene?" The *trans* alkene can be prepared by partial reduction of the corresponding alkyne, and the alkyne intermediate is suitable for making the required C–C bond disconnection:

Nucleophile Electrophile

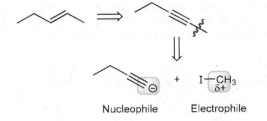

The retrosynthesis concludes by recognizing the required alkyne can be prepared from the given dihalide:

Now let's draw the forward process. To achieve the desired transformation, the geminal dihalide starting material can be converted into a terminal alkyne upon treatment with excess sodium amide (followed by water to protonate the alkynide ion). The next part of the synthesis (alkylation of the alkyne) can be accomplished via treatment with $NaNH_2$ followed by methyl iodide. Finally, the alkyne is treated with sodium in liquid ammonia to give the *trans* alkene (dissolving metal reduction).

1) excess $NaNH_2$
2) H_2O
3) $NaNH_2$
4) MeI
5) Na, NH_3

Note: The alkynide anion formed in step 1 can be used directly for the alkylation process (the neutral alkyne produced after step 2 does not need to be isolated and purified). In other words, step 2 (protonation) and step 3 (deprotonation) can be omitted, like this:

1) excess $NaNH_2$
2) MeI
3) Na, NH_3

(d) The desired transformation is a functional group interconversion (there is no change to the carbon skeleton). To begin the retrosynthesis we ask, "What starting material is needed to make a ketone?" Recall that ketones can be prepared by hydration of an alkyne, and the required alkyne can be prepared from the given dihalide:

Now let's draw the forward process. The geminal dihalide starting material can be converted into a terminal alkyne upon treatment with excess sodium amide (followed by water to protonate the alkynide ion). The resulting terminal alkyne can then undergo acid-catalyzed hydration in the presence of mercuric sulfate to give a methyl ketone:

1) excess $NaNH_2$
2) H_2O
3) H_2SO_4, H_2O, $HgSO_4$

(e) The desired transformation is a functional group interconversion (there is no change to the carbon skeleton). To begin the retrosynthesis we ask, "What starting material is needed to make a vinyl dibromide (with the specific stereochemistry shown)?" The target molecule can be prepared from an alkyne, and the required alkyne can be prepared from the given dihalide:

Now let's draw the forward process. The geminal dihalide starting material can be converted into a terminal alkyne upon treatment with excess sodium amide (followed by water to protonate the alkynide ion). The terminal alkyne can then be treated with one equivalent of bromine to give the desired product:

1) excess $NaNH_2$
2) H_2O
3) Br_2 (1 eq)

(f) The desired transformation is a functional group interconversion (there is no change to the carbon skeleton). To begin the retrosynthesis we ask, "What starting material is needed to make an alcohol?" Selecting an alkene is a logical choice, because the alkene can be prepared from an alkyne that, in turn, can be prepared from the given dihalide starting material:

Now let's draw the forward process. The synthesis begins by converting the dihalide into a terminal alkyne. This alkyne can then be reduced (in the presence of a poisoned catalyst) to give an alkene. This alkene can then be treated with dilute aqueous acid to give Markovnikov addition of H and OH, resulting in the desired product:

1) excess $NaNH_2$
2) H_2O
3) H_2, Lindlar's catalyst
4) dilute H_2SO_4

1) xs $NaNH_2$
2) H_2O

H_2, Lindlar's catalyst

dilute H_2SO_4

9.49. Treatment of 1,2-dichloropentane with excess sodium amide (followed by water to protonate the alkynide ion) gives 1-pentyne (compound **X**).

1) xs $NaNH_2$
2) H_2O

Compound **X**

Compound **X** undergoes acid-catalyzed hydration to give a methyl ketone, shown below:

H_2SO_4, H_2O
$HgSO_4$

9.50. Acetic acid has two carbon atoms, and carbon dioxide has one, so our starting alkyne has three carbon atoms:

$$H_3C{-}C{\equiv}C{-}H \xrightarrow[2)\ H_2O]{1)\ O_3} H_3C{-}COOH + CO_2$$

9.51. If two products are obtained, then the alkyne must be internal and unsymmetrical. There is only one such alkyne with the molecular formula C_5H_8:

H_2SO_4, H_2O
$HgSO_4$

9.52.

(a) As seen in Figure 9.7, a double bond can be converted into a triple bond via bromination, followed by two successive elimination reactions (with excess $NaNH_2$). Under these conditions, the resulting terminal alkyne is deprotonated, so water is introduced to protonate the alkynide ion, generating the alkyne:

1) Br_2
2) excess $NaNH_2$
3) H_2O

Br_2

(racemic)

1) xs $NaNH_2$
2) H_2O

(b) The desired transformation must lengthen the carbon chain by two carbon atoms, and we need to make a *trans* alkene. To begin the retrosynthesis we ask, "What starting material can be used to make a *trans* alkene?" The *trans* alkene can be prepared by partial reduction of the corresponding alkyne, and the alkyne intermediate is suitable for making the required C–C bond disconnection:

Nucleophile Electrophile

Now let's draw the forward process. The first part of the synthesis (alkylation of the alkyne) can be accomplished via treatment with $NaNH_2$ followed by ethyl iodide. After the alkylation process, a dissolving metal reduction will convert the alkyne into the desired *trans* alkene:

1) $NaNH_2$
2) EtI
3) Na, NH_3 (*l*)

1) $NaNH_2$
2) EtI

Na, NH_3 (*l*)

(c) When we compare the starting and final compounds, we observe two differences: the triple bond is gone and the carbon chain in the product is longer by three carbon atoms. The retrosynthetic analysis of the target molecule asks, "What starting material can be used to make an alkane?" One reaction we've seen that generates a product *with no functional groups* is catalytic hydrogenation of an alkyne.

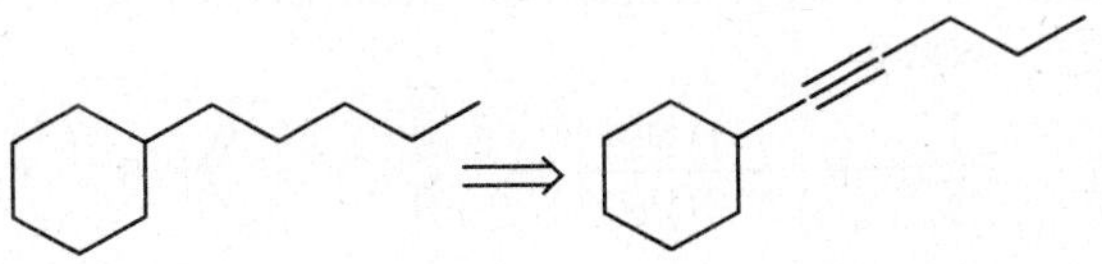

The alkyne intermediate is suitable for making the required C–C bond disconnection, so that is how the retrosynthesis continues:

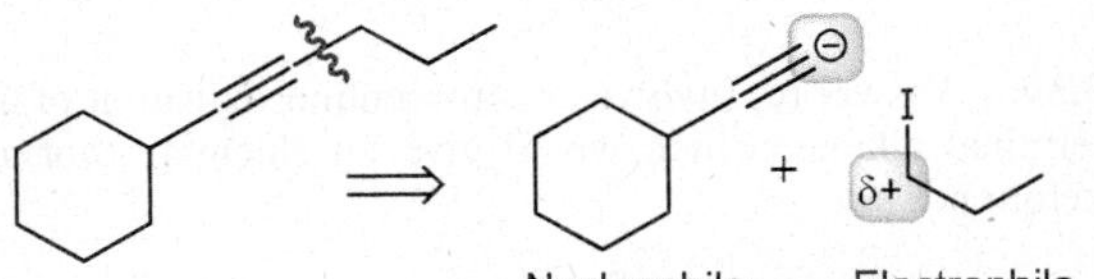

Nucleophile Electrophile

Now let's draw the forward process. The first part of the synthesis (alkylation of the alkyne) can be accomplished via treatment with $NaNH_2$ followed by *n*-propyl iodide. After the alkylation process, the resulting internal alkyne can be converted to the desired alkane upon hydrogenation:

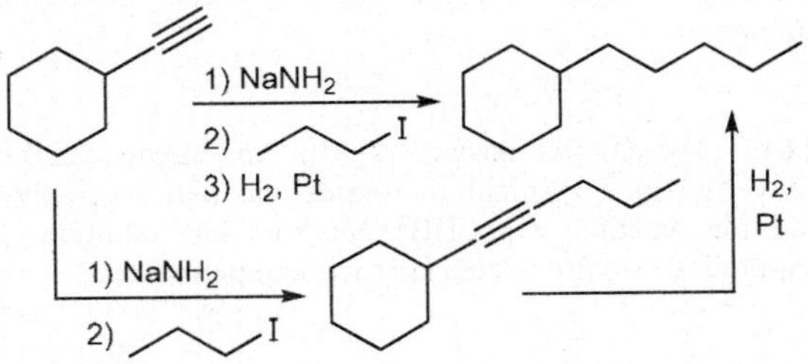

1) $NaNH_2$
2) I
3) H_2, Pt

1) $NaNH_2$
2) I

H_2, Pt

9.53. The starting material can either be a geminal dichloride or a vicinal dichloride:

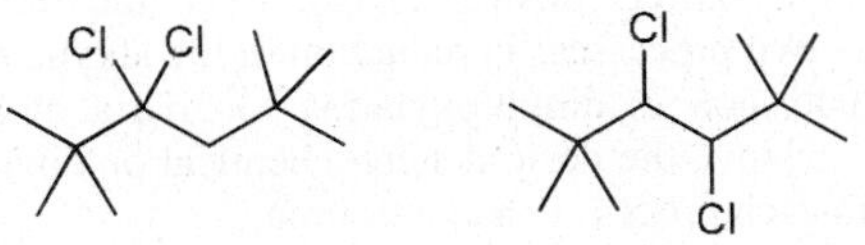

Cl Cl Cl Cl

Due to the symmetry of the compound, there are no other dichlorides that can be used to prepare the indicated alkyne via elimination.

9.54. In order to determine the structures of compounds **A-D**, we must work backwards. The last step is an alkylation process that installs an ethyl group, so compound **D** must be an alkynide ion, which is prepared from the corresponding terminal alkyne (compound **C**). The alkyne is made from the dibromide (compound **B**), which is formed in the first step when an alkene (compound **A**) is treated with Br_2:

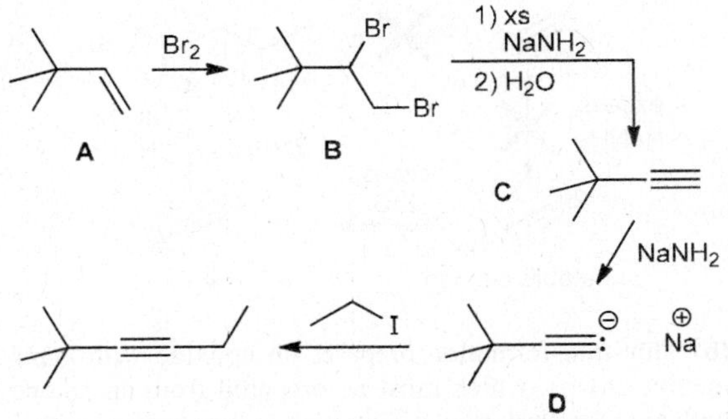

Br_2 → Br Br

1) xs $NaNH_2$
2) H_2O

A B C D

$NaNH_2$

I

Na

9.55. Terminal alkynes have the structure R–C≡C–H. The molecular formula indicates six carbon atoms, so the R group must be comprised of four carbon atoms. There are four different ways to connect these carbon atoms. They can be connected in a linear fashion, like this:

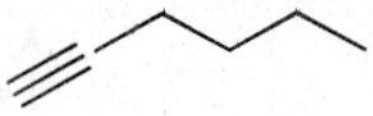

1-Hexyne

or there can be one methyl branch, which can be placed in either of two locations (C3 or C4), shown here:

3-Methyl-1-pentyne 4-Methyl-1-pentyne

or there can be two methyl branches, as shown here:

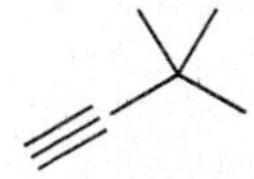

3,3-Dimethyl-1-butyne

9.56. Formation of 2,2-dimethyl-3-octyne (shown below) from acetylene would require two alkylation processes. That is, we must install an *n*-butyl group and a *tert*-butyl group. Installation of the *n*-butyl group can be readily achieved, but the *tert*-butyl group cannot, because installation of a *tert*-butyl group would require the use of a tertiary substrate, which will not undergo an S_N2 process:

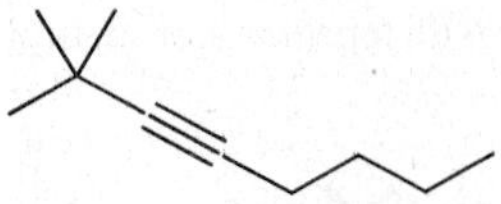

2,2-Dimethyl-3-octyne

9.57.

(a) This transformation requires the installation of two alkyl groups, but these groups must be added one at a time. The given synthesis fails because it attempts to add both ethyl groups at once, which would require a highly unstable dianion.

(b) This transformation prepares an epoxide with *trans* methyl groups, which must be prepared from an alkene with *trans* methyl groups. Catalytic hydrogenation with a poisoned catalyst produces a *cis* alkene, so the epoxide product would have the wrong stereochemistry.

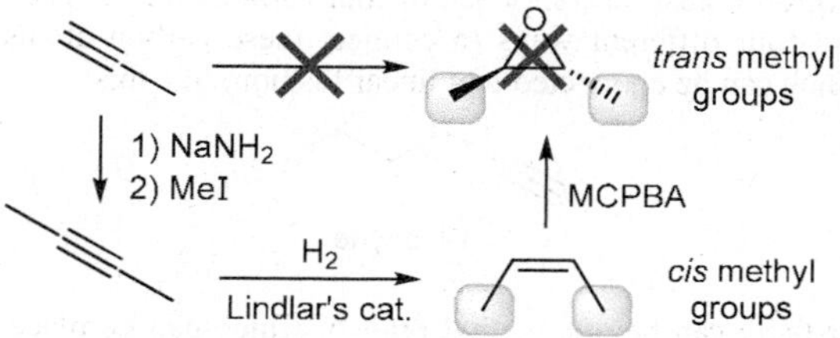

(c) This transformation requires the installation of an alkyl group, and the reduction of an alkyne to an alkene. The given synthesis does not work because it attempts to alkylate an *alkene* which fails because the proton attached to an sp^2 hybridized carbon atom is not acidic. Only a terminal *alkyne* can be deprotonated.

(d) This transformation requires Markovnikov addition of HBr to an alkene, the formation of a new C–C bond (via installation of an alkyl group on a terminal alkyne), and the hydration of an alkyne to furnish a ketone. The alkylation step fails because the acetylide anion is a strong base, and E2 elimination is favored for a secondary alkyl halide substrate. Alkylation of an alkyne is efficient only with a primary or methyl halide.

9.58. The products are carbon dioxide and a carboxylic acid, so we must be dealing with a terminal alkyne:

In this case, R = CH_3, so the correct answer is (a):

9.59. Answer (c) involves hydroboration-oxidation of a terminal alkyne, which would give an aldehyde (not a ketone):

9.60. The correct answer is (c). The reaction shows the conversion of an internal alkyne into a *trans* alkene. This can be accomplished via a dissolving metal reduction, which employs sodium metal in liquid ammonia.

9.61. The correct answer is (d). An alkyne can be converted to a geminal dibromide via two successive addition reactions with HBr. Markovnikov addition is required, so we use excess HBr without peroxides.

9.62. The correct answer is (b). This transformation requires two processes: 1) reduction of the alkyne to give an alkene, and 2) dihydroxylation to give a diol. One way to achieve the desired stereochemical outcome is to perform each process in a *syn* fashion.

9.63. The correct answer is (b). The most logical disconnection of the alkyne target molecule takes place at the C–C bond on either side of the alkyne. The *sp* hybridized carbon atom served as the nucleophile, so we draw a negative charge on the carbon atom to give a recognizable nucleophile (alkynide ion). The other carbon atom (at the disconnected bond) must have started out as an electrophile, so we draw a leaving group (such as Cl, Br, or I) at that position.

was a nucleophile
was an electrophile
What starting materials are needed?
Br
$\delta+$
Nucleophile
Electrophile

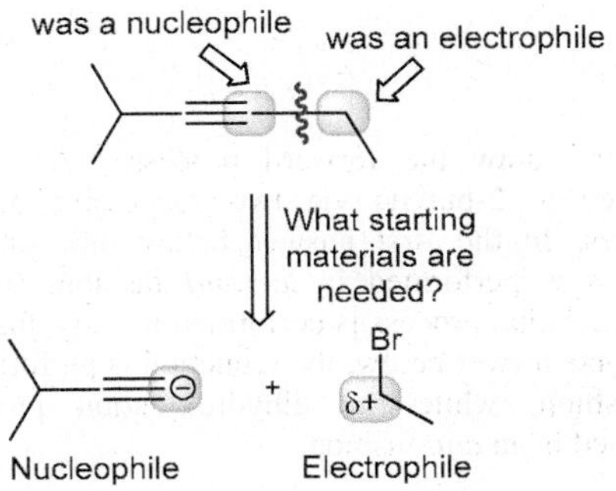

9.64. The correct answer is (c). A logical disconnection of the alkyne target molecule takes place at the C–C bond on either side of the alkyne. We must avoid making a disconnection at the sp^2 hybridized carbon atom on the benzene ring, because a leaving group at that position cannot undergo an S_N2 substitution.

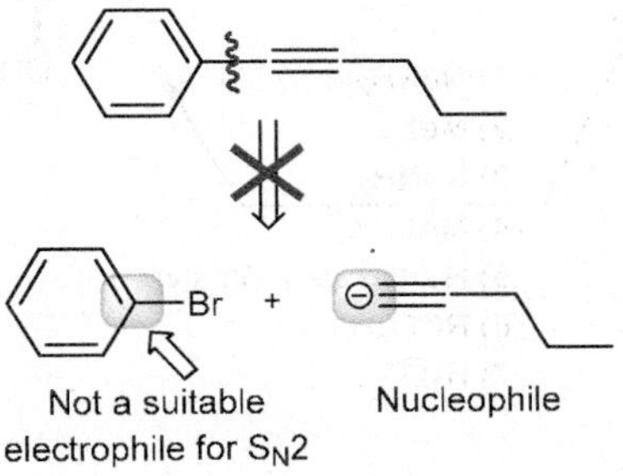

Br
Not a suitable electrophile for S_N2
Nucleophile

Therefore, the most logical disconnection is at the other side of the alkyne. The *sp* hybridized carbon atom served as the nucleophile, so we draw a negative charge on the carbon atom to give a recognizable nucleophile (alkynide ion). The other carbon atom (at the disconnected bond) must have started out as an electrophile, so we draw a leaving group (such as Cl, Br, or I) at that position. With this combination of an alkynide nucleophile and an electrophile with a leaving group on a primary carbon, we expect an S_N2 process to proceed smoothly to give the desired target molecule.

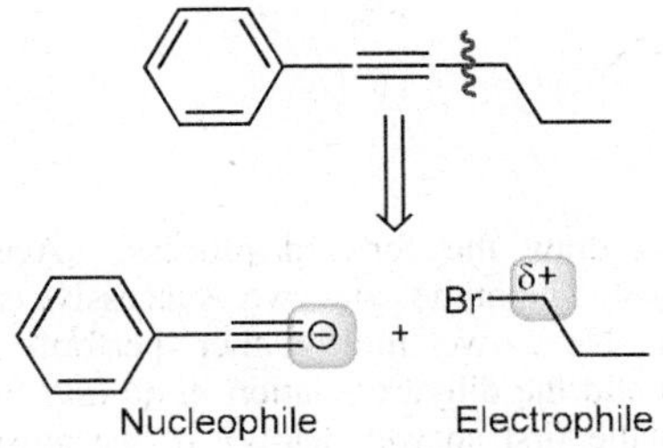

Br
$\delta+$
Nucleophile
Electrophile

9.65.
The correct answer is (d). These reagents indicate an alkylation process. In step 1, the alkyne is deprotonated by the strong base (H_2N^-) to give an alkynide ion. In step 2, this alkynide ion is used as a nucleophile to attack the given alkyl chloride (at the highlighted carbon atom, in an S_N2 reaction), thereby forming a new C–C bond:

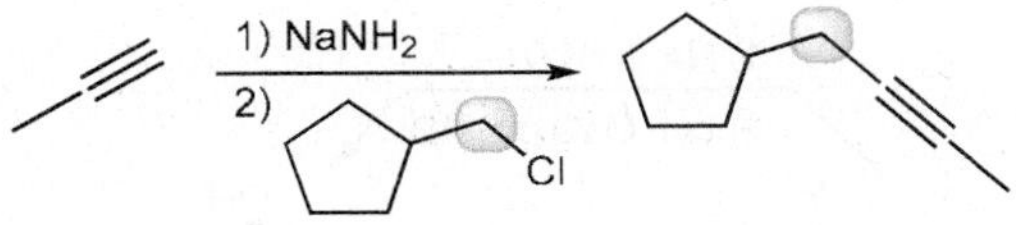

1) $NaNH_2$
2)
Cl

9.66.
(a) The alkyne can be reduced to an alkene, followed by bromination. Since the last step (bromination) proceeds via an *anti* addition, the desired stereoisomer can only be obtained if the previous step (reduction of the alkyne) is performed in an *anti* fashion as well. That is, we must perform a dissolving metal reduction:

1) Na, NH_3 (*l*)
2) Br_2
Br
Br
Na, NH_3 (*l*)
Br_2

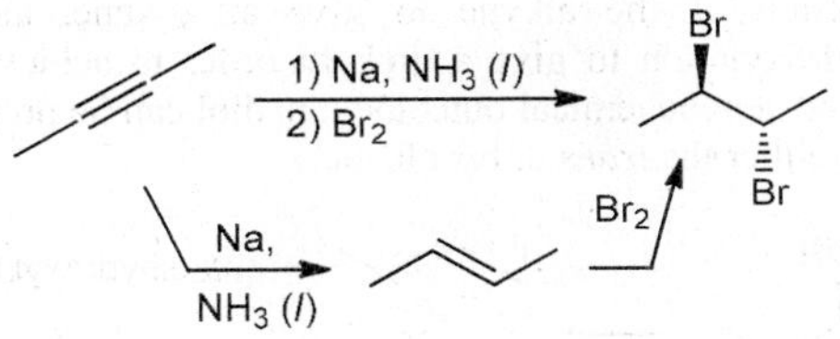

(b) The alkyne can be reduced to an alkene, followed by bromination. Since the last step (bromination) proceeds via an *anti* addition, the desired stereoisomer can only be obtained if the previous step (reduction of the alkyne) is performed in a *syn* fashion. That is, we must perform a hydrogenation reaction with a poisoned catalyst, rather than using a dissolving metal reduction.

1) H_2, Lindlar's catalyst
2) Br_2
Br
Br
+ En
H_2, Lindlar's cat.
Br_2

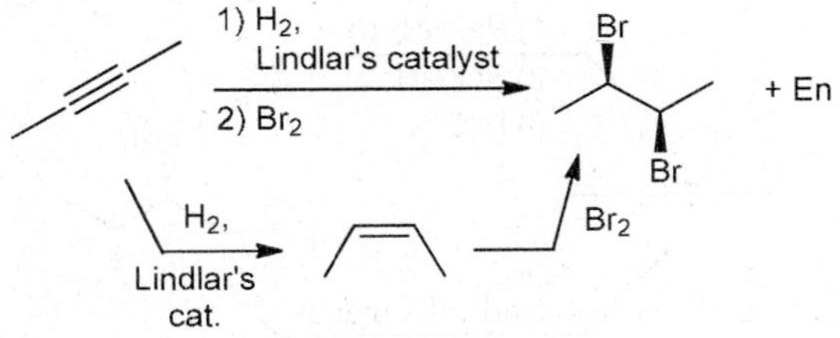

(c) This transformation requires two processes: 1) reduction of the alkyne to give an alkene, and 2) dihydroxylation to give a diol. In order to achieve the desired stereochemical outcome, the diol can be prepared from either the *trans* or *cis* alkene:

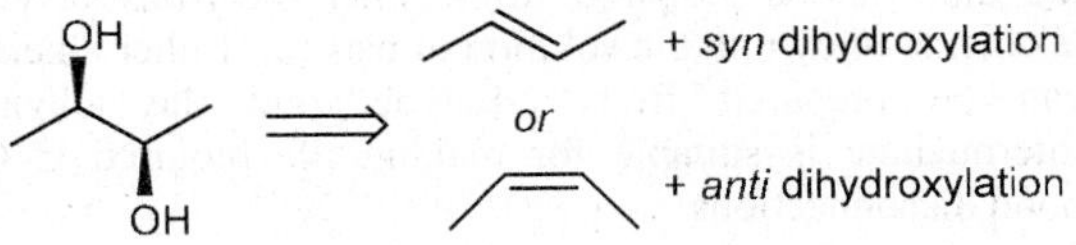

OH
OH
+ *syn* dihydroxylation
or
+ *anti* dihydroxylation

Either alkene can be prepared from the given starting material (2-butyne), and both answers are acceptable. In the first answer below, the reduction is performed in an *anti* fashion, while the dihydroxylation process is performed in a *syn* fashion. In the second answer below, the reduction is performed in a *syn* fashion, while the dihydroxylation process is performed in an *anti* fashion.

1) Na, NH_3 (*l*)
2) cat. OsO_4, NMO

OH

+ En

OH

1) H_2, Lindlar's Catalyst
2) RCO_3H
3) H_3O^+

(d) This transformation requires two processes: 1) reduction of the alkyne to give an alkene, and 2) dihydroxylation to give a diol. In order to achieve the desired stereochemical outcome, the diol can be prepared from either the *trans* or *cis* alkene:

OH

+ *anti* dihydroxylation

or

+ *syn* dihydroxylation

OH

Either alkene can be prepared from the given starting material (2-butyne), and both answers are acceptable. We must either perform both the reduction and the dihydroxylation in an *anti* fashion (as shown in the first answer below), or we must perform each process in a *syn* fashion (as shown in the second answer below).

1) Na, NH_3 (*l*)
2) RCO_3H
3) H_3O^+

OH

OH

1) H_2, Lindlar's Catalyst
2) cat. OsO_4, NMO

(e) This transformation requires three processes: 1) lengthening the carbon chain, 2) reduction of the alkyne to give an alkene, and 3) dihydroxylation to give a diol. In order to achieve the desired stereochemical outcome, the diol can be prepared from either the *trans* or *cis* alkene, as shown in the solution to part (c). Either alkene can be prepared from 2-butyne, and the alkyne intermediate is suitable for making the required C–C bond disconnections:

or

Now let's draw the forward process. Acetylene is converted to 2-butyne via two successive alkylation processes. In the first answer below, the subsequent reduction is performed in an *anti* fashion, while the dihydroxylation process is performed in a *syn* fashion. In the second answer below, the reduction is performed in a *syn* fashion, while the dihydroxylation process is performed in an *anti* fashion.

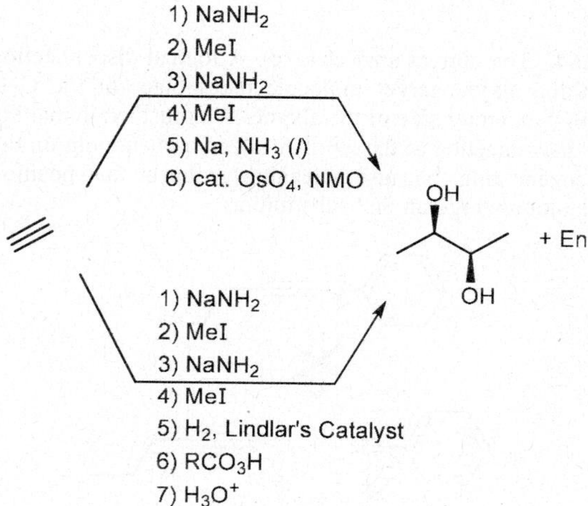

(f) This transformation requires three processes: 1) lengthening the carbon chain, 2) reduction of the alkyne to give an alkene, and 3) dihydroxylation to give a diol. In order to achieve the desired stereochemical outcome, the diol can be prepared from either the *trans* or *cis* alkene, as shown in part (d). Either alkene can be prepared from 2-butyne, and the alkyne intermediate is suitable for making the required C–C bond disconnections:

or

Now let's draw the forward process. Acetylene is converted to 2-butyne via two successive alkylation processes. Next, we must either perform both the reduction and the dihydroxylation in an *anti* fashion (as shown in the first answer below), or we must perform each process in a *syn* fashion (as shown in the second answer below).

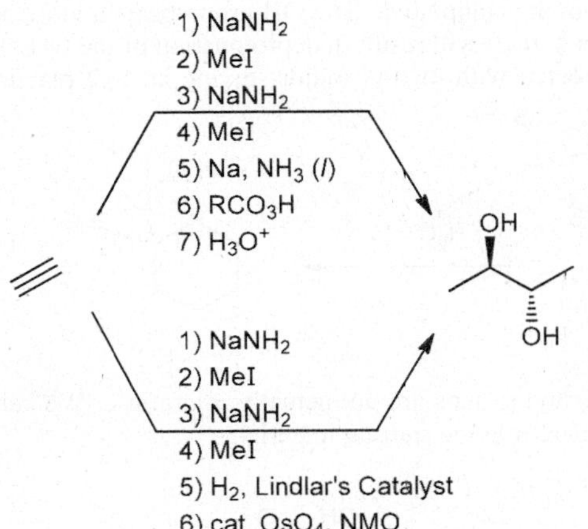

9.67. The replacement of a hydrogen atom with a deuterium can be accomplished with two "proton" transfer reactions. Deprotonation of the alkyne with sodium amide results in an alkynide ion. If the alkynide ion is treated with water (H_2O), the alkynide ion will be protonated again to regenerate the starting alkyne. If, however, the alkynide ion is treated with a source of deuterons (D_2O), then the desired deuterated compound will be produced:

1) $NaNH_2$
2) D_2O
$NaNH_2$
(base)
(acid)

9.68. The desired transformation is a functional group interconversion (there is no change to the carbon skeleton). To begin the retrosynthesis we ask, "What starting material can be used to make a ketone?" Recall that ketones can be prepared by hydration of an alkyne:

The required alkyne intermediate can be prepared from a dibromide that leads back to the given starting material:

Now let's draw the forward process. Conversion of the alkene starting material into the required alkyne can be accomplished via a two-step procedure. The alkene is treated with molecular bromine (Br_2) to give a vicinal dibromide, which is then treated with excess $NaNH_2$ (followed by water to protonate the alkynide ion) to give an alkyne. Finally, the alkyne can be converted into the desired methyl ketone via acid-catalyzed hydration in the presence of mercuric sulfate:

1) Br_2
2) xs $NaNH_2$
3) H_2O
4) H_2O, H_2SO_4, $HgSO_4$
Br_2
(racemic)
1) xs $NaNH_2$
2) H_2O
H_2O
H_2SO_4
$HgSO_4$

9.69. Since ozonolysis of the internal alkyne leads to only one carboxylic acid (rather than two), we can deduce that the internal alkyne must be symmetrical. If it has to be symmetrical, then it must be 4-octyne, since we installed a propyl group ourselves (that is, the starting alkyne already had one propyl group).

1) $NaNH_2$
2) I
4-octyne
1) O_3
2) H_2O
2 OH

9.70. The alkyne proton (pK_a ~ 25) is not the most acidic proton in the compound. The OH group bears a more acidic proton (pK_a ~ 18), so treatment with a strong base (such as sodium amide) will result in deprotonation of the OH group. The resulting alkoxide ion then serves as a nucleophile when treated with methyl iodide, giving an S_N2 reaction to generate the following product:

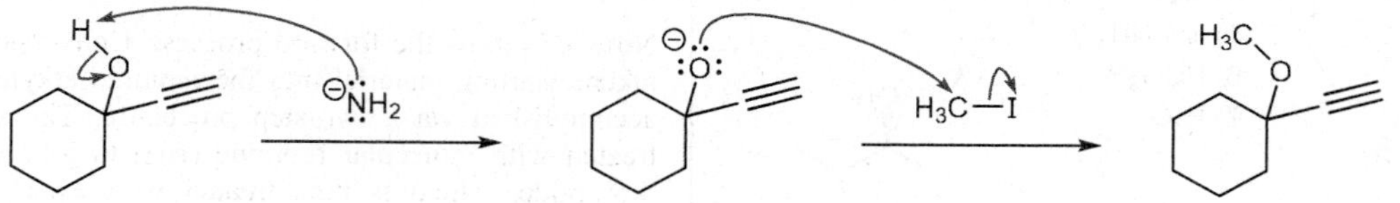

9.71. The key to solving this problem is to recognize that the methyl groups are not actually migrating. We can see this more clearly if we flip the product horizontally, and then compare it to the starting material:

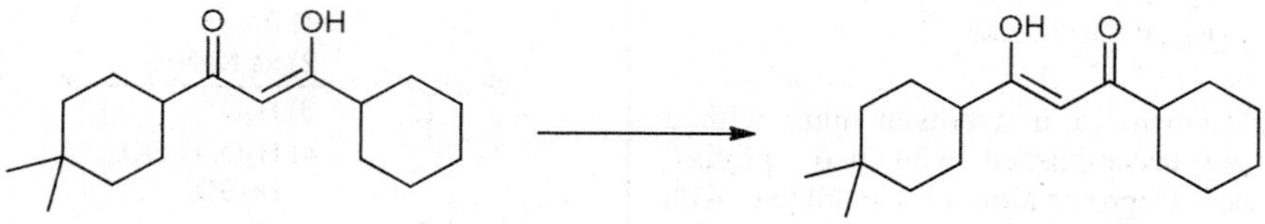

When drawn in this way, we can see more clearly that it is just a tautomerization process. Protonation of the C=O bond gives an intermediate that is highly resonance stabilized. Deprotonation then gives the product:

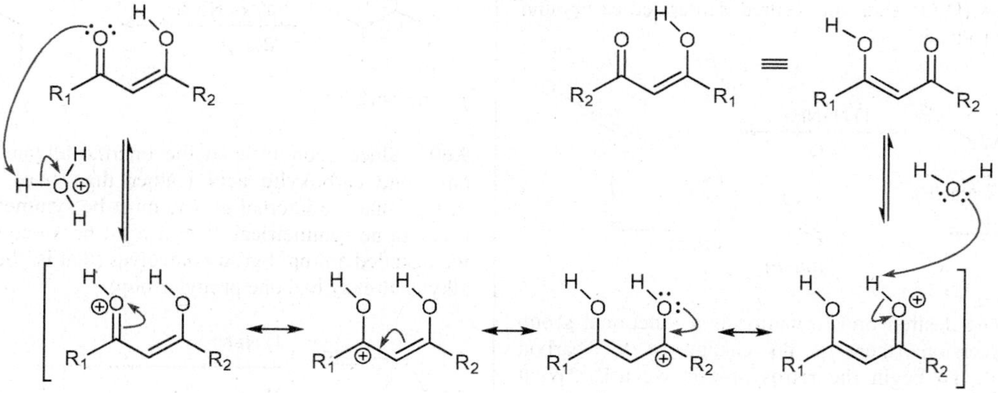

9.72. Much like an enol, the starting material in this case (called an enamine) also undergoes tautomerization via a similar mechanism. That is, the double bond is first protonated to generate a resonance-stabilized cation, which is then deprotonated. Notice that, in the first step, the double bond is protonated, rather than the nitrogen atom. Protonation of the nitrogen atom does not result in a resonance-stabilized cation.

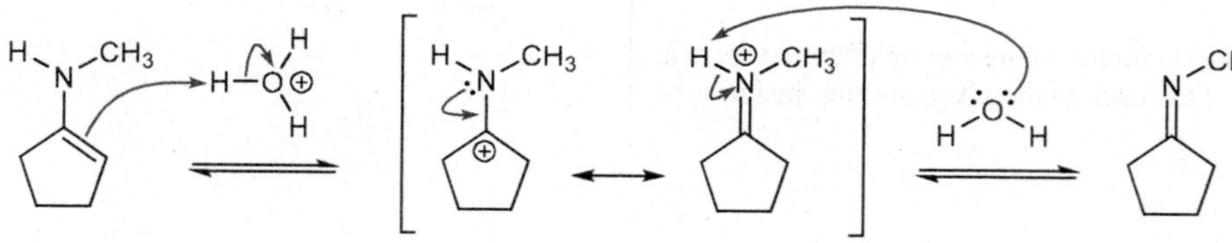

9.73.

(a) This transformation requires three processes: 1) lengthening the carbon chain, 2) reduction of the alkyne to give an alkene, and 3) epoxidation. In order to achieve the desired stereochemical outcome, the epoxide must be prepared from a *trans* alkene (specifically, *trans*-2-pentene), which can be prepared from 2-pentyne:

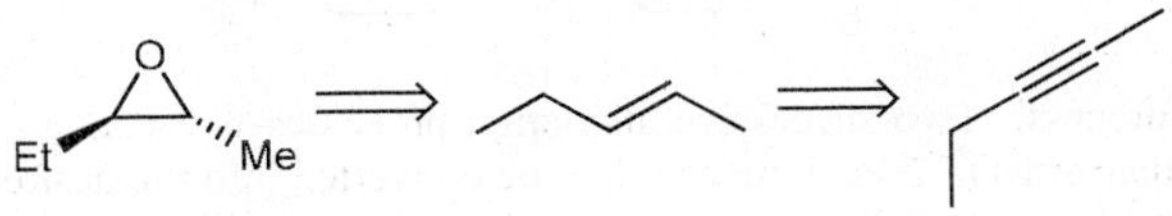

O, Et, Me ⟹ ⟹

We complete the retrosynthetic analysis by identifying that 2-pentyne is suitable for making the required C–C bond disconnections:

⟹ ⟹

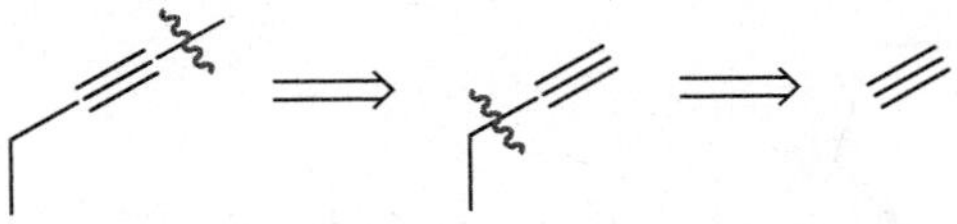

The retrosynthesis reveals the need for EtBr, but the only 2-carbon starting material that is allowed is acetylene, so EtBr must also be prepared from acetylene. An alkyl halide can be prepared from an alkene, which, in turn, can be prepared from acetylene:

Br ⟹ ⟹

Now let's draw the forward process. Two successive alkylation processes will convert acetylene into 2-pentyne (the alkylations can be done in either order). 2-Pentyne can then be converted into the desired epoxide target molecule via a dissolving metal reduction, followed by epoxidation of the resulting *trans* alkene, as shown here:

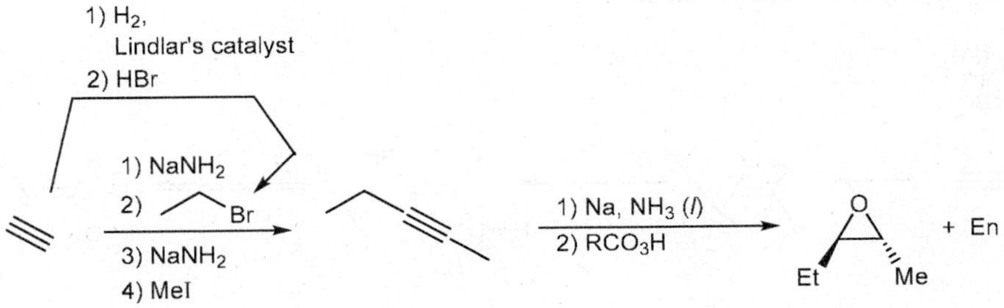

1) H_2, Lindlar's catalyst
2) HBr

1) $NaNH_2$
2) Br
3) $NaNH_2$
4) MeI

→

1) Na, NH_3 (*l*)
2) RCO_3H

→ O, Et, Me + En

(b) This transformation requires three processes: 1) lengthening the carbon chain, 2) reduction of the alkyne to give an alkene, and 3) epoxidation. In order to achieve the desired stereochemical outcome, the epoxide must be prepared from a *cis* alkene (specifically, *cis*-2-pentene), which can be prepared from 2-pentyne.

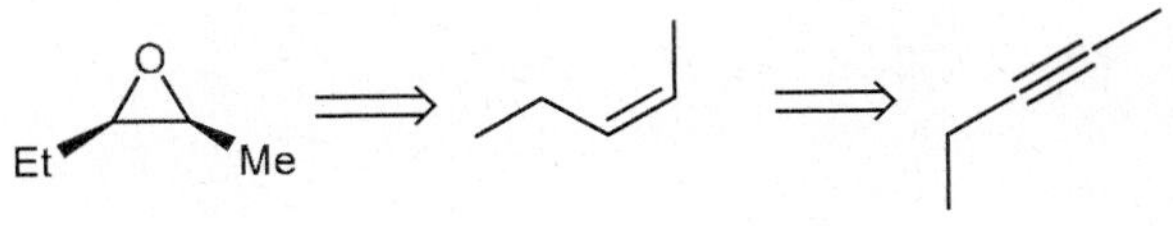

O, Et, Me ⟹ ⟹

We complete the retrosynthetic analysis by identifying that 2-pentyne is suitable for making the required C–C bond disconnections:

⟹ ⟹

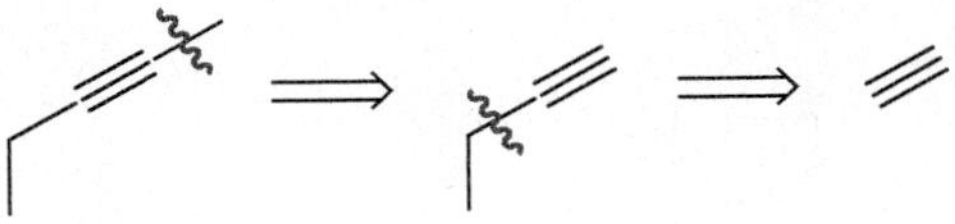

The retrosynthesis reveals the need for EtBr, but the only 2-carbon starting material that is allowed in acetylene, so EtBr must also be prepared from acetylene. An alkyl halide can be prepared from an alkene, which, in turn, can be prepared from acetylene:

Br ⟹ ⟹

Now let's draw the forward process. Two successive alkylation processes will convert acetylene into 2-pentyne (the alkylations can be done in either order). 2-Pentyne can then be converted into the desired epoxide target molecule via hydrogenation (with a poisoned catalyst), followed by epoxidation of the resulting *cis* alkene, as shown here:

1) H_2, Lindlar's catalyst
2) HBr

1) $NaNH_2$
2) Br
3) $NaNH_2$
4) MeI

1) H_2, Lindlar's catalyst
2) RCO_3H

O Et Me + En

9.74. The mechanism of alkyne halogenation is not entirely understood, but we can propose a mechanism for the addition of Br and OH across the triple bond. In the presence of Br_2, the alkyne is converted to a bromonium ion, which can be opened with water. The resulting oxonium ion is deprotonated to reveal an enol, which undergoes a two-step tautomerization to give the observed product:

Br :Br: R

⊕ Br R H O H R Br H–O ⊕ H H O H R Br H–O H H O ⊕ H

R Br O H O H R Br H–O ⊕ R ⊕ Br H–O:

9.75. D_3O^+ is directly analogous to H_3O^+, but the protons have been replaced with deuterons. In the presence of D_3O^+, tautomerization processes can occur. Below are two, successive tautomerization processes that can successfully explain formation of the deuterated product:

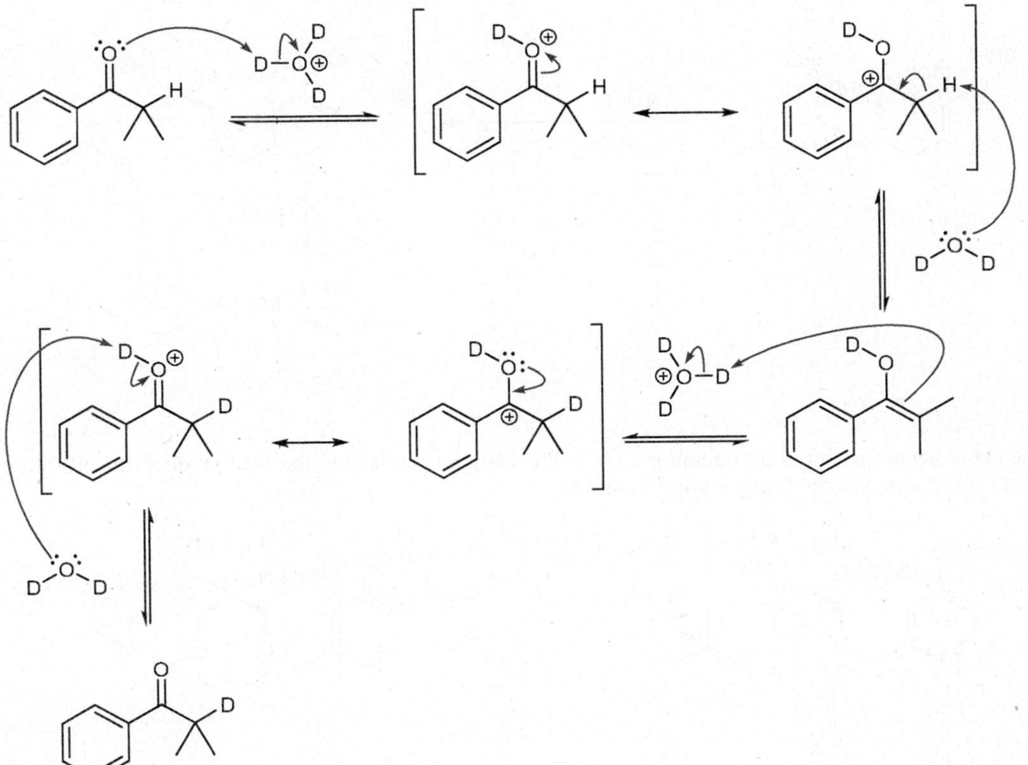

9.76.

(a) Compare the structures of the starting material and product, and identify the part of the structure that must be modified:

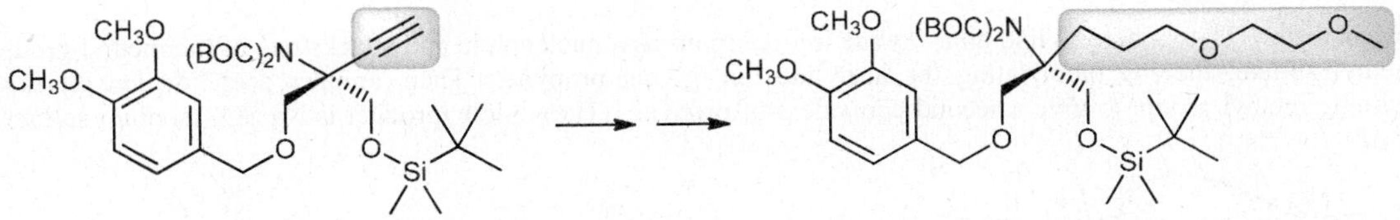

The rest of the structure remains unchanged. If we represent that part of the structure as an R group, the desired transformation can be shown as follows:

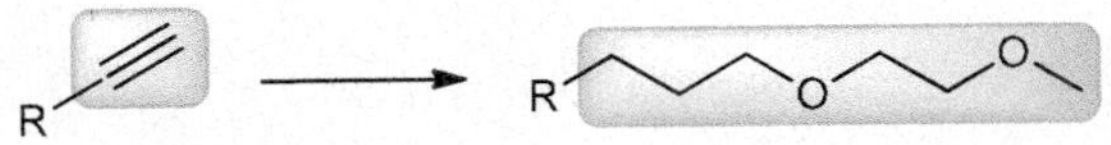

This transformation requires two processes: 1) lengthening the carbon chain, 2) reduction of the alkyne to give an alkane. We can number the alkyne carbon atoms in order to keep track of them throughout the retrosynthesis. The alkyl chain in the target molecule can be prepared from the alkyne, and this intermediate is suitable for making the required C–C bond disconnection:

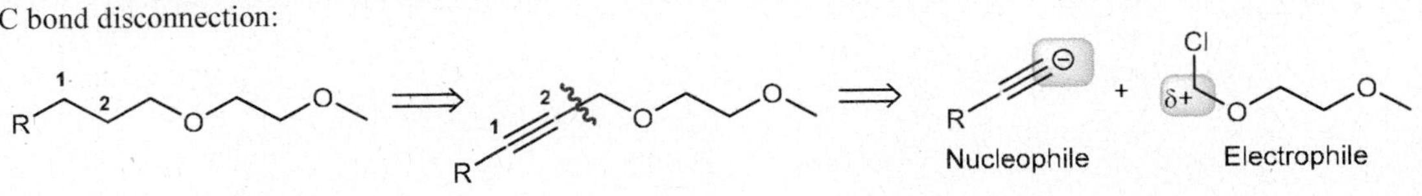

Nucleophile Electrophile

Now let's draw the forward process. The given transformation can be achieved by deprotonation of the terminal alkyne (with $NaNH_2$) and subsequent treatment with the required electrophile ($ClCH_2OCH_2CH_2OCH_3$ is called methoxy ethoxymethyl chloride or MEM chloride), followed by reduction of the alkyne group via hydrogenation.

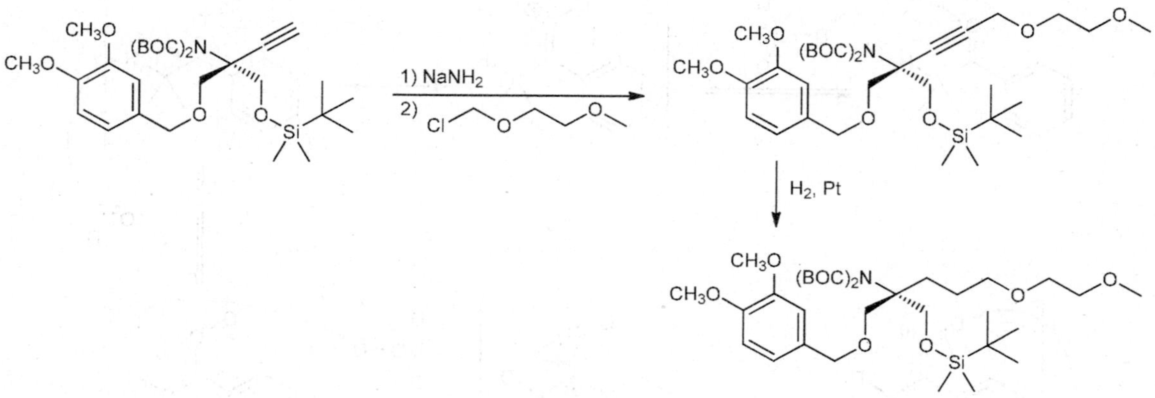

(b) The chiral center has the *S* configuration in both the starting material and the final product (the priorities, shown below, do not change as a result of the transformation).

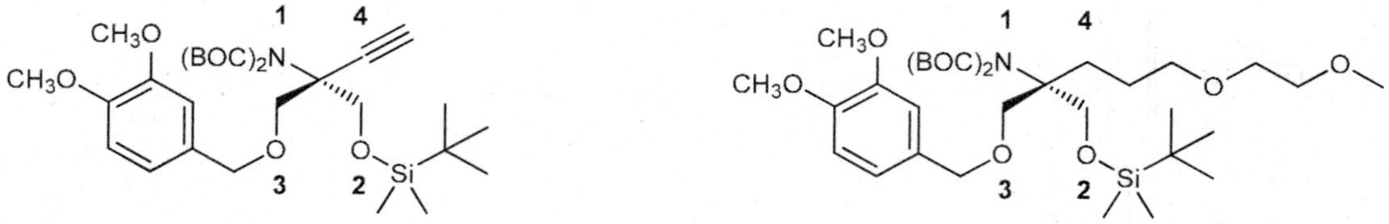

9.77.

(a) Dimethyl sulfate has two electrophilic methyl groups, each of which can be transferred to an acetylide ion:

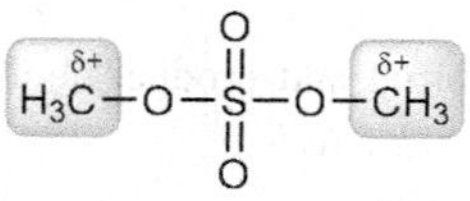

A mechanism is shown here. First, an acetylide ion functions as a nucleophile and attacks one of the methyl groups of dimethyl sulfate, thereby methylating the acetylide ion, giving propyne. Then, another acetylide ion attacks the remaining methyl group, to give a second equivalent of propyne. The ionic byproduct is Na_2SO_4 (sodium sulfate), as shown:

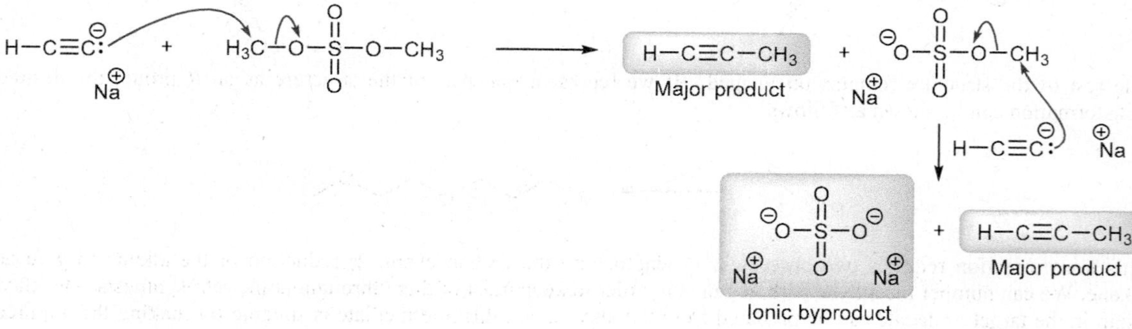

(b) During the course of the reaction, propyne molecules are being generated in the presence of acetylide ions. Under these conditions, propyne is deprotonated to give an alkynide ion which can then function as a nucleophile and attack a methyl group of dimethyl sulfate, giving 2-butyne:

$H_3C-C\equiv C-H$ (Major product) + $Na^{\oplus}$ $^{\ominus}:C\equiv C-H$ ⇌ $H_3C-C\equiv C:^{\ominus}$ $Na^{\oplus}$ + $H-C\equiv C-H$ (Acetylene)

$H_3C-C\equiv C:^{\ominus}$ + $H_3C-O-S(=O)_2-O-CH_3$ → $H_3C-C\equiv C-CH_3$ (2-Butyne) + $Na^{\oplus}$ $^{\ominus}O-S(=O)_2-O-CH_3$

Notice that in the first step of this process, an acetylide ion is converted to acetylene. This mechanism is therefore consistent with the observation that acetylene is present among the reaction products.

(c) With diethyl sulfate, an ethyl group is transferred in the alkylation reaction (rather than a methyl group). Therefore, the major product would be 1-butyne, as shown below:

$H-C\equiv C:^{\ominus}$ $Na^{\oplus}$ + $CH_3CH_2-O-S(=O)_2-O-CH_2CH_3$ → $H-C\equiv C-CH_2CH_3$ (1-butyne) + $Na^{\oplus}$ $^{\ominus}O-S(=O)_2-O-CH_2CH_3$

If the product undergoes further alkylation, the minor product would be 3-hexyne, as shown below:

$CH_3CH_2-C\equiv C-H$ (1-butyne) + $Na^{\oplus}$ $^{\ominus}:C\equiv C-H$ ⇌ $CH_3CH_2-C\equiv C:^{\ominus}$ $Na^{\oplus}$ + $H-C\equiv C-H$ (Acetylene)

$CH_3CH_2-C\equiv C:^{\ominus}$ + $CH_3CH_2-O-S(=O)_2-O-CH_2CH_3$ → $CH_3CH_2-C\equiv C-CH_2CH_3$ (3-Hexyne) + $Na^{\oplus}$ $^{\ominus}O-S(=O)_2-O-CH_2CH_3$

9.78. The alkyne isomerization is accomplished by a series of proton transfer reactions. Sodium amide is a strong base, so before the alkyne isomerization occurs, the alcohol group in **1** will be deprotonated to form an alkoxide ion (a negative charge on an oxygen atom). The alkoxide ion is carried through until the final reprotonation event (in step 2 when H_2O is introduced, which takes place *after* the transformation below is complete).

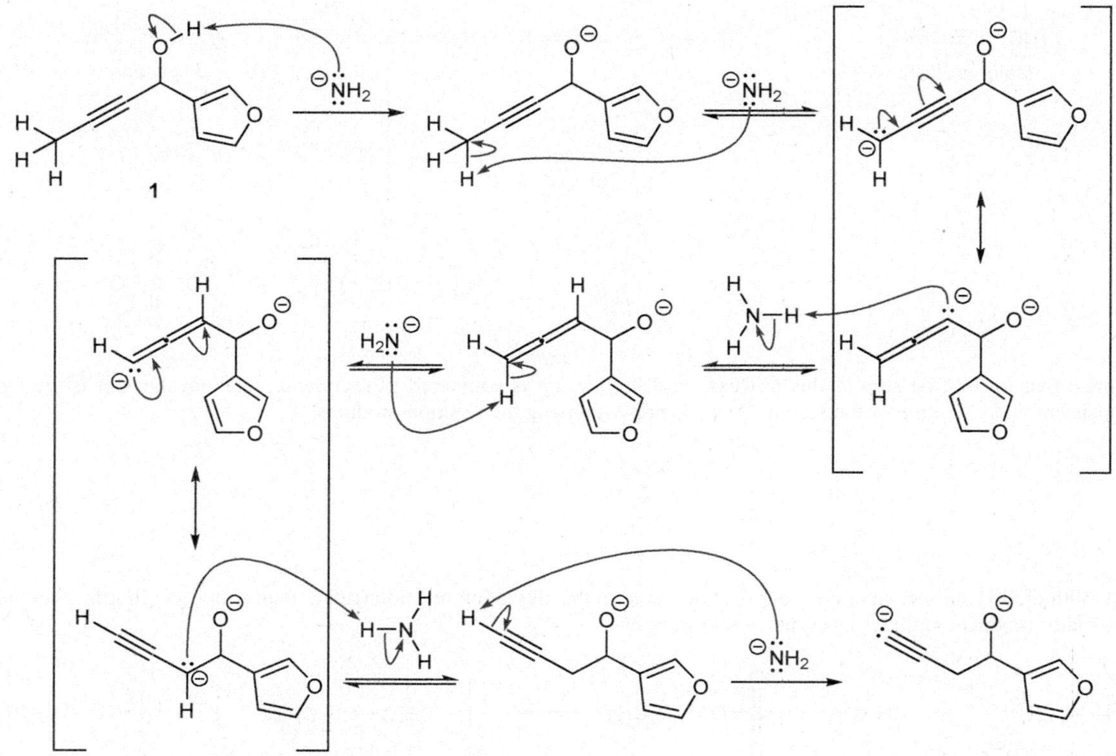

The formation of an alkynide ion (stabilized by having the lone pair on the carbon atom in an *sp* hybridized orbital) serves as a driving force for the reaction (see solution to problem 9.8). After the transformation above is complete, water is introduced as a proton source, giving **2**. The more strongly basic alkynide position is likely protonated first, followed by protonation of the alkoxide:

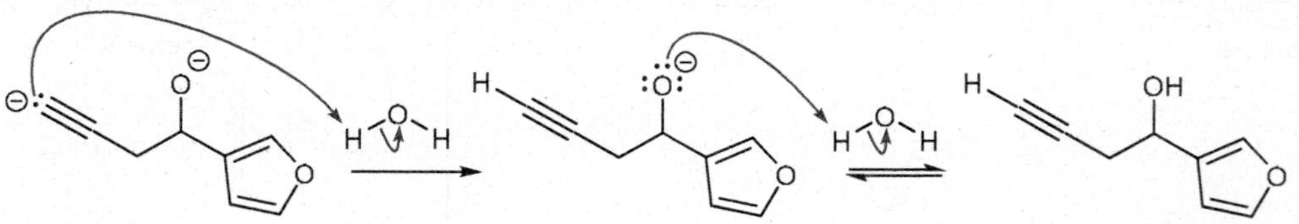

9.79.

(a) The problem statement indicates that the first mechanistic step for iodination of an alkyne is formation of a bridged iodonium intermediate, as shown below. This intermediate is analogous to the intermediate that we saw during halogenation of alkenes, although notice that the three-membered ring has a double bond (because the starting material was an alkyne, rather than an alkene).

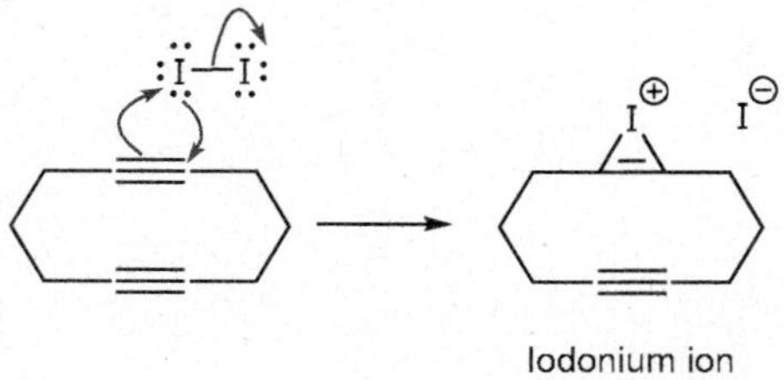

Iodonium ion intermediate

(b) As shown, the resulting carbocation would be vinylic, and we have seen that vinylic carbocations are generally too high in energy to form:

vinylic carbocation
(too high in energy to form)

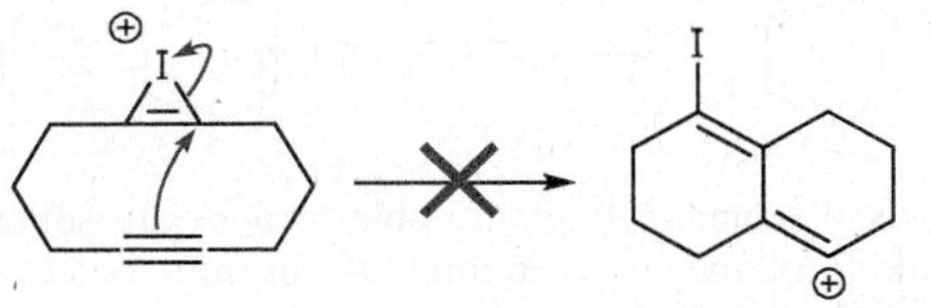

(c) The problem statement describes the following two-step mechanism, which avoids the formation of a vinylic carbocation intermediate. After the initial formation of the iodonium ion, the iodide anion can attack the second alkyne group, causing the π-electrons to attack the iodonium ion, giving the final product.

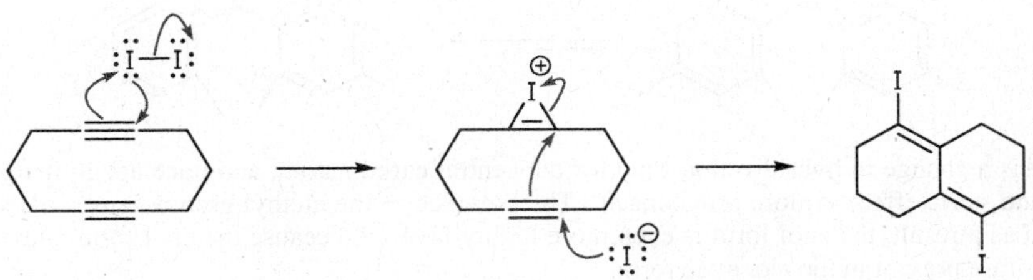

9.80.

(a) The highlighted carbon atom has four σ bonds, and is therefore sp^3 hybridized. As a result, the geometry around this carbon atom is expected to be tetrahedral, with approximate bond angles of 109.5°.

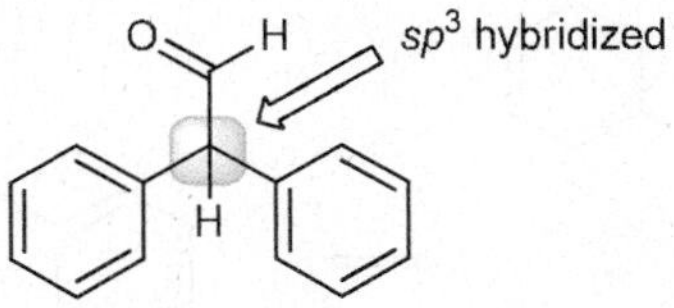

sp^3 hybridized

(a) Compound **1** is an aldehyde, and its enol is shown here:

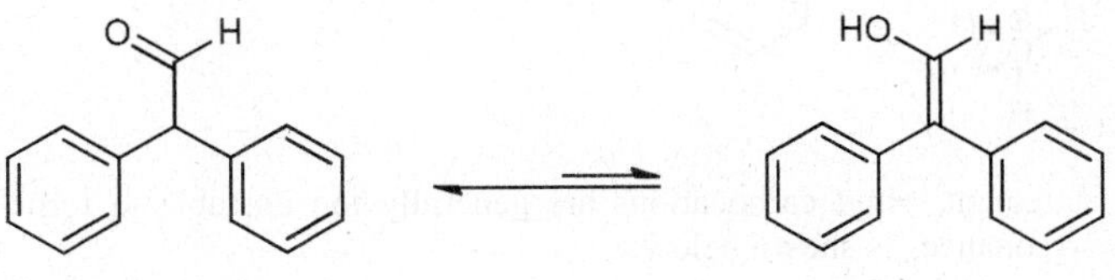

aldehyde enol

If we explore the same carbon atom that we analyzed in part (a), we find that this carbon atom (highlighted below) is sp^2 hybridized, with approximate bond angles of 120°.

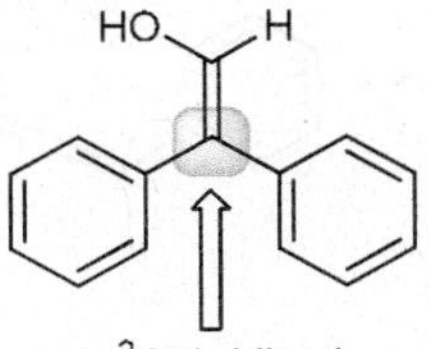

sp^2 hybridized

(c) As we noted in the solutions to parts (a) and (b), we expect the bond angles to change when compound **1** is converted into its enol form. Specifically, there is an increase in the bond angles from 109.5° to 120°, as shown:

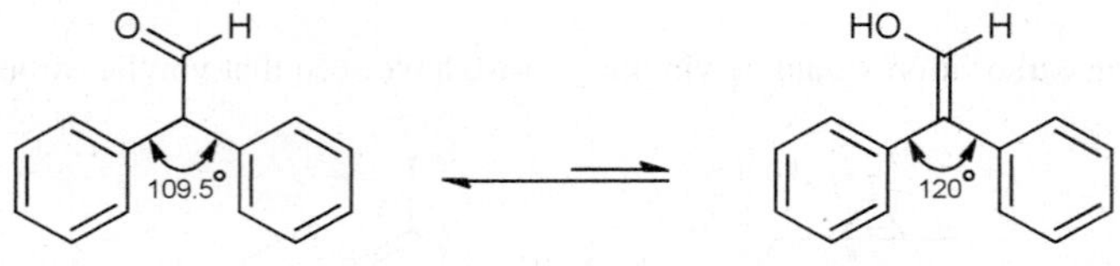

As such, the sterically demanding (bulky) aromatic rings are able to alleviate some steric strain in the enol form, relative to the aldehyde. In the aldehyde form, the aromatic rings are forced to be closer together in space. This steric effect causes the enol form to be particularly stable, and its concentration is significant (9.1% vs. the typical 0.014%).

(d) For compound **2**, the following equilibrium is established between the aldehyde and the enol:

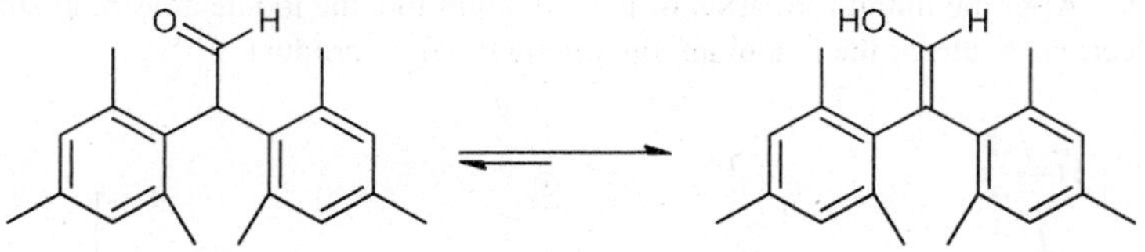

Once again, there is a change in hybridization state for the central carbon atom, and once again there is a steric effect. But in this case, the steric effect is more pronounced. The presence of the methyl groups causes the steric effect to be much greater, and as a result, the enol form is even more highly favored, because the enol form alleviates much of the significant steric strain present in the aldehyde form.

9.81.

(a) Begin by drawing the carbocation intermediate in this case:

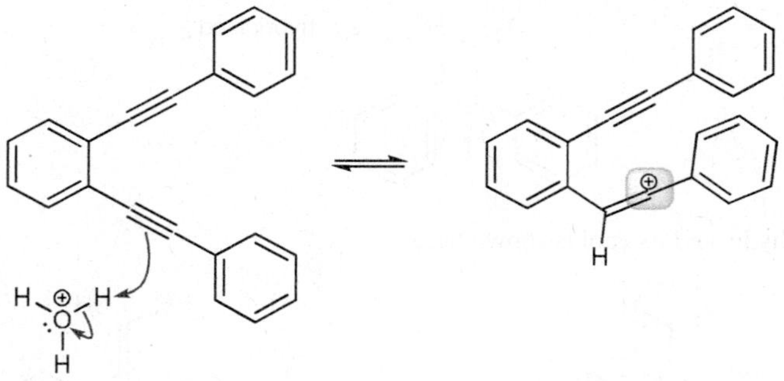

As noted in the problem statement, vinyl carbocations are generally too unstable to form, but this particular vinyl carbocation is stabilized by resonance, as shown below:

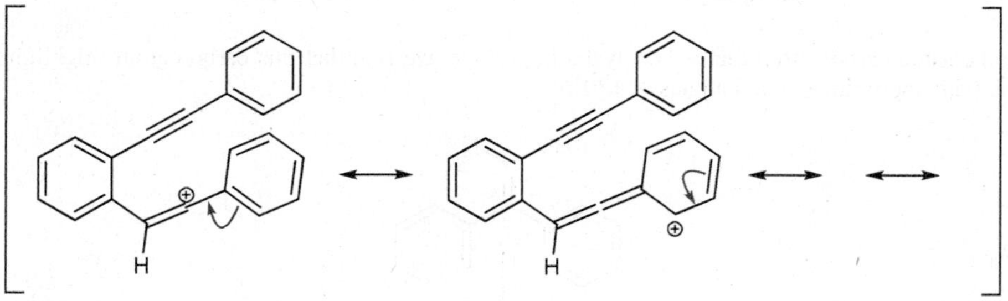

(b) The mechanism described in the problem statement is shown below. One of the alkyne groups is protonated resulting in the formation of a new C–H bond, and a resonance-stabilized vinyl carbocation. The π electrons from the other alkyne attack this carbocation, resulting in the formation of a new C–C bond, and a new vinyl carbocation.

Nucleophilic attack of water, followed by deprotonation, gives the enol, which tautomerizes to form the ketone, as shown:

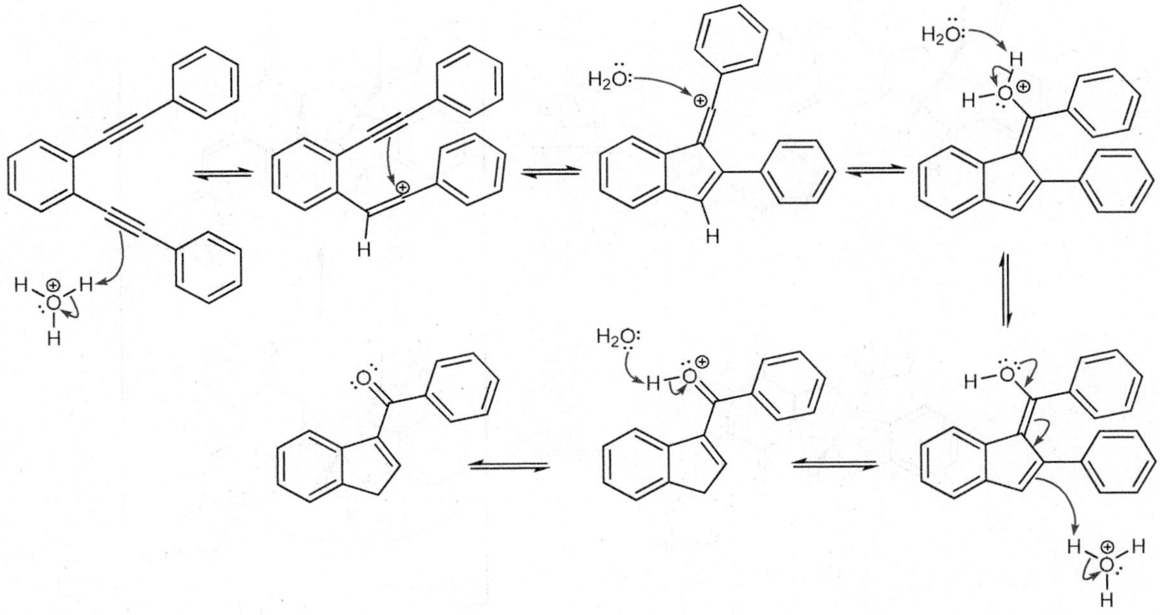

9.82.

(a) If the tautomerization process is base-catalyzed, then the first step will be deprotonation to give a resonance-stabilized anion, which is then protonated to give the tautomer.

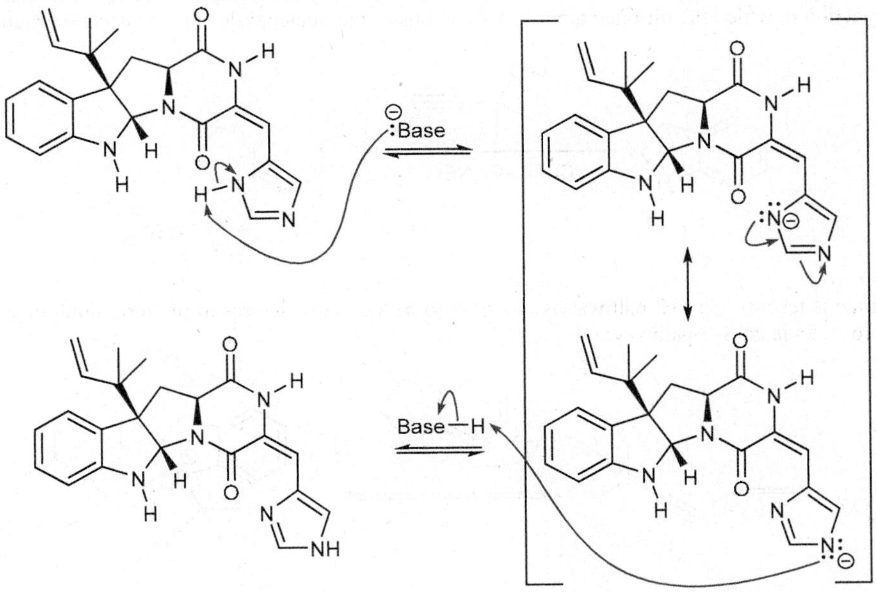

(b) If the tautomerization process is acid-catalyzed, then the first step will be protonation to give a resonance-stabilized cation, which is then deprotonated to give the tautomer.

H–A

A:

9.83.

(a) The molecular formula of compound **2** indicates that five carbon atoms have been installed (compound **1** contains only eight carbon atoms, while compound **2** contains thirteen carbon atoms). The lack of oxygen atoms in the molecular formula of compound **2** also indicates that the acetate group (OAc) has been completely removed, but the nitrogen atom is still present. These observations are consistent with the following structure, which can be formed via a substitution reaction in which the nitrogen atom in **1** functions as the nucleophile, and the acetate group functions as a leaving group.

AcO

CuCl, i-Pr_2NEt

1

2

Since the substrate is tertiary, an S_N2 pathway is too slow to be viable, as the result of steric hindrance. As such, the reaction must proceed via an S_N1 pathway:

OAc

AcO⊖

AcO⊖

2

i-Pr–N⊕–H AcO⊖ (i-Pr, Et)

(b) The reduction of a terminal alkyne to an alkene can be accomplished by treating compound **2** with H_2 and Lindlar's catalyst. Note: a terminal alkyne cannot be reduced via a dissolving metal reduction.

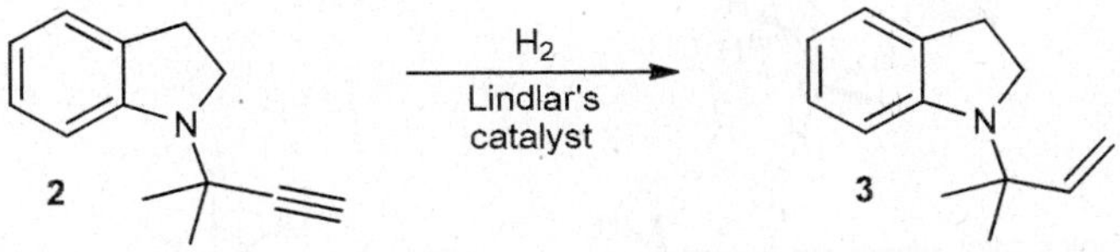

H_2
Lindlar's catalyst

9.84.

(a) The following is a wedge-and-dash structure for the *anti-anti* conformation of pentane. Notice that C1 and C4 are *anti* to each other (when looking down the C2-C3 bond), while C2 and C5 are also *anti* to each other (when looking down the C3-C4 bond).

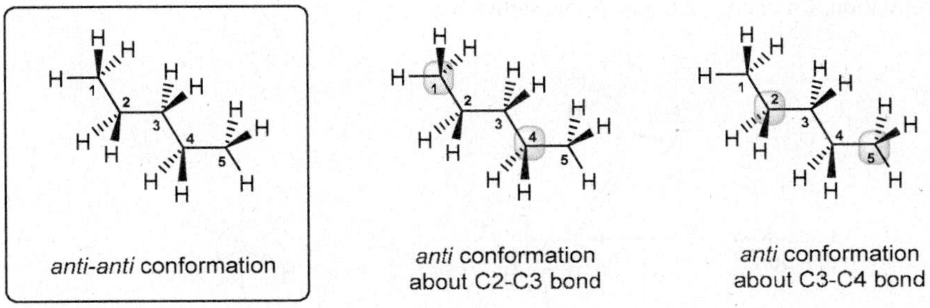

anti-anti conformation

anti conformation about C2-C3 bond

anti conformation about C3-C4 bond

The following is a wedge-and-dash structure for the *anti-gauche* conformation of pentane. Notice that C1 and C4 are *anti* to each other (when looking down the C2-C3 bond), while C2 and C5 are *gauche* to each other (when looking down the C3-C4 bond).

anti-gauche conformation

anti conformation about C2-C3 bond

gauche conformation about C3-C4 bond

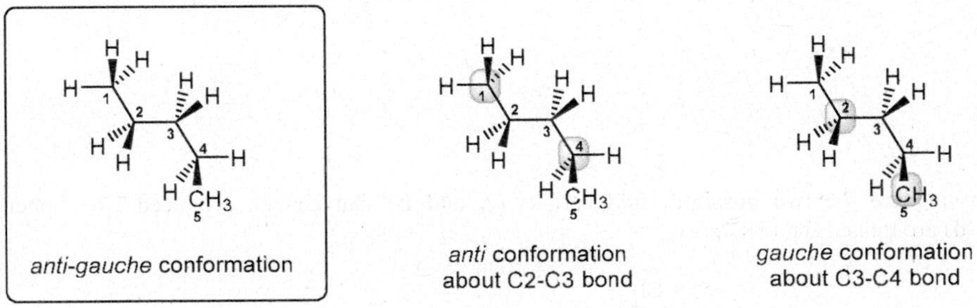

(b) The following conformer of 3-heptyne is analogous to the *anti-anti* conformer of pentane. Notice that C1 and C6 are *anti* to each other (when looking down the alkyne group), while C4 and C7 are also *anti* to each other (when looking down the C5-C6 bond).

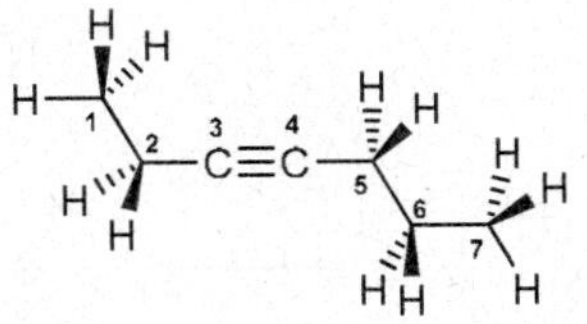

(c) In each of the lowest energy conformations of 3-heptyne, C1 and C6 are eclipsing each other (when looking down the alkyne group), as shown below. Note that the wedge-and-dash structure below is one of the two low-energy conformations, where C4 and C7 are *anti*:

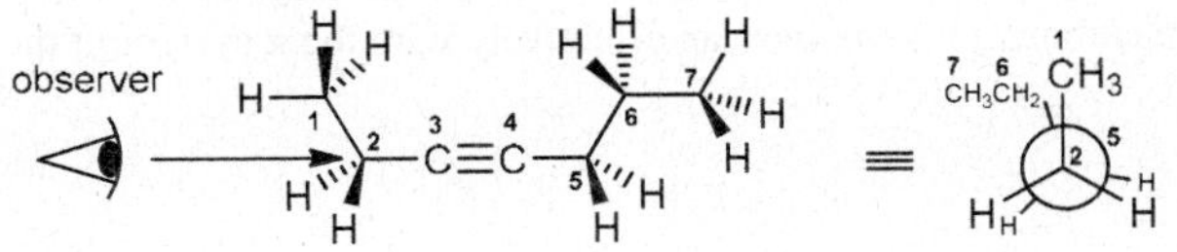

observer

(d) The difference between the two lowest energy conformations of 3-heptyne can be seen when looking down the C5-C6 bond. In one conformation, C4 and C7 are *anti* to each other, as shown below.

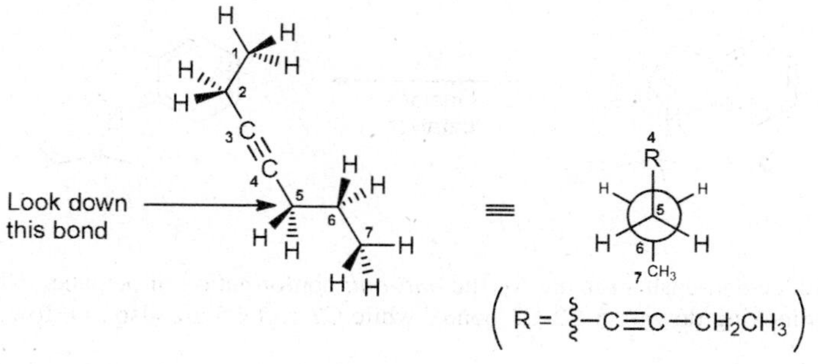

Look down this bond

$\left(R = \text{-}C{\equiv}C\text{-}CH_2CH_3 \right)$

In the other conformation, C4 and C7 are *gauche*, as shown here:

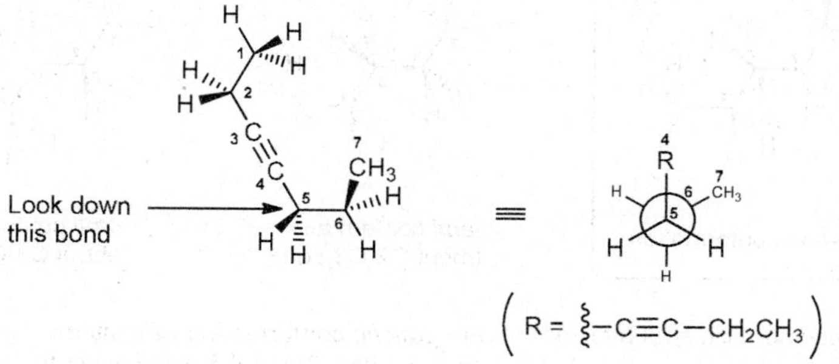

Look down this bond

$\left(R = \text{-}C{\equiv}C\text{-}CH_2CH_3 \right)$

9.85.

(a) The following are the two possible vinyl cations (**A** and **B**) that can be produced when phenyl-substituted acetylenes (**1a-d**) are treated with HCl:

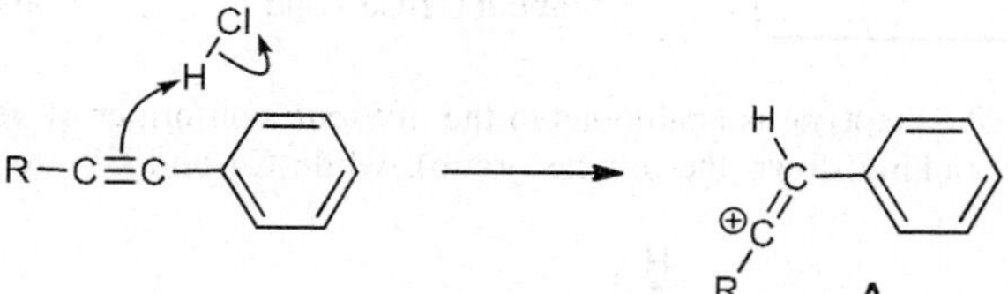

A

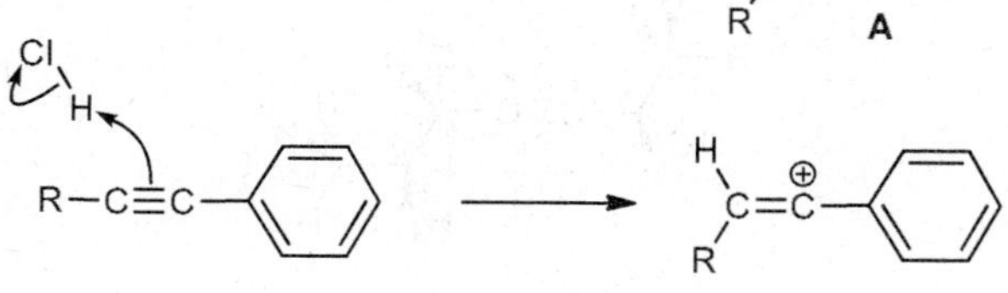

B

The empty 2*p* orbital (of the C+ atom) in cation **A** *cannot* be stabilized via resonance interaction with the aromatic ring, so we expect this cation to be highly unstable (as is the case for most vinyl carbocations). In contrast, the empty *p*-orbital of cation **B** (a benzylic carbocation) *can* overlap effectively with the π system of the aromatic ring:

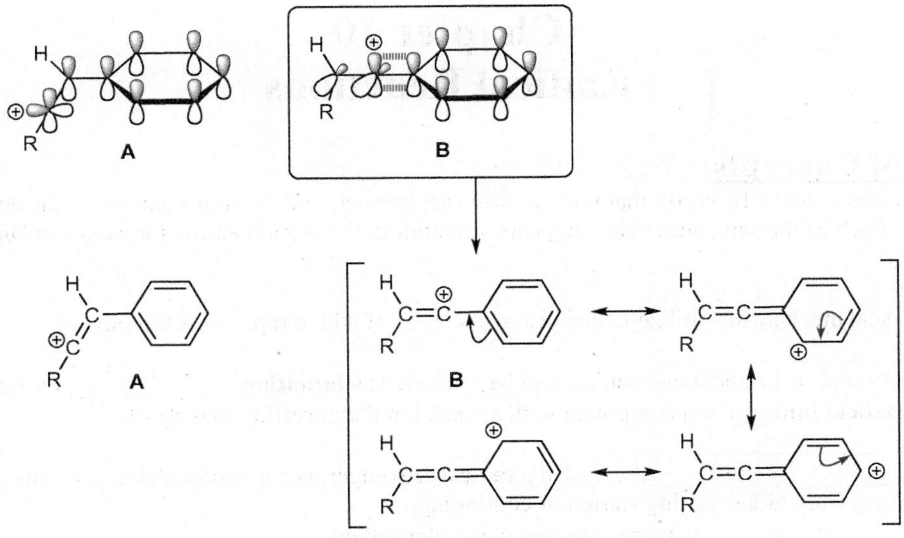

Since cation **B** is resonance-stabilized, it is much more stable than a typical vinyl carbocation. That is, cation **B** is more stable than cation **A**, explaining the regioselectivity observed in this series of hydrohalogenation reactions. This explains why Cl is installed at the benzylic position (the position next to the aromatic ring), because that is the location of the positive charge in the most stable carbocation.

(b) The stereoselectivity can be explained by considering the steric effects involved in the two competing transition states:

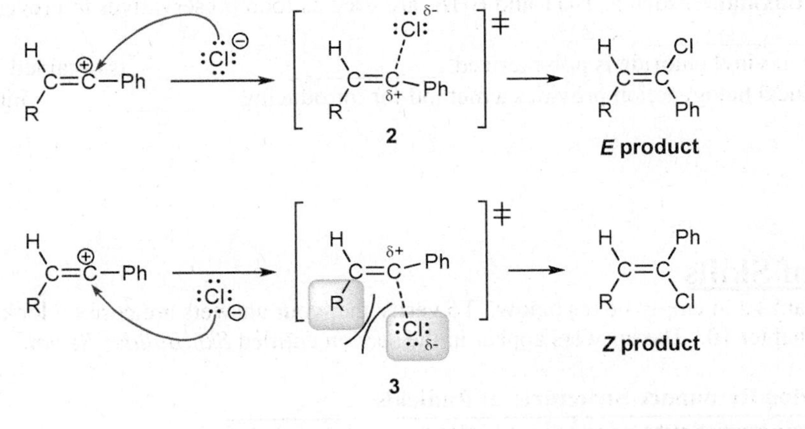

Attack of chloride on the vinyl carbocation via transition state **2** gives the *E* isomer, while transition state **3** results in the *Z* isomer. Transition state **3** involves a steric interaction between the alkyl group and the chloro group. For a small R group (such as R = Me), the preference for the *E* isomer is relatively small (70:30). As the size of the R group increases, there is increasing steric strain in transition state **3**, and the preference for the *E* isomer is enhanced. When R is the bulky *t*-butyl group, the *E* isomer is the exclusive product, via the lower energy transition state **2**.

Chapter 10
Radical Reactions

Review of Concepts

Fill in the blanks below. To verify that your answers are correct, look in your textbook at the end of Chapter 10. Each of the sentences below appears verbatim in the section entitled *Review of Concepts and Vocabulary*.

- Radical mechanisms utilize *fishhook arrows*, each of which represents the flow of ____________________.
- Every step in a radical mechanism can be classified as **initiation**, ____________, or **termination**.
- A **radical initiator** is a compound with a weak bond that readily undergoes ______________________.
- A ________________, also called a radical scavenger, is a compound that prevents a chain process from either getting started or continuing.
- ________________ is more selective than chlorination.
- When a new chiral center is created during a radical halogenation process, a ____________ mixture is obtained.
- _______________ can undergo **allylic bromination**, in which bromination occurs at the allylic position.
- Organic compounds undergo oxidation in the presence of atmospheric oxygen to produce **hydroperoxides**. This process, called ________________, is believed to proceed via a ____________ mechanism.
- **Antioxidants**, such as BHT and BHA, are used as food preservatives to prevent autooxidation of _______________ oils.
- When vinyl chloride is polymerized, ______________________ is obtained.
- Radical halogenation provides a method for introducing ______________ into an alkane.

Review of Skills

Fill in the blanks and empty boxes below. To verify that your answers are correct, look in your textbook at the end of Chapter 10. The answers appear in the section entitled *SkillBuilder Review*.

10.1 Drawing Resonance Structures of Radicals

Draw a resonance structure of the radical below, as well as the fishhook arrows that show its formation:

10.2 Identifying the Weakest C–H Bond in a Compound

Identify the weakest C–H bond in the following compound:

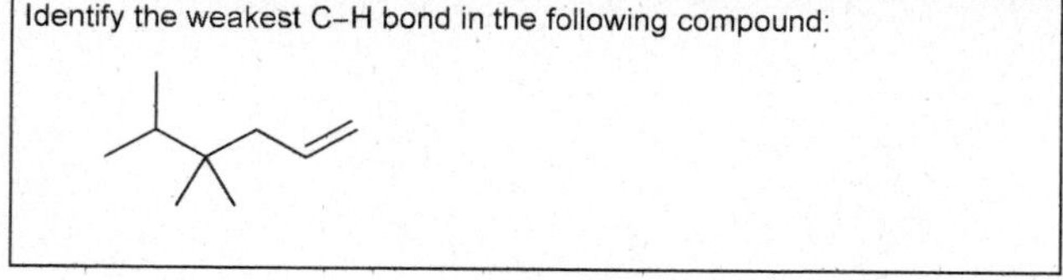

10.3 Identifying a Radical Pattern and Drawing Fishhook Arrows

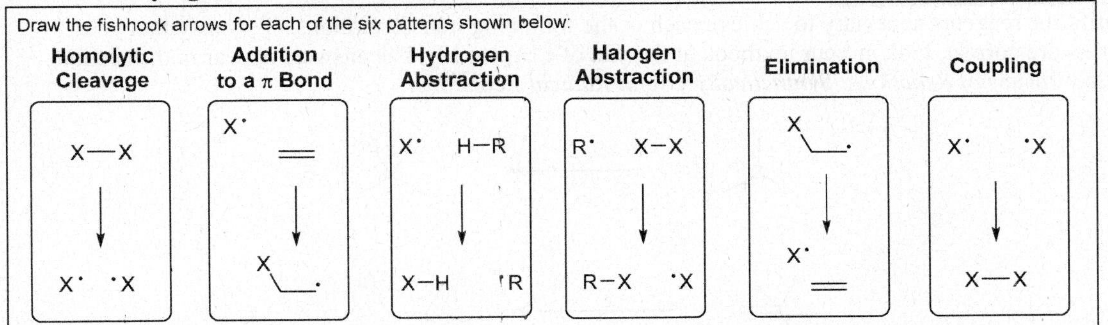

10.4 Drawing a Mechanism for Radical Halogenation

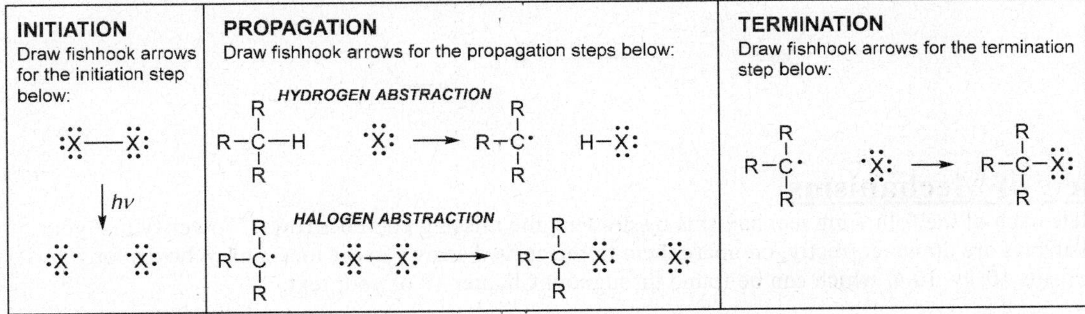

INITIATION
Draw fishhook arrows for the initiation step below:
hν
PROPAGATION
Draw fishhook arrows for the propagation steps below:
HYDROGEN ABSTRACTION
HALOGEN ABSTRACTION
TERMINATION
Draw fishhook arrows for the termination step below:

10.5 Predicting the Selectivity of Radical Bromination

Draw the expected major product of the following monobromination reaction:

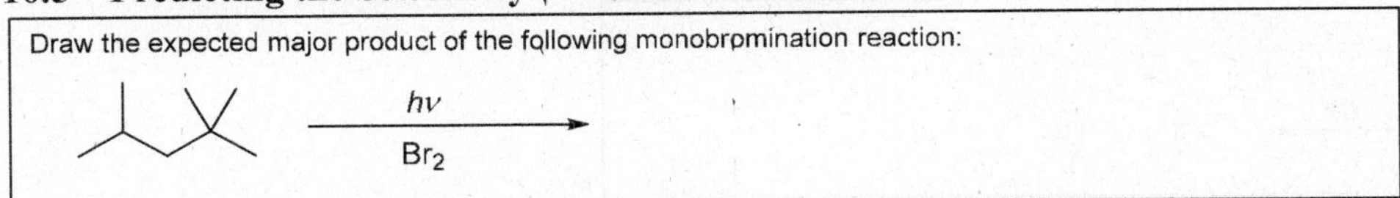

10.6 Predicting the Stereochemical Outcome of Radical Bromination

Draw the expected products of the following reaction:

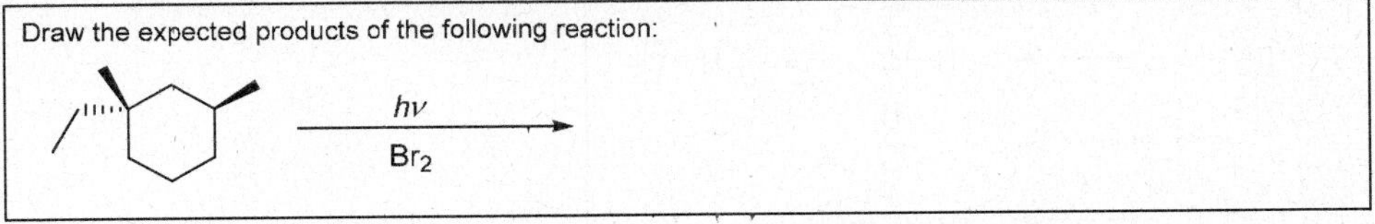

10.7 Predicting the Products of Allylic Bromination

Draw the expected products of the following reaction:

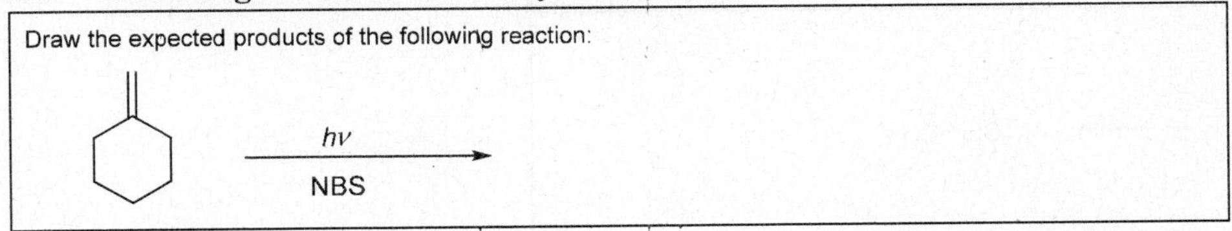

10.8 Predicting the Products for Radical Addition of HBr

Draw the expected products of the following reaction:

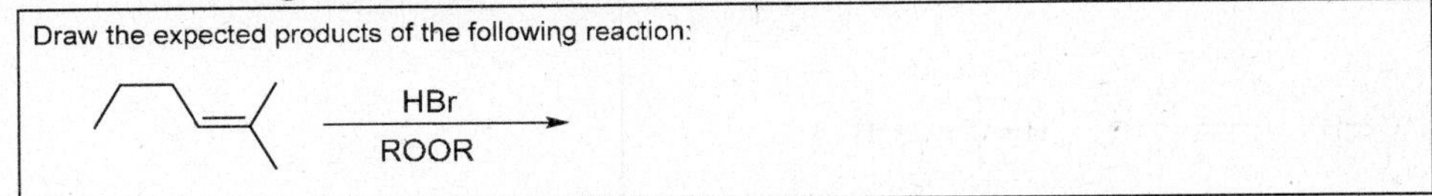

Review of Reactions

Identify the reagents necessary to achieve each of the following transformations. To verify that your answers are correct, look in your textbook at the end of Chapter 10. The answers appear in the section entitled *Review of Reactions: Synthetically Useful Radical Reactions.*

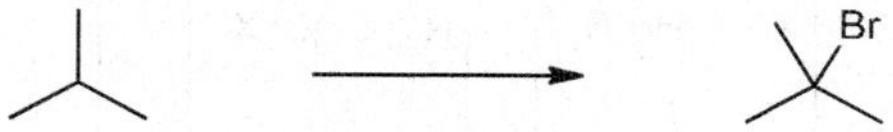

Br

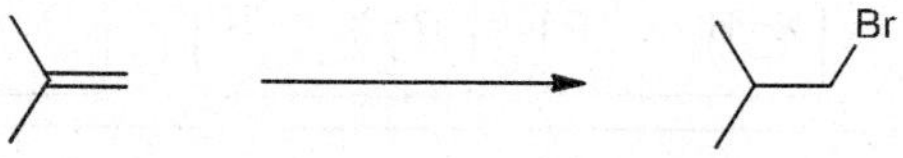

Br

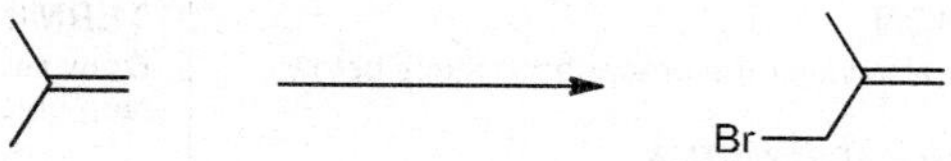

Br

Review of Mechanisms

Complete each of the following mechanisms by drawing the missing curved arrows. To verify that your curved arrows are drawn correctly, compare them to the curved arrows in the mechanism boxes for Mechanisms 10.1 – 10.4, which can be found throughout Chapter 10 of your text.

Mechanism 10.1 Radical Chlorination

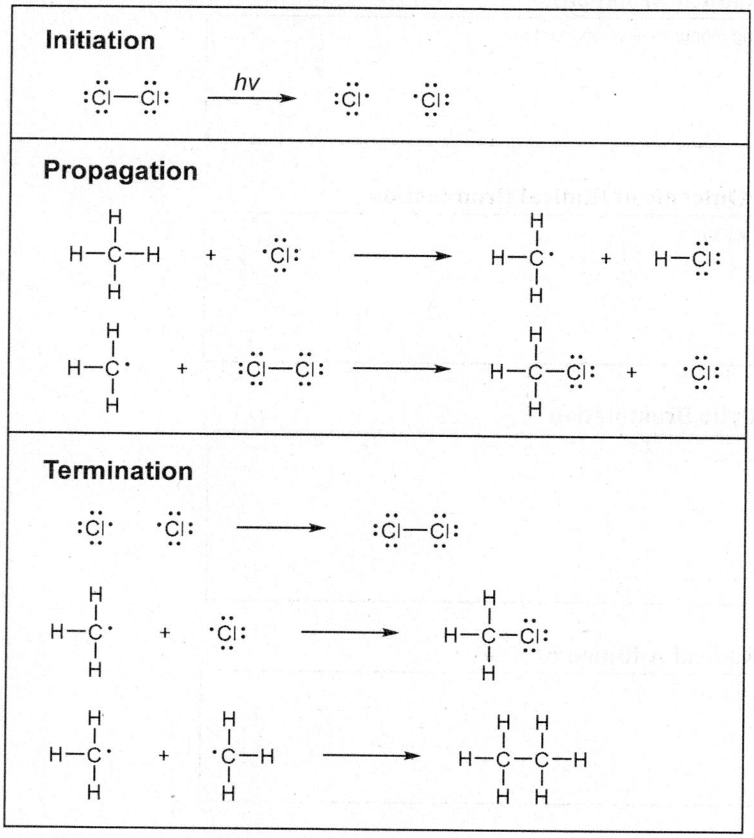

Initiation

:Cl—Cl: —hν→ :Cl· ·Cl:

Propagation

H—C(H)(H)—H + ·Cl: → H—C(H)(H)· + H—Cl:

H—C(H)(H)· + :Cl—Cl: → H—C(H)(H)—Cl: + ·Cl:

Termination

:Cl· ·Cl: → :Cl—Cl:

H—C(H)(H)· + ·Cl: → H—C(H)(H)—Cl:

H—C(H)(H)· + ·C(H)(H)—H → H—C(H)(H)—C(H)(H)—H

Mechanism 10.2 Autooxidation

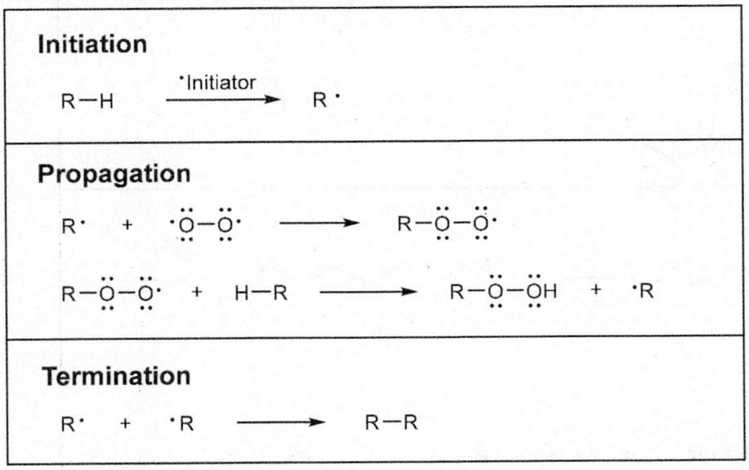

Initiation

R—H $\xrightarrow{\cdot\text{Initiator}}$ R·

Propagation

R· + ·O—O· ⟶ R—O—O·

R—O—O· + H—R ⟶ R—O—OH + ·R

Termination

R· + ·R ⟶ R—R

Mechanism 10.3 Radical Addition of HBr to an Alkene

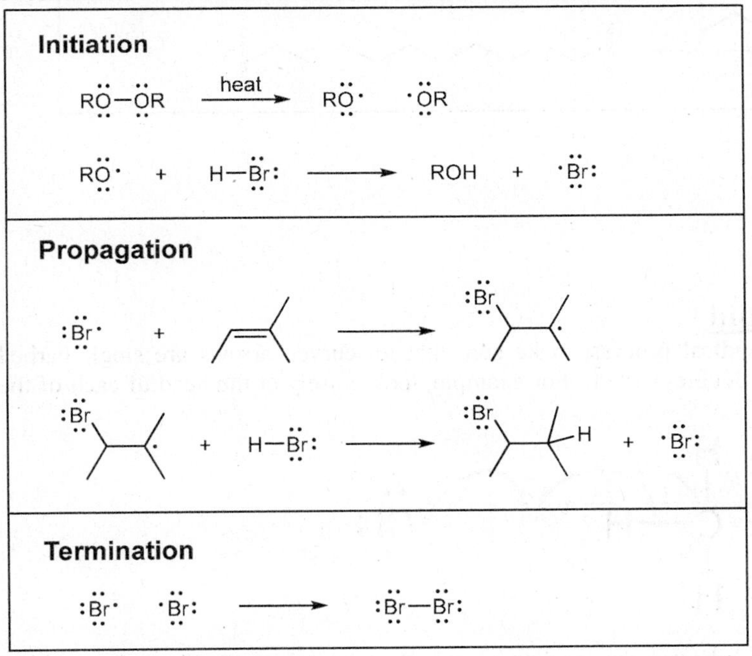

Initiation

RO—OR $\xrightarrow{\text{heat}}$ RO· ·OR

RO· + H—Br: ⟶ ROH + ·Br:

Propagation

:Br· + [alkene] ⟶ [:Br–substituted radical]

[:Br–substituted radical] + H—Br: ⟶ [:Br–substituted product]—H + ·Br:

Termination

:Br· ·Br: ⟶ :Br—Br:

Mechanism 10.4 Radical Polymerization

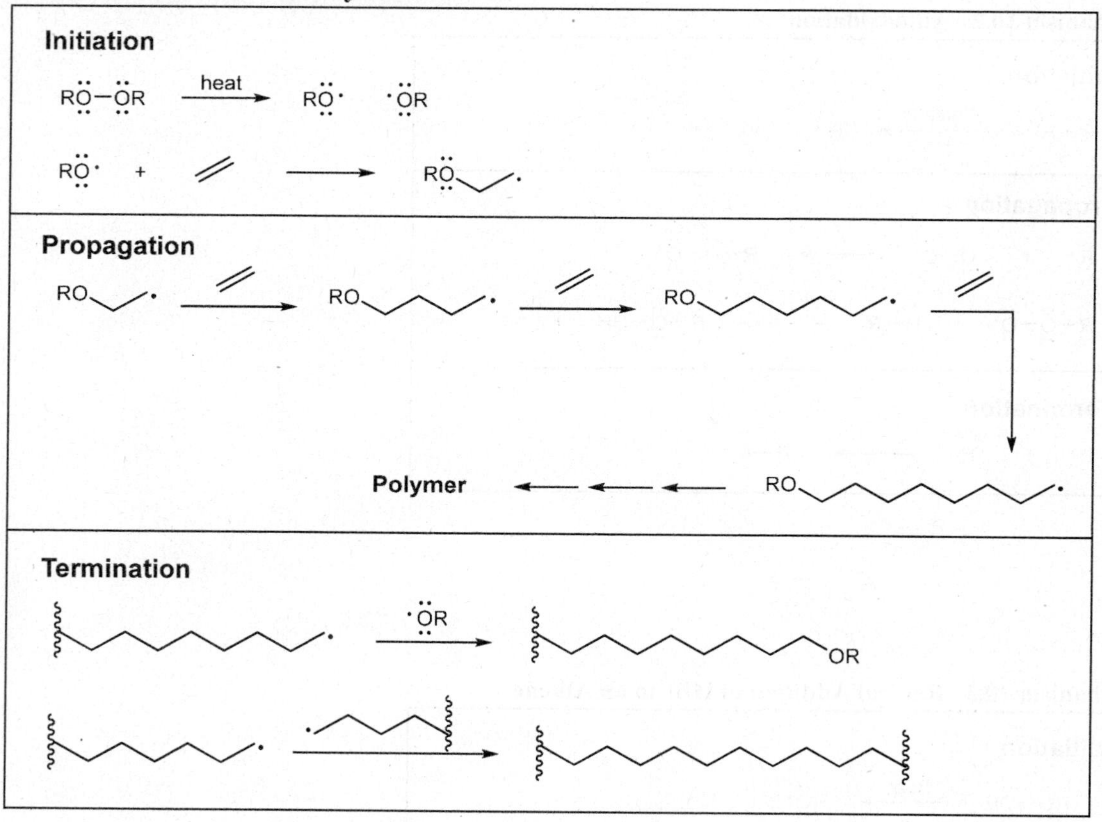

Initiation

RO–OR —heat→ RO· ·OR

RO· + ethylene → RO–CH₂CH₂·

Propagation

Polymer

Termination

OR

Common Mistakes to Avoid

When drawing a mechanism for a radical process, make sure that all curved arrows are single-barbed (called fishhook arrows), rather than double-barbed. For example, look closely at the head of each of the following fishhook arrows:

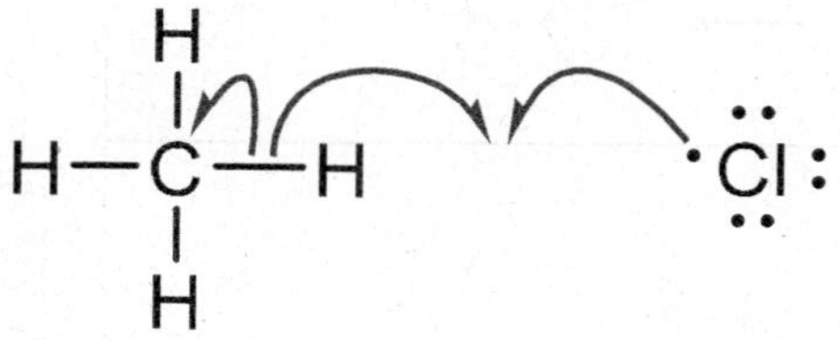

H–C(H)(H)–H ·Cl:

Each of these single-barbed arrows indicates the motion of one electron, while double-barbed arrows indicate the motion of two electrons. All of the mechanisms presented in this chapter utilize single-barbed arrows. Make sure to draw them properly.

Useful Reagents

The following is a list of common reagents covered in this chapter:

Reagents	Name of Reaction	Description of Reaction
Br_2, $h\nu$	Radical bromination	Under these conditions, an alkane undergoes bromination, with installation of Br (replacing a hydrogen atom) at the most substituted position.
Cl_2, $h\nu$	Radical chlorination	Under these conditions, an alkane undergoes chlorination. This reaction is less selective than bromination (but faster), and as such, it is generally most useful in situations where monochlorination results in only one regiochemical outcome (such as chlorination of cyclohexane or chlorination of 2,2-dimethylpropane).
HBr, ROOR	Hydrobromination	When treated with HBr in the presence of peroxides, an alkene undergoes *anti*-Markovnikov addition of H and Br.
NBS, $h\nu$	Allylic bromination	NBS, or *N*-bromosuccinimide, is a reagent that can be used to install a bromine atom at the allylic position of an alkene.

Solutions

10.1.

(a) The tertiary radical is the most stable, because alkyl groups stabilize the unpaired electron via a delocalization effect, called hyperconjugation. The primary radical is the least stable, because it lacks the stabilizing effect provided by multiple alkyl groups.

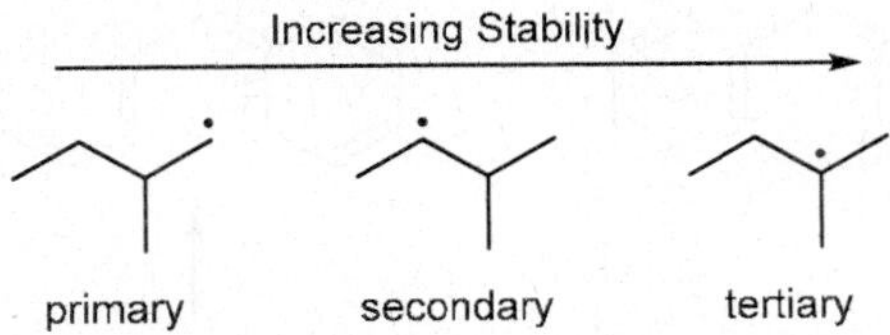

(b) The tertiary radical is the most stable, because alkyl groups stabilize the unpaired electron via a delocalization effect, called hyperconjugation. The primary radical is the least stable, because it lacks the stabilizing effect provided by multiple alkyl groups.

Increasing Stability

primary secondary tertiary

10.2.

(a) The unpaired electron occupies an allylic position, so it is resonance-stabilized. Three fishhook arrows are required, as shown:

(b) The unpaired electron occupies an allylic position, so it is resonance-stabilized. Three fishhook arrows are required, as shown below. The resulting resonance structure exhibits an unpaired electron that is allylic to another π bond, so again we draw three fishhook arrows to arrive at a third resonance structure:

(c) The unpaired electron occupies an allylic position, so it is resonance-stabilized. Three fishhook arrows are required, as shown below. The resulting resonance structure exhibits an unpaired electron that is allylic to another π bond, so again we draw three fishhook arrows to arrive at a third resonance structure:

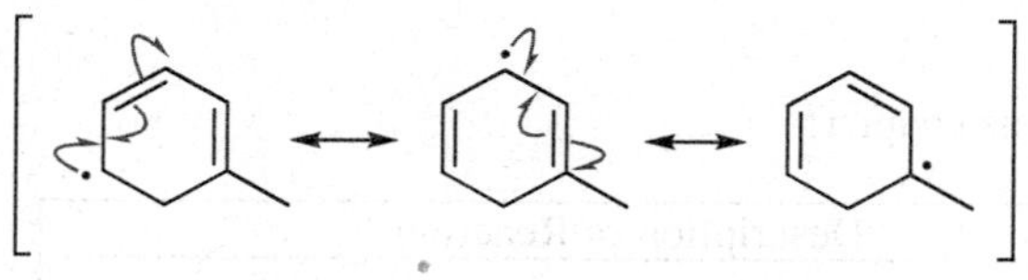

(d) The unpaired electron occupies an allylic position, so it is resonance-stabilized. Three fishhook arrows are required, as shown:

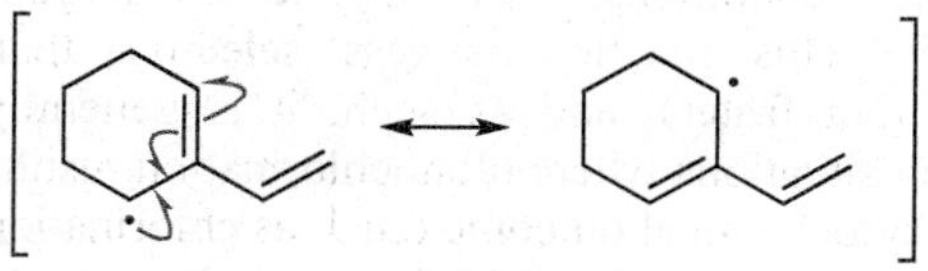

Note that the other π bond does not participate in resonance, because the unpaired electron is not allylic to that π bond.

10.3. The following hydrogen atom (highlighted) is connected to an allylic position, and its removal will generate a resonance-stabilized radical:

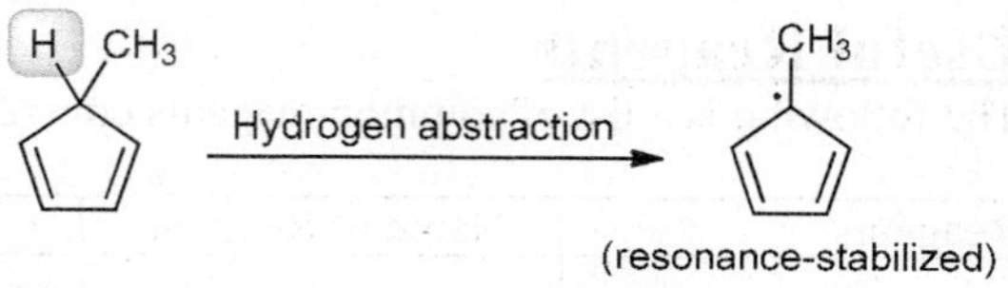

H CH3 Hydrogen abstraction CH3 (resonance-stabilized)

To draw the resonance structures for this radical, we begin by drawing three fishhook arrows. The resulting resonance structure also exhibits an unpaired electron that is allylic to a π bond, so again we draw three fishhook arrows as well as the resulting resonance structure. This process continues until we have drawn all five resonance structures, shown here:

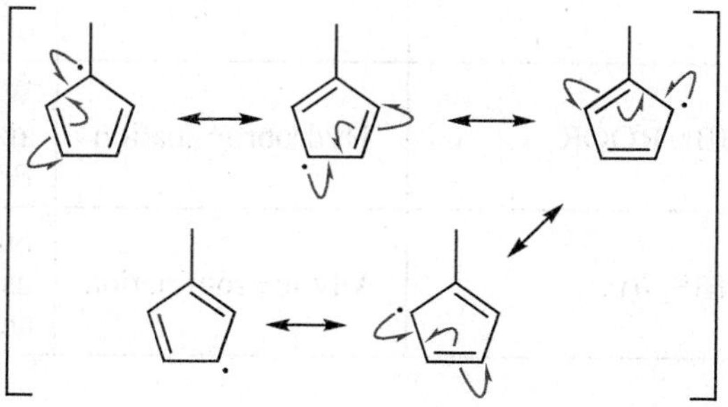

10.4. This radical is unusually stable, because it has a large number of resonance structures, so the unpaired electron is highly delocalized. As shown below, three fishhook arrows are required in order to draw each resonance structure:

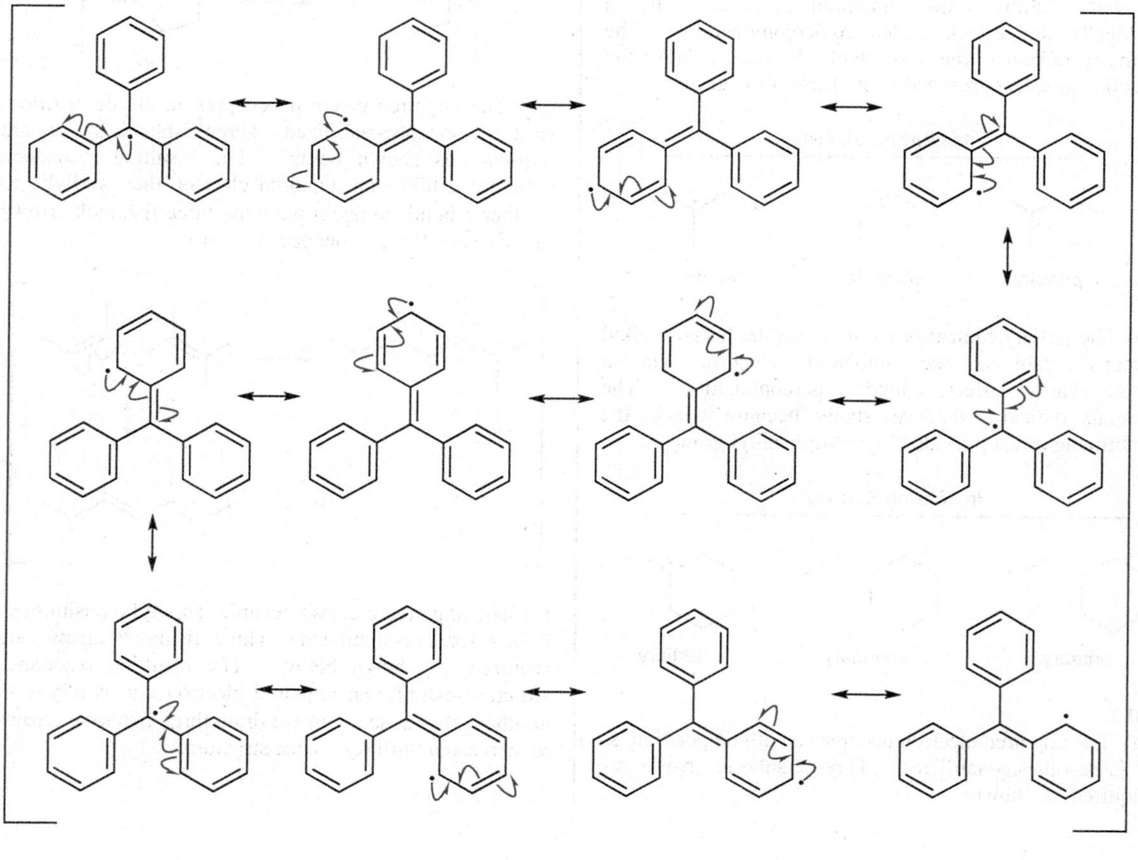

10.5.

(a) We consider each of the C—H bonds in the compound, and in each case, we imagine the radical that would result from homolytic cleavage of that C—H bond. Among all the C—H bonds, only one of them can undergo homolytic cleavage to generate a resonance-stabilized radical. Specifically, removal of the highlighted hydrogen atom will result in an allylic radical. And therefore, this C—H bond is the weakest C—H bond in the compound:

Hydrogen abstraction

(resonance-stabilized)

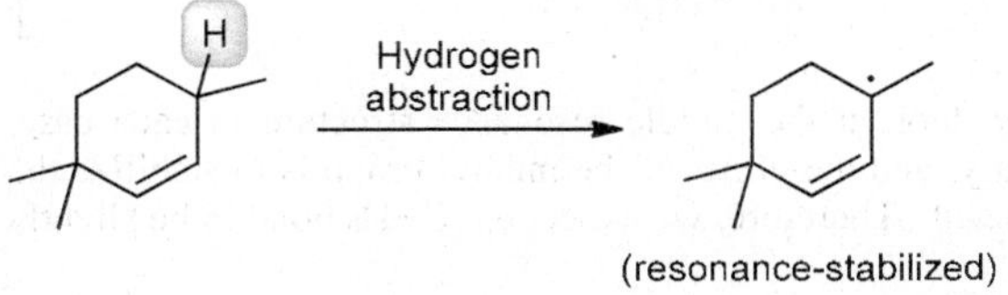

(b) We consider each of the C—H bonds in the compound, and in each case, we imagine the radical that would result from homolytic cleavage of that C—H bond. There are only two different kinds of C—H bonds that can undergo homolytic cleavage to generate a resonance-stabilized radical. Specifically, removing either of the highlighted hydrogen atoms will result in a resonance-stabilized intermediate (with three resonance structures):

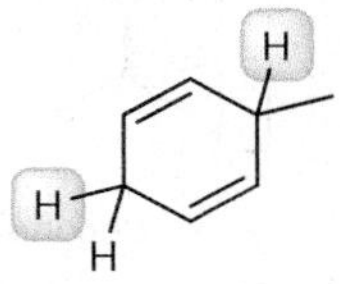

These two locations represent the two weakest C—H bonds in the compound. Between the two of them, the weaker C—H bond is the one that gives a tertiary allylic radical upon removal of the hydrogen atom (rather than a secondary allylic radical):

secondary allylic tertiary allylic

A tertiary allylic radical is more stable than a secondary allylic radical, so the following C—H bond is the weakest C—H bond in the compound:

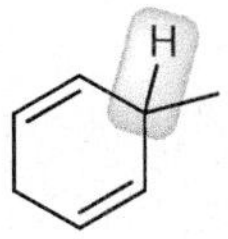

(c) We consider each of the C—H bonds in the compound, and in each case, we imagine the radical that would result from homolytic cleavage of that C—H bond. Among all the different types of C—H bonds, only one location can undergo homolytic cleavage to generate a resonance-stabilized radical. Specifically, removal of the highlighted hydrogen atom will result in an allylic radical. And therefore, this C—H bond is the weakest C—H bond in the compound:

Hydrogen abstraction

(resonance-stabilized)

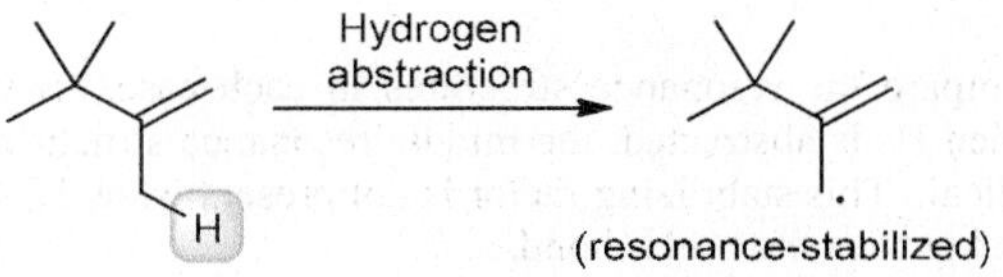

(d) We consider each of the C—H bonds in the compound, and in each case, we imagine the radical that would result from homolytic cleavage of that C—H bond. There are only two different kinds of C—H bonds that can undergo homolytic cleavage to generate a resonance-stabilized radical. Specifically, removing either of the highlighted hydrogen atoms will result in a resonance-stabilized intermediate (with two resonance structures):

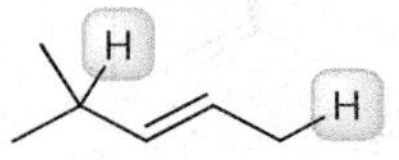

These two locations represent the two weakest C—H bonds in the compound. Between the two of them, the weaker C—H bond is the one that gives a tertiary allylic radical (rather than a primary allylic radical) upon removal of the hydrogen atom:

tertiary allylic primary allylic

Therefore, the following C—H bond is the weakest C—H bond in the compound:

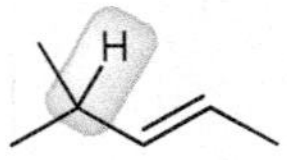

10.6. We begin by drawing the resonance structures of the radical that is formed when H_a is abstracted, as well as the resonance structures of the radical that is formed when H_b is abstracted:

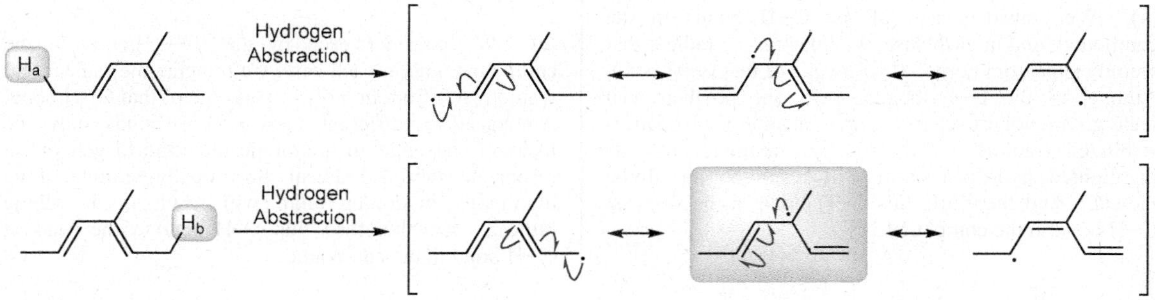

Compare the resonance structures in each case. Specifically, look at the middle resonance structure in each case. When H_b is abstracted, the middle resonance structure is tertiary, and the effect of the methyl group is to stabilize the radical. This stabilizing factor is not present when H_a is abstracted. Therefore, we expect the C—H_b bond to be slightly weaker than the C—H_a bond.

10.7. Imagine homolytically breaking each different type of C—H bond in the compound and then evaluate the stability of the resulting radical. There are ten locations to consider. Starting with the methyl group at the top, and proceeding methodically clockwise around the structure, gives the following analysis:

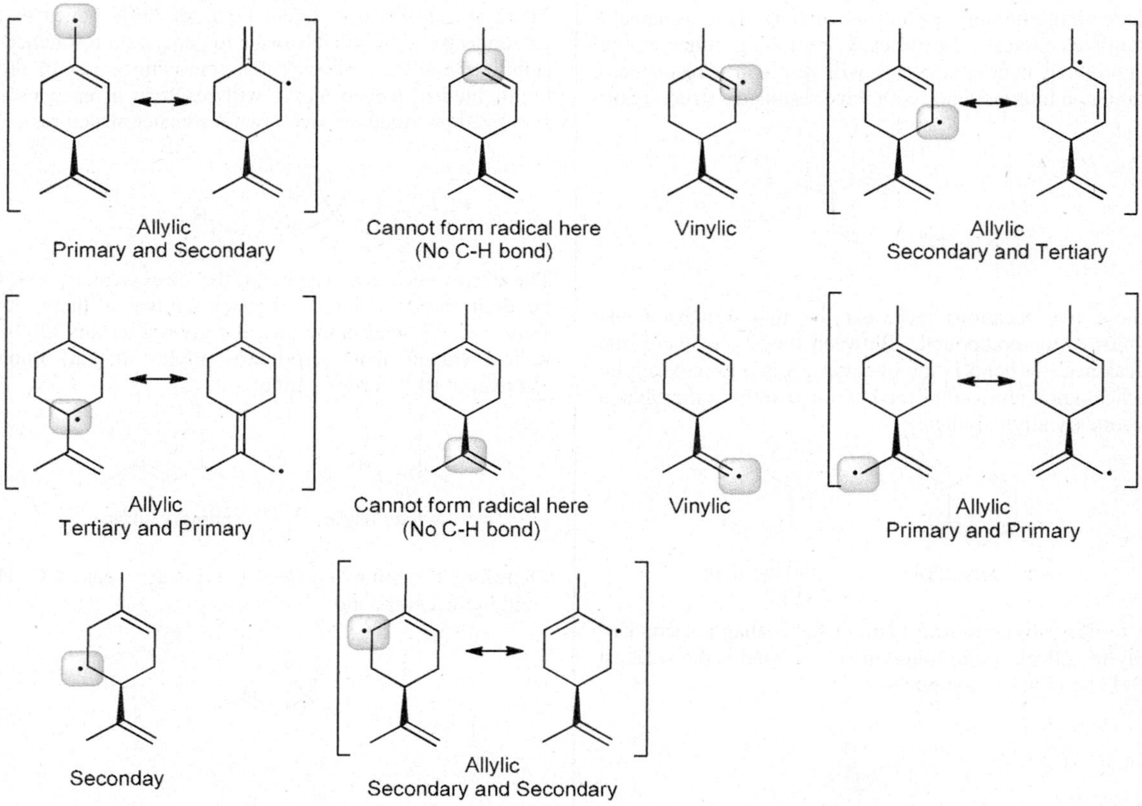

Two positions have no H atoms attached and therefore cannot react. Loss of a hydrogen atom at each of the remaining eight positions yields the eight radicals as shown. Two are vinylic, one is secondary and five are allylic. Each of the allylic radicals has two resonance forms and we must consider those forms when evaluating their relative stability. It appears that the most stable allylic radical has the unpaired electron delocalized over a secondary and a tertiary position, while the other allylic radicals are less substituted. Based on this, we conclude that the following C—H bonds are the weakest, since removal of a hydrogen atom from this location yields the most stable radical.

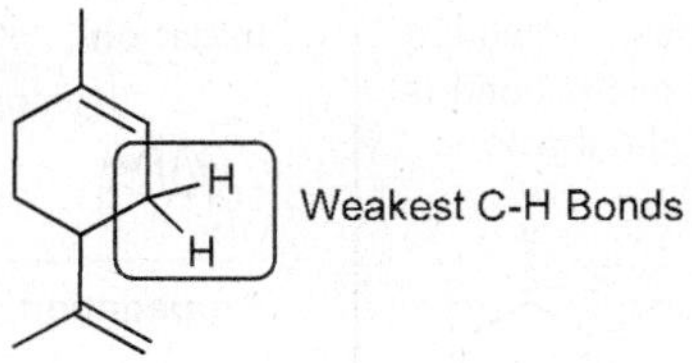

Weakest C-H Bonds

10.8.

(a) This step is a *hydrogen abstraction*, which requires a total of three fishhook arrows. In order to place the fishhook arrows properly, we must identify any bonds being formed (C–H) and any bonds being broken (H–Br). Formation of the C–H bond is shown with two fishhook arrows: one coming from the carbon radical and the other coming from the H–Br bond. The latter, together with the third fishhook arrow, shows the breaking of the H–Br bond:

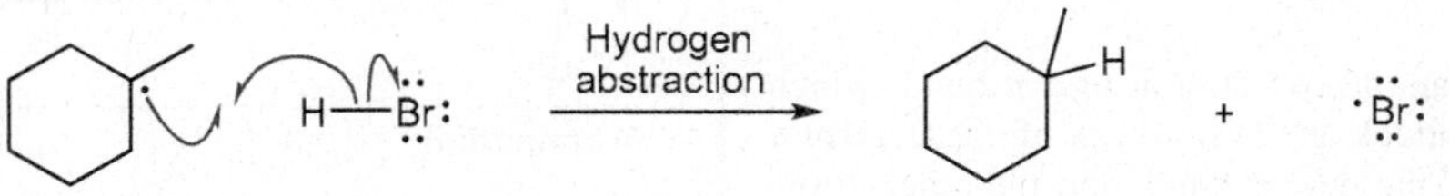

Hydrogen abstraction

(b) This step is *addition to a π bond*, which requires a total of three fishhook arrows. In order to place the fishhook arrows properly, we must identify any bonds being formed (C–Br) and any bonds being broken (C=C π bond). Formation of the C–Br bond is shown with two fishhook arrows: one coming from the bromine radical and the other coming from the π bond. The latter, together with the third fishhook arrow, shows the breaking of the C=C π bond:

Addition to a pi bond

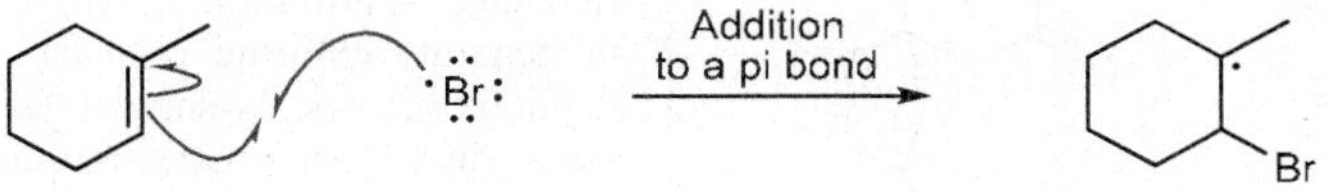

(c) This step is a *coupling* process, which requires a total of two fishhook arrows, showing formation of a bond (in this case, a C–C bond is formed):

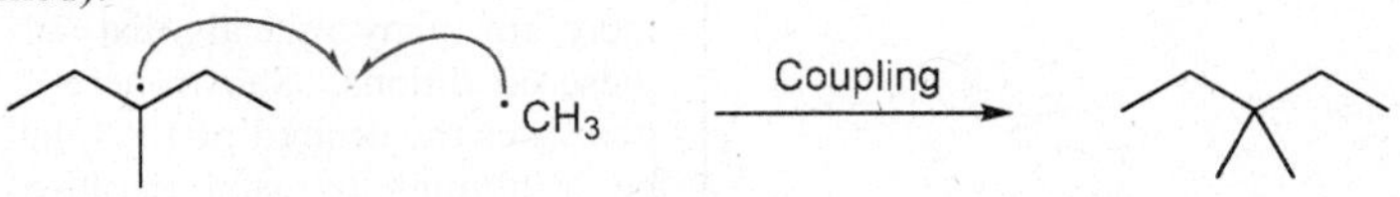

Coupling

(d) This step is a *hydrogen abstraction*, which requires a total of three fishhook arrows. In order to place the fishhook arrows properly, we must identify any bonds being formed (H–Br) and any bonds being broken (C–H). Formation of the H–Br bond is shown with two fishhook arrows: one coming from the bromine radical and the other coming from the C–H bond. The latter, together with the third fishhook arrow, shows the breaking of the C–H bond:

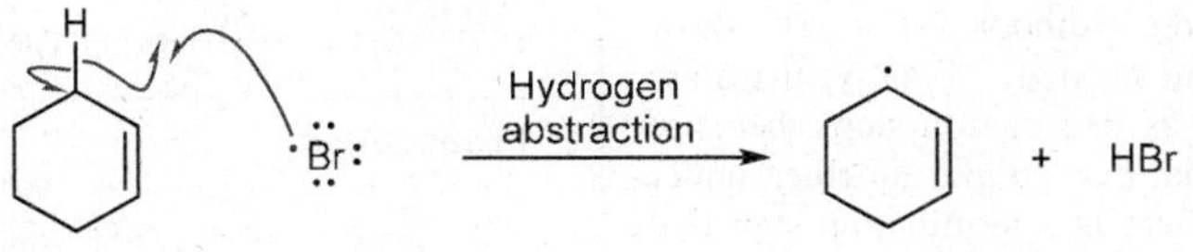

Hydrogen abstraction

(e) This step is an *elimination*, which requires a total of three fishhook arrows. In order to place the fishhook arrows properly, we must identify any bonds being formed (a C=C π bond) and any bonds being broken (C–C). Formation of the C=C π bond is shown with two fishhook arrows: one coming from the carbon radical and the other coming from the neighboring C–C bond. The latter, together with the third fishhook arrow, shows the breaking of the C–C bond:

Elimination

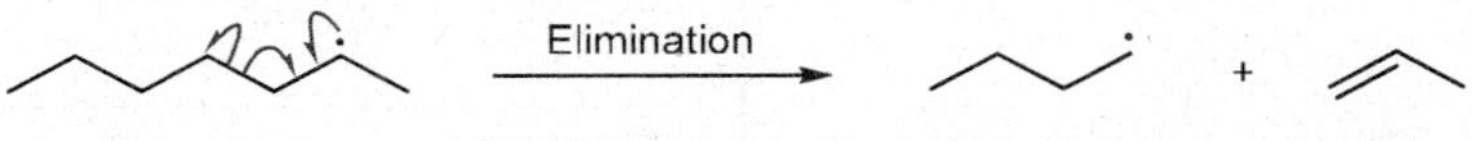

(f) This step is a *homolytic bond cleavage*, which requires a total of two fishhook arrows, showing breaking of a bond (in this case, an O–O bond is broken):

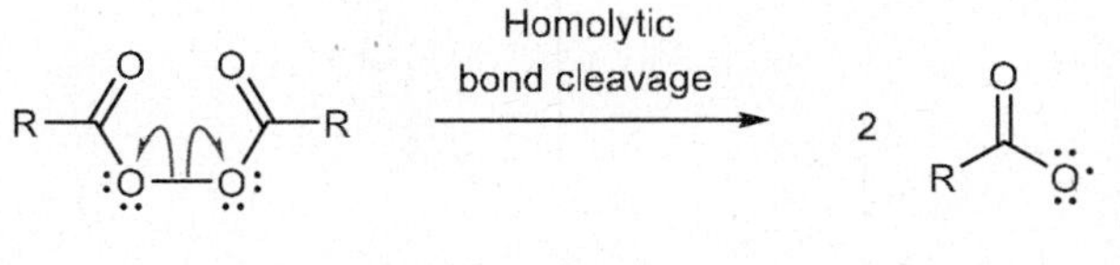

Homolytic bond cleavage

10.9. In this intramolecular process, a radical is reacting with a π bond. One π bond is broken and one σ bond is formed in this step – these bonds are highlighted below:

OEt EtO

π bond broken σ bond formed

This step is an example of addition to a π bond, which requires three fishhook arrows. Two of these arrows show formation of the new σ bond, and the other arrow shows where the unpaired electron will be located in intermediate **3**.

2 OEt EtO **3**

10.10.

(a) The mechanism will have three distinct stages. The first stage is initiation, in which the Cl–Cl bond is broken to generate chlorine radicals. This step requires two fishhook arrows, as shown. There are two propagation steps (hydrogen abstraction and halogen abstraction), each of which requires three fishhook arrows (to show the bonds being broken and formed). Finally, there are many steps that can serve as termination steps, because there are many radicals that can couple together under these conditions. Shown here is a termination step that generates the desired product, but it is still considered to be a termination step because this step reduces the number of radicals present in the reaction flask (two radicals are destroyed, without generating new radicals):

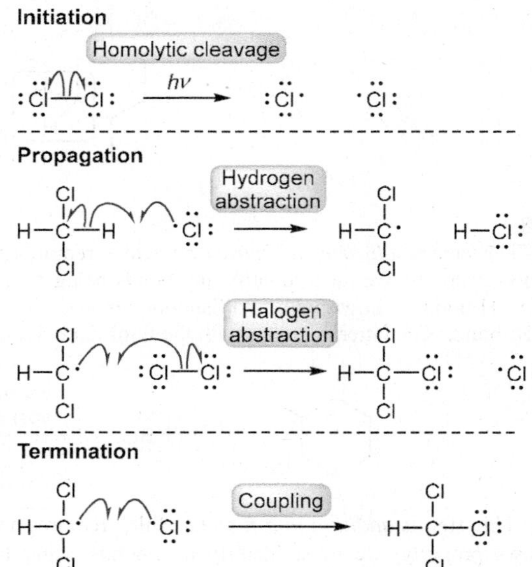

(b) The mechanism will have three distinct stages. The first stage is initiation, in which the Cl–Cl bond is broken to generate chlorine radicals. This step requires two fishhook arrows, as shown. There are two propagation steps (hydrogen abstraction and halogen abstraction), each of which requires three fishhook arrows (to show the bonds being broken and formed). Finally, there are many steps that can serve as termination steps, because there are many radicals that can couple together under these conditions. Shown here is a termination step that generates the desired product, but it is still considered to be a termination step because this step reduces the number of radicals present in the reaction flask (two radicals are destroyed, without generating new radicals):

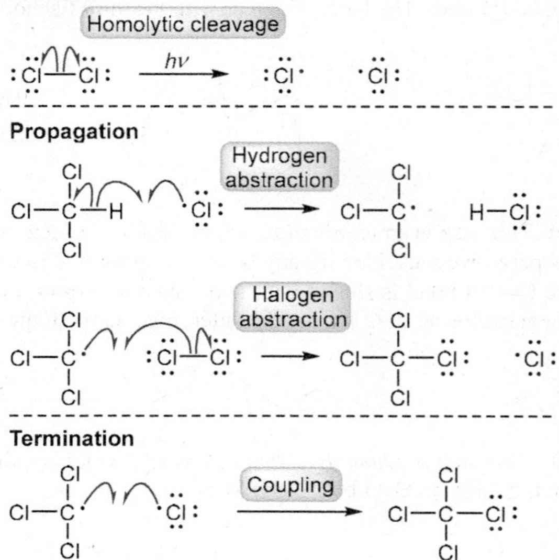

(c) The mechanism will have three distinct stages. The first stage is initiation, in which the Cl–Cl bond is broken to generate chlorine radicals. This step requires two fishhook arrows, as shown. There are two propagation steps (hydrogen abstraction and halogen abstraction), each of which requires three fishhook arrows (to show the bonds being broken and formed). Finally, there are many steps that can serve as termination steps, because there are many radicals that can couple together under these conditions. Shown below is a termination step that generates the desired product, but it is still considered to be a termination step because this step reduces the number of radicals present in the reaction flask (two radicals are destroyed, without generating new radicals):

Initiation

Homolytic cleavage

$$:\ddot{Cl}-\ddot{Cl}: \xrightarrow{h\nu} :\ddot{Cl}\cdot \quad \cdot\ddot{Cl}:$$

Propagation

Hydrogen abstraction

$$H_3C-CH_2-H \quad \cdot\ddot{Cl}: \longrightarrow H_3C-CH_2\cdot \quad H-\ddot{Cl}:$$

Halogen abstraction

$$H_3C-CH_2\cdot \quad :\ddot{Cl}-\ddot{Cl}: \longrightarrow H_3C-CH_2-\ddot{Cl}: \quad \cdot\ddot{Cl}:$$

Termination

Coupling

$$H_3C-CH_2\cdot \quad \cdot\ddot{Cl}: \longrightarrow H_3C-CH_2-\ddot{Cl}:$$

(d) The mechanism will have three distinct stages. The first stage is initiation, in which the Cl–Cl bond is broken to generate chlorine radicals. This step requires two fishhook arrows, as shown. There are two propagation steps (hydrogen abstraction and halogen abstraction), each of which requires three fishhook arrows (to show the bonds being broken and formed). Finally, there are many steps that can serve as termination steps, because there are many radicals that can couple together under these conditions. Shown below is a termination step that actually generates the desired product, but it is still considered to be a termination step because this step reduces the number of radicals present in the reaction flask (two radicals are destroyed, without generating new radicals):

Initiation

Homolytic cleavage

$$:\ddot{Cl}-\ddot{Cl}: \xrightarrow{h\nu} :\ddot{Cl}\cdot \quad \cdot\ddot{Cl}:$$

Propagation

Hydrogen abstraction

$$Cl_3C-CH_2-H \quad \cdot\ddot{Cl}: \longrightarrow Cl_3C-CH_2\cdot \quad H-\ddot{Cl}:$$

Halogen abstraction

$$Cl_3C-CH_2\cdot \quad :\ddot{Cl}-\ddot{Cl}: \longrightarrow Cl_3C-CH_2-\ddot{Cl}: \quad \cdot\ddot{Cl}:$$

Termination

Coupling

$$Cl_3C-CH_2\cdot \quad \cdot\ddot{Cl}: \longrightarrow Cl_3C-CH_2-\ddot{Cl}:$$

(e) The mechanism will have three distinct stages. The first stage is initiation, in which the Cl–Cl bond is broken to generate chlorine radicals. This step requires two fishhook arrows, as shown. There are two propagation steps (hydrogen abstraction and halogen abstraction), each of which requires three fishhook arrows (to show the bonds being broken and formed). Finally, there are many steps that can serve as termination steps, because there are many radicals that can couple together under these conditions. Shown here is a termination step that actually generates the desired product, but it is still considered to be a termination step because this step reduces the number of radicals present in the reaction flask (two radicals are destroyed, without generating new radicals):

Initiation

Homolytic cleavage

$$:\ddot{Cl}-\ddot{Cl}: \xrightarrow{h\nu} :\ddot{Cl}\cdot \quad \cdot\ddot{Cl}:$$

Propagation

Hydrogen abstraction

$$H_3C-CCl_2-CH_2-H \quad \cdot\ddot{Cl}: \longrightarrow H_3C-CCl_2-CH_2\cdot \quad H-\ddot{Cl}:$$

Halogen abstraction

$$H_3C-CCl_2-CH_2\cdot \quad :\ddot{Cl}-\ddot{Cl}: \longrightarrow H_3C-CCl_2-CH_2-\ddot{Cl}: \quad \cdot\ddot{Cl}:$$

Termination

Coupling

$$H_3C-CCl_2-CH_2\cdot \quad \cdot\ddot{Cl}: \longrightarrow H_3C-CCl_2-CH_2-\ddot{Cl}:$$

10.11.

(a) The mechanism for radical bromination of methane will have three distinct stages. The first stage is initiation, which creates bromine radicals. This initiation step involves homolytic bond cleavage (using Br_2 and either heat or light) and should employ two fishhook arrows.

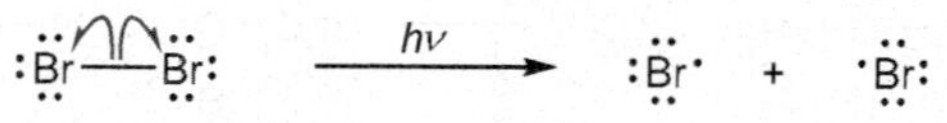

The next stage involves propagation steps. There are two propagation steps: hydrogen abstraction to remove a hydrogen atom, followed by halogen abstraction to attach a bromine atom. Each of these steps should have three fishhook arrows. These two steps together represent the core reaction. They show how the product is formed.

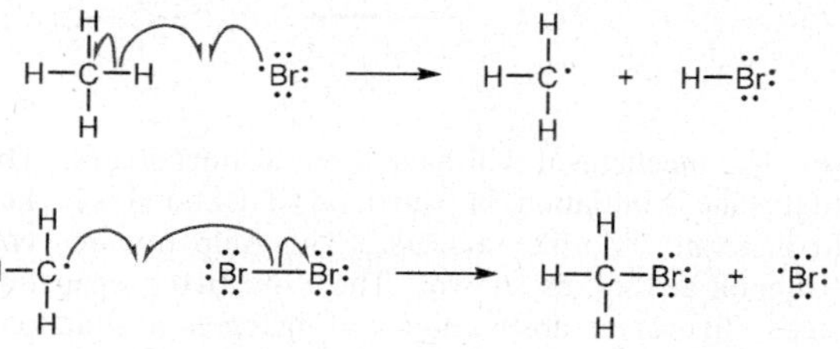

There are a number of possible termination steps to end the process. When drawing a mechanism for a radical reaction it is generally not necessary to draw all possible termination steps unless specifically asked to do so. It is sufficient to draw one termination step. Shown here is one termination step that also happens to produce the desired product.

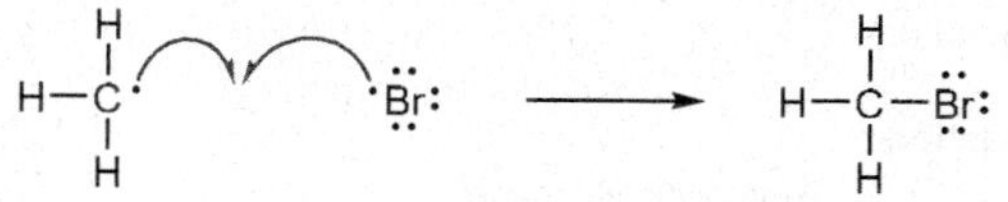

(b) During the bromination of methane, methyl radicals are produced. If two of these methyl radicals combine in a termination step, ethane is produced:

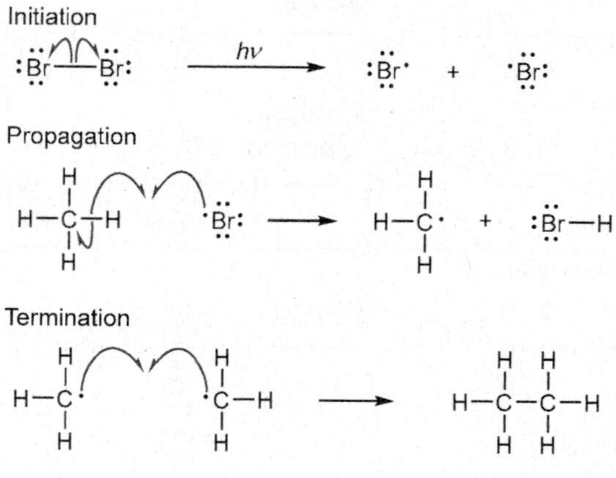

Initiation

Propagation

Termination

10.12.

(a) The tertiary position is expected to undergo selective bromination, giving the following alkyl bromide:

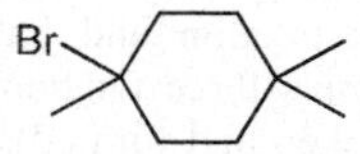

(b) The tertiary position is expected to undergo selective bromination, giving the following alkyl bromide:

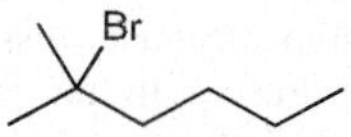

(c) The tertiary position is expected to undergo selective bromination, giving the following alkyl bromide:

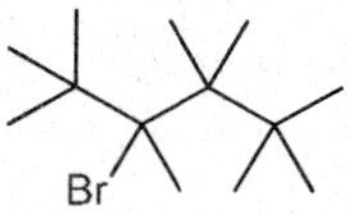

10.13. The tertiary allylic position is expected to undergo selective bromination, giving the following monobrominated compound as the major product:

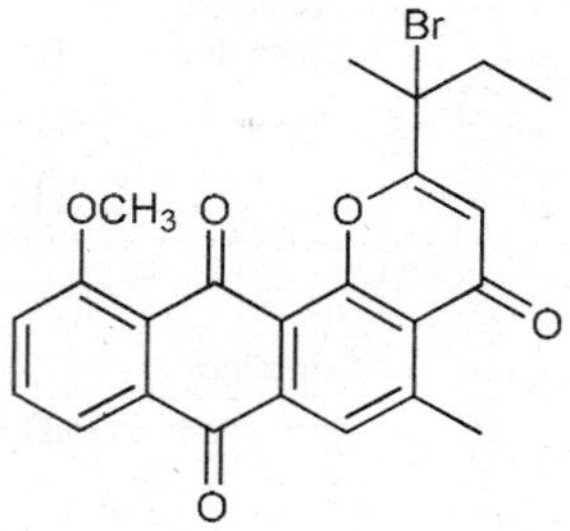

Major

Bromination at the benzylic position is also favorable because it proceeds via a resonance-stabilized, benzylic radical intermediate. The dibrominated minor product thus has a second bromine atom attached at the benzylic position:

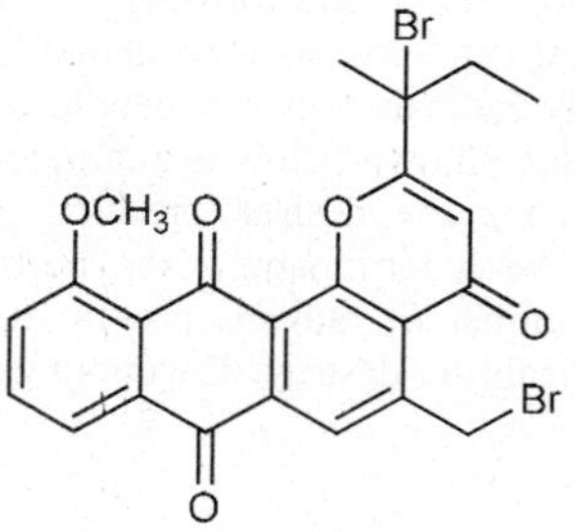

Minor

10.14.

(a) The starting compound contains only one tertiary position, so bromination occurs at this position. The product does not contain a chiral center:

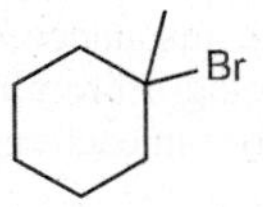

Br

no chiral center

(b) The starting compound contains only one tertiary position, so bromination occurs at this position. This position is an existing chiral center, so we expect loss of configuration to produce both possible enantiomers:

Br + Br

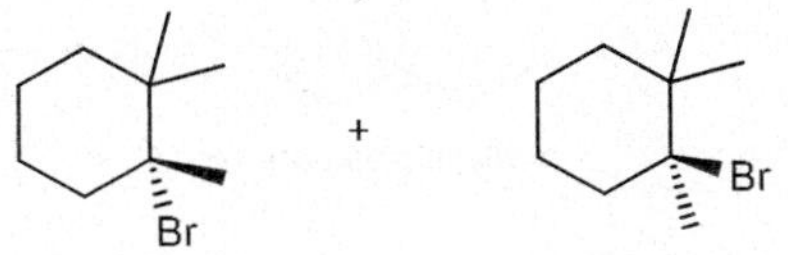

(c) The starting compound contains only one tertiary position, so bromination occurs at this position. This position is an existing chiral center, so we expect loss of configuration to produce both possible enantiomers:

Br + Br

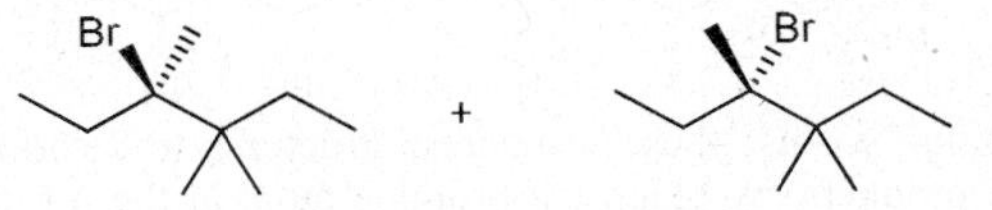

(d) The starting compound contains only one tertiary position, so bromination occurs at this position. This position is an existing chiral center, so we expect loss of configuration to produce both possible stereoisomers:

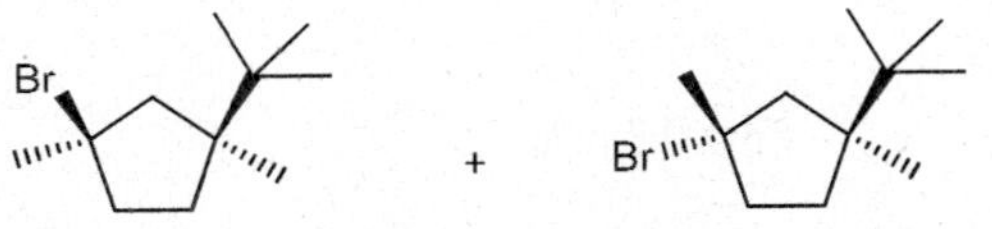

Br + Br

Notice that the other chiral center was unaffected by the reaction. The products are diastereomers.

10.15. The position undergoing halogenation is an existing chiral center, so we expect loss of configuration to produce both possible stereoisomers:

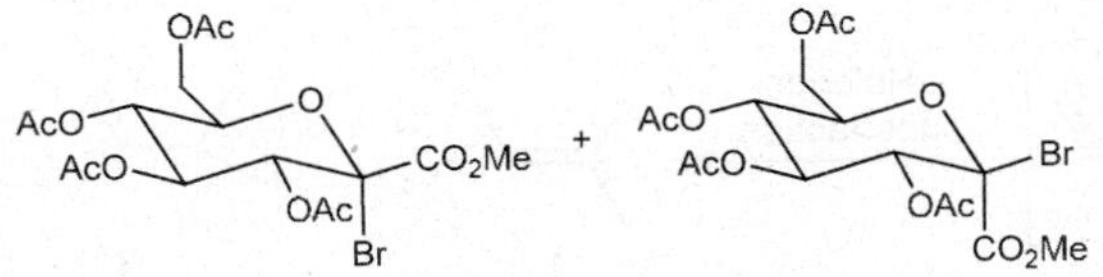

OAc, AcO, AcO, O, CO_2Me, OAc, Br + OAc, AcO, AcO, O, Br, OAc, CO_2Me

Notice that the other chiral centers are unaffected by the reaction. The products are diastereomers (rather than enantiomers), so they are not expected to be formed in equal amounts.

10.16.

(a) We begin by identifying allylic positions. There are four:

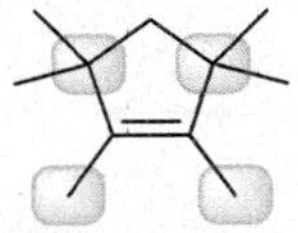

allylic positions

But two of these positions (the top two positions in the structure above) lack hydrogen atoms. So those positions cannot undergo radical bromination (a C–H bond is necessary because a key step in the mechanism is a hydrogen abstraction step). The remaining two allylic positions are identical (the molecule has a plane of symmetry that renders these two positions identical):

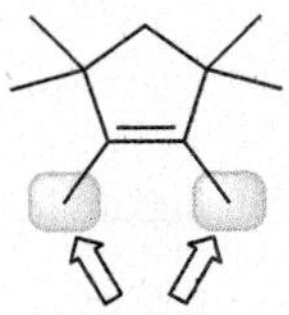

identical

So, we only need to consider allylic bromination occurring at one of these positions. To do that, we remove a hydrogen atom from the allylic position and draw the resonance structures of the resulting allylic radical:

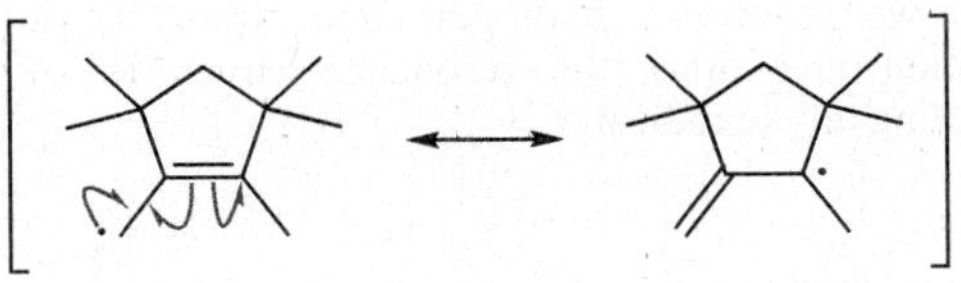

Finally, we use these resonance structures to determine the products, by placing a bromine atom at the position of the unpaired electron in each resonance structure:

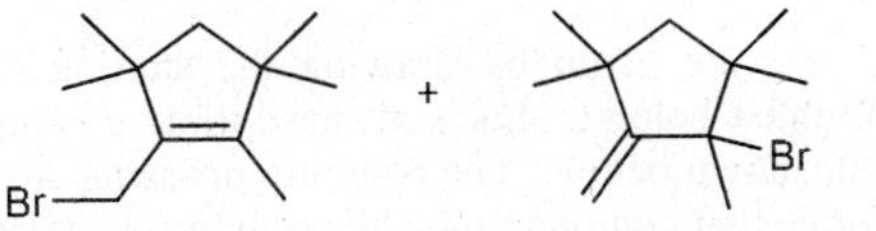

Br + Br

(b) We begin by identifying the allylic positions. There are two:

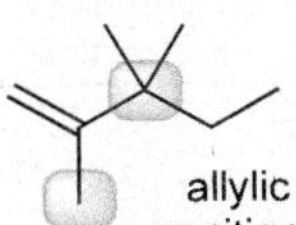

allylic positions

But one of these positions (the top position in the structure above) lacks a hydrogen atom. So this position cannot undergo radical bromination (a C–H bond is necessary because a key step in the mechanism is a hydrogen abstraction step). So, we only need to consider allylic bromination occurring at one position. To do that, we remove a hydrogen atom from the allylic position and draw the resonance structures of the resulting allylic radical:

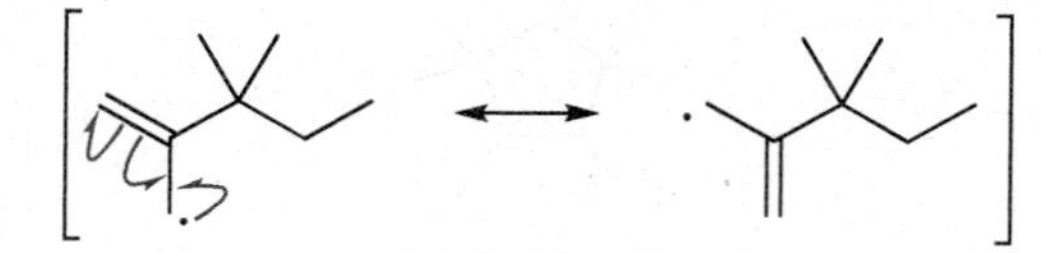

Finally, we use these resonance structures to determine the products, by placing a bromine atom at the position of the unpaired electron in each resonance structure. In this case, our methodical approach has produced two structures that are identical compounds, so this reaction has only one product:

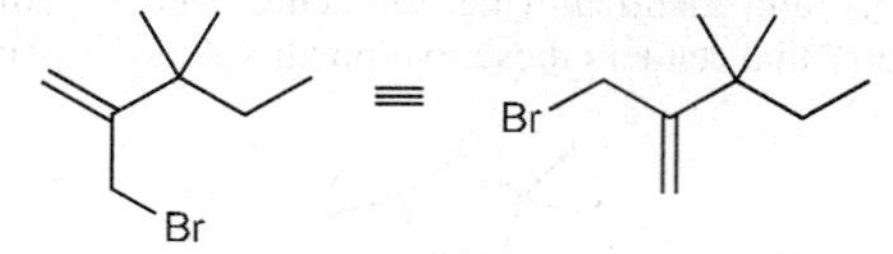

Br ≡ Br

(c) We begin by identifying allylic positions. There are two:

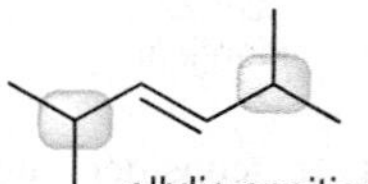

allylic positions

But these two positions are identical (due to molecular symmetry). So, we only need to consider allylic bromination occurring at one of these positions. To do that, we remove a hydrogen atom from the allylic position and draw the resonance structures of the resulting allylic radical:

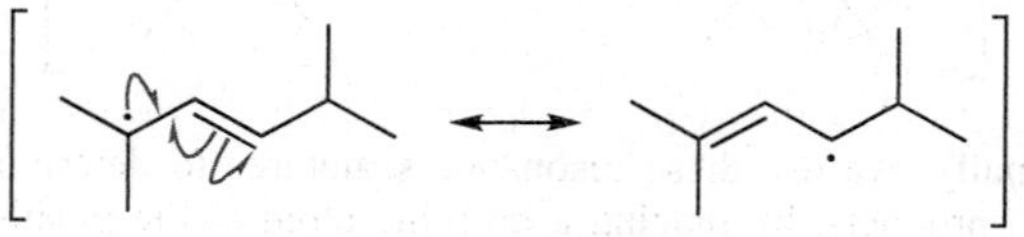

Finally, we use these resonance structures to determine the products, by placing a bromine atom at the position of the unpaired electron in each resonance structure:

Br + Br

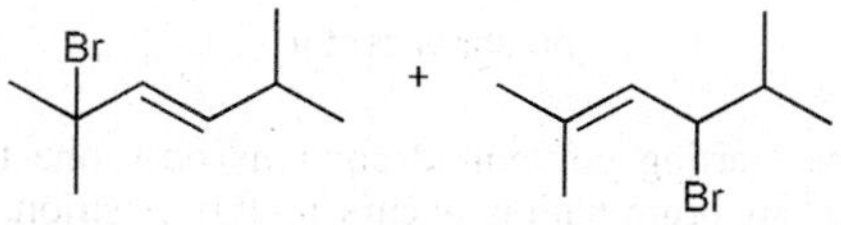

(d) We begin by identifying allylic positions. There is only one in this case:

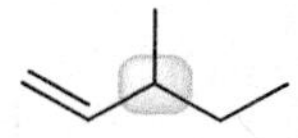

allylic position

Next, we remove a hydrogen atom from the allylic position and draw the resonance structures of the resulting allylic radical:

Finally, we use these resonance structures to determine the products, by placing a bromine atom at the position of the unpaired electron in each resonance structure:

Br + Br

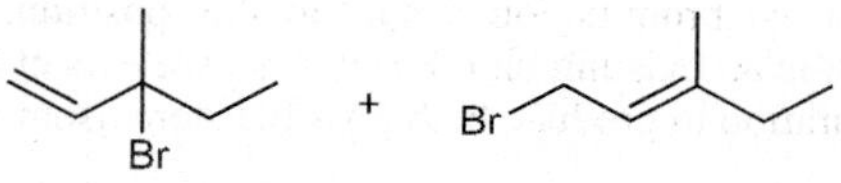

10.17. We begin by drawing the starting compound (2-methyl-2-butene) and identifying the allylic positions (highlighted below). Each of these three positions can undergo hydrogen abstraction to give a resonance-stabilized radical, shown below. The resulting products are also shown. Notice that there are a total of five products, labeled **1** – **5**. Notice that compound **4** is the only product that exhibits a chiral center. As such, a racemic mixture of compound **4** is expected.

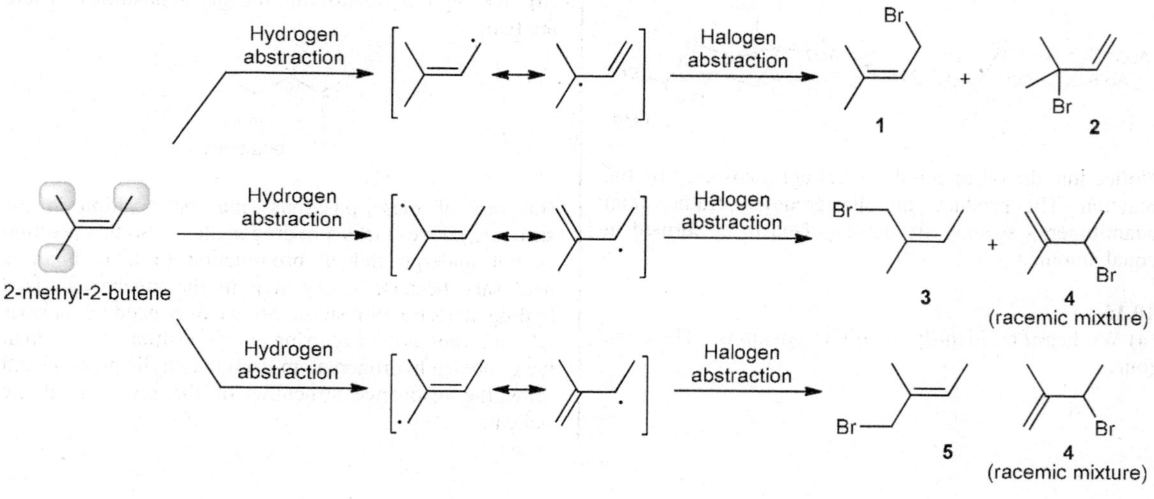

2-methyl-2-butene

Hydrogen abstraction → Halogen abstraction → Br **1** + Br **2**

Hydrogen abstraction → Halogen abstraction → Br **3** + Br **4** (racemic mixture)

Hydrogen abstraction → Halogen abstraction → Br **5** Br **4** (racemic mixture)

10.18. This compound has two allylic positions (highlighted below, A and B). Each allylic position has hydrogen atoms that can be abstracted, so we will need to consider products resulting from abstraction at each allylic position.

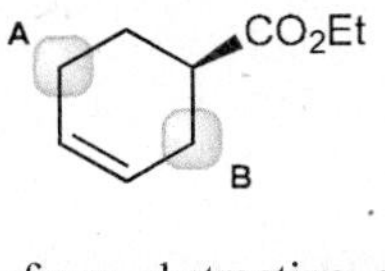

First, draw the resonance structures that result from abstracting a hydrogen atom from position **A** (in the first propagation step).

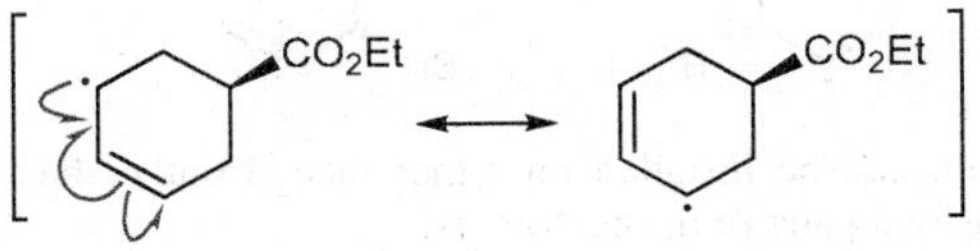

Then, halogen abstraction (in the second propagation step) can occur from either resonance structure, resulting in two possible regiochemical outcomes:

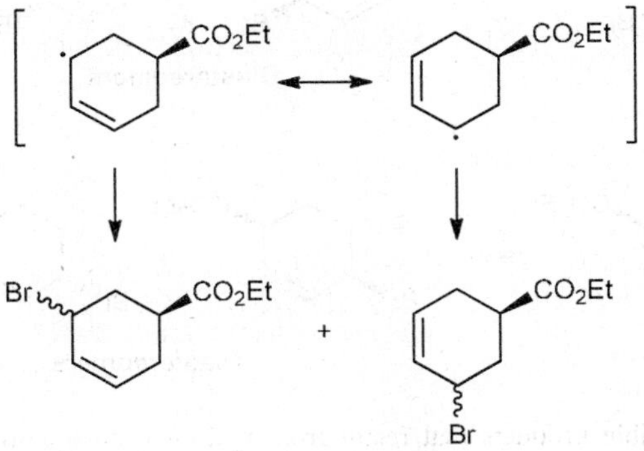

Now consider the stereochemical outcome in each of these cases. As described in Section 10.6, the bromine atom can be installed on either face (front or back) of the radical. This results in the formation of four products, as two pairs of diastereomers, as shown:

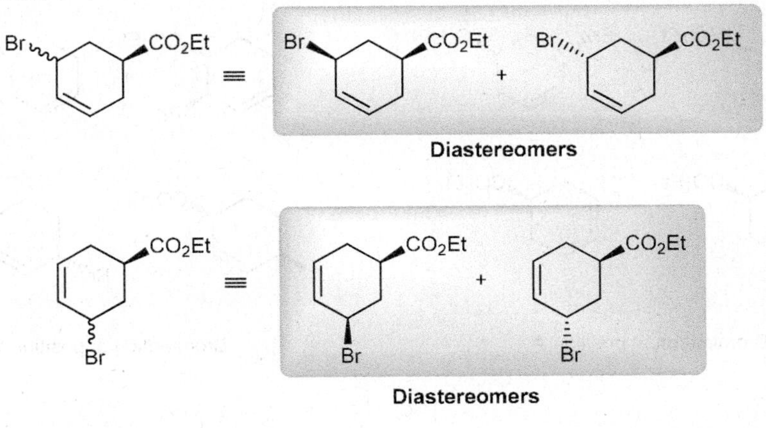

Diastereomers

Diastereomers

Now let's start over again, and abstract a hydrogen atom from the other allylic position (in the first propagation step):

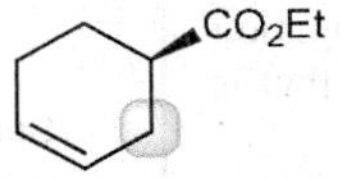

The resulting allylic radical is resonance-stabilized, so we draw both resonance structures. Once again, halogen abstraction (in the second propagation step) can occur from either resonance structure, resulting in two additional regiochemical outcomes:

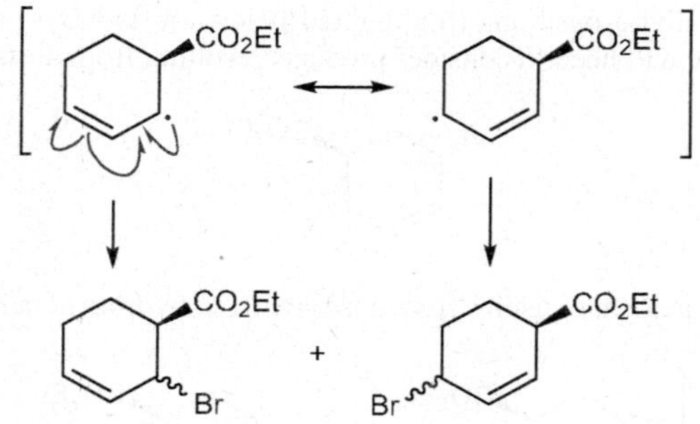

And once again, the bromine atom can be installed on either face (front or back) of the radical. This gives four additional bromination products, as two pairs of diastereomers:

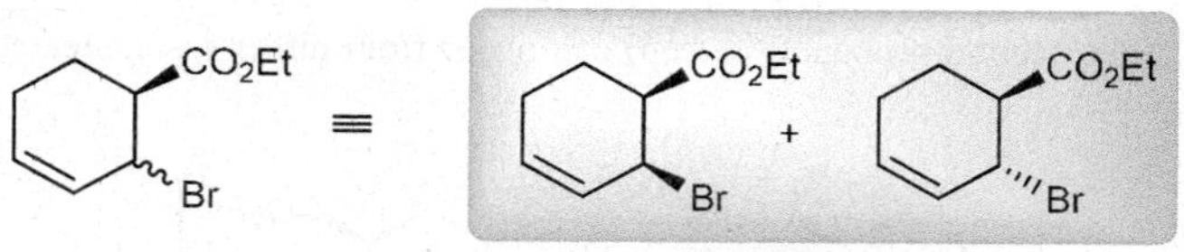

Diastereomers

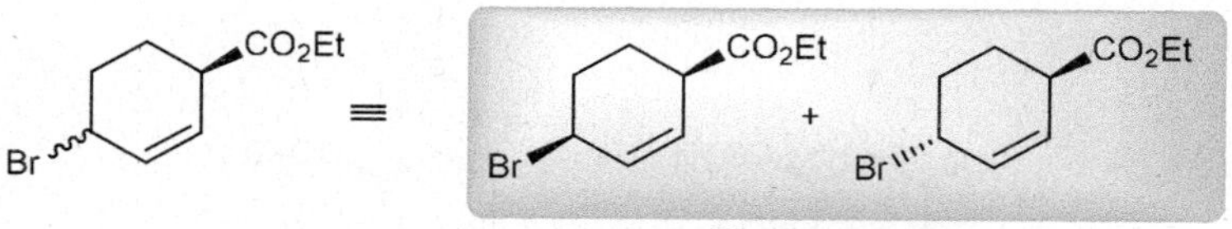

Diastereomers

In summary, there are four possible products that result from hydrogen abstraction at position **A** and four possible products that result from hydrogen abstraction at position **B**. This gives a total of eight products, which represent four pairs of diastereomers. Notice that in all eight isomers, the stereochemistry of the chiral carbon atom bonded to the ester remains unchanged, as no bonds to this atom change during the course of the reaction

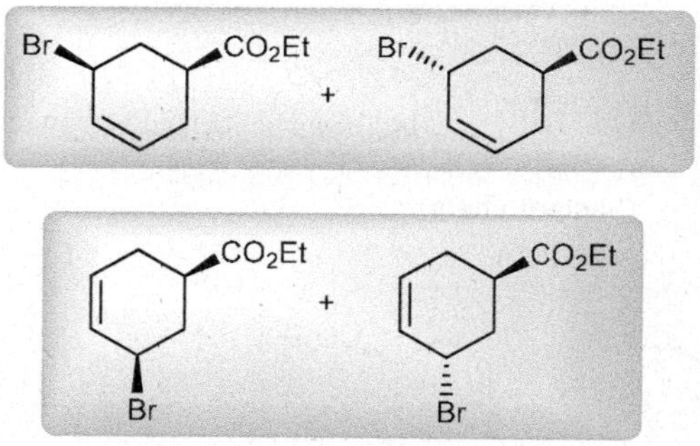

Bromination at position **A**

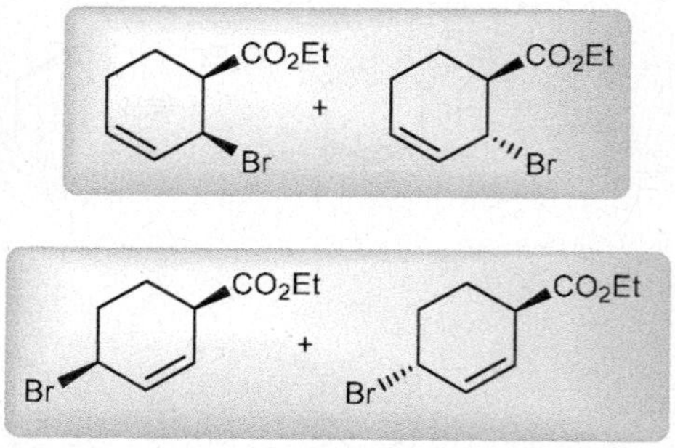

Bromination at position **B**

10.19. As seen in Section 10.8, destruction of ozone in the atmosphere occurs via the following two propagation steps (where R• represents a radical, such as Cl•, that is responsible for destroying ozone):

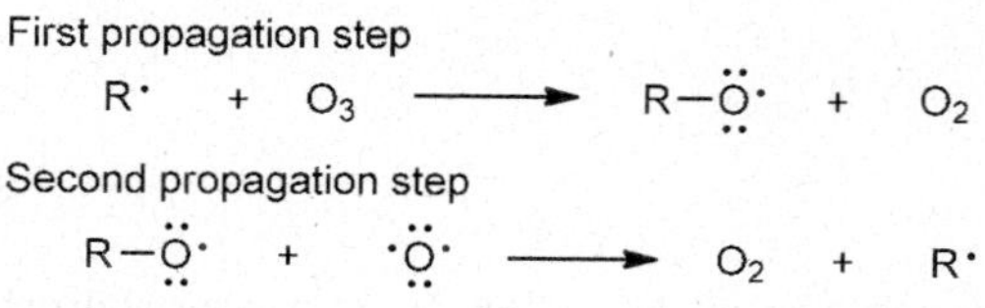

First propagation step

$$R\cdot + O_3 \longrightarrow R\text{—}\ddot{\underset{\cdot\cdot}{O}}\cdot + O_2$$

Second propagation step

$$R\text{—}\ddot{\underset{\cdot\cdot}{O}}\cdot + \cdot\ddot{\underset{\cdot\cdot}{O}}\cdot \longrightarrow O_2 + R\cdot$$

If we redraw these propagation steps with nitric oxide serving as the radical that destroys ozone, we get the following:

First propagation step

$\ddot{O}=\dot{N} + O_3 \longrightarrow \ddot{O}=\ddot{N}-\ddot{O}\cdot + O_2$

Second propagation step

$\ddot{O}=\ddot{N}-\ddot{O}\cdot + \cdot\ddot{O}\cdot \longrightarrow O_2 + \ddot{O}=\dot{N}$

Note that this is a chain process because the NO radical is regenerated in step 2. Since the NO radical is not used up in the overall process, it can continue reacting with more ozone.

10.20. Radicals will react with BHT and BHA because each of these compounds has a hydrogen atom that can be readily abstracted, thereby generating a resonance-stabilized radical. In each case, the phenolic hydrogen atom is abstracted (the hydrogen atom of the OH group connected to the aromatic ring). Similarly, vitamin E also exhibits a phenolic hydrogen atom. Abstraction of this hydrogen atom (highlighted below) gives a resonance-stabilized radical.

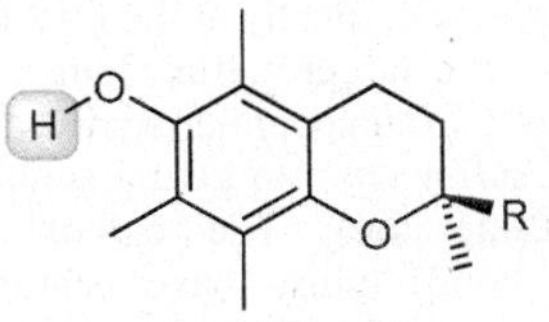

Vitamin E

10.21.

(a) The reagent (HBr) indicates that H and Br are added across the π bond. In the presence of peroxides (ROOR), we expect an *anti*-Markovnikov addition. That is, the bromine atom is installed at the less substituted position. In this case, one new chiral center is created, which results in a racemic mixture of the two possible enantiomers.

HBr / ROOR

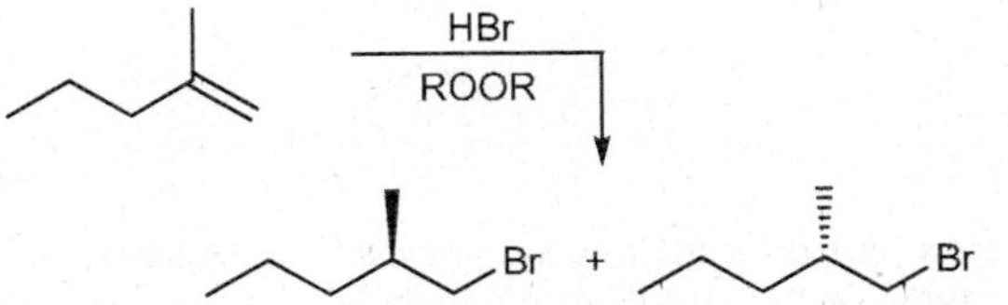

(b) The reagent (HBr) indicates that H and Br are added across the π bond. In the presence of peroxides (ROOR), we expect an *anti*-Markovnikov addition. That is, the bromine atom is installed at the less substituted position. In this case, no new chiral centers are formed:

HBr / ROOR

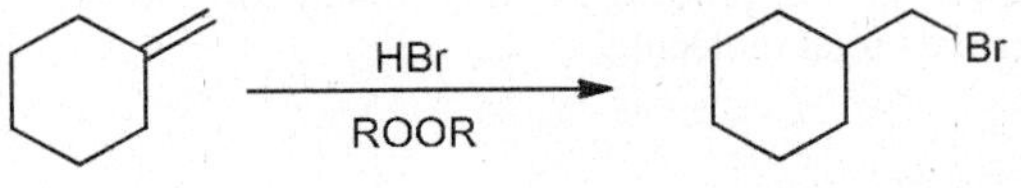

(no chiral centers)

(c) The reagent (HBr) indicates that H and Br are added across the π bond. In the presence of peroxides (ROOR), we expect an *anti*-Markovnikov addition. That is, the bromine atom is installed at the less substituted position. In this case, one new chiral center is created, which results in a racemic mixture of the two possible enantiomers.

HBr / ROOR

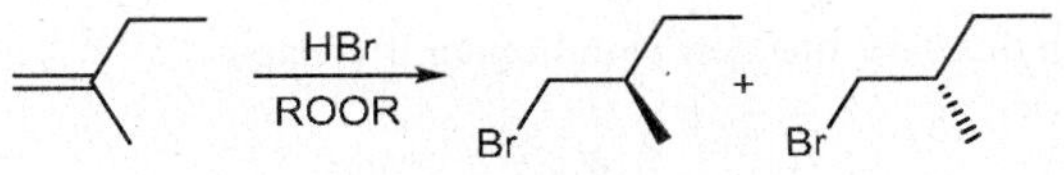

(d) The reagent (HBr) indicates that H and Br are added across the π bond. In the presence of peroxides (ROOR), we expect an *anti*-Markovnikov addition. That is, the bromine atom is installed at the less substituted position. In this case, one new chiral center is created, which results in a racemic mixture of the two possible enantiomers.

HBr / ROOR

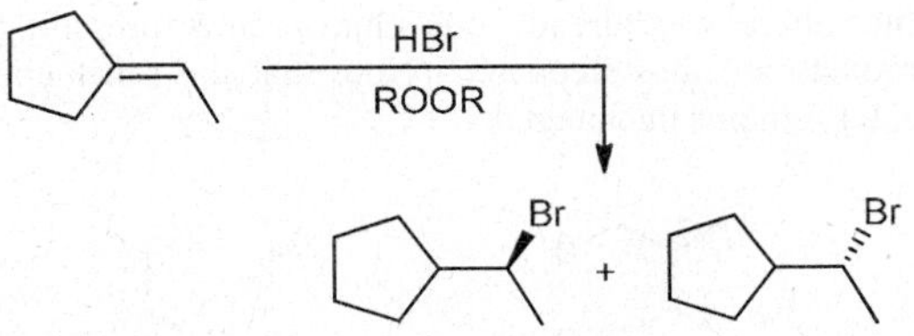

(e) The reagent (HBr) indicates that H and Br are added across the π bond. In the presence of peroxides (ROOR), we expect an *anti*-Markovnikov addition. That is, the bromine atom is installed at the less substituted position. In this case, no new chiral centers are formed:

HBr / ROOR

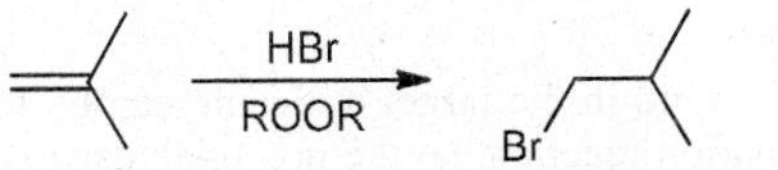

(no chiral centers)

(f) The reagent (HBr) indicates that H and Br are added across the π bond. In the presence of peroxides (ROOR), we expect an *anti*-Markovnikov addition, although that is irrelevant in this case, because the alkene is symmetrical. No new chiral centers are formed:

HBr / ROOR

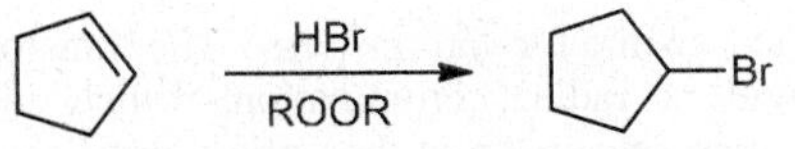

(no chiral centers)

10.22. The use of HBr indicates the addition of H and Br across the π bond. The problem statement indicates a radical process (caused by the use of light, rather than ROOR, in this case), so we expect *anti*-Markovnikov addition of HBr across the π bond. That is, the bromine atom will be installed at the less substituted position.

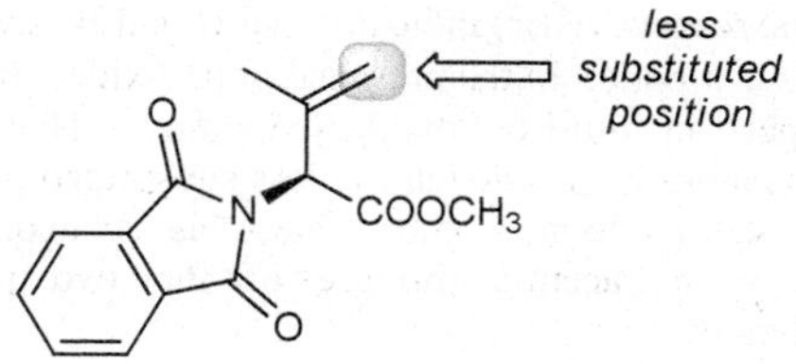

less substituted position

In this case, one new chiral center is created:

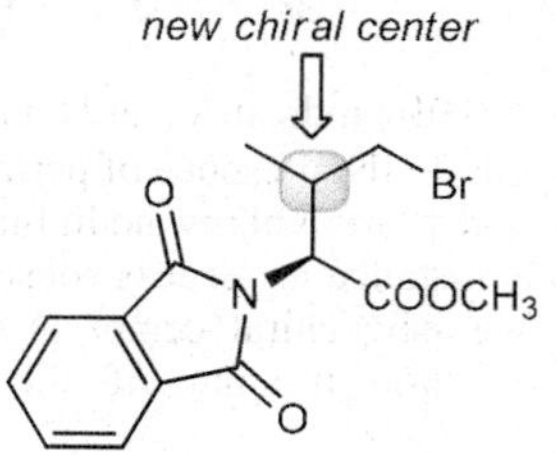

new chiral center

Since there was already one chiral center present, the products are diastereomers. Note that the other chiral center remains unchanged.

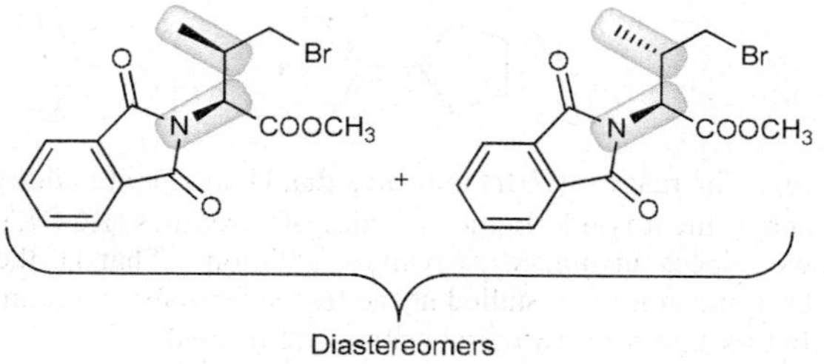

Diastereomers

10.23

(a) The π bond in the target molecule can be formed by an elimination reaction, so the potential starting material can be an alkyl halide. The tertiary bromide is an ideal intermediate, because it can be prepared by radical halogenation of the given alkane starting material, as shown in the following retrosynthetic analysis:

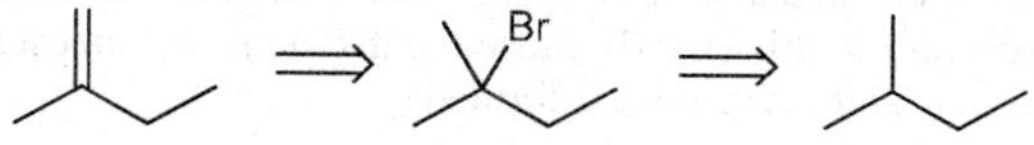

Now let's draw the forward process. The transformation begins with a radical bromination, which installs a bromine atom selectively at the tertiary position. For the subsequent E2 elimination, *tert*-butoxide would be the appropriate choice of strong base to give the target molecule (to give the less-substituted Hofmann product).

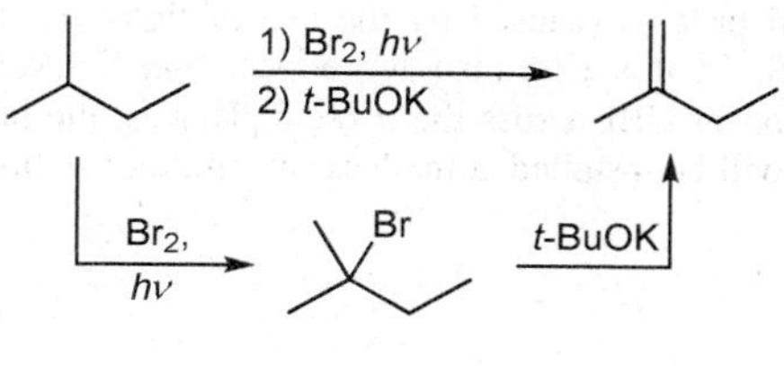

1) Br_2, *hν*
2) *t*-BuOK

Br_2, *hν*

t-BuOK

(b) The π bond in the target molecule can be formed by an elimination reaction, so the potential starting material can be an alkyl halide. The tertiary bromide is an ideal intermediate, because it can be prepared by radical halogenation of the given alkane starting material, as shown in the following retrosynthetic analysis:

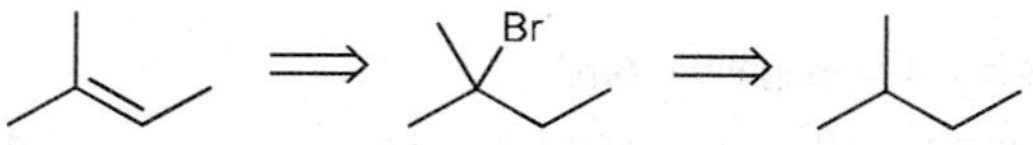

Now let's draw the forward process. The transformation begins with a radical bromination, which installs a bromine atom selectively at the tertiary position. For the subsequent E2 elimination, the more-substituted Zaitsev product is desired, so we must a use a strong base that is not sterically hindered, such as hydroxide, methoxide or ethoxide:

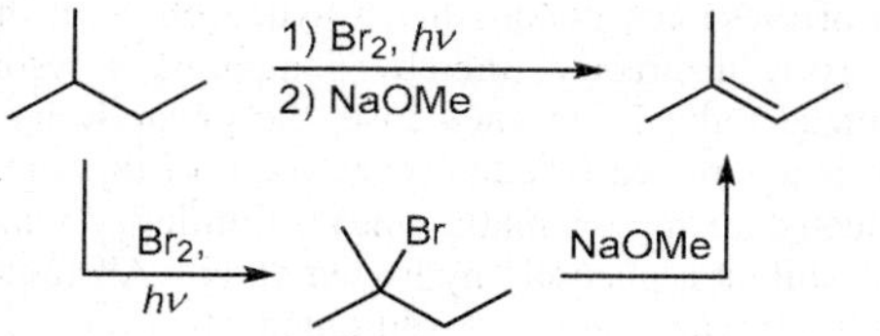

1) Br_2, *hν*
2) NaOMe

Br_2, *hν*

NaOMe

(c) Our retrosynthesis begins with a disconnection at the thiol group, more specifically at the C–S bond. The more electronegative and larger sulfur atom is more likely to have served as the nucleophile, so we draw a negative charge on the sulfur atom to give a suitable nucleophile (the hydrosulfide ion). The carbon atom (at the disconnected bond) must have started out as an electrophile, so we draw a leaving group at that position. We will select Cl or Br as the leaving group, because either of these halogens can be installed onto the starting alkane via radical halogenation of the given alkane starting material:

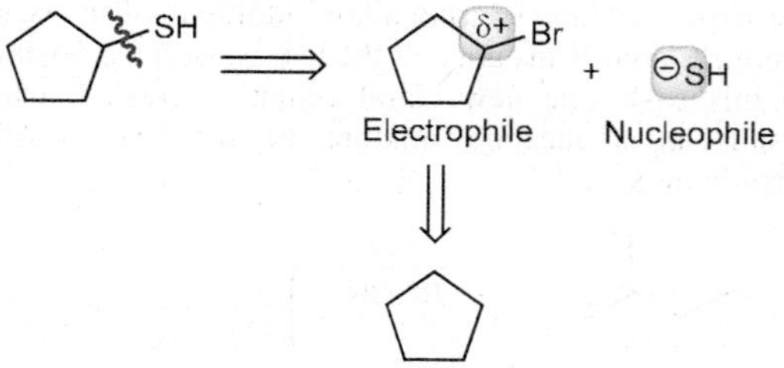

Electrophile Nucleophile

Now let's draw the forward process. The transformation begins with a radical bromination (or chlorination), which installs a bromine atom (or chlorine atom) at any of the equivalent cyclopentane carbon atoms. For the subsequent substitution reaction, the large hydrosulfide anion is a strong nucleophile but not a strong base, so the S_N2 pathway does not have any significant competition from the E2 pathway. With a secondary substrate, we expect an S_N2 process to proceed smoothly to give the desired target molecule.

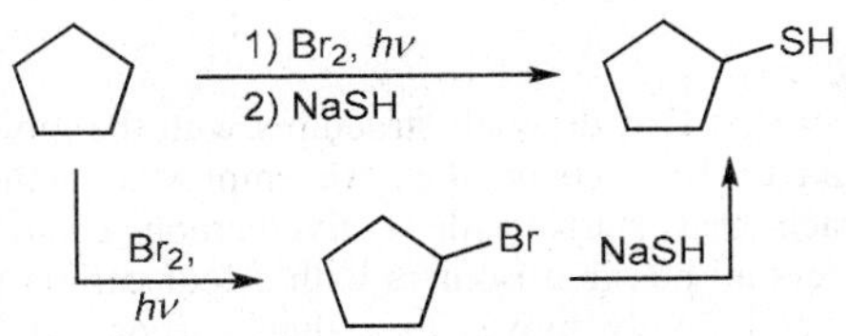

1) Br_2, $h\nu$
2) NaSH

Br_2, $h\nu$ — Br — NaSH — SH

(d) The retrosynthetic analysis of the target molecule asks, "What starting material can be used in order to make a *cis*-1,2-diol?" The two OH groups in the target molecule can be produced from an alkene:

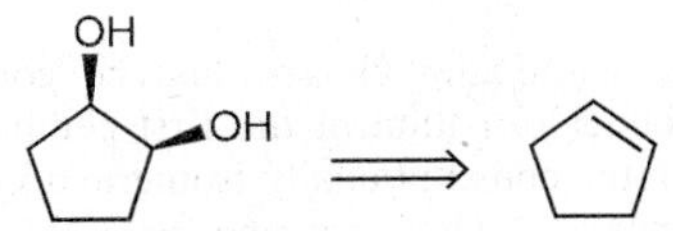

OH OH ⟹

This alkene can be produced from an alkyl halide that, in turn, can be prepared by radical halogenation of the starting alkane:

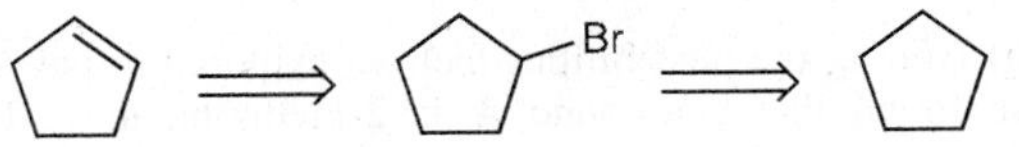

⟹ Br ⟹

Now let's draw the forward process. First, radical chlorination or bromination can be performed, thereby installing a leaving group, which then allows for an E2 process with a strong base to give an alkene. Any strong base can be used (there is only one regiochemical outcome, so a sterically hindered base is not required). It is acceptable to use hydroxide, methoxide or ethoxide for the E2 process. That being said, in this particular case, *tert*-butoxide will likely be more efficient, as it will suppress the competing S_N2 process (we would expect S_N2 to give a minor product if the base is not sterically hindered). Finally, *syn*-dihydroxylation is achieved by treatment with cold potassium permanganate:

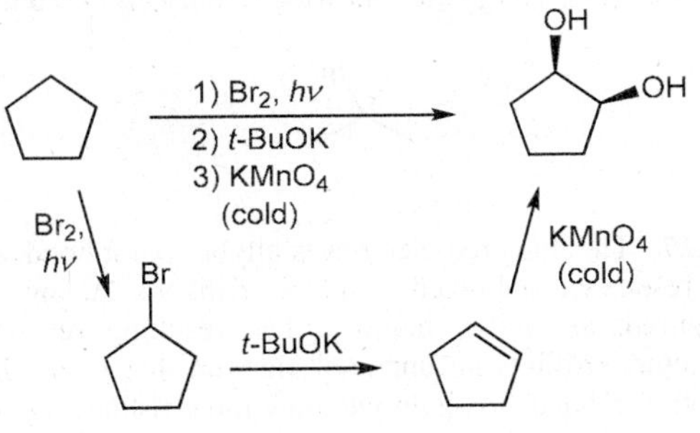

1) Br_2, $h\nu$
2) *t*-BuOK
3) $KMnO_4$ (cold)

OH OH

Br_2, $h\nu$ — Br — *t*-BuOK — $KMnO_4$ (cold)

10.24.

(a) The unpaired electron occupies an allylic position, so it is resonance-stabilized. Three fishhook arrows are required, as shown:

(b) The unpaired electron occupies an allylic position, so it is resonance-stabilized. Three fishhook arrows are required, as shown below. The resulting resonance structure exhibits an unpaired electron that is allylic to another π bond, so again we draw three fishhook arrows to arrive at a third resonance structure:

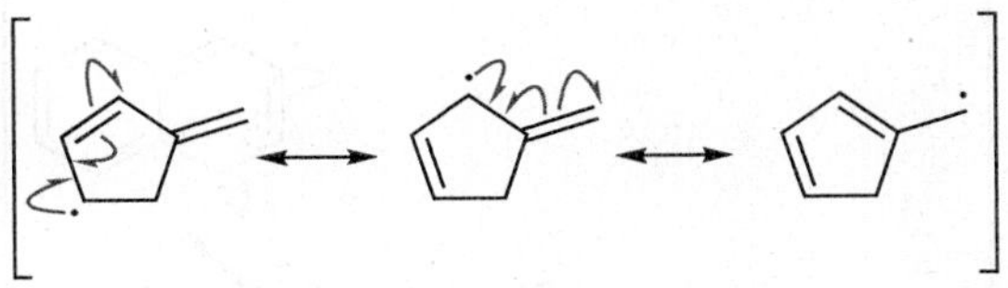

(c) The unpaired electron occupies an allylic position, so it is resonance-stabilized. Three fishhook arrows are required, as shown below. The resulting resonance structure exhibits an unpaired electron that is allylic to another π bond, so again we draw three fishhook arrows to arrive at a third resonance structure:

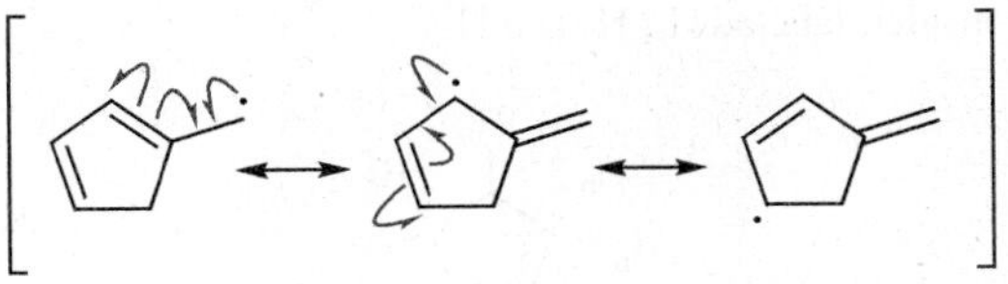

(d) The unpaired electron is allylic to a π bond, so it is resonance-stabilized. Three fishhook arrows are required, as shown below. The resulting resonance structure also exhibits an unpaired electron that is allylic to a π bond, so again we draw three fishhook arrows as well as the resulting resonance structure. This process continues until a total of five resonance structures are drawn, shown here:

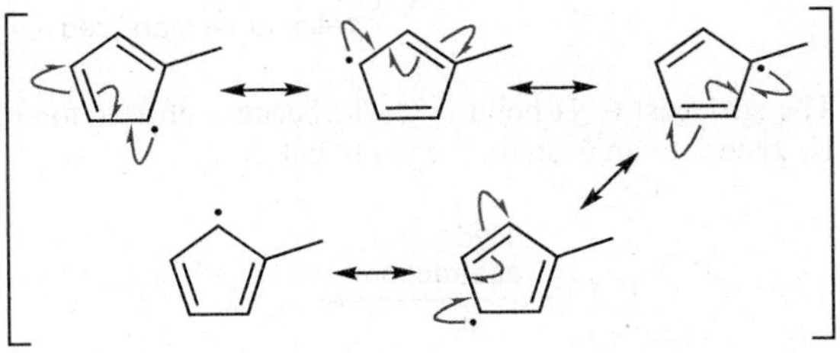

(e) The unpaired electron is allylic to a π bond, so it is resonance-stabilized. Three fishhook arrows are required, as shown below. The resulting resonance structure also exhibits an unpaired electron that is allylic to a π bond, so again we draw three fishhook arrows as well as the resulting resonance structure. This process continues until a total of five resonance structures are drawn, shown here:

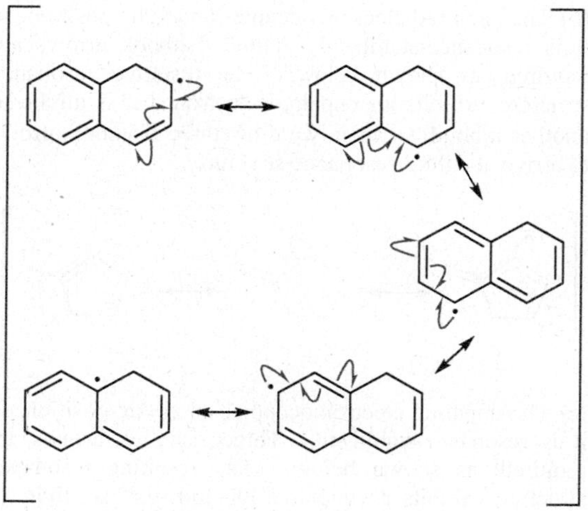

10.25. There are three different hydrogen atoms to consider, labeled H_a, H_b and H_c:

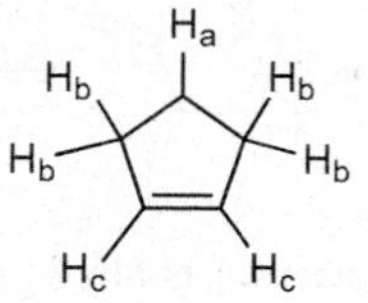

The weakest C–H bond is C–H_b, because abstraction of H_b generates a radical that is stabilized by resonance:

Hydrogen abstraction

resonance stabilized

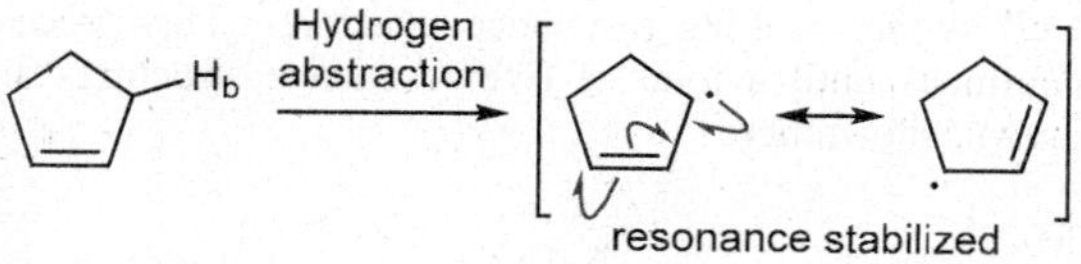

The strongest C–H bond is C–H_c, because abstraction of H_c generates an unstable vinyl radical:

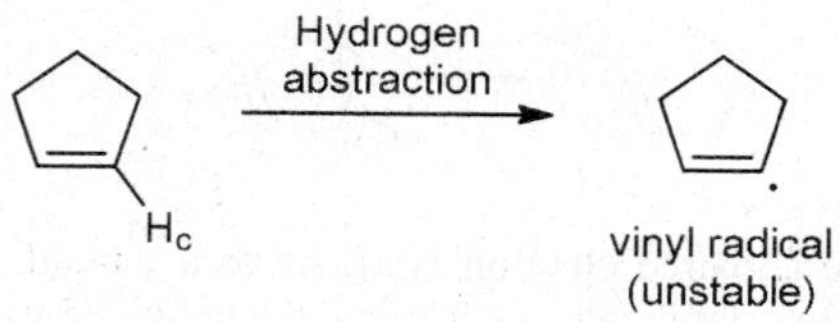

Hydrogen abstraction

vinyl radical (unstable)

The C–H_a bond strength is expected to fall in the middle of these two extremes, because abstraction of H_a generates a secondary radical that is not stabilized by resonance. We therefore expect the C–H bonds of cyclopentene to exhibit the following order of increasing bond strength:

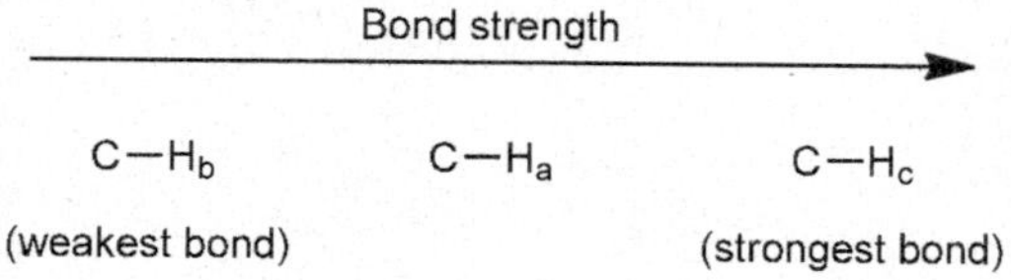

Bond strength

C–H_b (weakest bond) C–H_a C–H_c (strongest bond)

10.26.

(a) We must first draw all structures with the molecular formula C_5H_{12}. To do this, we employ a methodical approach that starts with a five-carbon chain, then considers all possible isomers with a four-carbon parent chain, and finally moves to a three-carbon parent (as seen in the solution to Problem 4.3). Such an approach gives the following three compounds:

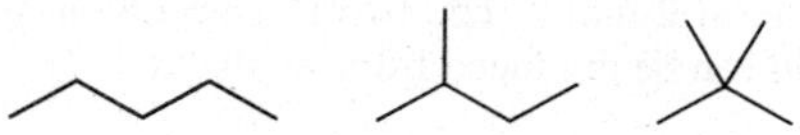

The middle compound above must be compound **A**, because monochlorination of the first compound above gives only three constitutionally isomeric alkyl chlorides (1-chloropentane, 2-chloropentane, or 3-chloropentane), while monochlorination of the last compound above gives only one regiochemical outcome. The middle compound is 2-methylbutane.

(b) Among the possibilities that we explored in part (a), we found that compound **A** is 2-methylbutane. This compound undergoes monochlorination to produce four constitutionally isomeric alkyl chlorides, shown here:

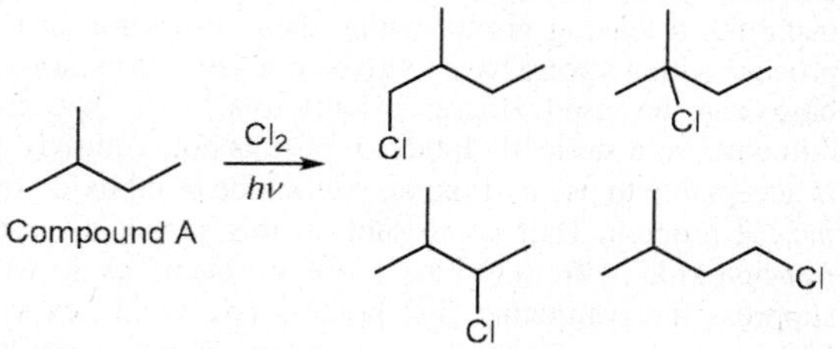

Compound A $\xrightarrow[h\nu]{Cl_2}$

(c) The tertiary position undergoes selective bromination, giving the following tertiary alkyl bromide:

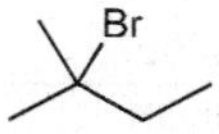

10.27. The unpaired electron is allylic to a π bond, so it is resonance-stabilized. Three fishhook arrows are required, as shown below. The resulting resonance structure exhibits an unpaired electron that is allylic to another π bond, so again we draw three fishhook arrows to arrive at a third resonance structure. This pattern appears one more time, giving a fourth resonance structure. This radical is particularly stable, because it has many resonance structures, so the unpaired electron is highly delocalized.

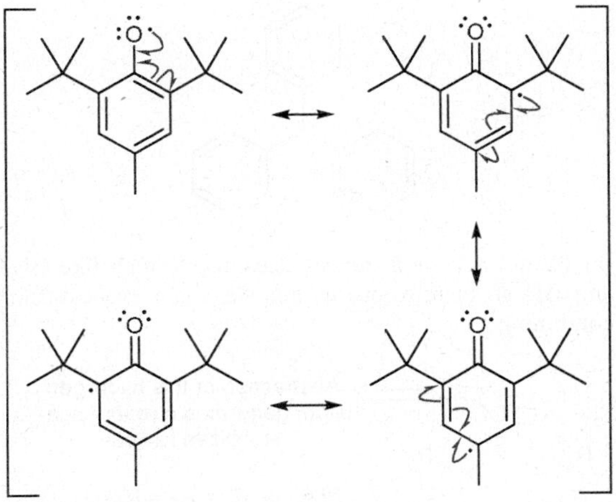

10.28. The benzylic hydrogen atom is the only hydrogen atom that can be abstracted to generate a resonance-stabilized radical. As such, the benzylic position is selectively brominated. The mechanism will have three distinct stages. The first stage is initiation, in which the N–Br bond (of NBS) is broken to generate a bromine radical. This step requires two fishhook arrows, as shown. There are two propagation steps (hydrogen abstraction and halogen abstraction), each of which requires three fishhook arrows (to show the bonds being broken and formed). The HBr produced in the first propagation step then reacts with NBS in an ionic reaction that produces Br_2. This Br_2 can be used in the second propagation step. Finally, there are many steps that can serve as termination steps, because there are many radicals that can couple together under these conditions. Shown here is a termination step that generates the desired product:

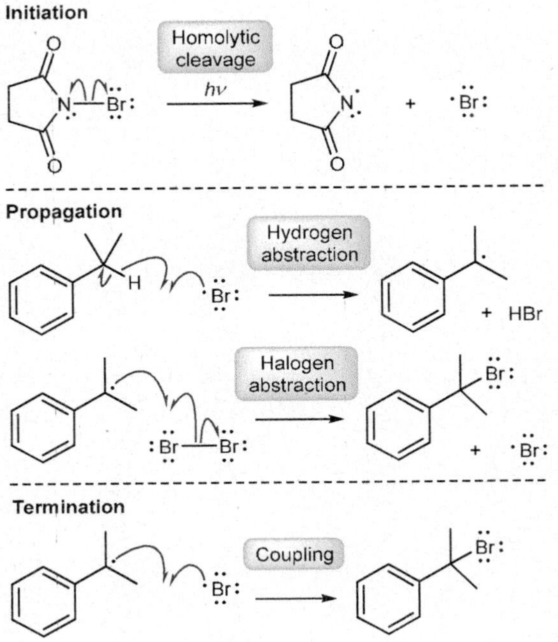

Initiation

Homolytic cleavage

hν

+ ·Br:

Propagation

Hydrogen abstraction

+ HBr

Halogen abstraction

+ ·Br:

Termination

Coupling

10.29. We must first draw all structures with the molecular formula C_5H_{12}. To do this, we employ the same methodical approach that was used in the solution to Problem 10.26, giving the following three compounds:

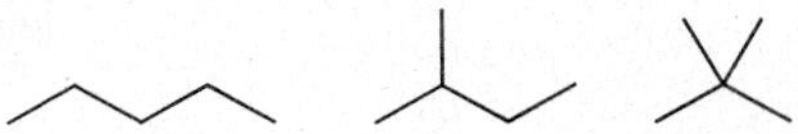

Monochlorination of the first compound above gives three constitutionally isomeric chloroalkanes (1-chloropentane, 2-chloropentane, or 3-chloropentane), while monochlorination of the second compound gives four constitutionally isomeric chloroalkanes, as detailed in the solution to problem **10.26**. The last compound above (2,2-dimethylpropane) gives only one monochlorination product, since all four CH_3 groups are equivalent:

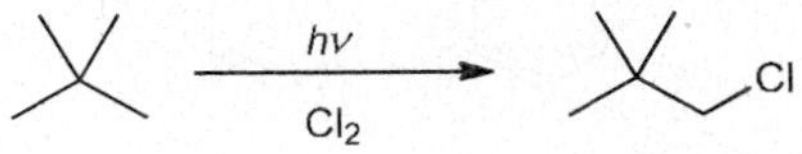

hν, Cl_2

10.30. Selective bromination at the benzylic position generates a new chiral center. The intermediate benzylic radical is expected to be attacked from either face of the planar radical with equal likelihood, giving rise to a racemic mixture of enantiomers:

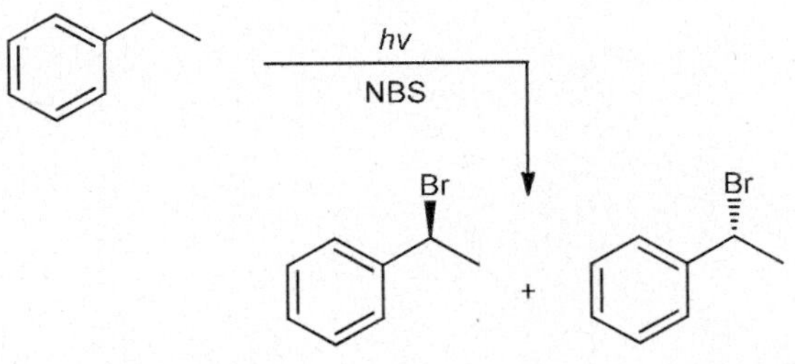

hν, NBS

10.31.

(a) Heating AIBN generates a radical that is resonance-stabilized (the unpaired electron is delocalized via resonance, which is a stabilizing effect):

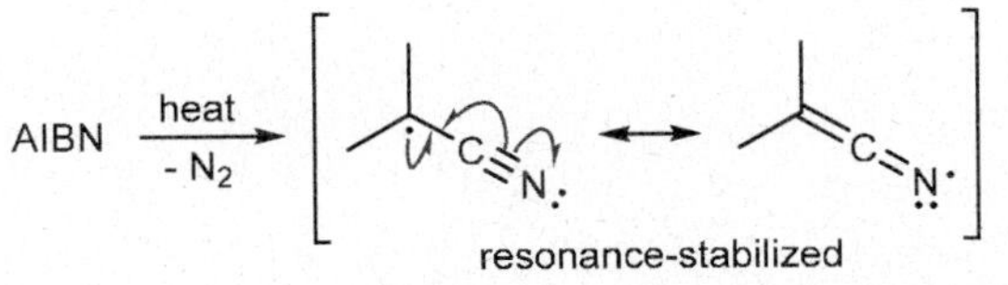

AIBN —heat, - N_2→

resonance-stabilized

In addition, the radical is also stable because it is tertiary. The methyl groups stabilize the unpaired electron via a delocalization effect, called hyperconjugation.

(b) Loss of nitrogen gas would result in the formation of vinyl radicals, which are too unstable to form under normal conditions:

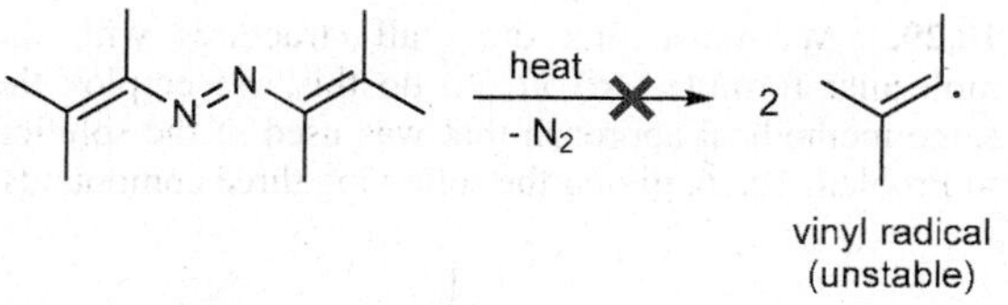

vinyl radical (unstable)

10.32.

(a) The central carbon atom is the only sp^3 hybridized carbon atom in the structure, so it is the only position that can readily undergo a hydrogen abstraction. Autooxidation at this location gives the following hydroperoxide:

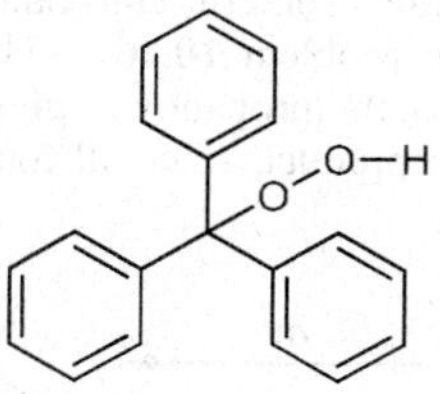

(b) The central carbon atom is benzylic to three aromatic rings. As such, that C–H bond is expected to be extremely weak. Hydrogen abstraction at this location generates a radical that is highly stabilized by resonance (see solution to Problem 10.4 for resonance structures):

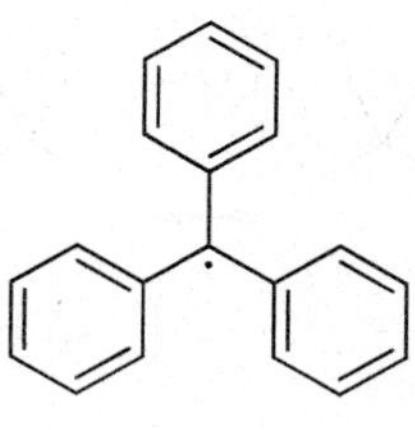

(c) Phenol acts as a radical scavenger (much like BHT and BHA), thereby preventing the chain process from continuing.

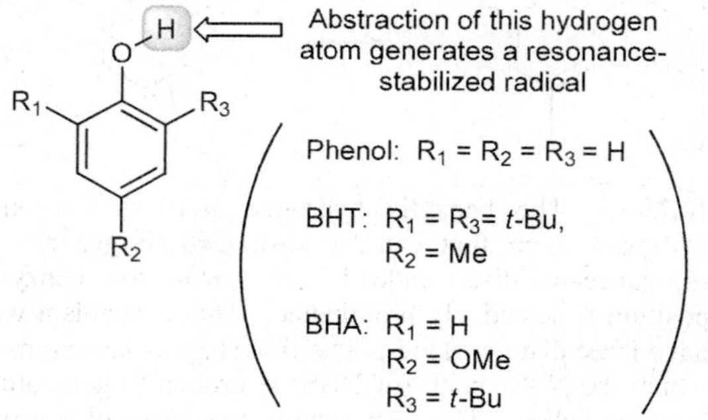

Abstraction of this hydrogen atom generates a resonance-stabilized radical

Phenol: $R_1 = R_2 = R_3 = H$

BHT: $R_1 = R_3 =$ *t*-Bu, $R_2 =$ Me

BHA: $R_1 = H$, $R_2 =$ OMe, $R_3 =$ *t*-Bu

10.33. A radical mechanism will have three distinct stages. The first stage is initiation, in which the N–Br bond (of NBS) is broken to generate a bromine radical. This step requires two fishhook arrows, as shown. The next stage is propagation. There are two propagation steps. The first is hydrogen abstraction, which requires three fishhook arrows and generates a resonance-stabilized radical. The HBr produced in the first propagation step then reacts with NBS in an ionic reaction that produces Br_2. This Br_2 can be used in the second propagation step. The second propagation step (halogen abstraction) also requires three fishhook arrows. Notice that this step can occur in either of two locations, as shown. The final stage is termination, and there are many steps that can serve as termination steps, because there are many radicals that can couple together under these conditions. Shown below are termination steps that generate the desired products:

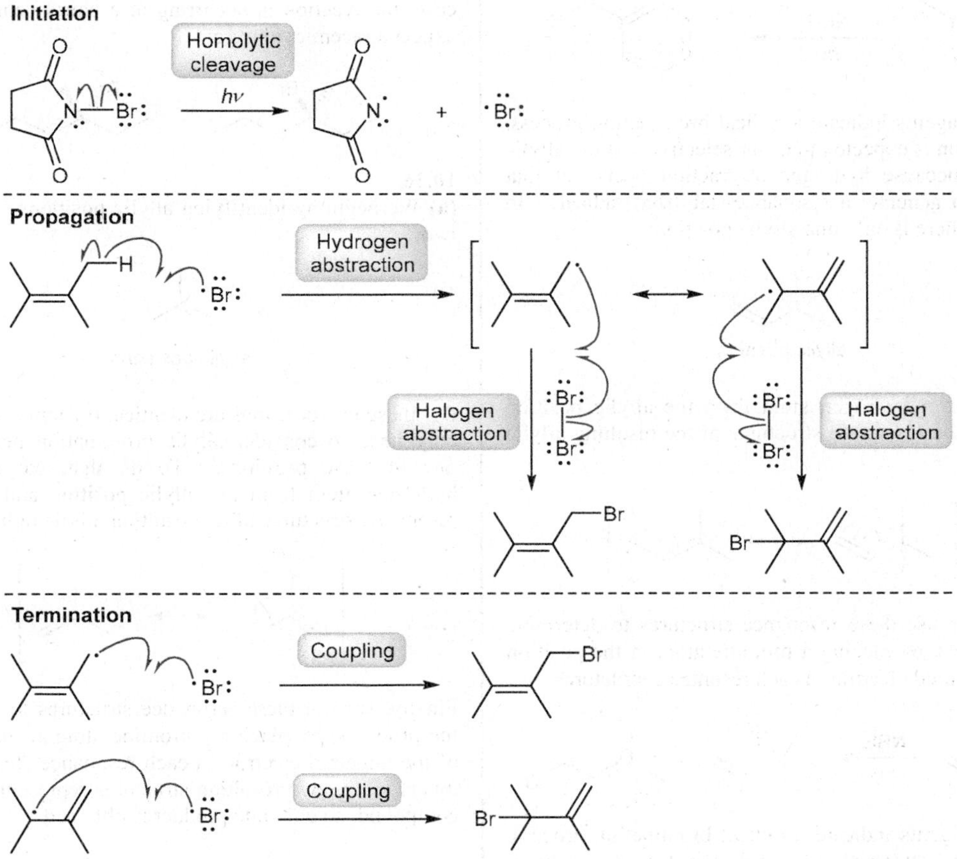

Initiation

Homolytic cleavage

$h\nu$

Propagation

Hydrogen abstraction

Halogen abstraction

Halogen abstraction

Termination

Coupling

Coupling

10.34.

(a) The reagents indicate a radical bromination process. The starting material has only one tertiary position, which undergoes selective bromination, giving the following tertiary alkyl bromide.

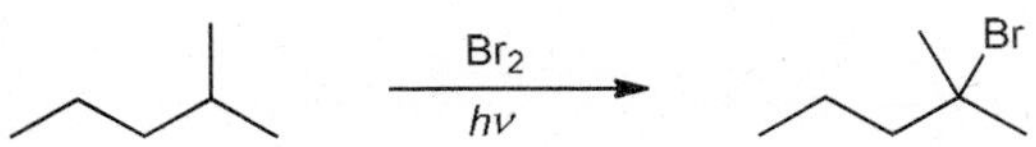

Br_2, $h\nu$

(b) As seen in Section 10.4, radical iodination is not thermodynamically favorable. We expect no reaction in this case:

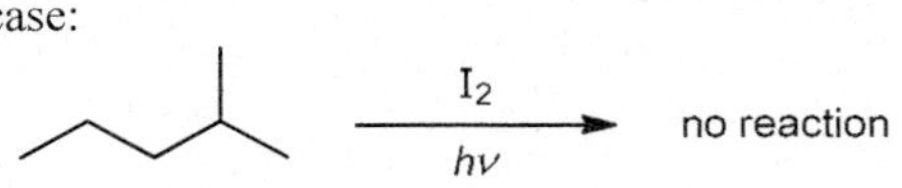

I_2, $h\nu$ → no reaction

(c) The reagents indicate a radical chlorination process. Radical chlorination is less selective than radical bromination, so we expect a mixture of products. That is, we expect chlorination to occur at each of the unique locations: C1, C2, or C3. Chlorination at C2 generates a product with a chiral center, so a racemic mixture is expected:

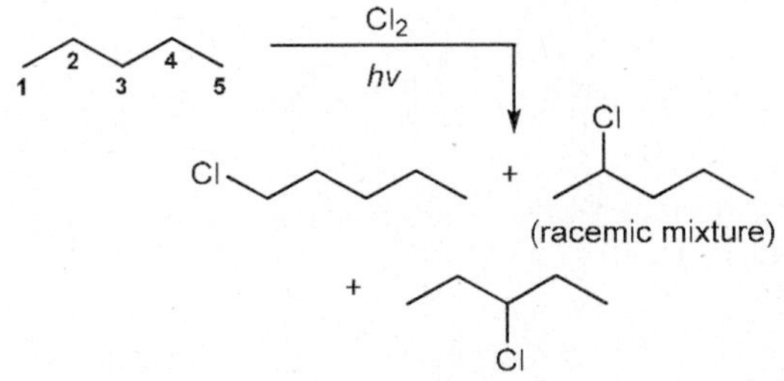

Cl_2, $h\nu$

(racemic mixture)

Chlorination at C4 yields the same product as chlorination at C2. Similarly, chlorination at C5 yields the same product as chlorination at C1.

(d) The reagents indicate a radical bromination process. Bromination is expected to occur at the benzylic position, as it is the only sp^3 hybridized carbon atom in the structure:

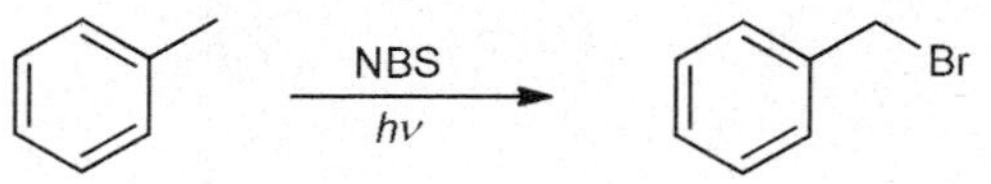

NBS, hν → Br

(e) The reagents indicate a radical bromination process. Bromination is expected to occur selectively at the allylic position (because hydrogen abstraction occurs at that position to generate a resonance-stabilized radical). In this case, there is only one allylic position:

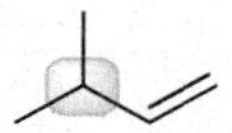

allylic position

We remove a hydrogen atom from the allylic position and draw the resonance structures of the resulting allylic radical:

Finally, we use these resonance structures to determine the products, by placing a bromine atom at the position of the unpaired electron in each resonance structure:

NBS, hν → Br + Br

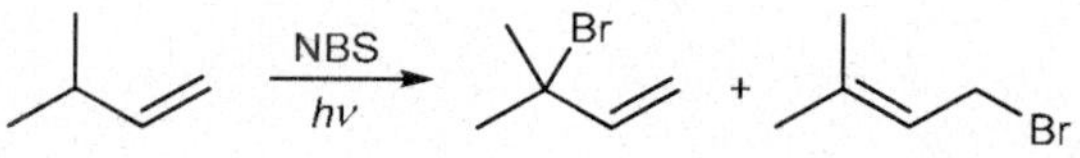

(f) The reagents indicate a radical bromination process. The starting material has only one tertiary position, which undergoes selective bromination, giving the following tertiary alkyl bromide.

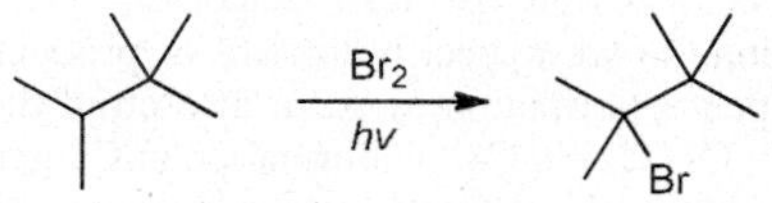

Br_2, hν → Br

10.35. We begin by drawing the starting material, (*S*)-3-methylhexane:

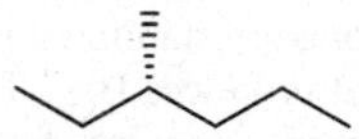

(*S*)-3-methylhexane

This compound has only one tertiary position, so bromination will occur selectively at that site. In this case, the reaction is occurring at a chiral center, so we expect a racemic mixture:

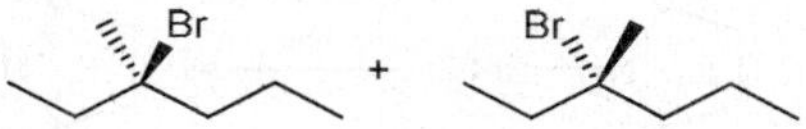

Br + Br

10.36.

(a) We begin by identifying allylic positions. There are two:

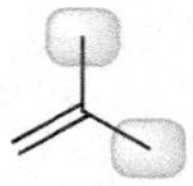

allylic positions

But these two positions are identical (symmetry). So, we only need to consider allylic bromination occurring at one of these positions. To do that, we remove a hydrogen atom from the allylic position and draw the resonance structures of the resulting allylic radical:

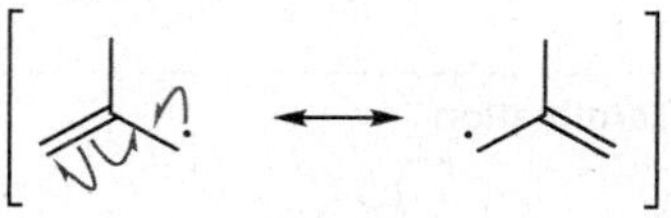

Finally, we use these resonance structures to determine the products, by placing a bromine atom at the position of the unpaired electron in each resonance structure. In this case, the two resulting structures represent the same compound, so only one product is obtained:

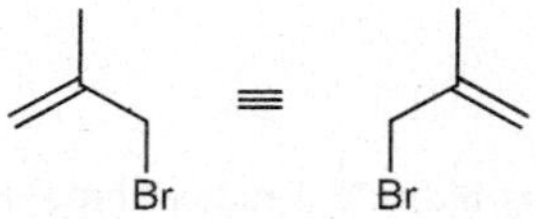

Br ≡ Br

(b) We begin by identifying the allylic positions. There are three of them (highlighted below), each of which can undergo hydrogen abstraction to give a resonance-stabilized radical, shown below. The resulting products are also shown. Notice that there are a total of five products, labeled **1** – **5**. Notice that many of the products exhibit a chiral center, and a racemic mixture is expected in each case.

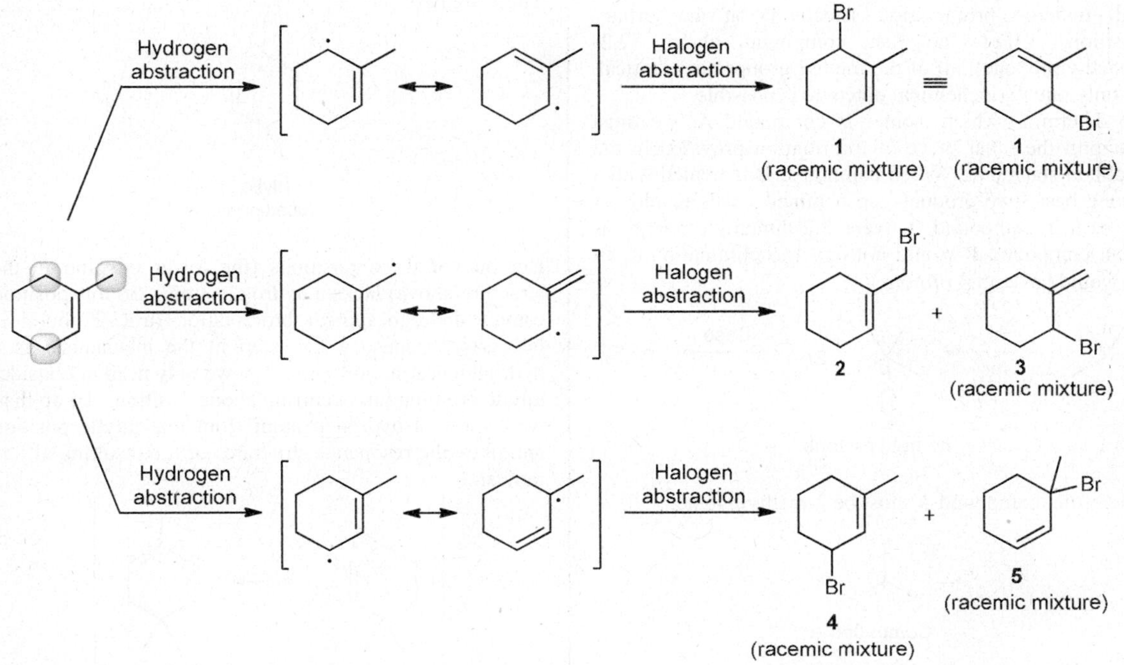

Hydrogen abstraction → Halogen abstraction → **1** (racemic mixture) ≡ **1** (racemic mixture)

Hydrogen abstraction → Halogen abstraction → **2** + **3** (racemic mixture)

Hydrogen abstraction → Halogen abstraction → **4** (racemic mixture) + **5** (racemic mixture)

10.37. As seen in Section 10.9, diethyl ether undergoes autooxidation to give the following hydroperoxide:

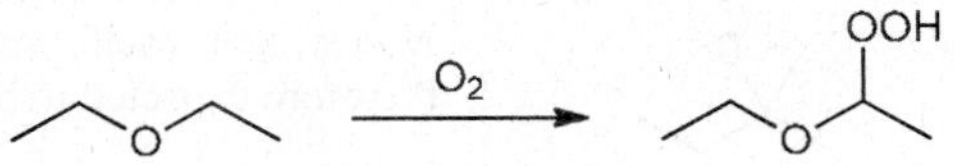

O_2 → OOH

As seen in Mechanism 10.2, autooxidation is believed to occur via two propagation steps. The first is a coupling step, and the second is a hydrogen abstraction, as shown here:

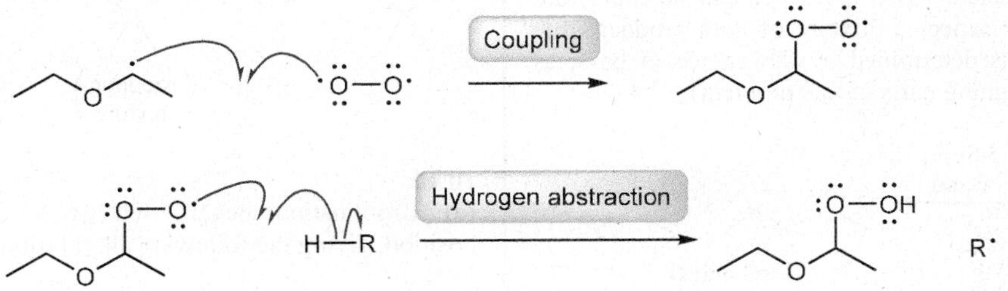

Coupling

Hydrogen abstraction

R·

10.38.

(a) Compound **A** has the molecular formula C_5H_{12}, so it must be one of the following three constitutional isomers:

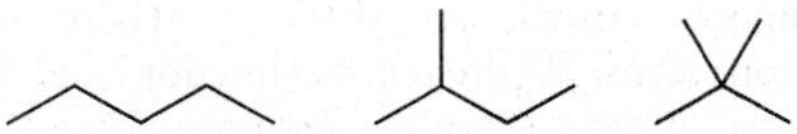

The first compound has no tertiary positions, so we would expect bromination to occur at one of the secondary positions. While there are three such positions in pentane (C2, C3, and C4), two of these positions are identical (C2 = C4). So there are two unique positions that are likely to be brominated: C2 and C3. That is, monobromination of pentane should produce a mixture of 2-bromopentane and 3-bromopentane. So, compound **A** cannot be pentane, because the problem statement indicates that monobromination would result in only one product. The other two constitutional isomers above are candidates, because each of them would give one alkyl halide as the product. The second compound above (2-methylbutane) will undergo bromination selectively at the tertiary position. For the last compound above (2,2-dimethylpropane), all of the methyl groups are identical, so only one regiochemical outcome is possible.

To determine which isomer is compound **A**, we must interpret the other piece of information provided in the problem statement. When compound **B** is treated with a strong base, two products are obtained. This would not be true if compound **A** were 2,2-dimethylpropane, as then, compound **B** would not undergo elimination at all (it would have no β protons):

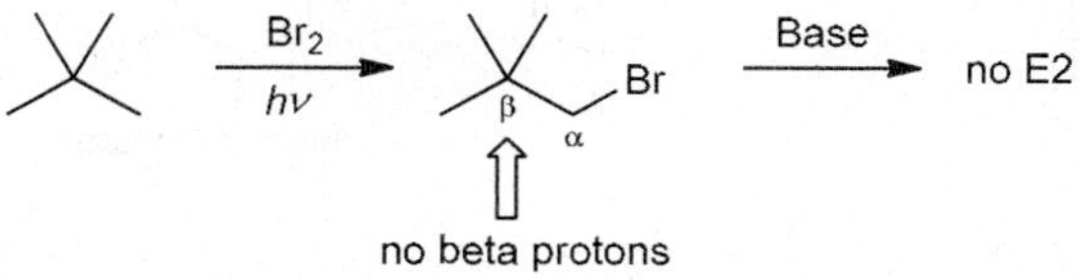

Br_2 / $h\nu$ — Br — Base — no E2

β α

no beta protons

Therefore, compound **A** must be 2-methylbutane:

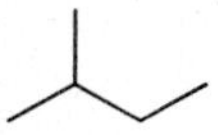

Compound **A**

As mentioned, compound **A** undergoes monobromination selectively at the tertiary position, giving the corresponding tertiary alkyl halide (compound **B**):

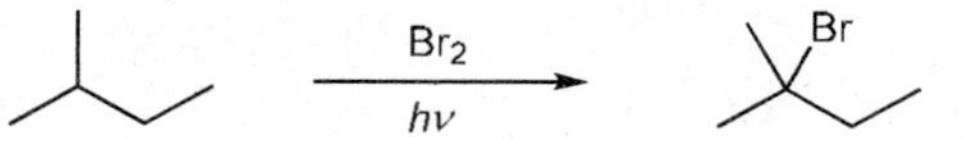

A — Br_2 / $h\nu$ — Br **B**

When compound **B** is treated with a strong base, an E2 reaction is expected. Two regiochemical outcomes are possible, so we expect a mixture of both products (the major product is determined by the choice of base, as seen in the remaining parts of this problem):

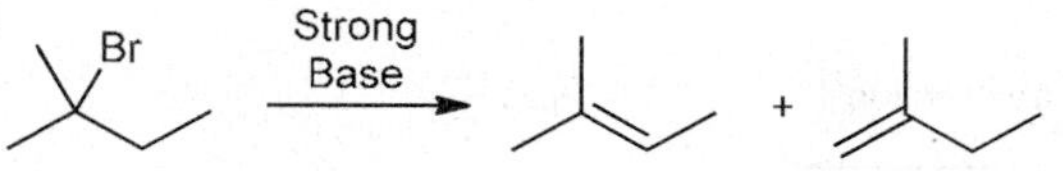

Br **B** — Strong Base — + **C** and **D**

(b) When compound **B** is treated with *tert*-butoxide (a sterically hindered base), the less-substituted (Hofmann) product is favored.

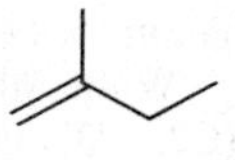

(c) When compound **B** is treated with sodium ethoxide, the more-substituted (Zaitsev) product is favored:

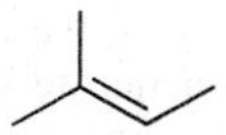

10.39. We begin by identifying the allylic positions. There are two:

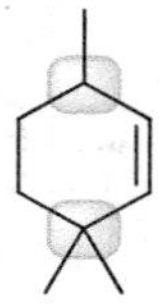

allylic positions

But one of these positions (the lower position in the structure above) lacks a hydrogen atom. So this position cannot undergo radical bromination (a C–H bond is necessary because a key step in the mechanism is a hydrogen abstraction step). So, we only need to consider allylic bromination occurring at one position. To do that, we remove a hydrogen atom from that allylic position and draw the resonance structures of the resulting allylic radical:

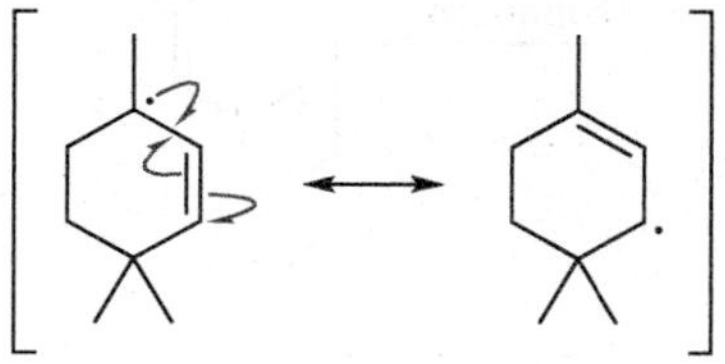

Finally, we use these resonance structures to determine the products, by placing a bromine atom at the position of the unpaired electron in each resonance structure. Notice that each product has a chiral center and is therefore expected to be produced as a racemic mixture:

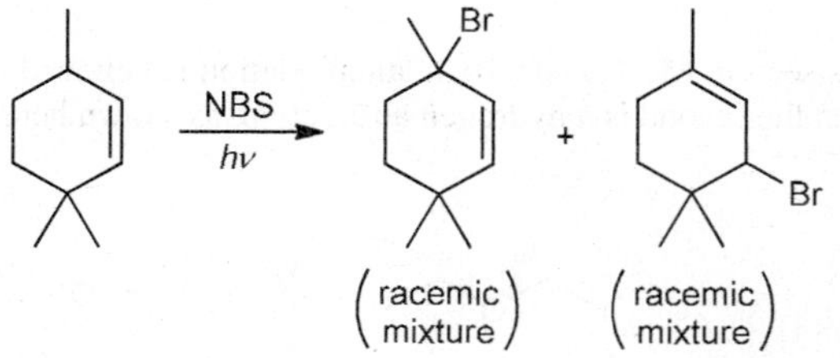

NBS / $h\nu$ — Br (racemic mixture) + Br (racemic mixture)

10.40.

(a) Bromination occurs selectively at the tertiary position, giving the following alkyl bromide:

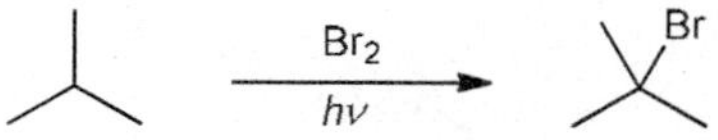

Br_2 / $h\nu$ — Br

(b) The mechanism will have three distinct stages. The first stage is initiation, in which the Br–Br bond is broken to generate bromine radicals. This step requires two fishhook arrows, as shown. There are two propagation steps (hydrogen abstraction and halogen abstraction), each of which requires three fishhook arrows (to show the bonds being broken and formed). In the first propagation step, a hydrogen atom is abstracted to give the more stable tertiary radical. In the second step, this radical undergoes halogen abstraction to give the product. There are many steps that can serve as termination steps, because there are many radicals that can couple together under these conditions. Shown below is a termination step that generates the desired product:

(c) If there was a minor product, it would only be formed via a primary radical, the formation of which is significantly slower under bromination conditions. The tertiary radical is selectively formed, which leads to the tertiary alkyl bromide as the product.

10.41.

(a) There are two tertiary positions in this case, highlighted here:

But these positions are identical, because they can be interchanged by an axis of symmetry, shown here (when we rotate 180 degrees about this axis, the same image is regenerated – you might want to build a molecular model to prove this to yourself):

So we only need to consider bromination at one of these positions (either one will lead to the same products). Since the reaction occurs at a chiral center, we expect monobromination to give both possible configurations (*R* and *S*) for that chiral center. The configuration of the other chiral center (not involved in the reaction) is retained. Therefore, the products are diastereomers:

(b) The starting compound has two tertiary positions, so dibromination will install one bromine atom at each of the tertiary positions:

Each chiral center can be produced with either the *R* or *S* configuration, so all possible stereoisomers are expected. With two chiral centers, we might expect four stereoisomers (2^n, where n = # of chiral centers = 2). However, in this case, there are only three stereoisomers, because one of them is a *meso* compound, shown below. (The *meso* compound has no enantiomer because it has an internal plane of symmetry and is therefore achiral. For a review of *meso* compounds, see Section 5.6)

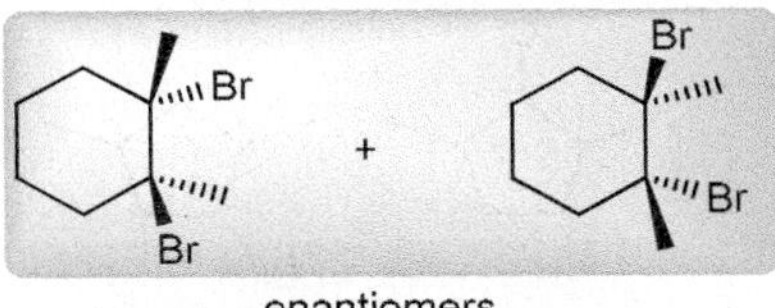

enantiomers

a *meso* compound

10.42. The correct answer is (c). When an aromatic compound bearing an alkyl side chain is treated with excess NBS, all benzylic hydrogen atoms will be replaced with bromine atoms.

10.43. The correct answer is (b). Iodination is the only reaction that is expected to be thermodynamically unfavorable (positive Δ*G*). Fluorination, chlorination and bromination are all expected to be thermodynamically favorable (negative Δ*G*), although fluorination will be too violent to be practically useful.

10.44. The correct answer is (c). Hydrogen abstraction is the first propagation step in radical bromination. The weakest C—H bond (the one with the smallest BDE) will be broken during this step. For a simple alkene, like 1-butene, an allylic C—H bond will be the weakest C—H bond, giving a stable, allylic radical intermediate after bond cleavage. The first two possibilities, (a) and (b), are vinylic radicals which are too high in energy to form. The fourth possibility (d) is a primary radical and is significantly less stable than an allylic radical.

10.45. The correct answer is (d). The tertiary radical is the most stable, because alkyl groups stabilize the unpaired electron via a delocalization effect, called hyperconjugation.

10.46. The correct answer is (b). This compound has seven carbon atoms, but there are only four unique positions (highlighted) where chlorination can occur:

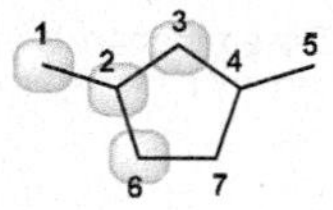

Chlorination at C1 produces the same result as chlorination at C5, chlorination at C2 produces the same result as chlorination at C4, and chlorination at C6 produces the same result as chlorination at C7. Therefore, the following four monochlorination products are expected:

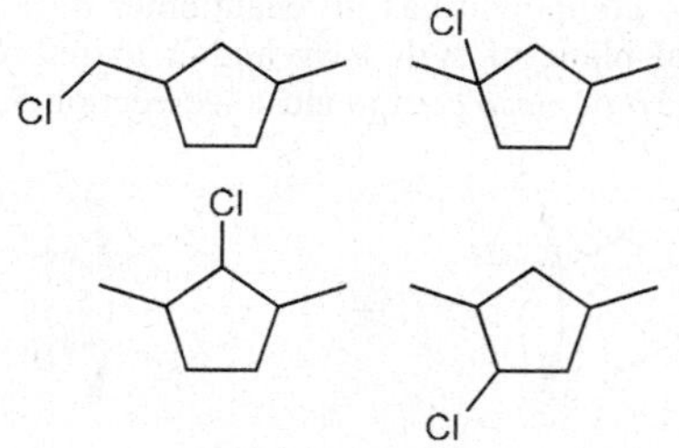

10.47. The correct answer is (a). After initiation of the radical mechanism, a bromine radical adds to the π bond, giving the more stable (secondary) carbon radical. This propagation step leads to the following intermediate:

10.48. The correct answer is (d). The mechanism for radical bromination of 2-methylpropane begins with initiation, which creates bromine radicals. This initiation step involves homolytic bond cleavage (using Br_2 and either heat or light) and should employ two fishhook arrows.

10.49. Methyl radicals are less stable than *tert*-butyl radicals so methyl radicals react with each other (couple) more rapidly. The tertiary *tert*-butyl radicals are more stable and are therefore less reactive. Also, methyl radicals are less hindered than *tert*-butyl radicals, so that further explains why methyl radicals couple more rapidly than *tert*-butyl radicals.

10.50.

(a) Monochlorination of cyclopentane gives only one product, because a reaction at any position generates the same product as a reaction at any other position:

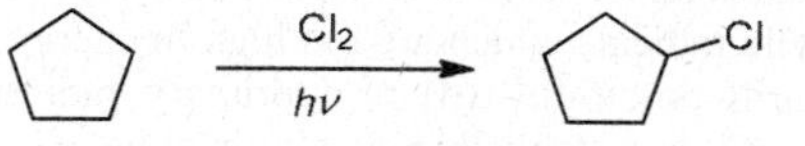

(b) This compound has six carbon atoms, but there are only four unique positions (highlighted) where chlorination can occur:

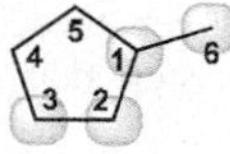

Chlorination at C4 produces the same result as chlorination at C3, and chlorination at C5 produces the same result as chlorination at C2. Therefore, the following four constitutional isomers are expected:

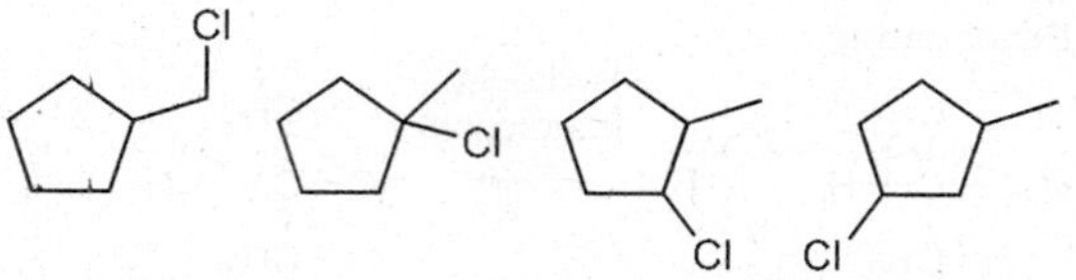

The last two structures have chiral centers and are therefore produced as mixtures of stereoisomers. But this can be ignored for purposes of solving this problem, because the problem statement asks for the number of *constitutional isomers* that are obtained.

(c) This compound has seven carbon atoms, but there are only three unique positions (highlighted) where chlorination can occur:

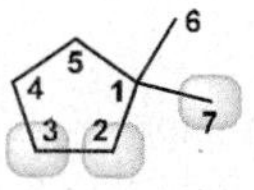

Chlorination cannot occur at C1, because that position does not have a C–H bond (which is necessary, because hydrogen abstraction is the first step of the chlorination process). Chlorination at C4 produces the same result as chlorination at C3, chlorination at C5 produces the same result as chlorination at C2, and chlorination at C6 produces the same result as chlorination at C7. Therefore, the following three constitutional isomers are expected:

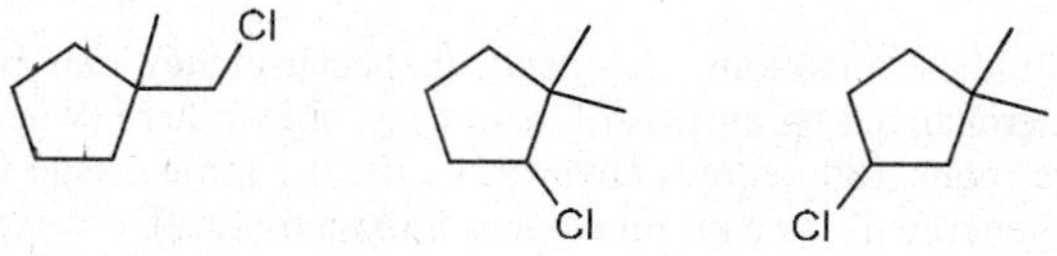

Each of the last two structures has a chiral center and is therefore produced as a mixture of enantiomers. But this can be ignored for purposes of solving this problem, because the problem statement asks for the number of *constitutional isomers* that are obtained.

(d) This compound has five carbon atoms, but there are only four unique positions (highlighted) where chlorination can occur:

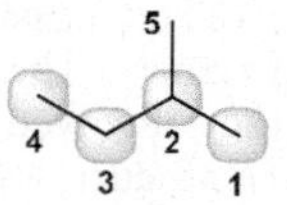

Chlorination at C5 produces the same result as chlorination at C1. Therefore, the following four constitutional isomers are expected:

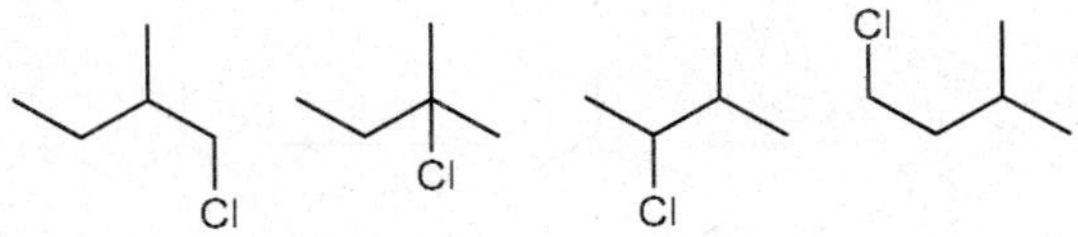

The first structure has a chiral center and is therefore produced as a racemic mixture of enantiomers. The same is true of the third structure above. But this can be ignored for purposes of solving this problem, because the problem statement asks for the number of *constitutional isomers* that are obtained.

(e) This compound has six carbon atoms, but there are only two unique positions (highlighted) where chlorination can occur:

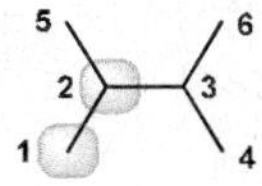

Chlorination at C3 produces the same result as chlorination at C2. Similarly, chlorination at C4, C5, or C6 produces the same result as chlorination at C1. Therefore, the following two constitutional isomers are expected:

The first structure has a chiral center and is therefore produced as a racemic mixture of enantiomers. But this can be ignored for purposes of solving this problem, because the problem statement asks for the number of *constitutional isomers* that are obtained.

(f) Monochlorination of cyclohexane gives only one product, because a reaction at any position generates the same product as a reaction at any other position:

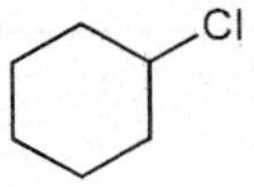

(g) This compound has seven carbon atoms, but there are only five unique positions (highlighted) where chlorination can occur:

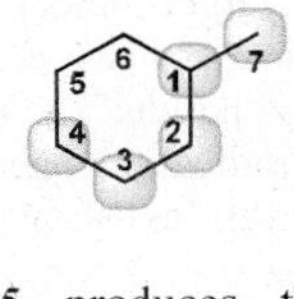

Chlorination at C5 produces the same result as chlorination at C3, and chlorination at C6 produces the same result as chlorination at C2. Therefore, the following five constitutional isomers are expected:

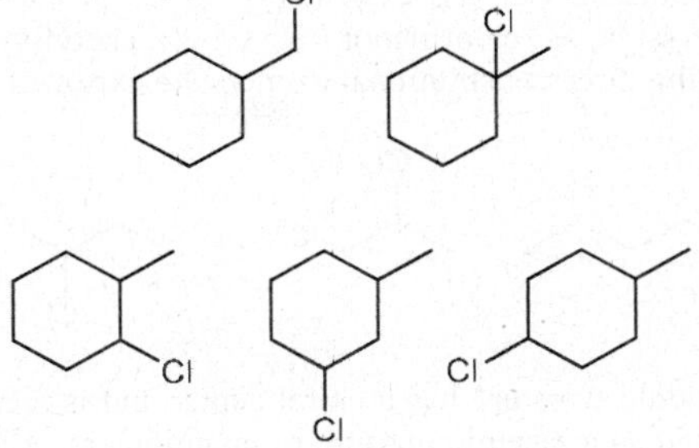

Each of the bottom three structures is produced as a mixture of stereoisomers. But this can be ignored for purposes of solving this problem, because the problem statement asks for the number of *constitutional isomers* that are obtained.

(h) This compound has eight carbon atoms, but two of them (the two central carbon atoms) cannot undergo chlorination because they lack a C–H bond. The remaining six carbon atoms (the methyl groups) all are identical, because a reaction at any position generates the same product as a reaction at any other position:

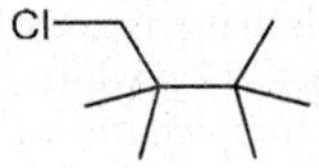

(i) This compound has six carbon atoms, but there are only three unique positions (highlighted) where chlorination can occur:

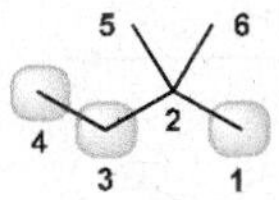

Chlorination cannot occur at C2, because that position lacks a C–H bond. Chlorination at C5 or C6 produces the same result as chlorination at C1. Therefore, the following three constitutional isomers are expected:

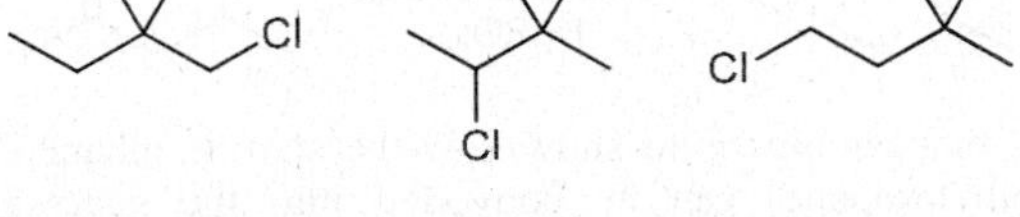

The middle structure has a chiral center and is therefore produced as a racemic mixture of enantiomers. But this can be ignored for purposes of solving this problem, because the problem statement asks for the number of *constitutional isomers* that are obtained.

(j) This compound has five carbon atoms, but there are only three unique positions (highlighted) where chlorination can occur:

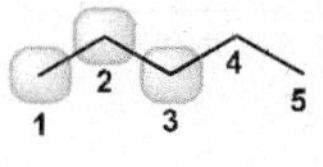

Chlorination at C4 produces the same result as chlorination at C2, and chlorination at C5 produces the same result as chlorination at C1. Therefore, the following three constitutional isomers are expected:

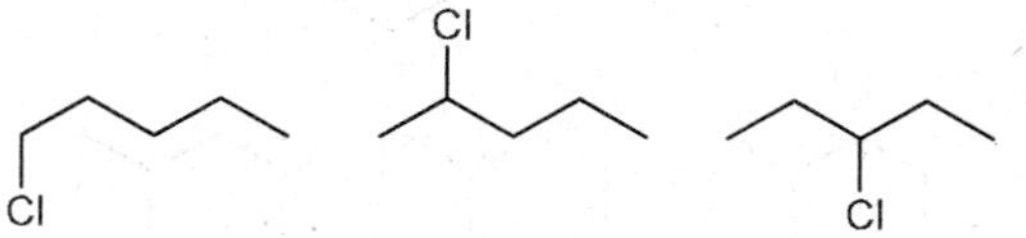

The middle structure has a chiral center and is therefore produced as a racemic mixture of enantiomers. But this can be ignored for purposes of solving this problem, because the problem statement asks for the number of *constitutional isomers* that are obtained.

10.51. Let's begin by drawing our product and starting materials, so that we can see them more clearly:

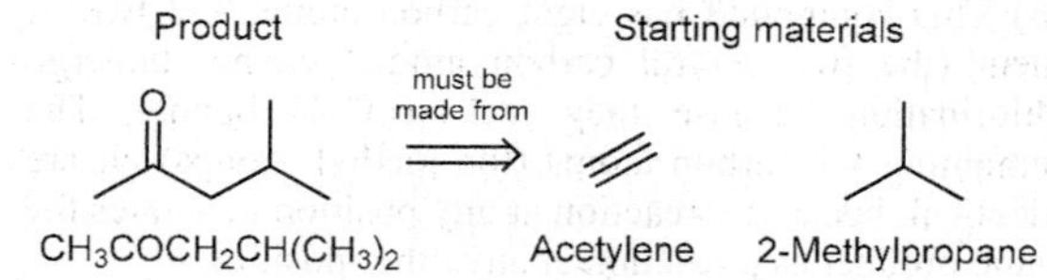

Product — must be made from — Starting materials

$CH_3COCH_2CH(CH_3)_2$ — Acetylene — 2-Methylpropane

The starting materials have two carbon atoms and four carbon atoms, respectively, while the product has six carbon atoms. Therefore, we must find a way to join the starting materials. So far, we have only seen one way to make a carbon-carbon bond. Specifically, acetylene can undergo alkylation, as shown here:

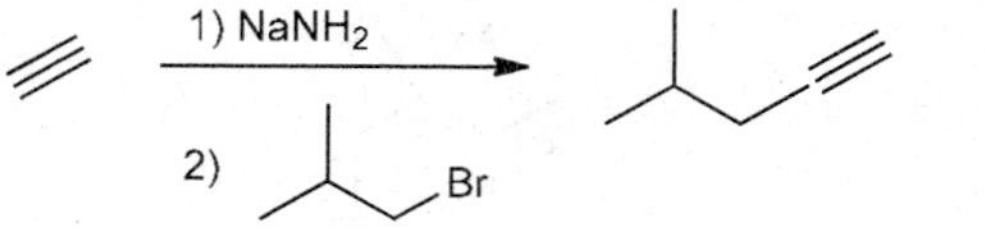

1) $NaNH_2$
2) Br

The resulting terminal alkyne can then be converted into the desired product via acid-catalyzed hydration in the presence of mercuric sulfate:

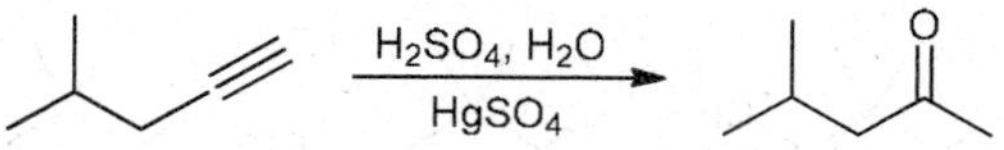

H_2SO_4, H_2O
$HgSO_4$

All that remains is to show how the starting alkane (2-methylpropane) can be converted into the necessary primary alkyl halide:

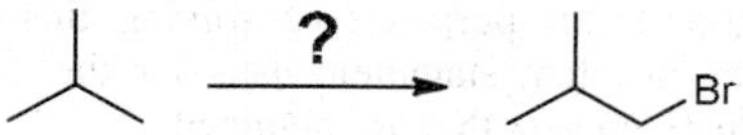

?

Radical bromination will indeed install a bromine atom, but this occurs selectively at the tertiary position:

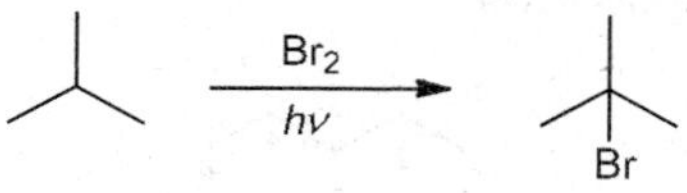

Br_2
$h\nu$

This tertiary alkyl bromide must now be converted into the desired primary alkyl bromide. That is, we must move the position of the bromine atom, which can be accomplished via a two-step process (elimination, followed by *anti*-Markovnikov addition of HBr):

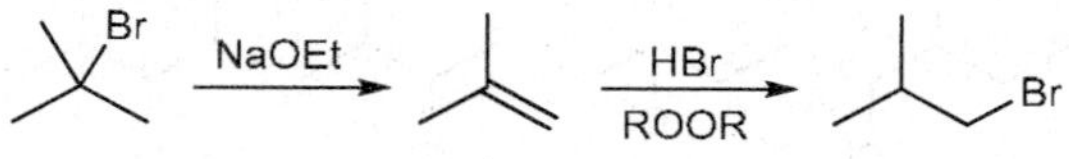

NaOEt — HBr, ROOR

In summary, the entire synthesis is shown here:

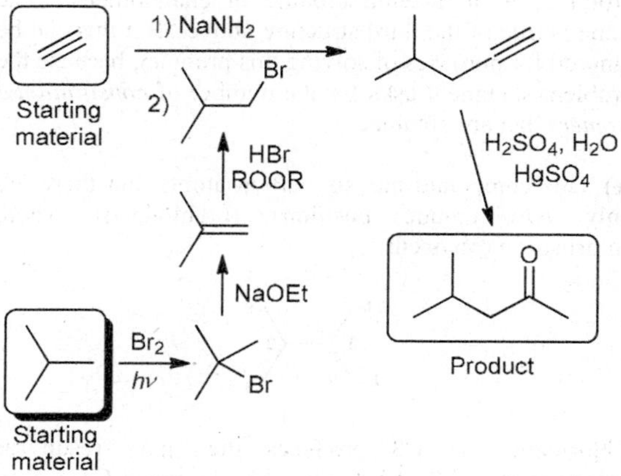

Starting material — 1) $NaNH_2$ 2) Br — HBr ROOR — NaOEt — Br_2, $h\nu$ — Starting material — H_2SO_4, H_2O $HgSO_4$ — Product

10.52.

(a) This transformation can be achieved via radical chlorination (under conditions that favor monochlorination):

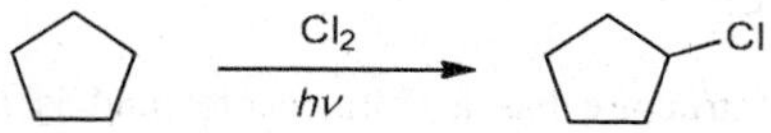

Cl_2
$h\nu$

(b) Radical iodination is not a feasible process (it is not thermodynamically favorable), so we cannot directly iodinate the starting cycloalkane. However, radical bromination (or alternatively, radical chlorination) can be performed, followed by an S_N2 reaction in which iodide replaces bromide:

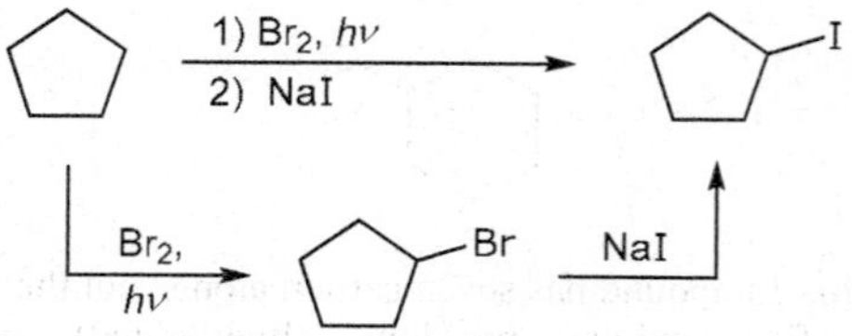

1) Br_2, $h\nu$
2) NaI

Br_2, $h\nu$ — Br — NaI

(c) The π bond in the target molecule can be formed by an elimination reaction, so we can make the target molecule from an alkyl halide. The alkyl halide, in turn, can be prepared by radical halogenation of the given alkane starting material, as shown in the following retrosynthesis:

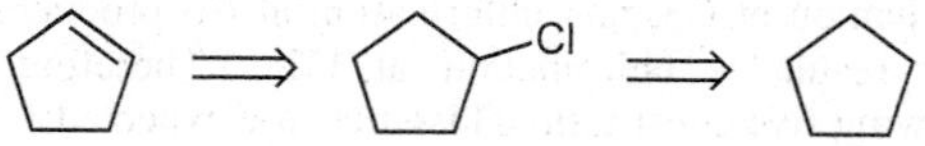

Now let's draw the forward process. First, radical chlorination or bromination can be performed, thereby installing a leaving group, which then allows for an E2 process with a strong base to give the desired product. Any strong base can be used (there is only one regiochemical outcome, so a sterically hindered base is not required). It is acceptable to use hydroxide, methoxide or ethoxide for the E2 process. That being said, in this particular case, *tert*-butoxide will likely be more efficient, as it will suppress the competing S_N2 process (we would expect S_N2 to give a minor product if the base is not sterically hindered).

1) Cl_2, $h\nu$
2) *t*-BuOK

Cl_2, $h\nu$ — Cl — *t*-BuOK

(d) A retrosynthetic analysis of the target molecule begins with the following question: "What starting material can be used to make a *trans*-1,2-dibromide?" The two Br atoms in the target molecule can be installed via bromination of alkene:

Br Br

This alkene can be produced from an alkyl halide that, in turn, can be prepared by radical halogenation of the given alkane starting material:

Br

Now let's draw the forward process. First, radical chlorination or bromination can be performed, thereby installing a leaving group, which then allows for an E2 process with a strong base to give cyclohexene. Any strong base can be used (there is only one regiochemical outcome, so a sterically hindered base is not required). It is acceptable to use hydroxide, methoxide or ethoxide for the E2 process. That being said, in this particular case, *tert*-butoxide will likely be more efficient, as it will suppress the competing S_N2 process. Finally, cyclohexene can be treated with Br_2 (*anti* addition of two bromine atoms) to give the target molecule. In summary, the desired transformation can be achieved via the following three-step synthesis:

1) Cl_2, $h\nu$
2) *t*-BuOK
3) Br_2

Br Br + En

Cl — *t*-BuOK — Br_2

(e) The π bond in the target molecule can be made from an alkyl halide via an elimination reaction. This alkyl halide is an ideal intermediate, because it can be prepared by radical halogenation of the given alkane starting material:

Br

Now let's draw the forward process. The transformation begins with a radical bromination, which installs a bromine atom selectively at the tertiary position. For the subsequent E2 elimination, the more-substituted Zaitsev product is desired, so we must a use a strong base that is not sterically hindered, such as hydroxide, methoxide or ethoxide:

1) Br_2, $h\nu$
2) NaOEt

Br_2, $h\nu$ — Br — NaOEt

10.53. Monochlorination of *cis*-1,2-dimethylcyclopentane generates twelve possible products, as six pairs of stereoisomers (one pair has an enantiomeric relationship, indicated below, and all other pairs have a diastereomeric relationship). In contrast, *trans*-dimethylcyclopentane produces only six different products, as shown below:

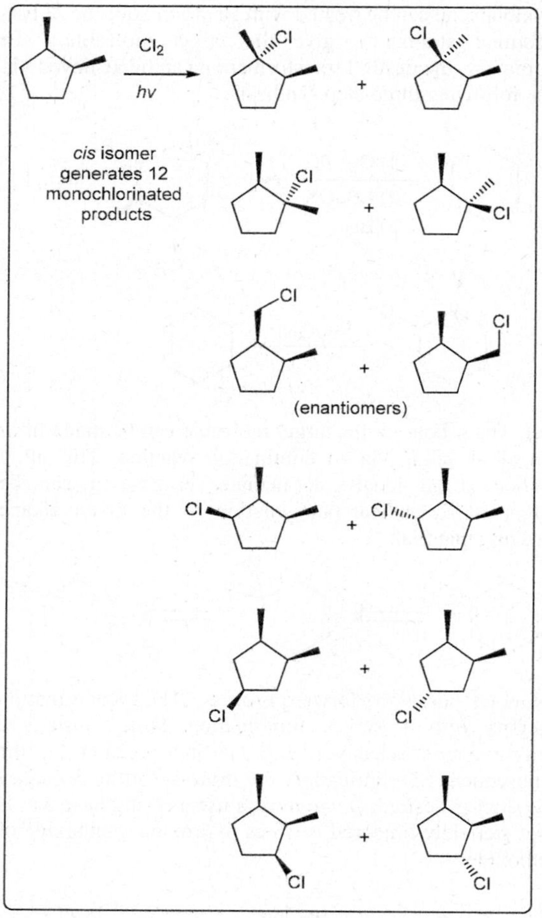

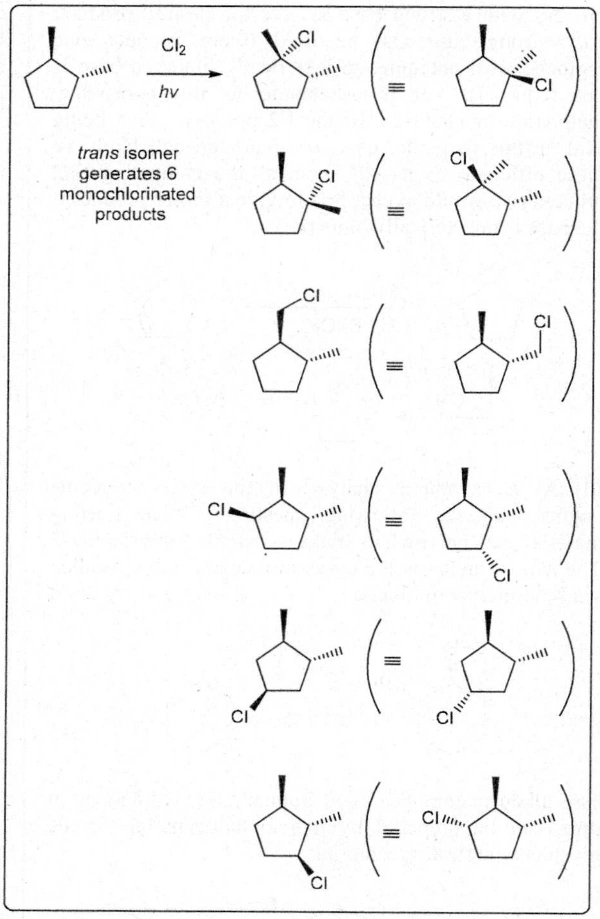

Cl_2, *hv*

cis isomer generates 12 monochlorinated products

(enantiomers)

Cl_2, *hv*

trans isomer generates 6 monochlorinated products

10.54.

(a) Recall that ΔG has two components: (ΔH) and ($-T\Delta S$). The first term is positive (recall that breaking a bond *consumes* energy), so the second term must have a large negative value in order for ΔG to be negative (which is necessary in order for the process to be thermodynamically favorable). This will be the case if ΔS and T are both large and positive (T cannot be negative, as it is in units of Kelvin). At high temperatures, both of these terms are indeed large. ΔS is large and positive because one chemical entity is being converted into two chemical entities, which significantly increases the entropy of the system.

(b) Recall that ΔG has two components: (ΔH) and ($-T\Delta S$). The magnitude of ($-T\Delta S$) is dependent on the temperature. As described in part (a), at high temperature, the ($-T\Delta S$) term dominates over (ΔH), and the reaction is thermodynamically favorable (ΔG is negative). However, at low temperature, the first term (enthalpy) dominates, and the reaction is no longer thermodynamically favored (ΔG is positive).

10.55. Bromination of compound **A** gives 2,2-dibromopentane:

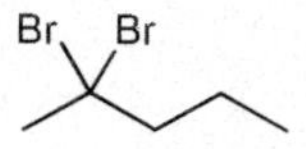

2,2-dibromopentane

So compound **A** must be 2-bromopentane:

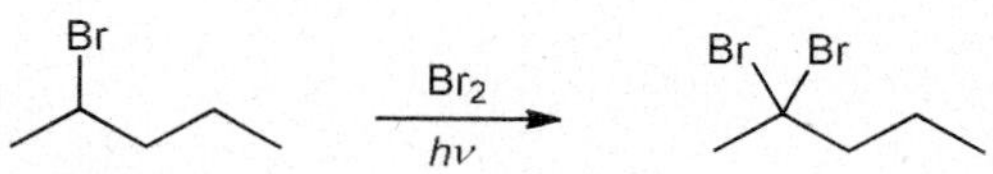

Compound **A** $\xrightarrow[hv]{Br_2}$

Compound **A** exhibits one chiral center. The problem statement provides the information that is necessary to determine the configuration of this chiral center. Specifically, we are told that treating compound **A** with a strong nucleophile (S_N2 conditions) results in a product with the *R* configuration. Since S_N2 reactions proceed via inversion of configuration, the chiral center must have the *S* configuration in compound **A**:

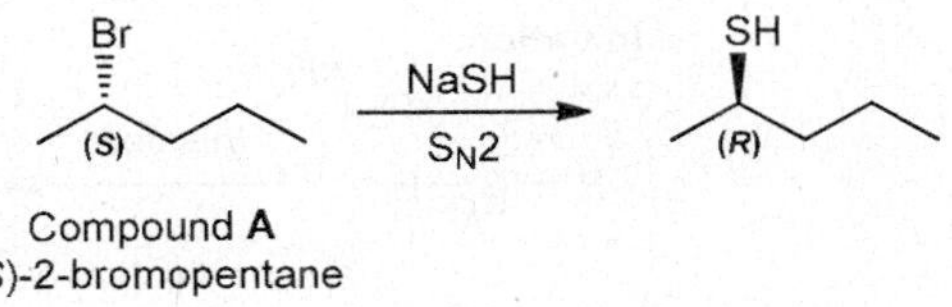

Br (S) NaSH S_N2 SH (R)

Compound **A**
(*S*)-2-bromopentane

10.56. The first propagation step in a bromination process is generally slow. In fact, this is the source of the selectivity for bromination processes. A pathway via a secondary radical will be significantly lower in energy than a pathway via a primary radical. As a result, bromination occurs predominantly at the more substituted (secondary) position. However, when chlorine is present, chlorine radicals can perform the first propagation step (hydrogen abstraction) very rapidly, and with little selectivity. Under these conditions, primary radicals are formed almost as readily as secondary radicals. The resulting radicals then react with bromine in the second propagation step to yield monobrominated products. Therefore, in the presence of chlorine, the selectivity normally observed for bromination is lost.

10.57. As shown in the problem statement, an acyl peroxide will undergo cleavage to give a radical that can then lose carbon dioxide:

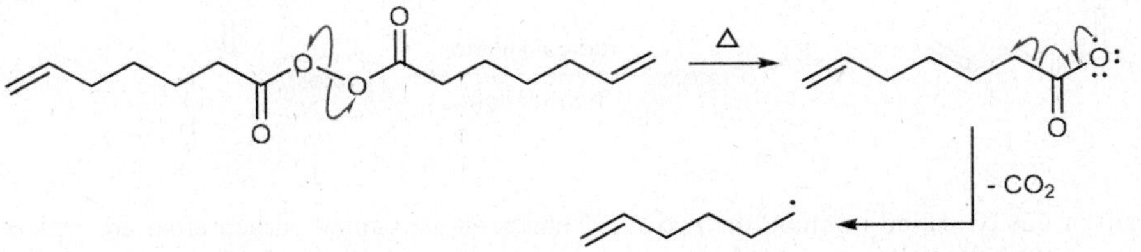

Δ

$- CO_2$

This radical is responsible for the formation of each of the reported products. The first product is formed when the radical undergoes hydrogen abstraction:

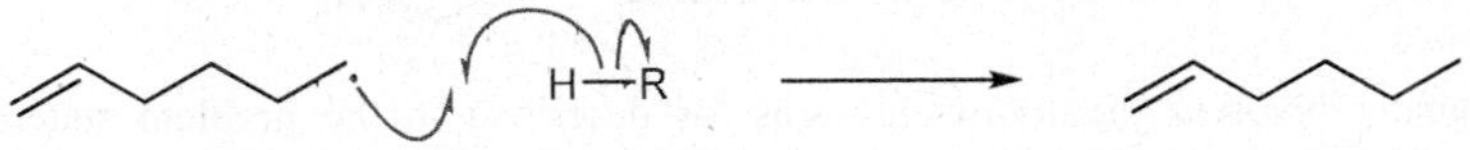

H—R

The second product is formed when two of the radicals couple with each other, as shown:

And finally, each of the cyclic products is formed via addition to a π bond (in an intramolecular fashion), followed by hydrogen abstraction:

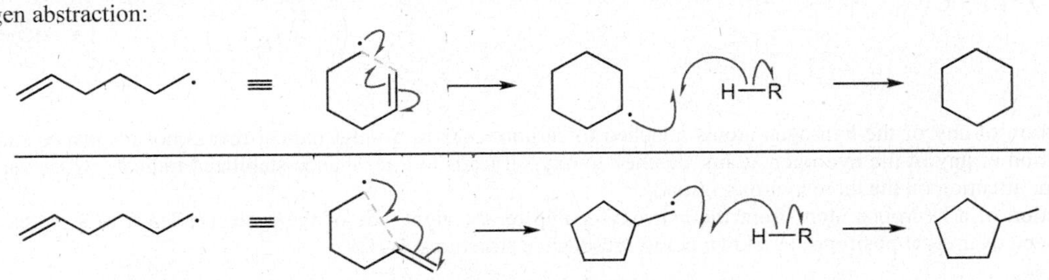

≡ H—R

≡ H—R

10.58.

(a) In this reaction, two groups (R′S and H) are being added across the π bond in an *anti*-Markovnikov fashion. This reaction resembles the *anti*-Markovnikov addition of HBr across a π bond (Section 10.10). Using that reaction as a guide, we can draw the following initiation and propagation steps. In the initiation step, an R′S radical is formed. The first propagation step is an addition to a π bond, and the second propagation step is a hydrogen abstraction to form the observed product and regenerate an R′S radical.

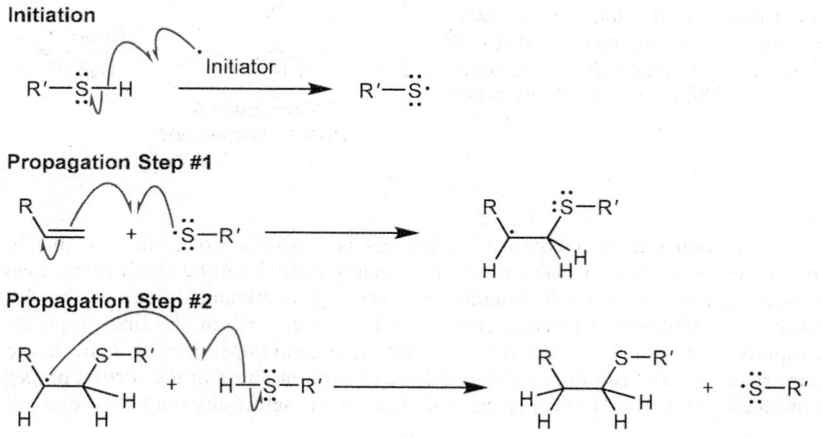

Initiation

Propagation Step #1

Propagation Step #2

(b) We expect an *anti*-Markovnikov addition of R′S and H across the π bond, as follows:

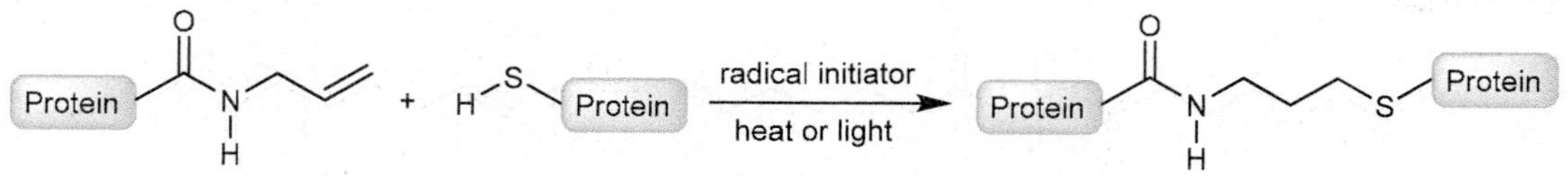

10.59. Overall, in this oxidation reaction, the two C—H bonds on the central carbon atom are replaced with a C=O double bond, as shown below. It is useful to keep this in mind as we draw our mechanism.

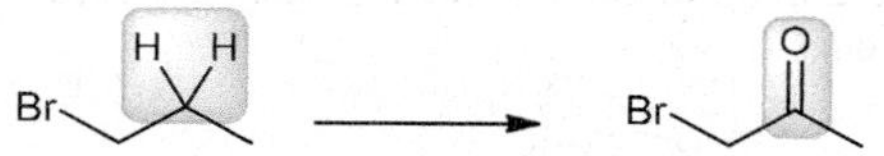

The following mechanism involves the following steps (as described in the problem statement): hydrogen atom abstraction, coupling with O_2, oxygen atom abstraction and hydrogen atom abstraction.

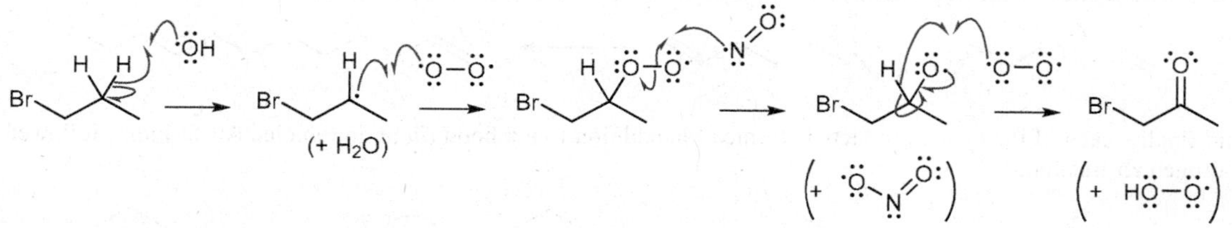

10.60.
Abstraction of any of the hydrogen atoms attached to carbon leads to a vinyl radical that is not resonance stabilized. Abstraction of any of the hydrogen atoms attached to oxygen leads to a resonance-stabilized radical. Thus, we should focus our attention on the three hydroxyl groups.
Abstraction of a hydrogen atom from the hydroxyl group on the right side of the molecule leads to a radical that is delocalized over eight positions, as shown below (resonance structures **A-H**):

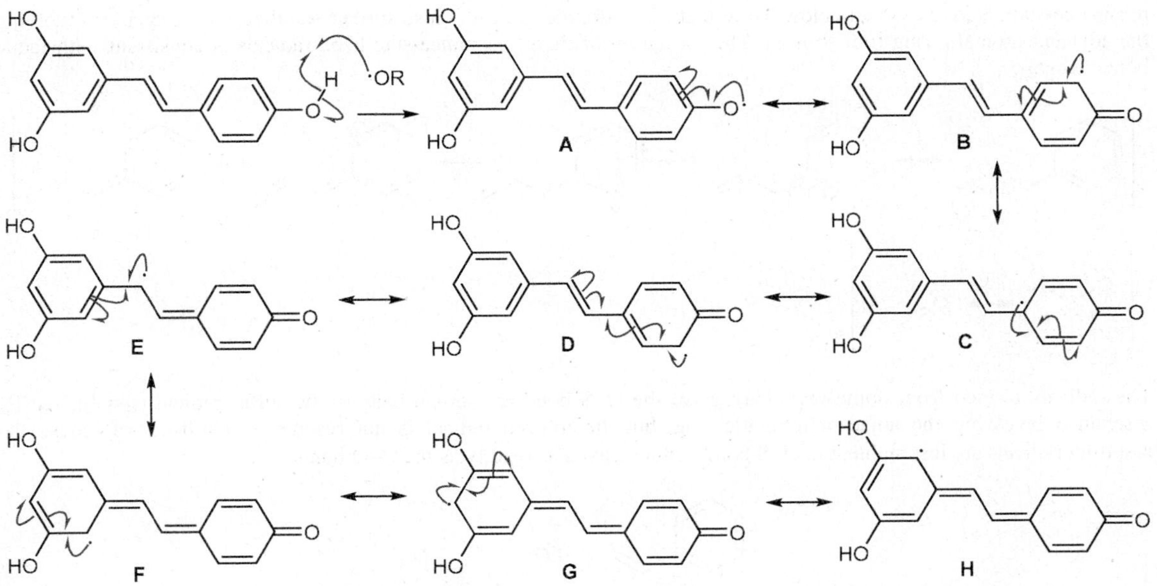

HO HO HO
H ·OR
HO HO A HO B
HO HO HO
HO E HO D HO C
HO HO HO
HO F HO G HO H

Abstraction of a hydrogen atom from either of the hydroxyl groups on the left side of the molecule leads to a radical that is delocalized over four positions, as shown here (resonance structures **I-L**).

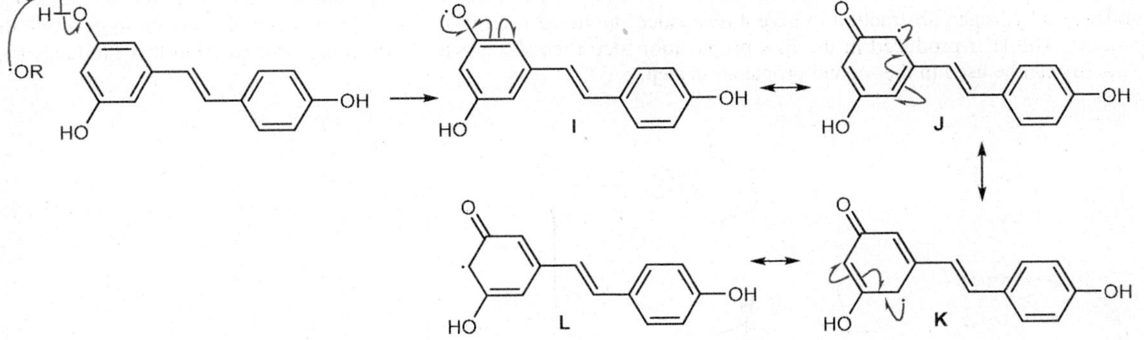

H-O ·OR OH HO
OH HO I
OH HO J
OH HO L
OH HO K

This suggests that the hydrogen atom of the hydroxyl group on the right side of the molecule is more susceptible to abstraction, because it leads to the formation of a more stable radical.

10.61.

(a) In the first step, the C–O bond undergoes homolytic cleavage, yielding the two radicals shown. Resonance structures of the sulfoxide radical demonstrate that the unpaired electron is delocalized over the oxygen and sulfur atoms, as shown. Recombination of the radicals in the second step generates the observed product, as shown.

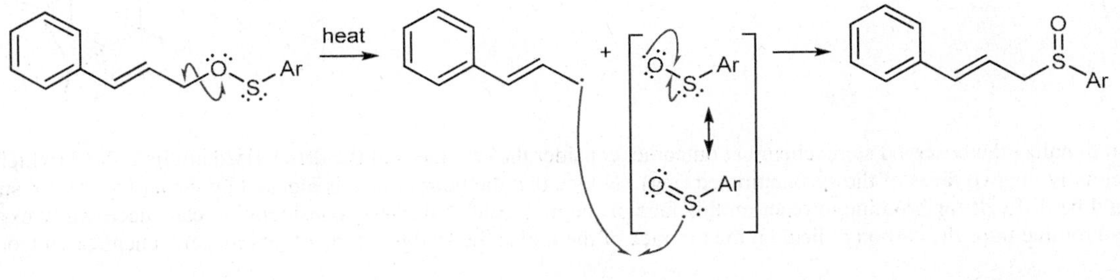

heat
Ar + Ar Ar
Ar

(b) To propose an explanation for the scission of the C–O bond over the O–S bond, we must analyze the radicals formed from each of these homolytic cleavages. Both of the radicals formed from cleavage of the C–O bond are resonance-stabilized, as shown below. Note that the sulfoxide radical is also further stabilized by resonance involving the adjacent aromatic ring (not shown). The formation of these resonance-stabilized radicals is consistent with facile bond cleavage.

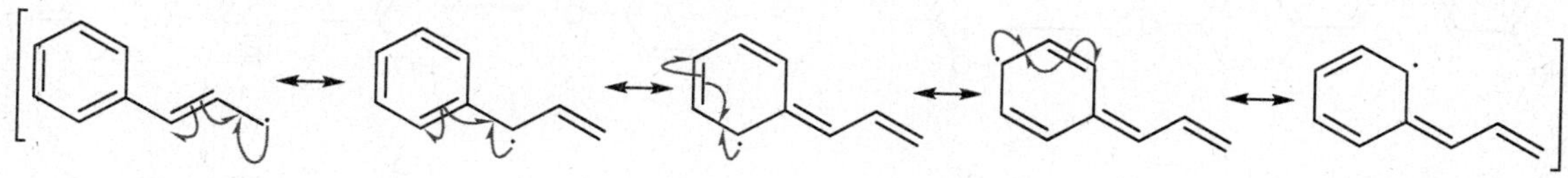

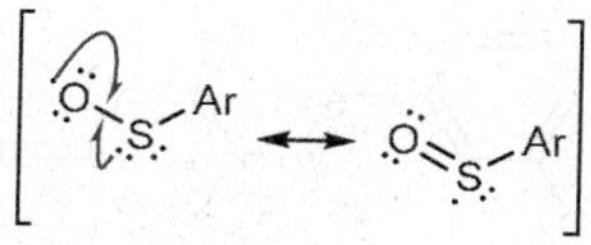

The radicals formed from homolytic cleavage of the O–S bond are shown below. The sulfur radical is stabilized by resonance involving the adjacent aromatic ring, but the oxygen radical is not resonance-stabilized. Because the resulting radicals are less stable, the O–S bond is not cleaved as readily as the C–O bond.

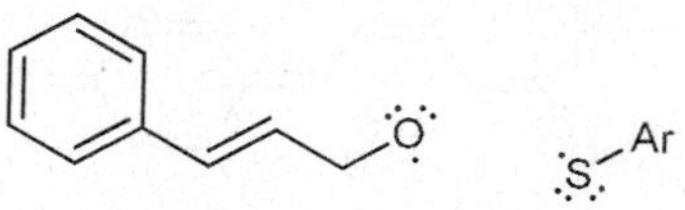

10.62.

In the first step, *tert*-butoxide is a strong, sterically hindered base, so we expect an E2 process, generating an alkene (note that there is only one regiochemical outcome for this E2 reaction, because there is only one beta proton that is anti-periplanar with the iodide leaving group). Then, when treated with NBS under radical conditions, the alkene undergoes hydrogen abstraction to give a resonance stabilized radical, followed by halogen abstraction to generate the product. The HBr produced in the first propagation step then reacts with NBS in an ionic reaction that produces Br_2. This Br_2 can be used in the second propagation step.

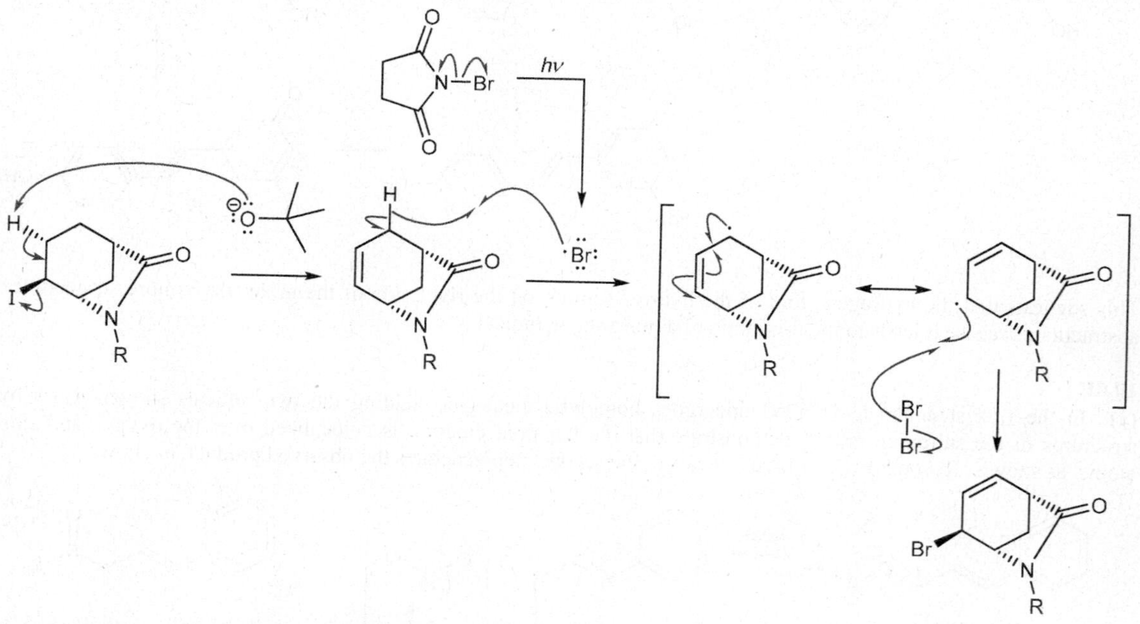

To rationalize the observed stereochemical outcome, consider the two faces of the delocalized allylic radical (which are essentially the two faces of the six-membered ring). Notice that the bottom face is blocked by the amide. As a result, it would be difficult for bromine to react on this face of the molecule. This steric consideration can successfully explain why bromine traps the carbon radical on the top face of the molecule, leading to the observed stereochemical outcome.

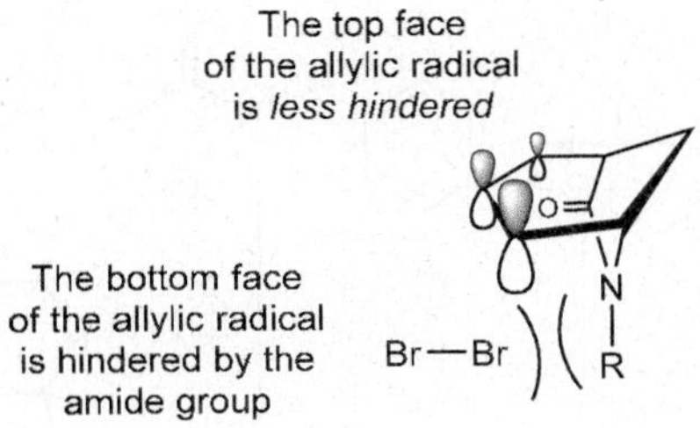

The top face of the allylic radical is *less hindered*

The bottom face of the allylic radical is hindered by the amide group

10.63. We know that a radical can undergo an addition reaction with a π bond, and if these two reactive sites are on the same carbon chain, then the resulting *intramolecular* addition reaction results in the formation of a ring. Both of the observed radicals are generated by such cyclization processes, occurring in a tandem (back-to-back) fashion. Depending on which π bond the initial radical attacks first, there are two different outcomes. The first tricyclic radical can be formed with the following two steps. First, the initial radical undergoes an intramolecular addition to a π bond, generating a 5-membered ring and a new radical. This radical once again undergoes addition to the remaining π bond, resulting in the formation of two 5-membered rings and a new radical.

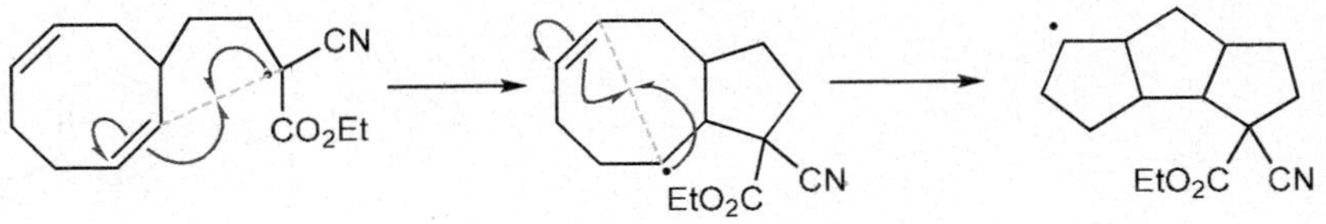

The second tricyclic radical can be formed if the initial radical undergoes addition with the π bond that is farther away (to form a six-membered ring), followed by a subsequent addition to the remaining π bond.

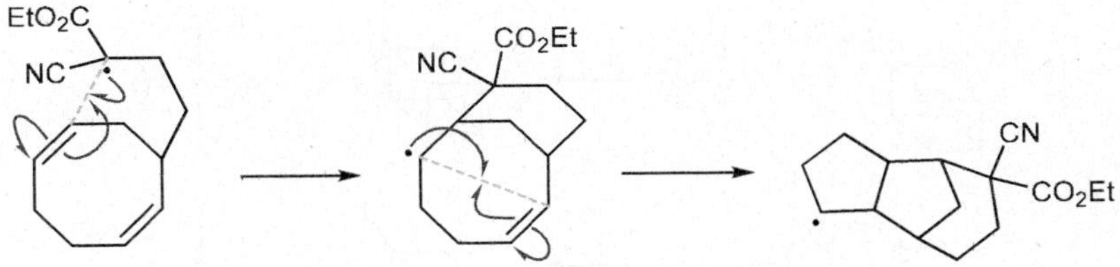

10.64. A comparison of the first product with the starting material reveals which bond in the starting material undergoes homolytic cleavage:

This bond undergoes homolytic cleavage

Starting material

Product

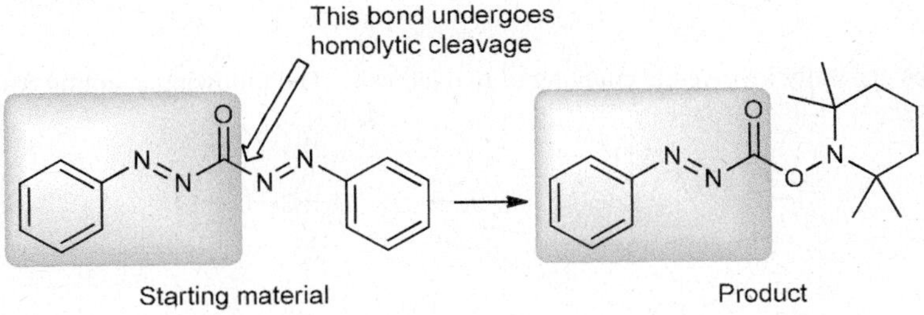

So we begin our mechanism by drawing homolytic cleavage of that bond, giving two radicals (**A** and **B**):

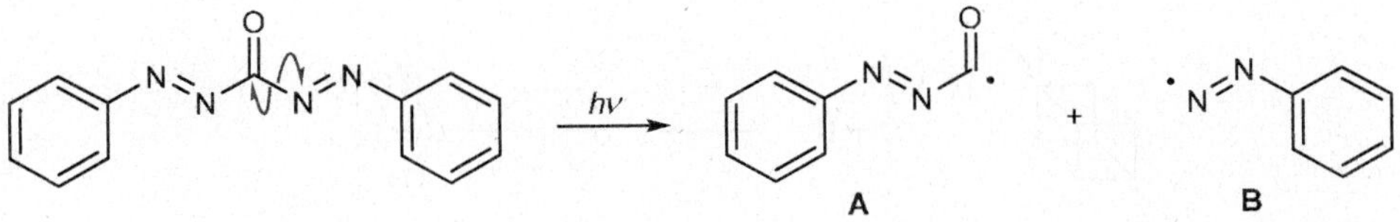

Radical **A** then couples with TEMPO to give the first product, while radical **B** undergoes further homolytic cleavage to give nitrogen gas and a phenyl radical, which couples with TEMPO to give the second product.

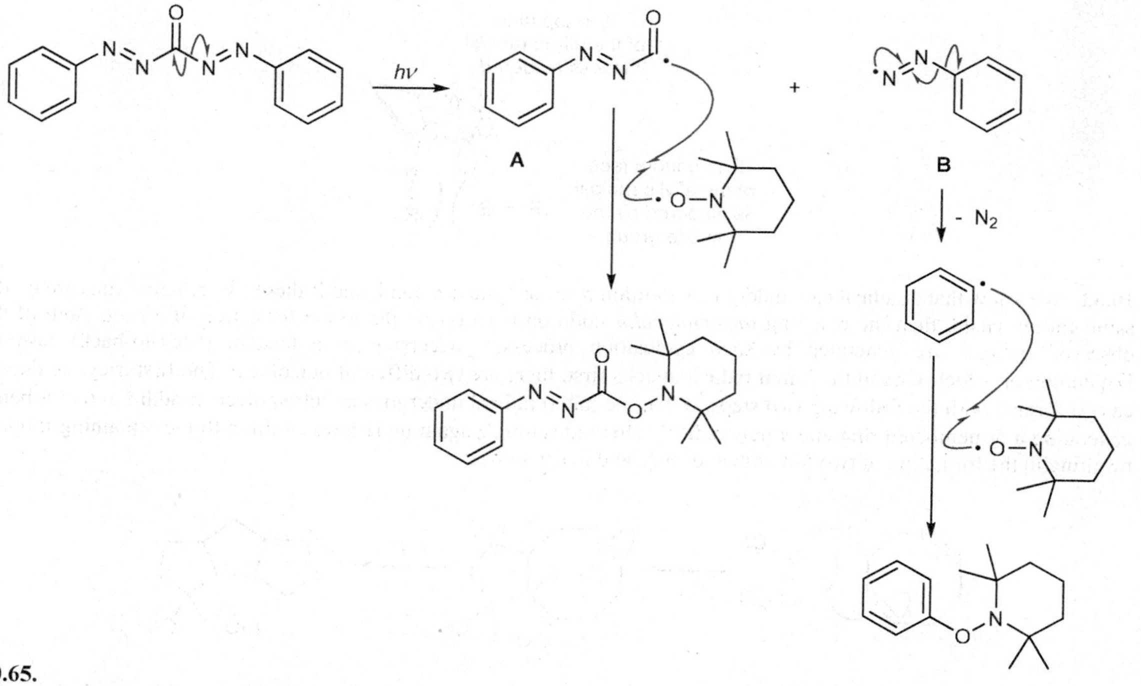

10.65.

(a) The first step of the propagation cycle is abstraction of a hydrogen atom from cubane, generating a cubyl radical. In the second step of the propagation cycle, halogen abstraction gives the product (iodocubane) and also regenerates the triiodomethyl radical.

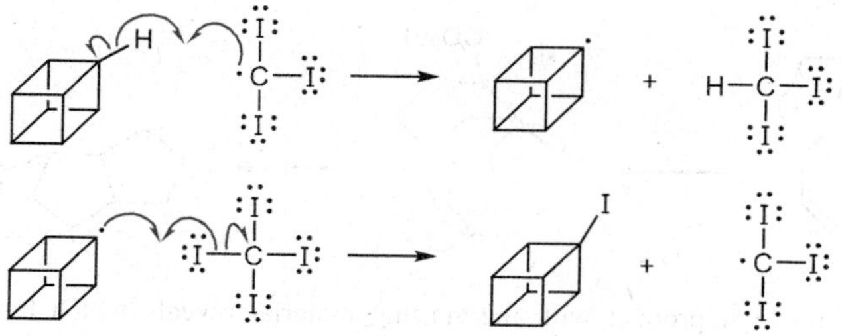

(b) Termination steps generally involve the coupling of two radicals. The following coupling reactions are all possible termination steps.

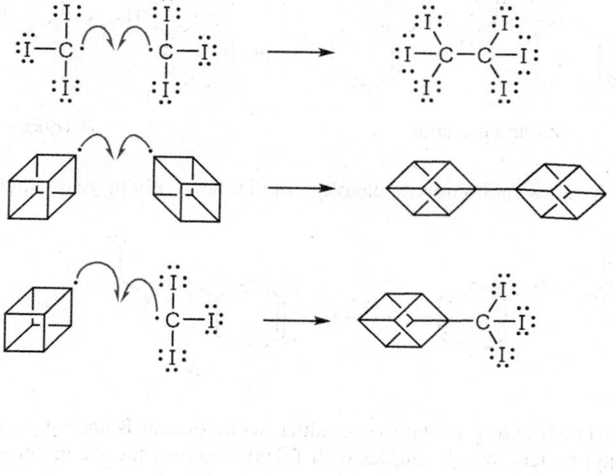

(c) Recall that for radical halogenation reactions, $\Delta G \approx \Delta H$ (Section 10.4). For the iodination of an alkane with I_2 (compound **2a**), ΔH will have a positive value:

$$(H_3C)C{-}H + I{-}I \longrightarrow (H_3C)C{-}I + H{-}I \quad \Delta H = +26 \text{ kJ/mol}$$

(H₃C)C–H: ~ 381 kJ/mol; I–I (**2a**): ~ 151 kJ/mol; (H₃C)C–I: ~ 209 kJ/mol; H–I (**4a**): ~ 297 kJ/mol

Since ΔH for the reaction is positive, ΔG for the reaction will also be positive. Therefore, the reaction is thermodynamically unfavorable. This argument is expected to hold true for cubane as well, assuming that the BDE for the C–H bond in cubane is not too different from the BDE for the C–H bond in $(CH_3)_3CH$. That is, iodination of cubane with molecular iodine (**2a**) is unfavorable.
In contrast, when **2b** is used as the reagent for iodination, ΔH for the reaction will be negative, which gives a negative value for ΔG. This can be rationalized by comparing the BDE values for **4a** and **4b**. The BDE of **4a** is 297 kJ/mol, while the BDE of **4b** is 423 kJ/mol. This significant difference in BDE values causes ΔH for the reaction to be negative when **2b** is used rather than **2a**, because a much stronger bond is being formed when **4b** is produced.

$$\text{cubane}{-}H + I{-}CI_3 \longrightarrow \text{cubane}{-}I + H{-}CI_3 \quad \Delta H = -59 \text{ kJ/mol}$$

cubane–H: ~ 381 kJ/mol; I–CI₃ (**2b**): ~ 192 kJ/mol; cubane–I: ~ 209 kJ/mol; H–CI₃ (**4b**): ~ 423 kJ/mol

Recall that bonds broken require an input of energy, contributing to a positive value of ΔH, while the formation of bonds releases energy, contributing to a negative value of ΔH.

10.66.
(a) The given peroxide is a radical initiator, and upon heating, it readily undergoes homolytic bond cleavage to generate resonance-stabilized radicals. These radicals can then abstract a halogen from CCl_4 in a halogen abstraction step to give a trichloromethyl radical ($Cl_3C\bullet$), shown below.

Homolytic Bond Cleavage

Halogen abstraction

trichloromethyl radical

There are two propagation steps, shown below. In the first propagation step, the π bond reacts with a trichloromethyl radical to give the more stable secondary radical (rather than the less stable primary radical). This step explains the regiochemical outcome of the process. Then, in the second propagation step, abstraction of a chlorine atom from CCl_4 gives the product and regenerates the reactive intermediate ($Cl_3C\bullet$), as expected for a propagation step.

Addition to a π Bond

Halogen abstraction

(b) The initiation steps are the same as those for a standard Kharasch reaction [see the two initiation steps in part (a) of this problem].

Based on our answer to part (a), we expect the first propagation step to involve addition of $Cl_3C\bullet$ across a π bond. In this case, there are two π bonds. The $Cl_3C\bullet$ radical can react with either π bond, but reaction with the more substituted alkene (disubstituted) leads to the observed product and produces the more stable 3° radical. This radical can then add to the other π bond (monosubstituted) in an intramolecular process to afford a primary radical that abstracts a chlorine atom from CCl_4 to give the desired product. Once again, notice that the final propagation step involves regeneration of the reactive intermediate, $Cl_3C\bullet$, as expected for a radical mechanism.

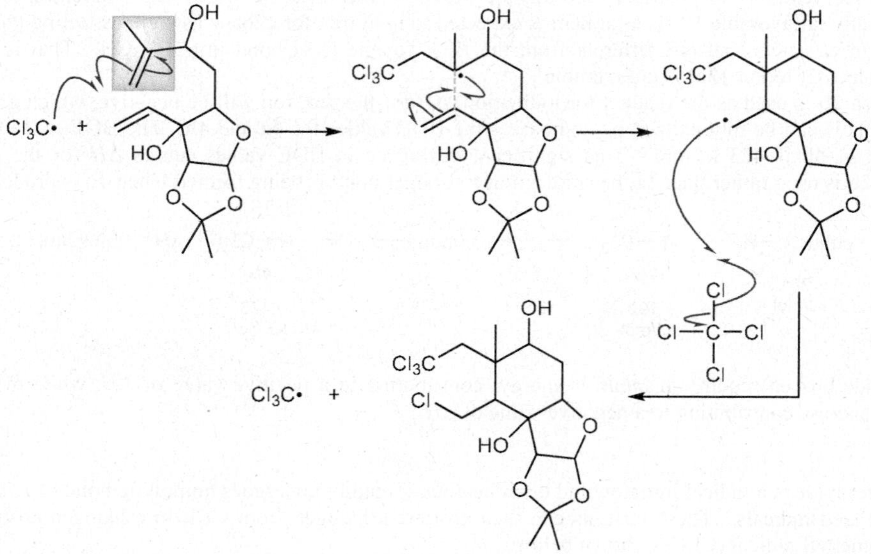

Chapter 11
Synthesis

Review of Concepts

Fill in the blanks below. To verify that your answers are correct, look in your textbook at the end of Chapter 11. Each of the sentences below appears verbatim in the section entitled *Review of Concepts and Vocabulary.*

- The position of a halogen can be moved by performing __________ followed by __________.
- The position of a π bond can be moved by performing __________ followed by __________.
- An alkane can be functionalized via radical __________.
- Every synthesis problem should be approached by asking the following two questions:
    1. Is there any change in the __________?
    2. Is there any change in the identity and/or location of the __________?
- In a __________ **analysis**, the last step of the synthetic route is first established, and the remaining steps are determined, working backward from the product.

Review of Skills

Fill in the blanks and empty boxes below. To verify that your answers are correct, look in your textbook at the end of Chapter 11. The answers appear in the section entitled *SkillBuilder Review.*

11.1 Changing the Identity or Position of a Functional Group

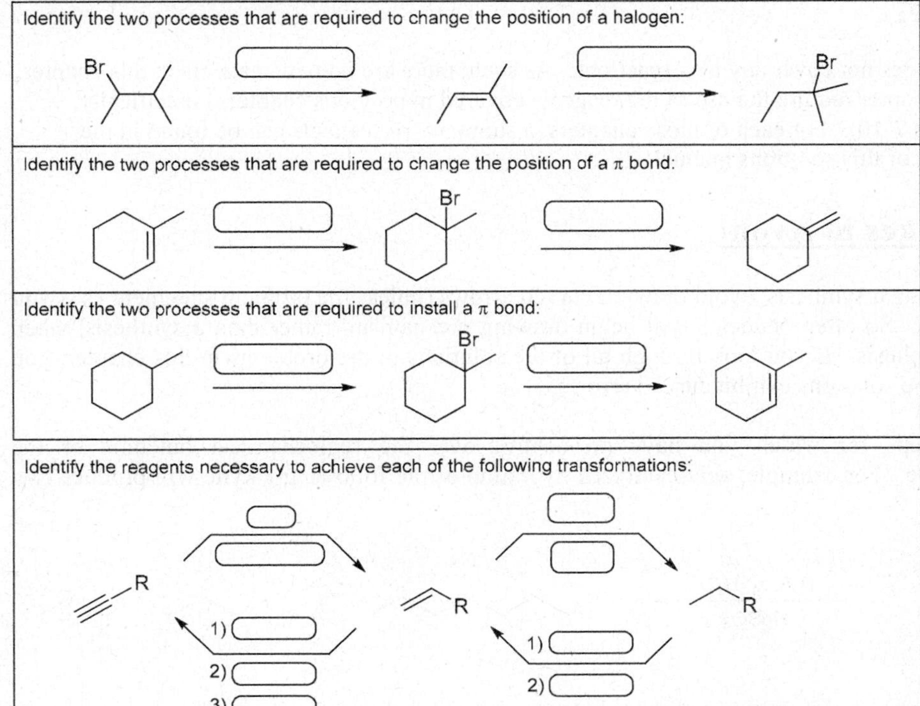

11.2 Changing the Carbon Skeleton

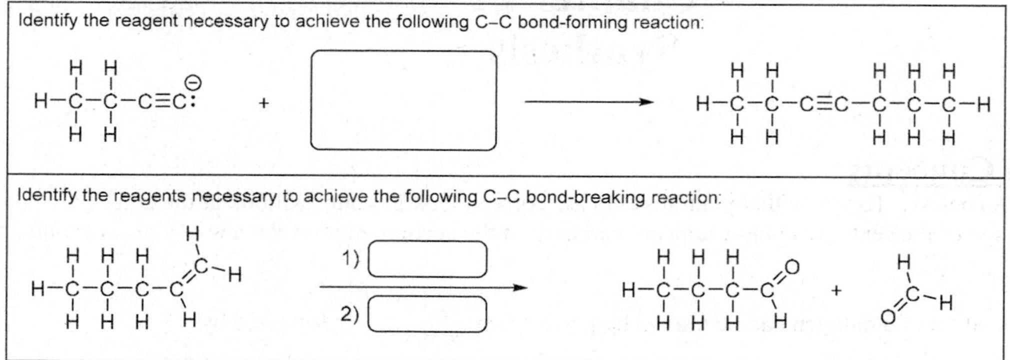

Identify the reagent necessary to achieve the following C–C bond-forming reaction:
Identify the reagents necessary to achieve the following C–C bond-breaking reaction:
1)
2)

11.3 Approaching a Synthesis Problem by Asking Two Questions

Identify the two questions to ask when approaching a synthesis problem:

1) ____?

2) ____?

11.4 Retrosynthetic Analysis

Complete the following retrosynthetic analysis by drawing the appropriate structures in the boxes provided:

Product ⟹ [ ] ⟹ [ ] ⟹ Starting material

Useful Reagents:

This chapter does not cover any new reactions. As such, there are no new reagents in this chapter. The problems in this chapter require the use of the reagents covered in previous chapters (specifically, Chapter 3 and Chapters 7-10). For each of those chapters, a summary of reagents can be found in the corresponding chapters of this solutions manual.

Common Mistakes to Avoid

When proposing a synthesis, avoid drawing curved arrows (unless the problem statement asks you to draw a mechanism). So often, students will begin drawing mechanism, rather than a synthesis, when asked to propose a synthesis. If you look through all of the solutions to the problems in this chapter, you will find that none of the solutions exhibit curved arrows.

Also, avoid using steps for which you have no control over the regiochemical outcome or the stereochemical outcome. For example, acid-catalyzed hydration of the following alkyne will produce two different ketones.

H_2SO_4, H_2O / $HgSO_4$ → ketone + ketone

This process should not be used if only one of these ketones is desired.

Solutions

11.1. The reagents for these reactions can be found in the summary material at the end of Chapter 8. They are shown again here:

Br_2, $h\nu$

NaOEt

HBr, ROOR

HBr

1) O_3 2) DMS

$KMnO_4$, NaOH, cold

1) RCO_3H 2) H_3O^+

Br_2, H_2O

Br_2

H_2, Pt

1) $BH_3 \cdot THF$ 2) H_2O_2, NaOH

H_3O^+

+ En

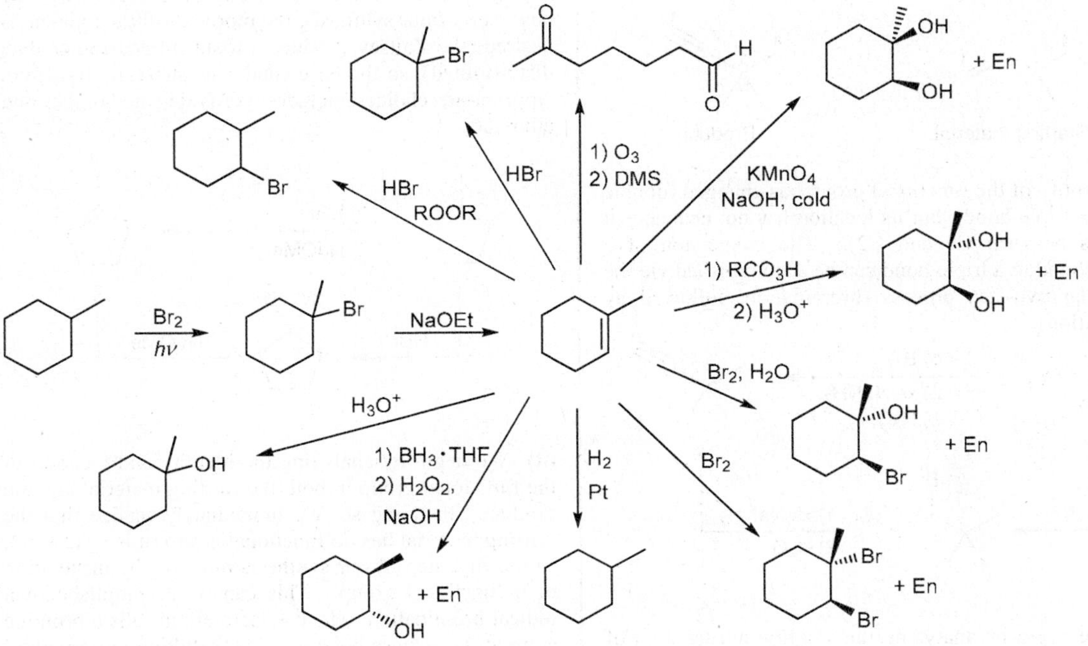

11.2. The reagents for these reactions can be found in the summary material at the end of Chapter 9. They are shown again here:

HBr

HBr

H_2SO_4, H_2O, $HgSO_4$

1) R_2BH 2) H_2O_2, NaOH

1) xs $NaNH_2$ 2) H_2O

1) xs $NaNH_2$ 2) H_2O

1) $NaNH_2$ 2) CH_3I

H_2, Lindlar's Catalyst

H_2, Pt

Na, $NH_3(l)$

Br_2

xs Br_2

1) O_3 2) H_2O

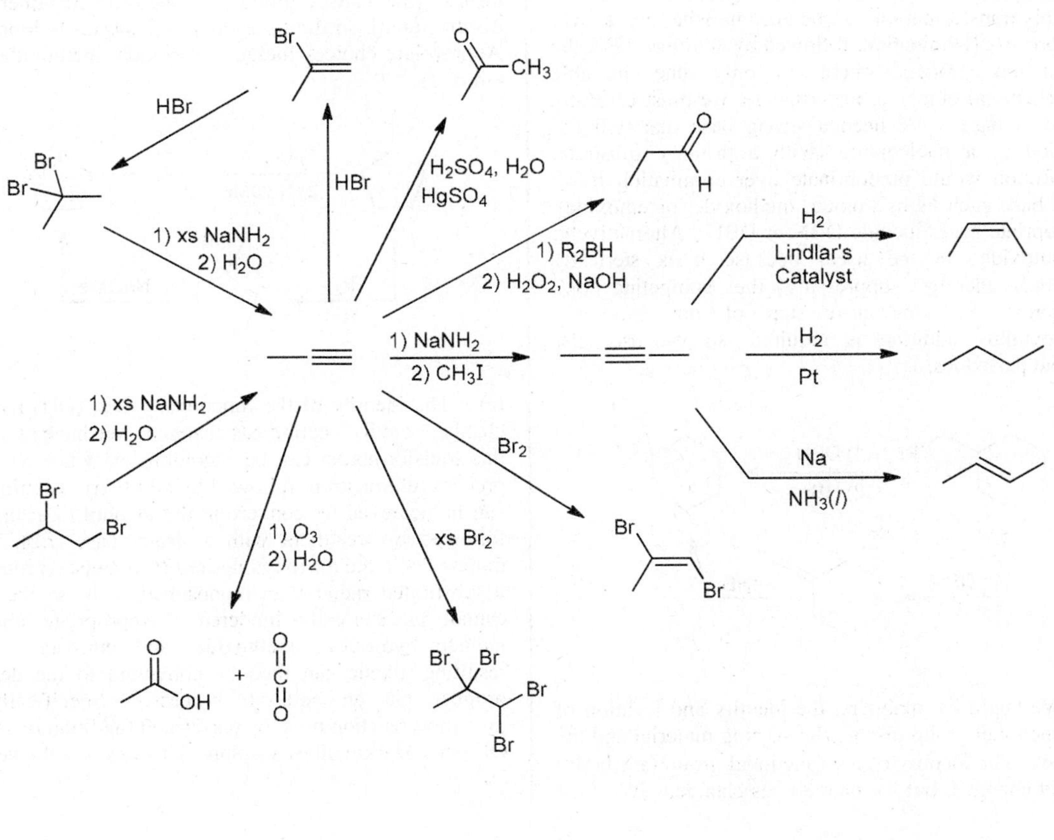

11.3.

(a) We begin by analyzing the identity and location of the functional group in both the starting material and the product:

Starting material Product

The identity of the functional group has changed (double bond → triple bond), but its location has not changed (it remains between C1 and C2). The conversion of a double bond to a triple bond can be accomplished via the following two-step process (bromination, followed by elimination):

1) Br_2
2) xs $NaNH_2$
3) H_2O

(b) We begin by analyzing the identity and location of the functional group in both the starting material and the product. The identity of the functional group (Br) has not changed, but its location has changed. We have seen that this transformation can be accomplished via a two-step process (elimination, followed by addition). For the elimination process, there is only one possible regiochemical outcome; nevertheless, we must carefully choose a base. We need a strong base that will not function as a nucleophile (with a primary substrate, substitution would predominate over elimination if we use a base such as hydroxide, methoxide, or ethoxide). Appropriate bases include DBN or DBU. Alternatively, *tert*-butoxide can be used, because it is sterically hindered, thereby suppressing the competing S_N2 reaction. For the next step of our synthesis, Markovnikov addition is required, so we use HBr without peroxides.

1) DBU
2) HBr

(c) We begin by analyzing the identity and location of the functional group in both the starting material and the product. The identity of the functional group (a π bond) has not changed, but its location has changed. We have seen that this transformation can be accomplished via a two-step process (addition, followed by elimination). For the first step of our synthesis, Markovnikov addition is required, so we use HBr without peroxides. For the next step of our synthesis, the more-substituted alkene is desired (the Zaitsev product – trisubstituted rather than disubstituted), so the base cannot be sterically hindered. Appropriate choices include hydroxide, methoxide, and ethoxide.

1) HBr
2) NaOMe

(d) We begin by analyzing the identity and location of the functional group in both the starting material and the product. In doing so, we immediately realize that the starting material has no functional group (it is an alkane), so the first step of our synthesis must be the installation of a functional group. This can be accomplished via radical bromination, which selectively installs a bromine atom at the tertiary position. The resulting tertiary alkyl halide can then be converted into the product via an elimination process. The more-substituted alkene is desired (the Zaitsev product – trisubstituted rather than disubstituted), so the base cannot be sterically hindered. Appropriate choices include hydroxide, methoxide, and ethoxide.

1) Br_2, *hv*
2) NaOMe

(e) The identity of the functional group (OH) has not changed, but its location has changed. We have seen that this transformation can be accomplished via a two-step process (elimination, followed by addition). Elimination can be achieved by converting the alcohol to a tosylate followed by treatment with a strong base (E2). The more-substituted alkene is desired (the Zaitsev product – trisubstituted rather than monosubstituted), so the base cannot be sterically hindered. Appropriate choices include hydroxide, methoxide, and ethoxide. The resulting alkene can then be converted to the desired product via an addition process. Specifically, a hydration reaction must be performed (addition of H and OH) in a Markovnikov fashion. This can be achieved by treating the alkene with dilute acid (acid-catalyzed hydration):

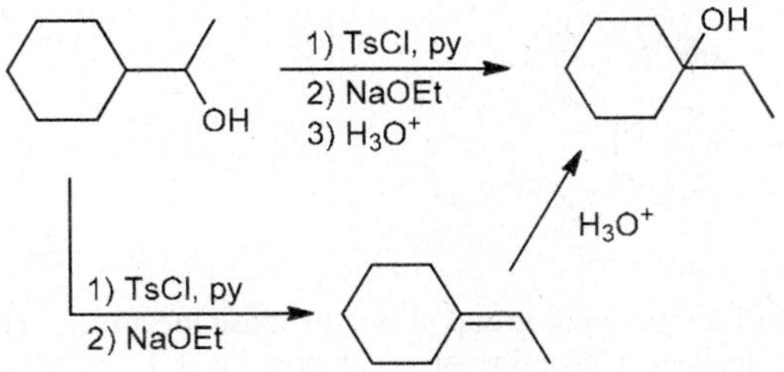

(f) We begin by analyzing the identity and location of the functional group in both the starting material and the product. The identity of the functional group has changed (Br → OH), and its location has changed as well. We have seen that this type of transformation can be accomplished via a two-step process (elimination of H and Br, followed by addition H and OH). For the first step of our synthesis, the more-substituted alkene is desired (the Zaitsev product – trisubstituted rather than monosubstituted), so the base cannot be sterically hindered. Appropriate choices include hydroxide, methoxide, and ethoxide. A hydration reaction must be performed (addition of H and OH) in a Markovnikov fashion. This can be achieved by treating the alkene with dilute acid (acid-catalyzed hydration):

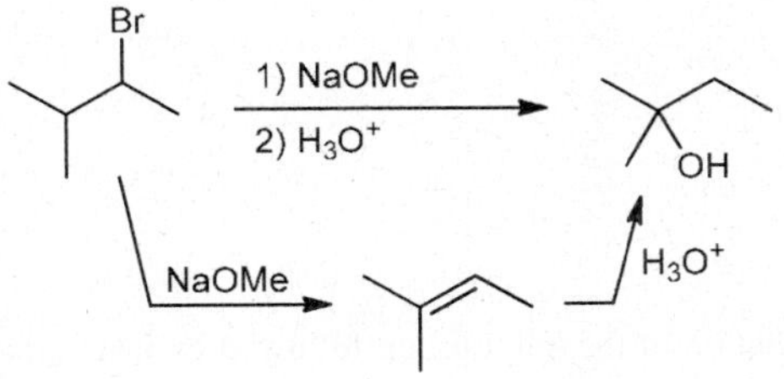

(g) We begin by analyzing the identity and location of the functional group in both the starting material and the product. The identity of the functional group has changed (OH → Br), and its location has changed as well. We have seen that this type of transformation can be accomplished via a two-step process (elimination of H and OH, followed by addition H and Br). For the first part of our synthesis (elimination), we must first convert the alcohol to a tosylate before treating with a strong base. The use of a sterically hindered base is not required (for a secondary substrate, E2 = major product, while S_N2 = minor product). Nevertheless, a sterically hindered base will be helpful here as it will suppress the competing S_N2 reaction. So *tert*-butoxide is used in the synthesis shown. Alternatively, DBU or DBN could be used. Note that an acid-catalyzed dehydration (E1 process) should be avoided, because heating the alcohol with concentrated sulfuric acid will likely involve a carbocation rearrangement (methyl shift). For the second part of our synthesis, we must perform an *anti*-Markovnikov addition of H and Br, so we use HBr in the presence of peroxides.

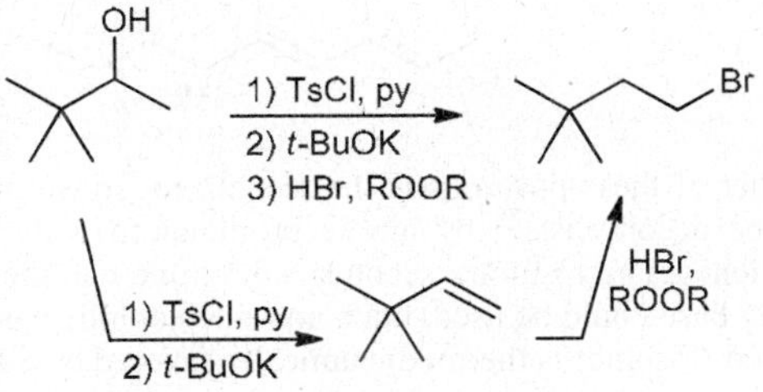

(h) We begin by analyzing the identity and location of the functional group in both the starting material and the product. The identity of the functional group (OH) has not changed, but its location has changed. We have seen that this type of transformation can be accomplished via a two-step process (elimination, followed by addition). For the first part of our synthesis (elimination), we must first convert the alcohol to a tosylate before treating with a strong base. The use of a sterically hindered base is not required (for a secondary substrate, E2 = major product, while S_N2 = minor product). Nevertheless, a sterically hindered base will be helpful here as it will suppress the competing S_N2 reaction. So *tert*-butoxide is used in the synthesis below. Alternatively, DBU or DBN could be used. Note that an acid-catalyzed dehydration (E1 process) should be avoided, because heating the alcohol with concentrated sulfuric acid will likely involve a carbocation rearrangement (methyl shift). For the second part of our synthesis, we must perform an *anti*-Markovnikov addition of H and OH. This can be achieved via hydroboration-oxidation, or alternatively, it can be achieved by *anti*-Markovnikov addition of HBr followed by an S_N2 process with hydroxide to give the desired alcohol:

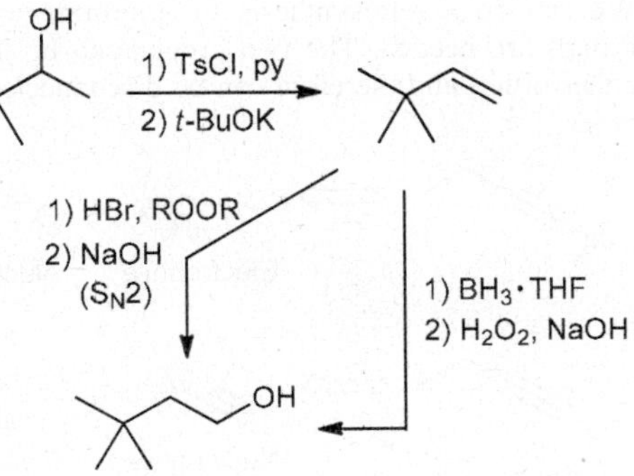

11.4. We begin by analyzing the identity and location of the functional groups in both the starting material and the product, and we look for a difference. In this case, a double bond is introduced at the locations labeled C1 and C2 below:

Neither of these positions is functionalized, so we must first install a functional group at one of these locations. This can be accomplished by allylic bromination with NBS, which installs a bromine atom at position C1 (an allylic position). The resulting secondary bromide can then be converted into the product via an elimination process. Any strong base could be used, but a non-nucleophilic base such as DBU is ideal to favor elimination over substitution. The desired *E* isomer is thermodynamically favored over the less stable *Z* isomer.

1) NBS, *hν*
2) DBU

NBS, *hν*

DBU

Br

11.5.

(a) The starting material (acetylene) has two carbon atoms, and the product has five carbon atoms. Therefore, we must change the carbon skeleton by forming carbon-carbon bonds with two successive alkylation processes. We can do a retrosynthesis to determine which alkyl groups are needed. The two groups can be installed in either order, and therefore can be disconnected in either order:

Electrophile Nucleophile

Nucleophile Electrophile

We must install a methyl group and an ethyl group. The ethyl group is installed by treating acetylene with sodium amide, followed by ethyl iodide. And the methyl group is installed in a similar way (upon treatment with sodium amide, followed by methyl iodide). The following shows installation of the ethyl group followed by installation of the methyl group:

1) $NaNH_2$
2) EtI
3) $NaNH_2$
4) MeI

1) $NaNH_2$
2) EtI

1) $NaNH_2$
2) MeI

(b) The desired transformation must lengthen the carbon chain by two carbon atoms, and there is a new triple bond. The alkyne target molecule is suitable for making the required C–C bond disconnection, so that is how the retrosynthesis begins:

Electrophile Nucleophile

A leaving group is needed for the substitution reaction with the acetylide nucleophile, but the OH group on the alcohol starting material is not a good leaving group. We must convert the alcohol to a tosylate before reaction with sodium acetylide:

1) TsCl, py
2) HC≡CNa

TsCl, py

OTs

H—C≡C:⊖

(c) When we compare the starting and final compounds, we observe three differences: the halide is gone, there is a ketone functional group, and the carbon chain in the product is *shorter* by one carbon atom (the starting material has eight carbon atoms, and the product has only seven carbon atoms). These changes can be accomplished via an ozonolysis process, which requires an alkene starting material. Numbering the carbon atoms is a helpful way to keep track of the original carbon skeleton:

The π bond in the required alkene can be formed by a Hofmann elimination of the given alkyl halide starting material, by treatment with a sterically hindered base, leading to the following synthesis:

1) *t*-BuOK
2) O_3
3) DMS

t-BuOK

1) O_3
2) DMS

(d) The starting material has one more carbon atom than the product (numbering the carbon atoms is a helpful way to keep track of the original carbon skeleton). Therefore, our synthesis must employ an ozonolysis process. Since the product is a carboxylic acid (rather than an aldehyde or ketone), we can conclude that the last step of our process must involve ozonolysis of an *alkyne*, rather than ozonolysis of an *alkene*:

The required alkyne can be prepared from a dibromide that leads back to the given starting material:

Now let's draw the forward process. The synthesis begins by converting the alkene to an alkyne – by first treating with Br_2, giving a dibromide, followed by treatment with excess sodium amide (and then water workup). At this point, ozonolysis of the alkyne produces the desired target molecule:

1) Br_2
2) xs $NaNH_2$
3) H_2O
4) O_3
5) H_2O

Br_2

1) xs $NaNH_2$
2) H_2O

1) O_3
2) H_2O

(e) The desired transformation must lengthen the carbon chain by two carbon atoms, and there is a new double bond. To begin the retrosynthesis, we ask, "What starting material can be used to make an alkene?" Table 11.1 helps us consider our options. A π bond can be made by an elimination reaction, or by partial reduction of an alkyne. Selecting an alkyne is a logical choice, because only an alkyne structure is suitable for making the required C–C bond disconnection:

Electrophile

Nucleophile

Now let's draw the forward process. The synthesis begins with an alkylation reaction, by treating the starting material with sodium acetylide. The resulting alkyne is treated with molecular hydrogen in the presence of a poisoned catalyst (such as Lindlar's catalyst) to produce the desired alkene target molecule:

1) HC≡CNa
2) H_2, Lindlar's cat.

H—C≡C:⊖

H_2
Lindlar's cat.

(f) The starting material has one more carbon atom than the product. Therefore, our synthesis must employ an ozonolysis process. Formation of an *aldehyde* product requires an *alkene* starting material. Numbering the carbon atoms is a helpful way to keep track of the original carbon skeleton:

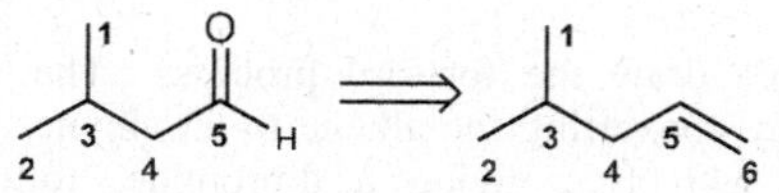

The required alkene can be formed by treatment of the alkyne starting material with molecular hydrogen in the presence of a poisoned catalyst (such as Lindlar's catalyst):

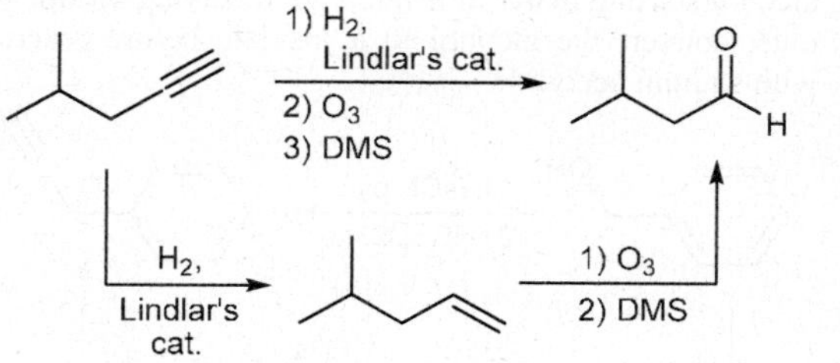

1) H_2, Lindlar's cat. 2) O_3 3) DMS

H_2, Lindlar's cat.

1) O_3 2) DMS

11.6. When we compare compounds **1** and **2**, we see that there are three carbon atoms in the side chain of compound **1**, and there are five carbon atoms in the side chain of compound **2**, so the carbon skeleton is changing. This requires a reaction in which a C–C bond is formed, thereby achieving the installation of a two-carbon fragment:

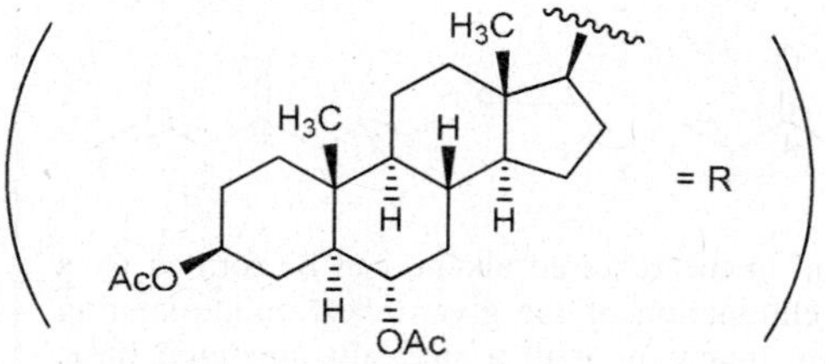

The only reaction we know (so far) that can install more than one carbon atom on an existing carbon skeleton is the reaction of an alkynide anion with an appropriate substrate bearing a good leaving group. An alcohol is not a good substrate for the required S_N2 reaction; however, conversion of the alcohol to a tosylate creates a good leaving group. Addition of an alkynide anion (specifically the acetylide anion) makes a new carbon-carbon bond and gives the desired five-carbon skeleton of the substituent. The resulting alkylation product (a terminal alkyne) can then be subjected to Markovnikov hydration to produce the desired methyl ketone.

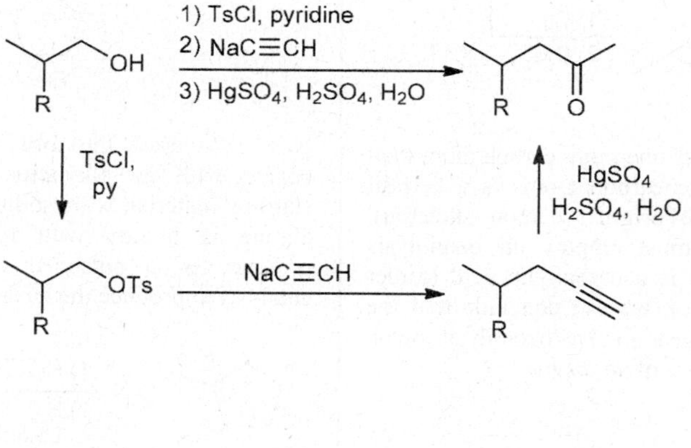

1) TsCl, pyridine 2) NaC≡CH 3) $HgSO_4$, H_2SO_4, H_2O

TsCl, py

$HgSO_4$ H_2SO_4, H_2O

NaC≡CH

11.7.

(a) The starting material has five carbon atoms, and the product has eight carbon atoms. So our synthesis must involve the installation of three carbon atoms. In addition, the identity of the functional group must be changed (from a triple bond to a *cis* double bond). To begin the retrosynthesis, we ask: "What starting material can be used to make a *cis* alkene?" The *cis* alkene can be prepared by partial reduction of the corresponding alkyne, and the alkyne structure is suitable for making the required C–C bond disconnection:

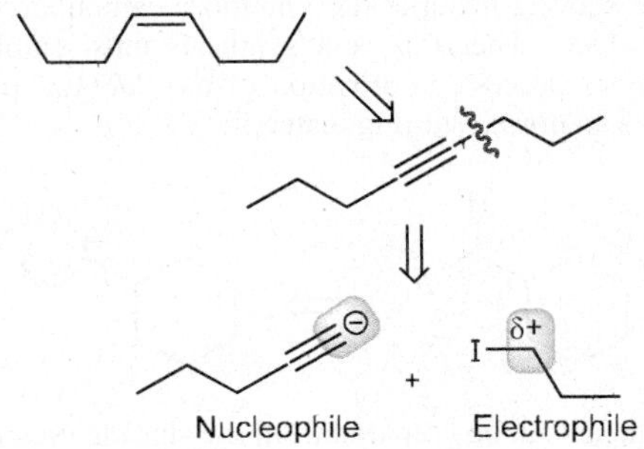

Nucleophile + Electrophile

Reduction of the alkyne (to give an alkene) must be the last step of our synthesis, because if we first reduce the triple bond, then we would not be able to perform the alkylation step. The alkylation step must be performed first, followed by reduction of the alkyne (with a poisoned catalyst), as shown here:

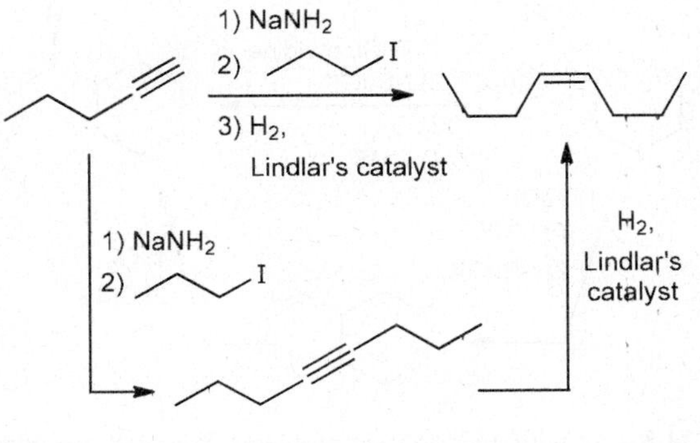

1) $NaNH_2$
2) I
3) H_2, Lindlar's catalyst

1) $NaNH_2$
2) I

H_2, Lindlar's catalyst

(b) The starting material has five carbon atoms, and the product has seven carbon atoms. The retrosynthesis of the alkyne target molecule begins by making a C–C bond disconnection adjacent to the triple bond. This disconnection is logical because it brings us to an alkyl halide that has five carbon atoms (a carbon chain that is the same length as the given starting material):

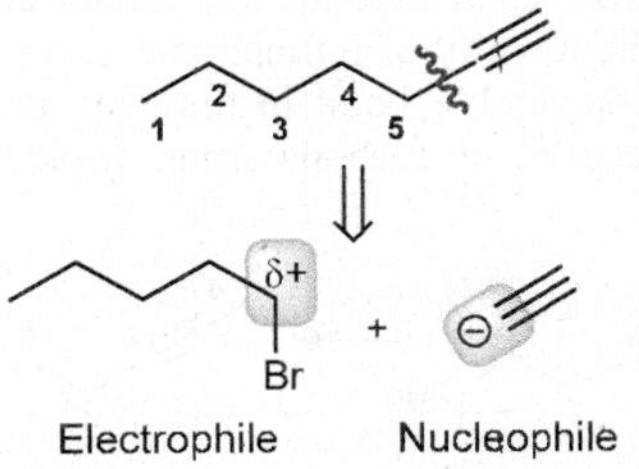

1 2 3 4 5

Br

Electrophile + Nucleophile

To continue the retrosynthesis, we ask: "What starting material can be used to make an alkyl halide?" Selecting an alkene is a logical choice, because the alkene, in turn, can be prepared from the given alkyne starting material:

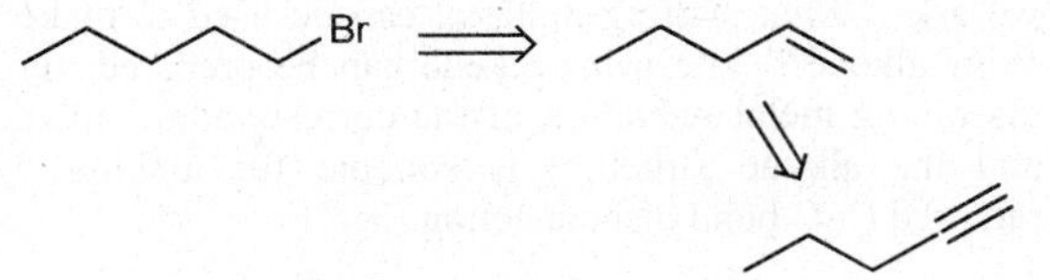

Br

To draw the forward process, we can convert the starting alkyne into an alkyl halide (via hydrogenation with a poisoned catalyst followed by an *anti*-Markovnikov addition of HBr). If this alkyl halide is treated with acetylide, the desired product is formed:

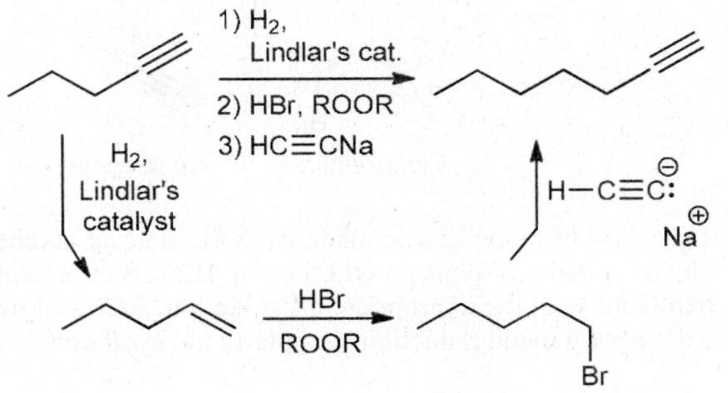

1) H_2, Lindlar's cat.
2) HBr, ROOR
3) HC≡CNa

H_2, Lindlar's catalyst

HBr, ROOR

Br

$H{-}C{\equiv}C{:}^{\ominus}$ $Na^{\oplus}$

(c) The starting material has one more carbon atom than the product. Therefore, our synthesis must employ an ozonolysis process. Since the product is a carboxylic acid (rather than an aldehyde or ketone), we can conclude that the last step of our process must involve ozonolysis of an *alkyne*, rather than ozonolysis of an *alkene*:

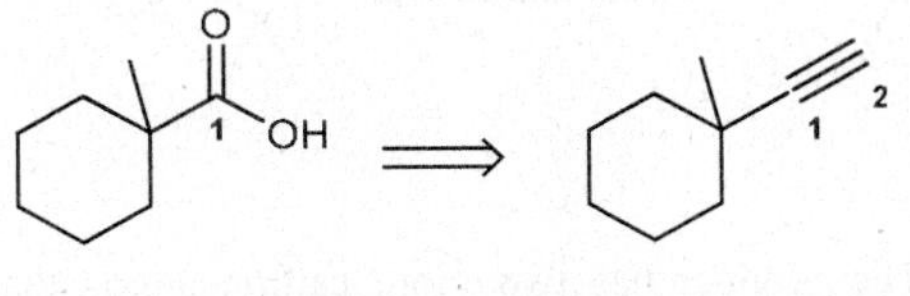

O, 1, OH ⟹ 1, 2

This alkyne can be prepared directly from the starting material upon treatment with excess sodium amide (followed by water workup). The complete synthesis is shown here:

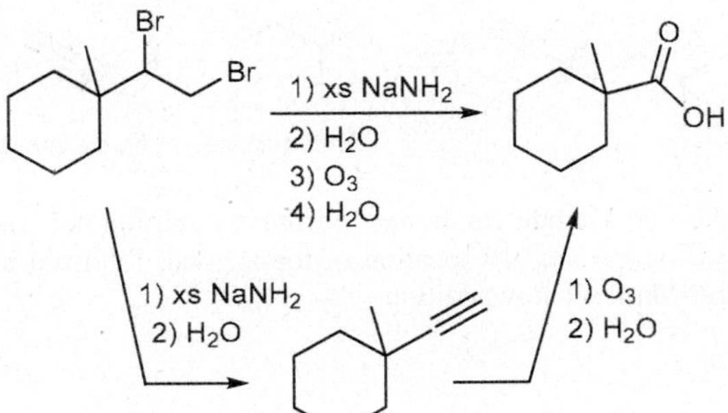

Br, Br

1) xs $NaNH_2$
2) H_2O
3) O_3
4) H_2O

O, OH

1) xs $NaNH_2$
2) H_2O

1) O_3
2) H_2O

(d) The starting material has six carbon atoms, and the product has nine carbon atoms. So our synthesis must involve the installation of three carbon atoms. Also, the location of the functional group has been changed. The product is a *trans* alkene, so to begin the retrosynthesis, we ask: "What starting material can be used to make a *trans* alkene?" The *trans* alkene can be prepared via a dissolving metal reduction of the corresponding alkyne, and the alkyne structure is suitable for making the required C–C bond disconnection:

δ+

Br

Electrophile Nucleophile

This alkyl bromide can be made from the starting alkene via an *anti*-Markovnikov addition of HBr. Subsequent treatment with the appropriate alkynide ion, followed by a dissolving metal reduction, completes the synthesis:

1) HBr, ROOR
2) $H_3C{-}C{\equiv}CNa$
3) Na, NH_3 (*l*)

HBr, ROOR

Na, NH_3 (*l*)

Br

$H_3C{-}C{\equiv}C{:}^{\ominus}$ $Na^{\oplus}$

(e) The product has two more carbon atoms than the starting material, and the location of the functional group has changed. The retrosynthesis of the alkyne target molecule begins by making a C–C bond disconnection adjacent to the triple bond:

δ+

Br

Electrophile Nucleophile

This alkyl halide can be made from the starting material by first moving the location of the π bond, followed by *anti*-Markovnikov addition:

1) HBr, ROOR
2) *t*-BuOK
3) HBr, ROOR
4) $HC{\equiv}CNa$

1) HBr, ROOR
2) *t*-BuOK

$H{-}C{\equiv}C{:}^{\ominus}$ $Na^{\oplus}$

HBr

ROOR

Br

(f) The starting material has one more carbon atom than the product. Therefore, our synthesis must employ an ozonolysis process. Formation of an *aldehyde* product requires an *alkene* starting material:

H 1 O ⟹ 1 2

This alkene can be prepared from the starting material by converting the alcohol to a tosylate, and then performing an E2 reaction with a sterically hindered base (giving the less-substituted, Hofmann alkene). Note that an acid-catalyzed dehydration (E1 process) would have given the more-substituted, Zaitsev product, along with rearrangement products. Finally, ozonolysis of the alkene produces the desired aldehyde target molecule:

OH

1) TsCl, pyridine
2) *t*-BuOK
3) O_3
4) DMS

H O

1) TsCl, pyridine
2) *t*-BuOK

1) O_3
2) DMS

11.8. As with all synthesis problems, we must consider the following two questions:
1) Is there is a change in the carbon skeleton?
2) Is there a change in the identity or location of any functional groups?
To answer the first question, we can assign numbers to the carbon atoms, and we see that one carbon atom (a methyl group) must be installed during this transformation. To answer the second question, there is a change in the identity of a functional group, because the carbon-carbon double bond in the starting material has been replaced by a carbon-carbon triple bond in the product.

OMe 5 3 1 6 4 2 OMe → 7 6 5 3 OMe 1 4 2 OMe

The retrosynthesis of the alkyne target molecule begins by making a C–C bond disconnection adjacent to the triple bond:

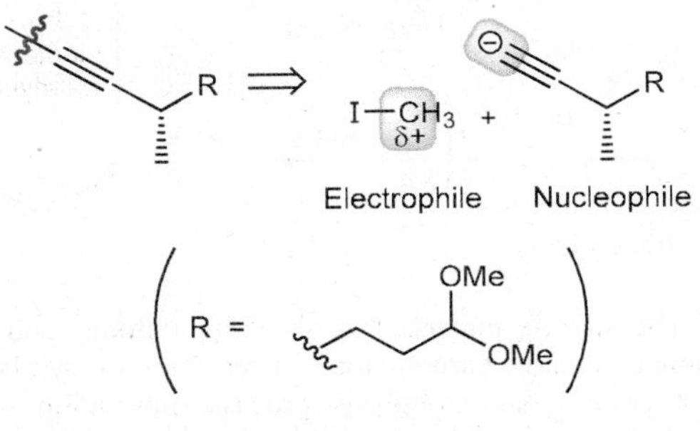

Electrophile Nucleophile

The required alkyne can be prepared from the alkene starting material using a two-step process: bromination followed by elimination (which requires aqueous workup to produce a neutral product). Alkylation of the resulting terminal alkyne can then be accomplished using a strong base and an appropriate methyl substrate, such as methyl iodide.

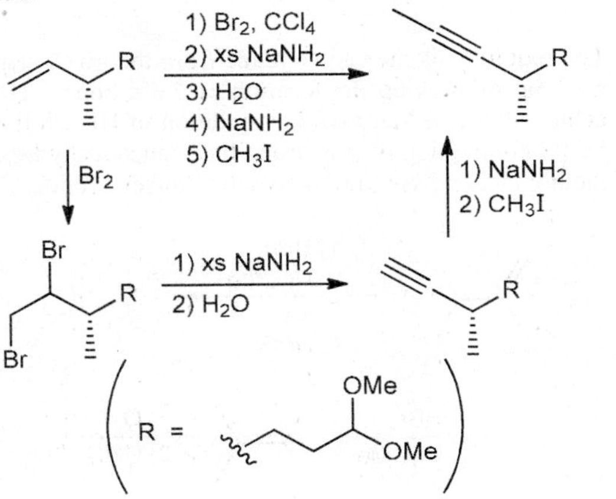

Notice that steps 3 and 4 can be omitted. The purpose of the water in step 3 is to protonate the alkynide ion (after both eliminations have occurred), but an alkynide ion is prepared in step 4 by deprotonation with $NaNH_2$. Thus, a more efficient synthesis would omit these steps, as shown here:

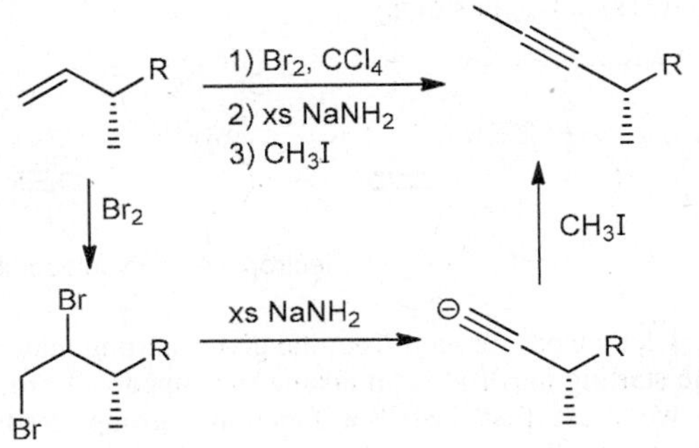

11.9.

(a) The product is an epoxide, which can be made from an alkene:

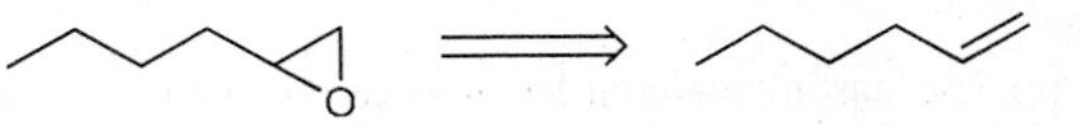

So the last step of our synthesis will likely be conversion of the alkene into the epoxide.

Now let's work forward from the starting material. The starting material has only four carbon atoms, while the product has six carbon atoms, so two carbon atoms must be installed. This can be achieved via an alkylation process:

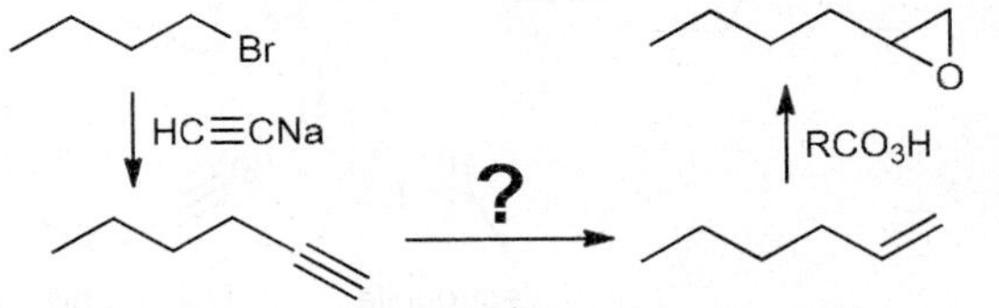

Now we must bridge the gap between the alkyne and the alkene. This can be achieved in one step, via hydrogenation with a poisoned catalyst. In summary, the desired transformation can be achieved with the following synthesis:

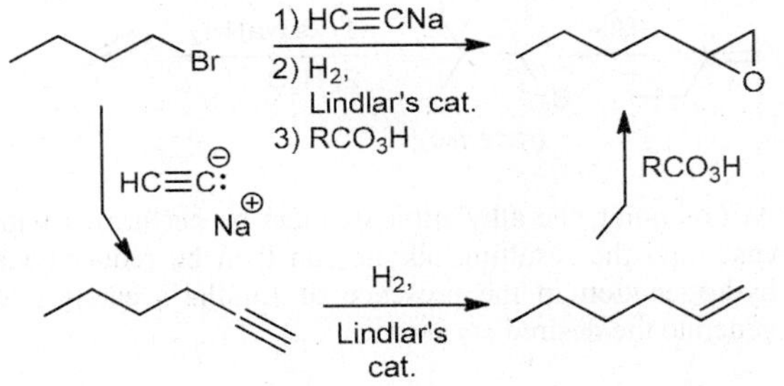

(b) The product is a methyl ketone, and we have seen that a methyl ketone can be prepared from an alkyne (via acid-catalyzed hydration):

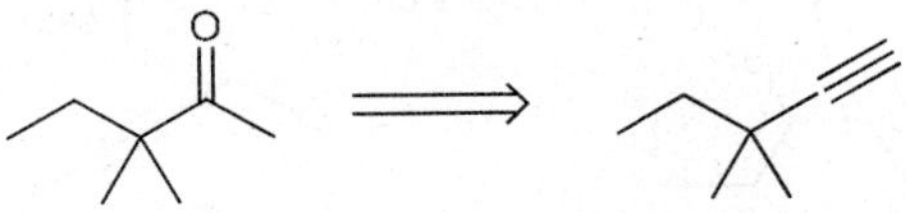

This alkyne can be prepared directly from the starting material upon treatment with excess sodium amide (followed by water workup). In summary, the desired transformation can be achieved with the following synthesis:

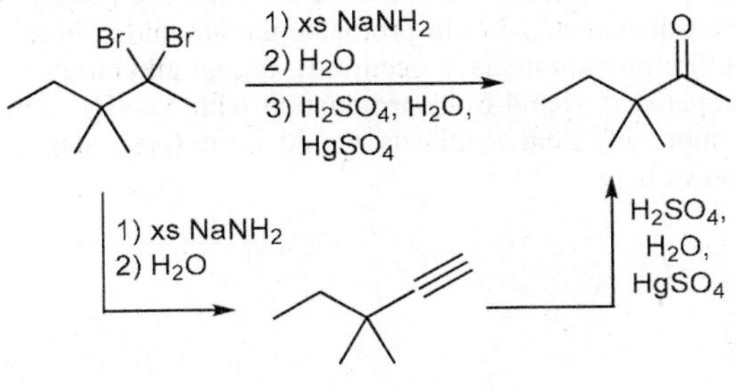

(c) The starting material has four carbon atoms, and the product has six carbon atoms. So our synthesis must involve the installation of two carbon atoms. To begin the retrosynthesis, we ask: "What starting material can be used to make a *cis* alkene?" The *cis* alkene can be prepared by partial reduction of the corresponding alkyne, and the alkyne structure is suitable for making the required C–C bond disconnection:

δ+ I + ⊖

Electrophile Nucleophile

The required alkyne can be prepared from the alkene starting material using a two-step process: bromination followed by elimination (which requires aqueous workup to produce a neutral product).

Br_2 → Br, Br (racemic) → 1) xs $NaNH_2$ 2) H_2O

At this point, the alkylation step can be performed with ease, and the resulting alkyne can then be reduced via hydrogenation in the presence of Lindlar's catalyst to generate the desired *cis* alkene:

1) Br_2
2) xs $NaNH_2$
3) H_2O
4) $NaNH_2$
5) EtI
6) H_2, Lindlar's cat.

Br_2; Br, Br (racemic); 1) xs $NaNH_2$ 2) H_2O; 1) $NaNH_2$ 2) EtI; H_2, Lindlar's catalyst

Notice that steps 3 and 4 can be omitted. The purpose of the water in step 3 is to protonate the alkynide ion (after both eliminations have occurred), but an alkynide ion is prepared in step 4 by deprotonation with $NaNH_2$. Thus, a more efficient synthesis would omit these steps, as shown here:

1) Br_2
2) xs $NaNH_2$
3) EtI
4) H_2, Lindlar's cat.

Br_2; Br, Br (racemic); 1) xs $NaNH_2$ 2) EtI; H_2, Lindlar's catalyst

(d) The starting material has six carbon atoms, and the product has three carbon atoms. Therefore, our synthesis must employ an ozonolysis process. Formation of a ketone product requires an alkene starting material, and ozonolysis of the alkene shown would produce *two* equivalents of the ketone target molecule:

O; 1 2 3 ⟹ 1 2 3, 1 2 3

The required alkene can be made from the given starting material by moving the location of the π bond. This is achieved with a Markovnikov addition of HBr, followed by E2 elimination with a strong base (such as hydroxide, methoxide, or ethoxide) to give the Zaitsev product.

1) HBr
2) NaOMe
3) O_3
4) DMS

1) HBr 2) NaOMe; 1) O_3 2) DMS

(e) Our retrosynthesis begins with a disconnection adjacent to the cyano group, at the C–C bond. Cyanide ($N{\equiv}C^-$) is a stable and familiar anion, so we will use cyanide as the nucleophile. The other carbon (at the disconnected bond) must have started out as an electrophile, so we draw a leaving group (such as Cl, Br, I or OTs) at that position.

C≡N ⟹ Br δ+ + ⊖C≡N

Electrophile Nucleophile

Now let's work forward from the given starting material. The starting material is an alkane (no functional group), so we must first install a functional group. Radical bromination will selectively install a bromine atom at a tertiary position:

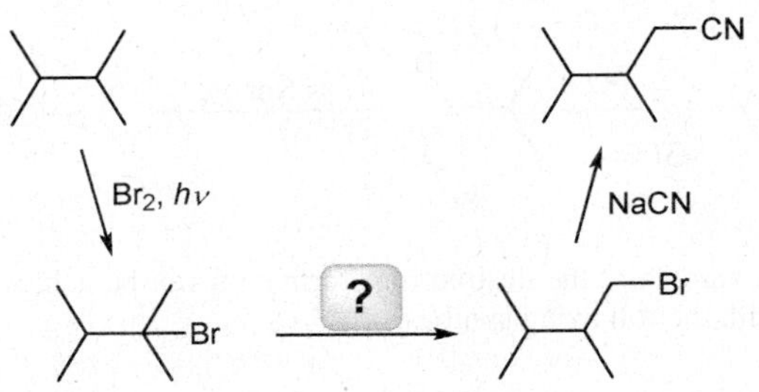

Now we must bridge the gap (by changing the position of the Br atom). This can be achieved in two steps: an E2 reaction with a strong, sterically hindered base (to give the Hofmann product), followed by an *anti*-Markovnikov addition of HBr. In summary, the desired transformation can be achieved with the following synthesis:

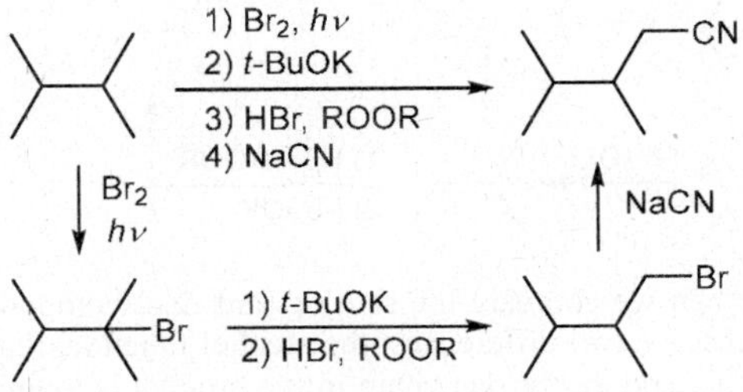

There are certainly other acceptable answers. For example, the *anti*-Markovnikov addition of HBr could be replaced with hydroboration-oxidation to give a primary alcohol. At that point, conversion of the alcohol into a tosylate makes a good leaving group before the final S_N2 process with cyanide as the nucleophile:

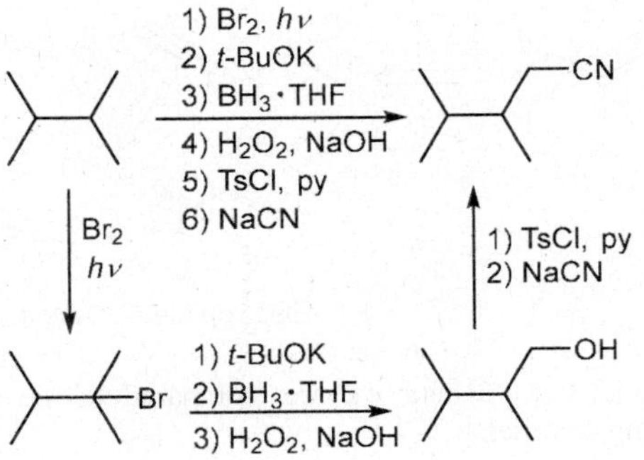

(f) In this transformation, the carbon chain has been extended, and a carboxylic acid functional group has been generated. Referring to Table 11.1, we can conclude that in order to make a carboxylic acid target molecule, the last step of our process must involve ozonolysis of an alkyne.

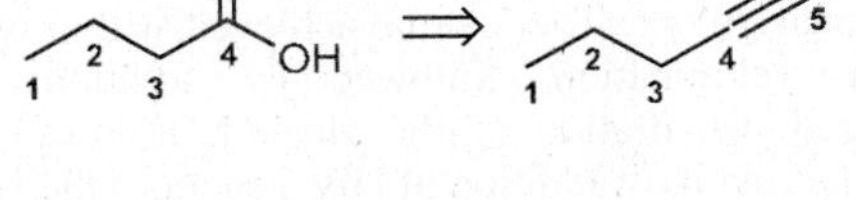

Continuing with our retrosynthetic analysis, the alkyne structure is suitable for making the required C–C bond disconnection:

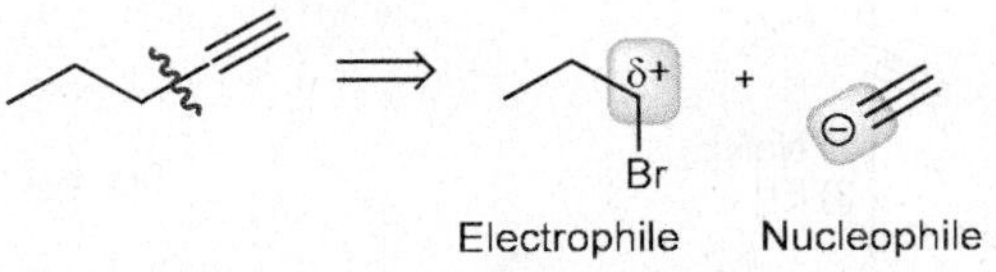

Now let's see where we are, compared to the given starting material:

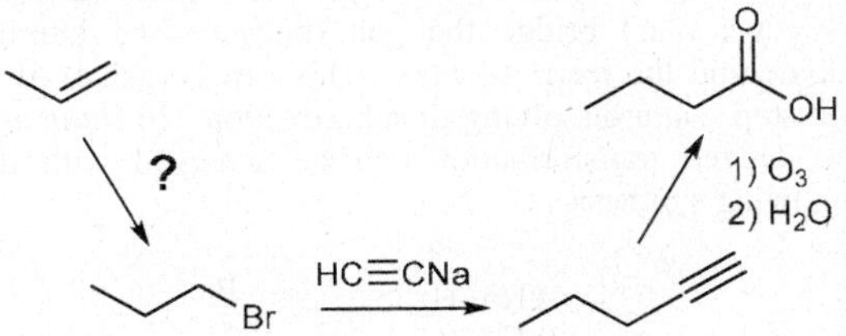

To bridge the gap, we must perform an *anti*-Markovnikov addition of HBr, which can be achieved in one step. In summary, the desired transformation can be achieved with the following synthesis:

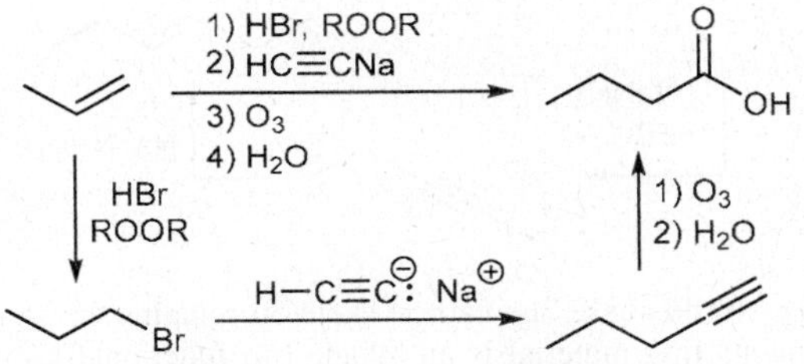

In this case, the carbon chain was lengthened by two carbon atoms (via an alkylation reaction), and then it was shortened by one carbon atom (via ozonolysis), resulting in one additional carbon atom in the target molecule.

(g) The product is a halohydrin, which can be made from the following *trans* alkene:

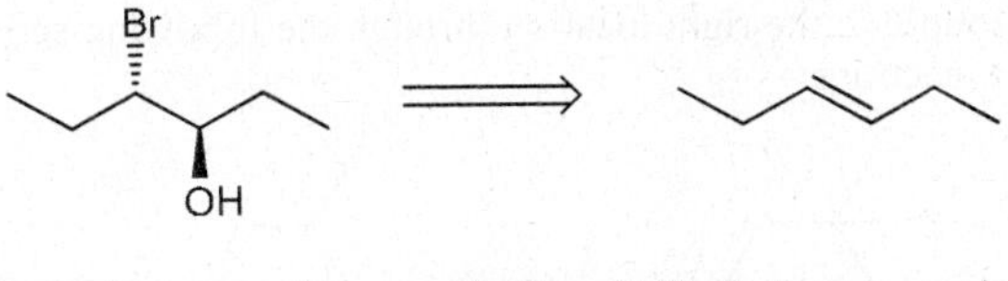

So the last step of our synthesis will likely be conversion of this alkene into the halohydrin.

Now let's work forward from the starting material. The starting material has only two carbon atoms, while the product has six carbon atoms. Two alkylation processes are required:

1) $NaNH_2$
2) EtI

1) $NaNH_2$
2) EtI

?

Br_2, H_2O

Br

OH + En

Now we must bridge the gap (between the internal alkyne and the *trans* alkene). This can be achieved in one step, via a dissolving metal reduction. In summary, the desired transformation can be achieved with the following synthesis:

1) $NaNH_2$
2) EtI
3) $NaNH_2$
4) EtI
5) Na, NH_3 (*l*)
6) Br_2, H_2O

Br

OH + En

1) $NaNH_2$
2) EtI

Br_2, H_2O

1) $NaNH_2$
2) EtI

Na, NH_3 (*l*)

(h) In this case, the carbon skeleton remains the same. The starting material is an alkane (no functional group) so we must begin our synthesis by installing a functional group. Radical bromination will selectively install a bromine atom at the tertiary position, giving the following tertiary alkyl halide:

Br_2
hν

Br

Now we must change both the location and the identity of the functional group. We can move the functional group into the right location through the following series of reactions:

Br

Elimination

Addition

Br

Elimination

And then finally, the double bond can be converted to the triple bond via the following two-step process:

Br_2

Br

Br

1) xs $NaNH_2$
2) H_2O

In summary, the desired transformation can be achieved with the following synthesis:

1) Br_2, *hν*
2) NaOEt
3) HBr, ROOR
4) *t*-BuOK
5) Br_2
6) xs $NaNH_2$
7) H_2O

Br_2, *hν*

1) Br_2
2) xs $NaNH_2$
3) H_2O

Br

NaOEt

1) HBr, ROOR
2) *t*-BuOK

(i) When we compare the starting and final compounds, we observe two differences: the alcohol functional group is gone, and the carbon chain in the product is longer by two carbon atoms. The retrosynthetic analysis of the target molecule asks: "What starting material can be used to make an alkane?" Referring to Table 11.1, we see that alkanes can be made by catalytic hydrogenation of an alkene or an alkyne. Selecting an alkyne is a logical choice, because only an alkyne structure is suitable for making the required C–C bond disconnection:

δ+

Br

+

Electrophile Nucleophile

Now let's see where we are, compared to the given starting material:

OH

?

H_2, Pt

Br

HC≡CNa

To bridge the gap, we must change the position of the functional group. That can be achieved with a two-step process (elimination, followed by addition). Acid-catalyzed dehydration of the alcohol, followed by an *anti*-Markovnikov addition of HBr generates the required alkyl bromide. In summary, the desired transformation can be achieved with the following synthesis:

1) conc. H_2SO_4, heat
2) HBr, ROOR
3) HC≡CNa
4) H_2, Pt

1) conc. H_2SO_4, heat
2) HBr, ROOR

H_2, Pt

H−C≡C:⊖ Na⊕

There are certainly other acceptable answers. For example, the *anti*-Markovnikov addition of HBr could be replaced with hydroboration-oxidation to give a primary alcohol. At that point, conversion of the alcohol into a tosylate makes a good leaving group before the SN2 process with acetylide as the nucleophile:

1) conc. H_2SO_4, heat
2) BH_3·THF
3) H_2O_2, NaOH
4) TsCl, py
5) HC≡CNa
6) H_2, Pt

1) conc. H_2SO_4, heat
2) BH_3·THF
3) H_2O_2, NaOH

H_2, Pt

1) TsCl, py
2) HC≡CNa

(j) The starting material has four carbon atoms, and the product has six carbon atoms, so our synthesis must involve the installation of two carbon atoms. The product is an aldehyde, so to begin the retrosynthesis, we ask: "What starting material can be used to make an aldehyde?" Referring to Table 11.1, we see than an aldehyde can be prepared by ozonolysis of an alkene, or by an *anti*-Markovnikov hydration of an alkyne. In this case, we are not shortening the carbon chain, so ozonolysis is not a reasonable option, and the alkyne starting material is the better choice. In addition, the corresponding alkyne structure is suitable for making the required C–C bond disconnection:

Electrophile Nucleophile

Now let's work forward from the given starting material. The starting material is an alkane (no functional group), so we must first install a functional group. Radical bromination will selectively install a bromine atom at a tertiary position:

Br_2, $h\nu$

1) R_2BH
2) H_2O_2, NaOH

?

HC≡CNa

To bridge the gap, we must change the position of the Br atom. That can be achieved with a two-step process (elimination, followed by addition). Treatment of the tertiary bromide with a strong base (such as hydroxide, methoxide, or ethoxide), followed by an *anti*-Markovnikov addition of HBr to the resulting alkene, generates the required alkyl bromide. In summary, the desired transformation can be achieved with the following synthesis:

1) Br_2, hv
2) NaOEt
3) HBr, ROOR
4) HC≡CNa
5) R_2BH
6) H_2O_2, NaOH

Br_2, $h\nu$

1) R_2BH
2) H_2O_2, NaOH

1) NaOEt
2) HBr, ROOR

H−C≡C:⊖

As demonstrated in part (i), there are other acceptable answers. For example, the *anti*-Markovnikov addition of HBr could be replaced with hydroboration-oxidation to give a primary alcohol. At that point, conversion of the alcohol into a tosylate makes a good leaving group suitable for the SN2 process with acetylide as the nucleophile.

11.10. Each of the following syntheses is one suggested synthetic pathway. There are other acceptable approaches that accomplish the same goals.

Each of the target compounds has a nine carbon linear chain, and our reagents must be alkenes with fewer than six carbon atoms. Thus, it is clear that we will be forming new C–C bonds in the course of this synthesis. The figure below outlines a retrosynthetic analysis for our target compound. An explanation of each of the steps (*a-h*) follows.

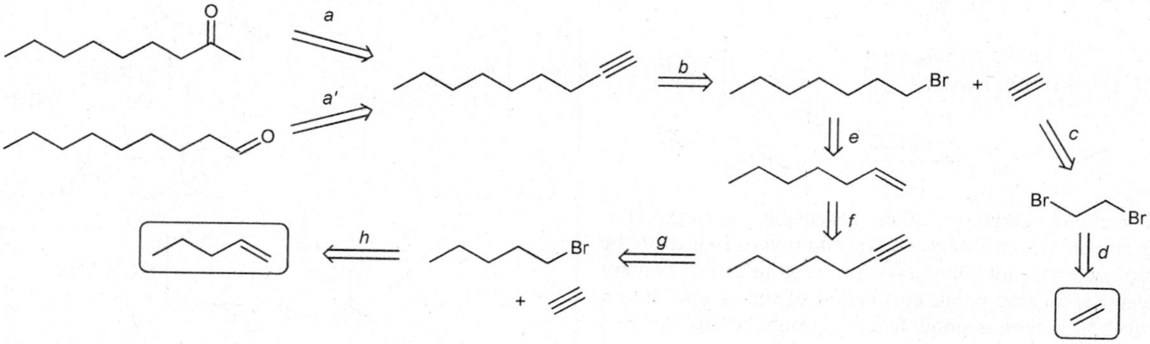

a. Either of the products can be prepared from a common synthetic intermediate, 1-nonyne, by hydration of the alkyne via (*a*) Markovnikov hydration, or (*a'*) *anti*-Markovnikov hydration.
b. The terminal alkyne can be prepared via reaction of 1-bromoheptane with an acetylide anion (formed by deprotonating acetylene).
c. Acetylene is prepared via a double elimination from 1,2-dibromoethane.
d. 1,2-Dibromoethane is prepared via bromination of ethylene (a suitable starting material).
e. 1-Bromoheptane is prepared via *anti*-Markovnikov addition of HBr across 1-heptene.
f. 1-Heptene is prepared via hydrogenation of 1-heptyne in the presence of a poisoned catalyst.
g. 1-Heptyne is prepared via reaction of 1-bromopentane with an acetylide anion.
h. 1-Bromopentane is prepared via *anti*-Markovnikov addition of HBr across 1-pentene (a suitable starting material).

Now, let's draw the forward scheme. In the presence of peroxides, the reaction of 1-pentene with HBr produces 1-bromopentane (via *anti*-Markovnikov addition). Subsequent reaction with acetylide [produced from ethylene as shown by bromination (Br_2), double elimination and deprotonation (excess $NaNH_2$)] provides 1-heptyne. Reduction to the alkene (H_2/Lindlar's catalyst) followed by *anti*-Markovnikov addition of HBr (HBr/peroxides) yields 1-bromoheptane. This primary alkyl bromide can then undergo an S_N2 reaction when treated with acetylide (prepared above), giving the common intermediate, 1-nonyne. A hydroboration-oxidation protocol (R_2BH then H_2O_2, NaOH) produces the target aldehyde, while an acid-catalyzed hydration (in the presence of mercuric sulfate) gives the target ketone.

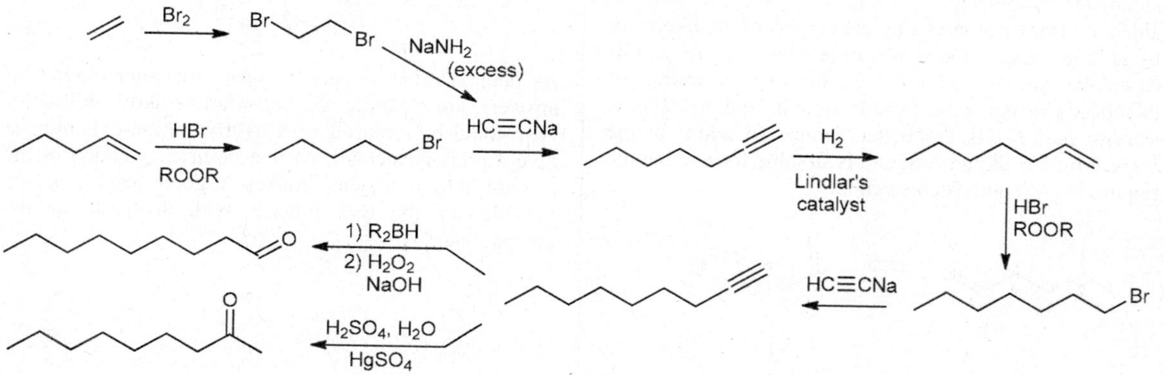

11.11. Most of the following reactions are addition reactions, which can be found in Chapter 8. The following reagents can be used to achieve each of the transformations shown:

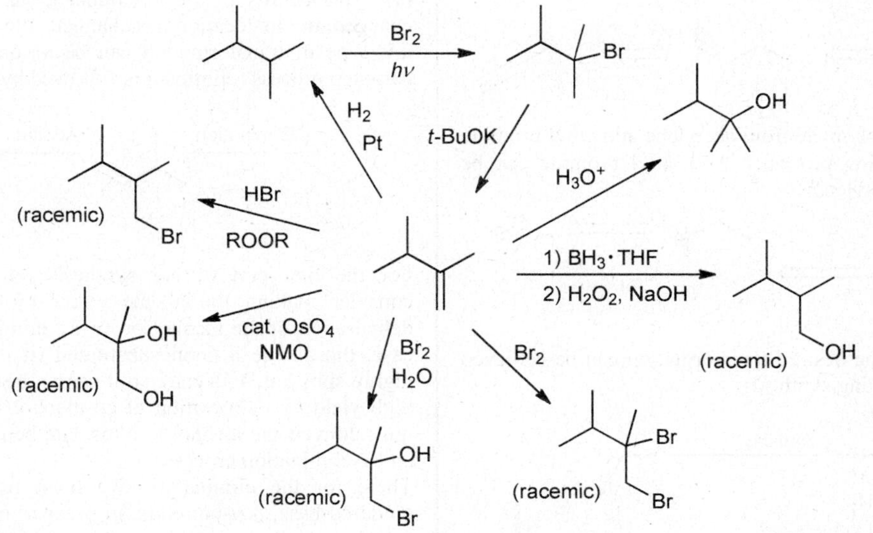

11.12. Most of the following reactions involve alkynes, which can be found in Chapter 9. The following reagents can be used to achieve each of the transformations shown:

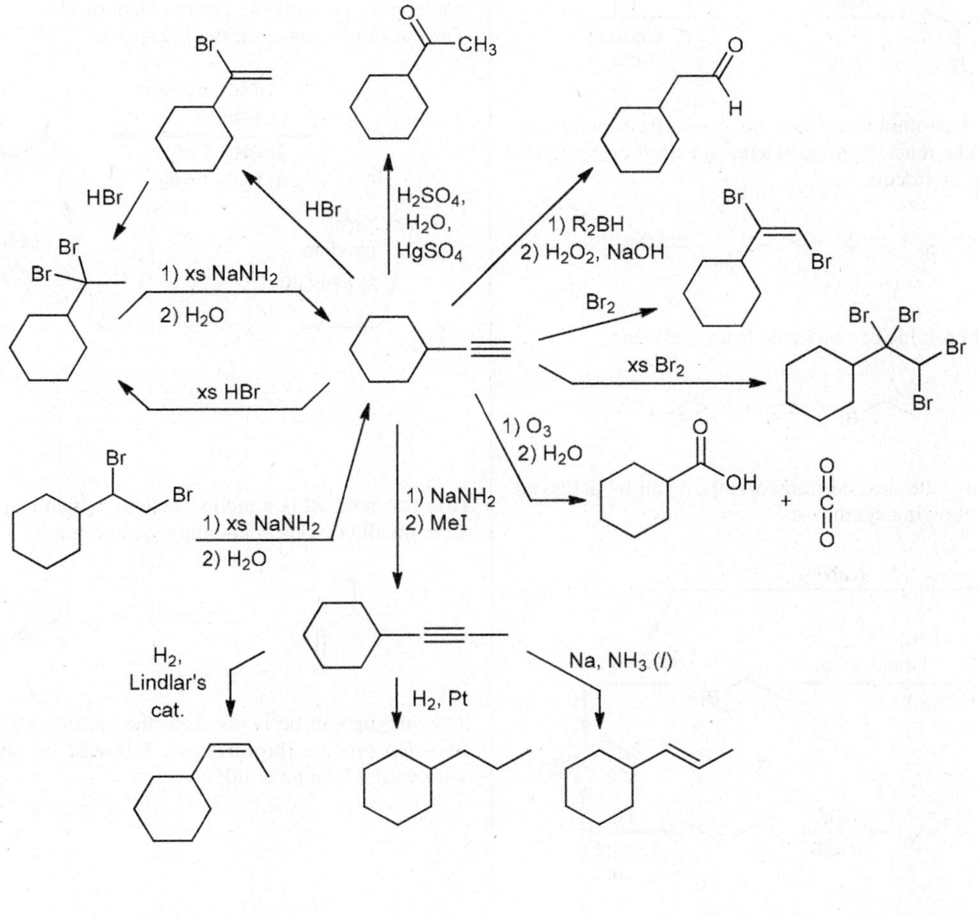

11.13. The product can be made from 1-butene, which can be made from 1-butyne:

1-Butyne can be made from acetylene and ethyl bromide via an alkylation process. And ethyl bromide can be made from acetylene:

In summary, the desired transformation can be achieved with the following synthesis:

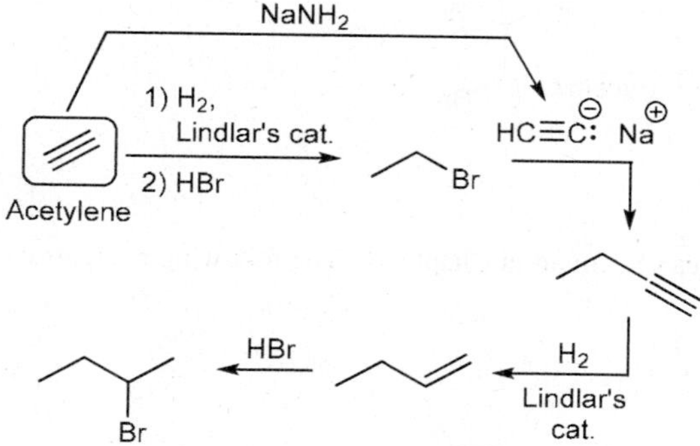

11.14. 1-Bromobutane can be made from 1-butyne, which can be made from acetylene and ethyl bromide via an alkylation process:

And ethyl bromide can be made from acetylene:

In summary, the desired transformation can be achieved with the following synthesis:

11.15.

(a) The identity of the functional group (OH) has not changed, but its location has changed. We have seen that this type of transformation can be accomplished via a two-step process (elimination, followed by addition):

For the first part of our synthesis (elimination), we consider making the alkene either by acid-catalyzed dehydration of the alcohol or by E2 elimination. In this case, the alkene is monosubstituted (it is not a stable, highly substituted alkene), so it cannot be synthesized in high yield by dehydration of an alcohol. Therefore, we must convert the alcohol to a tosylate before performing an E2 elimination process.

Then, for the elimination, we must use a sterically hindered base, *tert*-butoxide, in order to obtain the less-substituted alkene (Hofmann product). For the second part of our synthesis, we must perform an *anti*-Markovnikov addition of H and OH. This can be achieved via hydroboration-oxidation, or alternatively, it can be achieved by *anti*-Markovnikov addition of HBr followed by an S_N2 process (with hydroxide as a nucleophile) to give the desired alcohol.

The entire process is summarized here:

(b) The product is a methyl ketone, which can be made from an alkyne (via acid-catalyzed hydration):

This alkyne can be made from the starting alkene via a two-step process (bromination, followed by elimination with excess sodium amide):

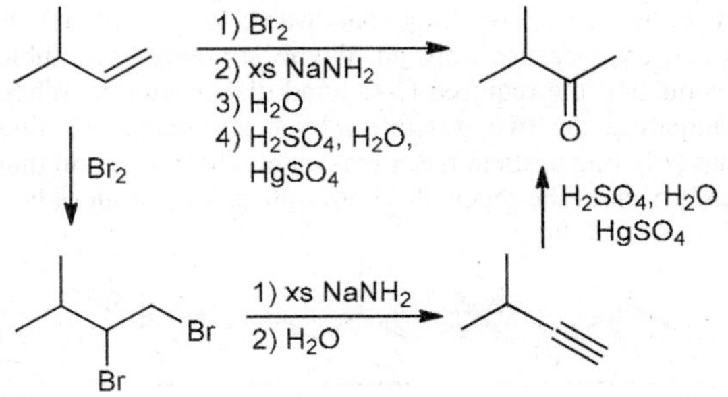

1) Br_2
2) xs $NaNH_2$
3) H_2O
4) H_2SO_4, H_2O, $HgSO_4$

Br_2

H_2SO_4, H_2O
$HgSO_4$

Br
Br

1) xs $NaNH_2$
2) H_2O

11.16. We must move the location of the π bond, and we have seen that this can be achieved via a two-step process (addition, followed by elimination). The addition step must occur in an *anti*-Markovnikov fashion, which can be achieved by treating the starting material with HBr in the presence of peroxides. The elimination process must give the less substituted alkene, so a sterically hindered base is required:

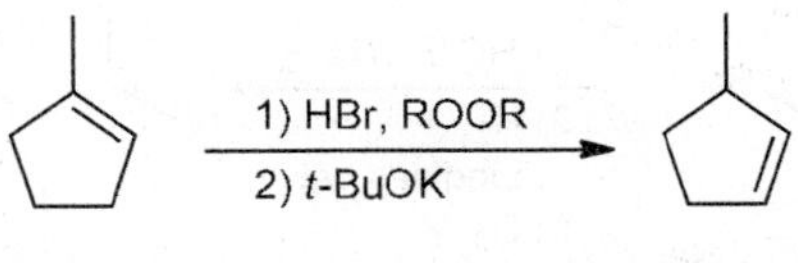

1) HBr, ROOR
2) *t*-BuOK

Alternatively, the addition of HBr can be replaced with hydroboration-oxidation (addition of H and OH), which is also an *anti*-Markovnikov addition. In that scenario, the resulting alcohol must first be converted to a tosylate before the elimination step can be performed.

11.17. Let's begin by drawing the starting material and the product, so that we can see the desired transformation more clearly:

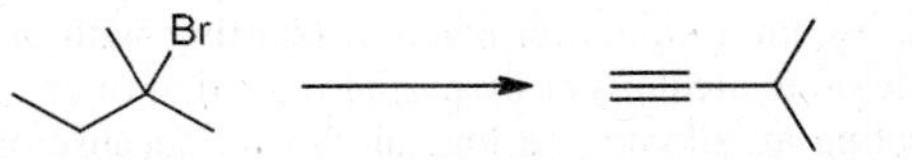

Br

2-bromo-
2-methylbutane

3-methyl-1-butyne

In this case, the identity and the location of the functional group have changed. During our synthesis, the functional group (Br) must be relocated, AND it must be converted into a triple bond. Moving the functional group can be achieved via elimination (to give the Zaitsev product) followed by addition (in an *anti*-Markovnikov fashion).

Br Elimination → Addition → Br

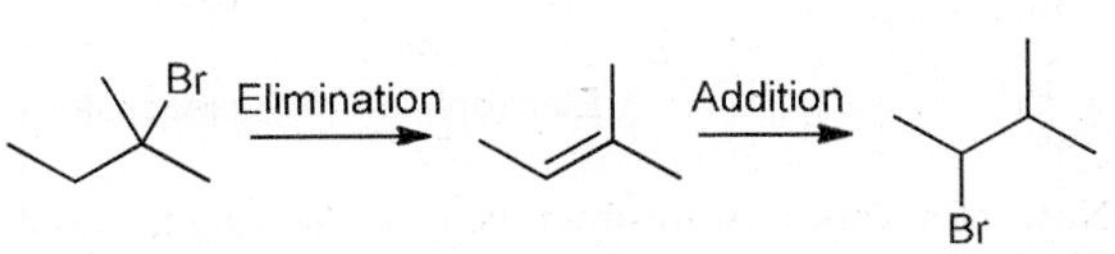

Then, changing the identity of the functional group can be achieved via the following three-step process:

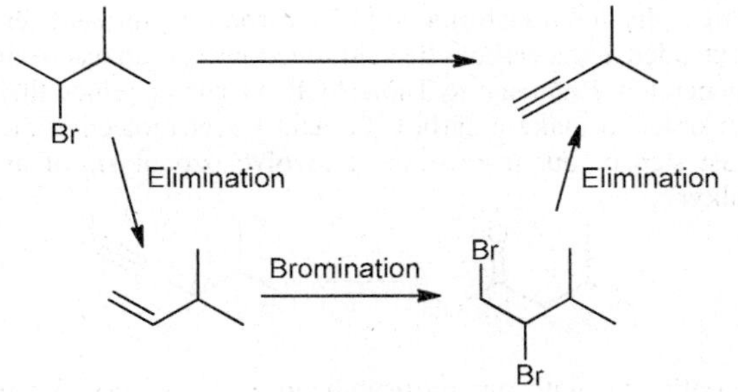

Br

Elimination

Elimination

Bromination

Br
Br

The necessary reagents are shown here:

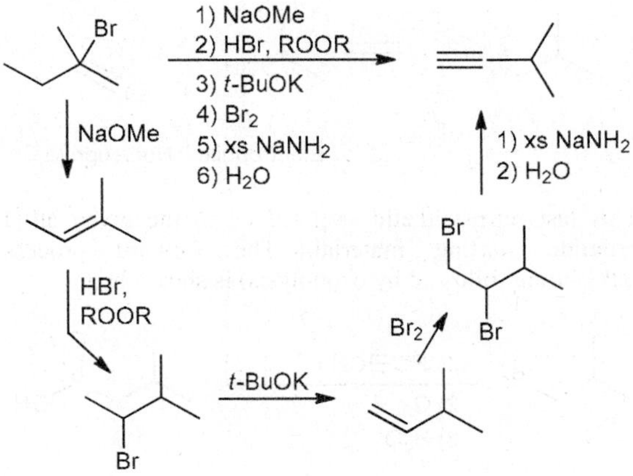

Br

1) NaOMe
2) HBr, ROOR
3) *t*-BuOK
4) Br_2
5) xs $NaNH_2$
6) H_2O

NaOMe

1) xs $NaNH_2$
2) H_2O

Br
Br

HBr,
ROOR

Br_2

t-BuOK

Br

11.18.

(a) The starting material has one more carbon atom than the product. Therefore, our synthesis must employ an ozonolysis process. Since the product is a carboxylic acid (rather than an aldehyde or ketone), we can conclude that the last step of our process must involve ozonolysis of an *alkyne*, rather than ozonolysis of an *alkene*:

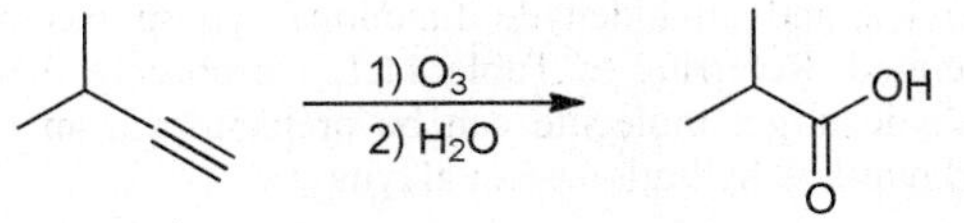

1) O_3
2) H_2O

OH
O

This alkyne can be made from the starting alkene via a two-step process (bromination, followed by elimination with excess sodium amide). The complete synthesis is shown here:

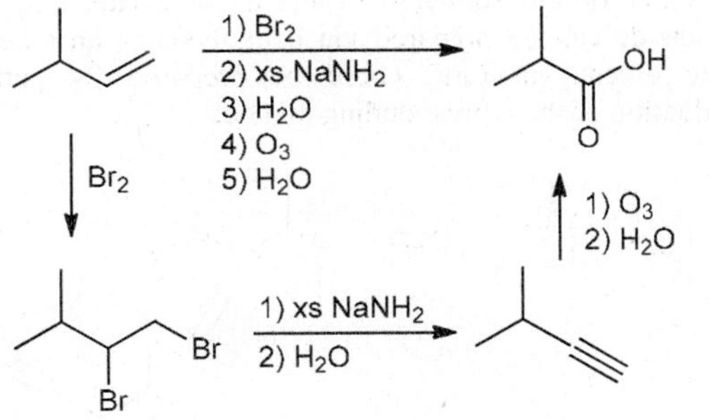

1) Br_2
2) xs $NaNH_2$
3) H_2O
4) O_3
5) H_2O

OH
O

Br_2

1) O_3
2) H_2O

Br
Br

1) xs $NaNH_2$
2) H_2O

(b) In this transformation, the carbon chain has been extended and a carboxylic acid functional group has been generated. Referring to Table 11.1, we can conclude that in order to make a carboxylic acid target molecule, the last step of our process must involve ozonolysis of an alkyne.

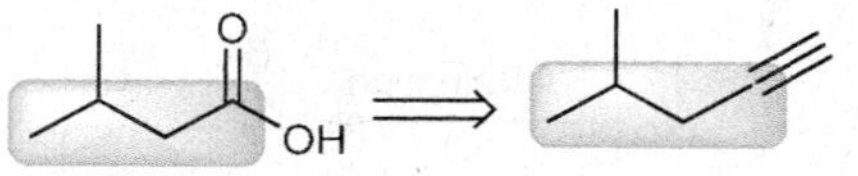

Continuing with our retrosynthetic analysis, the alkyne structure is suitable for making the required C–C bond disconnection:

Electrophile Nucleophile

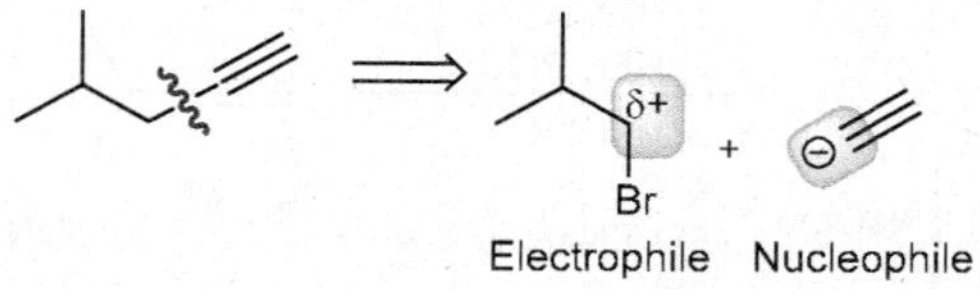

This last retrosynthetic step led us to the given alkyl bromide starting material. The forward process (alkylation, followed by ozonolysis) is shown here:

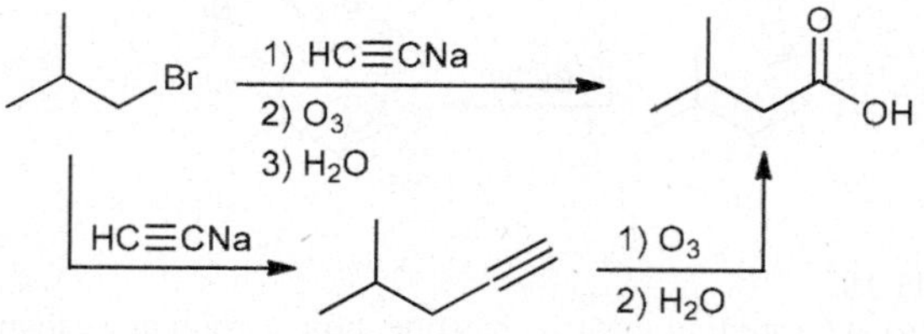

In this synthesis, the carbon chain was lengthened by two carbon atoms (via an alkylation reaction), and then it was shortened by one carbon atom (via ozonolysis), resulting in one additional carbon atom in the target molecule.

(c) In this transformation, the carbon chain has been extended and an aldehyde functional group has been generated. Referring to Table 11.1, we observe that an aldehyde target molecule can be prepared via an *anti*-Markovnikov hydration of an alkyne:

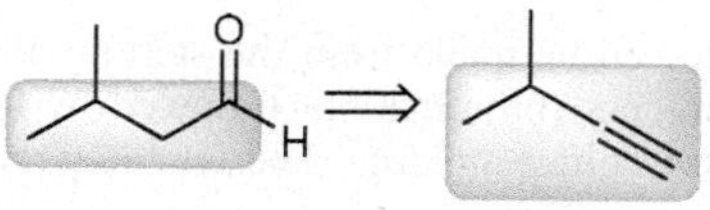

Another option shown in Table 11.1 reveals that the aldehyde can be prepared via ozonolysis of an alkene. The alkene, in turn, could be prepared by partial reduction of the corresponding alkyne:

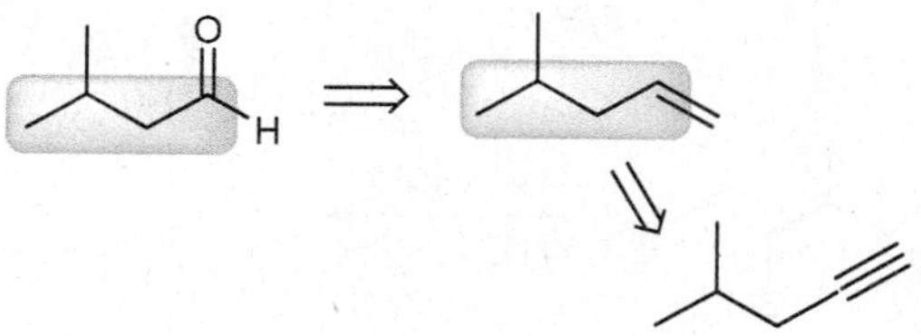

In either case, working backward to an alkyne is necessary, because only an alkyne structure is suitable for making the required C–C bond disconnection. When comparing the two possible alkyne structures, we find that only one of them has a reasonable disconnection that leads back to the given alkyl bromide starting material:

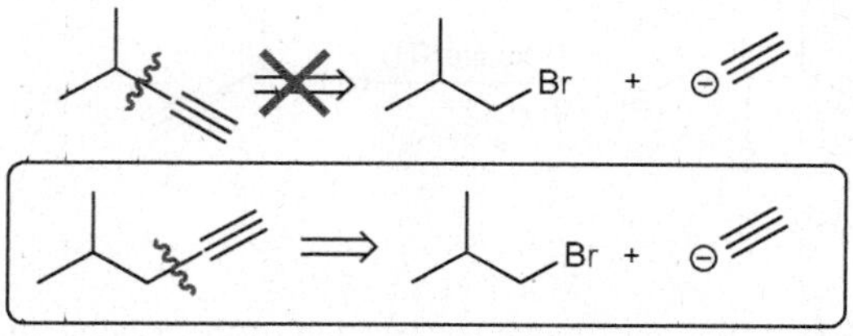

Therefore, the second retrosynthetic analysis we considered is the only feasible one. Based on that plan, the synthesis is achieved in three steps: alkylation with sodium acetylide, reduction of the alkyne to an alkene by hydrogenation in the presence of a poisoned catalyst, and ozonolysis to give the aldehyde target molecule:

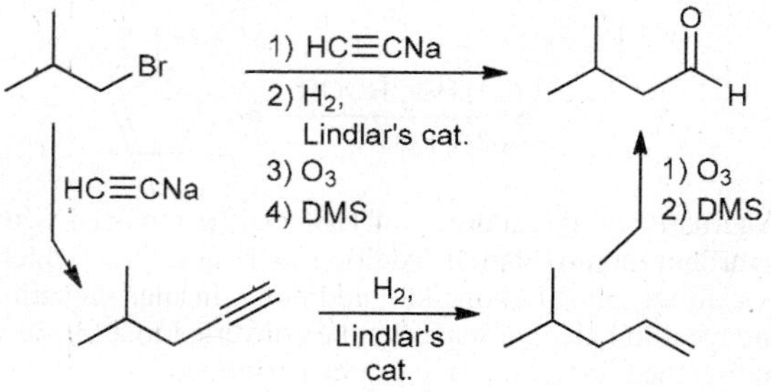

(d) The starting material has four carbon atoms, and the product has six carbon atoms, so our synthesis must involve the installation of two carbon atoms. The retrosynthetic analysis of the target molecule asks: "What starting material can be used to make an alkene?" Referring to Table 11.1, we see that an alkene can be made by an elimination reaction (starting with an alkyl halide or an alcohol) or by partial reduction of an alkyne. Selecting an alkyne is a logical choice, because only an alkyne structure is suitable for making the required C–C bond disconnection:

Electrophile Nucleophile

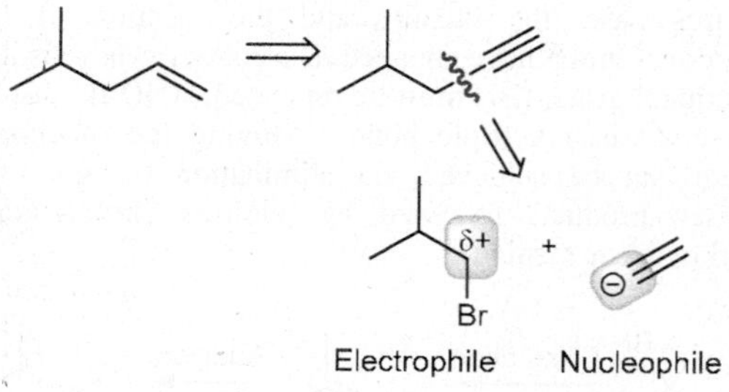

Now let's work forward from the given starting material. The starting material is an alkane (no functional group), so we must first install a functional group. Radical bromination will selectively install a bromine atom at a tertiary position:

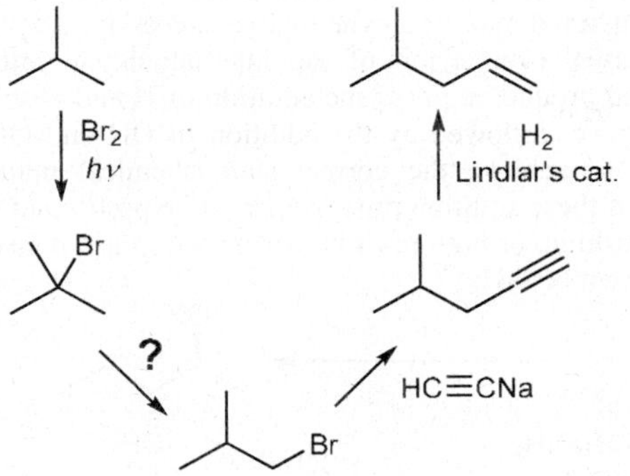

To bridge the gap, we must change the position of the Br atom. That can be achieved with a two-step process (elimination, followed by addition). Treatment of the tertiary bromide with a strong base (such as hydroxide, methoxide, or ethoxide), followed by an *anti*-Markovnikov addition of HBr to the resulting alkene, generates the required alkyl bromide. The primary alkyl halide is then treated with sodium acetylide to give an alkylation process. The resulting terminal alkyne can be hydrogenated in the presence of a poisoned catalyst to give the desired alkene:

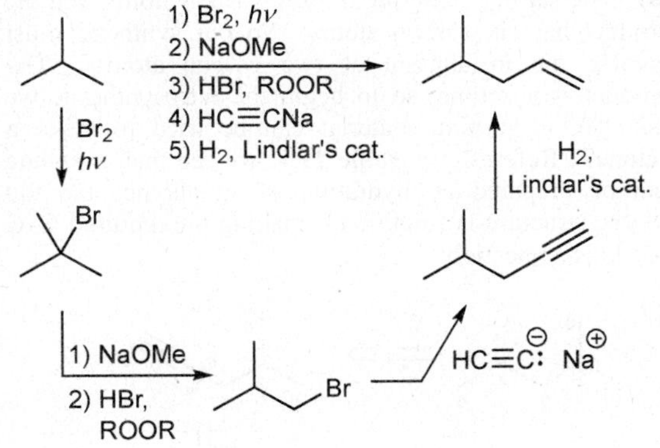

1) Br_2, $h\nu$
2) NaOMe
3) HBr, ROOR
4) HC≡CNa
5) H_2, Lindlar's cat.

As demonstrated in problem 11.9(i), there are other acceptable answers. For example, the *anti*-Markovnikov addition of HBr could be replaced with hydroboration-oxidation to give a primary alcohol. At that point, conversion of the alcohol into a tosylate makes a good leaving group before the S_N2 process with acetylide as the nucleophile.

(e) The starting material has three carbon atoms, and the product has five carbon atoms. We can install two carbon atoms by treating the starting material with sodium acetylide:

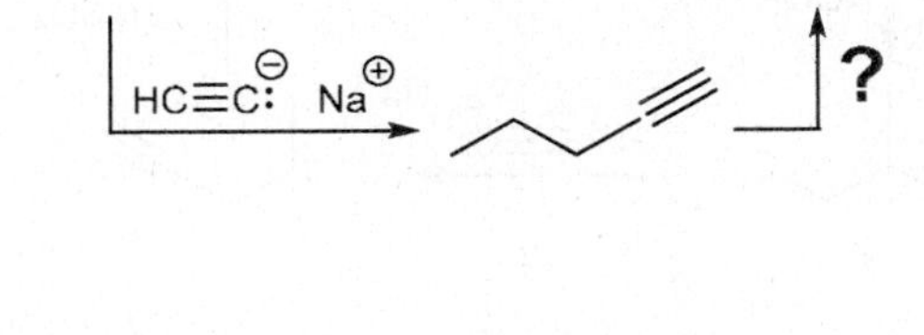

To bridge the gap, we observe that the triple bond is gone, two Br atoms have been added *and two H atoms have been added.* The geminal dihalide can be prepared in one step from the terminal alkyne, by treatment with excess HBr:

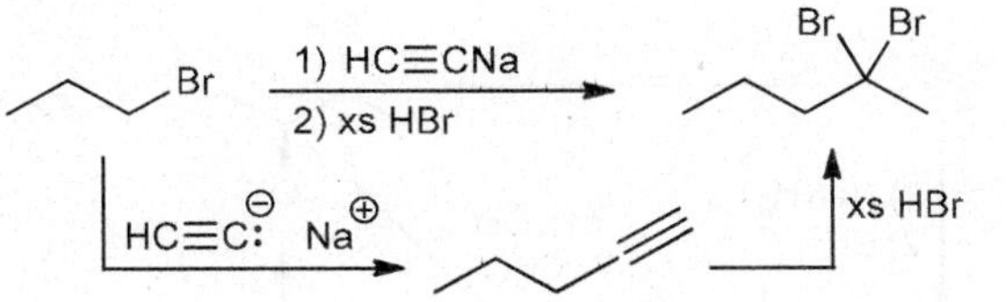

1) HC≡CNa
2) xs HBr

11.19. We begin by drawing the desired products, a pair of enantiomers:

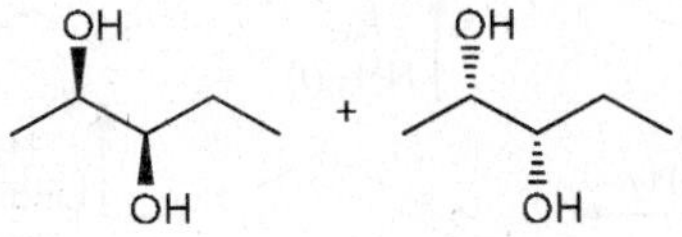

These compounds have five carbon atoms, but our starting materials can contain no more than two carbon atoms. So our synthesis must involve the formation of carbon-carbon bonds. This can be accomplished via the alkylation of acetylene (a compound with two carbon atoms). The retrosynthetic analysis of the target molecule asks: "What starting material can be used to make a 1,2-diol with the required stereochemistry?" The two OH groups in the target molecule can be prepared from either the *trans* or *cis* alkene:

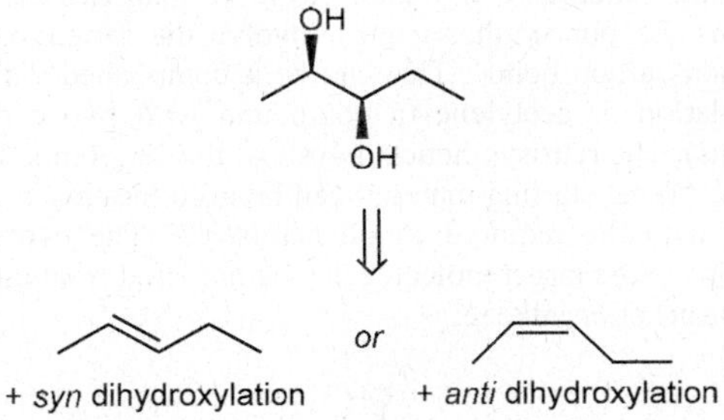

+ *syn* dihydroxylation or + *anti* dihydroxylation

Either alkene can be prepared from an alkyne, and the alkyne is suitable for making the required C–C bond disconnections:

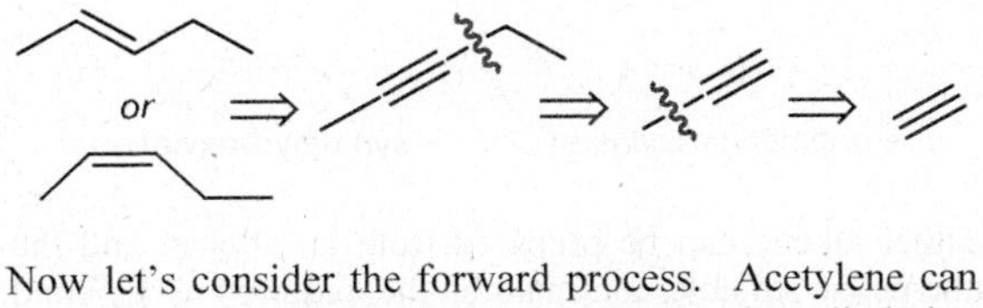

Now let's consider the forward process. Acetylene can be converted to 2-pentyne by two successive alkylation processes. Conversion of the internal alkyne into the desired product requires the addition of H and H to give an alkene, followed by the addition of OH and OH. In order to achieve the correct stereochemical outcome, one of these addition processes must be performed in a *syn* fashion, while the other must be performed in an *anti* fashion. In one answer shown below, the reduction is performed in an *anti* fashion, while the dihydroxylation process is performed with *syn* addition. In the other answer shown below, the reduction is performed in a *syn* fashion, while the dihydroxylation process is performed with *anti* addition:

1) $NaNH_2$ 2) EtI; 1) $NaNH_2$ 2) MeI; Na NH_3 (*l*); cat. OsO_4, NMO; H_2 Lindlar's cat.; 1) RCO_3H 2) H_3O^+; OH; OH; + En

11.20. We begin by drawing the desired products, a pair of enantiomers:

OH OH + OH OH

These compounds have five carbon atoms, but our starting materials can contain no more than two carbon atoms. So our synthesis must involve the formation of carbon-carbon bonds. This can be accomplished via the alkylation of acetylene (a compound with two carbon atoms). The retrosynthetic analysis of the target molecule asks: "What starting material can be used to make a 1,2-diol with the required stereochemistry?" The two OH groups in the target molecule can be prepared from either the *trans* or *cis* alkene:

OH OH

+ *anti* dihydroxylation *or* + *syn* dihydroxylation

Either alkene can be prepared from an alkyne, and the alkyne is suitable for making the required C–C bond disconnections:

or

Now let's consider the forward process. Acetylene can be converted into 2-pentyne by two successive alkylation processes. Conversion of the internal alkyne into the desired product requires the addition of H and H to give an alkene, followed by the addition of OH and OH. In order to achieve the correct stereochemical outcome, both of these addition processes must be performed in an *anti* fashion, or both must be performed in a *syn* fashion, as shown below:

1) $NaNH_2$ 2) EtI; 1) $NaNH_2$ 2) MeI; Na NH_3 (*l*); 1) RCO_3H 2) H_3O^+; H_2 Lindlar's cat.; cat. OsO_4, NMO; OH; OH; + En

11.21.

(a) The starting material has four carbon atoms, and the product has six carbon atoms. So our synthesis must involve the installation of two carbon atoms. The product is a ketone, so to begin the retrosynthesis, we ask: "What starting material can be used to make a ketone?" Referring to Table 11.1, we see than a ketone can be prepared by hydration of an alkyne, and the alkyne structure is suitable for making the required C–C bond disconnection:

O 1 2 3 4 5 6 ⟹ 1 2 3 4 5 6

δ+ Br + ⊖

Electrophile Nucleophile

The required alkyl bromide can be prepared from the given starting material by an *anti*-Markovnikov addition of HBr. In summary, the desired transformation can be achieved with the following synthesis:

1) HBr, ROOR 2) HC≡CNa 3) H_2SO_4, H_2O, $HgSO_4$

HBr, ROOR

Br

H−C≡C:⊖ Na⊕

H_2SO_4, H_2O $HgSO_4$

There are other acceptable answers. For example, the *anti*-Markovnikov addition of HBr could be replaced with hydroboration-oxidation to give a primary alcohol. At that point, conversion of the alcohol into a tosylate makes a good leaving group before the S_N2 process with acetylide as the nucleophile.

(b) The starting material has four carbon atoms, and the product has six carbon atoms, so our synthesis must involve the installation of two carbon atoms. The product is an aldehyde, so to begin the retrosynthesis we ask, "What starting material can be used to make an aldehyde?" Referring to Table 11.1, we see than an aldehyde can be prepared by ozonolysis of an alkene, or by an *anti*-Markovnikov hydration of an alkyne. In this case, we are not shortening the carbon chain, so ozonolysis is not a reasonable option, and the alkyne starting material is the better choice. In addition, the corresponding alkyne structure is suitable for making the required C–C bond disconnection:

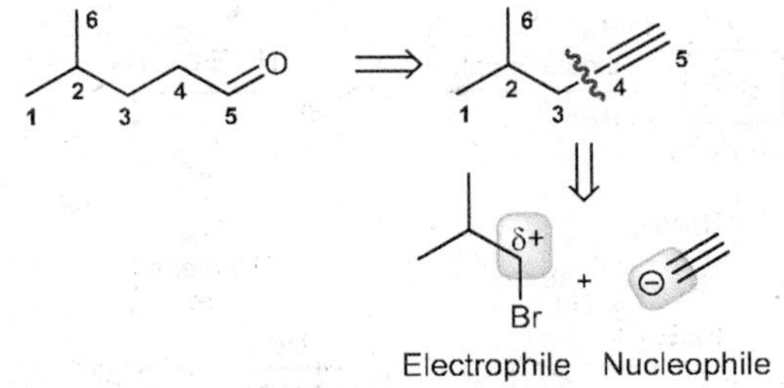

Electrophile Nucleophile

The required alkyl bromide can be prepared from the given starting material by an *anti*-Markovnikov addition of HBr. In summary, the desired transformation can be achieved with the following synthesis:

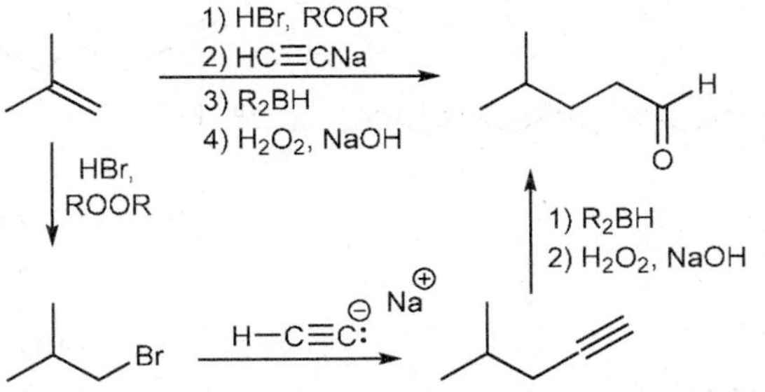

1) HBr, ROOR
2) HC≡CNa
3) R_2BH
4) H_2O_2, NaOH

HBr, ROOR

1) R_2BH
2) H_2O_2, NaOH

There are other acceptable answers. For example, the *anti*-Markovnikov addition of HBr could be replaced with hydroboration-oxidation to give a primary alcohol. At that point, conversion of the alcohol into a tosylate makes a good leaving group before the S_N2 process with acetylide as the nucleophile.

(c) In this transformation, the carbon chain has been extended (by one carbon atom), and a carboxylic acid functional group has been generated. Referring to Table 11.1, we can conclude that in order to make a carboxylic acid target molecule, the last step of our process must involve ozonolysis of an alkyne. Continuing with our retrosynthetic analysis, the alkyne structure is suitable for making the required C–C bond disconnection:

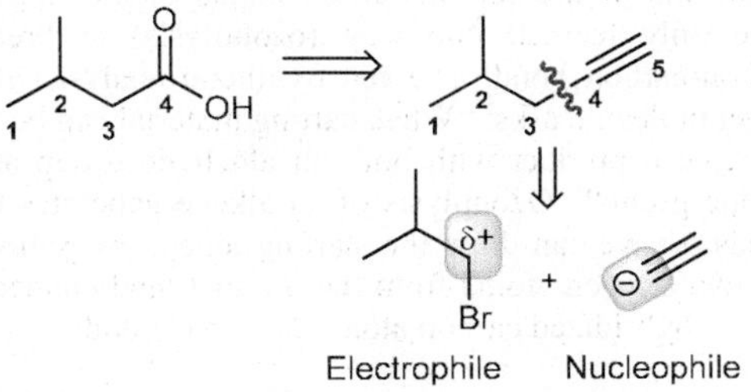

Electrophile Nucleophile

The required alkyl bromide can be prepared from the given starting material by an *anti*-Markovnikov addition of HBr. In summary, the desired transformation can be achieved with the following synthesis:

1) HBr, ROOR
2) HC≡CNa
3) O_3
4) H_2O

HBr, ROOR

1) O_3
2) H_2O

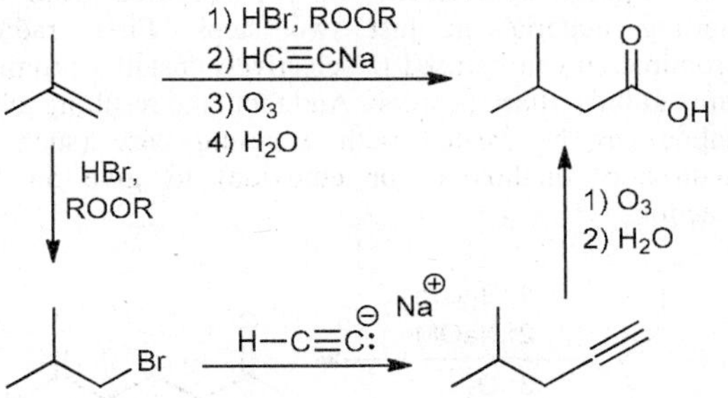

In this synthesis, the carbon chain was lengthened by two carbon atoms (via an alkylation reaction), and then it was shortened by one carbon atom (via ozonolysis), resulting in one additional carbon atom in the target molecule.

(d) The starting material has one more carbon atom than the product. Therefore, our synthesis must employ an ozonolysis process. Formation of an aldehyde product requires an alkene starting material. Numbering the carbon atoms is a helpful way to keep track of the original carbon skeleton:

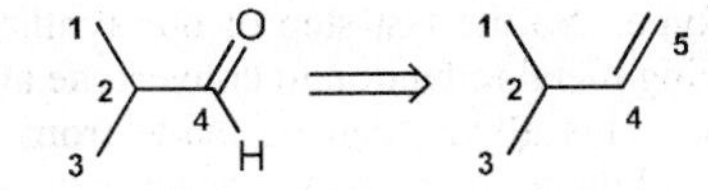

The required alkene can be formed by changing the position of the π bond in the given starting material (*anti*-Markovnikov addition of HBr, followed by elimination with a sterically hindered base to afford the Hofmann product). In summary, the desired transformation can be achieved with the following synthesis:

1) HBr, ROOR
2) *t*-BuOK
3) O_3
4) DMS

1) HBr, ROOR
2) *t*-BuOK

1) O_3
2) DMS

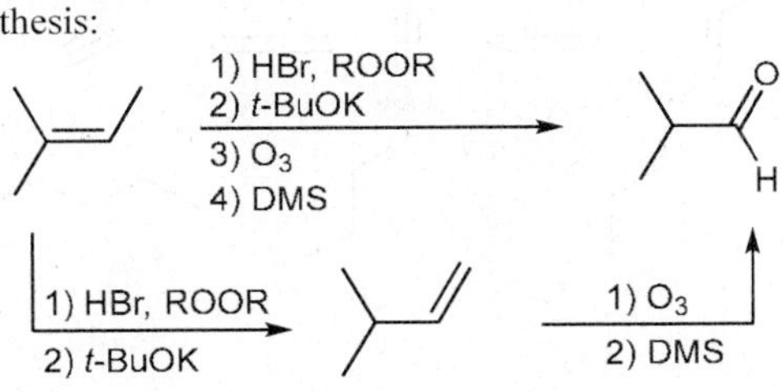

(e) The starting material is cyclic (it contains a ring) and the product is acyclic (it lacks a ring). So, we must break one of the carbon-carbon bonds of the ring. We have only learned one way (ozonolysis) to break a carbon-carbon bond. The retrosynthetic analysis of the target molecule asks: "What starting material can be used to make a product with both an aldehyde group and a ketone group?" Ozonolysis of an alkene generates C=O bonds, so we can draw the starting alkene by removing the two oxygen atoms from the product, and connecting the sp^2 hybridized carbon atoms as a C=C bond:

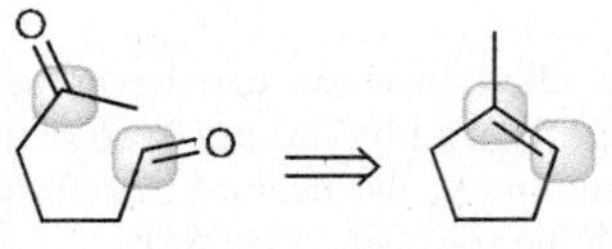

The required cycloalkene can be prepared from the starting material in just two steps. First, radical bromination can be used to selectively install a bromine atom at the tertiary position. And then, the resulting alkyl halide can be treated with a strong base (such as hydroxide, methoxide, or ethoxide) to give an E2 reaction:

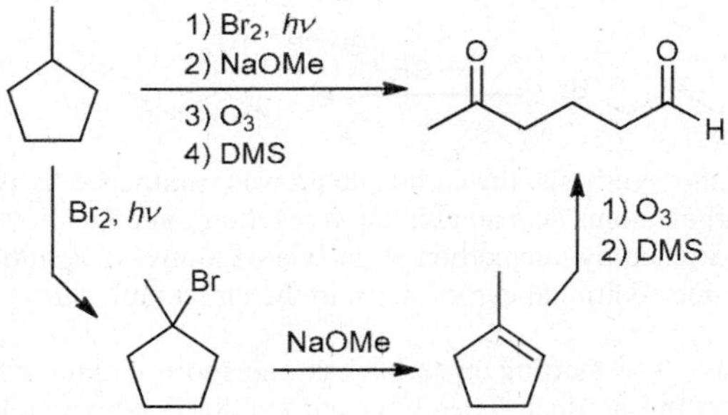

1) Br_2, *hν*
2) NaOMe
3) O_3
4) DMS

Br_2, *hν*

Br

NaOMe

1) O_3
2) DMS

H

11.22. The product is a *trans* alkene, which can be made from an alkyne. So the last step of our synthesis might be a dissolving metal reduction to convert the alkyne into the product. This alkyne can be made from acetylene and 1-bromobutane via two successive alkylation processes:

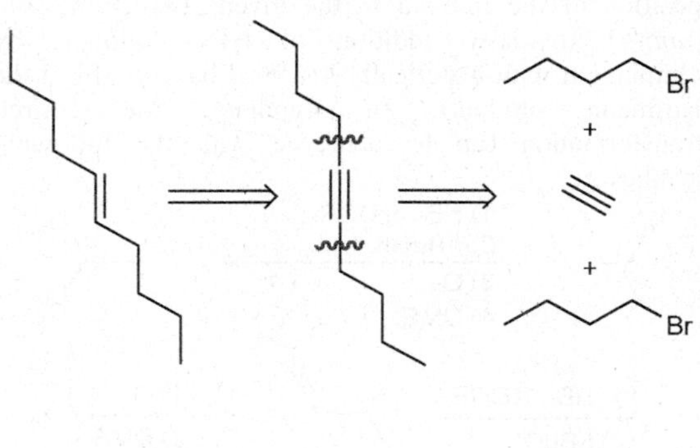

Br

\+

\+

Br

1-Bromobutane can be made from 1-butyne, which can be made from acetylene and ethyl bromide via an alkylation process:

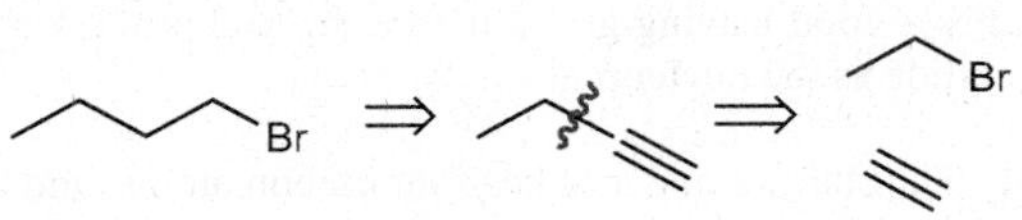

Br

Br

Finally, ethyl bromide can be made from acetylene:

Br

In summary, the desired transformation can be achieved with the following synthesis:

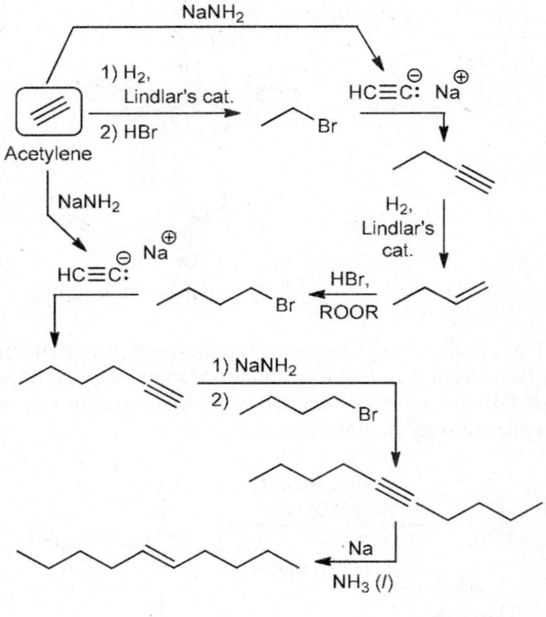

$NaNH_2$

1) H_2, Lindlar's cat.
2) HBr

Acetylene

Br

$HC{\equiv}C:^{\ominus}\ Na^{\oplus}$

H_2, Lindlar's cat.

$NaNH_2$

$HC{\equiv}C:^{\ominus}\ Na^{\oplus}$

HBr, ROOR

Br

1) $NaNH_2$
2) Br

Na
NH_3 (*l*)

11.23. We use the same approach taken in the previous problem. All carbon-carbon bonds are prepared via alkylation of an alkynide ion with the appropriate alkyl halide. Each alkyl halide must be prepared from acetylene. The last step of the synthesis is the reduction of an alkyne to a *cis* alkene via hydrogenation with a poisoned catalyst:

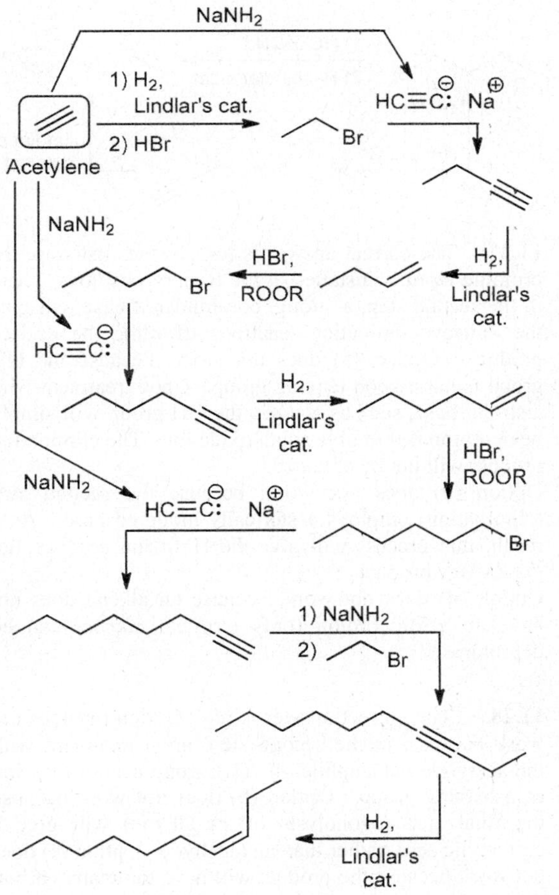

11.24. The starting material has two carbon atoms, and the product has five carbon atoms. So, we must join three fragments together (each of which has two carbon atoms), and then we must remove one of the carbon atoms. The latter process can be achieved via ozonolysis. Since the product is an aldehyde, it is reasonable to explore using ozonolysis as the last step of our synthesis:

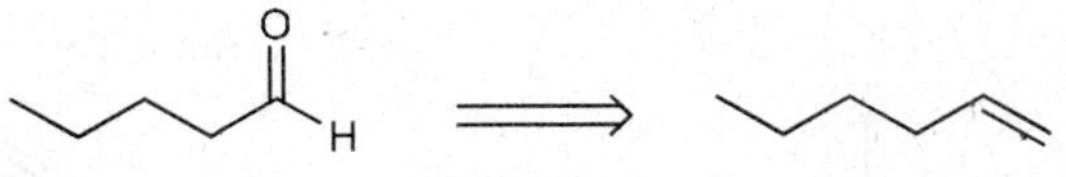

This alkene can be prepared from an alkyne, which can be prepared from acetylene and 1-bromobutane:

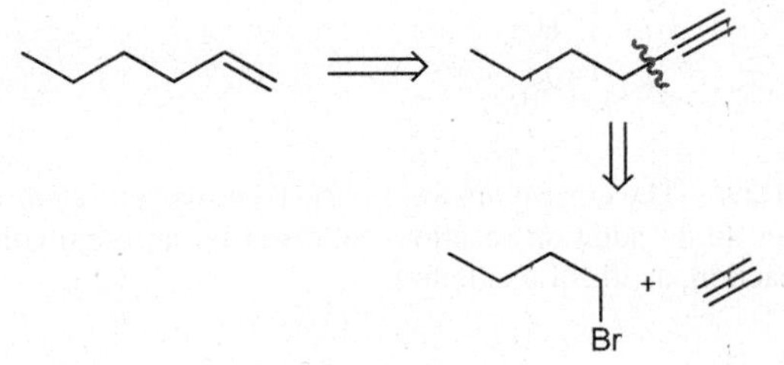

1-Bromobutane can be made from 1-butyne, which can be made from acetylene and ethyl bromide via an alkylation process:

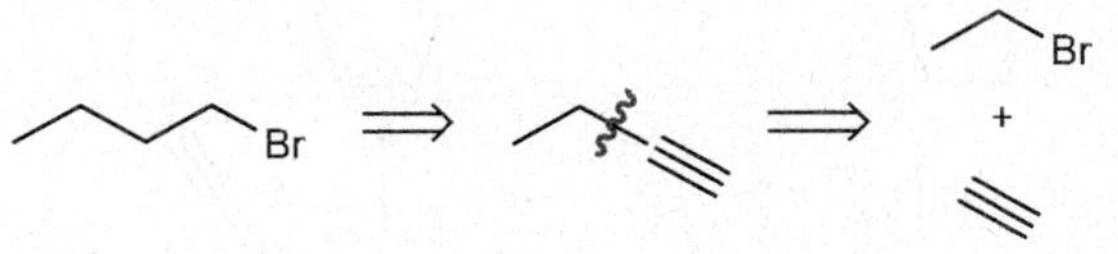

Finally, ethyl bromide can be made from acetylene:

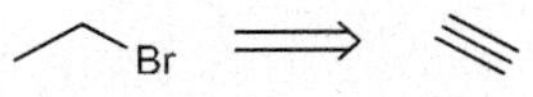

In summary, the desired transformation can be achieved with the following synthesis:

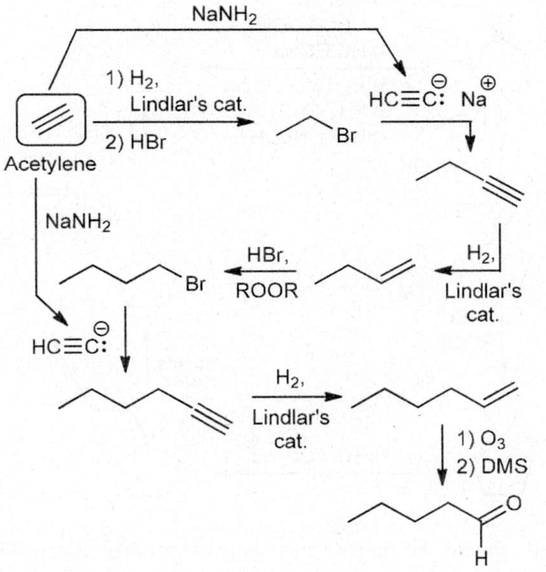

This synthesis represents just one correct answer to the problem. There are certainly other acceptable answers to this problem.

11.25. The desired transformation involves the installation of two carbon atoms, as well as a change in the location of the functional group. This can be achieved by converting the alcohol into a primary alkyl bromide, performing an alkylation process and then converting the triple bond into the desired alcohol:

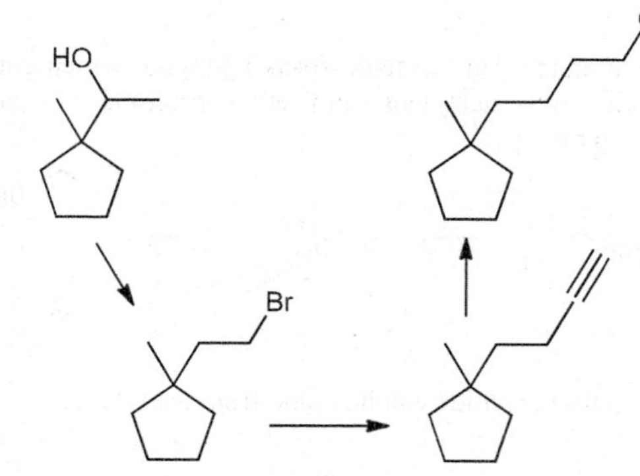
HO
OH
Br

The following reagents can be used to achieve the desired transformations:

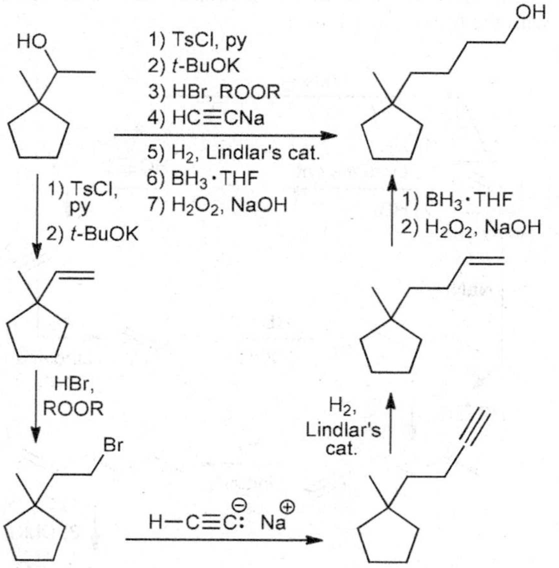
HO
1) TsCl, py
2) t-BuOK
3) HBr, ROOR
4) HC≡CNa
5) H2, Lindlar's cat.
6) BH3·THF
7) H2O2, NaOH
OH
1) TsCl, py
2) t-BuOK
1) BH3·THF
2) H2O2, NaOH
HBr, ROOR
H2, Lindlar's cat.
Br
H−C≡C: Na

It should be noted that there are other acceptable answers. As one example, the last part of our synthesis (hydroboration-oxidation) could be replaced with *anti*-Markovnikov addition of HBr to give a primary alkyl bromide, followed by an S_N2 process with hydroxide as the nucleophile:

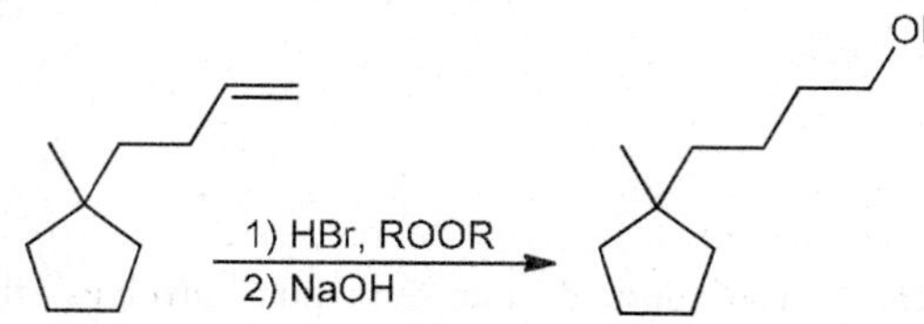
OH
1) HBr, ROOR
2) NaOH

11.26. The correct answer is (c). The starting material has four carbon atoms, and the product has six carbon atoms. The installation of two carbon atoms can be achieved by treating the starting primary alkyl bromide with sodium acetylide. The resulting alkylation product (a terminal alkyne) can then be reduced to an alkene via hydrogenation with a poisoned catalyst, such as Lindlar's catalyst:

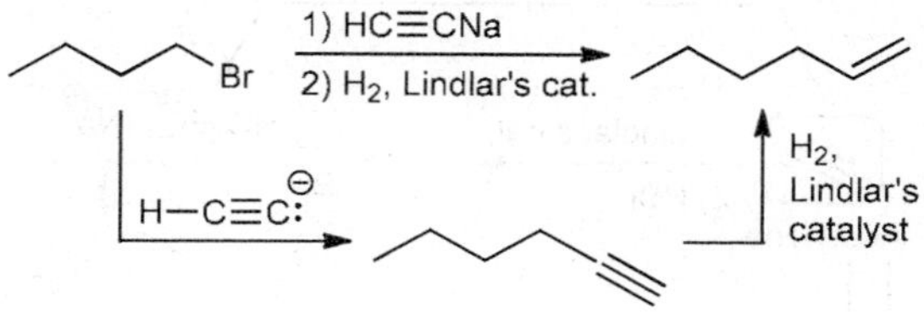
Br
1) HC≡CNa
2) H2, Lindlar's cat.
H−C≡C:
H2, Lindlar's catalyst

11.27. The correct answer is (c). In the first step, the bromine atom is installed at the tertiary position. Then, in the second step, a strong, non-hindered base will give the Zaitsev elimination reaction, affording the desired product. Option (a) does not work, because the OH group is not a good leaving group. Upon treatment with a strong base, such as NaOEt, the OH group will simply be deprotonated to give an alkoxide ion. The elimination product will not be obtained.
Option (b) does not work, because the second step (elimination) employs a sterically hindered base. As a result, that process will give the Hofmann product, not the Zaitsev product.
Option (d) does not work, because an alkene does not have an acidic proton (only terminal *alkynes* can be deprotonated).

11.28. The correct answer is (d). Option (a) does not work because, in the second step, upon treatment with the acetylide nucleophile, the OH group cannot function as a leaving group. Option (b) does not work because the final step (ozonolysis of an alkyne) will give a carboxylic acid, rather than an aldehyde. Option (c) does not work because the product will have too many carbon atoms (the starting material has six carbon atoms, and the product has seven carbon atoms, not eight). Option (d) is the correct answer, as shown below. *Anti*-Markovnikov addition of HBr gives a primary alkyl bromide, which is then converted into a terminal alkyne upon treatment with sodium acetylide. The terminal alkyne is then reduced to an alkene, followed by ozonolysis to give the desired aldehyde:

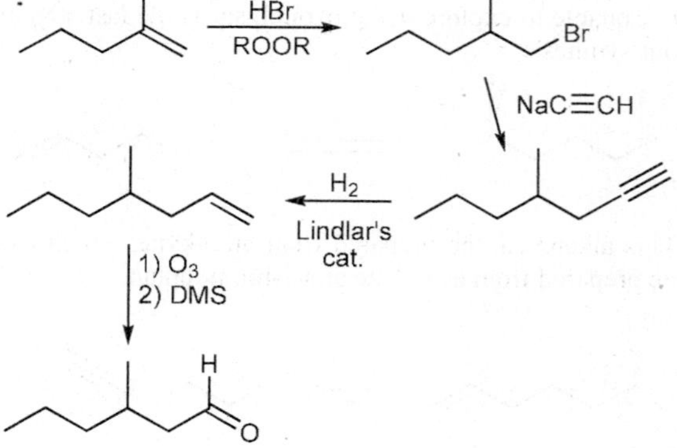
HBr
ROOR
Br
NaC≡CH
H2
Lindlar's cat.
1) O3
2) DMS
H
O

11.29. The correct answer is (d). To move a π bond, we can do an addition reaction, followed by an elimination reaction, as illustrated below:

1) HBr
2) *t*-BuOK

HBr
Addition

Br

t-BuOK
Elimination

11.30. The correct answer is (c). Ozonolysis causes oxidative cleavage of a C=C double bond or a C≡C triple bond. The loss of one or more carbon atoms from the alkene or alkyne starting material results in a product with a shorter carbon chain.

11.31. The correct answer is (b). This sequence begins with an elimination reaction, followed by an addition reaction, to move the position of the functional group. The Hofmann product is desired for the elimination reaction, so a strong, sterically hindered base is needed, and this is followed by an *anti*-Markovnikov addition of H and OH:

Br 1) *t*-BuOK
2) $BH_3 \cdot THF$
3) H_2O_2, NaOH

OH

t-BuOK

1) $BH_3 \cdot THF$
2) H_2O_2, NaOH

11.32. The correct answer is (a). As summarized in Table 11.1, a vicinal diol can be prepared from an alkene starting material, and a *trans* alkene can be prepared from an alkyne:

OH

OH

Finally, the most logical disconnection of the alkyne compound takes place at the C–C bond adjacent to the triple bond:

$\ominus$ + I–CH_3 ($\delta+$)

Nucleophile Electrophile

11.33. There are certainly many acceptable answers to this problem. The following retrosynthetic analysis employs the technique described in the problem statement:

OH

OH

Br + + O

This retrosynthetic analysis leads to the following synthesis:

Acetylene

OH

H_2, Pt

H_2,
Lindlar's
cat.

OH

Acetylene

1) $NaNH_2$

HBr

$NaNH_2$

2) O

3) H_2O

$HC{\equiv}C{:}^{\ominus}$ $Na^{\oplus}$

Br

H_2,
Lindlar's cat.

RCO_3H

Acetylene

11.34. The alkene in the target molecule can be prepared by partial reduction of the corresponding alkyne, and the alkyne structure is suitable for making the required C–C bond disconnection (employing the technique described in the problem statement that involves an electrophilic C=O group):

was an electrophile (C=O)

was a nucleophile

Nucleophile Electrophile

Acetylene is the only allowed source of carbon, so the aldehyde electrophile must be prepared from acetylene. Acetylene has two carbon atoms, and the aldehyde has only one carbon atom, so our synthesis must employ an ozonolysis process. Formation of an *aldehyde* product requires an *alkene* starting material:

This retrosynthetic analysis gives the following synthesis:

Acetylene

H_2, Lindlar's cat.

1) O_3 2) DMS

Acetylene

$NaNH_2$

1) HC≡CNa 2) H_2O

H_2, Lindlar's cat.

11.35. The key to solving this problem is recognizing that the cyclic product can be made from the following acyclic compound (which can be prepared from the starting material in just one step):

Upon treatment of this alkenyl alcohol with Br_2, a ring-forming reaction takes place that is similar to halohydrin formation:

Br_2

+ En

The π bond reacts with molecular bromine to give a bromonium ion, which is then attacked by the OH group in an intramolecular process. You may find it helpful to build molecular models to help visualize the stereochemistry of the ring-closing step.

According to the retrosynthetic analysis above, the desired transformation can be achieved in just two steps, shown here:

H_2 Lindlar's cat.

Br_2

+ En

11.36.

(a) The desired compound can be prepared from acetylene in just one step (via acid-catalyzed hydration):

H_2SO_4, H_2O $HgSO_4$

Alternatively, this transformation can also be achieved via hydroboration-oxidation of acetylene.

(b) The following synthesis represents just one correct answer to the problem. There are certainly other acceptable answers to this problem.

We have seen in previous problems that 1-butyne can be prepared from two equivalents of acetylene:

And 1-butyne can be converted into the aldehyde product in just two steps (partial hydrogenation, followed by ozonolysis):

H_2, Lindlar's cat.

1) O_3 2) DMS

In summary, the following synthesis can be used to make the desired compound from acetylene:

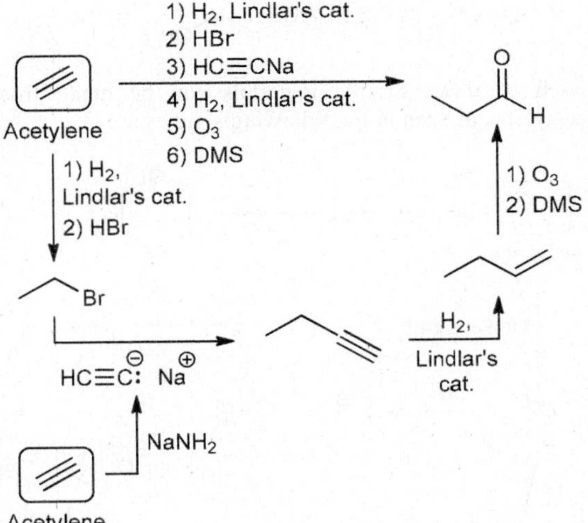

(c) The synthesis developed below is only one suggested synthetic pathway. There are likely other acceptable approaches that accomplish the same goal.

We have seen in previous problems that 1-butyne can be prepared from two equivalents of acetylene:

And 1-butyne can be converted into the aldehyde product via hydroboration-oxidation:

1) R_2BH
2) H_2O_2, NaOH

In summary, the following synthesis can be used to make the desired compound from acetylene:

1) H_2, Lindlar's cat.
2) HBr
3) HC≡CNa
4) R_2BH
5) H_2O_2, NaOH

(d) The synthesis developed below is only one suggested synthetic pathway. There are likely other acceptable approaches that accomplish the same goal.

The starting material has two carbon atoms, and the product has five carbon atoms. So, we must join three fragments together (each of which has two carbon atoms), and then we must remove one of the carbon atoms. The latter process can be achieved via ozonolysis. Since the product is an aldehyde, it is reasonable to explore using ozonolysis of an alkene as the last step of our synthesis:

This alkene can be prepared from an alkyne, which can be prepared from acetylene and 1-bromobutane:

1-Bromobutane can be made from 1-butyne, which can be made from acetylene and ethyl bromide via an alkylation process:

Finally, ethyl bromide can be made from acetylene:

In summary, the desired transformation can be achieved with the following synthesis:

11.37. Our retrosynthesis begins with a disconnection at the ether groups, at two of the C–O bonds, thereby breaking up the structure into two fragments, each with two carbon atoms. The more electronegative oxygen atom is more likely to have served as the nucleophile, so we draw a negative charge on the oxygen atom to give a suitable nucleophile (alkoxide ion). The carbon atom (at the disconnected bond) must have started out as an electrophile, so we draw a leaving group (such as Cl, Br, I or OTs) at that position:

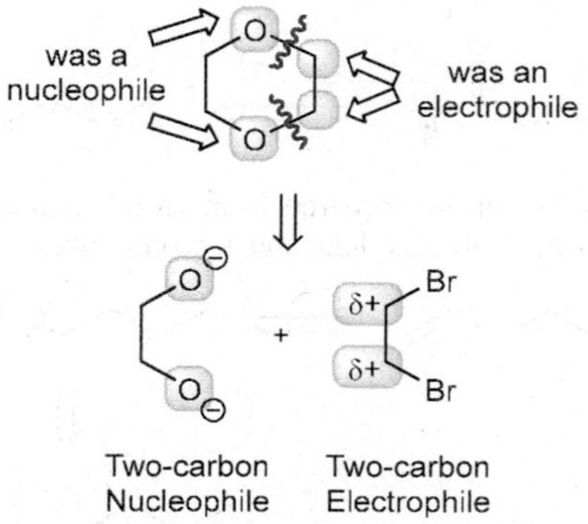

was a nucleophile

was an electrophile

Two-carbon Nucleophile

Two-carbon Electrophile

The dianion nucleophile can be prepared by deprotonation of the corresponding diol with excess NaH (as described in the problem statement), and the diol can be prepared from an alkene:

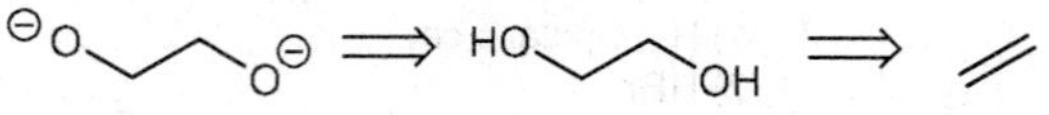

Each of these starting materials can be made from acetylene, as seen in the following synthesis:

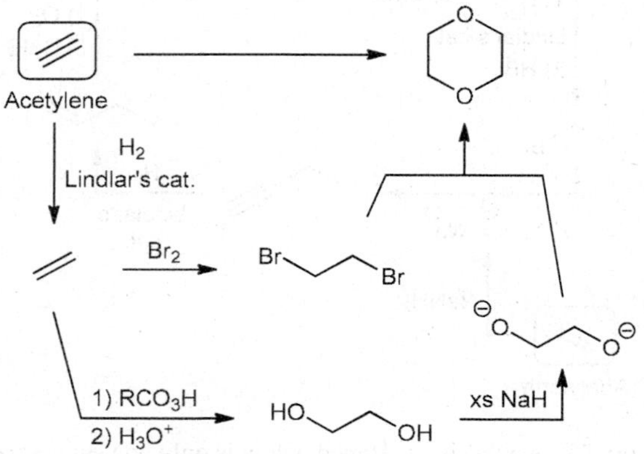

Acetylene

H_2 Lindlar's cat.

Br_2

1) RCO_3H 2) H_3O^+

xs NaH

11.38.

The synthesis developed below is only one suggested synthetic pathway. There are likely other acceptable approaches that accomplish the same goal.

An analysis of the structure of the product suggests the following origins of each of the carbon atoms in the product.

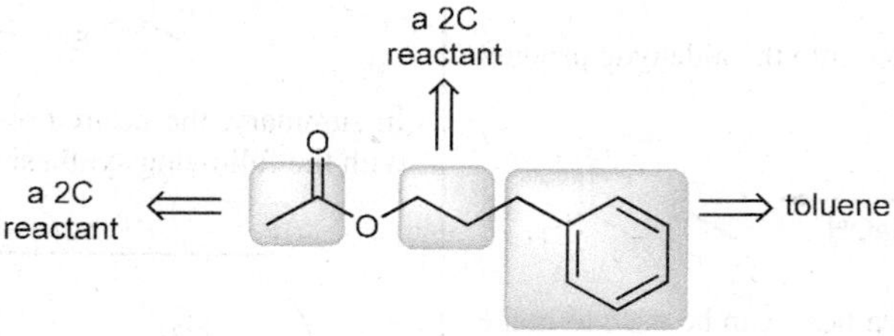

a 2C reactant

a 2C reactant

toluene

The following figure outlines a retrosynthetic analysis for our target molecule. An explanation of each of the steps (*a-e*) follows.

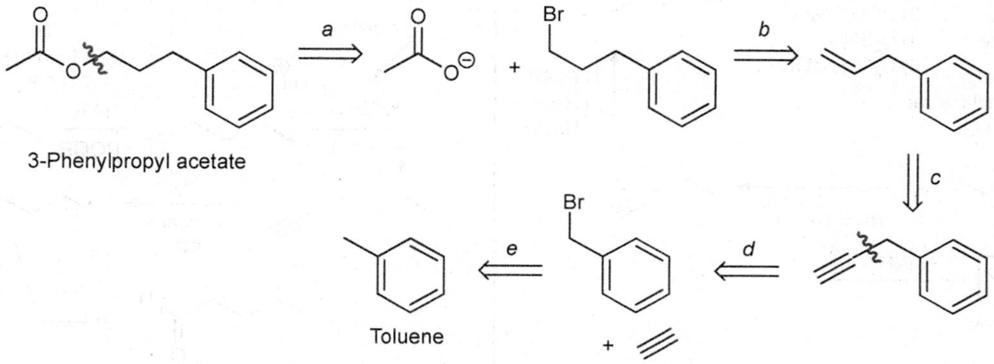

3-Phenylpropyl acetate

Toluene

a. The target molecule (3-phenylpropyl acetate) can be made via an SN2 reaction between a carboxylate ion and the primary alkyl bromide.
b. The primary alkyl bromide can be made by *anti*-Markovnikov addition of HBr to the monosubstituted alkene.
c. The alkene is made by reduction of the corresponding terminal alkyne.
d. The terminal alkyne is made by alkylating acetylene (using sodium amide to deprotonate) with benzyl bromide.
e. Benzyl bromide is made via radical bromination of toluene. The carbon atom adjacent to the aromatic ring (the benzylic position) is activated toward bromination due to the resonance-stabilized radical intermediate that forms.

Now, let's draw the forward scheme. Toluene is brominated using NBS and heat. Reaction with sodium acetylide (made by deprotonating acetylene with sodium amide) produces the terminal alkyne. The alkyne is reduced to the alkene using molecular hydrogen and Lindlar's catalyst. *Anti*-Markovnikov addition of HBr in the presence of peroxides produces the primary alkyl halide. SN2 substitution with the conjugate base of acetic acid (made by deprotonating acetic acid with sodium hydroxide) produces the desired product.

NaNH2
Br
NBS
heat
HC≡CNa
H2
Lindlar's catalyst
HBr
ROOR
Br
O
O
O
ONa
NaOH
O
OH

11.39.
The following synthesis is one suggested synthetic pathway. There are likely other acceptable approaches that accomplish the same goal.
Take note that the reactant and product each have six carbon atoms. This suggests that our synthetic plan will not necessarily involve any C–C bond-forming reactions. However, there is a change in the carbon skeleton, and we will need a C–C bond-*breaking* reaction to convert the cyclic starting material into an acyclic product.

O
6 carbon atoms
6 carbon atoms
O

The product contains two ketone groups, which is suggestive of an ozonolysis. The figure below outlines a retrosynthetic analysis for our target molecule. An explanation of each of the steps (*a-e*) follows.

O
a
b
Br
c
d
e
O
A
B
C
D
Br

a. The two ketone groups can be prepared via ozonolysis of 1,2-dimethylcyclobutene, **A** (we can draw the required alkene by removing the two oxygen atoms from the target molecule, and connecting the highlighted sp^2 hybridized carbon atoms as a C=C bond). When we compare alkene **A** with the given alkene starting material, we see that we must change the position of the π bond.
b. Alkene **A** is prepared via elimination from a suitable alkyl halide (**B**).

c. The tertiary alkyl bromide **B** is prepared via Markovnikov addition of HBr to alkene **C**.
d. Alkene **C** can be prepared via Zaitsev elimination of alkyl bromide **D**.
e. Alkyl bromide **D** can be made via addition of HBr to 3,4-dimethylcyclobutene (our given starting material).

Now, let's draw out the forward scheme. Treatment with HBr converts 3,4-dimethylcyclobutene to alkyl bromide **D**. Zaitsev elimination using a non-hindered base affords alkene **C**, which is subsequently converted to alkyl halide **B** using HBr (Markovnikov addition). Zaitsev elimination with a non-hindered base, followed by ozonolysis of the resulting alkene, gives the desired product, 2,5-hexanedione.

HBr → D (Br) ; NaOEt → C ; HBr → B (Br) ; NaOEt → A ; 1) O_3 2) DMS →

11.40.
The synthesis developed below is only one suggested synthetic pathway. There are likely other acceptable approaches that accomplish the same goal.

The following figure outlines a retrosynthetic analysis for our target molecule. An explanation of each of the steps (*a-h*) follows.

OH ⇒(a) OH ⇒(b) + O ⇒(c) ⇓(d) ⇐(e) Br ⇐(h) ≡ + Br ⇓(f) ⇓(g)

a. 1-Penten-3-ol can be made by reduction of the corresponding terminal alkyne.
b. The terminal alkyne can be made from acetylene (after deprotonation to form a nucleophile) and the aldehyde shown.
c. The aldehyde can be made by ozonolysis of *trans*-3-hexene.
d. *trans*-3-Hexene is prepared via a dissolving metal reduction of 3-hexyne.
e. 3-Hexyne is made by alkylation of 1-butyne with bromoethane.
f. Bromoethane is made by HBr addition to ethylene.
g. Ethylene is made by partial reduction of acetylene.
h. 1-Butyne is made by alkylation of acetylene using bromoethane (prepared as described above in steps *f-g*).

Now, let's draw the forward scheme. Acetylene is reduced to ethylene using molecular hydrogen and Lindlar's catalyst. Addition of HBr affords bromoethane. Reaction of the alkyl halide with sodium acetylide (prepared from acetylene and sodium amide) gives 1-butyne. Deprotonation of the terminal alkyne with sodium amide, followed by reaction with bromoethane produces 3-hexyne, which is subsequently reduced to *trans*-3-hexene using a dissolving metal reduction. (Alternatively, hydrogenation in the presence of Lindlar's catalyst provides the *cis* alkene.) Ozonolysis of 3-hexene (*trans* or *cis*) produces two equivalents of the desired aldehyde. Reaction of this electrophilic aldehyde with the sodium acetylide nucleophile, followed by treatment with water to protonate the alkoxide intermediate, forms a new C–C bond, and gives an alkynyl alcohol product. The alkyne is reduced to an alkene, using molecular hydrogen and Lindlar's catalyst, to produce the desired target molecule, 1-penten-3-ol.

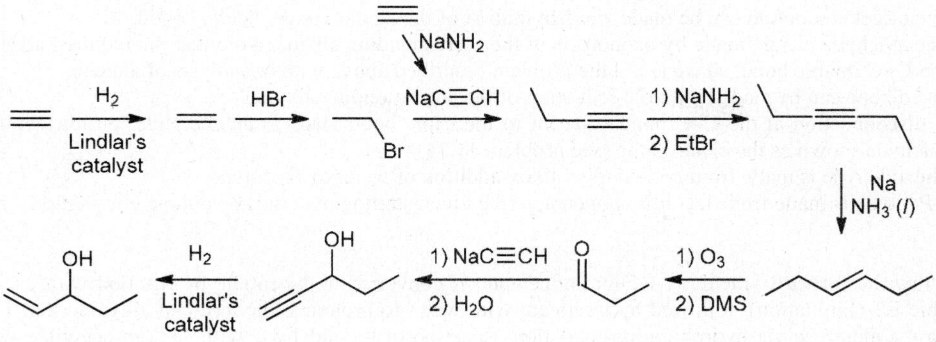

NaNH$_2$ · NaC≡CH · H$_2$, Lindlar's catalyst · HBr · Br · 1) NaNH$_2$ 2) EtBr · Na, NH$_3$ (*l*) · 1) O$_3$ 2) DMS · 1) NaC≡CH 2) H$_2$O · OH · H$_2$, Lindlar's catalyst

11.41.

The following synthesis is one suggested synthetic pathway. There are likely other acceptable approaches that accomplish the same goal.

The target molecule, (*E*)-2-hexenal, is bifunctional - containing both an alkene group and an aldehyde group. As summarized in Table 11.1, we have learned two ways to make aldehydes: (i) ozonolysis of alkenes and (ii) *anti*-Markovnikov hydration of terminal alkynes. In this case, it is not immediately apparent which of these methods we should use. For example, ozonolysis of the compound below would not be a good approach, as both C=C bonds are susceptible to cleavage, yielding the following three products.

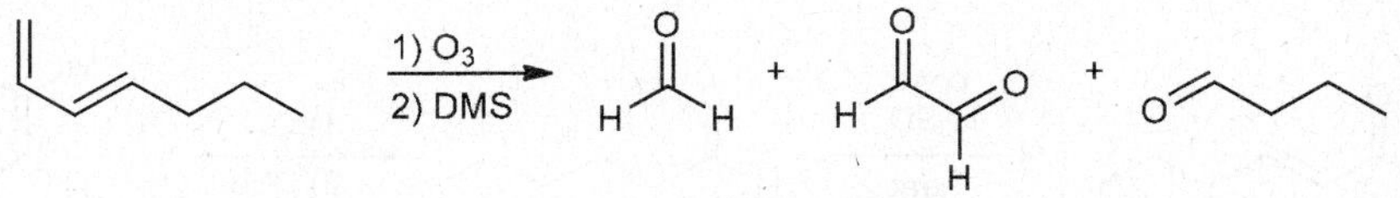

1) O$_3$ 2) DMS → + +

Anti-Markovnikov hydration of a terminal alkyne also does not appear to be a viable approach, as there is no obvious precursor that would allow installation of the C=C double bond adjacent to the aldehyde group.

Likewise, we know two ways to make an alkene: (i) reduction of an alkyne and (ii) elimination. We could potentially make the target molecule from the corresponding alkyne, but it is unclear what the next retrosynthetic step should be. (Note that in Chapter 13, we will learn reactions to make this approach possible.)

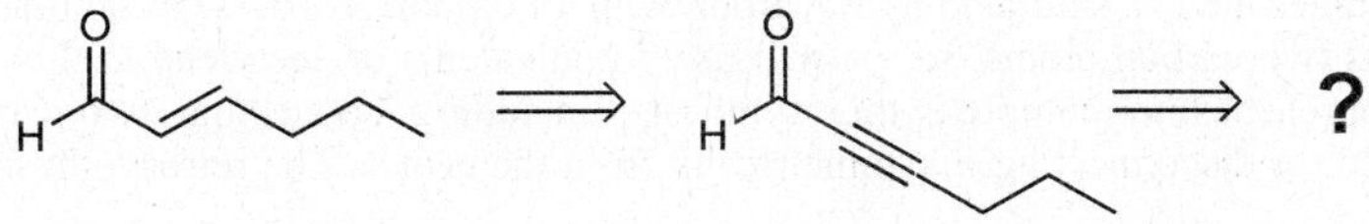

⟹ ⟹ **?**

Installing the alkene group by elimination turns out to be a viable approach in this case, as described below.

The figure below outlines a retrosynthetic analysis for our target molecule. An explanation of each of the steps (*a-f*) follows.

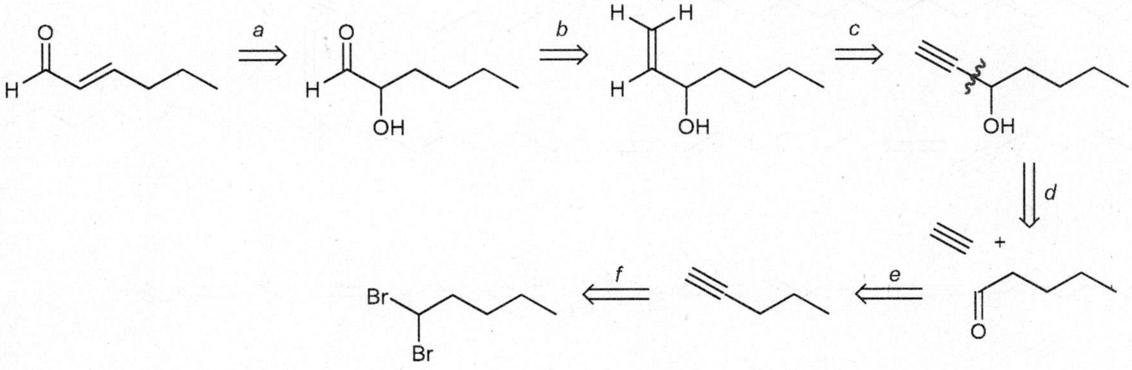

a ⟹ *b* ⟹ *c* ⟹ OH · *d* · *e* · *f* · Br Br

a. The target compound can be made via dehydration of the alcohol under acidic conditions.
b. The aldehyde can be made by ozonolysis of the corresponding alkene. Note that the required alkene has only one C=C double bond, so we avoid the problem described above with ozonolysis of a diene.
c. The alkene can be made by partial reduction of the corresponding alkyne.
d. A disconnection at the C–C bond adjacent to the triple bond leads to an acetylide ion nucleophile and the aldehyde shown as the electrophile (see problem 11.34).
e. The aldehyde is made from *anti*-Markovnikov addition of water to 1-pentyne.
f. 1-Pentyne is made from 1,1-dibromopentane (the given starting material) by double elimination.

Now let's draw the forward scheme. 1,1-Dibromopentane is converted to 1-pentyne by reaction with excess sodium amide (double E2 elimination), followed by treatment with water to protonate the terminal alkynide ion. 1-Pentyne is converted to the aldehyde via hydroboration-oxidation. Reaction of the aldehyde with sodium acetylide, followed by treatment with water to protonate the resulting alkoxide intermediate, forms a new C–C bond, and gives an alkynyl alcohol product. Reduction with H_2 and Lindlar's catalyst converts the alkyne group to an alkene group, and ozonolysis then converts the alkene to an aldehyde. Finally, treatment with concentrated acid and heat causes dehydration of the alcohol (E1 process), producing the target compound.

1) $NaNH_2$ (excess) 2) H_2O; 1) R_2BH 2) H_2O_2, NaOH; 1) HC≡CNa 2) H_2O; H_2, Lindlar's catalyst; 1) O_3 2) DMS; conc. H_2SO_4, heat

11.42.

The synthesis developed below is only one suggested synthetic pathway. There are likely other acceptable approaches that accomplish the same goal.

As seen in Table 4.1, tetradecane is a saturated hydrocarbon with 14 carbon atoms (with no branching), and our source of carbon (acetylene) has two carbon atoms, so we will use 7 equivalents of acetylene in this synthesis. There are a number of different approaches to complete this synthesis, including connecting two carbon atoms at a time sequentially from one end, or disconnecting it symmetrically from the center. The retrosynthesis below takes the latter of these two tactics.

The figure below outlines a retrosynthetic analysis for our target molecule. An explanation of each of the steps (*a-j*) follows.

a b c d e f g h i j Br + + Br Br + Br +

a. Tetradecane can be made via hydrogenation of 7-tetradecyne.
b. 7-Tetradecyne can be made by sequentially alkylating both sides of acetylene with 1-bromohexane.
c. 1-Bromohexane is made via an *anti*-Markovnikov addition of HBr across 1-hexene.
d. 1-Hexene is made by reduction of 1-hexyne using H_2 and Lindlar's catalyst.
e. 1-Hexyne can be produced from acetylene (after deprotonation to make a nucleophile) and 1-bromobutane.
f. 1-Bromobutane is made via an *anti*-Markovnikov addition of HBr across 1-butene.
g. 1-Butene is made by reduction of 1-butyne using H_2 and Lindlar's catalyst.
h. 1-Butyne is made from acetylene (after deprotonation) and 1-bromoethane.
i. 1-Bromoethane is made by addition of HBr across ethylene.
j. Ethylene is made by reduction of acetylene using H_2 and Lindlar's catalyst.

Now, let's draw out the forward scheme. Acetylene is reduced to ethylene using H_2 and Lindlar's catalyst. Addition of HBr, followed by S_N2 substitution with an acetylide nucleophile (made by deprotonation of acetylene with sodium amide) gives 1-butyne. Reduction of 1-butyne to 1-butene with H_2 and Lindlar's catalyst, followed by *anti*-Markovnikov addition of HBr in the presence of peroxide produces 1-bromobutane. A substitution reaction with sodium acetylide gives 1-hexyne. Another round of hydrogenation (H_2, Lindlar's catalyst), *anti*-Markovnikov addition (HBr, peroxide) and substitution (sodium acetylide) lengthens the chain by two more carbons, giving 1-octyne. Deprotonation of this terminal alkyne, followed by alkylation with a second equivalent of 1-bromohexane yields 7-tetradecyne. Finally, hydrogenation with H_2 and Pt produces the desired target molecule, tetradecane.

NaNH$_2$
H$_2$, Lindlar's catalyst
HBr
Br
NaC≡CH
H$_2$, Lindlar's catalyst
HBr, ROOR
Br
NaC≡CH
H$_2$, Lindlar's catalyst
HBr, ROOR
Br
NaC≡CH
1) NaNH$_2$
2) 1-bromohexane
H$_2$, Pt

11.43.
The synthesis developed below is only one suggested synthetic pathway. There are likely other acceptable approaches that accomplish the same goal.

Acetic acid and ethylene each have two carbon atoms, and our product has eight carbon atoms:

O
OH
+
O
O
2C reactant
2C reactant
8C product

A more detailed look at the product allows us to hypothesize where each of the two-carbon components will ultimately end up in the product (below). This is helpful in that it may allow us to determine which new bonds will be formed in the course of the reaction (*i.e.*, those connecting each of the 2C components).

2C
2C
2C
2C
O
O

By comparing the structures of the starting materials and product, we can also make an initial guess on the origins of each of the 2C components (shown below). It seems reasonable to assume that the ester will be derived from acetic acid (as both of these have a carbonyl flanked by a methyl group and an oxygen), and the other three 2C components will be derived from ethylene (with appropriate functional group modifications).

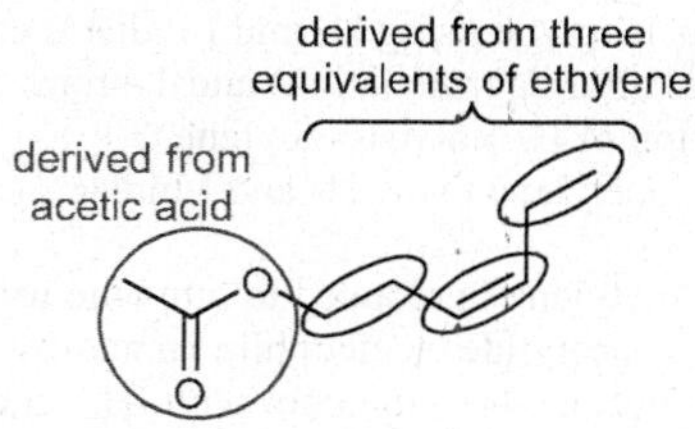

When considering which types of reactions to use, we will connect these pieces using a number of substitution reactions. Also, the only way we have learned to produce a *cis* alkene is via hydrogenation of an alkyne using H_2 and Lindlar's catalyst, so this will clearly be one of our steps.

The figure below outlines a retrosynthetic analysis for our target molecule. An explanation of each of the steps (*a-j*) follows.

a. Disconnection at the C–O bond leads to an S_N2 reaction between a carboxylate nucleophile (the conjugate base of a carboxylic acid) and an electrophilic substrate with an appropriate leaving group (*e.g.*, tosylate).
b. The carboxylate can be prepared from acetic acid (one of our given reactants) by treatment with a suitable base, such as NaOH.
c. The tosylate can be prepared from the corresponding alcohol.
d. The *cis* alkene can be produced from the corresponding alkyne (H_2 with Lindlar's catalyst).
e. This retrosynthetic step is the key disconnection that utilizes the reaction described in the problem statement. We can make this internal alkyne/alcohol by the reaction of an alkynide ion (formed by deprotonating 1-butyne) with an epoxide.
f. The epoxide is prepared from ethylene via epoxidation.
g. 1-Butyne is prepared by alkylating the conjugate base of acetylene with bromoethane.
h. Bromoethane is prepared via HBr addition to ethylene.
i. Acetylene is prepared via a double elimination from 1,2-dibromoethane.
j. 1,2-Dibromoethane is prepared via bromination of ethylene.

Now, let's draw out the forward scheme. This multistep synthesis uses three equivalents of ethylene (labeled **A**, **B**, **C** in the scheme shown) and one equivalent of acetic acid (labeled **D**). Ethylene (**A**) is converted to 1,2-dibromoethane upon treatment with bromine. Subsequent reaction with excess sodium amide produces an acetylide anion which is then treated with bromoethane [made from ethylene (**B**) and HBr] to produce 1-butyne. Deprotonation with sodium amide, followed by reaction with an epoxide [prepared by epoxidation of ethylene (**C**)] and treatment with water to protonate the alkoxide intermediate, produces a compound with an alkyne group and an alcohol group. Reduction of the alkyne to the *cis* alkene is accomplished with H_2 and Lindlar's catalyst, after which the alcohol is converted to a tosylate with tosyl chloride. Reaction with the conjugate base of acetic acid [produced by treating acetic acid (**D**) with NaOH] allows for an S_N2 reaction, thus yielding the desired product, *Z*-hexenyl acetate.

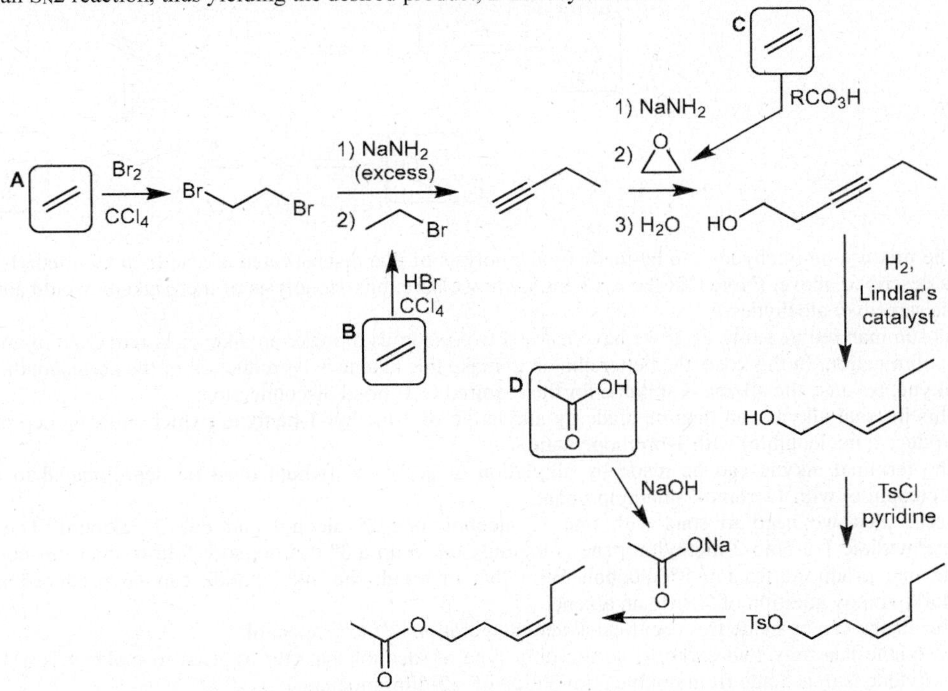

A, Br_2, CCl_4, Br, Br, 1) $NaNH_2$ (excess), 2) Br, HBr, CCl_4, B, 1) $NaNH_2$, 2) O, 3) H_2O, HO, C, RCO_3H, H_2, Lindlar's catalyst, HO, TsCl, pyridine, TsO, D, OH, O, NaOH, ONa, O, O, O

11.44.

The following synthesis is one suggested synthetic pathway. There are likely other acceptable approaches that accomplish the same goal.

As summarized in Table 11.1, we have learned two ways to make aldehydes: (i) *anti*-Markovnikov hydration of a terminal alkyne or (ii) ozonolysis of an alkene. However, in order to produce an equimolar mixture of both of these compounds from a single synthetic protocol (as described in the problem statement), ozonolysis of the following disubstituted alkene is the appropriate solution:

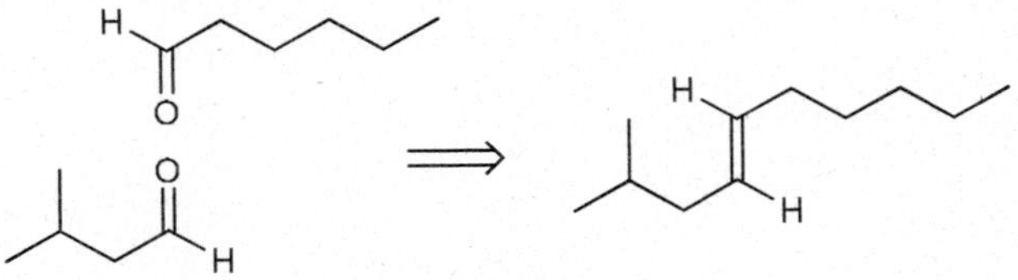

H, O, O, H, ⟹, H, H

The following figure outlines a retrosynthetic analysis for our target molecule. An explanation of each of the steps (*a*-*k*) follows.

a. The mixture of aldehydes can be made by ozonolysis of this disubstituted alkene (*trans*-2-methyl-4-decene), as described above. (Note that the *E* alkene is shown here, but ozonolysis of the *Z* alkene would also produce the same two aldehydes.)

b. As summarized in Table 11.1, we have learned two methods to make an alkene: 1) reduction of an alkyne or 2) elimination. In this case, the better choice to make this alkene is by reduction of the corresponding internal alkyne, because the alkyne is suitable for the required C–C bond disconnections.

c. This internal alkyne can thus be made by alkylation of 4-methyl-1-pentyne (which must be deprotonated to produce a nucleophile) with 1-bromopentane.

d. The terminal alkyne can be made by alkylation of acetylene (which must be deprotonated to produce a nucleophile) with 1-bromo-2-methylpropane.

e. Recall that we need to start with one 1° alcohol, one 2° alcohol and one 3° alcohol. The synthetic intermediate 1-bromo-2-methylpropane is the only one with a 3° carbon, so it follows that this compound is the one produced from a 3° alcohol. With this in mind, the alkyl halide can be produced from *anti*-Markovnikov addition of HBr to an alkene.

f. The alkene can be made from acid-catalyzed dehydration of the 3° alcohol.

g. Acetylene has only two carbons, so the only type of alcohol that can be used to make it is a 1° alcohol. Acetylene is thus made from double elimination of 1,2-dibromoethane.

h. 1,2-Dibromoethane is produced by bromination of ethylene.

i. Ethylene is produced by acid-catalyzed dehydration of ethanol, a 1° alcohol.

j. 1-Bromopentane is made by *anti*-Markovnikov addition of HBr to 1-pentene.

k. 1-Pentene is made from the 2° alcohol by tosylation followed by reaction with a bulky base to give the less substituted, Hofmann product.

Now let's draw the forward scheme. The 3° alcohol is converted to 2-methylpropene using strong, concentrated acid and heat (dehydration, an E1 process). *Anti*-Markovnikov addition of HBr (with peroxides) produces 1-bromo-2-methylpropane. Subsequent reaction with sodium acetylide (produced from the 1° alcohol by dehydration, bromination and double elimination, and deprotonation as shown) produces 4-methyl-1-pentyne. Deprotonation with sodium amide followed by reaction with 1-bromopentane (made from the 2° alcohol by tosylation, elimination and *anti*-Markovnikov addition of HBr) yields 2-methyl-4-decyne. Reduction using sodium metal in liquid ammonia produces the *E* alkene. Finally, ozonolysis followed by treatment with dimethyl sulfide produces an equimolar ratio of the two desired products, 3-methylbutanal and hexanal.

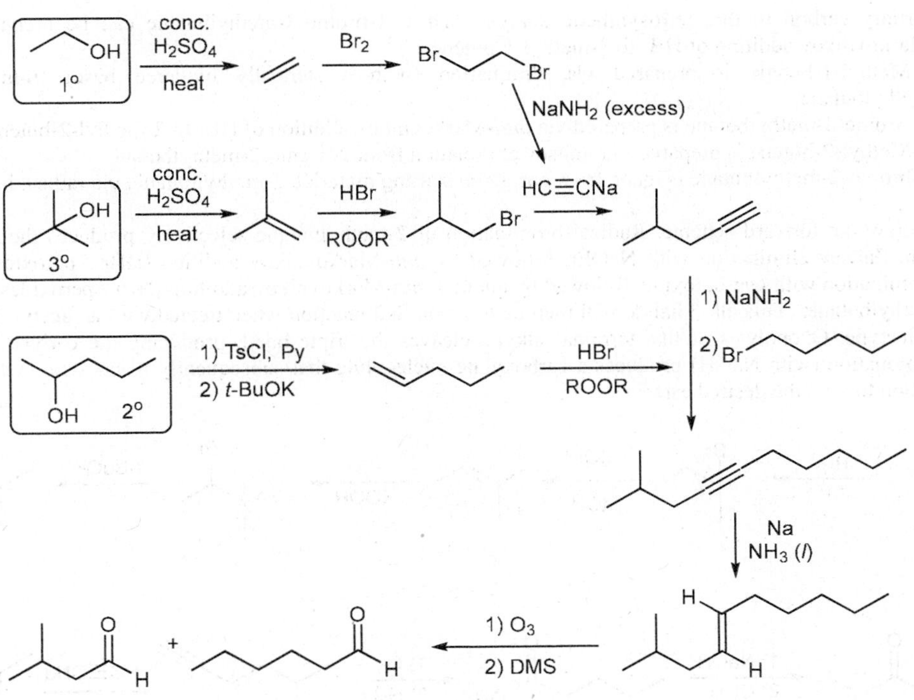

11.45.
The following synthesis is one suggested synthetic pathway. There are likely other acceptable approaches that accomplish the same goal.

An analysis of the structure of the product reveals that the five-carbon alkyl group (highlighted below) matches the skeletal structure of 2-methylbutane, the given starting material. This indicates which C–C bond must be made in the course of the synthesis.

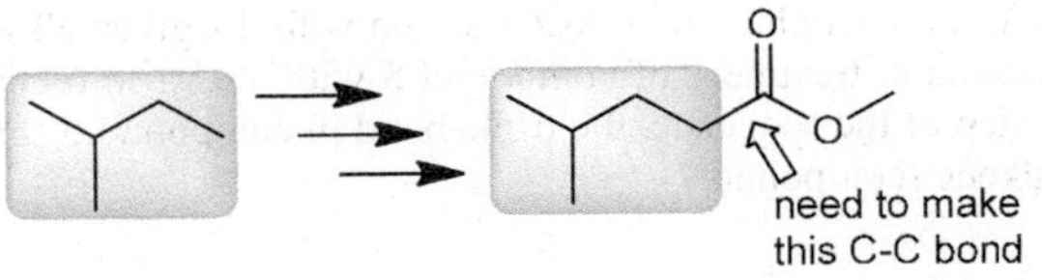

need to make this C-C bond

The figure below outlines a retrosynthetic analysis for our target compound. An explanation of each of the steps (*a-h*) follows.

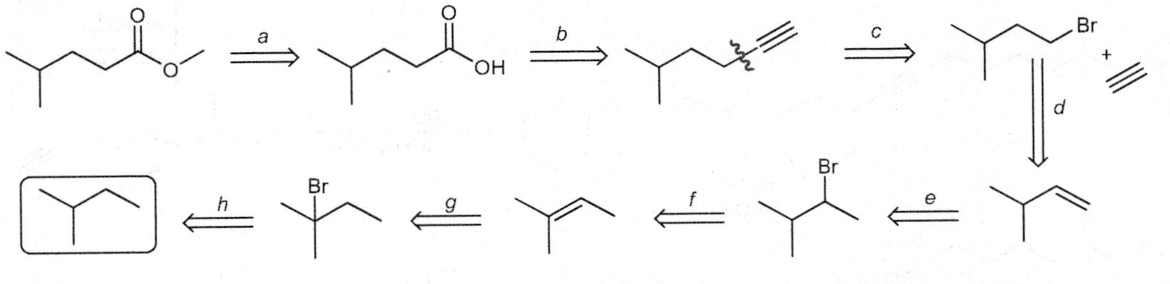

a. The only way we have learned to make an ester (so far) is via a reaction of a carboxylate nucleophile (the conjugate base of a carboxylic acid) and an alkyl halide (in this case, methyl iodide).

b. The only way we have learned to make a carboxylic acid (so far) is by ozonolysis of an alkyne.

c. The alkyne is prepared from the reaction of 1-bromo-3-methylbutane with an acetylide anion (formed by deprotonating acetylene). This alkyl halide has the same carbon skeleton as our given starting material (2-methylbutane), so our remaining steps primarily involve functional group manipulation.

d. Knowing that the first synthetic step must be radical halogenation of 2-methylbutane to produce the tertiary alkyl halide (the only useful reaction of alkanes), we need to migrate the functionality back toward the tertiary carbon in this retrosynthetic analysis. Thus, 1-bromo-3-methylbutane can be prepared via *anti*-Markovnikov addition of HBr to 3-methyl-1-butene.

e. 3-Methyl-1-butene is prepared via elimination (with a sterically hindered base) from 2-bromo-3-methylbutane.

f. 2-Bromo-3-methylbutane is prepared via *anti*-Markovnikov addition of HBr to 2-methyl-2-butene.

g. 2-Methyl-2-butene is prepared via Zaitsev elimination from 2-bromo-2-methylbutane.

h. 2-bromo-2-methylbutane is made from our given starting material, 2-methylbutane, via radical bromination.

Now, let's draw the forward scheme. Radical bromination of 2-methylbutane selectively produces the tertiary alkyl halide. Then, Zaitsev elimination with NaOEt, followed by *anti*-Markovnikov addition (HBr / peroxides), and then Hofmann elimination with *tert*-butoxide, followed by another *anti*-Markovnikov addition (HBr / peroxides) produces 1-bromo-3-methylbutane. This alkyl halide will then undergo an S_N2 reaction when treated with an acetylide ion to give 5-methyl-1-hexyne. Ozonolysis of this terminal alkyne cleaves the triple bond, producing the carboxylic acid (and CO_2). Deprotonation (with NaOH) produces a carboxylate nucleophile that subsequently reacts with methyl iodide in an S_N2 reaction to give the desired ester.

Br; Br_2; $h\nu$; NaOEt; EtOH; HBr; ROOR; Br; *t*-BuOK; HBr; ROOR; Br; HC≡CNa; 1) O_3; 2) H_2O; O; OH; 1) NaOH; 2) CH_3I; O; O

11.46. Hydroboration-oxidation of compound **1** affords a chiral organoborane (**2**). Compound **3** is an alkyl halide, which is expected to react with an acetylide nucleophile in an S_N2 reaction (as supported by the use of DMSO as the solvent), to give compound **4**. Upon treatment with the strong base BuLi, compound **4** is deprotonated to give an alkynide ion, which then serves as a nucleophile in an S_N2 reaction with the given alkyl iodide to produce compound **5**. As described in the problem statement, treatment of compound **5** with TsOH in methanol removes the THP group to give an alcohol (**6**). In the final step of the sequence, the triple bond in compound **6** is reduced by hydrogenation with a poisoned catalyst to give a *cis* alkene (compound **7**).

1; $BH_3 \cdot THF$; **2**; B; R; R; 1) Br_2; 2) NaOMe; **3**; Br; HC≡CLi; DMSO; **4**; 1) BuLi; 2) $I(CH_2)_7OTHP$; **5**; O; O; TsOH; MeOH; **6**; OH; H_2; Lindlar's catalyst; **7**; OH

Chapter 12
Alcohols and Phenols

Review of Concepts

Fill in the blanks below. To verify that your answers are correct, look in your textbook at the end of Chapter 12. Each of the sentences below appears verbatim in the section entitled *Review of Concepts and Vocabulary*.

- When naming an alcohol, the parent is the longest chain containing the __________ group.
- The conjugate base of an alcohol is called an ____________ ion.
- Several factors determine the relative acidity of alcohols, including ____________, ____________, and ______________________.
- The conjugate base of phenol is called a ____________, or _____________ ion.
- When preparing an alcohol via a substitution reaction, primary substrates will require S_N__ conditions, while tertiary substrates will require S_N__ conditions.
- Alcohols can be formed by treating a __________ **group** (C=O bond) with a ___________ **agent.**
- **Grignard reagents** are carbon nucleophiles that are capable of attacking a wide range of _______________, including the carbonyl group of ketones or aldehydes, to produce an alcohol.
- _____________ **groups**, such as the trimethylsilyl group, can be used to circumvent the problem of Grignard incompatibility and can be easily removed after the desired Grignard reaction has been performed.
- Tertiary alcohols will undergo an S_N__ reaction when treated with a hydrogen halide.
- Primary and secondary alcohols can undergo an S_N__ process when treated with either HX, $SOCl_2$, PBr_3, or when the hydroxyl group is converted into a tosylate group followed by nucleophilic attack.
- Tertiary alcohols undergo E1 elimination when treated with __________.
- Primary alcohols can undergo **oxidation** twice to give a ____________________.
- Secondary alcohols are oxidized only once to give a ___________
- PCC can be used to convert a primary alcohol into an ____________. Alternatively, primary alcohols can be converted into ________ with a __________ oxidation or a DMP-based oxidation.
- NADH is a biological reducing agent that functions as a ___________ delivery agent (very much like $NaBH_4$ or $LiAlH_4$), while NAD^+ is an ____________ agent.
- There are two key issues to consider when proposing a synthesis:
  1. A change in the _________________.
  2. A change in the ___________.

Review of Skills

Fill in the blanks and empty boxes below. To verify that your answers are correct, look in your textbook at the end of Chapter 12. The answers appear in the section entitled *SkillBuilder Review*.

12.1 Naming an Alcohol

Provide a systematic name for the following compound:

Cl Cl

OH

1) Identify the parent
2) Identify and name substituents
3) Assign locants to each substituent
4) Alphabetize
5) Assign Configuration

12.2 Comparing the Acidity of Alcohols

For each pair of compounds below, circle the compound that is more acidic:

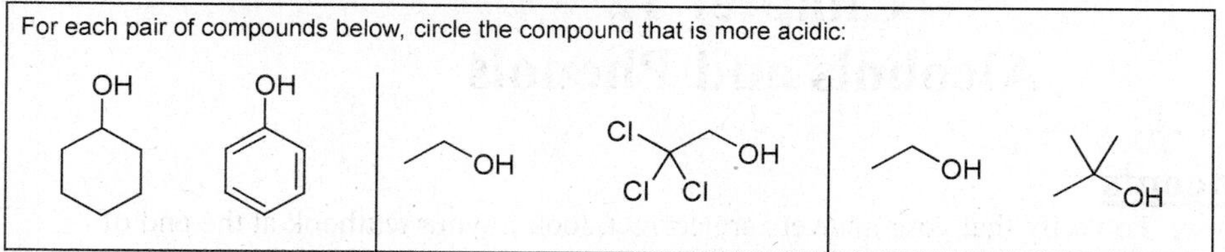

12.3 Identifying Oxidation and Reduction Reactions

In the following reaction, determine whether the starting material has been oxidized, reduced, or neither:

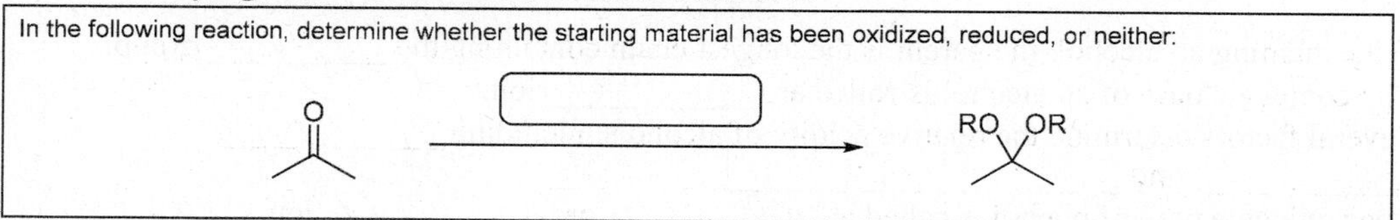

12.4 Drawing a Mechanism, and Predicting the Products of Hydride Reductions

Complete the mechanism below by drawing all curved arrows, intermediates and products.

H
H—Al—H
H

12.5 Preparing an Alcohol via a Grignard Reaction

Identify reagents that can achieve each of the following transformations:

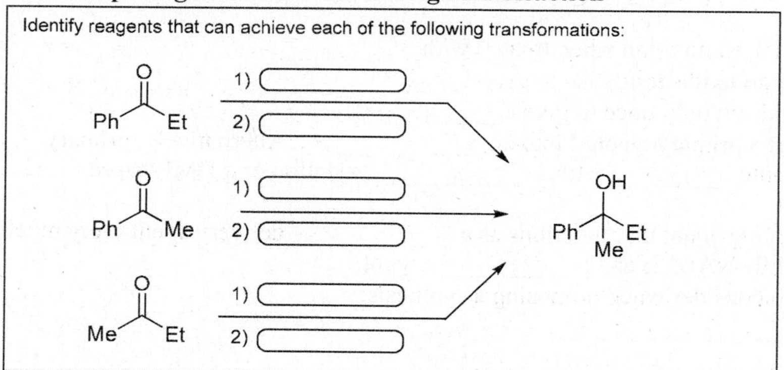

12.6 Proposing Reagents for the Conversion of an Alcohol into an Alkyl Halide

Identify reagents that can achieve each of the following transformations:

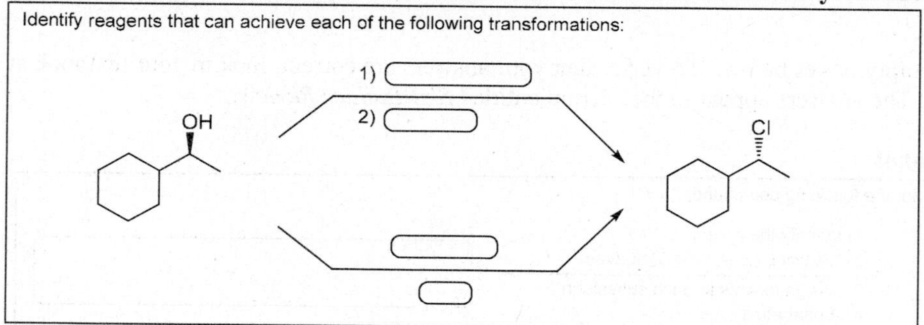

12.7 Predicting the Products of an Oxidation Reaction

Draw the expected product of the following reaction:

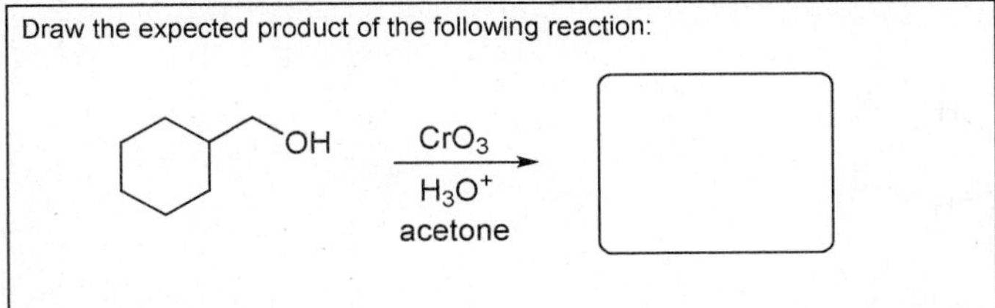

12.8 Converting Functional Groups

Identify reagents that can achieve each of the following functional group transformations:

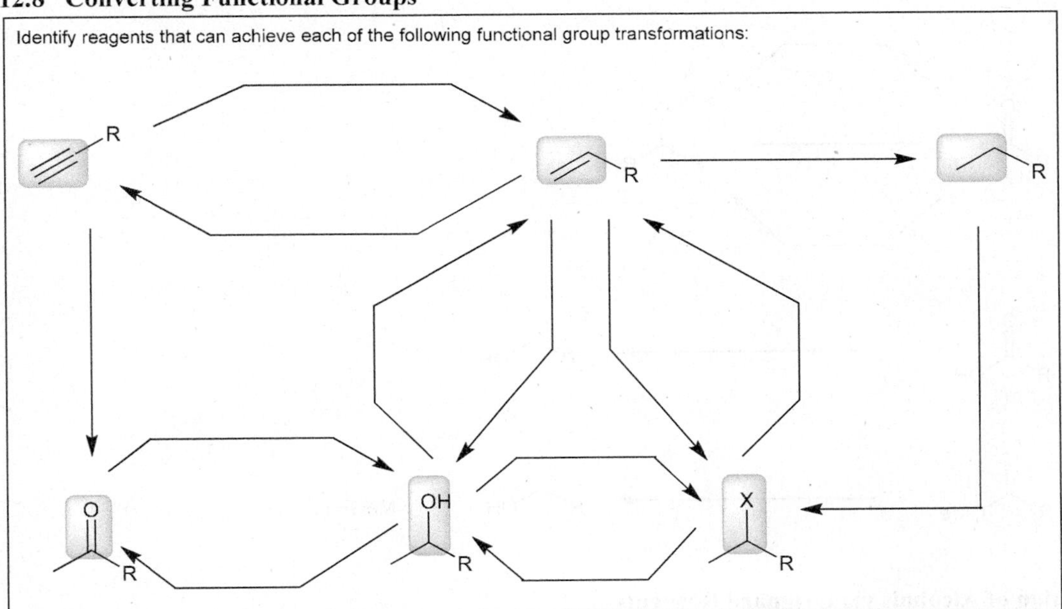

12.9 Proposing a Synthesis

Fill in the blanks:

As a guide for proposing a synthesis, ask the following two questions:

1) Is there a change in the ________________ skeleton?

2) Is there a change in the location or identity of the ____________________?

After proposing a synthesis, use the following two questions to analyze your answer:

1) Is the ____________________ outcome of each step correct?

2) Is the ____________________ outcome of each step correct?

Review of Reactions

Identify the reagents necessary to achieve each of the following transformations. To verify that your answers are correct, look in your textbook at the end of Chapter 13. The answers appear in the section entitled *Review of Reactions*.

Preparation of Alkoxides

$$ROH \longrightarrow RO^{\ominus} \quad Na^{\oplus}$$

Preparation of Alcohols via Reduction

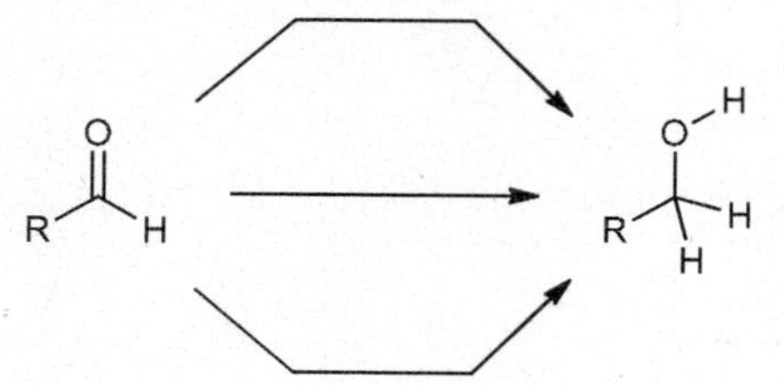

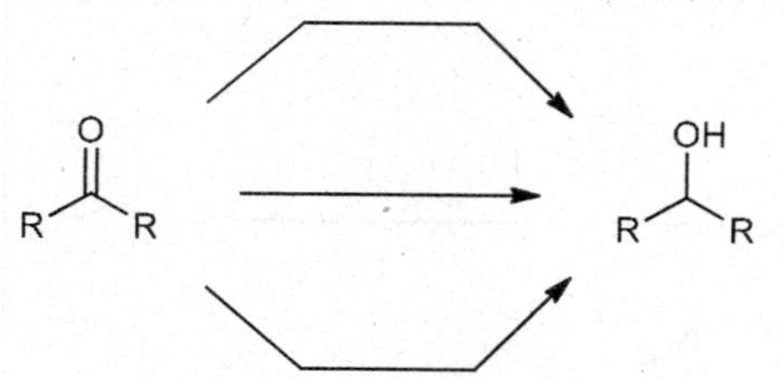

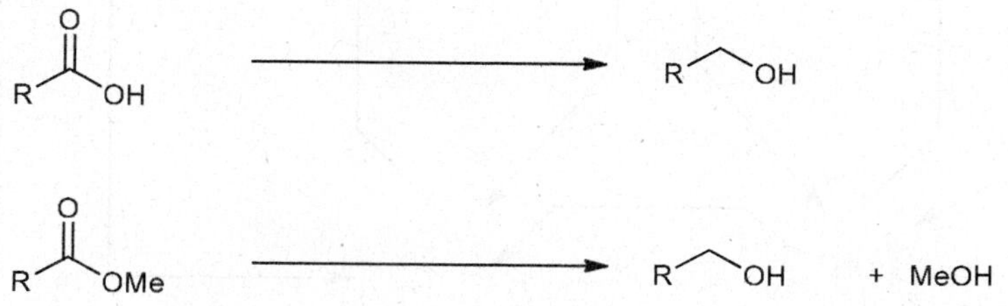

R–CHO → R–CH₂–OH

R–CO–R → R–CH(OH)–R

R–COOH → RCH_2OH

R–COOMe → RCH_2OH + MeOH

Preparation of Alcohols via Grignard Reagents

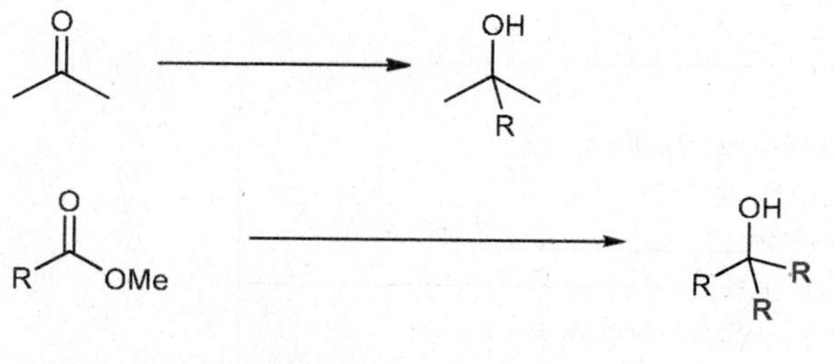

Protection and Deprotection of Alcohols

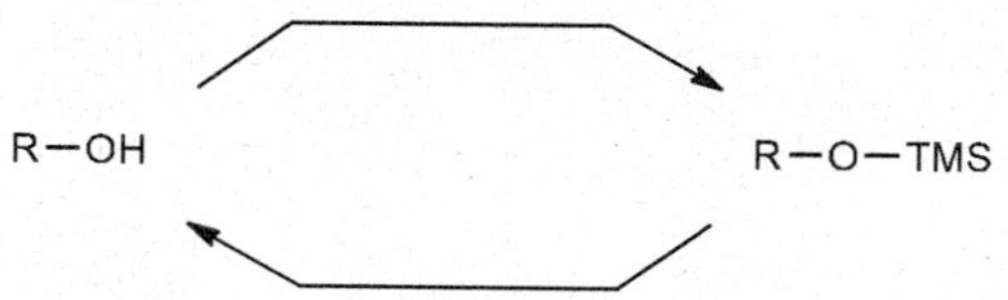

R–OH ⇌ R–O–TMS

SN1 Reactions with Alcohols

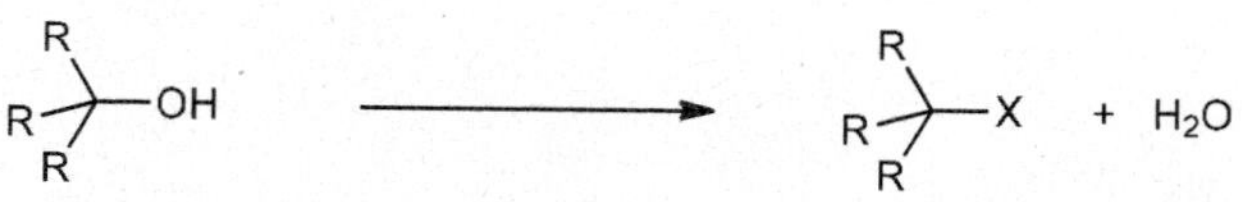

R_3C–OH → R_3C–X + H_2O

S$_N$2 Reactions with Alcohols

OH Br

OH Cl

E1 and E2 Reactions with Alcohols

OH + H_2O

OH OTs

Oxidation of Alcohols and Phenols

OH R R O R R

OH R O R OH

OH R O R H

OH O O

Review of Mechanisms

Complete each of the following mechanisms by drawing the missing curved arrows. To verify that your curved arrows are drawn correctly, compare them to the curved arrows in the mechanism boxes for Mechanisms 12.1 – 12.9, which can be found throughout Chapter 12 of your text.

Mechanism 12.1 Reduction of a Ketone or Aldehyde with $NaBH_4$

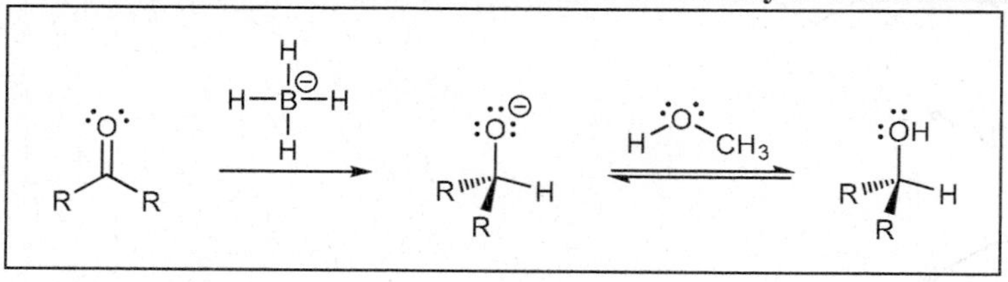

Mechanism 12.2 Reduction of a Ketone or Aldehyde with $LiAlH_4$

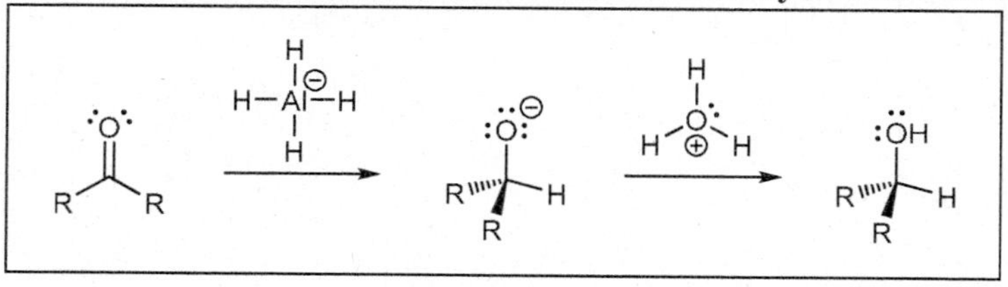

Mechanism 12.3 Reduction of an Ester with $LiAlH_4$

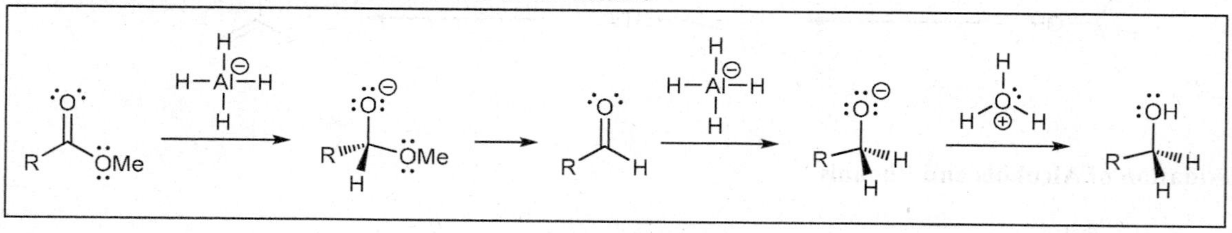

Mechanism 12.4 The Reaction Between a Grignard Reagent and a Ketone or Aldehyde

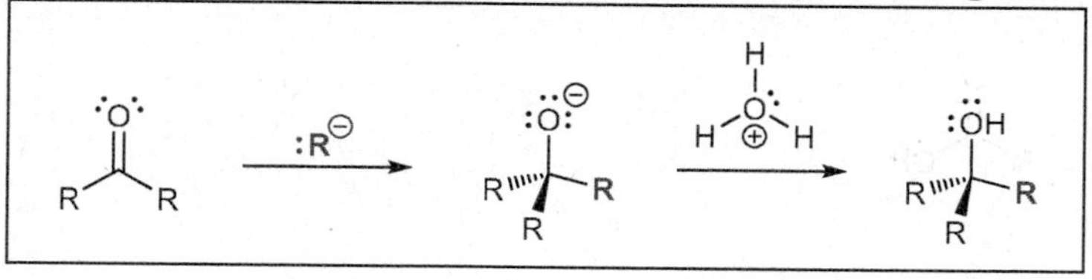

Mechanism 12.5 The Reaction Between a Grignard Reagent and an Ester

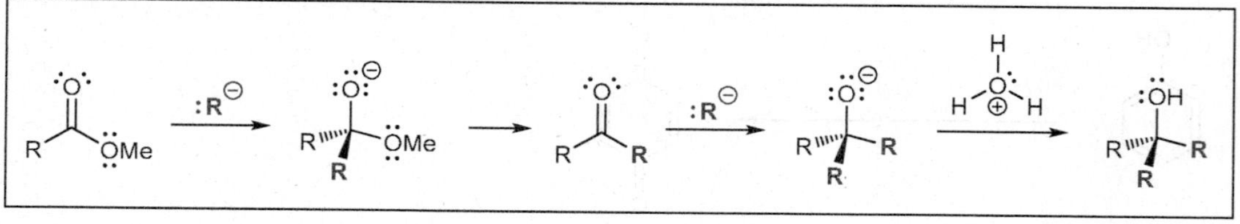

Mechanism 12.6 The Reaction Between $SOCl_2$ and an Alcohol

$SOCl_2$

py

Cl

Cl—S

O

H

Cl

Cl

H

Cl

:Cl:

Cl

O

O

O

:Cl: + SO_2 + :Cl:

Mechanism 12.7 The Reaction Between PBr_3 and an Alcohol

Br

P—Br

Br

H

Br + :Br: → Br: + $HOPBr_2$

Br

Mechanism 12.8 Oxidation of an Alcohol with Chromic Acid

Stage 1

R

H—OH + HO—Cr—OH ⇌ H—O—Cr—OH + H_2O

R

Stage 2

R

H—O—Cr—OH → E2 → R=O: + H_3O^+ + $HCrO_3^-$

R

Mechanism 12.9 Swern Oxidation

Stage 1

Cl O

H_3C

S—O:

H_3C

Cl O

Cl

+ :Cl:

CH_3

Cl—S⊕ + CO

CH_3 + CO_2

Stage 2

H_3C

S—Cl

H_3C

:OH

R—C—H

R

H_3C CH_3

O—H

Et_3N:

Et_3N:

O

R—C—R + CH_3—S—CH_3

Common Mistakes to Avoid

If you are proposing a synthesis that involves reduction of a ketone or aldehyde with $LiAlH_4$, make sure that the workup (with either H_3O^+ or H_2O) is shown as a separate step:

CORRECT

1) $LiAlH_4$ 2) H_3O^+ OH

INCORRECT

O $LiAlH_4$, H_3O^+ OH

(Don't make this mistake)

This is important because $LiAlH_4$ is incompatible with the proton source. In contrast, $NaBH_4$ is used in the presence of a proton source:

CORRECT

O $NaBH_4$ CH_3OH OH

So, when using $NaBH_4$ as a reducing agent, do not show the proton source (MeOH) as a separate step.

Also, when drawing a mechanism for the reduction of a ketone, aldehyde, or ester, make sure that the first curved arrow is placed on the Al–H bond, rather than on the negative charge:

CORRECT INCORRECT

In previous chapters, we have seen that it is generally acceptable to place the tail of a curved arrow on a negative charge, but this is an exceptional case. The negative charge in this case is not associated with a lone pair, so the tail of the curved arrow cannot be placed on the negative charge. It must be placed on the bond. This is true for reductions involving $NaBH_4$ as well:

CORRECT INCORRECT

Useful reagents

The following is a list of reagents that were new in this chapter:

Reagents	Function
NaH	A very strong base, used to deprotonate an alcohol to give an alkoxide ion.
Na	Will react with an alcohol to liberate hydrogen gas, giving an alkoxide ion.
$NaBH_4$, MeOH	A reducing agent (source of nucleophilic hydride). Can be used to reduce ketones or aldehydes to alcohols. Will not reduce esters or carboxylic acids.

1) $LiAlH_4$ 2) H_3O^+	A strong reducing agent (source of nucleophilic hydride). Can be used to reduce ketones, aldehydes, esters, or carboxylic acids to give an alcohol.
H_2, Pt	Reducing agent. Generally used to reduce alkenes and/or alkynes to alkanes, but in some cases, it can also be used to reduce ketones and/or aldehydes to alcohols.
Mg	Can be used to convert an organohalide (RX, where R = alkyl, aryl or vinyl group) into a Grignard reagent (RMgX).
RMgX	A Grignard reagent. Examples include MeMgBr, EtMgBr and PhMgBr. These reagents are very strong nucleophiles (and very strong bases as well), and they will react with aldehydes or ketones. Aldehydes are converted into secondary alcohols (except for formaldehyde which is converted to a primary alcohol), while ketones are converted to tertiary alcohols. Esters are converted to tertiary alcohols when treated with excess Grignard.
TMSCl, Et_3N	Trimethylsilyl chloride [$(CH_3)_3SiCl$], in the presence of a base (such as triethylamine), will protect an alcohol (ROH is converted to ROTMS).
TBAF	Tetrabutylammonium fluoride. Used for deprotection of alcohols with silyl protecting groups (ROTMS is converted to ROH).
HX	HBr and HCl are strong acids that also provide a source of a strong nucleophile. Can be used to convert an alcohol into an alkyl bromide or alkyl chloride.
TsCl, pyridine	Tosyl chloride (TsCl) will convert an alcohol into a tosylate. This is important because it converts a bad leaving group (HO^-) into a good leaving group (TsO^-).
PBr_3	Phosphorus tribromide (PBr_3) can be used to convert a primary or secondary alcohol into an alkyl bromide. If the OH group is connected to a chiral center, we expect inversion of configuration (typical for an S_N2 process).
$SOCl_2$, pyridine	Thionyl chloride ($SOCl_2$) can be used to convert a primary or secondary alcohol into an alkyl chloride. If the OH group is connected to a chiral center, reaction conditions can be selected for inversion of configuration (typical for an S_N2 process).
HCl, $ZnCl_2$	Can be used to convert an alcohol into an alkyl chloride.
$Na_2Cr_2O_7$, H_2SO_4, H_2O	A mixture of sodium dichromate ($Na_2Cr_2O_7$) and sulfuric acid (H_2SO_4) gives chromic acid, which is a strong oxidizing agent. Primary alcohols are oxidized to give carboxylic acids, while secondary alcohols are oxidized to give ketones. Tertiary alcohols are generally unreactive.
PCC, CH_2Cl_2	Pyridinium chlorochromate (PCC) is a mild oxidizing agent that will oxidize a primary alcohol to give an aldehyde, rather than a carboxylic acid. Secondary alcohols are oxidized to give ketones. Methylene chloride (CH_2Cl_2) is a solvent.
DMP, CH_2Cl_2	DMP (Des-Martin Periodinane) is a mild oxidizing agent that will oxidize a primary alcohol to give an aldehyde, rather than a carboxylic acid. Secondary alcohols are oxidized to give ketones.
1) DMSO, $(COCl)_2$, 2) Et_3N	Dimethyl sulfoxide (DMSO), oxalyl chloride [$(COCl)_2$], and triethyl amine (Et_3N) are the reagents for a Swern oxidation, which will oxidize a primary alcohol to give an aldehyde, rather than a carboxylic acid. Secondary alcohols are oxidized to give ketones.

Solutions

12.1.

(a) We begin by identifying the parent. The carbon atom connected to the OH group must be included in the parent. The longest chain that includes this carbon atom is six carbon atoms in length, so the parent is hexanol. There are three substituents (highlighted). Notice that the parent chain is numbered starting from the side that is closest to the OH group (the OH group is at C2 rather than C5). According to this numbering scheme, the methyl group is located at C2, and the bromine atoms are both at C5. Finally, we assemble the substituents alphabetically. The compound does not contain any chiral centers.

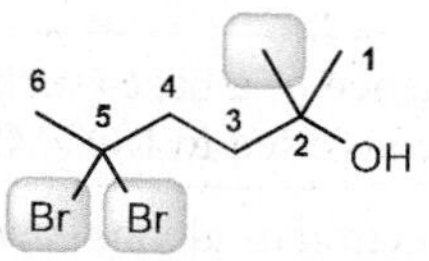

5,5-dibromo-2-methylhexan-2-ol
or
5,5-dibromo-2-methyl-2-hexanol

(b) We begin by identifying the parent. The carbon atom connected to the OH group must be included in the parent. The longest chain that includes this carbon atom is five carbon atoms in length, so the parent is pentanol. There are three substituents (highlighted), all of which are methyl groups. Notice that the parent chain is numbered starting from the side that is closest to the OH group (thereby placing the OH group at C1). According to this numbering scheme, the methyl groups are at C2, C3 and C4. We use the prefix "tri" to indicate three methyl groups. Finally, we assign a configuration to each chiral center.

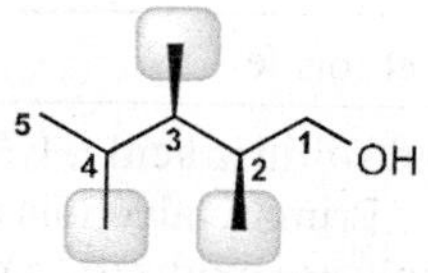

(2*S*,3*R*)-2,3,4-trimethylpentan-1-ol
or
(2*S*,3*R*)-2,3,4-trimethyl-1-pentanol

(c) We begin by identifying the parent (phenol). There are two substituents (highlighted), both of which are ethyl groups. Notice that the ring is numbered starting from the carbon atom bearing the OH group (it is not necessary to indicate a locant for the OH group, because in a ring, it is assumed to be at C1, by definition). According to this numbering scheme, the ethyl groups are at C2 and C6. We use the prefix "di" to indicate two ethyl groups. The compound does not contain any chiral centers.

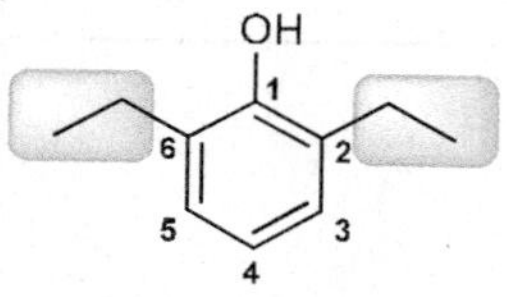

2,6-diethylphenol

(d) We begin by identifying the parent. The carbon atom connected to the OH group must be included in the parent. That carbon atom is part of a six-membered ring, so the parent is cyclohexanol. There are four substituents (highlighted), all of which are methyl groups. Notice that the parent chain is numbered starting from the carbon atom bearing the OH group (it is not necessary to indicate a locant for the OH group, because in a ring, it is assumed to be at C1, by definition). The numbers go counterclockwise, so as to give the lowest number to the first substituents (C2). According to this numbering scheme, the methyl groups are at C2, C2, C4 and C4. We use the prefix "tetra" to indicate four methyl groups. Finally, we assign a configuration to the chiral center.

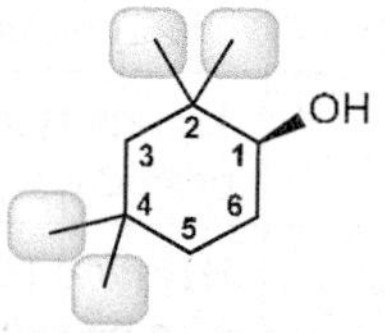

(*S*)-2,2,4,4-tetramethylcyclohexanol

12.2. The longest carbon chain contains 16 carbon atoms, and the chain is numbered to give the hydroxyl group the lowest number possible. Without the hydroxyl group, the alkene parent would be 2-hexadecene or hexadec-2-ene. To indicate the presence of the hydroxyl group, we name the compound as an alcohol, by dropping the "e" and adding the suffix "ol."

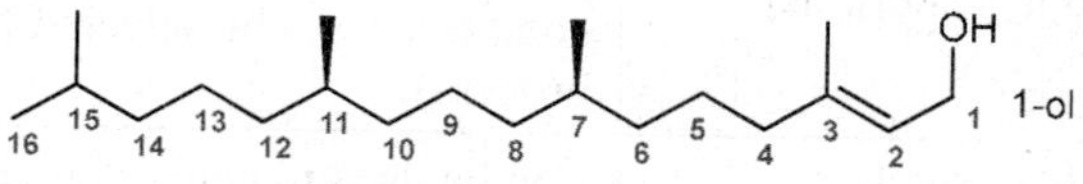

Hexadec-2-en-1-ol

Then identify the substituents and assign their locations.

3,7,11,15-tetramethyl

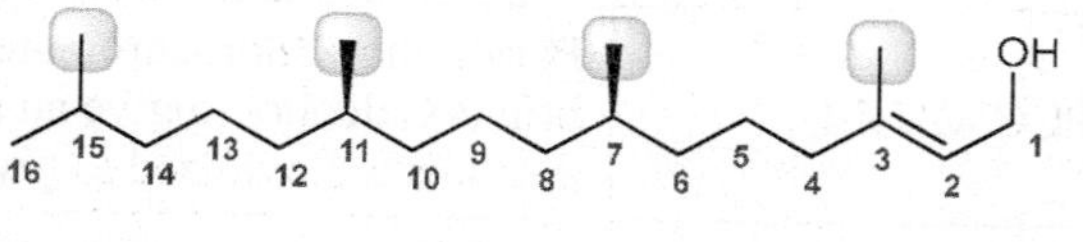

Finally, assign the configuration of each chiral center, as well as the configuration of the alkene unit:

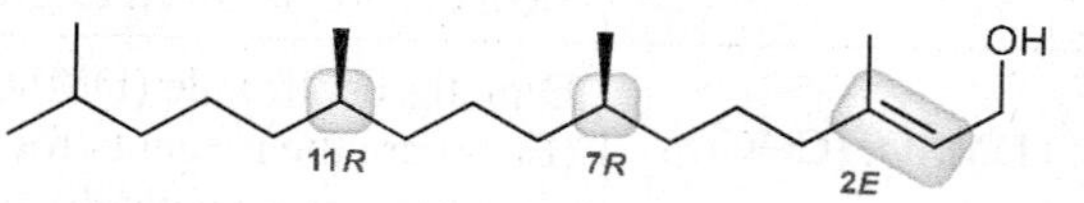

Putting it all together, the complete IUPAC name for phytol is:

(2*E*,7*R*,11*R*)-3,7,11,15-tetramethylhexadec-2-en-1-ol.

12.3. Nonyl mandelate has a longer alkyl chain than octyl mandelate and is therefore more effective at penetrating cell membranes, rendering it a more potent agent. Nonyl mandelate has a shorter alkyl chain than decyl mandelate and is therefore more water-soluble, enabling it to be transported through aqueous media and to reach its target destination more effectively.

12.4.

(a) When an alcohol is treated with elemental sodium (Na), the OH group is deprotonated, giving the corresponding alkoxide ion.

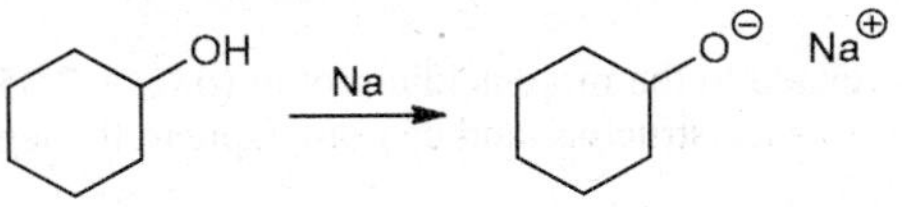

(b) When an alcohol is treated with sodium hydride (NaH), the OH group is deprotonated, giving the corresponding alkoxide ion.

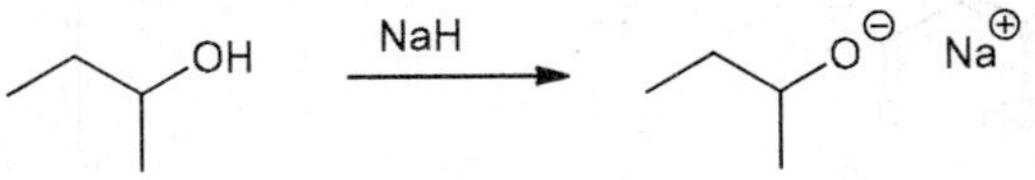

(c) When an alcohol is treated with elemental lithium (Li), the OH group is deprotonated, giving the corresponding alkoxide ion.

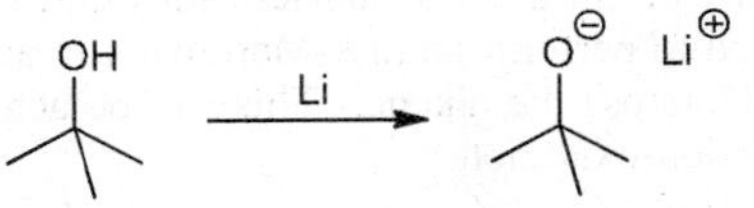

(d) When an alcohol is treated with sodium hydride (NaH), the OH group is deprotonated, giving the corresponding alkoxide ion.

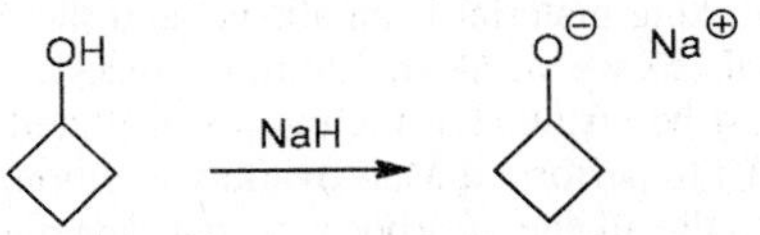

12.5.

(a) Upon treatment with aqueous acid, an alkoxide ion is protonated to give the corresponding alcohol:

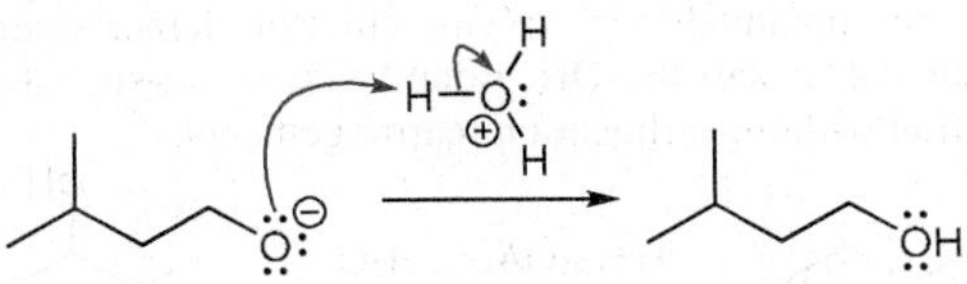

(b) Upon treatment with aqueous acid, an alkoxide ion is protonated to give the corresponding alcohol:

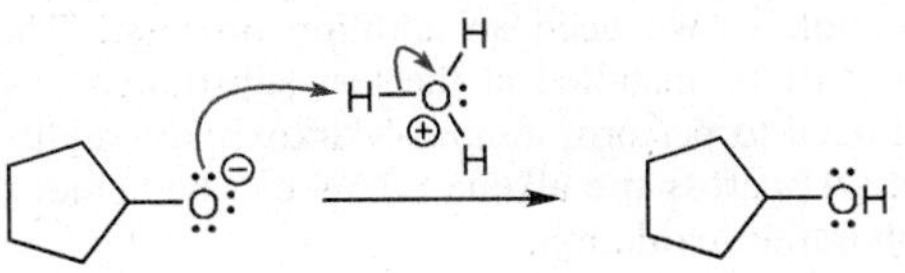

(c) Upon treatment with aqueous acid, an alkoxide ion is protonated to give the corresponding alcohol:

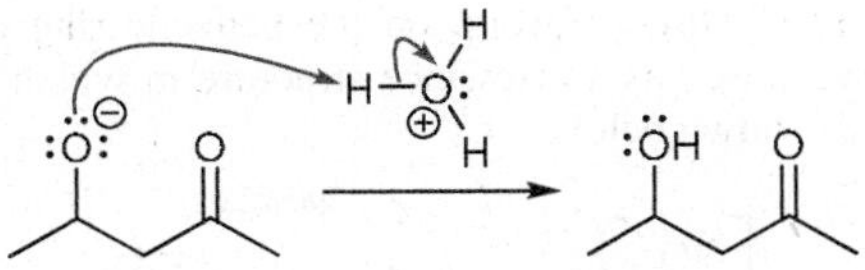

12.6.

(a) The first compound is more acidic because the conjugate base of a primary alcohol will be more easily solvated than the conjugate base of a tertiary alcohol. The better solvated, more stable conjugate base is the weaker conjugate base, and is related to the stronger parent acid:

(b) The first compound is more acidic because the electron-withdrawing effects of the chlorine atoms stabilize the conjugate base. The more stable conjugate base is the weaker conjugate base, and is related to the stronger parent acid:

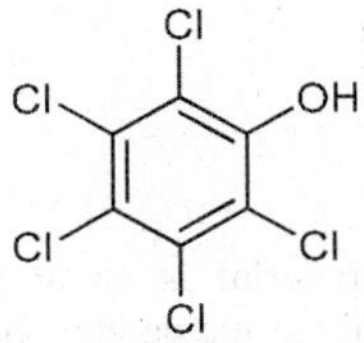

(c) The second compound is more acidic because its conjugate base is more stabilized by resonance, with the negative charge spread over two oxygen atoms, rather than just one oxygen atom. The more stable conjugate base is the weaker conjugate base, and is related to the stronger parent acid:

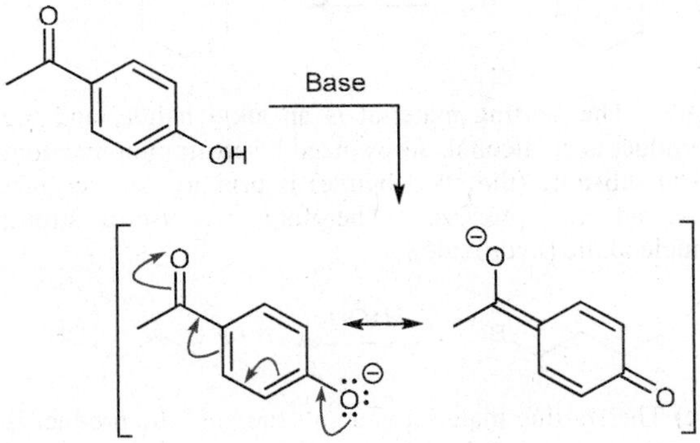

(d) The second compound (phenol) is more acidic because its conjugate base is stabilized by resonance. In contrast, the conjugate base of the first compound (*t*-BuOH) is not resonance-stabilized. The more stable conjugate base is the weaker conjugate base, and is related to the stronger parent acid:

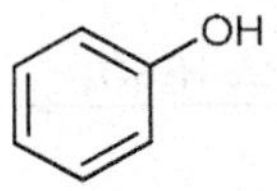

12.7. The hydroxyl group on the benzene ring on the left is expected to be more acidic (lower pK_a) because its conjugate base has a resonance structure in which the negative charge is spread onto an oxygen atom of the carbonyl group, as shown below:

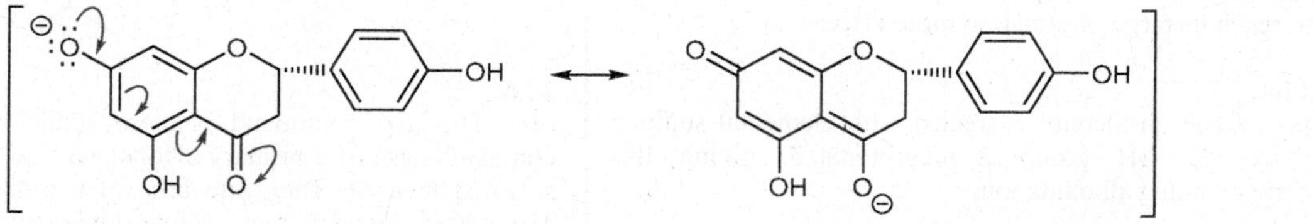

The more stable conjugate base is the weaker conjugate base, and is related to the more acidic proton (pK_{a1} = 7.05). In contrast, the hydroxyl corresponding to pK_{a2} does not have such a resonance structure and can only spread its negative charge onto carbon atoms.

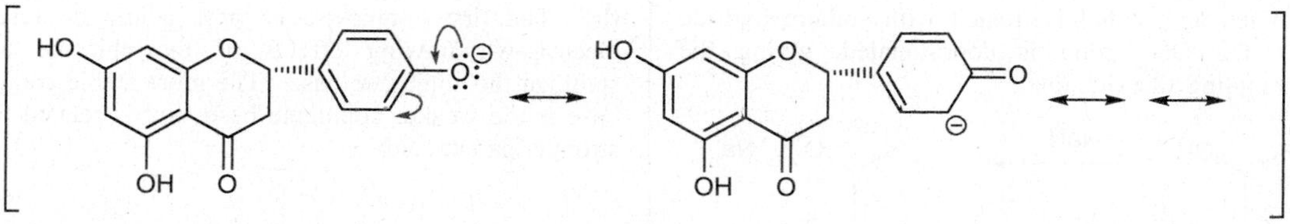

12.8.

(a) The starting material is an alkyl halide, and the product is an alcohol, so we need a substitution reaction. The substrate (the alkyl halide) is tertiary, so we must use an SN1 process. That is, we must use a weak nucleophile (water) rather than a strong nucleophile (hydroxide), because hydroxide is a strong base that would favor E2 over SN2.

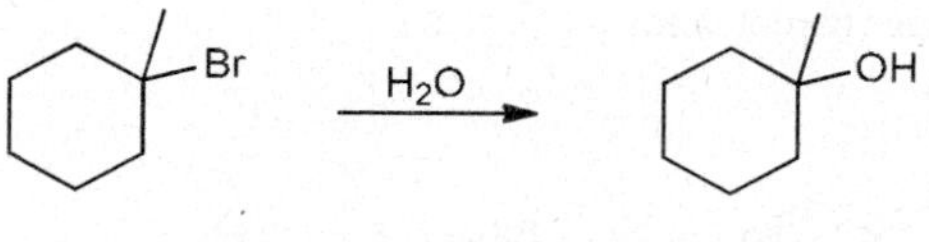

H_2O

(b) The starting material is an alkyl halide, and the product is an alcohol, so we need a substitution reaction. The substrate (the alkyl halide) is primary, so we must use an SN2 process. Therefore, we use a strong nucleophile (hydroxide).

NaOH

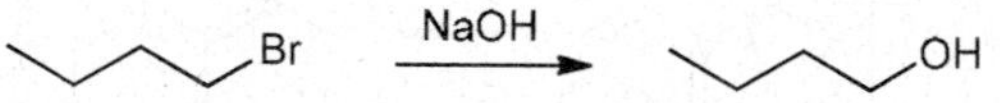

(c) The starting material is an alkene, and the product is an alcohol, so we need an addition process. The OH group must be installed at the more substituted position, so we need to perform a Markovnikov addition of H and OH across the alkene. Carbocation rearrangements are not a concern in this case (protonation of the alkene generates a tertiary carbocation which cannot rearrange), so acid-catalyzed hydration will give the desired product.

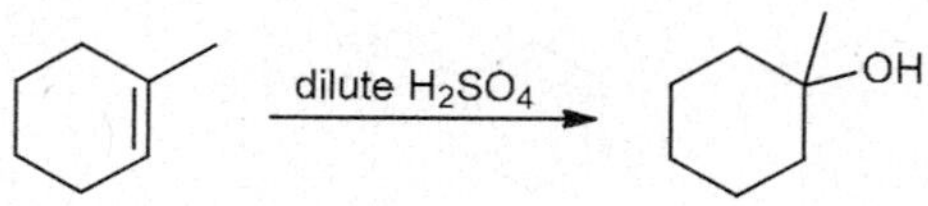

dilute H_2SO_4

(d) The starting material is an alkene, and the product is an alcohol, so we need an addition process. The OH group must be installed at the less-substituted position, so we need to perform an *anti*-Markovnikov addition of H and OH across the alkene. This can be achieved via hydroboration-oxidation.

1) $BH_3 \cdot THF$
2) H_2O_2, NaOH

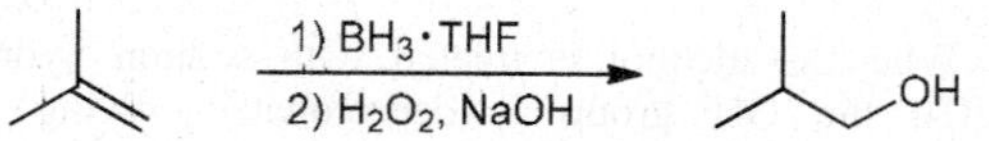

(e) The starting material is an alkene, and the product is an alcohol, so we need an addition process. The OH group must be installed at the more substituted position, so we need to perform a Markovnikov addition of H and OH across the alkene. Carbocation rearrangements are a concern in this case (protonation of the alkene generates a secondary carbocation which can rearrange to give a more stable, tertiary carbocation), so acid-catalyzed hydration cannot be used. Instead, the desired product can be obtained via oxymercuration-demercuration, which will install the OH group at the more-substituted position without carbocation rearrangements.

1) $Hg(OAc)_2$, H_2O
2) $NaBH_4$

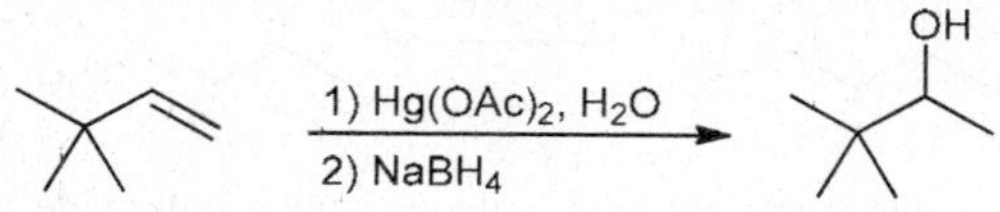

(f) The starting material is an alkene, and the product is an alcohol, so we need an addition process. The OH group must be installed at the less-substituted position, so we need to perform an *anti*-Markovnikov addition of H and OH across the alkene. This can be achieved via hydroboration-oxidation.

1) $BH_3 \cdot THF$
2) H_2O_2, NaOH

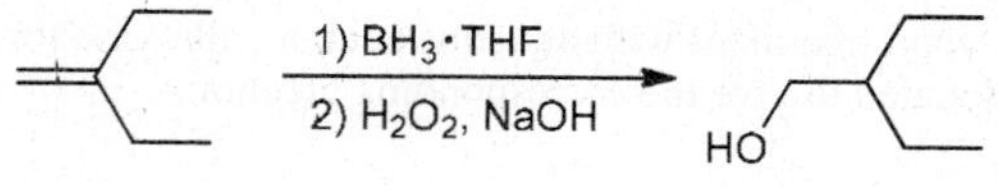

12.9.

(a) Let's begin by drawing the starting material and identifying the position that would lead to a primary alcohol.

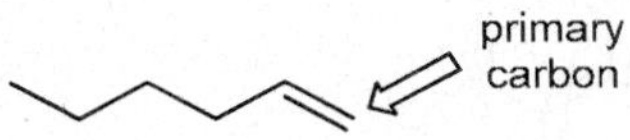

primary carbon

1-hexene

Addition of H and OH across this alkene will provide an alcohol. Markovnikov addition will give a secondary alcohol, so we must perform an *anti*-Markovnikov addition in order to obtain a primary alcohol. This can be achieved via hydroboration-oxidation.

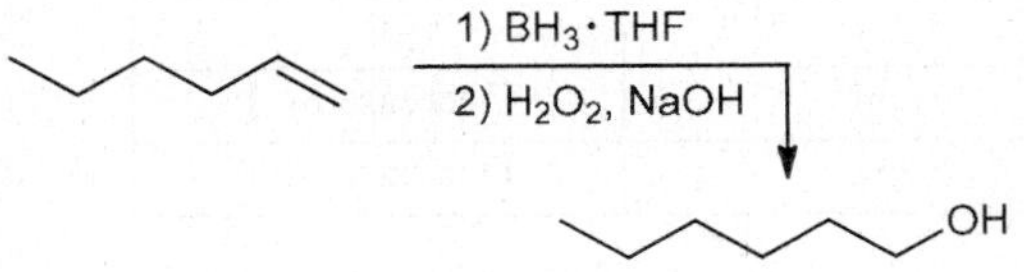

1) $BH_3 \cdot THF$
2) H_2O_2, NaOH

OH

(b) Let's begin by drawing the starting material and identifying the position that would lead to a secondary alcohol.

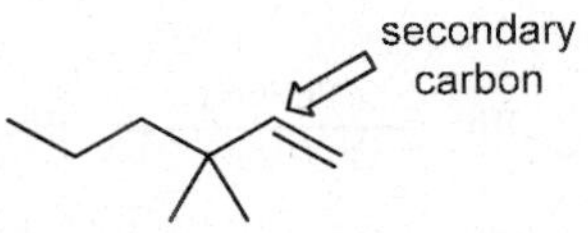

secondary carbon

3,3-dimethyl-1-hexene

Addition of H and OH across this alkene will provide an alcohol. Markovnikov addition will give a secondary alcohol, but we must be careful. Protonation of the alkene will generate a secondary carbocation which can rearrange (via a methyl shift) to give a more stable, tertiary carbocation. Therefore, acid-catalyzed hydration cannot be used. Instead, the desired product can be obtained via oxymercuration-demercuration, which will install the OH group at the more-substituted position without a carbocation rearrangement.

1) $Hg(OAc)_2$, H_2O
2) $NaBH_4$

OH

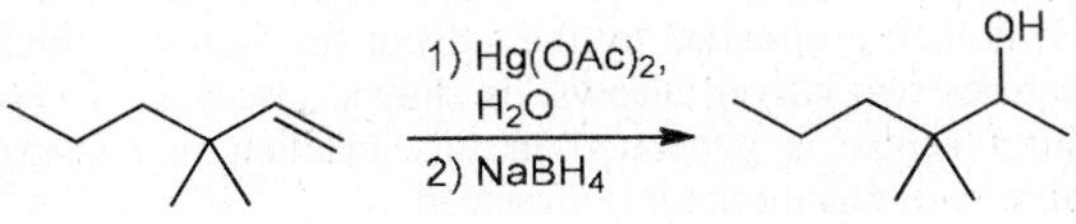

(c) Let's begin by drawing the starting material and identifying the position that would lead to a tertiary alcohol.

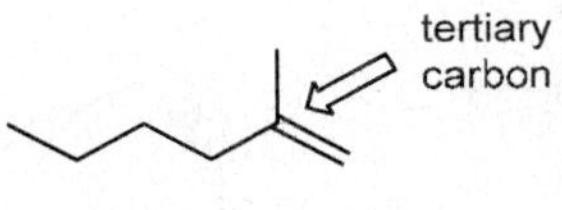

tertiary carbon

2-methyl-1-hexene

Addition of H and OH across this alkene will provide an alcohol. Specifically, Markovnikov addition will give a tertiary alcohol. Carbocation rearrangements are not a concern in this case (protonation of the alkene generates a tertiary carbocation which cannot rearrange), so acid-catalyzed hydration will give the desired product.

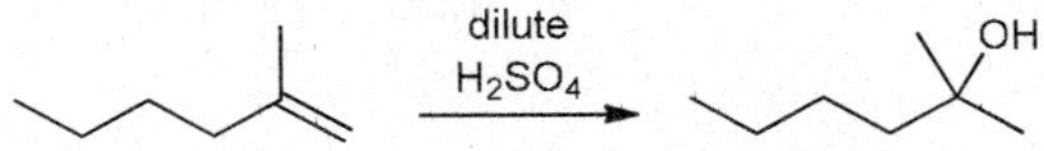

dilute H_2SO_4

OH

12.10.

(a) We focus on the carbon atom that undergoes a change in bonding as a result of the transformation. In the starting material, that carbon atom (the central carbon atom) has an oxidation state of +2. In the product, the same carbon atom has an oxidation state of +2. Since the oxidation state does not change, the starting material is neither oxidized nor reduced.

(b) We focus on the carbon atom that undergoes a change in bonding as a result of the transformation. In the starting material, that carbon atom has an oxidation state of +1. In the product, the same carbon atom has an oxidation state of +3. Since the oxidation state increases as a result of the transformation, the starting material is oxidized.

(c) We focus on the carbon atom that undergoes a change in bonding as a result of the transformation. In the starting material, that carbon atom has an oxidation state of +3. In the product, the same carbon atom has an oxidation state of -1. Since the oxidation state decreases as a result of the transformation, the starting material is reduced.

(d) We focus on the carbon atom that undergoes a change in bonding as a result of the transformation. In the starting material, that carbon atom has an oxidation state of +3. In the product, the same carbon atom has an oxidation state of +3. Since the oxidation state does not change, the starting material is neither oxidized nor reduced.

(e) We focus on the carbon atom that undergoes a change in bonding as a result of the transformation. In the starting material, that carbon atom has an oxidation state of 0. In the product, the same carbon atom has an oxidation state of +2. Since the oxidation state increases as a result of the transformation, the starting material is oxidized.

(f) We focus on the carbon atom that undergoes a change in bonding as a result of the transformation. In the starting material, that carbon atom has an oxidation state of +2. In the product, the same carbon atom has an oxidation state of +3. Since the oxidation state increases as a result of the transformation, the starting material is oxidized.

12.11. The carbon atoms in the ring have been numbered for ease of analysis. Let's consider each one separately.

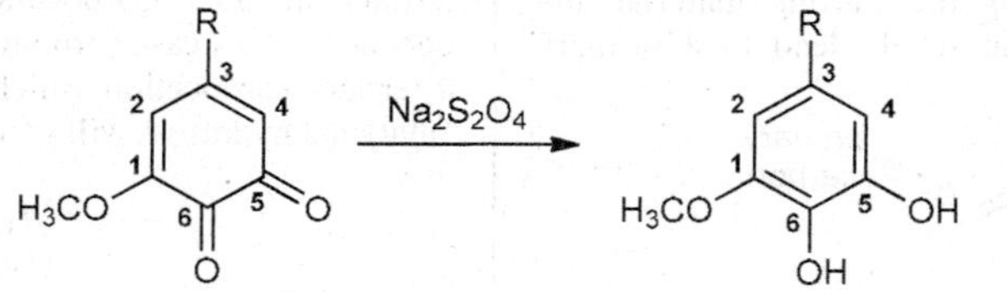

$\xrightarrow{Na_2S_2O_4}$

Carbon Atom	Starting Material Oxidation State	Product Oxidation State	Change in Oxidation State
1	+1	+1	0
2	-1	-1	0
3	0	0	0
4	-1	-1	0
5	+2	+1	Decrease by 1
6	+2	+1	Decrease by 1

Carbon atoms 1-4 do not undergo a change in oxidation state. Carbon atoms 5 and 6 both exhibit a decrease in oxidation state, so they are both reduced. This makes sense because both carbon atoms have lost a bond to oxygen.

12.12.

(a) Two curved arrows are used to show hydride delivery to the electrophilic carbon atom of the aldehyde carbonyl (C=O). Note that the tail of the first curved arrow is placed on the bond between Al and H (it is NOT placed on the negative charge). The resulting alkoxide ion is then protonated upon treatment with H_3O^+. This protonation step (called workup) requires two curved arrows, as shown.

1) $LiAlH_4$
2) H_3O^+

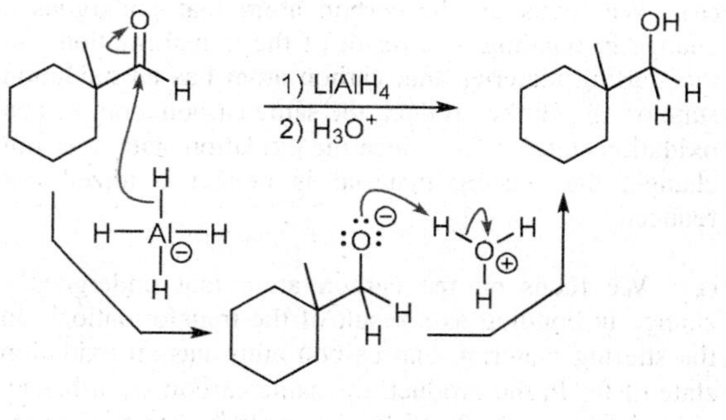

(b) Two curved arrows are used to show hydride delivery to the electrophilic carbon atom of the ketone carbonyl (C=O). Note that the tail of the first curved arrow is placed on the bond between Al and H (it is NOT placed on the negative charge). The resulting alkoxide ion is then protonated upon treatment with H_3O^+. This protonation step (called workup) requires two curved arrows, as shown.

1) $LiAlH_4$
2) H_3O^+

(Ph = phenyl)

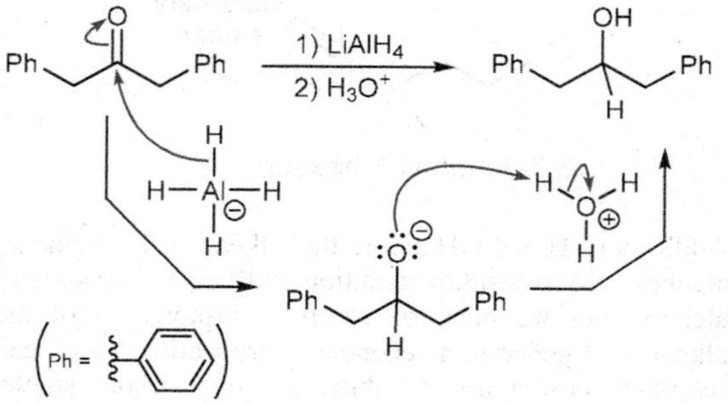

(c) Two curved arrows are used to show hydride delivery to the electrophilic carbon atom of the ketone carbonyl (C=O). Note that the tail of the first curved arrow is placed on the bond between B and H (it is NOT placed on the negative charge). The resulting alkoxide ion is then protonated by the solvent (methanol), which requires two curved arrows, as shown. Because a new chiral center is generated in this reaction, a racemic mixture of enantiomers is produced.

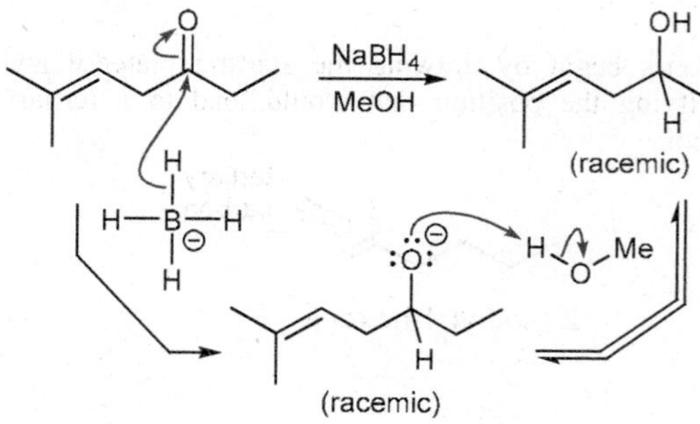

$NaBH_4$ / MeOH

(racemic)

(racemic)

(d) Two curved arrows are used to show hydride delivery to the electrophilic carbon atom of the ester carbonyl (C=O). Note that the tail of the first curved arrow is placed on the bond between Al and H (it is NOT placed on the negative charge). The resulting intermediate then ejects ethoxide as a leaving group, which requires two curved arrows. The resulting aldehyde is then further reduced by another equivalent of $LiAlH_4$. Once again, two curved arrows are used to show hydride delivery to the electrophilic carbon atom of the aldehyde carbonyl (C=O). The resulting alkoxide ion is then protonated upon treatment with H_3O^+. This protonation step (called workup) requires two curved arrows, as shown.

1) xs $LiAlH_4$
2) H_3O^+

- $EtO^{\ominus}$

12.13.

(a) Two curved arrows are used to show hydride delivery to the electrophilic carbon atom of the ketone carbonyl (C=O). Note that the tail of the first arrow is placed on the bond between B and H (it is NOT placed on the negative charge). The resulting alkoxide is then protonated, which also requires two curved arrows.

(b) The stereoisomer that is formed is the result of hydride delivery to the bottom face of the C=O bond, so that the newly installed hydrogen atom is on a dash, and the oxygen atom ends up on a wedge. This indicates that the borohydride ion approaches preferentially from the bottom face of the carbonyl group.

12.14.

(a) Treating the ketone with methyl magnesium bromide (followed by aqueous workup) gives a tertiary alcohol, with installation of a methyl group (highlighted):

1) MeMgBr
2) H_3O^+

(b) Treating the aldehyde with phenyl magnesium bromide (followed by aqueous workup) gives a secondary alcohol, with installation of a phenyl group (highlighted):

1) PhMgBr
2) H_3O^+

(racemic)

A new chiral center is generated in this reaction, and therefore, a 1:1 mixture of enantiomers is produced.

(c) Treating the ketone with the Grignard reagent shown (followed by aqueous workup) gives a tertiary alcohol, with installation of an alkyl group. The four-carbon alkyl group from the Grignard reagent is highlighted in the product:

1) MgBr
2) H_3O^+

(racemic)

A new chiral center is generated in this reaction, and therefore, a 1:1 mixture of enantiomers is produced.

(d) Treatment of the alkyl bromide with magnesium metal produces a Grignard reagent, which is then treated with formaldehyde to give a primary alcohol (after an aqueous workup):

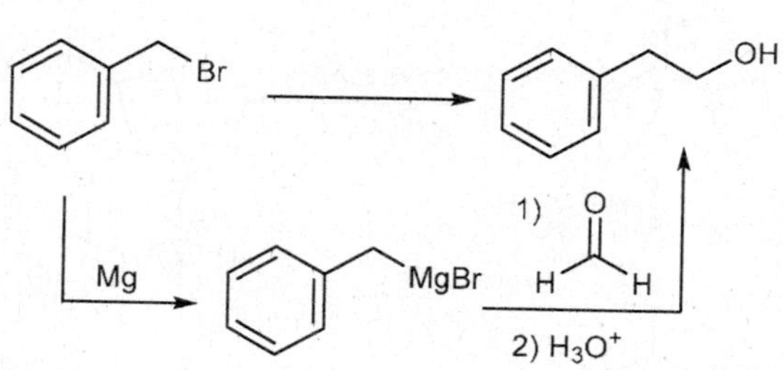

12.15.

(a) We begin by identifying the α position:

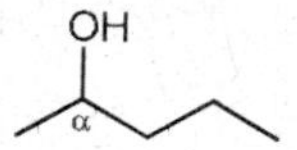

Next, we consider any possible C–C bond disconnections that can be made adjacent to the α position to give a suitable electrophile (C=O) and nucleophile (RMgBr). The target molecule has only two groups connected to the α position, so there are two possible retrosyntheses:

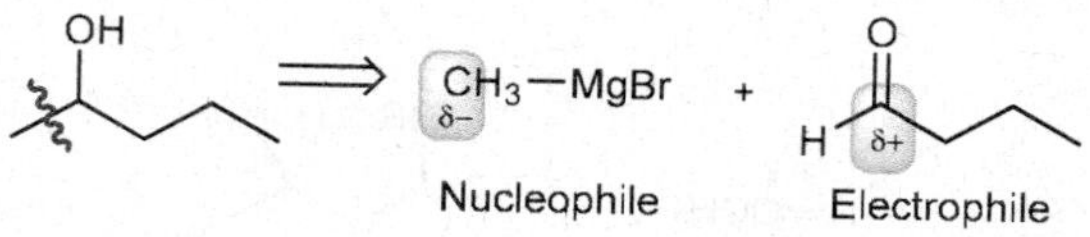

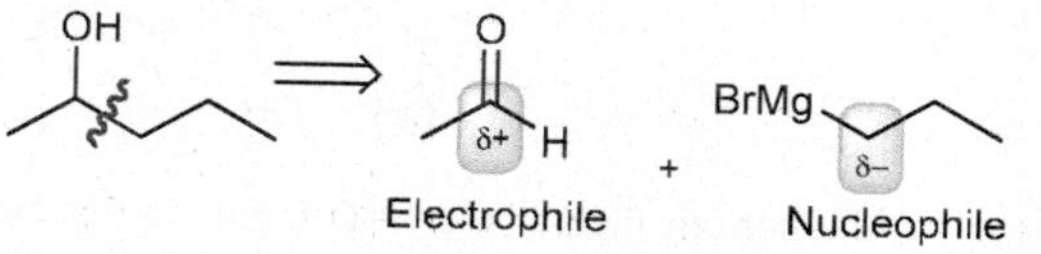

Either one of the groups attached to the α position could have been installed via a Grignard reaction with the appropriate aldehyde, as shown:

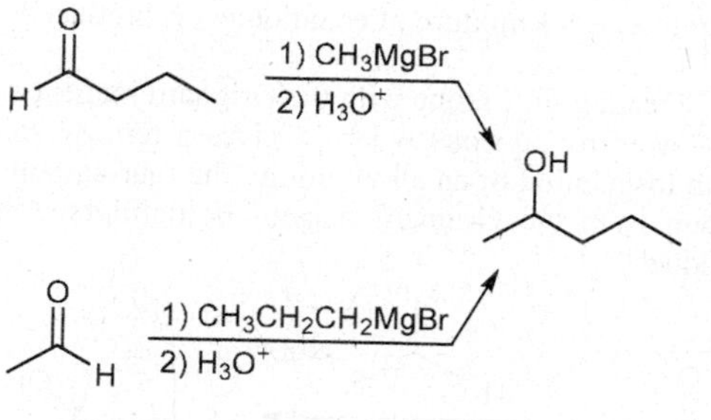

(b) We begin by identifying the α position:

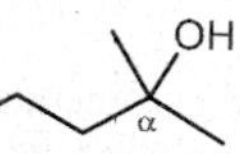

Next, we consider any possible C–C bond disconnections that can be made adjacent to the α position to give a suitable electrophile (C=O) and nucleophile (RMgBr). The target molecule has three groups connected to the α position, although two of them are identical. So there are only two different retrosyntheses:

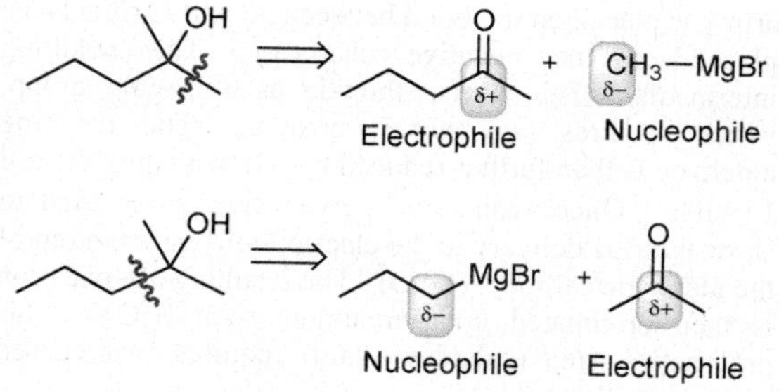

Either one of the unique groups attached to the α position could have been installed via a Grignard reaction, shown here:

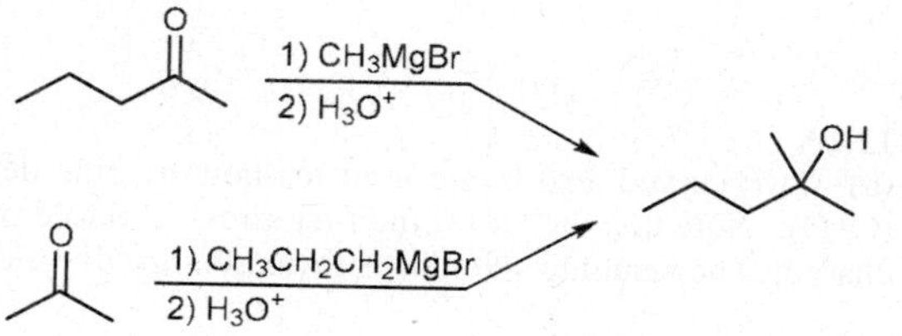

(c) We begin by identifying the α position:

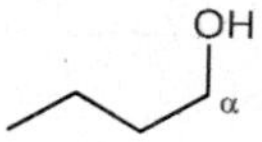

Next, we consider any possible C–C bond disconnections that can be made adjacent to the α position to give a suitable electrophile (C=O) and nucleophile (RMgBr). The target molecule has only one alkyl group connected to the α position, leading to a single possible retrosynthesis:

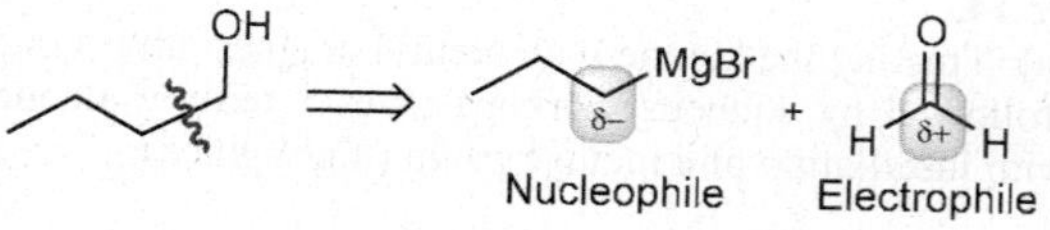

The *n*-propyl group attached to the α position could have been installed via a Grignard reaction with formaldehyde, as shown:

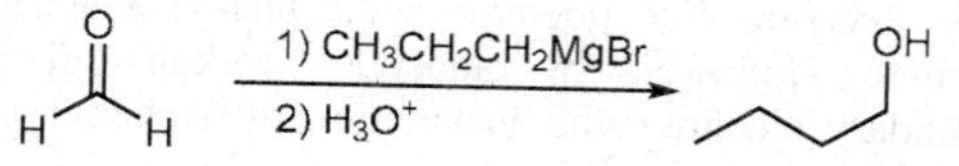

(d) We begin by identifying the α position:

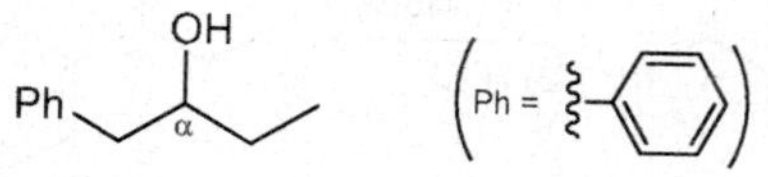

Next, we consider any possible C–C bond disconnections that can be made adjacent to the α position to give a suitable electrophile (C=O) and nucleophile (RMgBr). The target molecule has only two groups connected to the α position, so there are two possible retrosyntheses:

Electrophile Nucleophile

Nucleophile Electrophile

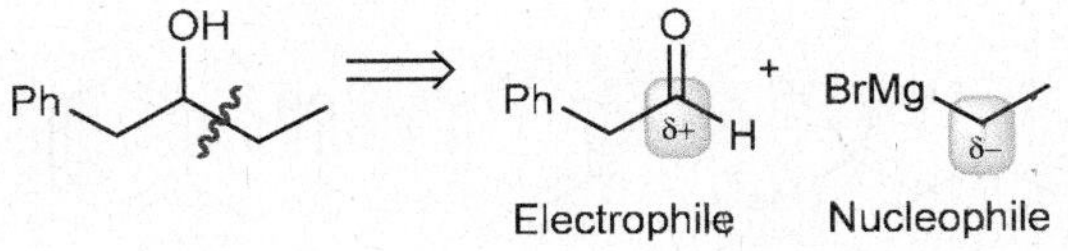

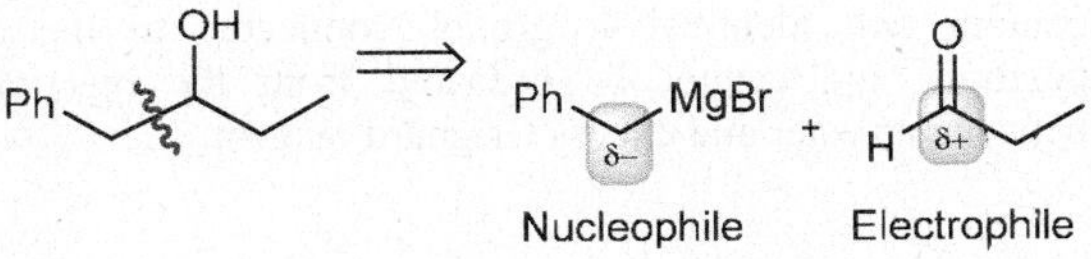

Either one of the groups attached to the α position could have been installed via a Grignard reaction with the appropriate aldehyde, as shown.

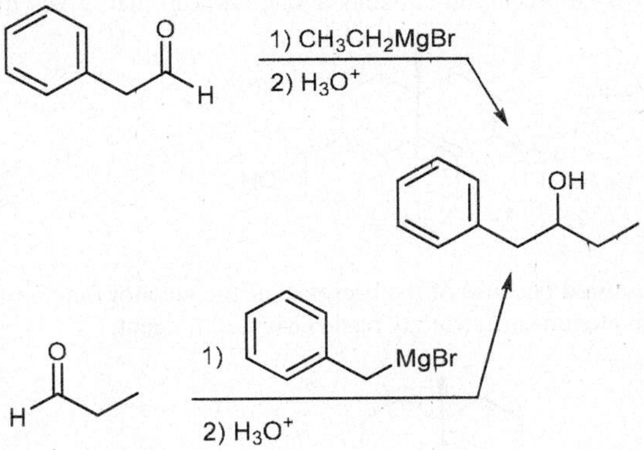

1) CH_3CH_2MgBr
2) H_3O^+

1) MgBr
2) H_3O^+

(e) We begin by identifying the α position:

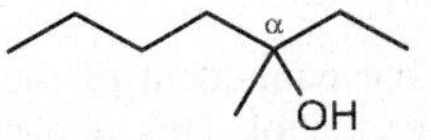

Next, we consider any possible C–C bond disconnections that can be made adjacent to the α position to give a suitable electrophile (C=O) and nucleophile (RMgBr). The target molecule has three different groups connected to the α position, so there are three possible retrosyntheses:

Electrophile Nucleophile

Electrophile Nucleophile

Nucleophile Electrophile

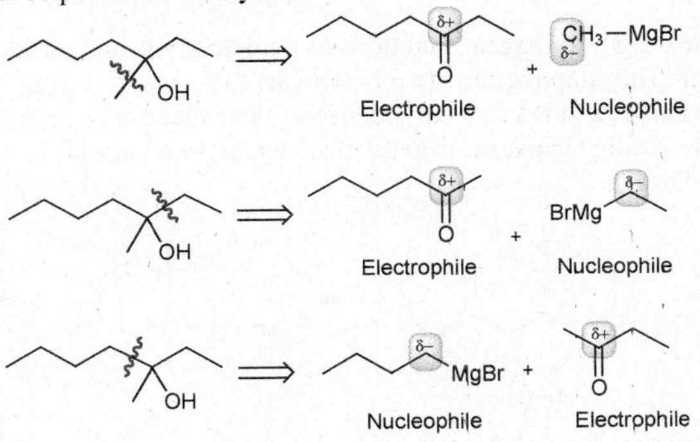

Any of these three groups attached to the α position could have been installed via a Grignard reaction with the appropriate ketone, as shown.

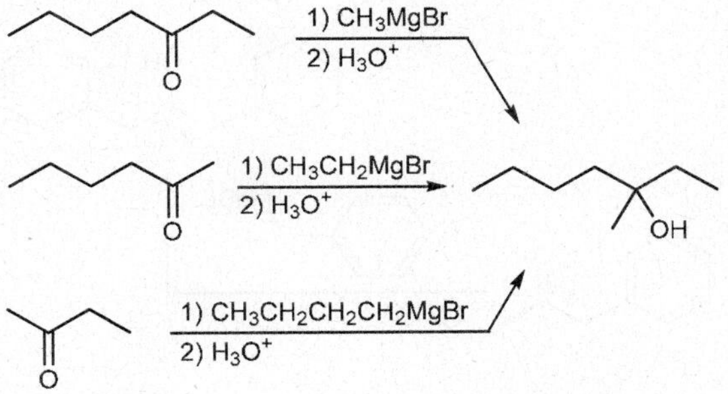

1) CH_3MgBr
2) H_3O^+

1) CH_3CH_2MgBr
2) H_3O^+

1) $CH_3CH_2CH_2CH_2MgBr$
2) H_3O^+

(f) We begin by identifying the α position:

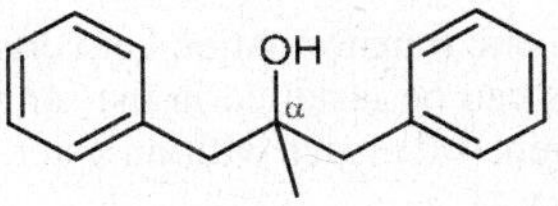

Next, we consider any possible C–C bond disconnections that can be made adjacent to the α position to give a suitable electrophile (C=O) and nucleophile (RMgBr). The target molecule has three groups connected to the α position, although two of them are identical. So there are only two different retrosyntheses:

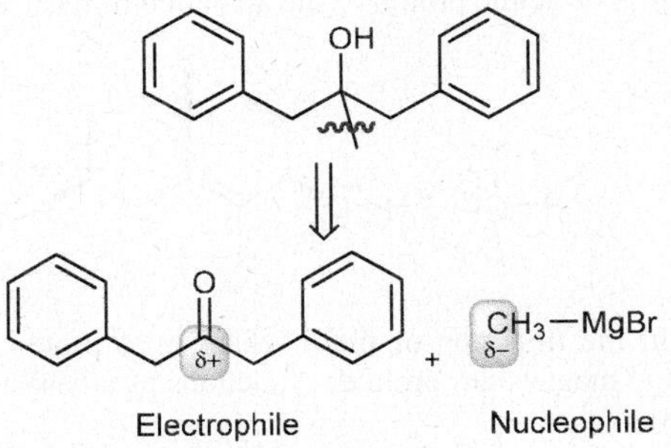

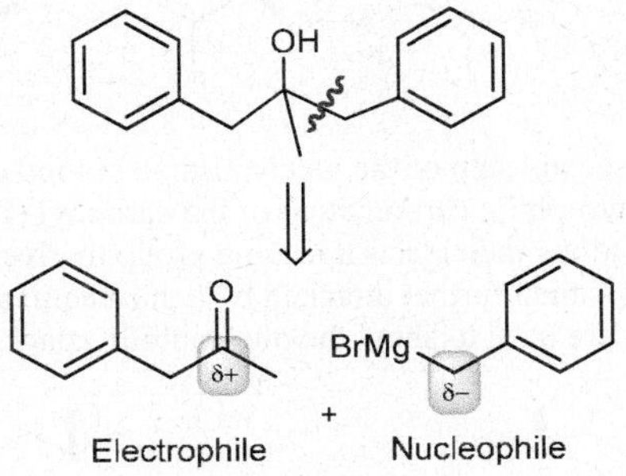

Electrophile Nucleophile

Electrophile Nucleophile

Either one of the unique groups attached to the α position could have been installed via a Grignard reaction, shown here.

1) CH_3MgBr
2) H_3O^+

OH

1) MgBr
2) H_3O^+

12.16. Each of the following two compounds can be prepared from the reaction between a Grignard reagent and an ester, because each of these compounds has two identical R groups connected to the α position:

Me OH Me

OH

The other four compounds from Problem 12.15 do not contain two identical R groups connected to the α position, and cannot be prepared from the reaction between an ester and excess Grignard reagent.

12.17.

(a) This synthetic transformation converts an ester to an alcohol with the installation of two new methyl groups. Both methyl groups can be installed in the same Grignard reaction, using CH_3MgBr (MeMgBr), since two equivalents of a Grignard reagent will react with an ester. After the reaction is complete, an aqueous acidic workup then gives the desired product.

HO CO₂Me **1** → 1) xs CH_3MgBr 2) H_3O^+ → HO OH **2**

(b) In this case, a third equivalent of the Grignard reagent is required because of the presence of the alcohol functional group. The acidic proton of the alcohol will react with one equivalent of the strongly basic Grignard reagent:

H–C(H)(H):⊖ + H–O ... CO_2Me → ⊖:O ... CO_2Me + CH_4

(c) In the first step of the mechanism, a proton transfer reaction occurs. One equivalent of the Grignard reagent (methyl magnesium bromide) functions as a base and removes the proton of the alcohol. This step requires two curved arrows.

H–C(H)(H):⊖ + H–O ... CO_2Me → ⊖:O ... CO_2Me + CH_4

In the second step of the mechanism, a second equivalent of the Grignard reagent functions as a nucleophile and attacks the electrophilic carbon atom of the carbonyl (C=O) of the ester. This step requires two curved arrows. The resulting intermediate then ejects a leaving group to give a ketone, which also requires two curved arrows. The electrophilic ketone is then further attacked by a third equivalent of the nucleophilic Grignard reagent. Once again, two curved arrows are used to show the nucleophilic attack, resulting in a dianion.

OMe → H_3C OMe → CH_3 → H_3C CH_3

Finally, the dianion is then protonated upon treatment with aqueous acid. There are two locations that are protonated, each of which requires two curved arrows, as shown. Notice that each anion is protonated in a separate step (this should not be drawn as one step with four curved arrows, because there are two distinct processes occurring, and it is unlikely that they occur precisely at the same moment).

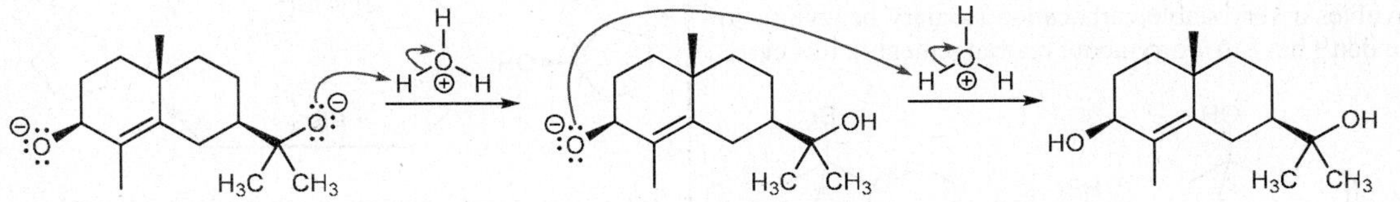

12.18.

(a) This type of transformation can be achieved via a Grignard reaction.

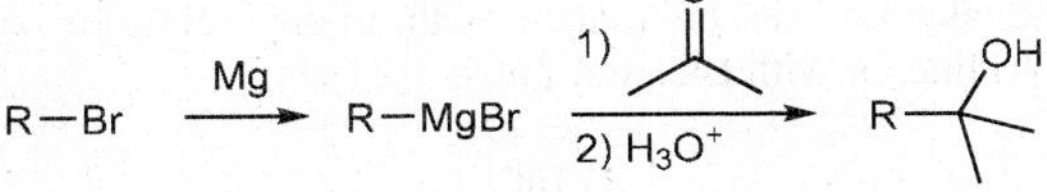

However, the starting halide has an acidic OH group, which is incompatible with a strongly basic Grignard reagent. To resolve this issue, we must first protect the OH group by converting it to a silyl ether (OTMS), and then we can perform the desired Grignard reaction. Deprotection with tetrabutylammonium fluoride (TBAF) then gives the desired product, as shown.

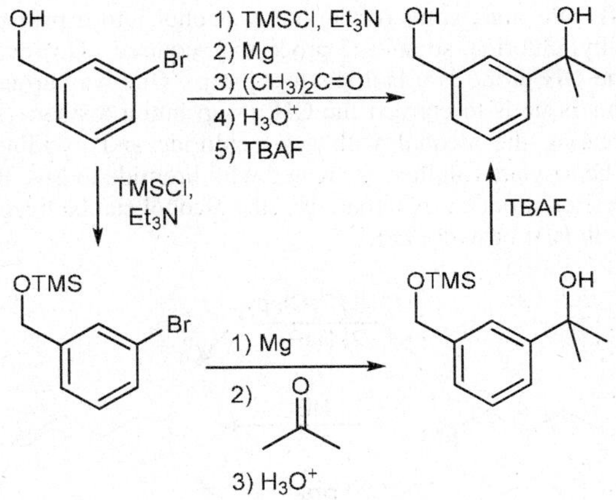

(b) This type of transformation can be achieved via a Grignard reaction in which the Grignard reagent is treated with 0.5 equivalents of an ester.

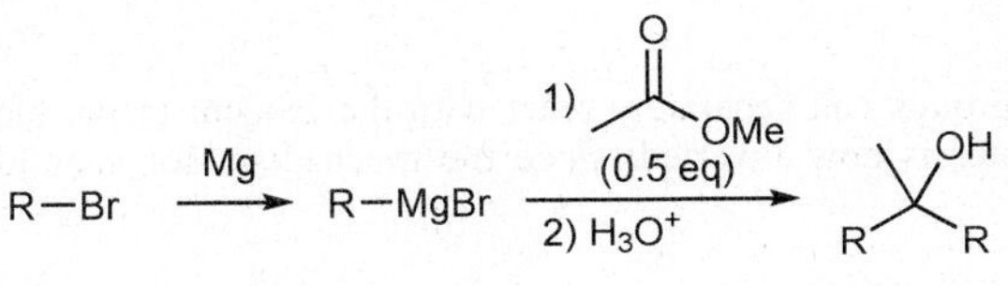

However, the starting halide has an acidic OH group, which is incompatible with a strongly basic Grignard reagent. To resolve this issue, we must first protect the OH group by converting it to a silyl ether (OTMS), and then we can perform the desired Grignard reaction. Deprotection with tetrabutylammonium fluoride (TBAF) then gives the desired product, as shown.

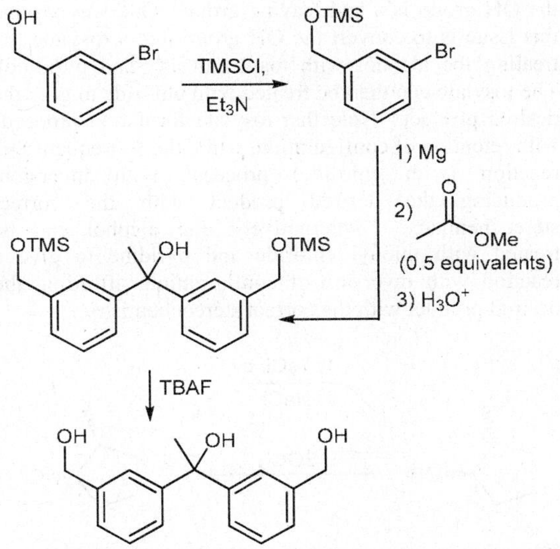

12.19.

(a) We must convert a secondary alcohol into a secondary alkyl bromide, with inversion of configuration, so an S_N2 process is required. However, the OH group is a bad leaving group. One way around this issue is to convert the OH group into a tosylate (by treating the alcohol with tosyl chloride and pyridine). The tosylate can then be treated with bromide to give the desired product. Note that tosylate formation proceeds with retention of configuration, while the subsequent S_N2 reaction (with bromide) proceeds with inversion, producing the desired product with the correct stereochemistry. Alternatively, the alcohol can be treated with PBr_3 to give a reaction with inversion of configuration, affording the desired product with the correct stereochemistry.

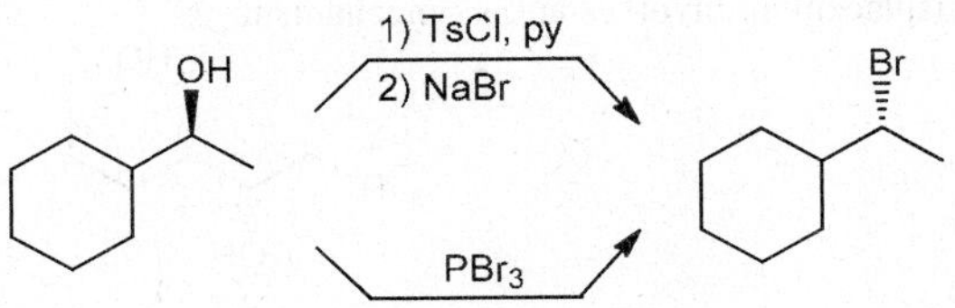

(b) We must convert a tertiary alcohol into a tertiary alkyl bromide. This can be achieved upon treatment with HBr, which provides both an acidic proton to convert the OH group into a good leaving group (water), as well as a nucleophilic bromide ion. This S_N1 process involves a very stable carbocation (tertiary benzylic), so we don't have to worry about rearrangement in this case.

OH HBr Br

(c) We must convert a secondary alcohol into a secondary alkyl chloride, with inversion of configuration, so an S_N2 process is required. However, the OH group is a bad leaving group. One way around this issue is to convert the OH group into a tosylate (by treating the alcohol with tosyl chloride and pyridine). The tosylate can then be treated with chloride to give the desired product. Note that tosylate formation proceeds with retention of configuration, while the subsequent S_N2 reaction (with chloride) proceeds with inversion, producing the desired product with the correct stereochemistry. Alternatively, the alcohol can be treated with thionyl chloride and pyridine to give a reaction with inversion of configuration, affording the desired product with the correct stereochemistry.

1) TsCl, py
2) NaCl
OH $SOCl_2$ py Cl

(d) We must convert a secondary alcohol into a secondary alkyl bromide, with inversion of configuration, so an S_N2 process is required. However, the OH group is a bad leaving group. One way around this issue is to convert the OH group into a tosylate (by treating the alcohol with tosyl chloride and pyridine). The tosylate can then be treated with bromide to give the desired product. Note that tosylate formation proceeds with retention of configuration, while the subsequent S_N2 reaction (with bromide) proceeds with inversion, producing the desired product with the correct stereochemistry. Alternatively, the alcohol can be treated with PBr_3 to give a reaction with inversion of configuration, affording the desired product with the correct stereochemistry.

1) TsCl, py
2) NaBr
OH Br
PBr_3

(e) We must convert a secondary alcohol into a secondary alkyl chloride, so a substitution process is required. However, the OH group is a bad leaving group. One way around this issue is to convert the OH group into a tosylate (by treating the alcohol with tosyl chloride and pyridine). The tosylate can then be treated with chloride to give the desired product. Alternatively, the alcohol can be treated with thionyl chloride and pyridine, or with HCl and $ZnCl_2$.

1) TsCl, py
2) NaCl
OH $SOCl_2$ py Cl
HCl
$ZnCl_2$

(f) We must convert a primary alcohol into a primary alkyl chloride, so an S_N2 process is required. However, the OH group is a bad leaving group. One way around this issue is to convert the OH group into a tosylate (by treating the alcohol with tosyl chloride and pyridine). The tosylate can then be treated with bromide to give the desired product. Alternatively, the alcohol can be treated with HBr or with PBr_3.

1) TsCl, py
2) NaBr
OH HBr Br
PBr_3

12.20.

(a) When diol **1** is treated with PBr_3, each of the two OH groups can separately react with the reagent to produce dibromide **2**. Notice that the configuration of the chiral center is now inverted, since the mechanism for bromide displacement involves an S_N2 mechanism.

HO R OH **1** PBr_3 Br R Br **2**

(b) The byproduct can be formed via the following process. First, the terminal alcohol is converted into a good leaving group upon treatment with PBr_3. However, before a bromide ion can attack this intermediate, an intramolecular S_N2-type reaction occurs – the internal OH group can displace the good leaving group of **4** to form cyclic ether **5** which can be deprotonated to produce by-product **3**. This mechanism agrees with the stereochemical outcome, because it does not involve the chiral center, and therefore the configuration of the chiral center is retained.

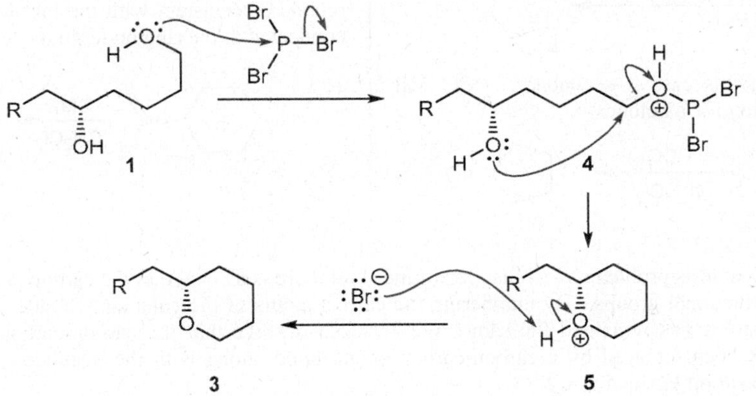

In the proposed mechanism shown above, the primary OH group is converted into a good leaving group, and the secondary OH group functions as a nucleophile. If instead the secondary OH group had reacted with PBr_3 (**6**), and the primary OH group had functioned as a nucleophile, then compound **7** would have been produced, which is not the by-product. Notice that with this mechanism, the configuration of the chiral center is now inverted, since the mechanism for bromide displacement involves an S_N2-type back-side attack. Compound **7** is the enantiomer of the observed by-product.

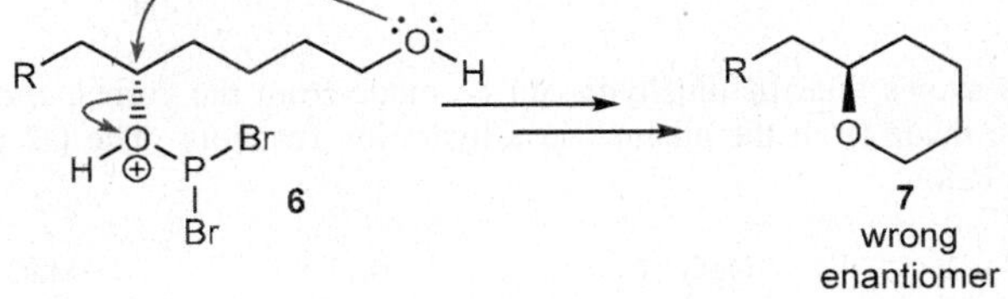

7
wrong enantiomer

12.21.

(a) An alcohol is converted to an alkene upon treatment with concentrated sulfuric acid. In this case, there are two possible regiochemical outcomes, and we expect that the more substituted alkene will be the major product.

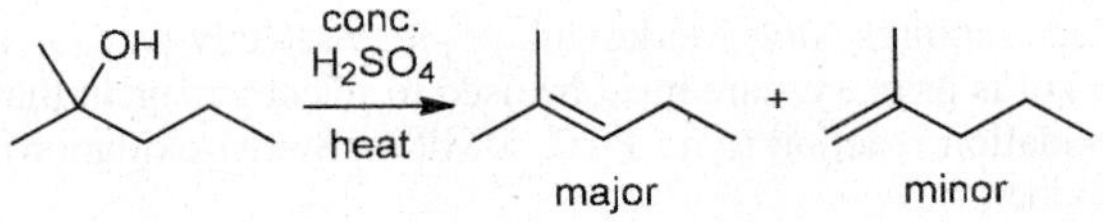

conc. H_2SO_4, heat — major + minor

(b) An alcohol is converted to a tosylate upon treatment with tosyl chloride and pyridine. This tosylate is a secondary substrate, and ethoxide is both a strong nucleophile and a strong base. Recall from Chapter 7 that a secondary substrate is expected to react with ethoxide via an E2 process to give the major product (while S_N2 gives the minor product).

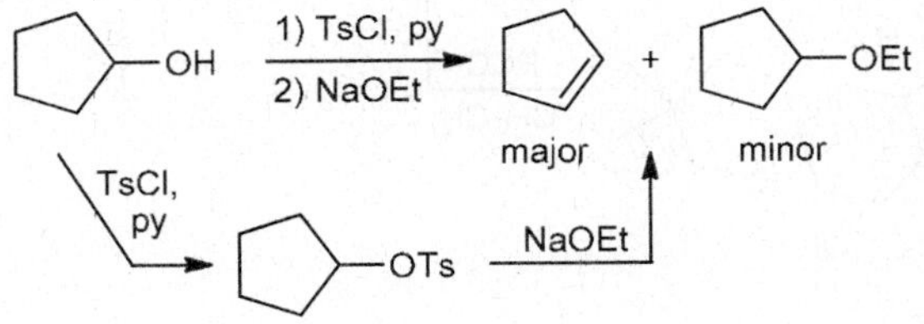

1) TsCl, py 2) NaOEt — major + minor (OEt); TsCl, py → OTs; NaOEt

12.22.

(a) The alcohol in this case is secondary. Chromic acid will oxidize the alcohol to give a ketone.

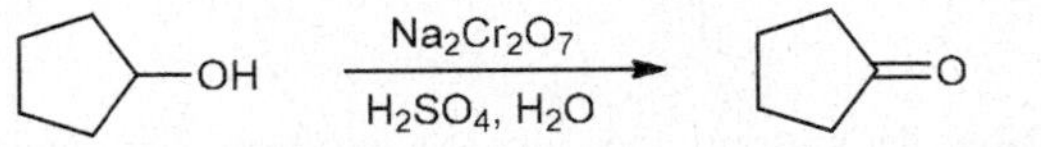

$Na_2Cr_2O_7$, H_2SO_4, H_2O

(b) The alcohol in this case is primary. Chromic acid will oxidize the alcohol to give a carboxylic acid.

$Na_2Cr_2O_7$, H_2SO_4, H_2O

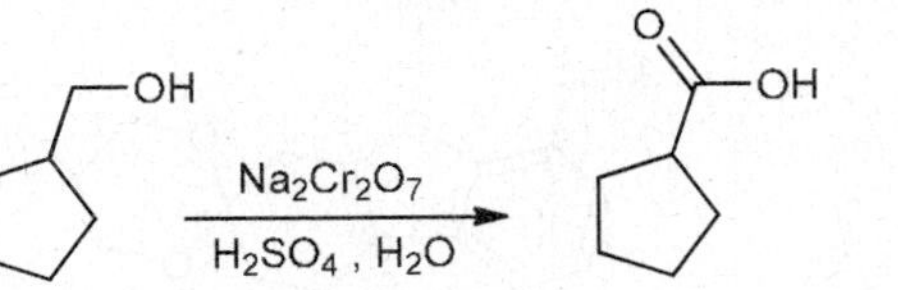

(c) The alcohol in this case is primary. This alcohol is oxidized upon treatment with chromic acid to give a carboxylic acid group, and the aldehyde group (already present) also undergoes oxidation to give a carboxylic acid group as well.

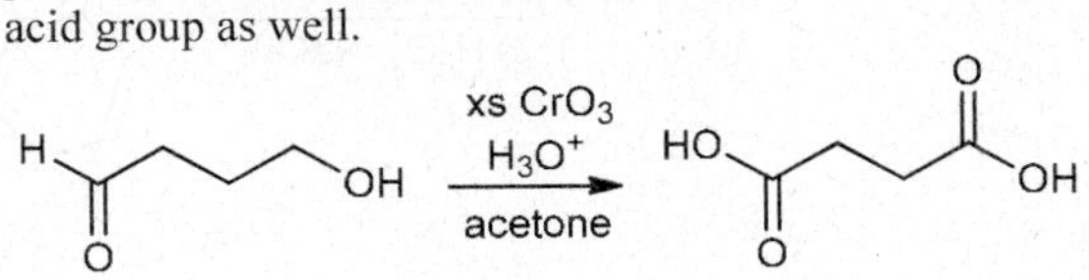

xs CrO_3, H_3O^+, acetone

(d) The alcohol in this case is primary. A Swern oxidation will oxidize the alcohol to give an aldehyde.

1) DMSO, $(COCl)_2$
2) Et_3N

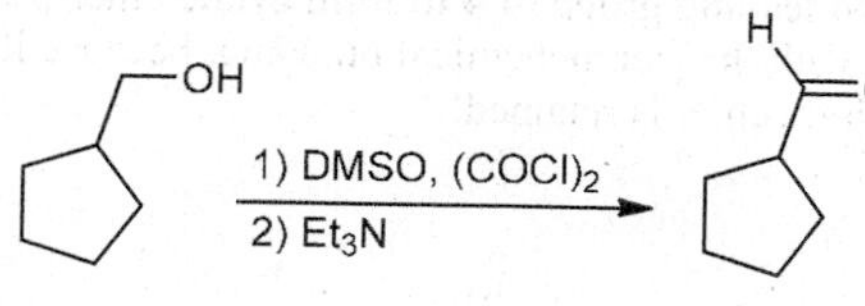

(e) The alcohol in this case is secondary. PCC will oxidize the alcohol to give a ketone.

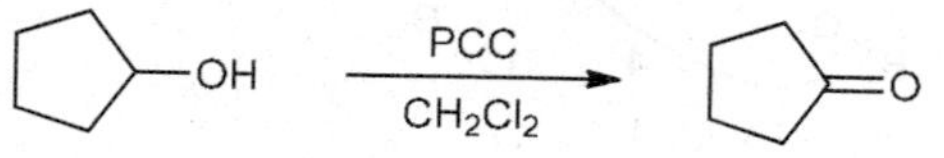

PCC
CH_2Cl_2

(f) The alcohol in this case is primary. Dess-Martin periodinane (DMP) will oxidize the alcohol to give an aldehyde, and the other aldehyde group (already present) is unreactive toward these conditions. Note that the *trans* orientation of the two groups on the ring is retained, consistent with the fact that none of the bonds connected to the ring undergo any changes.

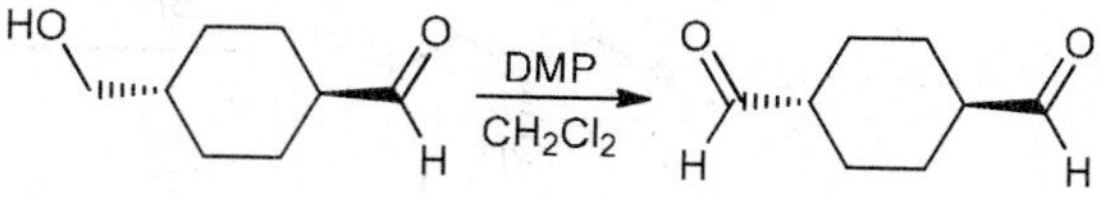

DMP
CH_2Cl_2

12.23. As with all synthesis problems, we must determine: 1) if there is a change in the carbon skeleton, and 2) if there is a change in the functional groups. By numbering the carbon atoms of the galantamine side chain, we see that the number of carbon atoms has remained the same. We can also observe that the carbon-carbon double bond of the starting material has been replaced by a carbon-carbon single bond, along with the installation of a carbon-oxygen double bond at the position labelled as C2.

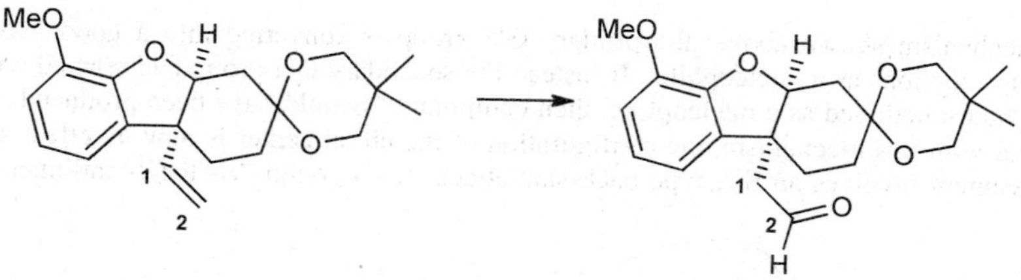

The following retrosynthesis shows that the aldehyde can be made from the corresponding alcohol via an oxidation reaction. This alcohol can be made from the alkene via a hydration reaction. The C2 position is highlighted in the retrosynthetic scheme shown below:

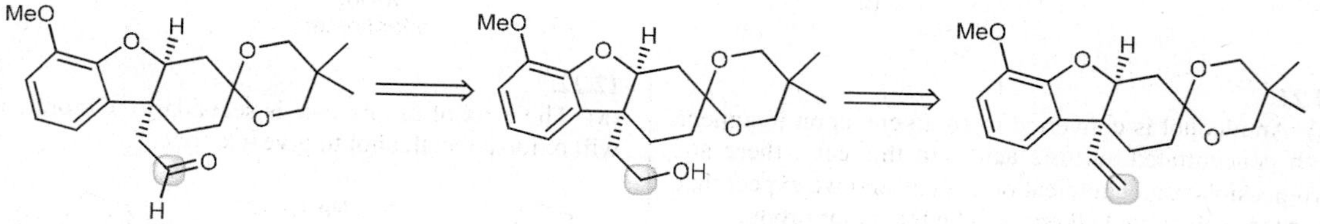

Now let's draw the forward process. The hydration step requires *anti*-Markovnikov regioselectivity, so a hydroboration-oxidation reaction is appropriate. Since the alcohol is primary, care must be used to select a reagent that will produce the aldehyde and not a carboxylic acid in the oxidation reaction (*e.g.*, PCC, DMP or Swern oxidation). The following reaction sequence would produce the desired product:

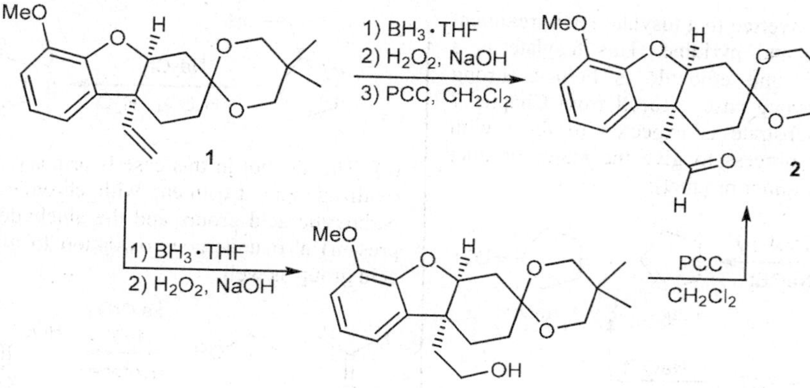

1) $BH_3 \cdot THF$
2) H_2O_2, NaOH
3) PCC, CH_2Cl_2

1) $BH_3 \cdot THF$
2) H_2O_2, NaOH

PCC
CH_2Cl_2

12.24.

(a) The retrosynthetic analysis of the target molecule asks, "What reaction have we seen that produces an alcohol product?" This analysis reveals two potential starting materials—an alkene or an aldehyde:

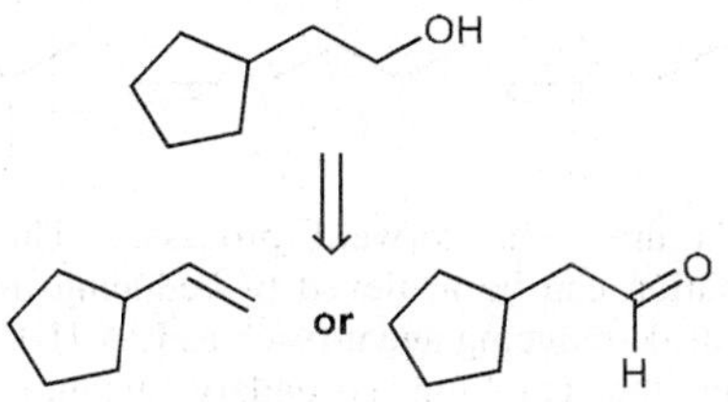

So we can see that there are at least two acceptable answers to this problem. In one option, the desired transformation is achieved by converting the alkyne into an alkene, followed by *anti*-Markovnikov addition of H and OH.

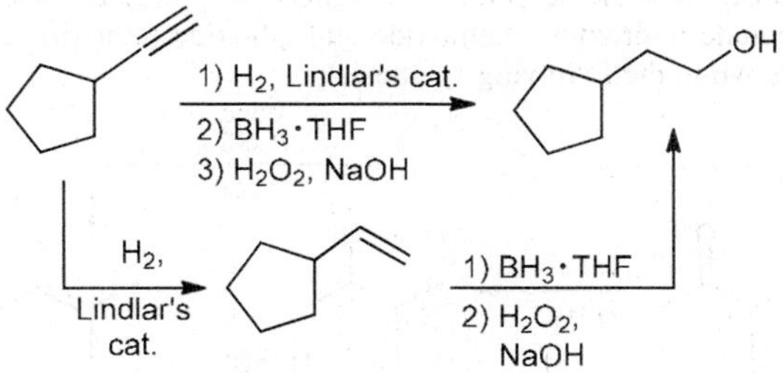

Another reasonable route involves conversion of the alkyne to an aldehyde via hydroboration-oxidation, followed by reduction with $LiAlH_4$.

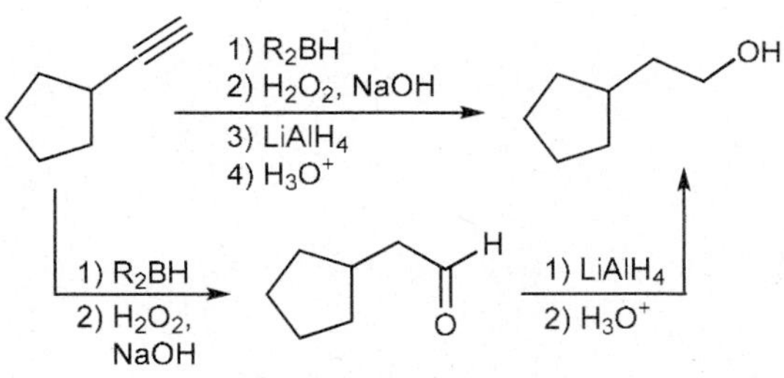

(b) The retrosynthetic analysis of the target molecule asks, "What reaction have we seen that produces an alkyne product?" This analysis reveals an alkene as a potential starting material:

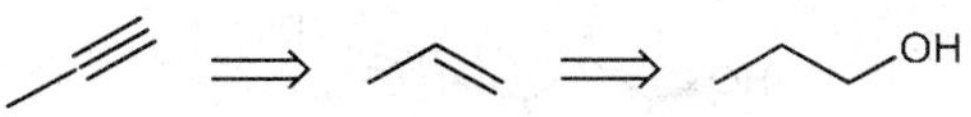

The required alkene can, in turn, be prepared from the given alcohol starting material.

Now let's draw the forward process. The desired transformation can be achieved by converting the alcohol into an alkene, either by acid-catalyzed dehydration (E1) or by E2 elimination of a good leaving group (e.g., tosylate). For the route involving an E2 elimination, it is important that the base is sterically hindered (*tert*-butoxide). With a primary substrate, S_N2 will likely predominate over E2 if the base is not sterically hindered. Alternatively, a non-nucleophilic base, such as DBU, can be used. Finally, the resulting alkene can be converted into the alkyne target molecule by addition of Br_2, followed by two successive elimination reactions, as shown.

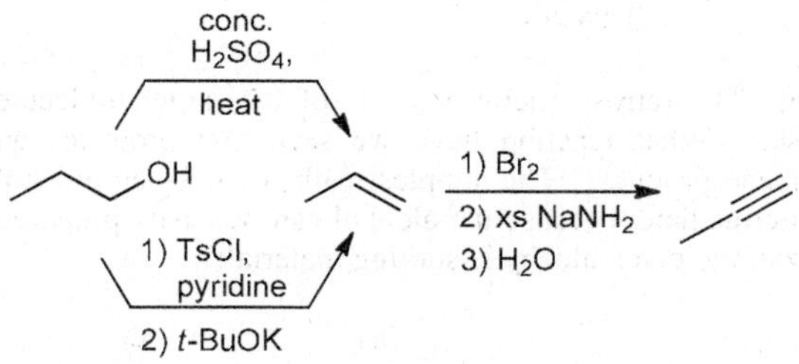

(c) The retrosynthetic analysis of the target molecule asks, "What reaction have we seen that produces an aldehyde product?" The simplest path involves an alcohol intermediate, because the alcohol can be easily prepared from the given alkene starting material:

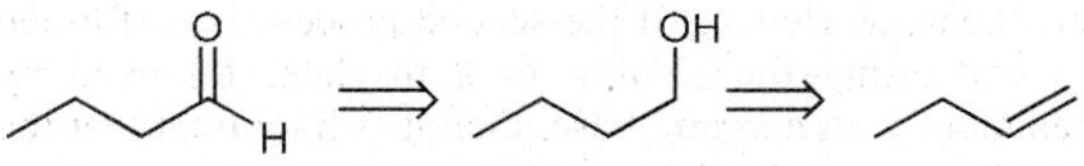

Now let's draw the forward process. The desired transformation can be achieved by converting the alkene into an alcohol via an *anti*-Markovnikov addition of H and OH, and then oxidizing the resulting primary alcohol with PCC to give the aldehyde, as shown.

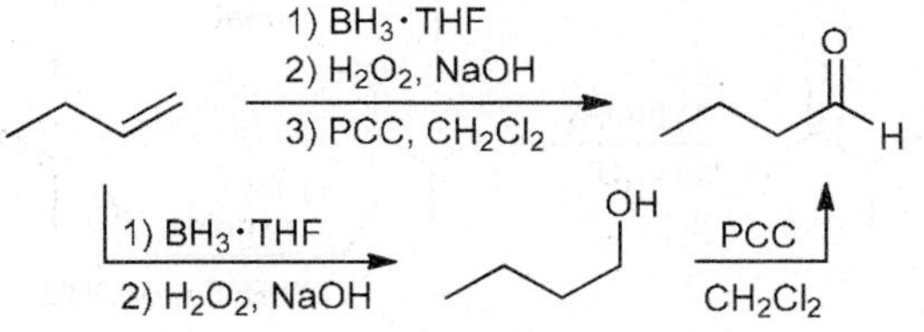

(d) The retrosynthetic analysis of the target molecule asks, "What starting material can be used to make an alkane?" One reaction we've seen that generates a product with no functional groups is catalytic hydrogenation of an alkene. The required alkene can be produced from the given alcohol:

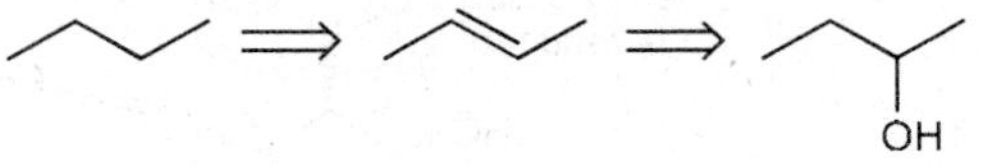

Now let's draw the forward process. The desired transformation can be achieved by converting the alcohol into an alkene, either by acid-catalyzed dehydration (E1) or by E2 elimination of a good leaving group (*e.g.*, tosylate). Finally, the alkene is reduced to the alkane target molecule by catalytic hydrogenation, as shown.

conc. H_2SO_4, heat

H_2 / Pt

OH

1) TsCl, py
2) NaOEt

(e) The retrosynthetic analysis of the target molecule asks, "What reaction have we seen that produces an alkene product?" The simplest path involves an alcohol intermediate, because the alcohol can be easily prepared from the given aldehyde starting material:

OH O H

Now let's draw the forward process. The desired transformation can be achieved by reducing the aldehyde, and then converting the resulting alcohol into an alkene, as shown. If the second process is performed by converting the alcohol to a tosylate, followed by treatment with a strong base, then it is important that the base is sterically hindered (*tert*-butoxide). With a primary substrate, S_N2 will likely predominate over E2 if the base is not sterically hindered. Alternatively, a non-nucleophilic base, such as DBU, can be used.

conc. H_2SO_4, heat

O H OH

1) $LiAlH_4$
2) H_3O^+

1) TsCl, pyridine
2) *t*-BuOK or DBU

(f) The retrosynthetic analysis of the target molecule asks, "What reaction have we seen that produces an alkene product?" The simplest path involves an alcohol intermediate, because the alcohol can be easily prepared from the given aldehyde starting material:

OH O

Now let's draw the forward process. The desired transformation can be achieved by reducing the ketone with a hydride reducing agent (such as $LiAlH_4$), and then converting the resulting secondary alcohol into an alkene, as shown. If the second process is performed by converting the alcohol into a tosylate, followed by treatment with a strong base, then it is important that the base is not sterically hindered. A sterically hindered base would give the less-substituted alkene as the major product (Hofmann product), but we need the more-substituted alkene (Zaitsev product). Appropriate bases include hydroxide, methoxide and ethoxide (ethoxide is shown in the following synthesis).

conc. H_2SO_4, heat

O OH

1) $LiAlH_4$
2) H_3O^+

1) TsCl, pyridine
2) NaOEt

12.25. Compound **1** contains both an ester group and an amide group. As described in the problem statement, treatment of compound **1** with $LiBH_4$ is expected to result in reduction of the ester group, while the amide group will remain unchanged. Notice that two equivalents of hydride are added to the ester carbon atom, reducing it to an alcohol.

ester group is reduced O OEt Ph HO

amide group is not reduced O N

OCH3 **1**

1) excess $LiBH_4$
2) H_2O

OH HO Ph O N OCH3 **2**

Treatment of diol **2** with excess NaH (a base to generate alkoxide ions), followed by excess benzyl bromide, converts both hydroxyl groups into ether groups. Finally, partial reduction of the alkyne by hydrogenation in the presence of a poisoned catalyst affords the *cis*-alkene (compound **3**). Note that the stereochemistry of all chiral centers is conserved throughout this synthetic sequence.

2 —(excess NaH, excess $PhCH_2Br$, DMF)→ … —(H_2, Lindlar's catalyst)→ 3

12.26.

(a) The product has more carbon atoms than the starting material, so we must form a carbon-carbon bond. This can be achieved with a Grignard reaction (using ethyl magnesium bromide to install an ethyl group). The resulting alcohol can then be oxidized to give the desired product. Any oxidizing agent (that we have seen) is suitable to convert this secondary alcohol to a ketone (*e.g.*, chromic acid, PCC, DMP or Swern oxidation).

1) EtMgBr
2) H_3O^+
3) $Na_2Cr_2O_7$, H_2SO_4, H_2O

1) EtMgBr
2) H_3O^+

$Na_2Cr_2O_7$, H_2SO_4, H_2O

(b) The product has one more carbon atom than the starting material, so we must form a carbon-carbon bond. This can be achieved with a Grignard reaction (using methyl magnesium bromide to install a methyl group). The resulting alcohol can then be oxidized to give the desired product. Any oxidizing agent (that we have seen) is suitable to convert this secondary alcohol to a ketone (*e.g.*, chromic acid, PCC, DMP or Swern oxidation).

1) MeMgBr
2) H_3O^+
3) $Na_2Cr_2O_7$, H_2SO_4, H_2O

1) MeMgBr
2) H_3O^+

$Na_2Cr_2O_7$, H_2SO_4, H_2O

12.27.

(a) The target molecule has one additional carbon atom, so the retrosynthesis requires a C—C bond disconnection. The disconnection is made at the α position of the alcohol to give a suitable electrophile (C=O) and nucleophile (RMgBr):

CH_3—MgBr

Electrophile Nucleophile

To complete the synthesis, we just need to bridge the gap between the given alcohol starting material and the required ketone. Conversion of a secondary alcohol to a ketone is an oxidation reaction, and we can use a wide variety of reagents for this step, such as chromic acid, PCC, DMP or Swern oxidation. Finally, a Grignard reaction with methyl magnesium bromide will give the desired target molecule:

1) $Na_2Cr_2O_7$, H_2SO_4, H_2O
2) MeMgBr
3) H_3O^+

$Na_2Cr_2O_7$, H_2SO_4, H_2O

1) MeMgBr
2) H_3O^+

(b) The target molecule has one additional carbon atom, so the retrosynthesis requires a C—C bond disconnection. The disconnection is made at the α position of the alcohol to give a suitable electrophile (C=O) and nucleophile (RMgBr):

CH_3—MgBr

Electrophile Nucleophile

To complete the synthesis, we just need to bridge the gap between the given alkyne starting material and the required aldehyde. The aldehyde can be made from a terminal alkyne via hydroboration-oxidation. Finally, a Grignard reaction with methyl magnesium bromide will give the desired target molecule:

1) R_2BH
2) H_2O_2, NaOH
3) MeMgBr
4) H_3O^+

OH

1) R_2BH
2) H_2O_2, NaOH

H

O

1) MeMgBr
2) H_3O^+

(c) The target molecule has two additional carbon atoms, so the synthesis requires a C—C bond-forming reaction. The retrosynthetic analysis of the target molecule asks, "What reaction have we seen that produces a ketone product?" The ketone can be prepared by oxidation of the corresponding alcohol, and the alcohol is suitable for making the required C—C bond disconnection (at the α position):

O

OH

O

δ+ H + BrMg δ−

Electrophile Nucleophile

Now let's draw the forward process. The aldehyde can be treated with a Grignard reagent (ethyl magnesium bromide), and the resulting alcohol can then be oxidized to give the desired target molecule. Conversion of a secondary alcohol to a ketone can be achieved with a wide variety of reagents, such as chromic acid, PCC, DMP or Swern oxidation.

O

H

1) EtMgBr
2) H_3O^+
3) $Na_2Cr_2O_7$, H_2SO_4, H_2O

O

1) EtMgBr
2) H_3O^+

OH

$Na_2Cr_2O_7$, H_2SO_4, H_2O

(d) The target molecule has one additional carbon atom, so the synthesis requires a C—C bond-forming reaction. The retrosynthetic analysis of the target molecule asks, "What reaction have we seen that produces an alkene product?" The alkene can be prepared by elimination of the corresponding alcohol, and the alcohol is suitable for making the required C—C bond disconnection (at the α position):

OH

CH_3—MgBr +
δ−

δ+

O

Nucleophile Electrophile

Now let's draw the forward process. The starting ketone can be treated with a Grignard reagent (methyl magnesium bromide) to give a Grignard reaction. The resulting alcohol can be converted to an alkene, either by acid-catalyzed dehydration (E1) or by E2 elimination of a good leaving group (*e.g.*, tosylate) to give the Zaitsev product:

conc. H_2SO_4 heat

O

1) MeMgBr
2) H_3O^+

OH

1) TsCl, py
2) NaOEt

(e) The target molecule has three additional carbon atoms, so the synthesis requires a C—C bond-forming reaction. The retrosynthetic analysis of the target molecule asks, "What reaction have we seen that produces an alkyl chloride product?" The chloride can be prepared by substitution of the corresponding alcohol, and the alcohol is suitable for making the required C—C bond disconnection (at the α position):

Cl

OH

δ−

MgCl

+

O

δ+

Nucleophile Electrophile

Now let's draw the forward process. The alkyl halide is first treated with magnesium metal to produce a Grignard reagent, followed by reaction with the required ketone (acetone). The resulting alcohol can then be converted to the desired target molecule by treatment with thionyl chloride (or $HCl/ZnCl_2$):

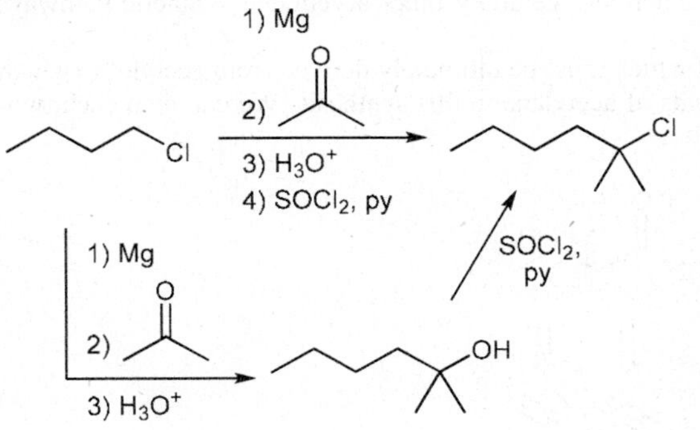

(f) The target molecule has two additional phenyl groups, so the retrosynthesis requires *two* C—C bond disconnections. The first disconnection in the retrosynthesis of the target molecule is made at the α position of the alcohol to give a suitable electrophile (C=O) and nucleophile (PhMgBr):

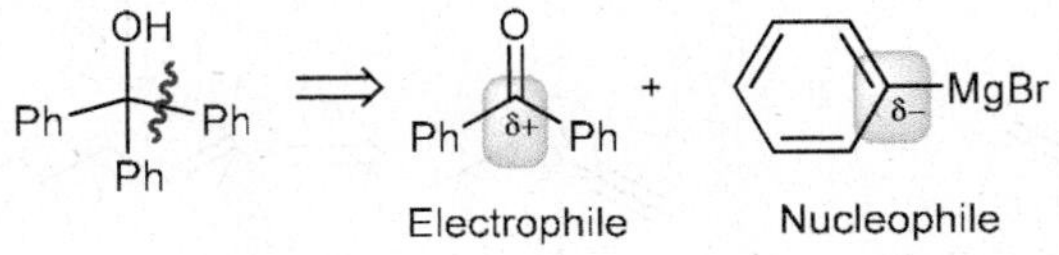

Electrophile Nucleophile

The retrosynthetic analysis continues by asking, "What reaction have we seen that produces a ketone product?" The ketone can be prepared by oxidation of the corresponding alcohol, and the alcohol is suitable for making the second required C—C bond disconnection (again at the α position):

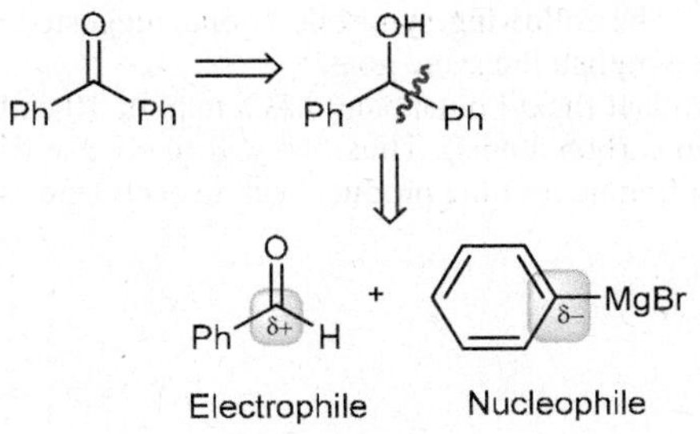

Electrophile Nucleophile

Now let's draw the forward process. The starting aldehyde is first treated with a Grignard reagent (phenyl magnesium bromide) to give a Grignard reaction. The resulting alcohol can be then converted to a ketone by oxidation with a variety of oxidizing agents, including chromic acid. Finally, a second Grignard reaction with phenyl magnesium bromide will give the desired target molecule:

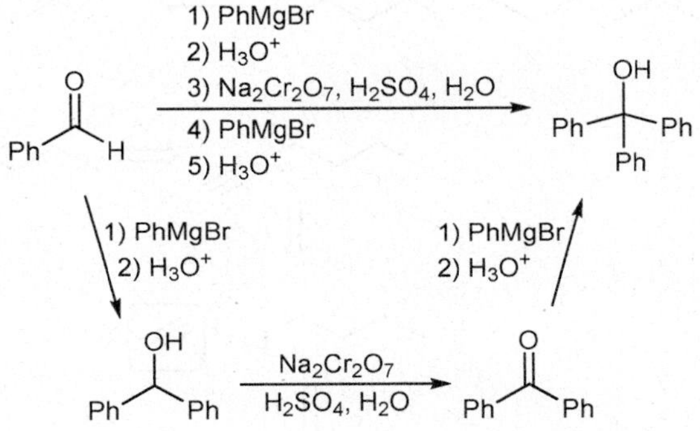

12.28. The following synthesis is one suggested approach. There are certainly other acceptable synthetic pathways that accomplish the same goal.

The product (hexyl butanoate) has a total of 10 carbons, all of which must be ultimately derived from acetylene (which has two carbon atoms). Thus, we will likely use five equivalents of acetylene in this synthesis. We can map each two-carbon fragment of the product back to acetylene as shown below.

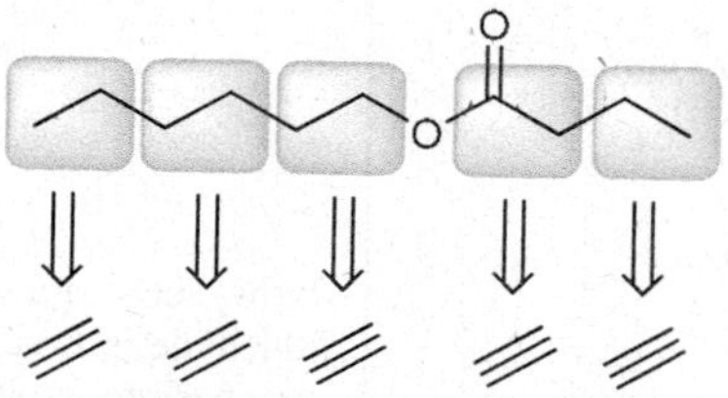

The figure below outlines a retrosynthetic analysis for our target molecule. An explanation of each of the steps (*a-l*) follows.

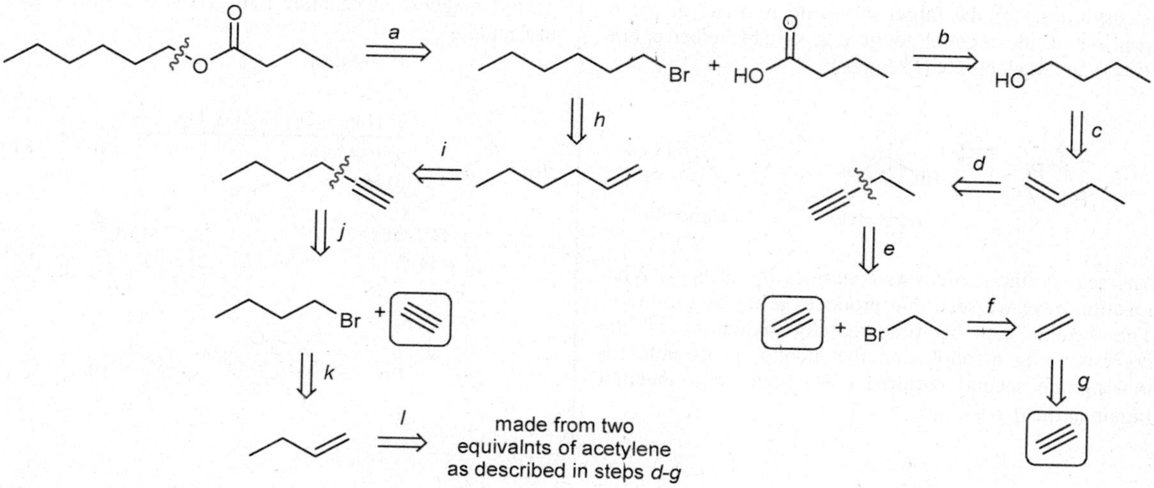

made from two equivalnts of acetylene as described in steps *d-g*

a. Hexyl butanoate is made from an S_N2 reaction between 1-bromohexane and the carboxylic acid shown (after deprotonation of the carboxylic acid to make a good nucleophile).

b. The carboxylic acid is made by oxidation of 1-butanol.

c. 1-Butanol is made by *anti*-Markovnikov hydration of 1-butene.

d. 1-Butene is made by partial reduction of 1-butyne.

e. 1-Butyne is made from an S_N2 reaction between acetylene (which must be deprotonated to form an acetylide ion) and bromoethane.

f. Bromoethane is made by addition of HBr to ethylene.

g. Ethylene is made by partial reduction of acetylene.

h. 1-Bromohexane is made by *anti*-Markovnikov addition of HBr to 1-hexene.

i. 1-Hexene is made by partial reduction of 1-hexyne.

j. 1-Hexyne is made from an S_N2 reaction between acetylene (which must be deprotonated to form an acetylide ion) and 1-bromobutane.

k. 1-Bromobutane is made by *anti*-Markovnikov addition of HBr to 1-butene.

l. 1-Butene is made as described in steps *d-g*.

Now let's draw the forward scheme. Acetylene is converted to bromoethane in two steps by hydrogenation with Lindlar's catalyst followed by addition of HBr to the resulting alkene. Reaction of bromoethane with sodium acetylide (made by deprotonation of acetylene with sodium amide) produces 1-butyne, which is then treated with H_2 and Lindlar's catalyst to furnish 1-butene. *Anti*-Markovnikov addition of water (via hydroboration-oxidation) gives 1-butanol, which is subsequently oxidized to the carboxylic acid using chromic acid. Deprotonation with sodium hydroxide followed by reaction with 1-bromohexane produces the product, hexyl butanoate. (1-Bromohexane is made in four steps from 1-butene, as shown. *Anti*-Markovnikov addition using HBr and peroxides gives 1-bromobutane. Reaction with sodium acetylide gives 1-hexyne, which is subsequently reduced to 1-hexene with H_2 and Lindlar's catalyst. *Anti*-Markovnikov addition using HBr and peroxides gives 1-bromohexane.)

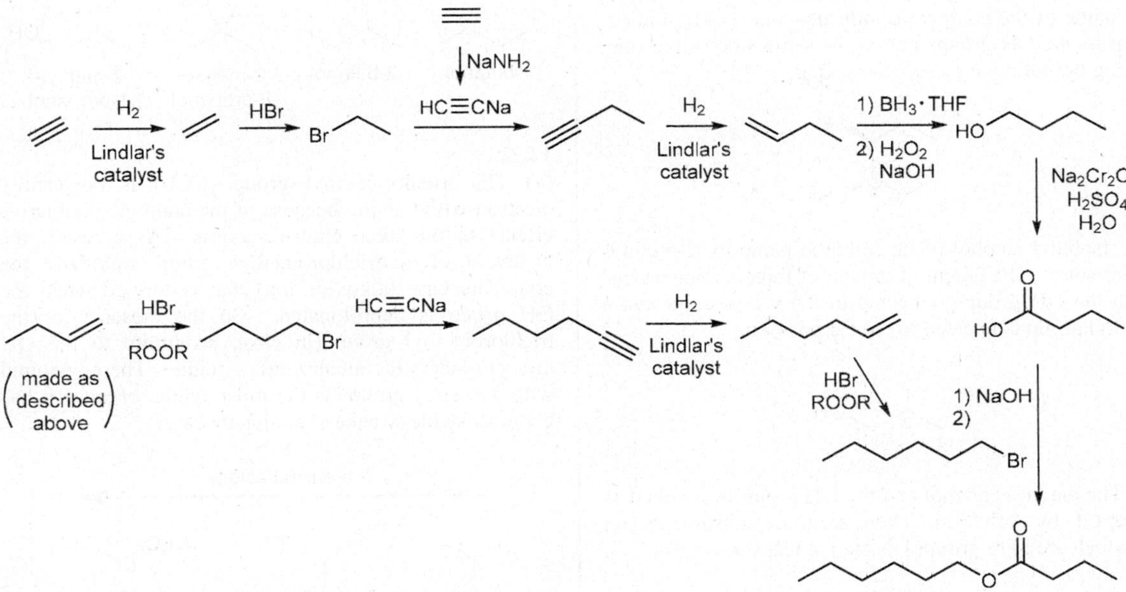

12.29.

(a) We begin by identifying the parent. The carbon atom connected to the OH group must be included in the parent. The longest chain that includes this carbon atom is five carbon atoms in length. So the parent is pentanol. There is only one substituent (highlighted) – a propyl group, located at C2. A locant is included to indicate the location of the OH group.

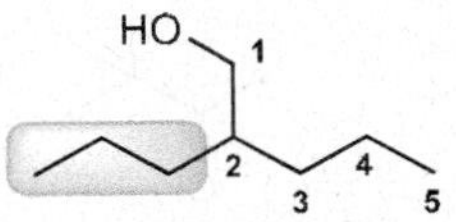

2-propyl-1-pentanol
or
2-propylpentan-1-ol

(b) We begin by identifying the parent. The carbon atom connected to the OH group must be included in the parent. The longest chain that includes this carbon atom is five carbon atoms in length. So the parent is pentanol. There is only one substituent (highlighted) – a methyl group, located at C4. Finally, we assign a configuration to the chiral center.

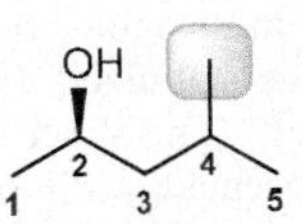

(*R*)-4-methyl-2-pentanol
or
(*R*)-4-methylpentan-2-ol

(c) We begin by identifying the parent (phenol). There are two substituents (highlighted) – a bromo group and a methyl group. Notice that the ring is numbered starting from the carbon atom bearing the OH group (it is not necessary to indicate a locant for the OH group, because in a ring, it is assumed to be at C1, by definition). We then number in the direction that gives the second substituent the lowest possible number (C2 rather than C4). According to this numbering scheme, the bromo group is at C2 and the methyl group is at C4. The substituents are arranged alphabetically in the name.

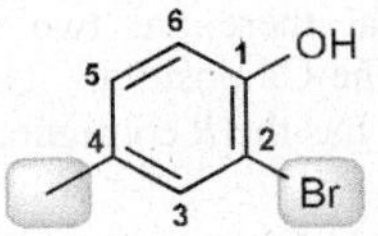

2-bromo-4-methylphenol

(d) We begin by identifying the parent. The carbon atom connected to the OH group must be included in the parent. That carbon atom is part of a six-membered ring, so the parent is cyclohexanol. There is only one substituent (highlighted) – a methyl group. Notice that the parent chain is numbered starting from the carbon atom bearing the OH group (it is not necessary to indicate a locant for the OH group, because in a ring, it is assumed to be at C1, by definition). The numbers go clockwise, so as to give the lowest number to the substituent (C2 rather than C6). Finally, we assign a configuration to each chiral center.

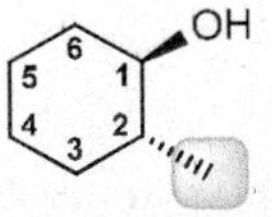

(1*R*,2*R*)-2-methylcyclohexanol

12.30.

(a) Cyclohexanediol is a six-membered ring containing two OH groups. The locants for the OH groups (C1 and

C2) indicate that they are on adjacent carbon atoms, and the name of the compound indicates a *cis* configuration. That is, the OH groups are on the same side of the ring, giving the following *meso* compound.

(b) Isobutyl alcohol is the common name for 2-methyl-1-propanol. The parent is a chain of three carbon atoms, with the OH group connected to the C1 position and a methyl group connected to the C2 position:

(c) The parent is phenol and the OH group is assumed to be at C1, by definition. There are three substituents (all of which are nitro groups) located at C2, C4 and C6.

(d) The parent (3-heptanol) is a chain of seven carbon atoms containing an OH group at the C3 position. The name indicates that there are two substituents (both methyl groups) at the C2 position. The C3 position is a chiral center, and it has the *R* configuration.

(e) Ethylene glycol is the common name for the following diol.

(f) The parent (1-butanol) is a chain of four carbon atoms with an OH group connected to the C1 position. The name indicates that there is one substituent (a methyl group) at C2. The C2 position is a chiral center, and it has the *S* configuration.

12.31. The solution to Problem 1.1d shows all constitutional isomers with the molecular formula $C_4H_{10}O$. Four of these isomers are alcohols, and have the following systematic names:

1-butanol, 2-butanol, 2-methyl-2-propanol, 2-methyl-1-propanol

12.32.

(a) The trichloromethyl group (CCl_3) is powerfully electron-withdrawing because of the combined inductive effects of the three chlorine atoms. As a result, the presence of a trichloromethyl group stabilizes the conjugate base (alkoxide ion) that is formed when the OH group is deprotonated. So the presence of the trichloromethyl group (in close proximity to the OH group) renders the alcohol more acidic. The compound with two such groups is the most acidic, because it has the most stable (weakest) conjugate base.

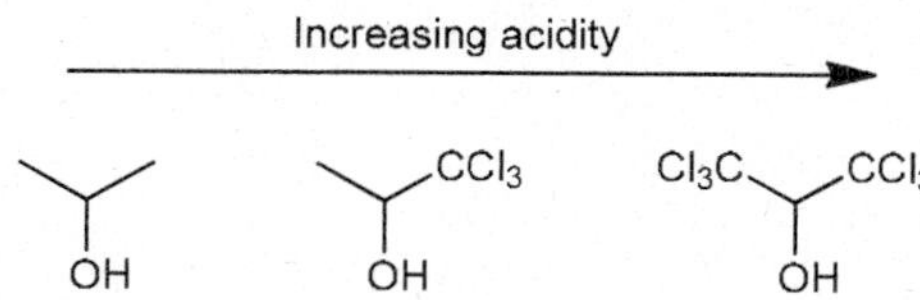

(b) The following order is based on a solvating effect, as described in Section 3.7. Specifically, the presence of *tert*-butyl groups will destabilize the conjugate base that is formed when the OH group is deprotonated. As such, the compound with two *tert*-butyl groups will be the least acidic, because it has the least stable (strongest) conjugate base.

Increasing acidity

(c) These compounds are expected to have the following relative acidity.

Increasing acidity

Cyclohexanol is the least acidic because its conjugate base is not resonance-stabilized (making it the least stable conjugate base). The other two compounds are much more acidic, because each of them generates a resonance-stabilized phenolate ion upon deprotonation. Among these two compounds, 2-nitrophenol is more acidic, because its conjugate base has an additional resonance structure in which the negative charge is placed on an oxygen atom of the nitro group (making it the most stable conjugate base).

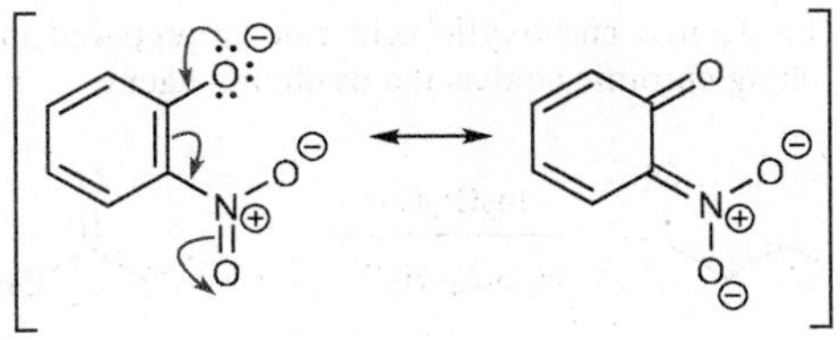

Note: This conjugate base has additional resonance structures not shown here.

12.33.

(a) This structure exhibits a lone pair next to a π bond, so we draw the two curved arrows associated with that pattern (see Section 2.10). The first curved arrow is drawn showing a lone pair becoming a π bond, while the second curved arrow shows a π bond becoming a lone pair. The resulting resonance structure also exhibits a lone pair next to a π bond, so again we draw the two curved arrows associated with that pattern. This is continued until we have drawn all of the resonance structures, shown here.

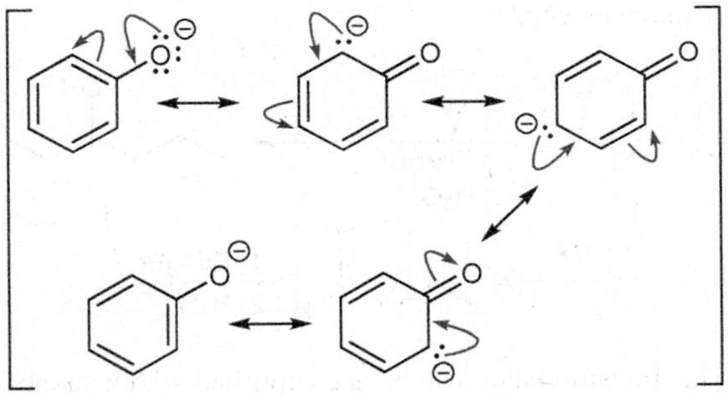

(b) This structure exhibits a lone pair next to a π bond, so we draw the two curved arrows associated with that pattern (see Section 2.10). The first curved arrow is drawn showing a lone pair becoming a π bond, while the second curved arrow shows a π bond becoming a lone pair.

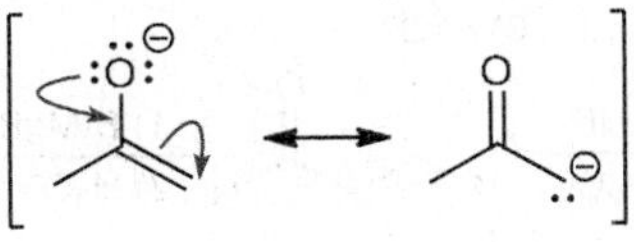

(c) This structure exhibits a lone pair next to a π bond, so we draw the two curved arrows associated with that pattern (see Section 2.10). The first curved arrow is drawn showing a lone pair becoming a π bond, while the second curved arrow shows a π bond becoming a lone pair. The resulting resonance structure also exhibits a lone pair next to a π bond, so again we draw the two curved arrows associated with that pattern, giving the third and final resonance structure.

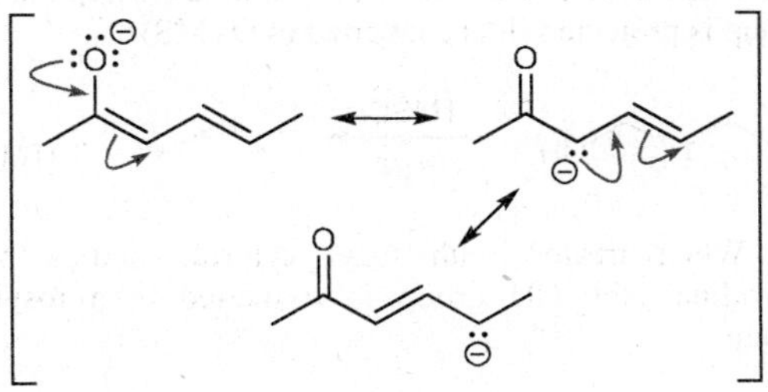

12.34.

(a) 1-Butanol is a primary alcohol. When treated with PBr_3, the OH group is replaced with Br, so the product is 1-bromobutane.

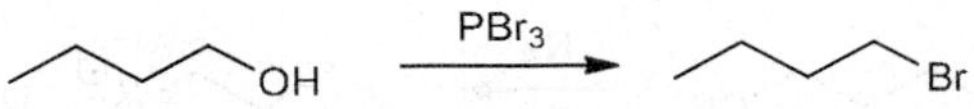

(b) 1-Butanol is a primary alcohol. When treated with $SOCl_2$ and pyridine, the OH group is replaced with Cl, so the product is 1-chlorobutane.

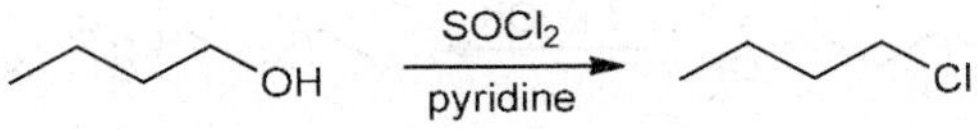

(c) 1-Butanol is a primary alcohol. When treated with HCl and $ZnCl_2$, the OH group is replaced with Cl, so the product is 1-chlorobutane.

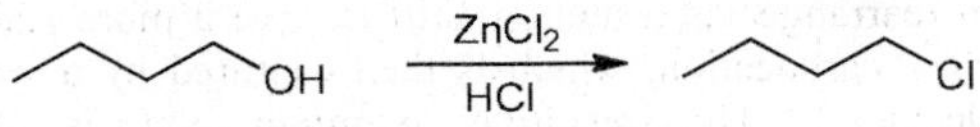

(d) When treated with Dess-Martin periodinane (DMP), a primary alcohol is oxidized to give an aldehyde (which is not further oxidized).

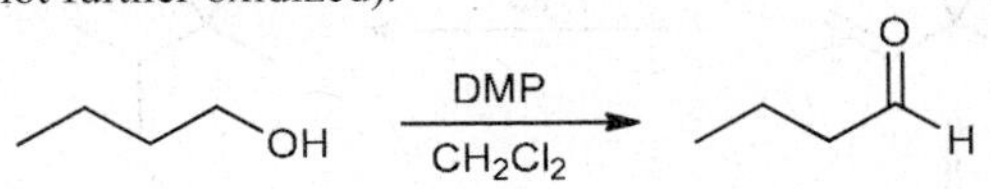

(e) When treated with chromic acid, a primary alcohol is oxidized to give a carboxylic acid.

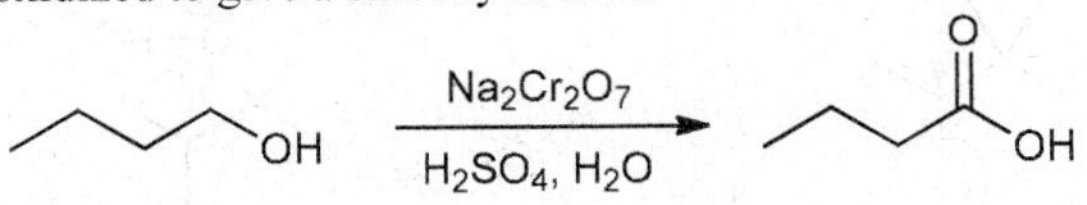

(f) When treated with lithium, an alcohol is converted to its conjugate base, an alkoxide ion.

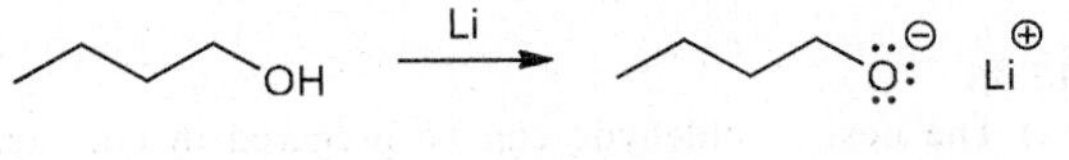

(g) When treated with sodium hydride (a strong base), an alcohol is deprotonated to give an alkoxide ion.

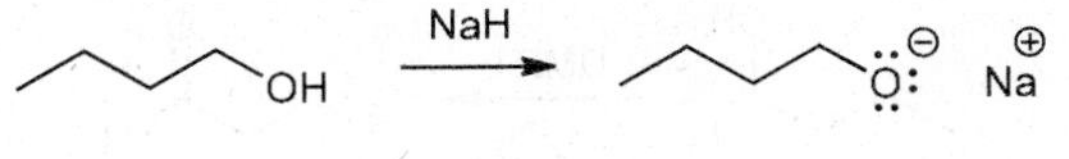

(h) When treated with TMSCl and a base (Et_3N), the OH group is protected (it is converted to OTMS).

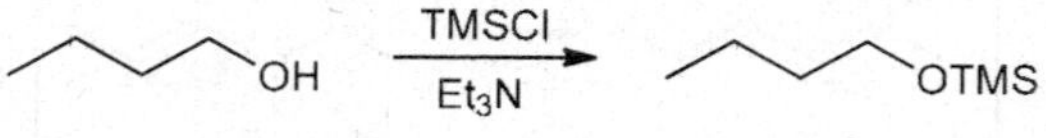

(i) When treated with tosyl chloride and a base (pyridine), the OH group is converted to a tosylate group.

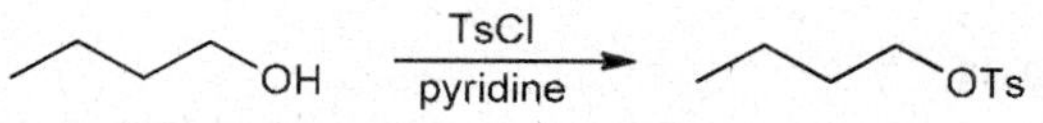

(j) When treated with sodium, an alcohol is converted to its conjugate base, an alkoxide ion.

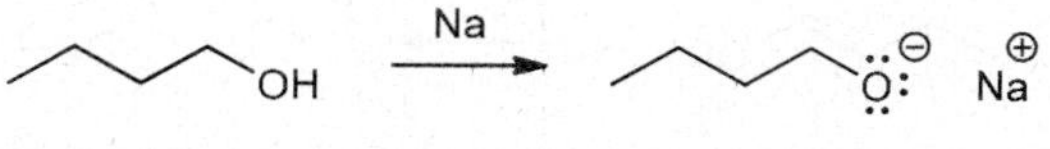

(k) When treated with *tert*-butoxide (a strong base), an alcohol is deprotonated to give an alkoxide ion.

12.35. When treated with aqueous acid, the π bond is protonated, giving a secondary carbocation (rather than a primary carbocation). This secondary carbocation can then rearrange via a methyl shift to give a more stable, tertiary carbocation, which is then captured by a water molecule. The resulting oxonium ion is then deprotonated by a molecule of water to give the product:

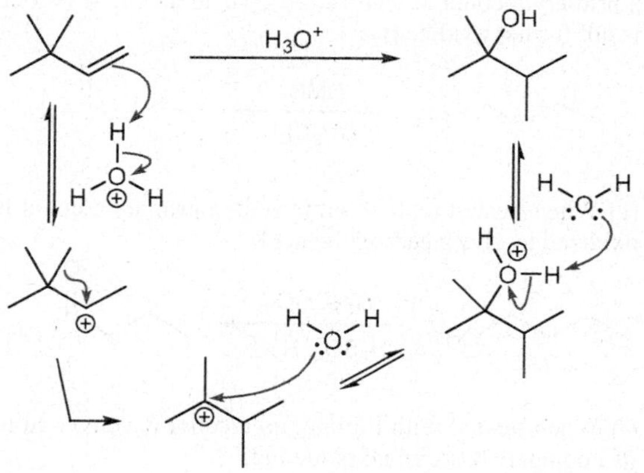

12.36.

(a) The desired aldehyde can be prepared in one step, using PCC, or DMP, or the Swern oxidation. One of these is shown below.

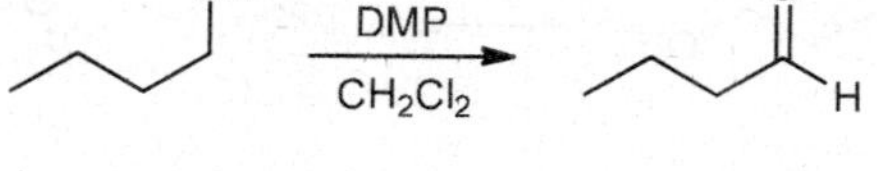

(b) The desired carboxylic acid can be prepared in one step, using chromic acid as the oxidizing agent.

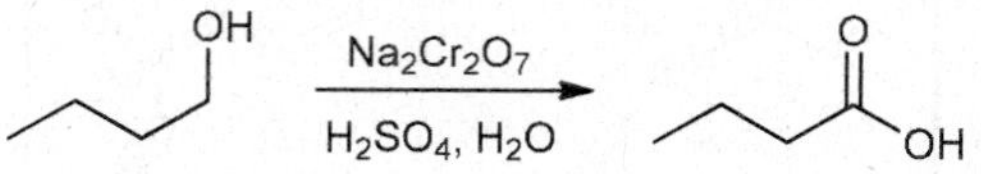

(c) The target molecule has two additional carbon atoms, so the retrosynthesis requires a C–C bond disconnection. The disconnection is made at the α position of the alcohol to give a suitable electrophile (C=O) and nucleophile (RMgBr):

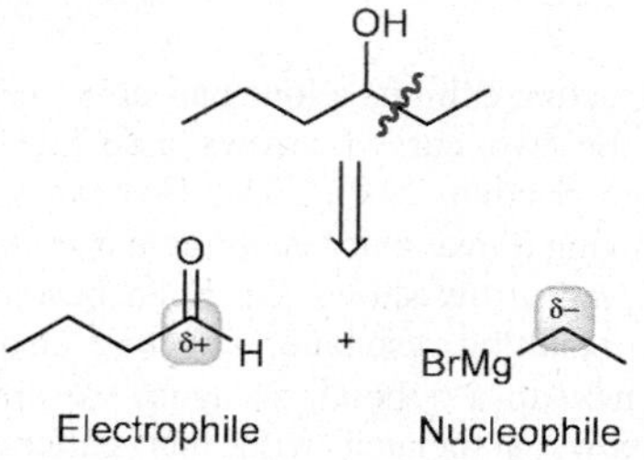

To complete the synthesis, we must bridge the gap between the given alcohol starting material and the required aldehyde:

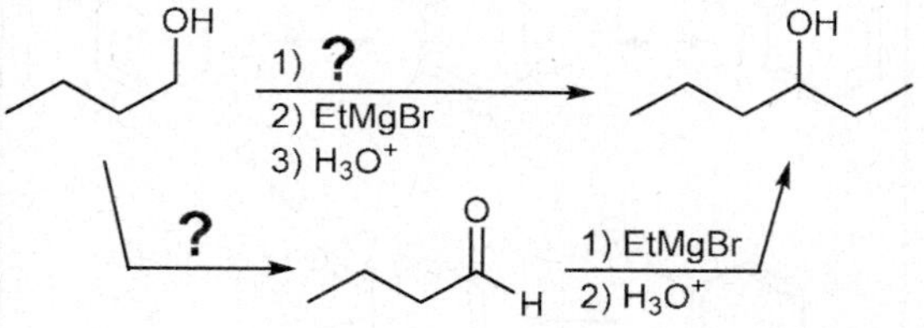

This transformation can be accomplished with a suitable oxidizing agent.

Now let's draw the forward process. The starting alcohol is first converted into an aldehyde (with DMP, or with PCC, or via a Swern oxidation) followed by a Grignard reaction with ethyl magnesium bromide.

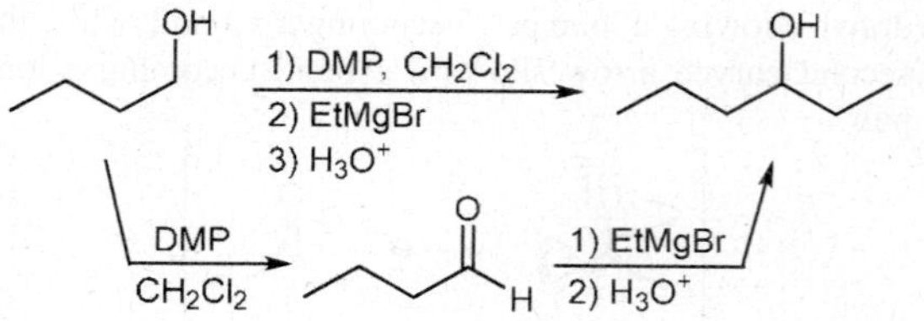

(d) The target molecule has two additional alkyl groups, so the retrosynthesis requires *two* C–C bond disconnections. The first disconnection in the retrosynthesis of the target molecule is made at the α position of the alcohol to give a suitable electrophile (C=O) and nucleophile (RMgBr):

Electrophile Nucleophile

The retrosynthetic analysis continues by asking, "What reaction have we seen that produces a ketone?" A ketone can be prepared by oxidation of the corresponding alcohol, and the alcohol is suitable for making the second required C–C bond disconnection (again at the α position):

Electrophile Nucleophile

To complete the synthesis, we must bridge the gap between the given alcohol starting material and the required aldehyde:

1) MeMgBr
2) H_3O^+

1) EtMgBr
2) H_3O^+

$Na_2Cr_2O_7$, H_2SO_4, H_2O

This transformation can be accomplished with a suitable oxidizing agent.

Now let's draw the forward process. The alcohol is first oxidized to give an aldehyde, which can then be treated with EtMgBr to give a Grignard reaction. The resulting alcohol is then oxidized to give a ketone that can undergo a second Grignard reaction. The first oxidation procedure can be performed with DMP (or PCC, or Swern oxidation) to give the aldehyde. The second oxidation procedure can also be achieved with a variety of oxidizing agents, including chromic acid.

1) DMP, CH_2Cl_2
2) EtMgBr
3) H_3O^+
4) $Na_2Cr_2O_7$, H_2SO_4, H_2O
5) MeMgBr
6) H_3O^+

DMP, CH_2Cl_2

1) MeMgBr
2) H_3O^+

1) EtMgBr
2) H_3O^+

$Na_2Cr_2O_7$, H_2SO_4, H_2O

(e) The target molecule has three additional carbon atoms, so the synthesis requires a C–C bond-forming reaction. The retrosynthetic analysis of the target molecule asks, "What reaction have we seen that produces a ketone?" A ketone can be prepared by oxidation of the corresponding alcohol, and the alcohol is suitable for making the required C–C bond disconnection (at the α position):

Electrophile Nucleophile

To complete the synthesis, we must bridge the gap between the given alcohol starting material and the required aldehyde:

1) $CH_3CH_2CH_2MgBr$
2) H_3O^+

$Na_2Cr_2O_7$, H_2SO_4, H_2O

This transformation can be accomplished with a suitable oxidizing agent.

Now let's draw the forward process. Oxidation with DMP (or PCC, or a Swern oxidation) gives an aldehyde, which can then be treated with a Grignard reagent to give a secondary alcohol. Oxidation of this alcohol gives the desired ketone, as shown.

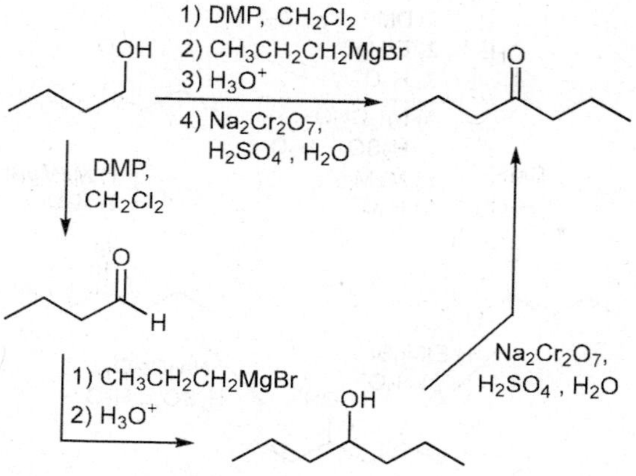

1) DMP, CH_2Cl_2
2) $CH_3CH_2CH_2MgBr$
3) H_3O^+
4) $Na_2Cr_2O_7$, H_2SO_4, H_2O

DMP, CH_2Cl_2

1) $CH_3CH_2CH_2MgBr$
2) H_3O^+

$Na_2Cr_2O_7$, H_2SO_4, H_2O

12.37.

(a) The desired product has only one group connected to the α position. That group could have been installed via a Grignard reaction with formaldehyde, as shown.

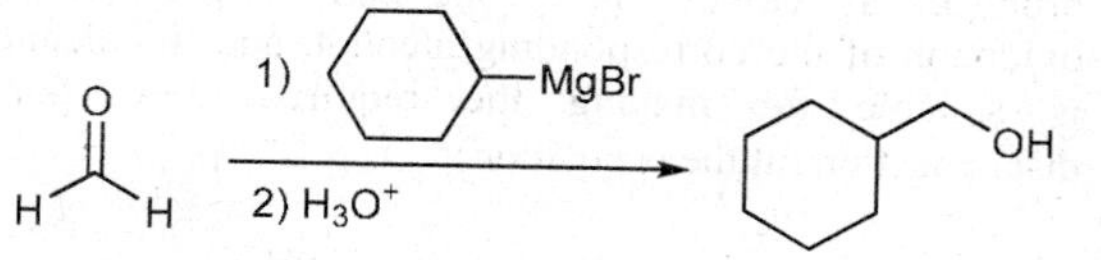

1) cyclohexyl–MgBr
2) H_3O^+

(b) The desired product has three groups connected to the α position. Any one of these groups could have been installed via a Grignard reaction with the appropriate ketone, as shown.

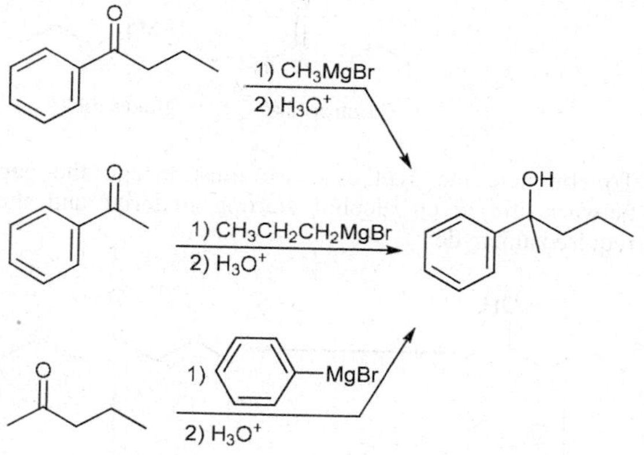

1) CH_3MgBr
2) H_3O^+

1) $CH_3CH_2CH_2MgBr$
2) H_3O^+

1) phenyl–MgBr
2) H_3O^+

(c) The desired product has only two groups connected to the α position. Either one of these groups could have been installed via a Grignard reaction with the appropriate aldehyde, as shown.

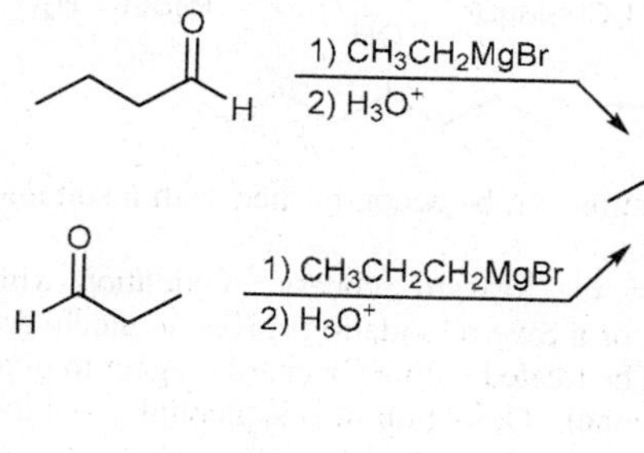

1) CH_3CH_2MgBr
2) H_3O^+

1) $CH_3CH_2CH_2MgBr$
2) H_3O^+

12.38.

(a) Reduction of the following aldehyde will afford the desired product.

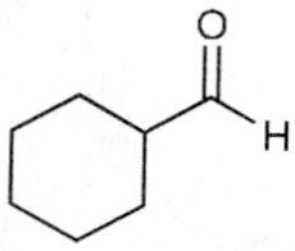

(b) Reduction of the following ketone will afford the desired product.

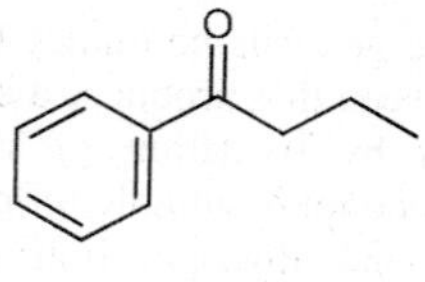

(c) Reduction of the following ketone will afford the desired product.

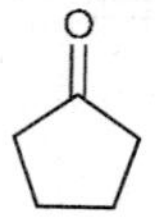

12.39.

(a) The target molecule has two additional carbon atoms, so the synthesis requires a C–C bond-forming reaction. The retrosynthetic analysis of the target molecule asks, "What reaction have we seen that produces a ketone product?" A ketone can be prepared via oxidation of the corresponding alcohol, and the alcohol is suitable for making the required C–C bond disconnection (at the α position):

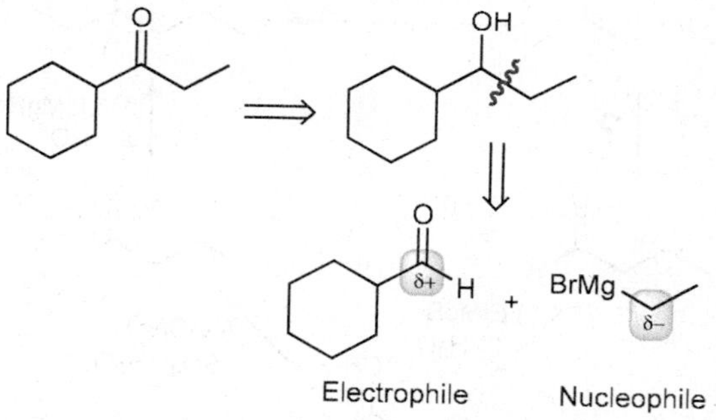

δ+ δ−

Electrophile Nucleophile

Now let's draw the forward process. The synthesis begins with a Grignard reaction with ethyl magnesium bromide, and the resulting alcohol can then be oxidized to give the desired target molecule:

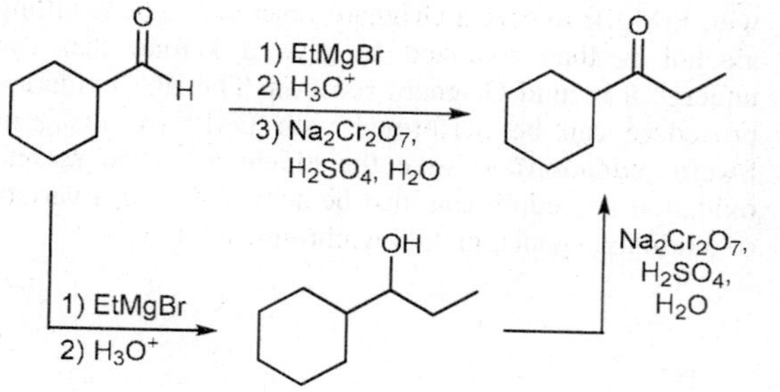

1) EtMgBr
2) H_3O^+
3) $Na_2Cr_2O_7$, H_2SO_4, H_2O

1) EtMgBr
2) H_3O^+

$Na_2Cr_2O_7$, H_2SO_4, H_2O

(b) Reduction of the aldehyde can be achieved with either $LiAlH_4$ or $NaBH_4$. This transformation cannot be achieved via catalytic hydrogenation, as that process would also reduce the carbon-carbon π bond.

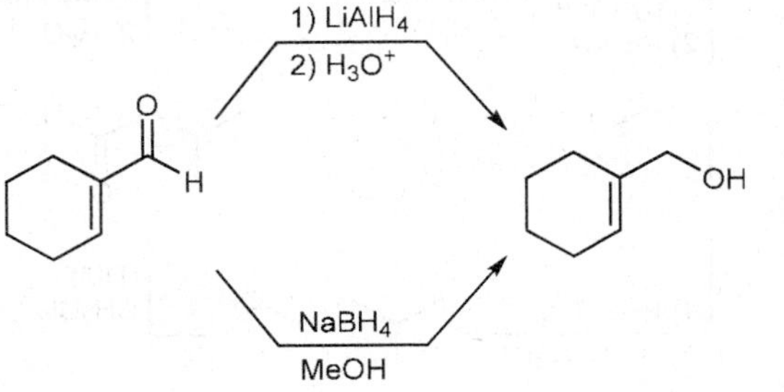

1) $LiAlH_4$
2) H_3O^+

$NaBH_4$
MeOH

(c) We begin by asking the following two questions:

1) *Is there a change in the carbon skeleton?* Yes, the carbon skeleton is increasing in size by one carbon atom.
2) *Is there a change in the functional groups?* Yes, the starting material has a carbon-carbon double bond, while the product has an OH group.

Now we must propose a strategy for achieving these changes.

The target molecule has one additional carbon atom, so the retrosynthesis requires a C—C bond disconnection. The disconnection is made at the α position of the alcohol to give a suitable electrophile (C=O) and nucleophile (RMgBr):

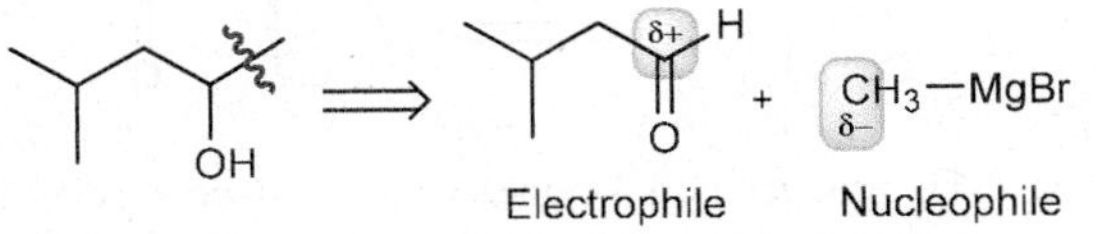

OH ⟹ δ+ H O + CH_3—MgBr δ−

Electrophile Nucleophile

To complete the synthesis, we need to convert the given alkene starting material into the required aldehyde. The aldehyde can be made by oxidation of the primary alcohol, or from the terminal alkyne via hydroboration-oxidation.

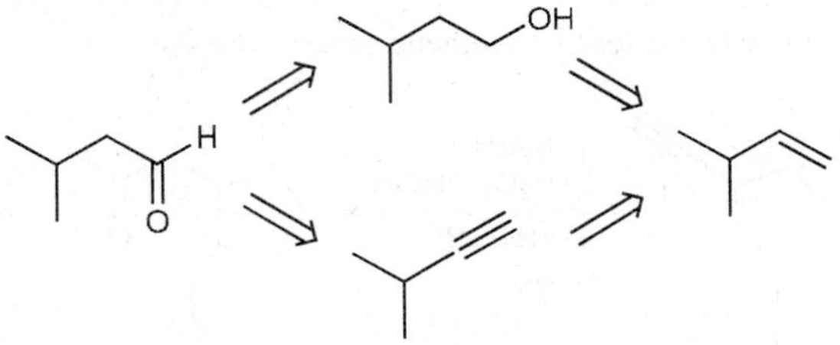

One solution for the synthesis is to convert the alkene into an alkyne (via bromination followed by double elimination) and then to perform hydroboration-oxidation to obtain the aldehyde. Another method is to perform hydroboration-oxidation with the starting alkene, followed by oxidation with PCC to give the aldehyde. Finally, treatment of the aldehyde with methyl magnesium bromide affords the desired target molecule.

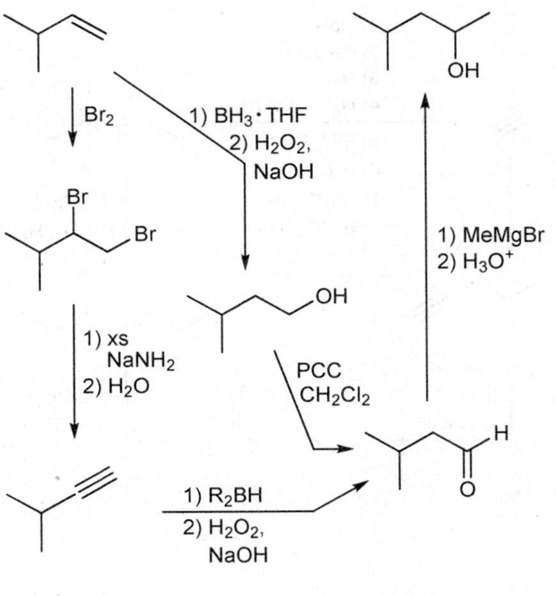

Br_2

1) $BH_3 \cdot THF$
2) H_2O_2, NaOH

1) xs $NaNH_2$
2) H_2O

PCC
CH_2Cl_2

1) R_2BH
2) H_2O_2, NaOH

1) MeMgBr
2) H_3O^+

(d) The starting material has two carbon atoms, while the product has seven carbon atoms, so we will need to create carbon-carbon bonds. The starting functional group is a triple bond, and the product is a ketone. There are certainly many acceptable answers to this problem. One such answer can be rationalized with the following retrosynthetic analysis.

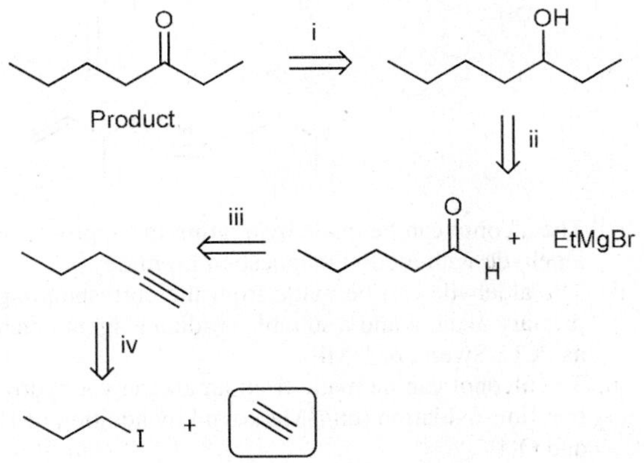

Product

\+ EtMgBr

\+ I

i. The product is a ketone, which can be made via oxidation of the corresponding secondary alcohol.
ii. The secondary alcohol can be made by treating the appropriate aldehyde with ethyl magnesium bromide.
iii. The aldehyde can be made from the corresponding terminal alkyne (via hydroboration-oxidation).
iv. The terminal alkyne can be made from the starting material via alkylation (to install a propyl group).

The reagents for this synthetic strategy are shown here.

1) $NaNH_2$
2) $CH_3CH_2CH_2I$
3) R_2BH
4) H_2O_2, NaOH
5) EtMgBr
6) H_3O^+
7) $Na_2Cr_2O_7$, H_2SO_4, H_2O

1) $NaNH_2$
2) I

1) R_2BH
2) H_2O_2, NaOH

1) EtMgBr
2) H_3O^+

$Na_2Cr_2O_7$, H_2SO_4, H_2O

(e) The starting material has five carbon atoms, while the product has six carbon atoms, so we will need to install a methyl group. There are certainly many acceptable answers to this problem. One such answer can be rationalized with the following retrosynthetic analysis.

i. The alcohol can be made by treating the appropriate aldehyde with methyl magnesium bromide.
ii. The aldehyde can be made from the corresponding primary alcohol and a suitable oxidizing agent (such as PCC, Swern, or DMP).
iii. The alcohol can be made from an alkene via hydroboration-oxidation (*anti*-Markovnikov addition of H and OH)
iv, The alkene can be made from the starting alcohol via an elimination process. Since OH is a bad leaving group, it must first be converted into a tosylate in order to perform an E2 reaction. A sterically hindered base is then used to favor the less-substituted alkene (Hofmann product).

Reagents for this synthetic strategy are shown here:

1) TsCl, py
2) *t*-BuOK
3) $BH_3 \cdot THF$
4) H_2O_2, NaOH
5) DMP, CH_2Cl_2
6) MeMgBr
7) H_3O^+

1) TsCl, pyridine
2) *t*-BuOK

1) $BH_3 \cdot THF$
2) H_2O_2, NaOH

DMP CH_2Cl_2

1) MeMgBr
2) H_3O^+

(f) The product has one more carbon atom than the starting material, so we will need to install a methyl group. The starting functional group is a triple bond, and the product is a ketone. There are certainly many acceptable answers to this problem. One such answer can be rationalized with the following retrosynthetic analysis.

i. The product is a ketone, which can be made via oxidation of the corresponding secondary alcohol.
ii. The secondary alcohol can be made by treating the appropriate aldehyde with methyl magnesium bromide.
iii. The aldehyde can be made from the corresponding terminal alkyne (via hydroboration-oxidation).

The reagents for this synthetic strategy are shown here.

1) R_2BH
2) H_2O_2, NaOH
3) MeMgBr
4) H_3O^+
5) $Na_2Cr_2O_7$, H_2SO_4, H_2O

1) R_2BH
2) H_2O_2, NaOH

1) MeMgBr
2) H_3O^+

$Na_2Cr_2O_7$, H_2SO_4, H_2O

12.40. Hydride functions as a base and removes a proton from the alcohol, giving an alkoxide ion. This intermediate has both a nucleophilic region (the negatively charged oxygen atom) and an electrophilic region (the position that is α to the bromine atom). As such, an intramolecular, S_N2-type process can occur, giving a cyclic product.

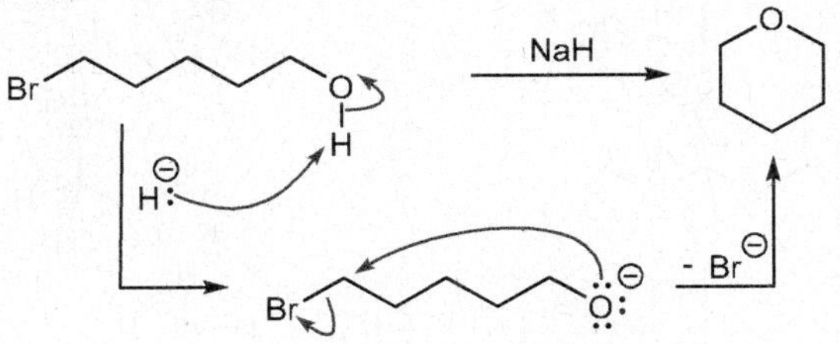

12.41. The major product is 1-methylcyclohexanol (resulting from Markovnikov addition of H and OH), which is a tertiary alcohol.

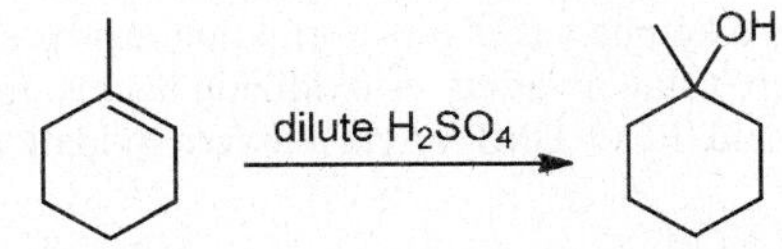

Tertiary alcohols do not generally undergo oxidation. In contrast, the minor product (2-methylcyclohexanol) is a secondary alcohol and can undergo oxidation to yield a ketone.

12.42. The conversion of compound **B** to compound **C** is achieved via a Grignard reaction that employs acetone as the electrophile. Therefore, the Grignard reagent (compound **B**) must be cyclohexyl magnesium bromide, as shown. Compound **A** is the alkyl bromide from which the Grignard reagent is prepared (bromocyclohexane).

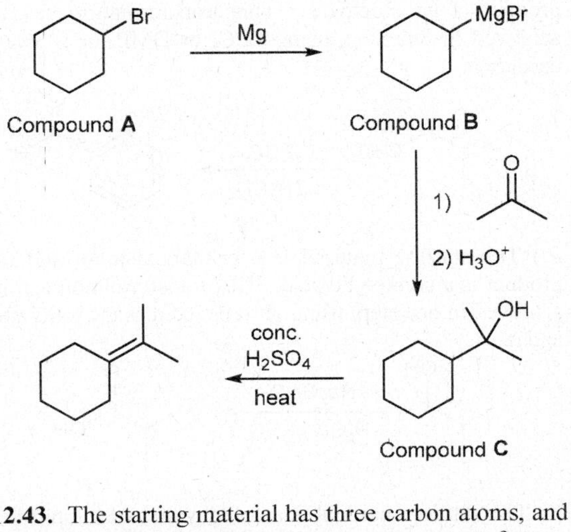

12.43. The starting material has three carbon atoms, and the product has six carbon atom, so we must form a carbon-carbon bond. This can be achieved with a Grignard reaction.

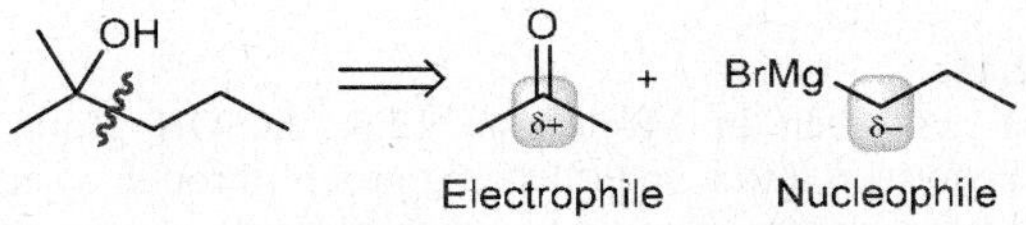

The reagents for this Grignard reaction (acetone and $CH_3CH_2CH_2MgBr$) can both be prepared from the starting alcohol, as shown here.

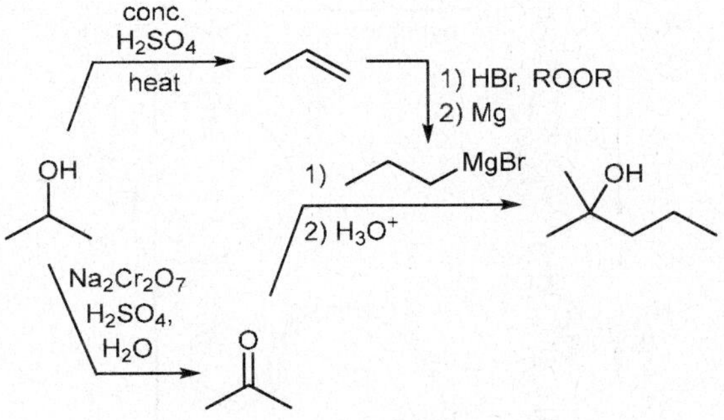

12.44.

(a) Two curved arrows are used to show hydride delivery to the electrophilic carbon atom of the ketone carbonyl (C=O). Note that the tail of the first curved arrow is placed on the bond between Al and H (it is NOT placed on the negative charge). The resulting alkoxide ion is then protonated upon treatment with aqueous acid. This protonation step (workup) requires two curved arrows, as shown.

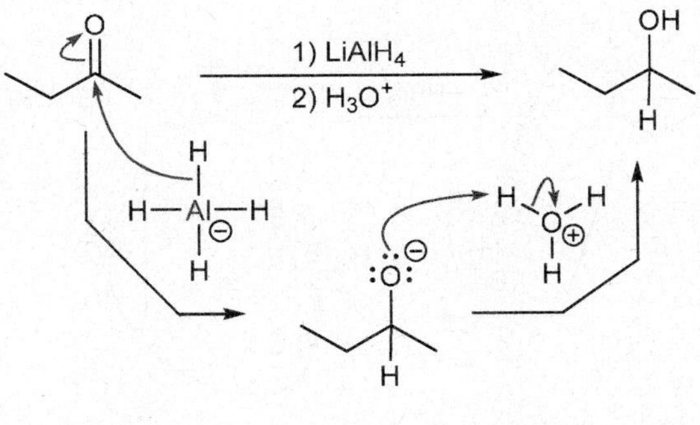

(b) Two curved arrows are used to show hydride delivery to the electrophilic carbon atom of the ketone carbonyl (C=O). Note that the tail of the first curved arrow is placed on the bond between B and H (it is NOT placed on the negative charge). The resulting alkoxide ion is then protonated by the protic solvent (methanol), which requires two curved arrows, as shown.

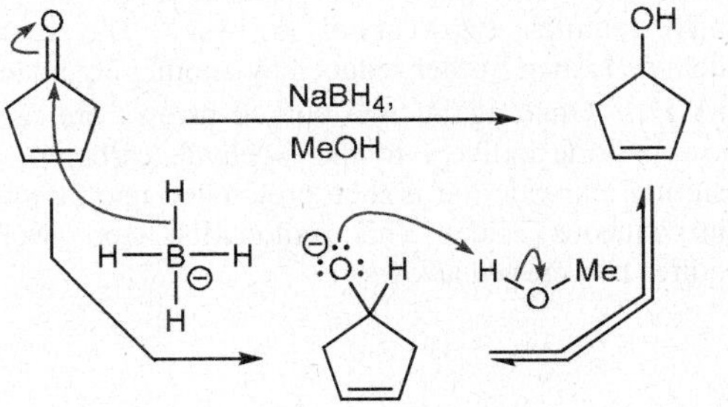

12.45.

(a) As seen in Mechanism 12.6, the OH group is transformed into a better leaving group (through a series of steps shown below), and then chloride attacks as a nucleophile and expels the leaving group (SN2 process).

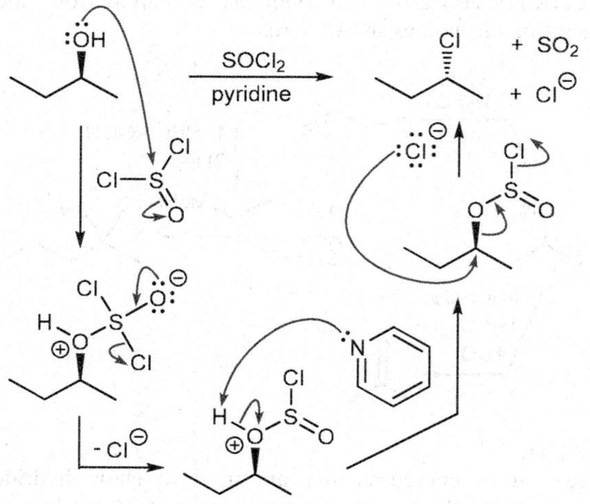

(b) As seen in Mechanism 12.7, the OH group is transformed into a better leaving group, and then bromide attacks as a nucleophile and expels the leaving group (SN2 process).

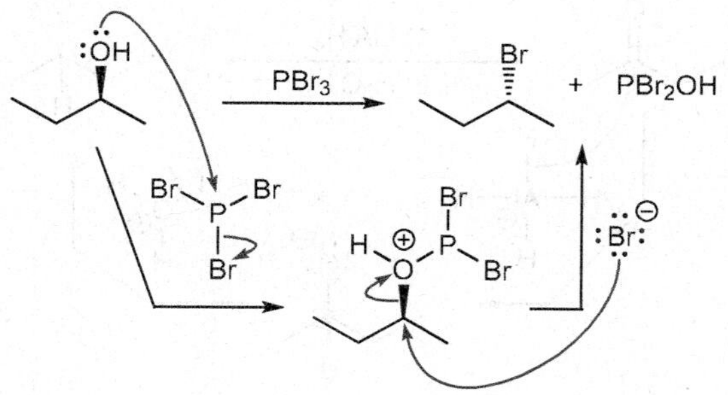

(c) Two curved arrows are used to show hydride delivery to the electrophilic carbon atom of the ester carbonyl (C=O). Note that the tail of the first curved arrow is placed on the bond between Al and H (it is NOT placed on the negative charge). The resulting intermediate then ejects methoxide as a leaving group, which requires two curved arrows. The resulting aldehyde is then further reduced by another equivalent of $LiAlH_4$. Once again, two curved arrows are used to show hydride delivery to the aldehyde carbonyl. The resulting alkoxide ion is then protonated upon treatment with aqueous acid. This protonation step (workup) requires two curved arrows.

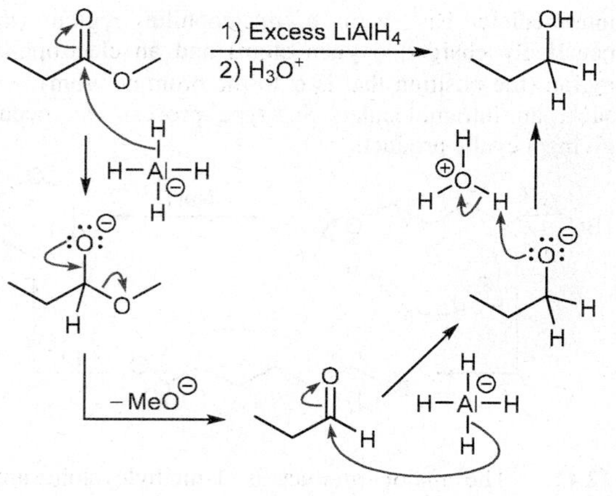

12.46.

(a) The starting material is a secondary alcohol, and the product is a ketone. This transformation can be achieved in one step, using a variety of oxidizing agents, including chromic acid, PCC, DMP or via a Swern oxidation.

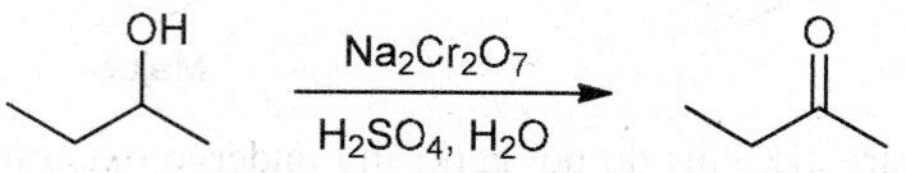

(b) The starting material is a primary alcohol and the product is an aldehyde. This transformation can be achieved in one step, using PCC, or DMP, or a Swern oxidation.

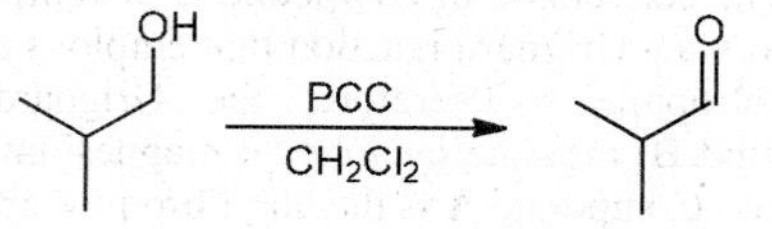

(c) The starting material is a primary alcohol, and the product is an aldehyde. This transformation can be achieved in one step, using PCC, or DMP, or a Swern oxidation.

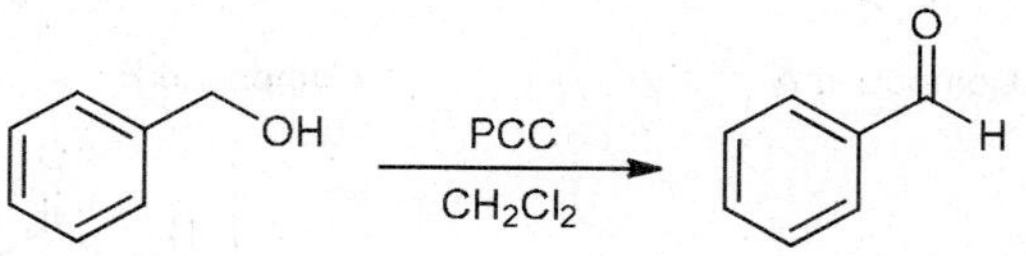

(d) The starting material is a primary alcohol and the product is a carboxylic acid. This transformation can be achieved in one step, using chromic acid as the oxidizing agent.

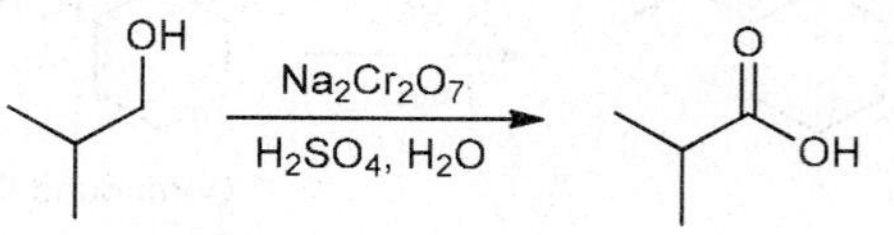

(e) The starting material is an aldehyde, and the product is an alcohol. This transformation can be achieved in one step, using $LiAlH_4$ or $NaBH_4$ as a reducing agent.

1) $LiAlH_4$
2) H_3O^+

(f) The starting material is a ketone and the product is an alcohol. This transformation can be achieved in one step, using $LiAlH_4$ or $NaBH_4$ as a reducing agent.

1) $LiAlH_4$
2) H_3O^+

12.47.

(a) Ozonolysis of the alkene gives a dialdehyde, which is then reduced when treated with excess $LiAlH_4$, followed by aqueous acidic workup, to give a diol.

1) O_3
2) DMS
3) xs $LiAlH_4$
4) H_3O^+

1) O_3
2) DMS

1) xs $LiAlH_4$
2) H_3O^+

(b) Ozonolysis of the alkene gives a dialdehyde, which is then reduced when treated with excess $LiAlH_4$, followed by aqueous acidic workup, to give a diol.

1) O_3
2) DMS
3) xs $LiAlH_4$
4) H_3O^+

1) O_3
2) DMS

1) xs $LiAlH_4$
2) H_3O^+

(c) Treating the aldehyde with ethyl magnesium bromide (followed by aqueous acidic workup) gives a secondary alcohol, which is then oxidized to give a ketone upon treatment with chromic acid. Finally, the ketone is converted to a tertiary alcohol when treated with ethyl magnesium bromide (followed by aqueous acidic workup).

1) EtMgBr
2) H_3O^+
3) $Na_2Cr_2O_7$, H_2SO_4, H_2O
4) EtMgBr
5) H_3O^+

1) EtMgBr
2) H_3O^+

$Na_2Cr_2O_7$, H_2SO_4, H_2O

1) EtMgBr
2) H_3O^+

(d) The aldehyde is reduced upon treatment with $LiAlH_4$, followed by aqueous acidic workup, to give an alcohol. Treating the alcohol with tosyl chloride and pyridine converts the alcohol into a tosylate.

1) $LiAlH_4$
2) H_3O^+
3) TsCl, pyridine

1) $LiAlH_4$
2) H_3O^+

TsCl, pyridine

(e) Acid-catalyzed hydration of the alkene gives a secondary alcohol, which is then oxidized to a ketone upon treatment with chromic acid. Finally, the ketone is converted to a tertiary alcohol when treated with a Grignard reagent (followed by aqueous acidic workup).

1) H_3O^+
2) $Na_2Cr_2O_7$, H_2SO_4, H_2O
3) PhMgBr
4) H_3O^+

H_3O^+

$Na_2Cr_2O_7$, H_2SO_4, H_2O

1) PhMgBr
2) H_3O^+

12.48.

(a) In the first step of the mechanism shown below, the Grignard reagent (methyl magnesium bromide) functions as a nucleophile and attacks the electrophilic carbon atom of the ketone carbonyl (C=O). This step requires two curved arrows. The resulting alkoxide ion is then protonated upon treatment with aqueous acid. This proton transfer step (workup) also requires two curved arrows, as shown.

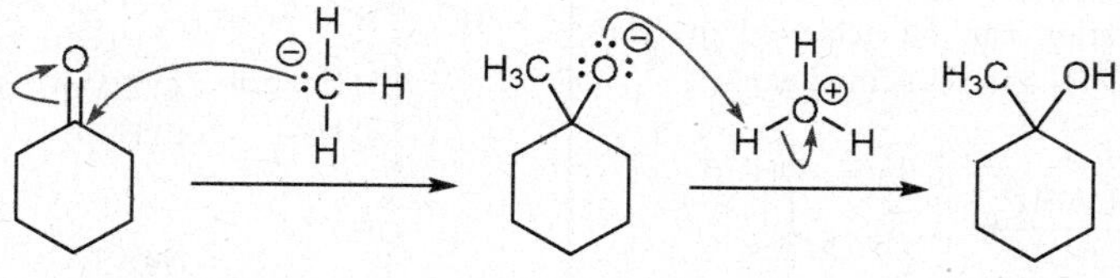

(b) In the first step of the mechanism shown below, the Grignard reagent (methyl magnesium bromide) functions as a nucleophile and attacks the electrophilic carbon atom of the ester carbonyl (C=O). This step requires two curved arrows. The resulting intermediate then ejects a leaving group to give a ketone, which requires two curved arrows. The ketone is then further attacked by another equivalent of the Grignard reagent. Once again, two curved arrows are used to show the nucleophilic attack, resulting in a dianion.

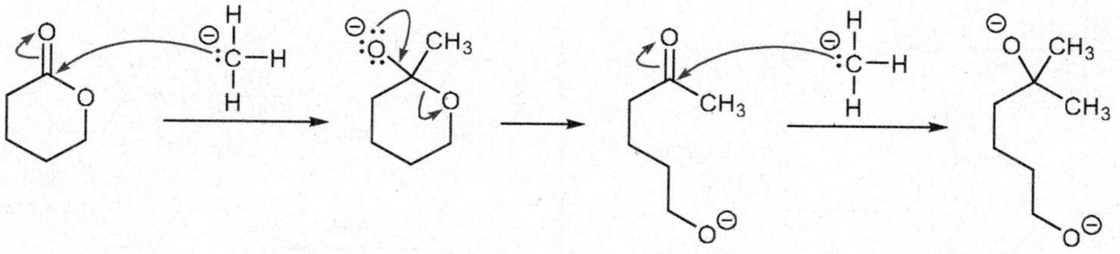

The resulting dianion is then protonated upon treatment with aqueous acid (workup). There are two locations that are protonated, each of which requires two curved arrows, as shown. Notice that each anion is protonated in a separate step (this should not be drawn as one step with four curved arrows, because there are two distinct processes occurring, and it is unlikely that they occur precisely at the same moment). The less-stable negative charge (the tertiary alkoxide) is the stronger base and is therefore protonated first.

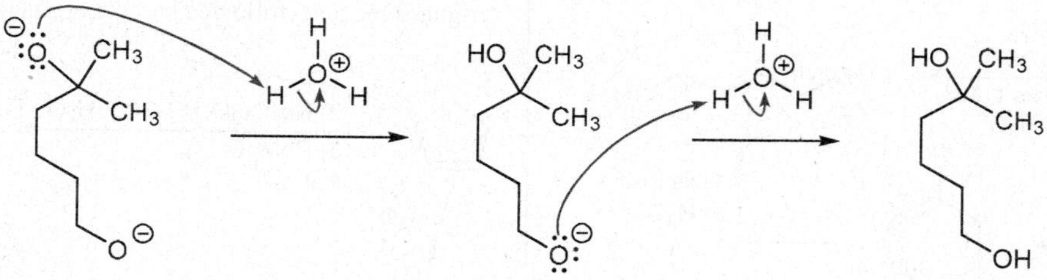

12.49. We begin by identifying the carbon atoms (highlighted below) that undergo a change in bonding as a result of the reaction:

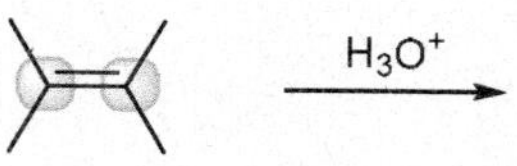

For each of these (highlighted) carbon atoms, we calculate its oxidation state in both the starting alkene and the alcohol product:

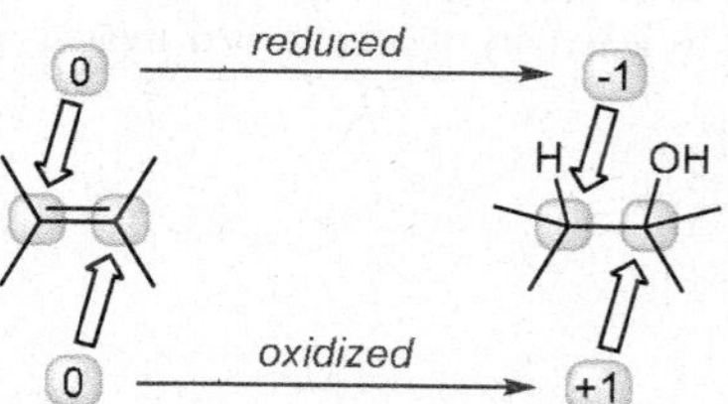

Notice that one carbon atom is *reduced* from an oxidation state of 0 to an oxidation state of -1, while the other carbon atom is *oxidized* from an oxidation state of 0 to an oxidation state of +1. Overall, the starting material does not undergo a net change in oxidation state and is, therefore, neither reduced nor oxidized.

12.50. We begin by identifying the carbon atoms (highlighted below) that undergo a change in bonding as a result of the reaction:

For each of these (highlighted) carbon atoms, we calculate its oxidation state in both the starting alkyne and the ketone product:

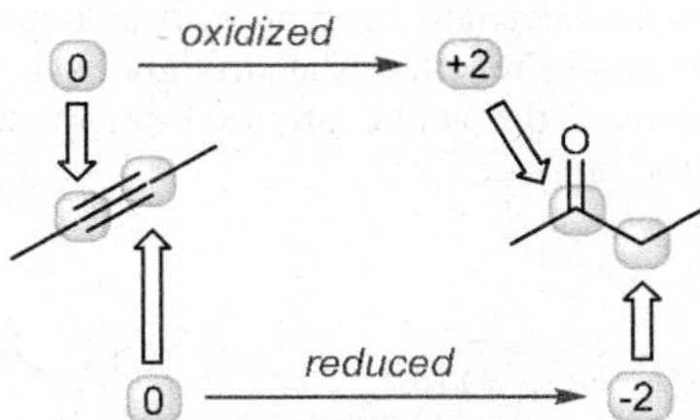

Notice that one carbon atom is *oxidized* from an oxidation state of 0 to an oxidation state of +2, while the other carbon atom is *reduced* from an oxidation state of 0 to an oxidation state of -2. Overall, the starting material does not undergo a net change in oxidation state and is, therefore, neither reduced nor oxidized.

12.51. Two curved arrows are used to show delivery of hydride to the electrophilic carbon atom of the ester carbonyl (C=O). Note that the tail of the first curved arrow is placed on the bond between Al and H (it is NOT placed on the negative charge). The resulting intermediate then ejects an alkoxide leaving group, which requires two curved arrows and opens the ring. The resulting aldehyde is then further reduced by another equivalent of $LiAlH_4$. Once again, two curved arrows are used to show hydride delivery, resulting in a dianion.

The dianion is then protonated upon treatment with aqueous acid (workup). There are two locations that are protonated, each of which requires two curved arrows, as shown. Notice that each protonation step is drawn separately. The less stable negative charge (the tertiary alkoxide) is the stronger base and is therefore protonated first.

12.52. In the first step of the mechanism, the Grignard reagent (methyl magnesium bromide) functions as a nucleophile and attacks the electrophilic carbon atom of the ester carbonyl (C=O). This step requires two curved arrows. The resulting intermediate then ejects an alkoxide leaving group to give a ketone, which requires two curved arrows and opens the ring. The ketone is then further attacked by another equivalent of the Grignard reagent. Once again, two curved arrows are used to show the nucleophilic attack, resulting in a dianion.

The resulting dianion is then protonated upon treatment with aqueous acid (workup). There are two locations that are protonated, each of which requires two curved arrows, as shown.

Notice that each protonation step is drawn separately. The less stable negative charge (the alkoxide) is the stronger base and is protonated first (rather than phenolate, which is a more stable (weaker) base because it is resonance-stabilized). In fact, the phenolate anion is more stable than a hydroxide ion, which explains why the proton source must be H_3O^+ rather than water. Water is not sufficiently acidic to protonate a phenolate ion (see Section 12.2, Acidity of Alcohols and Phenols).

12.53. The correct answer is (c). The starting material has two OH groups, one of which is primary, and the other is secondary. The secondary alcohol must be oxidized to a ketone, which can be achieved with a variety of reagents, but the primary alcohol must be oxidized to an aldehyde (and not a carboxylic acid). Only PCC will achieve the desired transformation.

12.54. The correct answer is (d). Options (a), (b), and (c) will achieve the desired synthesis of 2-methyl-2-pentanol via a Grignard reaction.

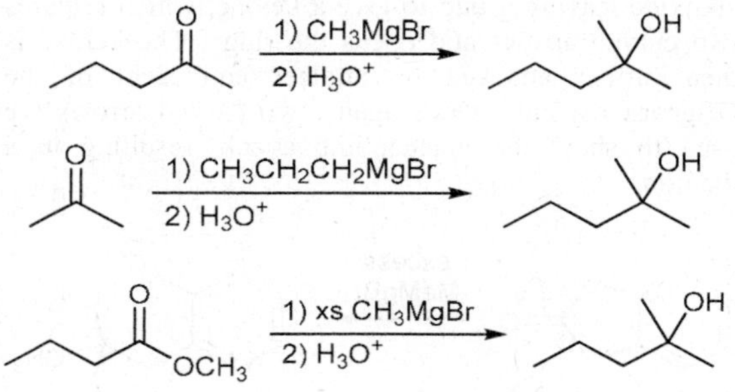
1) CH3MgBr
2) H3O+
OH
1) CH3CH2CH2MgBr
2) H3O+
OH
OCH3
1) xs CH3MgBr
2) H3O+
OH

Option (d) will NOT achieve the desired transformation. The strongly basic Grignard reagent will simply deprotonate the carboxylic acid, to give a carboxylate ion, thereby destroying the Grignard reagent.

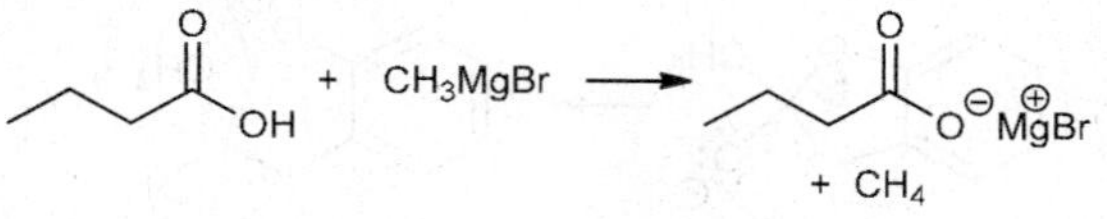
OH
+ CH3MgBr
MgBr
+ CH4

12.55. The correct answer is (b). Sodium borohydride is a reducing agent that will reduce ketones and aldehydes, but not esters. In this case, the ketone group will be reduced to a secondary alcohol, but the ester group will not undergo reduction.

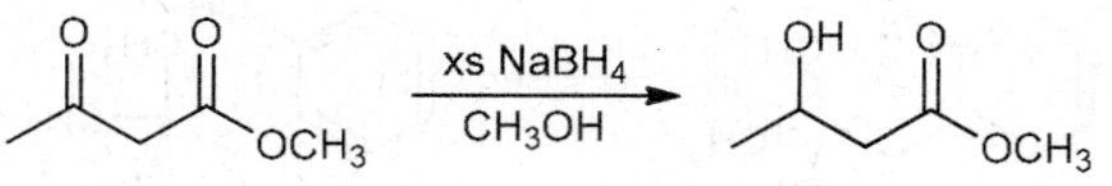
OCH3
xs NaBH4
CH3OH
OH
OCH3

12.56. The correct answer is (d). A logical retrosynthesis of the alcohol target molecule involves a disconnection at one of the C–C bonds adjacent to the C–OH bond. The Grignard reagent PhMgBr is a nucleophile that would afford the desired alcohol when combined with the aldehyde electrophile shown:

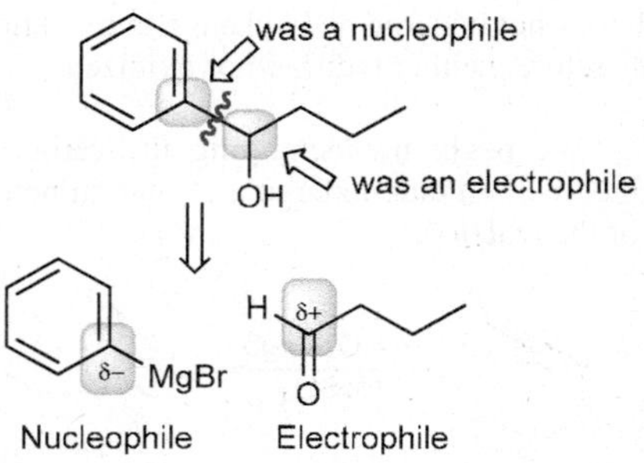
was a nucleophile
OH
was an electrophile
δ–
MgBr
H
δ+
O
Nucleophile
Electrophile

12.57. The correct answer is (c). The sequence begins with a substitution reaction to produce an alkyl bromide, followed by the formation of a Grignard reagent. Reaction of the Grignard reagent with methanol results in a proton transfer reaction (the strongly basic Grignard reagents removes the acidic proton from methanol), to give an alkane product:

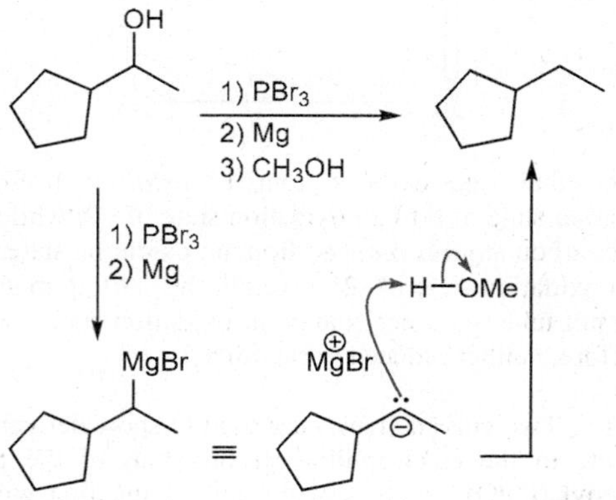
OH
1) PBr3
2) Mg
3) CH3OH
1) PBr3
2) Mg
H–OMe
MgBr
MgBr

12.58. The correct answer is (b). The starting material is a nucleophilic Grignard substrate, and the product is an alcohol, so the missing reagent must have an electrophilic carbonyl group. By locating the branched, four-carbon chain of the Grignard reagent within the product, is becomes clear that the carbon chain has been extended by one carbon atom. Therefore, the required reagent is the aldehyde shown (formaldehyde):

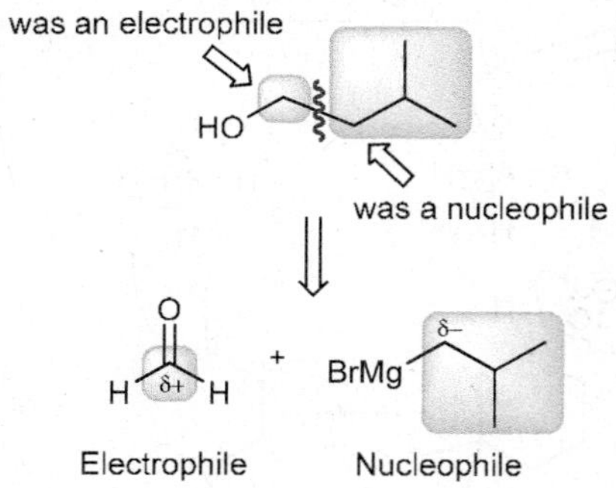
was an electrophile
HO
was a nucleophile
O
H
δ+
H
+
δ–
BrMg
Electrophile
Nucleophile

12.59. The correct answer is (a). The sequence begins with an ozonolysis reaction, producing a dialdehyde. Subsequent reduction of the dialdehyde with lithium aluminum hydride affords the corresponding diol, after workup. The starting alkene has five carbon atoms, so the product will have five carbon atoms as well.

1) O_3
2) DMS
3) $LiAlH_4$
4) H_3O^+

1) O_3
2) DMS

1) $LiAlH_4$
2) H_3O^+

12.60. The correct answer is (d). The highlighted atoms represent the tosylate leaving group (OTs). The given reaction is an S_N2 process, with the cyanide nucleophile displacing the tosylate leaving group:

NaCN
acetone

OTs leaving group

12.61. The correct answer is (a). This transformation represents an elimination reaction to give the less-substituted, Hofmann product, so dehydration of the alcohol is not a suitable approach (that would afford the more-substituted, Zaitsev product). Instead, a good leaving group is required (*e.g.*, Cl, Br, I, OTs) to facilitate an E2 elimination reaction. Therefore, the synthesis requires a two-step procedure: conversion of the OH group into a good leaving group, followed by treatment with a strong, sterically hindered base, such as potassium *tert*-butoxide, to generate the Hofmann product. Reaction of the alcohol with tosyl chloride and pyridine produces the tosylate leaving group:

1) TsCl, py
2) *t*-BuOK

TsCl
py

t-BuOK

12.62. Treating the aldehyde with methyl magnesium bromide (followed by aqueous acidic workup) gives the secondary alcohol, which can then be oxidized with a variety of oxidizing agents (*e.g.*, chromic acid, PCC, DMP or via Swern oxidation) to give a ketone. Treatment of the ketone with phenyl magnesium bromide gives a tertiary alcohol, after aqueous workup. Converting this alcohol to the less-substituted alkene (Hofmann product) requires that we first convert the OH group to a tosylate group, and then perform an E2 reaction with a sterically hindered base, such as *tert*-butoxide. Conversion of this alkene into a primary alcohol requires an *anti*-Markovnikov addition of H and OH across the alkene, which can be achieved via hydroboration-oxidation. Treatment with PBr_3 (or HBr) then converts the primary alcohol to a primary bromide. This alkyl bromide is then converted to a Grignard reagent (upon treatment with Mg) and then treated with formaldehyde to extend the carbon chain by one carbon atom (after aqueous workup). The resulting alcohol can then be oxidized to an aldehyde with PCC (or DMP, or Swern oxidation).

1) MeMgBr
2) H_3O^+

$Na_2Cr_2O_7$
H_2SO_4, H_2O

1) PhMgBr
2) H_3O^+

1) TsCl, pyridine
2) *t*-BuOK

1) $BH_3 \cdot THF$
2) H_2O_2, NaOH

PBr_3

1) Mg
2) $H_2C{=}O$
3) H_3O^+

PCC
CH_2Cl_2

12.63. 2-Nitrophenol is expected to be more acidic (lower pK_a) because its conjugate base has a resonance structure in which the negative charge is spread onto an oxygen atom of the nitro group:

Note that there are additional resonance structures (not shown), but the one drawn above is the key resonance structure that demonstrates the difference in stability between the isomers.

The more stable (weaker) conjugate base corresponds to the stronger acid. In contrast, the conjugate base of 3-nitrophenol does not have such a resonance structure, and the less stable (stronger) conjugate base corresponds to the weaker acid.

12.64. Most of the reagents for these transformations can be found in Figure 12.11. Note that some of these transformations have more than one suitable reagent from which to choose.

1) R_2BH 2) H_2O_2, NaOH

PCC CH_2Cl_2

1) $LiAlH_4$ 2) H_3O^+

1) Br_2 2) xs $NaNH_2$ 3) H_2O

H_2, Lindlar's cat.

1) TsCl 2) *t*-BuOK

1) $BH_3 \cdot THF$ 2) H_2O_2, NaOH

NaOH

PBr_3

HBr ROOR

t-BuOK

12.65.

(a) The product has one less carbon atom then the starting material, so we must break a carbon-carbon bond. Therefore, the last step of our synthesis must be ozonolysis of an alkene to give the desired ketone.

The retrosynthetic analysis continues by asking, "What reaction have we seen that produces an alkene product?" The simplest path involves an alcohol intermediate, because the alcohol can be easily prepared from the given aldehyde starting material:

Now let's draw the forward process. The desired transformation can be achieved by reducing the starting material (with $LiAlH_4$ or $NaBH_4$), followed by aqueous acidic workup, and then converting the OH group to a tosylate, and then treating the tosylate with a non-nucleophilic base (to favor E2 elimination over S_N2 substitution). Finally, ozonolysis of the alkene produces the desired ketone product.

1) $LiAlH_4$ 2) H_3O^+ 3) TsCl, pyridine 4) DBN 5) O_3 6) DMS

1) $LiAlH_4$ 2) H_3O^+

1) O_3 2) DMS

TsCl pyridine

DBN

(b) The target molecule has one additional carbon atom, so the synthesis requires a C–C bond-forming reaction. The retrosynthetic analysis of the target molecule asks, "What reaction have we seen that produces an alkene product?" The simplest path involves an alcohol intermediate, because the alcohol is suitable for making the required C–C bond disconnection (at the α position):

Electrophile

Nucleophile

Now let's draw the forward process. The synthesis begins with a Grignard reaction with methyl magnesium bromide, followed by aqueous acidic workup. The resulting alcohol can then be converted to the desired alkene via a two-step process. First, the alcohol is converted to a tosylate (because OH is a bad leaving group), and then the tosylate is treated with a sterically hindered base to give the less substituted alkene (the Hofmann product).

1) MeMgBr
2) H_3O^+
3) TsCl, pyridine
4) *t*-BuOK

1) MeMgBr
2) H_3O^+

t-BuOK

OH

OTs

TsCl
pyridine

(c) The starting material and the product have the same carbon skeleton. The identity of the functional group must be changed. The retrosynthetic analysis of the target molecule asks, "What reaction have we seen that produces an aldehyde product?" The simplest path involves an alcohol intermediate, because the alcohol can be easily prepared from the given alkyl chloride starting material:

O H ⟹ OH ⟹ Cl

Now let's draw the forward process. The alkyl chloride can first be converted to a primary alcohol upon treatment with hydroxide (via an S_N2 process), and then the alcohol can be oxidized with PCC (or DMP or Swern oxidation) to give the desired target molecule.

1) NaOH
2) PCC, CH_2Cl_2

NaOH

PCC
CH_2Cl_2

(d) The target molecule has one additional carbon atom, so the synthesis requires a C—C bond-forming reaction. The retrosynthetic analysis of the target molecule asks, "What reaction have we seen that produces a ketone product?" The ketone can be prepared by oxidation of the corresponding alcohol, and the alcohol is suitable for making the required C—C bond disconnection (at the α position):

$\delta+$ H + CH_3—MgBr $\delta-$

Electrophile Nucleophile

The required aldehyde can be prepared from the given alkyl chloride as seen in the previous problem:

Now let's draw the forward process. The alkyl chloride can first be converted to a primary alcohol via an S_N2 process, and then the alcohol can be oxidized with PCC (or DMP or Swern oxidation) to give the desired aldehyde. The aldehyde can undergo a Grignard reaction with methyl magnesium bromide, and oxidation of the resulting secondary alcohol produces the ketone target molecule. The reagents for this synthetic strategy are shown here.

1) NaOH
2) PCC, CH_2Cl_2
3) MeMgBr
4) H_3O^+
5) $Na_2Cr_2O_7$, H_2SO_4, H_2O

NaOH

$Na_2Cr_2O_7$, H_2SO_4, H_2O

PCC, CH_2Cl_2

1) MeMgBr
2) H_3O^+

(e) The starting material and the product have the same carbon skeleton. The identity of the functional group must be changed. The retrosynthetic analysis of the target molecule asks, "What reaction have we seen that produces a ketone product?" A reasonable choice is an alcohol intermediate, because the alcohol can be easily prepared from the given alkene starting material:

Now let's draw the forward process. The alkene is first converted to an alcohol via acid-catalyzed hydration, and then the alcohol can be oxidized with chromic acid to give the desired product.

1) dilute H_2SO_4
2) $Na_2Cr_2O_7$, H_2SO_4, H_2O

dilute H_2SO_4

$Na_2Cr_2O_7$
H_2SO_4, H_2O

(f) In the previous problem, we saw that the starting material can be converted into a ketone in just two steps.

1) dilute H_2SO_4
2) $Na_2Cr_2O_7$, H_2SO_4, H_2O

This ketone can be converted into the desired product with a Grignard reaction.

1) dilute H_2SO_4
2) $Na_2Cr_2O_7$, H_2SO_4, H_2O
3) MeMgBr
4) H_3O^+

dilute H_2SO_4

1) MeMgBr
2) H_3O^+

$Na_2Cr_2O_7$, H_2SO_4, H_2O

(g) In the previous problem, we saw that the starting material can be converted into a tertiary alcohol.

1) dilute H_2SO_4
2) $Na_2Cr_2O_7$, H_2SO_4, H_2O
3) MeMgBr
4) H_3O^+

This alcohol can be converted into the desired product in just one step (upon treatment with concentrated sulfuric acid).

1) dilute H_2SO_4
2) $Na_2Cr_2O_7$, H_2SO_4, H_2O
3) MeMgBr
4) H_3O^+
5) conc. H_2SO_4, heat

dilute H_2SO_4

conc. H_2SO_4, heat

$Na_2Cr_2O_7$, H_2SO_4, H_2O

1) MeMgBr
2) H_3O^+

(h) The target molecule has one additional carbon atom, so the retrosynthesis requires a C—C bond disconnection. The disconnection is made at the α position of the alcohol to give a suitable electrophile (C=O) and nucleophile (RMgBr):

OH ⟹ O δ+ + CH_3—MgBr δ−

Electrophile Nucleophile

In order to perform the desired Grignard reaction, we must first convert the starting alkyne into a ketone (acetone), which can be accomplished via acid-catalyzed hydration in the presence of mercuric sulfate. Reaction with methyl magnesium bromide, followed by aqueous acidic workup, produces the desired tertiary alcohol.

1) $HgSO_4$, H_2SO_4, H_2O
2) MeMgBr
3) H_3O^+

$HgSO_4$, H_2SO_4, H_2O

1) MeMgBr
2) H_3O^+

(i) The target molecule has one additional carbon atom, so the retrosynthesis requires a C—C bond disconnection. The disconnection is made at the α position of the alcohol to give a suitable electrophile (C=O) and nucleophile (RMgBr):

HO ⟹ δ+ O + CH_3—MgBr δ−

Electrophile Nucleophile

The retrosynthetic analysis continues by asking, "What reaction have we seen that produces a ketone product?" The simplest path involves an alcohol intermediate, because the alcohol can be easily prepared from the given alkene starting material:

O ⟹ OH ⟹

Now let's draw the forward process. We must first convert the starting alkene into a ketone, which can be accomplished via a two-step process (acid-catalyzed hydration, followed by oxidation). The resulting ketone can undergo a Grignard reaction with methyl magnesium bromide, followed by aqueous acidic workup, to give the desired target molecule.

1) dilute H_2SO_4
2) $Na_2Cr_2O_7$, H_2SO_4, H_2O
3) MeMgBr
4) H_3O^+

dilute H_2SO_4

1) MeMgBr
2) H_3O^+

$Na_2Cr_2O_7$, H_2SO_4, H_2O

(j) The desired transformation can be achieved in one step, using a Grignard reaction (followed by aqueous workup).

1) EtMgBr
2) H_3O^+

(k) The starting material and the product have the same carbon skeleton. The identity of the functional group must be changed. The retrosynthetic analysis of the target molecule asks, "What reaction have we seen that produces an alkene product?" The simplest path involves an alcohol intermediate, because the alcohol can be easily prepared from the given ketone starting material:

Now let's draw the forward process. The ketone is first converted to an alcohol via reduction (using either $LiAlH_4$ or $NaBH_4$), and then the alcohol can be treated with concentrated sulfuric acid to give the desired alkene (the more-substituted, Zaitsev product).

1) $LiAlH_4$
2) H_3O^+
3) conc. H_2SO_4, heat

1) $LiAlH_4$
2) H_3O^+

conc. H_2SO_4 heat

(l) The starting material and the product have the same carbon skeleton. The identity of the functional group must be changed. The retrosynthetic analysis of the target molecule asks, "What reaction have we seen that produces an alkene product?" The simplest path involves an alcohol intermediate, because the alcohol can be easily prepared from the given ketone starting material:

Now let's draw the forward process. The ketone is first converted to an alcohol via reduction (using either $LiAlH_4$ or $NaBH_4$), and then we must perform an elimination process. Since the less substituted alkene is desired (the Hofmann product), we must use a sterically hindered base. Since OH is a bad leaving group, it must first be converted to a good leaving group. This can be accomplished by treating the alcohol with tosyl chloride and the resulting tosylate can then be treated with *tert*-butoxide to give the desired target molecule.

1) $LiAlH_4$
2) H_3O^+
3) TsCl, py
4) *t*-BuOK

1) $LiAlH_4$
2) H_3O^+

TsCl, pyridine

t-BuOK

(m) The starting material and the product have the same carbon skeleton. The identity and location of the functional group must be changed. The answer to the previous problem allows us to convert the starting material into an alkene, which can be then converted into the alcohol product via hydroboration-oxidation (*anti*-Markovnikov addition of H and OH).

1) $LiAlH_4$
2) H_3O^+
3) TsCl, py
4) *t*-BuOK
5) $BH_3 \cdot THF$
6) H_2O_2, NaOH

1) $LiAlH_4$
2) H_3O^+

1) $BH_3 \cdot THF$
2) H_2O_2, NaOH

TsCl, pyridine

t-BuOK

(n) The target molecule has one additional carbon atom, so the synthesis requires a C–C bond-forming reaction. The retrosynthetic analysis of the target molecule asks, "What reaction have we seen that produces an alkene product?" The simplest path involves an alcohol intermediate, because the alcohol is suitable for making the required C–C bond disconnection (at the α position):

δ+ + CH_3–MgBr δ-

Electrophile Nucleophile

Now let's draw the forward process. The synthesis begins with a Grignard reaction with methyl magnesium bromide. The resulting alcohol can then be converted to the desired target molecule upon treatment with concentrated sulfuric acid (an E1 process that affords the more-substituted, Zaitsev product).

1) MeMgBr
2) H_3O^+
3) conc. H_2SO_4, heat

1) MeMgBr
2) H_3O^+

HO

conc. H_2SO_4, heat

(o) The desired transformation can be achieved in one step, using a Grignard reaction (followed by aqueous workup).

1) EtMgBr
2) H_3O^+

OH

(p) The starting material and the product have the same carbon skeleton. The identity of the functional group must be changed. The retrosynthetic analysis of the target molecule asks, "What reaction have we seen that produces an alkyl bromide product?" The simplest path involves an alcohol intermediate, because the alcohol can be easily prepared from the given aldehyde starting material:

Br ⟹ OH ⟹ H

Now let's draw the forward process. The desired transformation can be achieved by reducing the aldehyde (with either $LiAlH_4$ or $NaBH_4$) followed by treatment with PBr_3 to give the desired alkyl bromide.

1) $LiAlH_4$
2) H_3O^+
3) PBr_3

1) $LiAlH_4$
2) H_3O^+

OH

PBr_3

Br

(q) The target molecule has one additional carbon atom, so the retrosynthesis requires a C—C bond disconnection. The disconnection is made at the α position of the alcohol to give a suitable electrophile (C=O) and nucleophile (RMgBr):

OH ⟹ (δ+) H + CH_3—MgBr (δ−)

Electrophile Nucleophile

The retrosynthetic analysis continues by asking, "What reaction have we seen that produces an aldehyde product?" The simplest path involves an alcohol intermediate, because the alcohol can be easily prepared from the given alkene starting material:

H ⟹ OH ⟹

Now let's draw the forward process. We must first convert the starting alkene into an aldehyde, which can be accomplished via a two-step process: hydroboration-oxidation, followed by oxidation with PCC (or DMP or Swern oxidation). The resulting aldehyde can undergo a Grignard reaction with methyl magnesium bromide to give the desired target molecule, after aqueous workup.

1) $BH_3 \cdot THF$
2) H_2O_2, NaOH
3) PCC, CH_2Cl_2
4) MeMgBr
5) H_3O^+

OH

1) $BH_3 \cdot THF$
2) H_2O_2, NaOH

OH

PCC, CH_2Cl_2

H

1) MeMgBr
2) H_3O^+

12.66. The alcohol can be converted to an alkene via either of the two processes shown below. Notice that the second method involves converting the alcohol into a tosylate, followed by treatment with a strong base. If this method is used, it is important that the base is not sterically hindered, in order to favor the more-substituted alkene (Zaitsev product). Appropriate bases include hydroxide, methoxide and ethoxide (ethoxide is shown below).

OH

conc. H_2SO_4, heat

1) TsCl, pyridine
2) NaOEt

12.67. The desired transformation can be achieved by converting the alcohol to an alkene (via either of the two methods discussed in the previous problem), followed by *anti*-Markovnikov addition of H and OH to give the product (via hydroboration-oxidation).

conc. H_2SO_4, heat

OH

1) $BH_3 \cdot THF$
2) H_2O_2, NaOH

OH

1) TsCl, pyridine
2) NaOEt

There are certainly other acceptable answers. For example, conversion of the alkene to the desired alcohol can be achieved via *anti*-Markovnikov addition of HBr (in the presence of peroxides) followed by an S_N2 process in which hydroxide functions a nucleophile and replaces bromide.

12.68.

(a) The desired product can be made in just one step, via a Grignard reaction in which acetaldehyde (two carbon atoms) is treated with methyl magnesium bromide (one carbon atom), followed by aqueous acidic workup.

1) MeMgBr
2) H_3O^+

Acetaldehyde

(b) The desired transformation can be achieved via a Grignard reaction in which acetaldehyde (two carbon atoms) is treated with ethyl magnesium bromide (two carbon atoms), followed by aqueous acidic workup:

1) EtMgBr
2) H_3O^+

Acetaldehyde

Certainly, there are other acceptable answers, two of which are shown below. In both of these pathways, the starting materials are ethyl bromide (two carbon atoms) and sodium acetylide (two carbon atoms):

Br

$HC{\equiv}C:^{\ominus}\ Na^{\oplus}$

dilute H_2SO_4

H_2
Lindlar's catalyst

H_2SO_4, H_2O
$HgSO_4$

1) $LiAlH_4$
2) H_3O^+

(c) There are certainly many acceptable answers to this problem. One such answer can be rationalized with the following retrosynthetic analysis.

OH ⟹ (i) H; ⟹ (ii); Br + ⟸ (iii)

i. The product is an alcohol, which can be made from the corresponding aldehyde via reduction.
ii. The aldehyde can be made from a terminal alkyne via hydroboration-oxidation.
iii. The terminal alkyne can be made via an alkylation process.

Reagents for this synthetic strategy are shown here.

Br → 1) HC≡CNa; 2) R_2BH; 3) H_2O_2, NaOH; 4) $LiAlH_4$; 5) H_3O^+ → OH

HC≡CNa

1) R_2BH
2) H_2O_2, NaOH

1) $LiAlH_4$
2) H_3O^+

The following are alternative synthetic pathways that also achieve the desired transformation:

Br → OH

HC≡CNa

1) $BH_3 \cdot THF$
2) H_2O_2, NaOH

NaOH

Br

H_2
Lindlar's catalyst

HBr, ROOR

(d) There are certainly many acceptable answers to this problem. One such answer can be rationalized with the following retrosynthetic analysis.

OH ⟹ (i) H; ⟹ (ii); Br + ⟸ (iii)

i. The alcohol can be made by treating the appropriate aldehyde with ethyl magnesium bromide.
ii. The aldehyde can be made from a terminal alkyne (via hydroboration-oxidation).
iii. The terminal alkyne can be made via an alkylation process.

Reagents for this synthetic strategy are shown here.

1) HC≡CNa
2) R_2BH
3) H_2O_2, NaOH
4) EtMgBr
5) H_3O^+

HC≡CNa

1) EtMgBr
2) H_3O^+

1) R_2BH
2) H_2O_2, NaOH

The following is an alternative synthetic pathway that also achieves the desired transformation:

1) HC≡CNa
2) $NaNH_2$
3) EtBr
4) H_2SO_4, H_2O, $HgSO_4$
5) $LiAlH_4$
6) H_3O^+

HC≡CNa

1) $LiAlH_4$
2) H_3O^+

1) $NaNH_2$
2) EtBr

H_2SO_4, H_2O, $HgSO_4$

12.69. Two curved arrows are used to show hydride delivery to the electrophilic carbon atom of one of the ester carbonyl (C=O) groups. Note that the tail of the first curved arrow is placed on the bond between Al and H (it is NOT placed on the negative charge). The resulting intermediate then ejects an alkoxide leaving group, which requires two curved arrows and opens the ring. The resulting aldehyde is then further reduced by another equivalent of $LiAlH_4$. Once again, two curved arrows are used to show hydride delivery to the electrophilic carbonyl. This entire process is then repeated again for the other ester group, giving two equivalents of a dianion intermediate.

The resulting dianion is then protonated upon treatment with aqueous acid (workup). Notice that each protonation step is drawn separately (with two curved arrows for each step).

12.70. Note that the reaction mechanism is analogous to that of problem 12.70, except that the hydride nucleophile from problem 12.70 has been replaced with methyl magnesium bromide. Two curved arrows are used to show a Grignard reagent attacking the electrophilic carbon atom of one of the ester carbonyl (C=O) groups. The resulting intermediate then ejects an alkoxide leaving group, which requires two curved arrows and opens the ring. The resulting ketone is then further attacked by another equivalent of the Grignard reagent. Once again, two curved arrows are used to show the nucleophilic attack. This entire process is then repeated again for the other ester group, giving two equivalents of a dianion intermediate.

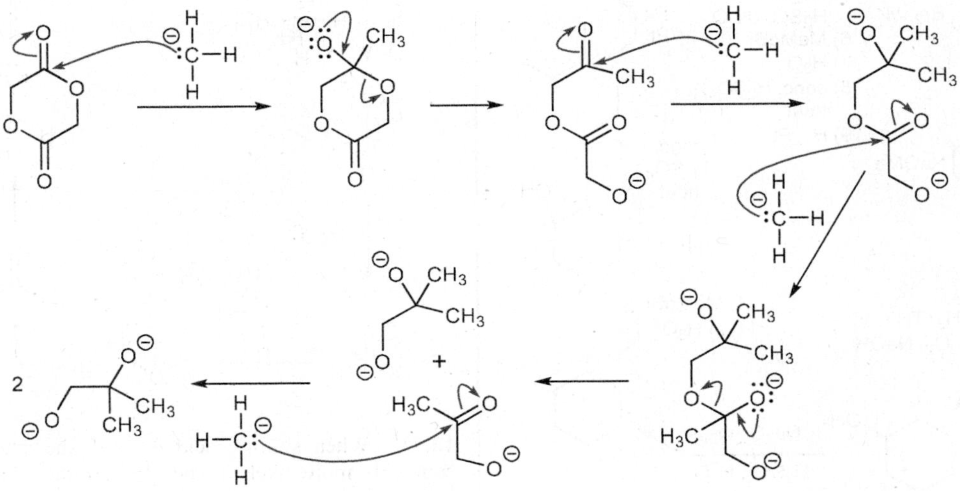

The resulting dianion is then protonated upon treatment with aqueous acid (workup). Notice that each protonation step is drawn separately (with two curved arrows for each step). The less-stable negative charge (the tertiary alkoxide) is the stronger base and is therefore protonated first.

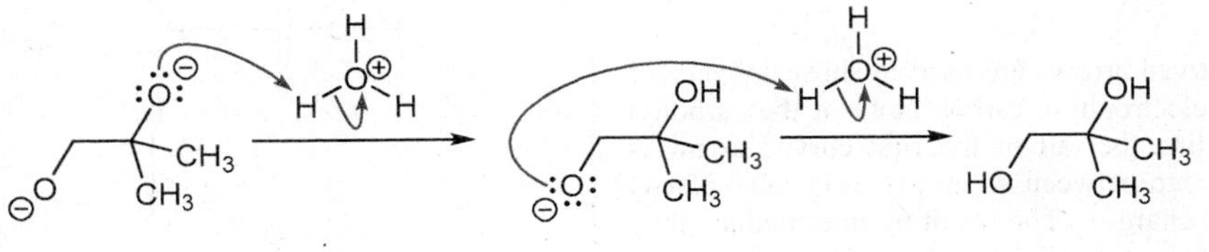

12.71. The product has one more carbon atom than the starting material. This extra methyl group can be installed using a Grignard reaction, as shown here:

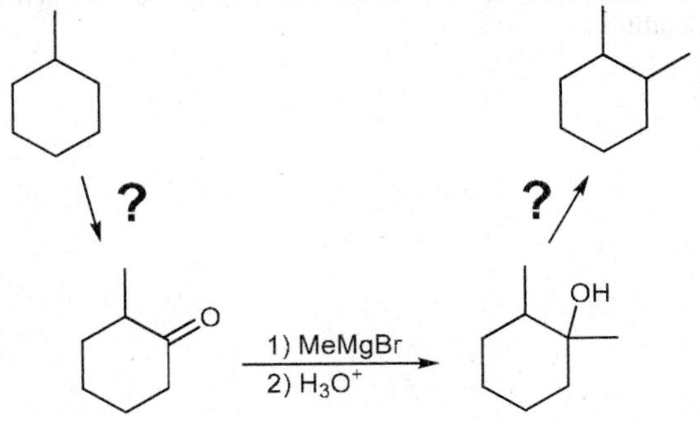

1) MeMgBr
2) H_3O^+

If we are going to use this Grignard reaction to form the critical carbon-carbon bond, then we must first convert the starting cycloalkane into the necessary ketone, AND we must convert the product of the Grignard reaction (a tertiary alcohol) into the desired product. There are certainly many acceptable synthetic routes that can be used. One such route is shown here. The starting material does not have a functional group, but one can be introduced via radical bromination. Elimination followed by hydroboration-oxidation gives an alcohol that can be oxidized to give the ketone necessary for the Grignard reaction. Then, after the Grignard reaction is complete, the resulting tertiary alcohol can be converted into the product in just two steps (E1 dehydration followed by catalytic hydrogenation):

1) Br_2, $h\nu$
2) NaOMe
3) $BH_3 \cdot THF$
4) H_2O_2, NaOH
5) $Na_2Cr_2O_7$, H_2SO_4, H_2O
6) MeMgBr
7) H_3O^+
8) conc. H_2SO_4, heat
9) H_2, Pt

Br$_2$, $h\nu$; NaOMe; 1) $BH_3 \cdot THF$ 2) H_2O_2, NaOH; $Na_2Cr_2O_7$, H_2SO_4, H_2O; 1) MeMgBr 2) H_3O^+; conc. H_2SO_4 heat; H_2 Pt

12.72. Two curved arrows are used to show delivery of hydride to the electrophilic carbon atom of the carbonyl (C=O). Note that the tail of the first curved arrow is placed on the bond between Al and H (it is NOT placed on the negative charge). The resulting intermediate then ejects methoxide as a leaving group, which requires two curved arrows. The resulting compound is then further reduced by another equivalent of $LiAlH_4$. Once again, two curved arrows are used to show hydride delivery to the ester carbonyl, resulting in an intermediate that ejects methoxide as a leaving group, which requires two curved arrows, to give formaldehyde. Formaldehyde is then reduced one last time by a third equivalent of $LiAlH_4$ (and once again, two curved arrows are required). The resulting alkoxide ion is then protonated upon treatment with aqueous acid (workup) to give the product (methanol). Note that three equivalents of methoxide are formed in this process, resulting in the formation of three equivalents of methanol after the workup.

1) Excess $LiAlH_4$
2) H_3O^+

- $MeO^{\ominus}$

- $MeO^{\ominus}$

12.73. When sulfuric acid is used, the actual proton source is more likely to be H_3O^+, rather than H_2SO_4, because aqueous sulfuric acid contains water (even when concentrated), and the equilibrium favors H_3O^+ over H_2SO_4, as shown here.

Therefore, our mechanism will show H_3O^+ as the proton source. In the first step of the mechanism, one of the OH groups is protonated to give a good leaving group. Loss of the leaving group then generates a tertiary carbocation. In this case, the tertiary carbocation can rearrange to produce a more stable cation that is resonance-stabilized. Water then functions as a base to remove a proton, giving the product. Notice that the last step employs water as the base. You cannot show hydroxide functioning as a base because the concentration of hydroxide is negligible in acidic conditions.

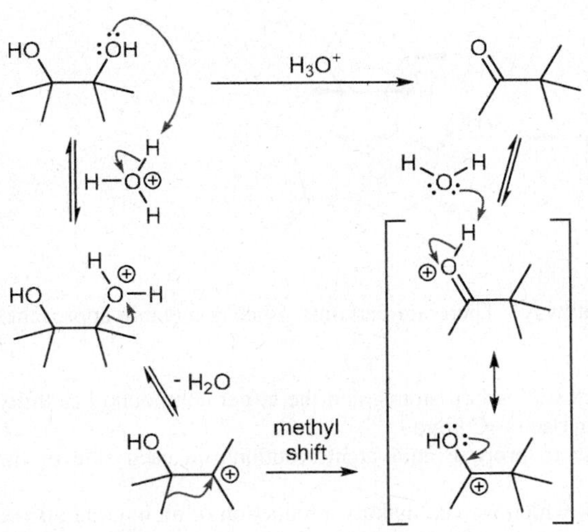

HO :ÖH H₃O⁺ H–O⊕ H H H–Ö:–H - H_2O methyl shift HÖ:

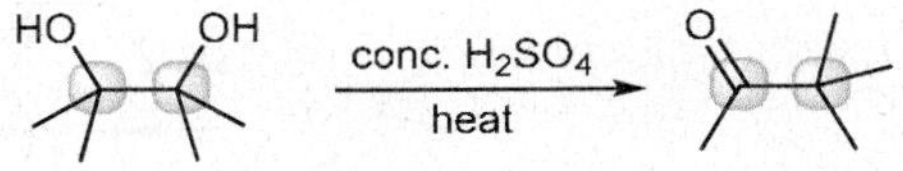

For each of these (highlighted) carbon atoms, we calculate its oxidation state in both the starting diol and the ketone product:

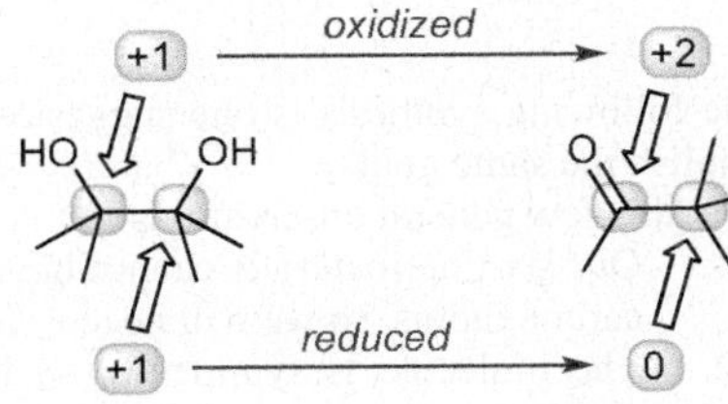

oxidized +1 +2 HO OH reduced +1 0

12.74. We begin by identifying the carbon atoms (highlighted below) that undergo a change in bonding as a result of the reaction:

HO OH conc. H_2SO_4 heat

Notice that one carbon atom is *oxidized* from an oxidation state of +1 to an oxidation state of +2, while the other carbon atom is *reduced* from an oxidation state of +1 to an oxidation state of 0. Overall, the starting material does not undergo a net change in oxidation state and is, therefore, neither reduced nor oxidized.

12.75. Analysis of the starting material and the product shows that the aldehyde and alkene groups have been reduced. Secondly, the aldehyde is selectively reduced in the presence of the ester. Selective reduction of the aldehyde with $NaBH_4$, followed by reduction of the alkene (H_2, Pt), affords the desired product. Note that $LiAlH_4$ cannot be used in this case, because $LiAlH_4$ would reduce the ester group as well.

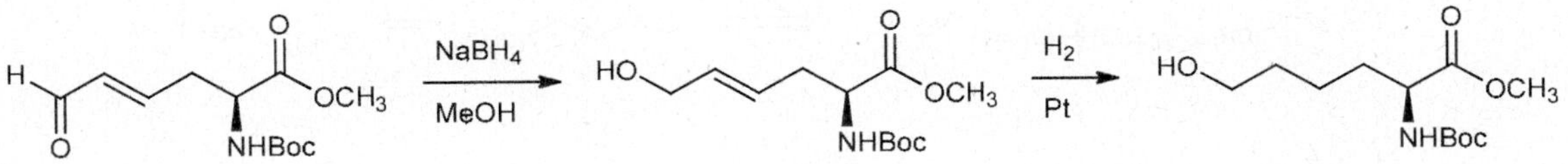

H OCH₃ NHBoc $NaBH_4$ MeOH HO OCH₃ NHBoc H_2 Pt HO OCH₃ NHBoc

12.76. We have seen that tetrabutylammonium fluoride (TBAF) can be used to remove silyl protecting groups. In this case, there are two such groups:

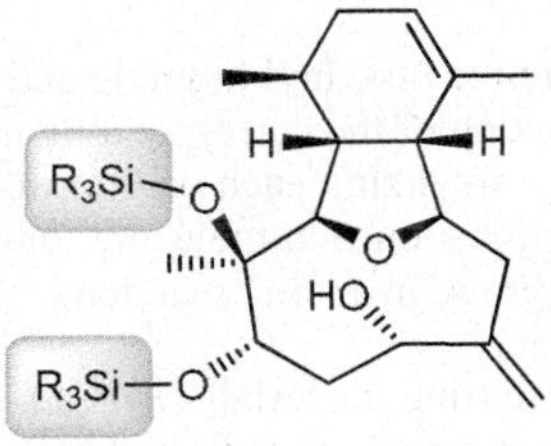

R₃Si–O R₃Si–O HO H H O

We learned about the trimethylsilyl protecting group (TMS), and the protecting groups employed in this case are very similar (in each case, one of the methyl (R) groups has been replaced with a *tert*-butyl group). This protecting group is called the *tert*-butyldimethylsilyl group (TBDMS, or just TBS for short), and it is removed upon treatment with TBAF, in much the same way that the TMS group is removed under similar conditions. The rest of the compound is expected to remain unchanged.

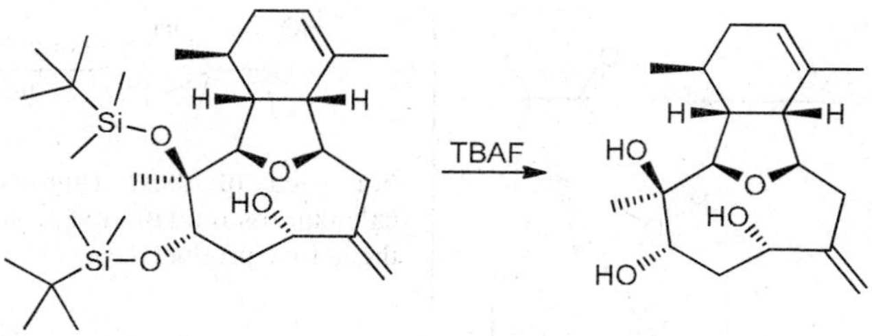

12.77. The following synthesis is one suggested synthetic pathway. There are certainly other acceptable approaches that accomplish the same goal.

Let's start with a few general observations:

1. Our starting materials cannot have more than eleven carbon atoms, and the target compound has thirty carbon atoms, so we will need to form more than one C—C bond.
2. The molecule is symmetric, so it makes sense to propose equivalent reactions on each side of the molecule.
3. The central C=C bond has the *cis* configuration, which we can make via reduction of an internal alkyne using H_2 and Lindlar's catalyst. This, however, cannot be the final step of our proposed synthesis, because these conditions would also reduce the two terminal alkyne groups. Thus, the two terminal alkyne groups need to be installed *after* reduction of the central alkyne group.

The figure below outlines a retrosynthetic analysis for our target molecule. An explanation of each of the steps (*a-d*) follows.

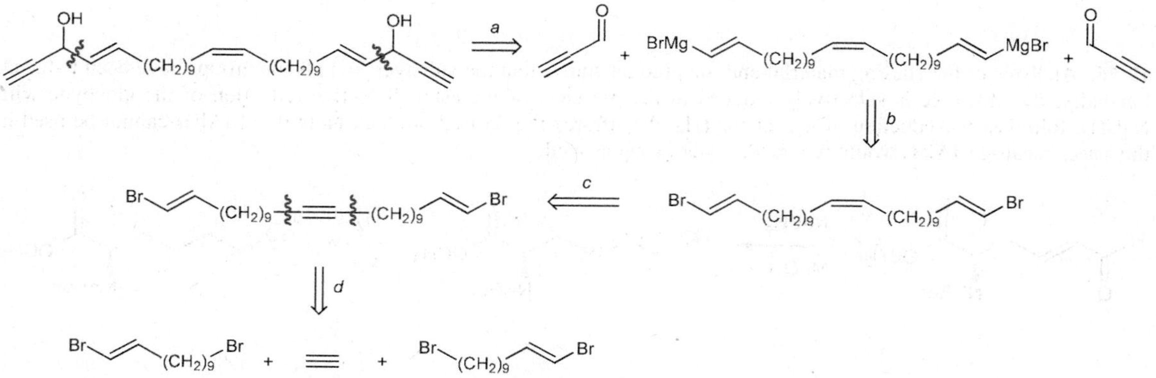

a. Duryne can be made from the reaction between the bis-vinyl Grignard reagent (a dianion) and two equivalents of the aldehyde shown.

b. The bis-vinyl Grignard is made from the bis-vinyl bromide and two equivalents of magnesium.

c. The *cis* alkene is produced via reduction of the corresponding internal alkyne.

d. The alkyne is made by sequentially alkylating each carbon of acetylene with (*E*)-1,11-dibromo-1-undecene. Note that while there are two bromines on this molecule, only the bromide attached to the sp^3 hybridized carbon atom can serve as a leaving group in an S_N2 reaction.

Now let's draw the forward scheme. The starting material, (*E*)-1,11-dibromo-1-undecene, is treated with sodium acetylide to produce a terminal alkyne. Deprotonation with sodium amide, followed by treatment with a second equivalent of (*E*)-1,11-dibromo-1-undecene gives the internal alkyne. Reduction of the alkyne with H_2 and Lindlar's catalyst affords the *cis* alkene. Further treatment with two equivalents of magnesium yields the bis-vinyl Grignard, which reacts with two equivalents of the aldehyde. Aqueous acidic workup produces the target molecule, duryne.

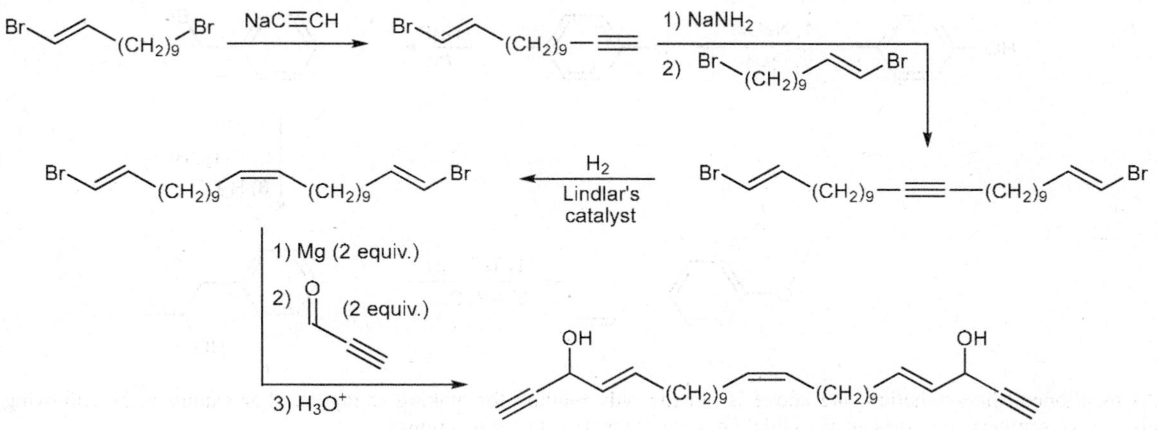

Br—CH=CH—$(CH_2)_9$—Br $\xrightarrow{NaC\equiv CH}$ Br—CH=CH—$(CH_2)_9$—C≡CH $\xrightarrow[\text{2) Br—(CH}_2)_9\text{—CH=CH—Br}]{\text{1) NaNH}_2}$ Br—CH=CH—$(CH_2)_9$—C≡C—$(CH_2)_9$—CH=CH—Br

$\xrightarrow[\text{Lindlar's catalyst}]{H_2}$ Br—CH=CH—$(CH_2)_9$—CH=CH—$(CH_2)_9$—CH=CH—Br

1) Mg (2 equiv.)
2) propynal (2 equiv.)
3) H_3O^+

→ HC≡C—CH(OH)—CH=CH—$(CH_2)_9$—CH=CH—$(CH_2)_9$—CH=CH—CH(OH)—C≡CH

12.78. The following synthesis is one suggested synthetic pathway. There are certainly other acceptable approaches that accomplish the same goal.
By comparing the structures of the starting material (4-methylphenol) and the product, it is clear that the following bonds (disconnected below) need to be made in this synthesis.

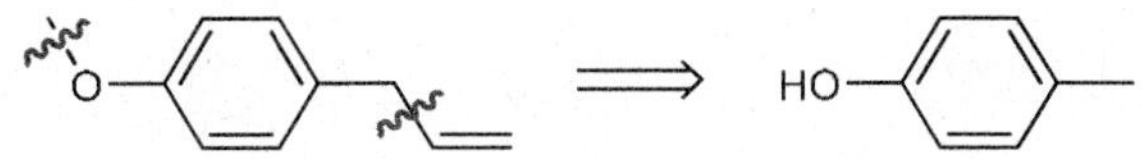

The left bond (C–O) can be made via an S_N2 process, while the right bond (C—C) can be made by either a Grignard reaction or by using an acetylide ion as a nucleophile. It is important that we make the ether bond early in our scheme to avoid an acid-base reaction between the phenolic proton ($pK_a \approx 10$) and the Grignard reagent or acetylide (both of which are strong bases). If the phenolic proton is subjected to a Grignard reagent, the latter would be destroyed via protonation. The same fate would occur for an acetylide ion that is treated with a compound bearing a phenolic proton.

The following is a retrosynthetic analysis for our target compound. An explanation of each of the steps (*a-e*) follows.

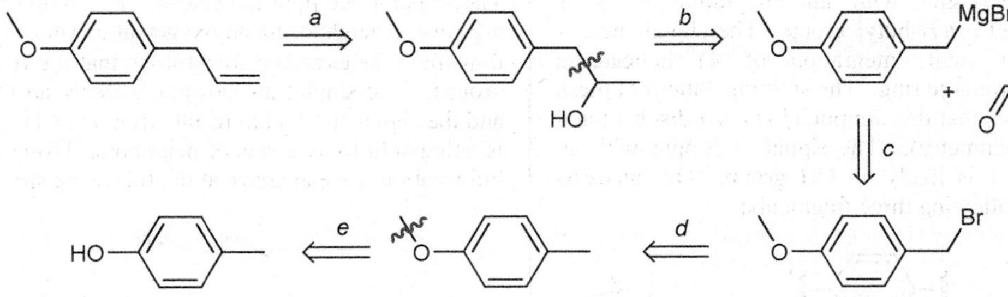

a. We can make the monosubstituted alkene by converting the OH group into a tosylate group and then performing an elimination reaction with a sterically hindered base.
b. The alcohol can be made via a Grignard reaction between the benzylic Grignard reagent and the aldehyde shown.
c. The Grignard reagent is made from the corresponding benzylic bromide.
d. The bromine atom can be installed via radical bromination at the benzylic position.
e. The methyl ether can be produced via an S_N2 reaction from the starting material.

Now let's draw the forward scheme. The starting material, 4-methylphenol, is deprotonated with sodium hydroxide and the resulting phenoxide serves as a nucleophile in an S_N2 reaction with bromomethane. The bromine atom is installed using *N*-bromosuccinimide and light. This compound is then converted into a Grignard reagent using magnesium. The Grignard reagent reacts with the appropriate aldehyde (CH_3CHO), followed by aqueous acidic workup to produce the alcohol. Conversion to the tosylate and subsequent elimination with *tert*-butoxide produces the less-substituted alkene (Hofmann product), estragole.

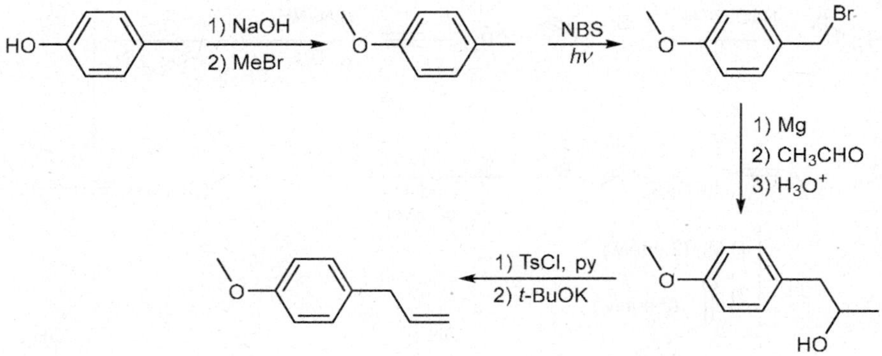

As mentioned, the synthetic route above is not the only method for making estragole. For example, the following alternative synthesis involves an acetylide ion, rather than a Grignard reaction:

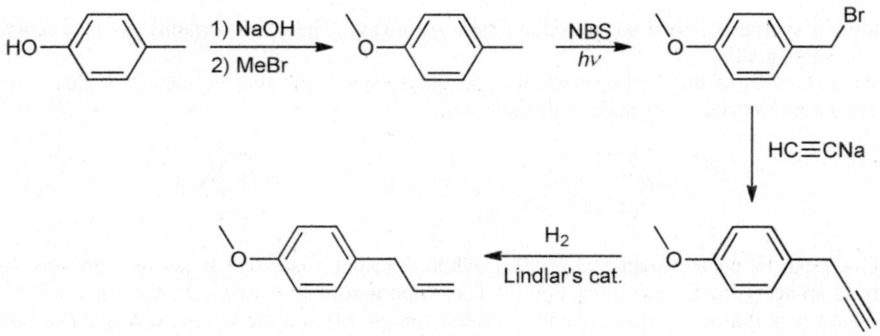

12.79. The singlet with an integration of 9 is characteristic of a *tert*-butyl group. The signals near 7 ppm (with a total integration of 4) indicate a disubstituted aromatic ring. The splitting pattern of these signals indicates that the compound is *para*-disubstituted (because of symmetry). The signal at 5 ppm with an integration of 1 is likely an OH group. Our analysis produces the following three fragments:

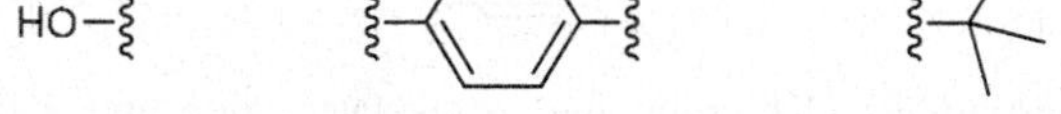

These three fragments can only be assembled in one way:

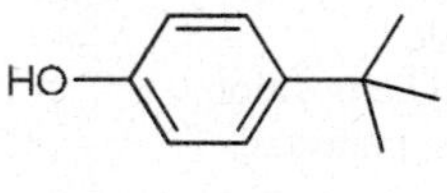

12.80. The ^{13}C NMR spectrum indicates that all three carbon atoms are in different environments. One of these signals appears between 50 and 100 ppm, indicating that one carbon atom is connected to an oxygen atom (the molecular formula indicates the presence of an oxygen atom). Now we turn to the ^{1}H NMR spectrum. The signal at 1 ppm has an integration of 3, indicating a CH_3 group. Since this signal is a triplet, the CH_3 group must be adjacent to a CH_2 group. The signal at 3.6 ppm indicates a CH_2 group (integration = 2) that is neighboring an oxygen atom (thus it is shifted downfield, as expected for protons that are α to an OH group). The singlet at 2.4 ppm is likely an OH group, and the signal at 1.6 ppm results from the CH_2 group that is being split by two sets of neighbors. Using all of this information, we can arrive at the following structure.

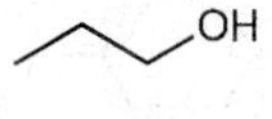

12.81. In the IR spectrum, the broad signal between 3200 and 3600 cm^{-1} indicates an OH group. The NMR spectrum indicates that there are only three different kinds of carbon atoms, yet the molecular formula indicates that the compound has five carbon atoms. Therefore, we must draw structures that possess enough symmetry such that there are only three unique kinds of carbon atoms. The following two structures are consistent with this analysis.

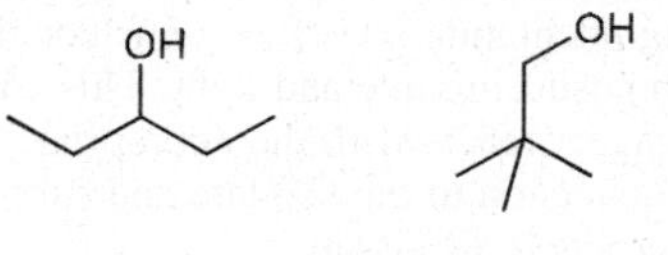

12.82. There is a multiplet just above 7 ppm, indicating an aromatic ring. Since the integration of this signal is 5, we expect the aromatic ring to be monosubstituted. The two triplets (at 2.8 ppm and 3.8 ppm) indicate two CH_2 groups that are neighboring each other, and the singlet at 2 ppm (with an integration of 1) is likely an OH group. Our analysis produces the following three fragments:

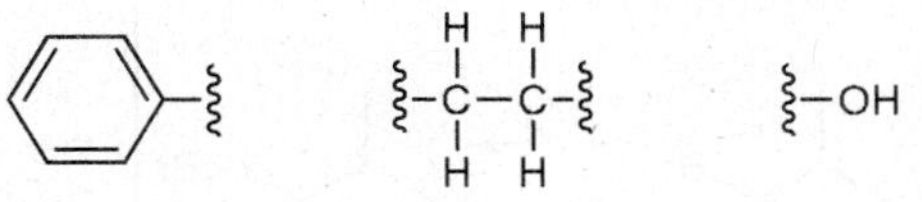

These three fragments can only be assembled in one way:

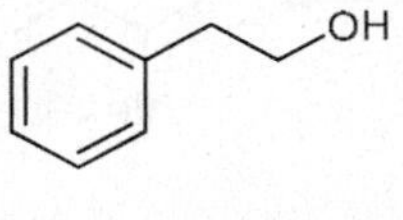

Notice that the signals for the CH_2 groups (the triplets) are shifted downfield. The signal at 3.8 ppm represents the CH_2 group next to the oxygen atom, and the signal at 2.8 ppm represents the CH_2 group next to the aromatic ring.

12.83. The following two syntheses are suggested synthetic pathways. There are certainly other acceptable approaches that accomplish the same goal.

Each of the target compounds has eight carbon atoms, suggesting the following disconnections, which break each carbon skeleton into two four-carbon fragments.

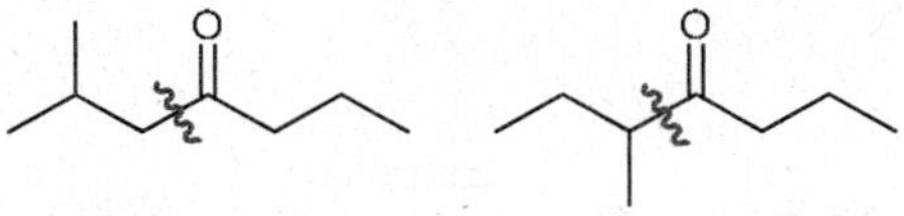

The following figure outlines a retrosynthetic analysis for our first target molecule. An explanation of each of the steps (*a-h*) follows.

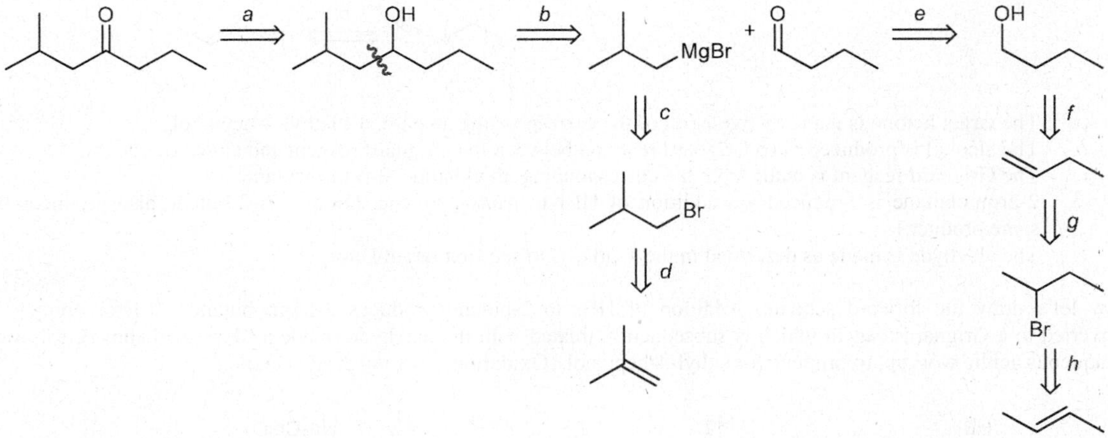

a. The target ketone is made by oxidation of the corresponding alcohol, 2-methyl-4-heptanol.
b. The alcohol is produced via a Grignard reaction between the Grignard reagent and aldehyde shown.
c. The Grignard reagent is made from the corresponding alkyl halide, 1-bromo-2-methylpropane.
d. 1-Bromo-2-methylpropane is produced via an *anti*-Markovnikov addition of HBr to 2-methylpropene.
e. The aldehyde is made from oxidation of 1-butanol.
f. 1-Butanol is made from 1-butene via hydroboration-oxidation.
g. 1-Butene is made from 2-bromobutane using a sterically hindered base to produce the less-substituted alkene.
h. 2-Bromobutane is made from HBr addition to *trans*-2-butene. (Note: *Cis*-2-butene also produces the same product.)

Now let's draw the forward scheme. Addition of HBr to 2-butene produces 2-bromobutane, which is subsequently treated with *tert*-butoxide to give 1-butene (Hofmann product). *Anti*-Markovnikov addition of water (via hydroboration-oxidation) followed by PCC (or DMP or Swern oxidation) gives the aldehyde. Reaction with the Grignard reagent (produced by *anti*-Markovnikov addition of HBr to 2-methylpropene, followed by treatment with magnesium, as shown) gives 2-methyl-4-heptanol after aqueous acidic workup. Oxidation produces the target ketone.

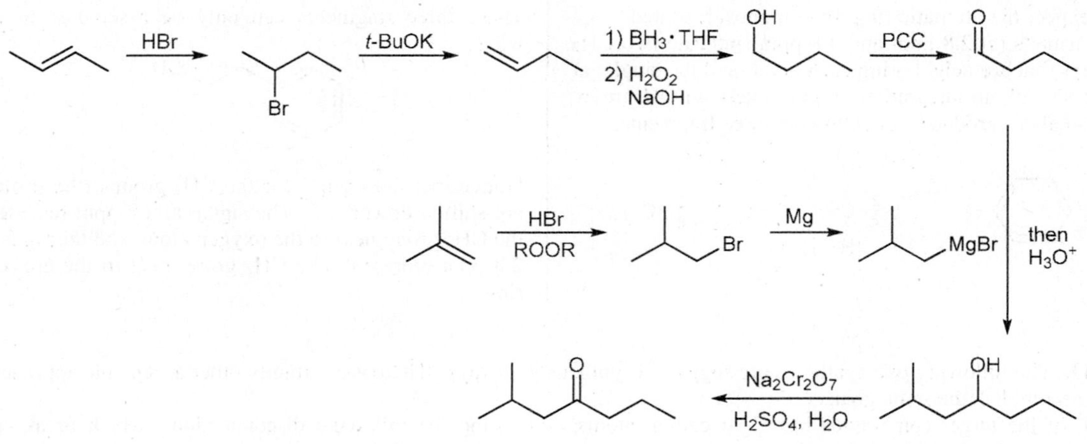

The following figure outlines a retrosynthetic analysis for our second target molecule. An explanation of each of the steps (*a-e*) follows.

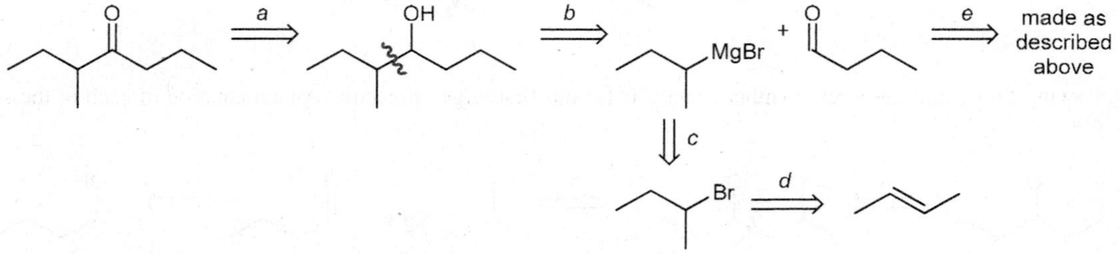

a. The target ketone is made by oxidation of the corresponding alcohol, 3-methyl-4-heptanol.
b. The alcohol is produced via a Grignard reaction between the Grignard reagent and aldehyde shown.
c. The Grignard reagent is made from the corresponding alkyl halide, 2-bromobutane.
d. 2-bromobutane is produced via addition of HBr to *trans*-2-butene. (Note: *Cis*-2-butene also produces the same product.)
e. The aldehyde is made as described in the synthesis of the first target ketone.

Now let's draw the forward scheme. Addition of HBr to 2-butene produces 2-bromobutane. 2-Bromobutane is converted to a Grignard reagent which is subsequently treated with the aldehyde (made as described above), followed by aqueous acidic workup, to produce 3-methyl-4-heptanol. Oxidation gives the target ketone.

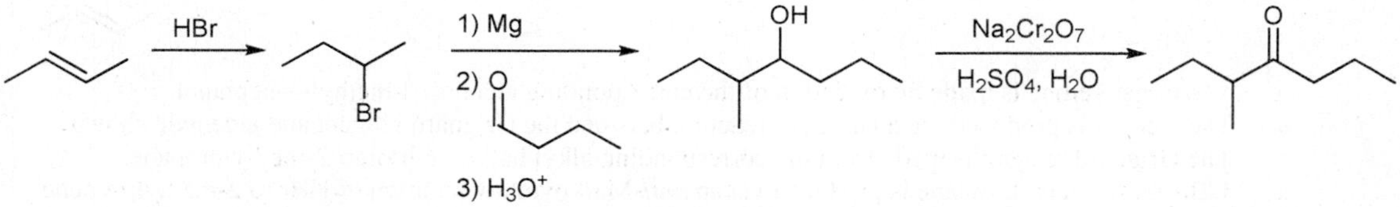

12.84.

(a) The alcohol is protonated by HBr, converting the OH group into a good leaving group. Water leaves, producing a tertiary carbocation intermediate. Bromide attacks the indicated carbon atom of the cyclopropyl group, thus opening up the ring and generating the product. The last step of the mechanism is aided by the relief of ring strain present in the three-membered ring.

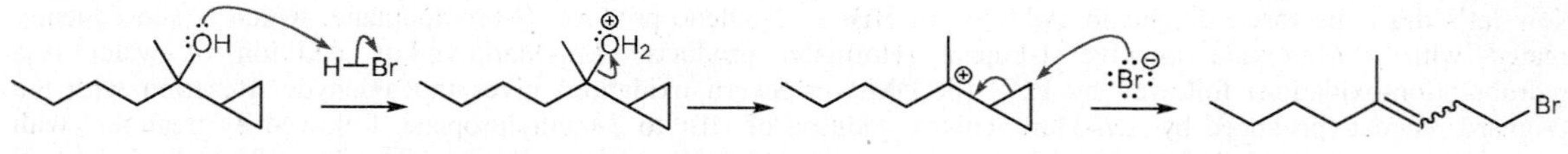

(b) The following synthesis is one suggested synthetic pathway. There are certainly other acceptable approaches that accomplish the same goal.

Let's start by looking at the product and attempting to map out the destinations of the carbon atoms in each of our given starting materials. The product is symmetric, which suggests that the groups from the left and right halves will have analogous origins. Compounds **A**, **B**, and **C** have four, five, and six carbon atoms, respectively. Both of the four-carbon linear termini of the product likely originate from the two equivalents of 1-bromobutane, which also has a linear four-carbon chain.

This leaves 16 carbon atoms in the central portion of the target structure (between the disconnections shown in the figure above). A careful analysis of this fragment (and keeping in mind the symmetry of the product) suggests that the six central carbon atoms are from **C**, while the two five-carbon units flanking this portion are derived from two equivalents of **B**.

Now let's consider the key step (described in the problem statement) and how it will fit into our synthesis. We need to fill in the gap between the starting materials and the key step, as well as the gap between the key step and the final product.

The tertiary alcohol above can be prepared via a Grignard reaction between butyl Grignard (prepared from **A**) and ketone **B**.

Now let's focus on the steps following the key step. The carbon atoms in the product can be mapped on to two equivalents of the alkenyl bromide and one equivalent of the diketone as shown in the figure below.

This suggests the following retrosynthetic approach.

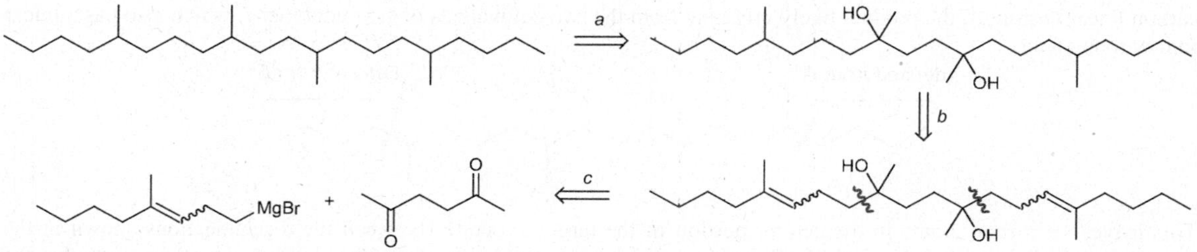

a. The product can be prepared in two steps from the diol: acid-catalyzed dehydration followed by hydrogenation of the resulting diene.
b. The alkyl chains can be prepared from alkenes via hydrogenation.
c. Disconnection at the indicated bonds indicates two equivalents of the Grignard reagent nucleophile and one equivalent of the diketone electrophile.

Now, let's draw out the forward scheme. 1-Bromobutane (**A**) is treated with magnesium to give a Grignard reagent, and subsequently reacted with ketone **B**. Aqueous acidic workup produces the tertiary alcohol. Reaction with HBr drives the ring-opening reaction to give the alkenyl bromide, as described in part (a). Conversion to the Grignard reagent by treatment with magnesium metal, followed by addition of 0.5 equivalents of diketone **C**, yields the dialkenyl diol after aqueous workup. This diene is reduced by catalytic hydrogenation. Acid-catalyzed dehydration of the diol, followed by hydrogenation of the resulting diene affords the desired product, 5,9,12,16-tetramethyleicosane.

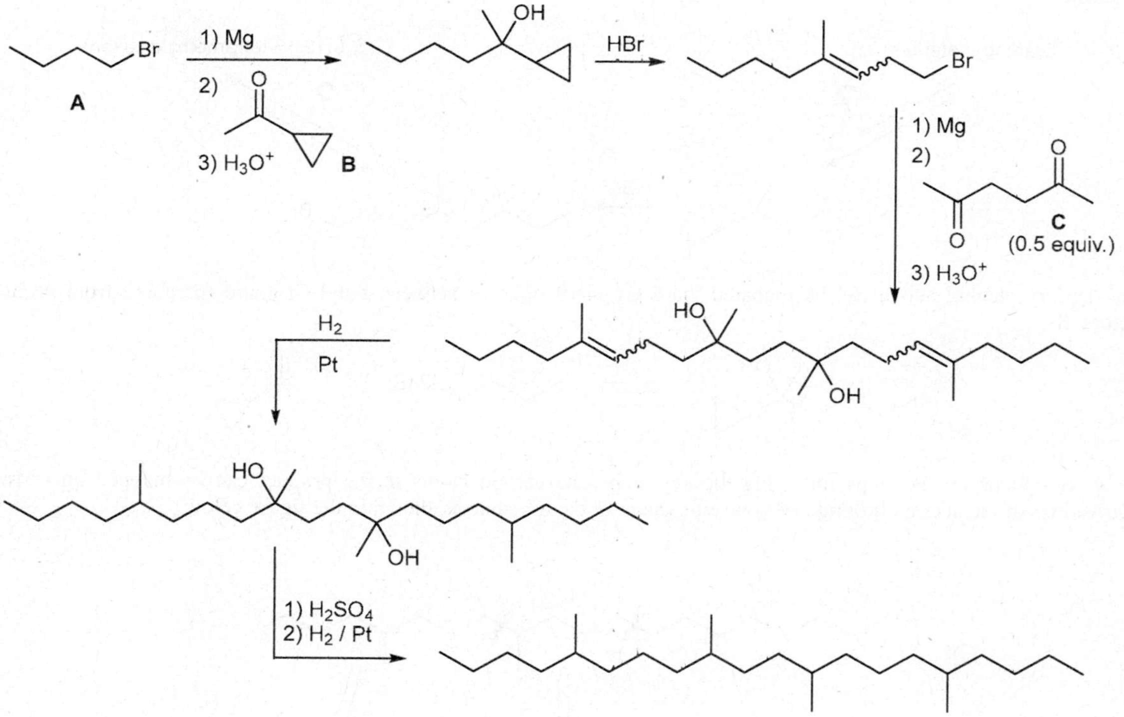

A: 1) Mg 2) B 3) H_3O^+; HBr; 1) Mg 2) C (0.5 equiv.) 3) H_3O^+; H_2 / Pt; 1) H_2SO_4 2) H_2 / Pt

12.85.

(a) TBAF (tetrabutylammonium fluoride) removes the silyl protecting group to give diol **3**, which then undergoes selective tosylation of the primary hydroxyl group (as described in the problem statement) to give tosylate **4**:

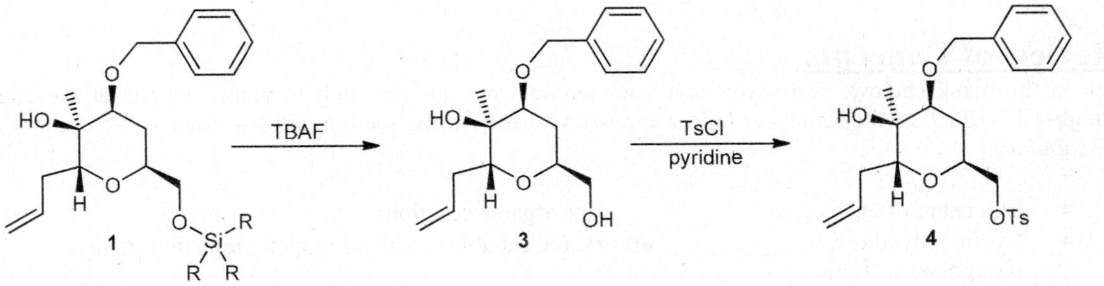

TBAF
TsCl
pyridine
1
3
4

(b) Methoxide functions as a base and deprotonates the tertiary hydroxyl group. The resulting alkoxide functions as a nucleophile and attacks the primary tosylate in an intramolecular, SN2-type process. Numbering the atoms in the six-membered ring can help us redraw the alkoxide ion in a conformation that is similar to the product.

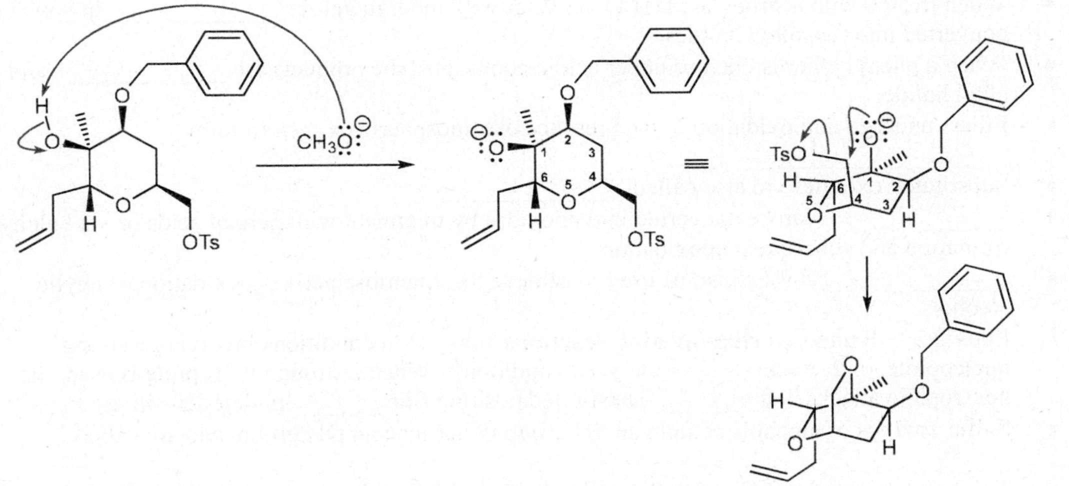

Note that these numbers do not necessarily adhere to IUPAC guidelines for assigning locants, but rather, they are simply tools that are used to verify that the alkoxide ion has been drawn correctly (it enables you to compare the configurations in each drawing, and prove to yourself that these drawings are the same). The use of locants can be especially helpful in ring-forming and ring-opening processes.

(c) The *cis* relationship places the negatively charged oxygen atom (nucleophile) in close proximity with the carbon atom bearing the tosylate group (electrophile), thereby enabling an intramolecular back-side attack. The *cis* relationship is necessary in order for the reaction to occur. In the following structure, where these two groups have a *trans* relationship, we can see that the reactive centers are too far apart for an intramolecular reaction to occur:

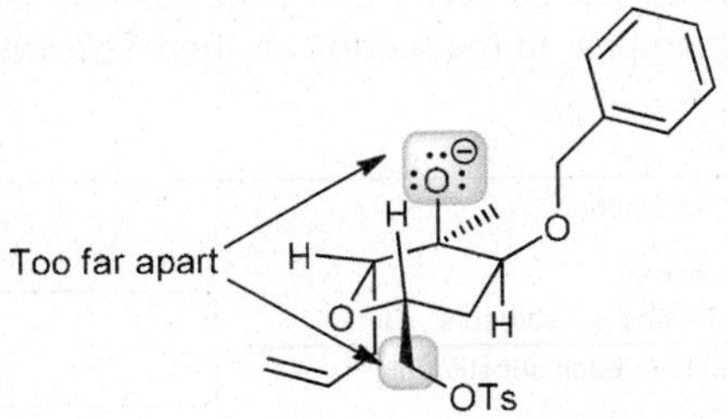

Too far apart

Chapter 13
Ethers and Epoxides; Thiols and Sulfides

Review of Concepts

Fill in the blanks below. To verify that your answers are correct, look in your textbook at the end of Chapter 13. Each of the sentences below appears verbatim in the section entitled *Review of Concepts and Vocabulary*.

- Ethers are often used as ____________ for organic reactions.
- Cyclic polyethers, or __________ **ethers**, are capable of solvating metal ions in organic (nonpolar) solvents.
- Ethers can be readily prepared from the reaction between an alkoxide ion and an ______________________, a process called a **Williamson ether synthesis**. This process works best for _________ or _____________ alkyl halides. _____________ alkyl halides are significantly less efficient, and _____________ alkyl halides cannot be used.
- When treated with a strong acid (HX), an ether will undergo **acidic** ____________ in which it is converted into two alkyl halides.
- When a phenyl ether is cleaved under acidic conditions, the products are ____________ and an alkyl halide.
- Ethers undergo autooxidation in the presence of atmospheric oxygen to form ________________.
- Substituted oxiranes are also called ______________.
- ______________ can be converted into epoxides by treatment with peroxy acids or via halohydrin formation and subsequent epoxidation.
- _____________ catalysts can be used to achieve the enantioselective epoxidation of allylic alcohols.
- Epoxides will undergo **ring-opening reactions** either 1) in conditions involving a strong nucleophile, or 2) under _____-catalyzed conditions. When a strong nucleophile is used, the nucleophile attacks at the ______-substituted position (the ______-hindered position).
- Sulfur analogs of alcohols contain an SH group rather than an OH group, and are called __________.
- Thiols can be prepared via an S_N2 reaction between sodium hydrosulfide (NaSH) and a suitable ____________________.
- The sulfur analogs of ethers (thioethers) are called ____________.
- Sulfides can be prepared from thiols in a process that is essentially the sulfur analog of the Williamson ether synthesis, involving a ___________ ion, rather than an alkoxide.

Review of Skills

Fill in the blanks and empty boxes below. To verify that your answers are correct, look in your textbook at the end of Chapter 13. The answers appear in the section entitled *SkillBuilder Review*.

13.1 Naming an Ether

Provide a systematic name for the following compound:

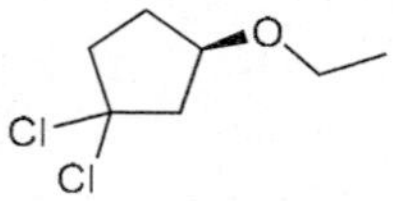

Cl Cl O

1) Identify the parent
2) Identify and name substituents
3) Assign locants to each substituent
4) Alphabetize
5) Assign configuration

13.2 Preparing an Ether via a Williamson Ether Synthesis

Identify reagents that can be used to achieve the following transformation:

OH 1) 2) O

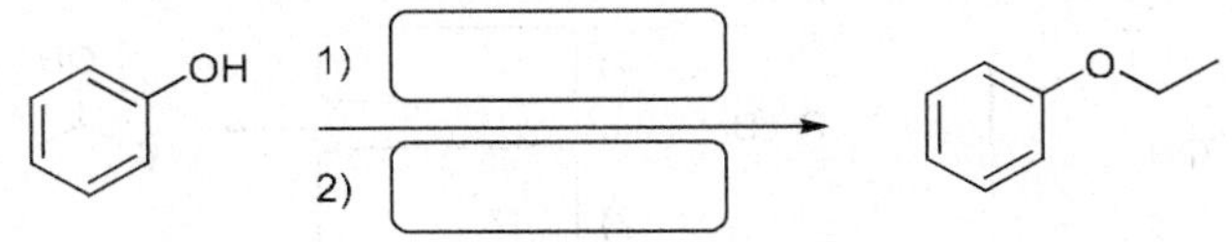

13.3 Preparing Epoxides

Identify reagents that can be used to achieve the following transformation:

Me Et Me 1) 2) Me Et Me + En

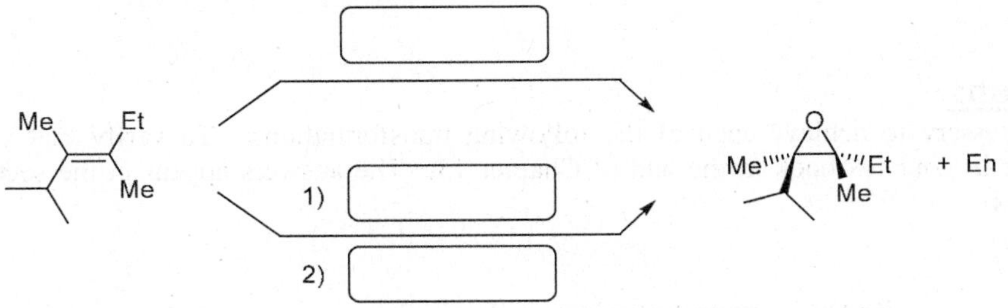

13.4 Drawing the Mechanism and Predicting the Product of the Reaction between a Strong Nucleophile and an Epoxide

Complete the mechanism below by drawing all curved arrows, intermediates and products.

O H :C≡N H–O⊕ H H

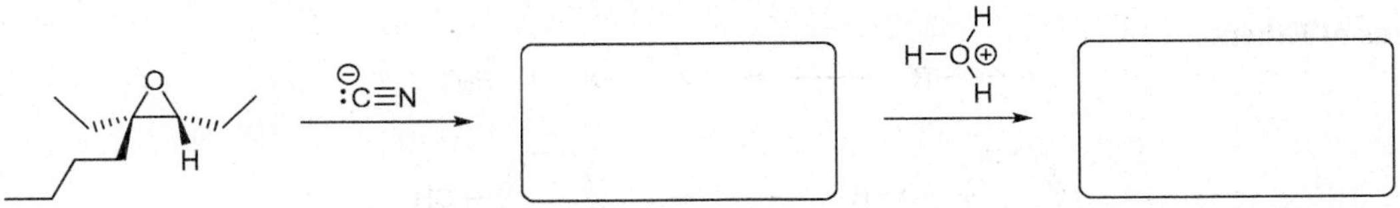

13.5 Drawing a Mechanism and Predicting the Product of Acid-Catalyzed Ring Opening

Complete the mechanism below by drawing all curved arrows, intermediates and products.

O Et Et Me H H–O⊕ H H EtOH EtOH

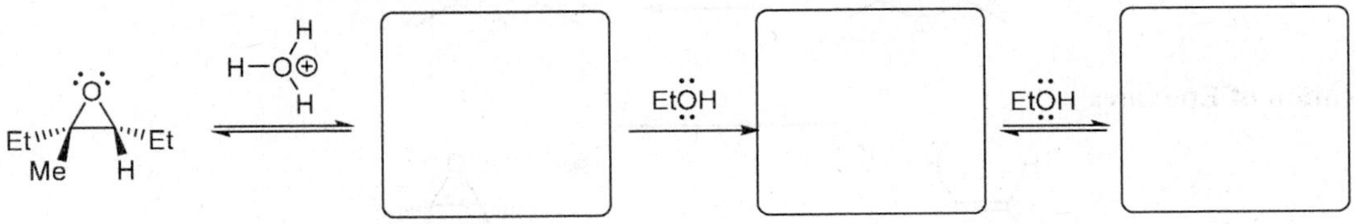

13.6 Installing Two Adjacent Functional Groups

Identify whether each ring-opening reaction below requires acidic conditions or basic conditions:

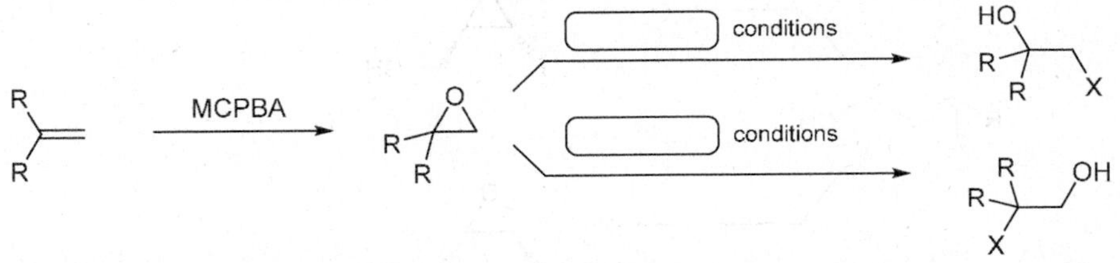

R R MCPBA O R R conditions HO R R X conditions R R OH X

13.7 Choosing the Appropriate Grignard Reaction

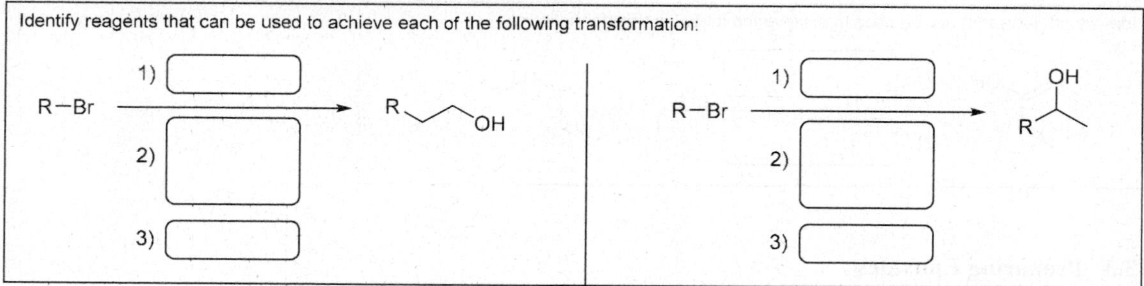

Review of Reactions

Identify the reagents necessary to achieve each of the following transformations. To verify that your answers are correct, look in your textbook at the end of Chapter 13. The answers appear in the section entitled *Review of Reactions.*

Preparation of Ethers

R—OH → R—O—R

Reactions of Ethers

R—O—R → 2 R—X + H_2O

+ R—X

OOH

Preparation of Epoxides

cis *cis*

Enantioselective Epoxidation

Ring-Opening Reactions of Epoxides

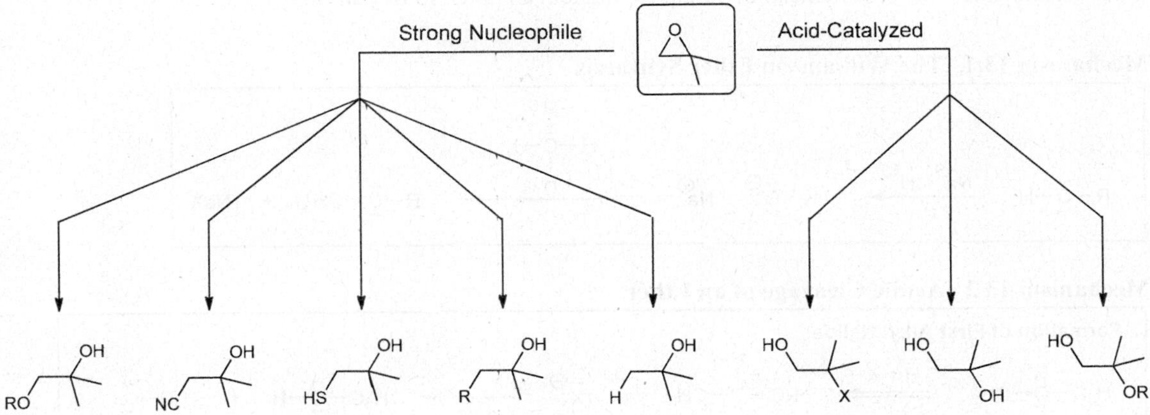

Thiols and Sulfides

Thiols

Br
R_1 R_2 ⟶ SH
R_1 R_2

2 RSH ⇌ R–S–S–R

Sulfides

R–SH ⟶ R–S–R

R–S–R ⟶ Me–S⊕(R)–R + $X^{\ominus}$

R–S–R ⟶ R–S(=O)–R ⟶ R–S(=O)(=O)–R

Review of Mechanisms

Complete each of the following mechanisms by drawing the missing curved arrows. To verify that your curved arrows are drawn correctly, compare them to the curved arrows in the mechanism boxes for Mechanisms 13.1 – 13.8, which can be found throughout Chapter 13 of your text.

Mechanism 13.1 The Williamson Ether Synthesis

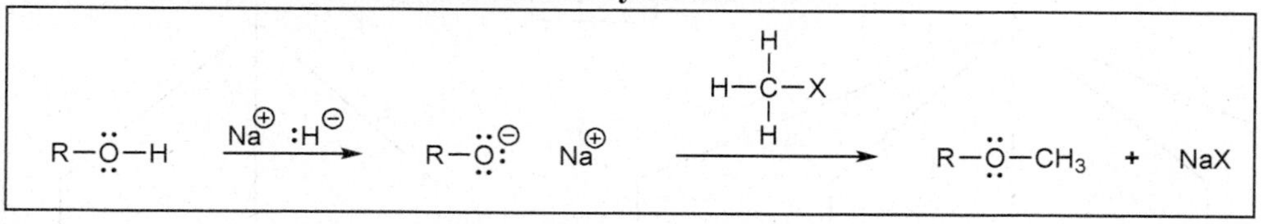

$R{-}\ddot{O}{-}H \xrightarrow{Na^{\oplus}\ :H^{\ominus}} R{-}\ddot{O}:^{\ominus}\ Na^{\oplus} \xrightarrow{H_3C{-}X} R{-}\ddot{O}{-}CH_3 + NaX$

Mechanism 13.2 Acidic Cleavage of an Ether

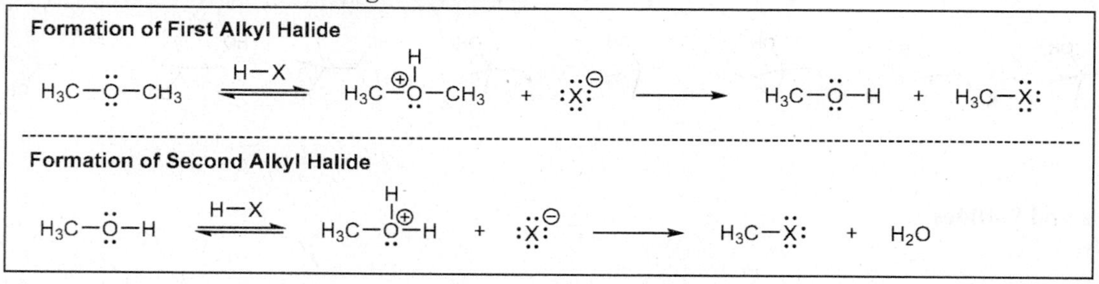

Formation of First Alkyl Halide

$H_3C{-}\ddot{O}{-}CH_3 \underset{}{\overset{H-X}{\rightleftharpoons}} H_3C{-}\overset{\oplus}{O}(H){-}CH_3 + :\ddot{X}:^{\ominus} \longrightarrow H_3C{-}\ddot{O}{-}H + H_3C{-}\ddot{X}:$

Formation of Second Alkyl Halide

$H_3C{-}\ddot{O}{-}H \underset{}{\overset{H-X}{\rightleftharpoons}} H_3C{-}\overset{\oplus}{O}(H){-}H + :\ddot{X}:^{\ominus} \longrightarrow H_3C{-}\ddot{X}: + H_2O$

Mechanism 13.3 Autooxidation of Ethers

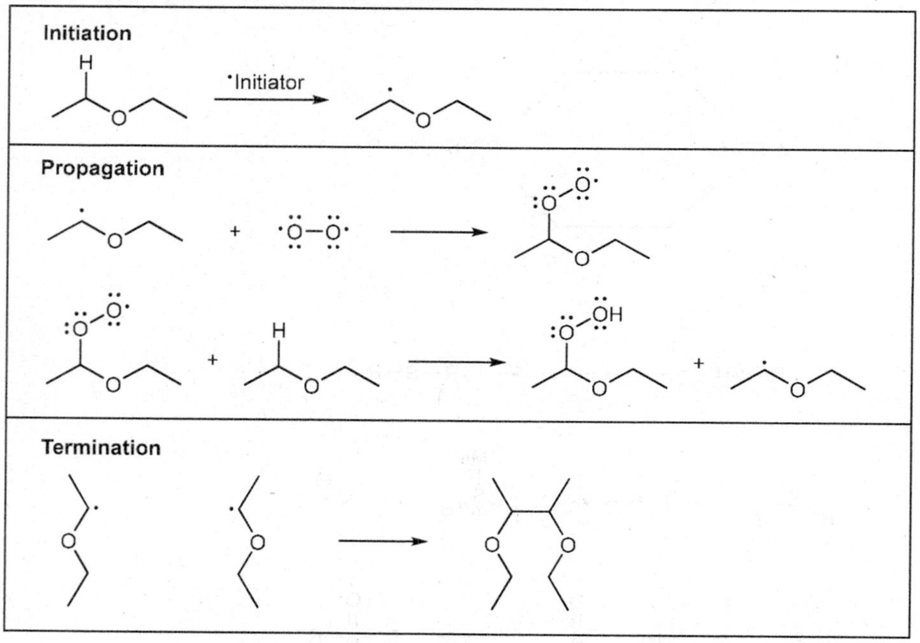

Initiation

·Initiator

Propagation

Termination

Mechanism 13.4 Epoxide Formation from Halohydrins

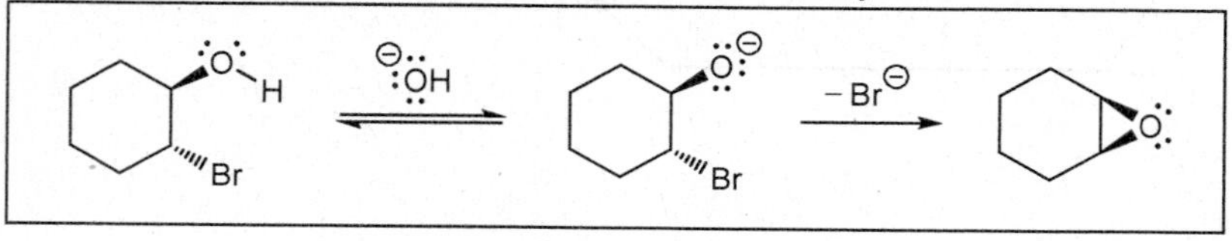

$:\ddot{O}H^{\ominus}$

$-\ Br^{\ominus}$

Mechanism 13.5 Epoxide Ring Opening with a Strong Nucleophile

Mechanism 13.6 Acid-Catalyzed Ring Opening of an Epoxide

Mechanism 13.7 Oxidation of Thiols

Deprotonation of the Thiol

$R{-}SH + Na^+ \; ^-OH \rightleftharpoons R{-}S^- \; Na^+ + H_2O$

Formation of the Disulfide

$R{-}S^- \; Na^+ \xrightarrow{Br{-}Br} R{-}S{-}Br \xrightarrow{Na^+ \; ^-S{-}R} R{-}S{-}S{-}R + NaBr$

Mechanism 13.8 Preparation of Sulfides from Thiols

$R{-}SH + Na^+ \; ^-OH \rightleftharpoons R{-}S^- \; Na^+ + H_2O \xrightarrow{R{-}X} R{-}S{-}R + NaX$

Common Mistakes to Avoid

We have seen that epoxides will react with a wide variety of nucleophiles (causing ring-opening reactions), either under acidic conditions or under basic conditions. If the attacking nucleophile is itself a strong base, such as a Grignard reagent, then acidic conditions cannot be used. That is, the following transformation is not possible to achieve, so avoid trying to do something like this:

[H^+] RMgBr

This doesn't work because Grignard reagents are not only strong nucleophiles, but they are also strong bases. And strong bases are incompatible with acidic conditions. If a Grignard reagent is subjected to a source of acid (even a relatively weak acid, such as H_2O), the Grignard reagent is irreversibly protonated to give an alkane, for example:

(MeMgBr) → Methane + HOMgBr

In summary, never use a Grignard reagent in the presence of an acid. The same rule applies to the use of $LiAlH_4$ (lithium aluminum hydride), which is both a strong nucleophile and a strong base. Therefore, much like a Grignard reagent, $LiAlH_4$ also cannot be used to open an epoxide under acidic conditions:

$[H^+]$ → ✗ $LiAlH_4$ →

Once again, this doesn't work because $LiAlH_4$ is incompatible with acidic conditions. Avoid making this mistake.

Useful reagents

The following is a list of commonly encountered reagents for reactions involving ethers, epoxides, thiols, and sulfides:

Reagents	Description
RX	An alkyl halide. Used for the alkylation of alcohols or thiols. First, the alcohol or thiol is deprotonated with a base, such as NaH or NaOH, and the resulting anion is then treated with the alkyl halide, thereby installing an alkyl group.
1) $Hg(OAc)_2$, ROH 2) $NaBH_4$	These reagents will achieve alkoxymercuration-demercuration of an alkene. This process adds H and OR in a Markovnikov fashion across the alkene.
HX	Will convert a dialkyl ether into two alkyl halides via cleavage of the C–O bonds. Will also react with an epoxide, thereby opening the ring, and installing a halogen at the more-substituted position.
MCPBA	*meta*-Chloroperoxybenzoic acid. An oxidizing agent that will convert an alkene into an epoxide.
RCO_3H	A peroxy acid. An oxidizing agent that will convert an alkene into an epoxide. MCPBA is an example of a peroxy acid.
1) Br_2, H_2O 2) NaOH	Alternative reagents for converting an alkene into an epoxide.
$(CH_3)_3COOH$, $Ti[OCH(CH_3)_2]_4$, (+)-DET or (–)-DET	Reagents for enantioselective (Sharpless) epoxidation.
NaOR (or RONa)	An alkoxide ion is both a strong nucleophile and a strong base. It can be used in a Williamson ether synthesis (reacts with an alkyl halide to form an ether), or to open an epoxide under basic conditions (the alkoxide ion attacks the less-substituted position).
NaCN	A good nucleophile that will react with an epoxide in a ring-opening reaction.
NaSH	A very strong nucleophile that will react with an epoxide in a ring-opening reaction. NaSH can also be used to prepare thiols from alkyl halides.
RMgBr	A Grignard reagent. A strong base and a strong nucleophile. Will react with an epoxide in a ring-opening reaction, to attack the less-substituted side (it is not possible to use acidic conditions and have the Grignard reagent attack the more-substituted side – see the previous section on common mistakes to avoid).

$LiAlH_4$	Lithium aluminum hydride is a source of nucleophilic hydride. It will react with an epoxide in a ring-opening reaction, to attack the less-substituted side (it is not possible to use acidic conditions and have a hydride ion attack the more-substituted side – see the previous section on common mistakes to avoid).
$[H^+]$, H_2O (or H_3O^+)	Aqueous acidic conditions. Under these conditions, an epoxide is opened to give a *trans* diol.
$[H^+]$, ROH	Under these conditions, an epoxide is opened, with a molecule of the alcohol attacking a protonated epoxide at the more-substituted position.
$NaOH/H_2O$, Br_2	Reagents for converting thiols into disulfides.
HCl, Zn	Reagents for converting disulfides into thiols.
H_2O_2	Strong oxidizing agent, used to oxidize sulfides to sulfoxides, and then further to sulfones.
$NaIO_4$	Oxidizing agent, used to oxidize sulfides to sulfoxides.

Solutions

13.1.

(a) The oxygen atom has two groups attached to it. One group has two carbon atoms and the other group has three carbon atoms. The parent is named after the larger group, so the parent is propane. The smaller group (together with the oxygen atom) is treated as an ethoxy substituent, and a locant is included to identify the location (C2) of the substituent on the parent chain.

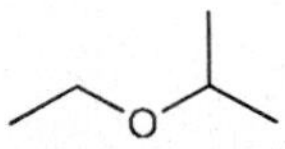

2-Ethoxypropane

(b) The oxygen atom has two groups attached to it. One group has three carbon atoms and the other group has two carbon atoms. The parent is named after the larger group, so the parent is propane. The smaller group (together with the oxygen atom) is treated as an ethoxy substituent, and a locant is included to identify the location (C1) of the substituent on the parent chain. The parent chain also has a chloro substituent, located at C2 of the parent. That position is a chiral center, so a stereodescriptor is required to identify the configuration (*S*). When assembling the name, the substituents are arranged alphabetically (chloro precedes ethoxy).

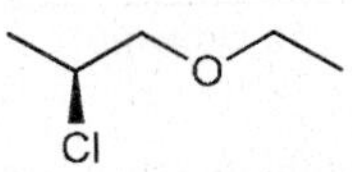

(*S*)-2-Chloro-1-ethoxypropane

(c) The oxygen atom has two groups attached to it. The parent is named after the larger group, so the parent is benzene. The smaller group (together with the oxygen atom) is treated as an ethoxy substituent, and a locant (C1) is included to identify the location of the substituent on the parent chain. The parent chain also has two chloro substituents, located at C2 and C4 of the parent. The parent is numbered to give the substituents the lowest possible numbers (1,2,4 instead of 1,3,4). When assembling the name, the substituents are arranged alphabetically (chloro precedes ethoxy).

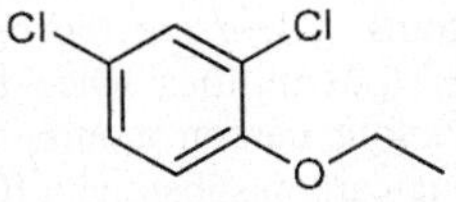

2,4-Dichloro-1-ethoxybenzene

(d) The parent is cyclohexanol (with OH located at C1, by definition), and the ethoxy group is listed as a substituent. A locant (C2) is included to indicate the location of the ethoxy group on the cyclohexanol parent, and stereodescriptors are included to indicate the configuration of each chiral center:

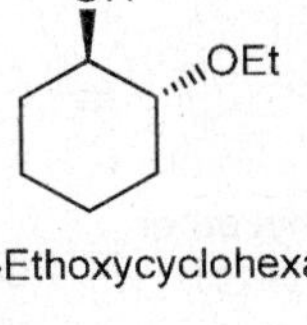

(*1R,2R*)-2-Ethoxycyclohexanol

(e) The parent is cyclohexene, and the ethoxy group is listed as a substituent. A locant (C1) is included to indicate the location of the ethoxy group on the cyclohexene parent:

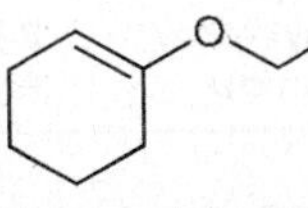

1-Ethoxycyclohexene

13.2.

(a) The parent (cyclobutane) is a four-membered ring, and there is an ethoxy group connected to the ring (which is defined as position C2 of the ring). This position is a chiral center, and it has the *R* configuration. There are two methyl groups, both located at C1.

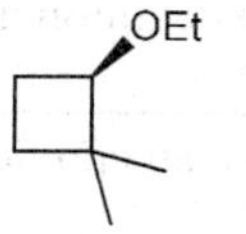

(*R*)-2-Ethoxy-1,1-dimethylcyclobutane

(b) This name has the format of a common name. It is an ether in which the oxygen atom is connected to a cyclopropyl group (a three-membered ring) and an isopropyl group (a three-membered chain, connected at the middle carbon atom).

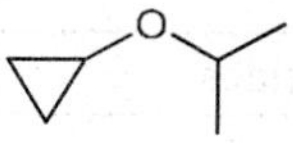

Cyclopropyl isopropyl ether

13.3. Recall that the general structure of an ether is R–O–R′, where R and R′ represent alkyl, aryl, or vinyl groups. For an ether with the molecular formula $C_5H_{12}O$, both groups must be alkyl groups. Further, if R has one carbon atom, then R′ must have four carbon atoms; and if R has two carbon atoms, then R′ must have three carbon atoms. These are the only options. Let's first consider the possibilities when R has one carbon atom and R′ has four carbon atoms. There is only one way to have a one-carbon substituent (CH_3), but there are four ways to assemble a four-carbon substituent. These options result in four possible ethers, shown below, along with their common and systematic names.

***n*-Butyl methyl ether** (common name)	**1-Methoxybutane** (systematic name)
Isobutyl methyl ether (common name)	**1-Methoxy-2-methylpropane** (systematic name)
***sec*-Butyl methyl ether** (common name)	**2-Methoxybutane** (systematic name)
***tert*-Butyl methyl ether** (common name)	**2-Methoxy-2-methylpropane** (systematic name)

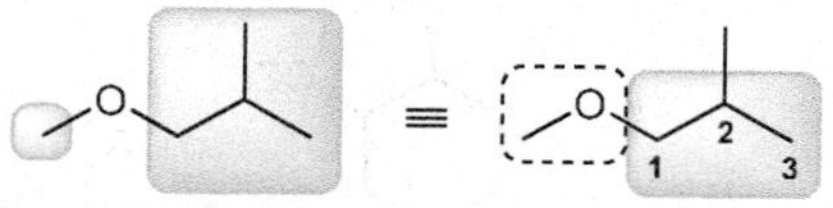

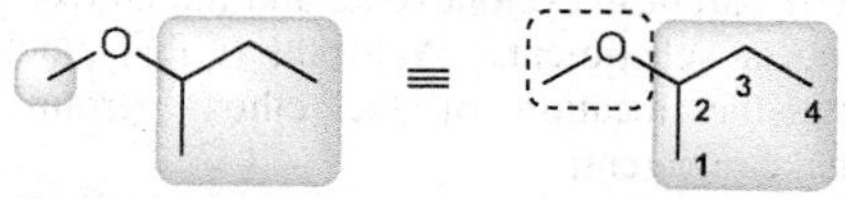

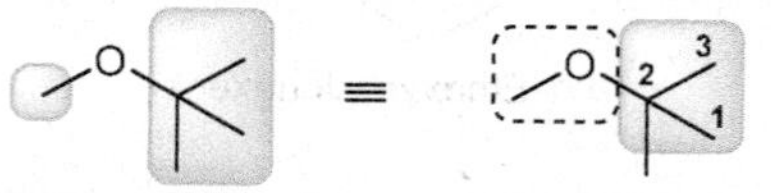

The common names of the four different types of butyl groups (*n*-butyl, isobutyl, *sec*-butyl, and *tert*-butyl) were discussed in Section 4.2. The systematic name of each compound uses the longest continuous carbon chain as the parent and treats the methoxy group as a substituent in all four cases.

Now let's consider the possibilities when R has two carbon atoms and R′ has three carbon atoms. There is only one way to have a two-carbon substituent, but there are two ways to assemble a three-carbon substituent (*n*-propyl or isopropyl). These options result in two possible ethers, shown below, along with their common and systematic names. The common names of the two types of propyl groups (*n*-propyl or isopropyl) were discussed in Section 4.2. Each systematic name uses the longest continuous carbon chain as the parent (propane) and treats the ethoxy group as a substituent.

Ethyl *n*-propyl ether (common name)	**1-Ethoxypropane** (systematic name)
Ethyl isopropyl ether (common name)	**2-Ethoxypropane** (systematic name)

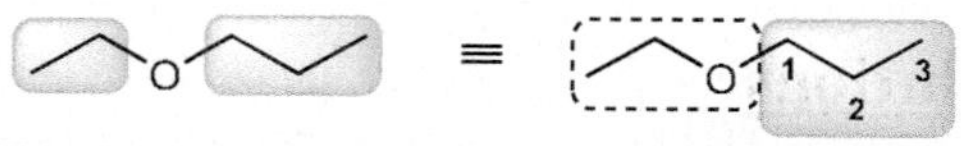

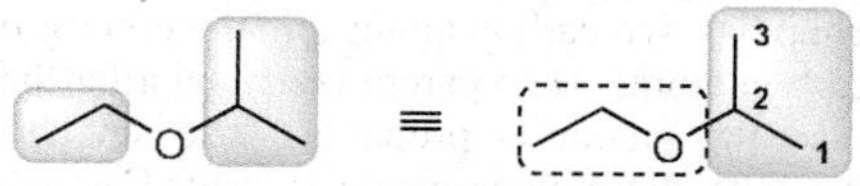

13.4.

(a) The cation is potassium, so we must use 18-crown-6, which solvates potassium ions.

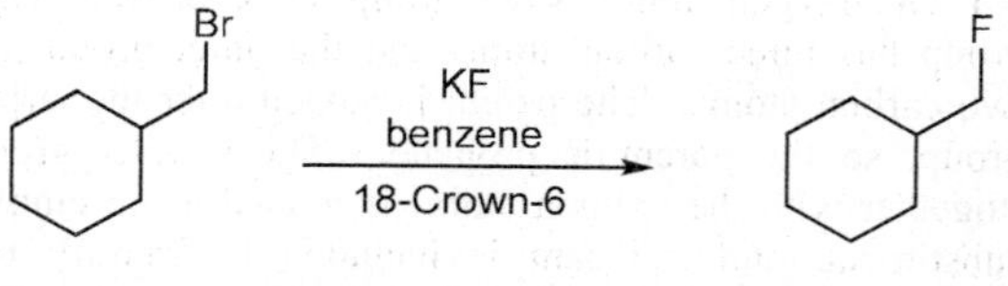

KF, benzene, 18-Crown-6

(b) The cation is sodium, so we must use 15-crown-5, which solvates sodium ions.

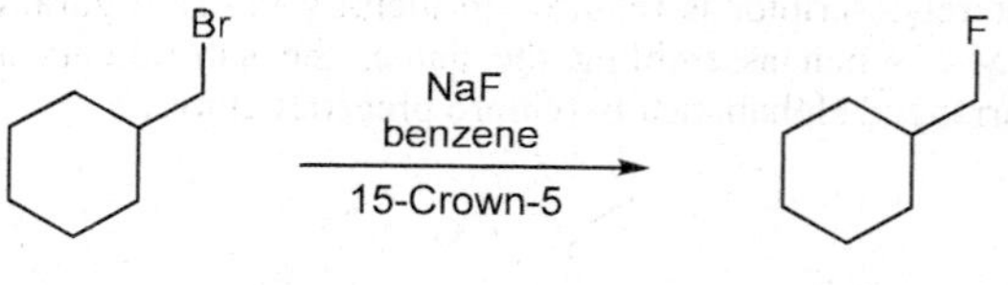

NaF, benzene, 15-Crown-5

(c) The cation is lithium, so we must use 12-crown-4, which solvates lithium ions.

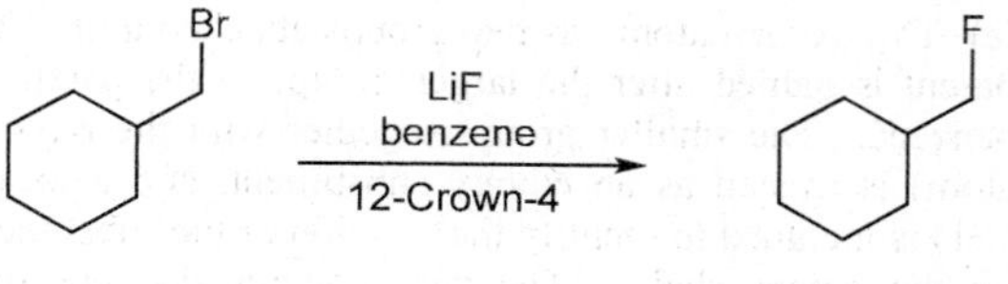

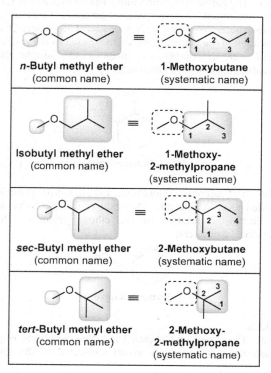

LiF, benzene, 12-Crown-4

(d) The cation is potassium, so we must use 18-crown-6, which solvates potassium ions.

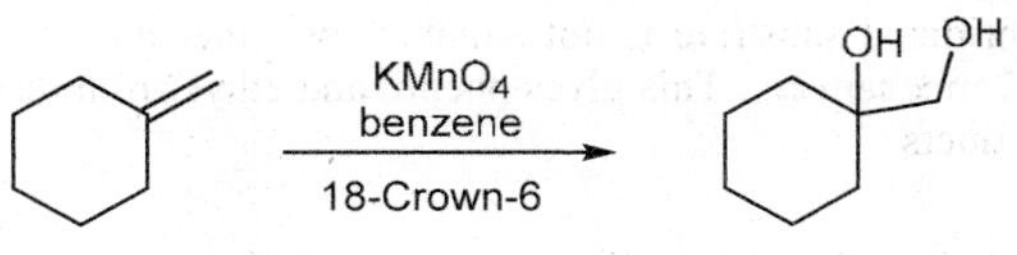

KMnO4 benzene 18-Crown-6

OH OH

13.5.

(a) A Williamson ether synthesis will be more efficient with a less sterically hindered substrate, since the process involves an SN2 reaction (at the highlighted position shown below). Therefore, in this case, it is better to start with a secondary alcohol and a primary alkyl halide, rather than a primary alcohol and a secondary alkyl halide:

OH 1) NaH 2) I

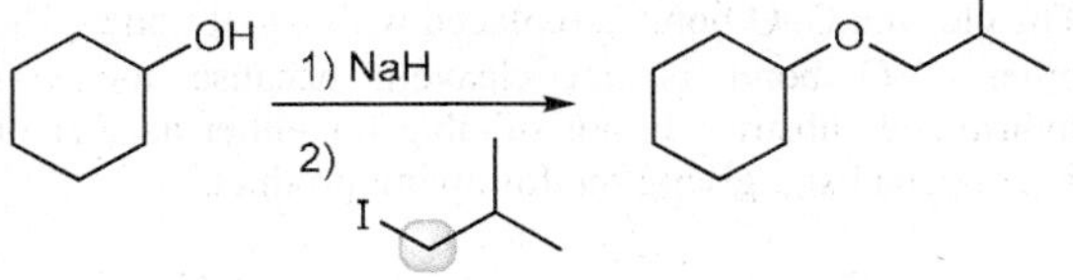

(b) A Williamson ether synthesis will be more efficient with a less sterically hindered substrate, since the process involves an SN2 reaction (at the highlighted position shown below). In this case, it is better to start with a secondary alcohol and a primary alkyl halide, rather than a primary alcohol and a secondary alkyl halide:

HO 1) NaH 2) I O

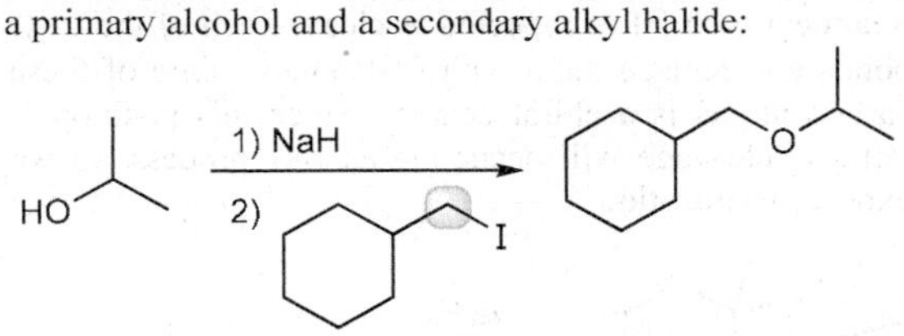

(c) A Williamson ether synthesis will be more efficient with a less sterically hindered substrate, since the process involves an SN2 reaction (at the highlighted position shown below). In this case, it is better to start with a tertiary alcohol and a methyl halide, rather than methanol and a tertiary alkyl halide:

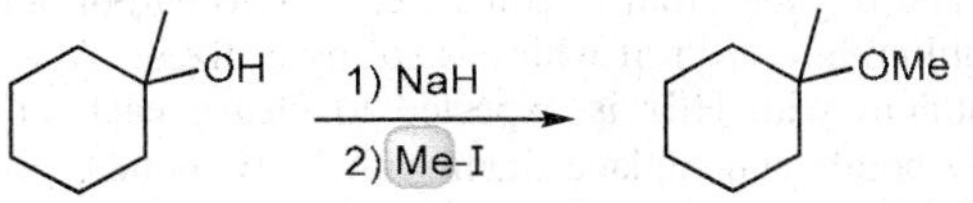

OH 1) NaH 2) Me-I OMe

(d) In order to perform an intramolecular Williamson ether synthesis, we must choose a starting compound that contains both an OH group and a halogen, as shown here:

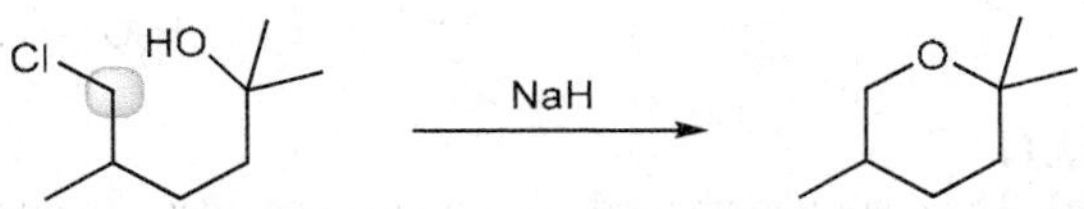

Cl HO NaH O

The OH group is deprotonated upon treatment with NaH (a strong base). The resulting alkoxide ion can then function as a nucleophile in an intramolecular, SN2-type reaction (back-side attack at the highlighted, primary position), expelling chloride as a leaving group, and giving a six-membered ring.

The alternative starting compound, shown below, cannot be used, because the leaving group is attached to a tertiary position, so an SN2-type process cannot occur at that location.

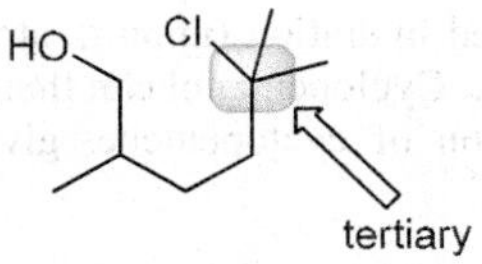

HO Cl tertiary

13.6. We begin by classifying the carbon atoms on either side of the oxygen atom. One is secondary and the other is primary:

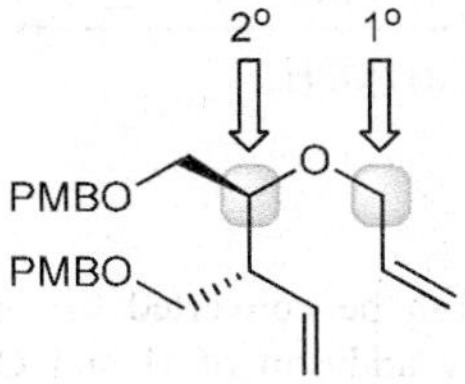

2° 1° PMBO PMBO O

A Williamson ether synthesis will be more efficient with a less sterically hindered substrate. Therefore, a primary halide and a secondary alcohol should be used, rather than a secondary halide and a primary alcohol:

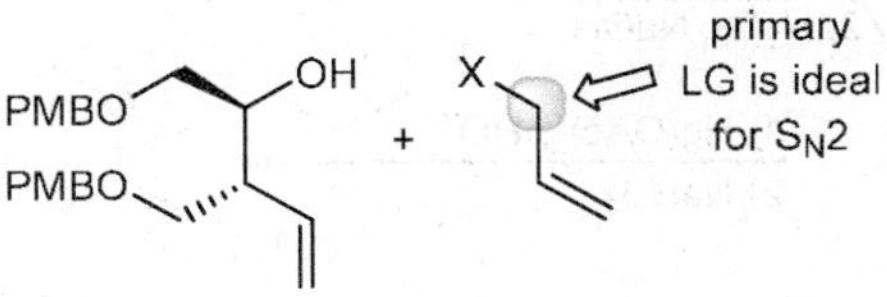

PMBO PMBO OH + X primary LG is ideal for SN2

The following reagents can be used to make the ether group of compound **1**:

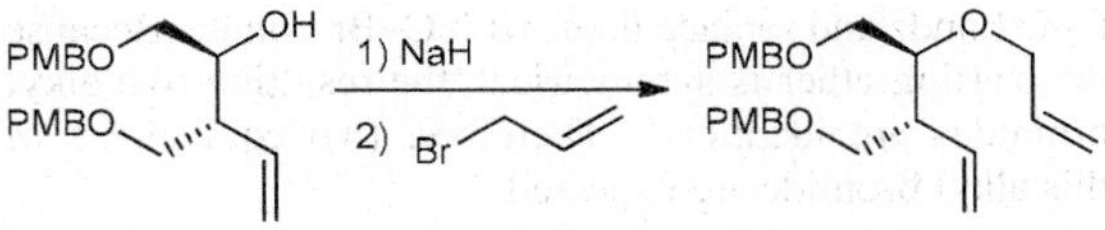

PMBO PMBO OH 1) NaH 2) Br PMBO PMBO O

13.7.

(a) The desired transformation involves the Markovnikov addition of H and OEt across the π bond. This can be accomplished via alkoxymercuration-demercuration, where EtOH is used in the first step of the process.

1) $Hg(OAc)_2$, EtOH 2) $NaBH_4$ OEt

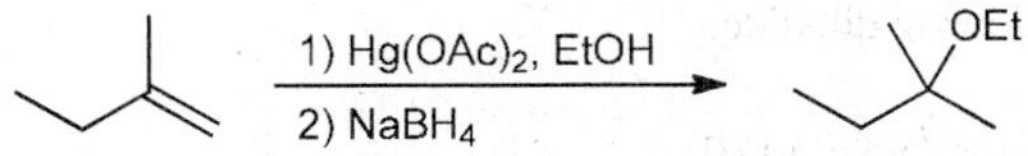

(b) The desired transformation involves the Markovnikov addition of H and OR across the π bond. This can be accomplished via alkoxymercuration-demercuration, where ROH (cyclobutanol) is used in the first step of the process.

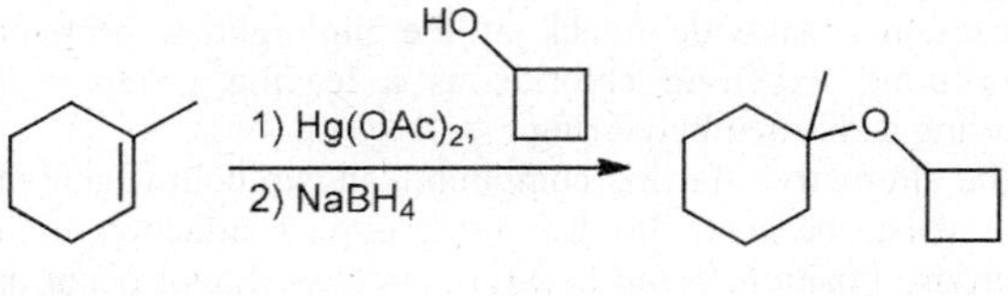

13.8. Cyclopentene can be converted to cyclopentanol via acid-catalyzed hydration (upon treatment with dilute aqueous H_2SO_4). Cyclopentanol can then be used for the alkoxymercuration of cyclopentene, giving the desired product.

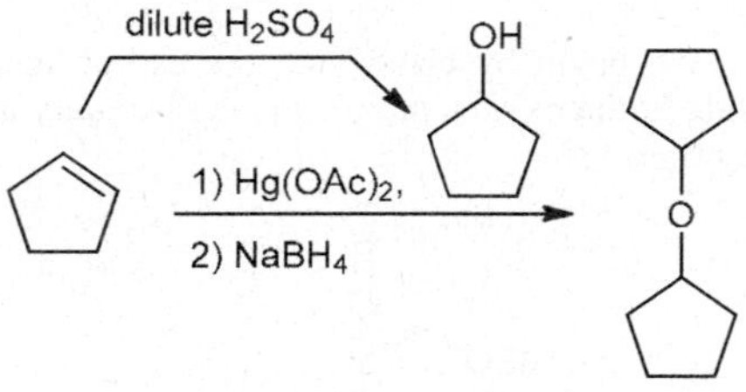

13.9. Propene can be converted to 1-propanol via an *anti*-Markovnikov addition of H and OH across the π bond, which can be achieved with hydroboration-oxidation. This alcohol can then be used for the alkoxymercuration of propene, giving the desired product via Markovnikov addition.

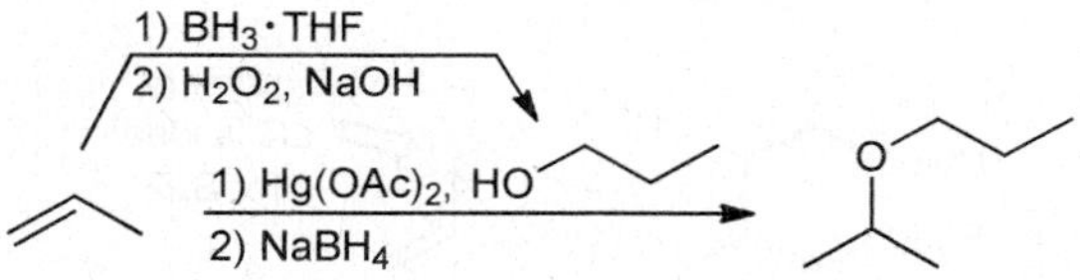

13.10.

(a) The oxygen atom is connected to two carbon atoms (highlighted), each of which is sp^3 hybridized. As such, treatment with HBr is expected to cleave each of the C–O bonds and replace them with C–Br bonds. Because the starting ether is symmetrical, the resulting two alkyl bromides are identical. Therefore, two equivalents of this alkyl bromide are expected.

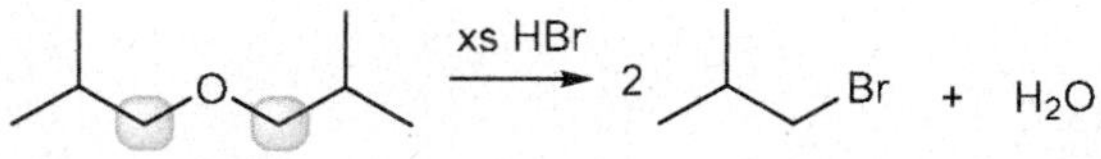

(b) The oxygen atom is connected to two carbon atoms (highlighted), each of which is sp^3 hybridized. As such, treatment with HI is expected to cleave each of the C–O bonds and replace them with C–I bonds, giving the following diiodide.

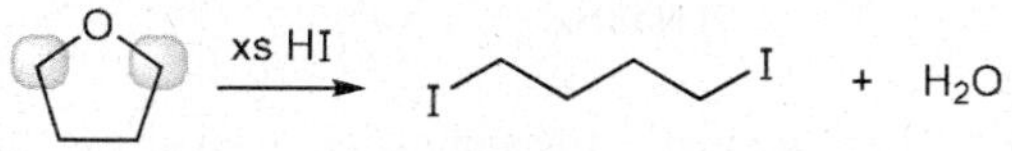

(c) The oxygen atom is connected to two carbon atoms. One of these carbon atoms (left) is sp^2 hybridized, while the other (right) is sp^3 hybridized. As such, treatment with HBr is expected to cleave only the C–O bond involving the sp^3-hybridized carbon atom (highlighted). The cleaved C–O bond is replaced with a C–Br bond. The other C–O bond is not cleaved, because the sp^2 hybridized substrate is not suitable for either an S_N1 or S_N2 mechanism. This gives phenol and ethyl bromide as products.

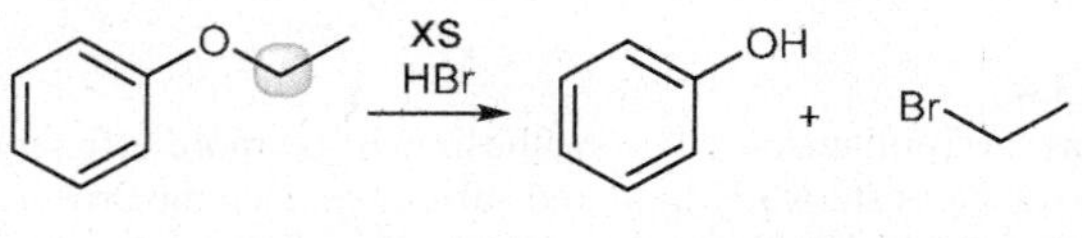

(d) The oxygen atom is connected to two carbon atoms. One of these carbon atoms (left) is sp^2 hybridized, while the other (right) is sp^3 hybridized. As such, treatment with HI is expected to cleave only the C–O bond involving the sp^3-hybridized carbon atom (highlighted). The cleaved C–O bond is replaced with a C–I bond. The other C–O bond is not cleaved, because the sp^2-hybridized substrate is not suitable for either an S_N1 or S_N2 mechanism, giving the following product.

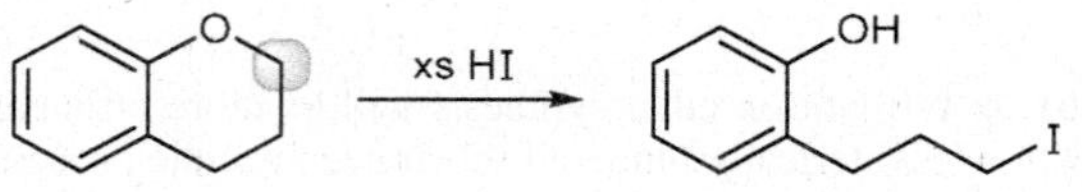

(e) The oxygen atom is connected to two carbon atoms (highlighted), each of which is sp^3 hybridized. As such, treatment with HI is expected to cleave each of the C–O bonds and replace them with C–I bonds. One of these carbon atoms is a chiral center. Since this position is tertiary, cleavage will occur via an S_N1 process, so we expect racemization.

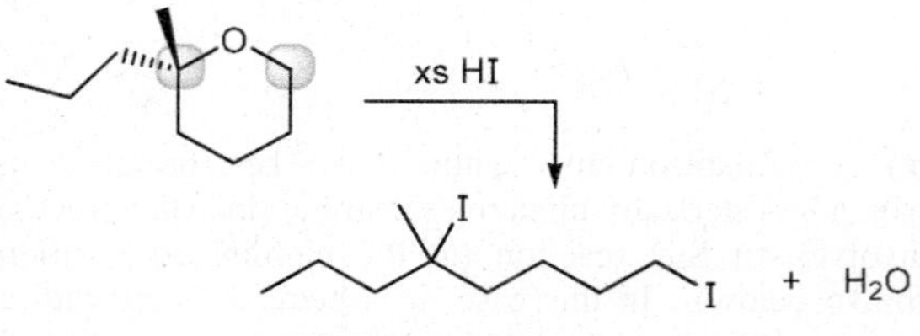

(f) The oxygen atom is connected to two carbon atoms (highlighted), each of which is sp^3 hybridized. As such, treatment with HBr is expected to cleave each of the C–O bonds and replace them with C–Br bonds, giving cyclohexyl bromide and ethyl bromide as products.

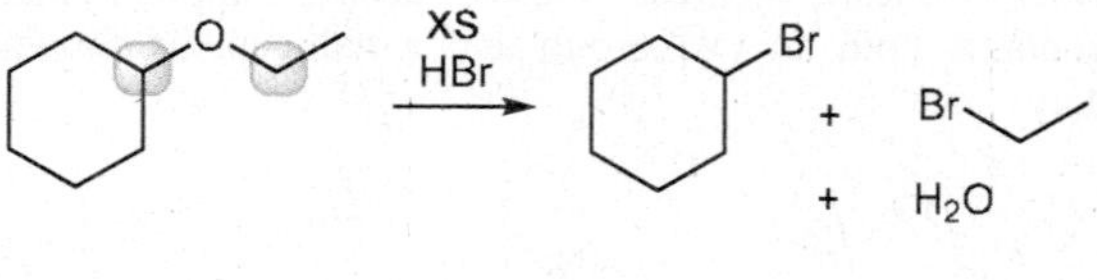

13.11.

(a) There are two methods for naming epoxides. In one method, the parent will be propane, and the oxygen atom is considered to be an epoxy substituent connected to the parent at C1 and C2. In addition, there is a methyl substituent at C2, and the substituents are alphabetized in the name, as shown below.

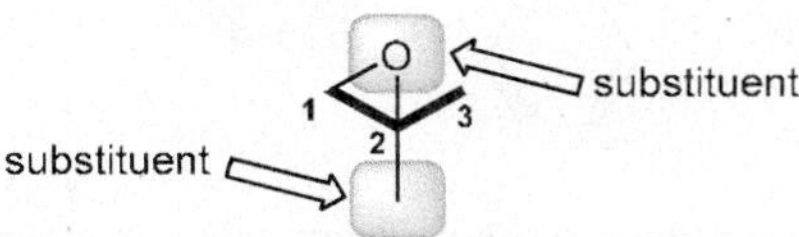

1,2-Epoxy-2-methylpropane

According to the second method for naming epoxides, the parent is considered to be the oxirane ring (with the O assigned as position 1, by definition). The ring has two methyl groups attached to it, both located at the 2 position.

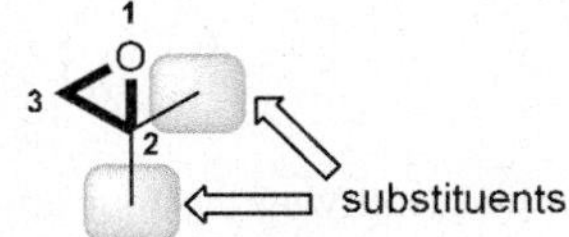

2,2-Dimethyloxirane

(b) There are two methods for naming epoxides. In one method, the parent will be ethane, and the oxygen atom is considered to be an epoxy substituent connected to the parent at C1 and C2. In addition, there are two phenyl substituents at C1, and the substituents are alphabetized in the name ("e" of epoxy comes before "p" of diphenyl), as shown below.

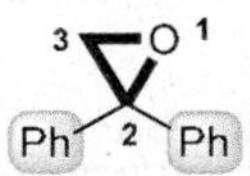

1,2-Epoxy-1,1-diphenylethane

According to the second method for naming epoxides, the parent is considered to be the oxirane ring (with the O assigned as position 1, by definition). The ring has two phenyl groups attached to it, both located at the 2 position.

2,2-Diphenyloxirane

(c) There are two methods for naming epoxides, although one of these methods will be less helpful because the two substituents (connected to the oxirane ring) are actually closed in a ring. This makes it difficult to name the compound as an oxirane. According to the first method for naming ethers, the parent is cyclohexane, and the oxygen atom is considered to be an epoxy substituent connected to the parent at C1 and C2.

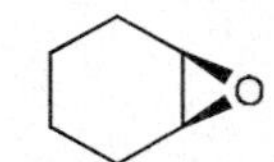
1,2-Epoxycyclohexane

13.12.

(a) There are two methods for naming epoxides. In one method, the parent will be propane, and the oxygen atom is considered to be an epoxy substituent connected to the parent at C1 and C2. In addition, there is a phenyl substituent at C2, and the substituents are alphabetized in the name, as shown below. Note that the configuration of the chiral center is indicated at the beginning of the name.

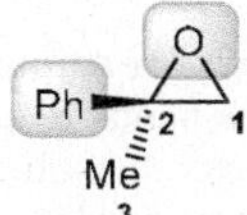

(*R*)-1,2-Epoxy-2-phenylpropane

According to the second method for naming epoxides, the parent is considered to be the oxirane ring (with the O assigned as position 1, by definition). The methyl group and the phenyl group are both considered to be substituents, and their locations are identified with locants. Finally, the configuration of the chiral center is indicated (at the beginning of the name).

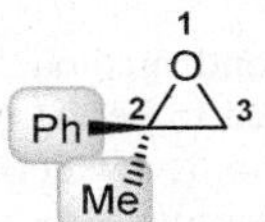

(*R*)-2-Methyl-2-phenyloxirane

In this case, the chiral center has the *R* configuration, as a result of the following prioritization scheme.

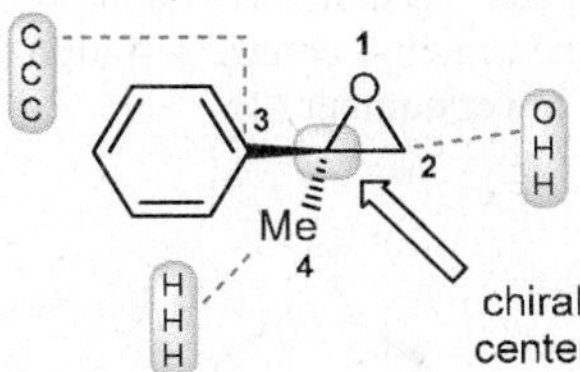

(b) There are two methods for naming epoxides. In one method, the parent will be heptane, and the oxygen atom is considered to be an epoxy substituent connected to the parent at C3 and C4. The configuration of each chiral center is indicated.

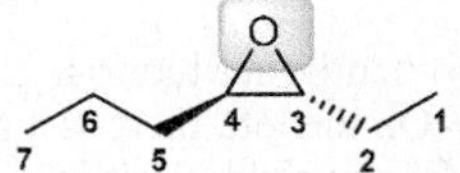

(3*R*,4*R*)-3,4-Epoxyheptane

According to the second method for naming epoxides, the parent is considered to be the oxirane ring (with the O assigned as position 1, by definition). The ring is connected to two substituents (a propyl group and an ethyl group). Their locations are identified with locants, and the configuration of each chiral center is indicated (at the beginning of the name).

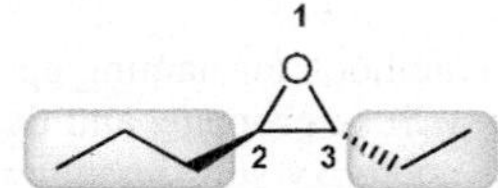

(2*R*,3*R*)-3-Ethyl-2-propyloxirane

Note that, since it is alphabetically first, the ethyl group will assume the C2 position.

(c) There are two methods for naming epoxides. In one method, the parent will be pentane, and the oxygen atom is considered to be an epoxy substituent connected to the parent at C2 and C3. There is also a methyl substituent located at C4, and the substituents are alphabetized in the name, as shown below. Note that the configuration of each chiral center is indicated at the beginning of the name.

(2*R*,3*S*)-2,3-Epoxy-4-methylpentane

According to the second method for naming epoxides, the parent is considered to be the oxirane ring (with the O assigned as position 1, by definition). The ring is connected to two substituents (a methyl group and an isopropyl group). Their locations are identified with locants, and the configuration of each chiral center is indicated (at the beginning of the name). If isopropyl is used as the substituent name, it comes first alphabetically and therefore is assigned the lower number (2). However, if the systematic name is used, (1-methylethyl), then methyl comes first alphabetically and is assigned the lower number (2).

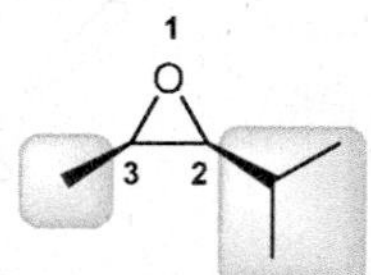

(2*S*,3*R*)-2-Isopropyl-3-methyloxirane
or
(2*R*,3*S*)-2-methyl-3-(1-methylethyl)oxirane

13.13.

(a) We begin by identifying the four groups attached to the epoxide ring. On the left, there is a methyl group and a phenyl group. On the right, there is a phenyl group and a hydrogen atom. Notice that the two phenyl groups are *trans* to each other:

These two groups must be *trans* to each other in the starting alkene.

MCPBA

+ En

(b) We begin by identifying the four groups attached to the epoxide ring. On the left, there is a cyclohexyl group and a hydrogen atom. On the right, there is a methyl group and a hydrogen atom. Notice that the two hydrogen atoms are *trans* to each other.

These two hydrogen atoms must be *trans* to each other in the starting alkene.

MCPBA

+ En

(c) The starting alkene must have the *E* configuration in order to obtain the desired epoxide, as shown.

MCPBA

+ En

(d) The starting alkene must have the *trans* configuration in order to obtain the desired epoxide, as shown.

MCPBA

+ En

13.14. Compounds **1** and **3** have the same carbon skeleton. Only the identity of a functional group has changed. The epoxide product is *cis*-disubstituted, so we can deduce that the carbon-carbon double bond of alkene **2** must also be *cis*-disubstituted (because the epoxidation process is stereospecific).

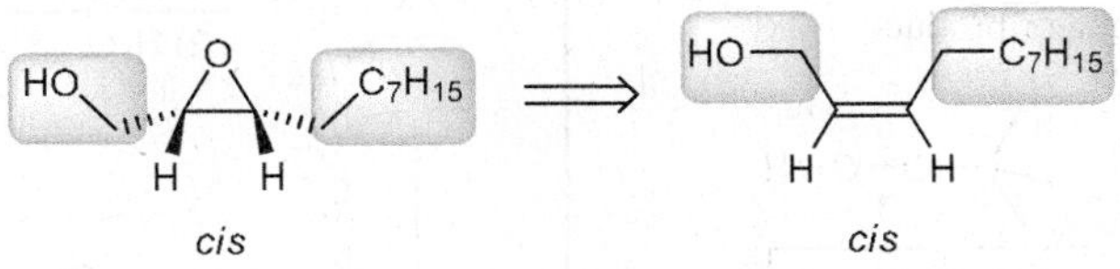

The carbon-carbon double bond of the alkene can be made via reduction of the alkyne.

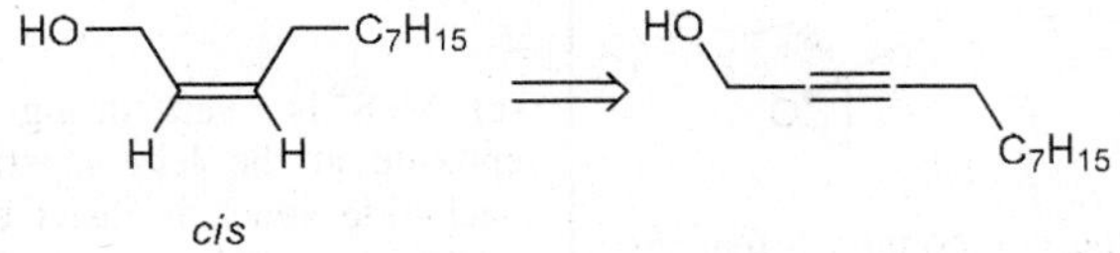

Now let's draw the forward process. Conversion of alkyne **1** to alkene **2** requires the use of H_2 and a poisoned catalyst to generate the *cis*-disubstituted alkene. In this particular case, the investigators used Ni_2B (also called the P-2 catalyst), but Lindlar's catalyst would also be acceptable. Conversion of alkene **2** into epoxide **3** can be accomplished using a peroxy acid, such as MCPBA, and results in a racemic mixture of epoxide **3**, because the epoxidation reaction can occur on either face (top or bottom) of the C=C bond.

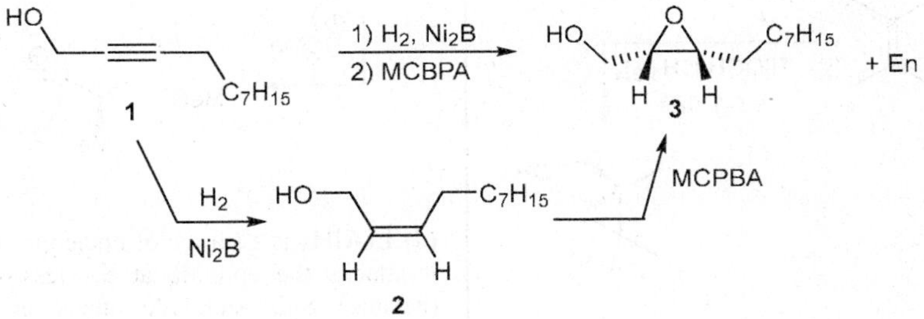

13.15.

(a) The allylic hydroxyl group appears in the upper right corner, so a Sharpless epoxidation with (+)-DET will generate an epoxide ring above the plane of the π bond, giving the following enantiomer as the major product.

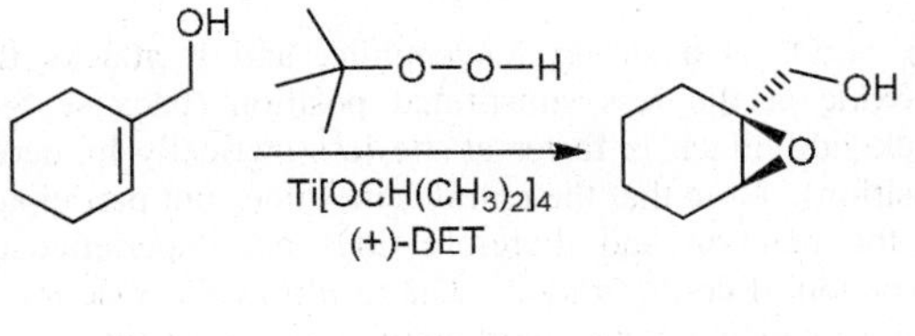

(b) We begin by redrawing the compound so that the allylic hydroxyl group appears in the upper right corner. A Sharpless epoxidation with (–)-DET will generate an epoxide ring below the plane of the π bond, giving the following enantiomer as the major product.

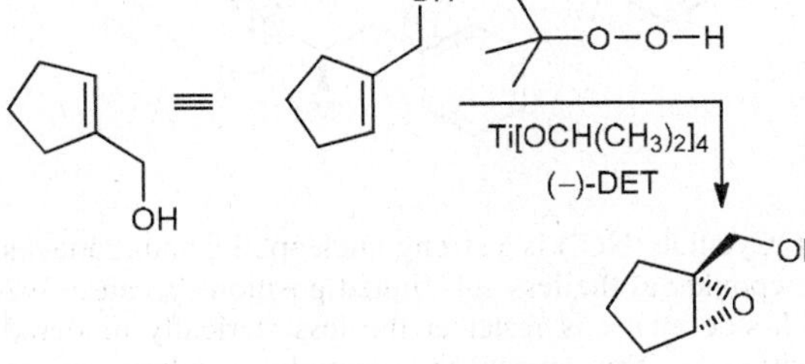

(c) We begin by redrawing the compound so that the allylic hydroxyl group appears in the upper right corner. A Sharpless epoxidation with (+)-DET will generate an epoxide ring above the plane of the π bond, giving the following enantiomer as the major product.

≡ Ti[OCH(CH$_3$)$_2$]$_4$ (+)-DET

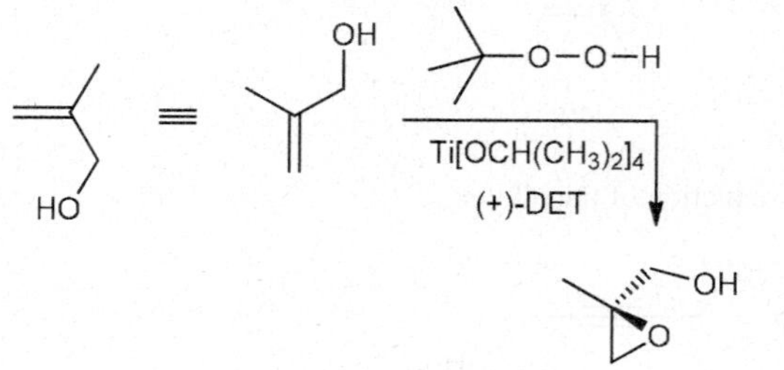

(d) We begin by redrawing the compound so that the allylic hydroxyl group appears in the upper right corner. A Sharpless epoxidation with (–)-DET will generate an epoxide ring below the plane of the π bond, giving the following enantiomer as the major product.

≡ Ti[OCH(CH$_3$)$_2$]$_4$ (–)-DET

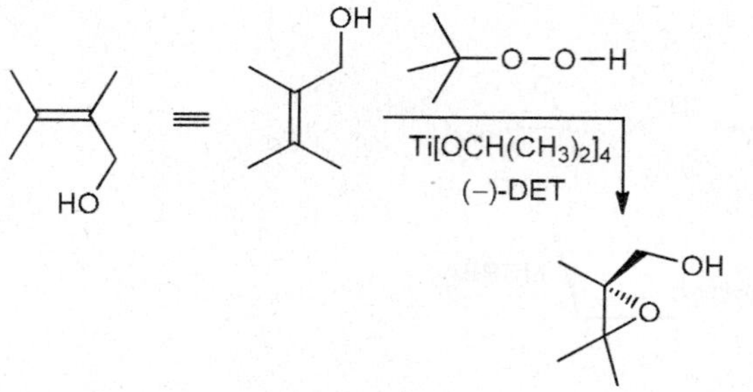

13.16.

(a) The Grignard reagent (PhMgBr) is a strong nucleophile, and it attacks the epoxide at the less substituted position (because S$_N$2 back-side attack is faster at the less sterically hindered position). The epoxide is opened, resulting in an alkoxide ion. This alkoxide is then protonated during workup with aqueous acid.

1) PhMgBr 2) H$_3$O$^+$

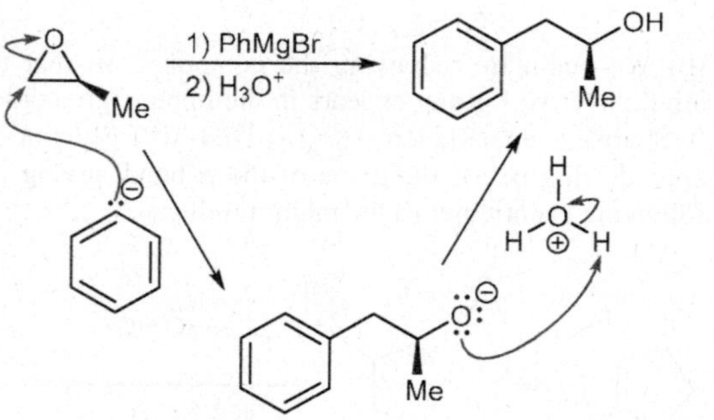

(b) Cyanide (NC$^-$) is a strong nucleophile, and it attacks the epoxide at the less substituted position (because S$_N$2 back-side attack is faster at the less sterically hindered position). This opens the epoxide, resulting in an alkoxide ion, which is then protonated during workup with aqueous acid.

1) NaCN 2) H$_3$O$^+$

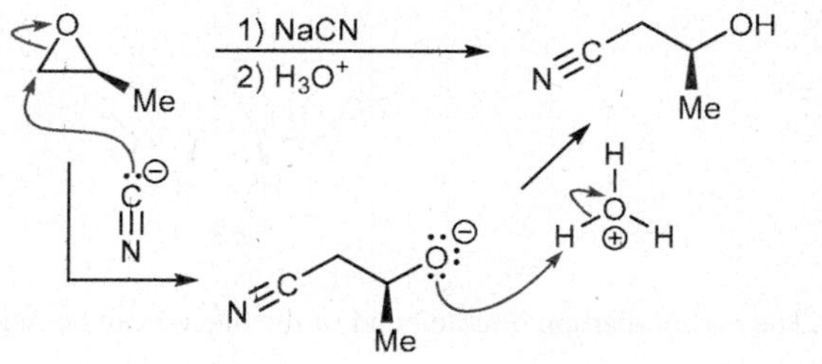

(c) MeS$^-$ is a very strong nucleophile, and it attacks the epoxide at the less substituted position (because S$_N$2 back-side attack is faster at the less sterically hindered position). This opens the epoxide, resulting in an alkoxide ion, which is then protonated during workup with aqueous acid.

1) NaSMe 2) H$_3$O$^+$

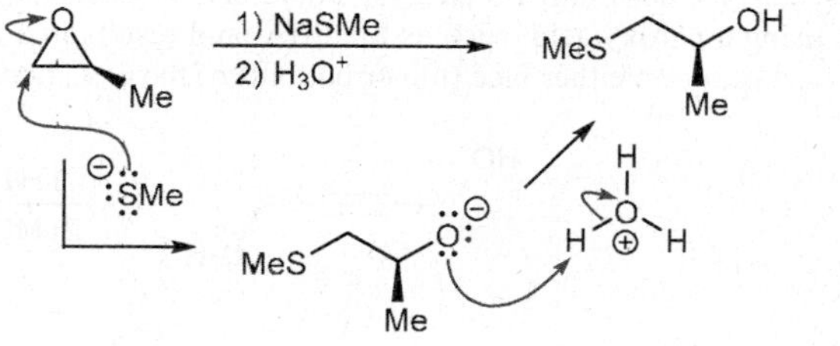

(d) LiAlH$_4$ is a source of nucleophilic hydride (H$^-$), and it attacks the epoxide at the less substituted position (because S$_N$2 back-side attack is faster at the less sterically hindered position). This opens the epoxide, resulting in an alkoxide ion, which is then protonated during workup with aqueous acid. Note that the product no longer has a chiral center.

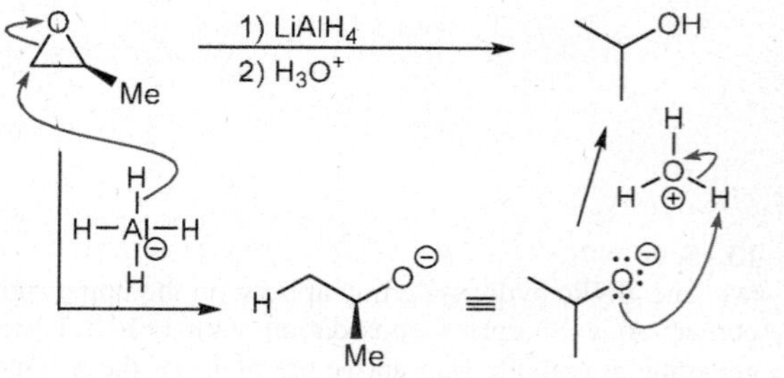

1) LiAlH$_4$ 2) H$_3$O$^+$

(e) EtO$^-$ is a strong nucleophile, and it attacks the epoxide at the less substituted position (because S$_N$2 back-side attack is faster at the less sterically hindered position). Note that the chiral center does not participate in the reaction and therefore does not experience an inversion of configuration. The resulting alkoxide ion is then protonated during workup with aqueous acid.

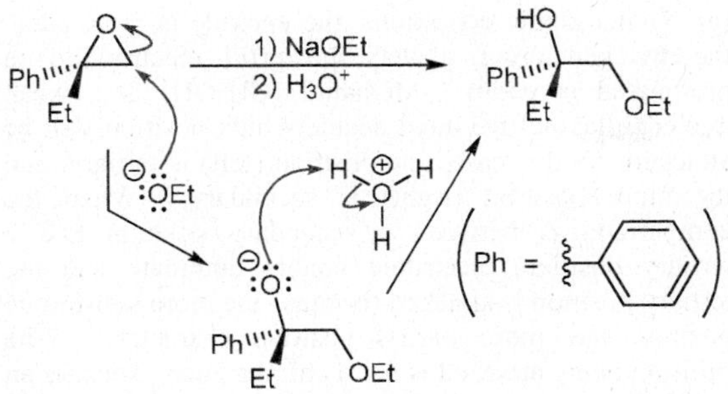

(f) $LiAlH_4$ is a source of nucleophilic hydride (H^-), and it attacks the epoxide at the less substituted position (because S_N2 back-side attack is faster at the less sterically hindered position). That position is a chiral center in the starting material, but in the product, it is no longer a chiral center. So even though the mechanism proceeds with back-side attack, there will not be an observable inversion of configuration at that location. The resulting alkoxide ion is then protonated during workup with aqueous acid.

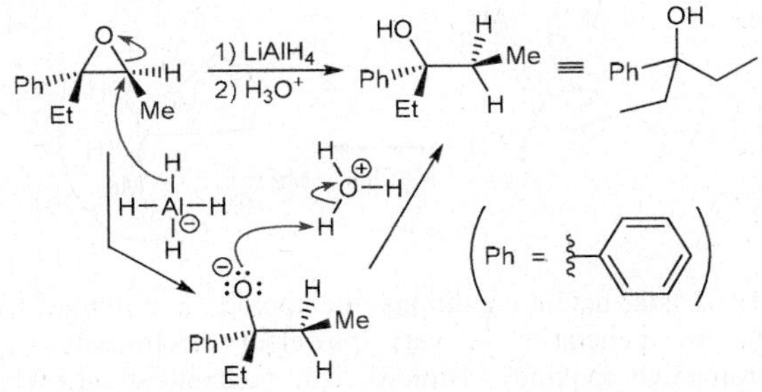

13.17. The reagent, lithium acetylide, has a C–Li bond. Since carbon is more electronegative than lithium, there is a partial negative (δ–) charge on the carbon atom. The difference in electronegativity is significant (carbon = 2.5 and lithium = 1.0), so the bond is sufficiently polar that it can be treated as ionic:

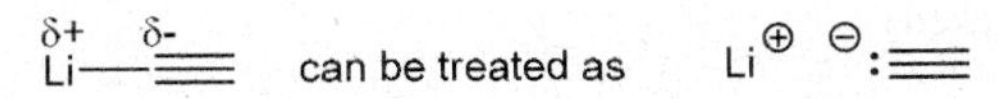

With this analysis, we can see the similarity between an organolithium reagent and a Grignard reagent (both are organometallic compounds, and both have similar reactivities). With negative character on the carbon atom, lithium acetylide is a strong nucleophile that will attack the less substituted side of the epoxide (highlighted below), because an S_N2 process will be faster at the less sterically hindered position. The less substituted side of the epoxide is not a chiral center, so even though the mechanism proceeds with back-side attack, there will not be an observable inversion of configuration at that location. The more substituted side will retain its original configuration since there are no bonds being made or broken at that position.

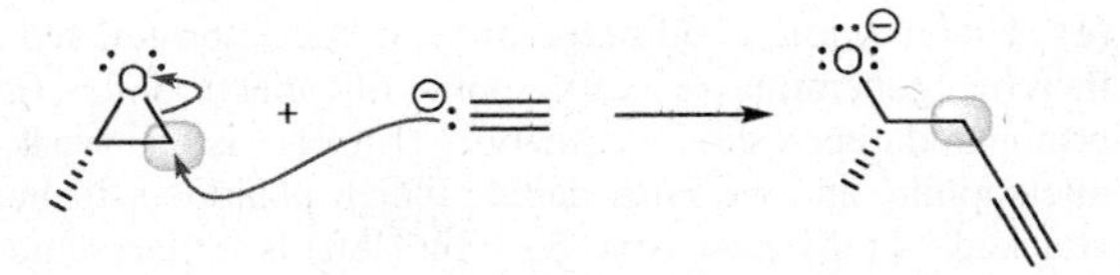

The resulting alkoxide ion is then protonated upon treatment with aqueous acid, to give the following alcohol (compound **2**):

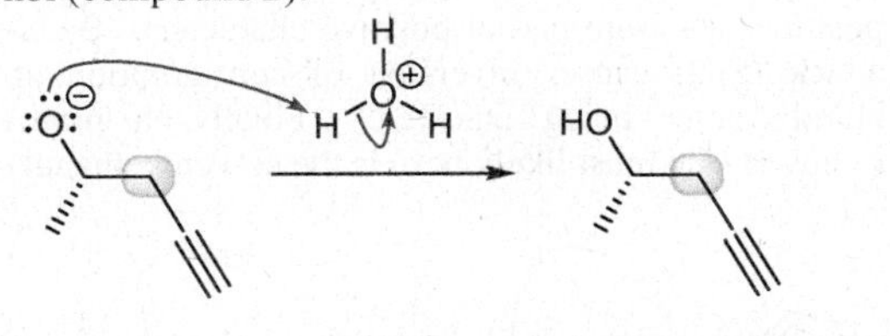

Compound 2
C_5H_8O

13.18.

(a) Under acidic conditions, the epoxide is protonated, thereby generating a very powerful electrophile (a protonated epoxide). Since the starting epoxide is symmetrical, regiochemistry is not a concern in this case. That is, the nucleophile can attack the epoxide at either position, giving the same product either way. Stereochemistry is also not an issue in this case, because the product does not contain any chiral centers.

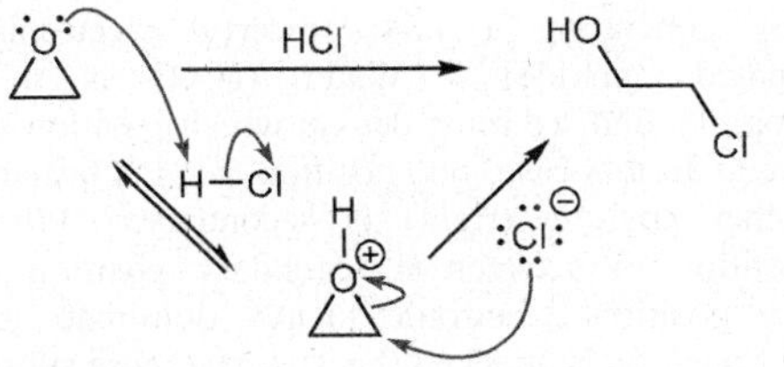

(b) Under acidic conditions, the epoxide is protonated, thereby generating a very powerful electrophile (a protonated epoxide). To determine where the nucleophile (bromide) attacks, we must decide whether steric or electronic effects dominate. In this case, one position (left) is primary and the other position (right) is secondary. Under these conditions, steric effects will dominate, and the attack is expected to occur at the less substituted position. The position being attacked is not a chiral center. There is an existing chiral center, but that center is not attacked, so we do not expect the configuration of that chiral center to change.

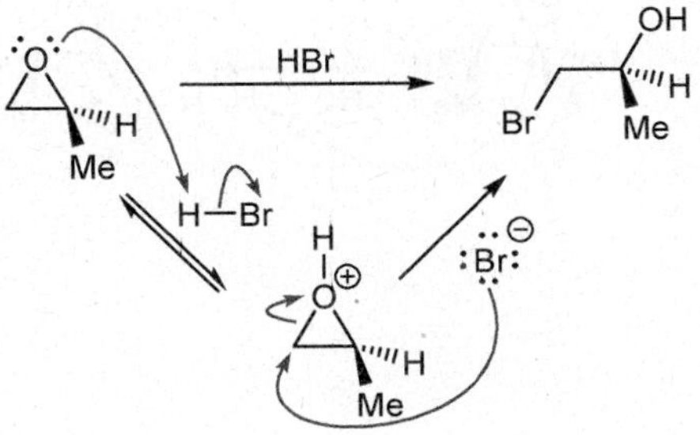

(c) Under acidic conditions, the epoxide is protonated, thereby generating a very powerful electrophile (a protonated epoxide). Ethanol (EtOH) is a weak nucleophile, and we must decide which position will be attacked. In this case, one position (left) is tertiary, and the other position (right) is secondary. When the competition is between a secondary position and a tertiary position, electronic factors dominate and the tertiary position is attacked (because the more substituted position has more partial positive character). Back-side attack (S_N2) causes inversion of configuration at the chiral center being attacked. Finally, a proton is removed (the most likely base is the solvent, ethanol).

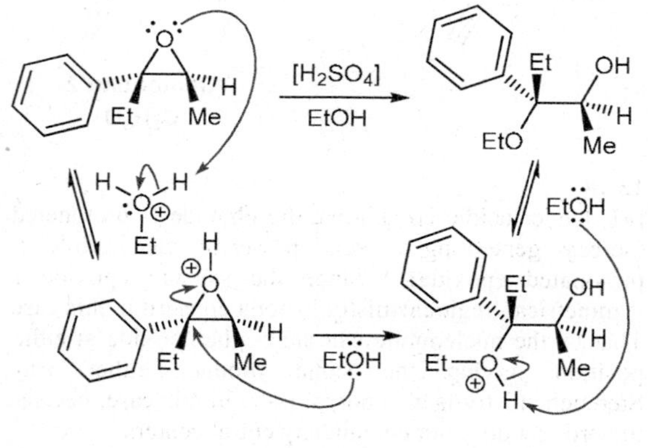

(d) Under acidic conditions, the epoxide is protonated, thereby generating a very powerful electrophile (a protonated epoxide). Water (H_2O) is a weak nucleophile, and we must decide which position will be attacked. In this case, one position (left) is tertiary, and the other position (right) is secondary. When the competition is between a secondary position and a tertiary position, electronic factors dominate and the tertiary position is attacked (because the more substituted position has more partial positive character). Back-side attack (S_N2) causes inversion of configuration at the chiral center being attacked. Finally, a proton is removed (the most likely base is the solvent, water).

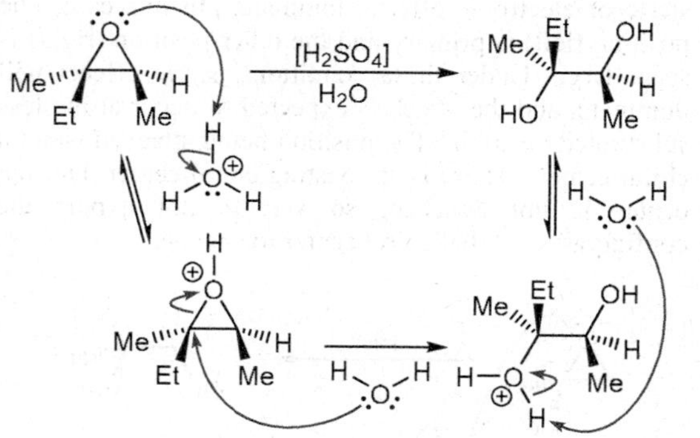

(e) Under acidic conditions, the epoxide is protonated, thereby generating a very powerful electrophile (a protonated epoxide). Methanol (CH_3OH) is a weak nucleophile, and we must decide which position will be attacked. In this case, one position (left) is tertiary, and the other position (right) is secondary. When the competition is between a secondary position and a tertiary position, electronic factors dominate and the tertiary position is attacked (because the more substituted position has more partial positive character). The position being attacked is not a chiral center. There is an existing chiral center, but that center is not attacked, so we do not expect the configuration of that chiral center to change. Finally, a proton is removed (the most likely base is the solvent, methanol).

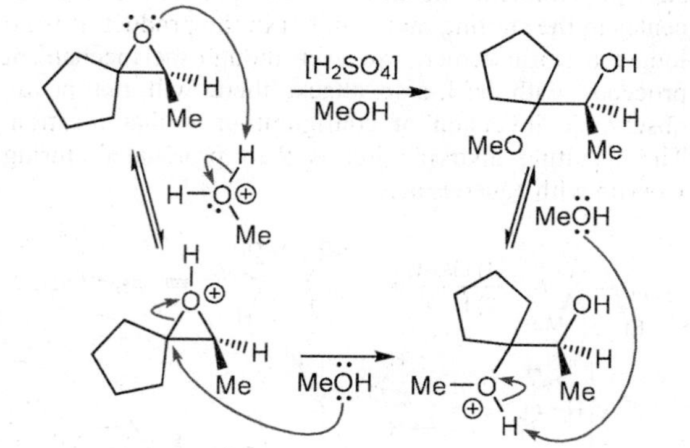

(f) Under acidic conditions, the epoxide is protonated, thereby generating a very powerful electrophile (a protonated epoxide). Bromide is a nucleophile, and we must decide which position will be attacked. In this case, one position (left) is tertiary, and the other position (right) is secondary. When the competition is between a secondary position and a tertiary position, electronic factors dominate and the tertiary position is attacked (because the more substituted position has more partial positive character). Back-side attack (S_N2) causes inversion of configuration at the chiral center being attacked.

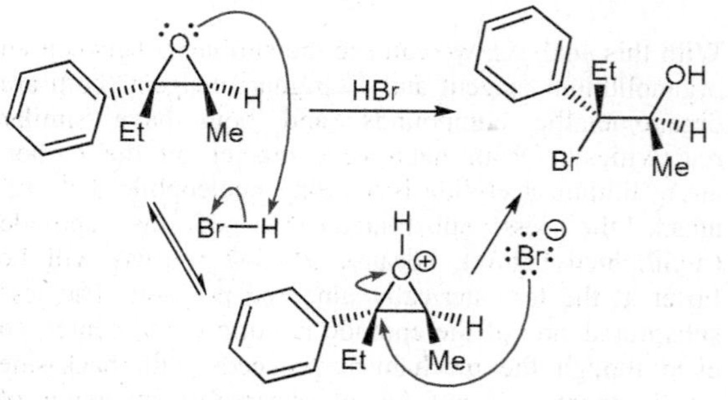

13.19. In this acid-catalyzed reaction, we expect the nucleophile (H_2O) to attack the more substituted (tertiary) position of the protonated epoxide (because the more substituted position has more partial positive character). The SN2 back-side attack would result in inversion of stereochemistry, but this is not a chiral center so no stereochemistry needs to be shown at the position where the SN2 attack occurred. The existing chiral center of the epoxide retains its configuration in the diol product.

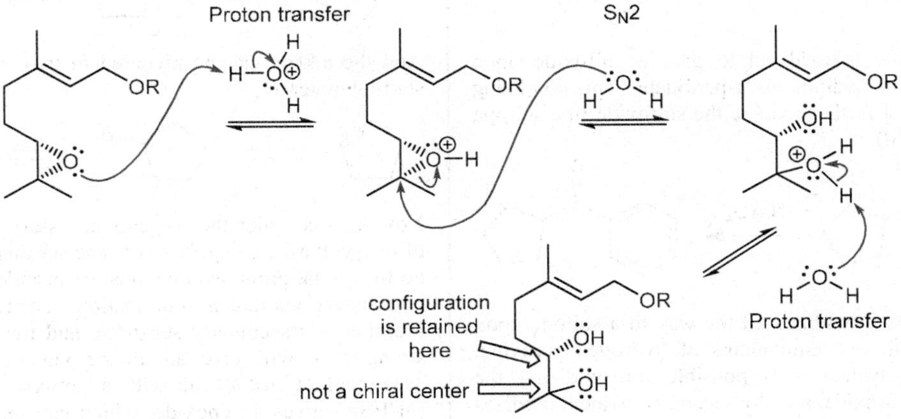

Proton transfer

SN2

configuration is retained here

not a chiral center

Proton transfer

The regiochemical outcome of the ring-opening reaction is determined at the SN2 step of the mechanism, so let's look closely at that step. Since the nucleophile (water) attacks the more substituted position in this acid-catalyzed reaction, the labeled oxygen of the epoxide remains attached to the less-substituted position, as shown here:

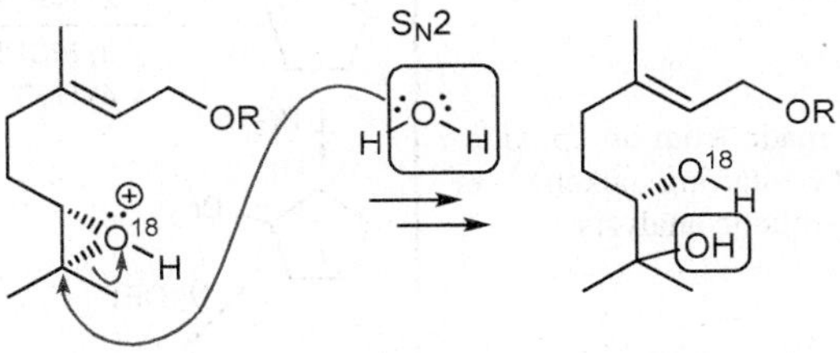

SN2

13.20.

(a) The desired thiol can be prepared by treating an appropriate alkyl halide (shown below) with sodium hydrosulfide (NaSH), to give an SN2 reaction.

NaSH

(b) The desired thiol can be prepared by treating an appropriate alkyl halide (shown below) with sodium hydrosulfide (NaSH), to give an SN2 reaction.

NaSH

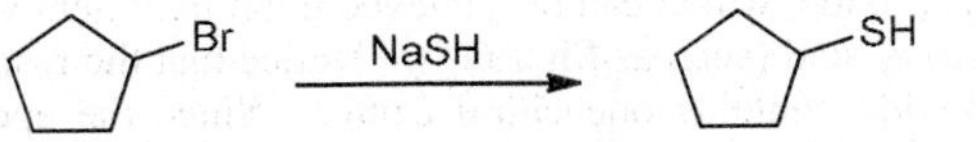

(c) The desired thiol can be prepared by treating an appropriate alkyl halide (shown) with sodium hydrosulfide (NaSH), to give an SN2 reaction. Notice that inversion of configuration is expected, so the starting alkyl halide must have the opposite configuration as the desired product.

NaSH

13.21.

(a) Treating a thiol with sodium hydroxide results in deprotonation of the thiol to give a thiolate ion (RS^-). This thiolate ion is a very strong nucleophile, and it will attack a primary alkyl halide to give an SN2 reaction. The product is a sulfide.

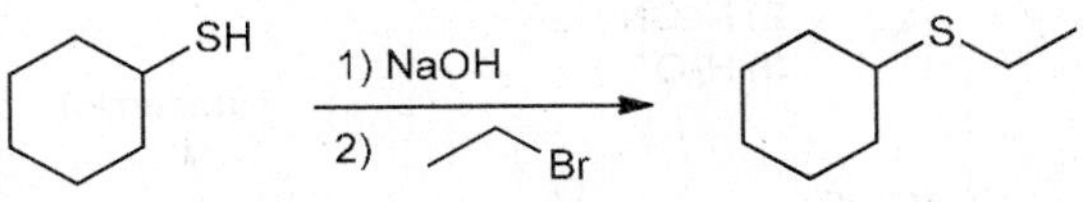

1) NaOH

2) Br

(b) A secondary alkyl bromide will serve as a substrate (electrophile) in an SN2 reaction, upon treatment with a thiolate ion (EtS^-). The reaction occurs at a chiral center, so we expect inversion of configuration, which is characteristic of S_N2 processes.

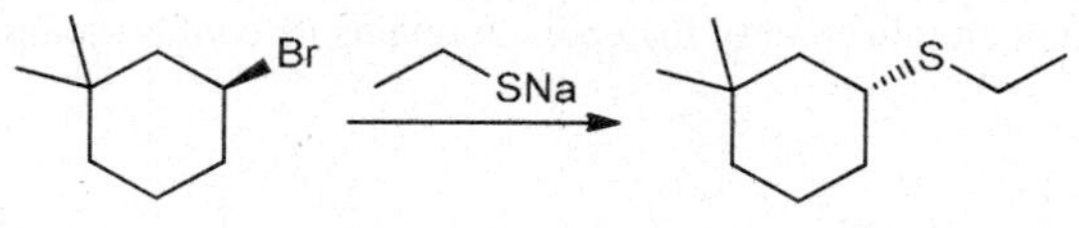

Br SNa S

(c) A sulfide is oxidized to give a sulfoxide upon treatment with sodium *meta*-periodate. This oxidizing agent does not further oxidize the sulfoxide (the sulfone is not obtained).

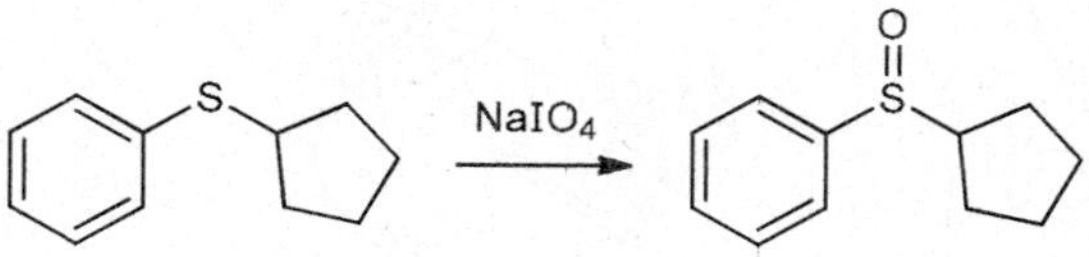

S $NaIO_4$ O S

(d) A sulfide is oxidized all the way to a sulfone upon treatment with two equivalents of hydrogen peroxide. The first equivalent is responsible for oxidizing the sulfide to a sulfoxide, and the second equivalent oxidizes the sulfoxide to a sulfone.

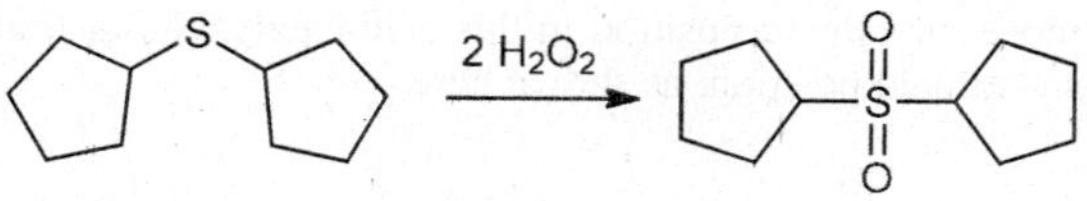

S $2\ H_2O_2$ O S O

13.22.

(a) The desired product can be made from an epoxide, which can be made from the starting alkene, as illustrated in the following retrosynthetic analysis:

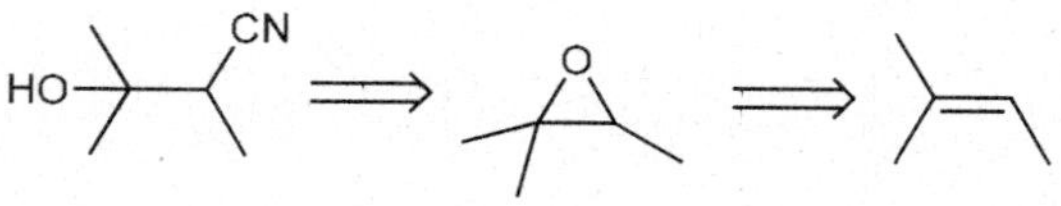

HO CN ⟹ O ⟹

Now let's consider the reagents necessary for each step of the synthesis. First, the alkene must be converted into an epoxide, which can be achieved upon treatment with a peroxy acid (such as MCPBA). Notice that the resulting epoxide contains one chiral center. Since the epoxide can form on either face of the π bond with equal likelihood, we expect a racemic mixture of the epoxide. Basic conditions are required (NaCN, rather than HCN) during the ring opening step, in order to ensure that the nucleophile (cyanide) attacks the less substituted position. If acidic conditions were employed (HCN), the nucleophile (cyanide) would attack the more substituted position.

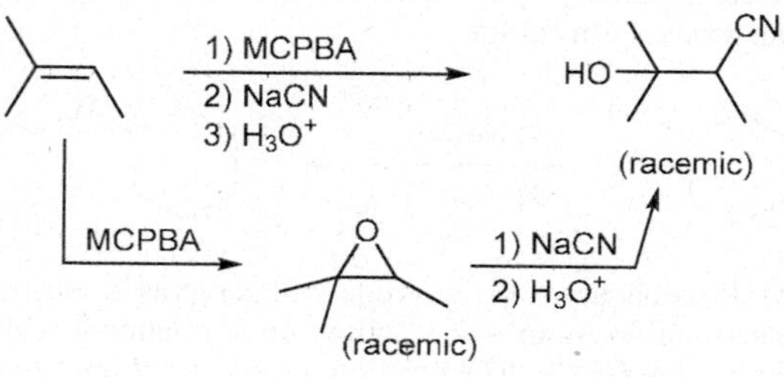

1) MCPBA 2) NaCN 3) H_3O^+ HO CN (racemic)

MCPBA O (racemic) 1) NaCN 2) H_3O^+

(b) The desired *trans*-diol can be made from an epoxide, which can be made from the corresponding alkene.

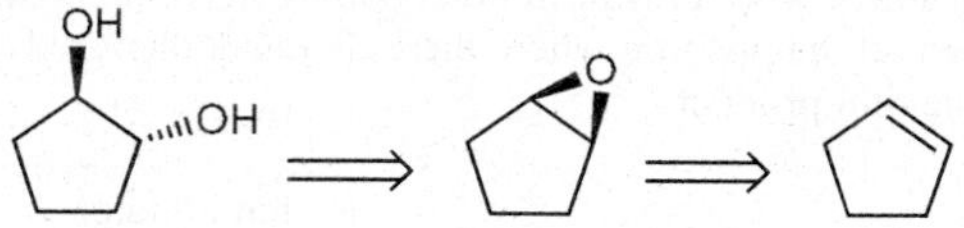

OH OH ⟹ O ⟹

And the alkene can be prepared in two steps from the starting material.

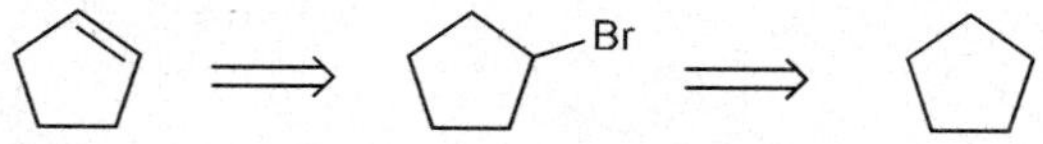

⟹ Br ⟹

Now let's consider the reagents necessary for each step of the synthesis, going forward. The starting material has no functional group, so one must be installed, which can be achieved via radical bromination. The resulting alkyl bromide is a secondary substrate, and treatment with a strong base will give an alkene via an E2 process. Treatment of the alkene with a peroxy acid (such as MCPBA) gives an epoxide, which can then be opened either under aqueous acidic conditions or under aqueous basic conditions to give the desired *trans*-diol.

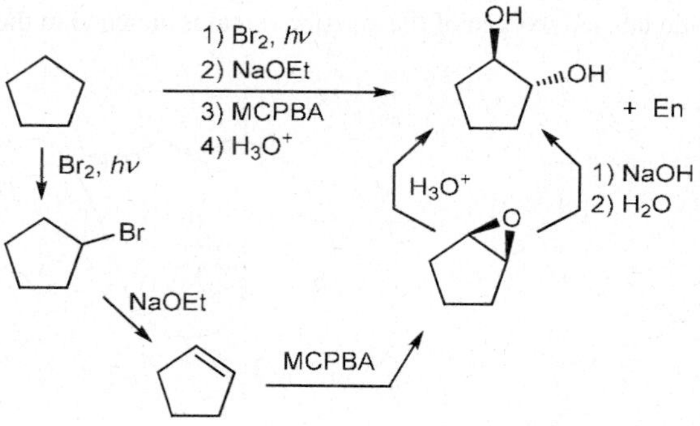

1) Br_2, $h\nu$ 2) NaOEt 3) MCPBA 4) H_3O^+ OH OH + En

Br_2, $h\nu$ Br NaOEt MCPBA O H_3O^+ 1) NaOH 2) H_2O

(c) The desired product can be made from an epoxide, which can be made from the starting alkene, as illustrated in the following retrosynthetic analysis:

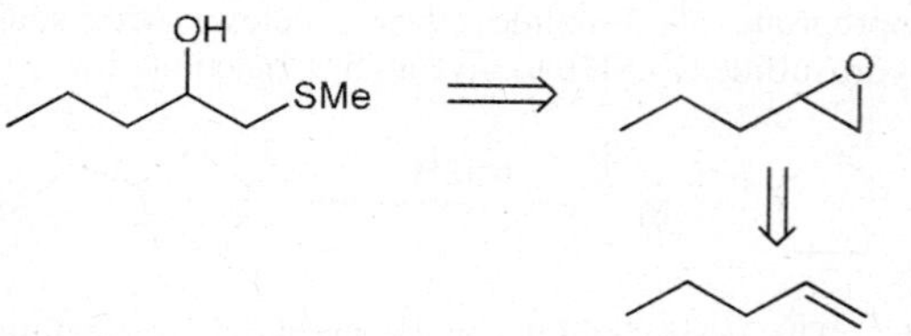

OH SMe ⟹ O ⟹

Now let's consider the reagents necessary for each step of the synthesis. First, the alkene must be converted into an epoxide, which can be achieved upon treatment with a peroxy acid (such as MCPBA). Notice that the resulting epoxide contains one chiral center. Since the epoxide can form on either face of the π bond with equal likelihood, we expect a racemic mixture of the epoxide. NaSMe must be used as the reagent (rather than MeSH and sulfuric acid) during the ring opening step, in order to ensure that the nucleophile (MeS^-) attacks the less substituted position.

1) MCPBA
2) NaSMe
3) H_3O^+
OH
SMe
(racemic)
MCPBA
O
(racemic)
1) NaSMe
2) H_3O^+

1) MCPBA
2) $[H_2SO_4]$, H_2S
SH
OH
(racemic)
MCPBA
O
(racemic)
$[H_2SO_4]$
H_2S

(d) The desired product can be made from an epoxide, which can be made from the starting alkene, as illustrated in the following retrosynthetic analysis:

SH
OH ⟹ O
⇓

Now let's consider the reagents necessary for each step of the synthesis. First, the alkene must be converted into an epoxide, which can be achieved upon treatment with a peroxy acid (such as MCPBA). Notice that the resulting epoxide contains one chiral center. Since the epoxide can form on either face of the π bond with equal likelihood, we expect a racemic mixture of the epoxide. Acidic conditions are required (H_2S and sulfuric acid, rather than NaSH) during the ring opening step, in order to ensure that the nucleophile (H_2S) attacks the more substituted position.

(e) The desired product (a ketone) can be made via oxidation of a secondary alcohol, which can be made from the starting epoxide, as illustrated in the following retrosynthetic analysis:

O ⟹ OH ⟹ O

Now let's consider the reagents necessary for each step of the synthesis. First, the epoxide must be converted into an alcohol, which can be achieved upon treatment with $LiAlH_4$, followed by water workup. The resulting alcohol is then oxidized to give the desired ketone.

O
(racemic)
1) $LiAlH_4$
2) H_3O^+
3) $Na_2Cr_2O_7$
H_2SO_4, H_2O
O
1) $LiAlH_4$
2) H_3O^+
OH
(racemic)
$Na_2Cr_2O_7$
H_2SO_4, H_2O

13.23. We always approach a synthesis problem by asking two questions.
1) Is there a change in the carbon skeleton? Yes, there is one more carbon atom in the product than in the starting material.
2) Is there a change in the functional group(s)? Yes, the starting material has an epoxide that is missing in the product. The product has two new functional groups, a ketone and a nitrile.

TIPS
OCH_3
O
⟶
TIPS
OCH_3
O
N

From a retrosynthetic perspective, the product ketone can be made via oxidation of a secondary alcohol, which can be made by opening the starting epoxide with a cyanide nucleophile:

TIPS
OCH_3
O
N
⟹
TIPS
OCH_3
OH
N
⟹
TIPS
OCH_3
O

Let's write out the steps in the forward direction. The epoxide-opening requires a strong nucleophile to achieve the desired regiochemistry (the less hindered position must be attacked). Use of NaCN, followed by workup to protonate the resulting alkoxide, accomplishes this and forms a new carbon-carbon bond in the process. Oxidation of the secondary alcohol to a ketone can be accomplished by a variety of reagents; one such option is shown below:

TIPS — OCH₃ epoxide → 1) NaCN 2) H_3O^+ 3) PCC → TIPS — OCH₃ ketone nitrile

1) NaCN 2) H_3O^+ → TIPS — OCH₃ — OH nitrile → PCC

13.24.

(a) A retrosynthetic analysis of the target molecule reveals that the required disconnection is not at the α position, but one bond farther away from the hydroxyl group. Therefore, the required disconnection leads to a Grignard reagent nucleophile and an *epoxide* electrophile.

Nucleophile Electrophile

Now let's draw the forward process. The desired transformation can be achieved by converting the starting alkyl halide into a Grignard reagent, and then treating the Grignard reagent with an epoxide (followed by aqueous workup).

1) Mg 2) epoxide 3) H_3O^+

Mg; MgBr; 1) epoxide 2) H_3O^+

(b) A retrosynthetic analysis of the target molecule reveals that the required disconnection is at the α position. Therefore, the required disconnection leads to a Grignard reagent nucleophile and an *aldehyde* electrophile.

Nucleophile Electrophile

Now let's draw the forward process. The desired transformation can be achieved by converting the starting alkyl halide into a Grignard reagent, and then treating the Grignard reagent with an aldehyde (followed by aqueous workup).

1) Mg 2) aldehyde 3) H_3O^+

(c) A retrosynthetic analysis of the target molecule reveals that the required disconnection is not at the α position, but one bond farther away from the hydroxyl group. Therefore, the required disconnection leads to a Grignard reagent nucleophile and an *epoxide* electrophile.

Nucleophile Electrophile

Now let's draw the forward process. The desired transformation can be achieved by converting the starting alkyl halide into a Grignard reagent, and then treating the Grignard reagent with an epoxide (followed by aqueous workup).

1) Mg 2) epoxide 3) H_3O^+

(d) The target molecule has four additional carbon atoms, so the synthesis requires a C–C bond-forming reaction. The retrosynthetic analysis of the target molecule asks, "What reaction have we seen that produces a ketone product?" The ketone can be prepared by oxidation of the corresponding alcohol, and the alcohol is suitable for making the required C–C bond disconnection (at the α position). Therefore, the required disconnection leads to a Grignard reagent nucleophile and an *aldehyde* electrophile.

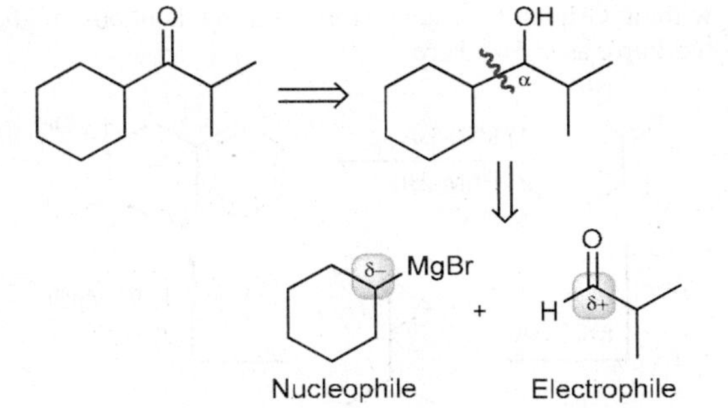

O
OH
α
MgBr
δ−
δ+
H
O
+
Nucleophile
Electrophile

Now let's draw the forward process. The desired transformation can be achieved by converting the starting alkyl halide into a Grignard reagent, treating the Grignard reagent with an aldehyde (followed by aqueous workup) to give an alcohol, and then oxidizing the alcohol to a ketone.

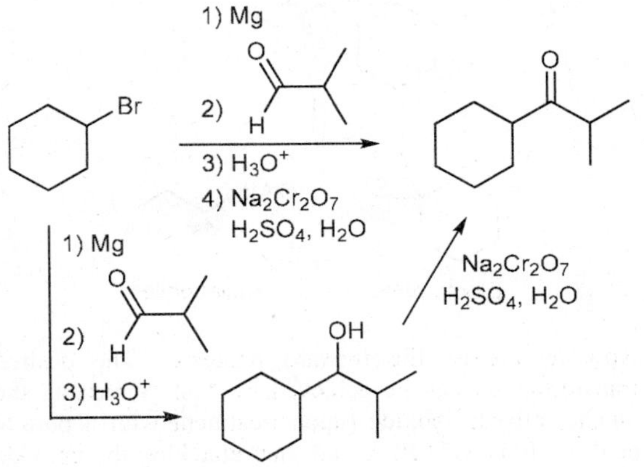

Br
1) Mg
2)
H
3) H3O+
4) Na2Cr2O7
H2SO4, H2O
O
1) Mg
2)
3) H3O+
OH
Na2Cr2O7
H2SO4, H2O

(e) The target molecule has two additional carbon atoms, so the synthesis requires a C–C bond-forming reaction. The retrosynthetic analysis of the target molecule asks, "What reaction have we seen that produces an alkyl chloride product?" The chloride can be prepared by substitution of the corresponding alcohol, and the alcohol is suitable for making the required C–C bond disconnection. In this case, the disconnection is not at the α position, but one bond farther away from the hydroxyl group. Therefore, the required disconnection leads to a Grignard reagent nucleophile and an *epoxide* electrophile.

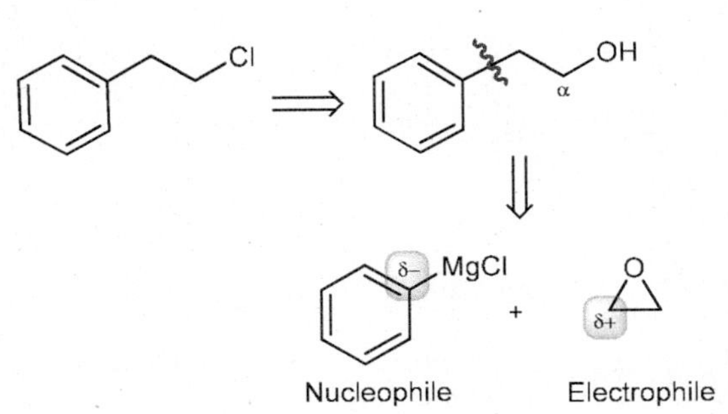

Cl
OH
α
MgCl
δ−
δ+
O
+
Nucleophile
Electrophile

Now let's draw the forward process. The desired transformation can be achieved by converting the starting aryl halide into a Grignard reagent, treating the Grignard reagent with an epoxide (followed by aqueous workup), and then converting the resulting alcohol into an alkyl chloride (via an S_N2 process).

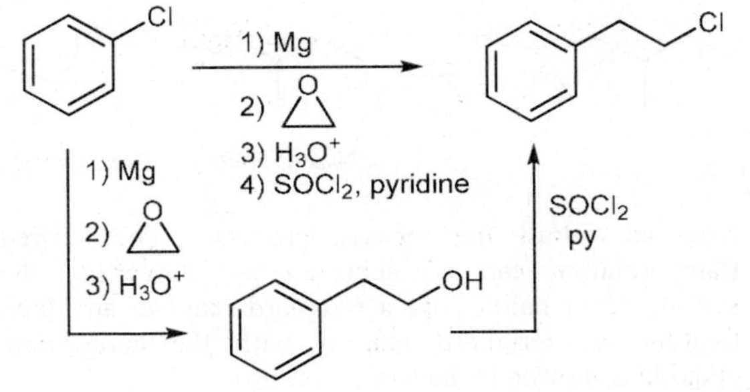

Cl
1) Mg
2)
3) H3O+
4) SOCl2, pyridine
Cl
1) Mg
2)
3) H3O+
OH
SOCl2
py

(f) The target molecule has two additional carbon atoms, so the synthesis requires a C–C bond-forming reaction. The retrosynthetic analysis of the target molecule asks, "What reaction have we seen that produces an alkyl chloride product?" The chloride can be prepared by substitution of the corresponding alcohol, and the alcohol is suitable for making the required C–C bond disconnection (at the α position). Therefore, the required disconnection leads to a Grignard reagent nucleophile and an *aldehyde* electrophile.

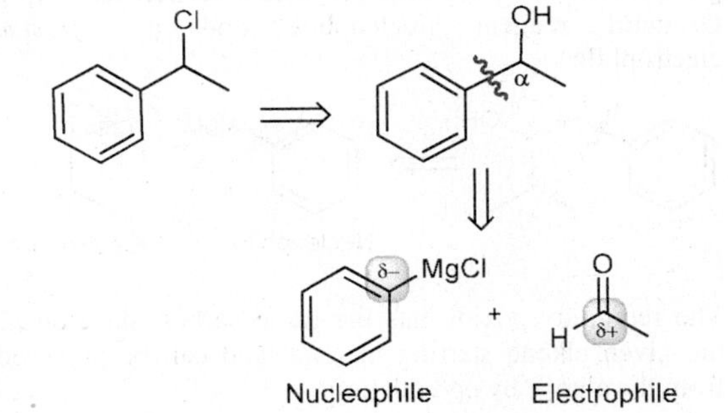

Cl
OH
α
MgCl
δ−
δ+
H
O
+
Nucleophile
Electrophile

Now let's draw the forward process. The desired transformation can be achieved by converting the starting aryl halide into a Grignard reagent, treating the Grignard reagent with an aldehyde (followed by aqueous workup), and then converting the resulting alcohol into an alkyl chloride (via an S_N2 process).

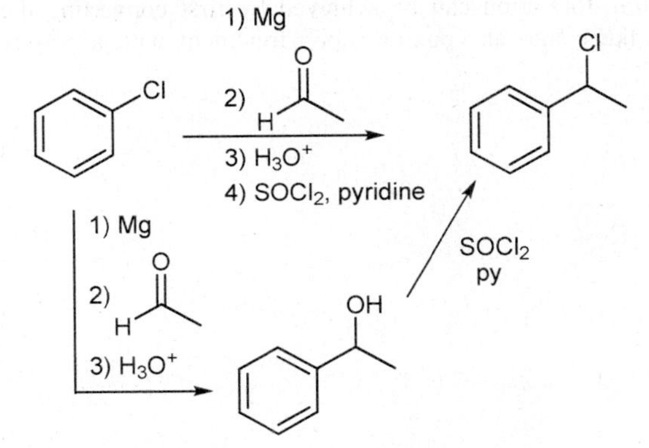

Cl
1) Mg
2)
H
3) H3O+
4) SOCl2, pyridine
Cl
1) Mg
2)
3) H3O+
OH
SOCl2
py

(g) A retrosynthetic analysis of the target molecule reveals that the required disconnection is not at the α position, but one bond farther away from the hydroxyl group. Therefore, the required disconnection leads to a Grignard reagent nucleophile and an *epoxide* electrophile.

OH ⟹ δ– MgCl + δ+ O

Nucleophile Electrophile

Now let's draw the forward process. The desired transformation can be achieved by converting the starting alkyl halide into a Grignard reagent, and then treating the Grignard reagent with the appropriate epoxide (followed by aqueous workup).

Cl → 1) Mg 2) epoxide 3) H_3O^+ → OH

(h) A retrosynthetic analysis of the target molecule reveals that the required disconnection is not at the α position, but one bond farther away from the hydroxyl group. Therefore, the required disconnection leads to a Grignard reagent nucleophile and an *epoxide* electrophile.

OH ⟹ δ– MgBr + δ+ O

Nucleophile Electrophile

The required epoxide has the same carbon skeleton as the given alkene starting material and can be prepared from the alkene by epoxidation.

Now let's draw the forward process. The desired transformation can be achieved by first converting the alkene into an epoxide (upon treatment with a peroxy acid, such as MCPBA), and then attacking the epoxide with a Grignard reagent (followed by aqueous acidic workup), as shown here:

1) MCPBA 2) PhMgBr 3) H_3O^+ → OH

MCPBA → epoxide → 1) PhMgBr 2) H_3O^+

(i) A retrosynthetic analysis of the target molecule reveals that the required disconnection is not at the α position, but one bond farther away from the hydroxyl group. Therefore, the required disconnection leads to a Grignard reagent nucleophile and an *epoxide* electrophile.

OH ⟹ O δ+ + BrMg δ–

Electrophile Nucleophile

Now let's draw the forward process. The desired transformation can be achieved by first converting the alkene into an epoxide (upon treatment with a peroxy acid, such as MCPBA), and then attacking the epoxide with a Grignard reagent (followed by aqueous acidic workup), as shown here:

1) MCPBA 2) $CH_3CH_2CH_2MgBr$ 3) H_3O^+ → OH

MCPBA → epoxide → 1) $CH_3CH_2CH_2MgBr$ 2) H_3O^+

13.25. A Grignard reagent with four carbon atoms is reacting with compound **2** to give compound **3** (which has eleven carbon atoms in the longest chain). Therefore, compound **2** must have seven carbon atoms in the longest chain. Additionally, the four-carbon chain is installed on the carbon atom that is adjacent to the carbon atom bearing the hydroxyl (OH) group (not attached to the alpha carbon); this is consistent with a reaction between a Grignard reagent and an epoxide. The configurations of the chiral centers in compounds **2** and **3** will remain the same, since these carbon atoms are not undergoing bond-making or bond-breaking.

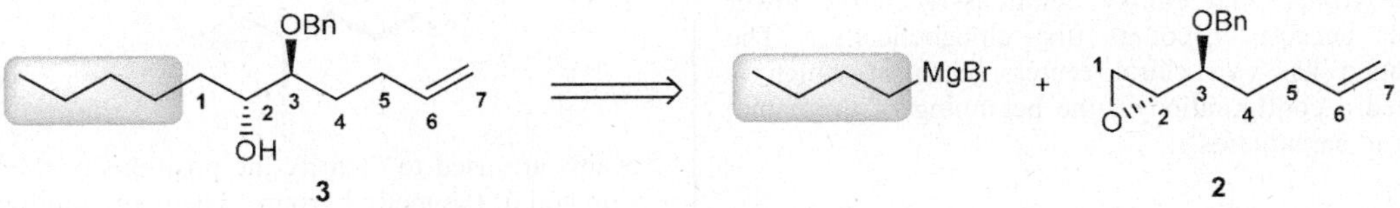

Now that we have drawn the structure of compound **2**, we need to provide mechanisms for each step of this overall transformation. Let's start with the conversion of compound **1** to compound **2**:

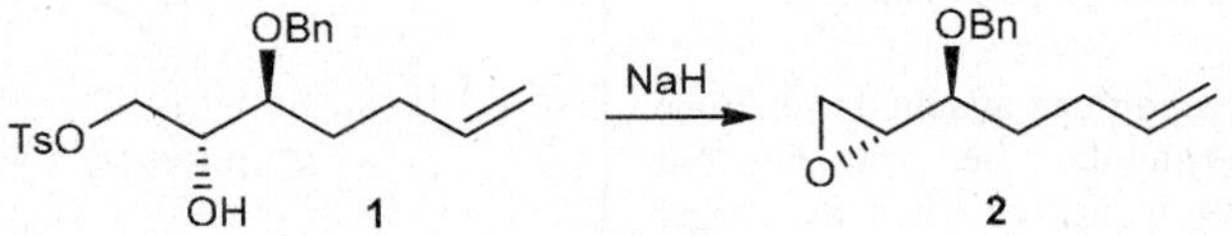

Recall that NaH is a strong, non-nucleophilic base. The mechanism begins with a proton transfer, as NaH deprotonates the alcohol. This is followed by an intramolecular S_N2-type reaction in which the tosylate group is the leaving group (this is an intramolecular Williamson ether synthesis). Note that the configuration at the alcohol carbon atom is unaffected in this reaction because no bonds are being made or broken with that carbon atom.

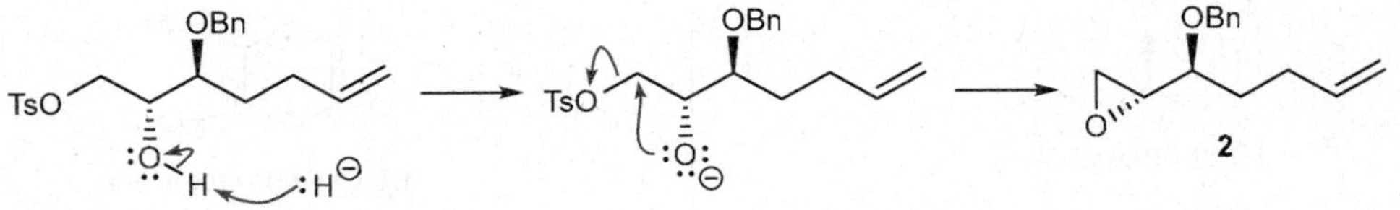

Now, let's consider the conversion of compound **2** to compound **3**:

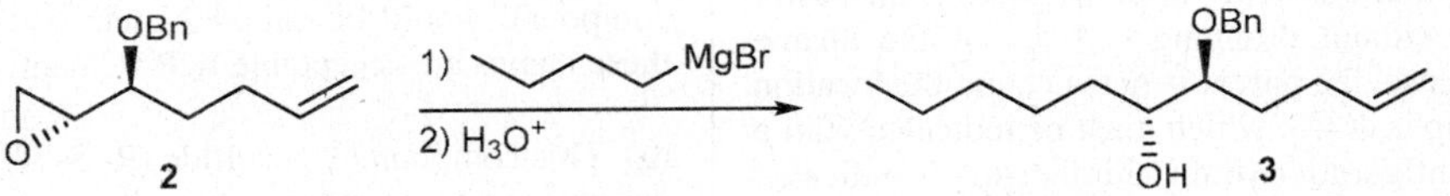

Recall that the Grignard reagent is characterized by a C-Mg bond that is very polar, and can be treated as an ionic bond. This carbanion is a strong nucleophile that attacks the less substituted carbon atom of the epoxide to form a new carbon-carbon bond and an alkoxide ion. Subsequent protonation of the alkoxide with aqueous acid produces an alcohol as the product. Note that the configuration at the more substituted carbon atom of the epoxide is unaffected in this reaction because no bonds are being made or broken with that carbon atom.

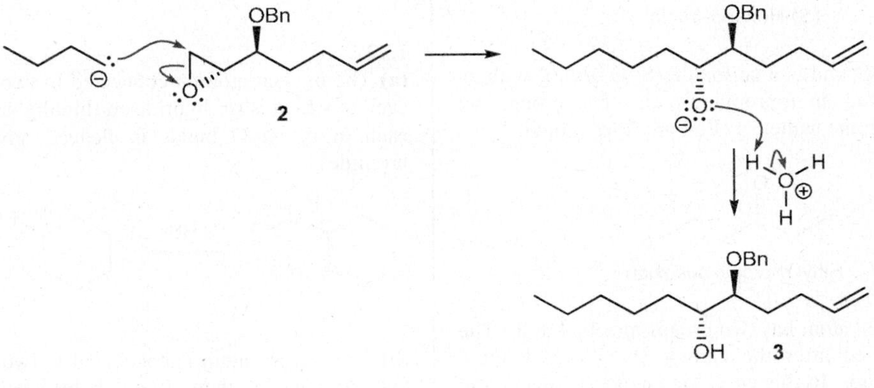

13.26.

(a) The oxygen atom has two groups attached to it. The parent is named after the larger group (cyclohexane). The smaller group (together with the oxygen atom) is treated as an ethoxy substituent. Locants are used to identify the positions of the ethoxy group and of the methyl group, with ethoxy being assigned the lower number because it comes first alphabetically. The compound has two chiral centers, each of which is assigned a configuration in the beginning of the name, placed in parentheses.

(1*S*,2*S*)-1-Ethoxy-2-methylcyclohexane

(b) The oxygen atom has two groups attached to it. One group has two carbon atoms and the other group has four carbon atoms. The parent is named after the larger group, so the parent is butane. The smaller group (together with the oxygen atom) is treated as an ethoxy substituent, and a locant (2) is included to identify the location of the substituent on the parent chain. The configuration of the chiral center is indicated at the beginning of the name, placed in parentheses.

(*R*)-2-Ethoxybutane

(c) The compound is named in the same way that we would name an alcohol (if the SH group were an OH group), except that the term "thiol" is used in the suffix of the name, without dropping the "e" of the alkane name. In this case, the parent is hexane, and the location of the SH group is at C3, which must be indicated with a locant. The configuration of the chiral center is indicated at the beginning of the name, placed in parentheses.

(*S*)-3-Hexanethiol

or

(*S*)-Hexane-3-thiol

(d) This compound is a sulfoxide (S=O group) with an ethyl group and an *n*-propyl group. The groups are alphabetized in the name, so ethyl precedes propyl.

Ethyl *n*-propyl sulfoxide

(e) The oxygen atom has two groups attached to it. The parent is named after the carbon chain containing a functional group. In this case, the parent is pentene (an alkene), and the smaller group (together with the oxygen atom) is treated as an ethoxy substituent.

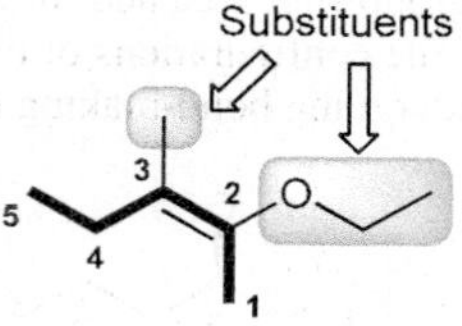

Locants are used to identify the positions of the ethoxy group and of the methyl group. The configuration of the alkene is indicated in the beginning of the name, placed in parentheses.

(*E*)-2-Ethoxy-3-methyl-2-pentene

or

(*E*)-2-Ethoxy-3-methylpent-2-ene

(f) The parent is benzene, and the two methoxy groups are treated as substituents. Locants are used to indicate the relative placement of the methoxy groups on the ring.

1,2-Dimethoxybenzene

Alternatively, the parent can be named as anisole (the common name for methoxybenzene), in which case the compound would be called 2-methoxyanisole. Both of these names are acceptable IUPAC names.

(g) This compound is a sulfide (R–S–R), and it is named very much like an ether. The sulfur atom is connected to two groups, ethyl and *n*-propyl, which are arranged alphabetically in the name.

Ethyl *n*-propyl sulfide

13.27.

(a) The oxygen atom is connected to two carbon atoms, each of which is sp^3 hybridized (highlighted). Therefore, each of the C–O bonds is cleaved, giving two alkyl bromides.

xs HBr → Br + CH_3Br + H_2O

(b) The oxygen atom is connected to two carbon atoms, but only one of them is sp^3 hybridized (highlighted).

Therefore, only one C–O bond is cleaved. The other C–O bond (where the C is sp^2 hybridized) is not cleaved, because the sp^2-hybridized substrate is not suitable for either an S_N1 or S_N2 mechanism. The products are phenol and methyl bromide.

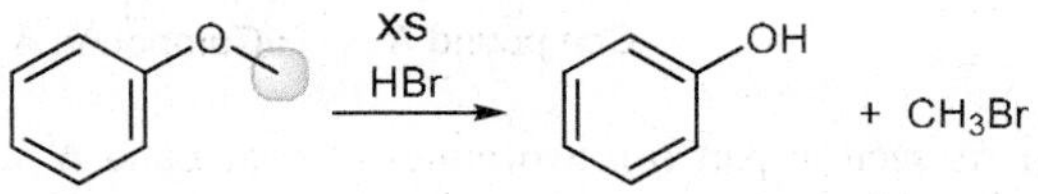

(c) The oxygen atom is connected to two carbon atoms, each of which is sp^3 hybridized (highlighted). Therefore, each of the C–O bonds is cleaved, giving two equivalents of 2-bromopropane.

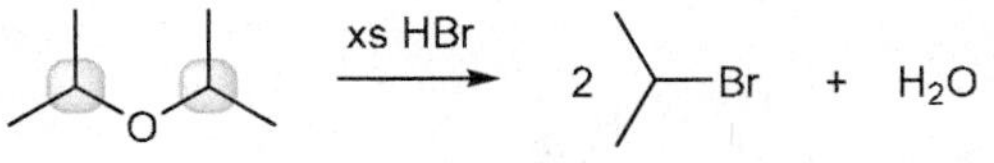

(d) The oxygen atom is connected to two carbon atoms, each of which is sp^3 hybridized (highlighted). Therefore, each of the C–O bonds is cleaved. As a result, the ring is opened, giving a dibromide, as shown.

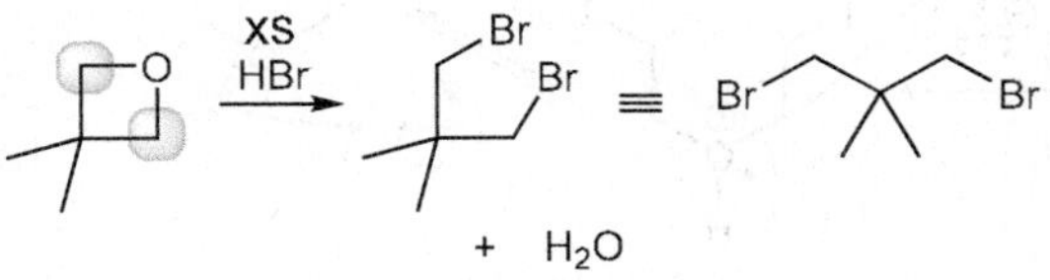

13.28. Ethers have the following structure:

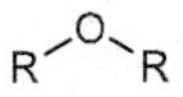

We are looking for ethers with the molecular formula $C_4H_{10}O$, which means that the four carbon atoms must be contained in the two R groups of the ether. One possibility is that each R group has two carbon atoms.

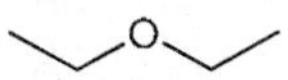

ethoxyethane
(diethyl ether)

Alternatively, one R group can have three carbon atoms and the other R group can have one carbon atom. The R group containing three carbon atoms can either be an *n*-propyl group or an isopropyl group, giving the following two additional isomers:

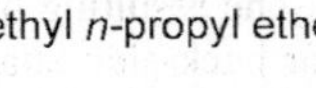

1-methoxypropane
(methyl *n*-propyl ether)

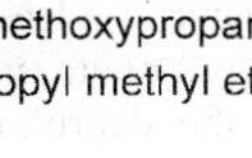

2-methoxypropane
(isopropyl methyl ether)

In total, there are three constitutionally isomeric ethers with the molecular formula $C_4H_{10}O$.

13.29.

(a) The desired transformation involves addition of H and OEt across a π bond. This can be achieved via alkoxymercuration-demercuration, where EtOH is used as the alcohol during the first step:

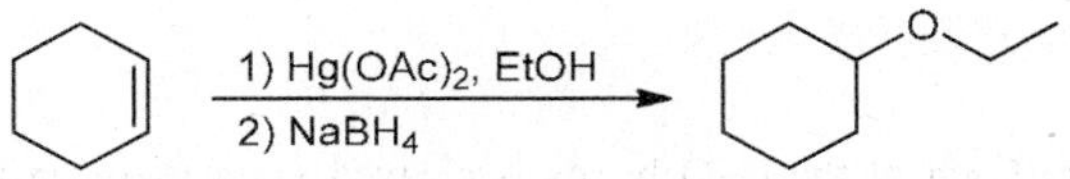

Alternatively, cyclohexene can be treated with dilute sulfuric acid to give cyclohexanol, which can then be used in a Williamson ether synthesis, together with ethyl iodide, to give the desired product:

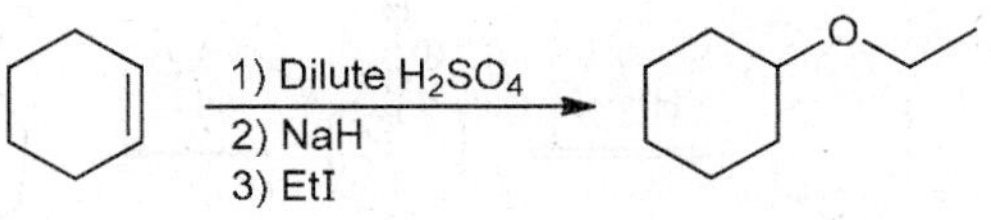

(b) The desired transformation involves an *anti* addition of OH and OMe across a π bond. This can be achieved by converting cyclohexene into an epoxide (upon treatment with a peroxy acid such as MCPBA), followed by a ring opening reaction. The second step of this process (opening the ring) can be performed under either acidic conditions or basic conditions.

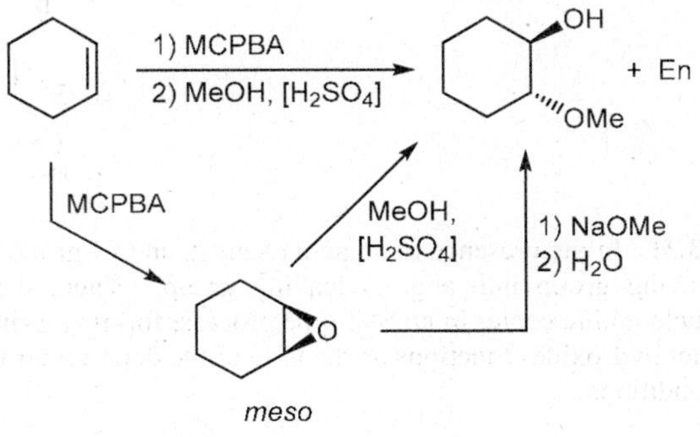

(c) The desired transformation involves addition of H and OR across a π bond. This can be achieved via alkoxymercuration-demercuration, where $(CH_3)_3COH$ is used as the alcohol during the first step.

13.30.

(a) The oxygen atoms are connected to four carbon atoms, each of which is sp^3 hybridized (highlighted). There are four C–O bonds, and each of them is cleaved under acidic conditions and replaced with a C–I bond, giving two moles of compound **A** and two moles of water.

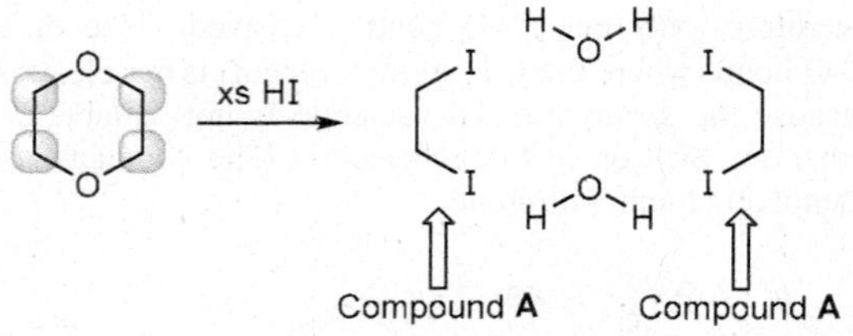

xs HI

Compound **A** Compound **A**

(b) As seen in part (a), two moles of compound **A** are produced for every one mole of 1,4-dioxane.

(c) Each of the C–O bonds is cleaved via a two-step process: (i) protonation of the oxygen atom to make a good leaving group (an oxonium ion), and (ii) an SN2 reaction, in which iodide functions as a nucleophile and attacks the primary substrate. Each of these two steps requires two curved arrows, as shown. Since there are four C–O bonds that undergo cleavage, our mechanism will have a total of eight steps, where each step utilizes two curved arrows, as shown.

- H_2O

- H_2O

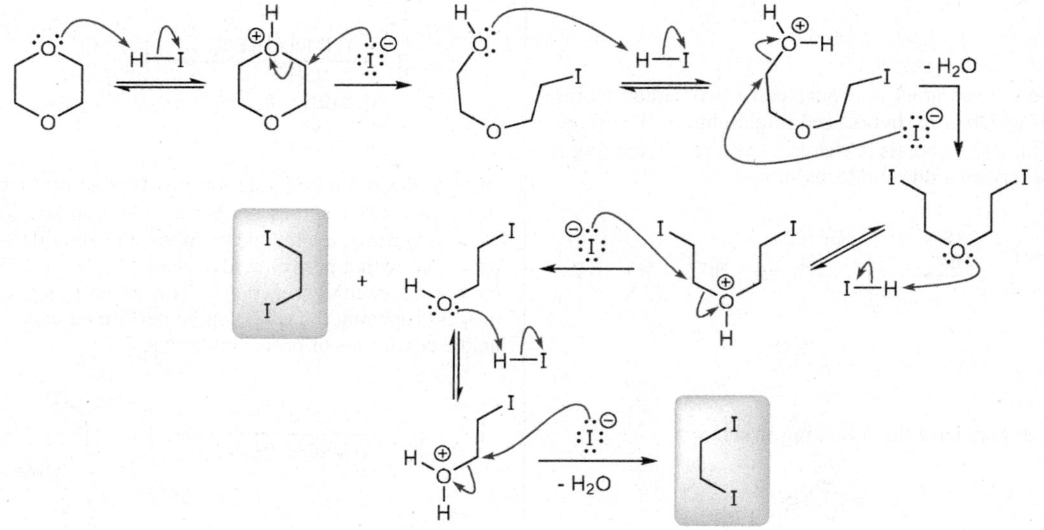

13.31. In the presence of an acid catalyst, an OH group is protonated to give an oxonium ion, thereby converting a bad leaving group into a good leaving group. Then, the other OH group (that was not protonated) functions as a nucleophilic center in an SN2-type process, forming a ring. Finally, deprotonation gives the product. Notice that water (not hydroxide) functions as the base in the deprotonation step (because there is virtually no hydroxide present in acidic conditions).

H_2SO_4

- H_2O

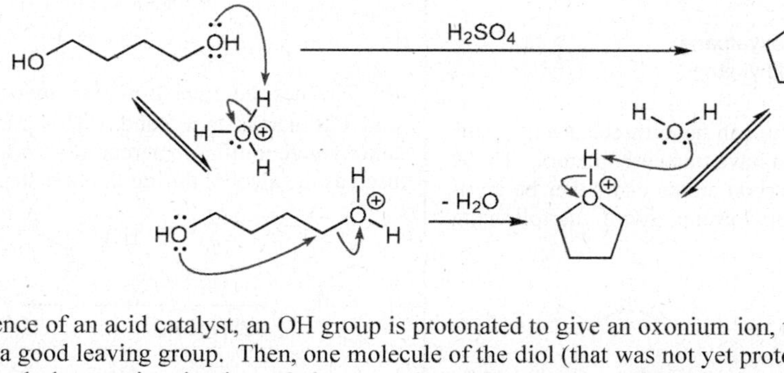

13.32. In the presence of an acid catalyst, an OH group is protonated to give an oxonium ion, thereby converting a bad leaving group into a good leaving group. Then, one molecule of the diol (that was not yet protonated) can function as a nucleophile and attack the oxonium ion in an SN2 process, expelling water as a leaving group. The resulting oxonium ion is then deprotonated. The previous three steps are then repeated (protonate, intramolecular back-side attack, and then deprotonate), giving the cyclic product. Notice that each of the deprotonation steps is shown with water functioning as the base (not hydroxide), because there is virtually no hydroxide present in acidic conditions.

13.33. Ethylene oxide has a high degree of ring strain, and readily functions as an electrophile in an S_N2 reaction. The reaction opens the ring and alleviates the ring strain. Oxetane has less ring strain and is, therefore, less reactive as an electrophile towards S_N2. The reaction can still occur, albeit at a slower rate, to alleviate the ring strain associated with the four-membered ring. THF has almost no ring strain and does not function as an electrophile in an S_N2 reaction. You might recall that THF is used as a polar aprotic solvent. The stability and low reactivity of THF contribute to making it suitable as a solvent.

13.34. Acetylene undergoes alkylation when treated with a strong base (such as $NaNH_2$) followed by methyl iodide. This process is then repeated to install a second methyl group, giving 2-butyne. This alkyne can be reduced either via hydrogenation with a poisoned catalyst to give a *cis* alkene, or via a dissolving metal reduction to give a *trans* alkene. Treatment of these alkenes with a peroxy acid (such as MCPBA) gives the epoxides shown.

1) $NaNH_2$ 2) MeI; 1) $NaNH_2$ 2) MeI; H_2, Lindlar's cat.; Na, NH_3; MCPBA; MCPBA

(meso) + Enantiomer

13.35. This transformation represents an intramolecular Williamson ether synthesis. Upon treatment with NaH (a strong base), the hydroxyl proton is removed, giving an alkoxide ion. This alkoxide ion (nucleophile) is tethered to an electrophilic carbon atom bearing a leaving group (bromide), so an intramolecular, S_N2-type process can occur, forming the cyclic ether product shown here.

NaH; $C_7H_{14}O$

13.36.

(a) Acetylene can be treated with $NaNH_2$ followed by $PhCH_2Br$ to install a benzyl group ($PhCH_2$). This process is then repeated (with methyl iodide as the alkyl halide) to install a methyl group. These two alkylation processes could also be performed in reverse order (with installation of the methyl group first, followed by installation of the benzyl group). The resulting internal alkyne can then be reduced via hydrogenation with a poisoned catalyst to give a *cis* alkene, which gives the desired epoxide (with *cis* alkyl groups) upon treatment with a peroxy acid, such as MCPBA.

1) $NaNH_2$ 2) $PhCH_2Br$ 3) $NaNH_2$ 4) MeI 5) H_2 Lindlar's cat. 6) MCPBA; + En; 1) $NaNH_2$ 2) $PhCH_2Br$; 1) $NaNH_2$ 2) MeI; H_2, Lindlar's cat.; MCPBA

(b) Acetylene can be treated with $NaNH_2$ followed by $PhCH_2Br$ to install a benzyl group ($PhCH_2$). This process is then repeated (with ethyl iodide as the alkyl halide) to install an ethyl group. These two alkylation processes could also be performed in reverse order (with installation of the ethyl group first, followed by installation of the benzyl group). The resulting internal alkyne can then be reduced via a dissolving metal reduction to give a *trans* alkene, which gives the desired epoxide (with *trans* alkyl groups) upon treatment with a peroxy acid (such as MCPBA).

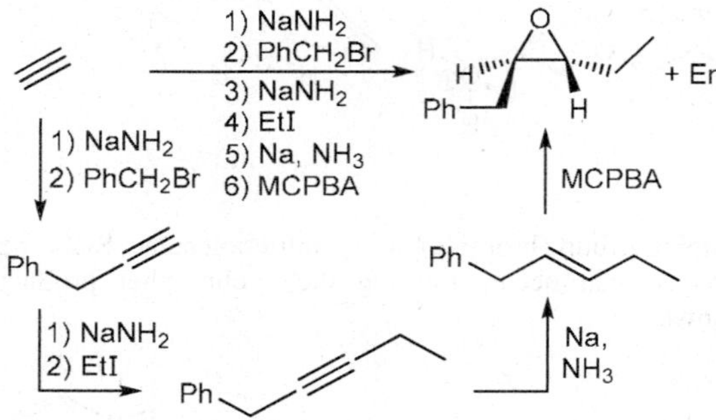

(c) Acetylene can be treated with $NaNH_2$ followed by ethyl iodide to install an ethyl group. This process is then repeated (with methyl iodide as the alkyl halide) to install a methyl group. These two alkylation processes could also be performed in reverse order (with installation of the methyl group first, followed by installation of the ethyl group). The resulting internal alkyne can then be reduced via hydrogenation with a poisoned catalyst to give a *cis* alkene, which gives the desired epoxide (with *cis* alkyl groups) upon treatment with a peroxy acid (such as MCPBA).

(d) Acetylene can be treated with $NaNH_2$ followed by ethyl iodide to install the first ethyl group. This process is then repeated (again with ethyl iodide as the alkyl halide) to install the second ethyl group. Notice that two alkylation processes are required, even though the same group is being installed on both sides of acetylene. The reagents ($NaNH_2$ followed by ethyl iodide) will install only one ethyl group. So these reagents must be repeated to install the second ethyl group (treating acetylene with two equivalents of $NaNH_2$ followed by two equivalents of ethyl iodide will NOT produce the desired internal alkyne). The resulting internal alkyne can then be reduced via a dissolving metal reduction to give a *trans* alkene, which gives the desired epoxide (with *trans* alkyl groups) upon treatment with a peroxy acid (such as MCPBA).

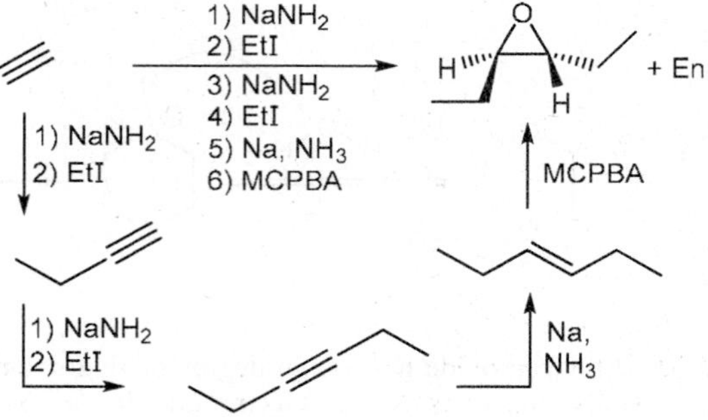

13.37.

(a) Treating the alkene with MCPBA (a peroxy acid) generates an epoxide. This epoxide has one chiral center, so we expect a racemic mixture of enantiomers (epoxide formation can occur on either face of the π bond with equal likelihood). Then, the epoxide is treated with methyl magnesium bromide, a strong nucleophile, resulting in a ring-opening reaction. Attack occurs at the less substituted position, because S_N2 back-side attack is faster at the less sterically hindered position. The resulting alkoxide ion is then protonated upon treatment with aqueous acid to give a secondary alcohol as the product. Notice that the epoxide intermediate is chiral and would be formed as a racemic mixture, but the final product does not have a chiral center.

(b) The starting material is an alkene, and the reagents indicate an alkoxymercuration-demercuration, resulting in the Markovnikov addition of H and OMe across the π bond. One chiral center is generated in the process, so we expect a racemic mixture of enantiomers.

(c) Treating the alkene with a peroxy acid generates an epoxide. This epoxide has one chiral center, so we expect a racemic mixture of enantiomers (epoxide formation can occur on either face of the π bond with equal likelihood). Then, the epoxide is treated with a strong nucleophile (MeS^-), resulting in a ring-opening reaction. Nucleophilic attack occurs at the less substituted position, because S_N2 back-side attack is faster at the less sterically hindered position. The resulting alkoxide ion is then protonated upon treatment with aqueous acid to give the product shown here. One chiral center is generated in the process, so we expect a racemic mixture of enantiomers.

1) RCO_3H
2) NaSMe
3) H_3O^+
SMe
OH
(racemic)
RCO_3H
1) NaSMe
2) H_3O^+
(racemic)

(d) This reaction sequence is a Williamson ether synthesis. Upon treatment with sodium hydride (NaH), an alcohol is deprotonated to give an alkoxide ion. The alkoxide then functions as a nucleophile when treated with ethyl chloride, giving an S_N2 reaction that affords the following ether:

OH 1) NaH 2) EtCl OEt

(e) Upon treatment with sodium hydride (NaH), an alcohol is deprotonated to give an alkoxide ion. The alkoxide then functions as a nucleophile and attacks the epoxide, giving a ring-opening reaction. Nucleophilic attack occurs at the less substituted position, because S_N2 back-side attack is faster at the less sterically hindered position. The resulting alkoxide ion is then protonated upon treatment with aqueous acid to give the following product:

OH 1) NaH 2) 3) H_3O^+ O OH

(f) Upon treatment with magnesium (Mg), the starting alkyl halide is converted into a Grignard reagent, which is a very strong nucleophile. When this Grignard reagent is treated with an epoxide, a ring-opening reaction occurs. The resulting alkoxide ion is then protonated upon treatment with aqueous acid to give the product shown:

Cl 1) Mg, diethyl ether 2) 3) H_3O^+ OH

13.38.

(a) Ethyl magnesium bromide is a strong nucleophile, so it can attack the epoxide (an electrophile) at the less substituted position, giving a ring-opening reaction. The resulting alkoxide ion is then protonated upon treatment with aqueous acid to give the tertiary alcohol as the product.

1) EtMgBr
2) H_3O^+
OH

(b) This is a Williamson ether synthesis. Sodium hydride is a strong base, and it will deprotonate an alcohol to give an alkoxide ion. This alkoxide ion will then function as a nucleophile when treated with a primary alkyl halide, giving an S_N2 reaction, as shown here:

1) NaH
2) EtI
OEt

(c) The acetylide ion is a strong nucleophile and it will attack the electrophilic epoxide at the less substituted position, giving a ring-opening reaction. The resulting alkoxide ion is then protonated upon treatment with aqueous acid, as shown:

1) HC≡CNa
2) H_3O^+
HO
HC≡C:

(d) Upon treatment with a strong acid, the epoxide is protonated. The resulting protonated epoxide is then attacked by MeSH (a good nucleophile). Nucleophilic attack occurs at the more substituted (tertiary) position, giving a ring-opening reaction. In the final step, a proton is removed to give the product shown:

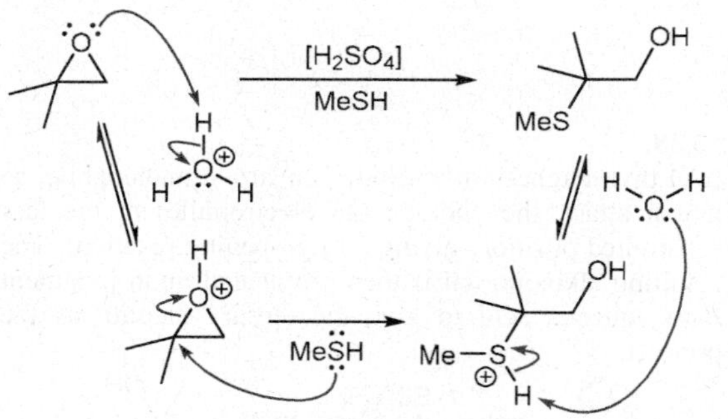

(e) This is an intramolecular Williamson ether synthesis. Sodium hydride is a strong base, and it will deprotonate the alcohol to give an alkoxide ion. This nucleophilic alkoxide ion contains a built-in electrophilic center (alkyl halide), so it can undergo an intramolecular S_N2-type process in which a chloride ion is ejected as a leaving group, giving the cyclic product shown:

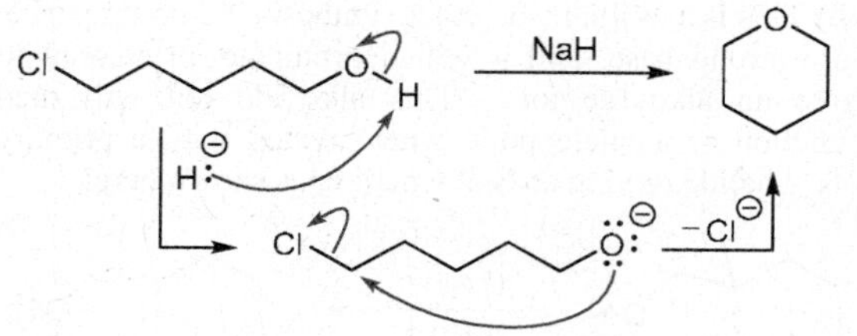

(f) Hydroxide functions as a nucleophile in an S_N2 reaction, ejecting chloride as a leaving group. The resulting alcohol is then deprotonated by another equivalent of hydroxide, giving an alkoxide ion. This nucleophilic alkoxide ion contains a built-in electrophilic center (alkyl halide), so it can undergo an intramolecular S_N2-type process in which a chloride ion is ejected as a leaving group, giving the cyclic product shown:

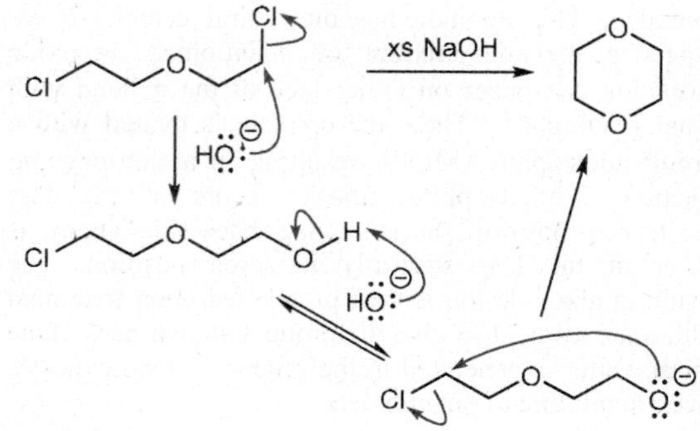

13.39. The starting material is a cyclic ether, in which the oxygen atom is connected to two carbon atoms, each of which is sp^3 hybridized (highlighted). As such, treatment with HBr is expected to cleave each of the C–O bonds and replace them with C–Br bonds, giving the following dibromide:

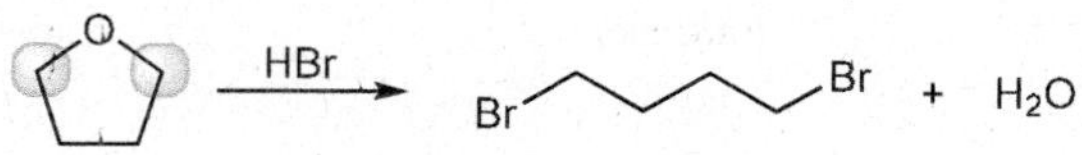

13.40. The starting material has six carbon atoms, and the product has six carbon atoms. So, the starting material must be some sort of cyclic ether, which opens to give a dibromide. By inspecting the dibromide product (*cis*-1,4-dibromocyclohexane), we can determine which two carbon atoms must have been connected to the oxygen atom in the starting cyclic ether (highlighted).

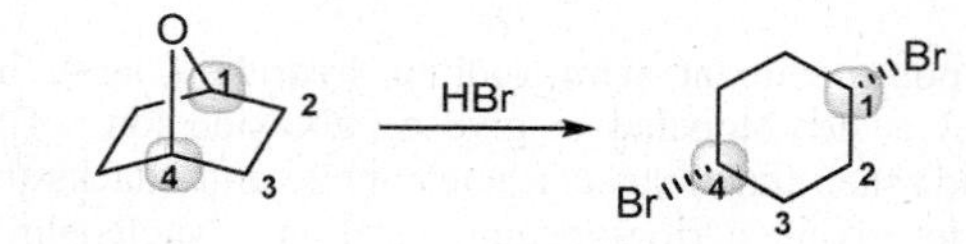

13.41. Ethyl magnesium bromide is a strong nucleophile, and can attack the epoxide at the less substituted position, in a ring-opening process, with inversion of configuration at the position that is attacked. The resulting alkoxide ion can then undergo an intramolecular S_N2-type process, expelling a chloride ion and generating a new epoxide. This epoxide can be attacked once again by ethyl magnesium bromide, once again at the less substituted position, and once again with inversion of configuration. The resulting alkoxide ion is then protonated upon treatment with aqueous acid, to give the observed product.

13.42.

(a) The starting material is an alkene, and the reagents indicate an alkoxymercuration-demercuration, resulting in the Markovnikov addition of H and OMe across the π bond.

1) $Hg(OAc)_2$, MeOH
2) $NaBH_4$

(b) The starting material is an alkene, and the reagents indicate an alkoxymercuration-demercuration, resulting in the Markovnikov addition of H and OR across the π bond.

1) $Hg(OAc)_2$,
2) $NaBH_4$

13.43.

(a) There are certainly many ways to prepare the target compound. The following strategy represents just one possible synthetic approach.

The central two carbon atoms (C3 and C4) come from acetylene, and the bonds at C2-C3 and C4-C5 are formed via the reaction between an acetylide ion and an epoxide. The forward process is shown below. Notice that after the first carbon-carbon bond is formed, the resulting compound is an alcohol. As such, the OH group must be protected before treating the terminal alkyne with a strong base (a strong base would simply deprotonate the alcohol, rather than the alkyne, if the OH group were not protected). The protecting group can be removed at any point after the second carbon-carbon bond is formed. Finally, the desired product is obtained upon reduction of the alkyne via hydrogenation in the presence of a poisoned catalyst, to give the *cis* alkene. Notice that the synthesis would still be successful if the order of the last two steps was reversed.

1) $NaNH_2$
2) epoxide
3) H_3O^+
4) TMSCl, Et_3N
5) $NaNH_2$
6) epoxide
7) H_3O^+
8) TBAF
9) H_2, Lindlar's cat.

1) $NaNH_2$ 2) epoxide 3) H_3O^+

TMSCl, Et_3N

1) $NaNH_2$ 2) epoxide 3) H_3O^+

TBAF

H_2, Lindlar's cat.

(b) There are certainly many ways to prepare the target compound. The following strategy represents just one possible synthetic approach.

The central two carbon atoms (C3 and C4) come from acetylene, and the bonds at C2-C3 and C4-C5 are formed via the reaction between an acetylide ion and an epoxide. The forward process is shown below. Notice that after the first carbon-carbon bond is formed, the resulting compound is an alcohol. As such, the OH group must be protected before treating the terminal alkyne with a strong base (a strong base would simply deprotonate the alcohol, rather than the alkyne, if the OH group were not protected). The protecting group can be removed after the second carbon-carbon bond is formed (either before or after the hydrogenation step). Finally, the desired product is obtained via hydrogenation of the alkyne followed by oxidation of the primary alcohol groups to aldehyde groups (using PCC or DMP or a Swern oxidation).

1) $NaNH_2$
2) [epoxide]
3) H_3O^+
4) TMSCl, Et_3N
5) $NaNH_2$
6) [epoxide]
7) H_3O^+
8) TBAF
9) H_2, Pt
10) PCC, CH_2Cl_2

1) $NaNH_2$ 2) [epoxide] 3) H_3O^+ — OH; TMSCl, Et_3N — OTMS; 1) $NaNH_2$ 2) [epoxide] 3) H_3O^+ — OH, OTMS; TBAF — OH, OH; H_2, Pt — OH, OH; PCC, CH_2Cl_2 — H, O, H, O

13.44. The following reagents can be used to achieve the desired transformations:

MCPBA; 1) EtMgBr 2) H_3O^+ — OH; $[H_2SO_4]$ MeOH — O–Me, OH; 1) NaOEt 2) H_3O^+ — OH, OEt; HBr — Br, OH; 1) NaCN 2) H_3O^+ — OH, CN; 1) $LiAlH_4$ 2) H_3O^+ — OH; 1) NaSMe 2) H_3O^+ — OH, SMe; 1) NaH 2) EtI — OEt, OEt; 1) NaH 2) CH_3I — O–Me, OMe

13.45. Alkoxymercuration-demercuration converts the alkene into an ether, which is then cleaved into two alkyl halides upon treatment with excess HI:

1) $Hg(OAc)_2$, EtOH 2) $NaBH_4$ — OEt; xs HI, heat — I + EtI + H_2O

Treatment of the alkene with a peroxy acid results in an epoxide, which is a *meso* compound in this case:

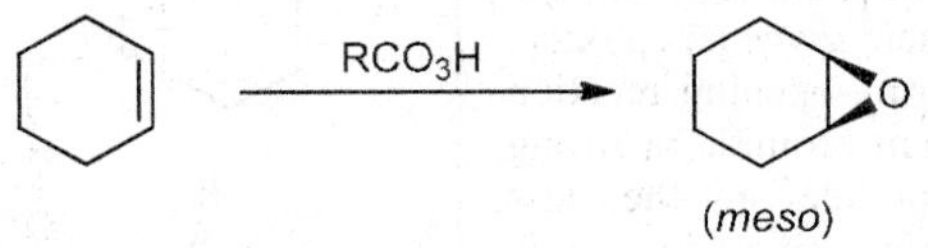

The epoxide can be opened in the presence of a variety of nucleophiles, with inversion of configuration, as shown:

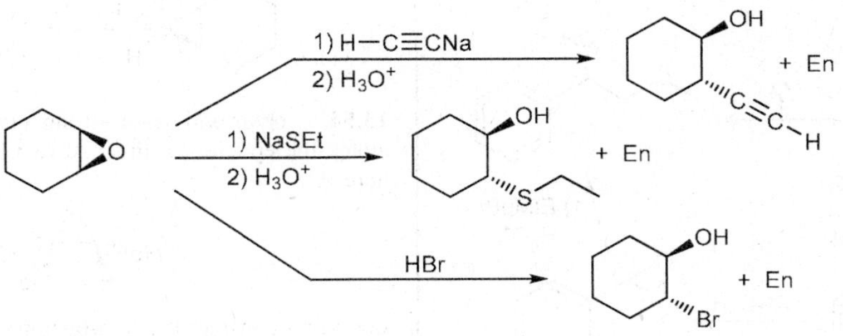

13.46. The correct answer is (b). The desired product is an ether (ROR) in which one R group is a phenyl group and the other R group is a tertiary alkyl group. A Williamson ether synthesis requires the use of an alkoxide ion and an alkyl halide, but the alkyl halide cannot be a tertiary alkyl halide or a phenyl halide. Therefore, this compound cannot be made via a Williamson ether synthesis, so options (a), (c), and (d) are not viable. The correct answer is option (b), which is an alkoxymercuration-demercuration. This process adds H and OPh across the alkene in a Markovnikov fashion, giving the desired product.

13.47. The correct answer is (d). Acidic options are employed, so none of the intermediates should contain alkoxide ions, which are strong bases and therefore inconsistent with acidic conditions. So options (b) and (c) are not correct. Option (a) suggests an S_N1 mechanism – loss of leaving group (opening of ring), followed by nucleophilic attack, but epoxide ring-opening reactions instead proceed via an S_N2 mechanism (concerted, back-side attack). The correct answer (d) is an example of an acid-catalyzed ring-opening reaction, with ethanol functioning as the nucleophile that opens the ring. In the first step of the mechanism, the epoxide is protonated, just as we would expect under acidic conditions. Then, this protonated epoxide is attacked by ethanol, opening the ring, and giving the intermediate shown in option (d). This intermediate is then deprotonated by ethanol to give the product.

13.48. The correct answer is (a). The starting material is an alkyl phenyl ether. The alkyl group is expected to be cleaved under these acidic conditions (the sp^3 center undergoing substitution is highlighted), giving an alkyl bromide. The other C–O bond is not cleaved, because the sp^2 hybridized substrate is not suitable for either an S_N1 or S_N2 mechanism.

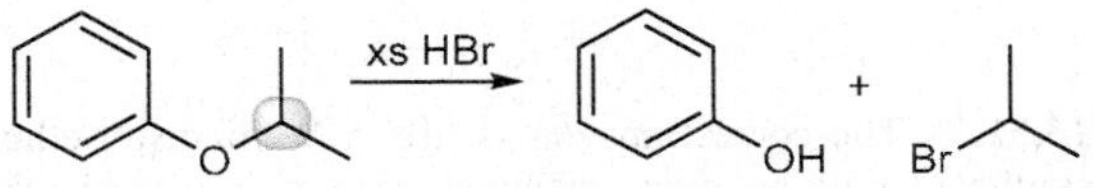

13.49. The correct answer is (a). The presence of two adjacent functional groups is the result of a ring-opening reaction of an epoxide. The hydroxyl group oxygen atom came from an epoxide electrophile, and the ethoxy group came from the attacking nucleophile (ethoxide), so the disconnection takes place at the bond connected to the ethoxy group, as shown.

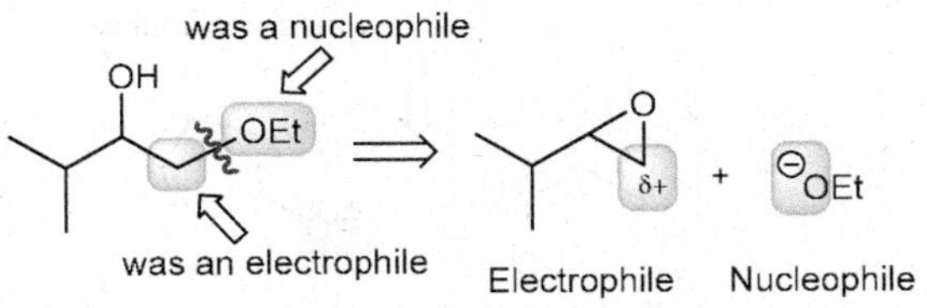

13.50. The correct answer is (c). The Grignard reagent (MeMgBr) is a strong nucleophile, and it attacks the epoxide at the less substituted position. The epoxide is opened, resulting in an alkoxide ion. This alkoxide is then protonated upon aqueous workup.

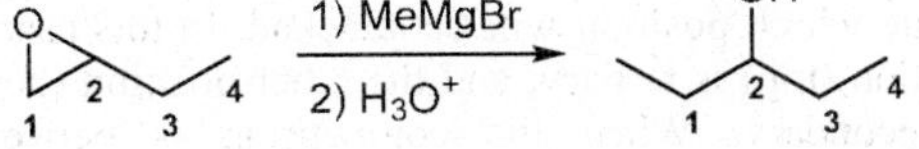

Note that the numbers shown above (in the product) do not follow IUPAC rules for numbering. Nevertheless, the practice of using numbers to count carbon atoms is a very effective tool that is being used in this case. It is OK to use numbers in this way, as long we are not using these numbers to name the product.

13.51. The correct answer is (b). The synthesis begins with an acid-catalyzed dehydration to produce an alkene. Treating the alkene with a peroxy acid gives an epoxide. Finally, this epoxide undergoes a ring-opening reaction when treated with methyl magnesium bromide (a strong nucleophile that attacks the epoxide at the less substituted position). The resulting alkoxide ion is protonated upon aqueous workup to give the desired product.

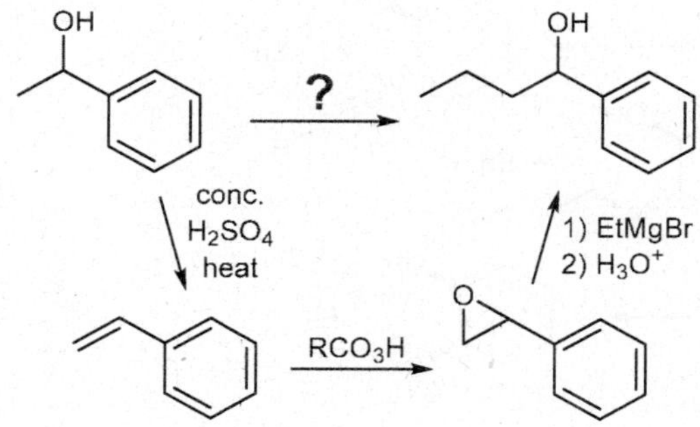

OH ? OH
conc. H_2SO_4 heat
1) EtMgBr 2) H_3O^+
RCO_3H
O

13.52. The correct answer is (d). A Williamson ether synthesis will be more efficient with a less sterically hindered substrate, since the process involves an S_N2 reaction. Therefore, the best retrosynthesis reveals a secondary alkoxide and a primary alkyl halide (rather than a primary alkoxide and a secondary alkyl halide):

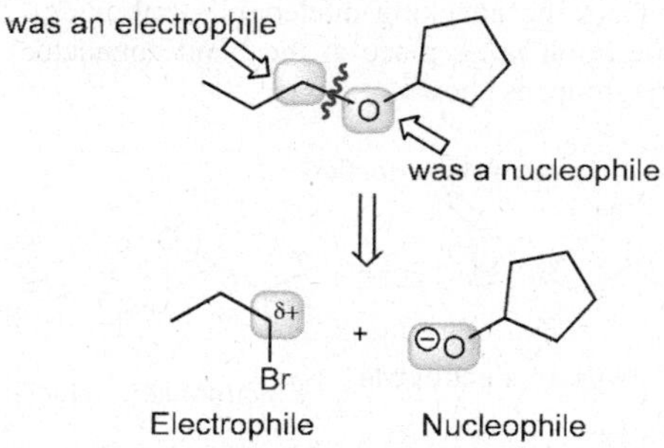

was an electrophile
O
was a nucleophile
δ+
Br
Electrophile
Nucleophile

13.53. The correct answer is (c). Under acidic conditions, the epoxide is protonated, thereby generating a very powerful electrophile (a protonated epoxide). Methanol (CH_3OH) is a weak nucleophile, and we must decide which position will be attacked. In this case, one position (top) is tertiary, and the other position (bottom) is secondary. When the competition is between a secondary position and a tertiary position, electronic factors dominate and the tertiary position is attacked (because the more substituted position has more partial positive character). Back-side attack causes inversion of configuration at the chiral center being attacked. Finally, a proton is removed (the most likely base is the solvent, methanol).

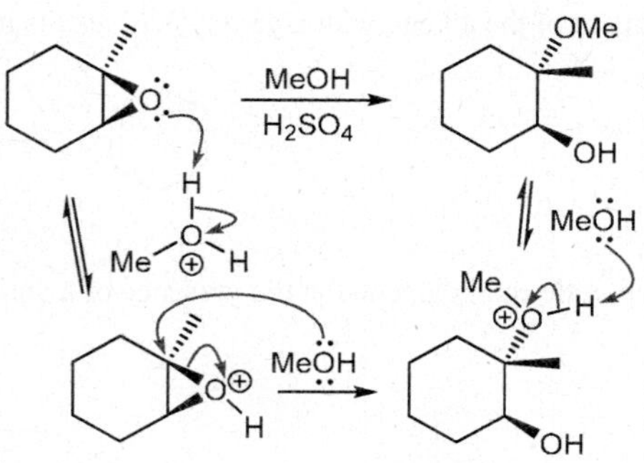

MeOH H_2SO_4 OMe OH
Me ⊕ H
MeOH
Me ⊕O H
MeOH
OH

13.54. Hydroxide is a strong nucleophile, and it can attack the epoxide at either of two locations, highlighted here:

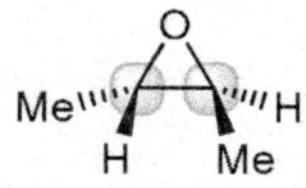

O
Me H
H Me

Let's first consider the ring-opening reaction occurring at the position on the left. In the presence of water, the resulting alkoxide ion is protonated to give a diol:

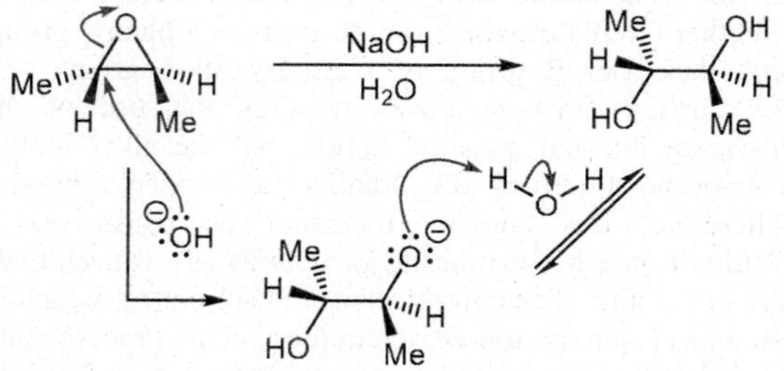

NaOH H_2O
Me OH H H HO Me
:OH
Me :O: H HO Me
H O H

This diol is a *meso* compound, which can be seen more clearly if we draw a Newman projection and then rotate about the central C–C bond:

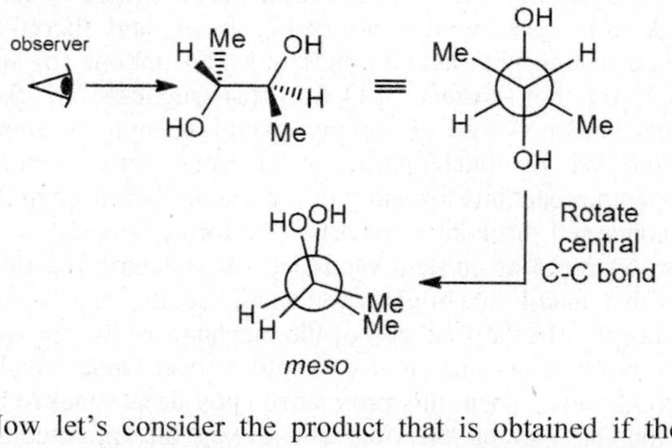

observer
H Me OH H HO Me
OH Me H H Me OH
Rotate central C-C bond
HO OH H H Me Me
meso

Now let's consider the product that is obtained if the ring-opening reaction occurs at the position on the right:

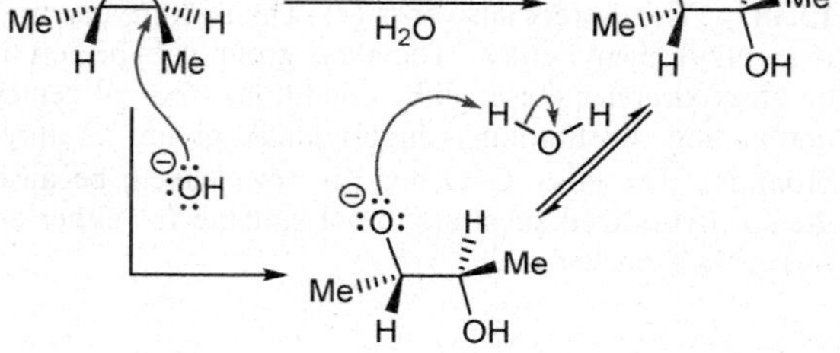

O
Me H H Me
NaOH H_2O
HO H Me Me H OH
:OH
H O H
:O: H Me Me H OH

Once again, notice that the resulting diol is a *meso* compound:

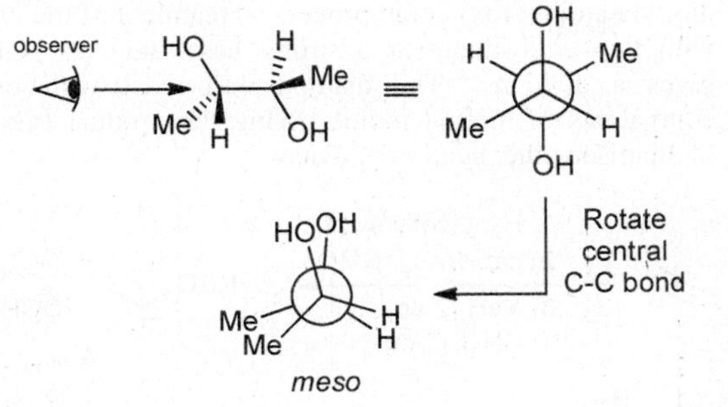

meso

In fact, this *meso* compound is the same *meso* compound that was obtained earlier (when the attack occurred at the position on the left).

meso *meso*

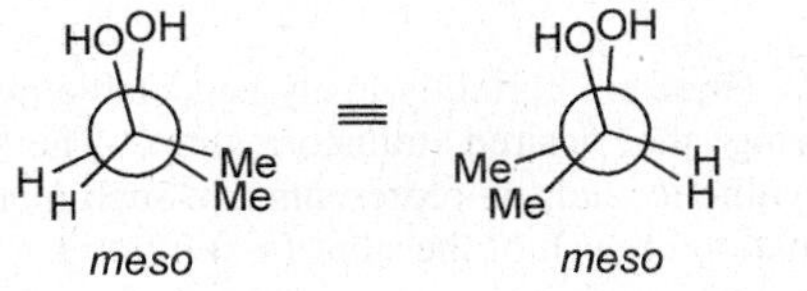

That is, the same product is obtained, regardless of which electrophilic position is attacked by hydroxide.

13.55. We begin by drawing the structure of *meso*-2,3-epoxybutane.

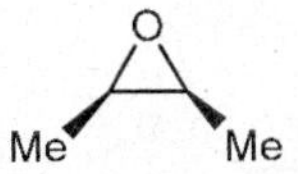

Hydroxide is a strong nucleophile, and it can attack the epoxide at either of two locations, just as we saw in the previous problem. Let's first consider the ring-opening reaction occurring at the position on the left. In the presence of water, the resulting alkoxide ion is protonated by water to give a diol:

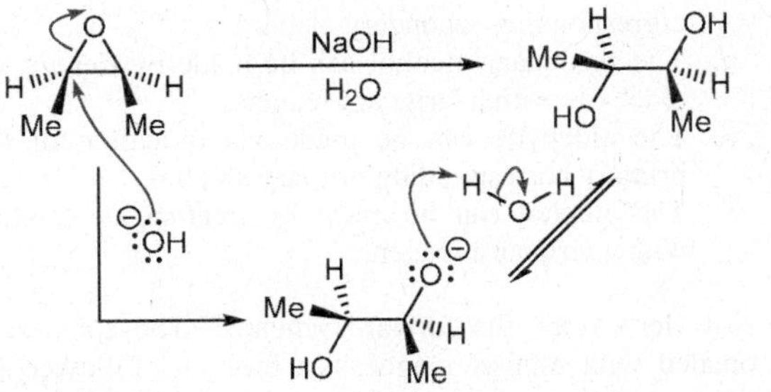

If we draw a Newman projection of the diol, as we did in the previous problem, we will see that this diol is **not** a *meso* compound.

not *meso*

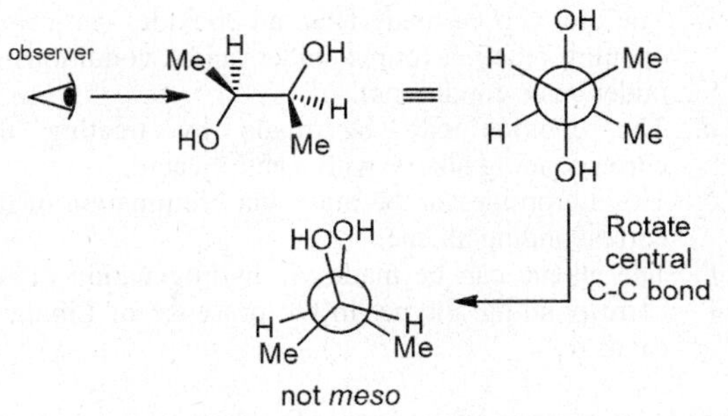

Now let's consider the product that is obtained if the ring-opening reaction occurs at the position on the right:

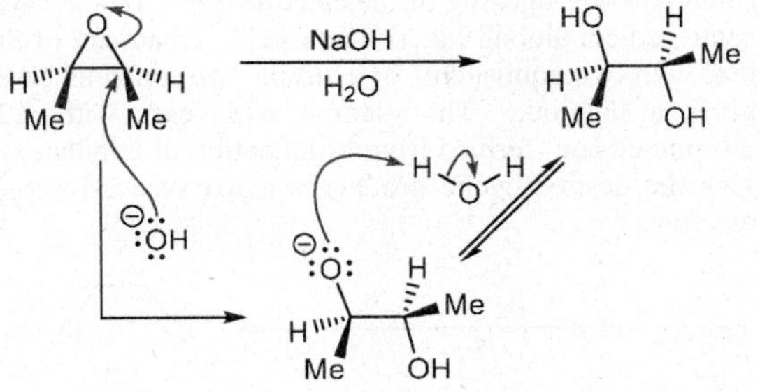

Once again, the resulting diol is ***not*** a *meso* compound. If we compare the two possible diols, we find that they are non-superimposable mirror images of each other, and therefore, they are enantiomers:

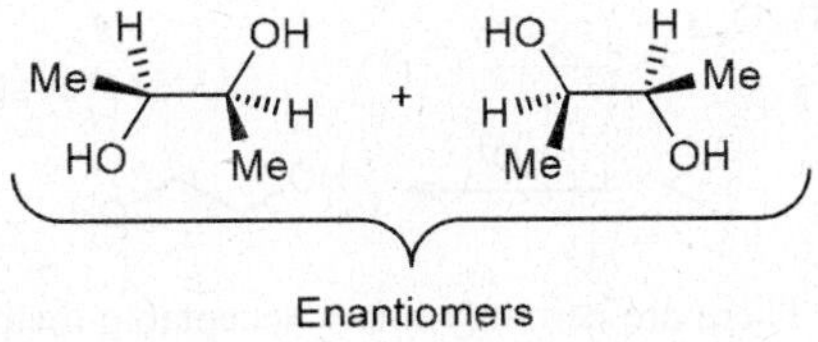

Enantiomers

13.56. There are certainly many acceptable methods for achieving the desired transformation. The following retrosynthetic analysis represents one such method. An explanation of each of the steps (*a-f*) follows.

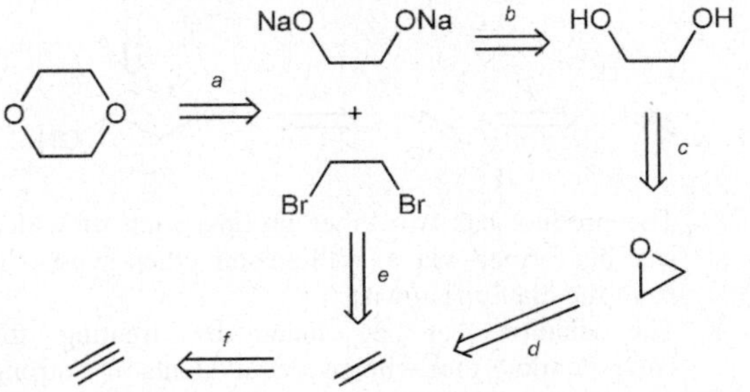

a. The cyclic product can be made by treating a dianion with a dihalide (via two successive S_N2 reactions).

b. The dianion can be made by treating the corresponding diol with two equivalents of a strong base (such as NaH).

c. The diol can be made from an epoxide, via a ring-opening reaction (either under acidic conditions or under basic conditions).
d. The epoxide can be made by treating the corresponding alkene with a peroxy acid.
e. The dibromide can be made via bromination of the corresponding alkene.
f. The alkene can be made via hydrogenation of the corresponding alkyne, in the presence of Lindlar's catalyst.

Now let's draw the forward scheme. Acetylene undergoes hydrogenation in the presence of Lindlar's catalyst to afford ethylene, which can be converted to an epoxide upon treatment with a peroxy acid. Acid-catalyzed ring-opening of the epoxide gives a diol (base-catalyzed conditions can also be used). Treatment of the diol with two equivalents of a strong base, such as NaH, gives a dianion. The dianion will react with 1,2-dibromoethane (formed from bromination of ethylene) to give the desired cyclic product via two successive S_N2 reactions.

H_2 Lindlar's catalyst; Br_2; Br Br; NaO ONa; RCO_3H; NaH (2 eq); H_3O^+; HO OH; O O

13.57. There are certainly many acceptable methods for achieving the desired transformation. The following retrosynthetic analysis represents one such method. An explanation of each of the steps (*a-d*) follows.

H_3CO OCH_3 ⇒ (a) NaO ONa ⇓ (b) HO OH ⇐ (c) ⇐ (d)

a. The product has two ether groups, each of which can be formed via a Williamson ether synthesis, from the dianion shown.
b. The dianion can be made by treating the corresponding diol with two equivalents of a strong base (such as NaH).
c. The diol can be made from an alkene, via a dihydroxylation process.
d. The alkene can be made via hydrogenation of the corresponding alkyne, in the presence of Lindlar's catalyst.

Now let's draw the forward scheme. Acetylene undergoes hydrogenation in the presence of Lindlar's catalyst to afford ethylene, which can be converted to a diol via a dihydroxylation process. Treatment of the diol with two equivalents of a strong base, such as NaH, gives a dianion. The dianion will react with two equivalents of methyl iodide giving the product (via a Williamson ether synthesis, twice).

1) H_2, Lindlar's cat.
2) cat. OsO_4, NMO
3) NaH (2 eq)
4) CH_3I (2 eq)

H_3CO OCH_3; CH_3I (2 eq); H_2 Lindlar's catalyst; NaO ONa; cat. OsO_4 NMO; HO OH; NaH (2 eq)

13.58. There are certainly many acceptable methods for achieving the desired transformation. The following retrosynthetic analysis represents one such method. An explanation of each of the steps (*a-f*) follows.

a b c d e f; O OH H

a. The product can be made via a Williamson ether synthesis, by treating the alcohol shown with base followed by ethyl iodide.
b. The alcohol can be made by treating a ketone with a Grignard reagent.
c. The ketone can be made via oxidation of the corresponding secondary alcohol.
d. The secondary alcohol can be made by treating an aldehyde with a Grignard reagent.
e. The aldehyde can be made via oxidation of the primary corresponding primary alcohol.
f. The alcohol can be made by treating an epoxide with a Grignard reagent.

Now let's draw the forward scheme. The epoxide is opened with methyl magnesium bromide, followed by aqueous acidic workup, to give 1-propanol. 1-Propanol is then oxidized to an aldehyde with PCC (or with DMP, or with a Swern oxidation). Treating the aldehyde with a Grignard reagent gives a secondary alcohol. Oxidation of the alcohol gives a ketone, which can be treated with ethyl magnesium bromide to give a tertiary alcohol. This alcohol is then deprotonated upon treatment with a strong base, such as NaH. The resulting anion then functions as a nucleophile in an S_N2 reaction with ethyl iodide to give the desired product.

1) MeMgBr
2) H_3O^+
3) PCC, CH_2Cl_2
4) EtMgBr
5) H_3O^+
6) $Na_2Cr_2O_7$, H_2SO_4, H_2O
7) EtMgBr
8) H_3O^+
9) NaH
10) EtI

1) MeMgBr
2) H_3O^+

PCC
CH_2Cl_2

1) EtMgBr
2) H_3O^+

$Na_2Cr_2O_7$, H_2SO_4, H_2O

1) EtMgBr
2) H_3O^+

1) NaH
2) EtI

13.59.

(a) A retrosynthetic analysis of the target molecule reveals that the required disconnection is not at the α position, but one bond farther away from the hydroxyl group. Therefore, the required disconnection leads to a Grignard reagent nucleophile and an epoxide electrophile.

δ+ + CH_3—MgBr δ−

Electrophile Nucleophile

Now let's draw the forward process. Treating the alkene with a peroxy acid (such as MCPBA) gives an epoxide. This epoxide undergoes a ring-opening reaction when treated with methyl magnesium bromide (a strong nucleophile), to give an alkoxide ion which is protonated upon treatment with aqueous acid to give the desired product.

1) MCPBA
2) MeMgBr
3) H_3O^+

MCPBA

1) MeMgBr
2) H_3O^+

(b) The following synthesis builds on the synthesis in the previous solution (**13.59a**). The product of that synthesis is treated with an oxidizing agent, such as chromic acid, to give the desired ketone:

1) MCPBA
2) MeMgBr
3) H_3O^+
4) $Na_2Cr_2O_7$, H_2SO_4, H_2O

MCPBA

1) MeMgBr
2) H_3O^+

$Na_2Cr_2O_7$, H_2SO_4, H_2O

(c) The target molecule has two additional carbon atoms, so the synthesis requires a C–C bond-forming reaction. The retrosynthetic analysis of the target molecule asks, "What reaction have we seen that produces an alkyl chloride product?" The chloride can be prepared by substitution of the corresponding alcohol, and the alcohol is suitable for making the required C–C bond disconnection. In this case, the disconnection is not at the α position, but one bond farther away from the hydroxyl group. Therefore, the required disconnection leads to a Grignard reagent nucleophile and an epoxide electrophile.

δ− MgCl + δ+

Nucleophile Electrophile

Now let's draw the forward process. The transformation can be achieved by converting the starting alkyl halide into a Grignard reagent, treating the Grignard reagent with ethylene oxide (followed by aqueous workup), and then converting the resulting alcohol into an alkyl chloride (via an S_N2 process).

1) Mg
2) ethylene oxide
3) H_3O^+
4) $SOCl_2$, pyridine

1) Mg
2) ethylene oxide
3) H_3O^+

$SOCl_2$, py

(d) The solution to this problem is a slight modification of the solution to the previous problem (**13.59c**). The only change is the structure of the epoxide:

1) Mg
2) propylene oxide
3) H_3O^+
4) $SOCl_2$, py

1) Mg
2) propylene oxide
3) H_3O^+

$SOCl_2$, py

(e) The starting material is an alcohol and the product is an ether. This transformation can be achieved via a Williamson ether synthesis. The alcohol is first deprotonated with a strong base (such as NaH) to give an alkoxide ion, which is then treated with ethyl iodide to give an S_N2 reaction (with iodide serving as the leaving group):

1) NaH
2) EtI

(f) A retrosynthetic analysis of the target molecule reveals that the required disconnection is not at the α position, but one bond farther away from the hydroxyl group. Therefore, the required disconnection leads to an alkoxide nucleophile and an epoxide electrophile.

Nucleophile Electrophile

Now let's draw the forward process. The alcohol is first deprotonated with a strong base (such as NaH) to give an alkoxide ion, which is then treated with ethylene oxide to give a ring-opening reaction. The resulting alkoxide ion is protonated upon aqueous workup to give the desired product:

1) NaH
2) (ethylene oxide)
3) H_3O^+

(g) A retrosynthetic analysis of the ether target molecule by disconnection of one of the C–O bonds leads to an alkoxide nucleophile and an alkyl halide electrophile (Williamson ether synthesis).

a: Electrophile + Nucleophile

b: Nucleophile + Electrophile

In this case, both possible disconnections lead to primary alkyl halides that are suitable for an S_N2 mechanism, so there are two reasonable approaches to this transformation. To put plan **a** into action, the starting alkene first undergoes *anti*-Markovnikov addition of H and Br (HBr with peroxides). Treatment of the resulting alkyl bromide with sodium ethoxide produces the desired ether target molecule.

1) HBr, ROOR
2) NaOEt

HBr, ROOR — NaOEt

For synthesis **b**, the starting alkene will undergo hydroboration-oxidation (*anti*-Markovnikov addition of H and OH) to give the primary alcohol. Deprotonation of this alcohol with a strong base (such as NaH) to give an alkoxide ion, followed by treatment with ethyl bromide produces the desired ether target molecule.

1) $BH_3 \cdot THF$
2) H_2O_2, NaOH
3) NaH
4) CH_3CH_2Br

1) $BH_3 \cdot THF$
2) H_2O_2, NaOH

1) NaH
2) CH_3CH_2Br

(h) A retrosynthetic analysis of the ether target molecule by disconnection of one of the C–O bonds leads to an alkoxide nucleophile and an alkyl halide electrophile (Williamson ether synthesis).

a: Electrophile + Nucleophile

b: Nucleophile + Electrophile

b is the better retrosynthesis

In this case, only one disconnection (**b**) leads to a primary alkyl halide that is suitable for an S_N2 mechanism. A synthesis using approach **a** would fail because E2 elimination would predominate in the reaction of a strong base (ethoxide) with a secondary alkyl halide:

E2 major + S_N2 minor

Now let's draw the forward process for synthesis **b**. The starting alkene will undergo acid-catalyzed hydration (Markovnikov addition of H and OH) to give the secondary alcohol. Deprotonation of this alcohol with a strong base (such as NaH) to give an alkoxide ion, followed by treatment with ethyl bromide produces the desired ether target molecule.

1) dilute H_2SO_4
2) NaH
3) EtBr

dilute H_2SO_4

OH

1) NaH
2) EtBr

(i) The following retrosynthesis illustrates the two steps that are required to achieve the desired transformation. The epoxide (target molecule) can be made from the corresponding alkene, which can be made from the given alkyne:

Now let's draw the forward process. Reduction of the alkyne in the presence of a poisoned catalyst affords an alkene, which is converted to the epoxide upon treatment with a peroxy acid (such as MCPBA).

1) H_2, Lindlar's catalyst
2) MCPBA

+ En

H_2, Lindlar's cat.

MCPBA

(j) The following retrosynthesis illustrates the two steps that are required to achieve the desired transformation. The epoxide (target molecule) can be made from the corresponding *cis*-alkene, which can be made from the given alkyne.

Note that in order to achieve the desired stereochemical outcome, the epoxide must be prepared from the *cis*-alkene.

Now let's draw the forward process. Reduction of the alkyne in the presence of a poisoned catalyst affords a *cis*-alkene, which is converted to the desired epoxide upon treatment with a peroxy acid (such as MCPBA).

1) H_2, Lindlar's catalyst
2) MCPBA

(meso)

H_2, Lindlar's cat.

MCPBA

(k) The following retrosynthesis illustrates the two steps that are required to achieve the desired transformation. The epoxide (target molecule) can be made from the corresponding *trans*-alkene, which can be made from the given alkyne.

Note that in order to achieve the desired stereochemical outcome, the epoxide must be prepared from the *trans*-alkene.

Now let's draw the forward process. Reduction of the alkyne via a dissolving metal reduction affords a *trans*-alkene, which is converted to the desired epoxide upon treatment with a peroxy acid (such as MCPBA).

1) Na, NH_3
2) MCPBA

+ En

Na, NH_3

MCPBA

(l) The starting material has five carbon atoms and the product has seven carbon atoms. The retrosynthetic analysis begins by identifying the position of the new C–C bond.

new C–C bond

1 2 3 4 5

1 2 3 4 5

OH

Note that the numbers shown above (in the product) do not follow IUPAC rules for numbering, Nevertheless, the practice of using numbers to count carbon atoms is a very effective tool that is being used in this case. It is OK to use numbers in this way, as long we are not using these numbers to name the product.

We can see that the required disconnection is not at the α position, but one bond farther away from the hydroxyl group, leading to a Grignard reagent nucleophile and an epoxide electrophile:

α

OH

MgBr

δ−

δ+

Nucleophile

Electrophile

Now let's draw the forward process. To prepare the epoxide, the alkyne starting material is first reduced in the presence of a poisoned catalyst to afford an alkene. The alkene is converted to the epoxide upon treatment with a peroxy acid (such as MCPBA), and then treatment with a Grignard reagent (and aqueous workup), gives the desired product.

1) H_2, Lindlar's cat.
2) MCPBA
3) EtMgBr
4) H_3O^+

OH

H_2, Lindlar's cat.

1) EtMgBr
2) H_3O^+

MCPBA

(m) We did not learn a C–C bond-forming reaction that results in an ether product. However, we did learn a way to cleave a phenyl ether to give phenol (using HX). A Williamson ether synthesis can then be used to reinstall a different alkyl group (this time an ethyl group, rather than a methyl group):

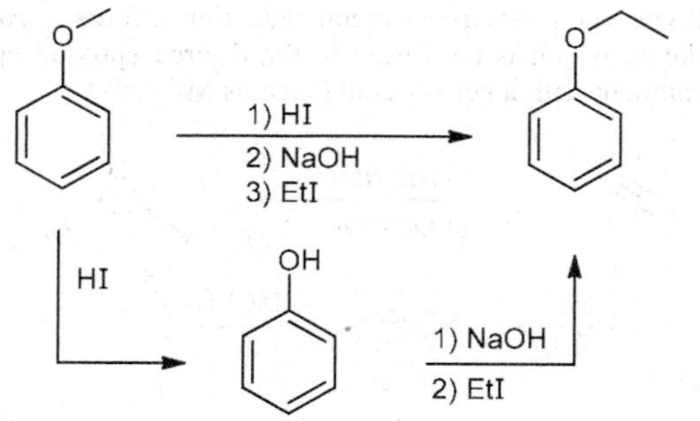

1) HI
2) NaOH
3) EtI

HI

1) NaOH
2) EtI

(n) A retrosynthetic analysis of the target molecule reveals that the required disconnection is not at the α position, but one bond farther away from the hydroxyl group. Therefore, the required disconnection leads to a Grignard reagent nucleophile and an epoxide electrophile.

Nucleophile Electrophile

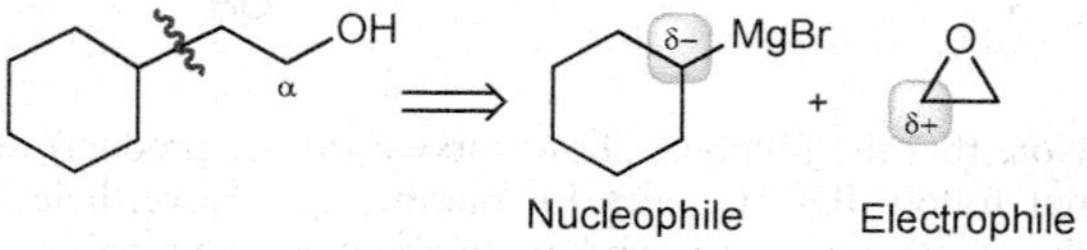

Now we have to bridge the gap between the given ether starting material and the alkyl bromide needed to prepare the Grignard reagent (bromocyclohexane). This transformation represents an ether cleavage, and it can be achieved by treatment with HBr. Conversion of bromocyclohexane into a Grignard reagent, followed by treatment with ethylene oxide (and aqueous workup), gives the desired product.

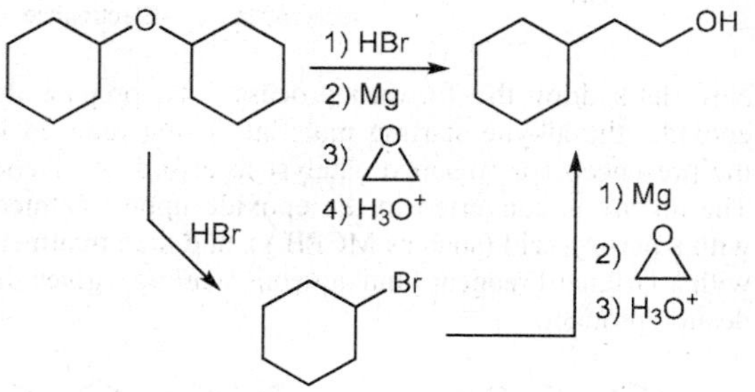

1) HBr
2) Mg
3) ethylene oxide
4) H_3O^+

HBr

1) Mg
2) ethylene oxide
3) H_3O^+

(o) A retrosynthetic analysis of the target molecule reveals that the required disconnection is not at the α position, but one bond farther away from the hydroxyl group. Therefore, the required disconnection leads to an alkoxide nucleophile and an epoxide electrophile.

Nucleophile Electrophile

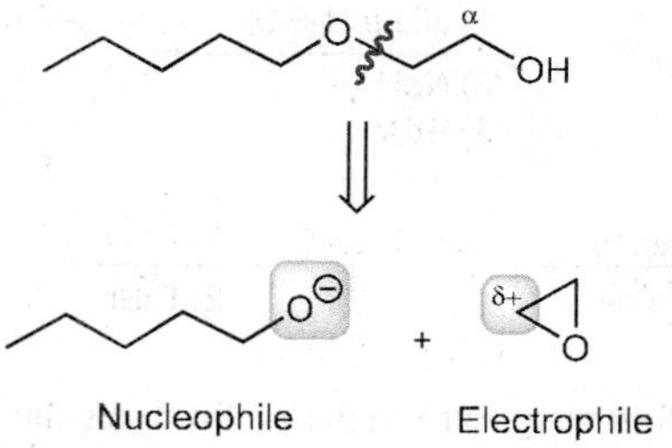

Now let's draw the forward process. The alcohol is first deprotonated with a strong base (such as NaH) to give an alkoxide ion, which is then treated with ethylene oxide to give a ring-opening reaction. The resulting alkoxide ion is protonated upon aqueous workup to give the desired product:

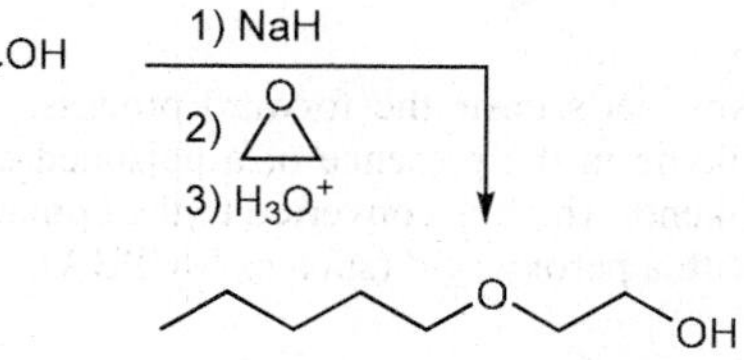

1) NaH
2) ethylene oxide
3) H_3O^+

(p) A retrosynthetic analysis of the target molecule reveals that the required disconnection is not at the α position, but one bond farther away from the hydroxyl group. Therefore, the required disconnection leads to a Grignard reagent nucleophile and an epoxide electrophile.

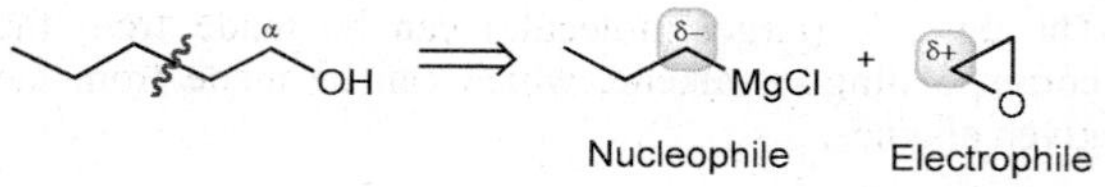

Nucleophile Electrophile

Now let's draw the forward process. The desired transformation can be achieved by converting the starting alkyl halide into a Grignard reagent, and then treating the Grignard reagent with an epoxide (followed by aqueous workup).

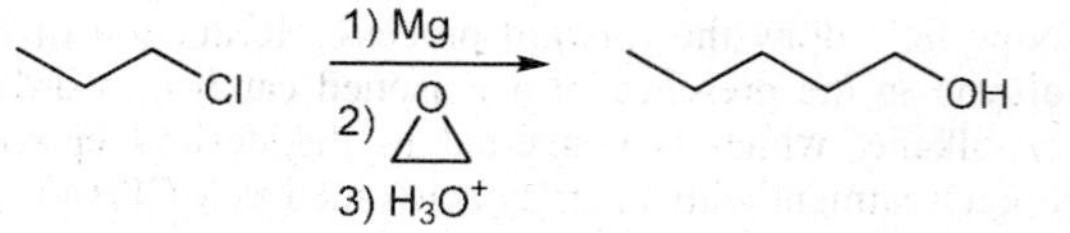

1) Mg
2) ethylene oxide
3) H_3O^+

(q) The retrosynthetic analysis of the target molecule asks, "What reaction have we seen that produces an aldehyde product?" The aldehyde can be prepared by oxidation of the corresponding primary alcohol, and the alcohol is suitable for making the required C–C bond disconnection. The disconnection is not at the α position, but one bond farther away from the hydroxyl group, leading to a Grignard reagent nucleophile and an epoxide electrophile.

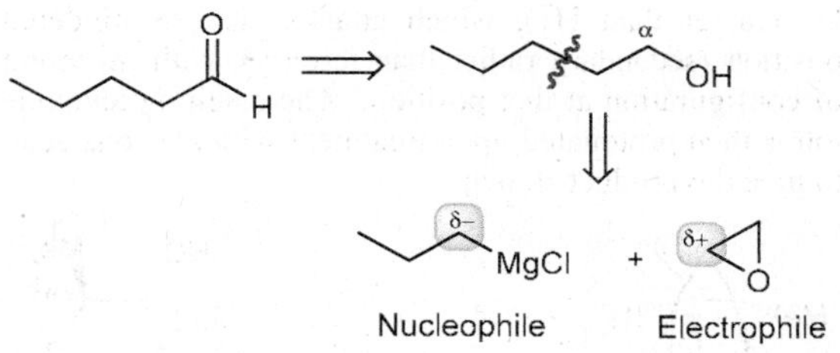

Nucleophile Electrophile

Now let's draw the forward process. The starting alkyl halide is converted into a Grignard reagent, which is then treated with ethylene oxide, followed by aqueous workup. Oxidation of the resulting primary alcohol with PCC (or DMP, or with a Swern oxidation) gives the desired aldehyde:

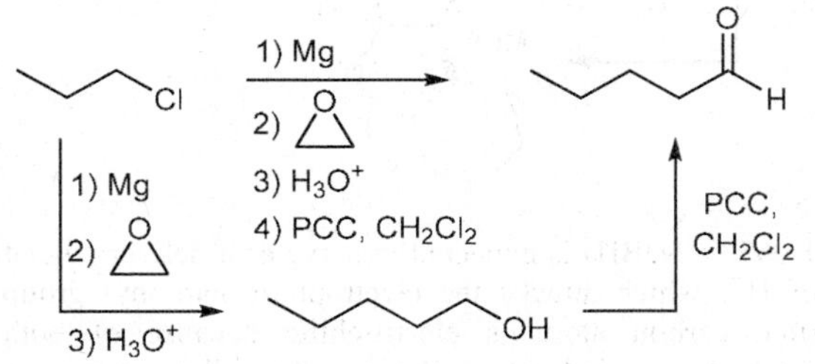

1) Mg
2) epoxide
3) H_3O^+
4) PCC, CH_2Cl_2

PCC, CH_2Cl_2

(r) This conversion can be achieved in one step, by treating the starting material with the epoxide shown, in the presence of acid catalysis. Under these conditions, the alcohol functions as a nucleophile and attacks a protonated epoxide to give a ring-opening reaction in which the nucleophile attacks the more substituted tertiary position (due to electronic effects).

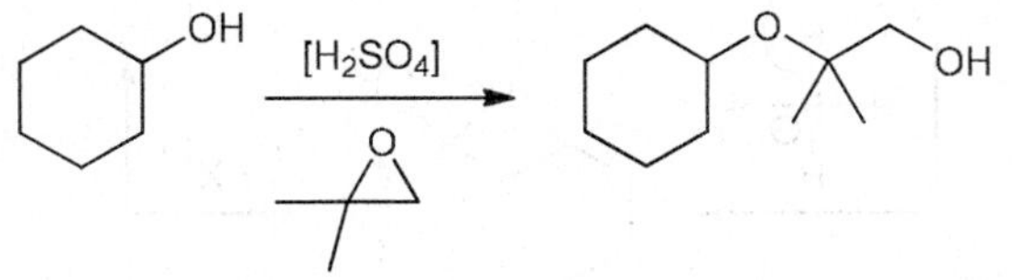

$[H_2SO_4]$

(s) This transformation is similar to the previous problem (**13.59r**), although in this case, the nucleophilic attack must occur at the less substituted position. This requires treating an epoxide with a strong nucleophile in basic conditions. The starting alcohol is first deprotonated with a strong base (such as NaH), and the resulting alkoxide ion is treated with the epoxide shown below. In this case, the nucleophile attacks the less substituted primary position (due to less steric hindrance). The resulting ring-opening reaction, followed by aqueous acidic workup, gives the desired product.

1) NaH
2) epoxide
3) H_3O^+

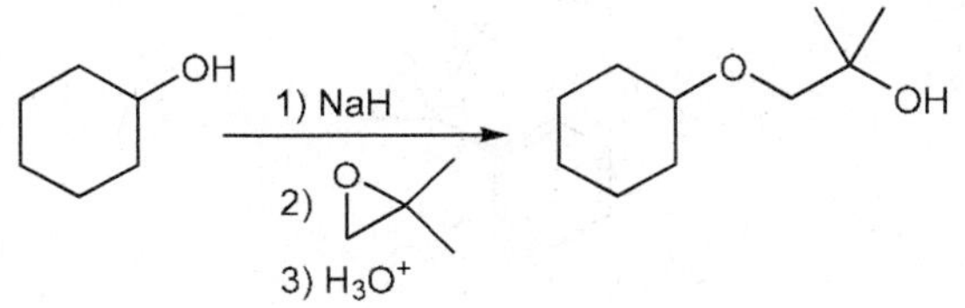

(t) The epoxide can be made from the following alkene, so we must find a way to make this alkene:

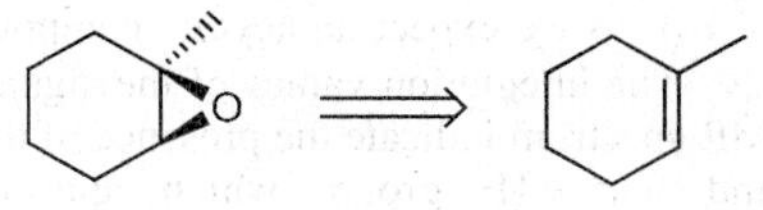

This alkene can be made from the starting alkane in two steps (radical bromination, followed by Zaitsev elimination with a non-bulky base), giving the following synthesis:

1) Br_2, $h\nu$
2) NaOEt
3) MCPBA

+ En

Br_2, $h\nu$

NaOEt

MCPBA

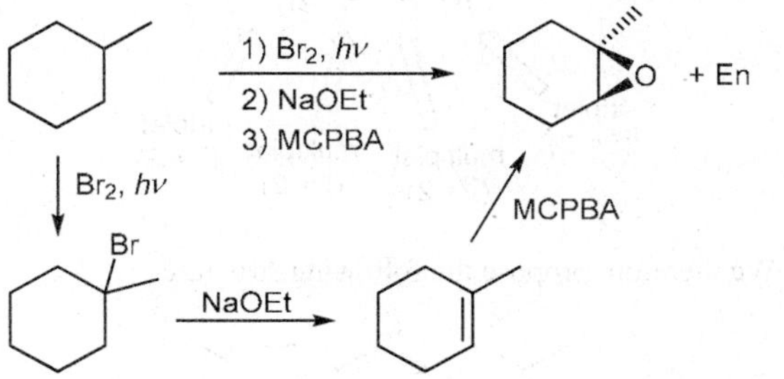

(u) The following synthesis builds on the synthesis in the previous solution (**13.59t**). The product of that synthesis is opened by a strong nucleophile (methoxide), which attacks the less substituted position. Aqueous acidic workup gives the desired product.

1) Br_2, $h\nu$
2) NaOEt
3) MCPBA
4) NaOMe
5) H_3O^+

+ En

Br_2, $h\nu$

1) NaOMe
2) H_3O^+

NaOEt

MCPBA

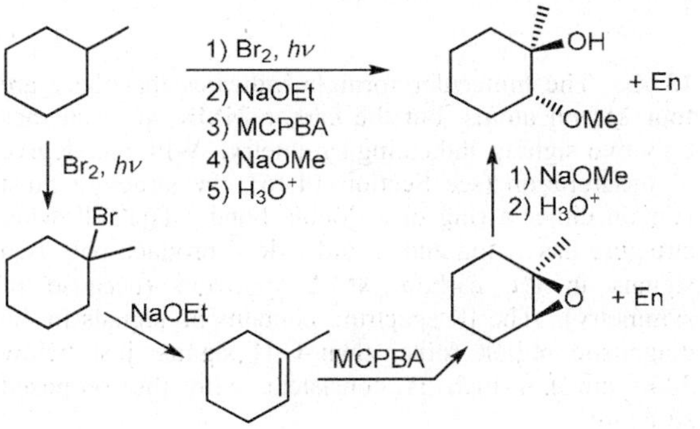

13.60. The molecular formula indicates that there are seven carbon atoms, but the spectrum has only five signals, indicating symmetry. With four degrees of unsaturation (see Section 14.16), we suspect an aromatic ring. There are four signals in the aromatic region of the spectrum, so we expect a monosubstituted ring, which explains the symmetry. The fifth signal appears above 50 ppm, indicating that it is next to an electronegative atom. We see in the molecular formula that there is an oxygen atom, so we propose the following structure, called methoxybenzene (also called anisole).

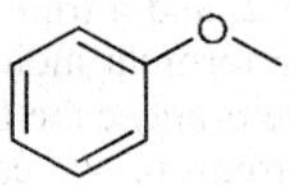

13.61. The ^{1}H NMR spectrum has four signals, and the total integration of those four signals is 2 + 2 + 2 + 3 = 9. However, the molecular formula indicates eighteen hydrogen atoms, so we expect a high degree of symmetry (one half of the molecule mirrors the other half). The compound has no degrees of unsaturation (see Section 14.16), so we expect an acyclic compound with no π bonds. The integration values of the signals in the proton NMR spectrum indicate the presence of three CH_2 groups and one CH_3 group, which appear to be connected to each other in a chain:

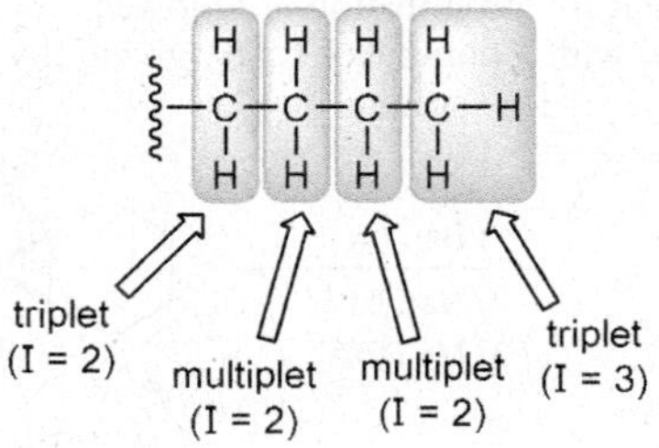

triplet (I = 2) multiplet (I = 2) multiplet (I = 2) triplet (I = 3)

We therefore propose the following structure:

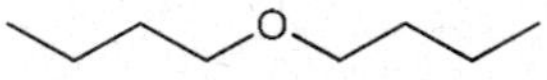

This structure is consistent with the carbon NMR spectrum as well. Notice that there are four different kinds of carbon atoms, thus giving rise to four signals. Only one of the four signals is above 50 ppm, indicating that it is next to an electronegative atom. This is consistent with the structure above.

13.62. The molecular formula indicates that there are four carbon atoms, but the carbon NMR spectrum has only two signals, indicating symmetry. With one degree of unsaturation (see Section 14.16), the structure must contain either a ring or a double bond. The following structure has a ring and would indeed produce only two signals in the carbon NMR spectrum (because of symmetry). The IR spectrum contains no signals in the diagnostic region (other than C-H signals just below 3000 cm^{-1}), which is consistent with the proposed structure.

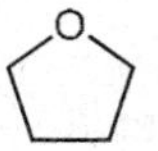

13.63. The spectrum has two signals, and the total integration of those two signals is 2 + 3 = 5. However, the molecular formula indicates ten hydrogen atoms, so we expect a high degree of symmetry (one half of the molecule mirrors the other half). The compound has no degrees of unsaturation (see Section 14.16), so we expect an acyclic compound with no π bonds. The signals in the spectrum are consistent with an ethyl group (a quartet with an integration of 2, and a triplet with an integration of 3). The molecular formula indicates the presence of an oxygen atom, so we propose the following structure in which the two ethyl groups mirror each other:

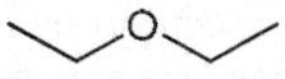

13.64. $LiAlD_4$ is expected to function very much like $LiAlH_4$. That is, it is expected to be a delivery agent of D^- (rather than H^-), which attacks the less hindered position (secondary rather than tertiary), with inversion of configuration at that position. The resulting alkoxide ion is then protonated upon treatment with aqueous acid, to give the product shown.

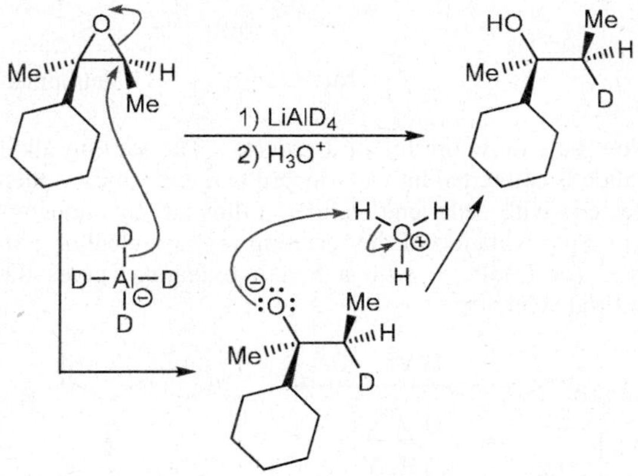

1) $LiAlD_4$
2) H_3O^+

13.65. $NaBH_4$ is expected to serve as a delivery agent of H^-, which attacks the electrophilic carbonyl group (that carbon atom is electrophilic because of both resonance and induction). The resulting alkoxide ion can then undergo an intramolecular S_N2-type reaction, expelling a halide as a leaving group, and generating the epoxide, as shown:

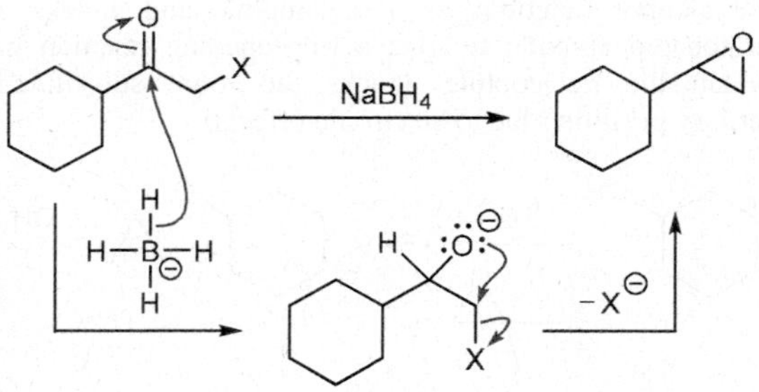

$NaBH_4$

$-X^\ominus$

13.66. When methyloxirane is treated with HBr, the regiochemical outcome is determined by a competition between steric and electronic factors, with steric factors prevailing – bromide attacks the less substituted position. However, when phenyloxirane is treated with HBr, electronic factors prevail in controlling the regiochemical outcome. Specifically, the position next to the phenyl group is a benzylic position and can stabilize a large partial positive charge.

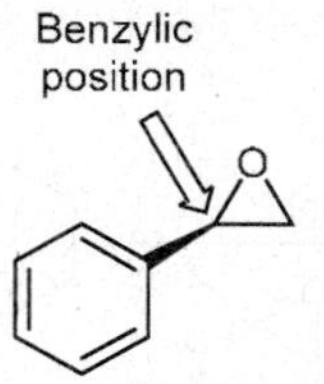

Benzylic position

In such a case, electronic factors are more powerful than steric factors, and bromide attacks the more substituted position.

13.67. This process for epoxide formation involves deprotonation of the hydroxyl group, followed by an intramolecular SN2-type attack. Recall that SN2 processes occur via back-side attack, which can only be achieved when both the hydroxyl group and the bromine occupy axial positions on the ring. Due to the steric bulk of a *tert*-butyl group, compound **A** spends most of its time in a chair conformation that has the *tert*-butyl group in an equatorial position. In this conformation, the OH and Br are indeed in axial positions, so the reaction can occur quite rapidly. In contrast, compound **B** spends most of its time in a chair conformation in which the OH and Br occupy equatorial positions. The SN2-type process cannot occur from this conformation.

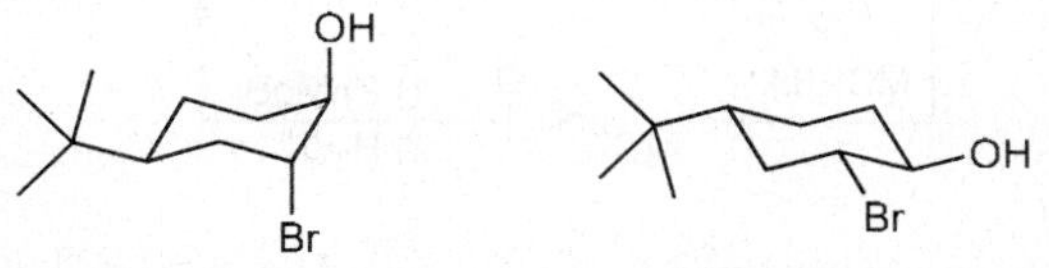

Most stable conformation of compound **A** (rapid back-side attack)

Most stable conformation of compound **B** (slow back-side attack)

When compound **A** is treated with NaOH, the hydroxyl group in compound **A** is deprotonated, giving an alkoxide ion which can serve as a nucleophile in an intramolecular SN2-type attack that gives an epoxide.

13.68. There are many acceptable methods for achieving the desired transformation. The following retrosynthetic analysis represents one such method. An explanation of each of the steps (*a-g*) follows.

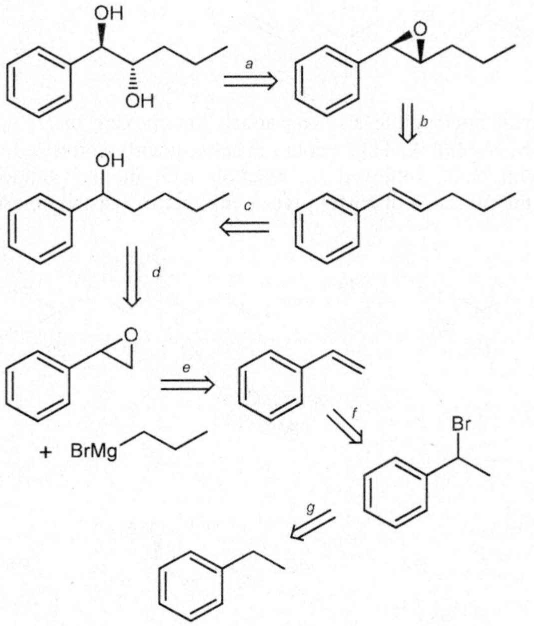

a. The diol can be made from an epoxide.
b. The epoxide can be made from the corresponding *trans*-alkene.
c. The *trans*-alkene can be made from an alcohol.
d. The alcohol can be made from an epoxide via a Grignard reaction.
e. The epoxide can be made from the corresponding alkene (called styrene).
f. Styrene can be made from the corresponding benzyl bromide.
g. The bromide can be made from ethyl benzene via radical bromination.

The forward synthetic scheme for the second pathway is illustrated here:

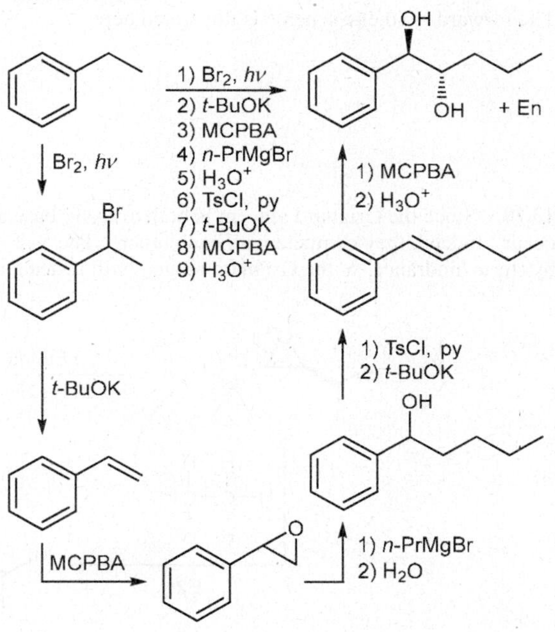

1) Br_2, *hν*
2) *t*-BuOK
3) MCPBA
4) *n*-PrMgBr
5) H_3O^+
6) TsCl, py
7) *t*-BuOK
8) MCPBA
9) H_3O^+

+ En

Br_2, *hν*

t-BuOK

MCPBA

1) *n*-PrMgBr
2) H_2O

1) TsCl, py
2) *t*-BuOK

1) MCPBA
2) H_3O^+

13.69. There are certainly many acceptable methods for achieving the desired transformation. The following retrosynthetic analysis represents one such method. An explanation of each of the steps (*a-e*) follows.

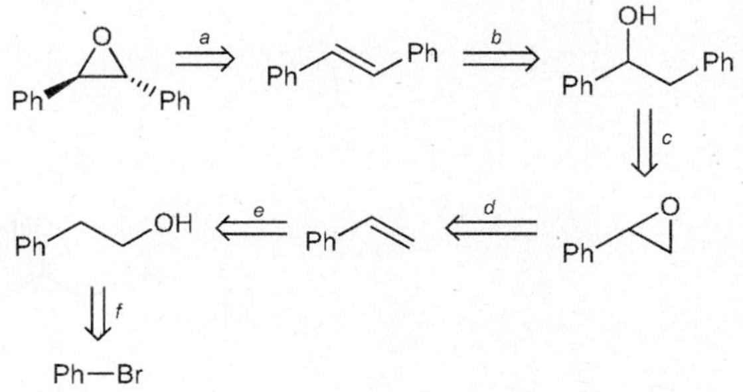

a. The epoxide can be made from the corresponding *trans* alkene.

b. The *trans* alkene can be made from the corresponding alcohol via a dehydration reaction (upon treatment with concentrated sulfuric acid).

c. The alcohol can be made from the reaction between a Grignard reagent (PhMgBr) and an epoxide, thereby installing a phenyl group.

d. The epoxide can be made from the corresponding alkene upon treatment with a peroxy acid.

e. The alkene can be made from the corresponding primary alcohol.

f. The primary alcohol can be made via Grignard reaction involving an epoxide.

The forward synthetic scheme is illustrated here:

Ph—Br → 1) Mg 2) epoxide 3) H_3O^+ 4) conc. H_2SO_4, heat 5) MCPBA 6) PhMgBr 7) H_3O^+ 8) conc. H_2SO_4, heat 9) MCPBA → epoxide + En

1) Mg 2) epoxide 3) H_3O^+; conc. H_2SO_4 heat; MCPBA; 1) PhMgBr 2) H_3O^+; conc. H_2SO_4 heat; MCPBA

13.70. Since the Grignard reagent is both a strong base and a strong nucleophile, substitution and elimination can both occur. Indeed, they compete with each other. The S_N2 mechanism (nucleophilic attack on the epoxide ring) is slowed by steric hindrance, so the Grignard reagent will instead act as a base, resulting in an E2 mechanism:

1) EtMgBr 2) H_3O^+; E2

13.71. In the first step, vinylmagnesium bromide is a powerful nucleophile and can attack the epoxide in a ring-opening reaction (attacking the less substituted carbon) to afford alcohol **A**. This alcohol is subsequently converted to ether **B** via a Williamson ether synthesis (deprotonation with NaH, followed by reaction with methyl iodide). Dihydroxylation of the terminal alkene with catalytic OsO_4 in the presence of NMO gives a mixture of diastereomeric diols **C** and **D**.

1) CH_2=CHMgBr 2) H_3O^+ → **A**; 1) NaH 2) CH_3I → **B**; cat. OsO_4, NMO → **C** + **D**

13.72. A new carbon-carbon bond must be made between C6 and C7:

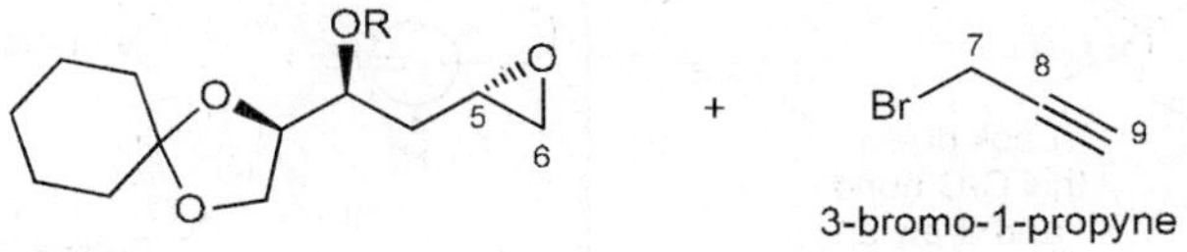

3-bromo-1-propyne

Epoxides are electrophilic functional groups and are subject to attack by a nucleophile. In order for the OH group to be ultimately positioned at C5, the nucleophile must attack the less substituted side of the epoxide (C6), which requires basic conditions (rather than acidic conditions, which often favors attack at the more substituted carbon).

The nucleophile for this reaction must be made from 3-bromo-1-propyne, which can be achieved by treatment with magnesium, thereby forming the Grignard reagent:

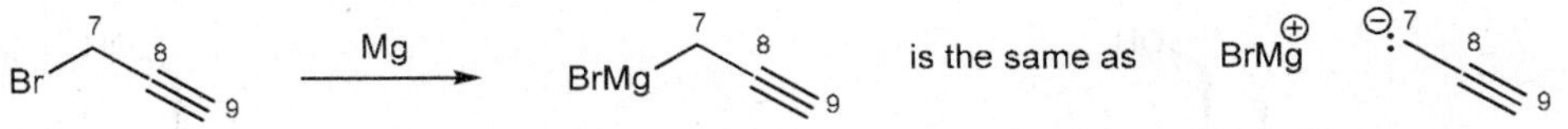

The desired transformation can therefore be achieved in the following way:

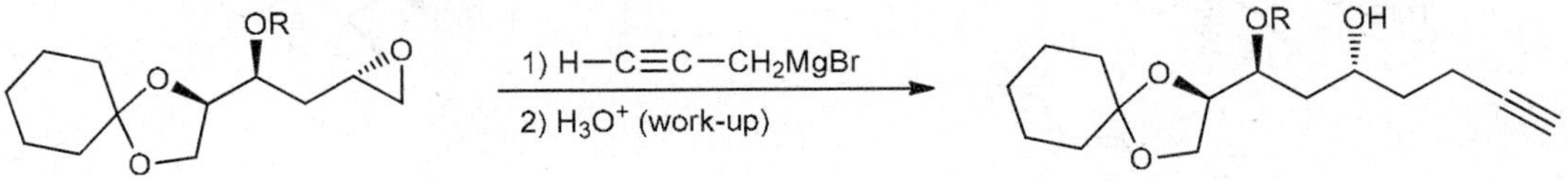

13.73. Let's begin by drawing the *trans*-decalin system in a way that illustrates the chair conformation of each six-membered ring. Assigning locants can be helpful in this situation, since they make it easier to place all of the substituents in the correct locations and in the correct configurations:

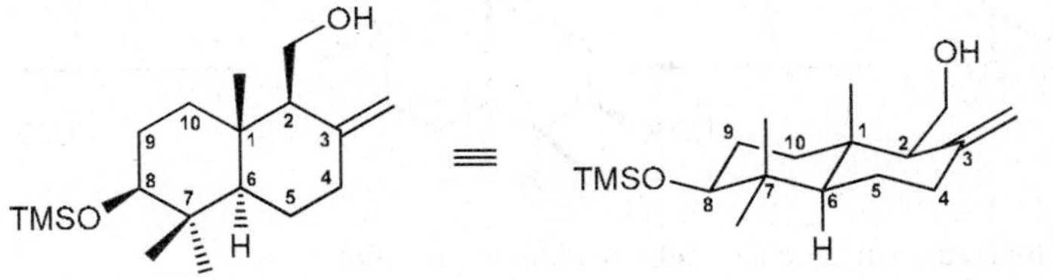

The *trans*-decalin structure imposes structural rigidity and limits the conformational freedom available to the compound.

Now let's consider the two faces of the π bond (the top face and the bottom face):

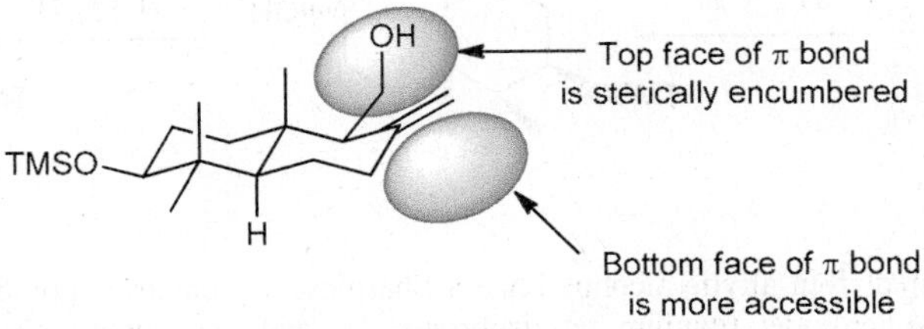

Top face of π bond is sterically encumbered

Bottom face of π bond is more accessible

These two faces of the π bond are not equally accessible. That is, the compound is confined to a conformation in which one face of the π bond is more sterically encumbered than the other. This steric consideration is difficult to see in the drawing above, and can be visualized more clearly if we draw a Newman projection, looking down the C2-C3 bond.

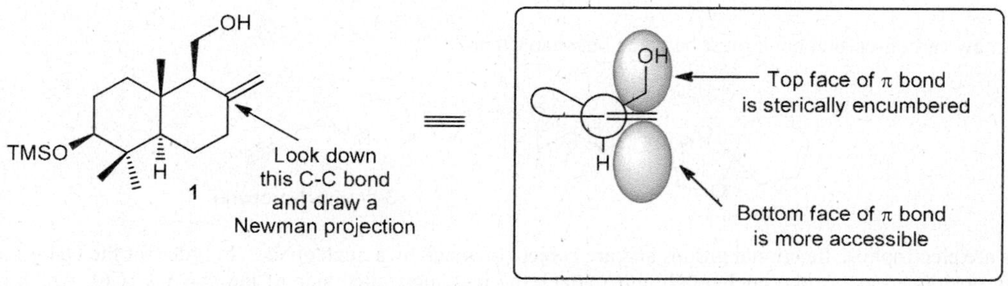

Since the bottom face of the π bond is more accessible, the epoxidation process will occur on that face, giving the following epoxide:

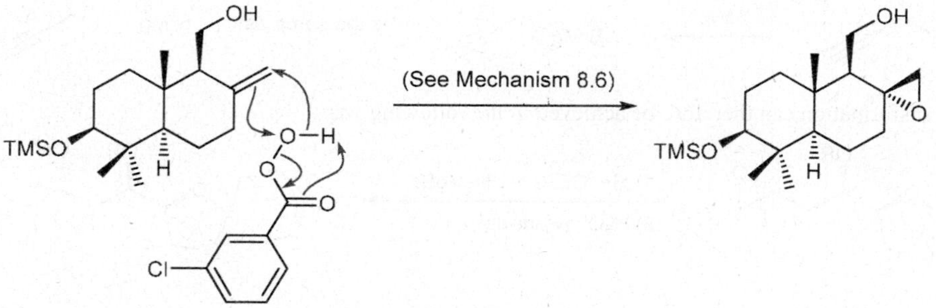

This epoxide is then opened with $LiAlH_4$, and under these conditions, it is likely that the primary OH group will be deprotonated, giving a dianion:

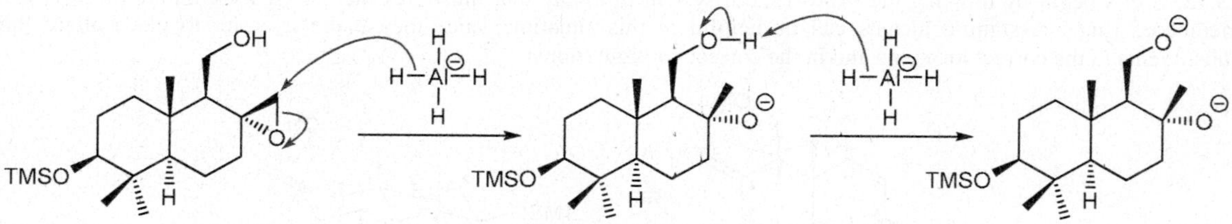

This dianion is then protonated upon aqueous acidic workup to give diol **2**:

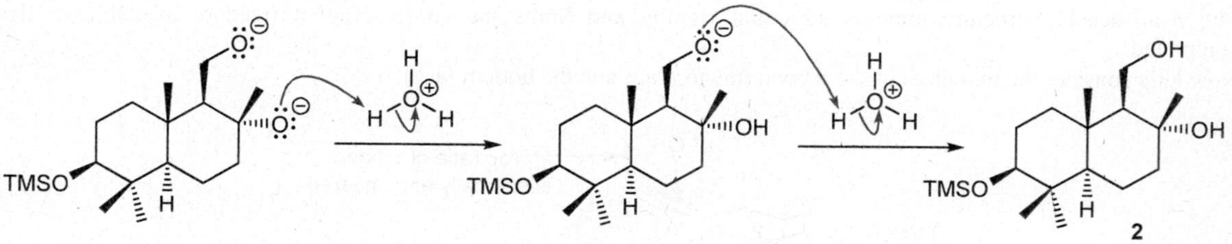

13.74.

(a) Epoxide **2** can be generated from allylic alcohol **1** via a Sharpless asymmetric epoxidation. The reagents for this reaction are: *tert*-butyl hydroperoxide, titanium tetraisopropoxide, and one enantiomer of diethyl tartrate (DET), depending on which epoxide enantiomer is required. In Figure 13.4, we saw a predictive tool for determining which enantiomer of DET would be required to afford epoxide **2**. Allylic alcohol **1** is orientated so that the allylic hydroxyl group appears in the upper right corner, and then we can see that (–)-DET is required in order for the epoxide ring to be formed on the bottom face of the molecule, corresponding to compound **2**.

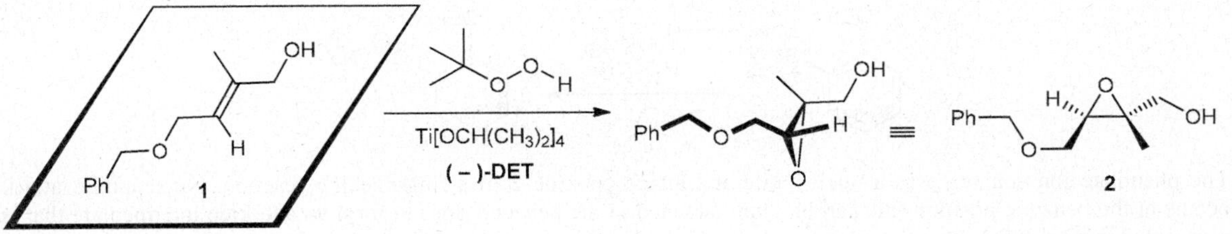

(b) In the first step, the epoxide group in compound **3** is protonated to form intermediate **5**. Notice that the epoxide is unsymmetrical; on the left side, the epoxide carbon is secondary, and on the right side, it is tertiary. Under acidic conditions, and when the epoxide has a tertiary carbon, the dominant effect is electronic (more important than steric considerations), and a nucleophile will attack at this site (Section 13.10). In this example, the nucleophile that opens the epoxide is the oxygen atom of the ester group, generating a resonance-stabilized intermediate (**6**) which exhibits a 5-membered ring. At this stage, the alcohol on the 2° carbon, which was generated from the epoxide opening, can attack to generate intermediate **7**. And finally, deprotonation of **7** generates compound **4**.

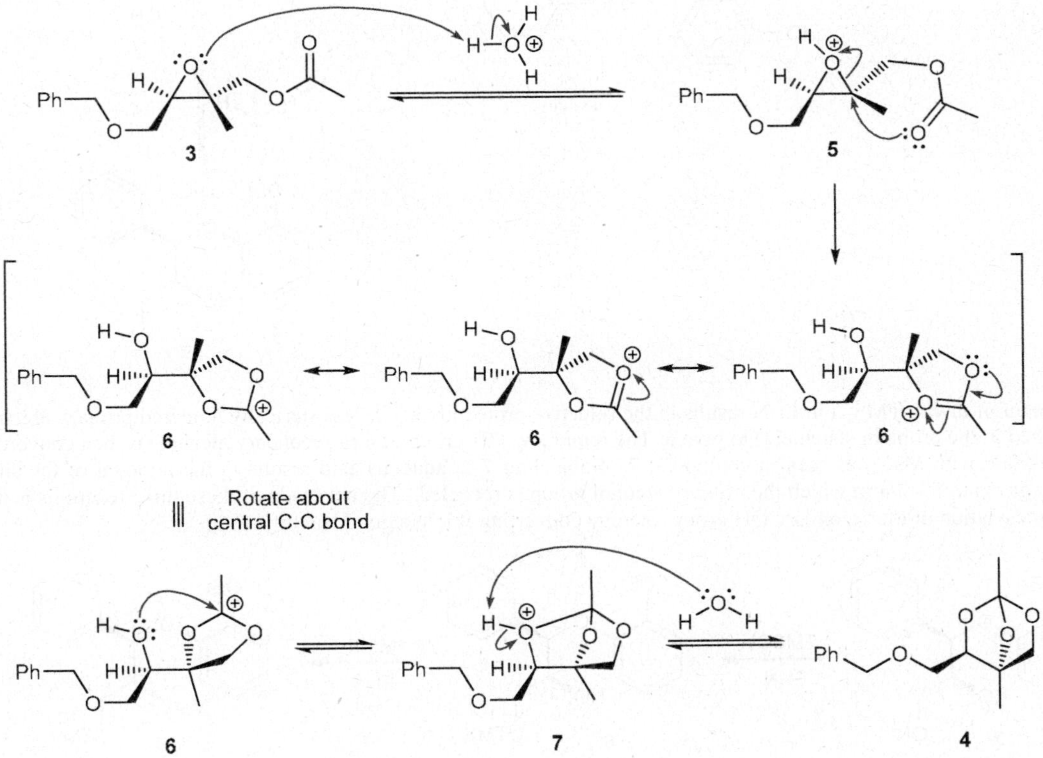

Rotate about central C-C bond

13.75.

(a) The conversion of **1** to **2** utilizes a Sharpless asymmetric epoxidation, so the product is expected to be an epoxide. The stereochemical outcome can be deduced by applying the paradigm provided in Figure 13.4. Since the chiral catalyst was formed using (–)-DIPT, which affords the same stereochemical outcome as (–)-DET, we can conclude that the epoxide forms "below the plane" when the alkene is drawn with the allylic hydroxyl group in the upper right corner:

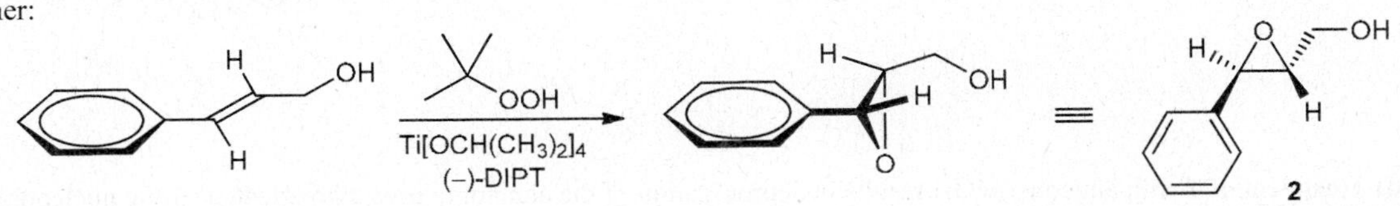

(b) First consider the reagents. Sodium hydroxide will deprotonate 2-ethoxyphenol to give a phenolate ion (shown below).

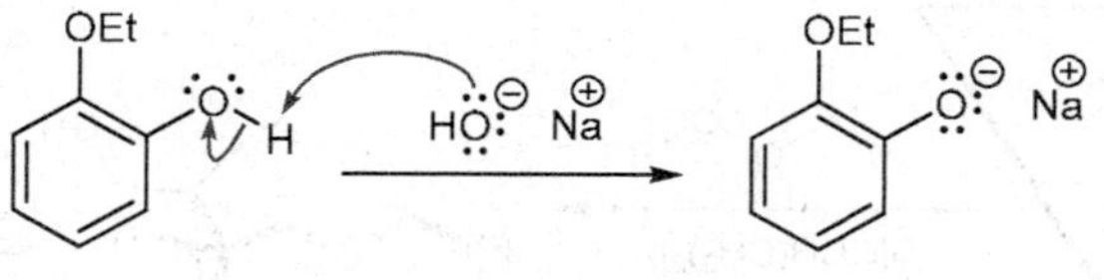

The phenolate ion can serve as a nucleophile and attack epoxide **2** in a ring-opening reaction. Nucleophilic attack occurs at the benzylic position (the carbon atom attached to the benzene ring) to form an alkoxide intermediate that is protonated to give diol **3**.

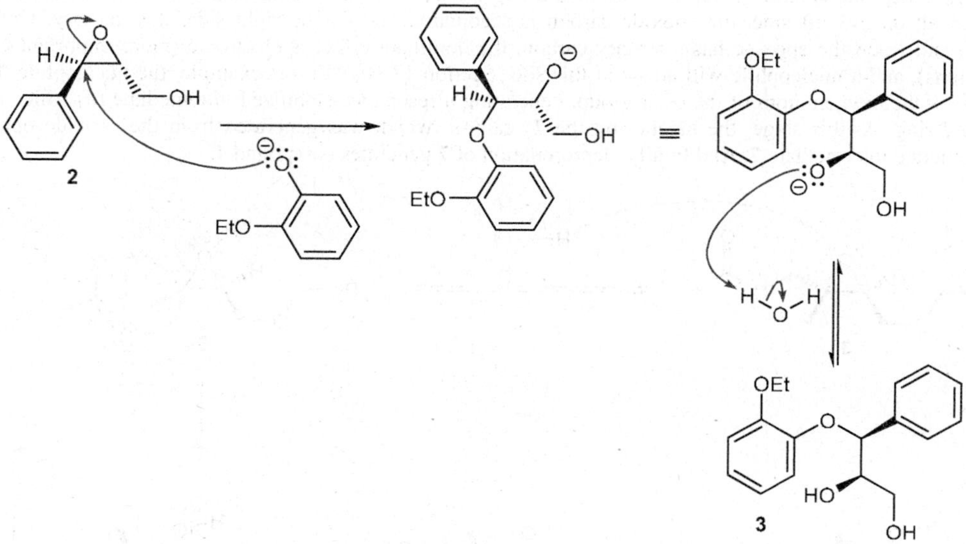

(c) Treatment of **3** with TMSCl in Et_3N results in the selective protection of the less sterically hindered primary alcohol (as described in the problem statement) to give **6**. The remaining OH group of **6** (a secondary alcohol) is then converted to the mesylate with MsCl, as seen in compound **7**. Subjecting **7** to aqueous acid results in the removal of the silyl protecting group to form **4** in which the primary alcohol group is revealed. The net result of these three reactions is the selective mesylation of the secondary OH group, thereby converting it into a good leaving group.

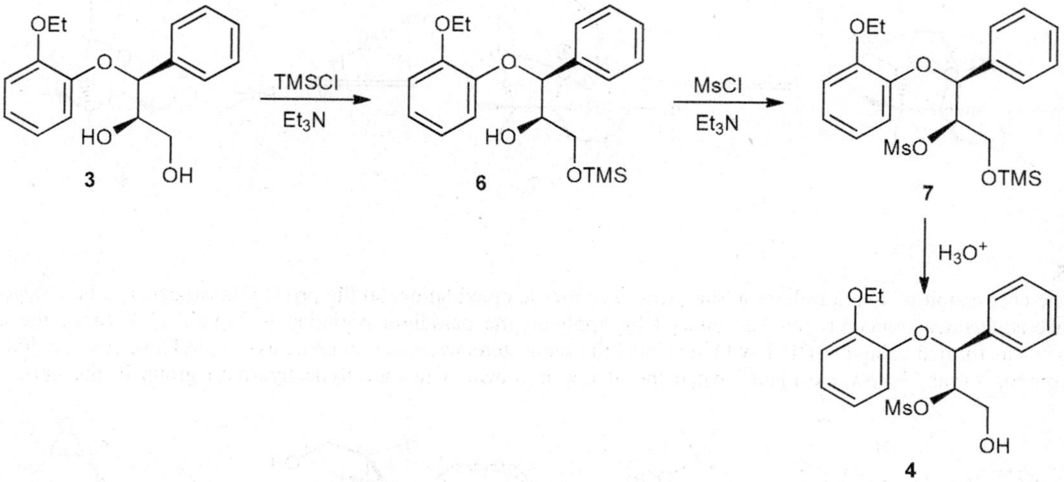

(d) Treatment of **4** with aqueous NaOH results in deprotonation of the alcohol to give alkoxide **8**, a strong nucleophile. This facilitates an intramolecular ring-forming reaction, via an S_N2-like back-side attack with inversion of configuration at the carbon atom bearing the mesylate leaving group, to give compound **5**.

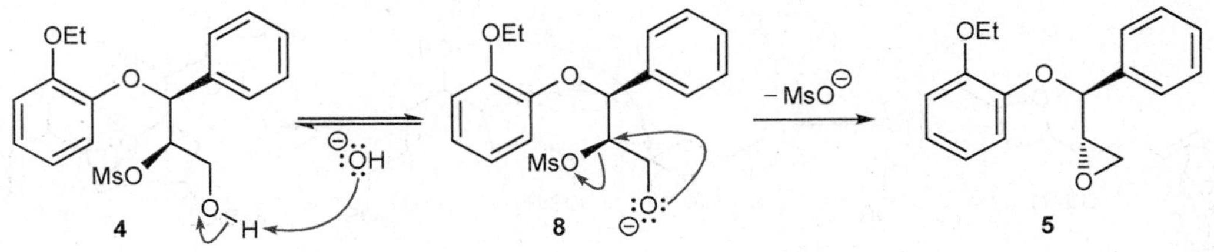

13.76.

(a) As described in the problem statement, the alcohol group is first deprotonated to give an alkoxide ion, which functions as a nucleophile and participates in an intramolecular SN2-type process in which the epoxide ring is opened and a new epoxide is formed (notice that back-side attack places the oxygen atom of the new epoxide on a dash). After protonation of the resulting alkoxide, the epoxide ring can then be opened with a thiophenolate ion (PhS^-) to give another alkoxide, which is then protonated to give the observed product.

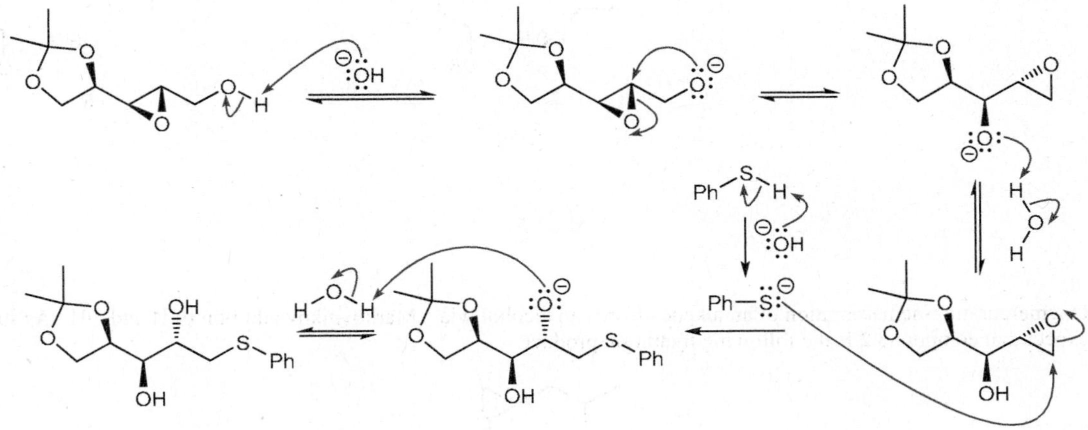

(b) The allylic alcohol undergoes a Sharpless asymmetric epoxidation to give the following epoxide, which then undergoes the transformation explored in part (a).

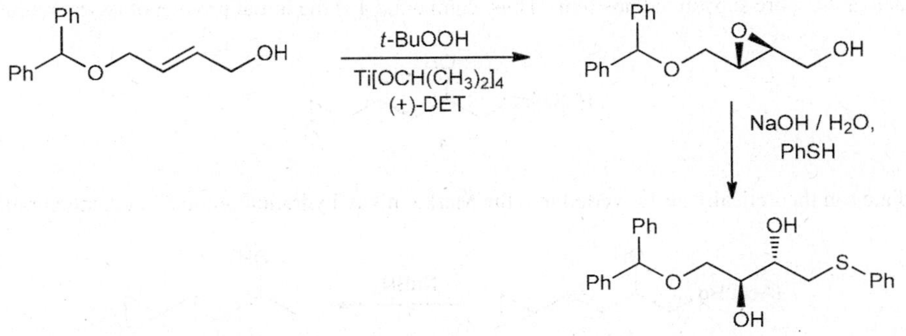

13.77. When compound **1** is treated with an electrophilic source of bromine, the alkene will readily react to form bromonium ion **3**. At this stage there are two possibilities for further reactivity. Typically, with an OH group present in the compound, the OH group can function as a nucleophile and attack the bromonium ion to form a cyclic ether. For example, consider path A. If this were to occur, an 8- or 9-membered ring would be formed, depending on which side of the bromonium ion is attacked by the alcohol. However, since we do not see the formation of a medium-sized ring in the product, a different reaction pathway must be occurring. Consider the lone pairs on the oxygen atom which is part of the 4-membered cyclic ether. If one of these lone pairs were to attack the bromonium ion (path B), the result would be a 5-membered ring fused to the 4-membered ring (compound **4**). However, this intermediate is not stable – the bridgehead oxygen atom has 3 bonds, making it positively charged and an excellent leaving group. At this stage, the nearby OH group reacts in an SN2-type fashion to open the 4-membered ring. Deprotonation of the resulting oxonium ion furnishes the observed product, epoxide **2**:

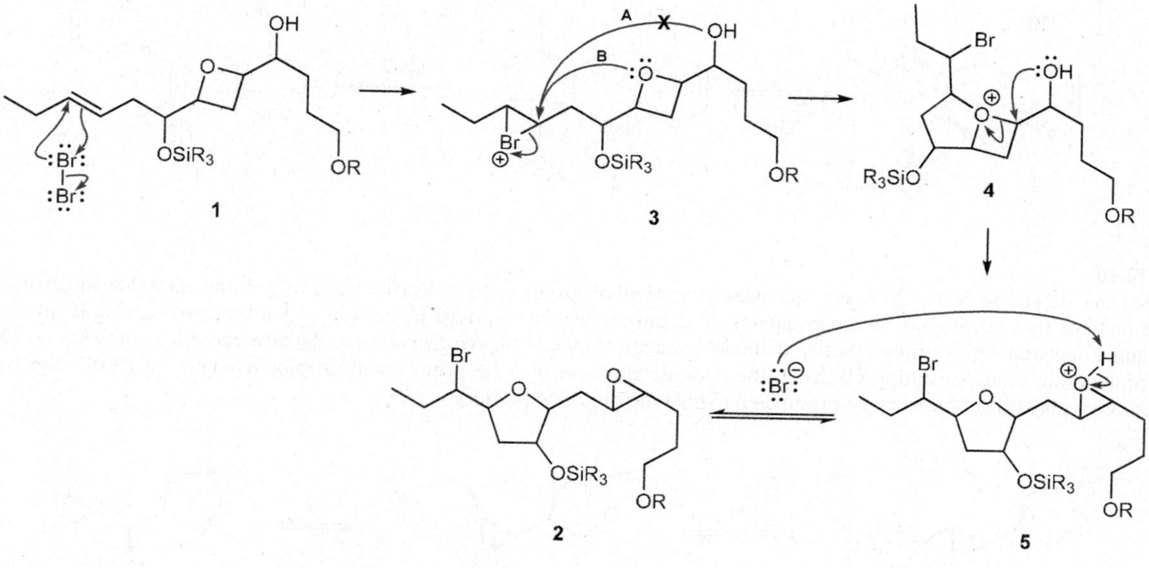

13.78.

(a) Oxymercuration-demercuration of an alkene affords an alcohol, via a Markovnikov addition of H and OH. As such, we expect that compound **2** is the following hydration product:

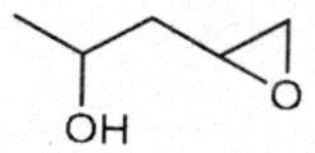

(b) Oxymercuration of compound **3** results in the addition of OH and HgOAc across the π bond, with the OH group being positioned at the more substituted position. Thus, compound **4** is the initial product of oxymercuration:

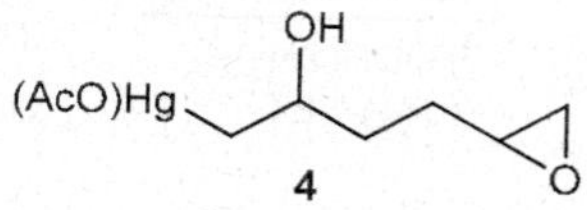

This intermediate can theoretically be converted into the Markovnikov hydration product via demercuration.

However, before $NaBH_4$ is introduced, **4** can also continue to react under the oxymercuration conditions, where $Hg(OAc)_2$ can behave as a Lewis acid. As shown, the $Hg(OAc)_2$ can interact with the oxygen atom of the epoxide group, much like an epoxide accepts a proton. This makes the epoxide ring more electrophilic and facilitates the intramolecular nucleophilic attack at the more substituted position by the nearby hydroxyl group. This intramolecular cyclization reaction forms a five-membered ring. Deprotonation produces compound **5**, and demercuration of **5** affords one of the major cyclization products, **8**.

The other cyclization product, **7**, is a result of the intramolecular epoxide ring-opening occurring at the less substituted position, forming a six-membered ring:

(c) Oxymercuration-demercuration of **1** can only result in a cyclic product if a four-membered ring is formed. Four-membered rings exhibit significant ring strain so their formation is slow under these conditions.

13.79. In the first step, iodine (the electrophile) adds to the terminal alkene to form an iodonium ion. Once formed, intramolecular attack of this intermediate by the methoxy group occurs, resulting in the formation of a 6-membered ring. This intermediate, however, is not stable and is quickly trapped by the iodide ion, resulting in ring opening to form a stable product. In the second step, thiophenol displaces the iodide in an S_N2 fashion, followed by deprotonation of the resulting intermediate using triethylamine as a base.

Step 1

Step 2

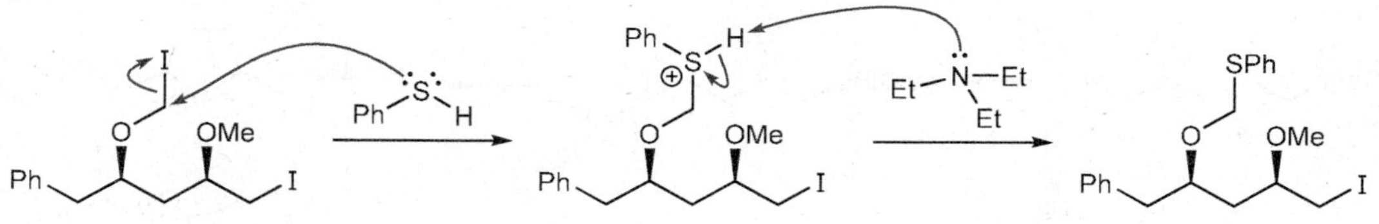

13.80. The epoxide in compound **1** has a tertiary carbon on the top side, and a primary carbon on the bottom side. Under acidic conditions, we have learned that electronic effects override steric effects when there is a tertiary carbon on one side of the epoxide. If this line of reasoning were to govern product formation in this example, we would expect the alcohol to attack the top carbon. However, we know from looking at the product of the ring-opening (compound **2**) that the alcohol must have attacked the primary carbon. Also of note is the stereochemistry at the ring opening site. After epoxide **1** is protonated, hexafluoro-2-propanol is expected to approach intermediate **3** from the bottom face of the molecule, resulting in compound **6**. Again, this is not what is observed:

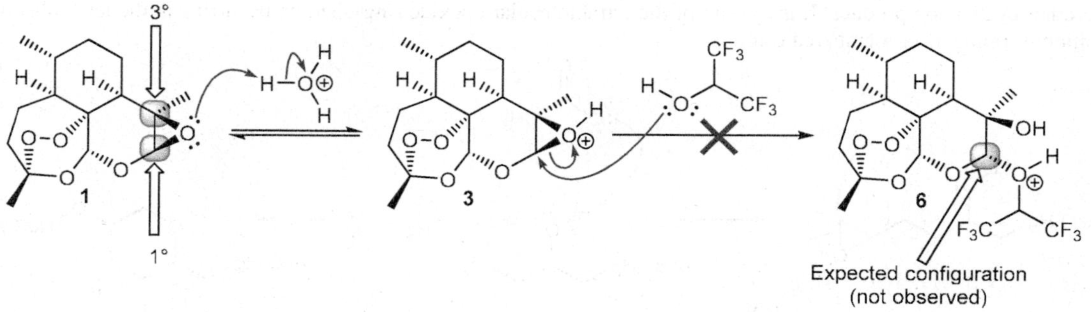

Expected configuration (not observed)

As it turns out, when this epoxide is protonated, the ring is able to open up (loss of leaving group), because it produces a carbocation that is resonance-stabilized (**4a**/**4b**). This cation can then be trapped by hexafluoro-2-propanol, which preferentially approaches from the top face of the molecule (the bottom face is blocked by the nearby methyl group and peroxide bridge), producing intermediate **5**. Finally, loss of a proton forms alcohol **2**. Overall, this nucleophilic substitution follows a stepwise path via a carbocation intermediate (SN1) rather than a concerted back-side attack (SN2). This explains the unexpected stereochemical outcome (no inversion of configuration).

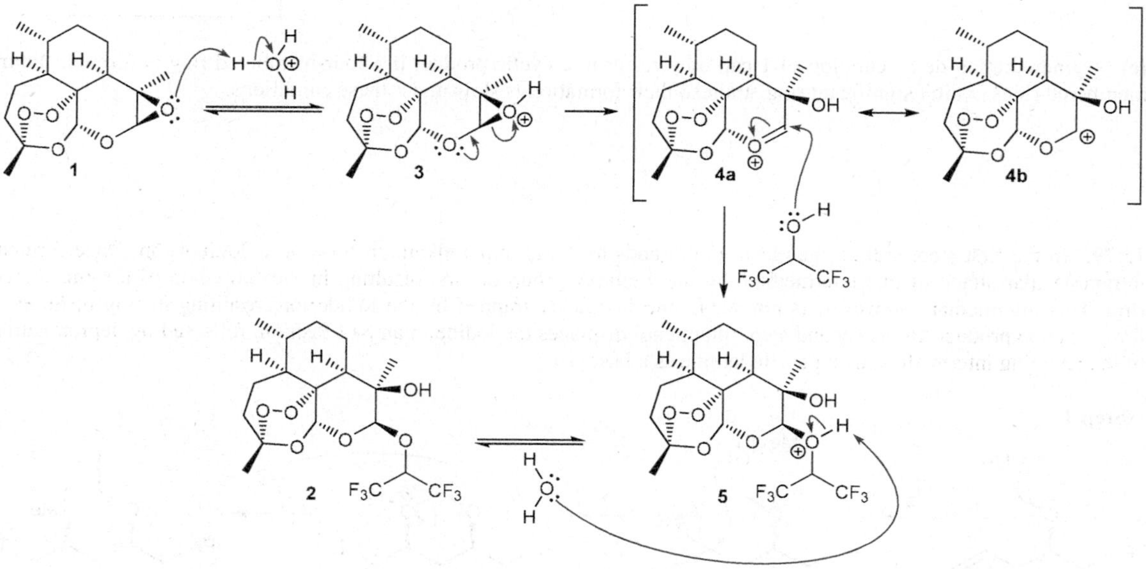

Notice that this example does not violate the principles underlying electronic trends (attack at secondary position rather than the tertiary position). Specifically, when there is an oxygen atom attached directly to one side of the epoxide, under acidic conditions, ring opening will predominantly occur at this site, *because it results in a highly stable (resonance-stabilized) cation.*

Chapter 14
Infrared Spectroscopy and Mass Spectrometry

Review of Concepts

Fill in the blanks below. To verify that your answers are correct, look in your textbook at the end of Chapter 14. Each of the sentences below appears verbatim in the section entitled *Review of Concepts and Vocabulary*.

- **Spectroscopy** is the study of the interaction between _______ and ________.
- The difference in energy (ΔE) between vibrational energy levels is determined by the nature of the bond. If a photon of light possesses exactly this amount of energy, the bond can absorb the photon to promote a _________________ **excitation**.
- IR spectroscopy can be used to identify which ________________ are present in a compound.
- The location of each signal in an IR spectrum is reported in terms of a frequency-related unit called _________________.
- The wavenumber of each signal is determined primarily by bond __________ and the __________ of the atoms sharing the bond.
- The intensity of a signal is dependent on the _______________ of the bond giving rise to the signal.
- ________________ C=C bonds do not produce signals.
- Primary amines exhibit two signals resulting from ___________ **stretching** and _____________ **stretching.**
- **Mass spectrometry** is used to determine the ____________________ and ________________________ of a compound.
- **Electron impact ionization (EI)** involves bombarding the compound with high energy _____________, generating a radical cation that is symbolized by $(M)^{+\bullet}$ and is called the **molecular ion,** or the __________ **ion.**
- Only the molecular ion and the cationic fragments are deflected, and they are then separated by their _________________ (*m/z*).
- The tallest peak in a mass spectrum is assigned a relative value of 100% and is called the __________ **peak.**
- The relative heights of the $(M)^{+\bullet}$ peak and the $(M+1)^{+\bullet}$ peak indicates the number of ___________________.
- A signal at M–15 indicates the loss of a _________ group; a signal at M–29 indicates the loss of an _________ group.
- ______________ alkanes have a molecular formula of the form C_nH_{2n+2}.
- Each double bond and each ring represents one **degree of** ______________.

Review of Skills

Fill in the blanks and empty boxes below. To verify that your answers are correct, look in your textbook at the end of Chapter 14. The answers appear in the section entitled *SkillBuilder Review*.

14.1 Analyzing an IR Spectrum

Step 1 Look for __________ bonds between 1600 and 1850 cm^{-1}	**Step 2** Look for __________ bonds between 2100 and 2300 cm^{-1}	**Step 2** Look for __________ bonds between 2750 and 4000 cm^{-1}
Guidelines: C=O bonds produce ________ signals C=C bonds generally produce ______ signals. Symmetrical C=C bonds do not appear at all.	**Guidelines:** ________ triple bonds do not produce signals	**Guidelines:** Draw a line at 3000 cm-1, and look for ________ or ________ C-H bonds to the left of the line. The shape of an O-H signal is affected by ____________ (due to H-bonding) Primary amines exhibit two N-H signals (__________ and _________ stretching)

14.2 Distinguishing Between Two Compounds Using IR Spectroscopy

Step 1 Work methodically through the expected ________________ of each compound.	**Step 2** Determine if any ____________ will be present for one compound but absent for the other.	**Step 3** For each expected signal, compare for any possible differences in ____________, ______________, or ______________.

14.3 Using the Relative Abundance of the $(M+1)^{+\bullet}$ Peak to Propose a Molecular Formula

Step 1 Fill in the boxes below to complete the formula that can be used to determine the number of carbon atoms in a compound using mass spectrometry: $$\frac{\left(\frac{\text{Abundance of } \square \text{ peak}}{\text{Abundance of } \square \text{ peak}}\right) \times 100\%}{1.1\ \%}$$	**Step 2** Analyze the mass of the molecular ion to determine if any ______________ are present.

14.4 Calculating HDI

Step 1 Rewrite the molecular formula "as if" the compound had no elements other than C and H, using the following rules: - Add one H for each ______________. - Ignore all ______________ atoms. - Subtract one H for each ______________.	**Step 2** Determine whether any H's are missing. Every two H's represent one degree of unsaturation: C_4H_9Cl ⟶ HDI = _____ C_4H_8O ⟶ HDI = _____ C_4H_9N ⟶ HDI = _____

Mistakes to Avoid

We have seen that the IR spectrum of an alkene will exhibit a signal near 1650 cm^{-1} (the characteristic signal for a C=C bond) if there is a dipole moment associated with the C=C bond. In such a case, the dipole moment changes as the C=C bond vibrates, creating an oscillating electric field that serves as an antenna to absorb the appropriate frequency of IR radiation. If the C=C bond does not have a dipole moment, then it cannot efficiently absorb IR radiation, and the signal near 1650 cm^{-1} will be absent. For example, consider the following two compounds:

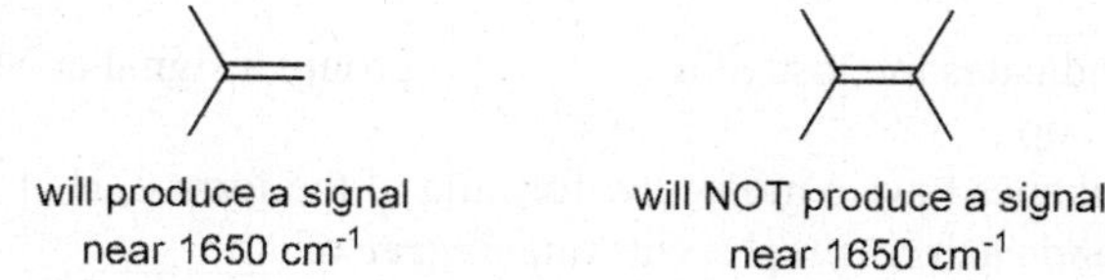

Don't be confused by the terms 'symmetrical alkene' and 'unsymmetrical alkene'. We might refer to the first compound as an unsymmetrical alkene (in reference to the dipole moment), but the truth is that this alkene still does possess some symmetry (an axis of symmetry, as well as two planes of symmetry).

Axis of symmetry

Plane of symmetry

Plane of symmetry

But these symmetry elements are not relevant for determining whether a C=C bond has a dipole moment, and therefore, they are not relevant for determining whether or not a C=C bond will produce a signal in an IR spectrum. When we refer to a symmetrical alkene, we are referring to the symmetry of the two vinylic positions:

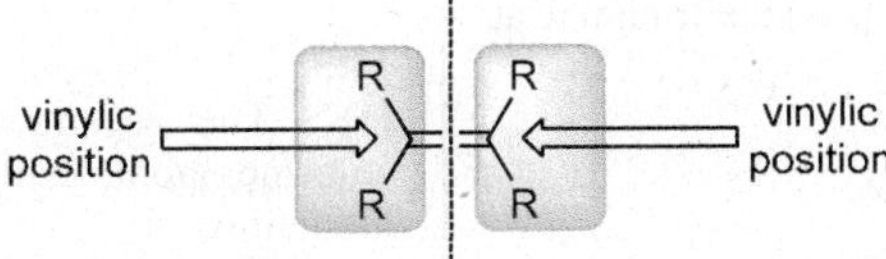

With this in mind, let's consider the following alkene:

While this molecule does possess some symmetry, you should avoid falling into the trap of calling it a symmetrical alkene and erroneously deciding that the C=C double bond will not produce a signal. In fact, the dipole moment for this C=C bond is expected to be quite large (because of the combined inductive effects of the chlorine atoms. One vinylic position (the one connected to the two chlorine atoms) is more electron-deficient (δ+) than the other vinylic position. As a result, this C=C bond is expected to produce a rather strong signal in the IR spectrum.

Solutions

14.1.

(a) The C–H bond is expected to produce the signal with the largest wavenumber, because bonds to H typically produce high-energy signals (due to the low mass of the hydrogen atom). Among the remaining two bonds, the triple bond is stronger than the double bond, so we expect the double bond to produce the signal with lowest wavenumber.

Increasing wavenumber

C–H C≡C C=C

(b) Each of the bonds in this case is a single bond. The C–H bond is expected to produce the signal with the larger wavenumber, because bonds to H typically produce high-energy signals (due to the low mass of the hydrogen atom).

Increasing wavenumber

C–H C–C

14.2.

(a) This compound exhibits an sp^2-hybridized carbon atom that is connected to a hydrogen atom. As such, this C–H bond (highlighted) should produce a signal above 3000 cm^{-1} (at approximately 3100 cm^{-1}).

> 3000 cm^{-1}

(b) This compound has three sp^2-hybridized carbon atoms, but none of them are connected to hydrogen atoms. And there are no *sp*-hybridized carbon atoms. Therefore, we do not expect a signal above 3000 cm^{-1}.

(c) This compound (an internal alkyne) has two *sp*-hybridized carbon atoms, but neither of them are connected to hydrogen atoms. And there are no sp^2-hybridized carbon atoms. Therefore, we do not expect a signal above 3000 cm^{-1}.

(d) This compound exhibits an sp^2-hybridized carbon atom that is connected to a hydrogen atom. As such, this C-H bond (highlighted) should produce a signal above 3000 cm^{-1} (at approximately 3100 cm^{-1}).

> 3000 cm^{-1}

(e) This compound has two sp^2-hybridized carbon atoms, but neither of them are connected to hydrogen atoms. And there are no *sp*-hybridized carbon atoms. Therefore, we do not expect a signal above 3000 cm^{-1}.

(f) This compound (a terminal alkyne) exhibits an *sp*-hybridized carbon atom that is connected to a hydrogen atom. As such, this C-H bond (highlighted) should produce a signal above 3000 cm^{-1} (at approximately 3300 cm^{-1}).

> 3000 cm^{-1}

14.3.

(a) Both carbonyl groups are ketones. One of the carbonyl groups (upper left) is not conjugated, so it is expected to produce a signal at approximately 1720 cm^{-1}. The other carbonyl group (bottom right) is conjugated to a C=C π bond, so it is expected to produce a signal at approximately 1680 cm^{-1}.

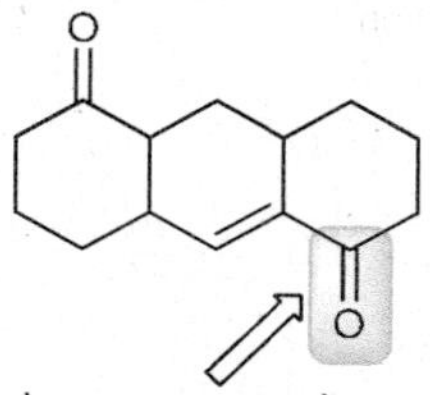

lower wavenumber
(carbonyl group is conjugated)

(b) Both carbonyl groups are esters. One of the ester groups (bottom right) is not conjugated, so it is expected to produce a signal at approximately 1735 cm^{-1}. The other carbonyl group (upper left) is conjugated to a C=C π bond, so it is expected to produce a signal at approximately 1710 cm^{-1}.

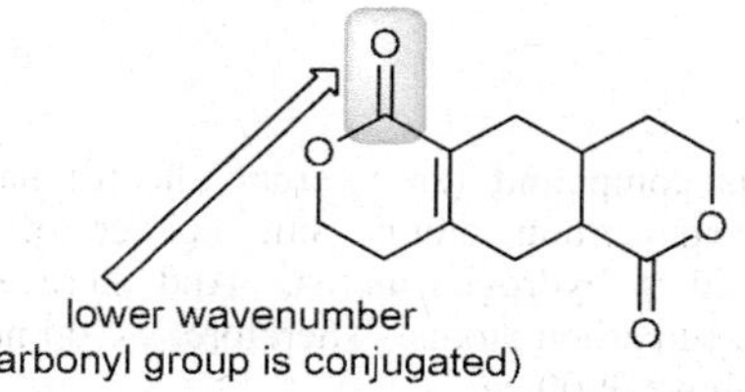

lower wavenumber
(carbonyl group is conjugated)

(c) The carbonyl group of a ketone is expected to produce a signal at approximately 1720 cm^{-1}, while the carbonyl group of an ester is expected to produce a signal at approximately 1735 cm^{-1}.

lower wavenumber
(a ketone)

higher wavenumber
(an ester)

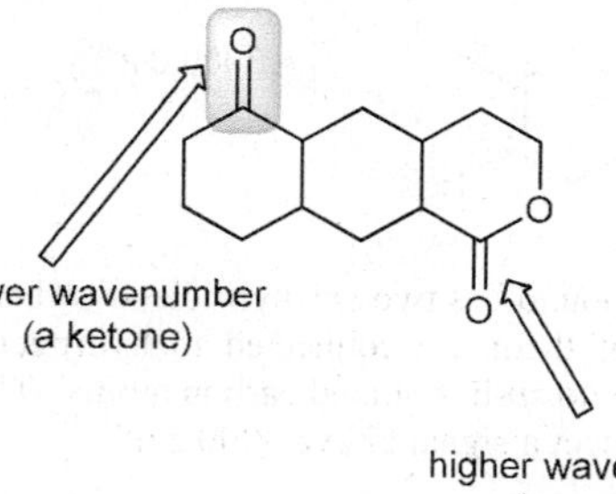

14.4. The C=C π bond in the conjugated compound produces a signal at lower wavenumber (1600 cm^{-1}) because it has some single-bond character, as seen in the third resonance structure below. This additional single bond character renders the conjugated C=C π bond weaker (relative to the nonconjugated C=C π bond of the other compound, which does not exhibit any single bond character).

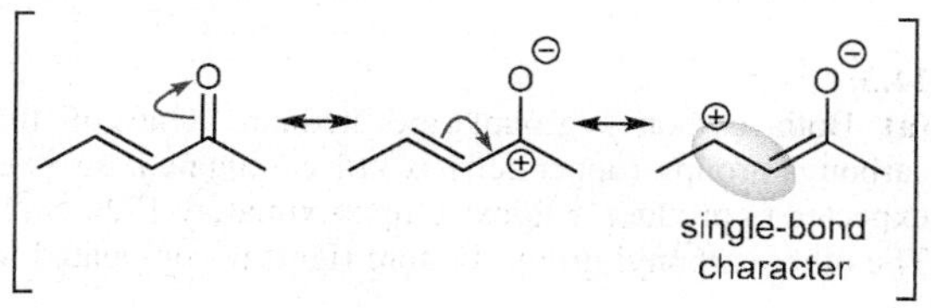

single-bond
character

14.5.

(a) The second compound has an electronegative chlorine atom, which withdraws electron density via induction.

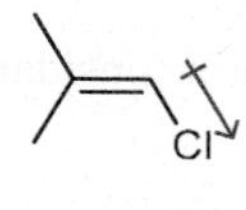

Therefore, the C=C bond in this compound has a larger dipole moment than the C=C bond in the other compound. As a result, we expect the chloroalkene to be more efficient at absorbing IR radiation (thereby producing a stronger signal).

(b) The C=C bond in the compound shown below will have a larger dipole moment because one carbon atom is connected to two chlorine atoms while the other carbon atom is not directly connected to any chlorine atoms. As a result, the two carbon atoms are in very different electronic environments, giving rise to a large dipole moment. We therefore expect this C=C bond to be more efficient at absorbing IR radiation (and therefore produce a stronger signal).

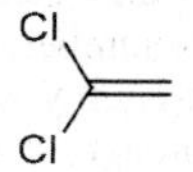

14.6. If we draw all significant resonance structures of 2-cyclohexenone, we see that one of the alkene carbon atoms is electron-deficient (highlighted in the third resonance structure):

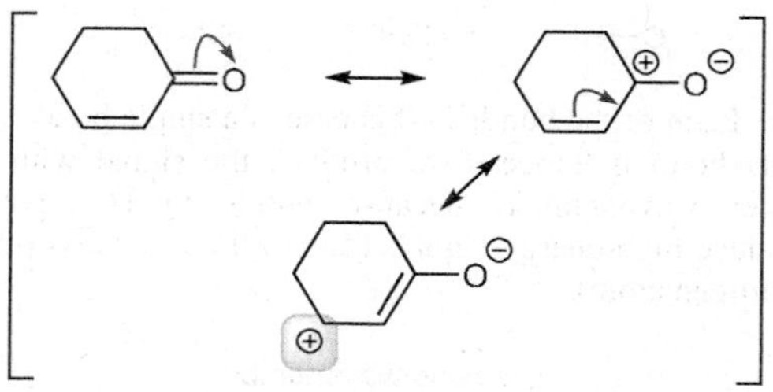

As a result, the two alkene carbon atoms experience very different electronic environments, giving rise to a large dipole moment. With a large dipole moment, this C=C bond is expected to be very efficient at absorbing IR radiation, thereby producing a strong signal.

14.7. The vinylic C–H bond should produce a signal at approximately 3100 cm^{-1}.

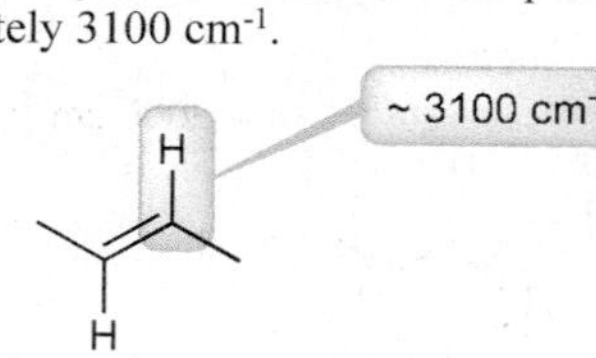

~ 3100 cm^{-1}

14.8. The narrow signal is produced by the O–H stretching in the absence of a hydrogen-bonding effect. The broad signal is produced by O–H stretching when hydrogen bonding is present. Hydrogen bonding effectively lowers the bond strength of the O–H bonds, because each hydrogen atom is slightly pulled away from the oxygen atom to which it is connected. A longer bond length (albeit temporary) corresponds with a weaker bond, which corresponds with a lower wavenumber.

14.9.

(a) The broad signal between 3200 and 3600 cm^{-1} is characteristic of an alcohol (ROH).

(b) This spectrum lacks broad signals above 3000 cm^{-1}, so the compound is neither an alcohol nor a carboxylic acid (both of which produce broad signals that reach as high as 3600 cm^{-1}).

(c) The extremely broad signal that extends from 2200 to 3600 cm^{-1} is characteristic of the O-H stretching of a carboxylic acid (RCO_2H). The signal just above 1700 cm^{-1} is also consistent with a carboxylic acid (for the C=O bond of the carboxylic acid).

(d) This spectrum lacks broad signals above 3000 cm^{-1}, so the compound is neither an alcohol nor a carboxylic acid (both of which produce broad signals that reach as high as 3600 cm^{-1}).

(e) The broad signal between 3100 and 3600 cm^{-1} is characteristic of an alcohol (ROH).

(f) The extremely broad signal that extends from 2200 to 3600 cm^{-1} is characteristic of the O–H stretching of a carboxylic acid (RCO_2H). The signal around 1700 cm^{-1} is also consistent with a carboxylic acid (for the C=O bond of the carboxylic acid).

14.10.

(a) The strong signal just above 1700 cm^{-1} is consistent with the stretching of the carbonyl group (C=O) of a ketone.

(b) The extremely broad signal that extends from 2200 to 3600 cm^{-1} is characteristic of the O–H stretching of a carboxylic acid (RCO_2H). The signal just above 1700 cm^{-1} is also consistent with a carboxylic acid (for the C=O bond of the carboxylic acid).

(c) The signal at approximately 3400 cm^{-1} is consistent with the stretching of the N–H bond of a secondary amine.

(d) The two signals at 3350 and 3450 cm^{-1} are consistent with the symmetric and asymmetric stretching of the N–H bonds of an NH_2 group. This indicates a primary amine.

(e) The strong signal just above 1700 cm^{-1} is consistent with the stretching of the carbonyl group (C=O) of a ketone.

(f) The broad signal between 3200 and 3600 cm^{-1} is characteristic of an alcohol (ROH).

14.11. The C_{sp^3}–H bonds of each methyl group can stretch symmetrically, asymmetrically, and in various combinations with respect to each other. Each one of these possible stretching modes is associated with a different wavenumber of absorption, giving a large number of overlapping peaks.

14.12.

(a) Begin by drawing a line at 1500 cm^{-1} and ignoring everything to the right (the fingerprint region). Then, look for any signals associated with double bonds (1600–1850 cm^{-1}) or triple bonds (2100–2300 cm^{-1}). In this case, there is a weak signal between 1600 and 1700 cm^{-1}, consistent with an alkene. Finally, we draw a line at 3000 cm^{-1}, and we look for signals to the left of this line. In this case, there is a signal at approximately 3100 cm^{-1}, which is consistent with a C_{sp^2}–H bond of an alkene. Among the possible structures, the alkene is the structure that is consistent with the signals in the spectrum.

(b) Begin by drawing a line at 1500 cm^{-1} and ignoring everything to the right (the fingerprint region). Then, look for any signals associated with double bonds (1600–1850 cm^{-1}) or triple bonds (2100–2300 cm^{-1}). In this case, there is a strong signal between 1700 and 1800 cm^{-1}, consistent with a carbonyl group. Among the possible structures, only two of them exhibit C=O bonds. One of these structures has carboxylic acid groups, and the spectrum does not match that compound (because that compound is expected to give a broad O–H stretch from 2200-3600 cm^{-1}, which is absent in our spectrum). The following structure (an ester) is consistent with the IR spectrum.

(c) Begin by drawing a line at 1500 cm^{-1} and ignoring everything to the right (the fingerprint region). Then, look for any signals associated with double bonds (1600–1850 cm^{-1}) or triple bonds (2100–2300 cm^{-1}). In this case, there are none. Next, we draw a line at 3000 cm^{-1}, and we look for signals to the left of this line. In this case, there are none. With no characteristic signals for any functional groups, this spectrum is consistent with an alkane.

(d) The broad signal between 3200 and 3600 cm^{-1} is characteristic of an alcohol (ROH). There is only one alcohol among the possible structures given.

(e) The extremely broad signal that extends from 2200 to 3600 cm^{-1} is characteristic of the O–H stretching of a carboxylic acid (RCO_2H). The signal just above 1700 cm^{-1} is also consistent with a carboxylic acid (for the C=O bond of a carboxylic acid group). Notice that this signal appears to be comprised of two overlapping signals, which can likely be attributed to symmetric and asymmetric stretching of the two carbonyl groups.

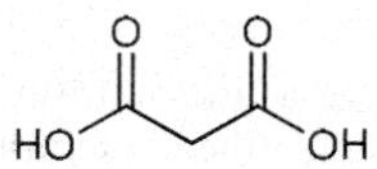

(f) The two signals at 3350 and 3450 cm^{-1} are consistent with the symmetric and asymmetric stretching of the N–H bonds of an NH_2 group. There is only one primary amine among the possible structures given.

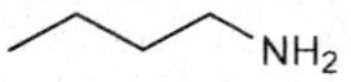

14.13. The following five signals are expected (presented in order of increasing wavenumber):
1) The C=C bond (expected to be ~ 1650 cm^{-1})
2) The C=O bond of the carboxylic acid group (expected to be ~ 1715 cm^{-1})
3) All C_{sp^3}–H bonds (expected to be < 3000 cm^{-1})
4) The C_{sp^2}–H bond (expected to be ~ 3100 cm^{-1})
5) The O–H bond of the carboxylic acid group (expected to be 2200 - 3600 cm^{-1})

14.14.

(a) The starting material is an alcohol and is expected to produce a typical signal for an O–H stretch (a broad signal between 3200 and 3600 cm^{-1}). In contrast, the product is a carboxylic acid and is expected to produce an even broader O–H signal (2200-3600 cm^{-1}) as a result of more extensive hydrogen bonding. In addition, the product can be differentiated from the starting material by looking for a signal at around 1715 cm^{-1}. The carboxylic acid product has a C=O bond and should exhibit this signal. The alcohol starting material lacks a C=O bond and will not show a signal at 1715 cm^{-1}.

(b) The starting material is an unsymmetrical alkyne and is expected to produce a signal at around 2200 cm^{-1}. In contrast, the product is an unsymmetrical alkene and is expected to produce a signal at around 1600 cm^{-1}. Also, the alkene product has C_{sp^2}–H bonds that are absent in the alkyne starting material. Therefore, the product is expected to have signals at around 3100 cm^{-1}, and the starting material will have no signal in that region.

(c) The C≡C bond in the starting material and the C=C bond in the product are each symmetrical and will not produce signals. However, the alkene product has C_{sp^2}–H bonds that are absent in the alkyne starting material. Therefore, the product is expected to have a signal at around 3100 cm^{-1}, and the starting material will have no signal in that region.

(d) The starting material will have two signals in the double-bond region (1600–1850 cm^{-1}): one for the C=O bond and one for the C=C bond. The product only has one signal in the double-bond region (1600–1850 cm^{-1}). It only has the signal for the C=O bond, which is expected to be at a higher wavenumber than the C=O signal in the starting material (because the C=O bond is not conjugated in the product).

14.15. The starting material has a cyano group (C≡N) and the triple bond is expected to produce a signal at approximately 2200 cm^{-1}. In contrast, the product is a carboxylic acid and is expected to produce a broad signal from 2200–3600 cm^{-1}, as well as a signal at 1715 cm^{-1} for the C=O bond.

14.16. The ketone starting material has a C=O signal at 1720 cm^{-1}, and that signal should be absent if the reaction was successful. The presence of the alcohol product can be confirmed by the appearance of an O–H signal at 3200 – 3600 cm^{-1}.

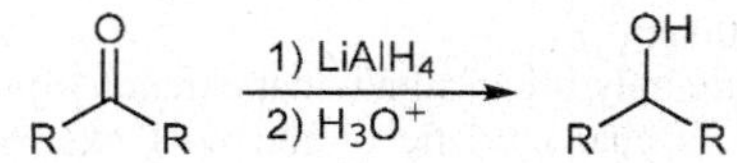

$$\text{R}_2\text{C=O} \xrightarrow[\text{2) } H_3O^+]{\text{1) } LiAlH_4} \text{R}_2\text{CH–OH}$$

14.17. 1-Chlorobutane is a primary substrate. When treated with sodium hydroxide, S_N2 substitution is expected to dominate over E2 elimination (see Chapter 7), but both products are expected to be obtained:

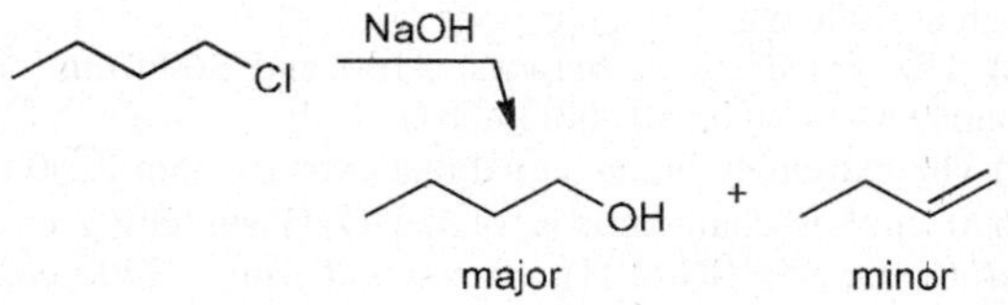

NaOH → OH (major) + (minor)

The substitution product is an alcohol and is expected to have a broad signal from 3200 to 3600 cm^{-1}. The elimination product is an unsymmetrical alkene and is expected to give a C=C signal at approximately 1650 cm^{-1}, as well as a C_{sp^2}–H signal at 3100 cm^{-1}.

14.18.

(a) These compounds have different molecular weights, so mass spectrometry can be used to distinguish between them. The first compound (called furan) has four carbon atoms, four hydrogen atoms and one oxygen atom, so it has the following molecular weight:

$$MW = (4\times12) + (4\times1) + (16) = 68$$

In contrast, the second compound (called cyclopentadiene) has five carbon atoms and six hydrogen atoms, so it has the following molecular weight:

$$MW = (5\times12) + (6\times1) = 66$$

Therefore, the mass spectrum of furan is expected to have a molecular ion peak at m/z = 68, while the mass spectrum of cyclopentadiene is expected to have a molecular ion peak at m/z = 66.

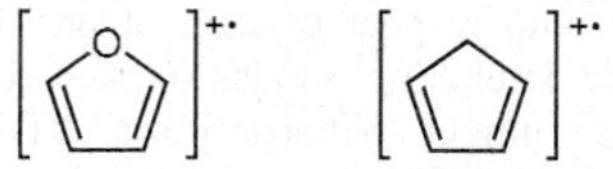

m/z = 68 m/z = 66

(b) The first compound (benzene) has six carbon atoms and six hydrogen atoms, so it has the following molecular weight:

$$MW = (6\times12) + (6\times1) = 78$$

In contrast, the second compound (pyridine) has five carbon atoms, five hydrogen atoms, and one nitrogen atom, so it has the following molecular weight:

$$MW = (5\times12) + (5\times1) + (14) = 79$$

These compounds have different molecular weights, so mass spectrometry can be used to distinguish between them. Specifically, the mass spectrum of benzene is expected to have a molecular ion peak at $m/z = 78$, while the mass spectrum of pyridine is expected to have a molecular ion peak at $m/z = 79$.

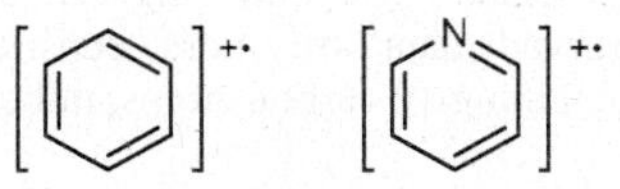

$m/z = 78$ $m/z = 79$

14.19.

(a) This compound does not have any nitrogen atoms. According to the nitrogen rule, this compound must have an even molecular weight.
With five carbon atoms, ten hydrogen atoms and one oxygen atom, the molecular weight is calculated as follows:

$$MW = (5\times12) + (10\times1) + (16) = 86$$

Therefore, we expect the molecular ion peak to appear at $m/z = 86$.

(b) This compound does not have any nitrogen atoms. According to the nitrogen rule, this compound must have an even molecular weight.
With five carbon atoms, eight hydrogen atoms and two oxygen atoms, the molecular weight is calculated as follows:

$$MW = (5\times12) + (8\times1) + (2\times16) = 100$$

Therefore, we expect the molecular ion peak to appear at $m/z = 100$.

(c) This compound has one nitrogen atom. According to the nitrogen rule, this compound must have an odd molecular weight, because it has an odd number of nitrogen atoms.
With six carbon atoms, fifteen hydrogen atoms and one nitrogen atom, the molecular weight is calculated as follows:

$$MW = (6\times12) + (15\times1) + (14) = 101$$

Therefore, we expect the molecular ion peak to appear at $m/z = 101$.

(d) This compound has two nitrogen atoms. According to the nitrogen rule, this compound must have an even molecular weight, because it has an even number of nitrogen atoms.
With five carbon atoms, fourteen hydrogen atoms and two nitrogen atoms, the molecular weight is calculated as follows:

$$MW = (5\times12) + (14\times1) + (2\times14) = 102$$

Therefore, we expect the molecular ion peak to appear at $m/z = 102$.

14.20.

(a) In order to determine the number of carbon atoms in the compound, we compare the relative heights of the $(M+1)^{+\bullet}$ peak and the $(M)^{+\bullet}$ peak, like this:

$$\frac{1.7\%}{38.3\%} \times 100\% = 4.4\%$$

This means that the $(M+1)^{+\bullet}$ peak is 4.4% as tall as the $(M)^{+\bullet}$ peak. Recall that each carbon atom in the compound contributes 1.1% to the height of the $(M+1)^{+\bullet}$ peak, so we must divide by 1.1% to determine the number of carbon atoms in the compound:

$$\text{Number of C} = \frac{4.4\%}{1.1\%} = 4$$

The compound contains four carbon atoms, which account for $4 \times 12 = 48$ amu (atomic mass units), but the molecular weight is 72 amu. So we must account for the remaining $72 - 48 = 24$ amu. The molecular formula cannot be C_4H_{24}, because a compound with four carbon atoms cannot have that many hydrogen atoms (we will see in Section 14.16 that if n represents the number of carbon atoms, the maximum number of hydrogen atoms is $2n+2$).
Therefore, there must be another element present. It cannot be a nitrogen atom as that would give an odd molecular weight (the nitrogen rule). So we try oxygen (16 amu), leaving only 8 amu for hydrogen atoms. This gives the following proposed molecular formula: C_4H_8O.

(b) In order to determine the number of carbon atoms in the compound, we compare the relative heights of the $(M+1)^{+\bullet}$ peak and the $(M)^{+\bullet}$ peak, like this:

$$\frac{4.3\%}{100\%} \times 100\% = 4.3\%$$

This means that the $(M+1)^{+\bullet}$ peak is 4.3% as tall as the $(M)^{+\bullet}$ peak. Recall that each carbon atom in the compound contributes 1.1% to the height of the $(M+1)^{+\bullet}$ peak, so we must divide by 1.1% to determine the number of carbon atoms in the compound:

$$\text{Number of C} = \frac{4.3\ \%}{1.1\ \%} = 3.9\ (\sim 4)$$

The compound cannot have 3.9 carbon atoms. It must have a whole number, so we round to the nearest whole number, which is 4.

The compound contains four carbon atoms, which account for 4 × 12 = 48 amu (atomic mass units), but the molecular weight is 68 amu. So we must account for the remaining 68 – 48 = 20 amu. The molecular formula cannot be C_4H_{20}, because a compound with four carbon atoms cannot have that many hydrogen atoms (we will see in Section 14.16 that if n represents the number of carbon atoms, the maximum number of hydrogen atoms is $2n+2$).

Therefore, there must be another element present. It cannot be a nitrogen atom as that would give an odd molecular weight (the nitrogen rule). So we try oxygen (16 amu), leaving only 4 amu for hydrogen atoms. This gives the following proposed molecular formula: C_4H_4O.

(c) In order to determine the number of carbon atoms in the compound, we compare the relative heights of the $(M+1)^{+\bullet}$ peak and the $(M)^{+\bullet}$ peak, like this:

$$\frac{4.6\%}{100\ \%} \times 100\ \% = 4.6\%$$

This means that the $(M+1)^{+\bullet}$ peak is 4.6% as tall as the $(M)^{+\bullet}$ peak. Recall that each carbon atom in the compound contributes 1.1% to the height of the $(M+1)^{+\bullet}$ peak, so we must divide by 1.1% to determine the number of carbon atoms in the compound:

$$\text{Number of C} = \frac{4.6\ \%}{1.1\ \%} = 4.15 \sim 4$$

The compound cannot have 4.15 carbon atoms. It must be a whole number, so we round to the nearest whole number, which is 4. That is, the compound contains four carbon atoms, which account for 4 × 12 = 48 amu (atomic mass units), but the molecular weight is 54 amu. So we must account for the remaining 54 – 48 = 6 amu. This indicates six hydrogen atoms, giving the following proposed molecular formula: C_4H_6.

(d) In order to determine the number of carbon atoms in the compound, we compare the relative heights of the $(M+1)^{+\bullet}$ peak and the $(M)^{+\bullet}$ peak, like this:

$$\frac{1.5\ \%}{19.0\ \%} \times 100\ \% = 7.9\ \%$$

This means that the $(M+1)^{+\bullet}$ peak is 7.9% as tall as the $(M)^{+\bullet}$ peak. Recall that each carbon atom in the compound contributes 1.1% to the height of the $(M+1)^{+\bullet}$ peak, so we must divide by 1.1% to determine the number of carbon atoms in the compound:

$$\text{Number of C} = \frac{7.9\ \%}{1.1\ \%} = 7.2 \sim 7$$

The compound cannot have 7.2 carbon atoms. It must be a whole number, so we round to the nearest whole number, which is 7. That is, the compound contains seven carbon atoms, which account for 7 × 12 = 84 amu (atomic mass units), but the molecular weight is 96 amu. So we must account for the remaining 96 – 84 = 12 amu. This indicates twelve hydrogen atoms, giving the following proposed molecular formula: C_7H_{12}.

14.21. The molecular formula of the first structure is $C_{10}H_{18}O$, while the molecular formula of the second structure is $C_9H_{14}O_2$. Each compound is expected to exhibit a molecular ion at m/z = 154, so the location of the parent signal cannot be used to differentiate these compounds. Instead, we can use the relative abundance of the $(M+1)^{+\bullet}$ peak. The following calculations show that the observed data are more consistent with a compound containing 10 carbon atoms, rather than 9.

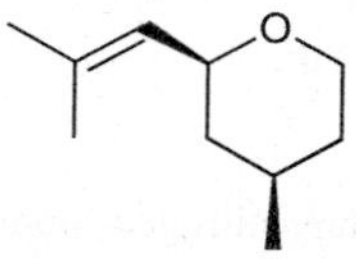

10 carbon atoms

expected abundance of $(M+1)^{+\bullet}$

$10 \times 1.1\% = 11\%$ of $(M)^{+\bullet}$

$(M+1)^{+\bullet} = 0.11 \times 17.8\%$

$(M+1)^{+\bullet} = 2.0\%$

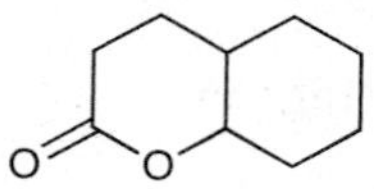

9 carbon atoms

expected abundance of $(M+1)^{+\bullet}$

$9 \times 1.1\% = 9.9\%$ of $(M)^{+\bullet}$

$(M+1)^{+\bullet} = 0.099 \times 17.8\%$

$(M+1)^{+\bullet} = 1.8\%$

The first structure is expected to give an $(M+1)^{+\bullet}$ peak with a relative abundance of 2.0%. Therefore, we conclude that the first structure is *cis*-rose oxide:

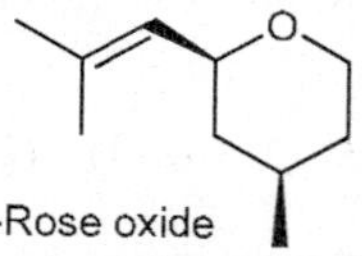

cis-Rose oxide

When comparing the relative abundance of the $(M+1)^{+\bullet}$ peaks in this particular case, the expected difference is quite small (1.8% vs. 2.0%), so other techniques will be more useful for differentiating these compounds, including high-resolution mass spectrometry (Section 14.13).

As a side note, these compounds can be easily distinguished by their strong odors: *cis*-rose oxide smells like roses, and the other compound (bicyclononalactone) smells like coconut!

14.22.

(a) The m/z = 77 fragment corresponds to M – 79, which is formed by loss of a Br atom. So the fragment does not contain Br. Also, the fragment has a smaller mass than a single ^{79}Br atom (77 < 79).

(b) Loss of a bromine radical from the molecular ion generates a phenyl cation, as shown:

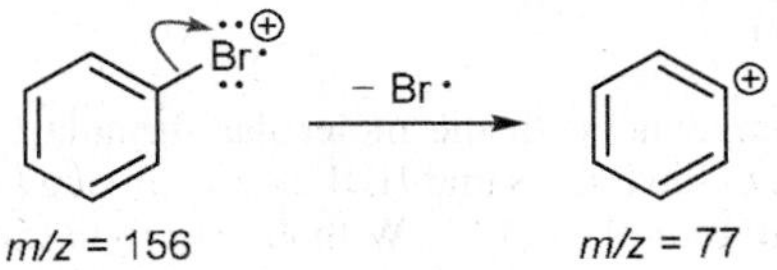

m/z = 156 — Br· → m/z = 77

Note: This type of fragmentation readily occurs under the high-energy conditions in a mass spectrometer, but this process is generally not otherwise observed, because phenyl carbocations are generally too high in energy to form at an appreciable rate during reactions performed in the laboratory.

14.23.
(a) There is not a significant $(M+2)^{+\bullet}$ peak, so neither bromine nor chlorine is present.
(b) There is not a significant $(M+2)^{+\bullet}$ peak, so neither bromine nor chlorine is present.
(c) The $(M+2)^{+\bullet}$ peak is approximately equal in height to the molecular ion peak, indicating the presence of a bromine atom.
(d) The $(M+2)^{+\bullet}$ peak is approximately one-third as tall as the molecular ion peak, indicating the presence of a chlorine atom.

14.24.
(a) A peak at M – 57 indicates the loss of a four-carbon radical fragment, which can result in the formation of a tertiary carbocation, as shown. This stable, tertiary carbocation is the fragment responsible for the base peak at M – 57:

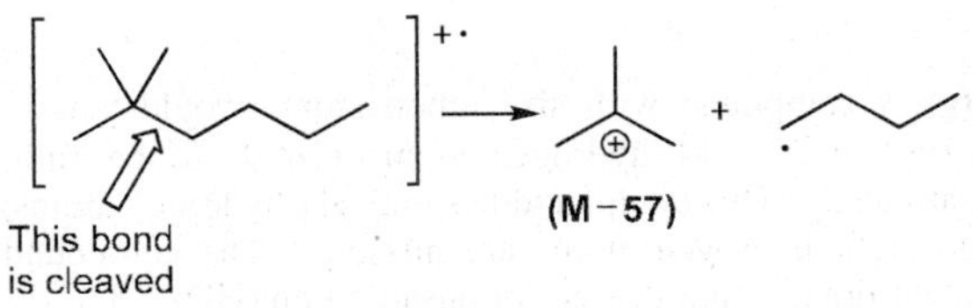

This bond is cleaved

(M – 57)

Remember that the uncharged fragment (the *n*-butyl radical) is not detected by the mass spectrometer.

(b) This carbocation is tertiary, and its formation is favored over the other possible (less stable) secondary and primary carbocations.
(c) They readily fragment to produce tertiary carbocations.
(d) M – 15 corresponds to loss of a methyl group. Indeed, loss of a methyl group would also produce a tertiary carbocation, but that pathway is less favorable because it involves formation of a methyl radical (which is less stable than a primary radical).

14.25. A fragment at M – 29 (loss of an ethyl group) corresponds to α cleavage of the molecular ion:

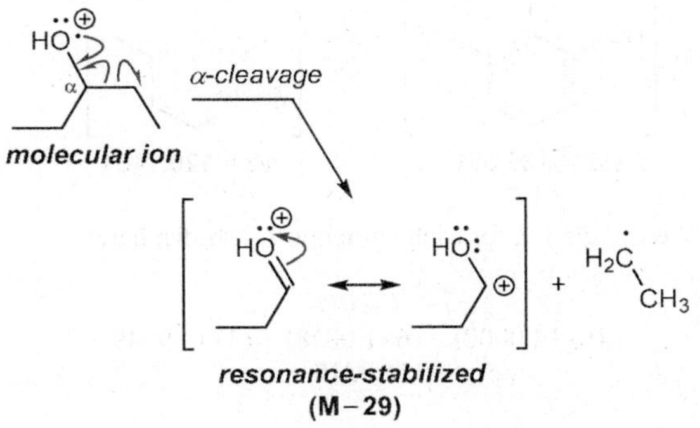

molecular ion — *α-cleavage* →

resonance-stabilized **(M – 29)** + H_2C·–CH_3

and a fragment at M – 18 (loss of water) should result from dehydration of the molecular ion:

molecular ion → **(M – 18)** + H_2O

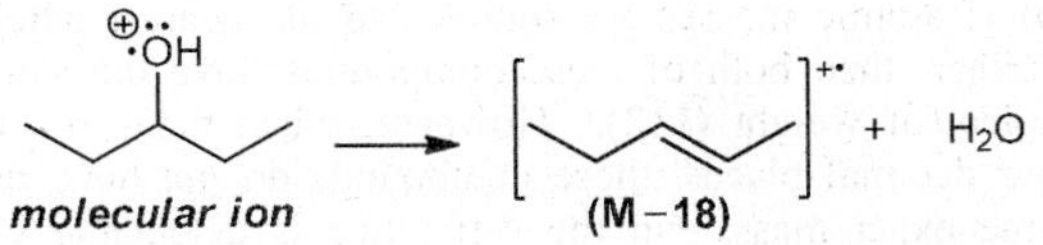

14.26. Fragmentation of the molecular ion, as shown below, results in the formation of a tertiary carbocation, as shown:

This bond is cleaved

(M – 43)

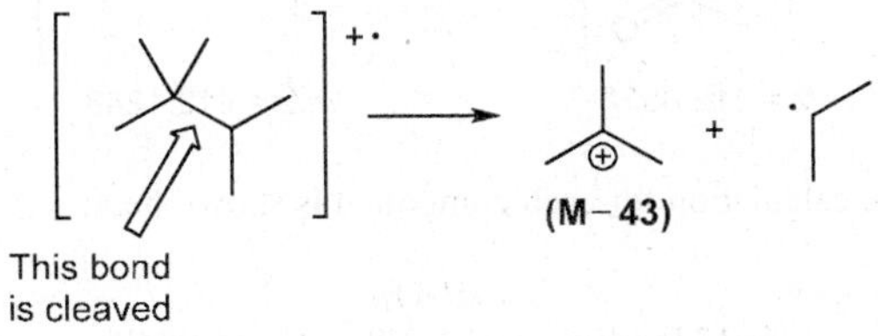

Notice that this fragmentation results in a tertiary carbocation as well as a secondary radical. No other possible fragmentation results in more stable fragments, so we expect this fragmentation to occur more often than any other possible fragmentation. As a result, the signal at M – 43 (corresponding to the tertiary carbocation) is expected to be the most abundant signal in the spectrum (the base peak).

14.27. In the first spectrum, the base peak appears at M – 29 (*m/z* = 83), signifying the loss of an ethyl group. Therefore, this spectrum is consistent with ethylcyclohexane. The second spectrum has a peak at M – 15 (*m/z* = 97), signifying the loss of a methyl group. So the second spectrum is consistent with 1,1-dimethylcyclohexane.

14.28.
(a) If atomic masses are rounded to the nearest whole number, then both of these compounds have the same molecular weight (126). However, when measured to four decimal places, these compounds do *not* have the same exact mass, and the difference is detectable via high-resolution mass spectrometry.

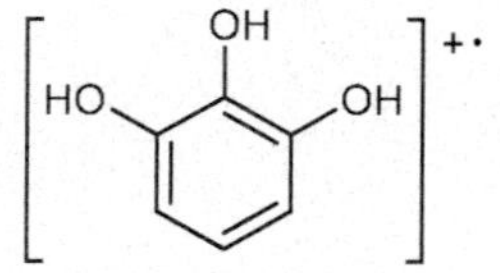

HO OH OH +· +·

m/z = 126.0315 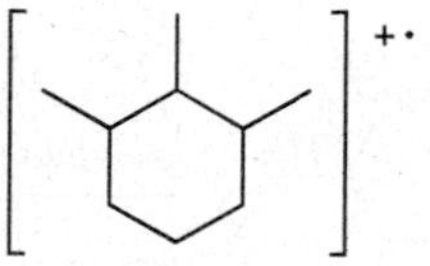 m/z = 126.1404

The calculation for each compound is shown here:

$C_6H_6O_3$
$(6\times12.0000) + (6\times1.0078) + (3\times15.9949)$
= **126.0315** amu

C_9H_{18}
$(9\times12.0000) + (18\times1.0078)$
= **126.1404** amu

(b) If atomic masses are rounded to the nearest whole number, then both of these compounds have the same molecular weight (112). However, when measured to four decimal places, these compounds do *not* have the same exact mass, and the difference is detectable via high-resolution mass spectrometry:

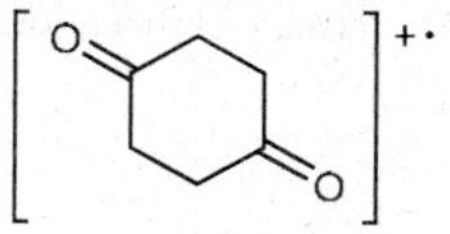

O O +· +·

m/z = 112.0522 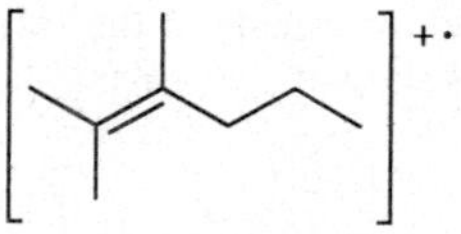 m/z = 112.1248

The calculation for each compound is shown here:

$C_6H_8O_2$
$(6\times12.0000) + (8\times1.0078) + (2\times15.9949)$
= **112.0522 amu**

C_8H_{16}
$(8\times12.0000) + (16\times1.0078)$
= **112.1248 amu**

14.29.
(a) The first compound (a phenol) should have a broad signal between 3200 and 3600 cm^{-1}, corresponding with O–H stretching. The second compound (an alkane) will not have such a signal.
(b) The first compound (a diketone) should have a strong signal around 1720 cm^{-1}, corresponding with C=O stretching. In contrast, the second compound (a tetrasubstituted alkene) will have a weak signal at around 1650 cm^{-1}, corresponding to the C=C bond. Note that this alkene does NOT have any C_{sp^2}–H bonds, so no signals at around 3100 cm^{-1} are expected in its IR spectrum.

14.30.
(a) A compound with six carbon atoms would require (2×6) + 2 = 14 hydrogen atoms (2*n*+2) to be fully saturated. This compound has only ten hydrogen atoms, so four hydrogen atoms are missing. The compound therefore has two degrees of unsaturation (HDI = 2).

(b) A compound with five carbon atoms would require (2×5) + 2 = 12 hydrogen atoms (2*n*+2) to be fully saturated (the presence of the oxygen atom does not affect this calculation). This compound has only ten hydrogen atoms, so two hydrogen atoms are missing. Therefore, the compound has one degree of unsaturation (HDI = 1).

(c) A compound with the molecular formula C_5H_9N is expected to have the same HDI as a compound with the molecular formula C_5H_8. With five carbon atoms, the compound would need to have twelve hydrogen atoms to be fully saturated (2*n*+2). With only eight hydrogen atoms, four are missing, representing two degrees of unsaturation (HDI = 2).

(d) A compound with the molecular formula C_3H_5ClO is expected to have the same HDI as a compound with the molecular formula C_3H_6 (Cl is treated like H, while O is ignored). With three carbon atoms, the compound would need to have eight hydrogen atoms to be fully saturated (2*n*+2). With only six hydrogen atoms, two are missing, representing one degree of unsaturation (HDI = 1).

(e) A compound with ten carbon atoms would require (2 × 10) + 2 = 22 hydrogen atoms (2*n*+2) to be fully saturated. This compound has only twenty hydrogen atoms, so two hydrogen atoms are missing. The compound therefore has one degree of unsaturation (HDI = 1).

(f) A compound with the molecular formula $C_4H_6Br_2$ is expected to have the same HDI as a compound with the molecular formula C_4H_8 (each Br is treated like an H). With four carbon atoms, the compound would need to have ten hydrogen atoms to be fully saturated (2*n*+2). With only eight hydrogen atoms, two are missing, representing one degree of unsaturation (HDI = 1).

(g) A compound with six carbon atoms would require (2×6) + 2 = 14 hydrogen atoms (2*n*+2) to be fully saturated. This compound has only six hydrogen atoms, so eight hydrogen atoms are missing. The compound therefore has four degrees of unsaturation (HDI = 4).

(h) A compound with the molecular formula C_2Cl_6 is expected to have the same HDI as a compound with the molecular formula C_2H_6 (each Cl is treated like an H). With two carbon atoms, the compound would need to have six hydrogen atoms to be fully saturated (2*n*+2). No hydrogen atoms are missing, representing zero degrees of unsaturation (HDI = 0).

(i) A compound with the molecular formula $C_2H_4O_2$ is expected to have the same HDI as a compound with the molecular formula C_2H_4 (the oxygen atoms can be ignored for purposes of calculating HDI). With two carbon atoms, the compound would need to have six hydrogen atoms to be fully saturated (2*n*+2). With only four hydrogen atoms, two are missing, representing one degree of unsaturation (HDI = 1).

(j) A compound with the molecular formula $C_{100}H_{200}Cl_2O_{16}$ is expected to have the same HDI as a compound with the molecular formula $C_{100}H_{202}$ (each Cl is treated like an H, and all of the oxygen atoms are ignored). With 100 carbon atoms, the compound would need to have 202 hydrogen atoms to be fully saturated ($2n$+2). No hydrogen atoms are missing, representing zero degrees of unsaturation (HDI = 0).

14.31. A compound with the molecular formula $C_3H_5ClO_2$ is expected to have the same HDI as a compound with the molecular formula C_3H_6 (the Cl is treated like an H, and both oxygen atoms are ignored). With three carbon atoms, the compound would need to have eight hydrogen atoms to be fully saturated ($2n$+2). Both $C_3H_5ClO_2$ and C_3H_6 are missing two hydrogen atoms, corresponding to one degree of unsaturation. The formula C_3H_8O is fully saturated, and the formula $C_3H_5NO_2$ has two degrees of unsaturation.

14.32. Strigol has four rings (numbered below) and five double bonds (highlighted below), for a total hydrogen deficiency index of nine:

This compound has 19 carbon atoms. If saturated, a compound with 19 carbon atoms would have 40 hydrogen atoms [(19 x 2) + 2], but an HDI of nine means that 18 hydrogen atoms are missing. So there must be 22 hydrogen atoms in the formula (40 – 18 = 22). Inspection of the structure also reveals six oxygen atoms, so the molecular formula of strigol must be $C_{19}H_{22}O_6$.

14.33. The range between 1600 and 1850 cm^{-1} is associated with the stretching of double bonds. Each of the compounds in this problem exhibits at least one double bond, if not two.

(a) The ketone carbonyl group is conjugated so it is expected to produce a signal near 1680 cm^{-1}. The C=C double bond is conjugated so it is expected to produce a signal near 1600 cm^{-1}.

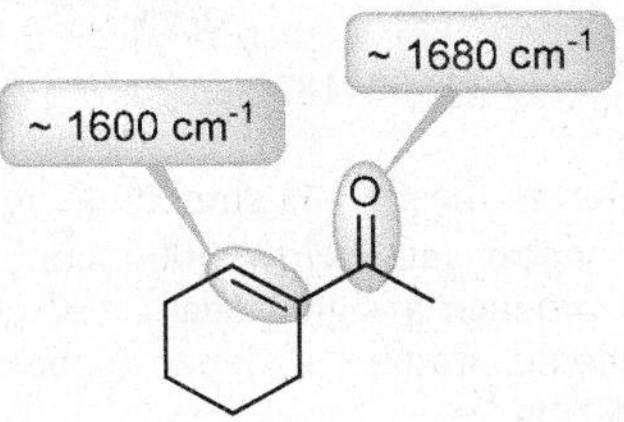

(b) The ketone carbonyl group is isolated (not conjugated) so it is expected to produce a signal near 1720 cm^{-1}. The C=C double bond is also isolated so it is expected to produce a signal near 1650 cm^{-1}.

(c) The carbonyl group of an ester will typically produce a signal near 1735 cm^{-1}.

(d) The carbonyl group of the ester is conjugated so it is expected to produce a signal near 1710 cm^{-1}. The C=C double bond is conjugated so it is expected to produce a signal near 1600 cm^{-1}.

14.34. The signals at highest wavenumber are expected to be associated with X–H bonds (because hydrogen has the lowest atomic mass). Among the list of bonds in the problem statement, there are three bonds that belong to this category (O–H, N–H, and C–H). These three bonds are expected to produce signals at approximately 3600, 3400, and 3100 cm^{-1}, respectively. Among the remaining bonds, the triple bond has the next highest wavenumber, followed by the double bonds (the C=O bond has a higher wavenumber than the C=C bond), followed by the C–O single bond.

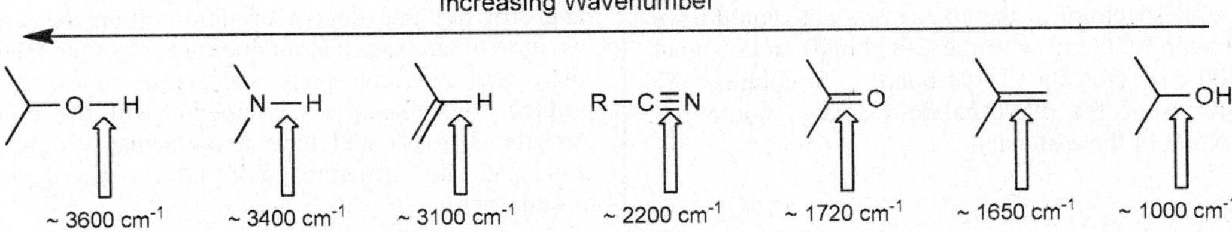

14.35.

(a) In addition to the C_{sp^3}–H stretching signals that are expected to appear just below 3000 cm^{-1}, the C=N bond and the C=O bond should each produce a signal in the double bond region (1600–1850 cm^{-1}).

(b) In addition to the C_{sp^3}–H stretching signals that are expected to appear just below 3000 cm^{-1}, the C_{sp^2}–H bond should produce a signal near 3100 cm^{-1}, and the C=C bond should produce a signal in the double bond region (~ 1650 cm^{-1}).

(c) In addition to the C_{sp^3}–H stretching signals that are expected to appear just below 3000 cm^{-1}, the C_{sp^2}–H bonds should produce a signal near 3100 cm^{-1}, and the C=C bond and the C=O bond should each produce a signal in the double bond region (1600–1850 cm^{-1}). The two C≡C bonds should also produce two signals around 2200 cm^{-1}, and the C_{sp}–H bond should produce a signal around 3300 cm^{-1}.

(d) In addition to the C_{sp^3}–H stretching signals that are expected to appear just below 3000 cm^{-1}, the C=O bond should produce a signal in the double bond region around 1715 cm^{-1}, and the O–H of the carboxylic acid group should produce a very broad signal from 2200–3600 cm^{-1}.

14.36.

(a) The IR spectrum of the alkene reactant should have a signal near 1650 cm^{-1} (for the C=C bond) and a signal near 3100 cm^{-1} (for the C_{sp^2}–H bond). In contrast, the IR spectrum of the alkane product should not exhibit either of these signals.

(b) The IR spectrum of the alcohol reactant should have a broad signal from 3200–3600 cm^{-1} (for the O–H bond), while the IR spectrum of the ketone product should not exhibit this signal, and instead should have a strong signal near 1720 cm^{-1} (for the C=O bond).

(c) The C=O bond of an ester should produce a signal at higher wavenumber (~1735 cm^{-1}) than the signal associated with the C=O bond of a ketone (~1720 cm^{-1}).

(d) The IR spectrum of the alkene reactant should have a signal near 1650 cm^{-1} (for the C=C bond) and a signal near 3100 cm^{-1} (for the C_{sp^2}–H bonds). In contrast, the IR spectra of the ketone and aldehyde products should not exhibit either of these signals. Instead, they should have signals near 1720–1730 cm^{-1} (for the C=O bonds).

(e) The IR spectrum of the alkene product should have a signal near 1650 cm^{-1} (for the C=C bond) and a signal near 3100 cm^{-1} (for the C_{sp^2}–H bond). In contrast, the IR spectrum of the alkyl halide reactant should not exhibit either of these signals.

14.37.

(a) The IR spectrum of this ketone is expected to exhibit C_{sp^3}–H signals just below 3000 cm^{-1}, and a C=O signal near 1720 cm^{-1}.

(b) The IR spectrum of this conjugated ketone is expected to exhibit C_{sp^3}–H signals just below 3000 cm^{-1}, a C_{sp^2}–H signal near 3100 cm^{-1}, a C=O signal near 1680 cm^{-1} (conjugated), and a C=C signal near 1600 cm^{-1} (conjugated).

(c) The IR spectrum of this ketone is expected to exhibit C_{sp^3}–H signals just below 3000 cm^{-1}, a C_{sp^2}–H signal near 3100 cm^{-1}, a C=O signal near 1720 cm^{-1} and a C=C signal near 1650 cm^{-1}.

(d) The IR spectrum of this alcohol is expected to exhibit C_{sp^3}–H signals just below 3000 cm^{-1}, a C_{sp^2}–H signal near 3100 cm^{-1}, a C=C signal near 1650 cm^{-1} and a broad O-H signal in the range of 3200–3600 cm^{-1}.

(e) The IR spectrum of this carboxylic acid is expected to exhibit C_{sp^3}–H signals just below 3000 cm^{-1}, a C=O signal near 1715 cm^{-1} and an extremely broad O-H signal in the range of 2200–3600 cm^{-1}.

(f) The IR spectrum of this keto-alcohol is expected to exhibit C_{sp^3}–H signals just below 3000 cm^{-1}, a C=O signal near 1720 cm^{-1} and a broad O-H signal in the range of 3200–3600 cm^{-1}.

14.38. The molecular ion peak appears at m/z = 86, so the compound must have a molecular weight of 86 amu. Each carbon atom contributes 12 amu, and each hydrogen atom contributes 1 amu. So, we begin by considering the following possibilities, each of which has one degree of unsaturation (as required by the problem statement).

$$C_4H_8 = (4\times12) + (8\times1) = 56 \text{ amu}$$
$$C_5H_{10} = (5\times12) + (10\times1) = 70 \text{ amu}$$
$$C_6H_{12} = (6\times12) + (12\times1) = 84 \text{ amu}$$

None of these possibilities exhibits the correct molecular weight. The closest is C_6H_{12}, but we are short by two amu (86 – 84 = 2 amu). If we add two hydrogen atoms (to give C_6H_{14}) then the compound will no longer have one degree of unsaturation (it would be fully saturated). So, we consider C_5H_{10}, which has a molecular weight of 70 amu. In order to get this number up to 86, we are missing 16 amu, which is exactly the weight of an oxygen atom. So, a compound with the molecular formula $C_5H_{10}O$ will have a molecular weight of 86 amu (and still has one degree of unsaturation, because the insertion of an oxygen atom does not affect the HDI). There are certainly many acceptable answers to this problem. For example, a compound with the molecular formula $C_4H_{10}N_2$ will have a molecular weight of 86 amu and its structure will have one degree of unsaturation.

14.39. The problem statement indicates that the $(M+1)^{+\bullet}$ peak is 10% as tall as the $(M)^{+\bullet}$ peak. Recall that each carbon atom in the compound contributes 1.1% to the height of the $(M+1)^{+\bullet}$ peak, so we must divide by 1.1% to determine the number of carbon atoms in the compound:

$$\text{Number of C} = \frac{10\ \%}{1.1\ \%} = 9.1 \sim 9$$

The compound cannot have 9.1 carbon atoms. It must be a whole number, so we round to the nearest whole number, which is nine. That is, the compound contains nine carbon atoms.

14.40. We begin by determining the molecular formula and molecular weight of each compound:

Cl	NH_2	OH	Br
C_6H_5Cl	C_6H_7N	C_6H_6O	C_6H_5Br
(MW = 112)	(MW = 93)	(MW = 94)	(MW = 156)

Each of these compounds has a different molecular weight, so we should be able to use match each compound with its spectrum if we focus on the molecular ion peak in each case.
Chlorobenzene (C_6H_5Cl) is consistent with spectrum (b) which has an $(M)^{+\bullet}$ peak at m/z = 112, as well as an $(M+2)^{+\bullet}$ peak that is one-third the height of the $(M)^{+\bullet}$ peak (characteristic of an organochloride).
Aniline (C_6H_7N) is consistent with spectrum (c) which has an $(M)^{+\bullet}$ peak at m/z = 93 (which is an odd number, consistent with the nitrogen rule).
Phenol (C_6H_6O) is consistent with spectrum (a) which has an $(M)^{+\bullet}$ peak at m/z = 94.
Bromobenzene (C_6H_5Br) is consistent with spectrum (d) which has an $(M)^{+\bullet}$ peak at m/z = 156, as well as an $(M+2)^{+\bullet}$ peak that is approximately the same height as the $(M)^{+\bullet}$ peak (characteristic of an organobromide).

14.41.
(a) The broad signal between 3200 and 3600 cm^{-1} indicates the presence of an OH group. The signals between 1600 and 1700 cm^{-1} indicate the presence of double bonds.
(b) In order to determine the number of carbon atoms in the compound, we compare the relative heights of the $(M+1)^{+\bullet}$ peak and the $(M)^{+\bullet}$ peak, like this:

$$\frac{3.9\ \%}{27.2\ \%} \times 100\ \% = 14.3\ \%$$

This means that the $(M+1)^{+\bullet}$ peak is 14.3% as tall as the $(M)^{+\bullet}$ peak. Recall that each carbon atom in the compound contributes 1.1% to the height of the $(M+1)^{+\bullet}$ peak, so we must divide by 1.1% to determine the number of carbon atoms in the compound:

$$\text{Number of C} = \frac{14.3\ \%}{1.1\ \%} = 13$$

Our analysis reveals that the compound likely has thirteen carbon atoms.

(c) The molecular ion peak appears at m/z = 196, so the compound must have a molecular weight of 196 amu. Each carbon atom contributes 12 amu, and each hydrogen atom contributes 1 amu. So, we begin by considering the following possibilities, each of which has two degrees of unsaturation (as required by the problem statement).

$$C_{12}H_{22} = (12\times12) + (22\times1) = 166 \text{ amu}$$
$$C_{13}H_{24} = (13\times12) + (24\times1) = 180 \text{ amu}$$
$$C_{14}H_{26} = (14\times12) + (26\times1) = 194 \text{ amu}$$

None of these possibilities exhibits the correct molecular weight. The closest is $C_{14}H_{26}$, but we are short by two amu (196 – 194 = 2 amu). If we add two hydrogen atoms (to give $C_{14}H_{28}$), then the compound will no longer have two degrees of unsaturation (it would only have one degree of unsaturation). So, we consider $C_{13}H_{24}$, which has a molecular weight of 180 amu. In order to get this number up to 196, we are missing 16 amu, which is exactly the weight of an oxygen atom. So, a compound the molecular formula $C_{13}H_{24}O$ will have a molecular weight of 196 amu (and still has two degrees of unsaturation, because the insertion of an oxygen atom does not affect the HDI).

14.42.
(a) Each of these compounds has the molecular formula C_6H_{12}.

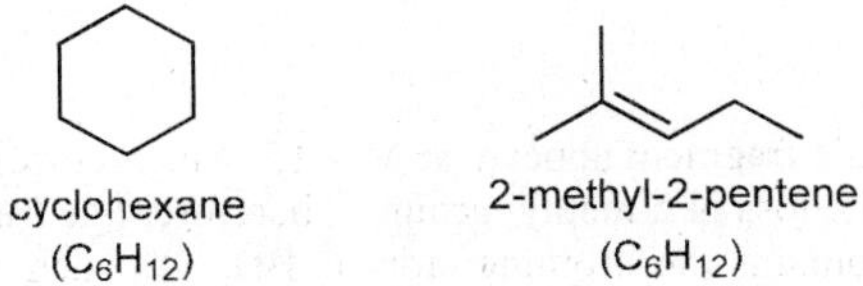

(b) A compound with six carbon atoms would require (6×2) + 2 = 14 hydrogen atoms ($2n$+2) to be fully saturated. Each of these compounds has only twelve hydrogen atoms, so each compound is missing two hydrogen atoms. Therefore, each of these compounds has one degree of unsaturation (HDI = 1). Alternatively, inspection of each structure also reveals the answer: cyclohexane has one ring, and 2-methyl-2-pentene has one π bond.

(c) No. Both compounds have exactly six carbon atoms and twelve hydrogen atoms, so each of these compounds will produce an $(M)^{+\bullet}$ peak with the same m/z, even with high resolution mass spectrometry.

$$C_6H_{12} = (6\times12.0000) + (12\times1.0078) = \mathbf{84.0936\ amu}$$

(d) The IR spectrum of the alkene would have a signal near 1650 cm^{-1} for the C=C bond and another signal near 3100 cm^{-1} for the C_{sp^2}–H bond. The IR spectrum of cyclohexane lacks both of these signals.

14.43. The parent ion for 1-ethyl-1-methylcyclohexane is expected to appear at m/z = 126 (calculation below).

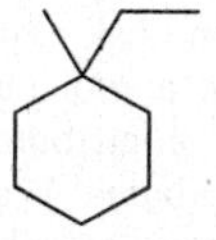

1-Ethyl-1-methylcyclohexane
(C_9H_{18})

$$\text{MW of } C_9H_{18} = (9\times12) + (18\times1) = 126 \text{ amu}$$

If the parent ion appears at m/z = 126, then the signal that appears at m/z = 111 is at (M – 15), which corresponds with the loss of a methyl group. The signal at m/z = 97 is at (M – 29), which corresponds with the loss of an ethyl group. Both fragmentations lead to a tertiary carbocation:

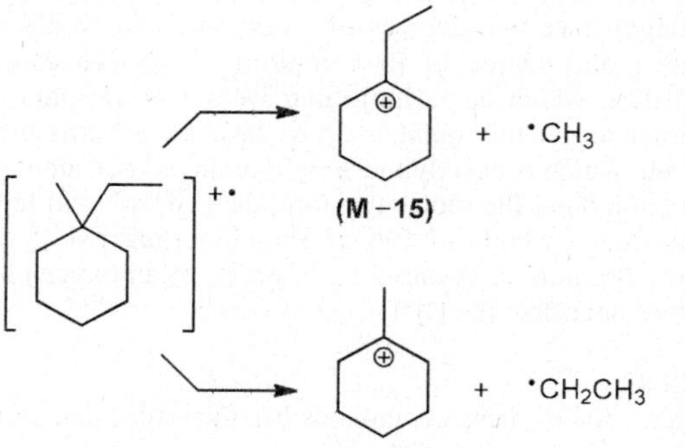

+ $^{\bullet}CH_3$

(M−15)

+ $^{\bullet}CH_2CH_3$

(M−29)

14.44.

(a) This fragment appears at M – 15, which corresponds with the loss of a methyl group. Therefore, this fragment still contains the bromine atom (^{79}Br). As such, we do expect a signal at M – 13 that is equal in height to the M – 15 peak (because there are two isotopes of bromine in roughly equal abundance). The M – 13 fragment arises when the M+2 molecular ion (which contains the ^{81}Br isotope) loses a methyl group.

(b) This fragment appears at (M – 29), which corresponds with the loss of an ethyl group. Therefore, this fragment still contains the bromine atom (^{79}Br). As such, we do expect a signal at M – 27 that is equal in height to the M – 29 peak (because there are two isotopes of bromine in roughly equal abundance). The M – 29 fragment arises when the M+2 molecular ion (which contains the ^{81}Br isotope) loses an ethyl group.

(c) This fragment appears at M – 79, which corresponds with the loss a bromine atom (^{79}Br). Therefore, the fragment no longer contains bromine. As such, we do not expect a signal at M – 77 (which we would only observe if the fragment still contained the bromine atom). If the M+2 molecular ion (which contains the ^{81}Br isotope) were to lose its bromine atom, it would produce the same fragment as when the molecular ion (which contains the ^{79}Br isotope) loses its bromine atom to produce the M – 79 peak.

14.45. As seen in Chapter 7, the choice of base affects the regiochemical outcome. Ethoxide favors the more-substituted (Zaitsev) product, while *tert*-butoxide favors the less-substituted (Hofmann) product.

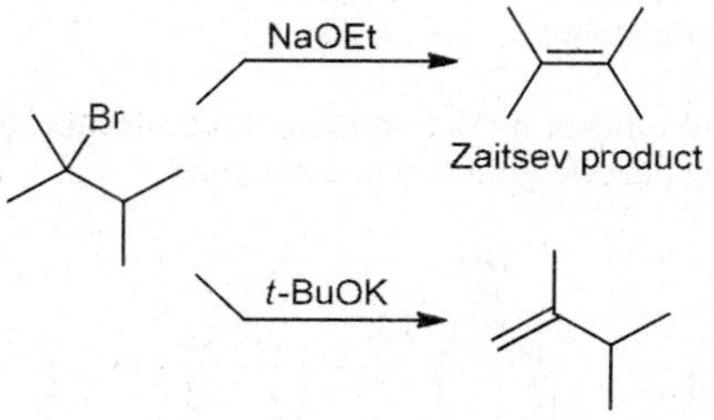

NaOEt

Br

Zaitsev product

t-BuOK

Hofmann product

The Zaitsev product is a symmetrical alkene, and as such, its IR spectrum will not exhibit a signal at 1650 cm^{-1} (where the signals for C=C bonds typically appear) or a signal at 3100 cm^{-1} (because this compound contains no C_{sp^2}–H bonds). In contrast, the Hofmann product will display both of these signals in its IR spectrum.

14.46.

(a)

The molecular ion peak appears at m/z = 66, so the compound must have a molecular weight of 66 amu. Each carbon atom contributes 12 amu, and each hydrogen atom contributes 1 amu. Five carbon atoms will account for 60 amu, leaving 6 amu for hydrogen atoms.

$$C_5H_6 = (5\times12) + (6\times1) = 66 \text{ amu}$$

The compound cannot have six carbon atoms, because that would exceed the known molecular weight (6×12 = 72 amu). And the compound cannot have four carbon atoms, because then we would not reach 66 amu even if the compound is fully saturated:

$$C_4H_{10} = (4\times12) + (10\times1) = 58 \text{ amu}$$

A hydrocarbon with four carbon atoms cannot have more than ten hydrogen atoms (2n+2).

(b) The molecular ion peak appears at m/z = 70, so the compound must have a molecular weight of 70 amu. Each carbon atom contributes 12 amu, and each hydrogen atom contributes 1 amu. The signal in the IR spectrum indicates the presence of a carbonyl group (C=O), so we must also account for at least one oxygen atom, which contributes 16 amu. If we start with four carbon atoms (4 × 12 = 48), and one oxygen atom (16), we get a total of 48 + 16 = 64 amu, which is just 6 amu short of the known molecular weight (70 amu). Six hydrogen atoms will supply the remaining 6 amu, giving the molecular formula C_4H_6O.

14.47.

(a) Octane is a saturated alkane with the molecular formula C_8H_{18}. The molecular weight of octane is calculated here:

$$C_8H_{18} = (8\times12) + (18\times1) = 114 \text{ amu}$$

Therefore, the molecular ion peak is expected to appear at m/z = 114, and indeed, that peak can be seen in the spectrum.

(b) The base peak is (by definition) the tallest peak in the spectrum, which appears at m/z = 43.

(c) The base peak (at m/z = 43) corresponds with the loss of a radical fragment that is 114 – 43 = 71 amu. That is, the base peak is at (M – 71), which corresponds with the loss of a pentyl radical to give a propyl cation, as shown here.

14.48.

(a) A compound with four carbon atoms would require (2×4) + 2 = 10 hydrogen atoms (2n+2) to be fully saturated. This compound has only six hydrogen atoms, so four hydrogen atoms are missing. The compound therefore has two degrees of unsaturation (HDI = 2).

(b) A compound with five carbon atoms would require (2×5) + 2 = 12 hydrogen atoms (2n+2) to be fully saturated. This compound has only eight hydrogen atoms, so four hydrogen atoms are missing. The compound therefore has two degrees of unsaturation (HDI = 2).

(c) A compound with 40 carbon atoms would require (2×40) + 2 = 82 hydrogen atoms (2n+2) to be fully saturated. This compound has only 78 hydrogen atoms, so four hydrogen atoms are missing. The compound therefore has two degrees of unsaturation (HDI = 2).

(d) A compound with 72 carbon atoms would require (2×72) + 2 = 146 hydrogen atoms (2n+2) to be fully saturated. This compound has only 74 hydrogen atoms, so 72 hydrogen atoms are missing. The compound therefore has 36 degrees of unsaturation (HDI = 36).

(e) A compound with the molecular formula $C_6H_6O_2$ is expected to have the same HDI as a compound with the molecular formula C_6H_6 (the oxygen atoms can be ignored for purposes of calculating HDI). With six carbon atoms, the compound would need to have fourteen hydrogen atoms to be fully saturated (2n+2). With only six hydrogen atoms, eight are missing, representing four degrees of unsaturation (HDI = 4).

(f) A compound with the molecular formula $C_7H_9NO_2$ is expected to have the same HDI as a compound with the molecular formula C_7H_8 (the oxygen atoms can be ignored, and we subtract one H because of the presence of an N). With seven carbon atoms, the compound would need to have sixteen hydrogen atoms to be fully saturated (2n+2). With only eight hydrogen atoms (in the equivalent C_7H_8), eight are missing, representing four degrees of unsaturation (HDI = 4).

(g) A compound with the molecular formula $C_8H_{10}N_2O$ is expected to have the same HDI as a compound with the molecular formula C_8H_8 (the oxygen atom can be ignored, and we subtract one H for each N). With eight carbon atoms, the compound would need to have eighteen hydrogen atoms to be fully saturated (2n+2). With only eight hydrogen atoms (in the equivalent C_8H_8), ten are missing, representing five degrees of unsaturation (HDI = 5).

(h) A compound with the molecular formula $C_5H_7Cl_3$ is expected to have the same HDI as a compound with the molecular formula C_5H_{10} (each Cl is treated like an H). With five carbon atoms, the compound would need to have twelve hydrogen atoms to be fully saturated (2n+2). With only ten hydrogen atoms (in the equivalent C_5H_{10}), two are missing, representing one degree of unsaturation (HDI = 1).

(i) A compound with the molecular formula C_6H_5Br is expected to have the same HDI as a compound with the molecular formula C_6H_6 (Br is treated like H). With six carbon atoms, the compound would need to have fourteen hydrogen atoms to be fully saturated (2n+2). With only six hydrogen atoms, eight are missing, representing four degrees of unsaturation (HDI = 4).

(j) A compound with the molecular formula $C_6H_{12}O_6$ is expected to have the same HDI as a compound with the molecular formula C_6H_{12} (the oxygen atoms can be ignored for purposes of calculating HDI). With six carbon atoms, the compound would need to have fourteen hydrogen atoms to be fully saturated (2n+2). With only twelve hydrogen atoms, two are missing, representing one degree of unsaturation (HDI = 1).

14.49. The compound contains only carbon atoms and hydrogen atoms, and there is a signal in the IR spectrum at 3300 cm^{-1}, indicating a C_{sp}–H bond. That is, the compound is a terminal alkyne. There are only two terminal alkynes with the molecular formula C_5H_8, shown here:

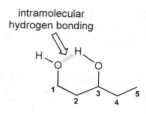

14.50. Limonene is a hydrocarbon, which means that it contains only carbon atoms and hydrogen atoms. That is, we must account for the entire molecular weight (136 amu) with only carbon atoms (12 amu) and hydrogen atoms (1 amu). With two double bonds and one ring, the compound has three degrees of unsaturation. So, we begin by considering the following possibilities, each of which has three degrees of unsaturation (as required by the problem statement).

$$C_9H_{14} = (9 \times 12) + (14 \times 1) = 122 \text{ amu}$$
$$C_{10}H_{16} = (10 \times 12) + (16 \times 1) = 136 \text{ amu}$$
$$C_{11}H_{18} = (11 \times 12) + (18 \times 1) = 150 \text{ amu}$$

The middle possibility ($C_{10}H_{16}$) matches exactly (136 amu).

14.51. Let's begin by drawing the structures of the compounds that must be distinguished:

1-Bromo-3-methyl-2-butene

2-Bromo-3-methyl-2-butene

Both of these compounds contain an unsymmetrical C=C bond that is expected to produce a signal in the range between 1600 and 1700 cm^{-1}, so we cannot use that signal to distinguish the compounds. Next we consider any other signals that are expected to appear in the diagnostic region (1600 – 4000 cm^{-1}). Neither compound has a triple bond so the region between 2100 and 2300 cm^{-1} won't be helpful. So we look to the region of the spectrum that has signals from X–H bonds. The first compound does indeed have a C_{sp^2}–H bond that is absent in the second compound:

C_{sp^2}–H

no C_{sp^2}–H

As such, the IR spectrum of the first compound is expected to exhibit a signal near 3100 cm^{-1}, while the IR spectrum of the second compound is not expected to have any signals above 3000 cm^{-1}. These compounds can therefore be distinguished by looking for a signal at 3100 cm^{-1} in the IR spectrum of each compound.

14.52. The compound can exhibit intramolecular hydrogen bonding even in dilute solutions, as shown here:

intramolecular hydrogen bonding

14.53. Each carbon atom contributes 1.1% to the $(M+1)^{+\bullet}$ peak, so with eight carbon atoms, we predict the $(M+1)^{+\bullet}$ peak to be 8.8% of the molecular ion peak. In addition, each nitrogen atom in the compound will contribute 0.37% to the $(M+1)^{+\bullet}$ peak. Three nitrogen atoms therefore contribute the same amount (1.1%) as one carbon atom. A compound with the molecular formula $C_8H_{11}N_3$ should have an $(M+1)^{+\bullet}$ peak that is 9.9% as tall as the molecular ion peak. If the molecular ion peak is 24% of the base peak, then the $(M+1)^{+\bullet}$ peak is expected to be 2.4% of the base peak.

14.54. A compound with the molecular formula C_4H_8O is expected to have the same HDI as a compound with the molecular formula C_4H_8 (the oxygen atom can be ignored for purposes of calculating HDI). With four carbon atoms, the compound would need to have ten hydrogen atoms to be fully saturated ($2n$+2). With only eight hydrogen atoms, two are missing, representing one degree of unsaturation (HDI = 1). A compound with one degree of unsaturation must contain either one double bond or one ring (but not both). The IR spectrum indicates the presence of a C=O bond, which accounts for the one degree of unsaturation (this means that the compound does NOT have a ring).
In the absence of a ring, there are only two ways to connect four carbon atoms (linear or branched):

Let's begin with the linear skeleton. There are two unique locations on the linear skeleton where a C=O bond can be placed (either at C1 or at C2):

Placing the C=O bond at C3 would be the same as placing it at C2. Similarly, placing the C=O bond at C4 would be the same as placing it at C1.
Next we turn to the branched skeleton. We cannot place a C=O bond at the central position, because that would violate the octet rule:

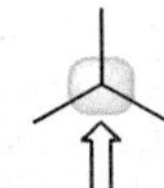

C=O bond can't be here,
because carbon can't have five bonds

And the remaining three positions are all identical. Placing the C=O bond at any of these three positions will give the same aldehyde:

In summary, there are only three possible structures that have the molecular formula C_4H_8O and contain a C=O group:

14.55. A compound with the molecular formula C_4H_8O is expected to have the same HDI as a compound with the molecular formula C_4H_8 (the oxygen atom can be ignored for purposes of calculating HDI). With four carbon atoms, the compound would need to have ten hydrogen atoms to be fully saturated ($2n+2$). With only eight hydrogen atoms, two are missing, representing one degree of unsaturation (HDI = 1). A compound with one degree of unsaturation must contain either one double bond or one ring (but not both). The IR spectrum does not have signals between 1600 and 1850 cm^{-1}, which indicates the absence of either a C=O bond or an unsymmetrical C=C bond. Therefore, the structure likely contains a ring. With four carbon atoms, there are only two different types of rings that are possible (a four-membered ring or a three-membered ring):

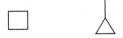

The broad signal between 3200-3600 cm^{-1} indicates an OH group, so our structure must include an OH group. There is only one unique location where an OH group can be placed on a four-membered ring:

Placement of the OH group at any other location would give the same compound (cyclobutanol). However, there are three unique locations where the OH group can be placed on the other skeleton (containing the three-membered ring):

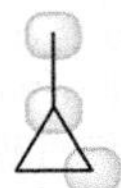

Our methodical analysis has revealed that there are four constitutional isomers with the molecular formula C_4H_8O that contain one ring and an OH group:

(Four possible stereoisomers)

14.56. A signal at 2200 cm^{-1} signifies the presence of a C≡C bond. There are only two possible constitutional isomers: 1-butyne or 2-butyne. 2-Butyne is symmetrical and would not produce a signal at 2200 cm^{-1}. The compound must be 1-butyne.

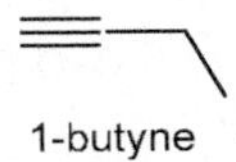

1-butyne

14.57. The correct answer is (a). The IR spectrum shows a very broad signal in the range of 2200–3600 cm^{-1}, as well as a strong signal at 1700 cm^{-1}, indicating the presence of a carboxylic acid. Carboxylic acids can only be made via oxidation of primary alcohols (not secondary or tertiary). Choice (d) is not correct because the compound may or may be acyclic or cyclic (we lack the information necessary to make such a determination).

14.58. The correct answer is (a). A peak at M – 29 corresponds with the loss of an ethyl group. Only option (a) can lose an ethyl group to give a stable carbocation (in this case, a benzylic carbocation that is stabilized by resonance):

$[C_6H_5CH_2CH_2CH_3]^{+\bullet} \rightarrow C_6H_5CH_2^{\oplus} + {}^{\bullet}CH_2CH_3$

A benzylic carbocation

14.59. The correct answer is (d). The IR spectrum has a narrow signal near 2100 cm^{-1}, which indicates the presence of a triple bond. In addition, the signal at 3300 cm^{-1} is consistent with a C_{sp}–H bond, which indicates that the compound is a terminal alkyne.

14.60. The correct answer is (c). The presence of a bromine atom in bromopropane gives rise to an $(M+2)^{+\bullet}$ peak that is approximately equivalent in height to the molecular ion peak (due to nearly equal abundance of ^{79}Br and ^{81}Br isotopes).

14.61. The correct answer is (d). Recall that each carbon atom in the compound contributes 1.1% to the height of the $(M+1)^{+\bullet}$ peak. A compound containing 17 carbon atoms is expected to have an M+1 signal that is approximately 19% of the molecular ion abundance (17 x 1.1% = 18.7%).

14.62. The correct answer is (d). The starting material is an alcohol and is expected to produce a typical signal for an O–H stretch (a broad signal between 3200 and 3600 cm^{-1}). In contrast, the ketone product has no O–H bond and will not show a broad signal between 3200 and 3600 cm^{-1} in its spectrum. In addition, the product can be differentiated from the starting material by looking for a signal at around 1720 cm^{-1}. The product has a C=O bond and should exhibit this signal. The starting material lacks a C=O bond and will not show a signal at 1720 cm^{-1}.

14.63. The correct answer is (c). The IR spectra for both a ketone and an ester will exhibit a signal for a C=O stretch, but only the spectrum of the ester exhibits a C–O stretch as well (~ 1100 cm^{-1}).

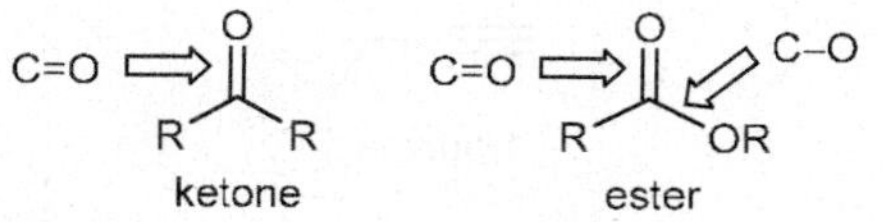

C=O ⟹ ketone (R–CO–R); C=O ⟹ ester (R–CO–OR) ⟸ C–O

ketone ester

14.64. The correct answer is (a). The mass spectrum shows a molecular ion at m/z 78. This corresponds with option (a) which has the molecular formula C_6H_6 (MW = 78). Option (b) has the molecular formula C_6H_8 (MW = 80), option (c) has the molecular formula C_6H_{12} (MW = 84), and option (d) has the molecular formula C_6H_4 (MW = 76). Only option (a) has the correct molecular weight.

14.65. The IR spectrum has a strong signal just above 1700 cm^{-1}, which indicates that the compound has a C=O bond. The mass spectrum has a molecular ion peak at m/z = 86, which indicates a molecular weight of 86 ($C_5H_{10}O$). The IR spectrum is consistent with a ketone rather than an aldehyde, because is lacks the aldehyde C–H peaks (weak signals at 2750 and 2850 cm^{-1}). The base peak is at M – 43, indicating the loss of a propyl group. The compound likely has a three carbon chain (either a propyl group or an isopropyl group) as shown in the following two structures.

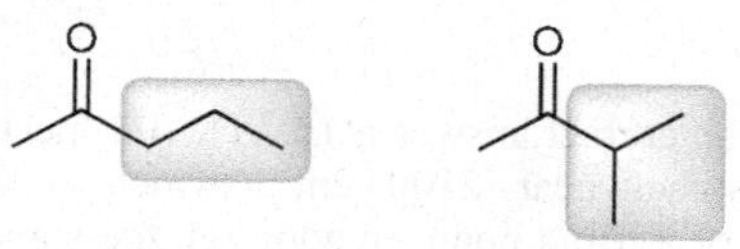

14.66. The IR spectrum has a signal near 2100 cm^{-1}, which indicates the presence of a triple bond. In addition, the narrow signal at 3300 cm^{-1} is consistent with a C_{sp}–H bond, which indicates that the compound is a terminal alkyne. The mass spectrum has a molecular ion peak at m/z = 68, which indicates a molecular weight of 68 amu. So we are looking for terminal alkynes with a molecular weight of 68 amu. Recall that each carbon atom contributes 12 amu, so five carbon atoms contribute 60 amu to the molecular weight, leaving just 8 amu left for hydrogen atoms (each hydrogen atom is 1 amu). That is, compounds with the molecular formula C_5H_8 will have a molecular weight of 68 amu. This molecular formula is consistent with two degrees of unsaturation (a triple bond).
There are two terminal alkynes with the molecular formula C_5H_8, shown here:

–≡–H –≡–H

14.67. All of the reactions in the following sequence were covered in previous chapters. The starting alkyl chloride is a tertiary substrate and will undergo an E2 reaction when treated with a strong base. With a base like ethoxide (which is not sterically hindered), the major product is the more-substituted alkene (the Zaitsev product), compound **A**. Hydroboration-oxidation of compound **A** gives an *anti*-Markovnikov addition of H and OH, affording alcohol **B**. When treated with tosyl chloride and pyridine, the alcohol is turned into the corresponding tosylate (compound **C**). Treatment of the tosylate with a sterically hindered base gives another E2 reaction, this time giving the less-substituted alkene (the Hofmann product), compound **D**. Treating **D** with a peroxy acid, such as MCPBA, gives epoxide **E**. When the epoxide is treated with a Grignard reagent (such as MeMgBr), followed by aqueous acidic workup, a ring-opening reaction occurs in which the Grignard reagent attacks the less-substituted side of the epoxide, to afford alcohol **F**. The last step of the sequence is a Williamson ether synthesis, in which a methyl group is installed to give compound **G**.

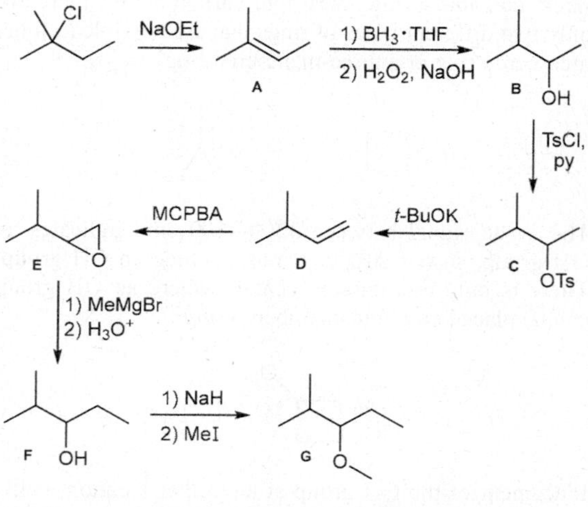

Cl → NaOEt → A → 1) BH_3·THF 2) H_2O_2, NaOH → B (OH) → TsCl, py → C (OTs) → t-BuOK → D → MCPBA → E (O) → 1) MeMgBr 2) H_3O^+ → F (OH) → 1) NaH 2) MeI → G (O)

(a) Compound **F** is an alcohol and its IR spectrum will exhibit a broad signal between 3200 and 3600 cm^{-1}. Compound **G** is an ether and its IR spectrum will not exhibit the same signal.

(b) Compound **D** is an alkene and its IR spectrum will exhibit a signal near 1650 cm^{-1} (for the C=C bond), as well as a signal near 3100 cm^{-1} (for the C_{sp^2}–H bond). Compound **E** is an epoxide, and its IR spectrum will not have these two signals.

(c) IR spectroscopy would not be helpful to distinguish these two compounds because they are both alcohols. Mass spectrometry could be used to differentiate these two compounds because they have different molecular weights.

(d) Compounds **A** and **D** have the same molecular formula, so mass spectrometry would only be useful if the fragmentation patterns were expected to be very different. In this case, it would be difficult to differentiate these compounds based on their fragmentation patterns, because both compounds will have a peak at M-15 (for loss of a methyl group). So mass spectrometry would not be helpful to distinguish compounds **A** and **D**.

14.68. The molecular formula C_4H_8 indicates an HDI = 1. As such, every constitutional isomer of C_4H_8 must contain either one ring or one double bond. The following structures are consistent with this description. Each of the first three constitutional isomers (shown here) exhibits a double bond, while each of the last two constitutional isomers exhibits a ring.

Among these isomers, only one of them (1-butene) can lose a methyl group to form a resonance-stabilized carbocation at (M – 15):

(M–15)
resonance-stabilized

$\cdot CH_3$

So we expect the signal at M – 15 to be particularly abundant in the mass spectrum of 1-butene.

14.69. As seen in the solution to Problem **2.51**, there are four constitutional isomers with the molecular formula C_4H_9Cl, shown here again:

Compound **A**

Only one of these four isomers, compound **A**, exhibits a chiral center. When compound **A** is treated with a strong base, the following three alkenes are formed (**B**, **C**, and **D**):

NaOEt
(racemic)
B **C** **D**

Compounds **B** and **C** are diastereomers (*cis* vs. *trans*), with the more stable *trans* alkene being favored as the major product. That leaves the minor product **D**, which is an unsymmetrical alkene. Accordingly, we do expect that the C=C bond will have a small dipole moment, so there should be a signal near 1650 cm^{-1} in the IR spectrum of compound **D**.

14.70. In order to determine the pattern of signals that we expect in the mass spectrum of a compound with two chlorine atoms, we must quickly review the reason for the pattern of signals that are observed for a compound with only one chlorine atom. Let's first imagine that all chlorine atoms had the same mass (35 amu). In such a hypothetical world, we would see the $(M)^{+\bullet}$ peak, but we would not observe a significant $(M+2)^{+\bullet}$ peak. However, all chlorine atoms are not the same mass, because there are two naturally abundant isotopes: ^{35}Cl and ^{37}Cl. The ^{37}Cl isotope represents approximately ¼ of all chlorine atoms in the world. As such, approximately ¼ of the molecular ions (for a compound with one chlorine atom) will contain an atom of ^{37}Cl instead of ^{35}C. This causes ¼ of the signal to be moved from $(M)^{+\bullet}$ to $(M+2)^{+\bullet}$.

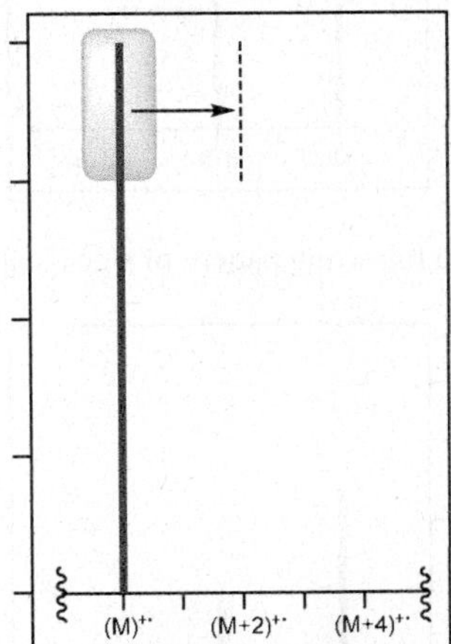

This gives the following familiar pattern of signals, in which the $(M+2)^{+\bullet}$ peak is approximately 1/3 the height of the $(M)^{+\bullet}$ peak:

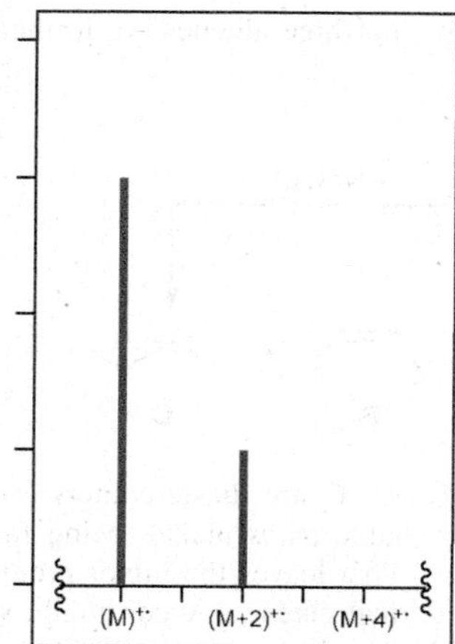

This is the characteristic pattern of signals for a compound with one chlorine atom.

Now let's consider a compound with *two* chlorine atoms. We will begin with the pattern of signals above, and determine how this pattern changes if the parent ion has two chlorine atoms, rather than just one. Once again, ¼ of each peak is moved up by two *m/z* units, because in ¼ of all of those ions, the second chlorine atom will be ^{37}Cl rather than ^{35}Cl.

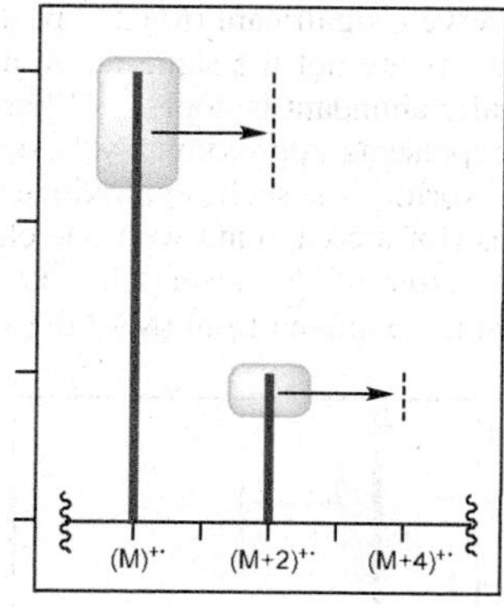

This gives the following pattern of signals at M, M+2, and M+4.

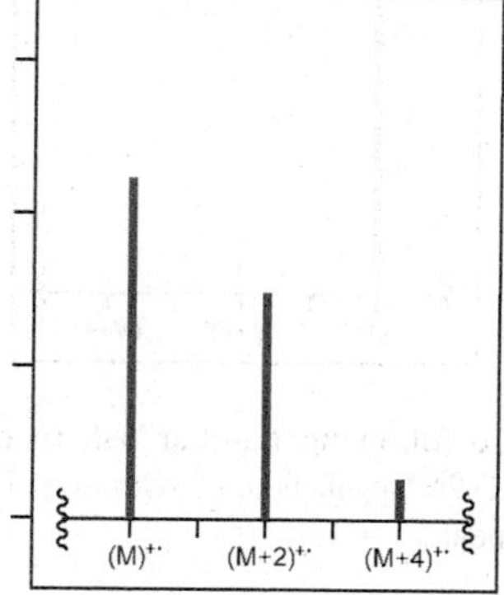

Below is a mathematical analysis that provides the same answer as shown above.

Each Cl atom has a 75% chance of being ^{35}Cl and a 25% chance of being ^{37}Cl. Therefore,

- the odds that both Cl atoms are the ^{35}Cl isotope (corresponding to the M peak) = (.75)(.75) = 56%
- the odds of having one Cl atom of each isotope (two possible combinations for M+2) = (.75)(.25) + (.25)(.75) = 38%
- the odds that both Cl atoms are the ^{37}Cl isotope (corresponding to the M+4 peak) = (.25)(.25) = 6%

The ratio of the three peaks (M : M+2 : M+4) will be 56:38:6 (approximately 9:6:1). This analysis matches the figure above.

14.71. The OH group in ephedrine can engage in intramolecular hydrogen bonding, even in dilute solutions.

intramolecular hydrogen bonding

14.72. In order to determine the pattern of signals that we expect in the mass spectrum of a compound with two bromine atoms, we must quickly review the reason for the pattern of signals that are observed for a compound with only one bromine atom. Let's first imagine that all bromine atoms had the same mass (79 amu). In such a hypothetical world, we would see the $(M)^{+\bullet}$ peak, but we would not observe a significant $(M+2)^{+\bullet}$ peak. However, all bromine atoms do not have the same mass, because there are two naturally abundant isotopes: ^{79}Br and ^{81}Br. Each isotope represents approximately half of all bromine atoms in the world. As such, approximately half of the parent ions (for a compound with one bromine atom) will contain an atom of ^{81}Br instead of ^{79}Br. This causes approximately half of the signal to be moved from $(M)^{+\bullet}$ to $(M+2)^{+\bullet}$.

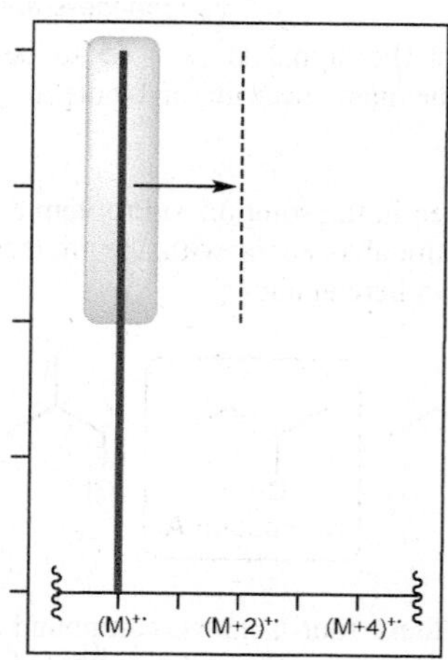

This gives the following familiar pattern of signals, in which the $(M+2)^{+\bullet}$ peak is approximately the same height as the $(M)^{+\bullet}$ peak:

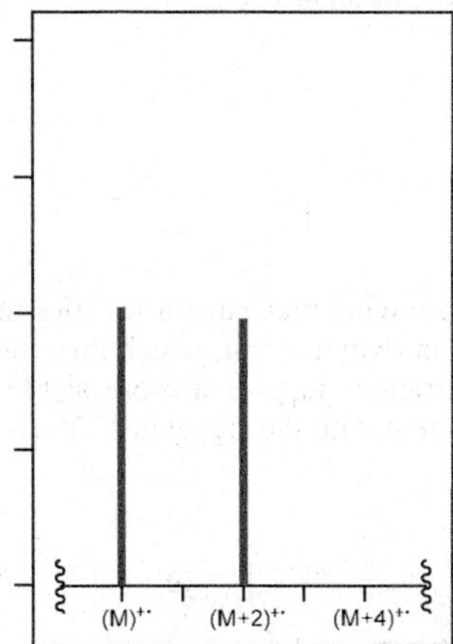

This is the characteristic pattern of signals for a compound with one bromine atom.

Now let's consider a compound with *two* bromine atoms. We will begin with the pattern of signals above, and determine how this pattern changes if the parent ion has two bromine atoms, rather than just one. Once again, half of each peak is moved up by two *m/z* units, because in half of all of those ions, the second bromine atom will be ^{81}Br rather than ^{79}Br.

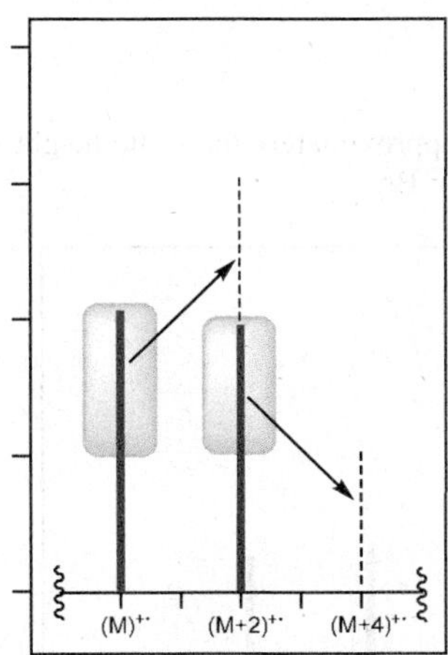

This gives the following pattern of signals at M, M+2, and M+4.

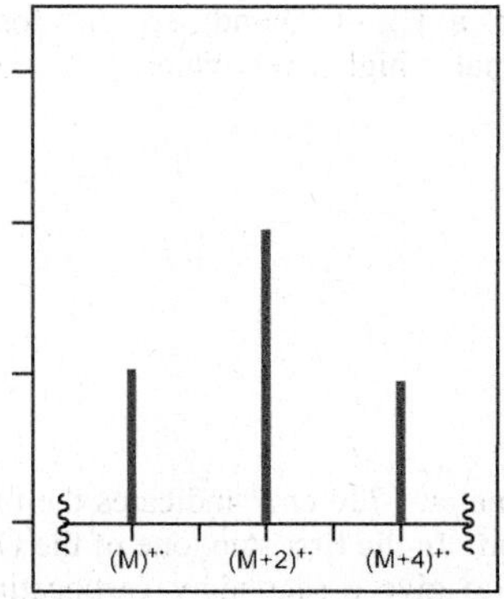

Below is a mathematical analysis that provides the same answer as shown above.

Each Br atom has (approximately) a 50% chance of being ^{79}Br and a 50% chance of being ^{81}Br. Therefore,

- the odds that both Br atoms are the ^{79}Br isotope (corresponding to the M peak) = (.50)(.50) = 25%
- the odds of having one Br atom of each isotope (two possible combinations for M+2) = (.50)(.50) + (.50)(.50) = 50%
- the odds that both Br atoms are the ^{81}Br isotope (corresponding to the M+4 peak) = (.50)(.50) = 25%

The ratio of the three peaks (M : M+2 : M+4) will be 25:50:25 (1:2:1). This analysis matches the figure above.

14.73.

Explanation #1 – Outside of the carbonyl group, one of the C–O bonds of the ester has some double bond character, as can be seen in the third resonance structure below:

This C–O bond is a stronger bond than the other C–O single bond, which does not have any double bond character. As a result, the stronger C–O bond (highlighted) produces a signal at higher wavenumber.

1300 cm^{-1} 1000 cm^{-1}

Explanation #2 – The C–O bond at 1300 cm^{-1} involves an sp^2-hybridized carbon atom, rather than an sp^3-hybridized carbon atom. The former has more *s*-character and holds its electrons closer to the positively charged nucleus. A C_{sp^2}–O bond is therefore shorter and stronger than a C_{sp^3}–O bond, so the former should produce a signal at higher wavenumber.

Both of these factors (resonance and hybridization) lead to the same conclusion that the C–O bond to the carbonyl carbon is stronger and shorter than the C–O bond to the sp^3 carbon atom.

14.74. A signal at 1720 cm^{-1} indicates the presence of a C=O bond. The following mechanism justifies the formation of a C=O bond. In the first step, one of the OH groups is protonated to give an oxonium ion, which then loses a leaving group (H_2O) to give a secondary carbocation. This carbocation then rearranges to give a more stable, resonance-stabilized cation, which is deprotonated to give the product (a ketone), consistent with the signal at 1720 cm^{-1}.

14.75. For a compound containing a bromine atom, the $(M+2)^{+\bullet}$ peak is approximately the same height as the $(M)^{+\bullet}$ peak, because approximately half of the bromine atoms are ^{81}Br, rather than ^{79}Br.

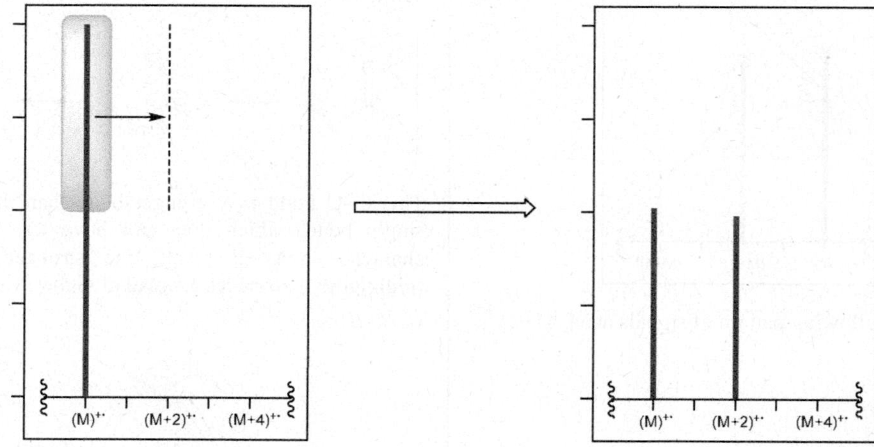

Now let's consider what happens if the parent ion also has a chlorine atom. In the solution to Problem **14.70**, we saw that the effect of a chlorine atom is to move ¼ of each peak by two *m/z* units, which gives the following pattern.

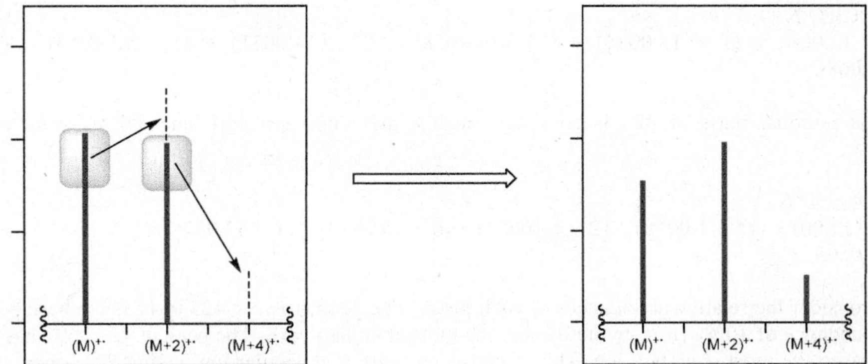

Incidentally, we would arrive at the same characteristic pattern if we had started our analysis by first considering the effect of the chlorine atom and only then considering the effect of the bromine atom, as done in Problem **14.72** and illustrated here:

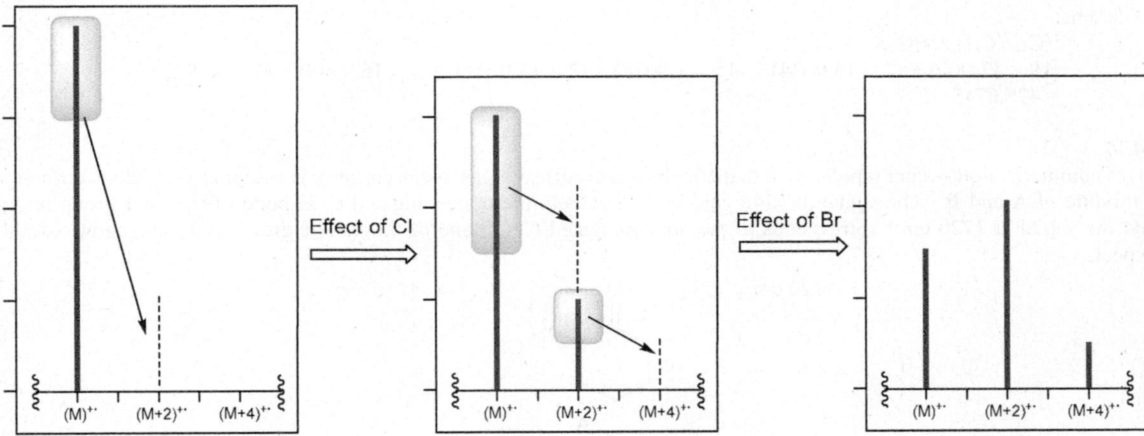

Below is a mathematical analysis that provides the same answer as shown above.
Each Br atom has (approximately) a 50% chance of being ^{79}Br and a 50% chance of being ^{81}Br. Furthermore, each Cl atom has a 75% chance of being ^{35}Cl and a 25% chance of being ^{37}Cl. Therefore,

- the odds of having ^{79}Br and ^{35}Cl (corresponding to the M peak) = (.50)(.75) = 37.5%
- the odds of having one light isotope and one heavy isotope (either ^{79}Br and ^{37}Cl, or ^{81}Br and ^{39}Cl), representing the M+2 peak = (.50)(.75) + (.50)(.25) = 50%
- the odds of having ^{81}Br and ^{37}Cl (corresponding to the M+4 peak) = (.50)(.25) = 12.5%

The ratio of the three peaks (M : M+2 : M+4) will be 37.5 : 50 : 12.5 (3:4:1). This analysis matches the figure above.

14.76.

(a) The formula for the dye is $C_{21}H_{15}N_2O_6S^-$. Using the values in Table 14.5 for C, H, N and O, and the values given in the question for S, we can calculate the expected mass for the parent ion using the most abundant isotope of each element (^{1}H, ^{12}C, ^{14}N, ^{16}O, ^{32}S) as follows.

$C_{21}H_{15}N_2O_6S^-$ =
(21 × 12.000) + (15 × 1.0078) + (2 × 14.0031) + (6 × 15.9949) + (1 × 31.9721)
= **423.0647**

This is consistent with reported base peak of the spectrum [(M^-), 100% relative intensity].
Now we must consider the other two peaks – the peak at 424.0681 is ~ 1 amu higher than the molecular ion, while the peak at 425.0605 is ~ 2 amu higher. Considering the relative natural isotopic abundances of C, H, N, O and S, the two elements that have the highest percentage of a second isotope are carbon (^{12}C = 98.93%, ^{13}C = 1.07%) and sulfur (^{32}S = 95.02%, ^{34}S = 4.21%). With this in mind, the peak at m/z = 424.0681 is consistent with a molecular ion where one of the ^{12}C atoms is replaced with a ^{13}C.

$^{12}C_{20}{}^{13}C_1H_{15}N_2O_6S^- =$
$(20 \times 12.000) + (1 \times 13.0034) + (15 \times 1.0078) + (2 \times 14.0031) + (6 \times 15.9949) + (1 \times 31.9721)$
$=$ **424.0681**

The peak with a nominal mass of 425 is consistent with a molecular ion with one ^{34}S, as demonstrated by the calculation below:

$C_{21}H_{15}N_2O_6{}^{34}S^- =$
$(21 \times 12.000) + (15 \times 1.0078) + (2 \times 14.0031) + (6 \times 15.9949) + (1 \times 33.9679)$
$=$ **425.0605**

Now, we must consider the relative abundances of each peak. The peak at $m/z = 423.0647$ is the base peak, because it has a relative abundance of 100% (it is, by definition, the most abundant ion). The peak at 424.0681 has an abundance of 22.5%, relative to the most abundant ion. This is consistent with a molecular ion having 21 carbon atoms, because the natural abundance of ^{13}C is 1.07%. [21 × 1.07% = 22.5%]
The peak at 425.0605 has an abundance of 4.21%, relative to the most abundant ion. This is consistent with a molecular ion having one sulfur atom, because the natural abundance of ^{34}S is 4.21%. [1 × 4.21% = 4.21%]

(b) The peak at $m/z = 425.0605$ is consistent with a molecular ion in which two of the ^{12}C atoms are replaced with two ^{13}C atoms:

$^{12}C_{19}{}^{13}C_2H_{15}N_2O_6S^- =$
$(19 \times 12.000) + (2 \times 13.0034) + (15 \times 1.0078) + (2 \times 14.0031) + (6 \times 15.9949) + (1 \times 31.9721)$
$=$ **425.0715**

14.77.
(a) Tautomerization occurs rapidly (and is difficult to prevent) so an IR spectrum of **A** is essentially an IR spectrum of a mixture of **A** and **B**. The signal at 1740 cm^{-1} corresponds to the nonconjugated C=O bond of the ester group in **A**, and the signal at 1720 cm^{-1} corresponds to the nonconjugated C=O bond of the ketone group in **A**, just as we would expect.

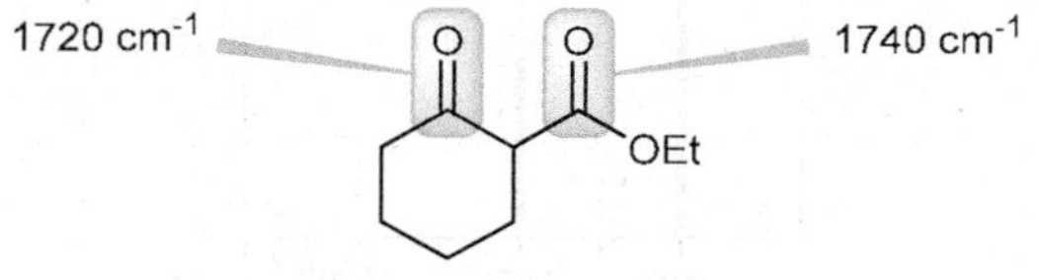

1720 cm^{-1} 1740 cm^{-1}
O O OEt
A

The C=O bond in compound **B** is part of a conjugated ester, so it appears at a lower wavenumber than a typical C=O bond of an ester. This must be the signal at 1660 cm^{-1}. This might be initially surprising, as we might have expected a conjugated ester to produce a signal around 1700 cm^{-1}. But on further inspection, we recognize that this conjugated system has an OH group, whose lone pairs are participating in resonance.

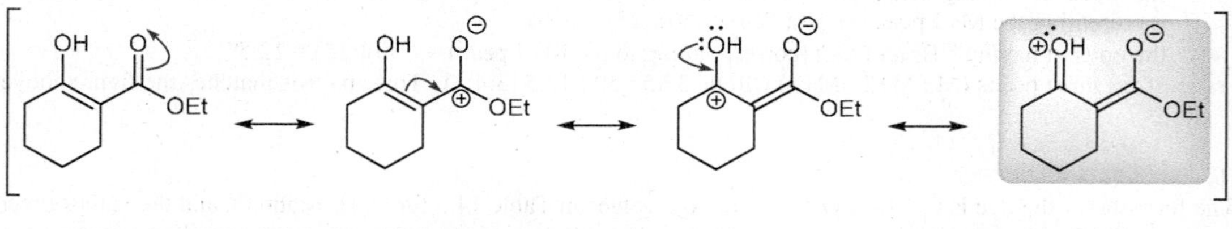

OH O OEt ⟷ OH O⊖ ⊕ OEt ⟷ :ÖH O⊖ ⊕ OEt ⟷ ⊕ÖH O⊖ OEt

As such, there is one additional resonance structure (highlighted above); this additional resonance structure gives the C=O bond additional single bond character (relative to other conjugated esters).
Finally, the signal at 1620 cm^{-1} corresponds with the conjugated C=C bond in **B**.

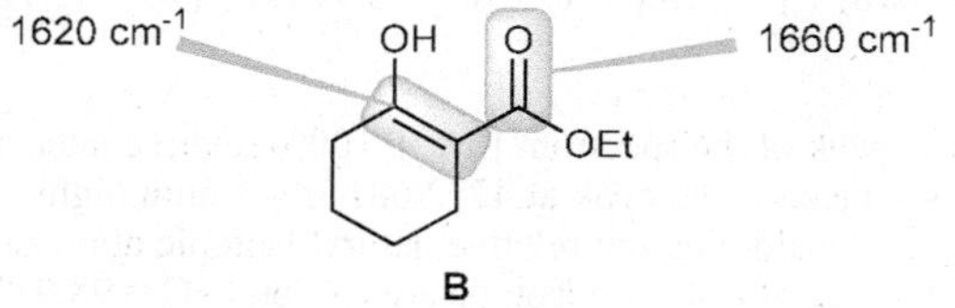

1620 cm^{-1} OH O 1660 cm^{-1}
OEt
B

(b) Compound **B** is capable of forming an intramolecular hydrogen bonding interaction, which lowers the energy of that tautomer. As such, it has a more significant presence at equilibrium.

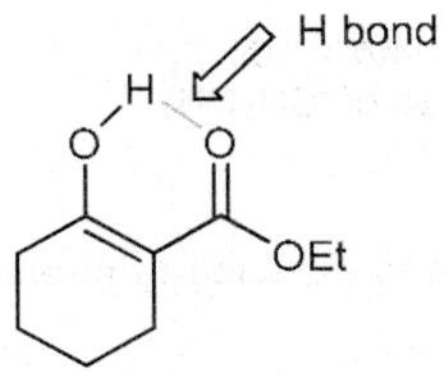

H bond

OEt

There is one additional factor that favors the enol in this case, which we will discuss in Chapter 16. Specifically, we will see that conjugation is a source of stabilization. Compound **B** possesses this type of stabilization, while compound **A** does not.

14.78.

The two compounds are isomers (both have the molecular formula C_8H_9BrO) so we expect the molecular ion from each of them to have the same *m/z* value. Recall that there are two isotopes of bromine with approximately equivalent natural abundance (^{81}Br and ^{79}Br), which results in two molecular ion peaks for each compound:

Molecular ions for **A**

m/z = 202 *m/z* = 200

Molecular ions for **B**

m/z = 202 *m/z* = 200

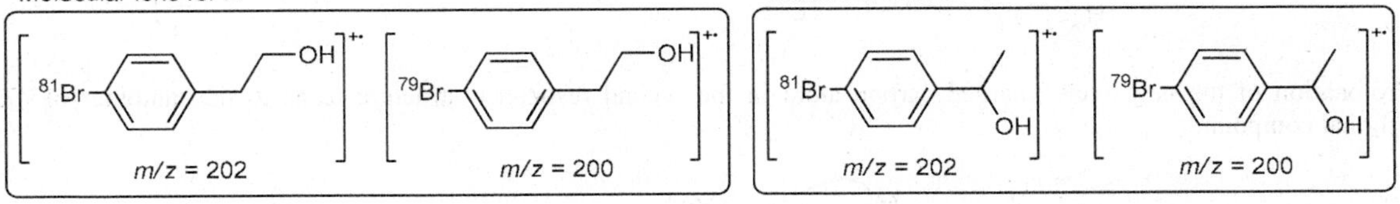

These molecular ions correspond to the two highest *m/z* peaks in each spectrum. Now we have to consider fragmentation patterns to match each spectrum to the correct isomer. Note that in each spectrum, there are several pairs of peaks separated by two mass units. This is consistent with the presence of bromine in these fragments, with the two peaks corresponding to fragments with ^{79}Br and ^{81}Br, respectively, just as observed for the molecular ion. In the peak list below, pairs separated by two mass units are shown in brackets.

Sample **X**, *m/z* = [202, 200], [187, 185], [159, 157], 121
Sample **Y**, *m/z* = [202, 200], [171, 169], 121

Focusing on sample **X**, the peaks at 187 and 185 represent a loss of 15 from the molecular ions 202 and 200, respectively. This is consistent with the loss of a $\bullet CH_3$ radical fragment. Compound **B** has a methyl group, while compound **A** does not, suggesting that this spectrum is from compound **B**. The peaks at 159 and 157 represent a loss of 43 from the molecular ions 202 and 200, respectively. This is consistent with the loss of a $\bullet CH(OH)CH_3$ radical fragment. The peak at 121 represents a loss of ^{81}Br• radical from the heavier molecular ion (202 – 81 = 121) or a loss of ^{79}Br• radical from the lighter molecular ion (200 – 79 = 121). These data are consistent with the structure of compound **B**, and the ions below.

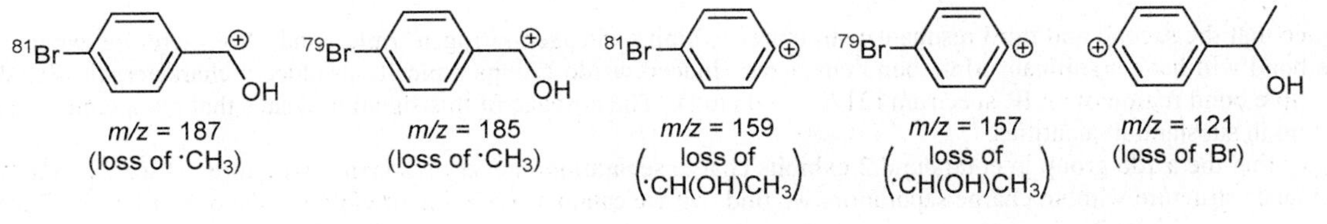

m/z = 187 (loss of ·CH_3)

m/z = 185 (loss of ·CH_3)

m/z = 159 (loss of ·CH(OH)CH_3)

m/z = 157 (loss of ·CH(OH)CH_3)

m/z = 121 (loss of ·Br)

Focusing on sample **Y**, the peaks at 171 and 169 represent a loss of 31 from the molecular ions 202 and 200, respectively. This is consistent with the loss of a $\bullet CH_2OH$ radical fragment. The peak at 121 represents a loss of ^{81}Br• radical from the heavier molecular ion (202 – 81 = 121) or a loss of ^{79}Br• radical from the lighter molecular ion (200 – 79 = 121), analogous to the corresponding peak in sample **X**. These data are consistent with the structure of compound **A** and with the ions shown below.

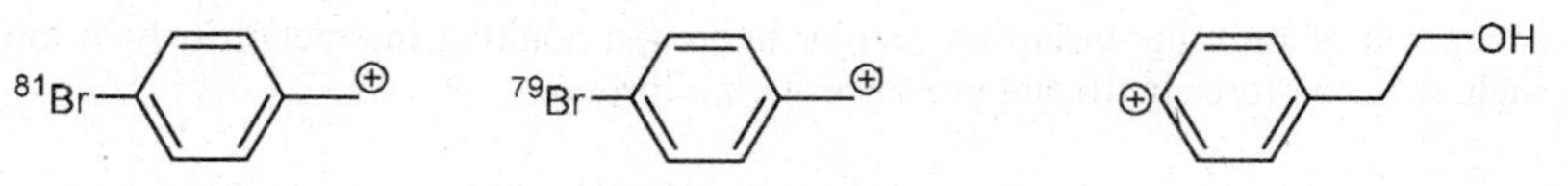

m/z = 171 (loss of ·CH_2OH)

m/z = 169 (loss of ·CH_2OH)

m/z = 121 (loss of ·Br)

14.79.

(a) The structure contains an N–H bond, which is expected to produce a signal above 3300 cm^{-1}, but that signal is absent from the spectrum.

(b) Recall that tautomers are constitutional isomers that can rapidly interconvert via a sequence of proton transfer reactions. We know that the tautomer of the proposed structure must not exhibit an N–H bond, so it is clear that this proton is involved in the tautomerization process. To determine the structure of the tautomer, we remove the proton (deprotonation is the first step when tautomerization occurs under base-catalyzed conditions) and draw the conjugate base, which is resonance-stabilized:

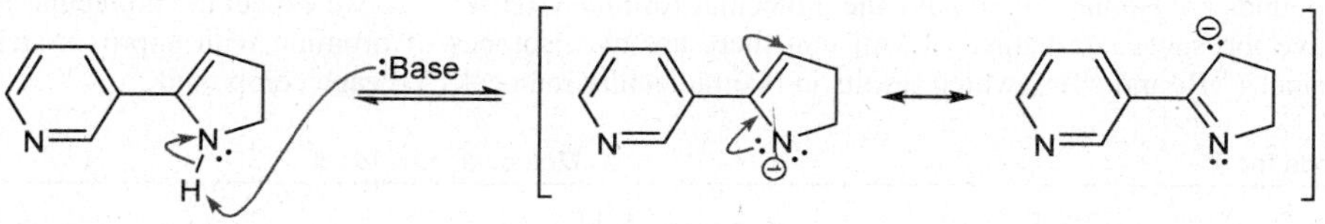

Protonation of the negatively charged carbon atom in the second resonance structure leads to the tautomer of the original compound:

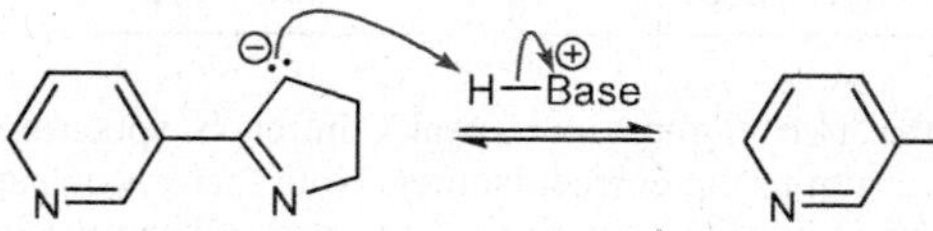

This compound lacks an N–H bond and is therefore consistent with the IR spectrum.

(c) The tautomer of the proposed structure exhibits a C=N bond, which is likely to be the source of the signal at 1621 cm^{-1} (this is the region of the spectrum where double bonds generally appear).

14.80. The azido group in compound **2** cannot be drawn without charge separation (much like a nitro group, which also cannot be drawn without charge separation). Nevertheless, the azido group is still stabilized by resonance, as seen here:

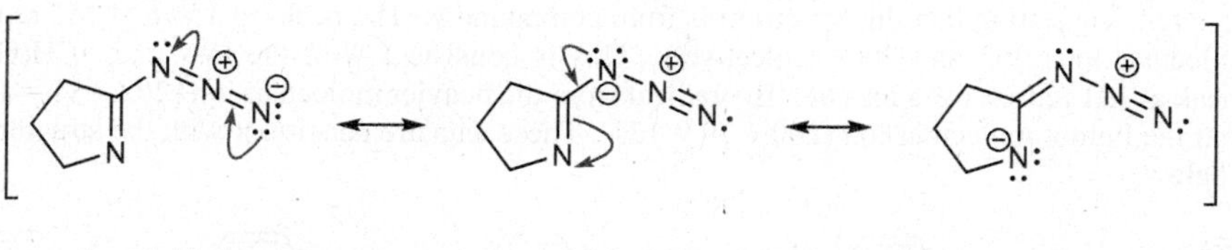

Notice that the second and third resonance structures exhibit a nitrogen-nitrogen triple bond. We therefore expect that this bond will have significant triple bond character. Indeed, azido groups typically produce a characteristic signal in the triple bond region of an IR spectrum (2120-2160 cm^{-1}). The absence of this signal indicates that compound **2** is not present in substantial quantities.

Notice that the azido group in compound **2** exhibits charge separation in every resonance structure. If we try to draw a resonance structure without charge separation, we find that we cannot do so without violating the octet rule:

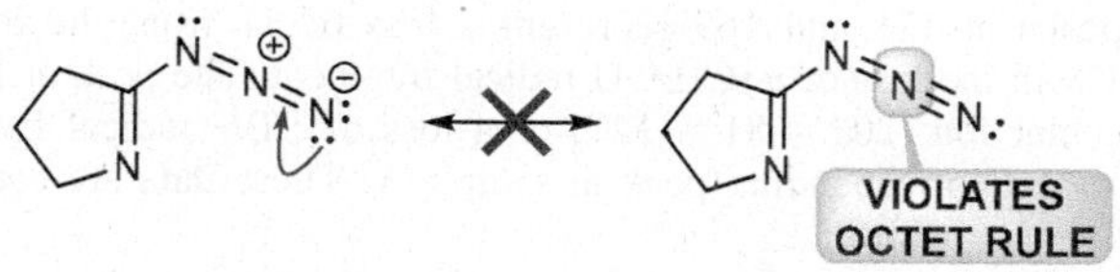

VIOLATES OCTET RULE

14.81. The molecular ion for cyclohexanone is expected at m/z = 98, so the two peaks at m/z = 55 must be due to cationic fragments. Considering the atoms present in cyclohexanone ($C_6H_{10}O$), we must consider all possible cationic formulas with m/z = 55.

One approach to this is to start with the heavier elements (carbon, oxygen) to come up with a formula close to (but not greater than) 55 and then adding in hydrogen atoms to make up the difference. For example, a cation with 4 carbon atoms (mass = 48) would require 7 hydrogen atoms (mass = 7) to give the correct total mass [48+7=55]. Alternatively, we can start with one oxygen atom (mass = 16), add three carbon atoms (mass = 36), and 3 hydrogen atoms (mass = 3) [16+36+3=55]. Thus, our two possible cationic formulas that fit are $C_3H_3O^+$ and $C_4H_7^+$.
Using the values in Table 14.5, we can calculate the precise mass for each formula:

$C_3H_3O^+$ = (3 × 12.000) + (3 × 1.0078) + (1 × 15.9949) = **55.0183** amu
$C_4H_7^+$ = (4 × 12.000) + (7 × 1.0078) = **55.0546** amu

The peak at m/z = 55.0183 (with a relative intensity of 86.7) is thus assigned to $C_3H_3O^+$, and the peak at m/z = 55.0546 (with a relative intensity of 13.3) is assigned to $C_4H_7^+$.

14.82. This amide has the structural feature required for a McLafferty rearrangement: a hydrogen atom that is gamma to a C=O group (analogous to the ketones and aldehydes discussed in section 14.12). The resulting ion (m/z = 113) and its mechanism of formation are presented below.

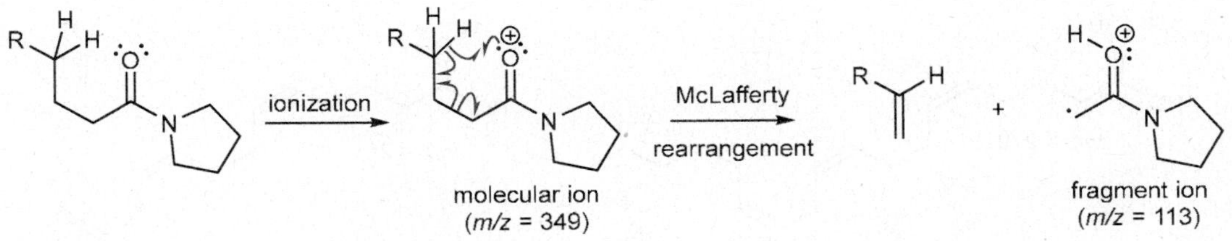

ionization
molecular ion
(m/z = 349)
McLafferty rearrangement
+
fragment ion
(m/z = 113)

14.83.
(a) The mass of pinolenic acid is 278 amu. The significant peak at 279 in positive ion mode is likely due to the addition of a proton, as this experiment is performed under conditions facilitating protonation. We can refer to this peak as the $(M+1)^+$ ion. Note that unlike molecular ions produced from electron ionization, this species is not a radical.

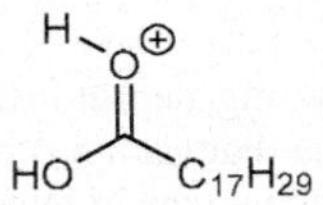

HO $C_{17}H_{29}$

Next, we need to determine the positions of the three *cis* double bonds. The peaks at 211, 171 and 117 (which only appear when ozone is present) are likely due to ozonolysis products of each of the three alkenes. In each case, the C=C is replaced with C=O bonds. These three peaks represent losses of 68, 108, and 162 from the $(M+1)^+$ ion. The change in mass between the $(M+1)^+$ ion and each of the ozonolysis products corresponds to the loss of carbon atoms and hydrogen atoms (from the chain terminus to each sp^2-hybridized carbon atom) as well as the gain of an oxygen atom. Accounting for the additional mass of oxygen (16) in each ionic fragment, the portions of the molecule cleaved by ozonolysis have masses of **84** (= 68+16), **124** (= 108+16) and **178** (= 162+16).

With this information, we can reconstruct the chain starting from the end distal to the carboxylic acid. A loss of 84 is consistent with the loss of a C_6H_{12} fragment.

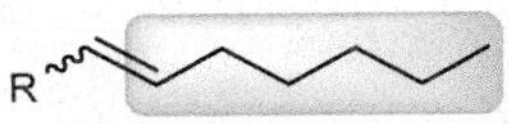

R
C_6H_{12}

Continuing up the chain, a loss of 124 is consistent with the loss of a C_9H_{16} fragment.

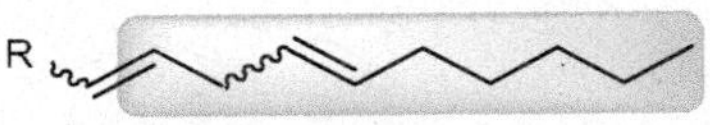

R
C_9H_{16}

A loss of 178 is consistent with the loss of a $C_{13}H_{22}$ fragment.

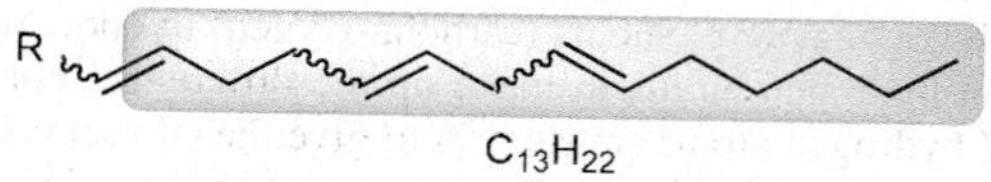

$C_{13}H_{22}$

With all of this information in hand, along with the knowledge that all of the double bonds have the *cis* configuration, we can propose the following structure for pinolenic acid:

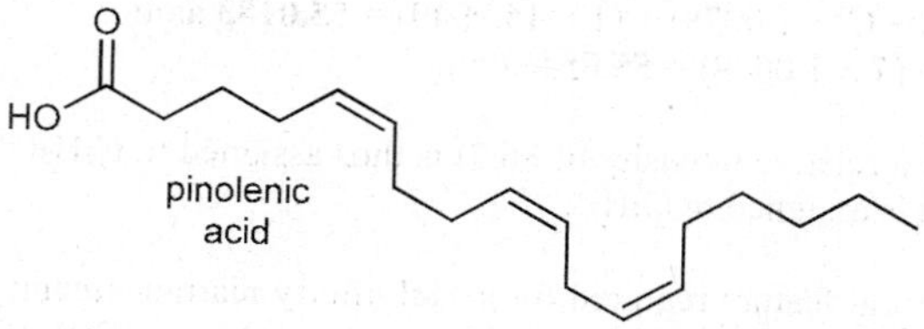

pinolenic acid

And we can propose the following structures for the four indicated fragments in positive ion mode.

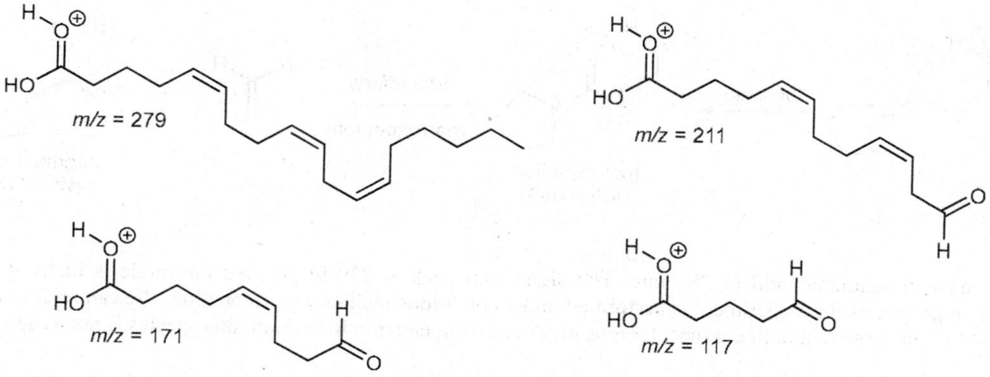

m/z = 279

m/z = 211

m/z = 171

m/z = 117

(b) In negative ion mode, under conditions facilitating deprotonation, we expect to form the conjugate base of the carboxylic acid, thus giving the following ions. Note that each has *two* mass units lower than the corresponding ions in positive ion mode, because each anionic structure has *two* fewer hydrogen atoms than each cationic structure.

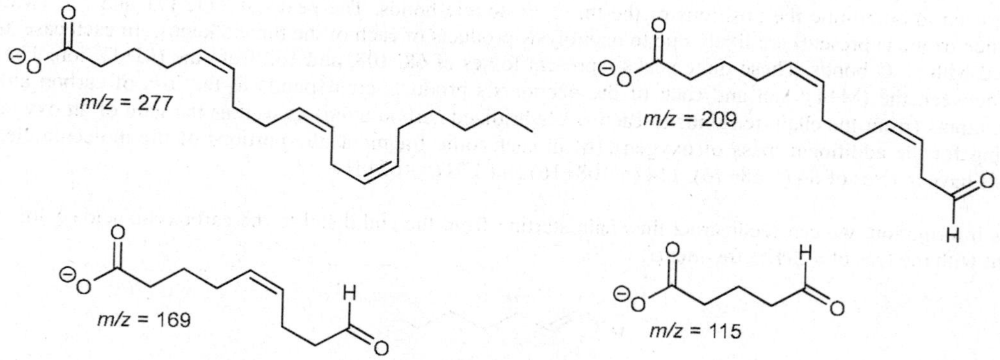

m/z = 277

m/z = 209

m/z = 169

m/z = 115

14.84.

(a) Each of these compounds has four rings, which we can label A-D:

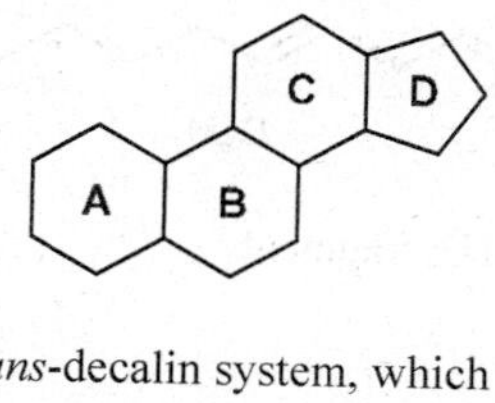

Notice that the A-B ring fusion resembles a *trans*-decalin system, which we explored at the end of Chapter 4:

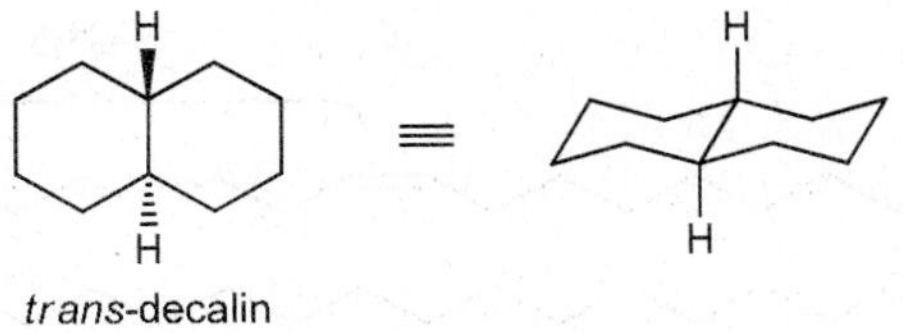

trans-decalin

Now let's consider the conformation of the four fused rings in compounds **1** and **2**. Building on the decalin system as a foundation, we can redraw these compounds (starting with the A-B ring fusion), in a way that shows the conformation of each ring:

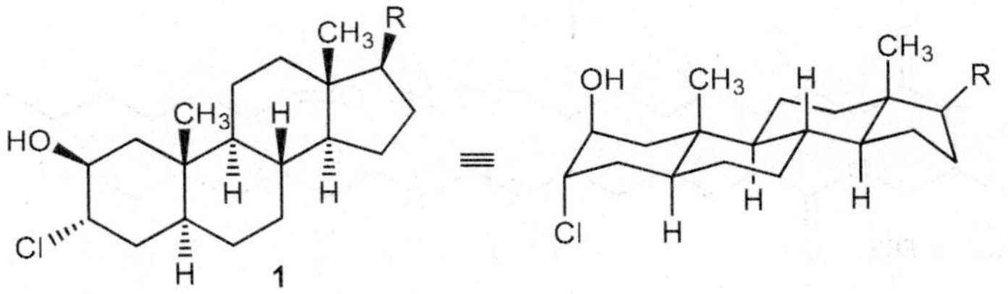

These polycyclic systems are conformationally rigid, because none of the six-membered rings can undergo a ring flip (you might find it helpful to build a model to prove to yourself that this is the case). Notice that in compound **2**, the OH group and the Cl group are in close proximity and can therefore participate in an intramolecular hydrogen-bonding interaction (which weakens the existing O-H bond in **2**, giving a lower wavenumber of absorption). In contrast, in compound **1** the OH group and the Cl group are far apart from each other, and cannot participate in an intramolecular hydrogen-bonding interaction.

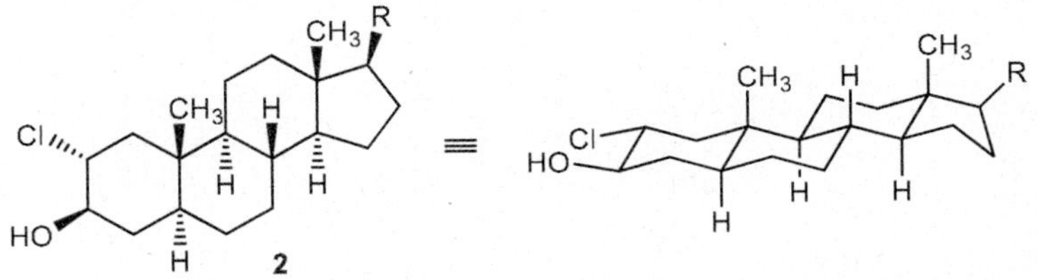

(b) The problem statement specifies that dilute solutions were investigated. As such, the effects of *inter*molecular hydrogen bonds are negligible and can be ignored. Compound **2** is capable of forming an *intra*molecular hydrogen bonding interaction (while compound **1** is not), so the signal for the OH group in compound **2** is expected to be broader.

14.85. The problem statement indicates that one of the two hydrocarbon chains contains a *cis* alkene group. To determine which of these two chains contains the double bond, consider the relative numbers of carbon atoms and hydrogen atoms in each chain. One of these chains ($C_{15}H_{31}$) is fully saturated, because there are two hydrogen atoms for every carbon atom in the chain, except for the terminal methyl group which has three hydrogen atoms. In contrast, the other chain ($C_{17}H_{33}$) must be unsaturated, because a saturated chain with 17 carbon atoms would be expected to have 35 hydrogen atoms (16 C with two H's and one C with three H's). Since it has only 33 hydrogen atoms, this chain must contain the double bond.

Now that we know which chain contains the double bond, we can determine the position of the double bond by analyzing the mass spectrometry data. The introduction of ozone allows for ozonolysis of the alkene, breaking the C=C bond and producing two new C=O bonds. The peak at m/z = 563 provides the information needed to determine the position of the double bond. The change in mass between the parent ion and the product of ozonolysis (110 mass units) corresponds to the loss of carbon atoms and hydrogen atoms (from the chain terminus to the first sp^2-hybridized carbon) as well as the gain of an oxygen atom. Accounting for the additional mass of oxygen (16), the portion of the molecule that was cleaved by ozonolysis has a mass of 126. This is consistent with the loss of a fragment with the formula C_9H_{18}. To summarize, 563 amu = 673 (molecular ion) – 126 (C_9H_{18}) + 16 (oxygen atom). The structure of the phospholipid, along with the ozonolysis product are shown here:

C_9H_{18}

HO–P(=O)(O⊖)–O

m/z = 673

ozonolysis

(+ O=)

m/z = 563

Chapter 15
Nuclear Magnetic Resonance Spectroscopy

Review of Concepts

Fill in the blanks below. To verify that your answers are correct, look in your textbook at the end of Chapter 15. Each of the sentences below appears verbatim in the section entitled *Review of Concepts and Vocabulary*.

- A spinning proton generates a **magnetic** _____________, which must align either with or against an imposed external magnetic field.
- All protons do not absorb the same frequency because of _______________, a weak magnetic effect due to the motion of surrounding electrons that either **shield** or **deshield** the proton.
- _______________ solvents containing a small amount of TMS are generally used for acquiring NMR spectra.
- In a ^{1}H NMR spectrum, each signal has three important characteristics: ________, ________ and ___________.
- When two protons are interchangeable by rotational symmetry, the protons are said to be ___________.
- When two protons are interchangeable by reflectional symmetry, the protons are said to be ___________.
- The left side of an NMR spectrum is described as _____ **field**, and the right side is described as _____ **field**.
- In the absence of inductive effects, a methyl group (CH_3) will produce a signal near _____ppm, a **methylene group** (CH_2) will produce a signal near _____, and a __________ **group** (CH) will produce a signal near _____. The presence of nearby groups increases these values somewhat predictably.
- The _______________, or area under each signal, indicates the number of protons giving rise to the signal.
- _______________ represents the number of peaks in a signal. A __________ has one peak, a __________ has two, a __________ has three, a __________ has four, and a __________ has five.
- Multiplicity is the result of **spin-spin splitting**, also called __________, which follows the ***n* + 1 rule**.
- When signal splitting occurs, the distance between the individual peaks of a signal is called the **coupling constant**, or _____ **value**, and is measured in hertz.
- Complex splitting occurs when a proton has two different kinds of neighbors, often producing a __________.
- ^{13}C is an __________of carbon, representing _____% of all carbon atoms.
- All ^{13}C-^{1}H splitting is suppressed with a technique called **broadband** _____________, causing all of the ^{13}C signals to collapse to ___________.

Review of Skills

Fill in the blanks and empty boxes below. To verify that your answers are correct, look in your textbook at the end of Chapter 15. The answers appear in the section entitled *SkillBuilder Review.*

15.1 Determining the Relationship between Two Protons in a Compound

For each of the following compounds, identify the relationship between the two indicated protons (are they homotopic, enantiotopic or diastereotopic?) and determine whether they are chemically equivalent.

Relationship:

Chemically equivalent?

15.2 Identifying the Number of Expected Signals in a ^{1}H NMR Spectrum

For each of the following compounds, determine whether the indicated protons are chemically equivalent.

Chemically equivalent?

15.3 Predicting Chemical Shifts

For each of the following compounds, predict the expected chemical shift of the indicated protons.

ppm ppm ppm

15.4 Determining the Number of Protons Giving Rise to a Signal

Step 1 Compare the relative ________ values, and choose the lowest number.	Step 2 Divide all integration values by the number from step 1, which gives the ratio of ________.	Step 3 Identify the number of protons in the compound (from the molecular formula) and then adjust the relative integration values so that the sum total equals the number of ________.

15.5 Predicting the Multiplicity of a Signal

Identify the expected multiplicity for each signal in the proton NMR spectrum of the following compound:

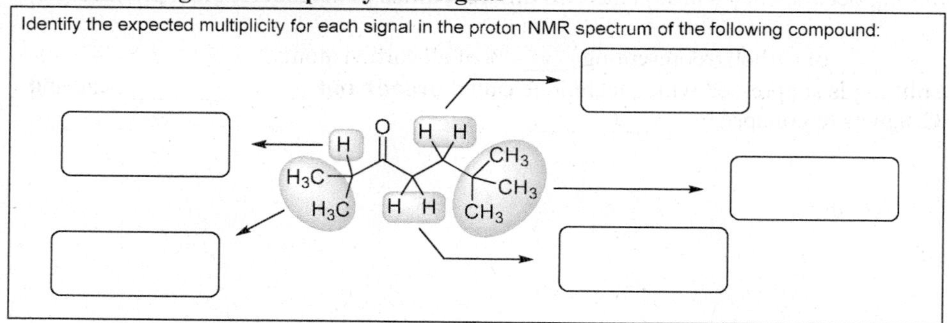

15.6 Drawing the Expected ^{1}H NMR Spectrum of a Compound

Step 1 Identify the number of ________.	Step 2 Predict the ________ of each signal.	Step 3 Determine the ________ of each signal by counting the number of ________ giving rise to each signal.	Step 4 Predict the ________ of each signal.	Step 5 Draw each signal.

15.7 Using ^{1}H NMR Spectroscopy to Distinguish Between Compounds

Step 1 Identify the number of ___________ that each compound will produce.	**Step 2** If each compound is expected to produce the same number of signals, then determine the ___________, ___________, and ___________ of each signal in both compounds.	**Step 3** Look for differences in the chemical shifts, multiplicities or integration values of the expected signals.

15.8 Analyzing a ^{1}H NMR Spectrum and Proposing the Structure of a Compound

Step 1 Use the ___________ to determine the ___________ HDI. An HDI of ____ indicates the possibility of an aromatic ring.	**Step 2** Consider the number of signals and integration of each signal (gives clues about the ___________ of the compound).	**Step 3** Analyze each signal (____, ____, and ____), and then draw fragments consistent with each signal. These fragments become our puzzle pieces that must be assembled to produce a molecular structure.	**Step 4** Assemble the fragments.	**Step 5** Verify that the proposed structure is consistent with all of the spectral data.

15.9 Predicting the Number of Signals and Approximate Location of Each Signal in a ^{13}C NMR Spectrum

sp^3 and *sp* hybridized carbon atoms produce signals below ______ ppm, while *sp*^{2} hybridized carbon atoms produce signals above ______ ppm. When *sp*^{2} hybridized carbon atoms are part of an aromatic ring, they produce signals between ____ and ____ ppm, while *sp*^{2} hybridized carbon atoms of carbonyl (C=O) groups produce weak signals above ______ ppm.

15.10 Determining Molecular Structure using DEPT ^{13}C NMR Spectroscopy

Complete the following chart by drawing the expected shape of each signal:

	CH_3	CH_2	CH	C
Broadband decoupled				
DEPT-90				
DEPT-135				

Mistakes to Avoid

In ^{1}H NMR spectroscopy, two neighboring CH_2 groups will only split each other if they are not chemically equivalent. For example compare the structures of butane and pentane, and in particular, compare the CH_2 group at the C2 position in each structure, highlighted below:

quartet — Butane: $CH_3-CH_2-CH_2-CH_3$ (H–C–C–C–C–H with all H shown)

sextet — Pentane: $CH_3-CH_2-CH_2-CH_2-CH_3$ (H–C–C–C–C–C–H with all H shown)

Butane Pentane

In the case of butane, the signal for the highlighted CH_2 group is expected to be a quartet as a result of the neighboring methyl group. Notice that the signal is not further split by the neighboring CH_2 group (C3), because the CH_2 groups at C2 and C3 are chemically equivalent and therefore, they don't split each other. In contrast, the signal for the highlighted CH_2 group does experience splitting. In the case of pentane, the C2 and C3 positions are not chemically equivalent. They occupy different electronic environments, and they are not interchangeable by symmetry. As a result, the signal for the highlighted CH_2 group (C2) is expected to be split into a sextet because it has 5 neighbors ($n + 1 = 6$). When analyzing a ^{1}H NMR spectrum, make sure to take this into account, as students will often misinterpret a spectrum as a result of failing to take this into account.

Solutions

15.1.

(a) When looking for symmetry, don't be confused by the position of the double bonds in the aromatic ring. Recall that we can draw the following two resonance structures:

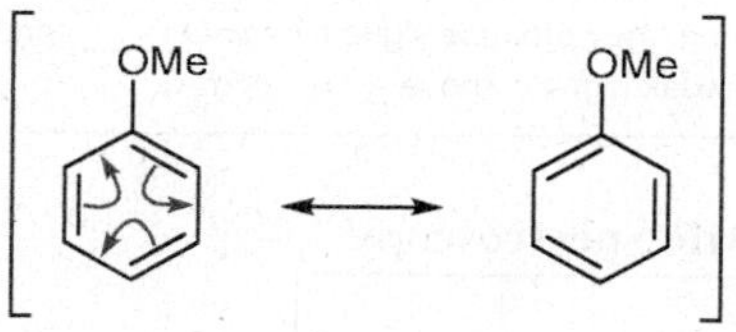

Neither resonance structure is more correct than the other. For purposes of looking for symmetry, it will be less confusing to draw the compound like this:

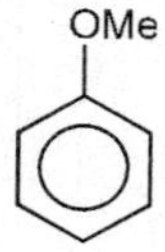

When drawn in this way, we can see that the two highlighted protons can be interchanged by rotational symmetry (the axis of symmetry is shown below). Therefore, the protons are homotopic.

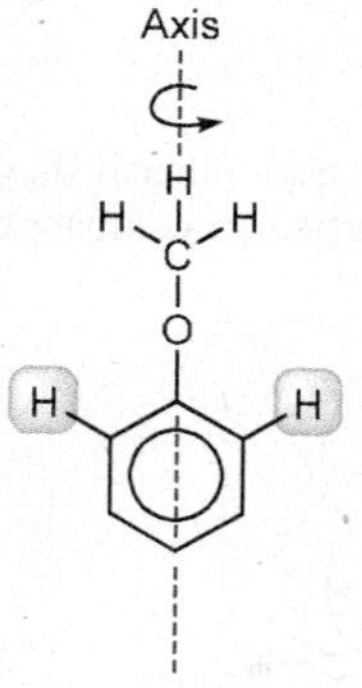

This conclusion can be verified by the replacement test. Specifically, each proton is replaced with deuterium, and the resulting compounds are found to be the same. Therefore, the protons are homotopic:

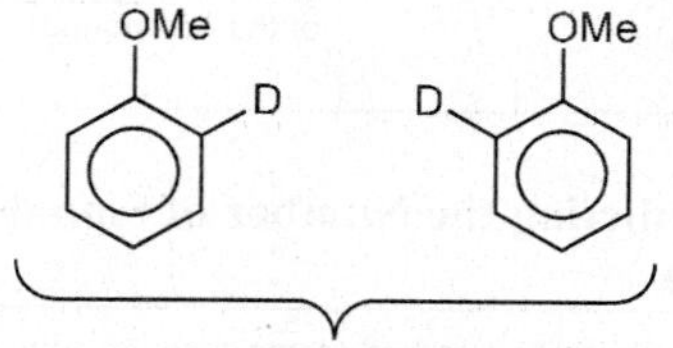

Same compound

(b) These protons cannot be interchanged by rotational symmetry, so they are not homotopic. They can be interchanged by reflectional symmetry (the plane of symmetry is the plane of the page). Therefore, the protons are enantiotopic.

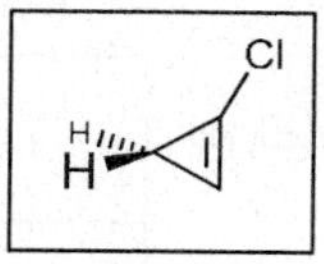

This conclusion can be verified by the replacement test. Specifically, each proton is replaced with deuterium, and the resulting compounds are found to be enantiomers. Therefore, the protons are enantiotopic:

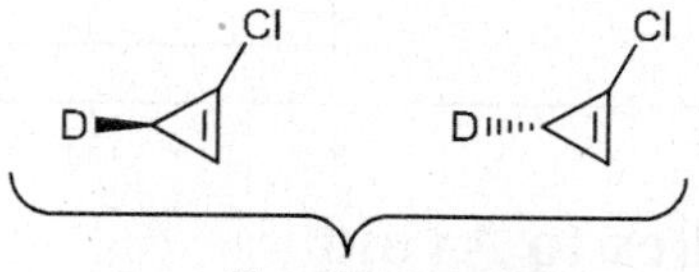

Enantiomers

(c) These protons cannot be interchanged by rotational symmetry, so they are not homotopic. They also cannot be interchanged by reflectional symmetry so they are not enantiotopic either. To determine if they are diastereotopic, we use the replacement test. Specifically, each proton is replaced with deuterium, and the resulting compounds are found to be diastereomers. Therefore, the protons are diastereotopic:

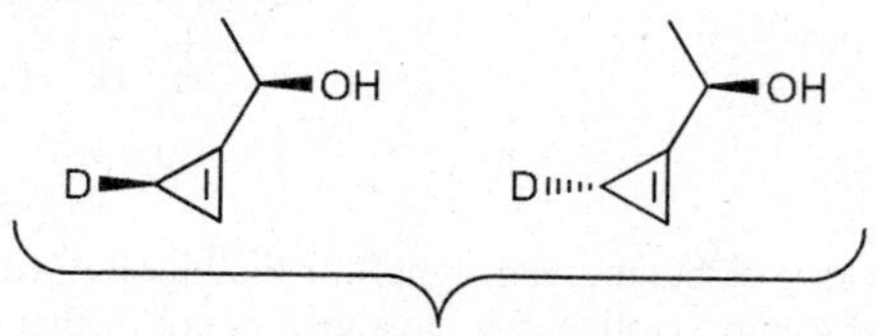

Diastereomers

(d) These protons cannot be interchanged by rotational symmetry, so they are not homotopic. They can be interchanged by reflectional symmetry (the plane of symmetry is the plane of the page). Therefore, the protons are enantiotopic.

This conclusion can be verified by the replacement test. Specifically, each proton is replaced with deuterium, and the resulting compounds are found to be enantiomers. Therefore, the protons are enantiotopic:

Enantiomers

(e) These protons can be interchanged by rotational symmetry (the axis of symmetry is shown below). Therefore, the protons are homotopic.

Axis

This conclusion can be verified by the replacement test. Specifically, each proton is replaced with deuterium, and the resulting compounds are found to be the same. Therefore, the protons are homotopic:

Same compound

15.2.

(a) All four protons shown in red can be interchanged either via rotation or reflection, so they are all chemically equivalent.

(b) The three protons of a methyl group are always equivalent (as we will soon see, immediately after the SkillBuilder), and in this case, the two methyl groups are equivalent to each other because they can be interchanged by rotation. Therefore, all six protons shown in blue are equivalent.

(c) Pentane has three different kinds of protons, shown here:

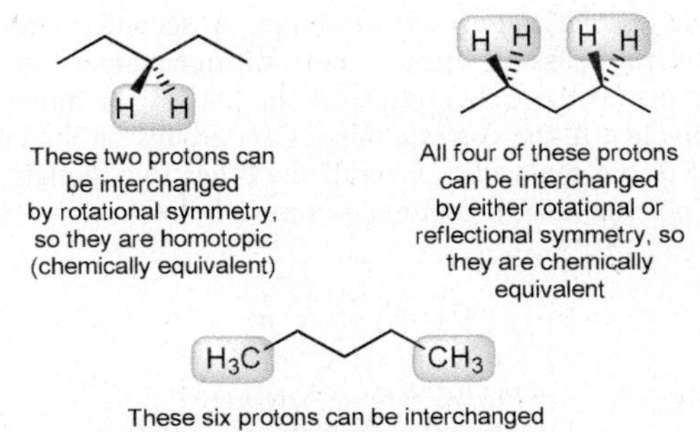
H H
These two protons can be interchanged by rotational symmetry, so they are homotopic (chemically equivalent)
H H H H
All four of these protons can be interchanged by either rotational or reflectional symmetry, so they are chemically equivalent
H3C
CH3
These six protons can be interchanged by either rotational or reflectional symmetry, so they are chemically equivalent

(d) Hexane has three different kinds of protons, shown here:

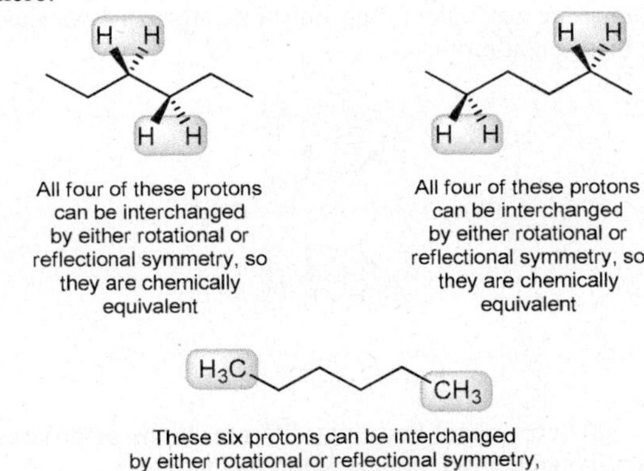
H H
H H
All four of these protons can be interchanged by either rotational or reflectional symmetry, so they are chemically equivalent
H H
H H
All four of these protons can be interchanged by either rotational or reflectional symmetry, so they are chemically equivalent
H3C
CH3
These six protons can be interchanged by either rotational or reflectional symmetry, so they are chemically equivalent

(e) The presence of a chlorine atom creates six different environments (in terms of proximity to the Cl), so there are six different kinds of protons, highlighted here:

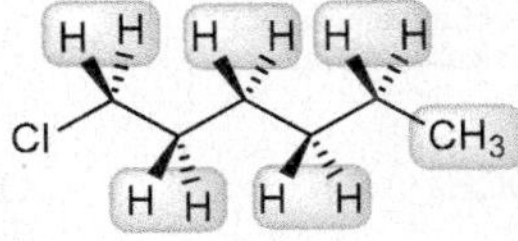
H H H H H H
Cl
CH3
H H H H

15.3. The compound must have a high degree of symmetry in order to have only one kind of proton. The molecular formula indicates that there are twelve protons. The equivalence of twelve protons can be achieved by having four methyl groups in identical environments. Four methyl groups account for four of the five carbon atoms in the compound. So, we can draw a structure in which the fifth carbon atom is connected to each of the methyl groups (shown below), providing the necessary symmetry:

15.4. In BZP, the two protons on each CH_2 group are equivalent, because they can be interchanged by reflectional symmetry (the plane of symmetry is the plane of the page). In addition, a second plane of symmetry passing through both nitrogen atoms in the ring makes the CH_2 groups on one half of the molecule equivalent to the corresponding CH_2 groups on the other half of the molecule. Overall, BZP has two signals for the piperazine CH_2 protons, as shown below:

DBZP has an additional plane of symmetry (slicing top to bottom in the structure shown), so all eight CH_2 protons are equivalent, and only one signal is expected for these eight protons.

15.5.

(a) This compound has eight different kinds of protons (highlighted below), giving rise to eight signals.

The three protons of a methyl group are always chemically equivalent

These two protons can be interchanged by reflectional symmetry, so they are chemically equivalent

The three protons of a methyl group are always chemically equivalent

These two protons can be interchanged by reflectional symmetry, so they are chemically equivalent

The three protons of a methyl group are always chemically equivalent

(b) This compound has four different kinds of protons (highlighted below), giving rise to four signals.

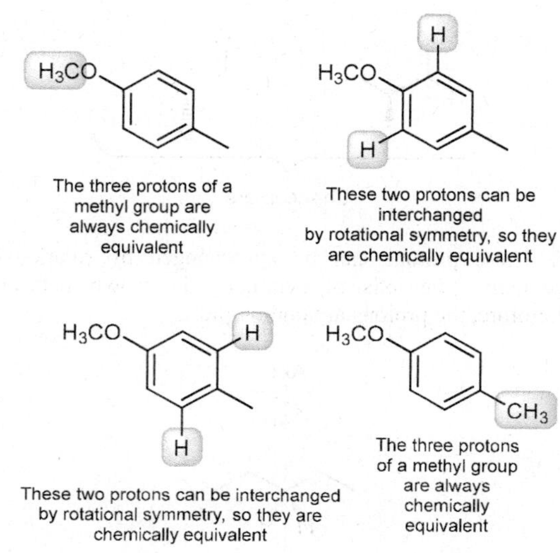
H₃CO
H₃CO
H
H
The three protons of a methyl group are always chemically equivalent
These two protons can be interchanged by rotational symmetry, so they are chemically equivalent
H₃CO
H
H
H₃CO
CH₃
These two protons can be interchanged by rotational symmetry, so they are chemically equivalent
The three protons of a methyl group are always chemically equivalent

(c) This compound has two different kinds of protons (highlighted below), giving rise to two signals.

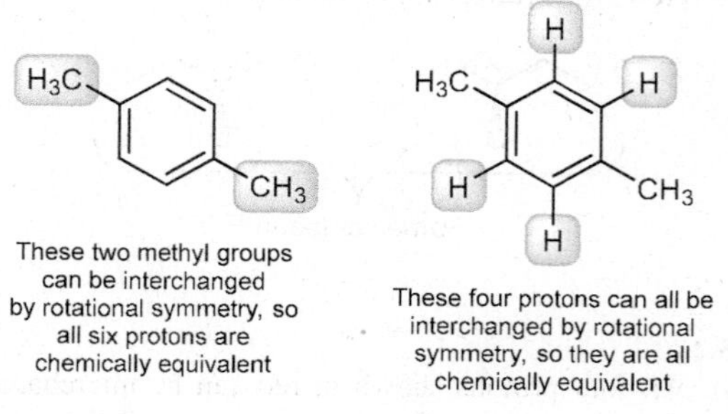
H₃C
CH₃
These two methyl groups can be interchanged by rotational symmetry, so all six protons are chemically equivalent
H₃C
H
H
H
H
CH₃
These four protons can all be interchanged by rotational symmetry, so they are all chemically equivalent

(d) This compound has three different kinds of protons (highlighted below), giving rise to three signals.

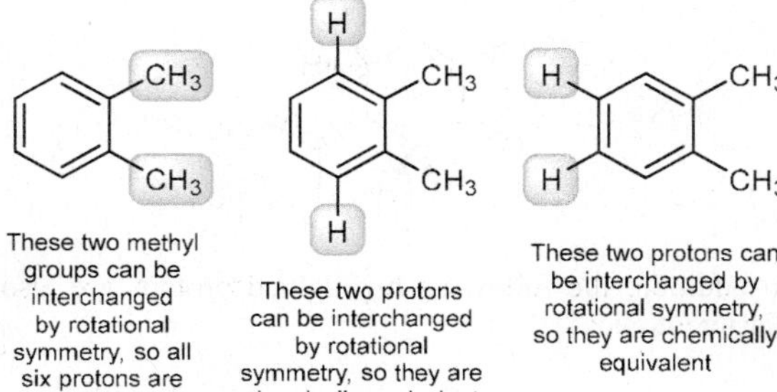

These two methyl groups can be interchanged by rotational symmetry, so all six protons are chemically equivalent

These two protons can be interchanged by rotational symmetry, so they are chemically equivalent

These two protons can be interchanged by rotational symmetry, so they are chemically equivalent

(e) This compound has five different kinds of protons (highlighted below), giving rise to five signals.

The three protons of a methyl group are always chemically equivalent

(f) This compound has three different kinds of protons (highlighted below), giving rise to three signals.

These two protons can be interchanged by rotational symmetry, so they are chemically equivalent

(g) This compound has four protons and none of them can be interchanged by rotational or reflectional symmetry. Each of the four protons occupies a unique electronic environment, giving rise to four signals.

(h) This compound has two different kinds of protons (highlighted below), giving rise to two signals.

These four methyl groups can be interchanged by rotational symmetry, so all twelve protons are chemically equivalent

These two protons can be interchanged by rotational symmetry, so they are chemically equivalent

(i) Due to the location of the two bromine atoms, each CH_2 group occupies a unique electron environment, giving rise to two separate signals (one for each CH_2 group). In addition, the two methyl groups are also in different electronic environments, giving two separate signals. In total, we expect four signals.

(j) Each of the protons in each CH_2 group is in a unique electronic environment, as a result of the presence of a chiral center. That is, each CH_2 group gives rise to two separate signals, so the two CH_2 groups collectively give rise to four different signals. The two methyl groups are also different from each other (because of their unequal proximity to the bromine atom), giving two more signals. In addition, there is one signal from the proton attached to the carbon bearing the bromine atom. In total, we expect seven signals.

(k) Heptane has four different kinds of protons, shown here:

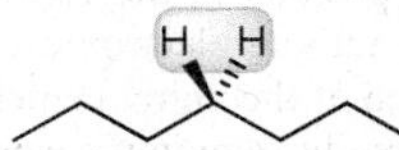

These two protons can be interchanged by rotational symmetry, so they are chemically equivalent

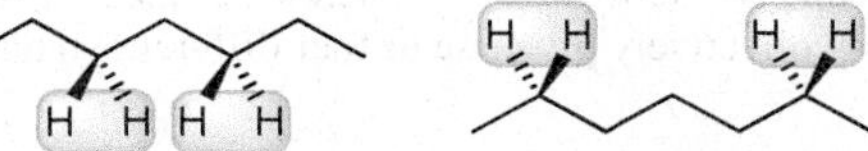

All four of these protons can be interchanged by either rotational or reflectional symmetry, so they are chemically equivalent

All four of these protons can be interchanged by either rotational or reflectional symmetry, so they are chemically equivalent

These six protons can be interchanged by either rotational or reflectional symmetry, so they are chemically equivalent

(l) Each of the three vinylic protons occupies a unique electronic environment, giving rise to three separate signals:

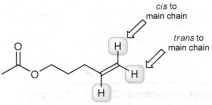

cis to main chain

trans to main chain

Each of these protons is unique. They are NOT chemically equivalent.

The two vinylic protons at the very end are different from each other because one is *cis* to the main chain and the other is *trans* to the main chain, as shown. Each of the CH_2 groups provides one signal (because each CH_2 group occupies a unique electronic environment), and the

CH_3 provides one more signal, giving a total of seven signals.

15.6. The presence of the bromine atom does not render C3 a chiral center because there are two ethyl groups connected to C3. Nevertheless, the presence of the bromine atom does prevent the two protons at C2 from being interchangeable by reflection. The replacement test gives a pair of diastereomers, so the protons are diastereotopic (which means that they are not chemically equivalent).

Br Br D D

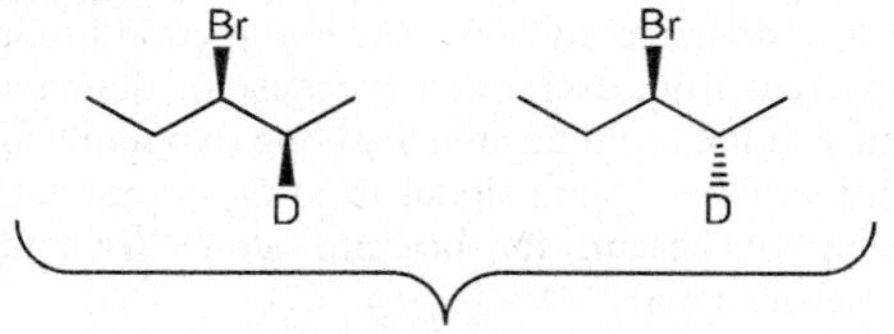

Diastereomers

15.7. Each of the protons in the following highlighted CH_2 groups is in a unique electronic environment, as a result of the presence of the chiral center. The protons in each CH_2 group are diastereotopic and not chemically equivalent. That is, each CH_2 group will give rise to two separate signals, because one H is on the same face as (*cis* to) the propenyl substituent, while the other H is further away from it, on the opposite face as (*trans* to) the propenyl substituent. Therefore, these two CH_2 groups collectively give rise to four different signals:

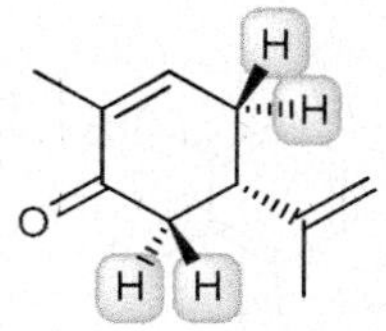

In addition, the following highlighted protons are also diastereotopic:

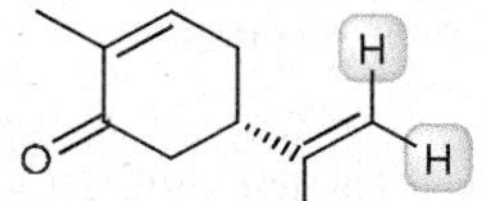

Furthermore, there are four more signals arising from the two methyl groups and the two CH groups, as shown:

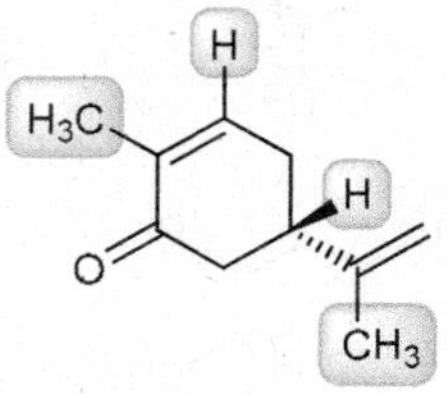

Therefore, we have seen that (*S*)-carvone has a total of ten different kinds of protons, giving rise to ten signals.

15.8.

(a) The ^1H NMR spectrum of this compound is expected to exhibit five signals. The calculation for the estimated chemical shift of each signal is shown here:

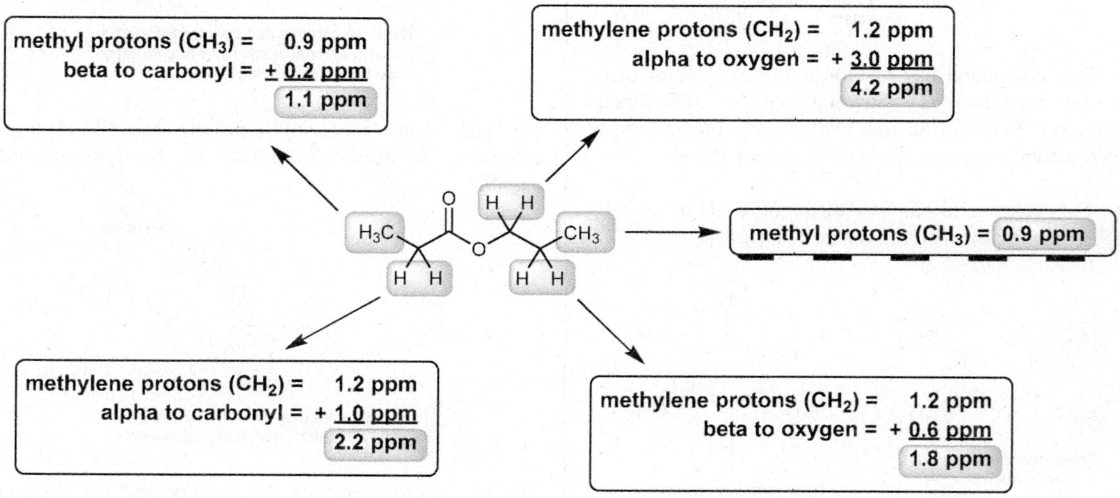

methyl protons (CH_3) = 0.9 ppm
beta to carbonyl = + 0.2 ppm
1.1 ppm

methylene protons (CH_2) = 1.2 ppm
alpha to oxygen = + 3.0 ppm
4.2 ppm

methyl protons (CH_3) = 0.9 ppm

methylene protons (CH_2) = 1.2 ppm
alpha to carbonyl = + 1.0 ppm
2.2 ppm

methylene protons (CH_2) = 1.2 ppm
beta to oxygen = + 0.6 ppm
1.8 ppm

(b) The ^{1}H NMR spectrum of this compound is expected to exhibit three signals. The calculation for the estimated chemical shift of each signal is shown here:

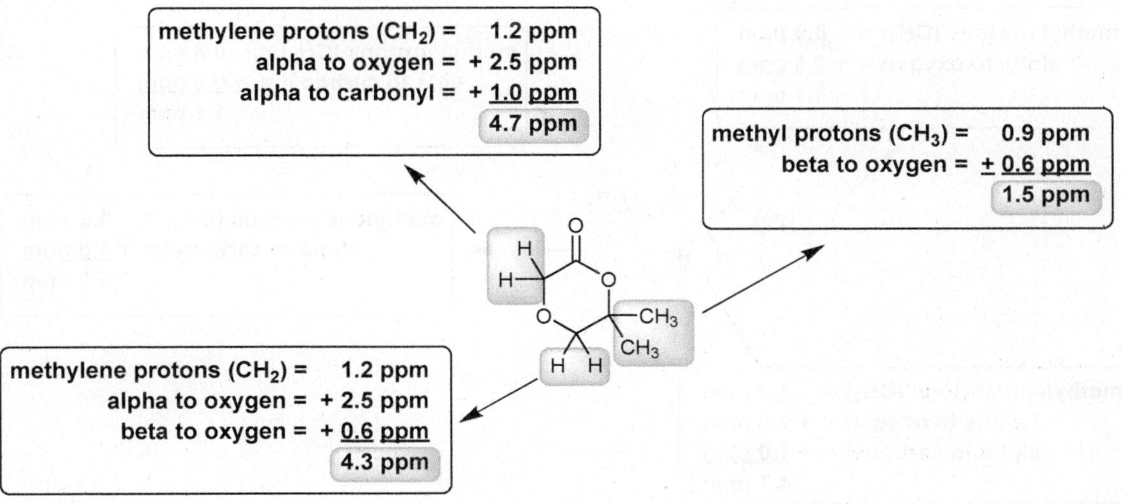

(c) The ^{1}H NMR spectrum of this compound is expected to exhibit four signals. The calculation for the estimated chemical shift of each signal is shown here:

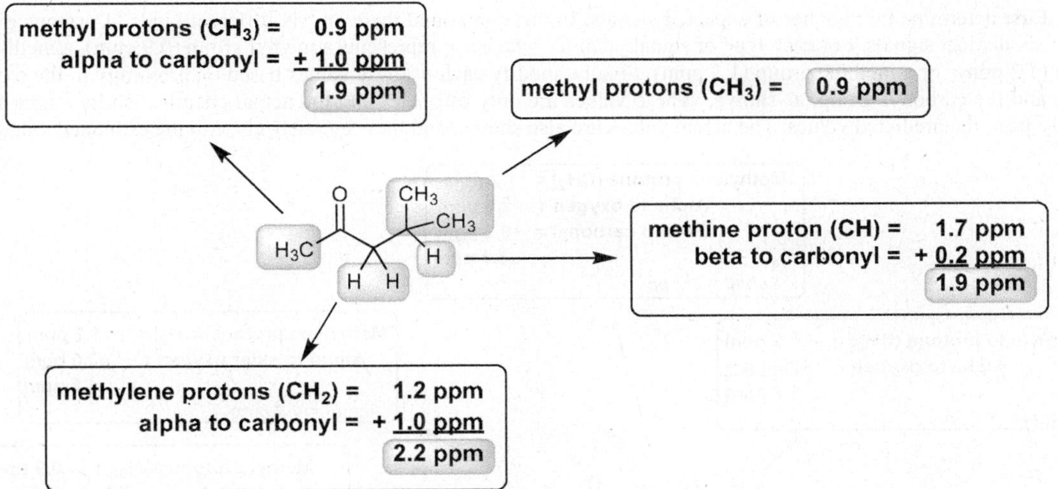

(d) The ^{1}H NMR spectrum of this compound is expected to exhibit four signals. The calculation for the estimated chemical shift of each signal is shown here:

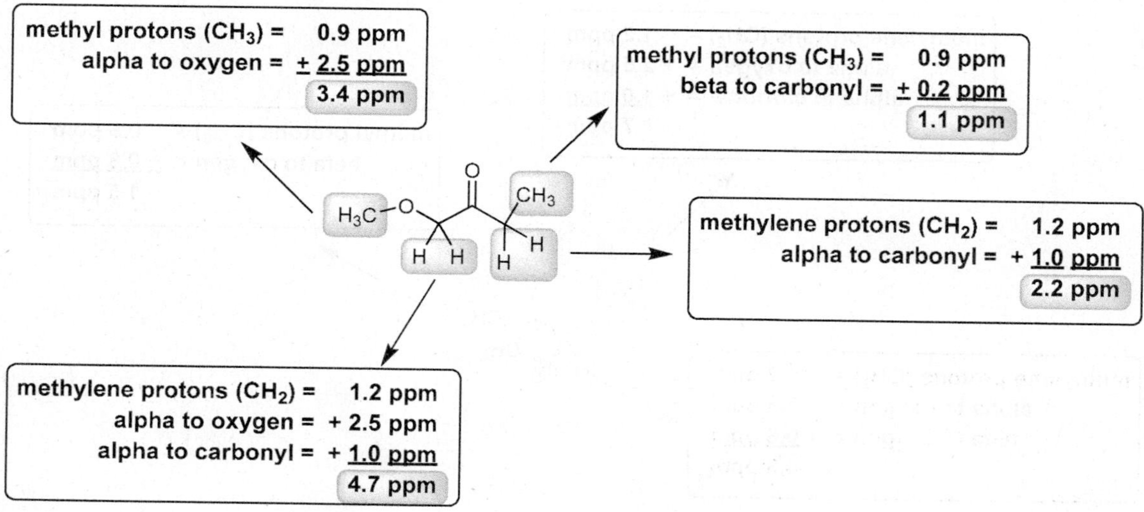

(e) All four methylene groups are equivalent, so the compound will have only one signal in its ^{1}H NMR spectrum. That signal is expected to appear at approximately (1.2 + 2.5 + 0.5) = 4.2 ppm.

methylene protons (CH_2) =	1.2 ppm
alpha to oxygen =	+ 2.5 ppm
beta to oxygen =	+ 0.5 ppm
	4.2 ppm

15.9. First determine the number of expected signals. In this compound there are six different kinds of protons giving rise to six distinct signals. For each type of signal, identify whether it represents a methyl group (0.9 ppm), a methylene group (1.2 ppm), or a methine group (1.7 ppm). Finally, modify each of these values based on proximity to the oxygen atoms and the carbonyl group, as shown. These values are only estimates and the actual chemical shifts might differ slightly from the predicted values. The actual values are also shown, and they are fairly close to the estimated values.

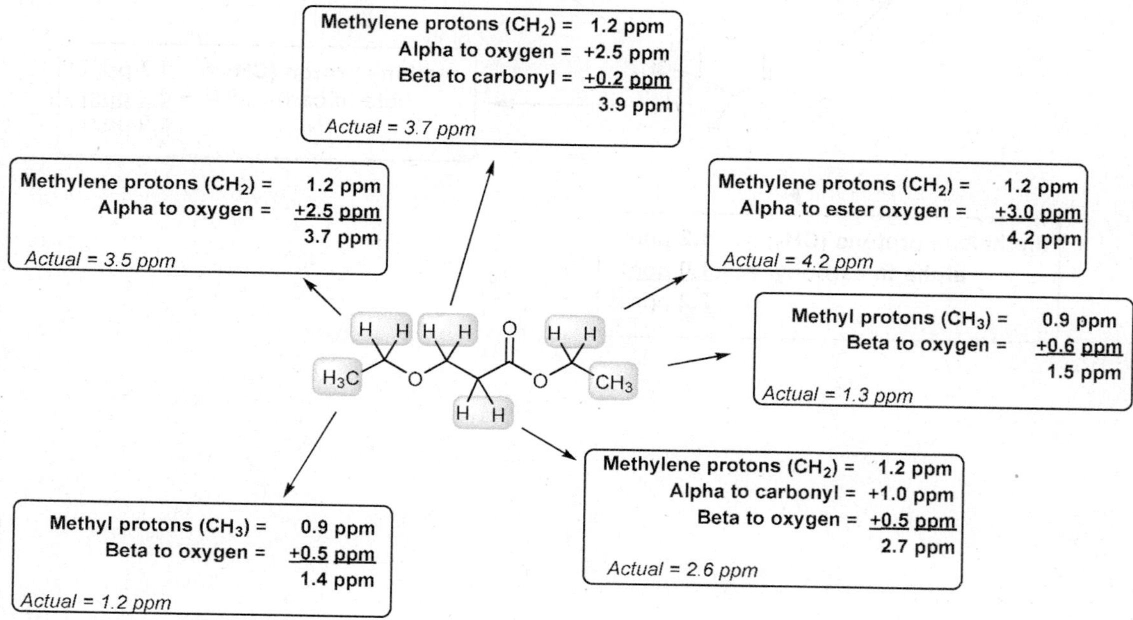

15.10.

(a) Using the values provided in Tables 15.1 and 15.2, we expect the following chemical shifts:

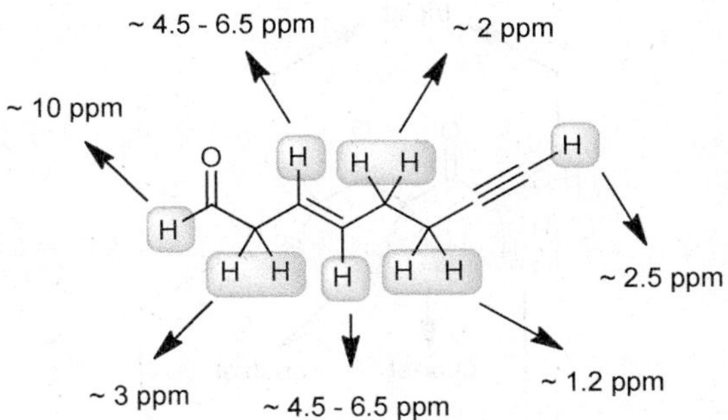

(b) Using the values provided in Tables 15.1 and 15.2, we expect the following chemical shifts:

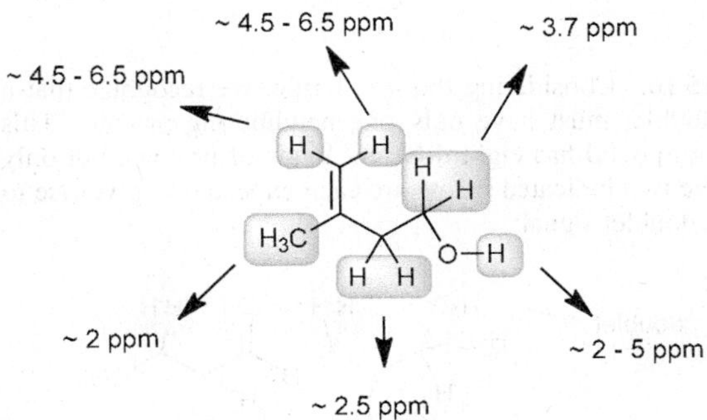

(c) Using the values provided in Tables 15.1 and 15.2, we expect the following chemical shifts:

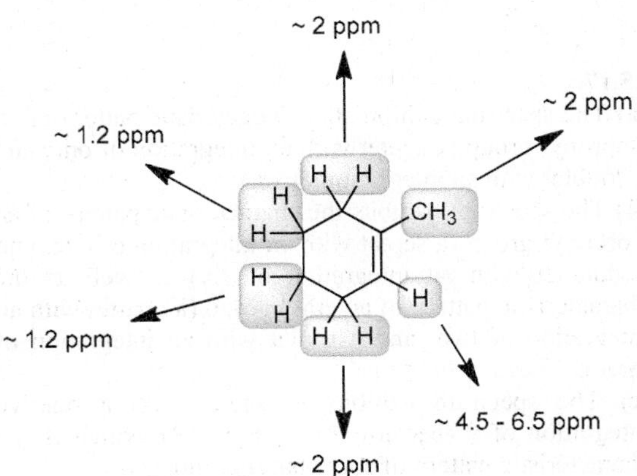

(d) Using the values provided in Tables 15.1 and 15.2, we expect the following chemical shifts:

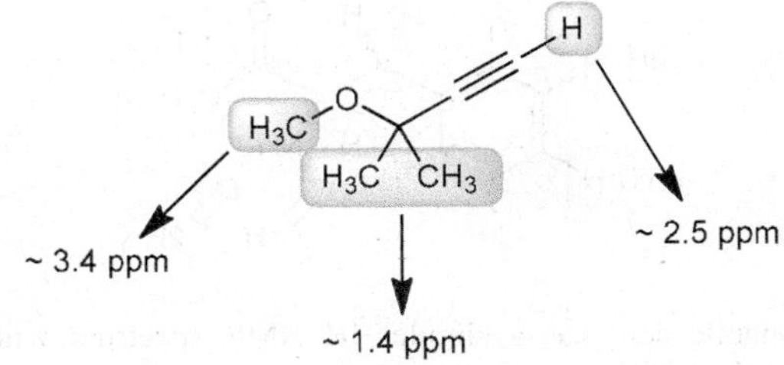

15.11. Among the integration values provided, the lowest number is 33.2, so we divide all integration values by 33.2, giving the following ratio:

1 : 1.5 : 1 : 1.5

Since there is no such thing as a half of a proton, these numbers must represent 2, 3, 2, and 3 protons, respectively. This is confirmed by the molecular formula, which indicates that the compound has ten hydrogen atoms.

Therefore,

The signal at 4.0 ppm represents two protons.
The signal at 2.0 ppm represents three protons.
The signal at 1.6 ppm represents two protons.
The signal at 0.9 ppm represents three protons.

15.12. Among the integration values provided, the lowest number is 17.1, so we divide all integration values by 17.1, giving the following ratio:

1 : 5 : 1 : 3

The molecular formula indicates that the compound has ten hydrogen atoms, so the numbers above are not only relative values, but they are also exact values.

Therefore,

The signal at 9.6 ppm represents one proton.
The signal at 7.5 ppm represents five protons.
The signal at 7.3 ppm represents one proton.
The signal at 2.1 ppm represents three protons.

15.13. Among the integration values provided, the lowest number is 18.92, so we divide all integration values by 18.92, giving the following ratio:

1 : 1 : 1

The molecular formula indicates that the compound has six hydrogen atoms (not just three), so the numbers above are only relative values. Each signal must actually represent two protons.

15.14. We expect a total of eight signals in the ^{1}H NMR spectrum of phenylalanine; note the symmetry in the benzene ring and the nonequivalent diastereotopic protons next to the chiral center bearing the NH_2 group.

Cinnamic acid has a simpler ^{1}H NMR spectrum with only six signals.

15.15.

(a) This compound has four different kinds of protons, highlighted here. In each case, we apply the $n + 1$ rule, giving the multiplicities shown:

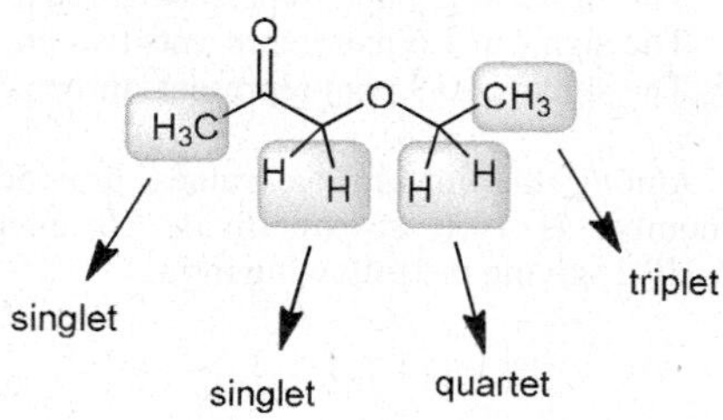

(b) This compound has four different kinds of protons, highlighted here. In each case, we apply the $n + 1$ rule, giving the multiplicities shown:

(c) This compound has four different kinds of protons, highlighted here. In each case, we apply the $n + 1$ rule, resulting in all singlets:

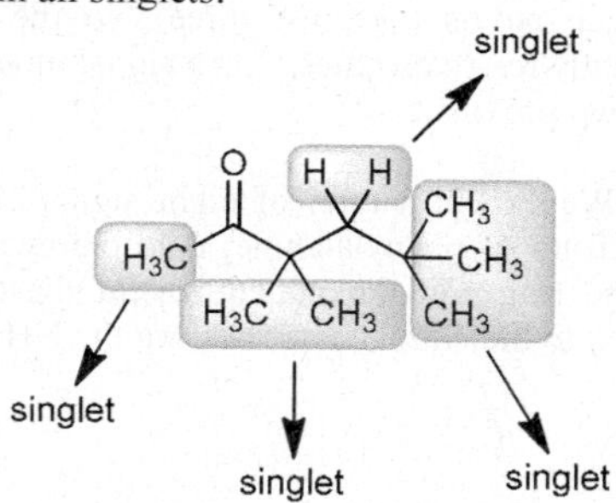

(d) This compound has six different kinds of protons, highlighted here. In each case, we apply the $n + 1$ rule, giving the multiplicities shown:

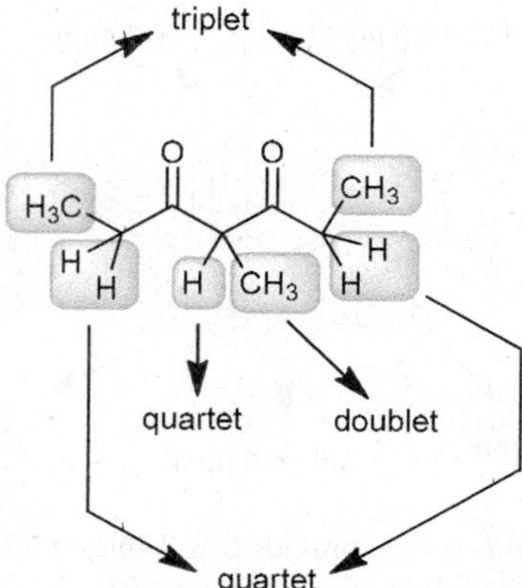

15.16. Considering the $n + 1$ rule, we recognize that a doublet must have only one neighboring proton. This compound has eight different kinds of protons, but only the two indicated below are each expected to give rise to a doublet signal:

15.17.

(a) The spectrum exhibits the characteristic pattern of an isopropyl group (a septet with an integration of one, and a doublet with an integration of six).

(b) The spectrum exhibits the characteristic pattern of an isopropyl group (a septet with an integration of one, and a doublet with an integration of six) as well as the characteristic pattern of an ethyl group (a quartet with an integration of two, and a triplet with an integration of three).

(c) The spectrum exhibits a singlet with a relative integration of 9 (because $55.0 / 6.0 \approx 9$), which is the characteristic pattern of a *tert*-butyl group.

(d) The spectrum does not exhibit the characteristic pattern of an ethyl group, an isopropyl group, or a *tert*-butyl group.

15.18. To determine the expected splitting pattern for the signal corresponding to H_a, we must consider the effects of the two non-equivalent neighbors, H_b and H_c, The former is coupled to H_a with coupling constant J_{ab} (11 Hz), and the latter is coupled to H_a with coupling constant J_{ac} (17 Hz). We begin with the larger coupling constant (J_{ac} in this case), which splits the signal into a doublet. Then, each peak of this doublet is then further split into a doublet because of the effect of H_b. The result is a doublet of doublets. It can be distinguished from a quartet because all four peaks should be approximately equal in height, as opposed to a quartet, in which the individual peaks have relative heights of 1 : 3 : 3 : 1.

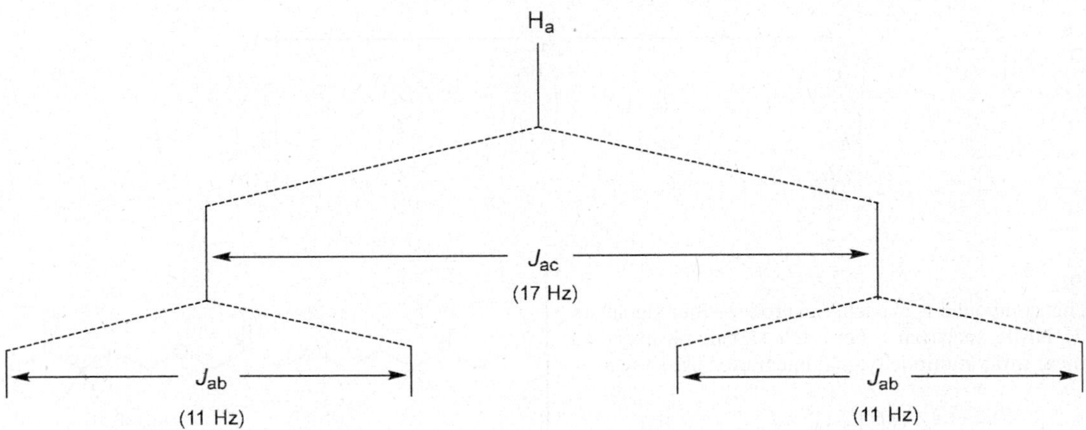

To determine the expected splitting pattern for the signal corresponding to H_b, we must consider the effects of the two non-equivalent neighbors, H_a and H_c, The former is coupled to H_b with a coupling constant J_{ab} (11 Hz), and the latter is coupled to H_b with a coupling constant J_{bc} (1 Hz). We begin with the larger coupling constant (J_{ab} in this case), which splits the signal into a doublet. Then, each peak of this doublet is then further split into a doublet because of the effect of H_c. The result is a doublet of doublets.

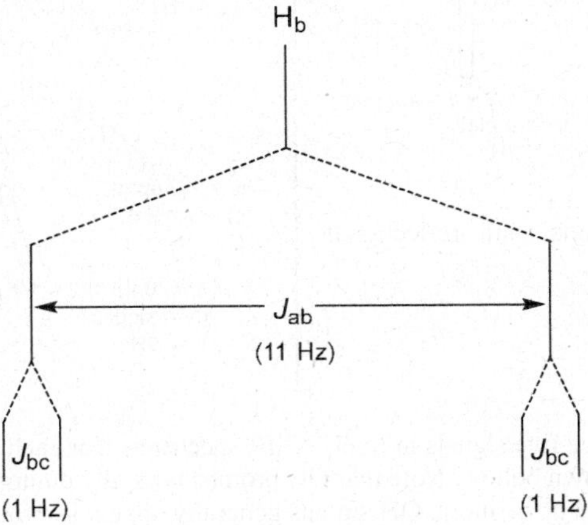

To determine the expected splitting pattern for the signal corresponding to H_c, we must consider the effects of the two non-equivalent neighbors, H_a and H_b, The former is coupled to H_c with a coupling constant J_{ac} (17 Hz), and the latter is coupled to H_c with a coupling constant J_{bc} (1 Hz). We begin with the larger coupling constant (J_{ac} in this case), which splits the signal into a doublet. Then, each peak of this doublet is then further split into a doublet because of the effect of H_b. The result is a doublet of doublets.

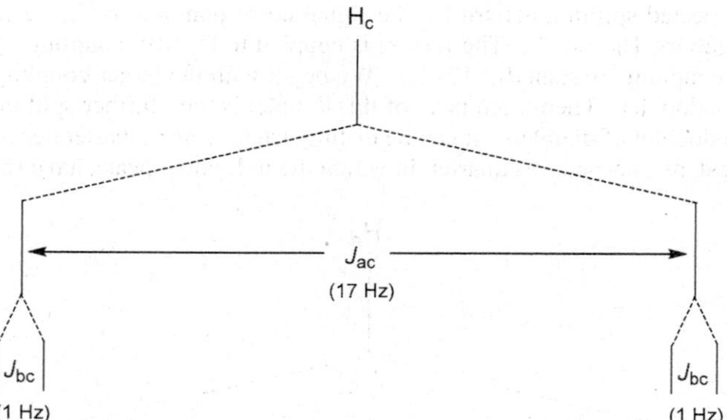

15.19.

(a) This compound is expected to produce four signals in its ^{1}H NMR spectrum. For each signal, its expected chemical shift, multiplicity, and integration are shown.

δ = 1.2 +1 = 2.2 ppm
m = triplet
I = 2H

δ = 1.2 + 0.2 = 1.4 ppm
m = quintet
I = 2H

δ = 1.2 + 0.6 = 1.8 ppm
m = quintet
I = 2H

δ = 1.2 + 3 = 4.2 ppm
m = triplet
I = 2H

(b) This compound has rotational and reflectional symmetry:

So we expect its ^{1}H NMR spectrum to exhibit only five signals, corresponding with the following highlighted protons. For each signal, its expected chemical shift, multiplicity, and integration are shown.

δ = 1.2 + 0.5 + 0.2 = 1.9 ppm
m = triplet
I = 2H

δ = 1.2 + 2.5 = 3.7 ppm
m = triplet
I = 2H

δ = 10 ppm
m = singlet
I = 1H

δ = 0.9 + 0.2 = 1.1 ppm
m = singlet
I = 6H

δ = 1.2 + 2.5 + 1.0 = 4.7 ppm
m = singlet
I = 2H

15.20. GHB is expected to produce five signals in its ^{1}H NMR spectrum. For each signal, its expected chemical shift, multiplicity, and integration are shown below. Note that OH protons typically do not couple with neighboring protons, and as a result, no splitting occurs. Furthermore, OH signals generally appear as broad singlets.

δ = 1.2 + 0.5 + 0.2 = 1.9 ppm
m = quintet
I = 2H

δ = ~2-5 ppm
m = singlet (broad)
I = 1H

δ = ~12 ppm
m = singlet (broad)
I = 1H

δ = 1.2 + 2.5 = 3.7 ppm
m = triplet
I = 2H

δ = 1.2 + 1.0 = 2.2 ppm
m = triplet
I = 2H

15.21.

(a) The first compound exhibits symmetry so it will have only three signals in its ^{1}H NMR spectrum, while the second compound will have six signals.

(b) Both compounds will exhibit ^{1}H NMR spectra with only two singlets. In each spectrum, the relative integration of the two singlets is 1:3. In the first compound, the singlet with the smaller integration value will be at approximately 2 ppm (alpha to a carbonyl). In the second compound, the singlet with the smaller integration value will be at approximately 4 ppm (alpha to the oxygen of an ester).

(c) The first compound will have only two signals in its ^{1}H NMR spectrum, while the second compound will have three signals.

(d) The first compound will have five signals in its ^{1}H NMR spectrum, while the second compound exhibits symmetry so it will have only three signals.

(e) The first compound exhibits symmetry so it will have only two signals in its ^{1}H NMR spectrum, while the second compound will have four signals.

(f) The first compound will have only one signal in its ^{1}H NMR spectrum (a singlet), while the second compound will have two signals (one signal will be a doublet and the other signal will have ten peaks).

15.22.

(a) Both compounds will have very similar ^{1}H NMR spectra, with the same number of signals, splitting patterns and integration values. The major difference will be the location of the 2H quartet signal. For the first structure, this signal is expected to appear near 2.2 ppm, because it is a methylene group (benchmark value = 1.2 ppm) next to a carbonyl group (+1). For the second structure, this signal will appear at approximately 4.2 ppm, because it is a methylene group next to an oxygen atom of an ester group (+3):

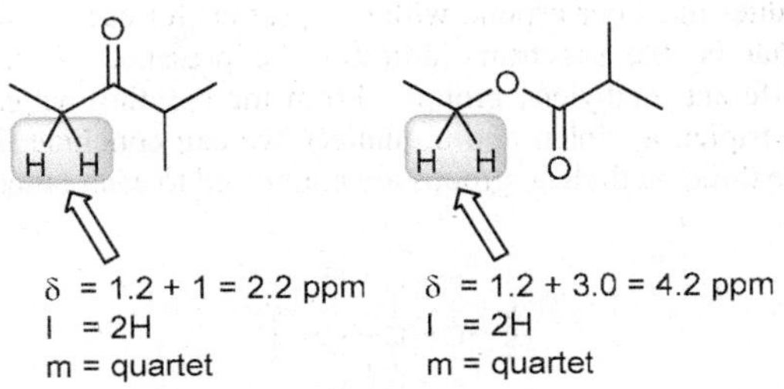

δ = 1.2 + 1 = 2.2 ppm
I = 2H
m = quartet

δ = 1.2 + 3.0 = 4.2 ppm
I = 2H
m = quartet

(b) These compounds are indeed expected to exhibit a different number of signals in their ^{1}H NMR spectra. However, most of the signals will appear in the region 1-2 ppm (in each spectrum), and may overlap, making it difficult to count the number of signals. The only distinctive peaks will be the vinyl and aldehyde signals. In fact, the presence or absence of a signal near 10 ppm is the easiest way to differentiate these compounds. The second compound is an aldehyde so it is expected to produce a signal near 10 ppm. The first compound will lack such a signal.

15.23.

(a) The molecular formula ($C_8H_{10}O$) indicates four degrees of unsaturation (see Section 14.16), which is highly suggestive of an aromatic ring. To determine the relative integration values, we divide each of the integration values by 9.1, giving a ratio of approximately 5 : 2 : 2 : 1. Since the compound has ten protons, the numbers above are not only relative values, but they are also exact values. Now let's analyze each of the signals individually.

The signal just above 7 ppm confirms our suspicion of an aromatic ring. This signal has an integration of 5H, indicating a monosubstituted aromatic ring:

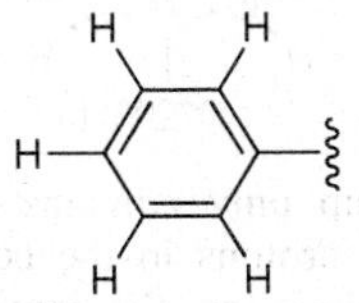

The spectrum also exhibits two triplets (just below 3 ppm and just below 4 ppm), indicating two methylene groups connected to each other:

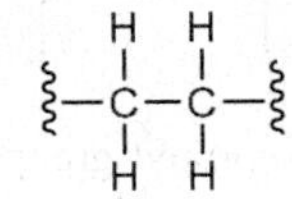

Each of these signals appears more downfield than we might expect for a methylene group (1.2 ppm), so each of these methylene groups must be connected to a group that causes a deshielding effect. This must be taken into account in our final structure.

If we inspect the two fragments that we have determined thus far (the monosubstituted aromatic ring and the methylene groups that neighbor each other), we will find that these two fragments account for nearly all of the atoms in the molecular formula ($C_8H_{10}O$). We only need to account for one more proton and one oxygen atom. The singlet at 2 ppm has an integration of 1, so this signal corresponds with only one proton (with no neighbors), so we conclude that the compound has an OH group.

There is only one way to assemble the three fragments:

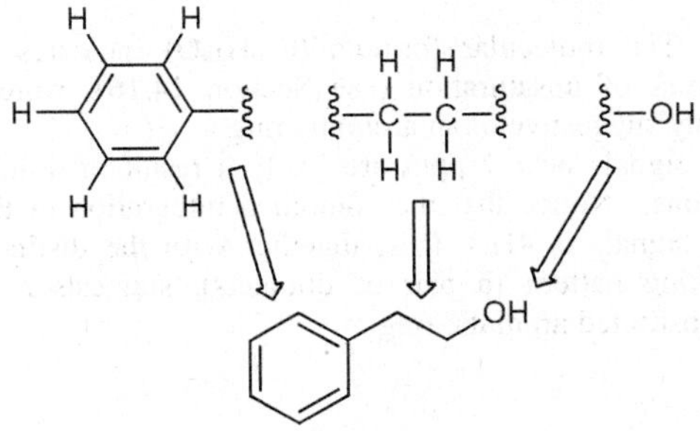

This structure is consistent with all of the signals, including the chemical shifts of the two triplets, which can be explained by the electron withdrawing effect of the oxygen atom as well as the local magnetic field established by the aromatic ring.

(b) The molecular formula ($C_7H_{14}O$) indicates one degree of unsaturation (see Section 14.16), which means that the compound must possess either a double bond or a ring. To determine the relative integration values, we divide each of the integration values by 10.8, giving a ratio of approximately 1 : 6. This spectrum has the characteristic pattern of an isopropyl group (a doublet with a relative integration of 6 and a septet with a relative integration of 1).

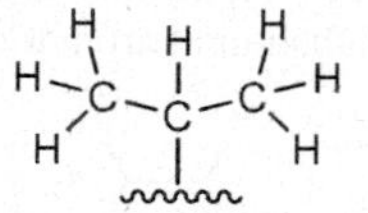

An isopropyl group only contains seven protons, but there are fourteen protons in the compound ($C_7H_{14}O$). We therefore conclude that the compound must contain two isopropyl groups, which are interchangeable by symmetry:

isopropyl ? isopropyl

Notice that the two isopropyl groups account for all but one of the carbon atoms in the compound. Therefore, there can only be one carbon atom in between the two isopropyl groups. This carbon atom cannot have any protons, since we don't see any other signals in the ^1H NMR spectrum. Also, we must still account for one oxygen atom ($C_7H_{14}O$), and we said that the compound must contain one degree of unsaturation. This all points to a carbonyl group at the central position:

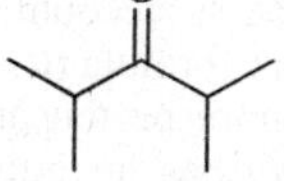

This structure is indeed consistent with the observed chemical shift at 2.7 ppm for the methine (CH) protons (1.7 + 1 = 2.7 ppm).

(c) The molecular formula ($C_{10}H_{14}O$) indicates four degrees of unsaturation (see Section 14.16), which is highly suggestive of an aromatic ring.
The signals near 7 ppm are likely a result of aromatic protons. Notice that the combined integration of these two signals is 4H. This, together with the distinctive splitting pattern (a pair of doublets), suggests a 1,4-disubstituted aromatic ring:

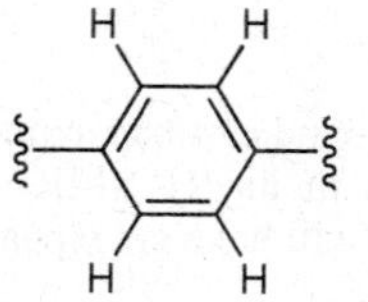

The spectrum also exhibits a singlet with an integration of 9H (at approximately 1.4 ppm) which is characteristic of a *tert*-butyl group.

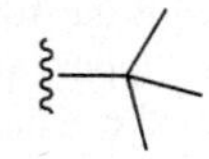

If we inspect the two fragments that we have determined thus far (the disubstituted aromatic ring and the *tert*-butyl group), we will find that these two fragments account for nearly all of the atoms in the molecular formula ($C_{10}H_{14}O$). We only need to account for one more proton and one oxygen atom. The peak just under 5 ppm has an integration of 1, so this signal corresponds with only one proton (with no neighbors), so we conclude that the compound has an OH group. This signal is broad, which is often (although not always) the case for signals arising from OH groups.
There is only one way to assemble the three fragments:

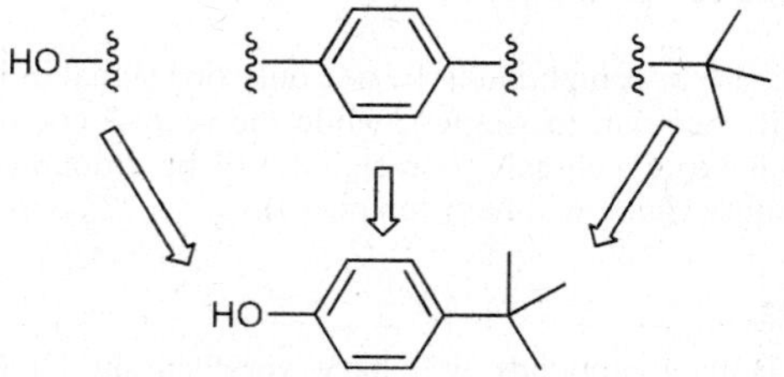

(d) The molecular formula ($C_4H_6O_2$) indicates two degrees of unsaturation (see Section 14.16), which means that the compound must possess either two double bonds, or two rings, or one ring and one double bond, or a triple bond.
To determine the relative integration values, we divide each of the integration values by 18.0, giving a ratio of approximately 1 : 1 : 1. The molecular formula indicates six protons (rather than three), so the relative integration values must correspond with two protons for each signal. That is, the spectrum indicates the presence of three different methylene groups. From the splitting patterns (a triplet, a triplet, and a quintet), we can conclude that the three methylene groups are connected to each other:

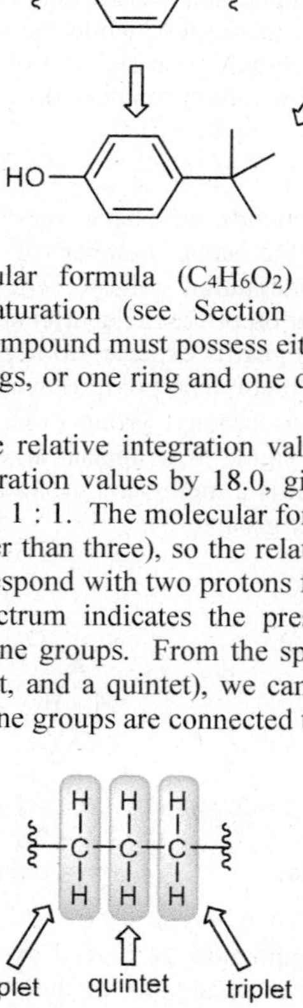

triplet quintet triplet

Now let's focus on the chemical shifts of the triplets (2.4 ppm and 4.3 ppm). Both signals are shifted downfield (relative to 1.2 ppm for a typical methylene group). One of these signals is significantly shifted downfield, perhaps because it is next to an oxygen atom (after all, the molecular formula indicates that there are two oxygen atoms in the compound):

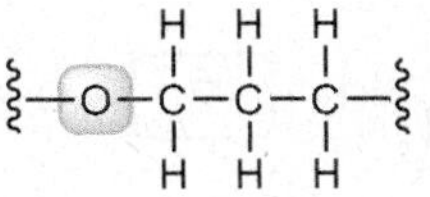

The other triplet is also shifted downfield, but the effect is weaker. This seems consistent with the effect of a carbonyl group:

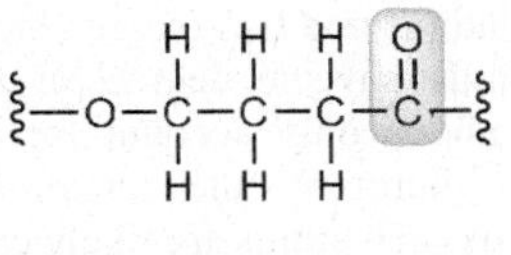

The central methylene group is beta to both the oxygen atom and the carbonyl group, and it feels the distant effects of both (explaining why that signal is shifted somewhat downfield itself).

The fragment above accounts for ALL of the atoms in the molecular formula, yet we are still missing one degree of unsaturation (the product must contain two degrees of unsaturation, but the fragment above has only one degree of unsaturation). Therefore, we close the ends together, giving the following structure:

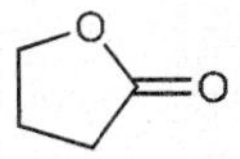

(e) The molecular formula ($C_9H_{10}O$) indicates five degrees of unsaturation (see Section 14.16), which is highly suggestive of an aromatic ring, in addition to either one double bond or one ring.

To determine the relative integration values, we divide each of the integration values by 13.9, giving a ratio of approximately 1 : 1.5 : 1 : 1.5. Since the compound has ten protons, the numbers above must correspond with:

2 : 3 : 2 : 3

Now let's analyze each of the signals individually.

The signals near 7 ppm are likely a result of aromatic protons. Notice that the combined integration of these signals is 5H, indicating a monosubstituted aromatic ring:

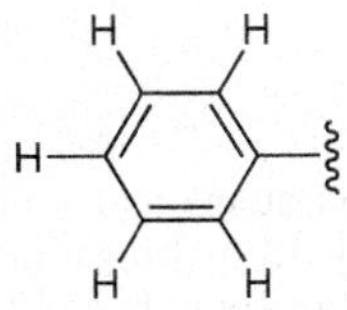

The spectrum also exhibits the characteristic pattern of an ethyl group (a quartet with an integration of 2 and a triplet with an integration of 3):

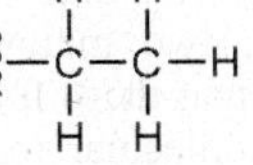

If we inspect the two fragments that we have determined thus far (the monosubstituted aromatic ring and the ethyl group), we will find that these two fragments account for nearly all of the atoms in the molecular formula ($C_9H_{10}O$). We only need to account for one more carbon atom and one oxygen atom. And let's not forget that our structure still needs one more degree of unsaturation, suggesting a carbonyl group:

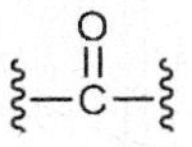

There is only one way to connect these three fragments.

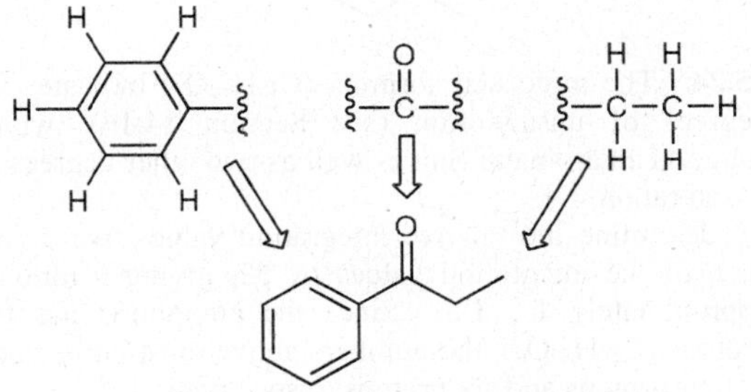

(f) The molecular formula ($C_5H_{12}O$) indicates no degrees of unsaturation (see Section 14.16), which means that the compound does not have a π bond or a ring.

To determine the relative integration values, we divide each of the integration values by 13.6, giving a ratio of approximately 1 : 2 : 6 : 3. Since the compound has twelve protons, the numbers above are not only relative values, but they are also exact values.

Now let's analyze each of the signals individually.

Let's begin with the two signals that represent the characteristic pattern for an ethyl group (a quartet with an integration of 2 and a triplet with an integration of 3).

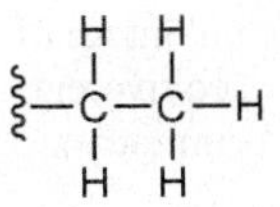

The singlet at 1.2 ppm has an integration of 6, indicating two methyl groups that can be interchanged via symmetry (and they cannot have any neighboring protons).

methyl H_3C ? CH_3 methyl

The singlet at 2.2 ppm has an integration of 1, so this signal corresponds with only one proton (with no neighbors), so we conclude that this likely represents an OH group (the molecular formula indicates the presence of an oxygen atom).

If we inspect the fragments that we have determined thus far (an ethyl group, two methyl groups and an OH group), we will find that these fragments account for all of the atoms in the molecular formula ($C_5H_{10}O$) except for one carbon atom. Indeed, this carbon atom is necessary to connect all of the fragments, as shown:

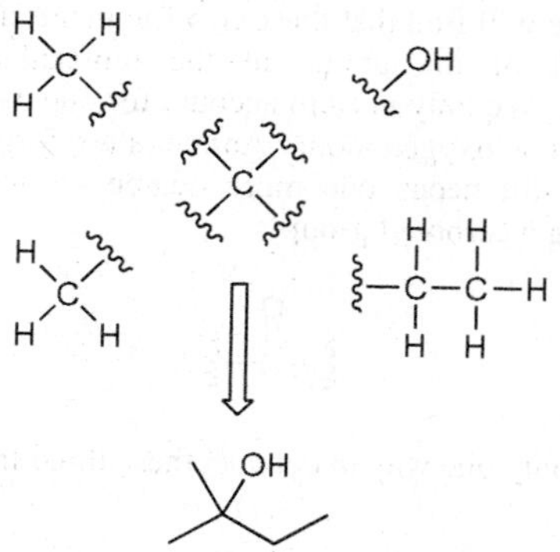

15.24. The molecular formula ($C_{10}H_{10}O_4$) indicates six degrees of unsaturation (see Section 14.16), which suggests an aromatic ring as well as two other degrees of unsaturation.
To determine the relative integration values, we divide each of the integration values by 52, giving a ratio of approximately 1 : 1.5. Since the compound has ten protons ($C_{10}H_{10}O_4$), the numbers above must correspond to four protons and six protons, respectively.
Now let's analyze each of the signals individually. The signal at 8.1 ppm is significantly downfield, and likely represents aromatic protons. Since it is a singlet with an integration of 4, it must correspond with a 1,4-disubstituted aromatic ring in which both substituents are identical (therefore rendering all four aromatic protons equivalent).

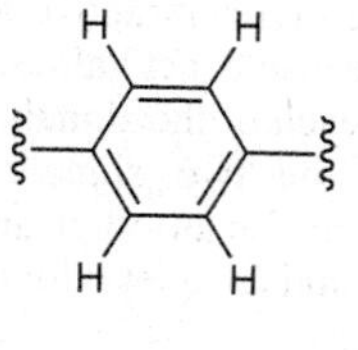

The other signal has an integration of 6, which likely represents two equivalent methyl groups (interchangeable by symmetry). Since the signal is a singlet, these methyl groups must not have any neighbors. The chemical shift is consistent with these methyl groups being connected to oxygen atoms:

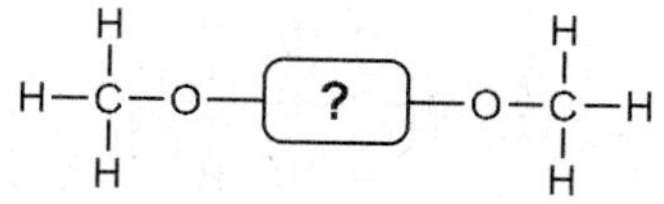

If we inspect the fragments that we have determined thus far (a 1,4-disubstituted aromatic ring, and two methoxy groups), we will find that these fragments account for all of the atoms in the molecular formula ($C_{10}H_{10}O_4$) except for two carbon atoms and two oxygen atoms. Recall that the compound must have six degrees of unsaturation, and the fragments above only account for four degrees of unsaturation. Therefore, the remaining two carbon atoms and two oxygen atoms are likely carbonyl groups:

There are certainly a few different ways to connect all of these fragments, but there is only one way to connect them without breaking the symmetry necessary to keep all four aromatic protons identical:

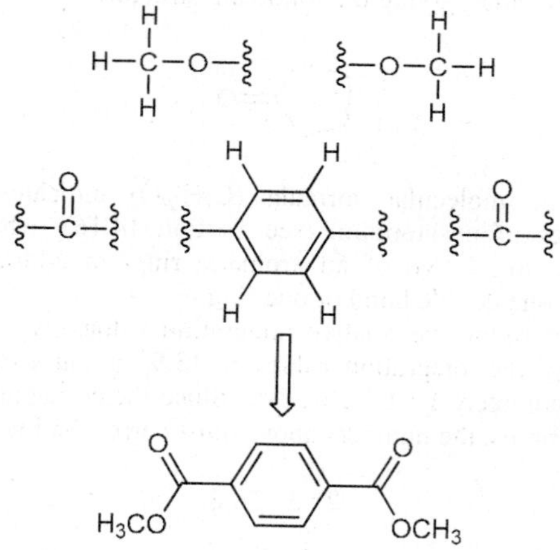

15.25. Begin by calculating the HDI. The molecular formula indicates 9 carbon atoms and 1 nitrogen atom. These would require 21 hydrogen atoms in order to be fully saturated. There are only 13 hydrogen atoms, so 8 hydrogen atoms are missing. Therefore, the HDI is 4. This is a relatively large number, and it would be inefficient to think about all of the possible ways to have 4 degrees of unsaturation. Any time we encounter an HDI of 4 or more, we should be on the lookout for an aromatic ring. Keep this in mind when analyzing the spectrum, which we expect to exhibit aromatic protons (near 7 ppm).
Next, consider the number of signals and the integration value for each signal. Be on the lookout for integration values that suggest the presence of symmetry. For example, a signal with an integration of 4 would suggest two equivalent CH_2 groups.

The spectral data indicates a total of 6 signals. Let's begin with the pair of triplets just below 3 ppm, each of which has an integration of 2. This suggests that there are two adjacent methylene groups.

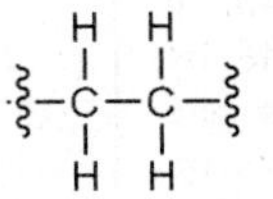

The singlet near 4 ppm has an integration of 3 which suggests an isolated methyl group, likely connected to an electronegative oxygen atom. A methoxy fragment, OCH_3, seems likely.

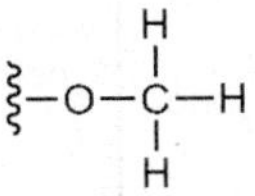

Moving downfield, there are 2 doublets near 7 ppm, each of which has an integration of 2. This pattern is characteristic of a 1,4-disubstituted benzene ring, bearing two different substituents (with different electronic demands). In such a case, there are two types of aromatic protons, each of which has an integration of 2 and is split into a doublet by its one neighbor:

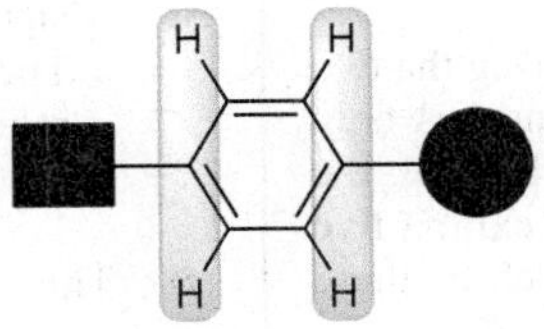

So far, we have the following three fragments:

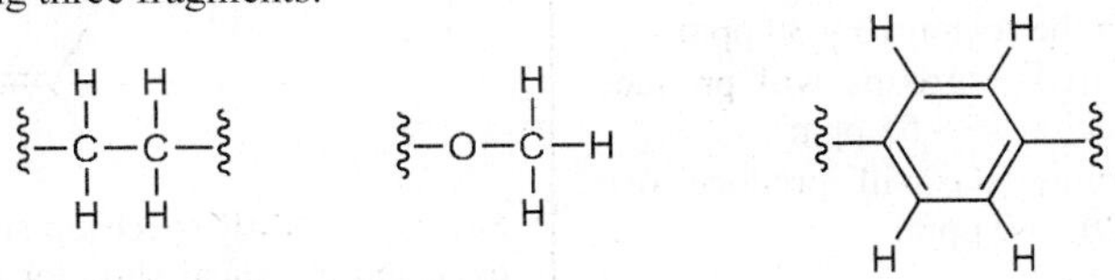

These fragments collectively account for 9 carbon atoms, 11 hydrogen atoms, and 1 oxygen atom. If we inspect the molecular formula, we see that we have accounted for all of the carbon atoms and the oxygen atom, but we must still account for 2 more hydrogen atoms and 1 nitrogen atom. Therefore, we conclude that the broad singlet near 1 ppm is likely to be an NH_2 group. We will see later (Chapter 22) that NH_2 protons generally appear as broad signals between 0.5 and 5.0 ppm.

Next, we assemble the fragments. There are two reasonable ways to connect the fragments in this case.

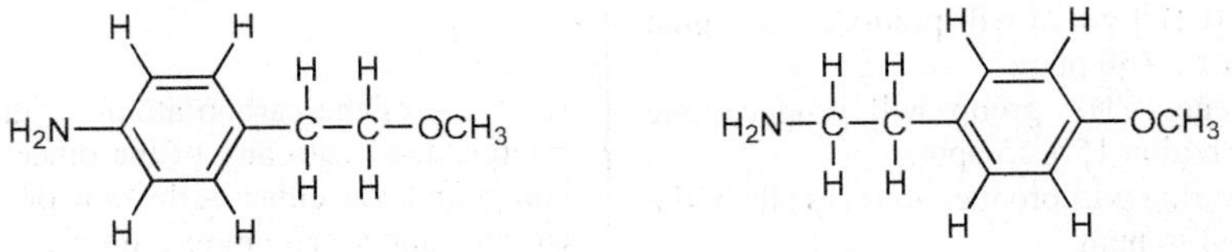

To distinguish between these two options, we consider chemical shifts. Note that the adjacent methylene groups appear fairly close to one another, near 3 ppm, indicating that neither is attached to the highly electronegative and deshielding oxygen atom. The first compound above shows a CH_2 group connected to oxygen, and we expect that compound to produce a triplet somewhere near 4 ppm, which is absent from the spectral data. The second compound above has the two CH_2 groups connected to an aromatic ring and an NH_2 group, respectively, which is consistent with the observed chemical shifts of these triplets. The structure is redrawn here in conventional bond-line notation.

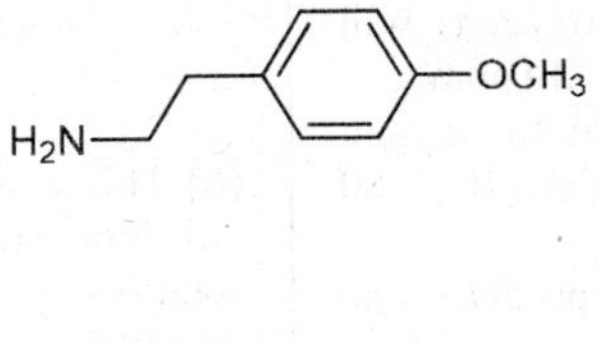

15.26.

(a) The two methyl groups occupy identical environments (there is a conformation of this molecule in which the two methyl groups are interchangeable by reflectional symmetry), so the methyl groups are chemically equivalent. All of the other carbon atoms are unique, giving rise to a total of four signals. The expected chemical shift for each of these signals (listed below) can be found in Table 15.4.

- The two, equivalent methyl (CH_3) groups will produce one signal in the region 10 – 30 ppm.
- The methylene (CH_2) group will produce one signal in the region 15 – 55 ppm.
- The methine (CH) group will produce one signal in the region 20 – 60 ppm.
- The carbonyl group will produce one weak signal in the region 185 – 220 ppm.

(b) This compound exhibits symmetry, rendering the two methyl groups equivalent. Similarly, the ring has only four unique signals, because of symmetry. In total, the ^{13}C NMR spectrum of this compound should exhibit five signals. The expected chemical shift for each of these signals (listed below) can be found in Table 15.4.

- The two, equivalent methyl (CH_3) groups will produce one signal in the region 10 – 30 ppm.
- The five methylene (CH_2) groups will produce three signals in the region 15 – 55 ppm.
- The quaternary carbon (C) will produce one signal in the region 20 – 60 ppm.

(c) Monosubstituted aromatic rings exhibit symmetry, giving only four unique carbon atoms in the aromatic ring. Each of the carbon atoms of the ethyl group is unique, giving a total of six signals. The expected chemical shift for each of these signals (listed below) can be found in Table 15.4.

- The methyl (CH_3) group will produce one signal in the region 10 – 30 ppm.
- The methylene (CH_2) group will produce one signal in the region 15 – 55 ppm.
- The aromatic ring will produce four signals in the region 110 – 170 ppm.

(d) This compound has no symmetry. Each of the carbon atoms is unique, giving a total of nine signals. The expected chemical shift for each of these signals (listed below) can be found in Table 15.4.

- One of the methyl (CH_3) groups (the methyl group that is NOT connected to oxygen) will produce one signal in the region 10 – 30 ppm.
- The methyl (CH_3) group connected to oxygen will produce one signal in the region 40 – 80 ppm.
- The methylene (CH_2) group will produce one signal in the region 15 – 55 ppm.
- The aromatic ring will produce six signals in the region 110 – 170 ppm.

(e) The aromatic ring is 1,4-disubstituted so it exhibits symmetry, giving four unique carbon atoms in the aromatic ring (four signals). In addition, there will be a signal for the carbon atom of the methoxy group, and finally, there will be two signals for the carbon atoms of the ethyl group. In total, there are seven signals. The expected chemical shift for each of these signals (listed below) can be found in Table 15.4.

- One of the methyl (CH_3) groups (the methyl group NOT connected to oxygen) will produce one signal in the region 10 – 30 ppm.
- The methyl group connected to oxygen will produce one signal in the region 40 – 80 ppm.
- The methylene (CH_2) group will produce one signal in the region 15 – 55 ppm.
- The aromatic ring will produce four signals in the region 110 – 170 ppm.

(f) This compound has symmetry, giving only five unique positions (highlighted):

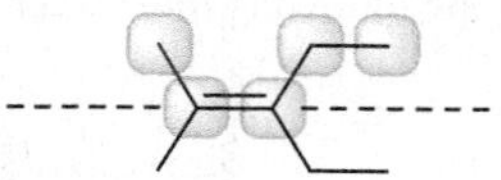

So the ^{13}C NMR spectrum should have five signals. The expected chemical shift for each of these signals (listed below) can be found in Table 15.4.

- The four methyl (CH_3) groups will produce two signals in the region 10 – 30 ppm.
- The two methylene (CH_2) groups will produce one signal in the region 15 – 55 ppm.
- The two vinylic carbon atoms will produce two signals in the region 100 – 150 ppm.

(g) None of the carbon atoms in this compound can be interchanged with any of the other carbon atoms in this compound via either reflection or rotation. All of the carbon atoms are unique, giving rise to seven signals. The expected chemical shift for each of these signals (listed below) can be found in Table 15.4.

- The four methyl (CH_3) groups will produce four signals in the region 10 – 30 ppm.
- The methylene (CH_2) group will produce one signal in the region 15 – 55 ppm.
- The two vinylic carbon atoms will produce two signals in the region 100 – 150 ppm.

(h) This compound exhibits a high degree of symmetry. All four methyl groups are interchangeable by either rotation or reflection, and therefore, all four methyl groups give rise to one signal. The two vinylic carbon atoms are also chemically equivalent, giving rise to one signal. In total, there are only two signals. The expected chemical shift for each of these signals (listed below) can be found in Table 15.4.

- The four methyl (CH_3) groups will produce one signal in the region 10 – 30 ppm.
- The two vinylic carbon atoms will produce one signal in the region 100 – 150 ppm.

(i) This compound exhibits a high degree of symmetry. All four carbon atoms are interchangeable by either rotation or reflection, and therefore, all four carbon atoms give rise to one signal, appearing in the region 40 – 80 ppm (as expected for an sp^3 hybridized carbon atom attached to an oxygen atom).

(j) None of the carbon atoms in this compound can be interchanged with any of the other carbon atoms in this compound via either reflection or rotation. All of the carbon atoms are unique, giving rise to five signals. The expected chemical shift for each of these signals (listed below) can be found in Table 15.4.

- The methyl (CH_3) group will produce one signal in the region 10 – 30 ppm.
- The methylene (CH_2) group next to oxygen will produce one signal in the region 40 – 80 ppm.
- The two vinylic carbon atoms will produce two signals in the region 100 – 150 ppm.
- The carbonyl group will produce one weak signal in the region 165 – 185 ppm.

15.27. Let's first determine which diastereomer has the *R* configuration and which has the *S* configuration. When we place the hydrogen on a dash, the three substituents will be arranged such that the sequence of priorities (1–2–3) is clockwise; therefore, compound **3** (having the *R* configuration) will have the H on a dash. Since compound **4** has the *S* configuration, we would simply need to invert the configuration at that center, meaning the H should be placed on a wedge.

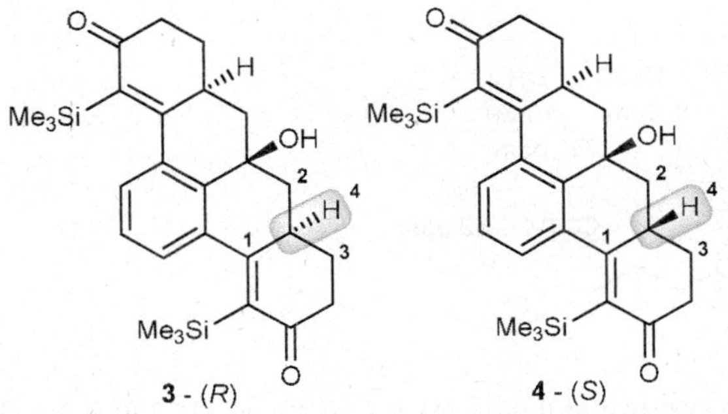

3 - (*R*) 4 - (*S*)

Upon close inspection, compound **3** has a plane of symmetry that compound **4** lacks. In fact, compound **3** is a *meso* compound. As a result, compound **3** should exhibit only thirteen signals in its ^{13}C NMR spectrum, while compound **4** should have twenty-three signals in its ^{13}C NMR spectrum. This would be an easy way to distinguish between these compounds.

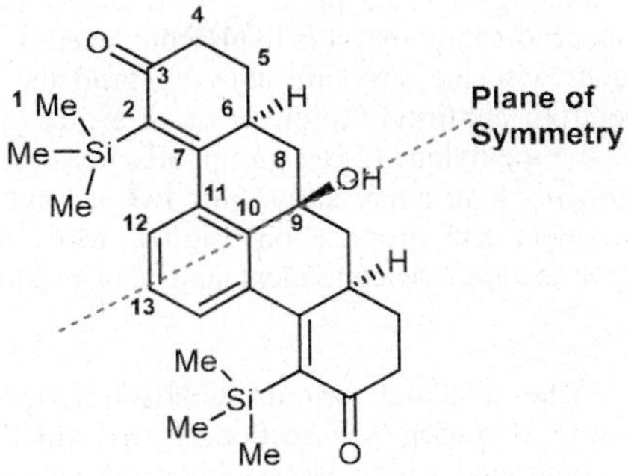

Plane of Symmetry

3 - (*R*)

15.28. The molecular formula ($C_5H_{10}O$) indicates one degree of unsaturation (see Section 14.16), which means that the compound must either have a double bond or a ring. The signal above 200 ppm (in the broadband-decoupled spectrum) indicates the source of the degree of unsaturation (C=O). The four signals in the range 10 – 60 ppm represent four unique sp^3 hybridized carbon atoms. Two of them are upside-down in the DEPT-135 spectrum, indicating that they are methylene groups. Based on the number of protons in the compound ($C_5H_{10}O$), the other two signals must be methyl groups (to give a total of ten protons).
Now we must connect a carbonyl group, two methylene groups and two methyl groups. There are only two ways to do that:

The first possibility cannot be correct because it has symmetry and would have only three signals in its broadband-decoupled ^{13}C spectrum. The latter structure is the only structure that is consistent with the data.

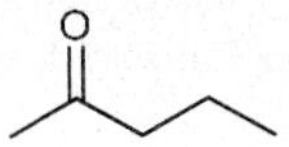

15.29. The molecular formula ($C_5H_{12}O$) indicates no degrees of unsaturation (see Section 14.16), which means that the compound cannot have either a double bond or a ring. The signal at 73.8 ppm must be produced by the carbon atom that is connected to the oxygen atom. Notice that only one signal can be found in the range 40 – 80 ppm, which is consistent with the compound being an alcohol, as indicated in the problem statement. There are only two other signals, indicating symmetry. Both of those signals represent sp^3 hybridized carbon atoms. The following structure meets these requirements:

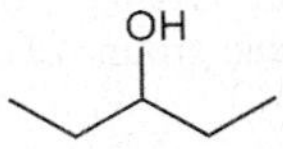

This structure is consistent with the information presented in the DEPT spectra. The DEPT-90 spectrum indicates there is only one CH group, and the chemical shift of that signal (73.8 ppm) is fairly downfield for a CH group, indicating that it is likely connected to oxygen (consistent with the structure above). And the DEPT-135 spectrum confirms the presence of a methyl (CH_3) group and a methylene (CH_2) group, also consistent with the symmetrical structure above (the two methyl groups are equivalent and produce one signal, while the two methylene groups are equivalent and also produce one signal)

15.30. The molecular formula ($C_7H_{14}O$) indicates one degree of unsaturation (see Section 14.16), which means that the compound must possess either a double bond or a ring. The signal above 200 ppm (in the broadband-decoupled spectrum) indicates the source of the degree of unsaturation (C=O). The five signals in the range 20 – 40 ppm represent five unique sp^3 hybridized carbon atoms. As such, the spectrum has a total of six signals for a compound with seven carbon atoms. That means that one of the signals in the range 20 – 40 ppm must represent two carbon atoms (for example, two equivalent methyl groups – the ^{1}H NMR data confirms the two equivalent methyl groups). Also notice that two of the signals are upside-down in the DEPT-135 spectrum, indicating that they are methylene groups. Based on the DEPT-90 spectrum, we can see that one of the signals is a CH group, while the last two signals must be methyl groups. There are several possible structures that are consistent with the information above, but only one of them is consistent with the ^{1}H NMR spectrum. Specifically, the singlet at 1.9 ppm (with an integration of 3) indicates a methyl ketone.

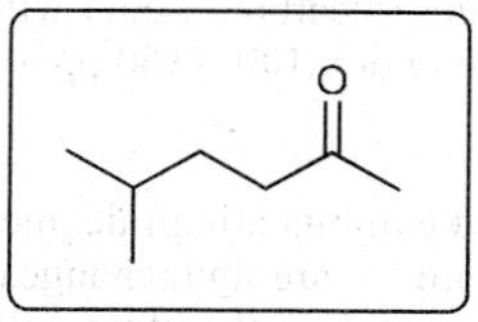

Consistent with ^{1}H NMR and ^{13}C NMR spectra

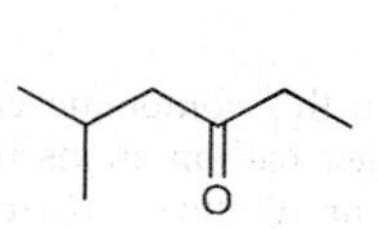

Consistent only with ^{13}C NMR spectrum

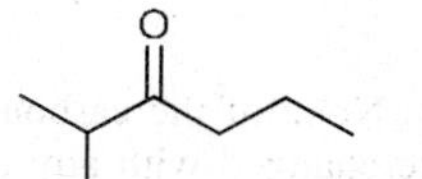

Consistent only with ^{13}C NMR spectrum

15.31. Diisobutyl phthalate has an internal plane of symmetry, so the two isobutyl groups are chemically equivalent (since they can be interchanged by rotational symmetry); the symmetry also bisects the aromatic ring. Although there are 16 carbon atoms in DIBP, only 7 signals are thus expected in its broadband-decoupled ^{13}C NMR spectrum. Using the values provided in Table 15.4, we expect the following approximate chemical shifts:

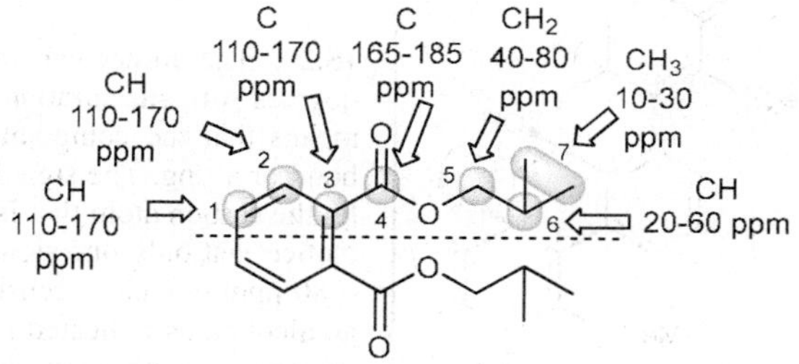

C 110-170 ppm
C 165-185 ppm
CH_2 40-80 ppm
CH_3 10-30 ppm
CH 110-170 ppm
CH 110-170 ppm
CH 20-60 ppm

The DEPT-90 spectrum exhibits three signals for the CH groups: two in the sp^2 hybridized region 110-170 (C1 and C2), and one in the 20-60 ppm region (C6). The DEPT-135 spectrum exhibits five signals (only the quaternary carbon atoms, C3 and C4, are missing); there is one signal in the 40-80 ppm region that is a negative signal, indicating the presence of a methylene group (CH_2) attached to an oxygen atom, C5.

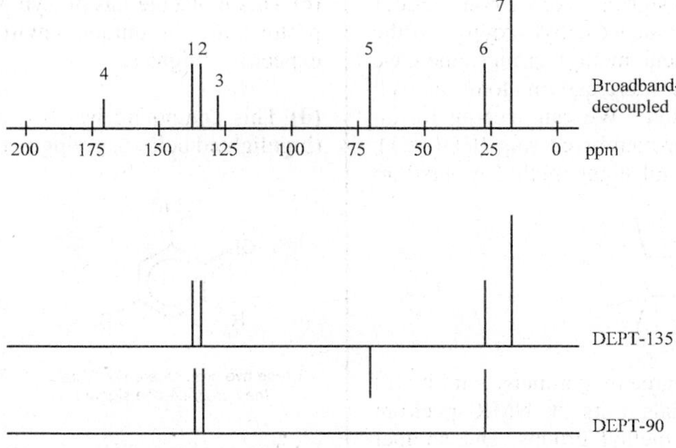

15.32.

(a) The molecular formula (C_5H_{10}) indicates one degree of unsaturation (see Section 14.16), which means that the compound must possess either a double bond or a ring. With only one signal in the spectrum, the structure must have a high degree of symmetry, such that all ten protons are equivalent. This is indeed the case for a five-membered ring (cyclopentane). There is no alkene with the molecular formula C_5H_{10} in which all ten protons occupy identical electronic environments. So cyclopentane is the only structure consistent with the spectral data:

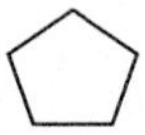

(b) The molecular formula ($C_5H_8Cl_4$) indicates no degrees of unsaturation (see Section 14.16), which means that the compound cannot have any double bonds, triple bonds, or rings. That is, the structure must be acyclic and cannot have any π bonds. With only one signal in the spectrum, the structure must have a high degree of symmetry, such that all eight protons are equivalent. This can be achieved with four equivalent methylene (CH_2) groups. Four CH_2 groups will be chemically equivalent if they are all attached to the same carbon atom (provided that all four CH_2 groups are connected to identical groups):

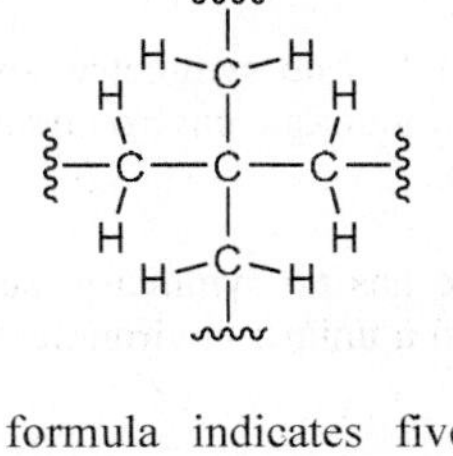

The molecular formula indicates five carbon atoms (which are now all accounted for) and four chlorine atoms, which can serve to cap the four loose ends shown above, like this:

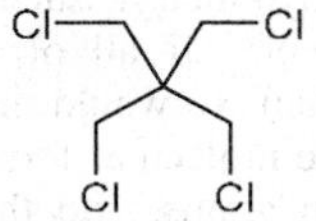

(c) The molecular formula ($C_{12}H_{18}$) indicates four degrees of unsaturation (see Section 14.16), which is highly suggestive of an aromatic ring. With only one signal in the spectrum, the structure must have a high degree of symmetry, such that all eighteen protons are equivalent. This can be achieved with six equivalent methyl groups, as seen in the following structure:

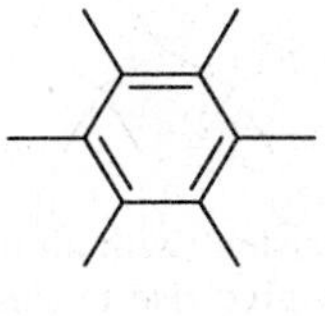

15.33. The molecular formula ($C_{12}H_{24}$) indicates one degree of unsaturation (see Section 14.16), which means that the compound must possess either a double bond or a ring. With only one signal in the ^{1}H NMR spectrum, the structure must have a high degree of symmetry, such that all twenty-four protons are equivalent. This can be accomplished with either twelve equivalent methylene (CH_2) groups or eight equivalent methyl (CH_3) groups. Since the former would use up all of the carbon atoms in the structure (all twelve), it is tempting to explore that possibility first. Indeed, a twelve-membered ring is comprised of twelve equivalent methylene groups, which should give rise to one signal in the ^{1}H NMR spectrum. And this compound (cyclododecane) also exhibits the correct degree of unsaturation (HDI = 1). However, this structure would give only one signal in its ^{13}C NMR spectrum, and the problem statement indicates that the

^{13}C NMR spectrum has two signals. So, we consider our other alternative (eight equivalent methyl groups). If the structure has eight equivalent methyl groups, then we still need to account for four more carbon atoms, as well as one degree of unsaturation. We can account for all four carbon atoms in a four-membered ring (HDI = 1). In the following structure, all eight methyl groups are equivalent.

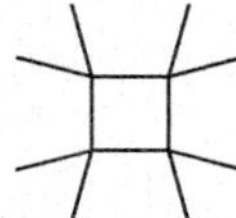

This structure has a high degree of symmetry, and it will indeed give rise to two signals in its ^{13}C NMR spectrum (one signal for all of the methyl groups, and another signal for all four carbon atoms of the ring).

15.34. The molecular formula ($C_{17}H_{36}$) indicates no degrees of unsaturation (see Section 14.16), which means that the compound cannot have any double bonds, triple bonds, or rings. That is, the structure must be acyclic and cannot have any π bonds. With only one signal in the ^{1}H NMR spectrum, the structure must have a high degree of symmetry, such that all thirty-six protons are equivalent. If all of these protons were methylene groups, then we would need at least eighteen carbon atoms, but the molecular formula indicates fewer than eighteen carbon atoms. So the thirty-six protons must be twelve equivalent methyl groups. Twelve methyl groups cannot all be attached to the same carbon atom, because the central carbon atom cannot have twelve bonds. However, twelve methyl groups will be equivalent if they comprise four equivalent *tert*-butyl groups, attached to one carbon atom.

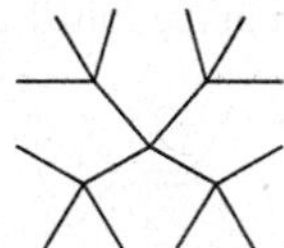

This structure is consistent with all of the available data, and it is expected to give rise to three signals in its ^{13}C NMR spectrum (one signal for the central carbon atom, another signal for the four carbon atoms connected to the central carbon atom, and then one last signal for all twelve, equivalent methyl groups).

15.35.
(a) All six methyl groups are equivalent (giving rise to one signal), while all four aromatic protons are also equivalent (giving rise to another signal). In total, we expect two signals.

(b) This molecule has no symmetry, so all four aromatic protons are in unique environments. Therefore, we expect four signals.

(c) This molecule has no symmetry, so all four aromatic protons are in unique environments. Therefore, we expect four signals.

(d) This compound has two different kinds of protons (highlighted below), giving rise to two signals.

These two protons are homotopic (they produce one signal)

These two protons are homotopic (they produce one signal)

(e) The methine (CH) proton gives one signal, while the two methyl groups collectively give one signal (with an integration of 6). In total, we expect two signals.

(f) The replacement test indicates that the two protons of the methylene group are diastereotopic. Therefore, each of these protons will produce its own signal. That is, the methylene group will give rise to two signals (because of the presence of the chiral center). The structure also has two methyl groups which are not equivalent to each other (because of their proximity to the Cl) so they also produce two different signals. Finally, the methine (CH) proton gives a signal, for a total of five signals.

15.36.
(a) This compound has four different kinds of carbon atoms (highlighted below), giving rise to four signals.

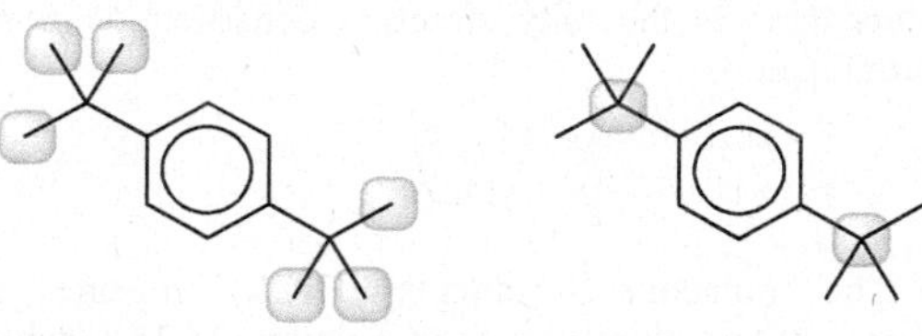

These six methyl groups are all equivalent (giving one signal)

These two carbon atoms are equivalent (giving one signal)

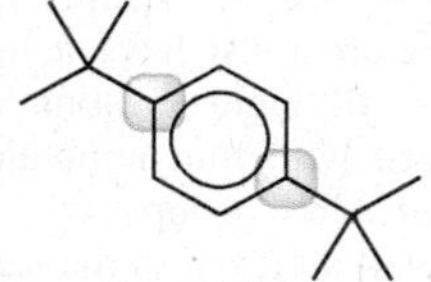

These two carbon atoms are equivalent (giving one signal)

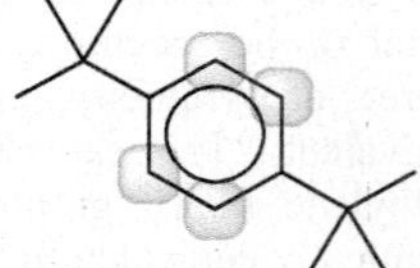

These four carbon atoms are all equivalent (giving one signal)

(b) This molecule has no symmetry, so each of the six carbon atoms is in a unique environment. Therefore, we expect six signals.

(c) This molecule has no symmetry, so each of the six carbon atoms is in a unique environment. Therefore, we expect six signals.

(d) This compound has four different kinds of carbon atoms (highlighted below), giving rise to four signals.

This carbon atom gives one signal

This carbon atom gives one signal

These two carbon atoms are equivalent (giving one signal)

These two carbon atoms are equivalent (giving one signal)

(e) The carbon atom of the methine (CH) group gives one signal, and the two methyl groups give one signal, for a total of two signals.

(f) Each of the four carbon atoms is in a unique environment, because of its substitution and its proximity to the chlorine atom. Therefore, we expect four signals.

15.37. The first compound exhibits symmetry that causes some of the carbon atoms to be equivalent (similar to the symmetry present in 15.36d). As such, the first compound will have five signals in its ^{13}C NMR spectrum. In contrast, the second compound lacks this symmetry. Each carbon atom occupies a unique environment, so the second compound is expected to produce seven signals in its ^{13}C NMR spectrum.

15.38. This compound has six different kinds of protons, highlighted here. In each case, we apply the $n + 1$ rule, giving the multiplicities shown:

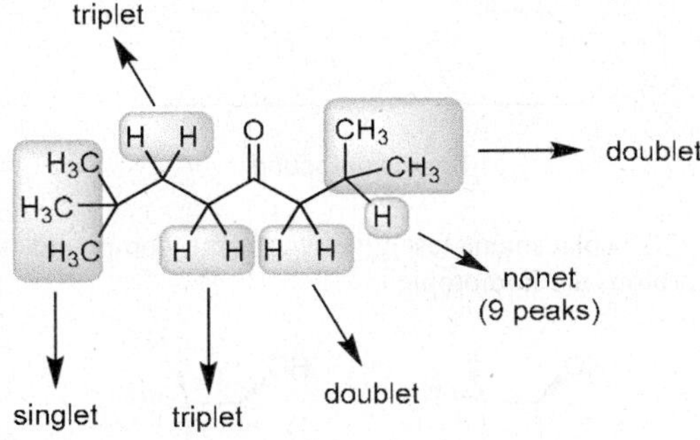

15.39.
(a) The first compound has a very high degree of symmetry, and will produce only four only signals in its ^{13}C NMR spectrum, while the second compound will produce twelve signals.
Also, the first compound will produce only two signals in its ^{1}H NMR spectrum, while the second compound will produce eight signals.

(b) The first compound is a *meso* compound. Two of the protons are enantiotopic (the protons that are alpha to the chlorine atoms) and are therefore chemically equivalent. As such, the first compound will only have two signals in its ^{1}H NMR spectrum, while the second compound will have three signals. For a similar reason, the first compound will only have two signals in its ^{13}C NMR spectrum, while the second compound will have three signals.

(c) The ^{13}C NMR spectrum of the second compound will have one more signal than the ^{13}C NMR spectrum of the first compound. The ^{1}H NMR spectra will differ in the following way: the OH group in the first compound will produce a singlet somewhere between 2 and 5 ppm with an integration of 1, while the methoxy group in the second compound will produce a singlet at approximately 3.4 ppm with an integration of 3.

(d) The first compound has symmetry that is not present in the second compound. As such, the first compound will have three signals in its ^{13}C NMR spectrum, while the second compound will have five signals.
For similar reasons, the first compound will have two signals in its ^{1}H NMR spectrum, while the second compound will have four signals.

15.40. The molecular formula (C_8H_{18}) indicates no degrees of unsaturation (see Section 14.16), which means that the compound does not have a π bond or a ring. With only one signal in its ^{1}H NMR spectrum, the structure must have a high degree of symmetry, such that all eighteen protons are equivalent. This can be achieved with six equivalent methyl groups, which account for six of the eight carbon atoms. The remaining two carbon atoms can be placed at the center of the structure, rendering all six methyl groups equivalent, like this:

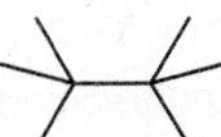

This compound will exhibit two signals in its ^{13}C NMR spectrum (one signal for the two central carbon atoms, and another signal for the six methyl groups).

15.41.
(a) The replacement test gives the same compound, so the protons are homotopic:

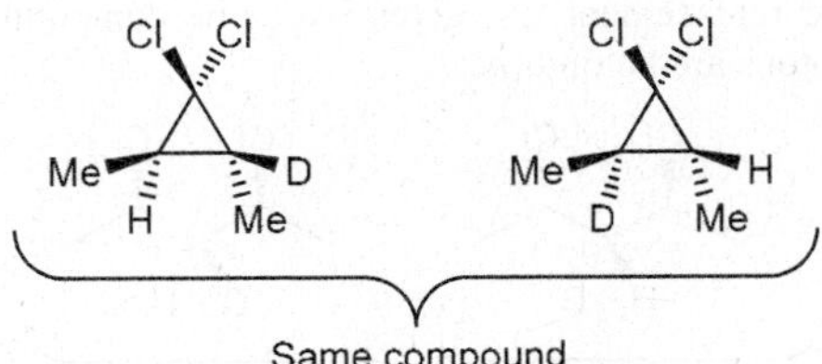

(b) The replacement test gives enantiomers, so the protons are enantiotopic:

Enantiomers

(c) The replacement test gives enantiomers, so the protons are enantiotopic:

Enantiomers

(d) The replacement test gives the same compound, so the protons are homotopic:

Same compound

(e) The replacement test gives diastereomers, so the protons are diastereotopic:

Diastereomers

These compounds are diastereomers because they are stereoisomers that are not mirror images of each other. If it seems to you like they should be enantiomers, keep in mind that each of these compounds has three chiral centers, and these compounds differ only in the configuration of one of the three chiral centers (each of the bridgehead positions is a chiral center).

(f) The replacement test gives the same compound, so the protons are homotopic:

Same compound

(g) The replacement test gives diastereomers, so the protons are diastereotopic:

Diastereomers

(h) The replacement test gives diastereomers, so the protons are diastereotopic:

Diastereomers

(i) The replacement test gives the same compound, so the protons are homotopic:

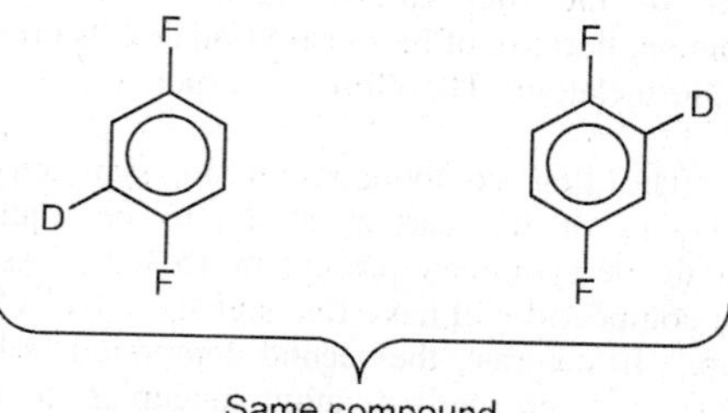

(j) The replacement test gives the same compound, so the protons are homotopic:

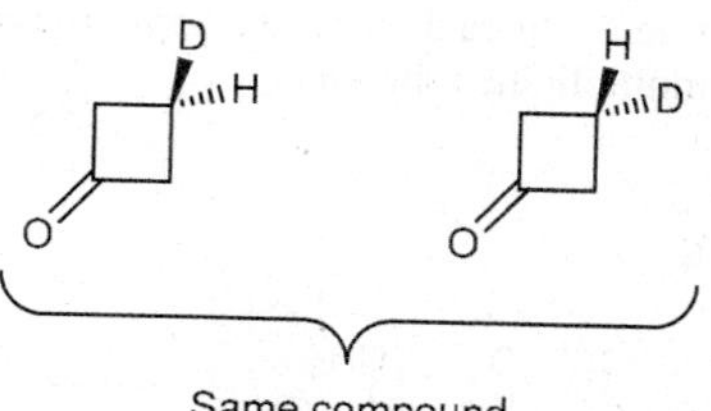

(k) The replacement test gives the same compound, so the protons are homotopic:

Same compound

(l) The replacement test gives diastereomers, so the protons are diastereotopic:

Diastereomers

(m) The replacement test gives enantiomers, so the protons are enantiotopic:

Enantiomers

(n) The replacement test gives diastereomers, so the protons are diastereotopic:

Diastereomers

(o) The replacement test gives the same compound, so the protons are homotopic:

Same compound

15.42. This compound has four aromatic protons. Because of their relationship with the respect to the ring (symmetry), we expect two types of protons, giving rise to a pair of doublets between 7 and 8 ppm:

The structure also has an ethyl group, so we expect the characteristic pattern of signals for an ethyl group. Specifically, we expect a triplet with an integration of 3 (corresponding to the CH_3 of the ethyl group) and a quartet with an integration of 2 (corresponding to the CH_2 of the ethyl group).

The signal for the CH_2 of the ethyl group is expected to appear at 1.2 + 1 = 2.2 ppm, while the signal for the CH_3 of the ethyl group is expected to appear at 0.9 + 0.2 = 1.1 ppm.
Finally, there are two neighboring methylene groups, shown here:

Each of these signals is expected to be a triplet with an integration of 2. And each of these signals is expected to be shifted downfield, because of the electron-withdrawing effects of the chlorine atom and oxygen atom. As seen in Section 16.5, each Cl adds approximately +2, and oxygen adds approximately +2.5 ppm. The CH_2 next to the oxygen atom should produce a signal near 1.2 + 2.5 + 0.4 = 4.1 ppm. The last term (+0.4) was for the effect of the distant Cl (one-fifth of 2.0). The CH_2 next to the Cl should produce a signal near 1.2 + 2.0 + 0.5 = 3.7 ppm.
The following hand-drawn spectrum shows all of the signals described above.

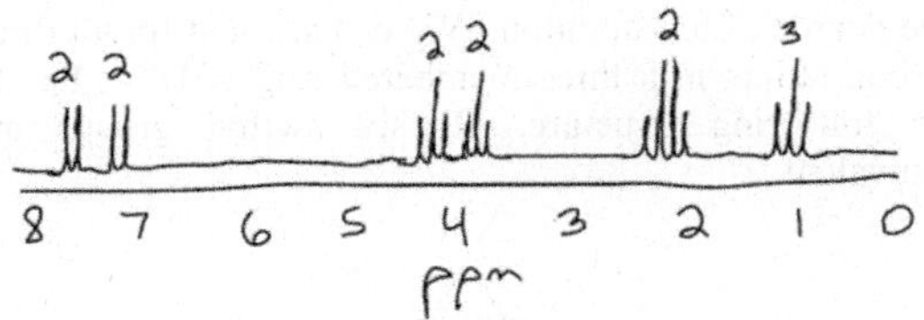

15.43.
(a) Each of the protons occupies a unique environment, and therefore, we expect four signals in the ^{1}H NMR spectrum of this compound.
(b) All of the halogens withdraw electron density from the neighboring proton, causing a downfield shift. But the effect will be strongest for fluorine (the most electronegative) and weakest for iodine (the least electronegative of the halogens in this compound).

Increasing chemical shift

$H_a > H_b > H_c > H_d$

(c) Each of the carbon atoms occupies a unique environment, and therefore, we expect four signals in the ^{13}C NMR spectrum of this compound.

(d) The carbon atoms follow the same trend exhibited by the protons (the carbon atom connected to fluorine will produce the signal that is farthest downfield).

15.44. The molecular formula (C_9H_{18}) indicates one degree of unsaturation (see Section 14.16), which means that the compound must possess either a double bond or a ring. With only one signal in the ^{1}H NMR spectrum, the structure must have a high degree of symmetry, such that all eighteen protons are equivalent. This can be accomplished with either nine equivalent methylene (CH_2) groups or six equivalent methyl (CH_3) groups. Since the former would use up all of the carbon atoms in the structure (all nine), it is tempting to explore that possibility first. Indeed, a nine-membered ring is comprised of nine equivalent methylene groups, which should give rise to one signal in the ^{1}H NMR spectrum:

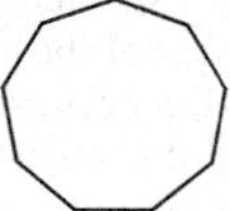

And this compound (cyclononane) also exhibits the correct degree of unsaturation (HDI = 1). However, this structure would give only one signal in its ^{13}C NMR spectrum, and the problem statement indicates that the ^{13}C NMR spectrum has two signals. So, we consider our other alternative (six equivalent methyl groups). If the structure has six equivalent methyl groups, then we still need to account for three more carbon atoms, as well as one degree of unsaturation. We can account for all three carbon atoms in a three-membered ring (HDI = 1). In the following structure, all six methyl groups are equivalent.

This structure has a high degree of symmetry, and it will indeed give rise to two signals in its ^{13}C NMR spectrum (one signal for all of the methyl groups, and another signal for all three carbon atoms of the ring).

15.45.

(a) This compound has three different methylene (CH_2) groups, giving rise to three separate signals.

In addition, each of the three methyl groups gives its own unique signal, as none of the methyl groups are in identical electronic environments (the two methyl groups on the left side of the structure are not in identical environments, because one is *trans* to the main chain, while the other is *cis* to the main chain):

Each of the vinylic CH groups are different, giving rise to two more signals:

And finally, the proton of the OH group gives one last signal, for a total of 3 + 3 + 2 + 1 = 9 signals.

(b) The methyl group gives one signal, and then each of the remaining protons gives rise to its own signal, for a total of six signals:

Note that each of the vinylic protons is unique. For example, the following two vinylic protons are different from each other even though they are connected to the same carbon atom:

15.46. Four CH_2 groups can be chemically equivalent if they are all attached to the same carbon atom (provided that the four CH_2 groups are all connected to identical groups:

The molecular formula indicates a total of nine carbon atoms and twenty hydrogen atoms, but the structure above only accounts for five carbon atoms and eight hydrogen atoms. We must still account for another four carbon atoms and twelve hydrogen atoms. This can be accomplished if we simply connect a methyl group to each of the CH_2 groups, giving the following structure:

The four methyl groups are chemically equivalent, giving rise to only signal. As such, the ^{1}H NMR spectrum of this compound is expected to exhibit only two signals (one for the CH_2 groups and the other for the CH_3 groups).

15.47. Below are the expected chemical shifts for each of the seven signals in this compound:

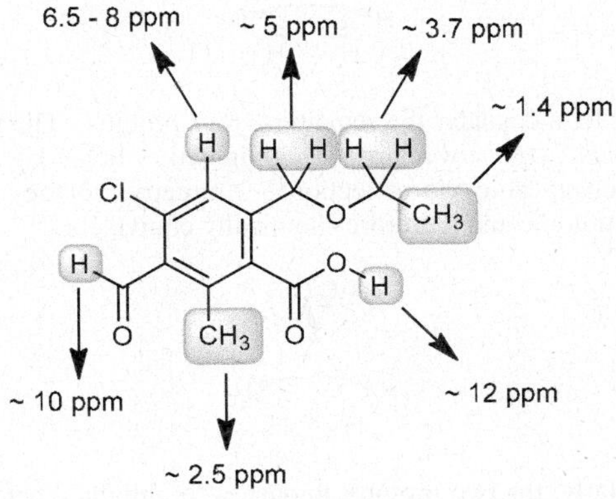

6.5 - 8 ppm ~ 5 ppm ~ 3.7 ppm ~ 1.4 ppm ~ 10 ppm ~ 12 ppm ~ 2.5 ppm

15.48.

(a) Symmetry in the ring gives four different signals for the carbon atoms of the ring 110 – 170, in addition to two signals for the vinylic carbon atoms. So in total, we expect six signals, all of which result from sp^2 hybridized carbon atoms, and therefore, we expect all six signals to appear in the region 110 – 170 ppm.

(b) Each of the carbon atoms of the ring occupies a unique environment, giving six signals. The two methyl groups occupy identical environments (they are interchangeable by reflectional symmetry), so they produce one signal. This can be seen more clearly if we draw wedges and dashes to illustrate the 3D orientation of the methyl groups:

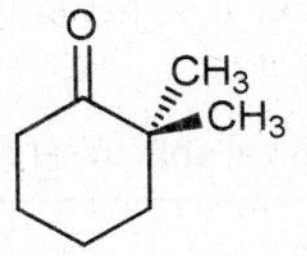

This gives a total of seven signals. The signal resulting from the carbon atom of the ketone carbonyl group is expected to appear in the region 185 – 220 ppm, while the remaining six signals should appear in the region 10 – 60 ppm.

(c) The compound is symmetrical, so we only need to consider half of the structure. We expect a total of four signals, corresponding with the following unique positions:

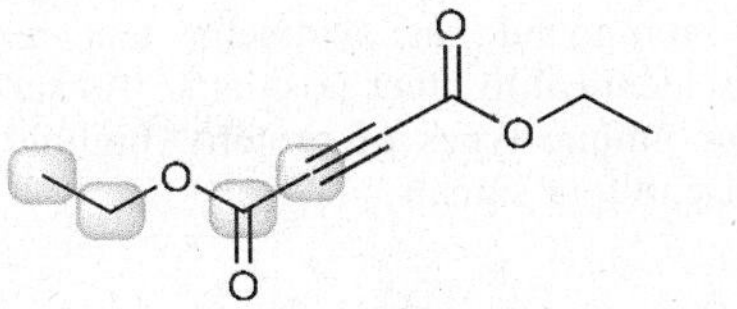

The signal from the carbon atom of the methyl group (sp^3 hybridized) will appear in the region 10 – 30 ppm. The carbon atom of the methylene (CH_2) group is also sp^3 hybridized, but it is next to an oxygen atom. So we expect that signal to appear in the region 40 – 80 ppm, together with the signal from the *sp* hybridized carbon. Finally, the signal from the ester carbonyl group is expected to appear in the region 165 – 185 ppm.

15.49. Let's begin by drawing the reaction described in the problem statement:

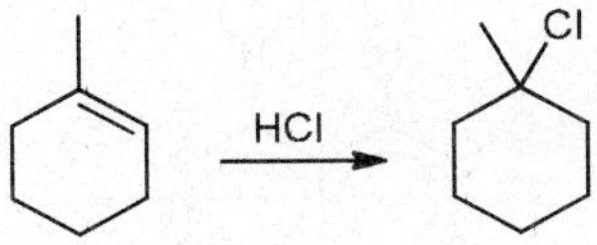

HCl

The Markovnikov product has symmetry that the *anti*-Markovnikov product lacks. As such, a ^{1}H NMR spectrum of the Markovnikov product should have fewer signals than a ^{1}H NMR spectrum of the *anti*-Markovnikov product.

15.50.

(a) The compound has a high degree of symmetry, and there are only two unique aromatic protons, highlighted below, giving rise to two signals.

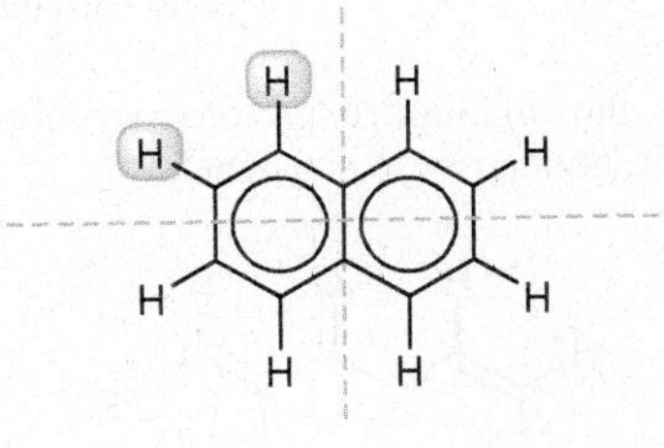

Each of the remaining aromatic protons can be interchanged with one of these positions (via either rotational or reflectional symmetry).

(b) The presence of the methyl group renders all of the aromatic protons different from each other (because of their proximity to the methyl group). As such, we expect eight signals:

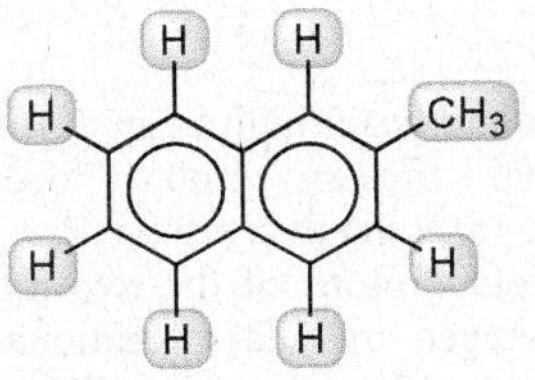

(c) The compound has symmetry that renders some positions identical to other positions. As such, there are only four unique types of protons, highlighted below, giving rise to four signals.

(d) The compound has a high degree of symmetry that renders some positions identical to other positions. As such, there are only two unique types of protons, highlighted below, giving rise to two signals.

(e) Each of the three vinylic protons is in a unique environment, so we expect three signals.

trans to nitro group

cis to nitro group

(f) Each of the highlighted protons occupies a unique environment, giving rise to six signals:

Note that the following positions are different from each other:

trans to oxygen atom

cis to oxygen atom

(g) The compound has a high degree of symmetry. As such, the two methyl groups occupy identical environments and collectively give rise to one signal. Similarly, all four protons of the two methylene (CH_2) can be interchanged by either rotational or reflection symmetry, so these four protons will collectively give rise to one signal. In total, we expect only two signals:

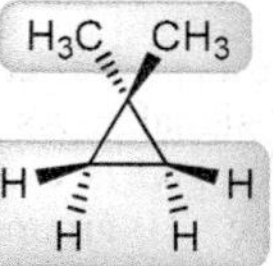

(h) The methyl group will produce one signal.

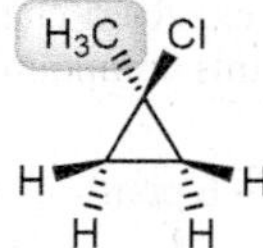

Now let's consider the remaining four protons. The two protons on wedges (highlighted below) are interchangeable via reflectional symmetry, so they are enantiotopic and therefore chemically equivalent.

Similarly, the two protons on dashes (highlighted below) are also interchangeable via reflectional symmetry, so they too are enantiotopic and therefore chemically equivalent.

In summary, we expect this compound to produce three signals in its ^1H NMR spectrum.

15.51. Among these three compounds, the first one (benzene) has aromatic protons, which are expected to produce a signal the farthest downfield (between 6.5 and 8 ppm). Acetylenic protons give signals that are relatively upfield (near 2.5 ppm) while vinylic protons are expected to produce a signal in the range of 4.5 – 6.5 ppm.

Increasing chemical shift in ^1H NMR spectroscopy

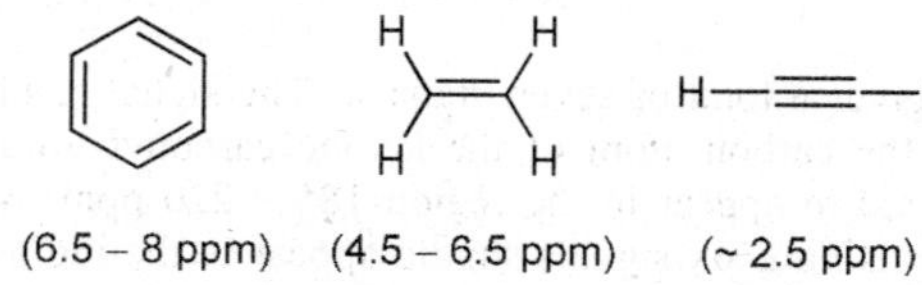

15.52. In Section 15.5, the term "chemical shift" was defined in the following way:

$$\delta = \frac{\text{observed shift from TMS (in Hz)}}{\text{operating frequency of the instrument (in Hz)}}$$

The problem statement indicates that the chemical shift of the proton is 1.2 ppm and the operating frequency of the spectrometer is 300 MHz. We then plug these values into the equation above, as shown:

$$1.2 \times 10^{-6} = \frac{\text{observed shift from TMS (in Hz)}}{300 \times 10^{6}\ \text{Hz}}$$

which gives the following observed shift from TMS (in Hz):

$$\text{observed shift from TMS (in Hz)} = \left(1.2 \times 10^{-6}\right) \times \left(300 \times 10^{6}\ \text{Hz}\right) = 360\ \text{Hz}$$

15.53. The molecular formula ($C_{13}H_{28}$) indicates no degrees of unsaturation (see Section 14.16), which means that the compound does not have a π bond or a ring. The ^{1}H NMR spectrum exhibits the characteristic pattern of an isopropyl group (a septet with an integration of 1 and a doublet with an integration of 6):

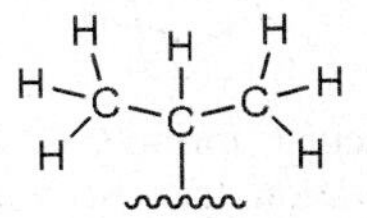

However, there are no other signals in this spectrum, and the molecular formula indicates that there are 28 protons (not just 7 protons, as we would expect for an isopropyl group). So the compound must be highly symmetrical, with four equivalent isopropyl groups (to account for all 28 protons). This also accounts for 12 of the 13 carbon atoms in this compound. The remaining carbon atom must be at the center, connected to all four isopropyl groups:

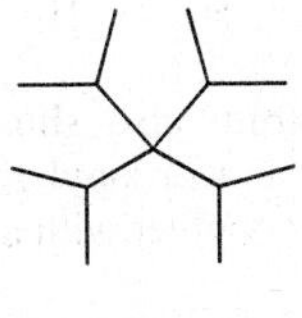

15.54. The molecular formula (C_8H_{10}) indicates four degrees of unsaturation (see Section 14.16), which is highly suggestive of an aromatic ring. This accounts for six of the eight carbon atoms in the structure. The other two carbon atoms must be connected to the ring, either as an ethyl group or as two methyl groups. Ethylbenzene would give an ^{1}H NMR spectrum with five signals and a ^{13}C NMR spectrum with six signals. The problem statement indicates fewer signals in each of these spectra, which means that the compound must have more symmetry than ethylbenzene. If we explore the three possible ways to connect two methyl groups to a ring (1,2- or 1,3- or 1,4-), we will find that only 1,4-dimethylbenzene has the necessary symmetry to give only two signals in the ^{1}H NMR spectrum and three signals in the ^{13}C NMR spectrum.

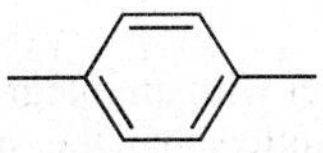

15.55. The molecular formula (C_3H_8O) indicates no degrees of unsaturation (see Section 14.16), which means that the compound does not have a π bond or a ring. The broad signal between 3200 and 3600 cm^{-1} indicates the presence of an OH group. The molecular formula indicates that the structure has three carbon atoms, yet the ^{13}C NMR spectrum exhibits only two signals (not three), indicating the presence of symmetry. This is only true for 2-propanol (not for 1-propanol):

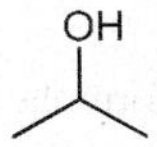

15.56. The molecular formula ($C_4H_6O_4$) indicates two degrees of unsaturation (see Section 14.16), which means that the compound must possess either two double bonds, or two rings, or one ring and one double bond, or a triple bond. The very broad signal (2500 – 3600 cm^{-1}) in the IR spectrum indicates the presence of a carboxylic acid group. The ^{1}H NMR spectrum has only two signals, with a total integration of 3, however the molecular formula indicates the presence of 6 protons. Therefore, the actual integration values for the signals are 2H and 4H, respectively. The broad singlet at 12.1 ppm is characteristic of a carboxylic acid group (COOH), as suggested by the IR spectrum, and since this signal has an integration value of 2H, we conclude that the compound must have two carboxylic acid groups. This accounts for both degrees of unsaturation, which means that the compound does not possess a ring. In order for the remaining four protons to be identical, they must be interchangeable by symmetry, which is indeed the case when we place two methylene (CH_2) groups in between the two carboxylic acid groups, like this:

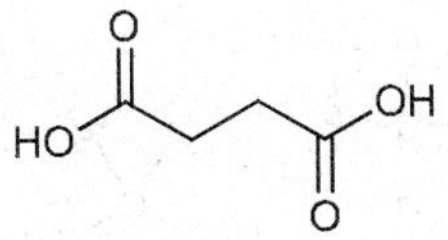

Note that the CH_2 groups do not split each other (they appear as singlets) because they are chemically equivalent. The $n + 1$ rule refers to n as the number of nonequivalent neighboring protons.

15.57.

(a) The molecular formula ($C_5H_{10}O$) indicates one degree of unsaturation (see Section 14.16), which means that the compound must possess either a double bond or a ring. The ^{1}H NMR spectrum exhibits the characteristic pattern of an isopropyl group (a doublet with an integration of 6, and a septet with an integration of 1):

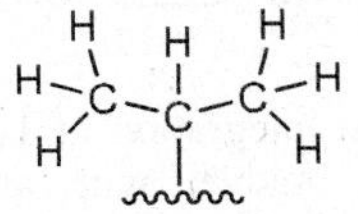

There is also a singlet with an integration of 3, indicating a methyl group. Notice that the signal for the methyl group appears at 2.12 ppm rather than 0.9 ppm, so it has been shifted downfield by just over 1 ppm, which is consistent with being adjacent to a C=O group (accounting for the one degree of unsaturation). The same downfield shift is true for the chemical shift of the CH of the isopropyl group. This gives the following structure:

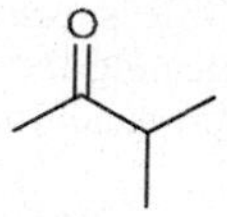

(b) The molecular formula ($C_5H_{12}O$) indicates no degrees of unsaturation (see Section 14.16), which means that the compound does not have a π bond or a ring. The ^{1}H NMR spectrum exhibits the characteristic pattern of an ethyl group (a quartet with an integration of 2, and a triplet with an integration of 3):

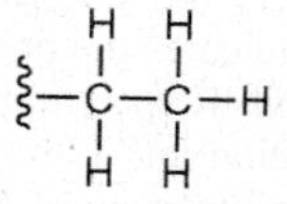

In addition, the singlet with an integration of 6 indicates two equivalent methyl groups:

methyl H_3C ? CH_3 methyl

And the singlet with an integration of 1 is suggestive of an OH group. These pieces account for all of the atoms in the compound except for one carbon atom, which we place in between the two methyl groups. The three fragments can then only be connected in one way, giving the following structure:

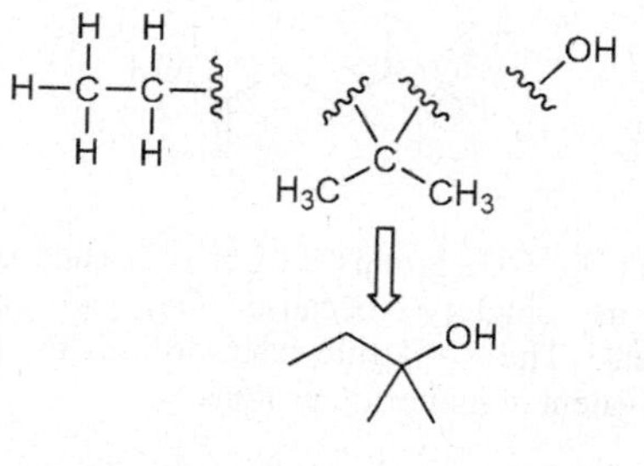

(c) The molecular formula ($C_4H_{10}O$) indicates no degrees of unsaturation (see Section 14.16), which means that the compound does not have a π bond or a ring. The ^{1}H NMR spectrum exhibits a signal with an integration of 6, which indicates two equivalent methyl groups:

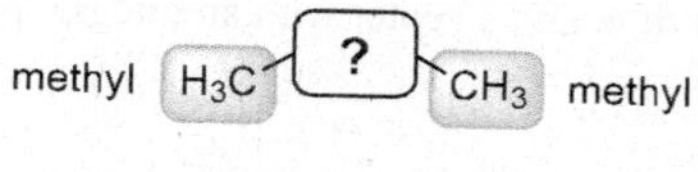

methyl H_3C ? CH_3 methyl

The singlet with an integration of 1 is suggestive of an OH group. The doublet with an integration of 2 indicates a methylene group with only one neighboring proton:

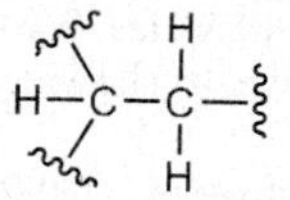

The CH (methine) proton is responsible for the last remaining signal, which is described as a multiplet (actually a nonet), indicating that this CH group is adjacent to the CH_2 group as well as the methyl groups, like this:

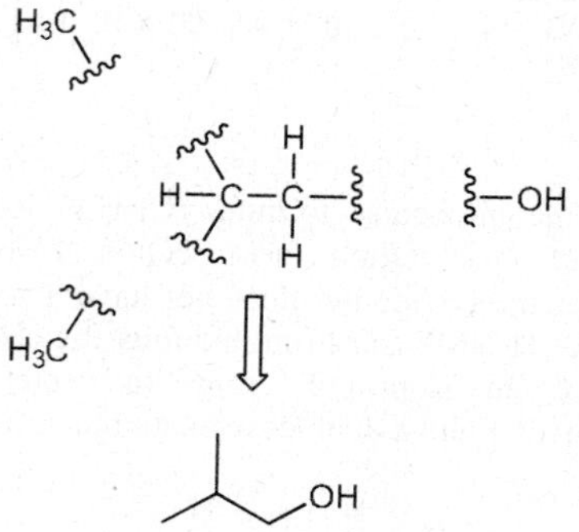

15.58. The molecular formula (C_9H_{12}) indicates four degrees of unsaturation (see Section 15.16), which is highly suggestive of an aromatic ring. This is confirmed by the multiplet just above 7 ppm in the ^{1}H NMR spectrum. This signal has an integration of 5, indicating that the ring is monosubstituted:

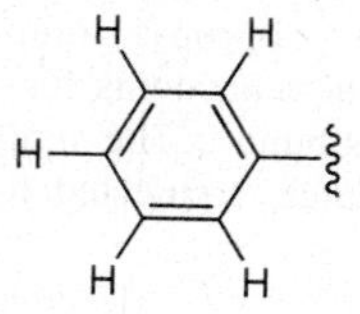

The ^{1}H NMR spectrum also shows the characteristic pattern of signals for an isopropyl group (a septet with an integration of 1 and a doublet with an integration of 6):

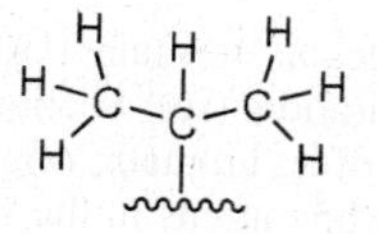

These two fragments (the monosubstituted aromatic ring and the isopropyl group) account for the entire structure:

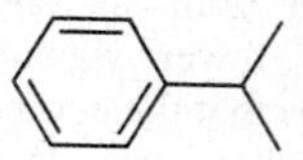

15.59. The molecular formula ($C_9H_{10}O_2$) indicates five degrees of unsaturation (see Section 14.16), which is highly suggestive of an aromatic ring, in addition to either one double bond or one ring. The multiplet just above 7 ppm in the ^{1}H NMR spectrum corresponds with aromatic protons, which confirms the presence of an aromatic ring. The integration of this signal is 5, which indicates that the ring is monosubstituted:

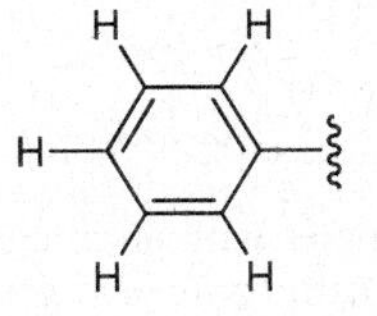

The presence of a monosubstituted ring is confirmed by the four signals between 110 and 170 ppm in the ^{13}C NMR spectrum (the region associated with sp^2 hybridized carbon atoms), just as expected for a monosubstituted aromatic ring:

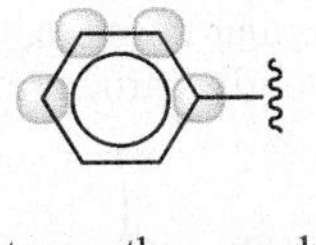

In the ^1H NMR spectrum, the very broad signal near 12 ppm (with an integration of 1) indicates the presence of a carboxylic acid group, which is confirmed by the signal near 180 ppm in the ^{13}C NMR spectrum (the region associated with carbonyl groups):

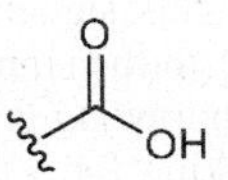

In the ^1H NMR spectrum, the pair of triplets (each with an integration of 2) indicates a pair of neighboring methylene groups:

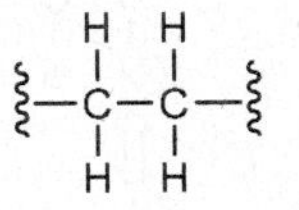

These methylene groups account for the two signals between 15 and 55 ppm in the ^{13}C NMR spectrum.
We have now analyzed all of the signals in both spectra, and we have uncovered three fragments, which can only be connected to each other in the following way:

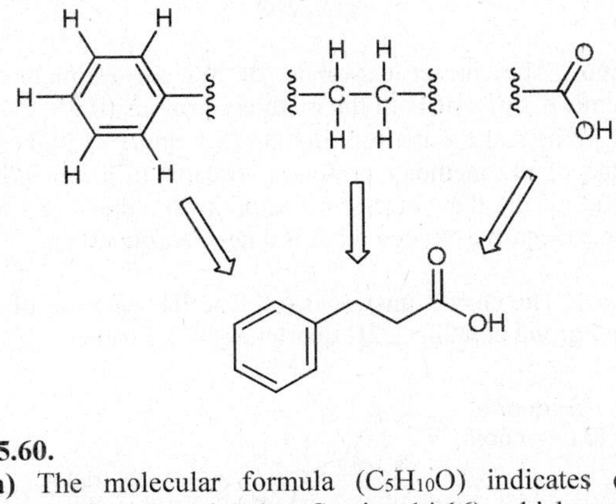

15.60.
(a) The molecular formula ($C_5H_{10}O$) indicates one degree of unsaturation (see Section 14.16), which means that the compound must possess either a double bond or a ring. One of the signals in the ^{13}C NMR spectrum appears above 200 ppm, indicating the presence of a carbonyl group (C=O), which accounts for the one degree of unsaturation.
In total, the ^{13}C NMR spectrum exhibits only three signals, while the molecular formula indicates the presence of five carbon atoms. Therefore, the structure must possess symmetry, giving only three different kinds of carbon atoms (one of which is a carbonyl group), as seen in the following structure:

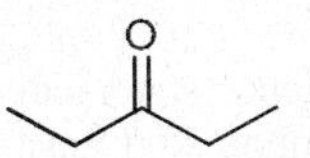

(b) The molecular formula ($C_6H_{10}O$) indicates two degrees of unsaturation (see Section 14.16), which means that the compound must possess either two double bonds, or two rings, or one double bond and one ring, or a triple bond. The ^{13}C NMR spectrum exhibits only three signals, while the molecular formula indicates the presence of six carbon atoms. Therefore, the structure must possess symmetry, giving only three different kinds of carbon atoms. Two of these types of carbon atoms must be sp^2 hybridized (because two signals appear between 100 and 150 ppm), while the third signal indicates a carbon atom attached to an oxygen atom. The following structure accounts for all of the observations above:

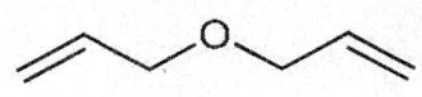

15.61. The problem statement indicates that the compound is an alcohol, so it must contain an OH group. The molecular formula ($C_4H_{10}O$) indicates no degrees of unsaturation (see Section 14.16), which means that the compound cannot have any double bonds, triple bonds, or rings. That is, the structure must be acyclic and cannot have any π bonds. The broadband decoupled spectrum has four signals, one of which appears above 60 ppm (this signal accounts for the carbon atom attached directly to the OH group). The other three signals in the broadband decoupled spectrum are all below 60 ppm, indicating that all of the carbon atoms are sp^3 hybridized (although we already knew that because HDI = 0). The DEPT-90 spectrum has only one signal, which means the compound has only one methine (CH) group. Furthermore, this signal is above 60 ppm, which indicates that this CH group is connected to directly to the OH group:

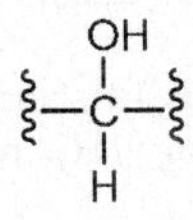

The DEPT-135 spectrum indicates that the other three signals (below 60 ppm) correspond with one methylene group (upside-down signal) and two methyl groups (right-side up signals that did not appear in the DEPT-90).
We have now analyzed all of the signals in all of the spectra, and we have uncovered four fragments, which can only be connected to each other in the following way:

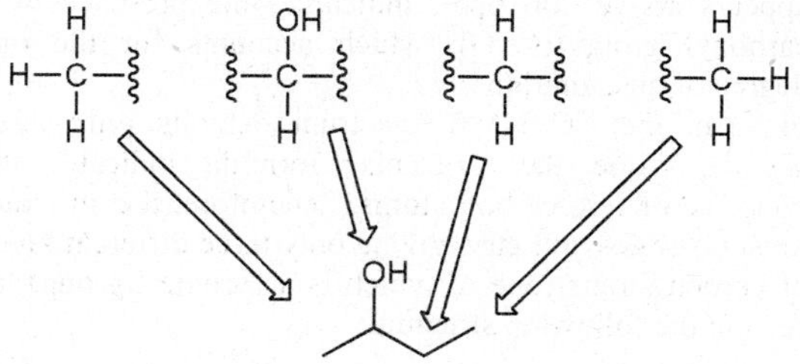

15.62. The problem statement indicates that the compound is an alcohol, so it must contain an OH group. The molecular formula ($C_6H_{14}O$) indicates no degrees of unsaturation (see Section 14.16), which means that the compound cannot have any π bonds or rings. That is, the structure must be an acyclic alcohol. The DEPT-135 spectrum exhibits five signals that are upside-down, indicating the presence of five methylene groups. One of these methylene groups must be connected to the OH group, because one of the upside-down signals appears above 60 ppm. The five methylene groups and the OH group account for 11 of the 14 protons in this compound. Therefore, the signal pointing up must correspond with a methyl group (rather than a CH group). There is only one way to connect a methyl group, five methylene groups, and an OH group, as shown here:

15.63. The correct answer is (a). The two signals farthest upfield are a quartet with an integration of 2 and a triplet with an integration of 3. Together, these two signals indicate the presence of an ethyl group:

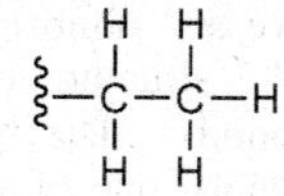

The signals near 7 ppm represent aromatic protons. Note that the total integration of the aromatic region is 2H + 2H = 4H, which means that the ring is disubstituted. Furthermore, the aromatic signals are a pair of doublets, indicating symmetry, which is achieved with a 1,4-disubstituted aromatic ring.

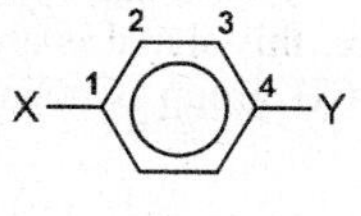

Therefore, the answer is (a), and the signal at 4.7 ppm (with an integration of 1H) represents the phenolic proton:

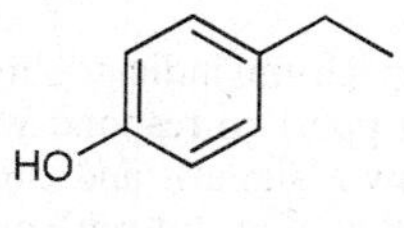

15.64. The correct answer is (b). The molecular formula ($C_4H_8O_2$) indicates one degree of unsaturation, which means that the compound must possess either a double bond or a ring. The ^1H NMR spectrum exhibits the characteristic pattern of an isopropyl group (a doublet with an integration of 6, and a septet with an integration of 1):

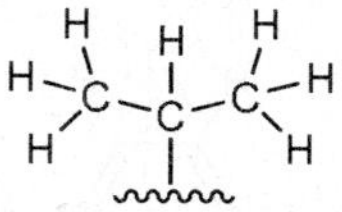

There is also a singlet with an integration of 1 at 11.38 ppm, indicating a carboxylic acid group (which accounts for the one degree of unsaturation).

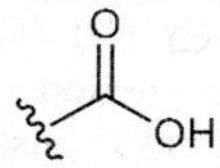

These two fragments account for all of the atoms that appear in the molecular formula, so we connect these fragments together to give structure (b):

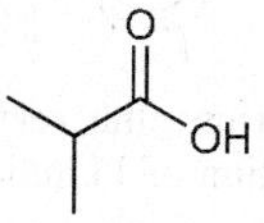

15.65. The correct answer is (b). A triplet indicates protons that are adjacent to a methylene (CH_2) group. The chemical shift of this signal (4 ppm) indicates that the protons giving rise to the signal are adjacent to some powerful electron-withdrawing group, such as an oxygen atom. So, we are looking for a compound that contains the following structural features:

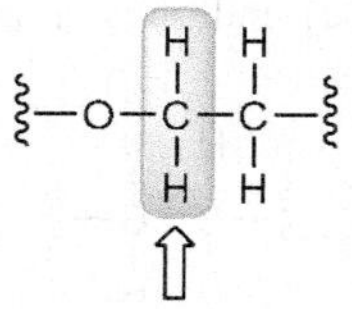

A triplet near 4 ppm

Only structure (b) has these features:

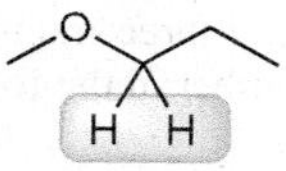

15.66. The correct answer is (d). We can estimate the chemical shifts of both the methoxy protons (0.9 + 2.5 = 3.4 ppm) and the methyl protons (0.9 ppm). A higher δ value of the methoxy protons correlates to a *downfield* shift, and the electron withdrawal due to the electronegative oxygen atom is a *deshielding* effect.

15.67. The correct answer is (a). The ^1H spectrum of an ethyl group contains a 2H quartet and a 3H triplet.

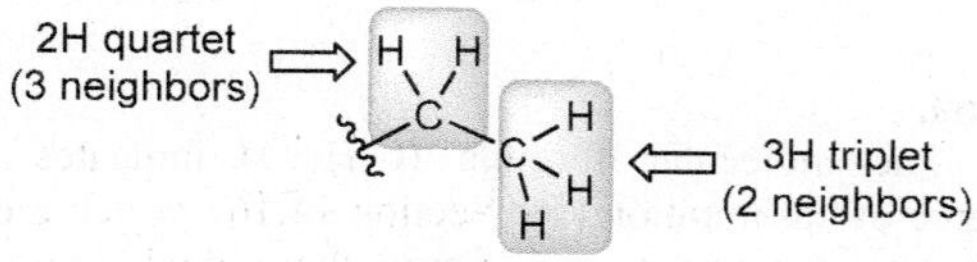

2H quartet (3 neighbors)

3H triplet (2 neighbors)

15.68. The correct answer is (d). The given compound has a total of six signals in its ^{13}C NMR and five signals in its ^{1}H NMR spectrum.

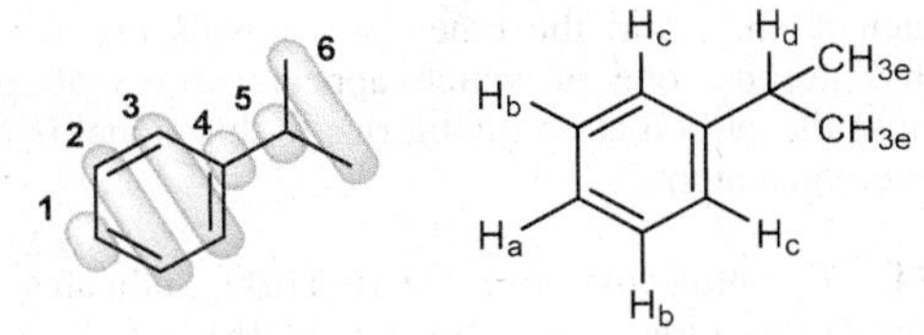

15.69. The correct answer is (d). The compound shown below is expected to produce four signals in its ^{1}H NMR spectrum. For each signal, its expected chemical shift, multiplicity, and integration closely correlate to the given spectrum.

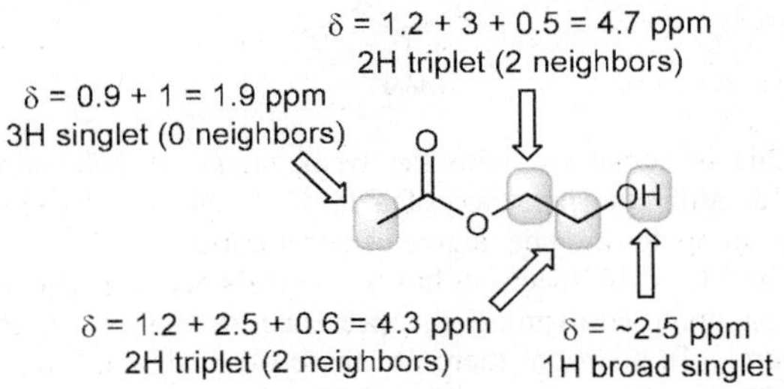

δ = 1.2 + 3 + 0.5 = 4.7 ppm
2H triplet (2 neighbors)

δ = 0.9 + 1 = 1.9 ppm
3H singlet (0 neighbors)

δ = 1.2 + 2.5 + 0.6 = 4.3 ppm
2H triplet (2 neighbors)

δ = ~2-5 ppm
1H broad singlet

15.70. The correct answer is (c). The compound shown below is expected to produce five signals in its ^{13}C NMR spectrum: one signal in the 185-220 ppm region (ketone C=O), three signals in the 110-170 ppm region (sp^2 hybridized aromatic), and one signal in the 10-60 ppm region (sp^3 hybridized).

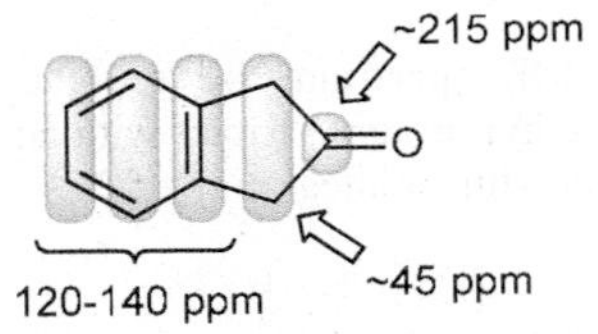

~215 ppm

120-140 ppm

~45 ppm

15.71. The correct answer is (c). Each of the given compounds has at least one internal plane of symmetry. The compound inside the box below exhibits a total of four signals in its ^{13}C NMR spectrum, while the other compounds each exhibit three signals in their ^{13}C NMR spectra.

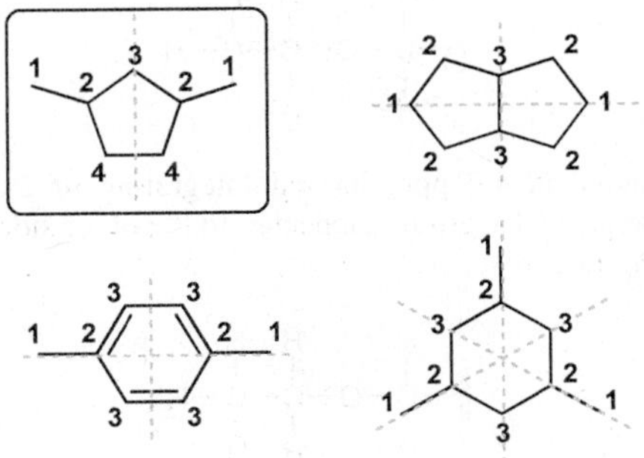

15.72. The molecular formula ($C_6H_{14}O_2$) indicates no degrees of unsaturation (see Section 14.16), which means that the compound cannot have any double bonds, triple bonds, or rings. That is, the structure must be acyclic and cannot have any π bonds. The IR spectrum has a broad signal between 3200 and 3600 cm^{-1}, indicating the presence of an OH group. The ^{13}C NMR spectrum has six signals, and the molecular formula indicates there are six carbon atoms, which means that the compound lacks symmetry that would interchange any of the carbon atoms. Since the compound has no degrees of unsaturation, all of the signals must arise from sp^3 hybridized carbon atoms. Indeed, three of the signals appear between 10 and 60 ppm, as expected for sp^3 hybridized carbon atoms. But the other three signals appear between 60 and 80 ppm, indicating that three carbon atoms are connected to an oxygen atom. We can therefore draw the following fragments (since the compound has an OH group, and since the molecular formula indicates only two oxygen atoms):

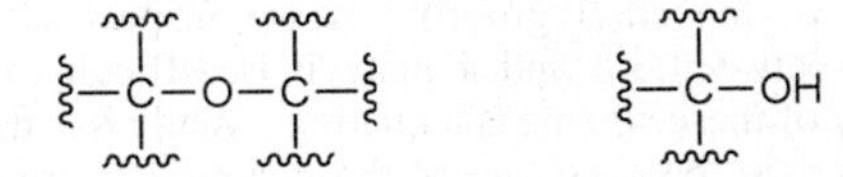

Now we explore the ^{1}H NMR spectrum. Let's begin with the signals downfield. There are three signals that appear between 3.5 and 4 ppm. One of these signals is clearly a triplet, but the other two signals are overlapping so it is difficult to determine their multiplicity (perhaps they are doublets that appear very close to each other, or perhaps they are overlapping triplets). We will revisit the multiplicity of these signals later. For now, let's focus on the chemical shifts and integration values for these signals. These signals are certainly from the three groups connected to oxygen atoms (because of their chemical shifts), and we notice that each of these signals has an integration of 2H, which allows us to modify the fragments above as follows:

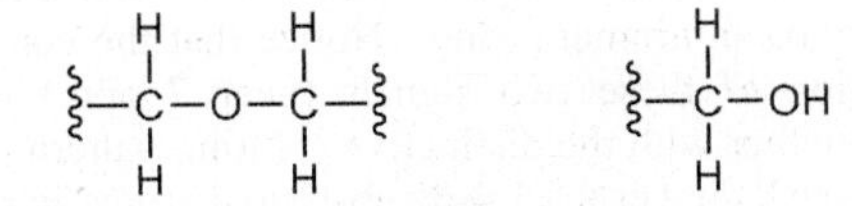

The singlet at 2.4 ppm has an integration of 1, which can be attributed to the OH group. Each of the multiplets near 1.5 ppm has an integration of 2H, indicating methylene groups that have complex splitting:

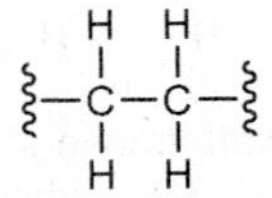

The triplet at 1.0 ppm has an integration of 3H, indicating a methyl group that is connected to a neighboring CH_2 group.

In summary, we have the following fragments:

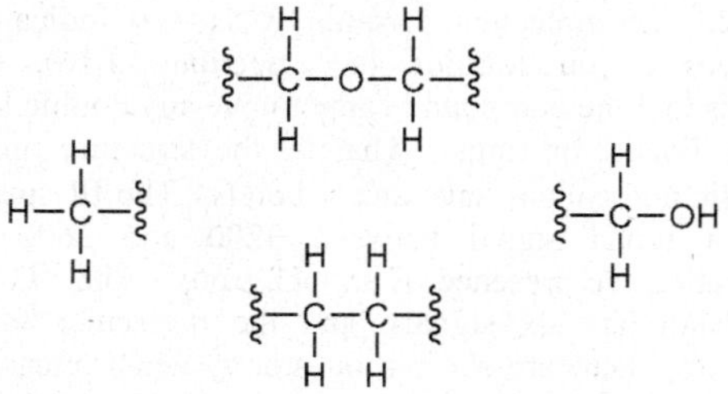

There are only two ways to connect these four fragments:

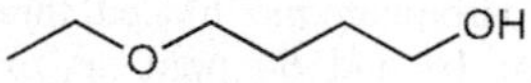

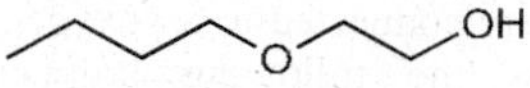

The first structure contains an ethyl group, which should produce a quartet with an integration of 2 (for the CH_2 portion of the ethyl group). If we inspect the three signals between 3.5 and 4 ppm, it is difficult to argue that any of these signals is a quartet. While it is difficult to be certain, because two of these signals overlap with each other, it looks more like each of these signals is a triplet, which would be consistent with the second structure:

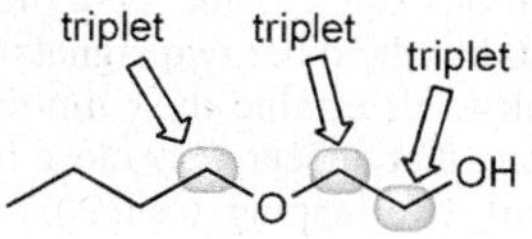

triplet triplet triplet

15.73. The molecular formula ($C_8H_{10}O$) indicates four degrees of unsaturation (see Section 14.16), which is highly suggestive of an aromatic ring. This is confirmed by the presence of signals just above 3000 cm^{-1} in the IR spectrum, and signals at approximately 1600 cm^{-1}. In the ^1H NMR spectrum, the signals near 7 ppm are likely a result of aromatic protons, which also confirms the presence of an aromatic ring. Notice that the combined integration of these two signals (near 7 ppm) is 4H. This, together with the distinctive splitting pattern (a pair of doublets), suggests a 1,4-disubstituted aromatic ring:

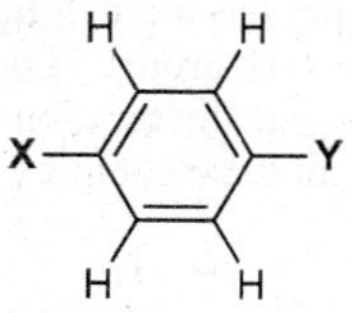

The spectrum also exhibits two singlets, each of which has an integration of 3H, indicating methyl groups. Notice that both signals are shifted downfield (relative the benchmark value for a methyl group of 0.9 ppm). One of them is shifted much more than other, indicating that it is likely next to an oxygen atom. This gives the following structure:

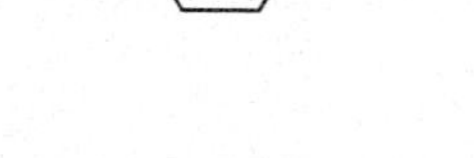

This structure is consistent with the ^{13}C NMR spectrum. Specifically, there are four signals for the aromatic ring, one of which is shifted downfield because it is next to an oxygen atom. And the other two signals are for the methyl groups, one of which appears above 60 ppm because the carbon atom giving rise to this signal is next to an oxygen atom.

15.74. The molecular formula ($C_5H_{10}O$) indicates one degree of unsaturation (see Section 14.16), which means that the compound must either have a double bond or a ring. The IR spectrum has a signal at approximately 3100 cm^{-1}, indicating the presence of a Csp^2–H bond:

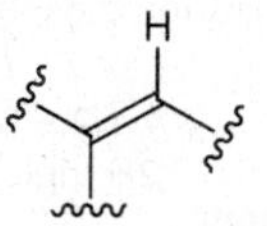

This is consistent with the weak signal at 1600 cm^{-1}, indicating the presence of a C=C double bond (which accounts for the one degree of unsaturation).

The ^{13}C NMR spectrum has two signals between 100 and 150 ppm, confirming the presence of a C=C double bond. In addition, there are three other signals, two of which appear between 40 and 80 ppm. These latter two signals are characteristic of carbon atoms connected to an oxygen atom. Since there is only one oxygen atom in the structure ($C_5H_{10}O$), these two carbon atoms must be connected to it:

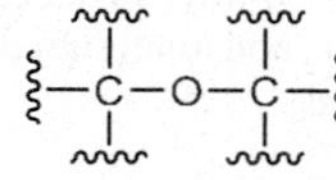

In the ^1H NMR spectrum, we see the characteristic pattern of an ethyl group (a triplet with an integration of 3 and a quartet with an integration of 2):

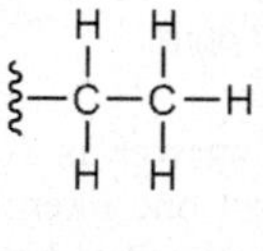

Notice that the quartet appears at 3.5 ppm, indicating that the ethyl group is one of the two groups that is connected to the oxygen atom:

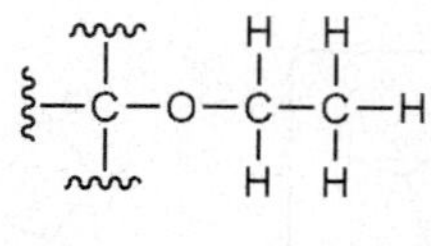

The signal at 4.0 ppm has an integration of 2, which represents a CH_2 group connected to the other side of the oxygen, like this:

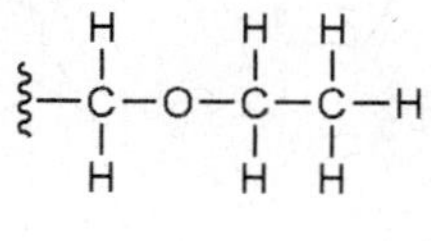

Remember that our structure must include a double bond, which completes the structure, and accounts for all of the atoms in the molecular formula ($C_5H_{10}O$):

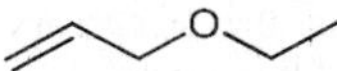

The signals above 5.0 correspond with the vinylic protons. There are three signals, and they are all splitting each other. Two of them are overlapping, to give a combined integration of 2.

15.75. The molecular formula ($C_8H_{14}O_3$) indicates that the compound has two degrees of unsaturation, so the structure must either have two rings, or two double bonds, or a ring and a double bond, or a triple bond.

In the IR spectrum, there are two signals that are suggestive of C=O groups, which would account for both degrees of unsaturation.

In the ^1H NMR spectrum, there are three signals. By comparing the heights of the S-curves, we can see that the *relative* integration values are 2:2:3. With a total of 14 protons (as seen in the molecular formula), these relative integration values must correspond to 4:4:6. This indicates a high level of symmetry in the compound.

The signal at approximately 1 ppm has an integration of 6, indicating two methyl groups that are identical because of symmetry. This signal is a triplet, which means that each of these CH_3 groups is neighboring a methylene (CH_2) group.

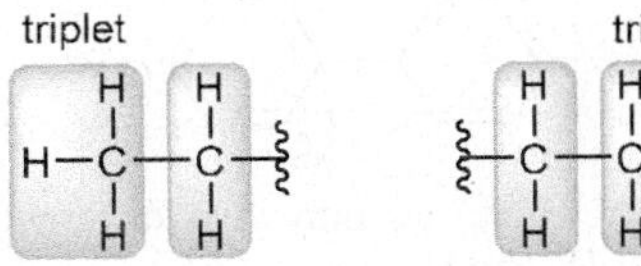

Based on the integration and multiplicities of the remaining two signals (at 1.7 ppm and 2.5 ppm), we can assemble the following two fragments, which must be identical by symmetry:

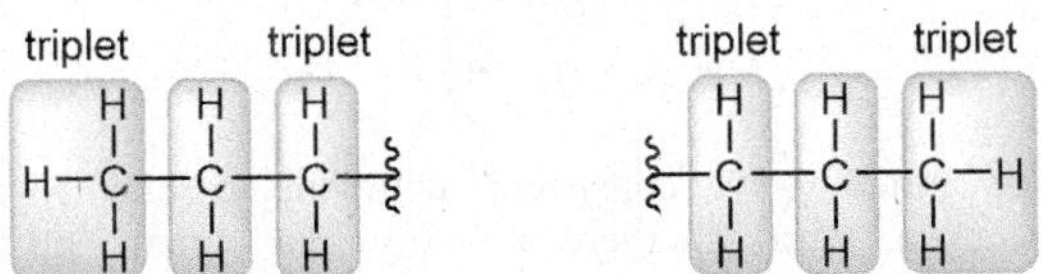

Each of the central methylene groups is being split by a neighboring methyl group and a neighboring methylene group. This could lead to a complex splitting pattern (either a triplet of quartets or a quartet of triplets). In this case, we are not seeing such a complex pattern. As mentioned in the textbook, this can happen when the *J*-values are fortuitously similar. In such a case, the system behaves as if it has five neighbors, and the $n + 1$ rule gives a sextet, which is what we see in this case.

Thus far, we have identified the presence of two, symmetrically positioned propyl groups, and there are two C=O bonds. Since the molecular formula indicates that there are three oxygen atoms in this compound, we can deduce the following structure:

The IR spectrum is consistent with this structure, in that the two signals (at 1755 and 1820 cm^{-1}) represent symmetrical and unsymmetrical stretching of two C=O bonds of the anhydride unit.

The ^{13}C NMR spectrum is also consistent with this structure. We see a carbon atom (of a C=O bond) at 180 ppm - notice that there is only one signal at 180 ppm, because the two carbonyl groups are equivalent by symmetry. As expected, three signals appear between 10 and 60 ppm, corresponding to the three carbon atoms of the propyl group (once again, the two propyl groups are identical to each other, giving rise to three signals rather than six). We can see from the DEPT-135 spectrum that two of these signals are methylene groups (because the signals are upside down), which is also consistent with the structure that we deduced above.

15.76. A compound with the molecular formula $C_6H_{10}O_4$ has two degrees of unsaturation, so any proposed structure must either have two rings, or two double bonds, or a ring and a double bond, or a triple bond.

The signal at 1747 cm^{-1} is likely a C=O bond of an ester, which accounts for one of the degrees of unsaturation.

In the proton NMR spectrum, there are four signals. The signal just above 5 ppm has an integration of 1, indicating a methine proton (CH), and it is a quartet, which indicates a neighboring methyl group. The signal for the methyl group should be a doublet (since it is next to the methine proton). That signal appears at 1.5ppm.

The quartet is significantly downfield (5 ppm), which indicates that it is neighbored by an oxygen atom, as well as some other group which can also shift the signal further downfield. The remaining two signals, just above 2 ppm and just below 4 ppm, are methyl groups (each has an integration of 3). Based on their chemical shifts, the former is likely to have a neighboring C=O group, while the latter is likely to be next to an oxygen atom:

The carbon NMR spectrum reveals that there are actually two carbonyl groups in this compound, and the molecular formula indicates four oxygen atoms, so it is likely that there are two ester groups that are overlapping in the IR spectrum (both at approximately 1747 cm^{-1}). This gives us three pieces that must be assembled:

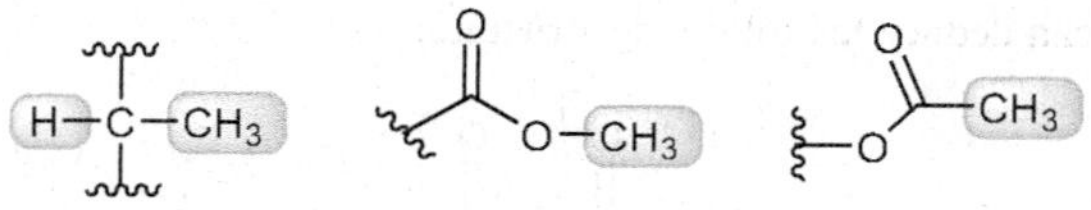

The following structure can be assembled from these three pieces:

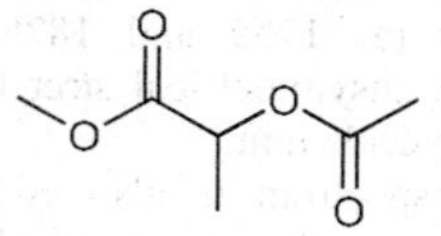

so far, we have only accounted for one degree of unsaturation (C=O) and only three oxygen atoms. The fourth oxygen atom, as well as the extra degree of unsaturation can be accounted for in the following structure which has the necessary symmetry, and is consistent with all of the spectra:

Notice that, because of symmetry, there are only four different kinds of carbon atoms in this compound, giving four signals in the carbon NMR spectrum. Two of these signals correspond to methylene (CH_2) groups, as confirmed by the presence of two upside-down signals in the DEPT-135 spectrum.

15.77. The molecular formula ($C_8H_{14}O_4$) indicates that the compound has two degrees of unsaturation, so the structure must either have two rings, or two double bonds, or a ring and a double bond, or a triple bond.
The signal at 1736 cm^{-1} is likely a C=O bond of an ester, which accounts for one of the degrees of unsaturation.
In the proton NMR spectrum, there are three signals. By comparing the heights of the S-curves, we can see that the *relative* integration values are 2:2:3. With a total of 14 protons (as seen in the molecular formula), these relative integration values must correspond to 4:4:6. This indicates a high level of symmetry in the compound.
The signal just above 4 ppm has an integration of 4, indicating two methylene (CH_2) groups that are identical because of symmetry. This signal is a quartet, which means that each of these two CH_2 groups is neighboring a methyl group. The signal for these methyl groups should be a triplet with an integration of 6. That signal appears just above 1 ppm.

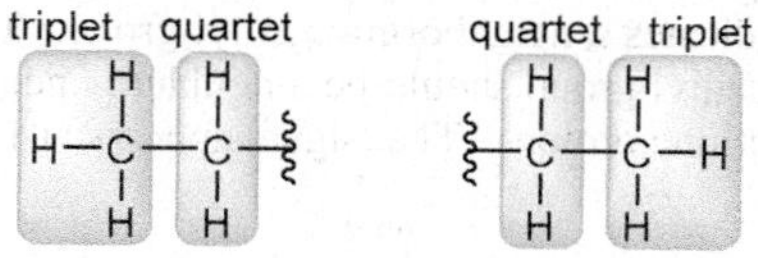

triplet quartet quartet triplet

The quartet is significantly downfield (4 ppm), which indicates that it is neighboring an oxygen atom.

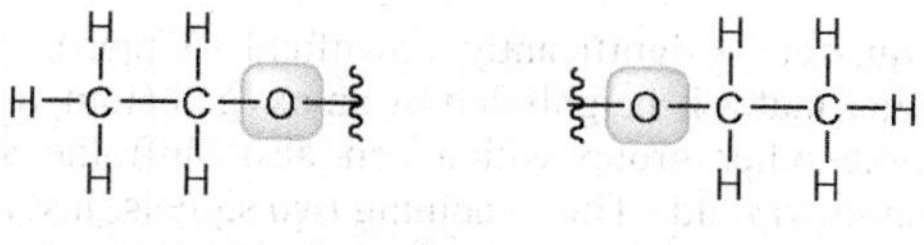

These two groups cannot be connected to each other, as that would not give us an opportunity to connect the remaining atoms in the compound.
The remaining signal, just below 3 ppm, is a singlet with an integration of 4, indicating two methylene groups that are equivalent by symmetry, with no neighboring protons:

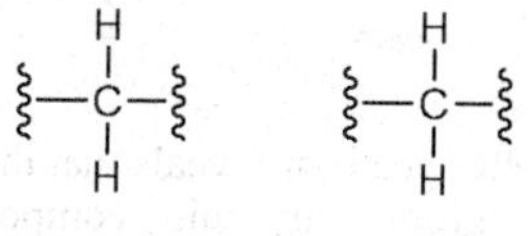

These two groups may or may not be connected to each other, but we know that they must be equivalent by symmetry, and they cannot have any neighbors.
The molecular formula indicates two degrees of unsaturation and the presence of four oxygen atoms, but so far, we have only accounted for one degree of unsaturation (C=O) and only three oxygen atoms. The fourth oxygen atom, as well as the extra degree of unsaturation can be accounted for in the following structure which has the necessary symmetry, and is consistent with all of the spectra:

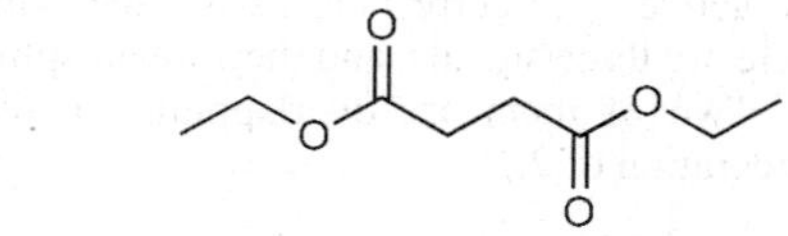

Notice that, because of symmetry, there are only four different kinds of carbon atoms in this compound, giving four signals in the carbon NMR spectrum. Two of these signals correspond to methylene (CH_2) groups, as confirmed by the presence of two upside-down signals in the DEPT-135 spectrum.

15.78. The molecular formula ($C_{12}H_8Br_2$) indicates that the compound has eight degrees of unsaturation, so the structure likely contains two aromatic rings (each of which represents four degrees of unsaturation).
In the proton NMR spectrum, there are only two signals, with the same relative integration. Since the molecule has eight protons (as seen in the molecular formula), we must conclude that each signal corresponds to four protons. Each of these signals is a doublet indicating only one neighboring proton. The following structure is consistent with this information:

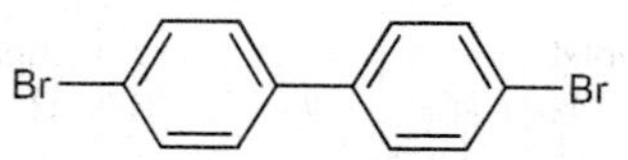

In this structure, there are only two different kinds of protons, labeled H_a and H_b:

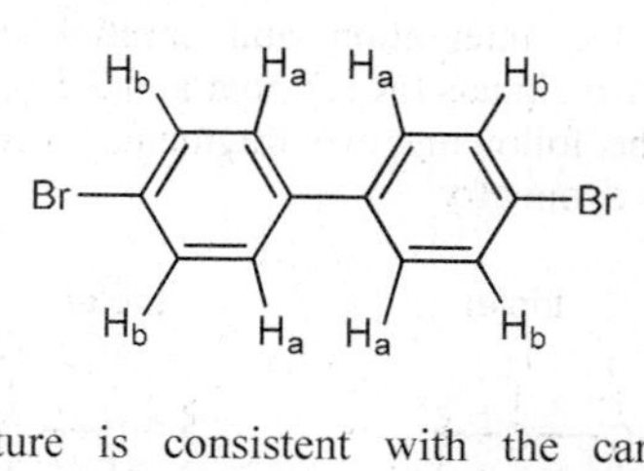

This structure is consistent with the carbon NMR spectrum, in which there are only four signals, all of which are aromatic:

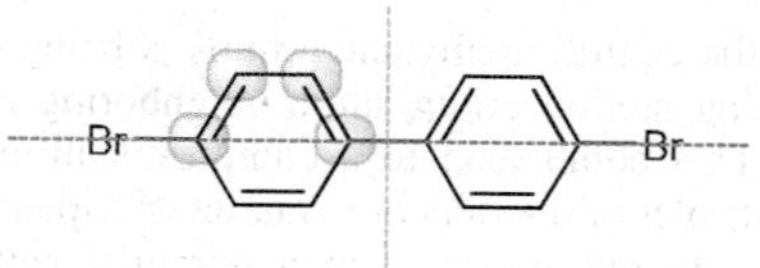

15.79.
(a) The height of the $(M+1)^{+\cdot}$ peak indicates that the compound has four carbon atoms (1.1% for each carbon atom). Since the molecular ion appears at m/z = 104, we know that nearly half of the molecular weight is due to carbon atoms (12 x 4 = 48). The rest of the molecular weight (104 – 48 = 56) must be attributed to oxygen

atoms and hydrogen atoms. There is a limit to how many hydrogen atoms there can be, since there are only four carbon atoms. Even if the compound is fully saturated, it could not have more than 10 protons ($2n+2$, where n is the number of carbon atoms). Oxygen has an atomic weight of 16. Therefore, the compound must contain at least three oxygen atoms (otherwise we could not account for the rest of the compound using hydrogen atoms alone). But the compound cannot contain *more* than three oxygen atoms, because four oxygen atoms have a combined atomic weight of 64, which already blows our budget, even before we place any hydrogen atoms (remember that our budget for O and H atoms is a total mass of 56). So, we conclude that the compound must have exactly three oxygen atoms. The remaining weight is accounted for with hydrogen atoms, giving the following molecular formula: $C_4H_8O_3$.

(b) The molecular formula ($C_4H_8O_3$) indicates that the compound has one degree of unsaturation, so the structure must either have a ring or a double bond.
In the IR spectrum, the broad signal between 3200 and 3600 cm^{-1} is characteristic of an O-H bond, and the signal at 1742 cm^{-1} is likely a C=O group of an ester.
In the proton NMR spectrum, there are four signals. By comparing the heights of the S-curves, we can see that the relative integration values are 2 : 2 : 1 : 3. With a total of eight protons (as seen in the molecular formula), these relative integration values correspond precisely to the number of protons giving rise to each peak. The signal just above 1 ppm corresponds to three protons, and is therefore a methyl group. Since this signal is a triplet, it must be next to a methylene (CH_2) group. The signal for that methylene group appears as a quartet, just as expected (since it is next to the methyl group) above 4 ppm. The location of this signal indicates that the methylene group is likely next to an oxygen atom (since it is shifted downfield). So far, we have the following fragment:

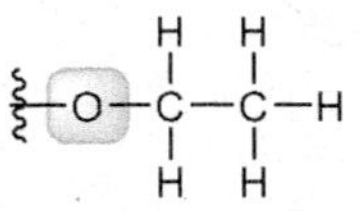

Now let's explore the remaining two signals in the proton NMR spectrum. The signal at approximately 3.6 ppm (with an integration of 1) vanishes in D_2O, indicating that it is the proton of the OH group (confirming our analysis of the IR spectrum). The singlet just above 4 ppm has an integration of 2, and therefore corresponds with an isolated methylene group (no neighbors). This methylene group is shifted significantly downfield, and our structure will have to take this into account.
In summary, we have the following fragments, which must be assembled.

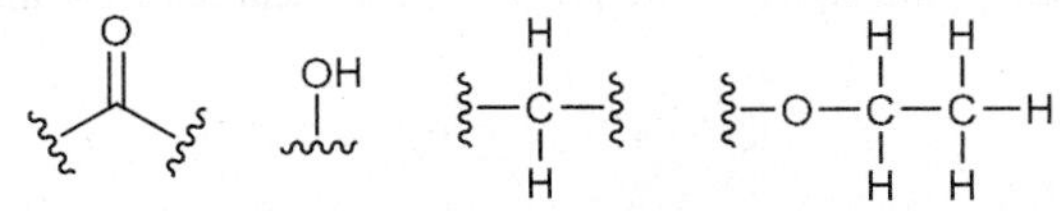

Since the C=O bond is likely part of an ester (based on the IR spectrum), we can redraw the following three fragments:

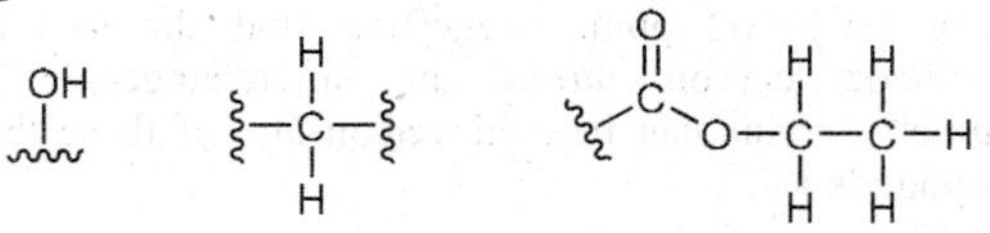

These fragments can only be assembled in one way:

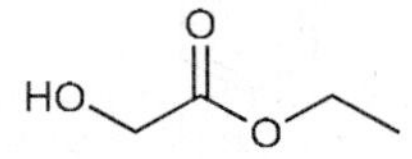

This structure is consistent with the carbon NMR spectrum, in which there is one signal for the C=O unit, two signals between 40 and 80 ppm (both of which must be methylene groups, based on the DEPT spectrum), and one signal between 10 and 30 ppm (representing the methyl group).

15.80. *N,N*-dimethylformamide (DMF) has three resonance structures:

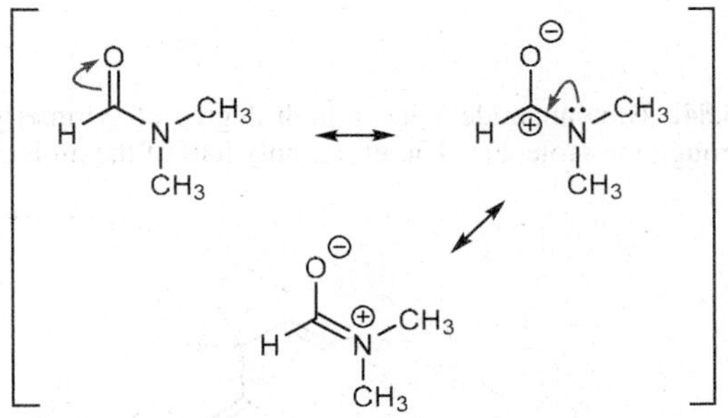

Consider the third resonance structure, in which the C-N bond is a double bond. This indicates that this bond is expected to have some double bond character. As such, there is an energy barrier associated with rotation about this bond, such that rotation of this bond occurs at a rate that is slower than the timescale of the NMR spectrometer. Therefore, the two methyl groups will appear as distinct signals in a ^{1}H NMR spectrum and in a ^{13}C NMR spectrum. At high temperature, more molecules will have the requisite energy to undergo free rotation about the C-N bond, so the process can occur on a timescale that is faster than the timescale of the NMR spectrometer. For this reason, the signals are expected to collapse into one signal at high temperature.

15.81. The first compound lacks a chiral center. The two methyl groups are enantiotopic and are therefore chemically equivalent. The second compound has a chiral center (the position bearing the OH group). As such, the two neighboring methyl groups are diastereotopic and are therefore not chemically equivalent. For this reason, the ^{13}C NMR spectrum of the second compound exhibits six signals, rather than five.

15.82. The molecular formula (C_8H_{10}) indicates four degrees of unsaturation (see Section 14.16), which is highly suggestive of an aromatic ring. The aromatic ring accounts for six of the eight carbon atoms, so we must account for the other two carbon atoms, which must be

sp^3 hybridized. However, only one signal appears in the region 10 – 60 ppm, suggesting that the two sp^3 hybridized carbon atoms are interchangeable by symmetry. This can be achieved in any of these three compounds:

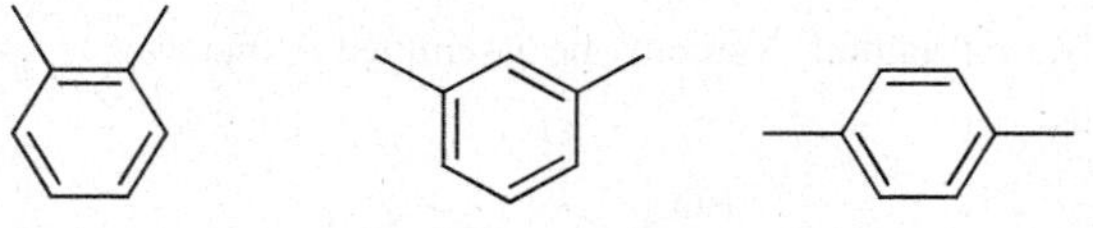

However, only one of these compounds, shown below, will exhibit four signals in the region 100 – 150 ppm.

15.83. The methyl group on the right side is located in the shielding region of the π bond, so the signal for this proton is moved upfield to 0.8 ppm.

15.84. Brevianamide S has a high degree of symmetry. In fact, there is a rotational axis of symmetry that runs right through the molecule. Therefore, only half of the molecule will need to be analyzed.

The compound lacks a chiral center and is achiral. As a result, the two protons for any methylene (CH_2) group will be equivalent to each other. There are two unique methylene groups (highlighted below), giving rise to two unique signals:

In addition, the methyl groups will all be equivalent, giving rise to one signal. The predicted chemical shift for all signals is shown below:

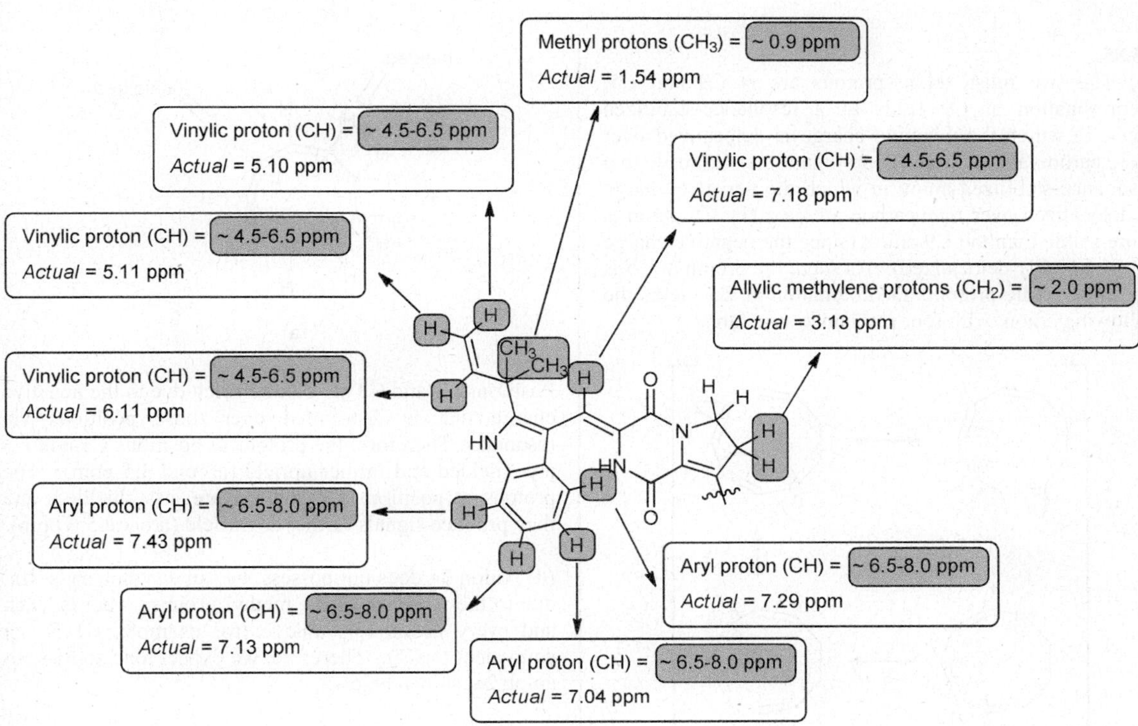

The multiplicity of each proton is determined with the $n + 1$ rule, as shown below. Note: complex splitting is observed for all three protons of the vinyl group.

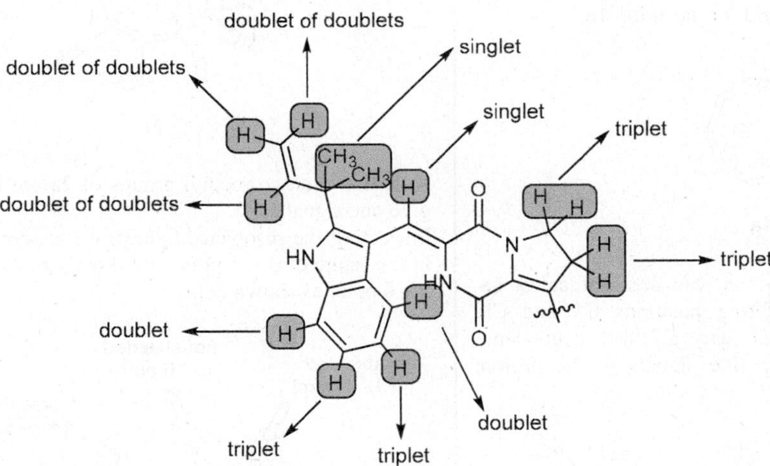

15.85.

(a) The two most acidic protons are at C9 and C5. Deprotonation at C9 leads to a resonance-stabilized anion in which the negative charge is delocalized over three carbon atoms, while deprotonation at C5 leads to a resonance-stabilized anion in which the negative charge is delocalized over *four* carbon atoms. The C5 anion is more stable than the C9 anion (since the negative charge is more highly delocalized). As such, the proton at C5 is the most acidic proton. Deprotonation at C5 yields the following anion, with four resonance structures:

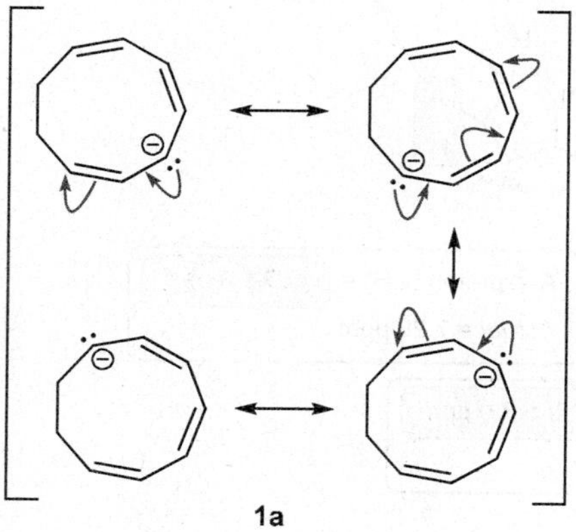

1a

(b) To look for symmetry and equivalent protons, let's consider the resonance hybrid for the anion **1a**:

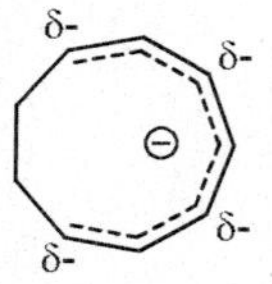

1a

When viewed in this way, we can see that anion **1a** possesses symmetry, rendering positions C8 and C9 equivalent. Similarly, C1 and C7 are equivalent. Therefore, we expect only five signals in the proton NMR spectrum of **1a**.

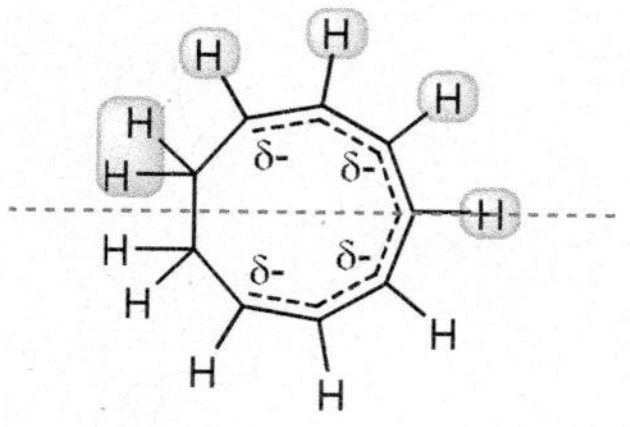

1a

(c) Once again, we can use the resonance hybrid to explain this effect:

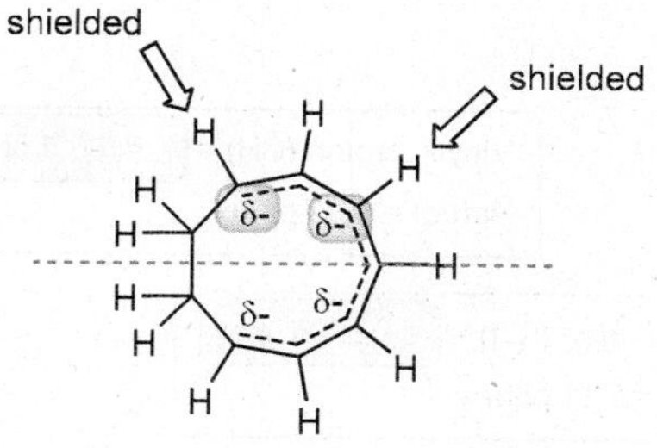

shielded

shielded

1a

Positions C1 and C3 are electron-rich due to the negative charge that is distributed over those positions via resonance Therefore, the protons at positions C1 and C3 are shielded and farther upfield (around 3.5 ppm). The protons at positions C2 and C4 are less shielded, and they produce signals farther downfield (around 5.6 ppm).

(d) Anion **2a** does not possess the same symmetry as **1a**, due to the presence of the methyl groups. That is, each and every position is unique (for example, C1 is not equivalent to C7). Therefore, we expect nine signals for anion **2a**, shown below:

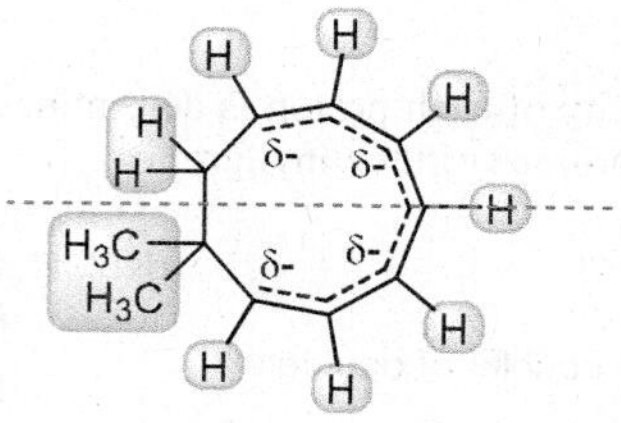

2a

Note that the two methyl groups of **2a** are identical and give one signal.

Inspecting the resonance hybrid, we expect four signals in the range of 3 – 4 ppm, and three signals in the range 5 – 6 ppm, as shown below:

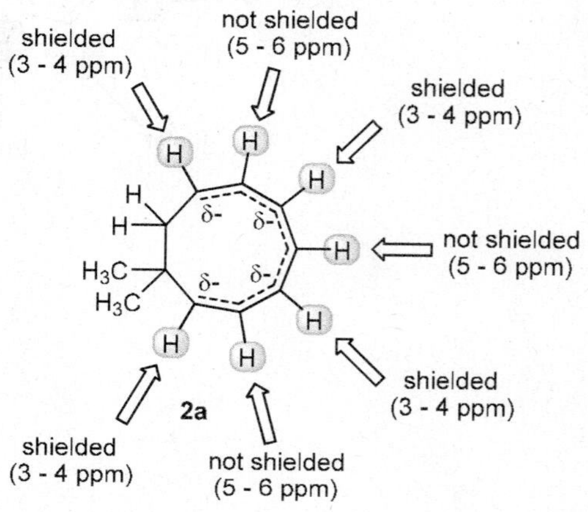

shielded (3 - 4 ppm)

not shielded (5 - 6 ppm)

shielded (3 - 4 ppm)

not shielded (5 - 6 ppm)

shielded (3 - 4 ppm)

shielded (3 - 4 ppm)

not shielded (5 - 6 ppm)

2a

The following are the actual data from the proton NMR spectrum of **2a**, which supports our prediction.

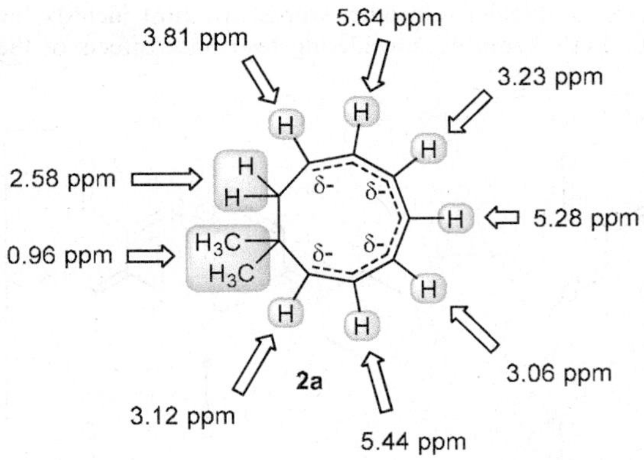

15.86.

(a) The nitro group is a powerful electron-withdrawing group (primarily due to resonance), which renders three positions on the ring electron-poor (highlighted below). This effect should cause H_a to be deshielded:

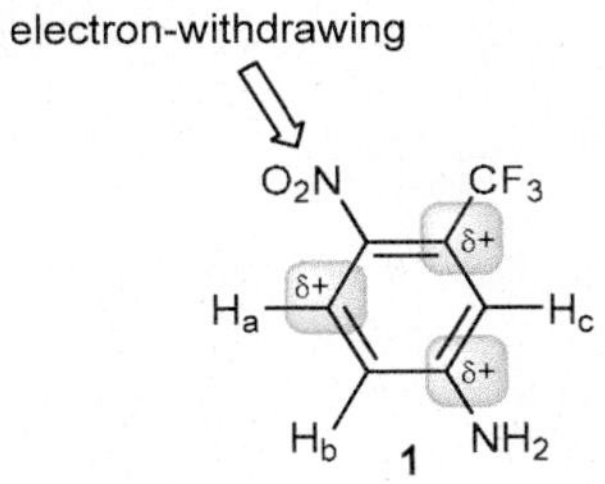

The amino group is a powerful electron-donating group (via resonance), which renders three positions on the ring electron-rich (highlighted below). This effect should cause H_b and H_c to be shielded:

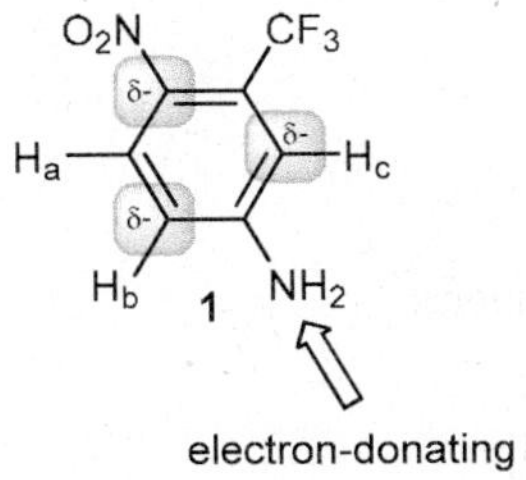

Based on these effects alone, we expect H_a to be downfield (deshielded) relative to H_b and H_c, and we expect that H_b and H_c will be upfield (shielded) relative to H_a. That is, based solely on the resonance effects of the nitro group and the amino group, we expect H_a to give the signal farthest downfield.

There is, however, one other group on the aromatic ring. The trifluoromethyl group is a powerful electron-withdrawing group, via induction (rather than resonance). As such, we expect its effect to diminish with distance, so H_a should be affected less than H_c. While this effect deshields H_c more than H_a, we would not expect this inductive effect to overwhelm the two resonance effects that suggest that H_a gives the most downfield signal. After all, there are two different resonance effects suggesting H_a is the most downfield signal, AND resonance is generally a stronger effect than induction.

(b) This transformation involves conversion of the amino group into an amide group. This has an impact on the lone pair on the nitrogen atom of the amine/amide; in **1** it is completely available for resonance into the ring, but in **2**, the adjacent carbonyl group involves this lone pair in resonance, pulling it away from the ring. As such, there are three positions on the ring that are the most affected, highlighted below:

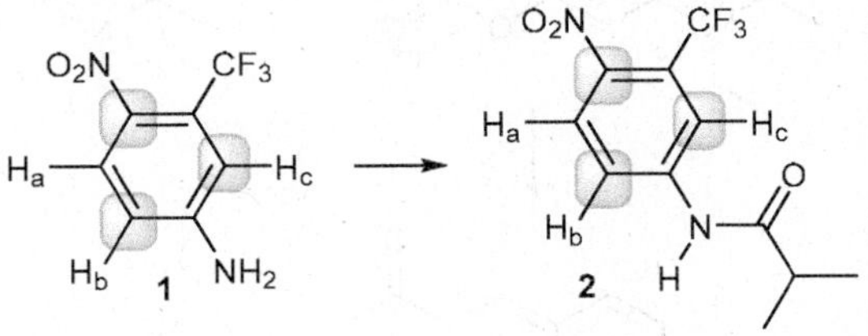

In both compound **1** and compound **2**, these three positions are shielded. But the shielding effect is greater in compound **1** than in compound **2**. In other words, these three positions become less shielded. As a result, we expect the signals for H_b and H_c to be more affected by this transformation than the signal for H_a. Specifically, H_b and H_c become less shielded as a result of this transformation, so we expect those two signals to move farther downfield.

The actual chemical shifts (shown below) support our predictions.

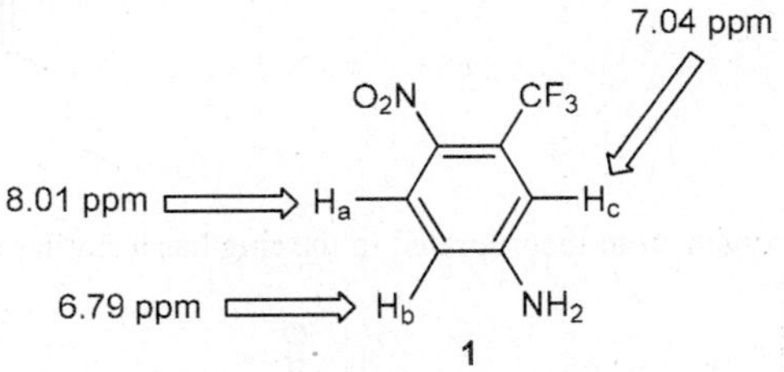

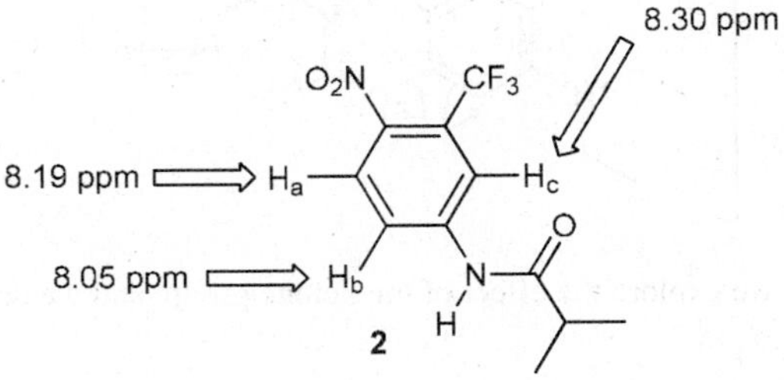

15.87.

(a) We are looking for the proton that is expected to be the most deshielded. In other words, we must identify the proton that resides in the most electron-deficient environment. Let's begin by considering resonance effects of the diethyl amino group:

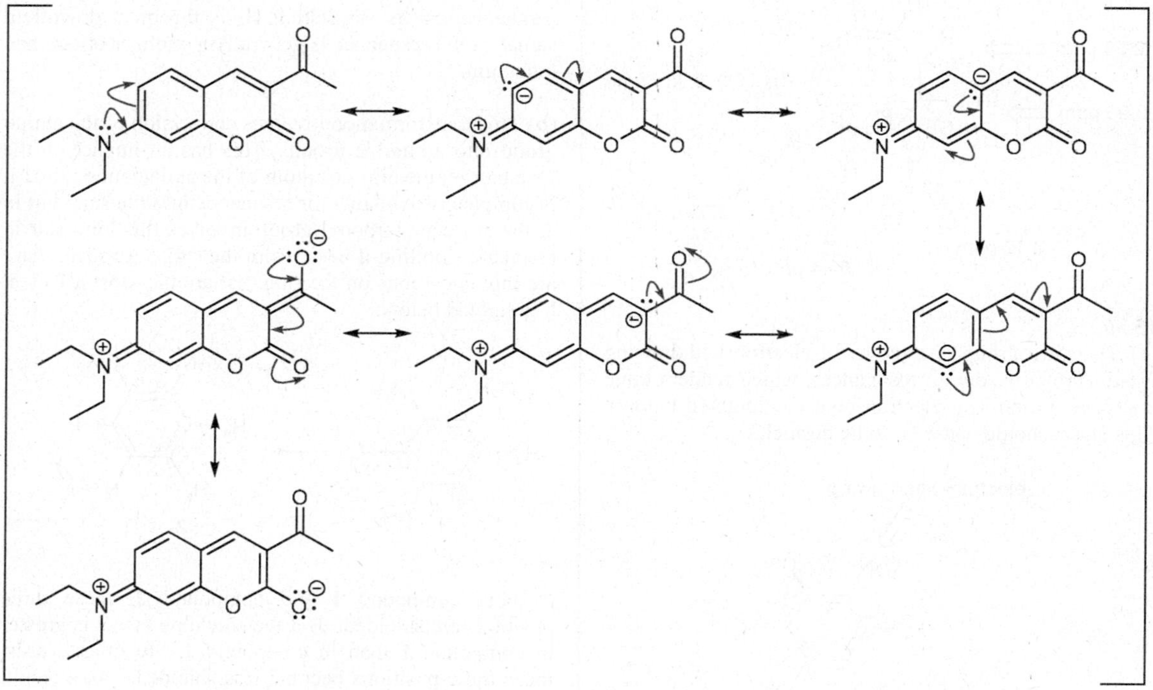

These resonance structures indicate that the diethylamino group donates electron density, so it has the opposite effect of what we are looking for; that is, the diethylamino group renders the following positions electron-rich:

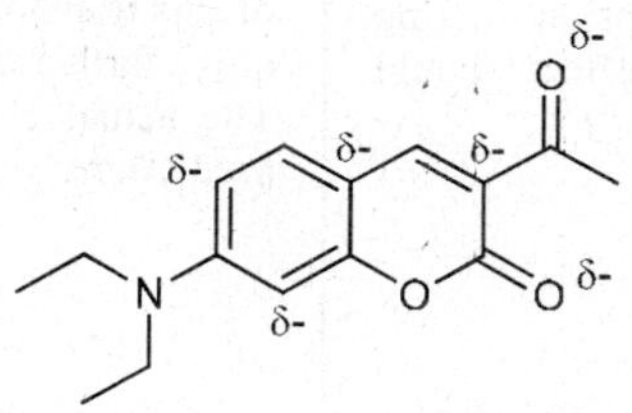

The oxygen atom incorporated in the ring has a similar effect on the same positions:

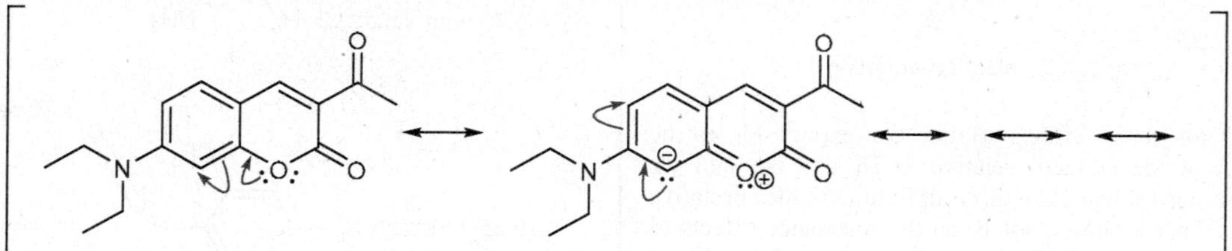

So we explore the effect of the ketone group, and we draw the following resonance structures:

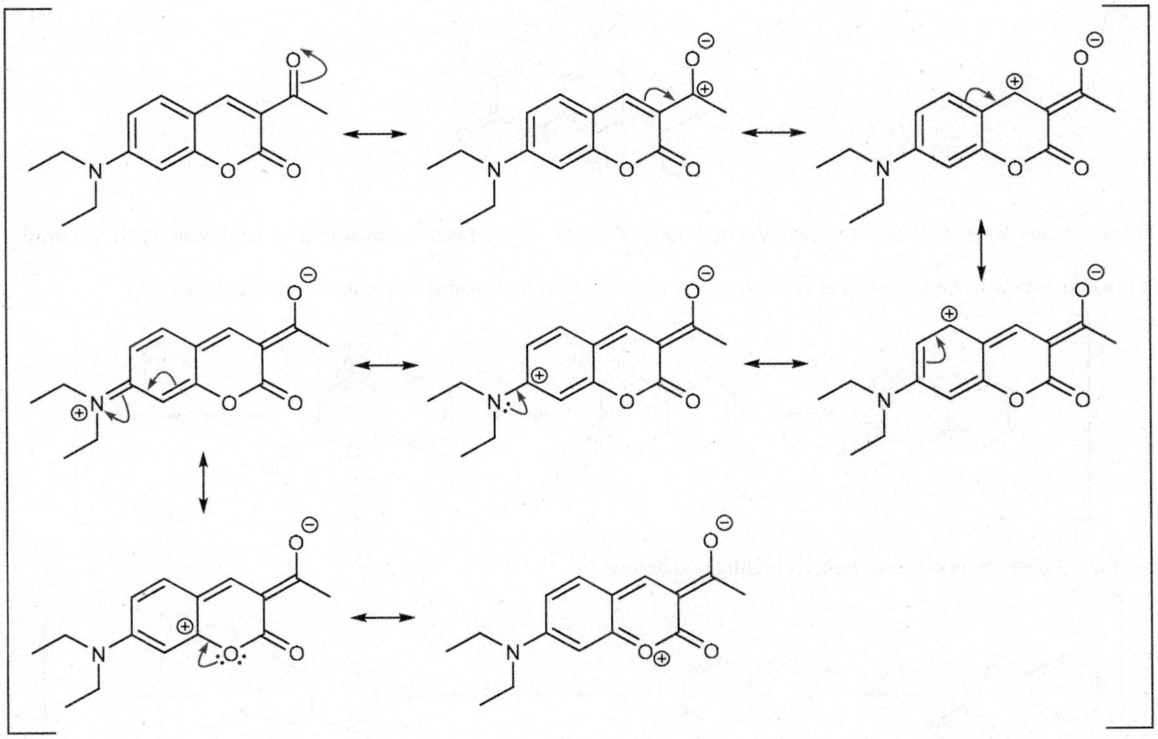

Notice that these resonance structures indicate that several positions are electron-deficient.

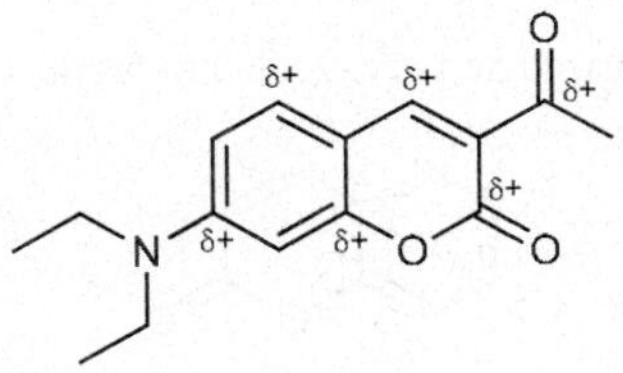

A similar effect is expected from the ester group:

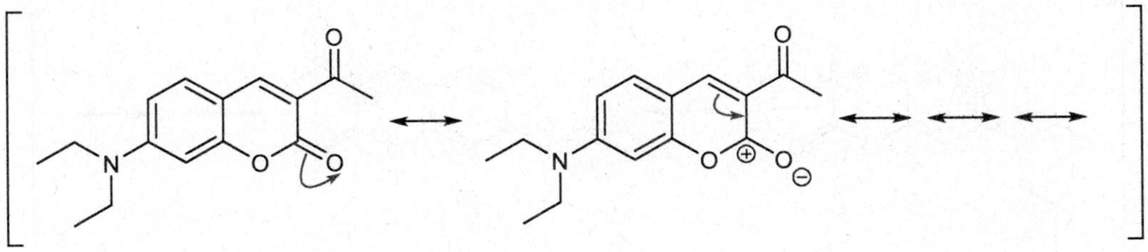

In summary, if we take into account *all* resonance effects (described above), the following picture emerges:

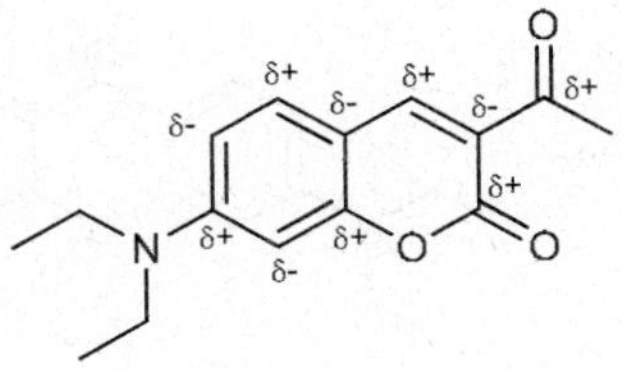

Notice that only two protons are directly attached to electron-deficient centers:

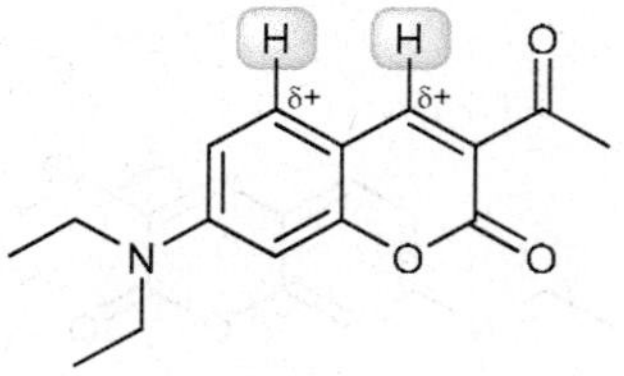

These are the two protons that are expected to produce the two signals farthest downfield in the proton NMR spectrum.

(b) In compound **1**, the C4 position is electron-deficient, as seen in the third resonance structure below:

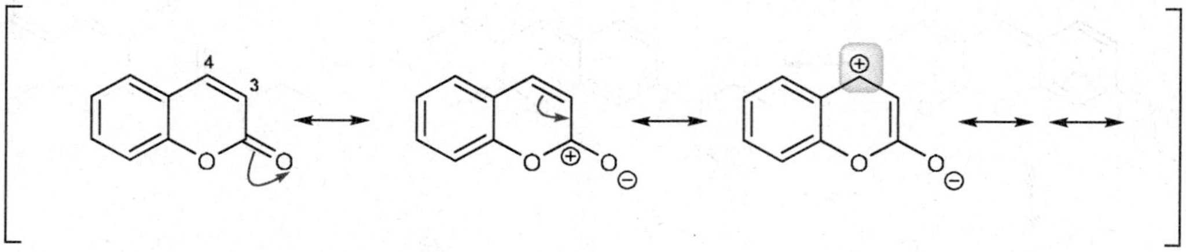

and the C3 position is electron-rich, as highlighted below:

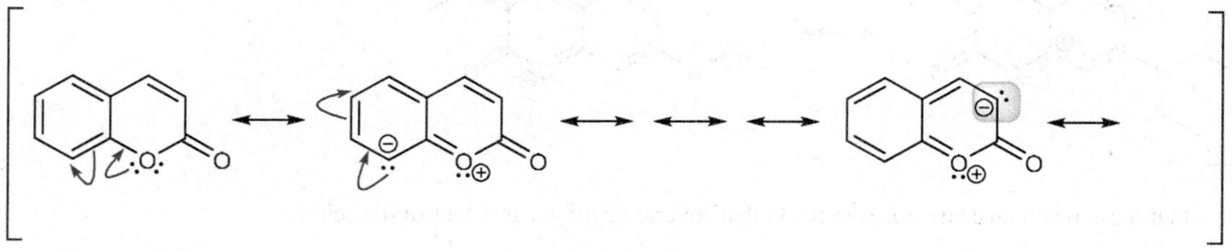

As a result, the C3–C4 π bond is highly polarized and is expected to produce a strong signal.

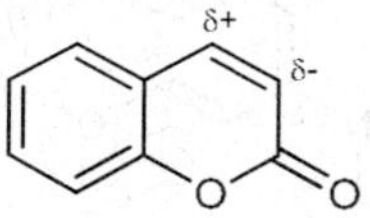

A similar effect is expected in compound **2**, but the effect should be much stronger, because the presence of the diethylamino group renders C3 even more electron-rich:

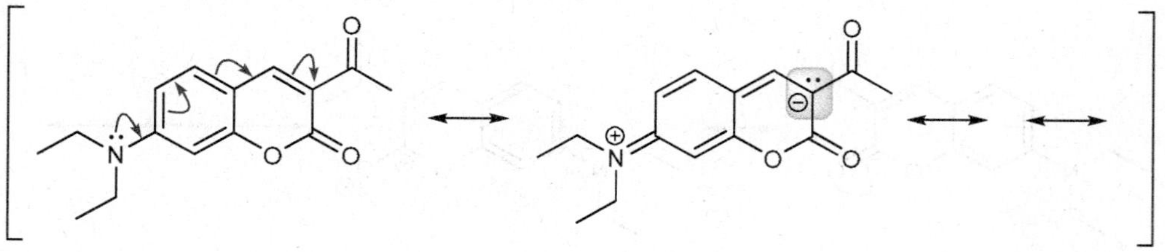

And the presence of the ketone renders C4 even more electron-deficient, as highlighted below:

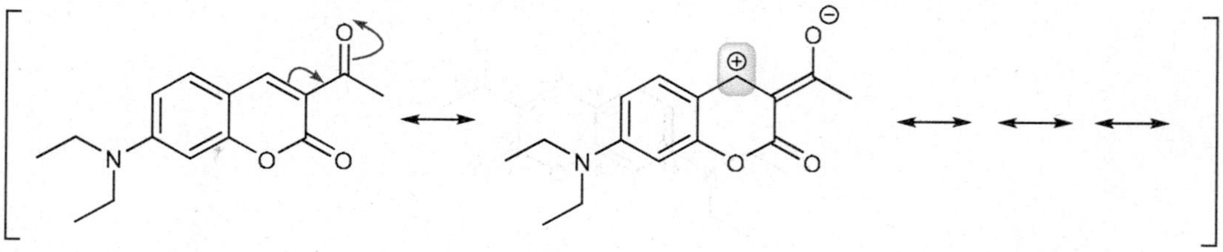

Therefore, the C3–C4 bond in compound **2** is expected to be even more polarized than the C3–C4 bond in compound **1**. So, compound **2** is expected to produce a stronger (more intense) signal.

Chapter 16
Conjugated Pi Systems and Pericyclic Reactions

Review of Concepts

Fill in the blanks below. To verify that your answers are correct, look in your textbook at the end of Chapter 16. Each of the sentences below appears verbatim in the section entitled *Review of Concepts and Vocabulary.*

- Conjugated dienes experience free-rotation about the C2-C3 bond, giving rise to two important conformations: *s*-________ and *s*-________. The ________ conformation is lower in energy.
- The ________ and ________ are referred to as **frontier orbitals**.
- An ________ **state** is produced when a π electron in the HOMO absorbs a photon of light bearing the appropriate energy necessary to promote the electron to a higher energy orbital.
- Reactions induced by light are called ______________ **reactions.**
- When butadiene is treated with one equivalent of HBr, two major products are observed, resulting from ______**-addition** and ______**-addition.**
- Conjugated dienes that undergo addition at low temperature are said to be under ___________ **control.**
- Conjugated dienes that undergo addition at elevated temperature are said to be under ________________ **control.**
- ______________ **reactions** proceed via a concerted process with a cyclic transition state, and they are classified as **cycloaddition reactions,** _____________ **reactions**, and **sigmatropic rearrangements.**
- The Diels–Alder reaction is a [_______] **cycloaddition** in which two C–C bonds are formed simultaneously.
- High temperatures can often be used to achieve the reverse of a Diels–Alder reaction, called a _________ **Diels–Alder**.
- The starting materials for a Diels–Alder reaction are a diene, and a ___________.
- The Diels–Alder reaction only occurs when the diene adopts an ____ conformation.
- When cyclopentadiene is used as the starting diene, a bridged bicyclic compound is obtained, and the _____ cycloadduct is favored over the _____ cycloadduct.
- Conservation of orbital symmetry determines whether an electrocyclic reaction occurs in a ___________ fashion or a ____________ fashion.
- A [_________] sigmatropic rearrangement is called a **Cope rearrangement** when all six atoms of the cyclic transition state are carbon atoms.
- Compounds that possess a conjugated π system will absorb UV or visible light to promote an electronic excitation called a __________ transition.
- The most important feature of the absorption spectrum is the _________, which indicates the wavelength of maximum absorption.
- When a compound exhibits a λ_{max} between 400 and 700 nm, the compound will absorb __________ light, rather than UV light.

Review of Skills

Fill in the blanks and empty boxes below. To verify that your answers are correct, look in your textbook at the end of Chapter 16. The answers appear in the section entitled *SkillBuilder Review*.

16.1 Proposing a Mechanism and Predicting the Products of Electrophilic Addition to Conjugated Dienes

In the space provided, draw a mechanism for the reaction that is expected to occur when the following compound is treated with HBr. Make sure to draw all possible products.

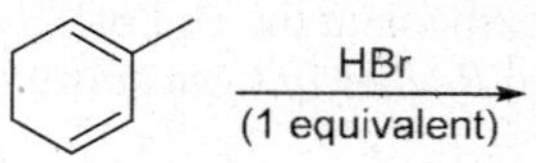

HBr
(1 equivalent)

16.2 Predicting the Major Product of an Electrophilic Addition to Conjugated Dienes

Draw the major product of the following reaction:

HBr (1 equivalent)
0°C

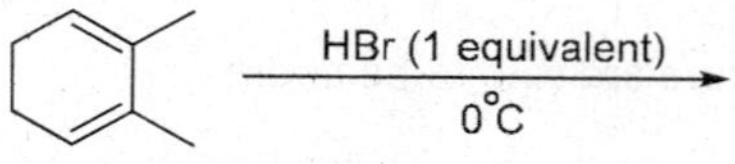

16.3 Predicting the Product of a Diels–Alder Reaction

Draw the major product of the following reaction:

NC CN

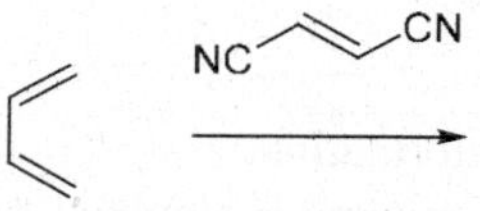

16.4 Predicting the Product of an Electrocyclic Reaction

Draw the major product of the following reaction:

Ph
hν
Ph

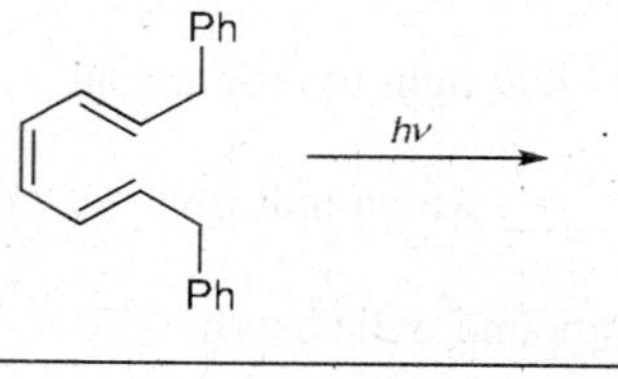

16.5 Using Woodward–Fieser Rules to Estimate λ_{max}

Use woodward-Fieser rules to estimate λ_{max} for the following compound:

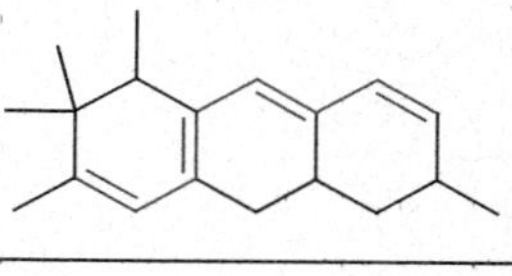

Base value =
Additional double bonds =
Auxochromic alkyl bonds =
Exocyclic double bond =
Homoannular diene =

Total =

Review of Reactions

Predict the Products for each of the following transformations. To verify that your answers are correct, look in your textbook at the end of Chapter 16. The answers appear in the section entitled *Review of Reactions*.

Preparation of Dienes

Br

t-BuOK

Br

t-BuOK

Br

Electrophilic Addition

HBr (1 equiv.)

Br_2 (1 equiv.)

Diels–Alder Reaction

Diels-Alder

X

+

Retro Diels-Alder

(very high temperature)

X

+

X

X

+

X

O OR

+

O OR

CHO

+

CHO

Electrocyclic Reactions

Sigmatropic Rearrangements

Cope Rearrangement

Claisen Rearrangement

Mistakes to Avoid

Many students have trouble drawing the product(s) of a Diels-Alder reaction when bicyclic structures are involved:

You might have trouble visualizing these structures, drawing them, or knowing where to place the substituents. The following are a few guidelines that might help you avoid making mistakes.
In a Diels-Alder reaction, an acyclic diene will give a product with a cyclohexene ring:

However, if the diene is cyclic (both π bonds are contained in the ring), then the product is a bicyclic structure:

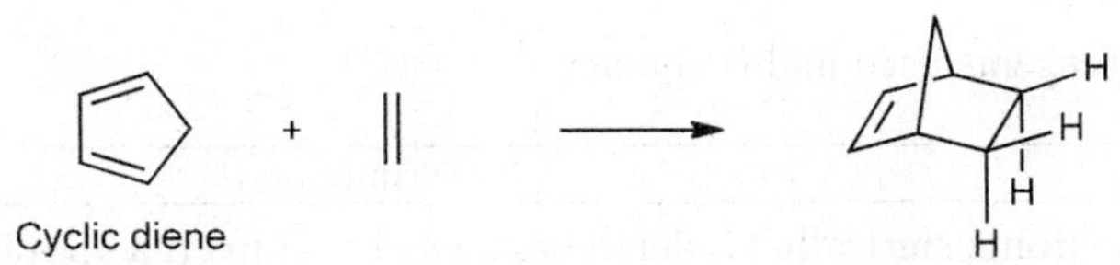

Formation of the bicyclic structure can be seen more clearly if we redraw the cyclic diene with distorted bond angles, like this:

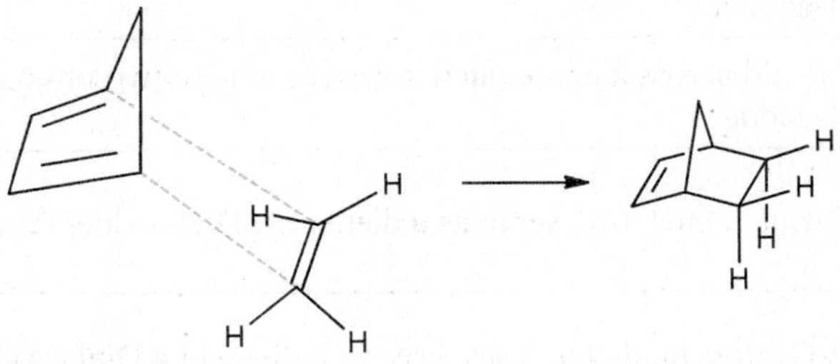

The dotted lines indicate the locations where σ bonds are forming as a result of the reaction. When the dienophile is monosubstituted or *cis*-disubstituted, the *endo* rule determines the product(s) obtained:

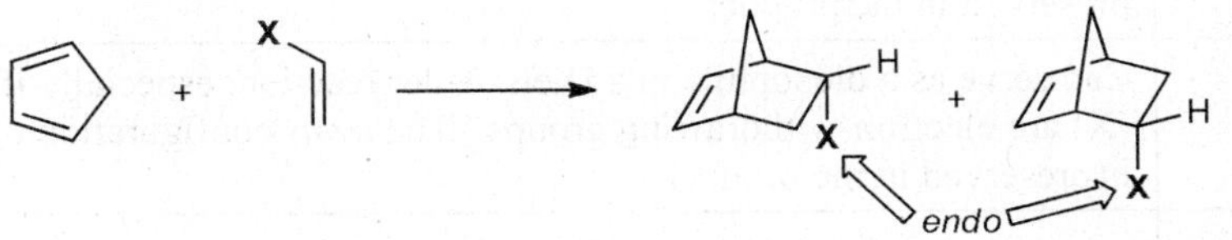

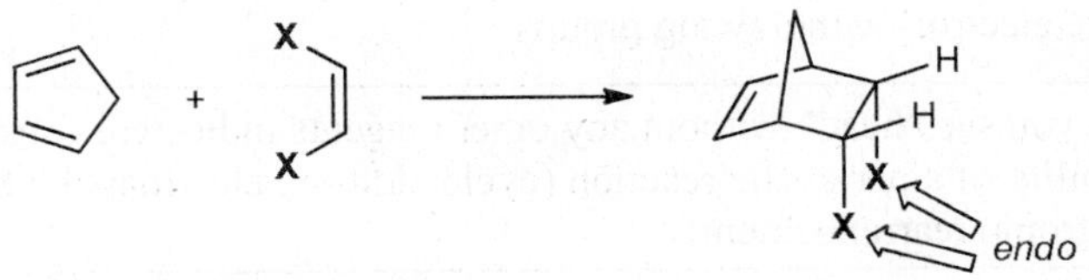

However, when a *trans*-disubstituted dienophile is used, the *endo* rule is not relevant. Two stereoisomeric products are obtained (in this case, enantiomers), and in each product, one group occupies an *endo* position while the other group occupies an *exo* position (because the *trans* configuration of the dienophile is preserved during a Diels-Alder reaction):

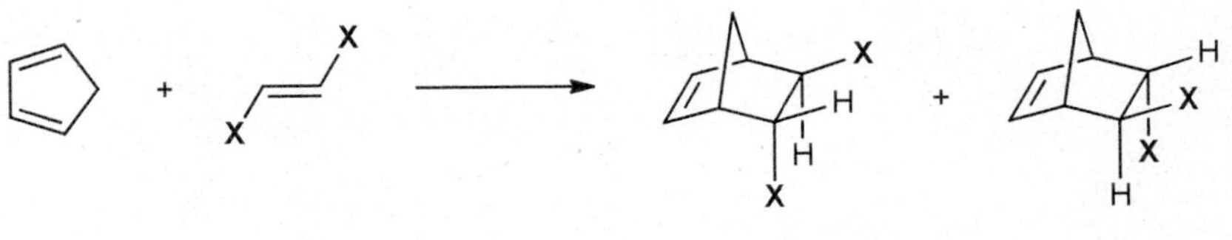

Useful reagents

The following is a list of reagents used in this chapter:

Reagents	Function
t-BuOK	A strong, sterically hindered base, used to convert a 1,2-dibromide or an allylic bromide into a conjugated diene.
HBr	Will add across a conjugated π system to give two products: a 1,2-adduct and a 1,4-adduct.
Br_2	Will add across a conjugated π system to give two products: a 1,2-adduct and a 1,4-adduct.
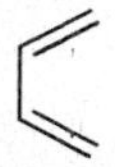	1,3-Butadiene. Can serve as a diene in a Diels-Alder reaction.
	1,3-Cyclopentadiene. Can serve as a diene in a Diels-Alder reaction.
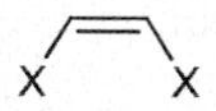 X X	Can serve as a dienophile in a Diels-Alder reaction, especially if the substituents (X) are electron-withdrawing groups. The *cis* configuration of the dienophile is preserved in the product.
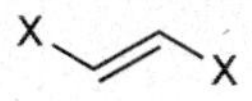 X X	Can serve as a dienophile in a Diels-Alder reaction, especially if the substituents (X) are electron-withdrawing groups. The *trans* configuration of the dienophile is preserved in the product.
 X X	Can serve as a dienophile in a Diels-Alder reaction, especially if the substituents (X) are electron-withdrawing groups.
heat	When you see "heat" without any other reagents indicated, consider the possibility of a pericyclic reaction (cycloaddition, electrocyclic reaction, or a sigmatropic rearrangement).
hν	When you see this term (pronounced *H-new*), or "light", without any other reagents indicated, consider the possibility of an electrocyclic reaction.

Solutions

16.1.

(a) The C=C bond in this compound is conjugated to two of the carboxylic acid groups:

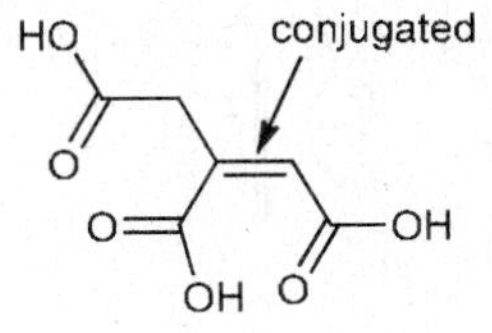

conjugated

(b) The C=C bond on the left is isolated, while the other two C=C bonds are conjugated to each other.

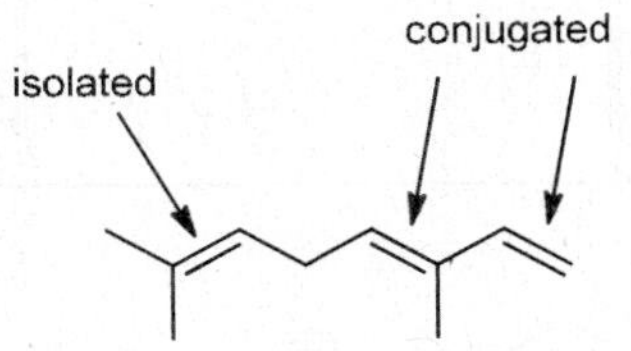

isolated conjugated

(c) One of C=C bonds is conjugated to the carbonyl (C=O) group, and the other C=C bond is isolated:

isolated conjugated

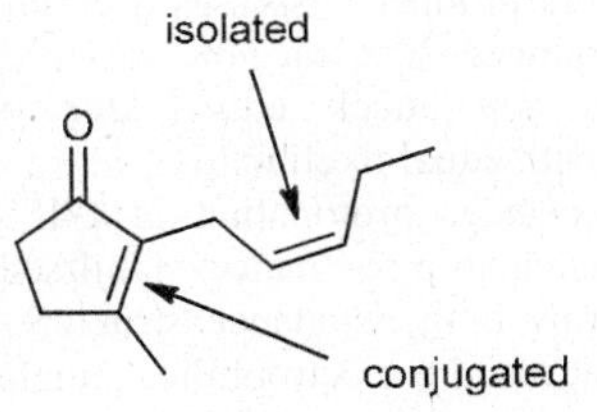

(d) One of C=C bonds is conjugated to the carbonyl group, and the other C=C bond is isolated:

isolated conjugated

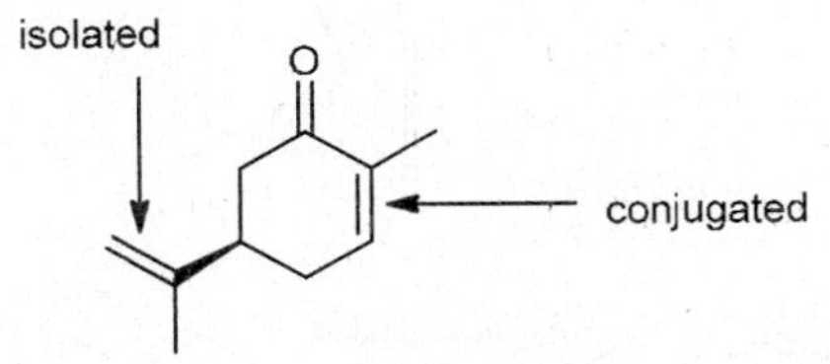

16.2. Radical bromination can be employed to install a functional group. The resulting alkyl halide can then be treated with a strong base to give an alkene. Bromination of the alkene will give a dibromide, which can then be converted into the product upon treatment with a strong, sterically hindered base, such as *tert*-butoxide. This final step involves two elimination reactions, giving the desired diene:

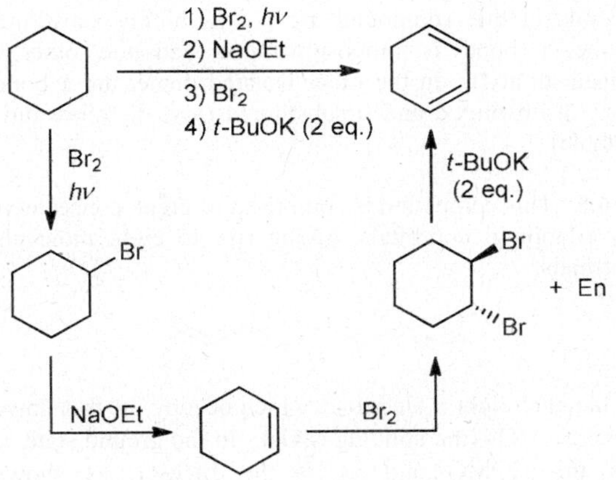

1) Br_2, $h\nu$
2) NaOEt
3) Br_2
4) *t*-BuOK (2 eq.)

Br_2, $h\nu$ → NaOEt → Br_2 → + En → *t*-BuOK (2 eq.)

16.3. The C1–C2 bond length is expected to be the shortest because it is a double bond (comprised of both a σ bond and a π bond):

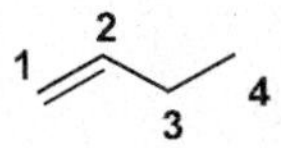

Among the other two bonds, C2–C3 is expected to be shorter than C3–C4 because the former is a $C_{sp^2}–C_{sp^3}$ bond, while the latter is a $C_{sp^3}–C_{sp^3}$ bond (sp^2 hybridized orbitals form shorter bonds because sp^2 hybridized orbitals are closer to the nucleus than sp^3 hybridized orbitals due to a greater amount of *s*-character).

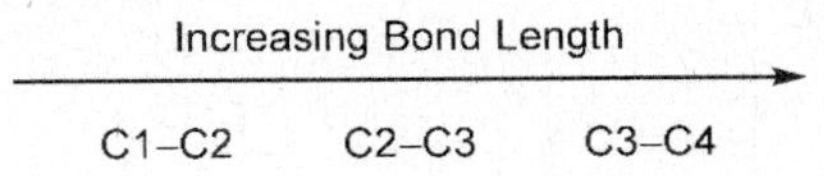

Increasing Bond Length

C1–C2 C2–C3 C3–C4

16.4.

(a) All three of these compounds will yield the same product (ethylcyclohexane) upon hydrogenation with two moles of hydrogen gas. Yet only one of these compounds is a conjugated diene, shown below.

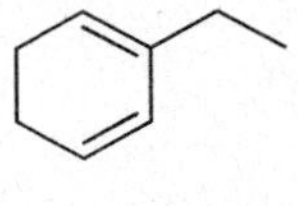

The other two compounds exhibit isolated π bonds. The conjugated diene will liberate the least heat because it is the most stable of the three compounds (lowest in energy).

(b) The following compound is expected to liberate the most heat upon hydrogenation with two moles of hydrogen gas:

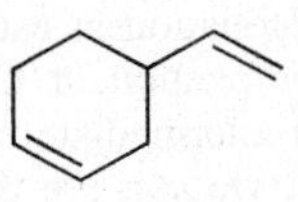

This isolated diene will liberate more heat than the other isolated diene, because it is the least stable diene. The π bonds in this compound are not as highly substituted (one π bond is monosubstituted and the other is disubstituted). In the other isolated diene, the π bonds are disubstituted and trisubstituted (and therefore more stable).

16.5. This compound is comprised of eight, consecutive, overlapping, *p* orbitals, giving rise to eight molecular orbitals:

There are eight π electrons, which occupy the four lower energy MOs (the bonding MOs). In the ground state, ψ_4 is the HOMO and ψ_5 is the LUMO, as shown. Photochemical excitation causes one electron to be promoted from ψ_4 to ψ_5. In the excited state, ψ_5 is the HOMO and ψ_6 is the LUMO:

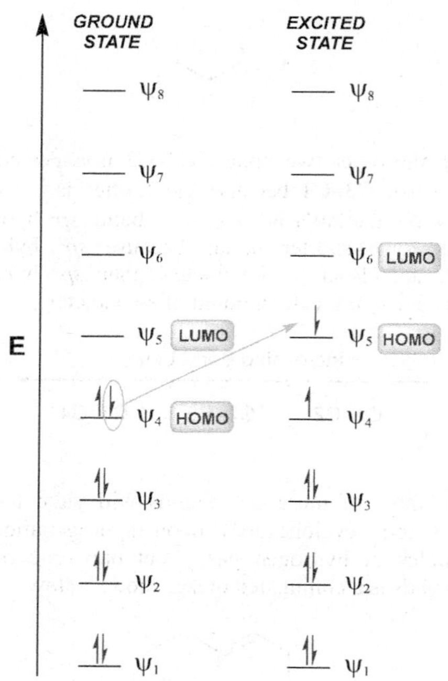

16.6.
(a) We first identify the locations where protonation can occur. There are four unique positions where protonation can occur, labeled C1 through C4:

Among these four positions, only protonation at C1 or at C4 will generate a resonance-stabilized carbocation. So, we must explore protonation at each of these positions. Let's begin with protonation at C1, which leads to a resonance–stabilized intermediate. When we draw both resonance structures, we can see that two positions are electrophilic. Nucleophilic attack can occur at either of these locations, although the product is the same in either case, leading to only one product:

(racemic) (racemic)

Notice that the product possesses a chiral center and is therefore produced as a racemic mixture (because the chloride ion can attack either face of the allylic carbocation with equal likelihood).

Now let's consider protonation at C4. Once again, protonation leads to a resonance–stabilized intermediate. When we draw both resonance structures, we can see that two positions are electrophilic. Nucleophilic attack can occur at either of these locations, giving two possible products, as shown:

(racemic)

In summary, we expect the following possible products:

(racemic) (racemic)

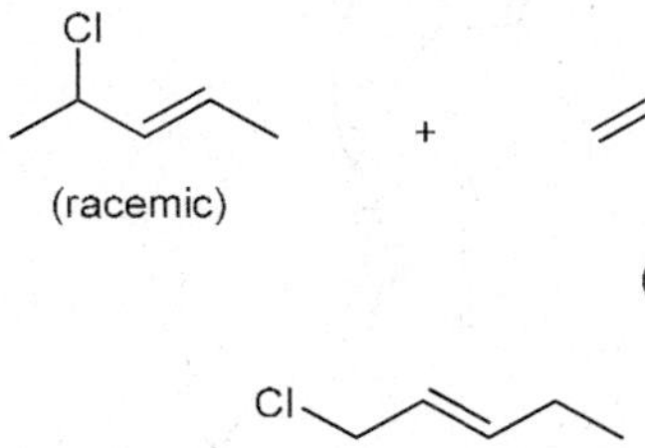

(b) We first identify the locations where protonation can occur. There are four unique positions where protonation can occur, labeled C2, C3, C4, and C5:

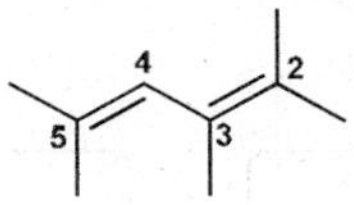

Among these four positions, only protonation at C2 or at C5 will generate a resonance-stabilized carbocation. So, we must explore protonation at each of these positions. Let's begin with protonation at C5, which leads to a resonance–stabilized intermediate. When we draw both resonance structures, we can see that two positions are electrophilic. Nucleophilic attack can occur at either of these locations, giving rise to two products:

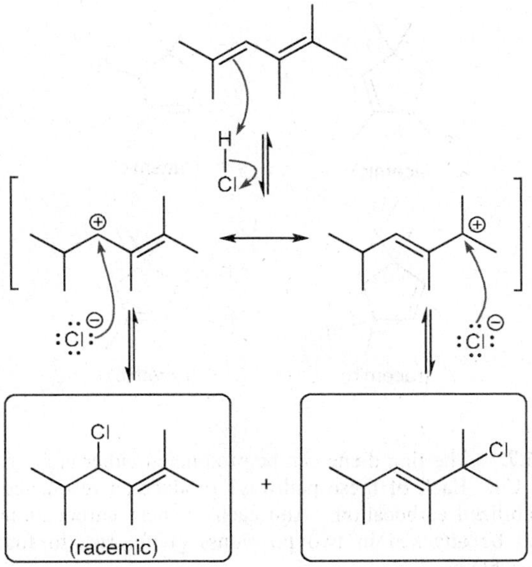

(racemic)

Notice that one of the products possesses a chiral center and is therefore produced as a racemic mixture (because the chloride ion can attack either face of the allylic carbocation with equal likelihood).

Now let's consider protonation at C2. Once again, protonation leads to a resonance–stabilized intermediate. When we draw both resonance structures, we can see that two positions are electrophilic. Nucleophilic attack can occur at either of these locations, giving two possible products, as shown:

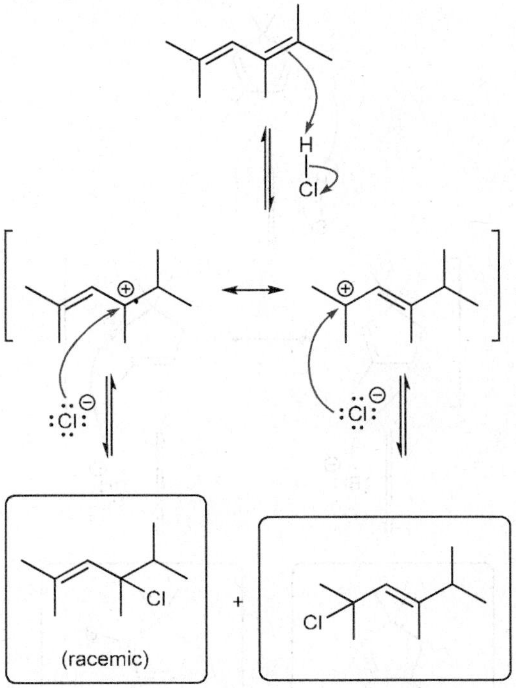

(racemic)

In summary, we expect the following possible products:

(racemic)

(racemic)

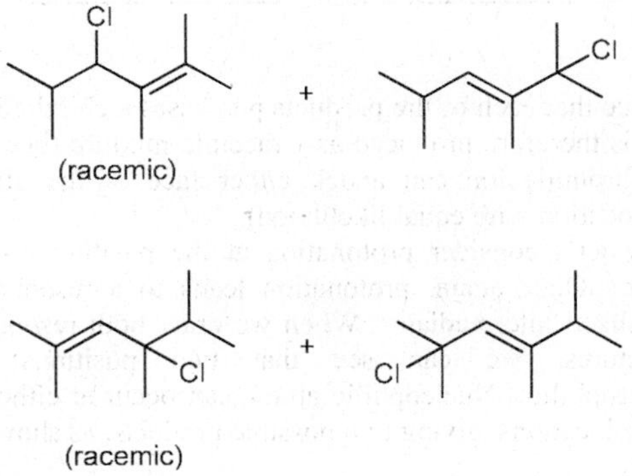

(c) We first identify the locations where protonation can occur. There are four unique positions where protonation can occur, although a resonance-stabilized intermediate can only be obtained upon protonation of one of the ends of the conjugated π system, highlighted below:

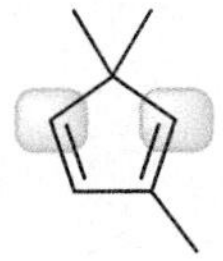

We must explore protonation at each of these positions. Let's begin with protonation of the position on the left, which leads to a resonance–stabilized intermediate. When we draw both resonance structures, we can see that two positions are electrophilic. Nucleophilic attack can occur at either of these locations, giving rise to two products:

(racemic) (racemic)

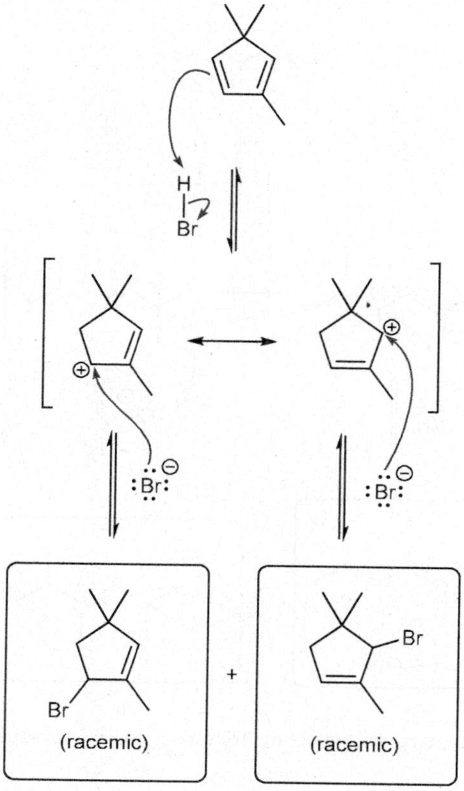

Notice that each of the products possesses a chiral center and is therefore produced as a racemic mixture (because the bromide ion can attack either face of the allylic carbocation with equal likelihood).

Now let's consider protonation at the position on the right. Once again, protonation leads to a resonance–stabilized intermediate. When we draw both resonance structures, we can see that two positions are electrophilic. Nucleophilic attack can occur at either of these locations, giving two possible products, as shown:

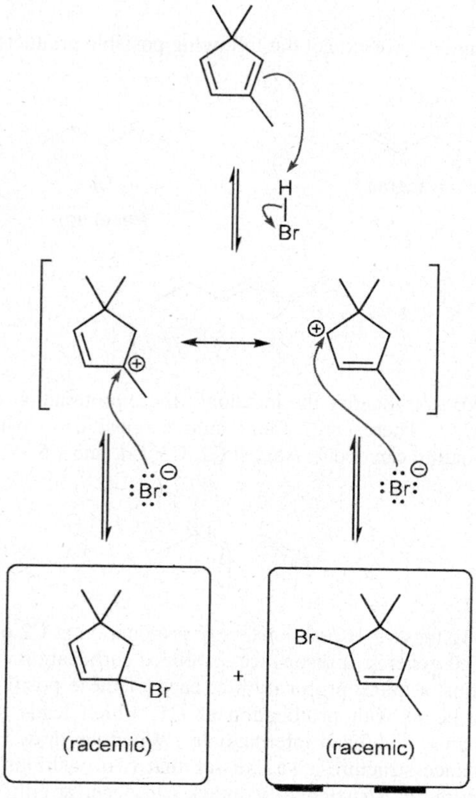

(racemic) (racemic)

In summary, we expect the following possible products:

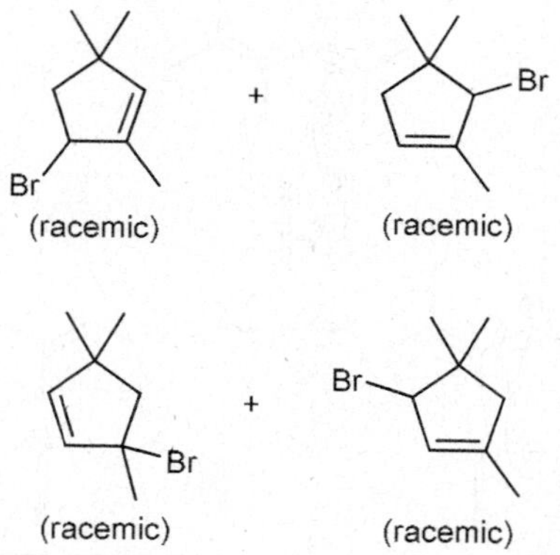

(racemic) (racemic)

(racemic) (racemic)

16.7. The first diene can be protonated either at C1 or at C4. Each of these pathways produces a resonance-stabilized carbocation. And each of these carbocations can be attacked in two positions, giving rise to four possible products.

In contrast, the second diene yields the same carbocation regardless of whether protonation occurs at C1 or at C4. This resonance-stabilized carbocation can be attacked in two positions, giving rise to two products (not four).

16.8. Recall that MCPBA produces an epoxide when it reacts with a carbon-carbon double bond. Compound **1** has two double bonds, so there are two possible epoxides that can be produced (not including stereoisomers). These constitutional isomers are compounds **2** and **3** (it does not matter which epoxide is labeled **2** and which is labeled **3**).

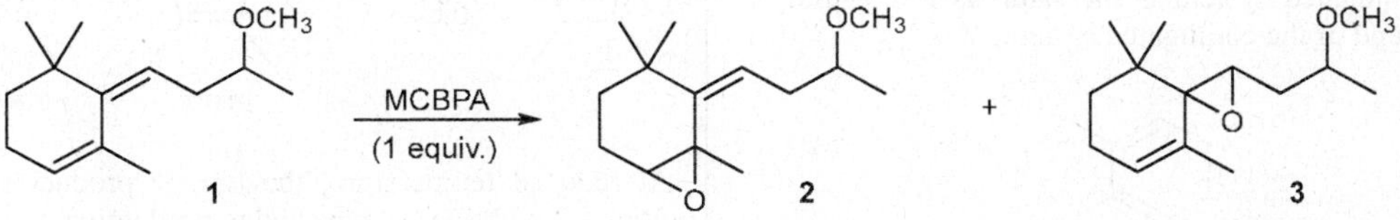

MCBPA (1 equiv.)

1 2 3

In the conversion of compound **2** into compound **4**, the mechanism begins with a proton transfer to give a protonated epoxide. Then, as indicated in the problem statement, the epoxide can open to give a resonance-stabilized allylic cation. After drawing the resonance structures of this allylic carbocation, nucleophilic attack by water, followed by a proton transfer, generates compound **4**.

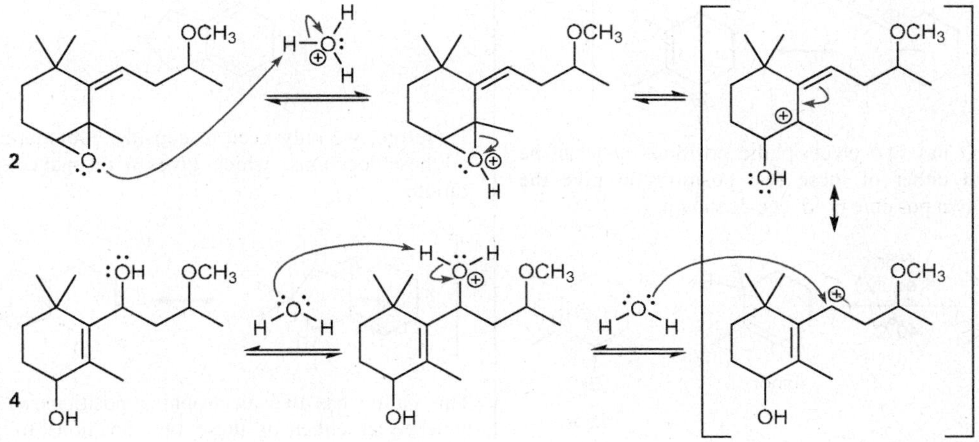

Likewise, the mechanism for the conversion of compound **3** into compound **4** begins with a proton transfer to create a protonated epoxide. The epoxide then opens to give a resonance-stabilized allylic cation. Nucleophilic attack by water, followed by a proton transfer, generates compound **4**.

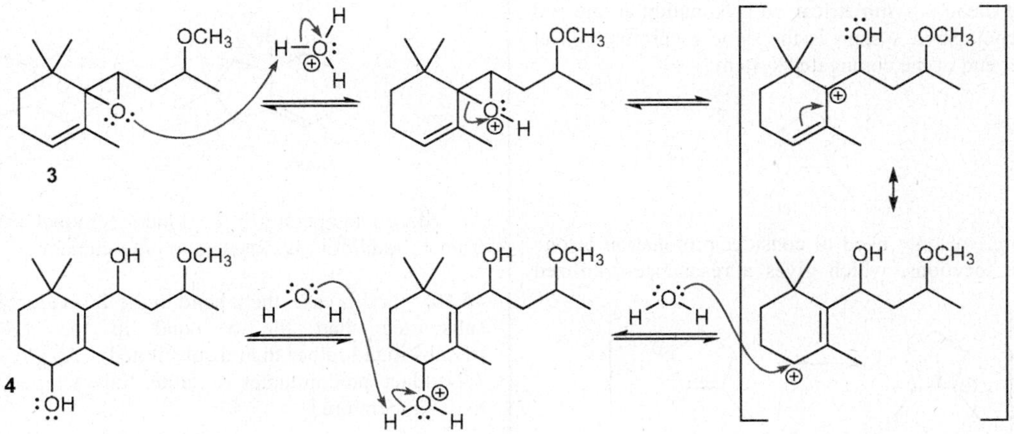

16.9.

(a) The diene is symmetrical, so protonation at one end of the conjugated system is the same as protonation at the other end of the conjugated system:

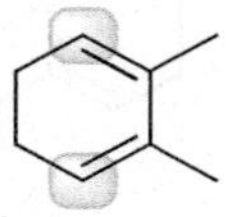

Therefore, we only need to consider protonation at one of these locations, which gives a resonance-stabilized cation:

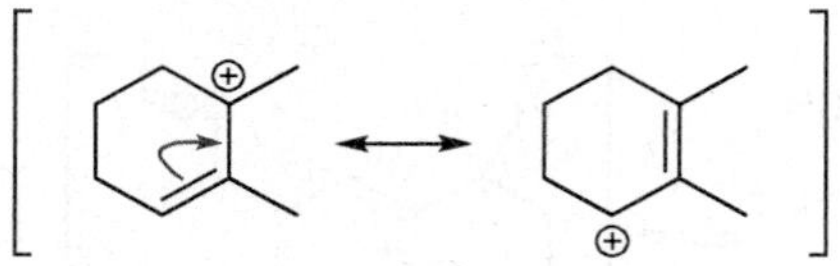

This cation has two electrophilic positions and can be attacked at either of these two positions to give the following two possible products, as shown.

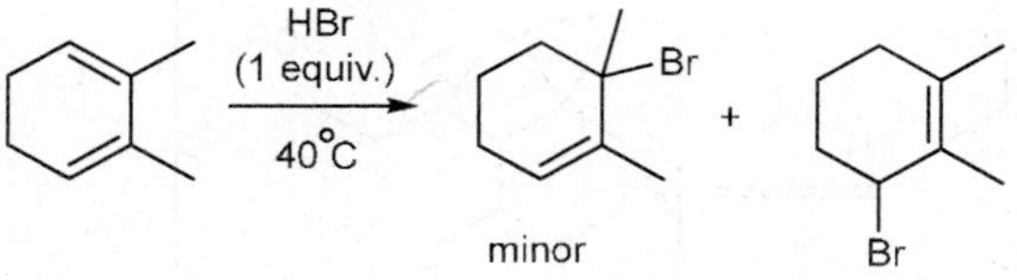

HBr (1 equiv.) 40°C — Br minor + Br major

At elevated temperature, the thermodynamic product (the compound with the more substituted π bond) is expected to predominate.

(b) The diene is symmetrical, so protonation at one end of the conjugated system is the same as protonation at the other end of the conjugated system:

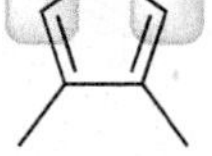

Therefore, we only need to consider protonation at one of these locations, which gives a resonance-stabilized cation:

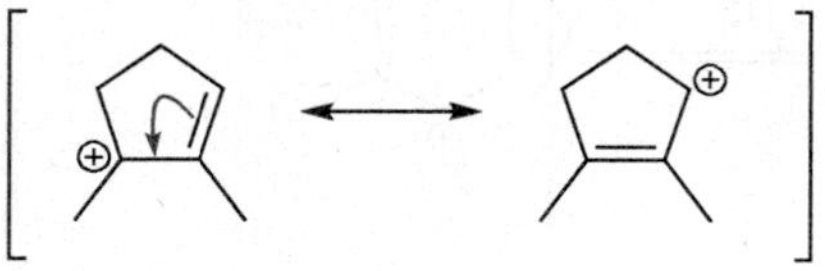

This cation has two electrophilic positions and can be attacked at either of these two positions to give the following two possible products, as shown.

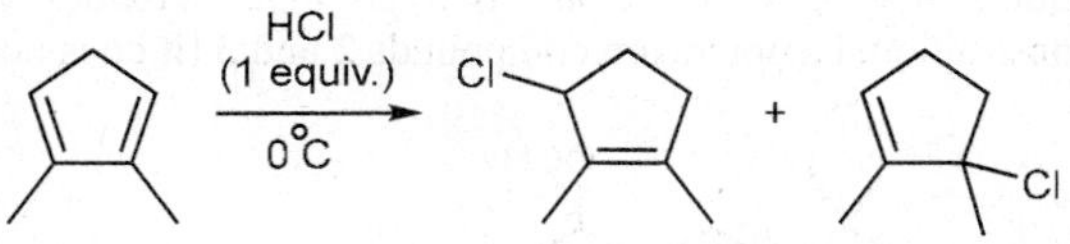

HCl (1 equiv.) 0°C — Cl minor + Cl major

At reduced temperature, the kinetic product (resulting from 1,2-addition) is expected to predominate.

(c) The diene is symmetrical, so protonation at one end of the conjugated system is the same as protonation at the other end of the conjugated system:

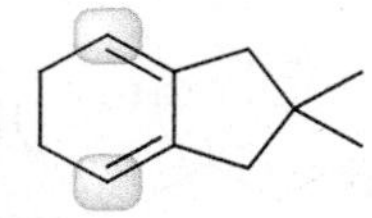

Therefore, we only need to consider protonation at one of these locations, which gives a resonance-stabilized cation:

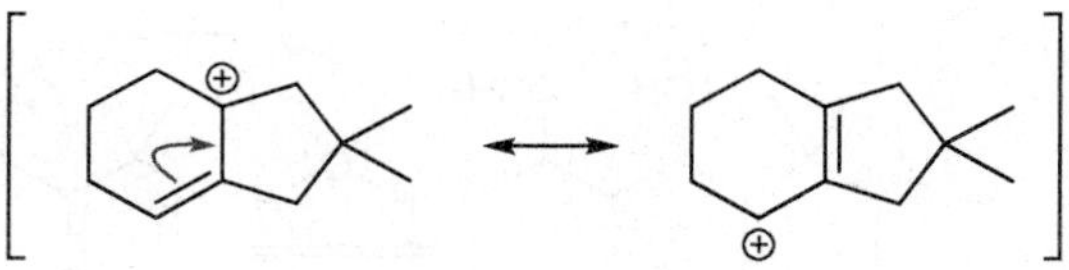

This cation has two electrophilic positions and can be attacked at either of these two positions to give the following two possible products, as shown.

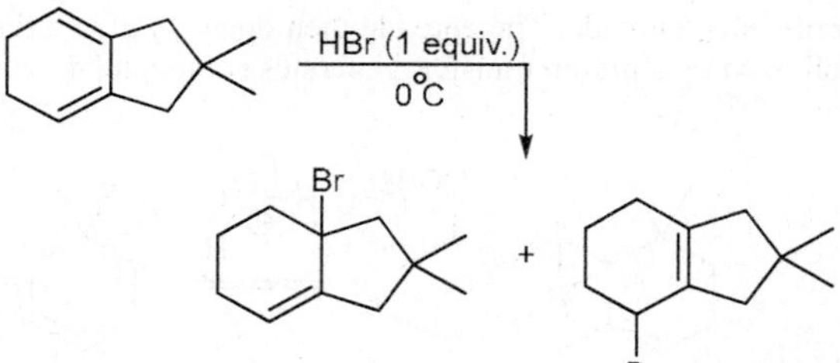

HBr (1 equiv.) 0°C — Br major + Br minor

At reduced temperature, the kinetic product (resulting from 1,2-addition) is expected to predominate.

16.10. In this case, the π bond in the 1,2-adduct is more substituted than the π bond in the 1,4-adduct (trisubstituted rather than disubstituted). As a result, the 1,2-adduct predominates at either low temperature or high temperature.

16.11. We first identify the locations where protonation can occur. There are four unique positions where protonation can occur, labeled C1 through C4:

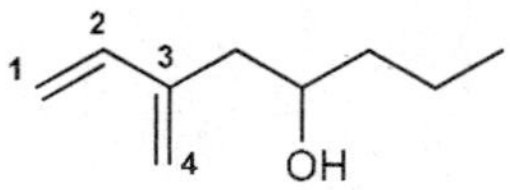

1 2 3 4 OH

Among these four positions, only protonation at C1 or at C4 will generate a resonance-stabilized carbocation. So, we must explore protonation at each of these positions. Protonation at C1 leads to a resonance-stabilized intermediate with carbocationic character shared between a secondary and a primary carbon:

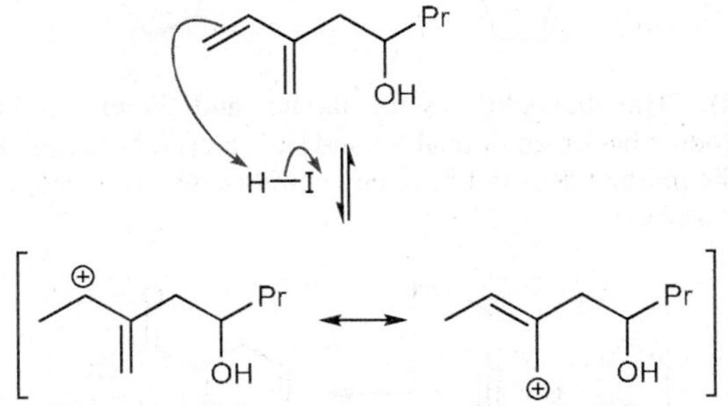

Secondary allylic

Primary allylic

Protonation at C4, however, leads to a resonance-stabilized intermediate with carbocationic character shared between a tertiary carbon and primary carbon:

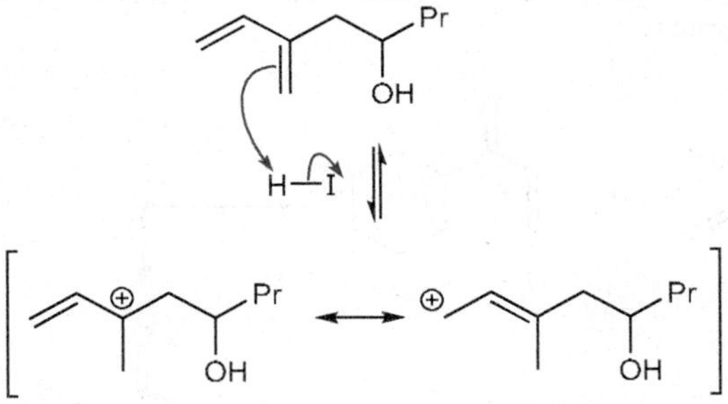

Tertiary allylic

Primary allylic

As observed with Markovnikov's Rule, protonation of a π bond will occur in the position that leads to the more stable carbocation intermediate, so the reaction begins with protonation at C4. When both resonance structures of the resulting carbocation intermediate are drawn, we can see that two positions are electrophilic. Nucleophilic attack of the iodide ion can occur at either of these locations, giving rise to two products. Under thermodynamic control, the major product is the more stable compound with the more substituted π bond, as shown below:

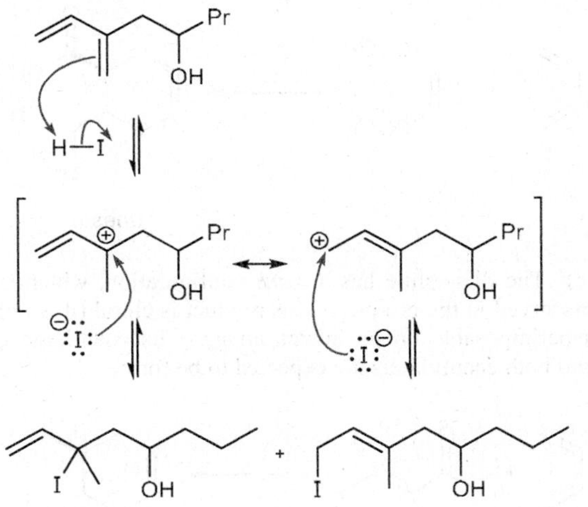

1,2-Adduct (kinetic)
Minor

1,4-Adduct (thermodynamic)
Major

16.12.

(a) This polymer is similar in structure to neoprene, but the chlorine atom has been replaced with a cyano group. So the starting material should be similar in structure to chloroprene (the monomer of neoprene), except that the chlorine atom is replaced with a cyano group:

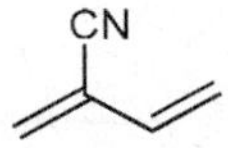

(b) This polymer is comprised of repeating units that bear two substituents (both fluorine atoms). The following monomer is necessary.

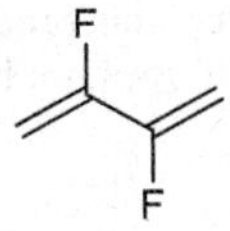

16.13.

(a) The dienophile has a *cis* configuration, which is preserved in the product. This product is superimposable on its mirror image, so it does not have an enantiomer. It is a *meso* compound.

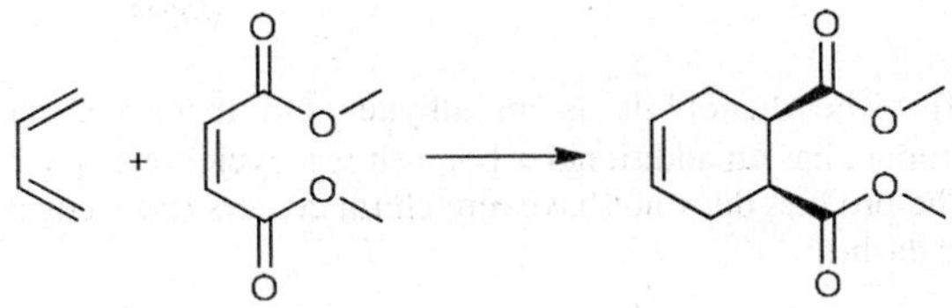

(*meso*)

(b) The dienophile has a *cis* configuration, which is preserved in the product. This product is superimposable on its mirror image, so it does not have an enantiomer. It is a *meso* compound.

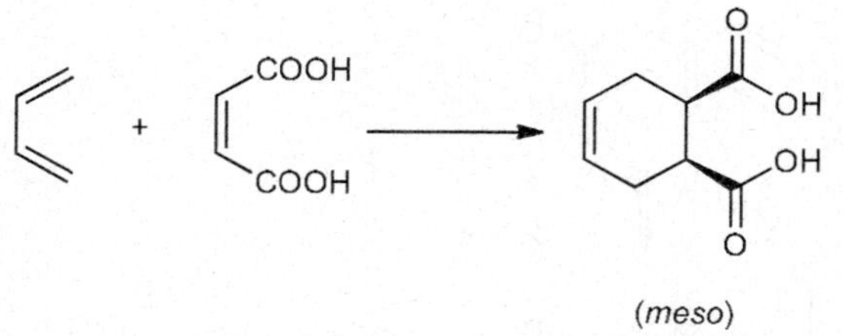

COOH
COOH
OH
OH
O
O

(meso)

(c) The dienophile has a *trans* configuration, which is preserved in the product. This product is chiral (it is not superimposable on its mirror image – its enantiomer), and both enantiomers are expected to be formed.

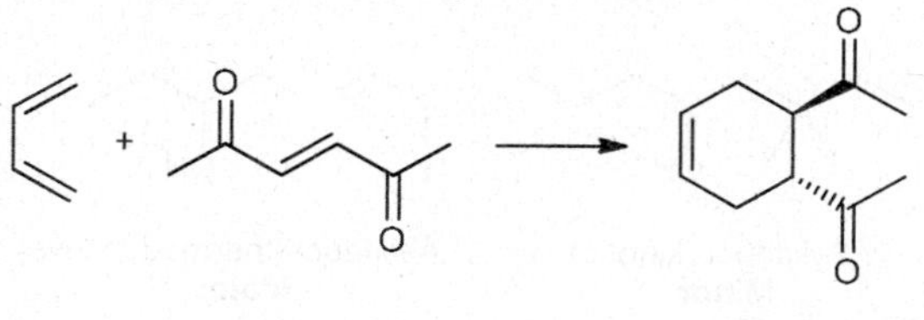

+ En

(d) The dienophile is an alkyne, and therefore, the product has an additional π bond (it is a cyclohexa*diene*). The product does not have any chiral centers (no wedges or dashes).

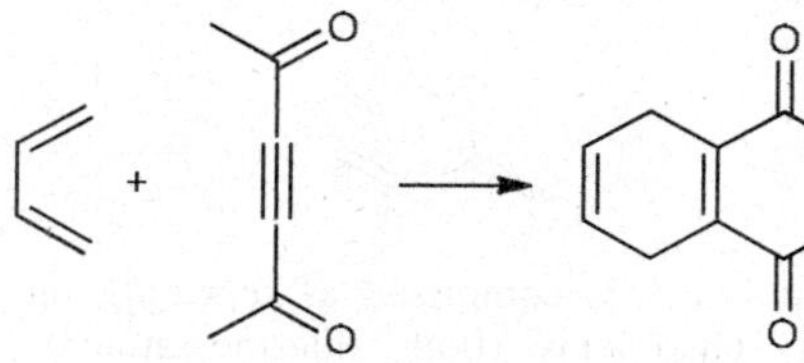

(e) The dienophile has a *cis* configuration, which is preserved in the product. This product is superimposable on its mirror image, so it does not have an enantiomer. It is a *meso* compound.

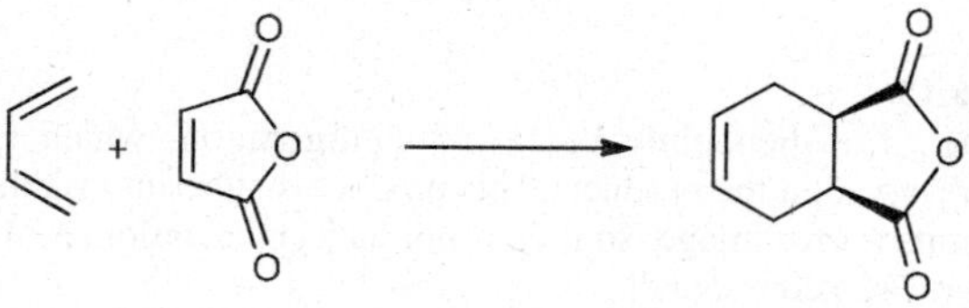

(meso)

(f) The dienophile is an alkyne, and therefore, the product has an additional π bond (it is a cyclohexa*diene*). The product does not have any chiral centers (no wedges or dashes).

CN
CN
CN
CN

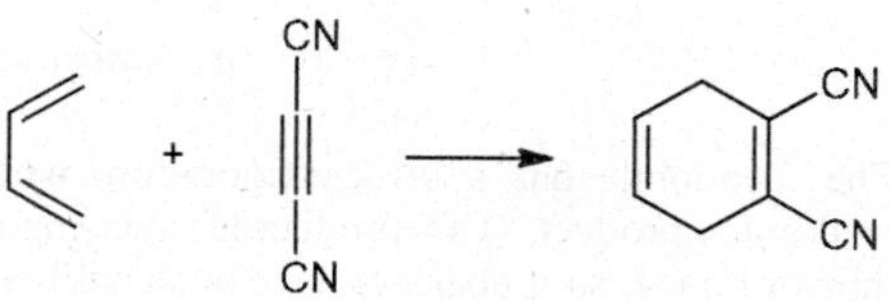

(g) The dienophile has a *cis* configuration, which is preserved in the product. This product is chiral (it is not superimposable on its mirror image – its enantiomer), and both enantiomers are expected to be formed.

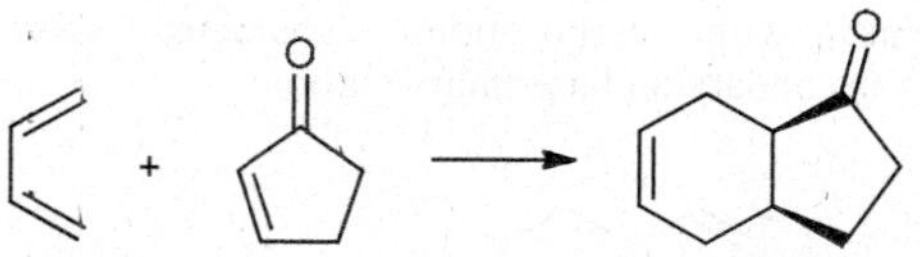

+ En

(h) The dienophile is an alkyne, and therefore, the product has an additional π bond (it is a cyclohexa*diene*). The product does not have any chiral centers (no wedges or dashes).

OH
OH

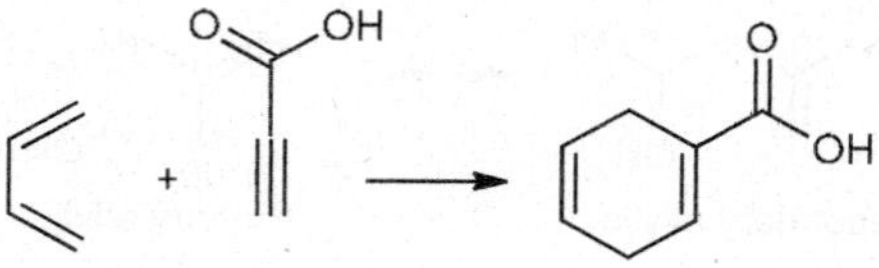

(i) The dienophile is a monosubstituted alkene, giving a cycloadduct with one chiral center. This product is chiral (it is not superimposable on its mirror image – its enantiomer), and both enantiomers are expected to be formed.

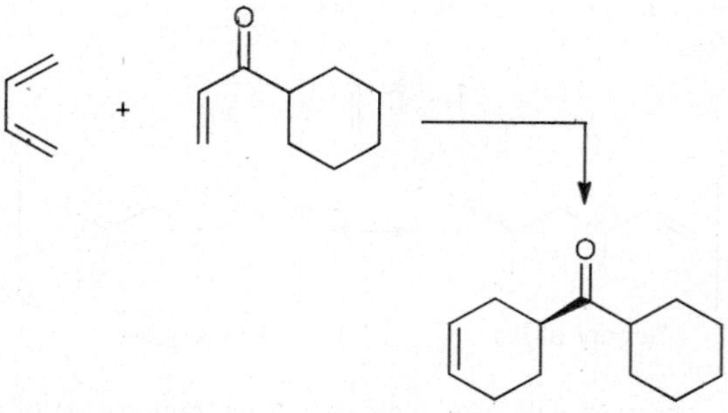

+ En

16.14.

(a) The Diels-Alder reaction forms two new C–C bonds, so the retrosynthesis of a Diels-Alder product requires the identification and disconnection of these two bonds. The product exhibits a *cis* orientation of the cyano groups, so the dienophile must have the *cis* configuration as well.

CN
CN
⟹
CN
CN

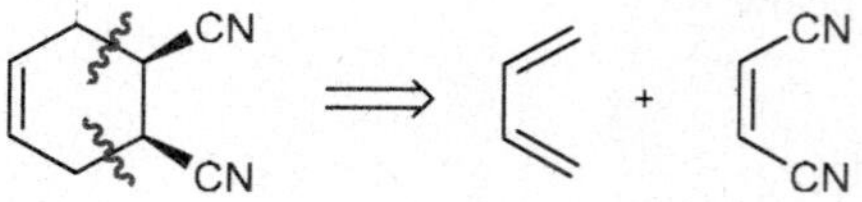

(b) The Diels-Alder reaction forms two new C–C bonds, so the retrosynthesis of a Diels-Alder product requires the identification and disconnection of these two bonds. The product exhibits a *cis* configuration, so the dienophile must have the *cis* configuration as well (although in this case, a *trans* configuration is not possible in a six-membered ring).

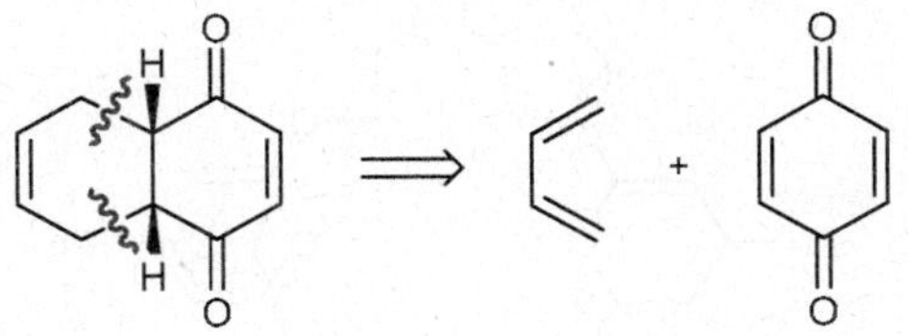

(c) The Diels-Alder reaction forms two new C–C bonds, so the retrosynthesis of a Diels-Alder product requires the identification and disconnection of these two bonds. In this case, there are no stereochemical outcomes to be accounted for (one new chiral center is generated, so the product is formed as a racemic mixture, as usual).

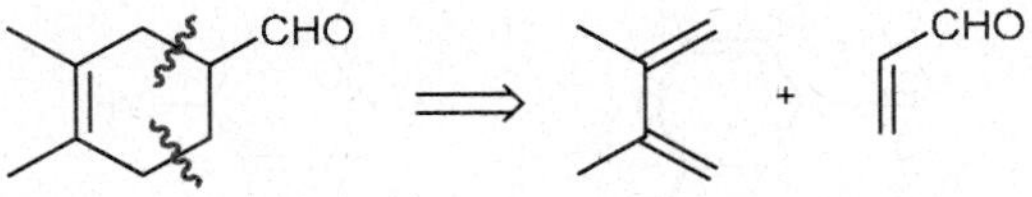

(d) The Diels-Alder reaction forms two new C–C bonds, so the retrosynthesis of a Diels-Alder product requires the identification and disconnection of these two bonds. The product exhibits a *trans* orientation of the ester groups, so the dienophile must have the *trans* configuration as well. Note that the retrosynthesis of either enantiomer leads to the same starting materials.

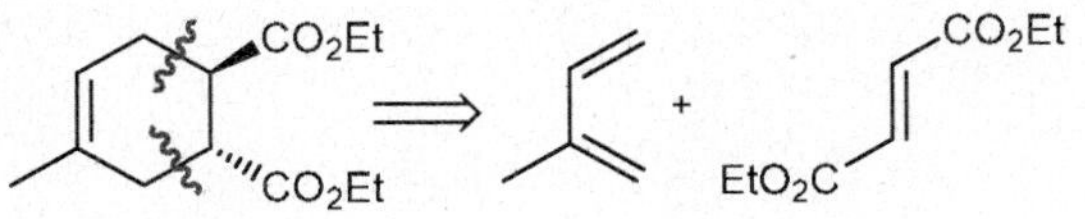

16.15. A Diels-Alder reaction produces a cyclohexene adduct. It follows that a hetero-Diels-Alder reaction should produce a cyclohexene-type product in which one or more of the carbon atoms of the cyclohexene ring are replaced with nitrogen or oxygen. Analysis of the macrocyclic product reveals two such moieties, highlighted below.

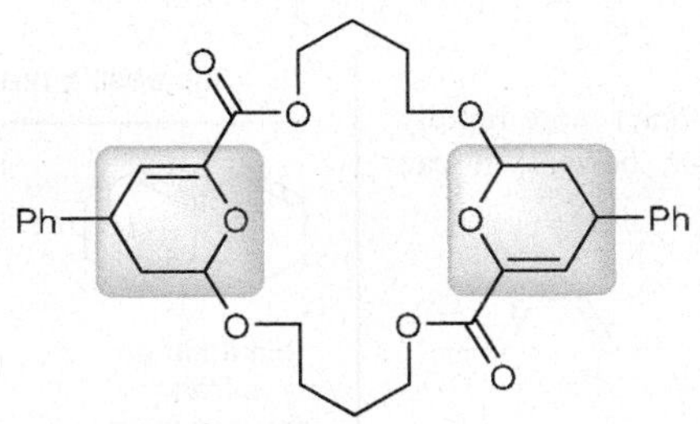

A retrosynthetic analysis, shown here, demonstrates that this product can be produced from two equivalents of the acyclic reactant, arranged head-to-tail, where the four disconnected bonds are produced in two hetero-Diels-Alder reactions.

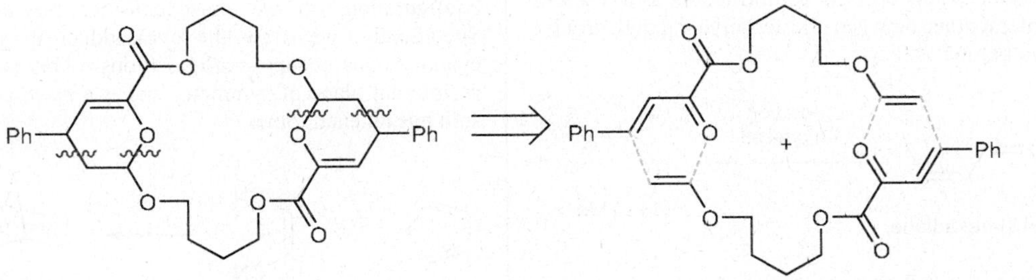

A mechanism consistent with this reaction is presented below. In the first step, the terminal alkene group of the bottom molecule serves as the "dienophile". On the top molecule, the alkene group next to the phenyl group and the adjacent, conjugated ketone serve the role of the "diene" (although in this case it is not formally a diene since one of the carbon atoms has been replaced by oxygen). This hetero-Diels-Alder reaction produces the first six-membered ring as shown. A second hetero-Diels-Alder reaction then proceeds at the other terminus of the molecule, thus producing the macrocyclic product.

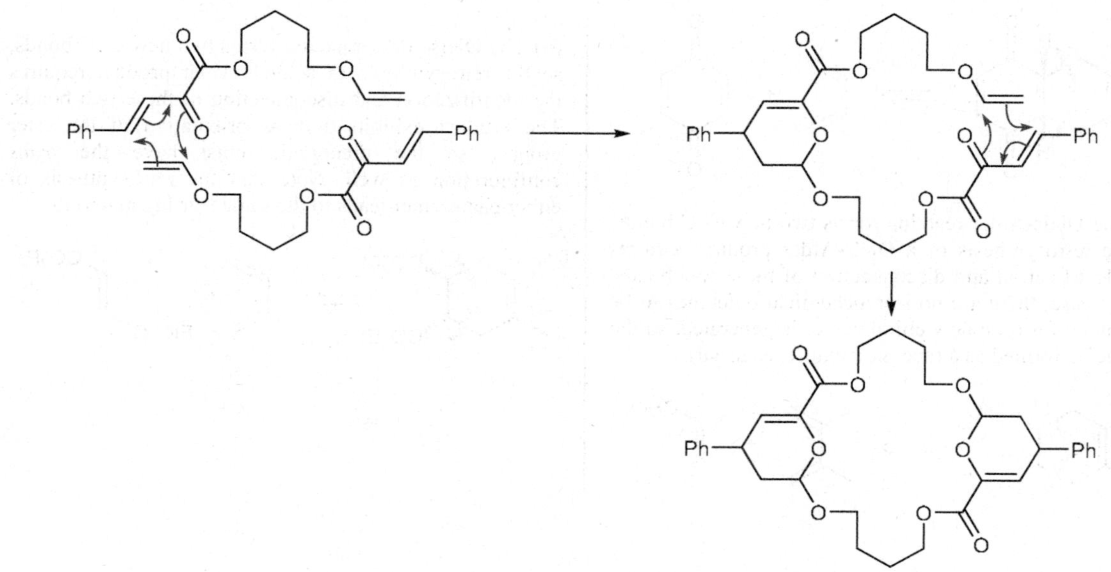

Ph O O Ph O O O O O O

Ph O O O Ph O O O O

Ph O O O O Ph O O O O

16.16.

(a) The 2*E*,4*E* isomer is expected to react more rapidly as a diene in a Diels–Alder reaction, because it can readily adopt an *s-cis* conformation.

rotate

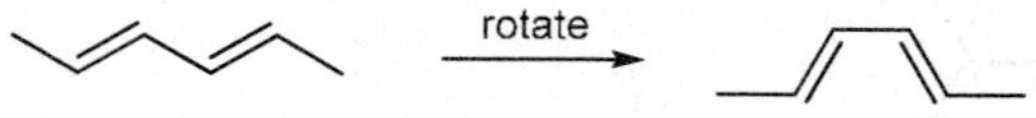

(2*E*,4*E*)-hexadiene

In contrast, the 2*Z*,4*Z* isomer is expected to react more slowly as a diene in a Diels–Alder reaction, because it cannot readily adopt an *s-cis* conformation, as a result of steric interactions between the terminal methyl groups, as shown below:

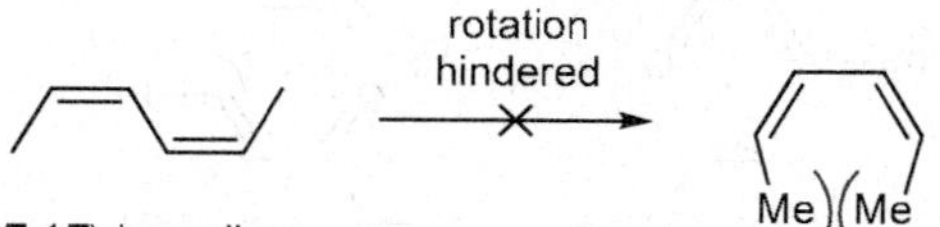

rotation hindered

(2*Z*,4*Z*)-hexadiene

Me Me

(b) One compound is locked in an *s-cis* conformation and will therefore be the most reactive in a Diels-Alder reaction. Another compound is locked in an *s-trans* conformation and will therefore be completely unreactive in a Diels-Alder reaction. The third molecule is drawn in an *s-trans* conformation but can adopt an *s-cis* conformation via rotation around the central C—C bond.

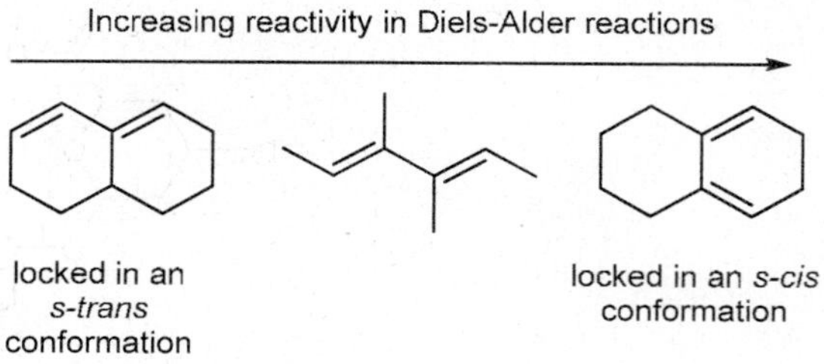

Increasing reactivity in Diels-Alder reactions

locked in an *s-trans* conformation

locked in an *s-cis* conformation

16.17.

(a) The diene is cyclic and the dienophile has a *cis* configuration, so we must consider the *endo* rule. Specifically, we draw the cycloadduct in which the cyano groups occupy *endo* positions. This product has an internal plane of symmetry and is a *meso* compound, so it has no enantiomer.

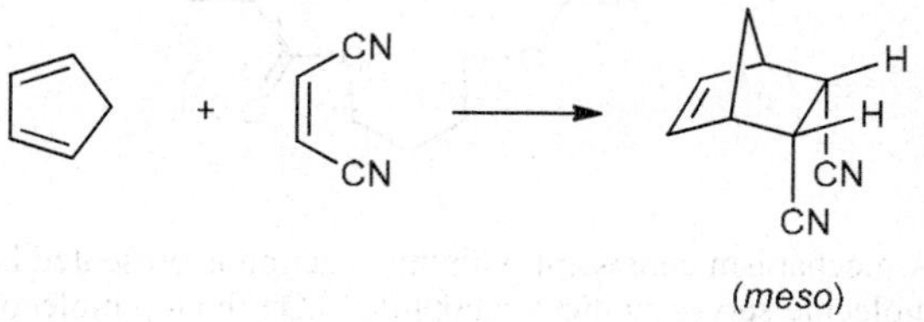

CN + CN H H CN CN

(*meso*)

If you have trouble seeing how the bicyclic framework is produced by this reaction, consider the following drawing in which the bond angles have been distorted to show how the bicyclic structure is formed:

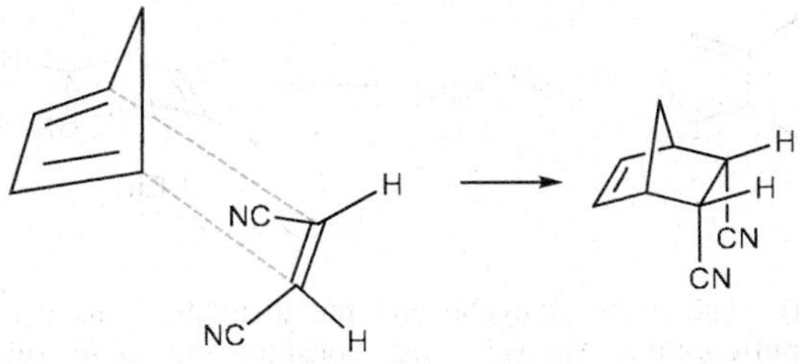

In this drawing, the dotted lines indicate the new σ bonds that are formed during the Diels-Alder reaction.

(b) The diene is cyclic, but the dienophile has a *trans* configuration. Therefore, the *endo* rule is not relevant in this case. Two products are expected, because the reaction can either occur like this,

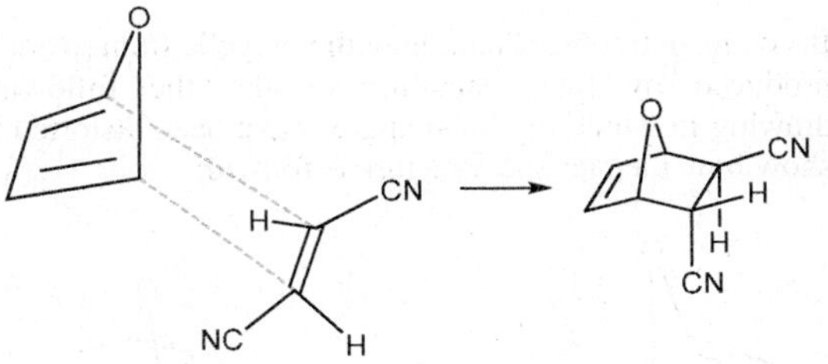

or the reaction can occur like this:

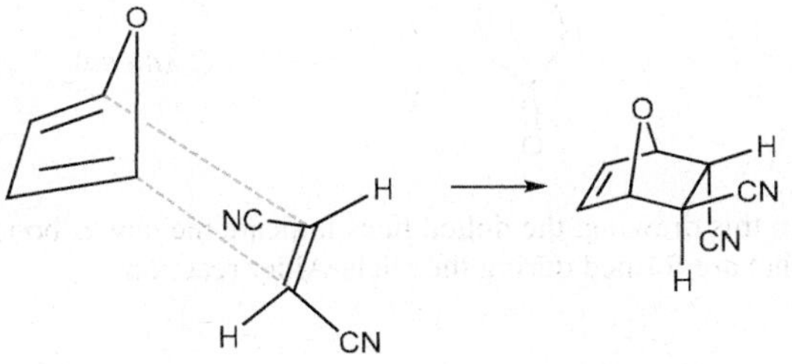

Either way, one group will occupy an *endo* position and the other group will occupy an *exo* position, because the configuration of the dienophile is preserved during a Diels-Alder reaction. Notice that the two possible products are non-superimposable mirror images of each other, so they represent a pair of enantiomers:

Enantiomers

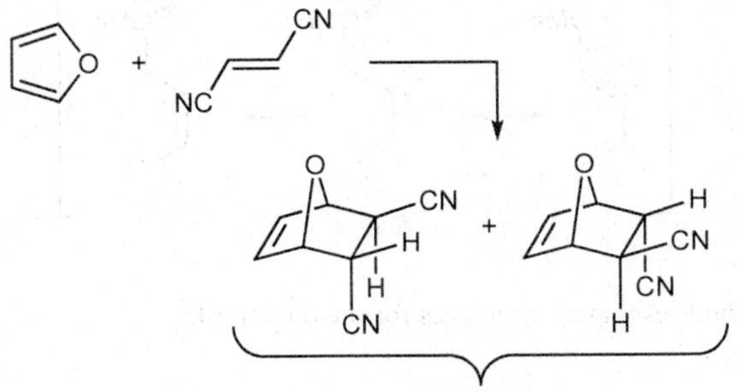

This can also be indicated in the following way:

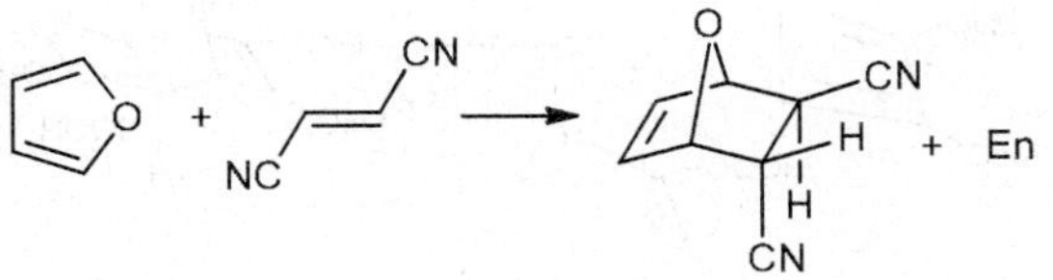

+ En

(c) The diene is cyclic, and the *endo* rule indicates that the substituent should occupy an *endo* position in the product. With this restriction in mind, the reaction can either occur like this,

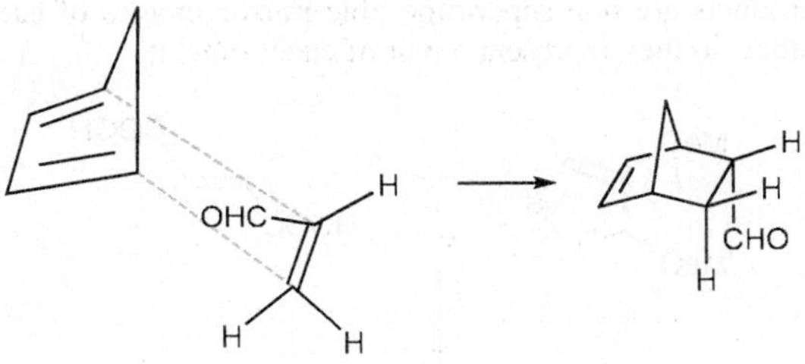

or the reaction can occur like this:

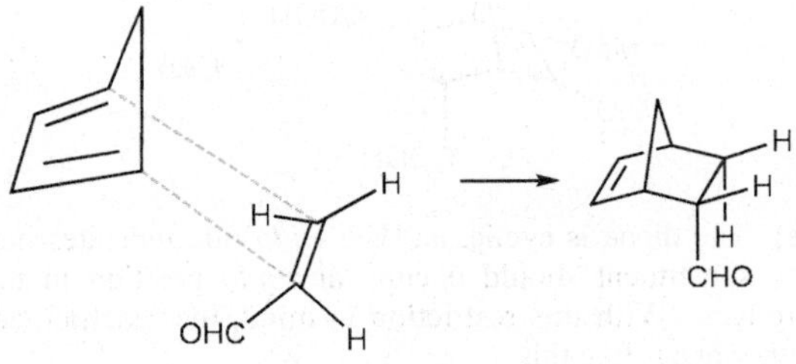

These two products represent a pair of enantiomers, and both are expected.

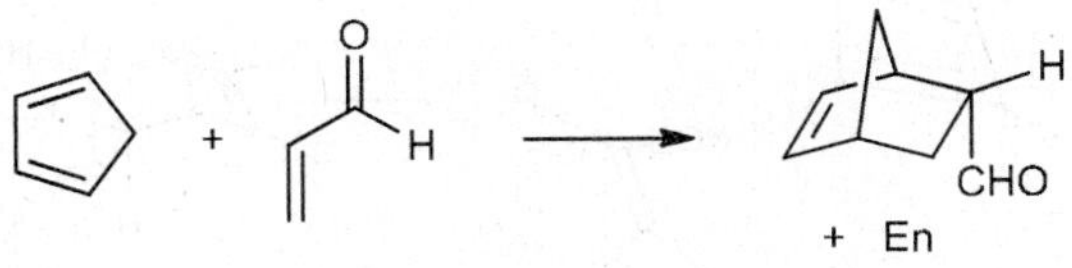

+ En

(d) The diene is cyclic, but the dienophile has a *trans* configuration. Therefore, the *endo* rule is not relevant in this case. Two products are expected, because the reaction can either occur like this,

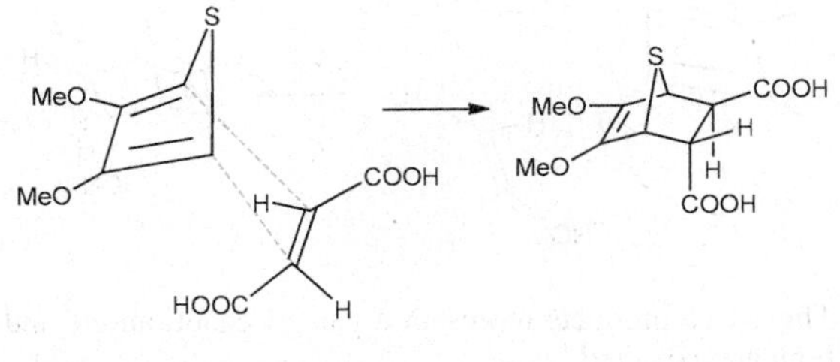

or the reaction can occur like this:

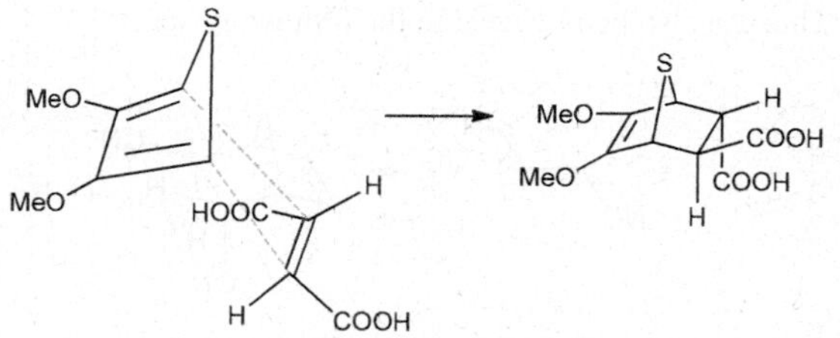

Either way, one group will occupy an *endo* position and the other group will occupy an *exo* position, because the configuration of the dienophile is preserved during a Diels-Alder reaction. Notice that the two possible products are non-superimposable mirror images of each other, so they represent a pair of enantiomers:

+ En

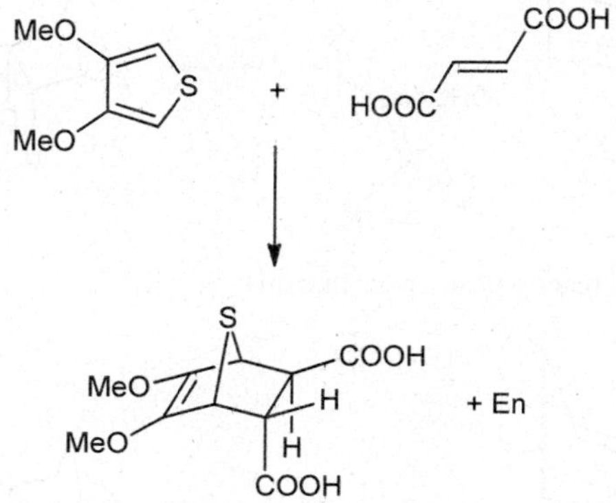

(e) The diene is cyclic, and the *endo* rule indicates that the substituent should occupy an *endo* position in the product. With this restriction in mind, the reaction can either occur like this,

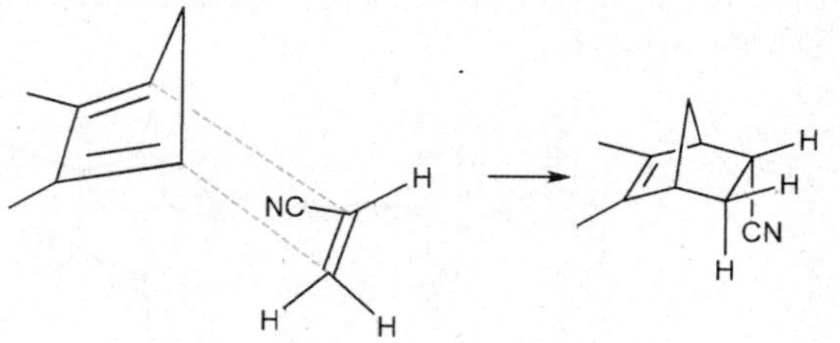

or the reaction can occur like this:

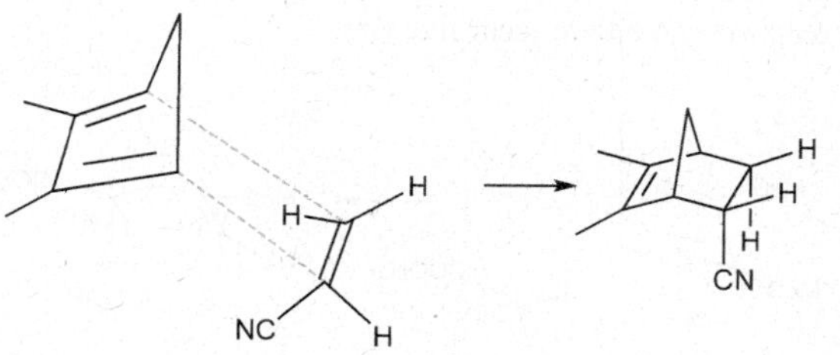

These two products represent a pair of enantiomers, and both are expected.

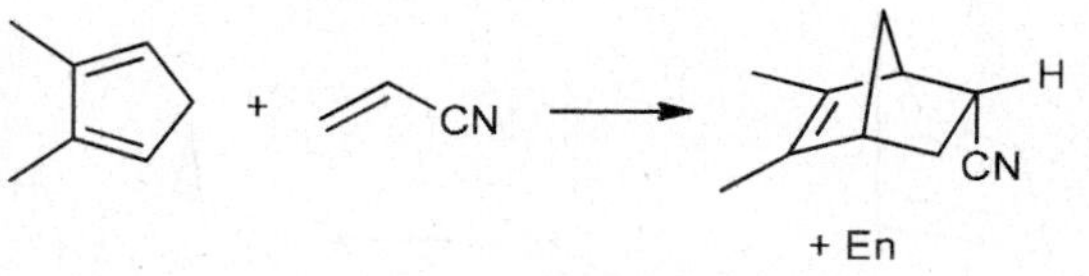

+ En

(f) The diene is cyclic and the dienophile has a *cis* configuration, so we must consider the *endo* rule. Specifically, we draw the *endo* product, which has an internal plane of symmetry in this case, so it is a *meso* compound (it has no enantiomer).

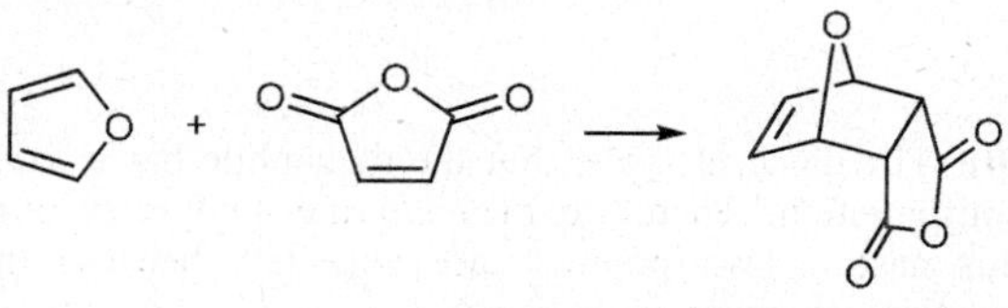

(*meso*)

If you have trouble seeing how the bicyclic framework is produced by this reaction, consider the following drawing in which the bond angles have been distorted to show how the bicyclic structure is formed:

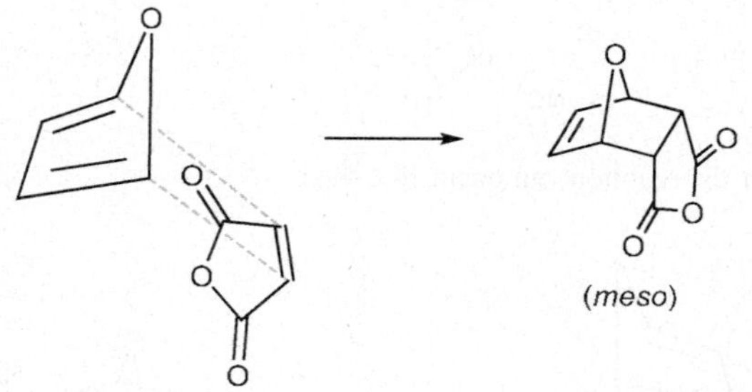

(*meso*)

In this drawing, the dotted lines indicate the new σ bonds that are formed during the Diels-Alder reaction.

16.18.

(a) Both the diene and the dienophile are unsymmetrical, so there are two possible regiochemical outcomes. The major product can be predicted by considering resonance structures for the diene,

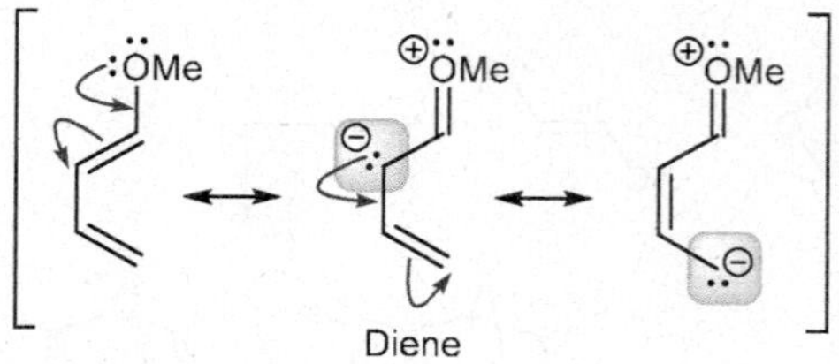

Diene

and resonance structures for the dienophile:

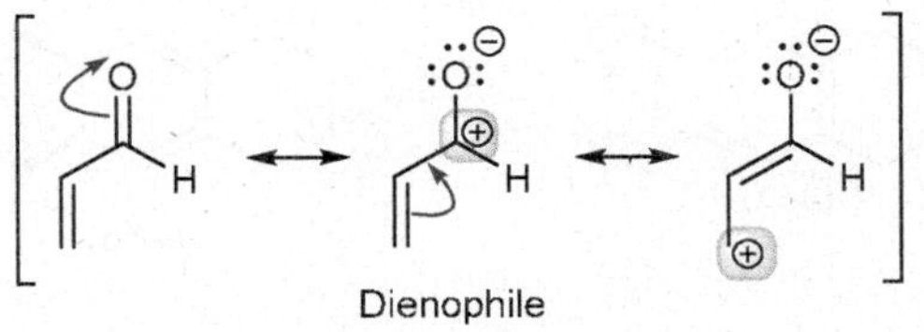

Dienophile

Notice that the diene is electron-rich, because of the electron-donating effect of the methoxy group; while the dienophile is electron-poor, because of the electron-withdrawing effect of the aldehyde group. The major product results when the regions of δ+ and δ– (highlighted) are aligned:

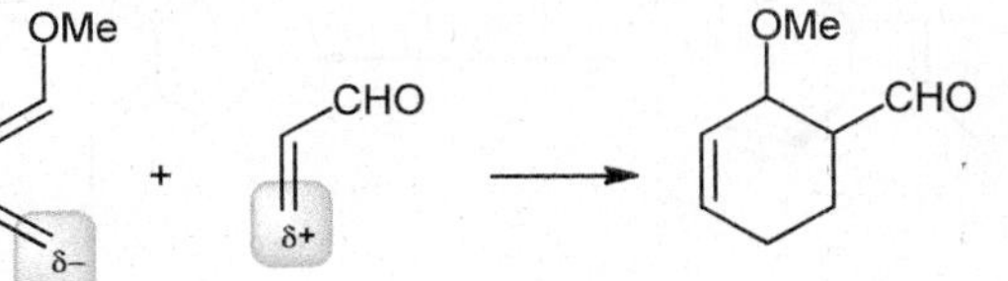

(b) Both the diene and the dienophile are unsymmetrical, so there are two possible regiochemical outcomes. The major product can be predicted by considering resonance structures for the diene,

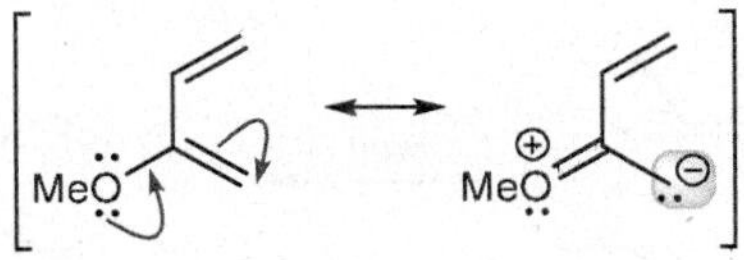

Diene

and resonance structures for the dienophile:

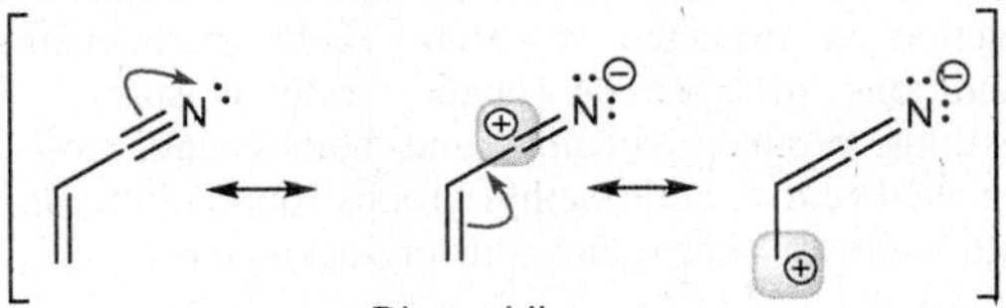

Dienophile

Notice that the diene is electron-rich, because of the electron-donating effect of the methoxy group; while the dienophile is electron-poor, because of the electron-withdrawing effect of the cyano group. The major product results when the regions of δ+ and δ– (highlighted) are aligned:

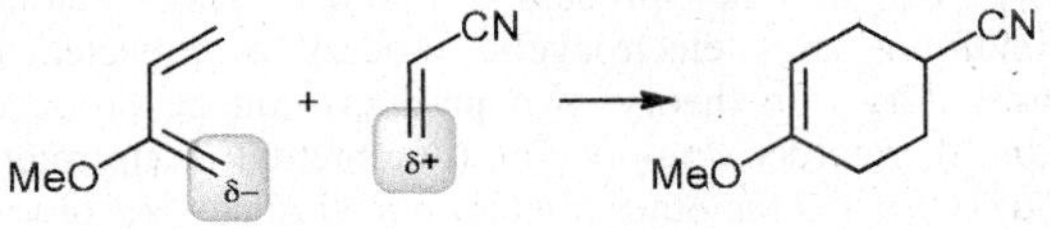

(c) Both the diene and the dienophile are unsymmetrical, so there are two possible regiochemical outcomes. The major product can be predicted by considering resonance structures for the diene,

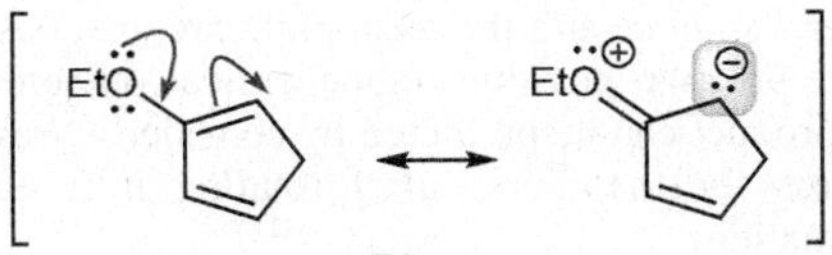

Diene

and resonance structures for the dienophile:

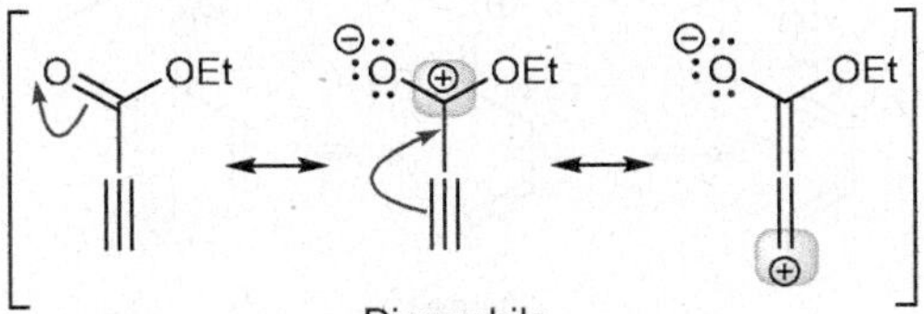

Dienophile

Notice that the diene is electron-rich, because of the electron-donating effect of the ethoxy group; while the dienophile is electron-poor, because of the electron-withdrawing effect of the ester group. The major product results when the regions of δ+ and δ– (highlighted) are aligned:

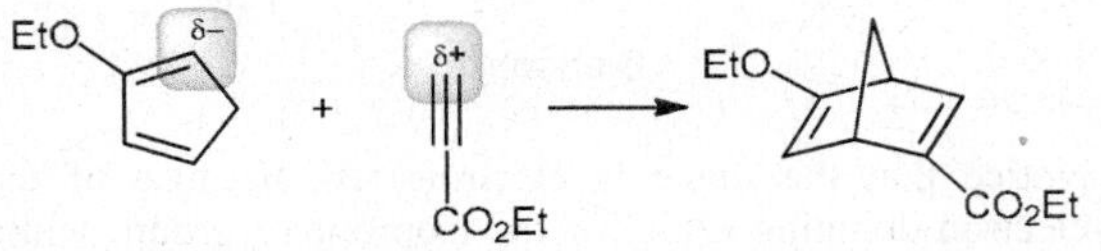

(d) Both the diene and the dienophile are unsymmetrical, so there are two possible regiochemical outcomes. The major product can be predicted by considering resonance structures for the diene,

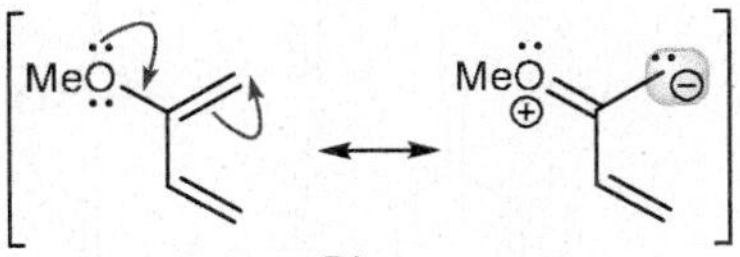

Diene

and resonance structures for the dienophile:

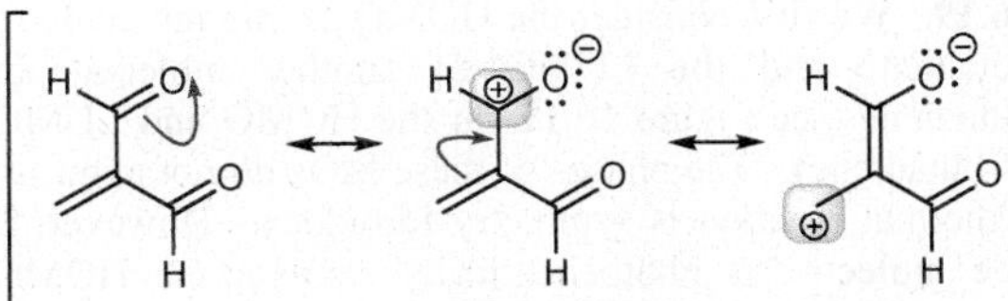

Dienophile

Notice that the diene is electron-rich, because of the electron-donating effect of the methoxy group; while the dienophile is electron-poor, because of the electron-withdrawing effects of both aldehyde groups. The major product results when the regions of δ+ and δ– (highlighted) are aligned:

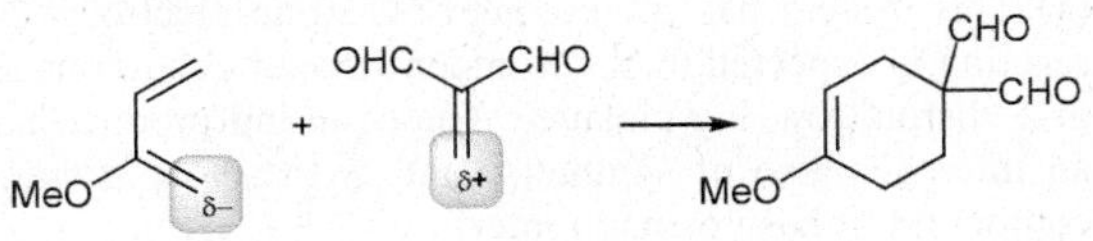

(e) Both the diene and the dienophile are unsymmetrical, so there are two possible regiochemical outcomes. The major product can be predicted by considering resonance structures for the diene, after rotating it to the *s-cis* conformation:

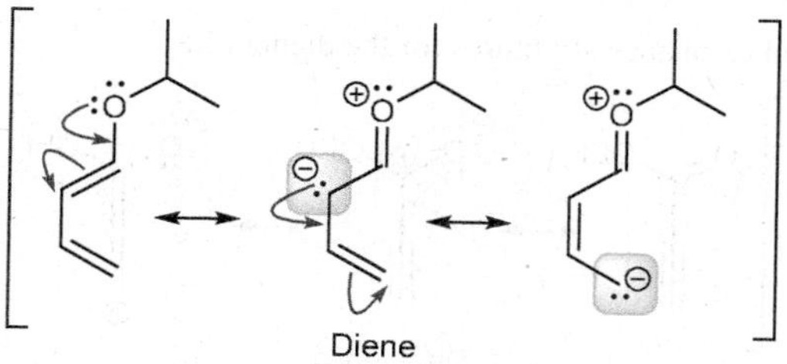

Diene

and resonance structures for the dienophile:

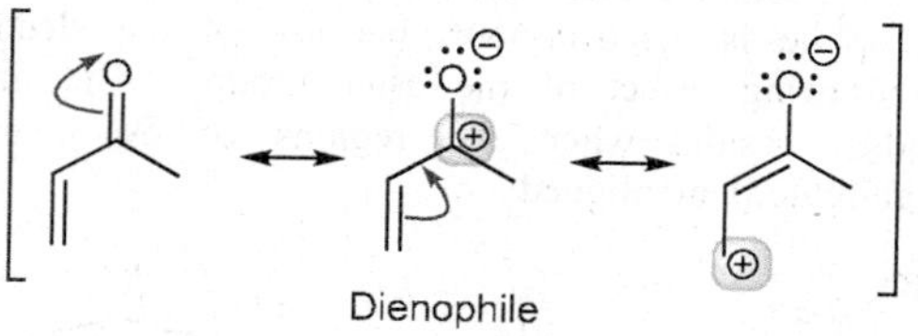

Dienophile

Notice that the diene is electron-rich, because of the electron-donating effect of the isopropoxy group; while the dienophile is electron-poor, because of the electron-withdrawing effect of the ketone group. The major product results when the regions of δ+ and δ– (highlighted) are aligned:

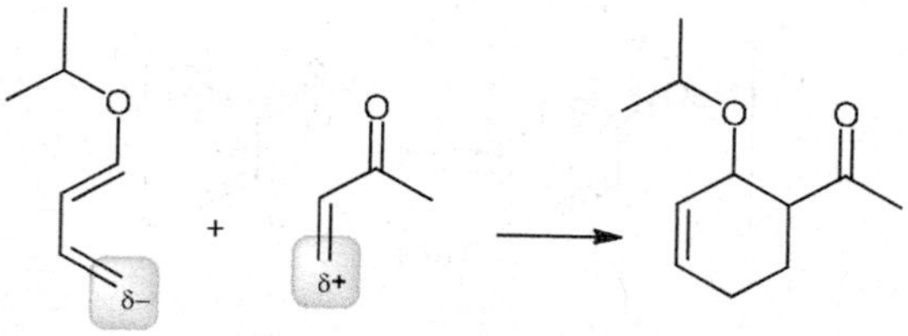

δ– + δ+

16.19. We first consider the HOMO of one molecule of butadiene and the LUMO of another molecule of butadiene (see Figure 16.17 for the HOMO and LUMO of butadiene). The phases of these MOs do not align, so a thermal reaction is symmetry-forbidden. However, if one molecule is photochemically excited, the HOMO and LUMO of the excited molecule are redefined. The phases of the frontier orbitals will align under these conditions (the HOMO of the excited molecule interacting with the LUMO of the ground-state molecule), so the reaction is expected to occur photochemically.

16.20.

(a) This system has six π electrons, so an electrocyclic reaction is expected to occur under thermal conditions to give disrotatory ring closure. The resulting product has an internal plane of symmetry and is therefore a *meso* compound (it has no enantiomer).

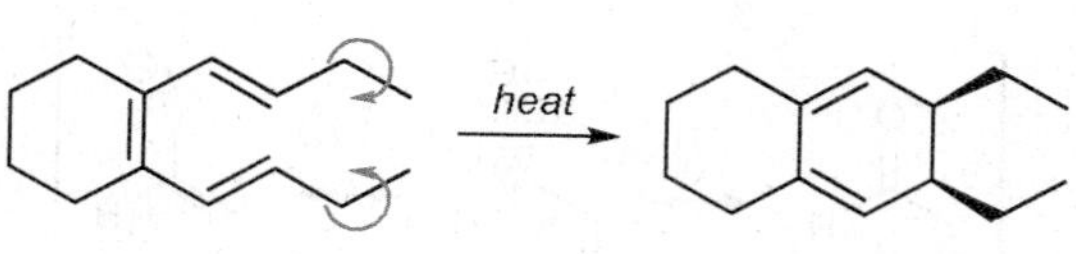

heat

(*meso*)

(b) In this reaction, the four-membered ring is opening (this is the reverse of an electrocyclic ring closure). If we look at the product, we see that four π electrons are involved in the process. Under thermal conditions, this electrocyclic process is expected be conrotatory, giving the following product.

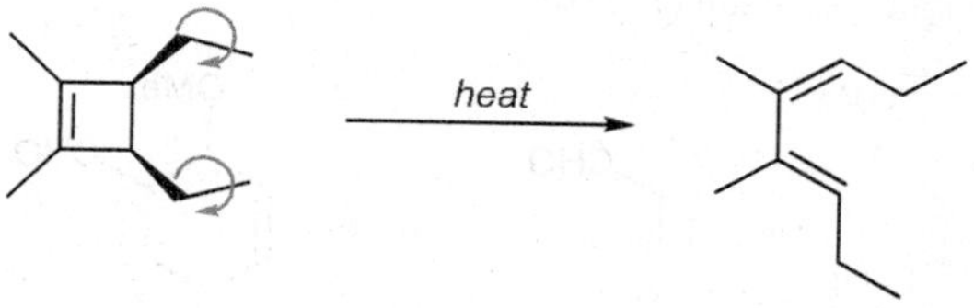

heat

(c) This system has six π electrons, so an electrocyclic reaction is expected to occur under thermal conditions to give disrotatory ring closure. The resulting product is chiral, and both enantiomers are expected (one ethyl group rotates clockwise and the other rotates counterclockwise, or vice versa, leading to both possible enantiomers).

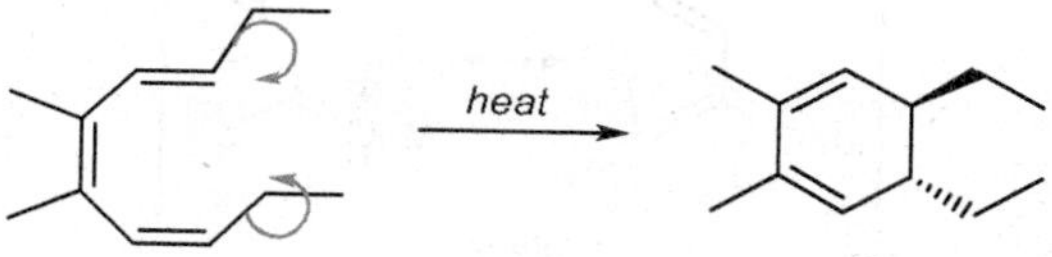

heat

+ En

16.21.

(a) This system has six π electrons, so an electrocyclic reaction is expected to occur under photochemical conditions to give conrotatory ring closure. The resulting product is chiral, and both enantiomers are expected (either both methyl groups rotate clockwise or both methyl groups rotate counterclockwise).

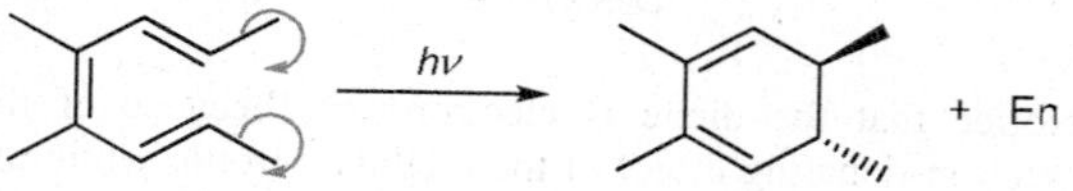

hν

+ En

(b) In this reaction, the four-membered ring is opening (this is the reverse of an electrocyclic ring closure). If we look at the product, we see that four π electrons are involved in the process. Under photochemical conditions, this electrocyclic process is expected be disrotatory. In theory, two products can be produced from disrotatory ring-opening (one methyl group rotates clockwise and the other rotates counterclockwise, or vice versa, leading to two possible products):

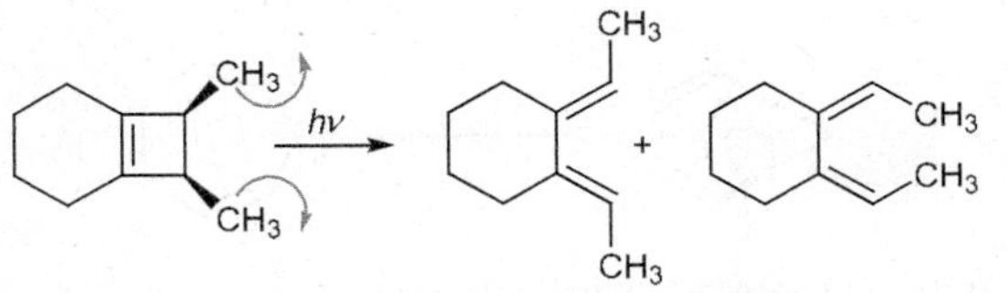

But the second product exhibits a severe steric interaction (the protons of the methyl groups are forced to occupy the same region of space), and this product is therefore not likely to be formed in substantial quantities. We therefore predict the following product for this reaction.

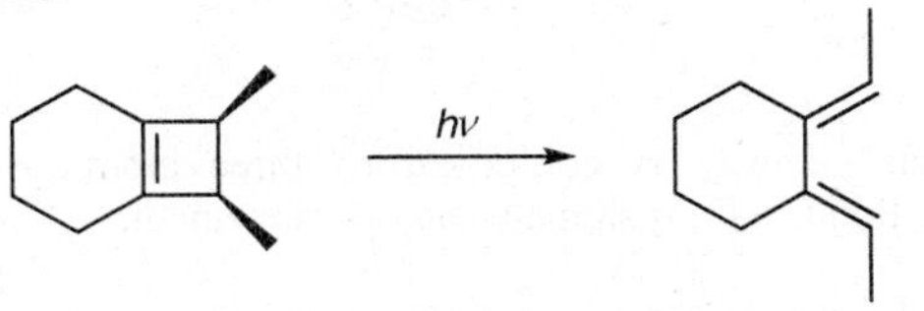

(c) This system has six π electrons, so an electrocyclic reaction is expected to occur under thermal conditions to give disrotatory ring closure. The resulting product is chiral, and both enantiomers are expected (one methyl group rotates clockwise and the other rotates counterclockwise, or vice versa, leading to both possible enantiomers).

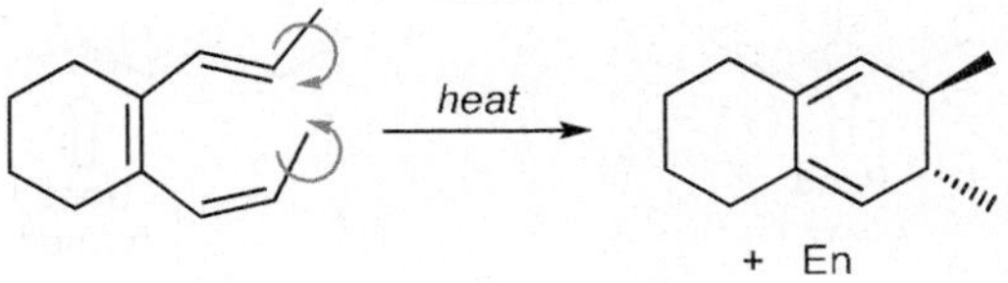

16.22. We begin by determining the number of π electrons that are involved in the reaction. Electrocyclic ring-opening and ring-closing reactions are equilibrium processes; in this case, a cyclobutene ring is in equilibrium with a diene. (In fact, ring opening is favored because it alleviates ring strain). From this equilibrium, we can see that 4 π electrons are involved in this electrocyclic reaction.

Since the reaction was done under thermal conditions (heat), the Woodward-Hoffmann rules predict a conrotatory ring opening. Thus, both ends of the π system will rotate clockwise, or both will rotate counterclockwise. If they both rotate clockwise, the initial *trans* stereochemistry of the ring will lead to double bonds that both have the *E* configuration.

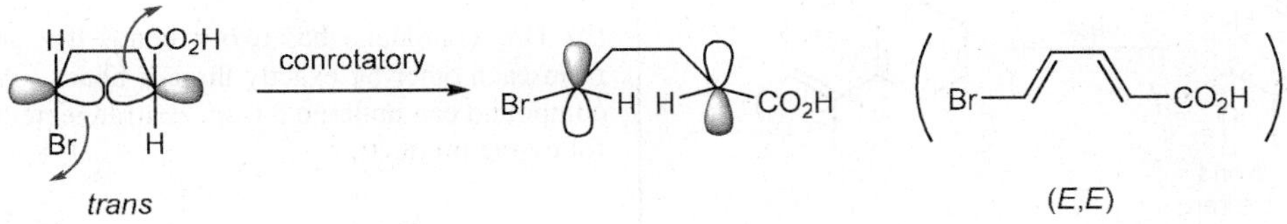

If they both rotate counterclockwise, the initial *trans* stereochemistry of the ring will lead to double bonds that both have the *Z* configuration.

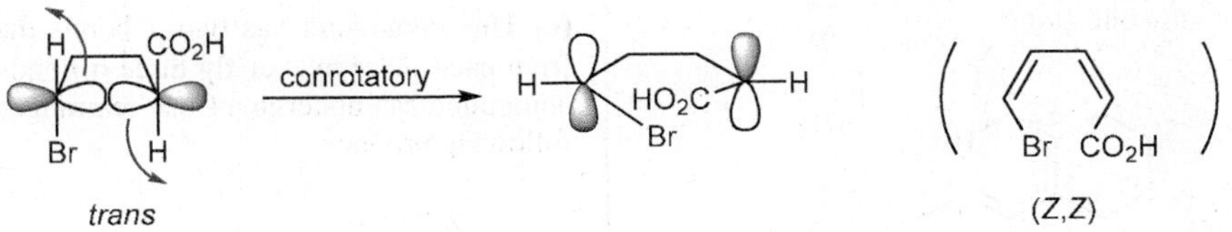

The Woodward-Hoffmann rules allow either the *E,E* isomer or the *Z,Z* isomer. It is not possible to get the *E,Z* or *Z,E* isomer under these conditions. Practically, only the *E,E* isomer is formed because the steric strain between the Br and CO_2H groups in the *Z,Z* isomer prevent its formation.

16.23.

(a) We begin by identifying the σ bond that is broken and the σ bond that is formed, highlighted here:

In the transition state, the bond that is breaking and the bond that is forming are separated by two different pathways, each of which is comprised of three atoms:

Therefore, this reaction is a [3,3] sigmatropic rearrangement.

(b) We begin by identifying the σ bond that is broken and the σ bond that is formed, highlighted here:

In the transition state, the bond that is breaking and the bond that is forming are separated by two different pathways: one is comprised of five atoms and the other is comprised of only one atom:

Therefore, this reaction is a [1,5] sigmatropic rearrangement.

16.24.

(a) This transformation can be achieved via the following sigmatropic rearrangement:

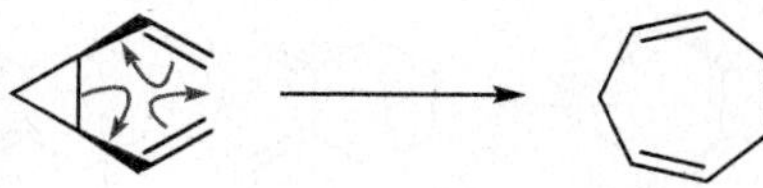

(b) We identify the σ bond that is broken and the σ bond that is formed, and then determine the pathways (highlighted below) that separate these bonds.

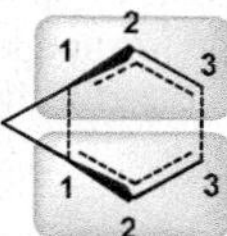

Each pathway is comprised of three atoms, so the reaction is a [3,3] sigmatropic rearrangement.

(c) In the sigmatropic rearrangement shown in part (a), we see that the ring strain associated with the three-membered ring is alleviated. The reverse process would involve forming a high-energy, three-membered ring. The equilibrium disfavors the reverse process.

16.25.

(a) This compound is an allylic vinylic ether, and it can therefore undergo a Claisen rearrangement to give the following product:

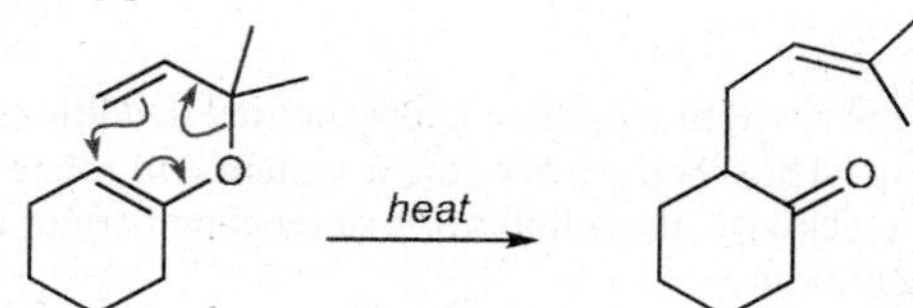

(b) This compound has two π bonds that are separated from each other by exactly three σ bonds. As such, this compound can undergo a Cope rearrangement to give the following product:

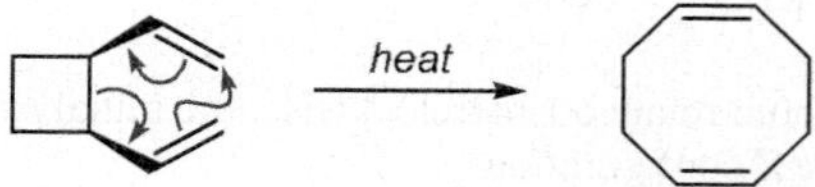

(c) This compound has two π bonds that are separated from each other by exactly three σ bonds. As such, this compound can undergo a Cope rearrangement to give the following product:

(d) This compound is an allylic phenyl ether. One of the π bonds in the phenyl group can serve an analogous role to the π bond in a vinylic group, so this molecule effectively behaves like an allylic vinylic ether. It can therefore undergo a Claisen rearrangement to give a ketone, which tautomerizes to give the enol, thereby re-establishing the aromatic ring:

16.26. This compound has two π bonds that are separated from each other by exactly three σ bonds. As such, this compound can undergo a Cope rearrangement. To draw the product of this reaction, it is helpful to redraw the starting material, as shown below, so that it is easier to see the ring of electrons responsible for the transformation:

16.27.

(a) This conjugated system (comprised of two π bonds) has five auxochromic alkyl groups, highlighted here:

and one exocyclic double bond:

The calculation for λ_{max} is shown here:

Base	=	217
Additional double bonds	=	0
Auxochromic alkyl groups	=	+25
Exocyclic double bond	=	+5
<u>Homoannular diene</u>	=	0
Total	=	247 nm

(b) This conjugated system (comprised of three π bonds) has five auxochromic alkyl groups, highlighted here:

and one exocyclic double bond:

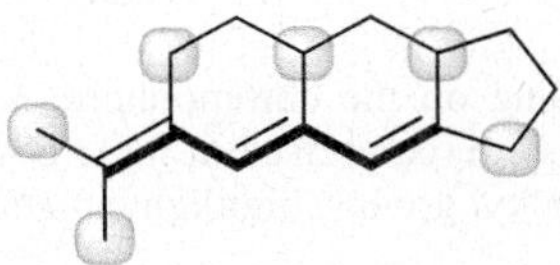

The calculation for λ_{max} is shown here:

Base	=	217
Additional double bonds	=	+30
Auxochromic alkyl groups	=	+25
Exocyclic double bond	=	+5
<u>Homoannular diene</u>	=	0
Total	=	277 nm

(c) This conjugated system (comprised of three π bonds) has six auxochromic alkyl groups, highlighted here:

and three exocyclic double bonds (each of the double bonds is exocyclic to a ring), shown here:

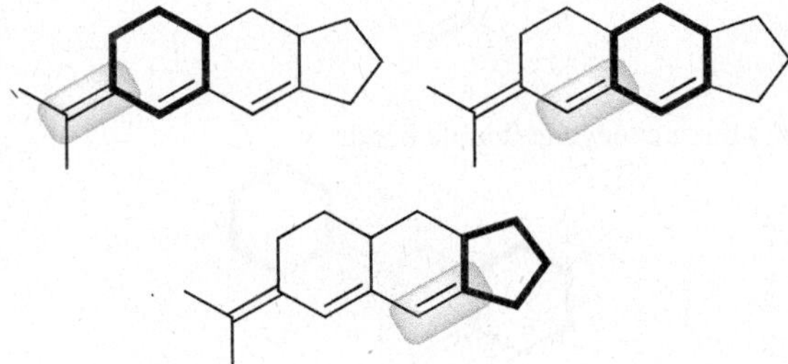

The calculation for λ_{max} is shown here:

Base	=	217
Additional double bonds	=	+30
Auxochromic alkyl groups	=	+30
Exocyclic double bonds	=	+15
<u>Homoannular diene</u>	=	0
Total	=	292 nm

(d) This conjugated system (comprised of three π bonds) has two of the double bonds in the same ring (homoannular):

In addition, there are seven auxochromic alkyl groups, highlighted here:

and one exocyclic double bond:

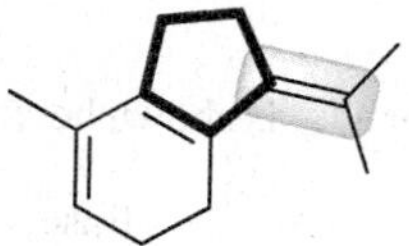

The calculation for λ_{max} is shown here:

Base	=	217
Additional double bonds	=	+30
Auxochromic alkyl groups	=	+35
Exocyclic double bonds	=	+5
Homoannular diene	=	+39
Total	=	326 nm

16.28. Focusing on the chromophore, we see that the conjugated system (comprised of three π bonds) has four auxochromic alkyl groups, highlighted here:

and three exocyclic double bonds:

We therefore predict that this compound will have a λ_{max} near 282 nm, as shown in the calculation below:

Base	=	217
Additional double bonds (1)	=	+ 30
Auxochromic alkyl groups (4)	=	+ 20
Exocyclic double bond (3)	=	+ 15
Homoannular dienes (0)	=	+ 0
Total	=	282 nm

Incidentally, the actual value is 265 nm. This demonstrates that our prediction was not exact, but it is nevertheless reasonably close to the actual value.

16.29.

(a) As seen in Figure 16.37 (the color wheel), the complementary color of orange is blue. Therefore, a compound that absorbs orange light will appear to be blue.

(b) As seen in Figure 16.37 (the color wheel), the complementary color of blue-green is red-orange. Therefore, a compound that absorbs blue-green light will appear to be red-orange.

(c) As seen in Figure 16.37 (the color wheel), the complementary color of orange-yellow is blue-violet. Therefore, a compound that absorbs orange-yellow light will appear to be blue-violet.

16.30.

(a) The parent ("cyclohex") indicates a six-membered ring, and the suffix ("diene") indicates the presence of two C=C bonds. The locants (1 and 4) indicate the positions of the two double bonds.

(b) The parent ("cyclohex") indicates a six-membered ring, and the suffix ("diene") indicates the presence of two C=C bonds. The locants (1 and 3) indicate the positions of the two double bonds.

(c) The parent ("pent") indicates a five-carbon chain, and the suffix ("diene") indicates the presence of two C=C bonds. The locants (1 and 3) indicate the positions of the two double bonds, and the stereodescriptor (*Z*) indicates the configuration of the C=C bond between C3 and C4.

(d) The parent ("hept") indicates a seven-carbon chain, and the suffix ("diene") indicates the presence of two C=C bonds. The locants (2 and 4) indicate the positions of the two double bonds, and the stereodescriptors (2*Z*, 4*E*) indicate the configurations of the double bonds.

(e) The parent ("but") indicates a four-carbon chain, and the suffix ("diene") indicates the presence of two C=C bonds. The locants (1 and 3) indicate the positions of the two double bonds. There are two methyl substituents, at positions C2 and C3.

16.31. Each of the highlighted compounds possesses a conjugated π system, which is shown with darker bonds.

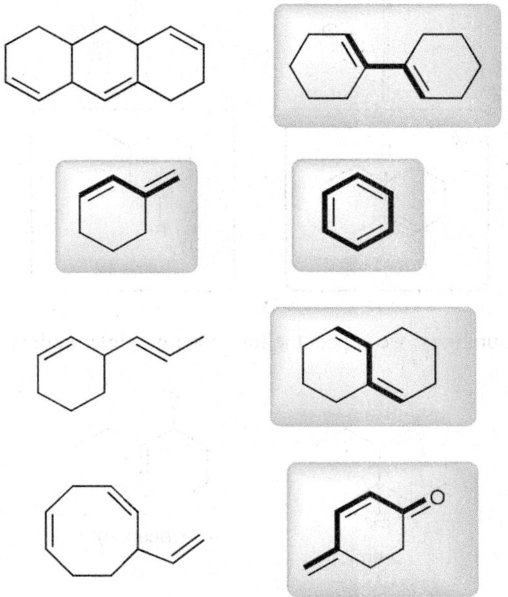

16.32. Potassium tert-butoxide is a strong, sterically hindered base, and the starting material will react with two equivalents of this base to undergo two successive elimination (E2) reactions, producing a conjugated, homoannular diene, as shown.

Br Br *t*-BuOK

16.33.

(a) The non-conjugated isomer (shown below) is less stable and higher in energy than the conjugated isomer.

As a result, this compound will liberate more heat upon hydrogenation.

(b) The non-conjugated isomer (shown below) is less stable and higher in energy than the conjugated isomer.

As a result, this compound will liberate more heat upon hydrogenation.

16.34. Treatment of 1,3-cyclohexadiene with one equivalent of HBr produces only one product (because both 1,2 addition and 1,4 addition yield the same product).

HBr (1 equiv.) Br (racemic)

16.35. The diene is protonated to give a resonance stabilized cation, which can then be attacked in one of two locations, leading to the 1,2-adduct and the 1,4-adduct. At low temperature, the kinetic product (the 1,2-adduct) dominates.

H—Br :Br: Br (racemic) MAJOR + Br

16.36. The diene is protonated to give a resonance stabilized cation, which can then be attacked in one of two locations, leading to the 1,2-adduct and the 1,4-adduct. At elevated temperature, the thermodynamic product (the 1,4-adduct) dominates.

H—Br :Br: Br (racemic) + Br MAJOR

16.37. We first identify the locations where protonation can occur. There are four unique positions where protonation can occur, although a resonance-stabilized intermediate can only be obtained upon protonation of one of the ends of the conjugated π system, highlighted below:

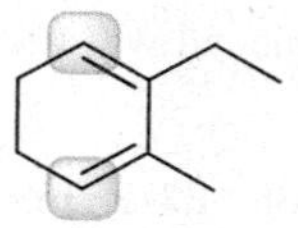

We must explore protonation at each of these positions. Let's begin with protonation of the bottom position, which leads to a resonance–stabilized intermediate. When we draw both resonance structures, we can see that two positions are electrophilic. Nucleophilic attack can occur at either of these locations, giving rise to two products:

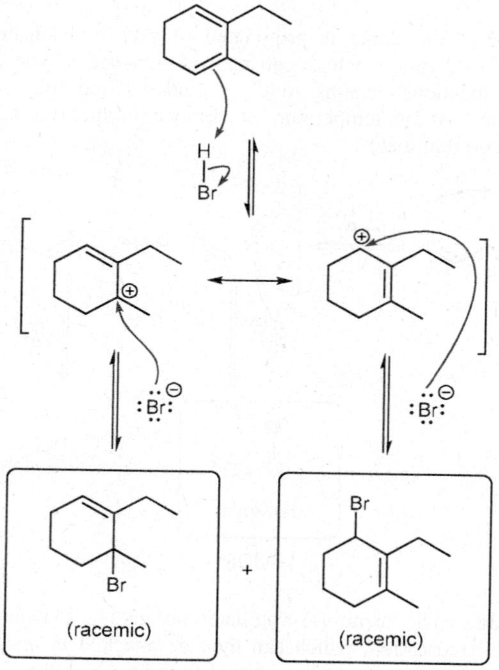

H–Br

Br

(racemic) + Br (racemic)

Notice that each of the products possesses a chiral center and is therefore produced as a racemic mixture (because the bromide ion can attack either face of the allylic carbocation with equal likelihood).

Now let's consider protonation at the position on top. Once again, protonation leads to a resonance–stabilized intermediate. When we draw both resonance structures, we can see that two positions are electrophilic. Nucleophilic attack can occur at either of these locations, giving two possible products, as shown:

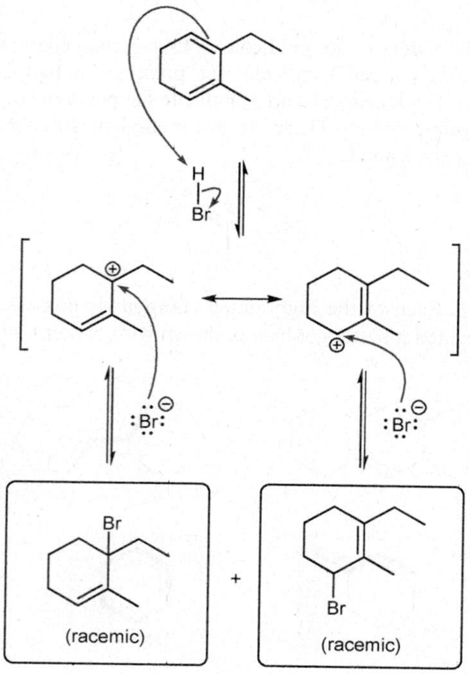

H–Br

Br (racemic) + Br (racemic)

In summary, we expect the following possible products:

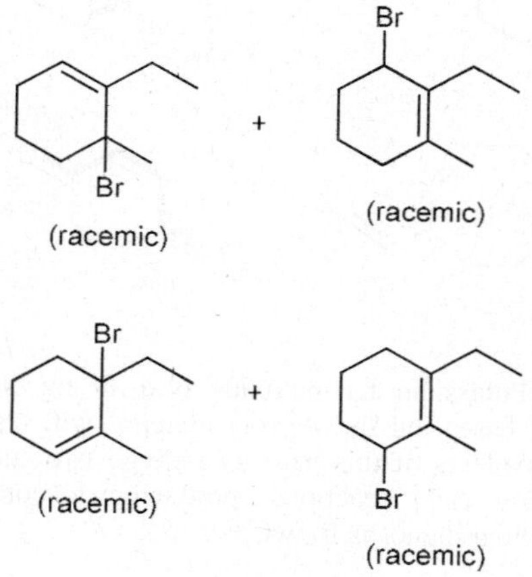

Br (racemic) + Br (racemic)

Br (racemic) + Br (racemic)

16.38. An increase in temperature allowed the system to reach equilibrium concentrations, which are determined by the relative stability of each product. Under these conditions, the 1,4-adducts predominate (the two tetrasubstituted alkenes shown in the solution to Problem 16.37, shown here again):

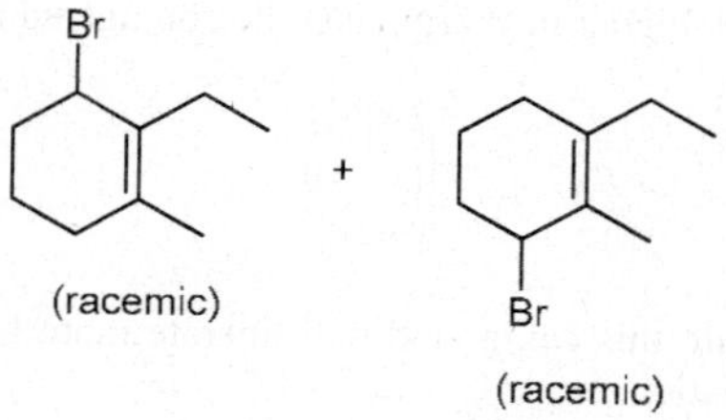

Br (racemic) + Br (racemic)

Once at equilibrium, lowering the temperature will not cause a decrease in the concentration of the 1,4-adducts.

16.39.

(a) The *tert*-butyl groups provide significant steric interactions that prevent the compound from adopting an *s-cis* conformation.

(b) This diene is not conjugated.

(c) The methyl groups provide a significant steric interaction in the *s-cis* conformation that prevents the compound from adopting this conformation.

(d) This diene cannot adopt an *s-cis* conformation.

16.40. The most reactive dienophile is the one connected to two electron-withdrawing C=O groups. Then, the dienophile with only one C=O group is the next most reactive. And finally, the least reactive dienophile is the one that lacks an electron-withdrawing substituent altogether.

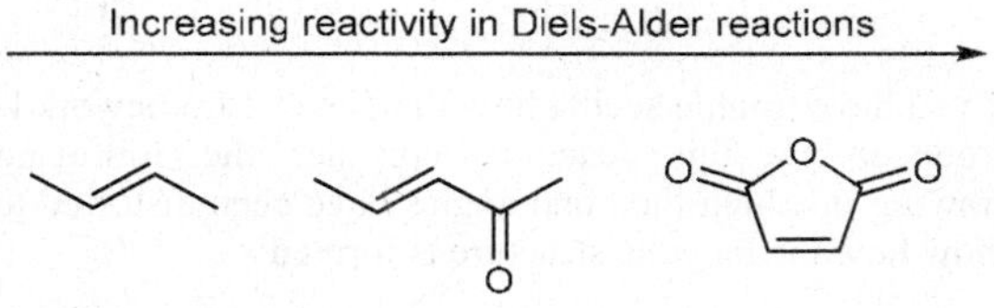

Increasing reactivity in Diels-Alder reactions

16.41. The π bonds in 1,2-butadiene are not conjugated, and λ_{max} is therefore lower than 217 nm. In fact, it is below 200 nm, which is beyond the range used by most UV-Vis spectrometers.

16.42.

(a) In a Diels-Alder reaction, the configuration of the dienophile (in this case, *trans*) is preserved in the product. The product is chiral, so it is formed as a pair of enantiomers:

HOOC COOH

COOH H H COOH + En

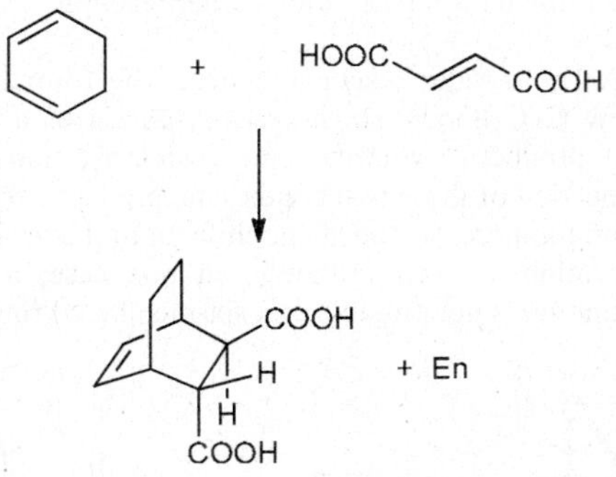

(b) In a Diels-Alder reaction, the configuration of the dienophile (in this case, *cis*) is preserved in the product. The product is chiral, so it is formed as a pair of enantiomers. Notice that both substituents occupy *endo* (rather than *exo*) positions:

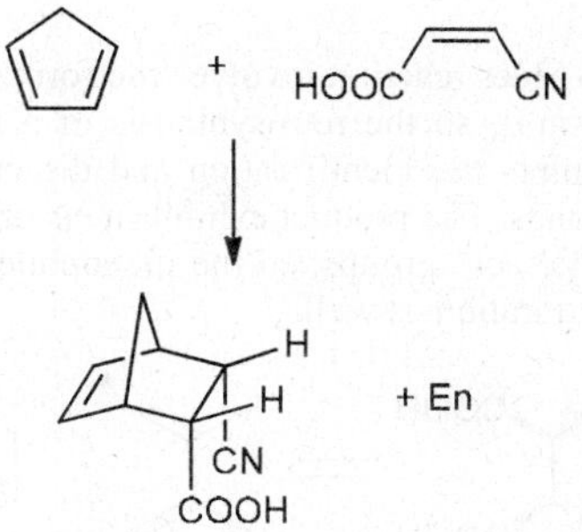

HOOC CN

H H CN COOH + En

(c) In a Diels-Alder reaction, the configuration of the dienophile (in this case, *trans*) is preserved in the product. The product is chiral, so it is formed as a pair of enantiomers:

+ En

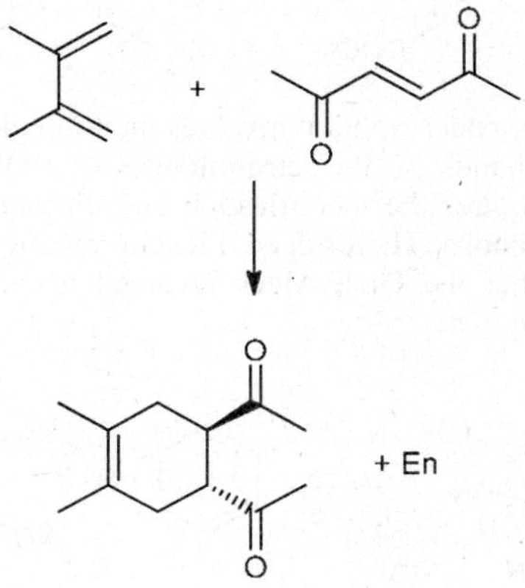

(d) The dienophile is an alkyne, and the product does not have any chiral centers (no wedges or dashes).

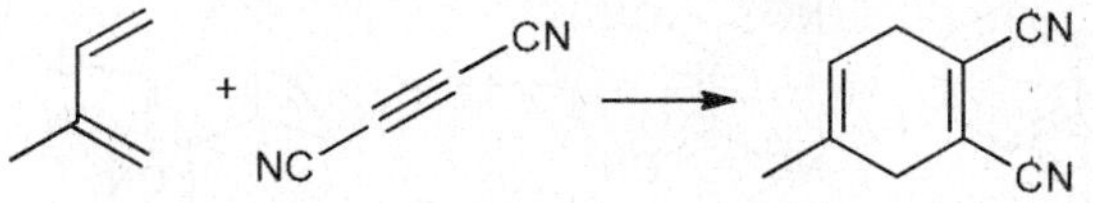

CN NC CN CN

(e) The starting material is a cyclic diene, but the *endo* rule is not relevant in this case, because an alkyne is used as the dienophile, so there are no *endo* positions. The product is a *meso* compound, and therefore has no enantiomer.

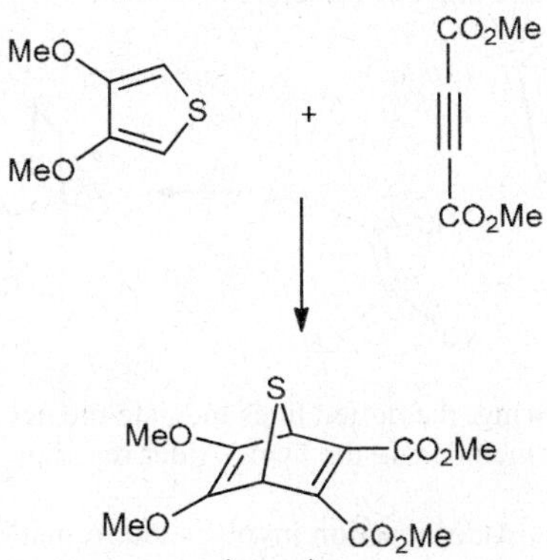

MeO MeO S + CO_2Me CO_2Me

S MeO MeO CO_2Me CO_2Me (*meso*)

16.43.

(a) A Diels-Alder reaction involves the formation of two new C–C bonds, so the retrosynthesis of a Diels-Alder product requires the identification and disconnection of these two bonds. The product exhibits a *cis* orientation of the carboxylic acid groups, so the dienophile must have the *cis* configuration as well.

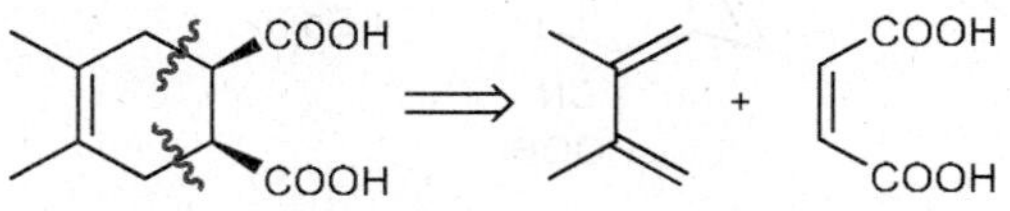

The forward process is shown here:

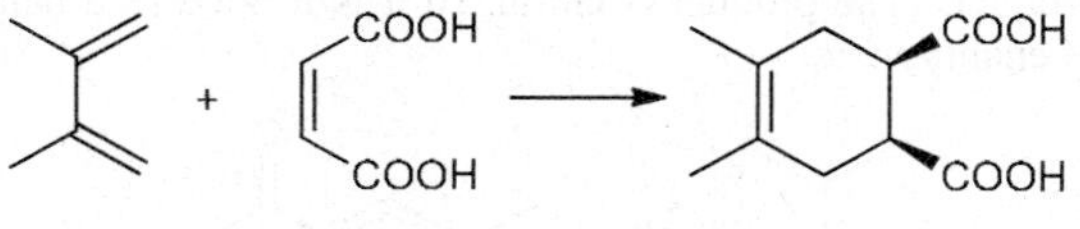

(b) A Diels-Alder reaction involves the formation of two new C–C bonds, so the retrosynthesis of a Diels-Alder product requires the identification and disconnection of these two bonds. The bridged, bicyclic product gives us the clue that the Diels-Alder involved a cyclic diene starting material.

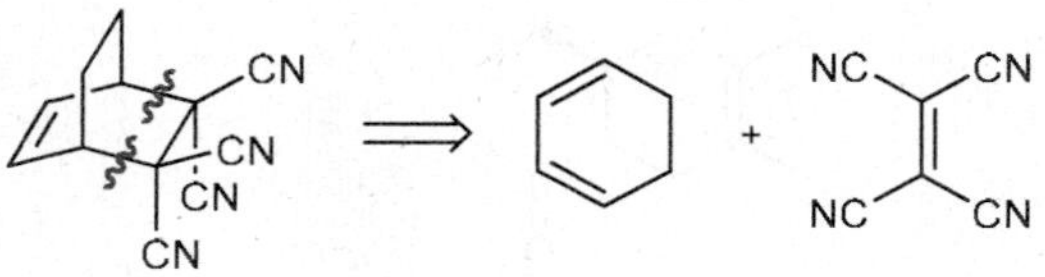

The forward process is shown here:

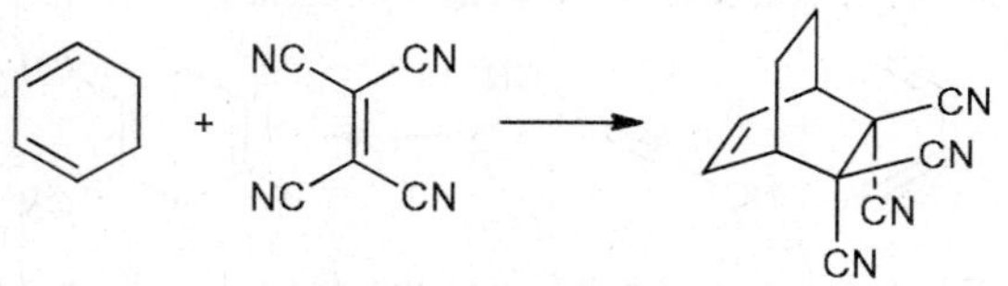

If you have trouble seeing how the bicyclic framework is produced by this reaction, consider the following drawing in which the bond angles have been distorted to show how the bicyclic structure is formed:

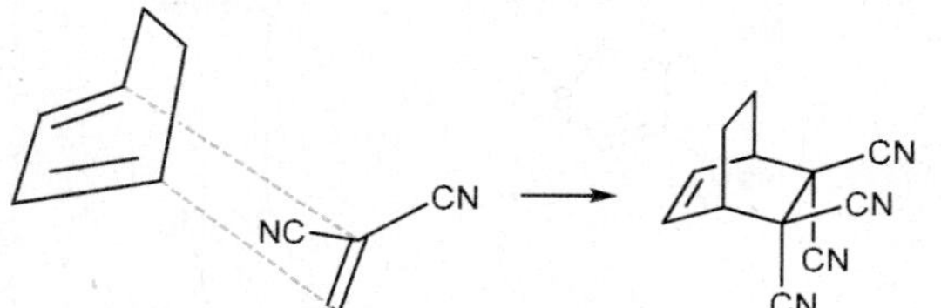

In this drawing, the dotted lines indicate the new σ bonds that are formed during the Diels-Alder reaction.

(c) A Diels-Alder reaction involves the formation of two new C–C bonds, so the retrosynthesis of a Diels-Alder product requires the identification and disconnection of these two bonds. The bridged, bicyclic product gives us the clue that the Diels-Alder involved a cyclic diene starting material.

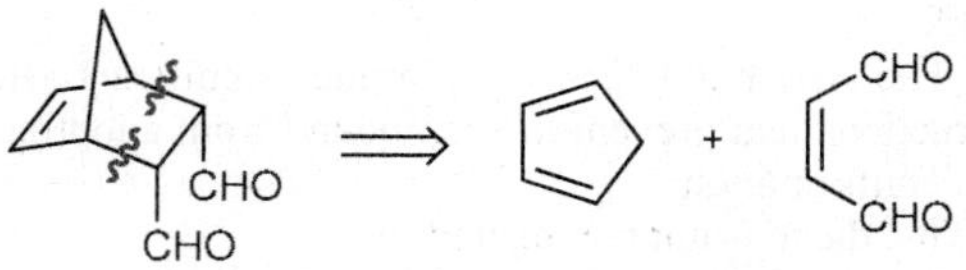

Notice that the starting dienophile must have the *cis* configuration, because the desired product exhibits a *cis* orientation of the aldehyde (CHO) groups.
The forward process is shown here (notice that the *endo* rule ensures formation of the desired product):

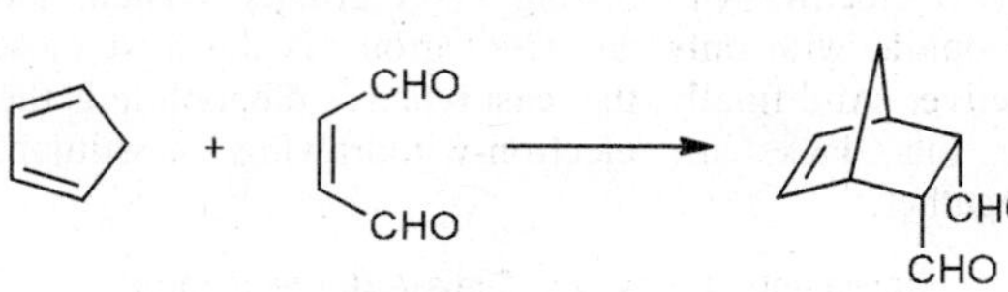

If you have trouble seeing how the bicyclic framework is produced by this reaction, consider the following drawing in which the bond angles have been distorted to show how the bicyclic structure is formed:

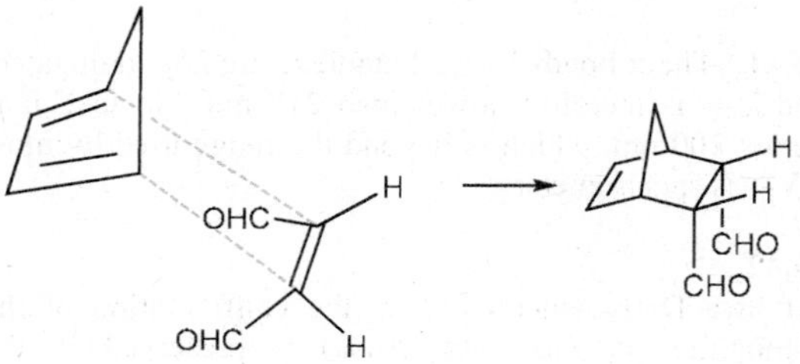

In this drawing, the dotted lines indicate the new σ bonds that are formed during the Diels-Alder reaction.

(d) A Diels-Alder reaction involves the formation of two new C–C bonds, so the retrosynthesis of a Diels-Alder product requires the identification and disconnection of these two bonds. The product exhibits a *cis* configuration, so the dienophile must have the *cis* configuration as well (although in this case, a *trans* configuration is not possible in a six-membered ring).

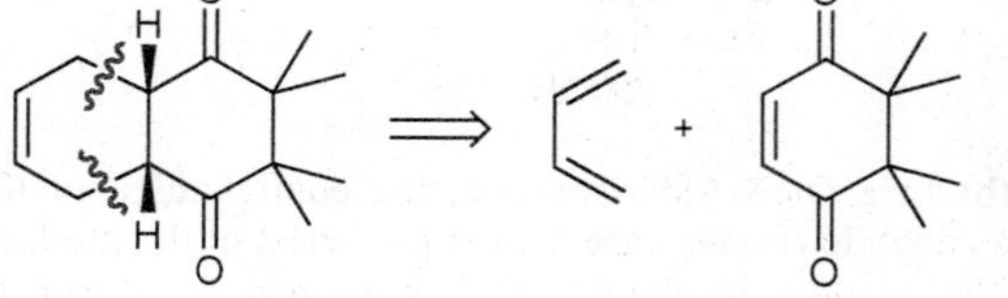

The forward process is shown here:

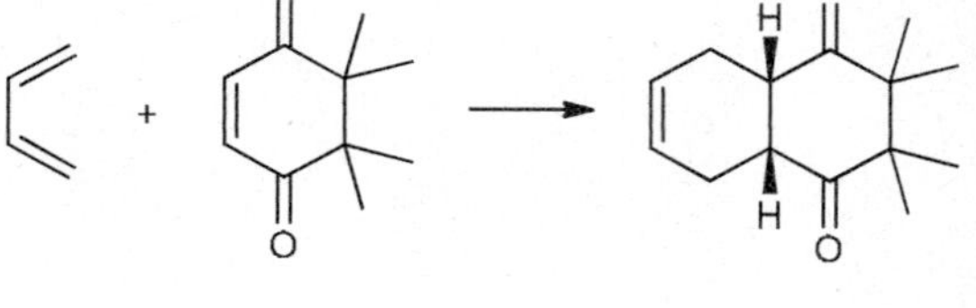

(e) A Diels-Alder reaction involves the formation of two new C–C bonds, so the retrosynthesis of a Diels-Alder product requires the identification and disconnection of these two bonds. The product exhibits a *cis* configuration, so the dienophile must have the *cis* configuration as well (although in this case, a *trans* configuration is not possible in a six-membered ring).

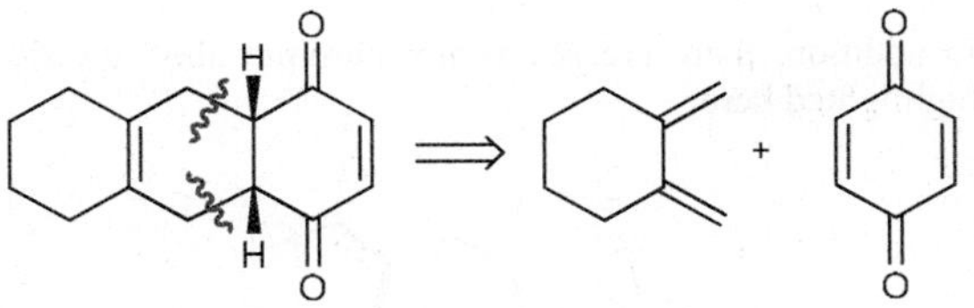

The forward process is shown here:

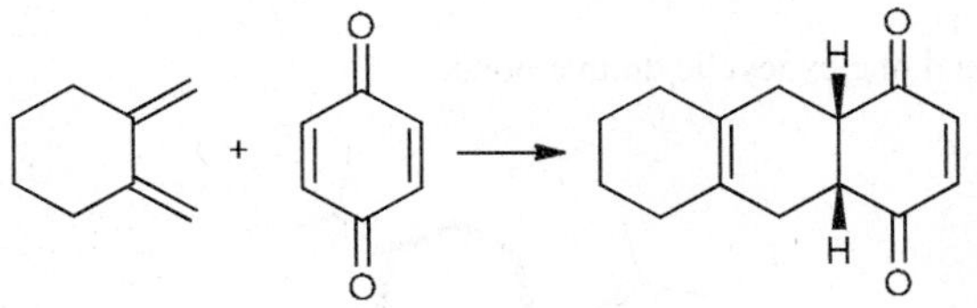

(f) A Diels-Alder reaction involves the formation of two new C–C bonds, so the retrosynthesis of a Diels-Alder product requires the identification and disconnection of these two bonds. The product exhibits a *trans* orientation of the ester groups, so the dienophile must have the *trans* configuration as well. Note that the retrosynthesis of either enantiomer leads to the same starting materials.

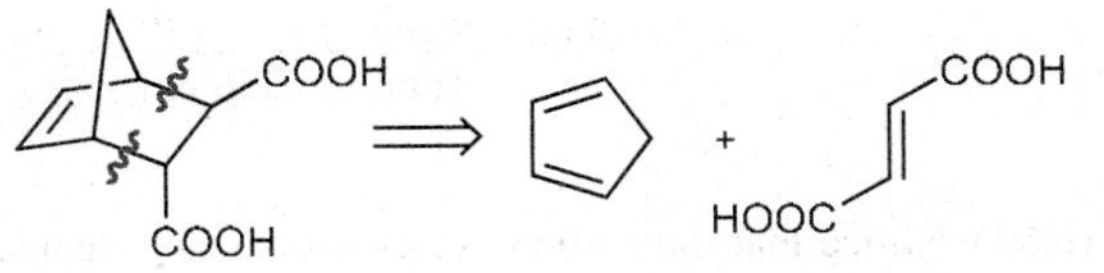

COOH COOH + HOOC COOH

The forward process is shown here:

COOH + HOOC COOH H H COOH + En

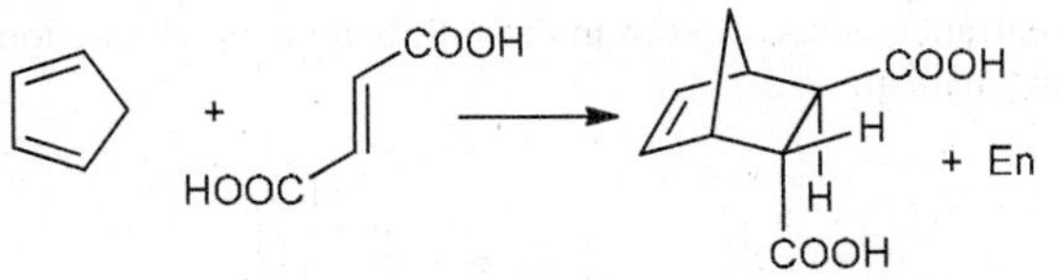

If you have trouble seeing why a pair of enantiomers is produced, see the solution to Problem **16.17b**, which is very similar to this problem (just replace the cyano groups with carboxylic acid groups).

(g) A Diels-Alder reaction involves the formation of two new C–C bonds, so the retrosynthesis of a Diels-Alder product requires the identification and disconnection of these two bonds. The product has a cyclohexadiene ring, so the starting dienophile must be an alkyne:

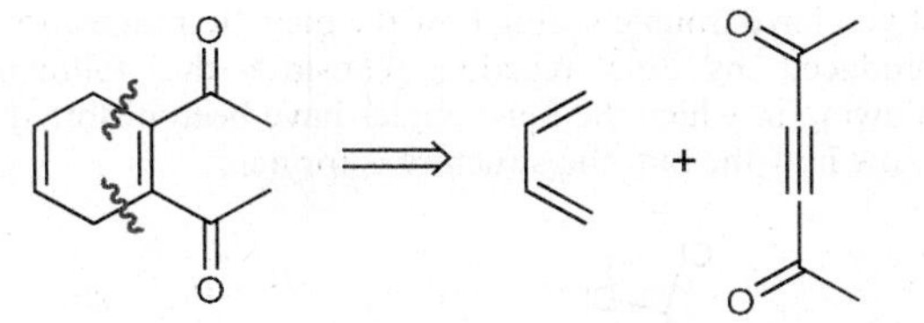

The forward process is shown here:

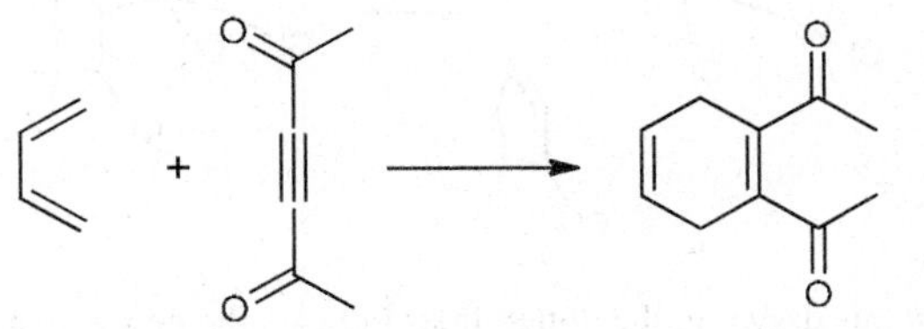

16.44. A product with the molecular formula $C_{14}H_{12}O_6$ can be formed via two successive Diels-Alder reactions, as shown here:

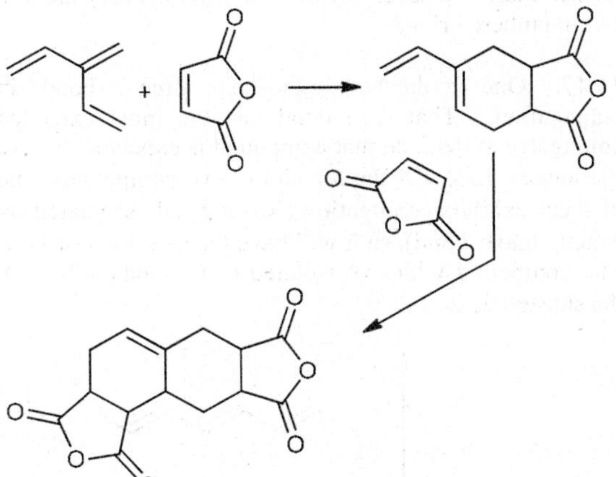

($C_{14}H_{12}O_6$)

16.45. The following diene and dienophile would be necessary to produce the desired compound via a Diels-Alder reaction:

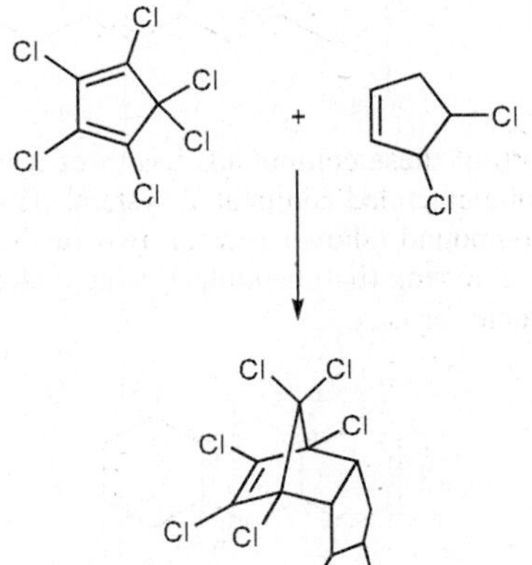

Chlordane

If you have trouble seeing how the bicyclic framework is produced by this reaction, consider the following drawing in which the bond angles have been distorted to show how the bicyclic structure is formed:

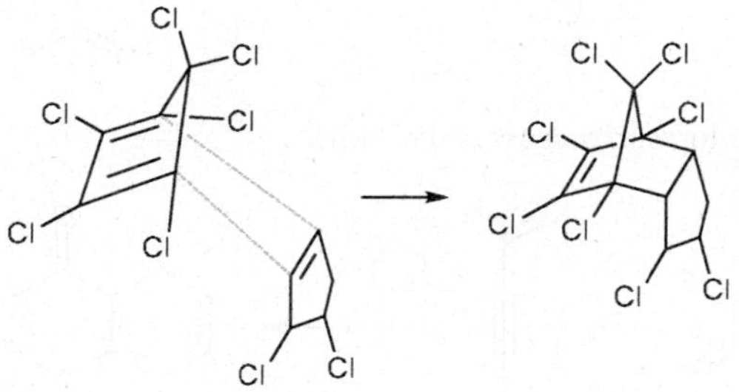

In this drawing, the dotted lines indicate the new σ bonds that are formed during the Diels-Alder reaction.

16.46. The two ends of the conjugated system are much farther apart in a seven-membered ring than they are in a five-membered ring.

16.47. One of the compounds has three π bonds in conjugation. That compound has the most extended conjugated system, so that compound is expected to have the longest λ_{max}. Of the remaining two compounds, one of them exhibits conjugation (two π bonds separated by exactly one σ bond), so it will have the next longest λ_{max}. The compound with two isolated C=C bonds will have the shortest λ_{max}:

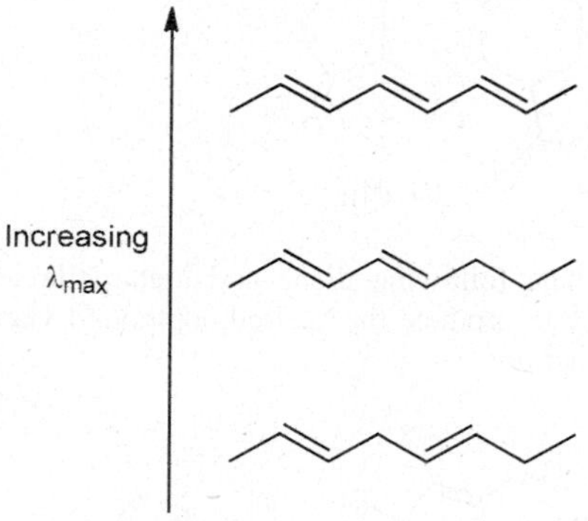

Increasing λ_{max}

16.48. Each of these compounds has three π bonds that comprise one extended conjugated system. However, in the first compound (shown below), two of the π bonds are in the same ring (homoannular), which adds +39 nm to the estimate for λ_{max}.

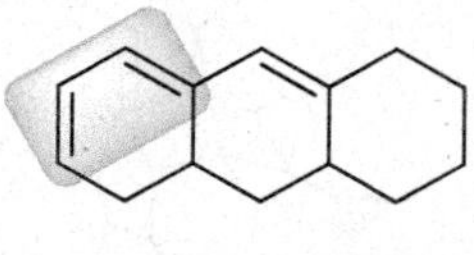

We therefore expect this compound to have the longer λ_{max}.

16.49. This conjugated system (comprised of four π bonds) has two of the double bonds in the same ring (homoannular):

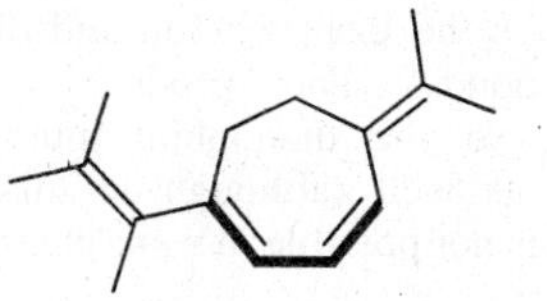

In addition, there are seven auxochromic alkyl groups, highlighted here:

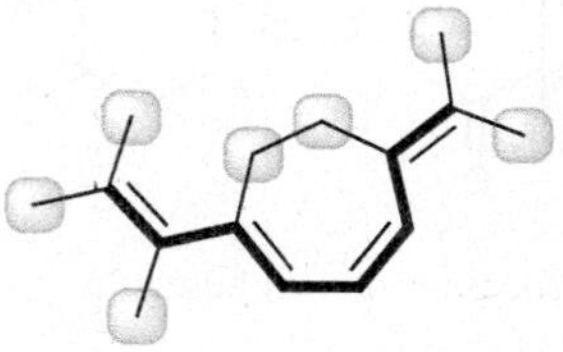

and one exocyclic double bond:

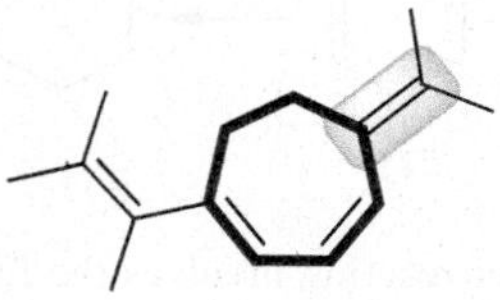

The calculation for λ_{max} is shown here:

Base	=	217
Additional double bonds	=	+60
Auxochromic alkyl groups	=	+35
Exocyclic double bonds	=	+5
Homoannular diene	=	+39
Total	=	356 nm

16.50. Notice that the carbon skeleton does not change during these reactions. It is the location of the deuteron, as well as the location of the π bonds, that changes. This is indeed characteristic of [1,5] sigmatropic rearrangements, as seen in the following general reaction mechanism:

Notice that the position of the proton changes, as well as the position of the π bonds. And we can certainly envision this process occurring with a deuteron, in place of the proton:

Much like the previous example, the position of the deuteron changes, as well as the position of the π bonds. This is exactly the type of transformation taking place in the reactions shown in the problem statement. It is therefore reasonable to explain each of these transformations with a [1,5] sigmatropic rearrangement, as shown:

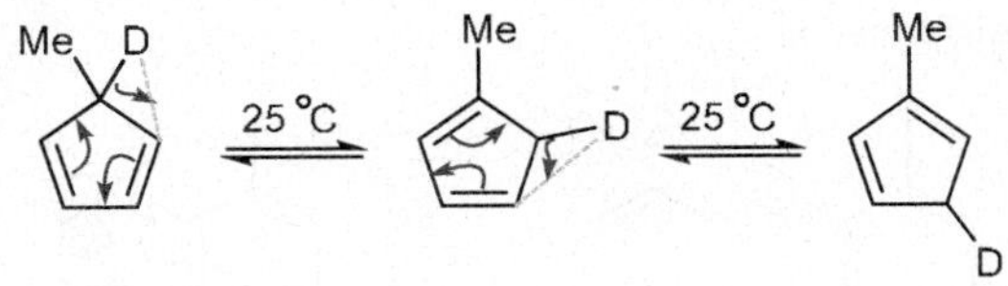

Me D 25 °C Me D 25 °C Me D

16.51. Notice that the carbon skeleton does not change during these reactions. It is the location of a deuteron, as well as the location of the π bonds, that changes. As described in the solution to the previous problem, these changes are characteristic of a [1,5] sigmatropic rearrangement:

D D heat D D

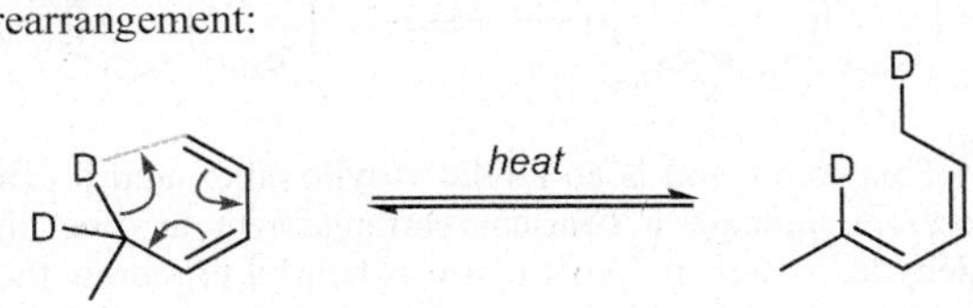

16.52.

(a) This system has six π electrons, so an electrocyclic reaction is expected to occur under thermal conditions to give disrotatory ring closure. The resulting product is a *meso* compound (has a plane of symmetry) and therefore has no enantiomer:

heat

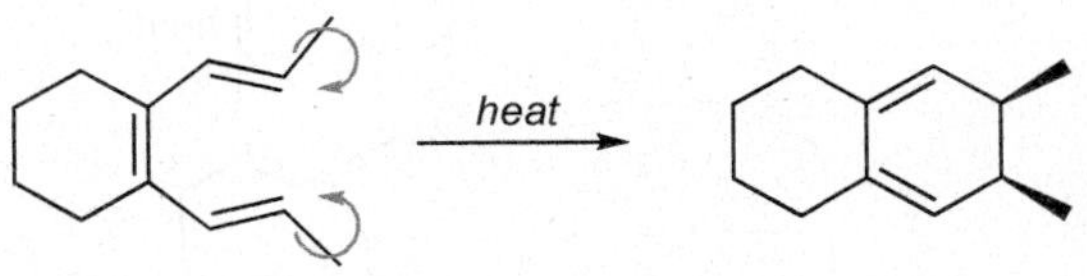

(*meso*)

(b) This system has six π electrons, so an electrocyclic reaction is expected to occur under photochemical conditions to give conrotatory ring closure. The resulting product is chiral, and both enantiomers are expected (the methyl groups can both rotate in a clockwise fashion, or they can both rotate in a counterclockwise fashion, giving both possible enantiomers).

hν

+ En

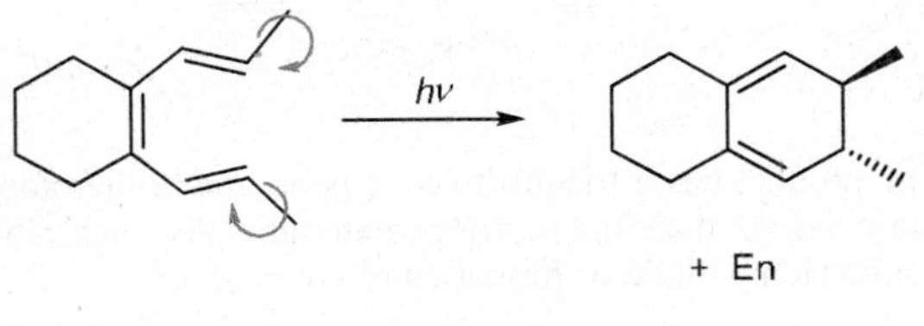

(c) This system has eight π electrons, so an electrocyclic reaction is expected to occur under thermal conditions to give conrotatory ring closure. The resulting product is chiral, and both enantiomers are expected (the methyl groups can both rotate in a clockwise fashion, or they can both rotate in a counterclockwise fashion, giving both possible enantiomers).

heat

+ En

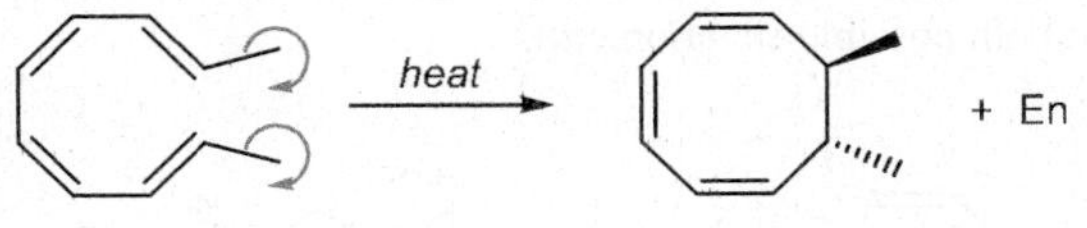

(d) This system has eight π electrons, so an electrocyclic reaction is expected to occur under photochemical conditions to give disrotatory ring closure. The resulting product is a *meso* compound (it has a plane of symmetry) and therefore has no enantiomer:

hν

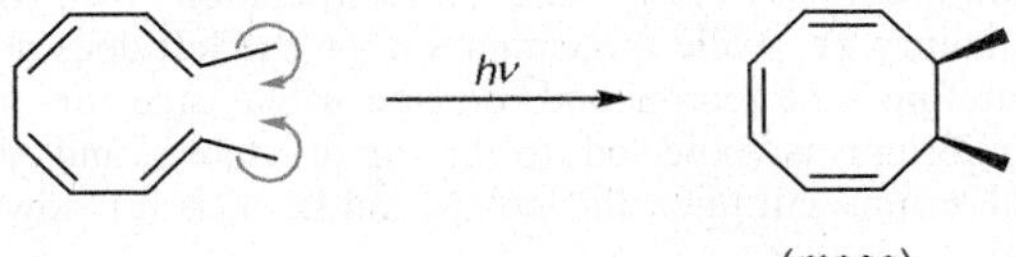

(*meso*)

16.53.

(a) This system has six π electrons, so an electrocyclic reaction is expected to occur under photochemical conditions to give conrotatory ring closure. The resulting product is a *meso* compound (has a plane of symmetry) and therefore has no enantiomer:

hν

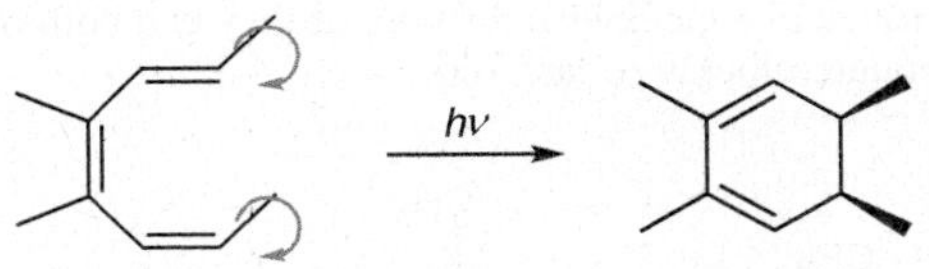

(*meso*)

This product is obtained whether both methyl groups rotate in a clockwise fashion, or whether both methyl groups rotate in a counterclockwise fashion:

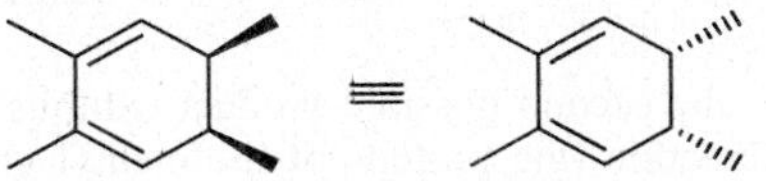

≡

(b) This system has six π electrons, so an electrocyclic reaction is expected to occur under thermal conditions to give disrotatory ring closure. The resulting product is chiral, and both enantiomers are expected (one methyl group rotates clockwise and the other rotates counterclockwise, or vice versa, leading to both possible enantiomers):

heat

+ En

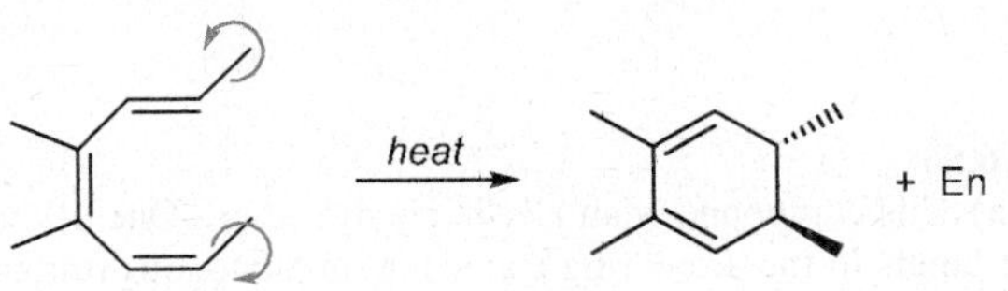

(c) This system has six π electrons, so an electrocyclic reaction is expected to occur under photochemical conditions to give conrotatory ring closure. The resulting product is chiral, and both enantiomers are expected (either both methyl groups can rotate clockwise, or both can rotate counterclockwise, leading to both possible enantiomers):

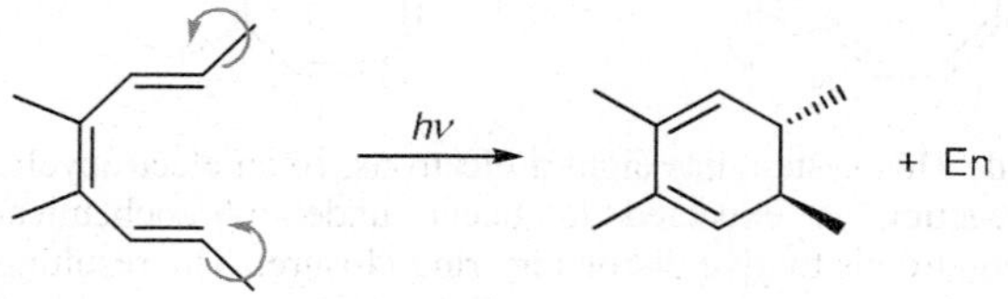

hv

+ En

16.54. The compound on the right side of the equilibrium has a π bond in conjugation with the aromatic ring, while the compound on the left does not. Therefore, the compound on the right side of the equilibrium is expected to be more stable, and the equilibrium will favor this compound because it is lower in energy.

16.55. In this reaction, the four-membered ring is opening (this is the reverse of an electrocyclic ring closure). If we look at the product, we see that four π electrons are involved in the process. Under thermal conditions, this electrocyclic process is expected be conrotatory. In theory, two products can be produced from conrotatory ring-opening (the methyl groups can both rotate in a clockwise fashion, or they can both rotate in a counterclockwise fashion):

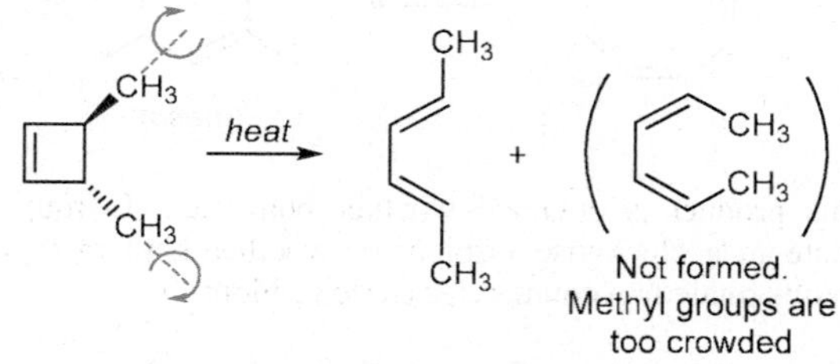

CH_3 CH_3 *heat* CH_3 CH_3 + CH_3 CH_3

Not formed.
Methyl groups are too crowded

However, the second possible product exhibits a severe steric interaction (the protons of the methyl groups are forced to occupy the same region of space), and this product is therefore not likely formed in substantial quantities. We therefore predict the following product for this reaction.

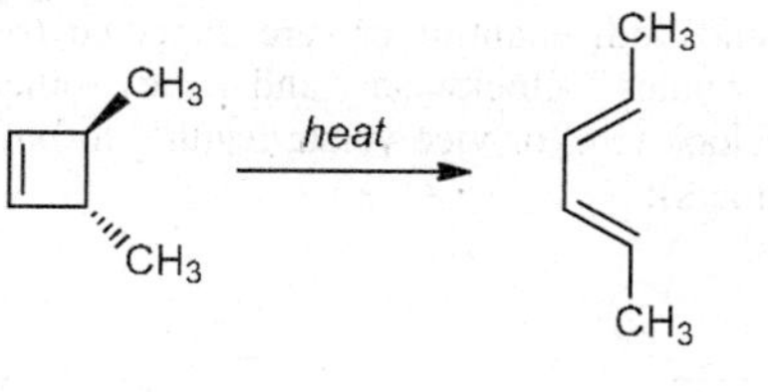

CH_3 CH_3 *heat* CH_3 CH_3

16.56.
(a) This compound is an allylic phenyl ether. One of the π bonds in the arene ring can serve an analogous role to the π bond in a vinylic group, so this molecule effectively behaves like an allylic vinylic ether. It can therefore undergo a Claisen rearrangement to give a ketone, which tautomerizes to give the enol, thereby re-establishing aromaticity. Note that it is helpful to redraw the starting material, as shown below, so that it is easier to see the ring of electrons responsible for the transformation:

O ≡ O *heat* OH *taut.* O

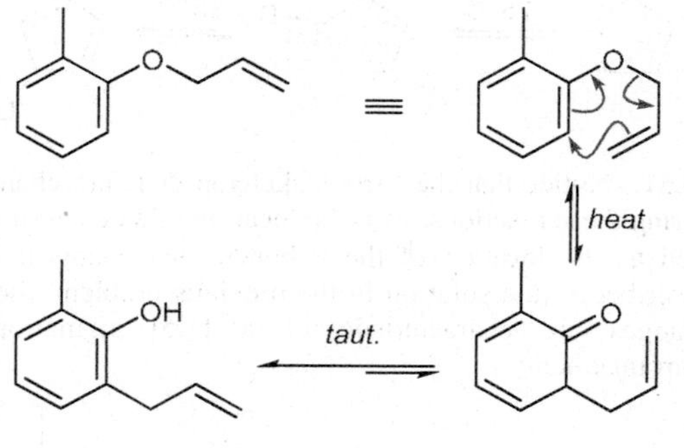

(b) This compound is an allylic vinylic ether, and it can therefore undergo a Claisen rearrangement to give an aldehyde, as shown. Note that it is helpful to redraw the starting material, as shown below, so that it is easier to see the ring of electrons responsible for the transformation:

O ≡ O *heat* O

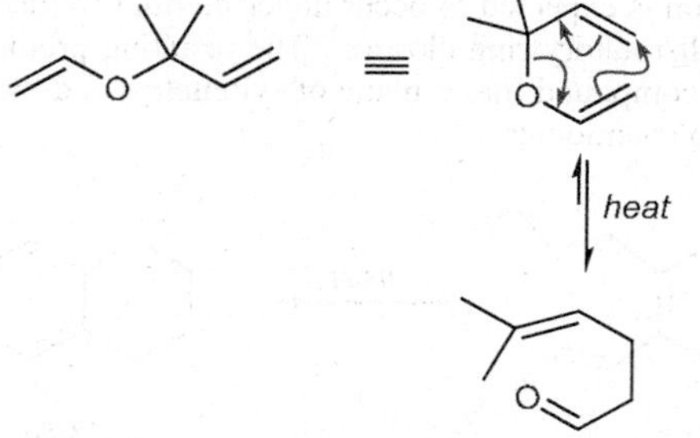

(c) This compound has two π bonds that are separated from each other by exactly three σ bonds. As such, this compound can undergo a Cope rearrangement. To draw the product of this reaction, it is helpful to redraw the starting material, as shown below, so that it is easier to see the ring of electrons responsible for the transformation:

≡ *heat*

The product has a trisubstituted π bond and is therefore more stable than the starting material. As such, the equilibrium will favor formation of the product.

16.57. Benzoquinone has two C=C π bonds, each of which can function as a dienophile in a Diels-Alder reaction. Two successive Diels-Alder reactions will afford a tricyclic structure. The tricyclic products are diastereomers and are formed because the second Diels-Alder reaction need not occur on the same face as the first Diels-Alder reaction.

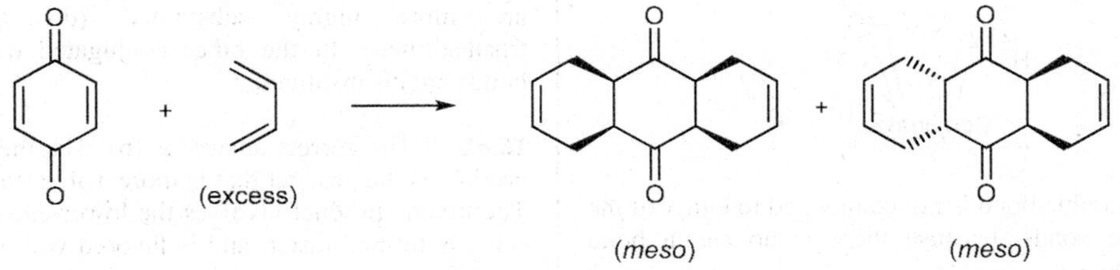

(excess)

(*meso*) (*meso*)

16.58. This transformation can occur via an intramolecular Diels-Alder reaction, in which a portion of the compound functions as the diene, while another portion of the compound functions as the dienophile. The result is a polycyclic compound (which is expected to be formed as a racemic mixture).

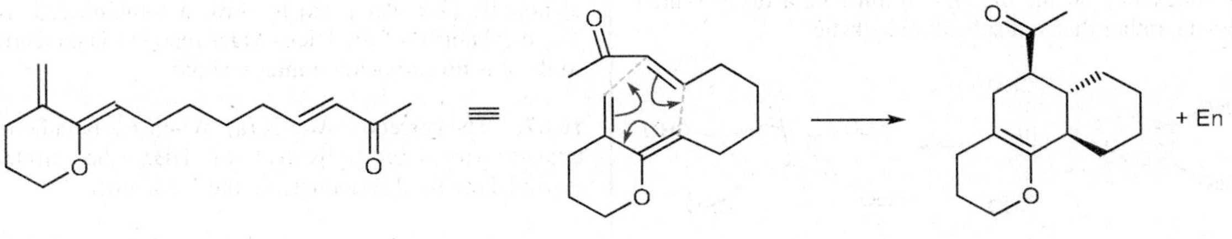

+ En

16.59. First count the number of conjugated double bonds. This compound has three. Two of them count toward the base value of 217 nm and the other will add +30.

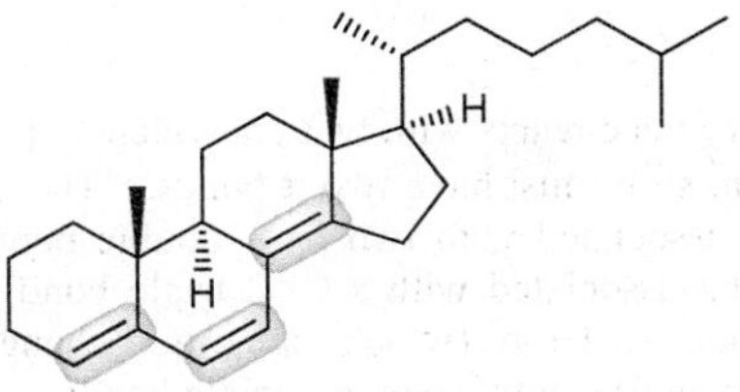

Next, look for any auxochromic alkyl groups. These are the carbon atoms connected directly to the chromophore. This compound has five auxochromic alkyl groups each of which adds +5, giving another +25.

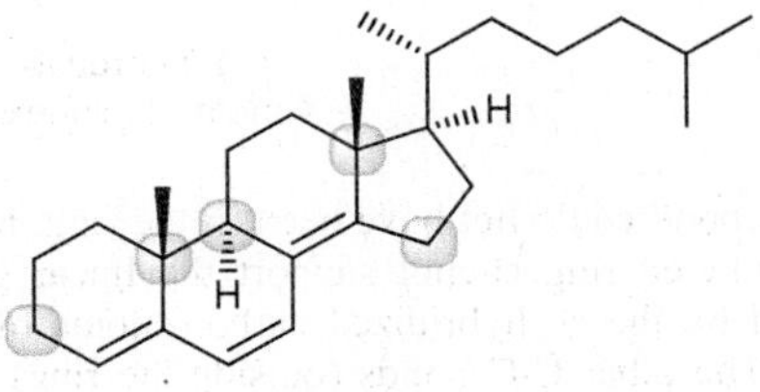

Next, look for exocyclic double bonds. In this case, two double bonds (highlighted below) are exocyclic. One of the double bonds is exocyclic to two rings giving a total of three exocyclic relationships that each add +5 for a total of +15.

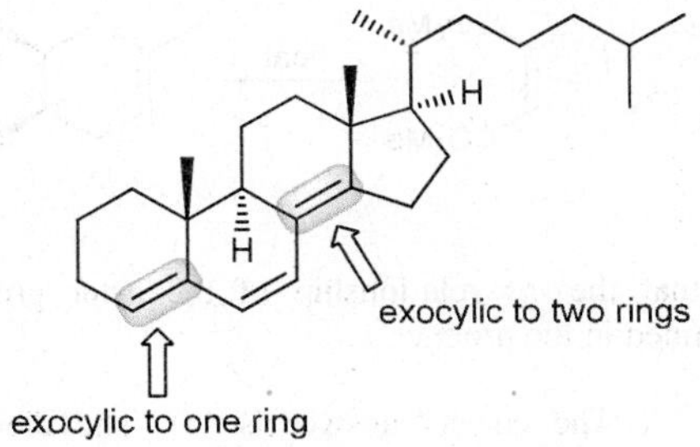

exocylic to two rings

exocylic to one ring

Finally, look for a homoannular diene. There appears to be no ring that has two double bonds contained wholly within it, so no adjustment is required for this compound. In summary, the following calculation predicts a λ_{max} of approximately 287 nm:

Base =	217
Additional double bond (1) =	+30
Auxochromic alkyl groups (5) =	+25
Exocyclic double bonds (3) =	+15
Homoannular diene (0) =	0
Total =	287 nm

This prediction is in fairly good agreement with the experimentally observed value of 283 nm.

16.60. The correct answer is (c). Recall that two double bonds are conjugated if they are separated from each other by exactly one sigma bond (no more and no less). The double bonds at the periphery are separated by a C–C bond that has one sigma bond and one π bond. Therefore, these two double bonds are conjugated:

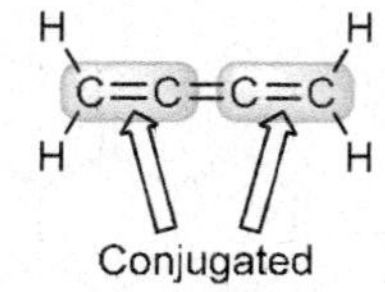

The central double bond is not conjugated to either of the other double bonds, because there is no sigma bond separating them.

16.61. The correct answer is (d). A Diels-Alder reaction would be ideal to prepare the target. The product is a substituted cyclohexadiene (rather than a substituted cyclohexene), so the dienophile must be a disubstituted alkyne, rather than a disubstituted alkene:

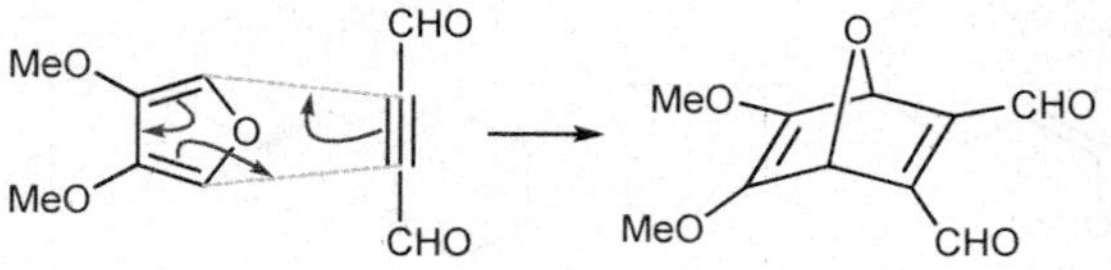

16.62. The correct answer is (b). This is a Diels-Alder reaction. We begin by redrawing the diene in its *s-cis* conformation, and aligning it with the dienophile, as shown here:

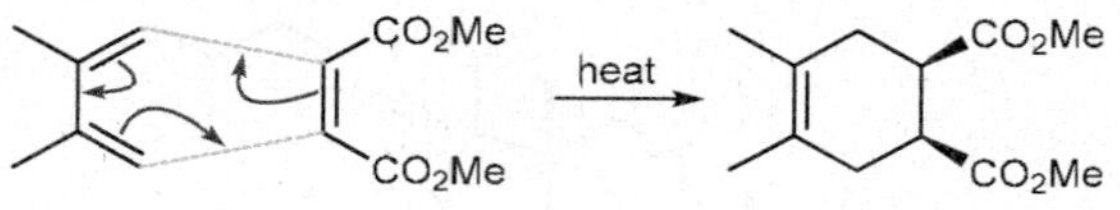

Note that the *cis* relationship of the ester groups is maintained in the product.

16.63. The correct answer is (d). The diene and dienophile combine to form a new cyclohexene ring. There are no new chiral centers, and there is no regiochemistry to consider when predicting the product of this Diels-Alder reaction.

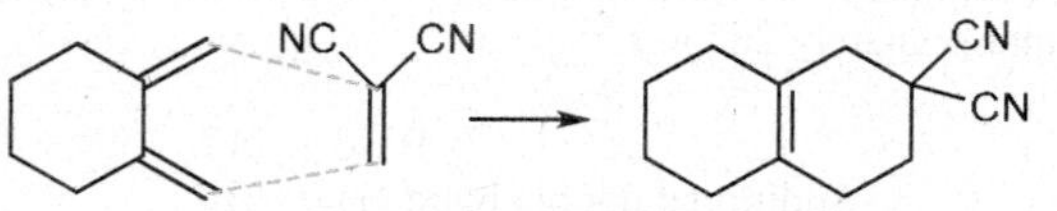

16.64. The correct answer is (c). Two of the given compounds are conjugated dienes:

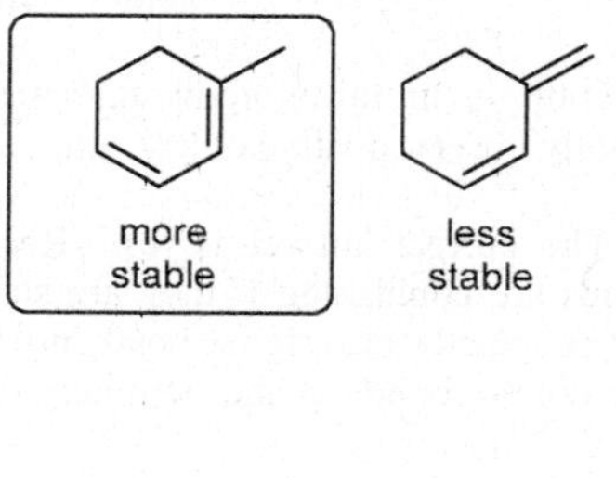

The other two compounds exhibit isolated π bonds and are therefore less stable than the two shown. The conjugated diene in the box is more stable than the other conjugated diene, because the π bonds in this compound are more highly substituted (disubstituted and trisubstituted). In the other conjugated diene, both π bonds are disubstituted.

16.65. The correct answer is (b). The thermodynamic product is the product that is more stable (lower energy). The kinetic product involves the lower-energy transition state, is formed faster, and is favored with cold reaction conditions.

16.66. The correct answer is (c). The Diels-Alder reaction is described as a [4+2] cycloaddition. A diene (4 π electrons) in the *s-cis* conformation reacts with a dienophile (2 π electrons) to form a 6-membered ring. The mechanism of the Diels-Alder reaction is concerted, so there is no carbocation intermediate.

16.67. The correct answer is (a). When 1,3-butadiene is treated with one equivalent of HBr, the products expected are the 1,2-adduct and the 1,4-adduct.

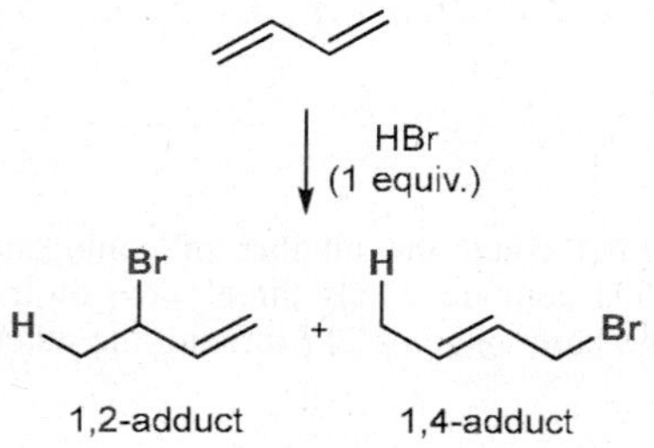

16.68.

(a) α-Terpinene reacts with two equivalents of molecular hydrogen, so it must have two π bonds. These π bonds must be associated with two C=C double bonds (rather than being associated with a C≡C triple bond), because the carbon skeleton (which does not change during hydrogenation) cannot support a triple bond:

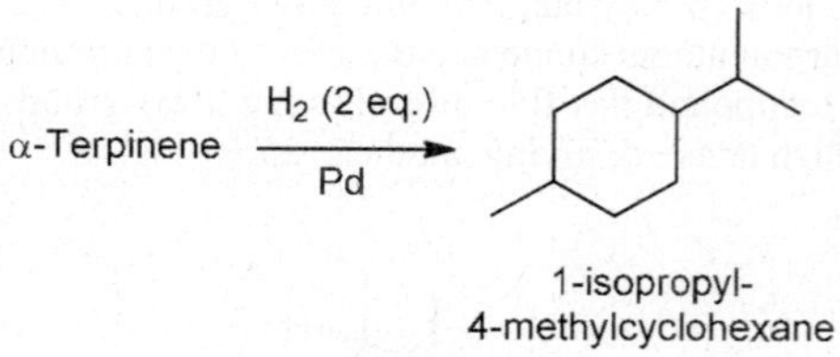

A triple bond could not have been in the ring, because a six-membered ring cannot support the linear geometry required by the *sp* hybridized carbon atoms of a triple bond. The other C-C bonds (outside the ring) can also not support a triple bond (without violating the octet rule by giving a carbon atom with five bonds). Therefore, α-terpinene must have two double bonds.

(b) The ozonolysis products indicate how the molecule must have been constructed, because ozonolysis breaks C=C bonds into C=O bonds:

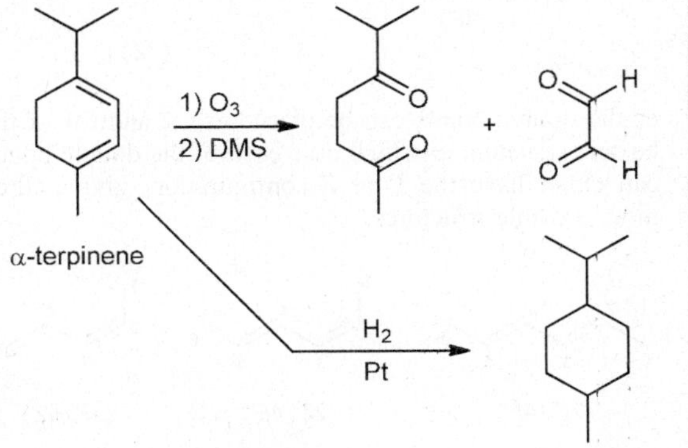

1) O_3
2) DMS

α-terpinene

H_2
Pt

(c) This conjugated system is a homoannular diene:

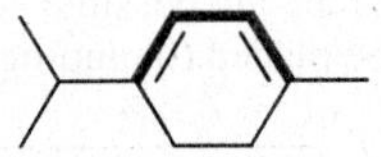

In addition, there are four auxochromic alkyl groups, highlighted here:

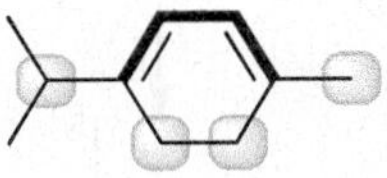

The calculation for λ_{max} is shown here:

Base	=	217
Additional double bonds	=	0
Auxochromic alkyl groups	=	+20
Exocyclic double bonds	=	0
Homoannular diene	=	+39
Total	=	276 nm

16.69. The molecular formula (C_7H_{10}) indicates three degrees of unsaturation (see Section 14.16). The problem statement indicates that compound **A** will react with two equivalents of molecular hydrogen (H_2). Therefore, we can conclude that compound **A** has two π bonds, which accounts for two of the three degrees of unsaturation. The remaining degree of unsaturation must be a ring. Ozonolysis yields two products, which together account for all seven carbon atoms:

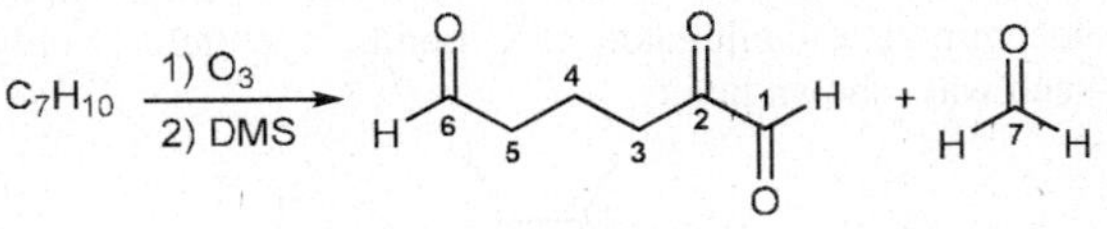

C_7H_{10} 1) O_3 2) DMS

Focus on the product with three carbonyl groups. Two of them must have been connected to each other in compound **A** (as a C=C bond), and the third carbonyl group must have been connected to the carbon atom of formaldehyde (CH_2O). This gives two possibilities:

Either C1 was connected to C6 (and C2 was connected to C7):

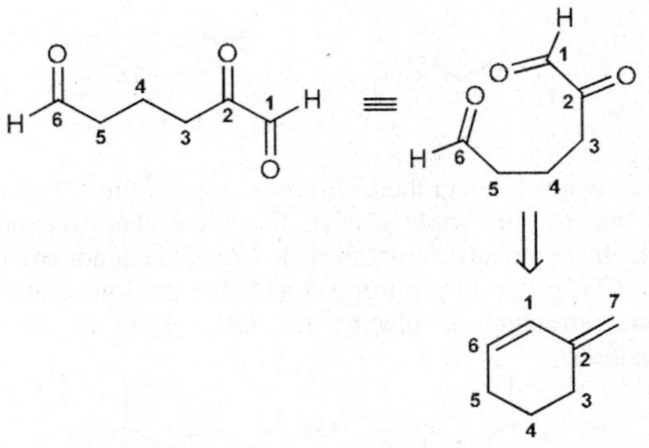

or C2 was connected to C6, and C1 was connected to C7:

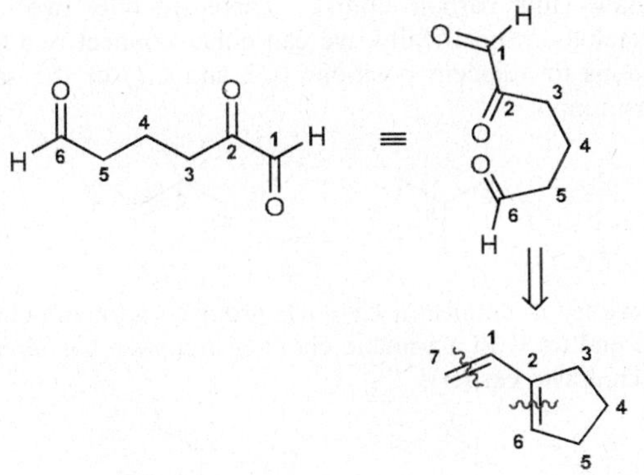

In summary, we have found two possible structures for compound **A**, both of which are conjugated dienes:

16.70. The molecular formula (C_6H_{10}) indicates two degrees of unsaturation (see Section 14.16). The problem statement indicates that all proposed structures must be conjugated dienes, which accounts for both degrees of unsaturation. That is, the proposed structures cannot have any rings. All structures must be acyclic.

To draw all possible conjugated dienes with the molecular formula C_6H_{10}, we must first consider all of the different ways in which six carbon atoms can be connected to each other.

We begin with a linear chain (parent = hexane):

Next, we look for any skeletons where the parent is pentane (five carbon atoms). There are only two such skeletons. Specifically, we can either connect the extra CH_3 group to positions C2 or C3 of the pentane chain:

We cannot connect the CH_3 group to positions C1 or C5, as that would simply give us the linear chain (hexane), which we already drew (above). We also cannot connect the CH_3 group to position C4 as that would generate the same structure as placing the CH_3 group at the C2 position:

Next, we look for any skeletons where the parent is butane (four carbon atoms). There are only two such skeletons. Specifically, we can either connect two CH_3 groups to adjacent positions (C2 and C3) or the same position:

If we try to connect a CH_3CH_2 group to a butane chain, we end up with a pentane chain (which was has already been drawn earlier):

In summary, there are five different ways in which six carbon atoms can be connected:

Parent = hexane

Parent = pentane

Parent = butane

For each one of these skeletons, we must consider all of the different unique positions where the double bonds can be placed (keeping in mind that they must remain conjugated). For the first skeleton (hexane), there are many different locations where the double bonds can be placed. For example, the double bonds can be at C1 and C3 (giving two stereoisomeric options, because the double bond at C3 can have either the *E* configuration or the *Z* configuration):

(3*E*) (3*Z*)

or the double bonds can be placed at C2 and C4 of the hexane skeleton, in which case each of the double bonds can either have the *E* or *Z* configuration, giving three more possible structures:

(2*E*, 4*E*) (2*Z*, 4*E*) (2*Z*, 4*Z*)

Next we move on to the skeletons that have only five carbon atoms in a linear chain, and for each of these skeletons, we consider all possible locations where the double bonds can be placed (including stereoisomers):

Next we move on to the skeletons that have only four carbon atoms in a linear chain. There are two such skeletons, and one of them cannot accommodate two π bonds (without violating the octet rule by giving more the four bonds to a carbon atom). The other skeleton with only four carbon atoms in a linear chain CAN accommodate conjugated C=C bonds, but there is only such way, shown here:

In summary, we have revealed twelve different conjugated dienes with the molecular formula C_6H_{10}, shown here:

Parent = hexadiene

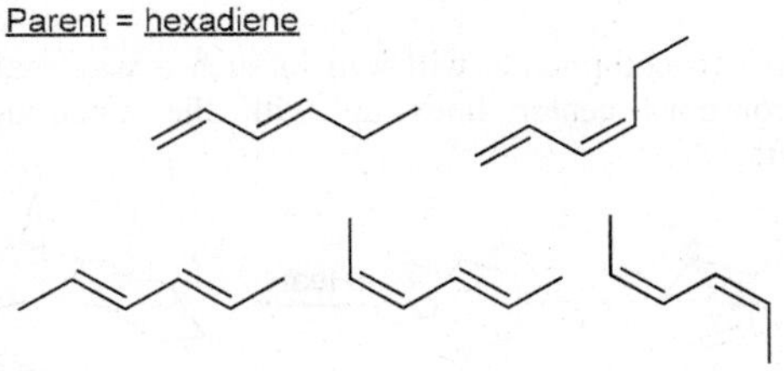

Parent = pentadiene

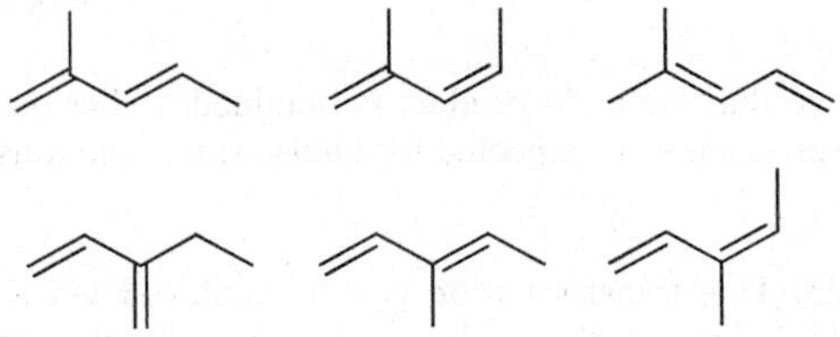

Parent = butadiene

16.71. Treating 1,3-butadiene with HBr at elevated temperature gives the 1,4-adduct, which can be treated with sodium hydroxide to give an S_N2 process in which Br is replaced with a hydroxyl group. Oxidation with PCC (or DMP or Swern) converts the alcohol into an aldehyde, which can then be treated with another equivalent of 1,3-butadiene to give a Diels-Alder reaction that affords the product.

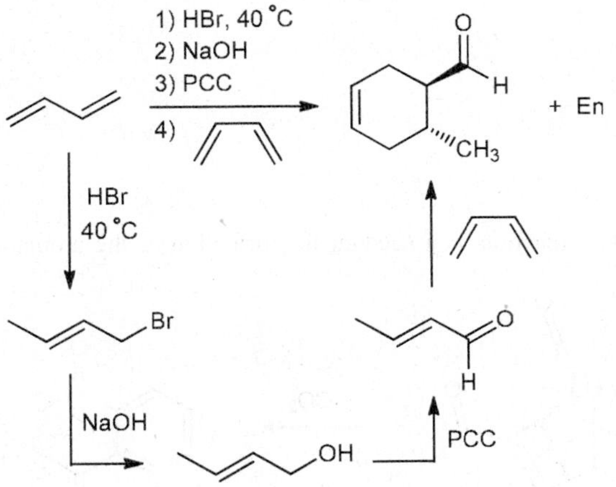

1) HBr, 40 °C
2) NaOH
3) PCC
4)

+ En

HBr
40 °C

NaOH

PCC

16.72. We have seen that electron-poor dienophiles react faster in a Diels-Alder reaction. Nitroethylene should be more reactive than ethylene in a Diels–Alder reaction, because the nitro group is electron-withdrawing, via resonance:

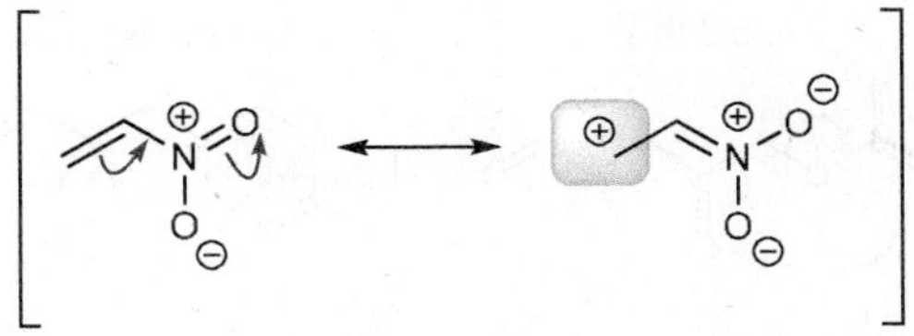

16.73. The starting material is an allylic vinylic ether, so it can undergo a Claisen rearrangement.

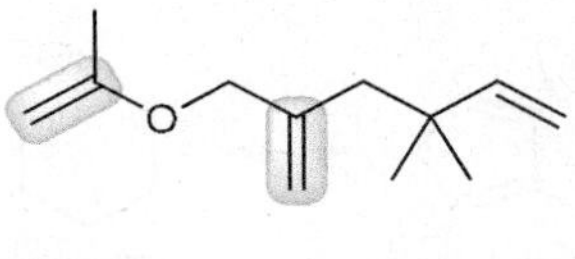

Alternatively, we also note that the starting material has two C=C bonds that are separated from each other by exactly three σ bonds, so it can undergo a Cope rearrangement:

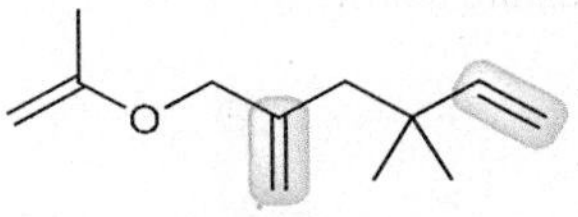

We will ultimately end up drawing both processes, and it does not matter the order in which we draw these two processes. Below, the Claisen rearrangement is drawn first, followed by the Cope rearrangement. If instead, the Cope rearrangement was drawn first, followed by the Claisen rearrangement, the same product would be obtained.

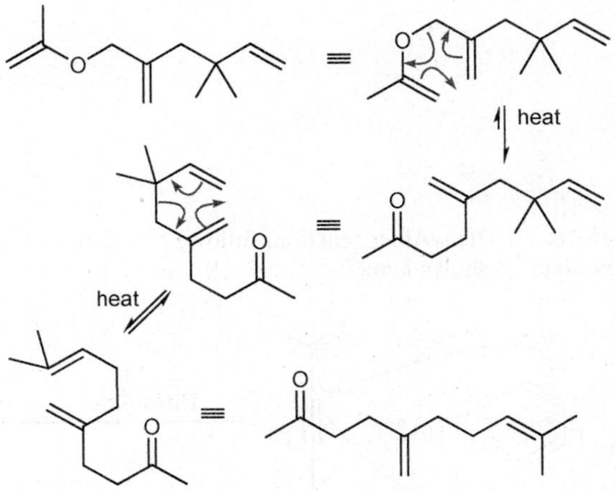

heat

heat

Note that for each of the sigmatropic processes above (the Claisen rearrangement and the Cope rearrangement), the compound is redrawn in such a way that enables us to clearly see the motion of the electrons that cause the reaction. You are likely to make a mistake if you try to draw the curved arrows without first redrawing the structure:

Avoid this:

Draw it like this:

16.74. First we determine the regiochemical outcome. The diene is electron-rich, as seen in the second resonance structure below:

The diene is electron-rich in this location

And the dienophile is electron-poor, as seen in the third resonance structure below:

The dienophile is electron-poor in this location

These two compounds will join in such a way that the electron-poor center lines up with the electron-rich center:

Heat

+ En

Notice that the endo product is obtained, rather than the exo product, as is expected for Diels-Alder reactions.

16.75. This transformation can be achieved via a retro Diels-Alder reaction (shown below), which requires elevated temperature, as described in Section 16.7.

heat

16.76. A Diels-Alder reaction, followed by a retro Diels-Alder reaction, can account for formation of the aromatic product, as shown here:

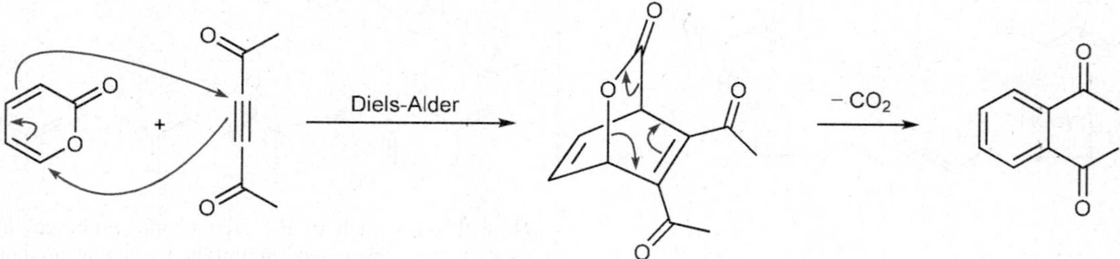

16.77. The nitrogen atom in divinyl amine is sp^2 hybridized. The lone pair is delocalized and joins the two neighboring π bonds into one conjugated system. As such, the compound absorbs light above 200 nm (UV light). In contrast, 1,4-pentadiene has two isolated double bonds and therefore does not absorb UV light in the region between 200 and 400 nm.

16.78. Notice that compound **1** contains a strained, four-membered ring. When this compound is heated to 120 °C, it will undergo a thermal electrocyclic reaction to form compound **2**, which possesses significantly less ring strain. The newly generated diene can then undergo a thermal, intramolecular Diels-Alder reaction with the alkyne, re-establishing the aromatic ring, and forming the hexacyclic product.

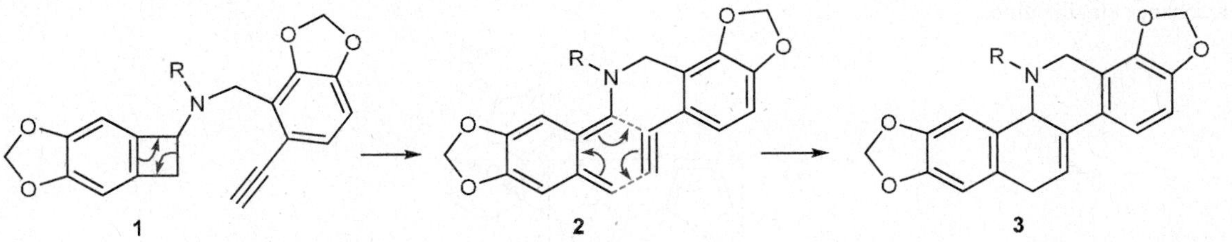

16.79. The first step involves a 6-π electrocyclic reaction that closes the first ring to form a cyclohexadiene. When this intermediate is redrawn, we can clearly see that the pendant alkene is in close enough proximity with the newly generated diene to induce an intramolecular [4+2] Diels-Alder cycloaddition, which will result in the formation of the tricyclic product.

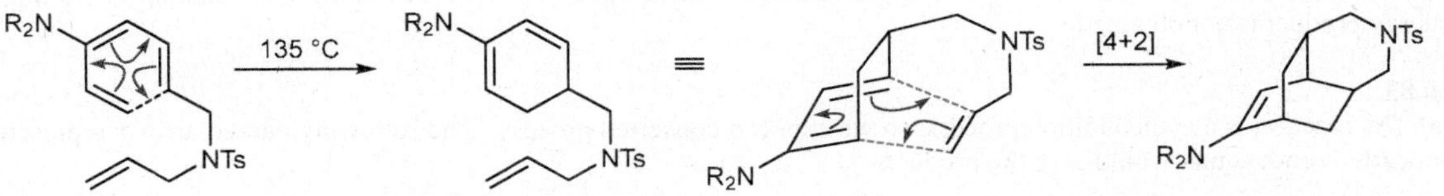

16.80. The divinylcyclopropane unit is converted to a cycloheptadiene in this [3,3] sigmatropic rearrangement. Interestingly, this rearrangement also opens one ring (cyclopropane) and forms a new ring (cycloheptadiene).

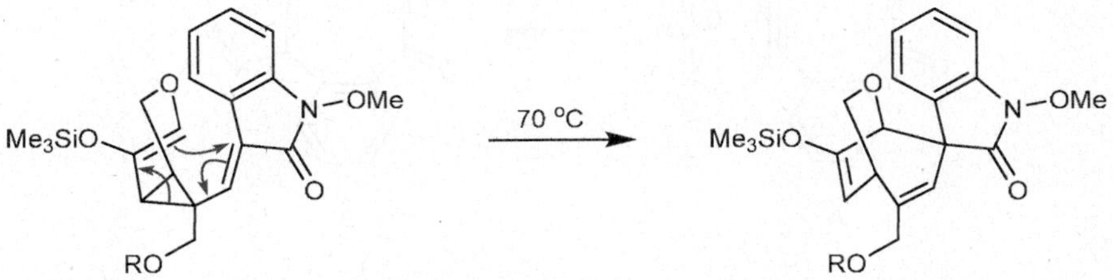

16.81. Compound **1** undergoes an electrocyclic reaction involving 6π-electrons. The reaction occurs with light, rather than heat, so we expect ring-closure to occur in a conrotatory fashion:

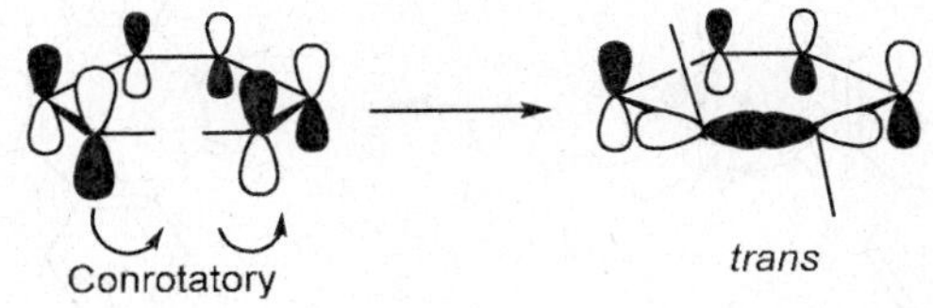

We therefore expect the two methyl groups to be *trans* to each other in the product:

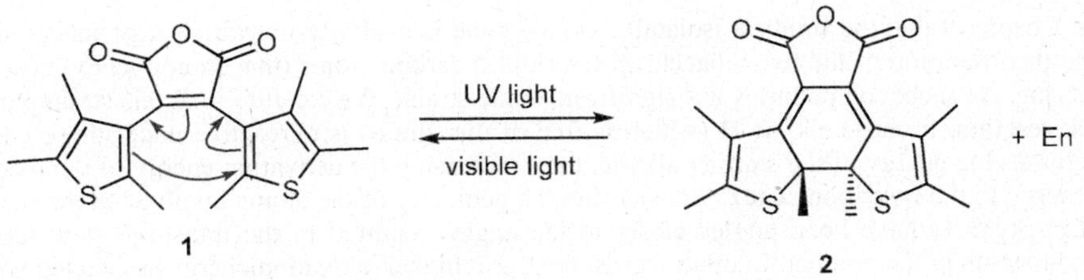

Compound **2** has rotational symmetry, but it lacks reflectional symmetry (see Section 5.6). As such, it is chiral, and it will be formed as a pair of enantiomers, with the other enantiomer being formed when both methyl groups rotate in the other direction.

16.82. Oligofurans are highly conjugated materials. When treated with maleimide, only one product results – the [4+2] addition to the terminal furan ring. Notice that this product is still conjugated – there are two furan rings that share resonance stabilization.

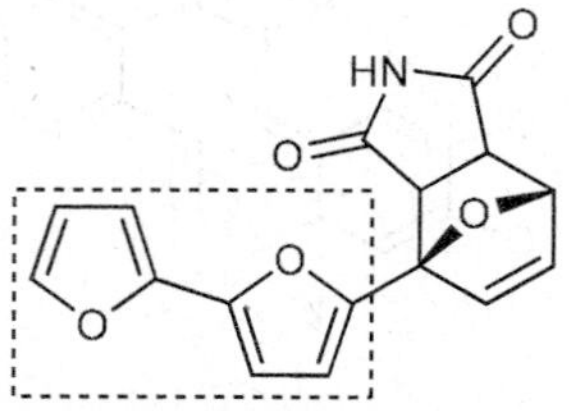

If the internal furan ring were to react, the conjugation between the remaining furan rings would be lost, leaving behind two isolated furan rings. This molecule would be much higher in energy because of the loss of extended conjugation. Since the product distribution of a Diels-Alder reaction is determined by thermodynamic considerations, the higher energy product is not obtained.

16.83.

(a) The reaction is a cycloaddition process, so we expect a concerted process. The following curved arrows represent a concerted process that would give the product:

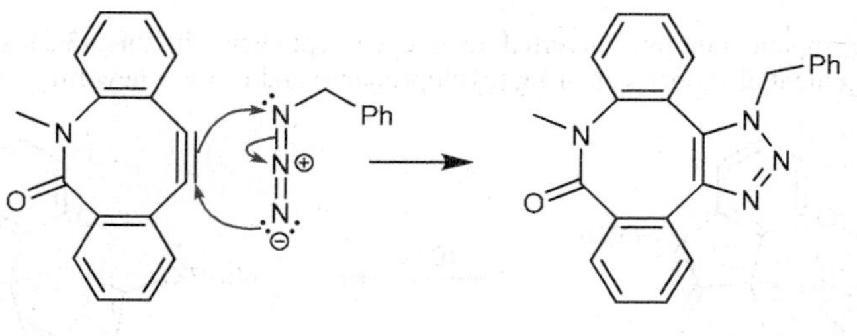

(b) Depending on the relative orientation of the azide and alkyne during the reaction, the following compound can also be formed.

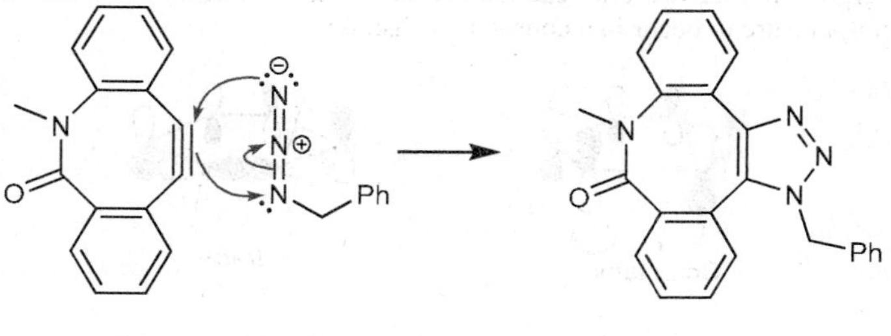

(c) Recall from Chapter 9 that the smallest isolatable cycloalkyne is cyclooctyne, which experiences significant angle strain due to the incorporation of the two adjacent *sp* hybridized carbon atoms (that should have linear geometry) into an 8-membered ring. As such, compound **A** has significant angle strain. We can infer that this strain plays an important role in the click reaction, because alkyne **D** (which is free of this strain) is unreactive under these conditions. This angle strain increases the energy of the starting alkyne, thus decreasing the activation energy of the reaction (since it is now closer in energy to the transition state). Considering the geometry of the atoms involved in the reaction, the angle strain forces the alkyne to have bond angles closer to the angles required in the transition state leading to the sp^2 hybridized carbon atoms in the product. In other words, there is a higher activation energy associated with distorting an unstrained alkyne (180°) to an alkene (120°), compared to the analogous conversion of a strained alkyne (<180°) to an alkene (120°).

Alkyne **E** is less reactive because the non-alkyne atoms in the ring are all sp^3 hybridized; this eases some of the angle strain due to their relative conformational flexibility. The multiple sp^2 hybridized atoms in the ring of alkyne **A** increase the strain on the alkyne, thus increasing the reaction rate.

Alkyne **F** is less reactive because one of the methyl groups (on the aromatic ring) sterically hinders the approach of the benzyl azide.

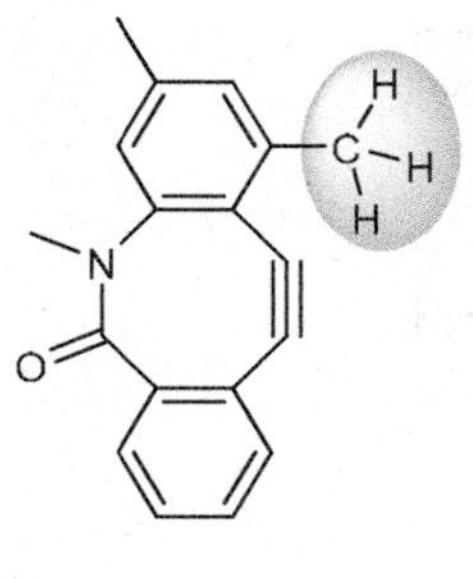

16.84.

(a) The following electrocyclic reaction will produce an eight-membered ring. Since the reaction involves eight π electrons, under thermal conditions, we expect conrotatory ring-closure, producing the *trans*-configuration seen in compound **2**. This ring-closing reaction can occur if either: (1) both ends of the forming ring rotate clockwise, or 2) both ends rotate counterclockwise. These two possibilities can occur with equal likelihood, resulting in a racemic mixture of compound **2**.

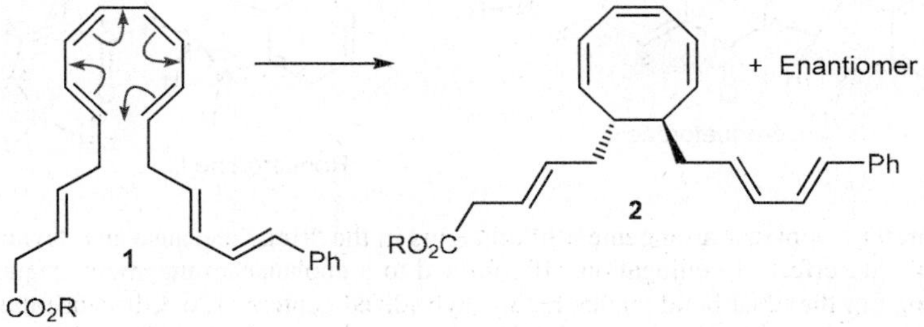

(b) The conversion of compound **2** to endiandric acid G occurs via an electrocyclic reaction involving six π electrons.

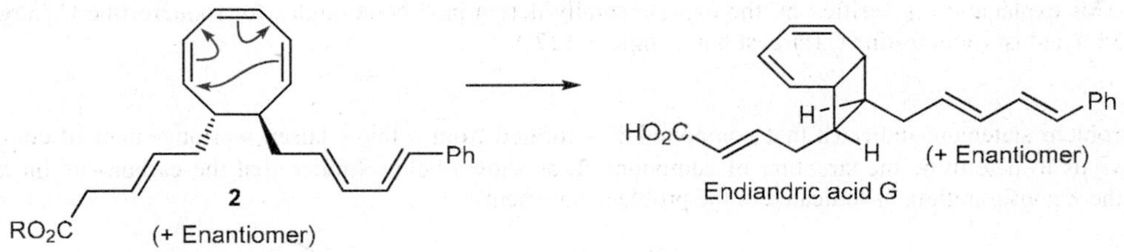

(c) Endiandric acid G contains two diene groups, but only one of them is in close proximity with the dienophile. An intramolecular Diels-Alder reaction, as shown below, gives endiandric acid C:

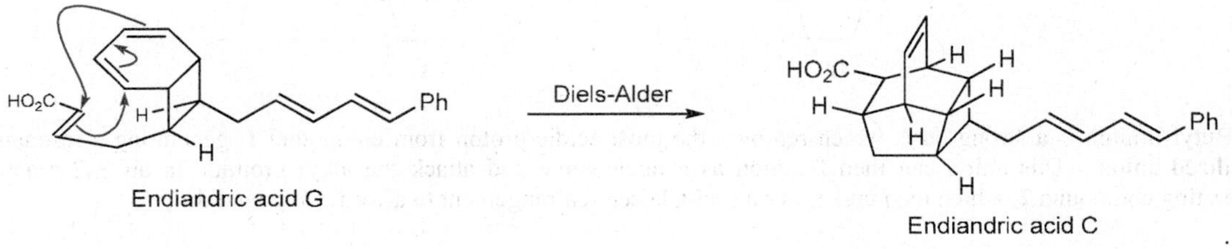

16.85. Both compounds exhibit intramolecular hydrogen bonding interactions, as shown below:

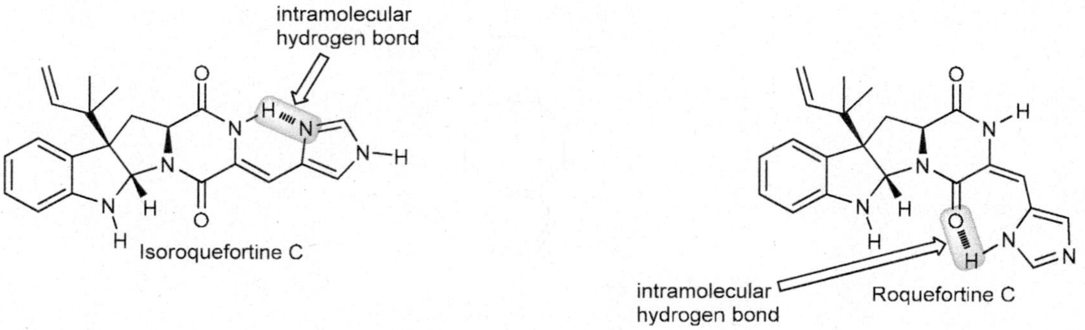

However, the hydrogen bonding interaction in isoroquefortine C involves a "ring" comprised of six atoms, while the hydrogen bonding interaction in roquefortine C involves a "ring" comprised of seven atoms.

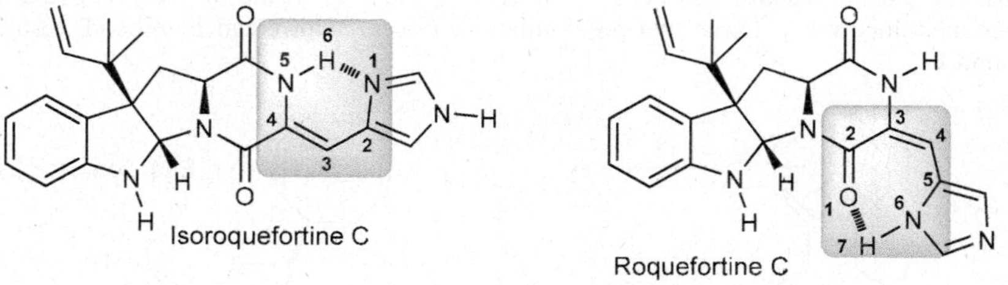

Both are expected to prefer a coplanar arrangement of all atoms in the "ring," because any deviations from coplanarity will diminish the stabilizing effect of conjugation. If confined to a coplanar arrangement, roquefortine C will have a more difficult time adopting the ideal bond angles for sp^2 hybridized centers (120°), because there are seven atoms in the "ring" rather than six. In contrast, isoroquefortine C (with its six atoms in the "ring") will not experience as much strain as it attempts to adopt the ideal bond angles in a coplanar arrangement. Deviation from the ideal bond angles or from coplanarity will cause an increase in energy. While both compounds will deviate (to a certain extent) from the ideal bond angles and from coplanarity, the deviation is expected to be more significant for roquefortine C. Therefore, roquefortine C is higher in energy than isoroquefortine C.

This explanation is verified by the experimentally determined bond angles for roquefortine C (largest bond angle is 134°) and isoroquefortine C (largest bond angle is 127°).

16.86.

(a) The problem statement indicates that compound **3** is formed from a thio-Claisen rearrangement of compound **2**. This allows us to determine the structure of compound **2**, as shown below (notice that the carbon-carbon π bond is shown in the *Z* configuration, as indicated in the problem statement):

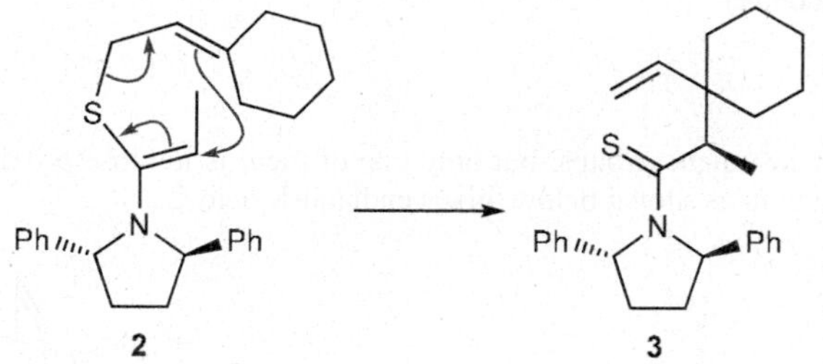

(b) Butyl lithium is a strong base, which removes the most acidic proton from compound **1**, generating a resonance-stabilized anion. This anion can then function as a nucleophile and attack the alkyl bromide in an SN2 reaction, generating compound **2**, which then undergoes a thio-Claisen rearrangement to afford compound **3**:

(c) The existing chiral centers have an influence on the stereochemical outcome of this reaction. Specifically, one of the phenyl groups provides steric hindrance that forces the reaction to occur in a stereospecific manner, so that the phenyl group and the cyclohexyl group remain far apart from one other:

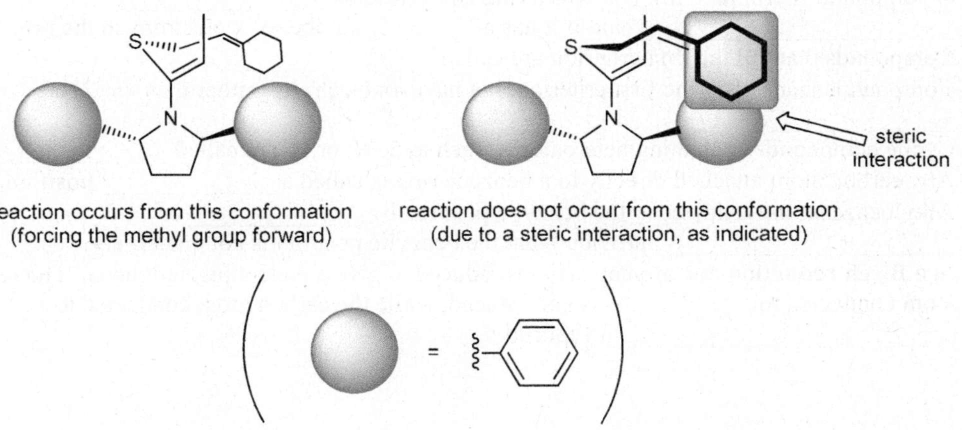

reaction occurs from this conformation
(forcing the methyl group forward)

reaction does not occur from this conformation
(due to a steric interaction, as indicated)

The reaction will proceed via the lowest energy transition state, which is most likely achieved from the first conformation above, rather than from the second conformation, in which there is a large steric interaction (highlighted).

Chapter 17
Aromatic Compounds

Review of Concepts

Fill in the blanks below. To verify that your answers are correct, look in your textbook at the end of Chapter 17. Each of the sentences below appears verbatim in the section entitled *Review of Concepts and Vocabulary*.

- When a benzene ring is a substituent, it is called a _________ **group**.
- Disubstituted derivatives of benzene can be differentiated by the use of the descriptors _______, ***meta*** and _______, or by the use of locants.
- Benzene is comprised of a ring of six identical C-C bonds, each of which has a bond order of _____.
- The **stabilization energy** of benzene can be measured by comparing _________ of hydrogenation.
- The stability of benzene can be explained with MO theory. The six π electrons all occupy ___________ MOs.
- The presence of a fully conjugated ring of π electrons is not the sole requirement for aromaticity. The requirement for an odd number of electron pairs is called ____________ **rule.**
- **Frost circles** accurately predict the relative energy levels of the ________ in a conjugated ring system.
- A compound is aromatic if it contains a ring comprised of _________________ ________________________ and if it has a ________ number of π electrons in the ring.
- Compounds that fail the first criterion are called ______________.
- Compounds that satisfy the first criterion, but have $4n$ electrons (rather than $4n+2$) are ______________.
- Cyclic compounds containing heteroatoms, such as S, N, or O, are called ________________.
- Any carbon atom attached directly to a benzene ring is called a _____________ **position.**
- Alkylbenzenes are oxidized at the benzylic position by _____________ or __________________, provided that the benzylic position is not quaternary.
- In a **Birch reduction**, the aromatic ring is reduced to give a nonconjugated diene. The carbon atom connected to ___________ is not reduced, while the carbon atom connected to _________________________ is reduced.

Review of Skills

Fill in the blanks and empty boxes below. To verify that your answers are correct, look in your textbook at the end of Chapter 17. The answers appear in the section entitled *SkillBuilder Review*.

17.1 Naming a Polysubstituted Benzene

Provide a systematic name for the following compound:

OH
Cl
Br

1) Identify the parent
2) Identify and name substituents
3) Assign locants to each substituent
4) Alphabetize

17.2 Determining Whether a Structure is Aromatic, Nonaromatic, or Antiaromatic

Identify each compound or ion below as aromatic, antiaromatic, or nonaromatic:

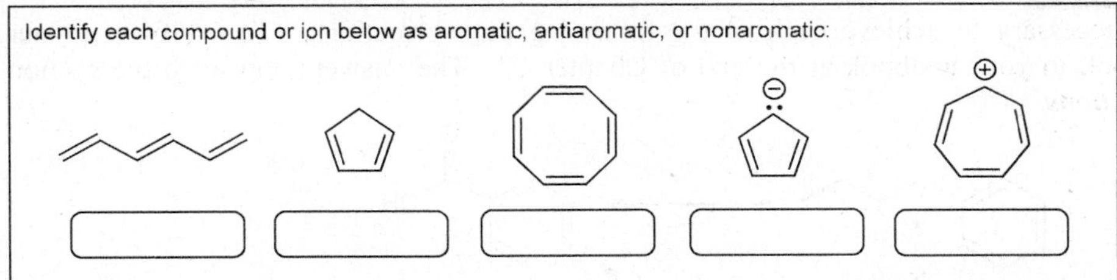

17.3 Determining Whether a Lone Pair Participates in Aromaticity

In the following compound, identify whether each lone pair participates in aromaticity:

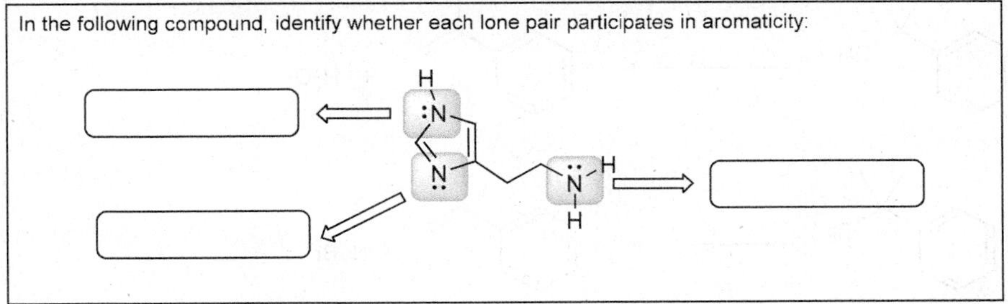

17.4 Manipulating the Side Chain of an Aromatic Compound

For each transformation below, identify the type of reaction that could be used (S_N2, S_N1, E2, etc.)

Br

OH

O

OH

OH

Br

17.5 Predicting the Product of a Birch Reduction

Predict the major product of the following reaction:

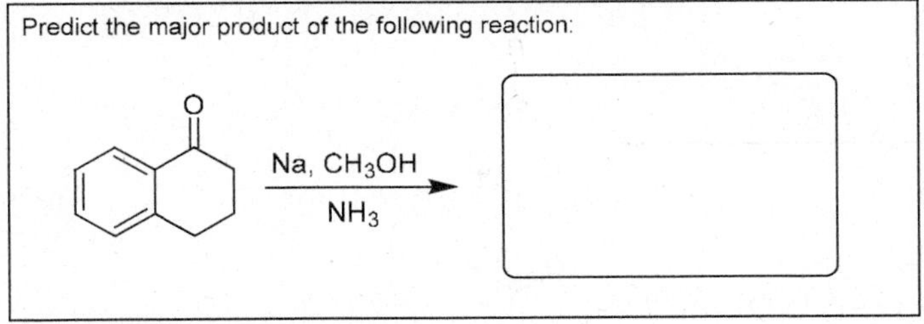

Review of Reactions

Identify the reagents necessary to achieve each of the following transformations. To verify that your answers are correct, look in your textbook at the end of Chapter 17. The answers appear in the section entitled *Review of Reactions*.

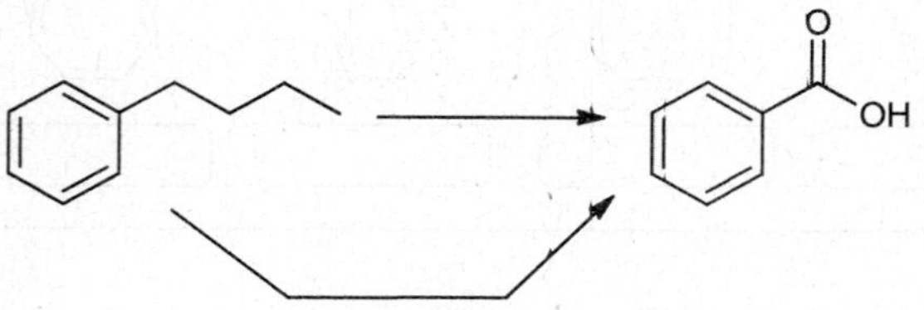

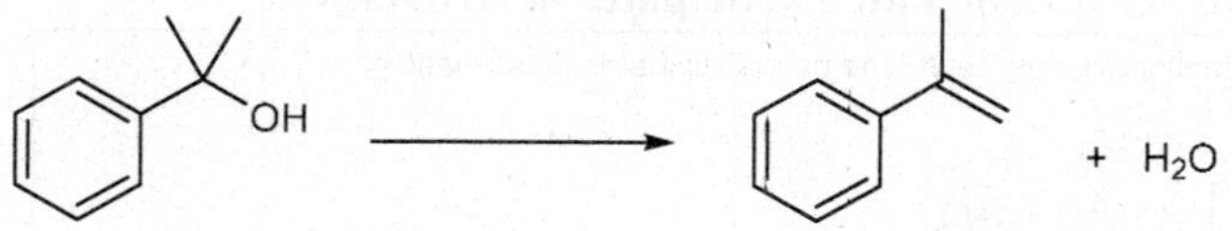

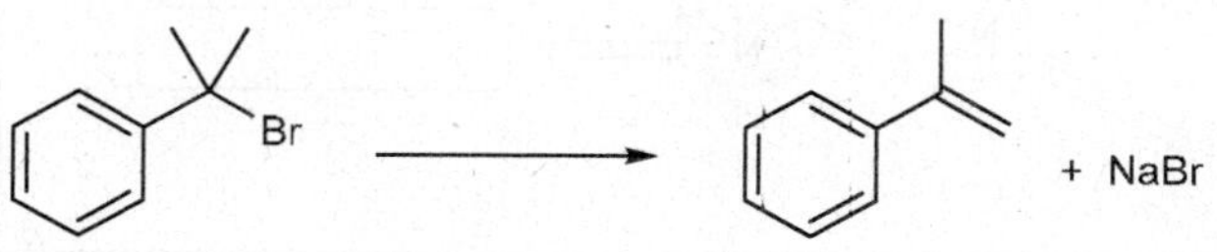

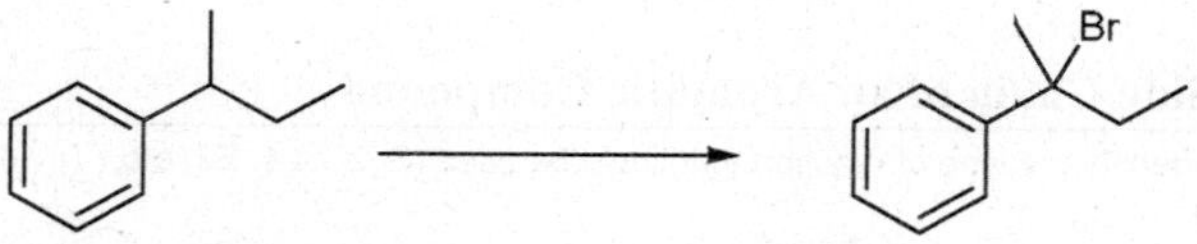

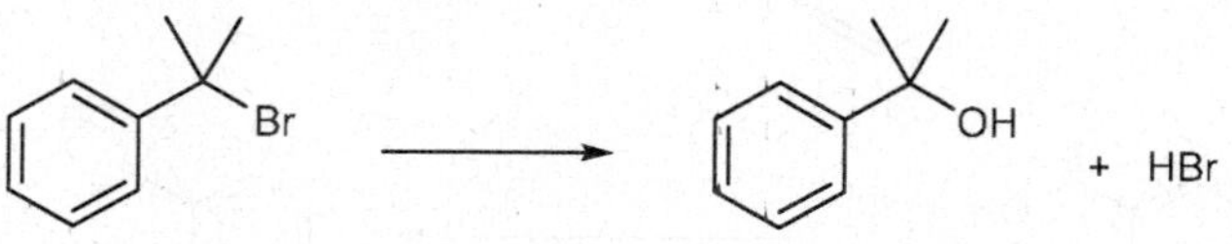

OH

OH

+ H_2O

Br

+ NaBr

Br

Br

OH

+ HBr

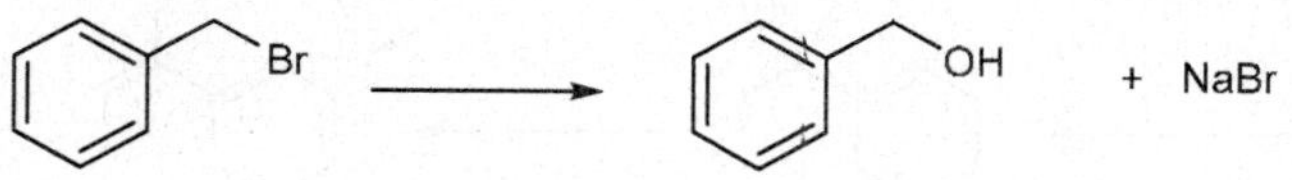

Br

OH

+ NaBr

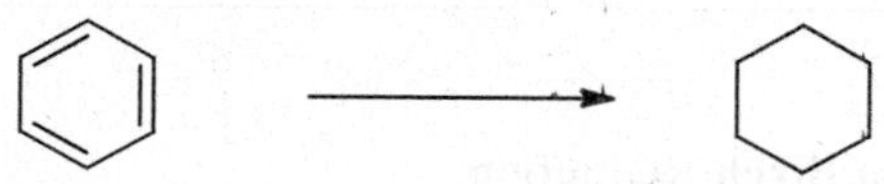

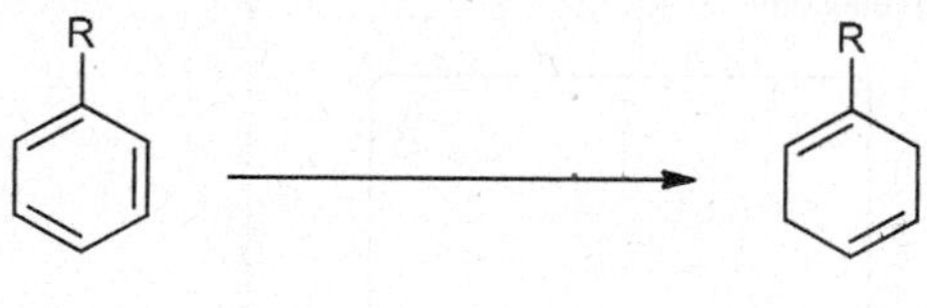

R

R

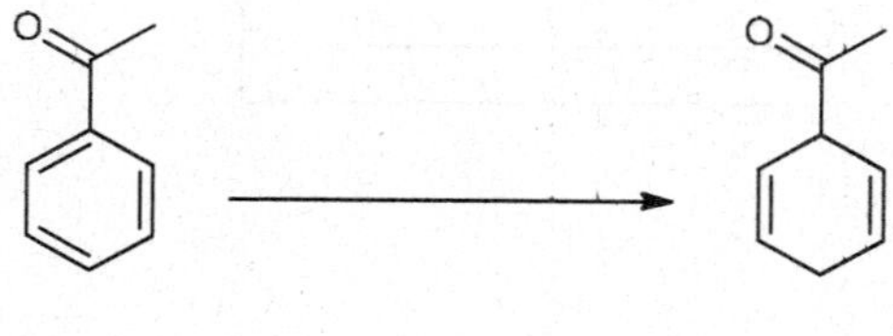

O

O

Review of Mechanisms

Complete the following mechanism by drawing the missing curved arrows. To verify that your curved arrows are drawn correctly, compare them to the curved arrows in Mechanism 17.1, which can be found in Section 17.7 of your text.

Mechanism 17.1 The Birch Reduction

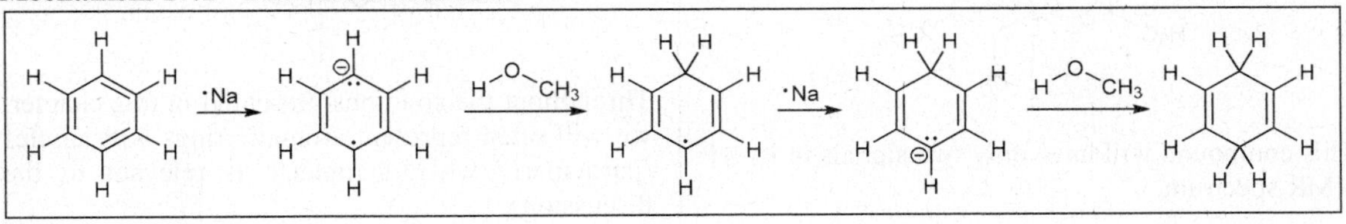

Mistakes to Avoid

When analyzing an NMR spectrum (^1H or ^{13}C), we must always consider the role of symmetry in affecting the number of signals in the spectrum. For example, consider the number of signals in the ^1H NMR spectrum of each of the following constitutional isomers:

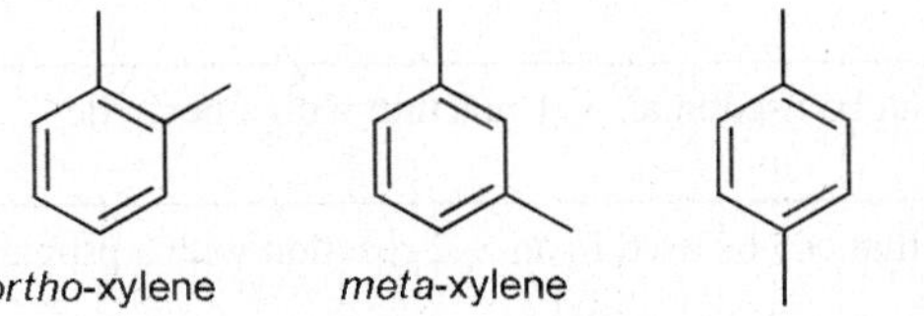

ortho-xylene *meta*-xylene *para*-xylene

Each of these compounds exhibits a different number of signals in its ^1H NMR spectrum (3 signals, 4 signals, and 2 signals, respectively). Similarly, each of these compounds exhibits a different number of signals in its ^{13}C NMR spectrum (4 signals, 5 signals, and 3 signals, respectively).

When analyzing an NMR spectrum of an aromatic compound, always make sure that the substitution pattern (*i.e.*, monosubstituted, *ortho*-disubstituted, etc.) is fully consistent with all signals in the spectrum (taking symmetry into account). When doing so, avoid being distracted by the positions of the double bonds. After all, the π electrons are delocalized, and there are resonance structures in which the double bonds are drawn in different locations. The best way to avoid being distracted is to draw a circle to represent the aromatic ring, like this:

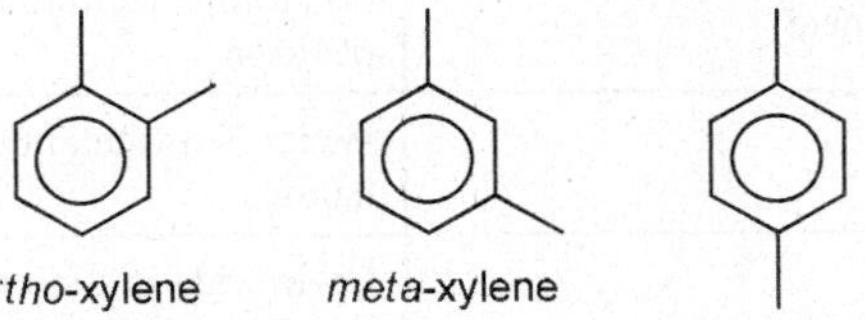

ortho-xylene *meta*-xylene *para*-xylene

To see the value of this approach, consider the following structure and determine how many signals are expected in the ^1H NMR and ^{13}C NMR spectra of this compound:

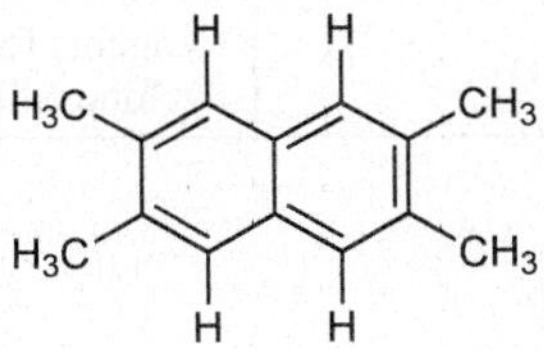

By redrawing the structure with circles to represent aromatic rings (rather than alternating single and double bonds), it becomes easier to see that there are only two unique kinds of protons in this structure (highlighted below):

This compound will have only two signals in its ^1H NMR spectrum.

Similarly, the compound will have only four signals in its ^{13}C NMR spectrum:

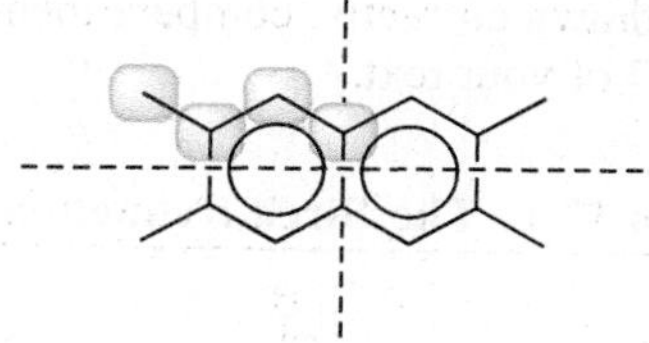

Throughout the solutions presented in this chapter, we will often represent aromatic rings with circles (particularly when symmetry is relevant to the discussion).

Useful reagents

The following is a list of reagents encountered in this chapter, as well as their specific function in the context of this chapter:

Reagents	Description
$Na_2Cr_2O_7$, H_2SO_4, H_2O	Sodium dichromate and sulfuric acid give chromic acid, which is a strong oxidizing agent that can be used to oxidize a benzylic position, provided that the benzylic position is not quaternary. The alkyl group (connected to the aromatic ring) is converted into a carboxylic acid group.
1) $KMnO_4$, H_2O, heat 2) H_3O^+	Potassium permanganate. A strong oxidizing agent that can be used to oxidize a benzylic position, provided that the benzylic position is not quaternary. The alkyl group (connected to the aromatic ring) is converted into a carboxylic acid group.
NBS, heat	N-Bromosuccinimide. A reagent that is used for radical bromination at the benzylic position.
H_2O	Water is a weak nucleophile that can be used in an S_N1 reaction with a benzylic halide.
NaOH	Hydroxide is a strong nucleophile that can be used in an S_N2 reaction with a primary benzylic halide.
conc. H_2SO_4	A strong acid that can be used to achieve acid-catalyzed dehydration to give an alkene.
NaOEt	Sodium ethoxide is a strong base that can be used to convert secondary or tertiary halides into alkenes (via an E2 process).
3 H_2, 100 atm, 150°C	Conditions for complete hydrogenation of benzene to give cyclohexane.
Na, CH_3OH, NH_3	Reagents for a Birch reduction, which reduces a benzene ring to give a 1,4-cyclohexadiene ring.

Solutions

17.1.

(a) The parent is benzaldehyde, shown in bold. An isopropyl group is located at C3 (also called the *meta* position), so the compound is 3-isopropylbenzaldehyde, (also called *meta*-isopropylbenzaldehyde), or 3-(1-methylethyl)benzaldehyde:

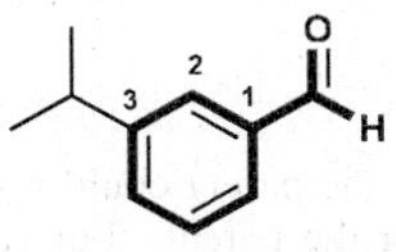

(b) The parent is toluene, shown in bold. A bromo group is located at C2 (also called the *ortho* position), so the compound is 2-bromotoluene (also called *ortho*-bromotoluene):

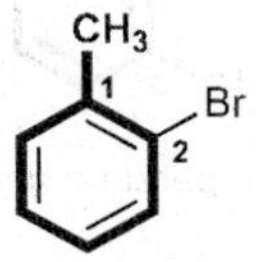

(c) The parent is phenol, shown in bold. Two nitro groups are located at C2 and C4, so the compound is called 2,4-dinitrophenol:

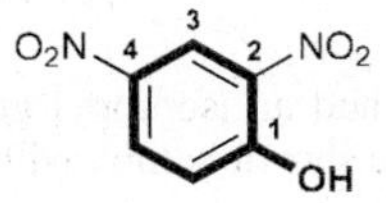

(d) The parent is benzene, shown in bold. There are three substituents (two isopropyl groups and one ethyl group). The assignment of locants, shown below, achieves the lowest numbers for all three substituents:

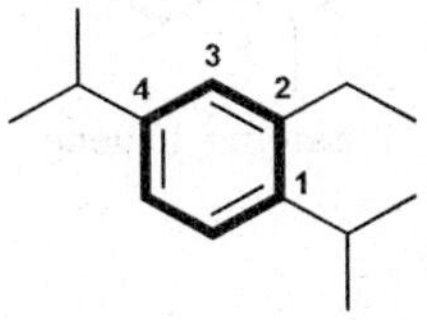

Therefore, the name of the compound is:
2-ethyl-1,4-diisopropylbenzene

(e) The parent is phenol, shown in bold. The substituents are each named and assigned a locant, as shown below. The locants start at the carbon atom connected to the OH group, and are then assigned counter-clockwise. This way, the ethyl group ("e") is assigned a lower locant than the isopropyl group ("i"), since there are no other differentiating factors that would allow us to decide which way to assign the locants.

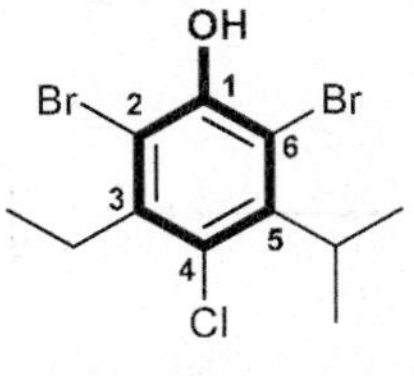

2,6-dibromo-4-chloro-3-ethyl-5-isopropylphenol, or
2,6-dibromo-4-chloro-3-ethyl-5-(1-methylethyl)phenol

17.2. When naming this compound, there are three choices for the parent. The compound can be named as a disubstituted phenol, as a disubstituted toluene, or as a trisubstituted benzene. For each of these possibilities, the assignment of locants is shown:

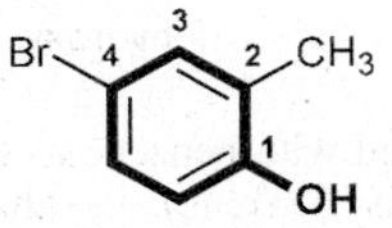

Parent = phenol

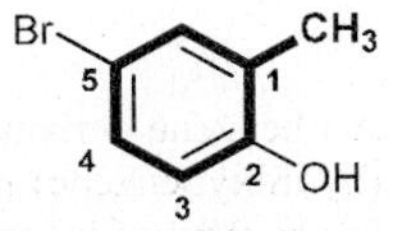

Parent = toluene

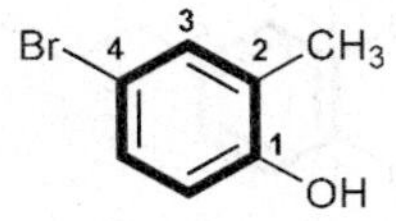

Parent = benzene

This gives the following three possible names:

(a) 4-bromo-2-methylphenol
(b) 5-bromo-2-hydroxytoluene
(c) 4-bromo-1-hydroxy-2-methylbenzene

17.3.

(a) The parent is anisole (methoxybenzene), and there are three other substituents (Br, Br, and Cl), in the following locations:

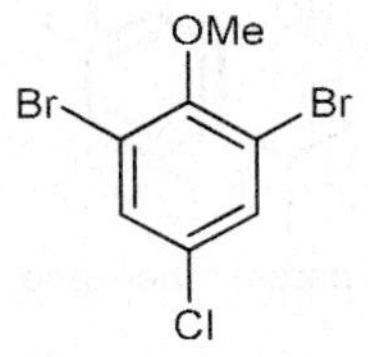

(b) The parent is phenol (hydroxybenzene), and there is one nitro group, located at C3 (the *meta* position):

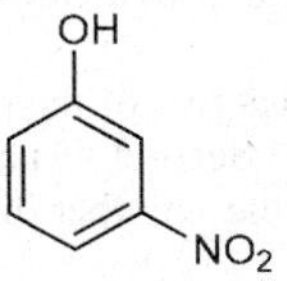

17.4. When naming this compound, there are three choices for the parent. The compound can be named as a disubstituted benzene, as a monosubstituted toluene, or as xylene. This gives the following possible names, all of which are acceptable IUPAC names:

(a) *meta*-xylene
(b) *meta*-dimethylbenzene
(c) 1,3-dimethylbenzene
(d) *meta*-methyltoluene
(e) 3-methyltoluene

17.5.

(a) Eugenol contains three substituents: an allyl group, a hydroxy group, and a methoxy group.

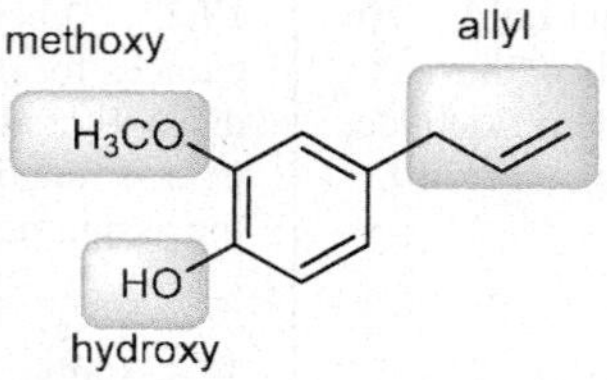

methoxy allyl H_3CO HO hydroxy

All benzene derivatives can be named with benzene as the parent. In this case, the parent could also be either phenol (hydroxybenzene) or anisole (methoxybenzene). So there are three choices for the parent. For each possibility, the assignment of locants is shown by numbering to give the lowest possible number to each consecutive substituent:

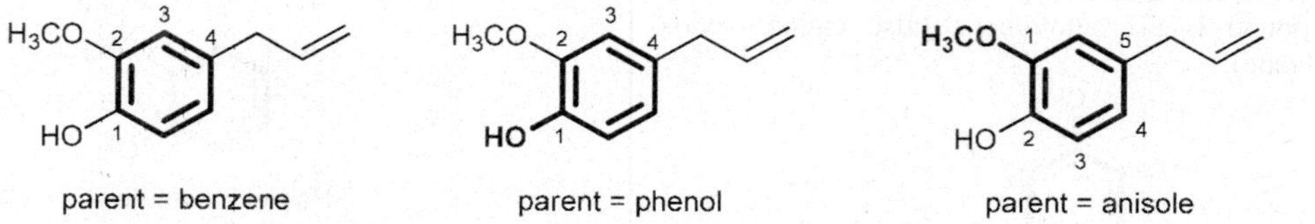

parent = benzene parent = phenol parent = anisole

The last step is to arrange the substituents alphabetically so the three names for eugenol are as follows:
As a benzene derivative: 4-allyl-1-hydroxy-2-methoxybenzene
As a phenol derivative: 4-allyl-2-methoxyphenol
As an anisole derivative: 5-allyl-2-hydroxyanisole

Thymol contains three substituents as well: a methyl group, a hydroxy group, and an isopropyl group. Each can be named based on benzene, phenol, or toluene as the parent. These possibilities are shown below, with the corresponding locants shown:

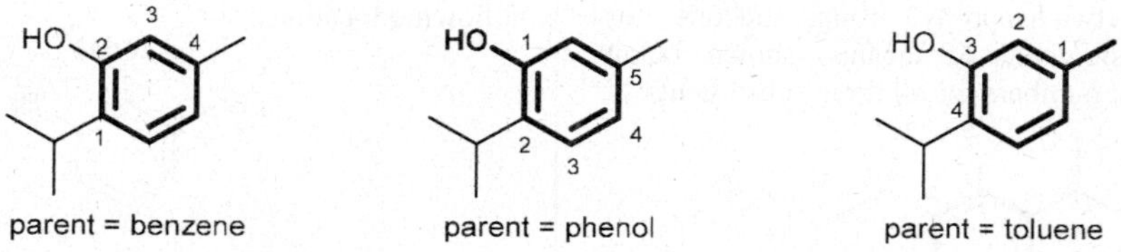

parent = benzene parent = phenol parent = toluene

Arranging the substituents we obtain the following three possible IUPAC names:
As a benzene derivative: 2-hydroxy-1-isopropyl-4-methylbenzene
As a phenol derivative: 2-isopropyl-5-methylphenol
As a toluene derivative: 3-hydroxy-4-isopropyltoluene

(b) This name violates two of the rules for naming polysubstituted aromatic compounds. First, the substituents are not listed alphabetically. Second, if named with benzene as the parent, the methyl group needs to have the first locant in order to minimize the number of the third locant. Thus, a correct name would be 2-hydroxy-4-isopropyl-1-methylbenzene.

17.6. The molecular formula (C_8H_8) indicates five degrees of unsaturation (see Section 14.16), four of which are accounted for by the aromatic ring, so the structure must also contain either a ring or a double bond (in addition to the aromatic ring). If compound **A** undergoes bromination to produce a dibromide, then it must contain an alkenyl C=C double bond (the aromatic ring does not react with Br_2). This C=C double bond accounts for the last degree of unsaturation, described above. The structure of compound **A**, as well as the structure of the resulting dibromide (compound **B**) are shown here:

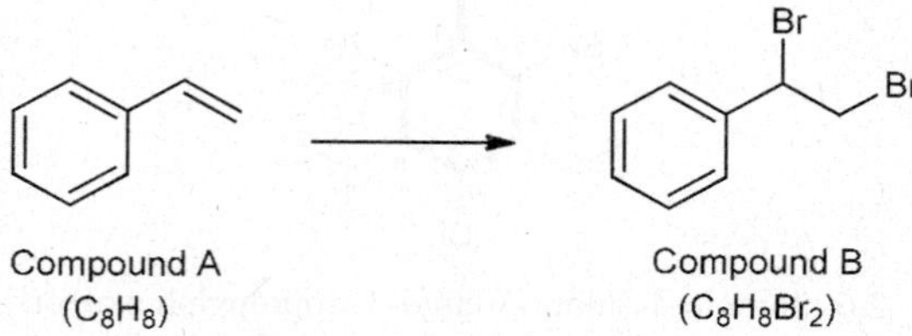

Compound A (C_8H_8) Compound B ($C_8H_8Br_2$)

17.7.

(a) According to Figure 17.1, the conversion from cyclohexane to cyclohexene is uphill in energy (ΔH has a positive value). This should make sense, because this conversion involves the breaking of two C–H bonds, and the formation of a H–H bond and a C=C π bond. Overall, there is a net cleavage of a strong σ bond and the formation of a weak π bond, which is thermodynamically disfavored.

(b) According to Figure 17.1, the conversion from cyclohexene to benzene is uphill in energy (ΔH has a positive value). This should make sense, because this conversion involves the breaking of four C–H bonds, and the formation of two H–H bonds and two C=C π bonds. Overall, there is a net cleavage of two strong σ bonds and the formation of two weak π bonds, which is thermodynamically disfavored.

(c) According to Figure 17.1, the conversion from cyclohexadiene to benzene is downhill in energy (ΔH has a negative value). This transformation is analogous to part (a), but in this case, it is thermodynamically favorable because of the gain in aromaticity (making the products more stable than the starting materials).

17.8.

(a) This compound has six π bonds, for a total of twelve π electrons. Twelve is not a Hückel number (instead, it is $4n$), so the compound is not expected to be aromatic.

(b) This compound has seven π bonds, for a total of fourteen π electrons. Fourteen is a Hückel number ($4n$+2). In Section 17.5, we saw that [14]annulene is somewhat destabilized by a steric interaction between the hydrogen atoms positioned inside the ring. Although [14]annulene is nonplanar, it does indeed exhibit aromatic stabilization, because the deviation from planarity is not too great.

(c) This compound has eight π bonds, for a total of sixteen π electrons. Sixteen is not a Hückel number (instead, it is $4n$), so the compound is not expected to be aromatic.

17.9. We draw a circle and inscribe a triangle inside the circle, with one of the points of the triangle at the bottom of the circle. Each location where the triangle touches the circle represents an energy level. The energy level on the bottom of the circle is a bonding MO, and the two energy levels on top are antibonding MOs. With one bonding MO, we expect that two π electrons are required in order to achieve aromaticity. The structure indeed has two π electrons, which both occupy the bonding MO. This Frost circle indicates that there is only one bonding MO and it is filled, while the antibonding MOs are empty. This structure is expected to exhibit aromatic stabilization.

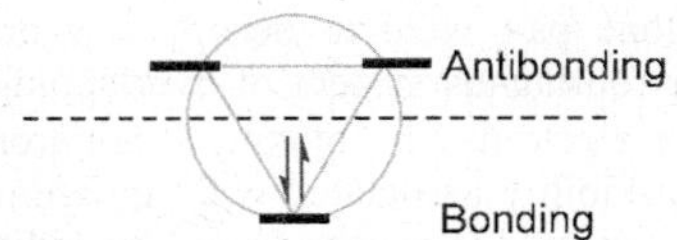

17.10. This structure exhibits a ring of continuously overlapping p orbitals, and there are 22 π electrons (a Hückel number). In addition, the ring is of sufficient size to accommodate the hydrogen atoms that are positioned inside the ring. Therefore, the compound is expected to exhibit aromatic stabilization.

17.11.

(a) This structure exhibits a ring of continuously overlapping p orbitals, but there are four π electrons ($4n$), rendering the structure antiaromatic. This anion is expected to be very high in energy (very unstable).

(b) This structure exhibits a ring of continuously overlapping p orbitals (C+ represents a carbon atom with an empty p orbital), and there are two π electrons ($4n$+2), rendering the structure aromatic.

(c) This structure exhibits a ring of continuously overlapping p orbitals, but there are four π electrons ($4n$), rendering the structure antiaromatic.

(d) This structure exhibits a ring of continuously overlapping p orbitals (the lone pair occupies a p orbital because it is resonance stabilized). There are ten π electrons ($4n$+2), rendering the structure aromatic.

17.12. In order to predict whether this dianion is aromatic, we must determine if the following two criteria have been met:

1. Does the compound contain a ring comprised of continuously overlapping p orbitals?

2. Is there a Hückel number of π electrons in the ring?

The lone pairs can occupy p orbitals, providing for continuous overlap of p orbitals around the ring, so the first criterion has been met. To determine if the second criterion has been met, we must count the number of π electrons. Each π bond counts as two electrons, and each lone pair counts as two electrons, for a total of 10 π electrons. Ten is a Hückel number ($4n$+2), so the dianion is aromatic. All ten π electrons are thus completely delocalized around the eight-membered ring.

17.13.

(a) One of the lone pairs occupies a p orbital, thereby rendering the compound aromatic (six π electrons). The other lone pair will occupy an sp^2-hybridized orbital, in the plane of the ring, extending away from the ring.

(b) One of the lone pairs occupies a p orbital, thereby rendering the compound aromatic (six π electrons). The other lone pair will occupy an sp^2-hybridized orbital, in the plane of the ring, extending away from the ring.

(c) If the lone pair were to occupy a p orbital, there would be a continuous system of overlapping p orbitals with eight π electrons ($4n$, making it antiaromatic). To avoid the instability associated with antiaromaticity, the lone pair is expected to occupy an sp^3-hybridized orbital.

(d) One of the lone pairs on the sulfur atom occupies a p orbital, thereby rendering the compound aromatic (six π electrons). The lone pair on the nitrogen atom occupies an sp^2-hybridized orbital that is in the plane of the ring (because the nitrogen atom is already using a p orbital as part of a π bond to establish aromaticity).

(e) There is only one lone pair (on oxygen) and it is not participating in aromaticity. That oxygen atom is already using a p orbital as part of a π bond to establish aromaticity.

(f) Each nitrogen atom has one lone pair, and neither is participating in aromaticity. In each case, the nitrogen atom is already using a p orbital as part of a π bond to establish aromaticity.

(g) The compound is not aromatic. In order to achieve a continuous system of overlapping p orbitals, each oxygen atom would need to contribute a lone pair in a p orbital, and that would give 8 π electrons (not a Hückel number).

(h) One of the lone pairs on the oxygen atom occupies a p orbital, thereby rendering the compound aromatic (six π electrons). The lone pair on the nitrogen atom occupies an sp^2-hybridized orbital that is in the plane of the ring (because the nitrogen atom is already using a p orbital as part of a π bond to establish aromaticity).

17.14. In addition to the aromatic rings indicated in the structures of the best-selling drugs (Section 17.1), each of the highlighted rings also has a continuous system of overlapping p orbitals, with six π electrons, and is therefore aromatic:

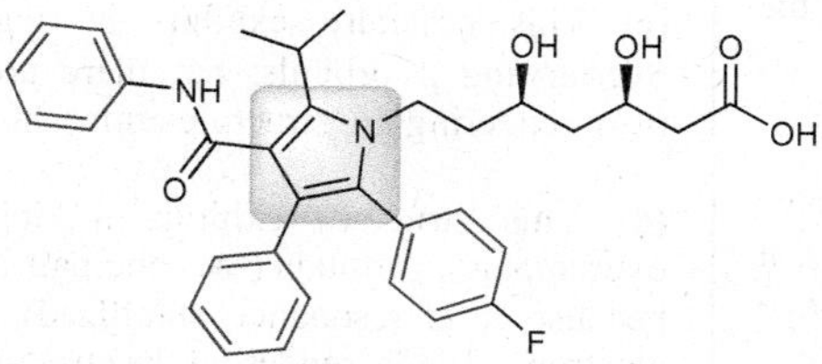

Lipitor

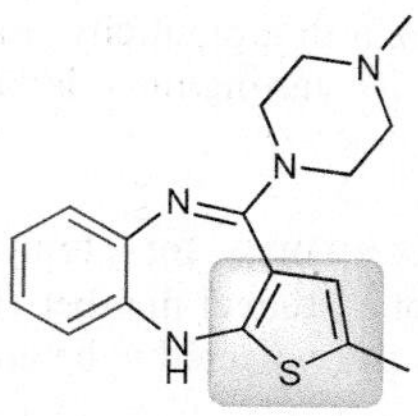

Zyprexa

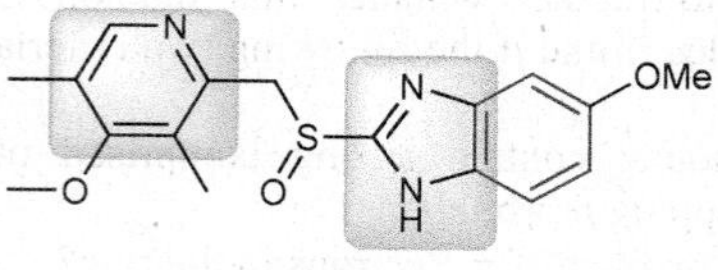

Prilosec

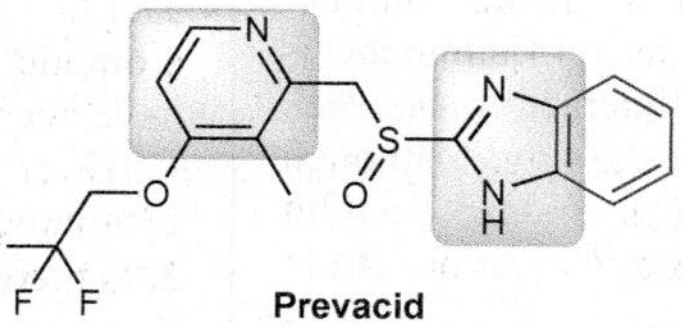

Prevacid

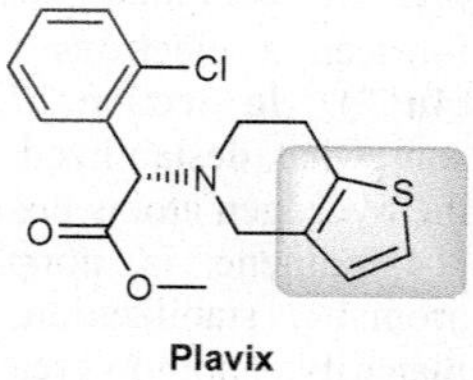

Plavix

17.15. The first compound is expected to be more acidic (is expected to have a lower pK_a), because deprotonation restores aromaticity to the ring.

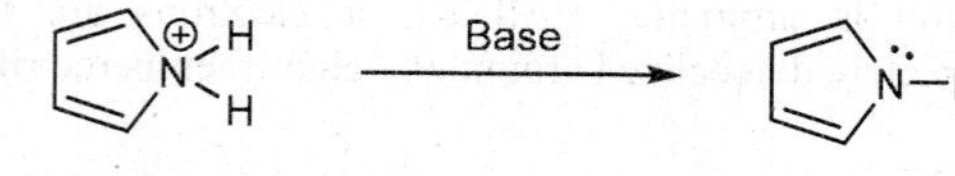

Base

not aromatic → aromatic

The second compound is already aromatic, even before deprotonation.

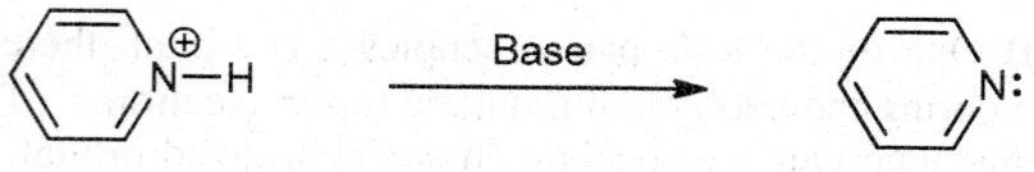

Base

aromatic → aromatic

In this case, the driving force for losing the proton is not as great.

17.16.

(a) Yes, it has the required pharmacophore (two aromatic rings separated by one carbon atom, and a tertiary amine).

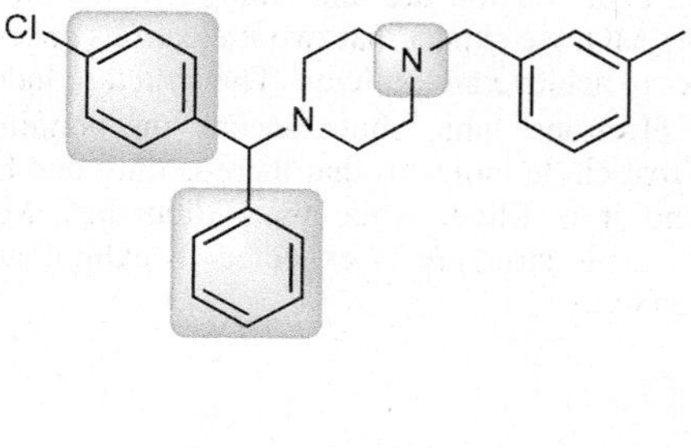

(b) Sedation occurs because meclizine is mostly nonpolar, so it crosses the blood-brain barrier and binds with receptors in the central nervous system.

(c) To modify the sedative properties of meclizine, we could introduce polar functional groups (such as OH groups or COOH groups) that reduce the ability of the compound to cross the blood-brain barrier.

17.17.

(a) The benzylic position undergoes oxidation, giving a carboxylic acid:

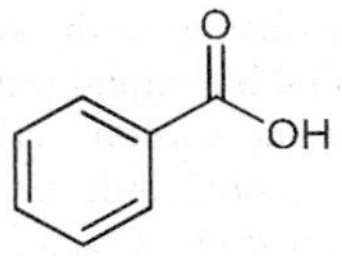

O
OH

(b) Each of the benzylic positions undergoes oxidation. Notice that oxidation of the ethyl group involves cleavage of a C-C bond to give a diacid with one less carbon atom than the starting material.

O O
HO OH

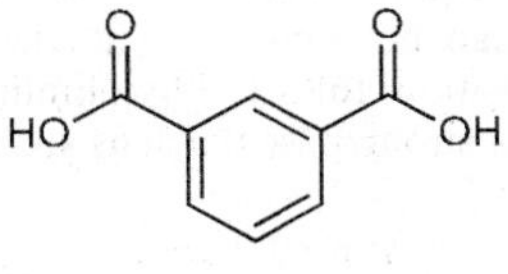

(c) There are three benzylic positions, but one of these positions (the *tert*-butyl group) is quaternary, and as such, that position does not undergo oxidation. The other two benzylic positions undergo oxidation, giving the following diacid:

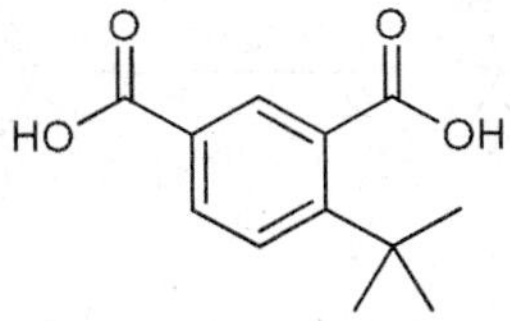

O O
HO OH

17.18.

(a) There are two questions to ask when approaching a synthesis problem.

1) Is there a change in the carbon skeleton?
2) Is there a change in the position or identity of the functional groups?

In this case, there is no change in the carbon skeleton:

OH

but there is a change in the functional group. Specifically, an OH group must be installed in the benzylic position. We did not encounter a one-step method for achieving this transformation. However, we can achieve the desired transformation in two steps (bromination at the benzylic position, followed by an S_N1 process in which water is used as a nucleophile).

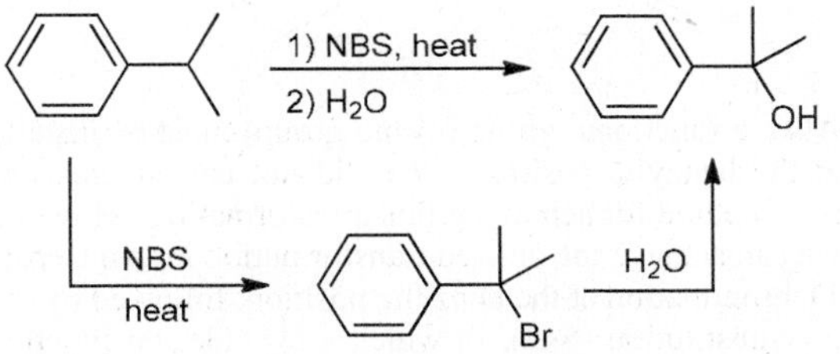

1) NBS, heat
2) H_2O
OH
NBS
heat
Br
H_2O

Note that the leaving group is on a tertiary position, so S_N2 substitution with NaOH would fail (E2 would predominate). Instead, a weak nucleophile is employed (H_2O) to initiate an S_N1 process.

(b) There are two questions to ask when approaching a synthesis problem.

1) Is there a change in the carbon skeleton?
2) Is there a change in the position or identity of the functional groups?

In this case, there is no change in the carbon skeleton.

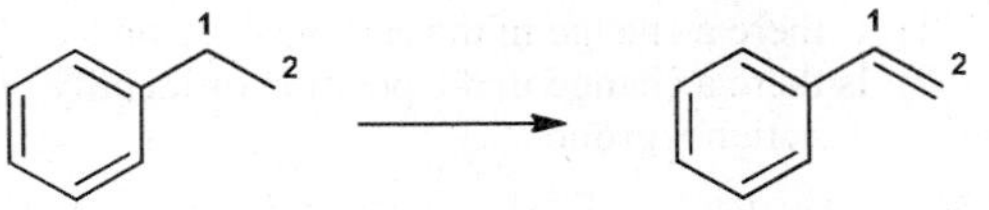

1 2 1 2

but there is a change in the functional group. Specifically, a π bond must be installed. We did not encounter a one-step method for achieving this transformation. However, we can achieve the desired transformation in two steps:
1) bromination at the benzylic position, followed by
2) elimination (E2) upon treatment with a strong base. A sterically hindered base, such as *tert*-butoxide, is required to promote elimination over substitution at the benzylic position:

1) NBS, heat
2) *t*-BuOK
Br
NBS
heat
t-BuOK

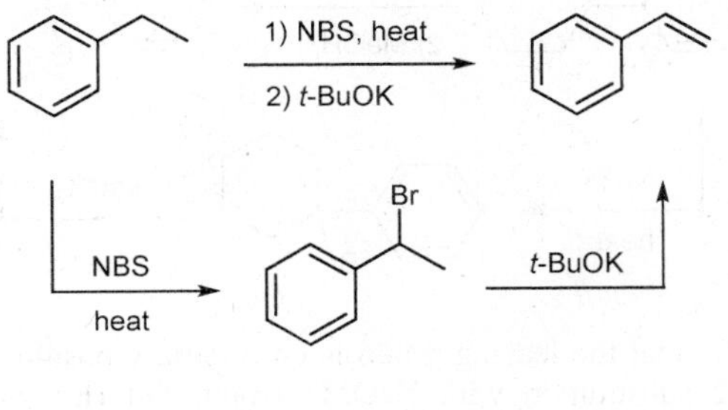

(c) There are two questions to ask when approaching a synthesis problem.

1) Is there a change in the carbon skeleton?
2) Is there a change in the position or identity of the functional groups?

In this case, there is a change in the carbon skeleton. The product has one more carbon atom than the starting material:

Also, a functional group (cyano group) must be installed at the benzylic position. We did not encounter a one-step method for achieving this transformation. However, we can achieve the desired transformation in two steps: 1) bromination at the benzylic position, followed by 2) substitution (S_N2), in which a cyanide ion functions as the nucleophile:

1) NBS, heat
2) NaCN

NBS
heat

NaCN

(d) There are two questions to ask when approaching a synthesis problem.

1) Is there a change in the carbon skeleton?
2) Is there a change in the position or identity of the functional groups?

In this case, there is no change in the carbon skeleton, but there is a change in the functional group. Specifically, a methoxy group (OMe) must be installed in the benzylic position. We did not encounter a one-step method for achieving this transformation. However, we can achieve the desired transformation in two steps:

1) bromination at the benzylic position, followed by
2) substitution (S_N1), in which methanol functions as the nucleophile:

1) NBS, heat
2) MeOH

NBS
heat

MeOH

Note that the leaving group is on a tertiary position, so S_N2 substitution with NaOMe would fail (E2 would predominate). Instead, a weak nucleophile is employed (MeOH) to initiate an S_N1 process.

(e) There are two questions to ask when approaching a synthesis problem.

1) Is there a change in the carbon skeleton?
2) Is there a change in the position or identity of the functional groups?

In this case, there is a change in the carbon skeleton. The product has one less carbon atom than the starting material:

This carbon atom must be removed

Also, a functional group (a C=O bond) must be installed. Let's begin with the change in the carbon skeleton. We have learned one suitable method (ozonolysis) to remove a carbon atom from the carbon skeleton and produce a ketone. Ozonolysis would indeed install the necessary functional group in the correct location:

1) O_3
2) DMS

This strategy requires that we first form the alkene above, which can be accomplished via bromination of the benzylic position, followed by elimination (E2) upon treatment with a strong base (such as sodium ethoxide):

1) NBS, heat
2) NaOEt
3) O_3
4) DMS

NBS,
heat

NaOEt

1) O_3
2) DMS

(f) There are two questions to ask when approaching a synthesis problem.

1) Is there a change in the carbon skeleton?
2) Is there a change in the position or identity of the functional groups?

In this case, there is a change in the carbon skeleton. The product has two additional carbon atoms that are absent in the starting material:

Also, a functional group (an OH group) must be installed at the benzylic position. Both of these goals (the change in carbon skeleton and installation of the functional group) can be accomplished via a Grignard reaction. That is, the following Grignard reaction can create the necessary C–C bond, while leaving an OH group in the desired location.

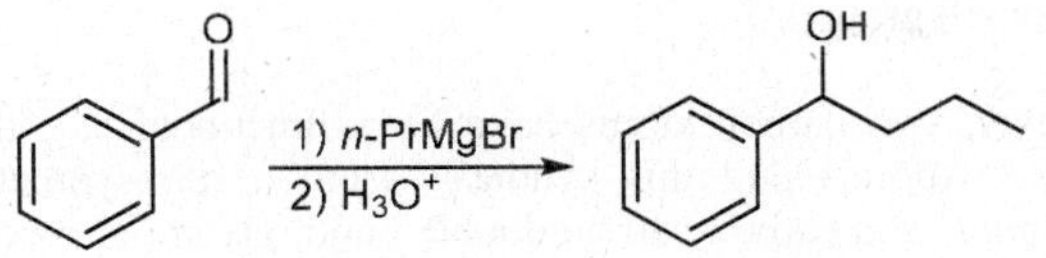

1) *n*-PrMgBr
2) H_3O^+

However, the necessary aldehyde does not have the same number of carbon atoms as the starting material:

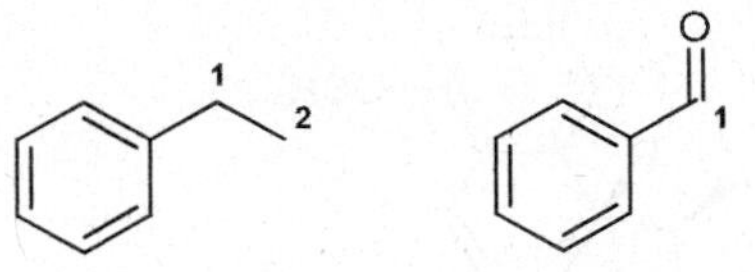

Nevertheless, we can easily convert the starting material into the desired aldehyde via the following three step process:

1) bromination at the benzylic position,
2) elimination (E2) upon treatment with a strong, sterically hindered base (to minimize S_N2 competition),
3) ozonolysis

This strategy is shown here:

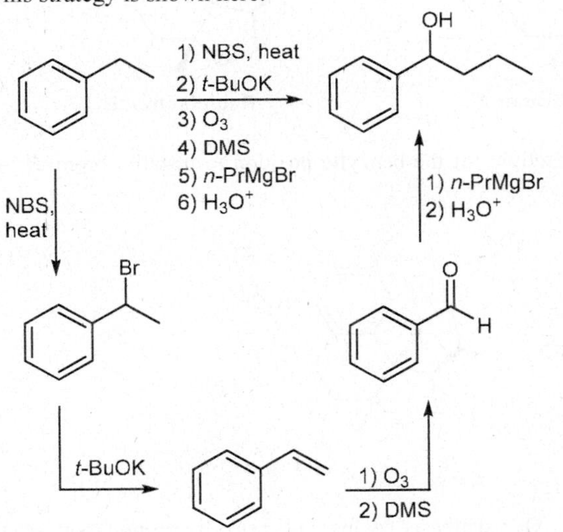

1) NBS, heat
2) *t*-BuOK
3) O_3
4) DMS
5) *n*-PrMgBr
6) H_3O^+

NBS, heat

1) *n*-PrMgBr
2) H_3O^+

t-BuOK

1) O_3
2) DMS

There are certainly other acceptable solutions to this problem. The solution above represents just one approach. For example, the following is another acceptable approach:

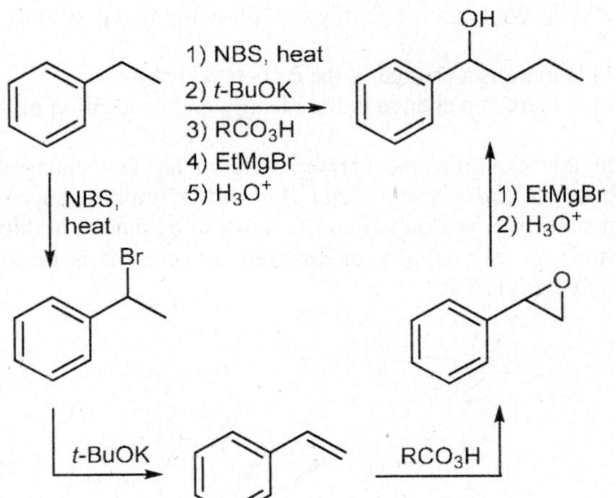

1) NBS, heat
2) *t*-BuOK
3) RCO_3H
4) EtMgBr
5) H_3O^+

NBS, heat

1) EtMgBr
2) H_3O^+

t-BuOK

RCO_3H

And here is yet one further way to achieve the desired transformation. In the following proposed synthesis, the regiochemical outcome of the final step is controlled by the preference for formation of a benzylic carbocation intermediate, and the regiochemical outcome of the penultimate step is controlled by the preference for formation of a conjugated π bond.

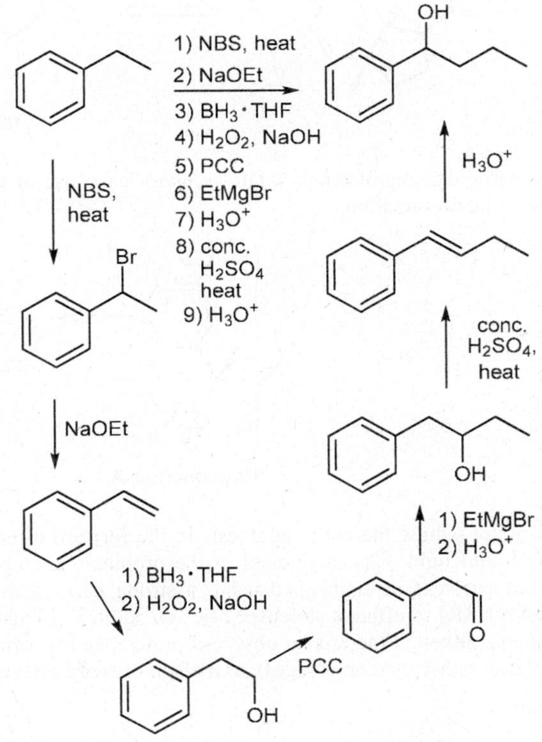

1) NBS, heat
2) NaOEt
3) $BH_3 \cdot THF$
4) H_2O_2, NaOH
5) PCC
6) EtMgBr
7) H_3O^+
8) conc. H_2SO_4 heat
9) H_3O^+

NBS, heat

H_3O^+

conc. H_2SO_4, heat

NaOEt

1) EtMgBr
2) H_3O^+

1) $BH_3 \cdot THF$
2) H_2O_2, NaOH

PCC

17.19. We begin by asking the following two questions:

1) Is there is a change in the carbon skeleton?
2) Is there is a change in the identity and/or location of the functional group(s)?

In this example, the carbon skeleton has not changed; however, two carbon atoms have been functionalized (the benzylic carbon atom and the carbon atom adjacent to it). Approaching this synthesis from a retrosynthetic perspective, we know that the diol can be made via dihydroxylation of a carbon-carbon double bond; given our recent study of the stability of aromatic systems, it is reasonable to expect that the aromatic π bonds will not undergo dihydroxylation:

OCH3 H OH OH ⟹ OCH3 H

The carbon-carbon double bond can be made via an elimination reaction, using a suitable leaving group, from either of two regioisomers:

OCH3 H ⟹ OCH3 H X or OCH3 H X

Regioisomer A **Regioisomer B**

In particular, regioisomer A allows us to make use of the reactivity of the benzylic position to install a bromide via benzylic bromination.

OCH3 H Br ⟹ OCH3 H

Regioisomer A

Now let's draw the entire synthesis in the forward direction. Our synthesis begins with benzylic bromination, using NBS and light. As mentioned in the problem statement, formation of the secondary benzylic bromide is favored. Elimination is accomplished using a strong base, such as DBU. The desired *cis*-diol can then be obtained using OsO_4/NMO to affect a stereospecific *syn* addition. This addition reaction can theoretically occur on either face of the alkene, although there is an observed preference for formation of the diastereomer shown (with the two OH groups on dashes, rather than on wedges), as a result of steric effects:

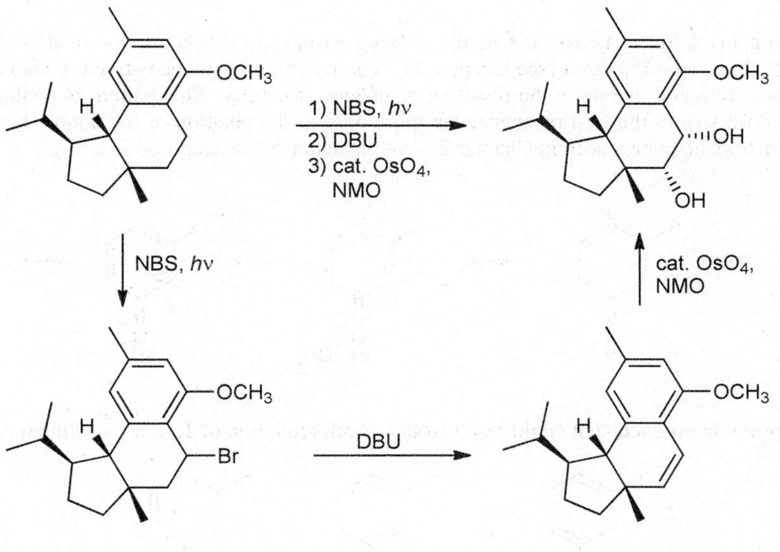

NBS, $h\nu$

1) NBS, $h\nu$
2) DBU
3) cat. OsO_4, NMO

DBU

cat. OsO_4, NMO

17.20.

(a) The ring bears one substituent (an alkyl group), which is electron donating. The carbon atom next to the alkyl group will not be reduced. In a Birch reduction, only two positions are reduced, and they must be 1,4 to each other. This gives the following product (note: the two structures below represent the same compound):

(b) The ring bears two substituents (alkyl groups), which are both electron donating. The carbon atoms next to the alkyl groups will not be reduced. In a Birch reduction, only two positions are reduced, and they must be 1,4 to each other. This gives the following product (note: the two structures below represent the same compound):

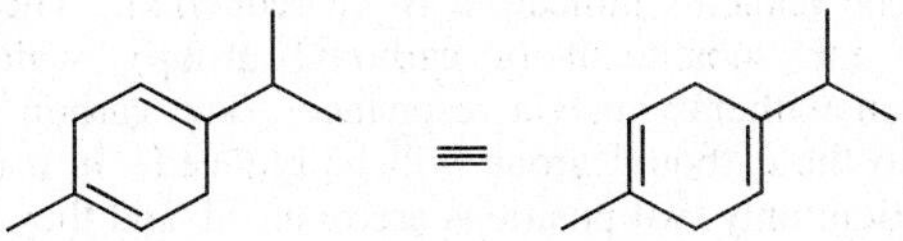

(c) The ring bears two substituents (alkyl groups), which are both electron donating. The carbon atoms next to the alkyl groups will not be reduced. In a Birch reduction, only two positions are reduced, and they must be 1,4 to each other. This gives the following product:

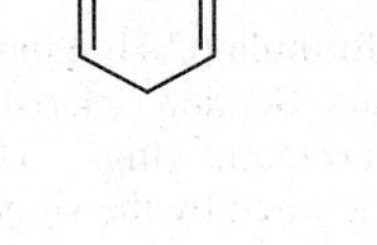

(d) The ring bears two substituents (alkyl groups), which are both electron donating. The carbon atoms next to the alkyl groups will not be reduced. In a Birch reduction, only two positions are reduced, and they must be 1,4 to each other. This gives the following product:

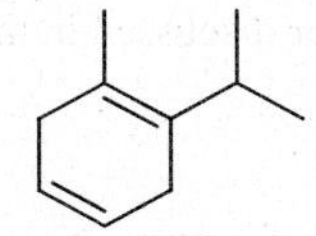

(e) The ring bears one substituent (a carboxylic acid group), which is electron withdrawing. The carbon atom next to the carboxylic acid group will be reduced. In a Birch reduction, only two positions are reduced, and they must be 1,4 to each other. This gives the following product:

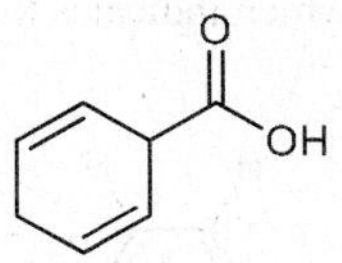

(f) The ring bears two substituents. The alkyl group is electron donating, so the carbon atom next to the alkyl group is not reduced. The carboxylic acid group is electron withdrawing, so the carbon atom next to the carboxylic acid group is reduced. In a Birch reduction, only two positions are reduced, and they must be 1,4 to each other. This gives the following product:

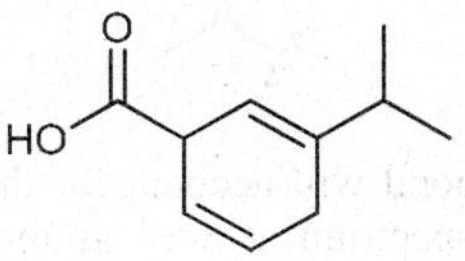

17.21.

(a) Compound **1** contains a lone pair (on the methoxy oxygen) next to a π-bond, so we draw two curved arrows associated with that characteristic pattern (see Chapter 2). The resulting resonance structure also exhibits a lone pair next to a π-bond, so once again, we draw the resulting resonance structure. This pattern is continued, until the final resonance structure differs from the first resonance structure only in the position of the double bonds (that is, in fact, one of the other characteristic patters seen in Chapter 2 – conjugated π-bonds enclosed in a ring).

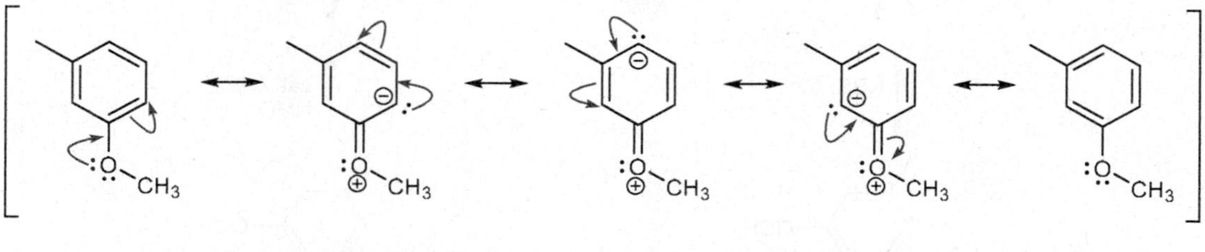

(b) There are three possible products that could result from a Birch reduction of **1**. Let's call them **A**, **B**, and **C**.

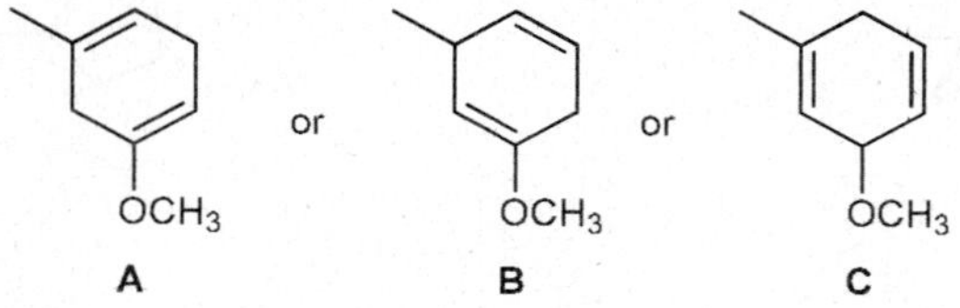

We know that alkyl groups are electron donating and they direct the reduction away from the point of attachment of the alkyl group. Thus, option **B** is unlikely because the carbon atom attached to the methyl group is reduced. If the methoxy substituent was predominantly electron-donating, **A** would be the major product; and if the methoxy substituent was predominantly electron-withdrawing, then **C** would be the major product. Since **A** is the major product, we deduce that the methoxy group must be electron-donating, and resonance must be the major factor. Indeed, this fact will be further discussed in the next chapter.

17.22.

(a) The molecular formula (C_8H_8O) indicates five degrees of unsaturation (see Section 14.16), which is highly suggestive of an aromatic ring, in addition to either one double bond or one ring. The aromatic ring is confirmed by the presence of a signal just above 3000 cm^{-1} in the IR spectrum, as well a signal at 1646 cm^{-1}. In the ^1H NMR spectrum, the multiplet at 7.5 ppm also confirms the aromatic ring. Notice that this multiplet has an integration of 5, which indicates that the aromatic ring is monosubstituted:

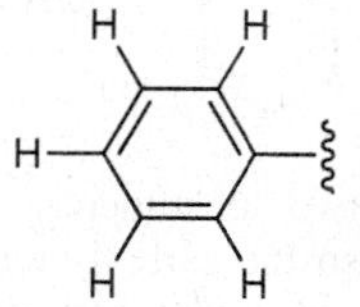

The singlet with an integration of 3 indicates a methyl group, and its chemical shift suggests it might be near a C=O bond:

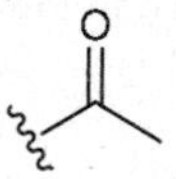

Indeed, a C=O bond will account for the signal at 1686 cm^{-1} in the IR spectrum as well as the fifth degree of unsaturation described earlier. The signal at 1686 cm^{-1} is consistent with a conjugated carbonyl group, as seen in the following structure:

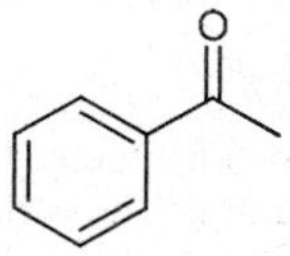

(b) As seen in Section 17.2, the common name for this compound is acetophenone.

(c) The reagents indicate a Birch reduction. The ring bears one substituent (a carbonyl group), which is electron withdrawing via resonance. The carbon atom next to the carbonyl group will be reduced. In a Birch reduction, only two positions are reduced, and they must be 1,4 to each other. This gives the following product:

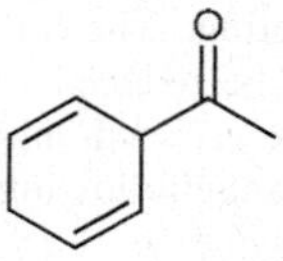

17.23.

(a) The molecular formula (C_8H_{10}) indicates four degrees of unsaturation (see Section 14.16), which is highly suggestive of an aromatic ring. The presence of an aromatic ring is confirmed by the signals just above 3000 cm^{-1} in the IR spectrum, as well as the signal near 1600 cm^{-1}. In the ^{1}H NMR spectrum, the multiplet at 7.1 ppm further confirms the presence of an aromatic ring. The integration of this signal is 4, indicating that the aromatic ring is disubstituted. The aromatic ring accounts for six of the eight carbon atoms in the compound (C_8H_{10}), so there must be two methyl groups attached to the ring (dimethyl benzene). To determine the substitution pattern (*ortho*, *meta* or *para*), we note that the ^{13}C NMR spectrum exhibits three signals in the region of 100-150 ppm, which indicates the *ortho* substitution pattern:

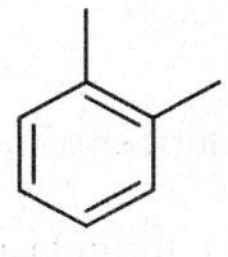

(b) As seen in Section 17.2, the common name for *ortho*-dimethylbenzene is *ortho*-xylene.

(c) When treated with chromic acid (a strong oxidizing agent), both benzylic positions are oxidized, giving the following diacid:

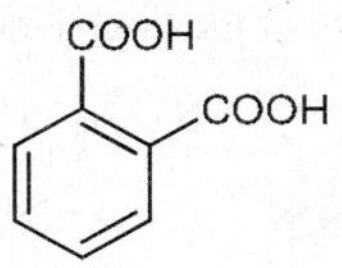

COOH

COOH

17.24.

(a) The parent is benzoic acid. An ethyl group is located at C4 (also called the *para* position), so the compound is 4-ethylbenzoic acid (also called *para*-ethylbenzoic acid):

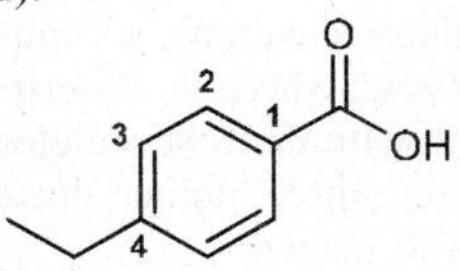

O 2 1 3 OH 4

(b) The parent is phenol. A bromine atom is located at C2 (also called the *ortho* position), so the compound is 2-bromophenol (also called *ortho*-bromophenol):

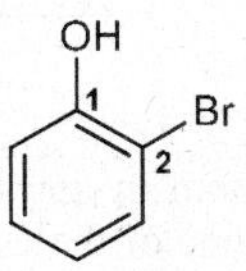

OH 1 Br 2

(c) The parent is phenol. A chlorine atom is located at C2 and a nitro group is located at C4, so the compound is 2-chloro-4-nitrophenol:

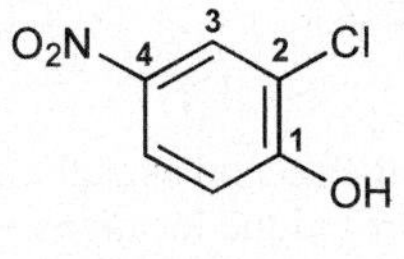

O_2N 4 3 2 Cl 1 OH

(d) The parent is benzaldehyde. A bromine atom is located at C2 and a nitro group is located at C5 (as shown), so the compound is 2-bromo-5-nitrobenzaldehyde:

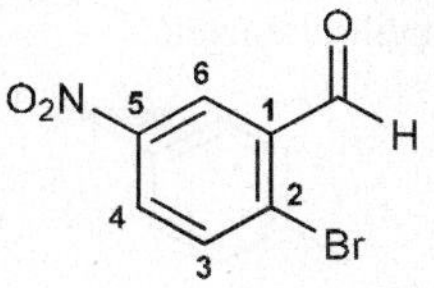

O 6 O_2N 5 1 H 2 4 Br 3

(e) The parent is benzene. There are two substituents, both isopropyl groups, located at C1 and C4, so the compound is 1,4-diisopropylbenzene (also called *para*-diisopropylbenzene):

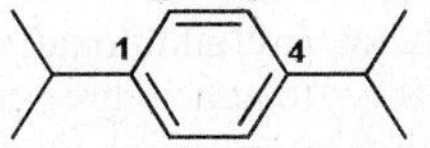

2 3 1 4

17.25.

(a) The parent is benzene, and there are two chlorine atoms that are *ortho* to each other (C1 and C2).

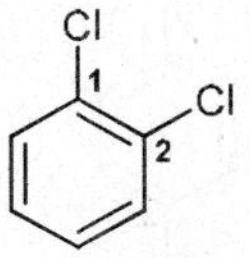

Cl 1 Cl 2

(b) As seen in Section 17.2, anisole is the common name for methoxybenzene:

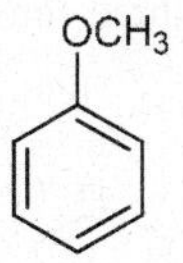

OCH_3

(c) The parent is toluene (methylbenzene), and there is a nitro group in the *meta* position (C3).

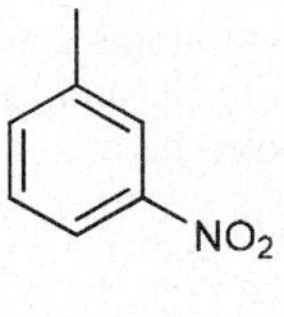

NO_2

(d) As seen in Section 17.2, aniline is the common name for aminobenzene:

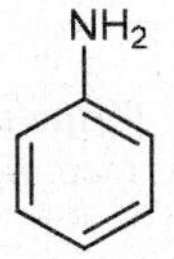

NH_2

(e) The parent is phenol (hydroxybenzene), and there are three bromine atoms, connected to positions C2, C4, and C6.

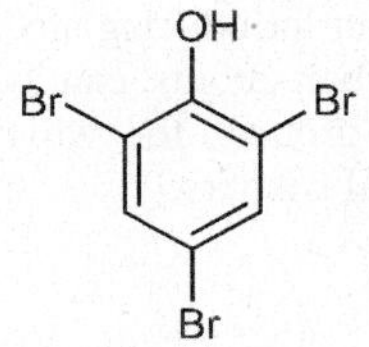

OH Br Br Br

(f) As seen in Section 17.2, *para*-xylene is the common name for *para*-dimethylbenzene:

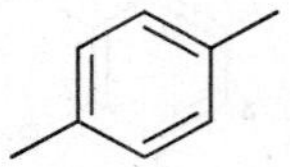

17.26. The molecular formula (C_9H_{12}) indicates four degrees of unsaturation (see Section 14.16), and the benzene ring accounts for all four degrees of unsaturation. The ring contains six carbon atoms, so the remaining three carbon atoms must be attached to the aromatic ring without any additional π bonds or rings. There are several ways to attach three carbon atoms to an aromatic ring without introducing any additional π bonds or rings. The carbon atoms can be attached as three individual methyl groups, for which there are three unique constitutional isomers:

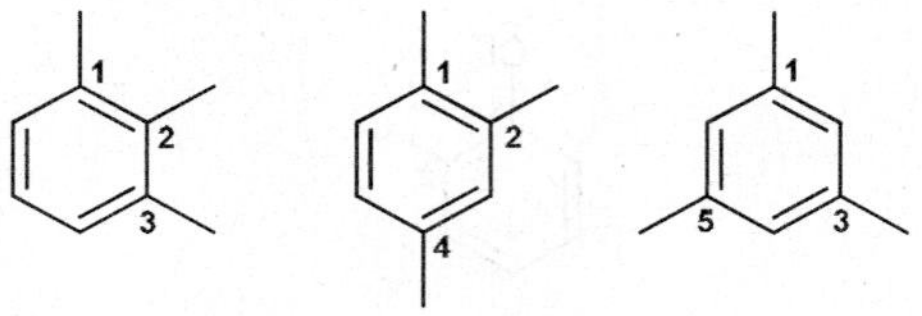

or there can be one ethyl group and one methyl group, for which there are also three constitutional isomers:

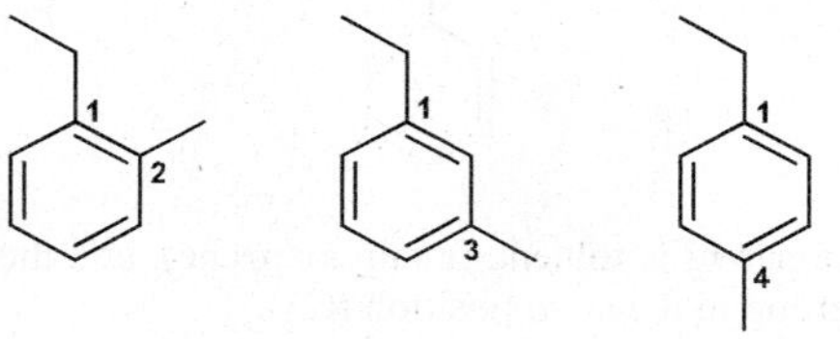

Finally, all three carbon atoms can be connected to the ring as a single substituent, and there are two ways to accomplish that, as shown here:

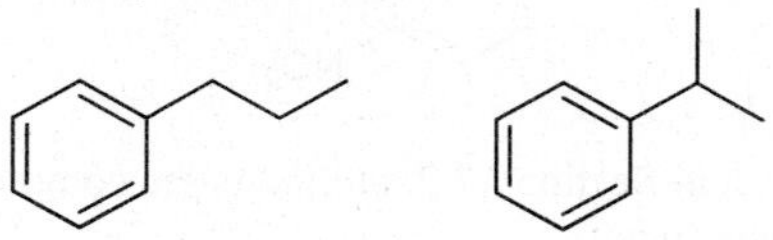

17.27. The molecular formula (C_8H_{10}) indicates four degrees of unsaturation (see Section 14.16), and the aromatic ring accounts for all four degrees of unsaturation. The ring contains six carbon atoms, so the remaining two carbon atoms must be attached to the aromatic ring without any additional π bonds or rings. There are several ways to attach two carbon atoms to an aromatic ring without introducing any additional π bonds or rings. The carbon atoms can be attached as two individual methyl groups, for which there are three unique constitutional isomers:

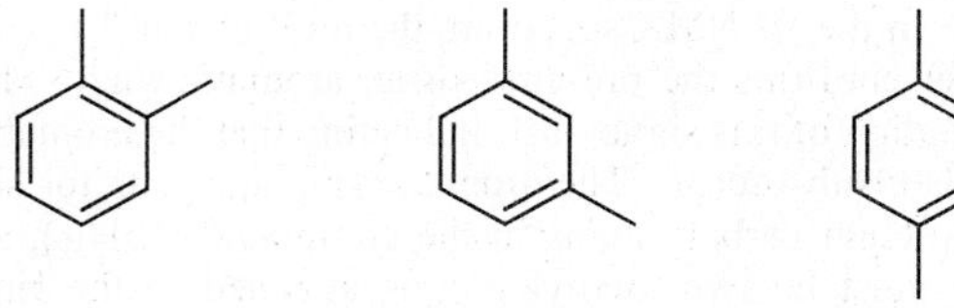

or there can simply be one ethyl group:

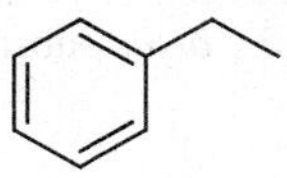

In total, we have seen four constitutional isomers.

17.28. The molecular formula (C_8H_9Cl) is similar to the molecular formula in the previous problem (C_8H_{10}), but one of the hydrogen atoms has been replaced with a chlorine atom. So, we begin our analysis by redrawing all four skeletons shown in the solution to the previous problem. In each case, we will draw a circle for the aromatic ring, rather than drawing π bonds, as it will be easier to identify symmetry without being distracted by the arbitrary locations of the double bonds:

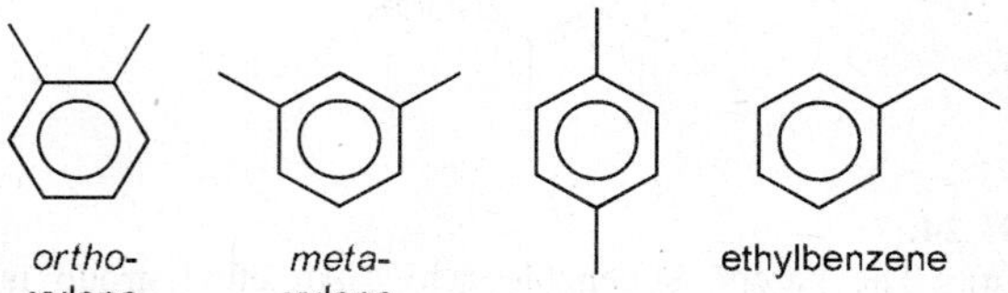

ortho-xylene *meta*-xylene *para*-xylene ethylbenzene

For each one of these skeletons, we must consider all of the unique locations where a chlorine atom can be placed. Let's begin with the first skeleton (*ortho*-xylene), for which there are three unique locations where the chlorine atom can be placed:

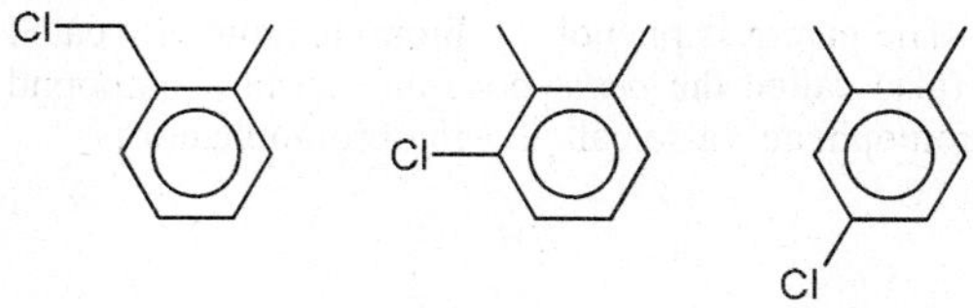

Placing the chlorine atom in any other location on this skeleton will result in one of the compounds above. For example:

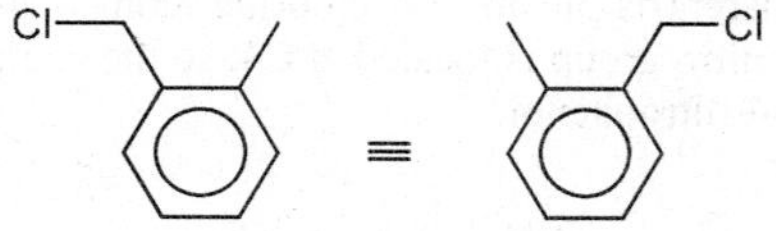

Next, we consider the second skeleton (*meta*-xylene), for which there are four unique locations where the chlorine atom can be placed:

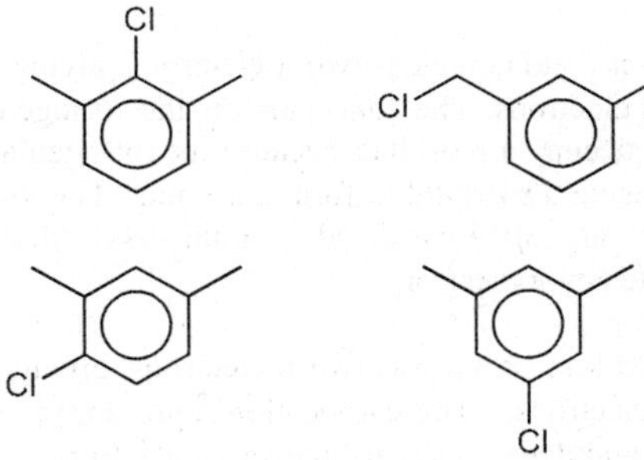

Next, we consider the third skeleton (*para*-xylene), for which there are only two unique locations where the chlorine atom can be placed:

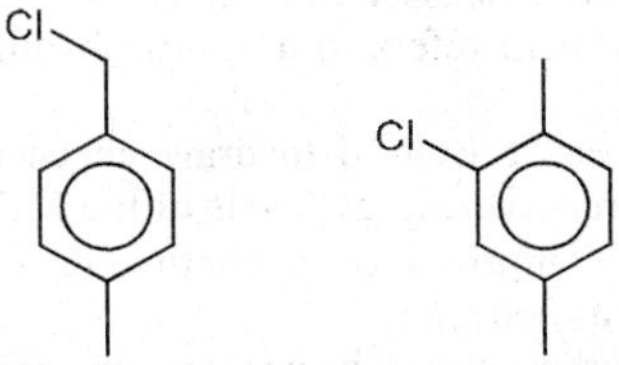

And finally, we consider the last skeleton (ethylbenzene), for which there are five unique locations where the chlorine atom can be placed:

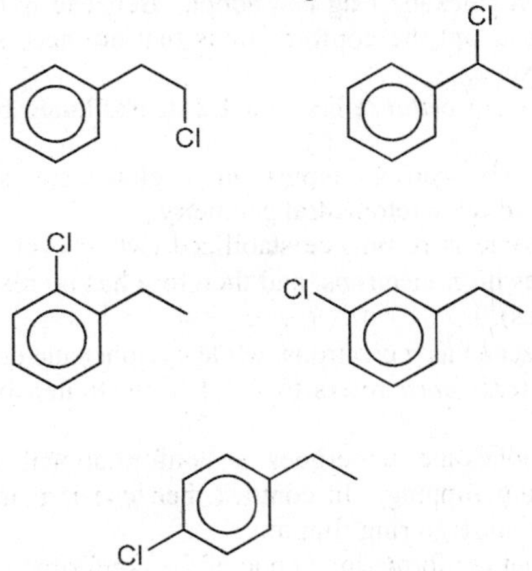

In total, we have seen fourteen constitutional isomers.

17.29. We begin by placing the first two nitro groups at positions C2 and C3:

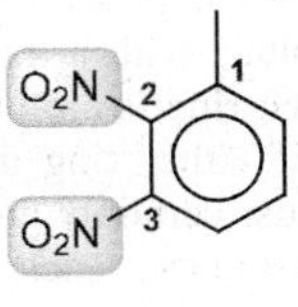

There are now three choices of where to place the third nitro group (C4, C5, or C6), giving rise to the following three constitutional isomers:

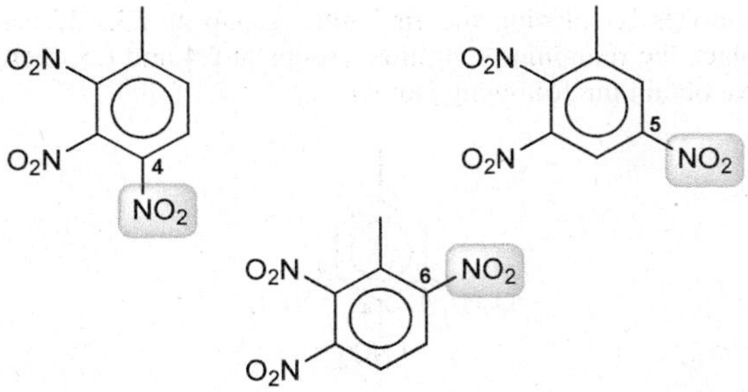

We have now exhausted all possibilities in which the first two nitro groups occupy positions C2 and C3. Next, we place the first two nitro groups at C2 and C4:

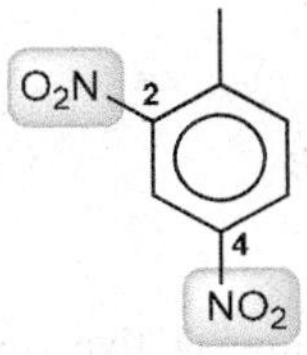

There are now two choices of where to place the third nitro group (C5 or C6), giving rise to the following two constitutional isomers, although the second isomer is the one described in the problem statement (so we will not count that one):

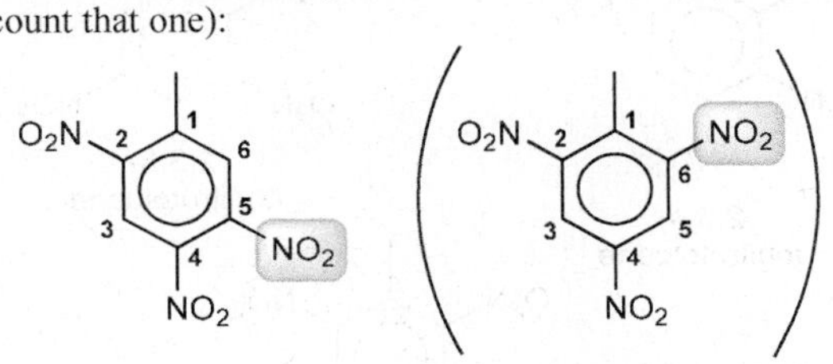

We do not count this isomer, as per the instructions in the problem statement

We have now exhausted all possibilities in which the first two nitro groups occupy positions C2 and C4. Next, we place the first two nitro groups at C2 and C5:

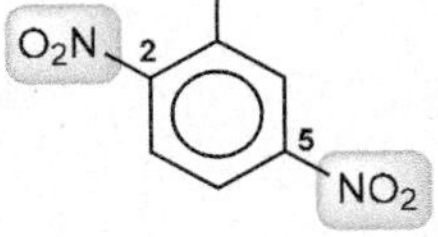

This leaves only one option for the third nitro group (C6), and this isomer has already been drawn earlier:

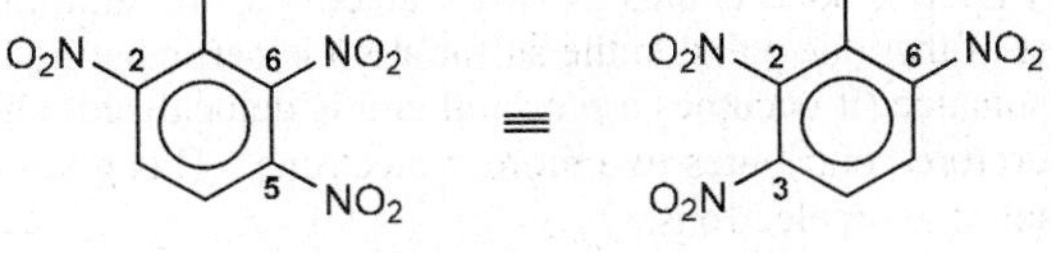

We have now considered all options for which the first nitro group is placed at C2. Now we continue our

analysis by placing the first nitro group at C3. If we place the remaining two nitro groups at C4 and C5, then we obtain the following isomer:

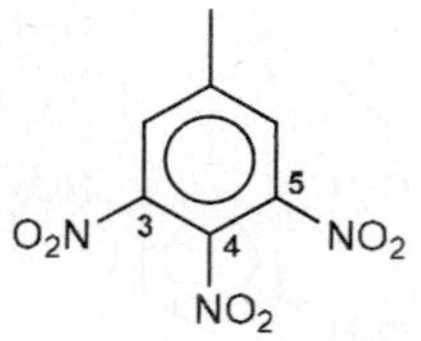

Any other arrangement of nitro groups will be identical to one of the arrangements already drawn. For example:

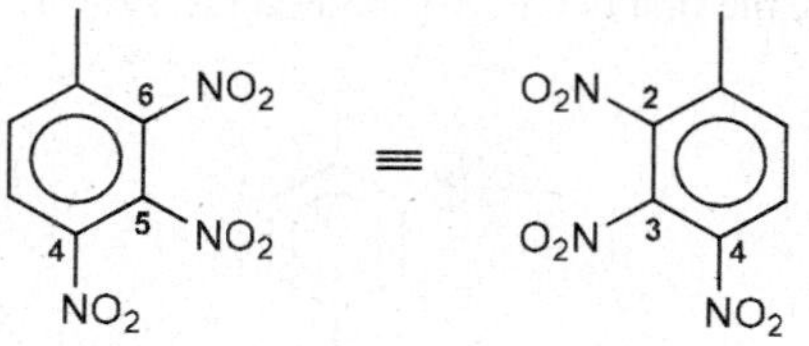

In summary, the following five isomers are consistent with the requirements described in the problem statement:

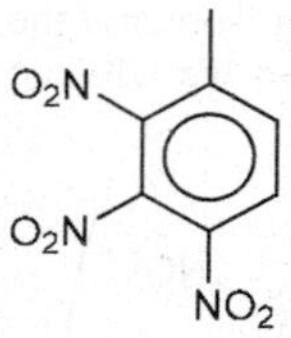

2,3,4-trinitrotoluene

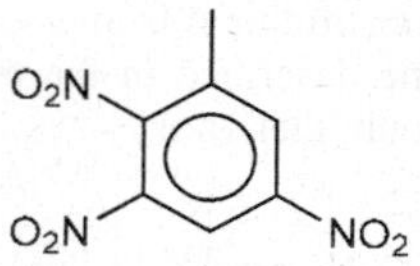

2,3,5-trinitrotoluene

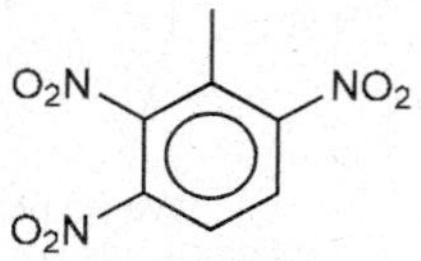

2,3,6-trinitrotoluene

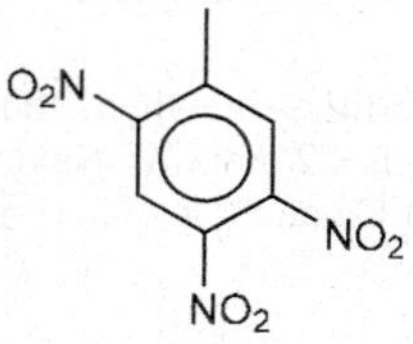

2,4,5-trinitrotoluene

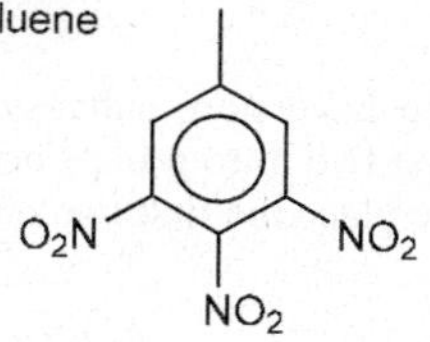

3,4,5-trinitrotoluene

17.30.

(a) Each π bond counts as two π electrons, and the lone pair also counts as two π electrons, giving a total of ten π electrons.

(b) Each π bond counts as two π electrons. In addition, one of the lone pairs on the sulfur atom is participating in resonance (it occupies a *p* orbital and is delocalized) and therefore contributes two more π electrons. This gives a total of six π electrons.

(c) Each π bond counts as two π electrons, giving a total of ten π electrons.

(d) Each π bond counts as two π electrons, giving a total of four π electrons. The lone pair on the nitrogen atom does not occupy a *p* orbital, because the nitrogen atom is already using a *p* orbital to form a π bond. The lone pair occupies an sp^2-hybridized orbital and does not contribute any π electrons.

(e) Each π bond counts as two π electrons, giving a total of six π electrons. The carbocation represents an empty *p* orbital and does not contribute any π electrons.

17.31. The terms used in reference to cyclohexane were discussed in Chapter 4, while the terms used in reference to benzene were discussed in Chapter 17.

(a) The term *meta* refers to a 1,3-disubstituted benzene ring.

(b) A Frost circle is used to draw an energy diagram showing the relative energy levels of the MOs associated with a ring comprised of a continuous conjugated π system, such as benzene.

(c) All six carbon atoms in benzene are sp^2 hybridized, with trigonal planar geometry.

(d) A chair conformation is one of the conformations that a cyclohexane ring can adopt. Benzene is flat and does not adopt the conformations that are accessible to cyclohexane.

(e) The term *ortho* refers to a 1,2-disubstituted benzene ring.

(f) All six carbon atoms in cyclohexane are sp^3 hybridized, with tetrahedral geometry.

(g) Benzene is resonance stabilized (while cyclohexane possesses no π electrons, and therefore has no resonance structures).

(h) Benzene has π electrons, while cyclohexane does not.

(i) The term *para* refers to a 1,4-disubstituted benzene ring.

(j) Cyclohexane undergoes a conformational change called ring flipping. In contrast, benzene is planar and does not undergo ring flipping.

(k) A boat conformation is one of the conformations that a cyclohexane ring can adopt. Benzene is flat and does not adopt the conformations that are accessible to cyclohexane.

17.32.

(a) Each of the rings is comprised of a continuous system of overlapping *p* orbitals with a total of six π electrons ($4n+2$). As such, each ring is aromatic. The compound would be aromatic if either ring alone were aromatic. This compound is most certainly aromatic. Notice that, in determining aromaticity, each ring is considered individually.

(b) As described in Section 17.5, the hydrogen atoms positioned inside the ring experience a steric interaction that forces this compound, called [10]annulene, out of planarity. Since the molecule cannot adopt a planar conformation, the *p* orbitals cannot continuously overlap with each other to form one system, and as a result,

[10]annulene does not meet the criteria for aromaticity. It is not aromatic.

(c) In order for this five-membered ring to exhibit a continuous system of overlapping *p* orbitals, each oxygen atom would have to be sp^2 hybridized (thereby placing a lone pair in a *p* orbital). If this were in fact to be the case, the ring would have eight π electrons (two π electrons for the π bond and two π electrons for each lone pair that occupies a *p* orbital). This would make the ring antiaromatic (with $4n$ π electrons). Therefore, not all of the oxygen atoms will adopt sp^2 hybridization. At least one of them will adopt sp^3 hybridization to avoid antiaromaticity. As such, the compound does not have a continuous system of *p* orbitals and is therefore not aromatic.

(d) One of the lone pairs on the oxygen atom occupies a *p* orbital, thereby rendering the compound aromatic (six π electrons). The lone pair on the nitrogen atom occupies an sp^2-hybridized orbital that is in the plane of the ring (because the nitrogen atom is already using a *p* orbital in a π bond to establish aromaticity).

(e) One of the carbon atoms in the ring is sp^3 hybridized (highlighted):

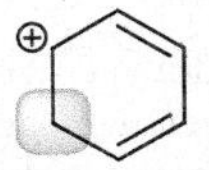

Therefore, the structure does not possess a continuous system of overlapping *p* orbitals, and it is not aromatic.

17.33.

(a) The six-membered ring is aromatic because it is comprised of a continuous system of overlapping *p* orbitals with six π electrons:

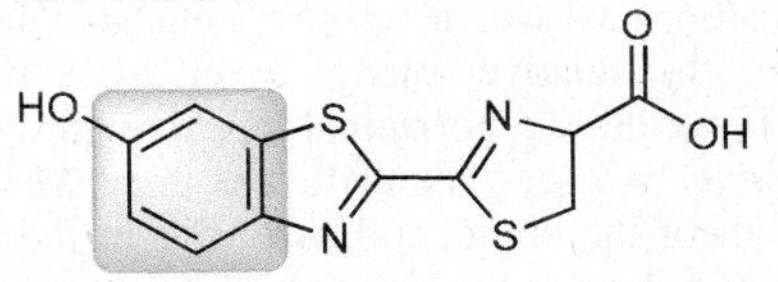

In addition, the adjacent five-membered ring is also aromatic:

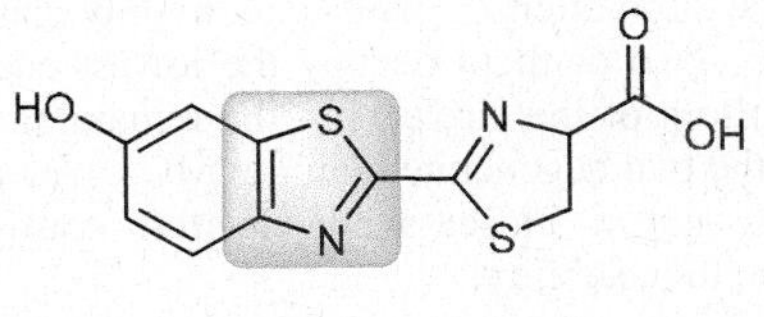

In this five-membered ring, one of the lone pairs on the sulfur atom occupies a *p* orbital, thereby rendering the ring aromatic (six π electrons). The lone pair on the nitrogen atom occupies an sp^2-hybridized orbital that is in the plane of the ring (because the nitrogen atom is already using a *p* orbital in a π bond to establish aromaticity).

The other five-membered ring (not highlighted above) is not aromatic because the ring contains sp^3-hybridized atoms and is therefore not a continuous system of overlapping *p* orbitals:

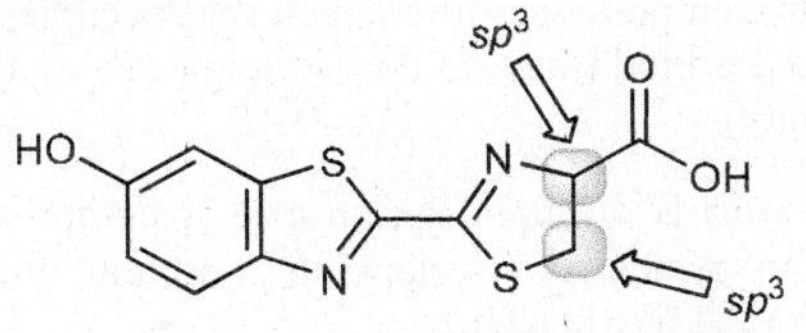

(b) Many of the lone pairs are delocalized by resonance (one lone pair on each of the sulfur atoms, as well as one lone pair on each of the OH groups). However, only one of these lone pairs (on the atom highlighted below) is participating in establishing aromaticity, as explained in the solution to part (a):

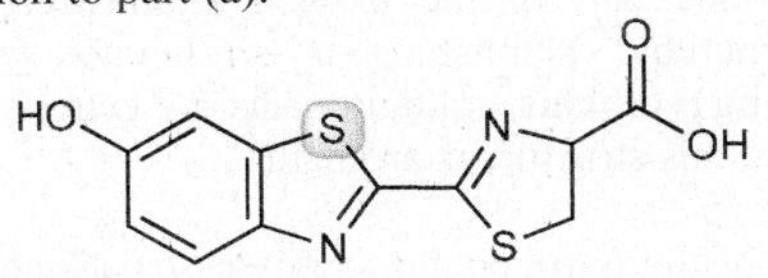

17.34.

(a) If the oxygen atom adopts sp^2 hybridization so that the lone can occupy a *p* orbital (to give a continuous system of overlapping *p* orbitals), then there would be a total of eight π electrons (six π electrons from the double bonds and two π electrons from the lone pair that occupies a *p* orbital). This compound would therefore be antiaromatic (with $4n$ π electrons) and extremely high in energy. To avoid the huge energy cost associated with antiaromaticity, the oxygen atom likely adopts sp^3 hybridization, so that the lone pair occupies an sp^3-hybridized orbital. This hybridization state destroys the continuous system of overlapping *p* orbitals, rendering the compound nonaromatic. As such, the lone pairs on the oxygen atom are expected to be localized (to avoid antiaromaticity).

(b) If the nitrogen atom adopts sp^2 hybridization so that the lone can occupy a *p* orbital (to give a continuous system of overlapping *p* orbitals), then there would be a total of eight π electrons (six π electrons from the double bonds and two π electrons from the lone pair that occupies a *p* orbital). This compound would therefore be antiaromatic (with $4n$ π electrons) and extremely high in energy. To avoid the huge energy cost associated with antiaromaticity, the nitrogen atom likely adopts sp^3 hybridization, so that the lone pair occupies an sp^3-hybridized orbital. This hybridization state destroys the continuous system of overlapping *p* orbitals, rendering the compound nonaromatic. As such, the lone pair on the nitrogen atom is expected to be localized (to avoid antiaromaticity).

(c) This compound is aromatic. One of the lone pairs of the sulfur atom occupies a *p* orbital, thereby establishing a continuous system of overlapping *p* orbitals, containing six π electrons. The lone pair on the nitrogen atom occupies an sp^2-hybridized orbital that is in the plane of the ring (because the nitrogen atom is already using a *p* orbital in a π bond to establish aromaticity).

(d) This compound is aromatic, much like benzene. Each nitrogen possesses a lone pair that occupies an sp^2-hybridized orbital (they do not participate in establishing aromaticity).

(e) The ring is aromatic because it is comprised of a continuous system of overlapping *p* orbitals with six π electrons, just like benzene.

(f) The nitrogen atom is sp^3 hybridized and does not have a *p* orbital, so there is not a continuous system of overlapping *p* orbitals. Therefore, this structure is nonaromatic.

(g) The lone pair of the oxygen atom occupies a *p* orbital, thereby establishing a continuous system of overlapping *p* orbitals, containing six π electrons. Therefore, this structure is aromatic.

(h) Each of the nitrogen atoms possesses a lone pair. If each of these nitrogen atoms adopts sp^2 hybridization so that each lone pair can occupy a *p* orbital, then there would be a continuous system of overlapping *p* orbitals, with a total of six π electrons (two π electrons from the double bond, and then two more π electrons from each lone pair that occupies a *p* orbital). Therefore, this compound is expected to exhibit aromatic stabilization.

17.35.

(a) The following compound is expected to undergo an SN1 reaction most readily, because loss of the leaving group generates an aromatic cation, as shown:

aromatic

This cation is aromatic because the oxygen atom can adopt sp^2 hybridization, thereby placing a lone pair in a *p* orbital. The positive charge represents an empty *p* orbital, so there is a continuous system of overlapping *p* orbitals. Each π bond contributes two π electrons, and the oxygen atom contributes two more π electrons, for a total of six π electrons.

(b) The following compound is expected to undergo an SN1 reaction least readily, because loss of the leaving group generates an antiaromatic cation (with 4*n* π electrons occupying a continuous system of overlapping *p* orbitals):

antiaromatic

The remaining compound is expected to have an intermediate rate of an SN1 reaction, because it involves a resonance-stabilized carbocation that is neither aromatic nor antiaromatic.

nonaromatic

17.36. The following compound is the most acidic among the group:

because deprotonation of this compound generates an aromatic ion (a continuous system of overlapping *p* orbitals containing six π electrons).

17.37. The second compound is a stronger base, because the lone pair on the nitrogen atom is localized and available to function as a base. In contrast, the nitrogen atom in the first compound is delocalized and is participating in aromaticity. This lone pair is unavailable to function as a base, because protonation of that nitrogen atom would destroy aromaticity.

17.38. In order to draw a Frost circle, we begin by drawing a circle, and then we place a pentagon inside the circle, with all five corners touching the circle, such that one of the corners points to the bottom of the circle, as shown here:

Each position where a corner touches the circle represents the relative energy level of a molecular orbital. If we draw a horizontal line through the center of the circle, we see that there are three MOs in the bottom half of the circle, and two MOs in the top half. The former represent the bonding MOs, while the latter represent the antibonding MOs. With three bonding MOs, we expect that six π electrons are required in order to achieve aromaticity. This cation only has four π electrons. Two of them occupy the lowest energy MO (at the bottom of the circle), but the remaining are split between the two degenerate bonding MOs. As such, this system is not a closed-shell electron configuration, giving rise to instability.

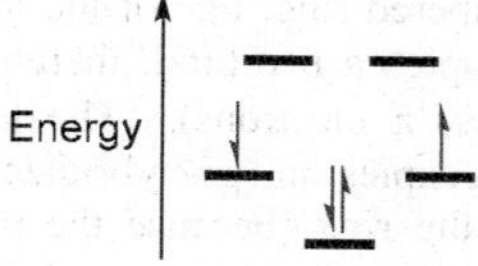

17.39. Cyclopentadiene is more acidic because its conjugate base is highly stabilized. Deprotonation of cyclopentadiene generates an anion that is aromatic, because it is a continuous system of overlapping *p*

orbitals containing six π electrons ($4n+2$). In contrast, deprotonation of cycloheptatriene gives an anion with eight π electrons ($4n$), expected to be highly unstable (antiaromatic).

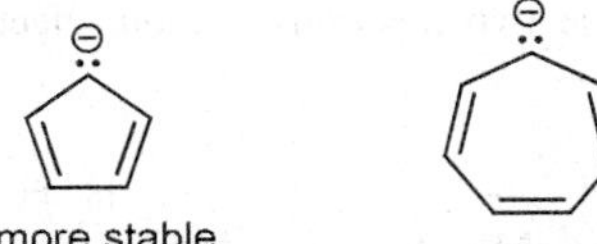

17.40. Yes, this compound is expected to be aromatic. The lone pairs on the nitrogen atoms do not contribute to aromaticity. They occupy sp^2-hybridized orbitals. One of the lone pairs on the oxygen atom (in the ring) occupies a *p* orbital, giving a continuous system of overlapping *p* orbitals containing six π electrons (two π electrons from each double bond, as well as two more π electrons from the lone pair that occupies a *p* orbital).

17.41. Benzene does not have three C–C single bond and three C=C double bonds. In fact, the pi bonds are delocalized by aromatic resonance, so all six C–C bonds of the ring have the same bond order and are the same length. However, cyclooctatetraene has four isolated π bonds. The molecule adopts a tub shape to avoid antiaromaticity. Some of the C–C bonds are double bonds (shorter in length), and some of the C–C bonds are single bonds (longer in length). Therefore, the two methyl groups can either be separated by a C–C single bond or a C=C double bond. And those two possibilities represent different compounds.

17.42.

(a) The starting material has a benzylic position that is not quaternary, so treatment with NBS is expected to result in bromination at the benzylic position to give the following product.

NBS, heat or light → (product with Br)

(b) The starting material has a benzylic position that is not quaternary, so treatment with chromic acid is expected to result in oxidation of the benzylic position. Note that only the benzylic carbon atom remains, while the other carbon atoms of the alkyl substituent are removed (characteristic of oxidation reactions at benzylic positions).

$Na_2Cr_2O_7$, H_2SO_4, H_2O → (product with O, OH)

(c) The starting material is an alcohol in which the OH group occupies a benzylic position. Upon treatment with sulfuric acid, we expect a dehydration reaction via an E1 process to form a conjugated π bond, as shown:

(HO) H_2SO_4, heat →

(d) The starting material is a secondary, benzylic bromide. Upon treatment with a non-nucleophilic base, such as DBU, we expect an E2 process to form a conjugated π bond, as shown:

(Br) DBU →

17.43.

(a) A monosubstituted aromatic ring is expected to produce four signals in the region between 110 and 170 ppm in the ^{13}C NMR spectrum. In addition, the isopropyl group is expected to produce two signals (because the methyl groups are equivalent) in the region between 10 and 60 ppm. In total, we expect six signals in the ^{13}C NMR spectrum.

(b) A *para*-disubstituted aromatic ring (where the two substituents are different from each other) is expected to produce four signals in the region between 110 and 170 ppm in the ^{13}C NMR spectrum. In addition, the methyl group is expected to produce one more signal in the region between 10 and 60 ppm. In total, we expect five signals in the ^{13}C NMR spectrum.

(c) The aromatic ring is 1,3,5-trisubstituted, and all three substituents are identical (methyl groups). Because of symmetry, there are only two unique kinds of aromatic carbon atoms, giving rise to two signals in the region between 110 and 170 ppm in the ^{13}C NMR spectrum. In addition, the three methyl groups are all equivalent and will collectively give rise to one signal in the region between 10 and 60 ppm. In total, we expect three signals in the ^{13}C NMR spectrum.

(d) The aromatic ring is 1,2,4-trisubstituted, and it lacks the symmetry necessary to render any of the carbon atoms identical. All six carbon atoms of the ring occupy unique environments, and similarly, all three methyl groups also occupy unique environments. We therefore expect six signals in the region between 110 and 170 ppm, and we expect three more signals in the region between 10 and 60 ppm, for a total of nine signals in the ^{13}C NMR spectrum.

17.44. The reagents indicate a Birch reduction, which is believed to occur via a four-step mechanism (as seen in Mechanism 17.1). In the first step, an electron is transferred to the aromatic ring, generating a radical anion, which is then protonated (with methanol functioning as the proton source). Another electron transfer process, followed by a proton transfer, results in the formation of 1,2,4,5-tetramethyl-1,4-cyclohexadiene, as shown. Notice that the four positions bearing the electron-donating alkyl groups are not reduced. The two positions without electron-donating substituents are reduced.

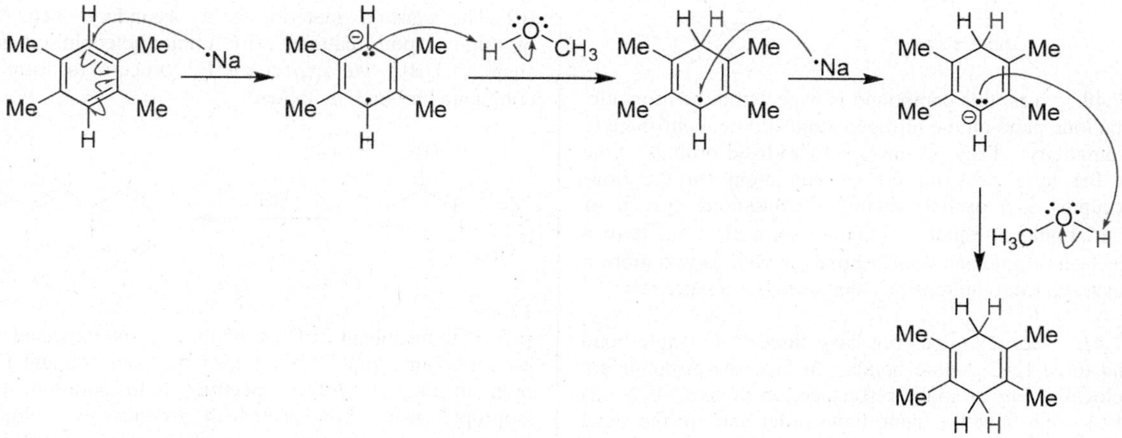

17.45. The reagents indicate a Birch reduction, and *meta*-xylene is expected to undergo a Birch reduction to produce a compound that will exhibit five signals in its ^{13}C NMR spectrum. The five unique carbon atoms are highlighted in the product:

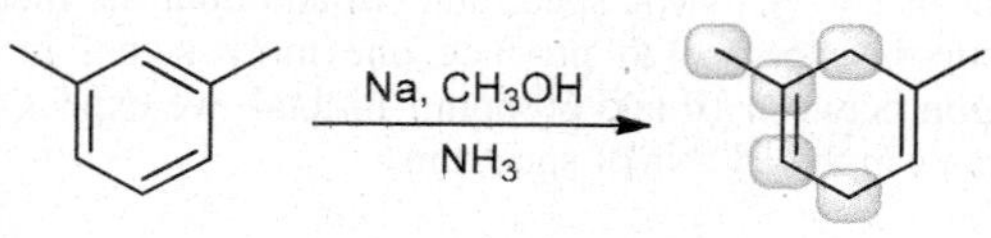

Na, CH_3OH / NH_3

The answer cannot be *ortho*-xylene, because a Birch reduction of *ortho*-xylene would produce a compound that exhibits only four signals in its ^{13}C NMR spectrum:

Na, CH_3OH / NH_3

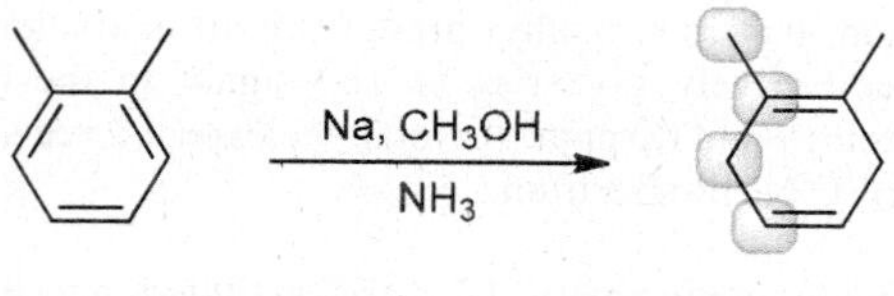

And the answer cannot be *para*-xylene, because a Birch reduction of *para*-xylene produces a compound that also exhibits only four signals in its ^{13}C NMR spectrum:

Na, CH_3OH / NH_3

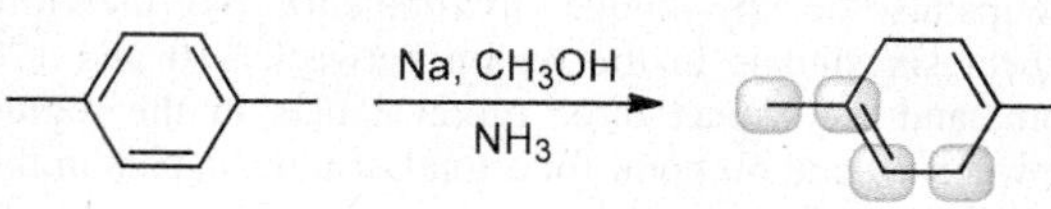

17.46.

(a) The first compound would lack Csp^2–H stretching signals just above 3000 cm^{-1}, while the second compound will have Csp^2–H stretching signals just above 3000 cm^{-1}.

(b) The ^{1}H NMR spectrum of the first compound will have only one signal, while the ^{1}H NMR spectrum of the second compound will have two signals.

(c) The ^{13}C NMR spectrum of the first compound will have only two signals, while the ^{13}C NMR spectrum of the second compound will have three signals.

17.47. When either compound is deprotonated, an aromatic anion is generated, which can be drawn with five resonance structures. The resulting anion is the same in either case:

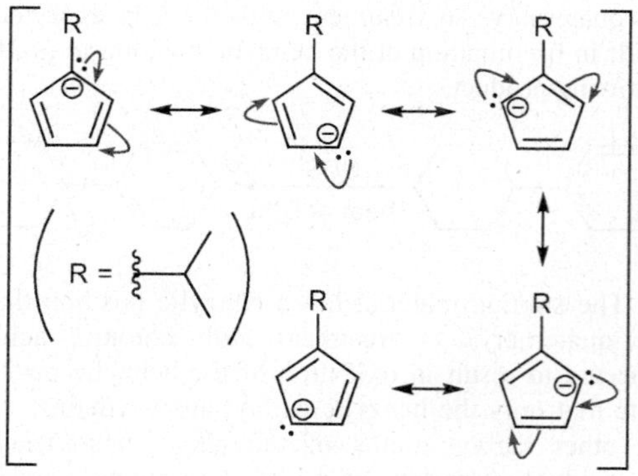

17.48. The correct answer is (a). In structure (a), the nitrogen atom has four sigma bonds, so it must be sp^3 hybridized. It does not have a *p* orbital, so the ring is NOT a continuous system of overlapping *p* orbitals. Therefore, structure (a) is nonaromatic. Structures (b), (c), and (d) are all aromatic.

17.49. The correct answer is (d). Structure (a) is a tertiary benzylic bromide, and it is therefore expected to undergo an SN1 reaction in ethanol, because the intermediate carbocation is highly stabilized:

Tertiary benzylic
carbocation intermediate

Similarly, structure (b) is also expected to undergo an S_N1 reaction, because the intermediate carbocation is aromatic and therefore highly stabilized:

Aromatic
carbocation intermediate

Structure (c) is also expected to undergo an S_N1 reaction, because the intermediate carbocation is tertiary and allylic (twice), so it is highly stabilized as well:

Tertiary allylic
carbocation intermediate

Only structure (d) gives an unstable carbocation upon loss of a leaving group:

Antiaromatic

This cation is unstable because it is antiaromatic (with $4n$ π electrons occupying a continuous system of overlapping *p* orbitals). Therefore, structure (d) is not expected to undergo an S_N1 reaction.

17.50. The correct answer is (b). In the first step, one equivalent of NBS will install one Br (not two) at the benzylic position:

NBS (one equiv.)
heat

The resulting secondary benzylic halide is expected to undergo an E2 reaction upon treatment with a strong, sterically hindered base, giving structure (b).

t-BuOK

17.51. The correct answer is (c). The benzylic position undergoes oxidation with chromic acid. Notice that oxidation of the ethyl group involves cleavage of a C–C bond to give a carboxylic acid with one less carbon atom than the starting material.

$Na_2Cr_2O_7$
H_2SO_4, H_2O

17.52. The correct answer is (d). [10]Annulene is not aromatic because the hydrogen atoms positioned inside the ring (shown in red) experience a steric interaction that forces the compound out of planarity:

17.53. The correct answer is (a). The lone pair on nitrogen atom **2** is delocalized by aromatic resonance. Since this lone pair is delocalized, it is less available to function as a base, making nitrogen atom **1** the stronger base. This prediction is confirmed by comparing the two conjugate acids. Protonation of nitrogen atom **1** does not have a significant effect on the stability of the molecule, because both rings are still aromatic. However, protonation of nitrogen atom **2** results in the loss of aromaticity in the 5-membered ring. Conjugate acid **2** is the less stable, stronger acid, making nitrogen atom **2** the weaker base.

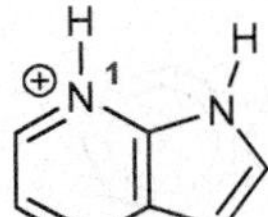

Conjugate acid **1**
more stable conj. acid
(aromaticity retained)

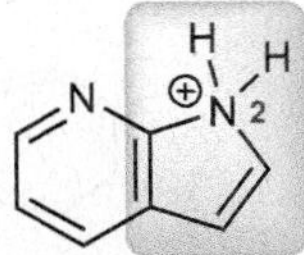

Conjugate acid **2**
less stable conj. acid
(aromaticity partially lost)

17.54. The correct answer is (c). The cycloheptatrienyl cation, also called the tropylium cation, is aromatic.

Because the compound is a 7-membered ring, its molecular orbital diagram has seven MOs. The three bonding MOs are filled (with six electrons from the three π bonds), as expected for an aromatic system.

E

17.55. In cycloheptatrienone, the resonance structures with C^+ and O^- contribute significant character to the overall resonance hybrid, because these forms are aromatic:

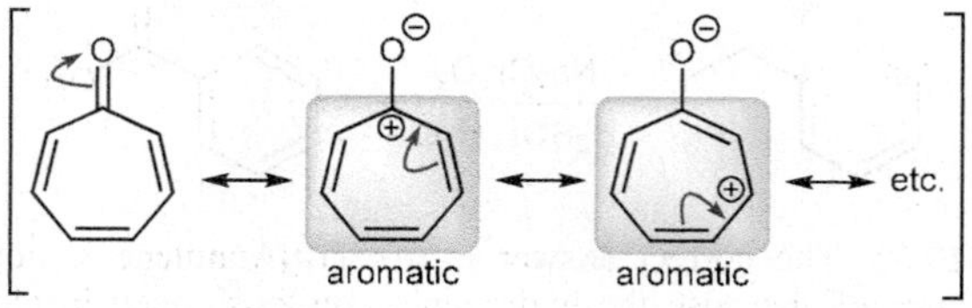

aromatic aromatic

Therefore, the oxygen atom of this C=O bond is particularly electron rich. A similar analysis of cyclopentadienone reveals resonance structures with antiaromatic character ($4n$ π electrons):

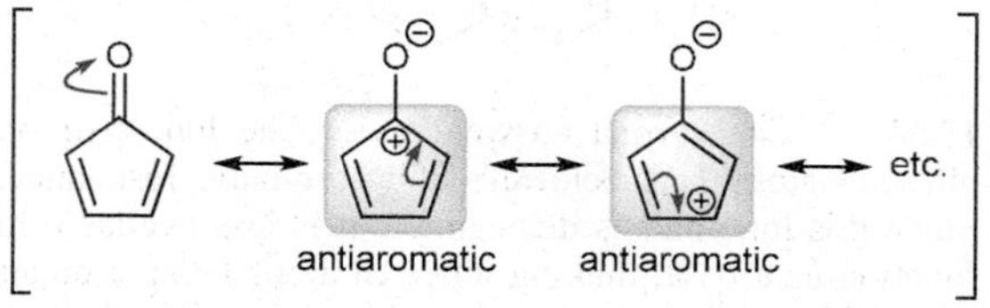

antiaromatic antiaromatic

These resonance structures contribute very little character to the overall resonance hybrid, and as a result, the oxygen atom of this C=O bond is less electron rich as compared with most C=O bonds.

17.56.

(a) Each of the rings in the following resonance structure is aromatic.

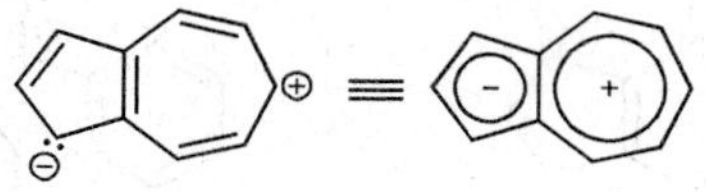

Therefore, this resonance structure contributes significant character to the overall resonance hybrid, which gives azulene a considerable dipole moment.

(b) The following compound has resonance structures in which both rings are aromatic, as shown here:

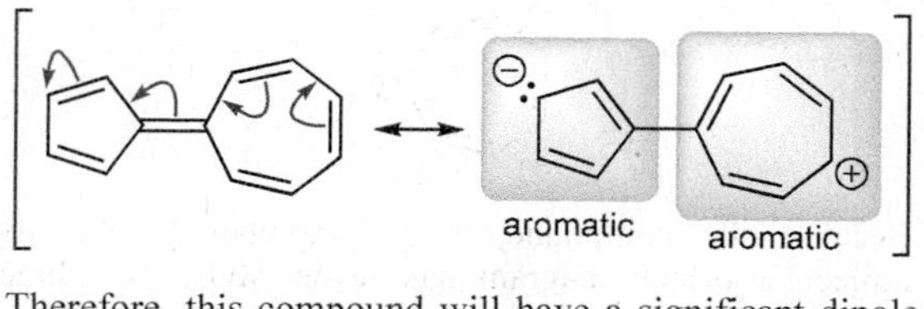

aromatic aromatic

Therefore, this compound will have a significant dipole moment.

In contrast, the other compound does not have resonance structures in which both rings are aromatic, and is not expected to exhibit a dipole moment.

17.57. There are two questions to ask when approaching a synthesis problem.

1) Is there a change in the carbon skeleton?
2) Is there a change in the position or identity of the functional groups?

In this case, there is a change in the carbon skeleton. The product has two more carbon atoms than the starting material:

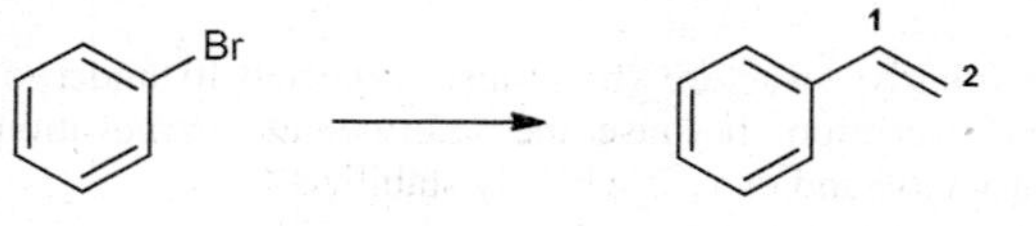

Also, a functional group (a π bond) must be installed. Let's begin with the change in the carbon skeleton. In order to increase the size of the carbon skeleton, we must form a C–C bond. A Grignard reaction will allow us to form the critical C–C bond, and the resulting alcohol can then be converted into the desired alkene via an E1 dehydration process, as shown below.

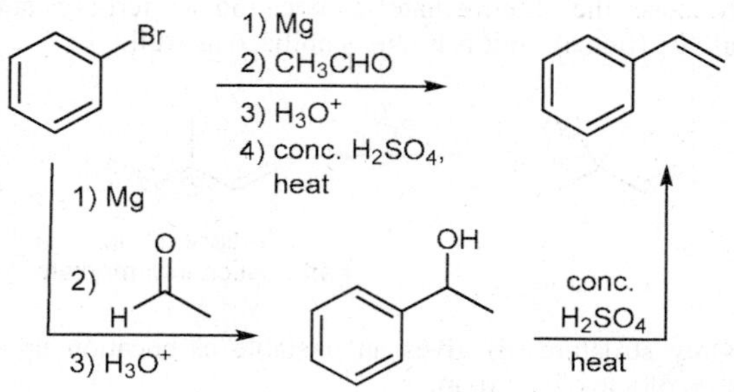

1) Mg
2) CH_3CHO
3) H_3O^+
4) conc. H_2SO_4, heat

1) Mg
2) H(C=O)CH₃
3) H_3O^+

conc. H_2SO_4 heat

17.58. In acidic conditions, the OH group is protonated, thereby generating a better leaving group (H_2O). Loss of the leaving group generates a tertiary carbocation, which can rearrange via a hydride shift to generate a resonance-stabilized benzylic carbocation. This benzylic carbocation is then captured by a chloride ion (generated in the first step of the mechanism) to give the product, as shown.

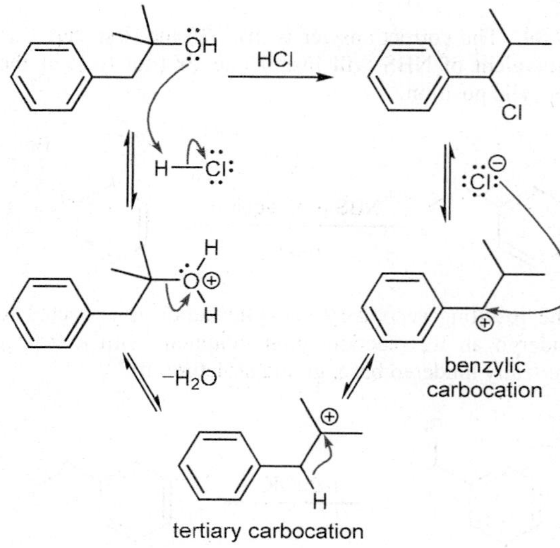

HCl

$-H_2O$

benzylic carbocation

tertiary carbocation

17.59. The molecular formula ($C_9H_{10}O_2$) indicates five degrees of unsaturation (see Section 14.16), which is strongly suggestive of an aromatic ring, as well as one additional double bond or ring. The signal just above 3000 cm^{-1} in the IR spectrum confirms the aromatic ring, as does the signal just above 1600 cm^{-1}. The ^1H NMR spectrum exhibits two doublets between 6.9 and 7.9 ppm, each with an integration of 2. This is the characteristic pattern of a disubstituted aromatic ring, in which the two substituents are different from each other:

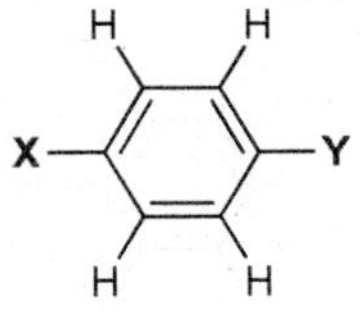

The singlet at 3.9 ppm (with an integration of 3) represents a methyl group. The chemical shift is downfield from the expected benchmark value of 0.9 ppm for a methyl group, indicating that it is likely next to an oxygen atom:

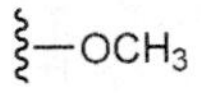

The singlet at 2.6 ppm (with an integration of 3) represents an isolated methyl group. The chemical shift of this signal suggests that the methyl group is neighboring a carbonyl group:

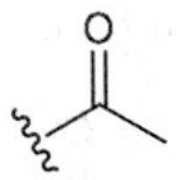

The carbonyl group accounts for one degree of unsaturation, and together with the aromatic ring, this would account for all five degrees of unsaturation. The presence of a carbonyl group is also confirmed by the signal at 196.6 ppm in the ^{13}C NMR spectrum (consistent with a ketone functional group).
We have uncovered three pieces, which can only be connected in one way, as shown:

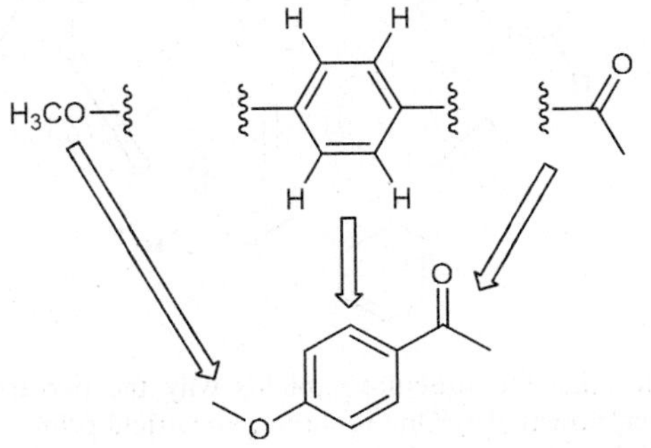

This structure is consistent with the ^{13}C NMR data: four signals for the sp^2-hybridized carbon atoms of the aromatic ring, and two signals for the sp^3-hybridized carbon atoms (one of which is above 50 ppm because it is next to an oxygen atom).

Also notice that the carbonyl group is conjugated to the aromatic ring, which explains why the signal for the C=O bond in the IR spectrum appears at 1676 cm^{-1}, rather than 1720 cm^{-1}.

17.60.

(a) If we functionalize the benzylic position (via bromination), the product can be obtained in just one more step, via a Williamson ether synthesis, as shown:

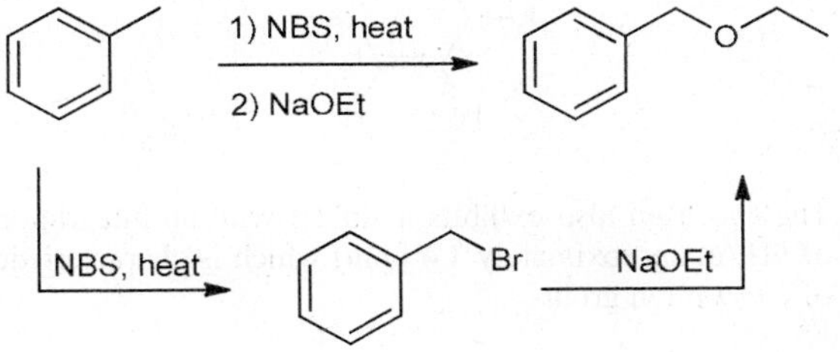

(b) This transformation requires that we move the position of the OH group. This was a strategy that was covered in Chapter 11. Specifically, we saw that the location of an OH group can be moved via elimination followed by addition. The first step is to convert the alcohol into an alkene via acid-catalyzed dehydration (E1). Then, hydroboration-oxidation will convert the alkene into the desired alcohol, via an *anti*-Markovnikov addition of H and OH across the π bond:

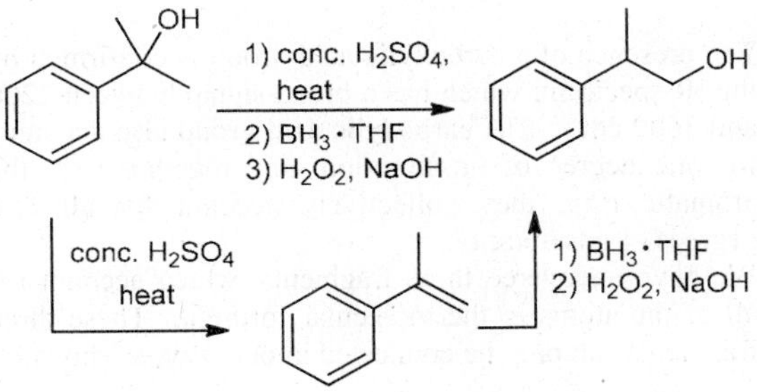

(c) The carbon skeleton is getting larger, so we must form a C–C bond. In addition, we must install a triple bond. Both of these goals can be achieved via bromination at the benzylic position, followed by an S_N2 reaction in which an acetylide ion is used as the nucleophile, as shown:

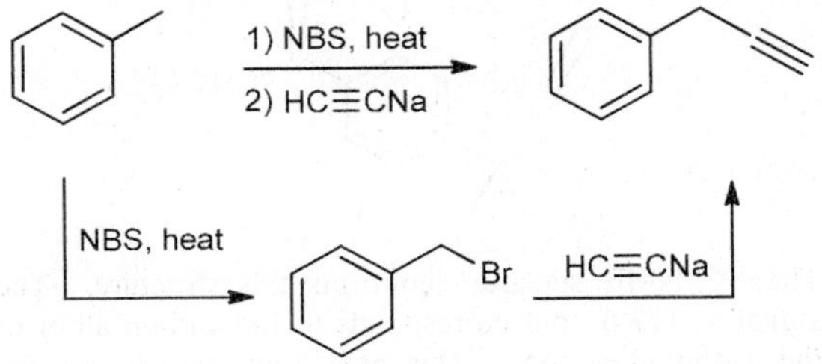

17.61. The molecular formula ($C_{11}H_{14}O_2$) indicates five degrees of unsaturation (see Section 14.16), which is highly suggestive of an aromatic ring, as well as either one double bond or one ring.

In the ^{1}H NMR spectrum, the signals near 7 ppm are likely a result of aromatic protons. Notice that the combined integration of these two signals is 4H. This, together with the distinctive splitting pattern (a pair of doublets), suggests a 1,4-disubstituted aromatic ring, in which the two substituents are different:

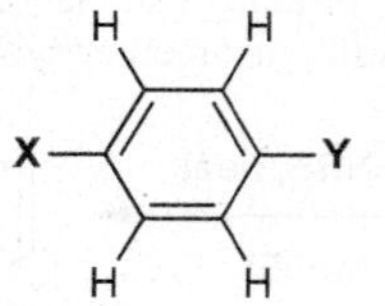

The spectrum also exhibits a singlet with an integration of 9H (at approximately 1.4 ppm) which is characteristic of a *tert*-butyl group.

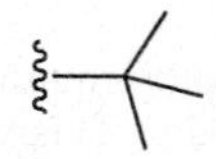

The ^{1}H NMR spectrum has only one other signal, very far downfield, characteristic of a carboxylic acid:

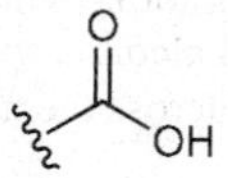

The presence of a carboxylic acid group is confirmed by the IR spectrum, which has a broad signal between 2200 and 3600 cm^{-1}. The carboxylic acid group also accounts for one degree of unsaturation, and together with the aromatic ring, they collectively account for all five degrees of unsaturation.
We have uncovered three fragments, which account for all of the atoms in the molecular formula. These three fragments can only be connected in one way, as shown:

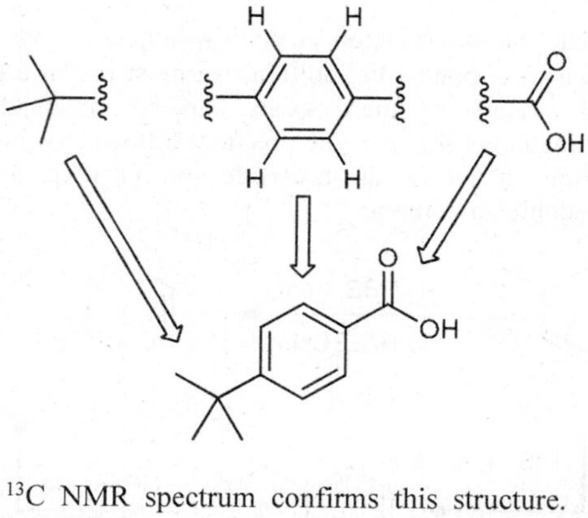

The ^{13}C NMR spectrum confirms this structure. The signal at 172.6 ppm corresponds to the carbon atom of the carbonyl group. The next four signals on the spectrum (above 100 ppm) are characteristic of a disubstituted aromatic ring, and finally, there are two signals between 10 and 60 ppm for the *tert*-butyl group (all three methyl groups are equivalent and give one signal, in addition to the signal from the benzylic carbon atom).

17.62. The molecular formula ($C_9H_{10}O$) indicates five degrees of unsaturation (see Section 14.16), which is highly suggestive of an aromatic ring, as well as either one double bond or one ring.
In the ^{1}H NMR spectrum, the multiplet near 7 ppm is likely a result of aromatic protons. The integration of this multiplet is 5H, indicating a monosubstituted aromatic ring:

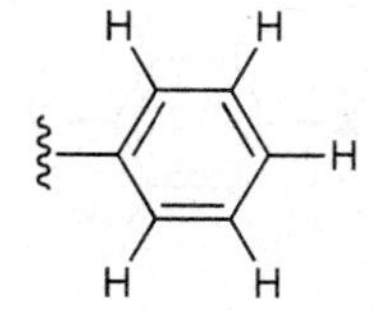

The spectrum also exhibits two triplets (near 3 ppm), indicating two methylene groups connected to each other:

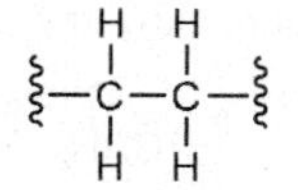

Each of these signals appears more downfield than we might expect for a methylene group (1.2 ppm), so each of these methylene groups must be connected to a group that causes a deshielding effect. This must be taken into account in our final structure.
The singlet near 10 ppm is characteristic of an aldehyde group:

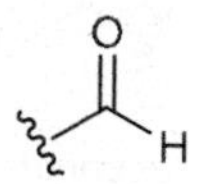

We have uncovered three fragments, which account for all of the atoms in the molecular formula. These three fragments can only be connected in one way, as shown:

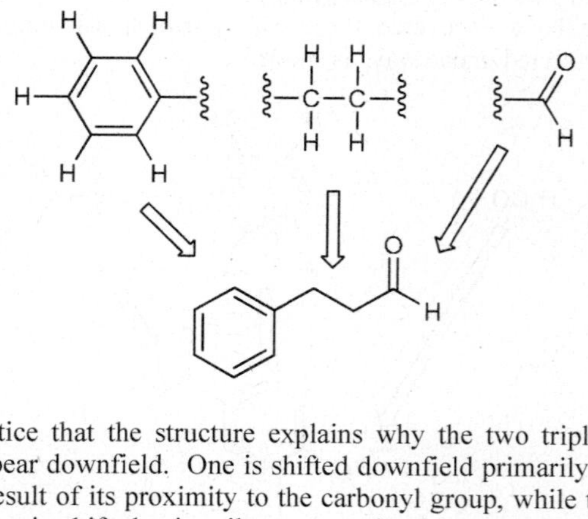

Notice that the structure explains why the two triplets appear downfield. One is shifted downfield primarily as a result of its proximity to the carbonyl group, while the other is shifted primarily as a result of its proximity to the aromatic ring.
The ^{13}C NMR spectrum confirms the structure above. The signal near 200 ppm corresponds to the carbon atom of the carbonyl group. The next four signals on the spectrum (above 100 ppm) are characteristic of a

monosubstituted aromatic ring, and finally, there are two signals between 10 and 60 ppm for the methylene groups. In addition, the IR spectrum also confirms this structure. The strong signal near 1700 cm^{-1} corresponds with the C=O bond, and the signals between 3000 and 3100 cm^{-1} correspond with the aromatic ring.

17.63.
(a) The second compound holds greater promise as a potential antihistamine, because it possesses two planar aromatic rings separated from each other by one carbon atom. The first compound has only one aromatic ring. The ring with the oxygen atom is not aromatic and not planar.
(b) The second compound is expected to exhibit sedative properties, because it lacks polar functional groups that would prevent it from crossing the blood-brain barrier, as described in the BioLinks application (that appears at the end of Section 17.5).

17.64. No, this compound possesses an allene group (C=C=C). The *p* orbitals of one C=C bond of the allene group do not overlap with the *p* orbitals of the other C=C bond (the two sets of *p* orbitals are orthogonal). This prevents the compound from having one continuous system of overlapping (*i.e.*, parallel) *p* orbitals.

17.65. The molecular formula (C_8H_{10}) indicates four degrees of unsaturation (see Section 14.16), consistent with an aromatic ring. The ^{1}H NMR spectrum of compound **A** has a multiplet near 7 ppm with an integration of 5H, indicating a monosubstituted aromatic ring:

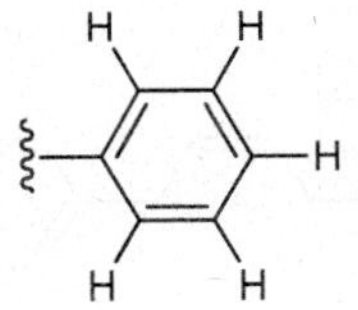

The aromatic ring accounts for six of the eight carbon atoms in the structure. If the ring is monosubstituted, then the other two carbon atoms must be an ethyl group, which is consistent with the two upfield signals in the ^{1}H NMR spectrum:

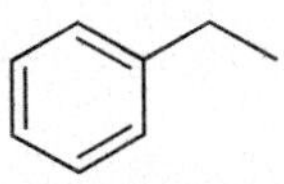

Compound **A**

Compound **B** is a constitutional isomer of **A** and exhibits four signals in its ^{13}C NMR spectrum, which is consistent with *ortho*-xylene (*meta*-xylene would produce five signals, while *para*-xylene would produce three signals in its ^{13}C NMR spectrum):

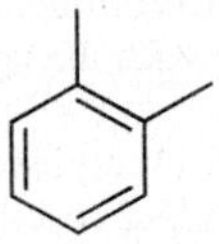

Compound **B**

Compound **C** is a constitutional isomer of **A** and **B**, and it exhibits three signals in its ^{13}C NMR spectrum, which is consistent with *para*-xylene. Compound **D** is a constitutional isomer of **A**, **B**, and **C**, and it must be *meta*-xylene:

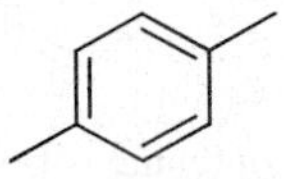

Compound **C** 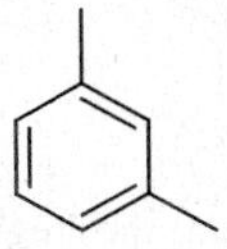 Compound **D**

17.66. The lone pair on the nitrogen atom in compound **A** is localized and is not participating in resonance. It is free to function as a base. The lone pair on the nitrogen atom in compound **B** is delocalized, allowing the compound to achieve aromatic stabilization via resonance:

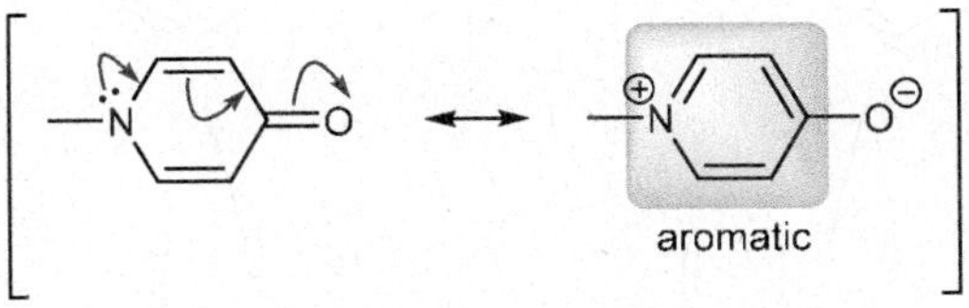

aromatic

In compound **B**, the lone pair on the nitrogen atom is participating in establishing aromaticity, and therefore, it is not available to function as a base.

17.67. There are two questions to ask when approaching a synthesis problem.

1) Is there a change in the carbon skeleton?
2) Is there a change in the position or identity of the functional groups?

In this case, there is a change in the carbon skeleton. The product has one less carbon atom than the starting material:

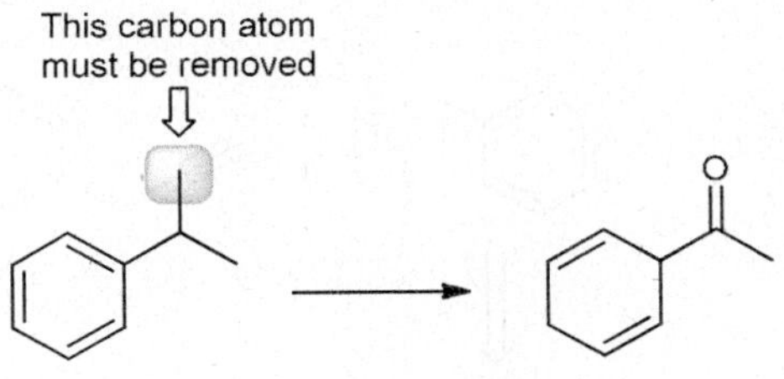

This carbon atom must be removed

Also, a functional group (a C=O bond) must be installed, and the aromatic ring must be reduced to a diene.
Let's begin with the change in the carbon skeleton. We have learned one suitable method (ozonolysis) to remove a carbon atom from the carbon skeleton and produce a ketone. Ozonolysis would indeed install the necessary functional group in the correct location:

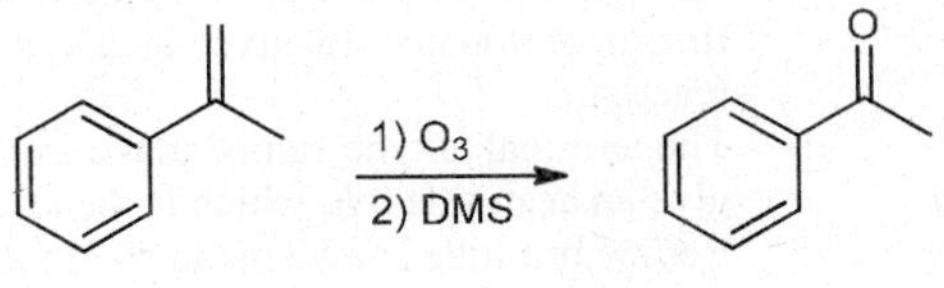

1) O_3
2) DMS

And the ketone above can be treated with Birch conditions to give the desired product:

Na, CH_3OH
NH_3

Notice that the regiochemical outcome of the Birch reduction is controlled by the carbonyl group, giving the desired product.
The strategy above requires that we first form the alkene from the starting material:

?

This can be accomplished via bromination of the benzylic position, followed by elimination (E2) upon treatment of the tertiary bromide with any strong base (such as sodium ethoxide). The entire synthesis is summarized here:

1) NBS, heat
2) NaOEt
3) O_3
4) DMS
5) Na, NH_3
CH_3OH

NBS, heat

Br

Na, NH_3
CH_3OH

NaOEt

1) O_3
2) DMS

17.68. The following retrosynthetic analysis (and subsequent explanation) represents one strategy for preparing the desired compound from the given starting materials (acetylene and toluene):

a

b

Br

\+

c

e

d

e

Br

\+

a. The *cis* epoxide can be made from the corresponding *cis* alkene, via epoxidation with a peroxy acid.

b. The *cis* alkene can be made from the corresponding alkyne, via a *syn* addition of molecular hydrogen (H_2) in the presence of a poisoned catalyst, such as Lindlar's catalyst.

c. The internal alkyne can be made from a terminal alkyne and benzyl bromide. The terminal alkyne is first deprotonated to give an alkynide ion, which is then treated with benzyl bromide (in an S_N2 reaction).

d. The terminal alkyne can be made from acetylene and benzyl bromide. Acetylene is first deprotonated to give an acetylide ion, which is then treated with benzyl bromide (in an S_N2 reaction).

e. Benzyl bromide can be made from toluene via radical bromination at the benzylic position.

The forward process is shown here:

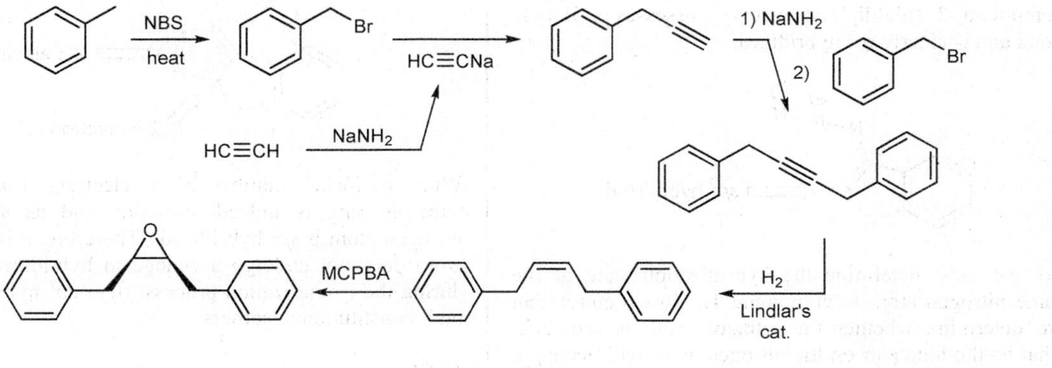

17.69.
(a) Each of the double bonds and triple bonds contributes two π electrons, and each of the unpaired electrons counts as one electron. This gives a total of 14 π electrons ($4n+2$), so the large ring in **B** is aromatic.
(b) Each of the negative charges is delocalized over three positions, as seen in the following three resonance structures:

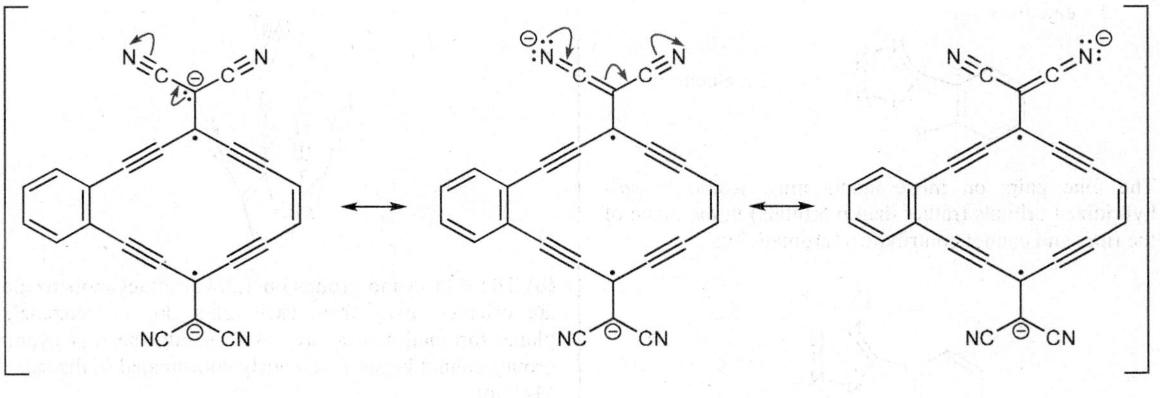

(c) Each of the unpaired electrons is highly delocalized, as seen in the following resonance structures:

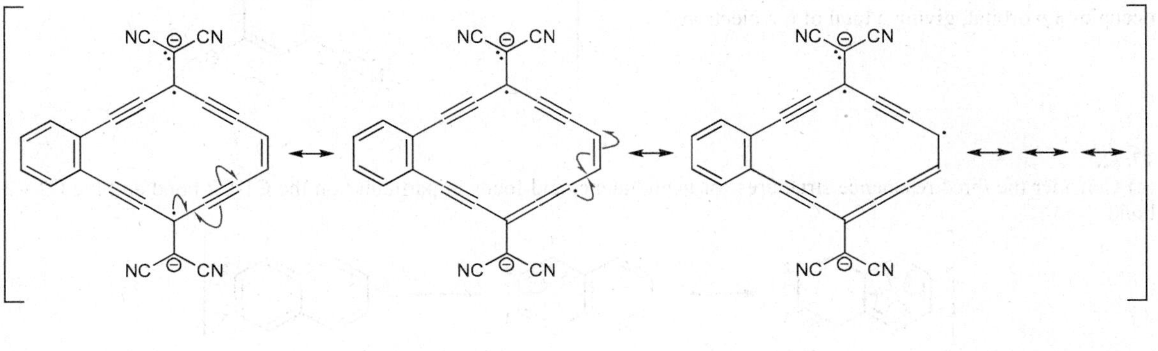

17.70. The nitrogen atom incorporated in the ring in compound **2** (highlighted below) possesses a double bond and is clearly sp^2 hybridized.

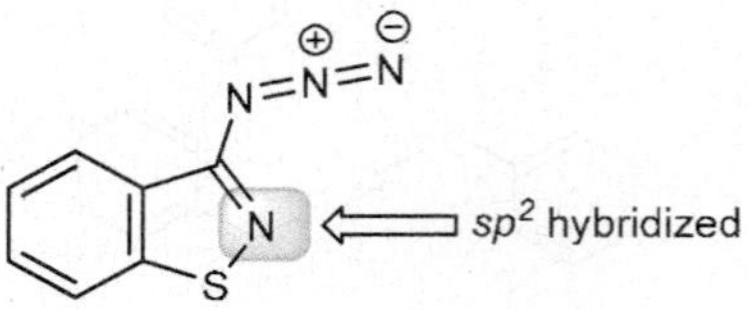

sp² hybridized

So, we must determine the hybridization state of the same nitrogen atom in compound **1**. This requires that we determine whether the tetrazole ring is aromatic. That is, the lone pair on the nitrogen atom will occupy a *p* orbital (rendering the nitrogen atom sp^2 hybridized) if doing so establishes aromaticity. However, if doing so establishes antiaromaticity, then the nitrogen atom would adopt sp^3 hybridization to avoid antiaromaticity.
To determine if the tetrazole ring is aromatic, we count the π electrons. The atoms highlighted below contribute a total of four π electrons (two for each π bond).

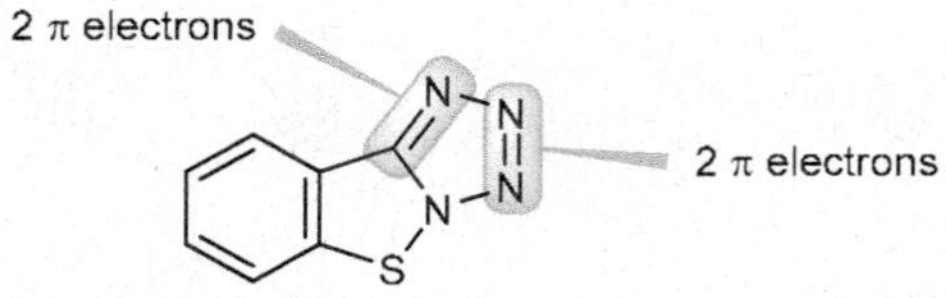

2 π electrons

2 π electrons

The lone pairs on those atoms must reside in sp^2-hybridized orbitals (rather than *p* orbitals) in the plane of the ring, and cannot contribute to aromaticity:

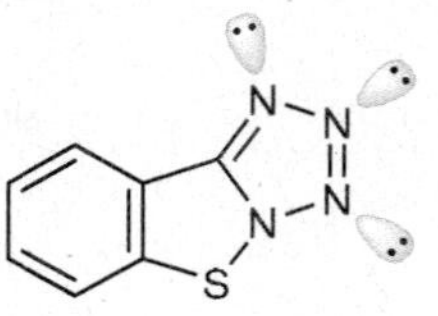

The lone pair on the remaining nitrogen atom therefore occupies a *p* orbital, giving a total of 6 π electrons.

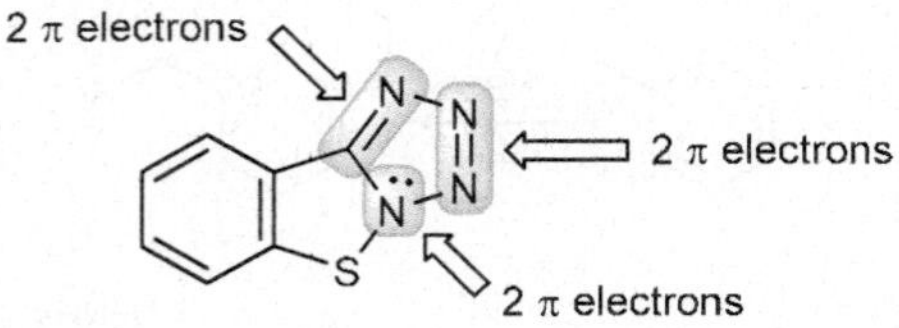

2 π electrons

2 π electrons

2 π electrons

With a Hückel number of π electrons (4*n*+2), the tetrazole ring is indeed aromatic, and as such, this nitrogen atom is sp^2 hybridized. Therefore, this nitrogen atom does not undergo a change in hybridization state during the isomerization process. It is sp^2 hybridized in both constitutional isomers.

17.71.
(a) If planar, the cyclooctatetraene ring would be antiaromatic (4*n* π electrons). It instead adopts a more stable tub conformation that is nonplanar, thus avoiding the energetic cost of antiaromaticity. The tub conformation allows all four cyano groups to be oriented approximately in the same direction such that they can coordinate to a single $Mg^{+\bullet}$ ion as shown.

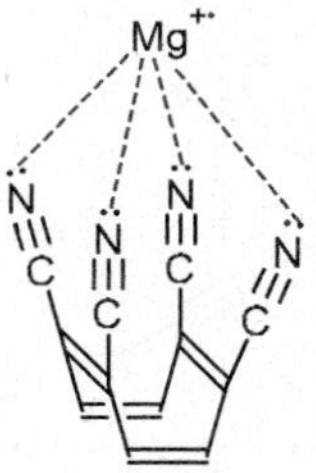

(b) The four cyano groups on 1,2,4,5-tetracyanobenzene are oriented away from each other due to benzene's planar (aromatic) structure. As a result, the four cyano groups cannot be simultaneously coordinated to the same $Mg^{+\bullet}$ ion.

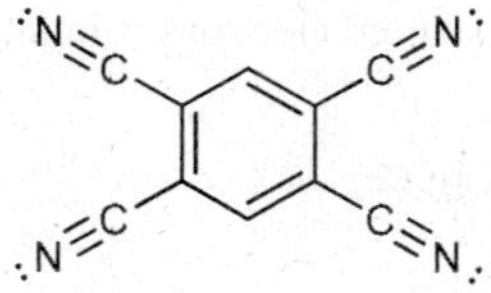

17.72.
(a) Consider the three resonance structures for naphthalene, and focus in particular on the C1-C2 bond and the C2-C3 bond:

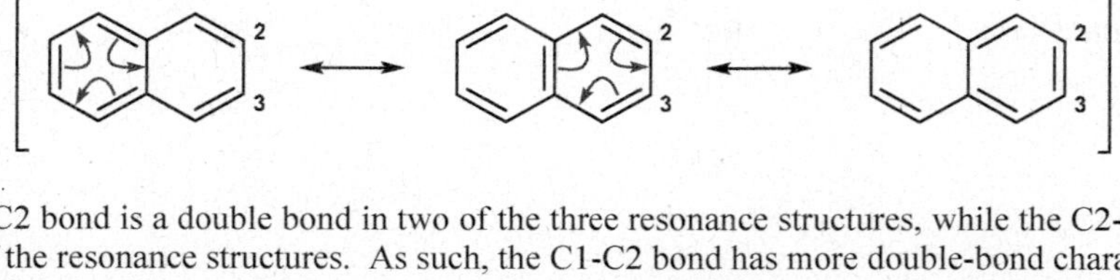

Notice that the C1-C2 bond is a double bond in two of the three resonance structures, while the C2-C3 bond is a double bond in only one of the resonance structures. As such, the C1-C2 bond has more double-bond character than the C2-C3 bond, so the C1-C2 bond will have a shorter bond length. Recall that double bonds are shorter than single bonds (see Table 1.2).

(b) Begin by drawing all of the resonance structures:

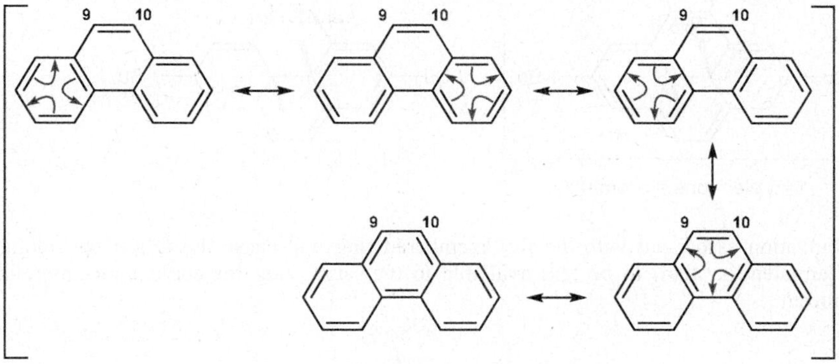

Notice that the C9-C10 bond is a double bond in four of the five resonance structures. No other bond in phenanthrene has this feature. As such, the C9-C10 bond is expected to have the most double-bond character, and consequently, the shortest bond length.

(c) In part (b), we determined that the C9-C10 bond has the most double-bond character, and in fact, this bond functions very much like a regular π bond. It will undergo addition with Br_2 to give a dibromide which retains aromaticity on two of the rings:

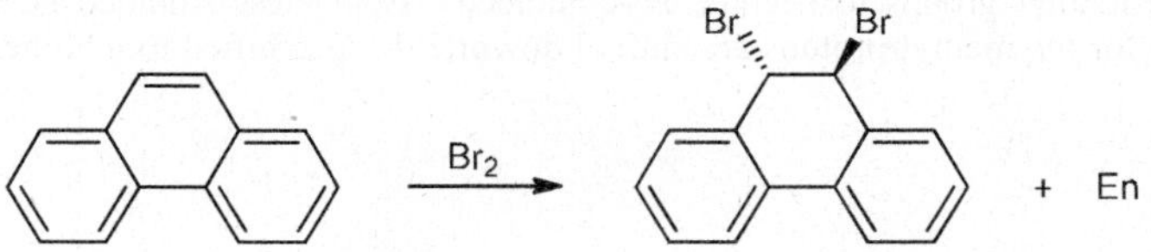

Br_2 + En

17.73.

(a) We know that *tert*-butyl groups are sterically bulky, so it is reasonable to suggest that the presence of three *tert*-butyl groups provides enough steric hindrance to slow the rate of the Diels-Alder process substantially, allowing the compound to be stable at room temperature.

(b) With these particular substituents, it is possible to draw resonance structures in which the double bonds are not positioned inside the ring, thereby diminishing the destabilizing effect of antiaromaticity.

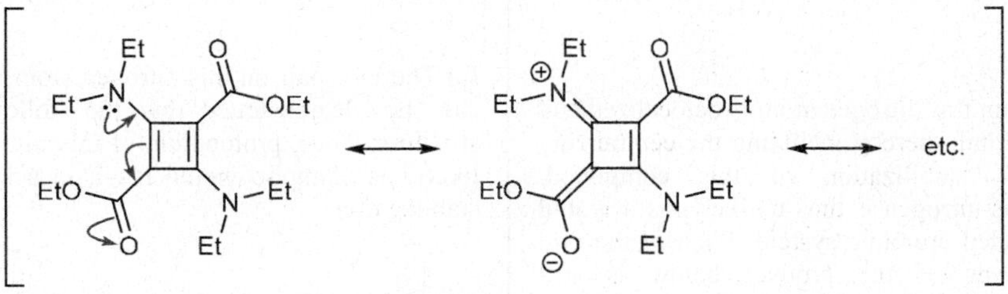

etc.

(c) Unlike systems with 4*n* π electrons, which exhibit destabilization associated with antiaromaticity, systems with (4*n*+2) π electrons exhibit aromatic stabilization. For these systems, the presence of resonance structures that diminish the aromatic stabilization will have a destabilizing effect.

17.74. In Chapter 15, we learned about magnetic anisotropy, and how aromatic ring currents have a shielding effect on substituents located at the interior of large aromatic rings. The starting material has an aromatic [14]annulene ring, which causes the two methyl groups to be shielded from the external magnetic field (even more so than the protons in TMS!), hence their chemical shift of –3.9 ppm. When treated with a base, deprotonation occurs to generate a new aromatic system – notice that the five-membered ring now has 6 π electrons:

t-Bu *t*-Bu *t*-Bu *t*-Bu

6 π electrons = aromatic

14 π electrons = aromatic

The aromatic stabilization associated with the five membered ring will cause the following highlighted π electrons (fused to the [14]annulene system) to be less available to the larger ring for conjugation, therefore resulting in a diminished ring current:

t-Bu *t*-Bu

As a result, the local magnetic field established by the [14]annulene system opposes the external magnetic less significantly, causing the methyl groups to become less shielded. Being less shielded is equivalent to being more deshielded, so the signals for the methyl protons are shifted downfield (*i.e.*, shifted to a higher ppm) from –3.9 ppm to –1.8 ppm.

17.75. First, let's consider the relative basicity of each nitrogen atom in the structure.

(a) The lone pair on this nitrogen atom is delocalized and occupying a *p* orbital, thereby including the central ring in the aromatic stabilization of this compound. Protonation of this nitrogen is thus unlikely, as it would disrupt this extended aromatic system. Each of the two six-membered rings of the product below is still aromatic, but the extended aromatic system would be lost.

Extended aromatic system

(b) The lone pair on this nitrogen atom occupies an sp^2-hybridized orbital that is orthogonal to the conjugated π system of the ring. Protonation of this nitrogen is thus reasonable, as the extended aromaticity of the resulting conjugate acid remains intact.

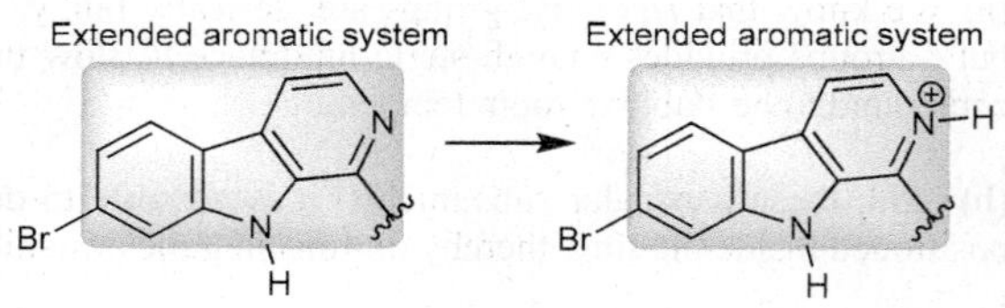

(c) The lone pair on this nitrogen atom is delocalized, as can be demonstrated by the following resonance structure. Thus, protonation of this nitrogen atom is not likely, as doing so would result in a loss of resonance stabilization.

(d) The lone pair on this nitrogen atom is localized, and not involved in any resonance structures. Protonation of this site is therefore reasonable.

Furthermore, protonation at this position leads to a resonance-stabilized conjugate acid in which the positive charge is delocalized over all three nitrogen atoms.

(e) The lone pair on this nitrogen atom is delocalized in a fashion analogous to location **c**. Protonation of this site is thus unlikely.

Based on the above analysis, treatment of the natural product with two equivalents of HCl is expected to afford the following conjugate acid:

17.76. In order to propose a structure for the fragment at m/z = 39, we should consider all of the possible combinations of atoms that will provide a mass-to-charge ratio of 39. Considering the atoms present (C, H, N) in each of the analytes, there are two possible ionic formulas that fit: C_2NH^+ or $C_3H_3^+$. Considering the fact that the fragment is very common (found in 17 of 20 samples), and relatively abundant (5-84% of the base peak), it is reasonable to assume that this fragment is a relatively stable cation. While there is no obvious stable cation with a formula of C_2NH^+, we can draw an aromatic cation with a formula of $C_3H_3^+$, called a cyclopropenium cation.

Now we must consider why the compounds in the top row produce this fragment, while those in the bottom row do not. Each of the compounds in the top row contains at least three contiguous carbon atoms with at least three hydrogen atoms among them (these six atoms are necessary for formation of the cyclopropenium cation), as highlighted below.

This is a structural feature that the three compounds in the bottom row lack. Thus, it is reasonable to propose that the cyclopropenium fragment results from rearrangements of the fragments from the molecules in the top row.

17.77. Many resonance structures can be drawn for dication **2**. The following are just a few of them:

Notice that only resonance structure **2a** has four complete aromatic rings. The extra stabilization associated with this structure means that this resonance structure will contribute more character to the overall resonance hybrid, and as a result, the two neighboring central carbon atoms in dication **2** will both exhibit a large degree of cationic character.

17.78.

(a) Sodium methoxide is a strong base, which removes the most acidic proton in the starting material. The resulting resonance-stabilized anion can then function as a nucleophile and attack 3-bromo-1-propyne in an S_N2 process, giving alkyne **A**.

F3C O N N H :ÖMe F3C O N N: Br F3C O N N

resonance-stabilized

A

(b) As mentioned in the solution to part (a), the anionic intermediate is resonance-stabilized, so we draw the resonance structures:

Among these resonance structures, the last one is expected to contribute the most character to the overall resonance hybrid, for two reasons: 1) the negative charge is placed on an oxygen atom (rather than on the less electronegative nitrogen atoms or carbon atom), and 2) this resonance structure exhibits a benzene-like ring. In fact, this anionic intermediate is indeed aromatic, and we expect the oxygen atom to bear the majority of the delocalized negative charge. Based on this, we can propose the following structure and mechanism of formation for product **B**:

F3C O N N: Br F3C O N N:

B

(c) In addition to the analysis in part (b) that explains why the oxygen atom has more ionic character than the nitrogen atom (making it more nucleophilic), compound **B** also has aromatic stabilization that is absent in **A**. As such, **B** is expected to be lower in energy, so the product mixture will have more of **B** than **A**.

not aromatic

A

aromatic

B

Chapter 18
Aromatic Substitution Reactions

Review of Concepts

Fill in the blanks below. To verify that your answers are correct, look in your textbook at the end of Chapter 18. Each of the sentences below appears verbatim in the section entitled *Review of Concepts and Vocabulary*.

- In the presence of iron, an _____________ **aromatic substitution** reaction is observed between benzene and bromine.
- Iron tribromide is a __________ acid that interacts with Br_2 and generates Br^+, which is sufficiently electrophilic to be attacked by benzene.
- Electrophilic aromatic substitution involves two steps:
    - Formation of the __________ **complex**, or **arenium ion.**
    - Deprotonation, which restores _________________.
- Sulfur trioxide (SO_3) is a very powerful ____________ that is present in fuming sulfuric acid. Benzene reacts with SO_3 in a reversible process called __________.
- A mixture of sulfuric acid and nitric acid produces the **nitronium ion** (NO_2^+). Benzene reacts with the nitronium ion in a process called ____________.
- A nitro group can be reduced to an ______group.
- **Friedel–Crafts alkylation** enables the installation of an alkyl group on ____________________. When choosing an alkyl halide, the carbon atom connected to the halogen must be ______hybridized.
- When treated with a Lewis acid, an acyl chloride will generate an ___________ **ion**, which is resonance stabilized and not susceptible to __________ rearrangements.
- When a Friedel–Crafts acylation is followed by a **Clemmensen reduction**, the net result is the installation of an ______ group.
- A methyl group is said to **activate** an aromatic ring and is an ______-______ **director**.
- All activators are ______-______directors.
- A nitro group is said to **deactivate** an aromatic ring and is a ______director.
- Most deactivators are ______directors.
- **Strong activators** are characterized by the presence of a _________________ immediately adjacent to the aromatic ring.
- **Strong deactivators** are powerfully electron withdrawing, either by ____________or ____________.
- When multiple substituents are present, the more powerful _________________ dominates the directing effects.
- In a **nucleophilic aromatic substitution** reaction, the aromatic ring is attacked by a ____________. This reaction has three requirements:
    - the ring must contain a powerful electron-withdrawing group (typically a ______group)
    - the ring must contain a ____________
    - the leaving group must be either ______or ______to the electron-withdrawing group.
- An **elimination-addition** reaction occurs via a ___________intermediate.

Review of Skills

Fill in the blanks and empty boxes below. To verify that your answers are correct, look in your textbook at the end of Chapter 18. The answers appear in the section entitled *SkillBuilder Review*.

18.1 Identifying the Effects of a Substituent

Place each of the following groups in the correct category below:

$—NR_3^{\oplus}$ —OR   —C(=O)R $—NH_2$ —O—C(=O)R   —C≡N   —Cl   —Br   —OH $—CX_3$ —I $—NO_2$ —R

Activators			Deactivators		
Strong	Moderate	Weak	Weak	Moderate	Strong

18.2 Identifying Directing Effects for Disubstituted and Polysubstituted Benzene Rings

In the following compound, identify the position that is most reactive towards electrophilic aromatic substitution.

CH_3 HO NO_2

18.3 Identifying Steric Effects for Disubstituted and Polysubstituted Aromatic Benzene Rings

In the following compound, identify the position that is most reactive towards electrophilic aromatic substitution.

NO_2

18.4 Using Blocking Groups to Control the Regiochemical Outcome of an Electrophilic Aromatic Substitution Reaction

Identify reagents that can be used to achieve the following transformation:

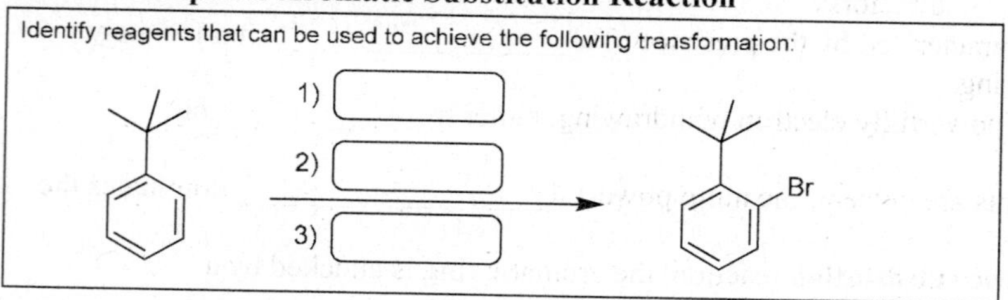

18.5 Proposing a Synthesis for a Disubstituted Benzene Ring

Identify reagents that can be used to achieve the following transformation:

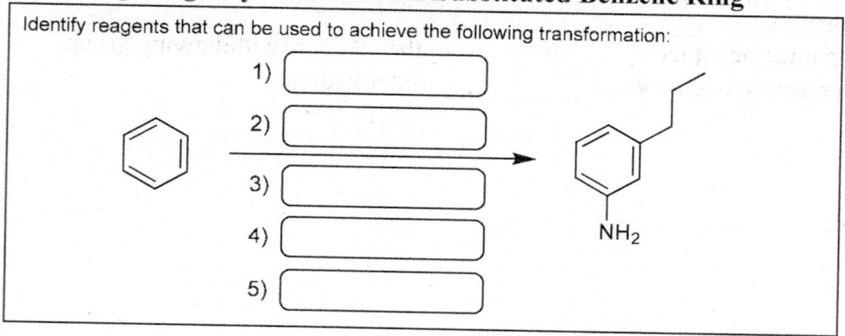

18.6 Proposing a Synthesis for a Polysubstituted Benzene Ring

Complete the following retrosynthetic analysis:

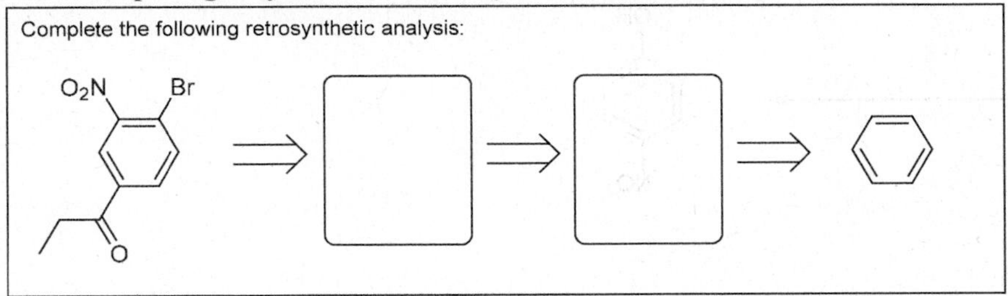
O2N
Br
O

18.7 Determining the Mechanism of an Aromatic Substitution Reaction

Indicate the mechanism that operates in each of the three scenarios shown in the following decision tree:

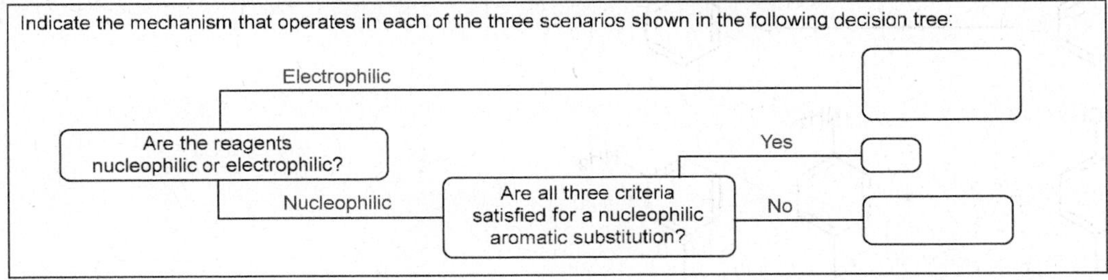
Electrophilic
Are the reagents nucleophilic or electrophilic?
Nucleophilic
Are all three criteria satisfied for a nucleophilic aromatic substitution?
Yes
No

Review of Reactions

Identify the reagents necessary to achieve each of the following transformations. To verify that your answers are correct, look in your textbook at the end of Chapter 18. The answers appear in the section entitled *Review of Reactions*.

Electrophilic Aromatic Substitution

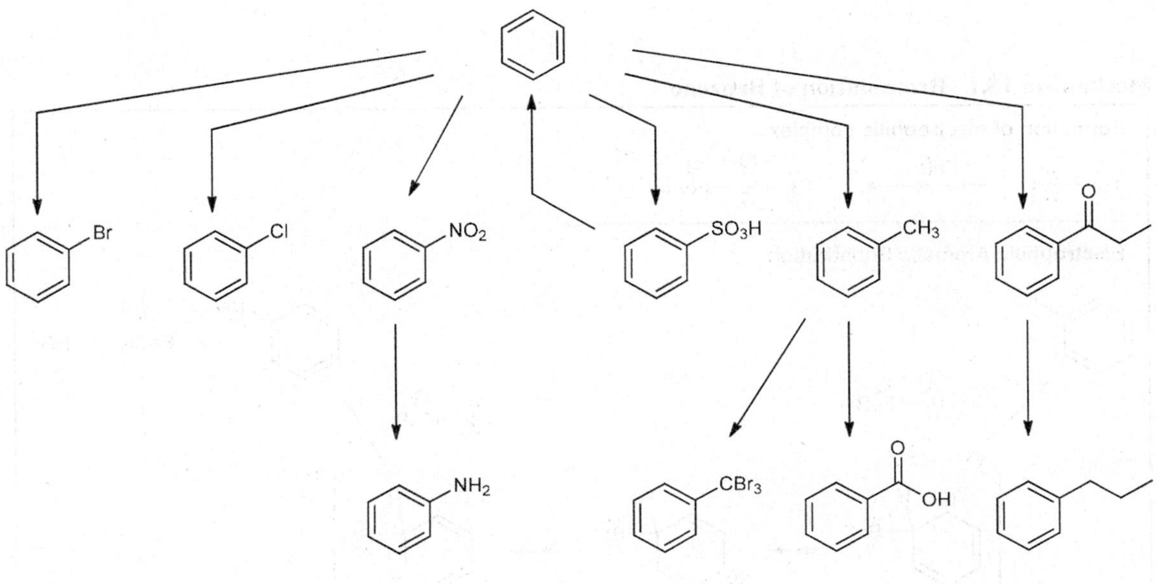

Nucleophilic Aromatic Substitution

Br
NO2
OH
NO2

Elimination-Addition

Cl
OH

Cl
NH2
+
NH2

Review of Mechanisms

Complete each of the following mechanisms by drawing the missing curved arrows. To verify that your curved arrows are drawn correctly, compare them to the curved arrows in the mechanism boxes for Mechanisms 18.1 – 18.9, which can be found throughout Chapter 18 of your text.

Mechanism 18.1 Bromination of Benzene

Formation of electrophilic complex

:Br—Br: —FeBr3→ :Br—Br—FeBr3

Electrophilic Aromatic Substitution

H
:Br—Br—FeBr3
:Br—FeBr3
Br:
+ FeBr3 + HBr
H Br: ⟷ H Br: ⟷ H Br:

Mechanism 18.2 Chlorination of Benzene

Formation of electrophilic complex

:Cl—Cl: $\xrightarrow{AlCl_3}$:Cl—Cl⊕—AlCl$_3$⊖

Electrophilic Aromatic Substitution

:Cl—Cl⊕—AlCl$_3$⊖

:Cl—AlCl$_3$⊖

+ $AlCl_3$ + HCl

Mechanism 18.3 A General Mechanism for Electrophilic Aromatic Substitution

E⊕

:Base

Mechanism 18.4 Sulfonation of Benzene

Mechanism 18.5 Nitration of Benzene

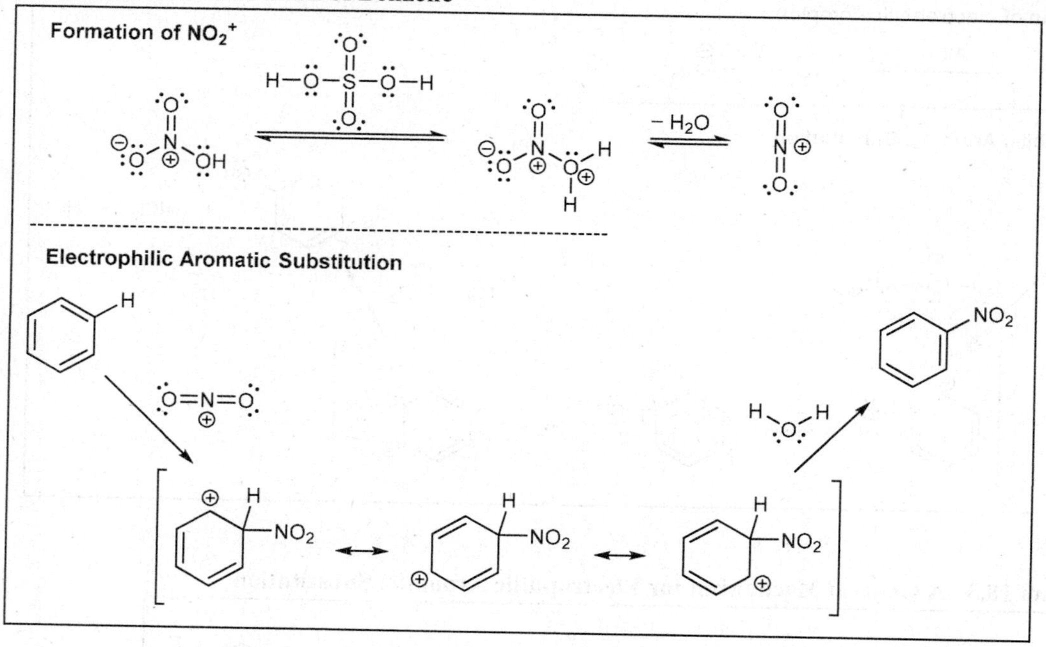

Formation of NO_2^+

H_2SO_4

$- H_2O$

Electrophilic Aromatic Substitution

H

NO_2

NO_2

NO_2

NO_2

Mechanism 18.6 Friedel-Crafts Alkylation

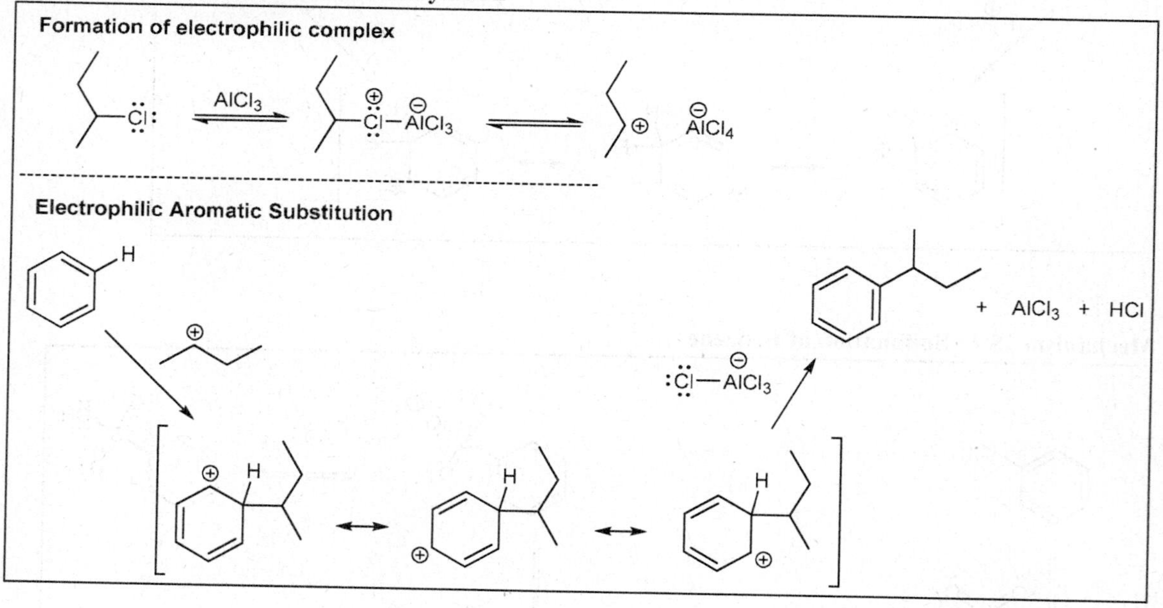

Formation of electrophilic complex

$AlCl_3$

$Cl-AlCl_3$

$AlCl_4$

Electrophilic Aromatic Substitution

H

$+ AlCl_3 + HCl$

$Cl-AlCl_3$

H H H

Mechanism 18.7 Friedel-Crafts Acylation

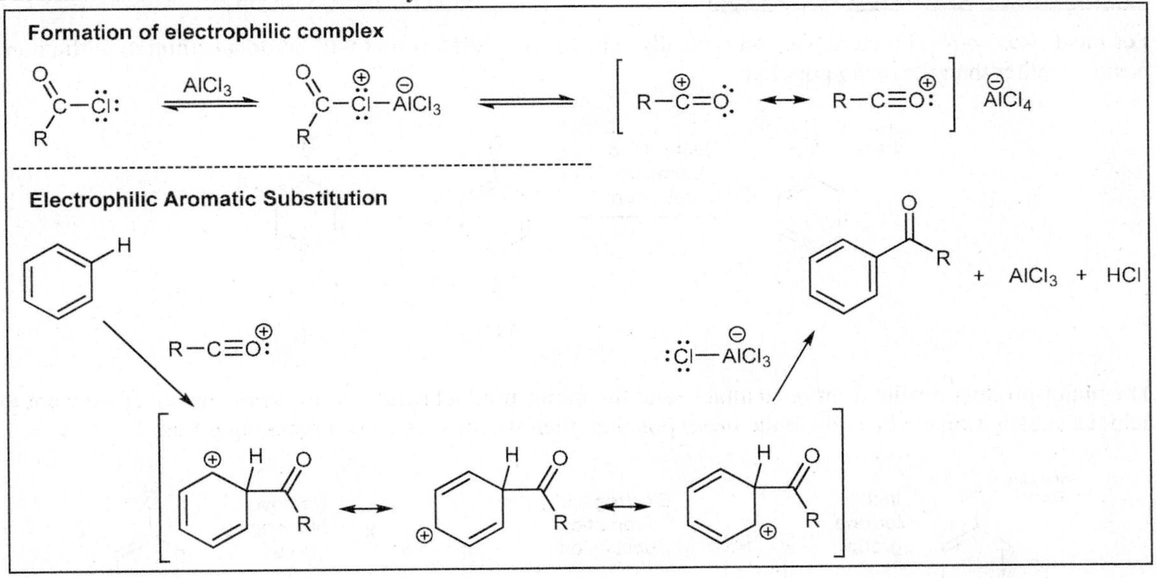

Formation of electrophilic complex

$AlCl_3$

R—C≡O: ⟷ R—C≡O: $AlCl_4^-$

R

Electrophilic Aromatic Substitution

H

R—C≡O:

:Cl—$AlCl_3$

R + $AlCl_3$ + HCl

H O R ⟷ H O R ⟷ H O R

Mechanism 18.8 Nucleophilic Aromatic Substitution (SNAr)

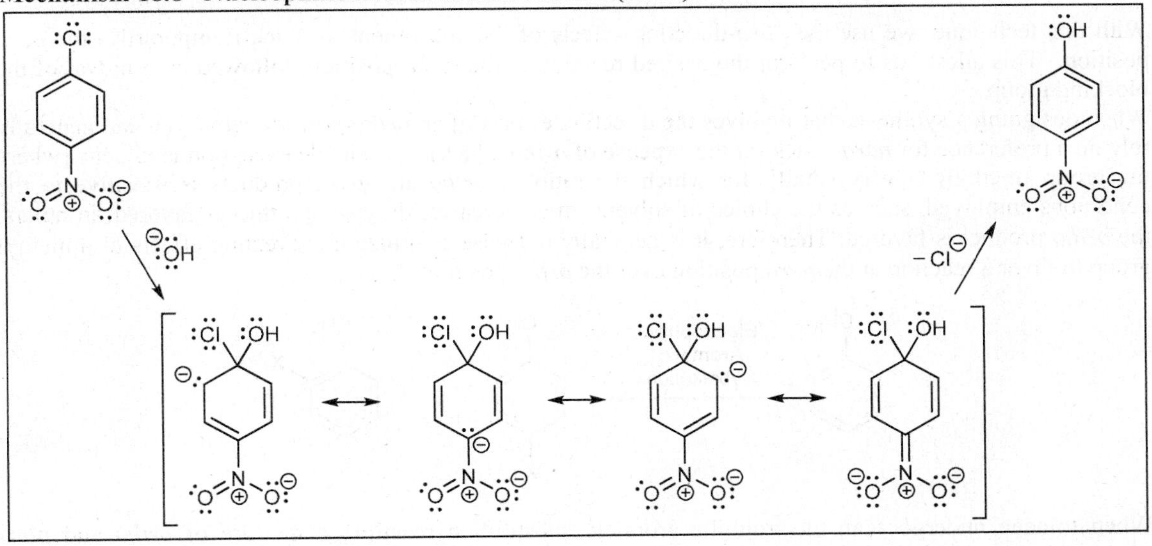

:Cl:

:OH

:OH

− Cl

:Cl :OH ⟷ :Cl :OH ⟷ :Cl :OH ⟷ :Cl :OH

Mechanism 18.9 Elimination-Addition

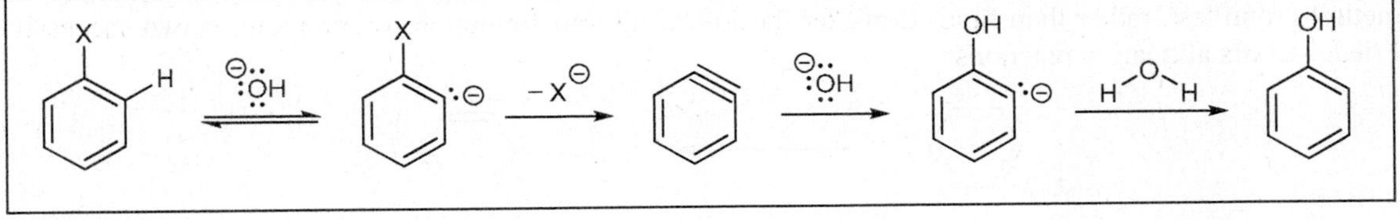

X H :OH X − X OH :OH H O H OH

Common Mistakes to Avoid

For most *ortho-para* directors, we can typically rely on steric effects that will favor substitution at the *para* position rather than the *ortho* position:

ortho-para director R — Electrophilic Aromatic Substitution → R, X (Major) + R, X (Minor)

The major product results from *para* attack, and the minor product results from *ortho* attack. If we want to achieve substitution exclusively at the *ortho* position, then we must employ a blocking group:

ortho-para director R — Install blocking group → R, SO_3H — Electrophilic Aromatic Substitution → R, X, SO_3H — Remove blocking group → R, X

With this technique, we use the *para*-directing effects of the substituent to block (temporarily) the *para* position. This allows us to perform the desired reaction at the *ortho* position, followed by removal of the blocking group.

When designing a synthesis that involves the directing effects of an *ortho-para* director, you can generally rely on a preference for *para* attack (at the expense of *ortho* attack). A notable exception is toluene [where the group (methyl) is very small], for which the ratio of *ortho* and *para* products is sensitive to the conditions employed, such as the choice of solvent. In some cases, the *para* product is favored; in others, the *ortho* product is favored. Therefore, it is generally not wise to utilize the directing effects of a methyl group to favor a reaction at the *para* position over the *ortho* position:

CH_3 — Electrophilic Aromatic Substitution → CH_3, X + CH_3, X

When toluene undergoes an electrophilic aromatic substitution reaction, a mixture of *ortho* and *para* products is unavoidable. Therefore, it would be inefficient to rely on the directing effects (to favor *para* attack over *ortho* attack) of a methyl group. Avoid making this mistake, as there are many ways that this mistake can manifest itself in a synthesis problem. For example, if you are designing a synthesis for a polysubstituted aromatic ring (where one of the substituents is a methyl group), consider installing the methyl group last, rather than first. Consider the following transformation, which requires two successive Friedel-Crafts alkylation reactions:

This transformation is best achieved if the *tert*-butyl group is installed prior to the methyl group. After the *tert*-butyl group is installed, subsequent methylation is forced to occur almost exclusively at the *para* position (because of the steric bulk of the *tert*-butyl group). In contrast, if methylation is performed first, then installation of the *tert*-butyl group will not proceed with the same degree of regioselectivity.

Useful reagents

The following is a list of reagents encountered in this chapter:

Reagents	Name of Reaction	Description
Br_2, $AlBr_3$ or $FeBr_3$	Bromination	Installation of a bromine atom on an aromatic ring.
Br_2	Bromination	Installation of a bromine atom on a moderately or strongly activated aromatic ring.
Cl_2, $AlCl_3$ or $FeCl_3$	Chlorination	Installation of a chlorine atom on an aromatic ring.
Cl_2	Chlorination	Installation of a chlorine atom on a moderately or strongly activated aromatic ring.
HNO_3, H_2SO_4	Nitration	Installation of a nitro group on an aromatic ring. Cannot be performed if an amino group is already present on the aromatic ring.
Fuming H_2SO_4	Sulfonation	Installation of a sulfonic acid group, often used as a blocking group.
Dilute H_2SO_4	Desulfonation	Removes a sulfonic acid group.
RCl, $AlCl_3$	Friedel-Crafts alkylation	Installation of an alkyl group on an aromatic ring. This process is limited to R groups that are not susceptible to carbocation rearrangements. Unless otherwise indicated, assume conditions will favor monoalkylation. This reaction cannot be performed on a moderately or strongly deactivated aromatic ring.
RC(=O)Cl , $AlCl_3$	Friedel-Crafts acylation	Installation of an acyl group on an aromatic ring. This reaction cannot be performed on a moderately or strongly deactivated aromatic ring.
1) Fe or Zn, HCl 2) NaOH	Reduction	Reduction of a nitro group to give an amino group.
Excess NBS	Benzylic bromination	Exhaustive bromination of the benzylic position.
1) $KMnO_4$, H_2O, heat 2) H_3O^+	Benzylic oxidation	Oxidation of the benzylic position to give a carboxylic acid.
Zn(Hg), HCl, heat	Clemmensen reduction	Reduction of a ketone or aldehyde carbonyl group to give a methylene (CH_2) group (from a ketone) or a methyl (CH_3) group (from an aldehyde).
NaOH	S_NAr or elimination-addition	A strong nucleophile used in nucleophilic aromatic substitution reactions as well as elimination-addition reactions.
$NaNH_2$	elimination-addition	A very strong base (and strong nucleophile) used in elimination-addition reactions.

Solutions

18.1. In the first step, the aromatic ring functions as a nucleophile and attacks the electrophile (I^+), giving a resonance-stabilized intermediate (called a sigma complex). The sigma complex then loses a proton to restore aromaticity, affording the product:

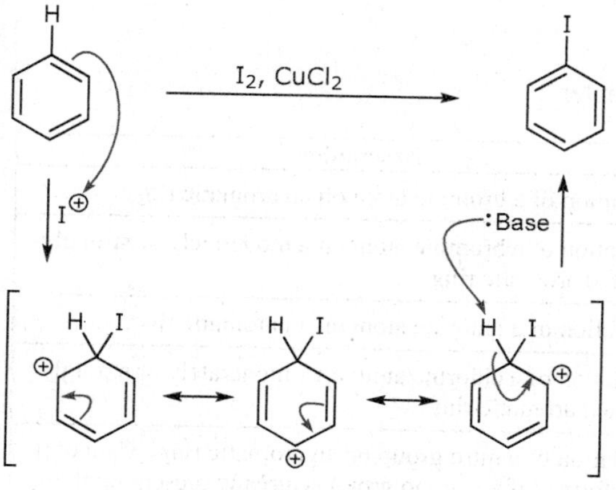

The problem statement has not provided enough information for us to determine the identity of the base that removes the proton in the last step of the process. This type of situation is extremely rare. In nearly all cases that you encounter throughout this course, you should be able to determine the identity of the base.

18.2. If we consider the fact that desulfonation is the reverse process of sulfonation, then the first step of desulfonation should be a proton transfer (since that was the last step of sulfonation). The aromatic ring is then protonated to give a resonance-stabilized intermediate (sigma complex), which then loses SO_3 to regenerate aromaticity.

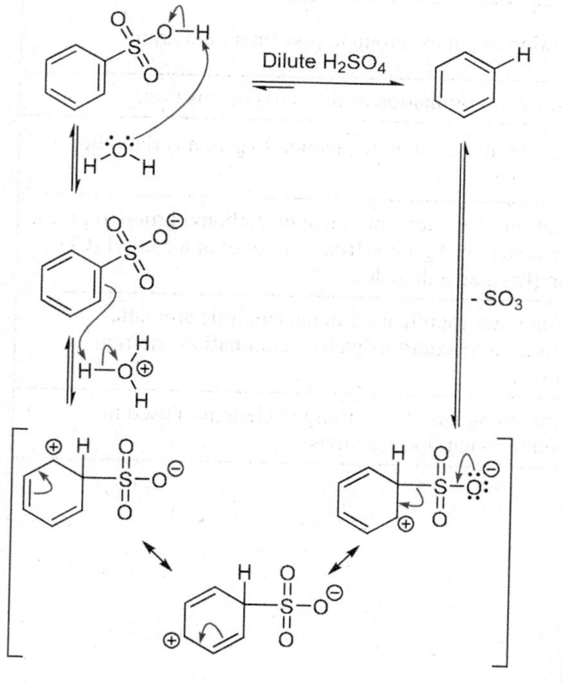

18.3. The aromatic ring functions as a base and removes a deuteron from D_2SO_4, giving an intermediate sigma complex, which then loses a proton to restore aromaticity.

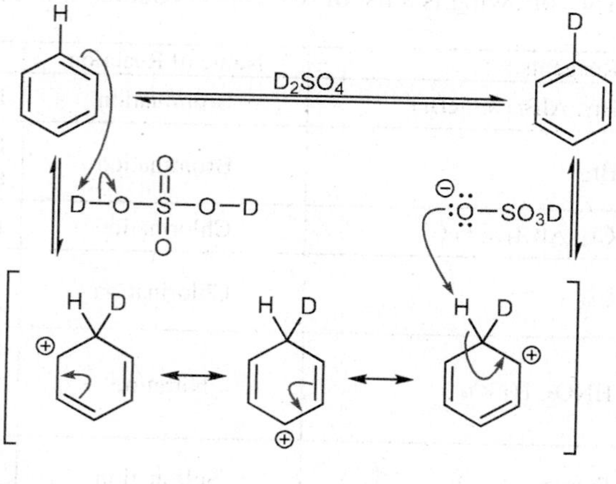

18.4.
In the presence of sulfuric acid, nitric acid can be protonated, followed by loss of water to give a nitronium ion (highlighted):

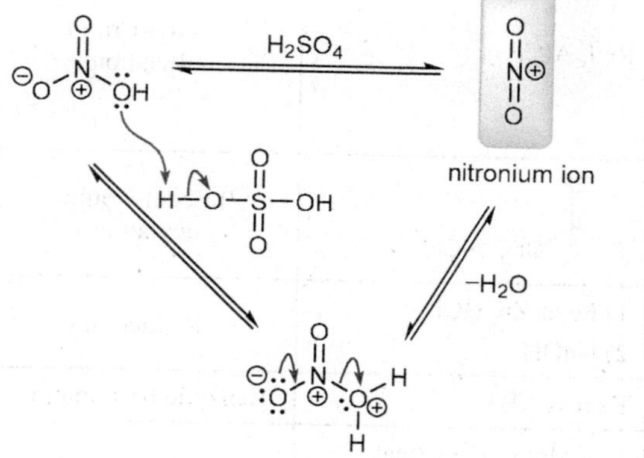

The aromatic ring then functions as a nucleophile and attacks the nitronium ion to give an intermediate sigma complex, which then loses a proton to restore aromaticity. Water is the likely base for the final deprotonation step, since water is present in these mixtures of concentrated aqueous acids (and there are no strong bases present in strongly acidic conditions).

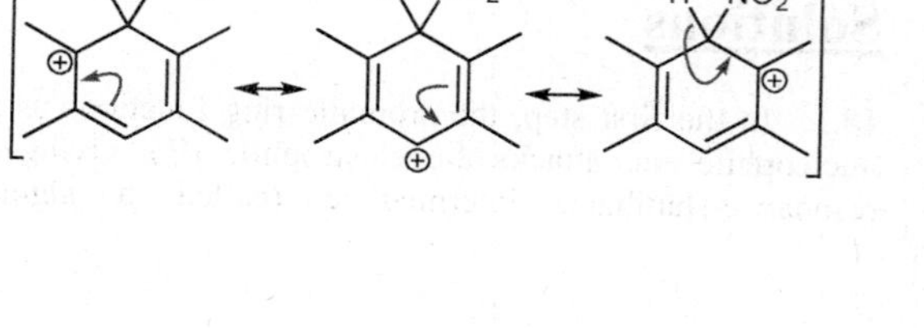

18.5.

(a) When benzene is treated with R–Cl in the presence of a Lewis acid, an R group is installed on the aromatic ring. In this case, the R group is a cyclohexyl group, so we expect a cyclohexyl group to be installed on the aromatic ring, giving the following product:

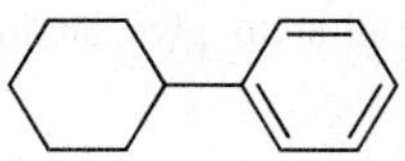

(b) When benzene is treated with R–Cl in the presence of a Lewis acid, an R group is installed on the aromatic ring. In this case, the R group is a *tert*-butyl group, so we expect a *tert*-butyl group to be installed on the aromatic ring. No rearrangement is expected, since the carbocation formed upon chloride departure is tertiary.

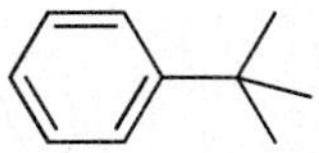

(c) When benzene is treated with R–Cl in the presence of a Lewis acid, an R group is installed on the aromatic ring. In this case, consider the complex that is formed when R–Cl interacts with the Lewis acid:

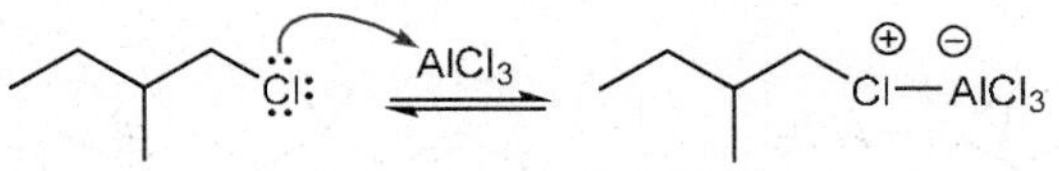

Cl: $AlCl_3$ ⇌ Cl–$AlCl_3$

This complex can either be attacked by benzene,

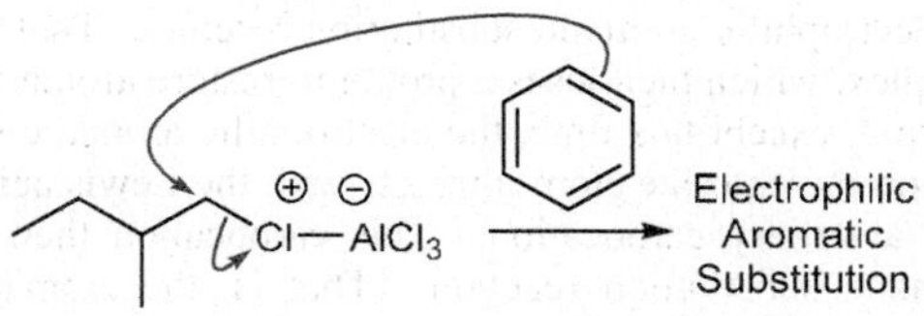

Cl–$AlCl_3$ → Electrophilic Aromatic Substitution

or the complex can rearrange, via a hydride shift, to give a tertiary carbocation, which can then react with benzene:

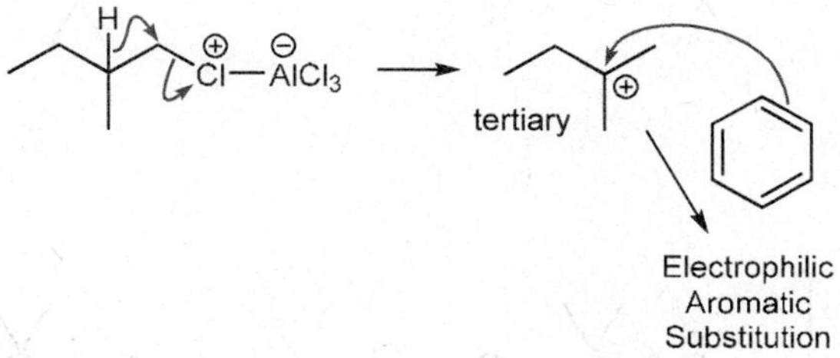

H Cl–$AlCl_3$ → tertiary

Electrophilic Aromatic Substitution

As such, there are two possible products for this electrophilic aromatic substitution reaction:

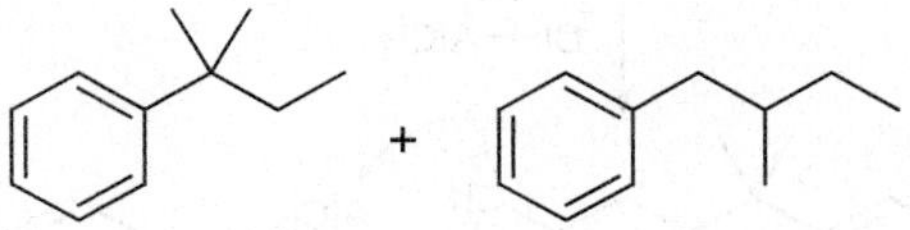

+

18.6. An interaction between a lone pair (on one of the chlorine atoms) and the Lewis acid generates a complex that can lose a leaving group ($AlCl_4^-$) to give a tertiary carbocation. This carbocation then functions as an electrophile in an electrophilic aromatic substitution reaction. That is, benzene attacks the carbocation to give an intermediate sigma complex, which then loses a proton to restore aromaticity. All of the steps above are then repeated once again to close the ring, except this time, the electrophilic aromatic substitution reaction occurs in an intramolecular fashion. A lone pair on the chlorine atom interacts with the Lewis acid to generate a complex that can lose a leaving group ($AlCl_4^-$) to give a tertiary carbocation. This carbocation then functions as an electrophile in an *intramolecular* electrophilic aromatic substitution reaction. That is, the aromatic ring attacks the carbocation to give an intermediate sigma complex, which then loses a proton to restore aromaticity:

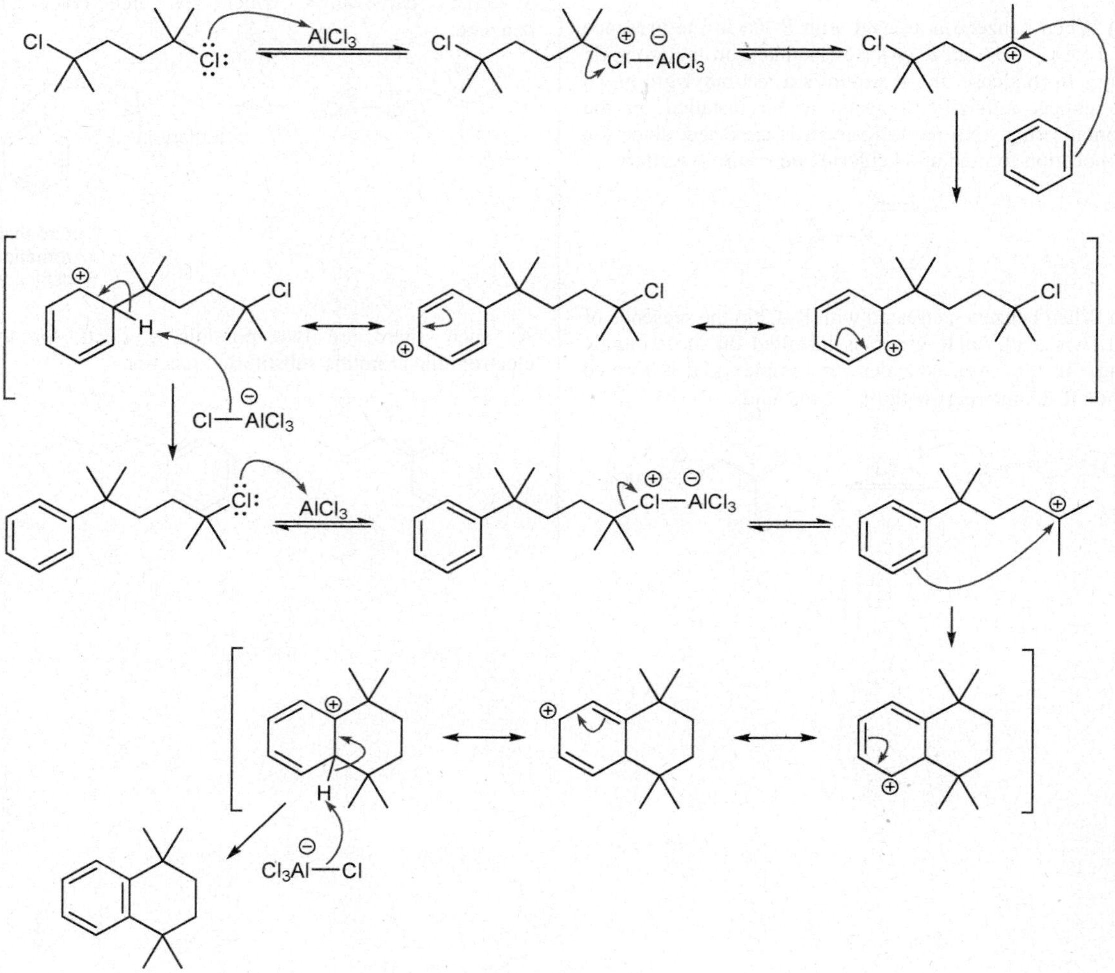

18.7. Sulfuric acid is used as an aqueous solution, so we indicate H_3O^+ as the proton source in our mechanism, rather than H_2SO_4 (because H_2SO_4 is a stronger acid than H_3O^+, so mixing H_2SO_4 and water causes the protons to be transferred from H_2SO_4 to H_2O, giving H_3O^+). In these acidic conditions, the alkene is protonated to give a carbocation. This carbocation then serves as an electrophile in an electrophilic aromatic substitution reaction. The aromatic ring attacks the electrophile to give an intermediate sigma complex, which is then deprotonated to restore aromaticity, thereby giving the product. Water is the likely base for the final deprotonation step, since water is present in a solution of sulfuric acid (and there are no strong bases present in strongly acidic conditions).

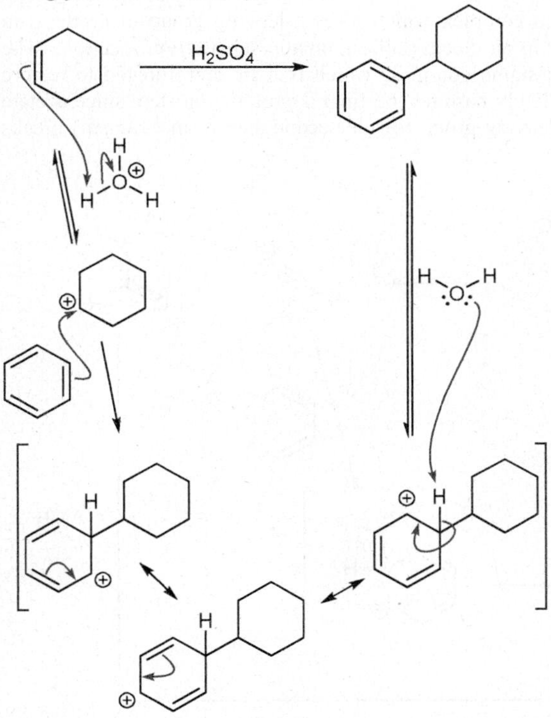

H_2SO_4

18.8.

(a) Making the desired product via a Friedel–Crafts alkylation process would require the use of the following type of alkyl halide:

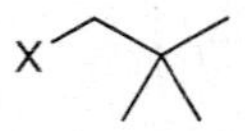

This alkyl halide will interact with a Lewis acid to give a complex that is likely to undergo a methyl shift:

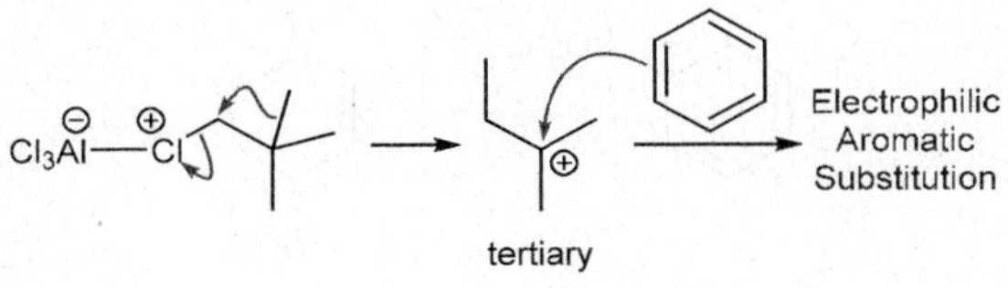

Cl_3Al — Cl → tertiary → Electrophilic Aromatic Substitution

Therefore, it is necessary to perform an acylation followed by a Clemmensen reduction to avoid carbocation rearrangements.

(b) Making the desired product via a Friedel–Crafts alkylation process would require the use of the following type of alkyl halide:

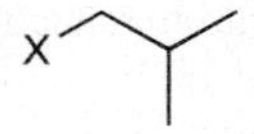

This alkyl halide will interact with a Lewis acid to give a complex that is likely to undergo a hydride shift:

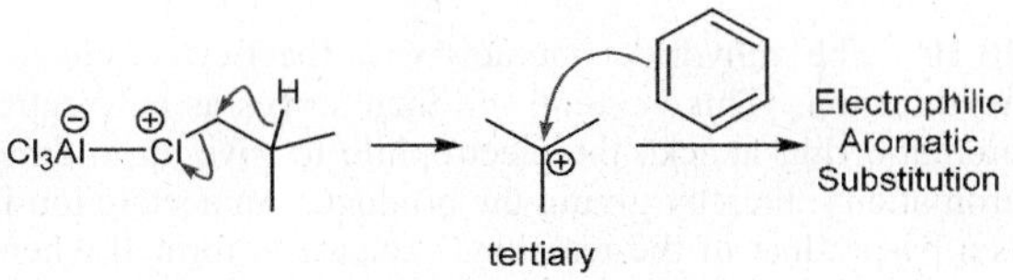

Cl_3Al — Cl → tertiary → Electrophilic Aromatic Substitution

Therefore, it is necessary to perform an acylation followed by a Clemmensen reduction to avoid carbocation rearrangements.

(c) Making the desired product via a Friedel–Crafts alkylation process would require the use of the following type of alkyl halide:

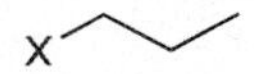

This alkyl halide will interact with a Lewis acid to give a complex that is likely to undergo a hydride shift:

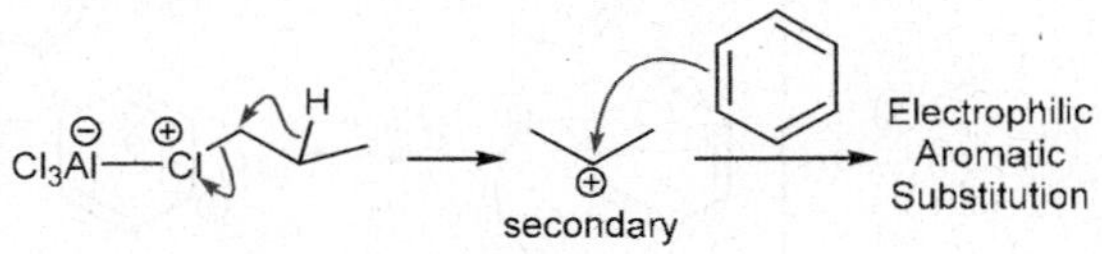

Cl_3Al — Cl → secondary → Electrophilic Aromatic Substitution

Therefore, it is necessary to perform an acylation followed by a Clemmensen reduction to avoid carbocation rearrangements.

(d) Making the desired product via a Friedel–Crafts alkylation process would require formation of a Lewis acid/base complex that cannot be stabilized by a rearrangement. Therefore, the compound can be made using a direct Friedel–Crafts alkylation.

18.9. Making the desired product via a Friedel–Crafts alkylation process would require the use of the following type of alkyl halide:

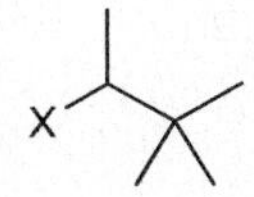

This alkyl halide will interact with a Lewis acid to give a complex that is likely to undergo a methyl shift:

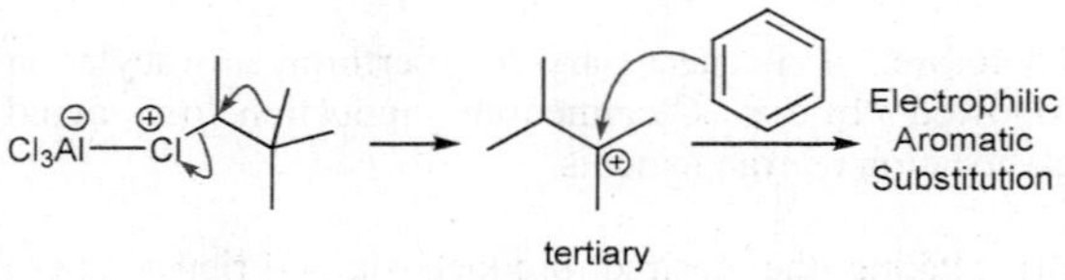

The desired product also cannot be made via acylation followed by a Clemmensen reduction, because the product of a Clemmensen reduction has two benzylic protons:

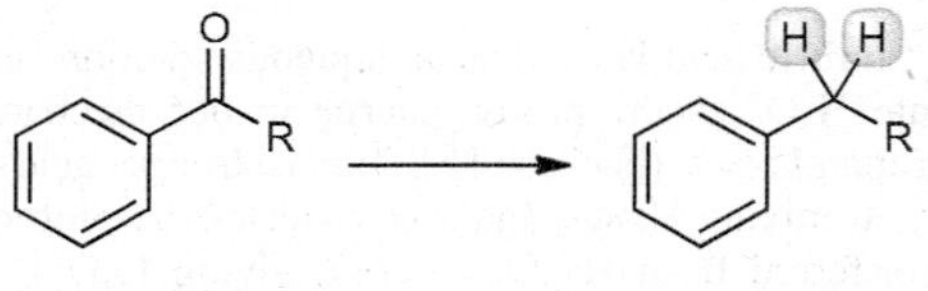

The desired product has only one benzylic proton, which means that it cannot be made via a Clemmensen reduction.

18.10. The anhydride interacts with the Lewis acid to give a complex which loses a leaving group to produce an acylium ion. This acylium ion then serves as an electrophile in an electrophilic aromatic substitution reaction. The aromatic ring attacks the electrophile to give an intermediate sigma complex, which is then deprotonated to restore aromaticity, thereby giving the product. An acetate ion is the likely base for the final deprotonation step, since acetate is a by-product of the reaction. Acetate is formed when the leaving group (of the second step of mechanism) breaks apart into $AlCl_3$ and acetate (as shown).

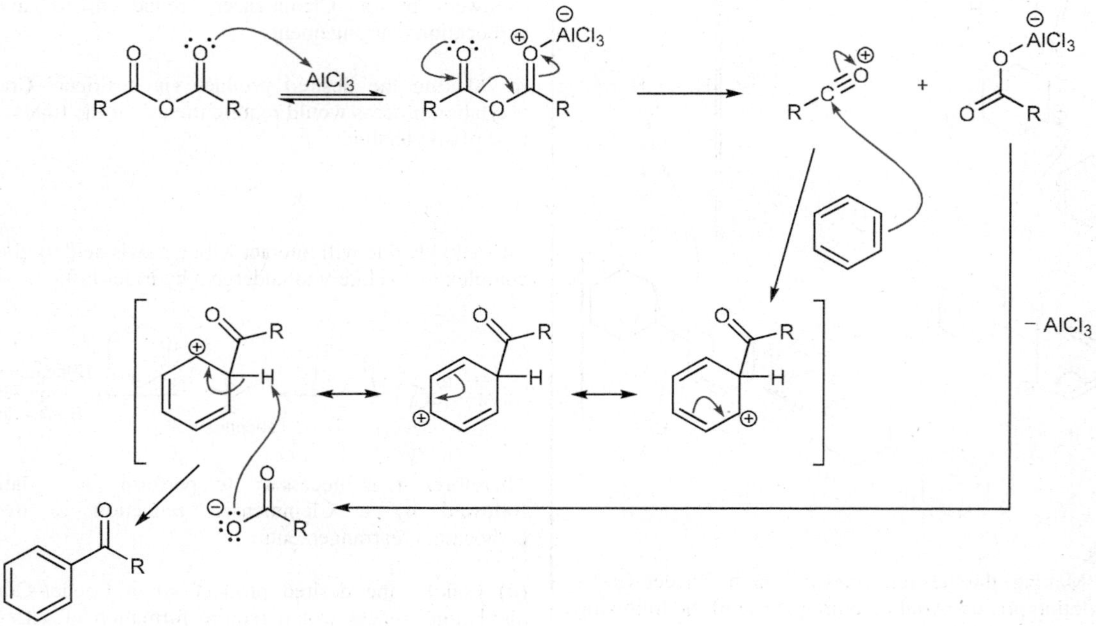

18.11. The methyl group is an activator, which directs the incoming electrophile (Br^+) to the *ortho* and *para* positions:

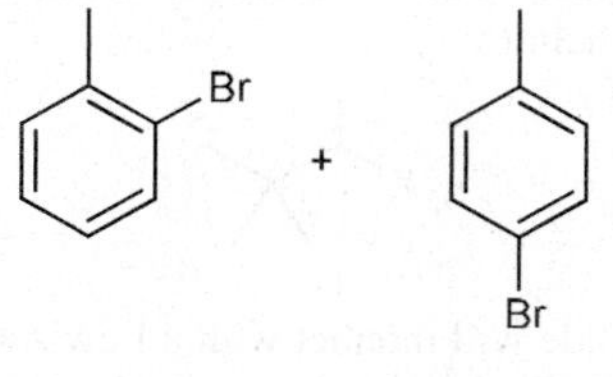

18.12.

(a) The ethoxy group is an activator, which directs the incoming electrophile (NO_2^+) to the *ortho* and *para* positions:

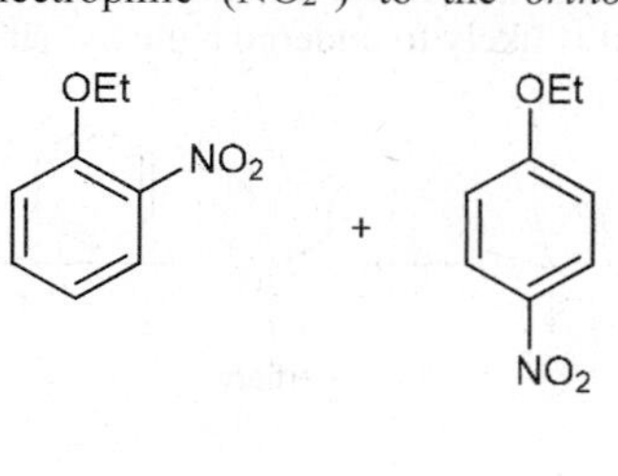

(b) In the presence of sulfuric acid, nitric acid can be protonated, followed by loss of water to give a nitronium ion:

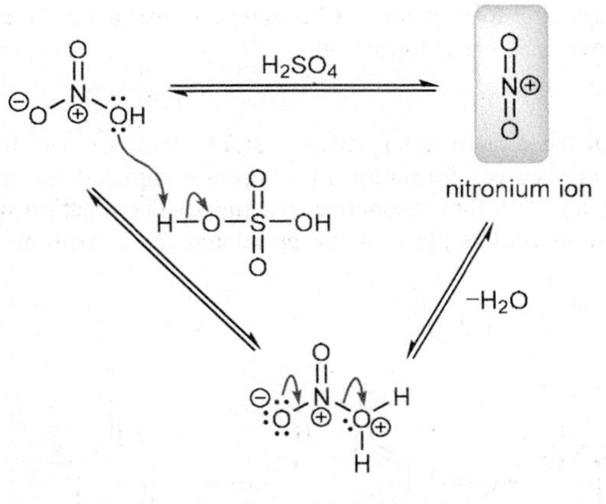

H_2SO_4

nitronium ion

$-H_2O$

The aromatic ring then functions as a nucleophile and attacks the nitronium ion. We will draw the attack occurring at the *para* position because the problem statement asked for a mechanism of formation of the major product (the *para* product predominates over the *ortho* product as a result of steric factors). The resulting intermediate sigma complex then loses a proton to restore aromaticity. Note that the signa complex is stabilized by the resonance effect of the ethoxy group, as shown in the fourth resonance structure:

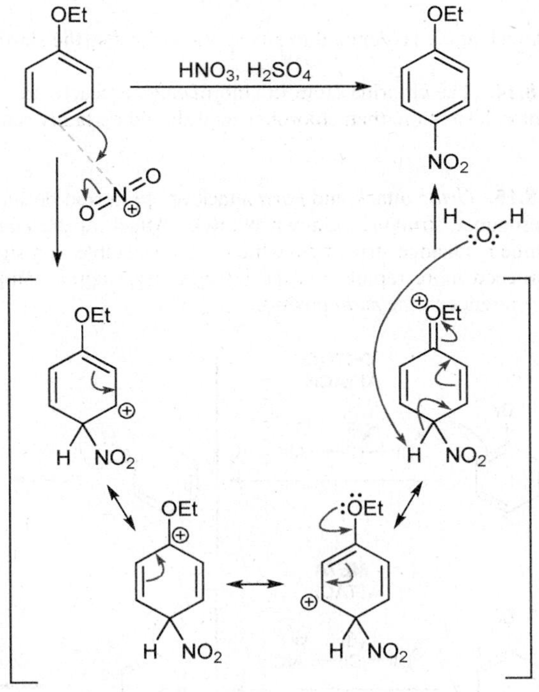

HNO_3, H_2SO_4

Water is the likely base for the final deprotonation step, since water is present (and there are no strong bases present in strongly acidic conditions).

18.13. As shown below, attack at C4 or C6 produces a sigma complex in which two of the resonance structures have a positive charge directly adjacent to an electron-withdrawing group (NO_2). These resonance structures are less contributing to the resonance hybrid, thereby destabilizing the sigma complex. In contrast, attack at C5 produces a sigma complex for which none of the resonance structures have a positive charge next to a nitro group.

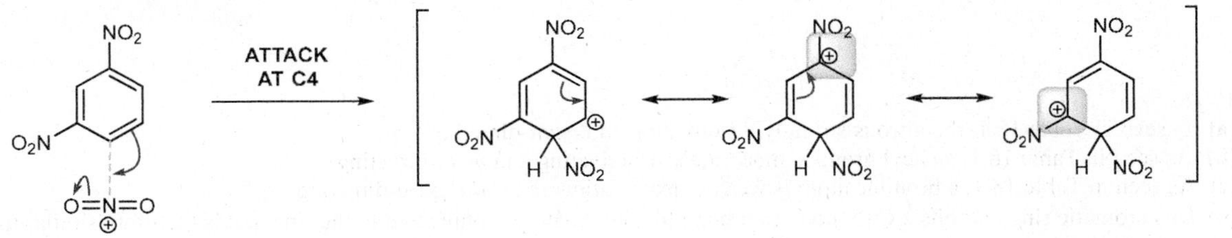

ATTACK AT C4

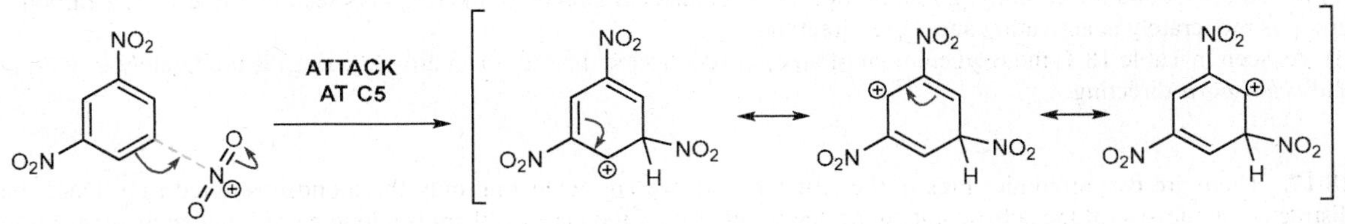

ATTACK AT C5

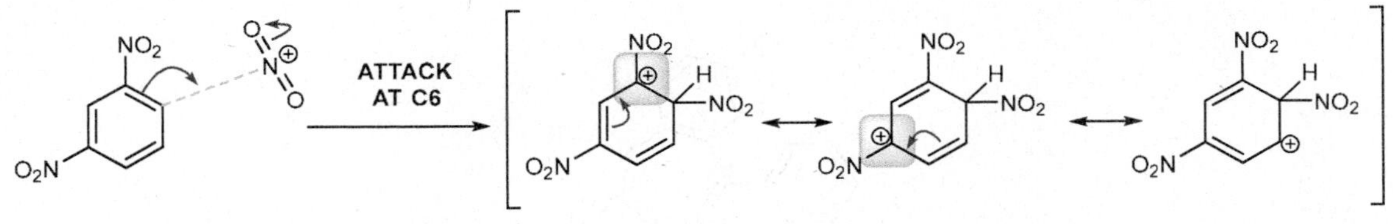

ATTACK AT C6

Attack at C2 is identical to attack at C4 because the starting material has symmetry that renders C2 and C4 identical.

18.14. The chlorine atom in chlorobenzene deactivates the ring relative to benzene. If benzene requires a Lewis acid for chlorination, than chlorobenzene should certainly require a Lewis acid for chlorination.

18.15. *Ortho* attack and *para* attack are preferred because each of these pathways involves a sigma complex with four resonance structures (shown below). Attack at the *meta* position involves formation of a sigma complex with only three resonance structures, which is not as stable as a sigma complex with four resonance structures. The reaction will proceed more rapidly via the lower energy sigma complex, so attack takes place at the *ortho* and *para* positions in preference to the *meta* position.

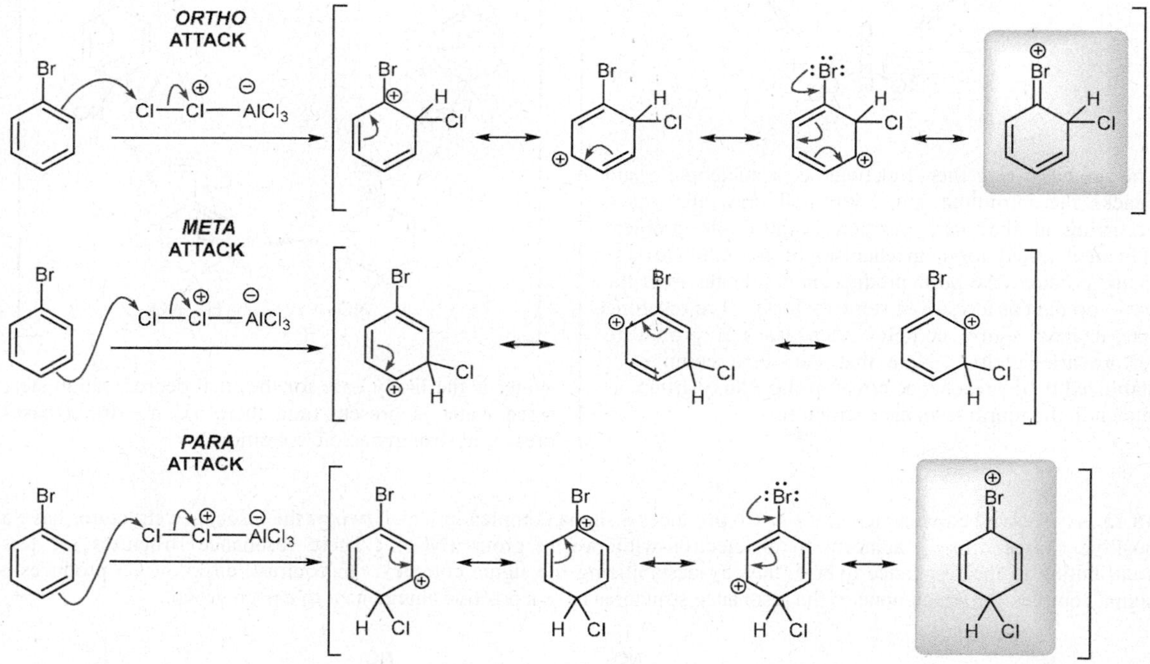

ORTHO ATTACK

META ATTACK

PARA ATTACK

18.16.
(a) As seen in Table 18.1, the nitro is strongly deactivating and *meta*-directing.
(b) As seen in Table 18.1, an acyl group is moderately deactivating and *meta*-directing.
(c) As seen in Table 18.1, a bromine atom is weakly deactivating and *ortho-para*-directing.
(d) This aromatic ring exhibits a C=N bond (a π bond to a heteroatom) conjugated to the ring and is therefore similar to a C=O bond that is conjugated to the ring. This group is moderately deactivating and *meta*-directing.
(e) In this compound the carbonyl group of the ester is connected directly to the ring. As seen in Table 18.1, a carbonyl group is moderately deactivating and *meta*-directing.
(f) As seen in Table 18.1, the oxygen atom of an ester (connected directly to an aromatic ring) is moderately activating and *ortho-para*-directing.

18.17. There are two aromatic rings in the structure, but we are evaluating only the monosubstituted ring. Don't be distracted by the size of the substituent connected to that ring. First check if there is a lone pair immediately adjacent to the ring. In this case, there isn't one. The atom directly attached to the benzene ring is a carbon atom.

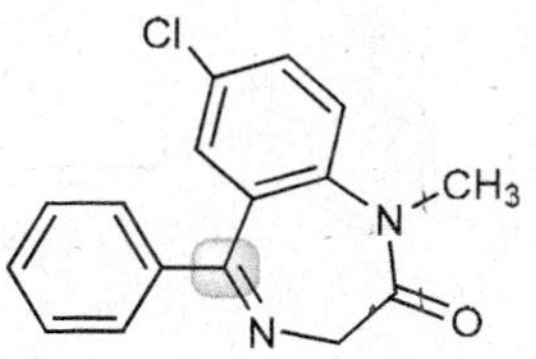

Next, check for a π bond to an electronegative atom, where the π bond is conjugated with the ring. The C=N bond is conjugated with the ring allowing for electron withdrawal via resonance, as shown here:

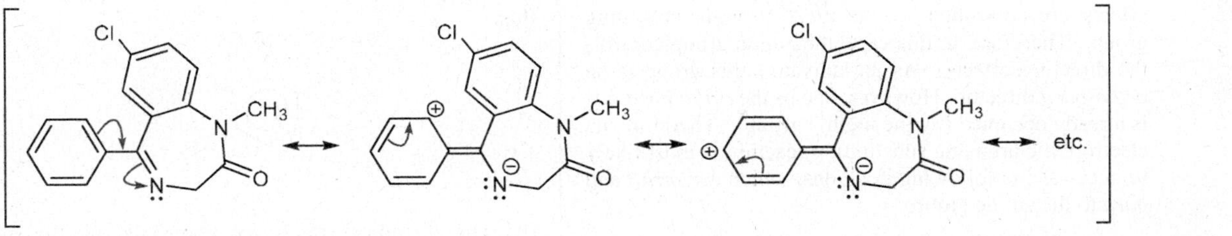

We therefore expect the monosubstituted ring to be moderately deactivated. Note the effect of the substituent. The resonance structures indicate that the *ortho* and *para* positions should be electron-poor (δ+). We therefore predict those positions to be the least likely to react with an electrophile. The deactivating substituent is (as usual) a *meta* director.

18.18.

(a) The methyl group is weakly activating and the nitro groups are strongly deactivating. The directing effects are controlled by the most strongly activating group. Therefore, in this case, the methyl group controls the directing effects. As an activator, the methyl group is an *ortho-para* director. However, the two *ortho* positions are both already occupied (by the nitro groups). Therefore, an electrophilic aromatic substitution reaction is most likely to occur at the position that is *para* to the methyl group, shown here:

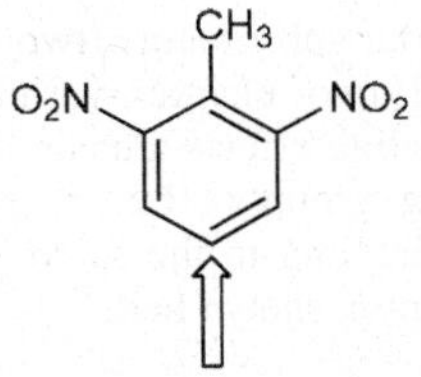

(b) The methyl group is weakly activating and the nitro groups are strongly deactivating. The directing effects are controlled by the most strongly activating group. Therefore, in this case, the methyl group controls the directing effects. As an activator, the methyl group is an *ortho-para* director. However, the *para* position and one of the *ortho* positions are already occupied (by the nitro groups). Therefore, an electrophilic aromatic substitution reaction is most likely to occur at the position that is *ortho* to the methyl group, shown here:

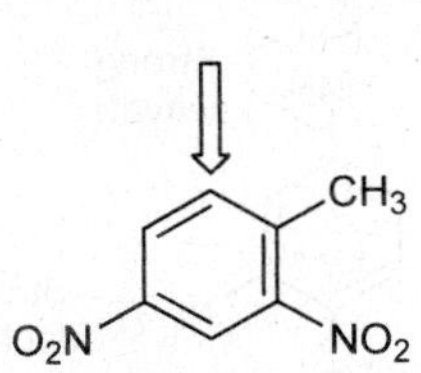

(c) This aromatic ring has three substituents. The methyl group is weakly activating. The other two groups are both esters, but they differ in the way they are connected to the ring. One group is moderately activating (the oxygen atom connected to the ring) and the other is moderately deactivating (the carbonyl group connected to the ring). The directing effects are controlled by the most strongly activating group. Therefore, in this case, the ester group (connected to the ring by its oxygen atom) controls the directing effects. As an activator, this group is an *ortho-para* director. However, the *para* position and one of the *ortho* positions are already occupied (by the other substituents). Therefore, an electrophilic aromatic substitution reaction is most likely to occur at the position that is *ortho* to the moderately activating group, shown here:

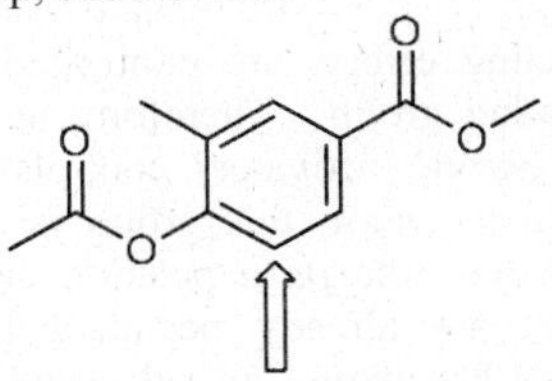

(d) The methyl groups are weakly activating and the cyano group is moderately deactivating. The directing effects are controlled by the most strongly activating group(s). Therefore, in this case, the methyl groups control the directing effects. Both of the methyl groups are directing to the same two locations (*ortho* and *para* to the methyl groups), shown here.

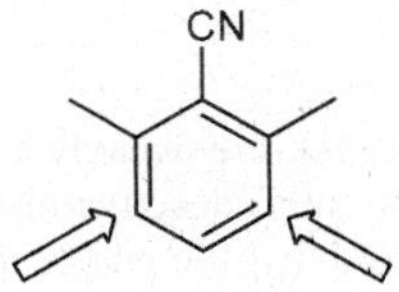

These two locations are identical because of symmetry, so only one major product is expected. Don't be confused by the arbitrary location of the double bonds. Perhaps it is less distracting if we draw the structure like this:

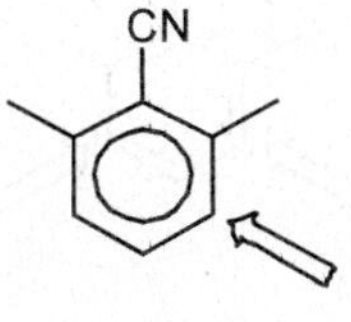

(e) This aromatic ring has three substituents. The methyl group is weakly activating. The carbonyl group (C=O) is moderately deactivating, and the amide group is moderately activating (see Table 18.1). The directing effects are controlled by the most strongly activating group. Therefore, in this case, the amide group controls the directing effects. As an activator, this group is an *ortho-para* director. However, one of the *ortho* positions is already occupied (by the methyl group). Therefore, an electrophilic aromatic substitution reaction is most likely to occur at the following positions, which are *ortho* and *para* to the amide group:

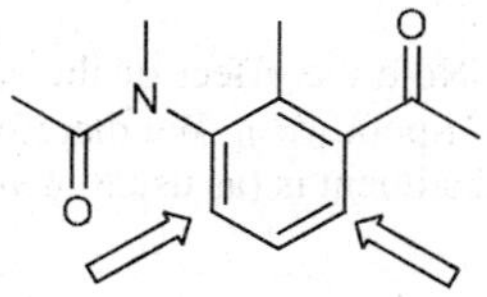

Note that each of the indicated positions is *ortho* to one substituent (and those substituents are similar in steric bulk). Thus, we don't expect steric considerations to play a significant role in differentiating these two positions. We will explore the impact of steric considerations in the section immediately following this SkillBuilder.

(f) The directing effects are controlled by the most strongly activating group. Therefore, in this case, the OH group (a strong activator) controls the directing effects. As an activator, this group is an *ortho-para* director. However, the *para* position and one of the *ortho* positions are already occupied (by the other substituents). Therefore, an electrophilic aromatic substitution reaction is most likely to occur at the position that is *ortho* to the OH group, shown here:

Br OH

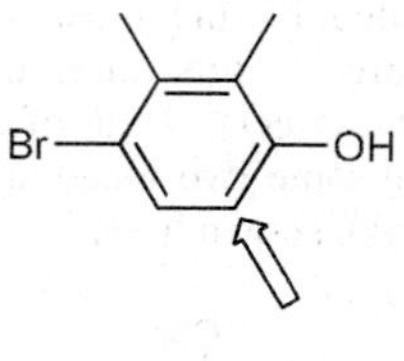

(g) The methyl groups are weakly activating and the bromo group is weakly deactivating. The directing effects are controlled by the most strongly activating group(s). Therefore, in this case, the methyl groups control the directing effects. Both of the methyl groups are directing to the same two locations (*ortho* and *para* to the methyl groups), shown here.

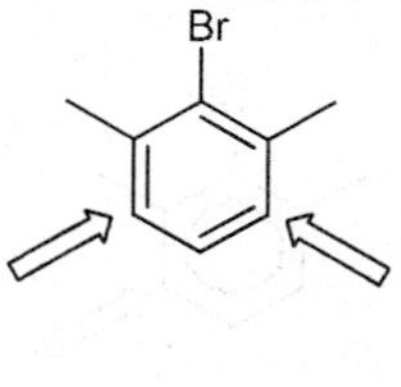

Br

These two locations are identical because of symmetry, so there is only one major product expected. Don't be confused by the arbitrary location of the double bonds. It may be less distracting if we draw the structure like this:

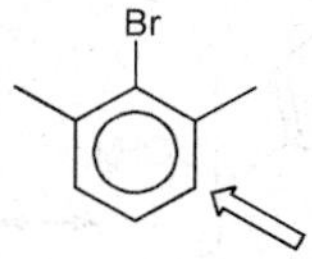

Br

(h) The directing effects are controlled by the most strongly activating group. Therefore, in this case, the OH group (a strong activator) controls the directing effects (rather than the methoxy group, which is only a moderate activator). As an activator, the OH group is an *ortho-para* director. However, the *para* position and one of the *ortho* positions are already occupied (by the other substituents). Therefore, an electrophilic aromatic substitution reaction is most likely to occur at the position that is *ortho* to the OH group, shown here:

OMe HO Br

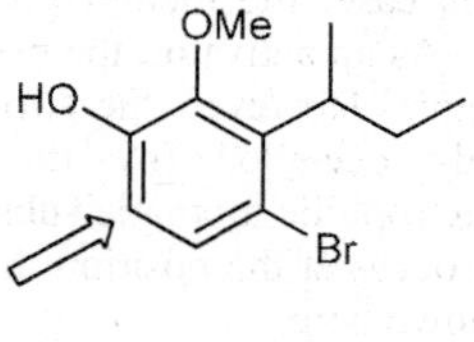

(i) This ring has four substituents (two electron-donating alkyl groups and two electron-withdrawing carbonyl groups). The directing effects are controlled by the most strongly activating group(s), but in this case, all four substituents are directing to the same location (*ortho* to the alkyl substituents), shown here.

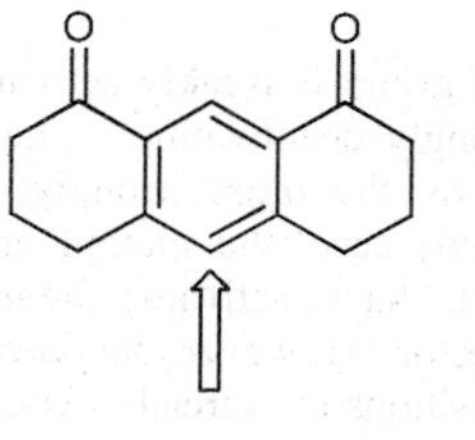

18.19. Begin by identifying the effect of each group on the aromatic ring.

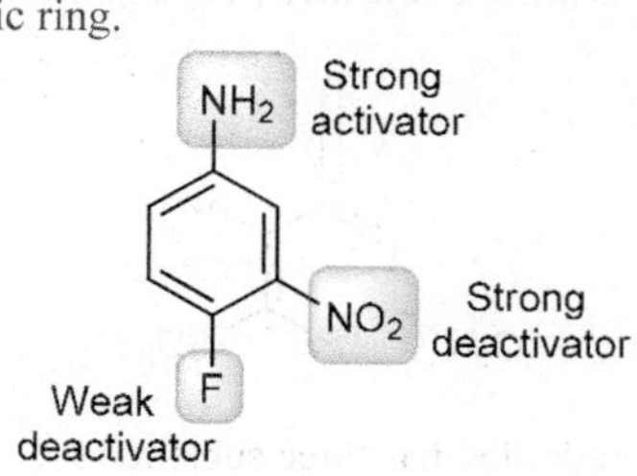

NH_2 Strong activator

NO_2 Strong deactivator

F Weak deactivator

The most powerful activator will dominate the directing effects. In this case the NH_2 group is the strongest activator so we consider the positions that are *ortho* and *para* to the NH_2 group.

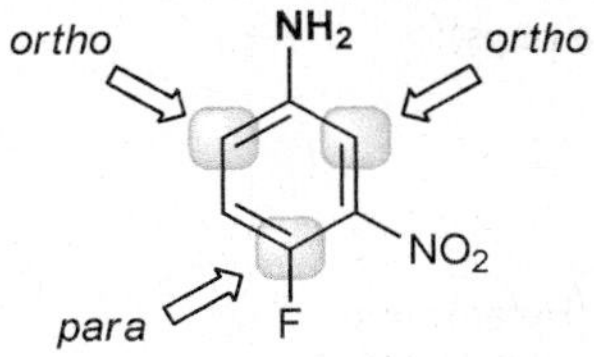

ortho NH2 ortho
NO2
para F

The *para* position is already occupied by F. There are two positions that are *ortho* to the NH_2 group, and both are available. Therefore, we expect that the reaction should occur primarily at these two positions:

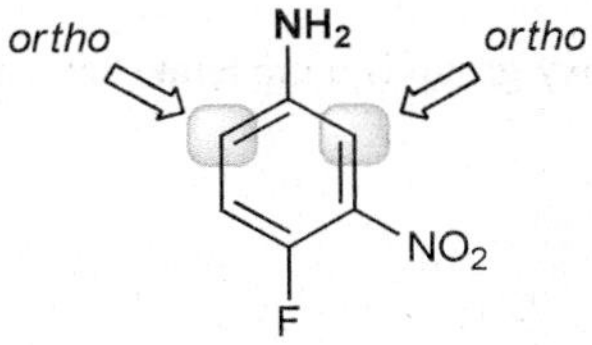

ortho NH2 ortho
NO2
F

However, we will soon see (immediately after this SkillBuilder) that the *ortho* position on the left is more accessible for steric reasons, so substitution at this position leads to the major product.

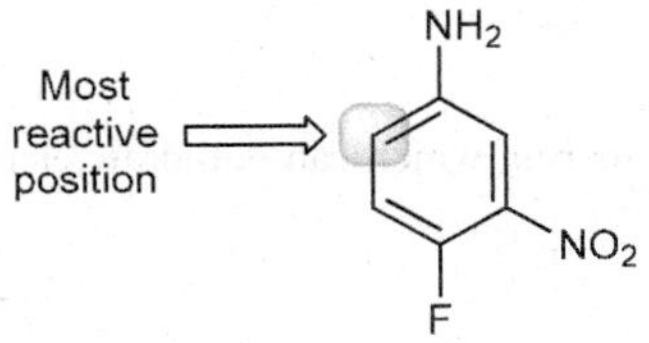

NH2
Most reactive position
NO2
F

18.20.

(a) The directing effects are controlled by the most strongly activating group. Therefore, in this case, the OH group (a strong activator) controls the directing effects. As an activator, the OH group is an *ortho-para* director. However, the *para* position is already occupied (by the methyl group). Therefore, an electrophilic aromatic substitution reaction should occur at an *ortho* position. Among the two *ortho* positions, one of them (top) is sterically hindered and we do not expect the reaction to occur at that location. Therefore, the reaction is most likely to occur at the following position:

Sterically hindered
HO NO2
Reaction occurs here

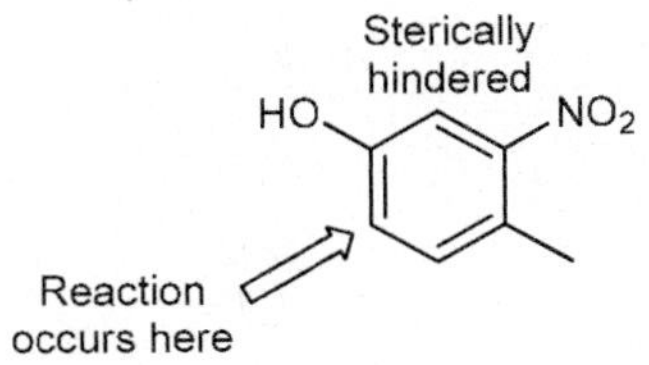

(b) The directing effects are controlled by the most strongly activating group. Therefore, in this case, the methoxy group (a moderate activator) controls the directing effects. As an activator, the methoxy group is an *ortho-para* director. However, the *para* position is already occupied (by the bromine atom). Therefore, an electrophilic aromatic substitution reaction should occur at an *ortho* position. Among the two *ortho* positions, one of them (top) is sterically hindered and we do not expect the reaction to occur at that location. Therefore, the reaction is most likely to occur at the following position:

Sterically hindered
O2N OMe
Br
Reaction occurs here

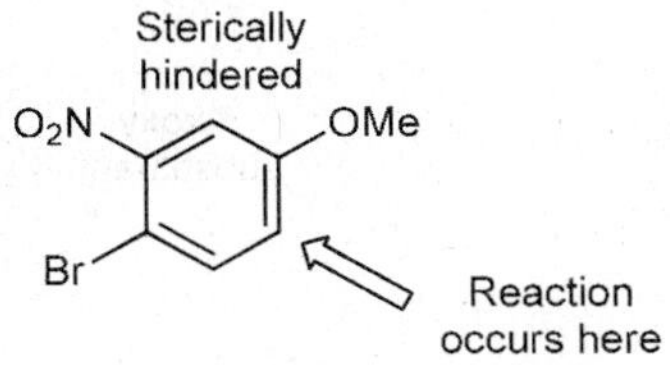

(c) This aromatic ring lacks an activating substituent. Both substituents are deactivators, and they are competing with each other (each group directs to the positions that are *meta* to itself). So electronically, the four aromatic positions are equally likely to undergo an electrophilic aromatic substitution reaction. The positions are differentiated from each other when we consider steric effects. The larger group (left) is more sterically bulky, and therefore blocks the positions closest to it. The reaction therefore occurs primarily *ortho* to the aldehyde group.

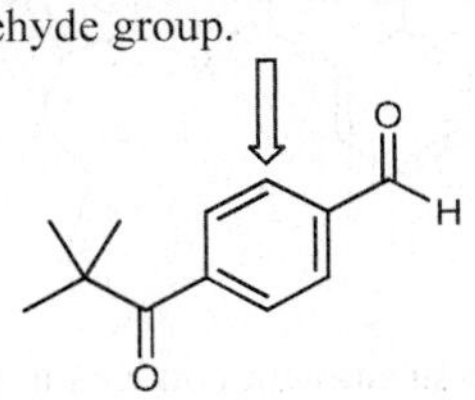

O
H
O

The two positions *ortho* to the aldehyde are identical because of symmetry, so a single major product is expected.

(d) The directing effects are controlled by the most strongly activating group(s). Therefore, in this case, the OH groups (strong activators) control the directing effects. However, each OH group directs to different locations (*ortho* and *para* to itself), so all available positions are activated. To differentiate between them, we note that the following position is too sterically hindered for a reaction to occur at this location:

OH
HO
Too sterically hindered

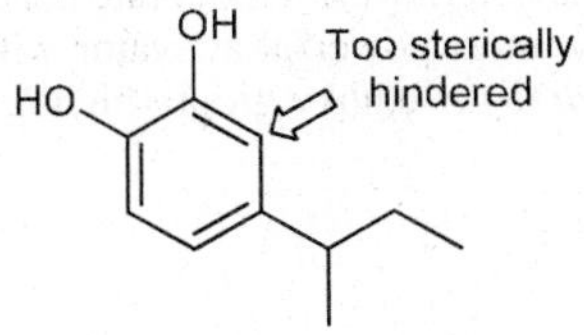

Among the other two locations, one of them is more sterically accessible than the other (being *ortho* to an OH group, as opposed to being *ortho* to a more bulky sec-butyl group), so the reaction is expected to occur at this location:

OH
HO

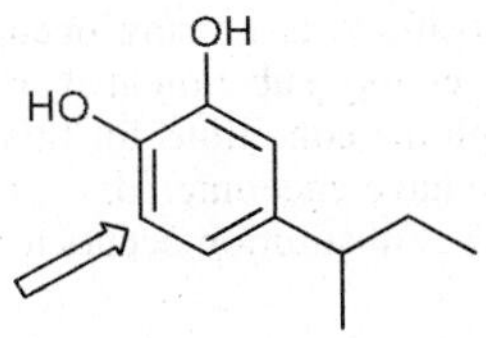

18.21. As mentioned in the problem statement, the benzene ring has two substituents. One is an alkoxy group, and the other is a heterocyclic substituent (it contains nitrogen atoms incorporated in rings):

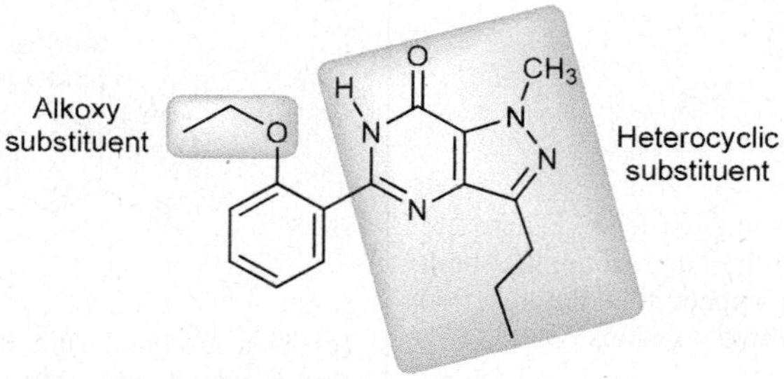

Alkoxy substituent

Heterocyclic substituent

Let's begin with the ethoxy group. With its lone pairs on oxygen, the ethoxy group is a moderate activator capable of donating electrons toward the ring, as expected for an alkoxy substituent:

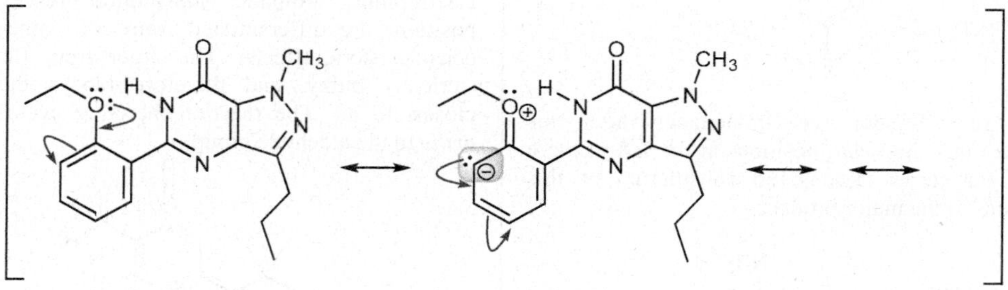

The heterocyclic substituent contains a π-bond conjugated with the benzene ring which can withdraw electron density from the ring via resonance:

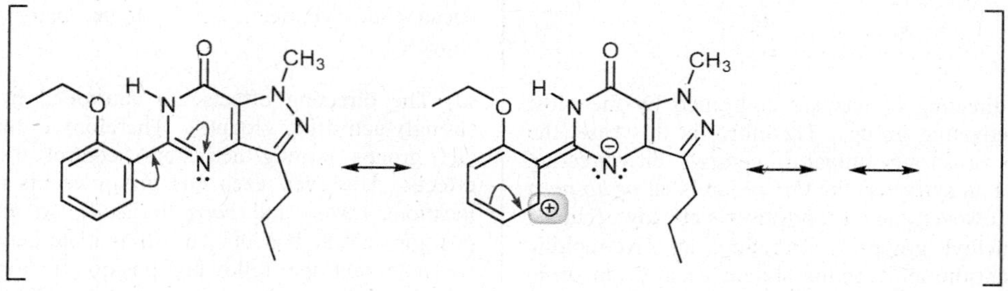

This makes the heterocycle a moderate deactivator.
Recall that the most powerful activator will dominate the directing effects, so we must consider the positions that are *ortho* and *para* to the ethoxy group (highlighted):

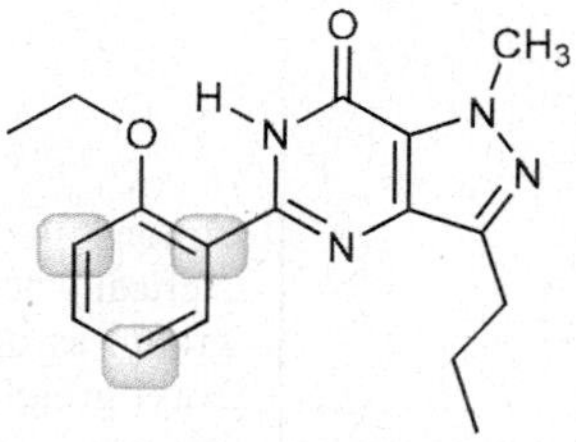

One of these positions is already occupied, leaving only two choices. Since the *ortho* position is more sterically hindered by the ethoxy substituent than the *para* position, we expect substitution to take place primarily at the *para* position. Though the conditions for this reaction are new, the resulting electrophilic aromatic substitution is similar to the reactions we have encountered. Specifically, the site of attack exchanges its proton for the electrophilic group, in this case $-SO_2Cl$. Substitution occurs to afford the following product:

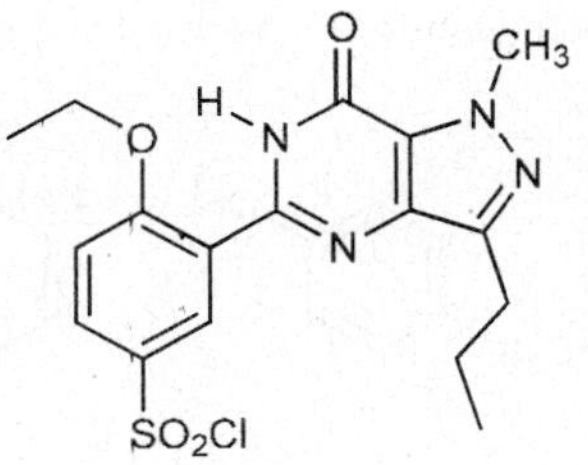

18.22.

(a) This aromatic ring has only one substituent (an alkyl group). Alkyl groups are weak activators and therefore *ortho-para* directors. An isopropyl group is sterically bulky, and we therefore expect a nitration reaction to occur predominantly at the *para* position. In order to install a nitro group at the *ortho* position, a blocking group will be required.

(b) This aromatic ring has only one substituent (an alkyl group). Alkyl groups are weak activators and therefore *ortho-para* directors. An isopropyl group is sterically bulky, and we therefore expect a bromination reaction to occur predominantly at the *para* position. Therefore, a blocking group is not required. Direct bromination will give primarily the desired product.

(c) The aromatic ring has two substituents: an amino group and a carbonyl group. The former is a strong activator and the latter is a moderate deactivator. The directing effects are typically controlled by the activating group, but in this case, both substituents direct to the same positions. As an activator, the amino group is an *ortho-para* director. However, one of the *ortho* positions is already occupied (by the other substituent). Therefore, there are only two locations where bromination can occur (*ortho* or *para* to the amino group):

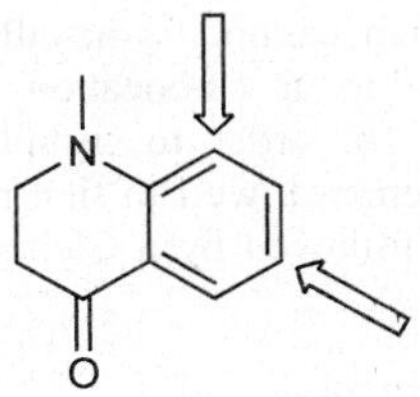

Among these two locations, one of them (top) is more sterically hindered than the other. Bromination is therefore expected to occur at the location that is more sterically accessible.

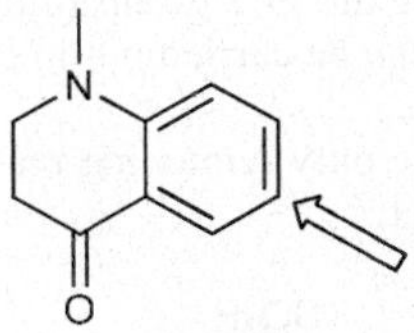

In order to install the bromine atom at the other location (*ortho* rather than *para*), a blocking group will be required.

(d) The directing effects for the desired nitration process are controlled by the most strongly activating group. Therefore, in this case, the OH group (a strong activator) controls the directing effects. As an activator, this group is an *ortho-para* director. However, one of the *ortho* positions is already occupied (by a methyl group). This leaves only two locations where the reaction can occur:

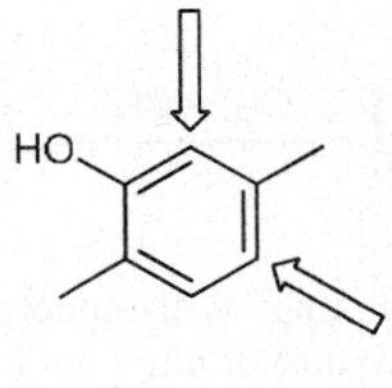

One of these locations (bottom right) is more accessible than the other, and this is exactly the position where a nitro group must be installed. Therefore, a blocking group will not be required.

18.23. Analyze the starting material. The OH group is a strong activator and an *ortho-para* director. Indeed, it is such a strong activator that bromination conditions will likely give a tribrominated product (with bromination occurring at the *para* position and both *ortho* positions), even at room temperature. Also, even if monobromination could be carried out cleanly, the product mixture would likely contain only some of the more sterically hindered (desired) product along with a substantial amount of the *para* isomer that would likely be the major product. Our strategy will be to block some reactive positions, brominate, and then unblock. As we have seen, reactive positions can be blocked via sulfonation, as shown here:

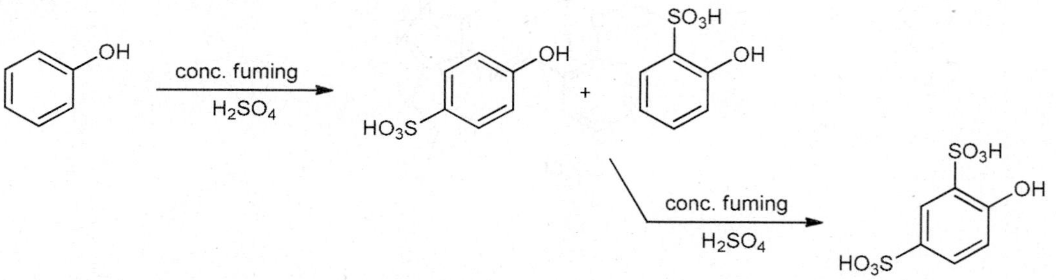

The first sulfonic acid group is installed quickly, and it deactivates the ring somewhat, so the second sulfonation is expected to occur more slowly (and possibly require a higher temperature). A third sulfonation would require an even higher temperature, so there is likely an optimum temperature at which the disulfonated product can be obtained in high yield (and indeed, this is the case!). The regiochemistry of the second sulfonation is controlled by the *ortho-para*-directing effect of the OH group (although the *meta*-directing SO_3H group also directs to the same location). This double blocking can be carried out in one synthetic step with excess fuming sulfuric acid, but is shown stepwise above for clarity.

Bromination at the only remaining reactive site, followed by desulfonation to remove the two blocking groups, yields the desired product:

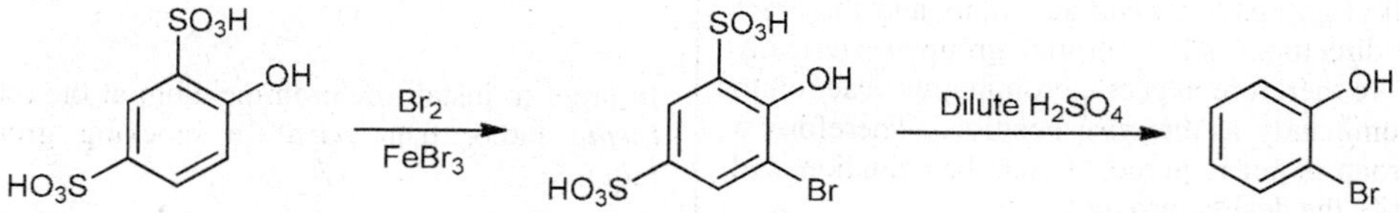

18.24.

(a) The aromatic ring will undergo chlorination upon treatment with molecular chlorine (Cl_2) and a Lewis acid ($AlCl_3$).

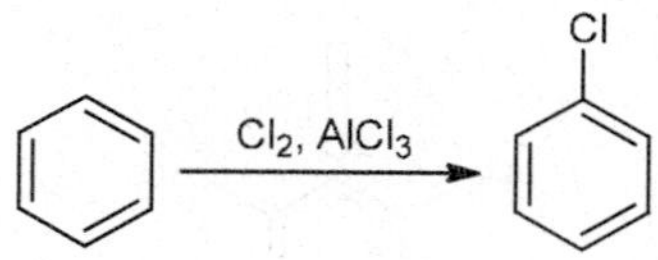

(b) The aromatic ring will undergo nitration upon treatment with a mixture of nitric acid and sulfuric acid.

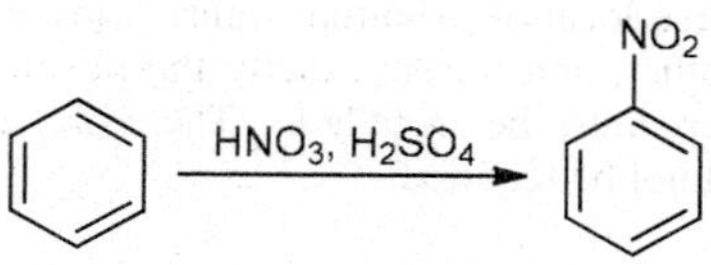

(c) The aromatic ring will undergo bromination upon treatment with molecular bromine (Br_2) and a Lewis acid ($AlBr_3$ or $FeBr_3$).

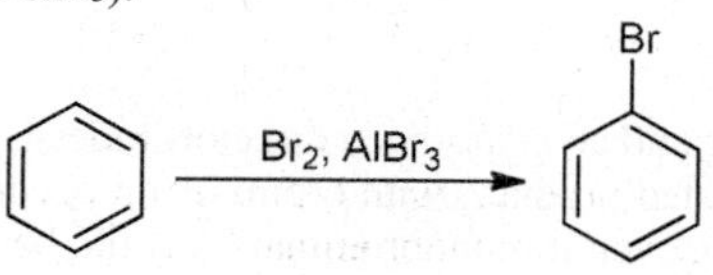

(d) The aromatic ring will undergo a Friedel-Crafts alkylation upon treatment with ethyl chloride and a Lewis acid ($AlCl_3$), to give ethylbenzene.

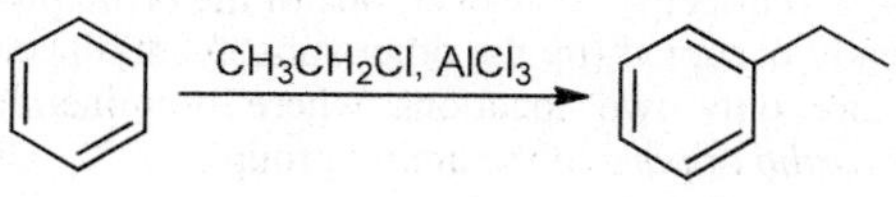

(e) A propyl group cannot be installed via a Friedel-Crafts alkylation, as a carbocation rearrangement is likely to occur. In order to install a propyl group (without rearrangement), we can first perform a Friedel-Crafts acylation, followed by a Clemmensen reduction, as shown:

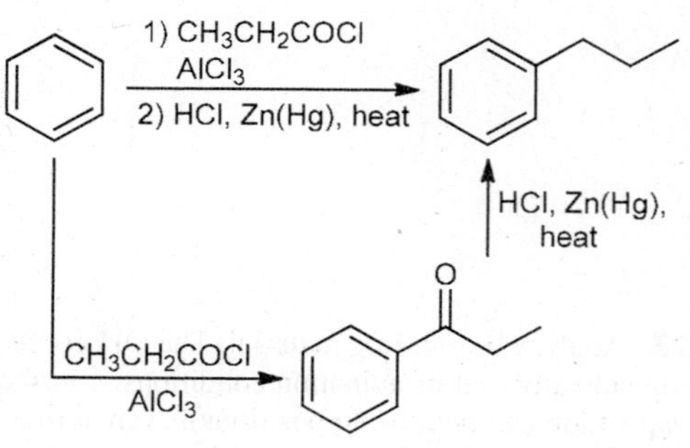

(f) An isopropyl group can be installed via a Friedel-Crafts alkylation in which isopropyl chloride is used as the alkyl halide:

$(CH_3)_2CHCl$, $AlCl_3$

Notice that we use isopropyl chloride to install an isopropyl group, rather than using *n*-propyl chloride and relying on a carbocation rearrangement (which would be inefficient because it would likely yield a mixture of products).

(g) We did not learn a one-step method for installing an amino group on an aromatic ring, but we did learn a two-step method for achieving that transformation. Specifically, we first install a nitro group, and then we reduce the nitro group to an amino group, as shown here:

1) HNO_3, H_2SO_4
2) Zn, HCl
3) NaOH

NH_2

HNO_3, H_2SO_4

NO_2

1) Zn, HCl
2) NaOH

(h) We did not learn a one-step method for installing a carboxylic acid group on an aromatic ring, but we did learn a multi-step method for achieving that transformation. Specifically, we first install a methyl group, and then we oxidize the methyl group to give the desired product:

1) CH_3Cl, $AlCl_3$
2) $KMnO_4$, H_2O, heat
3) H_3O^+

O OH

CH_3Cl, $AlCl_3$

1) $KMnO_4$, H_2O, heat
2) H_3O^+

(i) The aromatic ring will undergo a Friedel-Crafts alkylation upon treatment with methyl chloride and a Lewis acid ($AlCl_3$), to give toluene.

CH_3Cl, $AlCl_3$

18.25.

(a) These reagents will install a sulfonic acid group on the aromatic ring (a process called sulfonation), giving benzenesulfonic acid:

SO_3H

(b) These reagents will install a nitro group on the aromatic ring (a process called nitration), giving nitrobenzene:

NO_2

(c) These reagents will install a chlorine atom on the aromatic ring (a process called chlorination), giving chlorobenzene:

Cl

(d) These reagents will install an ethyl group on the aromatic ring (via a Friedel-Crafts alkylation), giving ethylbenzene:

(e) These reagents will install a bromine atom on the aromatic ring (a process called bromination), giving bromobenzene:

Br

(f) These reagents will achieve the installation of a nitro group on the aromatic ring, followed by reduction of the nitro group to give aminobenzene (also called aniline):

NH_2

18.26.

(a) Installation of the amino group requires a two-step process (nitration, followed by reduction), while installation of the bromine atom can be achieved in just one step. Therefore, our synthesis must have at least those three steps.

Now let's consider the order of events. These two groups must be installed in an *ortho* fashion. Both groups are *ortho-para* directing, so we could theoretically install either one first, and then use a blocking technique to install the second group in the correct location (*ortho*, rather than *para*). However, if we install the amino group first, there is a concern of polybromination, since aminobenzene (also called aniline) is highly activated. To avoid this problem, we first install the bromine atom (which does not activate the ring). Then, we perform a sulfonation to block the *para* position. Nitration then installs the nitro group in the desired location (*ortho*). Then, after removing the blocking group, the nitro group is reduced to give the desired product.

1) Br_2, $AlBr_3$
2) Fuming H_2SO_4
3) HNO_3, H_2SO_4
4) dilute H_2SO_4
5) Zn, HCl
6) NaOH

Br_2, $AlBr_3$

1) Zn, HCl
2) NaOH

Fuming H_2SO_4

dilute H_2SO_4

HNO_3, H_2SO_4

(b) Installation of the nitro group can be achieved in just one step (nitration), and installation of the chlorine atom can also be achieved in just one step.

Now let's consider the order of events. These two groups must be installed in a *meta* fashion. Only the nitro group is a *meta* director, so it must be installed first, followed by chlorination in the presence of a Lewis acid.

1) HNO_3, H_2SO_4
2) Cl_2, $AlCl_3$

HNO_3, H_2SO_4

Cl_2, $AlCl_3$

(c) Installation of the amino group requires a two-step process (nitration, followed by reduction), and installation of the propyl group also requires a two-step process (Friedel-Crafts acylation, followed by Clemmensen reduction), because direct alkylation would result in a mixture of products (due to carbocation rearrangements).

Now let's consider the order of events. These two groups must be installed in a *para* fashion. Both groups are *ortho-para* directors, so we might think that either group could be installed first. However, to favor the *para* product over the *ortho* product, we must capitalize on the steric bulk of the propyl group. That is, we install the propyl group first, followed by installation of the amino group. Also, the amino group is too highly activating and cannot be installed first.

1) CH_3CH_2COCl, $AlCl_3$
2) HCl, Zn(Hg), heat
3) HNO_3, H_2SO_4
4) Zn, HCl
5) NaOH

CH_3CH_2COCl, $AlCl_3$

1) Zn, HCl
2) NaOH

HCl, Zn(Hg), heat

HNO_3, H_2SO_4

(d) Installation of the amino group requires a two-step process (nitration, followed by reduction), while installation of the chlorine atom can be achieved in just one step.

Now let's consider the order of events. These two groups must be installed in a *meta* fashion, but both groups are *ortho-para* directors. Installation of these two groups (in either order) does not appear to give the desired *meta* substitution pattern. However, recall that installation of the amino group requires two steps. The first step is nitration, and a nitro group is a *meta*-director. So, we can achieve the desired transformation by performing the chlorination process after the nitration process but before the reduction process, as shown here:

1) HNO_3, H_2SO_4
2) Cl_2, $AlCl_3$
3) Zn, HCl
4) NaOH

HNO_3, H_2SO_4

1) Zn, HCl
2) NaOH

Cl_2, $AlCl_3$

(e) Installation of the tribromomethyl group requires a two-step process (Friedel-Crafts alkylation, followed by benzylic bromination), while installation of the bromine atom can be achieved in just one step.

Now let's consider the order of events. These two groups must be installed in a *meta* fashion. Only the tribromomethyl group is a *meta* director, so it must be installed first (via the required two-step process), followed by bromination in the presence of a Lewis acid:

1) CH_3Cl, $AlCl_3$
2) excess NBS, heat
3) Br_2, $AlBr_3$

CH_3Cl, $AlCl_3$

excess NBS, heat

Br_2, $AlBr_3$

(f) Installation of the tribromomethyl group requires a two-step process (Friedel-Crafts alkylation, followed by benzylic bromination), while installation of the bromine atom can be achieved in just one step.

Now let's consider the order of events. These two groups must be installed in a *para* fashion. Only the bromo group is an *ortho-para* director, so we install that group first, followed by installation of the tribromomethyl group (which requires two steps), as shown:

1) Br_2, $AlBr_3$
2) CH_3Cl, $AlCl_3$
3) excess NBS, heat

Br_2, $AlBr_3$

CH_3Cl, $AlCl_3$

excess NBS, heat

(g) This aromatic ring has two substituents, both of which are alkyl groups, although only the methyl group can be installed via a Friedel-Crafts alkylation process. The other alkyl group must be installed via a two-step process (Friedel-Crafts acylation, followed by Clemmensen reduction), because direct alkylation would result in a mixture of products (due to carbocation rearrangements).

Now let's consider the order of events. These two groups must be installed in an *ortho* fashion. Both groups are *ortho-para* directing, so we could theoretically install either one first, and then use a blocking technique to install the second group in the desired location (*ortho*, rather than *para*). However, if we install the methyl group first, then subsequent sulfonation is likely to produce a mixture of *ortho* and *para* products, since the methyl group is not very sterically demanding. To circumvent this lack of regiochemical control, it is best to install the bulky alkyl group first (using a Friedel-Crafts acylation, followed by a Clemmensen reduction), and then continue with a sulfonation step. This provides for much better selectivity during the sulfonation step, and therefore, a better yield of the final product.

1) $AlCl_3$,
2) HCl, Zn(Hg), heat
3) Fuming H_2SO_4
4) CH_3Cl, $AlCl_3$
5) Dilute H_2SO_4

$AlCl_3$,

HCl, Zn(Hg), heat

Fuming H_2SO_4

CH_3Cl, $AlCl_3$

Dilute H_2SO_4

(h) Installation of the nitro group can be achieved in just one step (nitration), and installation of the isopropyl group can also be achieved in just one step (Friedel-Crafts alkylation with isopropyl chloride and aluminum trichloride).

Now let's consider the order of events. These two groups must be installed in an *ortho* fashion. Only the alkyl group is an *ortho-para* director, so it must be installed first. Then, in order to install a nitro group in the *ortho* position (rather than the *para* position), a blocking group must be used. After the blocking group is installed, nitration will install a nitro group in the desired location. Desulfonation then affords the desired product.

1) $(CH_3)_2CHCl$, $AlCl_3$
2) Fuming H_2SO_4
3) HNO_3, H_2SO_4
4) Dilute H_2SO_4

$(CH_3)_2CHCl$, $AlCl_3$

Fuming H_2SO_4

HNO_3, H_2SO_4

Dilute H_2SO_4

(i) Installation of the propyl group requires a two-step process (Friedel-Crafts acylation, followed by Clemmensen reduction), because direct alkylation would result in a mixture of products (due to carbocation rearrangements). Installation of the chlorine atom can be achieved in just one step.

Now let's consider the order of events. These two groups must be installed in a *meta* fashion, but both groups are *ortho-para* directors. Installation of these two groups (in either order) does not appear to give the desired *meta* substitution pattern. However, as mentioned above, installation of the propyl group requires two steps. The first step is acylation, and the resulting acyl group is a *meta*-director. So, we can achieve the desired transformation by performing the chlorination process after the acylation process but before the reduction process, as shown here:

1) CH_3CH_2COCl, $AlCl_3$
2) Cl_2, $AlCl_3$
3) HCl, Zn(Hg), heat

CH_3CH_2COCl, $AlCl_3$

HCl, Zn(Hg), heat

Cl_2, $AlCl_3$

(j) Installation of each propyl group requires a two-step process (Friedel-Crafts acylation, followed by Clemmensen reduction), because direct alkylation would result in a mixture of products (due to carbocation rearrangements). These two groups must be installed in a *para* fashion. Propyl groups are *ortho-para* directors, and their steric bulk is expected to cause a preference for *para*-substitution. So we simply install one propyl group (via Friedel-Crafts acylation followed by reduction), followed by the second propyl group, as shown here:

1) CH_3CH_2COCl, $AlCl_3$
2) HCl, Zn(Hg), heat
3) CH_3CH_2COCl, $AlCl_3$
4) HCl, Zn(Hg), heat

CH_3CH_2COCl, $AlCl_3$

HCl, Zn(Hg), heat

CH_3CH_2COCl, $AlCl_3$

HCl, Zn(Hg), heat

18.27. The synthesis of 2-nitroaniline from benzene requires the installation of a nitro group and an amino group on a benzene ring. The nitro group can be installed in one step,

HNO_3, H_2SO_4

NO_2

while installation of the amino group can be achieved in two steps (nitration followed by reduction), as shown:

HNO_3, H_2SO_4

NO_2

1) Fe or Zn, HCl
2) NaOH

NH_2

Notice that each of these groups is installed via nitration of benzene, so nitration must be the first step of the synthesis. We must therefore decide whether to perform the second nitration before or after the reduction. Consider what will happen if the second nitration is attempted before the reduction. Since the existing nitro group is *meta*-directing, the 1,3-regioisomer will be produced, leading to 3-nitroaniline as the product. Additionally, the subsequent reduction may affect either one or both of the nitro groups.

HNO_3 / H_2SO_4

NO_2

HNO_3 / H_2SO_4

NO_2 NO_2

wrong regioisomer

1) Fe or Zn , HCl
2) NaOH

NO_2 NH_2 + NH_2 NH_2

wrong regioisomer, possible mixture of products

Now consider what will happen if the reduction is performed before the second nitration. The reduction will likely proceed smoothly; however, the following nitration step will be problematic. Under the acidic conditions employed during nitration, the amino group (an *ortho/para*-directing group) will be protonated to give an ammonium group, which is a *meta*-directing group. Therefore, nitration of aniline is not expected to give the desired regiochemical outcome

18.28.

(a) The following retrosynthetic analysis represents one strategy for preparing the desired compound. An explanation of each of the steps (*a-c*) follows.

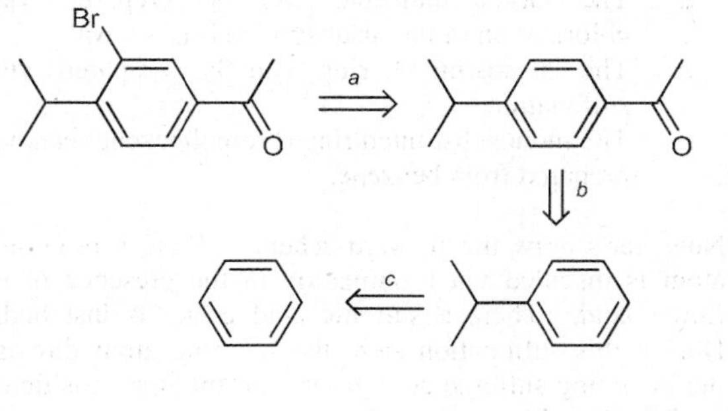

a. The target molecule can be prepared via bromination of the disubstituted ring shown.

b. The disubstituted ring can be prepared via a Friedel-Crafts acylation.

c. The monosubstituted ring (isopropylbenzene) can be prepared via a Friedel-Crafts alkylation.

Now, let's draw the forward scheme. First, an isopropyl group is installed via a Friedel-Crafts alkylation. Then, an acyl group is installed via a Friedel-Crafts acylation. During this acylation step, the isopropyl group directs the incoming acyl group into the *para* position. And finally, bromination will install the bromine atom in the correct location (*ortho* to the isopropyl group).

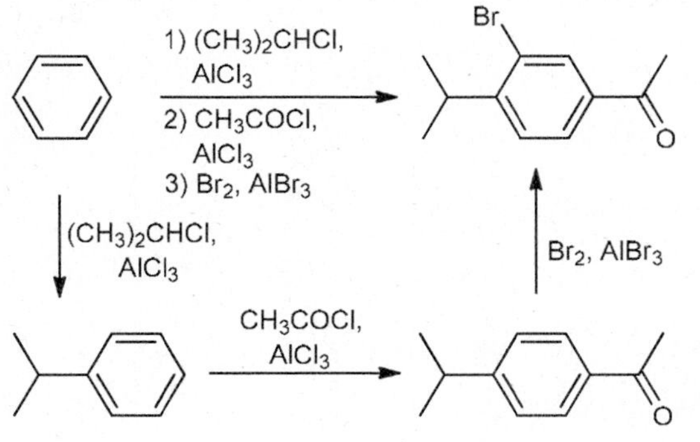

1) $(CH_3)_2CHCl$, $AlCl_3$
2) CH_3COCl, $AlCl_3$
3) Br_2, $AlBr_3$

$(CH_3)_2CHCl$, $AlCl_3$

CH_3COCl, $AlCl_3$

Br_2, $AlBr_3$

This strategy represents just one method for making the desired the compound. There are often other acceptable solutions. For example, after installation of the isopropyl group, the acyl group and the bromo group can be installed in the opposite order (first the bromo group and then the acyl group), although this would require a blocking group (to help direct the incoming bromo group into the *ortho* position). The first method (shown above) avoids the need for a blocking group.

(b) The following retrosynthetic analysis represents one strategy for preparing the desired compound. An explanation of each of the steps (*a-d*) follows.

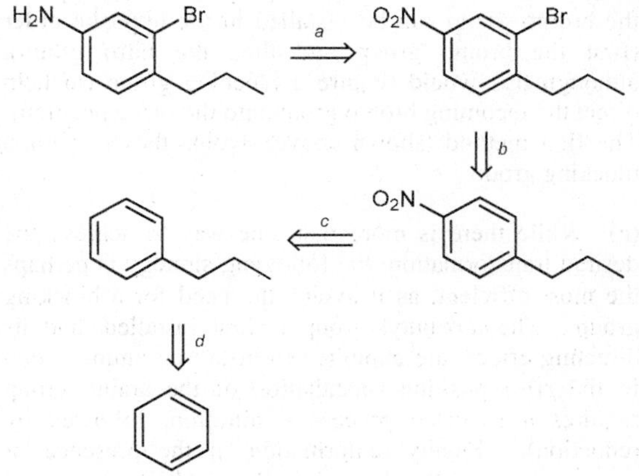

a. The target molecule can be prepared via reduction of a nitro group.

b. The bromo group can be installed in the correct location based on the directing effects of the propyl group.

c. The nitro group is installed in the *para* position via nitration.

d. A propyl group is installed onto the ring via Friedel-Crafts acylation, followed by reduction.

Now, let's draw the forward scheme. First, a propyl group is installed via a Friedel-Crafts acylation, followed by a Clemmensen reduction. This transformation could not have been achieved via direct alkylation (due to the problem of carbocation rearrangements). Then, a nitro group is installed in the *para* position because of the directing effects (and steric effects) of the propyl group. Bromination then installs the bromine atom in the correct location (*ortho* to the activating group), and finally, reduction of the nitro group gives the product.

1) CH_3CH_2COCl, $AlCl_3$
2) HCl, Zn(Hg), heat
3) HNO_3, H_2SO_4
4) Br_2, $AlBr_3$
5) Zn, HCl
6) NaOH

1) CH_3CH_2COCl, $AlCl_3$
2) HCl, Zn(Hg), heat

HNO_3, H_2SO_4

Br_2, $AlBr_3$

1) Zn, HCl
2) NaOH

The strategy above represents just one method for making the desired the compound. There are often other acceptable solutions. For example, the nitro group and the bromo group can be installed in the opposite order (first the bromo group and then the nitro group), although this would require a blocking group (to help direct the incoming bromo group into the *ortho* position). The first method (shown above) avoids the need for a blocking group.

(c) While there is more than one way to achieve the desired transformation, the following strategy is perhaps the most efficient, as it avoids the need for a blocking group. The *tert*-butyl group is first installed, and its directing effects are exploited to install an amino group in the *para* position (installation of the amino group requires a two-step process – nitration, followed by reduction). Finally, chlorination in the presence of excess Cl_2 gives the desired product ($AlCl_3$ is not used because the ring is highly activated). The regiochemical outcome of the final step is controlled by the electronic effects of the amino group (in concert with the steric effects of the *tert*-butyl group).

1) $(CH_3)_3CCl$, $AlCl_3$
2) HNO_3, H_2SO_4
3) Zn, HCl
4) NaOH
5) excess Cl_2

$(CH_3)_3CCl$, $AlCl_3$

1) HNO_3, H_2SO_4
2) Zn, HCl
3) NaOH

excess Cl_2

(d) The following retrosynthetic analysis represents one strategy for preparing the desired compound. An explanation of each of the steps (*a-c*) follows.

a. The target molecule can be prepared via chlorination of the disubstituted ring shown.
b. The disubstituted ring can be prepared via sulfonation.
c. The monosubstituted ring (bromobenzene) can be prepared from benzene.

Now, let's draw the forward scheme. First, a bromine atom is installed via bromination in the presence of a Lewis acid. Then, a sulfonic acid group is installed. During this sulfonation step, the bromine atom directs the incoming sulfonic acid group into the *para* position. And finally, chlorination will install the bromine atom in the correct location (*ortho* to the *ortho-para* directing bromine atom).

1) Br_2, $AlBr_3$
2) Fuming H_2SO_4
3) Cl_2, $AlCl_3$

Br_2, $AlBr_3$

Fuming H_2SO_4

Cl_2, $AlCl_3$

18.29. This transformation requires the installation of two groups, highlighted here:

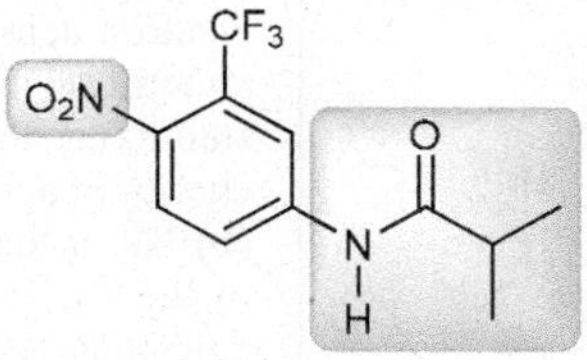

We begin by considering which of these two groups should be installed last. The nitro group is an electron-withdrawing substituent and a *meta* director. The amide group is an electron-donating substituent and an *ortho-para* director. Therefore, it makes sense for the last step of our synthesis to be nitration:

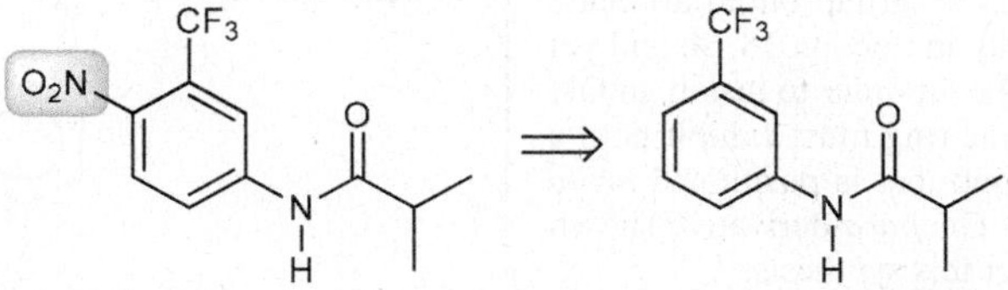

In this way, the amide group directs the final nitration step to occur in the desired location (*para* to the large amide group):

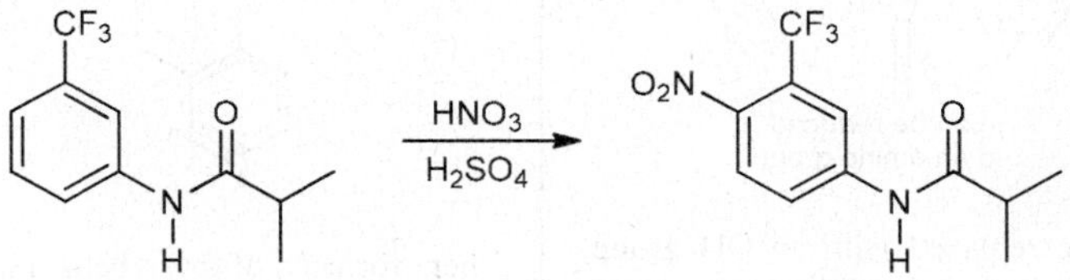

Continuing the retrosynthetic analysis, the amide group can be prepared from the corresponding aniline derivative which, in turn, can be made from reduction of a nitro group, as shown here:

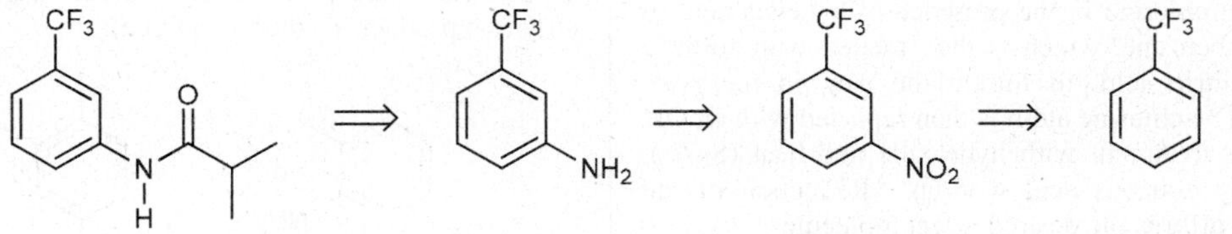

In summary, our forward synthesis has the following sequence of events.

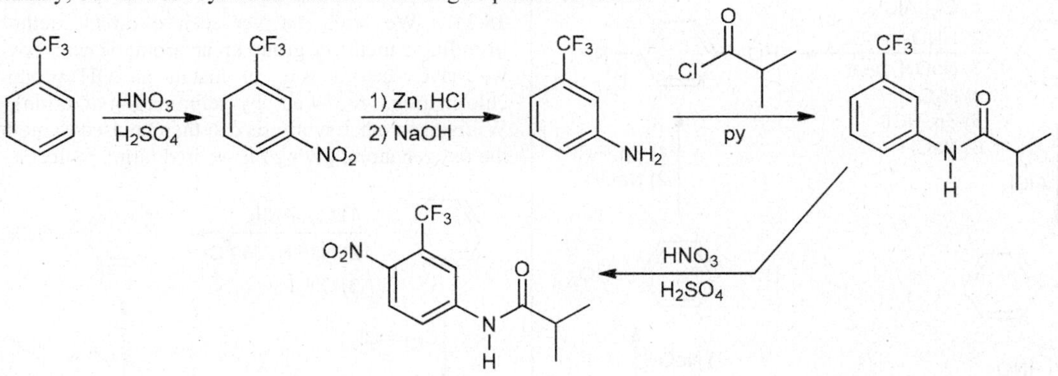

18.30. The ring contains a nitro group, as well as two leaving groups (chloride and bromide), and the reagent is a strong base (methoxide). As such, a nucleophilic aromatic substitution reaction can occur, but only if the leaving group is *ortho* or *para* to the nitro group. The bromine atom is *meta* to the nitro group, so the reaction does not occur there. Instead, chloride functions as the leaving group, because it is *para* to the nitro group.

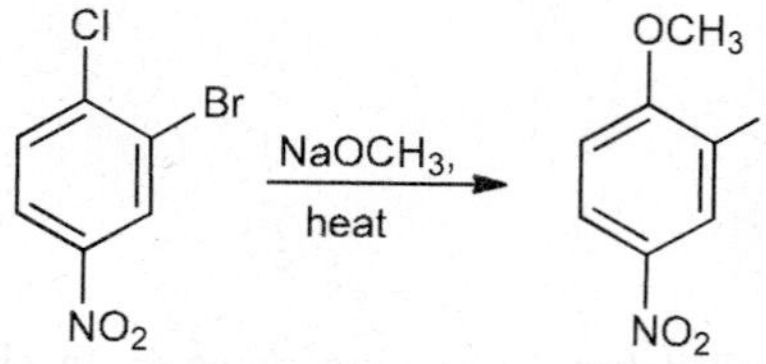

Cl Br NaOCH3, heat OCH3 Br NO2 NO2

18.31. This transformation involves the installation of two groups: NH_2 and OH. Thus far, we have only seen one way (SNAr) to install an OH group on an aromatic ring (we will see one other way in Section 18.14, and yet another method in Chapter 22). In order to install an OH group via an SNAr reaction, the ring must exhibit both a leaving group and a nitro group that is positioned *ortho* or *para* to the leaving group. The *para* derivative shown here would thus be suitable for this synthesis:

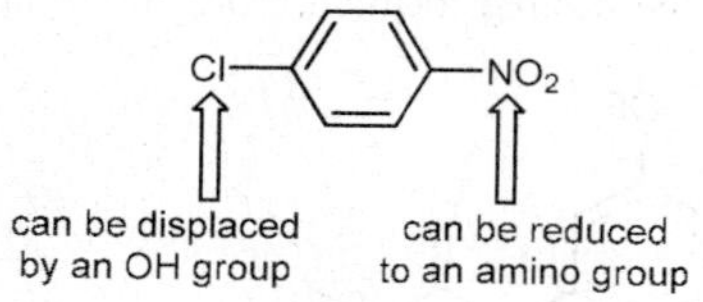

Cl NO2
can be displaced by an OH group
can be reduced to an amino group

The chlorine atom can be replaced with an OH group, and the nitro group can be reduced to an amino group, and the key intermediate above can be made in just two steps (chlorination, followed by nitration)

The entire synthesis is shown below. Benzene is first treated with chlorine in the presence of a Lewis acid to give chlorobenzene, which is then treated with sulfuric acid and nitric acid, to nitrate the ring in the *para* position. The chlorine atom is then replaced with an OH group upon treatment with hydroxide and heat (SNAr), followed by aqueous acid workup. Reduction of the nitro group affords the desired target molecule.

1) Cl_2, $AlCl_3$
2) HNO_3, H_2SO_4
3) NaOH, heat
4) H_3O^+
5) Zn, HCl
6) NaOH

HO NH2

Cl_2, $AlCl_3$

1) Zn, HCl
2) NaOH

Cl HO NO2

HNO_3, H_2SO_4

1) NaOH, heat
2) H_3O^+

Cl NO2

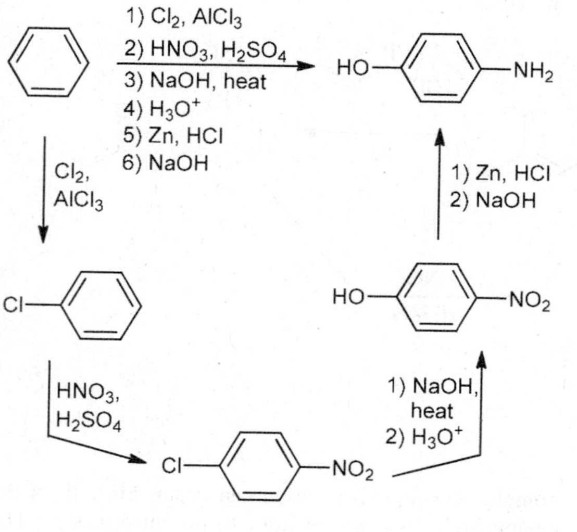

18.32.

(a) Each additional nitro group serves as a reservoir of electron density and provides for an additional resonance structure in the Meisenheimer complex, thereby stabilizing the Meisenheimer complex and lowering the energy of activation for the reaction.

(b) No, a fourth nitro group would not be *ortho* or *para* to the leaving group, and would not provide additional delocalization of the charge in the Meisenheimer complex. Therefore, we do not expect the temperature requirement to be significantly lowered.

18.33. The starting material is 4-chloro-2-methyltoluene, shown here:

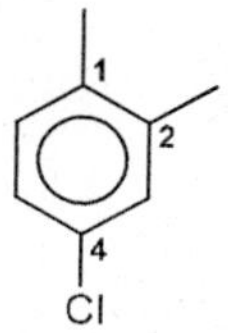

1 2 4 Cl

Upon treatment with sodium amide ($NaNH_2$), two possible benzyne intermediates can form:

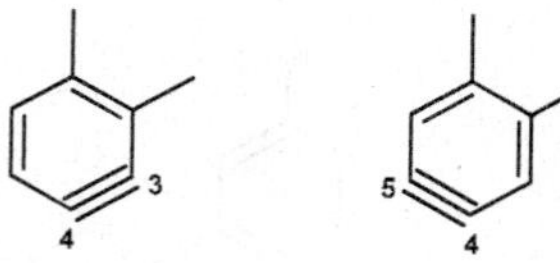

3 4 5 4

Then, for each of these benzyne intermediates, the amide ion can attack either side of the triple bond, which means that the NH_2 group is ultimately positioned at either C3, C4, or C5. Substitution at C4 yields the same product as substitution at C5 (because of symmetry), giving rise to only two products (rather than three):

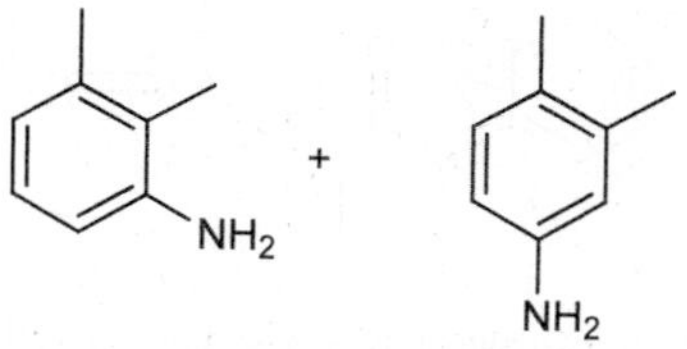

NH2 + NH2

18.34. We have not yet seen a direct method for installing a methoxy group on an aromatic ring, however, we have seen a way to install an OH group (via chlorination, followed by elimination-addition). A Williamson ether synthesis can then be used to methylate the oxygen atom, giving the desired target molecule.

1) Cl_2, $AlCl_3$
2) NaOH, 350°C
3) CH_3I

OCH3

Cl_2, $AlCl_3$

CH_3I

Cl NaOH, 350°C $O^{\ominus}$ $Na^{\oplus}$

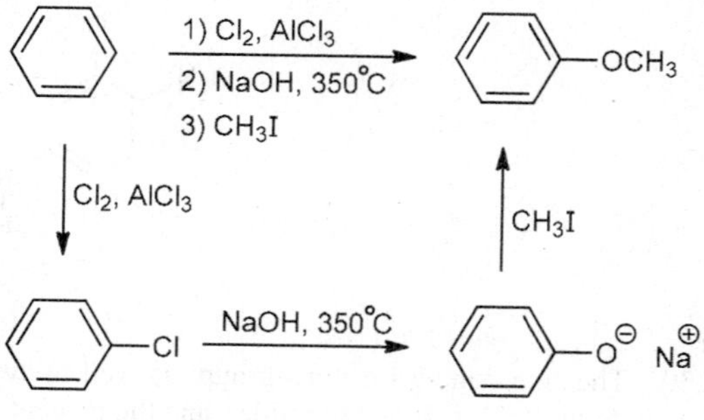

18.35.

(a) The reagent is hydroxide, which is both a strong nucleophile and a strong base. However, the three criteria for SNAr are not met (the aromatic ring lacks a nitro group), and therefore, the reaction must occur via an elimination-addition mechanism (as seen in Mechanism 18.9). The elimination-addition mechanism is also indicated by the high temperature (350°C).

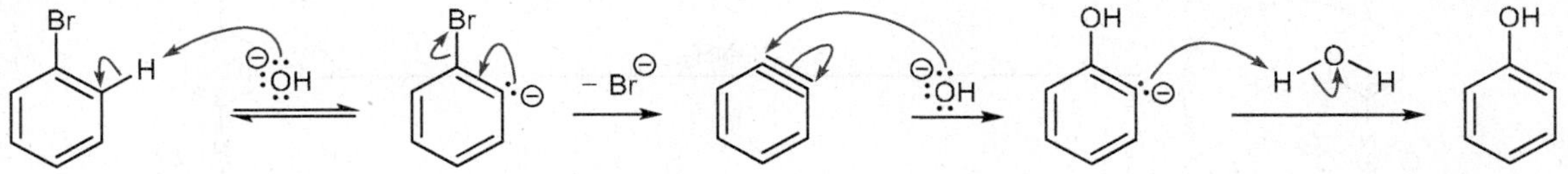

Under the basic conditions employed, the resulting product (phenol) is deprotonated to give a phenolate ion:

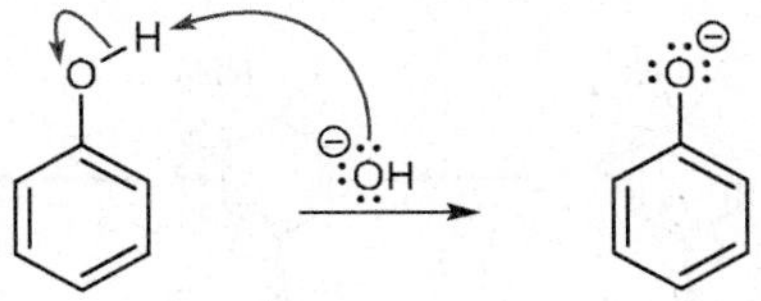

This is why aqueous acid must be introduced into the reaction flask after the reaction is complete, in order to serve as a proton source to regenerate phenol:

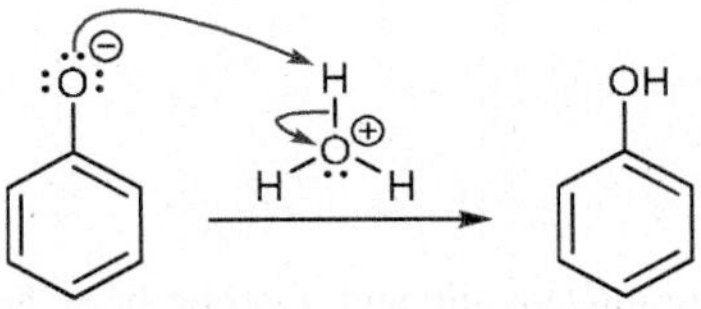

(b) These reagents provide a powerful electrophile, so the reaction must be an electrophilic aromatic substitution reaction. The first step of an electrophilic aromatic substitution reaction is formation of the electrophile that will react with the aromatic ring. This occurs when molecular bromine interacts with the Lewis acid to form a complex:

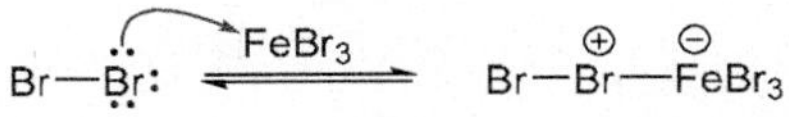

Then, one of the rings attacks the complex to give a resonance stabilized intermediate (a sigma complex), which then loses a proton to restore aromaticity (electrophilic aromatic substitution):

+ $FeBr_3$ + HBr

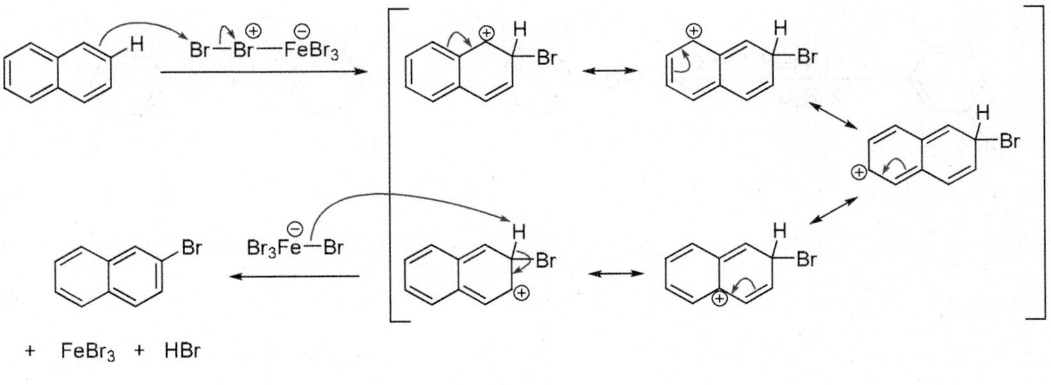

(c) The reagent (sodium amide) is a powerful nucleophile, and the three criteria for SNAr are met (a nitro group and a leaving group that are *para* to each other), so we expect an SNAr mechanism. In an SNAr reaction, the first step involves the nucleophile attacking the aromatic ring to generate a resonance-stabilized intermediate (a Meisenheimer complex). Loss of a leaving group then gives the product, as shown:

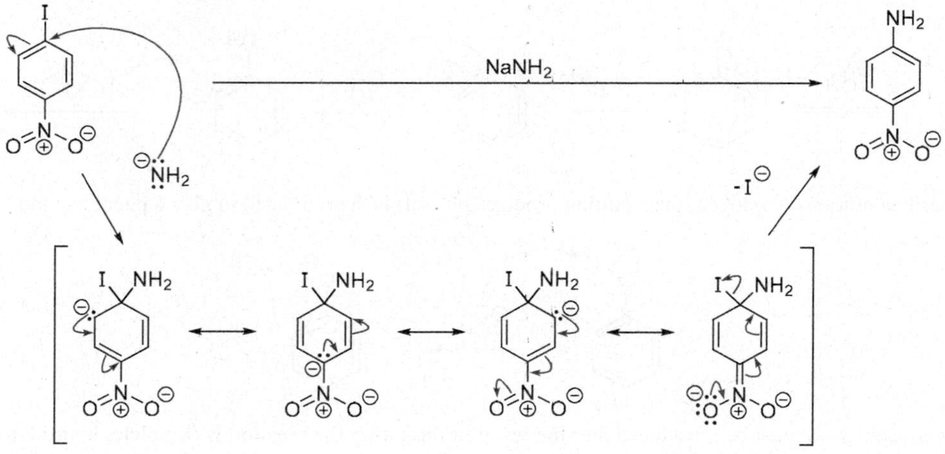

18.36. The reagent (hydroxide) is a strong nucleophile and a strong base, and the three criteria for SNAr are not met (the aromatic ring does not have a nitro group), so we expect an elimination-addition reaction. Upon treatment with sodium hydroxide, two possible benzyne intermediates are initially formed:

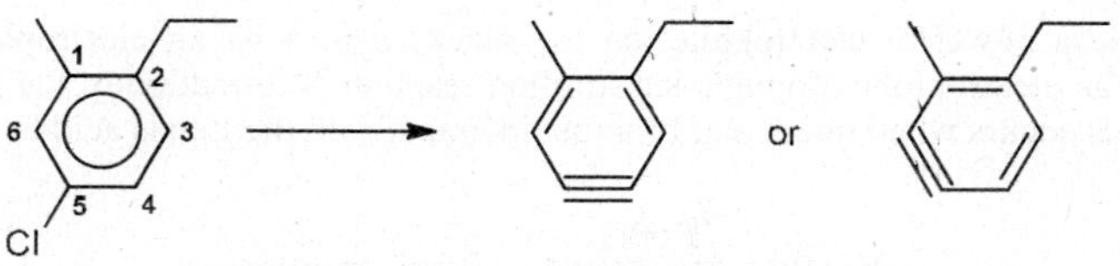

Then, for each of these benzyne intermediates, the hydroxide ion can attack either side of the triple bond, which means that the OH group is ultimately positioned at either C4, C5, or C6, giving rise to the following three products. Note that the second product can be made from either benzyne intermediate.

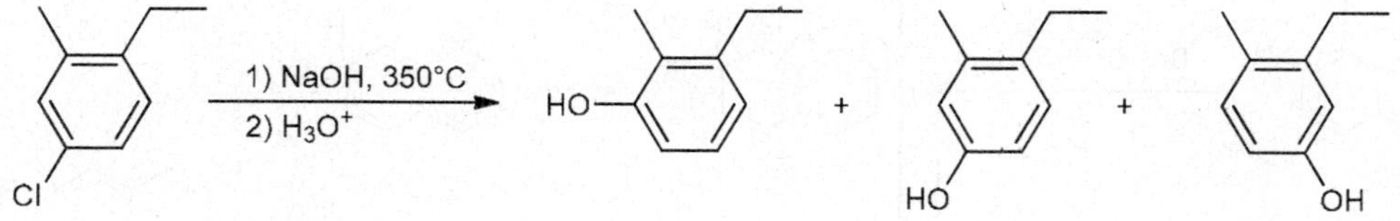

18.37.

(a) First look at the reagents. Both the acid chloride and $SnCl_4$ are electrophiles, so we can rule out S_NAr and elimination-addition. Additionally, the reaction involves the substitution of an H in an aromatic ring (pyrrole) with another group so this must be an electrophilic aromatic substitution. Since a methyl ketone results from the substitution, this is a Friedel-Crafts acylation.

Since the reaction is an electrophilic aromatic substitution, the intermediate formed must have a positive charge, as shown below.

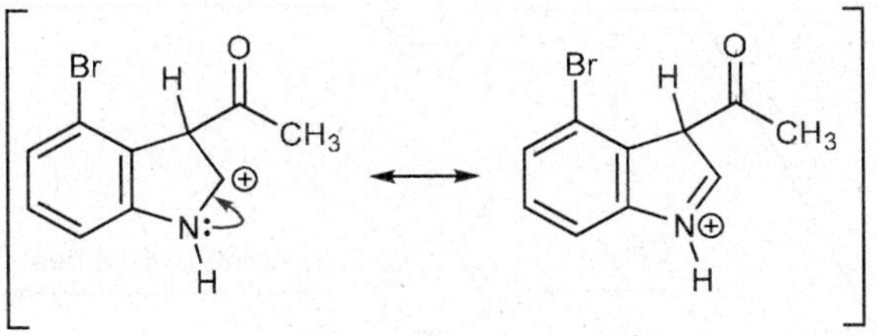

(b) The Friedel-Crafts acylation first involves the formation of an acylium ion when the acid chloride displaces a chloride anion from the Lewis acid followed by the loss of $SnCl_4$ leaving group.

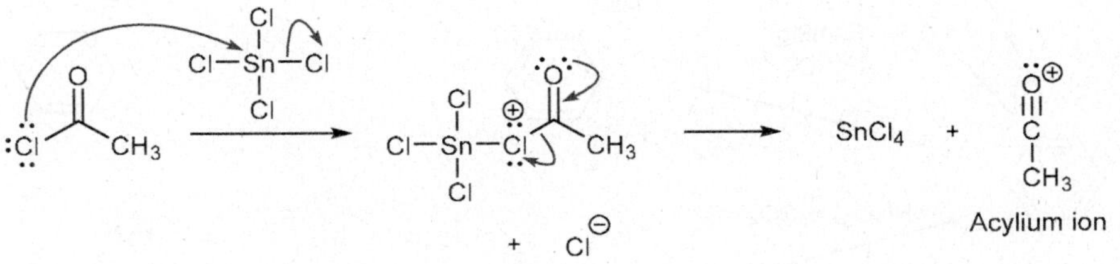

Acylium ion

Alternatively, the formation of the acylium could be drawn using a stepwise addition-elimination mechanism where the tin atom becomes pentavalent (which is reasonable for an element which is in the 5th row of the periodic table) and then loses a chloride:

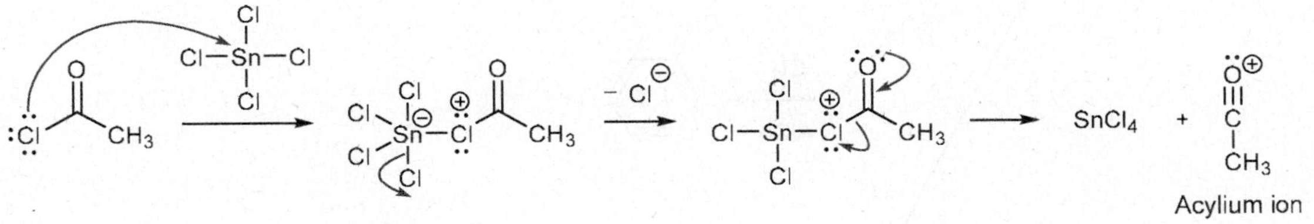

Acylium ion

This acylium ion is then attacked by the pyrrole ring to form a sigma complex:

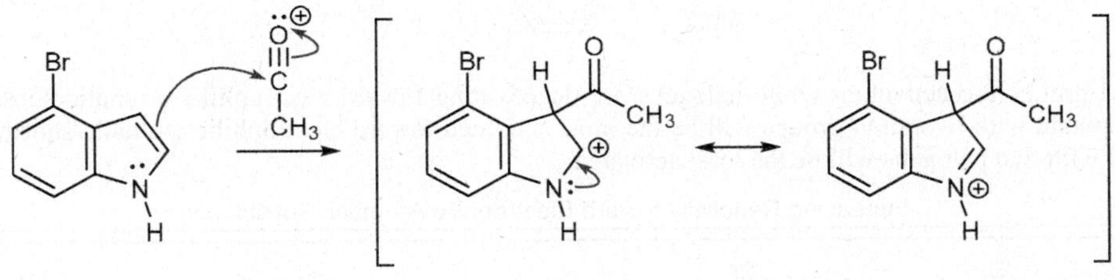

(c) Friedel-Crafts reactions frequently fail when amines are present due to the reaction of the nucleophilic amine and the Lewis acid. Since the lone pair of the nitrogen is part of an aromatic system (delocalized by resonance), it does not react with the Lewis acid.

18.38. The reagents for these reactions can be found in the Review of Reactions at the end of the chapter. They are shown again here:

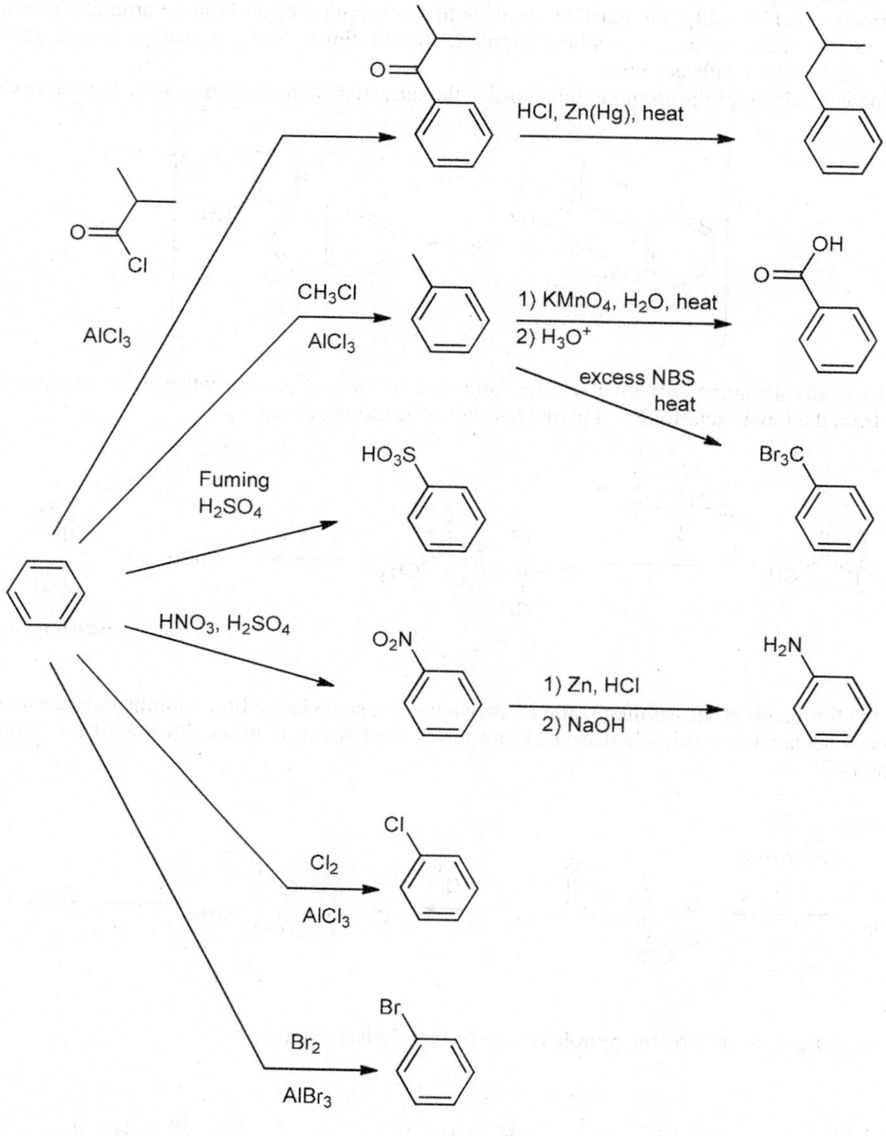

HCl, Zn(Hg), heat

OH

Cl

CH_3Cl

1) $KMnO_4$, H_2O, heat

$AlCl_3$

$AlCl_3$

2) H_3O^+

excess NBS

heat

HO_3S

Br_3C

Fuming

H_2SO_4

HNO_3, H_2SO_4

O_2N

H_2N

1) Zn, HCl

2) NaOH

Cl

Cl_2

$AlCl_3$

Br

Br_2

$AlBr_3$

18.39. Alkyl groups are activating while halogens are deactivating toward electrophilic aromatic substitution. As such, the compound with two alkyl groups will be the most activated toward electrophilic aromatic substitution, while the compound with two halogens will be the least activated:

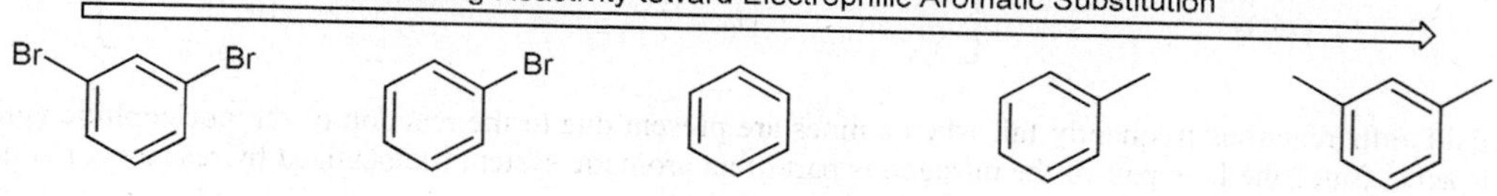

Increasing Reactivity toward Electrophilic Aromatic Substitution

Br Br Br

18.40. A nitro group is a strong deactivator. Therefore, among the four compounds, nitrobenzene is the least activated toward electrophilic aromatic substitution. The most activated ring is the one that exhibits two activators (OH and OMe).

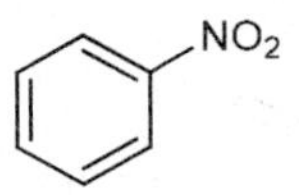

least activated 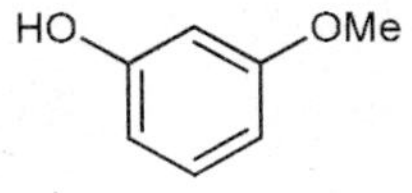 most activated

18.41.

(a) The reagents indicate a nitration reaction, so we must decide where the nitro group will be installed. The aromatic ring has one substituent (Br) which is an *ortho-para* director, so we expect nitration to occur at the *ortho* and *para* positions. Because of the size of the bromine atom, the *para* product is expected to predominate (a steric effect).

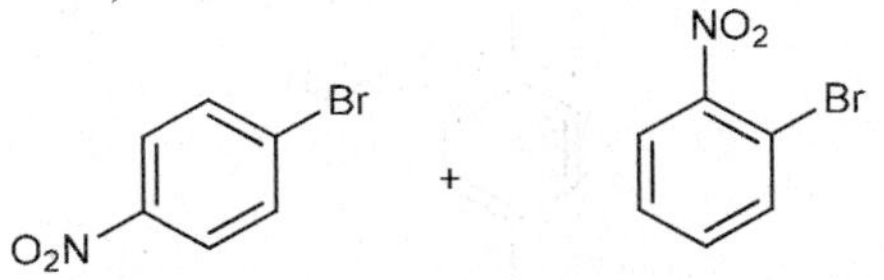

Major Minor

(b) The reagents indicate a nitration reaction, so we must decide where the nitro group will be installed. The aromatic ring has one substituent (an isopropyl group) which is an *ortho-para* director, so we expect nitration to occur at the *ortho* and *para* positions. Because of the size of the isopropyl group, the *para* product is expected to predominate (a steric effect).

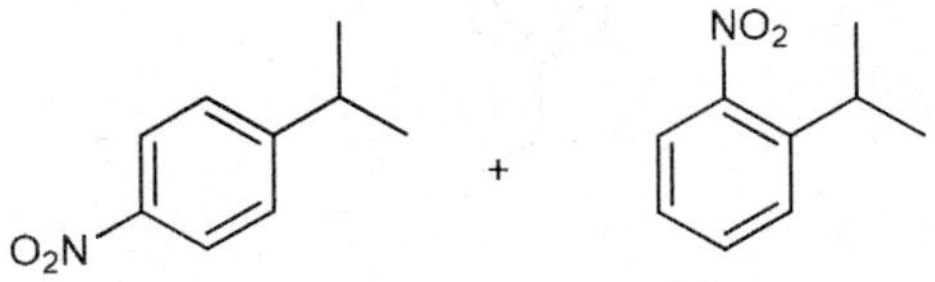

Major Minor

(c) The reagents indicate a nitration reaction, so we must decide where the nitro group will be installed. The aromatic ring has one substituent (a nitro group) which is a *meta* director, so we expect nitration to occur at the *meta* position.

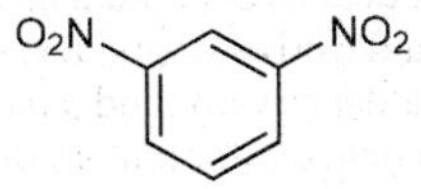

(d) The reagents indicate a nitration reaction, so we must decide where the nitro group will be installed. The aromatic ring has one substituent (a carbonyl group) which is a *meta* director, so we expect nitration to occur at the *meta* position.

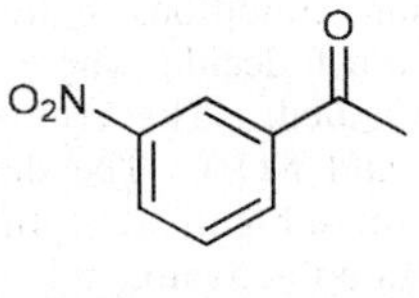

(e) The reagents indicate a nitration reaction, so we must decide where the nitro group will be installed. The aromatic ring has one substituent (a methoxy group) which is an *ortho-para* director, so we expect nitration to occur at the *ortho* and *para* positions. The *para* product is likely to predominate (because of a steric effect).

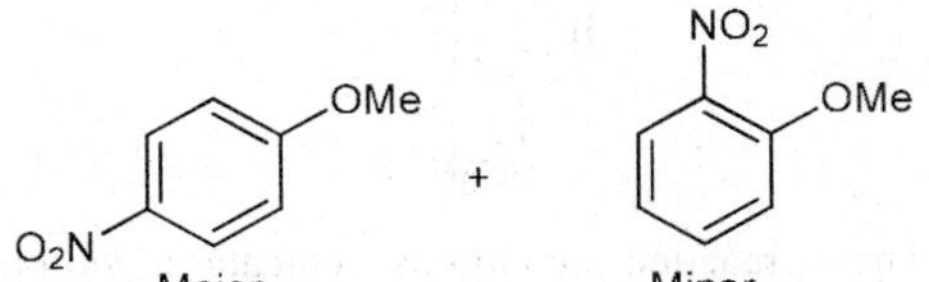

Major Minor

18.42.

(a) These reaction conditions indicate a sulfonation reaction, so we must decide where the sulfonic acid group will be installed. The aromatic ring has one substituent (Cl) which is an *ortho-para* director, so we expect sulfonation to occur at the *ortho* and *para* positions. For steric considerations, the *para* product (shown here) is likely to predominate over the *ortho* product.

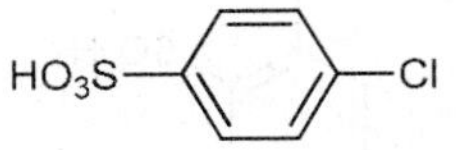

(b) These reaction conditions indicate a sulfonation reaction, so we must decide where the sulfonic acid group will be installed. The aromatic ring has one substituent (OH) which is an *ortho-para* director, so we expect sulfonation to occur at the *ortho* and *para* positions. For steric considerations, perhaps the *para* product (shown here) predominates over the *ortho* product, although it is a close call in this case. The OH group is similar in steric bulk to a methyl group, and we know that a methyl group provides little steric bulk (for example, sulfonation of toluene is expected to produce a mixture of *ortho* and *para* products, and the major product is difficult to predict).

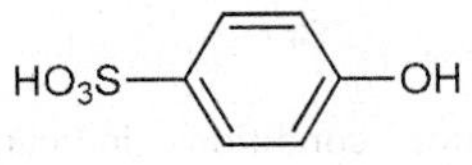

(c) These reaction conditions indicate a sulfonation reaction, so we must decide where the sulfonic acid group will be installed. The aromatic ring has one substituent (a carbonyl group) which is a *meta* director, so we expect sulfonation to occur at the *meta* position.

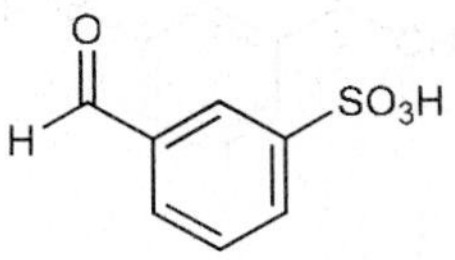

(d) These reaction conditions indicate a sulfonation reaction, so we must decide where the sulfonic acid group will be installed. The aromatic ring has two substituents (OH and NO_2). The directing effects are controlled by the more highly activating group, which is the OH group. As an activator, this group is an *ortho-para* director, so we expect sulfonation to occur at the *ortho* and *para* positions. For steric considerations, the *para* product (shown here) is likely to predominate over the *ortho* product.

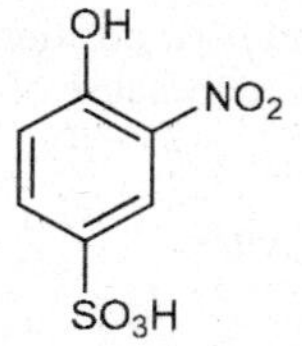

OH
NO_2
SO_3H

(e) These reaction conditions indicate a sulfonation reaction, so we must decide where the sulfonic acid group will be installed. The aromatic ring has two substituents (Br and Me). The directing effects are controlled by the more highly activating group, which is the methyl group. As an activator, this group is an *ortho-para* director, so we expect sulfonation to occur at the *ortho* and *para* positions. However, the *para* position is already occupied (by the bromine atom). Therefore, sulfonation can only take place at the *ortho* positions.

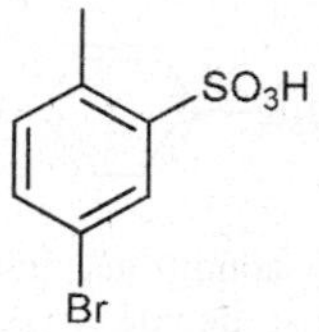

SO_3H
Br

In this case, sulfonation at either *ortho* position leads to the same product. The two *ortho* positions are identical because of symmetry. Don't be confused by the arbitrary location of the double bonds. It may be less distracting if we draw the structure like this:

SO_3H
Br

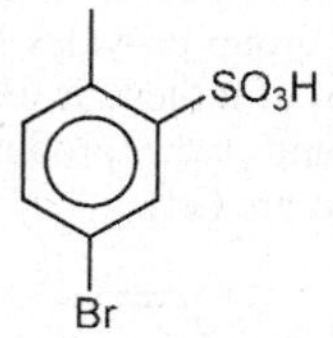

(f) These reaction conditions indicate a sulfonation reaction, so we must decide where the sulfonic acid group will be installed. The aromatic ring has one substituent (a carboxylic acid group) which is a *meta* director, so we expect sulfonation to occur at the *meta* position.

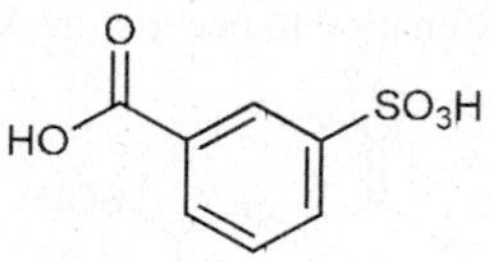

O
HO
SO_3H

(g) These reaction conditions indicate a sulfonation reaction, so we must decide where the sulfonic acid group will be installed. The aromatic ring has two substituents (methyl and ethyl). Both groups are weak activators, so all unoccupied aromatic positions are equally activated:

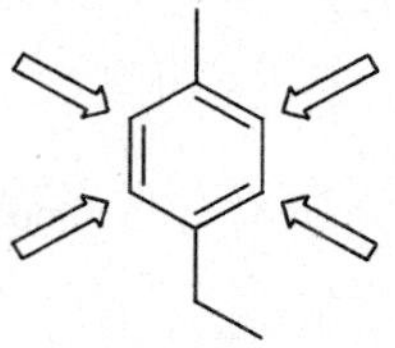

These positions are differentiated from each because of steric factors. The ethyl group is larger than the methyl group, so the reaction is expected to occur *ortho* to the methyl group, rather than *ortho* to the more sterically bulky ethyl group.

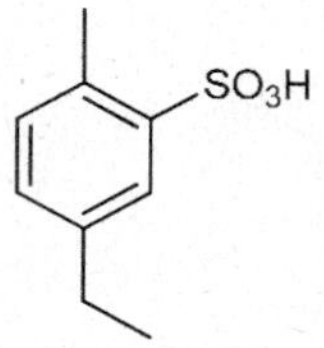

SO_3H

In this case, sulfonation at either *ortho* position leads to the same product. The two *ortho* positions are identical because of symmetry. Don't be confused by the arbitrary location of the double bonds. It may be less distracting if we draw the structure like this:

SO_3H

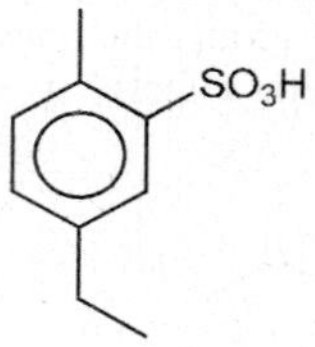

18.43. Each of these substituents appears in Table 18.1.
(a) This group is an activator and an *ortho-para* director.
(b) This group is an activator and an *ortho-para* director.
(c) This group is an activator and an *ortho-para* director.
(d) This group is a deactivator and an *ortho-para* director.
(e) This group is a deactivator and a *meta*-director.
(f) This group is a deactivator and a *meta*-director.
(g) This group is a deactivator and a *meta*-director.
(h) This group is a deactivator and a *meta*-director.
(i) This group is a deactivator and an *ortho-para* director.
(j) This group is a deactivator and a *meta*-director.

18.44. There are four aromatic rings, labeled **A** – **D**. Ring **A** is the most activated, because it exhibits a group that is a strong activator (an amino group). Ring **D** is moderately activated because it is connected to the oxygen atom of an ester. Ring **B** is weakly activated because it is connected to an alkyl group. And finally, ring **C** is deactivated because it is connected to a carbonyl group (a moderate deactivator).

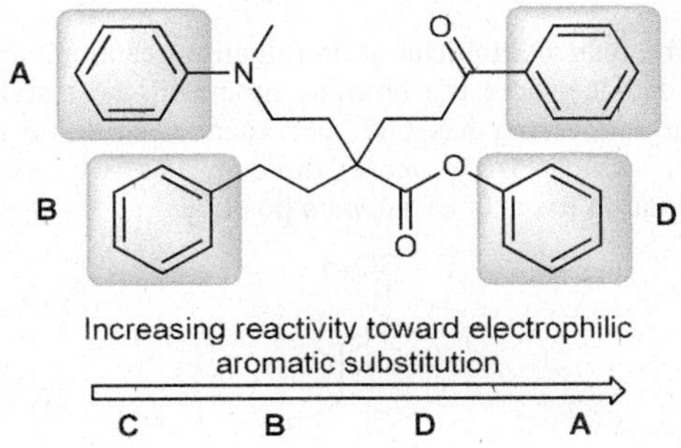

A B C D

Increasing reactivity toward electrophilic aromatic substitution

C B D A

18.45.

(a) The reagents indicate a Friedel-Crafts alkylation reaction (methylation), so we must decide where the methyl group will be installed. The aromatic ring has one substituent (Cl) which is an *ortho-para* director, so we expect methylation to occur at the *ortho* and *para* positions.

Cl + Cl

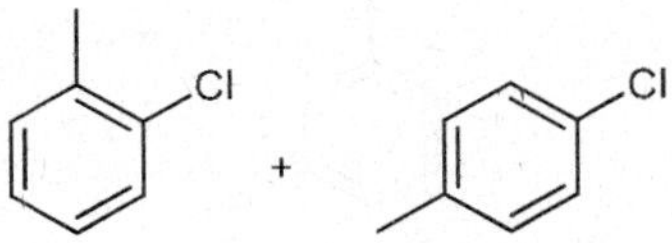

The *para* isomer is likely the major product because of steric considerations.

(b) This aromatic ring is moderately deactivated and will therefore not undergo a Friedel-Crafts alkylation.

(c) This aromatic ring is strongly deactivated and will therefore not undergo a Friedel-Crafts alkylation.

(d) The reagents indicate a Friedel-Crafts alkylation reaction (methylation), so we must decide where the methyl group will be installed. The aromatic ring has one substituent (ethyl) which is an *ortho-para* director, so we expect methylation to occur at the *ortho* and *para* positions.

+

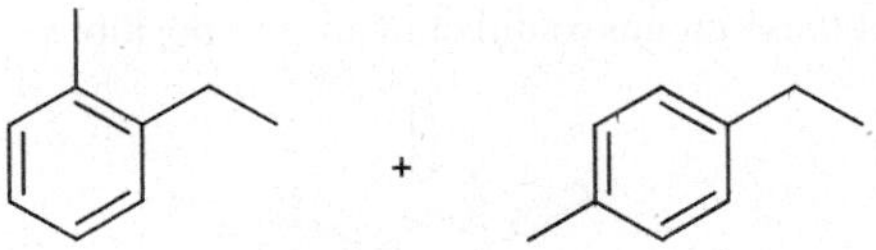

The *para* isomer is expected to be the major product because of steric considerations.

(e) The reagents indicate a Friedel-Crafts alkylation reaction (methylation), so we must decide where the methyl group will be installed. The aromatic ring has two substituents (I and methyl). The directing effects are controlled by the most strongly activating group. Therefore, in this case, the methyl group controls the directing effects. As an activator, this group is an *ortho-para* director. However, one of the *ortho* positions is already occupied (by the other substituent). Therefore, methylation will occur at the position that is *ortho* or *para* to the methyl group, shown here:

I + I

The first isomer is expected to be the major product because of steric considerations.

(f) The reagents indicate a Friedel-Crafts alkylation reaction (methylation), so we must decide where the methyl group will be installed. The aromatic ring has two substituents (propyl and methyl). Both groups are weak activators, and *ortho-para* directors. Therefore, we expect the following two products:

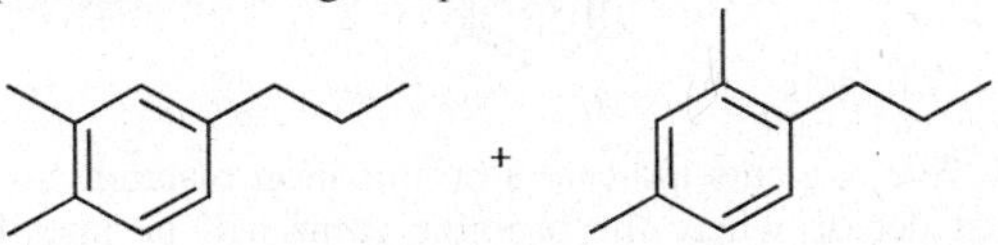

+

The first isomer is expected to be the major product because of steric considerations.

(g) This aromatic ring has two substituents, and both are moderately deactivating. As such, the ring is too deactivated to undergo a Friedel-Crafts alkylation.

(h) The reagents indicate a Friedel-Crafts alkylation reaction (methylation), so we must decide where the methyl group will be installed. The aromatic ring has two substituents (the oxygen atom of an ester, and an alkyl group). The directing effects are controlled by the most strongly activating group. Therefore, in this case, the ester group controls the directing effects. As an activator, this group is an *ortho-para* director. However, one of the *ortho* positions is already occupied (by the other substituent). Therefore, methylation will occur at the positions that are *ortho* and *para* to the ester group, shown here:

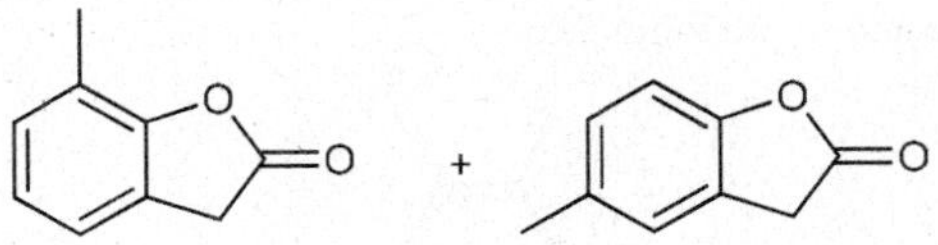

O =O + O =O

The second isomer is expected to be the major product because of steric considerations.

18.46.

(a) The reagents indicate a bromination reaction, so we must decide where the bromine atom will be installed. The aromatic ring has one substituent (Br) which is an *ortho-para* director, so we expect bromination to occur at the *ortho* and *para* positions. For steric considerations, the *para* product (shown here) is likely to predominate over the *ortho* product:

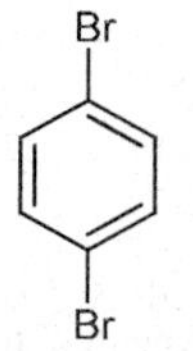

(b) The reagents indicate a bromination reaction, so we must decide where the bromine atom will be installed. The aromatic ring has one substituent (NO_2) which is a *meta* director, so we expect bromination to occur at the *meta* position:

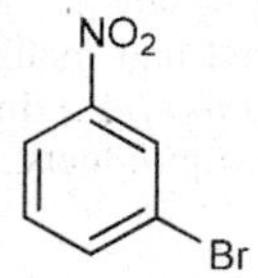

(c) The reagents indicate a bromination reaction, so we must decide where the bromine atom will be installed. The aromatic ring has two substituents (both methyl groups). Both groups are *ortho-para* directors, so all four unoccupied positions are activated:

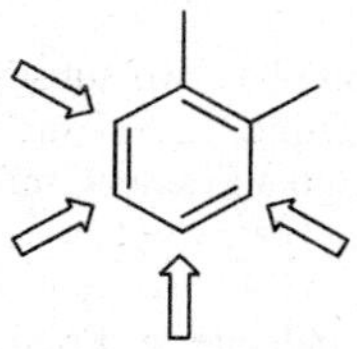

Because of symmetry, there are only two unique locations where bromination can occur:

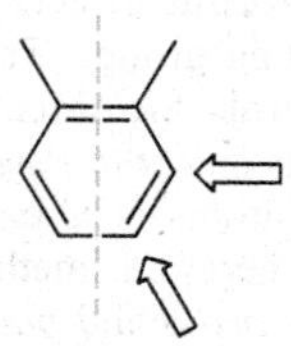

For steric considerations, the following product is expected to predominate.

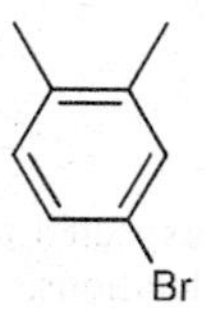

(d) The reagents indicate a bromination reaction, so we must decide where the bromine atom will be installed. The aromatic ring has one substituent (a *tert*-butyl group) which is an *ortho-para* director. The *tert*-butyl group is sterically bulky, so we expect the *para* product (shown here) to predominate over the *ortho* product.

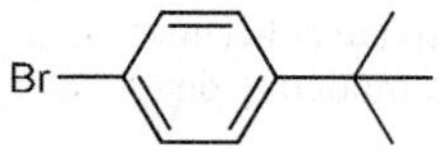

(e) The reagents indicate a bromination reaction, so we must decide where the bromine atom will be installed. The aromatic ring has one substituent (a sulfonic acid group) which is a *meta* director, so we expect bromination to occur at the *meta* position:

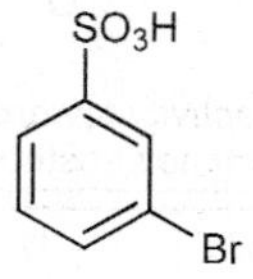

(f) The reagents indicate a bromination reaction, so we must decide where the bromine atom will be installed. The aromatic ring has one substituent (a carboxylic acid group) which is a *meta* director, so we expect bromination to occur at the *meta* position:

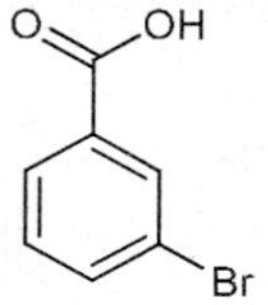

(g) The reagents indicate a bromination reaction, so we must decide where the bromine atom will be installed. The aromatic ring has one substituent (a carbonyl group) which is a *meta* director, so we expect bromination to occur at the *meta* position:

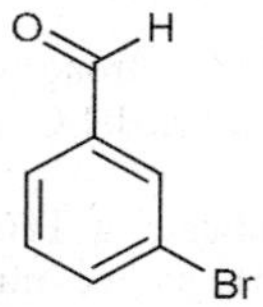

(h) The reagents indicate a bromination reaction, so we must decide where the bromine atom will be installed. The aromatic ring has two substituents (two bromine atoms). Both groups are *ortho-para* directors, so we expect these groups to direct to all four positions:

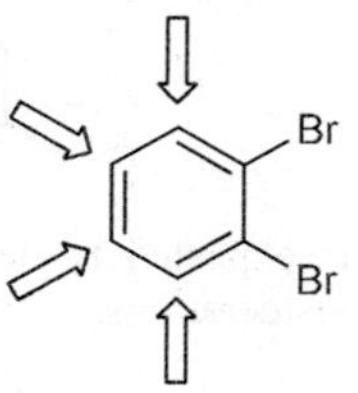

Because of symmetry, there are only two unique locations where bromination can occur:

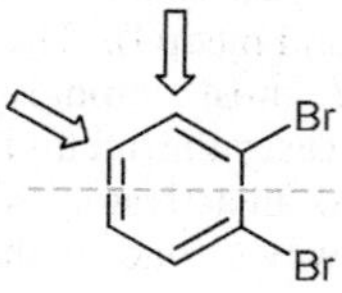

For steric considerations, the following product is expected to predominate.

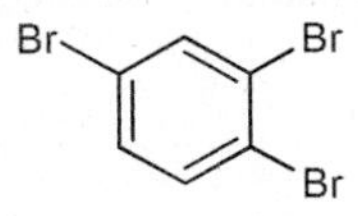

(i) The reagents indicate a bromination reaction, so we must decide where the bromine atom will be installed. The aromatic ring has two substituents: a methyl group (activator) and a nitro group (deactivator). The directing effects are controlled by the most strongly activating group. Therefore, in this case, the methyl group controls the directing effects. As an activator, this group is an *ortho-para* director, thereby activating the following three locations, one of which is sterically hindered:

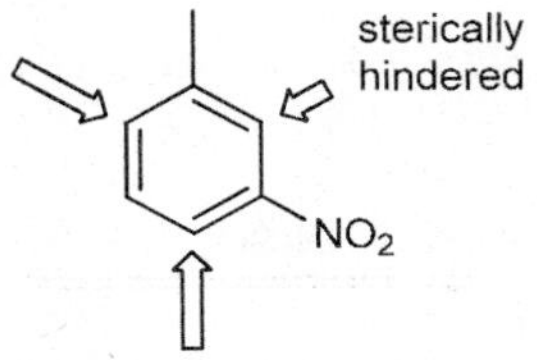

Among the remaining two positions, the position next to the nitro group is probably more hindered than the position next to the methyl group. So, we predict that bromination will occur predominantly near the methyl group:

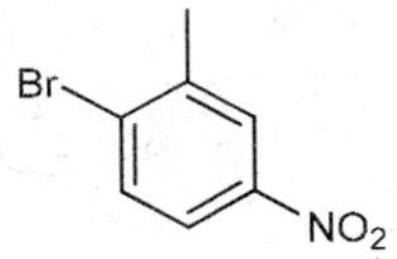

(j) The reagents indicate a bromination reaction, so we must decide where the bromine atom will be installed. The aromatic ring has two substituents (two bromine atoms). Both groups are *ortho-para* directors, so we expect these groups to direct to the following three positions, one of which is too sterically hindered:

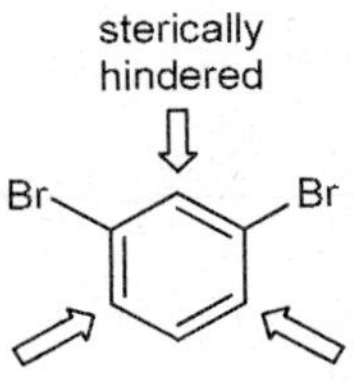

The remaining two positions are identical by symmetry. That is, bromination at either position will generate the same product:

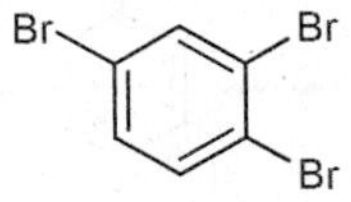

(k) The reagents indicate a bromination reaction, so we must decide where the bromine atom will be installed. The aromatic ring has two substituents (two bromine atoms), both of which are *ortho-para* directors, so we expect these groups to direct to all four positions (all of which of which are *ortho* to one of the halogens). Indeed, all four positions are equivalent by symmetry anyway. Bromination will install a third bromine atom which can be drawn in any of the four unoccupied positions (all are identical):

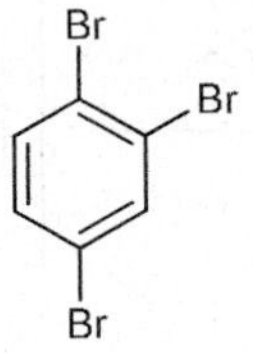

18.47. There are two aromatic rings, and each ring has two substituents. For each ring, the directing effects are controlled by the most strongly activating group. Therefore, in this case, the OR group (a moderate activator) controls the directing effects for each ring (rather than the *tert*-butyl groups, which are weak activators). As an activator, the OR group is an *ortho-para* director. However, in each ring, one of the *ortho* positions is already occupied (by the *tert*-butyl group). Therefore, nitration can occur at the following locations:

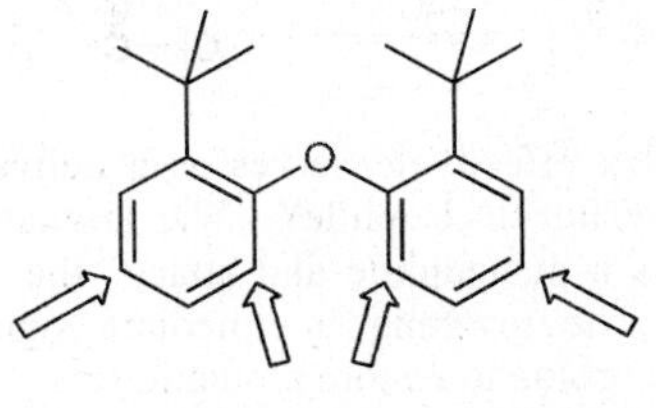

However, two of these locations are sterically hindered, so we only expect nitration at the remaining two (unencumbered) positions:

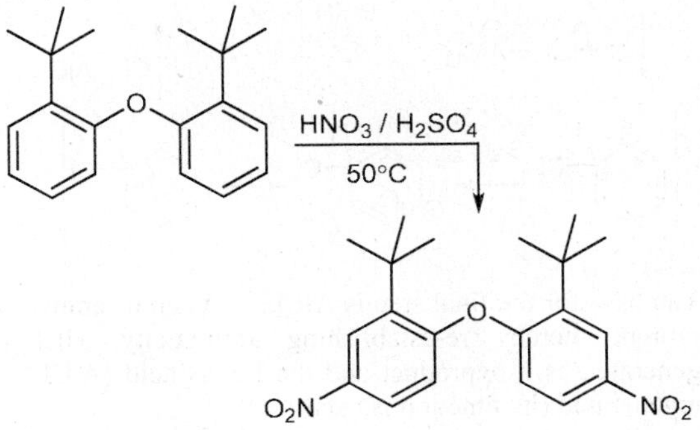

18.48. In acidic conditions, 2-methylpropene is protonated to give a tertiary carbocation, which can function as an electrophile in an electrophilic aromatic substitution reaction. The aromatic ring attacks the carbocation to give a resonance stabilized intermediate (sigma complex), followed by deprotonation which restores aromaticity. In aqueous acidic conditions, the proton source is H_3O^+ (in the beginning of the mechanism), and the base is H_2O (at the end of the mechanism).

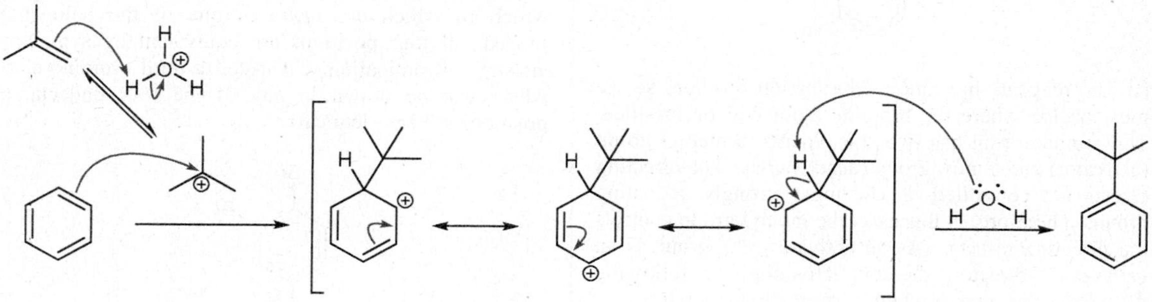

18.49.

(a) This is an electrophilic aromatic substitution reaction, and before we can draw the two steps of the process, we must begin by drawing formation of the electrophilic complex that will react with the aromatic ring. This occurs when molecular chlorine (Cl_2) attacks the Lewis acid, forming an electrophilic complex.

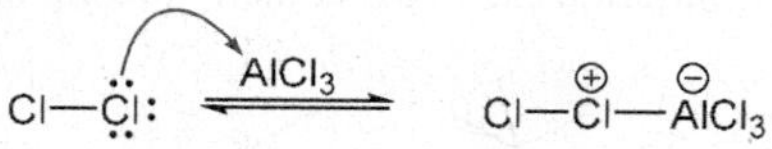

This complex effectively serves as a delivery agent of Cl^+ (a powerful electrophile). The aromatic ring then functions as a nucleophile and attacks the electrophilic complex. The resulting intermediate sigma complex then loses a proton to restore aromaticity.

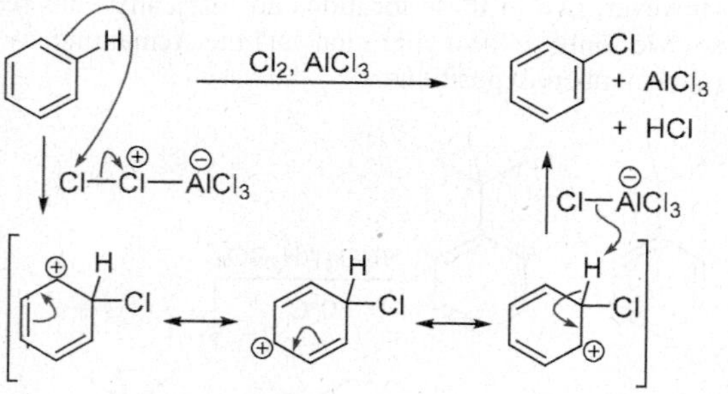

The base for the final step is $AlCl_4^-$. When it removes a proton, thereby re-establishing aromaticity, HCl is generated as a byproduct and the Lewis acid ($AlCl_3$) is regenerated (its function is catalytic).

(b) In the presence of sulfuric acid, nitric acid can be protonated, followed by loss of water to give a nitronium ion:

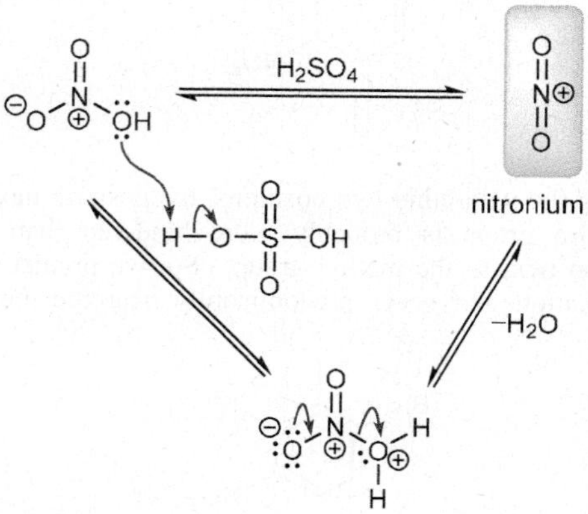

nitronium ion

The aromatic ring then functions as a nucleophile and attacks the nitronium ion. The resulting intermediate sigma complex then loses a proton to restore aromaticity. Water is the likely base for the final deprotonation step, since water is present in solutions of these strong acids (and there are no strong bases present in strongly acidic conditions).

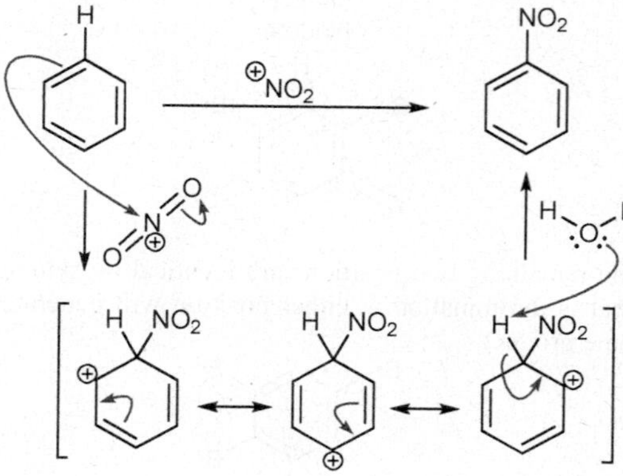

(c) This is an electrophilic aromatic substitution reaction (called sulfonation). As for all electrophilic aromatic substitution reactions, there are two key steps. First, the aromatic ring functions as a nucleophile and attacks the electrophilic species (SO_3). Then, the resulting intermediate sigma complex loses a proton to restore aromaticity. Notice that the product is an anion, which is protonated under these acidic conditions, as shown:

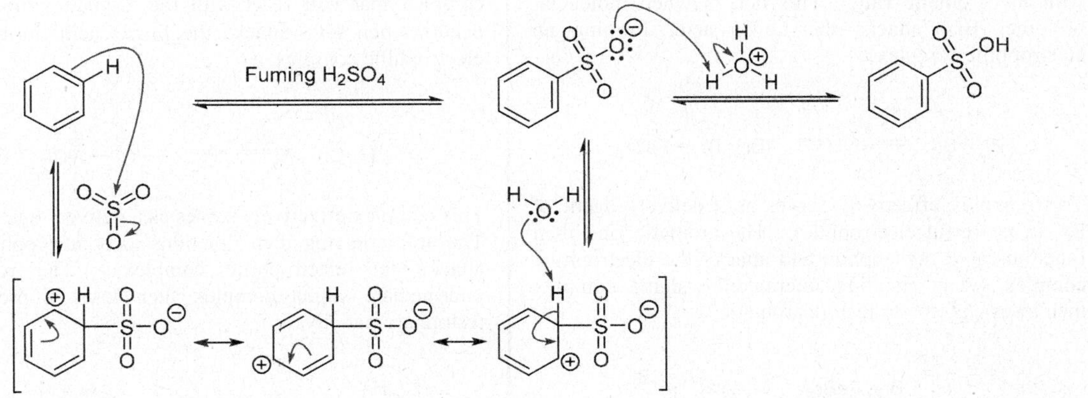

Fuming H_2SO_4

(d) This is a Friedel-Crafts alkylation process, which begins with the formation of the electrophilic complex that will react with the aromatic ring. Methyl chloride attacks the Lewis acid, forming an electrophilic complex. This complex effectively serves as a delivery agent of H_3C^+. The aromatic ring functions as a nucleophile and attacks the electrophilic complex. The resulting intermediate sigma complex then loses a proton to restore aromaticity, as shown:

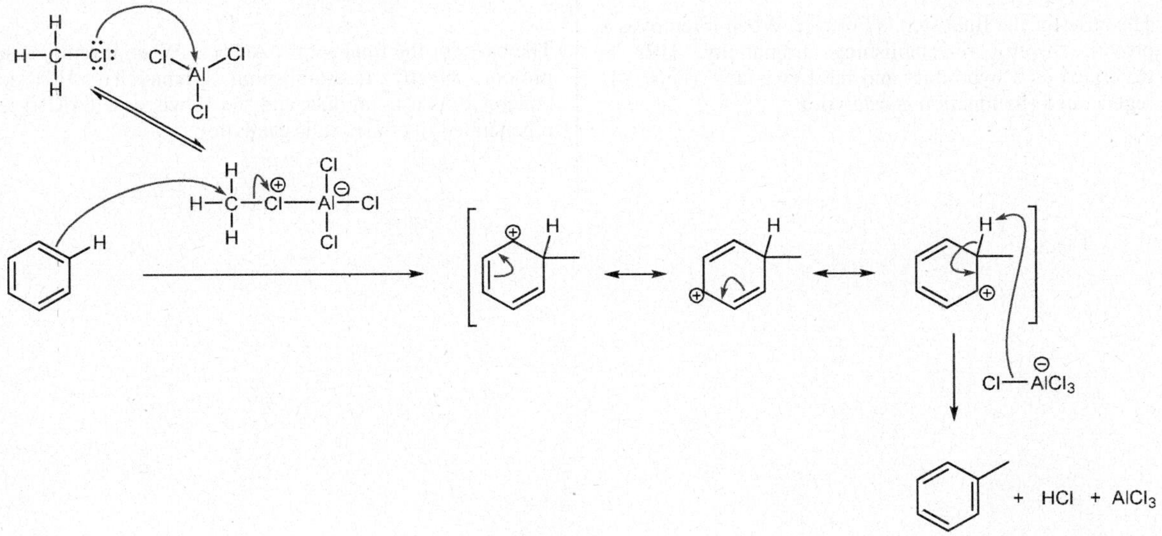

$Cl—AlCl_3$ (⊖)

+ HCl + $AlCl_3$

The base for the final step is $AlCl_4^-$. When it removes a proton, thereby re-establishing aromaticity, HCl is generated as a byproduct and the Lewis acid ($AlCl_3$) is regenerated (its function is catalytic).

(e) This is an electrophilic aromatic substitution reaction (called bromination), and before we can draw the two steps of the process, we must begin by drawing formation of the electrophilic complex that will react with the aromatic ring. This occurs when molecular bromine (Br_2) attacks the Lewis acid, forming an electrophilic complex.

$$Br{-}Br: \xrightleftharpoons{FeBr_3} Br{-}\overset{\oplus}{Br}{-}\overset{\ominus}{Fe}Br_3$$

This complex effectively serves as a delivery agent of Br^+ (a powerful electrophile). The aromatic ring then functions as a nucleophile and attacks the electrophilic complex. The resulting intermediate sigma complex then loses a proton to restore aromaticity.

Br₂, FeBr₃ → + FeBr₃ + HBr

The base for the final step is $FeBr_4^-$. When it removes a proton, thereby re-establishing aromaticity, HBr is generated as a byproduct and the Lewis acid ($FeBr_3$) is regenerated (its function is catalytic).

18.50.

(a) This is an electrophilic aromatic substitution reaction, and before we can draw the two steps of the process, we must begin by drawing formation of the electrophilic complex that will react with the aromatic ring. This occurs when I–Cl attacks the Lewis acid, forming an electrophilic complex.

$$I{-}Cl: \xrightleftharpoons{AlCl_3} I{-}\overset{\oplus}{Cl}{-}\overset{\ominus}{Al}Cl_3$$

This complex effectively serves as a delivery agent of I^+. The aromatic ring then functions as a nucleophile and attacks the electrophilic complex. The resulting intermediate sigma complex then loses a proton to restore aromaticity.

ICl, AlCl₃ → + AlCl₃ + HCl

The base for the final step is $AlCl_4^-$. When it removes a proton, thereby re-establishing aromaticity, HCl is generated as a byproduct and the Lewis acid ($AlCl_3$) is regenerated (its function is catalytic).

(b) This transformation involves two molecules of benzene reacting with one molecule of CH_2Cl_2, as indicated in the problem statement (0.5 equivalents of CH_2Cl_2). This process occurs via two successive electrophilic aromatic substitution reactions. The process begins with formation of the electrophilic complex that will react with the first aromatic ring. Methylene chloride (CH_2Cl_2) attacks the Lewis acid, forming an electrophilic complex. This complex effectively serves as a delivery agent of ClH_2C^+. The aromatic ring functions as a nucleophile and attacks the electrophilic complex. The resulting intermediate sigma complex then loses a proton to restore aromaticity, giving benzyl chloride. Then, all of the previous steps (formation of the electrophilic complex, followed by the two steps of an electrophilic aromatic substitution reaction) are repeated again, as shown, giving the product (diphenylmethane). Notice that, for each deprotonation step, the base is $AlCl_4^-$. When it removes a proton, HCl is generated as a byproduct and the Lewis acid ($AlCl_3$) is regenerated (its function is catalytic):

Cl—C(H)(H)—Cl: Cl—Al(Cl)—Cl

Cl—C(H)(H)—Cl⊕—Al⊖(Cl)(Cl)—Cl

Cl—AlCl$_3$ ⊖

Cl—Al(Cl)—Cl :Cl: + HCl + AlCl$_3$

⊖ AlCl$_3$ ⊕ Cl

Cl—AlCl$_3$ ⊖

18.51. The substituent (Cl) is replaced with a different substituent (OMe). Therefore, this is an aromatic substitution reaction. The reagent (methoxide) is nucleophilic, so we must determine whether this reaction occurs via an SNAr mechanism or via an elimination-addition mechanism. All three criteria for an SNAr mechanism are met: 1) there is a leaving group (chloride), and 2) there is an electron-withdrawing group (the carbonyl group is electron withdrawing via resonance), and 3) the leaving group is *para* to the electron-withdrawing group. Therefore, we draw an SNAr mechanism. In the first step, methoxide functions as a nucleophile and attacks the aromatic ring at the position bearing the leaving group, thereby forming an intermediate Meisenheimer complex. Then, a leaving group (chloride) is ejected, restoring aromaticity, and giving the product, as shown.

NaOMe

Cl OMe

$:\ddot{O}Me^{\ominus}$

Cl OMe

18.52.

(a) The reagents indicate the installation of a nitro group, followed by its reduction to an amino group. To determine where this group is installed, recall that the directing effects are controlled by the most strongly activating group. In this case, the methyl group is the only activating group (albeit weakly activating), so it controls the directing effects. As an activator, this group is an *ortho-para* director. However, the *para* position is already occupied (by the other substituent). Therefore, the group is installed in a position that is *ortho* to the methyl group.

1) HNO_3 / H_2SO_4
2) Zn, HCl
3) NaOH

Br NH_2 Br

HNO_3 / H_2SO_4 → NO_2, Br → 1) Zn, HCl 2) NaOH

In this case, both *ortho* positions are equivalent because of symmetry, so the group can be drawn in either ortho position (don't be distracted by the arbitrary placement of the double bonds in the aromatic ring):

H_2N … Br ≡ NH_2 … Br

(b) The starting material is *para*-xylene, in which the following four positions are identical (because of symmetry):

Therefore, installation of a third group leads to only one regiochemical outcome. Each group weakly activates the ring, so this compound will readily undergo a Friedel-Crafts acylation reaction, thereby installing an acyl group on the ring. Then, in the second step, the carbonyl group is reduced, giving the product shown:

1) $AlCl_3$, Cl(C=O)CH₂CH₂CH₃
2) Zn(Hg), HCl, heat

$AlCl_3$, Cl(C=O)… → Zn(Hg), HCl, heat

(c) The aromatic ring has only one substituent (an isopropyl group), which is an activator and therefore an *ortho-para* director. The *ortho* positions are sterically hindered (by their proximity to the bulky isopropyl group), so an electrophilic aromatic substitution reaction is expected to occur predominantly at the *para* position. In the first step, a Friedel-Crafts alkylation is used to install a methyl group in the *para* position. Then, both alkyl groups are oxidized to give a diacid, as shown:

1) CH_3Cl, $AlCl_3$
2) $KMnO_4$, H_2O, heat
3) H_3O^+

COOH COOH

CH_3Cl, $AlCl_3$ → 2) $KMnO_4$, H_2O, heat 3) H_3O^+

(d) The aromatic ring has only one substituent (a *tert*-butyl group), which is an activator and therefore an *ortho-para* director. The *ortho* positions are very sterically hindered (by their proximity to the bulky *tert*-butyl group), so an electrophilic aromatic substitution reaction is expected to occur almost exclusively at the *para* position. In the first step, a Friedel-Crafts alkylation is used to install a methyl group in the *para* position. Then, in the next step, the benzylic protons are replaced with bromine atoms. The *tert*-butyl group does not have any benzylic protons, so no bromine atoms are installed in that location. Only the methyl group undergoes benzylic bromination (exhaustively), as shown:

1) CH_3Cl, $AlCl_3$
2) excess NBS, heat

Br_3C

CH_3Cl, $AlCl_3$

excess NBS

18.53.

(a) Installation of the amino group requires a two-step process (nitration, followed by reduction), while installation of the bromine atom can be achieved in just one step.

Now let's consider the order of events. These two groups must be installed in a *meta* fashion, but both groups are *ortho-para* directors. Installation of these two groups (in either order) does not appear to give the desired *meta* substitution pattern. However, recall that installation of the amino group requires two steps. The first step is nitration, and a nitro group is a *meta*-director. So, we can achieve the desired transformation by performing the bromination process after the nitration process but before the reduction process, as shown here:

1) HNO_3, H_2SO_4
2) Br_2, $AlBr_3$
3) Zn, HCl
4) NaOH

H_2N Br

HNO_3, H_2SO_4

1) Zn, HCl
2) NaOH

O_2N Br_2, $AlBr_3$ O_2N Br

(b) Installation of the amino group requires two steps (nitration, followed by reduction), while installation of the ethyl group can be achieved in just one step (via a Friedel-Crafts alkylation). The substituents must be installed in a *meta* fashion, yet both substituents are *ortho-para* directors. To circumvent the problem, we consider utilizing the *meta*-directing effects of the nitro group to achieve the desired regiochemical outcome:

H_2N

O_2N O_2N

However, this strategy will not succeed because the middle step is flawed. Specifically, a Friedel-Crafts alkylation cannot be performed on a strongly deactivated ring (such as nitrobenzene). Therefore, we must find another way to install the two groups in a *meta* fashion. This can be accomplished if we install the ethyl group via a two-step process (Friedel-Crafts acylation, followed by reduction):

O

By installing the ethyl group in this way, we can capitalize on the directing effects of the carbonyl group, before reducing it. As seen below, a Friedel-Crafts acylation installs an acyl group, which is a *meta*-director. Subsequent nitration allows for the installation of a nitro group in the *meta* position. And finally, reduction (of both groups) gives the product, as shown.

1) CH_3COCl, $AlCl_3$
2) HNO_3, H_2SO_4
3) HCl, Zn(Hg), heat
4) NaOH

H_2N

CH_3COCl, $AlCl_3$

1) HCl, Zn(Hg), heat
2) NaOH

O O

HNO_3, H_2SO_4

O_2N

18.54.

(a) The second step of the synthesis will not work, because a strongly deactivated ring will not undergo a Friedel-Crafts alkylation. The product of the first step, nitrobenzene, will be unreactive in the second step.

(b) The second step of the synthesis will not efficiently install a propyl group, because a carbocation rearrangement is likely to occur, which can result in the installation of an isopropyl group. A mixture of products is expected.

(c) The second step of the synthesis will not install the acyl group in the *meta* position. It will be installed in a position that is either *ortho* or *para* to the bromine atom (with the *para* product likely to predominate, due to steric factors).

(d) The second step of the synthesis will not efficiently install the bromine atom in the *ortho* position, because of steric hindrance from the *tert*-butyl group. Bromination will occur primarily at the *para* position.

18.55.

(a) This compound has two aromatic rings, each of which is monosubstituted. One aromatic ring (left) is connected directly to a carbonyl group and is therefore deactivated. The other aromatic ring (right) is connected to a methylene group (CH_2) and is therefore activated (the substituent is treated like an alkyl group). So, we expect monobromination to occur on the activated ring. Since the substituent is an activator, it must also be an *ortho-para* director. The *ortho* positions are sterically hindered (by the large substituent), so the reaction is expected to occur predominantly at the *para* position, indicated here:

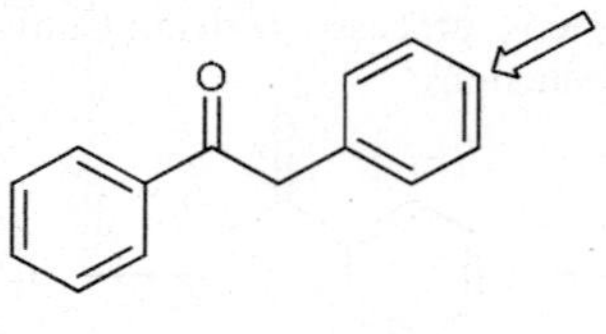

(b) This compound has two aromatic rings, each of which is monosubstituted. One aromatic ring (left) is connected directly to a carbonyl group and is therefore deactivated. The other aromatic ring (right) is connected to a nitrogen atom and is therefore activated. So, we expect monobromination to occur on the activated ring. Since the substituent is an activator, it must also be an *ortho-para* director. The *ortho* positions are sterically hindered (by the large substituent), so the reaction is expected to occur predominantly at the *para* position, indicated here:

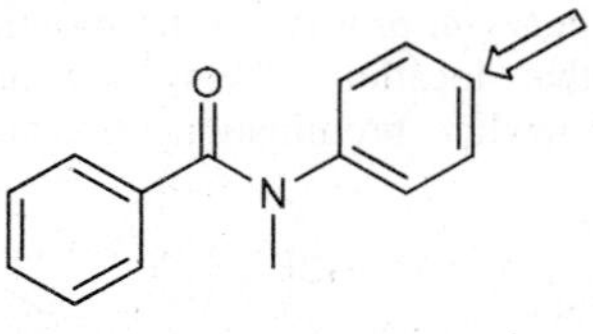

(c) This compound has two aromatic rings. The ring on the left has one only substituent, which is an activator. The ring on the right has two substituents: an activator and a deactivator. As such, we expect monobromination to occur on the left ring, which is more activated. Since the substituent (on the left ring) is an activator, it must also be an *ortho-para* director. The *ortho* positions are sterically hindered (by the large substituent), so the reaction is expected to occur predominantly at the *para* position, indicated here:

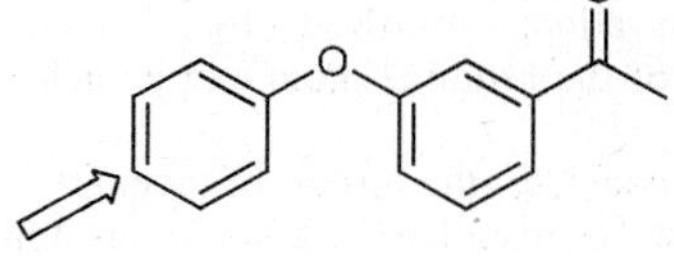

18.56. The reagent is sodium amide, which is strongly nucleophilic and strongly basic. The three criteria for SNAr are not met (the aromatic ring lacks a nitro group), and therefore, the reaction must occur via an elimination-addition mechanism (as seen in Mechanism 18.9), shown here. The benzyne intermediate can be attacked on either side of the triple bond, leading to the two products shown:

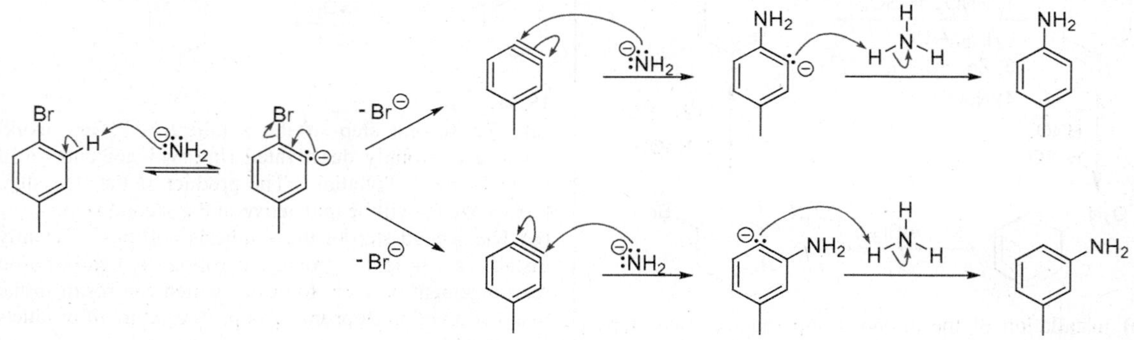

18.57. Phenol has only one substituent (OH) connected to the aromatic ring. This group is a strong activator, and is therefore an *ortho-para* director. If conditions are controlled such that three nitro groups are installed, then we expect the three nitro groups to be installed at the *ortho* and *para* positions, as shown:

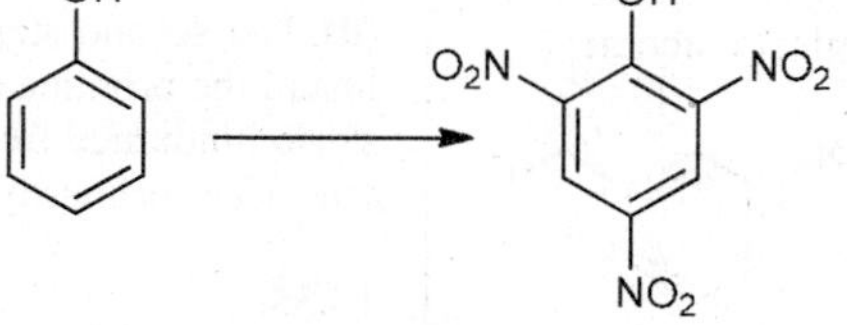

2,4,6-trinitrophenol
(picric acid)

18.58. The starting material is benzene, and the product exhibits a disubstituted aromatic ring. This will require two successive aromatic substitution reactions. The reagent (the diol) is not a very strong nucleophile, nor is it a very strong electrophile. However, in acidic conditions, the diol exists in equilibrium with a very powerful electrophile (a carbocation), as shown:

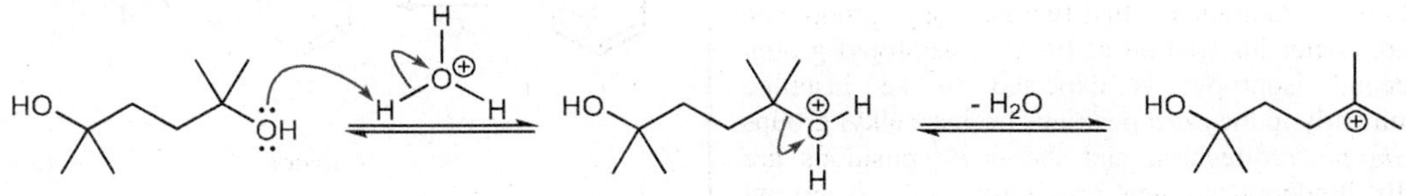

As seen above, one of the OH groups can be protonated, giving an excellent leaving group, which can leave to give a tertiary carbocation. Since the reagent is electrophilic, we expect that the reaction will proceed via electrophilic aromatic substitution. Accordingly, the aromatic ring attacks the carbocation to give an intermediate sigma complex, which then loses a proton to restore aromaticity:

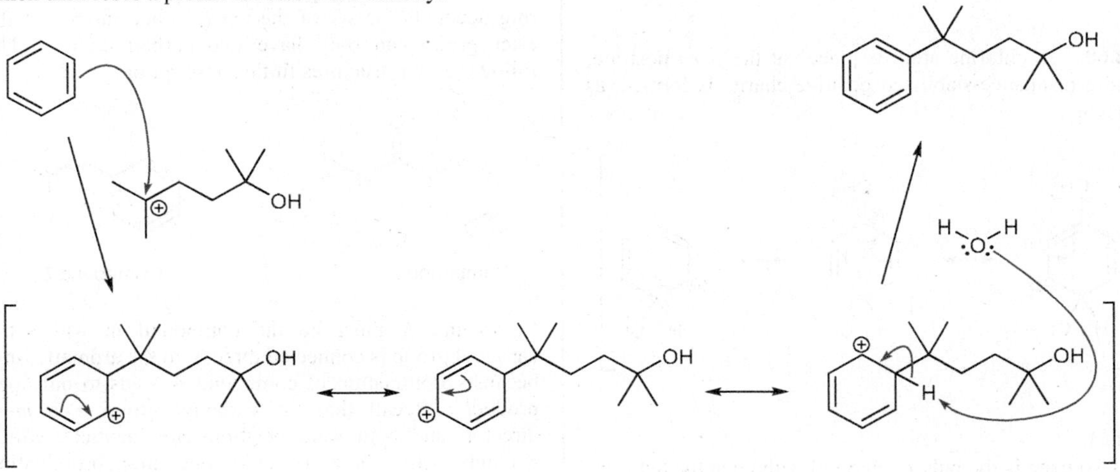

The entire process is then repeated again, except this time, the reaction will proceed in an *intramolecular* fashion, closing a ring. The OH group is protonated to give an excellent leaving group, which can leave to give a tertiary carbocation. The aromatic ring attacks the carbocation in an intramolecular process to give an intermediate sigma complex, which then loses a proton to restore aromaticity:

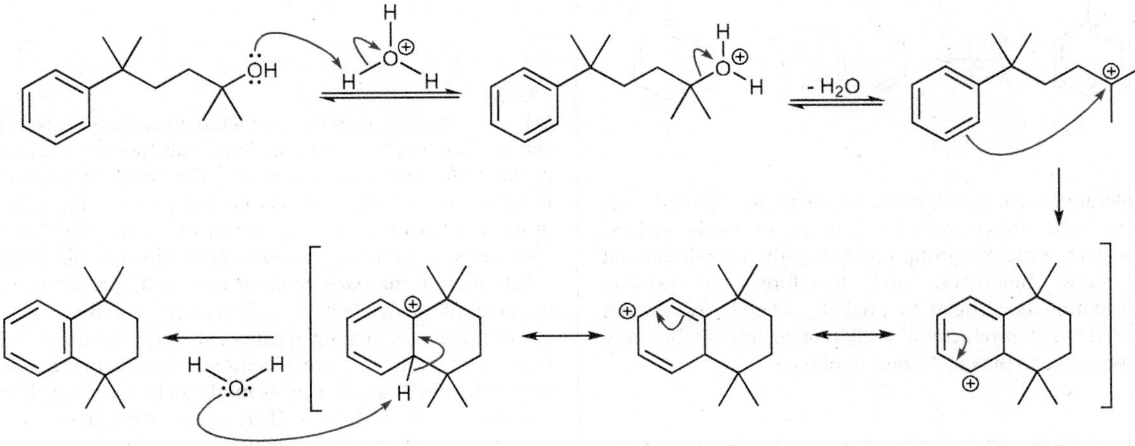

Since aqueous acidic conditions are employed (H_2SO_4 is generally dissolved in water), the proton source throughout the mechanism has been H_3O^+, and the base has always been H_2O, to be consistent with acidic conditions.

18.59. These reagents will achieve the installation of an isopropyl group on the aromatic ring. The problem statement indicates that conditions will favor dialkylation, which means that two isopropyl groups are installed. After installation of the first isopropyl group, the second isopropyl is expected to be installed predominantly in the *para* position (because alkyl groups are *ortho-para* directors, and the *ortho* positions are sterically hindered by their proximity to the isopropyl group):

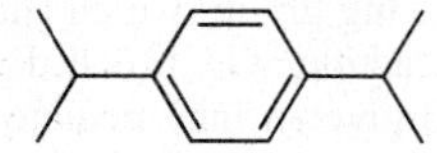

18.60. A chlorine atom is placed at the *para* position, and a resonance-stabilized positive charge is formed, as shown:

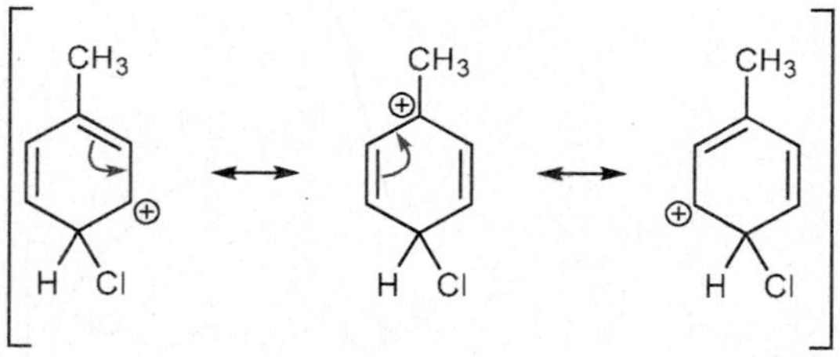

CH_3 H Cl ⟷ CH_3 H Cl ⟷ CH_3 H Cl

18.61.

(a) Toluene is the only compound with an activated ring, and it is expected to undergo a Friedel-Crafts reaction most rapidly. The methyl group is an *ortho-para* director, so we expect a mixture of *ortho* and *para* products (*ortho*-ethyltoluene and *para*-ethyltoluene):

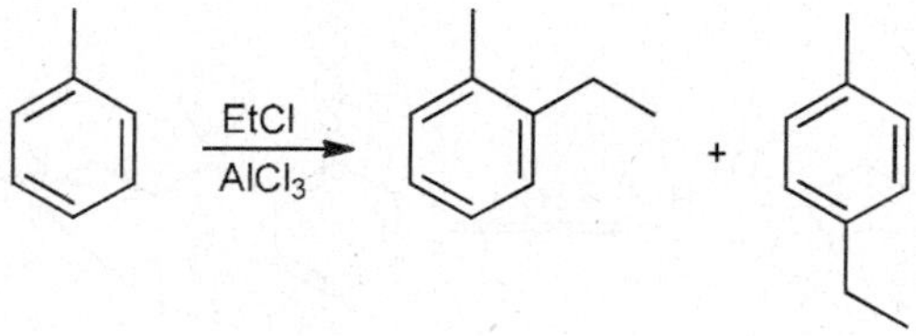

EtCl / $AlCl_3$ → +

Generally, *para* substitution products are favored over *ortho* substitution products because of steric factors, However, a methyl group provides only a small amount of steric hindrance, and therefore, the product distribution is difficult to predict. That is, we cannot predict which product will predominate, and it will likely be sensitive to the conditions employed.

(b) Anisole (methoxybenzene) exhibits the most activated ring, and it is therefore expected to undergo a Friedel-Crafts reaction most rapidly. The methoxy group is an *ortho-para* director, so we expect a mixture of *ortho* and *para* products, although the *para* product is expected to predominate because of steric factors:

OCH_3 EtCl / $AlCl_3$ → OCH_3 + OCH_3

minor major

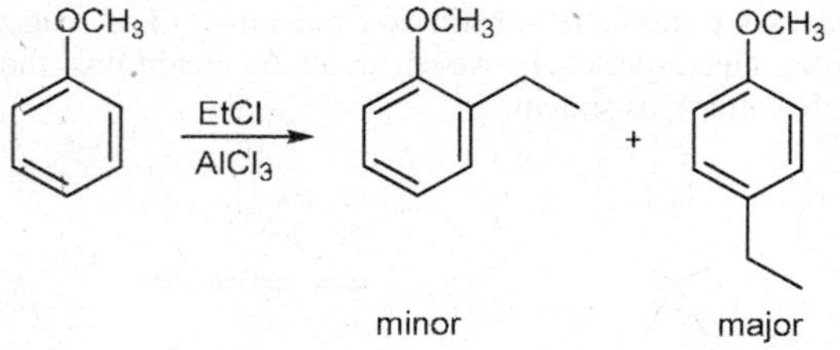

18.62. The molecular formula indicates five degrees of unsaturation, which accounts for the aromatic ring (four degrees) and the ester group (one degree). The aromatic ring accounts for six of the eight carbon atoms, so the ester group can only have two carbon atoms. The following two structures fit this description:

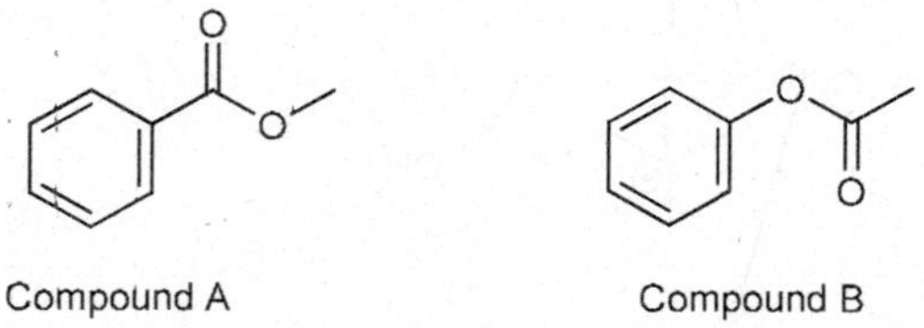

Compound A Compound B

Compound **A** must be the compound in which the carbonyl group is connected directly to the aromatic ring, because bromination of compound **A** leads to only one product. Recall that the carbonyl group is a *meta* director, and both *meta* positions are identical, which explains why there is only one monobromination product. In contrast, compound **B** exhibits an activating group, which is an *ortho-para* director. As such, we expect two monobromination products (bromination can occur at either the *ortho* position or the *para* position, with the *para* product expected to predominate due to the steric influence of the ester).

18.63.

(a) The starting material is methoxybenzene (anisole) and the product has two additional substituents, located at the *ortho* and *para* positions. The bromine atom is installed in just one step (bromination), and the nitro group can also be installed in just one step (nitration). The methoxy group is an *ortho-para* director, although substitution at the *para* position will likely predominate because of steric factors. Therefore, we must first perform the bromination reaction, thereby installing the bromine atom in the correct location. Note that no Lewis acid is required, as the ring is moderately activated; just one equivalent of Br_2 would be used. And finally, we perform a nitration reaction which installs the nitro group in the position that is *ortho* to the moderately activating methoxy group.

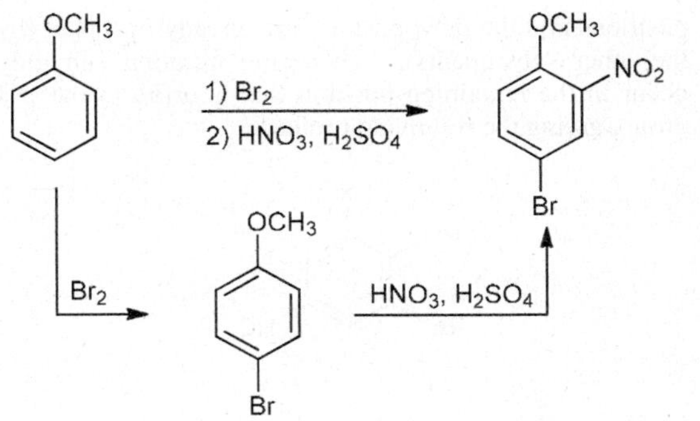

(b) The starting material is methoxybenzene (anisole) and the product has two additional substituents, located at the *ortho* and *para* positions. The bromine atom is installed in just one step (bromination), and the nitro group can also be installed in just one step (nitration). The methoxy group is an *ortho-para* director, although substitution at the *para* position will likely predominate because of steric factors. Therefore, we must first perform the nitration reaction, thereby installing the nitro group in the correct location. And then, we perform a bromination reaction which installs the bromine atom in the position that is *ortho* to the moderately activating methoxy group.

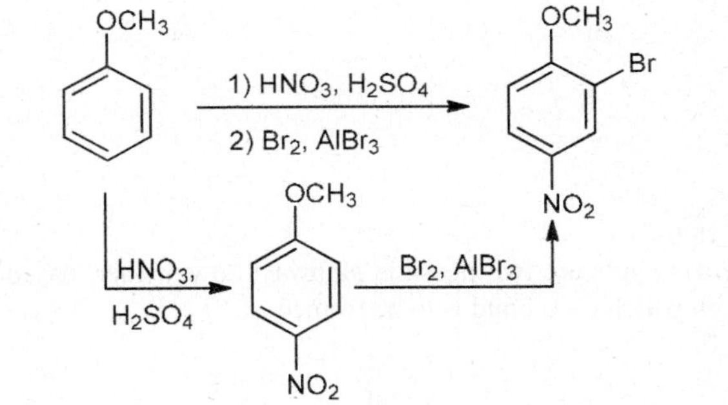

(c) The starting material exhibits an aromatic ring with only one substituent (an isopropyl group). In the product, the isopropyl group has been converted into a carboxylic acid, which can be achieved via benzylic oxidation with a strong oxidizing agent. In addition, a nitro group must be installed. Now we must determine the order of events. Both substituents (in the product) are *meta* directors, yet they are *ortho* to each other. Therefore, we must install the nitro group before the isopropyl group is oxidized, thereby relying on the *ortho-para* directing effects of an isopropyl group. A blocking group must be used because the isopropyl group provides steric bulk that would otherwise favor nitration at the *para* position, rather than the *ortho* position. By blocking the *para* position, we can perform a nitration reaction, thereby installing the nitro group in the desired location. Removing the blocking group, followed by oxidation of the benzylic position, gives the desired product.

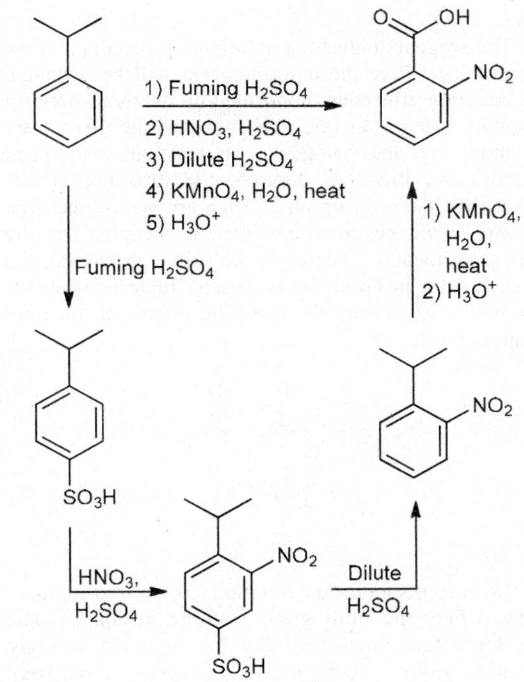

(d) The starting material exhibits an aromatic ring with only one substituent (an isopropyl group). In the product, the isopropyl group has been converted into a carboxylic acid, which can be achieved via benzylic oxidation with a strong oxidizing agent. In addition, a nitro group must be installed. Now we must determine the order of events. Both substituents (in the product) are *meta* directors, and they are indeed *meta* to each other. Therefore, we simply oxidize the benzylic position, and then perform a nitration reaction to give the desired product.

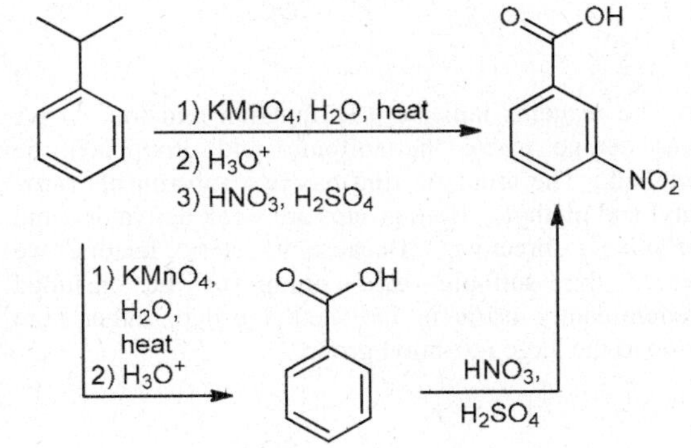

Note that an isopropyl group is an *ortho-para* director, but the desired nitration reaction must occur at the *meta* position, so nitration cannot be the first step of our synthesis. We must first convert the alkyl group (which is an *ortho-para* director) into a carboxylic acid group (which is a *meta* director).

18.64.

(a) The reagents indicate a chlorination reaction, so we must decide where the chlorine atom will be installed. The directing effects are controlled by the most strongly activating group. In this case, none of the groups are activators, but the halogen is at least an *ortho-para* director, and therefore controls the directing effects (rather than the ester groups, which are *meta*-directing). The two *ortho* positions are already occupied (by the other substituents). Therefore, chlorination is expected to occur at the position that is *para* to the bromine atom. This is consistent with the directing effects of the ester groups as well.

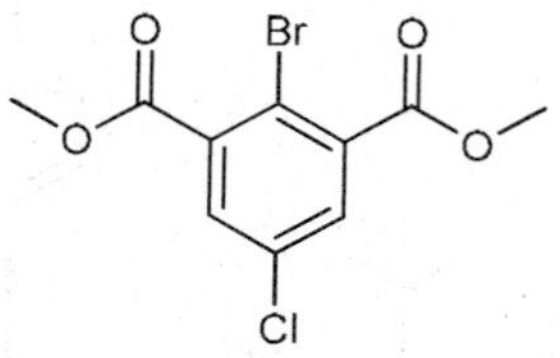

(b) The reagents indicate a nitration reaction, so we must decide where the nitro group will be installed. The directing effects are controlled by the most strongly activating group. Therefore, in this case, the methoxy group controls the directing effects. As an activator, this group is an *ortho-para* director. However, one of the *ortho* positions is already occupied (by a methyl substituent). That leaves two possible locations: the remaining *ortho* position and the *para* position. The former is too sterically hindered and the reaction is not likely to occur there. Therefore, we expect nitration to occur in the *para* position, giving the following product:

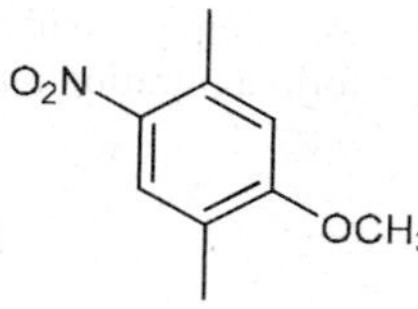

(c) The reagents indicate a sulfonation reaction, so we must decide where the sulfonic acid group will be installed. The aromatic ring has two substituents (*tert*-butyl and methyl). Both groups are weak activators, and *ortho-para* directors. Because of steric factors, we expect the sulfonic acid group to be installed predominantly *ortho* to the methyl group, rather than *ortho* to the large *tert*-butyl group.

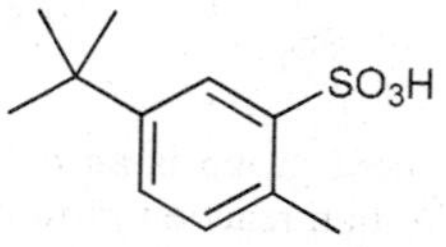

(d) The reagents indicate a nitration reaction, so we must decide where the nitro group will be installed. The directing effects are controlled by the most strongly activating group. Therefore, in this case, the OH group controls the directing effects. As an activator, this group is an *ortho-para* director. However, one of the *ortho* positions and the *para* position are already occupied (by the other substituents). Therefore, nitration can only occur in the remaining position that is *ortho* to the OH group, giving the following product.

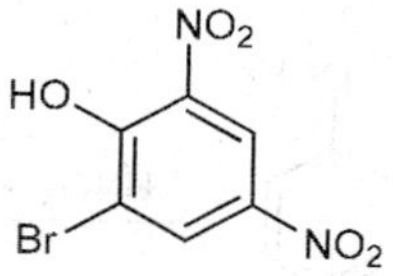

18.65. The reagents indicate a bromination reaction, so we must decide where the bromine atom will be installed. The directing effects are controlled by the most strongly activating group(s). The methyl groups are weakly activating and the bromo groups are weakly deactivating. Therefore, in this case, the methyl groups control the directing effects. Both of the methyl groups are directing to the same location (*ortho* and *para* to the methyl groups):

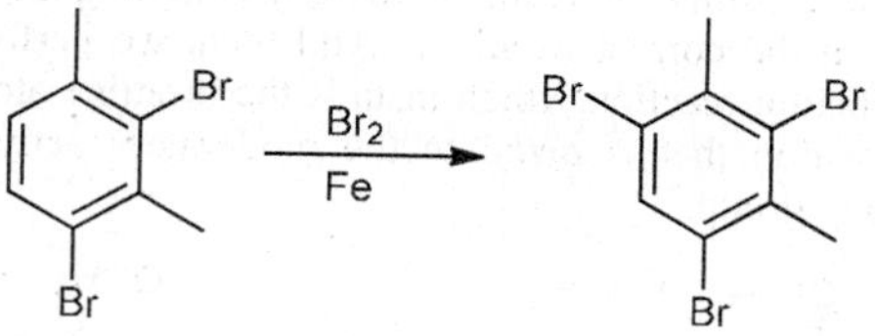

18.66.

(a) There are two possible pathways to consider, based on which C–C bond is to be formed:

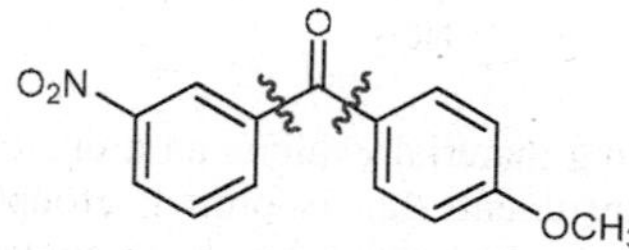

Let's consider one of these pathways (forming the bond on the left). Forming this bond via a Friedel-Crafts acylation would require the following starting materials:

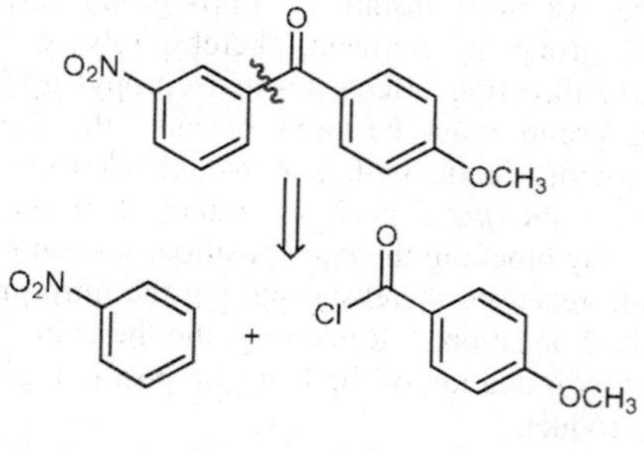

Indeed, the nitro group of nitrobenzene is *meta*-directing, so the regiochemistry is correct. However, we must remember that a Friedel-Crafts acylation cannot be performed on an aromatic ring that is strongly deactivated. Since nitrobenzene is strongly deactivated, this pathway is not feasible. So, we must explore the other possible pathway, forming this bond:

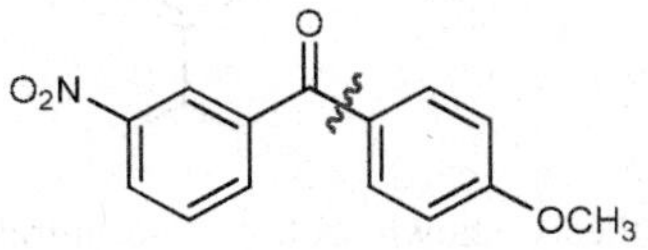

Forming this bond via a Friedel-Crafts acylation would require the following starting materials:

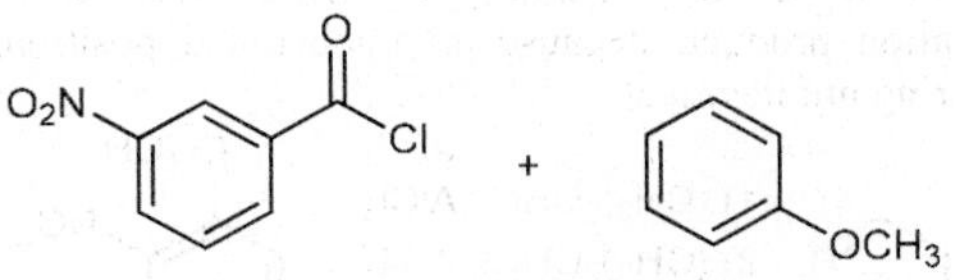

The methoxy group is an *ortho-para* director, and *para* substitution is expected to predominate (because of steric factors), giving rise to the desired product.

(b) There are two possible pathways to consider, based on which C–C bond is to be formed:

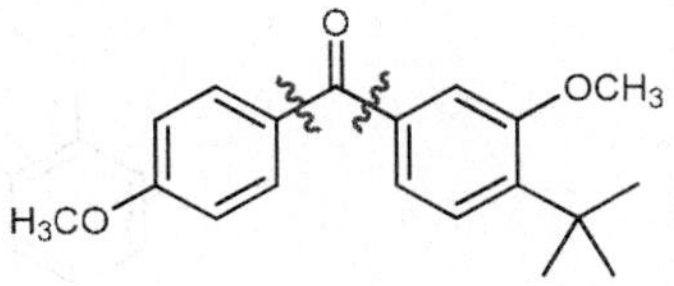

Let's consider one of these pathways (forming the bond on the right). Forming this bond via a Friedel-Crafts acylation would require the following starting materials:

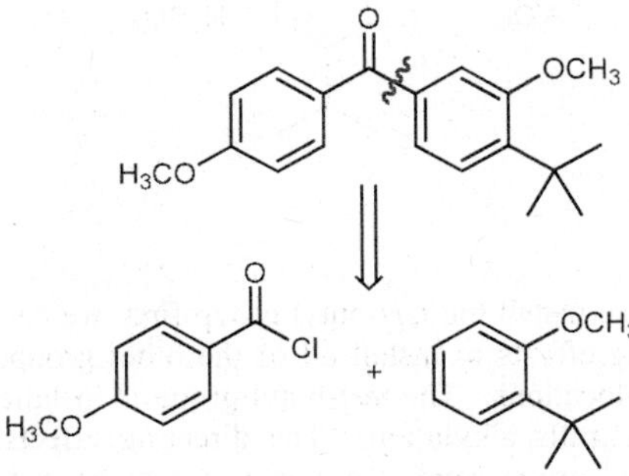

Let's consider the regiochemical outcome of a Friedel-Crafts acylation with these starting materials. The methoxy group controls the directing effects, so substitution is expected to occur at the position that is *para* to the methoxy group (not *meta* to the methoxy group).

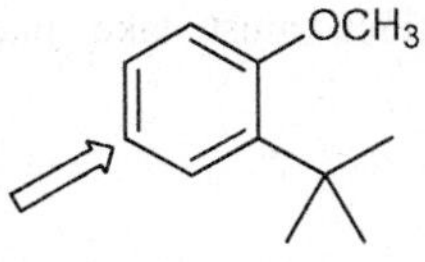

Therefore, this pathway is not feasible, because it would not give the correct product. So we must explore the other possible pathway, forming this bond:

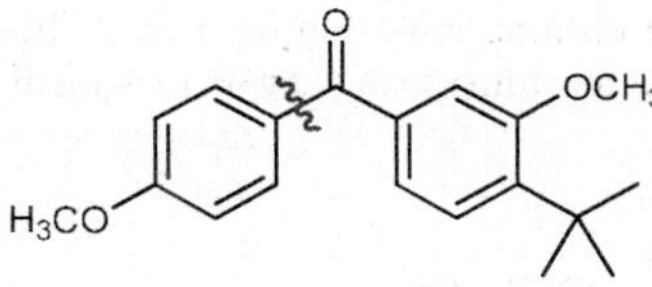

Forming this bond via a Friedel-Crafts acylation would require the following starting materials:

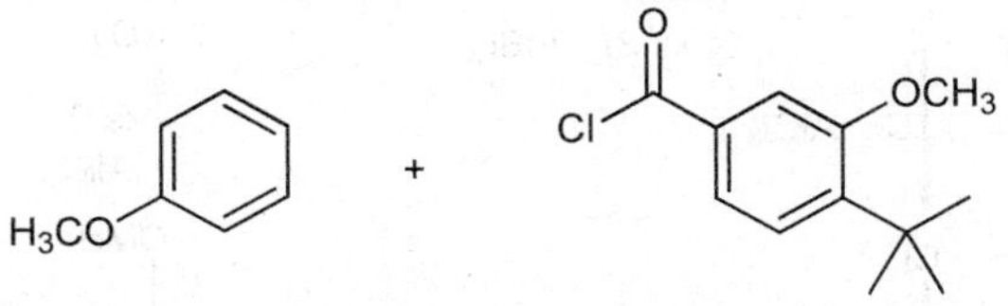

The methoxy group is an *ortho-para* director, and *para* substitution is expected to predominate (because of steric factors), giving rise to the desired product.

18.67.

(a) The product has four substituents. One of them is a methoxy group, which is an *ortho-para* director. The directing effects of this substituent, if installed first, can be used to install the other groups in the correct locations:

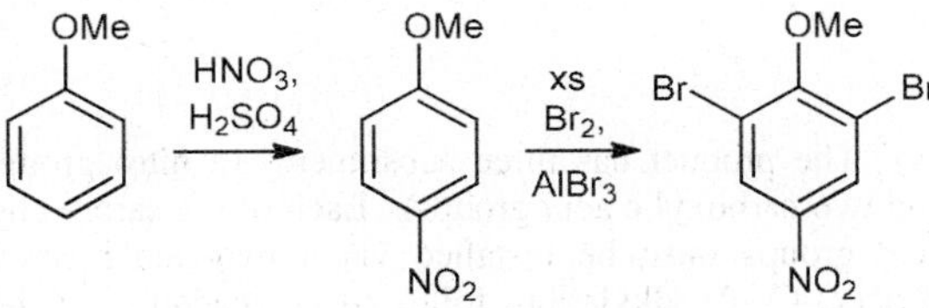

So the key to solving this problem is to convert benzene into methoxybenzene (anisole):

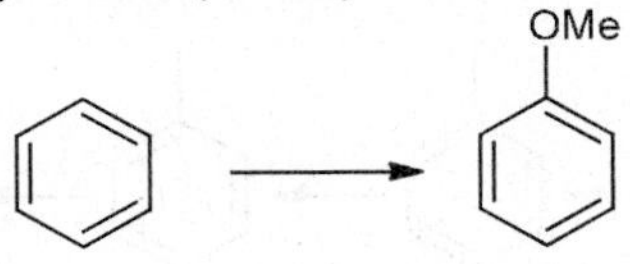

This can be achieved via the following three-step process.

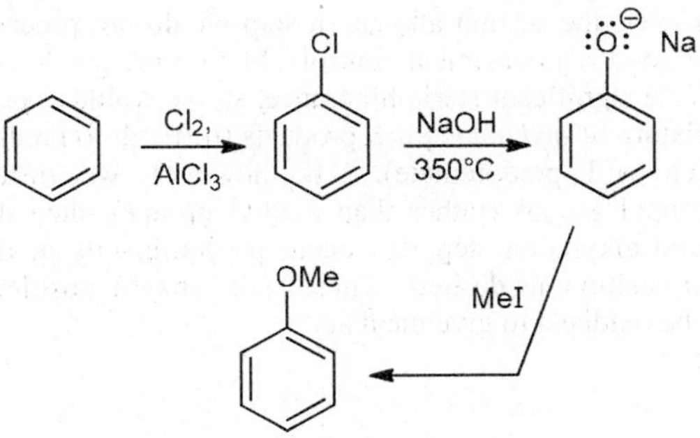

There are certainly other acceptable methods for achieving the desired transformation. Once such alternative is shown here, which uses an SNAr step in place of the elimination-addition step. This is possible by installing the nitro group prior to substitution of the Cl:

1) Cl_2, $AlCl_3$
2) HNO_3, H_2SO_4
3) NaOH
4) MeI
5) xs Br_2, $AlBr_3$

Cl_2, $AlCl_3$

HNO_3, H_2SO_4

NaOH

MeI

xs Br_2, $AlBr_3$

(b) The product has three substituents (a nitro group, and two carboxylic acid groups). Each of the carboxylic acid groups must be installed via a two-step method (Friedel-Crafts alkylation, followed by oxidation of the benzylic position). We could certainly install two methyl groups and then oxidize both of them:

However, the second alkylation step would not proceed with good regiochemical control. Methyl groups do not provide significant steric hindrance, so we would expect a mixture of *ortho* and *para* products (difficult to predict which will predominate). If, however, we install isopropyl groups (rather than methyl groups), then the second alkylation step will occur predominantly at the *para* position, as desired. Then, both benzylic positions can be oxidized to give the diacid:

In the synthesis shown below, the nitration step is performed prior to oxidation of the benzylic groups, because nitration of an activated ring is likely to proceed more readily than nitration of a deactivated ring. Nitration of *para*-diisopropylbenzene gives only one nitration product, because all unoccupied positions on the ring are identical.

1) $(CH_3)_2CHCl$, $AlCl_3$
2) $(CH_3)_2CHCl$, $AlCl_3$
3) HNO_3, H_2SO_4
4) $KMnO_4$, H_2O, heat
5) H_3O^+

$(CH_3)_2CHCl$, $AlCl_3$

$(CH_3)_2CHCl$, $AlCl_3$

HNO_3, H_2SO_4

1) $KMnO_4$, H_2O, heat
2) H_3O^+

(c) If we install the *tert*-butyl group first, we can use its directing effects to install all of the other groups in the correct locations. The *tert*-butyl group is installed via a Friedel-Crafts alkylation. The directing effects of the *tert*-butyl group cause the next group to be installed in the *para* position, so the next step of our synthesis must be a nitration step. Subsequent halogenation steps are forced to proceed at the positions that are *ortho* to the *tert*-butyl group. These positions are indeed sterically hindered, but the position that is *para* to the *tert*-butyl group is already occupied (by the nitro group). So any subsequent reactions must take place at the *ortho* positions.

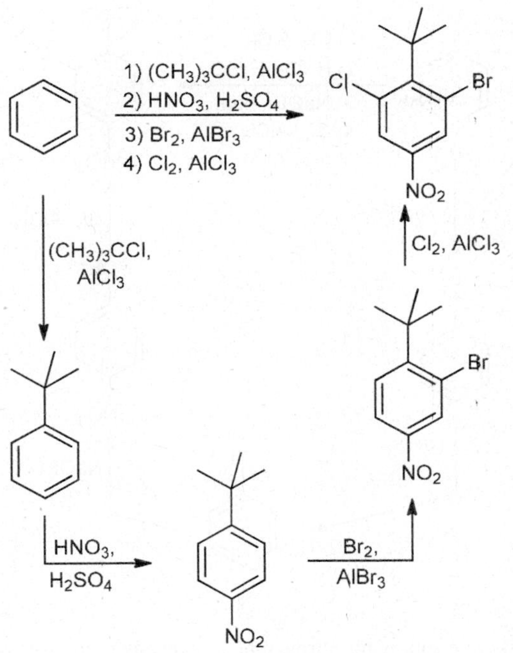
1) (CH3)3CCl, AlCl3
2) HNO3, H2SO4
3) Br2, AlBr3
4) Cl2, AlCl3
Cl
Br
NO2
Cl2, AlCl3
(CH3)3CCl, AlCl3
Br
NO2
HNO3, H2SO4
Br2, AlBr3
NO2

(d) We must install three groups (a chlorine atom, an acyl group, and a nitro group). Two of these groups are *meta* directors (the nitro group and the acyl group), while the other group (Cl) is an *ortho-para* director. The groups are all *meta* to each other in the product, so the Cl group must be installed last.

The first two steps of the synthesis must install the acyl group (via Friedel-Crafts acylation) and the nitro group (via nitration). These two steps cannot be performed in either order. The nitro group cannot be installed first, because nitrobenzene is too deactivated to undergo a Friedel-Crafts alkylation. Therefore, the order of events must be: 1) Friedel-Crafts alkylation, 2) nitration, and 3) chlorination. This synthesis is shown here:

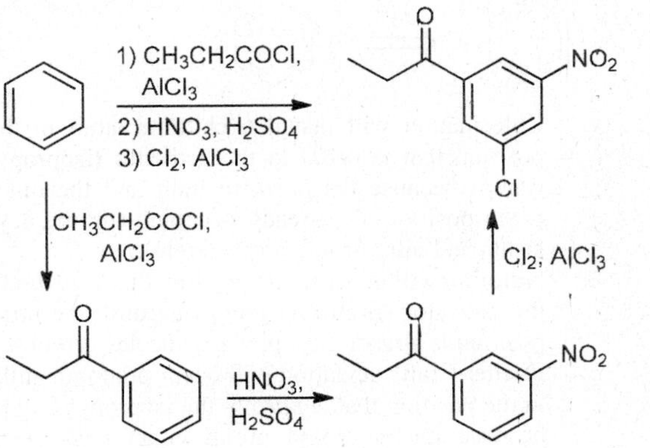
1) CH3CH2COCl, AlCl3
2) HNO3, H2SO4
3) Cl2, AlCl3
O
NO2
Cl
CH3CH2COCl, AlCl3
Cl2, AlCl3
O
O
NO2
HNO3, H2SO4

(e) There is likely more than one way to achieve the desired transformation. The following retrosynthetic analysis represents one strategy for preparing the desired compound. An explanation of each of the steps (*a-e*) follows.

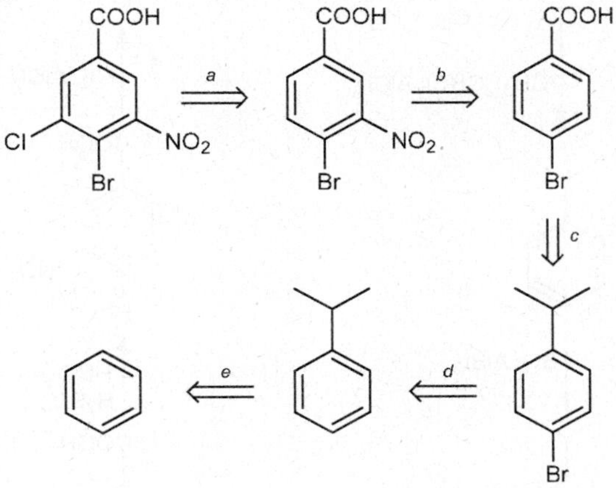
COOH
COOH
COOH
a
b
Cl
NO2
NO2
Br
Br
Br
c
e
d
Br

a. The product can be prepared by installing the Cl group last. None of the three groups (on the ring) are activators, but all three groups direct to the same location.

b. Installation of the nitro group would occur in the position that is *ortho* to the *ortho-para* directing group (and *meta* to the *meta*-directing group).

c. The carboxylic acid group can be made via oxidation of the benzylic position. A methyl group can certainly be oxidized to a carboxylic acid group, however, we choose a larger, sterically hindered alkyl group, as will soon be explained.

d. Bromination will occur predominantly in the *para* position, because the *ortho* positions are sterically hindered. This is indeed the reason why we chose an isopropyl group rather than a methyl group. With a methyl group, we would expect a mixture of *ortho* and *para* products.

e. Installation of an isopropyl group can be achieved via a Friedel-Crafts alkylation.

Now, let's draw the forward scheme. Note that in the first step (Friedel-Crafts alkylation), it would be inefficient to use 1-chloropropane and then rely on a carbocation rearrangement, as that would likely give a mixture of products (propylbenzene and isopropylbenzene).

1) $(CH_3)_2CHCl$, $AlCl_3$
2) Br_2, $AlBr_3$
3) $KMnO_4$, H_2O, heat
4) H_3O^+
5) HNO_3, H_2SO_4
6) Cl_2, $AlCl_3$

$(CH_3)_2CHCl$, $AlCl_3$

Br_2, $AlBr_3$

1) $KMnO_4$, H_2O, heat
2) H_3O^+

HNO_3, H_2SO_4

Cl_2, $AlCl_3$

COOH, Cl, NO_2, Br

(f) The following retrosynthetic analysis represents one strategy for preparing the desired compound. An explanation of each of the steps (*a-d*) follows.

OEt, Cl, NO_2

a. Chlorination will install a chlorine atom in the position that is *ortho* to the activator (OEt) because the *para* position is already occupied (by the nitro group).
b. An SNAr reaction will replace the chlorine atom with an ethoxy group.
c. Installation of the nitro group would occur predominantly in the position that is *para* to the *ortho-para* directing Cl group.
d. Chlorination gives chlorobenzene.

Now, let's draw the forward scheme, as described above:

1) Cl_2, $AlCl_3$
2) HNO_3, H_2SO_4
3) NaOEt, heat
4) Cl_2, $AlCl_3$

Cl_2, $AlCl_3$

HNO_3, H_2SO_4

NaOEt, heat

Cl_2, $AlCl_3$

OEt, Cl, NO_2

(g) The following retrosynthetic analysis represents one strategy for preparing the desired compound. An explanation of each of the steps (*a-d*) follows.

Cl, NO_2, O

a. Chlorination will install a chlorine atom in the position that is *ortho* to the activator (isopropyl group) because the *para* position and the other *ortho* position are already occupied (by the acyl group and nitro group, respectively).
b. Nitration will occur in the position that is *ortho* to the activator (isopropyl group) because the *para* position is already occupied (by the acyl group).
c. Friedel-Crafts acylation will occur predominantly in the position that is *para* to the isopropyl group, because the isopropyl group is an *ortho-para* director, and it is sterically bulky.
d. Friedel-Crafts alkylation can be used to install an isopropyl group.

Now, let's draw the forward scheme. Note that in the first step (Friedel-Crafts alkylation), it would be inefficient to use 1-chloropropane and then rely on a carbocation rearrangement, as that would likely give a mixture of products (propylbenzene and isopropylbenzene).

1) $(CH_3)_2CHCl$, $AlCl_3$
2) CH_3CH_2COCl, $AlCl_3$
3) HNO_3, H_2SO_4
4) Cl_2, $AlCl_3$

$(CH_3)_2CHCl$, $AlCl_3$

Cl_2, $AlCl_3$

CH_3CH_2COCl, $AlCl_3$

HNO_3, H_2SO_4

(**h**) The following retrosynthetic analysis represents one strategy for preparing the desired compound. An explanation of each of the steps (*a-c*) follows.

a. Nitration will occur in the correct location because of steric factors.

b. Bromination will install a bromine atom in the correct location since the branching on the non-aromatic ring provides significant steric bulk. Note that when considering electronic factors, all four open positions on the arene are similarly activated, each being either *ortho* or *para* to one of the alkyl substituents.

c. Two successive Friedel-Crafts alkylation reactions (the first intermolecular and the second intramolecular) will form the desired additional ring.

Now, let's draw the forward scheme, as described above:

1) Cl⁀⁀Cl (dichloride), $AlCl_3$
2) Br_2, $AlBr_3$
3) HNO_3, H_2SO_4

$AlCl_3$

Br_2, $AlBr_3$

HNO_3, H_2SO_4

18.68. These reagents will install a methyl group, and the problem statement indicates that three methyl groups are installed. Installation of the first methyl group gives toluene. Methylation of toluene gives a mixture of *ortho*-xylene and *para*-xylene.

Benzene → Toluene → *ortho*-Xylene + *para*-Xylene

ortho-Xylene undergoes methylation at C4 because of steric factors, while *para*-xylene undergoes methylation at C2 (because all four positions are equivalent). Note that when considering electronic factors on the *ortho*-dimethylbenzene intermediate, all four open positions on the arene are similarly activated, each being either *ortho* or *para* to one of the alkyl substituents. Either way, the product is the same: 1,2,4-trimethylbenzene

1,2,4-trimethyl benzene

18.69. All three available positions are sterically hindered, so bromination of the ring occurs very slowly.

18.70. The reagents indicate a bromination reaction, so we must decide where the bromine atom will be installed. The aromatic ring has two substituents: an alkyl group and a carbonyl group. The former is a weak activator and the latter is a moderate deactivator. The directing effects are controlled by the more strongly activating group, but in this case both groups direct to the same two psitions. As an activator, the alkyl group is an *ortho-para* director. However, one of the *ortho* positions is already occupied (by the other substituent). Therefore, there are only two locations where bromination can occur (*ortho* or *para* to the alkyl group):

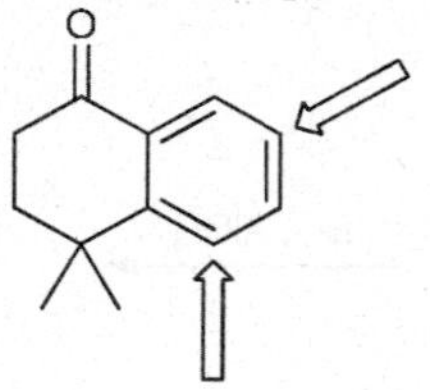

Among these two locations, one of them is more sterically hindered than the other. The reaction is expected to occur at the location that is more sterically accessible:

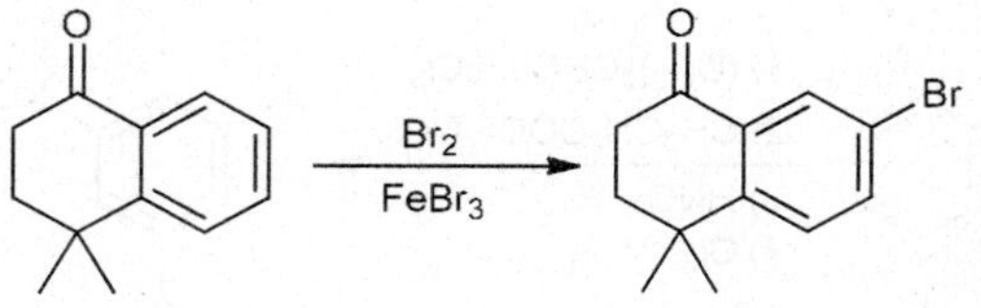

Br_2, $FeBr_3$

18.71. The reaction conditions (dilute H_2SO_4) will remove each of the sulfonic acid groups, giving toluene as the major product:

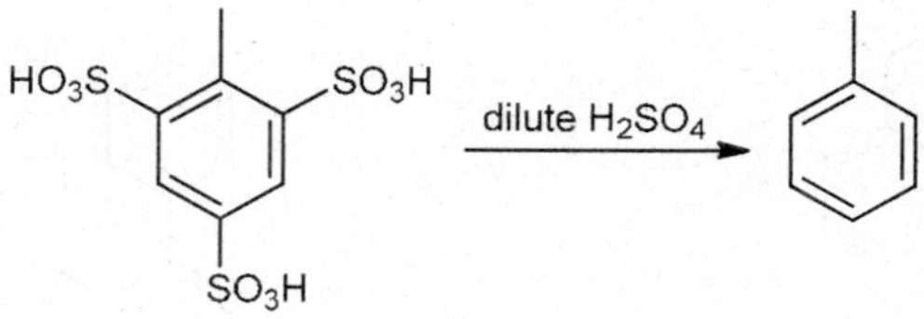

HO_3S, SO_3H, SO_3H — dilute H_2SO_4 →

18.72.

(a) The reagent (hydroxide) is a strong nucleophile, and the three criteria for SNAr are met (there is a nitro group and a leaving group that are *ortho* to each other), so we expect an SNAr mechanism. In an SNAr reaction, the nucleophile (hydroxide) replaces the leaving group (bromide) and maintains the same location:

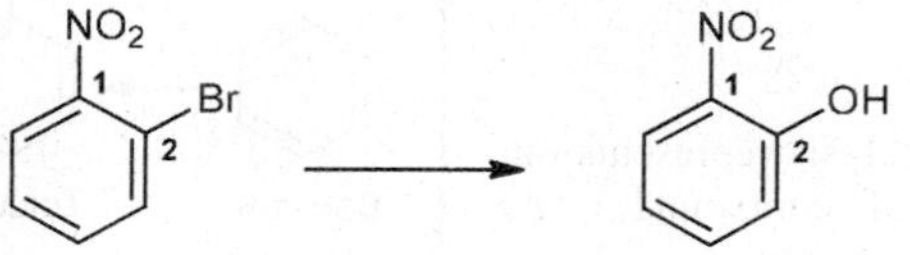

Under these basic conditions, the product (*ortho*-nitrophenol) is deprotonated upon its formation (because the resulting resonance stabilized phenoxide anion is more stable than hydroxide). In order to return the proton and regenerate *ortho*-nitrophenol, we must introduce a proton source into the reaction flask after the reaction is complete (aqueous acidic workup). The problem statement did not mention an acidic workup, so the product is the following salt (the conjugate base of *ortho*-nitrophenol):

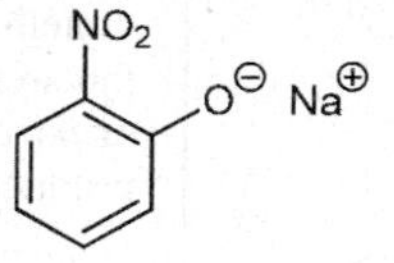

(b) In an SNAr mechanism, the nucleophile (hydroxide) attacks the ring at the position bearing the leaving group, giving the following resonance-stabilized intermediate (called a Meisenheimer complex):

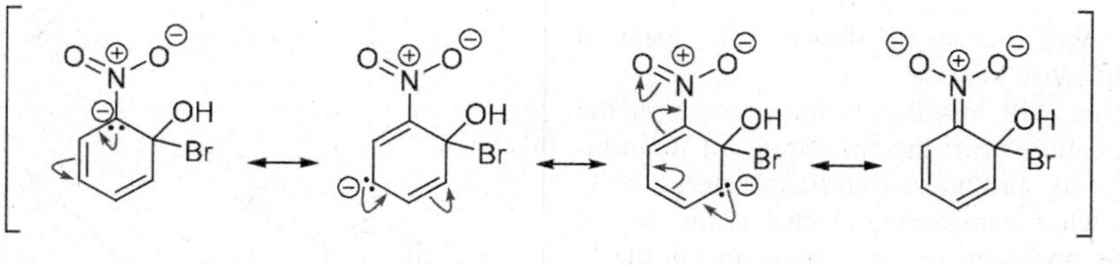

(c) No reaction is expected to take place. In order for an SNAr reaction to occur, the leaving group must be *ortho* or *para* to the nitro group. If the leaving group is *meta* to the nitro group, then the nitro group cannot function as a reservoir for electron density to stabilize the intermediate Meisenheimer complex via resonance delocalization.. The Meisenheimer complex is therefore too high in energy to form at an appreciable rate.

(d) Yes, this reaction is expected to proceed. This example meets all three criteria for an SNAr mechanism. The aromatic starting material exhibits a nitro group and a leaving group (bromide) that are *para* to each other.

18.73. The correct answer is (b). Friedel-Crafts reactions cannot be performed with aromatic rings that are strongly or moderately deactivated. Options (a) and (d) are both activated aromatic rings, and option (c) has a halogen, which is a weak deactivator. All three of these compounds will undergo Friedel-Crafts reactions. Only nitrobenzene is strongly deactivated and therefore, it will not undergo Friedel-Crafts reactions.

18.74. The correct answer is (c). One aromatic ring is connected to a carbonyl group (C=O), while the other aromatic ring is connected to an oxygen atom. The former is moderately deactivated, while the latter is moderately activated (as seen in Table 18.1).

This ring is moderately activated

As an activated ring, we expect *ortho-para* directing effects. The reaction will occur predominantly at the *para* position, rather than the *ortho* positions, because of steric factors.

18.75. The correct answer is (d). Option (a) would install a bromine substituent and an ethyl substituent, as desired, but the regiochemical outcome would not be correct. The two groups would be installed in a *para*-disubstituted fashion, rather than *meta*-disubstituted:

Br_2 / $AlBr_3$ → EtCl / $AlCl_3$ →

Similarly, option (b) would also install the two groups in a *para*-disubstituted fashion:

EtCl / $AlCl_3$ → Br_2 / $AlBr_3$ →

Option (c) employs a blocking technique, which would install the two groups in an *ortho*-disubstituted fashion (not *meta*, as desired):

EtCl / $AlCl_3$ → Fuming H_2SO_4 → Br_2 / $AlBr_3$ → Dilute H_2SO_4 →

Option (d) employs a Friedel-Crafts acylation, followed by bromination at the *meta* position (consistent with the carbonyl group being a *meta*-director), followed by a Clemmensen reduction, giving the desired product.

$AlCl_3$ → Br_2 / $AlBr_3$ → Zn(Hg) / HCl →

18.76. The correct answer is (b). The OH group donates electron density by resonance. This makes it an activating group for electrophilic aromatic substitution reactions and an *ortho-para* director.

18.77. The correct answer is (a). The aromatic ring has only one substituent (an isopropyl group), which is an electron-donating group and therefore an *ortho-para* director. The *ortho* positions are more sterically hindered, so the nitration reaction in the first step is expected to occur predominantly at the *para* position.

The second step is a chlorination reaction. To determine where the chlorine atom is installed, recall that the directing effects are controlled by the activating group. In this case, the isopropyl group is the only activating group (albeit weakly activating), so it controls the directing effects. The *para* position is already occupied (by the nitro group). Therefore, the chlorine atom is installed in a position that is *ortho* to the isopropyl group.

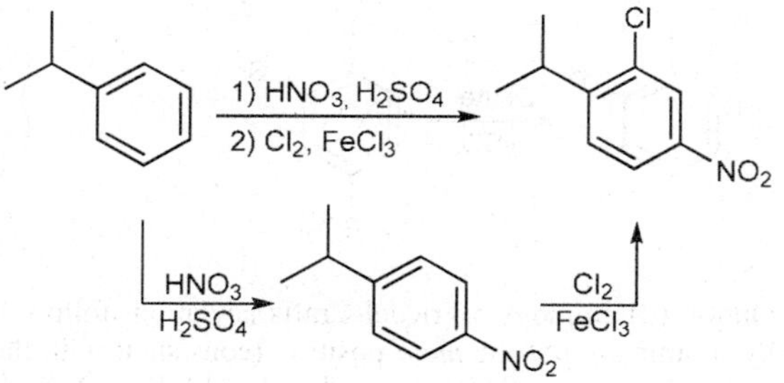

1) HNO_3, H_2SO_4
2) Cl_2, $FeCl_3$

18.78. The correct answer is (d). The reagent (sodium hydroxide) is a powerful nucleophile, and the three criteria for SNAr are met (a nitro group and a leaving group that are *para* to each other). To draw the product of an SNAr substitution mechanism, we replace the chloride leaving group with the hydroxide nucleophile.

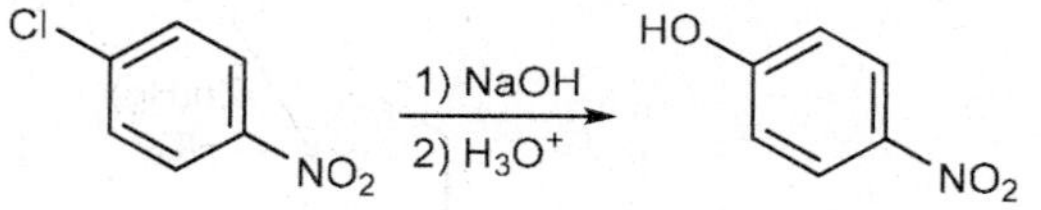

1) NaOH
2) H_3O^+

18.79. The correct answer is (a). The Friedel-Crafts acylation mechanism begins with the interaction of the acid chloride with the $AlCl_3$ Lewis acid. The resulting complex loses a leaving group to produce an acylium ion. The aromatic ring then functions as a nucleophile and attacks the acylium ion (we will draw the attack occurring at the *para* position because that leads to the given product). The resulting intermediate sigma complex then loses a proton to restore aromaticity.

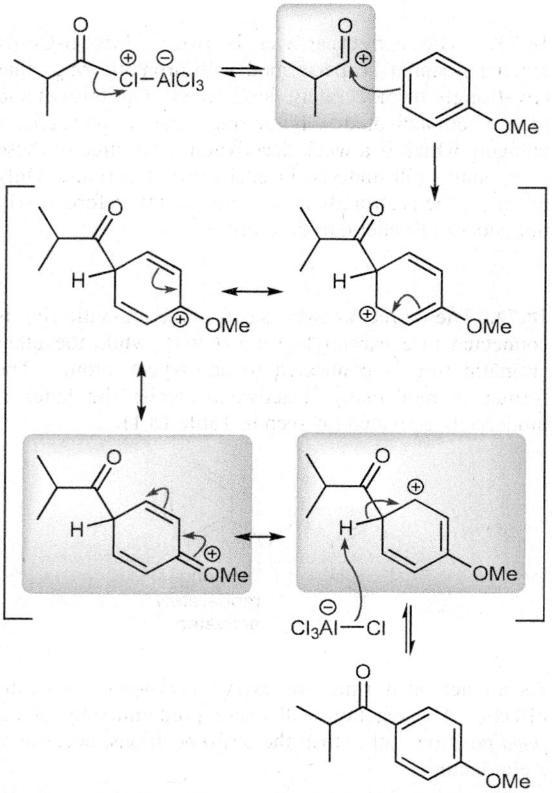

The three highlighted structures in the mechanism above were given as options in the multiple-choice problem. The following cation is not a possible intermediate because the sigma complex has a proton on the same carbon as the added electrophile, so that position cannot bear a positive charge:

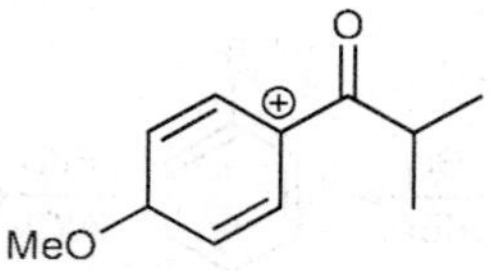

18.80.

(a) The nitroso group is a strong deactivator due to both resonance and induction. Considering induction, both the nitrogen and oxygen atoms withdraw electrons. More importantly, the pi-bond conjugated with the benzene ring withdraws electron density via resonance, moving electron density toward the highly electronegative oxygen as shown below.

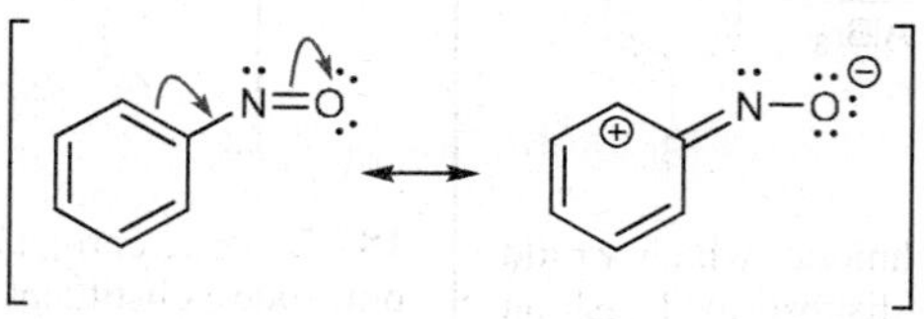

(b) Given the lone pair of electrons on the nitrogen atom, the sigma complex may be further stabilized through extra resonance when the electrophile adds to the *ortho* or *para* position. The *para*-substituted sigma-complex is shown below for a nitration reaction, with the participation of the nitrogen lone pair highlighted.

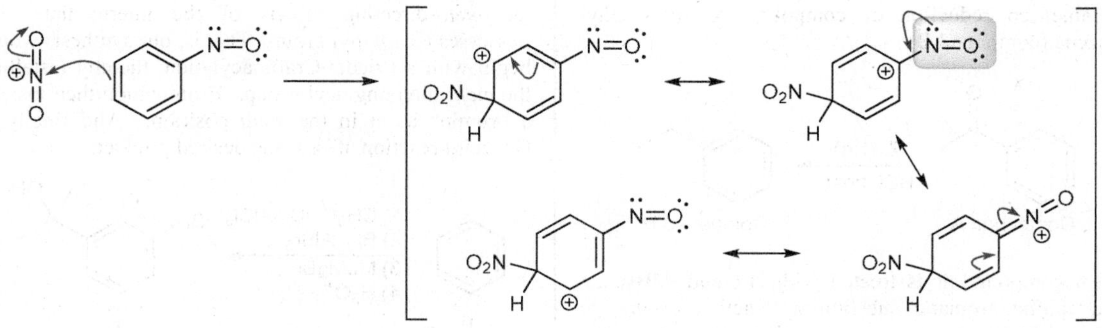

(c) Ring **A** is substituted with a nitrogen atom bearing a lone pair of electrons, and is therefore capable of donating electron density to the benzene ring via resonance. Thus, it is an activator and an *ortho-para* director. The nitrogen atom that is a substituent on ring **B** contains a positive charge along with a π bond to oxygen that strongly withdraws electrons from the benzene ring through both induction and resonance. This nitrogen atom on **B** is structurally analogous to a nitro group (positive charge on N) and would be expected to be a strong deactivator and *meta* director.

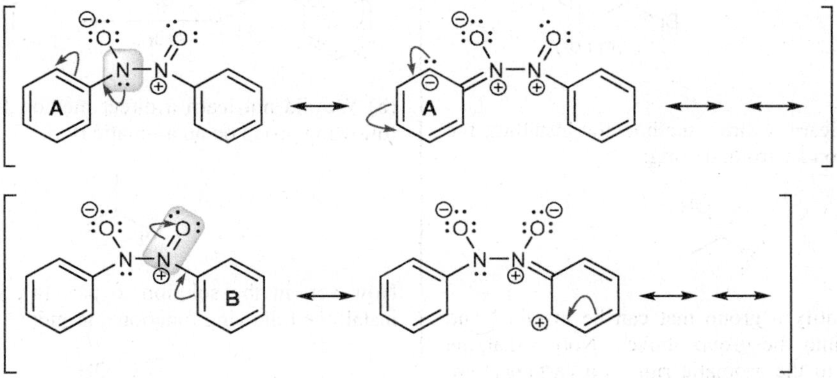

Since ring **A** is weakly activated and ring **B** is strongly deactivated, electrophilic aromatic substitution occurs on ring **A** in either the *ortho* or *para* position. Given that there is significant steric hindrance at the *ortho* position because of the large substituent, it makes sense that the major product occurs via *para* substitution on ring **A**.

18.81. The molecular formula indicates five degrees of unsaturation. One of these degrees of unsaturation must account for the signal in the double bond region of the IR spectrum (1680 cm^{-1}). The wavenumber of this signal indicates that it is likely a C=O bond that is conjugated:

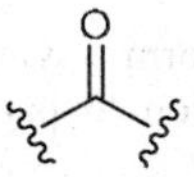

(conjugated)

In the ^{1}H NMR spectrum of compound **A**, the signals between 7.5 and 8 ppm indicate the presence of an aromatic ring, which account for the remaining four degrees of unsaturation. This signal has an integration of 5, indicating a monosubstituted aromatic ring:

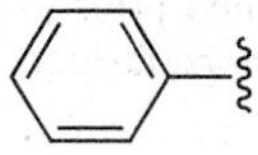

The upfield signal has an integration of 3, which indicates a methyl group:

$-CH_3$

These three fragments can only be connected in one way:

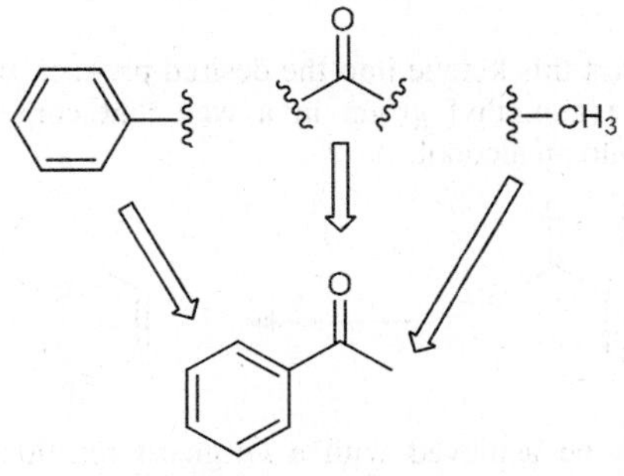

$-CH_3$

Compound **A**

Indeed, the C=O bond is conjugated in this structure. A Clemmensen reduction of compound **A** gives ethyl benzene (compound **B**):

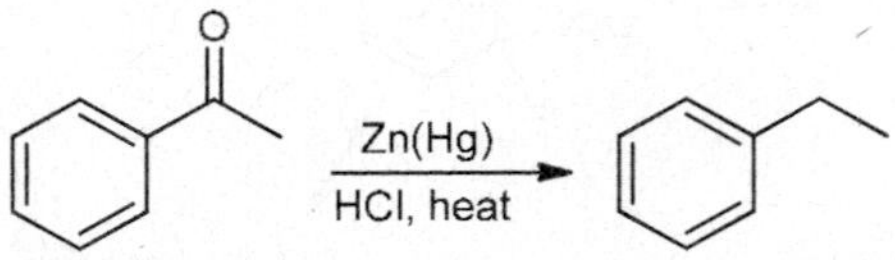

Zn(Hg) / HCl, heat

Compound **A** Compound **B**

When compound **B** is treated with Br_2 and $AlBr_3$, an electrophilic aromatic substitution reaction occurs, in which a bromine atom is installed on the ring. The reaction at the *para* position is expected to occur more rapidly because *ortho* attack is sterically hindered by the ethyl group:

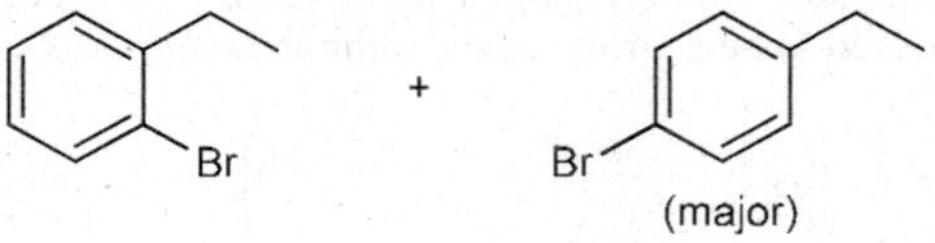

(major)

18.82.

(a) We did not learn a direct method for installing the following group on an aromatic ring:

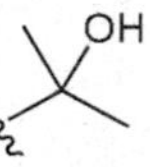

So we must identify a group that can be installed and then converted into the group above. Notice that the atom connected to the aromatic ring is a carbon atom, which means that the reaction requires the formation of a C–C bond. Therefore, we must use either a Friedel-Crafts alkylation or a Friedel-Crafts acylation. The alkylation reaction will install an alkyl group, *without* a functional group. In contrast, acylation will form a C–C bond *and* install a functional group in the desired location:

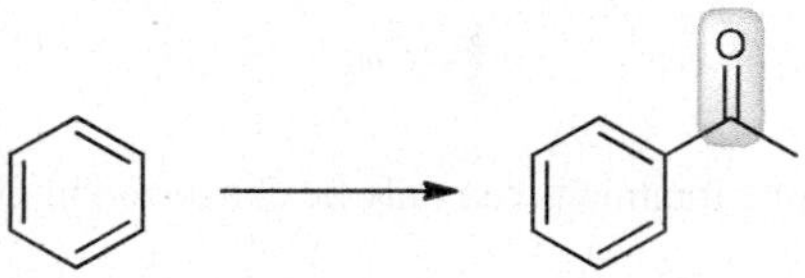

To convert this ketone into the desired product, we must introduce a methyl group in a way that converts the ketone into an alcohol:

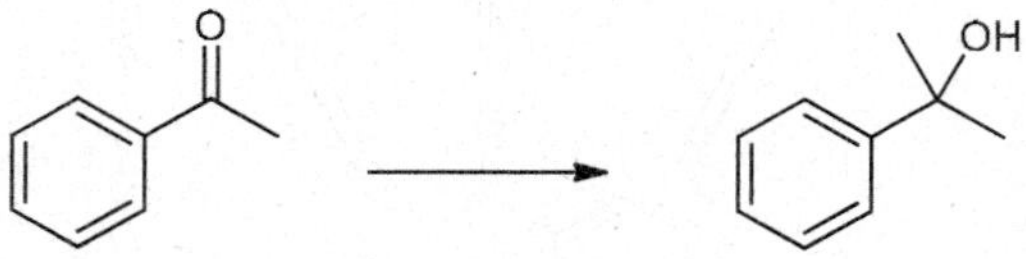

This can be achieved with a Grignard reaction, using methyl magnesium bromide.

Finally, we must consider the order of events. Notice that the product has two substituents that are *meta* to each other. This can be achieved if we take advantage of the *meta*-directing effects of the intermediate that possesses a carbonyl group. That is, our synthesis would begin with a Friedel-Crafts acylation, thereby installing the *meta*-directing acyl group. Bromination then installs a bromine atom in the *meta* position. And finally, a Grignard reaction affords the desired product:

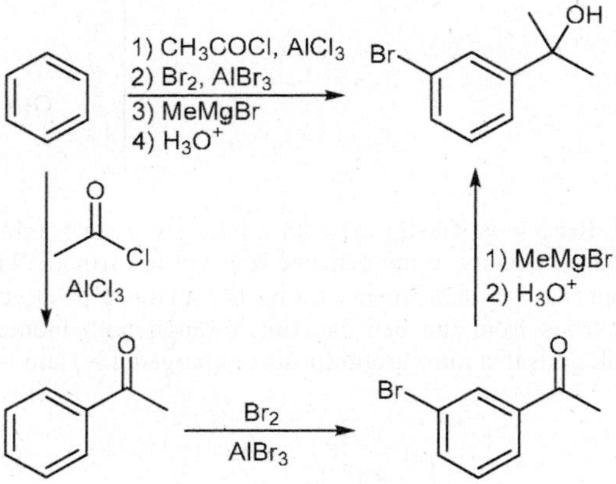

1) CH_3COCl, $AlCl_3$
2) Br_2, $AlBr_3$
3) MeMgBr
4) H_3O^+

$AlCl_3$ — Br_2 / $AlBr_3$ — 1) MeMgBr 2) H_3O^+

(b) We did not learn a direct method for installing the following group on an aromatic ring:

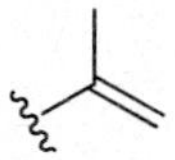

However, in the solution to part (a), we saw how to install the following functional group:

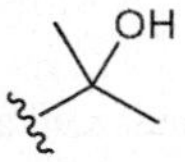

This functional group can be converted into the desired functional group in just one step – dehydration of the alcohol:

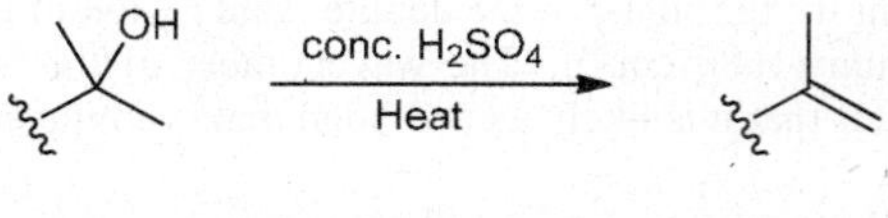

conc. H_2SO_4 / Heat

Therefore, we can perform a synthesis similar to the one described in the solution to part (a), followed by this reaction:

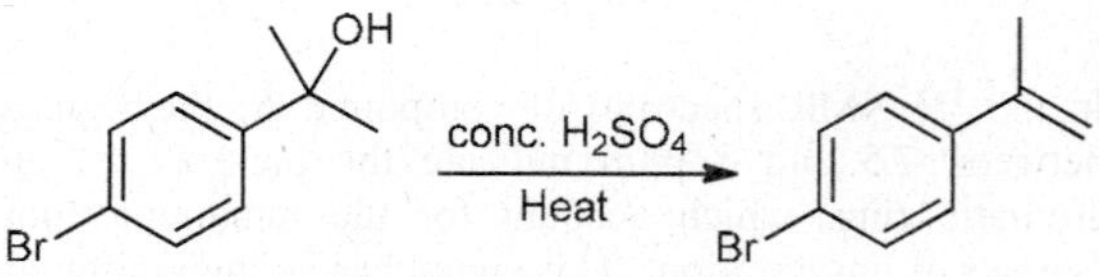

conc. H_2SO_4 / Heat

The complete synthesis is shown here. Notice that bromination is performed prior to acylation, in order to install the groups in a *para* fashion:

1) Br_2, $AlBr_3$
2) CH_3COCl, $AlCl_3$
3) MeMgBr
4) conc. H_2SO_4, heat

Br_2 $AlBr_3$

conc. H_2SO_4, heat

$AlCl_3$

MeMgBr

Alternatively, we can install the desired group in an entirely different way. Rather than performing a Friedel-Crafts acylation, we can perform a Friedel-Crafts alkylation to install an isopropyl group, which can then be functionalized via benzylic bromination, followed by elimination:

With this strategy, bromination of the aromatic ring is performed after installation of the isopropyl group, which allows us to employ the directing effects of the large isopropyl group to install the bromine atom primarily in the *para* position.
This alternative synthesis is summarized here:

1) $(CH_3)_2CHBr$, $AlBr_3$
2) Br_2, $AlBr_3$
3) NBS, heat
4) NaOEt

$(CH_3)_2CHBr$, $AlBr_3$

NaOEt

Br_2, $AlBr_3$

NBS, heat

18.83.
(a) The alkyl halide can interact with the Lewis acid to produce a carbocation:

$AlCl_3$ $-AlCl_4^{\ominus}$

The carbocation is planar, so the aromatic ring can attack either face of the carbocation, with equal likelihood. As a result, a Friedel-Crafts alkylation with this optically active alkyl halide is expected to produce a racemic pair of enantiomeric products:

(b) As explained in the solution to part (a), the reaction proceeds via a carbocation intermediate, which can be attacked from either face, leading to a racemic mixture. Therefore, the product mixture is optically inactive.

18.84.
The OH group activates the ring toward electrophilic aromatic substitution because the OH group donates electron density via resonance.

This effect gives electron density primarily to the *ortho* and *para* positions, as seen in the resonance structures above. These positions are shielded, and the protons at the *ortho* and *para* positions are expected to produce signals farther upfield (lower ppm) than protons at the *meta* position. Therefore, the *meta* protons correspond with the signal at 7.2 ppm.

18.85. The upfield signal has an integration of 3, and must correspond with the methyl group that is attached to the ring. The other signal (downfield) must correspond with aromatic protons. Since that signal has an integration of 2, the aromatic ring bears only two protons, which means that three nitro groups must have been installed.

The methyl group of toluene is an *ortho-para* director, and as such, nitration is expected to occur at the *ortho* and *para* positions. Each successive nitro group deactivates the ring, thereby requiring a higher temperature in order to install the next nitro group. At high enough temperature, all three positions will undergo nitration, giving 2,4,6-trinitrotoluene (also called TNT):

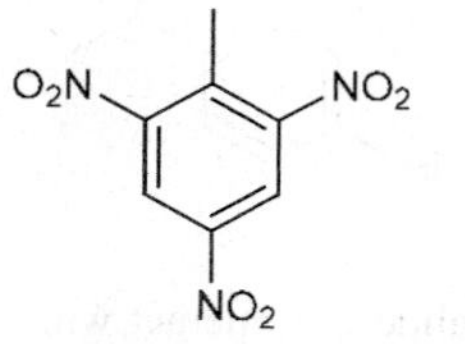

O_2N NO_2 NO_2

2,4,6-trinitrotoluene

Note that the two aromatic hydrogen atoms in the product are equivalent, consistent with the one downfield signal.

18.86.

(a) A phenyl group is an *ortho-para* director, because the sigma complex formed from *ortho* attack or *para* attack is highly stabilized by resonance (the positive charge is spread over both rings). The *ortho* position is sterically hindered while the *para* position is not, so we expect nitration to occur predominantly at the *para* position:

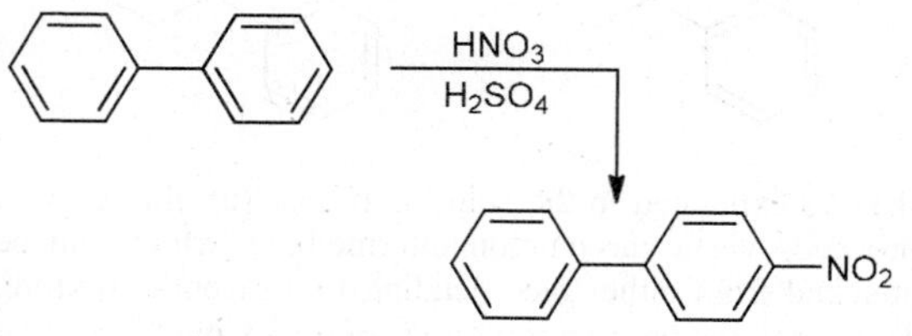

HNO_3 / H_2SO_4 → NO_2

(b) This carbonyl-bearing substituent withdraws electron density form the ring via resonance (three of the resonance structures of this compound exhibit a positive charge in the ring). As a result, this group is a moderate deactivator, and therefore a *meta* director:

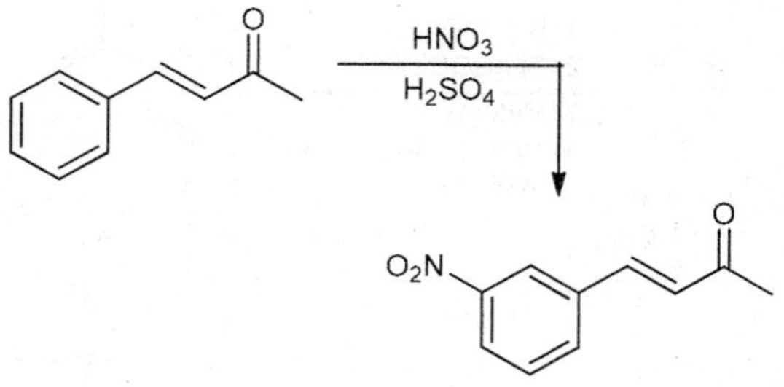

HNO_3 / H_2SO_4 → O_2N

18.87. Each of the chlorine atoms in the dichloride can interact with $AlCl_3$ and initiate a Friedel-Crafts alkylation process. During the first alkylation process, a tertiary benzylic carbocation can form via a hydride shift:

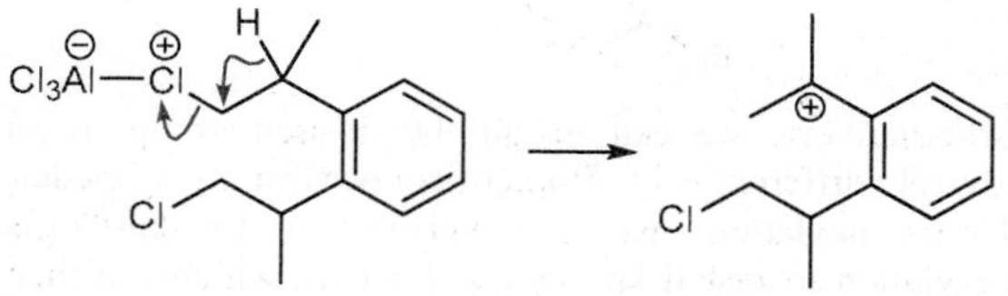

$Cl_3\overset{\ominus}{Al}—\overset{\oplus}{Cl}$ H Cl → Cl

This carbocation is then attacked by benzene in an electrophilic aromatic substitution reaction, giving the following:

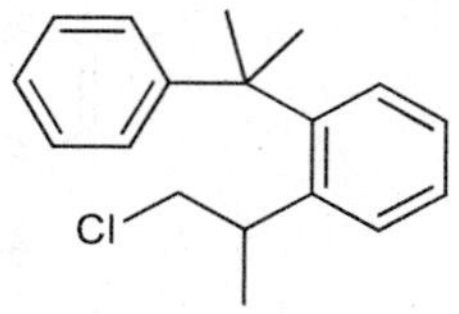

Cl

This compound still has one more chlorine atom, allowing for the second alkylation process. Once again, the second alkylation occurs in a similar fashion (via a tertiary, benzylic carbocation), but this time, the electrophilic aromatic substitution reaction proceeds in an intramolecular fashion:

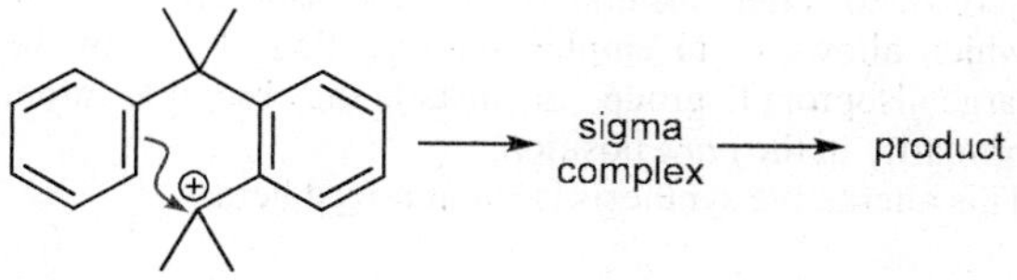

→ sigma complex → product

The following product is therefore expected:

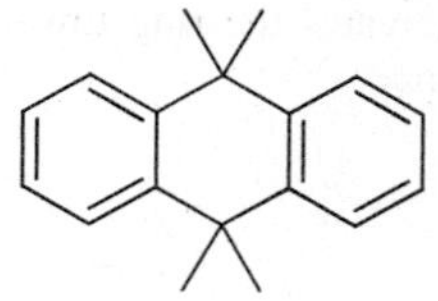

18.88. Formaldehyde is protonated to give a resonance-stabilized cation that can serve as an electrophile in an electrophilic aromatic substitution reaction. Phenol functions as the nucleophile and attacks the electrophile, giving a resonance-stabilized intermediate (sigma complex). Water can then serve as a base and remove a proton from the sigma complex, thereby restoring aromaticity:

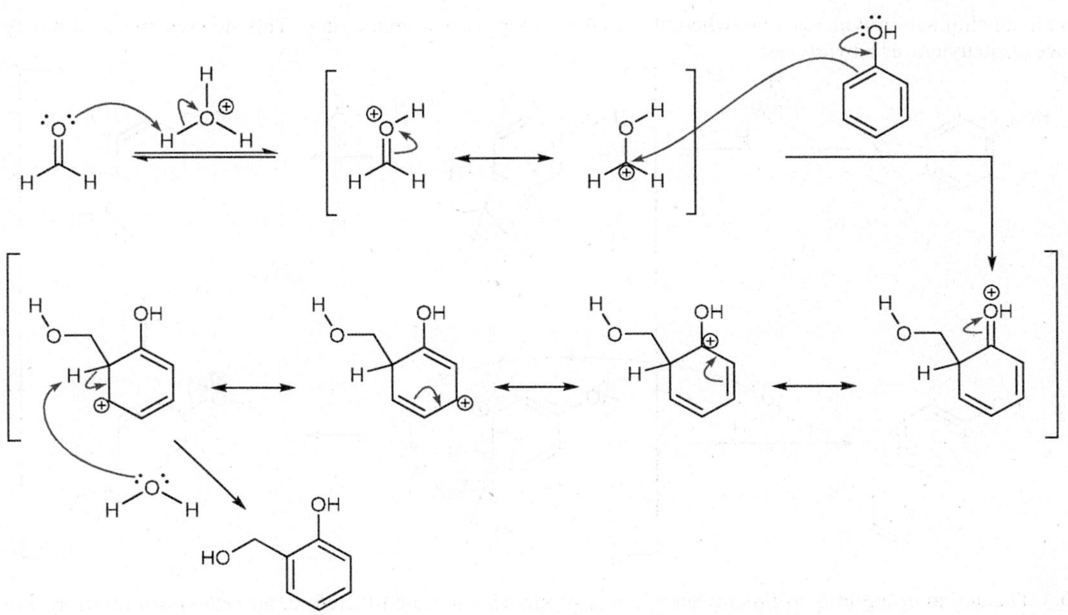

The resulting compound bears an OH group connected to a benzylic position. This OH can be protonated under acidic conditions, followed by loss of water as a leaving group to give a resonance-stabilized, benzylic carbocation. This carbocation can then serve as an electrophile in an electrophilic aromatic substitution reaction. Phenol functions as the nucleophile and attacks the electrophile, giving a resonance-stabilized intermediate (sigma complex). Water can then serve as a base and remove a proton from the sigma complex, thereby restoring aromaticity. This process continues, ultimately giving the structure of Bakelite

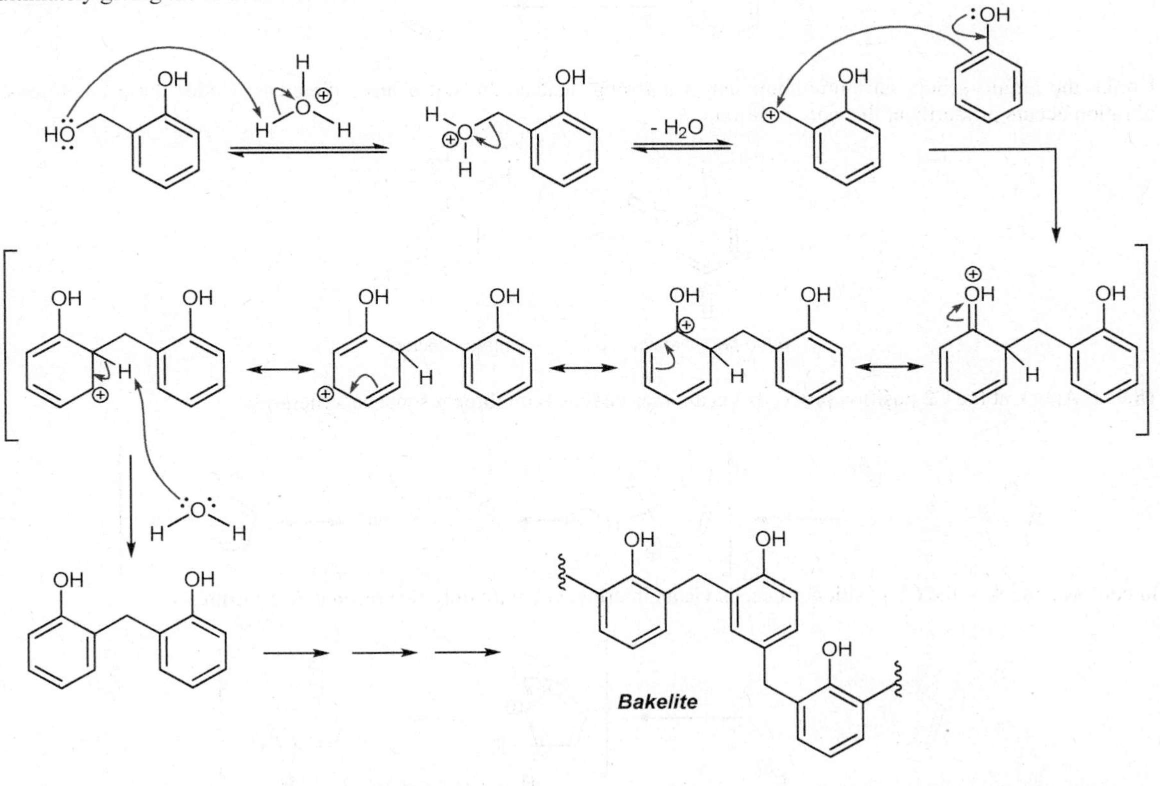

Bakelite

18.89. A *tert*-butyl group is removed from the ring and replaced with a proton, in what appears to be the reverse of an electrophilic aromatic substitution reaction. As such, we expect the reaction to proceed via an intermediate sigma complex. First, the aromatic ring is protonated to give a sigma complex. Water then functions as a base and removes a proton in an elimination-type reaction, where the leaving group is an aromatic ring. This step restores aromaticity, and produces isobutylene as a byproduct:

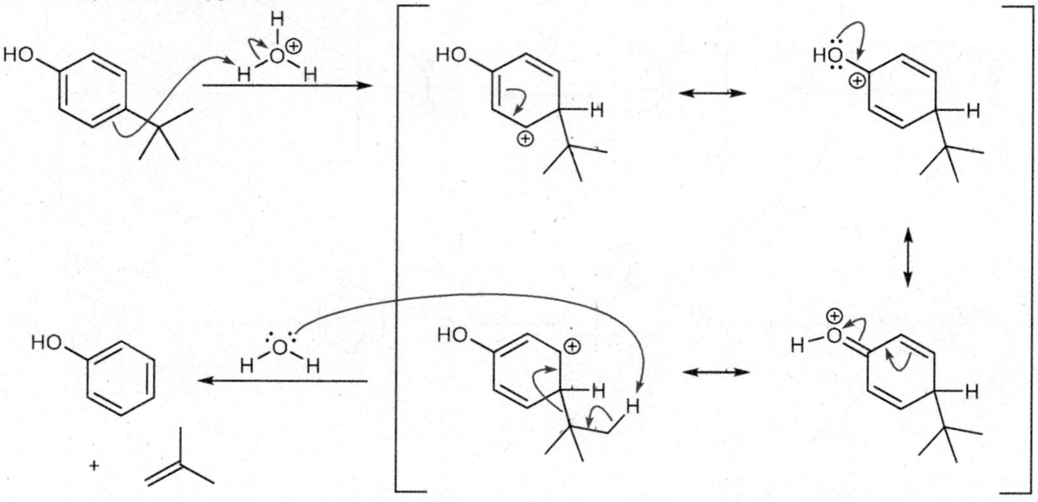

18.90. The amino group in *N,N*-dimethylaniline is a strong activator, and therefore, an *ortho-para* director. For this reason, bromination occurs at the *ortho* and *para* positions. However, in acidic conditions, the amino group is protonated to give an ammonium ion:

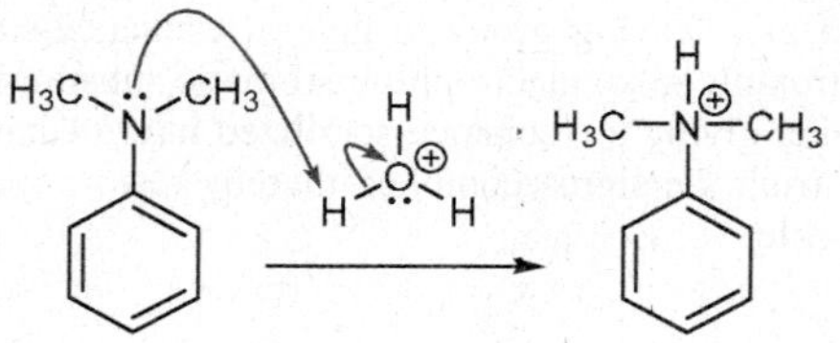

Unlike the amino group, an ammonium ion is a strong deactivator and a *meta* director. Under these conditions, nitration occurs primarily at the *meta* position

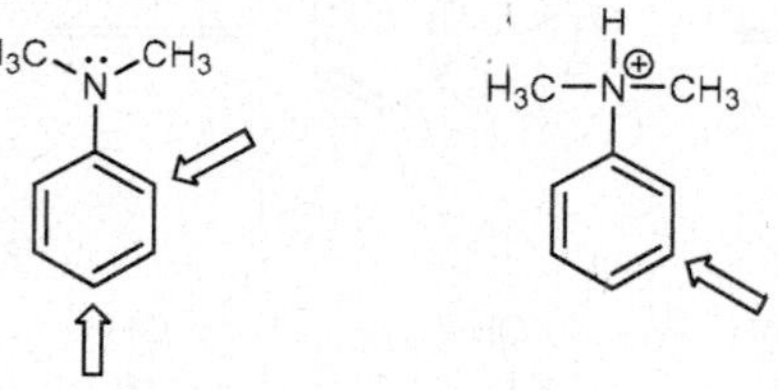

ortho-para director

meta-director

18.91. Attack at the C2 position proceeds via an intermediate with three resonance structures:

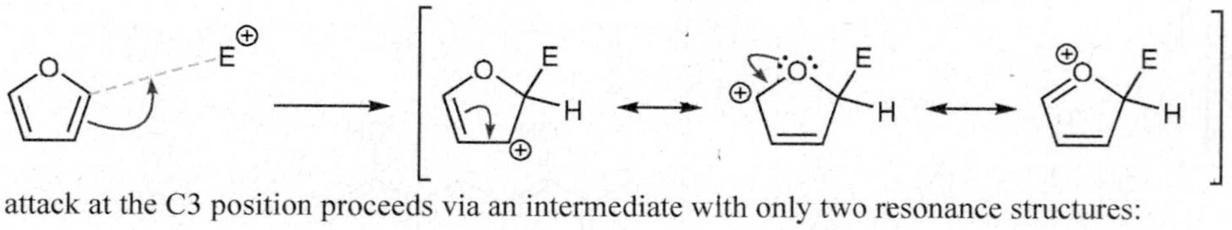

In contrast, attack at the C3 position proceeds via an intermediate with only two resonance structures:

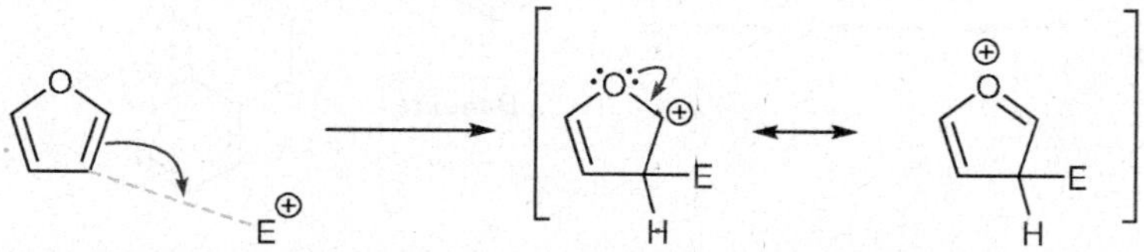

The intermediate for C2 attack is lower in energy than the intermediate for C3 attack. The transition state leading to the intermediate of C2 attack will therefore be lower in energy than the transition state leading to the intermediate of C3 attack. As a result, C2 attack occurs more rapidly.

18.92. In all cases, the X substituent is a methoxy group. It is the Y substituent that varies in identity. It is difficult to imagine that there could be a steric effect here, since the Y substituent is far away from the alkene π bond. The trend is more easily explained when we consider the electronic effect of the Y substituent. Specifically, the methoxy group (a powerful activator) can donate electron density via resonance into the alkene π bond, making it the most nucleophilic π bond among the disubstituted stilbenes. A methyl group is only weakly activating, so its effect should be smaller than the effect of a methoxy group (giving a smaller rate constant). A chloro group is a weak deactivator, which means that the net effect of the chloro group is to withdraw electron density from the aromatic ring and the conjugated alkene π bond. As a result, when Y = Cl, the rate constant is smaller than the rate constant when Y = Me. Finally, the nitro group is powerfully electron-withdrawing, via resonance, and it will significantly reduce the nucleophilicity of the alkene π bond, rendering it the least reactive double bond towards molecular bromine.

18.93.

(a) In aqueous acidic conditions, compounds **1** and **2** lose tritium via an electrophilic aromatic substitution reaction. First, the aromatic ring is protonated to give a resonance-stabilized sigma complex, which then undergoes loss of T^+ (rather than H^+) to re-establish aromaticity. Notice that the methoxy groups render the sigma complex particularly stable, because the positive charge is spread over the oxygen atoms as well (via resonance).

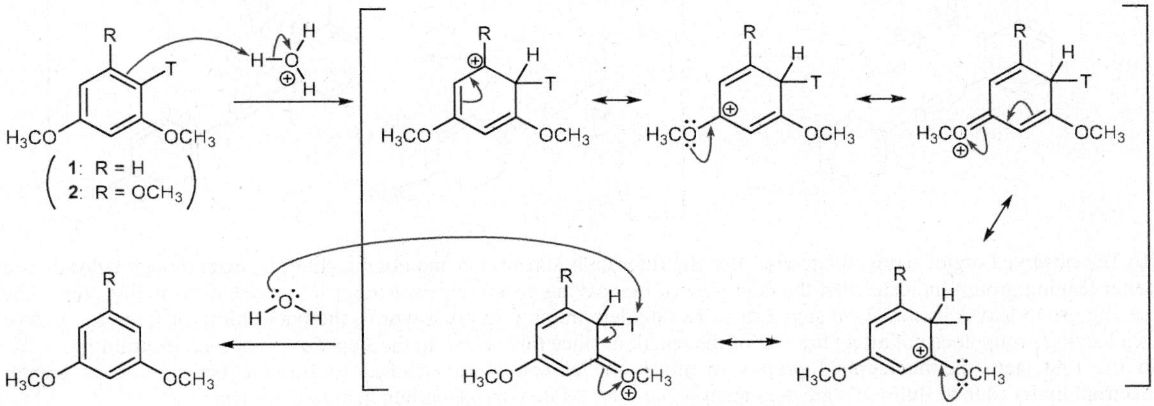

(b) Compound **2** has an extra methoxy group that is absent in compound **1**. This group further stabilizes the intermediate sigma complex by introducing one more resonance structure (highlighted below). That is, the positive charge is further delocalized:

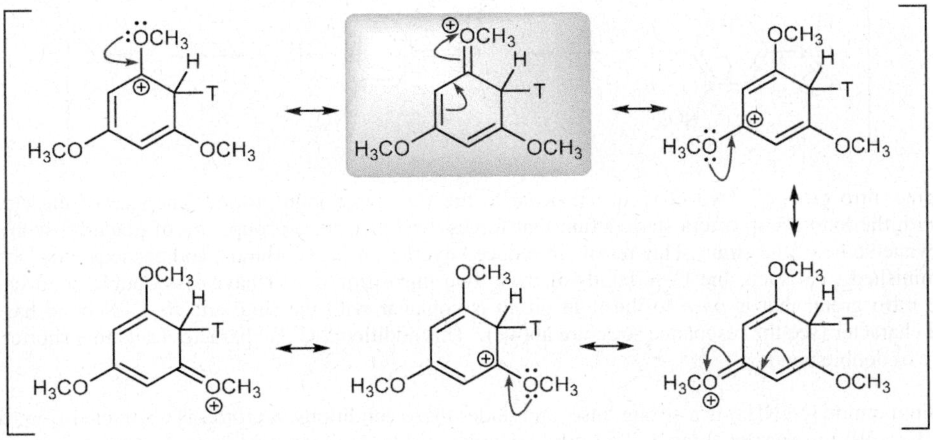

When compound **2** undergoes an electrophilic aromatic substitution reaction, the intermediate sigma complex exhibits a positive charge that is spread over three carbon atoms and three oxygen atoms. This intermediate is more stable than the intermediate formed when compound **1** undergoes an electrophilic aromatic substitution reaction. A more stable intermediate is associated with a lower-energy transition state, so it is therefore expected that compound **2** will lose tritium at a faster rate than compound **1**.

18.94.

(a) This is a reaction between a strong nucleophile (the terminal alkynide ion) and a nitro-substituted benzene ring, from which a halide is expelled. This satisfies the criteria for a nucleophilic aromatic substitution mechanism, as shown below. The alkynide ion attacks the carbon atom attached to fluorine, producing a resonance-stabilized Meisenheimer complex, which loses a fluoride leaving group to give the product.

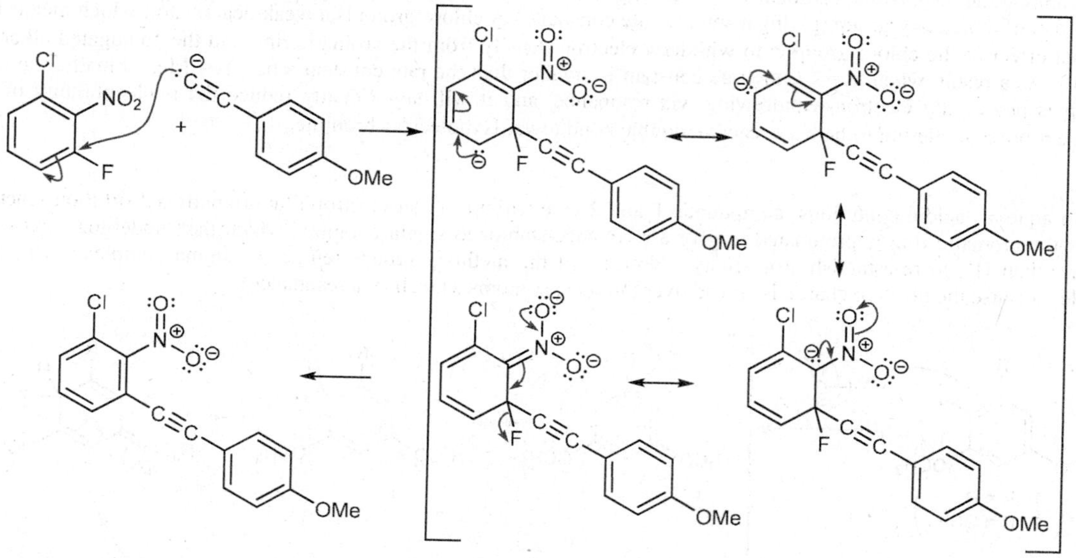

(b) The observed regiochemical outcome (that the fluoride is substituted and not the chloride, even though chloride is a better leaving group) indicates that the departure of the leaving group does not occur in the rate determining step. The leaving group leaves in step 2, so step 1 must be rate determining. In other words, the poor ability of fluoride to serve as a leaving group does not affect the rate of the reaction, since this occurs in the step that is not rate determining. In the first step, the nucleophile prefers to attack the carbon atom attached to fluorine because of its greater electrophilicity (due to fluorine's greater electronegativity) relative to the carbon next to chlorine.

18.95. Each of the three nitro groups is conjugated with the aromatic ring, and as a result, each of the C–N bonds has some double-bond character. The following resonance structure illustrates this for one of the nitro groups:

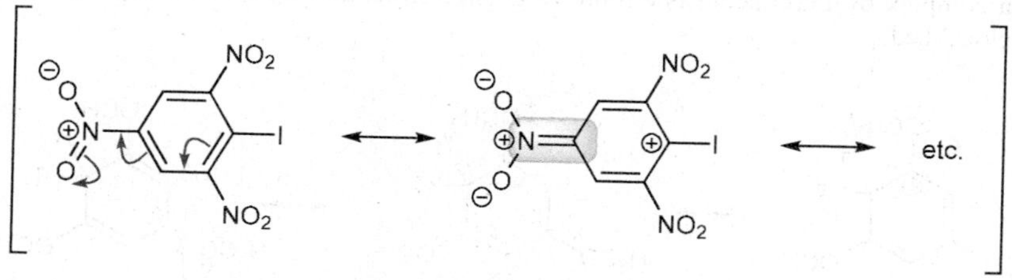

There are three nitro groups. Two of them are *ortho* to the very large iodo group. The size of the electron cloud associated with the iodo group offers steric strain that forces these two nitro groups out of planarity with the ring, in order to alleviate some of the strain. This results in reduced overlap of the *p* orbitals, and consequently, the resonance effect is diminished. As such, the C–N bonds of these two nitro groups will have less double-bond character. In contrast, the nitro group that is *para* to the iodo group is coplanar with the ring, and its C–N bond has significant double bond character (see the resonance structure above). This additional C=N character leads to a shorter bond, as is characteristic of double bonds.

18.96. Sodium amide ($NaNH_2$) is a strong base, and under these conditions, a proton is abstracted from the nitrogen atom of **1** ($pK_a \sim 30$) to generate anion **3**. This nitrogen atom will be re-protonated upon workup with water. It might seem unnecessary to draw this deprotonation step (only to protonate this same location at the end of the mechanism), but this step must be drawn to indicate that you understand that this proton is sufficiently acidic that it will not survive under these conditions (because anion **3** is resonance-stabilized). Next, deprotonation of the brominated benzene ring will produce **4**, which loses a bromide ion to form benzyne **5**. This intermediate is then trapped by the nearby aromatic ring to form anion **6**. Proton transfer will produce imine **7**, which can be rearomatized by the loss of a proton to form anion **8**. In the final step, protonation of **8** will form compound **2**. Note that the steps to convert intermediate **5** into intermediate **8** are consistent with two of the main reactions classes we have learned in this chapter: an electrophilic aromatic substitution reaction, and an elimination-addition.

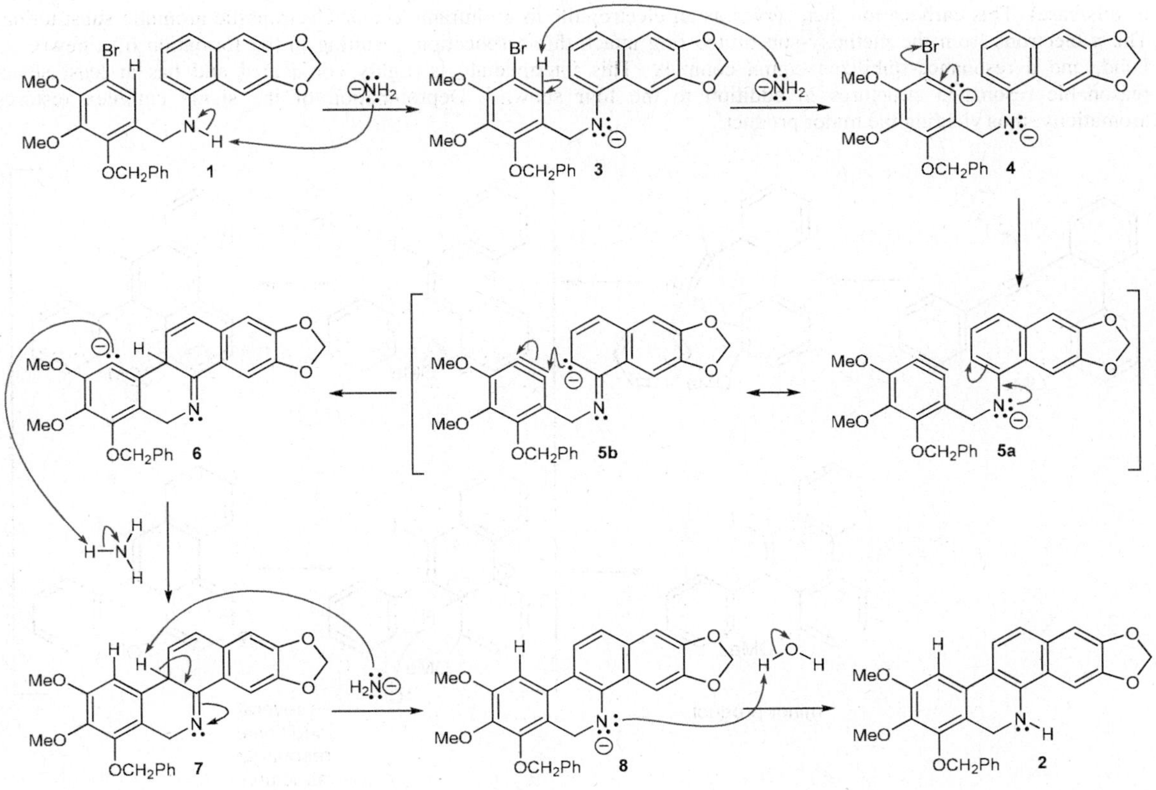

18.97.

(a) An analysis of the reactant and the minor product reveals that during the course of the reaction, a new C–C bond forms between the two carbon atoms indicated below.

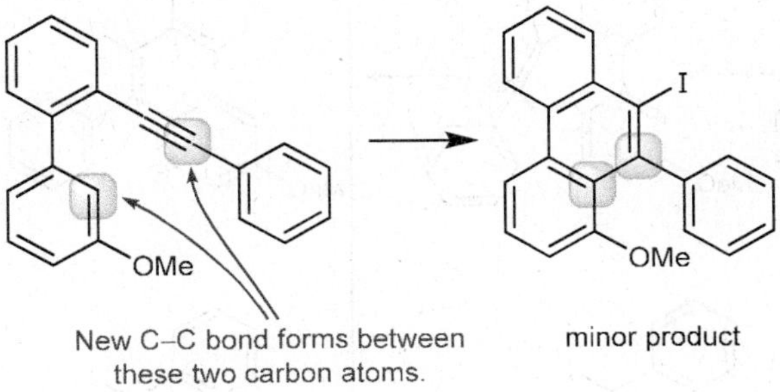

New C–C bond forms between these two carbon atoms.

minor product

Likewise, in the formation of the major product, an analogous new C–C bond forms after rotation around the C–C bond indicated below.

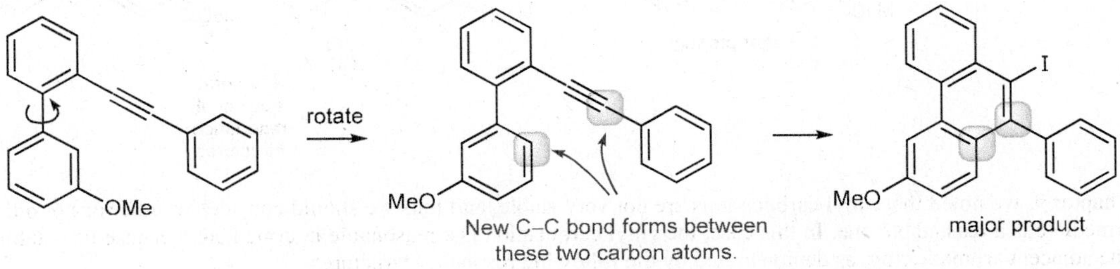

rotate

New C–C bond forms between these two carbon atoms.

major product

A mechanism for the formation of the minor product is presented below. The alkyne attacks ICl, resulting in the formation of a new C–I bond, and a vinyl carbocation (see below for a justification for formation of a vinyl carbocation in this case). This carbocation then serves as an electrophile in an intramolecular electrophilic aromatic substitution. The π electrons from the methoxy-substituted ring attack this carbocation, resulting in the formation of a new C–C bond, and a resonance-stabilized sigma complex. This intermediate is highly conjugated and has at least seven reasonable resonance structures in addition to the four shown. Deprotonation of the sigma complex restores aromaticity, thus yielding the minor product.

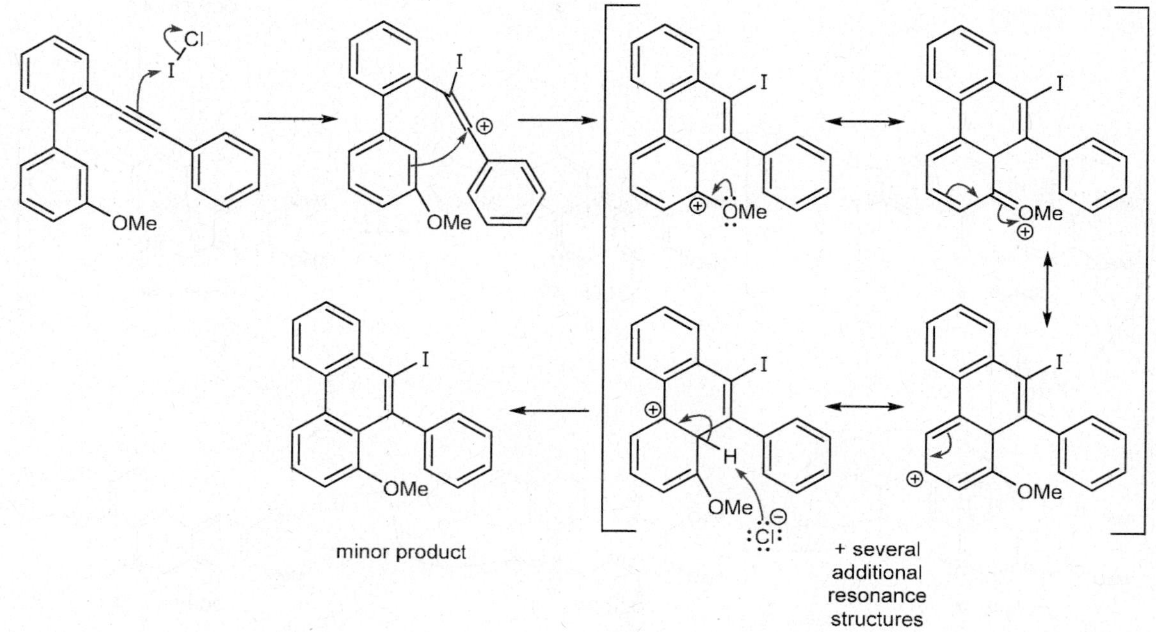

minor product

+ several additional resonance structures

A mechanism to account for the formation of the major product is given below. Each step is analogous to the mechanism presented above.

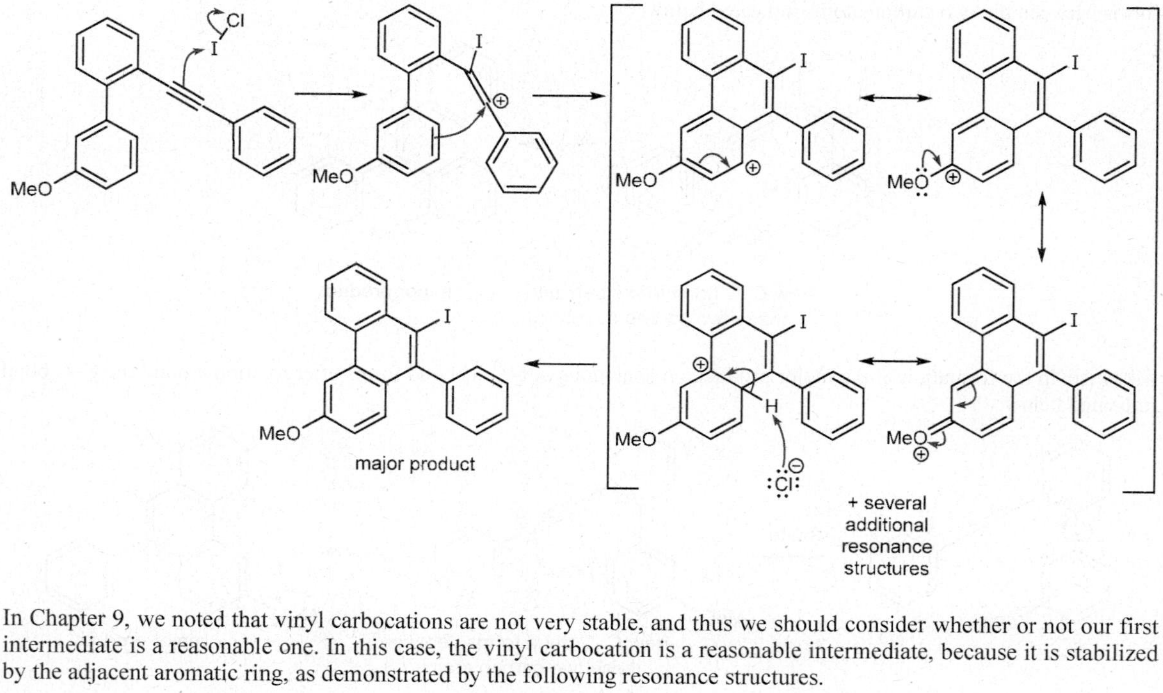

major product

+ several additional resonance structures

In Chapter 9, we noted that vinyl carbocations are not very stable, and thus we should consider whether or not our first intermediate is a reasonable one. In this case, the vinyl carbocation is a reasonable intermediate, because it is stabilized by the adjacent aromatic ring, as demonstrated by the following resonance structures.

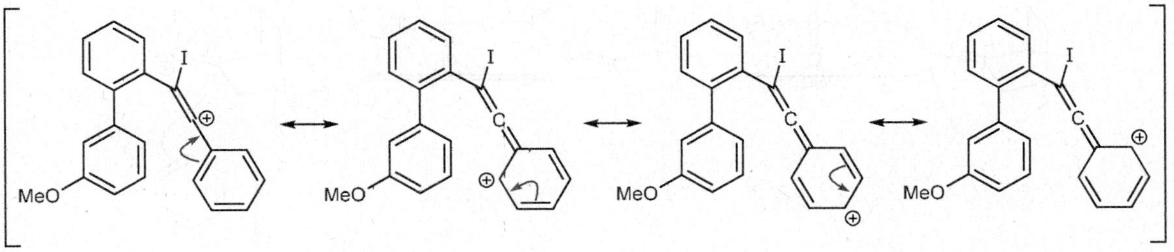

Alternatively, it would also be plausible to suggest a mechanism involving an iodonium ion, rather than a vinyl carbocation. In that case, the mechanism for formation of the major product would be as follows:

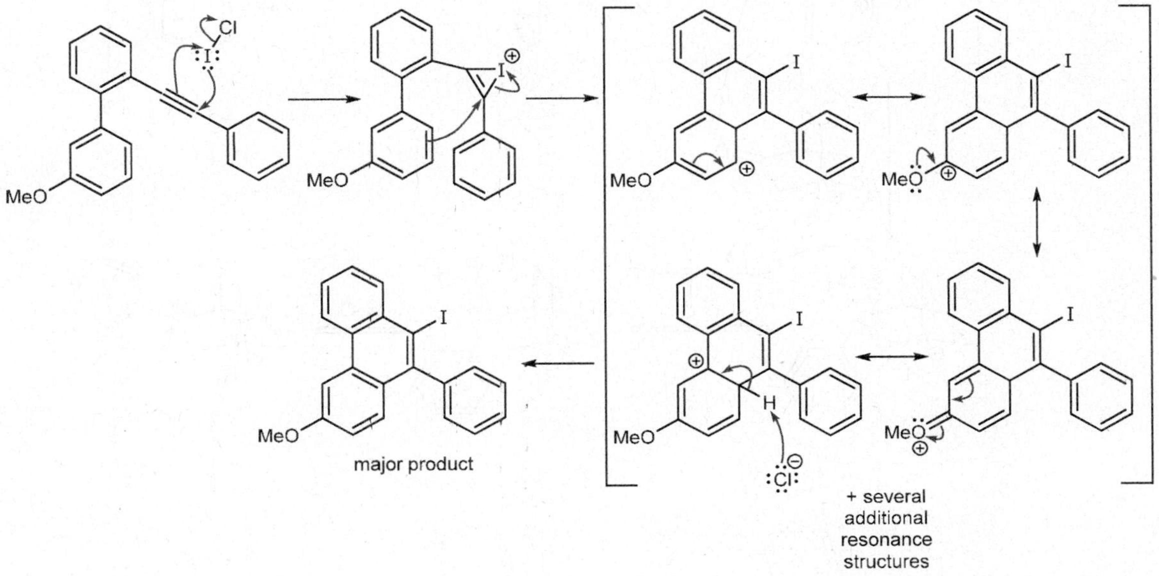

major product

+ several additional resonance structures

Formation of the minor product might also occur in this way, via an iodonium ion, rather than a vinyl carbocation.

(b) In the formation of each of the products, we should take into account both electronic and steric considerations. In the formation of the minor product, the new C–C bond is *ortho* to the methoxy group. In contrast, the new C–C bond is *para* to the methoxy in the formation of the major product. Both of these products are thus consistent with the methoxy group serving as an *ortho-para* director. The predominance of the major product can be explained by the increased steric accessibility of electrophilic carbocation to the position *para* to the methoxy group.

18.98. The first steps of the mechanism are two successive proton transfers to form the dication **2**. In each of these steps, triflic acid is the source of the proton. The amine is protonated first, to give an ammonium ion. If we compare the pK_a values of triflic acid (–14) and a typical ammonium ion (~10), we see that this proton transfer step is effectively irreversible, thus the irreversible reaction arrow for that step. Then, another proton transfer step generates the dication. Notice that protonation of the carbonyl group occurs as the second step, *after* protonation of the amine, because a protonated carbonyl group is a highly acidic species (p$K_a \approx -7$), relative to an ammonium ion.

Once the highly electrophilic dication intermediate is formed, an electrophilic aromatic substitution reaction occurs, in which benzene functions as the nucleophile. The resulting intermediate sigma complex is deprotonated to restore aromaticity. Under these acidic conditions, the hydroxyl group is protonated, generating an excellent leaving group (water), which leaves to give a resonance-stabilized benzylic carbocation (resonance structures not shown). This benzylic carbocation then functions as an electrophile in a second electrophilic aromatic substitution reaction. Finally, the resulting sigma complex is then deprotonated to restore aromaticity, thereby producing ammonium ion **3**.

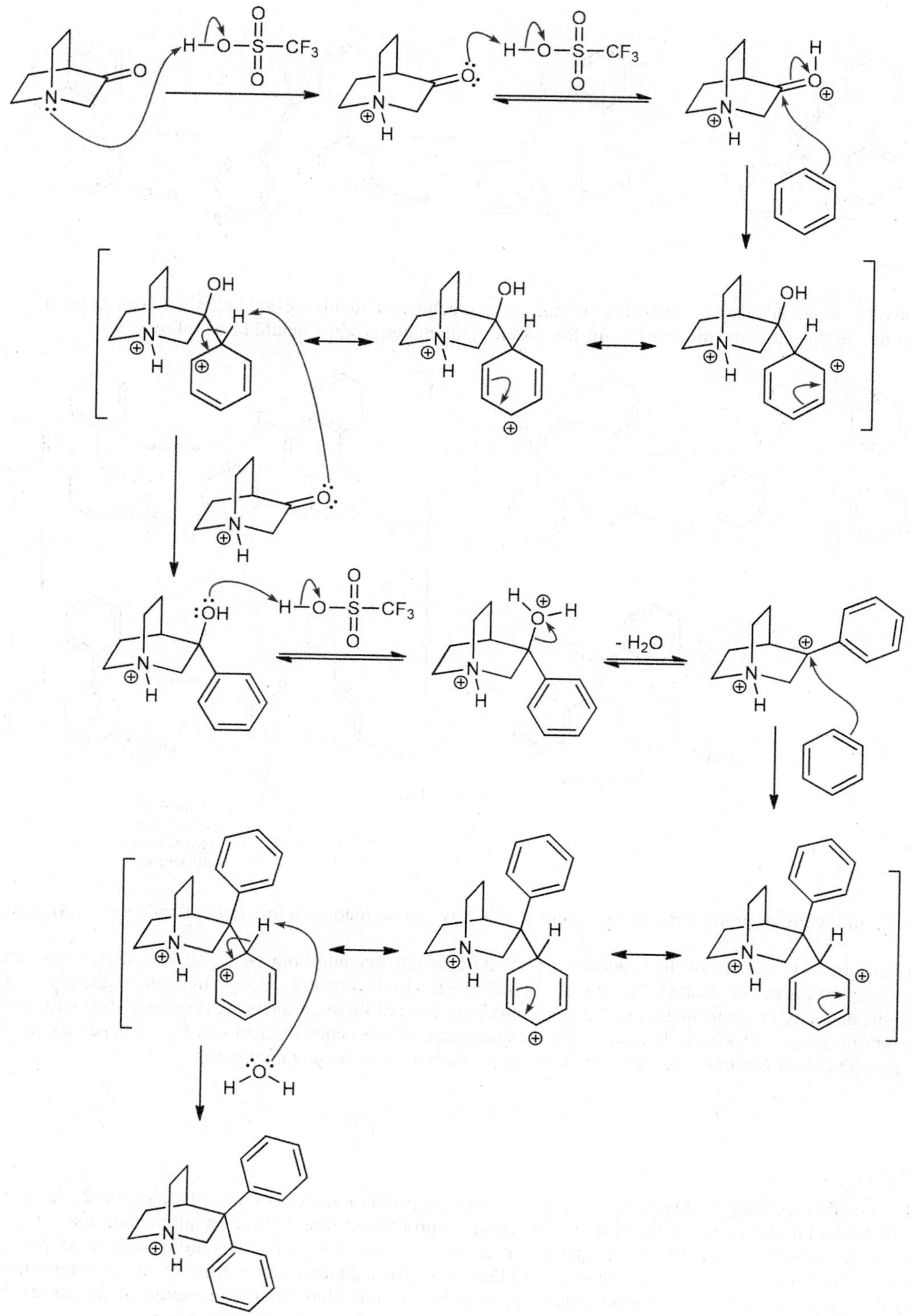

- H_2O

3

18.99.

(a) A reasonable mechanism for the conversion of **1** to **2** is shown. In the first step, *n*-butyllithium functions as a base and deprotonates the amine to produce a resonance-stabilized anion (resonance structures not shown). Boron trichloride serves as a Lewis acid in the next step, accepting an electron pair from the nucleophilic nitrogen atom, giving the Lewis acid – Lewis base complex. Subsequent loss of a chloride ion gives the compound **2**.

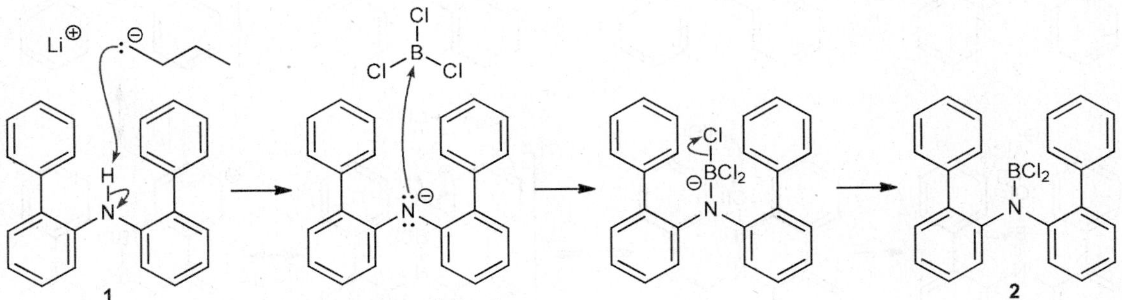

Li⊕ :⊖ H N Cl B Cl Cl N: ⊖ ⊖ Cl BCl2 N BCl2 N

1 resonance-stabilized intermediate 2

A mechanism for the final step is shown below. Overall, this step constitutes two subsequent electrophilic aromatic substitution reactions, resulting in the formation of two new B-C bonds. Addition of aluminum trichloride, a Lewis acid, serves to enhance the electrophilicity of the boron by forming a complex with one of the chlorine atoms. The boron is then attacked by one of the pendant arenes, with subsequent loss of $AlCl_4^-$ and formation of a resonance-stabilized sigma complex (several, but not all, of the many resonance structures are shown). Deprotonation with $AlCl_4^-$ restores aromaticity, giving a neutral intermediate. An identical sequence of mechanistic steps (complexation with $AlCl_3$, attack of the arene, loss of $AlCl_4^-$ and deprotonation) results in the formation of the second B-C bond, thus producing compound **3**.

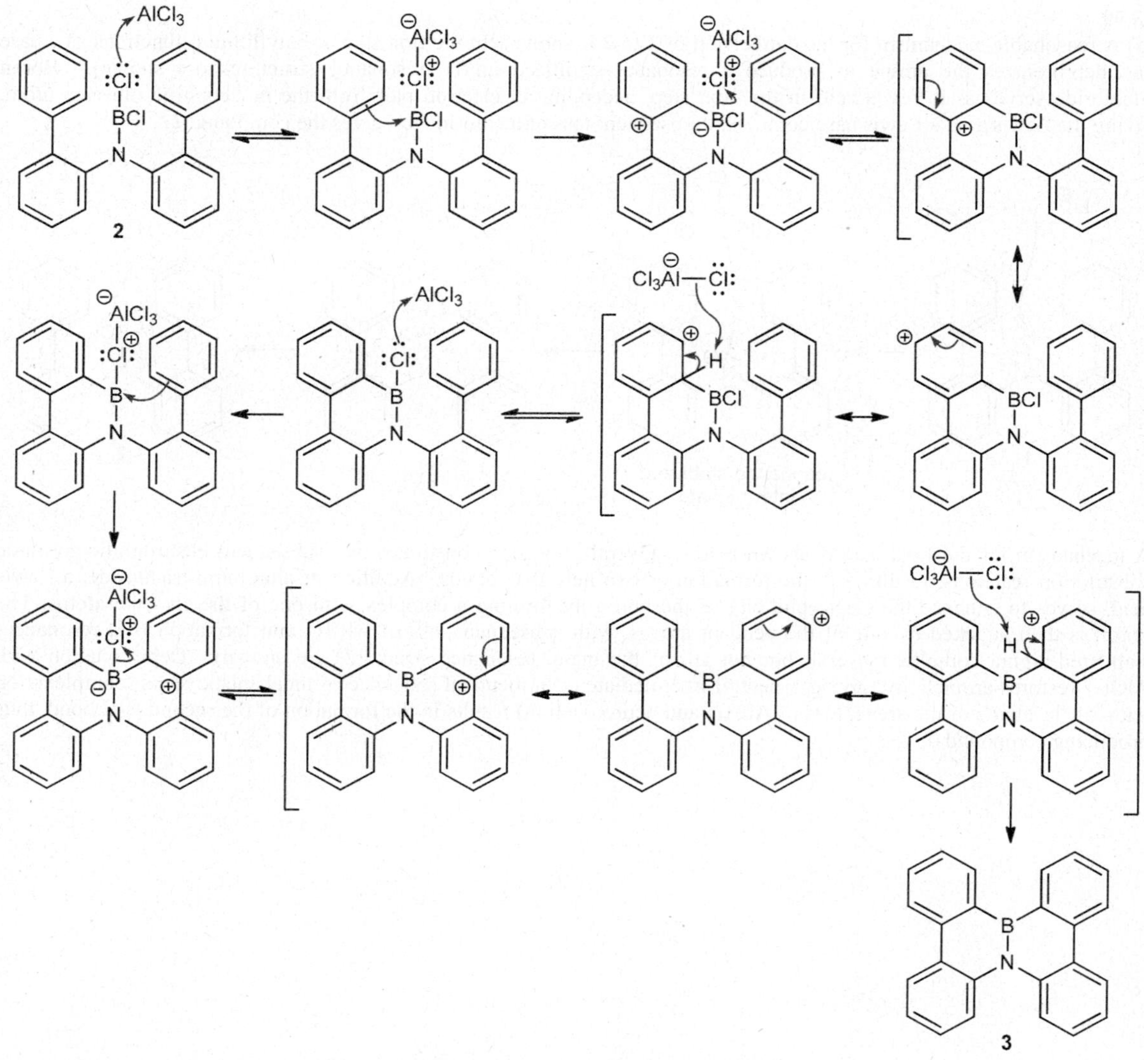

(b) Each of the four peripheral rings is clearly aromatic.

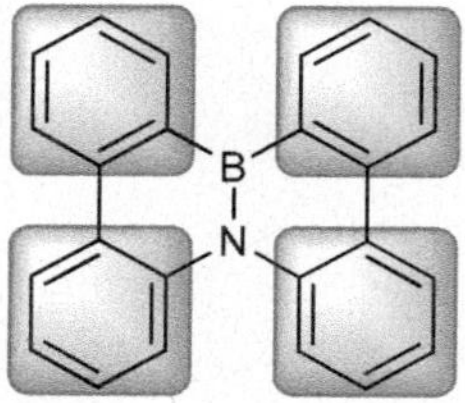

For the two central rings, the boron and nitrogen are both expected to be sp^2-hybridized, with boron having an empty *p*-orbital and nitrogen holding its lone pair in a *p*-orbital. As such, each central ring exhibits a continuous system of overlapping *p* orbitals with a total of 6 π electrons, thus fulfilling the criteria for aromaticity:

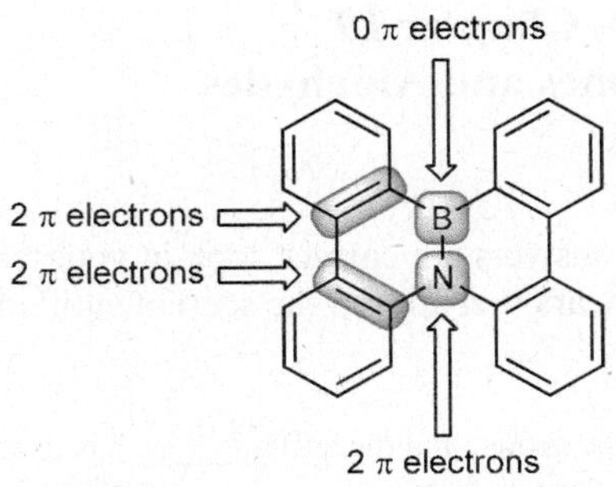

0 π electrons
2 π electrons
2 π electrons
2 π electrons

Therefore, all six rings are aromatic, providing for one extended aromatic system that is electronically similar to the following polycyclic aromatic compound:

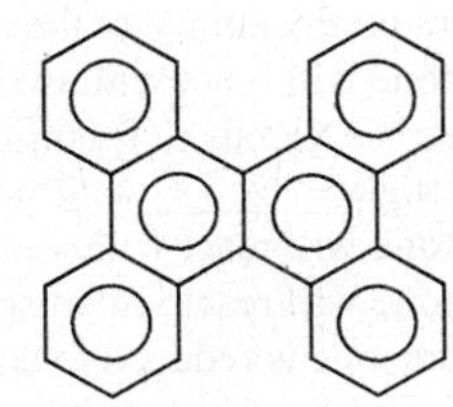

(c) In part b, we determined that we expect both central rings in **3** to be aromatic, and as such, we would expect the entire molecule to be planar. The observation that the central two rings are twisted (non-planar) indicates that the central rings actually have limited aromatic character, despite the fact that they have the right number of π electrons to be aromatic. There must be something causing the rings to twist out of planarity. This can be explained by the steric repulsion between the indicated hydrogen atoms shown in the structures below. A slightly twisted structure avoids this repulsive interaction.

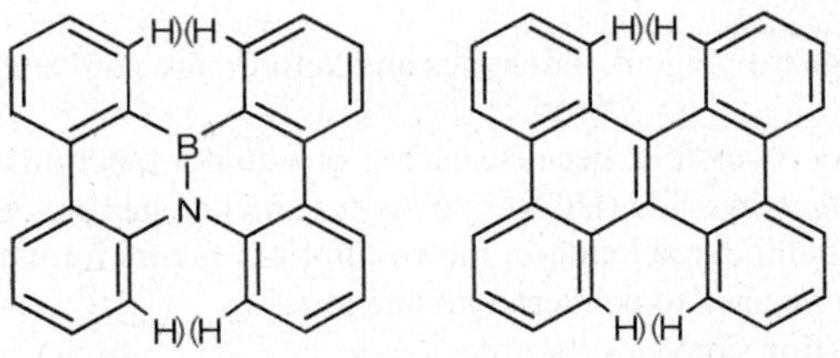

This steric interaction is very similar to the steric interaction in diphenylmethane, causing it to adopt a non-coplanar conformation, as seen in the BioLinks application at the end of section 17.5 (The Development of Non-Sedating Antihistamines).

Chapter 19
Ketones and Aldehydes

Review of Concepts

Fill in the blanks below. To verify that your answers are correct, look in your textbook at the end of Chapter 19. Each of the sentences below appears verbatim in the section entitled *Review of Concepts and Vocabulary*.

- The suffix "_____" indicates an aldehydic group, and the suffix "_____" is used for ketones.
- The electrophilicity of a carbonyl group derives from _____________ effects, as well as __________ effects.
- A general mechanism for nucleophilic addition under basic conditions involves two steps:
    1) nucleophilic attack
    2) _______________
- The position of equilibrium is dependent on the ability of the nucleophile to function as a ______________.
- In acidic conditions, an aldehyde or ketone will react with two molecules of alcohol to form an __________.
- The reversibility of acetal formation enables acetals to function as _______________ groups for ketones or aldehydes. Acetals are stable under strongly _______ conditions.
- In acidic conditions, an aldehyde or ketone will react with a primary amine to form an _________.
- In acidic conditions, an aldehyde or ketone will react with a secondary amine to form an __________.
- In the **Wolff-Kishner reduction**, a hydrazone is reduced to an _________ under strongly basic conditions.
- ______________ of acetals, imines, and enamines under acidic conditions produces ketones or aldehydes.
- In acidic conditions, an aldehyde or ketone will react with two equivalents of a thiol to form a ________.
- When treated with Raney nickel, thioacetals undergo **desulfurization** to yield a ____________ group.
- When treated with a hydride reducing agent, such as lithium aluminum hydride ($LiAlH_4$) or sodium borohydride ($NaBH_4$), aldehydes and ketones are reduced to ______________.
- The reduction of a carbonyl group with $LiAlH_4$ or $NaBH_4$ is not a reversible process, because hydride does not function as a _______________.
- When treated with a Grignard reagent, aldehydes and ketones are converted into alcohols, accompanied by the formation of a new __________ bond.
- Grignard reactions are not reversible, because carbanions do not function as ______________.
- When treated with hydrogen cyanide (HCN), aldehydes and ketones are converted into _____________. For most aldehydes and unhindered ketones, the equilibrium favors formation of the ________________.
- The **Wittig reaction** can be used to convert a ketone to an _________.
- A **Baeyer-Villiger oxidation** converts a ketone to an __________ by inserting __________________ next to the carbonyl group. Cyclic ketones produce cyclic esters, called ______________.

Review of Skills

Fill in the blanks and empty boxes below. To verify that your answers are correct, look in your textbook at the end of Chapter 19. The answers appear in the section entitled *SkillBuilder Review*.

19.1 Naming Aldehydes and Ketones

Provide a systematic name for the following compound:

O

1) Identify the parent
2) Identify and name substituents
3) Assign locants to each substituent
4) Alphabetize
5) Assign configuration

19.2 Drawing the Mechanism of Acetal Formation

Draw a mechanism for the acid-catalyzed conversion of a ketone to a hemiacetal. Make sure to draw all curved arrows and intermediates.

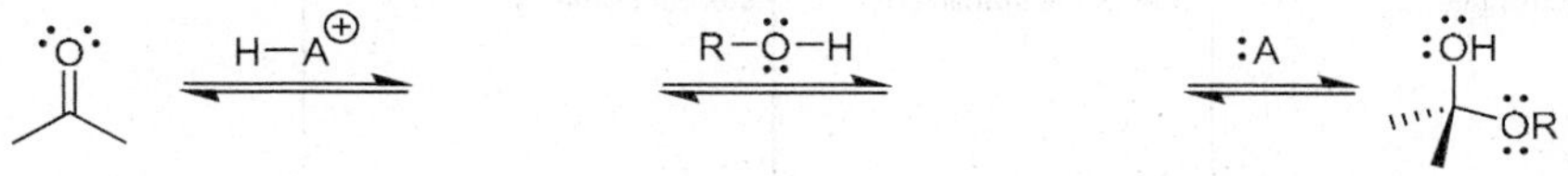

H–A⊕ ⇌ R–Ö–H ⇌ :A ⇌

Draw a mechanism for the acid-catalyzed conversion of a hemiacetal to an acetal. Make sure to draw all curved arrows and intermediates.

H–A⊕ ⇌ $-H_2O$ ⇌ R–Ö–H ⇌ :A ⇌

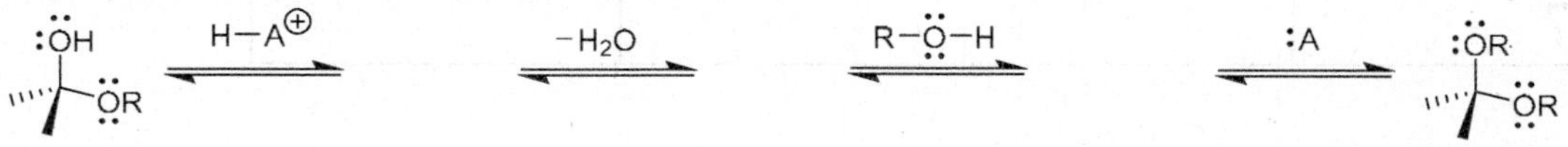

19.3 Drawing the Mechanism of Imine Formation

Draw a mechanism for the acid-catalyzed conversion of a ketone to a carbinolamine. Make sure to draw all curved arrows and intermediates.

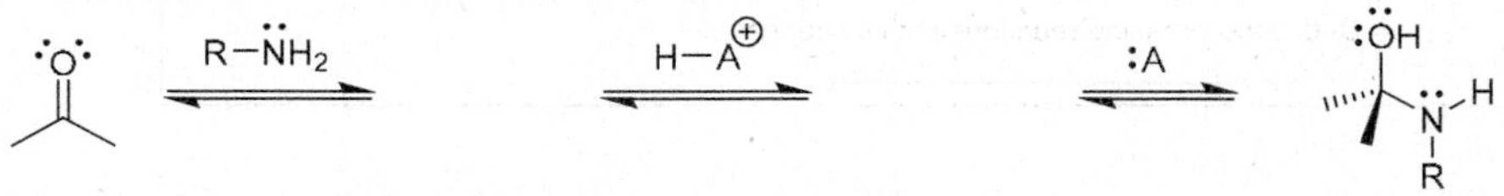

$R-NH_2$ ⇌ H–A⊕ ⇌ :A ⇌

Draw a mechanism for the acid-catalyzed conversion of a carbinolamine to an imine. Make sure to draw all curved arrows and intermediates.

H–A⊕ ⇌ $-H_2O$ ⇌ $R-NH_2$ ⇌

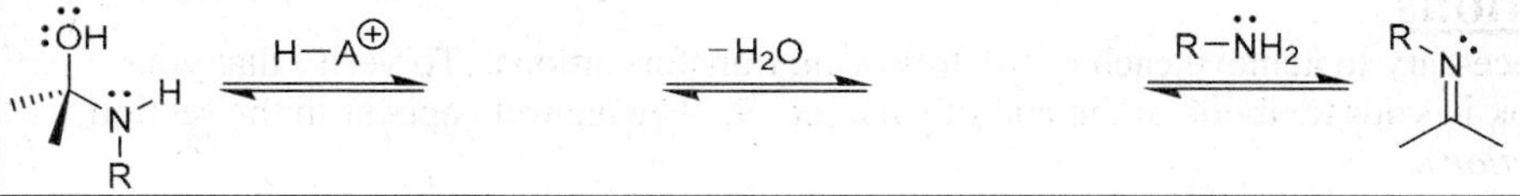

19.4 Drawing the Mechanism of Enamine Formation

Draw a mechanism for the acid-catalyzed conversion of a ketone to a carbinolamine. Make sure to draw all curved arrows and intermediates.

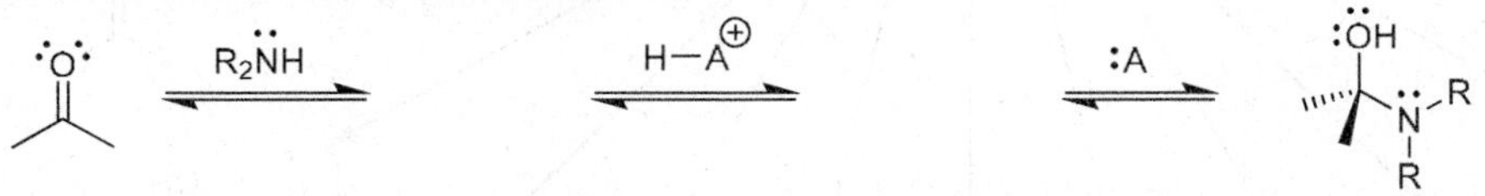

R_2NH ⇌ H–A⊕ ⇌ :A ⇌

Draw a mechanism for the acid-catalyzed conversion of a carbinolamine to an enamine. Make sure to draw all curved arrows and intermediates.

H–A⊕ ⇌ $-H_2O$ ⇌ R_2NH ⇌

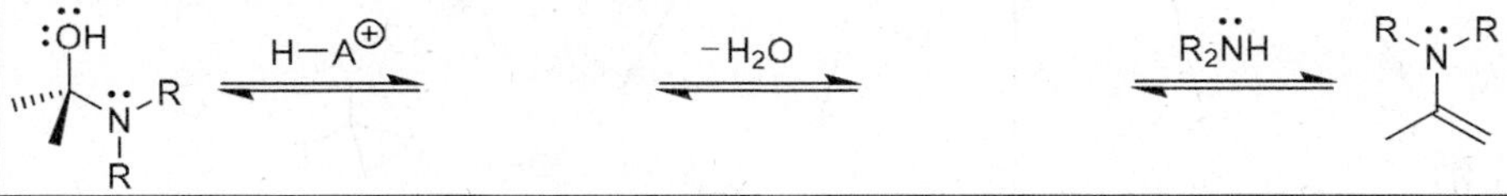

19.5 Drawing the Products of a Hydrolysis Reaction

Draw the products expected when the following compound is treated with aqueous acid. 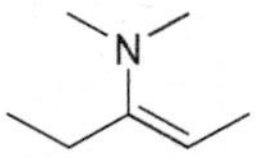	**Step 1** Identify the bond(s) expected to undergo cleavage.	**Step 2** Identify the carbon atom that will become a carbonyl group.	**Step 3** Determine the identity of the other fragment(s).

19.6 Predicting the Major Product of a Wittig or HWE Reaction

Example	**Step 1** Identify the nucleophile and electrophile.	**Step 2** Determine whether the Wittig reagent is stabilized.	**Step 3** Draw the alkene product with the appropriate stereochemistry.
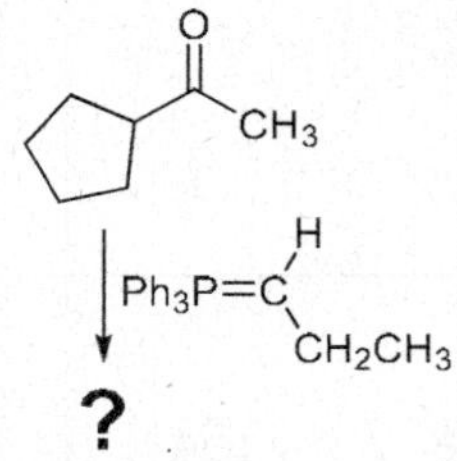 O; CH_3; H; $Ph_3P{=}C$; CH_2CH_3; ?			

19.7 Proposing a Synthesis

Begin by asking the following two questions: 1) Is there a change in the ________? 2) Is there a change in the ________?	Perform a retrosynthetic analysis, considering all of the C–C bond-forming reactions and all of the C–C bond-breaking reactions that you have learned so far. C–C bond-forming reactions in this chapter: - ________ - ________ - ________ C–C bond-breaking reactions in this chapter: - ________

Review of Reactions

Identify the reagents necessary to achieve each of the following transformations. To verify that your answers are correct, look in your textbook at the end of Chapter 19. The answers appear in the section entitled *Review of Reactions*.

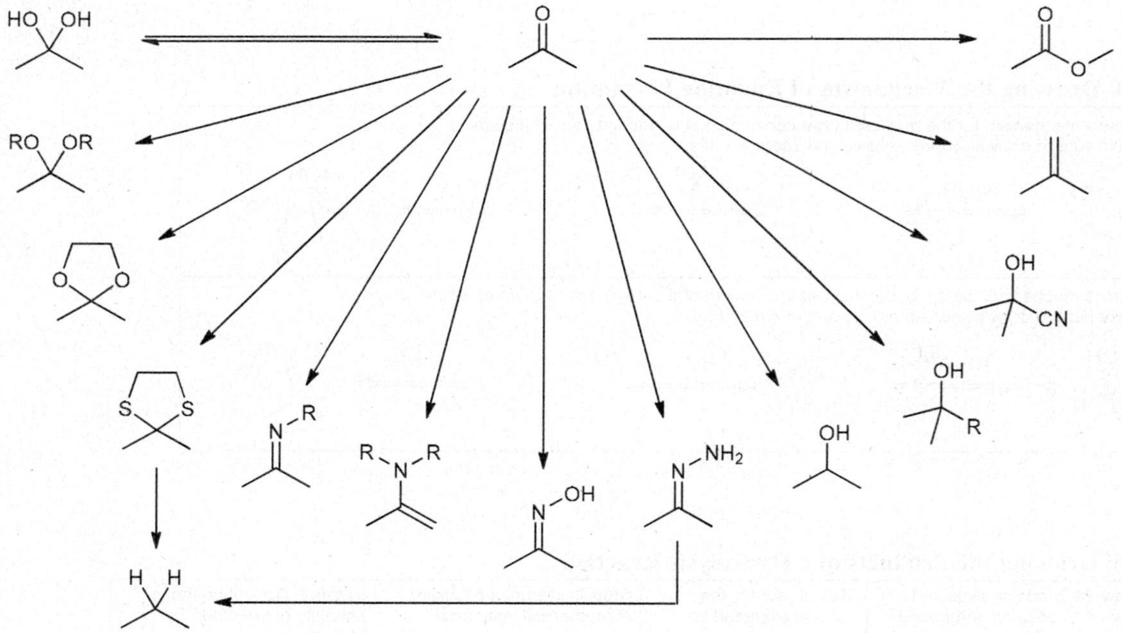

HO OH

O

O

RO OR

O O

OH

CN

OH

S S

R

N R

R R

N

OH

N OH

N NH_2

H H

Review of Mechanisms

Complete each of the following mechanisms by drawing the missing curved arrows. To verify that your curved arrows are drawn correctly, compare them to the curved arrows in the mechanism boxes for Mechanisms 19.1 – 19.14, which can be found throughout Chapter 19 of your text.

Mechanism 19.1 Nucleophilic Addition under Basic Conditions

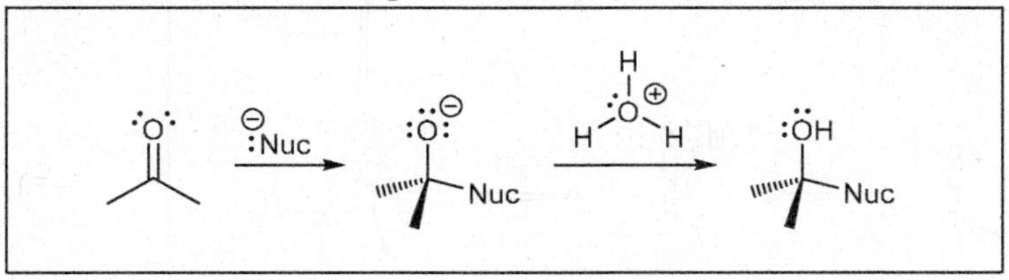

:Nuc⊖ → ; H_3O^+ →

Mechanism 19.2 Nucleophilic Addition under Acidic Conditions

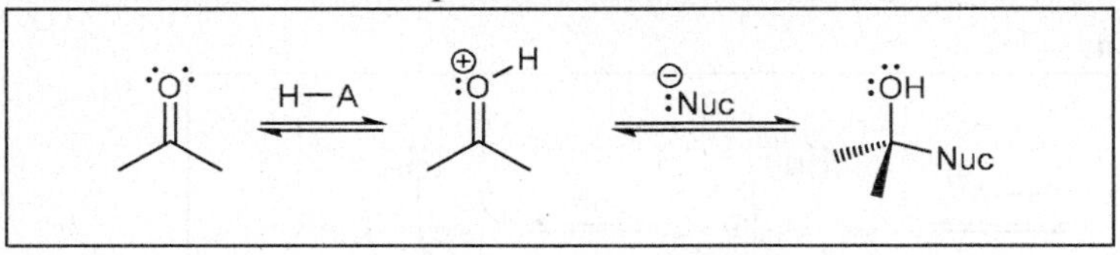

H—A ⇌ ; :Nuc⊖ ⇌

Mechanism 19.3 Base-Catalyzed Hydration

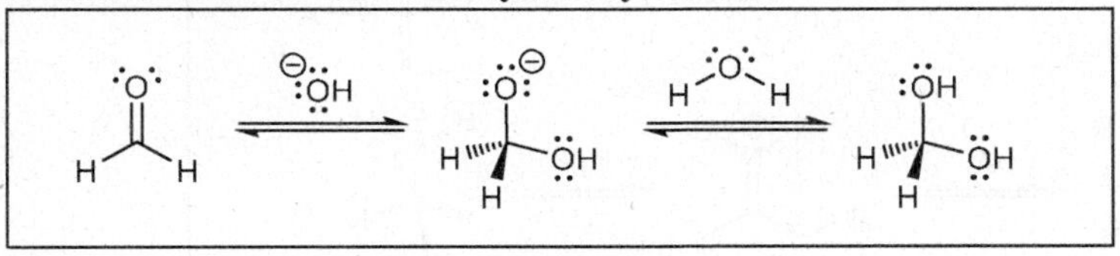

:OH⊖ ⇌ ; H_2O ⇌

Mechanism 19.4 Acid-Catalyzed Hydration

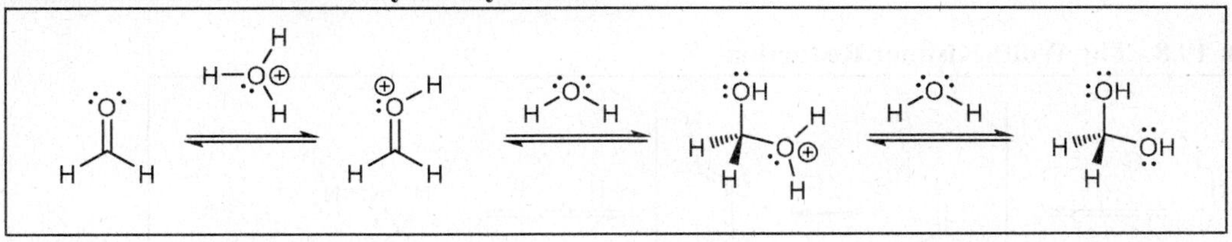

H_3O^+ ⇌ ; H_2O ⇌ ; H_2O ⇌

Mechanism 19.5 Acetal Formation

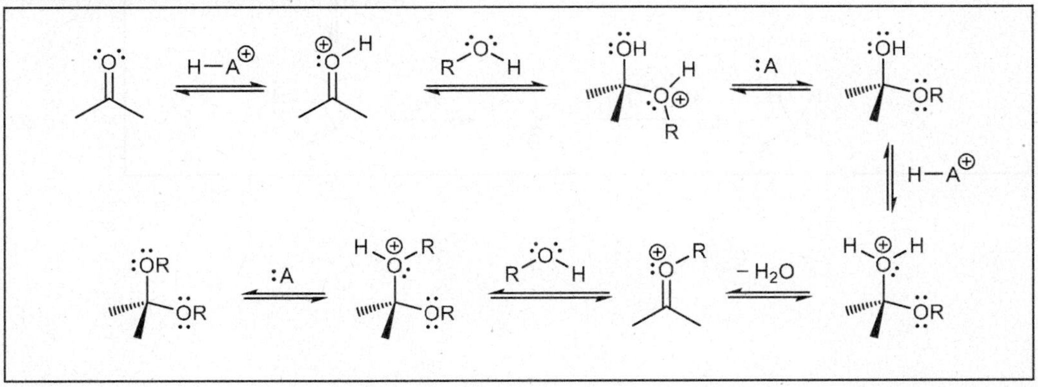

H—A⊕ ⇌ ; ROH ⇌ ; :A ⇌ ; H—A⊕ ⇌ ; – H_2O ⇌ ; ROH ⇌ ; :A ⇌

Mechanism 19.6 Imine Formation

H–A⊕ :A H–A⊕ – H_2O :A

Mechanism 19.7 Enamine Formation

H–A⊕ :A H–A⊕ – H_2O :A

Mechanism 19.8 The Wolff-Kishner Reduction

⊖:OH H–O–H ⊖:OH – N_2 H–O–H

Mechanism 19.9 Hydrolysis of Acetals

RÖ: :ÖR H–O⊕ H H RÖ: :O⊕–R H – ROH R–O⊕: H–O–H RÖ: :O⊕–H H H–O–H

RÖ: :ÖH H–O⊕–H H ⊕O: :ÖH R H – ROH H–O⊕: H–O–H :O:

Mechanism 19.10 The Reduction of Ketones or Aldehydes with Hydride Agents

:O: R R H–Al⊖–H (H, H) → :O:⊖ R H R → (H–O⊕–H, H) → :ÖH R H R

Mechanism 19.11 The Reaction Between a Grignard Reagent and a Ketone or Aldehyde

:O: R R :R⊖ → :O:⊖ R R R → (H–O⊕–H, H) → :ÖH R R R

Mechanism 19.12 Cyanohydrin Formation

:O: ⇌ (⊖:C≡N:) :O:⊖ CN ⇌ (H–C≡N:) :ÖH CN

Mechanism 19.13 The Wittig Reaction

:O: → (Ph–P⊕(Ph)(Ph)–C⊖:(H)(H)) → [Ph Ph Ph P O C H H] → H H + O=P(Ph)(Ph)Ph

Mechanism 19.14 The Baeyer-Villiger Oxidation

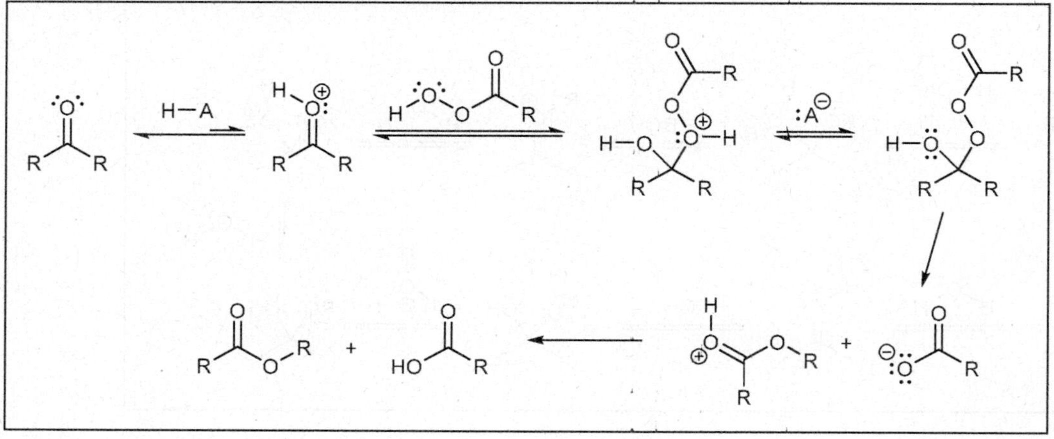

Common Mistakes to Avoid

This chapter covers many reactions, including the formation of acetals:

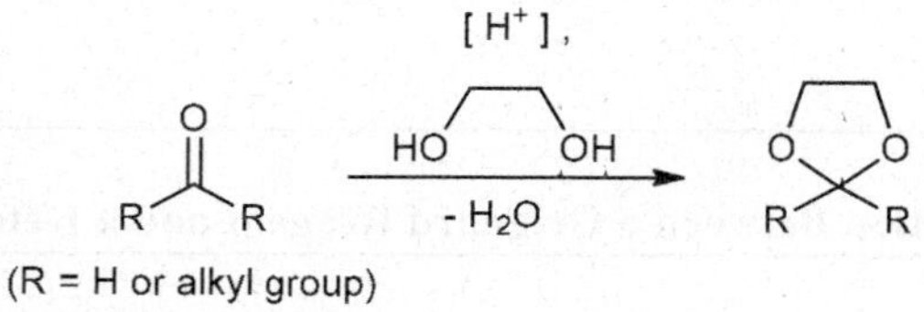

(R = H or alkyl group)

This reaction is covered in the context of the reactivity of ketones and aldehydes. The same reaction does not occur for other compounds containing a carbonyl group, such as esters:

an ester

Students often make this mistake, assuming that acetal formation will work for esters, just as it does for ketones. It does not. This is a common mistake, and it should be avoided.

As another example of a reaction that is often applied in the wrong context, consider the Wittig reaction.

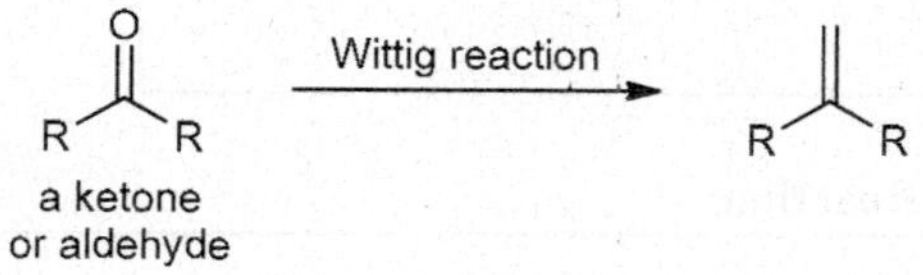

Wittig reaction

a ketone or aldehyde

Once again, Chapter 19 covers this reaction in the context of the reactivity of ketones and aldehydes. You cannot assume that the same reaction will occur for other compounds containing a carbonyl group, such as esters:

an ester

Students often make this type of mistake, by applying a reaction outside of the scope in which it was discussed. Try to avoid doing this. Whenever we cover a reaction that applies to ketones and aldehydes, you cannot assume that it will apply to esters (or any other functional group, for that matter).

Useful reagents

The following is a list of reagents encountered in this chapter:

Reagents	Type of Reaction	Description
H_3O^+	Hydrolysis of acetals imines, or enamines	Treating an acetal, an imine, or an enamine with H_3O^+ generates a carbonyl group via hydrolysis.
$[H^+]$, 2 ROH, ($-$ H_2O)	Acetal formation	Conversion of an aldehyde or ketone into an acetal. The acetal group can be used to protect aldehydes and ketones. The acetal group is stable to basic conditions but is removed when subjected to aqueous acidic conditions to regenerate the carbonyl group (a process called hydrolysis).
$[H^+]$, $HOCH_2CH_2OH$, $-$ H_2O	Cyclic acetal formation	Ethylene glycol can be used to convert an aldehyde or ketone into an acetal. The acetal group can be used to protect aldehydes and ketones. The acetal group is stable to basic conditions but is removed when subjected to aqueous acidic conditions to regenerate the carbonyl group (a process called hydrolysis).
$[H^+]$, $HSCH_2CH_2SH$, ($-$ H_2O)	Cyclic thioacetal formation	Ethylene thioglycol can be used to convert an aldehyde or ketone into a cyclic thioacetal.
Raney nickel	Desulfurization	Converts a thioacetal (or cyclic thioacetal) to an alkane.
$[H^+]$, RNH_2, ($-$ H_2O)	Imine formation	A primary amine (or ammonia) converts an aldehyde or ketone into an imine. The imine group is removed when subjected to aqueous acidic conditions to regenerate the carbonyl group (a process called hydrolysis).
$[H^+]$, R_2NH, ($-$ H_2O)	Enamine formation	A secondary amine converts an aldehyde or ketone into an enamine. The enamine group is removed when subjected to aqueous acidic conditions to regenerate the carbonyl group (a process called hydrolysis).
$[H^+]$, NH_2OH, ($-$ H_2O)	Oxime formation	Converts an aldehyde or ketone into an oxime.
$[H^+]$, NH_2NH_2, ($-$ H_2O)	Hydrazone formation	Converts an aldehyde or ketone into a hydrazone.
NaOH, H_2O, heat	Wolff-Kishner reduction	Reduces a hydrazone to an alkane.

1) $LiAlH_4$ 2) H_3O^+	Reduction	Reduces an aldehyde or ketone to an alcohol.
1) RMgBr 2) H_3O^+	Grignard reaction	When an aldehyde or ketone is treated with a Grignard reagent (followed by aqueous acidic workup), a carbon-carbon bond-forming reaction occurs, giving an alcohol that exhibits the newly formed C–C bond.
HCN, KCN	Cyanohydrin formation	Converts an aldehyde or ketone into a cyanohydrin.
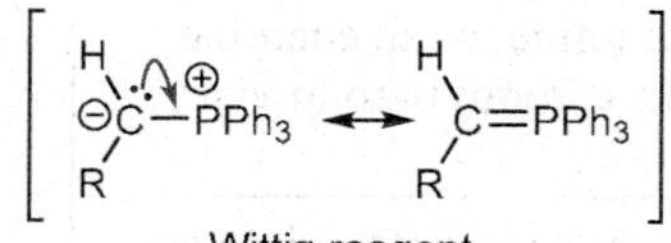 Wittig reagent	Wittig reaction	When an aldehyde or ketone is treated with a Wittig reagent, a carbon-carbon bond-forming reaction occurs, giving an alkene that exhibits the newly formed C=C double bond in the location of the former carbonyl group. If R is a simple alkyl group, such as methyl or ethyl, then the (*Z*) alkene is generally observed to be the major product.
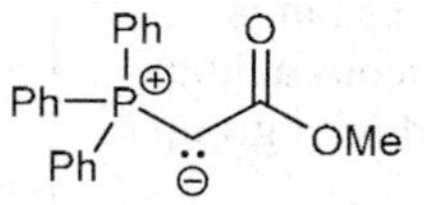	Wittig reaction	A stabilized Wittig reagent that will react with an aldehyde or ketone to give the (*E*) alkene as the major product.
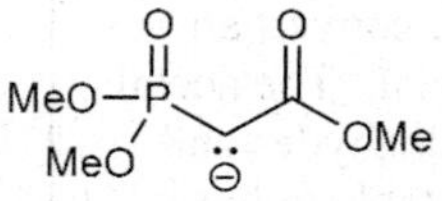	HWE reaction (Horner-Wadsworth-Emmons)	A HWE reagent functions like a stabilized Wittig reagent. It will react with an aldehyde or ketone to give the (*E*) alkene as the major product.
RCO_3H	Baeyer-Villiger oxidation	Converts a ketone into an ester (via insertion of an oxygen atom). Converts an aldehyde into a carboxylic acid (via insertion of an oxygen atom).

Solutions

19.1.

(a) We begin by identifying the longest chain that includes the carbon atom of the carbonyl group. In this case, the parent is a chain of six carbon atoms, with the carbonyl group at C1, so the parent is hexanal. Next, we identify the substituents. There are four: two methyl groups (both at C2) and two bromine atoms (both at C5). Finally, we assemble the substituents alphabetically, giving the following name:

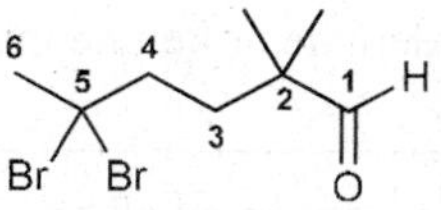

5,5-dibromo-2,2-dimethylhexanal

(b) We begin by identifying the longest chain that includes the carbon atom of the carbonyl group. In this case, the parent is a chain of six carbon atoms, with the carbonyl group at C2, so the parent is 2-hexanone. Next, we identify the substituents. There are three, all of which are methyl groups, located at C3, C4 and C5. We then assign a configuration to each of the chiral centers, giving the following name:

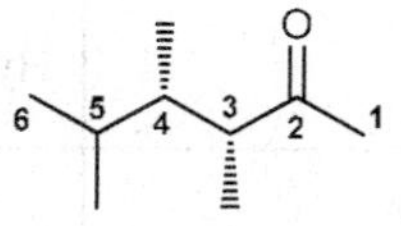

(3*R*,4*S*)-3,4,5-trimethyl-2-hexanone

(c) We begin by identifying the parent. In this case, the carbonyl group is part of a five-membered ring, so the parent is cyclopentanone. In a cyclic ketone, the carbonyl group is at C1 (by definition). Next, we identify the substituents. There are four, all of which are methyl groups (two at C2 and two at C5), giving the following name:

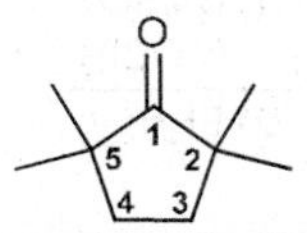

2,2,5,5-tetramethylcyclopentanone

(d) We begin by identifying the longest chain that includes the carbon atom of the carbonyl group. In this case, the parent is a chain of five carbon atoms, with the carbonyl group at C1, so the parent is pentanal. Next, we identify the substituents. There is only one substituent: a propyl group at C2, giving the following name:

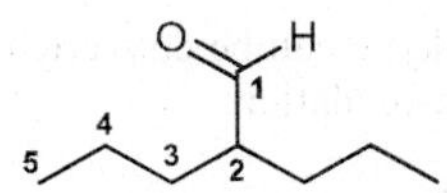

2-propylpentanal

(e) A cyclic compound containing an aldehyde adjacent to the ring is named as a cycloalkane carbaldehyde. In this case, the aldehyde group is connected to a four-membered ring (cyclobutane), giving the following name:

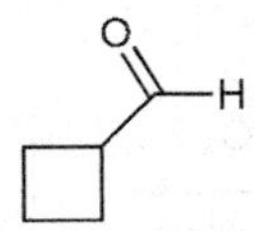

cyclobutanecarbaldehyde

19.2.

(a) The name (*S*)-3,3-dibromo-4-ethylcyclohexanone indicates that the parent (cyclohexanone) is a six-membered ring, with a carbonyl group incorporated in the ring (a cyclic ketone). There are three substituents: two bromine atoms (both at C3) and an ethyl group (at C4). The chiral center at C4 has the *S* configuration:

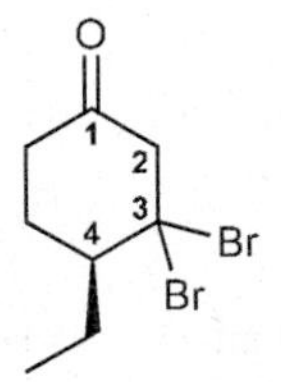

(b) The name 2,4-dimethyl-3-pentanone indicates that the parent (3-pentanone) is a chain of five carbon atoms, with the carbonyl group at C3. There are two substituents (both methyl groups), located at C2 and C4, as shown:

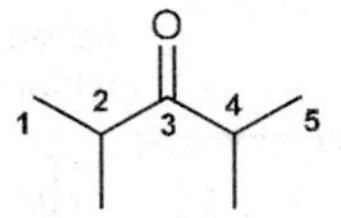

(c) The name (*R*)-3-bromobutanal indicates that the parent (butanal) is a chain of four carbon atoms, with the carbonyl group at C1 (by definition). There is only one substituent (Br), located at C3, which exhibits the *R* configuration, as shown:

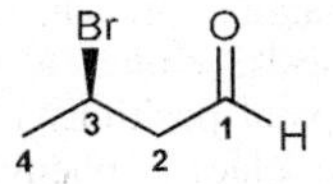

19.3.

(a) The parent is a six-membered ring with two carbonyl groups (a dione). The locants indicate the relative positions of the carbonyl groups (C1 and C3):

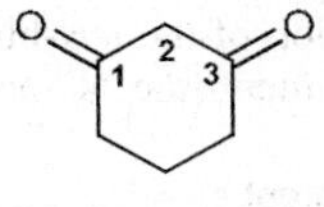

1,3-cyclohexanedione

(b) The parent is a six-membered ring with two carbonyl groups (a dione). The locants indicate the relative positions of the carbonyl groups (C1 and C4):

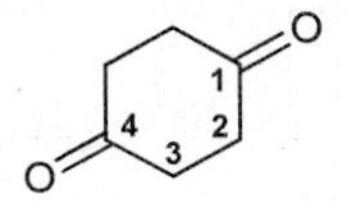

1,4-cyclohexanedione

(c) The parent is a chain of nine carbon atoms, for which three of the carbon atoms are carbonyl groups (a trione). The locants indicate the relative positions of the carbonyl groups (C2, C5, and C8):

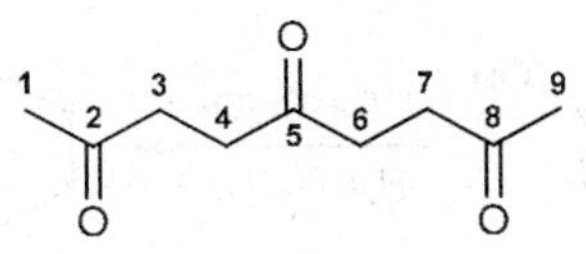

2,5,8-nonanetrione

19.4.

(a) To draw a complex molecule when given a complex name, it helps to start at the end and work backwards. First, draw the parent compound (methanone = a one-carbon ketone). Next, draw the substituents. There are two substituents in this case: phenyl and (2-hydroxy-4-methoxyphenyl). Draw the complex substituent by starting with a phenyl group and then adding the hydroxy and methoxy groups at the appropriate positions (recall that the point of attachment to the parent chain is defined as carbon #1 of the substituent). You may recognize that another possible name for this compound is 2-hydroxy-4-methoxybenzophenone.

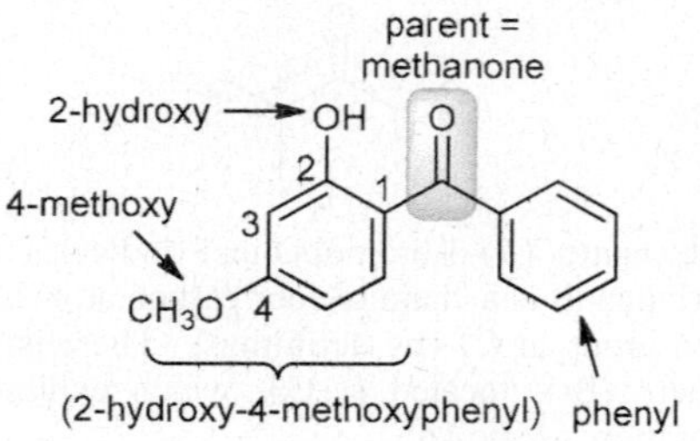

parent = methanone

2-hydroxy → OH

4-methoxy → CH_3O

(2-hydroxy-4-methoxyphenyl) phenyl

(b) Following the same approach, we begin by drawing the parent, propane-1,3-dione, a three-carbon ketone with carbonyl groups at positions 1 and 3. Next to one carbonyl group, we place a phenyl substituent with a methoxy group on carbon 4. Next to the other carbonyl group, we draw a phenyl substituent with a 1,1-dimethylethyl group (also known as a *t*-butyl group) at carbon 4. Note: the systematic use of the parentheses should clarify the position of the entire *tert*-butylphenyl substituent on the parent diketone compound (e.g. "3") as well as the position of the *tert*-butyl "sub-substituent" on the phenyl substituent (the "4" or *para* position).

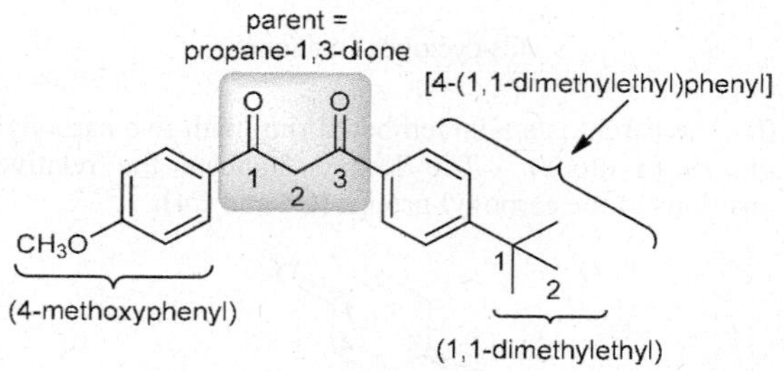

parent = propane-1,3-dione

[4-(1,1-dimethylethyl)phenyl]

CH_3O

(4-methoxyphenyl)

(1,1-dimethylethyl)

19.5.

(a) Oxidation of a secondary alcohol (to give a ketone) can be achieved with chromic acid (H_2CrO_4), which is prepared by mixing sodium dichromate ($Na_2Cr_2O_7$) and aqueous sulfuric acid. Alternatively, PCC, DMP, or Swern conditions can affect this transformation.

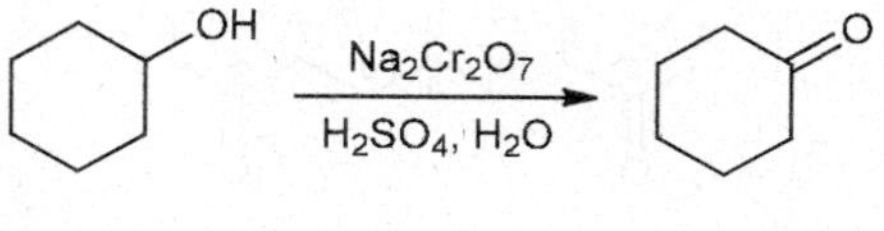

$Na_2Cr_2O_7$ / H_2SO_4, H_2O

(b) Oxidation of a primary alcohol (to give an aldehyde) can be achieved upon treatment with PCC, or upon treatment with DMP, or via a Swern oxidation:

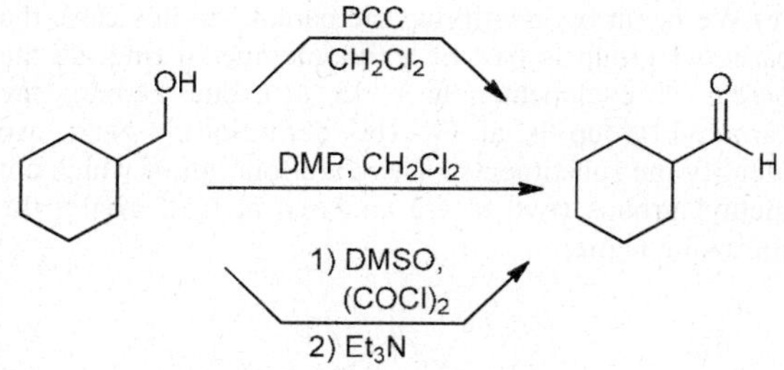

PCC, CH_2Cl_2

DMP, CH_2Cl_2

1) DMSO, $(COCl)_2$ 2) Et_3N

(c) A terminal alkyne can be converted into a methyl ketone upon acid-catalyzed hydration in the presence of mercuric sulfate:

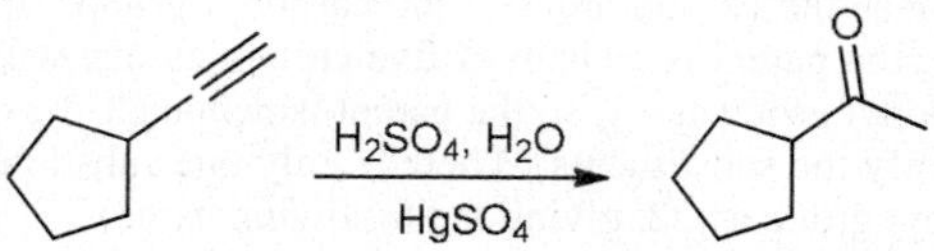

H_2SO_4, H_2O / $HgSO_4$

(d) A terminal alkyne can be converted into an aldehyde via hydroboration-oxidation:

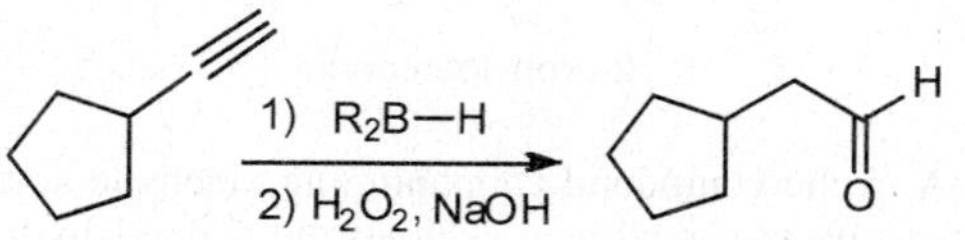

1) $R_2B{-}H$ 2) H_2O_2, NaOH

(e) A C=C bond can be cleaved into two carbonyl groups via ozonolysis. In this case, the ring is opened to give an acyclic product.

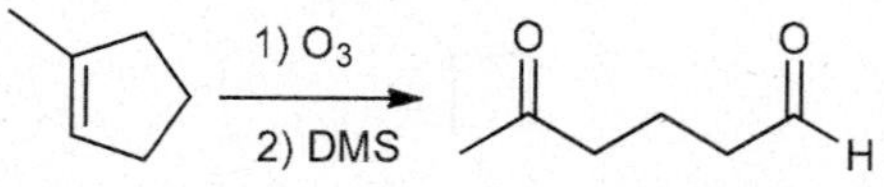

1) O_3 2) DMS

(f) An acyl group can be installed via a Friedel-Crafts acylation, using the appropriate acyl halide, as shown:

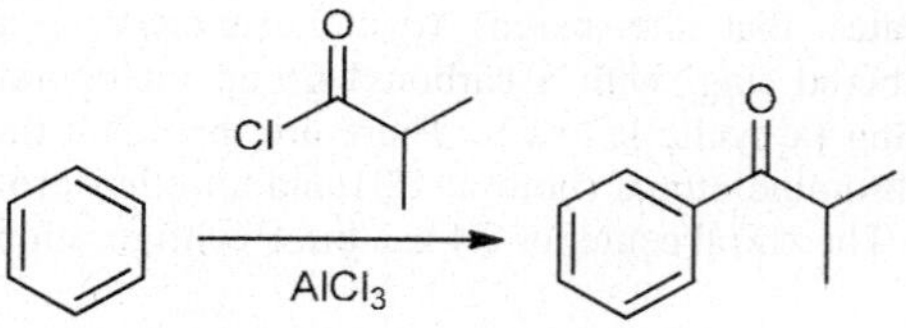

$AlCl_3$

19.6.

(a) As seen in Section 12.6, a Grignard reagent is a strong nucleophile that will attack a carbonyl group to give an alkoxide ion. After the reaction is complete, aqueous acid is introduced into the reaction flask, thereby protonating the alkoxide ion, giving an alcohol:

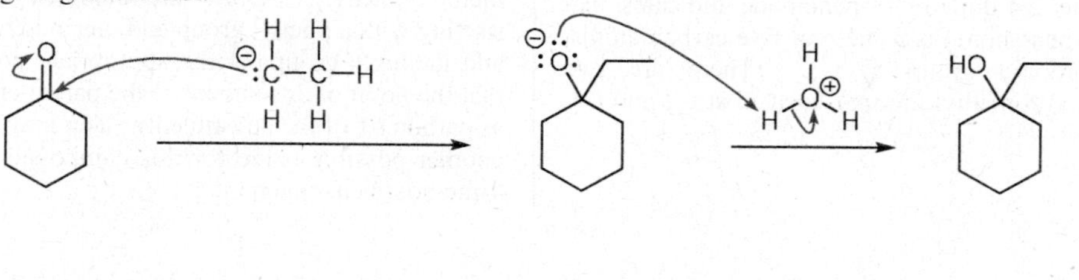

(b) Under acidic conditions, the carbonyl group is first protonated to give a resonance-stabilized cation. This cation is a strong electrophile that is then captured by a chloride ion to give the addition product shown. The equilibrium likely favors the ketone.

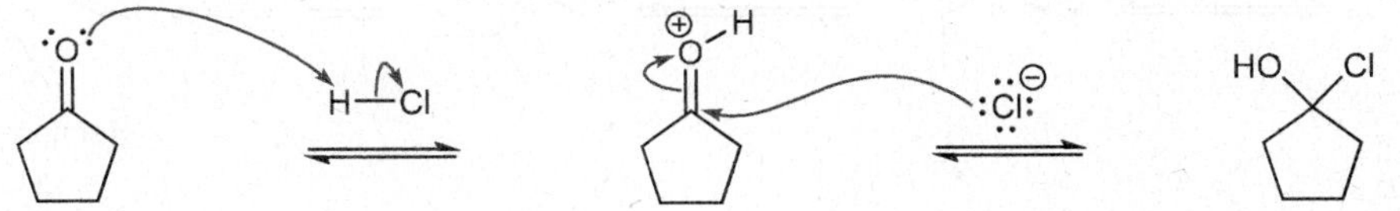

19.7. The carbonyl group in hexafluoroacetone is flanked by two very powerful electron-withdrawing groups (CF_3). These groups withdraw electron density from the carbonyl group, thereby increasing the electrophilicity of the carbonyl group. The resulting increase in energy of the reactant causes the equilibrium to favor the product (the hydrate, which is now lower in energy by comparison).

19.8.

(a) Acid-catalyzed acetal formation proceeds via a seven-step process, which can be divided into two parts:
Part one is conversion of the ketone into a hemiacetal. In the first step of the following mechanism, a proton transfer generates a protonated carbonyl group, which is an excellent electrophile and is subject to attack by a nucleophile. The alcohol functions as the nucleophile, attacking the protonated carbonyl group to give an oxonium ion that loses a proton to give the hemiacetal.
Part two of the mechanism is conversion of the hemiacetal into an acetal, which requires four steps. First a proton transfer step converts a bad leaving group (hydroxide) into an excellent leaving group (water). Loss of water gives an intermediate that can be attacked by another molecule of the alcohol. The resulting oxonium ion is then deprotonated to give an acetal. Note that in all steps of the mechanism, the proton source is a protonated alcohol, and the base is the alcohol (MeOH). No strong bases are employed or formed at any point during the mechanism.

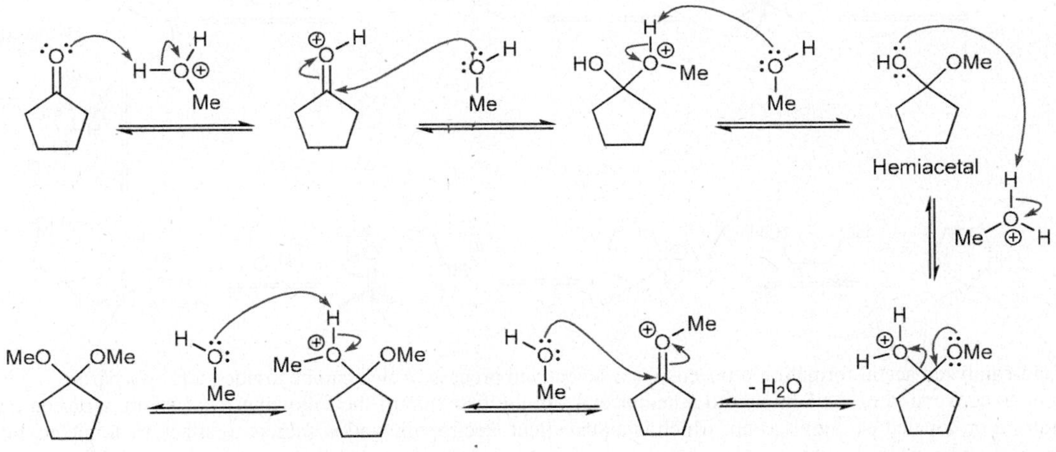

(b) Acid-catalyzed acetal formation proceeds via a seven-step process, which can be divided into two parts:
Part one is conversion of the ketone into a hemiacetal. In the first step of the following mechanism, a proton transfer generates a protonated carbonyl group, which is an excellent electrophile and is subject to attack by a nucleophile. The alcohol functions as the nucleophile, attacking the protonated carbonyl group to give an oxonium ion that loses a proton to give the hemiacetal.
Part two of the mechanism is conversion of the hemiacetal into an acetal, which requires four steps. First a proton transfer step converts a bad leaving group (hydroxide) into an excellent leaving group (water). Loss of water gives an intermediate that can be attacked by another molecule of the alcohol. The resulting oxonium ion is then deprotonated to give an acetal. Note that in all steps of the mechanism, the proton source is a protonated alcohol, and the base is the alcohol (EtOH). No strong bases are employed or formed at any point during the mechanism.

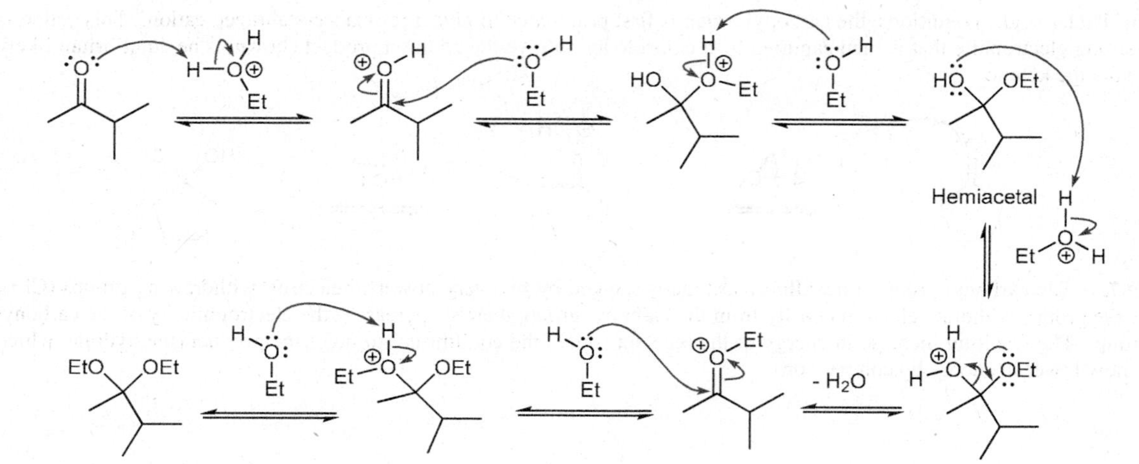

Hemiacetal

- H_2O

(c) Conversion of the ketone to an acetal occurs via a seven-step process. First the ketone is converted to a hemiacetal via the following three steps: 1) proton transfer, 2) nucleophilic attack, and then 3) proton transfer. And then the hemiacetal is converted into an acetal via the following four steps: 4) proton transfer, 5) loss of a leaving group, 6) nucleophilic attack, and 7) proton transfer. There is a subtle difference between this mechanism and the mechanisms shown in the solutions to part (a) and part (b) of this problem. The difference can be seen in step 6 (the second nucleophilic attack). In this case, step 6 occurs in an *intramolecular* fashion, giving a cyclic acetal:

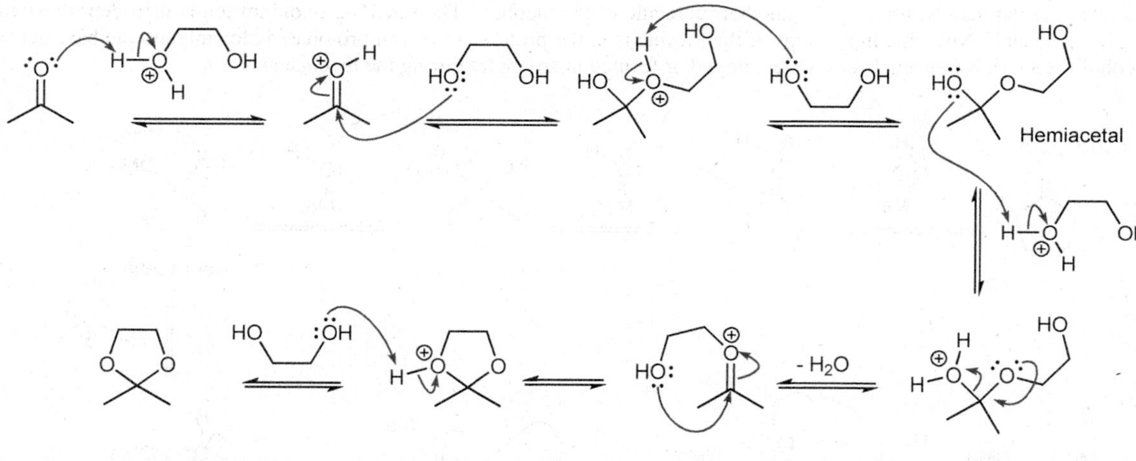

Hemiacetal

- H_2O

(d) Acid-catalyzed acetal formation proceeds via a seven-step process, which can be divided into two parts:
Part one is conversion of the ketone into a hemiacetal. In the first step of the following mechanism, a proton transfer generates a protonated carbonyl group, which is an excellent electrophile and is subject to attack by a nucleophile. In this case, the nucleophilic attack occurs in an *intramolecular* fashion to give a five-membered ring. This oxonium ion then loses a proton to give the hemiacetal.
Part two of the mechanism is conversion of the hemiacetal into an acetal, which requires four steps. First, a proton transfer step converts a bad leaving group (hydroxide) into an excellent leaving group (water). Loss of water gives an intermediate that can be attacked by another nucleophile. Once again, an intramolecular attack generates another five-membered ring. This oxonium ion is then deprotonated to give an acetal. Note that in all steps of the mechanism, the likely proton source is H_3O^+, and the base is water (H_2O). No strong bases are employed or formed at any point during the mechanism.

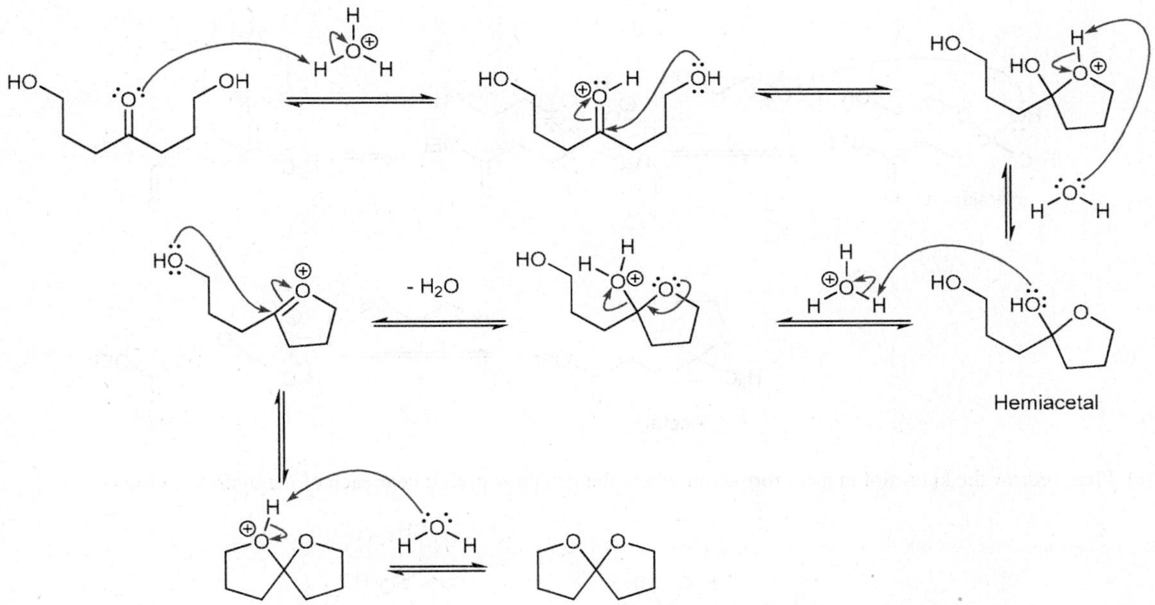

HO OH HO OH HO H HO HO H H - H_2O Hemiacetal

19.9.

(a) The reaction shows the formation of a cyclic acetal from a ketone. Since the reactant is a ketone, the reagent must be a diol and catalytic acidic conditions are required. Since the two oxygen atoms (in the acetal) are attached by two methylene (CH_2) groups, the diol reagent must also be attached by two methylene groups.

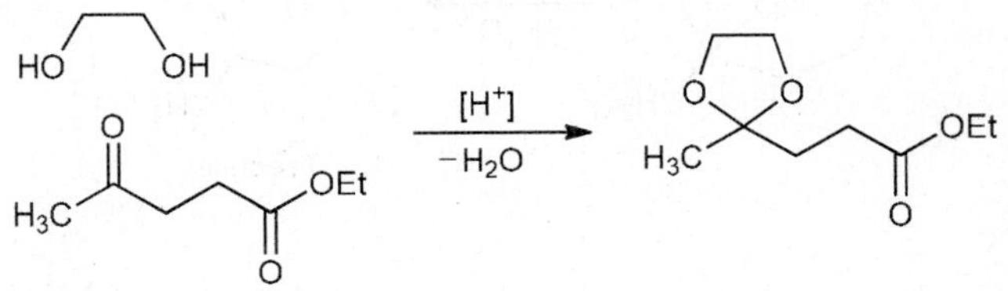

HO OH O H_3C OEt O $[H^+]$ $-H_2O$ O O H_3C OEt O

A strong acid must be used, typically H_2SO_4 or TsOH.

(b) The mechanism shown below is very similar to the mechanisms drawn in the solutions to Problems 19.8a-d. Specifically, it involves a seven-step process that can be divided into two parts: (1) formation of the hemiacetal (three steps) and (2) formation of the acetal (four steps).

Formation of the hemiacetal involves three mechanistic steps: 1) proton transfer, 2) nucleophilic attack, and then 3) proton transfer.

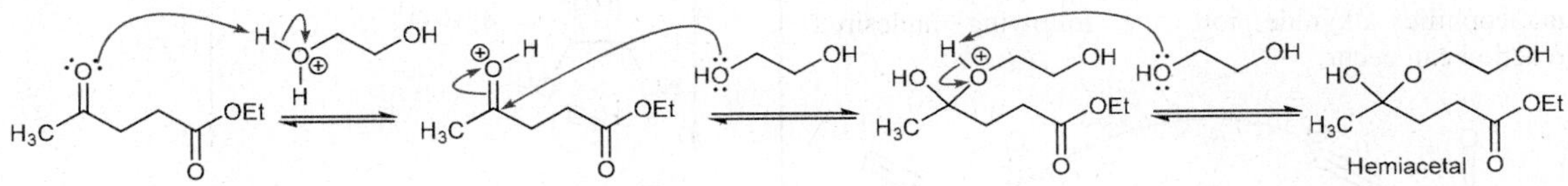

Hemiacetal

From the hemiacetal, the acetal forms in four mechanistic steps: 1) proton transfer, 2) loss of a leaving group, 3) nucleophilic attack, and then 4) proton transfer.

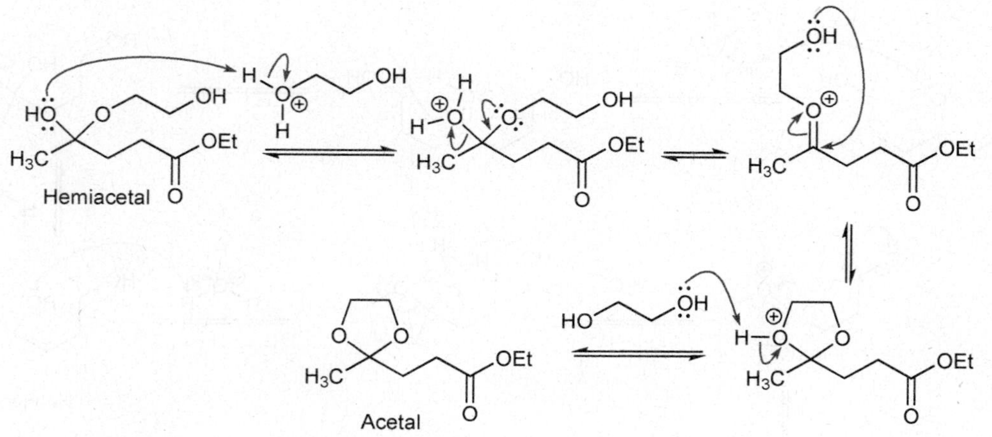

(c) First, redraw the keto-diol in a conformation where the ketone is placed near each of the hydroxyl groups.

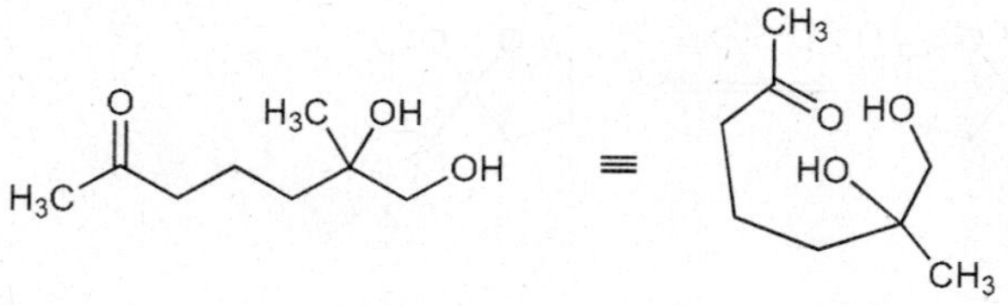

Next make a bond between the carbonyl carbon and each of the hydroxyl oxygen atoms. This produces a bicyclic acetal, as shown.

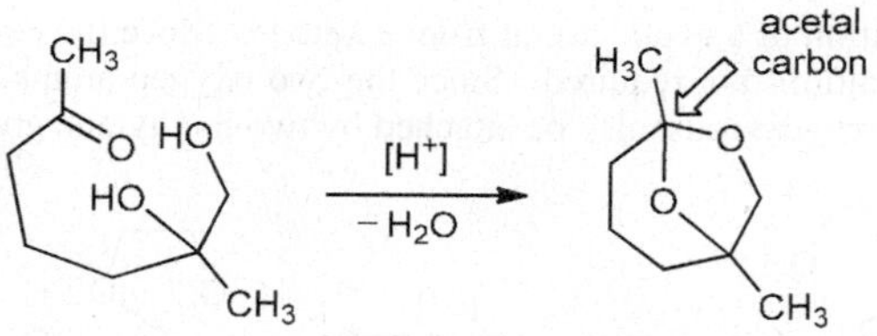

19.10.

(a) This type of transformation can be achieved via alkylation of a terminal alkyne, as seen in Section 9.10:

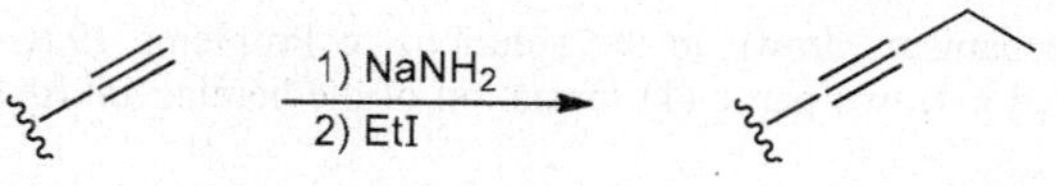

However, the starting material has a carbonyl group, which is a good electrophile. In the presence of the nucleophilic alkynide ion, the following undesired reaction can occur:

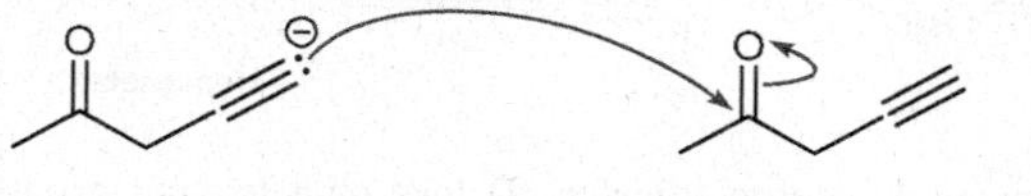

To avoid this undesired reaction, the carbonyl group is first converted into an acetal, which serves as a protecting group. The desired reaction is then performed, and finally, the protecting group (the acetal group) is removed with aqueous acid to give the product:

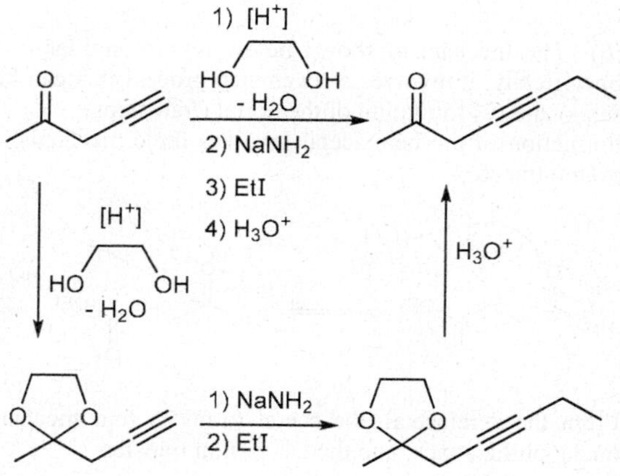

(b) This type of transformation can be achieved by treating the ester with excess phenyl magnesium bromide (see Mechanism 12.5):

1) xs PhMgBr
2) H_3O^+

However, the starting material has another carbonyl group, which is also a good electrophile. In the presence of a powerful nucleophile (the Grignard reagent), the following undesired reaction can occur:

To avoid this undesired reaction, the ketone is first converted into an acetal, which serves as a protecting group. Installation of the protecting group occurs exclusively at the carbonyl group of the ketone, rather than the ester group, because the ketone carbonyl is more reactive than the ester carbonyl group (esters are ***not*** converted to their corresponding acetals). Once the ketone has been protected, the desired reaction is then performed, and finally, the protecting group is removed with aqueous acid to give the product. This final step also protonates the alkoxide ion which was formed by the Grignard reaction:

1) [H^+], HO‿OH, - H_2O
2) xs PhMgBr
3) H_3O^+

[H^+]
HO‿OH
- H_2O

xs PhMgBr

H_3O^+

(c) This type of transformation can be achieved by treating the ester with excess $LiAlH_4$ (see Mechanism 12.3):

1) xs $LiAlH_4$
2) H_3O^+

However, the starting material has another carbonyl group, which is also a good electrophile. In the presence of $LiAlH_4$, the following undesired reaction can occur:

To avoid this undesired reaction, the ketone is first converted into an acetal, which serves as a protecting group. Installation of the protecting group occurs exclusively at the carbonyl group of the ketone, rather than the ester group, because the ketone carbonyl is more reactive than the ester carbonyl group (esters are ***not*** converted to their corresponding acetals). Once the ketone has been protected, the desired reaction is then performed, and finally, the protecting group is removed with aqueous acid to give the product. This final step also protonates the alkoxide ion which was formed by reduction of the ester group:

1) [H^+], HO‿OH
- H_2O
2) xs $LiAlH_4$
3) H_3O^+

[H^+]
HO‿OH
- H_2O

$LiAlH_4$
(2 eq.)

H_3O^+

19.11.

(a) Begin by identifying the carbon atom that is connected to two oxygen atoms (highlighted):

This carbon atom bears an acetal group, and will ultimately be converted into a carbonyl group in the product. The C–O bonds are cleaved, giving a ketone. Since these C–O bonds are contained in a ring, a diol is released.

H_3O^+

(b) Begin by identifying the carbon atom that is connected to two oxygen atoms (highlighted):

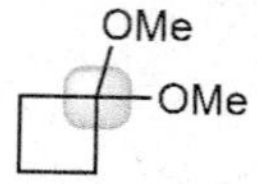

OMe
OMe

This carbon atom bears an acetal group, and will ultimately be converted into a carbonyl group in the product. The C-O bonds are cleaved, giving a ketone. Since these C-O bonds are not contained in a ring, two molecules of methanol are released.

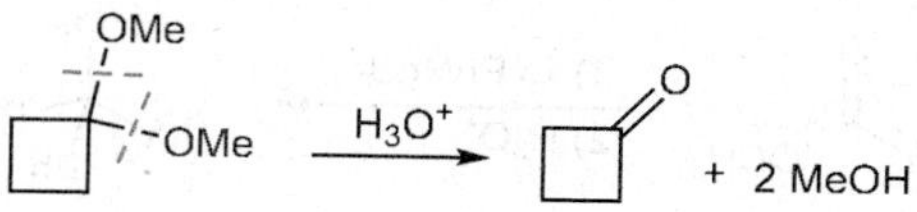

OMe
OMe
H_3O^+
O
+ 2 MeOH

19.12. The starting material contains both a carbonyl group and an OH group, and the product is a hemiacetal. As seen in Mechanism 19.5, formation of a hemiacetal requires three steps. First, the carbonyl is protonated, rendering it more electrophilic. Then, the alcohol group attacks the protonated carbonyl group (in an intramolecular fashion), giving an oxonium ion, which is deprotonated to give the hemiacetal, as shown:

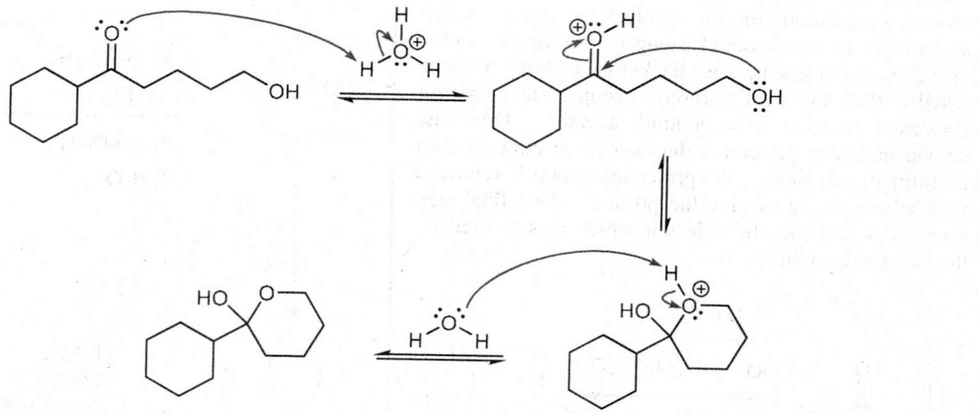

19.13. The molecular formula indicates two degrees of unsaturation. Since the product is a hemiacetal, the starting material must contain both a carbonyl group and an OH group. The carbon atom of the acetal group must have been the location of the carbonyl group. The OH group (that attacks the carbonyl group to form the hemiacetal) must be contained in the structure of compound A, which explains why a cyclic hemiacetal is formed. The two degrees of unsaturation in Compound A correlate to the ring and the π bond of the carbonyl group.

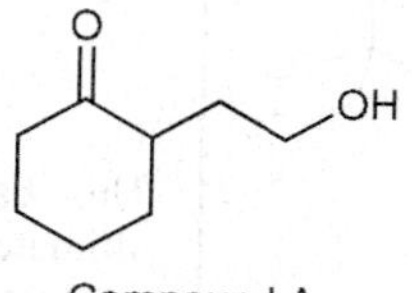

O
OH
Compound A

19.14.

(a) As seen in Mechanism 19.6, imine formation proceeds via six steps. First, the amine functions as a nucleophile and attacks the ketone. The resulting intermediate is then protonated to remove the negative charge (note that the likely proton source under these conditions is a protonated amine, called an ammonium ion). Proton transfer then gives a carbinolamine (note that the likely base for this step is a molecule of the amine). Protonation then converts the bad leaving group (hydroxide) into a good leaving group (water). Loss of water then gives an iminium ion, which is deprotonated to give an imine.

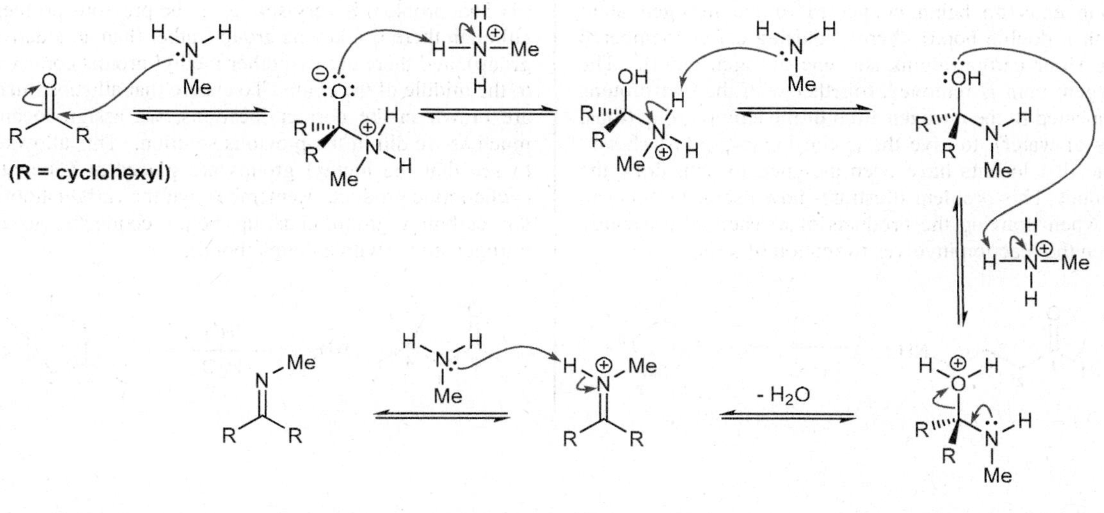

(R = cyclohexyl)

(b) As seen in the solution to part (a) of this problem, imine formation proceeds via six steps. The same six steps are drawn again here: 1) nucleophilic attack, 2) proton transfer, 3) proton transfer, 4) proton transfer, 5) loss of a leaving group, and 6) proton transfer.

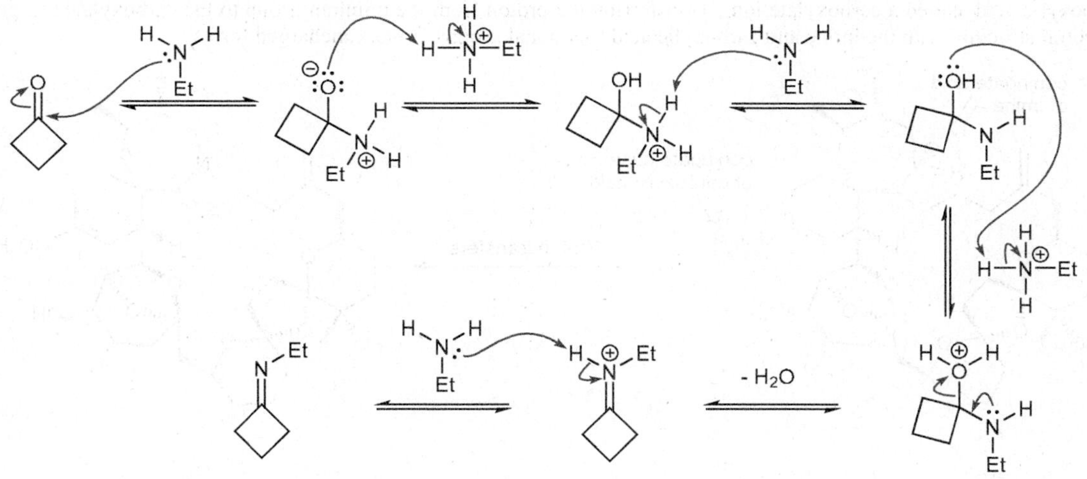

19.15.

(a) When treated with ammonia under acid-catalyzed conditions (with removal of water), a ketone is converted to an imine. In this case, ammonia (NH_3) is used as the nucleophile (rather than a primary amine, RNH_2), and as a result, the product will not have an R group connected to the nitrogen atom of the imine group. Instead, there will be a proton in that location (two of the protons from NH_3 are removed during imine formation, while the third proton remains connected to the nitrogen atom in the product).

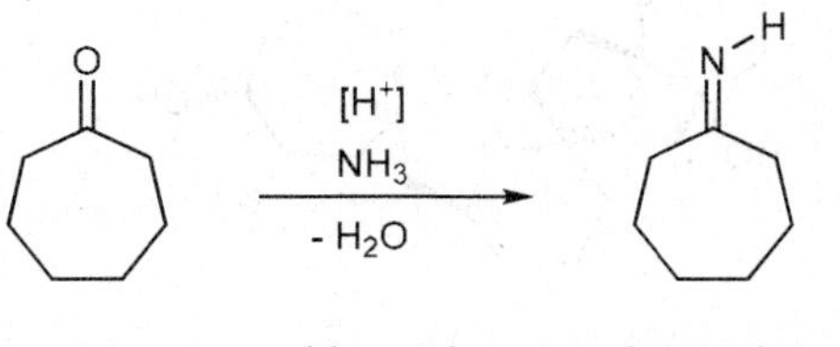

(b) Upon treatment with a primary amine under acid-catalyzed conditions (with removal of water), a ketone is converted to an imine. The carbon atom of the carbonyl group ends up being connected to the nitrogen atom (with a double bond). The oxygen atom is removed, together with the two protons connected to the nitrogen atom of the amine (thus, loss of water), to give the following imine product:

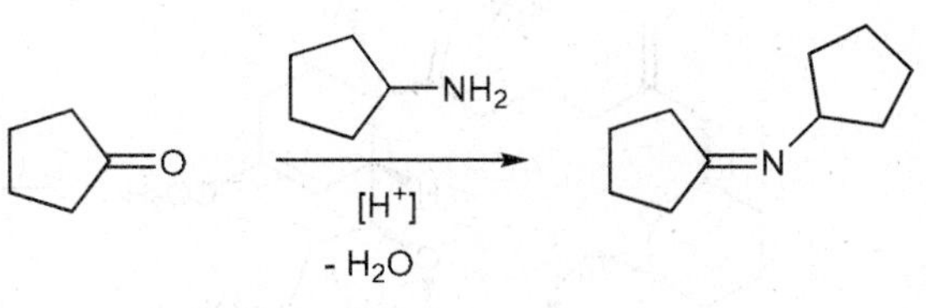

(c) Under acid-catalyzed conditions (with removal of water), an aldehyde will react with a primary amine to give an imine. In this case, the aldehyde group and the amino group are tethered to each other (in the same molecule), and the reaction therefore occurs in an intramolecular fashion. The carbon atom of the carbonyl group ends up being connected to the nitrogen atom (with a double bond), thereby closing a five-membered ring (four carbon atoms and one nitrogen atom). The oxygen atom is removed, together with the two protons connected to the nitrogen atom of the amino group (thus, loss of water), to give the cyclic imine product shown. Note that locants have been assigned to help draw the product. This problem illustrates how useful locants can be when drawing the products of a reaction, especially when the reaction involves formation of a ring.

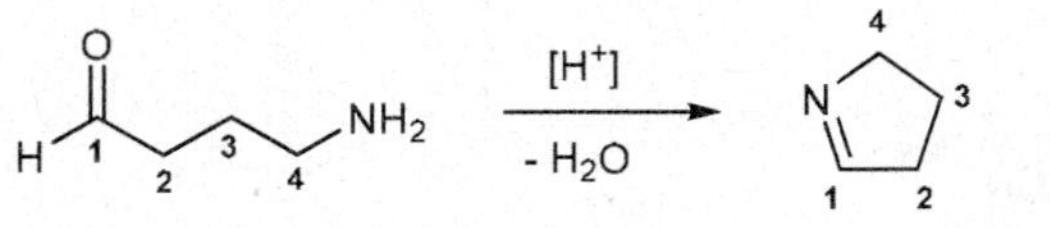

(d) This problem is very similar to the previous problem, although there is a ketone group (rather than an aldehyde group), and there are two other methyl groups connected to the middle of the chain. To ensure that all substituents are drawn in the correct locations, we assign locants, much as we did in the previous solution. This allows us to see that the methyl groups are placed at C4 in the cyclic imine product. Remember that the carbon atom of the carbonyl group ends up being connected to the nitrogen atom (with a double bond).

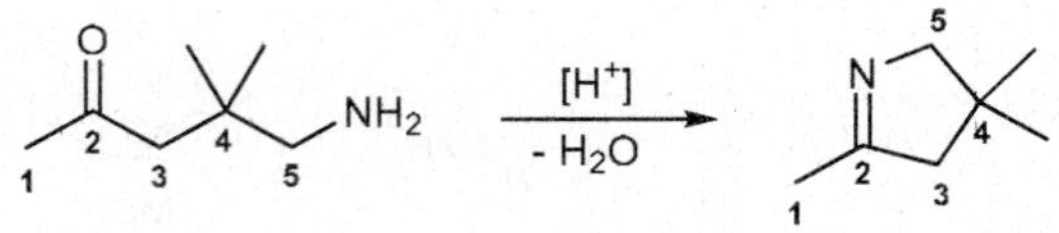

19.16.

(a) An iminium ion is the conjugate acid of an imine. Notice that pinnatoxin A also contains the conjugate base of a carboxylic acid, called a carboxylate ion. Transferring the proton from the iminium group to the carboxylate ion gives a neutral structure with the imine and carboxylic acid functional groups in their uncharged forms.

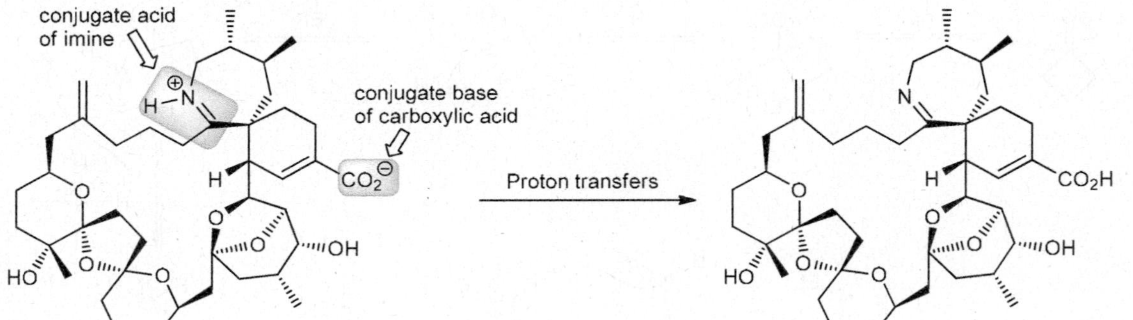

(b) To draw the amino ketone precursor, we consider the location of the C=N bond in the imine. The carbon atom of this bond must have been the carbonyl group in the starting ketone, and the nitrogen atom must have been connected to two protons in the starting amine. Thus, the following amino ketone reactant should give the desired imine (under acid-catalyzed conditions with the removal of water).

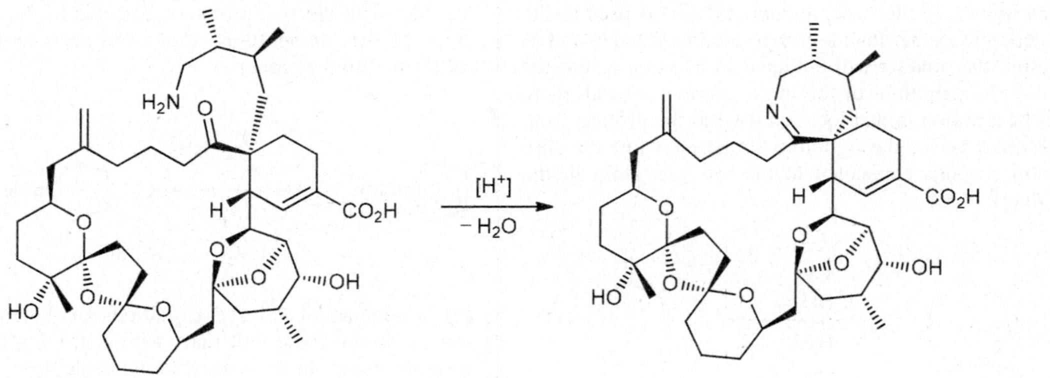

(c) Given the complexity of the structure, let's simplify the molecule by representing the large ring with R groups.

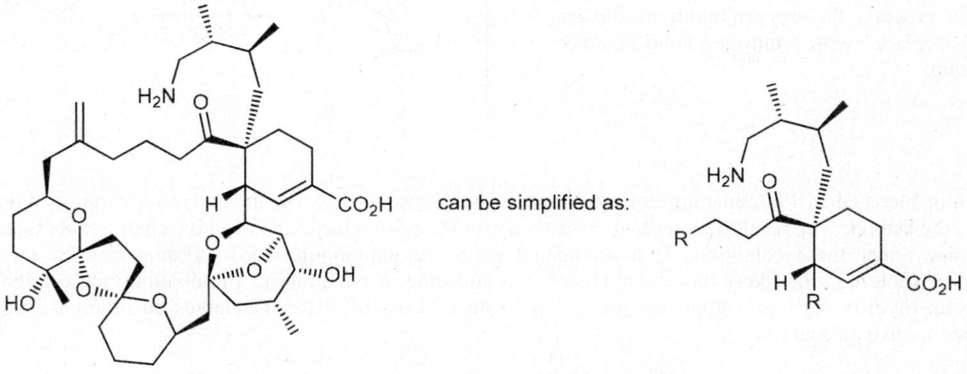

can be simplified as:

Imine formation proceeds via six steps. First, the amino group functions as a nucleophile and attacks the carbonyl group. The next three steps are a series of proton transfers assisted by a general acid ($R'—NH_3^+$) or base ($R'—NH_2$). For the entire mechanism, the general acid is likely the protonated amine of pinnatoxin (or related intermediates) and the general base is likely the free amine of pinnatoxin (or related intermediates). Expulsion of water affords the iminium ion that is deprotonated to give the imine.

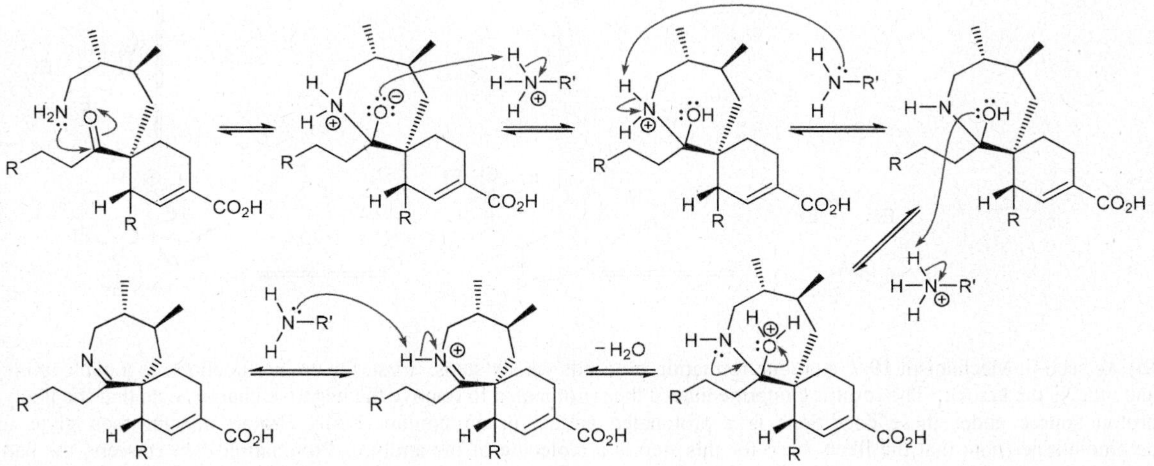

19.17.

(a) Treating a ketone with hydroxylamine under acidic conditions (with removal of water) gives an oxime ($R_2C{=}NOH$). During this process, the oxygen atom of the starting ketone is replaced with a nitrogen atom connected to an OH group:

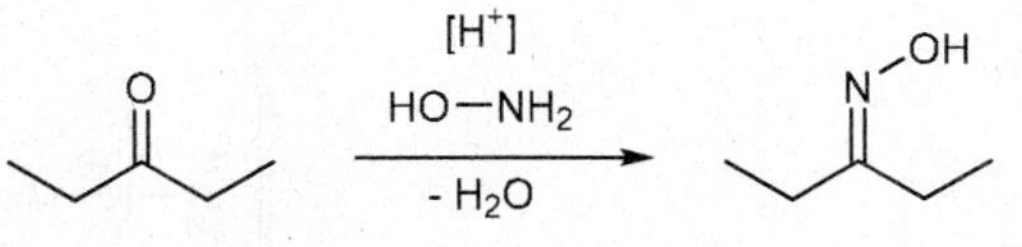

(b) Treating a ketone with hydrazine under acidic conditions (with removal of water) gives a hydrazone ($R_2C{=}NNH_2$). During this process, the oxygen atom of the starting ketone is replaced with a nitrogen atom connected to an NH_2 group:

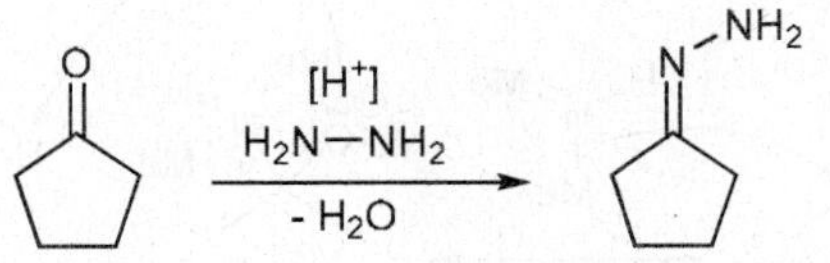

19.18.

(a) An oxime can be made by treating the corresponding ketone with hydroxylamine, in acid-catalyzed conditions (with removal of water), as shown.

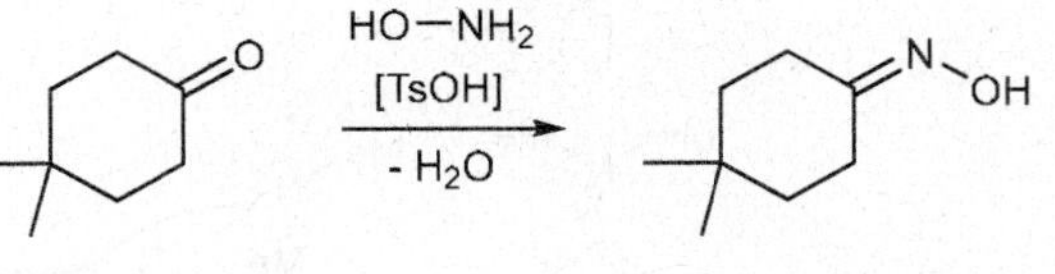

(b) Treating an aldehyde with hydrazine under acidic conditions (with removal of water) gives a hydrazone. During this process, the oxygen atom of the starting aldehyde is replaced with a nitrogen atom connected to an NH_2 group:

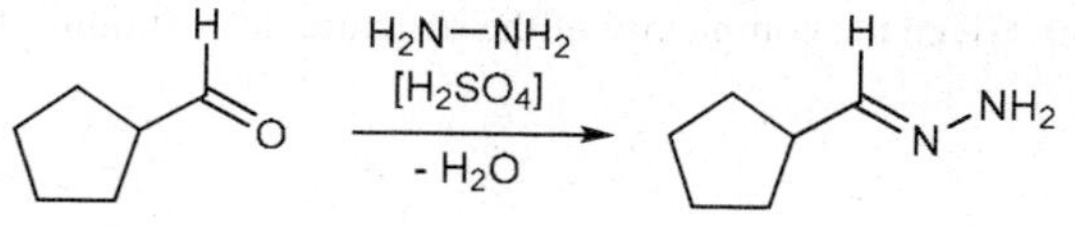

H2N—NH2
[H2SO4]
- H2O

19.19.

(a) As seen in Mechanism 19.7, enamine formation proceeds via six steps. First, the amine functions as a nucleophile and attacks the ketone. The resulting intermediate is then protonated to remove the negative charge (note that the likely proton source under these conditions is a protonated amine, or ammonium ion). Proton transfer then gives a carbinolamine (note that the likely base for this step is a molecule of the amine). Protonation then converts the bad leaving group (hydroxide) into a good leaving group (water). Loss of water then gives an iminium ion, which is deprotonated to give an enamine.

- H2O

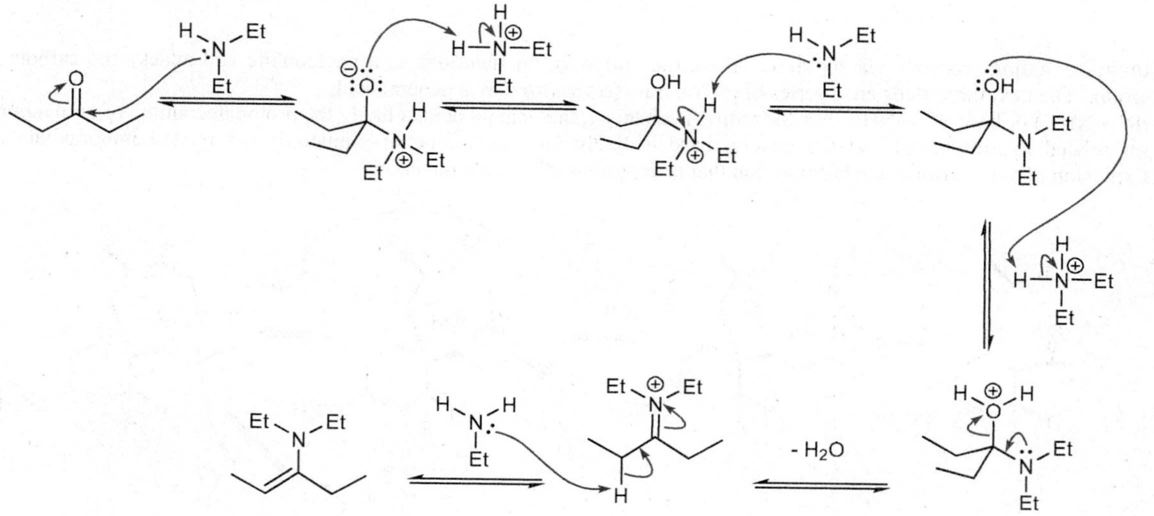

(b) As seen in Mechanism 19.7, enamine formation proceeds via six steps. First, the amine functions as a nucleophile and attacks the ketone. The resulting intermediate is then protonated to remove the negative charge (note that the likely proton source under these conditions is a protonated amine, or ammonium ion). Proton transfer then gives a carbinolamine (note that the likely base for this step is a molecule of the amine). Protonation then converts the bad leaving group (hydroxide) into a good leaving group (water). Loss of water then gives an iminium ion, which is deprotonated to give an enamine.

- H2O

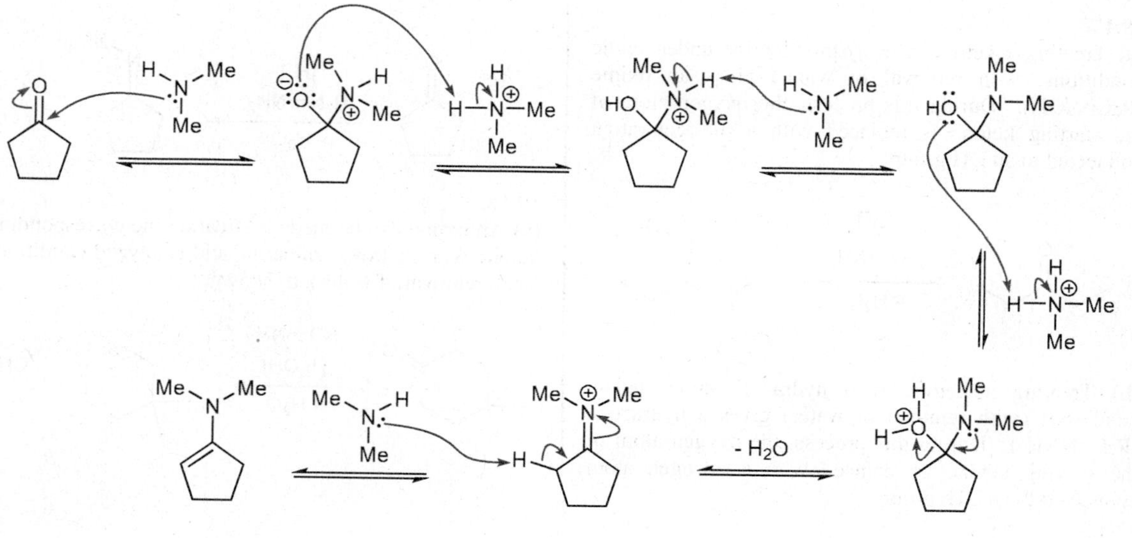

19.20.

(a) Upon treatment with a secondary amine under acid-catalyzed conditions (with removal of water), a ketone is converted to an enamine. The carbon atom of the carbonyl group ends up being connected to the nitrogen atom (with a single bond), and that same carbon atom ends up having a double bond to a neighboring carbon atom, as shown. In the process, the oxygen atom is removed altogether, as well as two protons (one from the nitrogen atom and the other from a carbon atom) to give loss of water.

[H⁺] N—H - H_2O N

(b) Upon treatment with a secondary amine under acid-catalyzed conditions (with removal of water), a ketone is converted to an enamine. The carbon atom of the carbonyl group ends up being connected to the nitrogen atom (with a single bond), and that same carbon atom ends up having a double bond to a neighboring carbon atom, as shown. In the process, the oxygen atom is removed altogether, as well as two protons (one from the nitrogen atom and the other from a carbon atom) to give loss of water.

H N O [H⁺] - H_2O N

(c) Under acid-catalyzed conditions (with removal of water), a ketone will react with a secondary amine to give an enamine. In this case, the carbonyl group and the amino group are tethered to each other (in the same molecule), and therefore, the reaction occurs in an intramolecular fashion, forming a five-membered ring (four carbon atoms and one nitrogen atom). The carbon atom of the carbonyl group ends up being connected to the nitrogen atom (with a single bond). That same carbon atom ends up having a double bond to a neighboring carbon atom. The oxygen atom is removed altogether, as well as two protons (one from the nitrogen atom and the other from a carbon atom) to give loss of water.

O 1 2 3 4 NH [H⁺] - H_2O N 4 3 2 1

Note that locants have been assigned to help draw the product. These locants do not have to conform to IUPAC rules, as they are just helpful tools that we are using to draw the product correctly. The starting material above is not numbered according to IUPAC rules but that is OK, as long as we don't name the starting material using these incorrect locants.

(d) Under acid-catalyzed conditions (with removal of water), a ketone will react with a secondary amine to give an enamine. In this case, the carbonyl group and the amino group are tethered to each other (in the same molecule), and therefore, the reaction occurs in an intramolecular fashion, forming a five-membered ring (four carbon atoms and one nitrogen atom). The carbon atom of the carbonyl group ends up being connected to the nitrogen atom (with a single bond). That same carbon atom ends up having a double bond to a neighboring carbon atom. The oxygen atom is removed altogether, as well as two protons (one from the nitrogen atom and the other from a carbon atom) to give loss of water. Locants have been assigned to help draw the product.

O 1 2 3 4 H N [H⁺] - H_2O N 4 3 2 1

19.21.

(a) Enamines are formed from secondary amines. Since compound **2** is an enamine, proline must be a secondary amine. If we analyze the structure of **2**, we can see that proline must contain a 5-membered ring, as shown:

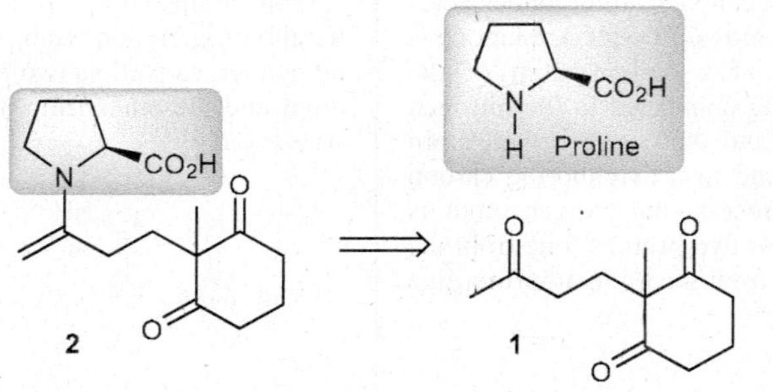

(b) The formation of the enamine intermediate proceeds via six steps. First the nitrogen atom in proline attacks a carbonyl group. The next three steps are a series of proton transfers assisted by a general acid or general base. Expulsion of water affords an iminium ion, followed by deprotonation to yield the enamine.

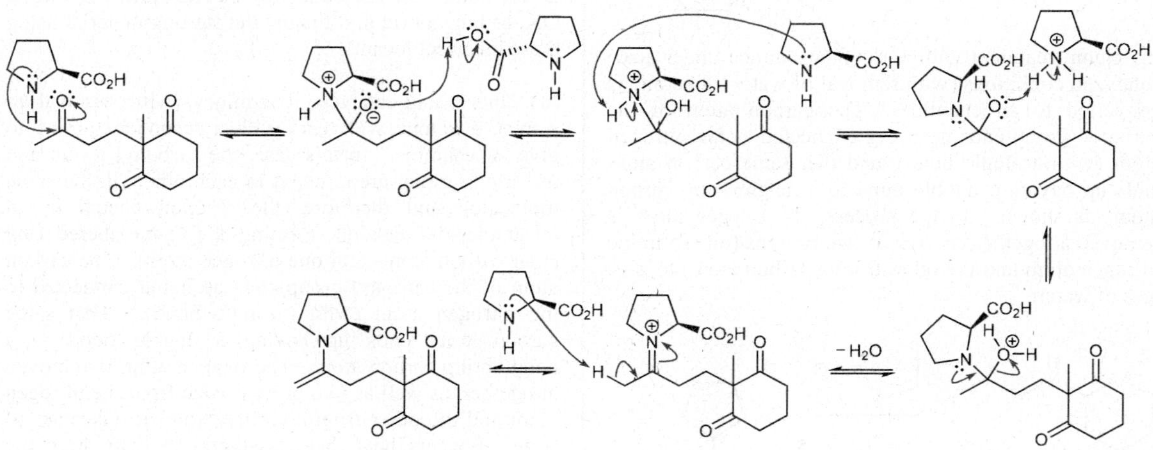

For the entire mechanism, the general acid is likely the carboxylic acid (R—CO_2H) or the protonated amine on proline ($R_2NH_2^+$) and the general base is likely the free amine on proline (R_2NH).

19.22. The starting material is a ketone, and the reagents indicate formation of a hydrazone, followed by a Wolff-Kishner reduction. The net result is complete reduction of the carbonyl group to a methylene group (CH_2):

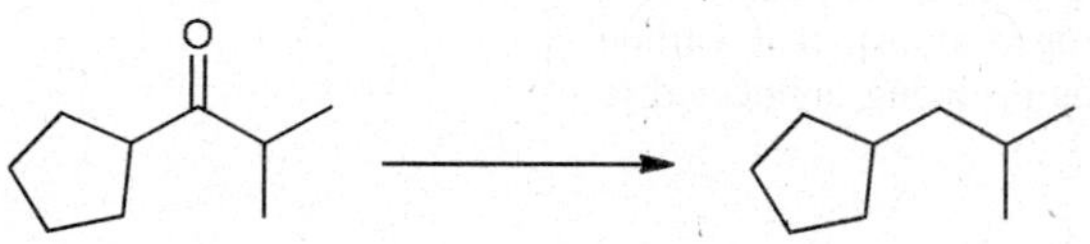

Formation of the hydrazone occurs via a six-step process (Mechanism 19.6). First, hydrazine functions as a nucleophile and attacks the ketone. The resulting intermediate is then protonated to remove the negative charge (note that the likely proton source under these conditions is a protonated hydrazine). Proton transfer then gives a carbinolamine (note that the likely base for this step is hydrazine). Protonation then converts the bad leaving group (hydroxide) into a good leaving group (water). Loss of water then gives an iminium ion, which is deprotonated to give the hydrazone.

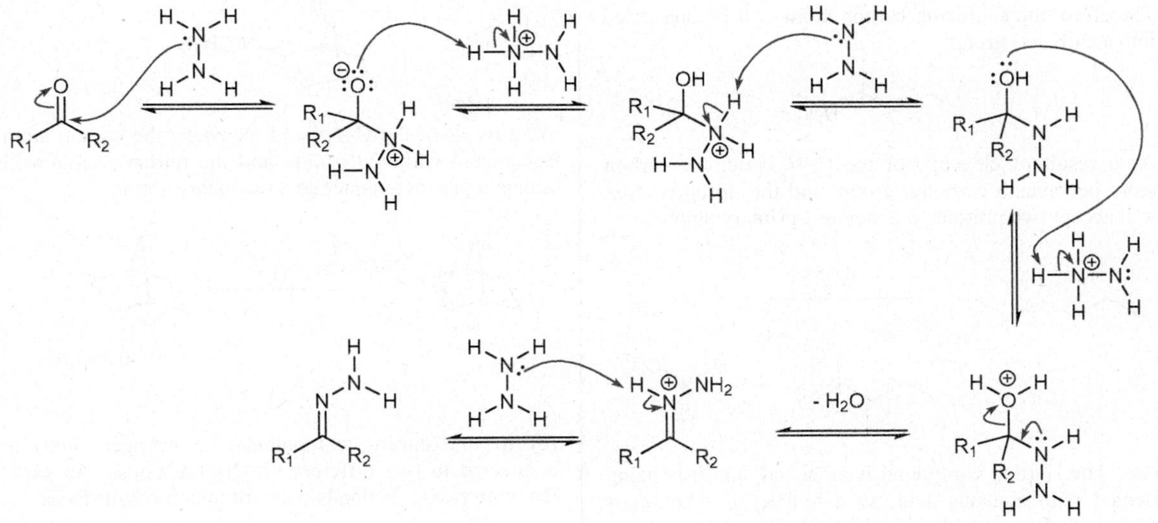

And then, under basic conditions, the hydrazone is reduced via the process shown here. As seen in Mechanism 19.8, three successive proton transfer steps, followed by loss of nitrogen gas, gives a carbanion, which is protonated by water to give the product:

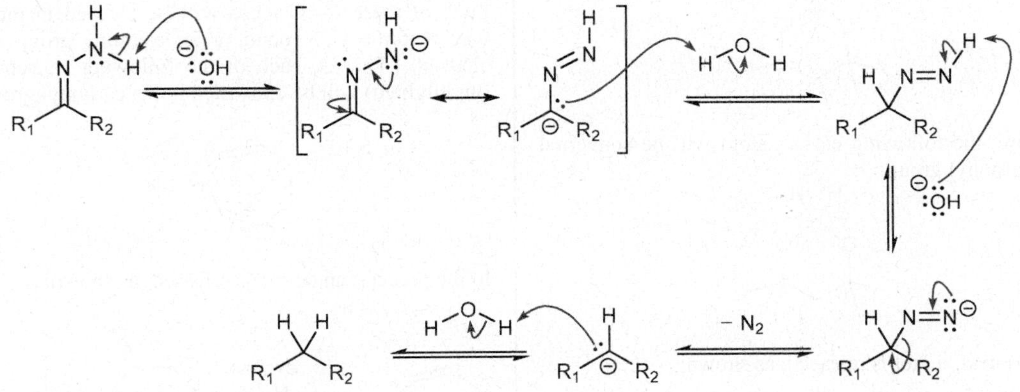

19.23.

(a) The starting compound is an acetal, and it is being treated with aqueous acid, so a hydrolysis reaction is expected. We first identify the bonds that will undergo cleavage. When an acetal undergoes hydrolysis, cleavage occurs for the C–O bonds of the acetal group:

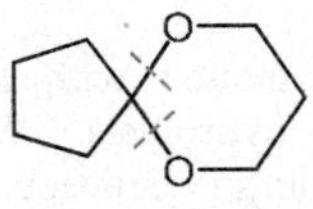

Therefore, the following carbon atom will be converted into a carbonyl group:

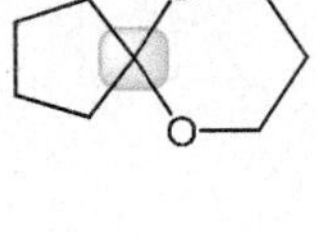

In the process, a diol is released, as shown:

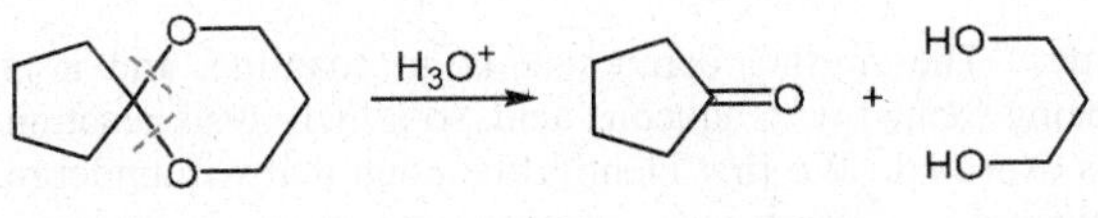

(b) The starting compound is an imine, and it is being treated with aqueous acid, so a hydrolysis reaction is expected. We first identify the bond that will undergo cleavage. When an imine undergoes hydrolysis, cleavage occurs for the C=N bond:

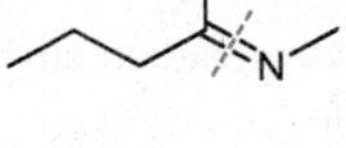

Therefore, the following carbon atom will be converted into a carbonyl group:

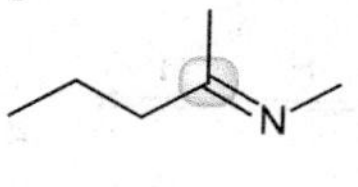

As a result of cleavage of the C=N bond, the carbon atom becomes a carbonyl group, and the nitrogen atom will accept two protons to generate a primary amine:

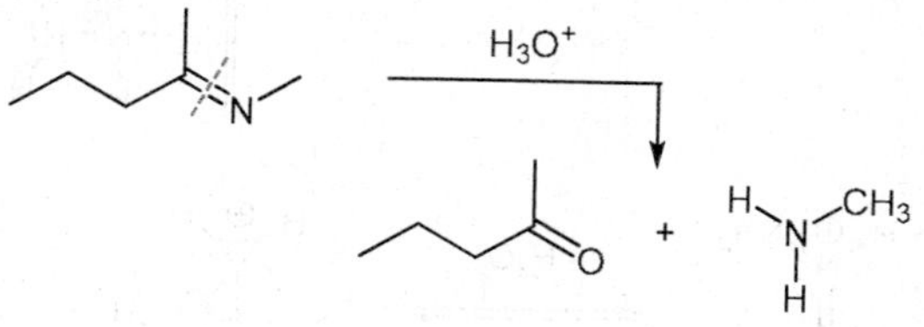

H_3O^+

(c) The starting compound is an acetal, and it is being treated with aqueous acid, so a hydrolysis reaction is expected. We first identify the bonds that will undergo cleavage. When an acetal undergoes hydrolysis, cleavage occurs for the C–O bonds of the acetal group:

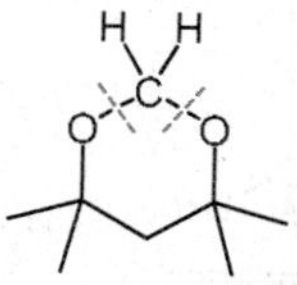

Therefore, the following carbon atom will be converted into a carbonyl group:

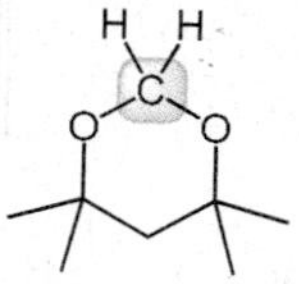

In the process, a diol is released, as shown:

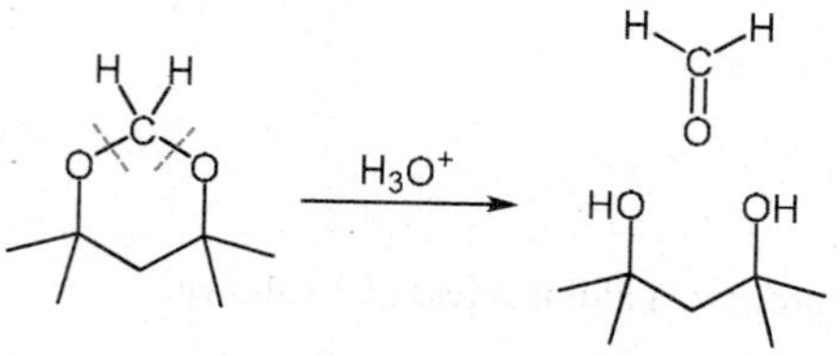

H_3O^+

(d) The starting compound is an enamine, and it is being treated with aqueous acid, so a hydrolysis reaction is expected. We first identify the bond that will undergo cleavage. When an enamine undergoes hydrolysis, cleavage occurs for the bond between the nitrogen atom and the sp^2-hybridized carbon atom to which it is attached:

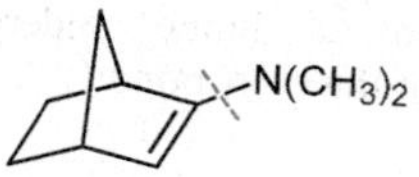

Therefore, the following carbon atom will be converted into a carbonyl group:

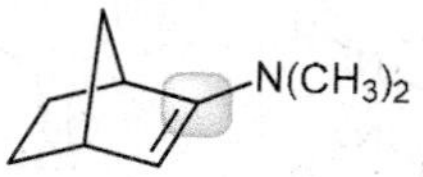

As a result of the C–N bond cleavage, the carbon atom becomes a carbonyl group, and the nitrogen atom will accept a proton to generate a secondary amine:

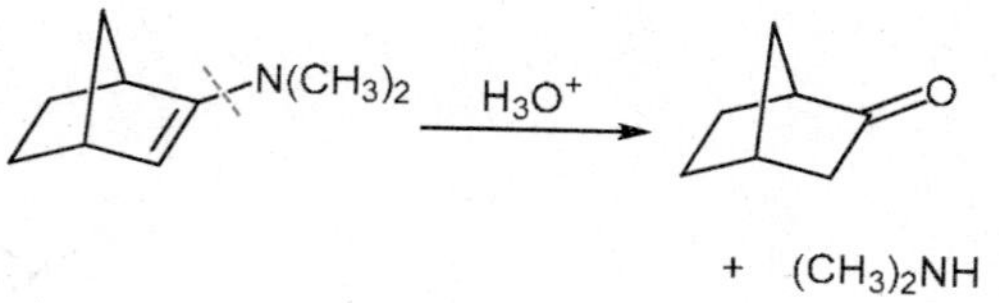

H_3O^+ + $(CH_3)_2NH$

(e) In the starting compound, the nitrogen atom is connected to two different vinylic positions. As such, there are two C–N bonds that will undergo hydrolysis:

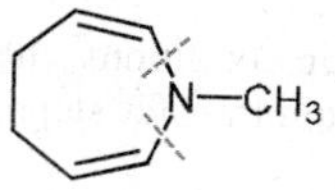

Each of these C–N bonds will be cleaved in the same way that the C–N bond of an enamine group can be cleaved. That is, each of the following carbon atoms (highlighted) will be converted into a carbonyl group:

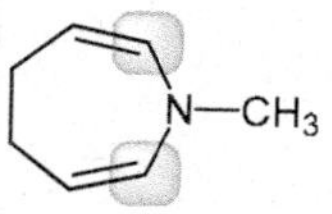

In the process, an amine is released, as shown:

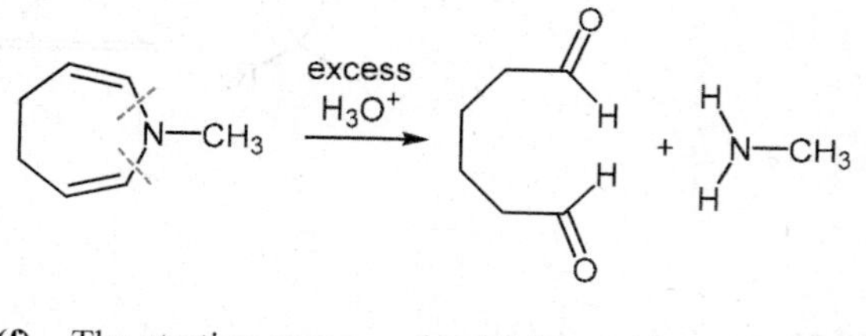

excess H_3O^+

(f) The starting compound exhibits a carbon atom that is connected to two oxygen atoms, as shown, and is therefore an acetal:

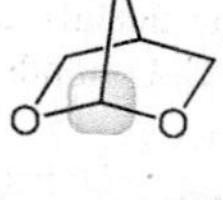

When an acetal is treated with aqueous acid, it is expected to undergo hydrolysis. We first identify the bonds that will undergo cleavage. When an acetal undergoes hydrolysis, cleavage occurs for the C–O bonds of the acetal group:

Each of these bonds is broken, thereby converting the carbon atom of the acetal group into a carbonyl group. In the process, each of the oxygen atoms will accept a proton to become an OH group, giving the following product, which exhibits a carbonyl group, as well as two OH groups:

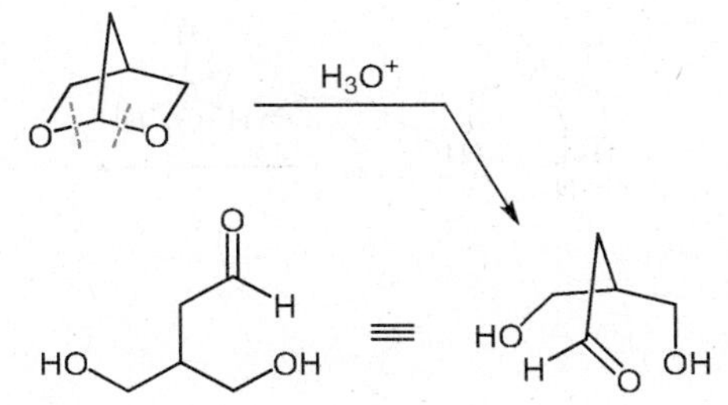

H_3O^+

HO OH H ≡ HO H O OH

19.24. First identify which bond(s) are susceptible to acid-catalyzed hydrolysis. Compound **1** contains a carbon atom flanked by two ether-like oxygen atoms and is therefore an acetal. These two C–O bonds can undergo hydrolytic cleavage upon reaction with excess water and catalytic acid.

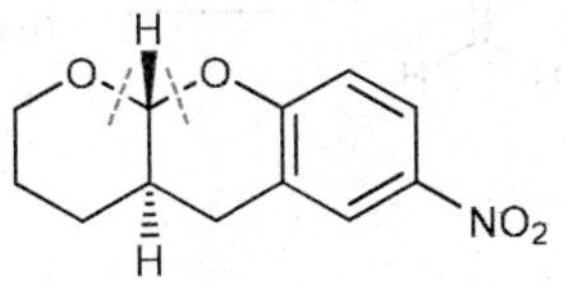

H O O H NO_2

Next, identify the carbon atom that is converted into a carbonyl group.

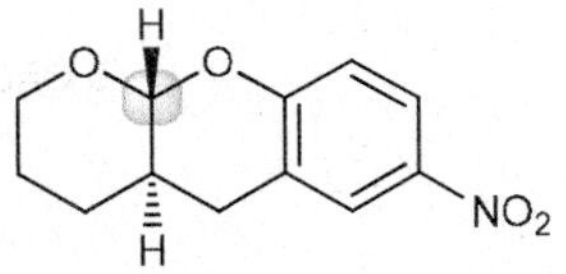

H O O H NO_2

Note that the acetal carbon bears a hydrogen atom and will therefore yield an aldehyde upon hydrolysis. Hydrolysis of this acetal leads to two ring-openings, so the "fragments" will still be tethered to the rest of the molecule in the final product. The two acetal oxygen atoms will become OH groups. The acetal carbon of **1** was a chiral center, and it has been converted into a planar aldehyde group (not a chiral center). The other chiral center in compound **1** was not affected by hydrolysis, so it retains its configuration (H is still on a dash).

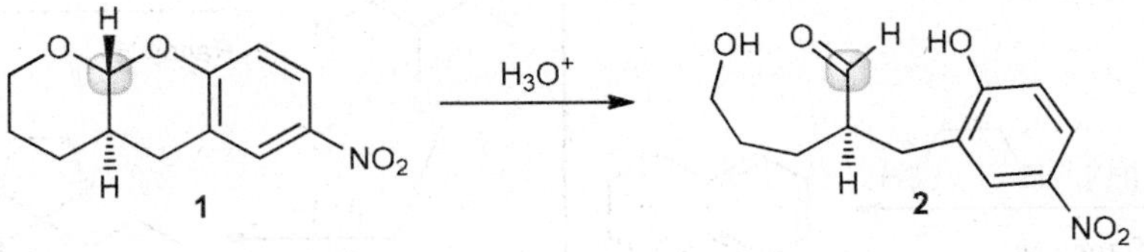

H O O H NO_2 **1** $\xrightarrow{H_3O^+}$ OH O H HO H **2** NO_2

19.25. In aqueous acidic conditions, one of the nitrogen atoms is protonated to give an ammonium ion. Loss of a leaving group then generates an iminium ion, which is attacked by water to give an oxonium ion. Loss of a proton then gives an alcohol (the likely base is water). Protonation of the neighboring nitrogen atom gives a leaving group, which leaves, thereby expelling a fragment with one carbon atom. A proton transfer provides the first equivalent of formaldehyde. The other equivalents are formed in a similar way.

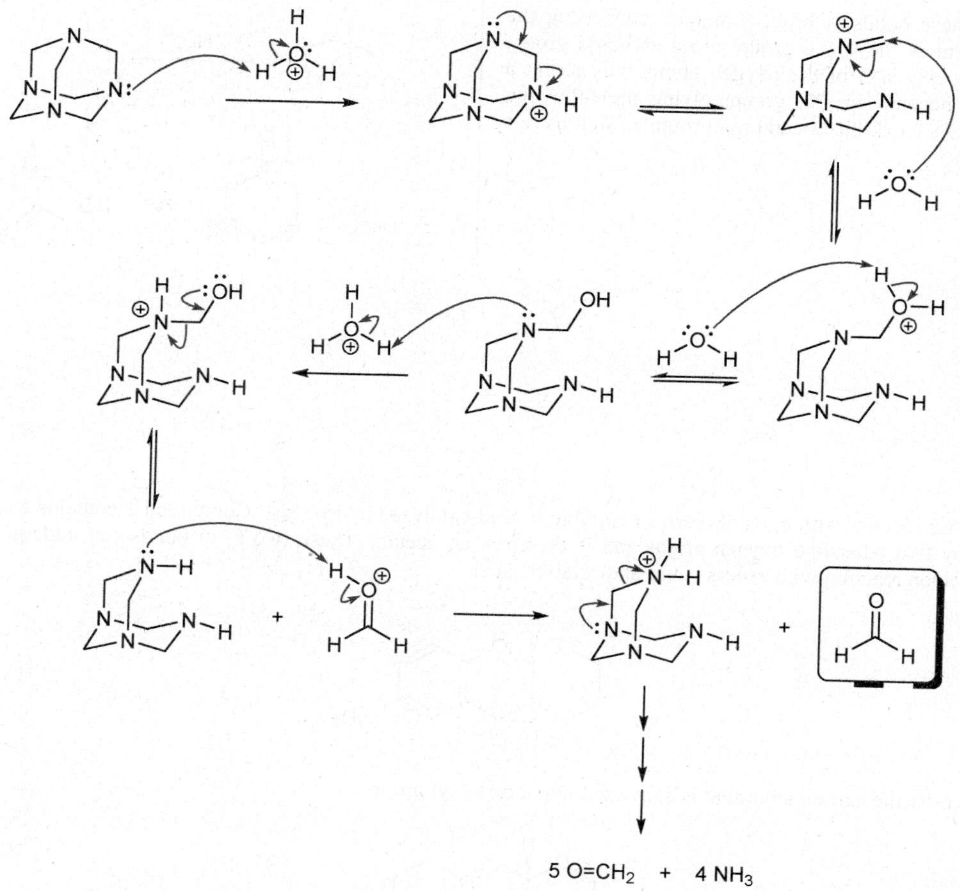

5 O=CH$_2$ + 4 NH$_3$

19.26.
(a) The starting material is a ketone, and the reagents indicate the formation of a cyclic thioacetal, followed by desulfurization with Raney nickel. The net result is conversion of the carbonyl group into a methylene (CH$_2$) group, as shown.

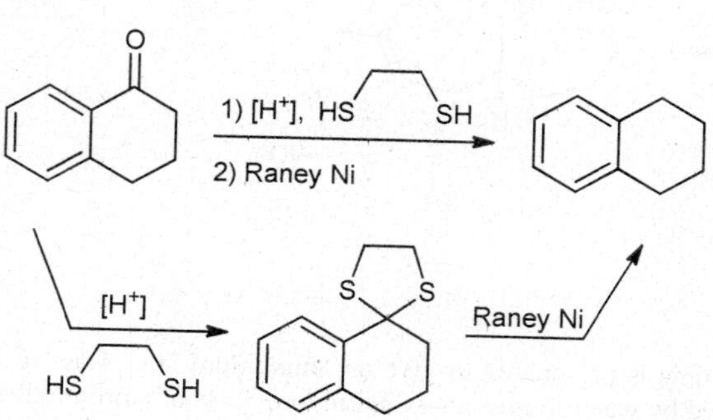

1) [H$^+$], HS SH
2) Raney Ni

[H$^+$]
HS SH
Raney Ni

(b) The starting material is an aldehyde, and the reagents indicate the formation of a cyclic thioacetal, followed by desulfurization with Raney nickel. The net result is the reduction of the carbonyl group to give an alkane.

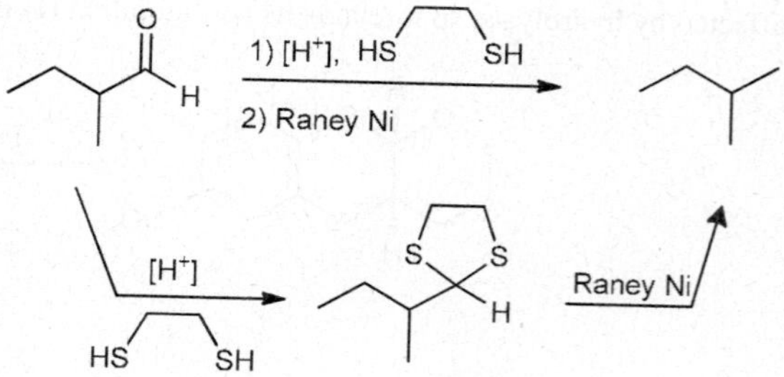

1) [H$^+$], HS SH
2) Raney Ni

[H$^+$]
HS SH
Raney Ni

19.27. When a ketone is treated with a dithiol in the presence of an acid catalyst, a cyclic thioacetal is formed. The carbonyl group is replaced with C–S bonds, as shown:

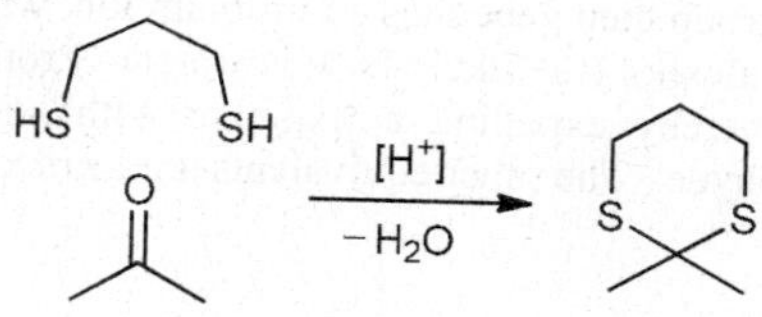

HS SH
[H$^+$]
– H$_2$O

19.28.

(a) The starting material is a ketone, and $LiAlH_4$ is a hydride reducing agent. Upon treatment with $LiAlH_4$, followed by aqueous acidic workup, the ketone is reduced to give the following secondary alcohol. A new chiral center is formed in the process, so we expect a racemic mixture of enantiomers:

1) $LiAlH_4$
2) H_3O^+

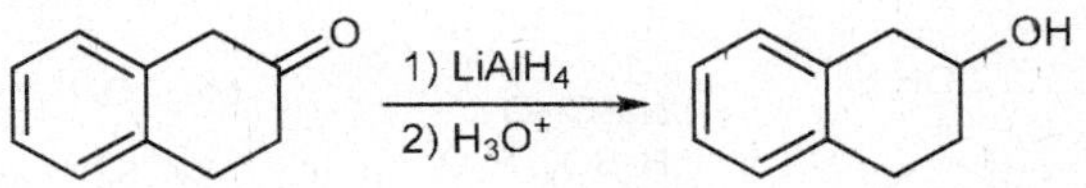

(b) The starting material is an aldehyde, and $NaBH_4$ is a hydride reducing agent. Upon treatment with $NaBH_4$ and methanol, the aldehyde is reduced to give the following primary alcohol:

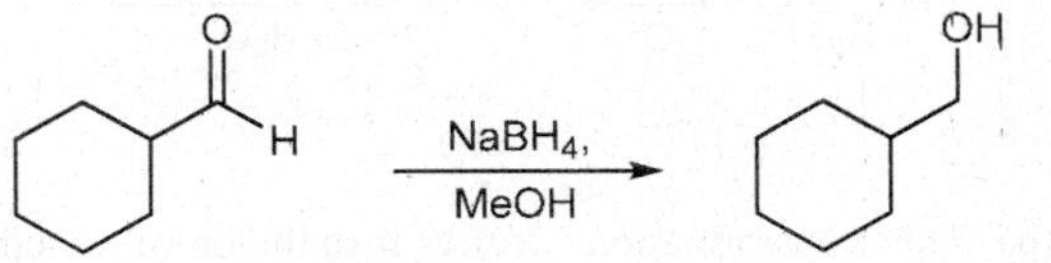

$NaBH_4$, MeOH

(c) The starting material is a ketone, and $LiAlH_4$ is a hydride reducing agent. Upon treatment with $LiAlH_4$, followed by aqueous acidic workup, the ketone is reduced to give the following secondary alcohol. A new chiral center is formed in the process, so we expect a racemic mixture of enantiomers:

1) $LiAlH_4$
2) H_3O^+

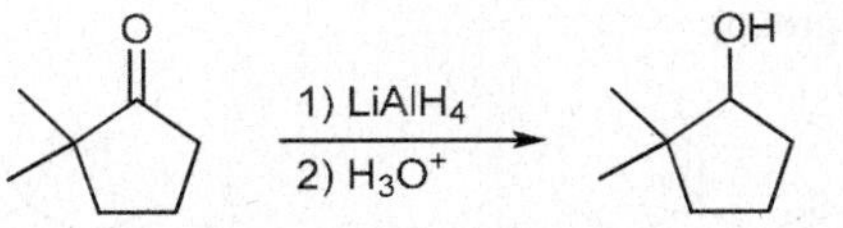

(d) The starting material is a ketone, and $NaBH_4$ is a hydride reducing agent. Upon treatment with $NaBH_4$ and methanol, the ketone is reduced to give the following secondary alcohol. A new chiral center is formed in the process, so we expect a racemic mixture of enantiomers:

$NaBH_4$, MeOH

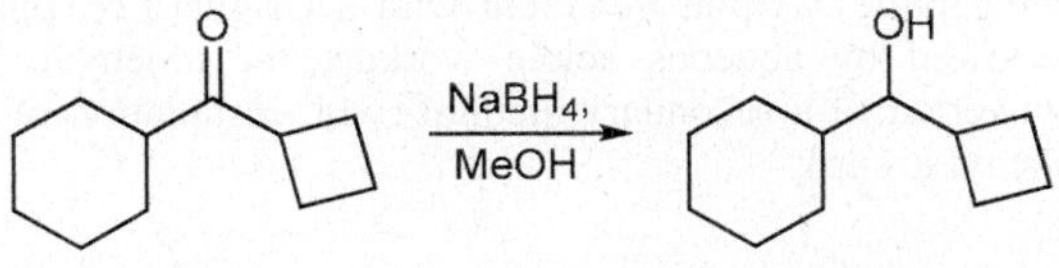

19.29.

(a) The following mechanism is consistent with the description given in the problem statement. After a hydroxide ion attacks one molecule of benzaldehyde, the resulting intermediate functions as a hydride delivery agent to attack another molecule of benzaldehyde, giving a carboxylic acid and an alkoxide ion. The alkoxide ion then deprotonates the carboxylic acid, generating a more stable carboxylate ion. When the reaction is complete, aqueous acid is added to the reaction flask in order to protonate the carboxylate ion, giving benzoic acid.

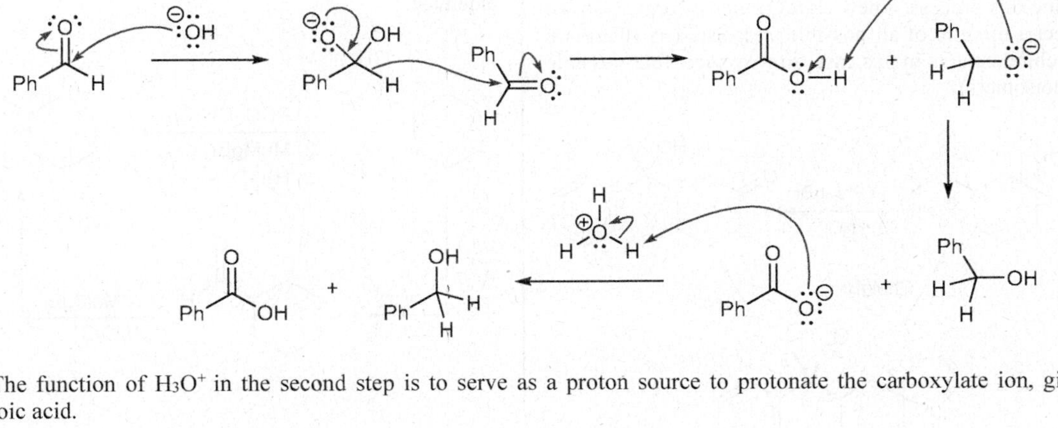

(b) The function of H_3O^+ in the second step is to serve as a proton source to protonate the carboxylate ion, giving benzoic acid.

(c) Water is only a weak acid (pK_a = 15.7) and is not sufficiently strong to serve as a proton source for a carboxylate ion (pK_a of $PhCO_2H$ is 4.2). See Section 3.5 for a discussion of this topic.

19.30.

(a) The starting material is a ketone, and ethyl magnesium bromide is a Grignard reagent (a strong nucleophile). Upon treatment with a Grignard reagent, followed by aqueous acidic workup, the ketone is converted to a tertiary alcohol (with installation of an ethyl group):

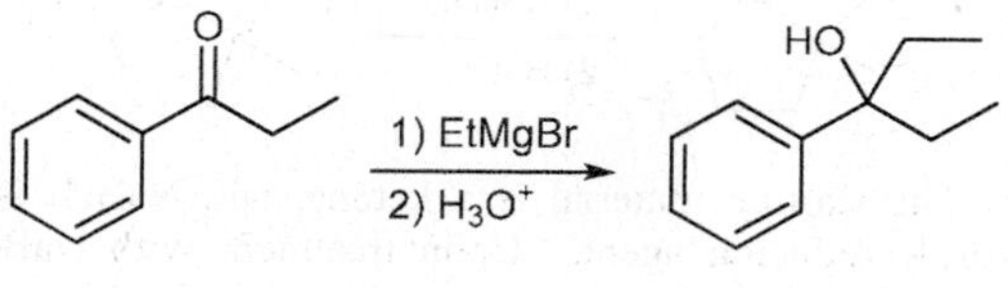

1) EtMgBr
2) H_3O^+

(b) The starting material is an aldehyde, and phenyl magnesium bromide is a Grignard reagent (a strong nucleophile). Upon treatment with a Grignard reagent, followed by aqueous acidic workup, the aldehyde is converted to a secondary alcohol (with installation of a phenyl group):

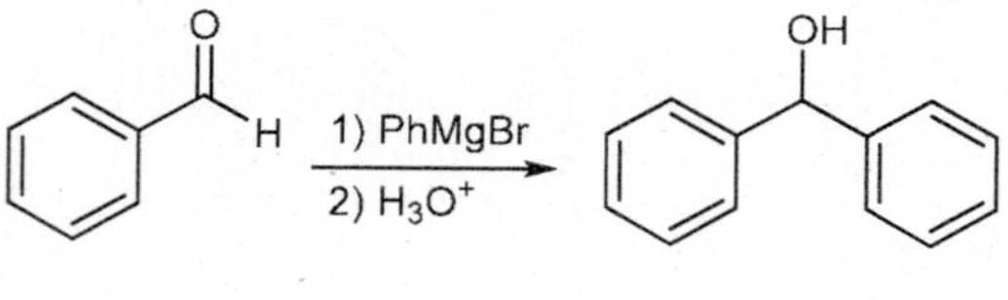

1) PhMgBr
2) H_3O^+

(c) The starting material is a ketone, and phenyl magnesium bromide is a Grignard reagent (a strong nucleophile). Upon treatment with a Grignard reagent, the ketone is attacked, giving the alkoxide ion shown. Treatment with aqueous acid then protonates the alkoxide ion (giving a tertiary alcohol) and also hydrolyzes the acetal group, giving a ketone group. During this process, a new chiral center is formed, so we expect a mixture of all possible stereoisomers (there are two chiral centers in this case, so we expect four possible stereoisomers):

1) PhMgBr
2) H_3O^+

PhMgBr

H_3O^+

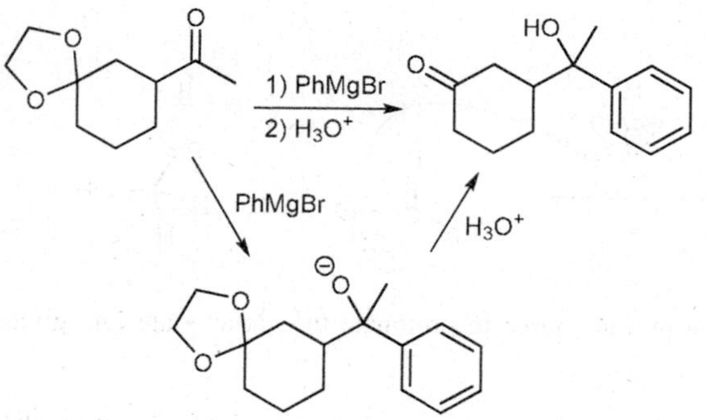

19.31.

(a) This transformation requires installation of a methyl group at the α position of an alcohol. We have not learned a direct way to achieve this transformation. However, if we first oxidize the alcohol to give a ketone, then treatment with methyl magnesium (followed by aqueous acidic workup) will install the methyl group and simultaneously reduce the ketone back to an alcohol. The net result is the conversion of a secondary alcohol into a tertiary alcohol, as shown. A variety of oxidizing agents can be used for the oxidation step, including chromic acid (formed by mixing sodium dichromate with aqueous sulfuric acid).

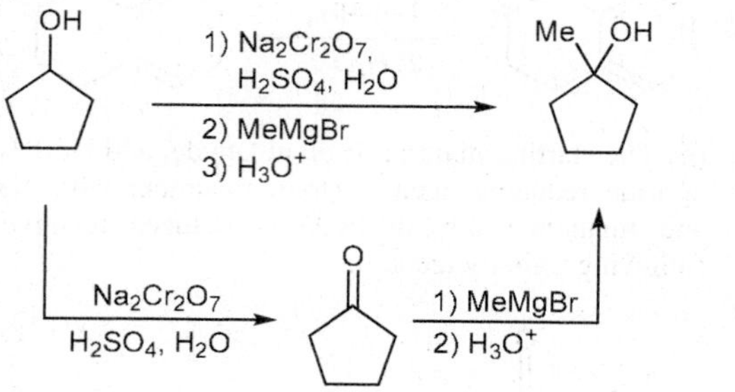

1) $Na_2Cr_2O_7$, H_2SO_4, H_2O
2) MeMgBr
3) H_3O^+

$Na_2Cr_2O_7$
H_2SO_4, H_2O

1) MeMgBr
2) H_3O^+

(b) This transformation requires installation of a methyl group at the α position of an alcohol. We have not learned a direct way to achieve this transformation. However, if we first oxidize the alcohol to give an aldehyde, then treatment with methyl magnesium (followed by aqueous acidic workup) will install the methyl group and simultaneously reduce the aldehyde back to an alcohol. This final step creates a new chiral center, so a racemic mixture of enantiomers is expected. The net result is the conversion of a primary alcohol into a secondary alcohol, as shown. The oxidation step can be performed with PCC (or DMP or Swern conditions), to give an aldehyde. The use of chromic acid would result in formation of a carboxylic acid, rather than an aldehyde.

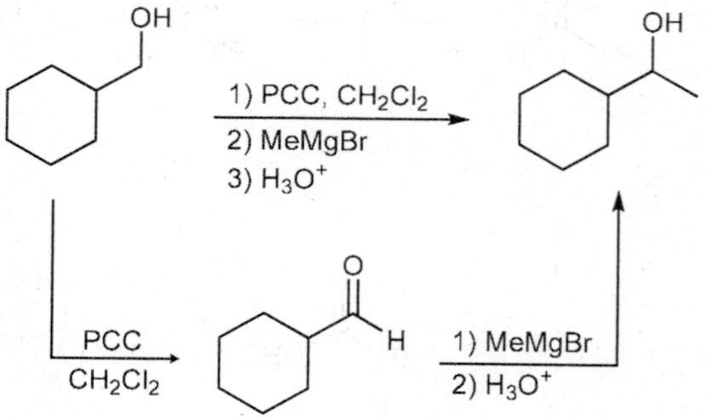

1) PCC, CH_2Cl_2
2) MeMgBr
3) H_3O^+

PCC
CH_2Cl_2

1) MeMgBr
2) H_3O^+

19.32.

(a) The starting material is a ketone, and the reagents indicate cyanohydrin formation, followed by reduction to give an amino alcohol. During cyanohydrin formation, a new chiral center is formed, so a racemic mixture is expected.

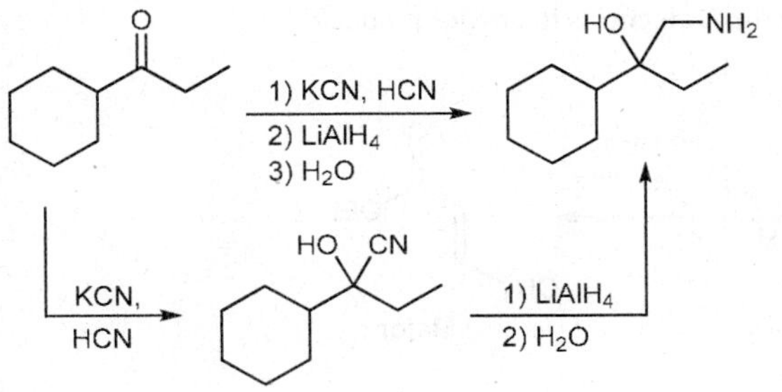

(b) The starting material is an aldehyde, and the reagents indicate cyanohydrin formation, followed by hydrolysis of the cyano group to give a carboxylic acid group. During cyanohydrin formation, a new chiral center is formed, so a racemic mixture is expected.

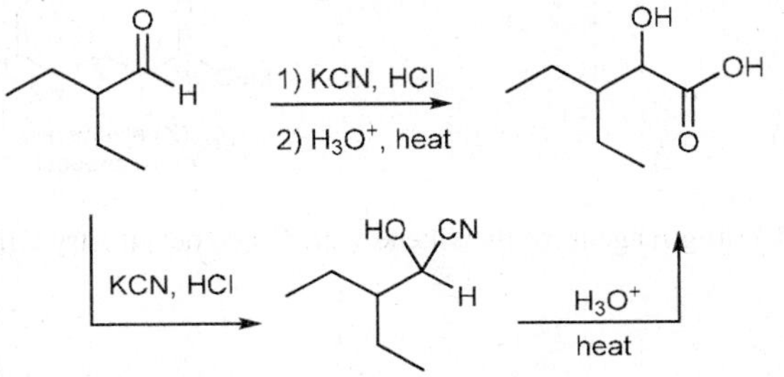

19.33.

(a) This transformation requires installation of a carboxylic acid group at the α position of an alcohol. We have not learned a direct way to achieve this transformation. However, if we first oxidize the alcohol to give a ketone, then treatment with KCN and HCl will convert the ketone into a cyanohydrin (thereby installing the extra carbon atom in the correct location, while simultaneously converting the carbonyl group back into an alcohol). Hydrolysis of the cyano group then gives the desired product. A variety of oxidizing agents can be used for the oxidation step, including chromic acid (formed by mixing sodium dichromate with aqueous sulfuric acid).

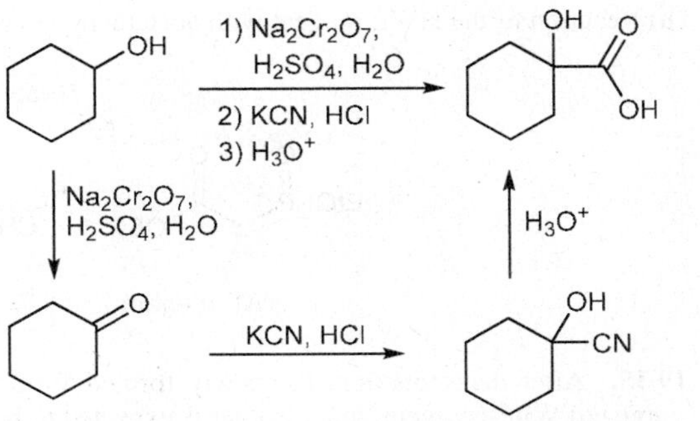

(b) This transformation requires installation of a CH_2NH_2 group at the α position of an alcohol. We have not learned a direct way to achieve this transformation. However, if we first oxidize the alcohol to give a ketone, then treatment with KCN and HCl will convert the ketone into a cyanohydrin (thereby installing the extra carbon atom in the correct location, while simultaneously converting the carbonyl group back into an alcohol). Reduction of the cyano group (with $LiAlH_4$, followed by aqueous acidic workup) then gives the desired product. A variety of oxidizing agents can be used for the oxidation step, including chromic acid (formed by mixing sodium dichromate with aqueous sulfuric acid).

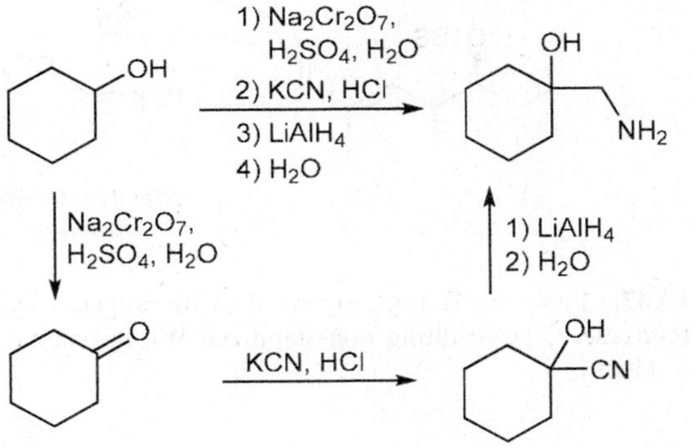

19.34.

(a) This is a reaction between a ketone and a non-stabilized Wittig reagent. Stereochemistry is not relevant in this case since there is only one stereoisomer possible, so the oxygen atom of the carbonyl group is replaced with the propyl group of the Wittig reagent, giving an alkene product. Numbering the carbon atoms on the Wittig reagent can help you draw the correct number of carbon atoms in the product.

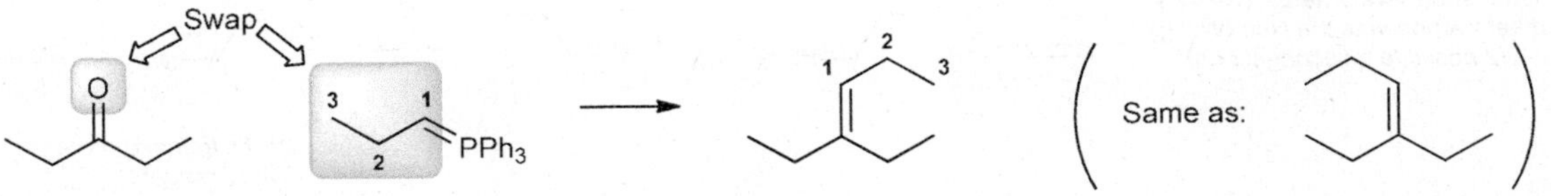

(b) Reaction of the HWE reagent with acetaldehyde will give the *E* alkene as the major product:

Swap

HWE reagent

HWE reaction

Major

19.35. After deprotonation, the newly formed Wittig reagent will react with the aldehyde. Because this is a non-stabilized Wittig reagent, the *Z* isomer is expected to be the major product.

Swap

Non-stabilized Wittig reagent

Wittig reaction

(*Z*)-alkene is major product

19.36. The presence of the ester group makes this a stabilized Wittig reagent, so the alkene with *E* stereochemistry will predominate as the major product.

Swap

Stabilized Wittig reagent

New C=C bond has E stereochemistry

19.37. First, the Wittig reagent must be prepared by deprotonation of the phosphonium salt with the given base, *t*-butoxide. The resulting non-stabilized Wittig reagent will react with the aldehyde to form a (*Z*) alkene as the major product.

Deprotonate

a non-stabilized Wittig reagent

Swap

Wittig

(Z) alkene is major product

19.38.

(a) There are two possible disconnections for the alkene target molecule.

One carbon was the E+ (C=O) & other carbon was the Nuc (Wittig) *(2 possible disconnections)*

or

1° RX to make Wittig reagent — Better (faster) S_N2

2° RX to make Wittig reagent — Slower S_N2

When comparing the two possible Wittig reagents, the one derived from the primary alkyl halide is preferred since it would be a faster and higher yielding S_N2 substitution with PPh_3. This non-stabilized Wittig reagent reacts with 2-butanone to give the target molecule, a *Z* alkene.

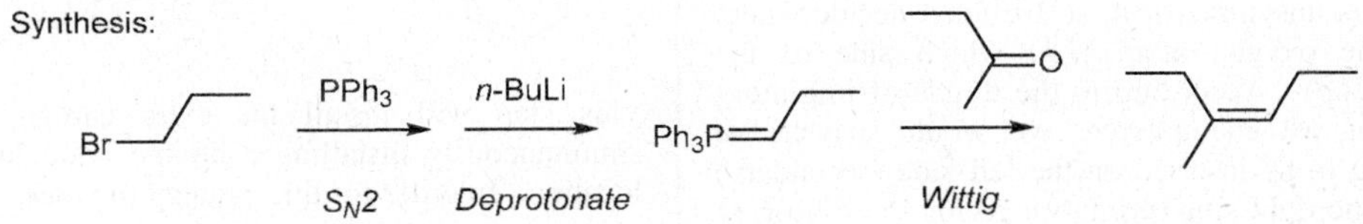

Synthesis:

Br — PPh3 (S_N2) → n-BuLi (Deprotonate) → Ph_3P= — Wittig →

(b) There is only one possible disconnection for this symmetrical alkene target molecule. The stabilized Wittig reagent (due to resonance with phenyl group) is prepared from benzyl bromide and reacts with benzaldehyde to give the target molecule, an *E* alkene.

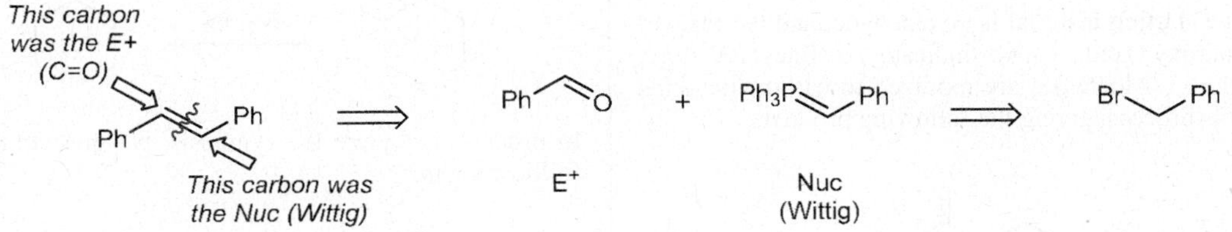

This carbon was the E+ (C=O)

Ph Ph

This carbon was the Nuc (Wittig)

⟹ Ph O (E+) + Ph_3P Ph (Nuc (Wittig)) ⟹ Br Ph

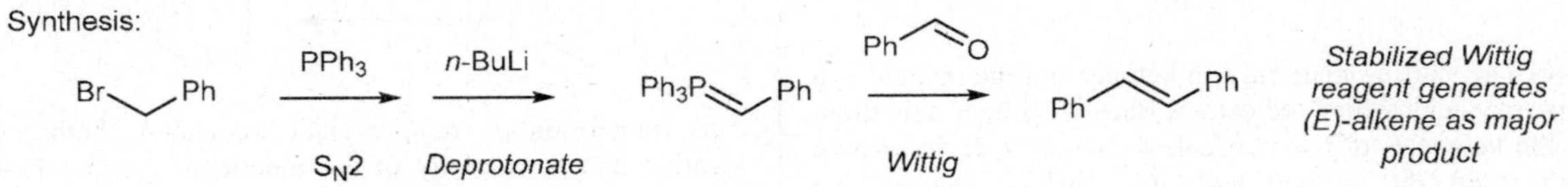

Synthesis:

Br Ph — PPh3 (S_N2) → n-BuLi (Deprotonate) → Ph_3P Ph — Ph O (Wittig) → Ph Ph

Stabilized Wittig reagent generates (E)-alkene as major product

(c) The best disconnection for this alkene target molecule is to align the ester group with the nucleophilic carbon (resulting in either a stabilized Wittig or HWE reagent); this will ensure the (*E*) configuration in the resulting alkene. The other component is a 5-carbon aldehyde (pentanal).

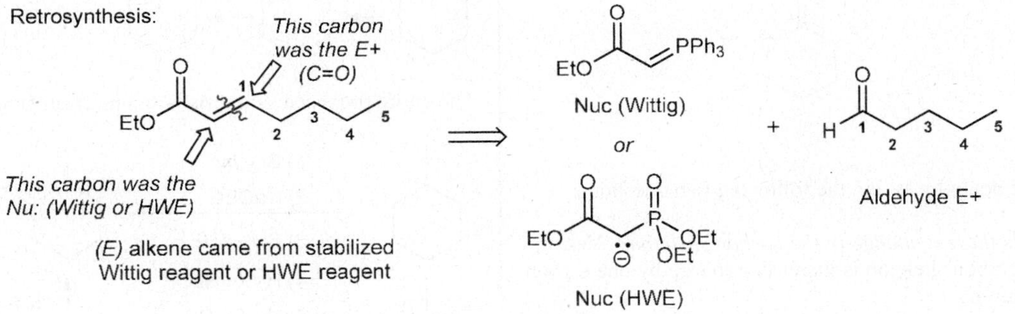

Retrosynthesis:

This carbon was the E+ (C=O)

EtO O 1 2 3 4 5

This carbon was the Nu: (Wittig or HWE)

(E) alkene came from stabilized Wittig reagent or HWE reagent

⟹ EtO O PPh3 — Nuc (Wittig)

or

EtO O O P OEt OEt — Nuc (HWE)

+ H O 1 2 3 4 5 — Aldehyde E+

Reaction of pentanal with either the Wittig or HWE reagent shown will afford the desired target molecule, with (*E*) stereochemistry.

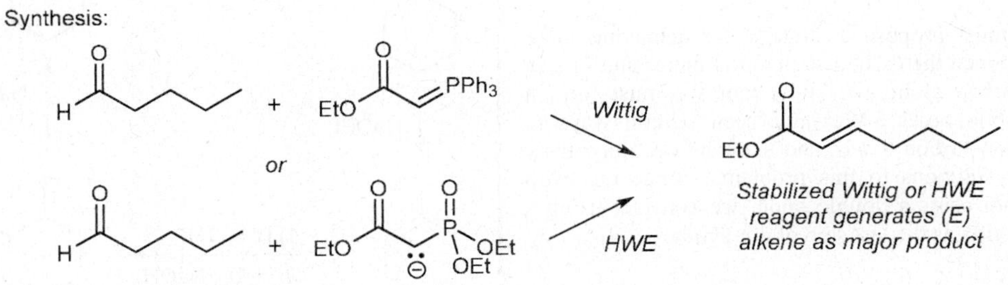

Synthesis:

H O + EtO O PPh3 — Wittig →

or

H O + EtO O O P OEt OEt — HWE →

EtO O

Stabilized Wittig or HWE reagent generates (E) alkene as major product

19.39.

(a) The starting material is a ketone, and the reagent is a peroxy acid, which indicates a Baeyer-Villiger oxidation. The ketone is unsymmetrical, so we must decide where to insert the oxygen atom (*i.e.*, which side of the carbonyl group). According to the trends of migratory aptitude that we encountered, we would expect the oxygen atom to be inserted on the left side (secondary) rather than the right side (primary), giving the following product:

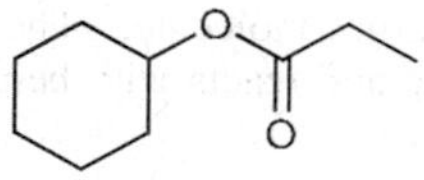

(b) The starting material is an aldehyde, and the reagent is a peroxy acid, which indicates a Baeyer-Villiger oxidation. Aldehydes are oxidized to carboxylic acids with this process, giving the following product:

OH

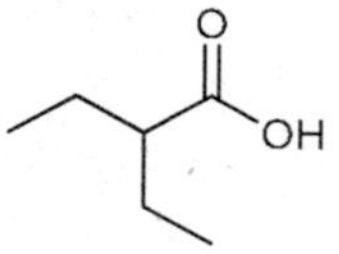

(c) The starting material is a ketone, and the reagent is a peroxy acid, which indicates a Baeyer-Villiger oxidation. The ketone is unsymmetrical, so we must decide where to insert the oxygen atom (*i.e.*, which side of the carbonyl group). According to the trends of migratory aptitude that we encountered, we would expect the oxygen atom to be inserted on the right side (tertiary) rather than the left side (secondary), giving the following product:

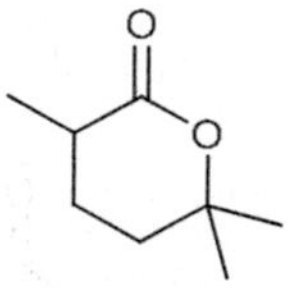

19.40.

(a) We begin by asking the following two questions:

1) *Is there a change in the carbon skeleton?* Yes, the carbon skeleton is increasing in size by one carbon atom.
2) *Is there a change in the functional groups?* Yes, the starting material lacks a functional group, and the product has an exocyclic double bond (extending from the ring).

Now we must propose a strategy for achieving these changes. Since the carbon skeleton is increasing in size by one carbon atom, we know that we must form a carbon-carbon bond. We have seen several ways to make carbon-carbon bonds, and there are certainly many acceptable solutions to this problem. Since the extra carbon atom bears a double bond, we consider using a Wittig reaction as the last step of our synthesis:

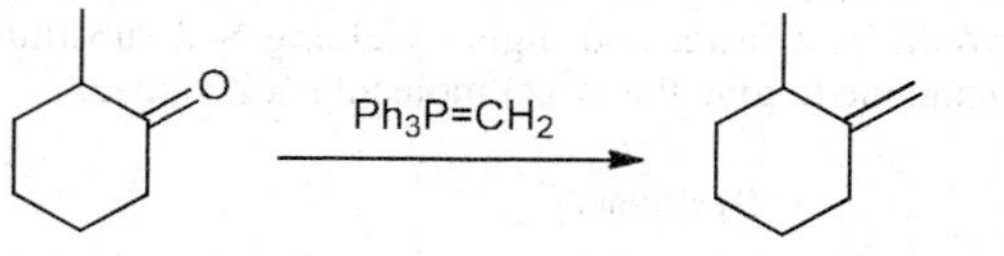

$Ph_3P{=}CH_2$

This step will install the extra carbon atom while simultaneously installing a double bond in the desired location. In order for this strategy to work, we must find a way to convert the starting material into the ketone above. Since the starting material lacks a functional group, the first step our synthesis must be a radical bromination:

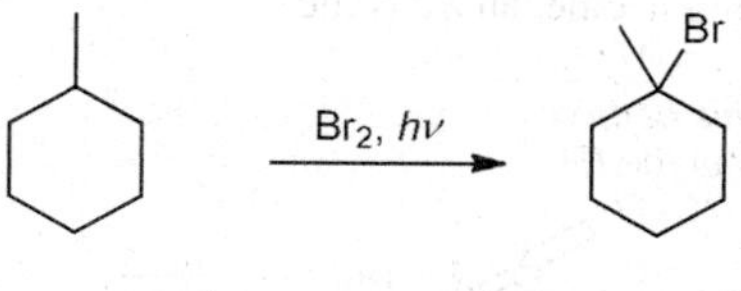

Br_2, $h\nu$

In order to complete the synthesis, we must bridge the following gap:

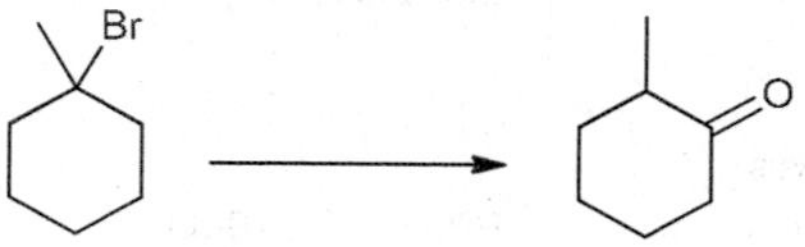

This transformation requires that we change both the location and the identity of the functional group. This can be achieved via elimination to give an alkene, followed by hydroboration-oxidation to give an alcohol, followed by oxidation:

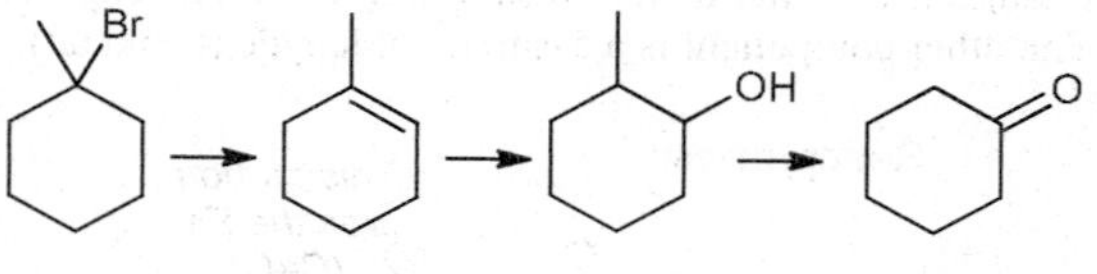

The entire proposed synthesis is summarized here:

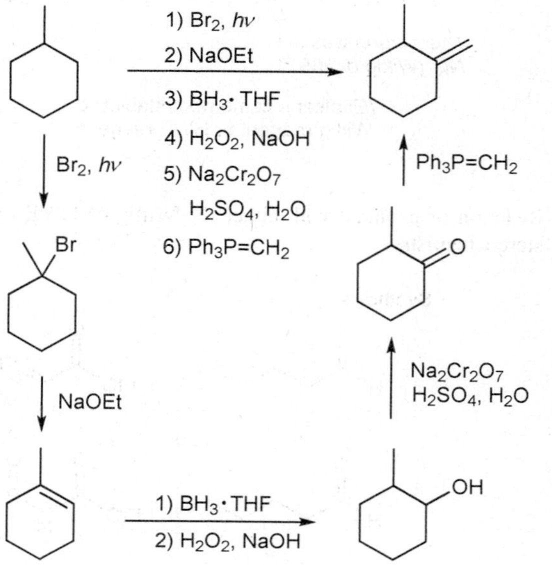

1) Br_2, $h\nu$
2) NaOEt
3) $BH_3 \cdot THF$
4) H_2O_2, NaOH
5) $Na_2Cr_2O_7$, H_2SO_4, H_2O
6) $Ph_3P{=}CH_2$

Br_2, $h\nu$

NaOEt

1) $BH_3 \cdot THF$
2) H_2O_2, NaOH

$Na_2Cr_2O_7$, H_2SO_4, H_2O

$Ph_3P{=}CH_2$

(b) We begin by asking the following two questions:

1) *Is there a change in the carbon skeleton?* Yes, the carbon skeleton is increasing in size by one carbon atom.
2) *Is there a change in the functional groups?* Yes, the starting material has a triple bond, and the product is an alcohol.

Now we must propose a strategy for achieving these changes. There are certainly many acceptable solutions to this problem. One such solution derives from the following retrosynthetic analysis. An explanation of each of the steps (*a-c*) follows.

OH a b c

a. The desired product is a primary alcohol, which can be made via hydroboration-oxidation of an alkene.
b. The alkene can be made from a ketone via a Wittig reaction.
c. The ketone can be made from the starting alkyne (via acid-catalyzed hydration).

Now let's draw the forward scheme. The starting alkyne is treated with aqueous acid in the presence of mercuric sulfate to give a hydration reaction (the initially formed enol rapidly tautomerizes to give a methyl ketone). This ketone is then treated with a Wittig reagent, thereby forming the crucial carbon-carbon bond. The resulting alkene is then expected to undergo hydroboration-oxidation to give *anti*-Markovnikov addition of H and OH across the double bond, affording the desired alcohol.

1) H_2SO_4, H_2O $HgSO_4$
2) $Ph_3P{=}CH_2$
3) $BH_3\cdot THF$
4) H_2O_2, NaOH

OH

H_2SO_4, H_2O $HgSO_4$

1) $BH_3\cdot THF$
2) H_2O_2, NaOH

O

$Ph_3P{=}CH_2$

(c) There are certainly many acceptable solutions to this problem. One such solution is virtually identical to the solution presented for Problem 19.40a, except that the last step has been replaced with a Grignard reaction:

1) Br_2, $h\nu$
2) NaOEt
3) $BH_3\cdot THF$
4) H_2O_2, NaOH
5) $Na_2Cr_2O_7$ H_2SO_4, H_2O
6) MeMgBr
7) H_3O^+

OH

Br_2, $h\nu$

1) MeMgBr
2) H_3O^+

Br

O

NaOEt

$Na_2Cr_2O_7$ H_2SO_4, H_2O

OH

1) $BH_3\cdot THF$
2) H_2O_2, NaOH

(d) We begin by asking the following two questions:

1) *Is there a change in the carbon skeleton?* Yes, the starting material is cyclic, and the product is acyclic. In addition, the starting material has seven carbon atoms, while the product has nine. Two carbon atoms must be installed and the ring must be opened.
2) *Is there a change in the functional groups?* Yes, the starting material lacks a functional group, and the product has two functional groups (C=O and OH).

Now we must propose a strategy for achieving these changes. The ring can be opened via ozonolysis, but only if we first install a double bond in the ring:

1) O_3
2) DMS

O H O

The necessary cycloalkene can be generated from the starting material in just two steps: 1) radical bromination installs a bromine atom in the tertiary position, and 2) subsequent treatment of the resulting tertiary alkyl halide with a strong base (such as sodium ethoxide) gives the cycloalkene above:

Br

This cycloalkene then undergoes ozonolysis to open the ring. After ozonolysis has been performed, the resulting dicarbonyl compound can be converted into the product in just two steps. First, treatment with excess methyl magnesium bromide (followed by aqueous acidic workup) installs two methyl groups, giving a diol. When treated with an oxidizing agent, the tertiary OH group is unaffected, while the secondary OH group is oxidized to a ketone.

1) Br_2, $h\nu$
2) NaOEt
3) O_3
4) DMS
5) excess MeMgBr
6) H_3O^+
7) $Na_2Cr_2O_7$, H_2SO_4, H_2O

Br_2, $h\nu$

NaOEt

1) O_3
2) DMS

1) excess MeMgBr
2) H_3O^+

$Na_2Cr_2O_7$, H_2SO_4, H_2O

(e) The product is an imine, which can be made from the corresponding aldehyde and primary amine:

This aldehyde can be made from the starting material via hydroboration-oxidation, followed by oxidation with PCC (or DMP or Swern conditions), as shown:

1) $BH_3 \cdot THF$
2) H_2O_2, NaOH
3) PCC, CH_2Cl_2
4) $[H_2SO_4]$, $- H_2O$

1) $BH_3 \cdot THF$
2) H_2O_2, NaOH

PCC, CH_2Cl_2

$[H_2SO_4]$, $- H_2O$

(f) While we certainly learned a way to install a bromine atom on a ring (bromination) we did not learn a way to install the other substituent (highlighted):

However, this substituent can be made from the following ketone, using a Wittig reaction:

And we have indeed seen a way to install an acyl group on a ring (Friedel-Crafts acylation). The two substituents are *meta* to each other, so we must utilize the directing effects of the acyl group to install the bromine atom in the correct location (Br is an *ortho-para* director, so it cannot be installed first). After the acyl group has been installed, the product can be obtained via bromination (which installs a bromine atom in the *meta* position), followed by a Wittig reaction, as shown:

1) $AlCl_3$
2) Br_2, $FeBr_3$
3) $Ph_3P{=}CH_2$

$AlCl_3$

Br_2, $FeBr_3$

$Ph_3P{=}CH_2$

(g) The desired product has a carbon atom (highlighted) connected to two oxygen atoms, so this compound is an acetal:

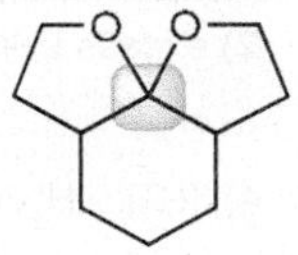

As such, the last step of our synthesis might be formation of the acetal group from the following precursor:

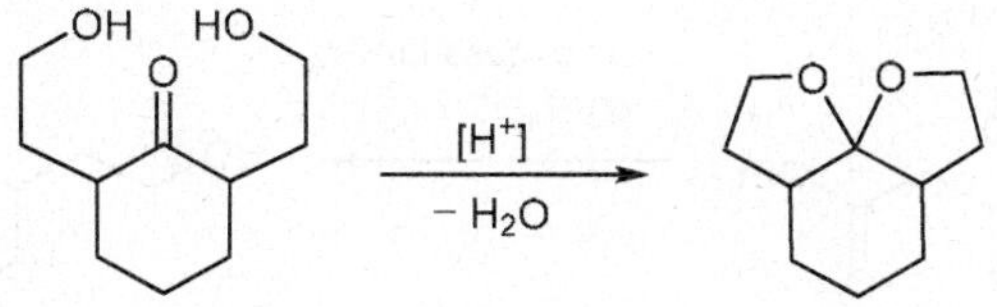

$[H^+]$

$- H_2O$

Notice that the reaction occurs in an intramolecular fashion because the two OH groups and the carbonyl group are tethered together.
Now we must determine how to make the precursor above from the starting material:

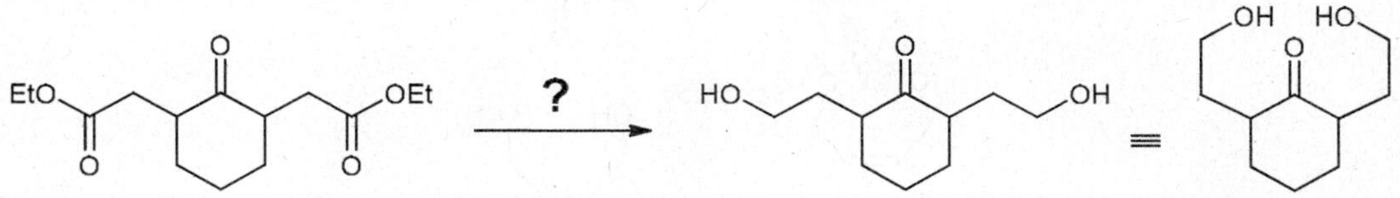

EtO OEt ? HO OH ≡ OH HO

It might be tempting to treat the starting material with a reducing agent, such as $LiAlH_4$ (xs), so it will reduce the two ester groups. Unfortunately, excess $LiAlH_4$ will also reduce the ketone:

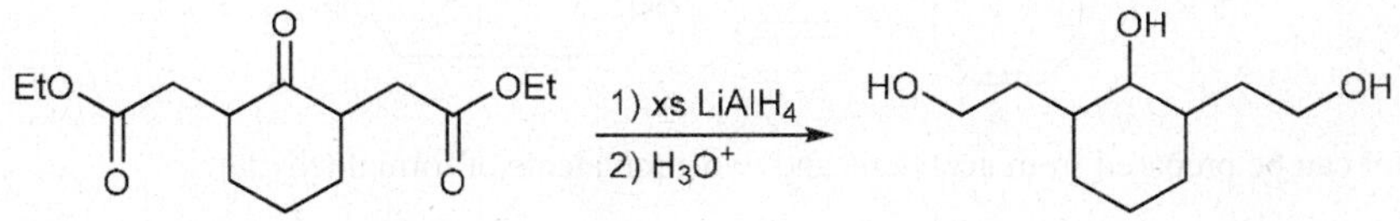

EtO OEt 1) xs $LiAlH_4$ 2) H_3O^+ HO OH OH

And once all three functional groups are reduced to a triol, we are stuck, because we have not learned a way to selectively oxidize the secondary OH group. This problem can be circumvented by using a protecting group first. That is, the ketone can be converted into an acetal group, thereby protecting it prior to the reduction step:

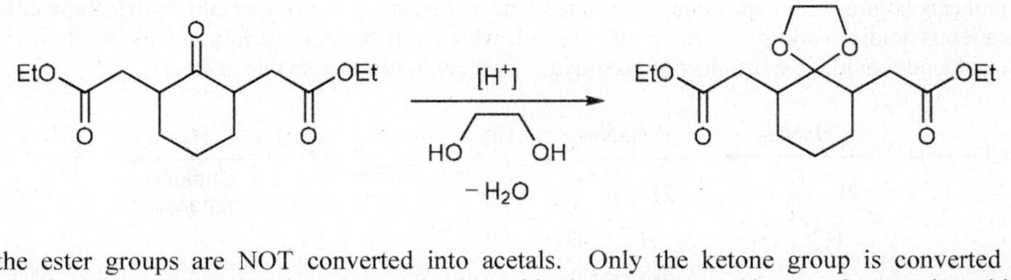

EtO OEt $[H^+]$ HO OH $- H_2O$ EtO OEt

Notice that the ester groups are NOT converted into acetals. Only the ketone group is converted to an acetal. Subsequent reaction with excess $LiAlH_4$ causes reduction of both ester groups (the acetal group is stable under these conditions). Treatment with aqueous acid will then protonate the dianion (each ester group is reduced to an alkoxide when treated with excess $LiAlH_4$). In addition, aqueous acid will also remove the acetal group, converting it back into a ketone. This allows us to perform the final step of the synthesis (intramolecular acetal formation) to give the product. The complete synthesis is shown here:

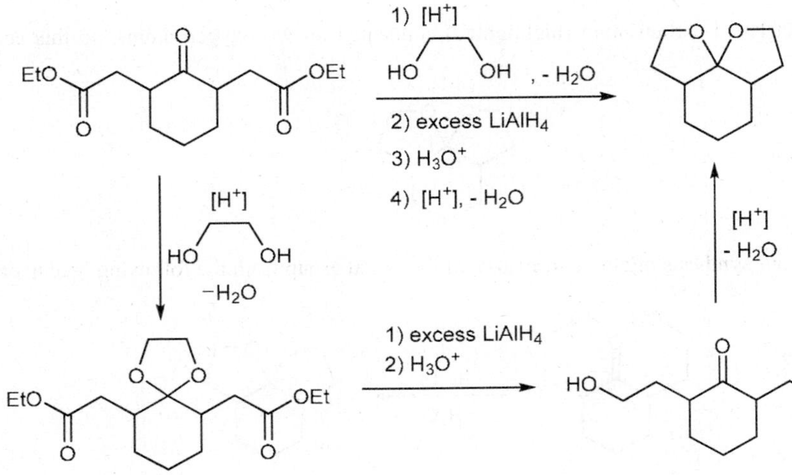

19.41. The product shown is a cyclic acetal, which can be made from a diol and acetone:

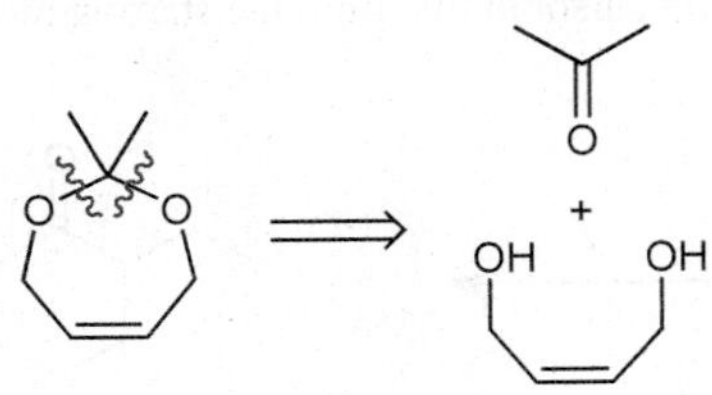

The alkene-diol can be prepared from the corresponding alkyne-diol:

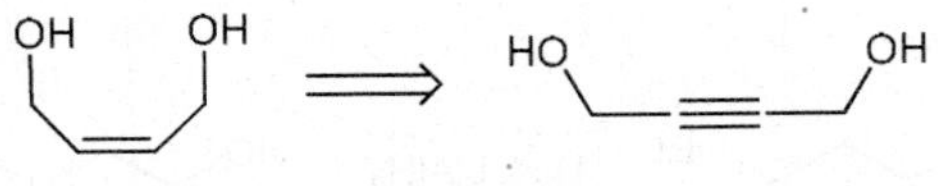

And the alkyne diol can be prepared from acetylene and two equivalents of formaldehyde:

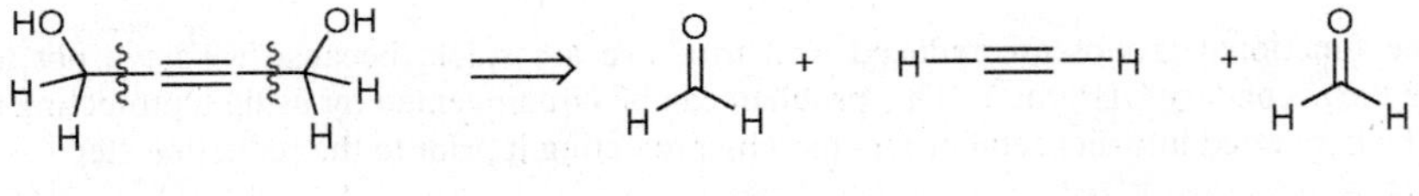

The forward synthesis begins with deprotonation of acetylene and treatment with formaldehyde. Repeating this process (followed by aqueous acidic workup) gives an alkyne-diol, which can be reduced to a *cis*-alkene-diol. That diol then reacts with acetone under acidic conditions (and removal of water) to give the cyclic acetal, **1**.

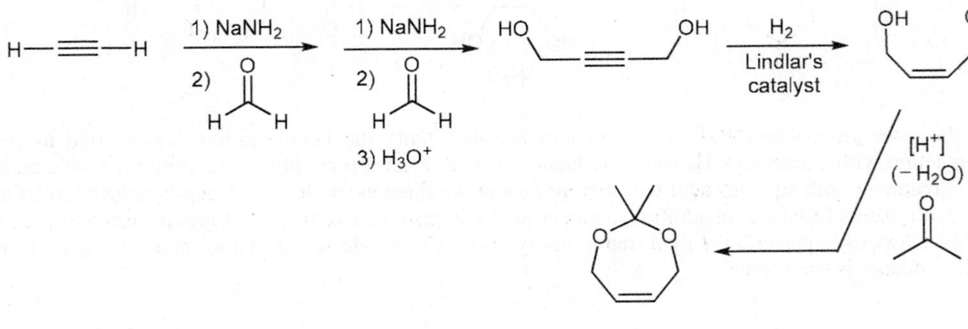

19.42. The signal at 1720 cm^{-1} indicates the presence of a carbonyl group. That carbonyl group is reduced to a methylene (CH_2) group upon treatment with 1,2-ethanedithiol followed by Raney nickel. Therefore, the position of the carbonyl group (in compound **A**) must correspond with one of the methylene groups in the product. Because of symmetry, there are only two unique methylene positions in the product.

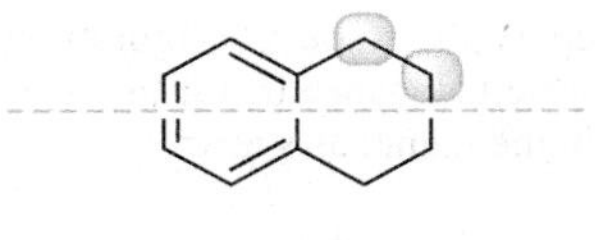

Therefore, compound **A** must be one of the following two structures:

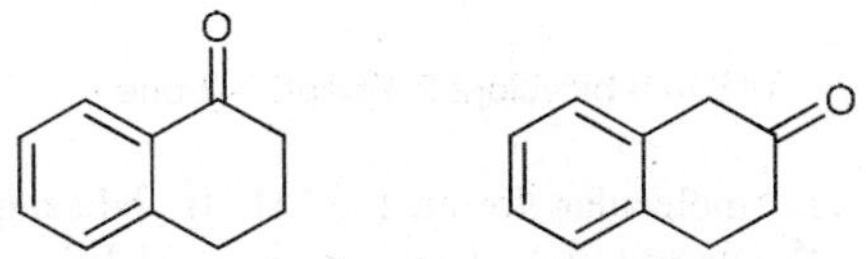

We can differentiate between these two structures based on the nature of the carbonyl group. In the first structure, the carbonyl group is conjugated to the aromatic ring. In the second structure, the carbonyl group is isolated. The signal at 1720 cm^{-1} indicates an isolated carbonyl group (a conjugated carbonyl group would produce a signal near 1680 cm^{-1}). Therefore, compound **A** must be the structure with the isolated carbonyl group.

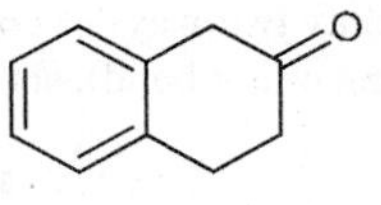

Compound **A**

19.43.

(a) We begin by identifying the parent. In this case, the carbonyl group is part of a five-membered ring, so the parent is cyclopentanone. In a cyclic ketone, the carbonyl group is at C1 (by definition). Next, we identify the substituents. There are two (a propyl group at C2 and a methyl group at C3), which are organized alphabetically in the name (methyl precedes propyl). We then assign a configuration to each of the chiral centers, giving the following name:

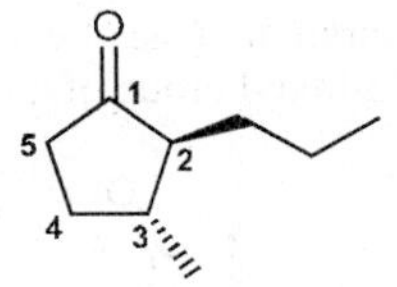

(2*S*,3*R*)-3-methyl-2-propylcyclopentanone

(b) A cyclic compound containing an aldehyde adjacent to the ring is named as a cycloalkane carbaldehyde. In this case, the aldehyde group is connected to a six-membered ring (cyclohexane), giving the following name:

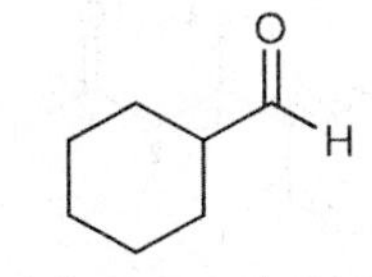

cyclohexanecarbaldehyde

(c) This compound is an aldehyde with a parent of four carbon atoms, so we might think the parent should be butanal. But this parent also contains a double bond, which is indicated by changing "an" to "en" (as in propane vs. propene). Therefore, in this case, it is 2-butenal. Notice that the location of the double bond (between C2 and C3) is indicated with a single locant (2). The location of the aldehyde group is at C1 (by definition) and need not be specified. The methyl group at C3 must be indicated in the name. The name does not have a stereodescriptor (*E* or *Z*) because the C=C bond is not stereoisomeric (one of the vinylic positions is connected to two identical groups).

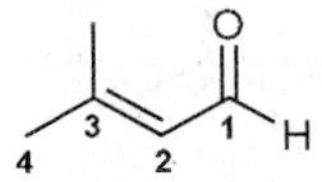

3-methyl-2-butenal

(d) The parent is a chain of six carbon atoms, with the carbonyl group at C3, so the parent is 3-hexanone. Next, we identify the substituents. There is only one: a methyl group at C4. We then assign a configuration to the chiral center, giving the following name:

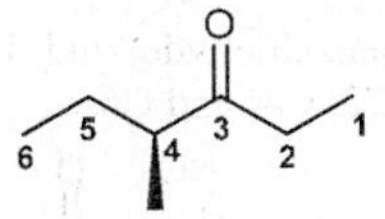

(*S*)-4-methyl-3-hexanone

19.44.

(a) The parent indicates three carbon atoms, and the suffix "dial" indicates two aldehyde groups at either end of the parent chain:

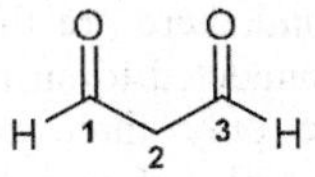

(b) The parent ("butanal") indicates four carbon atoms, and the suffix "al" indicates an aldehyde group (which is at C1, by definition). A phenyl substituent is located at C4:

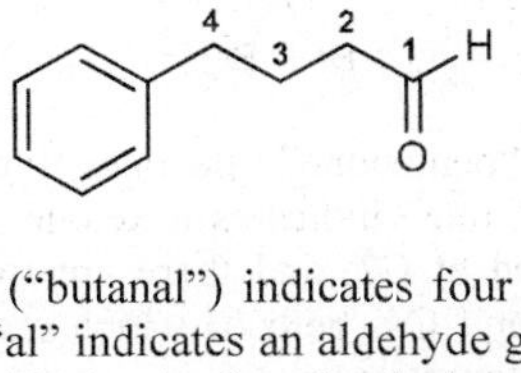

(c) The parent ("butanal") indicates four carbon atoms, and the suffix "al" indicates an aldehyde group (which is at C1, by definition). A phenyl substituent is located at C3, which is a chiral center with the *S* configuration:

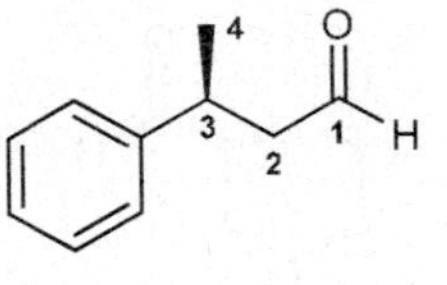

(d) The parent ("heptanone") indicates seven carbon atoms, and the suffix "one" indicates a ketone. The carbonyl group is located at C4, and there are four methyl groups (two at C3 and two at C5):

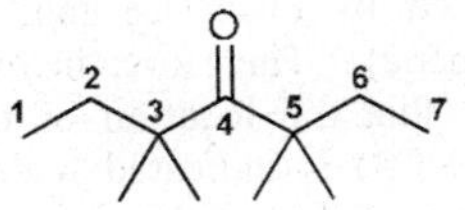

(e) The parent ("pentanal") indicates five carbon atoms, and the suffix "al" indicates an aldehyde group (which is at C1, by definition). An OH group is located at C3, which is a chiral center with the *R* configuration:

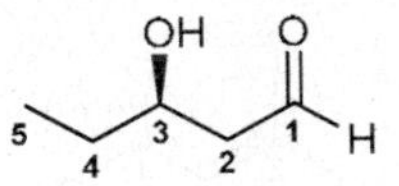

(f) The parent is acetophenone, and there is an OH group in the *meta* position (C3 of the ring):

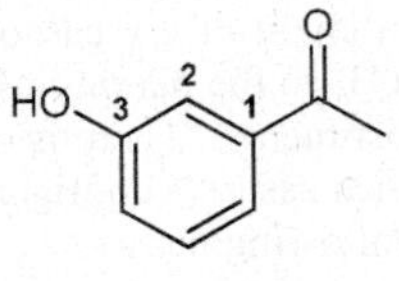

(g) The parent is benzaldehyde, and there are three nitro groups, located at C2, C4, and C6:

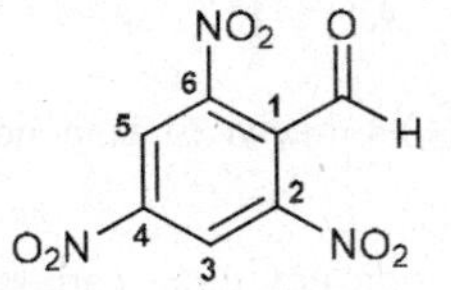

(h) The parent is acetaldehyde (which is the common name for ethanal), and there are three bromine atoms. Since C1 must be connected to an H (in order to be an aldehyde), there are only three positions where the bromine atoms can be placed, and they occupy all three of those positions. This is why locants were not necessary in the name.

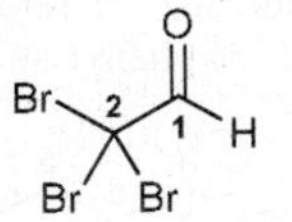

(i) The parent ("pentanone") indicates five carbon atoms, and the suffix "one" indicates a ketone. The carbonyl group is located at C2, and there are two OH groups, located at C3 and C4, both of which are chiral centers with the *R* configuration:

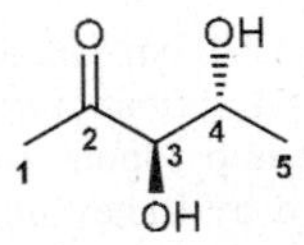

19.45. We begin by identifying the parent chain, which is a bicyclic structure. Using the skills covered in SkillBuilder 4.5, we assign the parent name for this structure and we assign a locant for the position of the carbonyl group (C2). The configurations of the two chiral centers (the bridgeheads, C1 and C4) are assigned and included in the name, as shown:

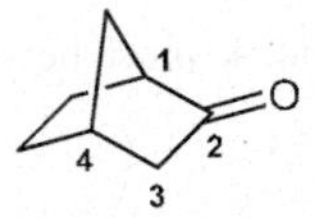

(1*S*,4*R*)-bicyclo[2.2.1]heptan-2-one

19.46. The molecular formula (C_4H_8O) indicates one degree of unsaturation (see Section 14.16), which accounts for the carbonyl group of the aldehyde:

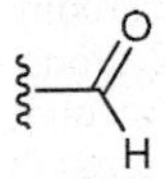

Since there is only one degree of unsaturation (which has now been accounted for), the remaining three carbon atoms do not comprise a ring and do not possess any π bonds. There are only two ways to connect three carbon atoms (without a ring or a π bond), shown here:

Therefore, there are only two aldehydes with the molecular formula C_4H_8O, shown here:

butanal

2-methylpropanal

19.47. The molecular formula ($C_5H_{10}O$) indicates one degree of unsaturation (see Section 14.16), which accounts for the carbonyl group of an aldehyde:

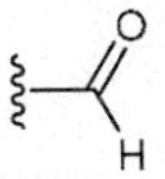

Since there is only one degree of unsaturation (which has now been accounted for), the remaining four carbon atoms do not comprise a ring and do not possess any π bonds. There are only four ways to connect four carbon atoms (without a ring or a π bond), shown here:

Linear Chain of 4

Linear Chain of 3

Linear Chain of 2

Therefore, there are only four aldehydes with the molecular formula $C_5H_{10}O$, shown here. One of these aldehydes exhibits a chiral center (highlighted):

pentanal 2-methylbutanal

3-methylbutanal 2,2-dimethylpropanal

19.48. The molecular formula ($C_6H_{12}O$) indicates one degree of unsaturation (see Section 14.16), which accounts for the carbonyl group of a ketone:

Since there is only one degree of unsaturation (which has now been accounted for), the remaining five carbon atoms do not comprise a ring and do not possess any π bonds. Now let's consider all of the different unique ways of connecting five carbon atoms around a carbonyl group. We can immediately rule out any isomers for which we place all five carbon atoms on one side of the carbonyl group, as that would generate an aldehyde, not a ketone. So, either there are four carbon atoms on one side of the carbonyl group and one carbon atom on the other side, OR, there are two carbon atoms on one side and three carbon atoms on the other side. We must explore each of these possibilities.

If we first consider having two carbon atoms on one side and three carbon atoms on the other side, there are only two such isomers, shown here:

Now we consider having four carbon atoms on one side of the carbonyl group and one carbon atom on the other side. The side with four carbon atoms can be arranged in one of four possible ways, shown here, giving rise to four more isomers:

Linear Chain of 4 Linear Chain of 3

Linear Chain of 2

In total, there are six different isomeric ketones with the molecular formula $C_6H_{12}O$, shown here:

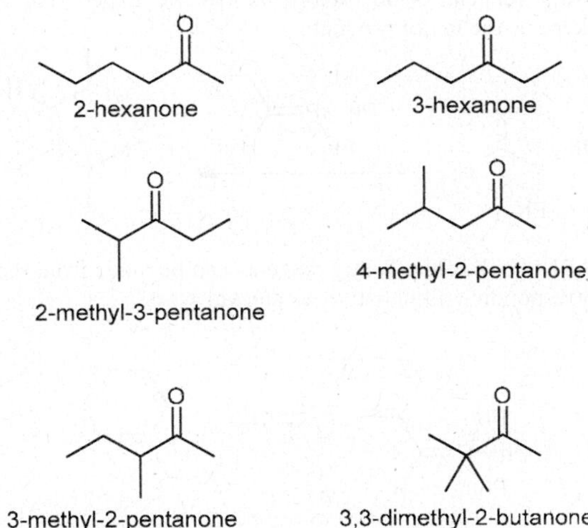

19.49.

(a) The following compound (benzaldehyde) is expected to react with a nucleophile more rapidly, because aldehydes are more reactive than ketones toward nucleophiles, as a result of electronic effects (aldehydes are more electrophilic than ketones) as well as steric effects (See Section 19.4).

(b) The following compound is expected to react with a nucleophile more rapidly, because each of the trifluoromethyl groups is very powerfully electron-withdrawing, thereby rendering the carbonyl group even more electrophilic.

F_3C CF_3

19.50.

(a) The Wittig reagent attacks the carbonyl group, thereby forming a C=C bond between the carbon atom of the carbonyl group and the carbon atom attached to phosphorus (in the Wittig reagent). In this case, the Wittig reagent is stabilized by the presence of the phenyl group, so we expect the product to have the *E* configuration.

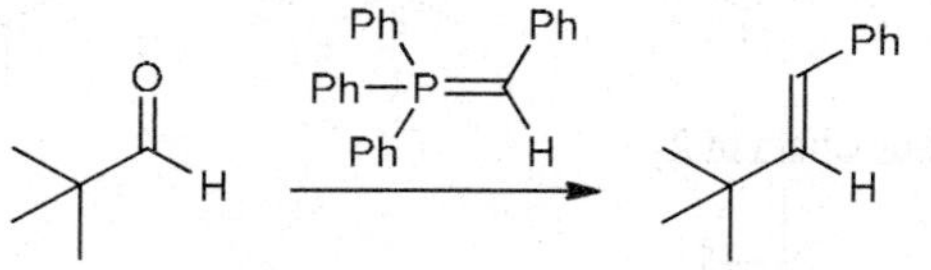

(b) The Wittig reagent attacks the carbonyl group, thereby forming a C=C bond between the carbon atom of the carbonyl group and the carbon atom attached to phosphorus (in the Wittig reagent). In this case, the Wittig reagent is not stabilized, so we expect the *Z* alkene as the major product:

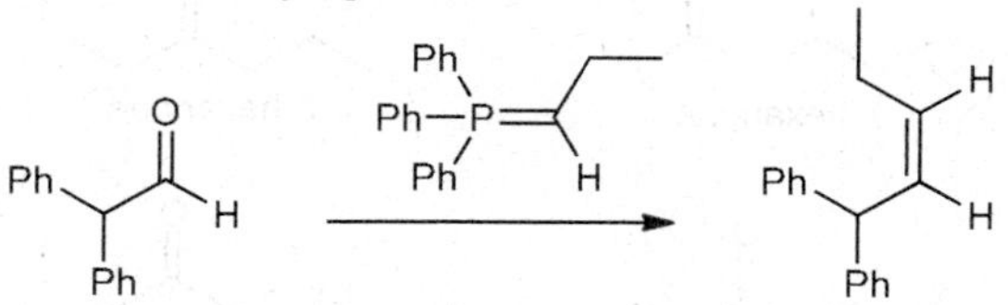

19.51. Each of the Wittig reagents can be made from the corresponding alkyl halide, as shown:

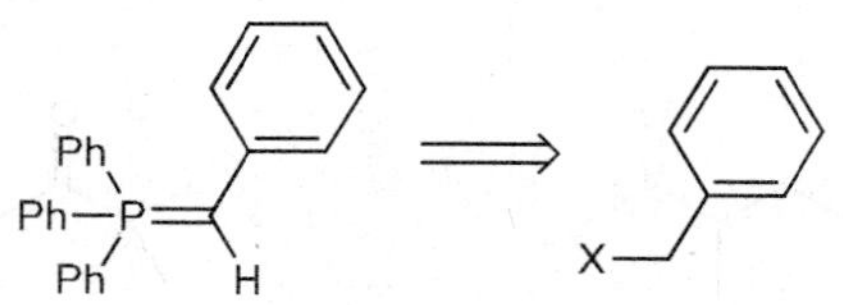

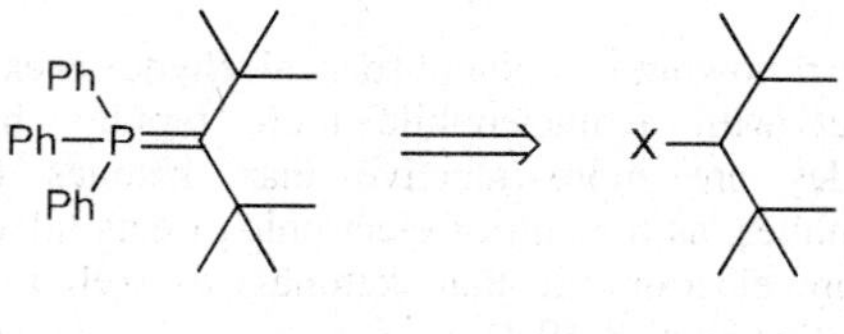

In the first case, the alkyl halide is primary, while in the second case, the alkyl halide is secondary (although this secondary alkyl halide is likely to behave much like a tertiary alkyl halide in that it exhibits significant steric hindrance). The latter alkyl halide will be more difficult to convert into a Wittig reagent, because it is too sterically hindered to undergo an S_N2 attack.

19.52.

(a) A Grignard reagent will attack the carbonyl group of a ketone, thereby forming a C–C bond and giving a tertiary alcohol. In this case, there are two possible routes that can be used to form the product via a Grignard reaction (the Grignard reaction can be used to install either a methyl group or an ethyl group), as shown:

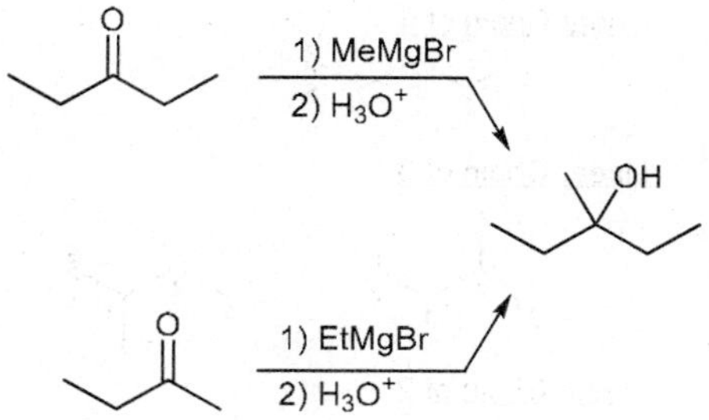

(b) A Grignard reagent will attack the carbonyl group of a ketone, thereby forming a C–C bond and giving an alcohol. In this case, treating cyclohexanone with ethyl magnesium bromide (followed by aqueous acidic workup) gives the desired product:

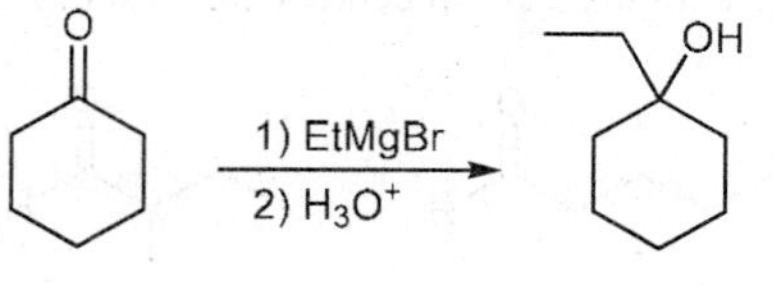

(c) Treating benzophenone with phenyl magnesium bromide (followed by aqueous acidic workup) will afford the desired product via a Grignard reaction:

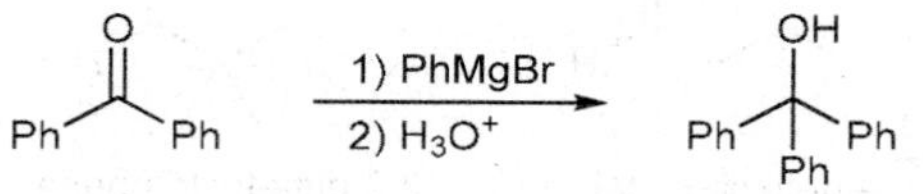

(d) A Grignard reagent will attack the carbonyl group of a ketone, thereby forming a C–C bond and giving a tertiary alcohol. In this case, there are two possible routes that can be used to form the product via a Grignard reaction (the Grignard reaction can be used to install either a phenyl group or a butyl group), as shown:

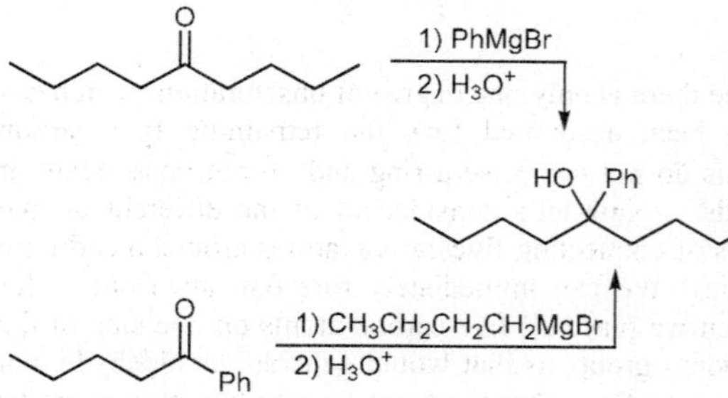

19.53. The following retrosynthetic analysis relies on a different strategy for opening the ring (rather than ozonolysis). Specifically, the final step might involve opening a cyclic ester (called a lactone) into the desired diol via reduction with $LiAlH_4$ (as seen in Section 12.4). This ester can be made from the starting material, as shown. An explanation of each of the steps (*a-d*) follows.

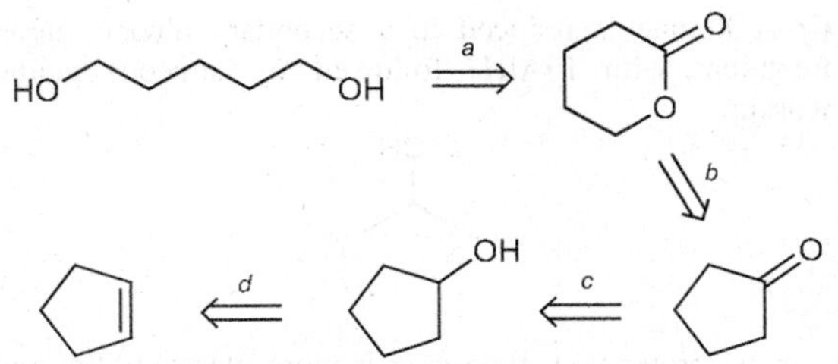

HO~~~OH ⟹(a) lactone ⟹(b) cyclopentanone ⟸(c) cyclopentanol ⟸(d) cyclopentene

a. The diol can be made from the lactone shown, upon reduction with excess $LiAlH_4$.

b. The lactone can be made from the corresponding ketone (cyclopentanone).

c. Cyclopentanone can be made via oxidation of the corresponding secondary alcohol (cyclopentanol).

d. Cyclopentanol can be made from cyclopentene via acid-catalyzed hydration.

Now let's draw the forward scheme. Cyclopentene is treated with aqueous acid to give cyclopentanol, which is then oxidized with chromic acid to give cyclopentanone. A Baeyer-Villiger oxidation converts the ketone into a lactone (a cyclic ester), which can then be converted into the product upon treatment with excess $LiAlH_4$ followed by aqueous acidic workup:

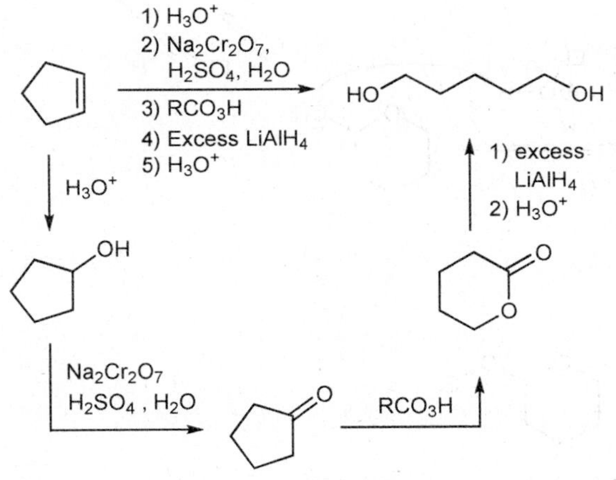

1) H_3O^+
2) $Na_2Cr_2O_7$, H_2SO_4, H_2O
3) RCO_3H
4) Excess $LiAlH_4$
5) H_3O^+

HO~~~OH

H_3O^+

$Na_2Cr_2O_7$
H_2SO_4 , H_2O

RCO_3H

1) excess $LiAlH_4$
2) H_3O^+

19.54.

(a) A ketone is converted into an imine upon treatment with ammonia in acid-catalyzed conditions (with removal of water).

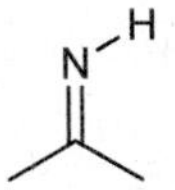

(b) A ketone is converted into an imine upon treatment with a primary amine in acid-catalyzed conditions (with removal of water).

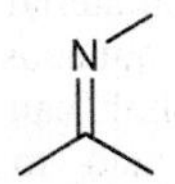

(c) A ketone is converted into an acetal upon treatment with two equivalents of an alcohol in acid-catalyzed conditions (with removal of water).

EtO OEt

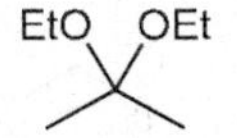

(d) A ketone is converted into an enamine upon treatment with a secondary amine in acid-catalyzed conditions (with removal of water).

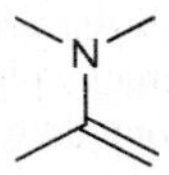

(e) A ketone is converted into a hydrazone upon treatment with hydrazine in acid-catalyzed conditions (with removal of water).

NH_2

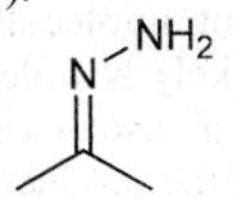

(f) A ketone is converted into an oxime upon treatment with hydroxylamine in acid-catalyzed conditions (with removal of water).

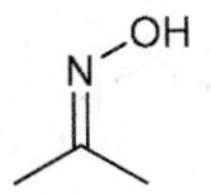

(g) A ketone is reduced to a secondary alcohol upon treatment with a reducing agent such as $NaBH_4$ (methanol serves as a proton source during the reaction).

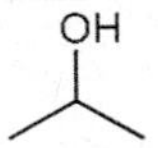

(h) A ketone is converted into an ester upon treatment with a peroxyacid. This process is called a Baeyer-Villiger oxidation.

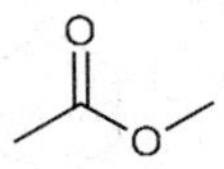

(i) A ketone is converted into a cyanohydrin upon treatment with HCN and KCN.

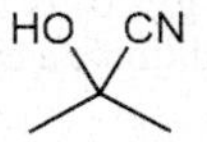

HO CN

(j) Ethyl magnesium bromide is a Grignard reagent (a very strong nucleophile) and it will attack a ketone to give a tertiary alcohol after aqueous acidic workup (with installation of the ethyl group at the α position to the OH group):

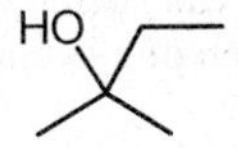

(k) A ketone is converted into an alkene when treated with a Wittig reagent. This particular Wittig reagent installs three carbon atoms, to give the following product:

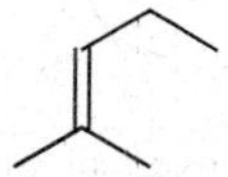

(l) A ketone is reduced to a secondary alcohol upon treatment with $LiAlH_4$ followed by aqueous acidic workup.

OH

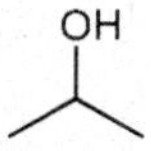

19.55. Under acid-catalyzed conditions, the carbonyl group can be protonated, rendering it more electrophilic, and subject to attack by a nucleophile. There are two options for the attacking nucleophile in the second step of the mechanism: 1) the protonated carbonyl group is tethered to an OH group that can function as a nucleophile and attack the protonated carbonyl group in an intramolecular fashion, or 2) the reagent (ethanol) can serve as a nucleophile and attack the protonated carbonyl group. Indeed, both of these processes will occur, as the product is an acetal. But we must decide which one is likely to occur first. As a general rule, intramolecular reactions will occur more rapidly than intermolecular reactions. As such, the second step of the mechanism will show the OH group attacking the protonated carbonyl group in an intramolecular fashion to give an oxonium ion, which then loses a proton to give a hemiacetal (the base for this step is likely a molecule of ethanol). Protonation of the hemiacetal gives a good leaving group (water) which leaves to give a resonance-stabilized cation that is attacked by ethanol. The resulting oxonium ion is then deprotonated to afford the product.

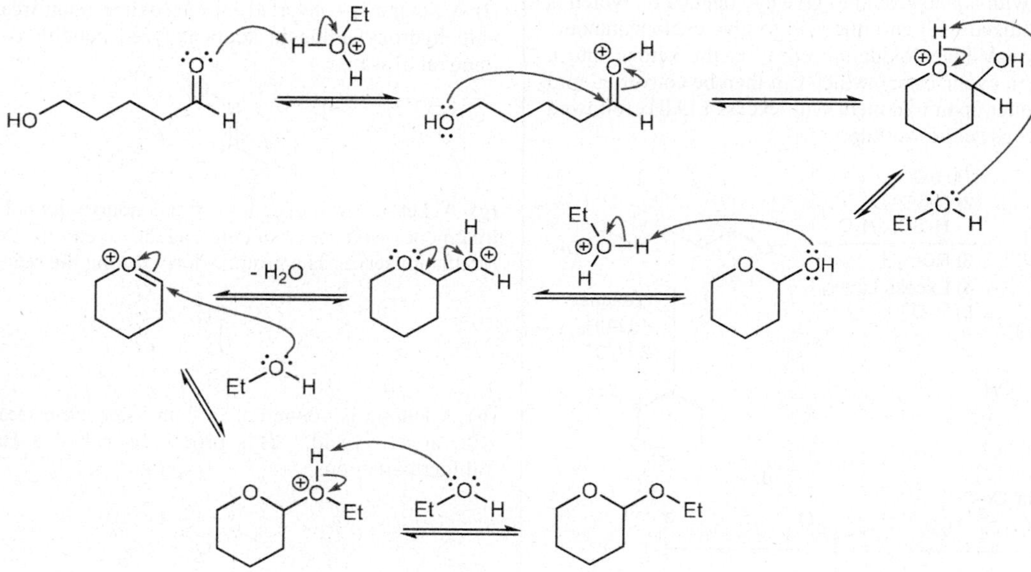

- H_2O

19.56. When ethylene glycol is treated with a ketone or aldehyde in acidic conditions, a cyclic acetal is formed:

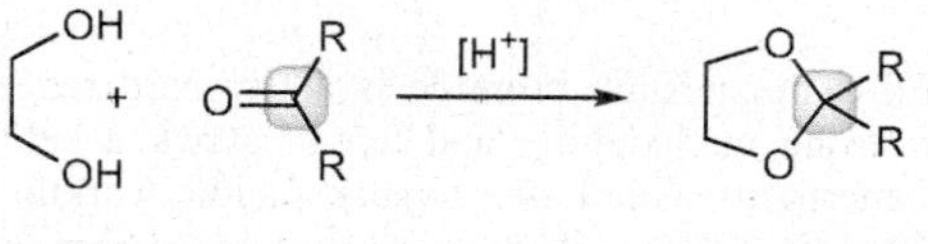

[H^+]

Catechol also has two OH groups (like ethylene glycol). So, when treated with a ketone or aldehyde in acidic conditions, catechol can behave much like ethylene glycol, giving formation of a cyclic acetal:

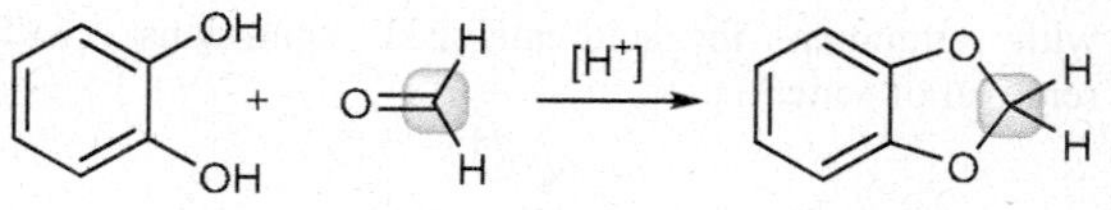

[H^+]

$C_7H_6O_2$

19.57.

(a) This transformation requires installation of a methyl group, which can be achieved with a Grignard reaction (by treating the starting material with methyl magnesium bromide, followed by aqueous acidic workup). The resulting tertiary alcohol can then be heated with concentrated sulfuric acid to give an E1 process, generating the desired alkene.

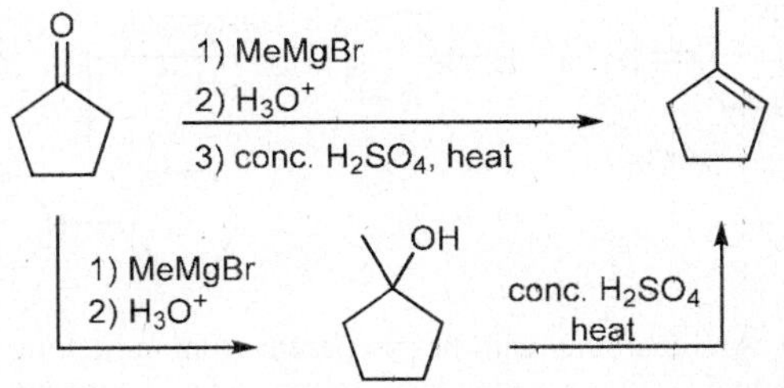

(b) This transformation requires installation of a carbon atom, while converting the carbonyl group into a C=C bond. This can be achieved in just one step, via a Wittig reaction:

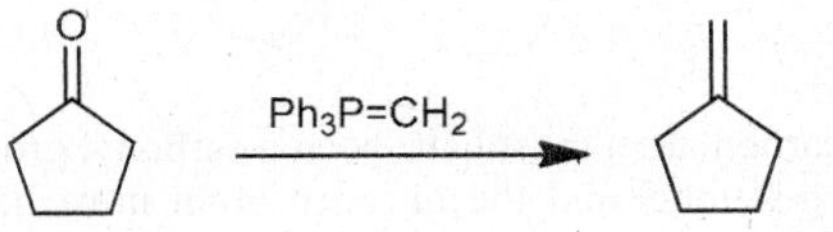

(c) This transformation can be achieved in just two steps: 1) conversion of the ketone to a cyanohydrin, followed by 2) hydrolysis of the cyano group to give a carboxylic acid:

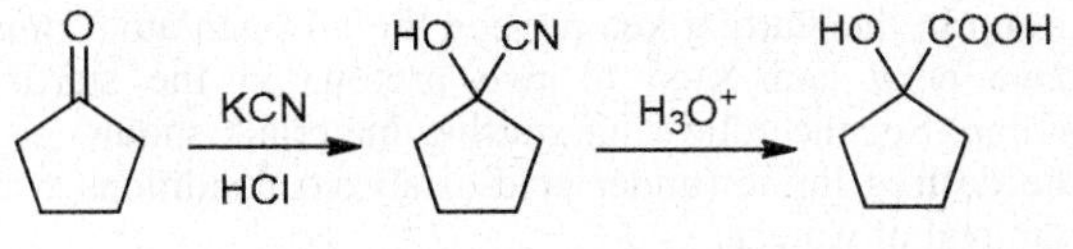

(d) This transformation involves insertion of an oxygen atom next to the carbonyl group of a ketone, thereby converting the ketone into a cyclic ester. This can be accomplished with a Baeyer-Villiger oxidation, which requires the use of a peroxyacid.

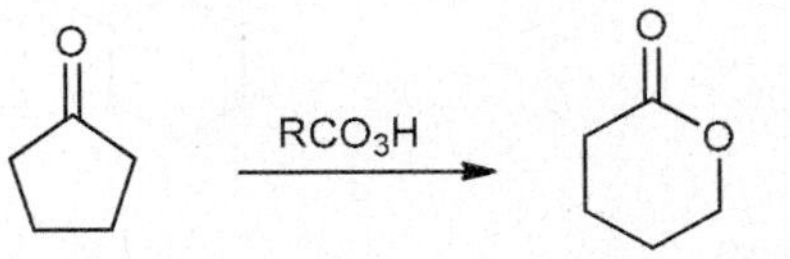

19.58. Under acid-catalyzed conditions, one of the carbonyl groups can be protonated, rendering it more electrophilic, and subject to attack by a nucleophile. Water serves as a nucleophile, attacking the protonated carbonyl group to give an oxonium ion, which then loses a proton (the base for this deprotonation step is likely a molecule of water). Under these conditions, the other carbonyl group can be protonated, rendering it more electrophilic (just as we saw with the first carbonyl group), and subject to attack by a nucleophile. One of the OH groups (present in the intermediate), can then function as a nucleophile and attack the protonated carbonyl group, giving an oxonium ion, which is deprotonated to give the product.

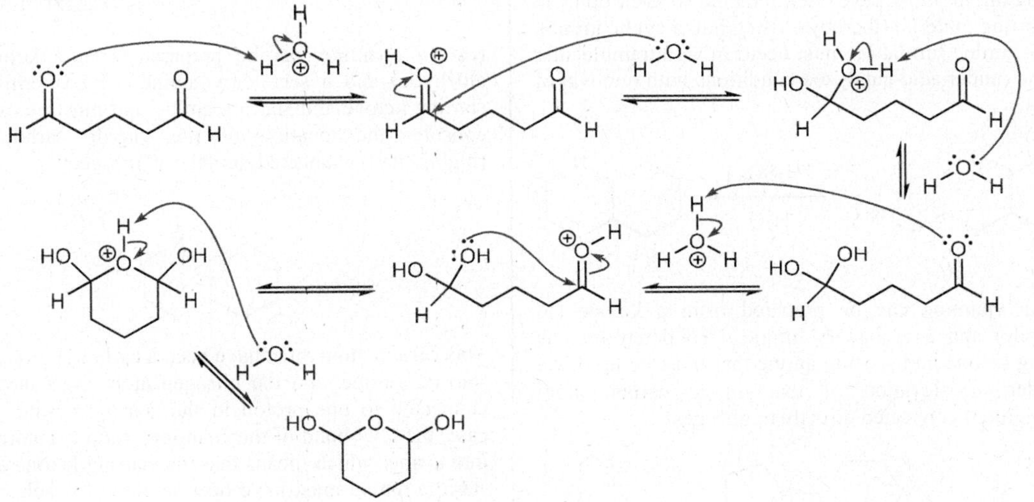

19.59.

(a) An imine can be prepared from a ketone (or aldehyde) and a primary amine. To determine the starting ketone and starting amine that must be used, we consider the location of the C=N bond in the imine. The carbon atom of this bond must have been a carbonyl group in the starting ketone, and the nitrogen atom must have been connected to two protons in the starting amine. So, the following starting materials should give the desired imine (under acid-catalyzed conditions with removal of water):

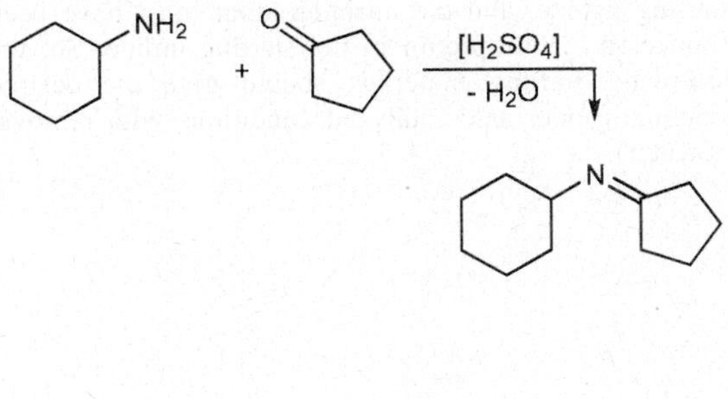

(b) An imine can be prepared from a ketone (or aldehyde) and a primary amine. To determine the starting ketone and starting amine that must be used, we consider the location of the C=N bond in the imine. The carbon atom of this bond must have been a carbonyl group in the starting ketone, and the nitrogen atom must have been connected to two protons in the starting amine. So, the following starting materials should give the desired imine (under acid-catalyzed conditions with removal of water):

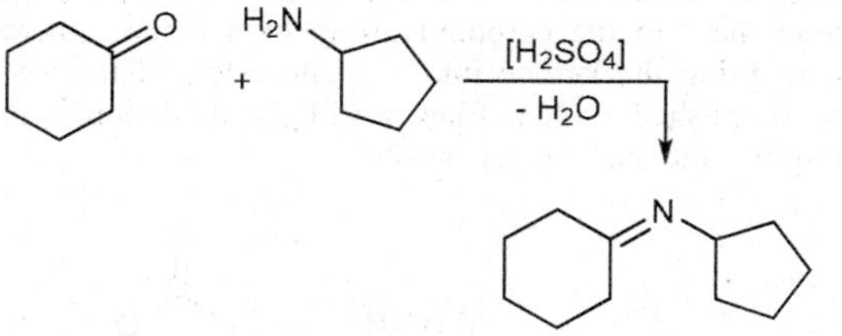

(c) An imine can be prepared from a ketone (or aldehyde) and a primary amine. To determine the starting ketone and starting amine that must be used, we consider the location of the C=N bond in the imine. The carbon atom of this bond must have been a carbonyl group in the starting ketone, and the nitrogen atom must have been connected to two protons in the starting amine. In this case, the C=N bond is incorporated into a ring, which means that the starting ketone and the starting amine must have been tethered to each other in the starting material (thereby closing into a cyclic imine). That is, imine formation must occur in an intramolecular fashion (under acid-catalyzed conditions with removal of water):

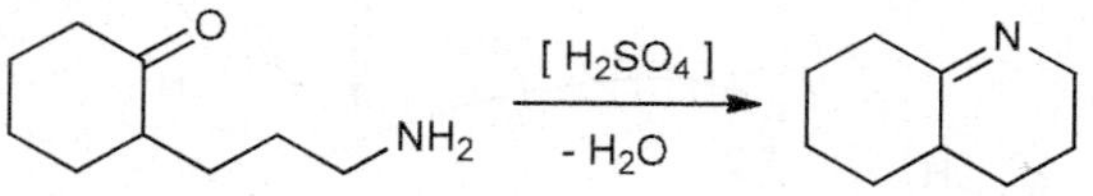

19.60.

(a) An enamine can be prepared from a ketone (or aldehyde) and a secondary amine. To determine the starting ketone and starting amine that must be used, we consider the location of the vinylic carbon atom (highlighted) connected directly to nitrogen:

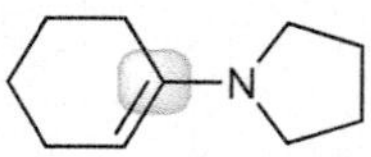

This carbon atom must have been a carbonyl group in the starting ketone, and the nitrogen atom must have been connected to one proton in the starting amine. So, the following starting materials should give the desired enamine (under acid-catalyzed conditions with removal of water):

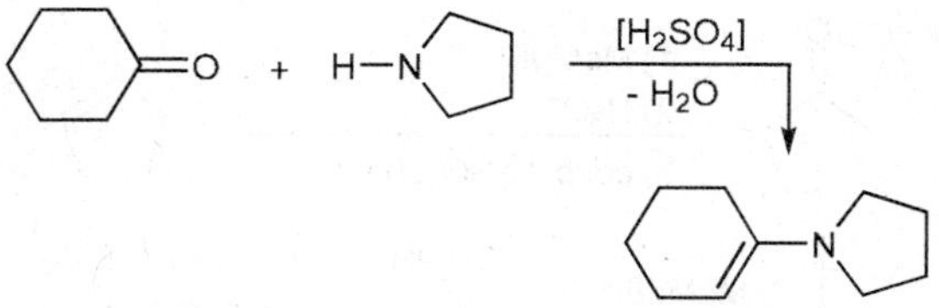

(b) An enamine can be prepared from a ketone (or aldehyde) and a secondary amine. To determine the starting ketone and starting amine that must be used, we consider the location of the vinylic carbon atom (highlighted) connected directly to nitrogen:

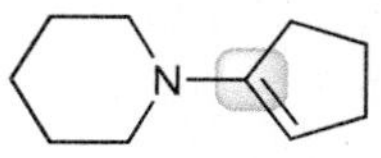

This carbon atom must have been a carbonyl group in the starting ketone, and the nitrogen atom must have been connected to one proton in the starting amine. So, the following starting materials should give the desired enamine (under acid-catalyzed conditions with removal of water):

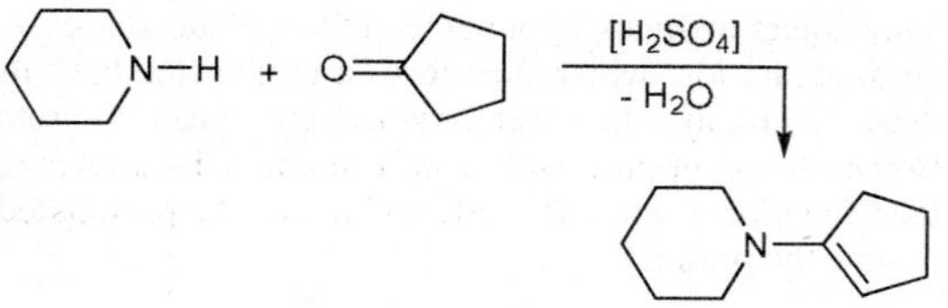

(c) An enamine can be prepared from a ketone (or aldehyde) and a secondary amine. To determine the starting ketone and starting amine that must be used, we consider the location of the vinylic carbon atom (highlighted) connected directly to nitrogen:

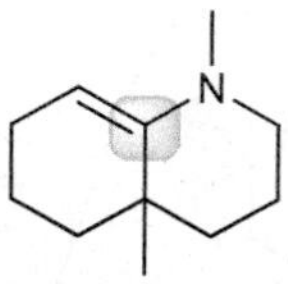

This carbon atom must have been a carbonyl group in the starting ketone, and the nitrogen atom must have been connected to one proton in the starting amine. In this case, the C-N bond of the enamine group is incorporated into a ring, which means that the starting ketone and the starting amine must have been tethered to each other in the starting material (thereby closing into a cyclic enamine). That is, enamine formation must occur in an intramolecular fashion (under acid-catalyzed conditions with removal of water):

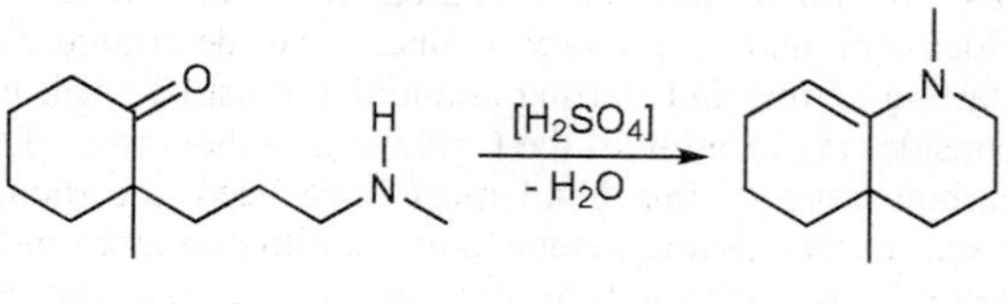

19.61.

(a) To determine the products of hydrolysis of an enamine, we must first identify the bond that will undergo cleavage. Bond cleavage is expected for the bond between the nitrogen atom and the sp^2-hybridized carbon atom to which it is attached:

Therefore, the following carbon atom will be converted into a carbonyl group:

As a result of the C-N bond cleavage, the carbon atom becomes a carbonyl group, and the nitrogen atom will accept a proton to generate a secondary amine (dimethyl amine):

(b) To determine the products of hydrolysis of an imine, we must first identify the bond that will undergo cleavage. Bond cleavage is expected for the C=N bond:

Therefore, the following carbon atom will be converted into a carbonyl group:

As a result of cleavage of the C=N bond, the carbon atom becomes a carbonyl group, and the nitrogen atom will accept two protons to generate a primary amine (methyl amine):

(c) To determine the products of hydrolysis of an acetal, we must first identify the bonds that will undergo cleavage. Bond cleavage is expected for the C-O bonds of the acetal group:

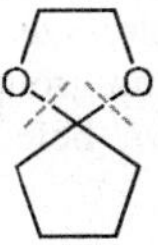

Therefore, the following carbon atom will be converted into a carbonyl group:

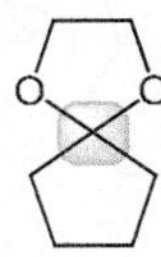

In the process, a diol is released, as shown:

(d) The starting compound exhibits a carbon atom that is connected to two oxygen atoms, as shown, and is therefore an acetal:

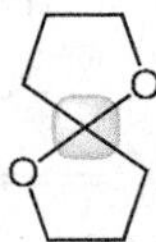

When an acetal is treated with aqueous acid, it will undergo hydrolysis. During acetal hydrolysis, cleavage occurs for the C-O bonds of the acetal group:

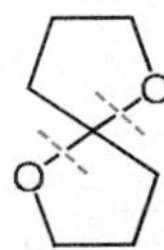

Each of these bonds is broken, thereby converting the carbon atom of the acetal group into a carbonyl group. In the process, each of the oxygen atoms will accept a proton to become an OH group, giving the following product, which exhibits a carbonyl group, as well as two OH groups:

19.62. This compound contains three different functional groups that will each undergo hydrolysis. The enamine (upper left corner) is hydrolyzed to give a ketone and a secondary amine. The imine (bottom left) is hydrolyzed to give an aldehyde and an amino group (tethered together). And finally, the cyclic acetal is hydrolyzed to give a ketone and ethylene glycol. Shown below, the three carbon atoms that will be converted into carbonyl groups have been highlighted:

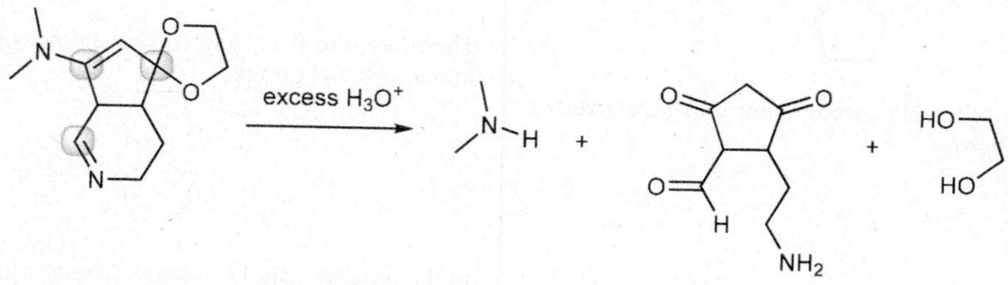

excess H_3O^+

19.63.

(a) This transformation involves hydrolysis of an enamine under acidic conditions, and therefore, we propose a mechanism that is the reverse of the mechanism for enamine formation (all of the same intermediates, but in reverse order). Under acid-catalyzed conditions, the π bond of the enamine can be protonated, giving an iminium ion which is then attacked by water to give an oxonium ion. Loss of a proton gives a carbinolamine intermediate. Protonation of the amino group gives a good leaving group, which leaves to give a resonance-stabilized cation that is deprotonated to give the product.

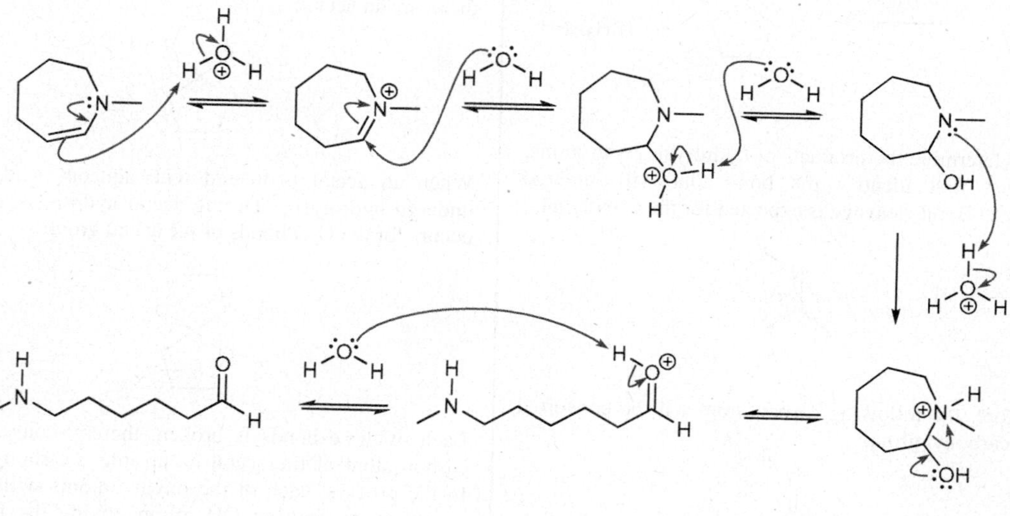

(b) This transformation involves hydrolysis of an imine under acidic conditions, and therefore, we propose a mechanism that is the reverse of the mechanism for imine formation (all of the same intermediates, but in reverse order). Under acid-catalyzed conditions, the imine is protonated, giving an iminium ion which is then attacked by water to give an oxonium ion. Loss of a proton gives a carbinolamine intermediate. Protonation of the amino group gives a good leaving group, which leaves to give a resonance-stabilized cation that is deprotonated to give the product.

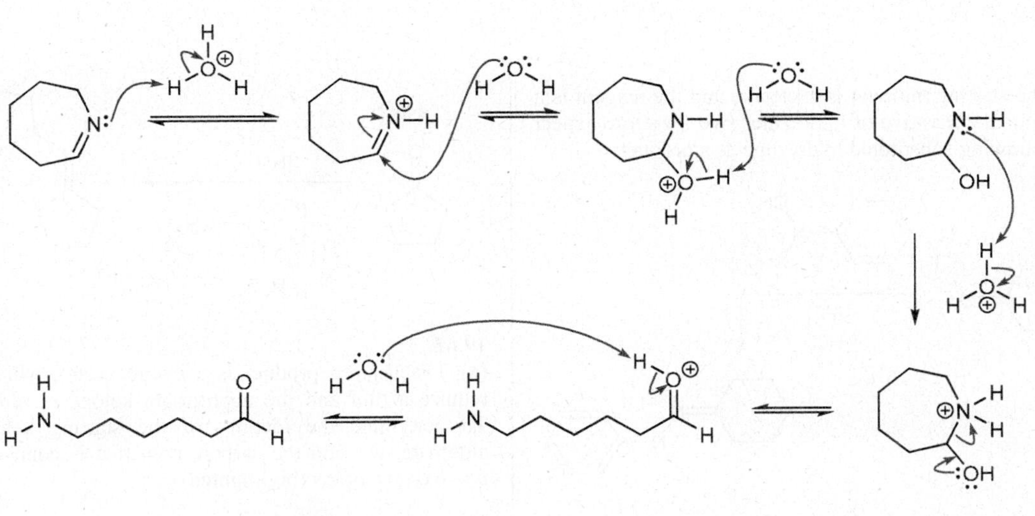

(c) This transformation involves hydrolysis of an acetal under acidic conditions, and therefore, we propose a mechanism that is the reverse of the mechanism for acetal formation (all of the same intermediates, but in reverse order). Under acid-catalyzed conditions, one of the oxygen atoms can be protonated (we can protonate either oxygen atom, and the end result will be the same), giving an oxonium ion. Loss of leaving group gives a resonance-stabilized cation that can be attacked by water to give another oxonium ion. Deprotonation gives a hemiacetal (water likely functions as the base for this step). Protonation, followed by loss of a leaving group gives a protonated aldehyde, which is deprotonated to give the product.

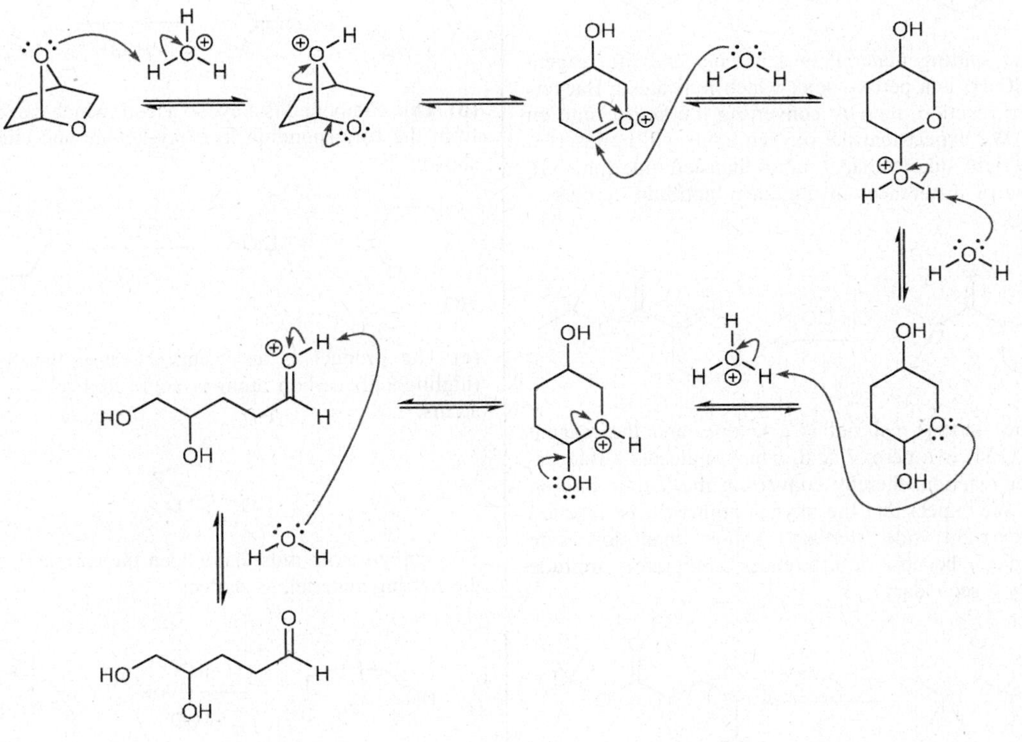

19.64.

(a) The starting material is a ketone, and the reagent is a substituted derivative of hydrazine. We therefore expect the following substituted hydrazone as a product:

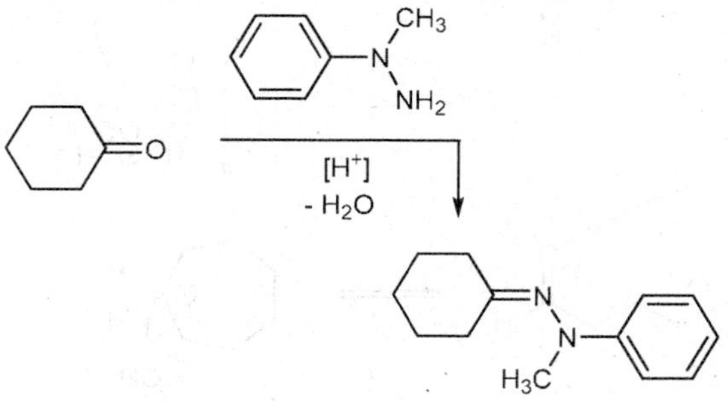

CH_3 N NH_2 O $[H^+]$ $- H_2O$ N N H_3C

(b) The starting material is a ketone, and the reagent is phenyl magnesium bromide (a Grignard reagent, which is a very strong nucleophile). The Grignard reagent will attack the ketone to give a tertiary alcohol (with installation of the phenyl group at the α position):

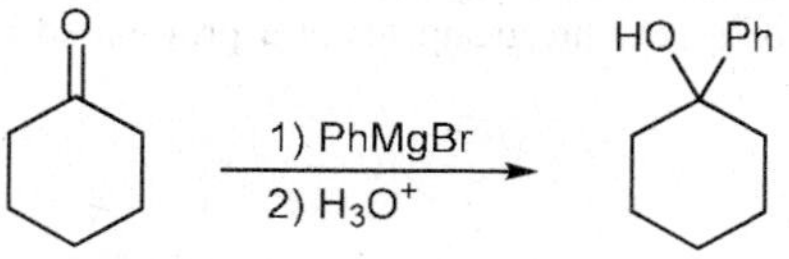

O 1) PhMgBr 2) H_3O^+ HO Ph

(c) The starting material is a ketone, and the reagent (CH_3CO_3H) is a peroxy acid, which indicates a Baeyer-Villiger reaction, thereby converting the ketone into an ester. We expect that the oxygen atom will be inserted on the right side (tertiary) rather than left side (phenyl), because of differences in migratory aptitude (tertiary > phenyl).

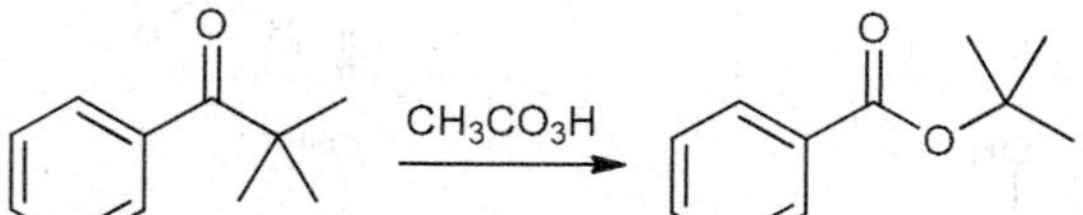

O CH_3CO_3H O O

(d) The starting material is a ketone, and the reagent (CH_3CO_3H) is a peroxy acid, which indicates a Baeyer-Villiger reaction, thereby converting the ketone into an ester. We expect that the oxygen atom will be inserted on the right side (tertiary) rather than left side (secondary), because of differences in migratory aptitude (tertiary > secondary).

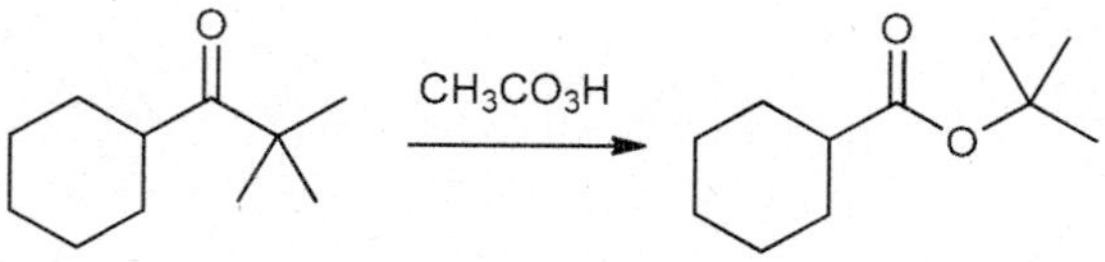

O CH_3CO_3H O O

(e) A ketone is converted into an imine upon treatment with a primary amine in acid catalyzed conditions (with removal of water). In the process, the C=O bond is replaced with a C=N bond.

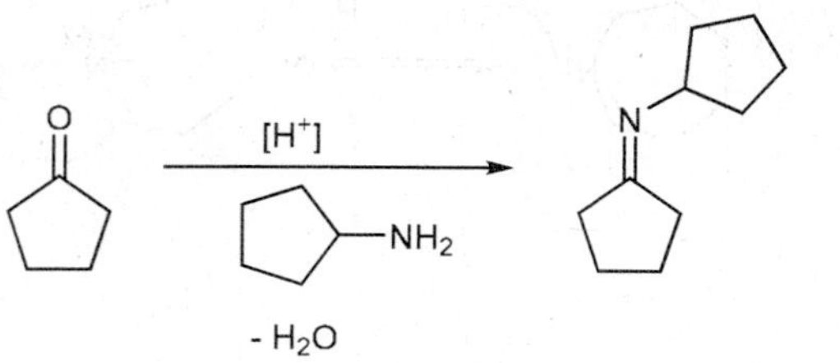

O $[H^+]$ NH_2 $- H_2O$ N

19.65.

(a) The desired product is a cyclic acetal, which will require a diol and the appropriate ketone or aldehyde. To determine the identity of the starting ketone or aldehyde, we find the carbon atom that is connected to two oxygen atoms (highlighted):

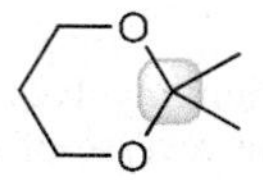

O O

This carbon atom bears the acetal group, so this carbon atom must have been the carbonyl group in the starting materials, as shown. The starting materials are 1,3-propanediol and acetone.

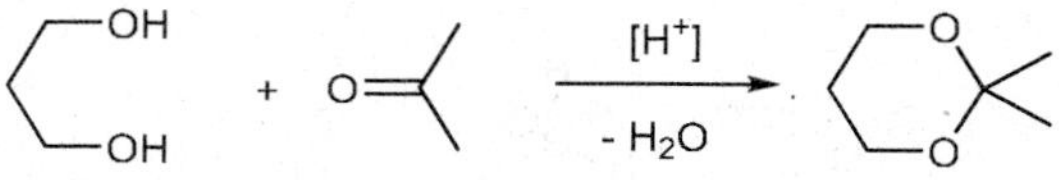

OH OH + O $[H^+]$ $- H_2O$ O O

(b) This compound is a cyclic acetal, which can be made from the corresponding hydroxy-ketone and ethanol, as shown:

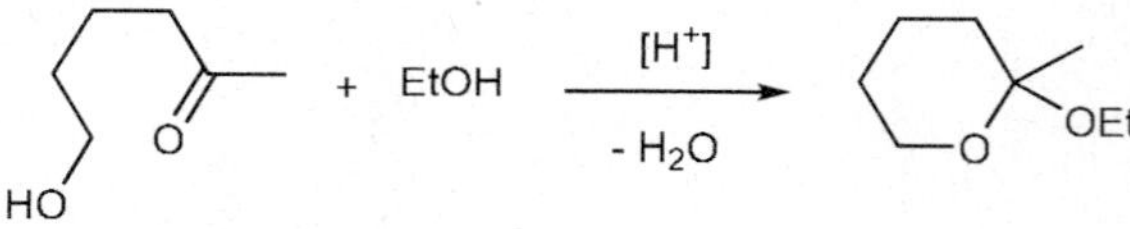

HO O + EtOH $[H^+]$ $- H_2O$ O OEt

(c) The product is an acetal, because the following (highlighted) carbon atom is connected to two oxygen atoms:

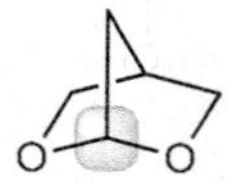

O O

This carbon atom must have been the carbonyl group in the starting material, as shown:

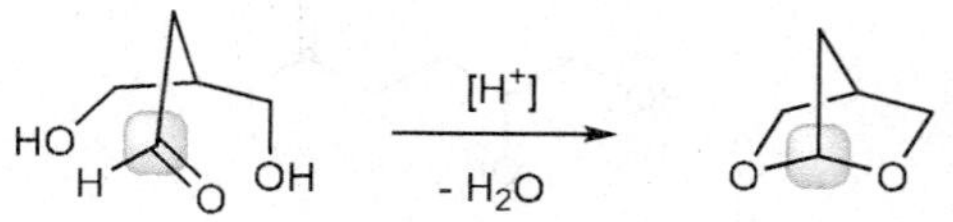

HO H O OH $[H^+]$ $- H_2O$ O O

The starting material exhibits a carbonyl group, as well as two OH groups, and can be redrawn like this:

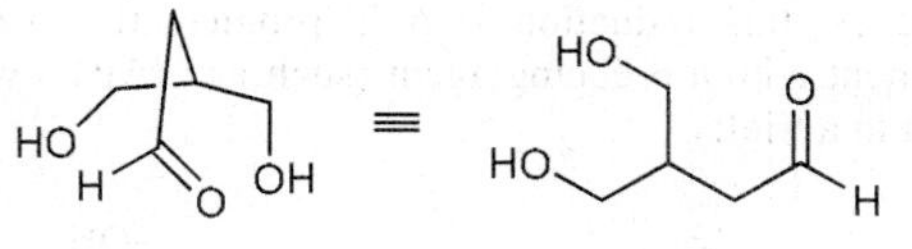

19.66. The desired product is an acetal, because the following (highlighted) carbon atom is connected to two oxygen atoms:

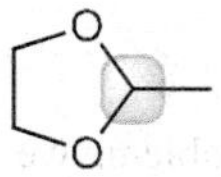

This carbon atom must have been the carbonyl group in the starting material, as shown:

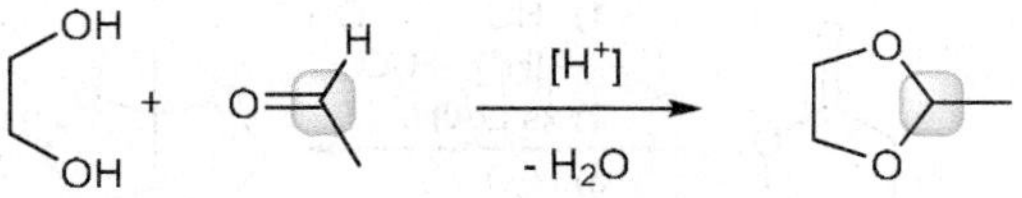

The diol and aldehyde shown above (ethylene glycol and acetaldehyde) can both be prepared from ethanol, as shown in the following retrosynthetic analysis. An explanation of each of the steps (*a-d*) follows.

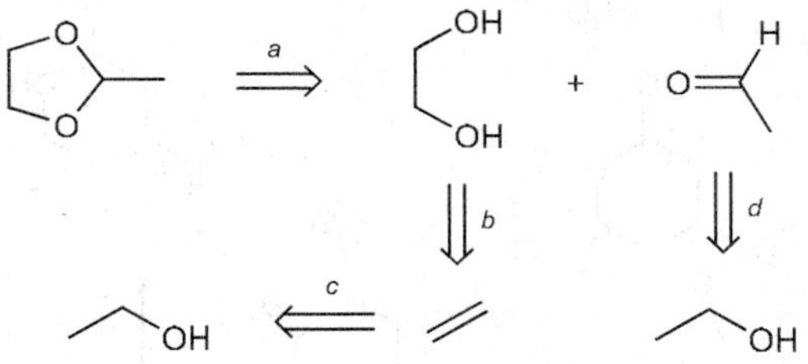

a. The acetal can be made from ethylene glycol and acetaldehyde.
b. Ethylene glycol can be made from ethylene via dihydroxylation.
c. Ethylene can be made from ethanol via acid-catalyzed dehydration.
d. Acetaldehyde can be prepared from ethanol via oxidation (with PCC)

Now let's draw the forward scheme. Ethanol is heated with concentrated sulfuric acid to give ethylene. Subsequent treatment with potassium permanganate (or osmium tetroxide) gives ethylene glycol.
Another equivalent of ethanol is oxidized with PCC (or DMP or Swern conditions) to give an aldehyde, which is then treated with ethylene glycol in acid-catalyzed conditions (with removal of water) to give the desired acetal, as shown.

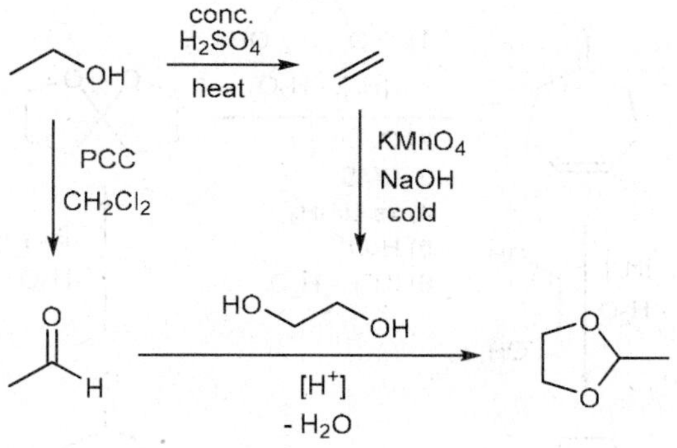

19.67.
(a) The desired product is an acetal, because the following (highlighted) carbon atom is connected to two oxygen atoms:

Therefore, this acetal can be made (via acetal formation) from the following dihydroxyketone:

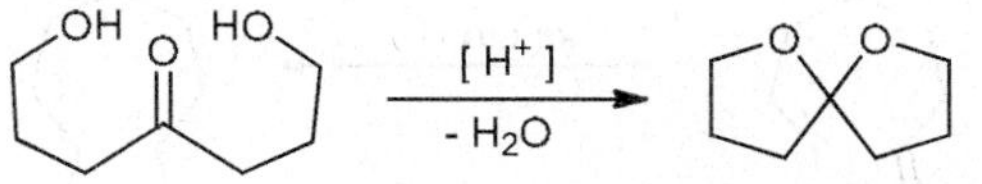

Now we must find a way to convert the starting material into the dihydroxyketone above. It might be tempting to perform ozonolysis on the starting material, followed by reduction:

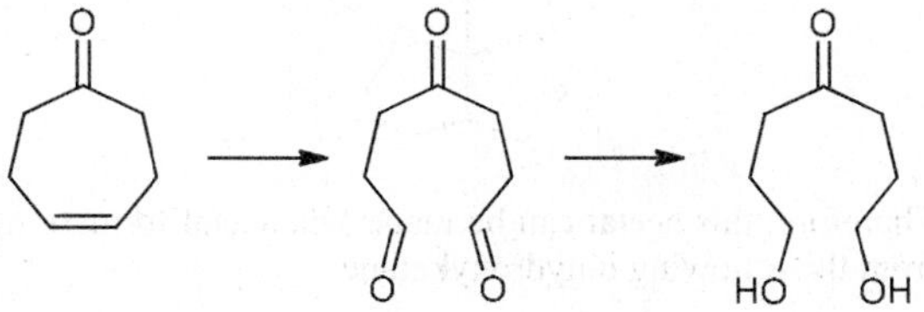

However, the reduction step is problematic, because we have not learned a way to selectively reduce the aldehyde groups in the presence of a ketone. Therefore, treatment with a reducing agent (such as excess $LiAlH_4$ or $NaBH_4$) would result in a triol:

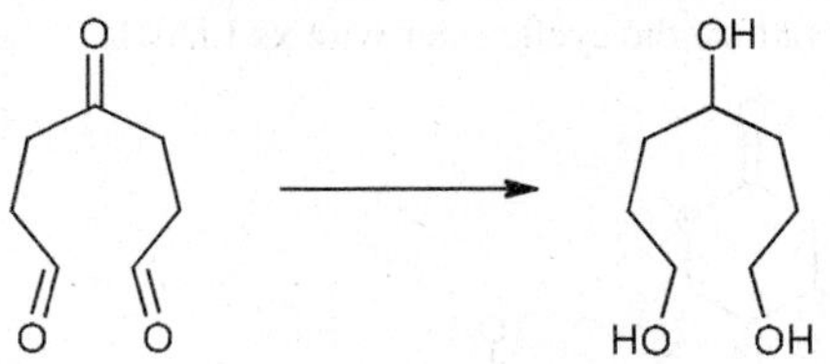

To circumvent this problem, we must first protect the ketone, before opening the ring with ozonolysis.
The entire synthesis is summarized here:

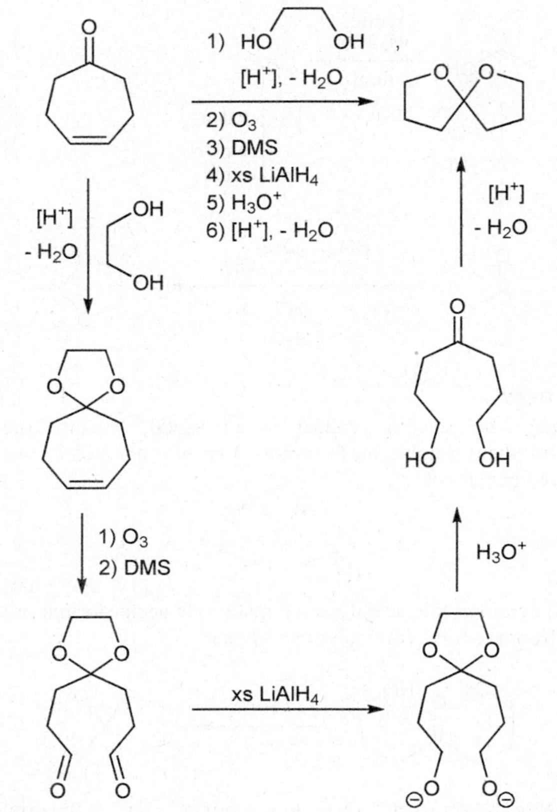

1) HO OH, $[H^+]$, - H_2O
2) O_3
3) DMS
4) xs $LiAlH_4$
5) H_3O^+
6) $[H^+]$, - H_2O

(b) The desired product is an acetal, because the following (highlighted) carbon atom is connected to two oxygen atoms:

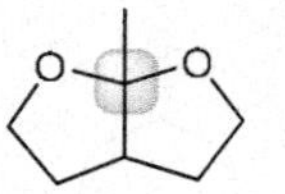

Therefore, this acetal can be made (via acetal formation) from the following dihydroxyketone:

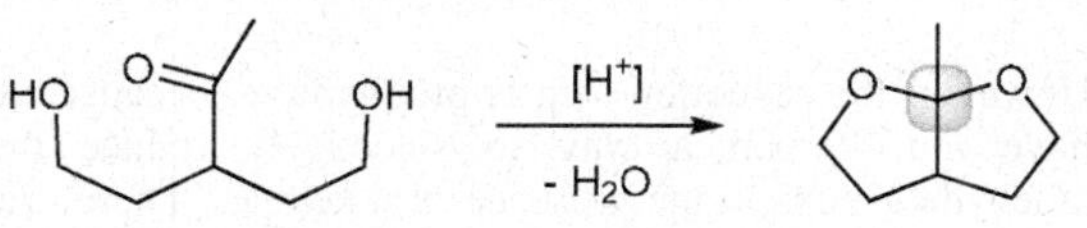

Now we must find a way to convert the starting material into the dihydroxyketone above. It might be tempting to simply reduce the cyclic ester with xs $LiAlH_4$:

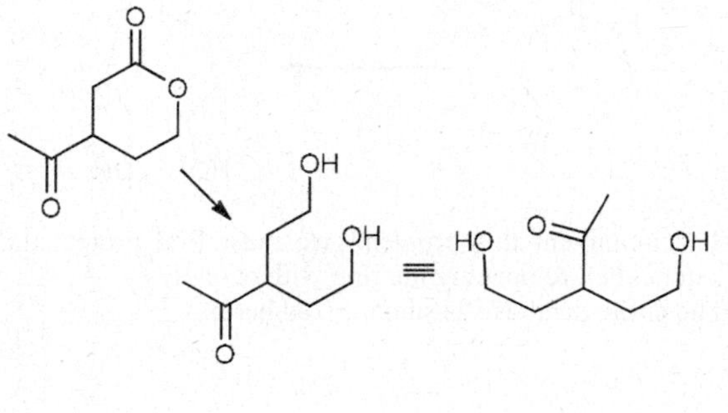

However, this reduction step is problematic, because treatment with a reducing agent (such as $LiAlH_4$) would result in a triol:

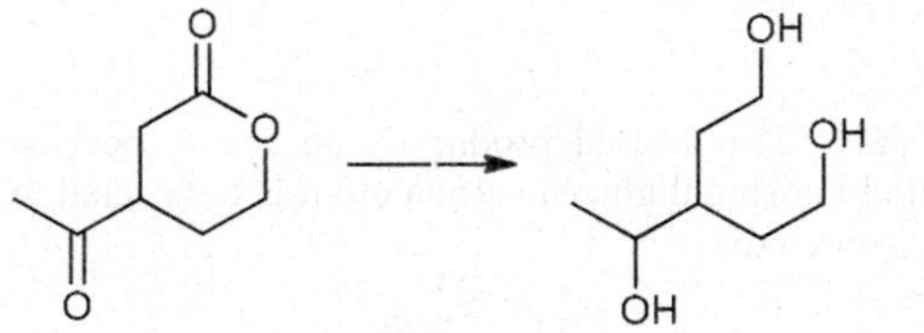

To circumvent this problem, we must first protect the ketone, before opening the ring with reduction. The entire synthesis is summarized here:

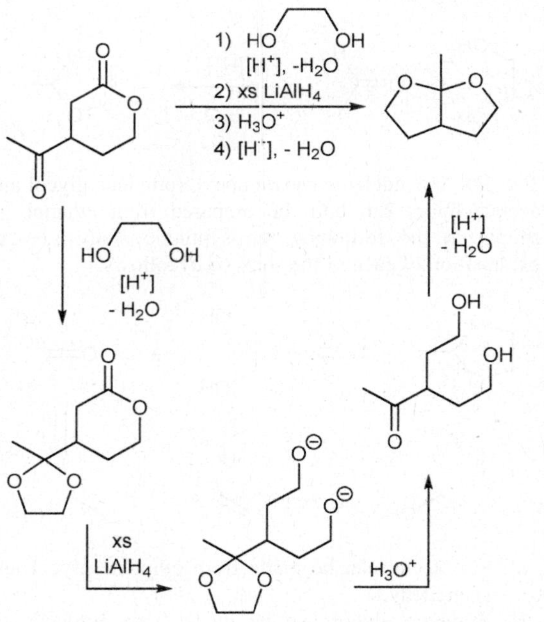

1) HO OH, $[H^+]$, -H_2O
2) xs $LiAlH_4$
3) H_3O^+
4) $[H^+]$, - H_2O

19.68. The starting compound exhibits a carbon atom that is connected to two oxygen atoms, as shown, and is therefore an acetal:

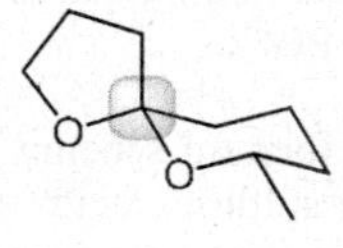

When an acetal is treated with aqueous acid, it is expected to undergo hydrolysis. We first identify the bonds that will undergo cleavage. When an acetal undergoes hydrolysis, cleavage occurs for the C-O bonds of the acetal group:

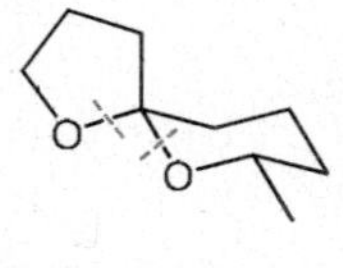

Each of these bonds is broken, thereby converting the carbon atom of the acetal group into a carbonyl group. In the process, each of the oxygen atoms will accept a proton to become an OH group, giving the following product, which exhibits a carbonyl group, as well as two OH groups:

H_3O^+

HO OH OH O HO

19.69.

(a) The product has three more carbon atoms than the starting material, which requires a C–C bond-forming reaction. Also, the position of the double bond must be moved. The extra three carbon atoms and the double bond can both be installed in the correct location if the last step of our synthesis is a Wittig reaction:

Ph_3P

This strategy requires converting the starting alkene (cyclohexene) into the ketone shown above (cyclohexanone), which can be achieved in two steps (via acid-catalyzed hydration, followed by oxidation with chromic acid). A Wittig reaction then gives the final product:

1) H_3O^+
2) $Na_2Cr_2O_7$, H_2SO_4, H_2O
3) PPh_3

H_3O^+ OH $Na_2Cr_2O_7$ H_2SO_4, H_2O O PPh_3

(b) The product is an acetal, which can be made from the corresponding ketone:

O [H⁺] 2 MeOH - H_2O MeO OMe

This ketone can be made from the starting material in just two steps. First, the dibromide is converted to an alkyne upon treatment with excess sodium amide (via two successive E2 reactions), followed by water work-up (to protonate the resulting alkynide ion). The terminal alkyne is then treated with aqueous acid, in the presence of mercuric sulfate, giving a hydration reaction. The initially formed enol will rapidly tautomerize to give a ketone. Conversion of the ketone to an acetal then gives the desired product.

Br Br
1) excess $NaNH_2$
2) H_2O
3) H_2SO_4, H_2O, $HgSO_4$
4) [H^+], 2 MeOH, - H_2O
MeO OMe

1) excess $NaNH_2$
2) H_2O

H_2SO_4, H_2O, $HgSO_4$

[H^+] 2 MeOH - H_2O

(c) The desired product is an acetal, because the following (highlighted) carbon atom is connected to two oxygen atoms:

Therefore, this acetal can be made (via acetal formation) from the following ketone and diol:

OH OH + O [H^+] - H_2O

Now we must find a way to convert the starting cycloalkene into the diol shown above. This can be achieved in just two steps. First, the ring is opened with an ozonolysis reaction to produce a dialdehyde. Then, the dialdehyde is reduced to a diol upon treatment with two equivalents of a reducing agent. Acetal formation then gives the final product.

1) O_3
2) DMS
3) $NaBH_4$ (2 eq.), MeOH
4) [H^+], acetone
(-H_2O)

1) O_3
2) DMS

H O O H

$NaBH_4$ (2 eq.) MeOH

OH OH

[H^+] acetone (-H_2O)

19.70. The steps of the following mechanism are based on the steps found in Mechanism 19.6 (imine formation). A lone pair on the nitrogen atom attacks one of the carbonyl groups, giving an intermediate that is protonated (under acidic conditions). The resulting cation is then deprotonated, followed by protonation of the OH group, thereby converting a bad leaving group into a good leaving group (water). Loss of the leaving group gives a cation, which is then deprotonated. The second carbonyl group is then attacked by a nucleophile, this time in an intramolecular fashion. The NH_2 group functions as a nucleophile and attacks the carbonyl group, closing a ring. The resulting intermediate is then protonated to remove the negative charge, followed by deprotonation. Protonation of the OH group converts a bad leaving group into a good leaving group (water), which then leaves. The resulting cation is then deprotonated to give the product.

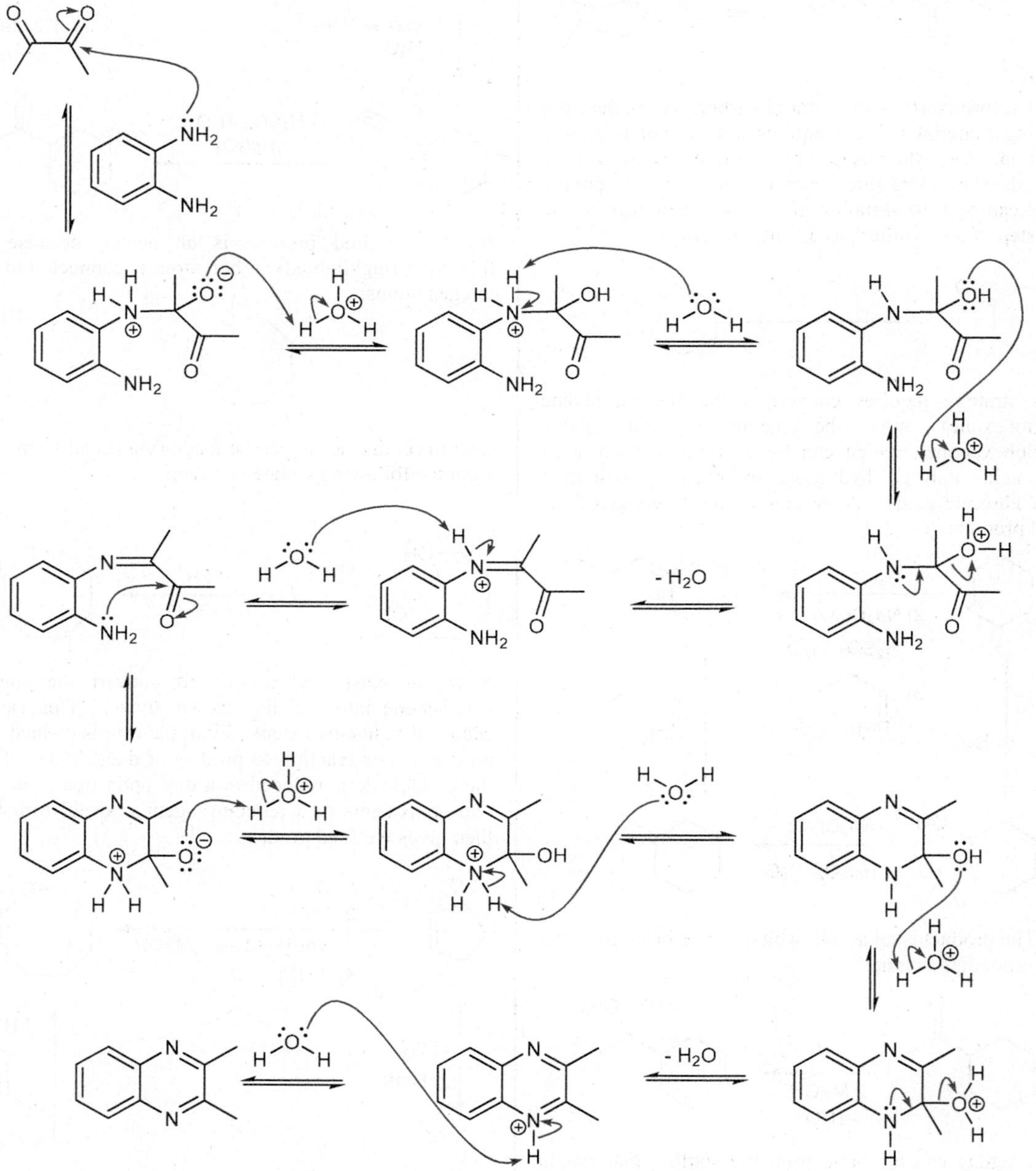

19.71. Cyclopropanone exhibits significant ring strain, with bond angles of approximately 60°. Some of this ring strain is relieved upon conversion to the hydrate, because an sp^2-hybridized carbon atom (that must be 120° to be strain-free) is replaced by an sp^3-hybridized carbon atom (that must be only 109.5° to be strain-free). In contrast, cyclohexanone is a larger ring and exhibits only minimal ring strain. Conversion of cyclohexanone to its corresponding hydrate does not alleviate a significant amount of ring strain.

19.72. 1,2-dioxane has two adjacent oxygen atoms and is therefore a peroxide. Like other peroxides, it is extremely unstable and potentially explosive.
1,3-dioxane has two oxygen atoms separated by one carbon atom. This compound is therefore an acetal. Like other acetals, it is only stable under basic conditions, but undergoes hydrolysis under mildly acidic conditions.
1,4-dioxane is neither a peroxide nor an acetal. It is therefore stable under basic conditions as well as mildly acidic conditions (like other ethers), and is used as a common solvent because of its inert behavior.

19.73.
(a) The desired product is an acetal, because the following (highlighted) carbon atom is connected to two oxygen atoms:

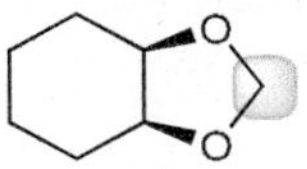

Therefore, this acetal can be made (via acetal formation) from the following diol and formaldehyde:

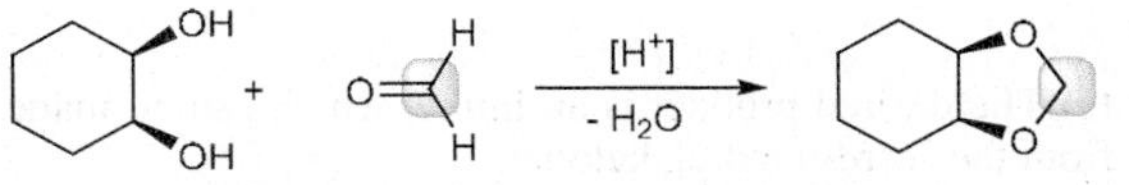

[H^+]
- H_2O

The diol can be made from the starting alkene in just one step, giving the following two-step synthesis:

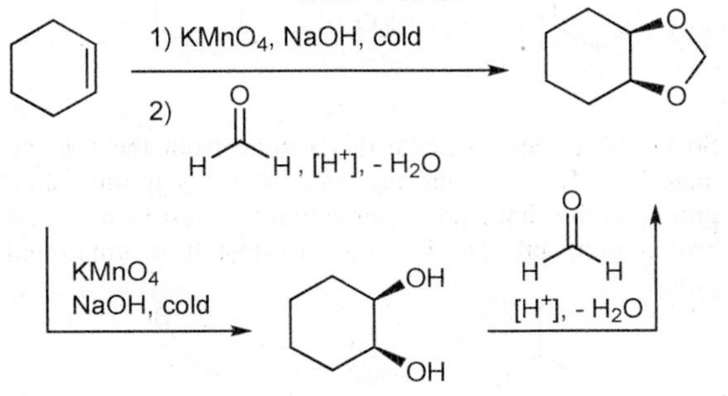

1) $KMnO_4$, NaOH, cold
2) H, [H^+], - H_2O

$KMnO_4$
NaOH, cold

[H^+], - H_2O

(b) There are certainly many acceptable solutions to this problem. One such solution derives from the following retrosynthetic analysis. An explanation of each of the steps (*a-c*) follows.

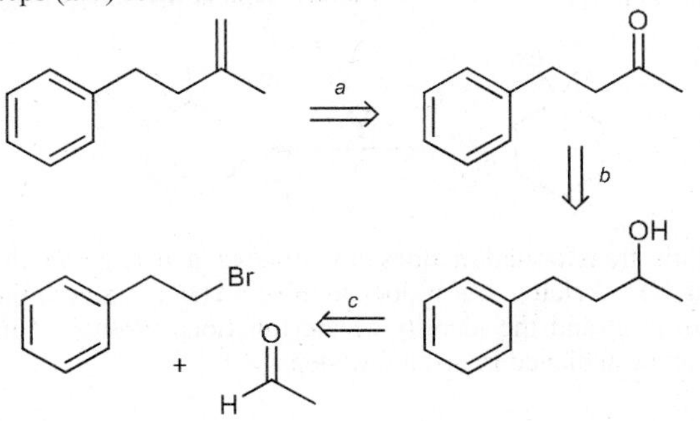

a. The desired alkene can be made from a ketone via a Wittig reaction.
b. The ketone can be made via oxidation of the corresponding secondary alcohol.
c. The alcohol can be made from the starting material via a Grignard reaction.

Now let's draw the forward scheme. Treating the starting material with magnesium gives a Grignard reagent, which is then treated with acetaldehyde (to give a Grignard reaction), followed by aqueous acidic workup, to give the alcohol. Conversion of the alcohol to a ketone requires an oxidation reaction, and we have chosen to use PCC as the oxidizing agent in this case, because chromic acid would likely oxidize the benzylic position. The resulting ketone is then converted into the desired product upon treatment with a Wittig reagent:

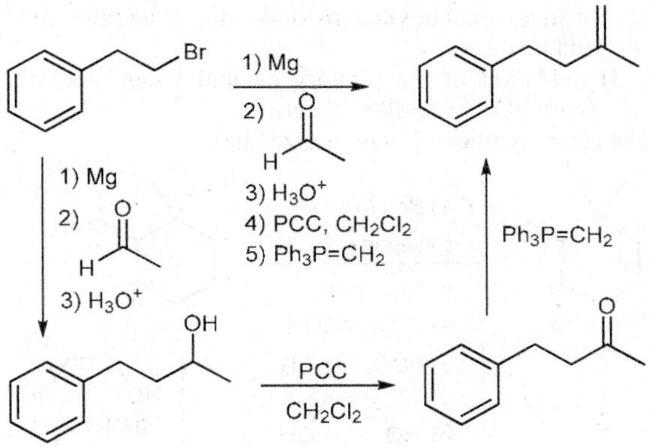

1) Mg
2) (acetaldehyde)
3) H_3O^+
4) PCC, CH_2Cl_2
5) $Ph_3P{=}CH_2$

1) Mg
2) (acetaldehyde)
3) H_3O^+

PCC
CH_2Cl_2

$Ph_3P{=}CH_2$

(c) The desired product is an acetal, because the following (highlighted) carbon atom is connected to two oxygen atoms:

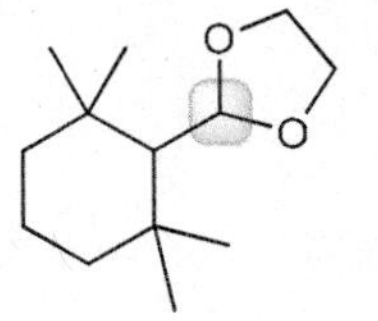

Therefore, this acetal can be made (via acetal formation) from the following aldehyde:

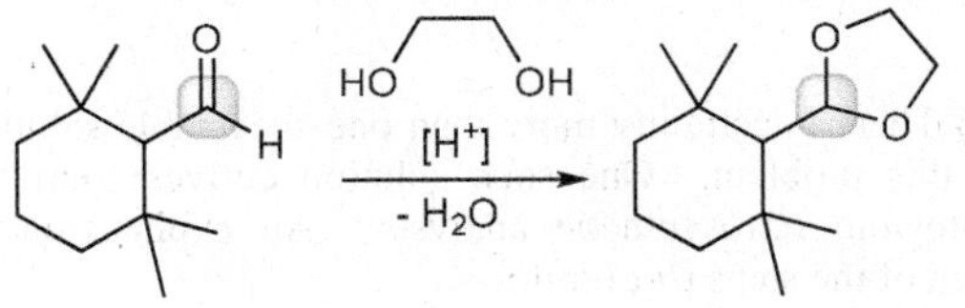

HO OH
[H^+]
- H_2O

So we will need to make this aldehyde from the starting material. But the starting material lacks a functional group, so the first step of our synthesis must be a radical bromination process in order to install a functional group:

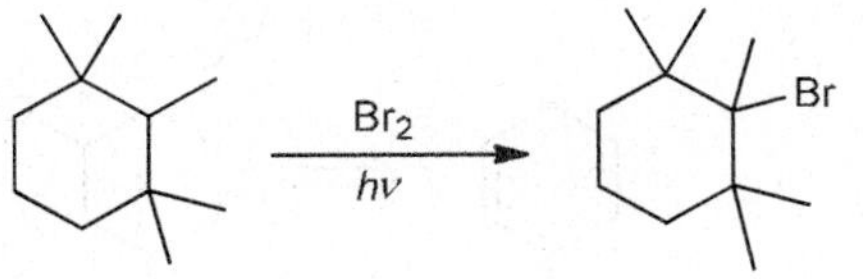

Br_2
$h\nu$

With the functional group installed, we must now bridge the gap between the first and last steps of the synthesis:

This transformation does not involve a change in the carbon skeleton, but it does involve a change in both the location and the identity of the functional group. This can be achieved in just a few steps:

1) elimination to give an alkene (upon treatment with a strong base),
2) hydroboration-oxidation to convert the alkene into a primary alcohol via *anti*-Markovnikov addition of H and OH, and
3) oxidation of the primary alcohol to an aldehyde (with PCC or DMP or Swern).

The entire synthesis is summarized here:

1) Br_2, $h\nu$
2) NaOEt
3) $BH_3 \cdot THF$
4) H_2O_2, NaOH
5) PCC, CH_2Cl_2
6) HO⌒OH, $[H^+]$, $- H_2O$

Br_2, $h\nu$ — NaOEt — 1) $BH_3 \cdot THF$ 2) H_2O_2, NaOH — PCC, CH_2Cl_2 — HO⌒OH, $[H^+]$, $- H_2O$

(d) There is certainly more than one acceptable solution to this problem. One such solution derives from the following retrosynthetic analysis. An explanation of each of the steps (*a-c*) follows.

a. The product is a cyanohydrin, which can be made from the corresponding ketone.
b. The ketone can be made via oxidation of the corresponding secondary alcohol.
c. The alcohol can be made from the starting alkene via acid-catalyzed hydration.

Now let's draw the forward scheme. The starting alkene is converted to an alcohol upon treatment with aqueous acid. Oxidation of the alcohol with chromic acid gives a ketone, which can then be converted into the desired cyanohydrin upon treatment with KCN and HCl:

1) H_3O^+
2) $Na_2Cr_2O_7$, H_2SO_4, H_2O
3) KCN, HCl

H_3O^+ — $Na_2Cr_2O_7$, H_2SO_4, H_2O — KCN, HCl

(e) The desired product is an imine, which can be made from the corresponding ketone:

$[H^+]$, NH_3, $- H_2O$

So we will need to make this ketone from the starting material. But the starting material lacks a functional group, so the first step of our synthesis must be a radical bromination process in order to install a functional group:

Br_2, $h\nu$

With the functional group installed, we must now bridge the gap between the first and last steps of the synthesis:

This transformation does not involve a change in the carbon skeleton, but it does involve a change in both the location and the identity of the functional group. This can be achieved in just a few steps:

1) elimination to give an alkene (upon treatment with a strong base),
2) hydroboration-oxidation to convert the alkene into a secondary alcohol via *anti*-Markovnikov addition of H and OH, and
3) oxidation of the secondary alcohol to a ketone (with chromic acid).

The entire synthesis is summarized here:

1) Br_2, $h\nu$
2) NaOEt
3) $BH_3 \cdot THF$
4) H_2O_2, NaOH
5) $Na_2Cr_2O_7$, H_2SO_4, H_2O
6) $[H^+]$, NH_3, $- H_2O$

Br_2, $h\nu$ — NaOEt — 1) $BH_3 \cdot THF$ 2) H_2O_2, NaOH — $Na_2Cr_2O_7$, H_2SO_4, H_2O — $[H^+]$, NH_3, $- H_2O$

(f) This transformation does not involve a change in the carbon skeleton, but it does involve a change in both the location and the identity of the functional group:

This can be achieved in just a few steps:
1) elimination with a strong, sterically hindered base to give the less-substituted alkene,
2) hydroboration-oxidation to convert the alkene into a primary alcohol via *anti*-Markovnikov addition of H and OH, and
3) oxidation of the primary alcohol to an aldehyde (with PCC or DMP or Swern conditions):

1) *t*-BuOK
2) $BH_3 \cdot THF$
3) H_2O_2, NaOH
4) PCC, CH_2Cl_2

t-BuOK — 1) $BH_3 \cdot THF$ 2) H_2O_2, NaOH — PCC, CH_2Cl_2

(g) The product is an enamine, which can be prepared from the corresponding ketone:

$[H^+]$, $(CH_3)_2NH$, $- H_2O$

This ketone can be made from benzene via a Friedel-Crafts acylation:

1) acetyl chloride, $AlCl_3$
2) $[H^+]$, $(CH_3)_2NH$, $- H_2O$

$AlCl_3$ — $[H^+]$, $(CH_3)_2NH$, $- H_2O$

(h) The desired product is an acetal, because the following (highlighted) carbon atom is connected to two oxygen atoms:

Therefore, this acetal can be made (via acetal formation) from the following diol and formaldehyde, as shown:

$[H^+]$, $- H_2O$

So we will need to make this diol from the starting material. This can be achieved in just two steps. First, the starting diyne is treated with sulfuric acid, in the presence of mercuric sulfate, to give a dione. Then, the dione can be reduced with two equivalents of $LiAlH_4$, followed by aqueous acidic workup, to give the diol, which is then converted into the desired acetal.

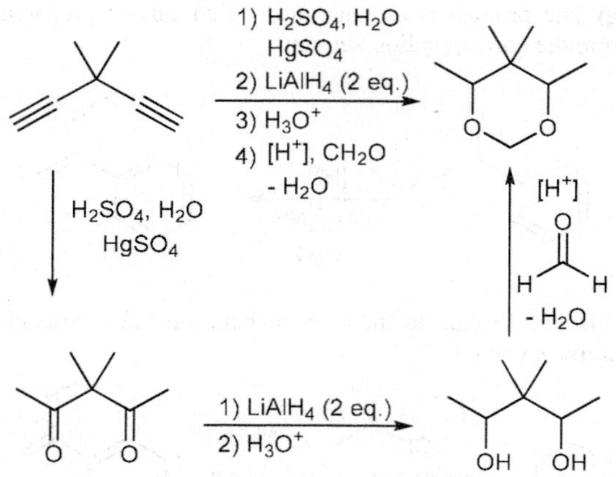

1) H_2SO_4, H_2O, $HgSO_4$ 2) $LiAlH_4$ (2 eq.) 3) H_3O^+ 4) [H^+], CH_2O - H_2O

H_2SO_4, H_2O $HgSO_4$

[H^+] - H_2O

1) $LiAlH_4$ (2 eq.) 2) H_3O^+

OH OH

19.74. The correct answer is (a). A ketone will react with a secondary amine (in the presence of an acid catalyst) to give an enamine.

19.75. The correct answer is (d). This process occurs under acid-catalyzed conditions. Under these conditions, option (d) is unlikely to form, because it has a negative charge on an oxygen atom, which is a strong base. Strong bases are unlikely to form in acidic conditions.

19.76. The correct answer is (b). The desired compound is an acetal, because the highlighted carbon atom is connected to two OR groups:

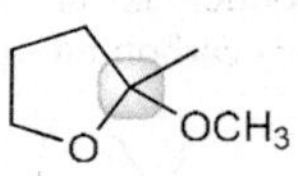

OCH_3

Acetal formation involves formation of the following bonds:

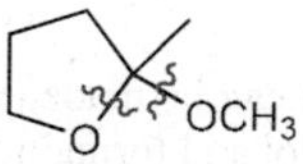

OCH_3

This acetal can be formed from the following reactants:

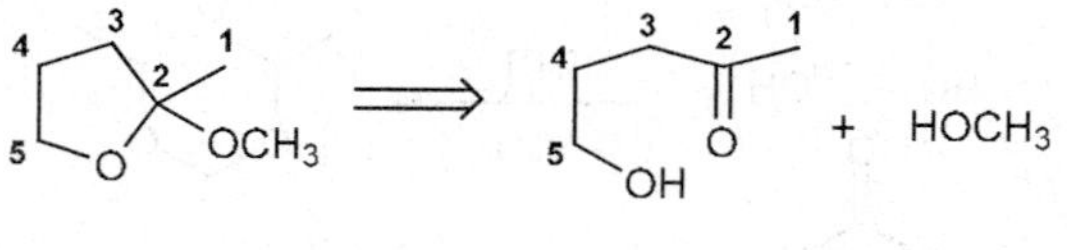

OCH_3 ⟹ OH + $HOCH_3$

19.77. The correct answer is (d). The parent is a chain of six carbon atoms, with the carbonyl group at C2, so the parent is 2-hexanone. Next, we identify the substituents. There is only one (a hydroxy group at C5):

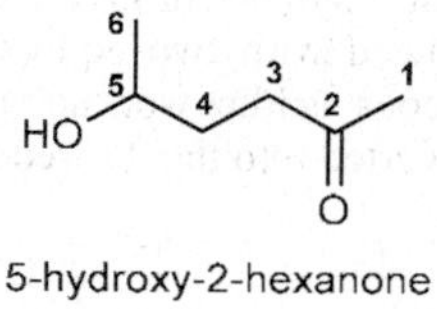

HO

5-hydroxy-2-hexanone

19.78. The correct answer is (c). To determine the products of hydrolysis of an acetal, we must first identify the bonds that will undergo cleavage. Bond cleavage is expected for the C–O bonds of the acetal group:

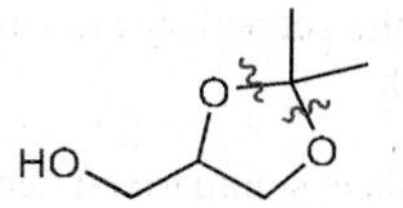

HO

Therefore, the following carbon atom (highlighted below) will be converted into a carbonyl group:

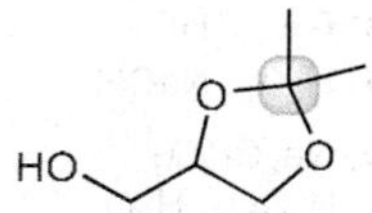

HO

In the process, a diol is released. In this case, an additional hydroxyl group is present, making one of the products a *triol*, rather than a diol, as shown:

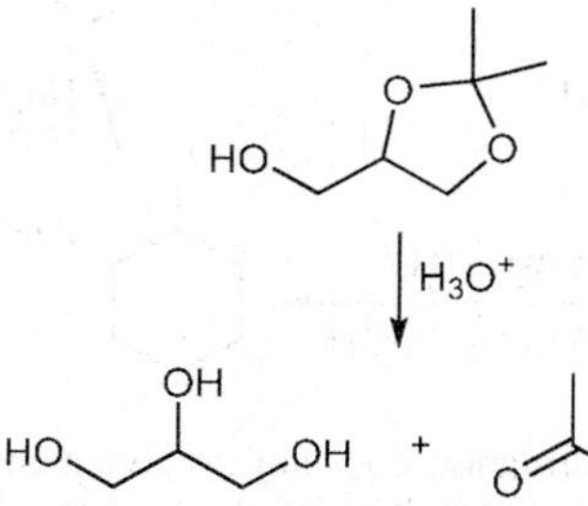

HO

H_3O^+

OH HO OH +

19.79. The correct answer is (b). The Wittig reagent attacks the carbonyl group, thereby forming a C=C bond between the carbon atom of the carbonyl group and the carbon atom attached to phosphorus (in the Wittig reagent). In this case, no stereoisomerism is possible in the product.

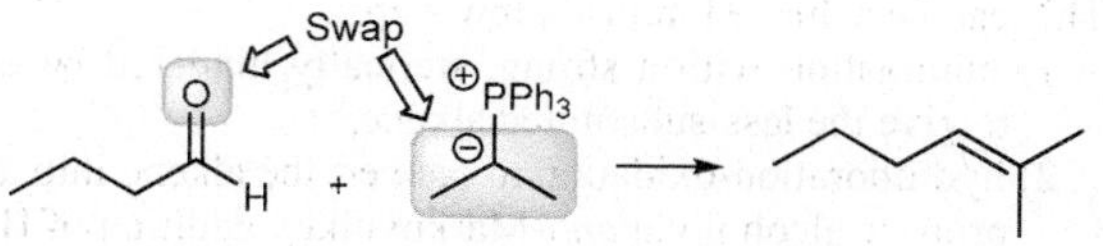

Swap

$^{\oplus}PPh_3$

19.80. The correct answer is (d). The starting material has both a ketone and an acetal, and the reagent is methyl magnesium bromide (a Grignard reagent, which is a very strong nucleophile). The Grignard reagent will not react with the acetal but will attack the electrophilic ketone to give a tertiary alkoxide (with installation of the methyl group at the α position).

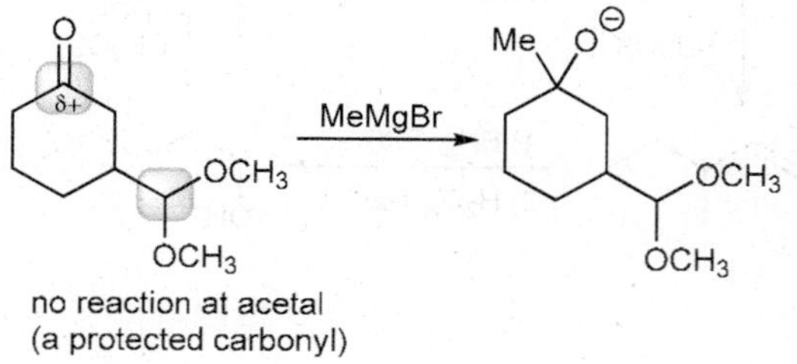

MeMgBr

Me

OCH_3 OCH_3

OCH_3 OCH_3

no reaction at acetal (a protected carbonyl)

The acidic aqueous workup (H_3O^+) serves to protonate the alkoxide intermediate. At the same time, the acetal protective group is removed. Hydrolysis of the acetal generates an aldehyde product.

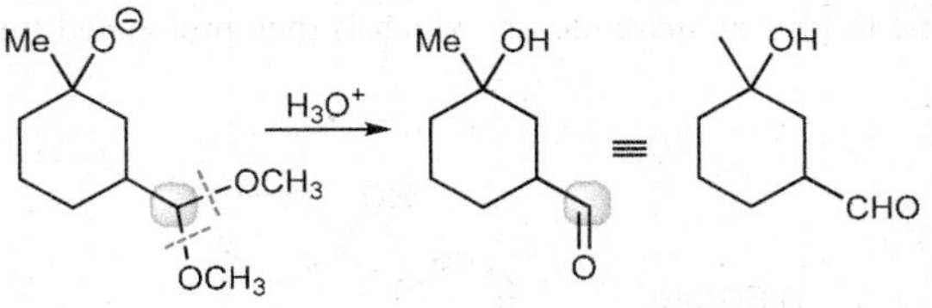

19.81. The correct answer is (c). We need to select a reagent that will react with the ketone, but not with the alkene. A ketone is reduced to a secondary alcohol upon treatment with $LiAlH_4$ followed by aqueous workup. The alkene functional group does not undergo reduction with hydride.

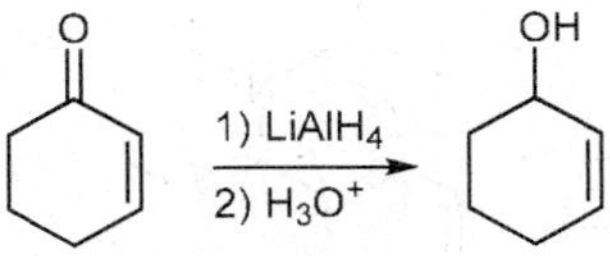

19.82. The molecular formula ($C_7H_{14}O$) indicates one degree of unsaturation (see Section 14.16). Therefore, the compound must have either one double bond or one ring. Treating compound **A** with a reducing agent ($NaBH_4$) gives an alcohol, so compound **A** is a ketone or aldehyde (accounting for the one degree of unsaturation). The ^1H NMR spectrum of compound **A** exhibits only two signals (for 14 protons). Therefore, the structure must be symmetrical. The two signals in the ^1H NMR spectrum are characteristic of an isopropyl group, which means that compound **A** must be diisopropyl ketone (or 2,4-dimethyl-3-pentanone). Conversion of compound **A** into a thioacetal, followed by desulfurization with Raney nickel, gives compound **B**. Notice that the carbonyl group has been reduced to a methylene (CH_2) group.

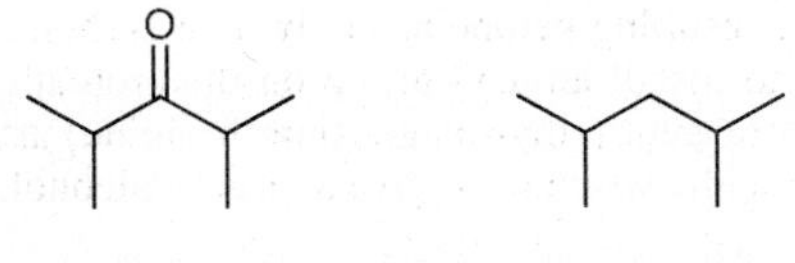

Compound **A** Compound **B**

(a) Compound **B** is symmetrical, much like compound **A**, giving rise to only three signals in its ^1H NMR spectrum, corresponding with the following protons:

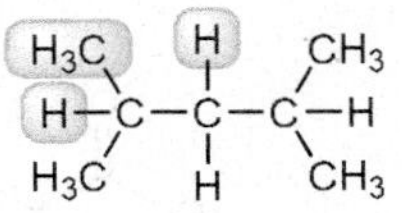

Compound **B**

(b) Compound **B** is symmetrical, giving rise to only three signals in its ^{13}C NMR spectrum, corresponding with the following three unique locations:

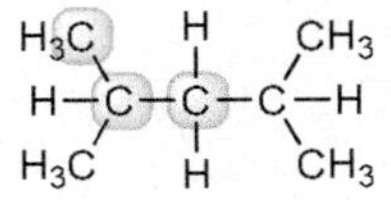

Compound **B**

(c) Compound **A** is a ketone, while compound **B** is an alkane. Therefore, compound **A** will exhibit a strong signal near 1715 cm^{-1}, while compound **B** will not exhibit a signal in the same region.

19.83. Compound **C** is converted to an enamine upon treatment with a secondary amine under acid-catalyzed conditions, so compound **C** must be the corresponding ketone. Once we know the structure of compound **C**, the other structures can be identified. Compound **A** must be an alkene, because ozonolysis gives a ketone (we know that only one carbon atom is lost during this process, because the molecular formula of compound **A** indicates 10 carbon atoms, while the resulting ketone has only nine carbon atoms). Compound **B** must be an acyl halide with three carbon atoms, in order to install an acyl group on the aromatic ring via a Friedel-Crafts acylation. Compound **D** is formed when a Grignard reagent (ethyl magnesium bromide) attacks the ketone to give an alkoxide ion which is then protonated (via aqueous acidic workup) to give a tertiary alcohol, as shown:

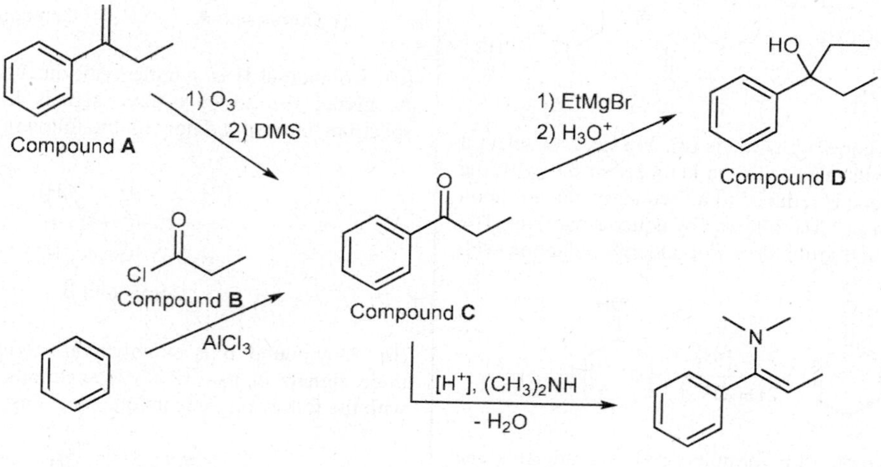

Compound A — 1) O_3 2) DMS → Compound C — 1) EtMgBr 2) H_3O^+ → Compound D (HO)

Compound B (Cl, O) + benzene, $AlCl_3$ → Compound C

Compound C — $[H^+]$, $(CH_3)_2NH$, $- H_2O$ → (enamine, N)

19.84. Cyclohexene is converted to cyclohexanol upon treatment with aqueous acid (acid-catalyzed hydration). Cyclohexanol is oxidized to cyclohexanone upon treatment with a strong oxidizing agent. Upon treatment with hydrazine in acid-catalyzed conditions, cyclohexanone is converted into the corresponding hydrazone. A Wolff-Kishner reduction then gives cyclohexane.
The conversion of cyclohexene to cyclohexane can be achieved more directly, in one step, via catalytic hydrogenation:

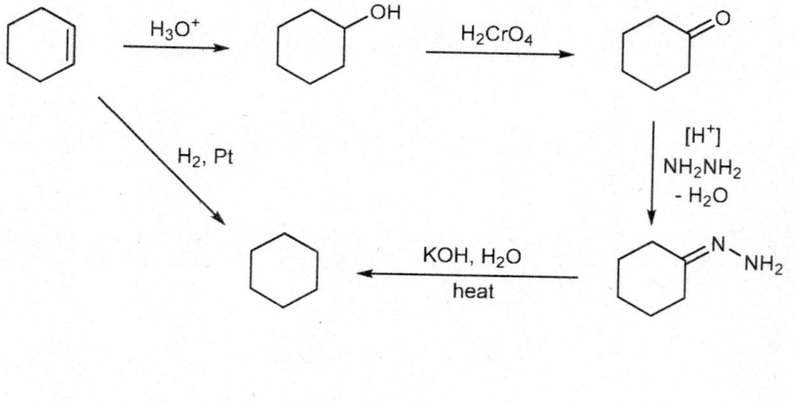

Cyclohexene — H_3O^+ → cyclohexanol (OH) — H_2CrO_4 → cyclohexanone (O) — $[H^+]$, NH_2NH_2, $- H_2O$ → hydrazone (N, NH_2) — KOH, H_2O, heat → cyclohexane

Cyclohexene — H_2, Pt → cyclohexane

19.85. Benzene undergoes an electrophilic aromatic substitution reaction when treated with Br_2 and a Lewis acid, such as $FeBr_3$, to give bromobenzene. Subsequent treatment with magnesium gives a Grignard reagent (phenyl magnesium bromide), which reacts with formaldehyde to give benzyl alcohol (after aqueous acidic workup). This alcohol is oxidized by PCC to give benzaldehyde, which is then converted into an acetal upon treatment with ethylene glycol under acid-catalyzed conditions.

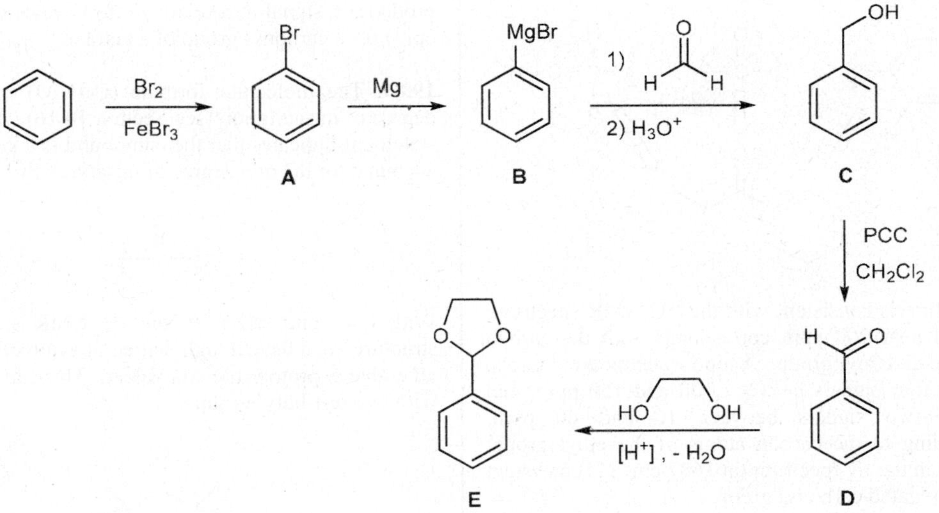

Br, Br_2, $FeBr_3$, A, MgBr, Mg, B, 1) H(C=O)H, 2) H_3O^+, C, OH, PCC, CH_2Cl_2, D, HO OH, $[H^+]$, $-H_2O$, E

19.86.

(a) The problem statement indicates that the compound is an aldehyde. The molecular formula (C_4H_6O) indicates two degrees of unsaturation (see Section 14.16), but the aldehyde group only accounts for one of the degrees of unsaturation. The signal at 1715 cm^{-1} (stretching of the carbonyl group) indicates that the carbonyl group is NOT conjugated. If it were conjugated, the signal would be expected to appear at lower wavenumber (below 1700 cm^{-1}).

The following two structures are consistent with the requirements described above. The first aldehyde is acyclic and exhibits a double bond that is not conjugated to the carbonyl group, while the second aldehyde is cyclic:

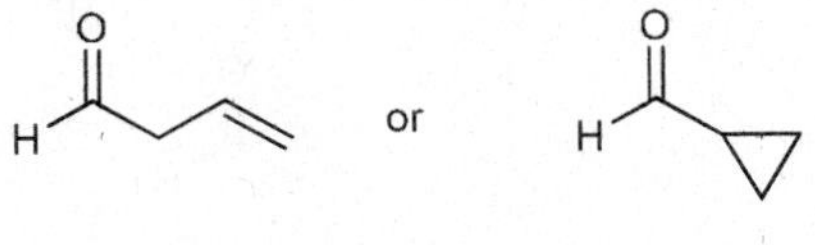

O, H, or, O, H

(b) The acyclic aldehyde would exhibit four signals in its ^{13}C NMR spectrum, while cyclopropyl carbaldehyde would exhibit only three signals in its ^{13}C NMR spectrum (two of the carbon atoms of the cyclopropyl group are identical). Furthermore, we would expect the chemical shifts to be markedly different.

19.87. The molecular formula ($C_9H_{10}O$) indicates five degrees of unsaturation (see Section 14.16), which is highly suggestive of an aromatic ring, in addition to either one double bond or one ring.

In the ^1H NMR spectrum, the signals near 7 ppm are likely a result of aromatic protons. Notice that the combined integration of these signals is 5H, indicating a monosubstituted aromatic ring:

H H H H H

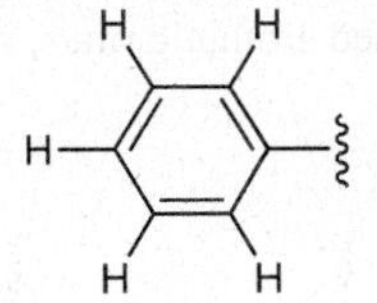

The spectrum also exhibits the characteristic pattern of an ethyl group (a quartet with an integration of 2 and a triplet with an integration of 3):

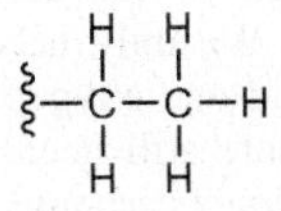

H H, C–C–H, H H

If we inspect the two fragments that we have determined thus far (the monosubstituted aromatic ring and the ethyl group), we will find that these two fragments account for nearly all of the atoms in the molecular formula ($C_9H_{10}O$). We only need to account for one more carbon atom and one oxygen atom. And let's not forget that our structure still needs one more degree of unsaturation, suggesting a carbonyl group:

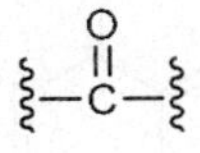

O, –C–

There is only one way to connect these three fragments.

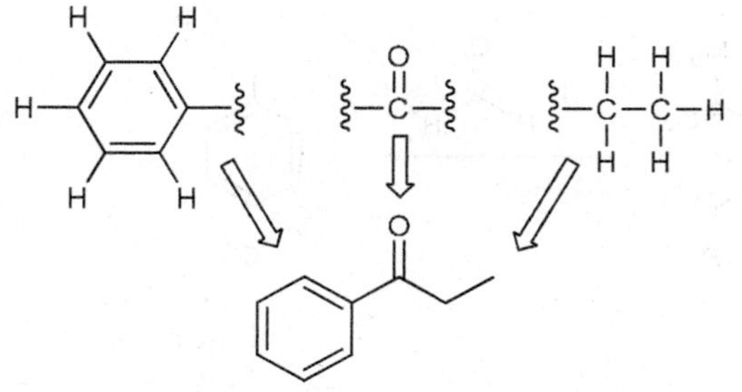

This structure is consistent with the ^{13}C NMR spectrum. The signal near 200 ppm corresponds with the carbon atom of the carbonyl group. A monosubstituted aromatic ring gives four signals between 100 and 150 ppm, and there are two signals between 10 and 60 ppm, corresponding to the carbon atoms of the ethyl group. The signal in the IR spectrum (at 1687 cm^{-1}) is consistent with a conjugated carbonyl group.

19.88. The molecular formula ($C_{13}H_{10}O$) indicates nine degrees of unsaturation (see Section 14.16), which is highly suggestive of two aromatic rings, in addition to either one double bond or one ring. The ^{13}C NMR spectrum exhibits only five signals, which must account for all thirteen carbon atoms in the compound. Therefore, many of the carbon atoms are identical, as a result of symmetry. There are four signals between 100 and 150 ppm, suggesting a monosubstituted aromatic ring. To account for so many degrees of unsaturation, as well as the symmetry that must be present, we propose two monosubstituted aromatic rings, rather than just one:

If we inspect these two fragments, we will find that they account for nearly all of the atoms in the molecular formula ($C_{13}H_{10}O$). We only need to account for one more carbon atom and one oxygen atom. And let's not forget that our structure still needs one more degree of unsaturation, suggesting a carbonyl group:

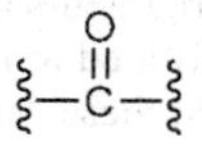

The signal near 200 ppm in the ^{13}C NMR spectrum corresponds with the carbon atom of the carbonyl group. There is only one way to connect these three fragments, as shown:

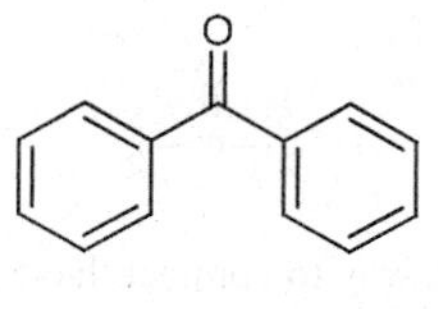

The carbonyl group in this compound (benzophenone) is conjugated to each of the rings, which explains why it produces a signal at a relatively low wavenumber (1660 cm^{-1}) for a carbonyl group of a ketone.

19.89. The molecular formula ($C_9H_{18}O$) indicates one degree of unsaturation (see Section 14.16). The problem statement indicates that the compound is a ketone, which accounts for the one degree of unsaturation:

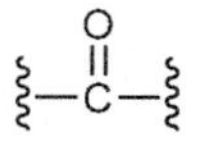

With only one signal in the ^1H NMR spectrum, the structure must have a high degree of symmetry, such that all eighteen protons are equivalent. This can be achieved with two tert-butyl groups:

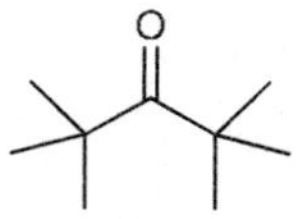

This compound is a ketone with a parent chain of five carbon atoms. The carbonyl group is located at C3, and there are four methyl groups (two at C2 and two at C4):

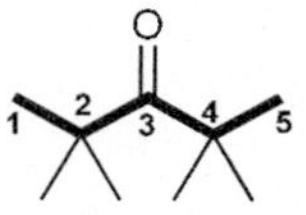

2,2,4,4-Tetramethyl-3-pentanone

19.90.

(a) There are certainly many acceptable solutions to this problem. One such solution derives from the following retrosynthetic analysis. An explanation of each of the steps (*a*-*c*) follows.

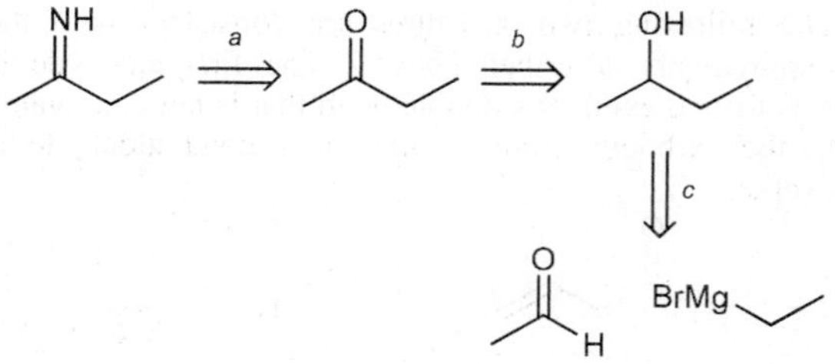

a. The desired product is an imine, which can be made from the corresponding ketone.

b. The ketone can be made via oxidation of the corresponding secondary alcohol.

c. The secondary alcohol can be made via a Grignard reaction from compounds containing no more than two carbon atoms.

Now let's draw the forward scheme. Ethyl bromide is converted into ethyl magnesium bromide (a Grignard reagent), which is then treated with acetaldehyde (to give a Grignard reaction), followed by aqueous acidic workup, to give 2-butanol. Oxidation gives 2-butanone, which can then be converted into the desired imine upon treatment with ammonia in acid-catalyzed conditions (with removal of water):

1) Mg
2) O, H
3) H_3O^+
4) $Na_2Cr_2O_7$, H_2SO_4, H_2O
5) $[H_2SO_4]$, NH_3, - H_2O

NH

Br

Mg

BrMg

$[H_2SO_4]$
NH_3
- H_2O

O

1) O, H
2) H_3O^+

OH

$Na_2Cr_2O_7$
H_2SO_4, H_2O

(b) There are certainly many acceptable solutions to this problem. One such solution derives from the following retrosynthetic analysis. An explanation of each of the steps (*a-e*) follows.

N ⟹ *a* O

b OH

c

O + CH_3MgBr ⟸ *e* OH ⟸ *d* O, H + BrMg

a. The desired product is an enamine, which can be made from the corresponding ketone (3-pentanone).
b. The ketone can be made via oxidation of the corresponding secondary alcohol (3-pentanol).
c. The secondary alcohol can be made via a Grignard reaction between ethyl magnesium bromide and an aldehyde (propanal).
d. Propanal can be made via oxidation of the corresponding alcohol (1-propanol).
e. 1-Propanol can be made by treating ethylene oxide with methyl magnesium bromide.

Now let's draw the forward scheme. Ethylene oxide is treated with methyl magnesium bromide, followed by aqueous acidic workup to give 1-propanol. This alcohol is then oxidized to an aldehyde with PCC (or DMP or Swern conditions), and the resulting aldehyde is then treated with ethyl magnesium bromide, followed by aqueous acidic workup, to give 3-pentanol. Oxidation with chromic acid gives 3-pentanone, which is then converted to the corresponding enamine upon treatment with dimethylamine and acid catalysis (with removal of water):

1) MeMgBr
2) H_3O^+
3) PCC, CH_2Cl_2
4) EtMgBr
5) H_3O^+
6) $Na_2Cr_2O_7$, H_2SO_4, H_2O
7) $[H_2SO_4]$, $(CH_3)_2NH$, - H_2O

O

N

1) MeMgBr
2) H_3O^+

OH

$[H_2SO_4]$
$(CH_3)_2NH$
- H_2O

O

PCC
CH_2Cl_2

$Na_2Cr_2O_7$
H_2SO_4
H_2O

O, H

1) EtMgBr
2) H_3O^+

OH

(c) The product is a cyclic acetal, which can be made from the corresponding ketone:

O → O O

This ketone can be prepared in a variety of ways, so there are certainly many acceptable solutions to this problem. One such solution derives from the following retrosynthetic analysis. An explanation of each of the steps (*a-e*) follows.

O ⟹ *a* OH

b

O, H + MgBr

c

BrMg + O ⟸ *e* OH ⟸ *d* Br

a. The ketone can be made via oxidation of the corresponding secondary alcohol (2-hexanol).
b. 2-Hexanol can be made via a Grignard reaction between acetaldehyde and *n*-butyl magnesium bromide.
c. *n*-Butyl magnesium bromide can be made from 1-bromobutane.
d. 1-Bromobutane can be made from 1-butanol via a substitution process.
e. 1-Butanol can be made by treating ethylene oxide with ethyl magnesium bromide.

Now let's draw the forward scheme. Bromoethane is converted into a Grignard reagent and then treated with ethylene oxide, followed by aqueous acidic workup to give 1-butanol. This alcohol is then converted to the corresponding alkyl bromide upon treatment with PBr_3. The alkyl bromide is then converted into a Grignard reagent (via insertion of Mg), and then treated with acetaldehyde, followed by aqueous acidic workup to give 2-hexanol. Oxidation with chromic acid gives 2-hexanone, which is then converted to the corresponding acetal upon treatment with ethylene glycol and acid catalysis (with removal of water):

1) Mg
2) ethylene oxide
3) H_3O^+
4) PBr_3
5) Mg
6) acetaldehyde
7) H_3O^+
8) $Na_2Cr_2O_7$, H_2SO_4, H_2O
9) HO–CH2CH2–OH, [H_2SO_4], - H_2O

Mg; 1) ethylene oxide 2) H_3O^+; PBr_3; 1) Mg 2) acetaldehyde 3) H_3O^+; $Na_2Cr_2O_7$, H_2SO_4, H_2O; HO–CH2CH2–OH, [H_2SO_4], - H_2O

(d) The product is a cyclic acetal, which can be made from the corresponding ketone:

This ketone can be prepared in a variety of ways, so there are certainly many acceptable solutions to this problem. One such solution derives from the following retrosynthetic analysis. An explanation of each of the steps (*a-d*) follows.

a, *b*, *c*, *d*; + MgBr; + MgBr

a. The ketone can be made via oxidation of the corresponding secondary alcohol (3-hexanol).
b. 3-Hexanol can be made via a Grignard reaction between butanal and ethyl magnesium bromide.
c. Butanal can be via oxidation of 1-butanol.
d. 1-Butanol can be made by treating ethylene oxide with ethyl magnesium bromide.

Now let's draw the forward scheme. Bromoethane is converted into a Grignard reagent and then treated with ethylene oxide, followed by aqueous acidic workup to give 1-butanol. This alcohol is then oxidized with PCC (or DMP or Swern) to give butanal, which is then treated with ethyl magnesium bromide (followed by aqueous acidic workup) to give 3-hexanol. Oxidation with chromic acid gives 3-hexanone, which is then converted to the corresponding acetal upon treatment with ethylene glycol and acid catalysis (with removal of water):

1) Mg
2) ethylene oxide
3) H_3O^+
4) PCC, CH_2Cl_2
5) EtMgBr
6) H_3O^+
7) $Na_2Cr_2O_7$, H_2SO_4, H_2O
8) HO–CH2CH2–OH, [H_2SO_4], - H_2O

Mg; MgBr; 1) ethylene oxide 2) H_3O^+; PCC, CH_2Cl_2; 1) EtMgBr 2) H_3O^+; $Na_2Cr_2O_7$, H_2SO_4, H_2O; HO–CH2CH2–OH, [H_2SO_4], - H_2O

(e) There are certainly many acceptable solutions to this problem. One such solution derives from the following retrosynthetic analysis. An explanation of each of the steps (*a-e*) follows.

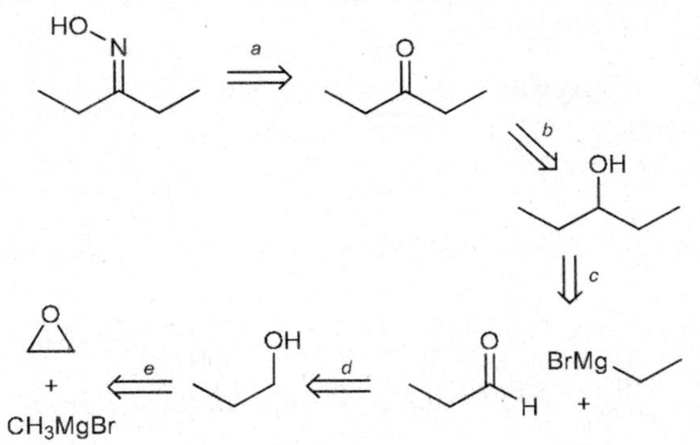

a. The desired product is an oxime, which can be made from the corresponding ketone (3-pentanone).
b. The ketone can be made via oxidation of the corresponding secondary alcohol (3-pentanol).
c. The secondary alcohol can be made via a Grignard reaction between ethyl magnesium bromide and an aldehyde (propanal).
d. Propanal can be made via oxidation of the corresponding alcohol (1-propanol).
e. 1-Propanol can be made by treating ethylene oxide with methyl magnesium bromide.

Now let's draw the forward scheme. Ethylene oxide is treated with methyl magnesium bromide, followed by aqueous acidic workup to give 1-propanol. This alcohol is then oxidized to an aldehyde with PCC (or DMP or Swern conditions), and the resulting aldehyde is then treated with ethyl magnesium bromide, followed by aqueous acidic workup, to give 3-pentanol. Oxidation with chromic acid gives 3-pentanone, which is then converted to the corresponding oxime upon treatment with hydroxylamine and acid catalysis (with removal of water):

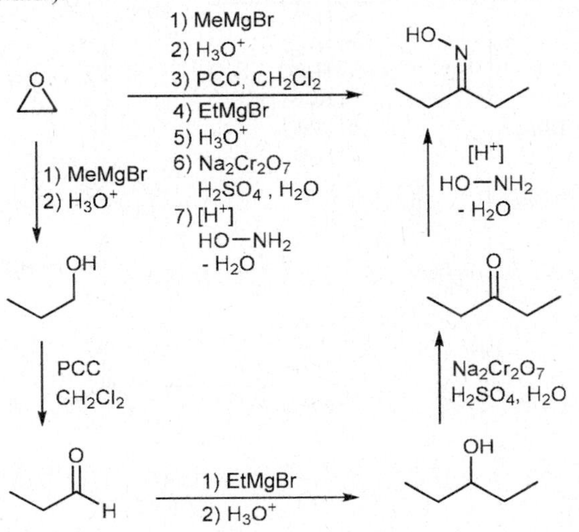

(f) There are certainly many acceptable solutions to this problem. One such solution derives from the following retrosynthetic analysis. An explanation of each of the steps (*a-e*) follows.

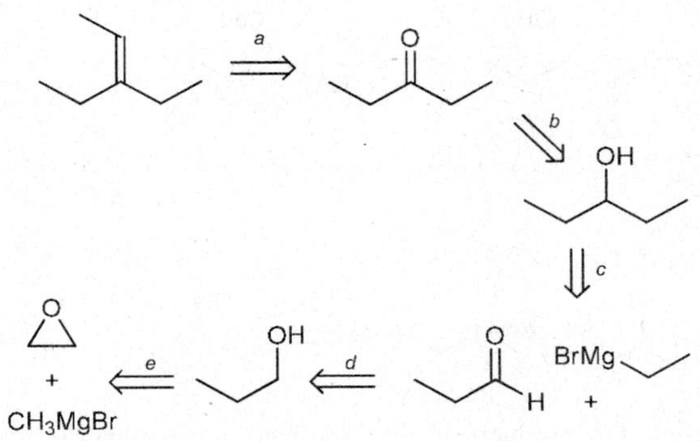

a. The desired product is an alkene, which can be made from a ketone (3-pentanone) via a Wittig reaction.
b. The ketone can be made via oxidation of the corresponding secondary alcohol (3-pentanol).
c. The secondary alcohol can be made via a Grignard reaction between ethyl magnesium bromide and an aldehyde (propanal).
d. Propanal can be made via oxidation of the corresponding alcohol (1-propanol).
e. 1-Propanol can be made by treating ethylene oxide with methyl magnesium bromide.

Now let's draw the forward scheme. Ethylene oxide is treated with methyl magnesium bromide, followed by aqueous acidic workup to give 1-propanol. This alcohol is then oxidized to an aldehyde with PCC (or DMP or Swern conditions), and the resulting aldehyde is then treated with ethyl magnesium bromide, followed by aqueous acidic workup, to give 3-pentanol. Oxidation with chromic acid gives 3-pentanone, which is then converted to the product upon treatment with the appropriate Wittig reagent, thereby installing two carbon atoms and a double bond in the correct location:

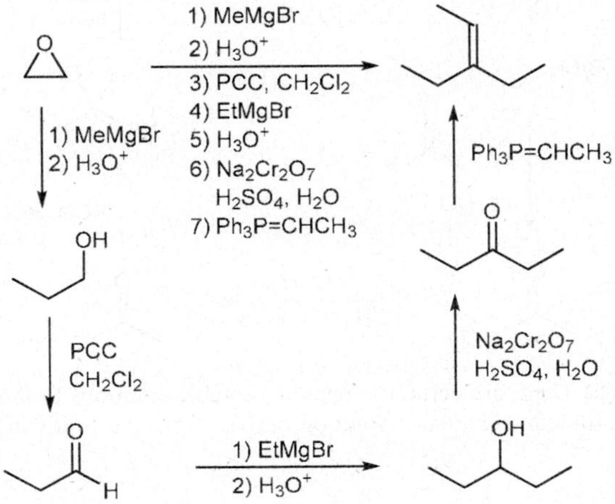

(g) There are certainly many acceptable solutions to this problem. One such solution derives from the following retrosynthetic analysis. An explanation of each of the steps (*a-d*) follows.

a. The product can be made from a cyanohydrin, via hydrolysis of the cyano group.
b. The cyanohydrin can be made from the corresponding ketone (2-butanone).
c. 2-Butanone can be made via oxidation of the corresponding secondary alcohol (2-butanol).
d. 2-Butanol can be made via a Grignard reaction between acetaldehyde and ethyl magnesium bromide.

Now let's draw the forward scheme. Ethyl bromide is converted into ethyl magnesium bromide, which is then treated with acetaldehyde (to give a Grignard reaction), followed by aqueous acidic workup, to give 2-butanol. Oxidation of 2-butanol with chromic acid gives 2-butanone, which can then be converted into a cyanohydrin upon treatment with KCN and HCl. And finally, hydrolysis of the cyano group gives the desired product:

1) Mg
2) acetaldehyde
3) H_3O^+
4) $Na_2Cr_2O_7$, H_2SO_4, H_2O
5) KCN, HCl
6) H_3O^+, heat

(h) There are certainly many acceptable solutions to this problem. One such solution derives from the following retrosynthetic analysis. An explanation of each of the steps (*a-d*) follows.

a. The product can be made from a cyanohydrin, via reduction of the cyano group.
b. The cyanohydrin can be made from the corresponding ketone (2-butanone).
c. 2-Butanone can be made via oxidation of the corresponding secondary alcohol (2-butanol).
d. 2-Butanol can be made via a Grignard reaction between acetaldehyde and ethyl magnesium bromide.

Now let's draw the forward scheme. Ethyl bromide is converted into ethyl magnesium bromide, which is then treated with acetaldehyde (to give a Grignard reaction), followed by aqueous acidic workup, to give 2-butanol. Oxidation of 2-butanol with chromic acid gives 2-butanone, which can then be converted into a cyanohydrin upon treatment with KCN and HCl. And finally, reduction of the cyano group gives the desired product:

1) Mg
2) acetaldehyde
3) H_2O
4) $Na_2Cr_2O_7$, H_2SO_4, H_2O
5) KCN, HCl
6) $LiAlH_4$
7) H_2O

19.91.

(a) In acid-catalyzed conditions, the starting material is protonated. There are two locations where protonation can occur (the lone pair of the nitrogen atom, or a lone pair of the oxygen atom). The nitrogen atom is more likely protonated first, because it is a stronger base (a protonated amine, called an ammonium ion, is a much weaker acid than a protonated ether, called an oxonium ion, as seen in the pK_a table on the inside cover of the textbook). Loss of a leaving group (dimethyl amine) gives a resonance-stabilized cation, which is then attacked by water. The resulting oxonium ion is then deprotonated to give a cyclic hemiacetal. Protonation, followed by loss of a leaving group gives a protonated carbonyl group, which then loses a proton to give the product. Notice that water functions as the base in each deprotonation step.

- $(CH_3)_2NH$

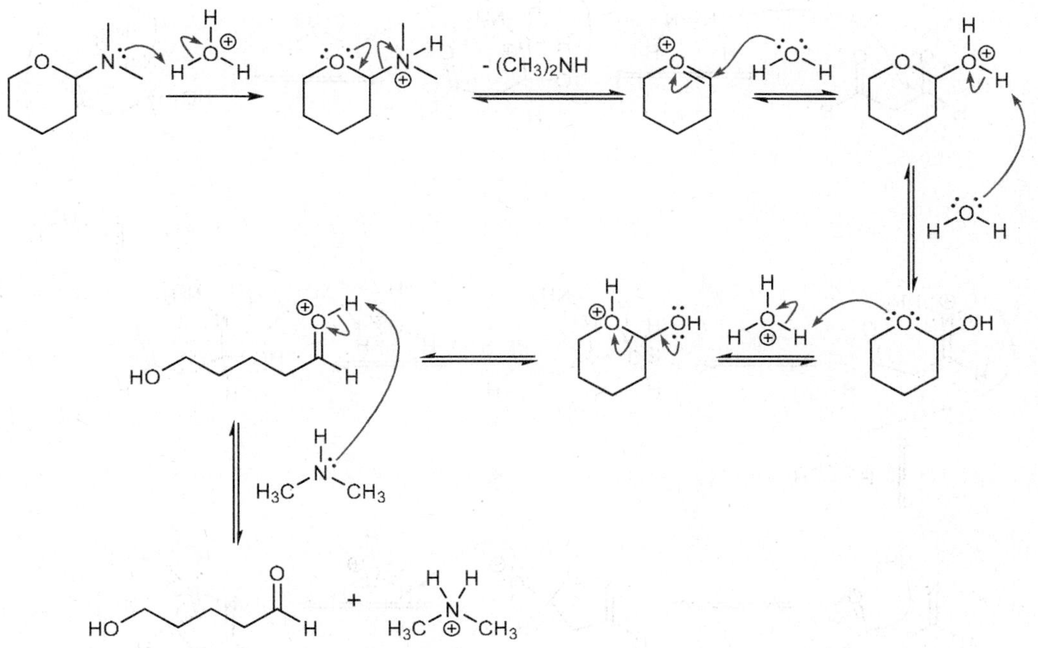

(b) The starting material is a vinyl ether, and it is being subjected to aqueous acidic conditions. This indicates that protonation is likely the first step of the mechanism. There are two locations to consider for protonation: the oxygen atom or the π bond. Protonation of the oxygen atom does not result in a resonance-stabilized cation, while protonation of the π bond does indeed result in a resonance-stabilized cation. As such, the first step is protonation of the π bond. The resulting intermediate is then attacked by water to give an oxonium ion, which is then deprotonated to give a hemiacetal. Protonation, followed by loss of a leaving group gives a protonated carbonyl group, which then loses a proton to give the product. Notice that water functions as the base in each deprotonation step.

- CH_3OH

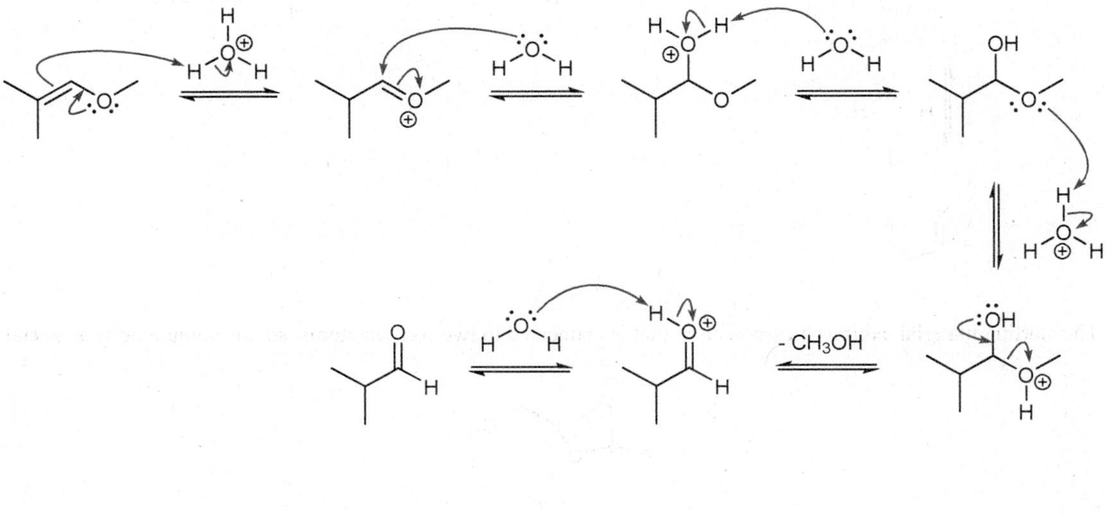

(c) Hydrazine is sufficiently nucleophilic to attack a carbonyl group directly (without prior activation of the carbonyl group via protonation). The resulting intermediate undergoes two successive proton transfer steps, giving an intermediate that is free of formal charges. Protonation of the OH group converts a bad leaving group into a good one (water). Loss of the leaving group gives a resonance-stabilized cation, which then loses a proton to give a hydrazone. An intramolecular nucleophilic attack gives an intermediate that undergoes two proton transfer steps to give an intermediate free of formal charges. Protonation of the OH group converts a bad leaving group into a good one (water). Loss of the leaving group gives a resonance-stabilized cation, which then loses a proton to give the product, which is aromatic.

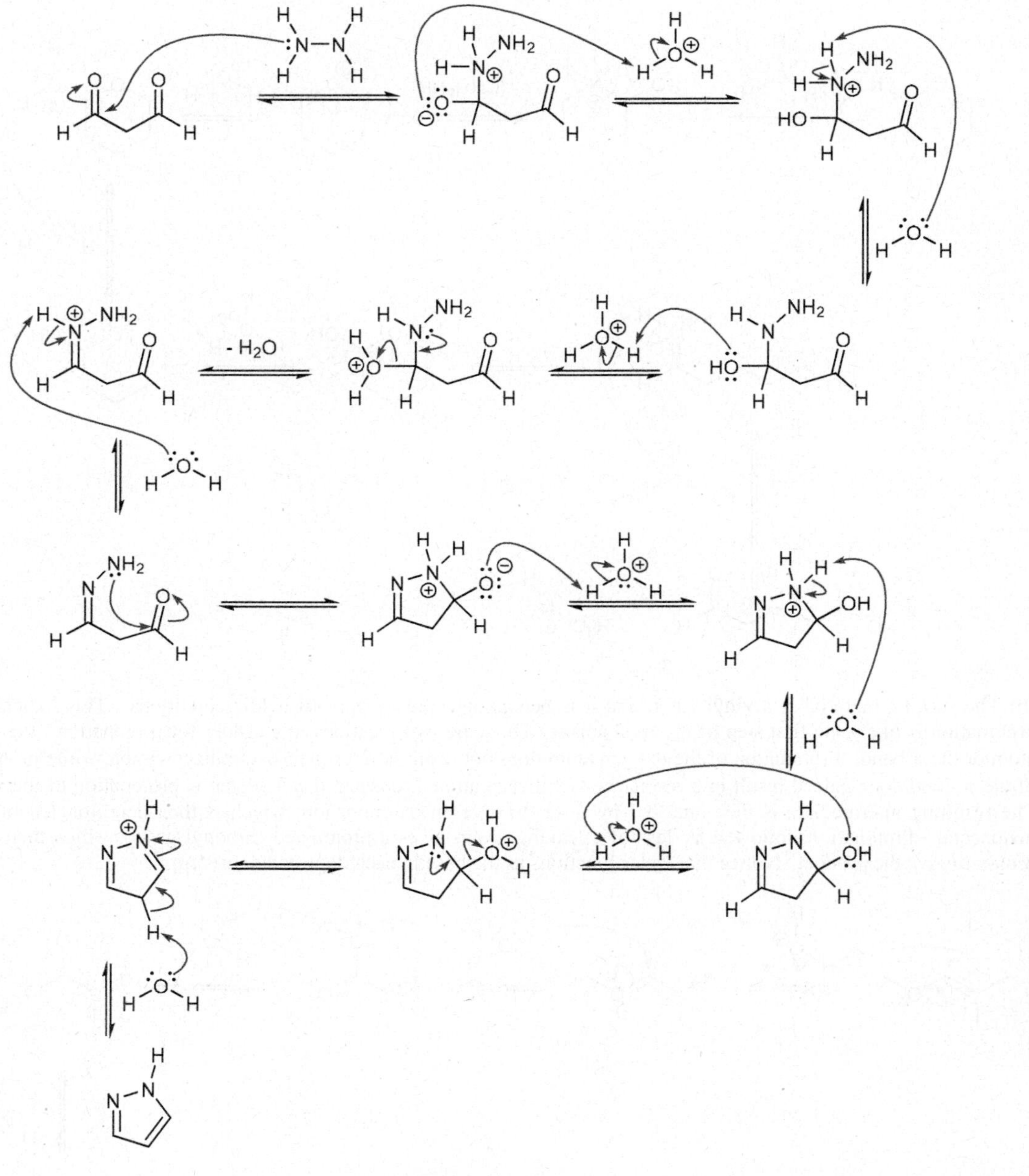

- H_2O

(d) The starting material exhibits a carbon atom that is connected to two oxygen atoms, so this compound is an acetal:

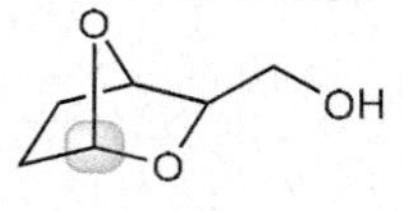

Protonation of the acetal (specifically at the oxygen atom in the bottom right corner of the structure) results in an oxonium ion that can lose a leaving group to give a resonance-stabilized cation. This intermediate has two OH groups. If the more distant OH group attacks the C=O bond, the resulting oxonium ion can lose a proton to give the product. Notice that the base for the deprotonation step is water, which was liberated in the first step.

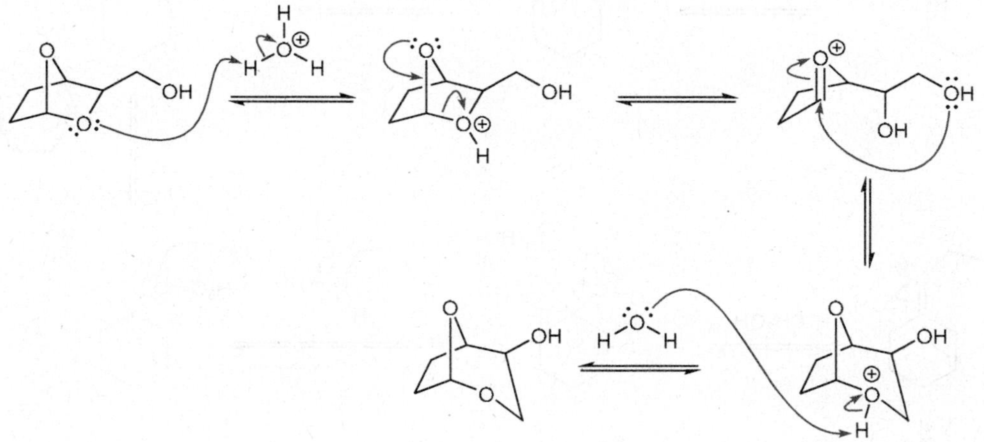

(e) The starting material exhibits a carbon atom that is connected to two oxygen atoms, so this compound is an acetal. Protonation of the acetal (specifically at the oxygen atom on the left) results in an oxonium ion that can lose a leaving group to give a resonance-stabilized cation. This intermediate has two OH groups. If the one on the right side attacks the C=O bond, the resulting oxonium ion can lose a proton to give the product. Notice that the base for the deprotonation step is water.

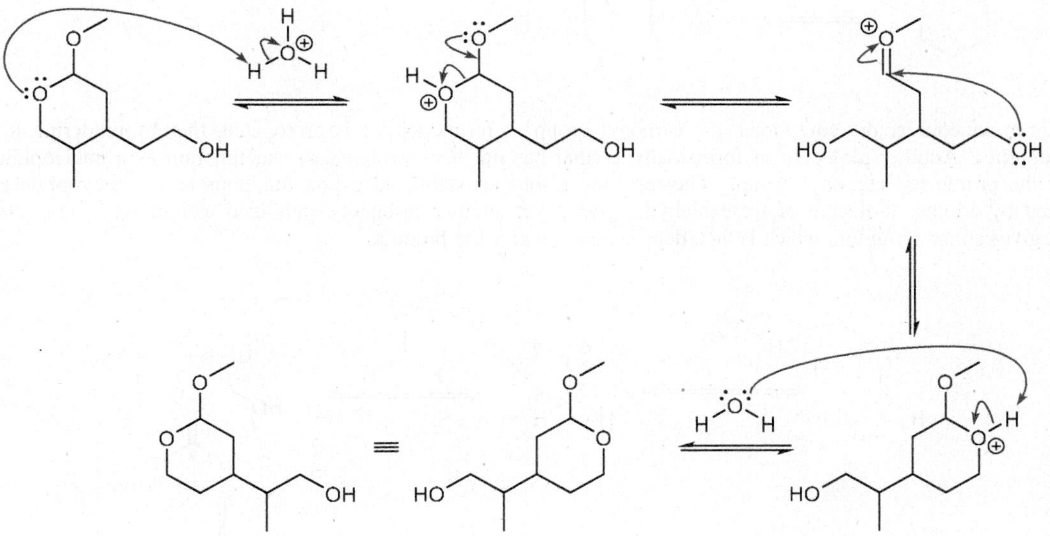

(f) The starting material is a vinyl ether, and it is being subjected to acidic conditions. This indicates that protonation is likely the first step of the mechanism. There are two locations to consider for protonation: the oxygen atom or the π bond. Protonation of the oxygen atom does not result in a resonance-stabilized cation, while protonation of the π bond does indeed result in a resonance-stabilized cation. As such, the first step is protonation of the π bond. The likely proton source is the conjugate acid of ethylene glycol, which received its proton from tosic acid (TsOH). Protonation results in a resonance-stabilized cation, which is then attacked by ethylene glycol to give an oxonium ion, followed by deprotonation to give a hemiacetal. Protonation of the methoxy group, followed by loss of a leaving group (methanol) gives a protonated carbonyl group, which then functions as an electrophile in an intramolecular nucleophilic attack. Deprotonation then gives the product.

Notice that ethylene glycol functions as the base in each deprotonation step.

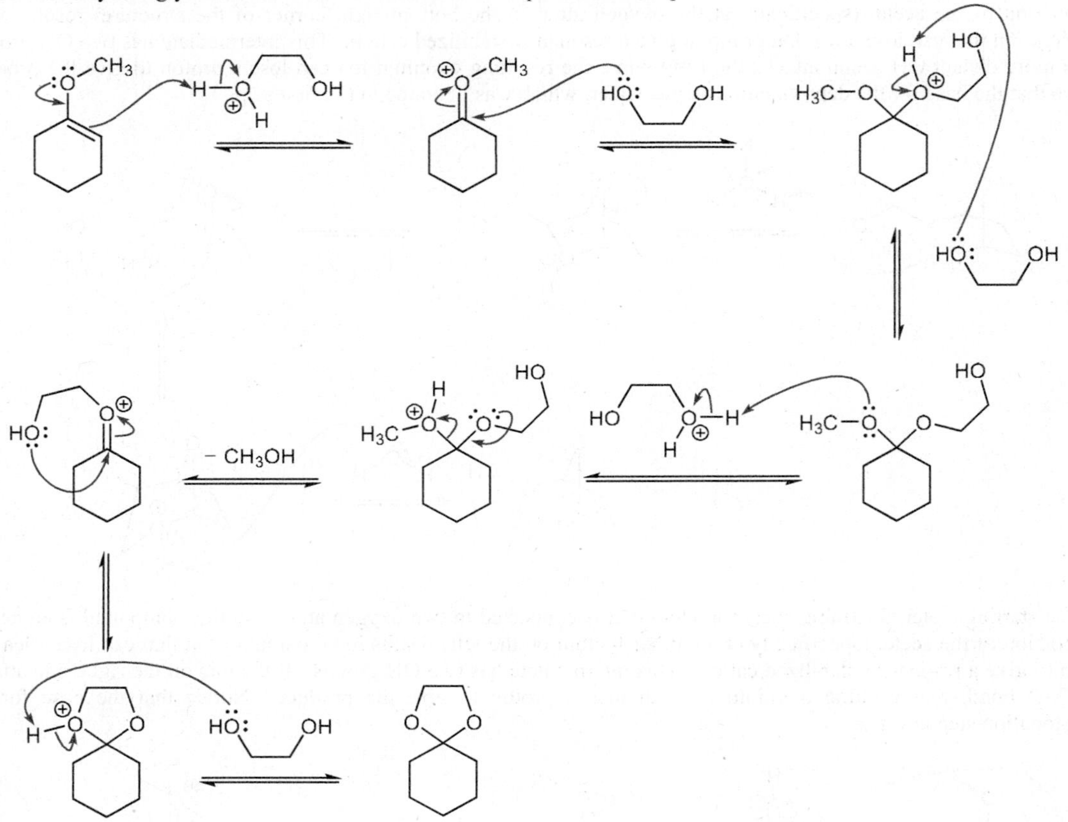

19.92. In aqueous acidic conditions, the carbonyl group of formaldehyde is protonated, thereby rendering it more electrophilic. Another molecule of formaldehyde (that has not been protonated) can function as a nucleophile and attack the protonated carbonyl group. The resulting resonance-stabilized cation functions as an electrophile and is attacked by another molecule of formaldehyde, giving yet another resonance-stabilized cation. An intramolecular attack gives an oxonium ion, which is then deprotonated to give the product.

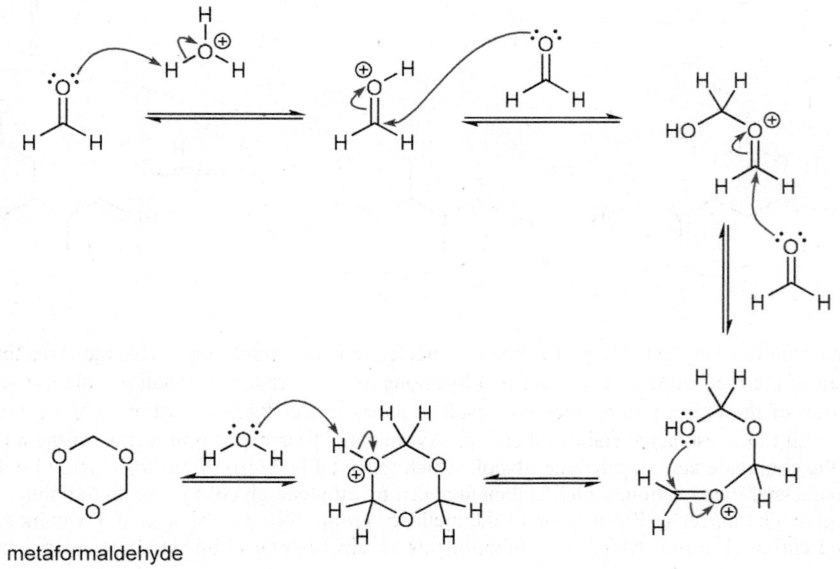

metaformaldehyde

19.93. The ketone group (the more reactive carbonyl) must first be protected by converting it into an acetal. Then, the ester can be reduced with excess $LiAlH_4$ to give an alcohol (if the ketone had not been protected, it would have also been reduced by $LiAlH_4$). Mild oxidation of the alcohol with PCC (or DMP or Swern conditions) gives an aldehyde, which is then converted to the corresponding thioacetal.

19.94.

(a) Compound **1** is a ketone, while compound **2** is an ester. This transformation from **1** to **2** can be achieved with a peroxy acid (RCO_3H). Then, compound **2** can be converted into diol **3** upon treatment with excess lithium aluminum hydride. Finally, after the conversion of diol **3** to tosylate **4**, compound **5** can be obtained if compound **4** is treated with a strong base, such as NaH. Under these conditions, the hydroxyl group is deprotonated to give an alkoxide ion, which functions as a nucleophilic center in an intramolecular S_N2-type process, giving an ether (compound **5**):

(b) Compound **1** has rotational symmetry, which compound **5** lacks. Therefore, compound **1** has fewer signals in its NMR spectrum.

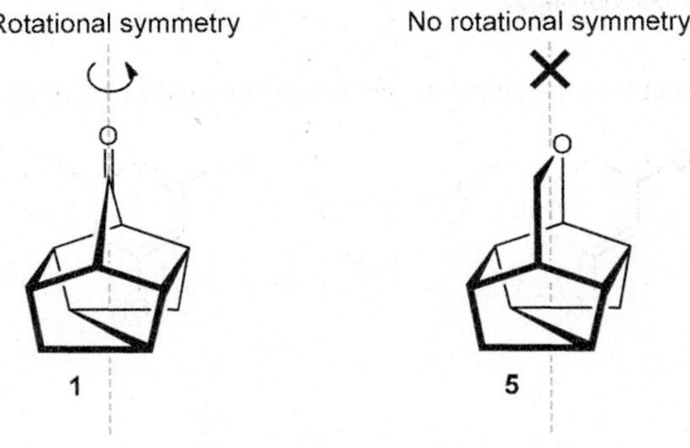

19.95. First, two curved arrows are used to show delivery of hydride to the ester. In the next step, two curved arrows show collapse of the charged tetrahedral intermediate and ejection of the leaving group, which in this case is ethoxide. The resulting aldehyde is then further reduced by another equivalent of hydride from $LiAlH_4$. Once again, two curved arrows are used to show hydride delivery. Finally, the resulting alkoxide ion is protonated upon treatment with aqueous acid, and two curved arrows are needed to show this proton transfer step. Therefore, the net reaction is the reduction of the ester functional group to generate an alcohol.

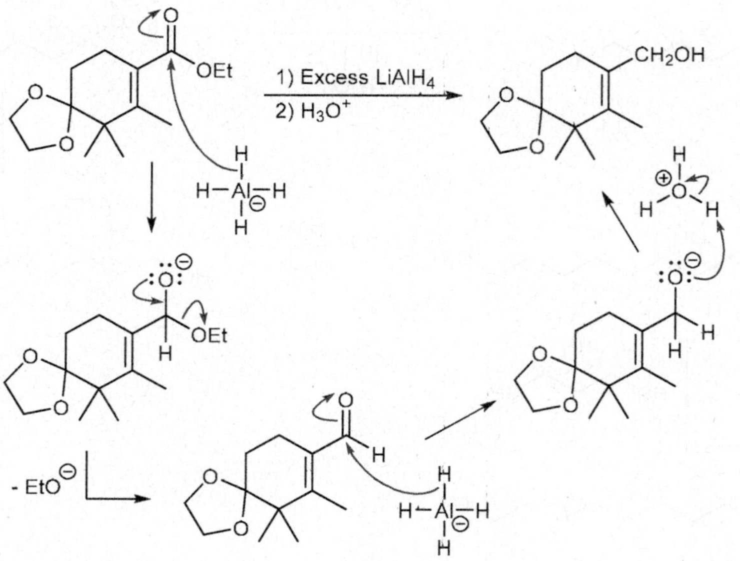

1) Excess $LiAlH_4$
2) H_3O^+

- $EtO^{\ominus}$

19.96. The carbon chain must be extended by adding a single additional carbon atom, and we are given a clue that a Wittig reaction is involved, so we know there is an alkene intermediate at some point. Using a retrosynthetic analysis, the target molecule (compound **2**) can be made from an alkene, via an *anti*-Markovnikov addition of H and OH:

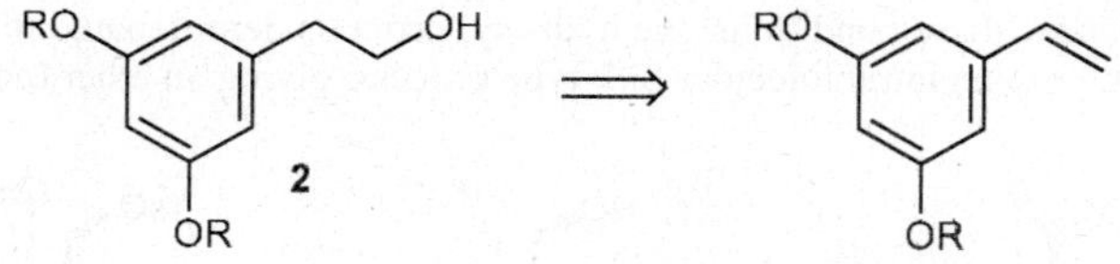

This alkene can be made from an aldehyde via a Wittig reaction:

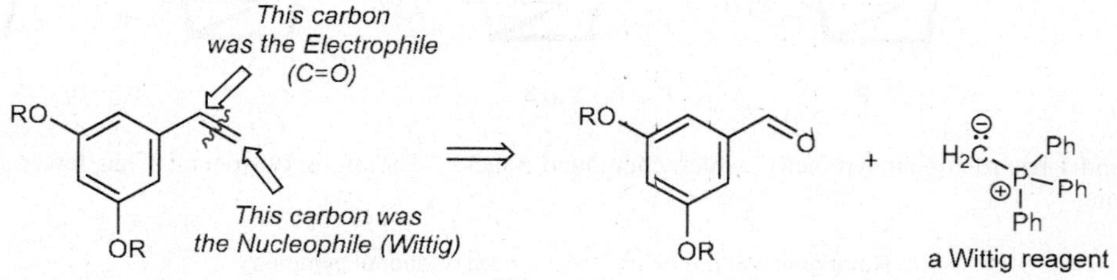

This carbon was the Electrophile (C=O)

This carbon was the Nucleophile (Wittig)

a Wittig reagent

And the aldehyde can be made from the corresponding alcohol (compound **1**) via oxidation:

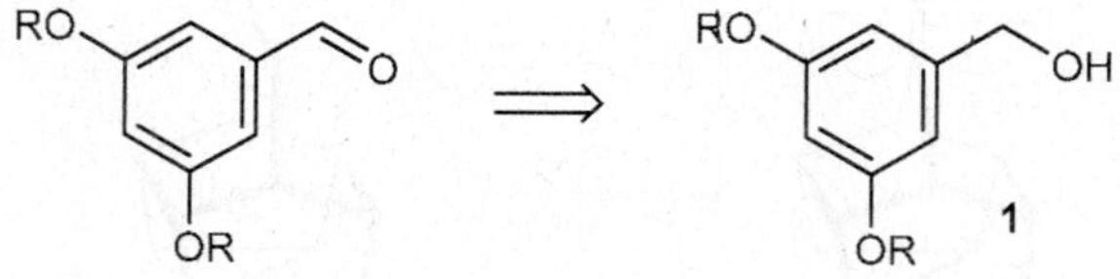

Now that we have completed the retrosynthetic planning, we can draw the forward process, as shown below. PCC oxidation of the primary alcohol gives an aldehyde, and reaction with a Wittig reagent affords the alkene. Hydroboration-oxidation adds H and OH in an *anti*-Markovnikov fashion, giving the desired alcohol.

1) PCC, CH_2Cl_2
2) $Ph_3P{=}CH_2$
3) $BH_3 \cdot THF$
4) H_2O_2, NaOH

PCC
CH_2Cl_2

$Ph_3P{=}CH_2$

1) $BH_3 \cdot THF$
2) H_2O_2, NaOH

19.97. As with all synthesis problems, we must determine 1) if there is a change in the carbon skeleton, and 2) if there is a change in identity or location of the functional group(s). Numbering the carbon atoms from left to right, we see that two carbon atoms are being introduced during this transformation. We also see that the alcohol functional group (OH) is in the same location in both the starting material and the product.

Importantly, a new C–C bond must be formed at the same location as the OH group (carbon 5 in our labeling scheme above, also the highlighted carbon below). This suggests a synthetic route involving a Grignard reagent attacking a C=O bond. That is, the product can be made from the following aldehyde, as shown in the following retrosynthetic scheme:

The starting material does not have the aldehyde group needed to react with a Grignard reagent. However, the required aldehyde can be made via oxidation of the primary alcohol, as shown in the following retrosynthetic scheme:

Now let's draw the forward process. There are many reagents that can be used to convert a primary alcohol into an aldehyde; one such reagent is PCC. The resulting aldehyde can then be treated with vinyl magnesium bromide ($CH_2{=}CHMgBr$), followed by workup with aqueous acid, to produce the desired product:

1) PCC, CH_2Cl_2
2) vinyl MgBr
3) H_3O^+

PCC
CH_2Cl_2

1) vinyl MgBr
2) H_3O^+

19.98. In all three reaction sequences, the first two steps involve the addition of a Grignard reagent to the ketone, followed by alcohol dehydration using concentrated sulfuric acid and heat. The products of each of these operations are shown below:

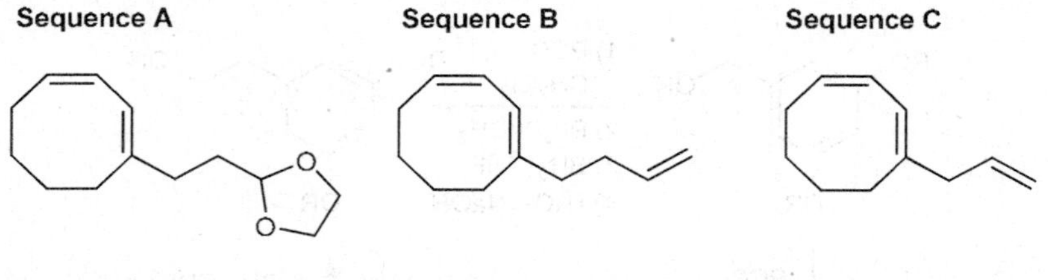

Sequence A

Sequence B

Sequence C

In sequence A, the final step is acetal deprotection using acetic acid and water, which will produce the desired product, aldehyde **2**. Under these conditions no undesired side reactions are expected to occur. In sequence B, ozonolysis of the terminal alkene will certainly produce an aldehyde, however, the molecule also contains two additional alkenes within the 8-membered ring. Since ozone is a non-selective oxidant, if this compound were subjected to O_3/DMS, all three alkene groups would react! Therefore, the major product of the reaction will not be aldehyde **2**, but a compound with three different carbonyl groups (as well as two other fragments). In sequence C, hydroboration/oxidation will convert the terminal alkene to a primary alcohol, which will then be transformed into the desired product via oxidation with PCC. However, just like we saw in sequence B, the other two π bonds will react under the hydroboration/oxidation conditions; the product isolated will not be aldehyde **2**. In conclusion, after a thorough analysis, only sequence A will lead to the desired product.

19.99. The product of the Wittig reaction is alkene **2**. The mechanism for acid-catalyzed hydration of this alkene begins with protonation of the π bond to generate **3**, a resonance-stabilized intermediate. Note that this intermediate is similar to the type of intermediate that we encountered during acetal formation/cleavage. Water then attacks to generate a charged tetrahedral intermediate (**4**) which can be deprotonated to form hemiacetal **5**. Protonation, followed by regeneration of the carbonyl group via loss of methanol, will produce protonated aldehyde **7**, which is deprotonated in the final step to afford aldehyde **8**.

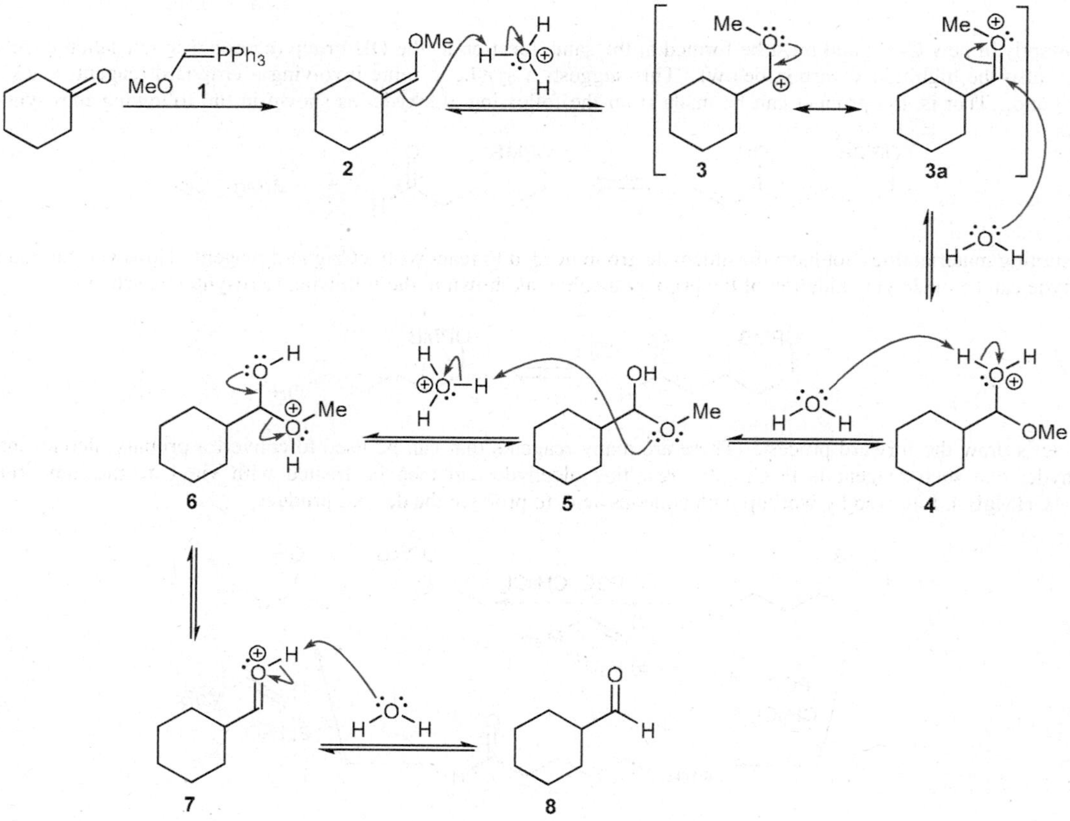

MeO=PPh3 1

2

3

3a

4

5

6

7

8

19.100. In the first step, the acid chloride reacts with $AlCl_3$ to form a resonance-stabilized acylium ion (**4**). In the absence of an aromatic ring, the C=C π bond will function as a nucleophile and trap the acylium ion to produce a carbocation (**5**). This carbocation is transformed into compound **1** if $AlCl_4^-$ transfers a chloride ion to the carbocation (path **A**). Alternatively, carbocation **5** is transformed into compound **2** via a 1,2-hydride shift to form tertiary carbocation **6** (path **B**), followed by deprotonation.

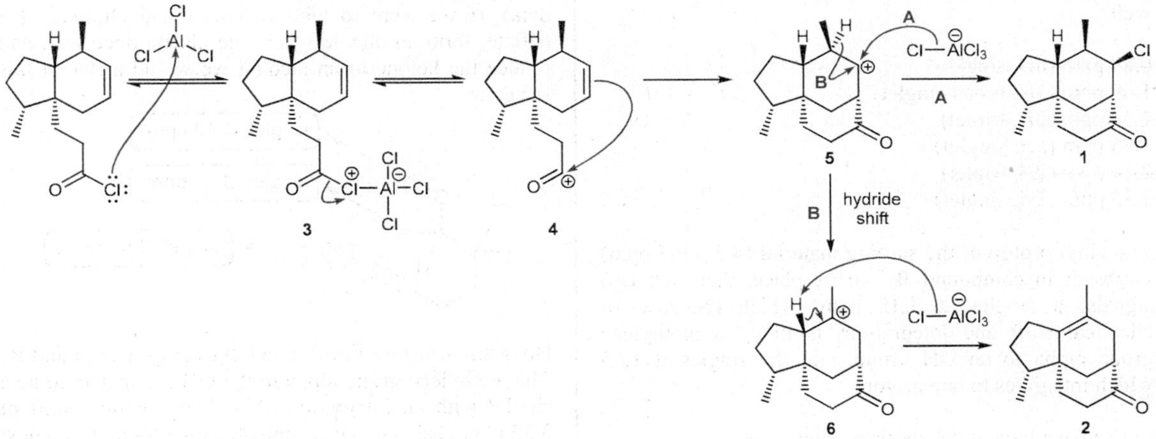

19.101.
(a) We begin by determining the HDI for compound **B**, as shown here:

$$HDI = \frac{1}{2}(2C + 2 + N - H - X)$$
$$= \frac{1}{2}(2 \cdot 8 + 2 + 0 - 14 - 0)$$
$$= \frac{1}{2}(4)$$
$$= 2$$

Therefore, compound **B** has two degrees of unsaturation. Now let's consider the IR data. Only two peaks are given, but they are important bands: a broad band around 3300 cm^{-1} is typical for an alcohol group, and 2117 cm^{-1} is diagnostic for a triple bond – either an alkyne (C≡C) or a nitrile (C≡N). Since nitrogen is absent from the molecular formula, the band at 2117 cm^{-1} must correspond with a C≡C triple bond. This accounts for both degrees of unsaturation.
At this stage, we consider the ^1H NMR data. Also, take into consideration the structure of the starting material; we know that the product should have at least some degree of similarity. The ^1H NMR data shows six types of protons; the easiest to assign is the broad singlet at 1.49 ppm which is likely a proton of an alcohol; the presence of an alcohol functional group is further confirmed by the IR absorption at 3305 cm^{-1}.
To assign the remaining signals, let's look at the starting material and consider what its ^1H NMR must look like.

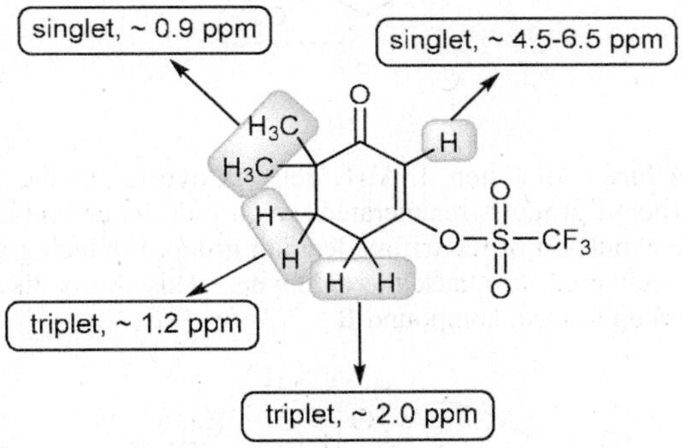

singlet, ~ 0.9 ppm
singlet, ~ 4.5-6.5 ppm
triplet, ~ 1.2 ppm
triplet, ~ 2.0 ppm

The starting material has no chiral centers, and only four unique types of protons. The two methyl groups will appear as a singlet near 0.9 ppm, the allylic methylene group is expected to appear as a triplet near 2.0 ppm, the methylene next to the gem-dimethyl group is expected to be a triplet near 1.2 ppm, and finally, the vinyl proton should be a singlet in the range 4.5-6.5 ppm.

Now let's take another look at the ^1H NMR data for compound **B**.

0.89 ppm (6H, singlet)
~~1.49 ppm (1H, broad singlet)~~
1.56 ppm (2H, triplet)
1.95 ppm (1H, singlet)
2.19 ppm (2H, triplet)
3.35 ppm (2H, singlet)

We've already assigned the 1.49 ppm broad singlet as an OH proton, so we can cross it off. We can also make the assumption that the 0.89 ppm singlet that integrates for 6 protons must be the gem-dimethyl group, and the two triplets at 1.56 ppm and 2.19 ppm are the two methylene groups from the starting material. Let's cross them out as well.

~~0.89 ppm (6H, singlet)~~
~~1.49 ppm (1H, broad singlet)~~
~~1.56 ppm (2H, triplet)~~
1.95 ppm (1H, singlet)
~~2.19 ppm (2H, triplet)~~
3.35 ppm (2H, singlet)

The vinyl proton of the starting material (4.5 – 6.5 ppm) is absent in compound **B**. In its place, there are two signals: a singlet at 3.35 ppm, which (because of chemical shift and integration) is likely a methylene group alpha to an OH group, and the singlet at 1.95 which integrates to one proton.

Let's take a look at the starting material again:

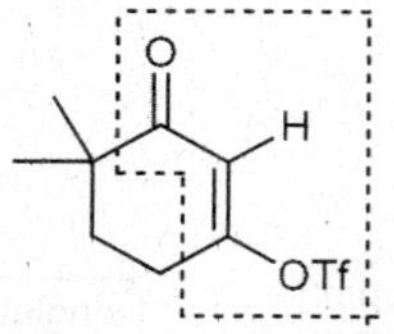

It seems that the left half is represented in compound **B**, however the right half has changed dramatically. From the molecular formula we know that there is no longer any sulfur in the molecule, so the triflate group must be gone. We also know that Compound B must exhibit an OH group and a C≡C triple bond (from the IR spectral data). If we were to take an eraser and eliminate the triflate, form an alkyne where the alkene once was, and reduce the ketone to an alcohol we would arrive at this structure:

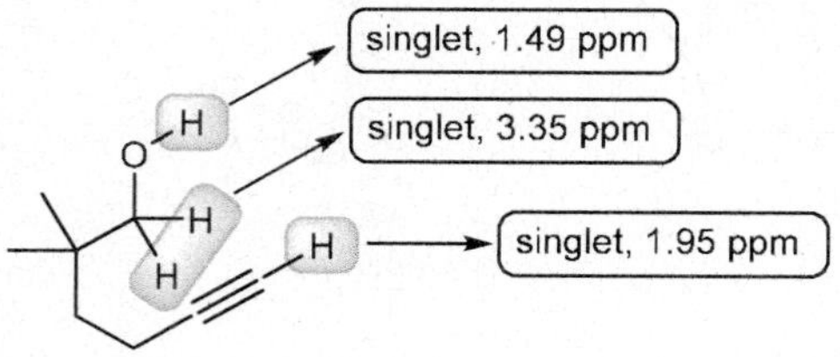

Does this structure fit the data left over for compound **B**? The methylene group alpha to the OH group should be a singlet with an integration of 2. The chemical shift of 3.35 ppm is also in the acceptable range for a proton next to oxygen. And the proton of the alkyne should also be a singlet, with an integration of 1. The chemical shift is slightly less than what the table in Chapter 15 shows (~2.5 ppm), but it is certainly within a margin of error. The molecular formula also matches.

(b) In the first step, $LiAlH_4$ delivers hydride to the carbonyl group to form the alkoxide intermediate **2**. Next, the carbonyl group is regenerated which induces carbon-carbon bond cleavage to simultaneously generate the alkyne via the expulsion of the triflate leaving group, producing aldehyde **3**. In the presence of $LiAlH_4$, another hydride ion can be delivered, to attack the aldehyde. This forms alkoxide intermediate **4**, which is protonated upon aqueous acidic workup to form compound **B**.

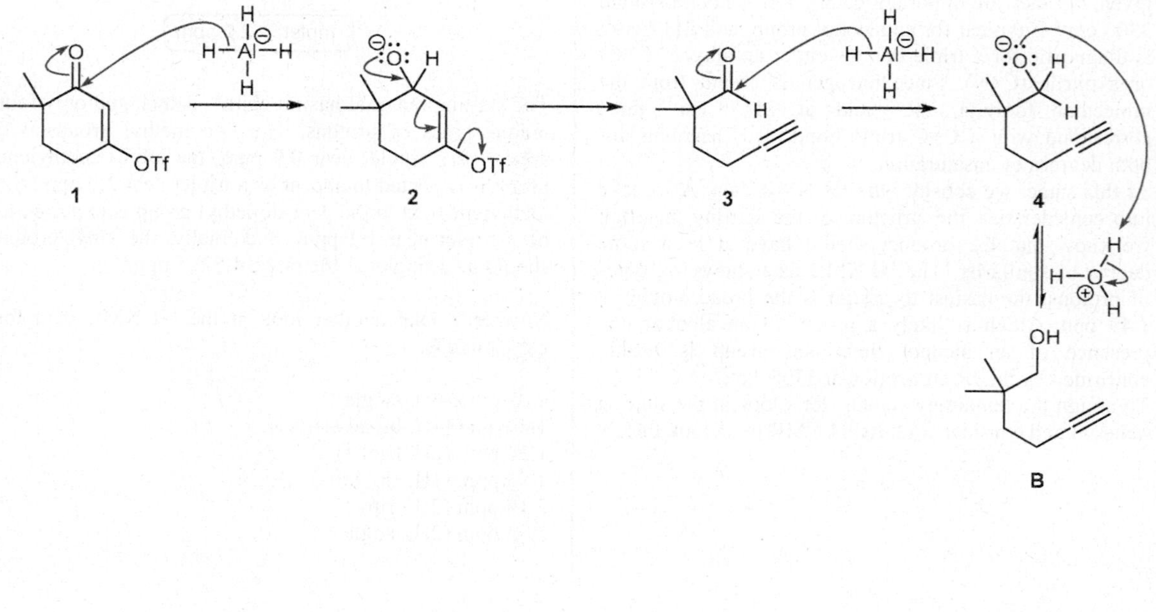

Chapter 20
Carboxylic Acids and Their Derivatives

Review of Concepts

Fill in the blanks below. To verify that your answers are correct, look in your textbook at the end of Chapter 20. Each of the sentences below appears verbatim in the section entitled *Review of Concepts and Vocabulary.*

- Treatment of a carboxylic acid with a strong base yields a _______________ salt.
- The pK_a of most carboxylic acids is between ____ and _____.
- Using the **Henderson-Hasselbalch equation**, it can be shown that carboxylic acids exist primarily as ______________________ at **physiological pH**.
- Electron-_____________ substituents can increase the acidity of a carboxylic acid.
- When treated with aqueous acid, a nitrile will undergo _______________, yielding a carboxylic acid.
- Carboxylic acids are reduced to ____________ upon treatment with lithium aluminum hydride or borane.
- Carboxylic acid derivatives exhibit the same _____________ state as carboxylic acids.
- Carboxylic acid derivatives differ in reactivity, with ________________ being the most reactive and ________________ the least reactive.
- When drawing a mechanism, avoid formation of a strong ___________ in acidic conditions, and avoid formation of a strong ____________ in basic conditions.
- When a nucleophile attacks a carbonyl group to form a tetrahedral intermediate, always reform the carbonyl group if possible, but avoid expelling _____ or ______.
- When treated with an alcohol, acid chlorides are converted into _________________.
- When treated with ammonia, acid chlorides are converted into _____________.
- When treated with a _____________ reagent, acid chlorides are converted into alcohols with the introduction of two alkyl groups.
- The reactions of anhydrides are the same as the reactions of ________________ except for the identity of the leaving group.
- When treated with a strong base followed by an alkyl halide, carboxylic acids are converted into __________.
- In a process called the **Fischer esterification**, carboxylic acids are converted into esters when treated with an ____________ in the presence of _______________.
- Esters can be hydrolyzed to yield carboxylic acids upon treatment with either aqueous base or aqueous ________. Hydrolysis under basic conditions is also called _______________.
- When treated with lithium aluminum hydride, esters are reduced to yield ___________. If the desired product is an aldehyde, then ________ is used as a reducing agent instead of $LiAlH_4$.
- When treated with a ____________ reagent, esters are reduced to yield alcohols, with the introduction of two alkyl groups.
- When treated with excess $LiAlH_4$, amides are converted into __________.
- Nitriles are converted to amines when treated with ___________.

Review of Skills

Fill in the blanks and empty boxes below. To verify that your answers are correct, look in your textbook at the end of Chapter 20. The answers appear in the section entitled *SkillBuilder Review.*

20.1 Drawing the Mechanism of a Nucleophilic Acyl Substitution Reaction

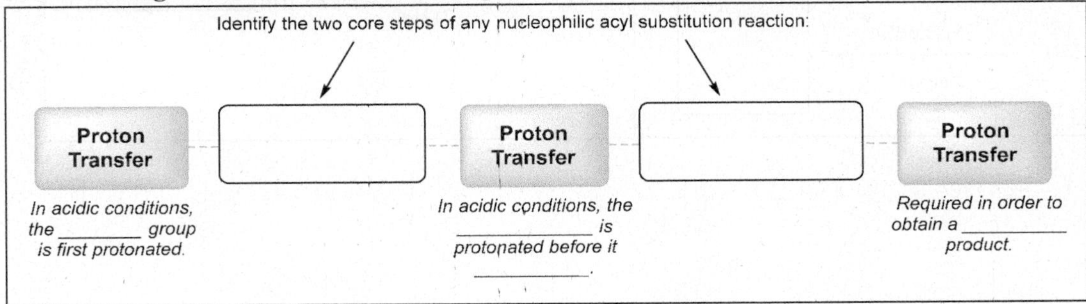

20.2 Interconverting Functional Groups

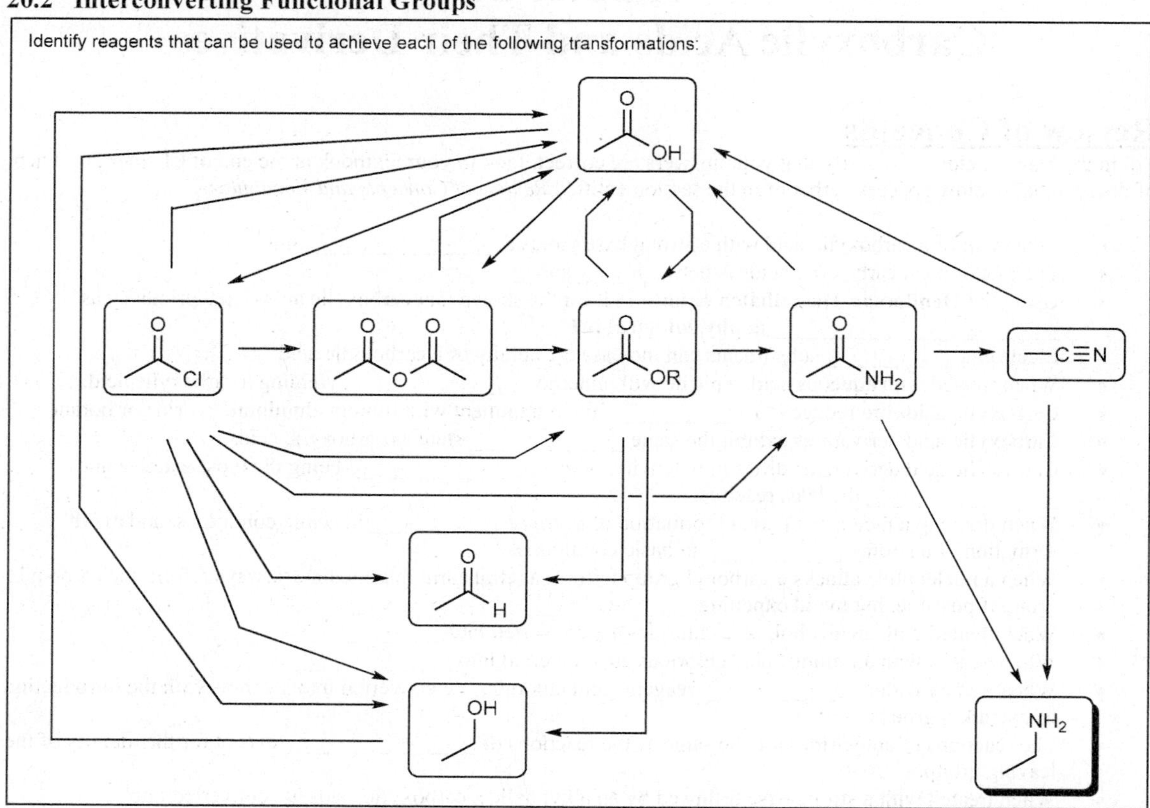

20.3 Choosing the Most Efficient C-C Bond-Forming Reaction

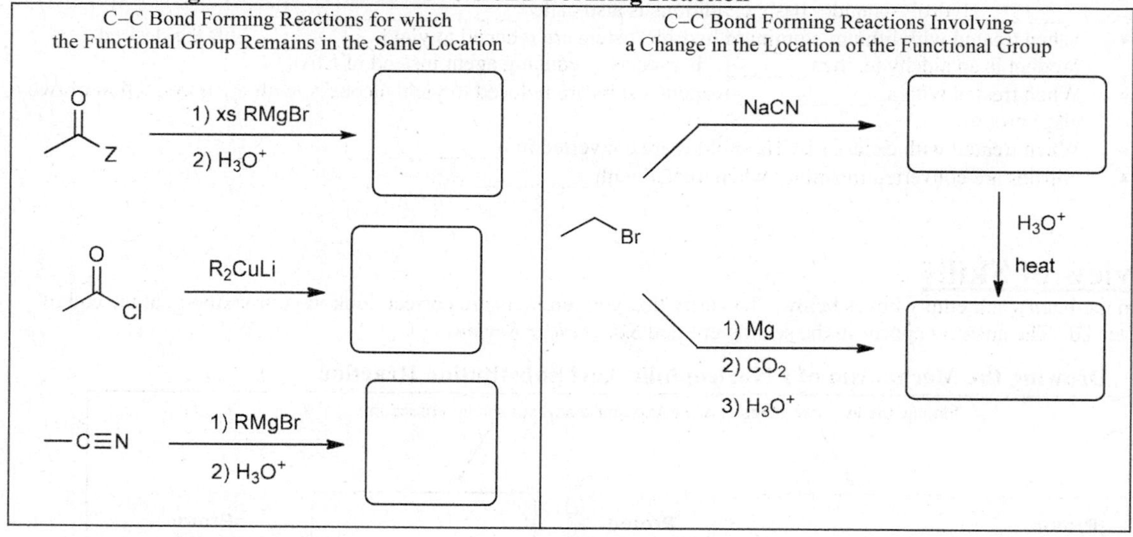

Review of Reactions

Identify the reagents necessary to achieve each of the following transformations. To verify that your answers are correct, look in your textbook at the end of Chapter 20. The answers appear in the section entitled *Review of Reactions*.

Preparation of Carboxylic acids

Reactions of Carboxylic Acids

Preparation and Reactions of Acid Chlorides

Preparation and Reactions of Acid Anhydrides

Preparation of Esters

Reactions of Esters

Preparation of Amides	*Reactions of Amides*
R–C(=O)–Cl ⟶ R–C(=O)–NH$_2$	R–C(=O)–NH$_2$ ⟶ R–C(=O)–OH (two routes); R–C(=O)–NH$_2$ ⟶ R–CH$_2$–NH$_2$

Preparation of Nitriles	*Reactions of Nitriles*
R–CH$_2$–Br ⟶ R–CH$_2$–C≡N; R–C(=O)–NH$_2$ ⟶ R–C≡N	R–C≡N ⟶ R–C(=O)–OH (two routes); R–C≡N ⟶ R–C(=O)–R; R–C≡N ⟶ R–CH$_2$–NH$_2$

Review of Mechanisms

Complete each of the following mechanisms by drawing the missing curved arrows. To verify that your curved arrows are drawn correctly, compare them to the curved arrows in the mechanism boxes for Mechanisms 20.1 – 20.15, which can be found throughout Chapter 20 of your text.

Mechanism 20.1 Hydrohalogenation

C(=O)–Z $\xrightarrow{:Nuc^{\ominus}}$ C(–O$^{\ominus}$)(Nuc)(Z) $\xrightarrow{-\,Z:^{\ominus}}$ C(=O)–Nuc

Mechanism 20.2 Preparation of Acid Chlorides via Thionyl Chloride

Part 1

$SOCl_2$ — Cl — Cl — HCl

Part 2

Cl — Cl — SO_2 + Cl^-

Mechanism 20.3 Hydrolysis of an Acid Chloride

$- Cl^-$

Mechanism 20.4 Reduction of an Acid Chloride with $LiAlH_4$

$- Cl^-$

Mechanism 20.5 The Reaction Between an Acid Chloride with a Grignard Reagent

$:R^-$ — $- Cl^-$ — $:R^-$

Mechanism 20.6 The Fischer Esterification Process

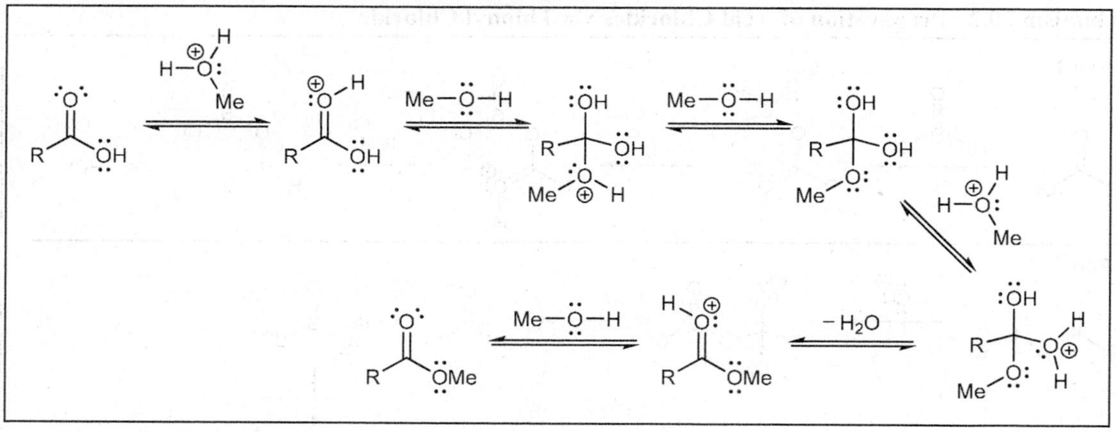

Mechanism 20.7 Base-Promoted Hydrolysis of Esters (Saponification)

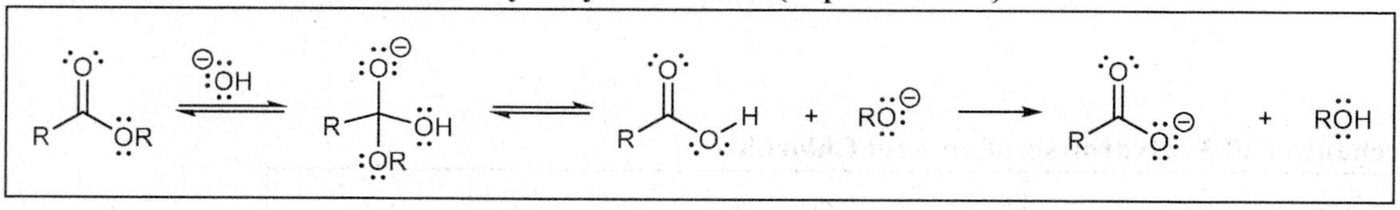

Mechanism 20.8 Acid-Catalyzed Hydrolysis of Esters

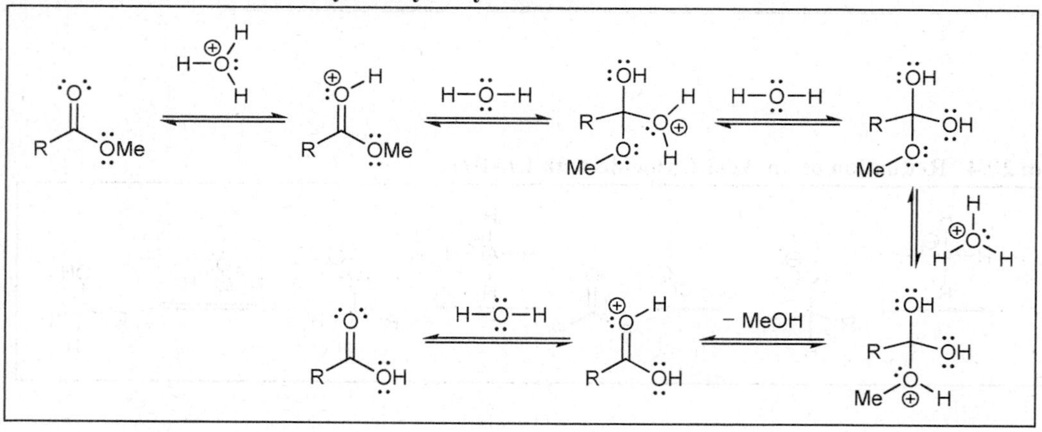

Mechanism 20.9 Reduction of an Ester with $LiAlH_4$

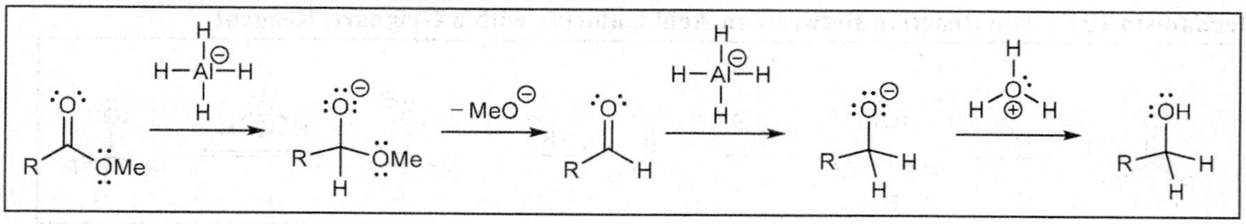

Mechanism 20.10 The Reaction Between an Ester and a Grignard Reagent

Mechanism 20.11 Acid-Catalyzed Hydrolysis of an Amide

Mechanism 20.12 Hydrolysis of Amides under Basic Conditions

Mechanism 20.13 Dehydration of Amides

Mechanism 20.14 Acid-Catalyzed Hydrolysis of Nitriles

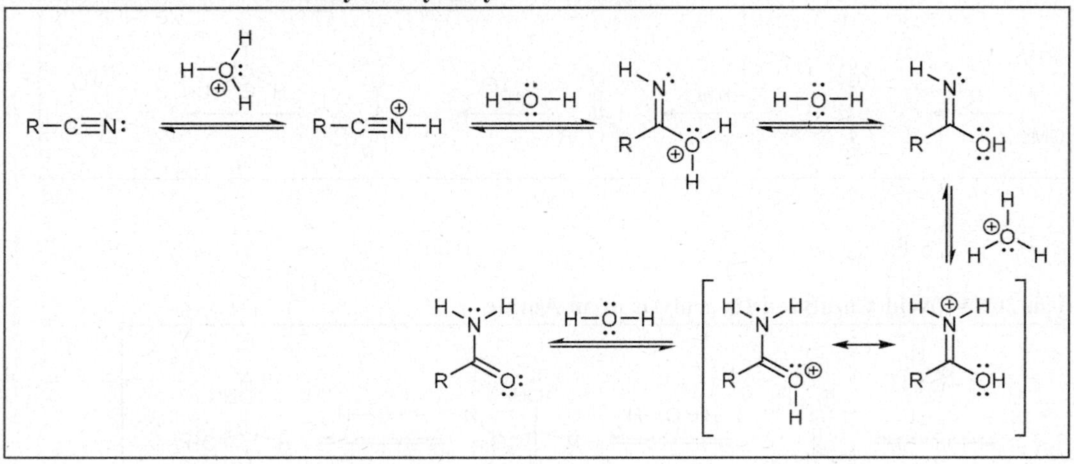

Mechanism 20.15 Base-Catalyzed Hydrolysis of Nitriles

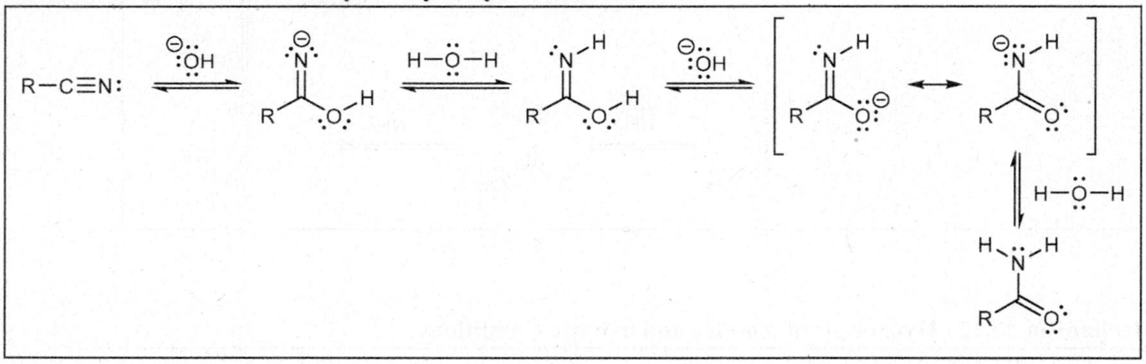

Common Mistake to Avoid

This chapter covers many reactions. One of these reactions is between an acid chloride and a lithium dialkylcuprate, giving a ketone as the product:

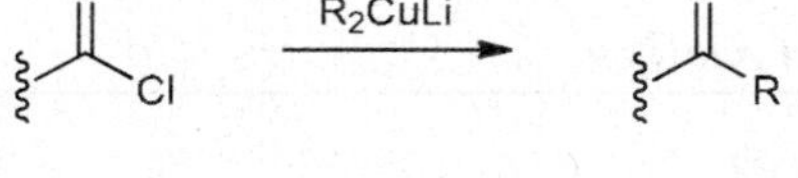

R_2CuLi

The resulting ketone is not further attacked by the lithium dialkylcuprate (unlike a Grignard reagent, which would attack the ketone). For some reason, students commonly propose a similar reaction between an ester and a lithium dialkylcuprate:

R_2CuLi

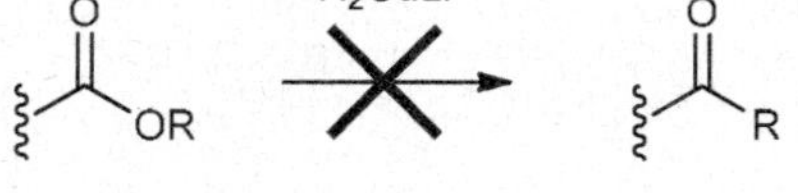

This reaction will not work. If a lithium dialkylcuprate will not attack a ketone, then it certainly won't attack an ester (which is less electrophilic than a ketone). Students often make this type of mistake, by applying a reaction outside of the scope in which it was discussed. Try to avoid doing this. Whenever we cover a reaction that applies to a particular functional group, you cannot assume that it will apply to other, less reactive, functional groups as well.

Useful reagents

The following is a list of reagents encountered in this chapter:

Reagents	Description
NaCN	This reagent will react with an alkyl halide to give a nitrile. Subsequent hydrolysis of the nitrile gives a carboxylic acid, with one more carbon atom than the starting alkyl halide.
1) Mg 2) CO_2 3) H_3O^+	These reagents can be used to convert an alkyl halide into a carboxylic acid, with the introduction of one carbon atom. Insertion of magnesium gives a Grignard reagent, which then attacks carbon dioxide to give a carboxylate ion, which is then protonated upon acidic workup.
1) $LiAlH_4$ 2) H_3O^+	Lithium aluminum hydride is a powerful hydride reducing agent. It will reduce ketones, aldehydes, esters, and carboxylic acids to give alcohols. Reduction of esters and carboxylic acids requires the use of excess $LiAlH_4$. Reduction of an amide (with $LiAlH_4$) gives an amine.
N	Pyridine is a weak base that is often used to neutralize for reactions that produce a strong acid as a by-product.
$SOCl_2$	Thionyl chloride can be used to convert a carboxylic acid into an acid halide. This reagent can also be used to dehydrate an amide to give a nitrile.
ROH	Alcohols are weak nucleophiles and weak bases. An alcohol can be used to convert an acid chloride or an acid anhydride into an ester.
NH_3	Ammonia is both a base and a nucleophile. Excess ammonia can be used to convert an acid chloride or an acid anhydride into an amide.
RNH_2	Primary amines are bases and nucleophiles. Excess amine can be used to convert an acid chloride or an acid anhydride into an amide.
R_2NH	Secondary amines are bases and nucleophiles. Excess amine can be used to convert an acid chloride or an acid anhydride into an amide.
1) xs RMgBr 2) H_3O^+	A Grignard reagent is a strong nucleophile. Two equivalents of a Grignard reagent will react with an acid chloride, with an anhydride, or with an ester, followed by aqueous acidic workup, to give an alcohol (with the introduction of two R groups). A Grignard reagent will also react with a nitrile, followed by hydrolysis, to give a ketone.
R_2CuLi	A lithium dialkylcuprate is a weak nucleophile. It will react with an acid chloride to give a ketone, but it will not react with ketones or esters.
1) $LiAl(OR)_3H$ 2) H_3O^+	Lithium trialkoxyaluminum hydrides are reducing agents that will convert an acid chloride or an acid anhydride into an aldehyde, without subsequent reduction of the resulting aldehyde.
1) DIBAH 2) H_3O^+	Diisobutylaluminum hydride is a hydride reducing agent that will convert an ester into an aldehyde.
H_3O^+	Aqueous acid will cause hydrolysis of an acid chloride, an anhydride, an ester, an amide, or a nitrile to give a carboxylic acid.
$[H^+]$, ROH	Under acidic conditions, an alcohol will react with a carboxylic acid via a Fischer esterification, giving an ester.

Solutions

20.1.

(a) In this molecule, the longest chain that contains both carboxylic acid groups is comprised of five carbon atoms, so the parent name (pentane) is given the suffix "–dioic acid", resulting in the IUPAC name pentanedioic acid. The common name is glutaric acid.

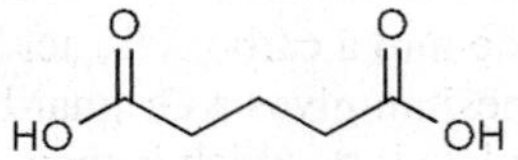

IUPAC name = pentanedioic acid
Common name = glutaric acid

(b) In this molecule, the longest chain that contains the carboxylic acid group is comprised of four carbon atoms, so the "e" in the parent name (butane) is replaced with the suffix "-oic acid", resulting in the IUPAC name butanoic acid. The common name is butyric acid.

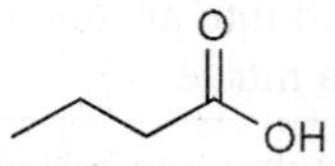

IUPAC name = butanoic acid
Common name = butyric acid

(c) The carboxylic acid group is attached to a ring, so the IUPAC name for this compound uses the name of the ring followed by the suffix "-carboxylic acid"; thus this compound is benzenecarboxylic acid. The common name is benzoic acid.

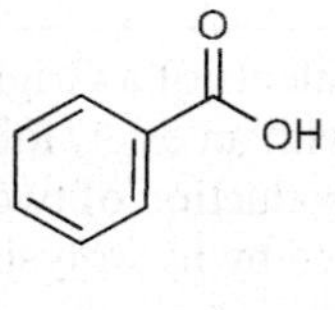

IUPAC name = benzenecarboxylic acid
Common name = benzoic acid

(d) In this molecule, the longest chain that contains both carboxylic acid groups is comprised of four carbon atoms, so the parent name (butane) is given the suffix "–dioic acid", resulting in the IUPAC name butanedioic acid. The common name is succinic acid.

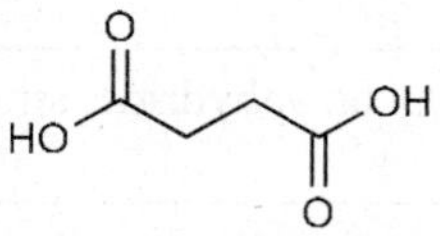

IUPAC name = butanedioic acid
Common name = succinic acid

(e) In this molecule, the parent chain is comprised of two carbon atoms, and the "e" in the parent name (ethane) is replaced with the suffix "-oic acid", resulting in the IUPAC name ethanoic acid. The common name is acetic acid.

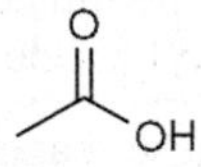

IUPAC name = ethanoic acid
Common name = acetic acid

(f) In this molecule, which contains only one carbon atom, the "e" in the parent name (methane) is replaced with the suffix "-oic acid", resulting in the IUPAC name methanoic acid. The common name is formic acid.

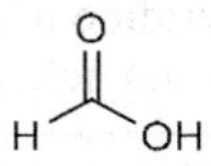

IUPAC name = methanoic acid
Common name = formic acid

20.2.

(a) This molecule has a four-membered ring (cyclobutane) connected to a carboxylic acid group.

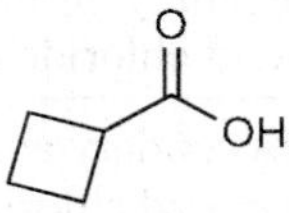

(b) The parent (butyric acid) has a four-carbon chain with a carboxylic acid at one terminus. The name indicates that there are two substituents (both chlorine atoms) at the C3 position.

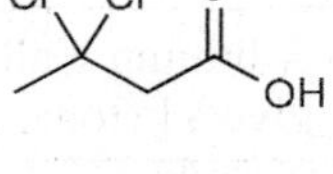

(c) The parent (glutaric acid) has a five-carbon chain with a carboxylic acid at each terminus. The name indicates that there are two substituents (both methyl groups) at the C3 position.

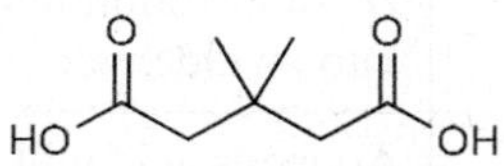

20.3.

(a) We begin by identifying the parent. The carbon atom of the carboxylic acid group must be included in the parent. The longest chain that includes this carbon atom is six carbon atoms in length, so the parent is hexanoic acid. There are four substituents (highlighted), all of which are methyl groups. Notice that the parent chain is numbered starting from the carbonyl carbon (defined as C1). According to this numbering scheme, two methyl groups are at C3, and two are at C4. We use the prefix "tetra" to indicate four methyl groups.

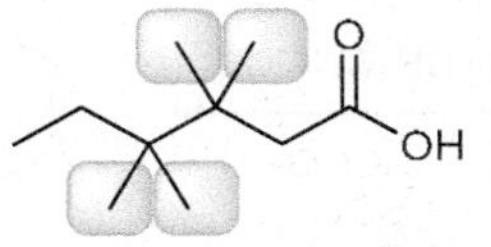

3,3,4,4-tetramethylhexanoic acid

(b) We begin by identifying the parent. The carbon atom of the carboxylic acid must be included in the parent. The longest chain that includes this carbon atom is five carbon atoms in length, so the parent is pentanoic acid. There is one substituent: a propyl group (highlighted). Notice that the parent chain is numbered starting from the carbonyl carbon (defined as C1). According to this numbering scheme, the propyl group is at C2.

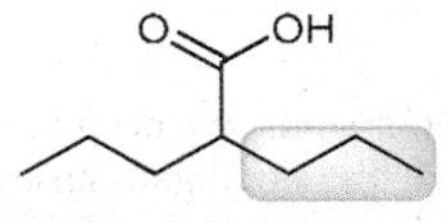

2-propylpentanoic acid

20.4. The conjugate base of the first compound is not resonance stabilized.

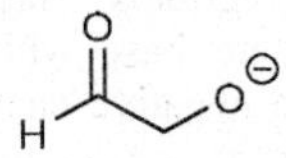

The second compound is more acidic because its conjugate base is resonance-stabilized:

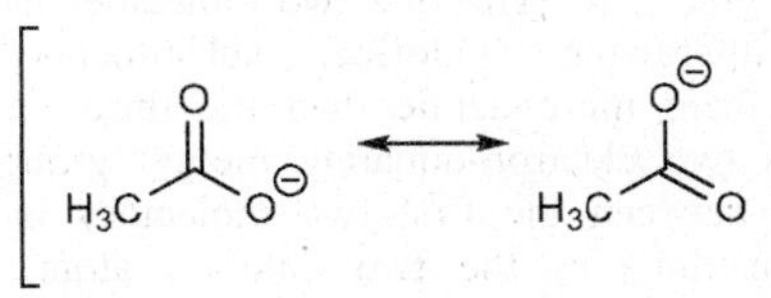

20.5. The conjugate base is resonance stabilized, with the negative charge spread over two oxygen atoms (much like the conjugate base of a carboxylic acid), as shown below. Note: there are three additional resonance structures that have not been drawn (each of which exhibits the negative charge on a carbon atom).

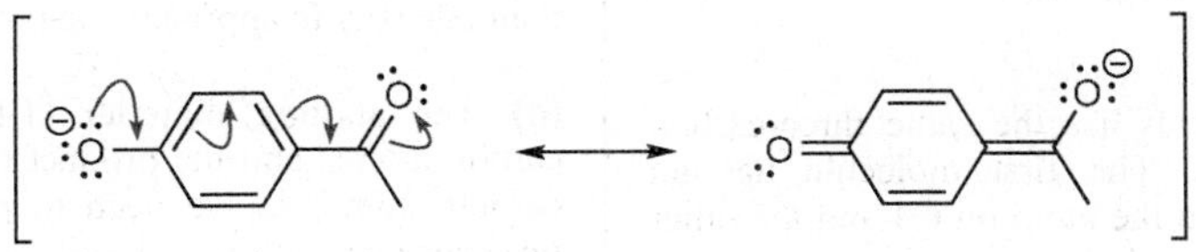

20.6. The first step is to draw the conjugate base of each molecule, including all relevant resonance structures. In the conjugate base of *meta*-hydroxyacetophenone, the negative charge is spread over only one oxygen atom (and three carbon atoms). In contrast, the conjugate base of *para*-hydroxyacetophenone has the negative charge spread over two oxygen atoms (and three carbon atoms), as shown in problem 20.5. The additional resonance structure of the *para* isomer's conjugate base renders it a more stable, weaker base, so *para*-hydroxyacetophenone is expected to be the stronger acid. *Meta*-Hydroxyacetophenone has the less stable, stronger conjugate base, so *meta*-hydroxyacetophenone is less acidic than the *para* isomer.

20.7. The first step is to identify the acid (formic acid) and base (hydroxide). When drawing a mechanism for an acid-base reaction, two curved arrows are required, as shown below. The tail of the first arrow is placed on a lone pair of the hydroxide oxygen atom, and the head is placed on the acidic hydrogen atom of formic acid. The tail of the second arrow is placed on the O-H bond, and the head is placed on the oxygen atom. The resulting carboxylate salt is named starting with the inorganic cation (potassium) followed by replacing the "-ic acid" suffix from formic acid to "-ate" giving the name potassium formate.

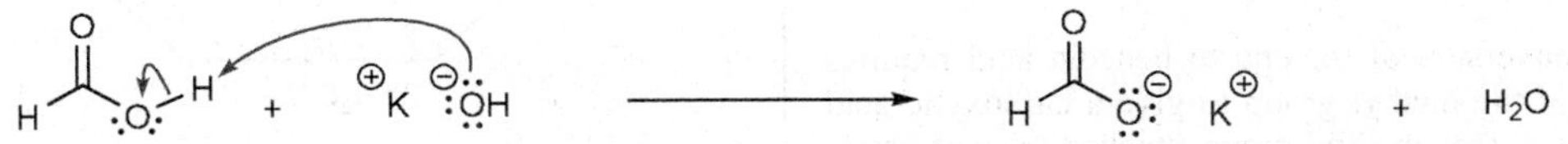

formic acid

potassium formate

20.8. To determine the relative amounts of acetic acid and its conjugate base (acetate), we enter the pK_a of acetic acid (4.76) and the pH (5.76) into the rearranged Henderson-Hasselbach equation as shown below:

$$\frac{[\text{conjugate base}]}{[\text{acid}]} = 10^{(\text{pH} - \text{p}K_a)} = 10^{(5.76 - 4.76)} = 10^1 = 10$$

The result shows that the conjugate base predominates under these conditions (at a ratio of 10:1).

20.9. For each set of acids, we need to assess the relative stability of each of the conjugate bases. A more stable, weaker conjugate base means that the corresponding acid is more acidic. When comparing the structures, we need to consider the nature of the substituents and their position relative to the carboxylic acid functional group. Electron-withdrawing groups stabilize the negative charge in the conjugate base while electron-donating groups destabilize the negative charge in the conjugate base. Substituents that are closer to the carboxylic acid have a greater effect on acidity.

(a) Each of the three acids has the same four-carbon parent: butyric acid. The first two molecules have two electron-withdrawing chlorine substituents each, rendering them more acidic than the third molecule, which has two electron-donating methyl groups. The difference between the first two molecules is in the relative positions of the two chlorine atoms. The compound with the two chlorine atoms on positions C2 and C3 is more acidic than the compound with the chlorine atoms on positions C2 and C4, because the chlorine atom on C3 has a greater effect than the chlorine atom on C4. The correct order, in increasing acidity, is thus: 3,3-dimethylbutyric acid < 2,4-dichlorobutyric acid < 2,3-dichlorobutyric acid.

(b) Each of the three acids has the same three-carbon parent: propionic acid. The first molecule has an electron-withdrawing bromine atom on C3 and the other two molecules have two bromine atoms each. The compound with two bromine atoms on C2 is the most acidic because the two bromine atoms are closest to the carboxylate anion in the conjugate base. The compound with two bromine atoms on C3 is more acidic than the compound with one bromine atom on C3 due to the additive effect of the two bromine atoms. The correct order, in increasing acidity, is thus: 3-bromopropionic acid < 3,3-dibromopropionic acid < 2,2-dibromo-propionic acid.

20.10.
(a) The starting material is an alcohol (ethanol) and the product is the corresponding carboxylic acid with two carbon atoms (acetic acid). Reagents that accomplish this oxidation reaction are: $Na_2Cr_2O_7$, H_2SO_4, H_2O.

(b) The conversion of toluene to benzoic acid requires oxidation of the methyl group to give a carboxylic acid group. Recall that an alkyl group attached to an aromatic ring is oxidized to a carboxylic acid by a strong oxidizing agent, as long as the alkyl group has at least one benzylic hydrogen atom. This conversion can thus be accomplished using: $Na_2Cr_2O_7$, H_2SO_4, H_2O.

(c) The conversion of benzene to benzoic acid requires the installation of a carbon atom on the carbon skeleton of the starting material. One approach to accomplish this transformation is to install a methyl group using a Friedel-Crafts alkylation (CH_3Cl, $AlCl_3$) followed by an oxidation of the methyl group using strongly oxidizing conditions, as shown.

1) CH_3Cl, $AlCl_3$
2) $Na_2Cr_2O_7$, H_2SO_4, H_2O

CH_3Cl, $AlCl_3$

$Na_2Cr_2O_7$, H_2SO_4, H_2O

OH

There are certainly other acceptable solutions. For example, bromination of benzene, followed by treatment of the resulting bromobenzene with magnesium, gives phenyl magnesium bromide. This Grignard reagent can then be treated with CO_2, followed by aqueous acidic workup, to give benzoic acid. This alternate solution illustrates an important point. For many of the synthesis problems that you will encounter, there may be more than one way to approach the problem.

(d) The starting material (1-bromobutane) has four carbon atoms, and the product (pentanoic acid) has five carbon atoms, so we need to propose a synthesis that involves installation of the fifth carbon atom. We can use NaCN to convert the four-carbon starting material to a five-carbon synthetic intermediate (a nitrile) via an S_N2 process. Hydrolysis of this intermediate (H_3O^+, heat) converts the nitrile to a carboxylic acid, giving the desired product. An alternate approach is to convert the starting alkyl halide to the corresponding Grignard reagent using Mg, followed by reaction with CO_2 to produce a carboxylate. Protonation with aqueous acid gives the desired carboxylic acid.

1) NaCN
2) H_3O^+, heat

Br

OH

1) Mg
2) CO_2
3) H_3O^+

(e) The conversion of ethylbenzene to benzoic acid requires an oxidation of the benzylic carbon atom of the starting material to form a carboxylic acid group. Recall that an alkyl group attached to an aromatic ring is oxidized to a carboxylic acid by a strong oxidizing agent, as long as the alkyl group has at least one benzylic hydrogen atom. This conversion can thus be accomplished using: $Na_2Cr_2O_7$, H_2SO_4, H_2O.

(f) The starting material (bromocyclohexane) has six carbon atoms, and the product (cyclohexanecarboxylic acid) has seven carbon atoms, so we need to propose a synthesis that includes the installation of the seventh carbon atom. We can use NaCN to convert the six-carbon starting material to a seven-carbon synthetic intermediate via an S_N2 process. Hydrolysis of this intermediate (H_3O^+, heat) converts the nitrile to a carboxylic acid, giving the desired product. An alternate approach is to convert the starting alkyl halide to the corresponding Grignard reagent using Mg, followed by reaction with CO_2 to produce a carboxylate. Protonation with aqueous acid gives the desired carboxylic acid.

1) NaCN
2) H_3O^+, heat

1) Mg
2) CO_2
3) H_3O^+

20.11.

(a) This synthesis requires the conversion of a six-carbon starting material (bromobenzene) to a seven-carbon product (benzyl alcohol), so we must include a reaction to form a new C–C bond. There are certainly multiple solutions to this problem. One such solution involves conversion of the starting material into a Grignard reagent, which can then be treated with formaldehyde, followed by acidic workup (H_3O^+) to produce the desired product (benzyl alcohol) as shown:

1) Mg
2) $H_2C{=}O$
3) H_3O^+

Mg

1) formaldehyde
2) H_3O^+

Alternatively, the starting material can be converted into a Grignard reagent, which can then be treated with CO_2 followed by aqueous acidic workup (H_3O^+) to produce benzoic acid. Reduction of the carboxylic acid using lithium aluminum hydride, followed by aqueous acidic workup, produces the desired product (benzyl alcohol) as shown:

1) Mg
2) CO_2
3) H_3O^+
4) $LiAlH_4$
5) H_3O^+

Mg

1) CO_2
2) H_3O^+

1) $LiAlH_4$
2) H_3O^+

(b) The starting material (toluene) and product (benzyl alcohol) have seven carbon atoms each. Oxidation of the starting material to benzoic acid can be accomplished with a suitable oxidizing agent to give benzoic acid. Reduction of the resulting carboxylic acid using lithium aluminum hydride, followed by aqueous acidic workup, produces the desired product (benzyl alcohol) as shown. Alternatively, a bromine atom can be installed in the benzylic position using NBS/heat. Subsequent reaction with NaOH produces the desired product via an S_N2 reaction, as shown:

1) $Na_2Cr_2O_7$, H_2SO_4, H_2O
2) xs $LiAlH_4$
3) H_3O^+

1) NBS, heat
2) NaOH

20.12.

(a) This symmetric anhydride is named by replacing "acid" from the corresponding carboxylic acid (propionic acid) with the suffix "anhydride", giving the name propionic anhydride (or propanoic anhydride).

propionic anhydride
or
propanoic anhydride

(b) This amide is named as a derivative of the carboxylic acid "propionic acid" by replacing the "-ic acid" suffix with "-amide". The two phenyl groups attached to the nitrogen atom are listed as substituents. Their position is indicated by the locant "*N*" thus giving the name *N,N*-diphenylpropionamide (or *N,N*-diphenylpropanamide).

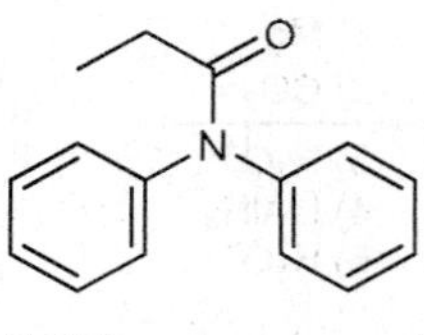

N,N-diphenylpropionamide
or
N,N-diphenylpropanamide

(c) This diester is named as a derivative of the parent dicarboxylic acid "succinic acid" by replacing the "-ic acid" suffix with "-ate". The two methyl groups attached to the oxygen atoms are indicated at the beginning, thus giving the name dimethyl succinate (or dimethyl butanedioate, based off the parent butanedioic acid).

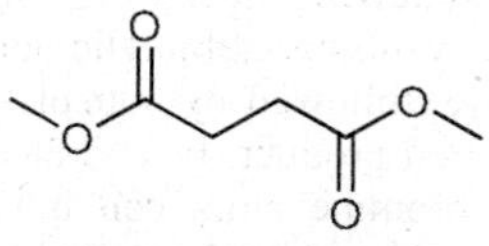

dimethyl succinate
or
dimethyl butanedioate

(d) This amide is named as a derivative of "cyclobutanecarboxylic acid" by replacing "-carboxylic acid" with "-carboxamide". The two alkyl groups attached to the nitrogen atom (ethyl and methyl) are listed in alphabetical order as substituents. Their position is indicated with the locant "*N*" thus giving the name *N*-ethyl-*N*-methylcyclobutanecarboxamide.

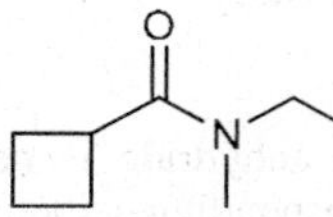

N-ethyl-N-methylcyclobutanecarboxamide

(e) This nitrile is named as a derivative of the carboxylic acid "butyric acid" by replacing the "-ic acid" suffix with "-onitrile" giving the name butyronitrile (or butanenitrile).

butyronitrile
or
butanenitrile

(f) This ester is named as a derivative of the parent carboxylic acid "butyric acid" by replacing the "-ic acid" suffix with "-ate". The propyl group attached to the oxygen atom is indicated at the beginning, thus giving the name propyl butyrate (or propyl butanoate).

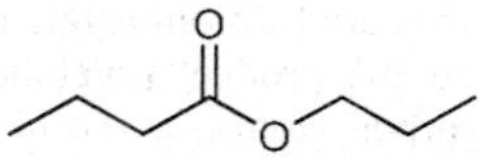

propyl butyrate
or
propyl butanoate

(g) This cyclic anhydride is named as a derivative of the parent dicarboxylic acid "succinic acid" by replacing "acid" with "anhydride", giving the name succinic anhydride (or butanedioic anhydride, based off the parent butanedioic acid).

succinic anhydride
or
butanedioic anhydride

(h) This ester is named as a derivative of the parent carboxylic acid "benzoic acid" by replacing the "-oic acid" suffix with "-ate". The methyl group attached to the oxygen atom is indicated at the beginning, thus giving the name methyl benzoate.

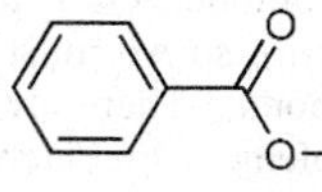

methyl benzoate

(i) This ester is named as a derivative of the parent carboxylic acid "acetic acid" by replacing the "-ic acid" suffix with "-ate". The phenyl group attached to the oxygen atom is indicated at the beginning, thus giving the name phenyl acetate (or phenyl ethanoate, based off the parent ethanoic acid).

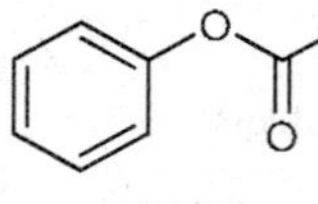

phenyl acetate
or
phenyl ethanoate

20.13.

(a) The parent (oxalic acid) is a dicarboxylic acid with two carbon atoms. The suffix "-ic acid" is replaced with "-ate", indicating that this is a diester. The name indicates that the two alkyl groups attached to the oxygen atoms are methyl groups.

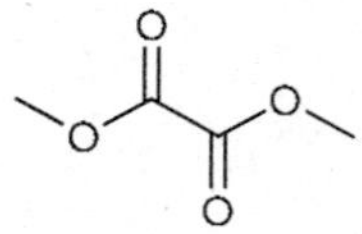

(b) The parent (cyclopentanecarboxylic acid) is a ring with five carbon atoms attached to a carboxylic acid. The suffix "-ic acid" is replaced with "-ate", indicating that this is an ester. The name indicates that the group attached to the oxygen atom is a phenyl group.

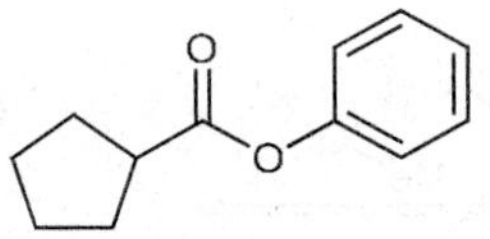

(c) The parent (propionic acid) is a carboxylic acid with three carbon atoms. The suffix "-ic acid" is replaced with "-amide", indicating that this compound is an amide. The name indicates that there is one methyl group attached to the nitrogen atom.

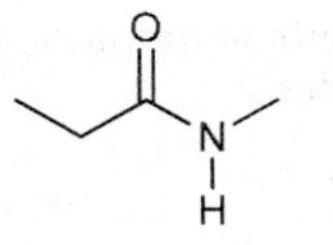

(d) The parent (propionic acid) is a carboxylic acid with three carbon atoms. The suffix "-ic acid" is replaced with "-yl chloride", indicating that this compound is an acid chloride.

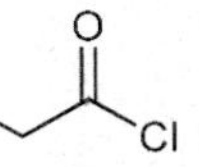

20.14.

(a) This mechanism has two steps: 1) nucleophilic attack, and 2) loss of a leaving group. The first step (nucleophilic attack), requires two curved arrows, which show the carboxylate ion functioning as a nucleophile and attacking the electrophilic carbonyl group, resulting in a tetrahedral intermediate. In step two (loss of a leaving group), the carbonyl group is reformed and chloride leaves, as shown with two curved arrows, resulting in the formation of an anhydride, as shown.

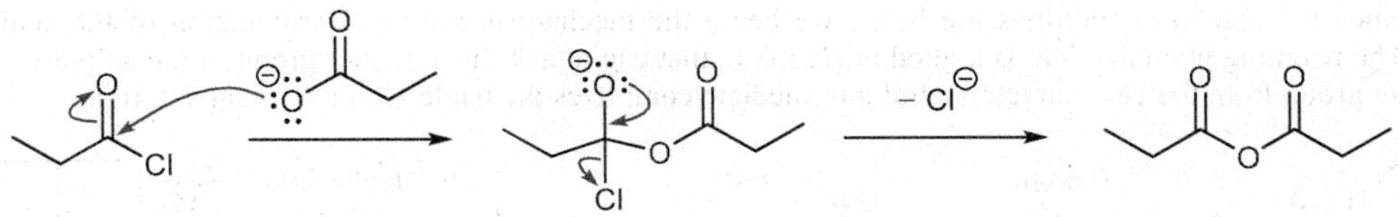

(b) This mechanism has three steps: 1) nucleophilic attack, 2) loss of a leaving group, and 3) proton transfer. The first step (nucleophilic attack) requires two curved arrows, which show ammonia functioning as a nucleophile and attacking the electrophilic carbonyl group, resulting in a tetrahedral intermediate. In step two (loss of a leaving group), the carbonyl group is reformed and chloride leaves, as shown with two curved arrows. In the third step (proton transfer), a second equivalent of ammonia serves as a base, deprotonating the cationic intermediate, resulting in the formation of an amide, as shown.

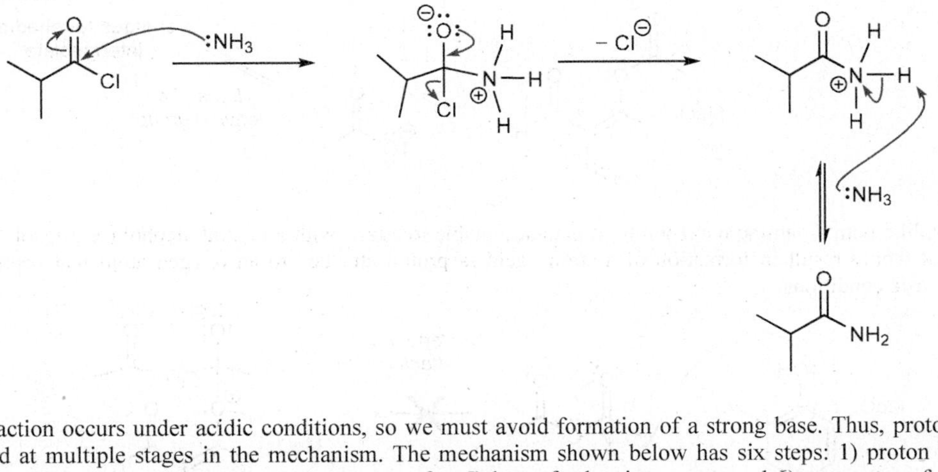

(c) This reaction occurs under acidic conditions, so we must avoid formation of a strong base. Thus, proton transfers are required at multiple stages in the mechanism. The mechanism shown below has six steps: 1) proton transfer, 2) nucleophilic attack, 3) proton transfer, 4) proton transfer, 5) loss of a leaving group, and 6) proton transfer. The first step (proton transfer) requires two curved arrows to show the transfer of a proton from $MeOH_2^+$ to the carbonyl group, resulting in the formation of an activated electrophile. In step two (nucleophilic attack), methanol serves as a nucleophile attacking the protonated carbonyl group, producing a cationic tetrahedral intermediate. We cannot immediately expel the OH group at this stage, as this would result in the formation of a strong base (hydroxide). This must be avoided in acidic conditions. In step 3 (proton transfer), methanol serves as a base, resulting in a neutral intermediate. Subsequently, in step 4 (proton transfer) a proton is transferred from $MeOH_2^+$ to the uncharged oxygen atom, as shown. In step 5 (loss of leaving group), the carbonyl group is reformed and water serves as the leaving group. In step 6 (proton transfer), methanol serves as a base which deprotonates the cationic intermediate, resulting in formation of the ester.

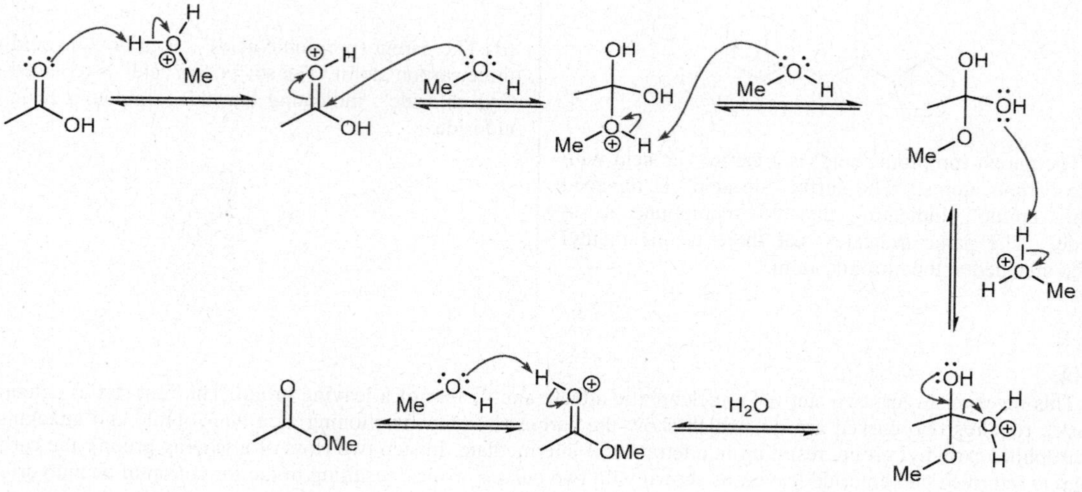

Me OH H Me H OH OMe OH Me H OH Me H H Me - H_2O OMe

20.15. Since the reaction conditions are basic, we begin the mechanism with a deprotonation of the acidic phenol proton. The resulting phenolate ion is a good nucleophile that can attack the carbonyl group of the anhydride. Loss of the leaving group from the charged tetrahedral intermediate completes the nucleophilic acyl substitution.

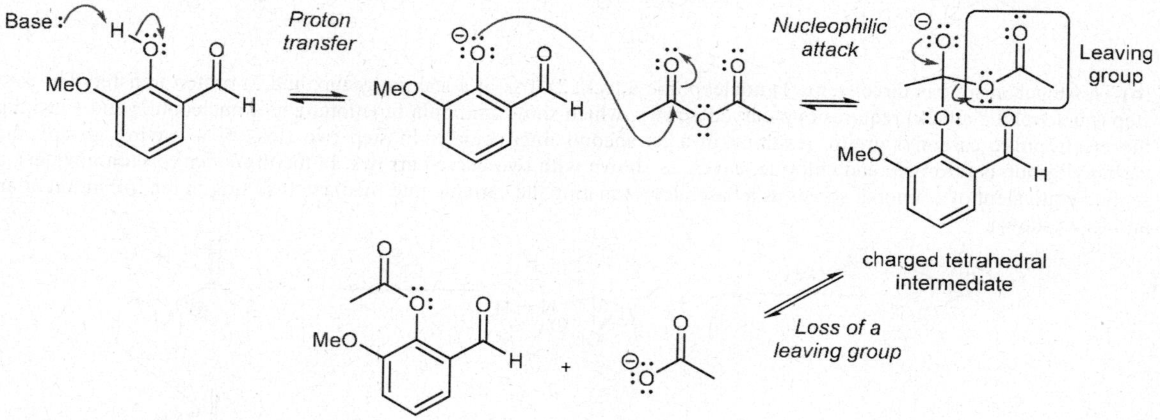

Base: *Proton transfer* *Nucleophilic attack* Leaving group

charged tetrahedral intermediate

Loss of a leaving group

Note that unlike with an amine nucleophile, it is unacceptable to attack with a neutral alcohol (or phenol) nucleophile because that would result in formation of a strong acid (a proton attached to an oxygen atom that bears a positive charge) in basic conditions.

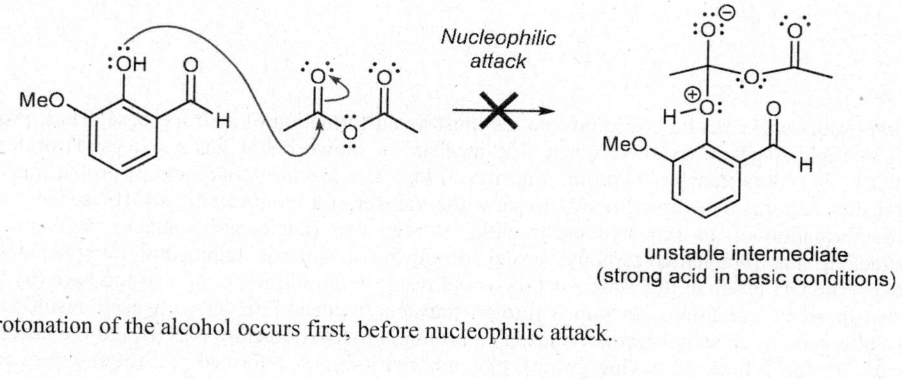

Nucleophilic attack

unstable intermediate
(strong acid in basic conditions)

Instead, deprotonation of the alcohol occurs first, before nucleophilic attack.

20.16.

(a) The reaction of an acid chloride with excess $LiAlH_4$, followed by aqueous acidic workup, results in the formation of the corresponding alcohol shown below.

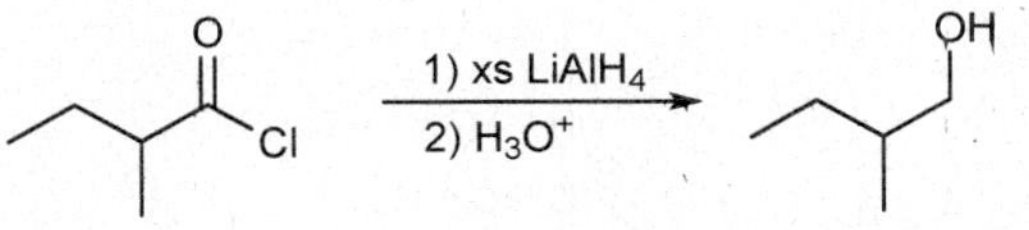
O
Cl
1) xs LiAlH4
2) H3O+
OH

(b) The reaction of an acid chloride with excess phenyl magnesium bromide, followed by aqueous acidic workup, results in the incorporation of two phenyl groups, giving a tertiary alcohol.

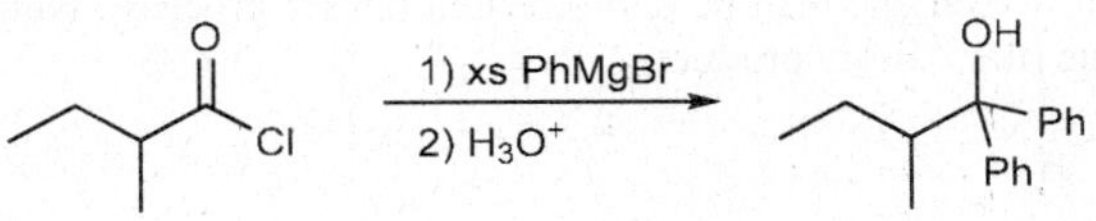
O
Cl
1) xs PhMgBr
2) H3O+
OH
Ph
Ph

(c) The reaction of an acid chloride with the selective hydride-reducing agent, $LiAl(OR)_3H$, produces an aldehyde. Subsequent reaction with a Grignard reagent, followed by aqueous acidic workup, gives a secondary alcohol.

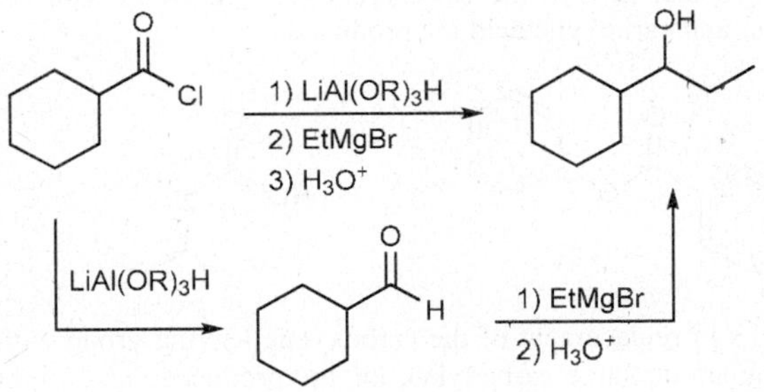
O
Cl
1) LiAl(OR)3H
2) EtMgBr
3) H3O+
OH
LiAl(OR)3H
O
H
1) EtMgBr
2) H3O+

(d) The reaction of an acid chloride with a lithium dialkylcuprate (a selective carbon nucleophile) produces a ketone. Subsequent reaction with $LiAlH_4$, followed by aqueous acidic workup, gives a secondary alcohol.

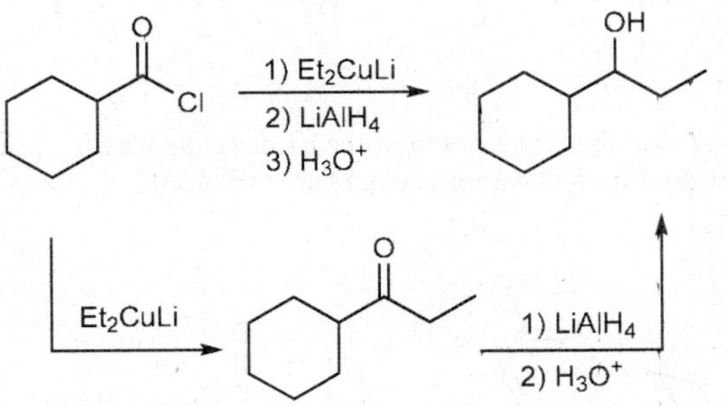
O
Cl
1) Et2CuLi
2) LiAlH4
3) H3O+
OH
Et2CuLi
O
1) LiAlH4
2) H3O+

(e) The reaction of an acid chloride with phenol (in the presence of pyridine) results in the replacement of the chlorine atom with the phenol oxygen atom, producing an ester, as shown.

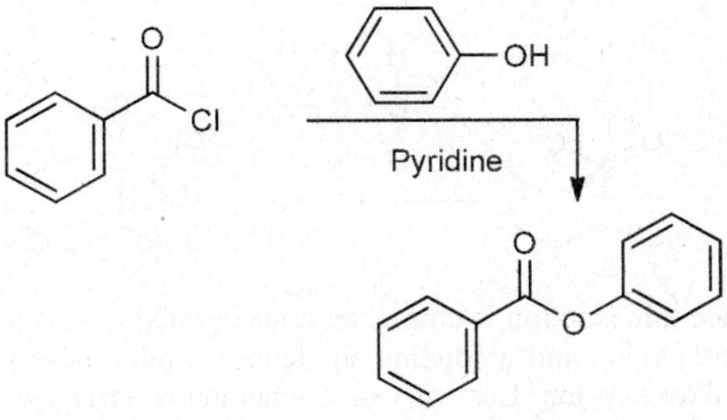
O
Cl
OH
Pyridine
O
O

(f) The reaction of an acid chloride with two equivalents of an amine results in the replacement of the chlorine atom with the amine nitrogen atom, producing an amide, as shown.

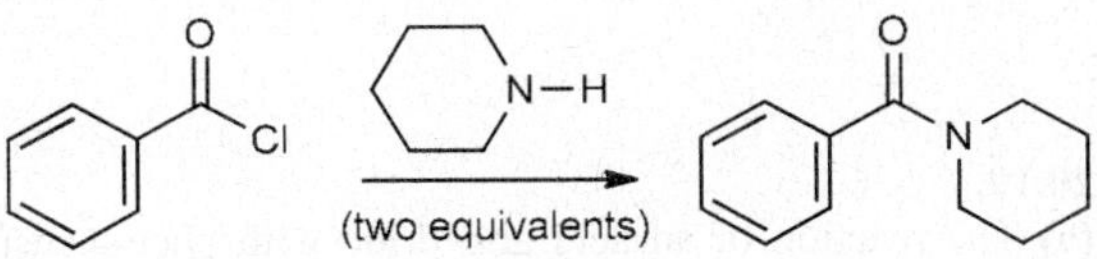
O
Cl
N—H
(two equivalents)
O
N

20.17. The conversion of benzyl alcohol to benzoyl chloride requires oxidation of the benzylic carbon atom. Subsequent reaction with thionyl chloride results in the conversion of the carboxylic acid to the desired acid chloride.

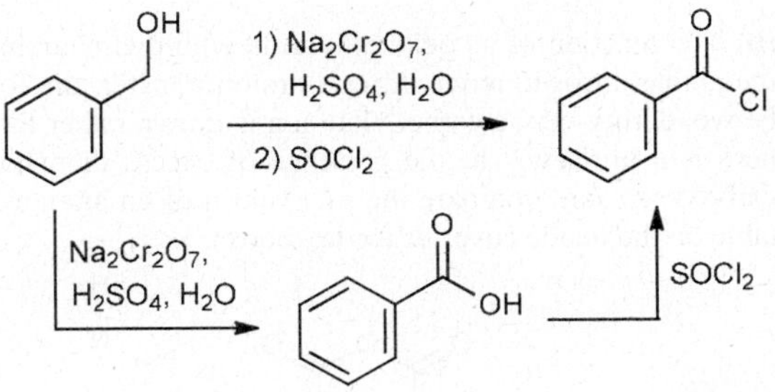
OH
1) Na2Cr2O7,
H2SO4, H2O
2) SOCl2
O
Cl
Na2Cr2O7,
H2SO4, H2O
O
OH
SOCl2

20.18. The reaction between an acid chloride and a Grignard reagent occurs via the following mechanism. In the first step (nucleophilic attack), the anionic carbon atom of the Grignard reagent serves as a nucleophile and attacks the electrophilic carbonyl group, resulting in a charged tetrahedral intermediate. In step two (loss of a leaving group), the carbonyl group is reformed and chloride leaves, shown with two curved arrows. In step three (nucleophilic attack), a second equivalent of the Grignard reagent attacks the carbonyl group of the intermediate ketone, resulting in the formation of another tetrahedral intermediate.

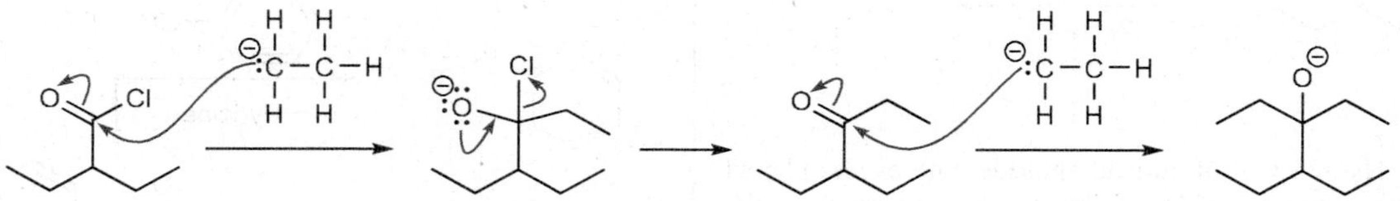

Once this reaction is complete, concentrated acid is added to the reaction flask. H_3O^+ serves as an acid, protonating the alkoxide ion and producing an alcohol. Under these strongly acidic conditions, the alcohol is further protonated, giving an oxonium ion. Loss of water generates a tertiary carbocation, which can then be deprotonated (an E1 process). Note that removal of this proton results in the more substituted alkene (the Zaitsev product).

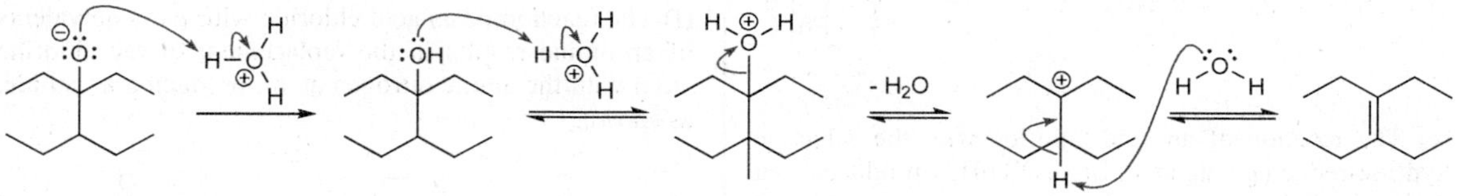

20.19.

(a) The reaction of an acid anhydride with phenol results in replacement of the carboxylate leaving group with the phenol oxygen atom. After a proton transfer, the following ester and carboxylic acid are produced.

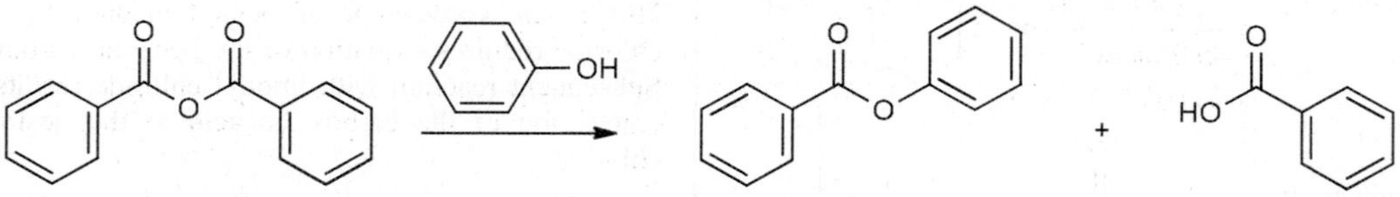

(b) The reaction of an acid anhydride with diethylamine results in replacement of the carboxylate leaving group with the amine nitrogen atom. After a proton transfer, the following amide and a carboxylate ion are produced. You might be wondering why a carboxylate ion is drawn rather than a carboxylic acid. This will be discussed in Chapter 23, but here is a preview: In the presence of excess diethylamine, the resulting carboxylic acid is deprotonated to give a carboxylate ion (compare the pK_a values of an ammonium ion and a carboxylic acid, which can be found in the pK_a table on the inside cover of the textbook).

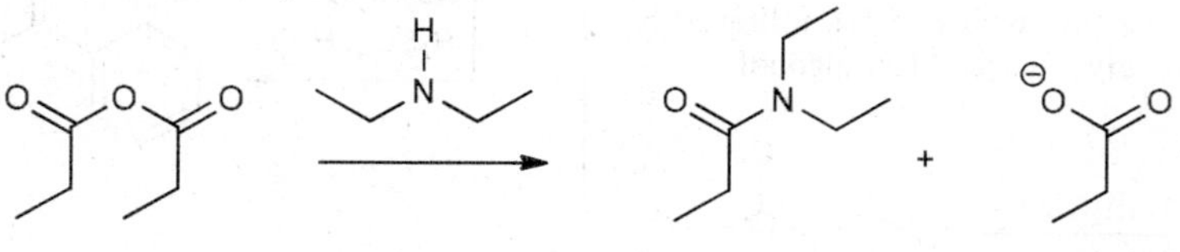

(c) The reaction of a phenol derivative with acetic anhydride results in the replacement of the carboxylate leaving group with the phenol oxygen atom. After a proton transfer, the following ester and carboxylic acid are produced.

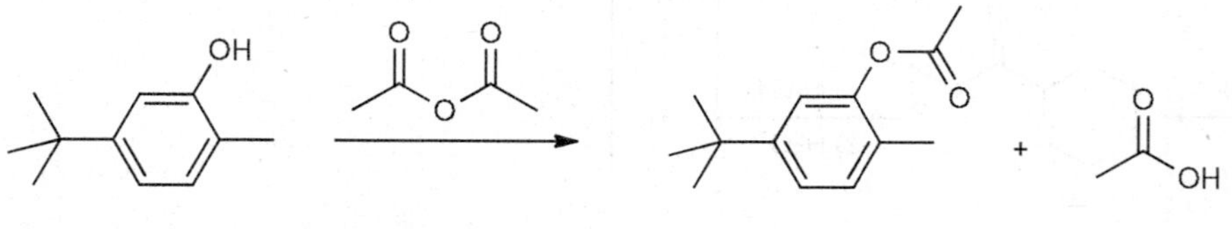

(d) The reaction of an acid anhydride with a cyclic secondary amine results in the replacement of the carboxylate leaving group with the amine nitrogen atom. After a proton transfer, the following amide and carboxylic acid are produced. You might be wondering why a carboxylate ion is drawn rather than a carboxylic acid. This will be discussed in Chapter 23, but here is a preview: In the presence of excess amine, the resulting carboxylic acid is deprotonated to give a carboxylate ion (compare the pK_a values of an ammonium ion and a carboxylic acid, which can be found in the pK_a table on the inside cover of the textbook).

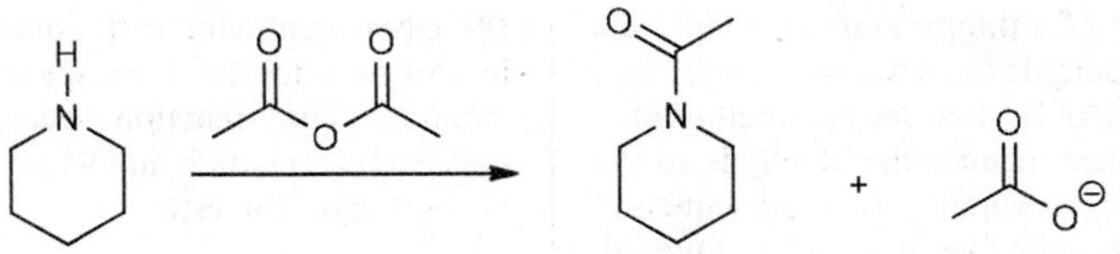

20.20. Three methods of converting benzoic acid to ethyl benzoate are shown below. In the first method, benzoic acid is deprotonated by NaOH. The intermediate salt (sodium benzoate) serves as a nucleophile in a subsequent S_N2 reaction with ethyl iodide. The second method is a Fischer esterification process, in which ethanol serves as both the solvent and a weak nucleophile. In the third method, benzoic acid is first converted to benzoyl chloride, and subsequently treated with ethanol (in the presence of pyridine) to produce the desired ester.

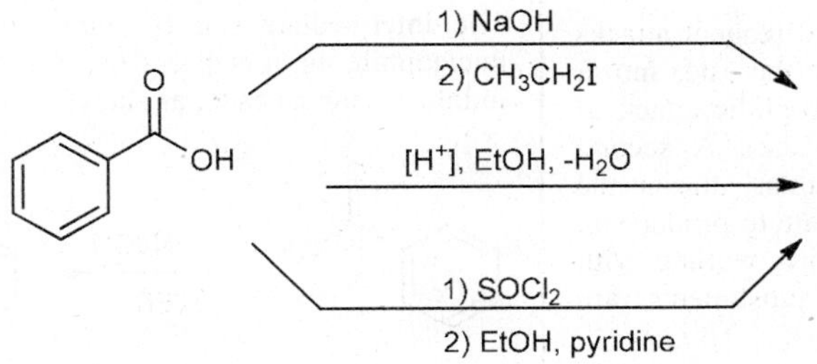

1) NaOH
2) CH_3CH_2I
$[H^+]$, EtOH, $-H_2O$
1) $SOCl_2$
2) EtOH, pyridine

20.21.

(a) Oxidation of benzyl alcohol to benzoic acid is accomplished using strongly oxidizing conditions, as shown below. Subsequent conversion to the corresponding ethyl ester can be accomplished by any of the three methods shown, as described in the solution to problem 20.20.

$Na_2Cr_2O_7$
H_2SO_4, H_2O
1) NaOH
2) CH_3CH_2I
$[H^+]$, EtOH, $-H_2O$
1) $SOCl_2$
2) EtOH, pyridine

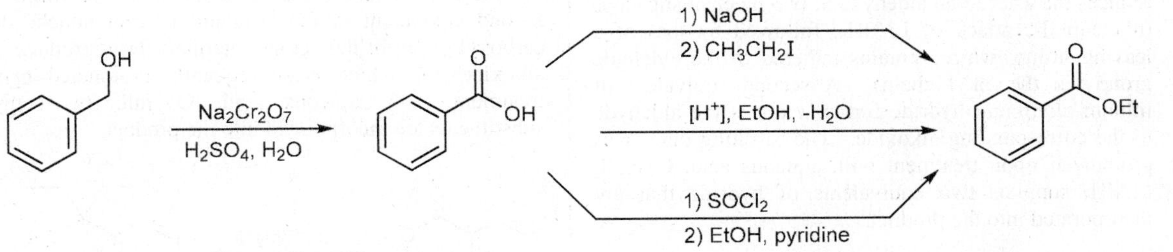

(b) Oxidation of styrene to benzoic acid is accomplished using strongly oxidizing conditions, as shown below. Subsequent conversion to the corresponding ethyl ester can be accomplished by any of the three methods shown, as described in the solution to problem 20.20.

$Na_2Cr_2O_7$
H_2SO_4, H_2O
1) NaOH
2) CH_3CH_2I
$[H^+]$, EtOH, $-H_2O$
1) $SOCl_2$
2) EtOH, pyridine

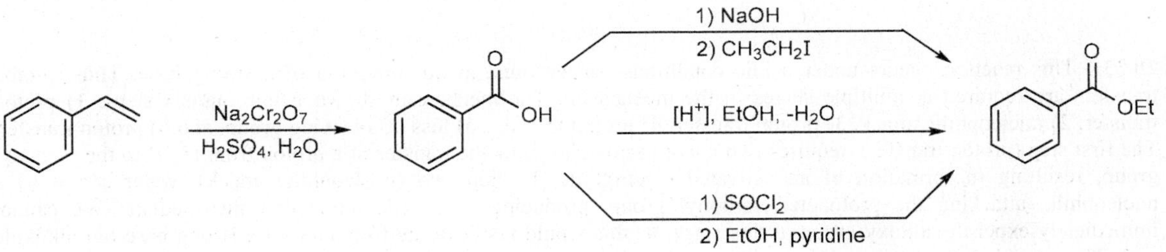

20.22.

(a) The first equivalent of lithium aluminum hydride reduces the ester to an aldehyde in two mechanistic steps (nucleophilic attack of $LiAlH_4$, then loss of methoxide). A second equivalent further reduces the aldehyde to the corresponding alkoxide, which is subsequently protonated upon workup with aqueous acid. Overall, $LiAlH_4$ supplies two equivalents of hydride that are incorporated into the product.

1) xs $LiAlH_4$ 2) H_3O^+ OMe OH + MeOH

(b) The first equivalent of the Grignard reagent attacks the carbonyl group, thereby converting the ester into a ketone in two mechanistic steps (nucleophilic attack of EtMgBr, followed by loss of methoxide). A second equivalent of the Grignard reagent then attacks the carbonyl group of the ketone intermediate to produce an alkoxide ion, which is protonated during workup with aqueous acid. Overall, two ethyl substituents are incorporated into the product.

1) xs EtMgBr 2) H_3O^+ OMe OH + MeOH

(c) The first equivalent of lithium aluminum hydride reduces the ester to an aldehyde in two mechanistic steps (nucleophilic attack of $LiAlH_4$, followed by loss of a leaving group, which remains tethered to the aldehyde group via the alkyl chain). A second equivalent of lithium aluminum hydride further reduces the aldehyde to the corresponding alkoxide. The resulting dianion is protonated upon treatment with aqueous acid. Overall, $LiAlH_4$ supplies two equivalents of hydride that are incorporated into the product.

1) xs $LiAlH_4$ 2) H_3O^+ HO OH

(d) Upon treatment with aqueous acid, an ethyl ester is hydrolyzed to the corresponding carboxylic acid and ethanol. The reaction occurs under acid-catalyzed conditions, in which an OH group ultimately replaces the OEt group of the ester.

OEt H_3O^+ OH + EtOH

(e) In step 1, benzoic acid is deprotonated by NaOH. The intermediate salt (sodium benzoate) serves as a nucleophile in a subsequent S_N2 reaction with ethyl iodide, giving an ester, as shown.

OH 1) NaOH 2) EtI

(f) The first equivalent of the Grignard reagent attacks the carbonyl group, thereby converting the cyclic ester into an acyclic ketone in two mechanistic steps (nucleophilic attack of EtMgBr, followed by loss of the leaving group, which remains tethered to the molecule via the alkyl chain attached to the aromatic ring). A second equivalent of the Grignard reagent attacks the carbonyl group of the ketone intermediate to produce an alkoxide ion, which is subsequently protonated upon treatment with aqueous acid. Overall, two ethyl substituents are incorporated into the product.

1) xs EtMgBr 2) H_3O^+ OH OH

20.23. This reaction occurs under acidic conditions, so we must avoid formation of a strong base. Thus, proton transfers are required at multiple stages in the mechanism. The mechanism shown below has six steps: 1) proton transfer, 2) nucleophilic attack, 3) proton transfer, 4) proton transfer, 5) loss of a leaving group, and 6) proton transfer. The first step (proton transfer), requires two curved arrows to show the transfer of a proton from H_3O^+ to the carbonyl group, resulting in formation of an activated electrophile. In step two (nucleophilic attack), water serves as a nucleophile, attacking the protonated carbonyl group, producing a cationic tetrahedral intermediate. We cannot immediately expel the alkoxy group at this stage, as this would result in the formation of a strong base (an alkoxide ion). This must be avoided in acidic conditions. This oxygen atom must first be protonated. However, protonation at this stage would result in an intermediate with two positive charges, which should be avoided, if possible. Therefore, in the next step (step 3), water serves as a base, resulting in a neutral intermediate. Subsequently, in step 4 (proton transfer) a proton is transferred from H_3O^+ to the oxygen atom, as shown. In step 5 (loss of leaving group), the carbonyl group is reformed and an alcohol serves as the leaving group, resulting in the opening of the ring. In step 6 (proton transfer), water serves as a base which deprotonates the cationic intermediate, resulting in formation of a bifunctional product.

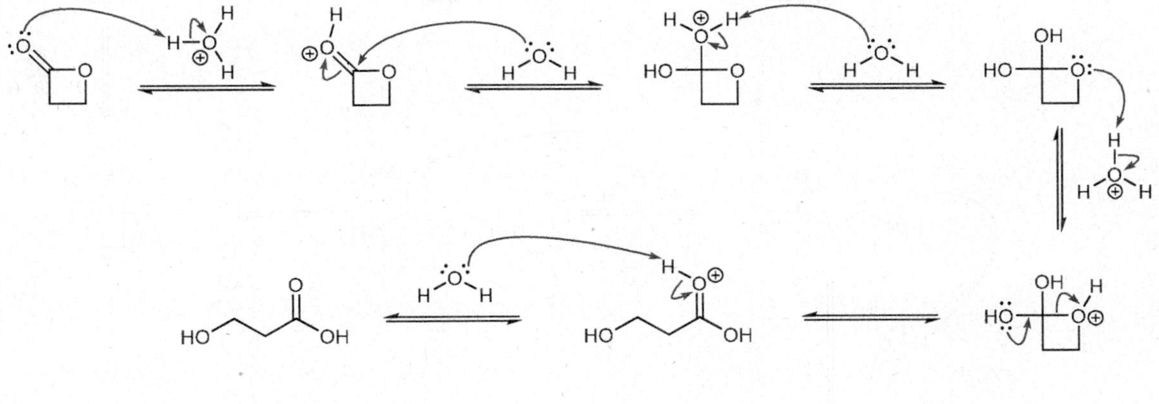

20.24.

(a) An amide is converted to the corresponding amine upon treatment with lithium aluminum hydride, followed by water workup.

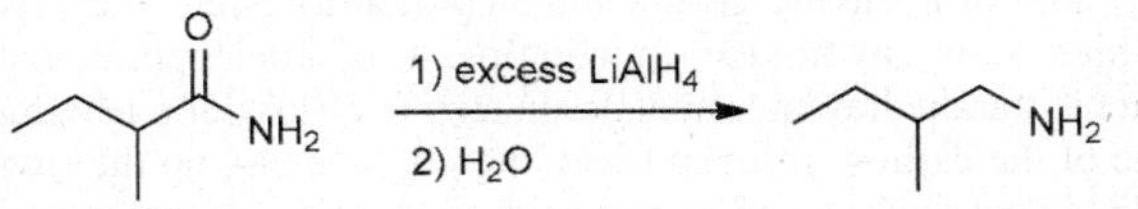

1) excess $LiAlH_4$
2) H_2O

(b) The reaction of an acid chloride with excess ammonia results in replacement of the chloride leaving group with an NH_2 group.

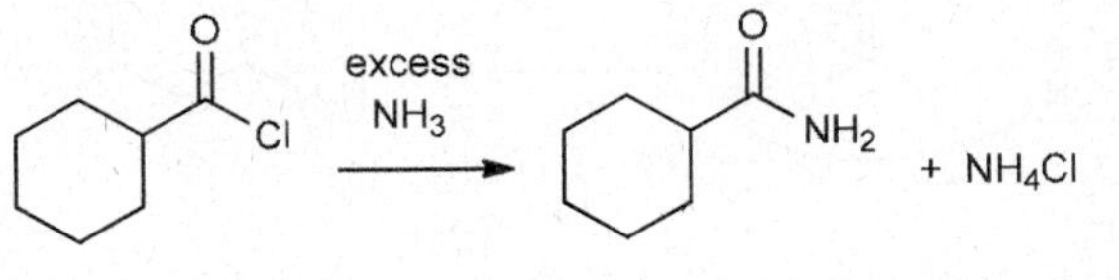

excess NH_3 + NH_4Cl

(c) An amide is hydrolyzed to give a carboxylic acid upon treatment with aqueous acid at elevated temperature. Under these acidic conditions, the by-product (ammonia) is protonated, resulting in formation of an ammonium ion.

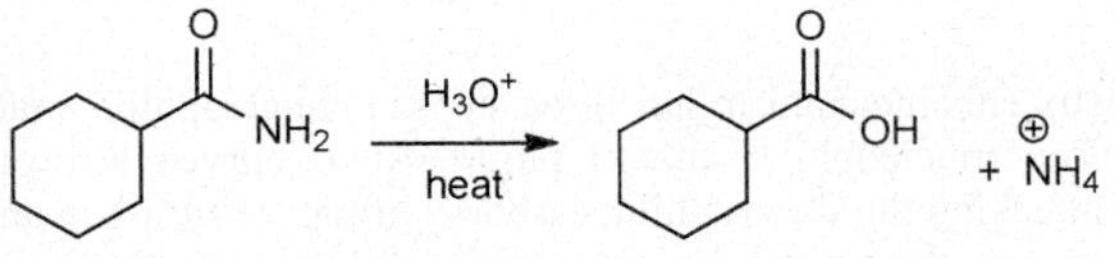

H_3O^+, heat; + $NH_4^{\oplus}$

20.25. An acid chloride can be converted to an amine in two steps. First, the acid chloride is treated with excess ammonia to produce the corresponding amide. Subsequent reduction with excess lithium aluminum hydride, followed by protonation with water, produces benzyl amine, as shown.

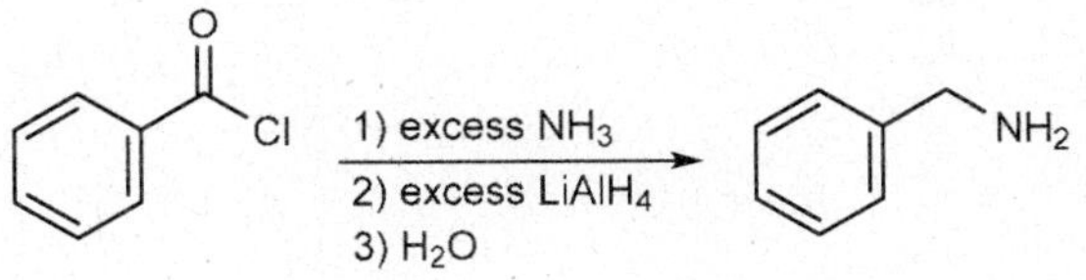

1) excess NH_3
2) excess $LiAlH_4$
3) H_2O

20.26.

(a) This reaction occurs under acidic conditions, so we must avoid formation of a strong base. Thus, proton transfers are required at multiple stages in the mechanism. The first step (proton transfer), requires two curved arrows to show the transfer of a proton from H_3O^+ to the carbonyl group, resulting in formation of an activated electrophile. In step two (nucleophilic attack), water serves as a nucleophile attacking the protonated carbonyl group, producing a cationic tetrahedral intermediate. We cannot immediately expel the amine group at this stage, as this would result in the formation of a strong base. This must be avoided in acidic conditions. This nitrogen atom must first be protonated. However, protonation at this stage would result in an intermediate with two positive charges, which should be avoided, if possible. Therefore, in the next step (step 3), water serves as a base, resulting in a neutral intermediate. Subsequently, in step 4 (proton transfer), a proton is transferred from H_3O^+ to the nitrogen atom, as shown. In step 5 (loss of leaving group), the carbonyl group is reformed and the amine serves as the leaving group, resulting in the opening of the ring. In step 6 (proton transfer), water serves as a base which deprotonates the cationic intermediate, resulting in formation of a bifunctional product. Under these acidic conditions, the amino group in the product is protonated to give an ammonium ion.

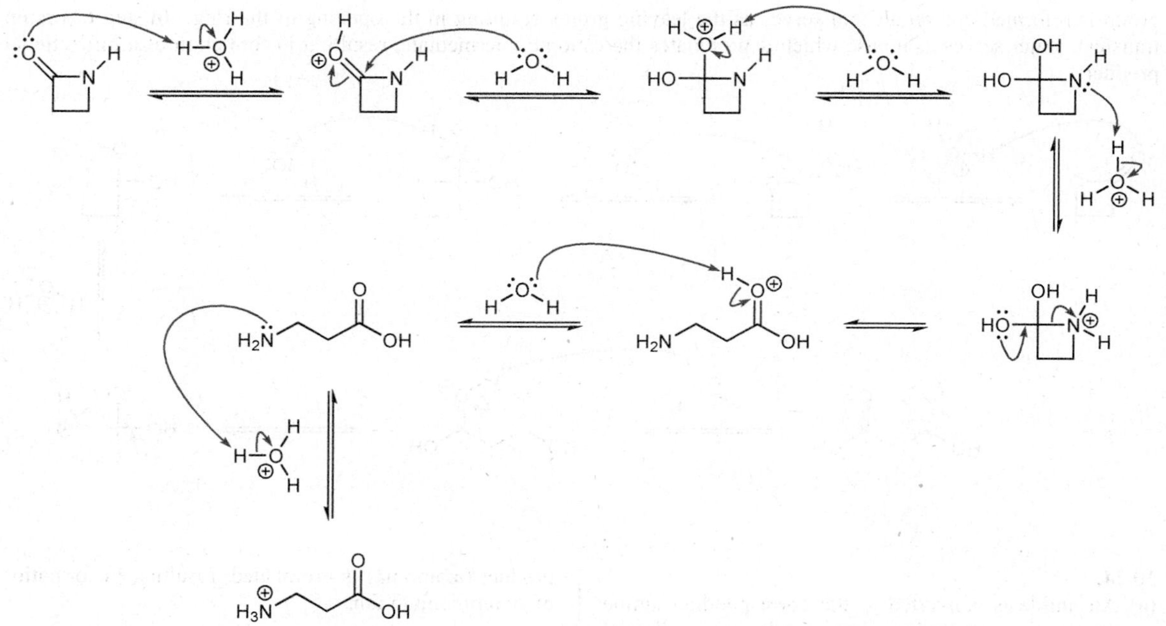

(b) This mechanism has three steps: 1) nucleophilic attack, 2) loss of a leaving group, and 3) proton transfer. The first step (nucleophilic attack) requires two curved arrows, which show hydroxide functioning as a nucleophile and attacking the electrophilic carbonyl group, resulting in an anionic tetrahedral intermediate. In step two (loss of a leaving group), the carbonyl group is reformed as a result of cleavage of the carbon-nitrogen bond, thereby opening up the ring and resulting in formation of a bifunctional anionic intermediate containing a carboxylic acid group and a deprotonated amine. In the third step (proton transfer) the anionic nitrogen atom serves as a base, deprotonating the carboxylic acid group in an intramolecular process, resulting in the formation of a carboxylate ion.

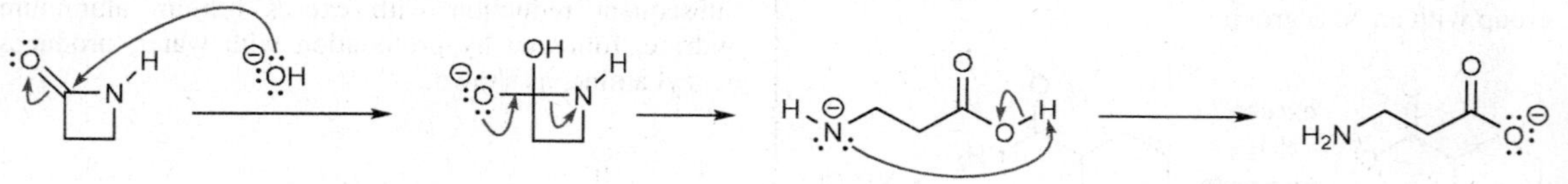

20.27.

(a) $LiAlH_4$ reduces the C≡N triple bond to a single bond via incorporation of two equivalents of hydride, producing the primary amine shown.

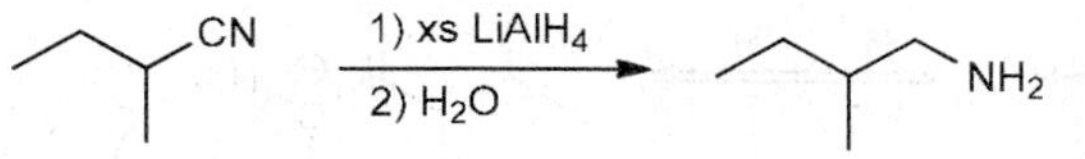

1) xs $LiAlH_4$
2) H_2O

(b) Treating benzyl bromide with sodium cyanide results in the formation of an intermediate nitrile (shown) via an S_N2 reaction. Subsequent attack by a Grignard reagent produces an anionic intermediate which is then protonated and hydrolyzed with H_3O^+ to form the corresponding ketone.

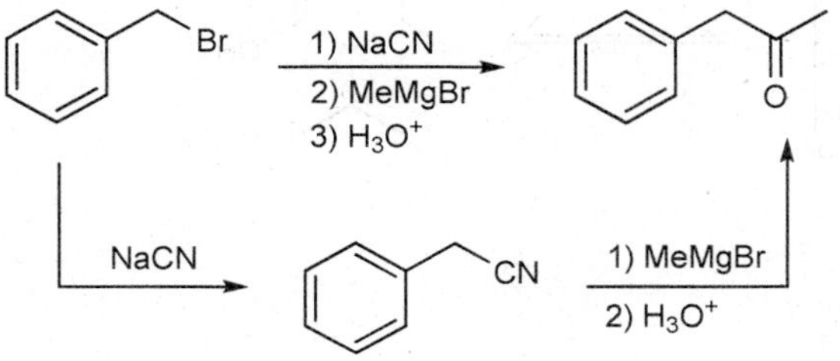

1) NaCN
2) MeMgBr
3) H_3O^+

NaCN

1) MeMgBr
2) H_3O^+

(c) Reaction of a nitrile with a Grignard reagent produces an anionic intermediate which is subsequently protonated and hydrolyzed with H_3O^+ to form a ketone, as shown. Reduction with $LiAlH_4$, followed by aqueous acidic workup, converts the ketone to the corresponding secondary alcohol.

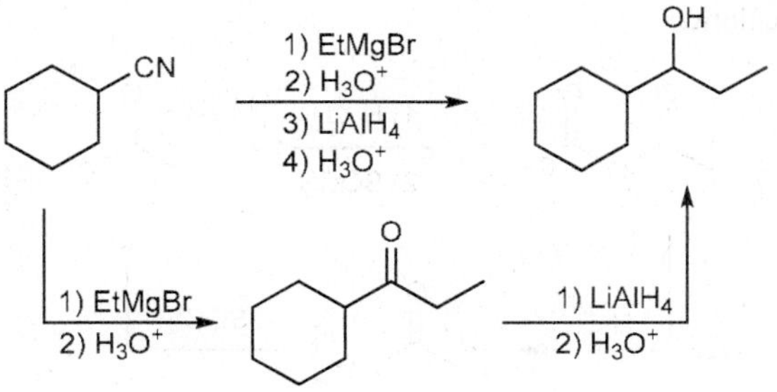

1) EtMgBr
2) H_3O^+
3) $LiAlH_4$
4) H_3O^+

1) EtMgBr
2) H_3O^+

1) $LiAlH_4$
2) H_3O^+

(d) A nitrile undergoes hydrolysis to give a carboxylic acid upon prolonged treatment with aqueous acid at elevated temperature.

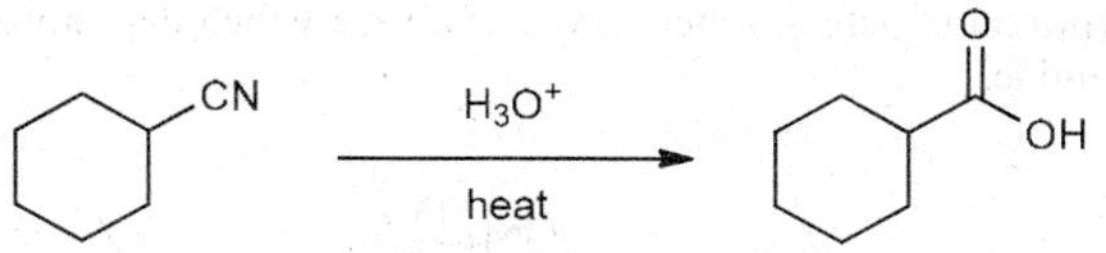

H_3O^+
heat

20.28.

(a) Reaction of the acid chloride with excess ammonia results in substitution of chloride with the nitrogen atom from ammonia. Thionyl chloride serves to dehydrate the resulting amide thus producing the desired nitrile as shown.

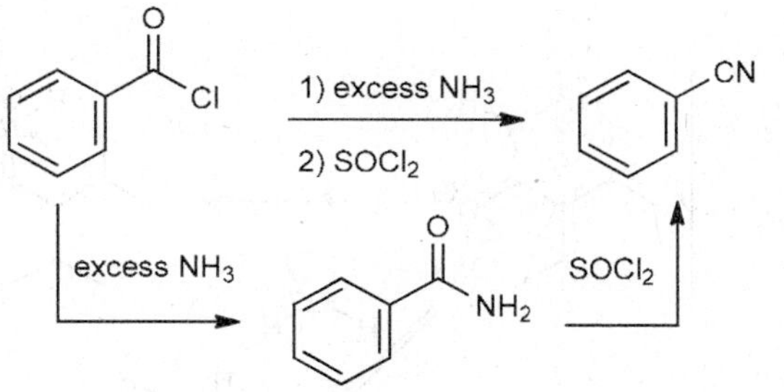

1) excess NH_3
2) $SOCl_2$

excess NH_3

$SOCl_2$

(b) Two approaches for this transformation are shown. Note that each approach incorporates an extra carbon atom to convert the starting material (which has seven carbon atoms) into the product (which has eight carbon atoms). In the first approach, sodium cyanide serves as a nucleophile in an S_N2 reaction, displacing the bromide to produce a nitrile. The nitrile is subsequently hydrolyzed upon heating with aqueous acid, to produce the desired carboxylic acid.
In the second approach, benzyl bromide is converted to benzyl magnesium bromide (a Grignard reagent), which serves as a nucleophile in a subsequent reaction with carbon dioxide. The resulting carboxylate ion is then protonated with H_3O^+ to produce the desired carboxylic acid.

1) NaCN
2) H_3O^+
heat

1) Mg
2) CO_2
3) H_3O^+

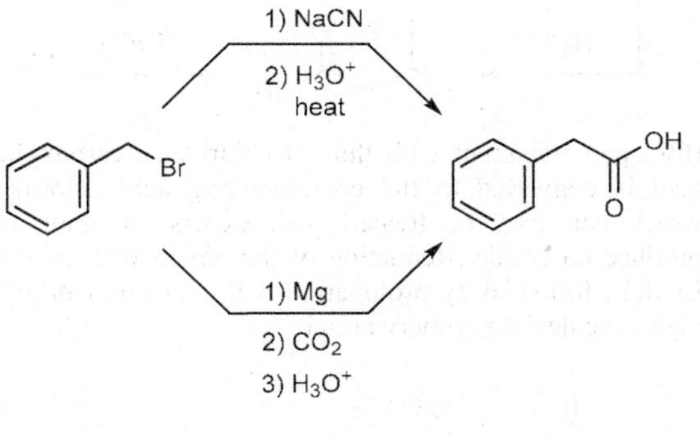

20.29. This reaction occurs under acidic conditions, so we must avoid formation of a strong base. Thus, proton transfers are required at multiple stages in this mechanism. The following mechanism has five steps: 1) proton transfer, 2) nucleophilic attack, 3) proton transfer, 4) proton transfer, and 5) proton transfer. The first step (proton transfer), requires two curved arrows to show the transfer of a proton from H_3O^+ to the nitrogen atom, resulting in the formation of an activated electrophile. In step two (nucleophilic attack), water serves as a nucleophile attacking the activated electrophilic carbon atom of the protonated nitrile, producing a cationic intermediate. In step 3 (proton transfer), water serves as a base, resulting in a neutral intermediate. Subsequently, in step 4 (proton transfer), a proton is transferred from H_3O^+ to the nitrogen atom to produce a cationic intermediate (two key resonance structures are shown). In step 5 (proton transfer), water serves as a base which deprotonates the cationic intermediate, resulting in the formation of an amide.

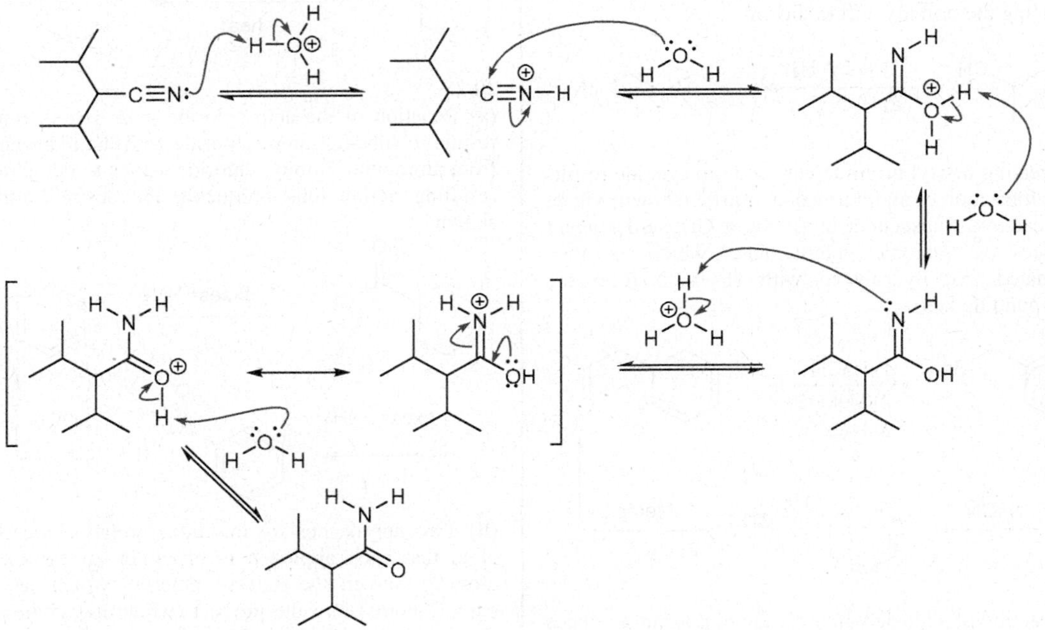

20.30.

(a) Acid-catalyzed hydrolysis of the ester produces the carboxylic acid, which can then be converted to the acid chloride upon treatment with thionyl chloride.

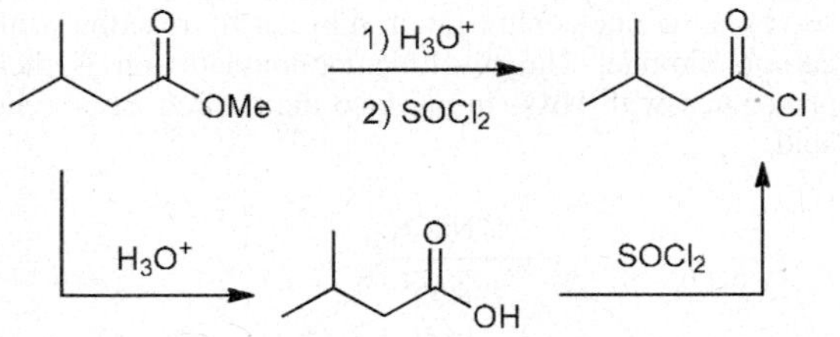

(b) Upon treatment with thionyl chloride, a carboxylic acid is converted to the corresponding acid chloride, which can then be treated with excess ammonia to produce an amide. Reduction of the amide with excess $LiAlH_4$, followed by protonation with aqueous workup, yields the desired primary amine.

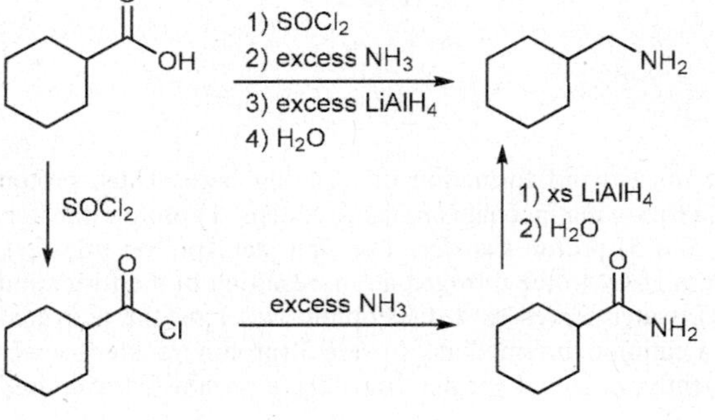

(c) Hydrolysis of acetic anhydride with water produces acetic acid, which can be converted to the desired product (acetyl chloride) upon treatment with thionyl chloride.

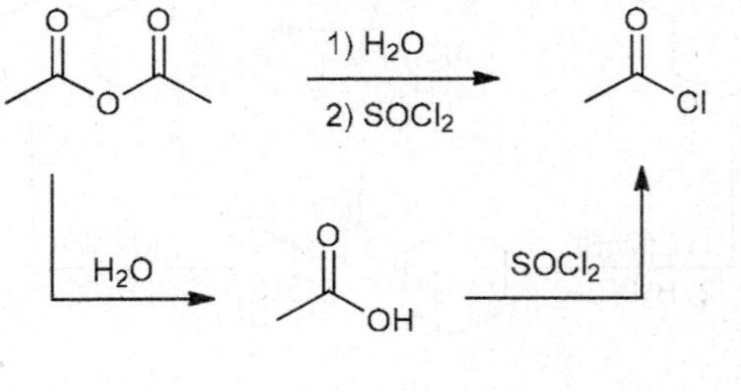

(d) The transformation requires conversion of an ester to an amine (with no change in the carbon skeleton). One way to achieve this transformation involves initial hydrolysis of the ester. The resulting carboxylic acid is converted to the acid chloride with thionyl chloride, which subsequently reacts with excess ammonia to produce the amide. Reduction of the amide with excess $LiAlH_4$, followed by aqueous workup, yields the desired primary amine.

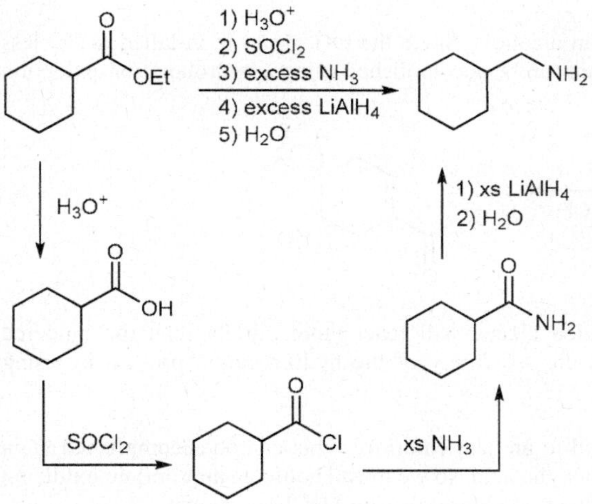

(e) A carboxylic acid can be converted to an acid chloride upon treatment with thionyl chloride. The acid chloride will then react with excess ammonia to produce an amide. Dehydration of the amide with thionyl chloride yields the desired nitrile.

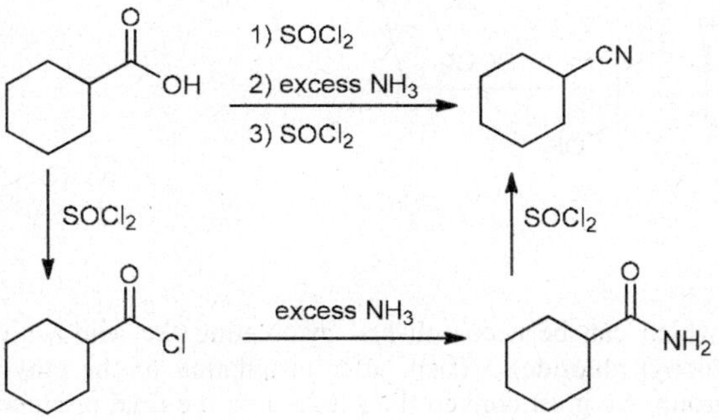

(f) This transformation involves hydrolysis of acetic anhydride to give acetic acid, which can be achieved in a single step, upon treatment with water.

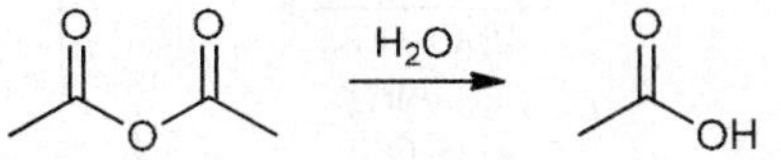

(g) An acid chloride can be converted to the corresponding ethyl ester in a single step, as shown.

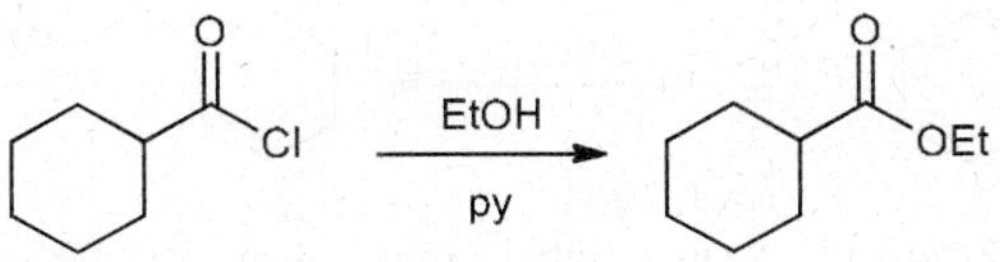

(h) Hydrolysis of the amide with H_3O^+ (and heat) produces a carboxylic acid, which can be converted to an acid chloride upon treatment with thionyl chloride. Subsequent reaction with a sterically hindered lithium trialkoxyaluminum hydride reagent, followed by aqueous acidic workup, produces the desired aldehyde.

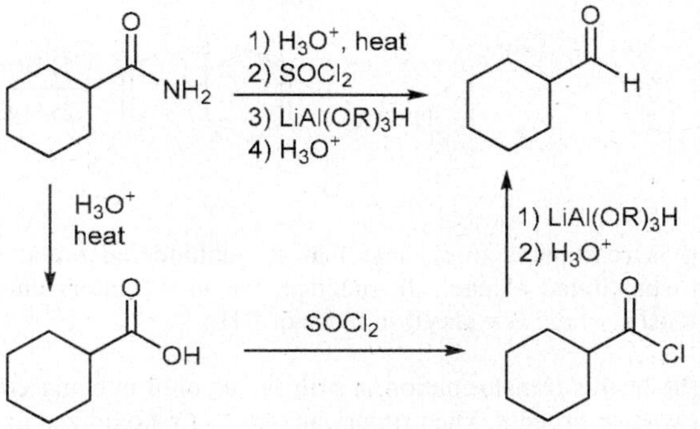

(i) Hydrolysis of the nitrile produces the carboxylic acid, which can be subsequently reduced to the primary alcohol with $LiAlH_4$, followed by aqueous acidic workup.

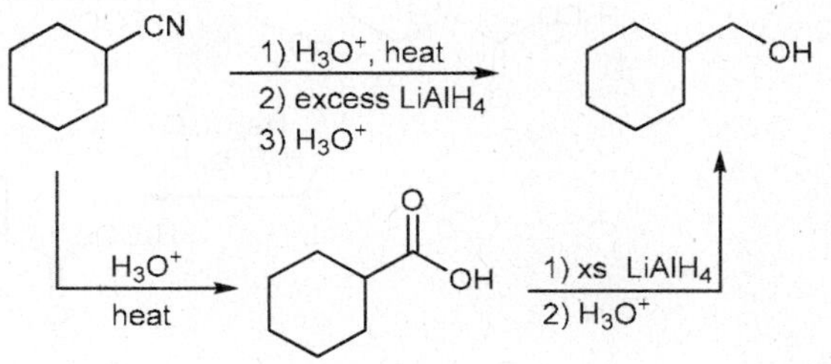

(j) Oxidation of a primary alcohol with a strong oxidizing agent yields a carboxylic acid, which can be subsequently converted to an acid chloride. Reaction with excess ammonia gives an amide, which can be dehydrated with thionyl chloride to give the desired nitrile.

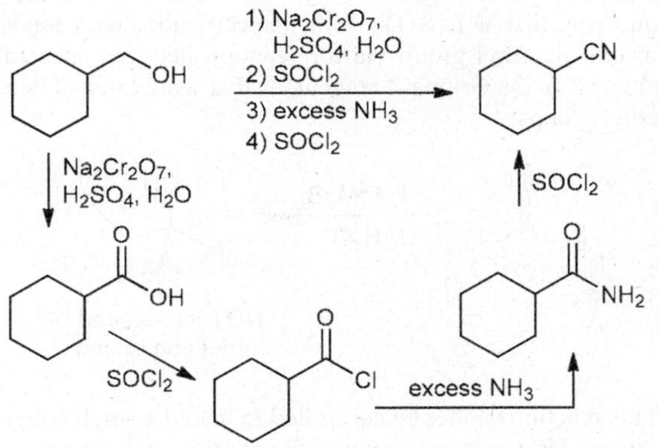

20.31.

(a) In this transformation, an alkene is being converted into an alcohol. Since the OH group is installed at the less substituted carbon, we need an *anti*-Markovnikov addition. This can be accomplished using a hydroboration-oxidation reaction sequence.

1) $BH_3 \cdot THF$
2) H_2O_2, NaOH

It is reasonable to assume that the unhindered monosubstituted alkene will react more rapidly than the hindered trisubstituted alkene. In practice, the investigators enhanced the selectivity of the hydroboration process by using R_2BH (where R = alkyl) in place of BH_3.

(b) In this transformation, a primary alcohol is being converted to an acid chloride. This can be accomplished via a two-step process. The primary alcohol is first oxidized to a carboxylic acid, so we must choose an appropriate oxidizing agent. This is followed by conversion of the carboxylic acid to the acid chloride, using $SOCl_2$, as shown.

1) $Na_2Cr_2O_7$, H_2SO_4, H_2O
2) $SOCl_2$

$Na_2Cr_2O_7$
H_2SO_4, H_2O

$SOCl_2$

20.32.

(a) The starting material has seven carbon atoms, and the product has nine carbon atoms. This requires installation of an ethyl group via a carbon-carbon bond-forming reaction. There are certainly several ways to achieve the installation of a single ethyl group. Let's first consider one way that will NOT work. Specifically, we cannot install the ethyl group via the reaction between an acid chloride and a Grignard reagent, as that would install two ethyl groups:

1) EtMgBr
2) H_3O^+

NOT the desired target compound

This reaction cannot be controlled to install a single ethyl group. However, a lithium dialkylcuprate will attack an acid chloride just once, installing just one ethyl group:

Et_2CuLi

In order to use this method to install an ethyl group, we must first convert the starting material into an acid halide (which can be accomplished by treating the acid with thionyl chloride). Then, after installation of the ethyl group, we must convert the ketone into the final product (which can be achieved via hydride reduction):

1) $SOCl_2$
2) Et_2CuLi
3) $LiAlH_4$
4) H_3O^+

$SOCl_2$

Et_2CuLi

1) $LiAlH_4$
2) H_3O^+

Alternatively, a single ethyl group can be installed via the reaction between an aldehyde and a Grignard reagent. This strategy gives the following synthesis: The carboxylic acid is first converted to an acid halide, followed by subsequent treatment with $LiAl(OR)_3H$ to give an aldehyde. The aldehyde can then be treated with ethyl magnesium bromide, followed by aqueous acidic workup, to give the desired product. This alternative strategy demonstrates that there may be more than one correct way to approach a synthesis problem.

(b) The most efficient solution to this problem is shown here. Bromobenzene is converted to phenyl magnesium bromide (a Grignard reagent) and then treated with formaldehyde, followed by aqueous acidic workup, to give benzyl alcohol, which serves as a nucleophile in a subsequent reaction with acetyl chloride to give the desired ester. Acetic anhydride is an equally effective reagent for the final esterification step.

1) Mg
2) CH_2O
3) H_3O^+
4) CH_3COCl, pyridine

Mg

1) $H_2C{=}O$
2) H_3O^+

pyridine

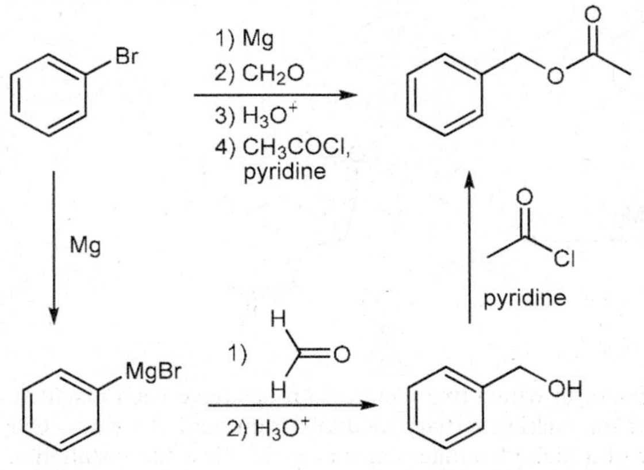

(c) There are certainly many acceptable solutions to this problem. One such solution derives from the following retrosynthetic analysis. An explanation of each of the steps (*a-d*) follows.

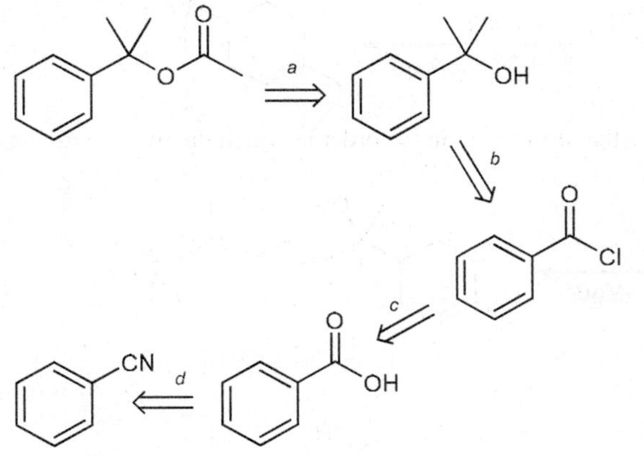

a. The desired product is an ester, which can be made via acetylation of the appropriate tertiary alcohol.
b. The tertiary alcohol can be made from an acid halide upon treatment with excess Grignard reagent.
c. The acid halide can be made from the corresponding carboxylic acid (benzoic acid).
d. Benzoic acid can be made from the starting material via hydrolysis.

Now let's draw the forward scheme. Hydrolysis of the nitrile to the carboxylic acid, followed by reaction with thionyl chloride, produces the acid chloride. Reaction with excess methyl magnesium bromide, followed by aqueous acidic workup, results in the formation of a tertiary alcohol, with the incorporation of two new methyl groups. The tertiary alcohol can then serve as a nucleophile in an acetylation reaction (upon treatment with acetyl chloride) to give the desired ester. Acetic anhydride is an equally effective reagent for the final esterification step.

1) H_3O^+, heat
2) $SOCl_2$
3) excess MeMgBr
4) H_3O^+
5) CH_3COCl, pyridine

H_3O^+ heat

pyridine

$SOCl_2$

1) xs MeMgBr
2) H_3O^+

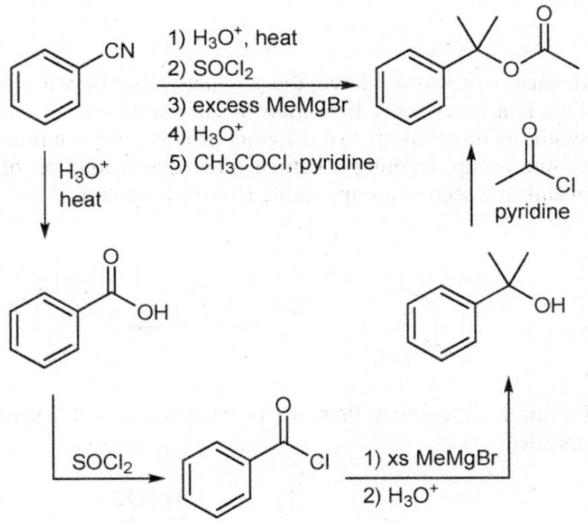

20.33. In this transformation, two new C–C bonds are being formed. Note that the two new groups are different. One is a simple methyl group and the other is a branched six-carbon group containing a double bond. The identity, *but not the position*, of the functional group is also changing.

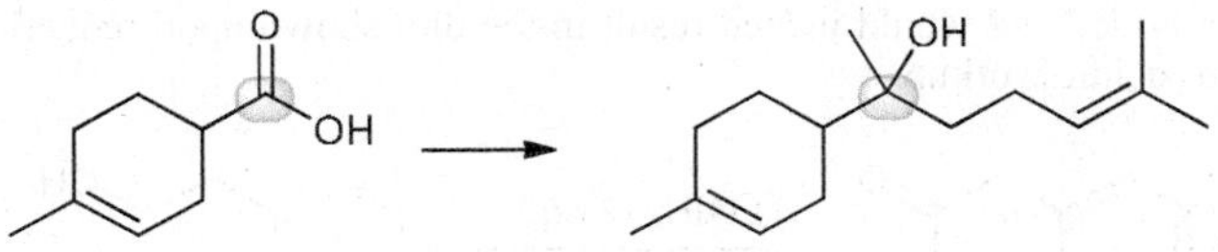

A reasonable first step is to convert the carboxylic acid into an acid chloride.

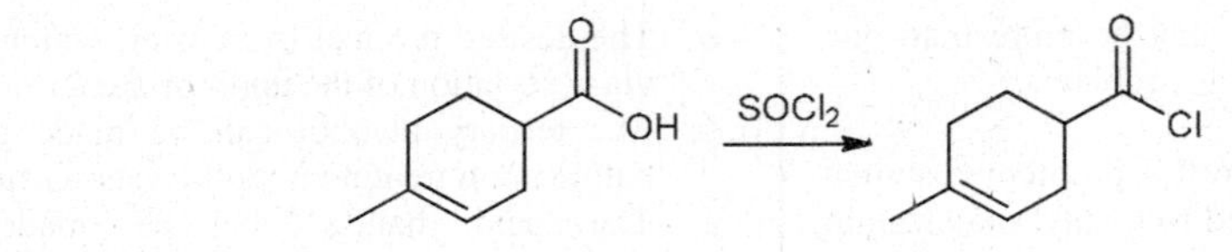
OH
SOCl2
Cl

The desired product is a tertiary alcohol, and we know that tertiary alcohols can be made from acid chlorides using excess Grignard reagent. This approach will not work in this case, because we must install two different groups. Consider what would happen if we treated the acid chloride above with excess Grignard reagent:

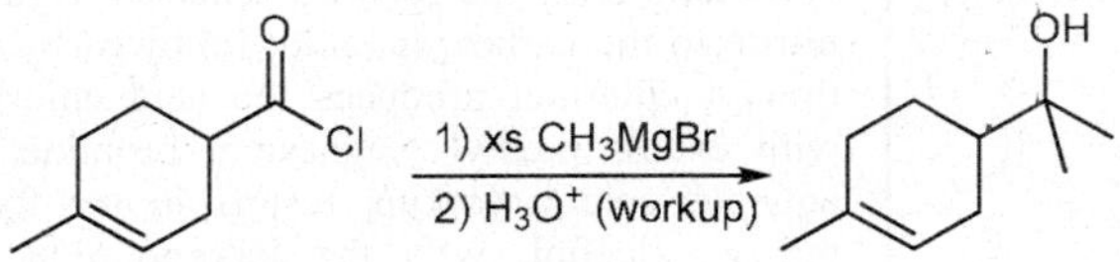
Cl
1) xs CH3MgBr
2) H3O+ (workup)
OH

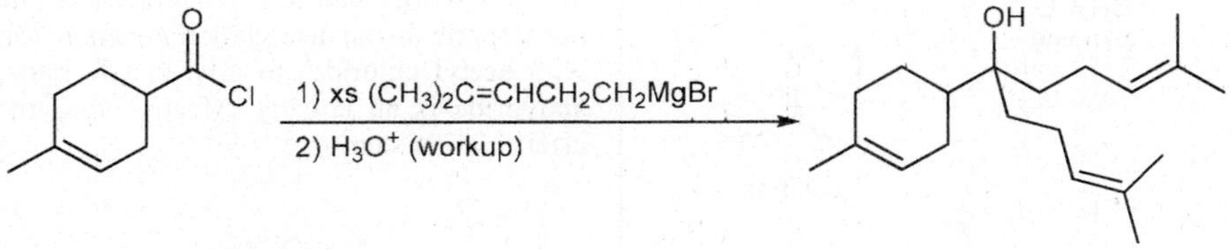
Cl
1) xs (CH3)2C=CHCH2CH2MgBr
2) H3O+ (workup)
OH

In each case shown above, the product would be a tertiary alcohol in which two identical groups have been installed. This is a noteworthy limitation to the use of excess Grignard for making tertiary alcohols from acid chlorides. Our synthesis must install two different groups, one at a time. Use of a dialkylcuprate reagent would allow the installation of one group, forming a ketone; subsequent addition of a Grignard reagent to the ketone, followed by protonation, installs the second group, giving a tertiary alcohol.

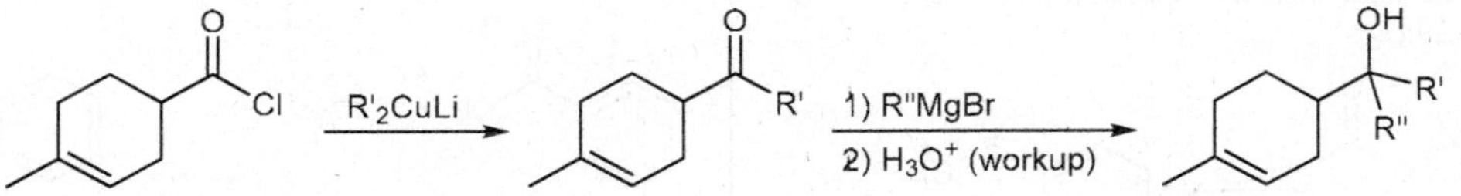
Cl
R'2CuLi
R'
1) R"MgBr
2) H3O+ (workup)
OH
R'
R"

Putting it all together, there are two variations of this synthesis that differ only in the order in which the two groups are installed:

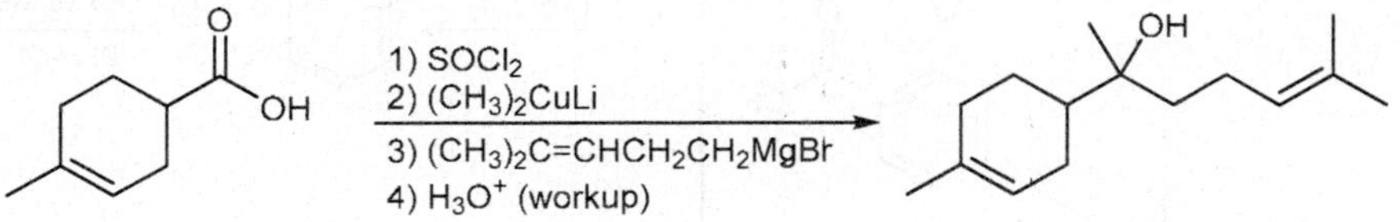
OH
1) SOCl2
2) (CH3)2CuLi
3) (CH3)2C=CHCH2CH2MgBr
4) H3O+ (workup)
OH

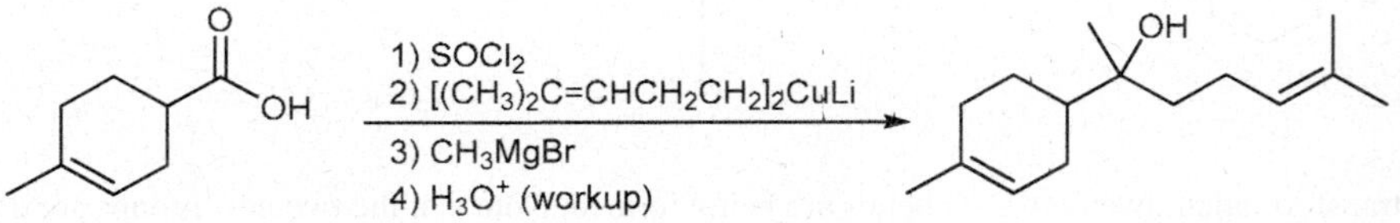
OH
1) SOCl2
2) [(CH3)2C=CHCH2CH2]2CuLi
3) CH3MgBr
4) H3O+ (workup)
OH

20.34. The signal at 1740 cm^{-1} indicates the presence of a carbonyl group (likely of an ester group) that is not conjugated with the aromatic ring (it would be at a lower wavenumber if it were conjugated). The cyclic ester (lactone) below fits the description provided and would indeed result in the diol shown upon reduction with two equivalents of $LiAlH_4$, followed by aqueous acidic workup.

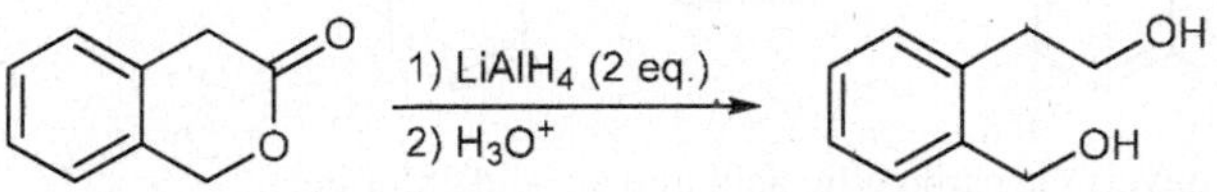
O
O
1) LiAlH4 (2 eq.)
2) H3O+
OH
OH

20.35.

(a) Each of these acids is a *para*-substituted benzoic acid. Relative acid strength depends on the electron-withdrawing (or electron-donating) capacity of the substituent. Electron-withdrawing groups pull electron density away from the ring, thus stabilizing the anionic charge on the conjugate base, thus giving a stronger acid. Electron-donating groups have the opposite effect: they donate electron density into the ring, thus destabilizing the anionic charge on the conjugate base, resulting in a weaker acid. Accordingly, the acids below are arranged in order of increasing acid strength. As seen in Table 18.1, a methoxy group is strongly electron-donating; a methyl group is weakly electron-donating; a bromine atom is weakly electron-withdrawing; a carbonyl group is a moderate electron-withdrawing group; a nitro group is a strong electron-withdrawing group.

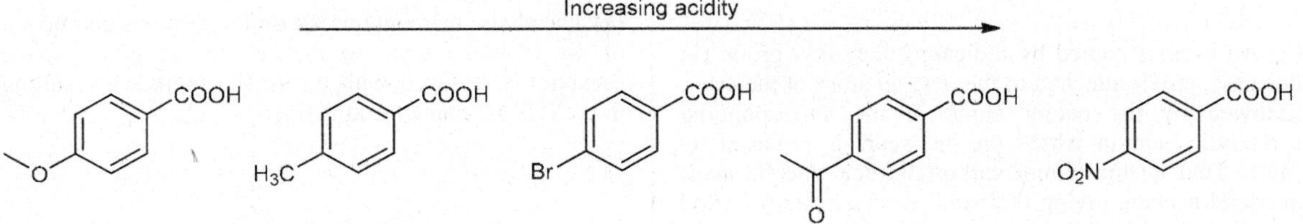

(b) When comparing the acids below, the difference in acidity is related to the proximity of the electron-withdrawing bromine atom to the carboxylic acid group. The closer the bromine atom is to the carboxylic acid group, the more the bromine atom stabilizes the negative charge on the conjugate base. Accordingly, the strongest acid in this series has the bromine atom immediately adjacent to the carboxylic acid (in the 1 position), followed by the 2-bromo isomer. The 3-bromo isomer is the weakest acid, as shown below.

Increasing acidity

20.36.

(a) The second carboxylic acid group is electron-withdrawing, and therefore stabilizes the negative charge in the conjugate base that is formed when the first proton is removed. This makes malonic acid a stronger acid than acetic acid.

(b) The carboxylate ion is electron-rich, and it destabilizes the conjugate base that is formed when the second proton is removed. This makes the second proton of malonic acid *less* acidic than acetic acid.

(c) Since both pK_{a1} and pK_{a2} are lower than 7.3, they are both expected to be largely deprotonated at physiological pH, resulting in the dianion shown below.

(d) The number of methylene (CH_2) groups separating the carboxylic acid groups is greater in succinic acid than in malonic acid. Therefore, the inductive effects described above in parts (a) and (b) are not as strong, and both pK_a values are closer to that of acetic acid.

20.37.

(a) When a carboxylic acid group is attached to a ring, it is named as an alkanecarboxylic acid. This compound (with a ring composed of five carbon atoms) is thus cyclopentanecarboxylic acid.

cyclopentanecarboxylic acid

(b) An amide is named by replacing the suffix "ic acid" or "oic acid" with "amide". The corresponding carboxylic acid is named cyclopentanecarboxylic acid. Replacement of "ic acid" with "amide" produces the name cyclopentanecarboxamide.

cyclopentanecarboxamide

(c) This acid chloride is named by replacing the "ic acid" from the parent (benzoic acid) with "yl chloride" to produce benzoyl chloride.

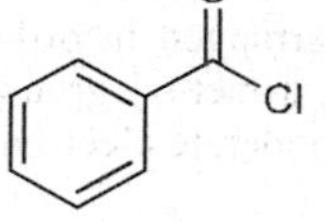

benzoyl chloride

(d) An ester is named by indicating the alkyl group (in this case, ethyl) attached to the oxygen atom of the ester, followed by the parent name of the corresponding carboxylic acid in which the "ic acid" is replaced by "ate". That is, the parent carboxylic acid (acetic acid) becomes acetate, giving the name ethyl acetate (or ethyl ethanoate, based off the name ethanoic acid).

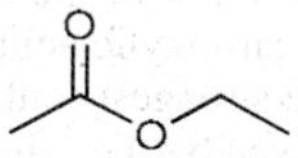

ethyl acetate
or
ethyl ethanoate

(e) The chain that contains the carboxylic acid group is comprised of six carbon atoms, so the "e" in the parent name (hexane) is replaced with the suffix "-oic acid", resulting in the IUPAC name hexanoic acid.

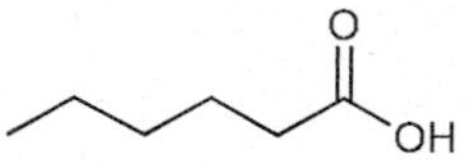

hexanoic acid

(f) The chain that contains the acid chloride group is comprised of five carbon atoms, so the "e" in the parent name (pentane) is replaced with the suffix "-oyl chloride", resulting in the IUPAC name pentanoyl chloride.

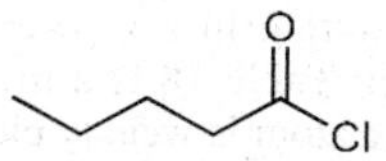

pentanoyl chloride

(g) The chain that contains the amide group is comprised of six carbon atoms, so the "e" in the parent name (hexane) is replaced with the suffix "-amide", resulting in the IUPAC name hexanamide.

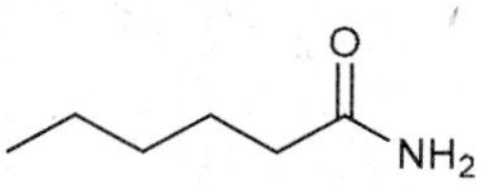

hexanamide

20.38. The common name for each molecule is shown below:

(a) acetic anhydride

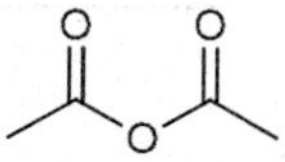

(b) benzoic acid

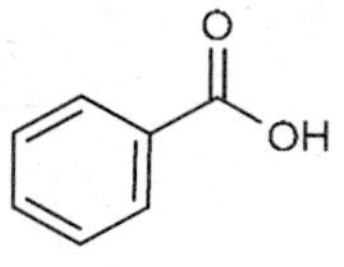

(c) formic acid

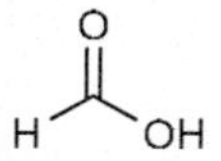

(d) oxalic acid

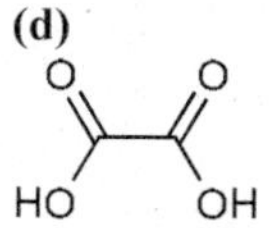

20.39. A molecular formula of $C_6H_{12}O_2$ corresponds with one degree of unsaturation (see Section 14.16), which accounts for the carboxylic acid group. Since there are no other degrees of unsaturation, all of the isomers must be acyclic, saturated carboxylic acids. There are eight isomers that fit this description, shown here. These isomers are identified by methodically considering each possible parent chain. There is only one isomer with a parent chain of six carbon atoms (the first isomer shown). Then, there are three isomers that have a parent chain of five carbon atoms (with one methyl substituent). The methyl group can be located at C2, C3 or C4 (it cannot be at C5, because that would simply generate a parent chain of six carbon atoms, and we have already accounted for that isomer). Then, there are several isomers with a parent chain of only four carbons (with either two methyl groups or with one ethyl group). Once again, these isomers are drawn methodically. The two methyl groups can both be at C2, or they can be at C2 and C3, or both can be at C3. And finally, there can be an ethyl group at C2. Notice that, for a four-carbon chain, an ethyl group cannot be placed at C3, as that would generate a structure with a parent of five carbon atoms (and we have already accounted for that isomer).

Each carboxylic acid is named by identifying the longest chain containing the carboxylic acid group and replacing the "e" at the end of the alkane name with "oic acid". Each chain is numbered with the carboxylic acid carbon atom being C1, and the substituents are identified accordingly. Three of the isomers exhibit chiral centers (highlighted).

chiral centers

hexanoic acid

2-methylpentanoic acid

3-methylpentanoic acid

4-methylpentanoic acid

2,2-dimethylbutanoic acid

2,3-dimethylbutanoic acid

3,3-dimethylbutanoic acid

2-ethylbutanoic acid

20.40. There are only two constitutional isomers, shown below. Each one is named by identifying the longest chain containing the acid chloride group and replacing the "e" in the parent name with the suffix "-oyl chloride". The first isomer is thus named butanoyl chloride. In the second (branched) isomer, the chain is numbered so that the carbonyl group is C1, which puts the methyl substituent on C2, resulting in the IUPAC name 2-methylpropanoyl chloride.

butanoyl chloride

2-methylpropanoyl chloride

20.41.

(a) Pentanoic acid is converted to 1-pentanol using a strong reducing agent ($LiAlH_4$), followed by aqueous acidic workup, as shown.

1) xs $LiAlH_4$ 2) H_3O^+

(b) Pentanoic acid is initially converted to 1-pentanol using a strong reducing agent ($LiAlH_4$), followed by aqueous acidic workup, as shown. Conversion to the tosylate followed by reaction with a strong, sterically hindered base produces 1-pentene via an E2 reaction. Note that a sterically hindered base is required in the last step because a non-sterically hindered base (*i.e.*, NaOEt) would result in the formation of the S_N2 product as the major product.

1) xs $LiAlH_4$ 2) H_3O^+ 3) TsCl, py 4) *t*-BuOK

1) xs $LiAlH_4$ 2) H_3O^+

TsCl, py

t-BuOK

(c) Conversion of pentanoic acid to hexanoic acid requires the installation of an extra carbon atom. Two efficient solutions to this problem involve new C–C bond formation by S_N2 with cyanide or by the reaction of a Grignard reagent with CO_2. The first solution derives from the following retrosynthetic analysis. An explanation of each of the steps (*a-d*) follows.

a. The desired product can be made via hydrolysis of a nitrile.

b. The nitrile can be made from 1-bromopentane (via an S_N2 process).

c. 1-Bromopentane can be made from 1-pentanol (upon treatment with PBr_3).

d. 1-Pentanol can be made from pentanoic acid via hydride reduction.

Now let's draw the forward scheme. Reduction of pentanoic acid gives 1-pentanol. Treating this alcohol with PBr_3 produces 1-bromopentane. Subsequent S_N2 substitution with sodium cyanide installs an extra carbon atom, producing a nitrile, which can be hydrolyzed to the desired product under acidic conditions;

1) xs $LiAlH_4$ 2) H_3O^+ 3) PBr_3 4) NaCN 5) H_3O^+, heat

1) xs $LiAlH_4$ 2) H_3O^+

H_3O^+ heat

OH CN OH

PBr_3 Br NaCN

20.42.

(a) *anti*-Markovnikov addition of water (via hydroboration-oxidation) produces 1-pentanol, which is subsequently oxidized to pentanoic acid using a strong oxidizing agent.

1) $BH_3 \cdot THF$ 2) H_2O_2, NaOH 3) $Na_2Cr_2O_7$, H_2SO_4, H_2O

1) $BH_3 \cdot THF$ 2) H_2O_2, NaOH

$Na_2Cr_2O_7$, H_2SO_4, H_2O

OH

(b) Conversion of 1-bromobutane to pentanoic acid requires the installation of an extra carbon atom. This extra carbon atom can be installed via an S_N2 process in which bromide is replaced with cyanide, thereby converting the alkyl halide into a nitrile. Subsequent acid-catalyzed hydrolysis produces the desired carboxylic acid, pentanoic acid.

Br 1) NaCN 2) H_3O^+, heat OH

NaCN CN H_3O^+ heat

Alternatively, treatment of the bromide with magnesium metal would generate a Grignard reagent. Subsequent reaction with CO_2, followed by aqueous workup, produces the desired pentanoic acid target compound.

20.43. As discussed in Chapter 18, the methoxy group is electron-donating via resonance, but electron-withdrawing via induction. The resonance donation effect is stronger, but only significantly affects the acidity when the methoxy group is in an *ortho* or *para* position relative to the carboxylic acid. Note that in the third resonance structure for the conjugate base of the *para* derivative below, there is a negative charge next to the carboxylate group (a destabilizing effect). In contrast, none of the resonance structures of the corresponding *meta* derivative have this destabilizing feature, and as such, the conjugate base of the *meta* derivative is more stable than the conjugate base of the *para* derivative. In fact, the conjugate base of the *meta* derivative is even more stable than the conjugate base of benzoic acid, because of the electron-withdrawing inductive effect of the methoxy group (in the absence of strong resonance effects that are present in the *ortho* or *para* derivative).

20.44.

(a) Reaction of hexanoyl chloride with an excess of ethyl amine produces the corresponding amide, where the chloride leaving group has been replaced with the nitrogen atom of the amine, as shown.

N H

(b) Reaction of hexanoyl chloride with an excess of $LiAlH_4$, followed by aqueous acidic workup, reduces the carboxylic acid group to the corresponding primary alcohol.

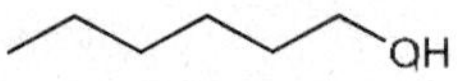

(c) Reaction of hexanoyl chloride with ethanol and pyridine produces the corresponding ester, where the chloride leaving group has been replaced with an ethoxy group.

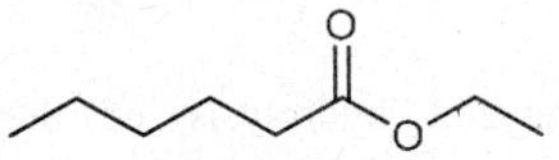

(d) Reaction of hexanoyl chloride with water produces the parent carboxylic acid, where the chloride leaving group has been replaced with an OH group.

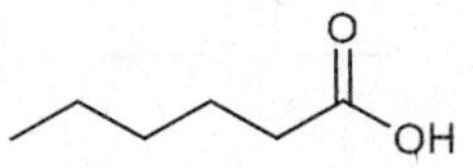

(e) Reaction of hexanoyl chloride with sodium benzoate produces the corresponding anhydride, where the chloride leaving group has been replaced with an oxygen atom of sodium benzoate, as shown.

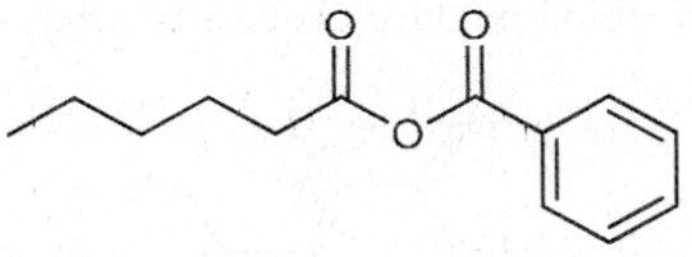

(f) Reaction of hexanoyl chloride with excess ammonia produces the corresponding amide, where the chloride leaving group has been replaced with an NH_2 group.

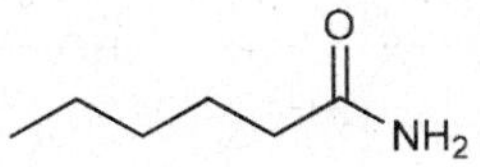

(g) Reaction of hexanoyl chloride with lithium diethylcuprate produces a ketone, where the chloride leaving group has been replaced with the ethyl group from the diethylcuprate, as shown.

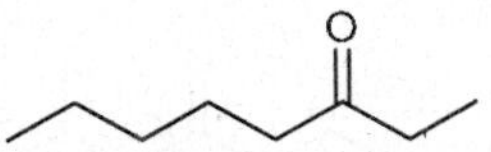

(h) Reaction of hexanoyl chloride with excess ethyl magnesium bromide, followed by aqueous acidic workup, produces a tertiary alcohol, where two new ethyl groups have been incorporated into the product. The first equivalent of ethyl magnesium bromide attacks the carbonyl group of the acid chloride to give a ketone (nucleophilic attack of EtMgBr, followed by loss of chloride). A second equivalent of ethyl magnesium bromide then attacks the carbonyl group of the ketone intermediate to produce an alkoxide ion, which is subsequently protonated during aqueous acidic workup, to give the following tertiary alcohol.

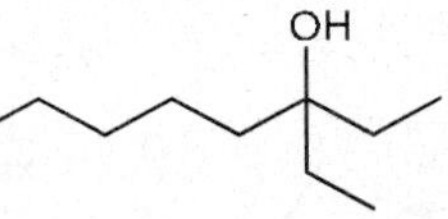

20.45.

(a) Reaction with thionyl chloride converts a carboxylic acid into the corresponding acid chloride, shown here.

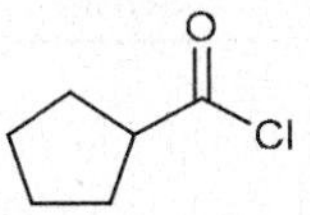

(b) A carboxylic acid is reduced to the corresponding alcohol upon treatment with $LiAlH_4$, followed by aqueous acidic workup.

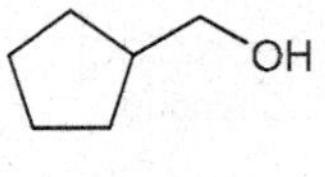

(c) Reaction with sodium hydroxide deprotonates the carboxylic acid to yield the sodium carboxylate.

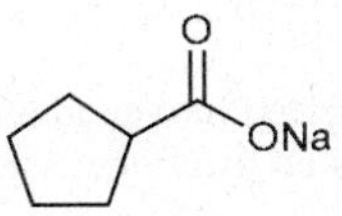

(d) Reaction with ethanol and catalytic acid converts the carboxylic acid to the ethyl ester shown, via a Fischer esterification.

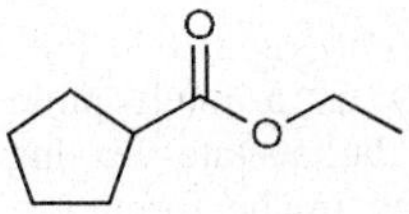

20.46.

(a) A carboxylic acid is reduced to the corresponding alcohol upon treatment with $LiAlH_4$, followed by aqueous acidic workup.

1) xs $LiAlH_4$
2) H_3O^+

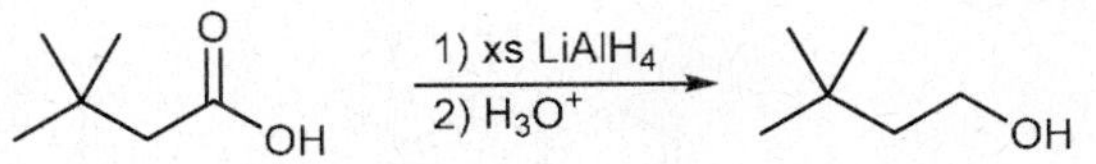

(b) Reaction with thionyl chloride converts the carboxylic acid to the corresponding acid chloride. Subsequent reaction with excess dimethylamine converts the acid chloride to the corresponding *N,N*-dimethylamide.

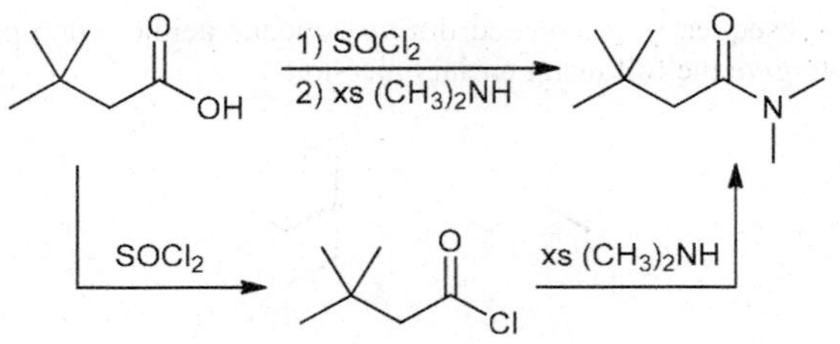

1) $SOCl_2$
2) xs $(CH_3)_2NH$

$SOCl_2$

xs $(CH_3)_2NH$

(c) An amide is converted into the corresponding nitrile upon treatment with thionyl chloride.

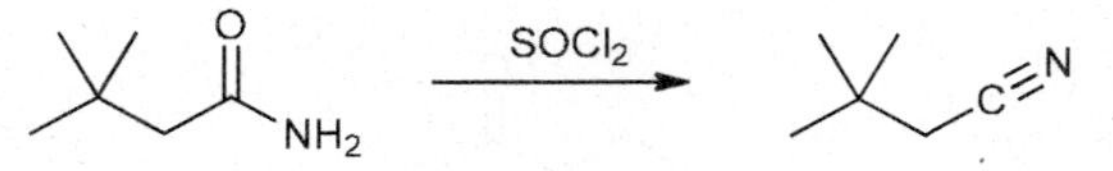

$SOCl_2$

(d) Acid catalyzed hydrolysis of an ester gives a carboxylic acid. Subsequent reaction with acetyl chloride and pyridine causes an acetylation reaction that gives the following anhydride as the product.

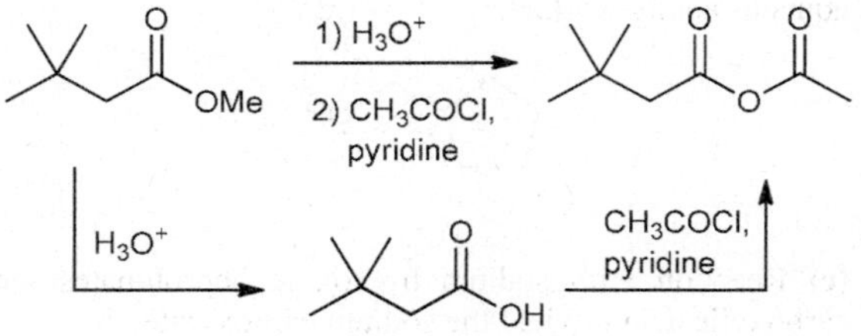

1) H_3O^+
2) CH_3COCl, pyridine

H_3O^+

CH_3COCl, pyridine

(e) An ester is converted into an aldehyde upon treatment with DIBAH, followed by aqueous acidic workup.

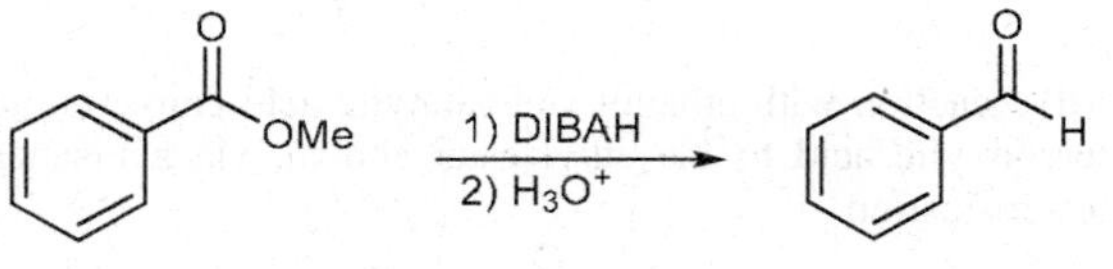

1) DIBAH
2) H_3O^+

(f) Phenol serves as a nucleophile in this reaction. Phenol replaces the acetate leaving group on acetic anhydride resulting in the formation of a phenyl ester and acetic acid, as shown. This process is called an acylation reaction.

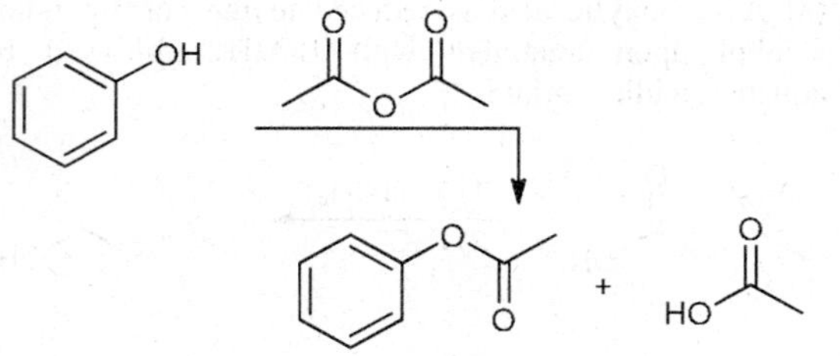

(g) Diphenylamine serves as a nucleophile in this reaction. Diphenylamine replaces the chloride leaving group on acetyl chloride resulting in the formation of the *N,N*-diphenylamide, as shown. This process is called an acylation reaction.

pyridine

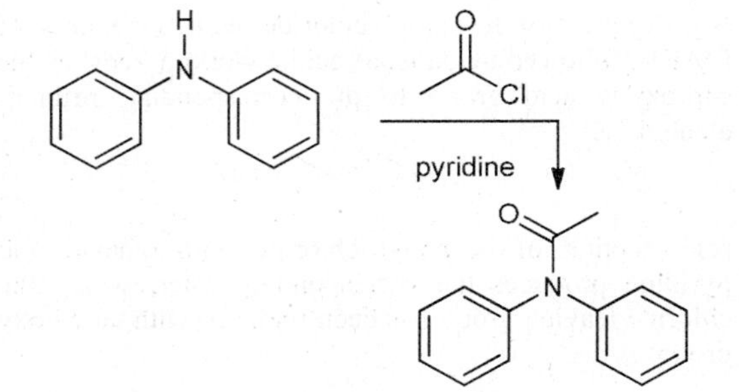

(h) Acid-catalyzed hydrolysis of the cyclic ester (lactone) results in formation of a bifunctional molecule containing both a carboxylic acid group and a phenol group.

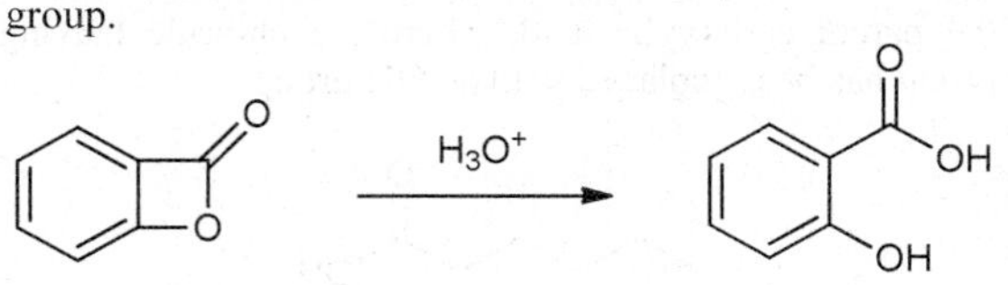

H_3O^+

20.47.
(a) The new bond that forms as a result of a Fischer esterification is the σ bond between the carbonyl group and the oxygen atom connected to it, as shown. Making this disconnection, it becomes evident that benzoic acid and phenol would produce the desired ester under acidic conditions.

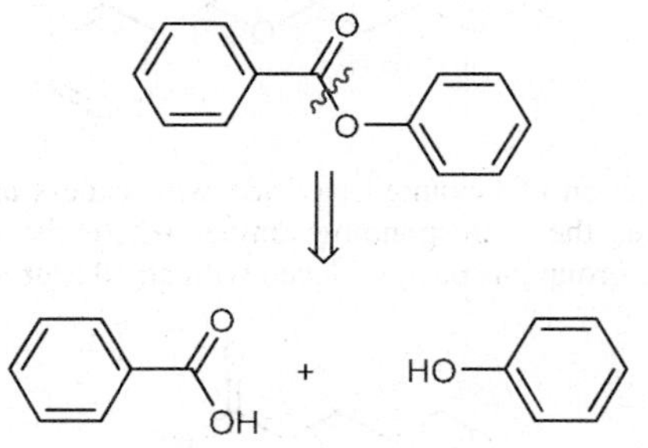

(b) The new bond that forms as a result of a Fischer esterification is the σ bond between the carbonyl group and the oxygen atom connected to it, as shown. Making this disconnection, it becomes evident that butyric acid and isopropanol would produce the desired ester under acidic conditions.

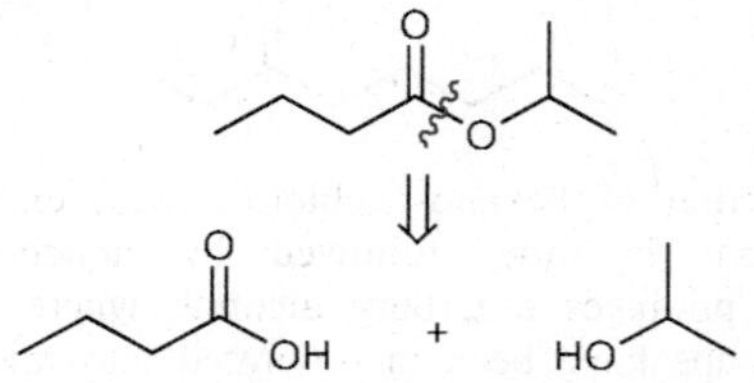

20.48. Oxidation of the primary alcohol gives the corresponding carboxylic acid (**A**). Reaction of **A** with thionyl chloride converts the carboxylic acid to the acid chloride (**B**). Reaction of **B** with excess ammonia produces the amide (**C**). Carboxylic acid (**A**) undergoes Fischer esterification upon reaction with ethanol and catalytic acid to produce the ethyl ester (**D**). Reaction of acid chloride **B** with a lithium trialkoxyaluminum hydride (followed by aqueous acidic workup) produces aldehyde **F**, which can also be made from ester **D** by reaction with DIBAH (**E**), followed by aqueous acidic workup.

$Na_2Cr_2O_7$, H_2SO_4, H_2O

[H^+] EtOH

A D

(E) 1) DIBAH 2) H_3O^+

$SOCl_2$

1) $LiAl(OR)_3H$ 2) H_3O^+

F B

xs NH_3

C

20.49.

(a) This transformation does not involve a change in the carbon skeleton. Only the identity of the functional group must be changed. To accomplish this transformation, the starting material can be treated with NaOH to give 1-pentanol via an S_N2 reaction. Subsequent oxidation of the primary alcohol gives the desired carboxylic acid.

1) NaOH 2) $Na_2Cr_2O_7$, H_2SO_4, H_2O

NaOH

$Na_2Cr_2O_7$ H_2SO_4, H_2O

(b) This transformation involves a change in the carbon skeleton (one extra carbon atom must be inserted), as well as a change in the identity and location of the functional group. To accomplish this transformation, the starting material can be treated with NaCN, thereby converting 1-bromopentane into hexanenitrile via an S_N2 reaction. Subsequent acid catalyzed hydrolysis of the nitrile (upon treatment with aqueous acid) gives hexanoic acid.

1) NaCN 2) H_3O^+, heat

NaCN

H_3O^+ heat

Alternatively, treatment of the bromide with magnesium metal would generate a Grignard reagent. Subsequent reaction with CO_2, followed by aqueous workup, produces the desired pentanoic acid target compound.

(c) First, 1-bromopentane can be converted into pentanoic acid in just two steps, as shown in the solution to part (a) of this problem. Then, treating this carboxylic acid with $SOCl_2$ will give the desired acid chloride.

1) NaOH 2) $Na_2Cr_2O_7$, H_2SO_4, H_2O 3) $SOCl_2$

NaOH

$SOCl_2$

$Na_2Cr_2O_7$ H_2SO_4, H_2O

(d) First, 1-bromopentane can be converted into hexanoic acid in just two steps, as shown in the solution to part (b) of this problem. Then, this carboxylic acid can be converted into an acid halide, followed by treatment with excess ammonia to give the desired product:

1) NaCN 2) H_3O^+, heat 3) $SOCl_2$ 4) xs NH_3

NaCN

xs NH_3

H_3O^+ heat

$SOCl_2$

(e) First, 1-bromopentane can be converted into pentanoic acid in just two steps, as shown in the solution to part (a) of this problem. This carboxylic acid can then be converted into an acid halide, followed by treatment with excess ammonia to give the desired product:

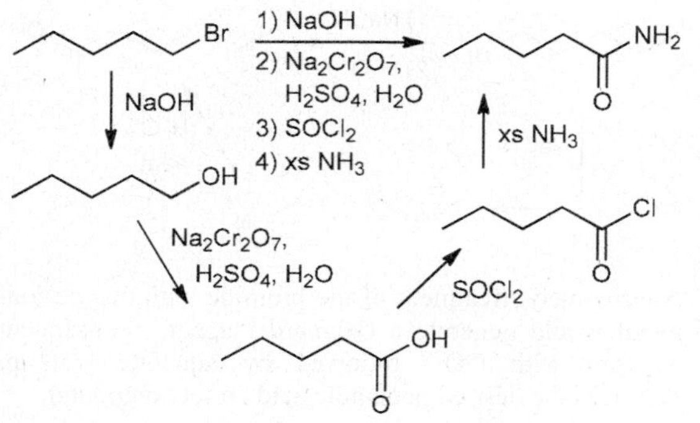

(f) First, 1-bromopentane can be converted into hexanoic acid in just two steps, as shown in the solution to part (b) of this problem. Upon treatment with ethanol in acid-catalyzed conditions, the carboxylic acid can be converted into an ester via a Fischer esterification process:

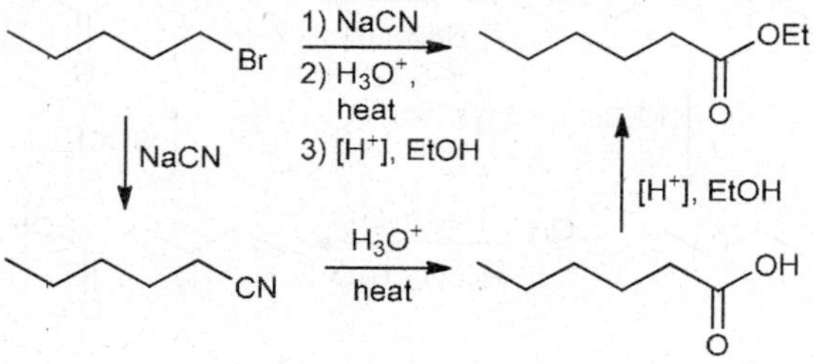

20.50.

(a) There are certainly many acceptable solutions to this problem. One such solution derives from the following retrosynthetic analysis. An explanation of each of the steps (*a-c*) follows.

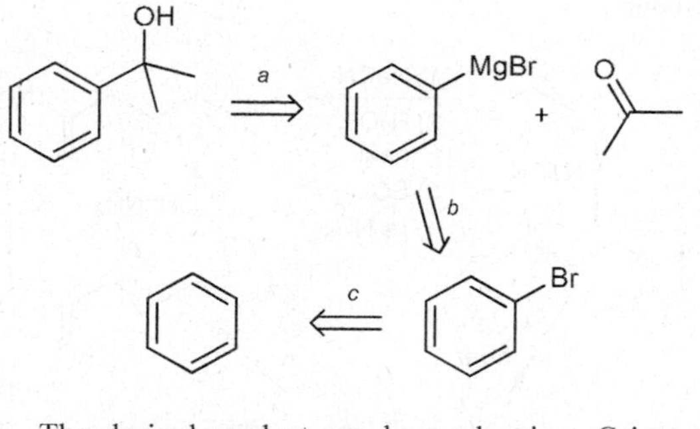

a. The desired product can be made via a Grignard reaction between phenyl magnesium bromide and acetone.

b. Phenyl magnesium bromide can be made from bromobenzene via insertion of magnesium.

c. Bromobenzene can be made from benzene via bromination.

Now let's draw the forward scheme. Benzene is converted into bromobenzene upon treatment with Br_2 and $AlBr_3$ (via an electrophilic aromatic substitution reaction). Bromobenzene is subsequently converted into phenyl magnesium bromide, which is then treated with acetone (in a Grignard reaction), followed by aqueous acidic workup, to give the desired tertiary alcohol.

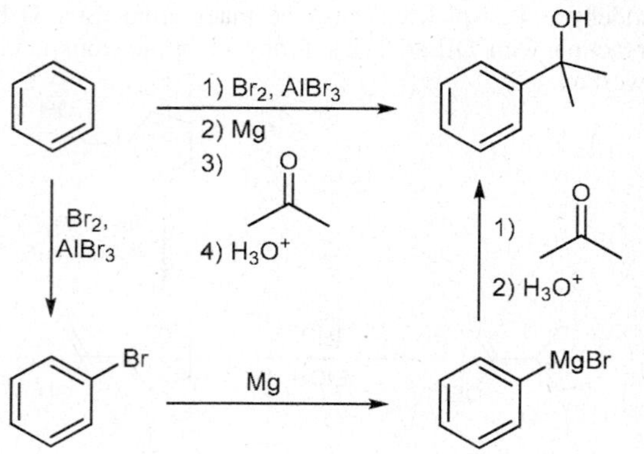

(b) There are certainly many acceptable solutions to this problem. One such solution derives from the following retrosynthetic analysis. An explanation of each of the steps (*a-e*) follows.

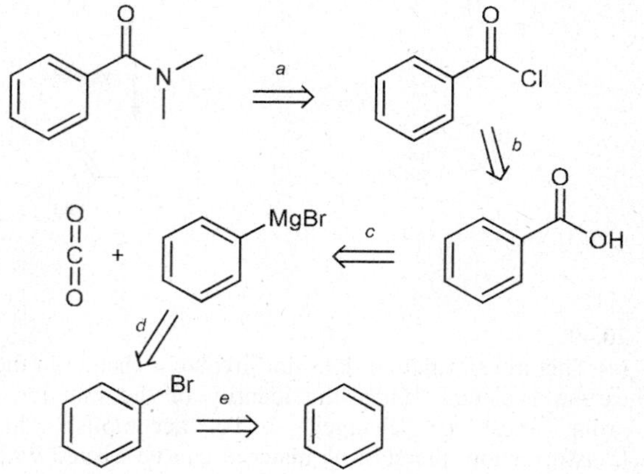

a. The desired amide can be made from the corresponding acid chloride (benzoyl chloride).

b. Benzoyl chloride can be made from benzoic acid upon treatment with thionyl chloride.

c. Benzoic acid can be made from the reaction between phenyl magnesium bromide and carbon dioxide.

d. Phenyl magnesium bromide can be made from bromobenzene via insertion of magnesium.

e. Bromobenzene can be made from benzene via bromination.

Now let's draw the forward scheme. Benzene is converted into bromobenzene upon treatment with Br_2 and $AlBr_3$ (via an electrophilic aromatic substitution reaction). Bromobenzene is subsequently converted into phenyl magnesium bromide, which is then treated with carbon dioxide, followed by aqueous acidic workup, to give benzoic acid. Conversion to the acid chloride, followed by reaction with excess dimethylamine, yields the desired amide.

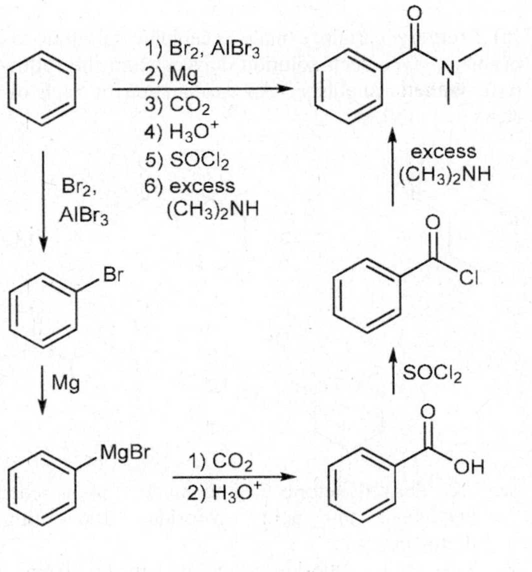

As mentioned, there are many alternative solutions. For example, benzoic acid can be made from benzene via Friedel-Crafts methylation, followed by benzylic oxidation.

(c) There are certainly many acceptable solutions to this problem. One such solution derives from the following retrosynthetic analysis. An explanation of each of the steps (*a-d*) follows.

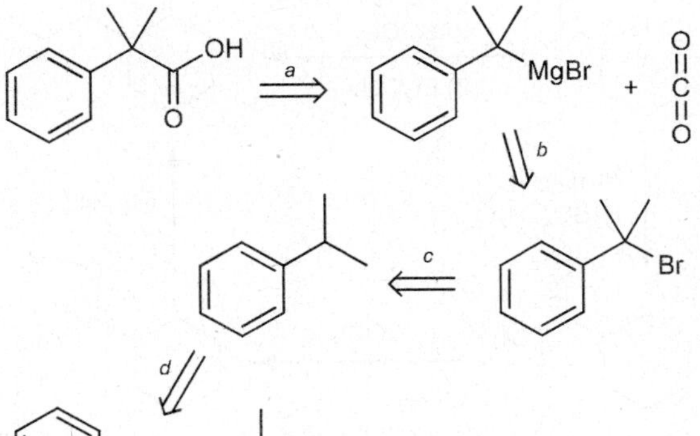

a. The desired carboxylic acid can be made from the reaction between a Grignard reagent and CO_2.
b. The Grignard reagent can be made from the corresponding tertiary benzylic halide.
c. The benzylic halide can be made from isopropyl benzene via benzylic bromination.
d. Isopropylbenzene can be made from benzene via a Friedel-Crafts alkylation.

Now let's draw the forward scheme. Benzene is converted into isopropyl benzene upon treatment with 2-chloropropane and a Lewis acid (via a Friedel-Crafts alkylation). Benzylic bromination replaces the benzylic hydrogen atom with a bromine atom. Conversion to a Grignard reagent, followed by reaction with carbon dioxide and subsequent workup, gives the desired carboxylic acid.

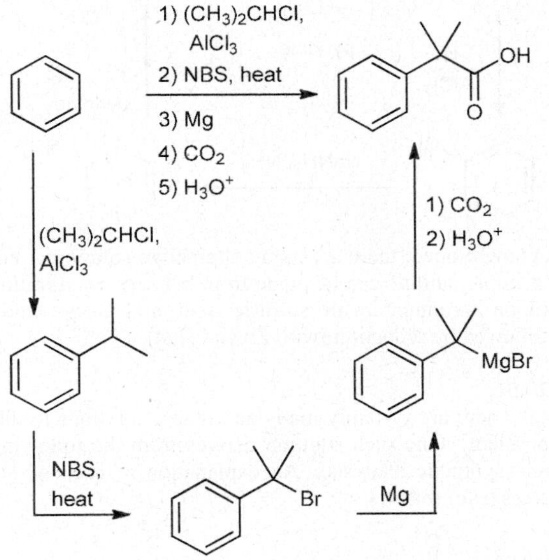

(d) There are certainly many acceptable solutions to this problem. One such solution derives from the following retrosynthetic analysis. An explanation of each of the steps (*a-c*) follows.

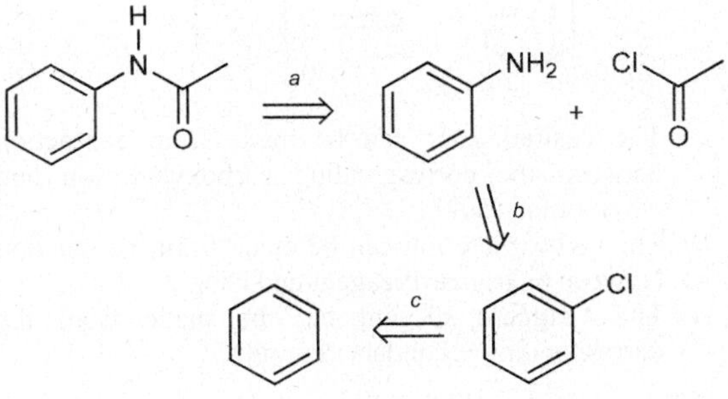

a. The desired amide can be made from aniline via an acylation reaction.
b. Aniline can be made from chlorobenzene via an elimination-addition process.
c. Chlorobenzene can be made from benzene via chlorination of the aromatic ring.

Now let's draw the forward scheme. Benzene is converted into chlorobenzene upon treatment with Cl_2 and $AlCl_3$ (via an electrophilic aromatic substitution reaction). Chlorobenzene is then converted to aniline via an elimination-addition reaction. Reaction with acetyl chloride (in the presence of pyridine) converts aniline to the desired amide.

1) Cl_2, $AlCl_3$
2) $NaNH_2$, NH_3
3) acetyl chloride, pyridine

Cl_2, $AlCl_3$

acetyl chloride, pyridine

$NaNH_2$, NH_3

As mentioned, there are many alternative solutions. For example, aniline can be made from benzene via nitration (upon treatment with sulfuric acid and nitric acid), followed by reduction (with Zn and HCl).

20.51.

(a) There are certainly many acceptable solutions to this problem. One such solution derives from the following retrosynthetic analysis. An explanation of each of the steps (*a-c*) follows.

a + EtI

b + CO_2

c

a. The desired ester can be made from a reaction between the corresponding carboxylate ion and ethyl iodide (SN2).

b. The carboxylate ion can be made from the reaction between a Grignard reagent and CO_2.

c. The Grignard reagent can be made from the corresponding secondary bromide.

Now let's draw the forward scheme. Bromocyclohexane is converted to a Grignard reagent, which subsequently reacts with carbon dioxide to produce a carboxylate ion. This anion then serves as a nucleophile in an SN2 reaction with iodoethane, giving the desired product.

1) Mg
2) CO_2
3) EtI

Mg

EtI

CO_2

MgBr

(b) There are certainly many acceptable solutions to this problem. One such solution derives from the following retrosynthetic analysis. An explanation of each of the steps (*a-c*) follows.

a + Et_2CuLi

b

c

a. The desired ketone can be made from a reaction between an acid chloride and lithium diethylcuprate.

b. The acid chloride can be made from the corresponding carboxylic acid.

c. The carboxylic acid can be made from the corresponding primary alcohol via oxidation.

Now let's draw the forward scheme. The primary alcohol is oxidized to the carboxylic acid, and subsequently converted to the acid chloride upon treatment with thionyl chloride. Reaction with lithium diethylcuprate then produces the desired ketone.

1) $Na_2Cr_2O_7$, H_2SO_4, H_2O
2) $SOCl_2$
3) Et_2CuLi

$Na_2Cr_2O_7$, H_2SO_4, H_2O

Et_2CuLi

$SOCl_2$

(c) There are certainly many acceptable solutions to this problem. One such solution derives from the following retrosynthetic analysis. An explanation of each of the steps (*a-f*) follows.

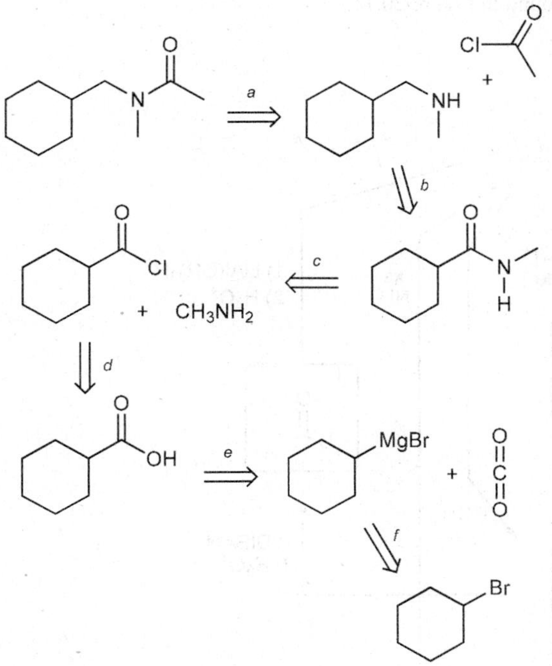

a. The desired amide can be made via acetylation of the corresponding secondary amine.
b. The secondary amine can be made via reduction of the corresponding amide.
c. The amide can be made from the corresponding acid halide (upon treatment with excess methyl amine).
d. The acid halide can be made from the corresponding carboxylic acid, upon treatment with thionyl chloride.
e. The carboxylic acid can be made from the reaction between a Grignard reagent and CO_2.
f. The Grignard reagent can be made from the corresponding secondary alkyl bromide.

Now let's draw the forward scheme. Bromocyclohexane is converted to a Grignard reagent, which subsequently reacts with carbon dioxide to produce the carboxylic acid (after aqueous acidic workup). Conversion to the acid chloride, followed by reaction with excess methylamine, yields an intermediate amide. This intermediate is then reduced to the corresponding amine. Finally, reaction with acetyl chloride (in the presence of pyridine) produces the desired product.

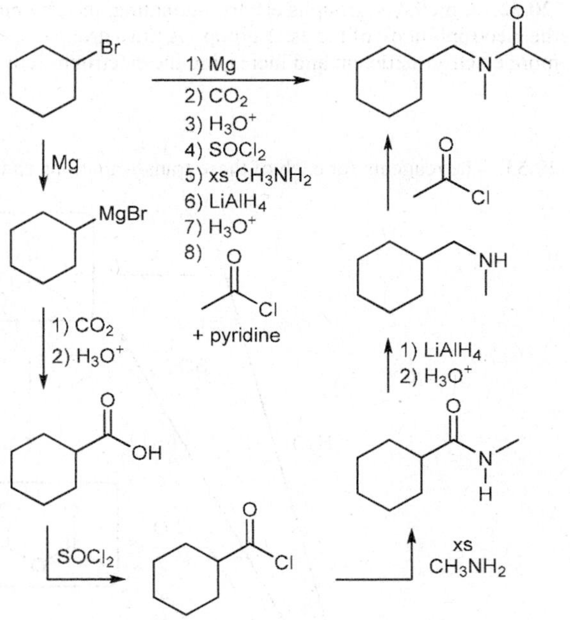

1) Mg
2) CO_2
3) H_3O^+
4) $SOCl_2$
5) xs CH_3NH_2
6) $LiAlH_4$
7) H_3O^+
8) acetyl chloride + pyridine

(d) There are certainly many acceptable solutions to this problem. One such solution derives from the following retrosynthetic analysis. An explanation of each of the steps (*a-b*) follows.

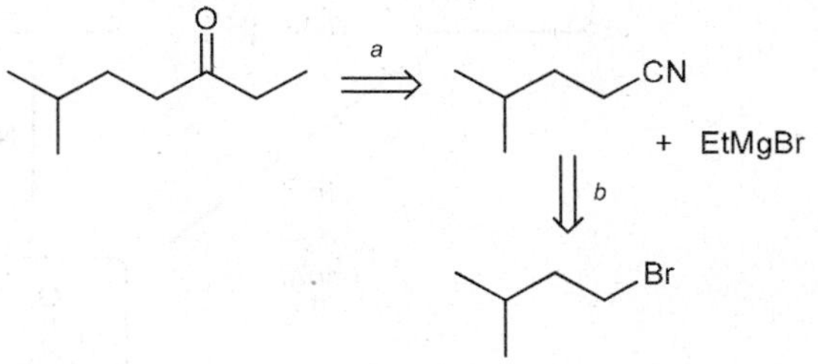

a. The desired ketone can be made from the reaction between a Grignard reagent and a nitrile.
b. The nitrile can be made from the corresponding primary bromide via an S_N2 process.

Now let's draw the forward scheme. Reaction of 1-bromo-3-methylbutane with sodium cyanide produces the nitrile, which is subsequently converted to the desired ketone upon treatment with ethyl magnesium bromide, followed by aqueous acid, as shown.

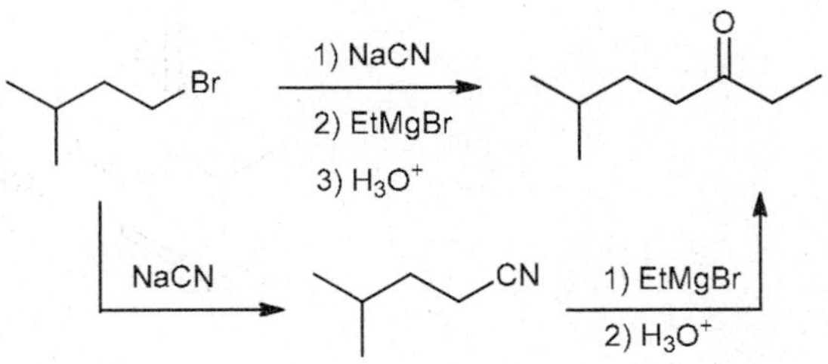

1) NaCN
2) EtMgBr
3) H_3O^+

20.52. A methoxy group is electron-donating, thereby making the carbonyl group more electron-rich and decreasing the electrophilicity of the ester group. A nitro group is electron-withdrawing, thereby making the carbonyl group even more electron-deficient and increasing the electrophilicity of the ester group.

20.53. The reagents for each of these transformations can be found in Figure 20.11.

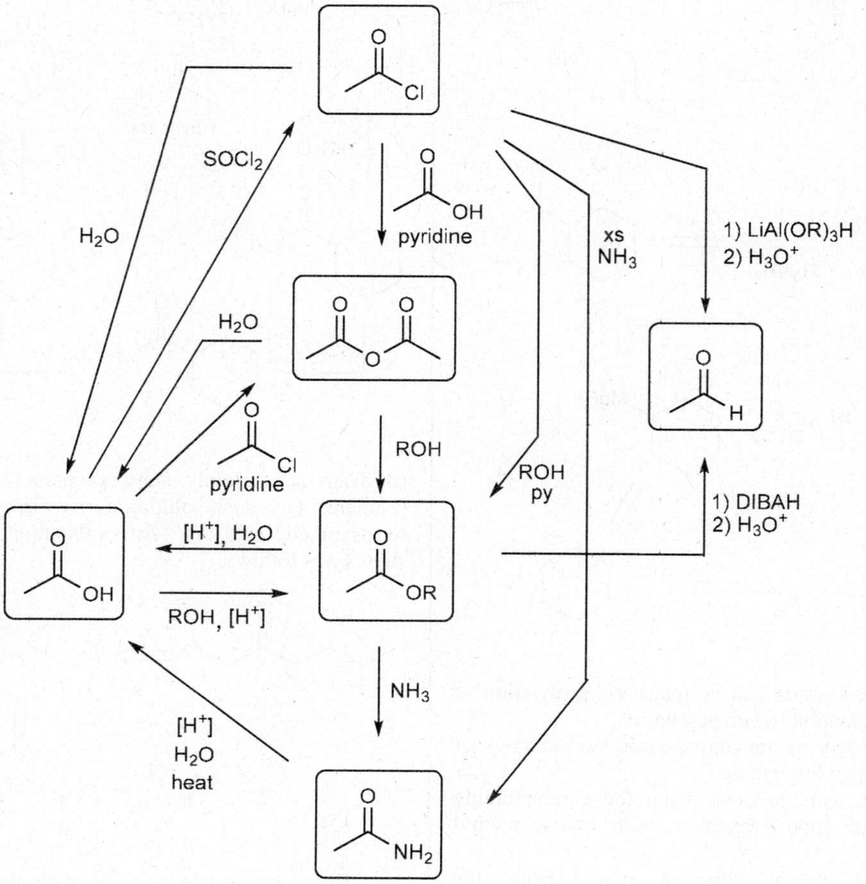

$SOCl_2$
pyridine
H_2O
H_2O
pyridine
$[H^+]$, H_2O
ROH, $[H^+]$
ROH
ROH
py
xs
NH_3
1) $LiAl(OR)_3H$
2) H_3O^+
1) DIBAH
2) H_3O^+
NH_3
$[H^+]$
H_2O
heat

20.54. There are certainly many acceptable solutions to this problem. One such solution derives from the following retrosynthetic analysis. An explanation of each of the steps (*a-d*) follows.

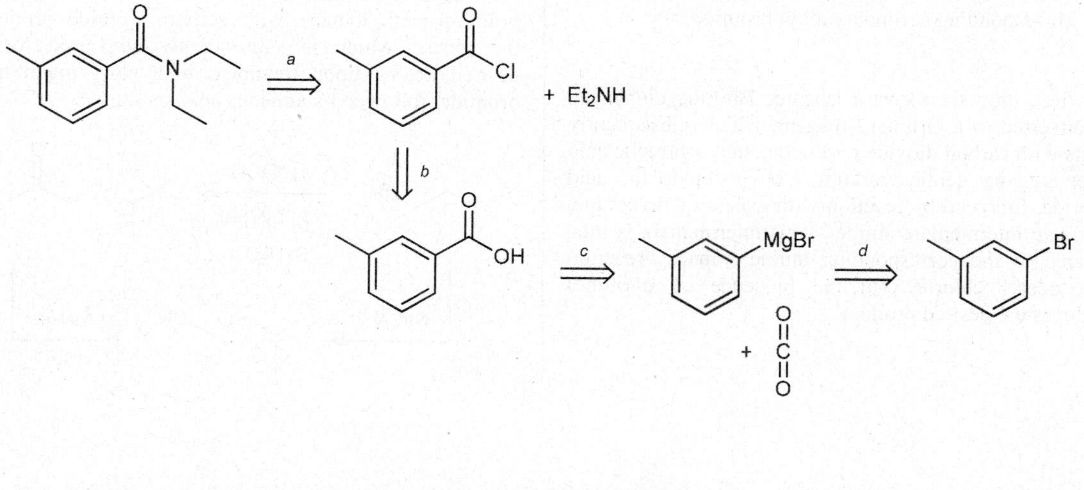

a
+ Et_2NH
b
c
d
MgBr
Br
+

a. The desired amide can be made from the corresponding acid halide (upon treatment with excess diethylamine).
b. The acid halide can be made from the corresponding carboxylic acid, upon treatment with thionyl chloride.
c. The carboxylic acid can be made from the reaction between a Grignard reagent and CO_2.
d. The Grignard reagent can be made from the starting material, via insertion of magnesium.

Now let's draw the forward scheme. Conversion of *meta*-bromotoluene to a Grignard reagent, followed by reaction with carbon dioxide and subsequent acidification, produces *meta*-methylbenzoic acid. The carboxylic acid is converted to the acid chloride, which then reacts with excess diethylamine to produce the desired amide, DEET.

Br

1) Mg
2) CO_2
3) H_3O^+
4) $SOCl_2$
5) xs Et_2NH

O
N

Mg

xs Et_2NH

MgBr

1) CO_2
2) H_3O^+

O
OH

$SOCl_2$

O
Cl

20.55. The carbonyl group of diphenyl carbonate has two phenoxide groups attached to it. Each of these groups can serve as a leaving group in a nucleophilic acyl substitution reaction. Accordingly, the first equivalent of methyl magnesium bromide replaces one phenoxide leaving group in the first two steps of the mechanism (nucleophilic attack, followed by loss of the leaving group) to produce an ester intermediate. The second equivalent of methyl magnesium bromide then replaces the second phenoxide group in an analogous manner (nucleophilic attack, followed by loss of the leaving group) to produce a ketone intermediate (acetone). A third equivalent of methyl magnesium bromide then attacks the carbonyl group of the ketone to produce *tert*-butoxide. Workup with aqueous acid (H_3O^+) protonates *tert*-butoxide, as well as the phenoxide ions, giving *tert*-butanol and two equivalents of phenol.

O O
O

1) excess MeMgBr
2) H_3O^+

OH

+ 2

OH

20.56. A mechanism for this reaction is shown below, in which the isotopically labeled ^{18}O atom is highlighted with a gray box. Protonation of the carbonyl group activates it toward nucleophilic attack by $^{18}OH_2$. Two successive proton transfers are followed by expulsion of the non-labeled oxygen atom (as a leaving group, H_2O). The loss of water from the charged tetrahedral intermediate is assisted by either one of the attached hydroxyl groups, leading to the labeled oxygen atom being in either position of the final product (after a final deprotonation step).

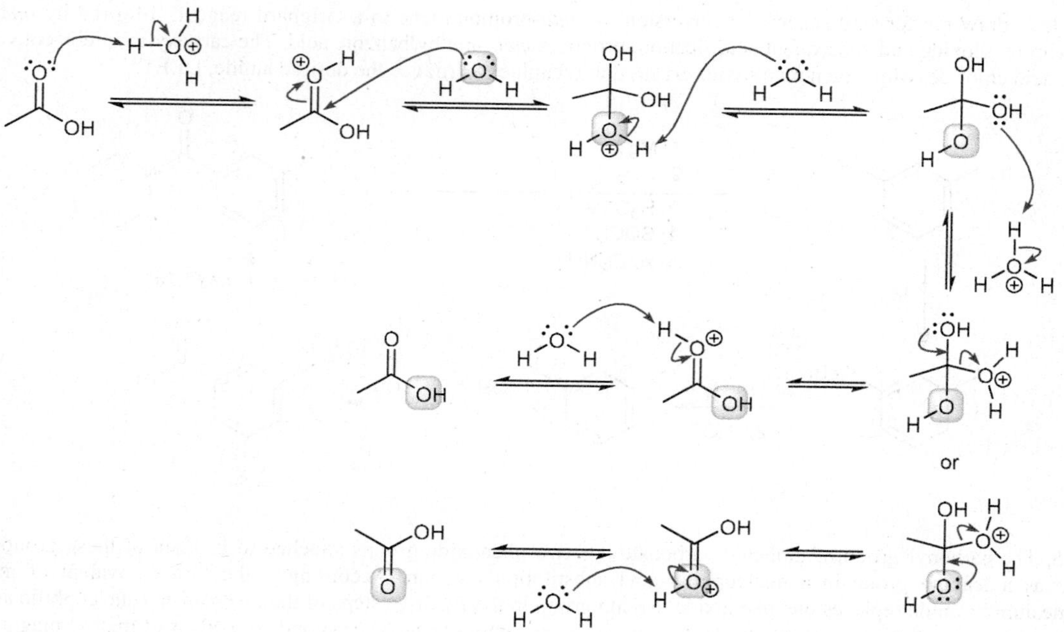

20.57.

(a) An alcohol (ROH) is used as a representative nucleophile in the mechanism below, which has six steps: 1) nucleophilic attack, 2) loss of a leaving group, 3) proton transfer, 4) nucleophilic attack, 5) loss of a leaving group and 6) proton transfer. ROH attacks the carbonyl group of phosgene to form a tetrahedral intermediate which subsequently expels a chloride ion. A proton transfer produces the neutral, monochlorinated intermediate shown below. A second round of these three steps (nucleophilic attack by ROH, loss of chloride and proton transfer) results in the overall substitution of the second chloride ion with the oxygen atom from the alcohol, thereby producing a carbonate ester, as shown.

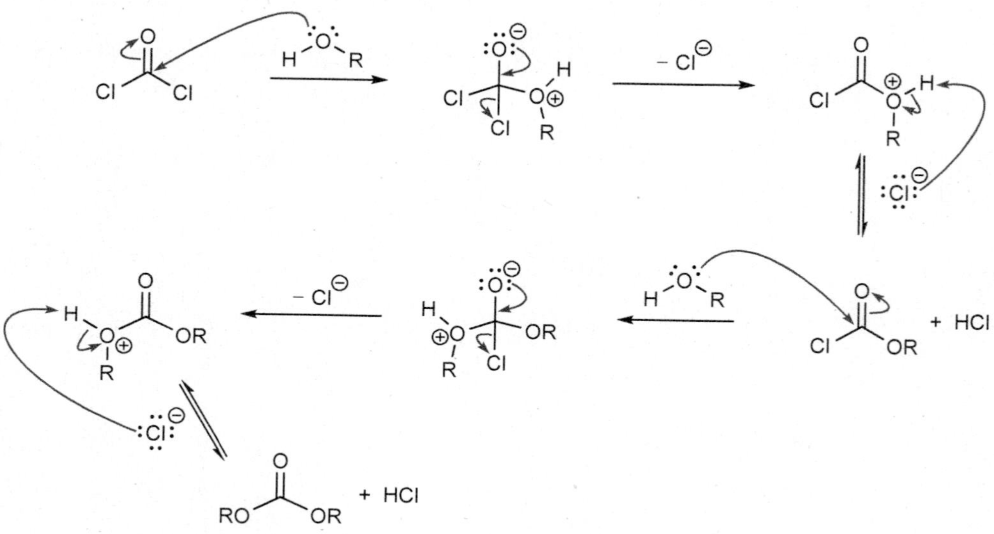

(b) The molecule below is produced from two acyl substitution reactions between phosgene and ethylene glycol via a mechanism analogous to the one in part (a) above. In this case, the second nucleophilic attack (step 4) is an intramolecular reaction, leading to the cyclic product shown here.

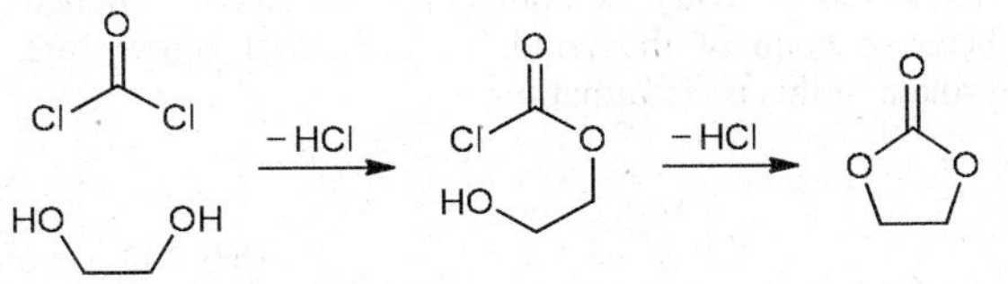

(c) Excess phenyl magnesium bromide reacts with phosgene to produce a tertiary alcohol (shown below), which results from the incorporation of three molar equivalents of the Grignard reagent. The first equivalent attacks the carbonyl group to produce a tetrahedral anionic intermediate; subsequent expulsion of a chloride ion gives an acid chloride intermediate. Likewise, a second equivalent of the Grignard reagent attacks the carbonyl group, producing a second anionic tetrahedral intermediate. Subsequent expulsion of a chloride leaving group gives a ketone intermediate, which is attacked by a third equivalent of phenyl magnesium bromide. Protonation of the resulting alkoxide ion (upon workup with aqueous acid) gives the tertiary alcohol, shown here.

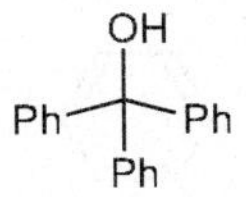

20.58.

(a) Hydrolysis of an ester group produces an alcohol and a carboxylic acid.

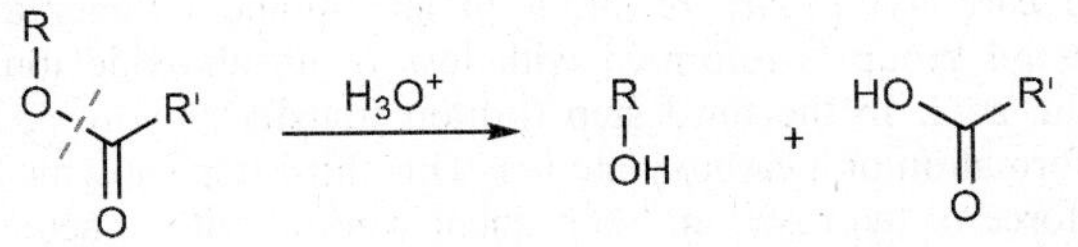

As such, hydrolysis of fluphenazine decanoate releases the hydrophobic chain as a carboxylic acid, giving the following primary alcohol:

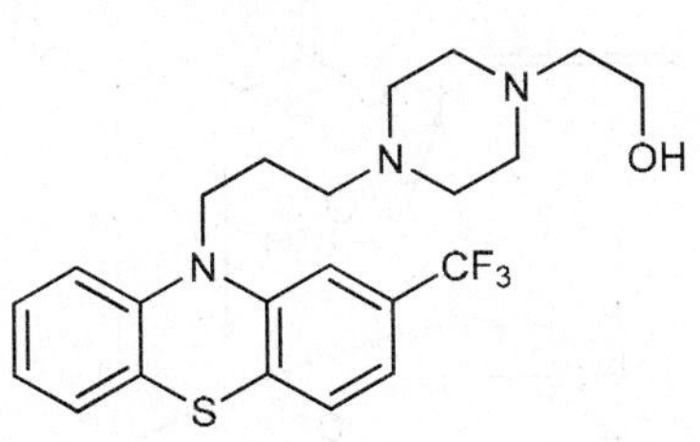

(b) The by-product of the reaction is a carboxylic acid containing ten carbon atoms. The "e" at the end of the parent alkane name (decane) is replaced with the suffix "-oic acid" to give the IUPAC name decanoic acid.

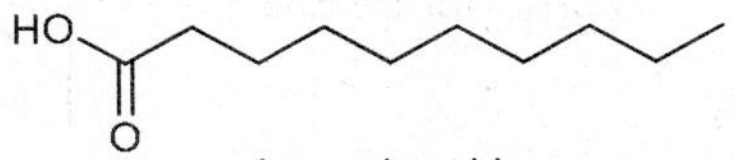

decanoic acid

20.59. There are certainly many acceptable solutions to this problem. One such solution derives from the following retrosynthetic analysis. An explanation of each of the steps (*a-d*) follows.

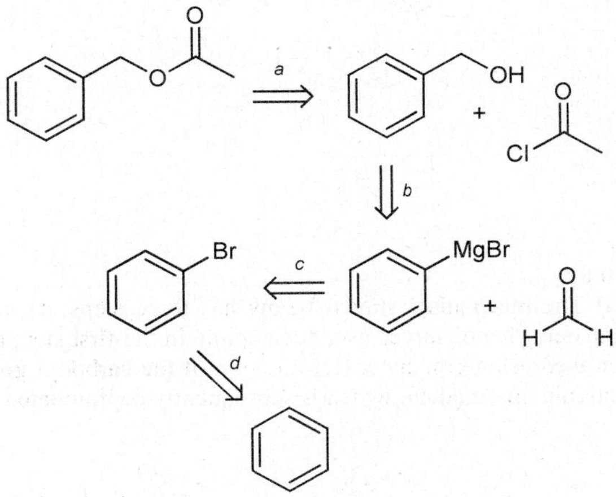

a. The ester can be made via acylation of the corresponding alcohol (benzyl alcohol).
b. The alcohol can be made from the reaction between phenyl magnesium bromide and formaldehyde.
c. Phenyl magnesium bromide can be made from bromobenzene, via insertion of magnesium.
d. Bromobenzene can be made from benzene via bromination of the aromatic ring.

Now let's draw the forward scheme. Benzene is converted into bromobenzene upon treatment with Br_2 and $AlBr_3$ (via an electrophilic aromatic substitution reaction). Bromobenzene is then converted to phenyl magnesium bromide (a Grignard reagent), which is then treated with formaldehyde, followed by aqueous acidic workup, to give benzyl alcohol. This alcohol then serves as a nucleophile in a subsequent acyl substitution reaction with acetyl chloride and pyridine to produce benzyl acetate (a process called acylation).

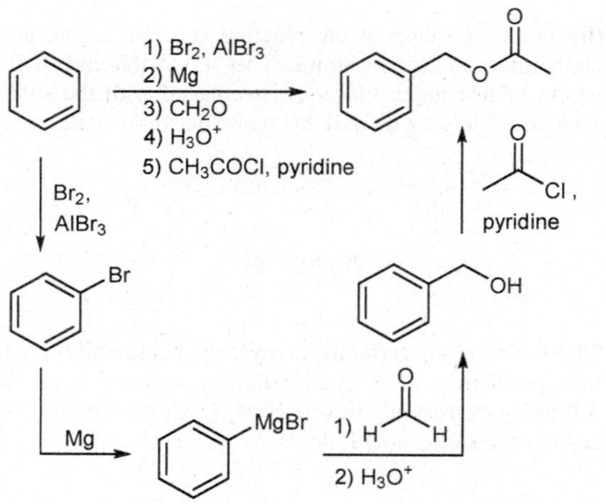

20.60. Hydrolysis of aspartame hydrolyzes both the amide group and the ester group in the molecule. Hydrolysis of the ester group produces methanol and the carboxylic acid group in phenylalanine, shown below. Hydrolysis of the amide group converts this group to an amine (highlighted in phenylalanine) and a carboxylic acid (highlighted in aspartic acid). Note that the stereochemistry at both chiral centers is retained because none of the bonds to the chiral centers are broken in this transformation.

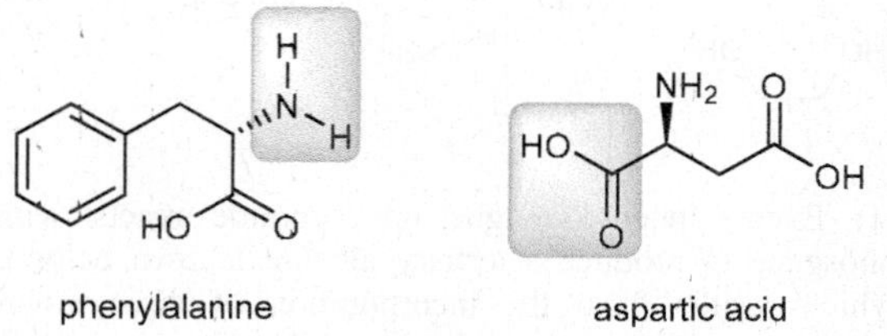

phenylalanine

aspartic acid

20.61.

(a) The mechanism shown below has three steps: 1) nucleophilic attack, 2) loss of a leaving group and 3) proton transfer. Phenol serves as a nucleophile in the first step, attacking the carbonyl group of the acid chloride to produce a tetrahedral intermediate. Reformation of the carbonyl group and expulsion of the leaving group (chloride) produces a cationic intermediate which is subsequently deprotonated by pyridine to give an ester.

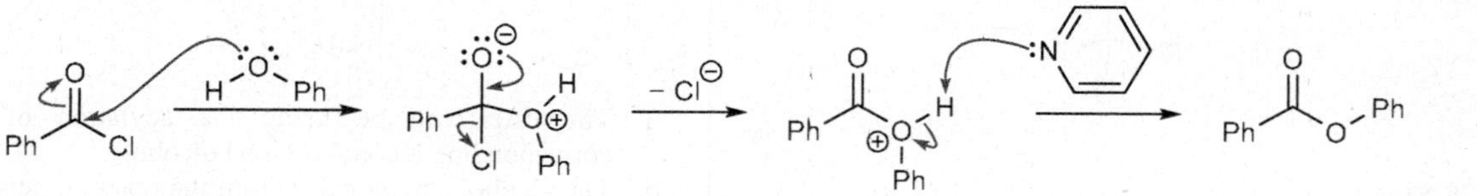

(b) The first part of this reaction (saponification) has three mechanistic steps: 1) nucleophilic attack, 2) loss of a leaving group and 3) proton transfer. The first step (nucleophilic attack) requires two curved arrows, which show hydroxide functioning as a nucleophile and attacking the electrophilic carbonyl group, resulting in an anionic tetrahedral intermediate. In step two (loss of a leaving group), the carbonyl group is reformed with loss of an alkoxide ion, resulting in opening of the ring and formation of a carboxylic acid. In the third step (proton transfer), hydroxide deprotonates the carboxylic acid intermediate, resulting in the formation of a carboxylate ion. This third step (in which a strong base deprotonates the carboxylic acid) is the driving force of the reaction. Subsequent workup with aqueous acid protonates the alkoxide and the carboxylate ion, regenerating the carboxylic acid.

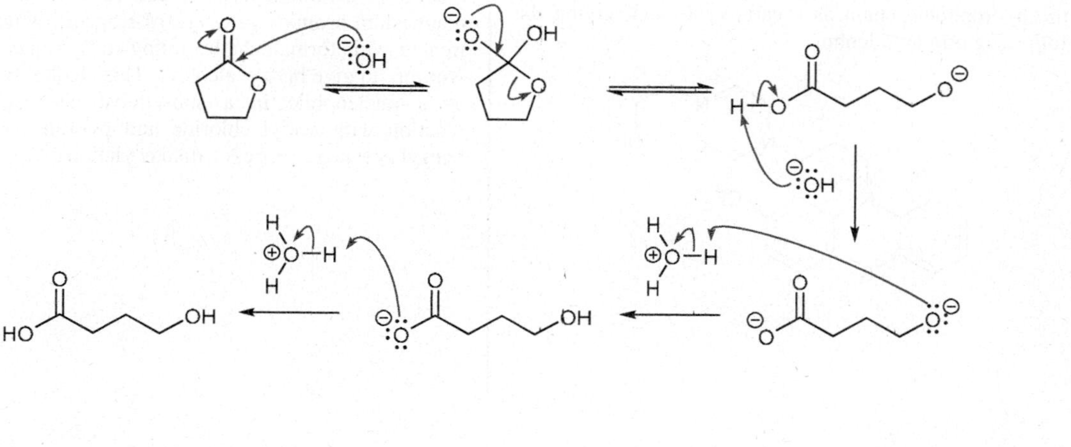

(c) The first part of this reaction (saponification) has three mechanistic steps: 1) nucleophilic attack, 2) loss of a leaving group and 3) proton transfer. The first step (nucleophilic attack) requires two curved arrows, which show hydroxide functioning as a nucleophile and attacking the electrophilic carbonyl group, resulting in an anionic tetrahedral intermediate. In step two (loss of a leaving group), the carbonyl group is reformed with loss of an alkoxide ion, resulting in opening of the ring and formation of a carboxylic acid. In the third step (proton transfer), hydroxide deprotonates the carboxylic acid intermediate, resulting in the formation of a carboxylate ion. This third step (in which a strong base deprotonates the carboxylic acid) is the driving force of the reaction. Subsequent workup with aqueous acid protonates the alkoxide and the carboxylate ion, regenerating the carboxylic acid.

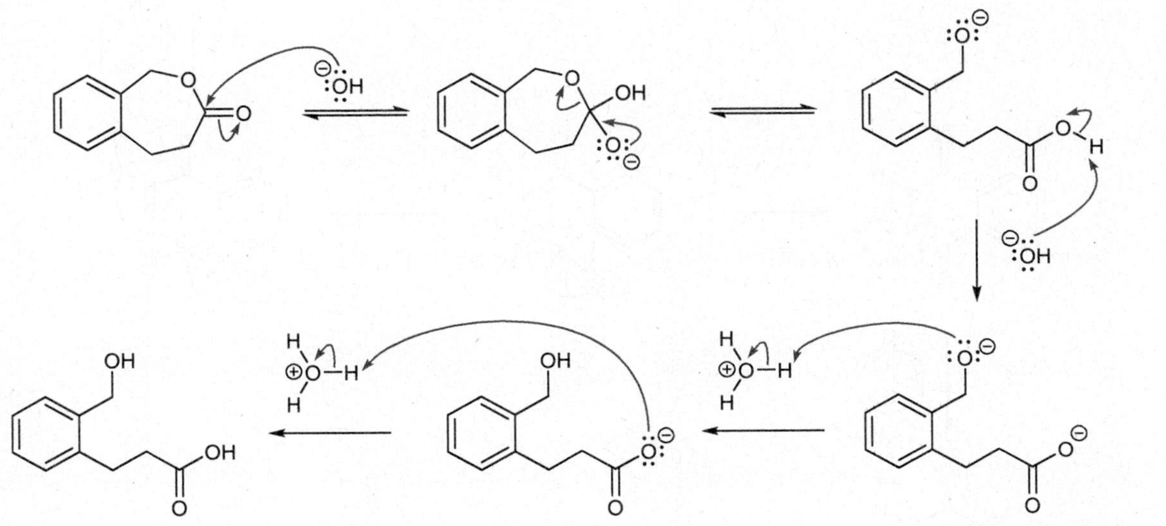

(d) The mechanism shown below has six steps: 1) nucleophilic attack, 2) loss of a leaving group, 3) proton transfer, 4) nucleophilic attack, 5) loss of a leaving group and 6) proton transfer. Hydrazine (NH_2NH_2) has two nucleophilic centers (each nitrogen atom has a lone pair), each of which can attack an acid chloride group. In the first step, hydrazine attacks one of the acid chloride groups to form a tetrahedral intermediate which subsequently expels a leaving group (chloride). Pyridine then functions as a base and removes a proton, giving an intermediate that bears no formal charges. Subsequent intramolecular nucleophilic attack by the second nitrogen atom on the second acid chloride group, followed by loss of chloride, results in an intermediate that is deprotonated to give the product. Once again, pyridine functions as the base for deprotonation.

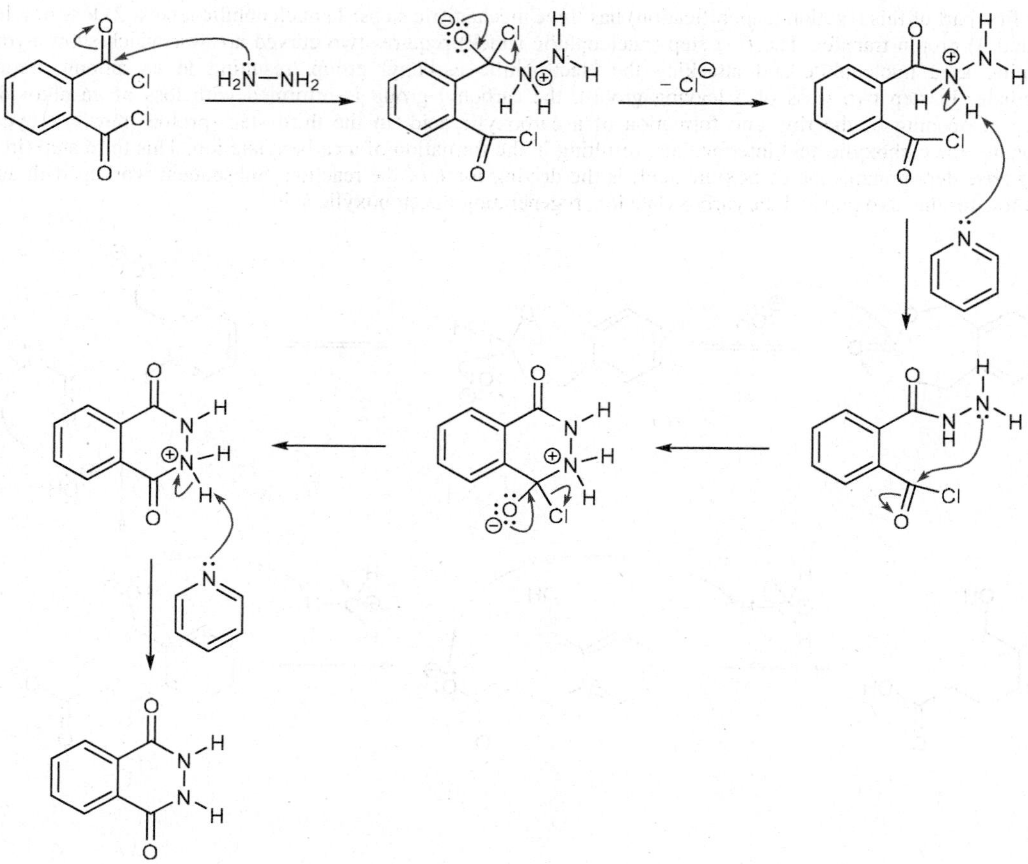

(e) The mechanism shown below has five steps: 1) nucleophilic attack, 2) loss of a leaving group, 3) nucleophilic attack, 4) proton transfer and 5) proton transfer. In the first step (nucleophilic attack), the anionic carbon atom of ethyl magnesium bromide serves as a nucleophile and attacks the electrophilic carbonyl group, resulting in a tetrahedral intermediate. In step two (loss of a leaving group), the carbonyl group is reformed with loss of an alkoxide ion, resulting in opening of the ring. In step three (nucleophilic attack) a second equivalent of the Grignard reagent attacks the carbonyl intermediate to produce a dianion.

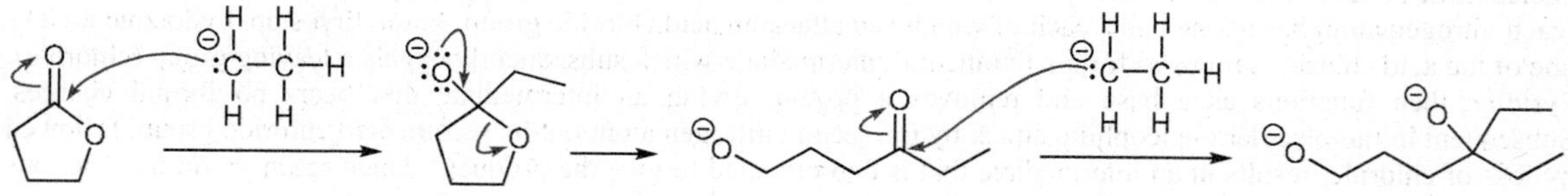

After the reaction is complete, a proton source is introduced into the reaction flask, thereby protonating the dianion. Each anion is protonated separately, so two separate steps are required.

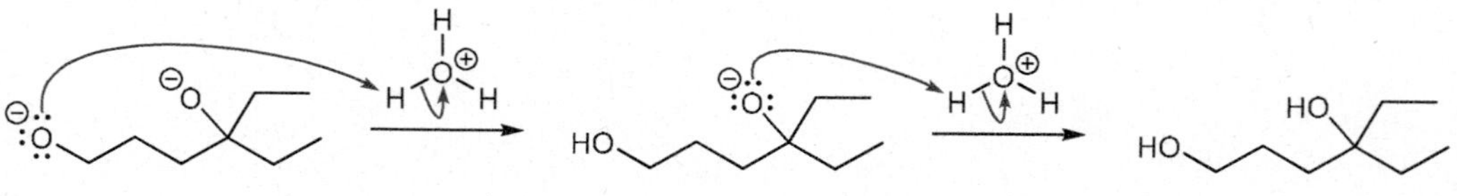

20.62. The three chlorine atoms withdraw electron density via induction. This effect renders the carbonyl group more electrophilic, and thus more reactive toward hydrolysis.

20.63. When treated with aqueous acid, each of the C–O bonds (on either side of the carbonyl group) is expected to undergo cleavage via an acid-catalyzed nucleophilic acyl substitution reaction. This produces a diol, shown below:

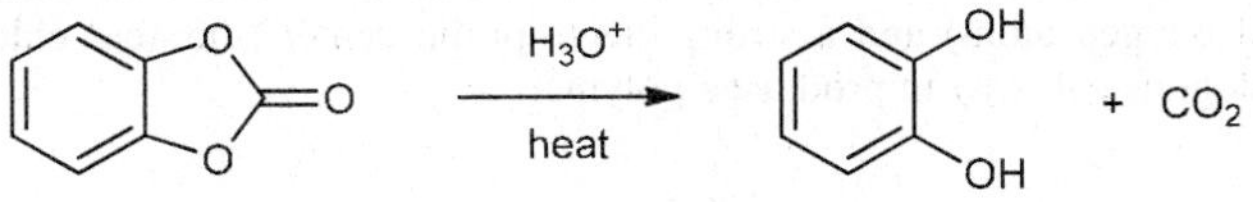

H_3O^+ / heat → OH, OH + CO_2

20.64.
(a) When treated with aqueous acid, the protecting group on the carboxylic acid is removed, giving the active drug shown below. Note that the configuration of each chiral center is conserved, because the bonds to those chiral centers were not involved in the reaction.

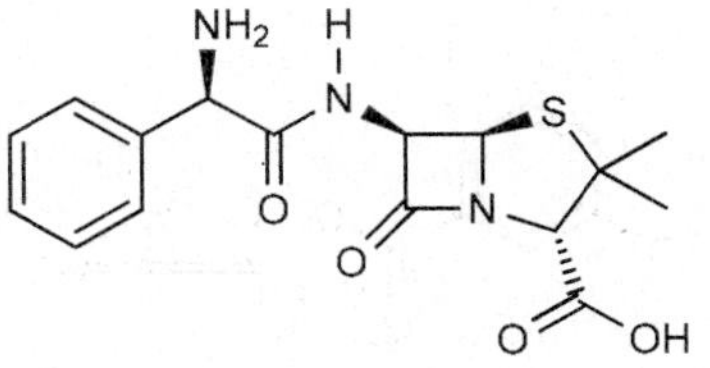

NH_2 H N S N O O O OH

(b) The active drug is ampicillin, as indicated in the BioLinks box.

20.65. Hydrolysis of the ester groups in Dexon results in cleavage of the bonds indicated by the arrows in the figure below. This results in the formation of glycolic acid. The IUPAC name is based on a parent carboxylic acid called acetic acid, with indication of an alcohol (hydroxy group) at C2. Since acetic acid only has one location that can bear a substituent, a locant is not required to indicate the position of the OH group. The IUPAC name is thus hydroxyacetic acid (or hydroxyethanoic acid).

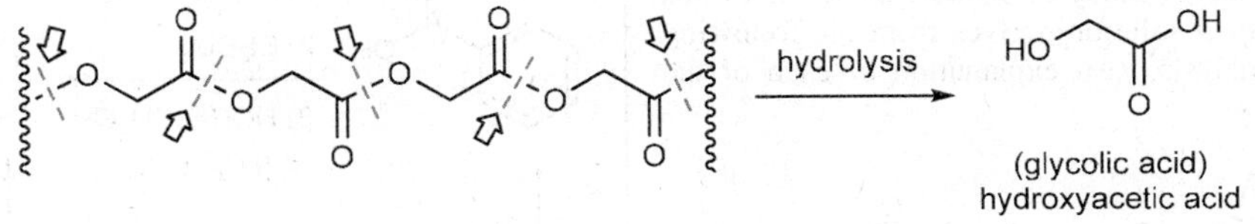

hydrolysis → HO OH O

(glycolic acid)
hydroxyacetic acid

20.66. Each of the monomers bears two identical functional groups. The first monomer has two acid chloride groups, each of which can serve as an electrophile in the polymerization reaction. The second monomer has two OH groups, each of which can serve as a nucleophile. Reaction of the two monomers thus produces the polymer below via a series of nucleophilic acyl substitution reactions. The monomeric origins of each section of the polyester polymer are highlighted below.

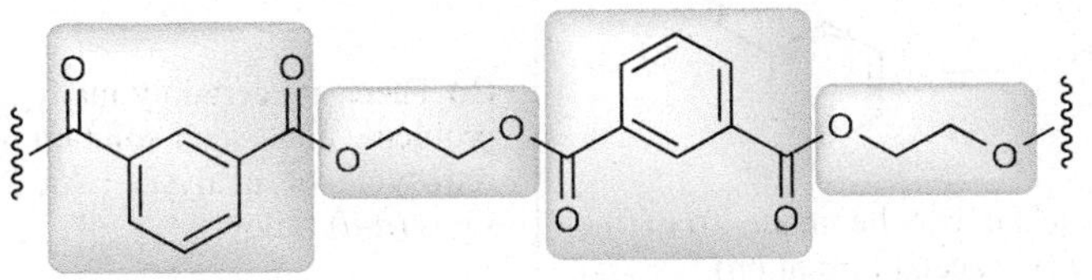

20.67. A retrosynthetic analysis of the polymer is shown below, where each of the bonds made during polymerization is indicated by an arrow. Each of these bonds can be made from a nucleophilic acyl substitution reaction between an amine and an acid chloride. Reaction of one monomer (bearing two electrophilic acid chloride groups) with the other monomer (bearing two nucleophilic amino groups) will produce the desired polymer.

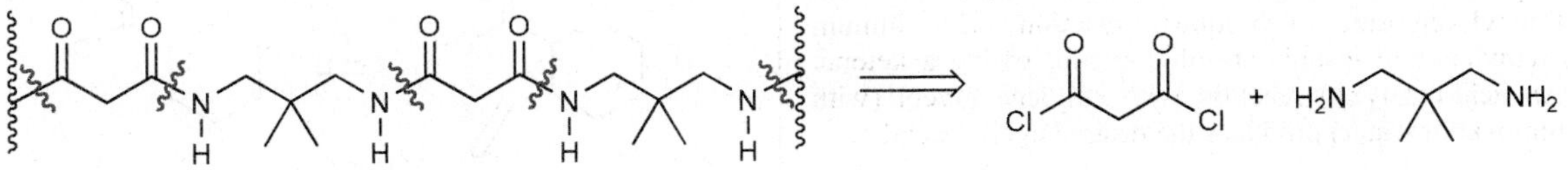

⟹ Cl Cl + H_2N NH_2

20.68. *meta*-Hydroxybenzoyl chloride (structure below) has a nucleophilic center (the OH group) as well as a strong electrophilic center (the acid chloride group) in a single molecule, thus making it susceptible to facile polymerization via the mechanism below. Each nucleophilic acyl substitution reaction has three steps: 1) nucleophilic attack, 2) loss of a leaving group and 3) proton transfer. In the first step (nucleophilic attack), the phenol oxygen atom of one molecule attacks the electrophilic carbonyl group on a second molecule resulting in the formation of a tetrahedral intermediate. A chloride leaving group is expelled in step 2 (loss of a leaving group), along with reformation of the carbonyl group. A proton transfer gives an intermediate that does not bear any formal charges. This intermediate (like the reactant) has a nucleophilic center (phenol oxygen atom) and a strong electrophilic center (the acid chloride group), thus allowing further reactions via an analogous pathway to produce a polymer.

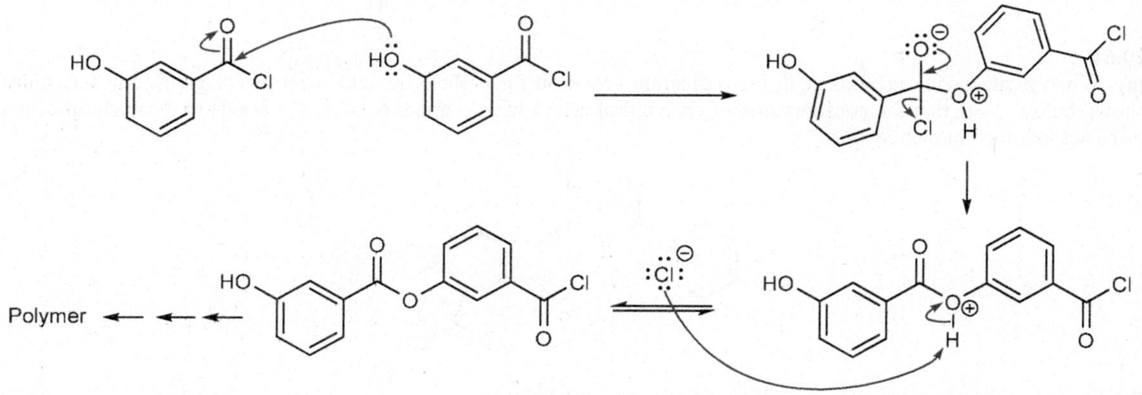

Polymer

20.69.

(a) There are certainly many acceptable solutions to this problem. One such solution derives from the following retrosynthetic analysis. An explanation of each of the steps (*a-c*) follows.

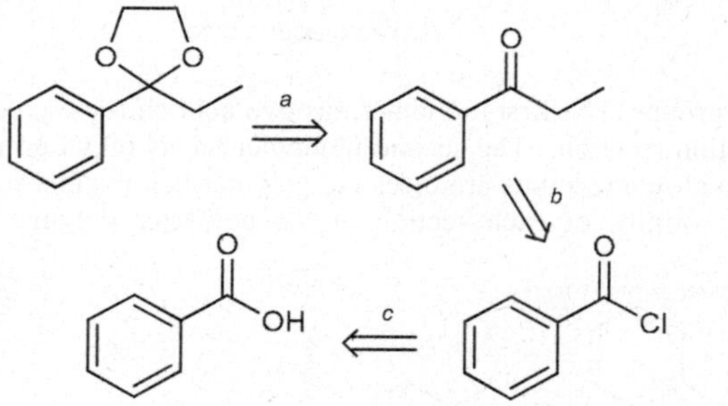

a. The desired cyclic acetal can be made from the corresponding ketone (via acetal formation).

b. The ketone can be made from benzoyl chloride, upon treatment with lithium diethylcuprate.

c. The acid halide can be prepared from benzoic acid.

Now let's draw the forward scheme. Benzoic acid is converted to benzoyl chloride upon treatment with thionyl chloride. Subsequent reaction with lithium diethylcuprate installs an ethyl group, giving a ketone. An acid-catalyzed reaction with ethylene glycol (with removal of water) produces the desired cyclic acetal.

1) $SOCl_2$
2) Et_2CuLi
3) $HOCH_2CH_2OH$, $[H^+]$, $-H_2O$

$SOCl_2$

Et_2CuLi

$HOCH_2CH_2OH$, $[H^+]$, $-H_2O$

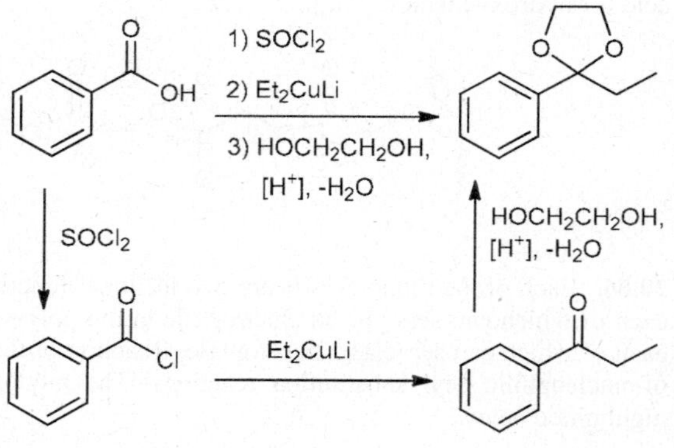

(b) There are certainly many acceptable solutions to this problem. One such solution derives from the following retrosynthetic analysis. An explanation of each of the steps (*a-d*) follows.

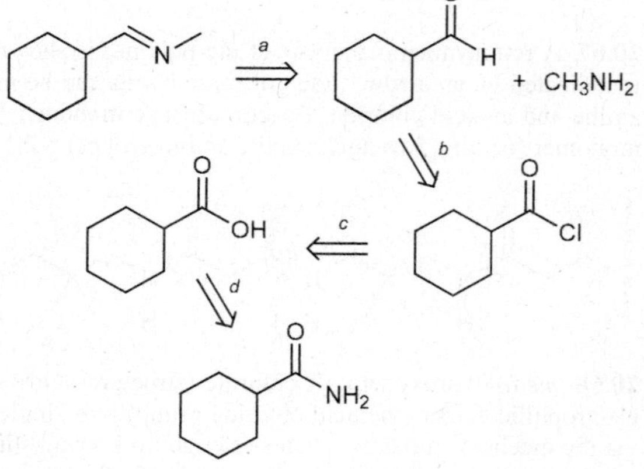

+ CH_3NH_2

a. The desired imine can be made from the corresponding aldehyde.
b. The aldehyde can be made from the corresponding acid chloride, upon treatment with $LiAl(OR)_3H$, followed by aqueous acidic workup.
c. The acid chloride can be prepared from the corresponding carboxylic acid, upon treatment with thionyl chloride.
d. The carboxylic acid can be made via hydrolysis of the starting amide.

Now let's draw the forward scheme. Acid catalyzed hydrolysis of the amide gives a carboxylic acid which is then converted to the acid chloride upon treatment with thionyl chloride. Reaction with a lithium trialkoxyaluminum hydride, followed by aqueous acidic workup, produces the aldehyde. Subsequent treatment of the aldehyde with methylamine under acid-catalyzed conditions (with removal of water) gives the desired imine.

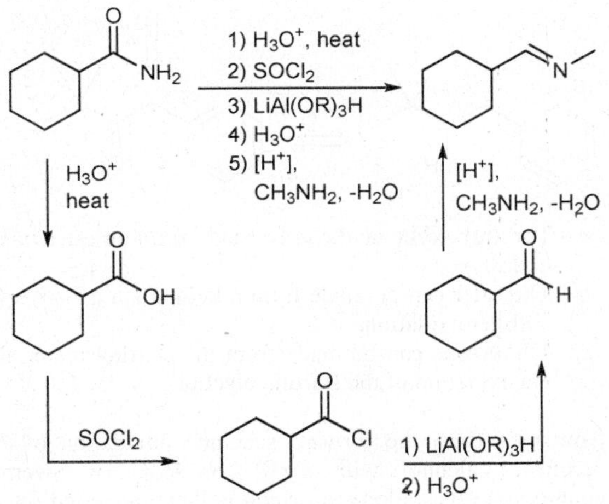

1) H_3O^+, heat
2) $SOCl_2$
3) $LiAl(OR)_3H$
4) H_3O^+
5) $[H^+]$, CH_3NH_2, $-H_2O$

H_3O^+ heat

$[H^+]$, CH_3NH_2, $-H_2O$

$SOCl_2$

1) $LiAl(OR)_3H$
2) H_3O^+

(c) There are certainly many acceptable solutions to this problem. One such solution derives from the following retrosynthetic analysis. An explanation of each of the steps (*a-d*) follows.

a b c d

+ $(CH_3)_2NH$

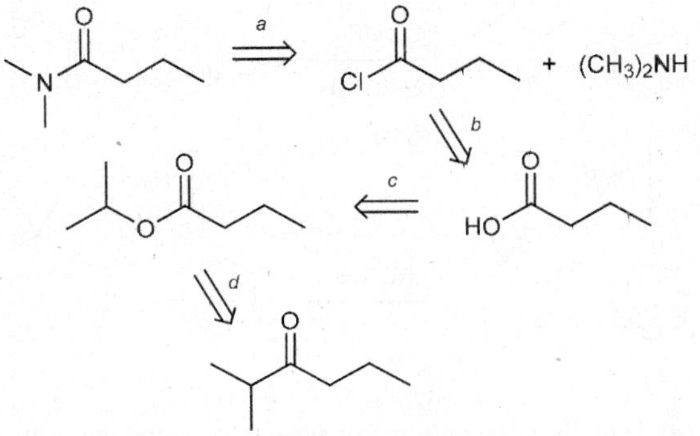

a. The desired amide can be made from the corresponding acid halide upon treatment with excess dimethylamine.
b. The acid halide can be made from the corresponding carboxylic acid, upon treatment with thionyl chloride.
c. The carboxylic acid can be prepared via hydrolysis of an ester.
d. The ester can be made from the starting ketone via a Baeyer-Villiger oxidation.

Now let's draw the forward scheme. Baeyer-Villiger oxidation of the starting ketone achieves the insertion of an oxygen atom between the carbonyl group and the more substituted alkyl group, thereby giving an ester. Acid-catalyzed hydrolysis of the ester gives butyric acid, which is then converted to the acid chloride upon treatment with thionyl chloride. The acid chloride is then converted into the desired product upon treatment with excess dimethylamine (via a nucleophilic acyl substitution reaction).

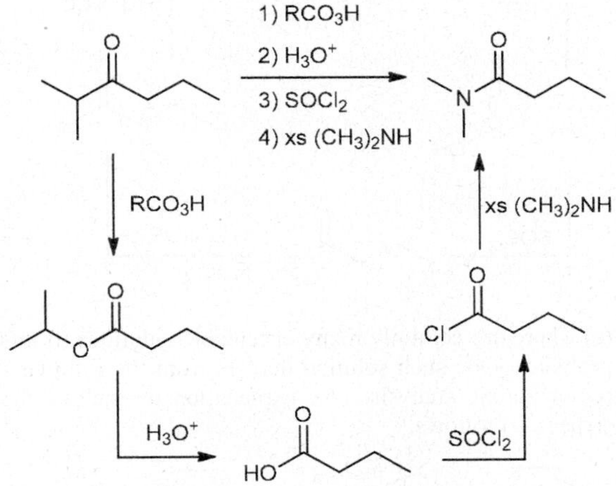

1) RCO_3H
2) H_3O^+
3) $SOCl_2$
4) xs $(CH_3)_2NH$

RCO_3H

xs $(CH_3)_2NH$

H_3O^+

$SOCl_2$

(d) There are certainly many acceptable solutions to this problem. One such solution derives from the following retrosynthetic analysis. An explanation of each of the steps (*a-d*) follows.

a b c d

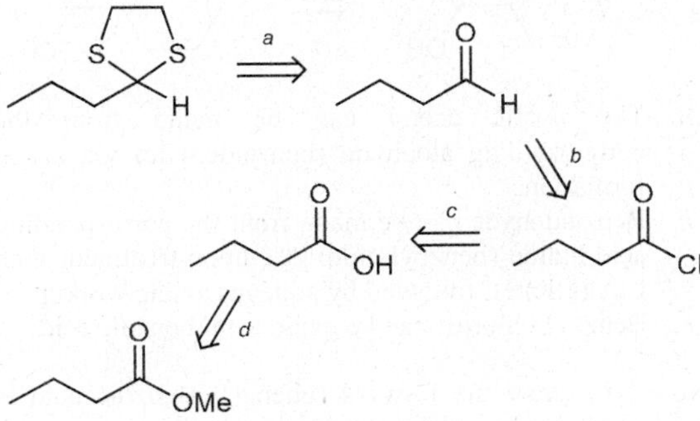

a. The cyclic thioacetal can be made from the corresponding aldehyde.
b. The aldehyde can be made from the corresponding acid halide, upon treatment with $LiAl(OR)_3H$, followed by aqueous acidic workup.

c. The acid halide can be made from the corresponding carboxylic acid, upon treatment with thionyl chloride.
d. The carboxylic acid can be made from the starting ester via hydrolysis.

Now let's draw the forward scheme. The ester undergoes hydrolysis upon treatment with aqueous acid, giving butyric acid, which is subsequently converted to an acid chloride upon treatment with thionyl chloride. Reaction with a lithium trialkoxyaluminum hydride produces an aldehyde, which is then converted into the desired cyclic thioacetal.

1) H_3O^+
2) $SOCl_2$
3) $LiAl(OR)_3H$
4) H_3O^+
5) $HSCH_2CH_2SH$, $[H^+]$, $-H_2O$

H_3O^+

$HSCH_2CH_2SH$, $[H^+]$, $-H_2O$

$SOCl_2$

1) $LiAl(OR)_3H$
2) H_3O^+

(e) There are certainly many acceptable solutions to this problem. One such solution derives from the following retrosynthetic analysis. An explanation of each of the steps (*a-c*) follows.

a. The cyclic acetal can be made from the corresponding aldehyde (benzaldehyde) via acetal formation.
b. Benzaldehyde can be made from the corresponding acid halide (benzoyl chloride), upon treatment with $LiAl(OR)_3H$, followed by aqueous acidic workup.
c. Benzoyl chloride can be made from benzoic acid.

Now let's draw the forward scheme. Benzoic acid is converted to benzoyl chloride upon treatment with thionyl chloride. Benzoyl chloride is then converted into an aldehyde upon treatment with a lithium trialkoxyaluminum hydride. The aldehyde is then treated with ethylene glycol under acid-catalyzed conditions (with removal of water), giving the desired cyclic acetal.

1) $SOCl_2$
2) $LiAl(OR)_3H$
3) H_3O^+
4) $HOCH_2CH_2OH$, $[H^+]$, $-H_2O$

$SOCl_2$

$HOCH_2CH_2OH$, $[H^+]$, $-H_2O$

1) $LiAl(OR)_3H$
2) H_3O^+

(f) There are certainly many acceptable solutions to this problem. One such solution derives from the following retrosynthetic analysis. An explanation of each of the steps (*a-c*) follows.

a. The carboxylic acid can be made from an ester, via hydrolysis.
b. The ester can be made from a ketone via a Baeyer-Villiger oxidation.
c. The ketone can be made from the starting material via oxidation of the starting alcohol.

Now let's draw the forward scheme. Oxidation of a secondary alcohol with DMP (or PCC or Swern conditions) gives a ketone, which is then converted into an ester via a Baeyer-Villiger oxidation (this process inserts an oxygen atom between the carbonyl group and the more substituted alkyl group). Finally, acid-catalyzed hydrolysis of the ester gives the desired carboxylic acid.

1) DMP
2) RCO_3H
3) H_3O^+

DMP

H_3O^+

RCO_3H

(g) There are certainly many acceptable solutions to this problem. One such solution derives from the following retrosynthetic analysis. An explanation of both steps follows.

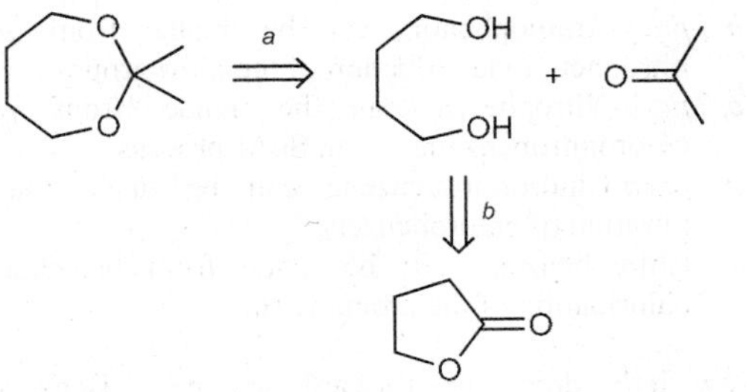

a. The product is a cyclic acetal, which can be made from a diol and a ketone (via acetal formation).
b. The diol can be made from the starting material via reduction with excess $LiAlH_4$.

Now let's draw the forward scheme. Reduction of the cyclic ester (lactone) with excess $LiAlH_4$, followed by aqueous acidic workup, causes the ring to open, giving 1,4-butanediol. This diol can then be treated with acetone under acid-catalyzed conditions (with removal of water) to give the desired cyclic acetal.

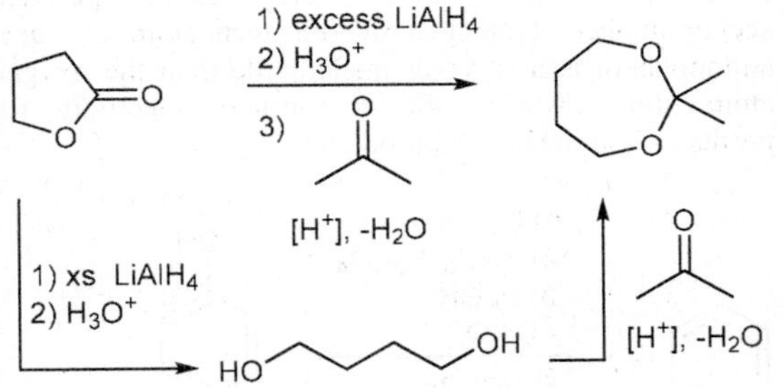

1) excess $LiAlH_4$
2) H_3O^+
3) [acetone], $[H^+]$, $-H_2O$

1) xs $LiAlH_4$
2) H_3O^+

$[H^+]$, $-H_2O$

20.70. The correct answer is (c). An ester will react with two equivalents of lithium aluminum hydride to give a diol, so we can rule out options (a) and (b). Between the remaining two options, (c) has the correct skeleton, which we can see more easily if we assign numbers to the skeleton, and then track the location of the methyl groups:

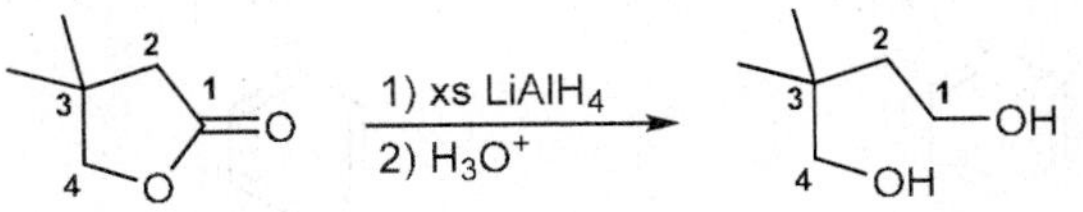

1) xs $LiAlH_4$
2) H_3O^+

20.71. The correct answer is (d). A Fischer esterification requires a carboxylic acid and an alcohol, so we can rule out options (a) and (c). Additionally, (b) has too many carbon atoms (3 + 5 = 8); the product has only seven carbon atoms, leading to the following starting materials:

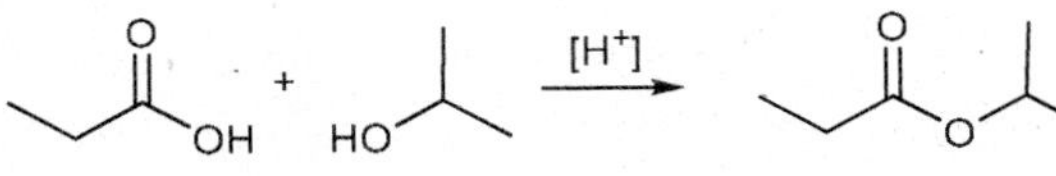

$[H^+]$

20.72. Option (d) is the correct answer because a negatively charged nitrogen atom is too strongly basic to be formed in acidic conditions. The other three structures (a-c) are all intermediates in the expected mechanism for an acid-catalyzed amide hydrolysis.

20.73. The correct answer is (b). An ester is named by indicating the alkyl group (in this case, 2-methylpropyl) attached to the oxygen atom of the ester, followed by the parent name of the corresponding carboxylic acid in which the "ic acid" is replaced by "ate". That is, the parent carboxylic acid (butanoic acid) becomes butanoate:

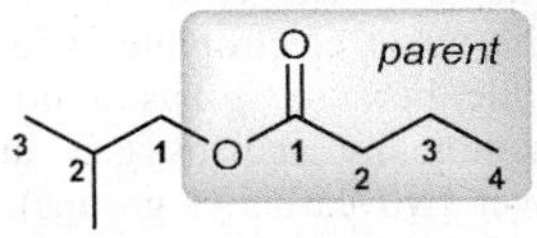

parent

2-methylpropyl butanoate

20.74. The correct answer is (a). The other three choices are intermediates in the mechanism of the reaction of MeMgBr with the given ester, as shown below:

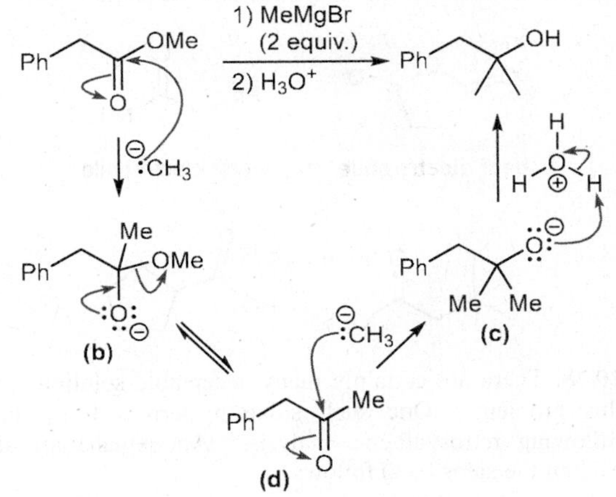

1) MeMgBr (2 equiv.)
2) H_3O^+

(b) (c) (d)

20.75. The correct answer is (c). The required transformation extends the carbon chain by one carbon atom, so synthesis II (epoxide ring opening reaction) fails because it adds two carbon atoms to the carbon chain. Synthesis I (Grignard reaction with carbon dioxide) and synthesis III (substitution with cyanide, followed by hydrolysis) both successfully install one additional carbon atom and give a carboxylic acid product, as shown here:

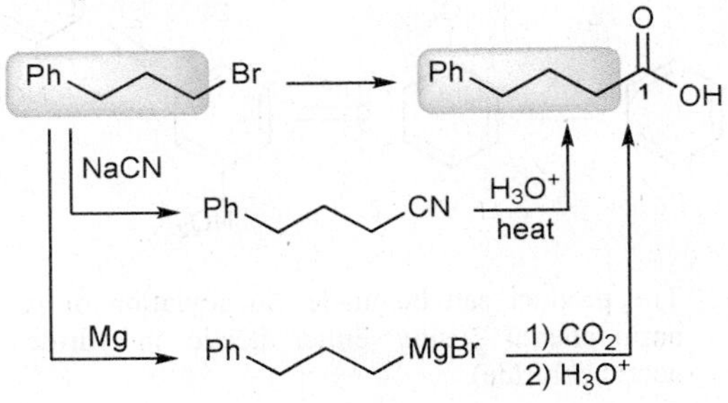

NaCN

H_3O^+, heat

Mg

1) CO_2
2) H_3O^+

20.76. The correct answer is (b). Reaction with thionyl chloride converts a carboxylic acid into the corresponding acid chloride, shown here.

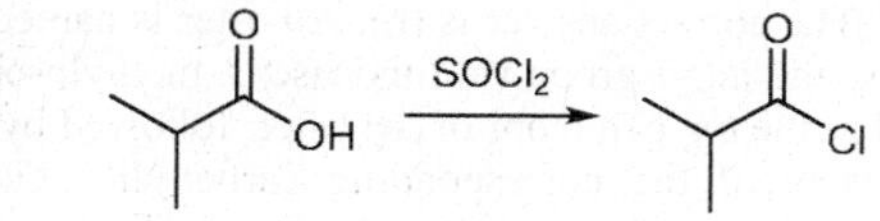

$SOCl_2$

OH Cl

20.77. The correct answer is (a). Of the functional groups given, the anhydride is the best electrophile, and the amide is the worst electrophile. The anhydride is a good electrophile because it has a minor amount of resonance donation by the oxygen atom lone pair (delocalized over two carbonyl groups), and it has an excellent leaving group (a resonance-stabilized carboxylate). The amide carbonyl is the least δ+ (and therefore the least electrophilic) because there is a significant amount of resonance donation involving the lone pair on the nitrogen atom. The reactivity of both the ester and the carboxylic acid falls somewhere in between the two extremes of the anhydride and the amide.

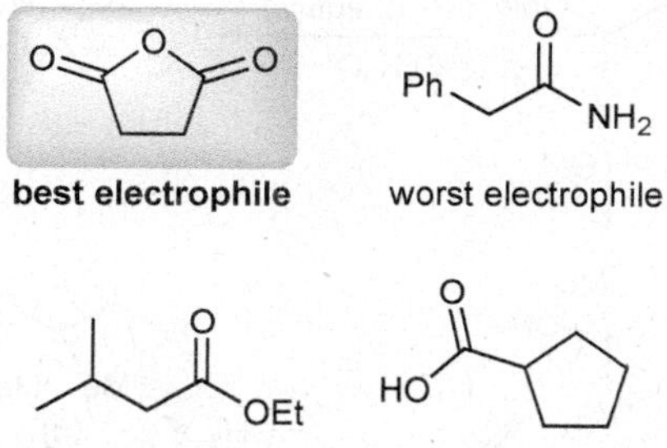

best electrophile worst electrophile

Ph NH_2 OEt HO

20.78. There are certainly many acceptable solutions to this problem. One such solution derives from the following retrosynthetic analysis. An explanation of each of the steps (*a-d*) follows.

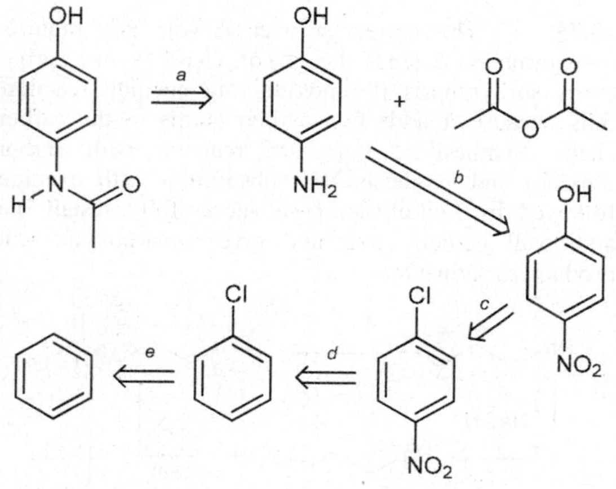

OH OH a + NH_2 H N b OH Cl Cl c e d NO_2 NO_2

a. The product can be made via acylation of *para*-aminophenol (using either acetic anhydride or acetyl chloride).

b. *para*-Aminophenol can be made from *para*-nitrophenol via reduction of the nitro group.

c. *para*-Nitrophenol can be made from *para*-chloronitrobenzene via an SNAr process.

d. *para*-Chloronitrobenzene can be made via the nitration of chlorobenzene.

e. Chlorobenzene can be made from benzene via chlorination of the aromatic ring.

Now let's draw the forward scheme. Benzene is converted into chlorobenzene via an electrophilic aromatic substitution (upon treatment with Cl_2 and $AlCl_3$). The chlorine substituent is an *ortho-para* director, thus allowing subsequent nitration to install a nitro group in the *para* position. This intermediate exhibits a leaving group (chloride) that is *para* to a strong electron withdrawing group (nitro), so this compound is susceptible to nucleophilic aromatic substitution upon treatment with hydroxide, to produce *para*-nitrophenol (after acid workup). Subsequent reduction of the nitro group with zinc and HCl gives *para*-aminophenol. Exposure to one molar equivalent of acetic anhydride produces the desired product, acetaminophen. Note that the nitrogen atom of *para*-aminophenol is a stronger nucleophile than the oxygen atom, thus allowing the appropriate selectivity for production of the desired product.

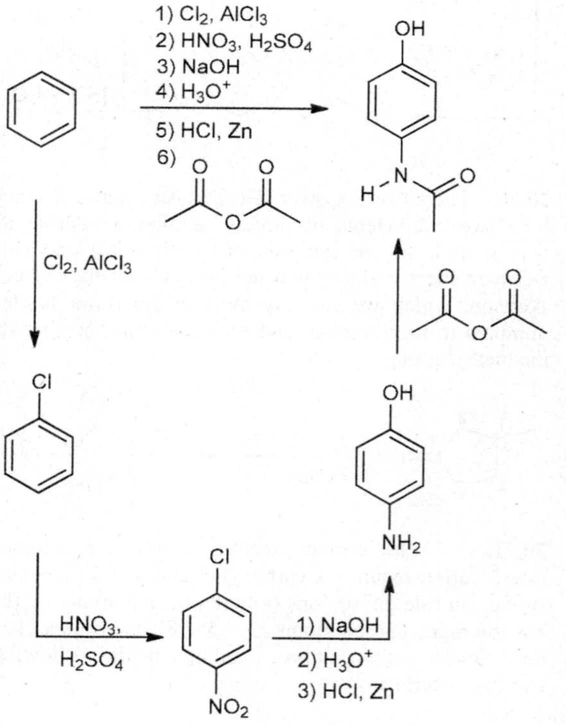

1) Cl_2, $AlCl_3$
2) HNO_3, H_2SO_4
3) NaOH
4) H_3O^+
5) HCl, Zn
6)

OH H N

Cl_2, $AlCl_3$

Cl OH NH_2 Cl

HNO_3, H_2SO_4

NO_2

1) NaOH
2) H_3O^+
3) HCl, Zn

20.79.

(a) There are certainly many acceptable solutions to this problem. One such solution derives from the following retrosynthetic analysis. An explanation of each of the steps (*a-f*) follows.

MgBr Br OH a b c + O d H e O f CH_3CN OH OH

a. The desired alcohol can be made from a Grignard reaction between acetaldehyde and ethyl magnesium bromide.
b. Ethyl magnesium bromide is made from ethyl bromide, by insertion of magnesium.
c. Ethyl bromide can be made from ethanol upon treatment with PBr_3.
d. Acetaldehyde can be made from ethanol via oxidation with PCC.
e. Ethanol can be made via reduction of acetic acid.
f. Acetic acid can be made via hydrolysis of acetonitrile.

Now let's draw the forward scheme. Hydrolysis of acetonitrile gives acetic acid which is subsequently reduced to ethanol upon treatment with excess $LiAlH_4$, followed by aqueous acidic workup. Upon treatment with PBr_3, ethanol is converted to ethyl bromide which is then converted to ethyl magnesium bromide (a Grignard reagent). A Grignard reaction with acetaldehyde (produced by oxidation of ethanol with PCC or DMP or Swern conditions), followed by aqueous acidic workup, produces the desired alcohol, 2-butanol.

PCC CH_3CN H_3O^+ O OH 1) xs $LiAlH_4$ 2) H_3O^+ OH PBr_3 Br 1) Mg 2) O H 3) H_3O^+ OH

(b) There are certainly many acceptable solutions to this problem. One such solution derives from the following retrosynthetic analysis. An explanation of each of the steps (*a-g*) follows.

BrMg a + OH H O b OH c OH O d MgBr + O=C=O e CH_3CN h O OH g OH f Br

a. The desired alcohol can be made from a Grignard reaction between propanal and ethyl magnesium bromide (formed in step *d*).
b. Propanal can be made from 1-propanol, upon treatment with PCC (or DMP or Swern conditions).
c. 1-Propanol can be made from propanoic acid, upon treatment with excess $LiAlH_4$, followed by aqueous acidic workup.
d. Propanoic acid can be made from a reaction between ethyl magnesium bromide and carbon dioxide.
e. Ethyl magnesium bromide is made from ethyl bromide, by insertion of magnesium.
f. Ethyl bromide can be made from ethanol upon treatment with PBr_3.
g. Ethanol can be made via reduction of acetic acid.
h. Acetic acid can be made via hydrolysis of acetonitrile.

Now let's draw the forward scheme. Hydrolysis of acetonitrile gives acetic acid which is subsequently reduced to ethanol upon treatment with excess $LiAlH_4$, followed by aqueous acidic workup. Upon treatment with PBr_3, ethanol is converted to ethyl bromide which is then converted to ethyl magnesium bromide (a Grignard reagent). A Grignard reaction with carbon dioxide, followed by protonation with H_3O^+, gives propanoic acid. This acid is converted into propanal via reduction (with excess $LiAlH_4$) followed by oxidation with PCC (or DMP or Swern conditions). Reaction with ethyl magnesium bromide (prepared as described above), followed by aqueous acidic workup, gives the desired alcohol, 3-pentanol.

CH_3CN —(H_3O^+, heat)→ acetic acid —(1) xs $LiAlH_4$ 2) H_3O^+)→ ethanol —(PBr_3)→ ethyl bromide —(1) Mg 2) CO_2 3) H_3O^+)→ propanoic acid —(1) xs $LiAlH_4$ 2) H_3O^+ 3) PCC)→ propanal —(1) EtMgBr 2) H_3O^+)→ 3-pentanol

(ethyl bromide —(Mg)→ EtMgBr)

(c) The desired product can be made from the product of **20.79(a)** in just two steps. Therefore, we would first perform the synthesis described in the solution to **20.79(a)**, followed by these two reactions:

2-butanol —($Na_2Cr_2O_7$, H_2SO_4, H_2O)→ 2-butanone —(1) EtMgBr 2) H_3O^+)→ 3-methyl-3-pentanol

(d) There are certainly many acceptable solutions to this problem. One such solution derives from the following retrosynthetic analysis. An explanation of each of the steps (*a*-*g*) follows.

3-ethyl-3-pentanol $\overset{a}{\Longrightarrow}$ propanoyl chloride + EtMgBr (2 equiv.)

propanoyl chloride $\overset{b}{\Longrightarrow}$ propanoic acid $\overset{c}{\Longrightarrow}$ EtMgBr + $O{=}C{=}O$

EtMgBr $\overset{d}{\Longrightarrow}$ ethyl bromide $\overset{e}{\Longrightarrow}$ ethanol $\overset{f}{\Longrightarrow}$ acetic acid $\overset{g}{\Longrightarrow}$ CH_3CN

a. The desired alcohol can be made from a Grignard reaction between propanoyl chloride and two equivalents of ethyl magnesium bromide (formed in step *d*).
b. Propanoyl chloride can be made from propanoic acid, upon treatment with thionyl chloride.
c. Propanoic acid can be made from a reaction between ethyl magnesium bromide and carbon dioxide.
d. Ethyl magnesium bromide is made from ethyl bromide, by insertion of magnesium.
e. Ethyl bromide can be made from ethanol upon treatment with PBr_3.
f. Ethanol can be made via reduction of acetic acid.
g. Acetic acid can be made via hydrolysis of acetonitrile.

Now let's draw the forward scheme. Hydrolysis of acetonitrile gives acetic acid which is subsequently reduced to ethanol upon treatment with excess $LiAlH_4$, followed by aqueous acidic workup. Upon treatment with PBr_3, ethanol is converted to ethyl bromide which is then converted to ethyl magnesium bromide (a Grignard reagent). A Grignard reaction with carbon dioxide, followed by protonation with H_3O^+, gives propanoic acid. Conversion to the acid chloride, followed by reaction with excess ethyl magnesium bromide (prepared as described above) produces the desired tertiary alcohol, 3-ethyl-3-pentanol.

CH_3CN —(H_3O^+, heat)→ acetic acid (OH) —(1) xs $LiAlH_4$; 2) H_3O^+)→ ethanol (OH) —(PBr_3)→ ethyl bromide (Br)

ethyl bromide —(1) Mg; 2) CO_2; 3) H_3O^+)→ propanoic acid (OH, O) —($SOCl_2$)→ propanoyl chloride (Cl, O)

ethyl bromide —(Mg)→ EtMgBr

propanoyl chloride —(1) xs EtMgBr; 2) H_3O^+)→ 3-ethyl-3-pentanol (OH)

20.80. The mechanism shown below has 12 steps: 1) proton transfer, 2) loss of a leaving group, 3) nucleophilic attack, 4) proton transfer, 5) proton transfer, 6) loss of a leaving group, 7) proton transfer, 8) nucleophilic attack, 9) proton transfer, 10) proton transfer, 11) loss of a leaving group and 12) proton transfer. In step 1 (proton transfer), one of the acetal oxygen atoms is protonated by H_3O^+, activating it as a leaving group (either oxygen atom can be protonated, which will ultimately lead to the same product). In step 2, an alcohol group leaves as a leaving group. Water then attacks the activated carbonyl group in step 3 (nucleophilic attack). In step 4 (proton transfer), water deprotonates the cationic oxygen atom, and in step 5 (proton transfer), the other oxygen atom in the hemiacetal is protonated by H_3O^+, producing a cationic intermediate. In step 6, an alcohol serves as a leaving group. Protonation of the carbonyl group (step 7, proton transfer) activates this carbonyl group for intramolecular attack by one of the tethered alcohol groups (step 8, nucleophilic attack), thus forming a ring (a five-membered ring is more likely formed than a more strained, four-membered ring). Deprotonation by water (step 9, proton transfer) followed by protonation by H_3O^+ (step 10, proton transfer) produces a cationic intermediate with an activated leaving group (water). In step 11, water serves as a leaving group. In step 12 (proton transfer), water serves as a weak base to deprotonate the cationic oxygen atom, resulting in the formation of the final product (a lactone).

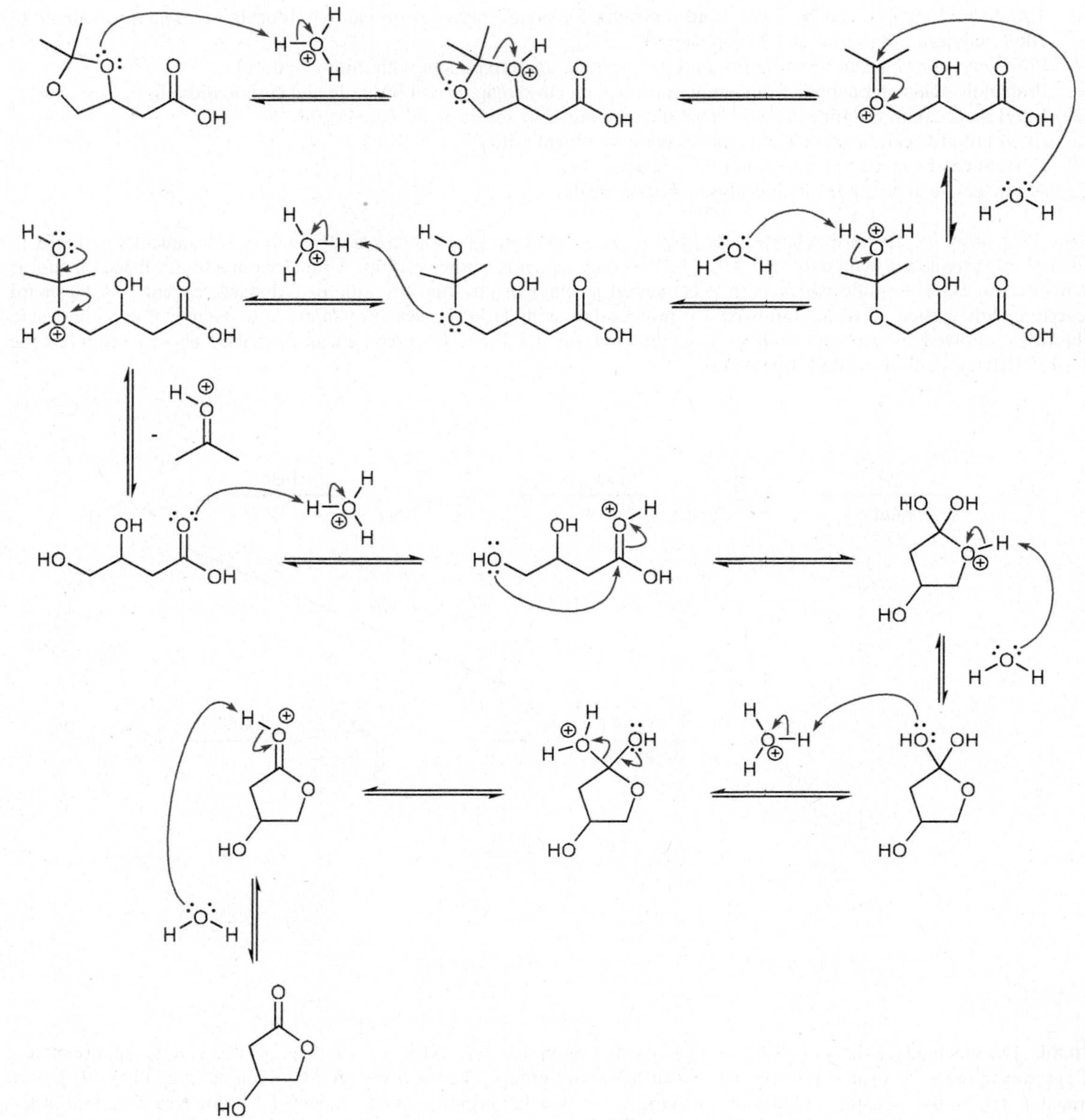

20.81. The carboxylic acid has a molecular formula of $C_5H_{10}O_2$. Thionyl chloride replaces the OH group on the carboxylic acid with a chlorine atom, thus the molecular formula of the resulting acid chloride is C_5H_9ClO. Considering the possible acid chloride isomers with this molecular formula, only one (compound **A**) gives a single signal in its ^{1}H NMR spectrum

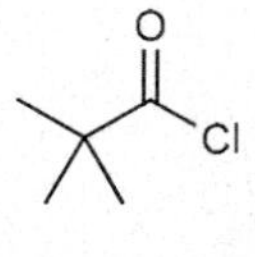

Compound **A**

When compound **A** is treated with excess ammonia, a nucleophilic acyl substitution reaction occurs, producing the amide shown.

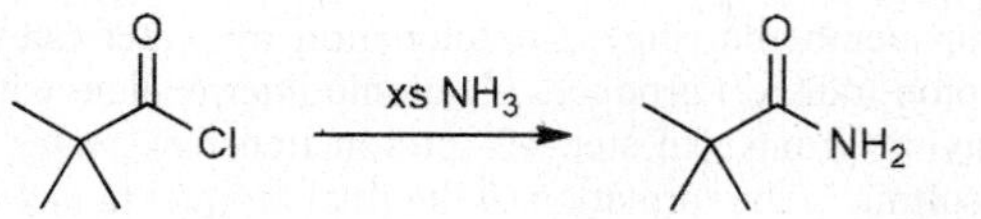

xs NH_3

20.82. The molecular formula ($C_{10}H_{10}O_4$) indicates six degrees of unsaturation (see Section 14.16), which is highly suggestive of an aromatic ring (which accounts for four degrees of unsaturation) plus two more degrees of unsaturation (either two rings, or two double bonds, or a ring and a double bond, or a triple bond).

There are two signals (both singlets) in the ^1H NMR spectrum, with integration values of 3H (4.0 ppm) and 2H (8.1 ppm). There are a total of 10 hydrogen atoms in the molecule, so the ratio of the integration values (3:2) must represent a 6H:4H ratio of the hydrogen atoms in the molecule (6H + 4H = 10H total). This indicates a high degree of symmetry in the structure.

The signal at 8.1 ppm (with an integration of 4H) is consistent with the chemical shift expected for aromatic protons. The integration (4H) indicates a disubstituted ring, and the multiplicity of this signal (it is a singlet) suggests a *para*-disubstituted aromatic ring with two equivalent substituents, thereby rendering all four aromatic protons equivalent (thus giving rise to a singlet):

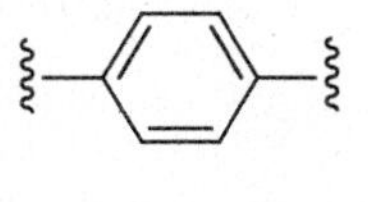

The signal at 4.0 ppm (with an integration of 6H) is consistent with two identical methyl groups, each of which must be next to an oxygen atom to justify the downfield shift of the signal:

MeO— —OMe

The two methoxy fragments and the aromatic ring account for all of the atoms in the molecular formula except for two carbon atoms and two oxygen atoms. Since we still need to account for two more degrees of unsaturation, and since the proposed structure must retain its high degree of symmetry, we propose the following structure:

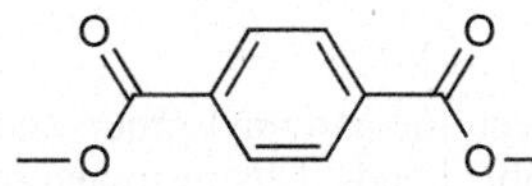

20.83. An IR spectrum of butyric acid should have a very broad signal between 2200 and 3600 cm^{-1} due to the O–H stretch of the carboxylic acid. An IR spectrum of ethyl acetate will not have this signal.

20.84. The ^1H NMR spectrum of *para*-chlorobenzaldehyde should have a signal at approximately 10 ppm corresponding to the aldehydic proton. The ^1H NMR spectrum of benzoyl chloride should not have a signal near 10 ppm.

20.85. The molecular formula ($C_8H_8O_3$) indicates five degrees of unsaturation (see Section 14.16), which is highly suggestive of an aromatic ring (which accounts for four degrees of unsaturation) plus either one ring or one double bond.

The broad signal between 2200 cm^{-1} and 3600 cm^{-1} in the IR spectrum is consistent with the O-H stretch of a carboxylic acid.

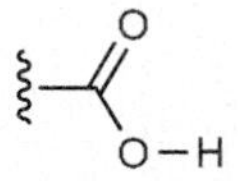

O
O–H

The ^1H NMR spectrum exhibits a signal at approximately 12 ppm, confirming the presence of a carboxylic acid group. The pair of doublets (with a combined integration of 4H) appearing between 7 and 8 ppm is characteristic of a *para*-disubstituted aromatic ring (with two different substituents):

X— —Y

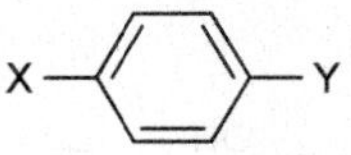

The singlet near 4 ppm has an integration of 3H, indicating a methyl group. The downfield chemical shift of this signal indicates that the methyl group is likely attached to an oxygen atom:

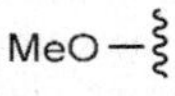

MeO—

There is only one way to connect the three fragments:

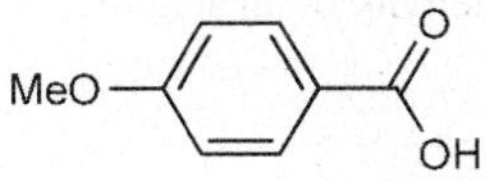

MeO— O OH

The ^{13}C NMR spectrum is consistent with this structure. The most downfield signal (~170 ppm) is consistent with the carbonyl group. A disubstituted aromatic ring (bearing two different substituents) is expected to produce four signals between 100 and 150 ppm. We do in fact see four signals, although one of them is above 150 ppm, which likely corresponds with the carbon atom connected to the methoxy group (an oxygen atom is electronegative, causing a deshielding effect). Finally, the signal between 50 and 100 ppm is consistent with the carbon atom of the methoxy group (an sp^3 hybridized carbon atom attached to an electronegative atom).

20.86. The mechanism shown below has 6 steps: 1) proton transfer, 2) nucleophilic attack, 3) proton transfer, 4) proton transfer, 5) loss of a leaving group and 6) proton transfer. In step 1 (proton transfer), the carbonyl group is protonated by H_3O^+, activating it as an electrophile. In step 2 (nucleophilic attack), the tethered alcohol group serves as a nucleophile in an intramolecular nucleophilic attack. In step 3 (proton transfer), water deprotonates the oxonium ion, and in step 4 (proton transfer), a different oxygen atom is protonated by H_3O^+, activating it as a leaving group. In step 5 (loss of a leaving group), an alcohol serves as a leaving group. Deprotonation (step 6) gives the final, rearranged product. If the oxygen atom of the OH group in the starting material is an isotopic label (as indicated by the highlighted boxes), then we would expect the label to be incorporated into the ring of the product, as shown.

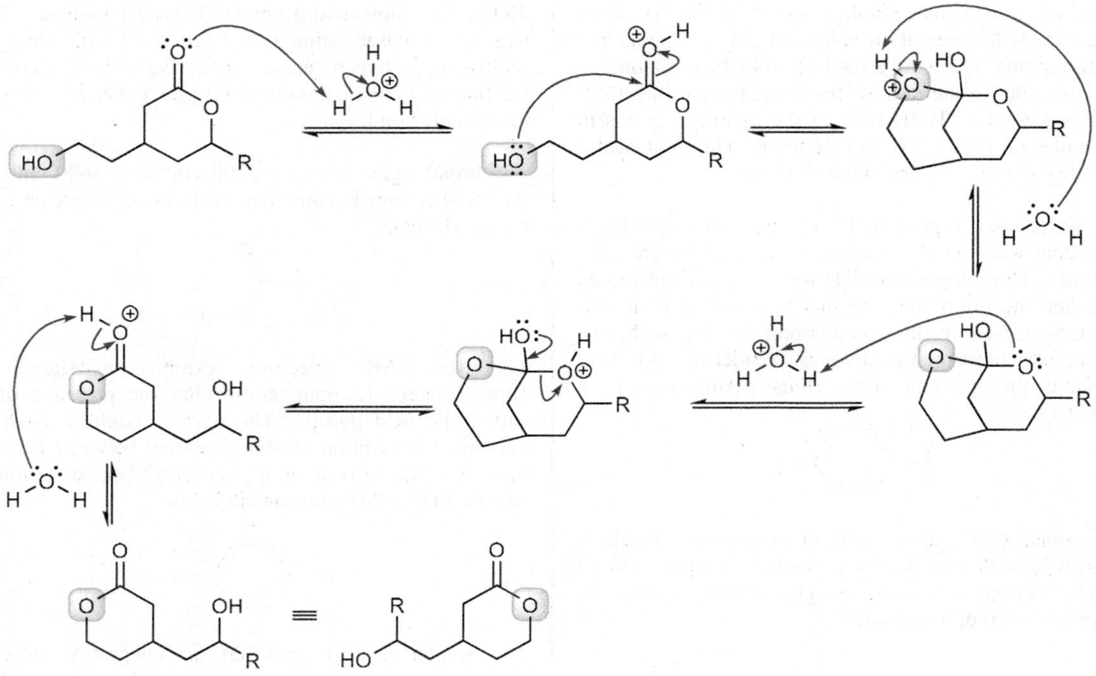

20.87. The lone pair of the nitrogen atom (of the amide group) is participating in aromaticity and is therefore less available to donate electron density into the carbonyl group. As a result, the carbonyl group is less electron-rich and more electrophilic than the carbonyl group of a regular amide (where the lone pair contributes significant electron density to the carbonyl group via resonance). Also, when this compound functions as an electrophile in a nucleophilic acyl substitution reaction, the leaving group is particularly stable because it is an aromatic anion in which the negative charge is spread over all five atoms of the aromatic ring. With such a good leaving group, this compound more closely resembles the reactivity of an acid halide than an amide.

20.88.

(a) DMF, like most amides, exhibits restricted rotation about the bond between the carbonyl group and the nitrogen atom, due to the significant contribution of the resonance form with a C=N double bond. This restricted rotation causes the methyl groups to be in different electronic environments (one methyl group is *cis* to the C=O and the other is *trans* to the C=O). They are not chemically equivalent and will therefore produce two different signals (in addition to the signal from the other proton in the compound). Upon treatment with excess $LiAlH_4$, followed by water workup, DMF is reduced to an amine:

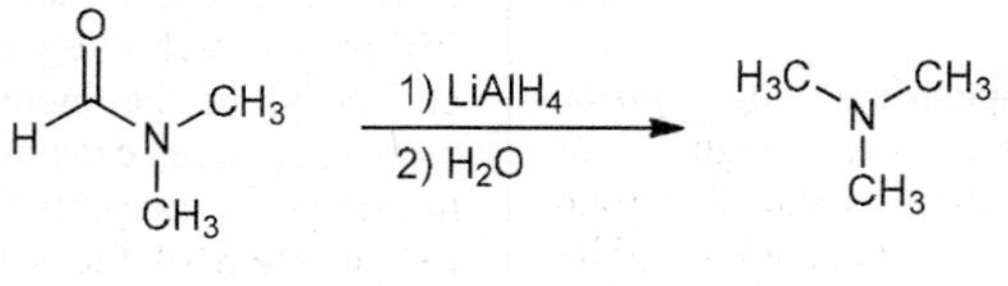

1) $LiAlH_4$
2) H_2O

This amine does not exhibit restricted rotation with respect to its C–N bonds. As such, all of the methyl groups are now chemically equivalent and will together produce only one signal.

(b) Restricted rotation causes the methyl groups to be in different electronic environments. As a result, the ^{13}C NMR spectrum of DMF should have three signals.

20.89. The first step of the synthesis involves deprotonation of the alcohol group in compound **1** using NaH, generating an alkoxide ion. This alkoxide ion is then treated with the chiral 2-bromo ester, to give an S_N2 reaction (note the inversion of configuration of the chiral center bearing the methyl group). Reduction of the ester with DIBAH provides an aldehyde, which is transformed into the terminal olefin (compound **2**) using a Wittig reaction.

Ph, OH — NaH, then Br–CH(CH₃)–CO₂Et → DIBAH → Ph₃P=CH₂ → 2

20.90. Notice that the product has one more carbon atom than the starting material, and therefore, we must form a new C–C bond. This can be accomplished by first installing a leaving group at one of the benzylic positions, followed by an S_N2 process in which the leaving group is replaced with cyanide. Acid-catalyzed hydrolysis of the resulting nitrile affords the desired carboxylic acid. This route is preferable to the formation of a Grignard reagent followed by condensation with CO_2, as there are two bromine atoms in the molecule.

NBS, heat → NaCN → H_3O^+, heat

20.91. A possible synthesis is shown below. When analyzing the starting material, **1**, and the product, **2**, it can be seen that the alcohol is transformed into an ether and that the ester is converted to a different ester. So first determine a method to make an ether; the Williamson ether synthesis (alkoxide + RX) is a convenient method for ether formation.

For the conversion of the ester to a different ester, notice that the carbonyl carbon of ester **1** is still part of the side chain, but it is now a methylene (CH_2) and therefore the ester must be reduced at some point ($LiAlH_4$ is known to reduce esters). Reduction of the ester affords a 1° alcohol, which can then be converted to the new ester:

1) xs $LiAlH_4$ 2) H_3O^+

Conversion of the alcohol into the ester could be achieved either via an acid-catalyzed Fischer esterification ($ROH + RCO_2H$) or the via addition of an alcohol to an acid chloride ($ROH + RCOCl$). The acid chloride method is preferable, because acidic conditions required for the Fischer esterification could produce undesired side reactions with other functional groups present.

Finally, the order of addition of these reagents is important so that you don't get an undesired product. See the correct order below, which shows the formation of ether **3** first (Williamson ether synthesis), followed by reduction of the ester to give alcohol **4**, and finally, conversion of alcohol **4** to the desired ester **2** using the corresponding acid chloride.

Williamson Ether Synthesis 1) NaH, THF 2) allyl bromide → **3** → 1) $LiAlH_4$ 2) H_2O → **4** → acryloyl chloride, pyridine → **2**

20.92.
The desired transformation of compound **1** to compound **2** can be achieved via reduction of the carboxylic acid, followed by substitution. Direct conversion of the resultant alcohol may be accomplished using PBr_3, or one can utilize a two-step method involving: 1) tosylate formation using TsCl and pyridine followed by, 2) S_N2 displacement using sodium bromide in DMSO:

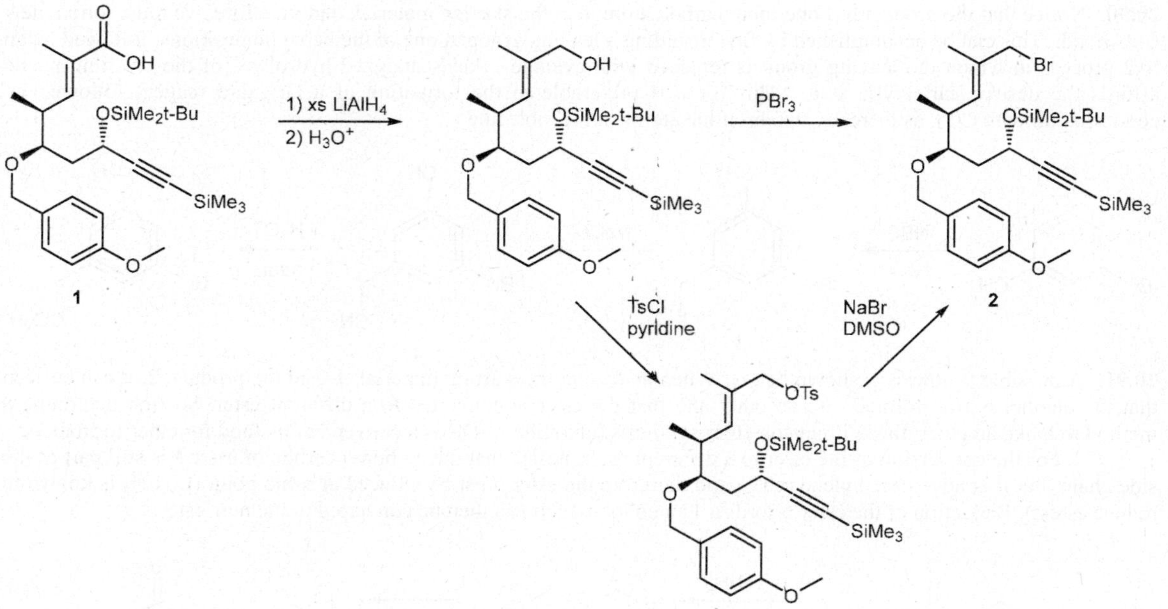
OH
OSiMe2t-Bu
SiMe3
1
1) xs LiAlH4
2) H3O+
OH
OSiMe2t-Bu
SiMe3
PBr3
Br
OSiMe2t-Bu
SiMe3
2
TsCl
pyridine
NaBr
DMSO
OTs
OSiMe2t-Bu
SiMe3

20.93.
(a) Compound **1** is a nitrile (it contains a cyano group), which can undergo hydrolysis with aqueous sodium hydroxide. This gives rise to three possible products, only one of which is the desired product:

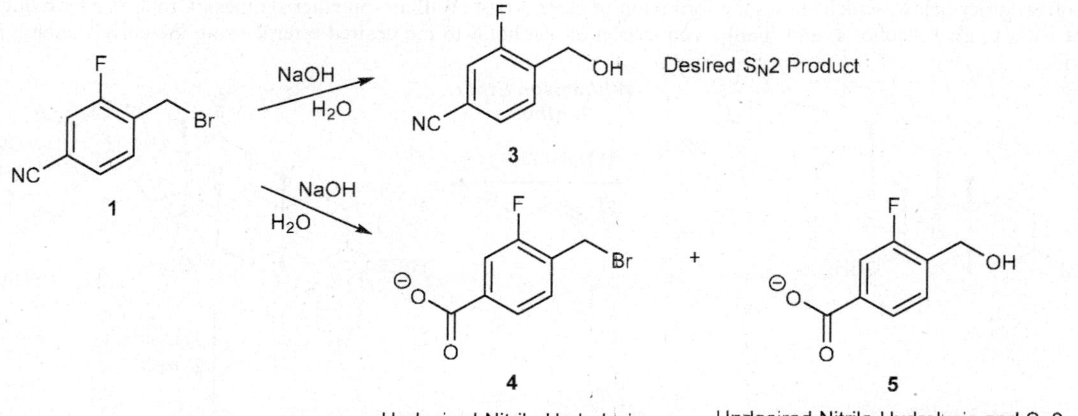
F
Br
NC
1
NaOH
H2O
F
OH
NC
3
Desired SN2 Product
NaOH
H2O
F
Br
4
Undesired Nitrile Hydrolysis
+
F
OH
5
Undesired Nitrile Hydrolysis and SN2

(b) A likely mechanism for the reaction sequence is shown here. The conversion of compound **1** to compound **2** is an S_N2 displacement of the bromide with acetate ($CH_3CO_2^-$). The reaction of the ester with NaOMe/MeOH is expected to proceed via an acyl substitution mechanism in which methoxide (CH_3O^-) attacks the carbonyl carbon of the ester (just as hydroxide would during hydrolysis). This generates a charged tetrahedral intermediate that collapses to give a new ester and eject the alkoxide leaving group, which is protonated by the solvent to give the desired alcohol, compound **3**.

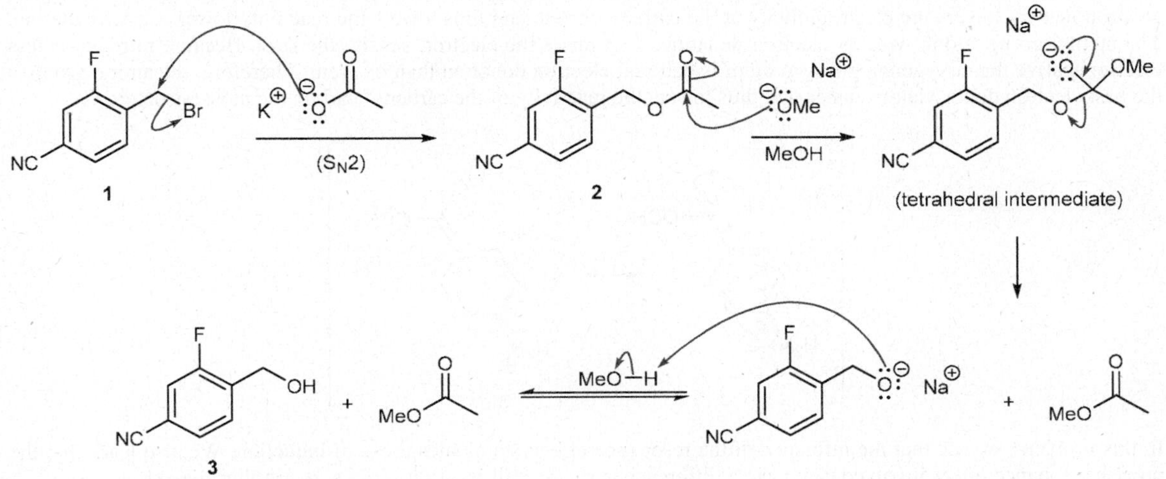

20.94. Note that methyl benzoate is the reference compound among the series of compounds examined. So, benzoates with rate constants larger in value than 1.7 $M^{-1}min^{-1}$ are more reactive while those with lower values are less reactive than the reference compound. The aromatic ring of methyl *p*-nitrobenzoate is considerably lower in electron density due to resonance interaction between the electron-withdrawing nitro group and the ring. However, in one of the resonance forms (structure **A** below), the electron-deficient carbon atom of the ring is adjacent to the carbonyl carbon atom. Thus, the electrophilicity at the carbonyl functional group is the largest, making its reactivity towards nucleophilic attack by hydroxide anion the fastest for this compound.

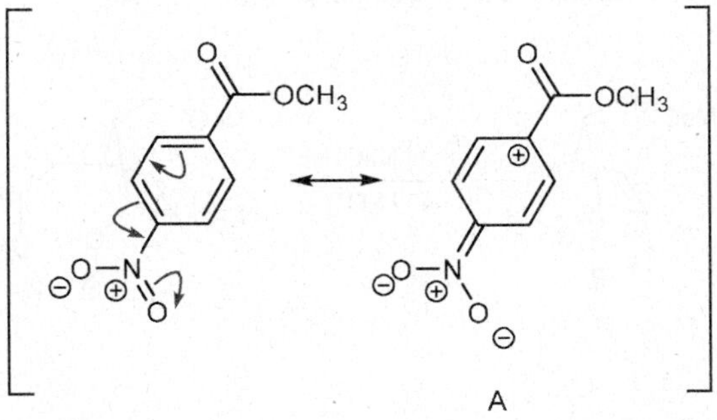

While the aromatic ring of methyl *m*-nitrobenzoate is also expected to be just as deficient in electron density as that of the *para* isomer, the positive charge on the ring never occupies the carbon atom adjacent to the carbonyl carbon atom (structure **B**). So, the electron deficiency (electrophilicity) at the carbonyl carbon atom is not as large as that of the *para* isomer. Thus, the reaction rate is slower than that of the *para* isomer.

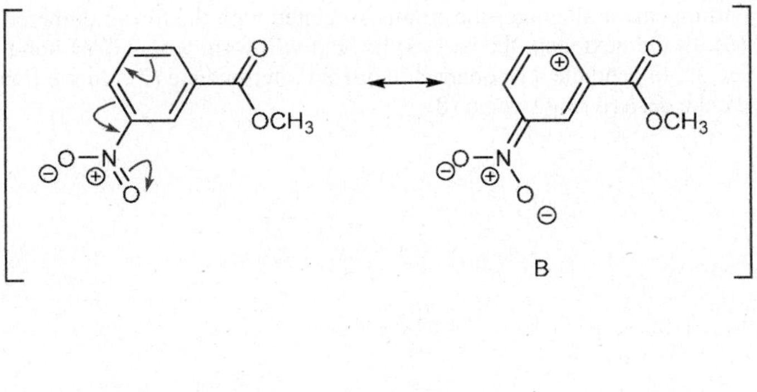

The main electronic interaction between the halogen atoms and the aromatic ring is induction by which both chlorine and bromine are expected to withdraw electron density from the ring. Since chlorine is more electronegative than bromine, the *m*-chlorophenyl ring is more electron-deficient than the *m*-bromophenyl ring, and thus more reactive.

The remaining three benzoates are less reactive than the reference compound. This suggests that the substituents found in these compounds are all electron-donating so as to increase the electron density at the carbonyl group. The increased electron density lowers the electrophilicity at the carbonyl group and thus lowers the reactivity towards hydroxide ion. The methyl group is only weakly electron-donating, so it raises the electron density the least. Because nitrogen is less electronegative than oxygen, a nitrogen atom is better at electron donation than oxygen. Therefore, the amino group is the best electron donor via resonance and thus lowers the reactivity of the carbonyl carbon the most (structure **C**).

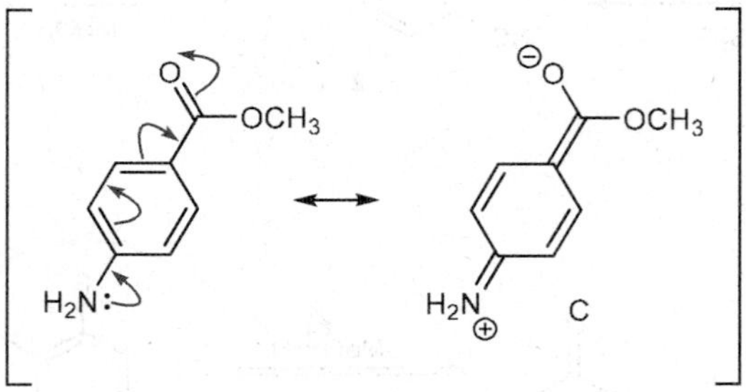

In this problem, we see that the influence of the resonance effects dominates those of induction. We also learn that the precise resonance forms involved can make a difference in the overall reactivity (*p*- vs. *m*-nitrobenzoates).

20.95. Compound **1** is a ketone, so treatment with a peroxy acid will give the corresponding lactone (cyclic ester). Notice that the oxygen atom is inserted between the carbonyl group and the bridgehead position (because that position is more substituted than the other side of the carbonyl group). Hydrolysis of the lactone gives compound **3**. In Corey's synthesis, this hydrolysis step was performed under basic conditions (saponification), and under those conditions, the product (compound **3**) would be deprotonated to give a carboxylate ion. Acid workup is necessary in order to protonate the carboxylate ion and regenerate the carboxylic acid (compound **3**).

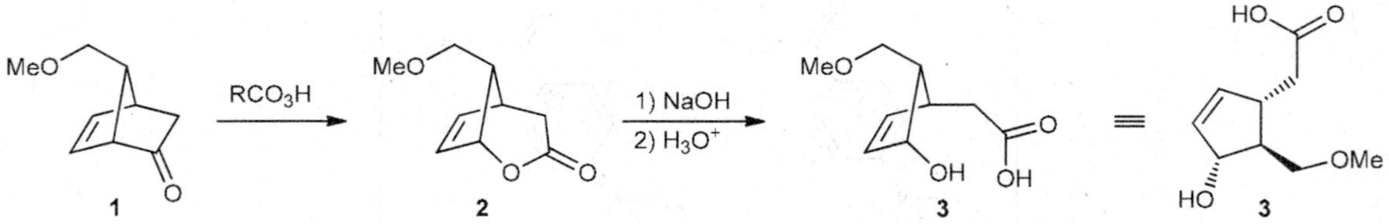

20.96. The first step of the synthesis involves deprotonation of the carboxylic acid with sodium hydride, followed by intramolecular esterification of **2** via loss of the mesylate group, giving **3**. Note that the newly formed ring is on the bottom face of the molecule. Next, protonation of the ester with aqueous acid will generate an activated carbonyl (**4**) that spontaneously leaves as a neutral carboxylic acid, generating a secondary carbocation (**5**). Next, a 1,2-carbon migration occurs so that the four-membered ring opens and the pair of electrons in the C–C σ-bond moves over one carbon (this carbocation rearrangement alleviates the strain associated with the four-membered ring), which generates a new, tertiary carbocation (**6**). In the next step, the carboxylic acid will form a new C–O bond with the carbocation, on the bottom face of the molecule, to generate a resonance-stabilized intermediate (**7**). In the final step of the mechanism, **7** is deprotonated to produce the desired ring system (**8**).

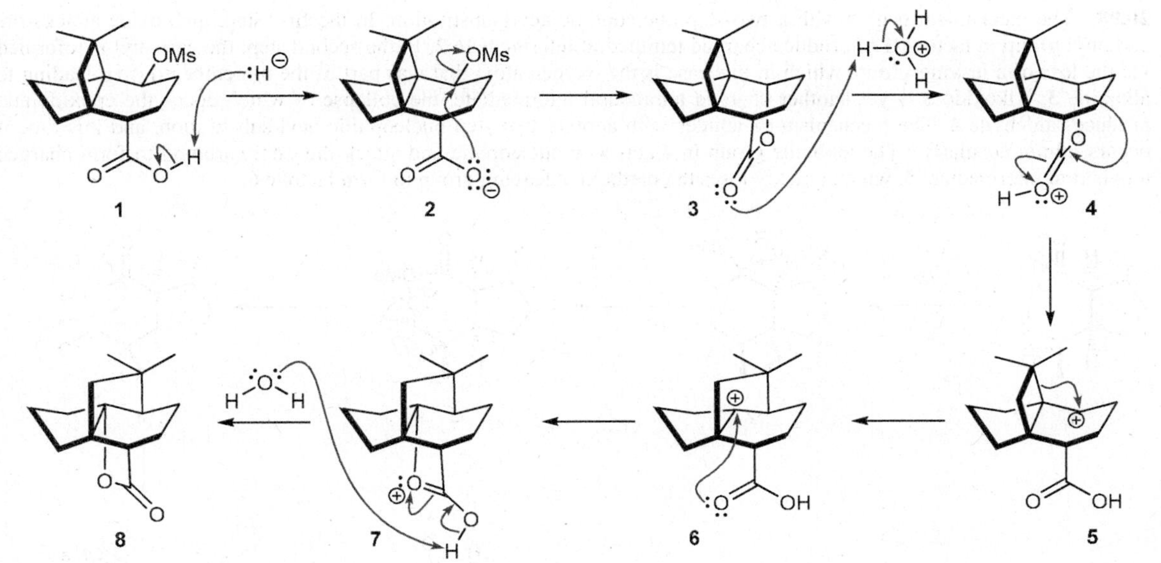

20.97. Deprotonation of alcohol **1** with sodium hydride will produce an alkoxide (**3**) that can easily react at the lactone carbonyl to form intermediate **4** (note: the bond lengths and bond angles in **4** have been exaggerated). Once formed, the unstable tetrahedral intermediate quickly decomposes to generate a new 5-membered lactone via the expulsion of an alkoxide leaving group. While it may be difficult to see in its current form, intermediate **5** can be redrawn as **5a**, which clearly shows the desired fused 5,5-ring system of the product with the proper stereochemistry at the ring junction. In the final step, the alkoxide in **5a** reacts at the carbon-bromine bond in an SN2 fashion to produce epoxide **2**.

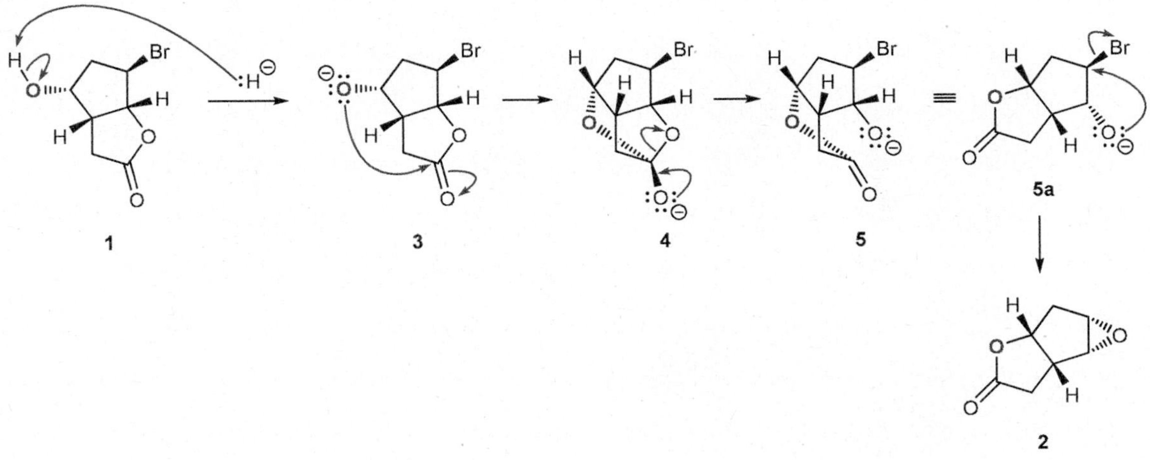

20.98. The mechanism begins with a two-step nucleophilic acyl substitution. In the first step, methoxide attacks the carbonyl group in lactone **1** to produce charged tetrahedral intermediate **2**. In the second step, the carbonyl is reformed via the loss of a leaving group, which in this case is the oxygen atom that was part of the 6-membered ring, leading to alkoxide **3**. Alkoxide **3** is yet another charged tetrahedral intermediate, the collapse of which opens the epoxide and produces aldehyde **4**. The mechanism concludes with another two-step nucleophilic acyl substitution, and this time it occurs intramolecularly. The alkoxide group in **4** acts as a nucleophile and attack the ester carbonyl to form charged tetrahedral intermediate **5**, which quickly loses the methoxide leaving group to form lactone **6**.

Chapter 21
Alpha Carbon Chemistry: Enols and Enolates

Review of Concepts

Fill in the blanks below. To verify that your answers are correct, look in your textbook at the end of Chapter 21. Each of the sentences below appears verbatim in the section entitled *Review of Concepts and Vocabulary*.

- In the presence of catalytic acid or base, a ketone will exist in equilibrium with an _____. In general, the equilibrium will significantly favor the ________.
- When treated with a strong base, the α position of a ketone is deprotonated to give an __________.
- ________________ or ______ will irreversibly and completely convert an aldehyde or ketone into an enolate.
- In the **haloform reaction**, a ________ ketone is converted into a carboxylic acid upon treatment with excess base and excess halogen followed by acid workup.
- When an aldehyde is treated with sodium hydroxide, an **aldol addition reaction** occurs, and the product is a ____________________________.
- For most simple aldehydes, the position of equilibrium favors the aldol product. For most ketones, the reverse process, called a ______**-aldol reaction** is favored.
- When an aldehyde is heated in aqueous sodium hydroxide, an **aldol** ___________ **reaction** occurs, and the product is an _______________________. Elimination of water occurs via an ______ **mechanism**.
- **Crossed aldol,** or **mixed aldol reactions** are aldol reactions that occur between different partners and are only efficient if one partner lacks __________ or if a **directed aldol addition** is performed.
- Intramolecular aldol reactions show a preference for formation of ______ and ____-membered rings.
- When an ester is treated with an alkoxide base, a **Claisen condensation reaction** occurs, and the product is a ______________.
- The α position of a ketone can be alkylated by forming an enolate and treating it with an ______________.
- For unsymmetrical ketones, reactions with _____ at low temperature favor formation of the kinetic enolate, while reactions with ______ at room temperature favor the thermodynamic enolate.
- When LDA is used with an unsymmetrical ketone, alkylation occurs at the ________________ position.
- The ____________________ **synthesis** can be used to prepare substituted derivatives of acetone, while the ____________________ **synthesis** can be used to prepare substituted derivatives of acetic acid.
- Aldehydes and ketones that possess _____-unsaturation are susceptible to nucleophilic attack at the β position. This reaction is called a ___________ **addition,** or *1,4-addition*, or a **Michael reaction**.

Review of Skills

Fill in the blanks and empty boxes below. To verify that your answers are correct, look in your textbook at the end of Chapter 21. The answers appear in the section entitled *SkillBuilder Review*.

21.1 Drawing Enolates

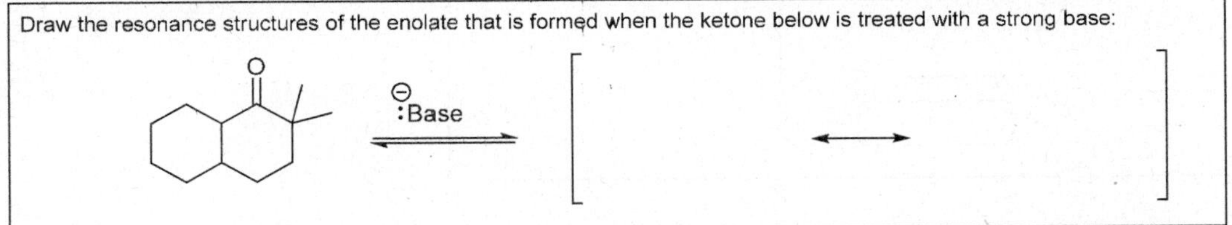

Draw the resonance structures of the enolate that is formed when the ketone below is treated with a strong base:
O
:Base

21.2 Predicting the Products of an Aldol Addition Reaction

Predict the product of the aldol addition reaction that occurs when the following aldehyde is treated with sodium hydroxide:

H — O — NaOH

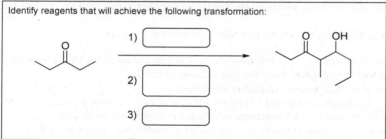

21.3 Drawing the Product of an Aldol Condensation

Predict the product of the aldol condensation reaction that occurs when the following compound is heated with sodium hydroxide:

O — NaOH, heat

21.4 Identifying the Reagents Necessary for a Crossed Aldol Reaction

Identify reagents that will achieve the following transformation:

O — 1) — 2) — 3) — O OH

21.5 Using the Acetoacetic Ester Synthesis

Identify reagents that will achieve the following transformation:

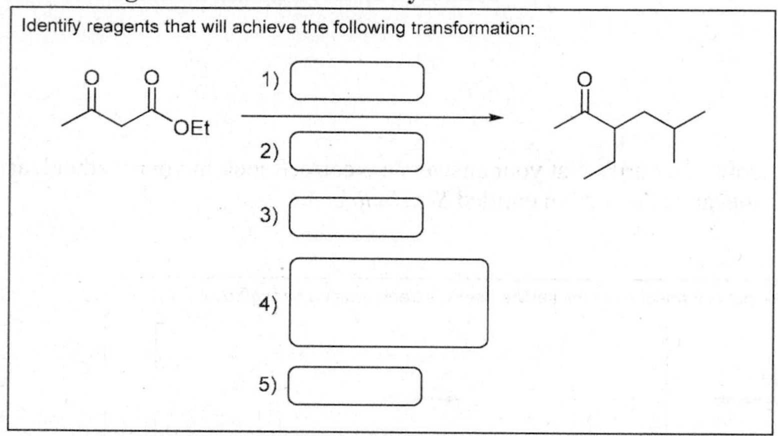

21.6 Using the Malonic Ester Synthesis

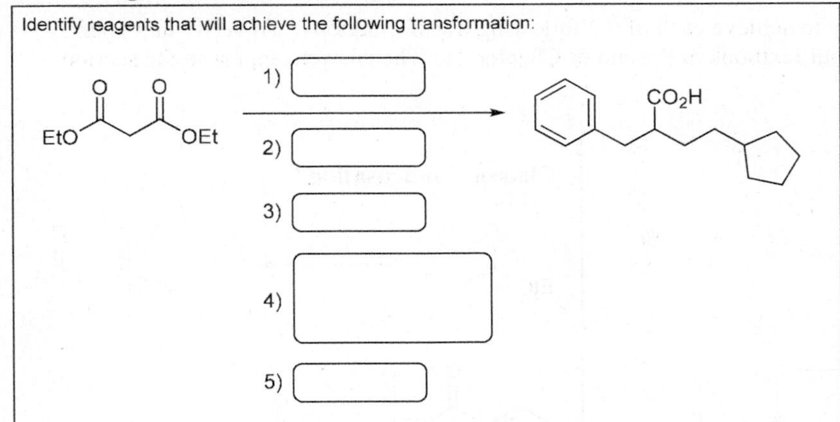

21.7 Determining When to Use a Stork Enamine Synthesis

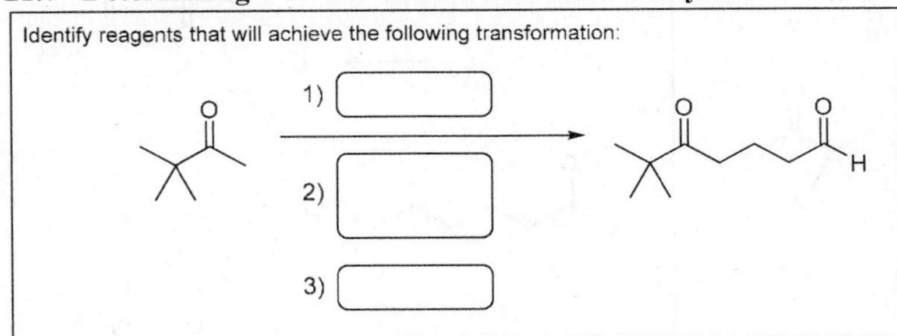

21.8 Determining which Addition or Condensation Reaction to Use

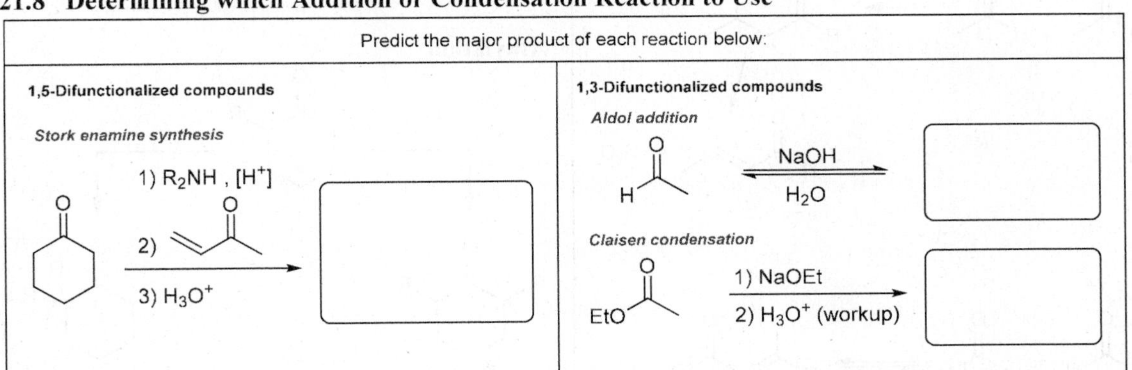

21.9 Alkylating the α and β Positions

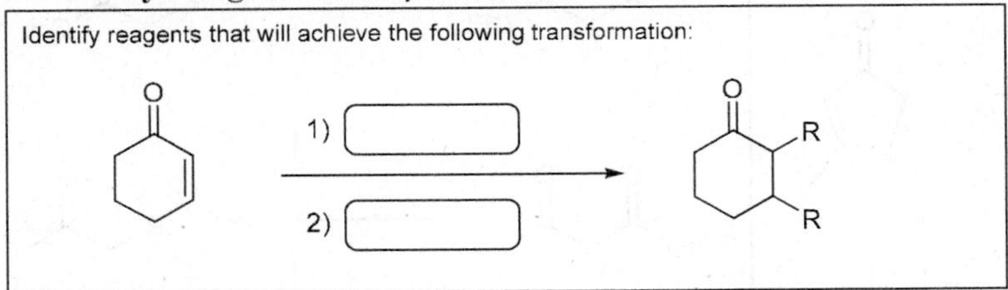

Review of Reactions

Identify the reagents necessary to achieve each of the following transformations. To verify that your answers are correct, look in your textbook at the end of Chapter 21. The answers appear in the section entitled *Review of Reactions*.

Alpha Halogenation

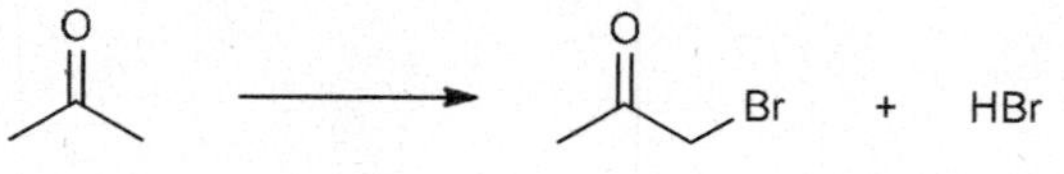

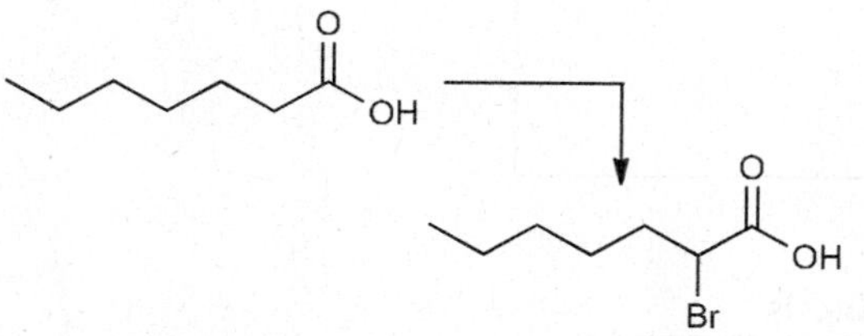

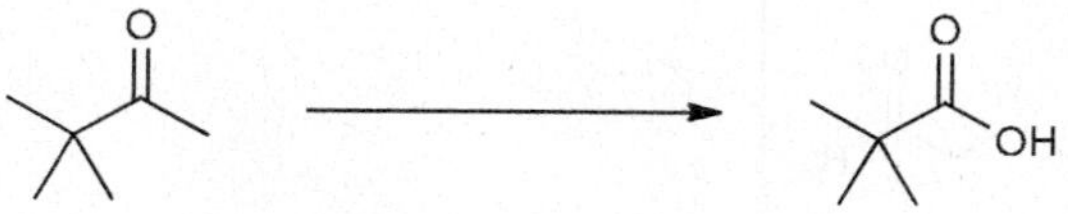

Br + HBr

OH

OH

Br

OH

Aldol Reactions

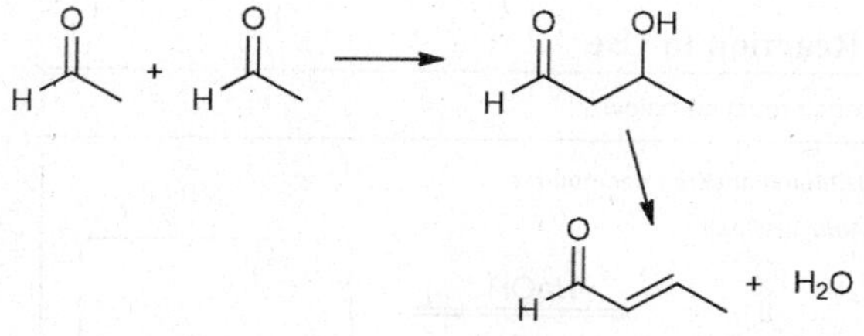

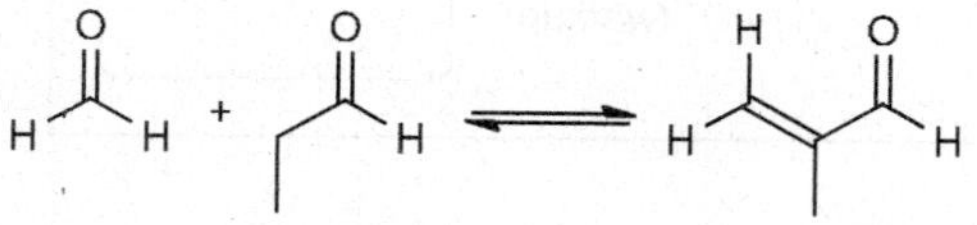

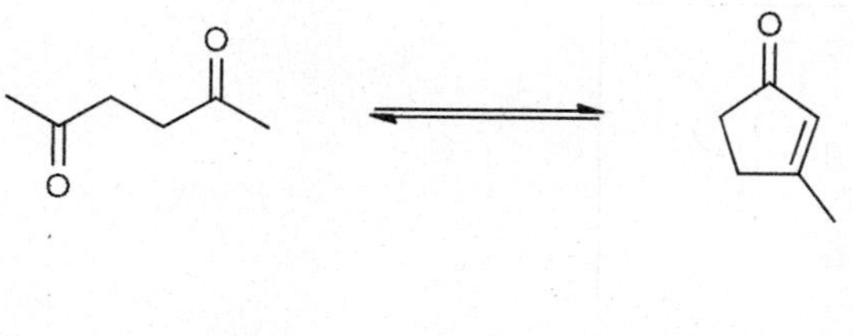

H + H → H OH

H + H_2O

H H + H ⇌ H H H

Claisen Condensation

EtO → EtO

OEt + OEt

OEt

EtO OEt

OEt

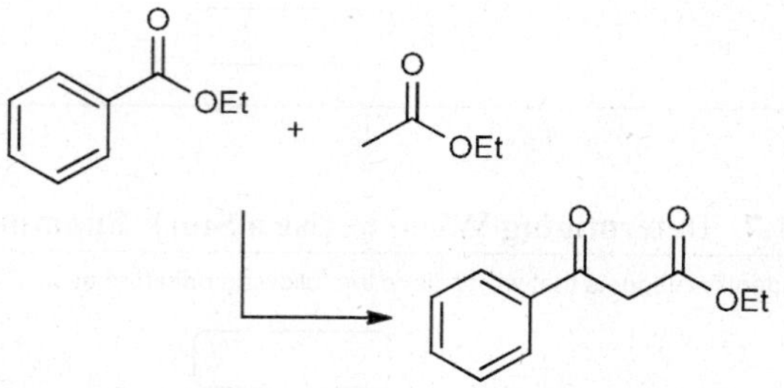

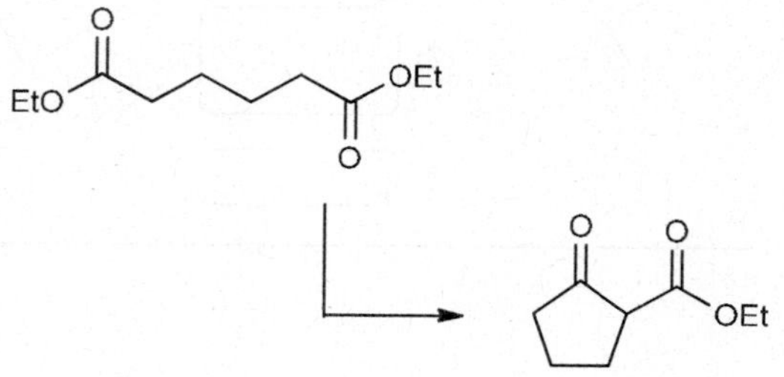

Alkylation

R

R R

R R

EtO OEt → R OH

EtO → R

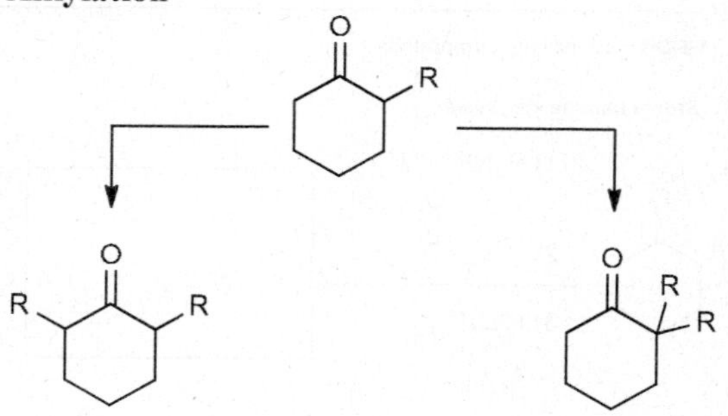

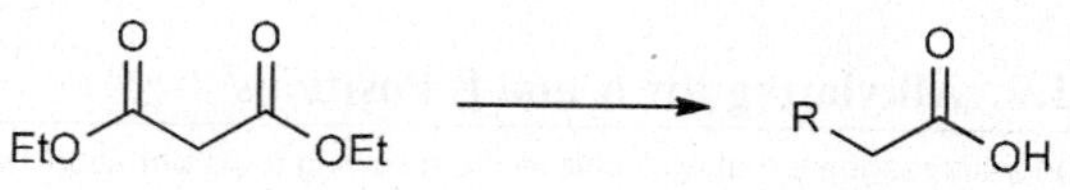

Michael Additions

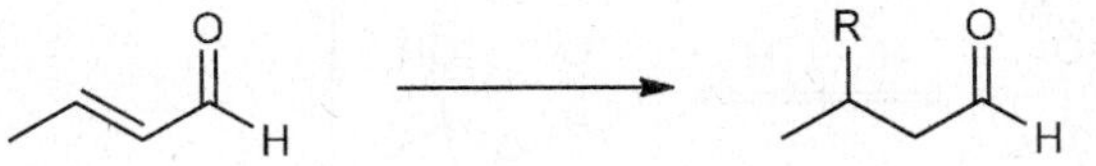

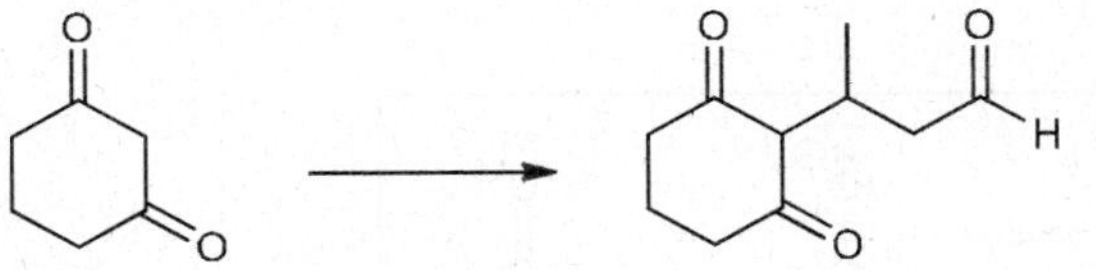

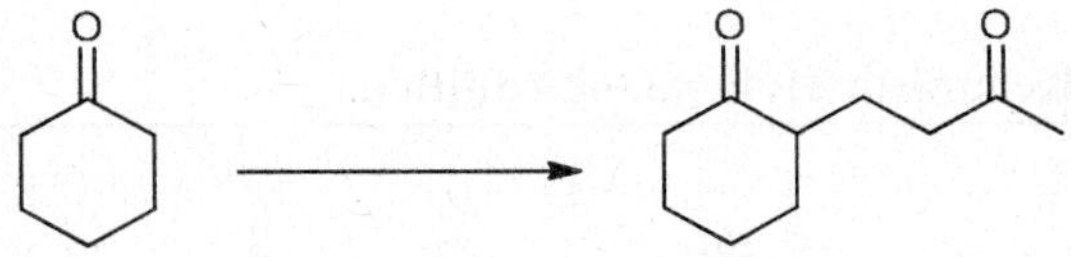

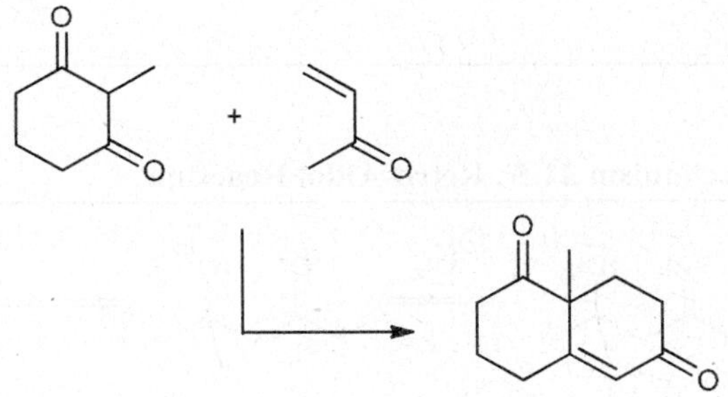

Review of Mechanisms

Complete each of the following mechanisms by drawing the missing curved arrows. To verify that your curved arrows are drawn correctly, compare them to the curved arrows in the mechanism boxes for Mechanisms 21.1 – 21.7, which can be found throughout Chapter 21 of your text.

Mechanism 21.1 Acid-Catalyzed Tautomerization

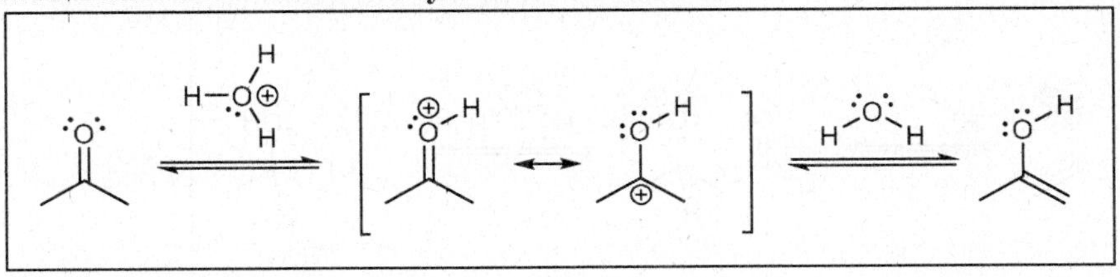

Mechanism 21.2 Base-Catalyzed Tautomerization

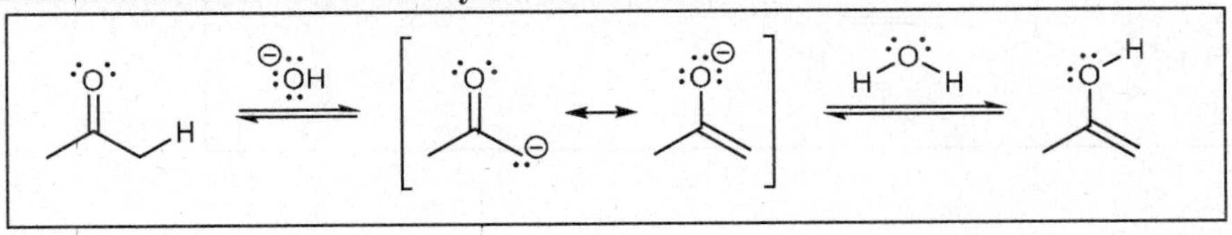

Mechanism 21.3 Acid-Catalyzed Halogenation of Ketones

Part 1: Enol Formation

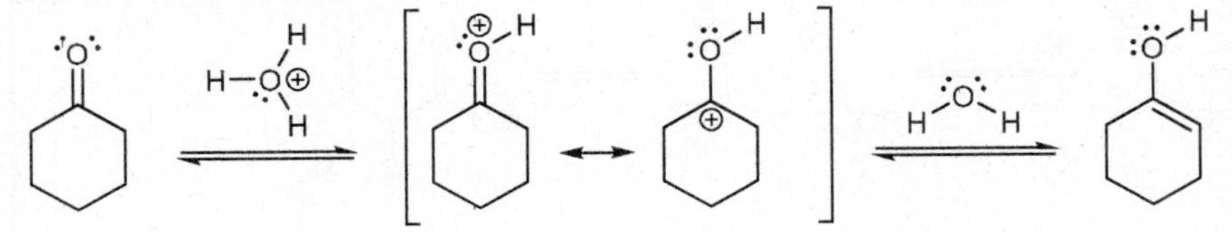

Part 2: Halogenation

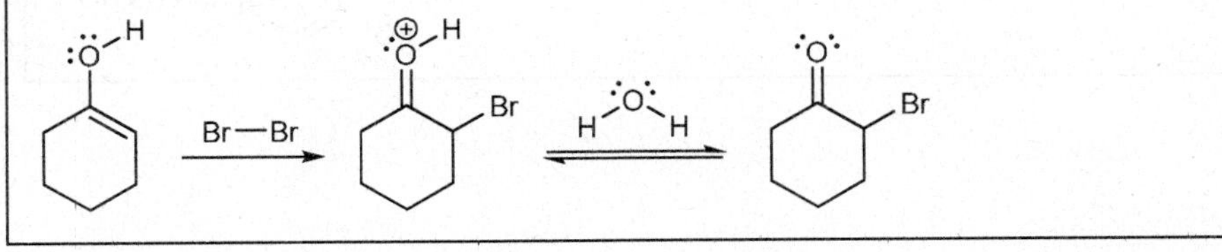

Br—Br

Mechanism 21.4 Aldol Addition

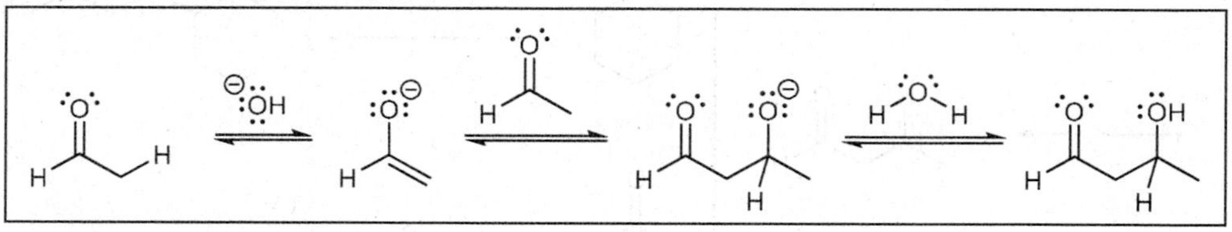

Mechanism 21.5 Retro-Aldol Reaction

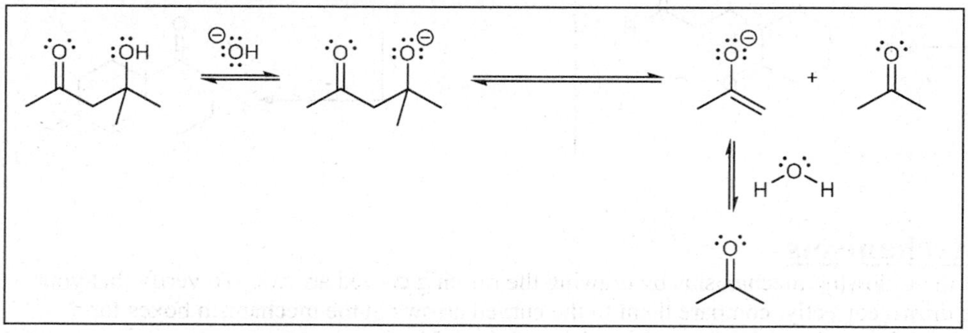

Mechanism 21.6 Aldol Condensation

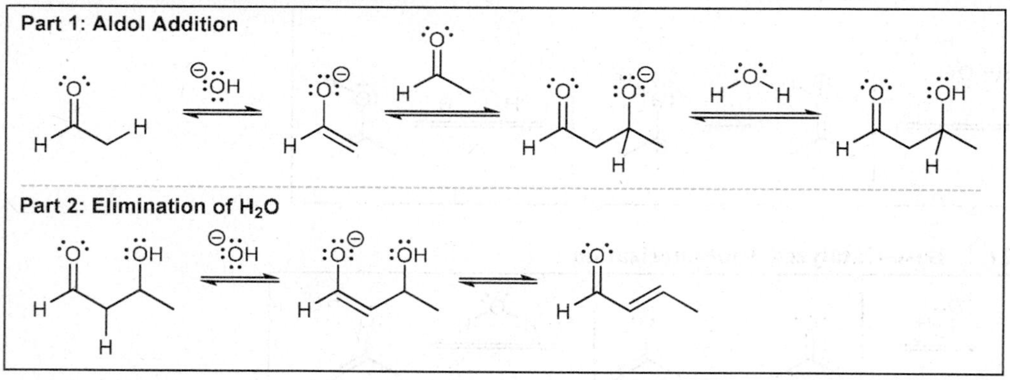

Part 1: Aldol Addition

Part 2: Elimination of H_2O

Mechanism 21.7 Claisen Condensation

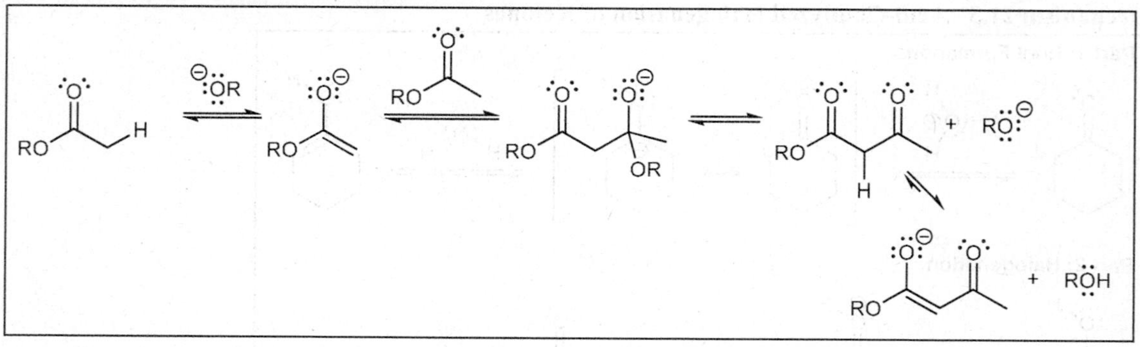

Common Mistake to Avoid

This chapter covers many reactions. Some of them require aqueous workup, some require aqueous *acidic* workup, while others require no workup at all. Students often confuse the appropriate workup conditions by associating the workup conditions with the reagents, rather than the reaction occurring. For example, consider the following reaction:

R_2CuLi

This reaction (covered in Section 20.8) involves a lithium dialkylcuprate being used as a nucleophile, giving a ketone as a product. In contrast, consider the following reaction (from Chapter 21), which also employs a lithium dialkylcuprate as a nucleophile:

1) R_2CuLi
2) H_3O^+

R_2CuLi H_3O^+

In this case, the initial product of the reaction is an enolate, which must be protonated, thereby requiring acidic workup. This was not the case for the reaction between an acid chloride and a lithium dialkylcuprate. It would therefore be a mistake to memorize that lithium dialkylcuprates always require aqueous acidic workup (or to memorize the opposite), because it depends on the situation. Rather than memorizing arbitrary rules that don't always apply, it would be wiser to focus on understanding *why* certain reactions require workup while others do not. Your understanding will be facilitated by a strong focus on reaction mechanisms.

Useful reagents

The following is a list of reagents encountered in this chapter:

Reagents	Type of Reaction	Description
$[H_3O^+]$, Br_2	α-Bromination	These reagents can be used to install a bromine atom at the α position of a ketone (or aldehyde). Subsequent treatment of the resulting α-bromoketone with pyridine gives an α,β-unsaturated ketone. This two-step process can be used to introduce α,β-unsaturation into a ketone or aldehyde.
1) Br_2, PBr_3 2) H_2O	Hell-Volhard-Zelinsky reaction	These reagents can be used to install a bromine atom at the α position of a carboxylic acid.
1) NaOH, Br_2 2) H_3O^+	Haloform reaction	These reagents can be used to convert a methyl ketone into a carboxylic acid. This process is most efficient when the other α position (of the starting ketone) bears no protons.
NaOH, H_2O	Aldol addition reaction	Aqueous sodium hydroxide will cause an aldol addition reaction between two equivalents of an aldehyde or ketone to give a β-hydroxyaldehyde (or a β-hydroxyketone).
NaOH, H_2O, heat	Aldol condensation	Aqueous sodium hydroxide and heat will cause an aldol condensation between two equivalents of an aldehyde or ketone to give an α,β-unsaturated aldehyde (or an α,β-unsaturated ketone).

1) NaOEt 2) H_3O^+	Claisen condensation	These reagents will cause two equivalents of an ester to undergo a condensation reaction, giving a β-ketoester. An acidic aqueous workup is required to afford the neutral product.
1) LDA, -78°C 2) RX	Alkylation	These conditions can be used to install an alkyl group at the less-substituted α position of an unsymmetrical ketone (via the kinetic enolate).
1) NaH, 25°C 2) RX	Alkylation	These conditions can be used to install an alkyl group at the more-substituted α position of an unsymmetrical ketone (via the thermodynamic enolate).
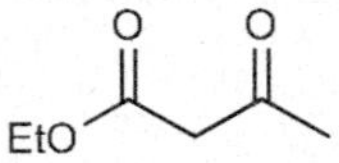 EtO (ethyl acetoacetate structure)	Acetoacetic ester synthesis	Ethyl acetoacetate can be converted into a derivative of acetone upon treatment with ethoxide, followed by an alkyl halide, followed by hydrolysis and decarboxylation with aqueous acid and heat.
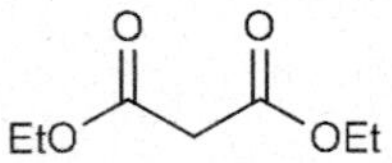 EtO, OEt (diethyl malonate structure)	Malonic ester synthesis	Diethyl malonate can be converted into a substituted carboxylic acid (more specifically, a derivative of acetic acid) upon treatment with ethoxide, followed by an alkyl halide, followed by hydrolysis and decarboxylation with aqueous acid and heat.
1) R_2CuLi 2) H_3O^+	Michael reaction	A lithium dialkylcuprate is a weak nucleophile and can serve as a Michael donor. It will undergo conjugate addition with a suitable Michael acceptor (see Table 21.2).

Solutions

21.1. Under acid-catalyzed conditions, the carbonyl group is first protonated, generating a resonance-stabilized intermediate, which is then deprotonated at the α position to give an enol. Notice that the acid for the protonation step is a hydronium ion, and the base for the deprotonation step is water, consistent with acidic conditions.

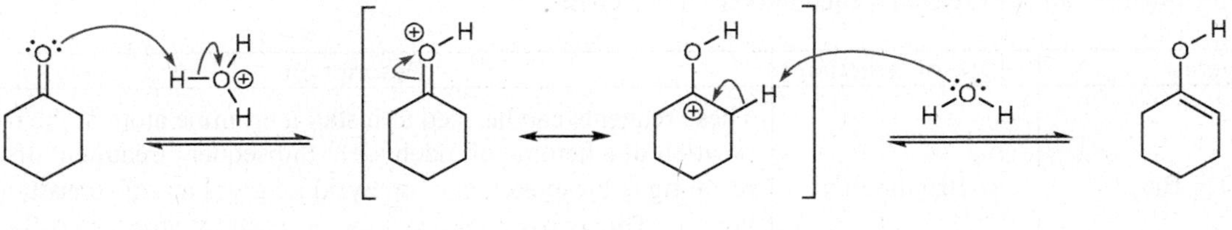

21.2. If we carefully inspect the solution to the previous problem, we find that the final step of the mechanism is deprotonation of the α position, thereby converting the resonance-stabilized cationic intermediate into an enol. Therefore, the reverse of this process must begin with protonation of the α position, thereby converting the enol into a resonance-stabilized cationic intermediate. Subsequent deprotonation of this intermediate gives the ketone. Notice that the acid for the protonation step is a hydronium ion, and the base for the deprotonation step is water, consistent with acidic conditions.

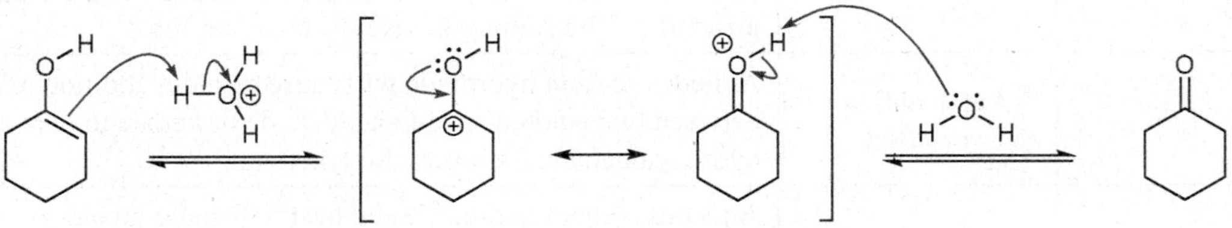

21.3. Under base-catalyzed conditions, the α position is first deprotonated, generating a resonance-stabilized anionic intermediate. The oxygen atom in this intermediate is then protonated to give an enol. Since the ketone is unsymmetrical, the two α positions are not equivalent. Therefore, the enol can be formed at either α position, as shown below. Notice that, in each case, the base for the deprotonation step is a hydroxide ion, and the acid for the protonation step is water, consistent with basic conditions.

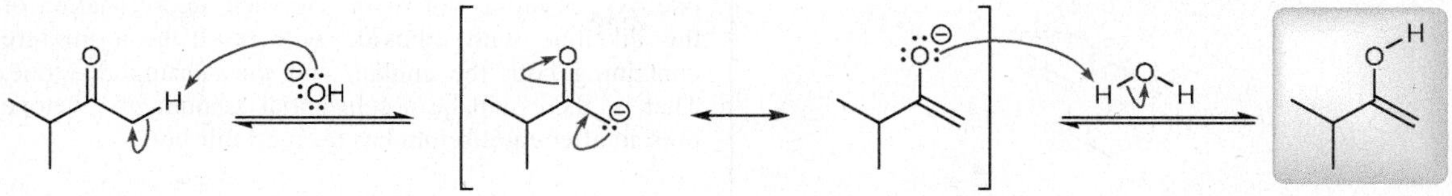

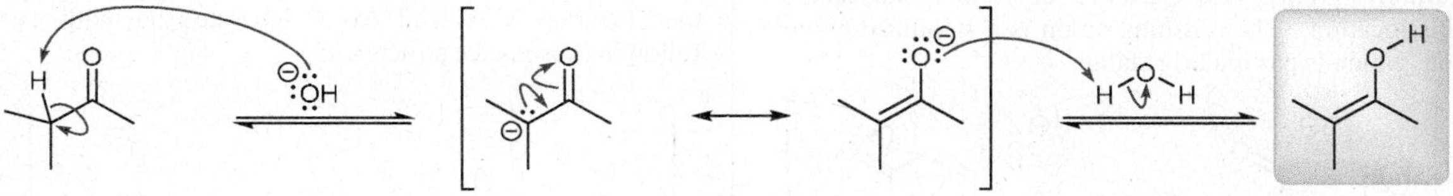

21.4.

(a) This compound has two α positions, although they are equivalent because the ketone is symmetrical. Deprotonation at either location will lead to the same enolate ion, which has the following resonance structures:

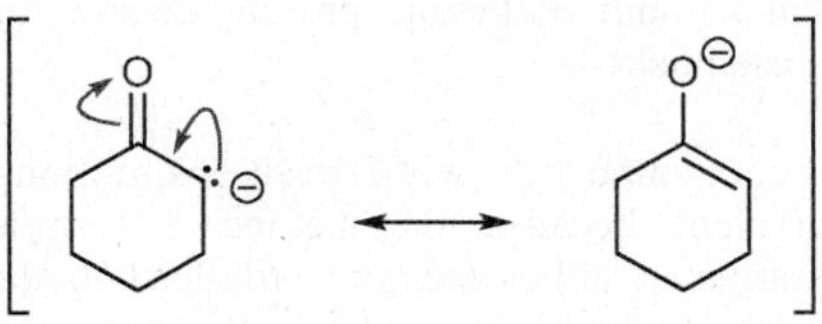

(b) This compound has two α positions, although only one of these positions bears protons. Deprotonation at that location will lead to an enolate ion with the following resonance structures:

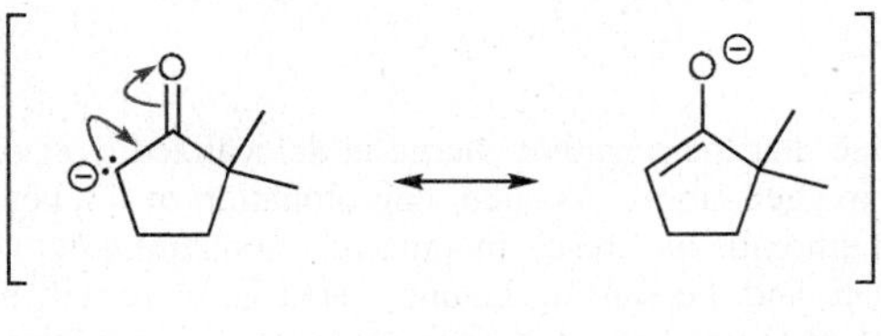

(c) This compound has two α positions, although only one of these positions bears protons. Deprotonation at that location will lead to an enolate ion with the following resonance structures:

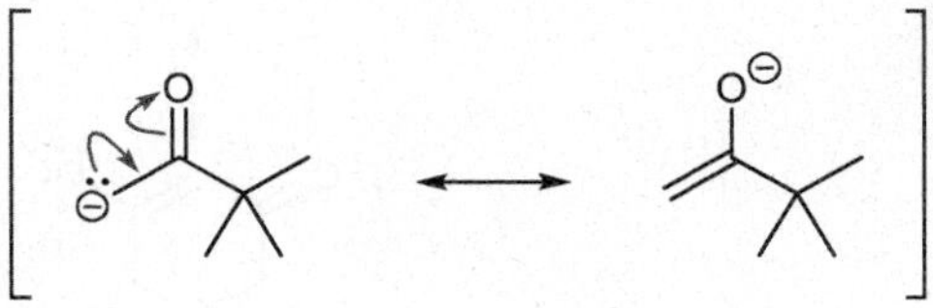

(d) This compound has two α positions, although only one of these positions bears protons. Deprotonation at that location will lead to an enolate ion with the following resonance structures:

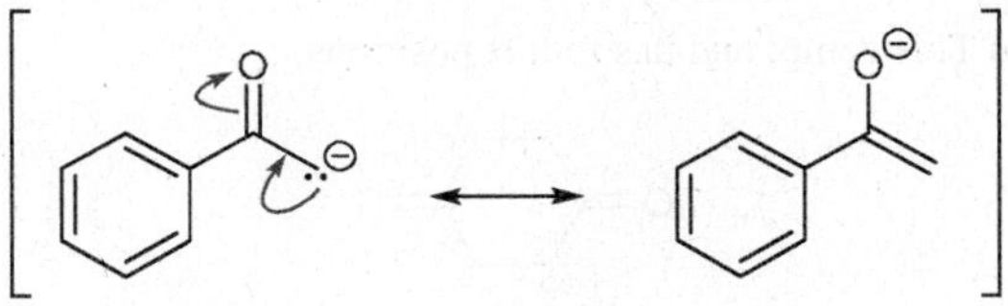

(e) This compound is an aldehyde and therefore has only one α position. Deprotonation at that location will lead to an enolate ion with the following resonance structures:

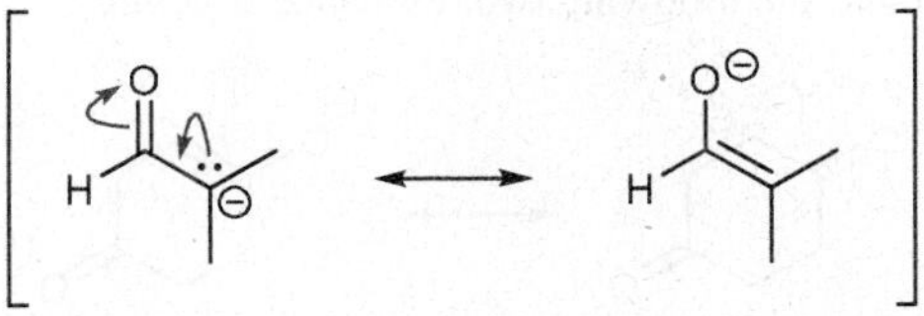

21.5. The ester has only one alpha carbon, and the proton at that position will be removed when treated with a strong base. Just as we would expect for the enolate of an aldehyde or a ketone, the resulting ester enolate has two resonance structures.

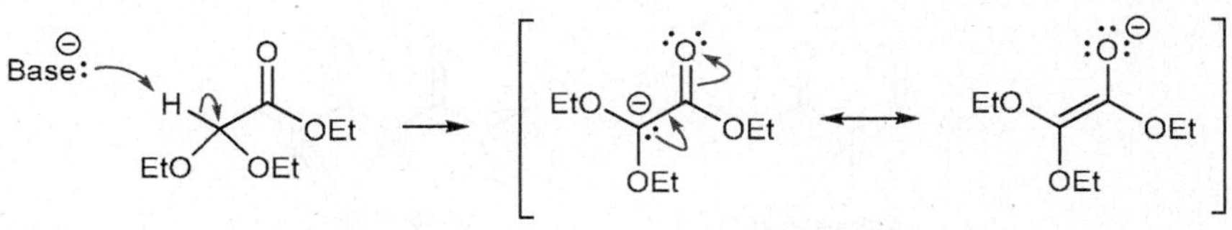

21.6.

(a) This compound has three α positions:

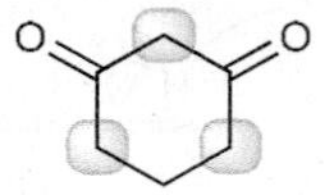

Among these positions, the central position is α to both carbonyl groups, and therefore, deprotonation occurs at this location. The resulting anion is a stabilized enolate ion, which is particularly stable:

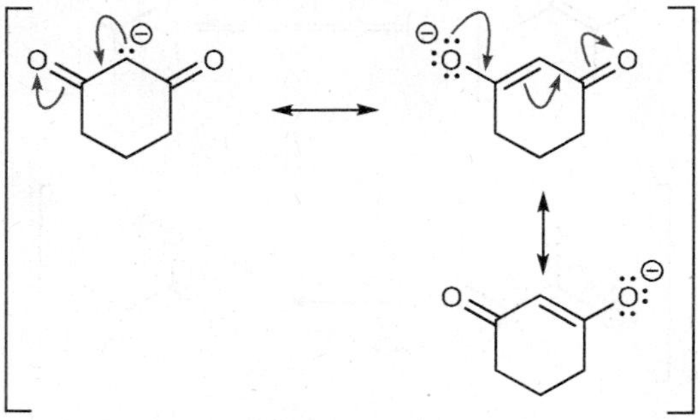

If ethoxide is used as the base to form the enolate, then enolate formation can be treated as nearly complete. That is, there will not be a substantial amount of diketone present after equilibrium has been established.

(b) This compound has four α positions:

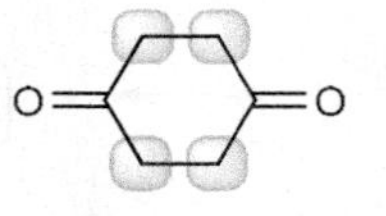

All four of these positions are equivalent because of symmetry (the structure has been rotated to make the symmetry more apparent). Therefore, deprotonation at any one of these positions results in the same enolate, which has the following two resonance structures:

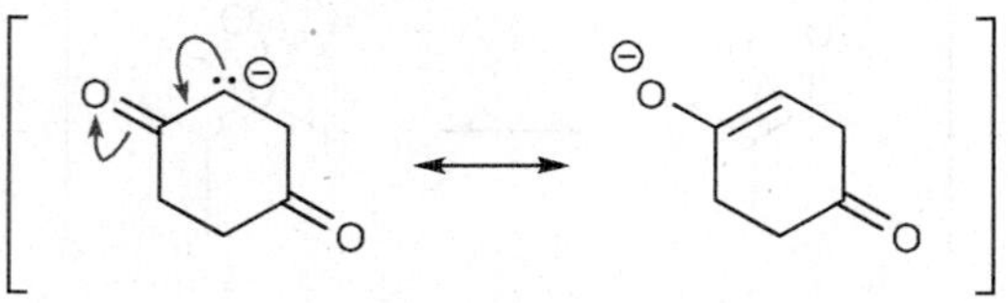

Notice that the negative charge is delocalized over only one oxygen atom (not two). As such, deprotonation of the diketone with ethoxide will result in a mixture containing both the enolate and the starting diketone. That is, there will be a substantial amount of diketone present after equilibrium has been established.

(c) This compound has two α positions, although only one of these positions bears protons. Deprotonation at that location will lead to an enolate ion with the following resonance structures:

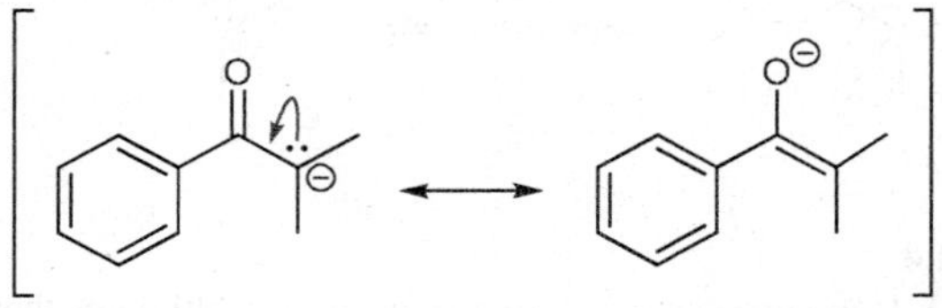

Notice that the negative charge is delocalized over only one oxygen atom. As such, deprotonation of the ketone with ethoxide will result in a mixture containing both the enolate and the starting ketone. That is, there will be a substantial amount of ketone present after equilibrium has been established.

(d) This compound has two α positions, although they are equivalent because the ketone is symmetrical. Deprotonation at either location will lead to the same enolate ion, which has the following resonance structures:

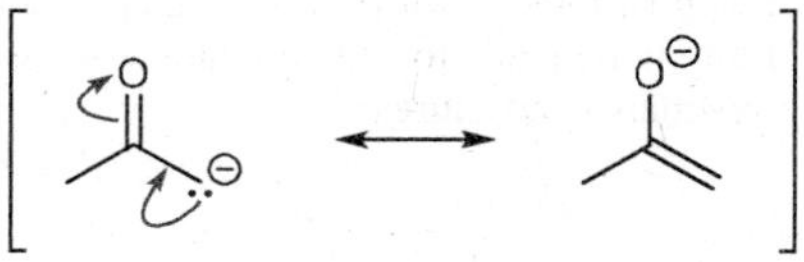

Notice that the negative charge is delocalized over only one oxygen atom. As such, deprotonation of the ketone with ethoxide will result in a mixture containing both the enolate and the starting ketone. That is, there will be a substantial amount of ketone present after equilibrium has been established.

21.7.

(a) 2,4-Dimethyl-3,5-heptanedione is more acidic because its conjugate base is a stabilized enolate (the anion is delocalized over *two* oxygen atoms). The other compound (4,4-dimethyl-3,5-heptanedione) cannot form a stabilized enolate because there are no protons connected to the carbon atom that is in between both carbonyl groups.

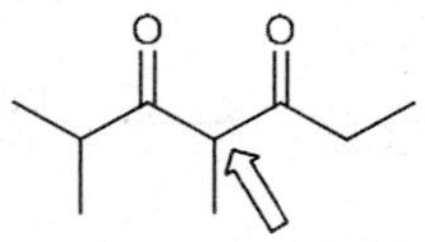

(more acidic)

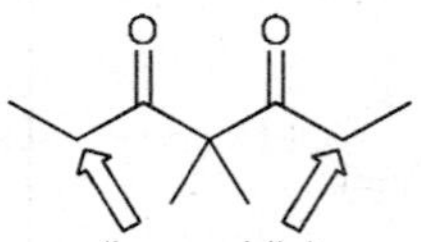

(less acidic)

(b) 1,3-Cyclopentanedione is more acidic because its conjugate base is a stabilized enolate (the anion is delocalized over *two* oxygen atoms). The other compound (1,2-cyclopentanedione) cannot form a stabilized enolate because the carbonyl groups are adjacent to each other.

(more acidic) (less acidic)

(c) Acetophenone is more acidic than benzaldehyde because acetophenone has α protons, while benzaldehyde does not. Deprotonation of acetophenone results in a resonance-stabilized conjugate base (an enolate ion).

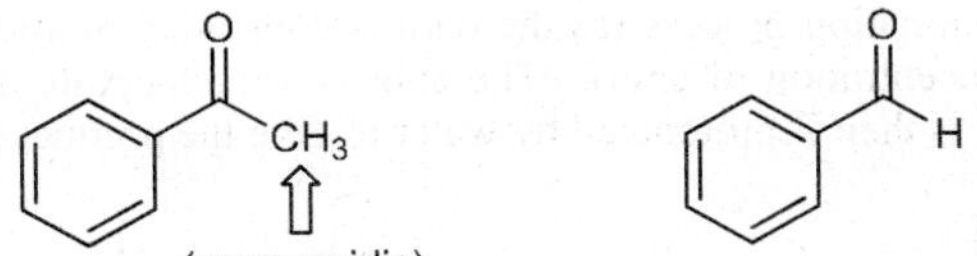

(more acidic)

21.8.

(a) These reagents indicate bromination at the α position, followed by elimination to give an α,β-unsaturated ketone:

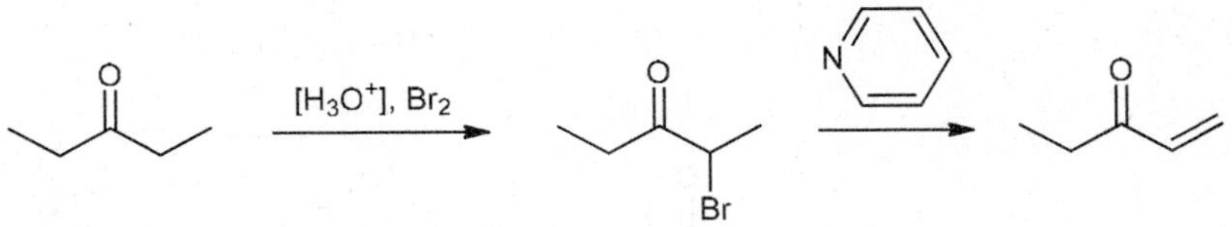

$[H_3O^+]$, Br_2

The following is a mechanism accounting for the entire transformation. Under acid-catalyzed conditions, the mechanism begins with a two-step tautomerization process to form an enol. The carbonyl group is protonated, giving a resonance-stabilized intermediate, which can then be deprotonated to give an enol. There is only a small amount of enol present at equilibrium, but its steady presence is responsible for the bromination process (as the enol is consumed by reacting with Br_2, the equilibrium is adjusted to replenish the small concentration of enol). The enol is a nucleophile and can attack molecular bromine (Br_2) to give an intermediate which is then deprotonated by water to give the product of the first reaction.

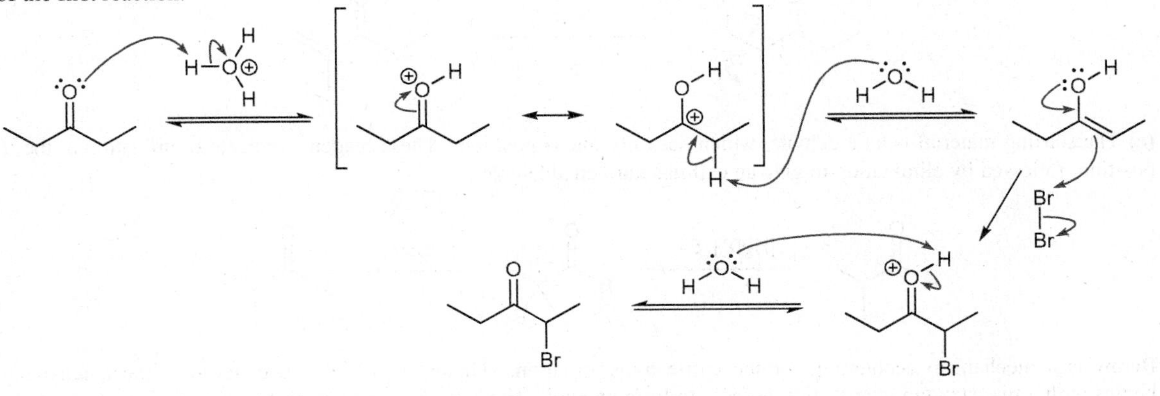

When this α-bromoketone is subsequently treated with pyridine, an E2 reaction gives the product:

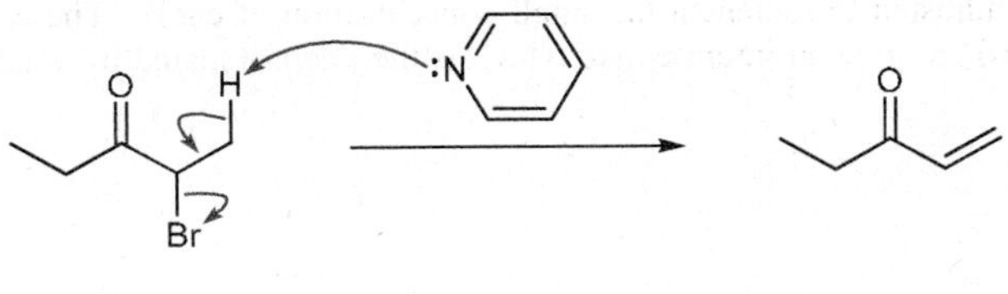

(b) These reagents indicate bromination at the more substituted α position, followed by elimination to give an α,β-unsaturated ketone:

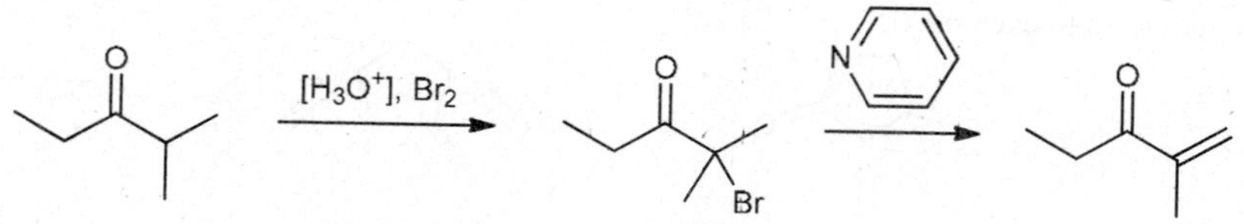

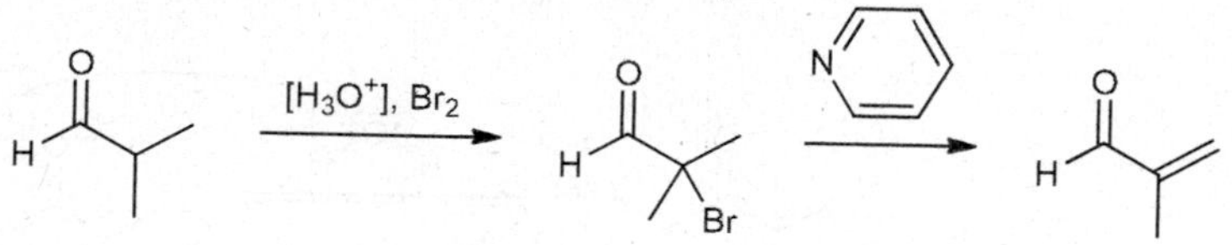

$[H_3O^+]$, Br_2

Br

Below is a mechanism accounting for the entire transformation. Under acid-catalyzed conditions, the mechanism begins with a two-step tautomerization process to form an enol. The carbonyl group is protonated, giving a resonance-stabilized intermediate, which can then be deprotonated to give an enol (the more substituted enol is favored over the less substituted enol at equilibrium). There is only a small amount of enol present at equilibrium, but its steady presence is responsible for the bromination process (as the enol is consumed by reacting with Br_2, the equilibrium is adjusted to replenish the small concentration of enol). The enol is a nucleophile and can attack molecular bromine (Br_2) to give an intermediate which is then deprotonated by water to give the product of the first reaction.

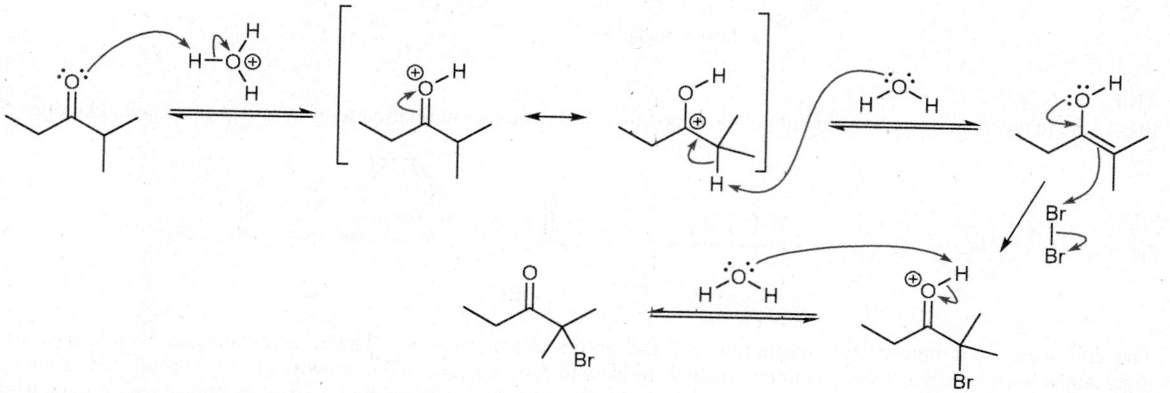

When this α-bromoketone is subsequently treated with pyridine, an E2 reaction gives the product:

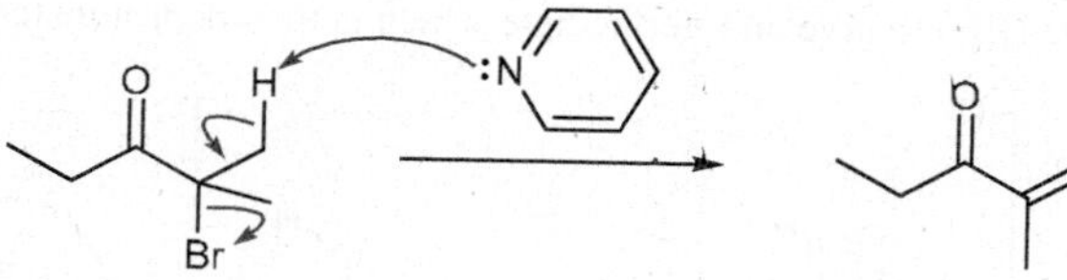

(c) The starting material is an aldehyde, which has only one α position. These reagents indicate bromination at the α position, followed by elimination to give an α,β-unsaturated aldehyde:

$[H_3O^+]$, Br_2

Below is a mechanism accounting for the entire transformation. Under acid-catalyzed conditions, the mechanism begins with a two-step tautomerization process to form an enol. The carbonyl group is protonated, giving a resonance-stabilized intermediate, which can then be deprotonated to give an enol. There is only a small amount of enol present at equilibrium, but its steady presence is responsible for the bromination process (as the enol is consumed by reacting with Br_2, the equilibrium is adjusted to replenish the small concentration of enol). The enol is a nucleophile and can attack molecular bromine (Br_2) to give an intermediate which is then deprotonated by water to give the product of the first reaction.

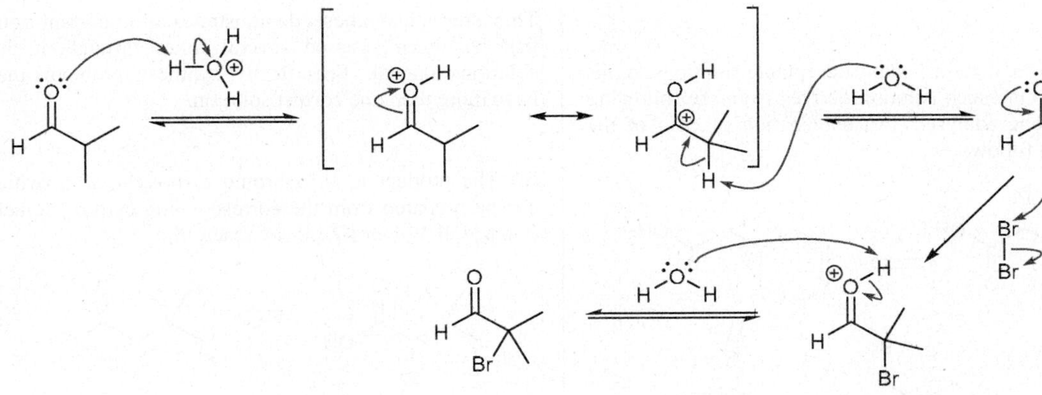

When this α-bromoaldehyde is subsequently treated with pyridine, an E2 reaction gives the product:

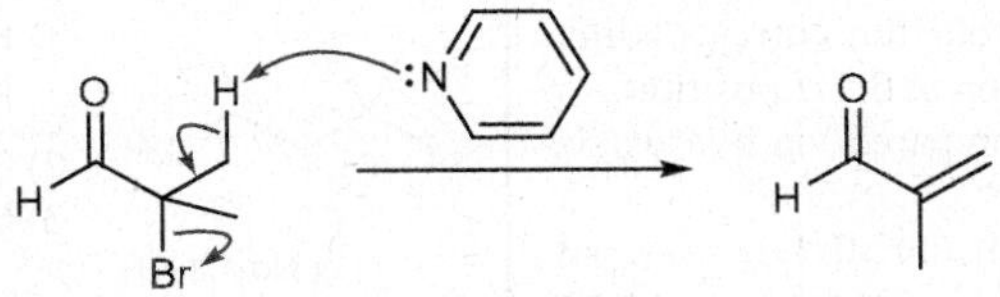

21.9.

(a) The product is an α,β-unsaturated ketone, which can be prepared from the corresponding saturated ketone (via acid-catalyzed halogenation followed by elimination),

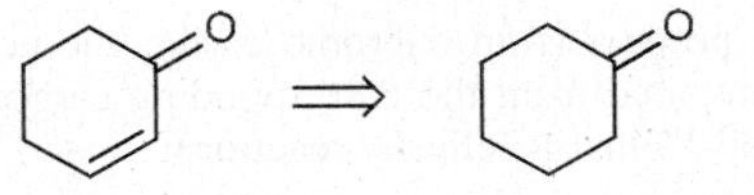

and this saturated ketone can be prepared from the starting secondary alcohol via oxidation.

The complete synthesis is shown here. The first step employs chromic acid as the oxidizing agent. Alternatively, PCC (or DMP or a Swern oxidation) can be used to affect the same transformation.

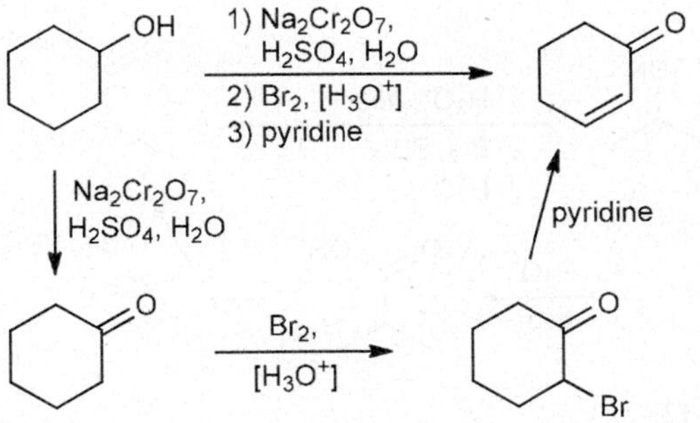

1) $Na_2Cr_2O_7$, H_2SO_4, H_2O
2) Br_2, $[H_3O^+]$
3) pyridine

$Na_2Cr_2O_7$, H_2SO_4, H_2O

Br_2, $[H_3O^+]$

pyridine

(b) The product is an α,β-unsaturated aldehyde, which can be prepared from the corresponding saturated aldehyde (via acid-catalyzed halogenation followed by elimination):

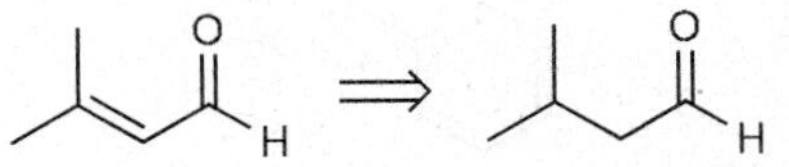

The saturated aldehyde can be prepared from the starting primary alcohol via oxidation with PCC (or with DMP or a Swern oxidation).

1) PCC, CH_2Cl_2
2) Br_2, $[H_3O^+]$
3) pyridine

PCC, CH_2Cl_2

Br_2, $[H_3O^+]$

pyridine

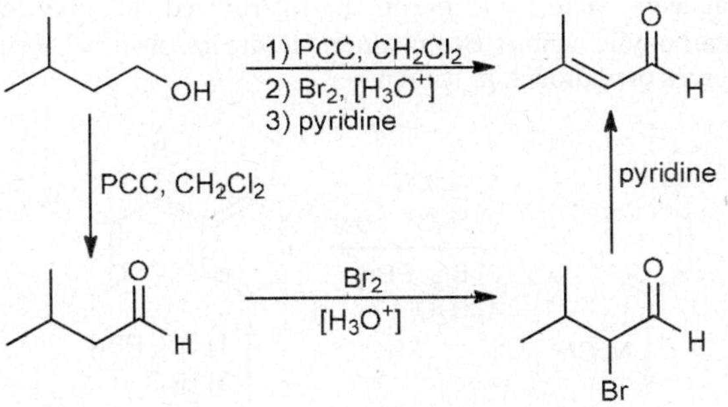

21.10.

(a) The starting material is a carboxylic acid, and the reagents indicate a Hell-Volhard-Zelinsky reaction. This process installs a bromine atom at the α position, as shown:

1) Br_2, PBr_3
2) H_2O

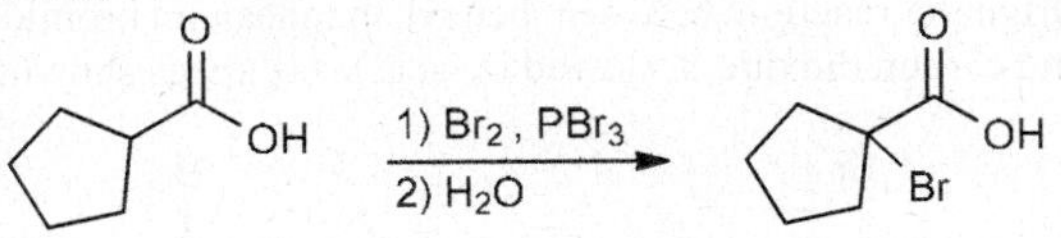

(b) The starting material is a carboxylic acid, and the reagents indicate a Hell-Volhard-Zelinsky reaction. This process installs a bromine atom at the α position, as shown:

1) Br_2, PBr_3
2) H_2O

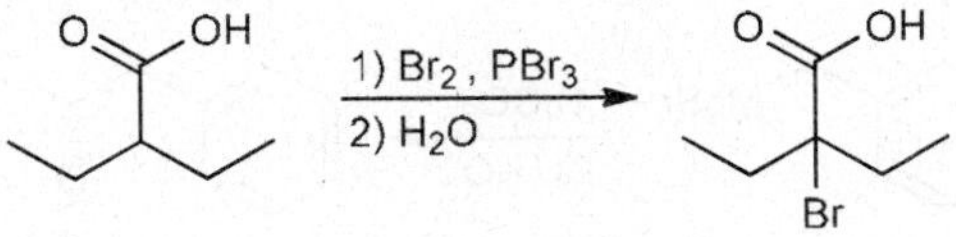

21.11.

(a) There are certainly many acceptable solutions to this problem. One such solution derives from the following retrosynthetic analysis. An explanation of each of the steps (*a-c*) follows.

Br OH O a OH O b Br c CN

a. The product can be made from the corresponding carboxylic acid via bromination at the α position.

b. The carboxylic acid can be prepared via hydrolysis of the corresponding nitrile.

c. The nitrile can be made from the starting material (benzyl bromide) via an S_N2 process in which cyanide is used as a nucleophile.

Now let's draw the forward scheme. Benzyl bromide is treated with sodium cyanide, giving an S_N2 reaction that results in formation of a nitrile. Upon treatment with aqueous acid, the nitrile is hydrolyzed to give a carboxylic acid. Bromination at the α position then gives the product, as shown.

Br 1) NaCN 2) H_3O^+, heat 3) Br_2, PBr_3 4) H_2O Br OH O

NaCN 1) Br_2, PBr_3 2) H_2O

CN H_3O^+ heat OH O

Alternatively, the carboxylic acid can be prepared via a Grignard reaction between benzyl magnesium bromide and carbon dioxide, followed by acid workup, as shown:

Br 1) Mg 2) CO_2 3) H_3O^+ 4) Br_2, PBr_3 5) H_2O Br OH O

Mg 1) Br_2, PBr_3 2) H_2O

MgBr 1) CO_2 2) H_3O^+ OH O

This alternate synthesis demonstrates an important point that has been stressed several times throughout this solutions manual. Specifically, synthesis problems may have more than one correct solution.

(b) The product is an α-bromo carboxylic acid, which can be prepared from the corresponding carboxylic acid (via a Hell-Volhard-Zelinsky reaction):

O OH Br ⟹ O OH

This carboxylic acid can be prepared from the starting primary alcohol via oxidation with chromic acid:

OH 1) $Na_2Cr_2O_7$, H_2SO_4, H_2O 2) Br_2, PBr_3 3) H_2O O OH Br

$Na_2Cr_2O_7$ H_2SO_4, H_2O O OH 1) Br_2, PBr_3 2) H_2O

(c) The product is an α-bromo carboxylic acid, which can be prepared from the corresponding carboxylic acid (via a Hell-Volhard-Zelinsky reaction):

O OH Br ⟹ O OH

This carboxylic acid can be prepared from the starting nitrile via hydrolysis with aqueous acid, as shown:

CN 1) H_3O^+, heat 2) Br_2, PBr_3 3) H_2O O OH Br

H_3O^+ heat O OH 1) Br_2, PBr_3 2) H_2O

21.12.

(a) The starting material is a methyl ketone, which is converted into the corresponding carboxylic acid (shown here) via the haloform reaction:

O OH

(b) The starting material is a methyl ketone, which is converted into the corresponding carboxylic acid (shown here) via the haloform reaction:

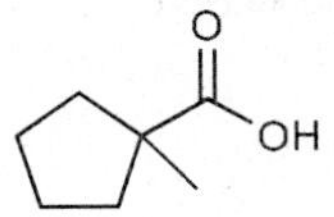

(c) The starting material is a methyl ketone, which is converted into the corresponding carboxylic acid (shown here) via the haloform reaction:

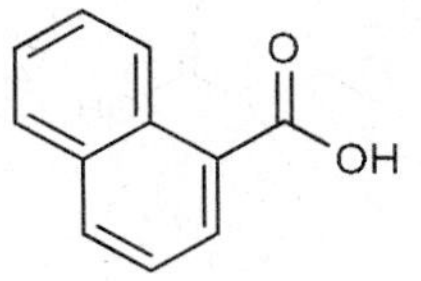

21.13.

(a) There are certainly many acceptable solutions to this problem. One such solution derives from the following retrosynthetic analysis. An explanation of each of the steps (*a-c*) follows.

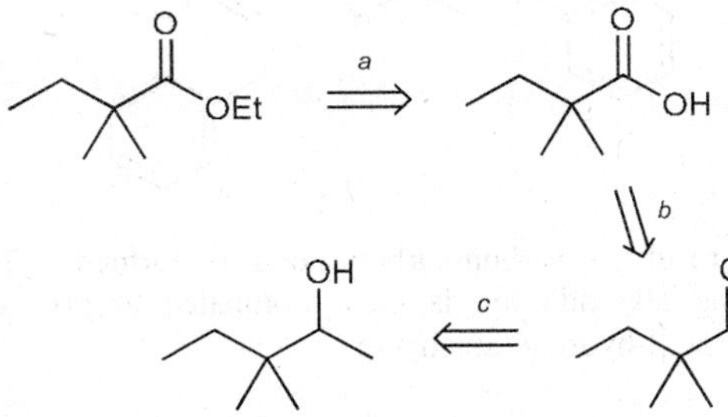

a. The product is an ester, so it can be made from the corresponding carboxylic acid via a Fischer esterification.

b. The carboxylic acid can be prepared from the corresponding methyl ketone via a haloform reaction.

c. The ketone can be made via oxidation of the corresponding secondary alcohol.

Now let's draw the forward scheme. The starting alcohol is oxidized upon treatment with chromic acid (alternatively, PCC or DMP can be used for this step). The resulting ketone is then treated with molecular bromine (Br_2) and sodium hydroxide, followed by aqueous acid, to give a carboxylic acid (via a haloform reaction). Finally, the carboxylic acid is treated with ethanol in the presence of an acid catalyst, giving the desired ester (via a Fischer esterification).

1) $Na_2Cr_2O_7$, H_2SO_4, H_2O
2) Br_2, NaOH
3) H_3O^+
4) $[H^+]$, EtOH

$Na_2Cr_2O_7$, H_2SO_4, H_2O

$[H^+]$, EtOH

1) Br_2, NaOH
2) H_3O^+

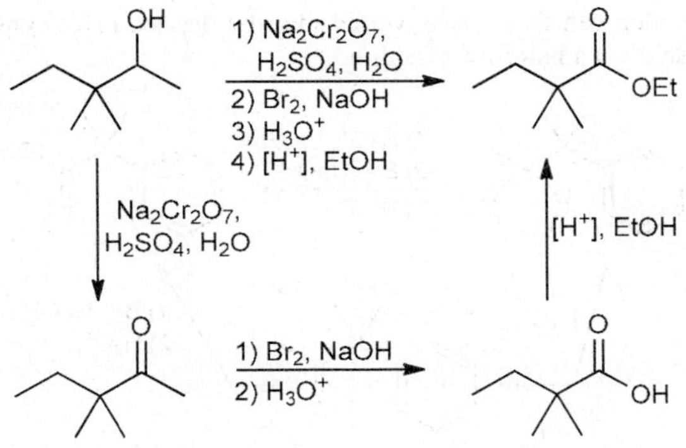

(b) An efficient solution to this problem derives from the following retrosynthetic analysis. An explanation of each of the steps (*a-c*) follows.

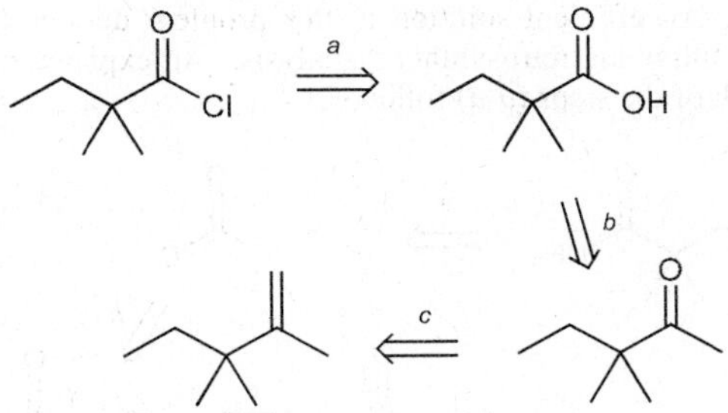

a. The product is an acid chloride which can be made from the corresponding carboxylic acid.

b. The carboxylic acid can be prepared from the corresponding methyl ketone via a haloform reaction.

c. The ketone can be made from the starting alkene via ozonolysis.

Now let's draw the forward scheme. Ozonolysis converts the starting alkene into a ketone (with loss of a carbon atom). The resulting ketone is then treated with molecular bromine (Br_2) and sodium hydroxide, followed by aqueous acid, to give a carboxylic acid (via a haloform reaction). Finally, the carboxylic acid is converted into an acid chloride upon treatment with thionyl chloride.

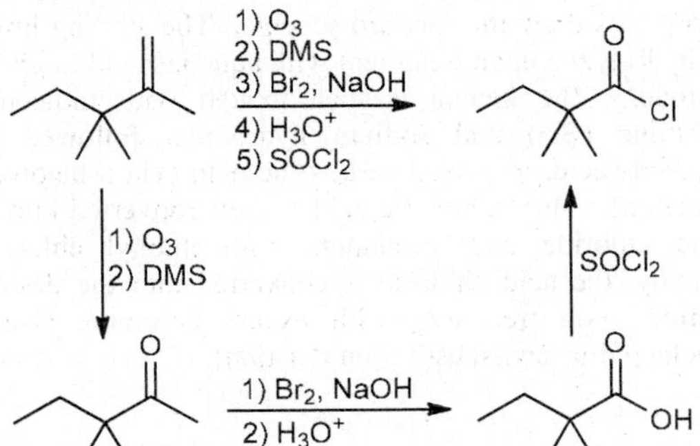

1) O_3
2) DMS
3) Br_2, NaOH
4) H_3O^+
5) $SOCl_2$

1) O_3
2) DMS

$SOCl_2$

1) Br_2, NaOH
2) H_3O^+

(c) The starting material is an acetal. Upon treatment with aqueous acid, the acetal is converted to a ketone, which can then be converted into the desired carboxylic acid via a haloform reaction.

1) H_3O^+
2) Br_2, NaOH
3) H_3O^+

H_3O^+

1) Br_2, NaOH
2) H_3O^+

(d) An efficient solution to this problem derives from the following retrosynthetic analysis. An explanation of each of the steps (*a-d*) follows.

a. The product is an amide, which can be made from the corresponding acid chloride.
b. The acid chloride can be made from the corresponding carboxylic acid upon treatment with thionyl chloride.
c. The carboxylic acid can be prepared from the corresponding methyl ketone via a haloform reaction.
d. The ketone can be made via hydrolysis of the starting imine.

Now let's draw the forward scheme. The starting imine is hydrolyzed upon treatment with aqueous acid to give a ketone. The ketone is then treated with molecular bromine (Br_2) and sodium hydroxide, followed by aqueous acid, to give a carboxylic acid (via a haloform reaction). The carboxylic acid is then converted into an acid chloride upon treatment with thionyl chloride. Finally, the acid chloride is converted into the desired amide upon treatment with excess ammonia (via a nucleophilic acyl substitution reaction).

1) H_3O^+
2) Br_2, NaOH
3) H_3O^+
4) $SOCl_2$
5) xs NH_3

H_3O^+

xs NH_3

1) Br_2, NaOH
2) H_3O^+

$SOCl_2$

21.14.

(a) The α position of one molecule of the aldehyde is deprotonated, and the resulting enolate functions as a nucleophile and attacks the electrophilic carbonyl group of another molecule of the aldehyde:

As a result, a carbon-carbon bond is formed. The resulting alkoxide ion is then protonated to give the following β-hydroxy aldehyde:

(b) The α position of one molecule of the aldehyde is deprotonated, and the resulting enolate functions as a nucleophile and attacks the electrophilic carbonyl group of another molecule of the aldehyde:

As a result, a carbon-carbon bond is formed. The resulting alkoxide ion is then protonated to give the following β-hydroxy aldehyde:

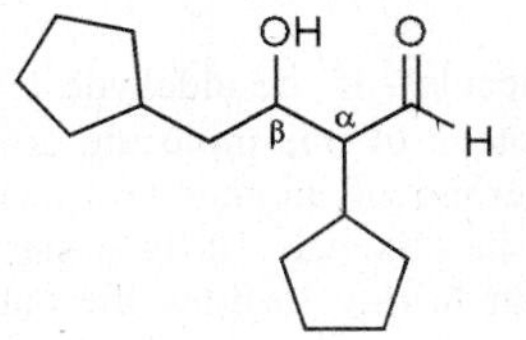

(c) The α position of one molecule of the aldehyde is deprotonated, and the resulting enolate functions as a nucleophile and attacks the electrophilic carbonyl group of another molecule of the aldehyde:

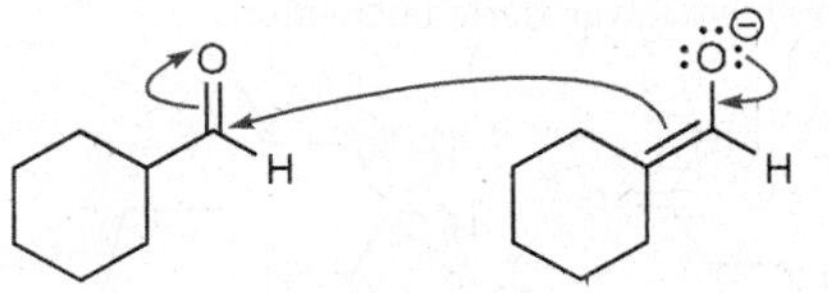

As a result, a carbon-carbon bond is formed. The resulting alkoxide ion is then protonated to give the following β-hydroxy aldehyde:

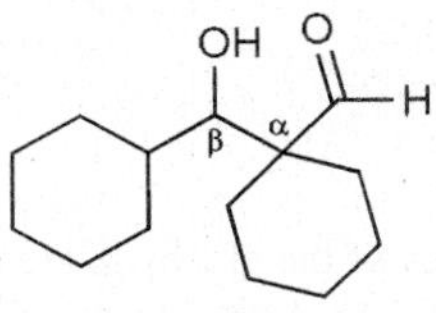

(d) The α position of one molecule of the aldehyde is deprotonated, and the resulting enolate functions as a nucleophile and attacks the electrophilic carbonyl group of another molecule of the aldehyde:

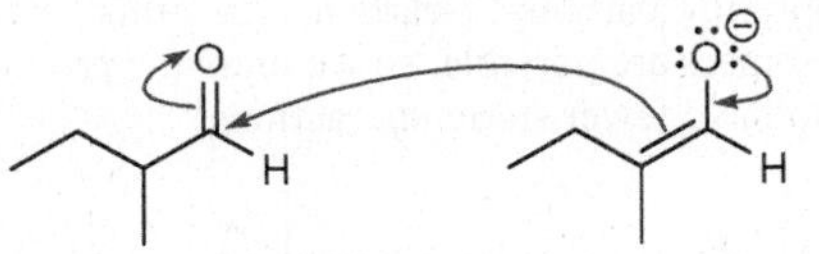

As a result, a carbon-carbon bond is formed. The resulting alkoxide ion is then protonated to give the following β-hydroxy aldehyde:

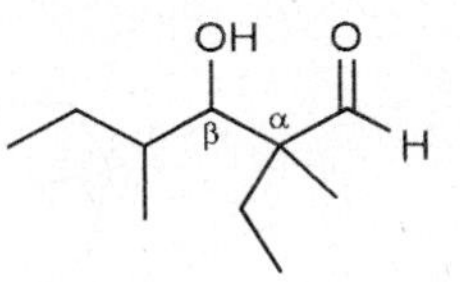

21.15.

(a) This compound has two α positions, although they are equivalent because the ketone is symmetrical. That is, deprotonation at either location will lead to the same enolate. This enolate can then function as a nucleophile and attack the electrophilic carbonyl group of another molecule of the ketone. As a result, a carbon-carbon bond is formed. The resulting alkoxide ion is then protonated to give a β-hydroxy ketone, as shown:

NaOH, H_2O

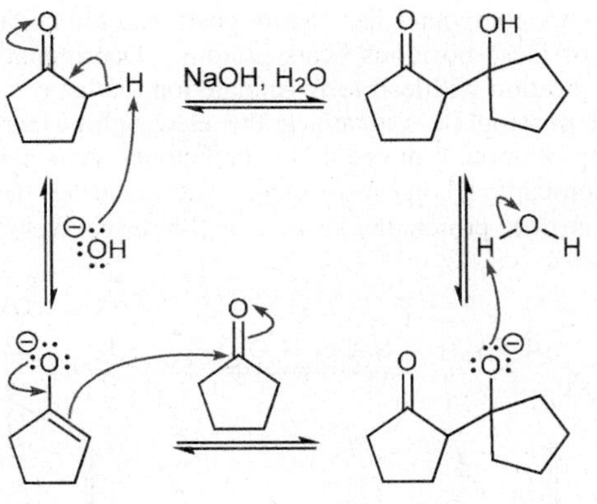

(b) This compound has two α positions, although only one of these positions bears protons. Deprotonation at that location will lead to an enolate ion that can function as a nucleophile and attack the electrophilic carbonyl group of another molecule of the ketone. As a result, a carbon-carbon bond is formed. The resulting alkoxide ion is then protonated to give a β-hydroxy ketone, as shown:

NaOH, H_2O

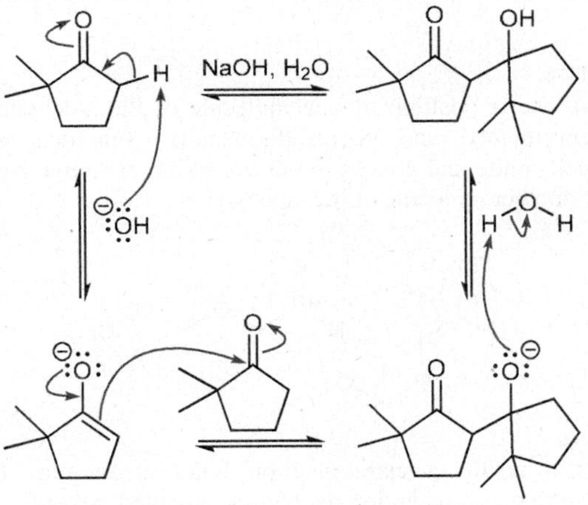

(c) This compound has two α positions, although they are equivalent because the ketone is symmetrical. That is, deprotonation at either location will lead to the same enolate. This enolate can then function as a nucleophile and attack the electrophilic carbonyl group of another molecule of the ketone. As a result, a carbon-carbon bond is formed. The resulting alkoxide ion is then protonated to give a β-hydroxy ketone, as shown:

NaOH, H_2O

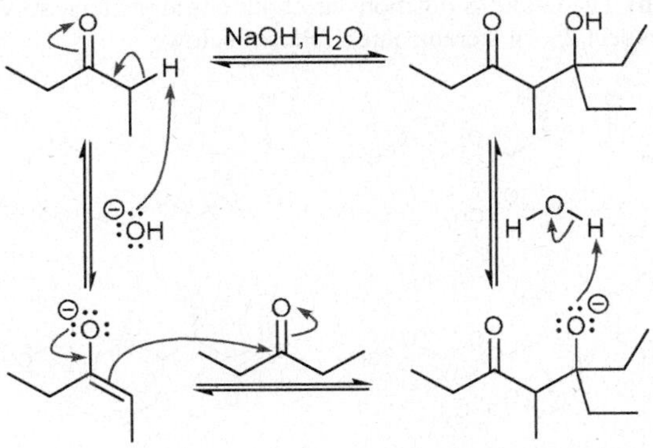

(d) This compound has two α positions, although only one of these positions bears protons. Deprotonation at that location will lead to an enolate ion that can function as a nucleophile and attack the electrophilic carbonyl group of another molecule of the ketone. As a result, a carbon-carbon bond is formed. The resulting alkoxide ion is then protonated to give a β-hydroxy ketone, as shown:

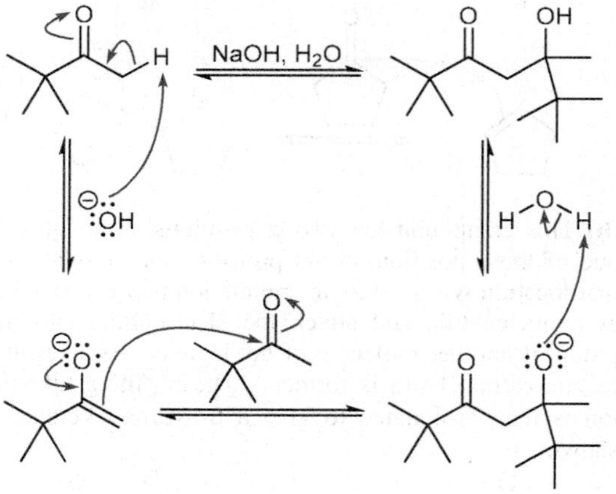

NaOH, H_2O

21.16.

(a) The α position of one molecule of the aldehyde is deprotonated, and the resulting enolate functions as a nucleophile and attacks the electrophilic carbonyl group of another molecule of the aldehyde:

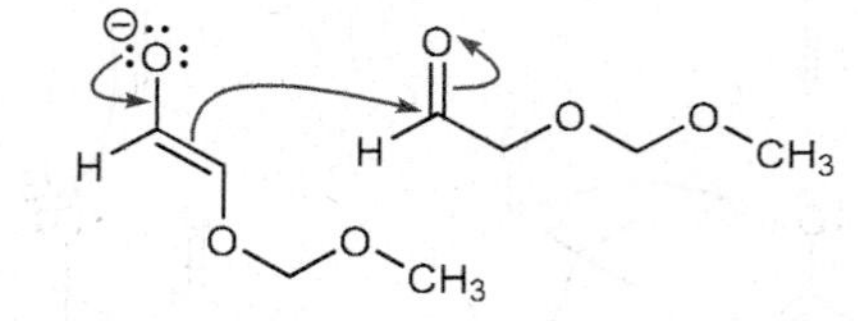

As a result, a carbon-carbon bond is formed. The resulting alkoxide ion is then protonated to give the following β-hydroxy aldehyde:

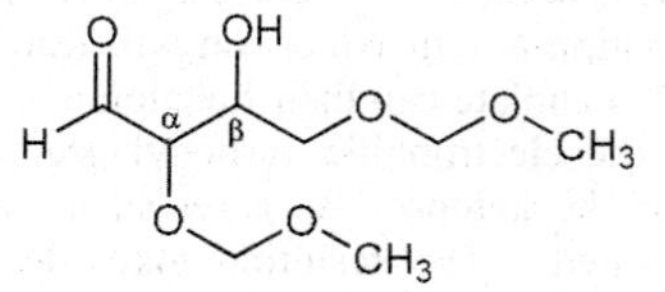

(b) The α and β positions are both chiral centers, so we expect $2^2 = 4$ stereoisomers, shown below:

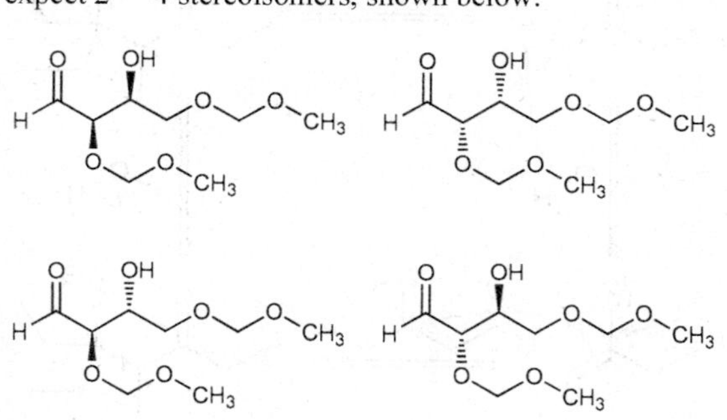

21.17.

(a) Two molecules of the aldehyde are redrawn such that two α protons of one molecule are directly facing the carbonyl group of another molecule (highlighted). When drawn in this way, it is easier to predict the product without having to draw the entire mechanism. We simply remove the two α protons and the oxygen atom, and we replace them with a double bond. The dashed line illustrates the new carbon-carbon bond being formed, between the nucleophilic alpha carbon (enolate) and the electrophilic carbonyl carbon. In this case, two stereoisomers are possible, so we draw the product that is likely to have fewer steric interactions:

$-H_2O$

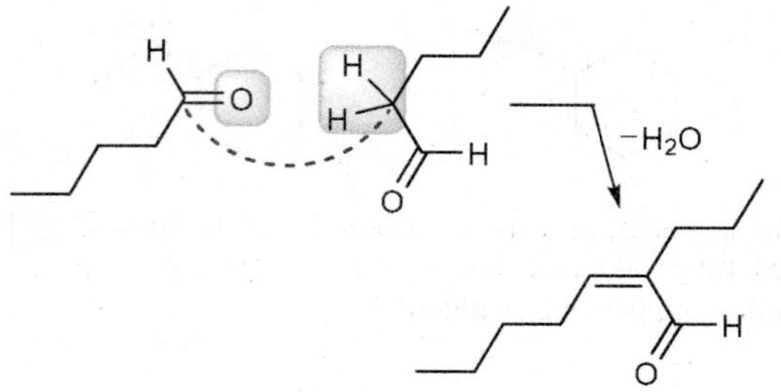

(b) Two molecules of the aldehyde are redrawn such that two α protons of one molecule are directly facing the carbonyl group of another molecule (highlighted). We then remove the two α protons and the oxygen atom, and we replace them with a double bond. The dashed line illustrates the new carbon-carbon bond being formed, between the nucleophilic alpha carbon (enolate) and the electrophilic carbonyl carbon. In this case, two stereoisomers are possible, so we draw the product that is likely to have fewer steric interactions:

$-H_2O$

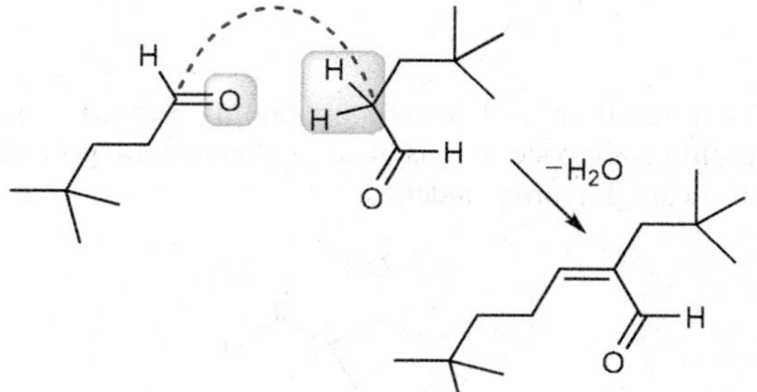

(c) Two molecules of the aldehyde are redrawn such that two α protons of one molecule are directly facing the carbonyl group of another molecule (highlighted). We then remove the two α protons and the oxygen atom, and we replace them with a double bond. The dashed line illustrates the new carbon-carbon bond being formed, between the nucleophilic alpha carbon (enolate) and the electrophilic carbonyl carbon. In this case, two stereoisomers are possible, so we draw the product that is likely to have fewer steric interactions:

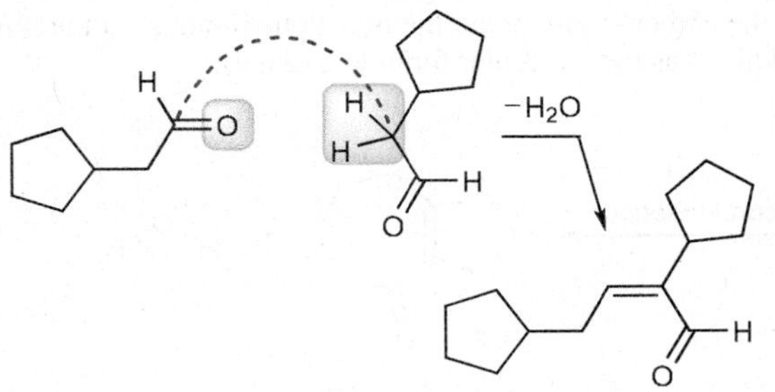

(d) This compound has two α positions, but only one of these positions has the required *two* α protons for the condensation reaction to occur.

Two molecules of the ketone are redrawn such that two α protons of one molecule are directly facing the carbonyl group of another molecule (highlighted). We then remove the two α protons and the oxygen atom, and we replace them with a double bond. The dashed line illustrates the new carbon-carbon bond being formed, between the nucleophilic alpha carbon (enolate) and the electrophilic carbonyl carbon. In this case, two stereoisomers are possible, so we draw the product in which the larger groups are farther apart from each other (circled below). This *E* isomer has fewer steric interactions:

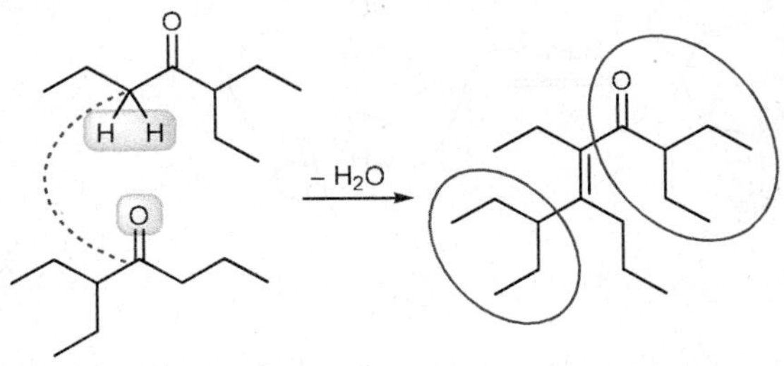

(e) This compound has two α positions (each with the required two α protons), although they are equivalent because the ketone is symmetrical, so we only need to consider the reaction occurring at one of these locations.

Two molecules of the ketone are redrawn such that two α protons of one molecule are directly facing the carbonyl group of another molecule (highlighted). We then remove the two α protons and the oxygen atom, and we replace them with a double bond. The dashed line illustrates the new carbon-carbon bond being formed, between the nucleophilic alpha carbon (enolate) and the electrophilic carbonyl carbon:

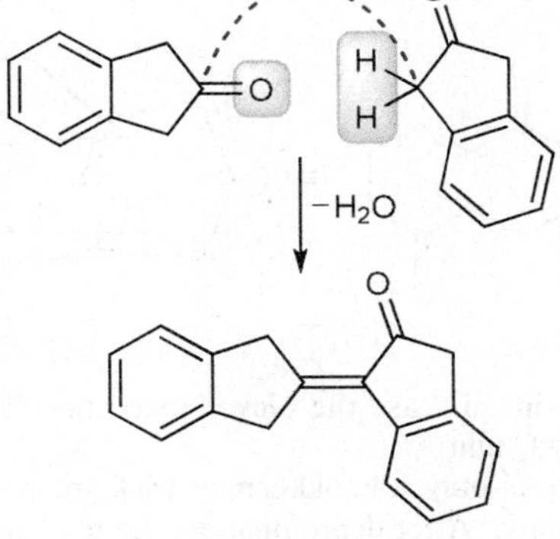

(f) This compound has two α positions, but only one of these positions has the required *two* α protons for the condensation reaction to occur.

Two molecules of the ketone are redrawn such that two α protons of one molecule are directly facing the carbonyl group of another molecule (highlighted). We then remove the two α protons and the oxygen atom, and we replace them with a double bond. The dashed line illustrates the new carbon-carbon bond being formed, between the nucleophilic alpha carbon (enolate) and the electrophilic carbonyl carbon. In this case, two stereoisomers are possible, so we draw the product with fewer steric interactions (the *E* isomer):

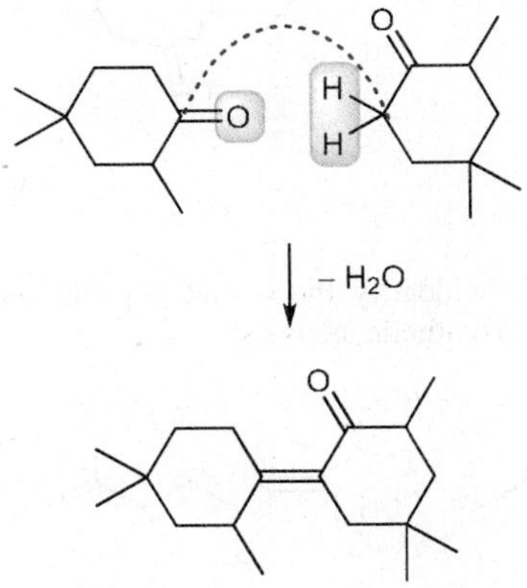

21.18. Compound **1** has 8 equivalent alpha protons, highlighted below:

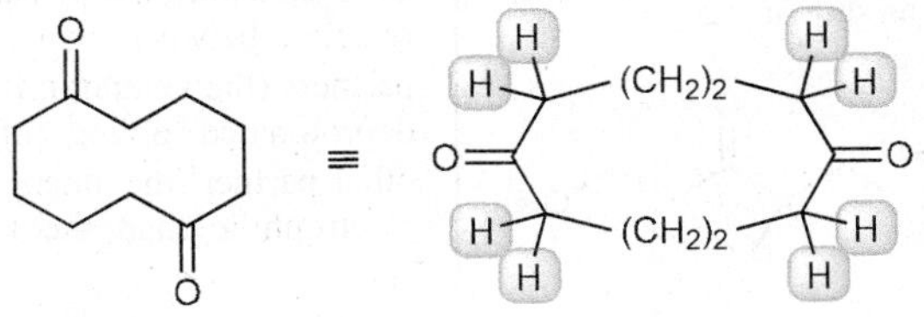

Removal of any one of these protons with a base will yield the same enolate and ultimately lead to the same final aldol condensation product. To draw the aldol condensation product, we redraw compound **1** to show one set of alpha protons directly facing the other carbonyl group. Removing the carbonyl group and the two alpha protons (in the box below) gives the structure of the final product, compound **2**, which has the molecular formula $C_{10}H_{14}O$.

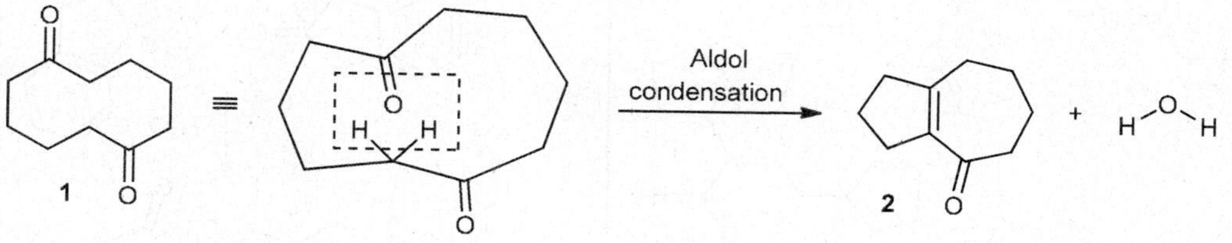

Aldol condensation

Note that in this case the bicyclic structure limits the product to only one possible stereoisomer around the newly-formed C=C unit.

The above is only a bookkeeping trick for predicting the product. It is beneficial to think about the mechanism as shown below. After deprotonation, the resulting enolate ion attacks the carbonyl group, giving an alkoxide ion. After protonation of the alkoxide, a base-catalyzed elimination process gives an α,β–unsaturated ketone:

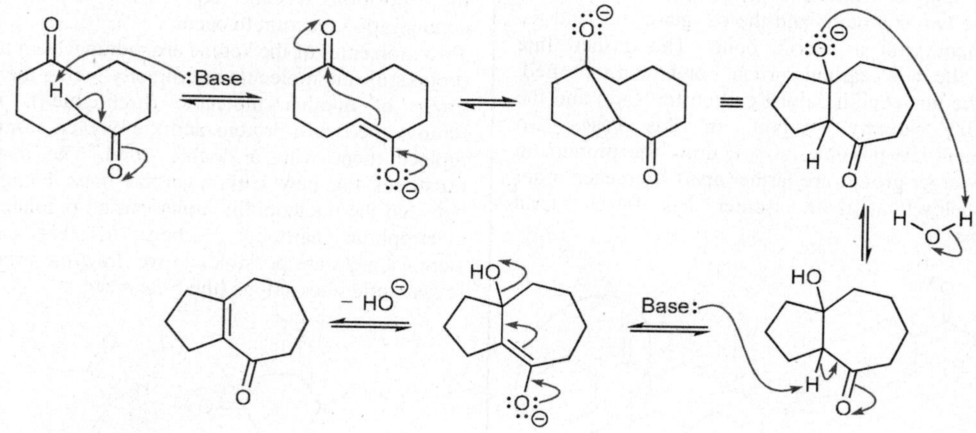

21.19.

(a) We first identify the α and β positions, and then apply a retrosynthetic analysis:

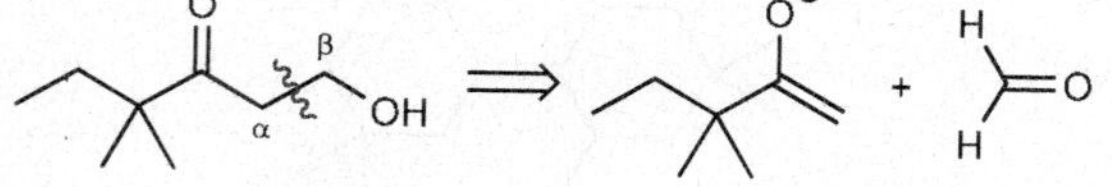

This transformation can be achieved with an aldol reaction between two different partners. One of the partners (the ketone) only has one α position that can be deprotonated (giving only one possible enolate), and the other partner (formaldehyde) lacks α protons and is more electrophilic than the ketone. As such, LDA is not required as a base for this directed aldol addition. Sodium hydroxide can be used, as shown:

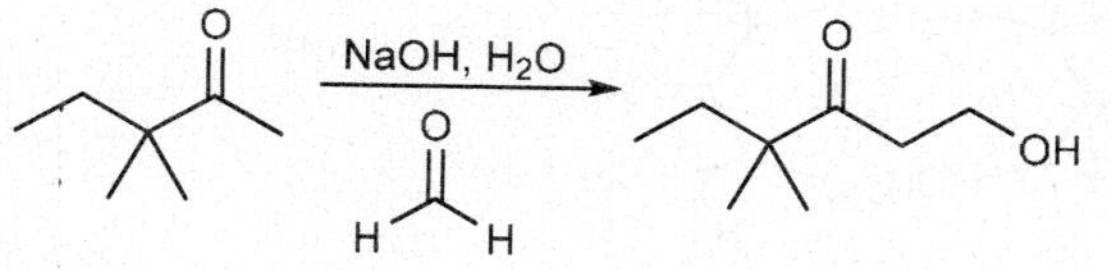

NaOH, H_2O

(b) We first identify the α and β positions, and then apply a retrosynthetic analysis:

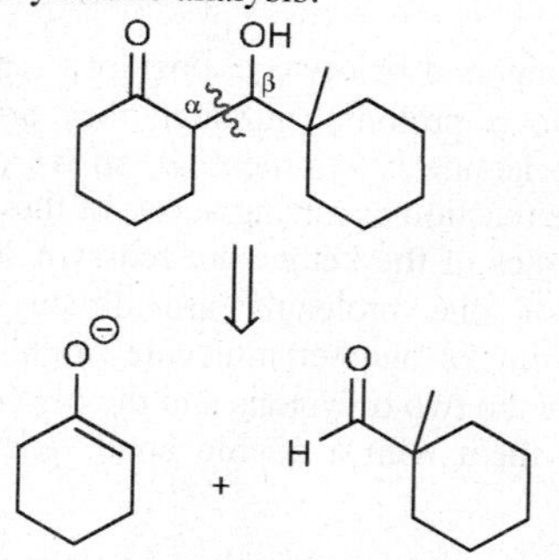

This transformation can be achieved with an aldol reaction between two different partners. One of the partners (the ketone) only has one α position that can be deprotonated (giving only one possible enolate), and the other partner (the aldehyde) lacks α protons and is more electrophilic than the ketone. As such, LDA is not required as a base for this directed aldol addition. Sodium hydroxide can be used, as shown:

NaOH, H_2O

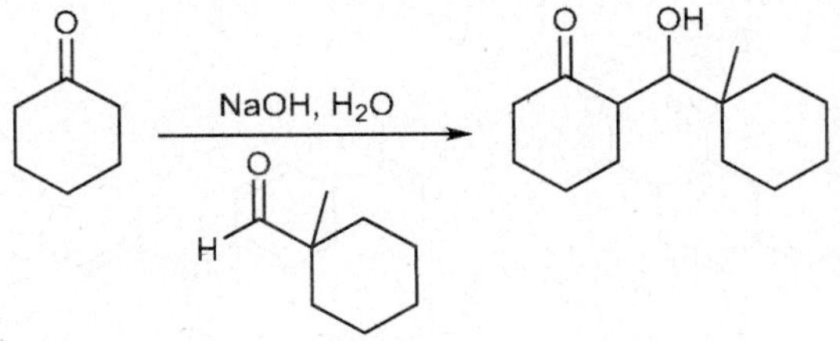

(c) We first identify the α and β positions, and then apply a retrosynthetic analysis:

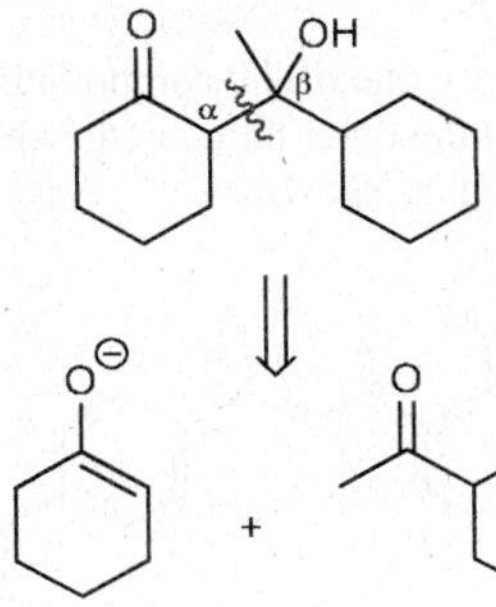

Two different partners are required, each of which can be deprotonated to give an enolate. Therefore, this transformation must be achieved with a directed aldol reaction, using LDA as the base, followed by aqueous acidic workup, as shown:

1) LDA

2)

3) H_3O^+

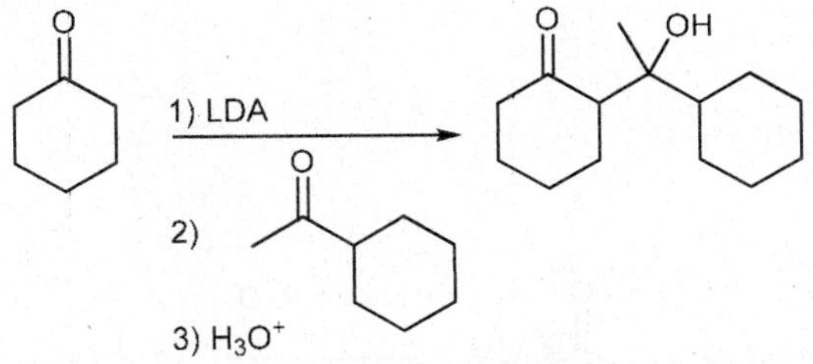

(d) We first identify the α and β positions, and then apply a retrosynthetic analysis to identify the starting materials:

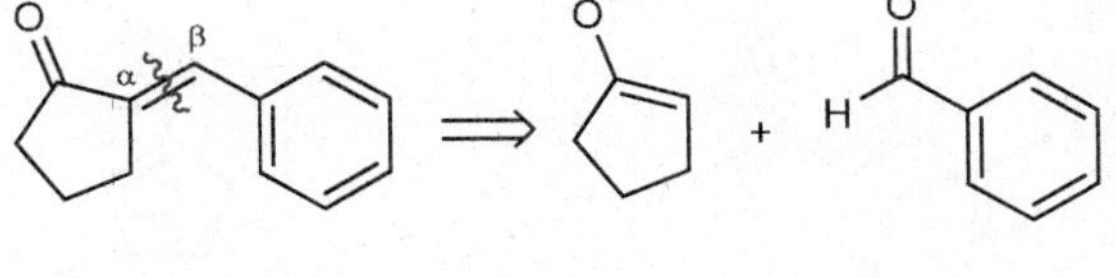

This transformation can be achieved with an aldol reaction between two different partners. One of the partners (the ketone) only has one α position that can be deprotonated (giving only one possible enolate), and the other partner (the aldehyde) lacks α protons and is more electrophilic than the ketone. As such, LDA is not required as a base for this directed aldol addition. Sodium hydroxide can be used, as shown:

NaOH, H_2O
heat

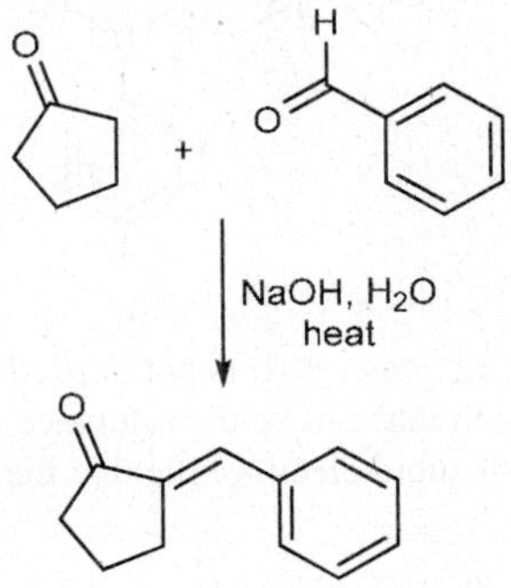

(e) We first identify the α and β positions, and then apply a retrosynthetic analysis:

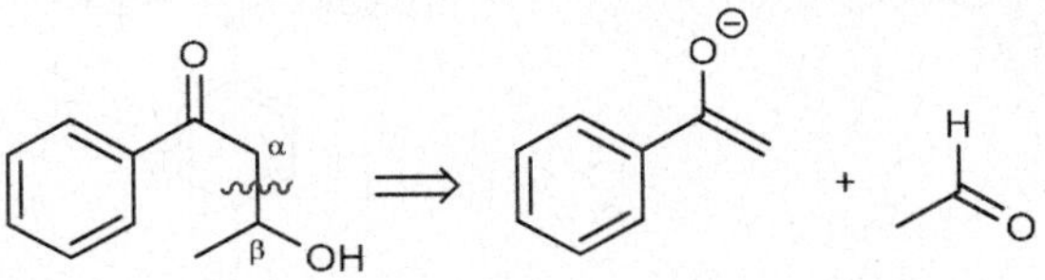

Two different partners are required, each of which can be deprotonated to give an enolate. Therefore, this transformation must be achieved with a directed aldol reaction, using LDA as the base, followed by aqueous acidic workup, as shown:

1) LDA

2)

3) H_3O^+

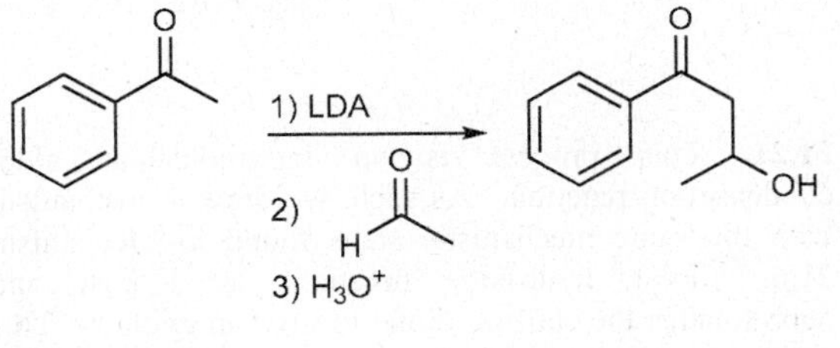

21.20. We first identify the part of the structure of DPHP that is produced by an aldol reaction, and look for either a β–hydroxy carbonyl or an α,β-unsaturated carbonyl. We do see in DPHP an α,β-unsaturated carbonyl. After identifying the α and β positions, we then apply a retrosynthetic analysis to identify the starting materials:

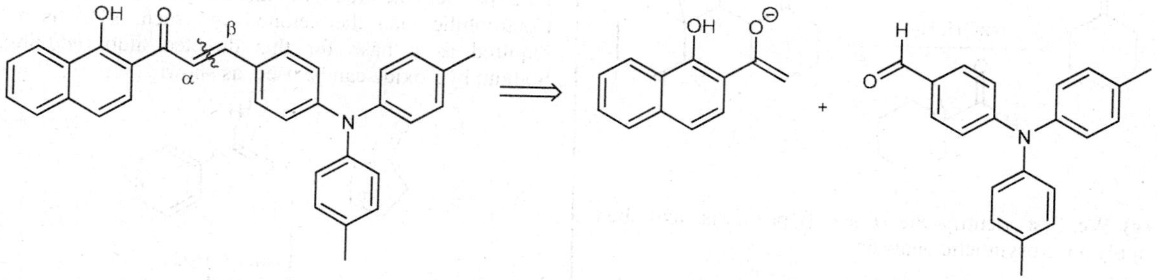

A strong base, such as LDA, is not needed for this crossed aldol reaction since only one of the compounds (the ketone) has an α position that can be deprotonated (giving only one possible enolate) and the other partner (the aldehyde) lacks α protons and is more electrophilic that the ketone. Sodium hydroxide can be used, as shown:

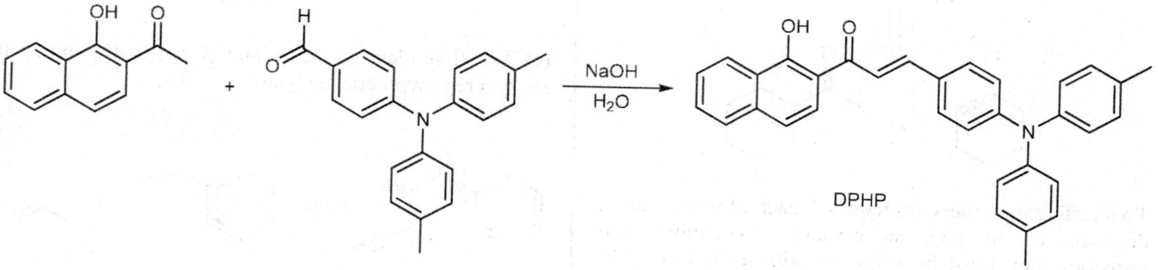

21.21. This process is an intramolecular aldol condensation reaction. As such, we draw a mechanism with the same mechanistic steps found in Mechanism 21.6. First, hydroxide functions as a base and deprotonates the starting dione to give an enolate. This enolate ion is a nucleophilic center, and it will attack the carbonyl group present in the same structure (an intramolecular process), thereby closing a five-membered ring. The resulting alkoxide ion is then protonated to give a β-hydroxy ketone. Hydroxide then functions as a base again, deprotonating the α position. The resulting enolate then ejects a hydroxide ion, giving the condensation product, as shown:

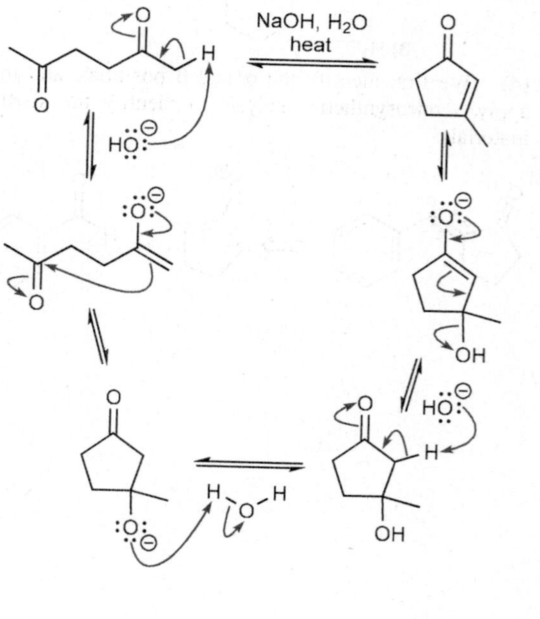

21.22. The following process is the reverse of the mechanism shown in the previous problem. As such, all of the intermediates are identical to the intermediates in the previous problem, but they appear in reverse order. First, hydroxide attacks the β position of the α,β-unsaturated ketone (conjugate addition), giving an enolate. The enolate is then protonated to give a β-hydroxyketone, which is subsequently deprotonated to give an alkoxide ion. The carbonyl group is then reformed, with loss of an enolate as a leaving group (this is a retro-aldol process). The resulting enolate is then protonated to give the dione. Notice that each of the protonation steps employs water as an acid, consistent with basic conditions (strong acids are not measurably present under these conditions).

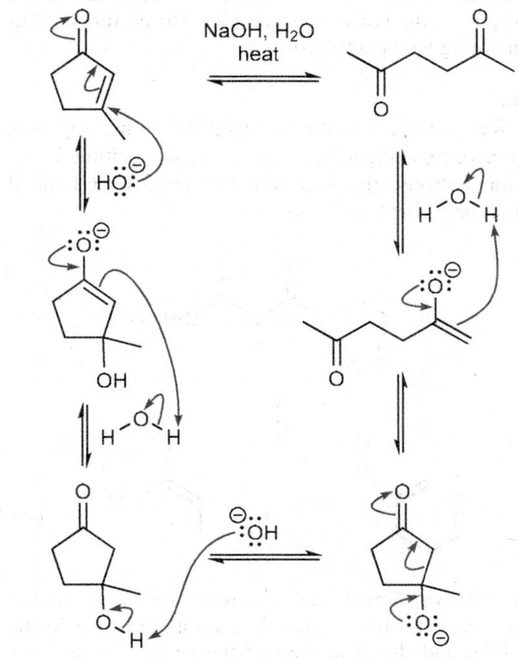

21.23. This process is an intramolecular aldol condensation reaction, and is similar to Problem **21.21**, with one additional methylene (CH_2) group in between the two carbonyl groups. As such, a six-membered ring is formed, rather than a five-membered ring. Other than this small difference, this mechanism is identical to the mechanism shown in the solution to Problem **21.21**.
First, hydroxide functions as a base and deprotonates the starting dione to give an enolate. This enolate ion is a nucleophilic center, and it will attack the carbonyl group present in the same structure (an intramolecular process), thereby closing a six-membered ring. The resulting alkoxide ion is then protonated to give a β-hydroxy ketone. Hydroxide then functions as a base again, deprotonating the α position. The resulting enolate then ejects a hydroxide ion, giving the condensation product, as shown:

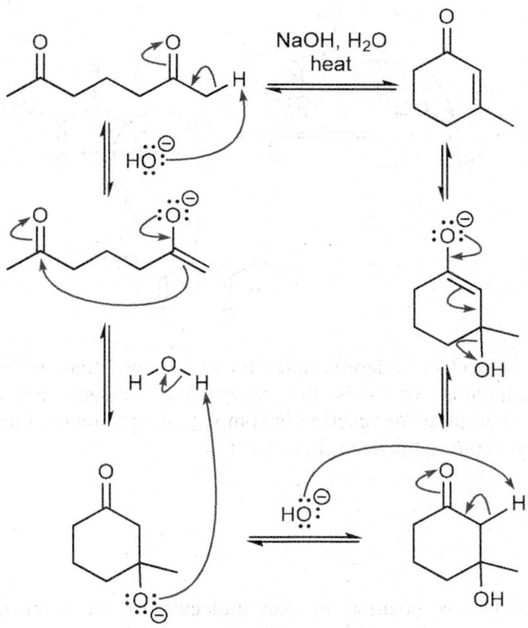

21.24.
(a) The starting material is an ester, and the product is a β-ketoester. Therefore, this process is a Claisen condensation. In this case, two identical partners will react with each other (this is not a crossed Claisen condensation). Since the alkoxy group of the ester is an ethoxy group, we must use sodium ethoxide as the base (NaOEt), in order to avoid hydrolysis or trans-esterification.

(b) The starting material is an ester, and the product is a β-ketoester. Therefore, this process is a Claisen condensation. In this case, two identical partners will react with each other (this is not a crossed Claisen condensation). Since the alkoxy group of the ester is a *tert*-butoxy group, we must use potassium *tert*-butoxide as the base (*t*-BuOK), in order to avoid hydrolysis or transesterification.

21.25.
(a) The α position of one molecule of the ester is deprotonated, and the resulting enolate functions as a nucleophile and attacks the electrophilic carbonyl group of another molecule of the ester. As a result, a carbon-carbon bond is formed, giving a charged tetrahedral intermediate. The carbonyl group is then reformed via loss of an ethoxide ion, affording a β-ketoester, as shown:

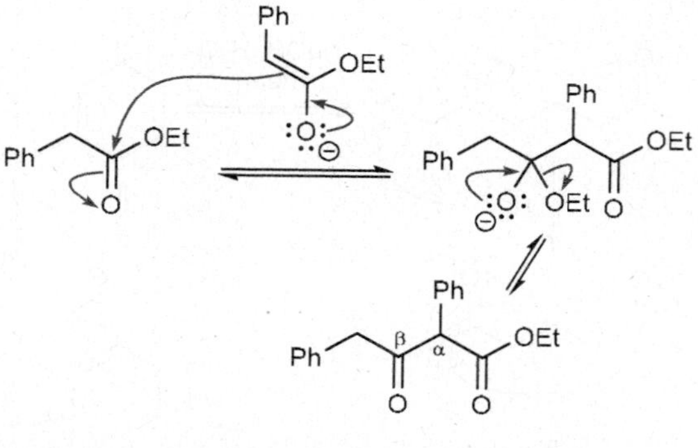

This product is deprotonated under the conditions of its formation, which is the reason for aqueous acidic workup after the reaction is complete (to protonate, thus regenerating the neutral product).

(b) The α position of one molecule of the ester is deprotonated, and the resulting enolate functions as a nucleophile and attacks the electrophilic carbonyl group of another molecule of the ester. As a result, a carbon-carbon bond is formed, giving a charged tetrahedral intermediate. The carbonyl group is then reformed via loss of a methoxide ion, affording a β-ketoester, as shown:

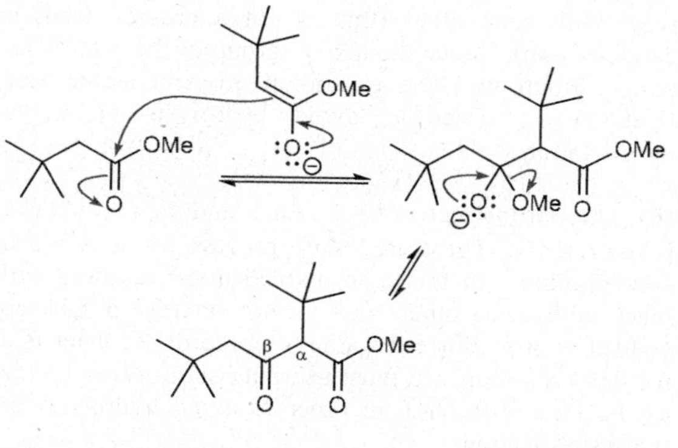

This product is deprotonated under the conditions of its formation, which is the reason for aqueous acidic workup after the reaction is complete (to protonate, thus regenerating the neutral product).

(c) The α position of one molecule of the ester is deprotonated, and the resulting enolate functions as a nucleophile and attacks the electrophilic carbonyl group of another molecule of the ester. As a result, a carbon-carbon bond is formed, giving a charged tetrahedral intermediate. The carbonyl group is then reformed via loss of an ethoxide ion, affording a β-ketoester, as shown:

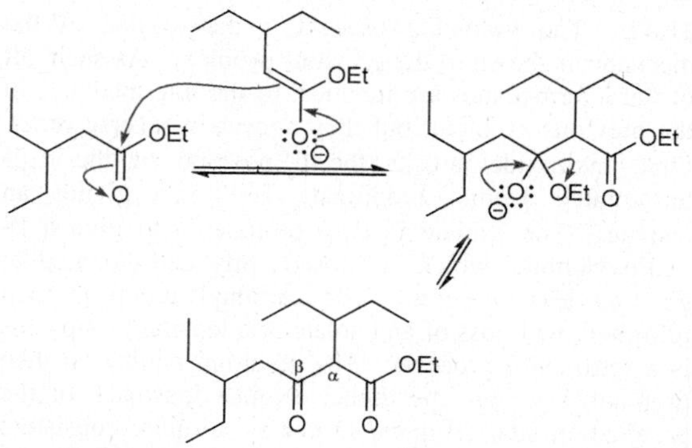

This product is deprotonated under the conditions of its formation, which is the reason for aqueous acidic workup after the reaction is complete (to protonate, thus regenerating the neutral product).

21.26.

(a) We first identify the α and β positions, and then apply a retrosynthetic analysis. The α position is the location between the two carbonyl groups, and the β position bears the keto group:

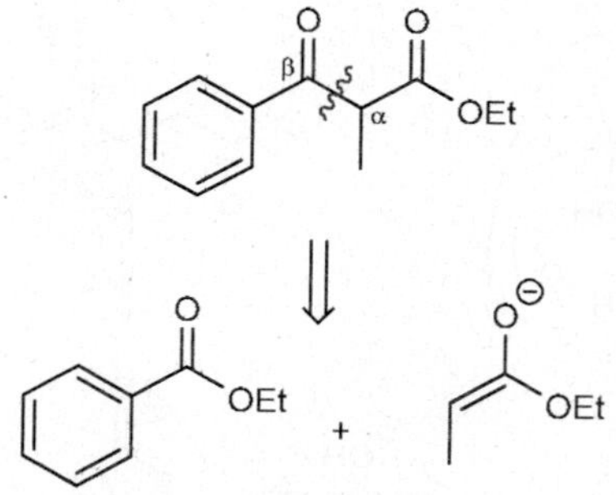

Since the two partners are different, we use a crossed Claisen condensation. LDA is used as the base in the first step, and the final step of the process is aqueous acidic workup, as shown:

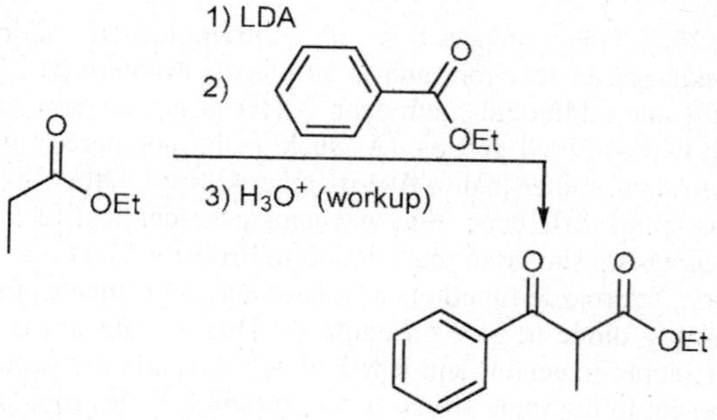

1) LDA

2)

3) H_3O^+ (workup)

(b) We first identify the α and β positions, and then apply a retrosynthetic analysis. The α position is the location between the two carbonyl groups, and the β position bears the keto group:

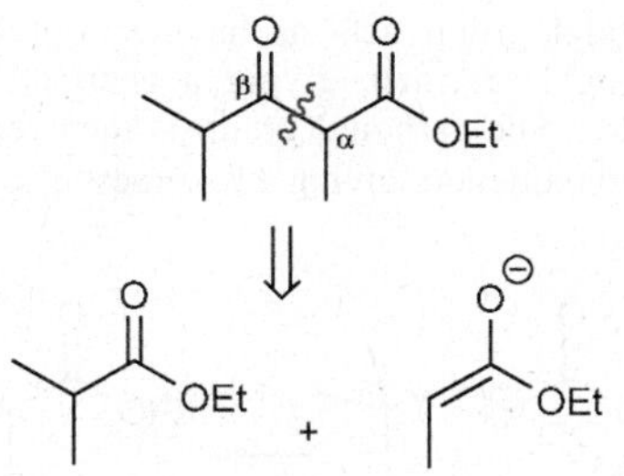

Since the two partners are different, we use a crossed Claisen condensation. LDA is used as the base in the first step, and the final step of the process is aqueous acidic workup, as shown:

1) LDA

2)

3) H_3O^+ (workup)

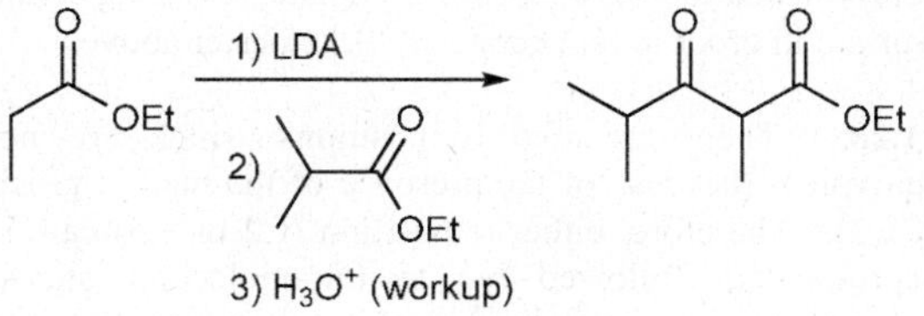

(c) We first identify the α and β positions, and then apply a retrosynthetic analysis. The α position is the location between the two carbonyl groups, and the β position bears the keto group:

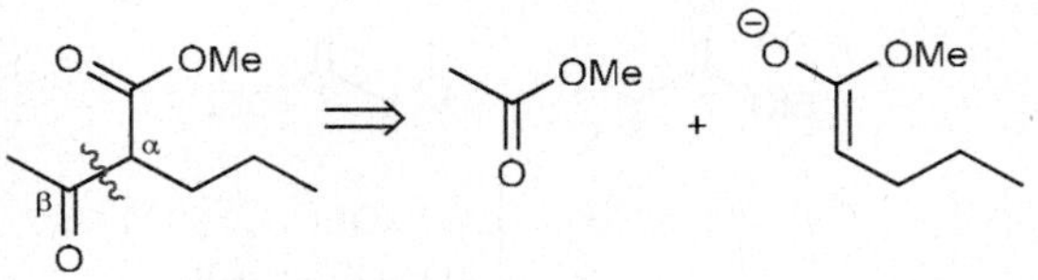

Since the two partners are different, we use a crossed Claisen condensation. LDA is used as the base in the first step, and the final step of the process is aqueous acidic workup, as shown:

1) LDA

2)

3) H_3O^+ (workup)

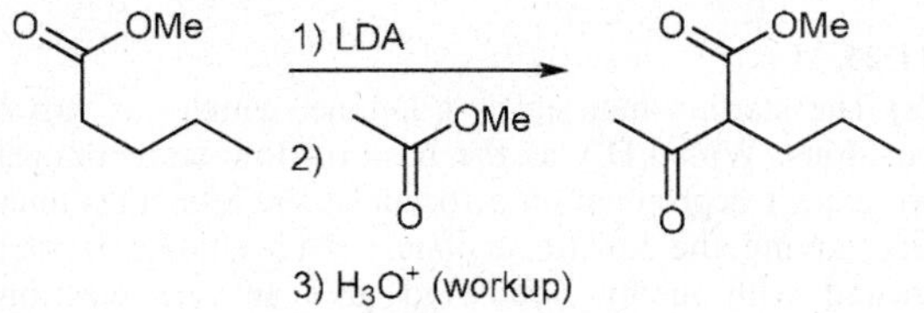

(d) We first identify the α and β positions, and then apply a retrosynthetic analysis. The α position is the location between the two carbonyl groups, and the β position bears the keto group:

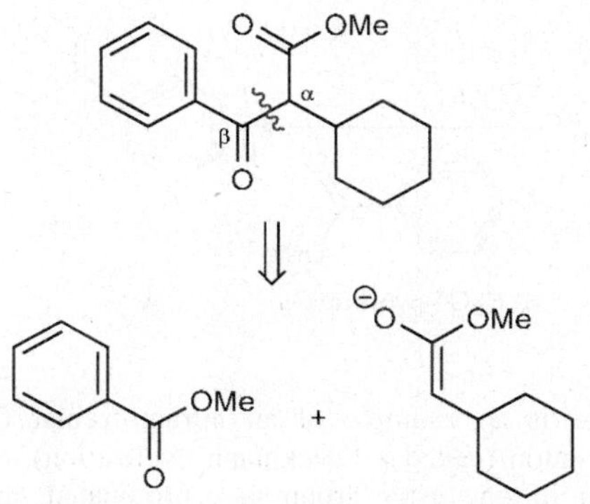

Since the two partners are different, we use a crossed Claisen condensation. LDA is used as the base in the first step, and the final step of the process is aqueous acidic workup, as shown:

1) LDA

2)

3) H_3O^+ (workup)

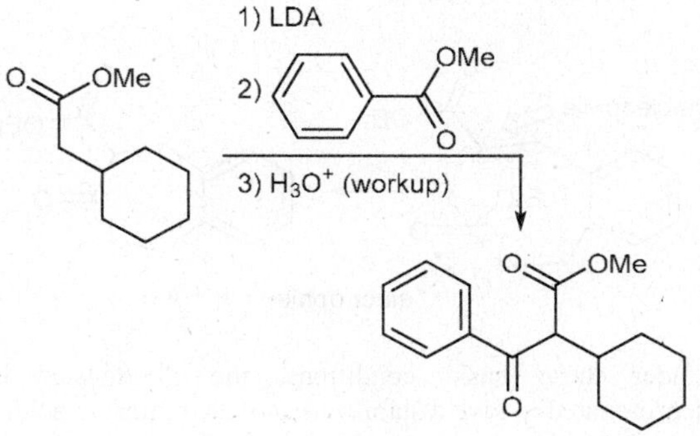

(e) We first identify the α and β positions, and then apply a retrosynthetic analysis. The α position is the location between the two carbonyl groups, and the β position bears the keto group:

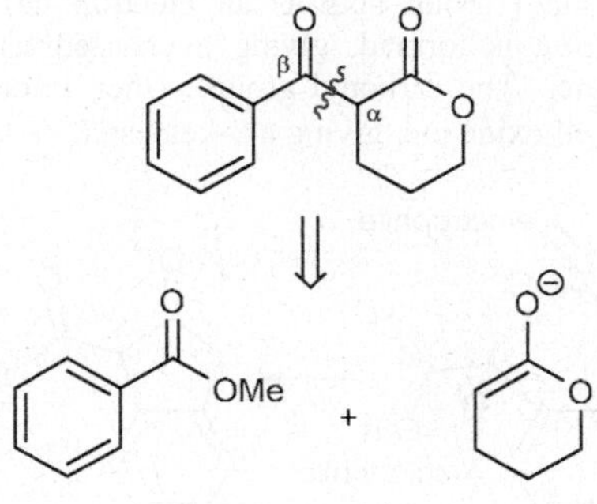

Since the two partners are different, we use a crossed Claisen condensation. LDA is used as the base in the first step, and the final step of the process is aqueous acidic workup, as shown:

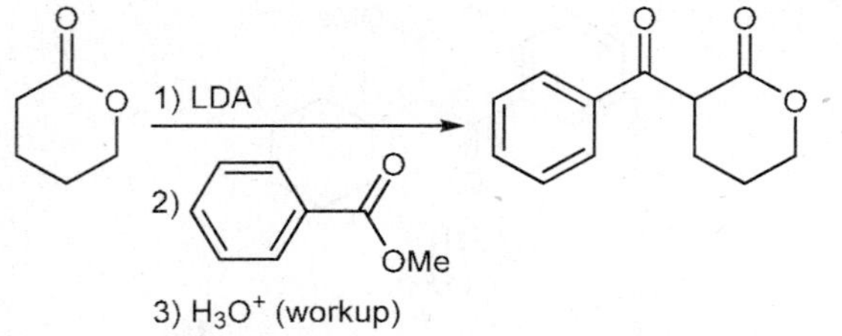

21.27.

(a) This is an example of an intramolecular Claisen condensation (called a Dieckmann cyclization). The α position of one ester group is deprotonated, and the resulting enolate functions as a nucleophile and attacks the other carbonyl group within the same structure (the other carbonyl group acts as an electrophile). As a result, a ring is formed, giving a charged tetrahedral intermediate. The carbonyl group is then reformed via loss of an ethoxide ion, giving a β-ketoester:

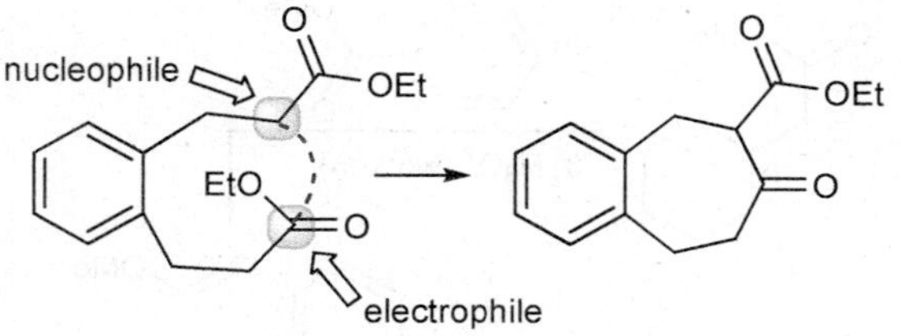

Under these basic conditions, the β-ketoester is deprotonated to give a stabilized enolate, requiring acidic workup in order to regenerate the β-ketoester above.

(b) This is an example of an intramolecular Claisen condensation (called a Dieckmann cyclization). The α position of one ester group is deprotonated, and the resulting enolate functions as a nucleophile and attacks the other carbonyl group within the same structure (the other carbonyl group acts as an electrophile). As a result, a ring is formed, giving a charged tetrahedral intermediate. The carbonyl group is then reformed via loss of an ethoxide ion, giving a β-ketoester:

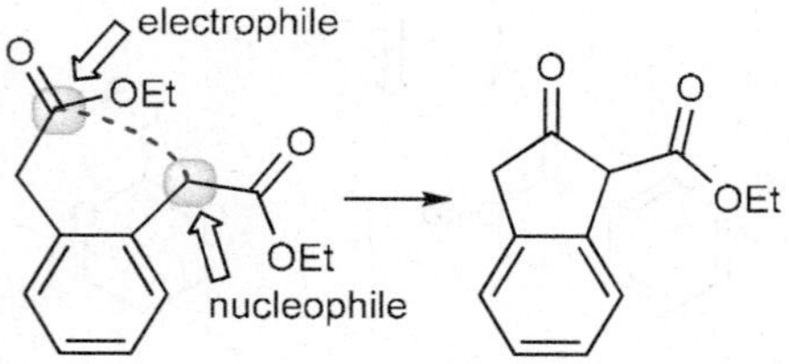

Under these basic conditions, the β-ketoester is deprotonated to give a stabilized enolate, requiring acidic workup in order to regenerate the β-ketoester above.

(c) This is an example of an intramolecular Claisen condensation (called a Dieckmann cyclization). The α position of one ester group is deprotonated, and the resulting enolate functions as a nucleophile and attacks the other carbonyl group within the same structure (the other carbonyl group acts as an electrophile). As a result, a ring is formed, giving a charged tetrahedral intermediate. The carbonyl group is then reformed via loss of an ethoxide ion, giving a β-ketoester:

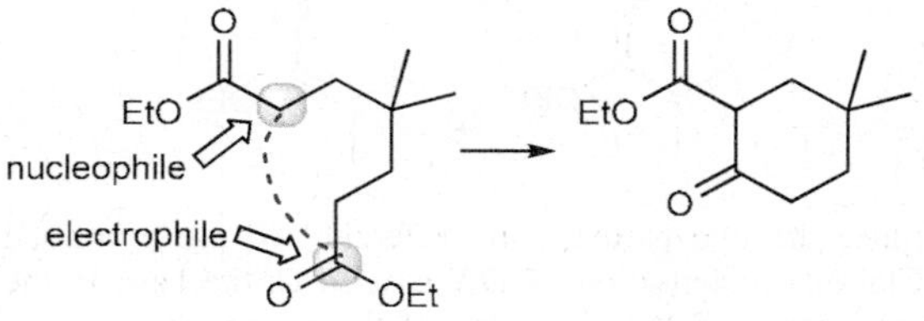

Under these basic conditions, the β-ketoester is deprotonated to give a stabilized enolate, requiring acidic workup in order to regenerate the β-ketoester above.

21.28. There are two α positions which are not equivalent (because of the presence of the methyl group at C3). Therefore, either α position (C2 or C6) can be deprotonated, followed by an intramolecular attack, leading to the following two possible condensation products. That is, the cyclization process can either result in a bond between C2 and C7 or between C6 and C1. Each product has the usual β-ketoester pattern of functional groups:

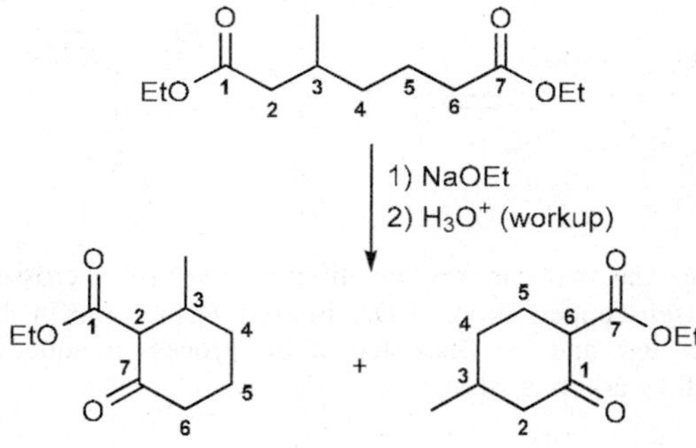

21.29.

(a) The starting material is a ketone, which has two α positions. With LDA as the base (at low temperature), we expect deprotonation to occur at the less substituted site, giving the kinetic enolate. This enolate is then treated with methyl iodide to give an S_N2 reaction, thereby installing the methyl group at the less substituted α position, as shown.

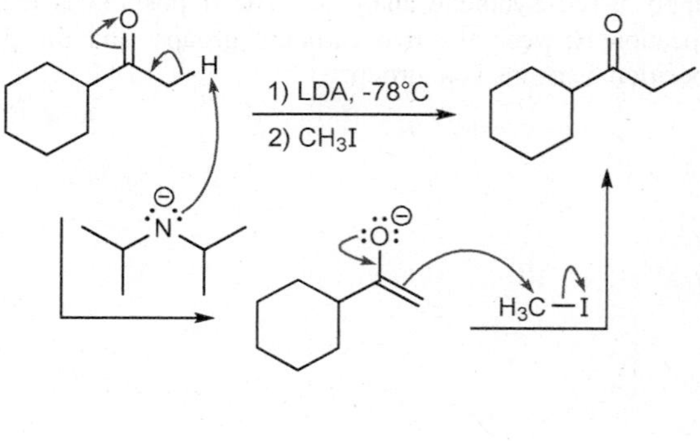

(b) The starting material is a ketone, which has two α positions. At room temperature, with NaH as the base, we expect deprotonation to occur at the more substituted site, giving the thermodynamic enolate. This enolate is then treated with benzyl bromide to give an S_N2 reaction, thereby installing the benzyl group at the more substituted α position, as shown.

1) NaH
2) $PhCH_2Br$

(c) The starting material is a ketone, which has two α positions, although they are equivalent because the ketone is symmetrical. Deprotonation at either location will lead to the same enolate ion. In the first step, LDA functions as a base and deprotonates the ketone to give an enolate. This enolate is then treated with ethyl iodide to give an S_N2 reaction, thereby installing an ethyl group. Subsequent treatment with LDA (at low temperature), followed by methyl iodide, installs a methyl group at the other (less substituted) α position via the kinetic enolate. The net result is the installation of an ethyl group at one α position and the installation of a methyl group at the other α position:

1) LDA, -78°C
2) EtI
3) LDA, -78°C
4) CH_3I

21.30. This transformation does not involve a change in the identity or location of the functional group (a hydroxyl group), but it does involve a change in the carbon skeleton:

An ethyl group must be installed, although we have not learned a way to do this in one step. A multistep strategy is necessary. One strategy for achieving this transformation derives from the following retrosynthetic analysis. An explanation of each of the steps (*a-c*) follows.

a. The alcohol can be made via reduction of the corresponding ketone.
b. The ethyl group can be installed through alkylation of 3-methyl-2-pentanone (via the kinetic enolate).
c. 3-Methyl-2-pentanone can be made from the starting alcohol via oxidation.

Now let's draw the forward scheme. The starting alcohol is oxidized with chromic acid to give 3-methyl-2-pentanone. Alternatively, PCC (or DMP) can be used to affect the same transformation. The ketone is then treated with LDA at low temperature to deprotonate the less substituted α position, giving the kinetic enolate. Upon treatment with ethyl iodide, an ethyl group is installed in the desired location. The resulting ketone is then reduced with $LiAlH_4$, followed by aqueous acidic workup, to give the product.

1) $Na_2Cr_2O_7$, H_2SO_4, H_2O
2) LDA, -78°C
3) EtI
4) $LiAlH_4$
5) H_3O^+

$Na_2Cr_2O_7$, H_2SO_4, H_2O

1) LDA, -78°C
2) EtI

1) $LiAlH_4$
2) H_3O^+

21.31.
(a) The product is a methyl ketone that has the following (highlighted) group connected to the α position:

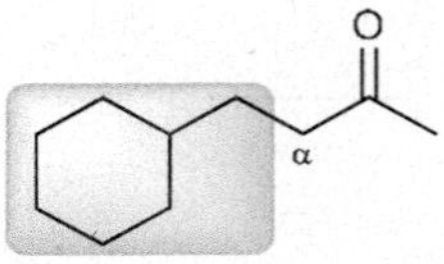

This group can be installed via an acetoacetic ester synthesis, using the following alkyl halide:

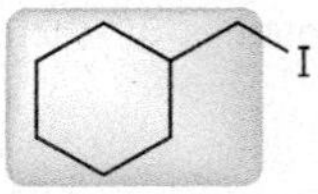

An acetoacetic ester synthesis begins with the deprotonation of ethyl acetoacetate (using ethoxide as a base). The resulting resonance-stabilized conjugate base is then treated with the alkyl halide above, thereby installing the alkyl group. Subsequent hydrolysis and decarboxylation give the product, as shown. The three carbon atoms in the ethyl acetoacetate starting material that remain in the product (after decarboxylation) have been highlighted:

1) NaOEt
2)
3) H_3O^+, heat

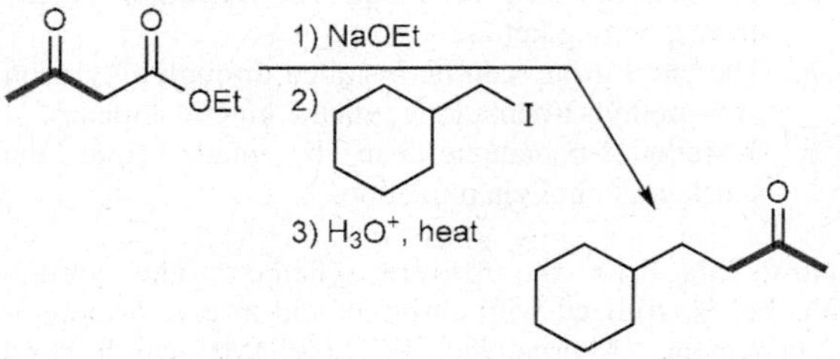

(b) The product is a methyl ketone that has the following two (highlighted) groups connected to the α position:

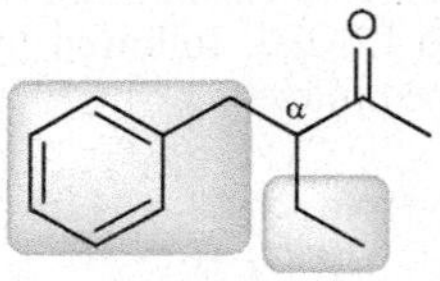

Each of these groups can be installed via an acetoacetic ester synthesis, using the following halides:

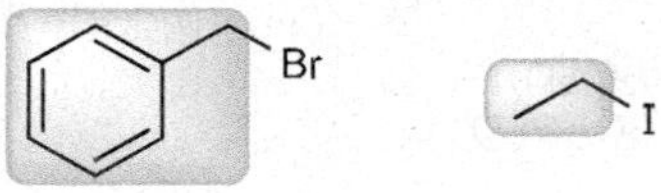

An acetoacetic ester synthesis begins with the deprotonation of ethyl acetoacetate (using ethoxide as a base). The resulting resonance-stabilized conjugate base is then treated with one of the halides above, to install one of the two groups. The other group is installed in a similar way (deprotonation with a base, followed by alkylation). Subsequent hydrolysis and decarboxylation give the product, as shown. The three carbon atoms in the ethyl acetoacetate starting material that remain in the product (after decarboxylation) have been highlighted:

1) NaOEt
2) EtI
3) NaOEt
4) $PhCH_2Br$
5) H_3O^+, heat

1) NaOEt
2) EtI

1) NaOEt
2) $PhCH_2Br$

H_3O^+, heat

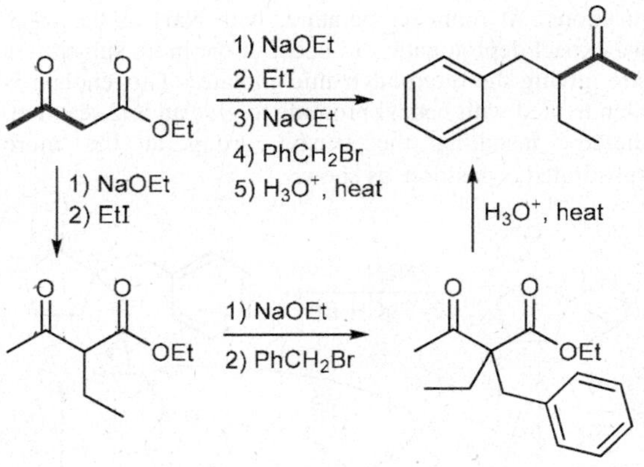

(c) The product is a methyl ketone that has the following two (highlighted) groups connected to the α position:

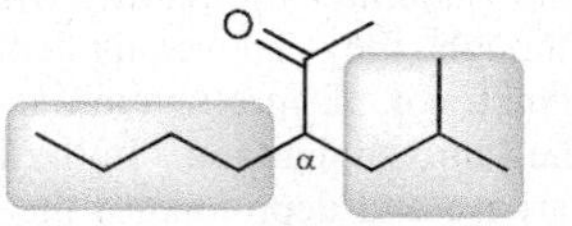

Each of these groups can be installed via an acetoacetic ester synthesis, using the following alkyl halides:

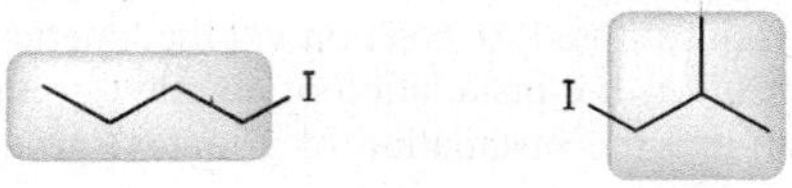

An acetoacetic ester synthesis begins with the deprotonation of ethyl acetoacetate (using ethoxide as a base). The resulting resonance-stabilized conjugate base is then treated with one of the alkyl halides above, thereby installing one of the two alkyl groups. The other alkyl group is installed in a similar way (deprotonation with a base, followed by alkylation). Subsequent hydrolysis and decarboxylation give the product, as shown. The three carbon atoms in the ethyl acetoacetate starting material that remain in the product (after decarboxylation) have been highlighted:

1) NaOEt
2)
3) NaOEt
4)
5) H_3O^+, heat

1) NaOEt
2)

H_3O^+, heat

3) NaOEt
4)

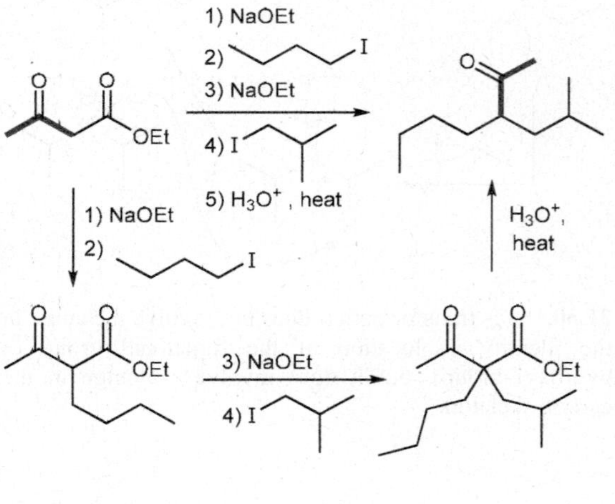

(d) The product is a methyl ketone that has the following two (highlighted) groups connected to the α position:

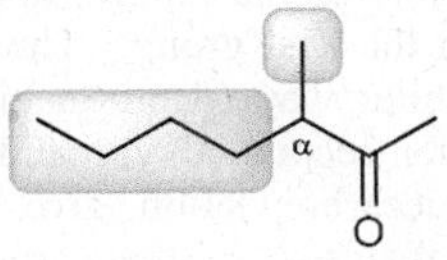

Each of these groups can be installed via an acetoacetic ester synthesis, using methyl iodide and *n*-butyl iodide, respectively.

An acetoacetic ester synthesis begins with the deprotonation of ethyl acetoacetate (using ethoxide as a base). The resulting resonance-stabilized conjugate base is then treated with one of the alkyl halides above, thereby installing one of the two alkyl groups. The other alkyl group is installed in a similar way (deprotonation with a base, followed by alkylation). Subsequent hydrolysis and decarboxylation give the product, as shown. The three carbon atoms in the ethyl acetoacetate starting material that remain in the product (after decarboxylation) have been highlighted:

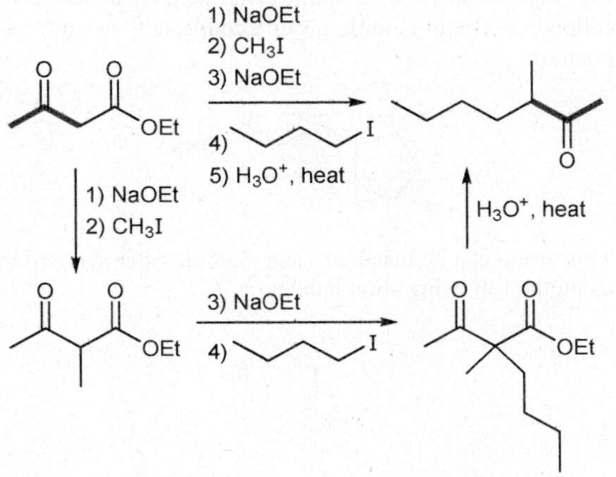

1) NaOEt
2) CH_3I
3) NaOEt
4) I
5) H_3O^+, heat

1) NaOEt
2) CH_3I

3) NaOEt
4) I

H_3O^+, heat

21.32. Comparison of carbon skeletons reveals that three carbon atoms have been introduced in this transformation; specifically, carbon atoms 8-10 look like acetone and can be installed using an acetoacetic ester synthesis.

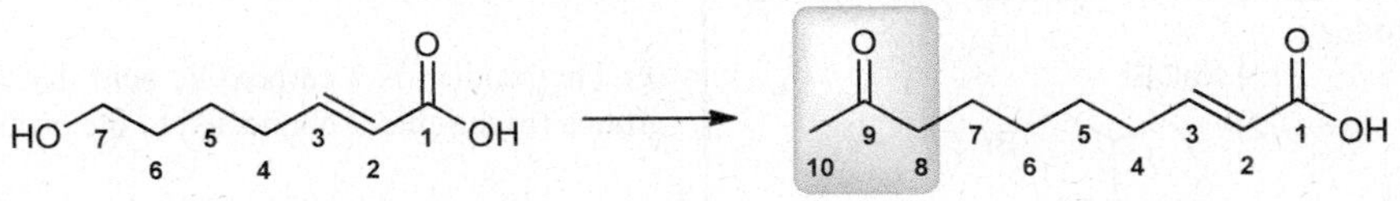

That makes carbon 8 the α carbon, so carbon 7 will need to have a leaving group. The necessary alkyl halide will be a primary halide and a good substrate for the acetoacetic acid synthesis, provided that the COOH group is deprotonated, as indicated in the problem statement.

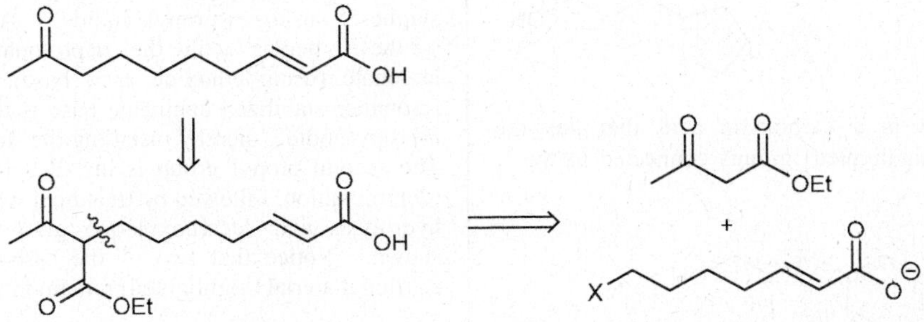

The alkyl halide must be made from the alcohol starting material. The mild reagent PBr_3 can be used to convert the primary alcohol to a primary bromide.

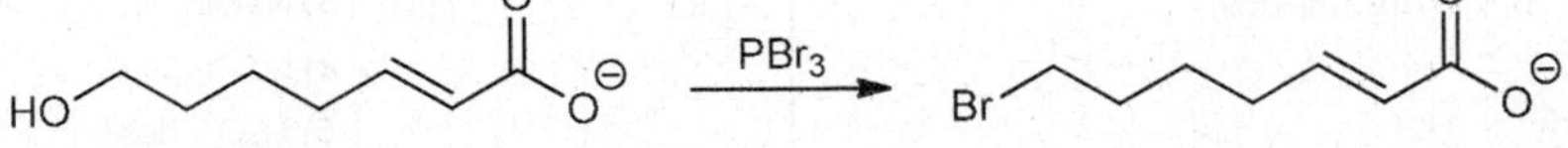

Putting it all together, we have the following sequence:

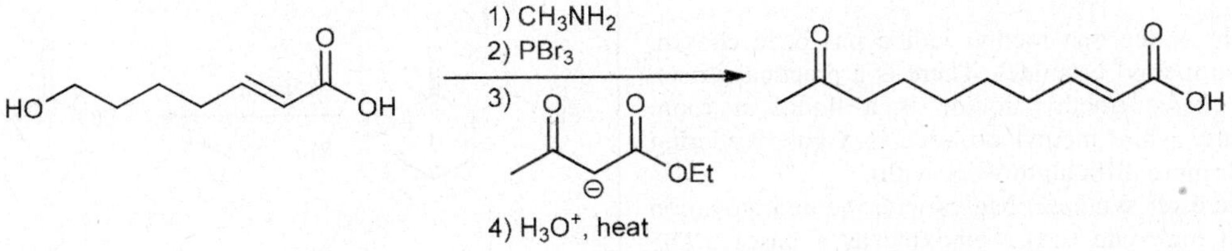

1) CH_3NH_2
2) PBr_3
3) (enolate of ethyl acetoacetate)
4) H_3O^+, heat

21.33.

(a) The product is a carboxylic acid that has the following (highlighted) group connected to the α position:

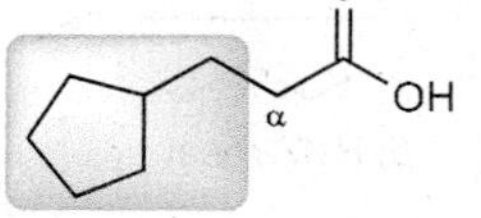

This group can be installed via a malonic ester synthesis, using the following alkyl halide:

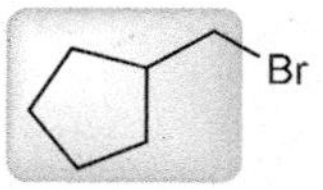

A malonic ester synthesis begins with the deprotonation of diethyl malonate (using ethoxide as a base). The resulting resonance-stabilized conjugate base is then treated with the alkyl halide above, thereby installing the alkyl group. Subsequent hydrolysis and decarboxylation give the product, as shown. Notice that two of the carbon atoms in the starting material (highlighted) remain in the product:

1) NaOEt
2)
3) H_3O^+, heat

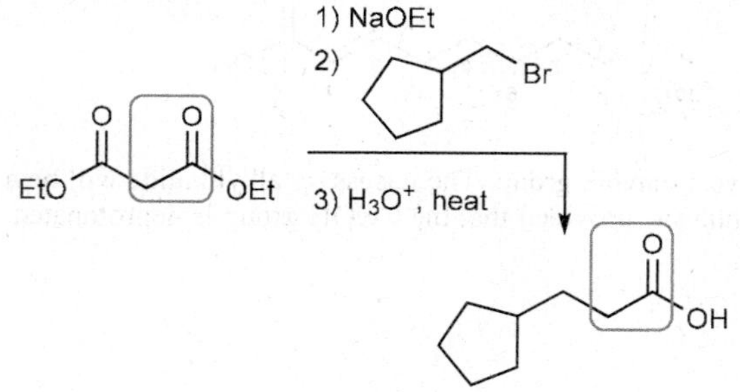

(b) The product is a carboxylic acid that has the following two (highlighted) groups connected to the α position:

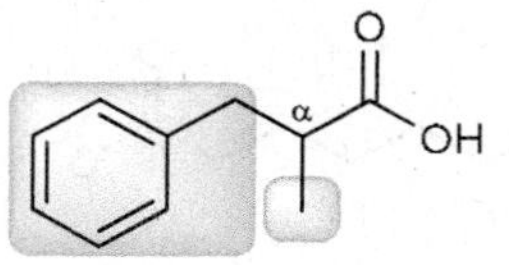

Both of these groups can be installed via a malonic ester synthesis, using the following halides:

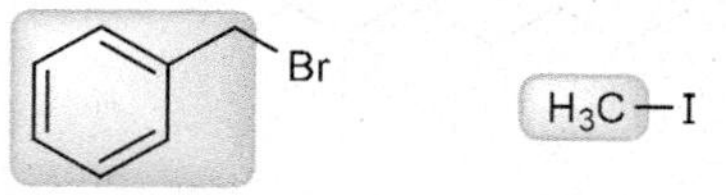

$H_3C—I$

You might notice that methyl iodide has been chosen, rather than methyl bromide. There is a practical reason for this choice (methyl iodide is a liquid at room temperature, while methyl bromide is a gas, rendering the CH_3Br more difficult to work with).

A malonic ester synthesis begins with the deprotonation of diethyl malonate (using ethoxide as a base). The resulting resonance-stabilized conjugate base is then treated with the one of the halides above (either one), thereby installing the first group. The second group is installed in a similar way (deprotonation, followed by treatment with the second alkyl halide). Subsequent hydrolysis and decarboxylation give the product, as shown. Notice that two of the carbon atoms in the starting material remain in the product (after decarboxylation), and these two carbon atoms have been highlighted:

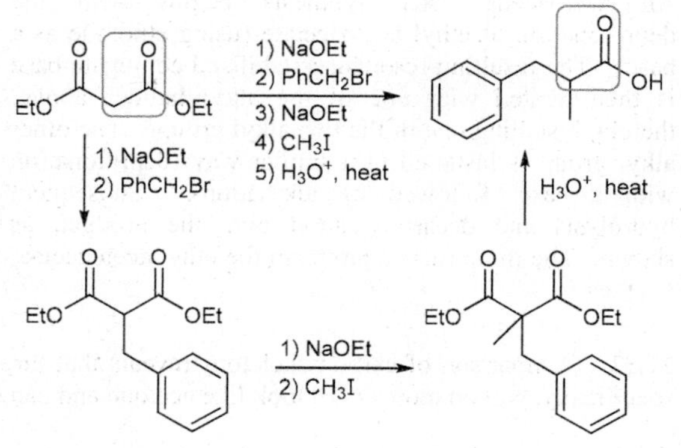

1) NaOEt
2) $PhCH_2Br$
3) NaOEt
4) CH_3I
5) H_3O^+, heat

1) NaOEt
2) $PhCH_2Br$

1) NaOEt
2) CH_3I

H_3O^+, heat

(c) The product is a carboxylic acid that has two propyl groups (highlighted) connected to the α position:

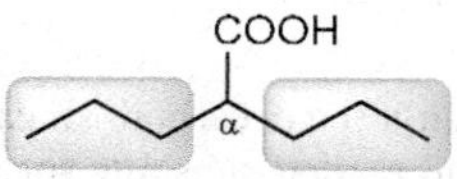

Both of these groups can be installed via a malonic ester synthesis, using *n*-propyl iodide. A malonic ester synthesis begins with the deprotonation of diethyl malonate (using ethoxide as a base). The resulting resonance-stabilized conjugate base is then treated with *n*-propyl iodide, thereby installing the first propyl group. The second propyl group is installed in a similar way (deprotonation, followed by treatment with). Subsequent hydrolysis and decarboxylation give the product, as shown. Notice that two of the carbon atoms in the starting material (highlighted) remain in the product:

1) NaOEt
2)
3) NaOEt
4)
5) H_3O^+, heat

1) NaOEt
2)

1) NaOEt
2)

H_3O^+, heat

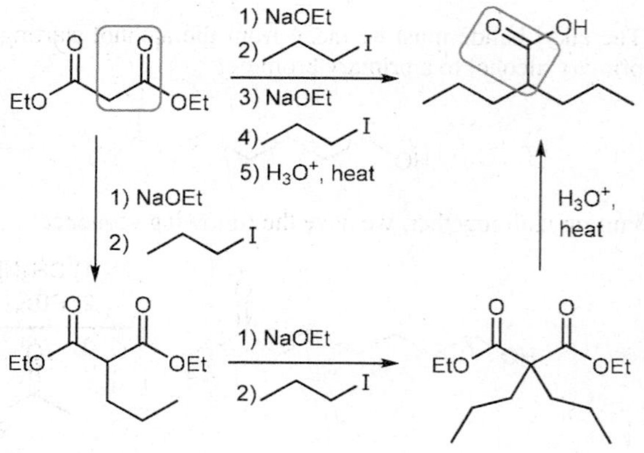

(d) The product is a carboxylic acid that has the following two (highlighted) groups connected to the α position:

Both of these groups can be installed via a malonic ester synthesis, using *n*-propyl bromide and methyl iodide, respectively. The reason for using methyl iodide (rather than methyl bromide) was discussed in the solution to part (b).

A malonic ester synthesis begins with the deprotonation of diethyl malonate (using ethoxide as a base). The resulting resonance-stabilized conjugate base is then treated with the one of the alkyl halides above (either one), thereby installing the first alkyl group. The second alkyl group is installed in a similar way (deprotonation, followed by treatment with the second alkyl halide). Subsequent hydrolysis and decarboxylation give the product, as shown. Notice that two of the carbon atoms in the starting material (highlighted) remain in the product:

1) NaOEt
2) CH_3I
3) NaOEt
4) Br
5) H_3O^+, heat

1) NaOEt
2) CH_3I

1) NaOEt
2) Br

H_3O^+, heat

(e) The product is a carboxylic acid that has the following two (highlighted) groups connected to the α position:

Both of these groups can be installed via a malonic ester synthesis, using ethyl iodide and isobutyl iodide, respectively.

A malonic ester synthesis begins with the deprotonation of diethyl malonate (using ethoxide as a base). The resulting resonance-stabilized conjugate base is then treated with the one of the alkyl halides above (either one), thereby installing the first alkyl group. The second alkyl group is installed in a similar way (deprotonation, followed by treatment with the second alkyl halide). Subsequent hydrolysis and decarboxylation give the product, as shown. Notice that two of the carbon atoms in the starting material (highlighted) remain in the product:

1) NaOEt
2) EtI
3) NaOEt
4) I
5) H_3O^+, heat

1) NaOEt
2) EtI

1) NaOEt
2) I

H_3O^+, heat

21.34. Product **1** is simply the alkylation product of diethylmalonate. Therefore, treatment of diethyl malonate with base (sodium ethoxide) followed by methyl iodide affords compound **1**.

1) NaOEt
2) CH_3I

The steps leading from **1** to **2** constitute a malonic ester synthesis, where an aryl halide is used instead of an alkyl halide, giving a nucleophilic aromatic substitution (rather than a simple S_N2 back-side attack). This effectively installs an aryl group onto the malonate derivative (**1**) rather than installing an alkyl group. Sodium ethoxide serves as the base to deprotonate **1** making a nucleophilic enolate.

NaOEt

1,2-Difluoro-4-nitrobenzene is a very electron-deficient arene that can be attacked by the enolate of **1**. Recall that for an SNAr reaction to take place, there must be (1) a powerful electron-withdrawing group on the aromatic ring (in this case the nitro group), (2) a leaving group (in this case a fluoride) and (3) the leaving group must be either *ortho* or *para* to the electron-withdrawing group. This third point dictates which fluoride is expelled (the one that is *para* to the nitro group):

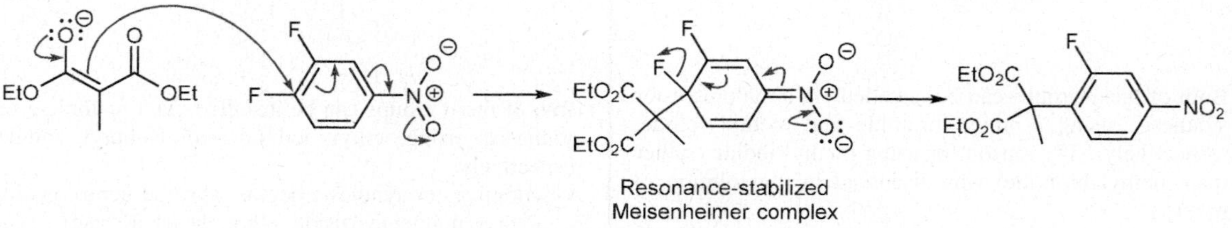

Resonance-stabilized Meisenheimer complex

You may note that fluoride is not typically a good leaving group. For the SNAr reaction, it turns out that the attack of the benzene ring (which breaks aromaticity) is the rate-determining step. For this reason, the leaving group ability is much less relevant for SNAr. In fact, the electron-withdrawing nature of fluoride actually *increases* the rate of SNAr (relative to Cl and Br) due to making the benzene ring more electrophilic.

Finally, the last step uses H_3O^+ and heat to invoke a hydrolysis and subsequent decarboxylation of the arylated malonate product. Product **2** is a carboxylic acid as shown below:

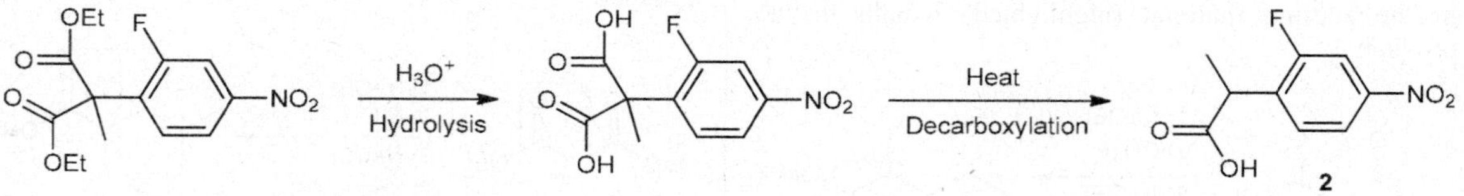

21.35.

(a) The starting material is an α,β-unsaturated ketone, which can function as a Michael acceptor (see Table 21.2), and the reagent (lithium diethylcuprate) can function as a Michael donor. As such, we expect a conjugate addition, thereby installing an ethyl group at the β position. Note that the α position is protonated by the aqueous workup to give a neutral product:

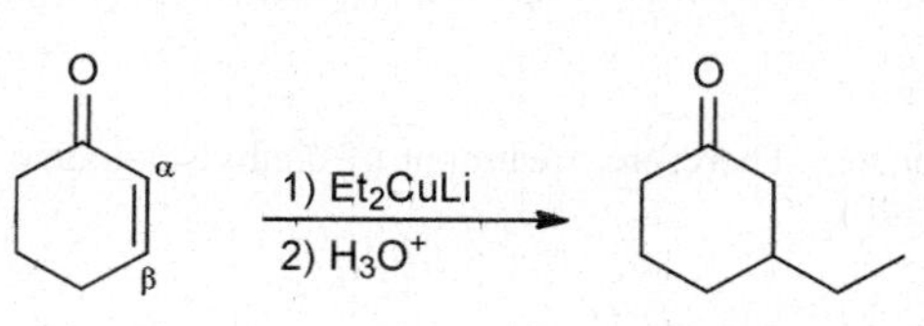

(b) The starting material is an α,β-unsaturated nitrile, which can function as a Michael acceptor (see Table 21.2), and the reagent (lithium diethylcuprate) can function as a Michael donor. As such, we expect a conjugate addition thereby installing an ethyl group at the β position. Note that the α position is protonated by the aqueous workup to give a neutral product:

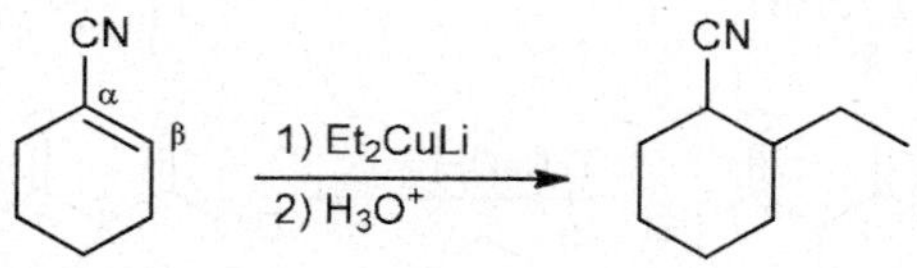

(c) The starting material is an α,β-unsaturated ester, which can function as a Michael acceptor (see Table 21.2), and the reagent (lithium diethylcuprate) can function as a Michael donor. As such, we expect a conjugate addition, thereby installing an ethyl group at the β position. Note that the α position is protonated by the aqueous workup to give a neutral product:

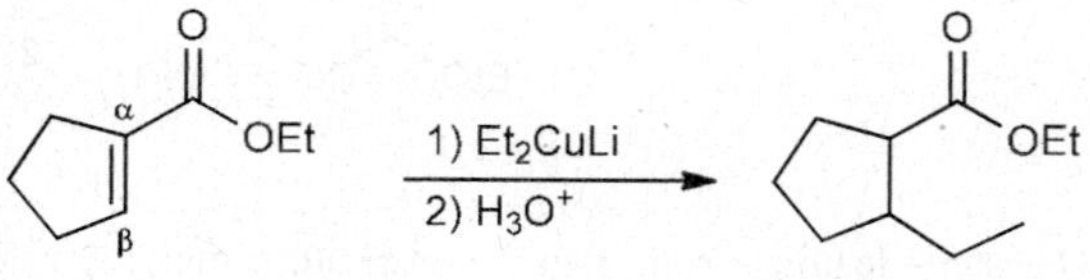

21.36. The starting material has an acidic proton, which is removed upon treatment with a strong base, such as hydroxide. The resulting resonance-stabilized conjugate base (a stabilized enolate) functions as a Michael donor and attacks the Michael acceptor (an α,β-unsaturated ketone). Subsequent acid workup causes protonation to give an enol, which tautomerizes to give a ketone, as shown.

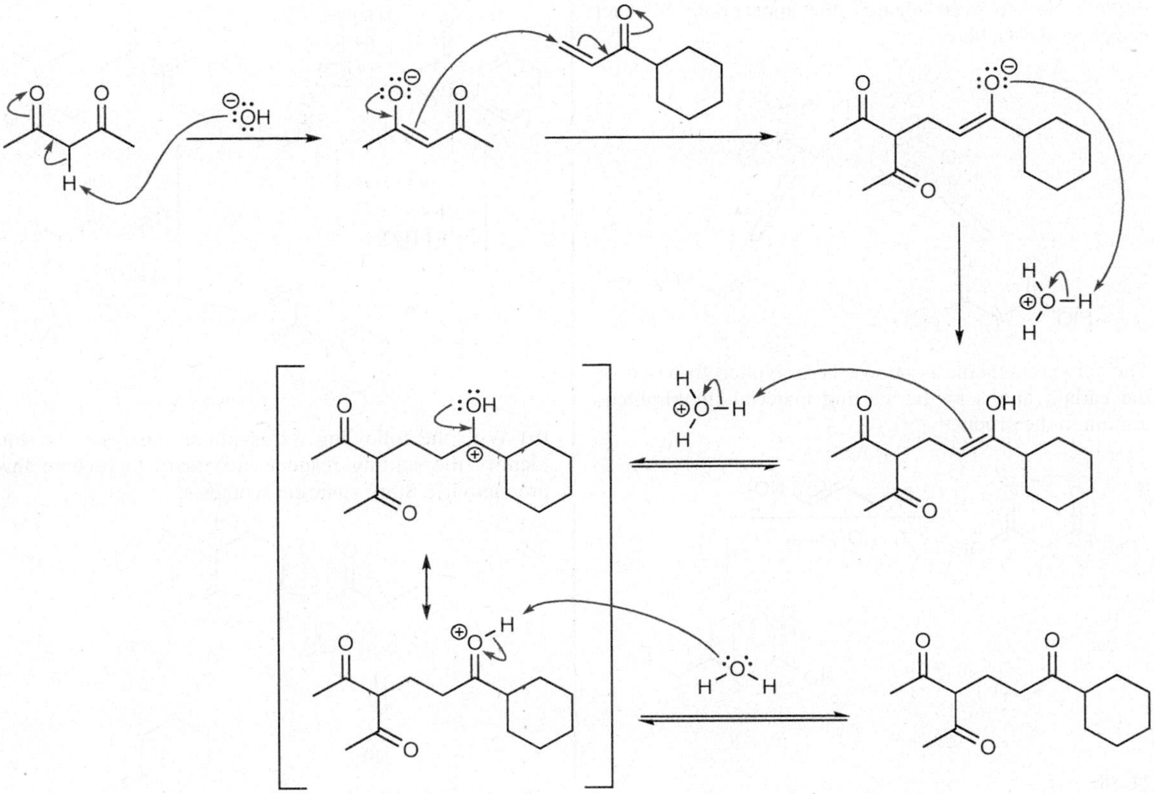

21.37.

(a) Recall that the malonic ester synthesis is useful for creating carboxylic acids that possess either one or two alkyl groups at the α position:

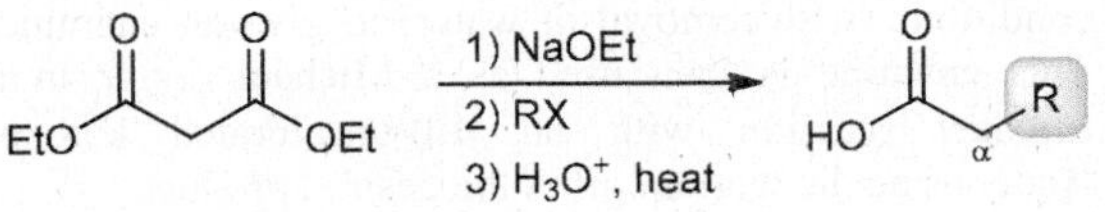

1) NaOEt
2) RX
3) H_3O^+, heat

Therefore, making the desired product via a malonic ester synthesis would require installation of the following highlighted group:

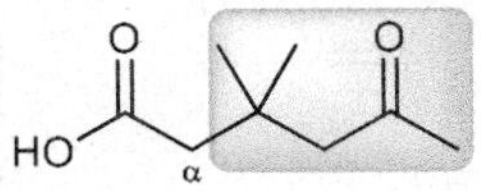

Installation of this group via alkylation (as seen in Section 21.5) would require a tertiary alkyl halide:

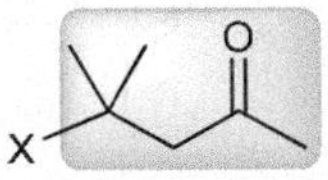

However, a tertiary alkyl halide will not undergo an S_N2 reaction because it is too sterically hindered. So this method will not work. The problem statement indicates that a Michael reaction can be used to achieve the desired transformation. That is, we would use an electrophile with the same carbon skeleton as the alkyl halide above, but the electrophilic position will be the β position of an α,β-unsaturated ketone:

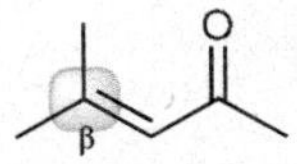

This compound can function as a Michael acceptor, thereby allowing the desired transformation, as seen in the following synthesis. Notice that two of the carbon atoms in the starting material (highlighted) remain in the product:

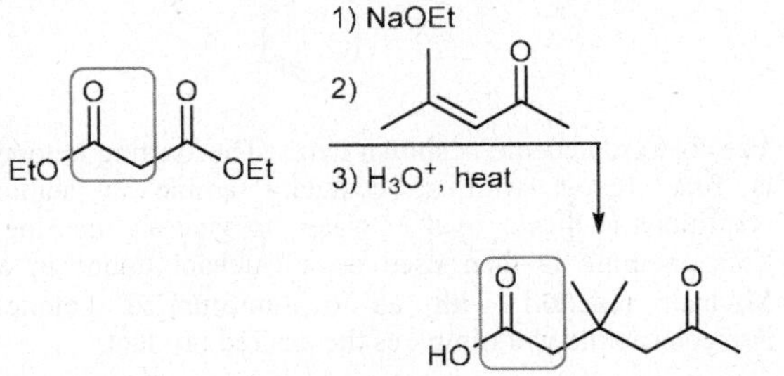

1) NaOEt
2)
3) H_3O^+, heat

(b) As described in the solution to part (a), the conjugate base of diethyl malonate can function as a Michael donor. So we must identify the appropriate Michael acceptor, shown here:

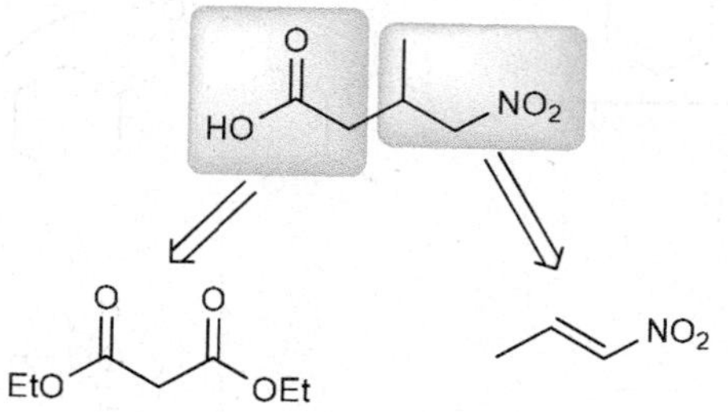
HO
NO2
EtO
OEt
NO2

The forward scheme is shown here. Notice that two of the carbon atoms in the starting material (highlighted) remain in the product:

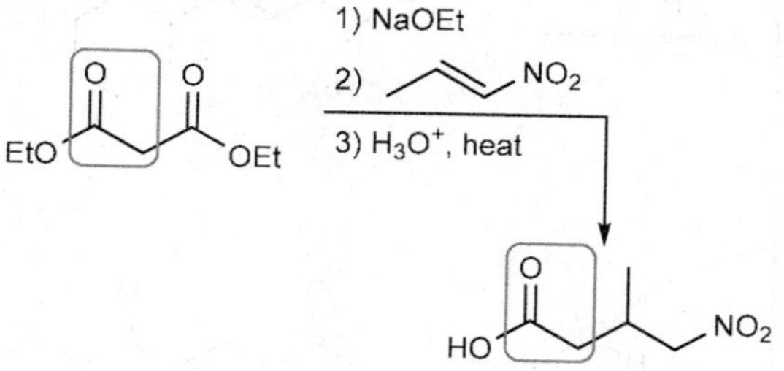
1) NaOEt
2)
NO2
3) H3O+, heat
EtO
OEt
HO
NO2

21.38.

(a) With the following retrosynthetic analysis, we can identify the starting reagents necessary to prepare this product via a Stork enamine synthesis:

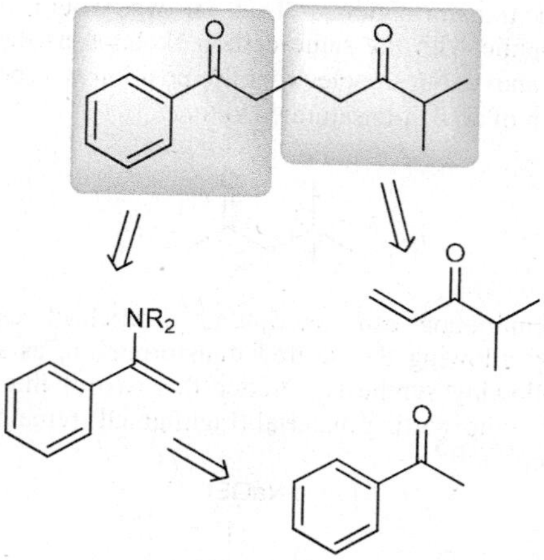
NR2

The forward scheme is shown here. The starting ketone is first treated with a secondary amine in acidic conditions (with removal of water) to give an enamine. This enamine is then used as a Michael donor in a Michael reaction with an α,β-unsaturated ketone. Aqueous acidic workup gives the desired product.

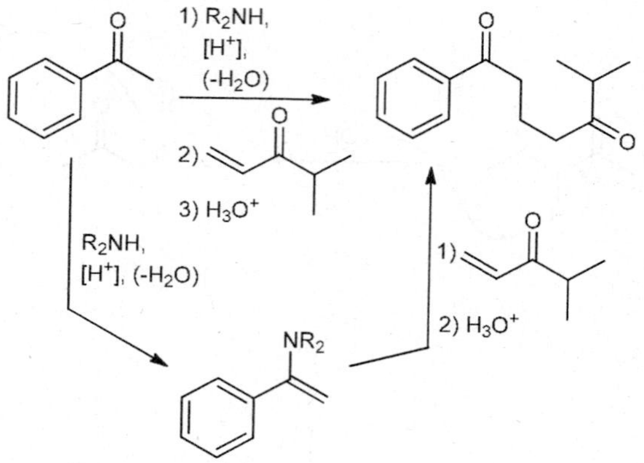
1) R2NH,
[H+],
(-H2O)
2)
3) H3O+
R2NH,
[H+], (-H2O)
1)
2) H3O+
NR2

(b) With the following retrosynthetic analysis, we can identify the starting reagents necessary to prepare this product via a Stork enamine synthesis:

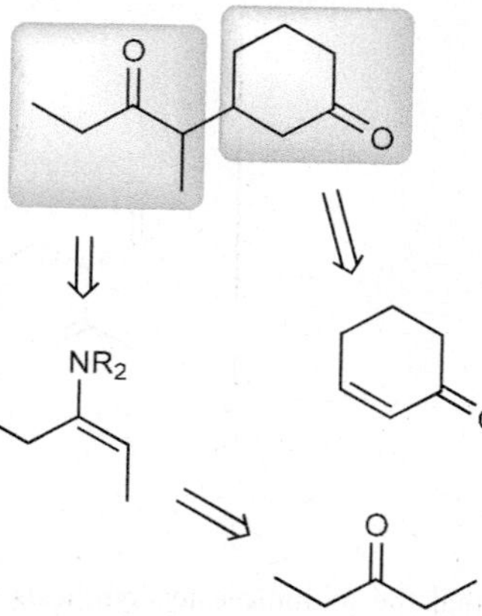
NR2

The forward scheme is shown here. The starting ketone is first treated with a secondary amine in acidic conditions (with removal of water) to give an enamine. This enamine is then used as a Michael donor in a Michael reaction with an α,β-unsaturated ketone. Aqueous acidic workup gives the desired product.

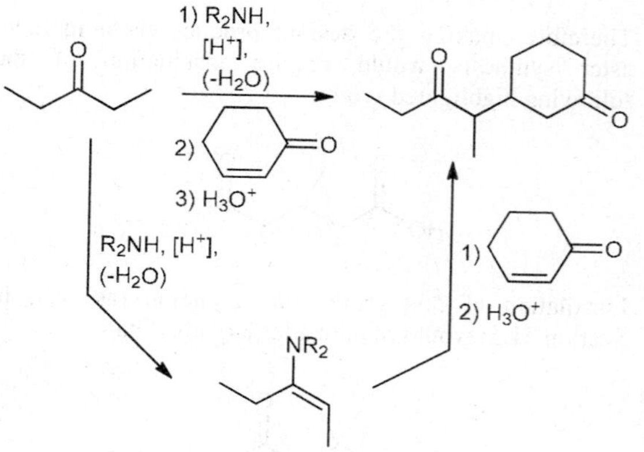
1) R2NH,
[H+],
(-H2O)
2)
3) H3O+
R2NH, [H+],
(-H2O)
1)
2) H3O+
NR2

(c) With the following retrosynthetic analysis, we can identify the starting reagents necessary to prepare this product via a Stork enamine synthesis:

The forward scheme is shown here. The starting ketone is first treated with a secondary amine in acidic conditions (with removal of water) to give an enamine. This enamine is then used as a Michael donor in a Michael reaction with an α,β-unsaturated ketone. Aqueous acidic workup gives the desired product.

1) R_2NH, $[H^+]$, $(-H_2O)$
2)
3) H_3O^+

R_2NH, $[H^+]$, $(-H_2O)$

1)
2) H_3O^+

21.39. With the following retrosynthetic analysis, we can identify the starting reagents necessary to prepare compound **2** via a Stork enamine synthesis:

Acrylonitrile

The forward scheme is shown here. The starting ketone **1** is first treated with a secondary amine in acidic conditions (with removal of water) to give an enamine. This enamine is then used as a Michael donor in a Michael reaction with an α,β-unsaturated compound, acrylonitrile, as the Michael acceptor. Aqueous acidic workup gives the desired product **2**.

1) R_2NH, $[H^+]$, $(-H_2O)$
2) CN
3) H_3O^+

R_2NH, $[H^+]$, $(-H_2O)$

1) CN
2) H_3O^+

Now let's consider the conversion of **2** to **3**. You might be wondering why a Stork enamine synthesis is used, rather than treating compound **2** with LDA, followed by an allyl halide. This won't work, because LDA can deprotonate the position that is alpha to the cyano group. Therefore, a Stork enamine synthesis is required to convert **2** to **3**. We can identify the starting reagents necessary to prepare compound **3** via a Stork enamine as a nucleophile and an alkyl halide as an electrophile.

The forward scheme is shown here. The starting ketone **2** is first treated with a secondary amine in acidic conditions (with removal of water) to give an enamine. Note that deprotonation at the less hindered alpha carbon is favored to give the less substituted enamine. This enamine is then used as a nucleophile in an S_N2 reaction with allyl iodide. Aqueous acidic workup gives the desired product **3**.

1) R_2NH, $[H^+]$, $(-H_2O)$
2) I
3) H_3O^+

R_2NH, $[H^+]$, $(-H_2O)$

1) I
2) H_3O^+

Ketone **3** can be converted into alkene **4** upon treatment with a Wittig reagent, and hydrolysis of the cyano group gives the desired carboxylic acid product **5**. In the literature synthesis, this hydrolysis step was performed under basic conditions (like saponification), and under those conditions, the product would be deprotonated to give a carboxylate ion. Acid workup is necessary in order to protonate the carboxylate ion and regenerate the carboxylic acid.

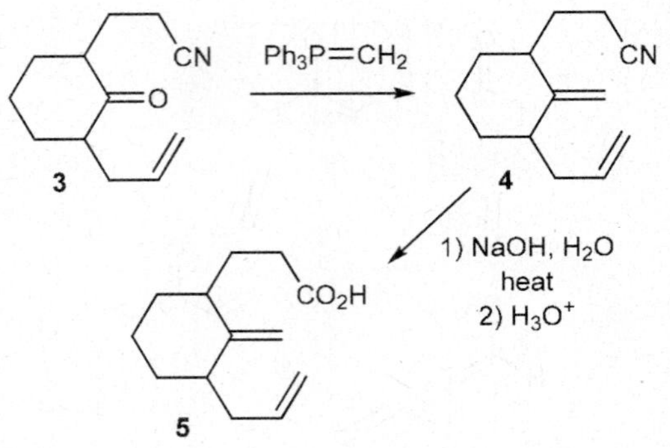

CN $Ph_3P{=}CH_2$ CN

3 4

1) NaOH, H_2O heat 2) H_3O^+

CO_2H

5

21.40. In the presence of a strong base, the β-dicarbonyl compound is deprotonated, giving a resonance-stabilized intermediate (stabilized enolate) which then functions as a Michael donor, attacking the β position of the α,β-unsaturated ketone in a Michael reaction. The resulting enolate is then protonated, and then subsequently deprotonated to give a different enolate (these two steps represent equilibration of the enolates). The new enolate then attacks one of the carbonyl groups to initiate an aldol condensation. The resulting alkoxide ion is then protonated by water. Subsequent deprotonation and loss of hydroxide gives the product. Notice that water functions as the acid for all protonation steps, consistent with basic conditions (strong acids are not measurably present under these conditions).

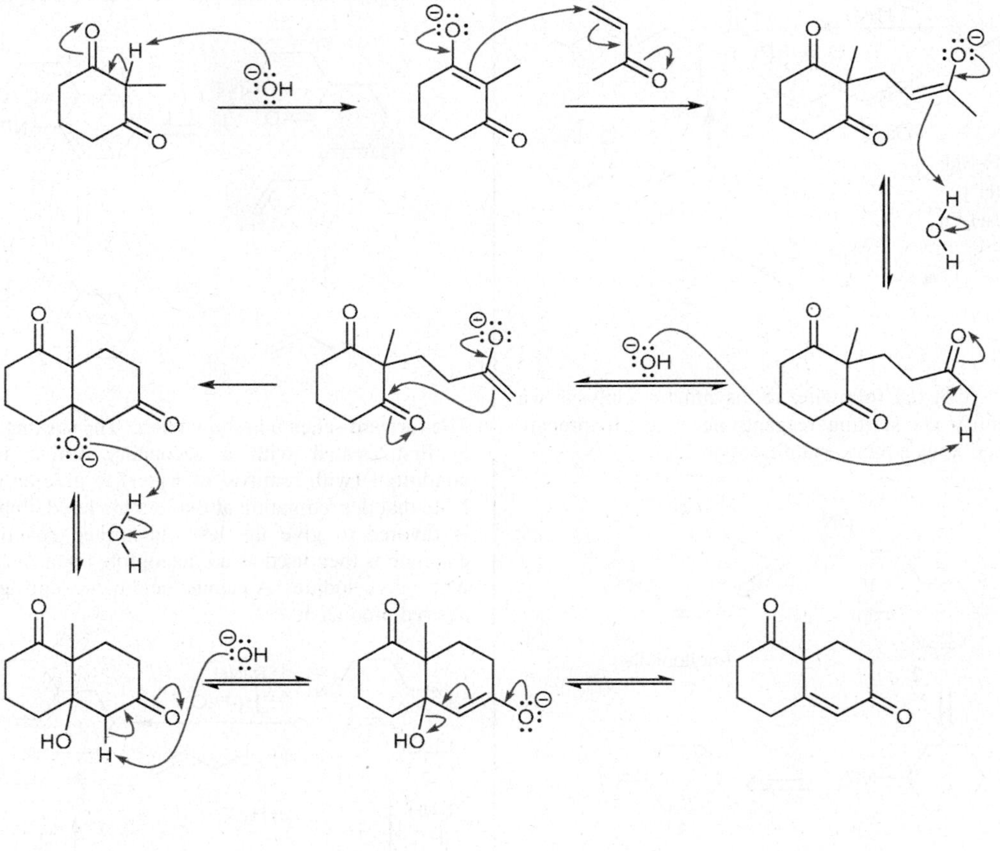

21.41. A Robinson annulation is comprised of a Michael reaction, followed by an intramolecular aldol condensation. To determine the starting materials necessary to prepare the desired product via a Robinson annulation, we draw the following retrosynthetic analysis:

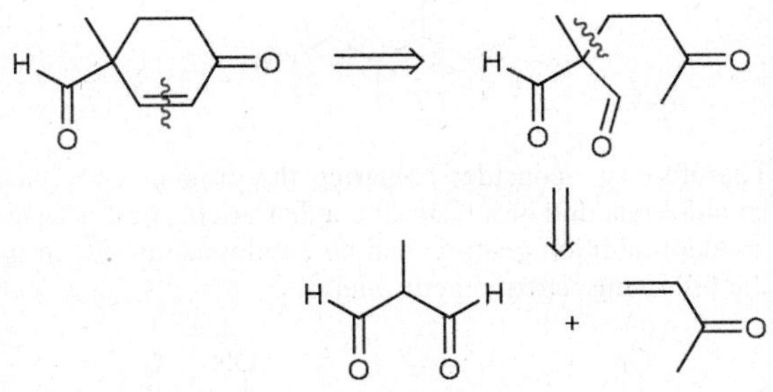

These two steps do not represent two separate reactions. A Robinson annulation can be performed in one reaction flask, as shown in the following forward scheme:

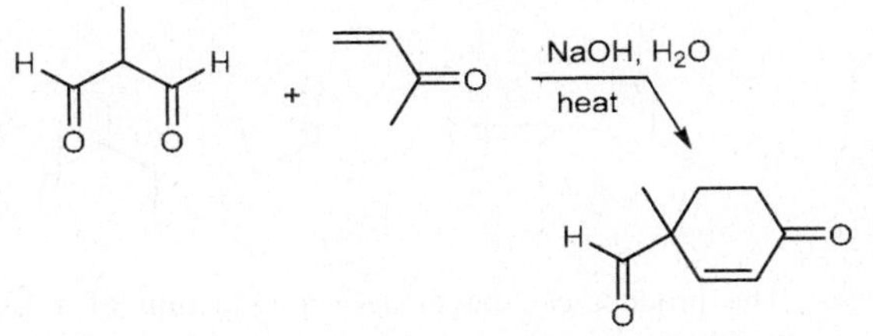

NaOH, H_2O
heat

21.42.
(a) The product is 1,5-difunctionalized:

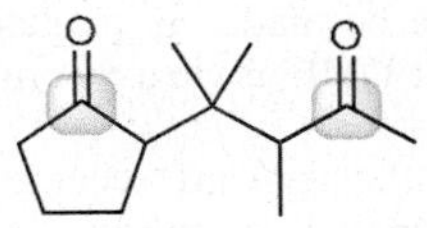

Therefore, we consider preparing the product via a Michael reaction, which would require the following starting materials:

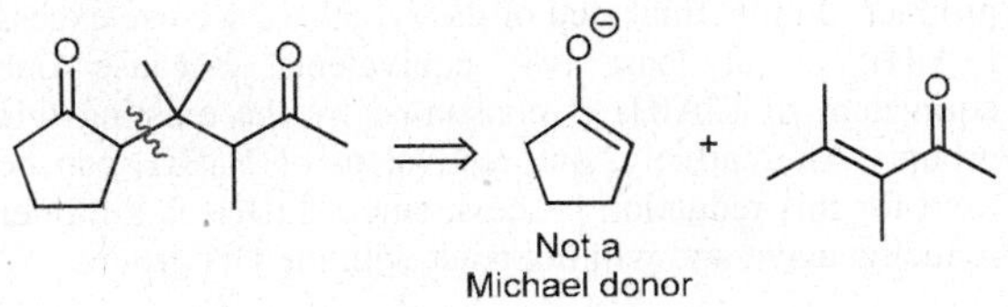

Not a Michael donor

Since enolates are not efficient Michael donors, we must consider a Stork enamine synthesis (in which we use an enamine, rather than an enolate, as a Michael donor). The enamine can be made directly from cyclopentanone.

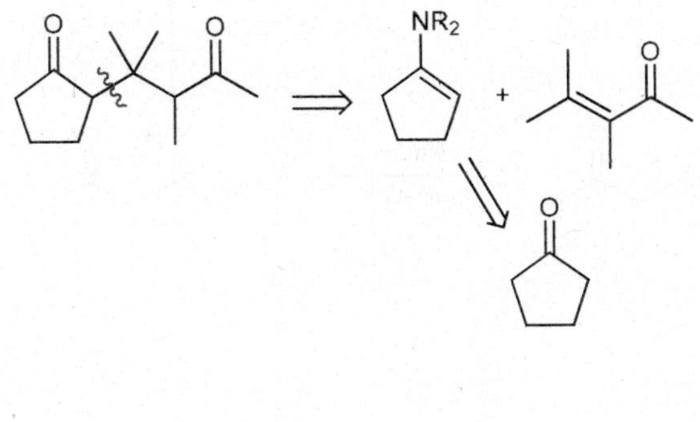

The forward scheme is shown here:

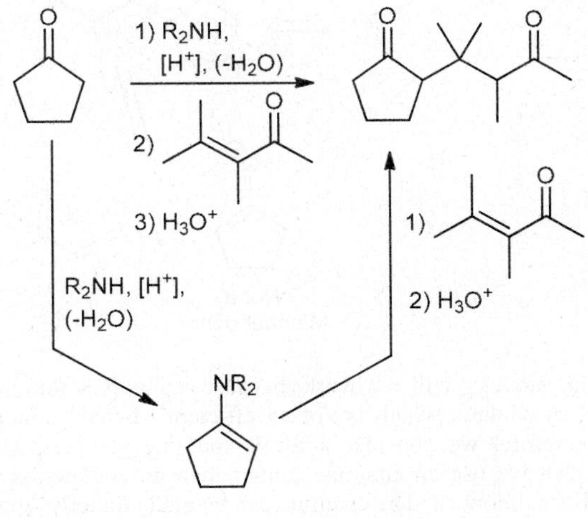

1) R_2NH, $[H^+]$, $(-H_2O)$
2)
3) H_3O^+

R_2NH, $[H^+]$, $(-H_2O)$

1)
2) H_3O^+

(b) The product is 1,3-difunctionalized:

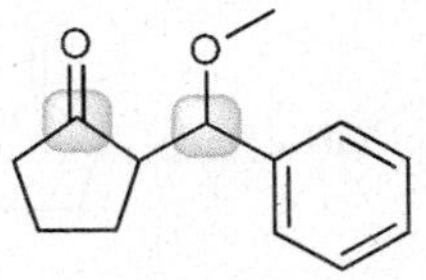

Therefore, we consider preparing the product via either an aldol reaction or a Claisen condensation. In this case, a directed aldol addition, followed by methylation of the resulting alkoxide ion (in a Williamson ether synthesis), gives the desired product:

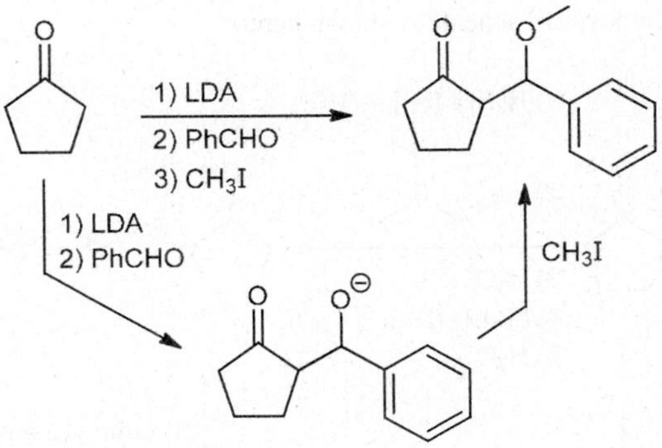

1) LDA
2) PhCHO
3) CH_3I

1) LDA
2) PhCHO

CH_3I

(c) The product is 1,5-difunctionalized:

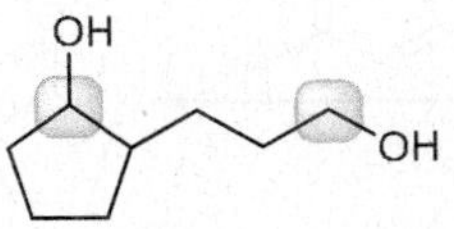

Therefore, we consider preparing the product via a Michael reaction:

Not a
Michael donor

This strategy will not work, because it involves the use of an enolate, which is not an efficient Michael donor. Therefore, we consider a Stork enamine synthesis (in which we use an enamine, rather than an enolate, as a Michael donor). The enamine can be made directly from cyclopentanone.

The forward scheme is shown here:

1) R_2NH, $[H^+]$, $(-H_2O)$
2) (propenal)
3) H_3O^+
4) $LiAlH_4$ (2 eq.)
5) H_3O^+

R_2NH, $[H^+]$, $(-H_2O)$

1) (propenal)
2) H_3O^+

1) $LiAlH_4$ (2 eq.)
2) H_3O^+

21.43.
(a) The product is 1,3-difunctionalized:

Therefore, we consider preparing the product via either an aldol reaction or a Claisen condensation. In this case, an aldol addition reaction can be employed, as shown in the following retrosynthetic analysis.

a. The product can be made via reduction of a β-hydroxyaldehyde.
b. The β-hydroxyaldehyde can be made via an aldol addition reaction between two molecules of propanal.
c. Propanal can be made via oxidation of 1-propanol with PCC (or DMP or via a Swern oxidation).

Now let's draw the forward scheme. Upon treatment with PCC, 1-propanol is oxidized to give propanal. Treating propanal with sodium hydroxide then gives a β-hydroxyaldehyde (via an aldol addition reaction between two molecules of propanal). Reduction with excess $LiAlH_4$, followed by aqueous acidic workup, gives the product. In this final step of the synthesis, we use excess $LiAlH_4$, or at least two equivalents, because one equivalent of $LiAlH_4$ is consumed by the existing OH group. Alternatively, one equivalent of $NaBH_4$ can be used for this reduction process, since $NaBH_4$ is a milder reducing agent and will not react with the OH group.

1) PCC, CH_2Cl_2
2) NaOH, H_2O
3) $LiAlH_4$
4) H_3O^+

PCC
CH_2Cl_2

1) xs $LiAlH_4$
2) H_3O^+

NaOH
H_2O

(b) The product is 1,5-difunctionalized:

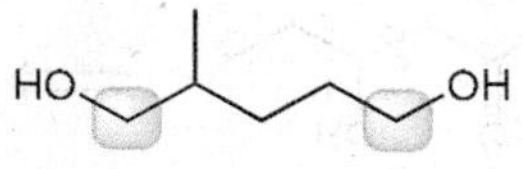

HO OH

Therefore, we consider preparing the product via a Michael reaction:

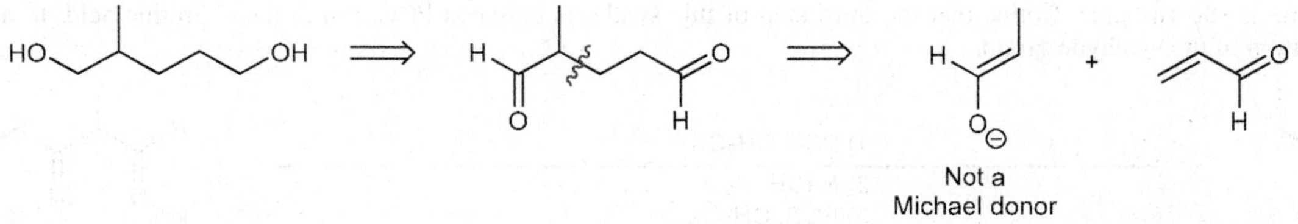

Not a Michael donor

This strategy will not work, because it involves the use of an enolate, which is not an efficient Michael donor. Therefore, we consider a Stork enamine synthesis (in which we use an enamine, rather than an enolate, as a Michael donor). Both the Michael donor and the Michael acceptor can be made from propanal, which can be made from 1-propanol via oxidation with PCC (or DMP or via a Swern oxidation):

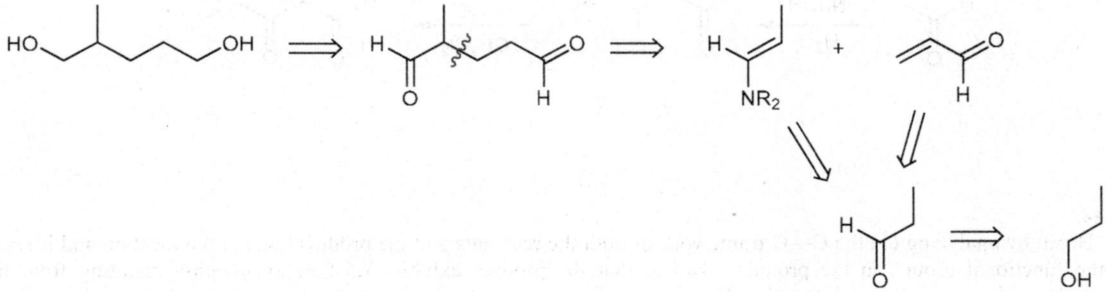

NR_2

The forward scheme is shown here:

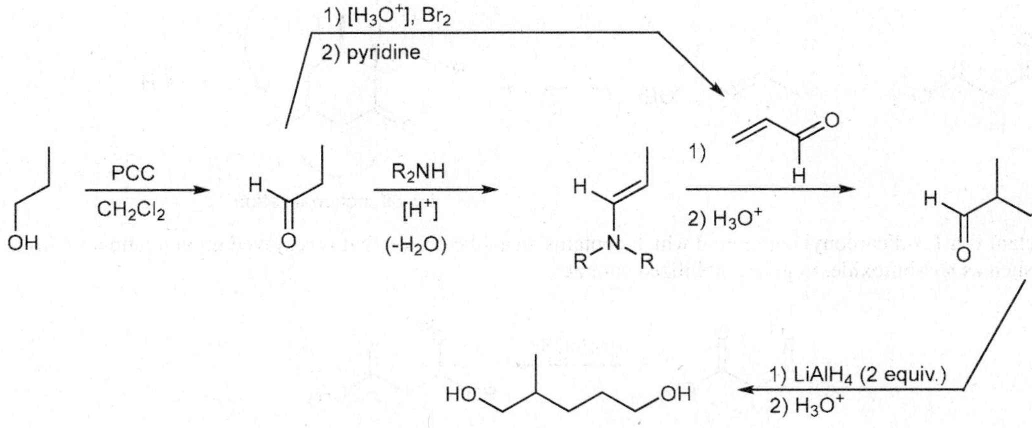

1) $[H_3O^+]$, Br_2
2) pyridine

PCC, CH_2Cl_2 → R_2NH, $[H^+]$, $(-H_2O)$ → 1) acrolein, 2) H_3O^+ → 1) $LiAlH_4$ (2 equiv.), 2) H_3O^+

(c) The product has two imine groups which can be made from the corresponding dicarbonyl compound upon treatment with ammonia in acid-catalyzed conditions (with removal of water):

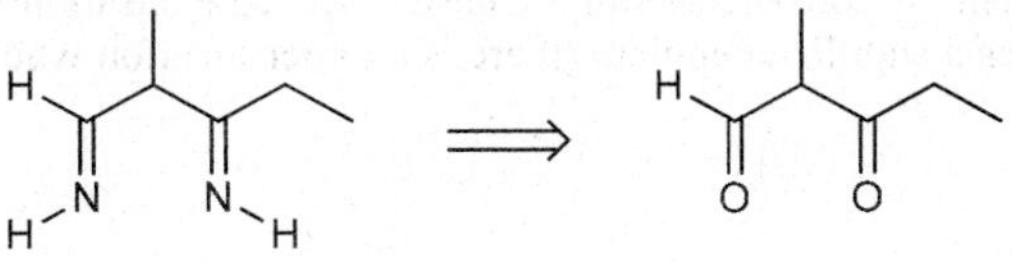

This dicarbonyl compound is 1,3-difunctionalized and can be made from a β-hydroxyaldehyde, which can be made from propanal via an aldol addition reaction:

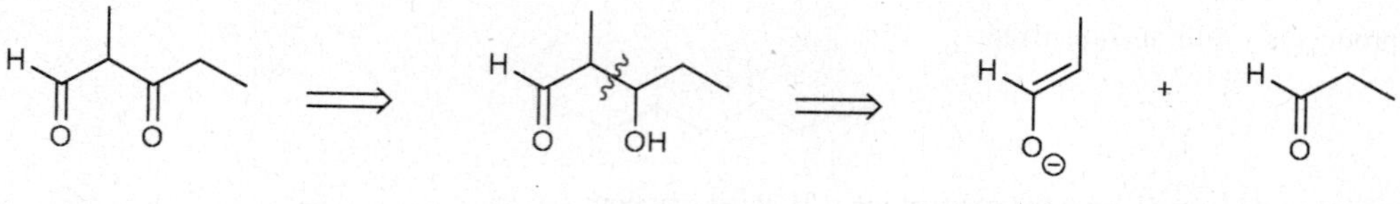

And propanal can be made from 1-propanol via oxidation with PCC (or DMP or via a Swern oxidation). The forward scheme is shown here. Notice that the third step of this synthesis employs PCC, rather than chromic acid, to avoid oxidation of the aldehyde group.

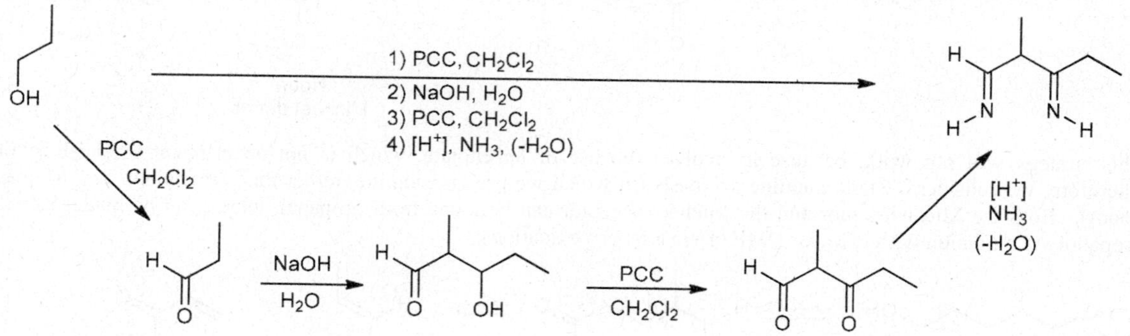

1) PCC, CH_2Cl_2
2) NaOH, H_2O
3) PCC, CH_2Cl_2
4) [H^+], NH_3, (-H_2O)

PCC CH_2Cl_2 — NaOH H_2O — PCC CH_2Cl_2 — [H^+] NH_3 (-H_2O)

21.44.

(a) Begin by analyzing (1) the C—C framework of both the reactants and the product and (2) the location and identity of the functional groups in the product. Notice that the product exhibits 1,5-functionalization resulting from the Michael reaction, as well as 1,3-functionalization resulting from the Claisen-type reaction.

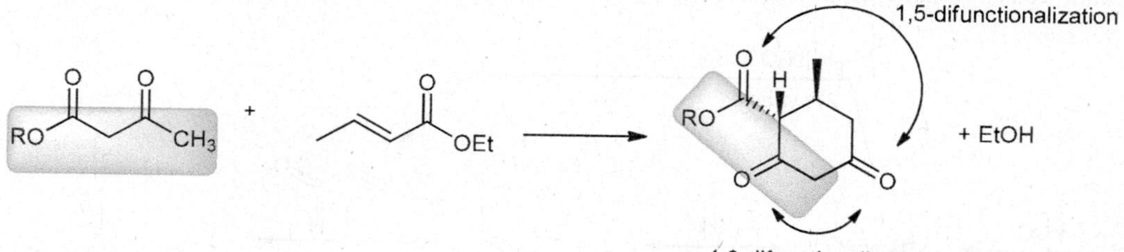

1,5-difunctionalization

+ EtOH

1,3-difunctionalization

The first reactant is a 1,3-dicarbonyl compound which contains an acidic proton that is removed upon treatment with a strong base, such as *tert*-butoxide, to give a stabilized enolate.

t-BuOK

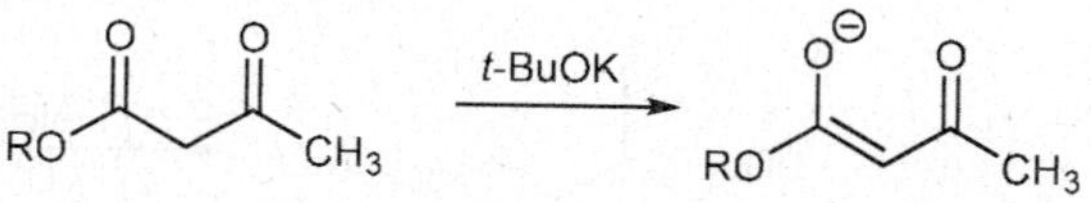

Due to the stability of this enolate, the first step is 1,4-addition (a Michael reaction). The other reactant contains an α,β-unsaturated ester capable of participating as a Michael acceptor, giving 1,5-difunctionalization, as shown below. Next, notice that a stable 6-membered ring could result if the ketone –CH_3 group were deprotonated and the resulting enolate attacked the ethyl ester. This is consistent with a Claisen-type condensation which would afford 1,3-diketone. As a final step, deprotonation gives a stabilized enolate (there is another location where deprotonation could also occur to give a stabilized enolate):

Then, upon aqueous acidic workup, the stabilized enolate is protonated to give the observed product:

(b) Claisen condensations are nucleophilic acyl substitution reactions where the nucleophile is an ester enolate and the electrophile is an ester; intramolecular versions are known as Dieckmann condensations. This reaction, however, involves a ketone enolate (rather than an ester enolate) attacking an ester in a nucleophilic acyl substitution, so neither of these names strictly apply.

21.45.

(a) This transformation requires the installation of two groups (one at the α position and the other at the β position). This can be achieved by treating the α,β-unsaturated ketone with lithium dimethylcuprate, thereby installing a methyl group and generating an enolate, which is then treated with benzyl iodide to install a benzyl group:

1) Me_2CuLi

2) I

(b) This transformation requires the installation of two methyl groups (one at the α position and the other at the β position), as well as reduction of the aldehyde group. Installation of the two methyl groups can be achieved by treating the α,β-unsaturated ketone with lithium dimethylcuprate, followed by methyl iodide. Reduction is then achieved with $LiAlH_4$, followed by aqueous acidic workup.

1) Me_2CuLi
2) MeI
3) $LiAlH_4$
4) H_3O^+

1) Me_2CuLi
2) MeI

1) $LiAlH_4$
2) H_3O^+

(c) The desired transformation requires the installation of two alkyl groups (one at the α position and the other at the β position), as well as conversion of the aldehyde group into an acid chloride. The acid chloride can be prepared from a carboxylic acid, which, in turn, can be prepared by oxidation of the corresponding aldehyde, as shown in the following retrosynthesis:

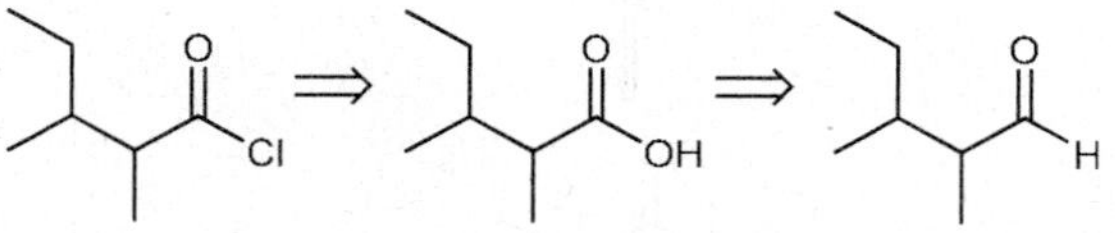

Continuing the retrosynthesis, this aldehyde can be made from the given starting, with the installation of two alkyl groups:

Et_2CuLi + MeI

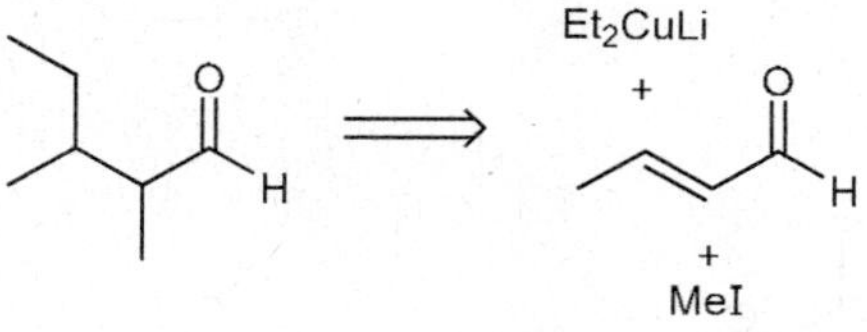

Now let's draw the forward process. Installation of the two alkyl groups can be achieved by treating the α,β-unsaturated aldehyde with lithium diethylcuprate, followed by methyl iodide. The aldehyde is then converted into an acid halide via oxidation with chromic acid (to give a carboxylic acid) followed by treatment with thionyl chloride.

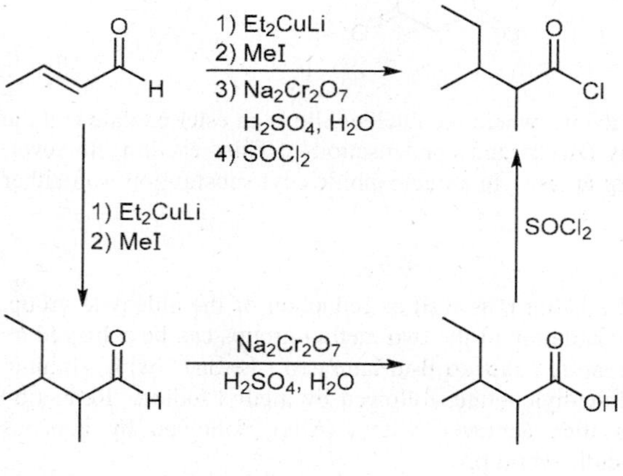

1) Et_2CuLi
2) MeI
3) $Na_2Cr_2O_7$, H_2SO_4, H_2O
4) $SOCl_2$

1) Et_2CuLi
2) MeI

$Na_2Cr_2O_7$, H_2SO_4, H_2O

$SOCl_2$

Note than the acid chloride group is formed at the end of the synthesis, because if it were formed in the beginning of the synthesis, then lithium diethylcuprate would react with the acid chloride group.

(d) If the acetal is hydrolyzed with aqueous acid, the resulting α,β-unsaturated ketone can be treated with lithium diethylcuprate, followed by ethyl iodide, to install the two ethyl groups in the correct locations. The ketone can then be converted back into an acetal upon treatment with ethylene glycol under acidic conditions (with removal of water).

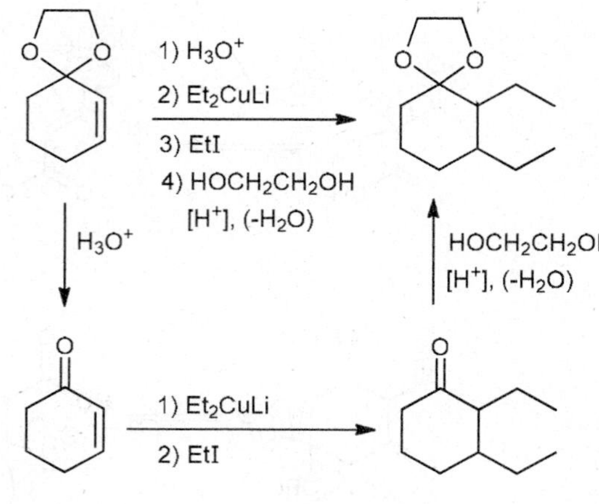

1) H_3O^+
2) Et_2CuLi
3) EtI
4) $HOCH_2CH_2OH$, $[H^+]$, $(-H_2O)$

H_3O^+

$HOCH_2CH_2OH$, $[H^+]$, $(-H_2O)$

1) Et_2CuLi
2) EtI

(e) The imine can be prepared from the corresponding aldehyde:

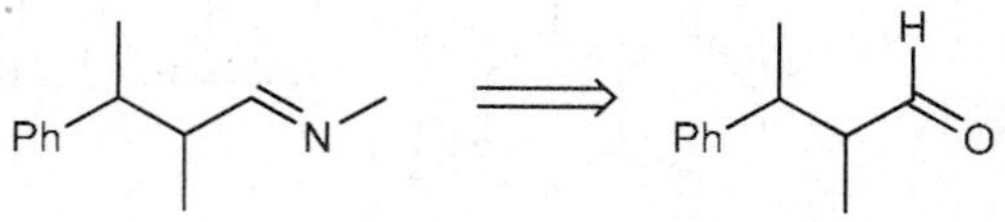

This aldehyde can be made from an α,β-unsaturated aldehyde, with installation of a phenyl group at the β position and installation of a methyl group at the α position):

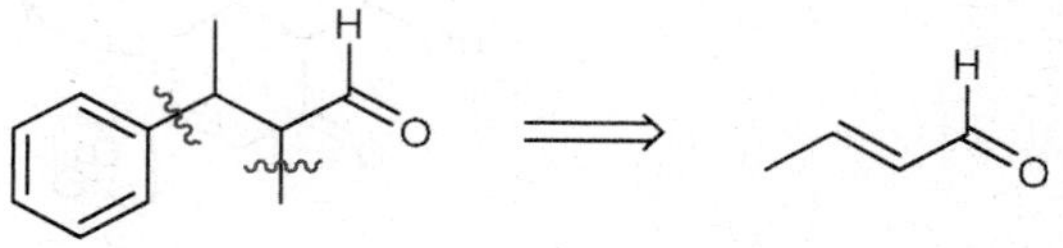

The α,β-unsaturated aldehyde can be made from the starting material via oxidation with PCC (or DMP or via a Swern oxidation). The entire synthesis is summarized here:

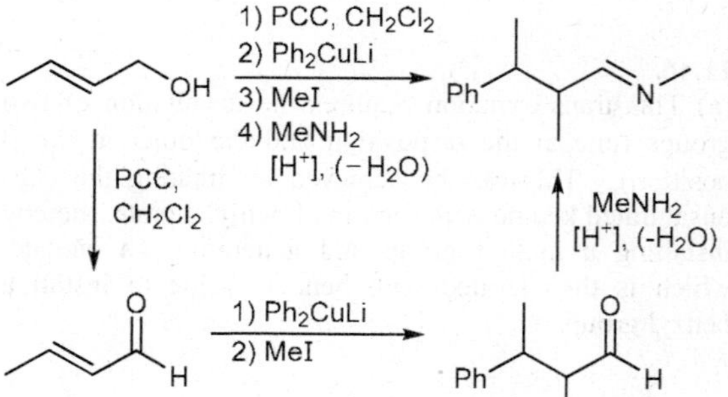

1) PCC, CH_2Cl_2
2) Ph_2CuLi
3) MeI
4) $MeNH_2$, $[H^+]$, $(-H_2O)$

PCC, CH_2Cl_2

$MeNH_2$, $[H^+]$, $(-H_2O)$

1) Ph_2CuLi
2) MeI

(f) This problem is similar to the previous problem, although the final step is reduction of the aldehyde with a Clemmensen reduction to give an alkane.

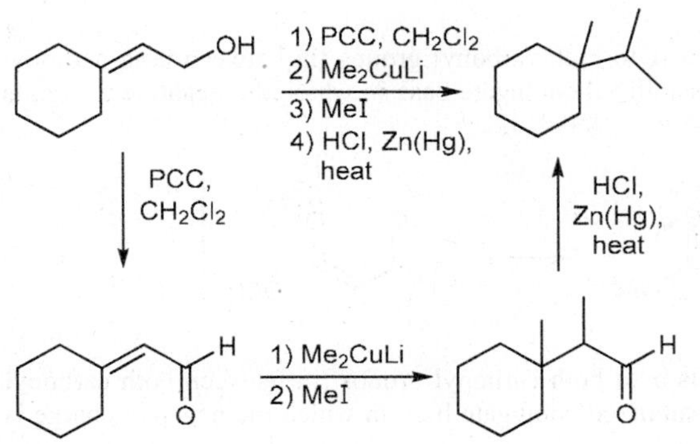

1) PCC, CH_2Cl_2
2) Me_2CuLi
3) MeI
4) HCl, Zn(Hg), heat

PCC, CH_2Cl_2

HCl, Zn(Hg), heat

1) Me_2CuLi
2) MeI

21.46. This transformation requires the installation of two groups (one at the α position and the other at the β position). This can be achieved by treating the α,β-unsaturated carbonyl with lithium dimethylcuprate, thereby installing a methyl group and generating an enolate, which can then be treated with allyl iodide to install an allyl group at the α position, as shown below:

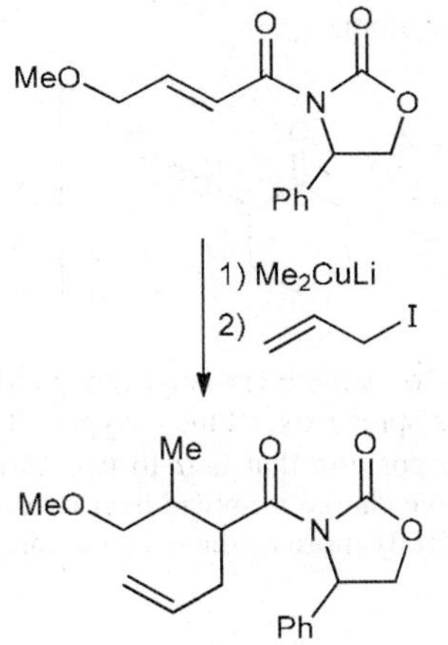

1) Me_2CuLi
2)

21.47.
(a) Deprotonation (of the highlighted proton) results in a resonance-stabilized enolate ion. Therefore, the highlighted proton is the most acidic proton (with a pK_a just below 20), because its removal leads to a stabilized conjugate base.

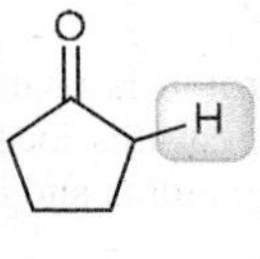

(b) This compound does not have an acidic proton, and is expected to have a pK_a above 20.

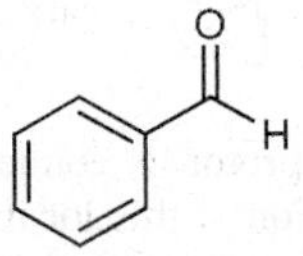

(c) Deprotonation (of the highlighted proton) results in a resonance-stabilized enolate ion. Therefore, the highlighted proton is the most acidic proton (with a pK_a just below 20), because its removal leads to a stabilized conjugate base.

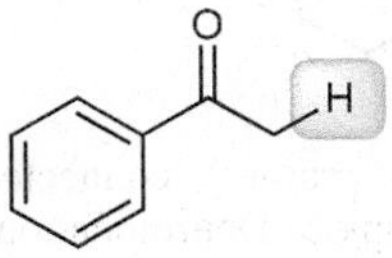

(d) Deprotonation (of the highlighted proton) results in a resonance-stabilized conjugate base (a stabilized enolate). Therefore, the highlighted proton is the most acidic proton (with a pK_a below 20), because its removal leads to the most stable conjugate base possible.

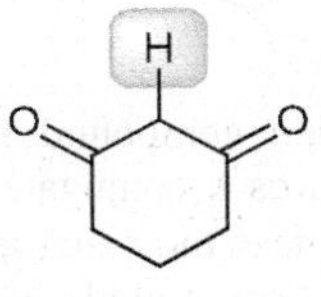

(e) Deprotonation (of the highlighted proton) results in an alkoxide ion. As such, the compound below is expected to have a pK_a lower than 20 (see the pK_a table at the beginning of the textbook).

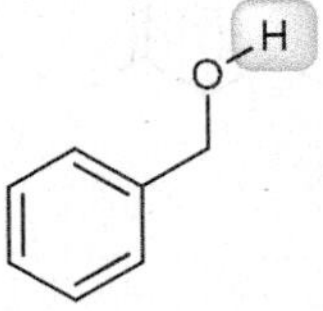

21.48. The most acidic proton is connected to the position that is α to two carbonyl groups (in between both carbonyl groups). Deprotonation at this location leads to a resonance-stabilized conjugate base in which the negative charge is spread over two oxygen atoms and one carbon atom. As such, the compound below is expected to have a pK_a lower than 10 (see the pK_a table at the beginning of the textbook)

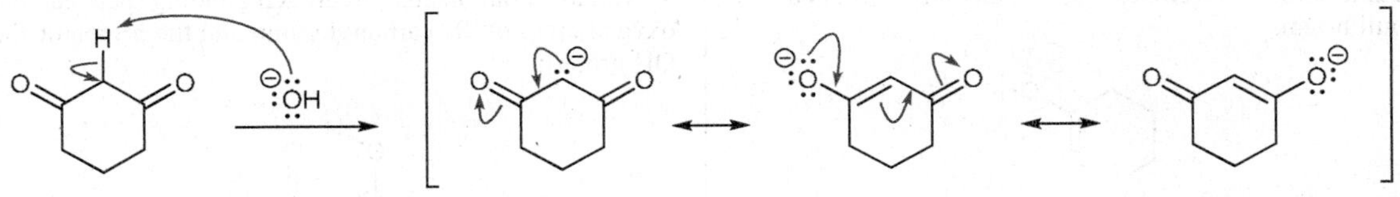

21.49.

(a) The most acidic proton is connected to the position that is α to both carbonyl groups (in between both carbonyl groups). Deprotonation at this location leads to a resonance-stabilized conjugate base in which the negative charge is spread over two oxygen atoms and one carbon atom).

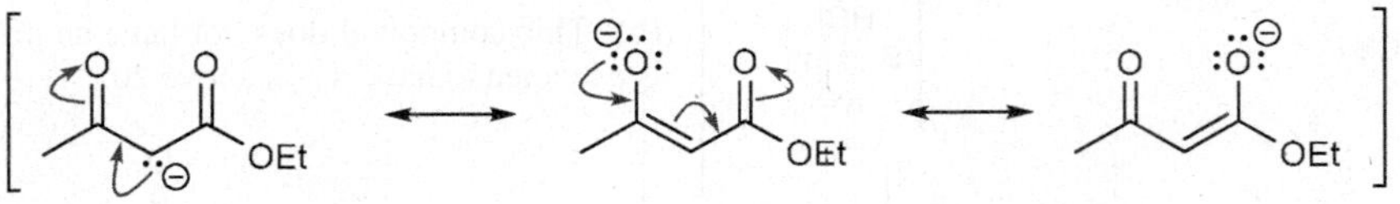

(b) The most acidic proton is connected to the position that is α to both carbonyl groups (in between both carbonyl groups). Deprotonation at this location leads to a resonance-stabilized conjugate base in which the negative charge is spread over two oxygen atoms and one carbon atom).

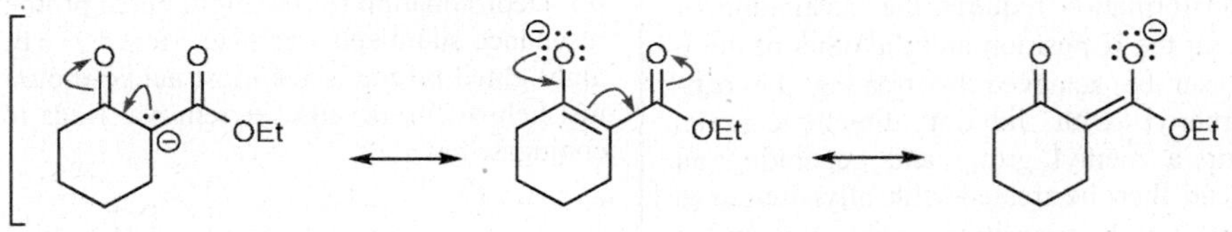

(c) The most acidic proton is connected to the position that is α to the carbonyl group as well as the cyano group (in between both groups). Deprotonation at this location leads to a resonance-stabilized conjugate base in which the negative charge is spread over an oxygen atom, a nitrogen atom and a carbon atom).

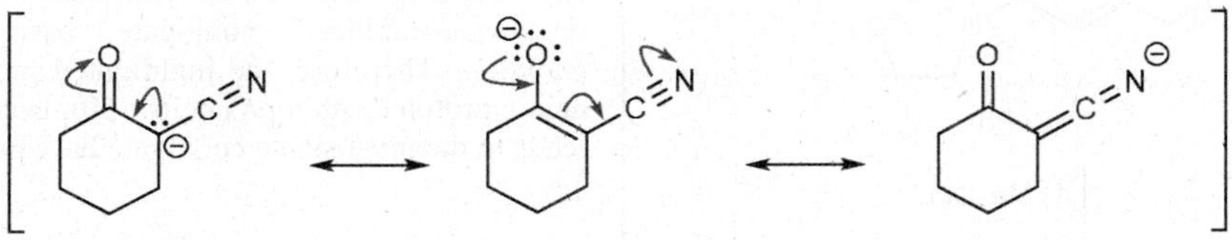

21.50. The most acidic compound is the one that exhibits a position that is α to three carbonyl groups (deprotonation of this compound gives a conjugate base in which the negative charge is spread over three oxygen atoms and one carbon atom). The next most acidic compound is the one that exhibits a position that is α to two carbonyl groups (deprotonation of this compound gives a conjugate base in which the negative charge is spread over two oxygen atoms and one carbon atom). Of the remaining two compounds, an alcohol is generally more acidic than a ketone (see the pK_a table at the beginning of the textbook).

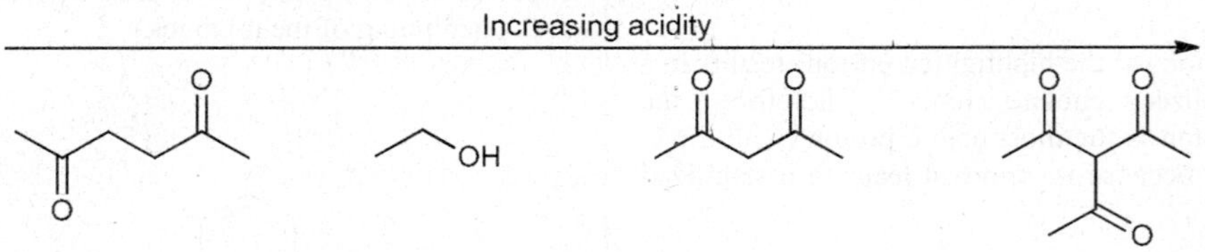

Increasing acidity

21.51.

(a) This enol does not exhibit a significant presence at equilibrium. Ketones are generally favored at equilibrium.

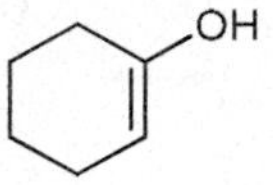

(b) The following enol does exhibit a significant presence at equilibrium because it exhibits conjugation as well as intramolecular hydrogen bonding (between the oxygen atom of the carbonyl group and the proton of the OH group):

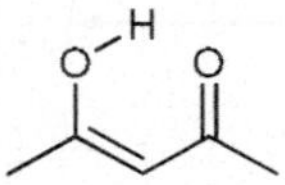

(c) The following enol does indeed exhibit a significant presence at equilibrium, because it is aromatic.

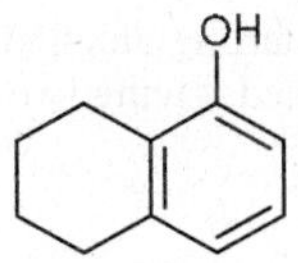

In fact, in this case, the ketone does not exhibit a significant presence at equilibrium. The aromatic ring is so strongly favored, that we cannot detect the ketone present in the mixture.

21.52. Ethyl acetoacetate has two carbonyl groups:

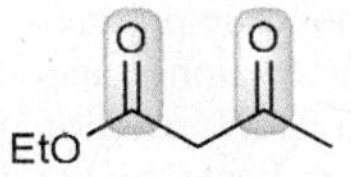

One of them (left) has only one α position and can therefore only form one enol:

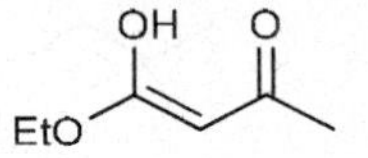

The other carbonyl group (right) has two α positions and can therefore form two different enols:

In total, there are three enol isomers.

21.53.

(a) This compound has only one α position. LDA is a strong base and it will deprotonate the compound (at the α position), resulting in the following enolate.

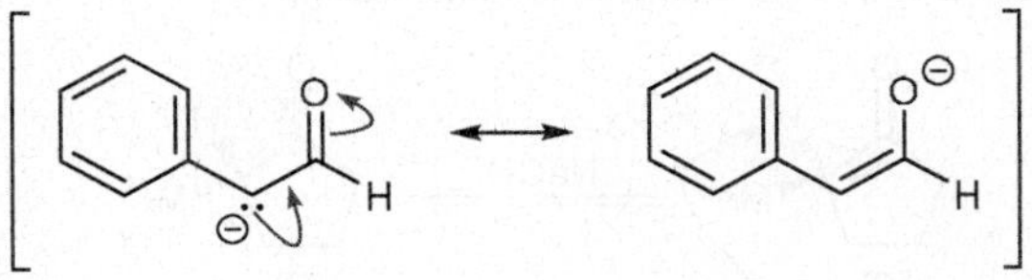

(b) This compound has only one α position. LDA is a strong base and it will deprotonate the compound (at the α position), resulting in the following enolate.

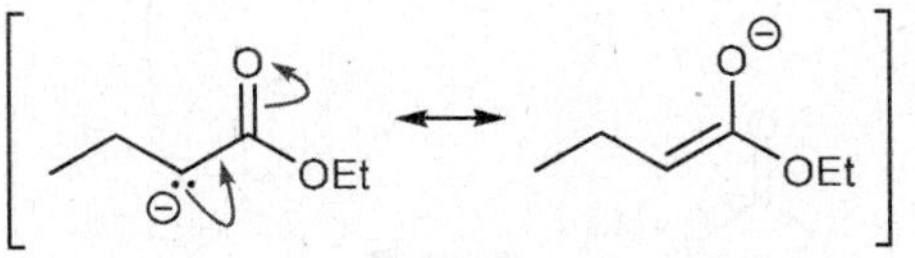

(c) This compound has two α positions. LDA is a strong, sterically hindered base, so deprotonation will occur at the less substituted position. Deprotonation at that position results in the following kinetic enolate:

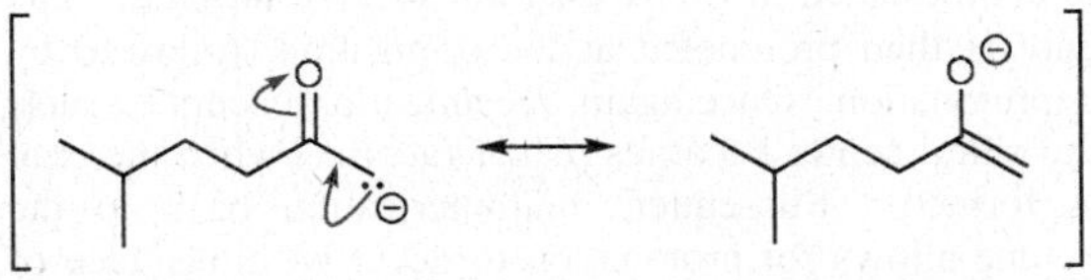

21.54. Deprotonation at the highlighted γ position results in an enolate anion that has three resonance structures. The negative charge is spread over one oxygen atom and two carbon atoms:

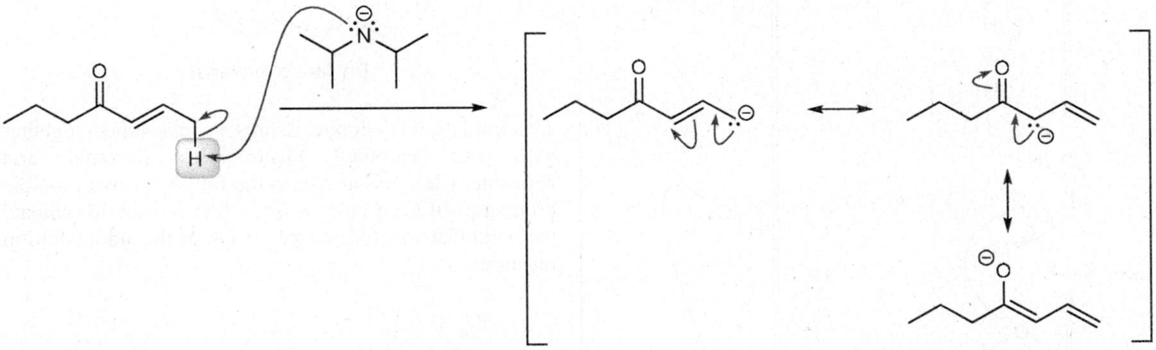

21.55. Deprotonation at the α carbon changes the hybridization state of the α carbon from sp^3 (tetrahedral) to sp^2 (planar). When the α position is protonated once again, the proton can be placed on either side of the planar α carbon, resulting in racemization:

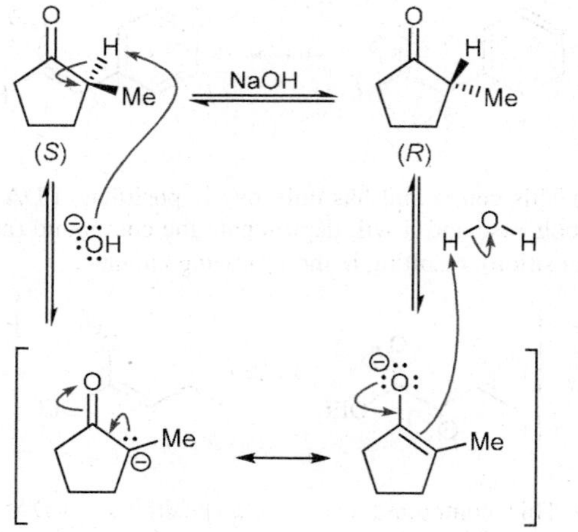

NaOH
(*S*) (*R*)

21.56. In acidic conditions, the carbonyl group is first protonated, resulting in a resonance-stabilized cation that is deprotonated at the α position to give an enol. The enol is then protonated at the α position, followed by deprotonation. Once again, racemization occurs because the chiral center becomes planar (achiral) when the enol is formed. Subsequent tautomerization back to the ketone allows for protonation to occur on either face of the planar enol, giving a racemic mixture.

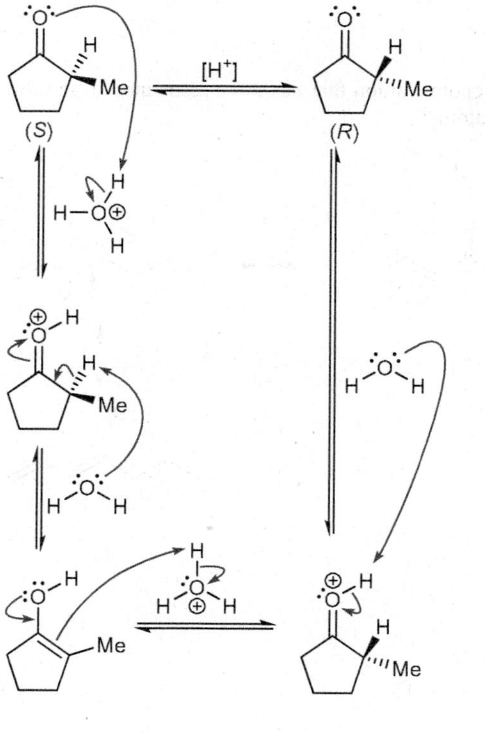

[H⁺]
(*S*) (*R*)

21.57. Each of the starting aldehydes has an α position that can be deprotonated, giving two possible enolates:

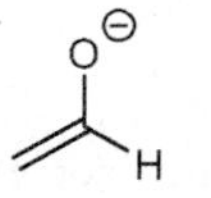

Enolate of acetaldehyde

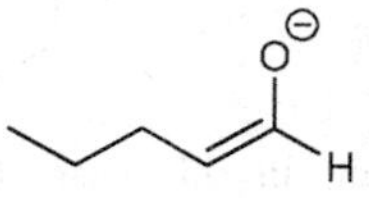

Enolate of pentanal

So, there are two nucleophiles in solution, as well as two electrophiles (acetaldehyde and pentanal), giving rise to the following four possible products (the top two are the result of self aldol reactions, and the bottom two are produced by mixed aldol reactions). In each case, a wavy line is used to indicate the bond that was formed as a result of the aldol addition reaction:

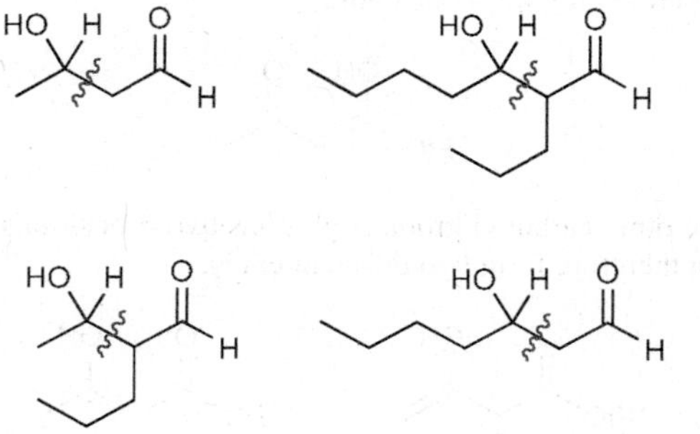

21.58. Hexanal has an α position that bears protons, but the α position of benzaldehyde does not bear any protons. As such, only one enolate can form under these conditions:

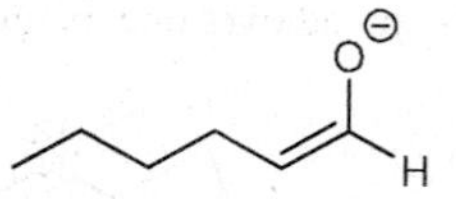

Enolate of hexanal

This enolate nucleophile is present in solution together with two carbonyl electrophiles (hexanal and benzaldehyde), giving rise to the following two possible products. In each case, a wavy line is used to indicate the bond that was formed as a result of the aldol addition reaction:

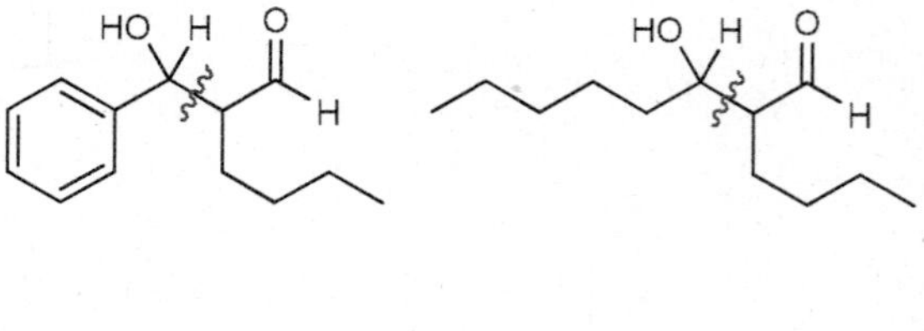

21.59. The mechanism begins with a two-step tautomerization process to form an enol. The carbonyl group is protonated, giving a resonance-stabilized intermediate that is then deprotonated to give an enol. A second, two-step tautomerization process results in the observed product. Protonation of the enol results in a resonance-stabilized, benzylic carbocation intermediate that is then deprotonated to give the product.

In the product, the carbonyl group and the aromatic ring are conjugated. However, in the starting material, the carbonyl group and the aromatic ring are *not* conjugated. Formation of conjugation results in stabilization (a decrease in energy) which serves as a driving force for formation of the product.

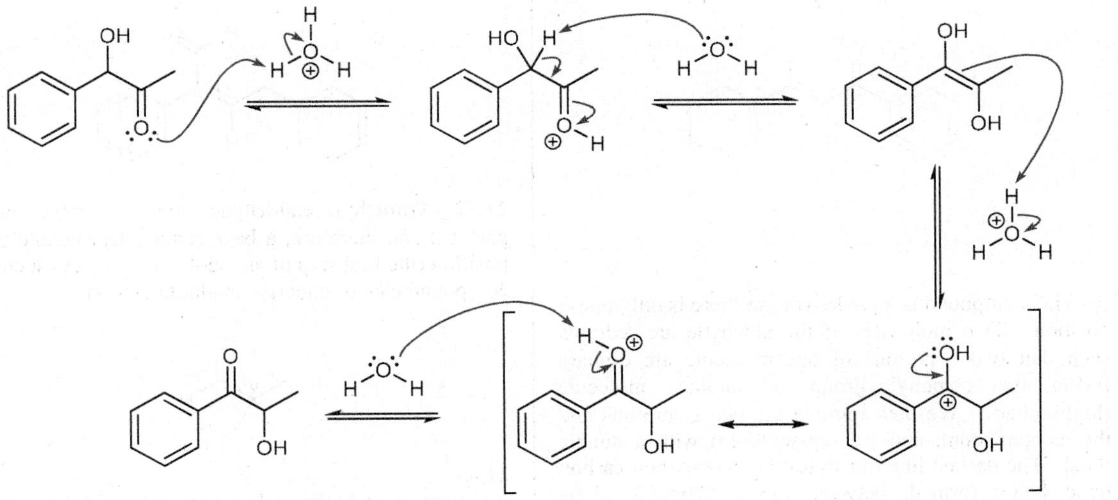

21.60. The mechanism begins with a two-step tautomerization process to form an enol. Hydroxide functions as a base and deprotonates the α position, giving a resonance-stabilized enolate (only the more significant resonance structure is drawn below) that is protonated to give an enol. A second, two-step tautomerization process results in the observed product. The enol is then deprotonated to give another enolate ion, which is then protonated to give the product.

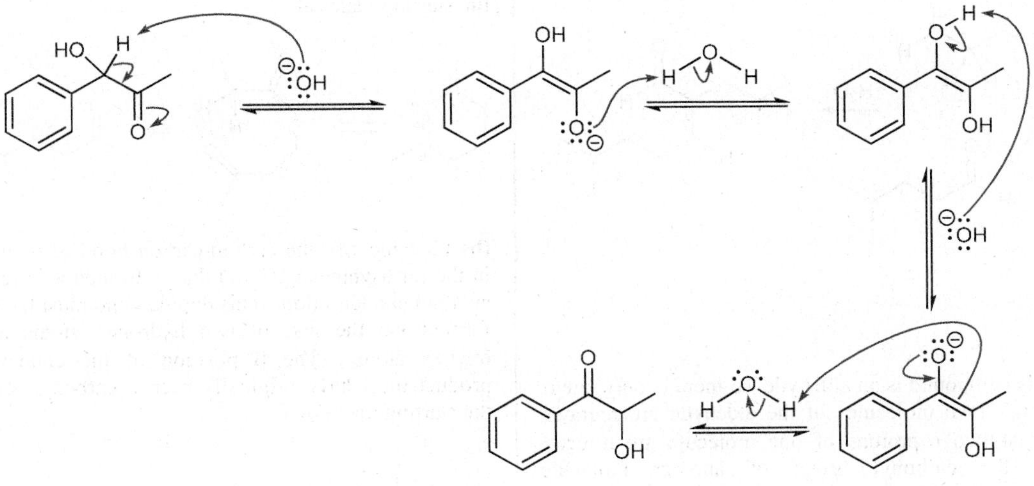

21.61.

(a) This compound (acetophenone) has only one α position that bears protons. Two molecules of the ketone are redrawn such that two α protons of one molecule are directly facing the carbonyl group of another molecule (highlighted). We then remove the two α protons and the oxygen atom, and we replace them with a double bond. The dashed line illustrates the new carbon-carbon bond being formed, between the nucleophilic alpha carbon (enolate) and the electrophilic carbonyl carbon. In this case, two stereoisomeric products are possible, and we draw the *E* isomer, rather than the *Z* isomer, because the former has fewer steric interactions.

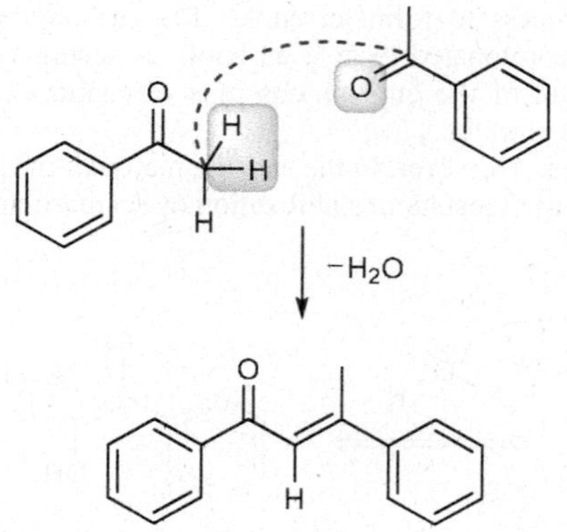

(b) This compound is an aldehyde, so there is only one α position. Two molecules of the aldehyde are redrawn such that two α protons of one molecule are directly facing the carbonyl group of another molecule (highlighted). We then remove the two α protons and the oxygen atom, and we replace them with a double bond. The dashed line illustrates the new carbon-carbon bond being formed, between the nucleophilic alpha carbon (enolate) and the electrophilic carbonyl carbon. In this case, two stereoisomeric products are possible, and we draw the isomer that exhibits fewer steric interactions.

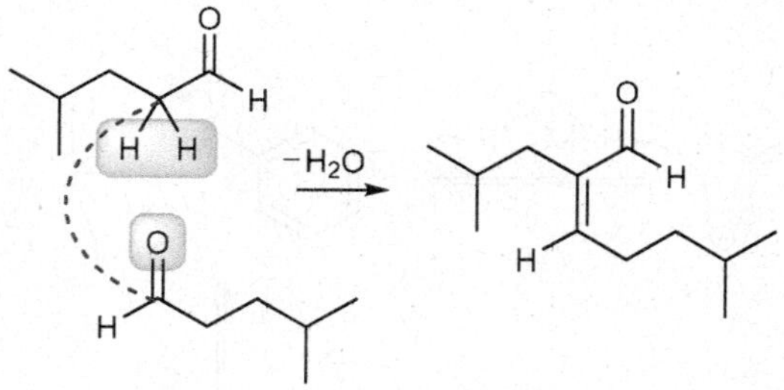

(c) This compound is an aldehyde, so there is only one α position. Two molecules of the aldehyde are redrawn such that two α protons of one molecule are directly facing the carbonyl group of another molecule (highlighted). We then remove the two α protons and the oxygen atom, and we replace them with a double bond. The dashed line illustrates the new carbon-carbon bond being formed, between the nucleophilic alpha carbon (enolate) and the electrophilic carbonyl carbon. In this case, two stereoisomeric products are possible, and we draw the isomer that exhibits fewer steric interactions.

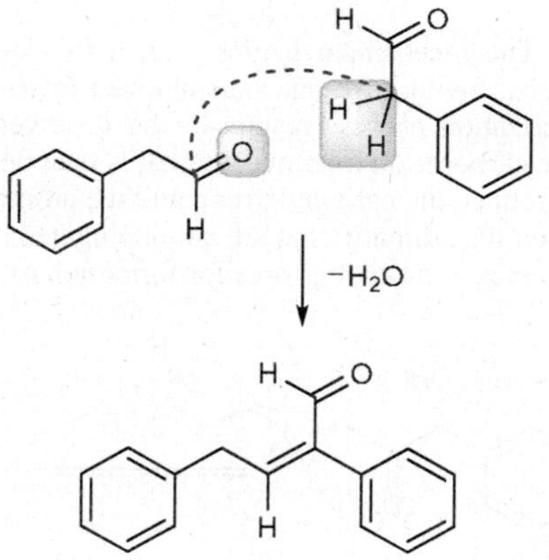

21.62. Trimethylacetaldehyde does not have any α protons, and therefore, a base cannot deprotonate the α position (the first step of an aldol reaction). As such, this compound cannot undergo an aldol reaction.

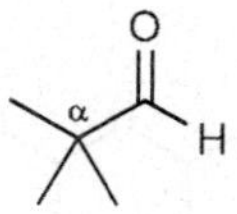

21.63.

(a) First identify the carbon-carbon bond (disconnected in the retrosynthesis below) that is formed as a result of an aldol condensation. This double bond must have been formed via the loss of two hydrogen atoms and an oxygen atom. The β position of the condensation product must have originally been a carbonyl group in the starting material:

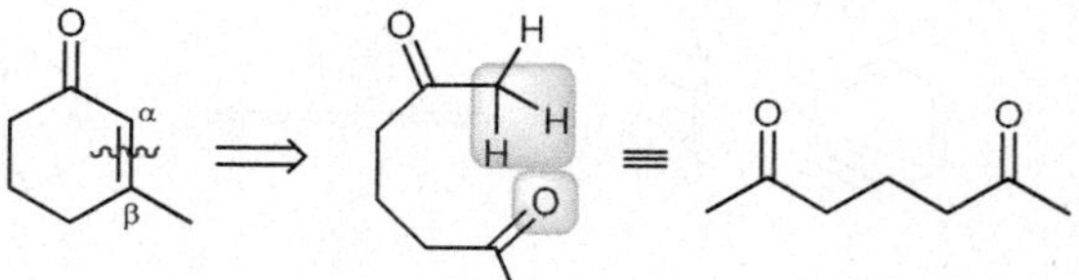

(b) First identify the carbon-carbon bond (disconnected in the retrosynthesis below) that is formed as a result of an aldol condensation. This double bond must have been formed via the loss of two hydrogen atoms and an oxygen atom. The β position of the condensation product must have originally been a carbonyl group in the starting material:

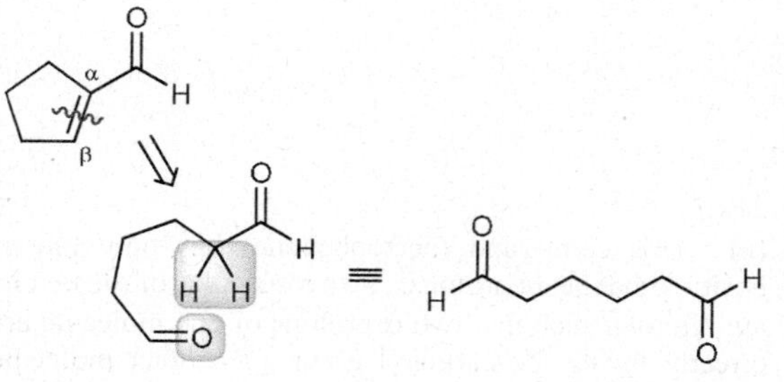

(c) First identify the carbon-carbon bond (disconnected in the retrosynthesis below) that is formed as a result of an aldol condensation. This double bond must have been formed via the loss of two hydrogen atoms and an oxygen atom. The β position of the condensation product must have originally been a carbonyl group in the starting material:

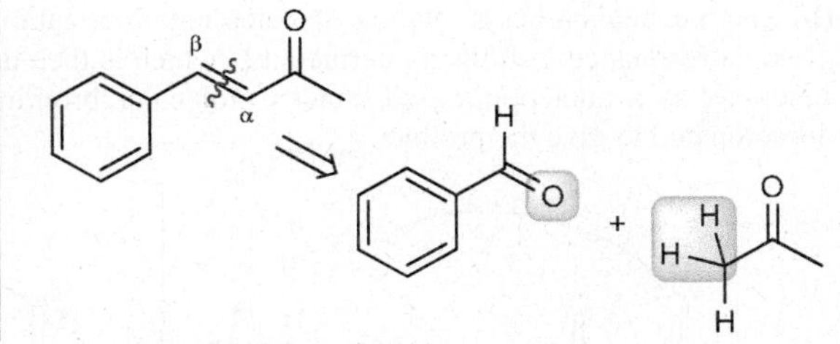

21.64. In acidic conditions, the nucleophilic agent must be an enol, rather than an enolate, because enolate ions are fairly basic and are therefore incompatible with acidic conditions. Therefore, the mechanism begins with a two-step tautomerization process. In the first step, the carbonyl group is protonated, giving a resonance-stabilized intermediate, which is then deprotonated at the α position to give an enol. The enol then functions as a nucleophile and attacks another protonated carbonyl group. The resulting resonance-stabilized cation is then deprotonated to give the aldol addition product.

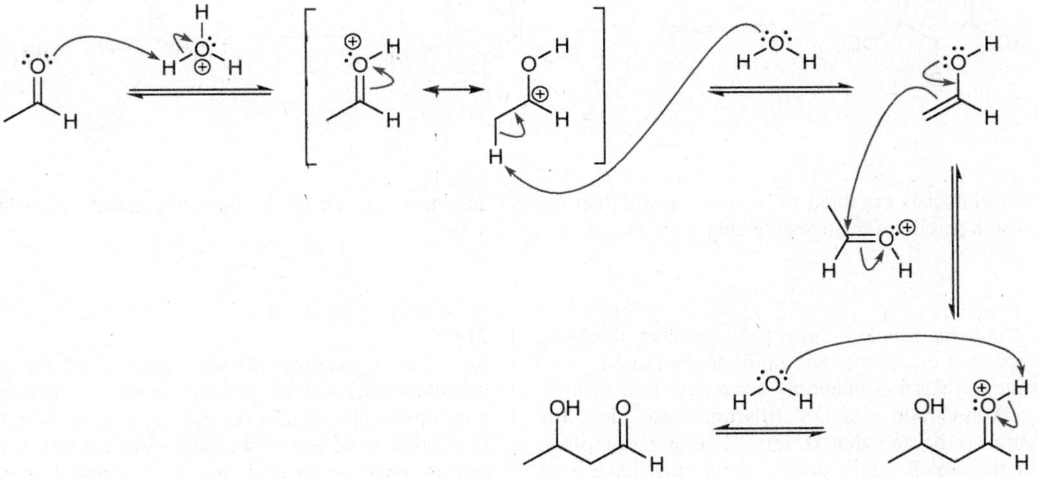

21.65.

(a) In acidic conditions, the nucleophilic agent must be an enol, rather than an enolate, because enolate ions are fairly basic and are therefore incompatible with acidic conditions. The reaction still occurs at the α position, installing a bromine atom at that position (α-halogenation), giving the following product.

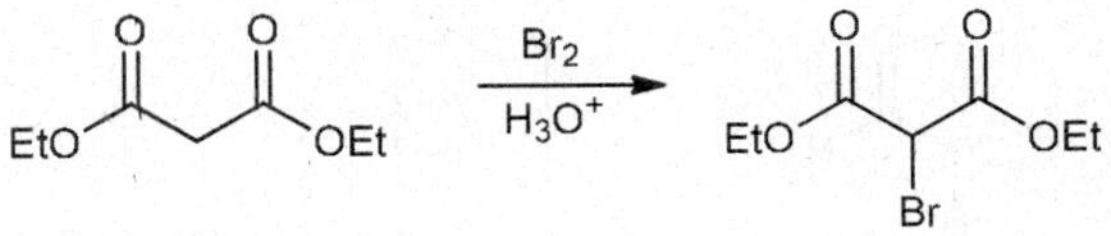

(b) The mechanism begins with a two-step tautomerization process. In the first step, the carbonyl group is protonated, giving a resonance-stabilized intermediate, which is then deprotonated at the α position to give an enol. The enol then functions as a nucleophile and attacks molecular bromine (Br_2). The resulting resonance-stabilized cation is then deprotonated to give the product.

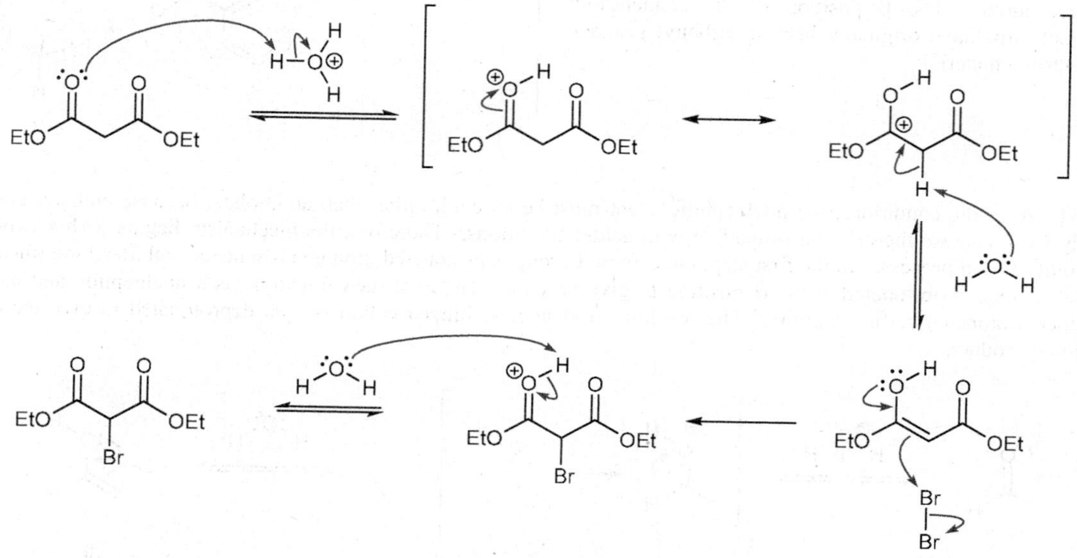

(c) The product is expected to be more acidic than diethyl malonate because of the inductive effect of the bromine atom, which stabilizes the negative charge in the conjugate base.

21.66. Cinnamaldehyde is an α,β-unsaturated aldehyde, so it can be made via an aldol condensation. To determine the starting materials necessary, first identify the carbon-carbon bond (disconnected in the retrosynthesis below) that is formed as a result of an aldol condensation. This double bond must have been formed via the loss of two hydrogen atoms and an oxygen atom. The β position of the condensation product must have originally been a carbonyl group in the starting material:

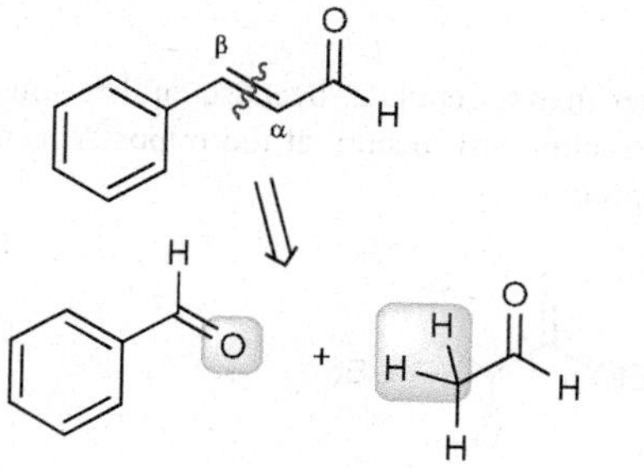

Therefore, cinnamaldehyde can be made from benzaldehyde and acetaldehyde, as shown:

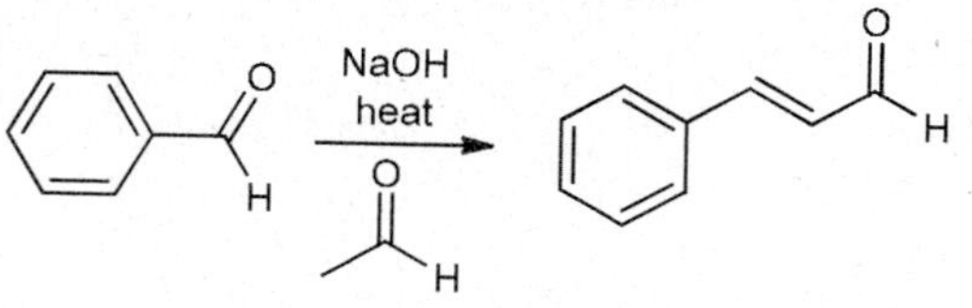

21.67.

(a) The α position of one molecule of the ester is deprotonated, and the resulting enolate functions as a nucleophile and attacks the electrophilic carbonyl group of another molecule of the ester. As a result, a carbon-carbon bond is formed, giving a charged tetrahedral intermediate. The carbonyl group is then reformed via loss of an ethoxide ion, affording a β-ketoester, as shown:

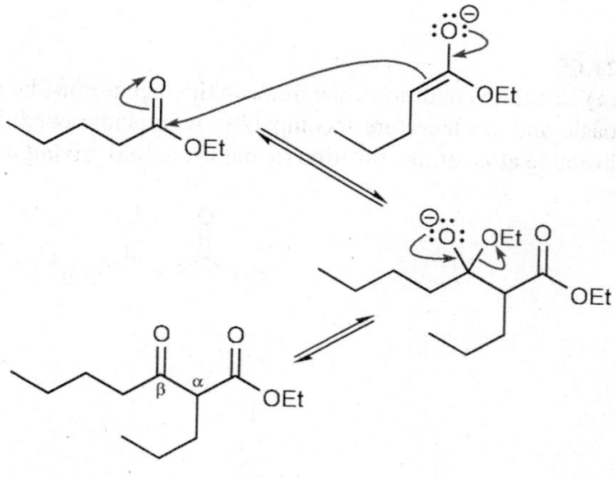

This product is deprotonated under the conditions of its formation, which is the reason for the acid workup after the reaction is complete (to regenerate the neutral product).

(b) The α position of one molecule of the ester is deprotonated, and the resulting enolate functions as a nucleophile and attacks the electrophilic carbonyl group of another molecule of the ester. As a result, a carbon-carbon bond is formed, giving a charged tetrahedral intermediate. The carbonyl group is then reformed via loss of an ethoxide ion, affording a β-ketoester, as shown:

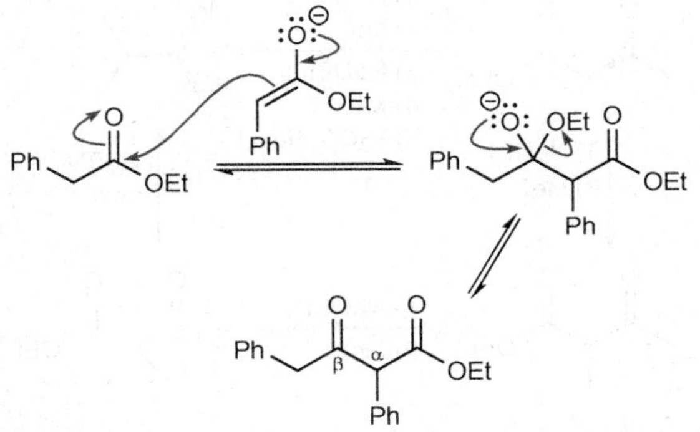

This product is deprotonated under the conditions of its formation, which is the reason for the acid workup after the reaction is complete (to regenerate the neutral product).

21.68.

(a) The product is a carboxylic acid that has the following (highlighted) group connected to the α position:

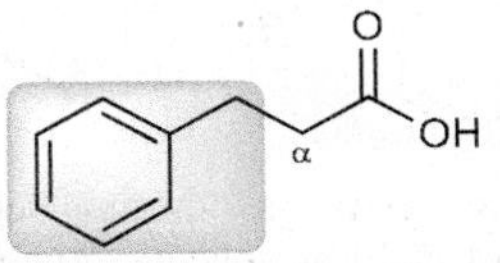

This group can be installed via a malonic ester synthesis, using benzyl bromide:

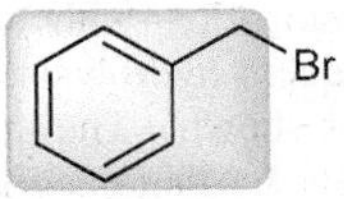

A malonic ester synthesis begins with the deprotonation of diethyl malonate (using ethoxide as a base). The resulting resonance-stabilized conjugate base is then treated with benzyl bromide, thereby installing the alkyl group. Subsequent hydrolysis and decarboxylation give the product, as shown. Notice that two of the carbon atoms in the starting material (highlighted) remain in the product:

1) NaOEt
2) $PhCH_2Br$
3) H_3O^+, heat

1) NaOEt
2) $PhCH_2Br$

H_3O^+, heat

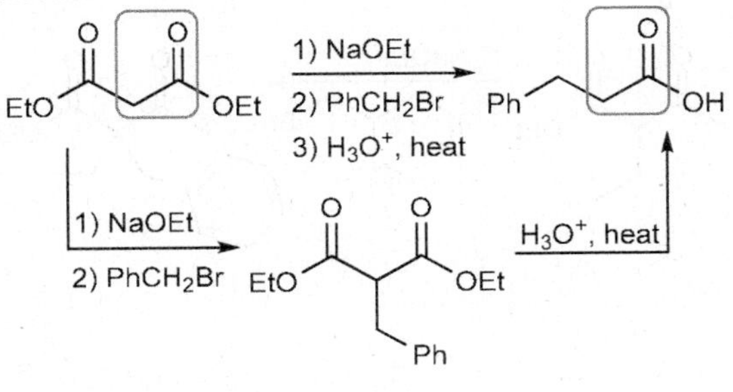

(b) The product is a carboxylic acid that has two methyl groups (highlighted) connected to the α position:

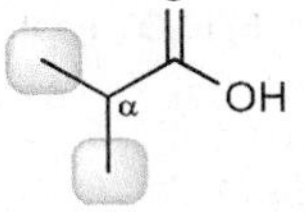

Each of these groups can be installed via a malonic ester synthesis, using methyl iodide. The reason for using methyl iodide (rather than methyl bromide) was discussed in the solution to Problem **21.31b**.

A malonic ester synthesis begins with the deprotonation of diethyl malonate (using ethoxide as a base). The resulting resonance-stabilized conjugate base is then treated with methyl iodide, thereby installing the first methyl group. The second methyl group is installed in a similar way (deprotonation, followed by treatment with methyl iodide). Subsequent hydrolysis and decarboxylation give the product, as shown. Notice that two of the carbon atoms in the starting material remain in the product (after decarboxylation), and these two carbon atoms have been highlighted:

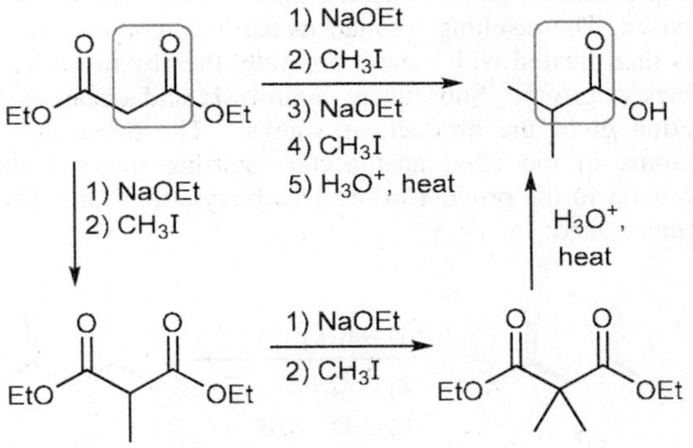

1) NaOEt
2) CH_3I
3) NaOEt
4) CH_3I
5) H_3O^+, heat

1) NaOEt
2) CH_3I

1) NaOEt
2) CH_3I

H_3O^+, heat

(c) The product is a carboxylic acid that has two benzyl groups (highlighted) connected to the α position:

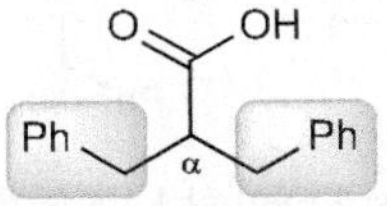

Each of these groups can be installed via a malonic ester synthesis, using benzyl bromide.

A malonic ester synthesis begins with the deprotonation of diethyl malonate (using ethoxide as a base). The resulting resonance-stabilized conjugate base is then treated with benzyl bromide, thereby installing the first benzyl group. The second benzyl group is installed in a similar way (deprotonation, followed by treatment with benzyl bromide). Subsequent hydrolysis and decarboxylation give the product, as shown. Notice that two of the carbon atoms in the starting material (highlighted) remain in the product:

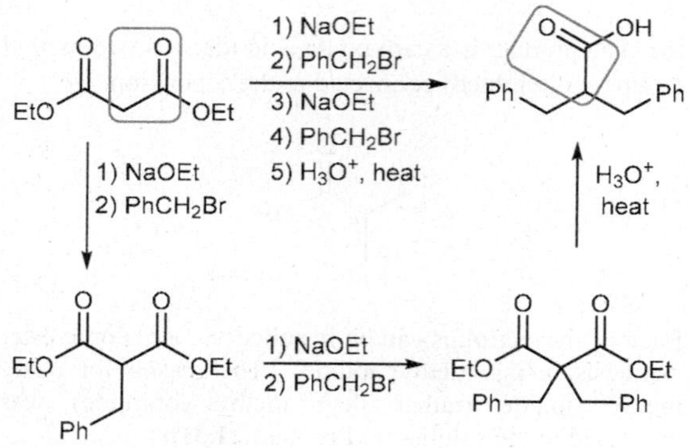

1) NaOEt
2) $PhCH_2Br$
3) NaOEt
4) $PhCH_2Br$
5) H_3O^+, heat

1) NaOEt
2) $PhCH_2Br$

H_3O^+, heat

1) NaOEt
2) $PhCH_2Br$

21.69.

(a) The product is a methyl ketone that has a benzyl group (highlighted) connected to the α position:

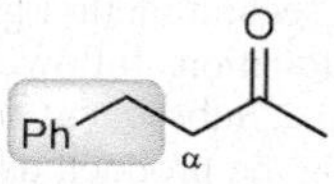

This group can be installed via an acetoacetic ester synthesis, using benzyl bromide.

An acetoacetic ester synthesis begins with the deprotonation of ethyl acetoacetate (using ethoxide as a base). The resulting resonance-stabilized conjugate base is then treated with benzyl bromide, thereby installing a benzyl group. Subsequent hydrolysis and decarboxylation gives the product, as shown. The three carbon atoms in the ethyl acetoacetate starting material that remain in the product (after decarboxylation) have been highlighted:

1) NaOEt
2) $PhCH_2Br$
3) H_3O^+, heat

1) NaOEt
2) $PhCH_2Br$

H_3O^+ heat

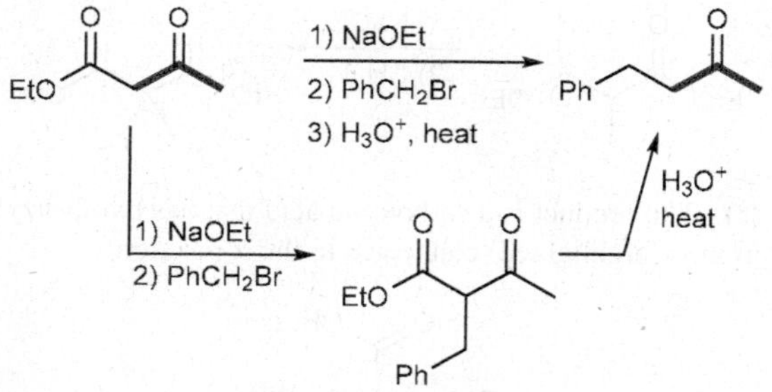

(b) The product is a methyl ketone that has two methyl groups (highlighted) connected to the α position:

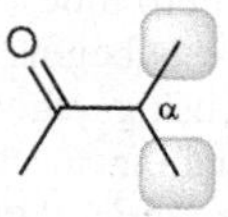

Each of these groups can be installed via an acetoacetic ester synthesis, using methyl iodide to install each methyl group.

An acetoacetic ester synthesis begins with the deprotonation of ethyl acetoacetate (using ethoxide as a base). The resulting resonance-stabilized conjugate base is then treated with methyl iodide, thereby installing the first methyl group. The second methyl group is installed in a similar way (deprotonation with a base, followed by alkylation). Subsequent hydrolysis and decarboxylation give the product, as shown. The three carbon atoms in the ethyl acetoacetate starting material that remain in the product (after decarboxylation) have been highlighted:

1) NaOEt
2) MeI
3) NaOEt
4) MeI
5) H_3O^+, heat

1) NaOEt
2) MeI

H_3O^+ heat

1) NaOEt
2) MeI

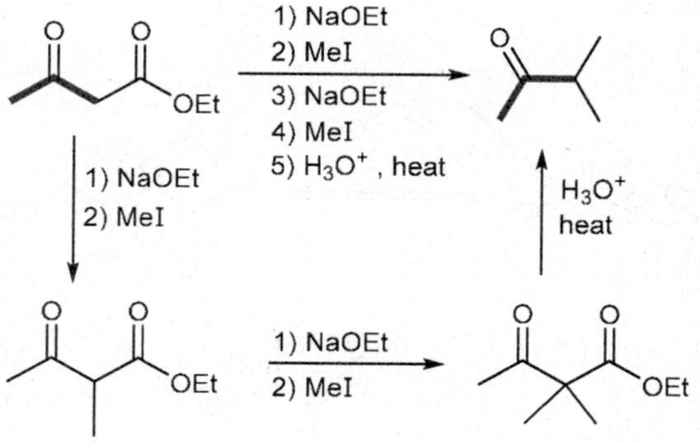

(c) The product is a methyl ketone that has the following two (highlighted) groups connected to the α position:

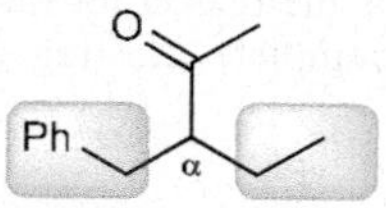

Each of these groups can be installed via an acetoacetic ester synthesis, using the following halides:

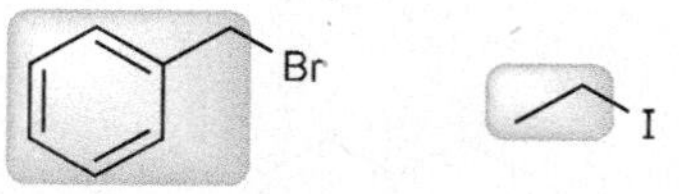

An acetoacetic ester synthesis begins with the deprotonation of ethyl acetoacetate (using ethoxide as a base). The resulting resonance-stabilized conjugate base is then treated with one of the halides above, to install one of the two groups. The other group is installed in a similar way (deprotonation with a base, followed by alkylation). Subsequent hydrolysis and decarboxylation give the product, as shown. The three carbon atoms in the ethyl acetoacetate starting material that remain in the product (after decarboxylation) have been highlighted:

1) NaOEt
2) EtI
3) NaOEt
4) $PhCH_2Br$
5) H_3O^+, heat

1) NaOEt
2) EtI

H_3O^+ heat

1) NaOEt
2) benzyl bromide

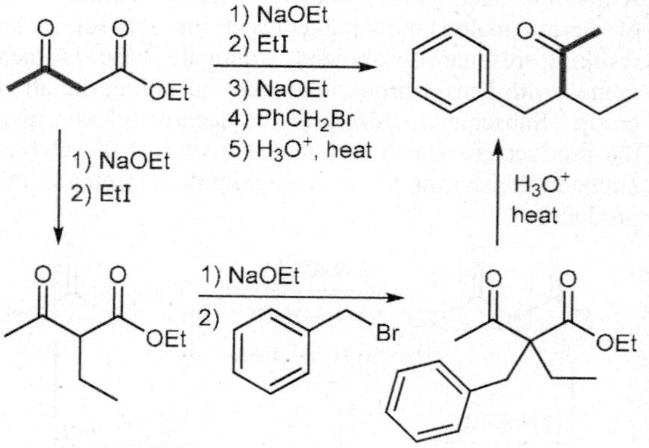

21.70. Protonation of the isolated π bond gives a tertiary carbocation, which is then deprotonated to give a fully conjugated system. Next, tautomerization of the ketone to an enol completes the mechanism (a two-step process). Protonation of the carbonyl group then gives a resonance-stabilized cation (there are two additional significant resonance structures that have not been drawn). This intermediate is then deprotonated to generate aromaticity, which is the driving force for this process.

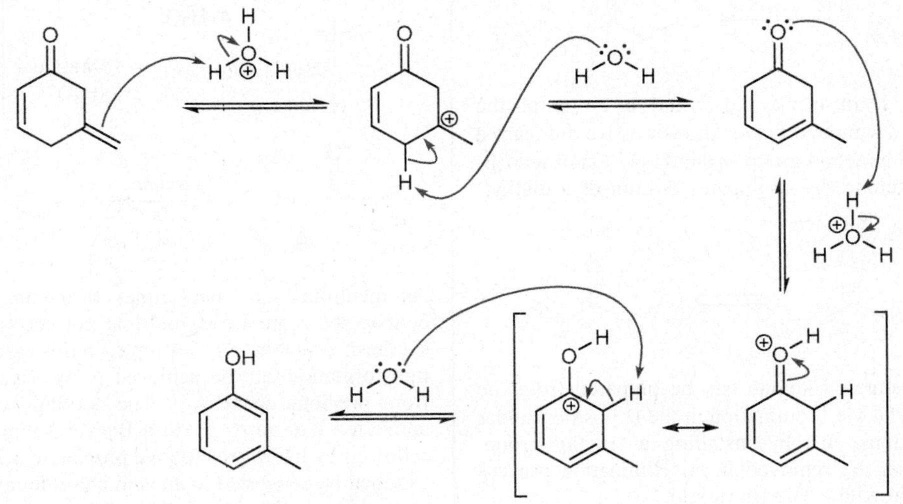

21.71. The reaction conditions suggest an aldol condensation. This compound has three α positions. During an aldol condensation, one of the α positions must be deprotonated to give an enolate, which will attack the other carbonyl group (in an intramolecular process). But we must decide which of the three possible enolates gives rise to the product, as all three possible enolates are expected to be present at equilibrium. If either of the interior α positions is deprotonated to give an enolate ion, the resulting intramolecular attack would generate a four-membered ring:

However, the third enolate can participate in an intramolecular attack that gives a six-membered ring:

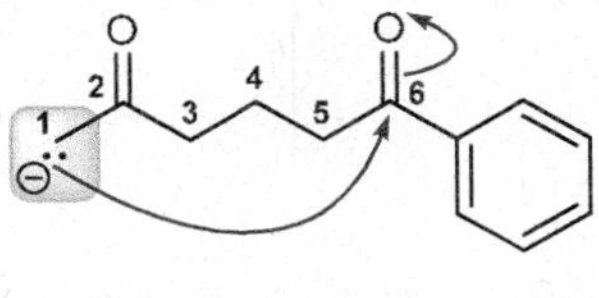

Formation of a six-membered ring is favored over formation of a four-membered ring, because a six-membered ring is relatively strain-free, while a four-membered ring is highly strained (and therefore higher in energy).
Now that we have identified which carbon-carbon bond will be formed during an intramolecular aldol condensation, we can draw the product by removing the following two highlighted α protons and the oxygen atom, and we replace them with a double bond, giving a product with the molecular formula $C_{12}H_{12}O$:

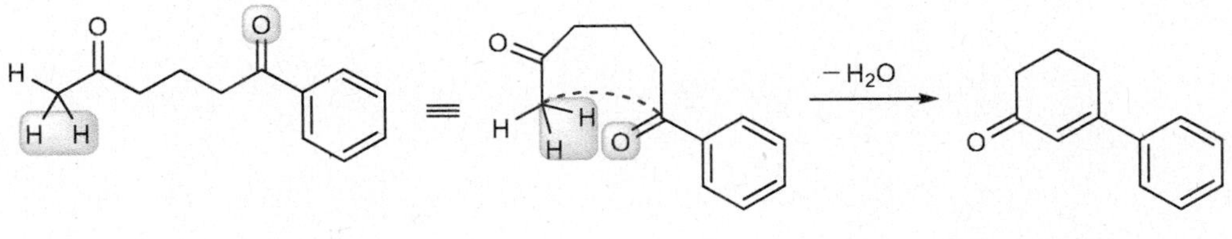

21.72. This transformation requires the installation of a methyl group at a β position:

We did not learn a way to install a group at the β position of a saturated ketone, however, we did learn a way to install a methyl group at the β position of an α,β-unsaturated ketone (by conjugate addition of a dialkylcuprate):

This α,β-unsaturated ketone can be prepared from the starting material via bromination at the α position under acidic conditions, thereby installing a leaving group, which can then be removed in an elimination process upon treatment with a base (pyridine):

Br

The forward scheme is shown here:

1) $[H_3O^+]$, Br_2
2) pyridine
3) Me_2CuLi
4) H_3O^+

$[H_3O^+]$, Br_2

1) Me_2CuLi
2) H_3O^+

Br

pyridine

As mentioned so many times throughout this entire course, there are often multiple correct solutions to a synthesis problem. For example, in this case, the desired transformation can be achieved using several reactions from previous chapters. The starting ketone can be converted into an ester via a Baeyer-Villiger oxidation, followed by hydrolysis to give propanoic acid. This acid can then be converted to an acid chloride upon treatment with thionyl chloride, followed by conversion to the product upon treatment with lithium di-*n*-propylcuprate:

1) RCO_3H
2) H_3O^+
3) $SOCl_2$
4) n-Pr_2CuLi

RCO_3H

n-Pr_2CuLi

Cl

$SOCl_2$

H_3O^+

OH

21.73. The ester is converted into a β-ketoester via a crossed Claisen condensation. Upon treatment with aqueous acid at elevated temperature, the ester group is hydrolyzed and the resulting β-ketoacid (not drawn below) undergoes decarboxylation to give acetophenone. A crossed aldol condensation (with formaldehyde) gives an α,β-unsaturated ketone. Treating the α,β-unsaturated ketone with lithium diethylcuprate, followed by water workup, installs an ethyl group at the β position, and removes the unsaturation between the α and β positions. Another crossed aldol condensation (with formaldehyde again) gives the final product.

LDA; 1) ; 2) H_3O^+; H_3O^+, heat; NaOH, H_2O, heat; 1) Et_2CuLi 2) H_3O^+; NaOH, H_2O, heat

21.74.

(a) The starting material is a methyl ketone, and the reagents indicate a haloform reaction, giving a carboxylic acid:

1) NaOH, xs I_2 2) H_3O^+ → + CHI_3

(b) The starting material is a ketone, and the reagents indicate α-bromination, which will only occur at an α position that bears protons. In this case, there is only one such position. Under these conditions, a bromine atom is installed at this α position, to give the following product:

Br_2, CH_3COOH

(c) The starting material is a ketone, and the reagents indicate an aldol condensation. The starting ketone has only one α position that bears protons, and the reaction occurs at this location. To draw the product, two molecules of the ketone are redrawn such that two α protons of one molecule are directly facing the carbonyl group of another molecule (highlighted). We then remove the two α protons and the oxygen atom, and we replace them with a double bond. In this case, two stereoisomers are possible, so we draw the product with fewer steric interactions:

$-H_2O$

21.75. This transformation represents a retro-aldol reaction, which occurs via a mechanism that is the reverse of an aldol condensation (all the same intermediates, but in reverse order). First a hydroxide ion functions as a nucleophile and attacks the electrophilic β position of the α,β-unsaturated ketone (a conjugate addition). The resulting enolate is then protonated to give a β-hydroxyketone. Deprotonation gives an alkoxide ion, which then reforms a carbonyl group by expelling an enolate ion as a leaving group. This enolate ion is then protonated to give cyclohexanone. Notice that water is the proton source for the protonation steps, consistent with basic conditions (strong acids, such as hydronium ions, are not measurably present under these conditions).

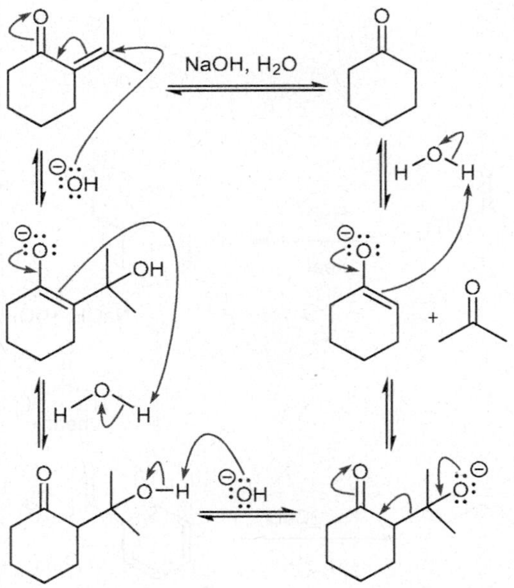

21.76.

(a) Upon treatment with aqueous acid, each of the four ester groups is hydrolyzed, giving a compound with four carboxylic acid groups, shown here:

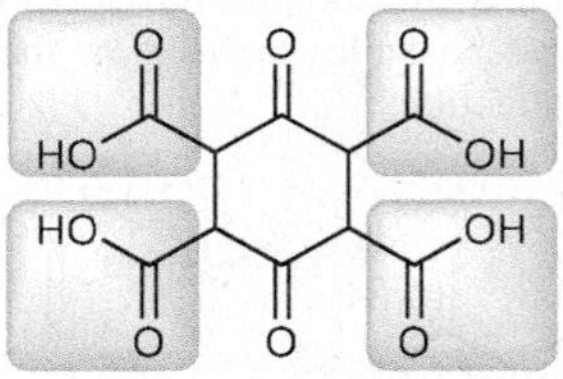

Each of these four carboxylic acid groups is β to a carbonyl group, and will therefore undergo decarboxylation upon heating. This gives the dione shown below, as well as four equivalents of ethanol (from hydrolysis) and four equivalents of carbon dioxide (from decarboxylation):

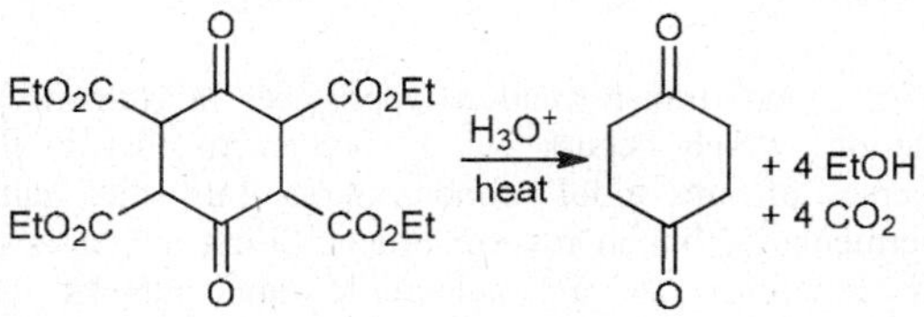

(b) The starting material is a diester. Upon treatment with ethoxide, an intramolecular Claisen condensation (followed by acid workup) gives a β-ketoester via formation of a ring. Upon heating with aqueous acid, the β-ketoester is hydrolyzed to a β-ketoacid (not shown), which then undergoes decarboxylation (under the conditions of its formation) to give the ketone shown below. Notice that the configuration of each chiral center remains unchanged because the chiral centers are not involved in the reaction.

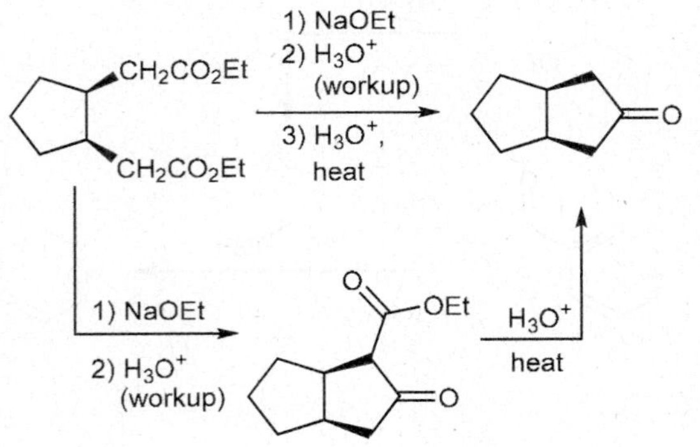

(c) The starting material is an unsymmetrical ketone. Treatment with bromine in aqueous acidic conditions gives α-bromination, which is expected to occur at the more substituted position. Subsequent treatment of the resulting α-bromoketone with pyridine (a base) gives an elimination reaction to afford an α,β-unsaturated ketone. Treatment of the α,β-unsaturated ketone with lithium diethylcuprate, followed by methyl iodide, achieves the installation of an ethyl group at the β position and a methyl group at the α position:

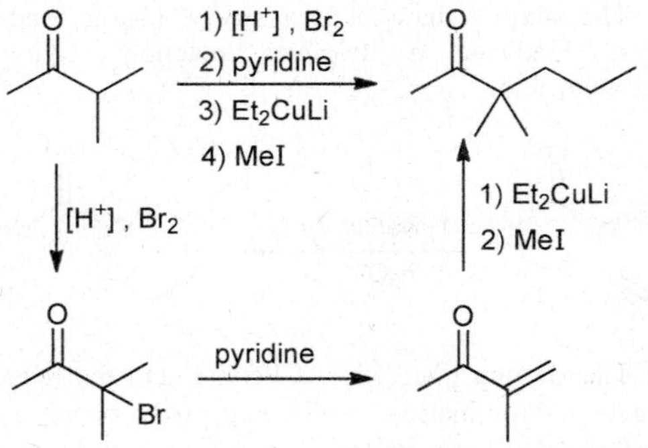

(d) The starting material is an unsymmetrical ketone, and LDA is a strong, sterically hindered base. At low temperature, LDA will irreversibly deprotonate the ketone at the less substituted α position to give the kinetic enolate. Subsequent treatment of the enolate with ethyl iodide will install an ethyl group at this α position:

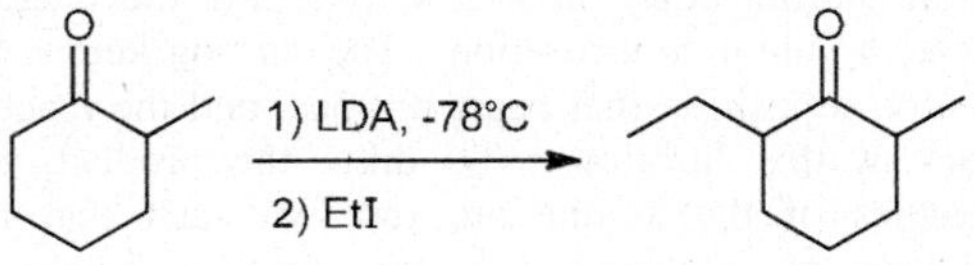

21.77.

(a) One strategy for achieving the desired transformation derives from the following retrosynthetic analysis. An explanation of each of the steps (*a-d*) follows.

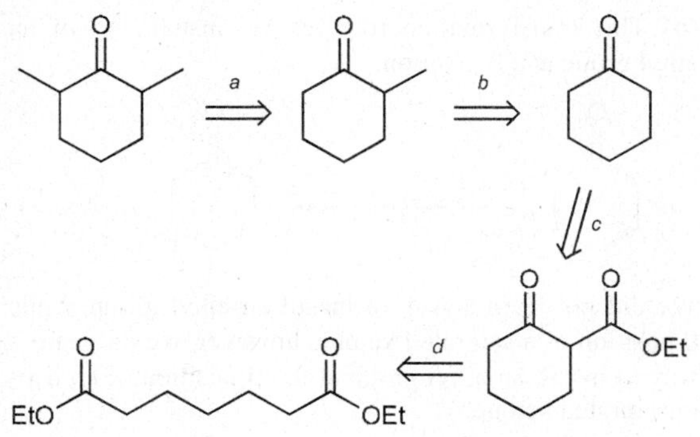

a
b
c
O
O
O
O
O
O
O
OEt
d
EtO
OEt

a. The product can be made from 2-methylcyclohexanone via alkylation of an α position (with LDA as the base, to control the regiochemical outcome).

b. 2-Methylcyclohexanone can be made from cyclohexanone via alkylation of an α position.

c. Cyclohexanone can be made from a β-ketoester via hydrolysis and subsequent decarboxylation.

d. The β-ketoester can be made from the starting material via a Dieckmann cyclization.

Now let's draw the forward scheme. Treating the starting diester with ethoxide, followed by acid workup, gives a β-ketoester via a Dieckmann cyclization. Upon treatment with aqueous acid and heat, the β-ketoester is hydrolyzed to give a β-keto acid, which then undergoes decarboxylation to give cyclohexanone. Two subsequent alkylation processes will install the two methyl groups. The choice of base in the first alkylation is not so critical, because cyclohexanone is symmetrical (both α positions are equivalent). But during the second alkylation process, LDA must be used at low temperature, in order to install the methyl group at the less substituted α position (via the kinetic enolate).

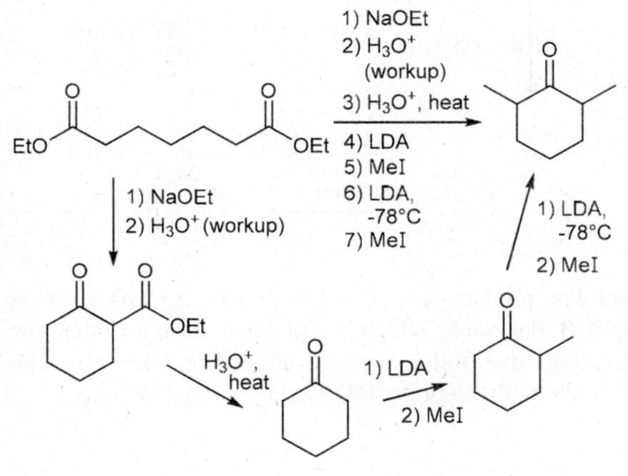

1) NaOEt
2) H_3O^+ (workup)
3) H_3O^+, heat
4) LDA
5) MeI
6) LDA, -78°C
7) MeI
EtO
OEt
1) NaOEt
2) H_3O^+ (workup)
OEt
H_3O^+, heat
1) LDA
2) MeI
1) LDA, -78°C
2) MeI

Alternatively, and perhaps more efficiently, installation of the first methyl group can be performed immediately after the Dieckmann cyclization (before hydrolysis and decarboxylation). In this way, the anionic product of the Dieckmann cyclization (without acid workup) is used as a nucleophile to attack methyl iodide in an S_N2 process, thereby installing the first methyl group, as shown.

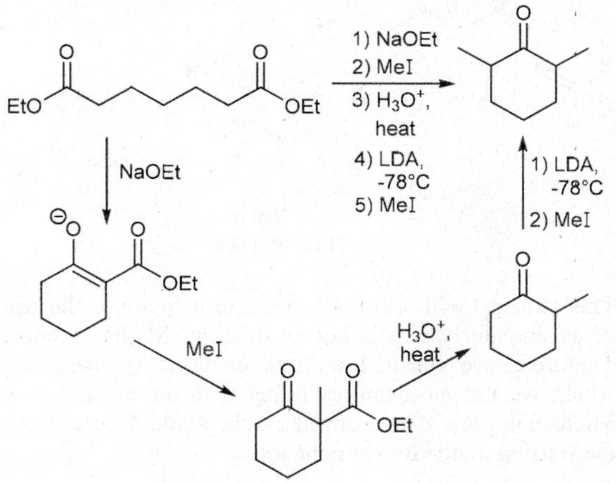

EtO
OEt
1) NaOEt
2) MeI
3) H_3O^+, heat
4) LDA, -78°C
5) MeI
NaOEt
1) LDA, -78°C
2) MeI
OEt
MeI
H_3O^+, heat
OEt

(b) The product can be made via two successive aldol condensation reactions, one of which is intramolecular and the other is intermolecular:

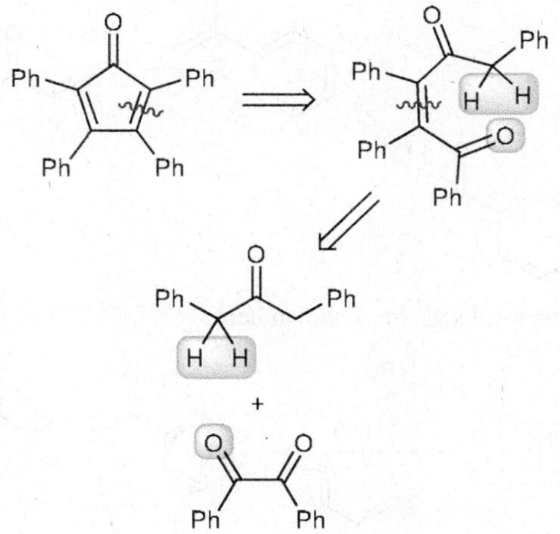

Ph
Ph
Ph
Ph
Ph
H
H
Ph
Ph
Ph
Ph
Ph
H H
+
Ph
Ph

The entire transformation can be achieved in one reaction flask, by treating the starting material with the dione above in basic conditions.

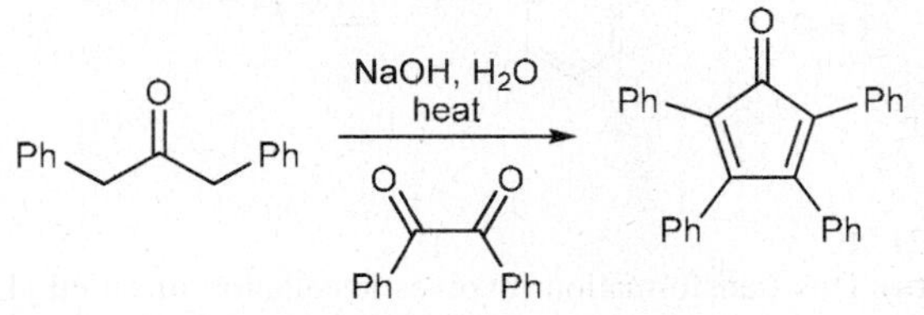

Ph
Ph
NaOH, H_2O
heat
Ph
Ph
Ph
Ph
Ph
Ph

(c) The product is 1,5-difunctionalized:

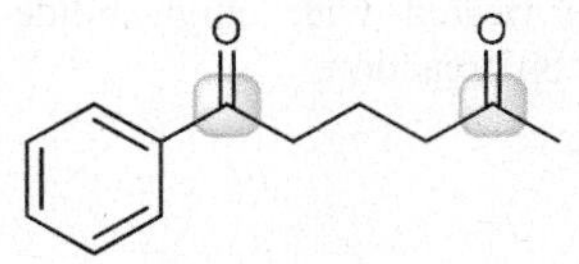

Therefore, we consider preparing the product via a Michael reaction:

Not a Michael donor

This strategy will not work, because it involves the use of an enolate, which is not an efficient Michael donor. Therefore, we consider a Stork enamine synthesis (in which we use an enamine, rather than an enolate, as a Michael donor). The enamine can be made directly from the starting material (acetophenone).

The forward scheme is shown here:

1) R_2NH, $[H^+]$, $(-H_2O)$
2)
3) H_3O^+

R_2NH, $[H^+]$, $(-H_2O)$

NR_2

1)
2) H_3O^+

21.78.

(a) This transformation involves installation of an ethyl group at the α position of a ketone. A strong base is used to deprotonate the α position, giving an enolate, which is then treated with ethyl iodide to give the product via an S_N2 reaction.

1) LDA
2) EtI

(b) This transformation requires the installation of an ethyl group at a β position:

We did not learn a way to install an alkyl group at the β position of a saturated ketone, however, we did learn a way to install an alkyl group at the β position of an α,β-unsaturated ketone:

This α,β-unsaturated ketone can be prepared from the corresponding α-bromoketone upon treatment with a base (pyridine). And the α-bromoketone can be made from the starting material via α-bromination under acidic conditions:

The forward scheme is shown here:

1) Br_2, $[H_3O^+]$
2) pyridine
3) Et_2CuLi
4) H_3O^+

Br_2, $[H_3O^+]$

1) Et_2CuLi
2) H_3O^+

pyridine

(c) The product is a ketone with alkyl groups at the α and β positions, which could have been installed by treating the following α,β-unsaturated ketone with lithium diethylcuprate, followed by methyl iodide:

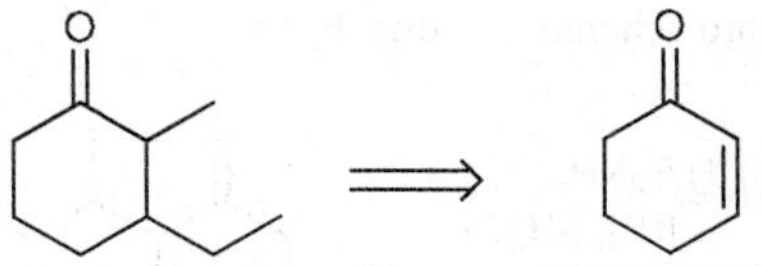

This α,β-unsaturated ketone can be prepared from the corresponding α-bromoketone upon treatment with a base (pyridine). And the α-bromoketone can be made from the starting material via α-bromination under acidic conditions:

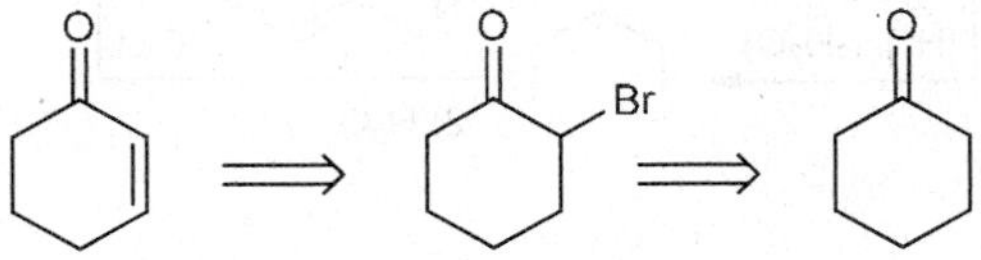

Br

The forward scheme is shown here:

1) Br_2, $[H_3O^+]$
2) pyridine
3) Et_2CuLi
4) MeI

Br_2, $[H_3O^+]$

1) Et_2CuLi
2) MeI

Br

pyridine

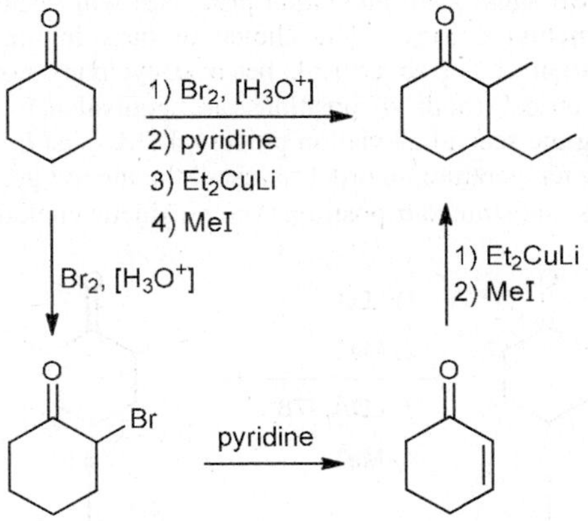

(d) The product is 1,5-difunctionalized:

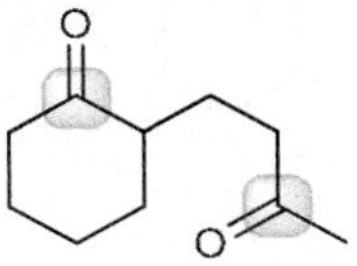

Therefore, we consider preparing the product via a Michael reaction:

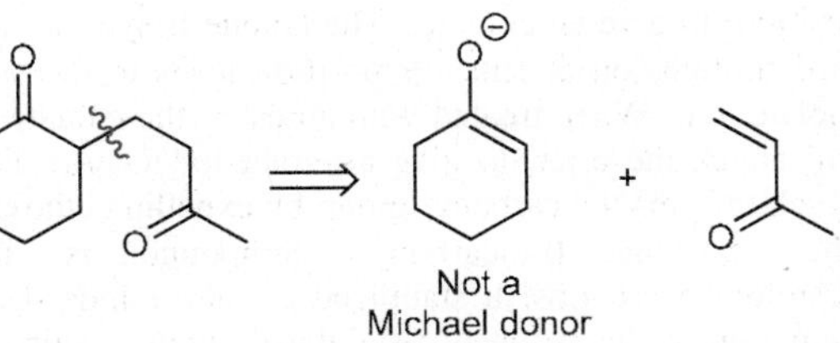

Not a Michael donor

This strategy will not work, because it involves the use of an enolate, which is not an efficient Michael donor. Therefore, we consider a Stork enamine synthesis (in which we use an enamine, rather than an enolate, as a Michael donor). The enamine can be made from the starting ketone upon treatment with a secondary amine under acid-catalyzed conditions (with removal of water):

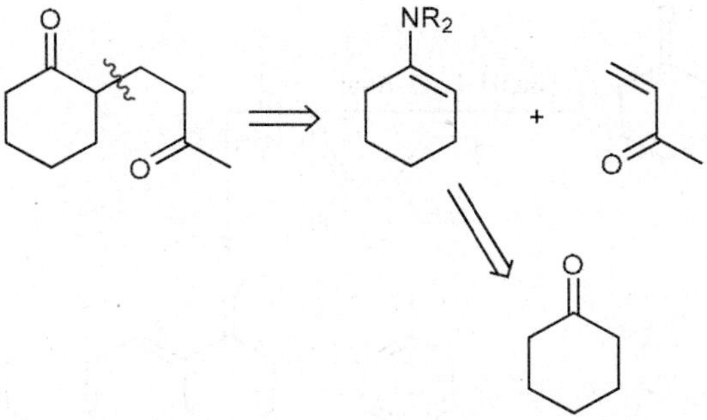

NR_2

The forward scheme is shown here:

1) R_2NH, $[H^+]$, $(-H_2O)$
2) [methyl vinyl ketone]
3) H_3O^+

R_2NH, $[H^+]$, $(-H_2O)$

NR_2

1) [methyl vinyl ketone]
2) H_3O^+

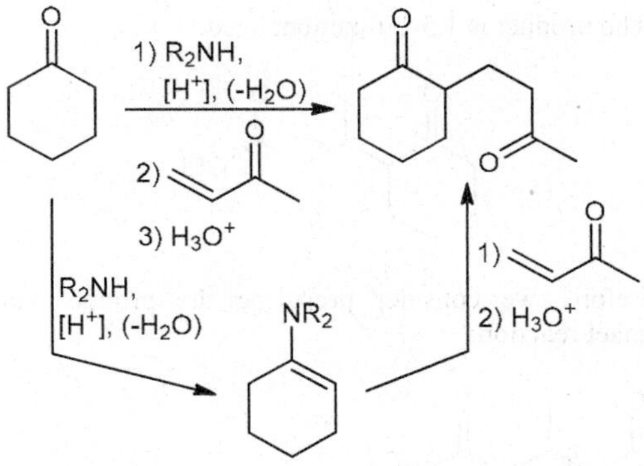

(e) The product is a ketone that exhibits α,β-unsaturation on either side of the carbonyl group. This suggests two aldol condensation reactions, one at each α position of the starting ketone (cyclohexanone):

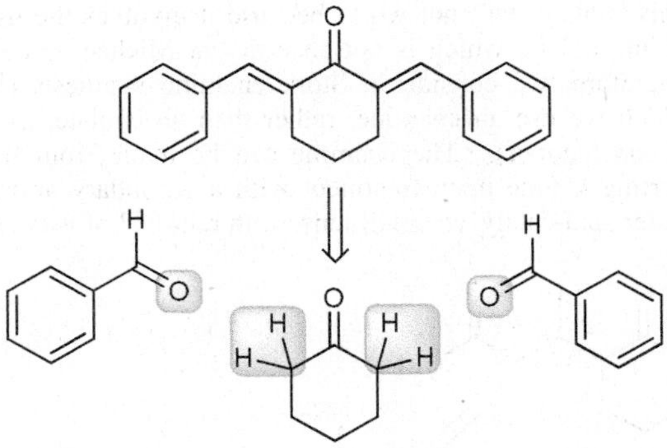

Both aldol condensation reactions can be performed in a single reaction flask, by treating cyclohexanone with excess benzaldehyde under basic conditions:

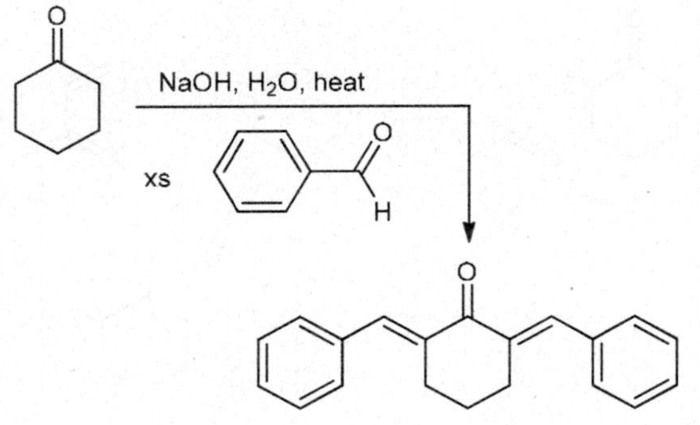

(f) The product is 1,5-difunctionalized:

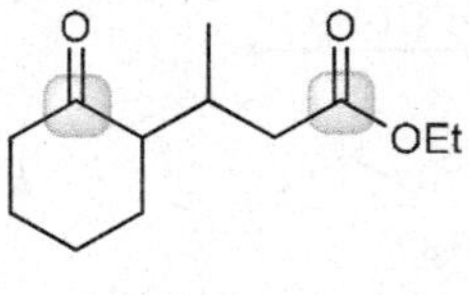

Therefore, we consider preparing the product via a Michael reaction:

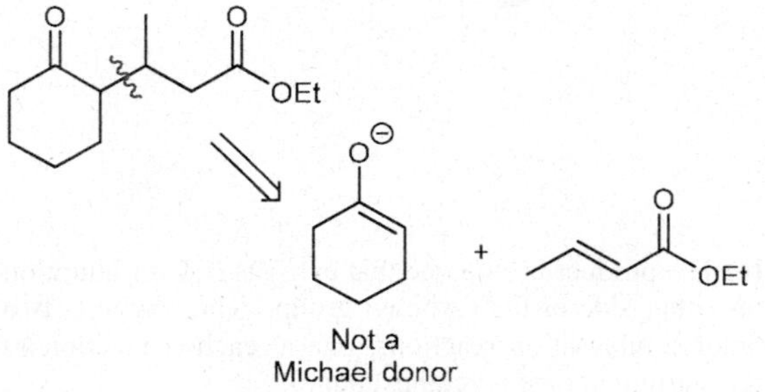

Not a Michael donor

This strategy will not work, because it involves the use of an enolate, which is not an efficient Michael donor. Therefore, we consider a Stork enamine synthesis (in which we use an enamine, rather than an enolate, as a Michael donor). The enamine can be made from the starting ketone upon treatment with a secondary amine under acid-catalyzed conditions (with removal of water):

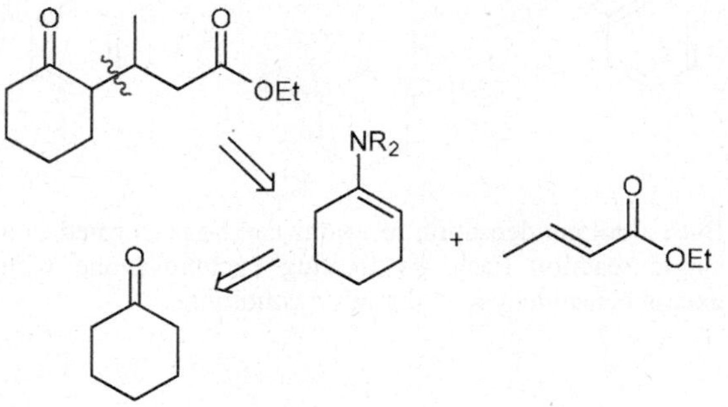

The forward scheme is shown here:

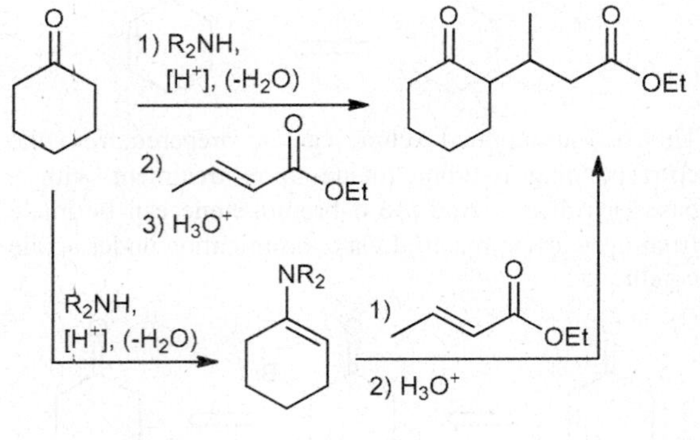

(g) Two subsequent alkylation processes will install the two methyl groups. The choice of base in the first alkylation is not so critical, because cyclohexanone is symmetrical (both α positions are equivalent). But during the second alkylation process, LDA must be used at low temperature, in order to install the methyl group at the less substituted α position (via the kinetic enolate).

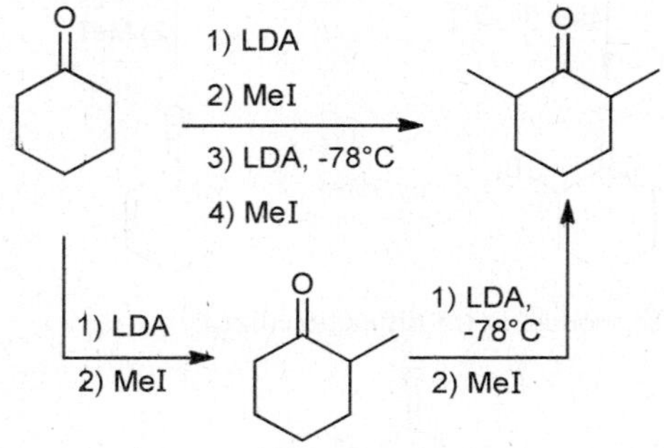

21.79. LDA is a strong, sterically hindered base, and it will irreversibly deprotonate the cyclohexanone (at the α position) to give an enolate. The ketone is symmetrical, so deprotonation at either α position leads to the same enolate ion. When treated with an ester, the enolate ion will attack the ester, to give a tetrahedral intermediate, which reforms the carbonyl group by expelling ethoxide. The resulting β-dicarbonyl compound is then deprotonated to give a stabilized enolate. Indeed, the formation of this resonance-stabilized anion is a driving force for this reaction. After the reaction is complete, an acid is required to protonate this anion.

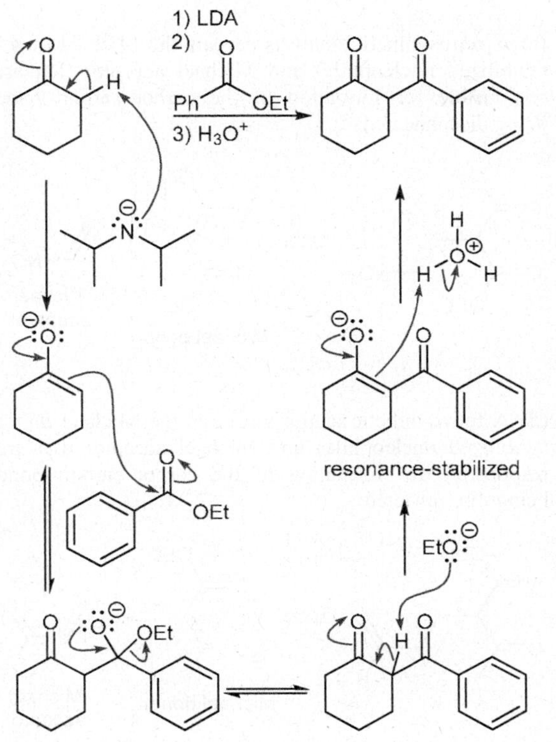

1) LDA
2) Ph OEt
3) H_3O^+

resonance-stabilized

1) LDA
2) EtO OEt
3) H_3O^+

21.80. LDA is a strong, sterically hindered base, and it will irreversibly deprotonate the cyclohexanone (at the α position) to give an enolate. The ketone is symmetrical, so deprotonation at either α position leads to the same enolate ion. When treated with a carbonate, the enolate ion will attack the carbonate, to give a charged tetrahedral intermediate, which reforms the carbonyl group by expelling ethoxide. The resulting β-ketoester is then deprotonated to give a stabilized enolate. Indeed, the formation of this resonance-stabilized anion is a driving force for this reaction. After the reaction is complete, an acid is required to protonate this anion.

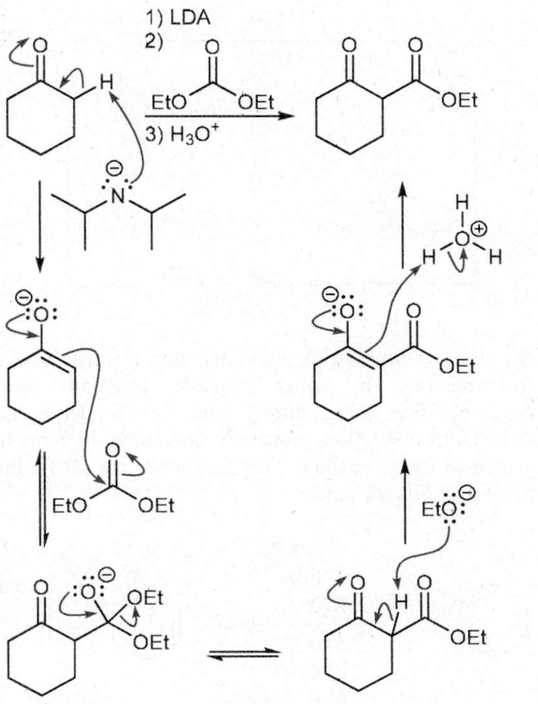

21.81. LDA is a strong, sterically hindered base, and it will irreversibly deprotonate the ester (at the α position) to give an ester enolate. When treated with a ketone, the enolate ion will attack the ketone, to give an alkoxide ion. After the reaction is complete, an acid is required to protonate this alkoxide ion.

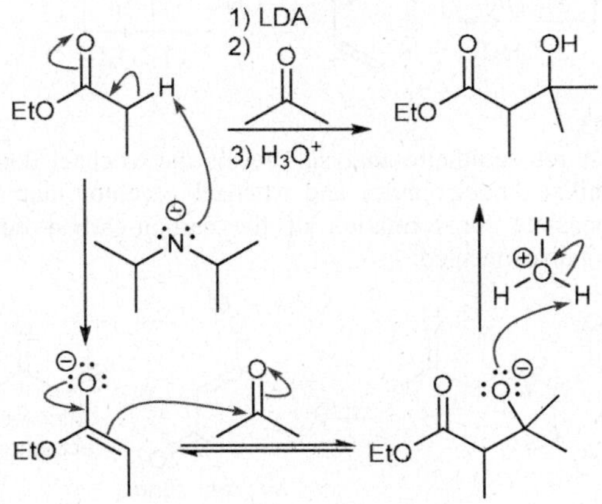

1) LDA
2)
3) H_3O^+

21.82.

(a) The following mechanism is consistent with the description in the problem statement.

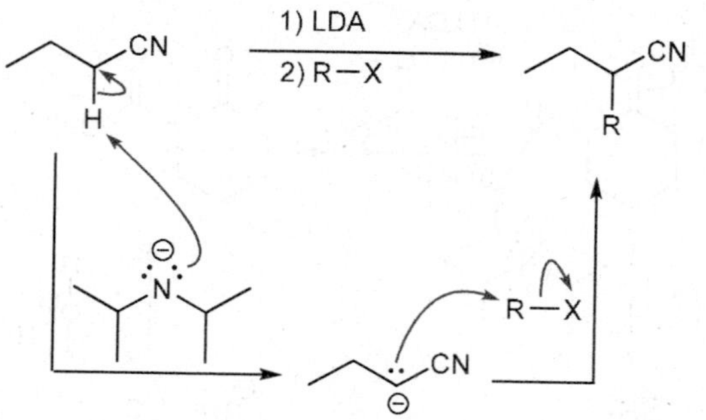

(b) Benzyl bromide is converted into a nitrile (via an S_N2 reaction in which cyanide functions as a nucleophile). This nitrile can then undergo two successive alkylation processes, installing two methyl groups at the α position. Hydrolysis of the nitrile then gives a carboxylic acid.

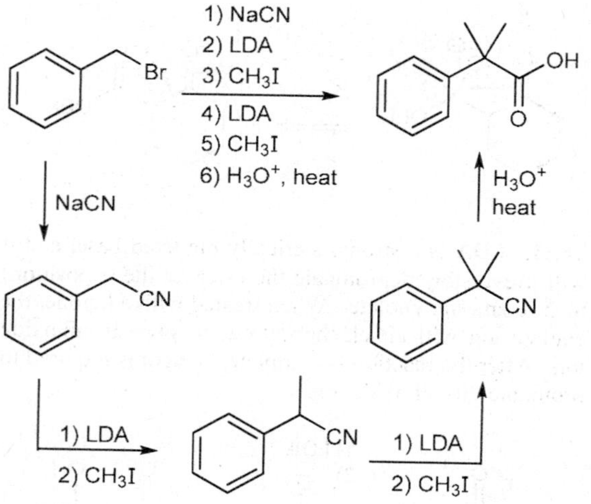

21.83.

(a) A retrosynthetic analysis reveals the Michael donor (stabilized nucleophile) and Michael acceptor that are responsible for formation of the carbon-carbon bond being disconnected:

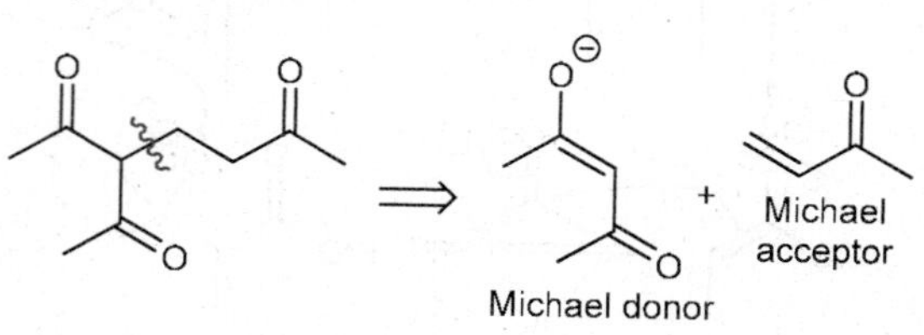

(b) A retrosynthetic analysis reveals the Michael donor (stabilized nucleophile) and Michael acceptor that are responsible for formation of the carbon-carbon bond being disconnected:

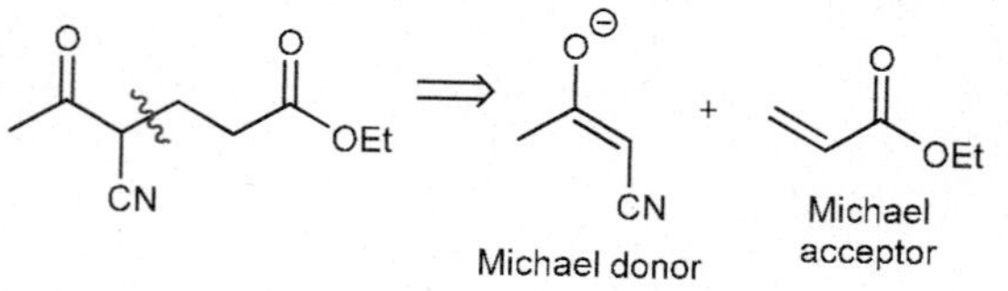

(c) A retrosynthetic analysis reveals the Michael donor (stabilized nucleophile) and Michael acceptor that are responsible for formation of the carbon-carbon bond being disconnected:

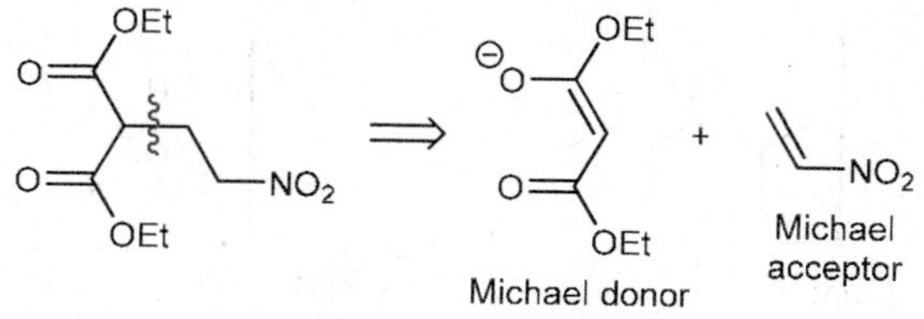

(d) A retrosynthetic analysis reveals the Michael donor (stabilized nucleophile) and Michael acceptor that are responsible for formation of the carbon-carbon bond being disconnected:

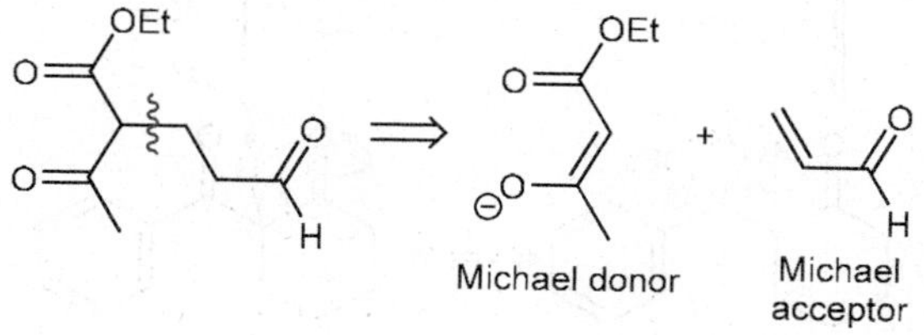

(e) A retrosynthetic analysis reveals the Michael donor (stabilized nucleophile) and Michael acceptor that are responsible for formation of the carbon-carbon bond being disconnected:

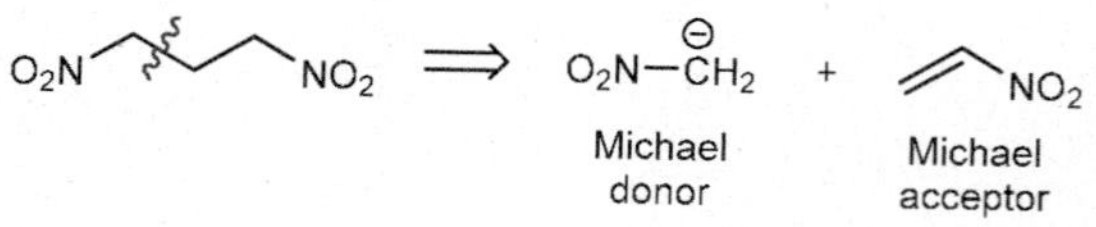

21.84.

(a) The conjugate base of diethyl malonate functions as a nucleophile and attacks *n*-propyl bromide in an S_N2 process, expelling bromide as a leaving group, and giving the following product:

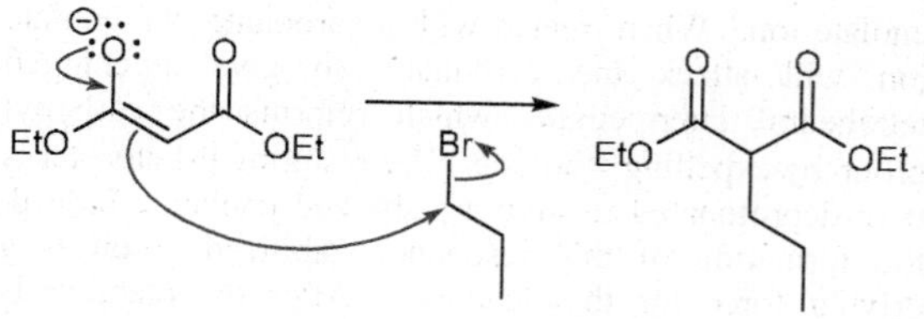

(b) The conjugate base of diethyl malonate functions as a nucleophile and attacks the epoxide at the less substituted (more accessible) position, thereby opening the epoxide and forming an alkoxide ion. Acid workup converts the alkoxide ion into an alcohol, as shown:

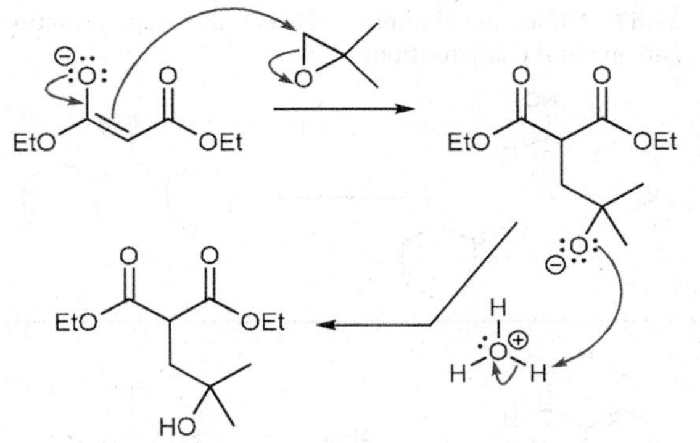

(c) The conjugate base of diethyl malonate functions as a nucleophile and attacks the acid chloride to give a charged tetrahedral intermediate, which expels a chloride ion to reform a carbonyl group (via a nucleophilic acyl substitution reaction):

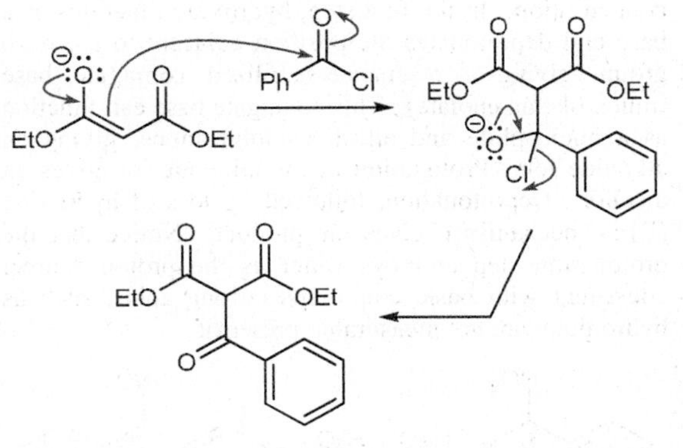

(d) The conjugate base of diethyl malonate functions as a nucleophile and attacks the β position of the α,β-unsaturated ketone (a Michael reaction). The resulting enolate is converted back into a ketone when protonated at the alpha carbon, upon treatment with aqueous acid:

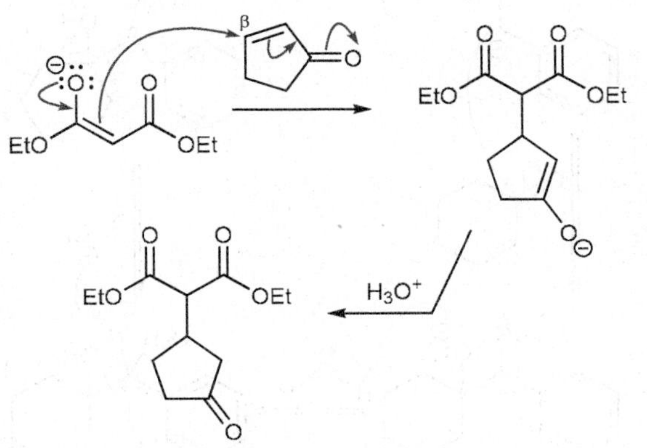

(e) The conjugate base of diethyl malonate functions as a nucleophile and attacks benzyl iodide in an S_N2 process, expelling iodide as a leaving group, and giving the following product:

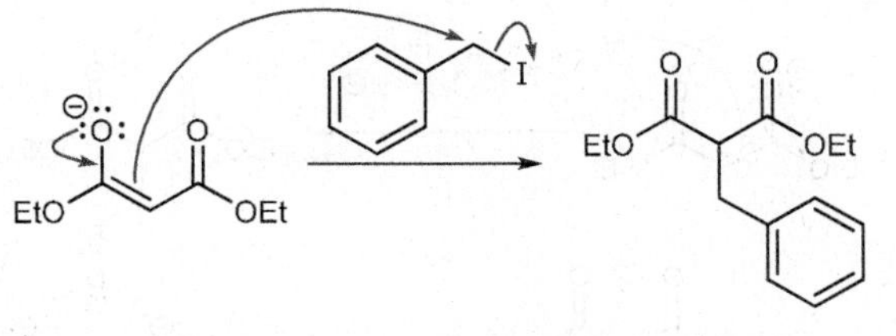

(f) The conjugate base of diethyl malonate functions as a nucleophile and attacks the β position of the α,β-unsaturated nitrile (a Michael reaction). The resulting intermediate is converted back into a nitrile when protonated at the alpha carbon, upon acid workup:

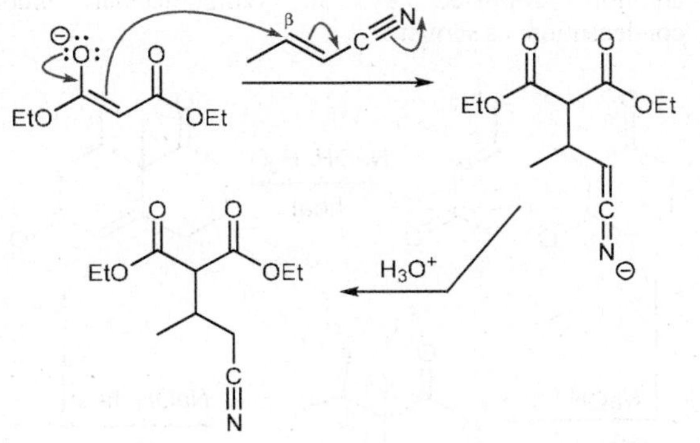

(g) The conjugate base of diethyl malonate functions as a nucleophile and attacks the acid anhydride to give a charged tetrahedral intermediate, which expels an acetate ion to reform a carbonyl group (via a nucleophilic acyl substitution reaction):

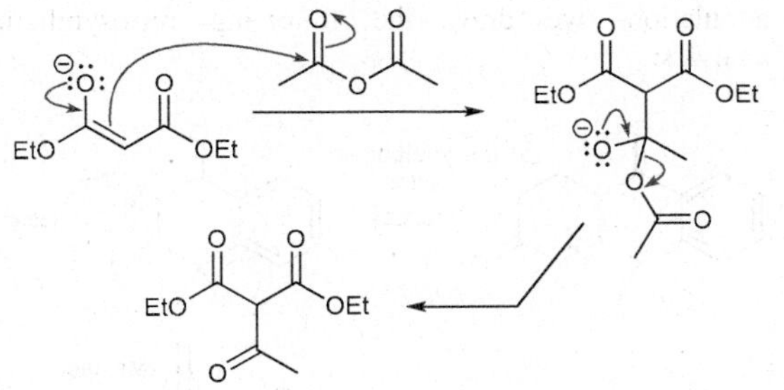

This compound is deprotonated to some extent under these conditions (by the acetate ion, which can function as a base), which serves as a driving forced to push the reaction to completion. Then, acid workup gives the proton back, regenerating the neutral product shown above.

(h) The conjugate base of diethyl malonate functions as a nucleophile and attacks the β position of the α,β-unsaturated nitro compound (a Michael reaction). The resulting intermediate is converted back into a nitro compound when protonated at the alpha carbon, upon acid workup:

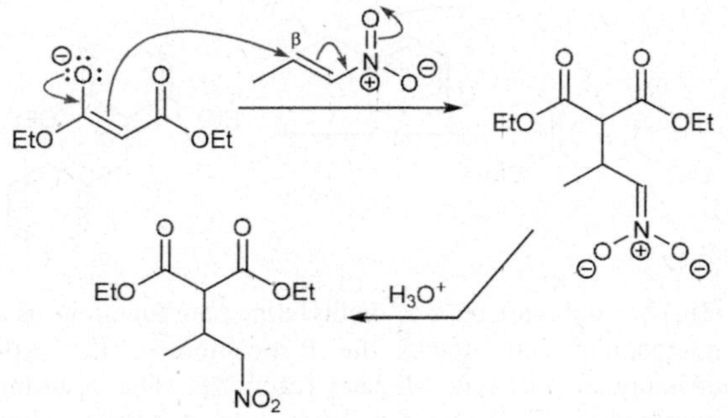

EtO OEt EtO OEt EtO OEt H_3O^+ NO_2

21.85. A Robinson annulation is comprised of a Michael addition, followed by an intramolecular aldol condensation, as shown:

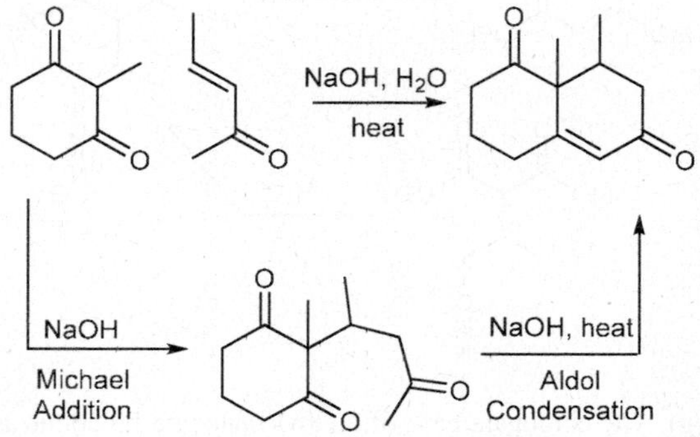

NaOH, H_2O heat

NaOH Michael Addition

NaOH, heat Aldol Condensation

21.86. A Robinson annulation is comprised of a Michael reaction, followed by an intramolecular aldol condensation. To determine the starting materials necessary to prepare the desired product via a Robinson annulation, we draw the following retrosynthetic analysis:

Intramolecular aldol

Michael

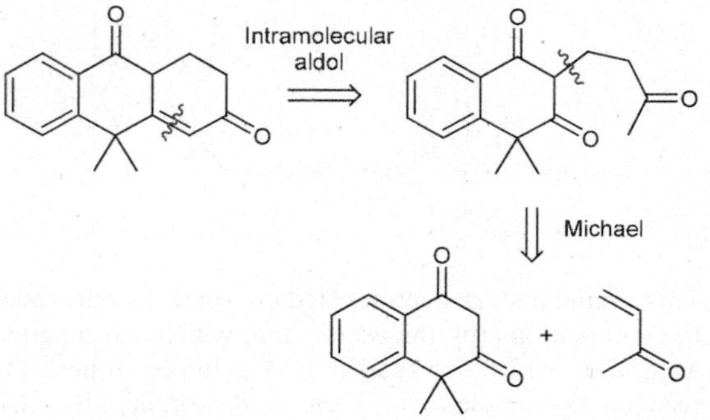

These two steps do not represent two separate reactions. A Robinson annulation can be performed in one reaction flask, as shown in the following forward scheme:

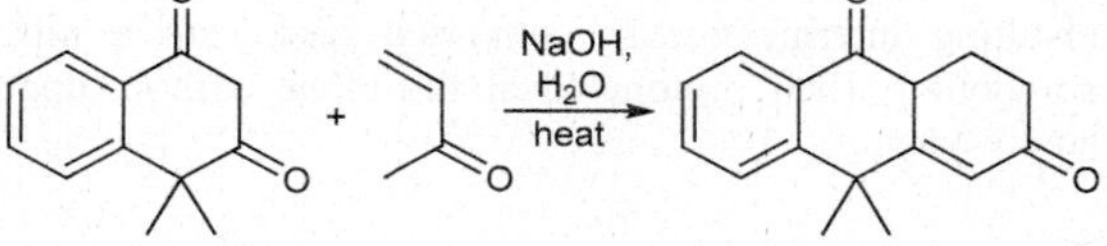

NaOH, H_2O heat

21.87. Notice the similarity between this transformation and an aldol condensation:

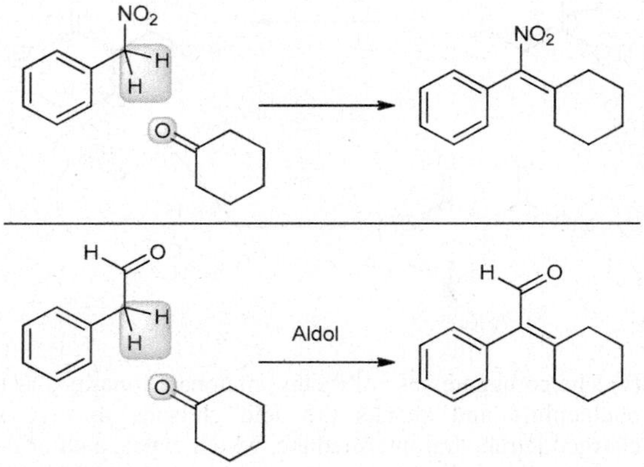

NO_2 H H NO_2

H O H H Aldol H O

Indeed, we will draw a mechanism (below) that is extremely similar to the mechanism of an aldol condensation. In the first step, hydroxide functions as a base and deprotonates the position adjacent to the nitro group, giving a resonance-stabilized conjugate base (much like an enolate). This conjugate base can function as a nucleophile and attack cyclohexanone, giving an alkoxide ion. Protonation of the alkoxide ion gives an alcohol. Deprotonation, followed by loss of hydroxide (E1cb mechanism), gives the product. Notice that the protonation step employs water as the proton source, consistent with basic conditions (strong acids, such as hydronium, are not measurably present):

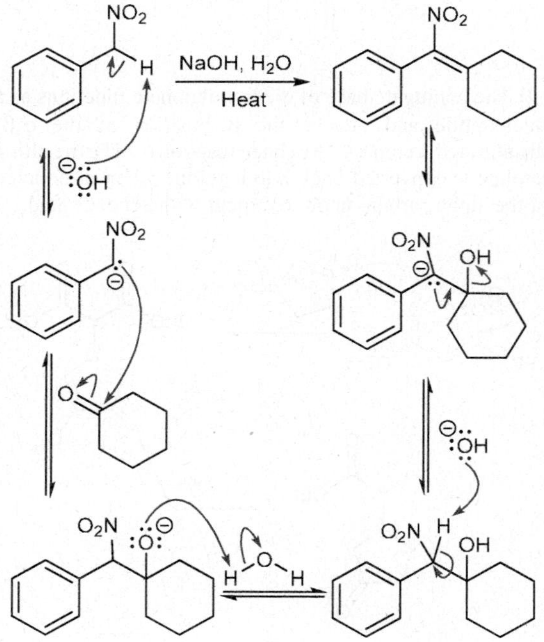

NO_2 H NaOH, H_2O Heat NO_2

:OH

NO_2 O_2N OH

:OH

O_2N H O H O_2N H OH

21.88. There are certainly many acceptable solutions to this problem. One such solution derives from the following retrosynthetic analysis. An explanation of each of the steps (*a-c*) follows.

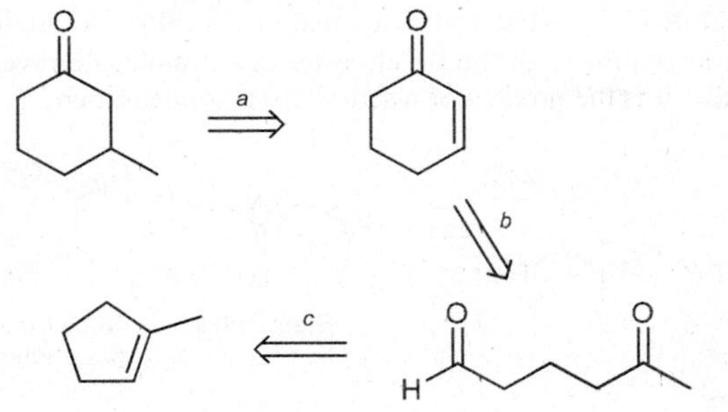

a. The product has seven carbon atoms, while the starting material only has six carbon atoms. The extra carbon atom can be installed in the last step of the synthesis, via a Michael addition (between a lithium dialkylcuprate and an α,β-unsaturated ketone).

b. The α,β-unsaturated ketone can be made via an intramolecular aldol condensation, starting the appropriate dicarbonyl compound.

c. The dicarbonyl compound can be made from the starting material via ozonolysis.

Now let's draw the forward scheme:

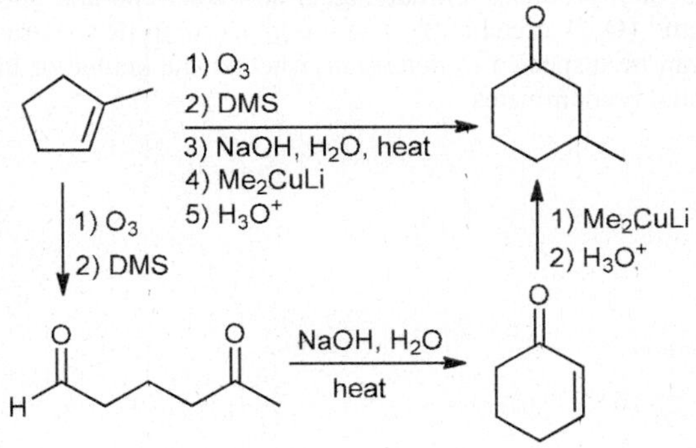

1) O_3
2) DMS
3) NaOH, H_2O, heat
4) Me_2CuLi
5) H_3O^+

1) O_3
2) DMS

NaOH, H_2O
heat

1) Me_2CuLi
2) H_3O^+

21.89. The correct answer is (b). Options (c) and (d) are enols, not enolates, so they can be ruled out. The base (diisopropylamide) is sterically hindered, so we expect formation of the less substituted (kinetic) enolate:

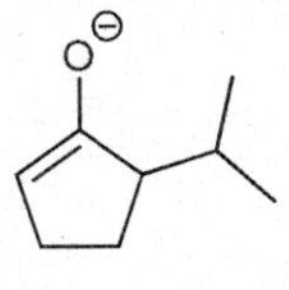

21.90. The correct answer is (c). The product is an α,β-unsaturated ketone:

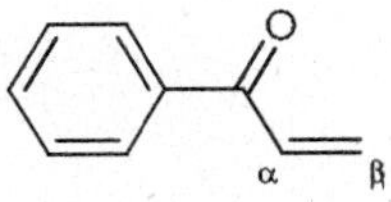

α β

Therefore, it can be prepared via an aldol condensation, using the following starting materials:

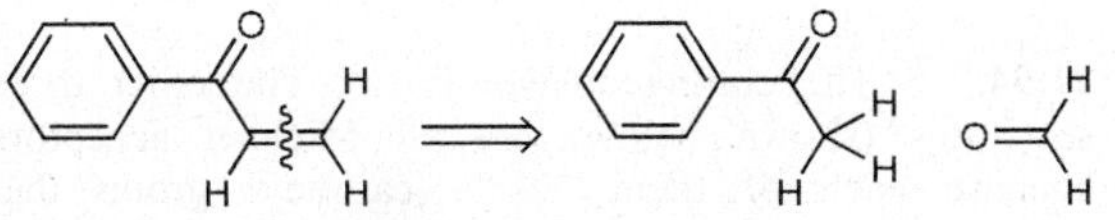

These compounds correspond with option (c).

21.91. The correct answer is (d). This process involves the installation of two alkyl groups bear a carbonyl group: an ethyl group at the alpha position and a methyl group at the beta position. This can be achieved by first installing the methyl group at the beta position, by treating the starting compound with Me_2CuLi, to give an enolate, followed by alkylation with ethyl iodide to give the desired product:

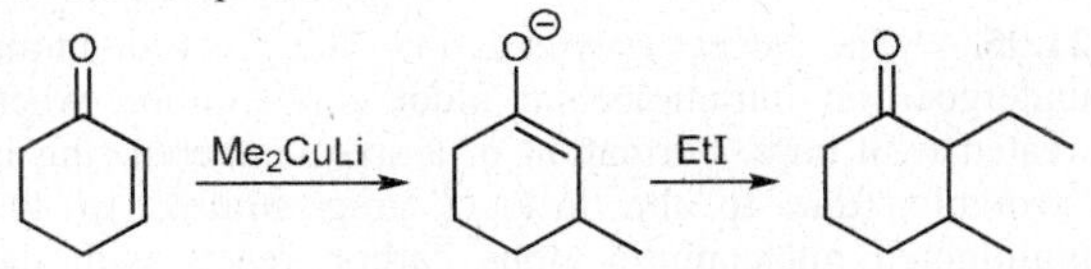

Me_2CuLi → EtI →

21.92. The correct answer is (c). The most acidic proton is connected to the position that is α to two carbonyl groups (in between both carbonyl groups). Deprotonation at this location leads to a resonance-stabilized conjugate base in which the negative charge is spread over two oxygen atoms and one carbon atom).

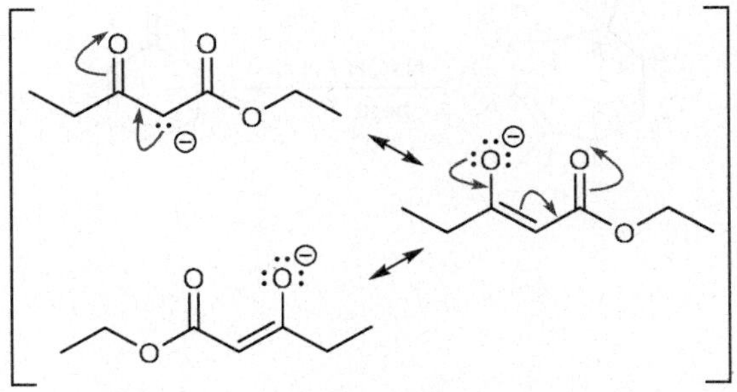

21.93. The correct answer is (d). This reaction sequence is an example of a malonic ester synthesis. It begins with the deprotonation of diethyl malonate (using ethoxide as a base). The resulting resonance-stabilized conjugate base is then treated with bromocyclopentane, thereby installing a cyclopentyl group. Subsequent hydrolysis and decarboxylation give a carboxylic acid product, as shown. Notice that two of the carbon atoms in the starting material (highlighted) remain in the product:

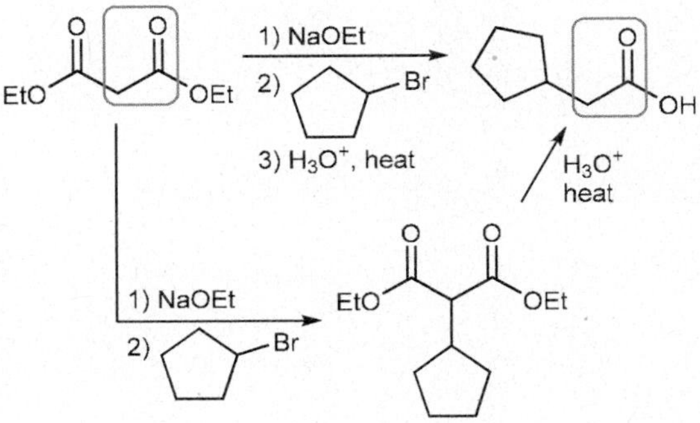

1) NaOEt
2) cyclopentyl-Br
3) H_3O^+, heat

1) NaOEt
2) cyclopentyl-Br

H_3O^+
heat

21.94. The correct answer is (a). The other three structures (shown below) are all Michael acceptors because each of them has a carbonyl group that possesses α,β unsaturation. The β position of the unsaturated ketone, ester and aldehyde structures are electrophilic, and would be expected to react with a Michael donor.

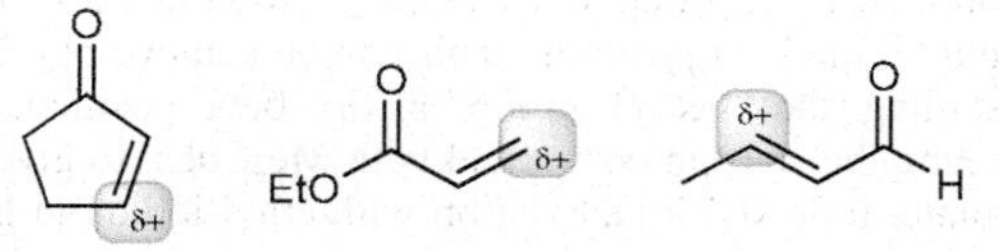

21.95. The correct answer is (a). The given diketone undergoes an intramolecular aldol condensation when treated with base. Formation of a six-membered ring is favorable (due to the lack of ring strain), so the highlighted nucleophilic alpha carbon reacts with the appropriately located electrophilic carbonyl.

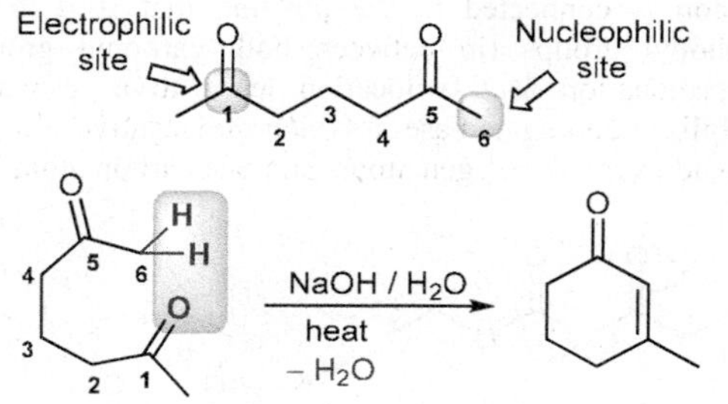

21.96. The correct answer is (b). A logical retrosynthesis of the β-keto ester target molecule reveals that it is the product of a self-Claisen condensation:

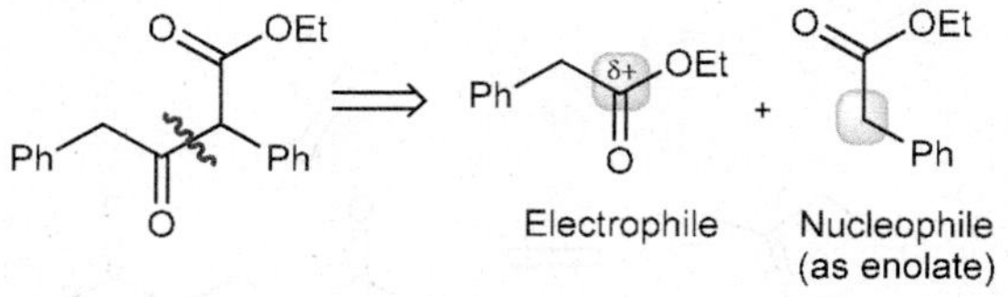

None of the other options would successfully prepare the desired target molecule. Recall that S_N2 processes are not possible with aryl halides, because backside attack is not possible if the leaving group is on an sp^2 hybridized carbon atom.

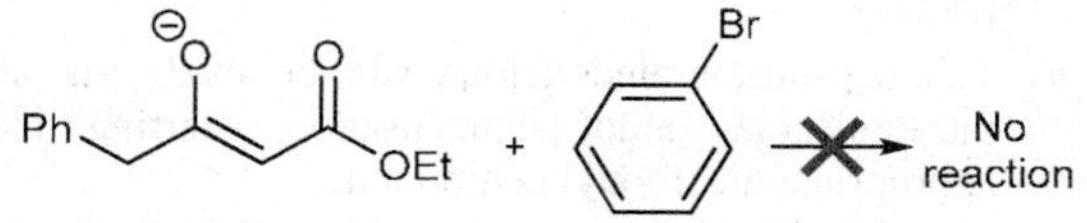

21.97. A ketone generally produces a strong signal at approximately 1720 cm^{-1} (C=O stretching), while an alcohol produces a broad signal between 3200 and 3600 cm^{-1} (O–H stretching). These regions of an IR spectrum can be inspected to determine whether the ketone or the enol predominates.

21.98. Under acidic conditions, one of the OH groups is protonated. If the middle OH group is protonated, the resulting leaving group (water) can leave to give a secondary carbocation. A hydride shift then gives a resonance-stabilized cation that is deprotonated to give a hydroxyaldehyde. Protonation of the OH group (to give a good leaving group), followed by an E2 process, gives the product. In this last step, an E2 process is more likely than an E1 process, because the latter would involve formation of a primary carbocation. Notice that water is the base for the deprotonation steps, consistent with acidic conditions (strong bases, such as hydroxide, are not measurably present under these conditions).

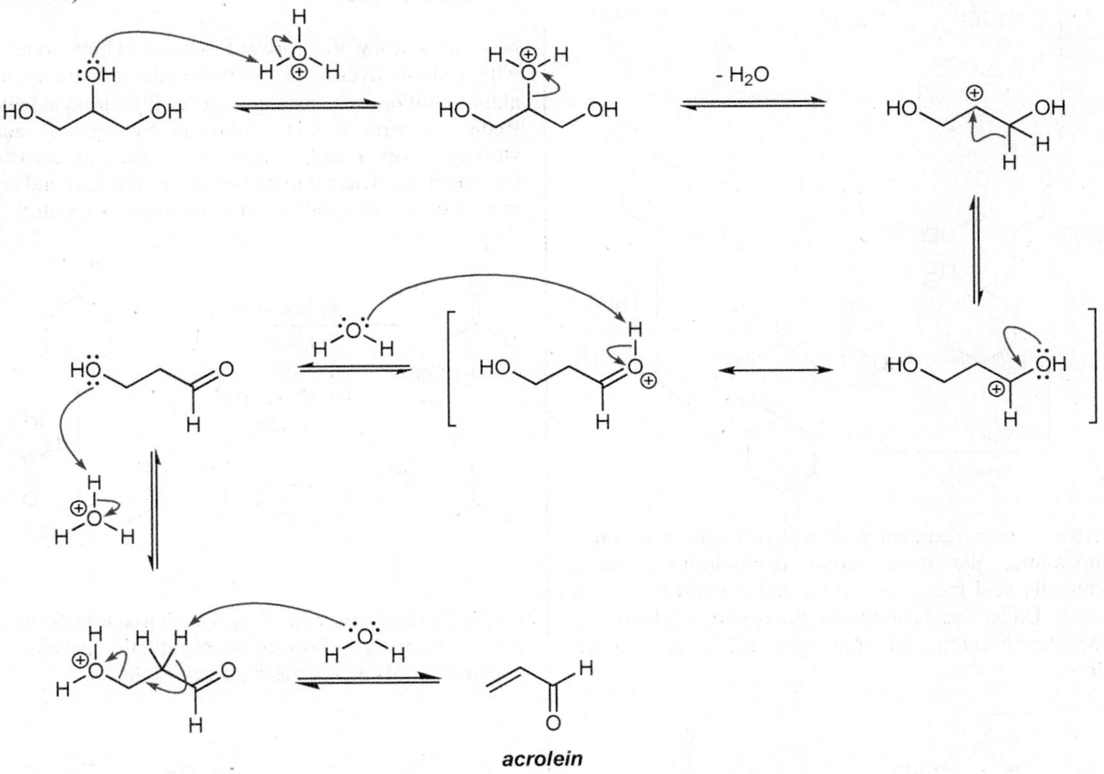

- H_2O

acrolein

21.99. Upon treatment with aqueous acid, the nitrile is hydrolyzed to give a β-ketoacid, which undergoes decarboxylation at elevated temperature to give the following ketone:

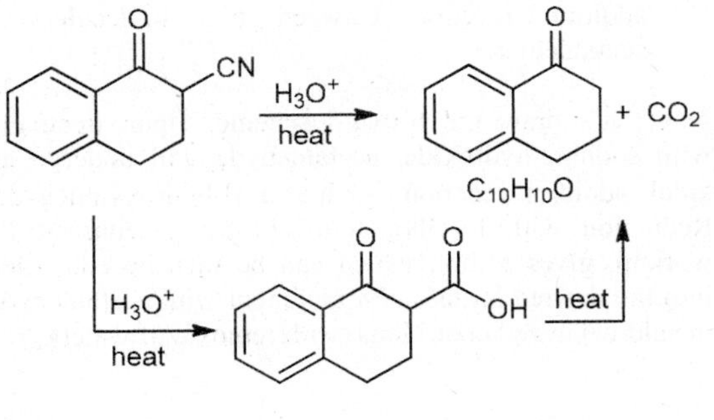

CN H_3O^+ heat + CO_2 $C_{10}H_{10}O$ H_3O^+ heat OH heat

21.100. Using the strategy described in the problem statement, the desired lactone can be made if we use the following epoxide, instead of ethylene oxide:

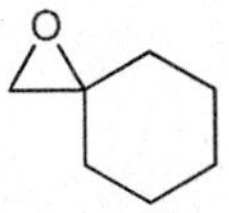

The synthesis, as described in the problem statement, is shown here (follow the location of the cyclohexyl group):

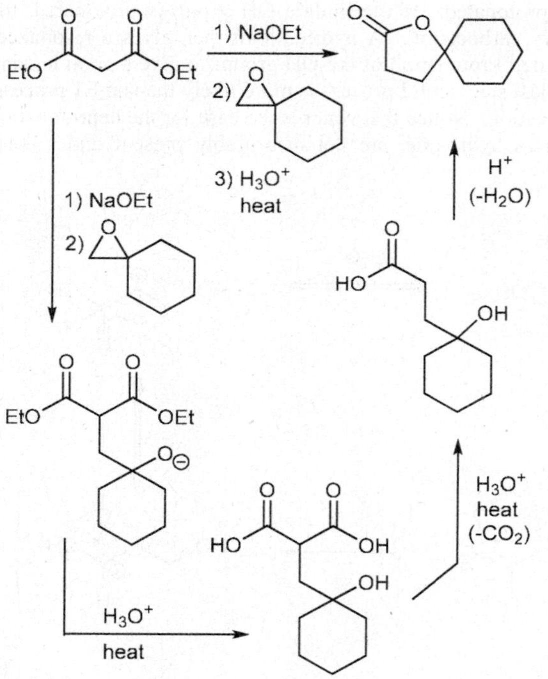

1) NaOEt
2) [epoxide]
3) H_3O^+
heat

H^+
($-H_2O$)

1) NaOEt
2) [epoxide]

H_3O^+
heat
($-CO_2$)

H_3O^+
heat

21.101. Upon treatment with aqueous acid at elevated temperature, the ester group is hydrolyzed to a carboxylic acid group, and the acetal is hydrolyzed to a ketone. Under these conditions, the resulting β-keto acid undergoes decarboxylation to give the ketone shown below:

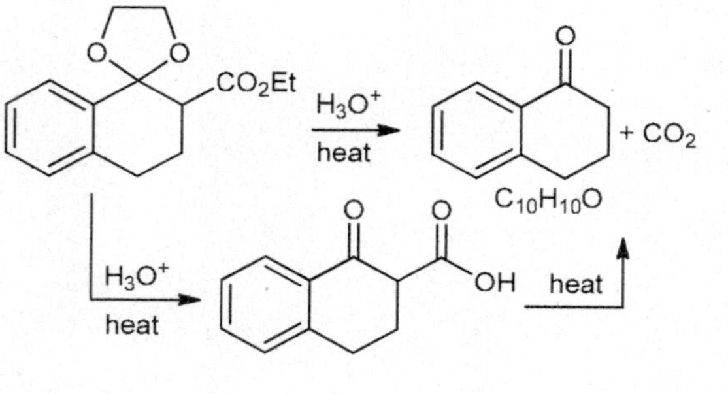

H_3O^+
heat

+ CO_2

$C_{10}H_{10}O$

H_3O^+
heat

heat

21.102.

(a) One strategy for achieving the desired transformation derives from the following retrosynthetic analysis. An explanation of each of the steps (*a-c*) follows.

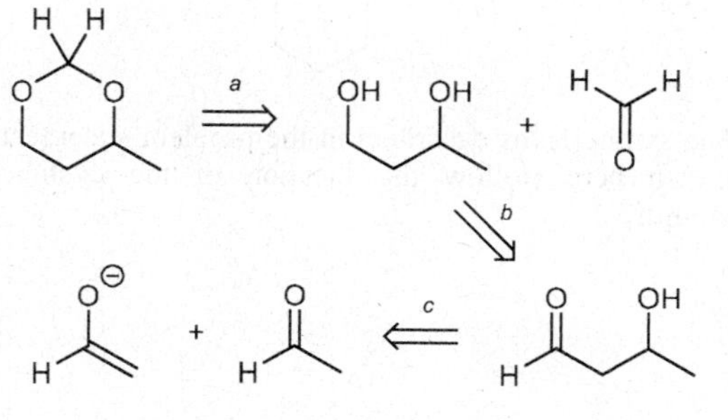

a. b. c.

a. The product is a cyclic acetal, which can be prepared from a diol and formaldehyde.
b. The diol can be made via reduction of a β-hydroxyaldehyde.
c. The β-hydroxyaldehyde can be made via an aldol addition reaction between two molecules of acetaldehyde.

Now let's draw the forward scheme. Upon treatment with sodium hydroxide, acetaldehyde will undergo an aldol addition reaction, giving a β-hydroxyaldehyde. Reduction with $LiAlH_4$, followed by aqueous acidic workup, gives a diol, which can be then be converted into the desired acetal upon treatment with formaldehyde in acid-catalyzed conditions (with removal of water).

1) NaOH, H_2O
2) $LiAlH_4$
3) H_3O^+
4) CH_2O, [H^+], $-H_2O$

NaOH
H_2O

1) $LiAlH_4$
2) H_3O^+

[H^+]
($-H_2O$)

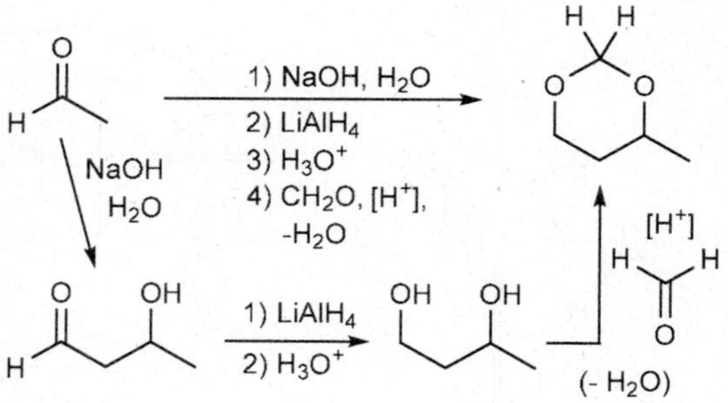

(b) One strategy for achieving the desired transformation derives from the following retrosynthetic analysis. An explanation of each of the steps (*a-c*) follows.

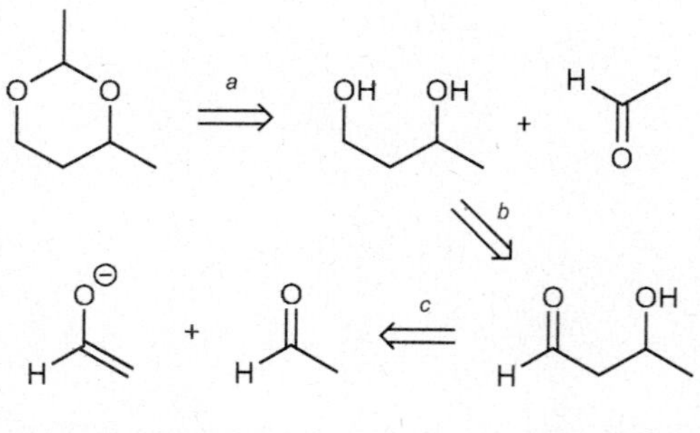

a. b. c.

a. The product is a cyclic acetal, which can be prepared from a diol and acetaldehyde.
b. The diol can be made via reduction of a β-hydroxyaldehyde.
c. The β-hydroxyaldehyde can be made via an aldol addition reaction between two molecules of acetaldehyde.

Now let's draw the forward scheme. Upon treatment with sodium hydroxide, acetaldehyde will undergo an aldol addition reaction, giving a β-hydroxyaldehyde. Reduction with $LiAlH_4$, followed by aqueous acidic workup, gives a diol, which can be then be converted into the desired acetal upon treatment with acetaldehyde in acid-catalyzed conditions (with removal of water).

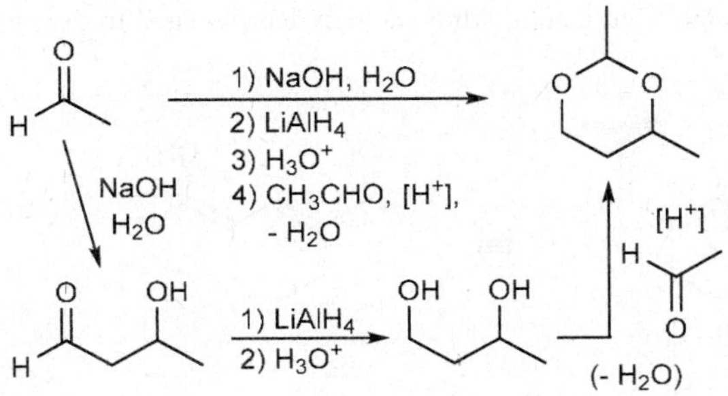

1) NaOH, H_2O
2) $LiAlH_4$
3) H_3O^+
4) CH_3CHO, $[H^+]$, $- H_2O$

NaOH, H_2O

1) $LiAlH_4$
2) H_3O^+

$[H^+]$

(- H_2O)

21.103.
When treated with aqueous acid, both compound **A** and compound **B** undergo racemization at the α position (via the enol as an intermediate). Each of these compounds establishes an equilibrium between *cis* and *trans* isomers. But the position of equilibrium is very different for compound **A** than it is for compound **B**. The equilibrium for compound **A** favors a *cis* configuration, because that is the configuration for which the compound can adopt a stable chair conformation in which both groups occupy equatorial positions. The equilibrium for compound **B** favors a *trans* configuration, because that is the configuration for which that compound can adopt a stable chair conformation in which both groups occupy equatorial positions.

21.104. The problem statement indicates that oxidation of the alcohol with PCC gives an aldehyde, which indicates that the alcohol must be primary. There are only two primary alcohols with the molecular formula $C_4H_{10}O$:

Oxidation of the first compound (1-butanol) gives 1-butanal, which is expected to produce four signals in its ^1H NMR spectrum. In contrast, oxidation of the second compound (2-methyl-1-propanol) is expected to produce an aldehyde that exhibits only three signals in its ^1H NMR spectrum:

PCC

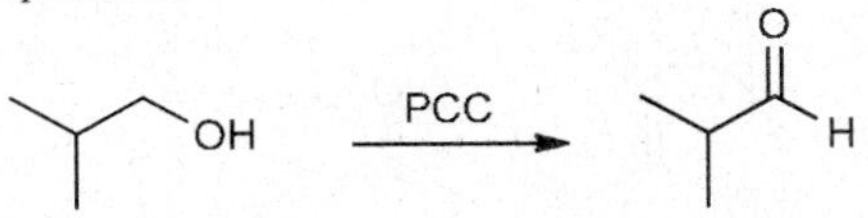

Upon treatment with aqueous sodium hydroxide, this aldehyde is deprotonated to give an enolate ion that can function as a nucleophile and attack the carbonyl group of another molecule of the aldehyde:

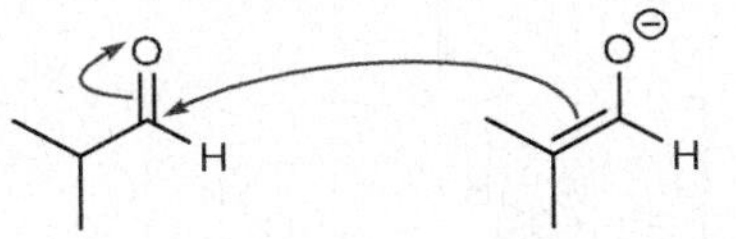

As a result, a carbon-carbon bond is formed. The resulting alkoxide ion is then protonated to give a β-hydroxy aldehyde, as shown:

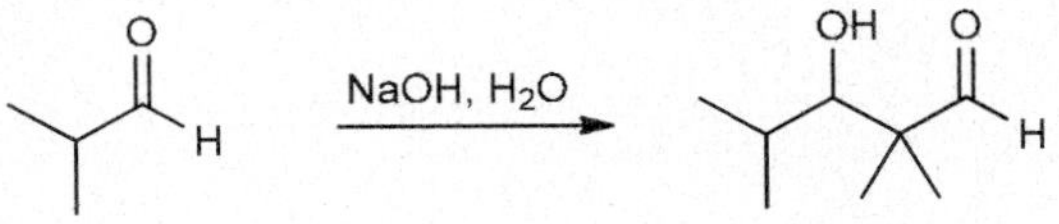

NaOH, H_2O

21.105. Treatment of acetaldehyde with aqueous sodium hydroxide results in a β-hydroxy aldehyde that can be converted into the desired diol via reduction with $LiAlH_4$, followed by aqueous acidic workup, as shown:

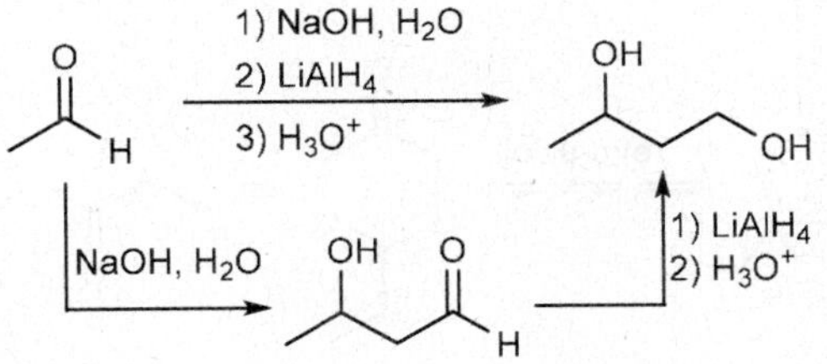

1) NaOH, H_2O
2) $LiAlH_4$
3) H_3O^+

NaOH, H_2O

1) $LiAlH_4$
2) H_3O^+

21.106. Protonation of the carbonyl group gives a resonance-stabilized cation, which is then deprotonated to give the product. This process is an example of tautomerization:

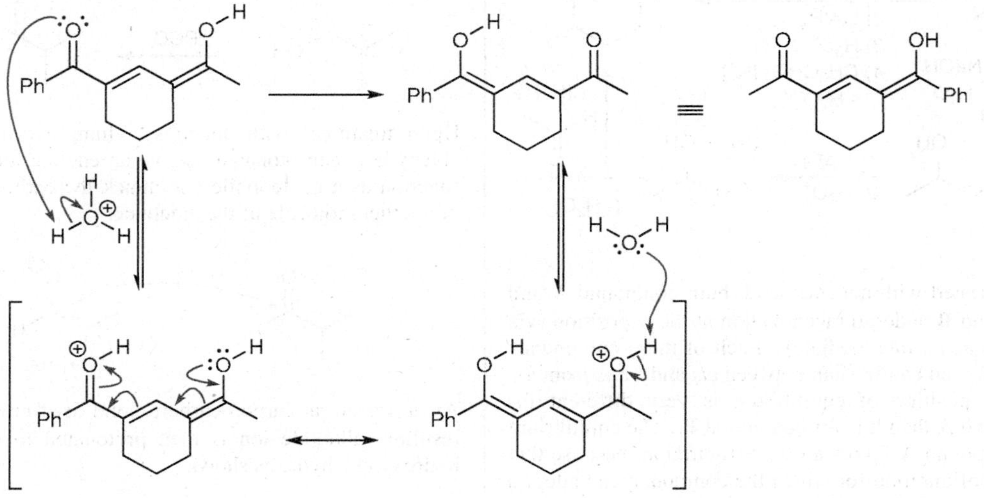

21.107. A retro-aldol reaction opens the ring into an acyclic diketone, which then closes up again via an intramolecular aldol condensation:

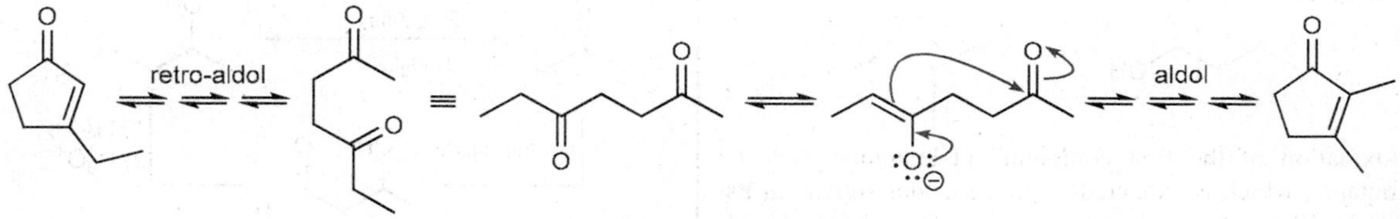

The retro-aldol process proceeds via a mechanism similar to the mechanism seen in the solution to Problem **21.22**. In the first step, hydroxide attacks the β position of the α,β-unsaturated ketone (a conjugate addition), giving an enolate. The enolate is then protonated to give a β-hydroxyketone, which is subsequently deprotonated at a different position to give an alkoxide ion. The carbonyl group is then formed, with loss of an enolate as a leaving group. The resulting enolate is then protonated to give the diketone.

Then, an intramolecular aldol reaction occurs, thereby closing a six-membered ring. First, hydroxide functions as a base, giving a new enolate. This enolate ion is a nucleophilic center, and it will attack the carbonyl group present in the same structure (an intramolecular process), thereby closing a five-membered ring. The resulting alkoxide ion is then protonated to give a β-hydroxy ketone. Hydroxide then functions as a base again, deprotonating the α position. The resulting enolate then ejects a hydroxide ion, giving the condensation product, as shown. Notice that each of the protonation steps employs water as the proton source, consistent with basic conditions (strong acids are not measurably present under these conditions).

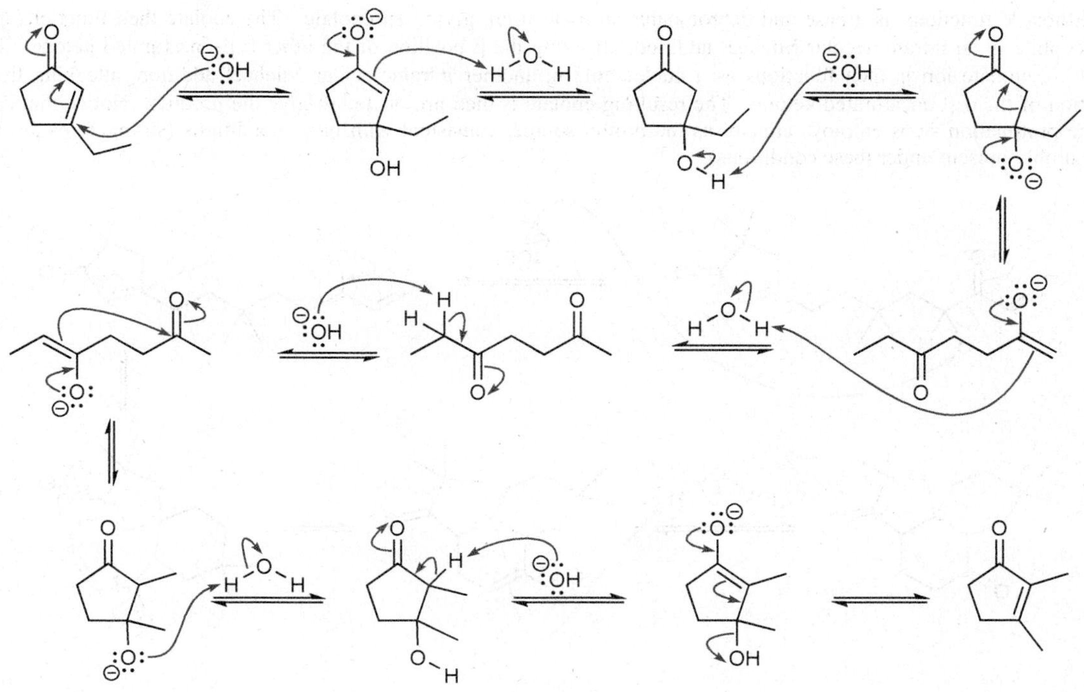

21.108.

(a) Hydroxide functions as a base and deprotonates the α position of the ketone, giving an enolate. The enolate then functions as a nucleophile in an intramolecular Michael addition, attacking the β position of the α,β-unsaturated ketone. The resulting enolate ion is then protonated to give a ketone, which is then further deprotonated to give a new enolate (all possible enolates are present at equilibrium). This enolate then attacks the other carbonyl group in an intramolecular attack, giving an alkoxide ion, which is then protonated to give the product. Notice that each of the protonation steps employs water as the proton source, consistent with basic conditions (strong acids are not measurably present under these conditions).

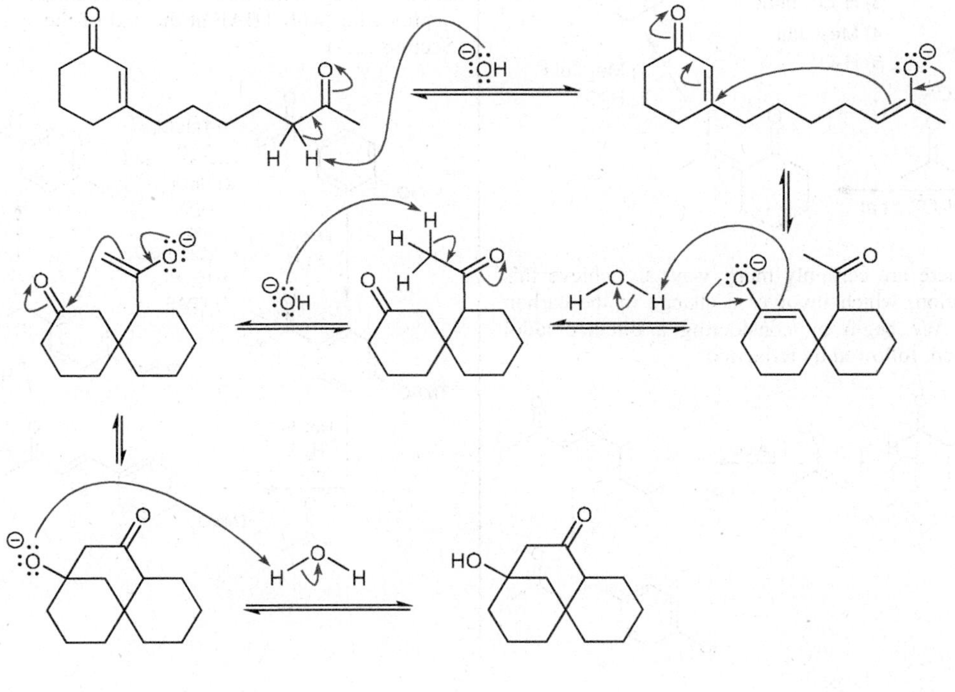

(b) Ethoxide functions as a base and deprotonates an α position, giving an enolate. The enolate then functions as a nucleophile in an intramolecular Michael addition, attacking the β position of the other α,β-unsaturated ketone. The resulting enolate ion is then functions as a nucleophile in another intramolecular Michael addition, attacking the β position of the α,β-unsaturated ketone. The resulting enolate is then protonated to give the product. Notice that each of the protonation steps employs ethanol as the proton source, consistent with basic conditions (strong acids are not measurably present under these conditions).

21.109. Direct alkylation would require performing an S_N2 reaction on a tertiary substrate, which will not occur. Instead the enolate would function as a base and E2 elimination would be observed instead of S_N2. The desired transformation can be achieved via a directed aldol condensation, followed by a Michael addition, as shown:

1) LDA
2) acetone
3) H_3O^+, heat
4) Me_2CuLi
5) H_3O^+

1) LDA
2) acetone
3) H_3O^+, heat

1) Me_2CuLi
2) H_3O^+

21.110. There are certainly many ways to achieve this transformation, which involves a change in the carbon skeleton. We begin by considering a directed aldol condensation, followed by reduction:

Aldol

This strategy suffers from a fatal flaw. The phenolic OH group is more acidic than acetone. Therefore, it is not possible to form the enolate of acetone without first deprotonating the phenolic OH group. And deprotonation in that location would generate a resonance-stabilized anion, in which the negative charge is spread over several positions including the oxygen atom of the carbonyl group, thereby deactivating the aldehyde group as an electrophile. This obstacle can be circumvented by protecting the OH group before the desired transformation is performed, and then deprotecting with TBAF at the end of the synthesis (see Section 12.7).

1) TMSCl, Et_3N
2) NaOH, H_2O, heat, $(CH_3)_2CO$
3) H_2, Pt
4) TBAF

TMSCl, Et_3N

NaOH, H_2O, heat, acetone

H_2, Pt

TBAF

21.111.

(a) The desired product is an ester, which can be made from the corresponding carboxylic acid via a Fischer esterification:

The carboxylic acid can be made with a malonic ester synthesis, as shown in the following scheme. Notice that two of the carbon atoms in the starting material (highlighted) remain in the product:

1) NaOEt
2) Br
3) H_3O^+, heat
4) [H^+], EtOH, (-H_2O)

1) NaOEt
2) Br
3) H_3O^+, heat

[H^+], EtOH, (-H_2O)

(b) The desired product is a primary alcohol, which can be made from the corresponding carboxylic acid via reduction:

This carboxylic acid can be made with a malonic ester synthesis, as shown in the following scheme. Notice that two of the carbon atoms in the starting material (highlighted) remain in the product:

1) NaOEt
2) $PhCH_2Br$
3) H_3O^+, heat
4) xs $LiAlH_4$
5) H_3O^+

1) NaOEt
2) $PhCH_2Br$
3) H_3O^+, heat

1) xs $LiAlH_4$
2) H_3O^+

(c) The desired product is an amide, which can be made from the corresponding carboxylic acid (via an acid halide), as shown:

This carboxylic acid can be made with a malonic ester synthesis, as shown in the following scheme. Notice that two of the carbon atoms in the starting material (highlighted) remain in the product:

1) NaOEt
2) R Br
3) H_3O^+, heat
4) $SOCl_2$
5) excess NH_3

1) NaOEt
2) R Br
3) H_3O^+, heat

$SOCl_2$

xs NH_3

21.112. When a dibromide is used (rather than two separate alkyl halides), a cyclic product is expected, as shown. Notice that two of the carbon atoms in the starting material (highlighted) remain in the product:

NaOEt

NaOEt

H_3O^+ heat

21.113.

(a) The desired product is an acetal, which can be made from the corresponding ketone:

This ketone can be made from ethyl acetoacetate via an acetoacetic ester synthesis, as shown in the following scheme. The three carbon atoms in the ethyl acetoacetate starting material that remain in the product (after decarboxylation) have been highlighted:

1) NaOEt
2) Br
3) H_3O^+, heat
4) $HOCH_2CH_2OH$, [H^+], (- H_2O)

1) NaOEt
2) Br

H_3O^+, heat

$HOCH_2CH_2OH$, [H^+], (-H_2O)

(b) The desired product is an alcohol, which can be made from the corresponding ketone via a reduction process:

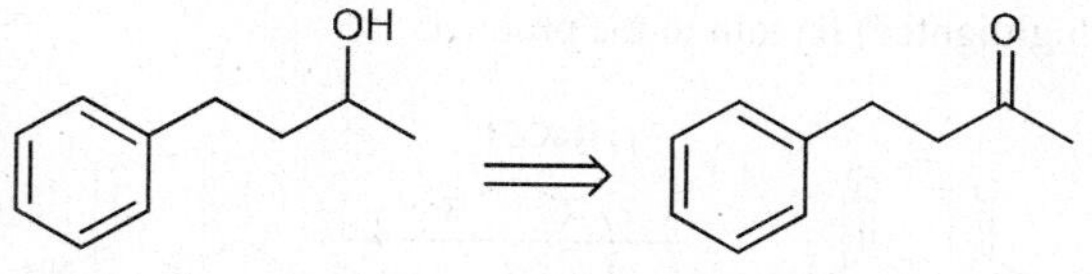

This ketone can be made from ethyl acetoacetate via an acetoacetic ester synthesis, as shown in the following scheme. The three carbon atoms in the ethyl acetoacetate starting material that remain in the product (after decarboxylation) have been highlighted:

1) NaOEt
2) $PhCH_2Br$
3) H_3O^+, heat
4) $LiAlH_4$
5) H_3O^+

OEt

OH

1) NaOEt
2) Br

1) $LiAlH_4$
2) H_3O^+

OEt

H_3O^+
heat

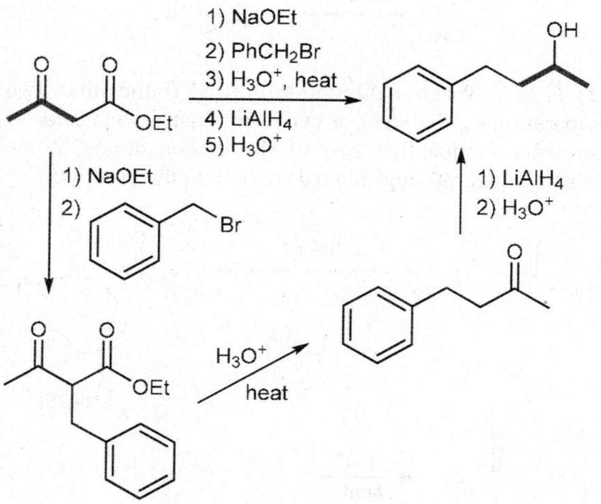

(c) The desired product is an imine, which can be made from the corresponding ketone:

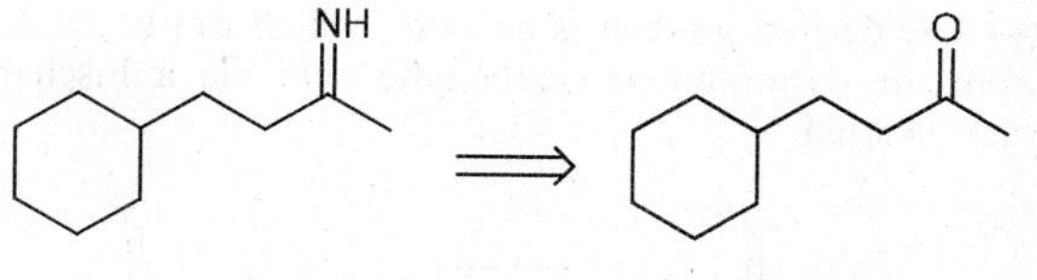

NH

This ketone can be made from ethyl acetoacetate via an acetoacetic ester synthesis, as shown in the following scheme. The three carbon atoms in the ethyl acetoacetate starting material that remain in the product (after decarboxylation) have been highlighted:

1) NaOEt
2) Br
3) H_3O^+, heat
4) $[H^+]$, NH_3, $(-H_2O)$

OEt

NH

1) NaOEt
2) Br

$[H^+]$, NH_3, $(-H_2O)$

OEt H_3O^+, heat

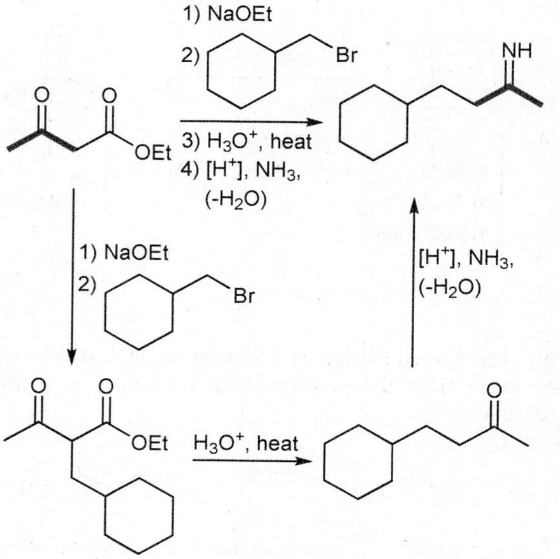

21.114. Using the approach described in the problem statement, the product can be made from an acyclic diester, as shown in the following retrosynthetic analysis:

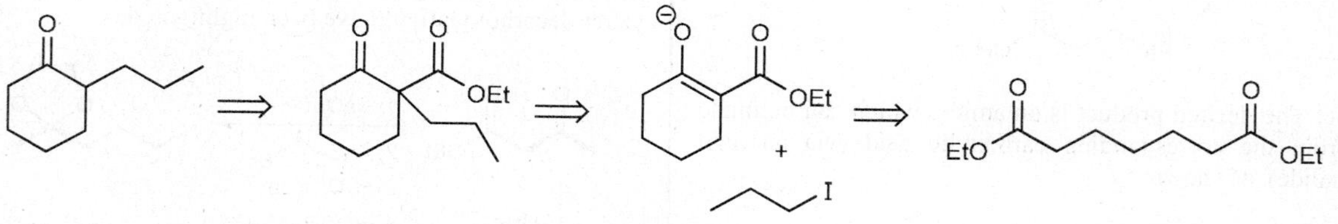

OEt

OEt

\+

I

EtO OEt

The diester can be made from the corresponding diacid (via Fischer esterification of both carboxylic acid groups), and the diacid can be made via oxidation of the starting diol:

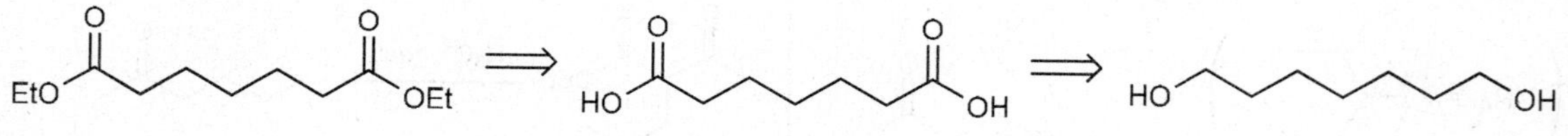

EtO OEt HO OH HO OH

Now let's draw the forward scheme. The starting diol is converted to a diacid upon treatment with chromic acid. This diacid is then converted to a diester upon treatment with ethanol and an acid catalyst (with removal of water). The resulting diester will undergo a Dieckmann cyclization upon treatment with sodium ethoxide, followed by aqueous acidic workup. Alkylation, followed by hydrolysis and decarboxylation, gives the product, as shown.

HO OH $Na_2Cr_2O_7$, H_2SO_4, H_2O HO OH

$[H^+]$, EtOH, $(-H_2O)$

EtO OEt

1) NaOEt
2) H_3O^+ (workup)

OEt

1) NaOEt
2) I
3) H_3O^+, heat

21.115.

(a) The compound possesses three functional groups, and can be assembled in a variety of ways. One method capitalizes on the 1,5-arrangement of two of the functional groups (highlighted):

OEt H

Therefore, we consider preparing the product via a Michael reaction. The following retrosynthetic analysis is based on assembly of the carbon skeleton via a Michael reaction, as well as an aldol condensation reaction to prepare the Michael acceptor, and a Claisen condensation to prepare the Michael donor:

Michael ⟹ OEt + H ; aldol ⟹ H + H

Claisen ⟹ OEt + OEt

OH ⟸ OH ⟸

OH

Now let's show the forward scheme for this strategy. One equivalent of ethanol is oxidized with PCC (or DMP or via a Swern oxidation) to give acetaldehyde, which is then heated with aqueous sodium hydroxide to give an α,β-unsaturated aldehyde (via an aldol condensation). Another equivalent of ethanol is oxidized with chromic acid to give a carboxylic acid, which is then treated with ethanol under acidic conditions to give an ester (via Fischer esterification). The ester is then converted into a β-ketoester (via a Claisen condensation). The β-ketoester is then deprotonated with ethoxide to give a stabilized enolate which then attacks the α,β-unsaturated aldehyde to give a Michael reaction:

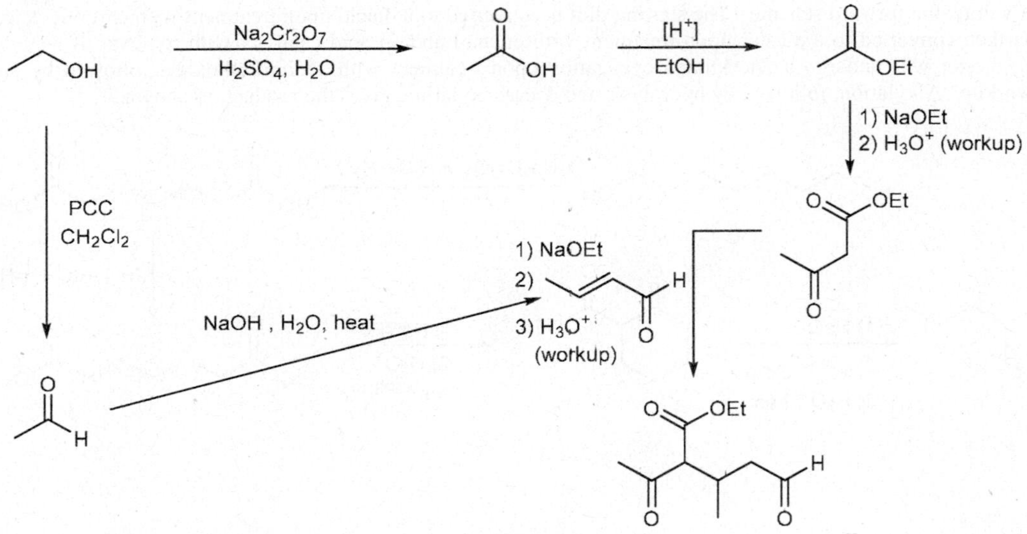

$Na_2Cr_2O_7$ / H_2SO_4, H_2O ; $[H^+]$ / EtOH ; 1) NaOEt 2) H_3O^+ (workup) ; PCC / CH_2Cl_2 ; NaOH , H_2O, heat ; 1) NaOEt 2) 3) H_3O^+ (workup)

(b) The product is 1,5-difunctionalized:

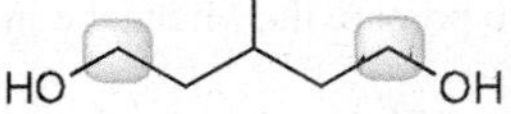

Therefore, we consider preparing the product via a Michael reaction:

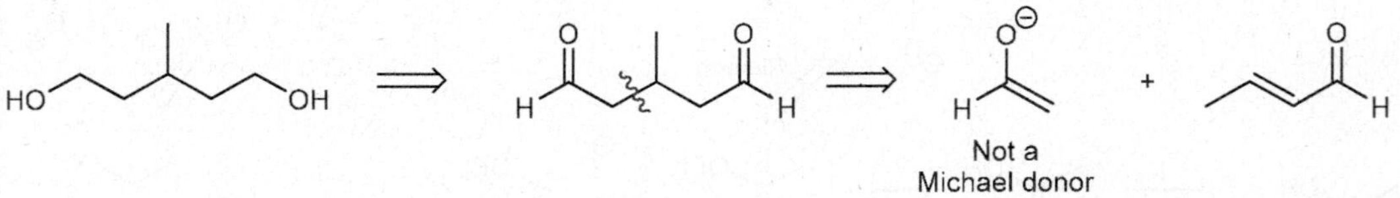

Not a Michael donor

This strategy will not work, because it involves the use of an enolate, which is not an efficient Michael donor. Therefore, we consider a Stork enamine synthesis (in which we use an enamine, rather than an enolate, as a Michael donor). The Michael donor can be made from acetaldehyde, which can be made from ethanol via oxidation with PCC (or DMP or via a Swern oxidation). The Michael acceptor can be made from two equivalents of acetaldehyde via an aldol condensation:

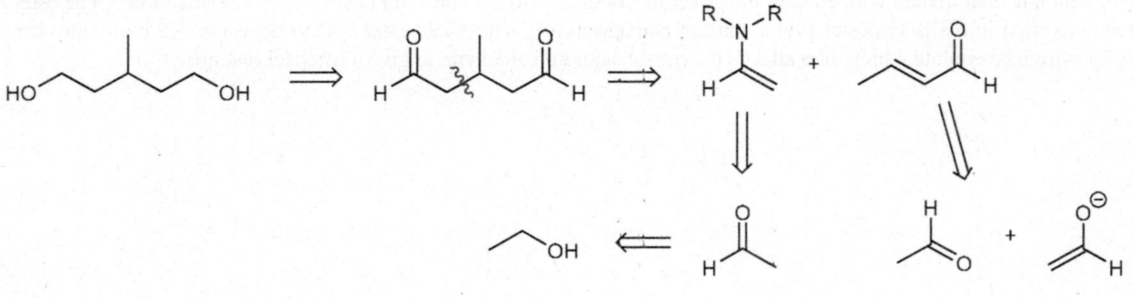

The forward scheme is shown here:

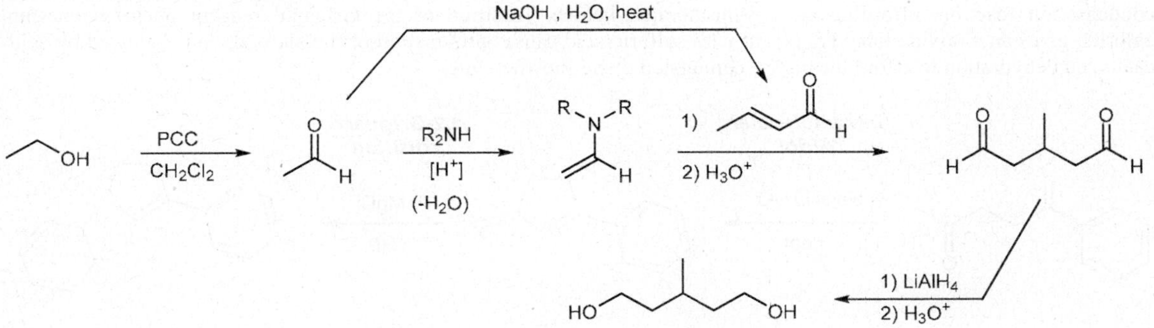

NaOH , H_2O, heat

PCC / CH_2Cl_2; R_2NH / $[H^+]$ / $(-H_2O)$; 1) (crotonaldehyde) 2) H_3O^+; 1) $LiAlH_4$ 2) H_3O^+

21.116. Intermediate **A** is the corresponding enamine, while the alkylation product **B** is 2-methylcyclohexanone:

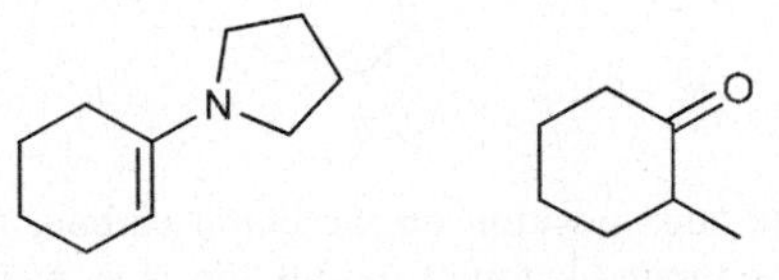

A B

Product **B** is a result of the hydrolysis of the iminium ion shown here, which forms after alkylation of enamine **A** by methyl iodide:

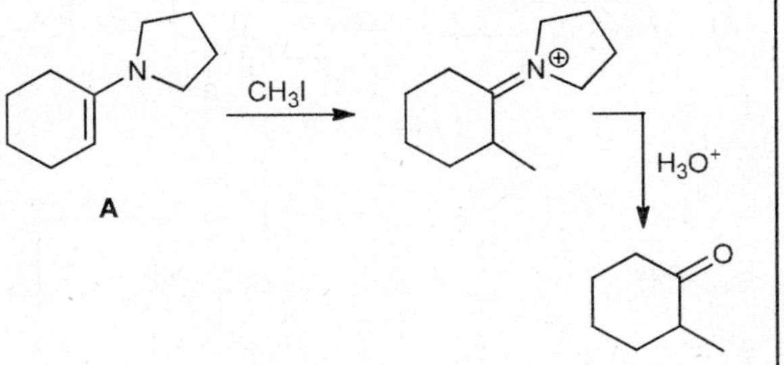

A CH_3I H_3O^+ B

21.117.
(a) Only one diastereomer of the enamine forms due to rotational symmetry possessed by the secondary amine. Due to this symmetry property, forming the π bond at either alpha carbon results in identical enamines.

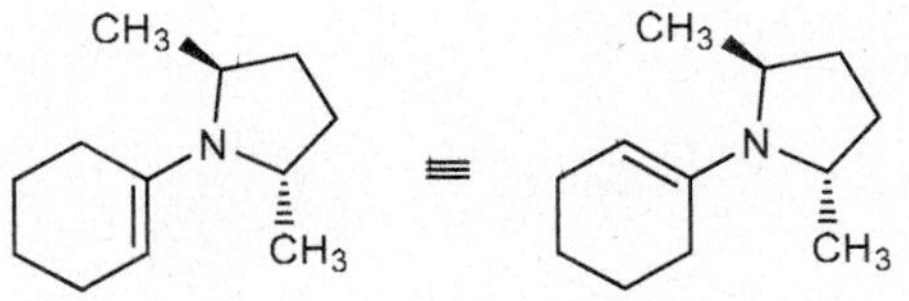

CH_3 CH_3 ≡ CH_3 CH_3

(b) When the enamine in part (a) reacts with methyl iodide, the methyl group closer to the α carbon atom of the enamine is oriented below the plane of the page.

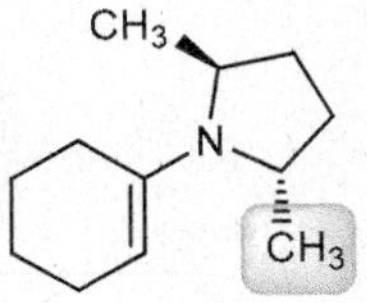

CH_3 CH_3

So, approach by the electrophile from that direction is hindered, resulting in (*R*)-2-methylcyclohexanone being the minor product. In contrast, since the approach of methyl iodide from the top face of the page is unimpeded by this substituent, (*S*)-2-methylcyclohexanone is the major product. Producing an excess of one enantiomer results in a high *ee* (rather than forming a racemic mixture).

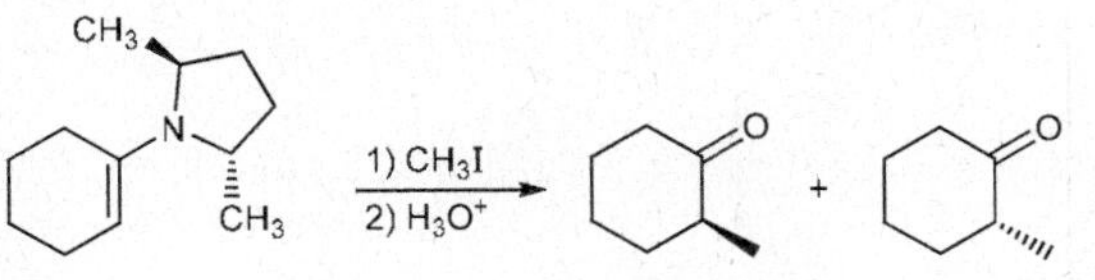

CH_3 CH_3 1) CH_3I 2) H_3O^+ +

S-enantiomer (major) *R*-enantiomer

21.118. While there are three alpha positions in the molecule, deprotonation at only one of them (the terminal methyl group) will lead to the formation of a 5-membered ring. The first reaction is a base-catalyzed, intramolecular aldol condensation reaction, affording an α,β-unsaturated ketone. Addition of the Grignard reagent phenylmagnesium chloride gives an alkoxide ion. Treatment with sulfuric acid causes protonation of the alkoxide ion, followed by acid-catalyzed dehydration to afford the highly conjugated diene shown below.

Intramolecular Aldol

$NaOH/H_2O$

heat

1,2-Grignard addition

PhMgCl

THF

Acid-Catalyzed Dehydration

H_2SO_4

21.119. In the starting material, the dimethoxybenzyl group is in the "up" position on the chiral carbon, from the perspective drawn. This group thus attacks the β carbon of the α,β-unsaturated ketone from the top face, pushing the other aromatic ring "down", so that it ends up *cis* to the hydrogen on the adjacent chiral center.

OMe MeO MeO H N R $F_3C-S-O-S-CF_3$

OMe MeO MeO H N R CF_3

OMe MeO MeO H N R CF_3

OMe H N R CF_3

OMe H N R CF_3

OMe H N R CF_3

OMe H N R CF_3

OMe H N R CF_3

N

OMe H N R CF_3

21.120. This transformation involves two successive Michael addition reactions. The sequence begins with the deprotonation of diethyl malonate with potassium carbonate to give a stabilized anion (Michael donor) that attacks the α,β-unsaturated ketone **1** (Michael acceptor) in a conjugate addition to afford enolate **3**. After protonation of enolate **3** to give compound **4**, the β-diester moiety is then deprotonated to give the resonance-stabilized enolate **5**. At this point a second Michael addition occurs as the anion (Michael donor) attacks the carbon containing the two thioethers (Michael acceptor) in a conjugate fashion to give enolate **6**. Finally, the electrons of the enolate come back down to make the ketone and the α,β-unsaturation by expelling the ethanethiolate anion as a leaving group to afford the final product **2**. Alternatively, the reverse order of these two Michael additions will likewise afford the product.

EtO OEt H H K⊕ OK

EtO OEt Cl N H EtS SEt 1 H₃C

Michael Addition

CH₃ K⊕ Cl EtO₂C EtO₂C N H EtS SEt 3

KO O H

CH₃ Cl EtO EtO H N H EtS SEt 4 K⊕ OK

CH₃ Cl EtO EtO N H EtS SEt K⊕ 5

Michael Addition

H₃C K⊕ EtO₂C EtO₂C EtS SEt HN 6 Cl

H₃C EtO₂C EtO₂C SEt HN 2 Cl + EtSK

21.121.

(a) Under strongly basic conditions, an enolate is formed. Then, under these conditions, an elimination-addition reaction can occur (Section 18.14), in which the nucleophilic enolate attacks the tethered benzyne in a ring-forming reaction, followed by protonation.

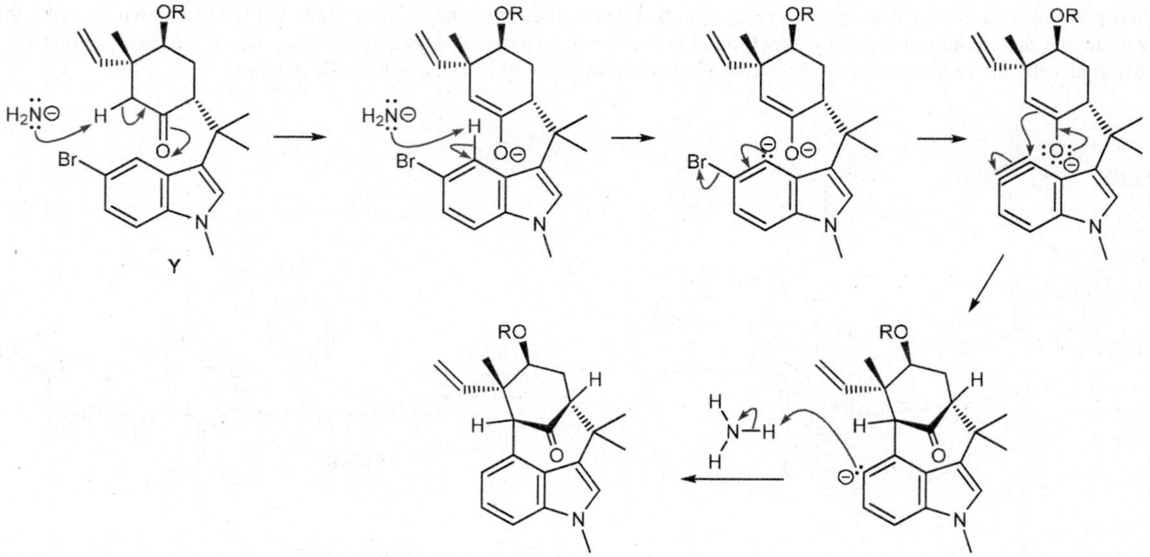

(b) The enolate ion has two nucleophilic centers, which means that it can attack from the oxygen atom, as well as from the alpha carbon atom. If the oxygen atom of the enolate functions as the nucleophile and attacks the benzyne unit, the following side product is formed:

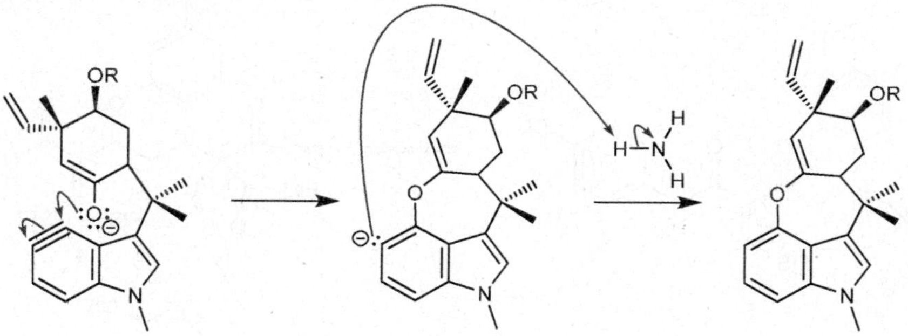

21.122.

(a) Cyclic hemiacetal **A** is in equilibrium with its open-chain form, which has an aldehyde group and a hydroxyl group. The aldehyde group can undergo a Wittig reaction when treated with a stabilized ylide ($Ph_3P{=}CHCO_2Et$), giving an α,β-unsaturated ester **B** (the stabilized Wittig reagent is expected to give the *trans* alkene as the major product). Next, the hydroxyl group is converted to a tosylate group upon treatment with tosyl chloride to afford compound **C**:

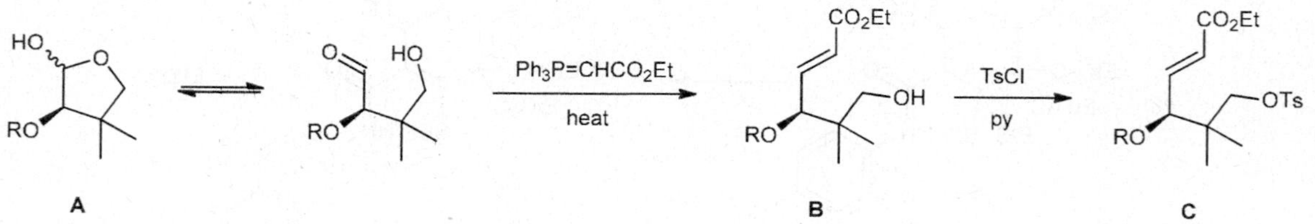

(b) The conversion of compound **C** to **D** begins with a conjugate addition reaction, in which Me_2CuLi functions as a Michael donor and attacks the α,β-unsaturated ester (the Michael acceptor) to afford an enolate. This enolate can then function as a nucleophile in an intramolecular S_N2-type process (shown below), in which the enolate undergoes α-alkylation to close the cyclopentyl ring system, giving compound **D**:

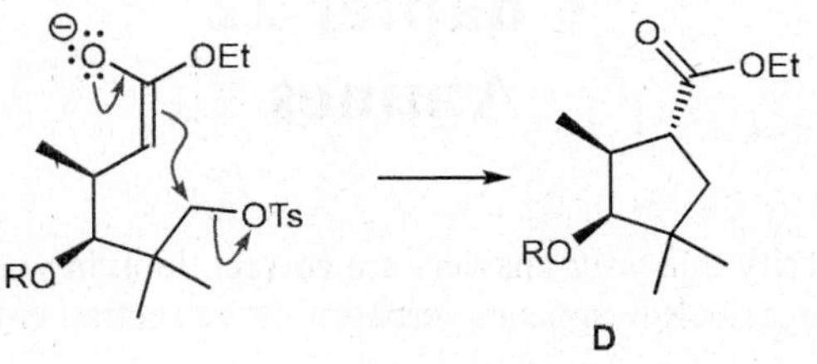

(c) The ester is first reduced to an alcohol (compound **E**), which can be acylated, using either acetic anhydride or acetyl chloride and pyridine:

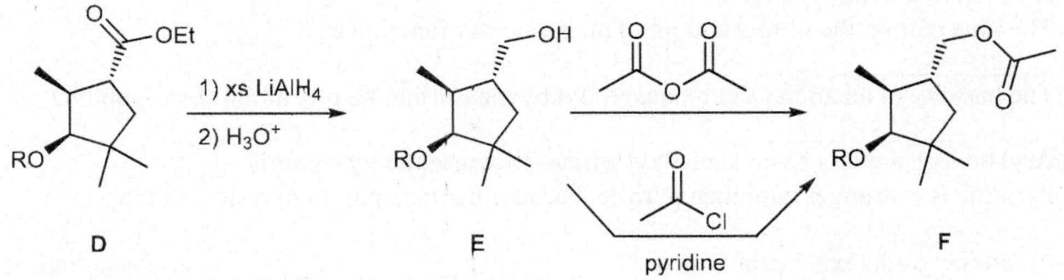

21.123. Methoxide functions as a base and deprotonates the most acidic position, leading to a stabilized enolate. This enolate then functions as a nucleophile and attacks the ester in an intramolecular nucleophilic acyl substitution reaction. The resulting charged tetrahedral intermediate loses methoxide to reform the carbonyl group. The OH group is then deprotonated to give an alkoxide ion, which then attacks the newly formed carbonyl group to give another charged tetrahedral intermediate, which is protonated to give the product.

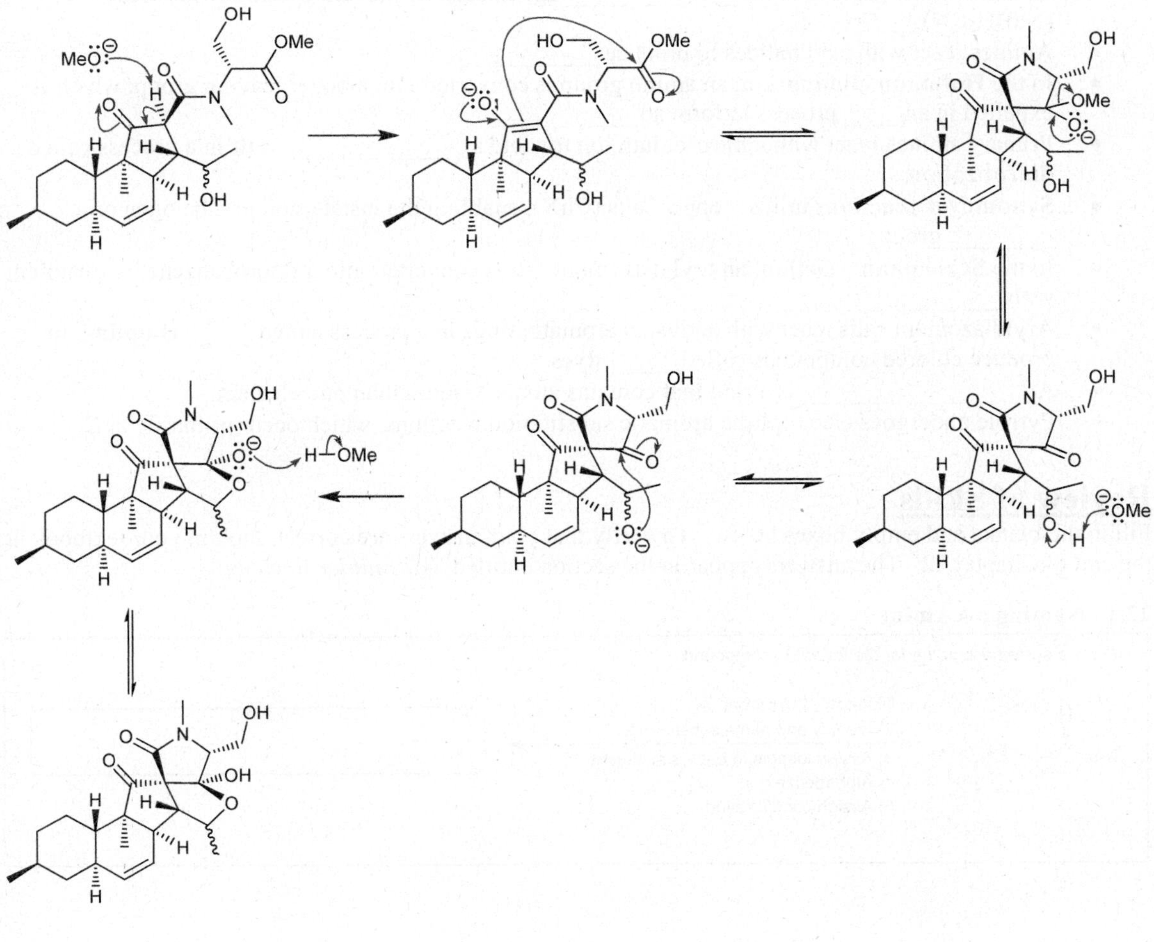

Chapter 22
Amines

Review of Concepts

Fill in the blanks below. To verify that your answers are correct, look in your textbook at the end of Chapter 22. Each of the sentences below appears verbatim in the section entitled *Review of Concepts and Vocabulary*.

- Amines are ____________, ____________, or ____________, depending on the number of groups attached to the nitrogen atom.
- The lone pair on the nitrogen atom of an amine can function as a ________ or __________________.
- The basicity of an amine can be quantified by measuring the pK_a of the corresponding __________________.
- Aryl amines are less basic than alkyl amines, because the lone pair is __________________.
- Pyridine is a stronger base than pyrrole, because the lone pair in pyrrole participates in _____________.
- An amine group exists primarily as ______________________________ at physiological pH.
- The **azide synthesis** involves treating an ________________ with sodium azide, followed by ________________.
- The ___________ **synthesis** generates primary amines upon treatment of potassium phthalimide with an alkyl halide, followed by hydrolysis or reaction with N_2H_4.
- Amines can be prepared via **reductive amination**, in which a ketone or aldehyde is converted into an imine in the presence of a ______________ agent, such as **sodium cyanoborohydride** ($NaBH_3CN$).
- Amines react with acyl halides to produce ___________.
- In the **Hofmann elimination**, an amino group is converted into a better leaving group, which is expelled in an _____ process to form an ____________.
- Primary amines react with a nitrosonium ion to yield a ______________ **salt** in a process called **diazotization**.
- **Sandmeyer reactions** utilize copper salts (CuX), enabling the installation of a halogen or a ________ group.
- In the **Schiemann reaction**, an aryl diazonium salt is converted into a fluorobenzene by treatment with _________________.
- Aryldiazonium salts react with activated aromatic rings in a process called _____ **coupling**, to produce colored compounds called _____ **dyes**.
- A __________________ is a ring that contains atoms of more than one element.
- Pyrrole undergoes electrophilic aromatic substitution reactions, which occur primarily at C__.

Review of Skills

Fill in the blanks and empty boxes below. To verify that your answers are correct, look in your textbook at the end of Chapter 22. The answers appear in the section entitled *SkillBuilder Review*.

22.1 Naming an Amine

Provide a systematic name for the following compound:

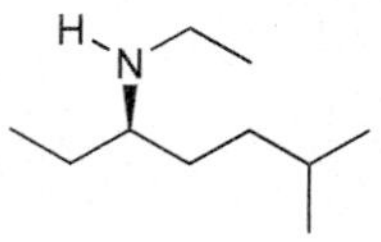

1) Identify the parent
2) Identify and name substituents
3) Assign locants to each substituent
4) Alphabetize
5) Assign configuration

22.2 Preparing a Primary Amine via the Gabriel Reaction

Identify reagents that will achieve the following transformation:

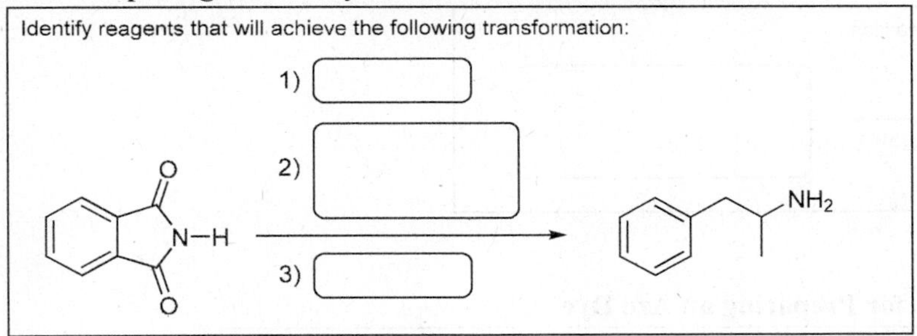

22.3 Preparing an Amine via a Reductive Amination

Identify reagents that will achieve each of the following transformations:

H
R O
CH3
R N
H
R NH2

22.4 Synthesis Strategies

Identify the reactant that is used as a nucleophile in each of the three processes shown below:

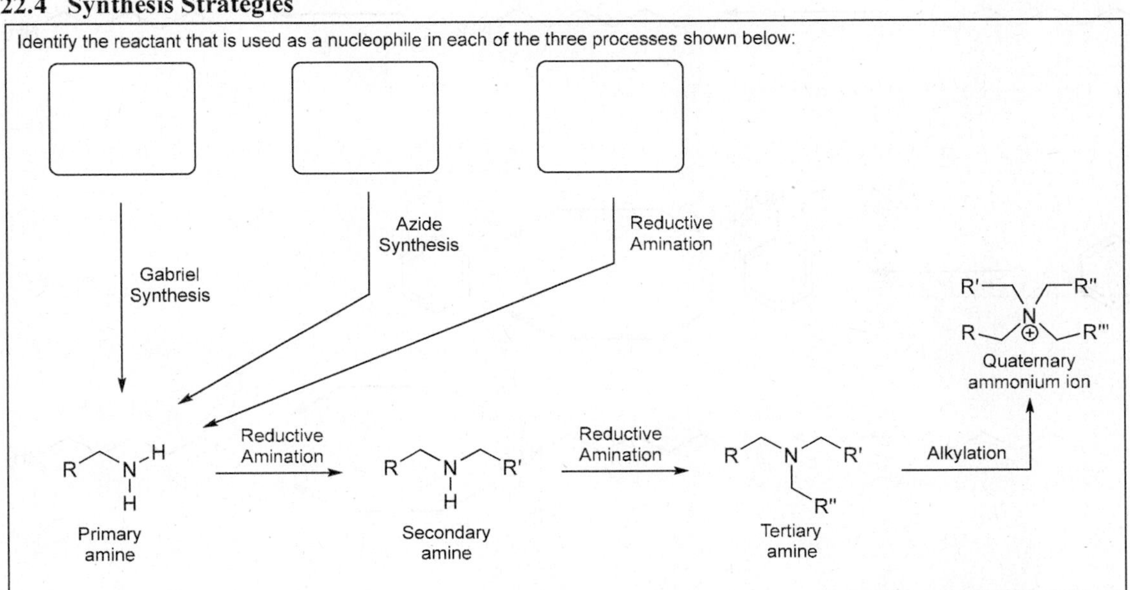

22.5 Predicting the Product of a Hofmann Elimination

Predict the major product of the following reaction:

NH_2

1) excess CH_3I

2) Ag_2O, H_2O, heat

22.6 Determining the Reactants for Preparing an Azo Dye

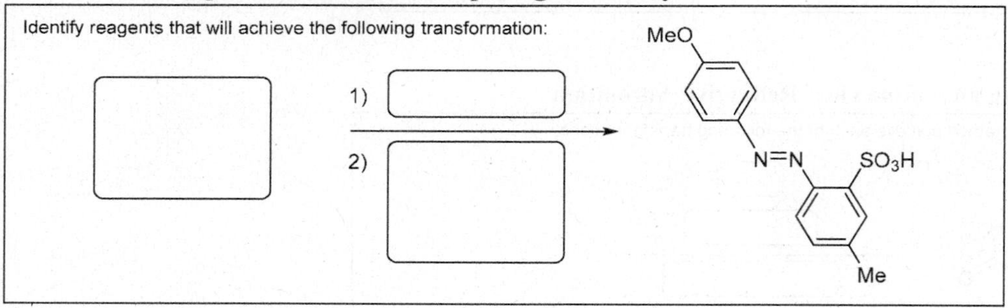
Identify reagents that will achieve the following transformation:
MeO
1)
2)
N=N
SO3H
Me

Review of Reactions

Identify the reagents necessary to achieve each of the following transformations. To verify that your answers are correct, look in your textbook at the end of Chapter 22. The answers appear in the section entitled *Review of Reactions*.

Preparation of Amines

Br → C≡N → NH_2

OH → NH_2 → NH_2

→ NO_2 → NH_2

X → N_3 → NH_2

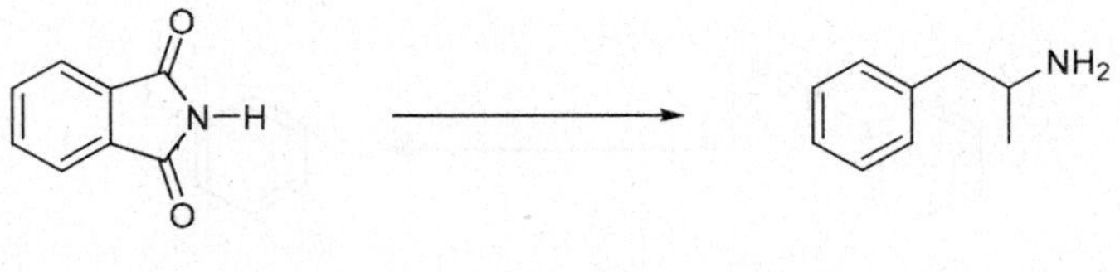

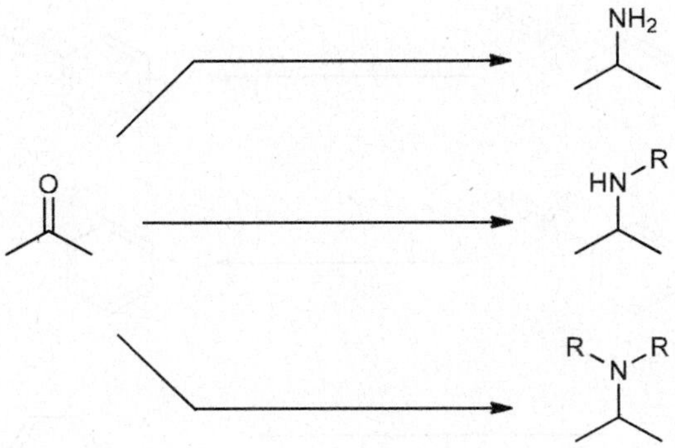

Reactions of Amines

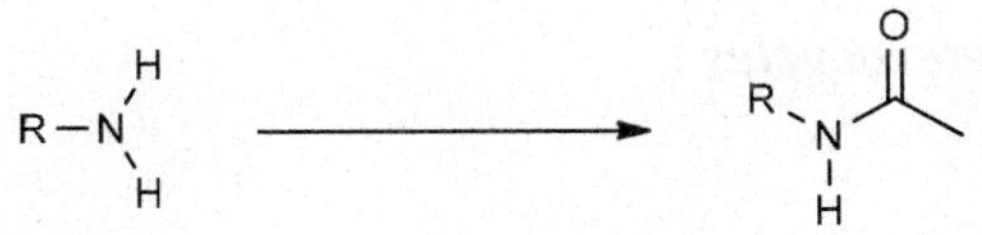

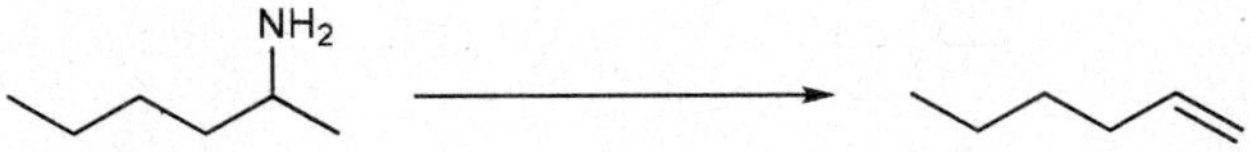

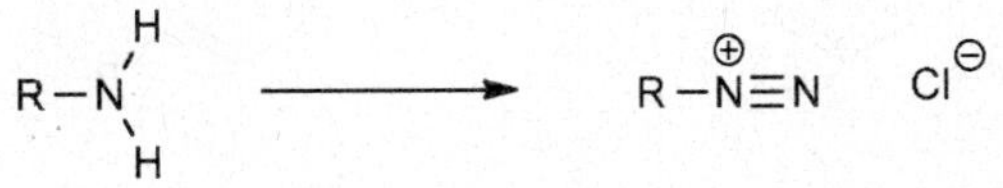

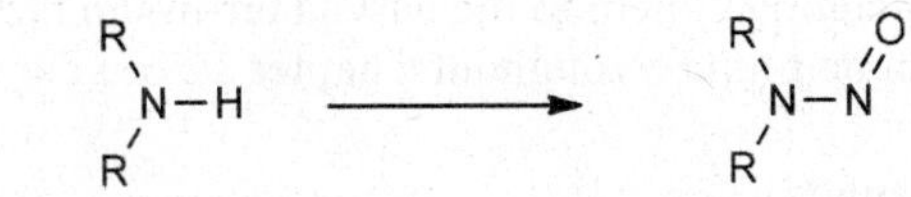

Reactions of Aryldiazonium Salts

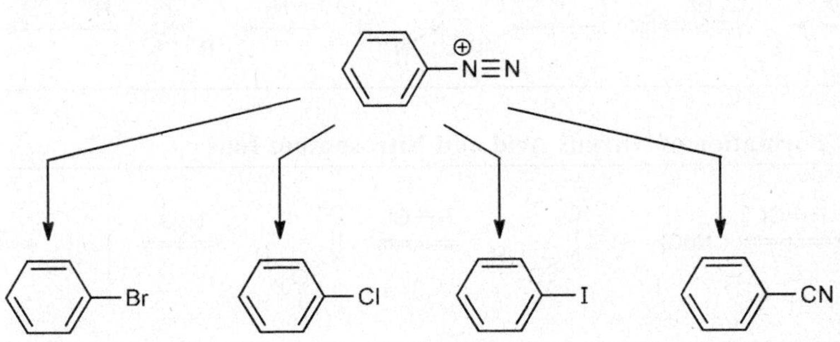

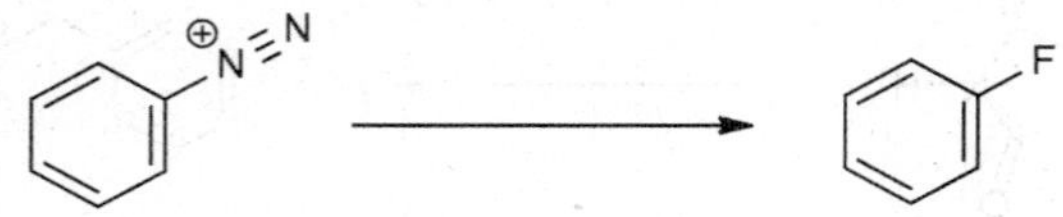

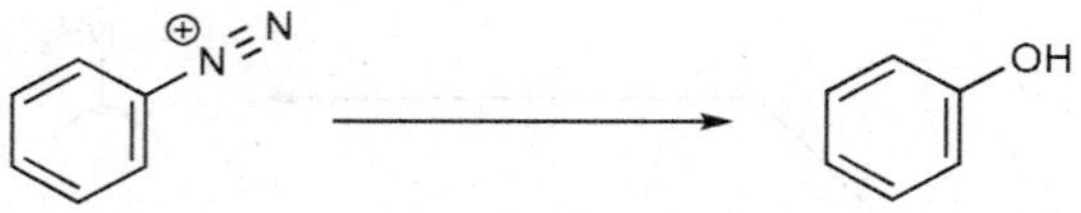

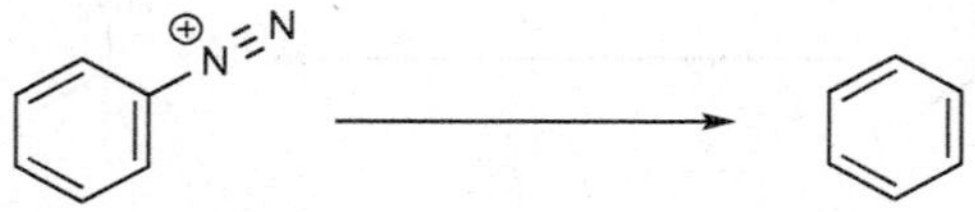

Reactions of Nitrogen Heterocycles

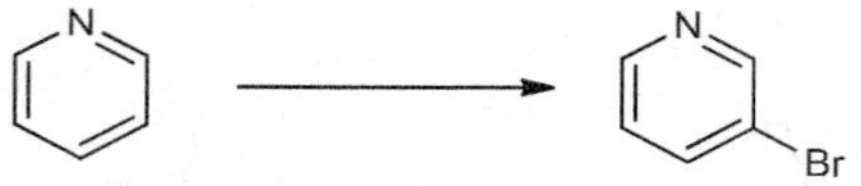

Review of Mechanisms

Complete each of the following mechanisms by drawing the missing curved arrows. To verify that your curved arrows are drawn correctly, compare them to the curved arrows in the mechanism boxes for Mechanisms 22.1 – 22.4, which can be found throughout Chapter 22 of your text.

Mechanism 22.1 The Azide Synthesis

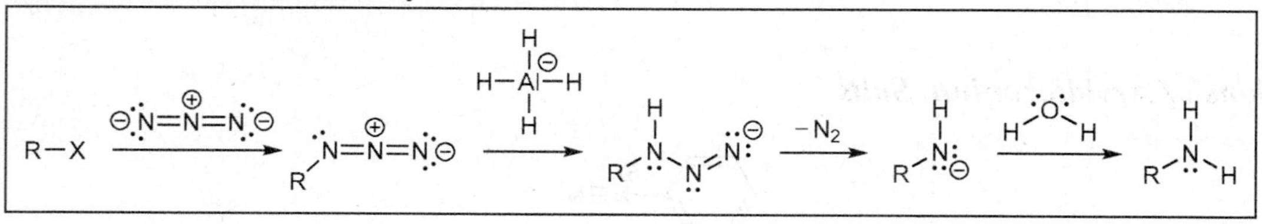

Mechanism 22.2 Formation of Nitrous Acid and Nitrosonium Ions

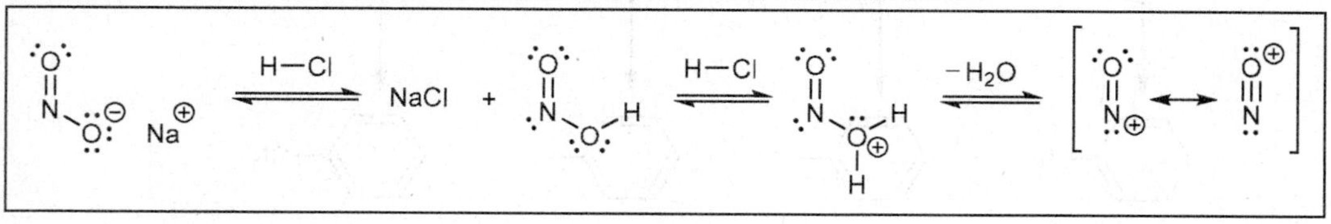

Mechanism 22.3 Formation of *N*-Nitrosamines

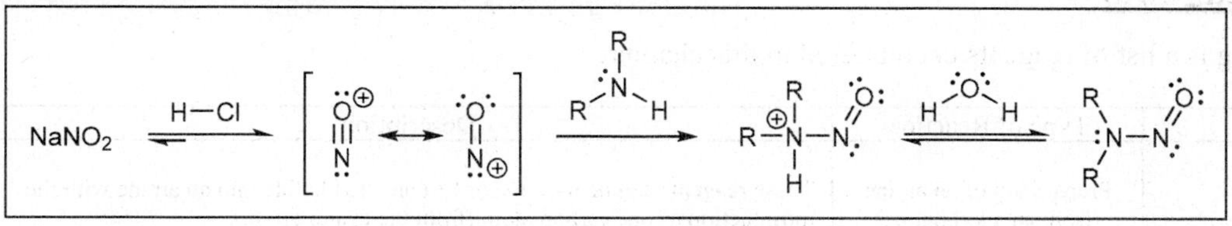

Mechanism 22.4 Diazotization

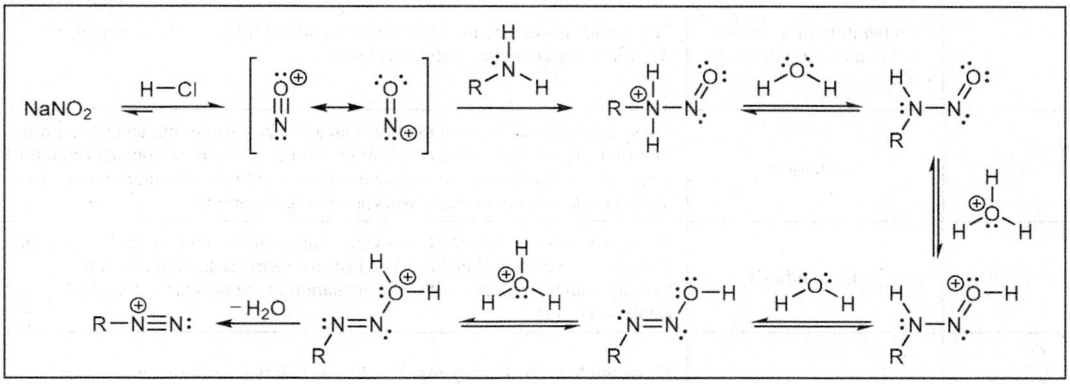

Common Mistake to Avoid

Whenever you learn a new reaction, pay close attention to any restrictions that may apply. For example, the Gabriel synthesis employs an S_N2 process to create the critical C–N bond of a primary amine:

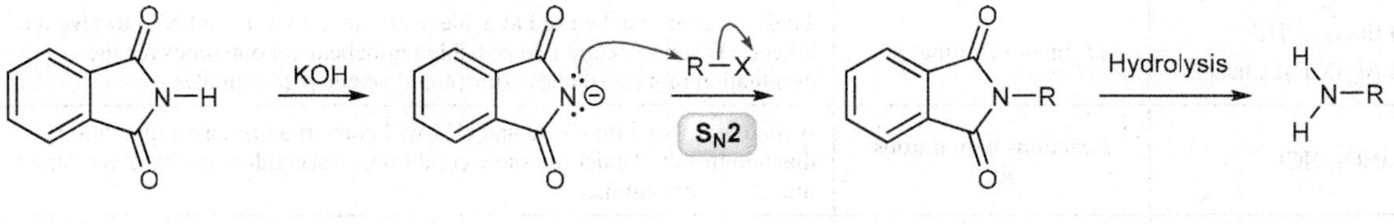

Since an S_N2 process is employed, a tertiary alkyl halide cannot be used, because tertiary alkyl halides are too sterically hindered to undergo an S_N2 process. It is a common mistake to attempt to use a tertiary alkyl halide in a Gabriel synthesis, because it is easy to forget the restrictions that apply. Keep this in mind for all reactions that you study. Make sure that you understand the circumstances under which each reaction can or cannot be used.

Useful reagents

The following is a list of reagents encountered in this chapter:

Reagents	Type of Reaction	Description
1) NaCN 2) xs $LiAlH_4$ 3) H_2O	Preparation of an amine (from an alkyl halide)	These reagents can be used to convert an alkyl halide into an amine with the introduction of one carbon atom (from the cyano group).
1) $SOCl_2$ 2) xs NH_3 3) xs $LiAlH_4$ 4) H_2O	Preparation of an amine (from a carboxylic acid)	These reagents can be used to convert a carboxylic acid into an amine, without a change in the carbon skeleton.
1) Fe, H_3O^+ 2) NaOH	Reduction	These reagents can be used to reduce an aryl nitro group into an amino group. The first step employs acidic conditions, so the amine is protonated (under the conditions of its formation) to give an ammonium ion. The ammonium ion is then deprotonated upon basic workup, giving the amine.
1) NaN_3 2) $LiAlH_4$ 3) H_2O	Azide synthesis	These reagents can be used to convert an alkyl halide into an amine, without a change in the carbon skeleton. The last two steps (reduction and water workup) can be replaced with hydrogenation in the presence of a metal catalyst (H_2, Pt).
O, N–H, O	Gabriel synthesis	Phthalimide is the starting material for the Gabriel synthesis, which can be used to prepare primary amines. Phthalimide is treated with KOH to give potassium phthalimide, which is then treated with an alkyl halide, giving an S_N2 reaction. The product of the S_N2 process is then hydrolyzed (upon treatment with hydrazine or aqueous acid) to release the amine.
$NaBH_3CN$	Reductive amination	In the presence of an acid catalyst, sodium cyanoborohydride can be used to achieve a reductive amination. The reaction occurs between a ketone (or aldehyde) and an amine (or ammonia). This process can be used to convert a primary amine into a secondary amine. Similarly, a secondary amine is converted into a tertiary amine.
O, Cl	Acetylation	An amine will undergo acetylation (giving an amide) when treated with acetyl chloride.
1) Excess CH_3I 2) Ag_2O, H_2O, heat	Hofmann elimination	These reagents can be used to achieve elimination of H and NH_2 to give an alkene. When there are two possible regiochemical outcomes for the elimination process, the less substituted alkene predominates.
$NaNO_2$, HCl	Reactions with nitrous acid	A mixture of sodium nitrite and HCl will convert a primary amine into a diazonium salt. Under the same conditions, a secondary amine is converted into an *N*-nitrosamine.
CuBr	Sandmeyer reaction	When an aryldiazonium salt is treated with CuBr, the diazonium group is replaced with a bromine atom.
CuCl	Sandmeyer reaction	When an aryldiazonium salt is treated with CuCl, the diazonium group is replaced with a chlorine atom.
CuI	Sandmeyer reaction	When an aryldiazonium salt is treated with CuI, the diazonium group is replaced with an iodine atom.
CuCN	Sandmeyer reaction	When an aryldiazonium salt is treated with CuCN, the diazonium group is replaced with a cyano group.
HBF_4	Fluorination (Schiemann reaction)	When an aryldiazonium salt is treated with HBF_4, the diazonium group is replaced with a fluorine atom.
H_2O, heat	Preparation of phenol	When an aryldiazonium salt is treated with water and heat, the diazonium group is replaced with an OH group.
H_3PO_2	Reduction	When an aryldiazonium salt is treated with H_3PO_2, the diazonium group is replaced with a hydrogen atom.

Solutions

22.1.

(a) This compound is an amine that has only one alkyl group connected to the nitrogen atom. Since this alkyl group is complex, we must name the compound as an alkanamine (rather than an alkyl amine). The parent is comprised of four carbon atoms (thus, butanamine), and the amino group is located at C1. There are two methyl groups, both located at C3.

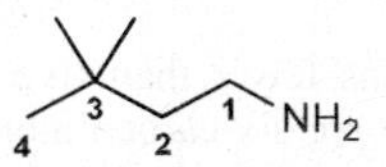

3,3-dimethyl-1-butanamine

(b) This compound is an amine that has only one simple alkyl group (a cyclopentyl group) connected to the nitrogen atom, so we can name this compound as an alkyl amine, in addition to naming it as an alkanamine. Therefore, this compound is cyclopentylamine.

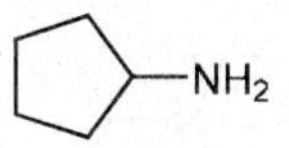

cyclopentylamine
or
cyclopentanamine

(c) This compound is an amine that has three simple alkyl groups (two methyl groups and a cyclopentyl group) connected to the nitrogen atom, so we can name this compound as a trialkyl amine, in addition to naming it as an alkanamine. The alkyl groups are listed in alphabetical order:

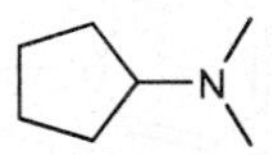

cyclopentyldimethylamine
or
N,N-dimethylcyclopentanamine

(d) This compound is an amine that has three simple alkyl groups (all ethyl groups) connected to the nitrogen atom, so we can name this compound as a trialkyl amine, in addition to naming it as an alkanamine:

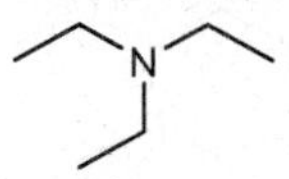

triethylamine
or
N,N-diethylethanamine

(e) This compound is an amine that has only one alkyl group connected to the nitrogen atom. Since this alkyl group is complex, we must name the compound as an alkanamine (rather than an alkyl amine). The parent is a six-membered ring (thus, cyclohexanamine), and there is an isopropyl group located at C3. There are two chiral centers, and the configuration of each is listed at the beginning of the name:

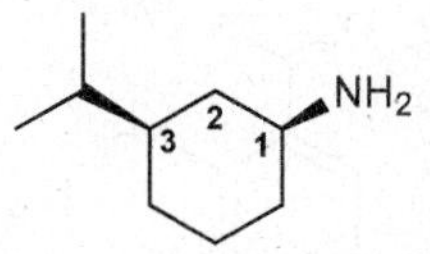

(1*S*,3*R*)-3-isopropylcyclohexanamine

(f) This compound has two functional groups (an OH group and an NH_2 group). The OH group takes priority, so the compound is named as an alcohol (cyclohexanol), with the amino group listed as a substituent, located at C3. There are two chiral centers, and the configuration of each is listed at the beginning of the name:

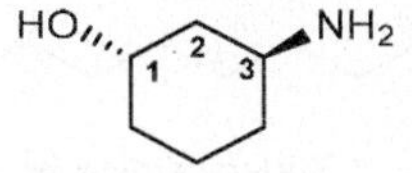

(1*S*,3*S*)-3-aminocyclohexanol

22.2.

(a) The name indicates a dialkyl amine, in which both alkyl groups are simple groups (a cyclohexyl group and a methyl group):

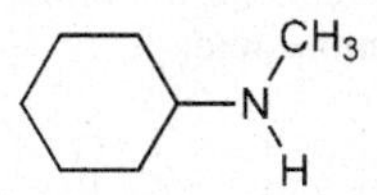

(b) The name indicates a trialkyl amine, in which all three alkyl groups are cyclobutyl groups:

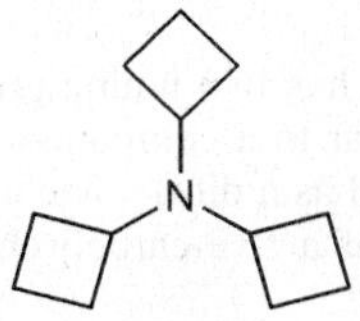

(c) The parent is aniline (or aminobenzene), and there are two ethyl groups (one at C2 and the other at C4):

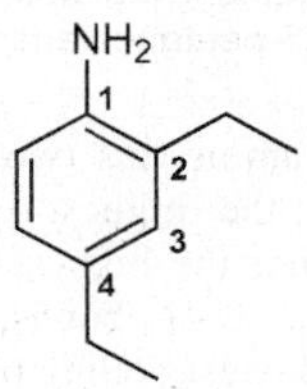

(d) The parent is a six-membered ring that bears an amino group (thus, cyclohexanamine). There is a methyl group at C2, and the configuration of each chiral center (C1 and C2) is indicated in the name:

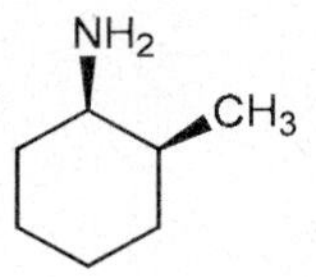

(e) The parent is benzaldehyde, and there is an amino substituent in the *ortho* position.

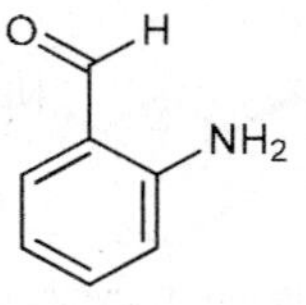

22.3.
(a) This compound is an amine that has only one simple alkyl group (a 13-carbon chain is a tridecyl group) connected to the nitrogen atom, so we can name this compound as an alkyl amine or an alkanamine. Therefore, this compound can be named tridecylamine or 1-tridecanamine.

Tridecylamine *or* 1-tridecanamine

(b) The parent is a four-carbon carboxylic acid (butanoic acid). The NH_2 group is called an amino group. The carboxylic acid functionality has a higher suffix priority than amines, therefore, this compound is named 4-aminobutanoic acid.

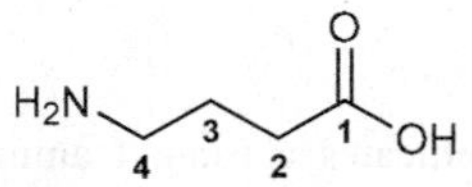

4-Aminobutanoic acid

(c) This compound has two amine groups and is named as a diamine (similar to a compound with two hydroxyl groups being named as a diol). The amino groups are at positions 1 and 5 of a five-carbon chain so it is named pentane-1,5-diamine.

Pentane-1,5-diamine or 1,5-pentanediamine

22.4. The primary amine has two N-H bonds and is expected to exhibit the highest extent of hydrogen bonding, and therefore, the highest boiling point. The tertiary amine lacks N-H bonds, and is therefore expected to have the lowest boiling point.

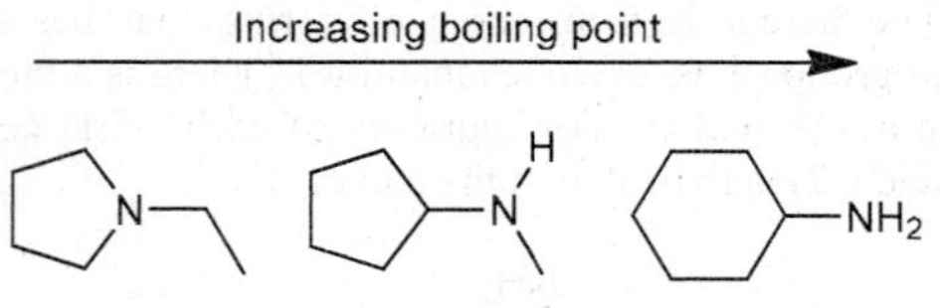

Increasing boiling point

22.5.
(a) This amine has more than five carbon atoms per amino group (there are eight carbon atoms and only one amino group). Therefore, this compound is not expected to be water soluble.

(b) This amine has fewer than five carbon atoms per amino group (there are only three carbon atoms and one amino group). Therefore, this compound is expected to be water soluble.

(c) This diamine has fewer than five carbon atoms per amino group (there are six carbon atoms and two amino groups). Therefore, this compound is expected to be water soluble.

22.6.
(a) The following compound is expected to be a stronger base because the lone pair is localized, and therefore more available to function as a base.

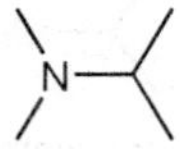

The other compound exhibits a delocalized lone pair, and is therefore a weaker base.

(b) The following compound is expected to be a stronger base because the lone pair is not participating in aromatic resonance (recall that 8 π electrons delocalized in a ring describes an unstable, *antiaromatic* system). It is available to function as a base.

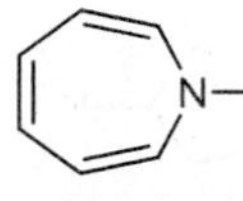

In contrast, the other compound is aromatic, and the lone pair is delocalized (in order to establish aromaticity), so it is much less available to function as base.

(c) The following compound is expected to be a stronger base because the nitrogen atom has a lone pair that is localized, and therefore more available to function as a base.

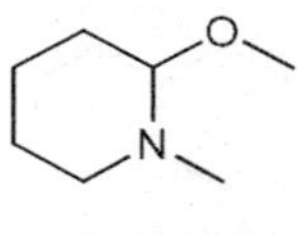

The other compound exhibits a nitrogen atom with a delocalized lone pair, and is therefore a weaker base.

(d) The following compound is expected to be a stronger base because the lone pair is not participating in aromaticity. The lone pair occupies an sp^2-hybridized orbital (directed away from the ring, in the plane of the ring) and is available to function as a base.

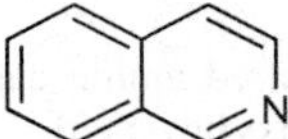

In contrast, the other compound (shown in the problem statement) exhibits a nitrogen atom with a lone pair that is highly delocalized (in order to establish aromaticity in the five-membered ring), so it is much less available to function as base.

22.7. In all of these compounds, the lone pair (on the nitrogen atom) is delocalized throughout two aromatic rings:

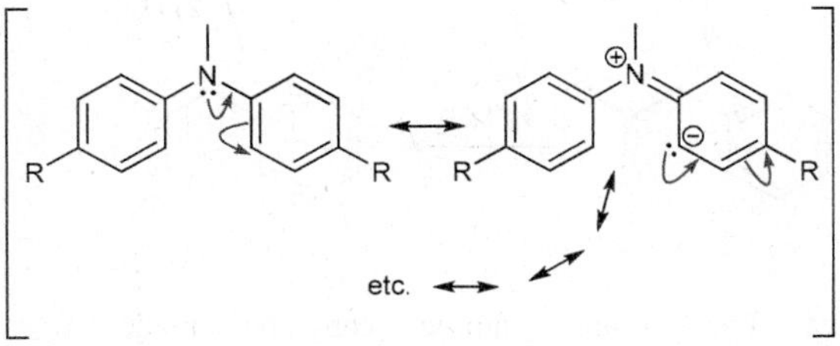

One of the three compounds (shown in the problem statement) has two methyl groups (electron-donating), which destabilize the delocalized lone pair:

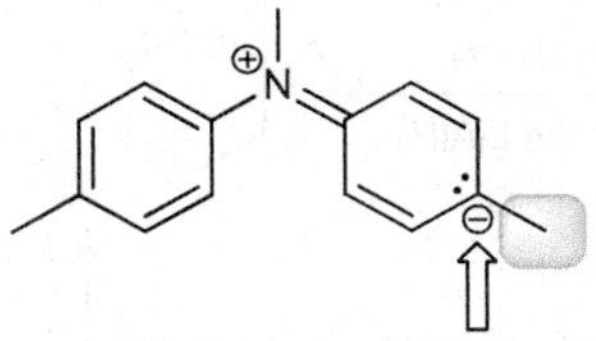

Charge is destabilized by methyl group

This compound is the strongest base, because the delocalization effect is diminished by the effect of the alkyl groups.
In contrast, the following compound has an aldehyde group. This group is electron-withdrawing, and a resonance structure can be drawn in which the nitrogen atom bears a positive charge, and the oxygen atom bears a negative charge:

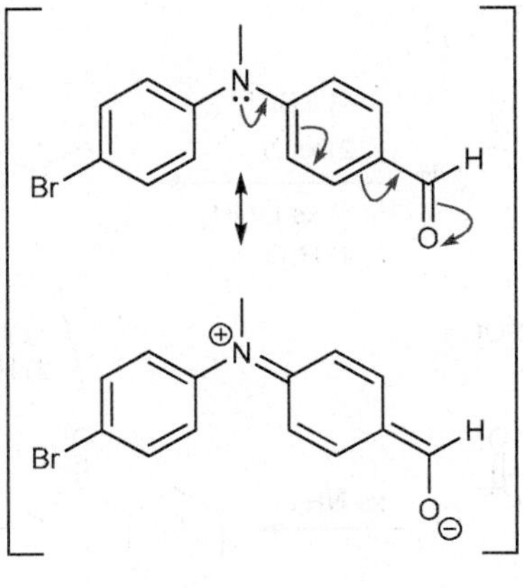

This resonance contributor is significant, and it renders the lone pair highly delocalized, and therefore a very poor base. In summary, we predict the following order of base strength:

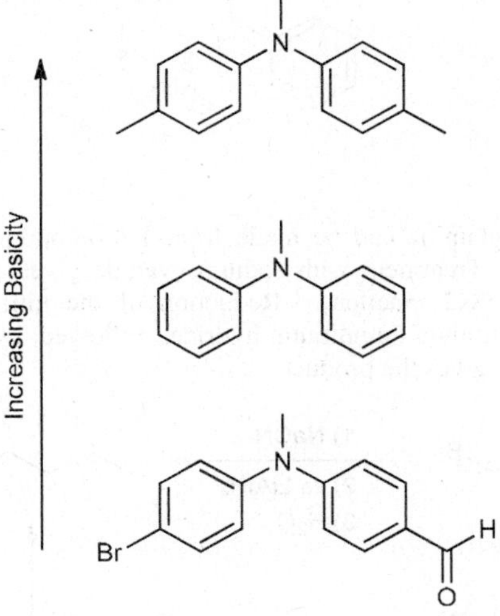

22.8. In the reactant, the lone pair of the amino group is delocalized via resonance. In the product, the lone pair of the amino group is localized, and is therefore more available to function as a base.

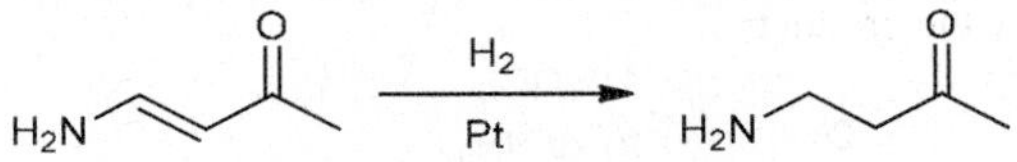

22.9.
(a) At physiological pH, the amino group exists primarily as a charged ammonium ion:

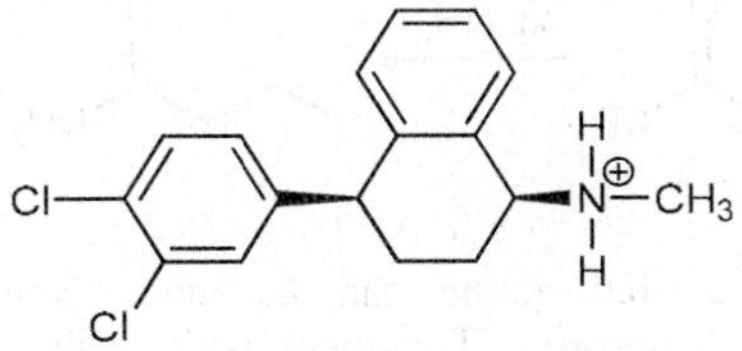

(b) At physiological pH, the amino group exists primarily as a charged ammonium ion:

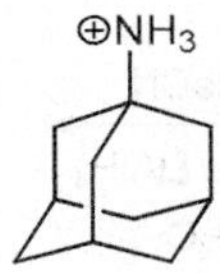

(c) At physiological pH, the amino group exists primarily as a charged ammonium ion:

OH

$\overset{\oplus}{N}H_3$

22.10.

(a) Butylamine can be made from 1-bromopropane, as shown. Treatment with sodium cyanide gives a nitrile (via an S_N2 reaction). Reduction of the nitrile with excess lithium aluminum hydride, followed by water workup, gives the product:

Br → 1) NaCN 2) xs $LiAlH_4$ 3) H_2O → NH_2

NaCN → CN → 1) xs $LiAlH_4$ 2) H_2O

Alternatively, butylamine can be made from butanoic acid, as shown. Treatment with thionyl chloride gives an acid chloride, which can be treated with excess ammonia to give an amide. The amide is then reduced with excess lithium aluminum hydride, followed by water workup, to give the product:

O, OH → 1) $SOCl_2$ 2) xs NH_3 3) xs $LiAlH_4$ 4) H_2O → NH_2

$SOCl_2$

1) xs $LiAlH_4$ 2) H_2O

O, Cl → xs NH_3 → O, NH_2

(b) The desired amine can be made from benzyl bromide, as shown. Treatment with sodium cyanide gives a nitrile (via an S_N2 reaction). Reduction of the nitrile with excess lithium aluminum hydride, followed by water workup, gives the product:

Br → 1) NaCN 2) xs $LiAlH_4$ 3) H_2O → NH_2

NaCN → CN → 1) xs $LiAlH_4$ 2) H_2O

Alternatively, the desired amine can be made from the corresponding carboxylic acid, as shown below. Treatment with thionyl chloride gives an acid chloride, which can then be treated with excess ammonia to give an amide. The amide is then reduced with excess lithium aluminum hydride, followed by water workup, to give the product:

OH, O → 1) $SOCl_2$ 2) xs NH_3 3) xs $LiAlH_4$ 4) H_2O → NH_2

$SOCl_2$

1) xs $LiAlH_4$ 2) H_2O

Cl, O → xs NH_3 → NH_2, O

(c) The desired amine can be made from bromocyclohexane, as shown. Treatment with sodium cyanide gives a nitrile (via an S_N2 reaction). Reduction of the nitrile with excess lithium aluminum hydride, followed by water workup, gives the product:

Br → 1) NaCN 2) xs $LiAlH_4$ 3) H_2O → NH_2

NaCN → CN → 1) xs $LiAlH_4$ 2) H_2O

Alternatively, the desired amine can be made from the corresponding carboxylic acid, as shown below. Treatment with thionyl chloride gives an acid chloride, which can then be treated with excess ammonia to give an amide. The amide is then reduced with excess lithium aluminum hydride, followed by water workup, to give the product:

O, OH → 1) $SOCl_2$ 2) xs NH_3 3) xs $LiAlH_4$ 4) H_2O → NH_2

$SOCl_2$

1) xs $LiAlH_4$ 2) H_2O

O, Cl → xs NH_3 → O, NH_2

22.11. This compound cannot be prepared from an alkyl halide or a carboxylic acid, using the methods described in this section, because both methods produce an amine with two alpha protons:

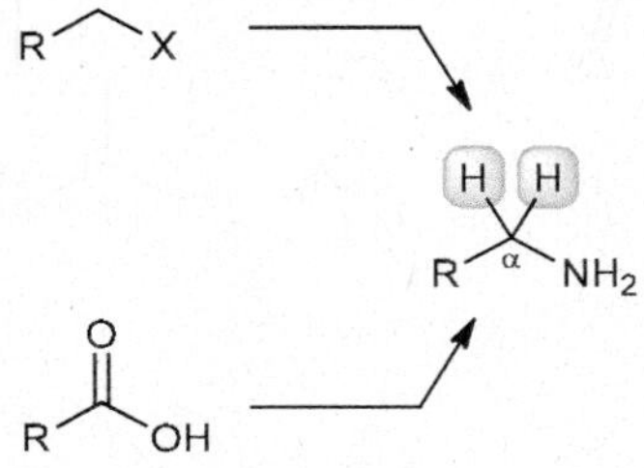

The desired product has two methyl groups at the alpha position:

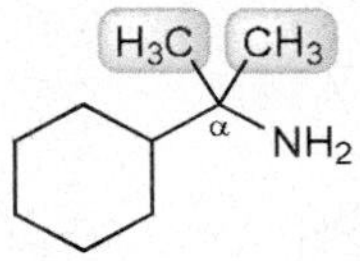

So this product cannot be made with either of the synthetic methods above.

22.12.

(a) We begin by identifying an alkyl halide that can serve as a precursor:

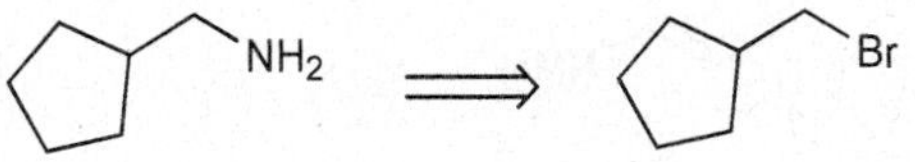

In the Gabriel synthesis, phthalimide is the starting material, and three steps are required. In the first step, phthalimide is deprotonated by hydroxide to give potassium phthalimide, which can serve as a nucleophile and attack the alkyl halide above in an S_N2 process. Subsequent treatment with hydrazine (or aqueous acid) releases the desired amine:

1) KOH
2)
3) H_2NNH_2

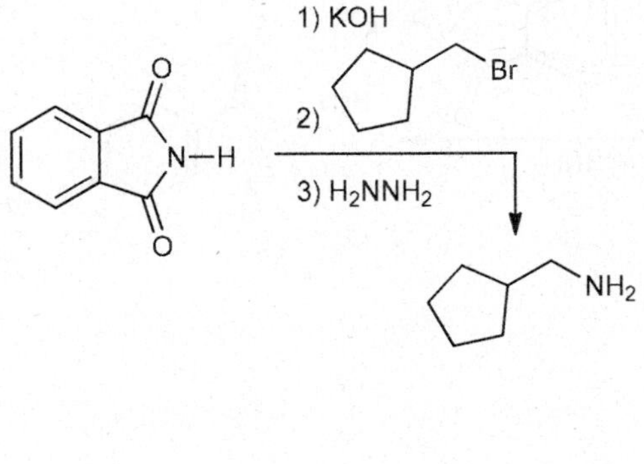

(b) We begin by identifying a halide that can serve as a precursor:

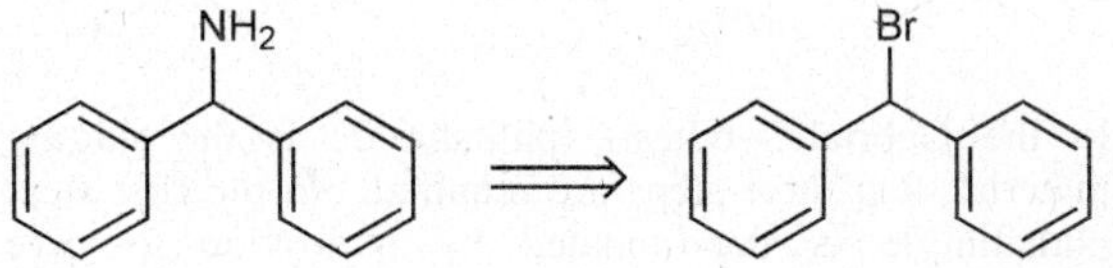

In the Gabriel synthesis, phthalimide is the starting material, and three steps are required. In the first step, phthalimide is deprotonated by hydroxide to give potassium phthalimide, which can serve as a nucleophile and attack the halide above in an S_N2 process. Subsequent treatment with hydrazine (or aqueous acid) releases the desired amine:

1) KOH
2)
3) H_2NNH_2

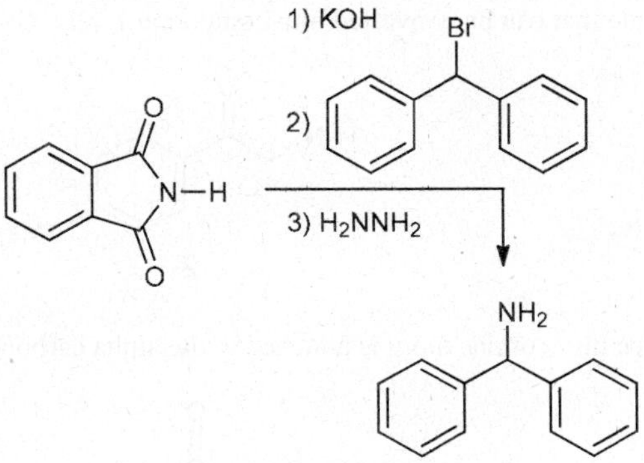

(c) We begin by identifying an alkyl halide that can serve as a precursor:

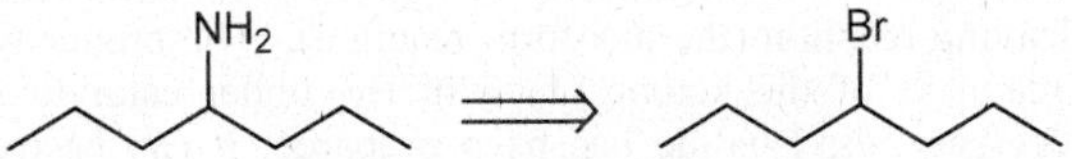

In the Gabriel synthesis, phthalimide is the starting material, and three steps are required. In the first step, phthalimide is deprotonated by hydroxide to give potassium phthalimide, which can serve as a nucleophile and attack the alkyl halide above in an S_N2 process. Subsequent treatment with hydrazine (or aqueous acid) releases the desired amine:

1) KOH
2)
3) H_2NNH_2

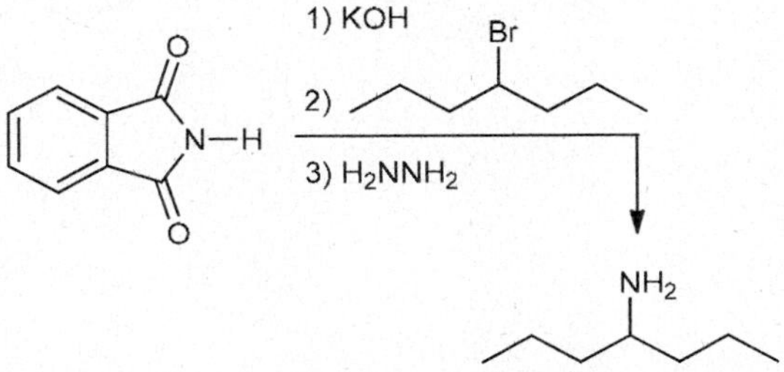

(d) We begin by identifying an alkyl halide that can serve as a precursor:

NH2 ⟹ Br

In the Gabriel synthesis, phthalimide is the starting material, and three steps are required. In the first step, phthalimide is deprotonated by hydroxide to give potassium phthalimide, which can serve as a nucleophile and attack the alkyl halide above in an SN2 process. Subsequent treatment with hydrazine (or aqueous acid) releases the desired amine:

1) KOH
2) Br
3) H_2NNH_2

N–H

NH2

22.13. The problem statement dictates that a Gabriel synthesis be employed, so we begin by identifying a suitable alkyl halide that can be converted into compound **2** via a Gabriel synthesis.

H_2N ... NMe_2 **2** ⟹ Br ... NMe_2

Since the bromine atom is attached to the alpha carbon, this halide can be installed by an alpha-halogenation of **1**.

Br ... NMe_2 ⟹ NMe_2 **1**

Enolization of **1** under basic conditions would lead to tribromination of the alpha carbon, followed by a C–C bond cleaving reaction (the haloform reaction). The proper way to mono-brominate **1** involves the use of acidic conditions. Treatment of the ketone (**1**) with Br_2 under catalytic acidic conditions affords the alpha bromo ketone. Once the necessary alkyl halide has been prepared, it can be treated with potassium phthalimide, followed by hydrazine (or aqueous acid), to give the desired amine as shown:

1 NMe_2 —$[H_3O^+]$, Br_2→ Br ... NMe_2 —1) $N^{\ominus}$ $K^{\oplus}$ 2) H_2NNH_2→ H_2N ... **2** NMe_2

22.14.

(a) The compound has two C–N bonds:

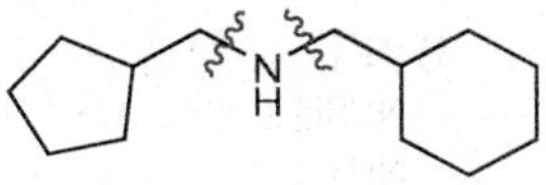

Each of these bonds can be made via a reductive amination, giving two possible synthetic routes, shown here:

[H⁺], NaBH₃CN

NH₂

[H⁺], NaBH₃CN

NH₂

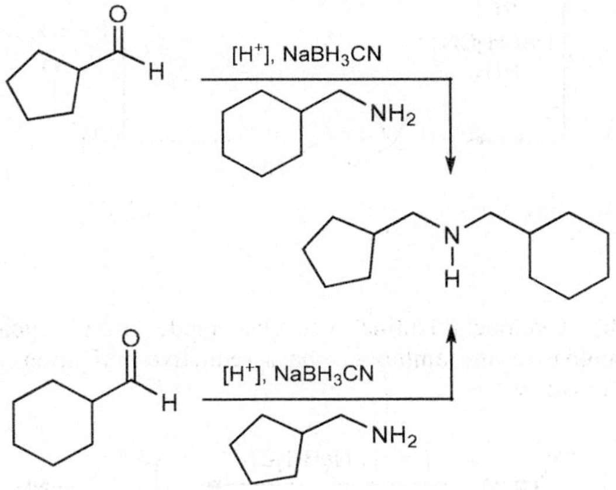

(b) The compound has three C-N bonds:

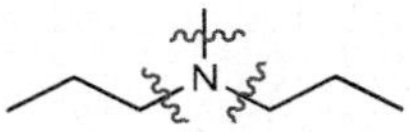

Each of these bonds can be made via a reductive amination. However, two of them are equivalent (because of symmetry), giving two possible synthetic routes, shown here:

[H⁺], NaBH₃CN

[H⁺], NaBH₃CN

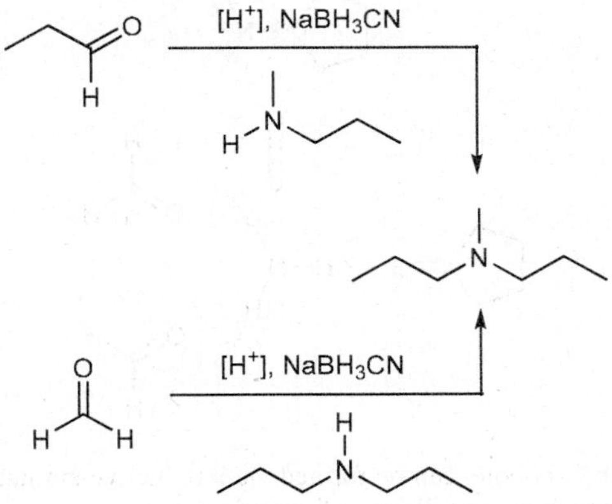

(c) The compound has two C-N bonds:

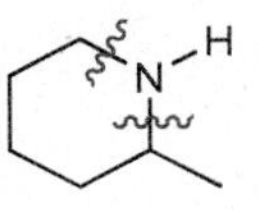

Each of these bonds can be made via a reductive amination, giving two possible synthetic routes, shown here:

[H⁺], NaBH₃CN

[H⁺], NaBH₃CN

NH₂

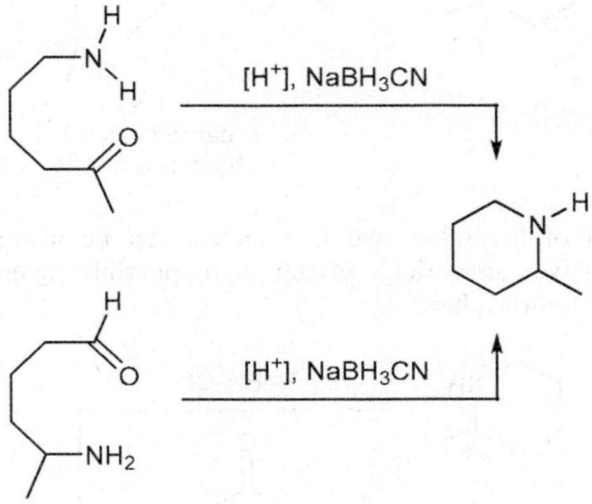

(d) The compound has three C-N bonds:

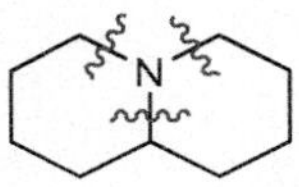

Each of these bonds can be made via a reductive amination. However, two of them are equivalent (because of symmetry), giving two possible synthetic routes, shown here:

[H⁺], NaBH₃CN

HN

[H⁺], NaBH₃CN

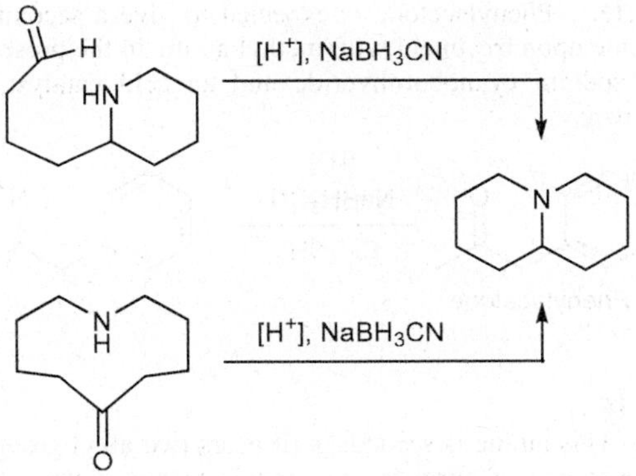

(e) The compound has three C-N bonds:

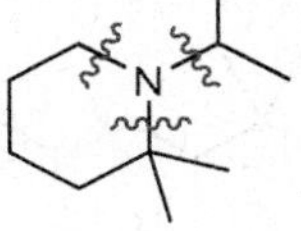

However, one of these bonds cannot be made via a reductive amination, because the starting material cannot have a pentavalent carbon atom:

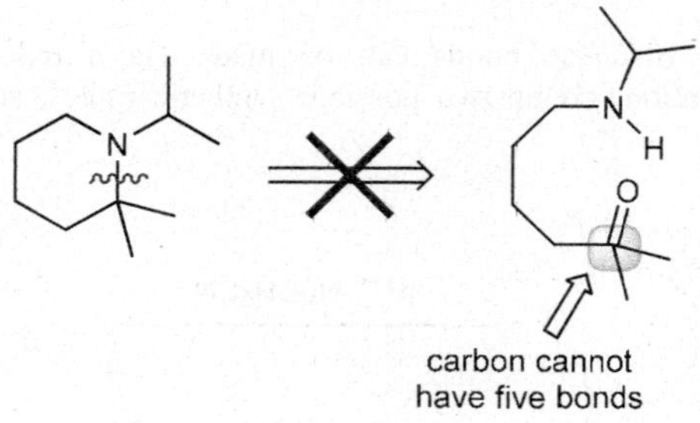

carbon cannot have five bonds

Each of the other two C-N bonds can be made via reductive amination, giving two possible synthetic routes, shown here:

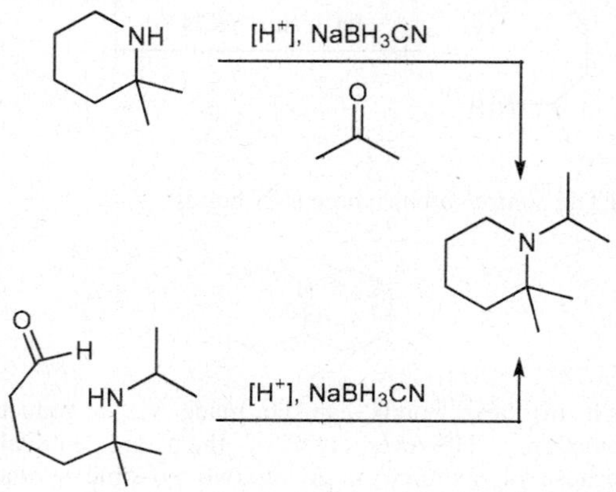

[H⁺], NaBH$_3$CN

[H⁺], NaBH$_3$CN

22.15. Phenylacetone is expected to give a secondary amine upon treatment with methyl amine in the presence of sodium cyanoborohydride and an acid catalyst, as shown:

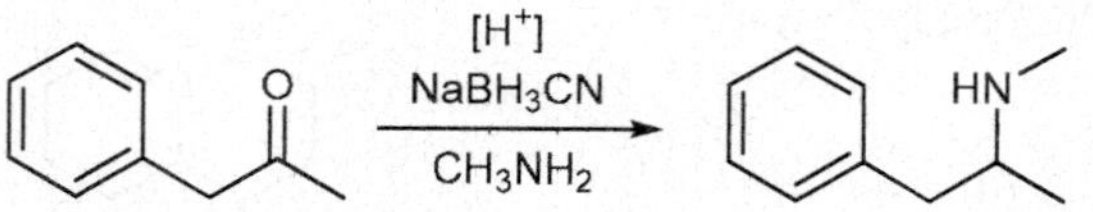

[H⁺] NaBH$_3$CN CH$_3$NH$_2$

Phenylacetone

22.16.

(a) This amine is secondary (it bears two alkyl groups). The source of nitrogen is ammonia, which dictates that each group must be installed via a reductive amination process. The following retrosynthetic analysis reveals the necessary starting materials:

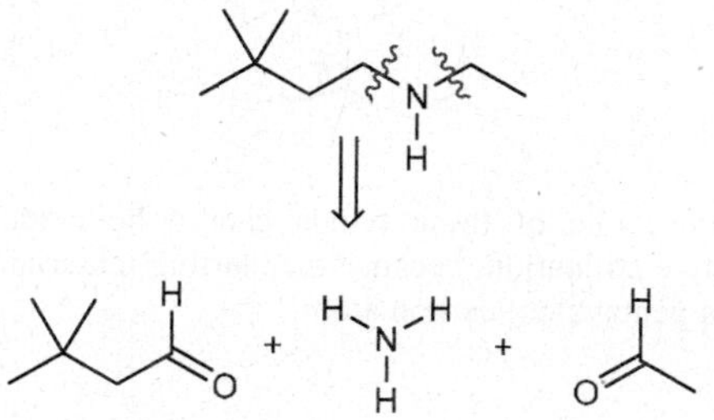

Each C-N bond can be formed via a reductive amination, as shown in the following forward scheme. Note that the two reductive amination reactions can be done in either order.

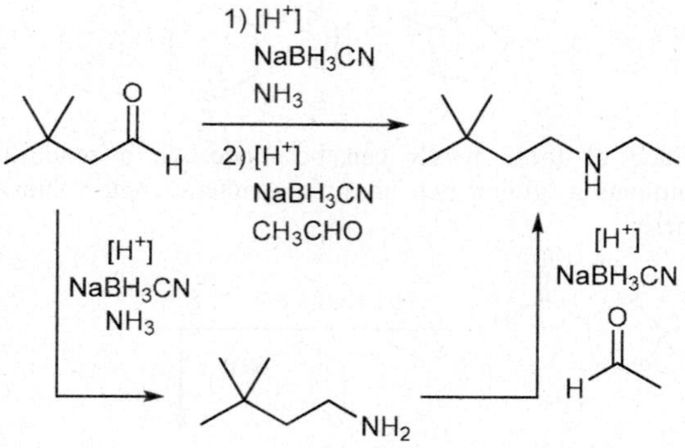

1) [H⁺] NaBH$_3$CN NH$_3$
2) [H⁺] NaBH$_3$CN CH$_3$CHO

[H⁺] NaBH$_3$CN NH$_3$

[H⁺] NaBH$_3$CN

(b) Cyclopentylamine can be made from cyclopentanone and ammonia, via a reductive amination, as shown:

[H⁺], NaBH$_3$CN NH$_3$

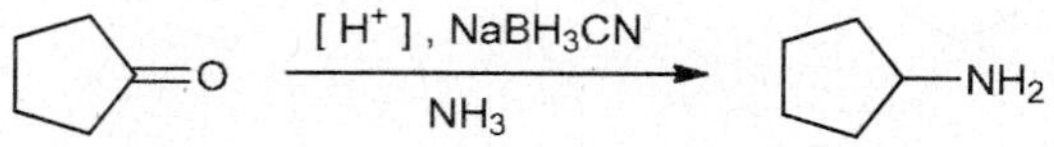

(c) This amine is tertiary (it bears three alkyl groups). The source of nitrogen is ammonia, which dictates that each group must be installed via a reductive amination process. The following retrosynthetic analysis reveals the necessary starting materials:

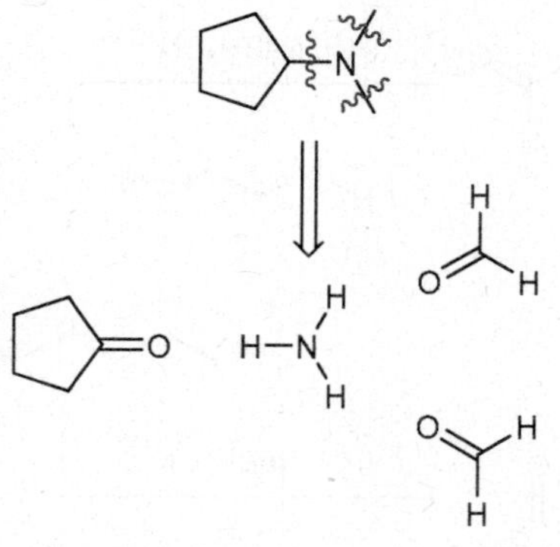

Each C-N bond can be formed via a reductive amination, as shown in the following forward scheme:

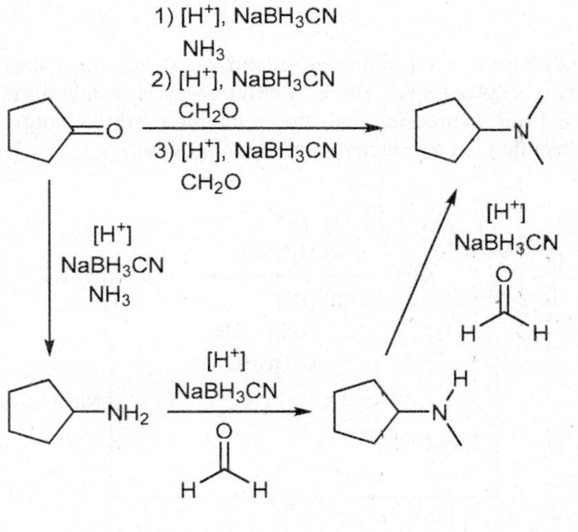

1) $[H^+]$, $NaBH_3CN$, NH_3
2) $[H^+]$, $NaBH_3CN$, CH_2O
3) $[H^+]$, $NaBH_3CN$, CH_2O

$[H^+]$, $NaBH_3CN$, NH_3

$[H^+]$, $NaBH_3CN$

$[H^+]$, $NaBH_3CN$

(d) This amine is secondary (it bears two ethyl groups). The source of nitrogen is ammonia, which dictates that each group must be installed via a reductive amination process. The following retrosynthetic analysis reveals the necessary starting materials:

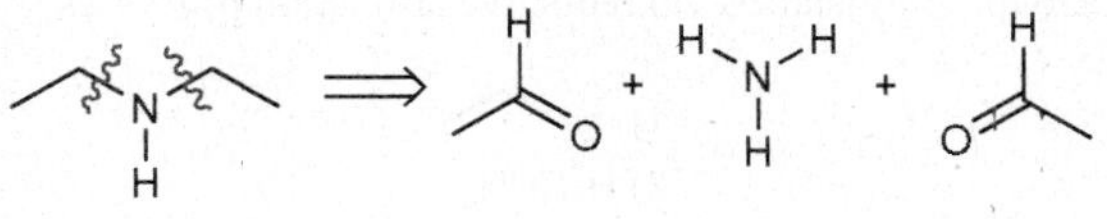

Each C-N bond can be formed via a reductive amination, as shown in the following forward scheme:

1) $[H^+]$, $NaBH_3CN$, NH_3
2) $[H^+]$, $NaBH_3CN$, CH_3CHO

$[H^+]$, $NaBH_3CN$, NH_3

$[H^+]$, $NaBH_3CN$

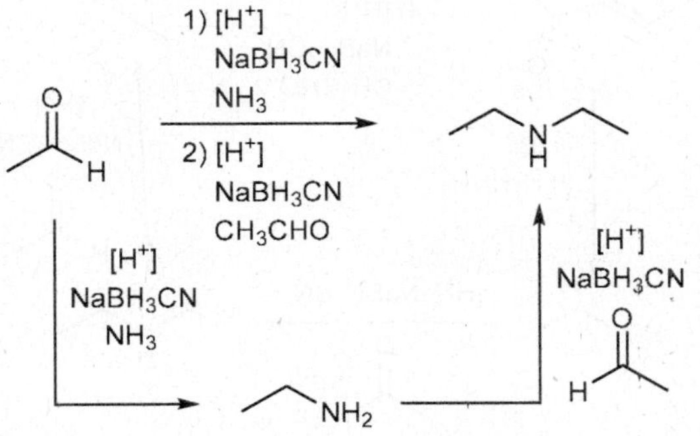

(e) This amine is tertiary (it bears three ethyl groups). The source of nitrogen is ammonia, which dictates that each group must be installed via a reductive amination process. The following retrosynthetic analysis reveals the necessary starting materials:

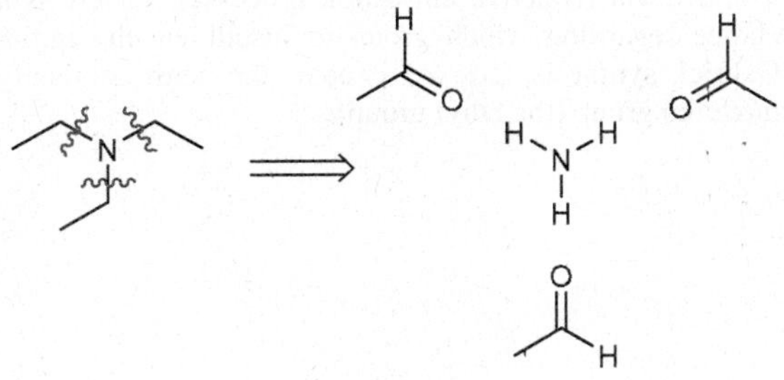

Each C–N bond can be formed via a reductive amination, as shown in the following forward scheme:

1) $[H^+]$, $NaBH_3CN$, NH_3
2) $[H^+]$, $NaBH_3CN$, CH_3CHO
3) $[H^+]$, $NaBH_3CN$, CH_3CHO

$[H^+]$, $NaBH_3CN$, NH_3

$[H^+]$, $NaBH_3CN$

$[H^+]$, $NaBH_3CN$

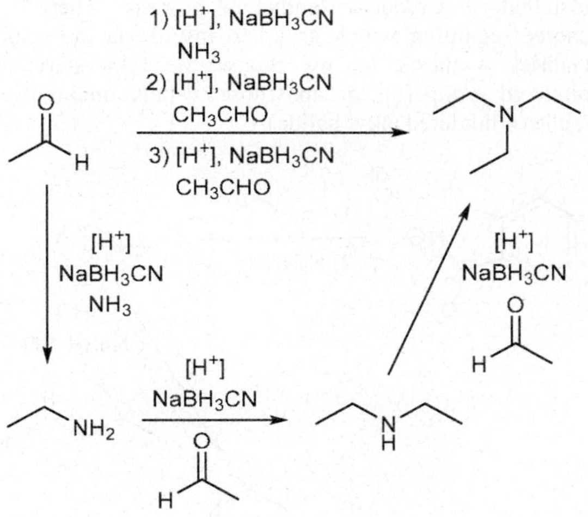

(f) This amine is tertiary (it bears three alkyl groups). The source of nitrogen is ammonia, which dictates that each group must be installed via a reductive amination process. The following retrosynthetic analysis reveals the necessary starting materials:

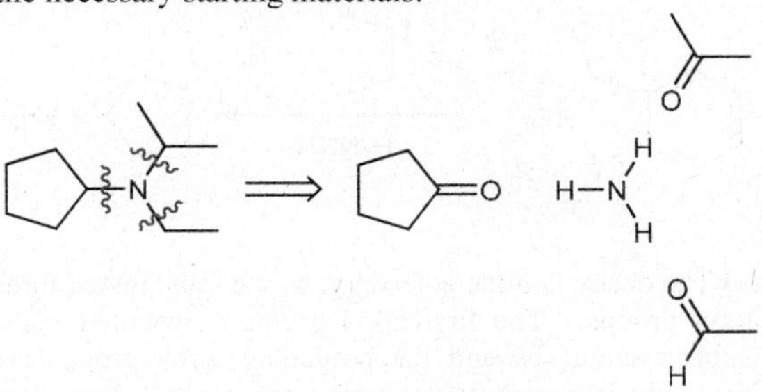

Each C-N bond can be formed via a reductive amination, as shown in the following forward scheme:

1) $[H^+]$, $NaBH_3CN$, NH_3
2) $[H^+]$, $NaBH_3CN$, CH_3CHO
3) $[H^+]$, $NaBH_3CN$, $(CH_3)_2CO$

$[H^+]$, $NaBH_3CN$, NH_3

$[H^+]$, $NaBH_3CN$

$[H^+]$, $NaBH_3CN$

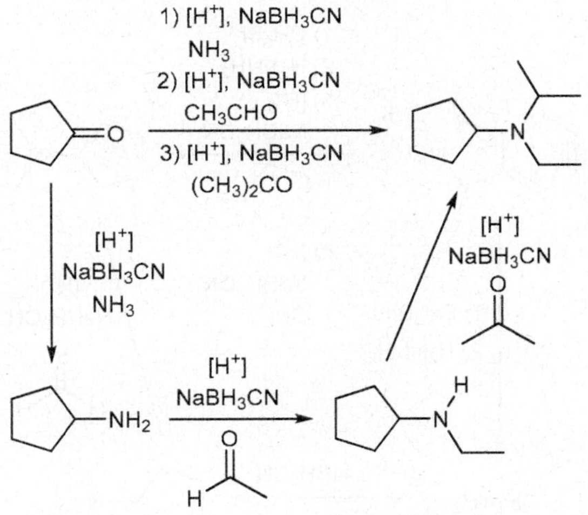

22.17.

(a) The desired amine is secondary, so we must install two alkyl groups. The first alkyl group is installed via a Gabriel synthesis, and the remaining alkyl group is installed via a reductive amination process. There is a choice regarding which group to install via the initial Gabriel synthesis, so we choose the least sterically hindered group (the group whose installation involves the least hindered alkyl halide):

1) Br
2) H_2NNH_2
H_2N
$[H^+]$
$NaBH_3CN$
O
H
N
H

(b) The desired amine is primary, so we only need to install one alkyl group, which can be achieved with a Gabriel synthesis, as shown:

1) Br
2) H_2NNH_2
NH_2

(c) The desired amine is tertiary, so we must install three alkyl groups. The first alkyl group is installed via a Gabriel synthesis, and the remaining alkyl groups are installed via reductive amination processes. There is a choice regarding which group to install via the initial Gabriel synthesis, so we choose the least sterically hindered group (one of the methyl groups):

1) CH_3Br
2) H_2NNH_2
3) $[H^+]$
$NaBH_3CN$
O
4) $[H^+]$
$NaBH_3CN$
CH_2O
N
1) CH_3Br
2) H_2NNH_2
$[H^+]$
$NaBH_3CN$
H H
CH_3NH_2
$[H^+]$, $NaBH_3CN$
O
H
N

(d) The desired amine is secondary, so we must install two alkyl groups. The first ethyl group is installed via a Gabriel synthesis, and the remaining ethyl group is installed via a reductive amination process:

1) Br
2) H_2NNH_2
3) $[H^+]$
$NaBH_3CN$
CH_3CHO
N
H
1) Br
2) H_2NNH_2
H_2N
$[H^+]$
$NaBH_3CN$
O
H

(e) The desired amine is tertiary, so we must install three alkyl groups (all ethyl groups). The first ethyl group is installed via a Gabriel synthesis, and the remaining ethyl groups are installed via reductive amination processes.

1) Br
2) H_2NNH_2
3) $[H^+]$
$NaBH_3CN$
CH_3CHO
4) $[H^+]$
$NaBH_3CN$
CH_3CHO
N
1) Br
2) H_2NNH_2
$[H^+]$
$NaBH_3CN$
O
H
H_2N
$[H^+]$, $NaBH_3CN$
O
H
N
H

(f) The desired amine is tertiary, so we must install three alkyl groups. The first alkyl group is installed via a Gabriel synthesis, and the remaining alkyl groups are installed via reductive amination processes. There is a choice regarding which group to install via the initial Gabriel synthesis, so we choose the least sterically hindered group (the ethyl group):

1) Br⌒ 2) H_2NNH_2 3) [H^+] $NaBH_3CN$ (cyclopentanone) 4) [H^+] $NaBH_3CN$ $(CH_3)_2CO$

1) Br⌒ 2) H_2NNH_2

[H^+] $NaBH_3CN$ (acetone)

H_2N⌒ — [H^+], $NaBH_3CN$ (cyclopentanone)

22.18.

(a) The desired amine is secondary, so we must install two alkyl groups. The first alkyl group is installed via an azide synthesis, and the remaining alkyl group is installed via a reductive amination process. There is a choice regarding which group to install via the initial azide synthesis, so we choose the least sterically hindered group (the group whose installation involves the least hindered alkyl halide):

Br⌒ — 1) NaN_3 2) H_2, Pt 3) [H^+], $NaBH_3CN$ $(CH_3)_3CCH_2CHO$

1) NaN_3 2) H_2, Pt

H_2N⌒ — [H^+], $NaBH_3CN$

(b) The desired amine is primary, so we only need to install one alkyl group, which can be achieved via an azide synthesis, as shown:

Br (cyclopentyl) — 1) NaN_3 2) H_2, Pt → NH_2

(c) The desired amine is tertiary, so we must install three alkyl groups. The first alkyl group is installed via an azide synthesis, and the remaining alkyl groups are installed via reductive amination processes. There is a choice regarding which group to install via the initial azide synthesis, so we choose the least sterically hindered group (one of the methyl groups):

CH_3Br — 1) NaN_3 2) H_2, Pt 3) [H^+], $NaBH_3CN$ (cyclopentanone) 4) [H^+], $NaBH_3CN$ CH_2O

1) NaN_3 2) H_2, Pt

[H^+] $NaBH_3CN$ (formaldehyde)

CH_3NH_2 — [H^+], $NaBH_3CN$ (cyclopentanone)

(d) The desired amine is secondary, so we must install two alkyl groups. The first ethyl group is installed via an azide synthesis, and the remaining ethyl group is installed via a reductive amination process:

Br⌒ — 1) NaN_3 2) H_2, Pt 3) [H^+], $NaBH_3CN$ CH_3CHO

1) NaN_3 2) H_2, Pt

[H^+] $NaBH_3CN$ (acetaldehyde)

H_2N⌒

(e) The desired amine is tertiary, so we must install three alkyl groups (all ethyl groups). The first ethyl group is installed via an azide synthesis, and the remaining ethyl groups are installed via reductive amination processes.

Br⌒ — 1) NaN_3 2) H_2, Pt 3) [H^+], $NaBH_3CN$ CH_3CHO 4) [H^+], $NaBH_3CN$ CH_3CHO

1) NaN_3 2) H_2, Pt

[H^+] $NaBH_3CN$ (acetaldehyde)

H_2N⌒ — [H^+], $NaBH_3CN$ (acetaldehyde)

(f) The desired amine is tertiary, so we must install three alkyl groups. The first alkyl group is installed via an azide synthesis, and the remaining alkyl groups are installed via reductive amination processes. There is a choice regarding which group to install via the initial azide synthesis, so we choose the least sterically hindered group (the ethyl group):

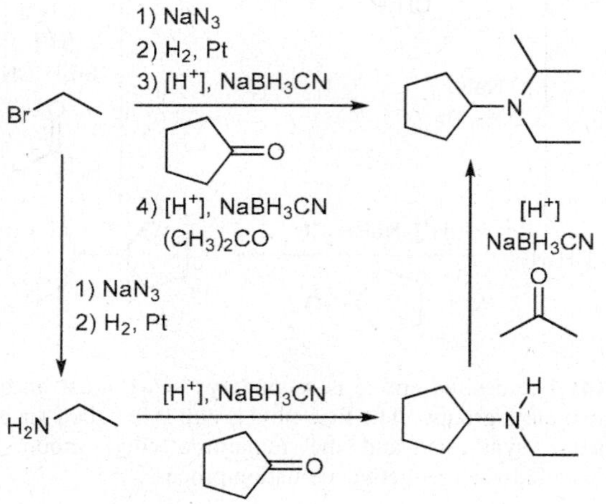

1) NaN_3
2) H_2, Pt
3) $[H^+]$, $NaBH_3CN$

4) $[H^+]$, $NaBH_3CN$
$(CH_3)_2CO$

$[H^+]$
$NaBH_3CN$

1) NaN_3
2) H_2, Pt

$[H^+]$, $NaBH_3CN$

22.19. For the conversion of **1** to **2**, the desired C-N bond can be installed via a reductive amination process, with the appropriate cyclic amine, as shown:

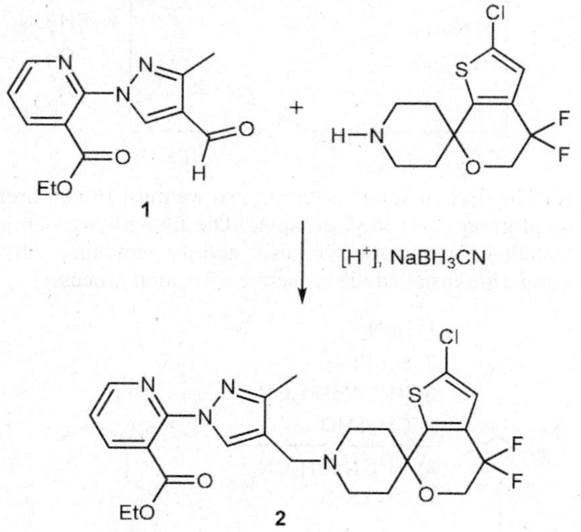

$[H^+]$, $NaBH_3CN$

Then, conversion of **2** to **3** requires reduction of an ester group, which can generally be achieved with excess $LiAlH_4$, followed by acidic workup:

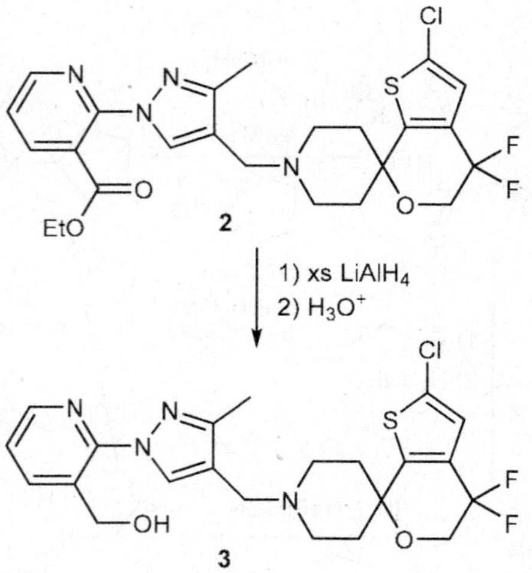

1) xs $LiAlH_4$
2) H_3O^+

22.20. Protection of the amino group (by acylation to give an amide) allows for direct nitration of the ring (in the *para* position). After nitration is complete, the acetyl group can be removed by hydrolysis of the amide in aqueous basic or acidic conditions:

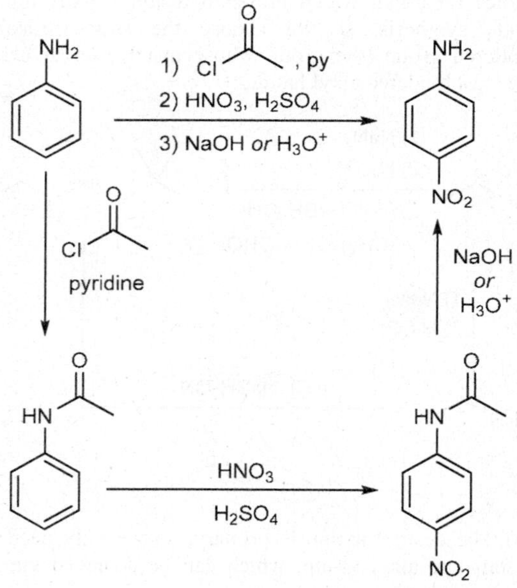

1) acetyl chloride, py
2) HNO_3, H_2SO_4
3) NaOH *or* H_3O^+

pyridine

NaOH *or* H_3O^+

HNO_3
H_2SO_4

22.21. Direct chlorination of nitrobenzene would result in a *meta*-disubstituted product (because the nitro group is *meta*-directing). So we must first reduce the nitro group into an amino group, thereby converting a *meta* director into an *ortho-para* director:

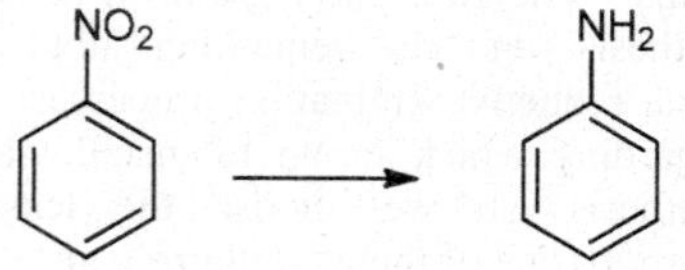

Monochlorination of aniline (in the *para* position) will then give the product. Unfortunately, aniline will not efficiently undergo monochlorination (the ring is too highly activated to install just one chlorine atom). However, the strongly activating effect of the amino group can be temporarily diminished via acylation to give the amide. Then, after monochlorination has been performed, the acetyl group can be removed by amide hydrolysis in aqueous basic or acidic conditions. The entire synthesis is shown here:

1) HCl, Zn
2) Cl(C=O)CH3, pyridine
3) Cl_2
4) NaOH *or* H_3O^+

HCl, Zn
pyridine
Cl_2
NaOH *or* H_3O^+

22.22. An efficient solution derives from the following retrosynthetic analysis. An explanation of each of the steps (*a*-*d*) follows.

a. The product can be made from ethyl amine and acetyl chloride, via a nucleophilic acyl substitution reaction.
b. Ethyl amine can be made from acetamide via reduction with $LiAlH_4$, followed by water workup.
c. Acetamide can be made from acetyl chloride, via a nucleophilic acyl substitution reaction.
d. Acetyl chloride can be made from acetic acid upon treatment with thionyl chloride.

Now let's draw the forward scheme. Acetic acid is treated with thionyl chloride to give acetyl chloride. One equivalent of acetyl chloride is converted into ethylamine (via amide formation, followed by reduction), which is then treated with another equivalent of acetyl chloride to give the product:

1) xs NH_3
2) xs $LiAlH_4$
3) H_2O

$SOCl_2$

(2 equiv.)

22.23.

(a) The starting material is an amine, and the reagents indicate a Hofmann elimination. There are two β positions, but they are equivalent because of symmetry. As such, there is only one possible regiochemical outcome for the elimination process:

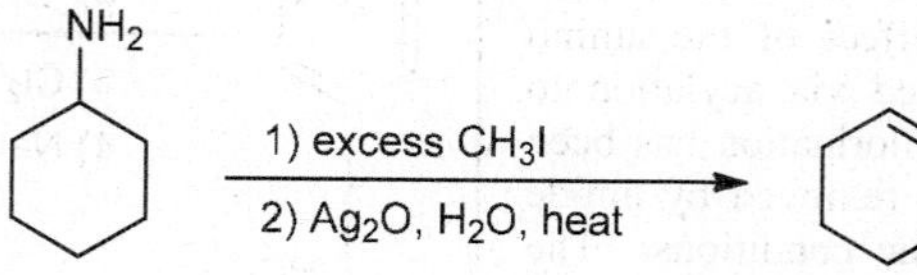

NH_2

1) excess CH_3I

2) Ag_2O, H_2O, heat

(b) The starting material is an amine, and the reagents indicate a Hofmann elimination. There are two β positions, and we expect elimination to occur at the β position that leads to the less substituted alkene:

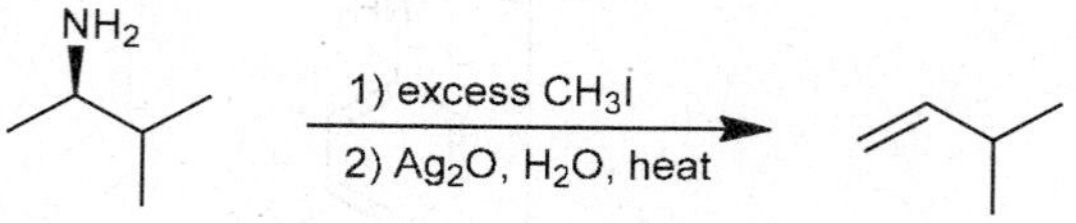

NH_2

1) excess CH_3I

2) Ag_2O, H_2O, heat

(c) The starting material is an amine, and the reagents indicate a Hofmann elimination. There are two β positions, and we expect elimination to occur at the β position that leads to the less substituted alkene:

N

1) excess CH_3I

2) Ag_2O, H_2O, heat

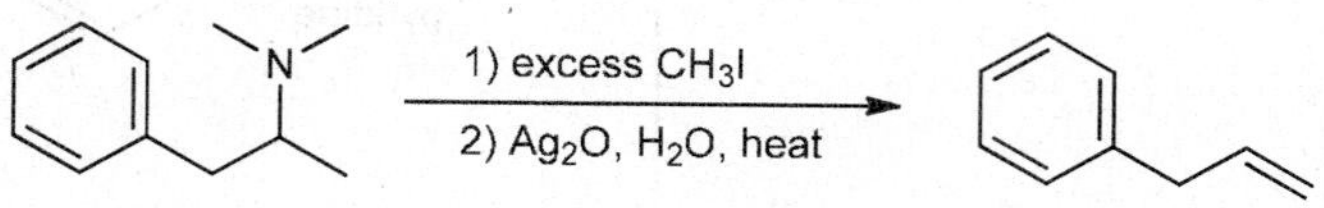

22.24. The third product is perhaps the most revealing. It indicates that the structure of PCP must contain an aromatic ring for which the benzylic position is connected to a nitrogen atom, and the same benzylic position is also part of a cyclohexyl ring. This justifies formation of the first product shown. The second product indicates that the nitrogen atom in PCP must be incorporated in a six-membered ring. This ring is opened during formation of the third product.

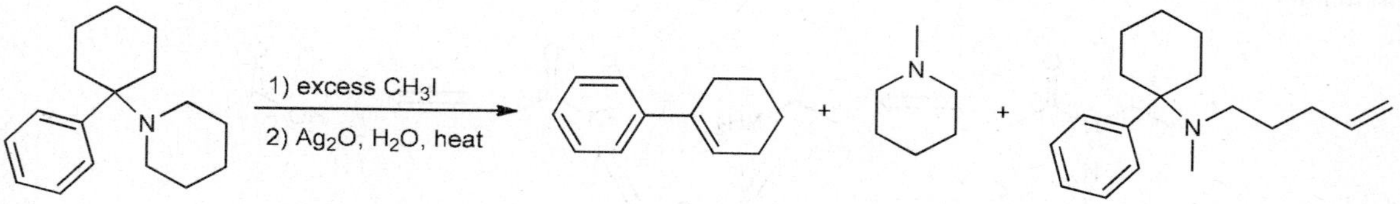

N

1) excess CH_3I

2) Ag_2O, H_2O, heat

N + N +

22.25.

(a) The starting material is a primary amine, so it is converted into the corresponding diazonium salt upon treatment with sodium nitrite and HCl:

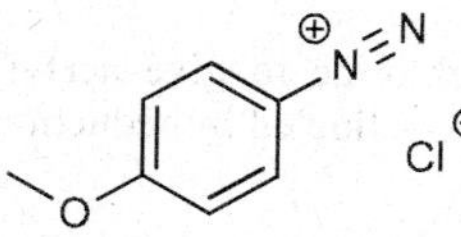

$\oplus$ N≡N

Cl $\ominus$

O

(b) The starting material is a secondary amine, so it is converted into the corresponding *N*-nitrosamine upon treatment with sodium nitrite and HCl:

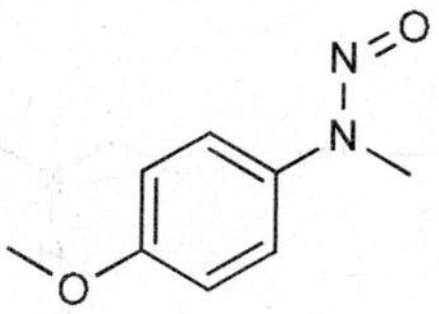

N=O

N

O

(c) The starting material is a secondary amine, so it is converted into the corresponding *N*-nitrosamine upon treatment with sodium nitrite and HCl:

O

N

N

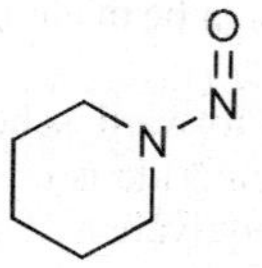

(d) The starting material is a secondary amine, so it is converted into the corresponding *N*-nitrosamine upon treatment with sodium nitrite and HCl:

O

N

N

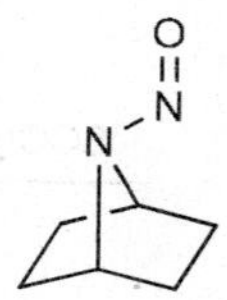

22.26.

(a) The desired transformation requires installation of an isopropyl group in the *para* position, as well as conversion of the NH_2 group into a cyano group. The addition of the alkyl group can be achieved via a Friedel-Crafts alkylation, while the cyano group can come from a Sandmeyer reaction (via a diazonium ion).

Now let's consider the order of events. An amino group is a strong activator, and therefore an *ortho-para* director, while a cyano group is a *meta* director. Therefore, in order to achieve *para*-disubstitution, the isopropyl group must be installed before conversion of the amino group into a cyano group:

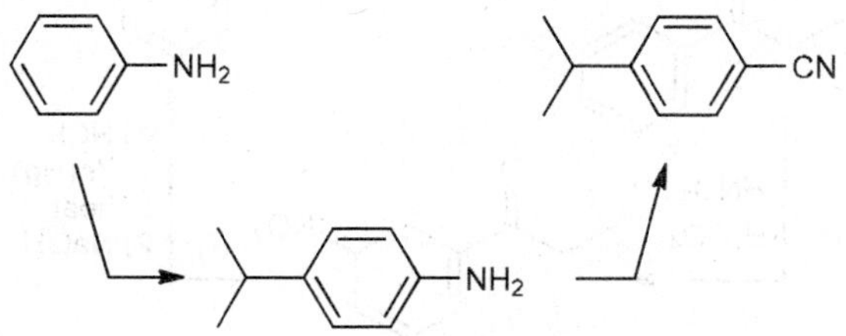

However, this strategy has one flaw. A Friedel-Crafts alkylation requires the use of a Lewis acid ($AlCl_3$), which can interact with the lone pair of the amino group, thereby converting the activating amino group into a deactivating group (see Section 22.8). As such, a Friedel-Crafts alkylation will not work. This issue can be avoided by protecting the amino group first as an amide, thereby reducing the electron-donating ability of the lone pair on the nitrogen atom. The desired Friedel-Crafts alkylation is then performed (installing an isopropyl group), followed by hydrolysis to restore the amino group:

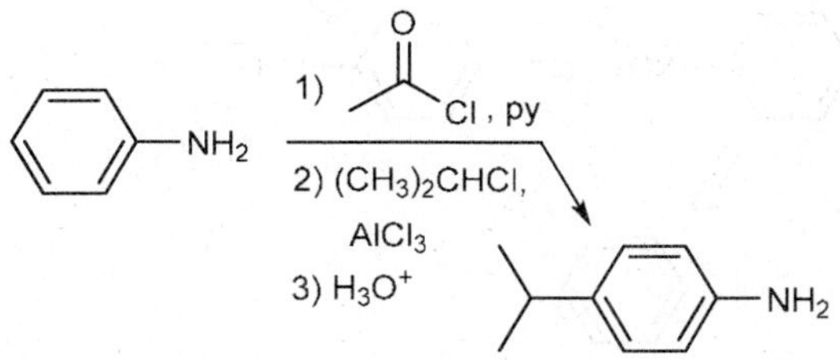

1) acetyl chloride, py
2) $(CH_3)_2CHCl$, $AlCl_3$
3) H_3O^+

And finally, conversion of the amino group into a cyano group is achieved in two steps. First, *para*-isopropylaniline is treated with sodium nitrite and HCl, giving an aromatic diazonium ion, which is then treated with CuCN to give a Sandmeyer reaction that affords the desired product:

1) $NaNO_2$, HCl
2) CuCN

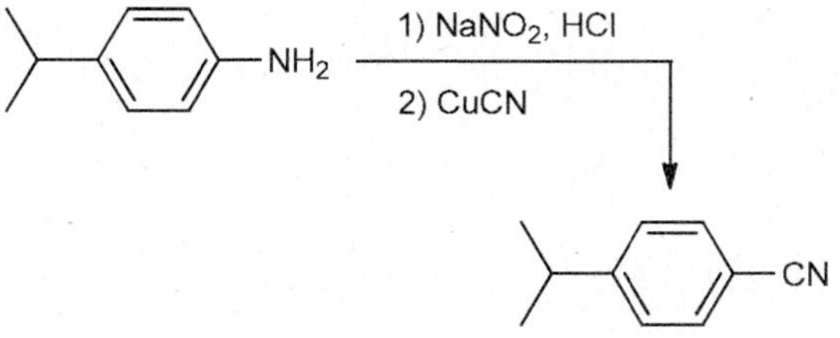

(b) The *meta*-directing effect of the nitro group enables installation of a bromine atom in the correct location (the *meta* position):

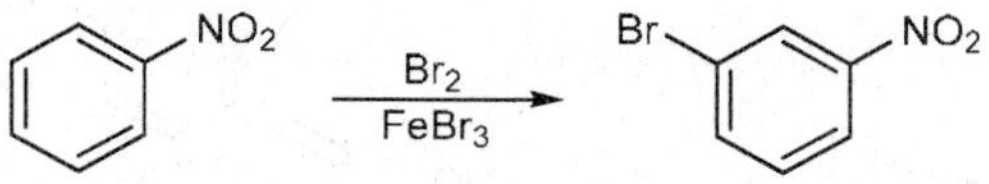

Br_2 / $FeBr_3$

Now we must replace the nitro group with a bromine atom. One method for accomplishing this transformation involves the use of a Sandmeyer reaction, as seen in the following retrosynthetic analysis:

Sandmeyer

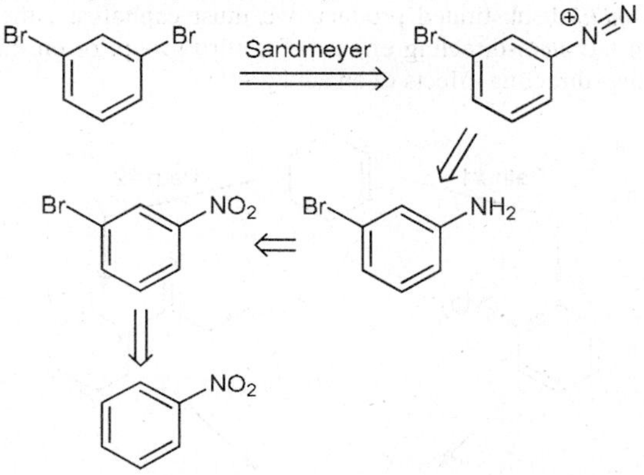

Now let's draw the forward scheme. The starting material is treated with Br_2 and a Lewis acid, thereby installing a bromine atom in the *meta* position. Reduction of the nitro group gives *meta*-bromoaniline, which is then converted into an aromatic diazonium ion upon treatment with sodium nitrite and HCl. The aromatic diazonium ion is then treated with CuBr to give a Sandmeyer reaction that affords the desired product, as shown:

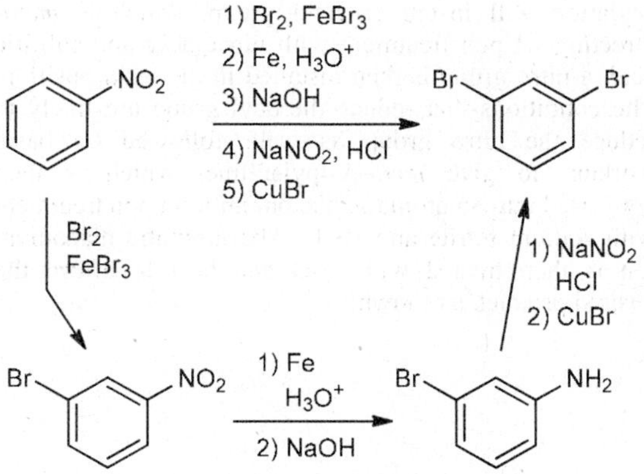

1) Br_2, $FeBr_3$
2) Fe, H_3O^+
3) NaOH
4) $NaNO_2$, HCl
5) CuBr

Br_2 / $FeBr_3$

1) Fe / H_3O^+
2) NaOH

1) $NaNO_2$ / HCl
2) CuBr

(c) The starting material is benzene and the product is disubstituted. Specifically, we must install a propyl group and a hydroxyl group. Installation of a propyl group *without rearrangement* can be achieved with a Friedel-Crafts acylation (followed by reduction). Installation of a hydroxyl group can be achieved via a diazonium ion, as shown in the following retrosynthetic analysis:

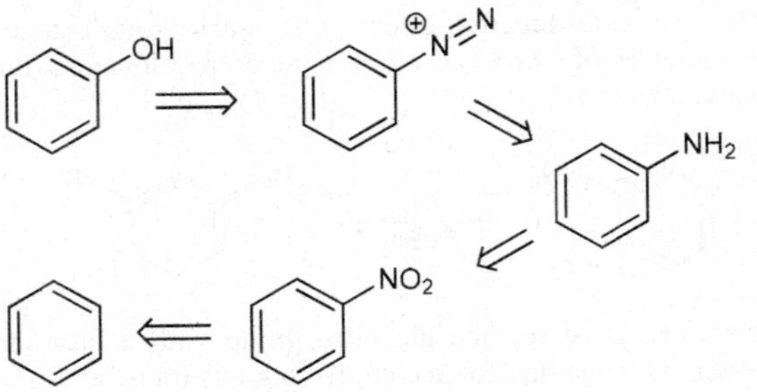
OH
⊕ N≡N
NH2
NO2

Now let's consider the order of events. In order to form a *meta*-disubstituted product, we must capitalize either on the *meta*-directing effects of a nitro group, or on the *meta*-directing effects of an acyl group:

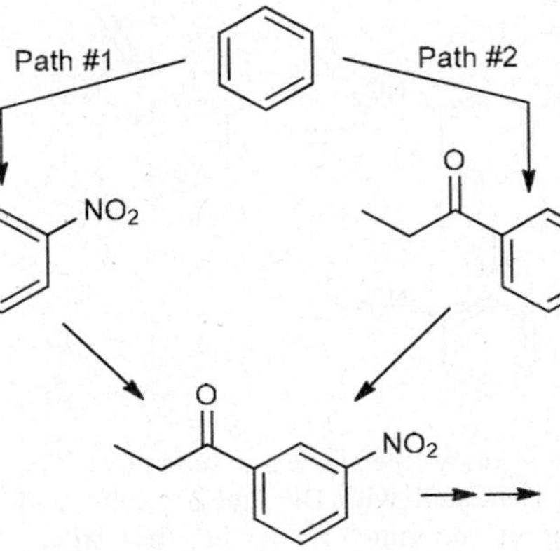
Path #1
Path #2
NO2
O
O
NO2

The first path is flawed, because it involves a Friedel-Crafts acylation process on a strongly deactivated ring, which will not occur. Therefore, only the second pathway is viable.

Now let's draw the forward scheme. A Friedel-Crafts acylation will install an acyl group, which is *meta*-directing. Upon treatment with nitric acid and sulfuric acid, a nitro group is then installed in the *meta* position. The conditions that reduce the acyl group are likely to reduce the nitro group as well, followed by basic workup, to give *meta*-propylaniline, which is then converted into an aromatic diazonium ion upon treatment with sodium nitrite and HCl. The aromatic diazonium ion is then treated with H_2O and heat to afford the desired product, as shown:

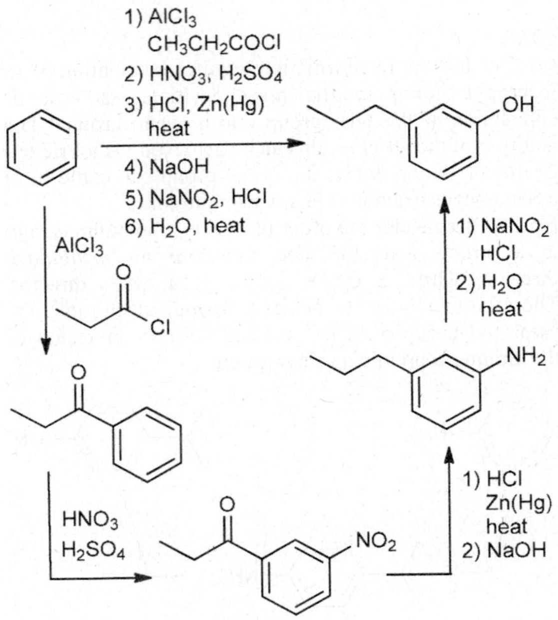
1) AlCl3
CH3CH2COCl
2) HNO3, H2SO4
3) HCl, Zn(Hg)
heat
4) NaOH
5) NaNO2, HCl
6) H2O, heat
OH
AlCl3
O
Cl
1) NaNO2
HCl
2) H2O
heat
NH2
O
1) HCl
Zn(Hg)
heat
2) NaOH
HNO3
H2SO4
O
NO2

(d) Installation of a *tert*-butyl group can be achieved with a Friedel-Crafts alkylation. The carboxylic acid group can come from hydrolysis of a cyano group, which can be installed via a diazonium ion, as shown in the following retrosynthetic analysis:

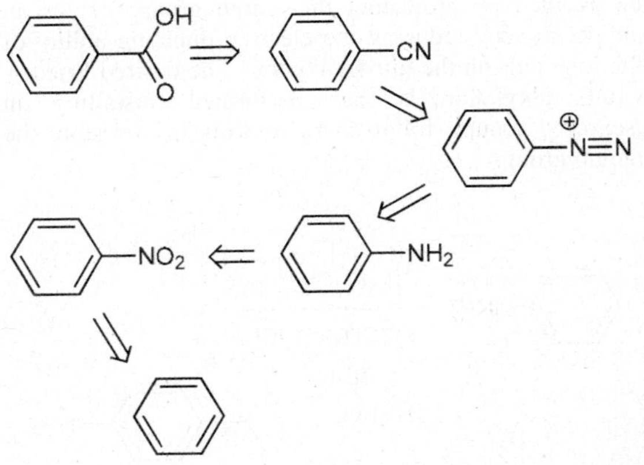
OH
O
CN
⊕
N≡N
NO2
NH2

Now let's consider the order of events. The desired product is *para*-disubstituted, which can be achieved by installing the *tert*-butyl group first. This group is very large and will favor nitration at the *para* position, as seen in the following forward scheme:

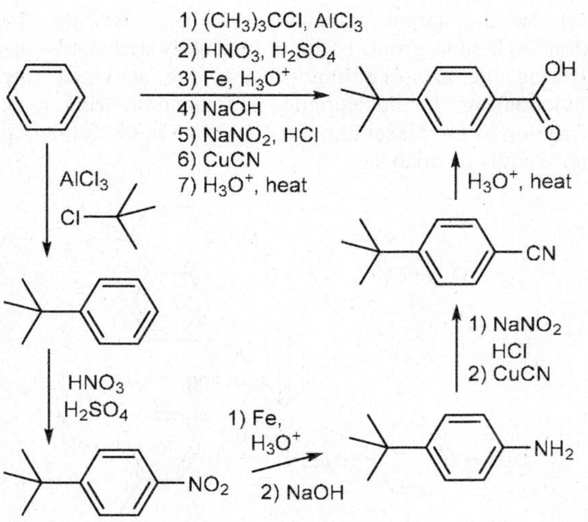

Alternatively, the first step of the synthesis above can be followed by a Friedel-Crafts alkylation (with MeI and $AlCl_3$), followed by oxidation of the benzylic position with chromic acid to give the desired carboxylic acid product.

(e) Installation of a chlorine atom can be achieved by treating benzene with Cl_2 and $AlCl_3$ (via an electrophilic aromatic substitution reaction). Installation of a fluorine atom cannot be achieved via a similar process, but it can be achieved via a diazonium ion, as shown in the following retrosynthetic analysis:

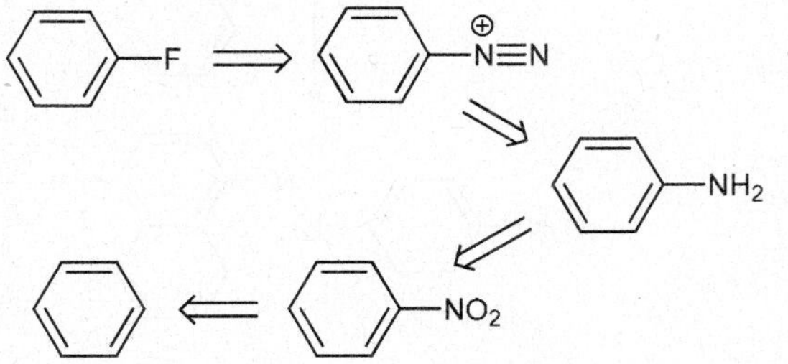

Now let's consider the order of events. Chlorine is larger than fluorine, so it is reasonable to install the chlorine atom first (thereby favoring the *para*-disubstituted product over the *ortho*-disubstituted product). In fact, we cannot install the fluorine atom first, because the directing effects of a fluorine substituent were not discussed in Chapter 18 (beyond scope of course).

Now let's draw the forward scheme. The starting material is treated with Cl_2 and a Lewis acid, thereby installing a chlorine atom. A nitro group is then installed in the *para* position, upon treatment with nitric acid and sulfuric acid. Reduction of the nitro group gives *para*-chloroaniline, which is then converted into an aromatic diazonium ion upon treatment with sodium nitrite and HCl. The aromatic diazonium ion is then treated with HBF_4 (a Schiemann reaction) to give the desired product, as shown:

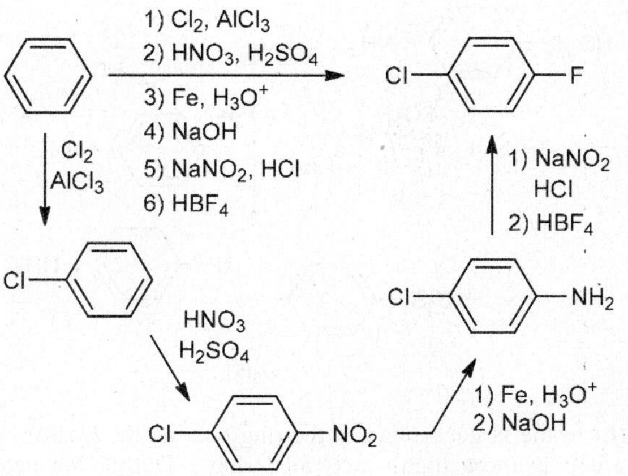

(f) This transformation can be achieved by replacing the chlorine atom with an amino group (via elimination-addition), followed by conversion of aniline into a diazonium ion, followed by subsequent treatment with CuBr (via a Sandmeyer reaction):

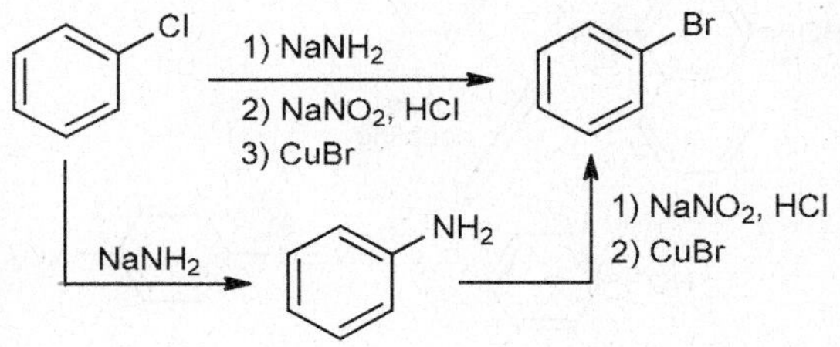

22.27.

(a) The target molecule has two aromatic rings. The ring bearing the amino group is more highly activated. During the azo coupling process, the activated ring functions as the nucleophile, and the other ring must function as the diazonium ion, as shown in the following retrosynthetic analysis:

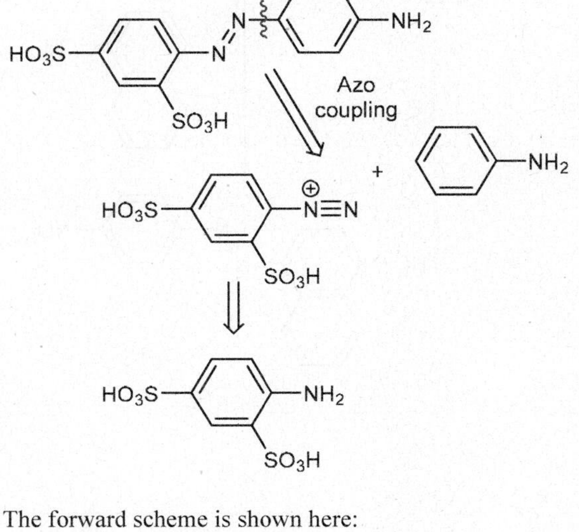

The forward scheme is shown here:

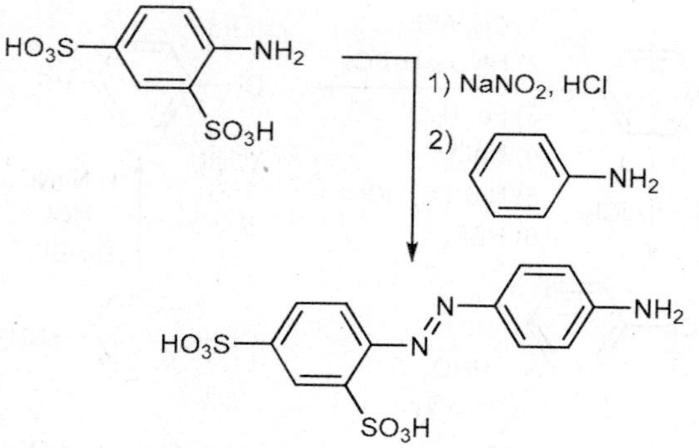

(b) In the target molecule, the ring bearing the hydroxyl group is more highly activated ring. During the azo coupling process, the activated ring functions as the nucleophile, and the other ring must function as the diazonium ion, as shown in the following retrosynthetic analysis:

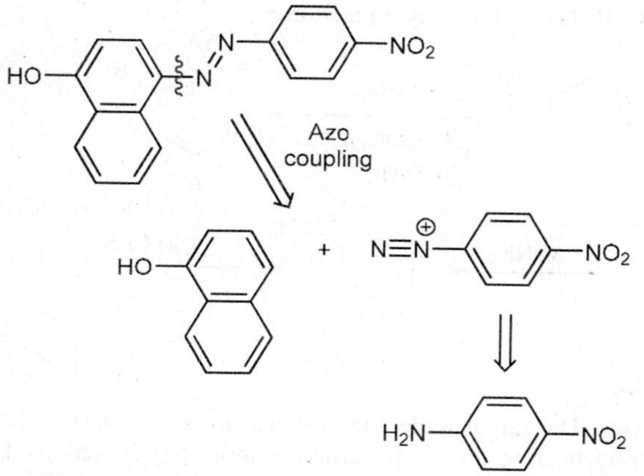

The forward scheme is shown here:

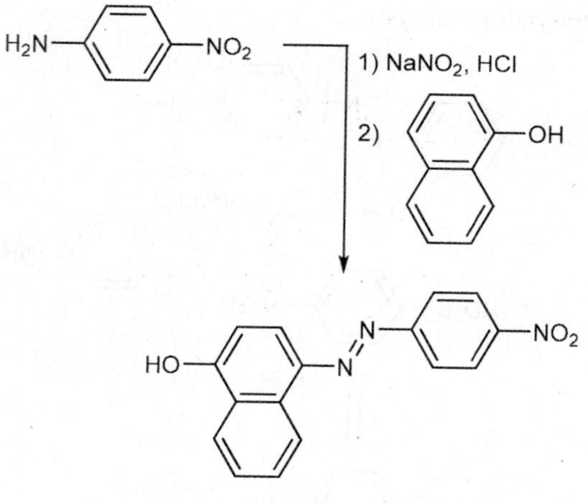

(c) In the target molecule, the ring bearing the dimethylamino group is the more highly activated ring. During the azo coupling process, the activated ring functions as the nucleophile, and the other ring must function as the diazonium ion, as shown in the following retrosynthetic analysis:

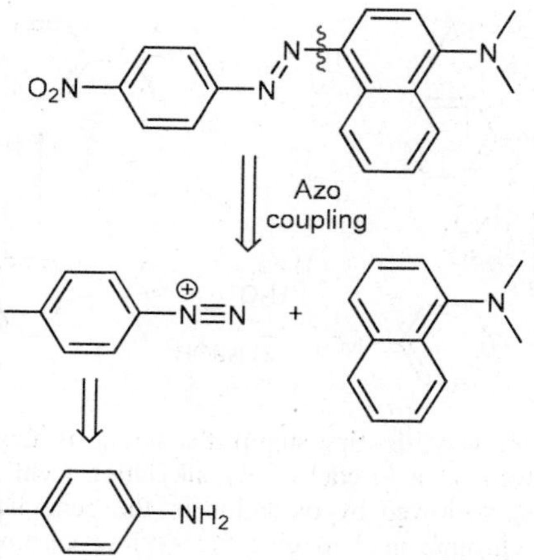

The forward scheme is shown here:

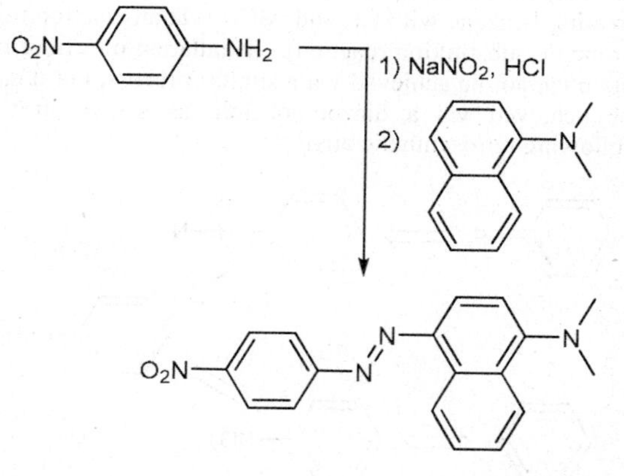

22.28.

(a) The retrosynthesis is guided by the instructions that the synthesis must begin with 3,4,5-trimethoxyaniline. Disconnection at the appropriate location leads to the required diazonium salt (prepared from 3,4,5-trimethoxyaniline) and the dihydroxybenzene derivative shown (called 2-hydroxyphenyl-*para*-toluenesulfonate).

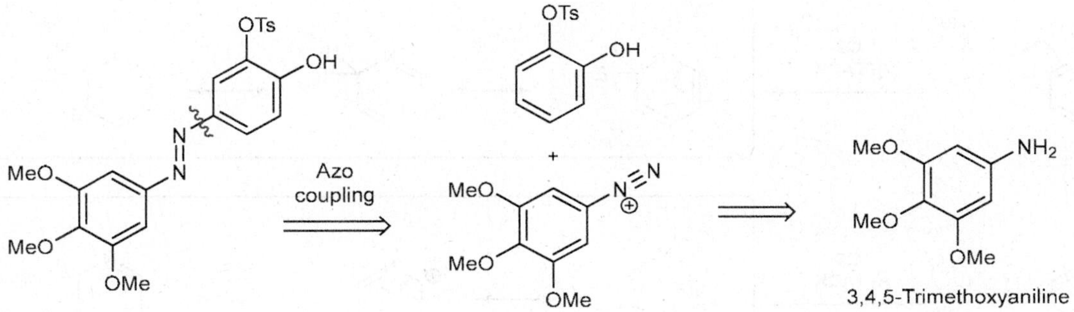

OTs OH MeO N N MeO OMe

Azo coupling

OTs OH + MeO N≡N ⊕ MeO OMe

MeO NH2 MeO OMe

3,4,5-Trimethoxyaniline

Note that the electrophilic aromatic substitution reaction takes place at the position that is *para* to the more strongly activating OH group (the sulfonate ester is only a moderate activator, similar to a regular ester). The forward scheme is shown here:

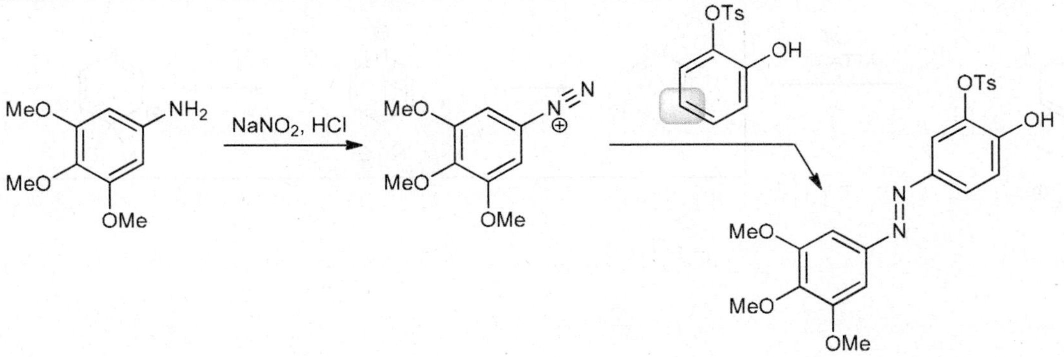

MeO NH2 MeO OMe

NaNO2, HCl

MeO N≡N ⊕ MeO OMe

OTs OH

OTs OH N N MeO MeO OMe

(b) The *trans* isomer has the two aromatic rings on opposite sides of the N=N double bond. To draw the *cis* isomer, we need to place both groups on the same side of the double bond:

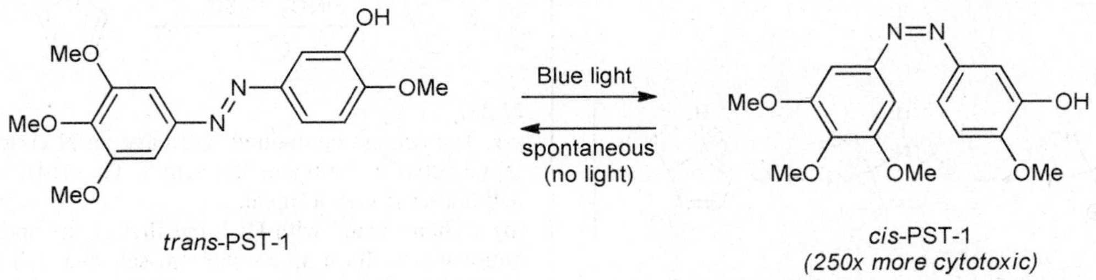

OH MeO N OMe MeO N MeO

trans-PST-1

Blue light

spontaneous (no light)

N=N MeO OH MeO OMe OMe

cis-PST-1

(250x more cytotoxic)

22.29. Attack at either C2 or C4 generates a sigma complex intermediate that exhibits a resonance structure with a nitrogen atom that bears a positive charge and lacks an octet (highlighted below). Attack at C3 generates a more stable intermediate (carbon is more electropositive than nitrogen so a carbon atom can better stabilize a positive charge associated with an unfilled octet):

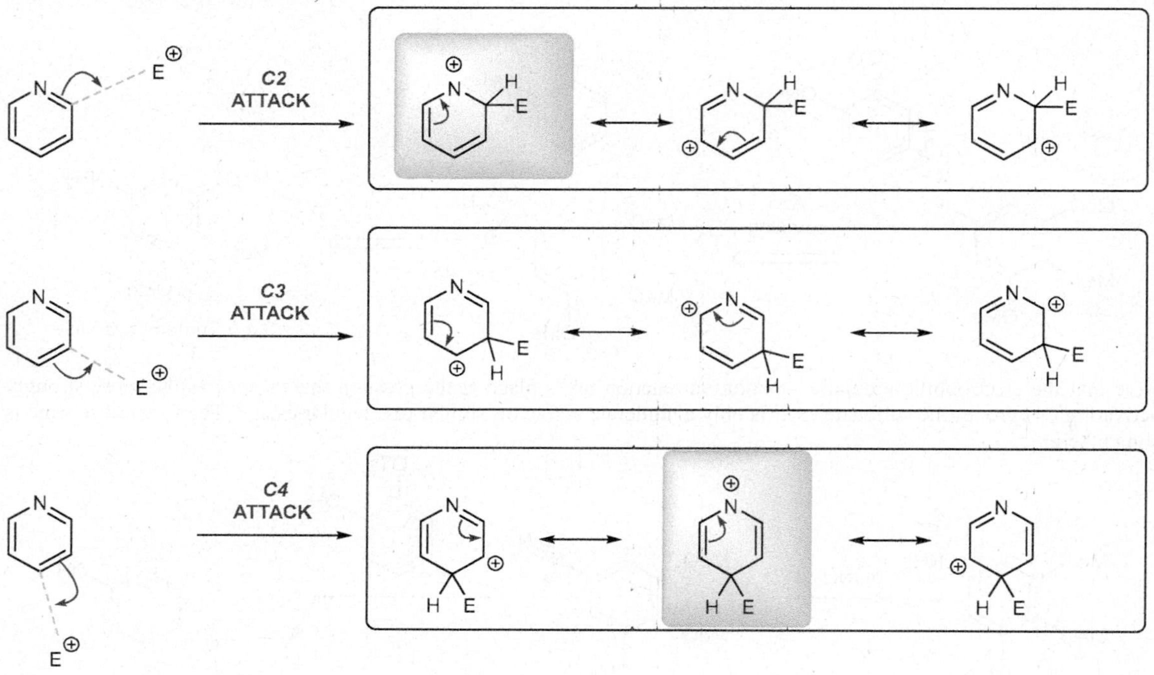

C2 ATTACK

C3 ATTACK

C4 ATTACK

22.30. Attack at the C2 position proceeds via an intermediate with three resonance structures:

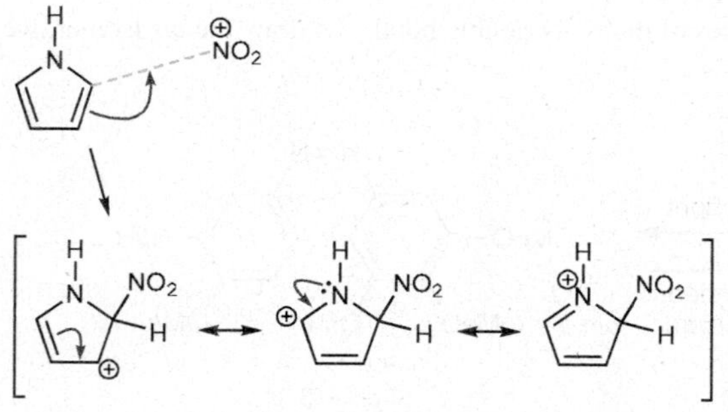

In contrast, attack at the C3 position proceeds via an intermediate with only two resonance structures:

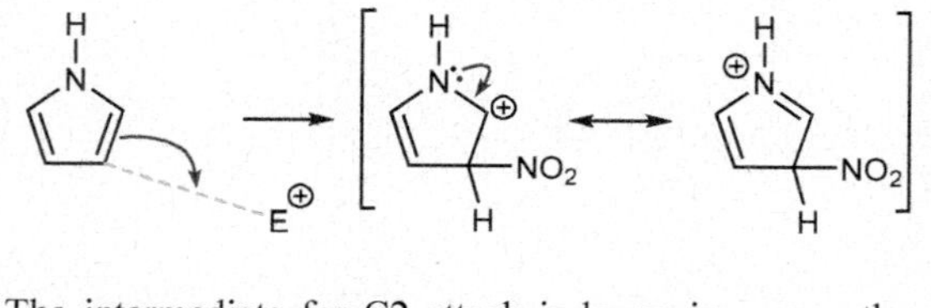

The intermediate for C2 attack is lower in energy than the intermediate for C3 attack. The transition state leading to the intermediate of C2 attack will therefore be lower in energy than the transition state leading to the intermediate of C3 attack. As a result, C2 attack occurs more rapidly, giving the following product:

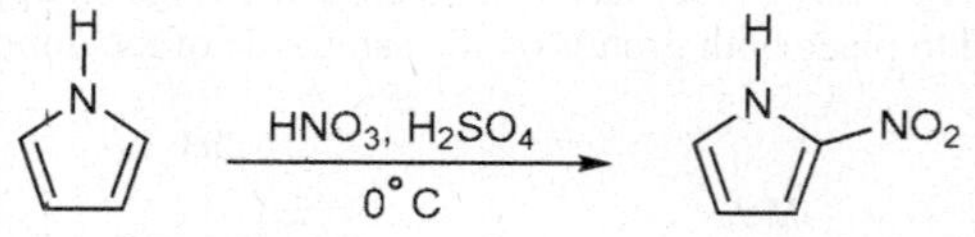

HNO_3, H_2SO_4, 0°C

22.31.
(a) The second compound will have an N-H stretching signal between 3300 and 3500 cm^{-1}. The first compound will not have such a signal.
(b) When treated with HCl, the first compound will be protonated to form an ammonium salt that will produce an IR signal between 2200 and 3000 cm^{-1}. The second compound is not an amine and will not exhibit the same behavior.

22.32.
(a) The ^{1}H NMR spectrum of the first compound will have a singlet resulting from the N-methyl group. The ^{1}H NMR spectrum of the second compound is not expected to exhibit any singlets.
(b) The ^{1}H NMR spectrum of the first compound will have six signals, while the ^{1}H NMR spectrum of the second compound will have only three signals.

22.33. The designation "primary" indicates that *one* carbon group, and two hydrogen atoms, are attached to the nitrogen atom, while the designation "secondary" indicates that *two* carbon groups and one hydrogen atom are attached to the nitrogen atom:

secondary

primary

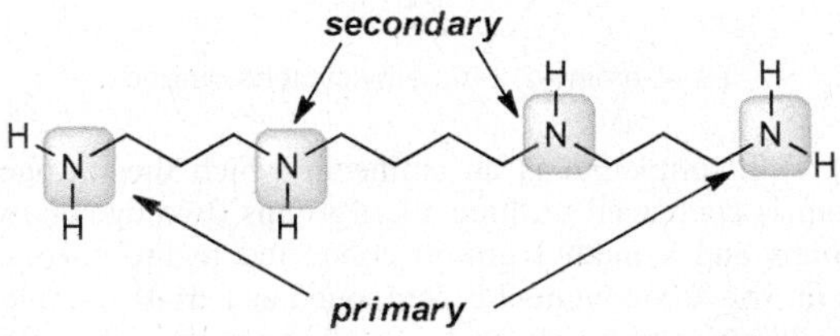

22.34.

(a) The lone pair that is farther away from the rings is the most basic, because that lone pair is localized. The lone pair of the other nitrogen atom is delocalized via resonance.

(b) The dimethylamino group exhibits a localized lone pair, and as such, it is expected to exist primarily as a charged ammonium ion (pK_a ~ 10; see pK_a values in Table 3.1) at physiological pH. In contrast, the other nitrogen atom exhibits a delocalized lone pair, and it is not expected to be protonated at physiological pH (see discussion of the Henderson-Hasselbalch equation in Section 20.3).

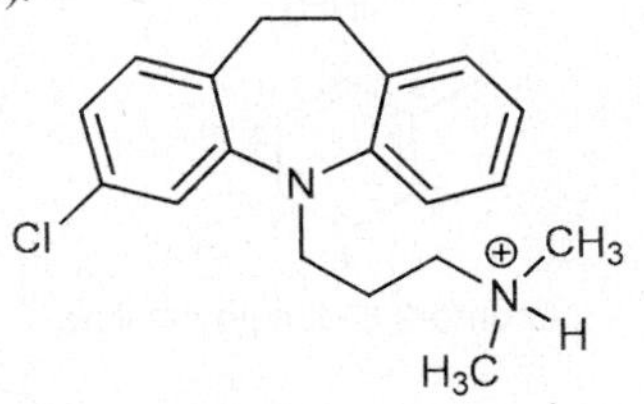

22.35. The nitrogen atom of the amide group exhibits a delocalized lone pair, so this lone pair will certainly not be the most basic. Indeed, amides do not function as effective bases. Each of the remaining two nitrogen atoms exhibits a localized lone pair. The nitrogen atom of the aromatic system has the localized lone pair in an sp^2-hybridized orbital (in the plane of the ring, and going away from the ring). In contrast, the other nitrogen atom (highlighted) has the localized lone pair in an sp^3-hybridized orbital. The sp^3-hybridized nitrogen atom (highlighted) is expected to be a better base than the sp^2-hybridized nitrogen atom, because the former has a lone pair that is farther away from the nucleus (held less tightly) and is therefore more available to function as a base. Also, a comparison of pK_a values (see Table 3.1) indicates that pyridine is a weaker base than triethyl amine (compare the pK_a values of the corresponding ammonium ions).

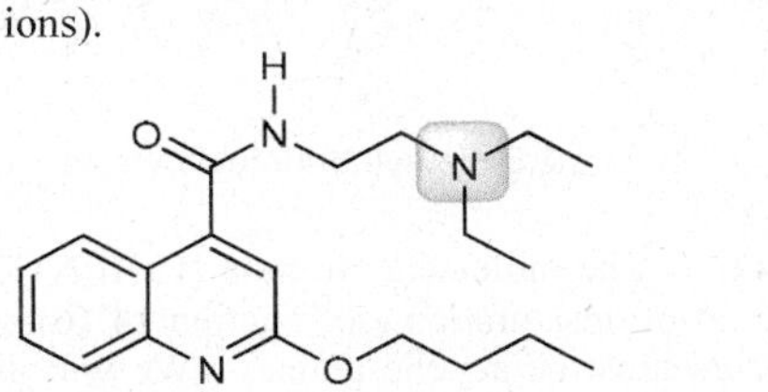

22.36.

(a) Pyridine is a weaker base than trimethylamine because the lone pair of pyridine occupies an sp^2-hybridized orbital, rather than an sp^3-hybridized orbital. By occupying an sp^2-hybridized orbital, the electrons of the lone pair have more *s* character and are therefore closer to the positively charged nucleus, rendering them less basic. As such, trimethylamine is a stronger base than pyridine:

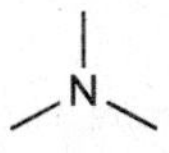

(b) The nitrogen atom of an amide group exhibits a lone pair that is highly delocalized and is therefore not expected to function as a base. Pyridine is a stronger base because the lone pair is localized (the lone pair occupies an sp^2-hybridized orbital):

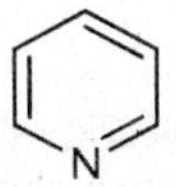

22.37.

(a) The parent is aniline (aminobenzene), and there are two alkyl groups connected to the nitrogen atom (an ethyl group and an isopropyl group):

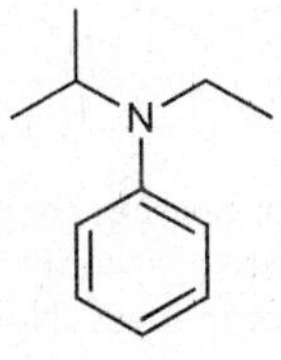

(b) The name indicates a three-membered ring connected to a nitrogen atom, as well as two substituents (both methyl groups) connected to the nitrogen atom.

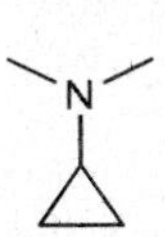

(c) The parent is a five-membered chain (pentane) that bears an amino group at C2. In addition, there is a dimethyl amino group located at C3. The configuration of each chiral center (C2 and C3) is indicated in the name:

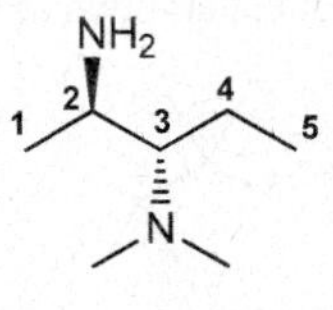

(d) The name indicates a primary amine in which the nitrogen atom is connected to a benzyl ($PhCH_2$–) group:

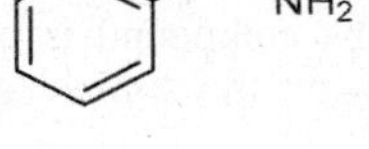

22.38. Only one of the nitrogen atoms (highlighted) has a localized lone pair.

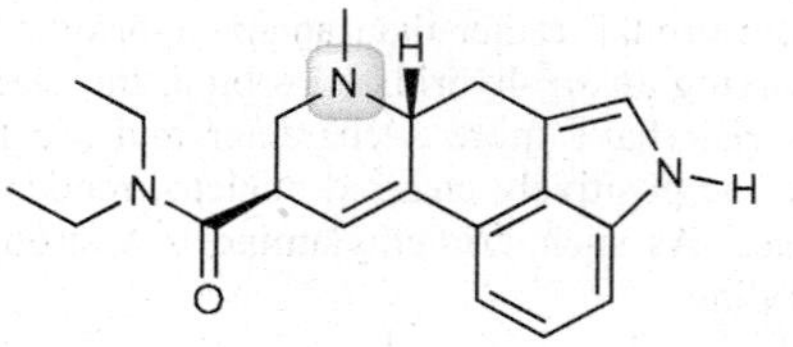

As such, this nitrogen atom is significantly more basic than the other two nitrogen atoms, each of which exhibits a highly delocalized lone pair. The nitrogen atom on the left is part of an amide group, and is not expected to function as a base, while the nitrogen atom on the right is using its lone pair to establish aromaticity. So that lone pair is also largely unavailable to serve as a base.

22.39.

(a) Recall that an atom bearing four different groups is a chiral center. There are two chiral centers (highlighted) in this compound. Notice that, in this case, the nitrogen atom is a chiral center because it is connected to four different groups (one of which is a lone pair).

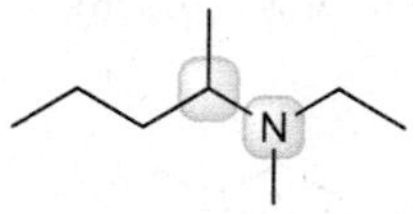

(b) Recall that an atom bearing four different groups is a chiral center. There is only one chiral center (highlighted) in this compound. Notice that, in this case, the nitrogen atom is not a chiral center because it is connected to two identical groups (two propyl groups).

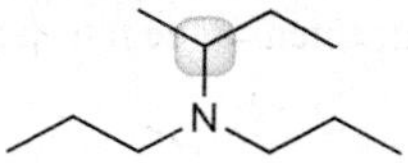

22.40.

(a) This compound is an amine that has only one alkyl group connected to the nitrogen atom. Since this alkyl group is complex, we must name the compound as an alkanamine (rather than an alkyl amine). The parent is comprised of six carbon atoms (thus, hexanamine), and the amino group is connected to C1. There are four methyl groups (two at C2 and two at C3), resulting in the following name:

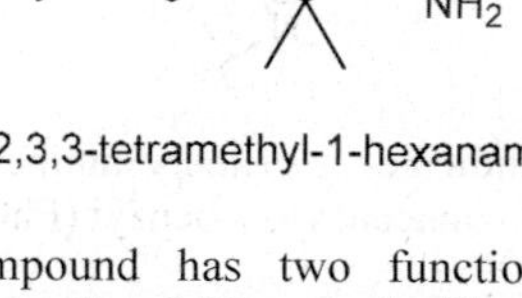

2,2,3,3-tetramethyl-1-hexanamine

(b) This compound has two functional groups (a carbonyl group and an NH_2 group). The carbonyl group takes priority, so the compound is named as a ketone (cyclohexanone), with the amino group listed as a substituent, located at C4. In addition, there are two methyl groups, both located at C2. The configuration of the chiral center is listed at the beginning of the name:

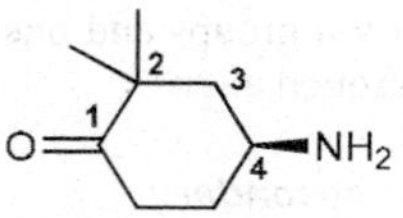

(*S*)-4-amino-2,2-dimethylcyclohexanone

(c) This compound is an amine in which the nitrogen atom is connected to three alkyl groups (two cyclobutyl groups and a methyl group) connected to the nitrogen atom, so we can name this compound as a trialkyl amine, in addition to naming it as an alkanamine. The alkyl groups are listed in alphabetical order:

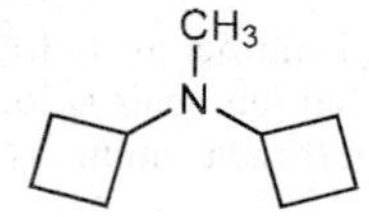

dicyclobutylmethylamine or
N-cyclobutyl-*N*-methylcyclobutanamine

(d) The parent is aniline, and there are two methyl groups (one at C2 and the other at C6), as well as one bromine atom located at C3:

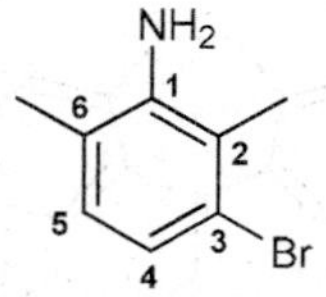

3-bromo-2,6-dimethylaniline

(e) The parent is aniline, and there are three substituents: a propyl group at C3 and two methyl groups (both connected to the nitrogen atom).

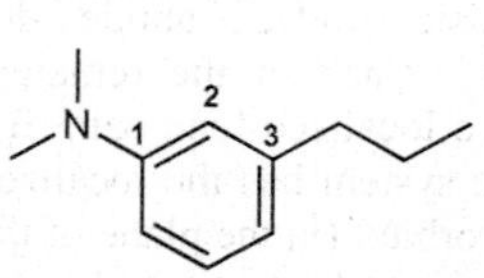

N,*N*-dimethyl-3-propylaniline

(f) The parent is pyrrole, which by definition places the nitrogen at position 1, and there are three substituents: a methyl group connected to the nitrogen atom, and two ethyl groups at C2 and C5.

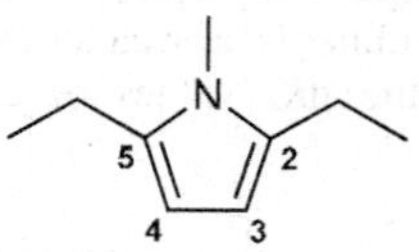

2,5-diethyl-*N*-methyl pyrrole

22.41. The molecular formula ($C_4H_{11}N$) indicates no degrees of unsaturation (see Section 14.16), so all of the isomers must be acyclic amines. We will methodically consider all possible primary, secondary, and tertiary amines. Let's begin our analysis with tertiary amines. There is only one isomer that is a tertiary amine:

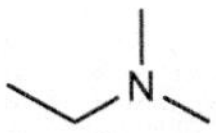

ethyldimethylamine or
N,N-dimethylethanamine

And there are three isomers that are all secondary amines:

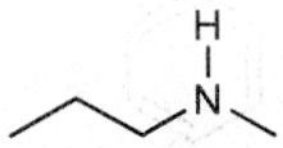

methylpropylamine or
N-methyl-1-propanamine

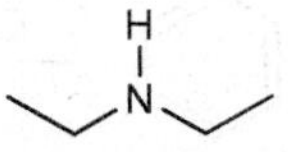

diethylamine or
N-ethylethanamine

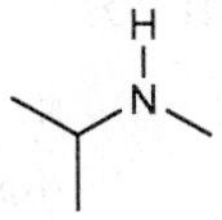

isopropylmethylamine or
N-methyl-2-propanamine

And finally, there are four isomers that are all primary amines:

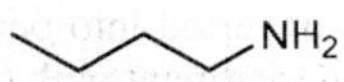

n-butylamine or
1-butanamine

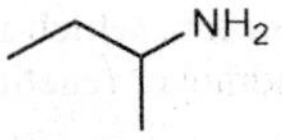

sec-butylamine or
2-butanamine

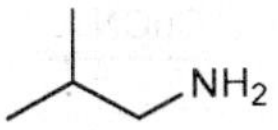

isobutylamine or
2-methyl-
1-propanamine

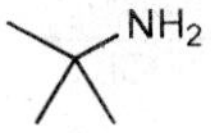

tert-butylamine or
2-methyl-
2-propanamine

In total, there are eight constitutional isomers with the molecular formula $C_4H_{11}N$.

22.42. The molecular formula ($C_5H_{13}N$) indicates no degrees of unsaturation (see Section 14.16), so all of the isomers must be acyclic amines. The following three isomers are all tertiary amines (acyclic and fully saturated), and none of them have a chiral center:

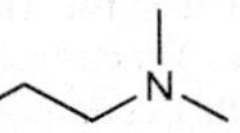

dimethylpropylamine or
N,N-dimethyl-1-propanamine

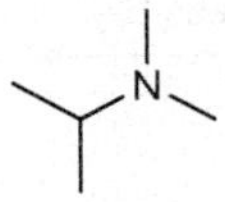

isopropyldimethylamine or
N,N-dimethyl-2-propanamine

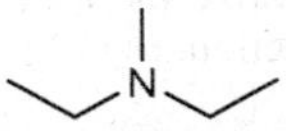

diethylmethylamine or
N-ethyl-*N*-methylethanamine

22.43.

(a) The lone pair on pyridine functions as a base and deprotonates acetic acid, giving a pyridinium ion and an acetate ion, as shown. Two curved arrows must be drawn. The first curved arrow shows the base attacking the proton, and the second curved arrow shows heterolytic cleavage of the O-H bond:

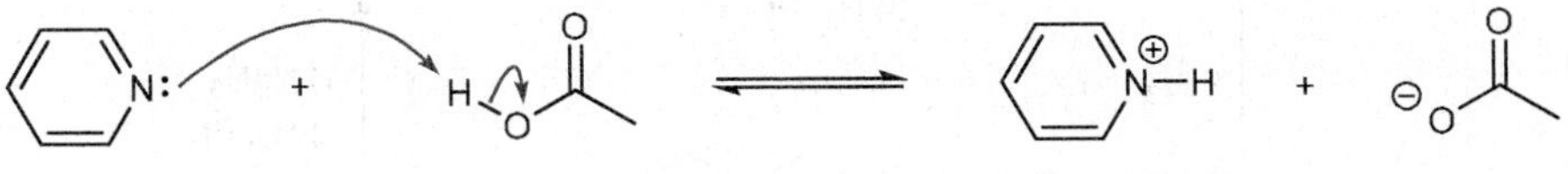

Base Acid

(b) The tertiary amine functions as a base and deprotonates the carboxylic acid (benzoic acid), giving an ammonium ion and a benzoate ion, as shown. Two curved arrows must be drawn. The first curved arrow shows the base attacking the proton, and the second curved arrow shows heterolytic cleavage of the O-H bond:

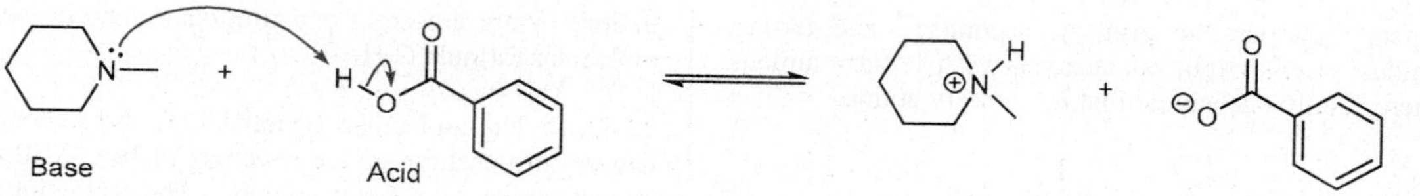
Base
Acid

22.44.

(a) The amino group of aniline is an *ortho-para* director and it strongly activates the ring toward electrophilic aromatic substitution. When treated with excess Br_2, we expect bromination to occur in the two *ortho* positions and the *para* position, giving 2,4,6-tribromoaniline:

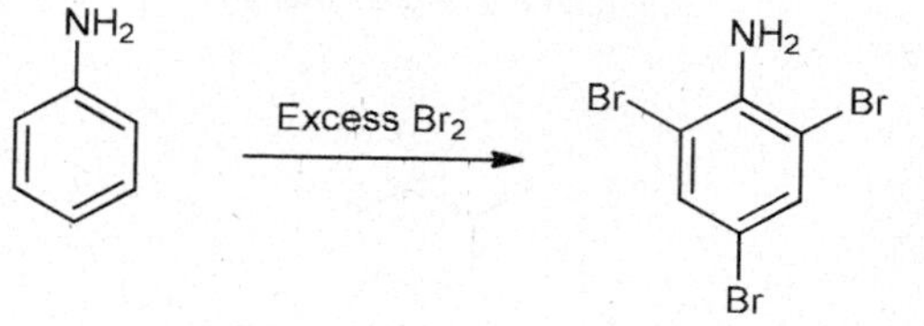
NH2
Excess Br2
NH2
Br
Br
Br

(b) Aniline is a strong nucleophile. When treated with an acid chloride, the amino group undergoes acylation, giving the following product. Pyridine functions as an acid sponge to neutralize the HCl that is produced as a by-product of the reaction.

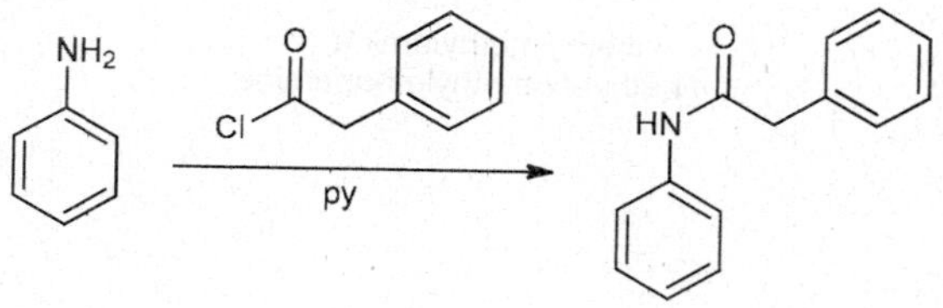
NH2
O
Cl
py
O
HN

(c) When treated with excess methyl iodide, the amino group of aniline undergoes exhaustive alkylation to give a quaternary ammonium salt:

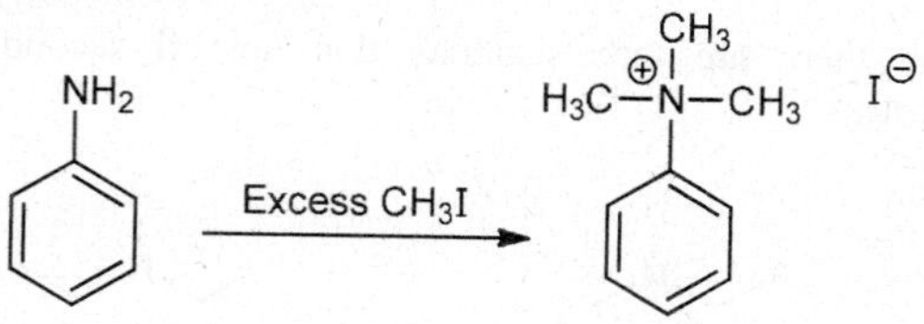
NH2
Excess CH3I
CH3
H3C—N—CH3
I

(d) Treating aniline with sodium nitrite and HCl gives a diazonium ion, which is then converted into benzene upon treatment with H_3PO_2:

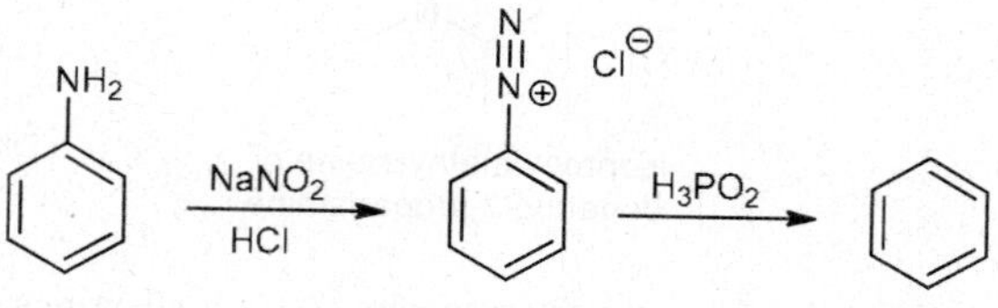
NH2
NaNO2
HCl
N
N
Cl
H3PO2

(e) Treating aniline with sodium nitrite and HCl gives a diazonium ion, which is then converted into benzonitrile via a Sandmeyer reaction (upon treatment with CuCN):

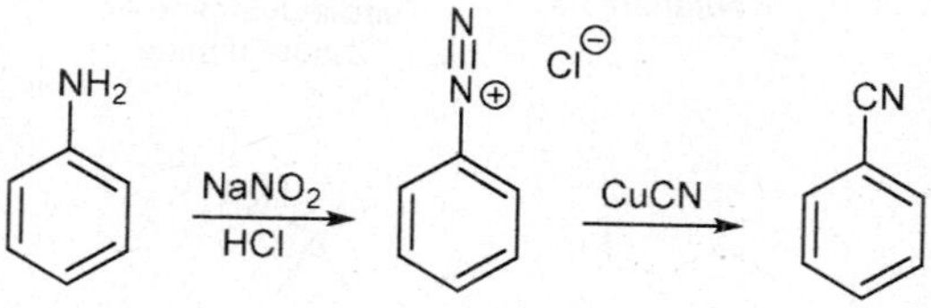
NH2
NaNO2
HCl
N
N
Cl
CuCN
CN

22.45.

(a) This transformation does not involve a change in the carbon skeleton. Only the identity of the functional group must be changed. This can be achieved by converting the alcohol into an alkyl halide (upon treatment with PBr_3), followed by an azide synthesis, as shown below. Alternatively, the alkyl halide can be converted into the desired amine via a Gabriel synthesis.

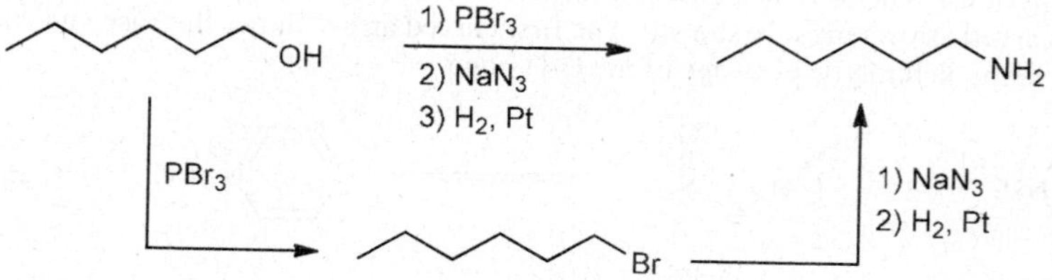
OH
1) PBr3
2) NaN3
3) H2, Pt
NH2
PBr3
Br
1) NaN3
2) H2, Pt

(b) This transformation involves a change in the carbon skeleton, as well as a change in the identity and location of the functional group. There are certainly many ways to install the extra carbon atom and manipulate the functional group as necessary. The most efficient method involves converting the alcohol into an alkyl halide (upon treatment with PBr_3), followed by an S_N2 reaction with cyanide as the nucleophile, thereby giving a nitrile and extending the chain by one carbon atom. Reduction of the nitrile with excess lithium aluminum hydride, followed by water workup, gives the product:

OH
1) PBr_3
2) NaCN
3) xs $LiAlH_4$
4) H_2O
NH_2
PBr_3
1) xs $LiAlH_4$
2) H_2O
Br
NaCN
CN

(c) The starting material has six carbon atoms, while the product has only five carbon atoms. In order to remove a carbon atom, a carbon-carbon bond must be broken, which can be accomplished via ozonolysis:

1) O_3
2) DMS
O
H
+
H
O
H

This strategy requires that we first convert the starting alcohol into the alkene above, which can be achieved by treating the alcohol with PBr_3, giving an alkyl halide, followed by elimination with a strong, sterically hindered base (such as *tert*-butoxide) to give an alkene. Ozonolysis of the alkene gives pentanal, which can be converted into the product via reductive amination, as shown:

OH
1) PBr_3
2) *t*-BuOK
3) O_3
4) DMS
5) $[H^+]$, $NaBH_3CN$, NH_3
NH_2
PBr_3
$[H^+]$, $NaBH_3CN$
NH_3
Br
O
t-BuOK
1) O_3
2) DMS

22.46.

(a) This transformation does not involve a change in the carbon skeleton. Only the identity of the functional group must be changed. This can be achieved via an azide synthesis, as shown below. Alternatively, the alkyl halide can be converted into the desired amine via a Gabriel synthesis.

Br
1) NaN_3
2) H_2, Pt
NH_2

(b) This transformation involves a change in the carbon skeleton, as well as a change in the identity and location of the functional group. There are certainly many ways to install the extra carbon atom and manipulate the functional group as necessary. The most efficient method involves an S_N2 reaction with cyanide as the nucleophile, thereby converting the alkyl bromide into a nitrile, and extending the chain by one carbon atom. Reduction of the nitrile with excess lithium aluminum hydride, followed by water workup, gives the product:

Br
1) NaCN
2) xs $LiAlH_4$
3) H_2O
NH_2
NaCN
CN
1) xs $LiAlH_4$
2) H_2O

(c) This transformation does not involve a change in the carbon skeleton, although the identity of the functional group must be changed. There are certainly many ways to change the identity of the functional group. One efficient method involves conversion of the carboxylic acid to the corresponding amide (upon treatment with thionyl chloride to give an acid chloride, followed by treatment with excess NH_3). Reduction of the amide with excess lithium aluminum hydride, followed by water workup, gives the product:

1) $SOCl_2$
2) xs NH_3
3) xs $LiAlH_4$
4) H_2O

$SOCl_2$

xs NH_3

1) xs $LiAlH_4$
2) H_2O

(d) This transformation does not involve a change in the carbon skeleton, although the identity of the functional group must be changed. This can be accomplished via reduction of the nitrile (upon treatment with excess lithium aluminum hydride, followed by water workup):

1) xs $LiAlH_4$
2) H_2O

22.47. Aziridine has significant ring strain, and this strain will increase significantly during pyramidal inversion (as the bond angle must increase during the geometric change associated with pyramidal inversion; see Figure 22.3). This provides a significant energy barrier for pyramidal inversion at room temperature.

22.48. The diethylamino group exhibits a localized lone pair, and as such, it is expected to exist primarily as a charged ammonium ion (pK_a ~ 10; see pK_a values in Table 3.1) at physiological pH. In contrast, the other nitrogen atom exhibits a highly delocalized lone pair, and it is not expected to be protonated at physiological pH (see discussion of the Henderson-Hasselbalch equation in Section 20.3).

22.49. The following mechanism is based on Mechanism 19.6 (imine formation), although the final step is reduction, rather than a proton transfer step. In the first step, ammonia is a strong nucleophile and will attack formaldehyde directly. The resulting intermediate is then protonated, followed by subsequent deprotonation to give a carbinolamine. Protonation of the carbinolamine gives an excellent leaving group (H_2O), which leaves to give an iminium ion. Finally, sodium cyanoborohydride acts as a delivery agent of a hydride ion, which reduces the iminium ion to give methyl amine, as shown.

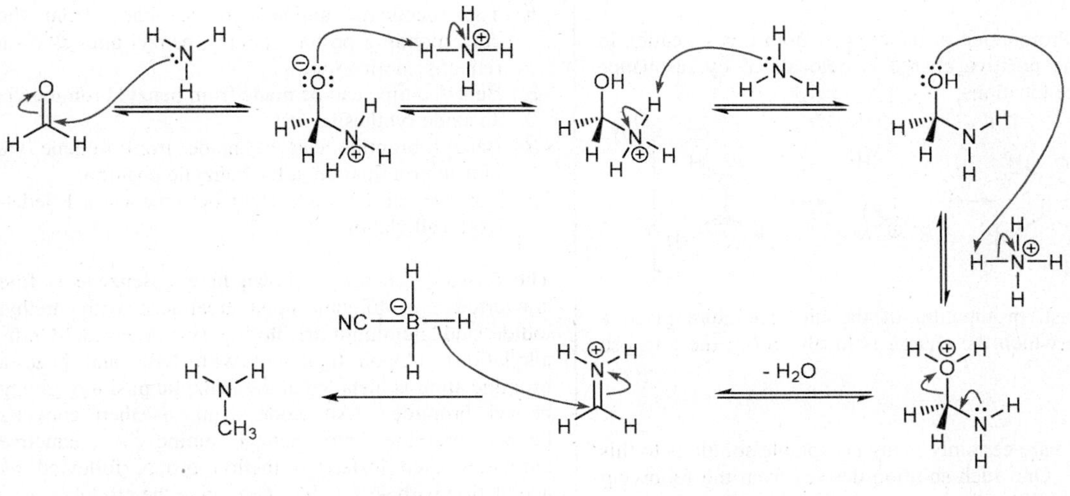

22.50. In acidic conditions, the amino group is protonated to give an ammonium ion. The ammonium group is a powerful deactivator and a *meta*-director.

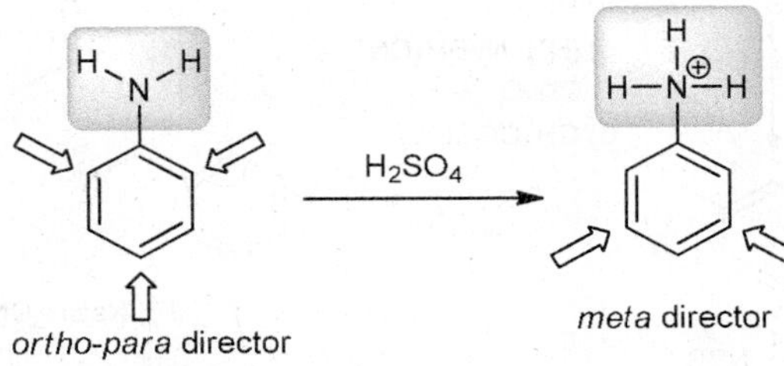

H_2SO_4

ortho-para director

meta director

22.51.

(a) In each compound, the lone pair of the amino group is delocalized because it is adjacent to the aromatic ring. However, the lone pair is more strongly delocalized for the compound that exhibits a nitro group in the *para* position (rather than the *meta* position). This extra delocalization is a result of the following additional resonance structure, in which electron density is delocalized onto the nitro group:

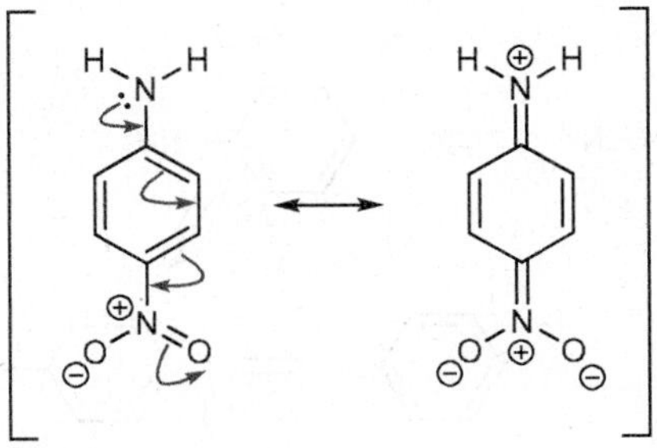

A similar resonance structure (in which electron density is delocalized onto the nitro group) cannot be drawn for *meta*-nitroaniline. As such, the lone pair in *meta*-nitroaniline is less delocalized and is therefore a stronger base. This explains why *para*-nitroaniline is a weaker base than *meta*-nitroaniline.

(b) The basicity of *ortho*-nitroaniline should be closer in value to *para*-nitroaniline, because a resonance structure can be drawn in which the lone pair is delocalized onto the nitro group:

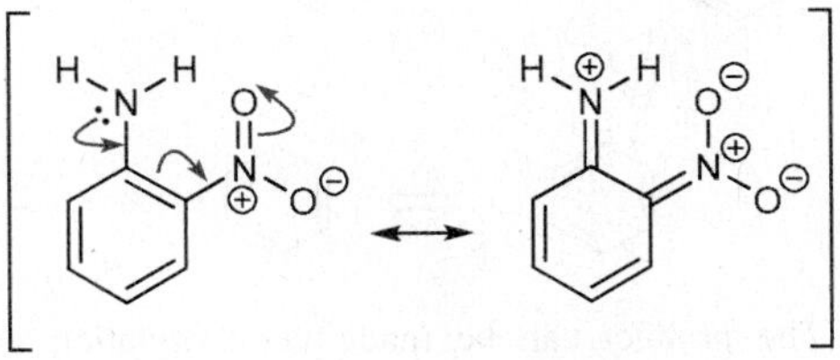

22.52. The starting amine exhibits two positions that are β to the dimethyl amino group:

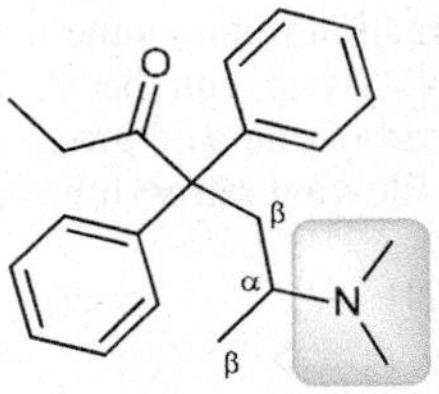

During a Hofmann elimination, the amino group is removed, and a double bond is formed between the α position and the less substituted β position, giving the following product:

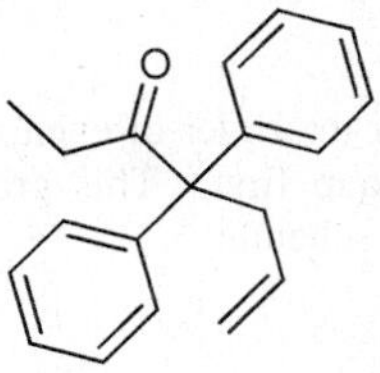

22.53. Protonation of the oxygen atom gives a cation in which the positive charge is delocalized by resonance over three locations:

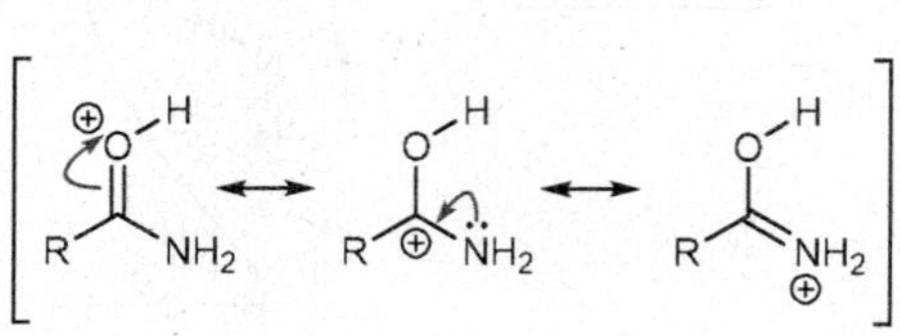

In contrast, protonation of the nitrogen atom gives a cation in which the charge is localized (on the nitrogen atom).

22.54.
(a) There are certainly many acceptable solutions to this problem. One such solution derives from the following retrosynthetic analysis. An explanation of each of the steps (*a-e*) follows.

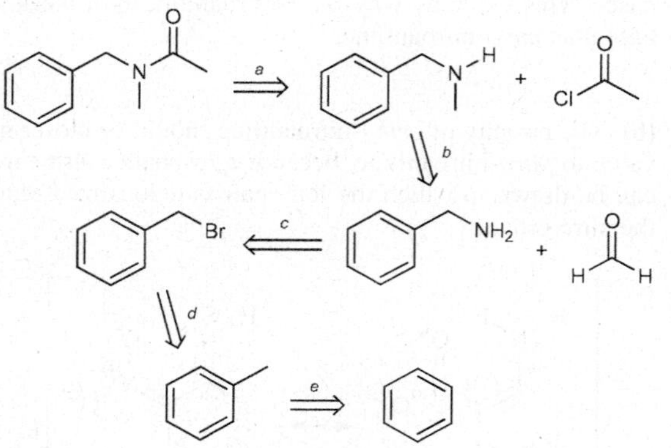

a. The product can be made via acetylation of the corresponding secondary amine.

b. The secondary amine can be made from the corresponding primary amine (benzyl amine) via a reductive amination.

c. Benzyl amine can be made from benzyl bromide via an azide synthesis.

d. Benzyl bromide can be made from toluene via radical bromination at the benzylic position.

e. Toluene can be made from benzene via a Friedel-Crafts alkylation.

The forward scheme is shown here. Benzene is first converted into toluene upon treatment with methyl iodide and aluminum trichloride (via a Friedel-Crafts alkylation). Upon treatment with NBS and heat, a bromine atom is installed at the benzylic position, giving benzyl bromide. An azide synthesis then converts benzyl bromide into benzyl amine. A reductive amination then installs a methyl group, followed by acetylation with acetyl chloride to give the product:

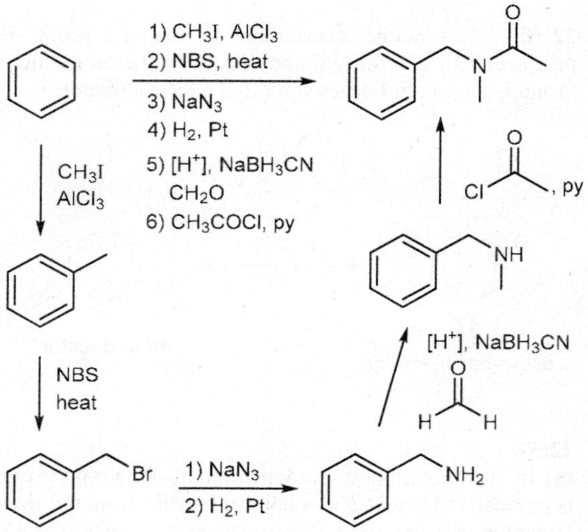

1) CH_3I, $AlCl_3$
2) NBS, heat
3) NaN_3
4) H_2, Pt
5) [H^+], $NaBH_3CN$, CH_2O
6) CH_3COCl, py

(b) This transformation requires the installation of a cyclohexyl group as well as a carboxylic acid group. Installation of the cyclohexyl group can be achieved with a Friedel-Crafts alkylation, using chlorocyclohexane and $AlCl_3$. Installation of a carboxylic acid group can be achieved by hydrolysis of a cyano group (installed via a diazonium ion), as shown in the following retrosynthetic analysis:

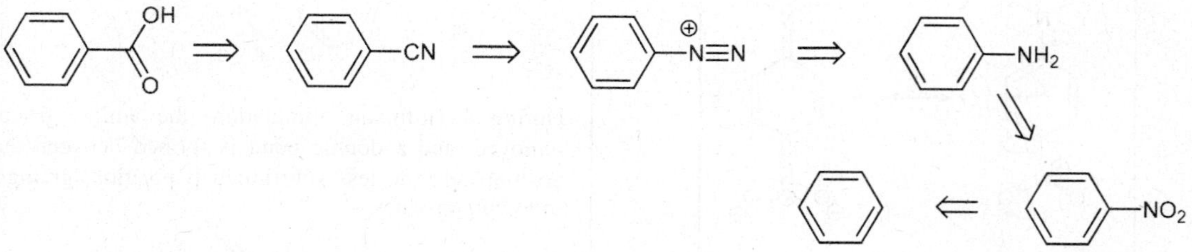

Now let's consider the order of events. The desired product is *para*-disubstituted, which can be achieved by installing the cyclohexyl group first. This group is very large and will favor nitration at the *para* position, as seen in the following forward scheme:

1) [cyclohexyl chloride] $AlCl_3$
2) HNO_3, H_2SO_4
3) Fe, H_3O^+
4) NaOH
5) $NaNO_2$, HCl
6) CuCN
7) H_3O^+, heat

Cl

$AlCl_3$

HNO_3
H_2SO_4

NO_2

1) Fe, H_3O^+
2) NaOH

NH_2

1) $NaNO_2$, HCl
2) CuCN

CN

H_3O^+, heat

COOH

22.55. In Chapter 20, we learned that an amide linkage can be hydrolyzed under aqueous basic conditions. In this case, the compound has two amide linkages, each of which is hydrolyzed via a nucleophilic acyl substitution process. Hydroxide functions as a nucleophile and attacks one of the carbonyl groups to give a charged tetrahedral intermediate that reforms the carbonyl group by expelling a negatively charged nitrogen atom as a leaving group. The resulting anion then undergoes an intramolecular proton transfer step, giving a more stable carboxylate ion. Then, the remaining amide group undergoes hydrolysis in a similar way. Specifically, hydroxide attacks the carbonyl group to give a charged tetrahedral intermediate that reforms the carbonyl group by expelling a negatively charged nitrogen atom as a leaving group. Under these conditions, the carboxylic acid group is deprotonated to give a carboxylate ion, and the amide ion is protonated (by water) to give an amine:

22.56. Hydrazine releases the amine via two successive nucleophilic acyl substitution reactions (as shown), giving the following by-product:

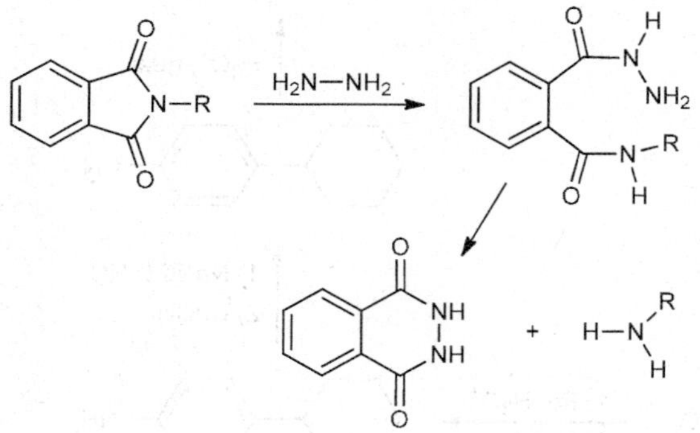

22.57.

(a) The starting material is a secondary amine and the reagents indicate a Hofmann elimination. In the first step, two methyl groups are installed, converting the secondary amine into a quaternary ammonium ion. Then, treatment with aqueous silver oxide and heat results in cleavage of a C-N bond (thereby opening the ring) and formation of a double bond, as shown:

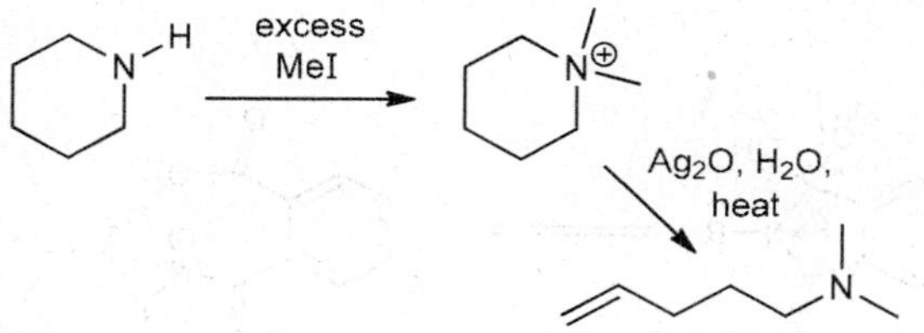

(b) This is a Gabriel synthesis. Since the alkyl halide is ethyl bromide, the product is ethylamine:

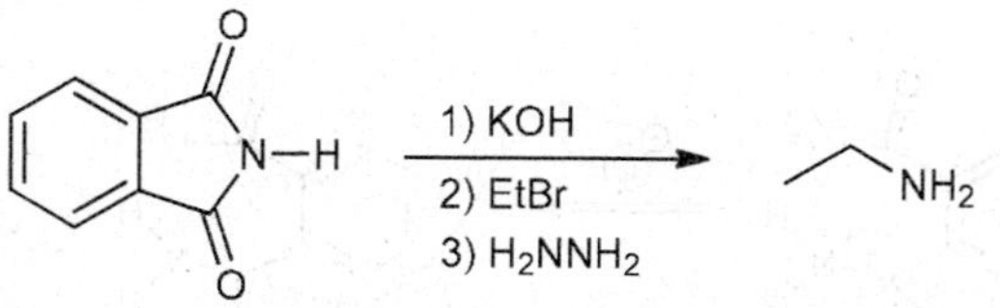

(c) The starting alkyl halide is converted into a nitrile upon treatment with sodium cyanide in a polar aprotic solvent (via an S_N2 process). Hydrolysis of the nitrile with aqueous acid and heat gives a carboxylic acid, which is then converted into an acid chloride upon treatment with thionyl chloride. The acid chloride is then converted into an amide upon treatment with excess ammonia, via a nucleophilic acyl substitution reaction.

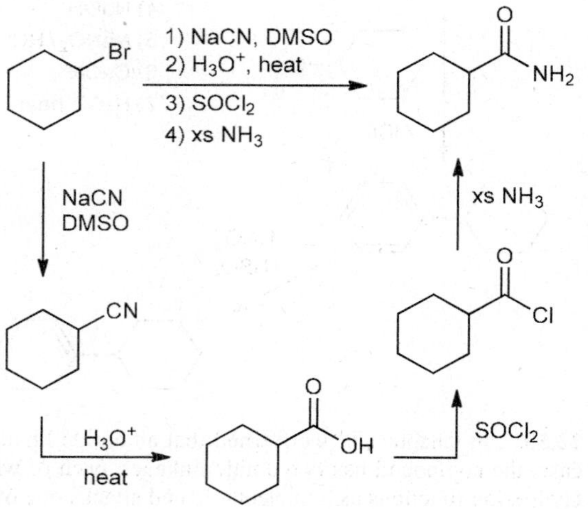

(d) Treating benzene with a mixture of nitric and sulfuric acid results in nitration of the aromatic ring, giving nitrobenzene. Subsequent treatment of nitrobenzene with iron in aqueous acid (followed by basic workup) results in reduction of the nitro group, giving aniline. Aniline is converted into a diazonium ion upon treatment with sodium nitrite and HCl, and the diazonium ion is converted into benzonitrile via a Sandmeyer reaction (with CuCN):

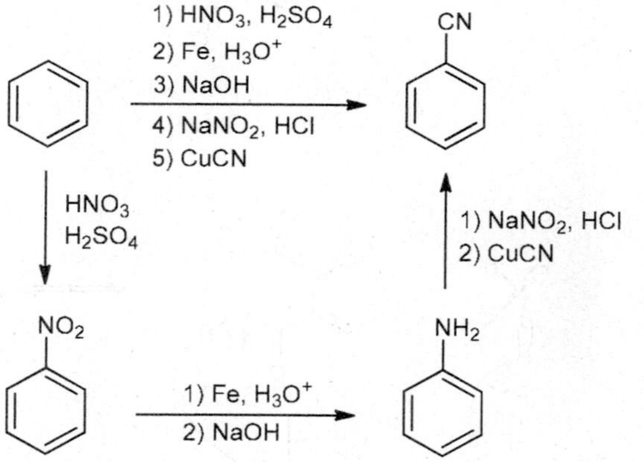

22.58. The carbonyl group is converted into a dimethyl amino group via a reductive amination process. A Hofmann elimination then gives an alkene. The double bond cannot be formed at a bridgehead position (Bredt's rule), so there is only one possible regiochemical outcome for the elimination process. Ozonolysis of the alkene gives a dialdehyde (in a *cis* configuration), which is then converted into a diamine via reductive amination of each carbonyl group.

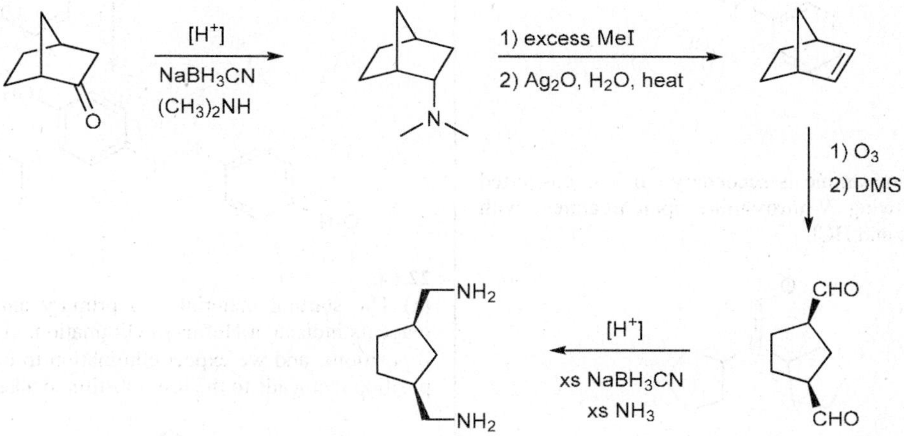

[H^+], $NaBH_3CN$, $(CH_3)_2NH$

1) excess MeI
2) Ag_2O, H_2O, heat

1) O_3
2) DMS

CHO
CHO

[H^+]
xs $NaBH_3CN$
xs NH_3

NH_2
NH_2

22.59. The conjugate base of pyrrole is highly stabilized because it is an aromatic anion and it is resonance stabilized, spreading the negative charge over all five atoms of the ring, as shown below. Pyrrole is relatively acidic (compared with other amines) because its conjugate base is so highly stabilized.

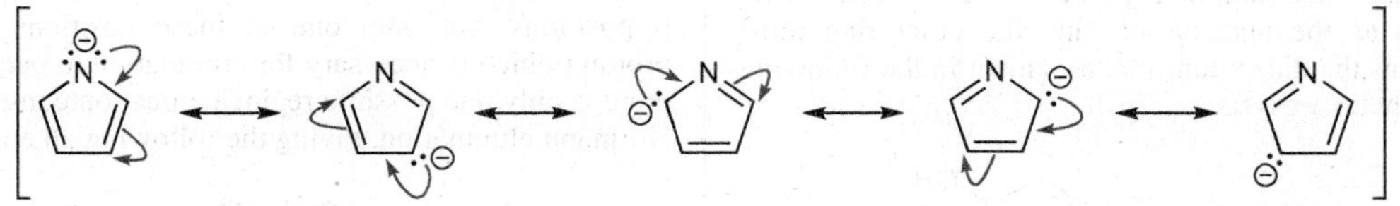

22.60. The desired amine is primary (it has only one C–N bond), and it can be made from the following ketone and ammonia via a reductive amination:

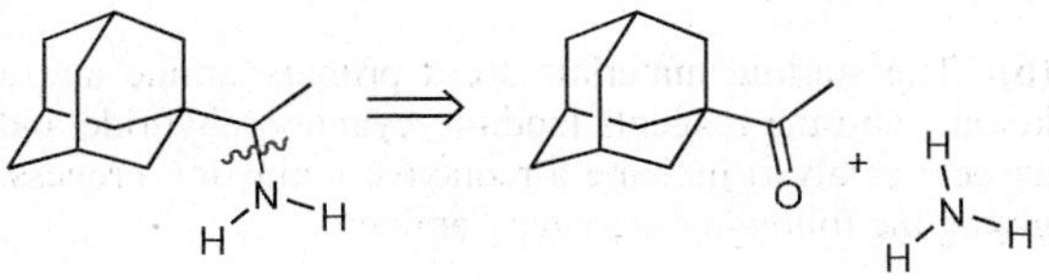

22.61.
The compound has three C-N bonds:

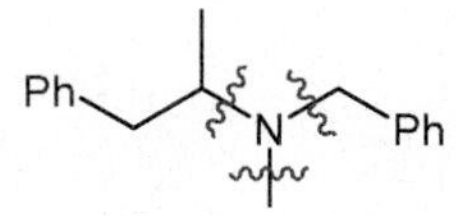

Each of these bonds can be made via a reductive amination, giving three possible synthetic routes. Two are shown here (the third route begins with formaldehyde):

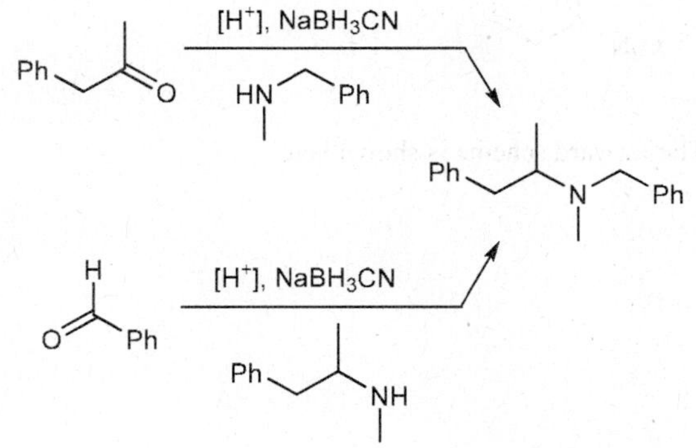

[H^+], $NaBH_3CN$

[H^+], $NaBH_3CN$

22.62.

(a) The starting amine is primary, so it is converted into the following diazonium salt upon treatment with sodium nitrite and HCl:

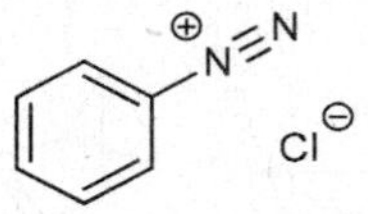

(b) The starting amine is secondary, so it is converted into the following *N*-nitrosamine upon treatment with sodium nitrite and HCl:

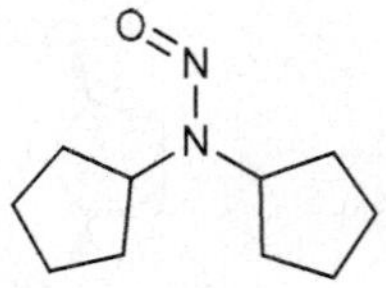

22.63. The target molecule has two aromatic rings. The ring bearing the hydroxyl group is more highly activated. During the azo coupling process, the activated ring functions as the nucleophile, and the other ring must function as the diazonium ion, as shown in the following retrosynthetic analysis:

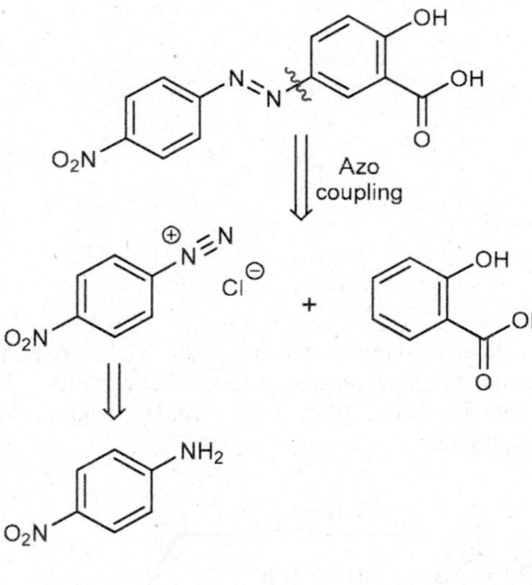

Azo coupling

The forward scheme is shown here:

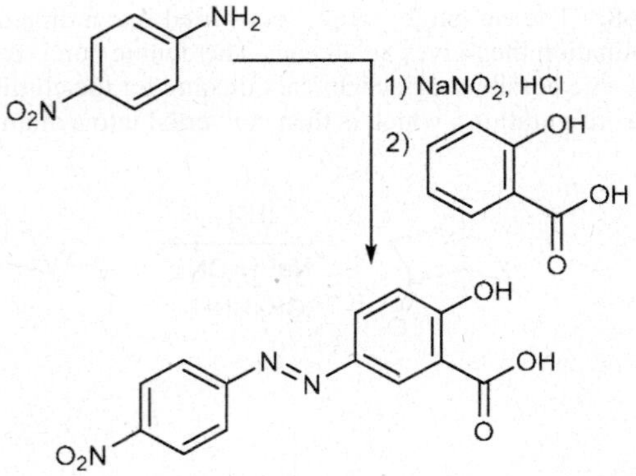

1) $NaNO_2$, HCl

2)

22.64.

(a) The starting material is a primary amine, and the reagents indicate a Hofmann elimination. There are two β positions, and we expect elimination to occur at the β position that leads to the less substituted alkene:

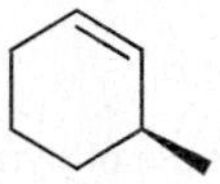

(b) The starting material is a primary amine, and the reagents indicate a Hofmann elimination. There are two β positions, but only one of these positions bears a proton (which is necessary for elimination to occur). So, there is only one possible regiochemical outcome for this Hofmann elimination, giving the following alkene:

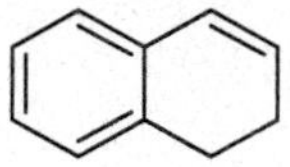

22.65.

(a) Reduction of the nitro group, followed by basic workup, gives *meta*-bromoaniline:

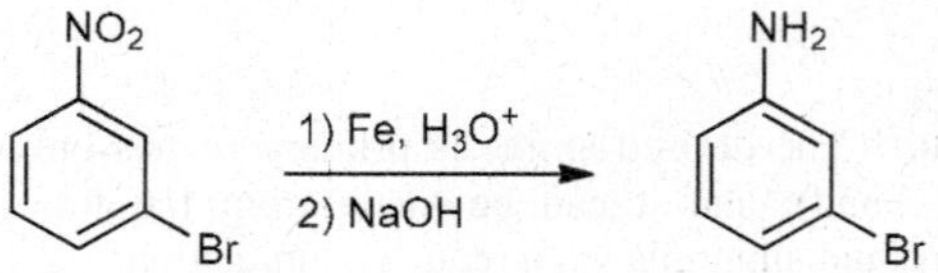

1) Fe, H_3O^+

2) NaOH

(b) The starting materials are a primary amine and a ketone, and the reagents (sodium cyanoborohydride and an acid catalyst) indicate a reductive amination process, giving the following secondary amine:

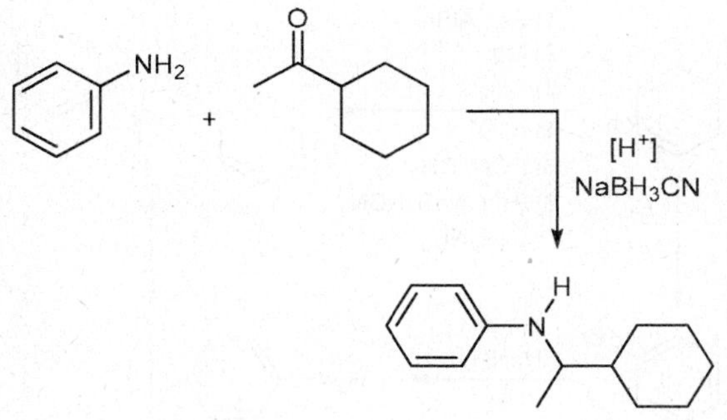

(c) A nitrile is reduced to an amine upon treatment with excess lithium aluminum hydride, followed by water workup:

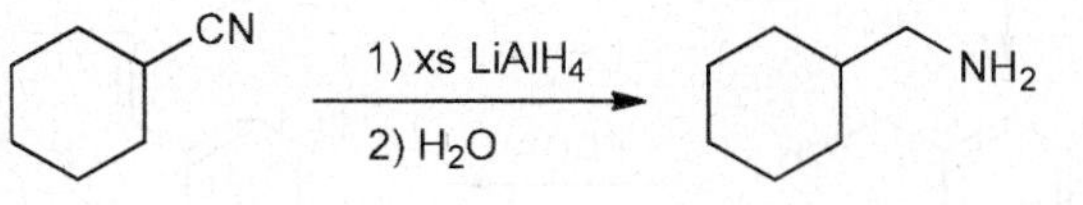

(d) An amide is reduced to an amine upon treatment with excess lithium aluminum hydride, followed by water workup:

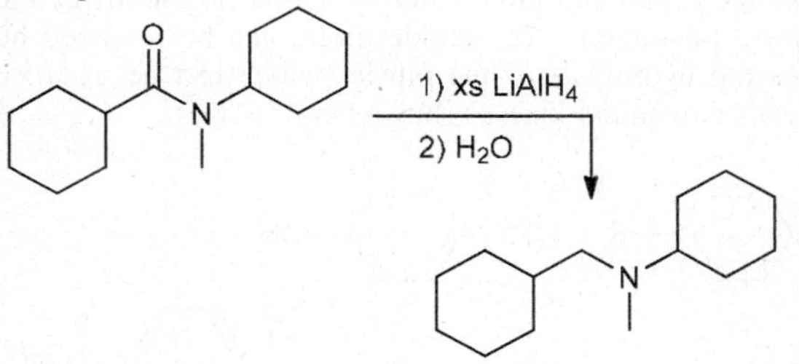

22.66.

(a) Upon treatment with water, the diazonium group is replaced with a hydroxyl group, giving *meta*-bromophenol:

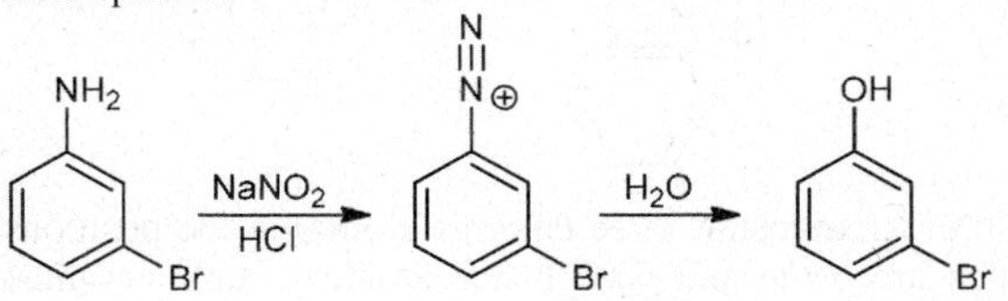

(b) Upon treatment with HBF_4, the diazonium group is replaced with a fluorine atom (via a Schiemann reaction), as shown:

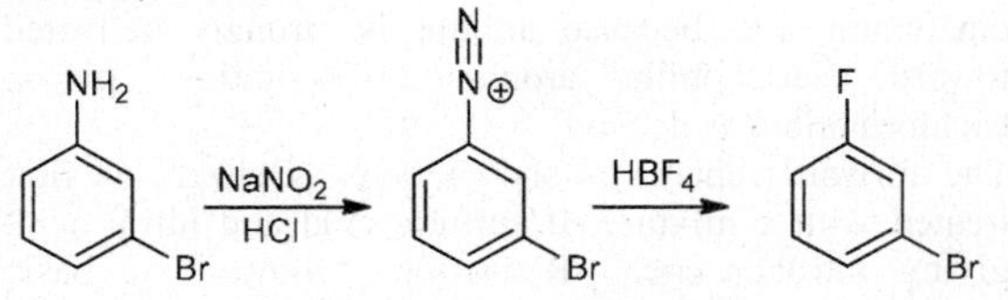

(c) Upon treatment with CuCN, the diazonium group is replaced with a cyano group via a Sandmeyer reaction:

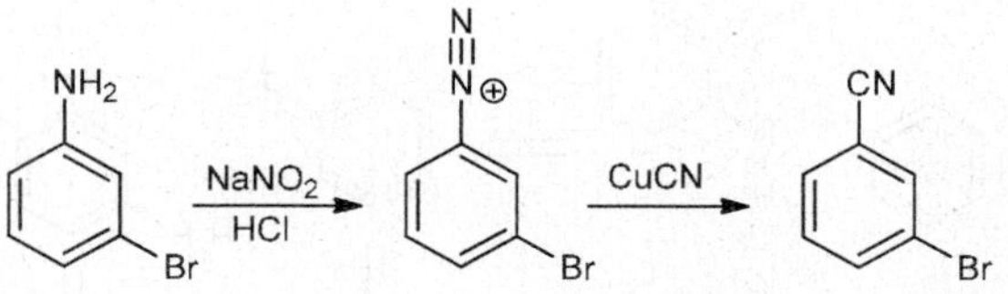

(d) Upon treatment with H_3PO_2, the diazonium group is replaced with a hydrogen atom, giving bromobenzene:

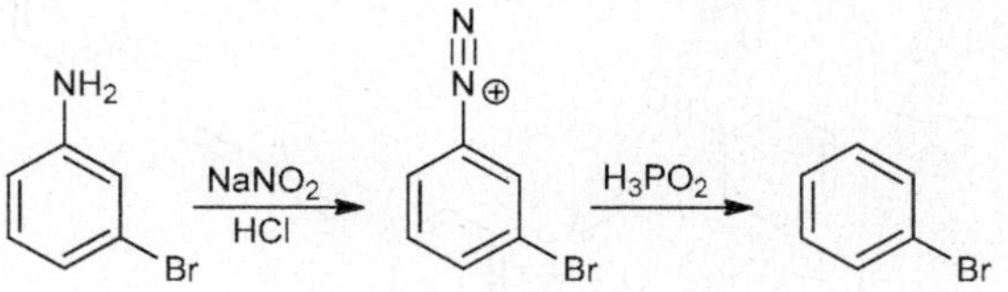

(e) Upon treatment with CuBr, the diazonium group is replaced with a bromine atom via a Sandmeyer reaction:

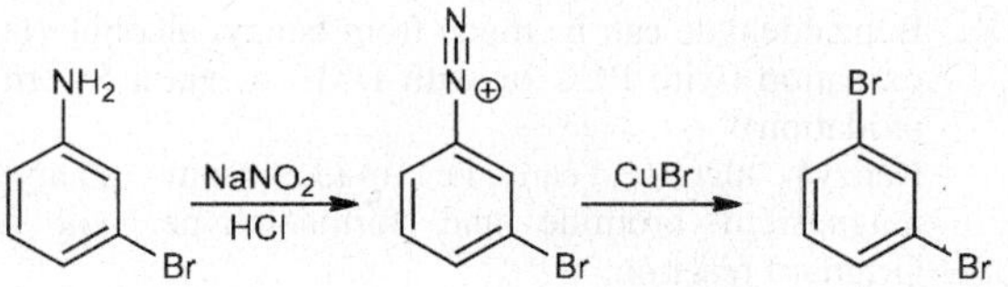

22.67. In this case, the carbonyl group and the amino group are tethered together (both functional groups are present in one compound), so we expect an intramolecular reductive amination to occur, thereby forming a new ring to give a bicyclic product, as shown:

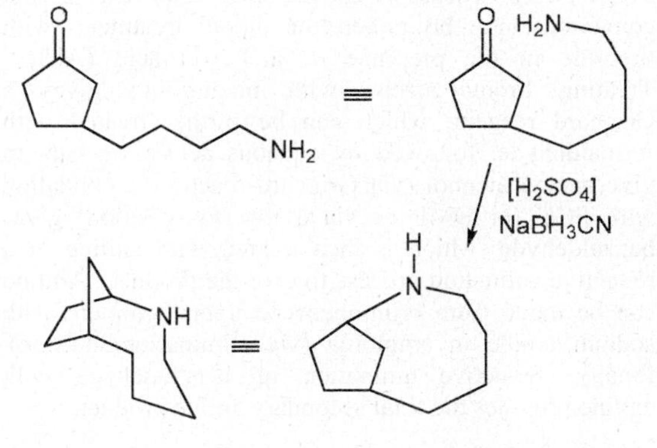

22.68.

(a) There are certainly many acceptable solutions to this problem. One such solution derives from the following retrosynthetic analysis. An explanation of each of the steps (*a-f*) follows.

a. The product is a secondary amine, which can be made from benzaldehyde and aniline via a reductive amination.

b. Benzaldehyde can be made from benzyl alcohol via oxidation (with PCC or with DMP or via a Swern oxidation).

c. Benzyl alcohol can be made from phenyl magnesium bromide and formaldehyde, via a Grignard reaction.

d. Phenyl magnesium bromide can be made from bromobenzene, upon treatment with magnesium.

e. Bromobenzene can be made from benzene via an electrophilic aromatic substitution reaction.

f. Aniline can be made from bromobenzene via elimination-addition.

The forward scheme is shown here. Benzene is first converted into bromobenzene upon treatment with bromine in the presence of a Lewis acid ($AlBr_3$). Treating bromobenzene with magnesium gives a Grignard reagent, which can be further treated with formaldehyde, followed by aqueous acidic workup, to give benzyl alcohol (via Grignard reaction). Oxidation with PCC (or DMP or via a Swern oxidation) gives benzaldehyde, which is then treated with aniline in a reductive amination process to give the product. Aniline can be made from bromobenzene upon treatment with sodium amide in ammonia (via elimination-addition). Finally, reductive amination of benzaldehyde with aniline provides the final secondary amine product.

1) Br_2, $AlBr_3$
2) Mg
3) CH_2O
4) H_3O^+
5) PCC, CH_2Cl_2
6) $[H^+]$, $NaBH_3CN$, $C_6H_5NH_2$

Br_2, $AlBr_3$

$NaNH_2$, NH_3

$[H^+]$, $NaBH_3CN$

1) Mg
2) $H_2C{=}O$
3) H_3O^+

PCC, CH_2Cl_2

(b) This transformation requires the installation of an amide group and three chlorine atoms (in the *ortho* and *para* positions). The amide group can be prepared by partial hydrolysis of the nitrile (which can be installed via a diazonium ion), as shown here:

In order to install three chlorine atoms (in the positions that are *ortho* and *para* to the amide group), we must perform chlorination (with excess chlorine) during a stage in the process when the ring is highly activated (thereby giving trichlorination). This can be accomplished immediately prior to making the diazonium ion, because aniline is strongly activated toward electrophilic aromatic substitution, giving trichlorination, as desired.

The forward scheme is shown here. Benzene is first treated with a mixture of sulfuric acid and nitric acid, giving nitrobenzene. Reduction, followed by basic workup, gives aniline, which will undergo trichlorination when treated with excess chlorine to give 2,4,6-trichloroaniline. Treatment with sodium nitrite and HCl converts the substituted aniline into a diazonium ion, which can then be treated with CuCN to give a nitrile (via a Sandmeyer reaction). Partial hydrolysis of the nitrile with aqueous acid then gives the amide product.

1) HNO_3, H_2SO_4
2) Fe, H_3O^+
3) NaOH
4) xs Cl_2
5) $NaNO_2$, HCl
6) CuCN
7) H_3O^+

HNO_3 H_2SO_4

1) Fe, H_3O^+
2) NaOH

xs Cl_2

1) $NaNO_2$, HCl
2) CuCN

H_3O^+

(c) There are certainly many acceptable solutions to this problem. One such solution derives from the following retrosynthetic analysis. An explanation of each of the steps (*a-f*) follows.

a. The product is an amide, which can be made from benzoyl chloride and aniline via a nucleophilic acyl substitution reaction.
b. Benzoyl chloride can be made from benzoic acid upon treatment with thionyl chloride.
c. Benzoic acid can be made from phenyl magnesium bromide and carbon dioxide, via a Grignard reaction.
d. Phenyl magnesium bromide can be made from bromobenzene, upon treatment with magnesium.
e. Bromobenzene can be made from benzene via an electrophilic aromatic substitution reaction.
f. Aniline can be made from bromobenzene via elimination-addition.

The forward scheme is shown here. Benzene is first converted into bromobenzene upon treatment with bromine in the presence of a Lewis acid ($AlBr_3$). Treating bromobenzene with magnesium gives a Grignard reagent, which can be further treated with carbon dioxide, followed by acidic workup, to give benzoic acid. Treating benzoic acid with thionyl chloride gives benzoyl chloride, which is then treated with aniline in a nucleophilic acyl substitution to give the product. Aniline can be made from bromobenzene upon treatment with sodium amide in ammonia (via elimination-addition):

1) Br_2, $AlBr_3$
2) Mg
3) CO_2
4) H_3O^+
5) $SOCl_2$
6) $C_6H_5NH_2$, pyridine

Br_2 $AlBr_3$

$NaNH_2$ NH_3

pyridine

1) Mg
2) CO_2
3) H_3O^+

$SOCl_2$

(d) There are certainly many acceptable solutions to this problem. One such solution derives from the following retrosynthetic analysis. An explanation of each of the steps (*a-f*) follows.

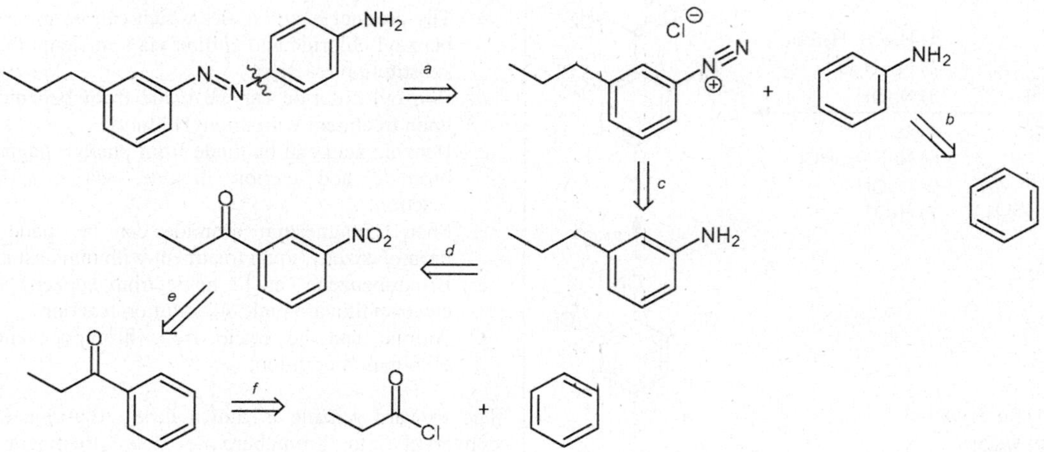

NH2
a
Cl
N≡N
NH2
b
c
NO2
d
NH2
e
f
Cl

a. The product is an azo dye, which can be made via azo coupling from a diazonium ion and an activated aromatic ring.
b. Aniline can be made from benzene via chlorination followed by treatment with sodium amide in liquid ammonia (elimination-addition).
c. The diazonium ion can be made from *meta*-propylaniline, upon treatment with sodium nitrite and HCl.
d. *meta*-Propylaniline can be made via reduction of the disubstituted ring shown.
e. The disubstituted ring can be prepared via nitration of an aromatic ketone (giving nitration at the *meta* position).
f. The aromatic ketone can be made via a Friedel-Crafts acylation.

The forward scheme is shown here. One equivalent of benzene is converted into aniline via chlorination (with Cl_2 and $AlCl_3$) followed by elimination-addition (with sodium amide in liquid ammonia). Another equivalent of benzene is subjected to a Friedel-Crafts acylation (thereby installing an acyl group), followed by nitration (in the *meta* position), followed by conditions that will reduce both the carbonyl group and the nitro group, giving *meta*-propylaniline. Treatment with sodium nitrite and HCl gives a diazonium ion, which is then treated with aniline (in an azo coupling process) to give the desired azo dye.

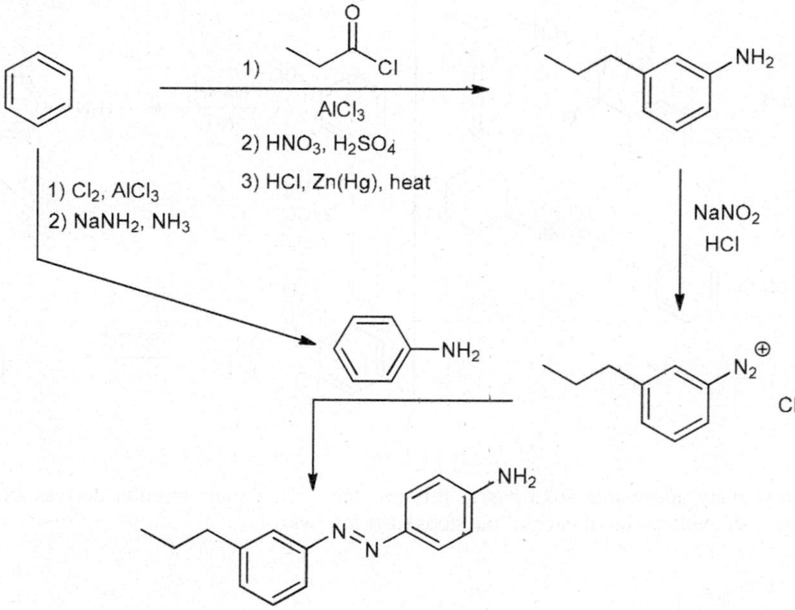

1)
Cl
AlCl3
2) HNO3, H2SO4
3) HCl, Zn(Hg), heat
NH2
1) Cl2, AlCl3
2) NaNH2, NH3
NaNO2
HCl
NH2
N2
Cl
NH2
N
N

22.69. The molecular formula ($C_6H_{15}N$) indicates no degrees of unsaturation (see Section 14.16), so all of the isomers must be saturated, acyclic amines. The IR data indicates that we are looking for structures that lack an N-H bond (i.e., tertiary amines). Let's first consider all tertiary amines in which the nitrogen atom is connected to two methyl groups. Since there must be a total of six carbon atoms in each structure (and the two methyl groups only account for two carbon atoms), we must consider all of the different ways in which the remaining four carbon atoms can be connected. There are four ways, shown here (*n*-butyl, isobutyl, *sec*-butyl, and *tert*-butyl):

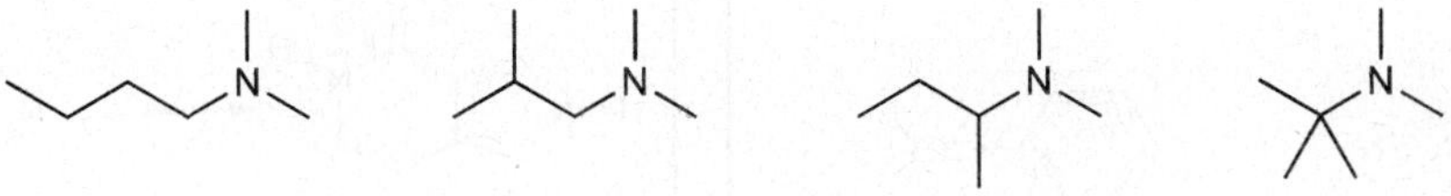

Now let's consider all isomers in which the nitrogen atom is connected to one methyl group and one ethyl group. Since there must be a total of six carbon atoms in each structure (while one methyl group and one ethyl group only account for three carbon atoms), we must consider all of the different ways in which the remaining three carbon atoms can be connected. There are only two ways, shown here (*n*-propyl and isopropyl):

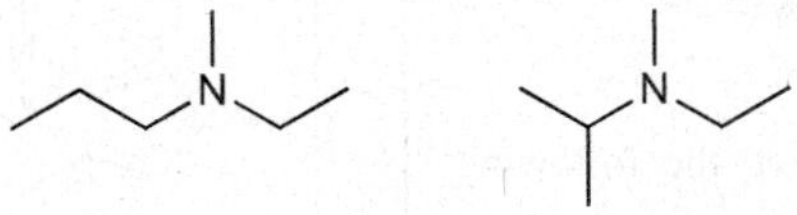

And finally, there is only one isomer in which the nitrogen atom has two ethyl groups. In this structure, the third group is also an ethyl group, giving triethylamine:

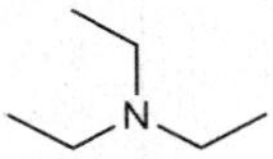

There are no other isomers possible. In total, we have seen seven different tertiary amines with the molecular formula $C_6H_{15}N$.

22.70. The compound has two nitrogen atoms. One of the nitrogen atoms (adjacent to the aromatic ring) exhibits a delocalized lone pair, while the other nitrogen atom (of the NH_2 group) exhibits a localized lone pair. The localized lone pair is more nucleophilic than the delocalized lone pair, so only the NH_2 group is converted into a quaternary ammonium ion, as shown here:

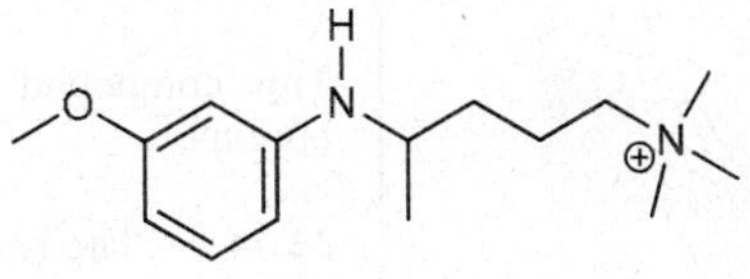

22.71. Pyrrole functions as a nucleophile (preferentially at C2, as discussed in Section 22.12) and attacks acetyl chloride, giving a charged tetrahedral intermediate that can then expel a chloride leaving group, thereby regenerating a carbonyl group. The resulting cation is resonance-stabilized, much like a sigma complex. Pyridine then functions as a base and removes a proton, thereby restoring aromaticity and generating the product:

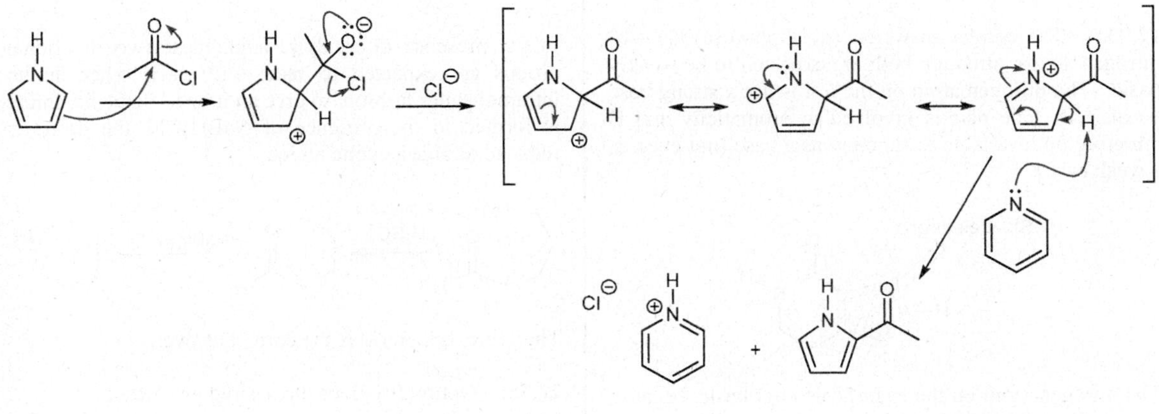

22.72.
(a) An azo coupling reaction will give the following product:

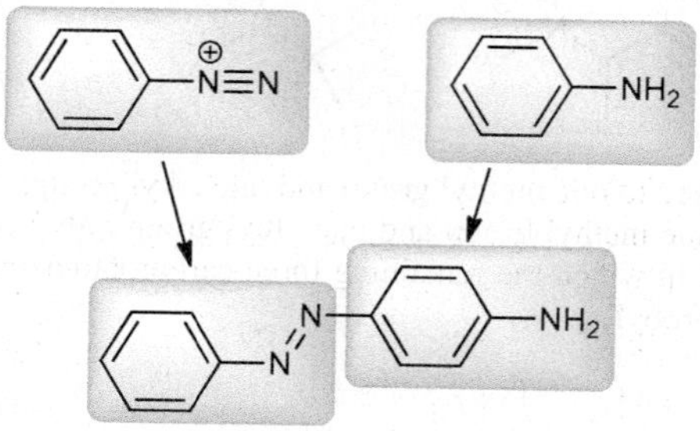

(b) An azo coupling reaction will give the following product:

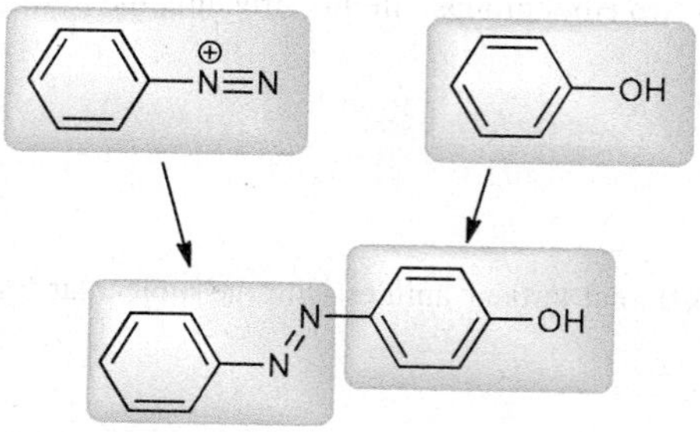

(c) An azo coupling reaction will give the following product:

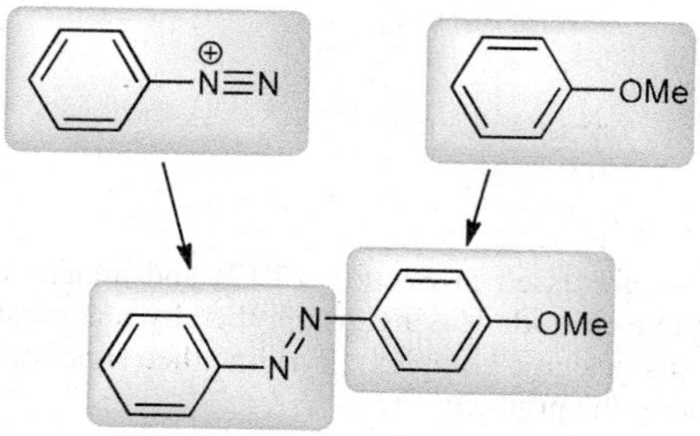

22.73. The correct answer is (d). Option (a) has two nitrogen atoms, although both are expected to be weakly basic. The nitrogen atom on the left is not a strong base because its lone pair is involved in aromaticity and is therefore not available to function as a base (not even as a weak base):

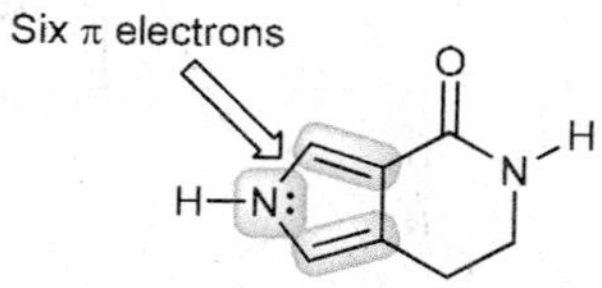

Six π electrons

The nitrogen atom on the right is also not basic, because its lone pair is delocalized by resonance:

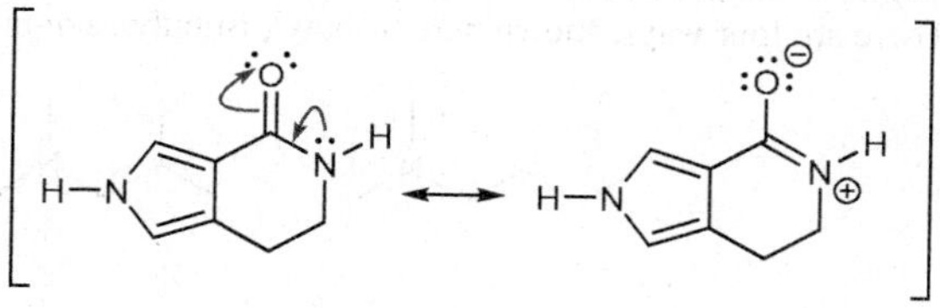

In option (b), the nitrogen atom on the right has a lone pair that is delocalized,

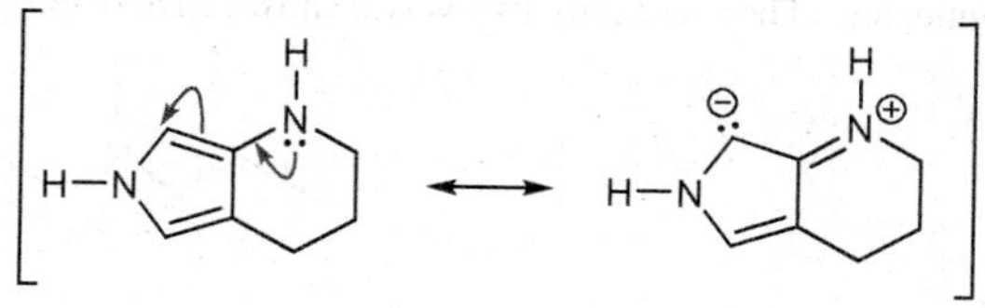

but the delocalization is not as pronounced as it is in the case of an amide (because a resonance structure with C^- is less significant than a resonance structure with O^-). This compound might function as a base, although it would not be expected to be a very strong base, so we continue our analysis of the remaining two options.
Option (c) has only one nitrogen atom, and its lone pair is involved in aromaticity.
Option (d) is the correct answer, because it has a nitrogen atom (on the right side) with a localized lone pair:

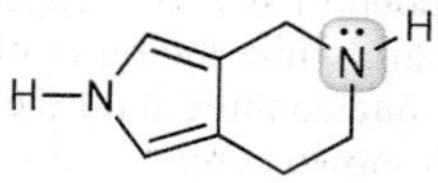

This compound is the strongest base among the four options.

22.74. The reactant has both an amino group and an aldehyde group:

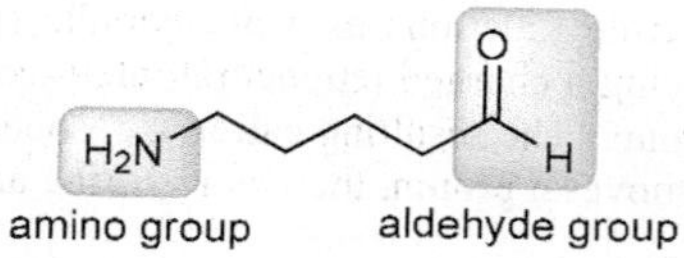

amino group aldehyde group

In the presence of catalytic acid, these two functional groups are expected to react with each other, in an intramolecular fashion, to give an imine. Since the imine is formed in the presence of $NaBH_3CN$, the imine is reduced to give a cyclic amine:

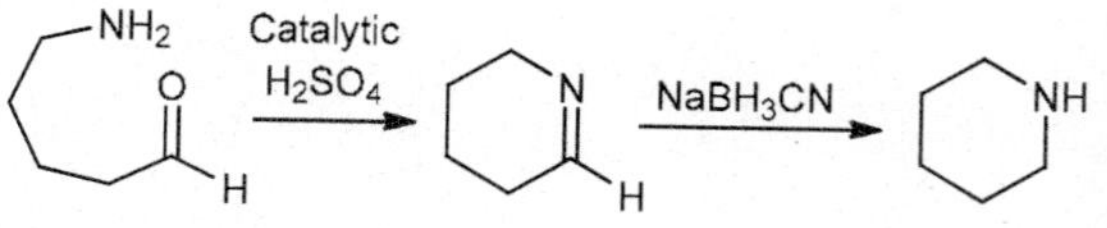

Catalytic H_2SO_4 → $NaBH_3CN$ →

Therefore, option (a) is the correct answer.

22.75. Option (b) gives the desired product:

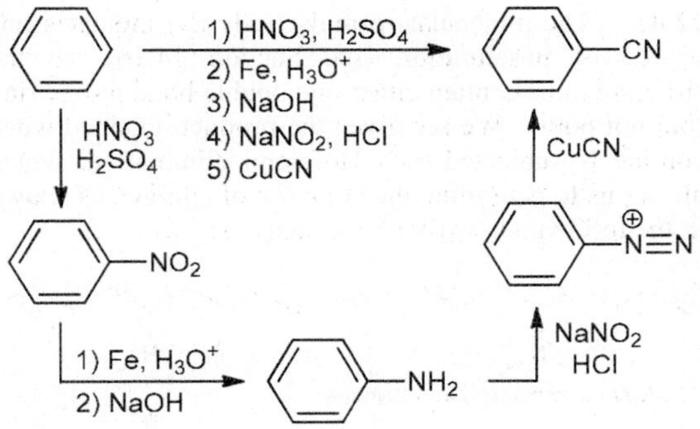

Nitration of the ring gives nitrobenzene, which is then reduced to give aniline. Aniline is then converted into a diazonium ion, followed by reaction with CuCN to give the product. In options (c) and (d), the second step (in each case) is not a viable reaction. Option (a) is not a known reaction. Benzene is not expected to react with HCN.

22.76. The correct answer is (a). The starting material is an amine, and the reagents indicate a Hofmann elimination. There are two β positions, but they are equivalent because of symmetry. As such, there is only one possible regiochemical outcome for the elimination process:

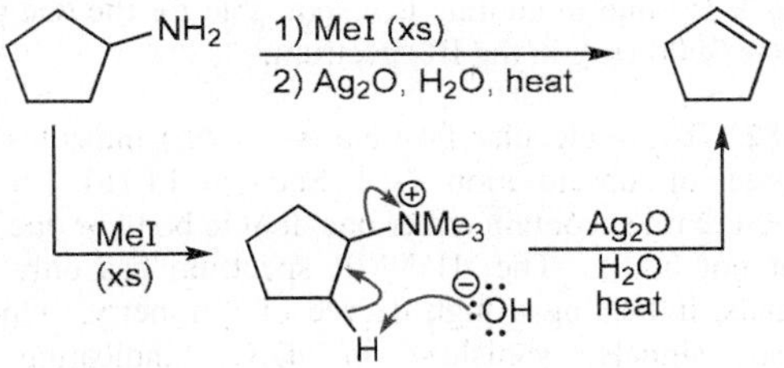

22.77. The correct answer is (c). The reaction sequence begins with acylation of the amino group. With the amino group protected as an amide, the aromatic ring can safely undergo nitration of the ring (in the *para* position). After nitration is complete, the acetyl group is removed by basic hydrolysis of the amide:

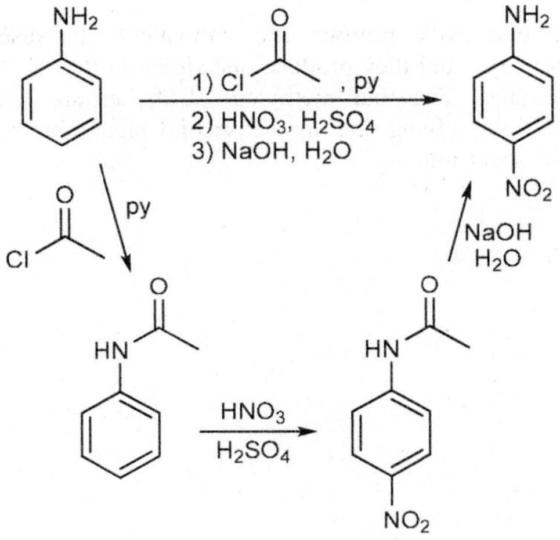

22.78. The correct answer is (c). Halides are *ortho-para* directing, but the bromine and chlorine atoms have a *meta* relationship in the product, so a *meta* directing group must have been present at some point during the synthesis. A nitro group can be used for its *meta*-directing effect, and the nitro group can be replaced with a halide by using a Sandmeyer reaction:

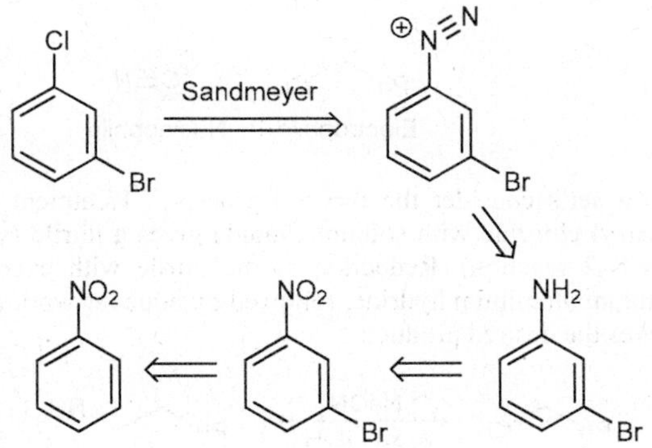

Now let's consider the forward process. The synthesis begins with nitration of benzene. Nitrobenzene is then treated with Br_2 and a Lewis acid, thereby installing a bromine atom in the *meta* position. Reduction of the nitro group gives *meta*-bromoaniline, which is then converted into an aromatic diazonium ion upon treatment with nitrous acid (produced *in situ* by combining sodium nitrite and HCl). The aromatic diazonium ion is then treated with CuCl to give a Sandmeyer reaction that affords the desired product, as shown:

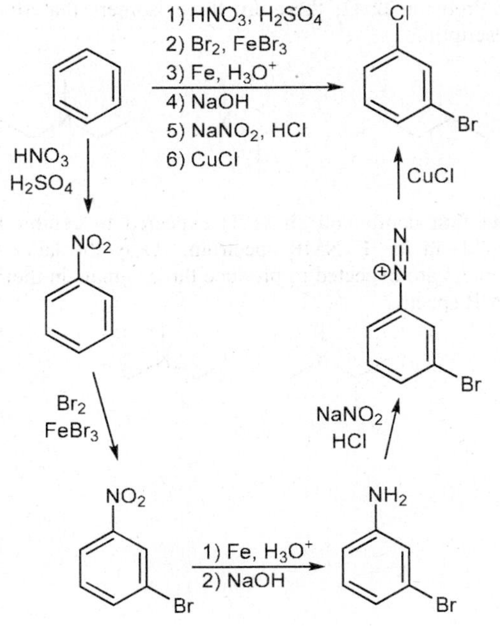

22.79. The correct answer is (a). The target molecule is an amine, and the carbon chain has been extended by one carbon atom. The amine can be prepared from the corresponding nitrile, which is suitable for the required C–C bond disconnection:

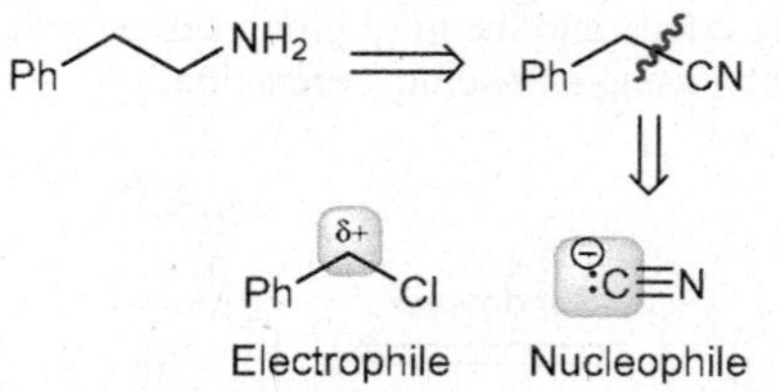

Ph NH$_2$ ⟹ Ph CN

Ph Cl (δ+) :C≡N (⊖)

Electrophile Nucleophile

Now let's consider the forward process. Treatment of benzyl chloride with sodium cyanide gives a nitrile (via an S$_N$2 reaction). Reduction of the nitrile with excess lithium aluminum hydride, followed by aqueous workup, gives the desired product:

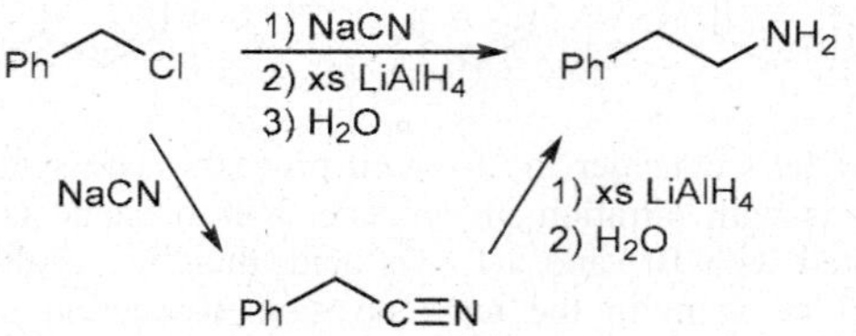

Ph Cl → 1) NaCN 2) xs LiAlH$_4$ 3) H$_2$O → Ph NH$_2$

NaCN ↘ Ph C≡N ↗ 1) xs LiAlH$_4$ 2) H$_2$O

22.80. The molecular formula ($C_5H_{13}N$) indicates no degrees of unsaturation (see Section 14.16), so all of the isomers must be saturated, acyclic amines. The IR data indicates that we are looking for structures that lack an N-H bond (*i.e.*, tertiary amines). As seen in the solution to Problem **22.42**, there are three isomers that fit this description:

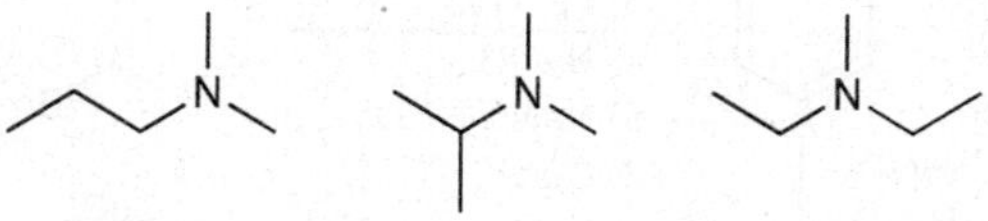

The first compound above is expected to exhibit four signals in its ^{1}H NMR spectrum. Only the latter two isomers are expected to produce three signals in their ^{1}H NMR spectra:

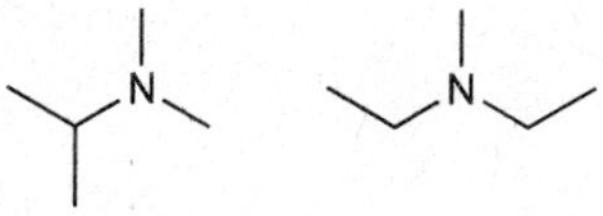

22.81. The molecular formula ($C_8H_{17}N$) indicates one degree of unsaturation (see Section 14.16), so the structure must contain either one double bond or one ring (but not both). We are given the product obtained when coniine is subjected to a Hofmann elimination, which allows us to determine the structure of coniine, as shown in the following retrosynthetic analysis:

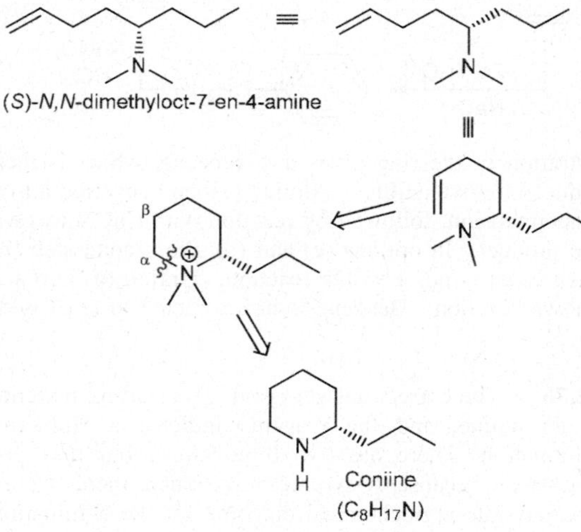

(*S*)-*N*,*N*-dimethyloct-7-en-4-amine

β α

Coniine ($C_8H_{17}N$)

The N-H bond in coniine is responsible for the one peak above 3000 cm^{-1} in the IR spectrum.

22.82. The molecular formula ($C_4H_{10}N_2$) indicates one degree of unsaturation (see Section 14.16), so the structure must contain either one double bond or one ring (but not both). The ^{1}H NMR spectrum has only two signals, indicating a high degree of symmetry. One of these signals vanishes in D_2O, indicating an exchangeable proton, consistent with an N-H bond. The other signal must account for all of the other protons. The following structure accounts for all of the observations:

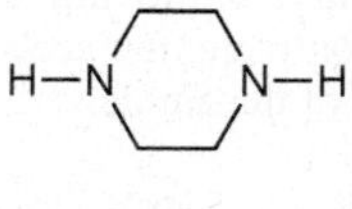

H—N N—H

The two N-H protons are equivalent (because of symmetry), and they produce one signal in the ^{1}H NMR spectrum. The four methylene (CH_2) groups are all equivalent, giving rise to the second signal in the ^{1}H NMR spectrum.

22.83. There are certainly many acceptable solutions to this problem. One such solution derives from the following retrosynthetic analysis, in which the final step of the synthesis employs the strategy described in the problem statement (opening an epoxide with an amine functioning as the nucleophile). An explanation of each of the steps (*a-f*) follows.

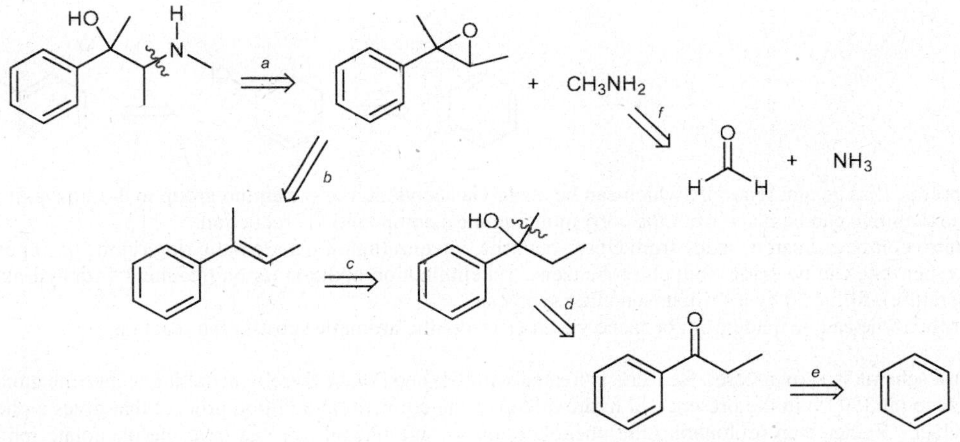

a. The product can be made by treating the appropriate epoxide with methyl amine, as described in the problem statement.
b. The epoxide can be made from the corresponding alkene, upon treatment with a peroxyacid.
c. The alkene can be made from an alcohol, via acid-catalyzed dehydration.
d. The alcohol can be made from a ketone, via a Grignard reaction (with methyl magnesium bromide, followed by aqueous acidic workup).
e. The ketone can be made from benzene via a Friedel-Crafts acylation.
f. Methylamine can be made from formaldehyde and ammonia via a reductive amination.

The forward scheme is shown here. Benzene is treated with an acyl chloride and $AlCl_3$, thereby installing an acyl group via a Friedel-Crafts acylation. The resulting ketone is then treated with methyl magnesium bromide, followed by aqueous acidic workup, to give a tertiary alcohol. This alcohol undergoes dehydration upon treatment with concentrated sulfuric acid and heat, giving the more substituted alkene (Zaitsev product). Treating the alkene with a peroxyacid gives an epoxide. The epoxide is then converted into the desired product upon treatment with methyl amine (which can be made from formaldehyde and ammonia via a reductive amination process).

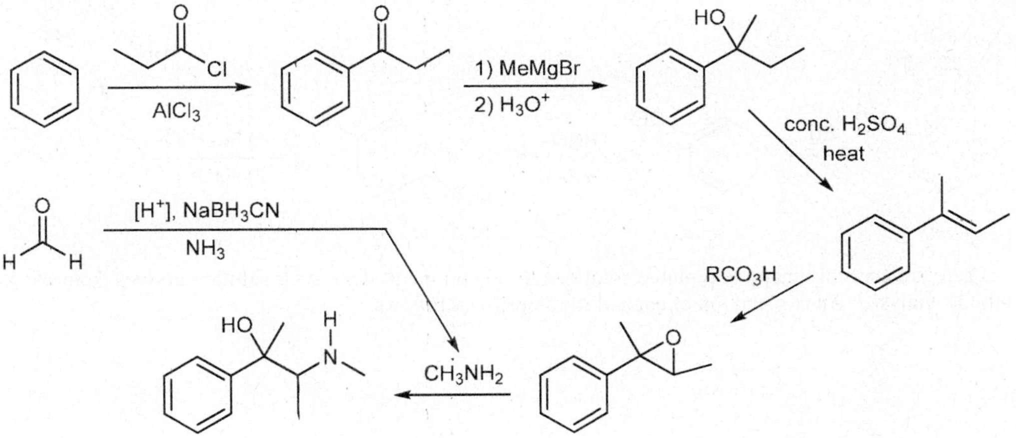

22.84. There are certainly many acceptable solutions to this problem. One such solution derives from the following retrosynthetic analysis. An explanation of each of the steps (*a-e*) follows.

a. The product has an amide group, which can be made via acetylation of the amino group in 4-ethoxyaniline.
b. 4-Ethoxyaniline can be made from the corresponding nitro compound via reduction.
c. The nitro compound can be made from ethoxybenzene, via nitration of the aromatic ring (in the *para* position).
d. Ethoxybenzene can be made from chlorobenzene, via elimination-addition (upon treatment with hydroxide at high temperature), followed by a Williamson ether synthesis.
e. Chlorobenzene can be made from benzene via an electrophilic aromatic substitution reaction.

The forward scheme is shown here. Benzene is treated with Cl_2 and $AlCl_3$, thereby installing a chlorine atom. Heating chlorobenzene (at 350°C) in the presence of hydroxide gives an elimination-addition process that gives a phenolate ion as the product. Rather than protonating the phenolate ion to give phenol, we can treat the phenolate ion with ethyl iodide, giving ethoxybenzene via an S_N2 process. When ethoxybenzene is treated with a mixture of sulfuric acid and nitric acid, nitration occurs at the *para* position (the ethoxy group is an *ortho-para* director, and the *para* position is favored because of steric factors). Reduction of the nitro group, followed by acetylation of the resulting amino group, gives the desired product.

1) Cl_2, $AlCl_3$
2) NaOH, 350°C
3) EtI
4) HNO_3, H_2SO_4
5) Fe, H_3O^+
6) NaOH
7) CH_3COCl, py

Cl_2 / $AlCl_3$; NaOH / 350°C; EtI; HNO_3 / H_2SO_4; 1) Fe, H_3O^+ 2) NaOH; Cl, py

22.85. There are certainly many acceptable solutions to this problem. One such solution derives from the following retrosynthetic analysis. An explanation of each of the steps (*a-g*) follows.

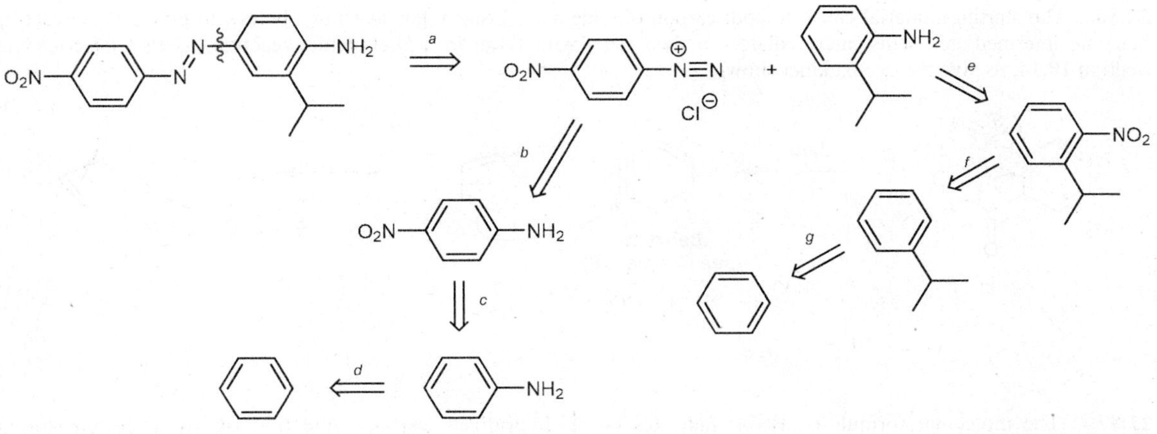

a. The product is an azo dye, which can be made via azo coupling from a diazonium ion and an activated aromatic ring.
b. The diazonium ion can be made from *para*-nitroaniline, upon treatment with sodium nitrite and HCl.
c. *para*-Nitroaniline can be made from aniline via nitration. This process requires the use of a protective group (via acylation) prior to nitration, and removal of the protective group after nitration (because aniline will not directly undergo nitration to give *para*-nitroaniline).
d. Aniline can be made from benzene via nitration followed by reduction of the nitro group.
e. The substituted aniline can be made via reduction of the corresponding nitro compound.
f. The nitro compound can be made via nitration. This process requires the use of a blocking group (via sulfonation) prior to nitration, so that nitration will occur at the *ortho* position (rather than the *para* position). Desulfonation is then required after nitration (to remove the sulfonic acid blocking group).
g. Isopropyl benzene can be made from benzene via a Friedel-Crafts alkylation.

The forward scheme is shown here:

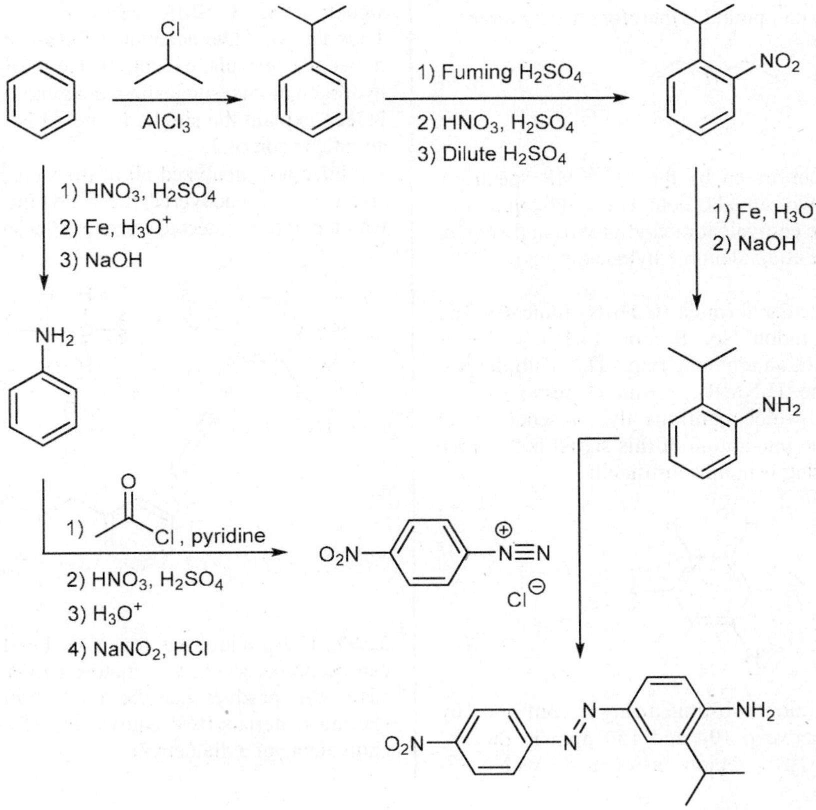

22.86. The starting material can lose both carbon dioxide and nitrogen gas, as shown below, to give a very reactive benzyne intermediate. This intermediate can then react with furan in a Diels-Alder reaction, as first described in Section 18.14, to give the cycloadduct shown.

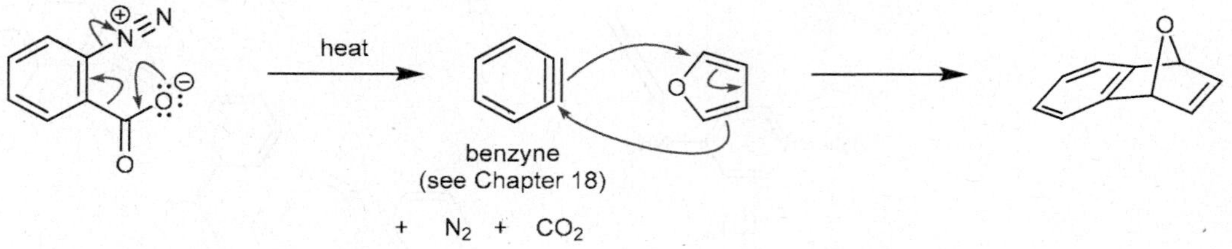

heat

benzyne
(see Chapter 18)

+ N_2 + CO_2

22.87. The molecular formula ($C_6H_{15}N$) indicates no degrees of unsaturation (see Section 14.16), so the structure does not contain a π bond or a ring. That is, the structure must be a saturated, acyclic amine. The ^1H NMR spectrum exhibits the characteristic pattern of an ethyl group (a quartet with an integration of 2 and a triplet with an integration of 3).

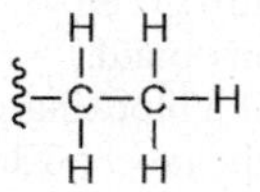

There are no other signals in the ^1H NMR spectrum, indicating a high degree of symmetry. That is, these two signals must account for all fifteen protons in the compound, indicating that there are three equivalent ethyl groups. The compound is therefore triethylamine:

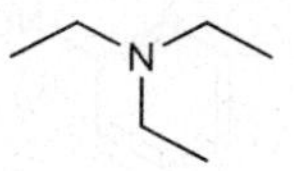

This analysis is confirmed by the ^{13}C NMR spectrum, which has only two signals, both below 60 ppm (one signal for the three equivalent methyl groups and another signal for the three equivalent methylene groups).

22.88. The molecular formula ($C_8H_{11}N$) indicates four degrees of unsaturation (see Section 14.16), which is highly suggestive of an aromatic ring. The multiplet just above 7 ppm in the ^1H NMR spectrum corresponds with aromatic protons, which confirms the presence of an aromatic ring. The integration of this signal is 5, which indicates that the ring is monosubstituted:

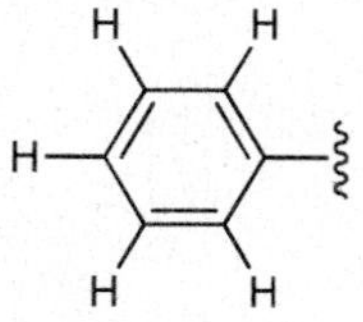

The presence of a monosubstituted ring is confirmed by the four signals between 100 and 150 ppm in the ^{13}C NMR spectrum (the region associated with sp^2-hybridized carbon atoms), just as expected for a monosubstituted aromatic ring:

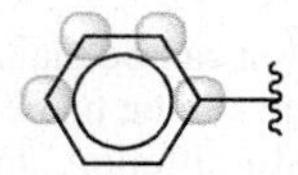

In the ^1H NMR spectrum, the pair of triplets (each with an integration of 2) indicates a pair of neighboring methylene groups:

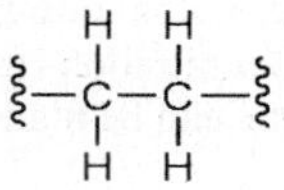

These methylene groups account for the two upfield signals in the ^{13}C NMR spectrum.

Thus far, we have accounted for all of the atoms in the molecular formula, except for one nitrogen atom and two hydrogen atoms, suggesting an amino group. This would indeed explain the singlet in the ^1H NMR spectrum with an integration of 2.

We have now analyzed all of the signals in both spectra, and we have uncovered the following three fragments, which can be connected to each other in only one way:

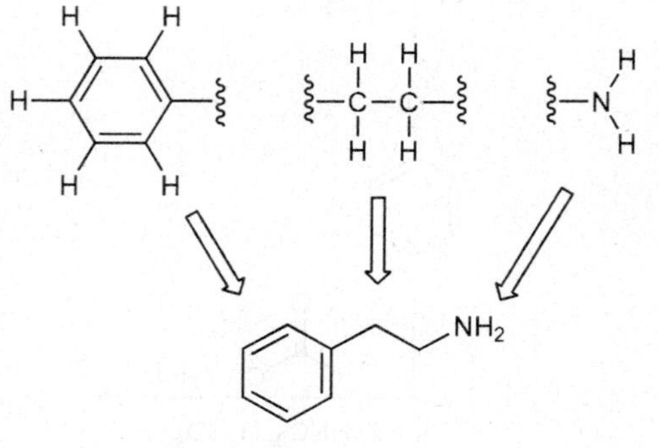

NH_2

22.89. The product has four C-N bonds, each of which can be prepared via a reductive amination process. As such, the product can be made from the following starting materials (two equivalents of ammonia and two equivalents of a dialdehyde):

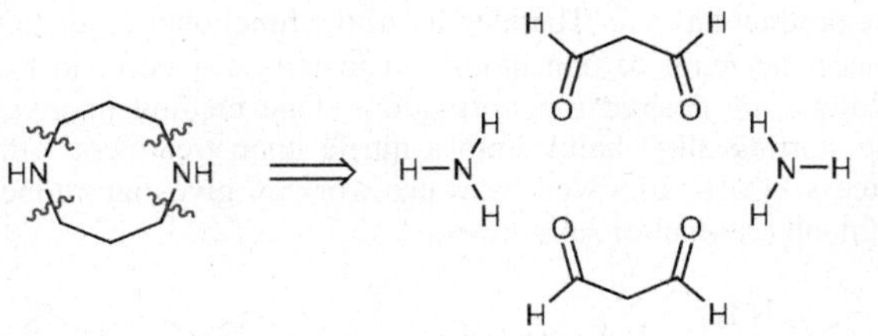

The necessary dialdehyde has only three carbon atoms, but the starting material (benzene) has six carbon atoms. This suggests that we must somehow break apart the aromatic ring into two fragments. This might seem impossible at first, as we have seen that aromatic rings are particularly stable. We did, however, cover a reaction that destroys aromaticity (a Birch reduction will convert benzene into 1,4-cyclohexadiene). If a Birch reduction is followed by ozonolysis, the resulting dialdehyde can then be treated with excess ammonia and excess sodium cyanoborohydride (with an acid catalyst) to give the product:

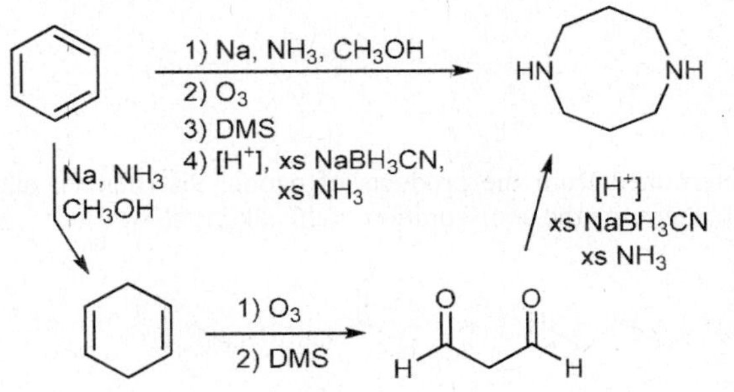

1) Na, NH_3, CH_3OH
2) O_3
3) DMS
4) [H^+], xs $NaBH_3CN$, xs NH_3

Na, NH_3, CH_3OH

1) O_3
2) DMS

[H^+]
xs $NaBH_3CN$
xs NH_3

22.90. The starting material exhibits a five-membered ring, while the product exhibits a six-membered ring that contains a nitrogen atom. Since we have not learned a way to insert a nitrogen atom into an existing ring, we must consider opening the ring, and then closing it back up again (in a way that incorporates the nitrogen atom into the ring). There are certainly many acceptable synthetic routes. One such route derives from the following retrosynthetic analysis. An explanation of each of the steps (*a-d*) follows.

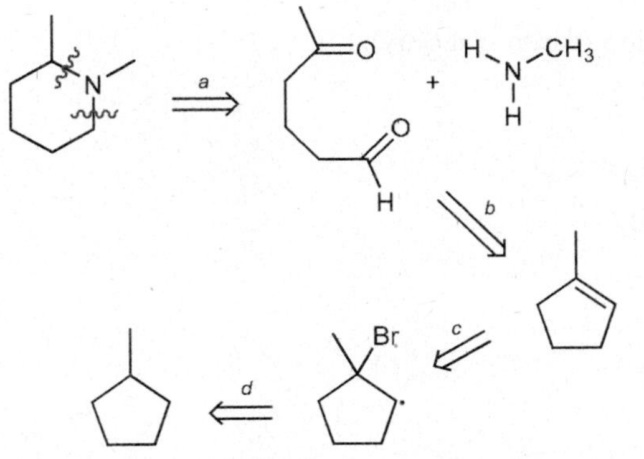

a. The product is a tertiary amine, and it can be made from a dicarbonyl compound and methyl amine, via two reductive amination processes.

b. The dicarbonyl compound can be made via ozonolysis of 1-methylcyclopentene.

c. 1-Methylcyclopentene can be made from 1-bromo-1-methylcyclopentane via an elimination reaction.

d. 1-Bromo-1-methylcyclopentane can be made from methylcyclopentane via radical bromination.

The forward scheme is shown here. Methylcyclopentane will undergo radical bromination selectively at the tertiary position, giving a tertiary alkyl bromide. This alkyl bromide will undergo an elimination reaction upon treatment with a non-hindered strong base, such as sodium ethoxide, to give the more substituted, Zaitsev product. Ozonolysis of the resulting alkene gives a dicarbonyl compound, which can then be converted into the product upon treatment with methyl amine and sodium cyanoborohydride (with acid catalysis):

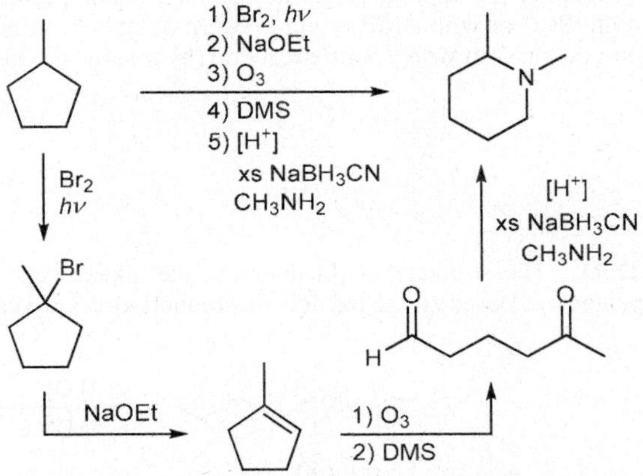

1) Br_2, $h\nu$
2) NaOEt
3) O_3
4) DMS
5) [H^+]
xs $NaBH_3CN$
CH_3NH_2

Br_2, $h\nu$

NaOEt

1) O_3
2) DMS

[H^+]
xs $NaBH_3CN$
CH_3NH_2

22.91. This amine is tertiary, and each C-N bond can be made via a reductive amination process, as shown in the following retrosynthetic analysis:

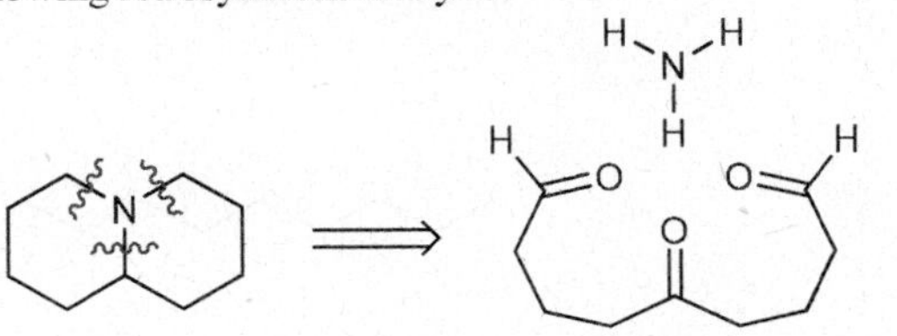

The necessary tricarbonyl compound can be made from the starting material via ozonolysis, as shown in the following forward scheme:

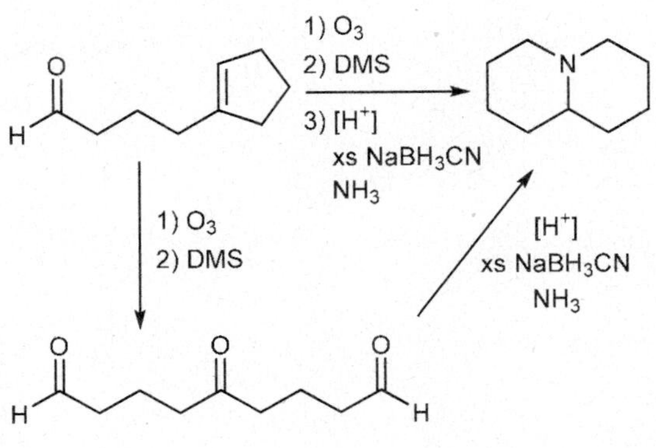

1) O_3
2) DMS
3) [H^+]
xs $NaBH_3CN$
NH_3

1) O_3
2) DMS

[H^+]
xs $NaBH_3CN$
NH_3

22.92. The starting material has nine carbon atoms, while the product has ten. The identity of the functional group has also changed, so we must propose a synthesis that introduces the tenth carbon atom and installs a π bond in the appropriate location. There are certainly many ways to achieve the desired transformation. One method involves introduction of the tenth carbon atom via conversion of the starting alkyl halide into a nitrile upon treatment with cyanide (an SN2 reaction). Reduction of the nitrile with excess $LiAlH_4$, followed by water workup, gives an amine, which can then be converted into the desired alkene via a Hofmann elimination, as shown:

Br
1) NaCN
2) xs $LiAlH_4$
3) H_2O
NH_2
1) excess CH_3I
2) Ag_2O, H_2O
heat
NaCN
CN
1) xs $LiAlH_4$
2) H_2O

As an alternate approach, the starting alkyl halide can be treated with NaOH to give an alcohol, which can be oxidized (with PCC or with DMP or via a Swern oxidation) to give an aldehyde. This aldehyde can then be converted directly into the product with a Wittig reaction (by treatment with $Ph_3P{=}CH_2$).

22.93. The structure of the intermediate alkene can be determined from the products of ozonolysis (butanal and pentanal). Based on the ozonolysis products alone, we cannot determine the configuration of the alkene (*E* or *Z*):

1) O_3
2) DMS
(*E* or *Z*)
H + H
O O

The *E* alkene can be made via a Hofmann elimination from two possible amines:

NH_2
NH_2

But only one of these amines lacks a chiral center, as shown below (due to symmetry):

NH_2
Compound **A**

22.94. Sodium nitrite is protonated upon treatment with HCl, giving nitrous acid (HONO), which can be further protonated under these acidic conditions. The resulting cation can lose water (an excellent leaving group), giving a nitrosonium ion, as shown:

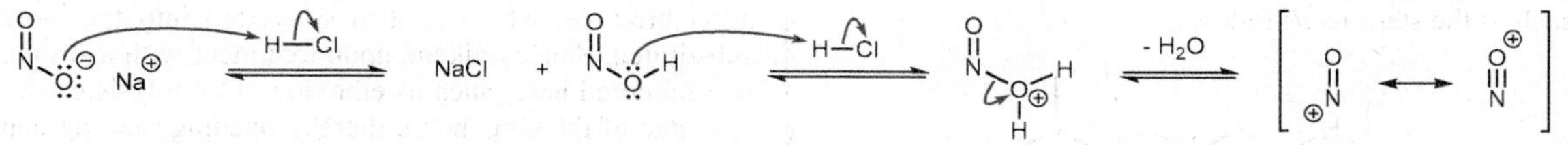

The amino group attacks the nitrosonium ion, giving a cation, which then loses a proton to give an intermediate *N*-nitrosamine. Protonation, followed by deprotonation, gives a tautomer of the *N*-nitrosamine. Protonation of this tautomer, followed by loss of a leaving group (water) gives a diazonium ion. Loss of the diazonium group (as N_2 gas) would generate a primary carbocation, which is unlikely to occur because of the high energy cost associated with primary carbocations. However, a hydride shift can occur at the same time as the leaving group leaves (see the discussion at the very end of Section 7.8), giving a secondary carbocation. A subsequent methyl shift generates a more stable, tertiary benzylic carbocation. Finally, deprotonation gives the product. Notice that in acidic conditions, water functions as the base for all deprotonation steps, rather than hydroxide (which is not measurably present in acidic conditions).

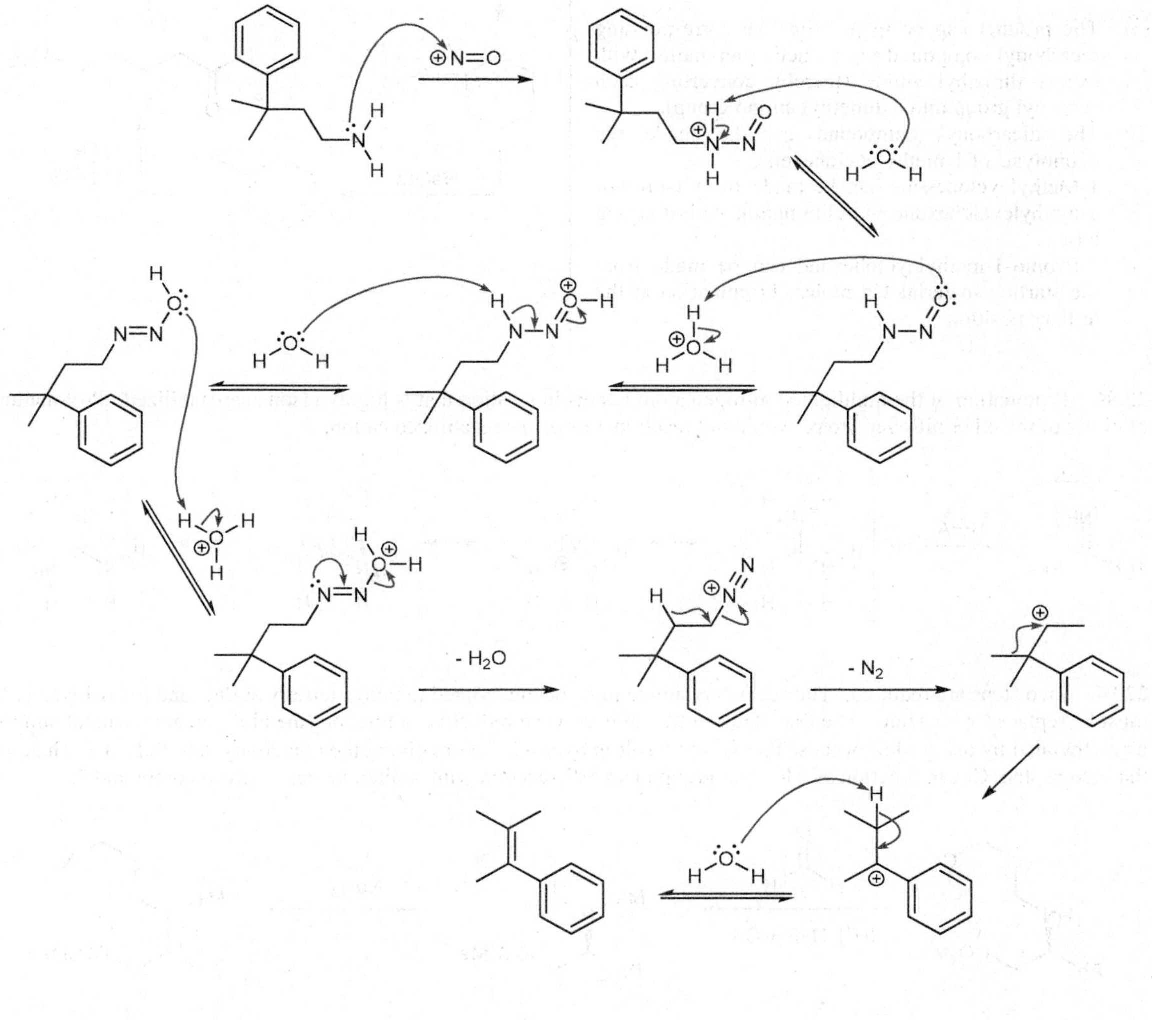

22.95. There are certainly many acceptable solutions to this problem. One such solution derives from the following retrosynthetic analysis. An explanation of each of the steps (*a-d*) follows.

a. The product can be made from the corresponding dicarbonyl compound via reductive amination with excess dimethyl amine (thereby converting each carbonyl group into a dimethyl amino group).

b. The dicarbonyl compound can be made via ozonolysis of 1-methylcyclohexene.

c. 1-Methylcyclohexene can be made from 1-bromo-1-methylcyclohexane via elimination with a strong base.

d. 1-Bromo-1-methylcyclohexane can be made from the starting material via radical bromination at the tertiary position.

Now let's draw the forward scheme. Radical bromination of the starting cycloalkane gives a tertiary alkyl bromide, which is then converted into the more substituted, Zaitsev alkene upon treatment with a strong, non-hindered base, such as ethoxide. Ozonolysis causes cleavage of the C=C bond, thereby opening the ring and giving a dicarbonyl compound, which can then be converted into the product via reductive amination, upon treatment with excess dimethylamine and sodium cyanoborohydride with acid catalysis.

1) Br_2, $h\nu$
2) NaOEt
3) O_3
4) DMS
5) $[H^+]$
xs $NaBH_3CN$
xs $(CH_3)_2NH$

Br_2, $h\nu$

Br

NaOEt

1) O_3
2) DMS

$[H^+]$
xs $NaBH_3CN$
xs $(CH_3)_2NH$

22.96. Protonation of the highlighted nitrogen atom results in a cation that is highly resonance-stabilized. Protonation of either of the other nitrogen atoms would not result in a resonance-stabilized cation.

22.97. Two steps are required. The secondary amine must be methylated to give a tertiary amine, and the halogen (Cl) must be replaced with azide. The first step can be achieved via a reductive amination (the nitrogen atom cannot simply be methylated by using MeI, because that would result in over-alkylation, giving the quaternary salt, R_4N^+ I^-). Then, in the second step, Cl can function as a leaving group in an S_N2 reaction with sodium azide, to afford compound **2**.

$[H^+]$, $NaBH_3CN$

NaN_3

1

2

22.98. We begin by identifying the bond that must be made:

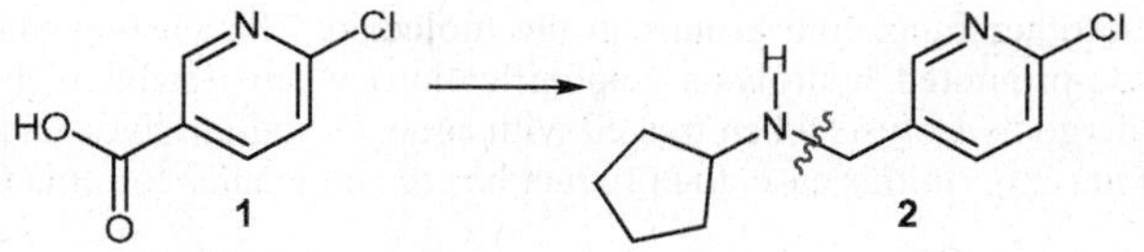

Forming this C-N bond will require the use of cyclopentylamine:

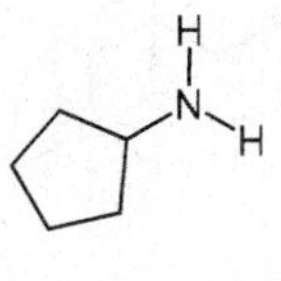

In order to form the desired C-N bond, the carboxylic acid can be reduced to an alcohol (which can be achieved with excess $LiAlH_4$ followed by acidic workup), but then we must decide what to do with the alcohol.

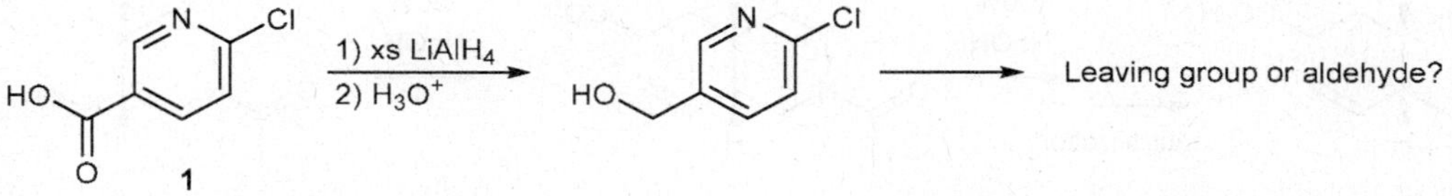

1) xs $LiAlH_4$
2) H_3O^+

Leaving group or aldehyde?

We can either convert the alcohol into a leaving group (OTs, Cl, Br) and then treat it with cyclopentylamine in an S_N2 reaction, or we can convert the alcohol into an aldehyde and then perform a reductive amination with cyclopentylamine:

S_N2

reductive amination

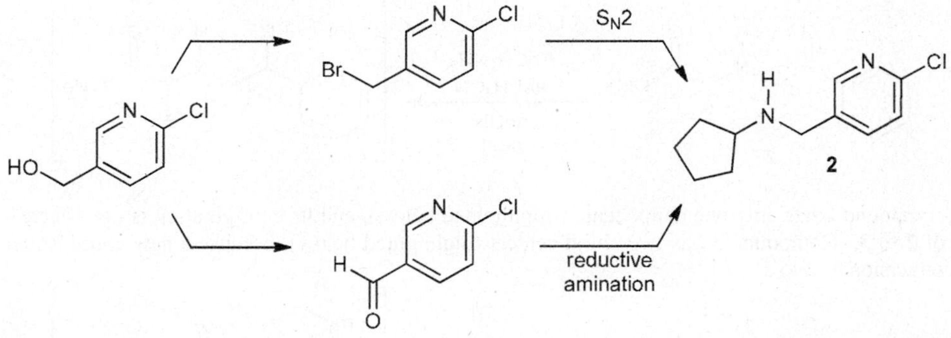

The first path is expected to be inefficient, because it will be difficult to achieve monoalkylation. More likely, polyalkylation will occur, especially with the activated benzylic leaving group. The second path is expected to be more efficient, which gives the following synthesis:

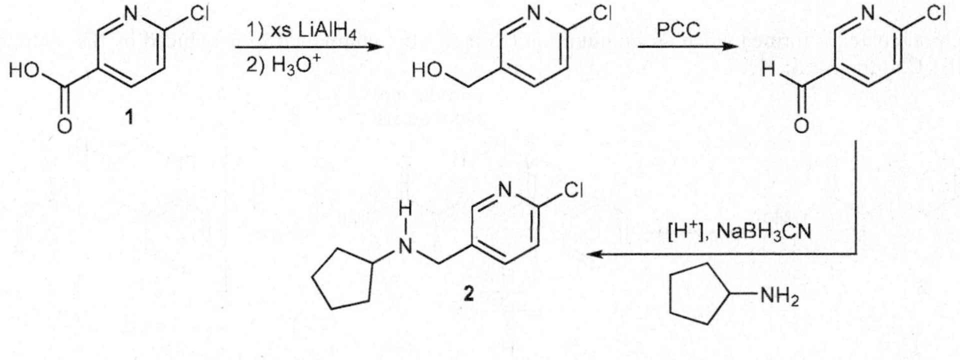

1) xs $LiAlH_4$
2) H_3O^+

PCC

[H^+], $NaBH_3CN$

22.99. The starting material has a primary amino group. When treated with excess methyl iodide, it will undergo exhaustive methylation to produce a quaternary ammonium salt. When this salt is treated with NaOH, first an anion exchange occurs, followed by an E2 elimination to produce an alkene (Hofmann elimination). At this stage it would be useful to take inventory of the other functional groups in the molecule. The starting material also contains an ethyl ester. Esters will undergo base-promoted hydrolysis (saponification) when treated with aqueous sodium hydroxide (Chapter 20). Amides also undergo hydrolysis when treated with aqueous sodium hydroxide, to afford a carboxylic acid as well as an amine (also Chapter 20). In this case, the product has the molecular formula $C_{10}H_{17}NO_2$.

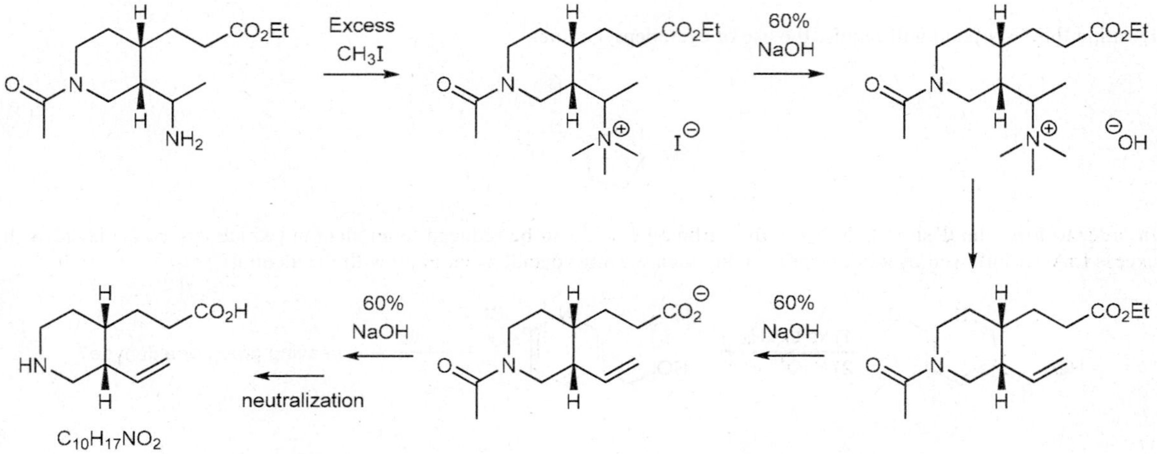

Excess CH_3I → 60% NaOH → 60% NaOH → 60% NaOH, neutralization → $C_{10}H_{17}NO_2$

22.100.

(a) The ketone group of compound **2** undergoes reductive amination (the ester group is unreactive under these conditions) to give compound **3**, as shown below:

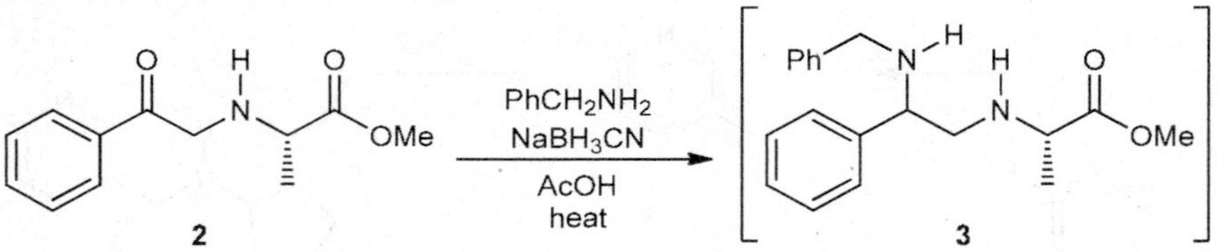

2 → ($PhCH_2NH_2$, $NaBH_3CN$, AcOH, heat) → **3**

Notice that compound **2** has only one chiral center (highlighted below), and its configuration is not affected during the conversion of **2** to **3**. Compound **3** has two chiral centers (highlighted below), because a new chiral center is created during the conversion of **2** to **3**:

2 → ($PhCH_2NH_2$, $NaBH_3CN$, AcOH, heat) → **3**

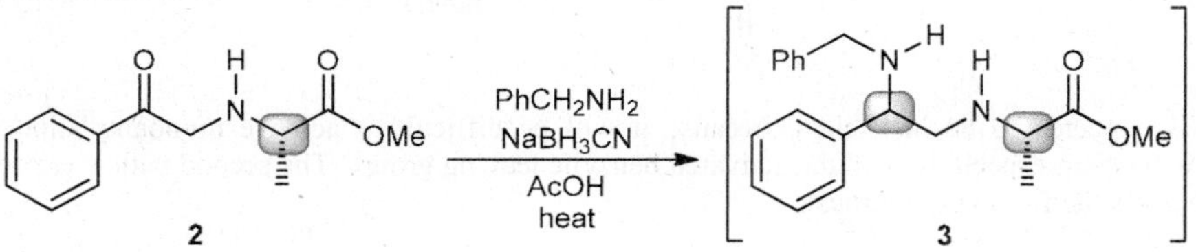

This new chiral center is formed when an iminium ion (formed from compound **2**) is reduced by the hydride reducing agent ($NaBH_3CN$) to give **3**:

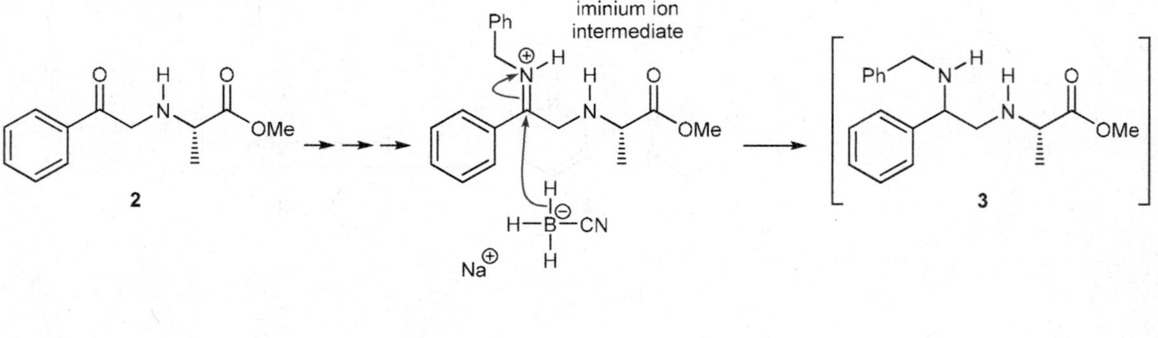

2 → → → iminium ion intermediate → **3**

Focus carefully on the reduction step above, in which the new chiral center is created. Since the reducing agent can approach from either face of the C=N π bond, we would expect the following two diastereomeric products:

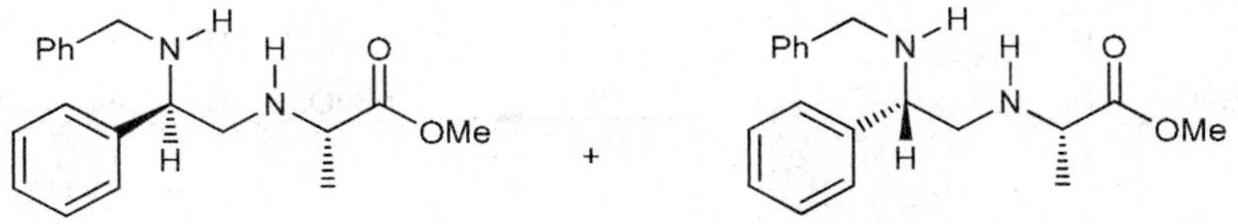

(b) The problem statement indicates that compound **3** undergoes a reaction that produces a cyclic amide. So we inspect the structure of **3** to determine which functional groups can react with each other to produce a cyclic amide. Compound **3** has two amine groups and one ester group,

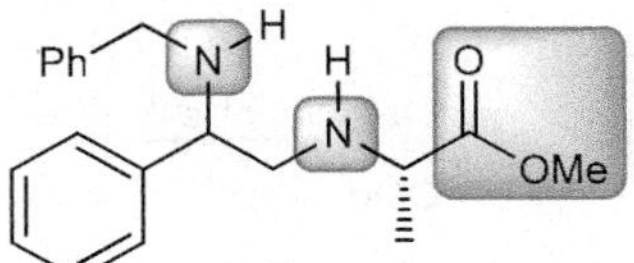

and we learned in Chapter 20 that an amide can be formed from an amine and an ester:

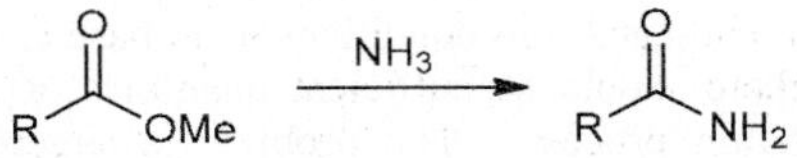

This type of reaction is generally slow and inefficient, but in our case, the reaction is intramolecular, so it can occur more rapidly. The reaction can occur in two possible ways, either forming a 3-membered ring or a 6-membered ring, depending on which amine group reacts with the ester group:

Formation of a 3-membered ring

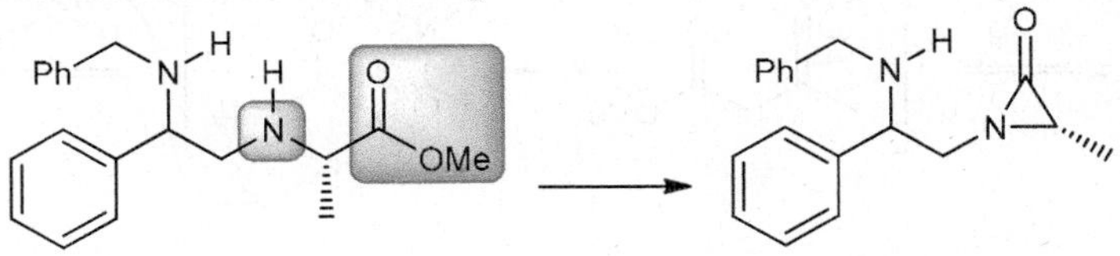

Formation of a 6-membered ring

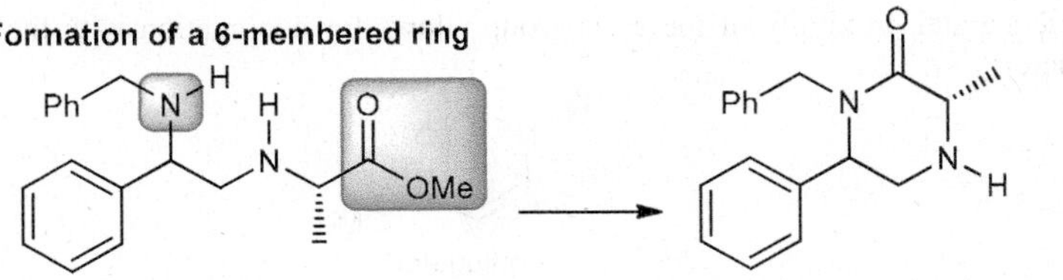

The 3-membered ring will be higher in energy than the 6-membered ring as a result of significant ring strain in the 3-membered ring. As such, the 6-membered ring will be formed as the product.

Since compound **3** was formed as a diastereomeric mixture, we expect that compound **4** will also be produced as a diastereomeric mixture:

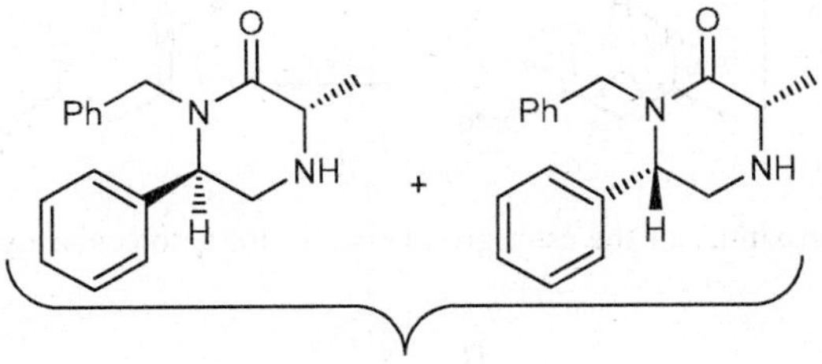

22.101.

(a) Hydrogenation of an azide is expected to give an amine.

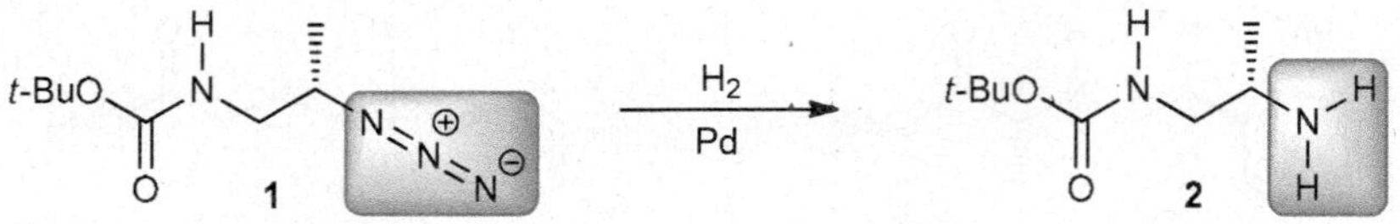

Compound **2** is a primary amine, and the problem statement indicates that it reacts with methyl acetoacetate to give an imine. We saw in Chapter 19 that an imine is the product generated when a primary amine is treated with a ketone. Methyl acetoacetate does have a ketone group, so we expect that compound **3** will have the following structure:

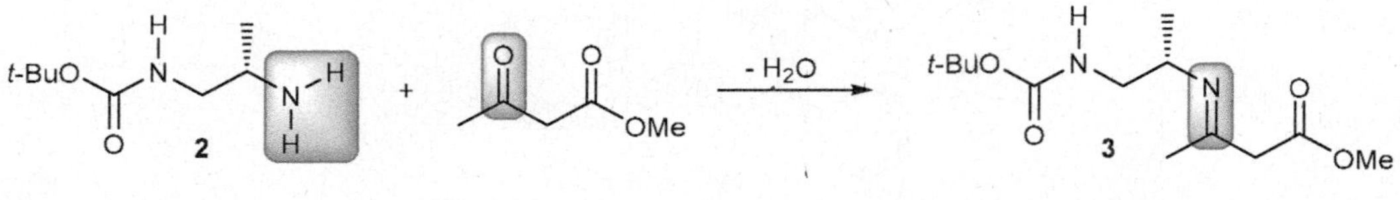

In Chapter 19, we saw that the process was catalyzed by acid, but in this case, the problem statement indicates that acid catalyst was not employed, and the reaction proceeded without it.

(b) Tautomerization can occur either in acid-catalyzed conditions or in base-catalyzed conditions. In fact, even if we do not introduce either acid or base, there should be sufficient quantities of either (adsorbed to the surface of the glassware) to catalyze the tautomerization process. The problem statement asks us to draw an acid-catalyzed mechanism. Much like keto-enol tautomerization, the process should require two steps (protonation and deprotonation). In acidic conditions, protonation occurs first, to give a resonance-stabilized cation, which then undergoes deprotonation:

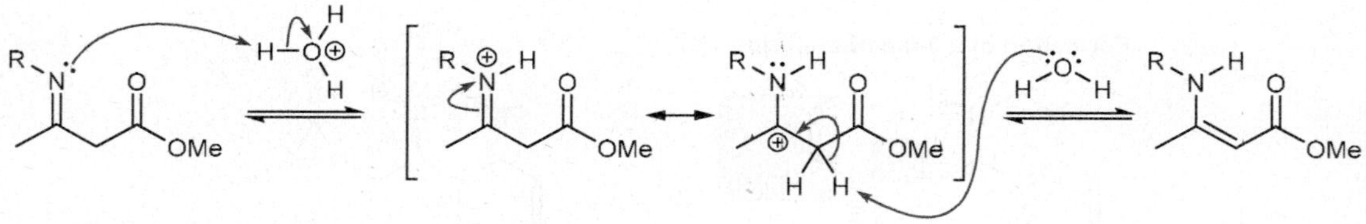

(c) The following are three reasons why the enamine is particularly stable in this case:

(1) The presence and proximity of the ester group allows for conjugation with the C=C bond, which is a stabilizing factor:

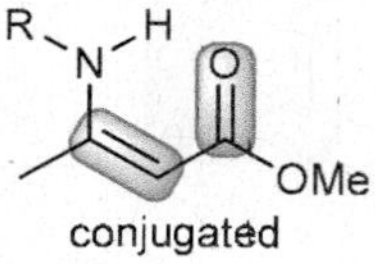

conjugated

(2) The lone pair of this enamine is particularly delocalized, as a result of the resonance structure shown below. This delocalization contributes to the stability of this enamine.

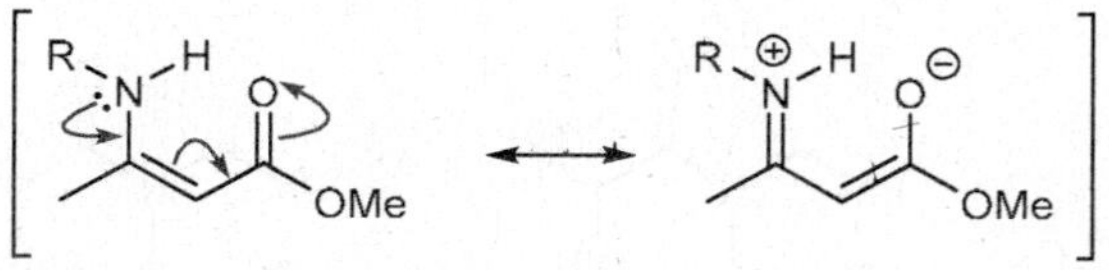

(3) The presence and proximity of the ester group enables intramolecular hydrogen bonding, which is also a stabilizing factor.

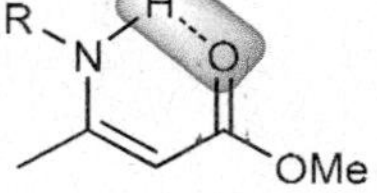

All three factors (described above) contribute to the enhanced stability of the enamine in this particular case, and as such, the equilibrium favors formation of the enamine.

22.102. The mechanism begins as we would expect for the formation of an imine by reaction of a primary amine and an aldehyde. The nucleophilic amino group of compound **1** attacks the electrophilic carbonyl group of the aldehyde to produce intermediate **3**. After three successive proton transfer steps, water is lost, by collapse of a charged tetrahedral intermediate, producing iminium ion **7**. The iminium ion is an electrophile that can be trapped by a nucleophile. In this example, the nucleophile is the attached aromatic ring. In Chapter 18, we learned about electrophilic aromatic substitution reactions (such as the Friedel-Crafts acylation); we know that the two methoxy groups in **7** sufficiently activate the aromatic ring, enabling the ring to function as a nucleophile and attack an electrophile. Nucleophilic attack by the aromatic ring will produce the sigma complex intermediate **8**, in which the electrophilic group has been added *para* to one of the methoxy groups. Next, deprotonation will restore aromaticity and produce compound **9**. This process is known as the Pictet-Spengler condensation reaction. In the final stage of the synthesis, the lone pair on the nitrogen atom will function as a nucleophile and initiate a back-side attack on the 1° alkyl chloride (S_N2-type process) to form the 5-membered ring. A final deprotonation produces desired product, compound **2**.

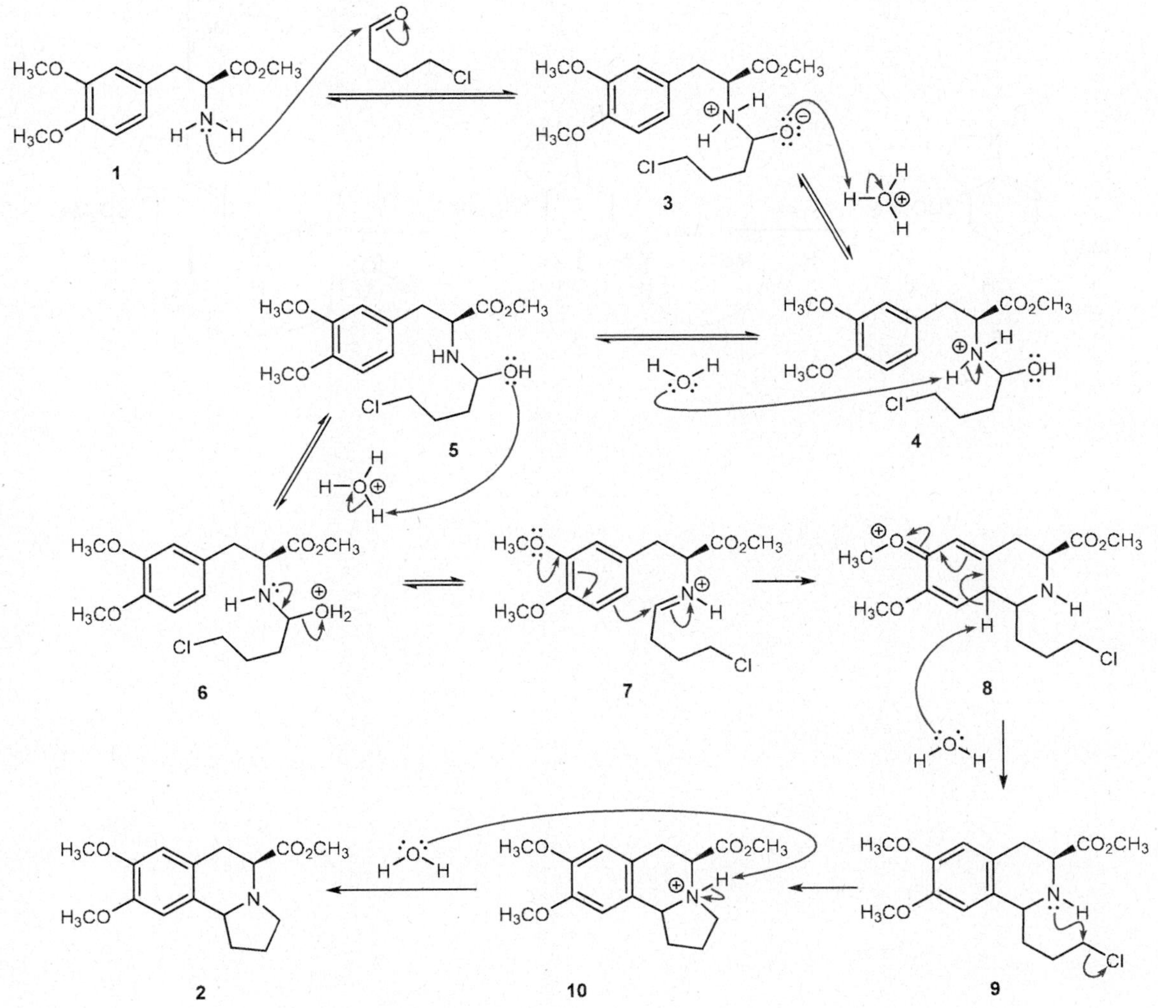

22.103. If we redraw the starting material so that the pendant π bond is in close proximity to the diene, we can envision a thermal [4+2] Diels-Alder reaction occurring. Notice that the dienophile will approach the diene, resulting in the oxo-bridge on one face and the ester group on the other face of the newly formed system. After cyclization, compound **2** can undergo a nitrogen-induced fragmentation that will result in the formation of a C-N π-bond via the breakage of the C-O σ-bond. Compound **3** has a highly electrophilic α,β-unsaturated iminium ion that can be trapped by methanol to produce intermediate **4**. Because the top face of the six-membered ring is hindered by the alkoxide, methanol will approach from the bottom face, thereby installing the methoxy group on a dash. The final two steps of the mechanism are proton transfer steps, producing the desired compound (**6**).

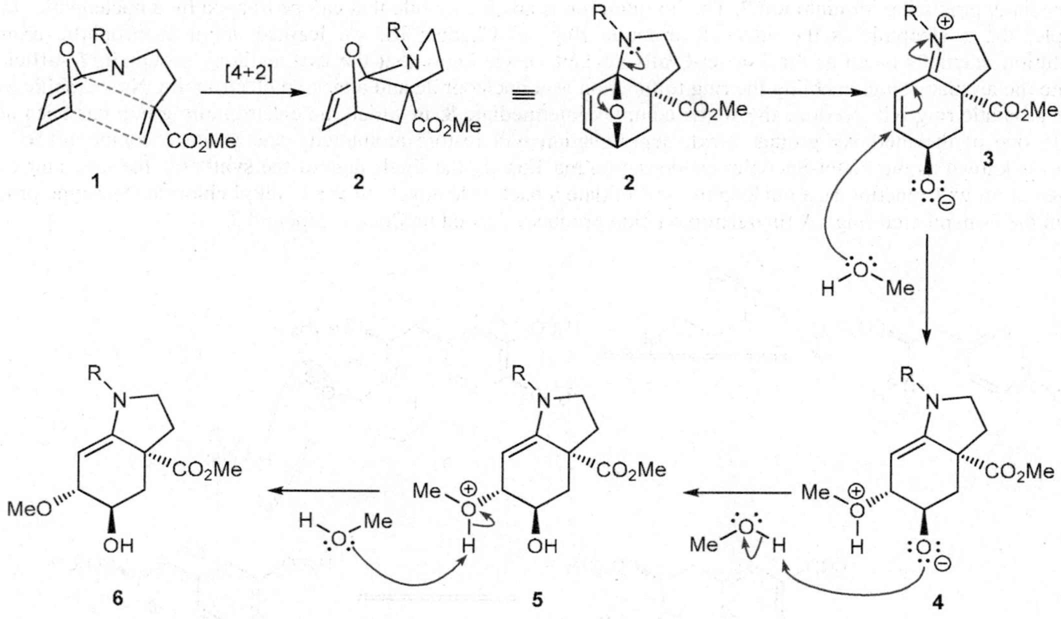

Chapter 23
Organometallic Compounds

Review of Concepts

Fill in the blanks below. To verify that your answers are correct, look in your textbook at the end of Chapter 23. Each of the sentences below appears verbatim in the section entitled *Review of Concepts and Vocabulary*.

- Organolithium (RLi) and organomagnesium (RMgX) compounds are strong bases and strong ____________.
- Alkenes will react with ICH_2ZnI to form a three-membered ring, in a process called ________________.
- Stille coupling, which occurs in the presence of a suitable palladium catalyst, is the reaction between an ________________ and an organic electrophile to form a new C–C σ bond. The coupling process is stereospecific. It is observed to proceed with ______________ of configuration at each of the C=C units.
- Suzuki coupling, which occurs in the presence of a suitable palladium catalyst and a base, is the reaction between an ______________________ and an organic electrophile to form a new C–C σ bond. The coupling process is stereospecific. It is observed to proceed with ______________ of configuration at each of the C=C units.
- Negishi coupling, which occurs in the presence of a suitable palladium catalyst, is the reaction between an organic electrophile (R′X) and an organometallic species containing ____________, to form a new C–C σ bond. The coupling process is stereospecific. It is observed to proceed with ________________ of configuration at each of the C=C units.
- The Heck reaction is a coupling reaction that occurs between an aryl, vinyl or benzyl halide (RX) and an ____________ in the presence of an appropriate Pd catalyst and a base.
- Alkene _________________ is a process that is characterized by the redistribution (changing of position) of carbon-carbon double bonds. When terminal alkenes are used, the evolution of ___________ gas drives the reaction toward the formation of one alkene with excellent yields.
- When the starting material is a diene and the reaction is conducted in dilute solutions (thereby favoring an intramolecular process over intermolecular processes), alkene metathesis can serve as a method for ring-formation. This process is called _________________ metathesis. Similarly, ring-opening metathesis can be achieved in the presence of _____________.

Review of Skills

Fill in the blanks and empty boxes below. To verify that your answers are correct, look in your textbook at the end of Chapter 23. The answers appear in the section entitled *SkillBuilder Review*.

SkillBuilder 23.1 Identifying the Partners for a Corey-Posner/Whitesides-House Coupling Reaction

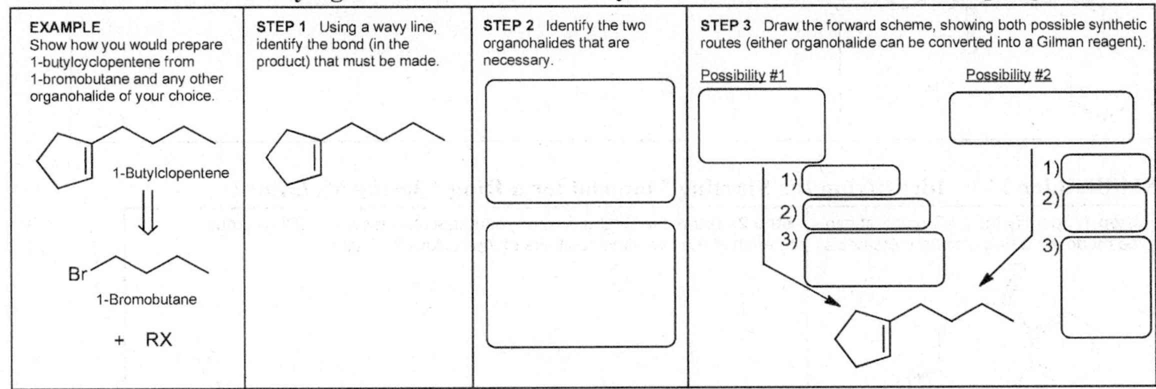

SkillBuilder 23.2 Predicting the Product of a Stille Coupling Reaction

Step 1 Consider a Stille coupling reaction between the following two coupling partners. Circle the carbon atoms that will be joined via a new σ bond.	Steps 2 and 3 Rotate the coupling partners and draw the product, making sure to retain the configuration of any C=C units.

SkillBuilder 23.3 Predicting the Product of a Suzuki Coupling Reaction

Step 1 Consider a Suzuki coupling reaction between the following two coupling partners. Circle the carbon atoms that will be joined via a new σ bond.	Steps 2 and 3 Rotate the coupling partners and draw the product, making sure to retain the configuration of any C=C units.

SkillBuilder 23.4 Predicting the Product of a Negishi Coupling Reaction

Step 1 Consider a Negishi coupling reaction between the following two coupling partners. Circle the carbon atoms that will be joined via a new σ bond.	Steps 2 and 3 Orient the coupling partners and draw the product, making sure to retain the configuration of any C=C units.

SkillBuilder 23.5 Identifying the Partners for a Heck Reaction

Step 1 Using a wavy line, identify the bond (in the following desired product) that can be made via a Heck reaction.	Step 2 Identify the organohalide and alkene necessary. Keep in mind that Heck reactions are stereoselective with respect to the starting alkene, and stereospecific with respect to the organohalide (if it is a vinyl halide, as in this case).

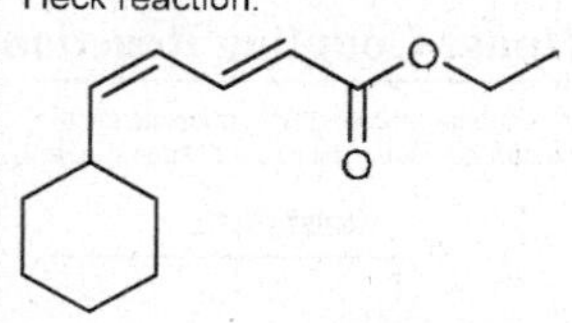

SkillBuilder 23.6 Identifying the Starting Material for a Ring-Closing Metathesis

Step 1 Identify the C=C unit that can be made via a ring-closing metathesis.	Step 2 Break the C=C unit apart, and attach a methylene (CH_2) group to each of the two vinyl positions of the broken C=C unit.

Review of Reactions

Identify the reagents necessary to achieve each of the following transformations. To verify that your answers are correct, look in your textbook at the end of Chapter 23. The answers appear in the section entitled *Review of Reactions*.

Preparation of Organolithium and Organomagnesium Compounds

R—X ⟶ R—Li + LiX

R—X ⟶ R—Mg—X

Preparation of Gilman Reagents

R—X ⟶ R—Li ⟶ R_2CuLi

Coupling Reaction of a Gilman Reagent with an Organohalide

R_2CuLi ⟶ R—R′

Simmons-Smith Reaction

Stille Coupling

R—X ⟶ RMgX ⟶ R—$SnBu_3$

R—$SnBu_3$ + R′—X ⟶ R—R′

Suzuki Coupling

Ph–Li → Ph–B(OMe)$_2$

R–BY$_2$ + R′–X → R–R′

Negishi Coupling

R–ZnX + R′–X → R–R′

Heck Reaction

R–X + alkene → R-substituted alkene

Alkene Metathesis

(E + Z isomers)

Common Mistake to Avoid

When a cyclic alkene is used in a Heck reaction, make sure to move the position of the C=C unit when drawing the product:

R–X + cyclohexene → (Pd(OAc)$_2$, PPh$_3$, Et$_3$N) R–cyclohexene

It is a common mistake to forget to move the C=C unit:

Incorrect

R−X + (cyclohexene) —[$Pd(OAc)_2$, PPh_3, Et_3N]→ R−(cyclohexene)

(Don't make this mistake)

Recall that the C=C unit moves because this is the only regiochemical outcome that accommodates *syn* elimination. This is the case whenever a cyclic alkene is used as a starting material in a Heck reaction, so make sure to take this into account when drawing the product of a Heck reaction.

Useful reagents

The following is a list of reagents that appear in this chapter:

Reagents	Function
Li (2 eq.)	Used to convert an organohalide into an organolithium compound.
RLi	An organolithium reagent. A very strong nucleophile and a very strong base.
Mg, Et_2O	Reagents for converting an organohalide into a Grignard reagent.
RMgX	A Grignard reagent. A very strong nucleophile and a very strong base.
CuX (0.5 eq.)	Used to convert an organolithium compound into a Gilman reagent.
CH_2I_2, Zn-Cu, Et_2O	Reagents for a Simmons-Smith cyclopropanation.
Bu_3SnCl	Used to convert an organolithium or organomagnesium compound into an organostannane.
$Pd(PPh_3)_4$	A catalyst used for coupling reactions, including Stille coupling, Suzuki coupling, Negishi coupling, and the Heck reaction.
$Pd(OAc)_2$	A catalyst used for coupling reactions, including Stille coupling, Suzuki coupling, Negishi coupling, and the Heck reaction.
H−B (catecholborane structure)	Catecholborane. Used to convert an alkyne into a vinyl boronic ester. Also used to convert an alkene into an alkyl boronic ester.
H−B (9-BBN structure)	9-BBN. Used to convert an alkyne into a vinyl borane. Also used to convert an alkene into an alkyl borane.
$B(OMe)_3$	Trimethylborate. Used to convert an aryllithium compound into an aryl boronic ester.
$ZnBr_2$	Zinc dibromide. Used to convert an organolithium or organomagnesium compound into an organozinc compound.
Zn, Et_2O	Zinc and diethyl ether. Used to convert an organolithium or organomagnesium compound into an organozinc compound.
Grubbs catalyst	Used to achieve alkene metathesis, including ring-opening metathesis and ring-closing metathesis.

Solutions

23.1. The first compound is an organomagnesium compound (C–Mg bond), while the second compound is an organozinc reagent (C–Zn bond). We must compare these two C–M bonds to determine which bond has greater ionic character. The difference in electronegativity between C (2.5) and Mg (1.2) is greater than the difference in electronegativity between C (2.5) and Zn (1.6). Therefore, the carbon atom of a C–Mg bond is expected to have a greater partial negative character, and thus be more nucleophilic than the carbon atom of the C–Zn bond. That is, organomagnesium compounds are expected be more nucleophilic than organozinc compounds. Indeed, they are.

23.2. Compound **A** is formed when the iodide group (in 4-iodotoluene) is replaced with a lithium atom, giving an organolithium compound:

Li (2 eq.) → A + LiI

Upon treatment with H_2O, compound **A** functions as a base and removes a proton from H_2O, giving toluene (compound **B**):

A —H_2O→ B + LiOH

Compound **C** is formed via insertion of magnesium, as shown:

Mg, Et_2O → C

Compound **C** is converted into compound **B** upon treatment with a proton source, such as H_2O (compound **D**):

C —H_2O (D)→ B + HOMgI

Upon treatment with D_2O, compound **C** functions as a base and removes a deuteron from D_2O, giving a deuterated product (compound **E**), as shown:

C —D_2O→ E + DOMgI

23.3. Compound **C** functions as a base and removes a deuteron, which requires two curved arrows, as shown:

C → E + DOMgI

23.4. Treating a nitrile with a Grignard reagent, followed by aqueous acid, gives ketone **A** (Section 20.13):

1) MeMgBr 2) H_3O^+ → A

When a ketone (compound **A**) is treated with PhMgBr, followed by aqueous acidic workup, a phenyl group is installed and the product is alcohol **B** (Section 12.6):

A —1) PhMgBr 2) H_3O^+→ B

When compound **A** is treated with EtMgBr, followed by aqueous acidic workup, an ethyl group is installed and the product is alcohol **C** (Section 12.6):

A —1) EtMgBr 2) H_3O^+→ C

The starting ester can be converted into compound **B** upon treatment with two equivalents of PhMgBr (compound **D**), followed by aqueous acidic workup, as described in Section 20.11:

1) D (2 eq.) 2) H_3O^+ → B

The starting ketone can be converted into compound **C** upon treatment with one equivalent of MeMgBr (compound **E**), followed by aqueous acidic workup, as described in Section 12.6:

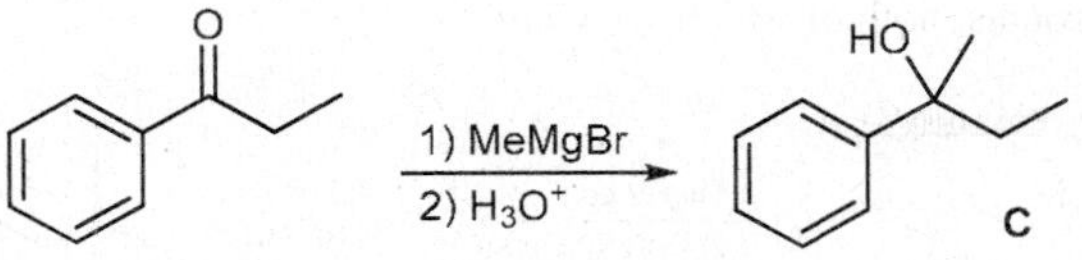

1) MeMgBr
2) H_3O^+

C

23.5. When the racemic epoxide is treated with ethyl magnesium bromide, followed by aqueous acidic workup, an ethyl group is installed at the less substituted side of the epoxide, and the three-membered ring is opened to give a racemic alcohol (compound **A**):

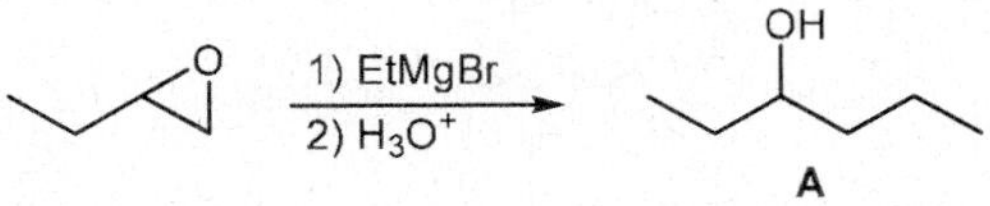

1) EtMgBr
2) H_3O^+

A

Propanal can be converted into compound **A** upon treatment with *n*-propyl magnesium bromide (compound **B**), followed by aqueous acidic workup, as shown:

1) BrMg
2) H_3O^+

A

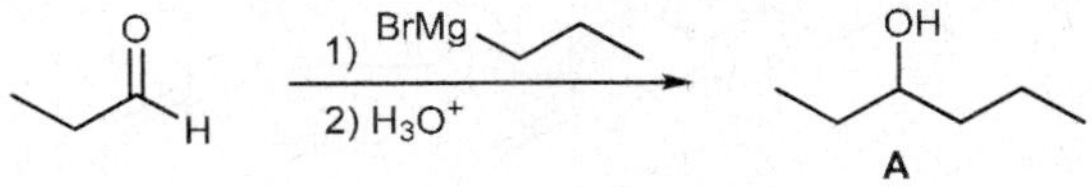

Butanal can be converted into compound **A** upon treatment with ethyl magnesium bromide (compound **C**), followed by aqueous acidic workup, as shown:

1) MgBr
2) H_3O^+

A

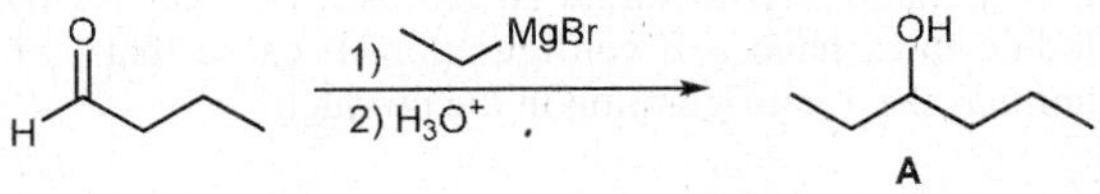

23.6. The first equivalent of the Grignard reagent attacks the carbonyl group, followed by loss of the leaving group to give a ketone (note that the leaving group is still tethered to the ketone). A second equivalent of the Grignard reagent attacks the carbonyl group of the ketone intermediate to produce a dianion, which is subsequently protonated during aqueous acidic workup to give a diol. Overall, the ring has been opened, and two methyl substituents are incorporated into the product:

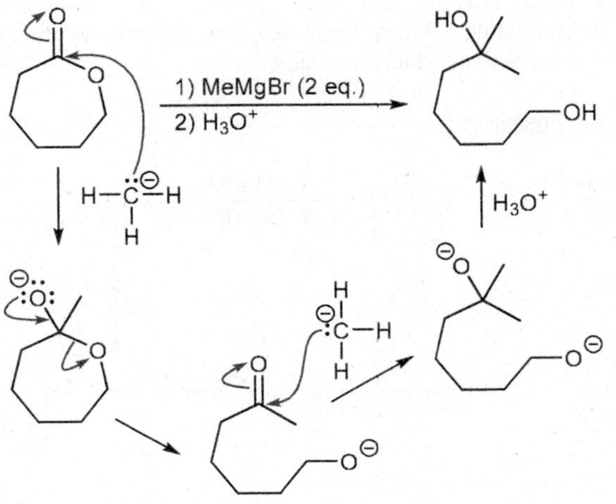

1) MeMgBr (2 eq.)
2) H_3O^+

H_3O^+

23.7.

(a) First determine which bond (in the product) will be made via a coupling process, considering the alkyl halide identified in the problem statement:

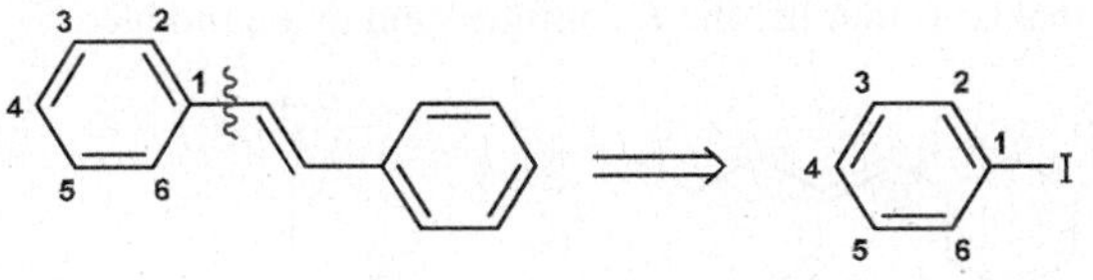

Next, draw the two organohalides that are necessary for the coupling process. One of the organohalides has already been identified in the problem statement (iodobenzene). The other organohalide (of our choice) must be a vinyl halide, so we draw an iodide (because iodides are more reactive than bromides or chlorides). This vinyl iodide must have an *E* configuration, because that C=C unit has the *E* configuration in the product:

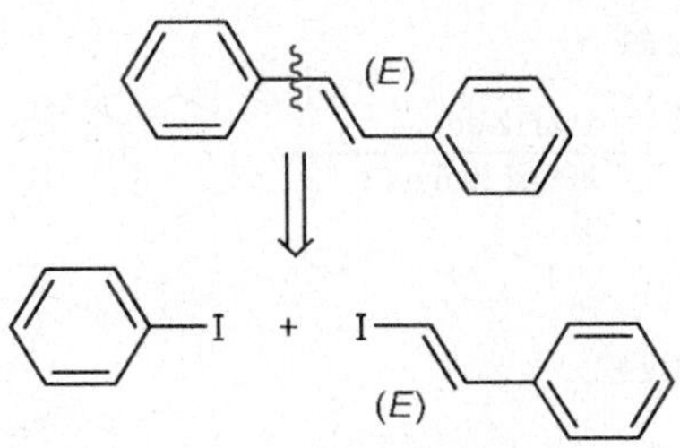

(*E*)

(*E*)

One of these organohalides must be converted into a Gilman reagent and then treated with the other organohalide. This leads to two possible synthetic routes, both of which are viable:

Possibility #1

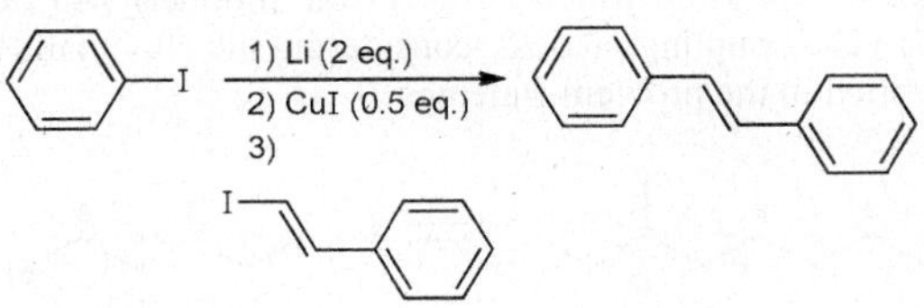

1) Li (2 eq.)
2) CuI (0.5 eq.)
3)

Possibility #2

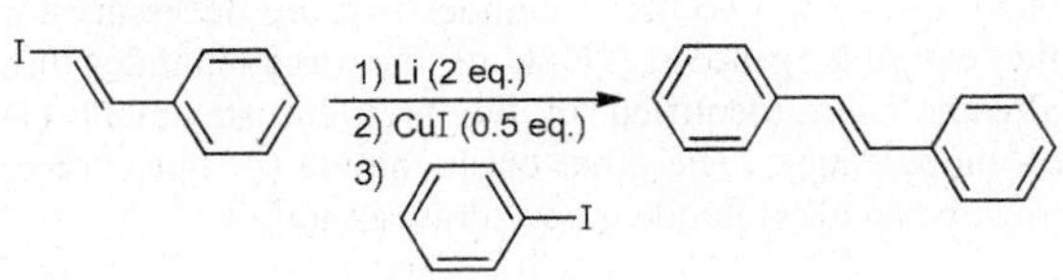

1) Li (2 eq.)
2) CuI (0.5 eq.)
3)

(b) First determine which bond (in the product) will be made via a coupling process, considering the alkyl halide identified in the problem statement:

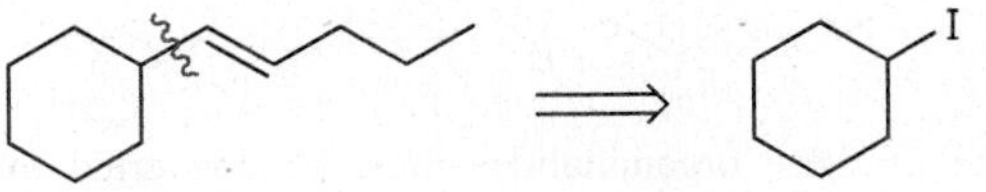

Next, draw the two organohalides that are necessary for the coupling process. One of the organohalides has already been identified in the problem statement (cyclohexyl iodide). The other organohalide (of our choice) must be a vinyl halide, so we draw an iodide. This vinyl iodide must have an *E* configuration, because that C=C unit has the *E* configuration in the product:

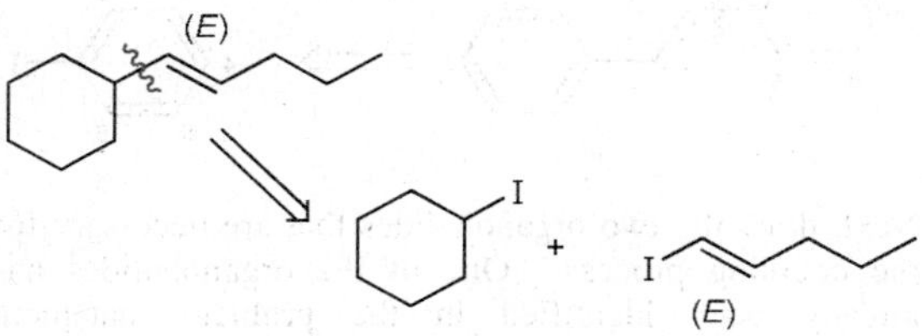

(E)

I + I (E)

One of these organohalides must be converted into a Gilman reagent and then treated with the other organohalide. This leads to two possible synthetic routes, both of which are viable:

Possibility #1

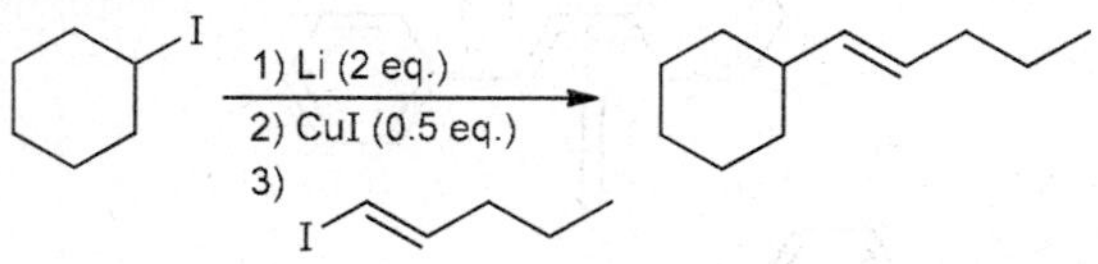

1) Li (2 eq.)
2) CuI (0.5 eq.)
3) I

Possibility #2

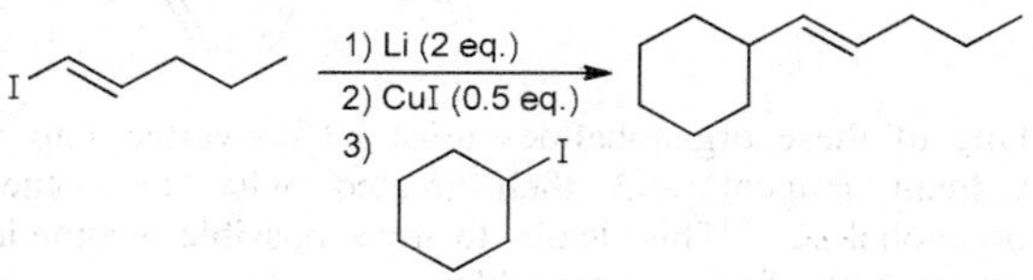

1) Li (2 eq.)
2) CuI (0.5 eq.)
3) I

(c) First determine which bond (in the product) will be made via a coupling process, considering the alkyl halide identified in the problem statement:

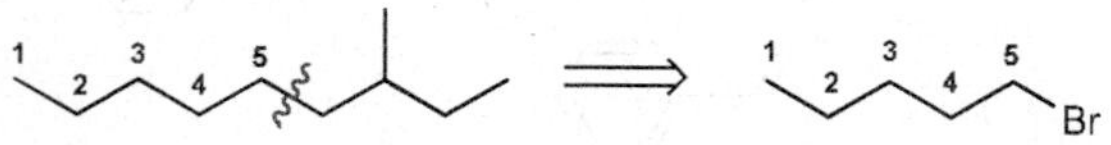

1 2 3 4 5 ⟹ 1 2 3 4 5 Br

Next, draw the two organohalides that are necessary for the coupling process. One of the organohalides has already been identified in the problem statement (1-bromopentane). The other organohalide (of our choice) must be an alkyl halide, so we draw an iodide:

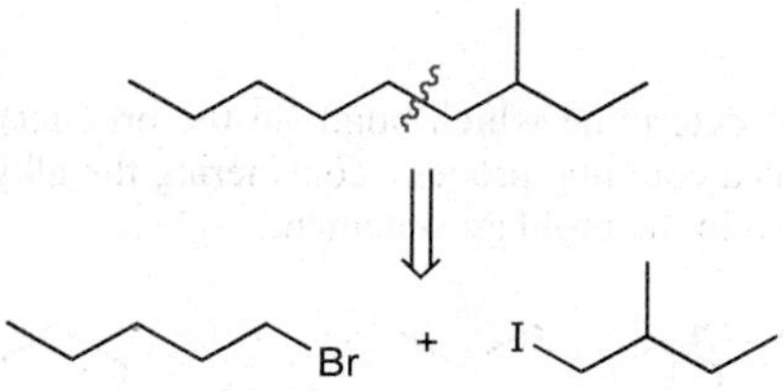

Br + I

One of these organohalides must be converted into a Gilman reagent and then treated with the other organohalide. This leads to two possible synthetic routes, both of which are viable:

Possibility #1

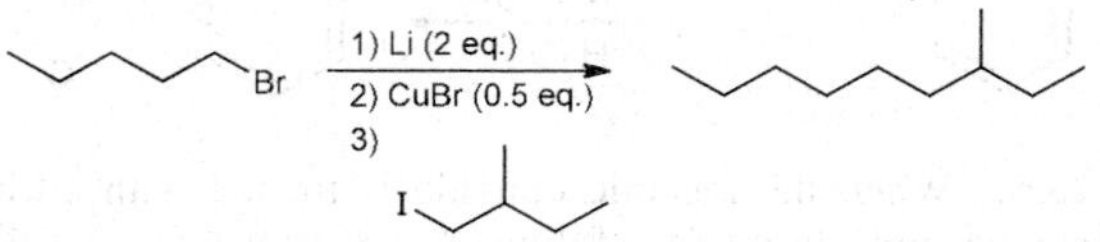

Br 1) Li (2 eq.)
2) CuBr (0.5 eq.)
3) I

Possibility #2

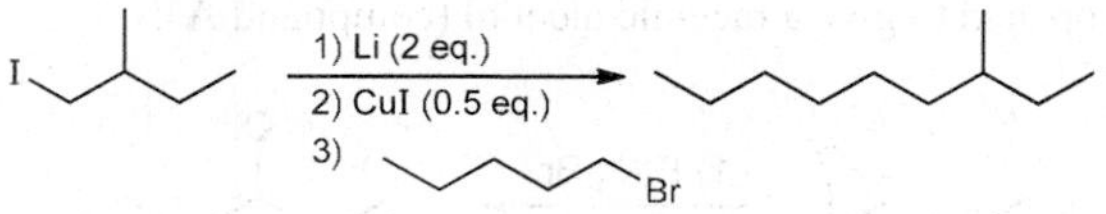

I 1) Li (2 eq.)
2) CuI (0.5 eq.)
3) Br

(d) First determine which bond (in the product) will be made via a coupling process, considering the alkyl halide identified in the problem statement:

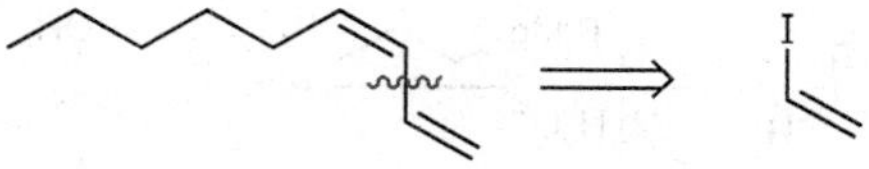

⟹ I

Next, draw the two organohalides that are necessary for the coupling process. One of the organohalides has already been identified in the problem statement (vinyl iodide). The other organohalide (of our choice) must be a vinyl halide, so we draw an iodide. This latter vinyl iodide must have a *Z* configuration, because that C=C unit has the *Z* configuration in the product:

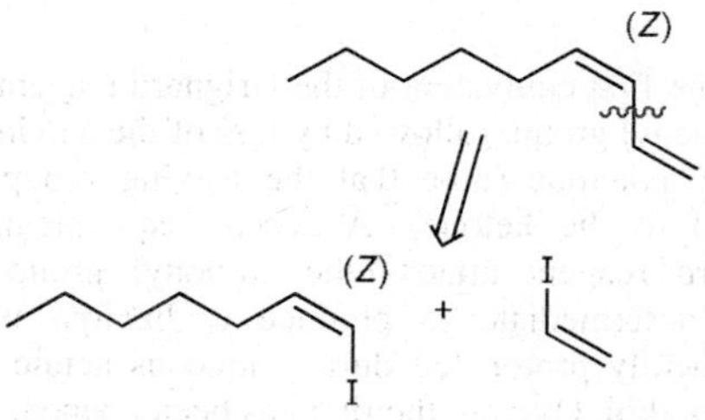

(Z)

(Z) I + I

One of these organohalides must be converted into a Gilman reagent and then treated with the other organohalide. This leads to two possible synthetic routes, both of which are viable:

Possibility #1

I 1) Li (2 eq.)
2) CuI (0.5 eq.)
3) I

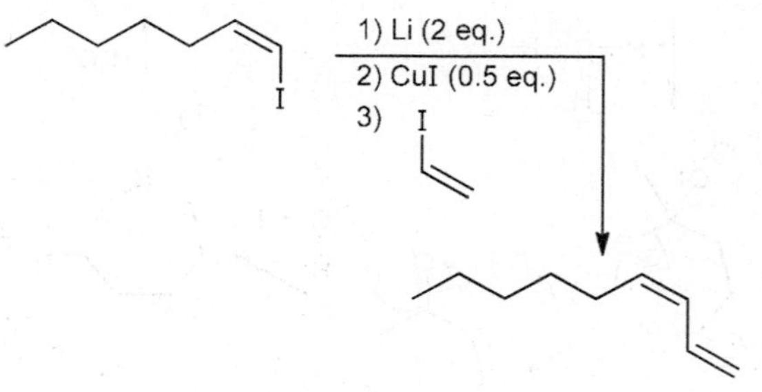

Possibility #2

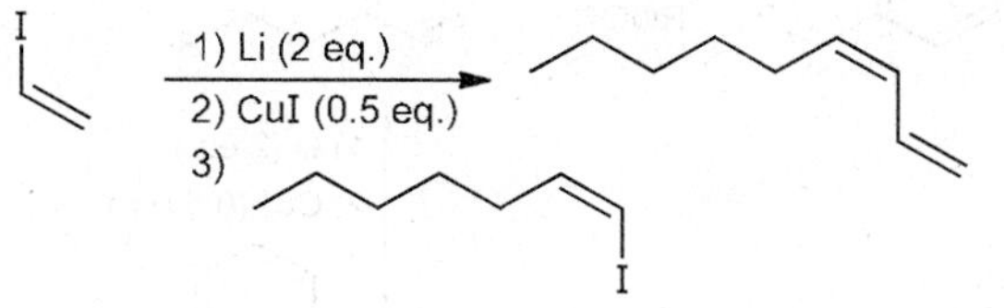

1) Li (2 eq.)
2) CuI (0.5 eq.)
3)

23.8.

(a) First determine which bond (in the product) will be made via a coupling process. The problem statement dictates that each organohalide must have no more than 6 carbon atoms. Since the product has twelve carbon atoms, that leaves us with only one choice, indicated below with a wavy line:

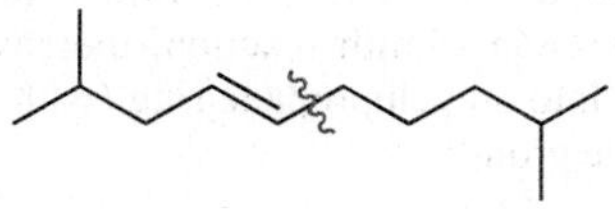

Next, draw the two organohalides that are necessary for the coupling process. In this case, we need an alkyl iodide and a vinyl iodide, as shown. Note that the vinyl iodide must have an *E* configuration, because that C=C unit has the *E* configuration in the product:

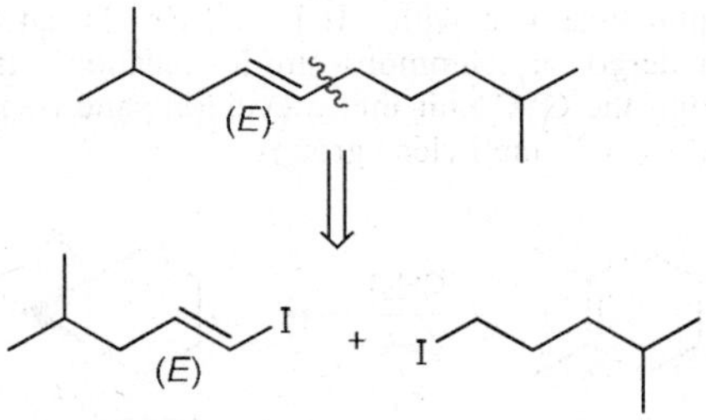

(*E*)

(*E*)

One of these organohalides must be converted into a Gilman reagent and then treated with the other organohalide. This leads to two possible synthetic routes, both of which are viable:

Possibility #1

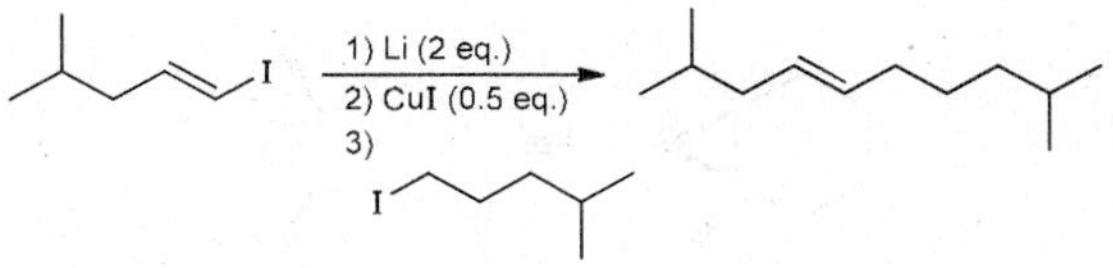

1) Li (2 eq.)
2) CuI (0.5 eq.)
3)

Possibility #2

1) Li (2 eq.)
2) CuI (0.5 eq.)
3)

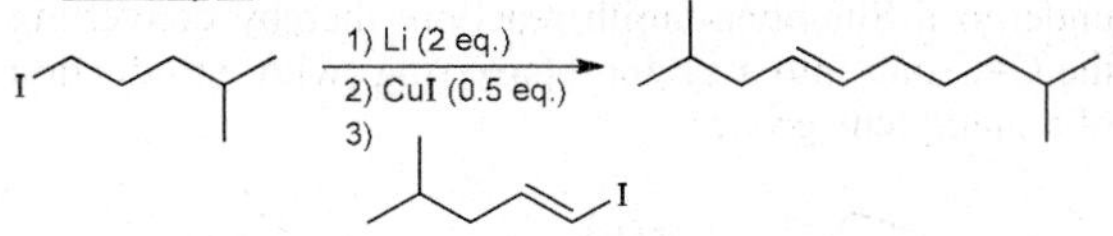

(b) First determine which bond (in the product) will be made via a coupling process. The problem statement dictates that each organohalide must have no more than 6 carbon atoms. Since the product has twelve carbon atoms, that leaves us with only one choice, indicated below with a wavy line:

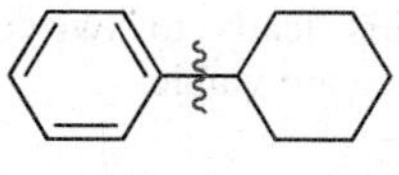

Next, draw the two organohalides that are necessary for the coupling process. In this case, we need cyclohexyl iodide and iodobenzene, as shown:

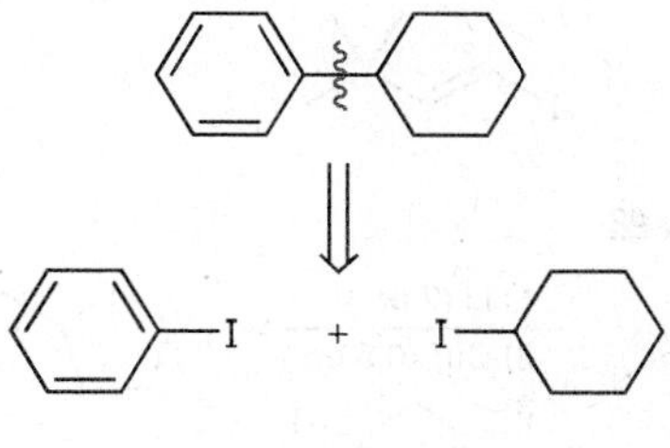

One of these organohalides must be converted into a Gilman reagent and then treated with the other organohalide. This leads to two possible synthetic routes, both of which are viable:

Possibility #1

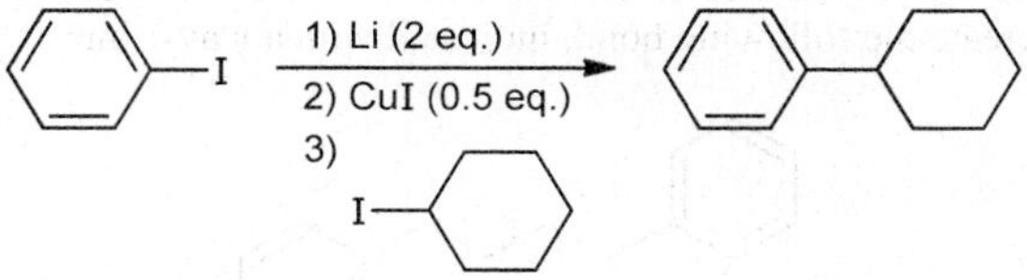

1) Li (2 eq.)
2) CuI (0.5 eq.)
3)

Possibility #2

1) Li (2 eq.)
2) CuI (0.5 eq.)
3)

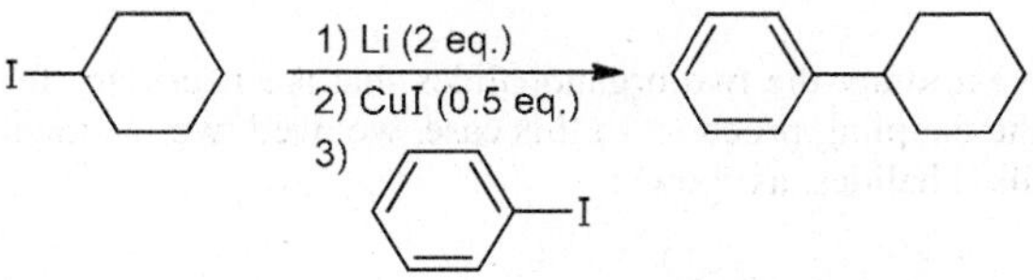

(c) First determine which bond (in the product) will be made via a coupling process. The problem statement dictates that each organohalide must have no more than 6 carbon atoms, so we select the following bond for a disconnection (six carbon atoms on one side and five on the other):

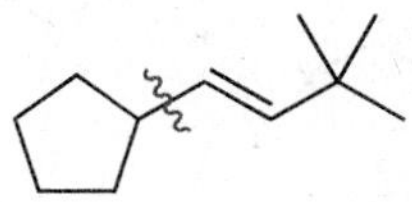

Next, draw the two organohalides that are necessary for the coupling process. In this case, we need cyclopentyl iodide and a vinyl iodide. Note that the vinyl iodide must have an *E* configuration, because that C=C unit has the *E* configuration in the product:

(*E*) (*E*)

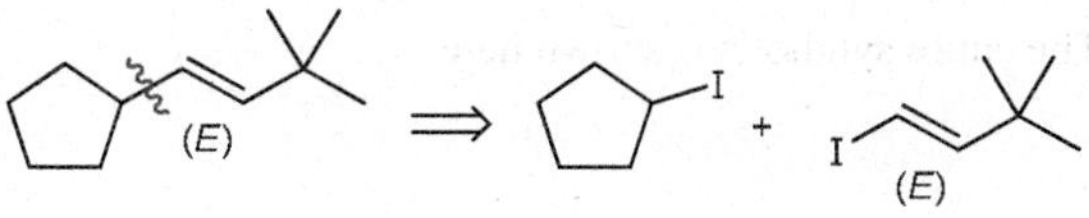

One of these organohalides must be converted into a Gilman reagent and then treated with the other organohalide. This leads to two possible synthetic routes, both of which are viable:

Possibility #1

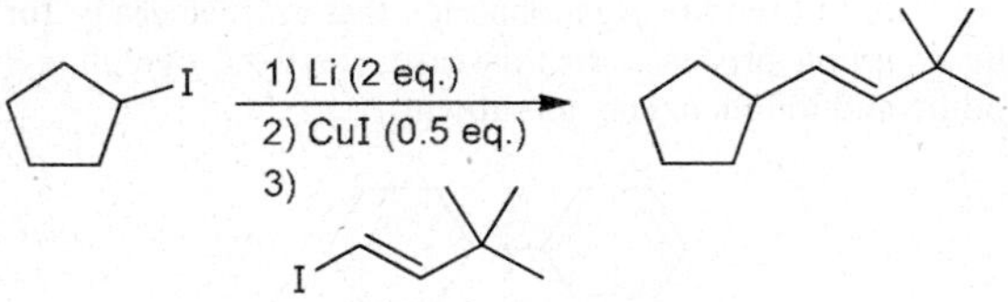

1) Li (2 eq.)
2) CuI (0.5 eq.)
3)

Possibility #2

1) Li (2 eq.)
2) CuI (0.5 eq.)
3)

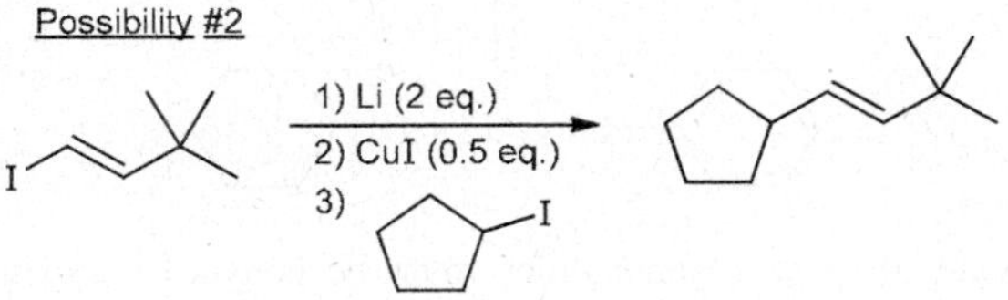

23.9.

(a) First determine which bond (in the product) will be made via a coupling process. The problem statement dictates that styrene (which has eight carbon atoms) must be the only source of carbon atoms. Therefore, we must create the following bond, indicated with a wavy line:

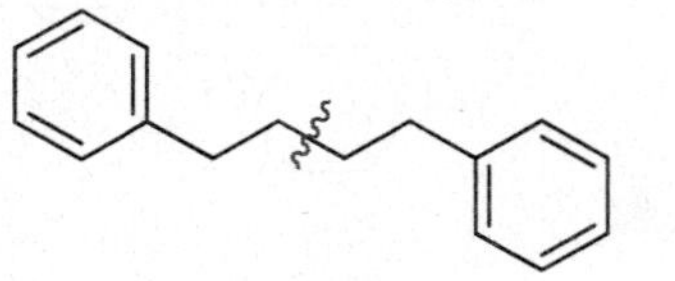

Next, draw the two organohalides that are necessary for the coupling process. In this case, we need two identical alkyl halides, as shown:

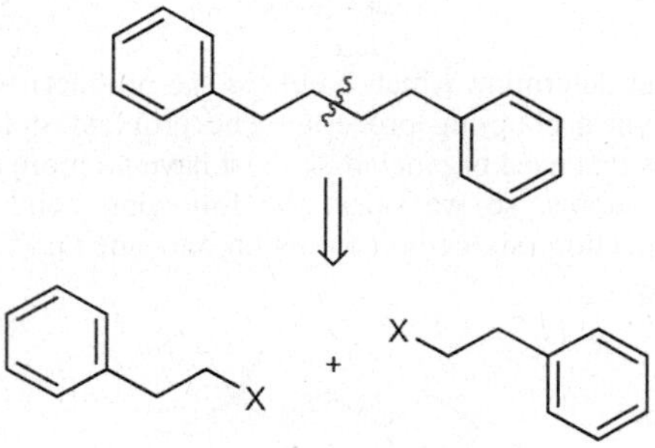

These alkyl halides can be made from styrene via the *anti*-Markovnikov addition of HBr (see Section 10.10):

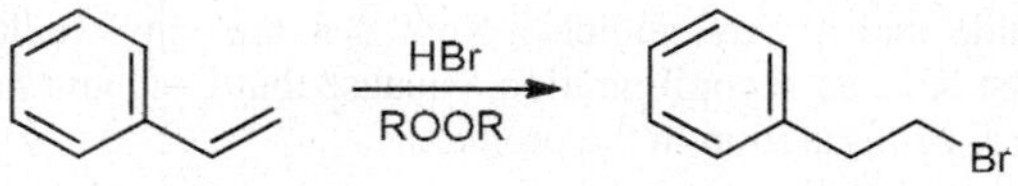

HBr
ROOR

The entire synthesis is shown here:

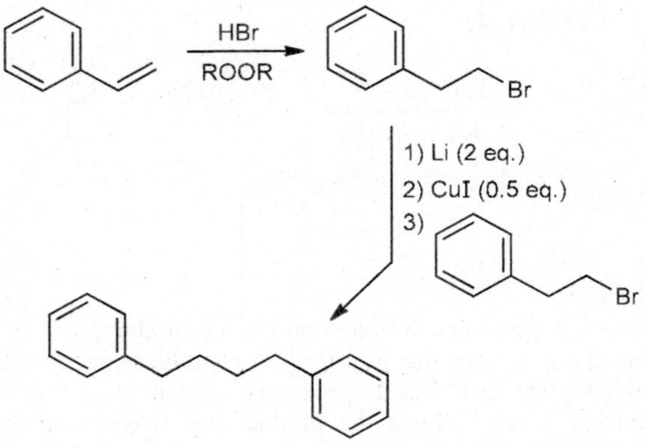

HBr
ROOR

1) Li (2 eq.)
2) CuI (0.5 eq.)
3)

23.10.

(a) Upon treatment with CH_2I_2 and Zn-Cu, an alkene will undergo a Simmons-Smith reaction, thereby converting the C=C unit into a cyclopropane ring (with introduction of a methylene group):

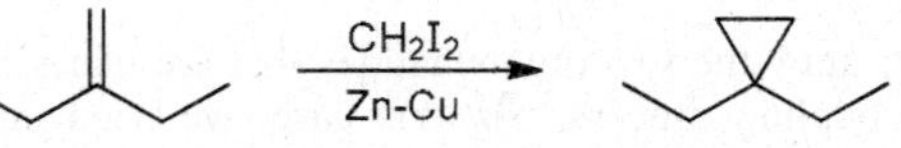

CH_2I_2
Zn-Cu

In this case, the product has no chiral centers, so wedges and dashes are not drawn.

(b) Upon treatment with CH_2I_2 and Zn-Cu, an alkene will undergo a Simmons-Smith reaction, thereby converting the C=C unit into a cyclopropane ring (with introduction of a methylene group):

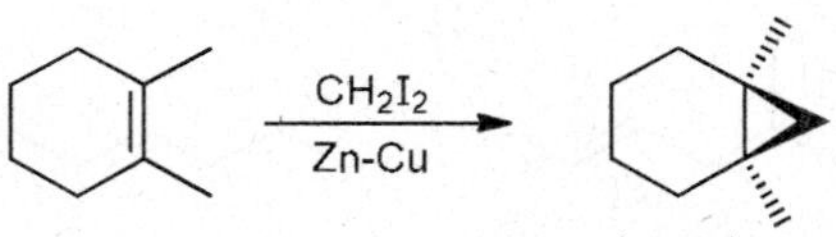

CH_2I_2
Zn-Cu

(*meso*)

In this case, the product has two chiral centers, and is a *meso* compound (it is not a chiral molecule and therefore it does not have an enantiomer). Introduction of the methylene group can occur on either face of the π bond, giving the same *meso* compound in either case:

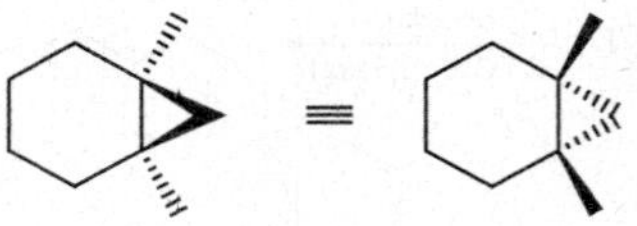

(c) Upon treatment with CH_2I_2 and Zn-Cu, an alkene will undergo a Simmons-Smith reaction, thereby converting the C=C unit into a cyclopropane ring (with introduction of a methylene group):

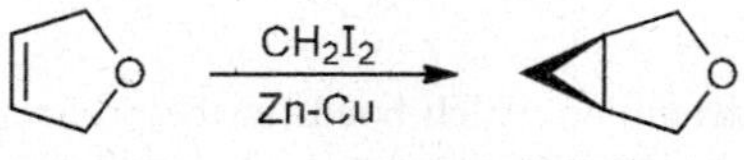

CH_2I_2
Zn-Cu

(*meso*)

In this case, the product has two chiral centers, and is a *meso* compound (it is not a chiral molecule and therefore it does not have an enantiomer). Introduction of the methylene group can occur on either face of the π bond, giving the same *meso* compound in either case:

(d) Upon treatment with CH_2I_2 and Zn-Cu, an alkene will undergo a Simmons-Smith reaction, thereby converting the C=C unit into a cyclopropane ring (with introduction of a methylene group):

CH_2I_2 / Zn-Cu; + En

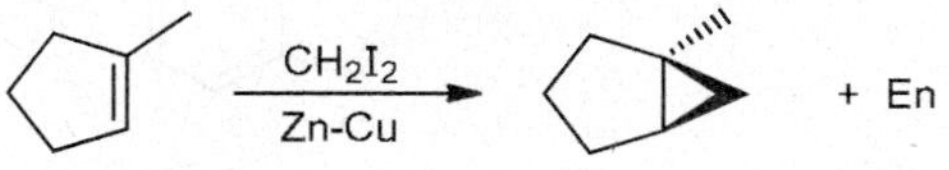

In this case, the product is a chiral compound. Since the methylene group can be installed on either face of the π bond, we expect a pair of enantiomers formed as a racemic mixture..

(e) Upon treatment with CH_2I_2 and Zn-Cu, an alkene will undergo a Simmons-Smith reaction, thereby converting the C=C unit into a cyclopropane ring (with introduction of a methylene group). In this case, the starting alkene has a *trans* configuration, which is preserved in the product. Introduction of the methylene group can occur on either face of the π bond, giving rise to a pair of enantiomers. The chiral product is thus formed as a racemate:

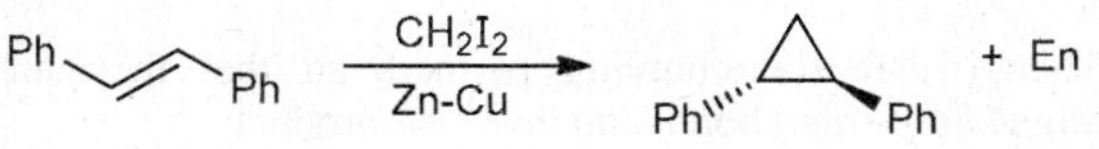

Ph, Ph; CH_2I_2 / Zn-Cu; Ph, Ph; + En

23.11.

As indicated in the problem statement, the geminal dimethyl group of spongian-16-one can be accessed via hydrogenolysis of a cyclopropane ring, and cyclopropane **2** can be prepared from compound **1** via the Simmons-Smith reaction, as shown in the following retrosynthetic analysis:

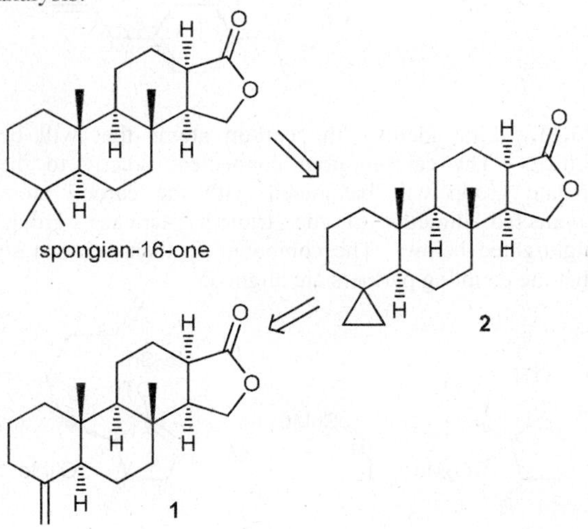

spongian-16-one; **2**; **1**

The forward reaction sequence is shown here:

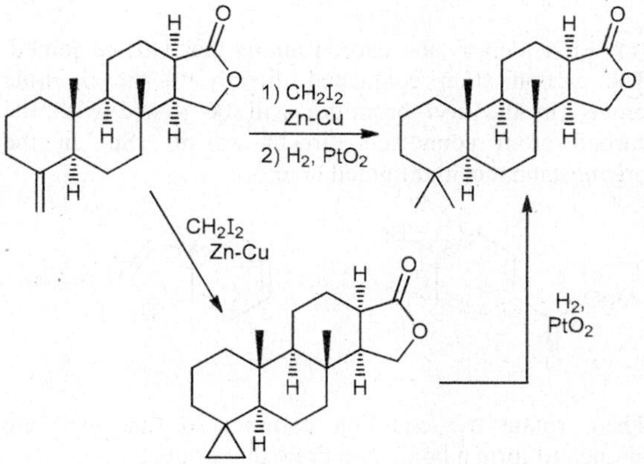

1) CH_2I_2, Zn-Cu; 2) H_2, PtO_2

CH_2I_2, Zn-Cu; H_2, PtO_2

23.12.

(a) First identify the carbon atoms that will be joined. The carbon atom connected directly to the bromide group (in the aryl bromide) will be joined with the carbon atom connected directly to tin (Sn) in the organostannane, highlighted here:

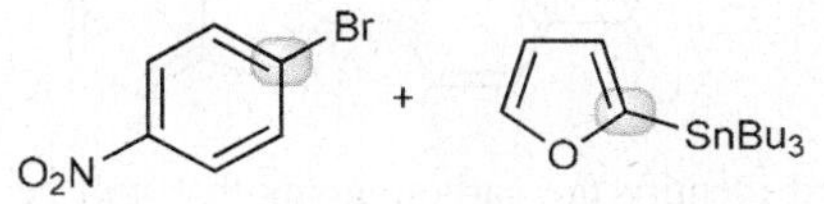

O_2N, Br; + ; $SnBu_3$

Then, rotate the coupling partners so that they are aligned to form a bond, and draw the product:

O_2N

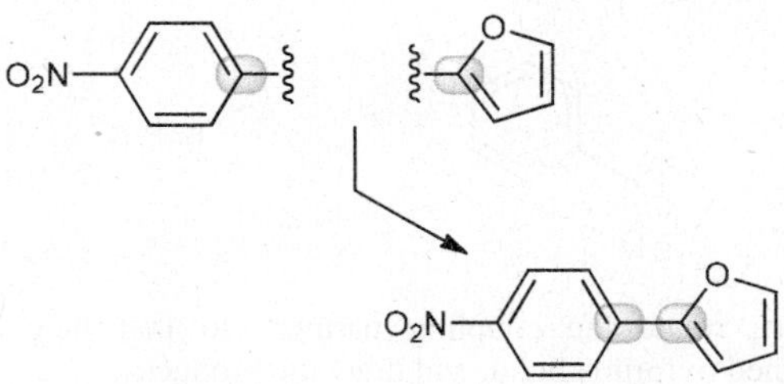

(b) First identify the carbon atoms that will be joined. The carbon atom connected directly to the iodide group (in the vinyl iodide) will be joined with the carbon atom connected directly to tin (Sn) in the organostannane, highlighted here:

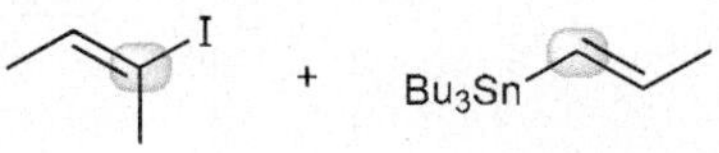

I; + ; Bu_3Sn

Then, rotate the coupling partners so that they are aligned to form a bond, and draw the product:

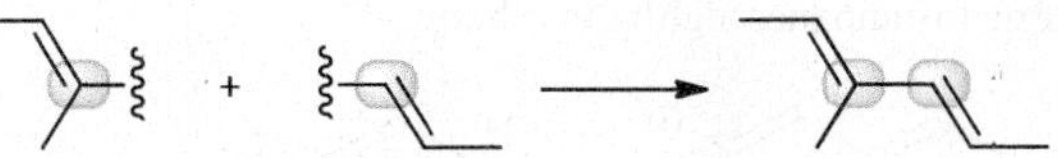

In this case, there are two C=C units (one in the organic electrophile and the other in the organostannane). So, when drawing the product, we must be careful that each of these C=C units maintains its *E* configuration.

(c) First identify the carbon atoms that will be joined. The carbon atom connected directly to the bromide group (in the aryl bromide) will be joined with the carbon atom connected directly to tin (Sn) in the organostannane, highlighted here:

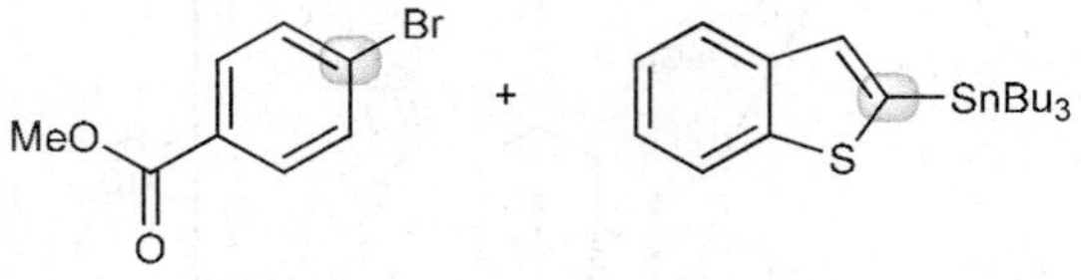

Then, rotate the coupling partners so that they are aligned to form a bond, and draw the product:

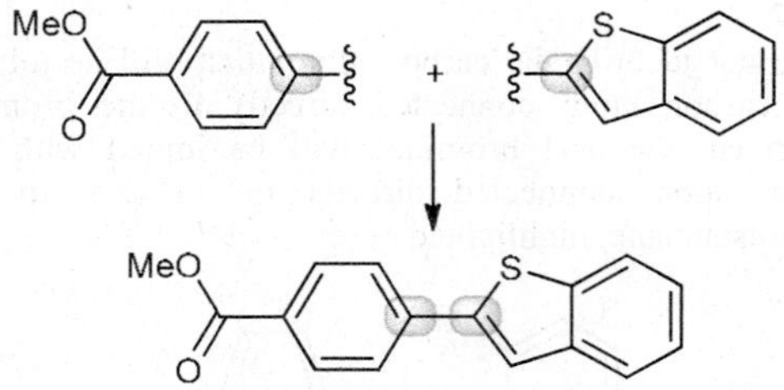

(d) First identify the carbon atoms that will be joined. The carbon atom connected directly to the iodide group (in the aryl iodide) will be joined with the carbon atom connected directly to tin (Sn) in the organostannane, highlighted here:

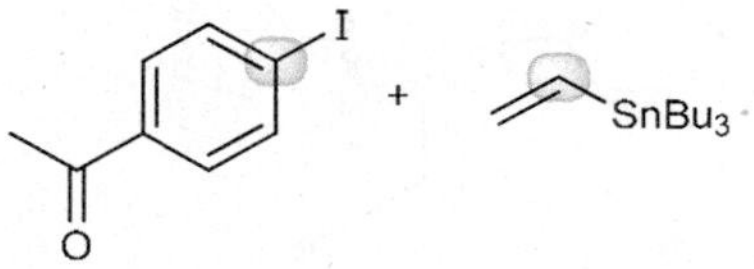

Then, rotate the coupling partners so that they are aligned to form a bond, and draw the product:

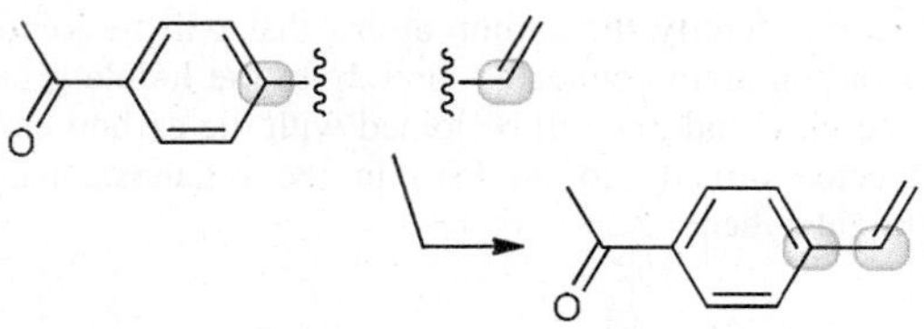

(e) First identify the carbon atoms that will be joined. The carbon atom connected directly to the bromide group (in the vinyl bromide) will be joined with the carbon atom connected directly to tin (Sn) in the organostannane, highlighted here:

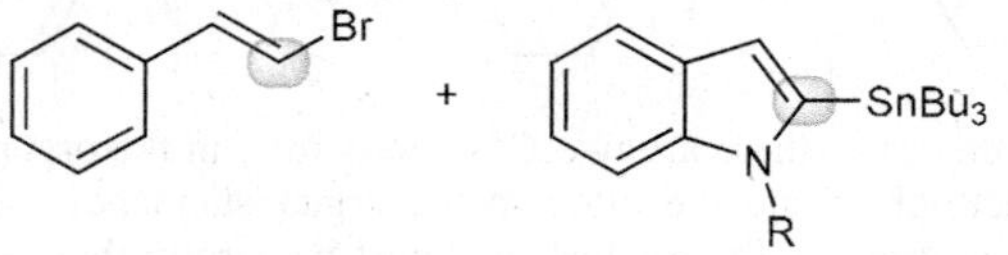

Then, rotate the coupling partners so that they are aligned to form a bond, and draw the product:

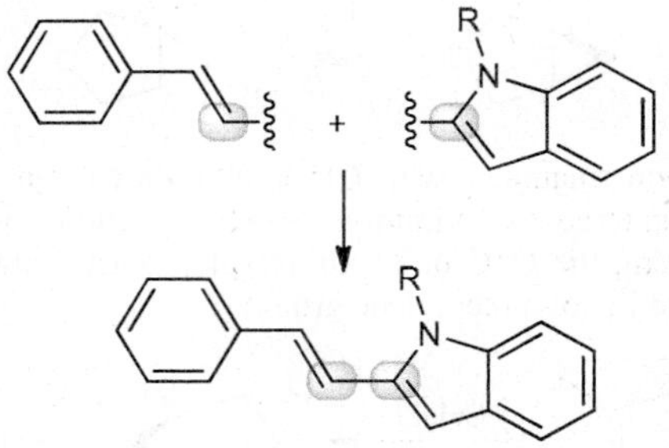

In this case, the organic electrophile has a C=C unit with the *E* configuration. So, when drawing the product, we must be careful that that this C=C unit maintains its *E* configuration.

(f) First identify the carbon atoms that will be joined. The carbon atom connected directly to the triflate group (in the vinyl triflate) will be joined with the carbon atom connected directly to tin (Sn) in the organostannane, highlighted here:

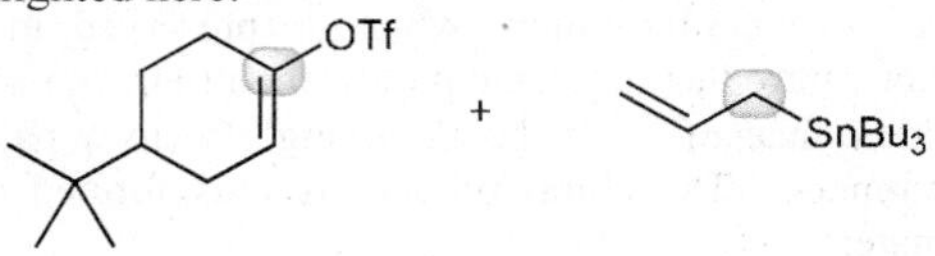

Then, rotate the coupling partners so that they are aligned to form a bond, and draw the product:

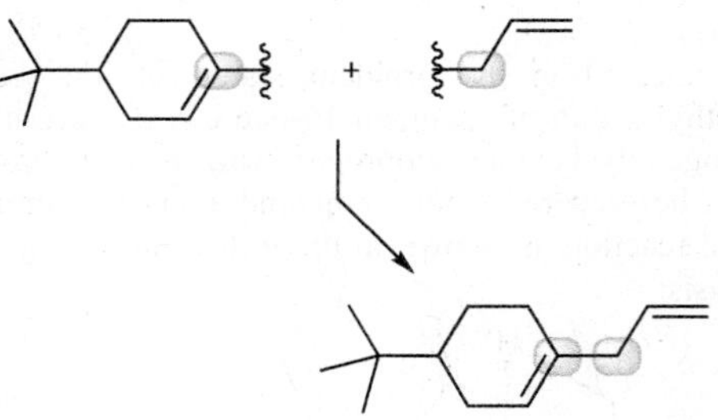

23.13. First identify the carbon atoms that will be joined. The carbon atom connected directly to the triflate group will be joined with the carbon atom connected directly to the trimethylstannane group, highlighted below. The compound is then redrawn so that the coupling partners are aligned:

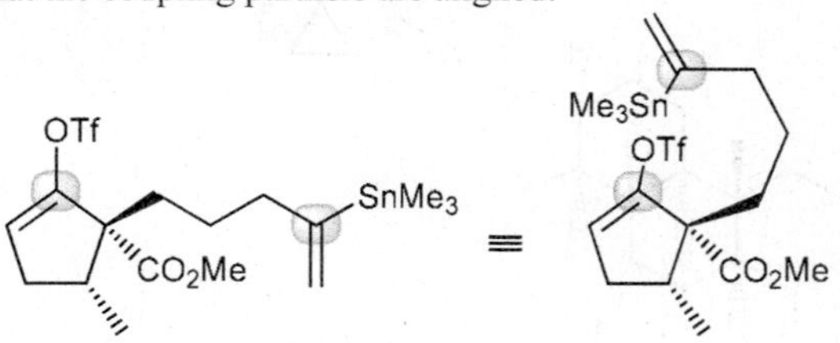

A new σ bond is then formed between the highlighted carbon atoms, giving the following diene:

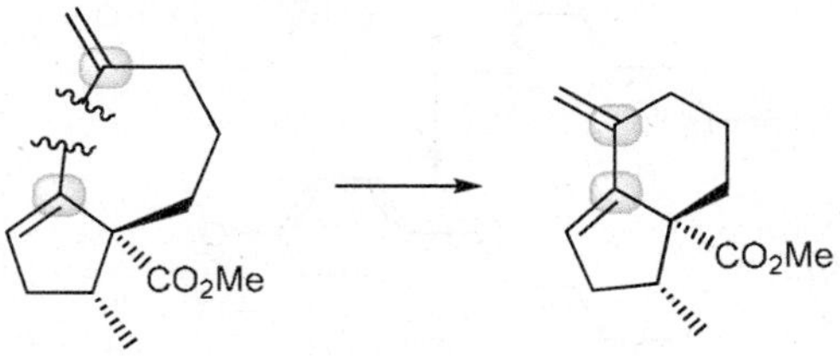

23.14. The problem statement indicates an intramolecular process, which means that both coupling partners are tethered within the same molecule. The carbon atom connected directly to the iodide group will be joined with the carbon atom connected directly to tin (Sn) in the organostannane, highlighted here:

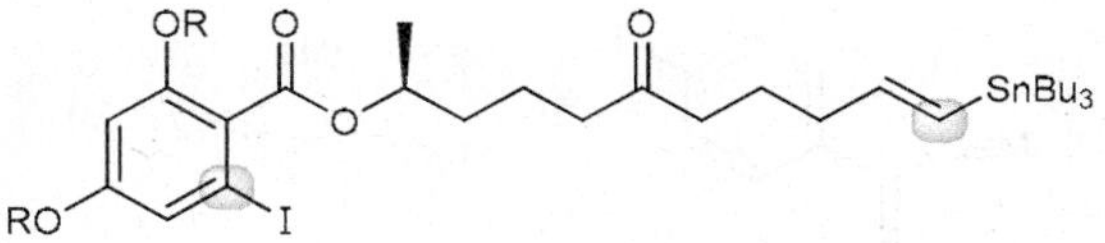

Next we align the coupling partners and draw the product of the intramolecular Stille coupling process:

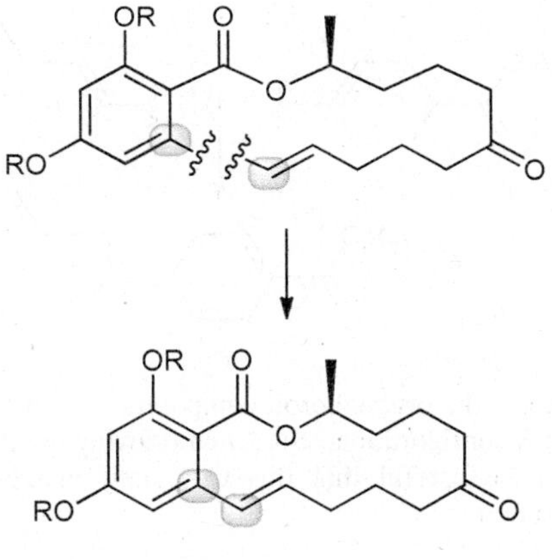

Then, in the final step of the synthesis, the coupling product is treated with HCl to remove the protecting groups, converting both OR groups into OH groups, giving (*S*)-zearalenone, as shown:

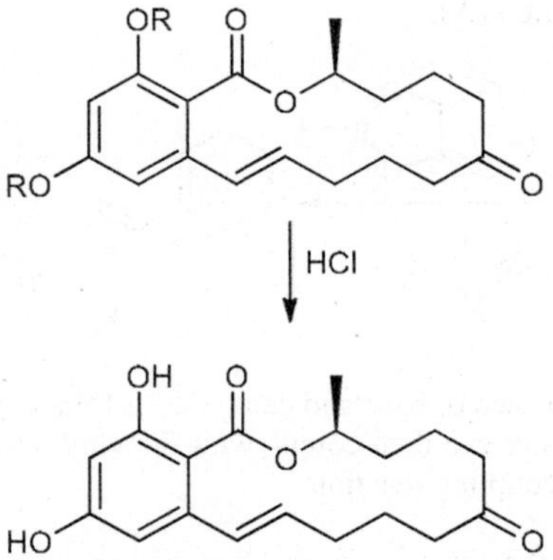

$C_{18}H_{22}O_5$

23.15. First identify the carbon atoms that will be joined. The carbon atom connected directly to the triflate group will be joined with the carbon atom connected directly to the trimethylstannane group, highlighted below:

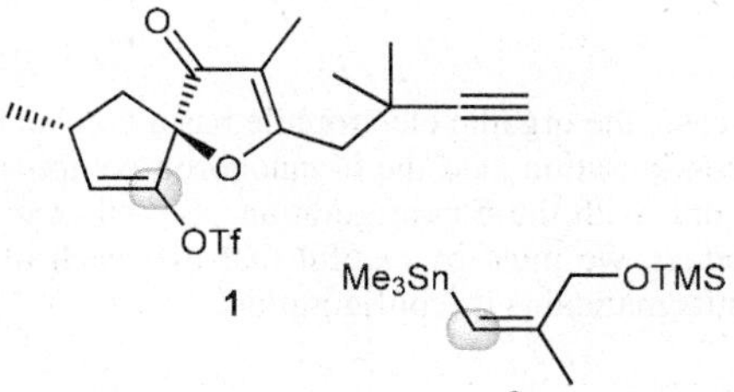

Then, redraw the coupling partners so that they are aligned to form a bond (neither of the coupling partners needs to be rotated in this case), and draw the product **3**:

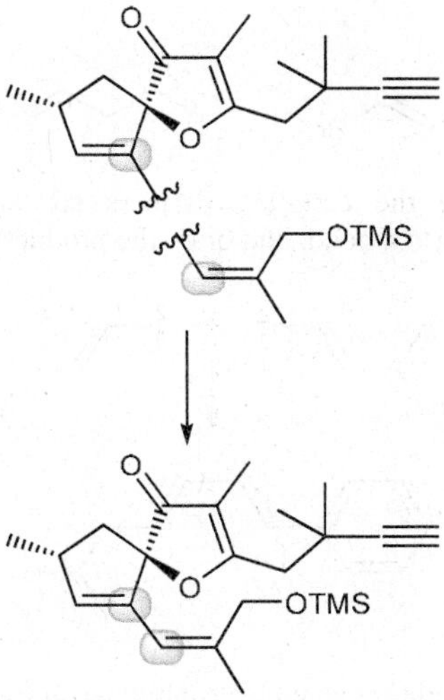

23.16.

(a) First identify the carbon atoms that will be joined. The carbon atom connected directly to the triflate group will be joined with the carbon atom connected directly to boron in the organoboron compound, highlighted here:

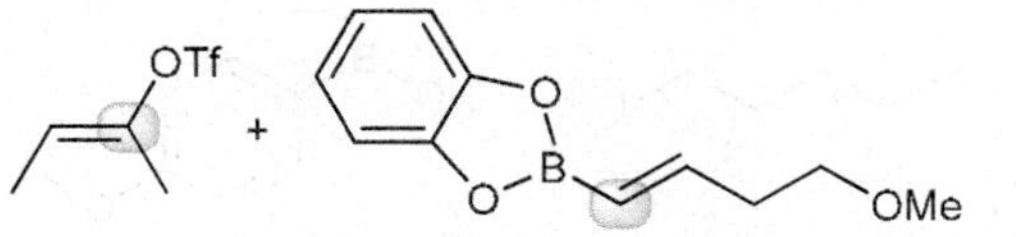

Then, rotate the coupling partners so that they are aligned to form a bond, and draw the product:

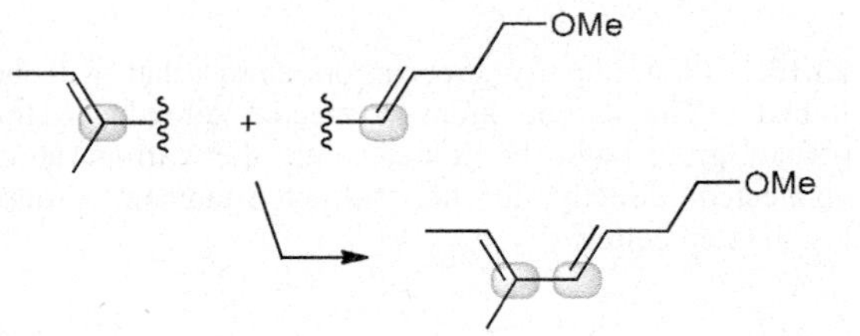

In this case, the organic electrophile has a C=C unit with the *E* configuration, and the organoboron compound has a C=C unit with the *E* configuration. So, when drawing the product, we must be careful that that each of these C=C units maintains its configuration.

(b) First identify the carbon atoms that will be joined. The carbon atom connected directly to the bromide will be joined with the carbon atom connected directly to boron in the organoboron compound, highlighted here:

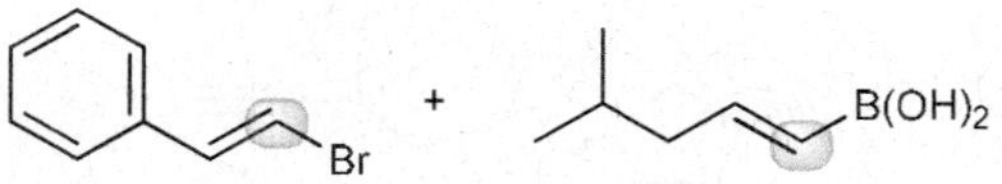

Then, rotate the coupling partners so that they are aligned to form a bond, and draw the product:

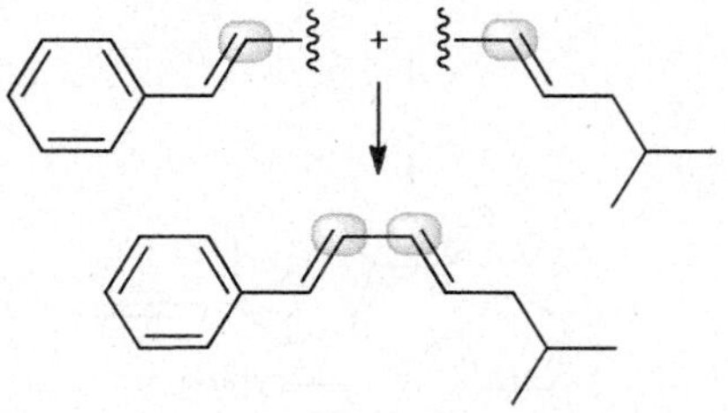

In this case, the organic electrophile has a C=C unit with the *E* configuration, and the organoboron compound has a C=C unit with the *E* configuration. So, when drawing the product, we must be careful that that each of these C=C units maintains its configuration.

(c) First identify the carbon atoms that will be joined. The carbon atom connected directly to the iodide will be joined with the carbon atom connected directly to boron in the organoboron compound, highlighted here:

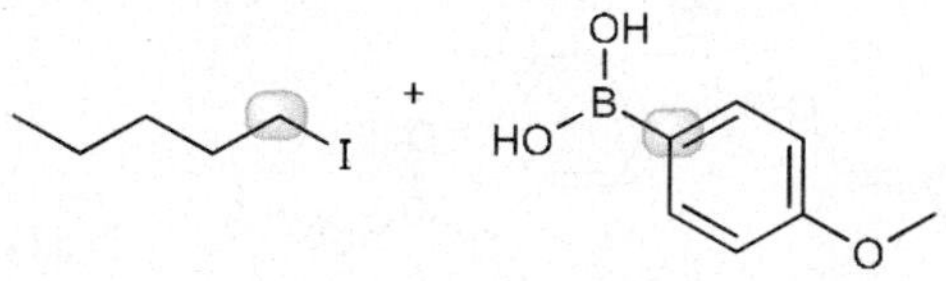

Then, rotate the coupling partners so that they are aligned to form a bond, and draw the product:

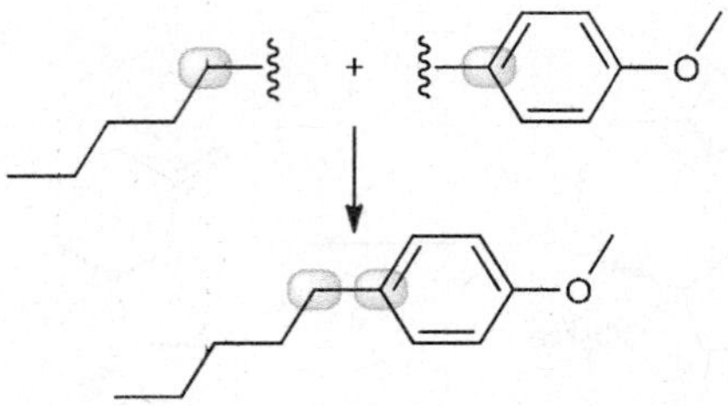

(d) First identify the carbon atoms that will be joined. The carbon atom connected directly to the iodide will be joined with the carbon atom connected directly to boron in the organoboron compound, highlighted here:

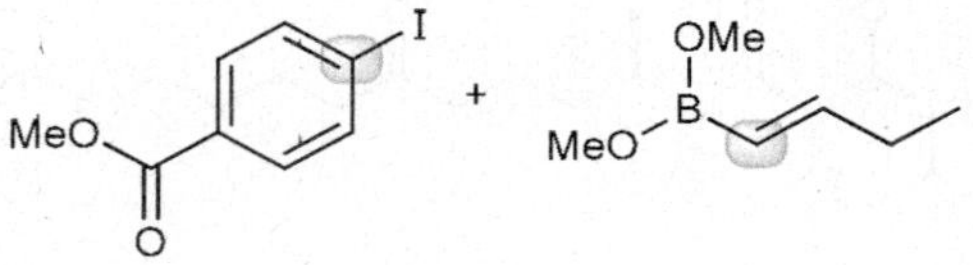

Then, rotate the coupling partners so that they are aligned to form a bond, and draw the product:

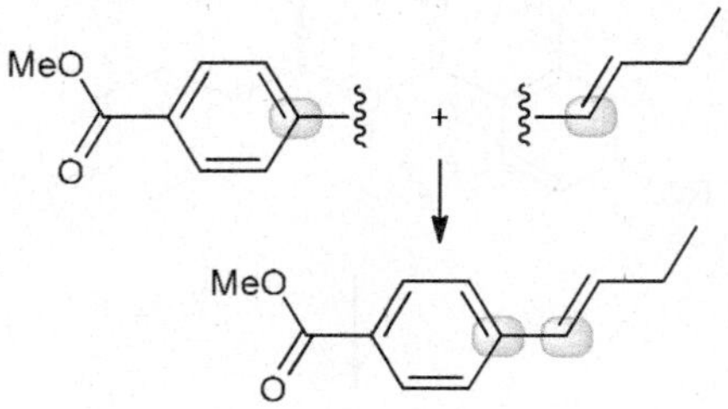

In this case, the organoboron compound has a C=C unit with the *E* configuration. So, when drawing the product, we must be careful that this C=C unit maintains its configuration.

23.17.

(a) Upon treatment with catecholborane, the alkyne is converted into a vinyl boronic ester with the *E* configuration (**A**):

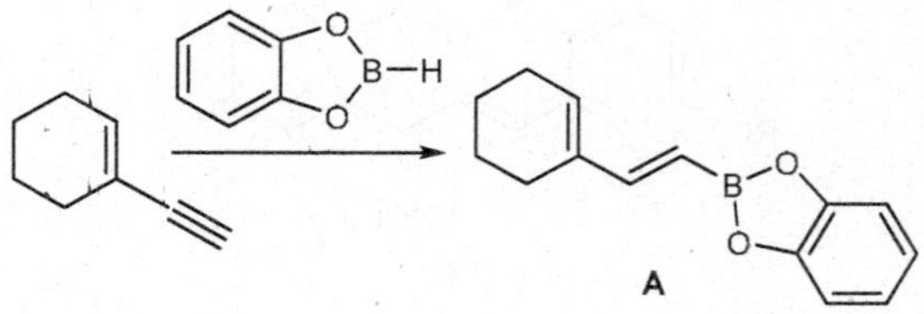

In the presence of base and catalytic $Pd(PPh_3)_4$, this vinyl boronic ester can then couple with the vinyl bromide via a Suzuki coupling reaction.

To draw the product of the Suzuki coupling reaction, first identify the carbon atoms that will be joined. The carbon atom connected directly to the bromide will be joined with the carbon atom connected directly to boron in the organoboron compound, highlighted here:

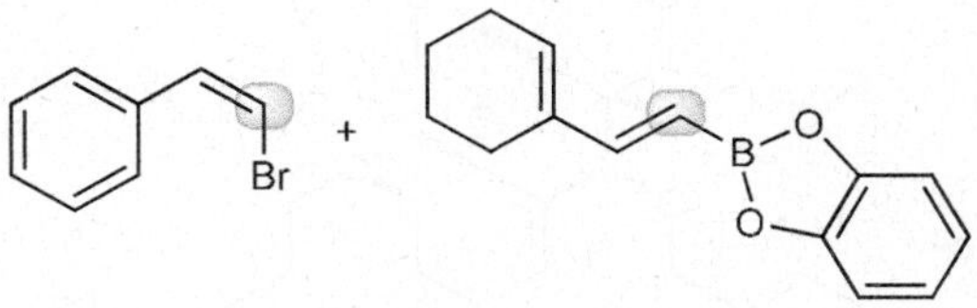

Then, rotate the coupling partners so that they are aligned to form a bond, and draw the product (**B**):

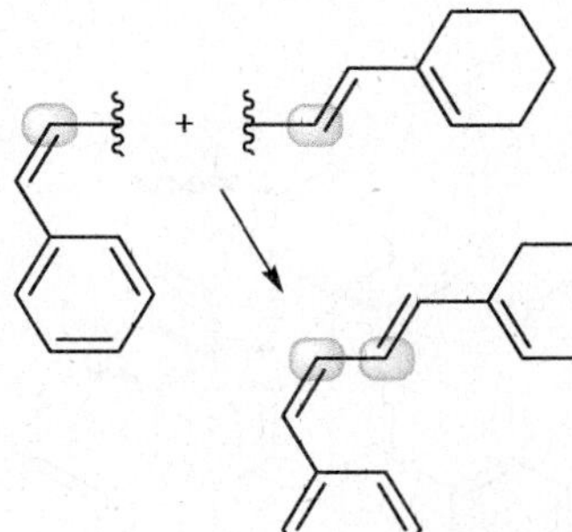

In this case, the organic electrophile has a C=C unit with the *Z* configuration, and the organoboron compound has a C=C unit with the *E* configuration. So, when drawing the product, we must be careful that that each of these C=C units maintains its configuration.

23.18. In the first step of the reaction sequence, 9-BBN adds across a π bond (hydroboration). Of the three π bonds indicated below, the one that is monosubstituted is the least sterically hindered and thus most susceptible to reaction with this bulky cyclic alkylborane.

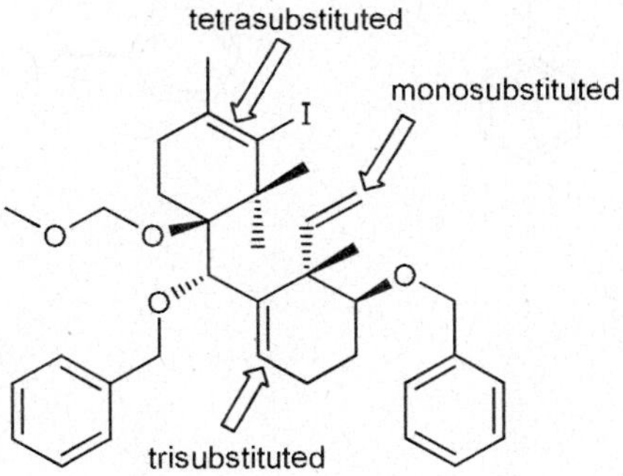

An additional hint suggesting reaction at the monosubstituted π bond comes from the wording of the problem statement (that this sequence makes an 8-membered ring). The first reaction installs a borane group, so the second reaction is likely an intramolecular Suzuki coupling between the carbon atom bonded to the boron after step 1 (C8 below) and the carbon atom bonded to iodine (C1 below). Bond formation between these two carbon atoms would indeed make a new 8-membered ring, as indicated by the numbered atoms below.

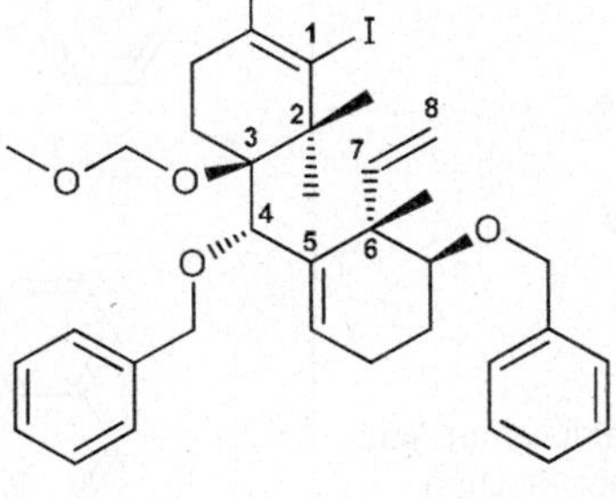

During step one of the process, 9-BBN adds across the monosubstituted π bond in an *anti*-Markovnikov addition, giving compound **2**.

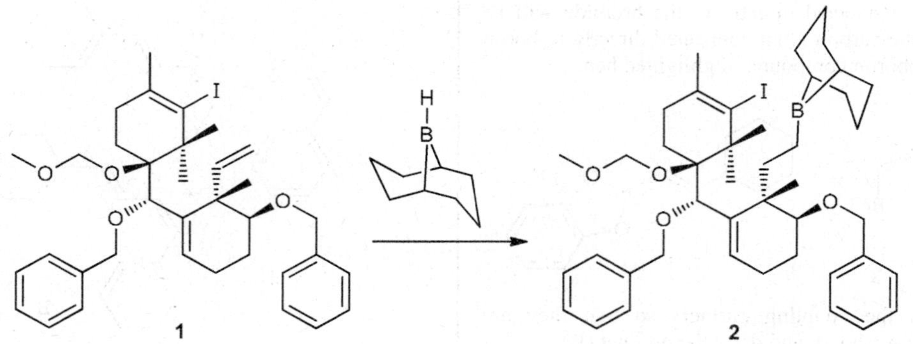

1

2

Compound **2** then undergoes an intramolecular Suzuki coupling reaction upon treatment with a palladium catalyst under basic conditions, giving compound **3**. The highlighted carbon atoms are joined as result of this process:

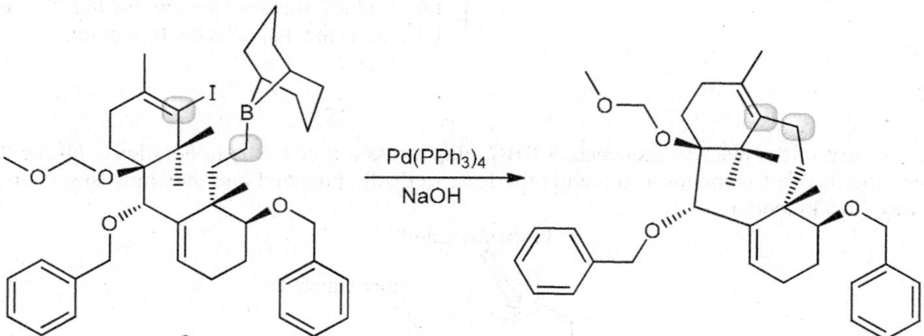

$Pd(PPh_3)_4$
NaOH

2

3

23.19.
(a) We begin by identifying the bonds in compound **1** that will be made, highlighted here:

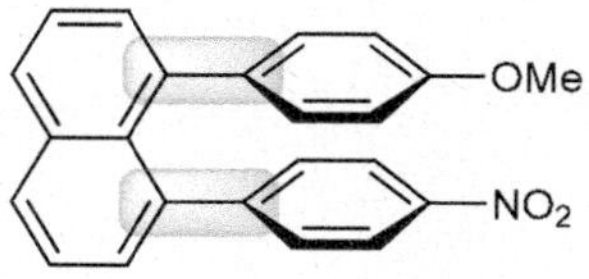

1

Compound **A** contains both an aryl bromide group and an aryl boronic acid group. So, disconnection of compound **1** at the top biaryl bond will reveal a *para*-methoxy boronic acid coupling partner, while disconnection at the bottom biaryl bond will reveal a *para*-nitro aryl iodide partner, shown below:

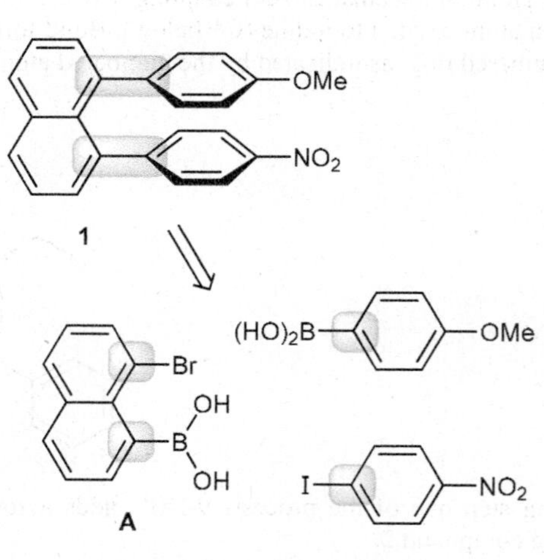

1

A

Now let's consider the forward reaction sequence. Compound **1** is prepared from compound **A** via two successive Suzuki cross-coupling reactions:

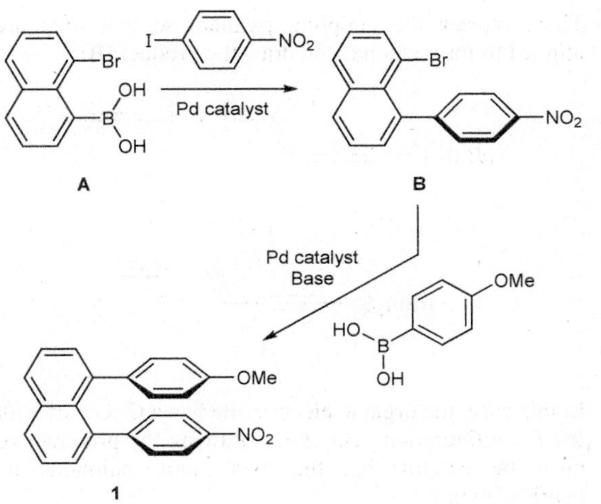

A

B

1

Compound **B** (shown above) is formed via the Suzuki cross coupling between compound **A** and the *para*-nitro aryl iodide.

(b) Compound **A** contains both an aryl bromide group and an aryl boronic acid group, yet it does not react with itself under these conditions. Rather, it prefers to cross couple with the *para*-nitro aryl iodide. This is likely due to the fact that aryl iodides participate much faster in Suzuki cross coupling reactions than aryl bromides.

23.20.

(a) First identify the carbon atoms that will be joined. The carbon atom connected directly to the bromide will be joined with the carbon atom connected directly to Zn in the organozinc, highlighted here:

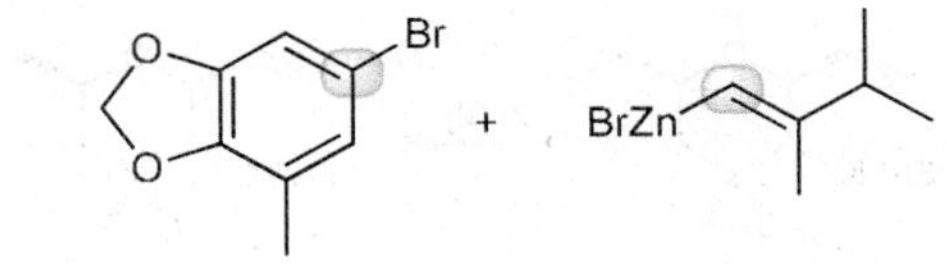

Then, realign the coupling partners so that they are aligned to form a bond, and draw the product:

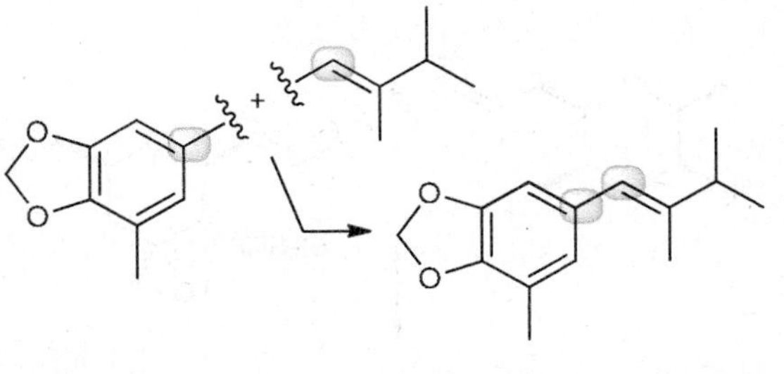

In this case, the organozinc has a C=C unit with the *E* configuration. So, when drawing the product, we must be careful that this C=C unit maintains its configuration.

(b) First identify the carbon atoms that will be joined. The carbon atom connected directly to the bromide will be joined with the carbon atom connected directly to zinc in the organozinc, highlighted here:

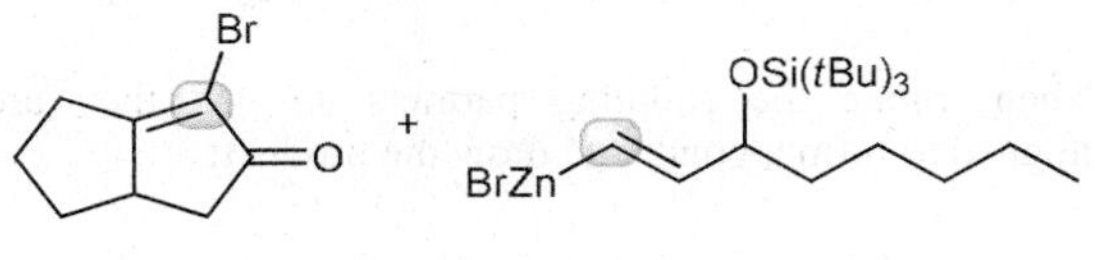

Then, rotate the coupling partners so that they are aligned to form a bond, and draw the product:

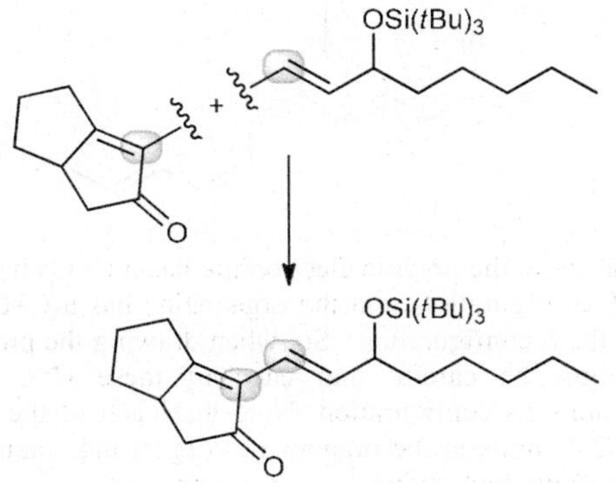

In this case, the organozinc has a C=C unit with the *E* configuration. So, when drawing the product, we must be careful that this C=C unit maintains its configuration.

(c) First identify the carbon atoms that will be joined. The carbon atom connected directly to the iodide will be joined with the carbon atom connected directly to zinc in the organozinc, highlighted here:

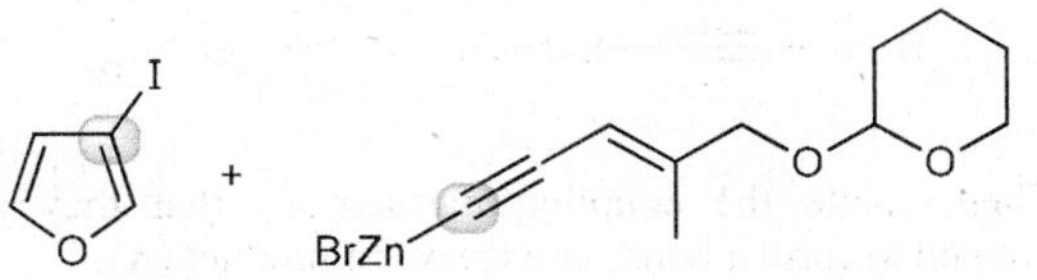

Then, rotate the coupling partners so that they are aligned to form a bond, and draw the product:

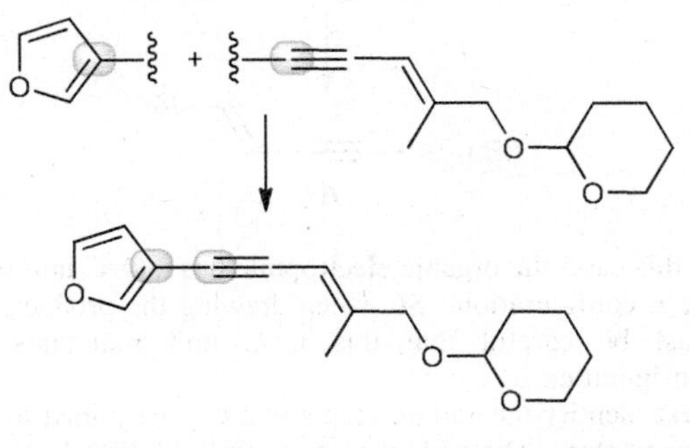

(d) First identify the carbon atoms that will be joined. The organic electrophile is a dibromide; and the more reactive vinyl bromide serves as the coupling partner. The sp^2-hybridized carbon atom connected directly to the bromide will be joined with the carbon atom connected directly to zinc in the organozinc, highlighted here:

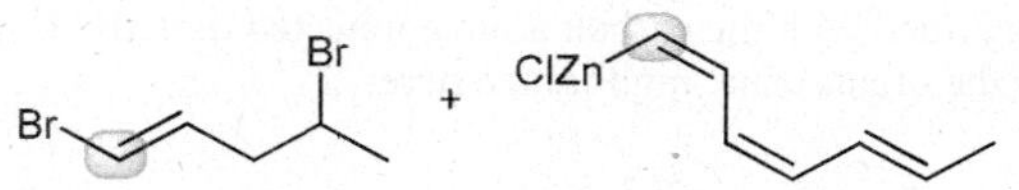

Then, rotate the coupling partners so that they are aligned to form a bond, and draw the product:

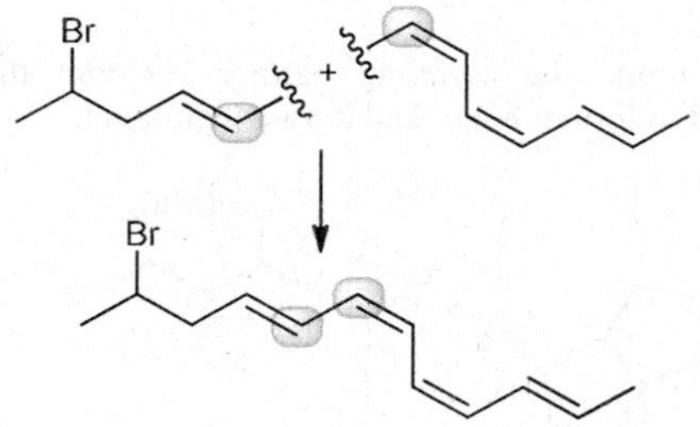

In this case, the organic electrophile has a C=C unit with the *E* configuration, and the organozinc has a C=C unit with the *Z* configuration. So, when drawing the product, we must be careful that each of these C=C units maintains its configuration. Note that each of the other two C=C units in the organozinc reagent also maintains its configuration giving the *E*,*Z*,*Z*,*E* product.

23.21. First identify the carbon atoms that will be joined in the first step. The organic electrophile is both a vinyl iodide and a vinyl bromide. The more reactive vinyl iodide serves as the coupling partner. The carbon atom connected directly to the iodide will be joined with the carbon atom connected directly to zinc in the organozinc, highlighted here:

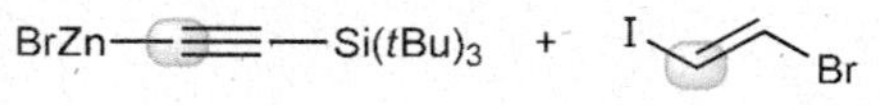

Then, rotate the coupling partners so that they are aligned to form a bond, and draw the product (**A**):

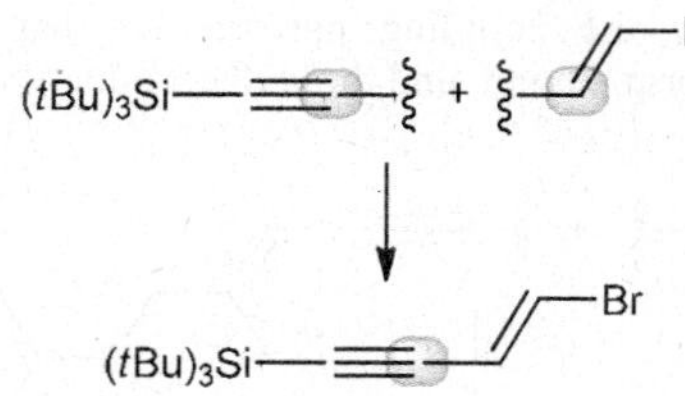

In this case, the organic electrophile has a C=C unit with the *E* configuration. So, when drawing the product, we must be careful that this C=C unit maintains its configuration.

Next identify the carbon atoms that will be joined in the second step. The carbon atom connected directly to the bromide in **A** will be joined with the carbon atom connected directly to zinc in the organozinc, highlighted here:

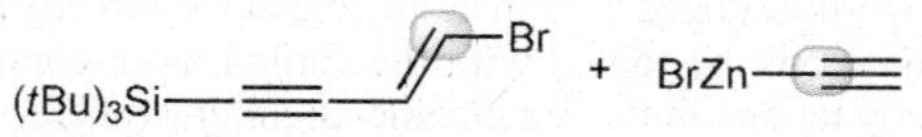

Then, redraw the coupling partners so that they are aligned to form a bond, and draw the product (**B**):

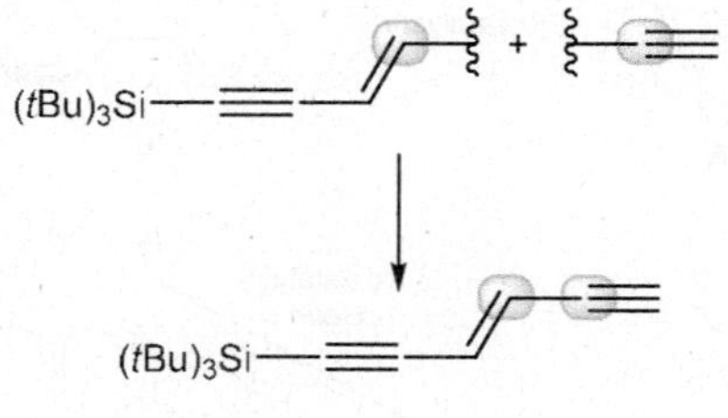

In this case, the organic electrophile has a C=C unit with the *E* configuration. So, when drawing the product, we must be careful that this C=C unit maintains its configuration.

23.22. First identify the carbon atoms that will be joined. The carbon atom connected directly to the iodide (of the vinyl iodide) will be joined with the carbon atom connected directly to Zr in the organozirconium, highlighted here:

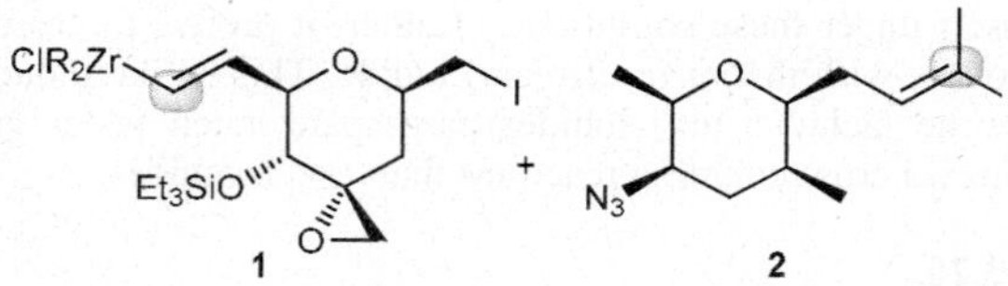

Note that $ZnCl_2$ is used in this reaction, so the active organometallic species is likely the organozinc compound resulting from a transmetallation from Zr to Zn:

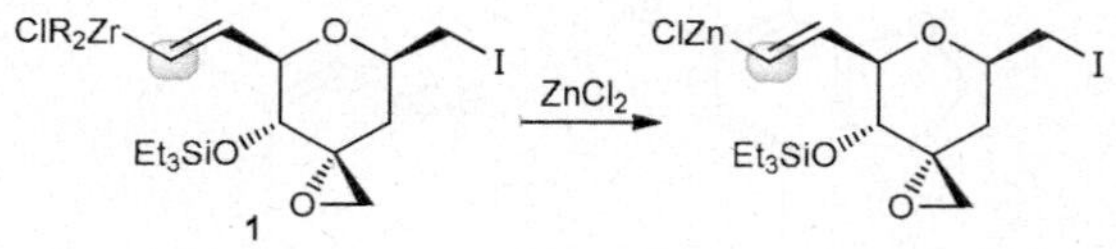

Next, realign the coupling partners so that they are aligned to form a bond, and draw the product:

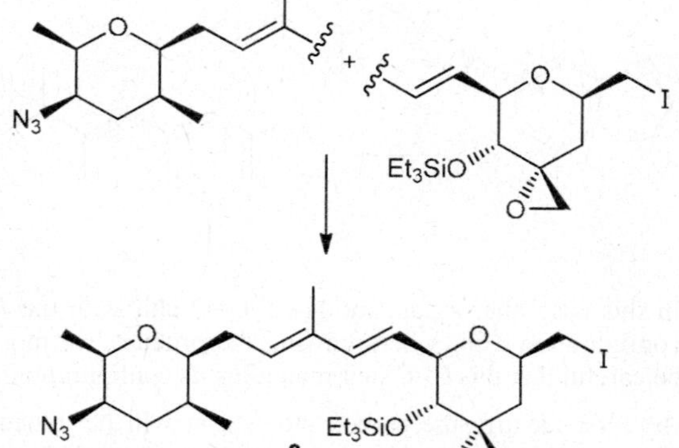

In this case, the organic electrophile has a C=C unit with the *E* configuration, and the organozinc has a C=C unit with the *E* configuration. So, when drawing the product, we must be careful that each of these C=C units maintains its configuration.

23.23.

(a) First identify the new C-C bond in compound **3** that was formed via Negishi coupling. Analysis of the structures of **1** and **3** leads to the identification of the newly formed bond in **3** as indicated here with a wavy line:

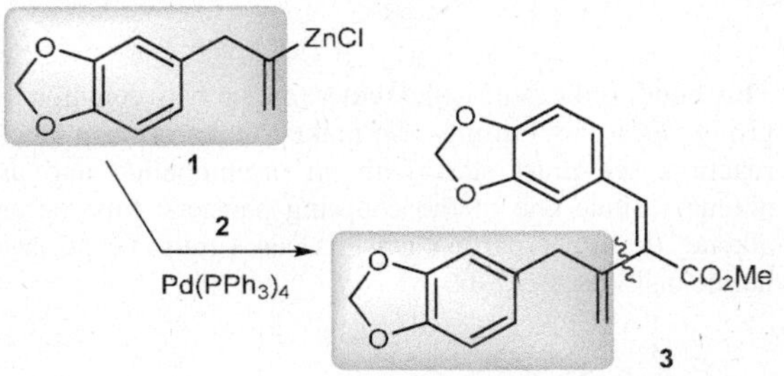

ZnCl

1

2

$Pd(PPh_3)_4$

CO_2Me

3

Next, identify the structure of organic electrophile **2** (a vinyl bromide) by disconnecting the C-C bond that is formed in the Negishi coupling:

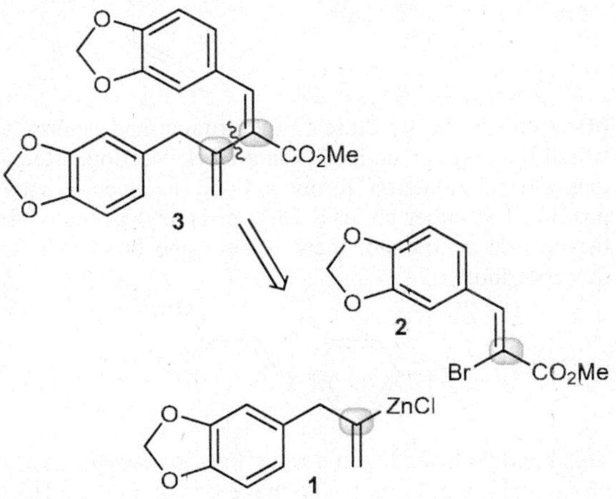

CO_2Me

3

2

Br CO_2Me

ZnCl

1

Note that at the C=C unit of **2** the arene and ester (CO_2Me) are *trans* to each other. This *trans* relationship is preserved in **3**.

(b) The question states that savinin and gadain are isomers of each other, differing only in the configuration of the C=C unit. Changing the configuration of the *E* alkene in savinin thus provides the structure of gadain, a *Z* alkene:

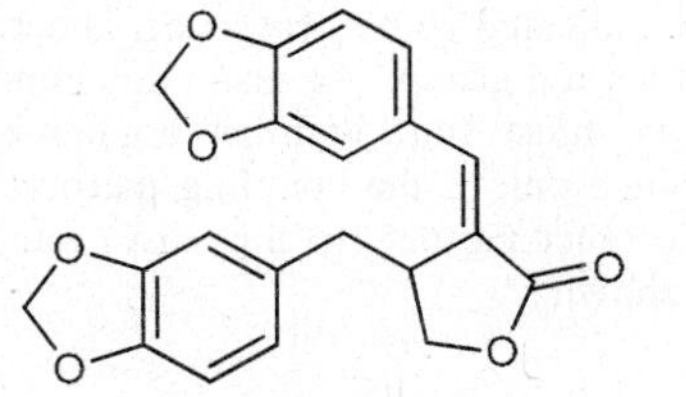

savinin

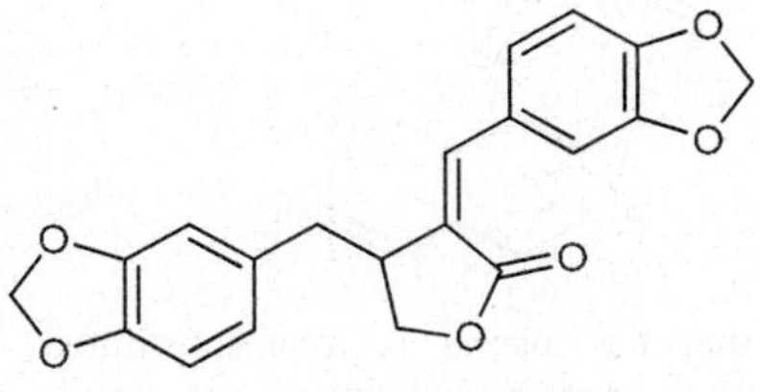

gadain

Changing the configuration of **2** gives the vinyl bromide that would be utilized to produce the intermediate that would be subsequently converted to gadain, as shown:

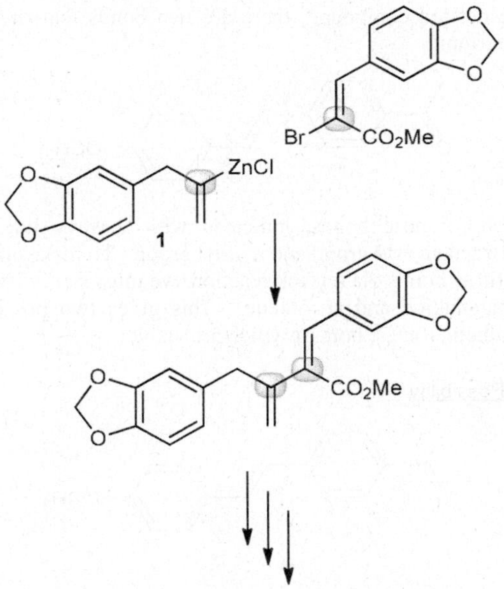

Br CO_2Me

ZnCl

1

CO_2Me

gadain

23.24.

(a) Analyze the structure of the product and determine which bond(s) can be made via a Heck reaction. Recall that a Heck reaction forms a bond between a vinyl position and either an aryl, vinyl or benzyl position. In the desired compound, there is only one bond that fits this criterion:

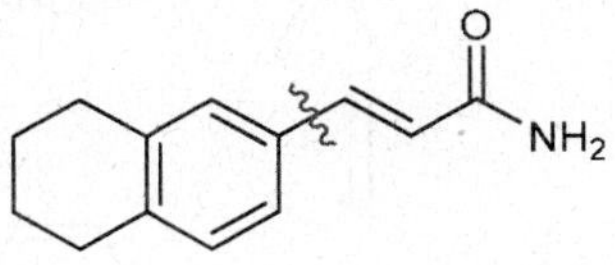

O

NH_2

This bond, indicated with a wavy line, is between an aryl group and a vinyl group. To make this bond via a Heck reaction, we must start with an organohalide and an alkene. Since one of the coupling partners must be an alkene, the other partner (in this case) must be an aryl halide, as shown:

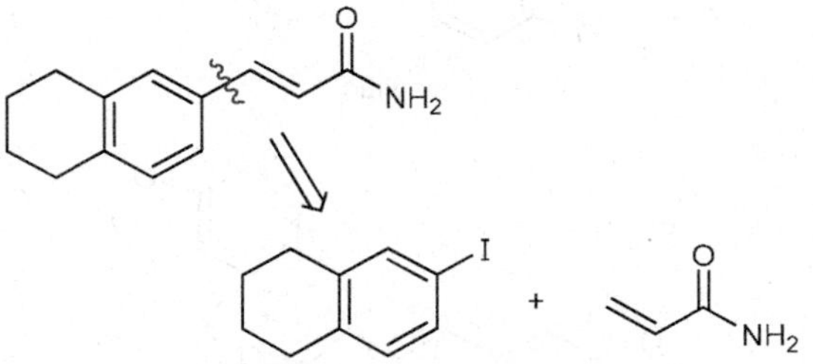

Notice that the alkene is monosubstituted, with an electron-withdrawing substituent, so we expect the reaction to be very effective.

(b) Analyze the structure of the product and determine which bond(s) can be made via a Heck reaction. Recall that a Heck reaction forms a bond between a vinyl position and either an aryl, vinyl or benzyl position. In the desired compound, there are two bonds that fit this criterion:

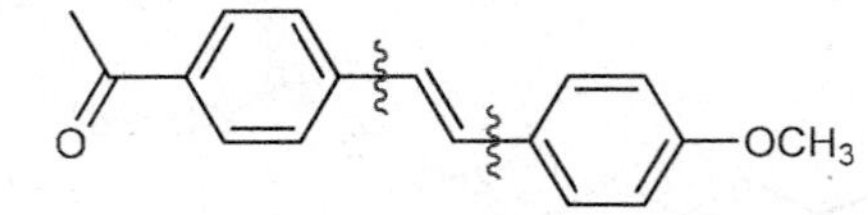

Each of these bonds, indicated with a wavy line, is between an aryl group and a vinyl group. To make either of these bonds via a Heck reaction, we must start with an organohalide and an alkene. This gives two possible synthetic routes, both of which are viable:

Possibility #1

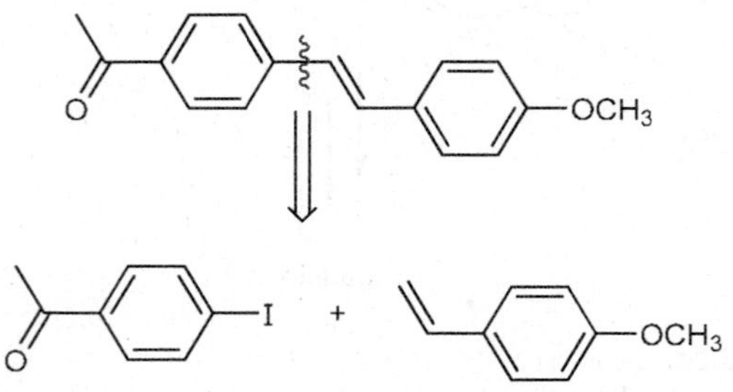

Possibility #2

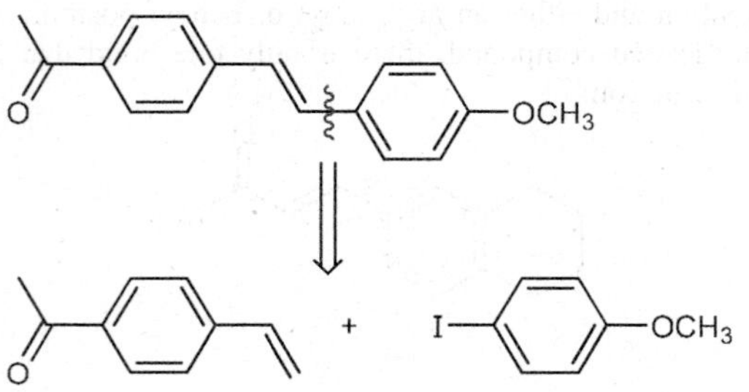

The second possibility has the added advantage that the starting monosubstituted alkene has an electron-withdrawing substituent (so we expect the reaction to be very effective).

(c) Analyze the structure of the product and determine which bond(s) can be made via a Heck reaction. Recall that a Heck reaction forms a bond between a vinyl position and either an aryl, vinyl or benzyl position. In the desired compound, there is only one bond that fits this criterion:

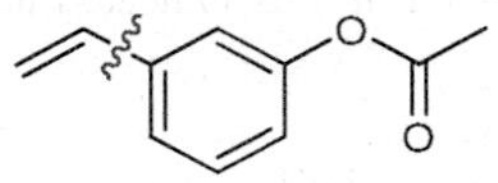

This bond, indicated with a wavy line, is between an aryl group and a vinyl group. To make this bond via a Heck reaction, we must start with an organohalide and an alkene. Since one of the coupling partners must be an alkene, the other partner (in this case) must be an aryl halide, as shown:

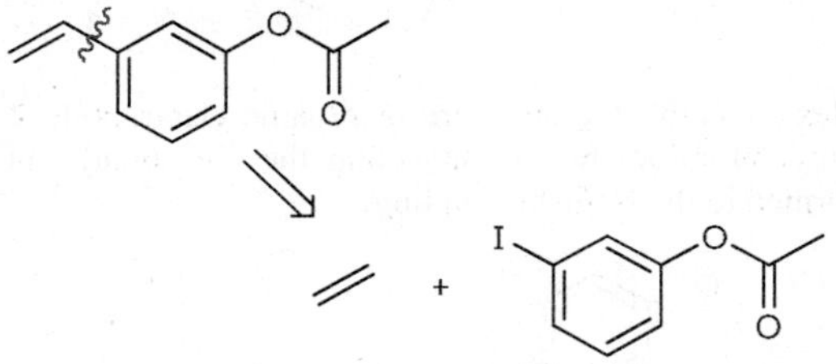

(d) Analyze the structure of the product and determine which bond(s) can be made via a Heck reaction. Recall that a Heck reaction forms a bond between a vinyl position and either an aryl, vinyl or benzyl position. In the desired compound, there is only one bond that fits this criterion:

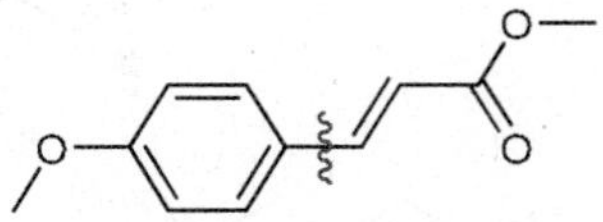

This bond, indicated with a wavy line, is between an aryl group and a vinyl group. To make this bond via a Heck reaction, we must start with an organohalide and an alkene. Since one of the coupling partners must be an alkene, the other partner (in this case) must be an aryl halide, as shown:

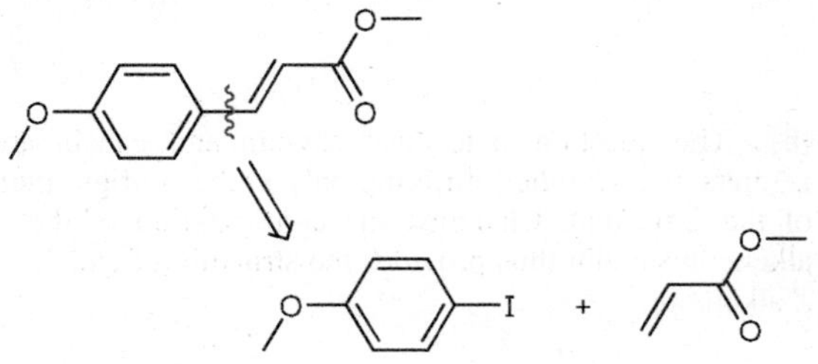

Notice that the alkene is monosubstituted, with an electron-withdrawing substituent, so we expect the reaction to be very effective.

(e) Analyze the structure of the product and determine which bond(s) can be made via a Heck reaction. Recall that a Heck reaction forms a bond between a vinyl position and either an aryl, vinyl or benzyl position. In the desired compound, there is only one bond that fits this criterion:

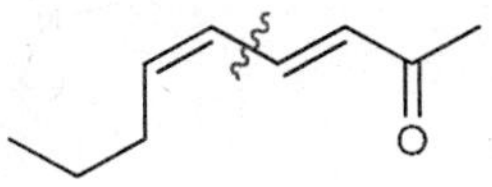

This bond, indicated with a wavy line, is between two vinyl groups. Since this bond must be made via a Heck reaction, we must consider the following two possibilities, each of which starts with an alkene and a vinyl iodide:

Possibility #1

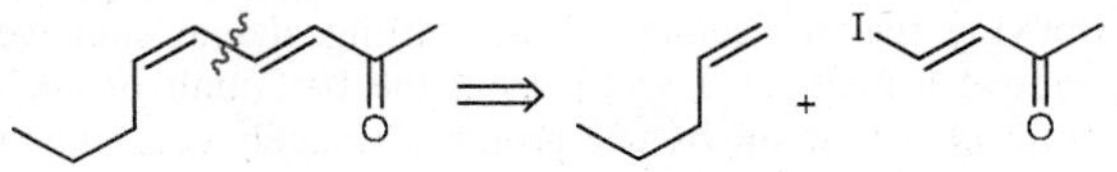

Possibility #2

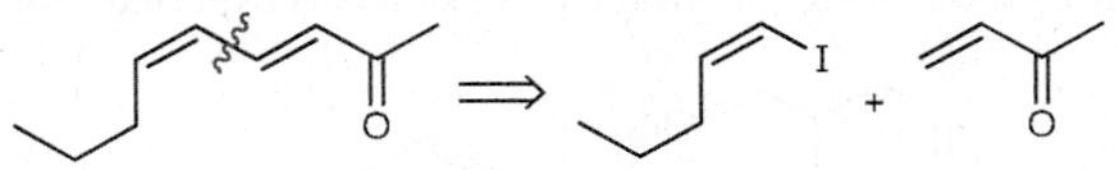

The first possibility is not viable because Heck reactions are stereoselective with respect to the starting alkene, and the *E* isomer (rather than the *Z* isomer) would be expected:

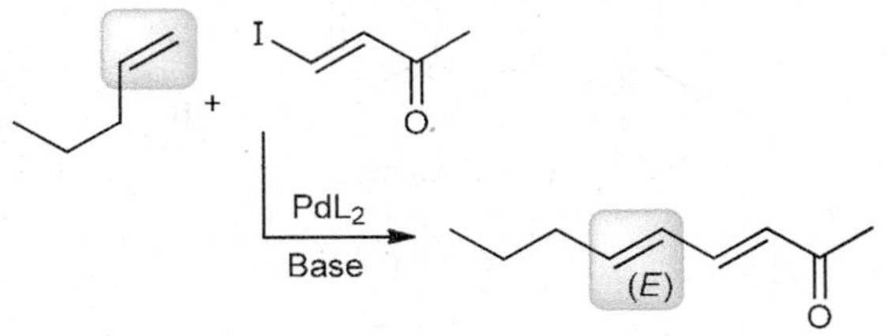

PdL_2

Base

(*E*)

This is not the desired stereochemical outcome, so we turn our attention to the second possibility. This possibility works, and it has the added advantage that the starting alkene is monosubstituted, with an electron-withdrawing substituent (so we expect the reaction to be very effective).

PdL_2, Base

(*Z*)

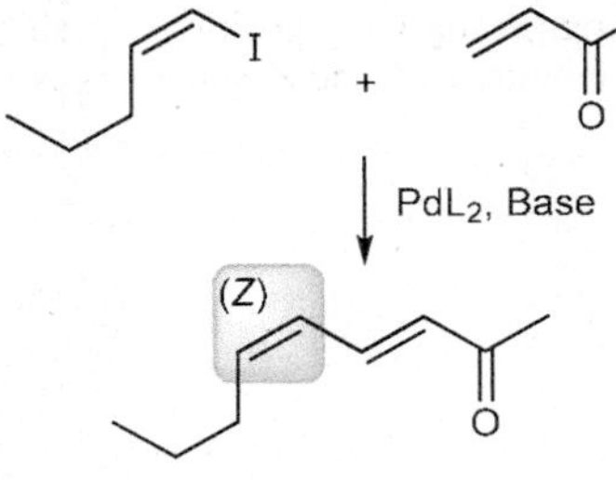

(f) Analyze the structure of the product and determine which bond(s) can be made via a Heck reaction. Recall that a Heck reaction forms a bond between a vinyl position and either an aryl, vinyl or benzyl position. In the desired compound, there is only one bond that fits this criterion:

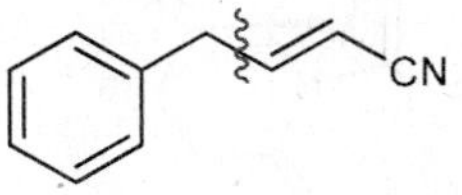

This bond, indicated with a wavy line, is between a benzyl group and a vinyl group. To make this bond via a Heck reaction, we must start with an organohalide and an alkene. Since one of the coupling partners must be an alkene, the other partner (in this case) must be a benzyl halide, as shown:

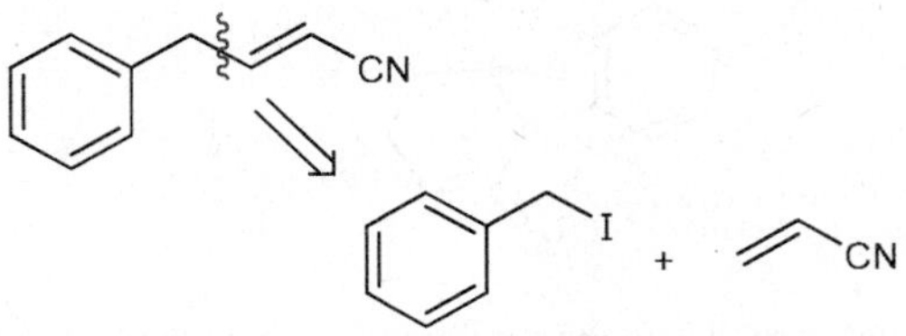

Notice that the alkene is monosubstituted, with an electron-withdrawing substituent, so we expect the reaction to be very effective.

23.25. In a Heck reaction, an alkene is coupled with an organohalide. In this case, the two coupling partners are tethered together in the same compound:

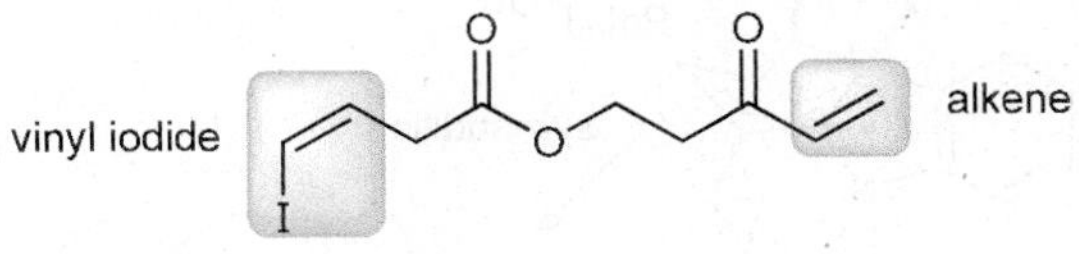

vinyl iodide

alkene

Upon treatment with a base and a suitable palladium catalyst, an intramolecular Heck reaction can occur. Notice that the *Z* configuration of the vinyl iodide is preserved. The new C-C bond is formed regioselectively and stereoselectively, with the *E* configuration being formed:

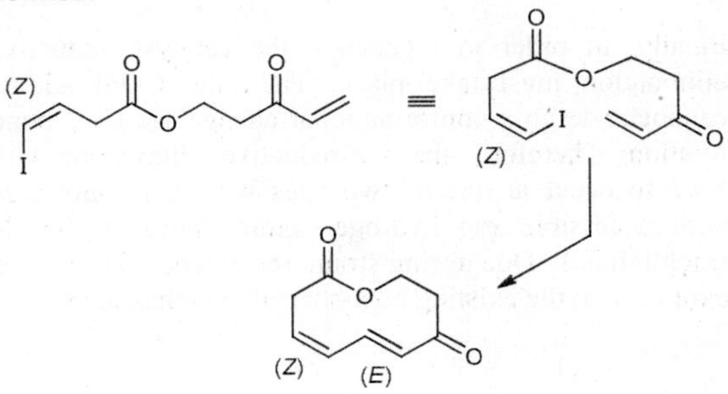

(*Z*)

(*Z*)

(*Z*) (*E*)

23.26. Begin by identifying the carbon atoms that will be joined. The carbon atom connected to the iodide will be joined directly to the proximal carbon of the alkene (a bond will be formed between the two highlighted positions):

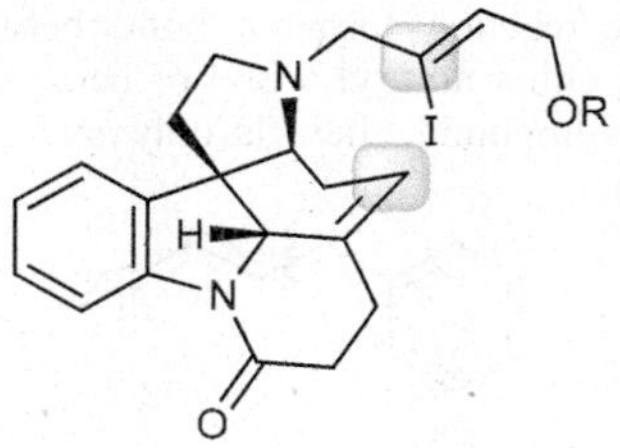

Next, consider the intermediate that forms when the palladium catalyst interacts with compound **1**.

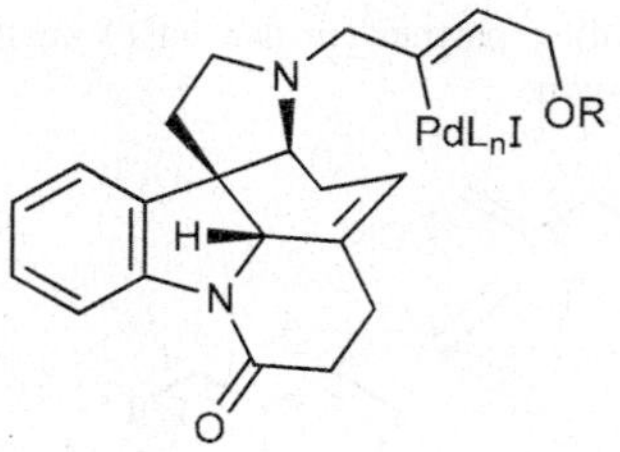

In this case, due to the configuration about the C–N bond (the N is on a wedge, coming out of the plane of the page), we expect oxidative addition to occur above (rather than below) the plane of the boat-shaped cyclohexene ring. When *syn*-addition occurs across the alkene, the new C–C bond forms the 6-membered ring with stereochemistry as shown.

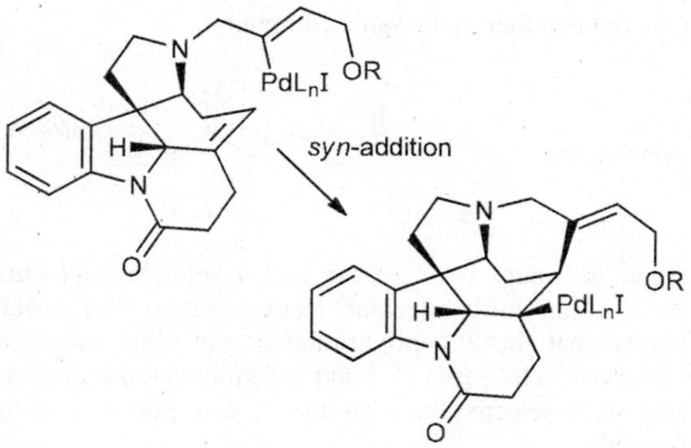

Finally, in order to regenerate the catalyst, reductive elimination must take place. The new C–Pd adduct cannot undergo a conformational change via C–C bond rotation. Therefore, the *syn*-reductive elimination will have to occur at one of two sites within the molecule with accessible *syn* hydrogen atoms *beta* to the Pd (highlighted). Due to ring strain, the alkene that forms is exocyclic to the existing boat-shaped cyclohexane ring.

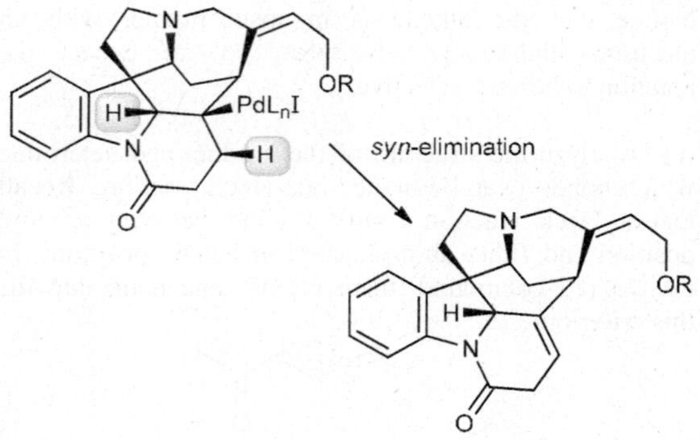

23.27.

(a) The starting material is a monosubstituted alkene, which is an excellent starting material for alkene metathesis. To draw the metathesis product, we begin by drawing two molecules of the starting alkene, and we remove a methylene group from the C=C unit of each molecule (these methylene groups ultimately combine to give ethylene gas). We then draw a double bond between the remaining vinyl positions, as shown. Don't forget that a mixture of *E* and *Z* isomers is expected.

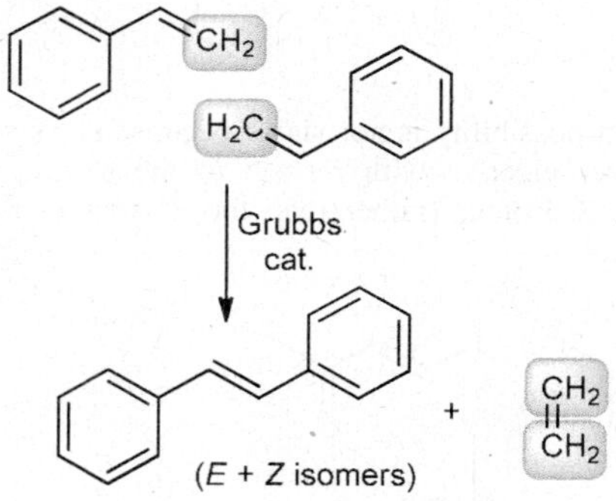

(b) The starting material is a 2,2-disubstituted alkene, which is an excellent starting material for alkene metathesis. To draw the metathesis product, we begin by drawing two molecules of the starting alkene, and we remove a methylene group from the C=C unit of each molecule (these methylene groups ultimately combine to give ethylene gas). We then draw a double bond between the remaining vinyl positions, as shown. Don't forget that a mixture of *E* and *Z* isomers is expected.

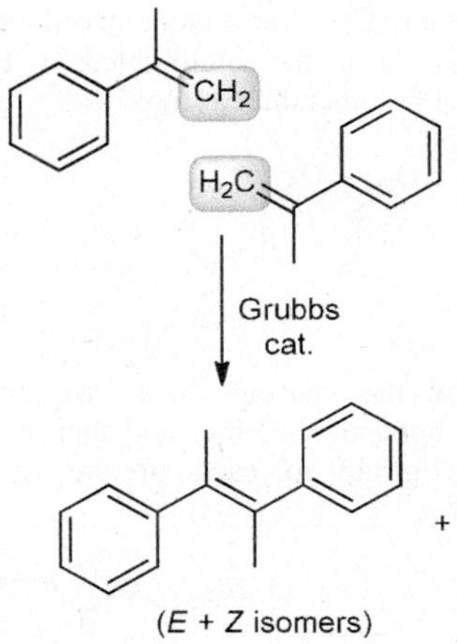

CH_2

H_2C

Grubbs cat.

CH_2 + || CH_2

(*E* + *Z* isomers)

(c) The starting material is a 2,2-disubstituted alkene, which is an excellent starting material for alkene metathesis. To draw the metathesis product, we begin by drawing two molecules of the starting alkene, and we remove a methylene group from the C=C unit of each molecule (these methylene groups ultimately combine to give ethylene gas). We then draw a double bond between the remaining vinyl positions, as shown.

CH_2 H_2C

Grubbs cat.

+ CH_2 || CH_2

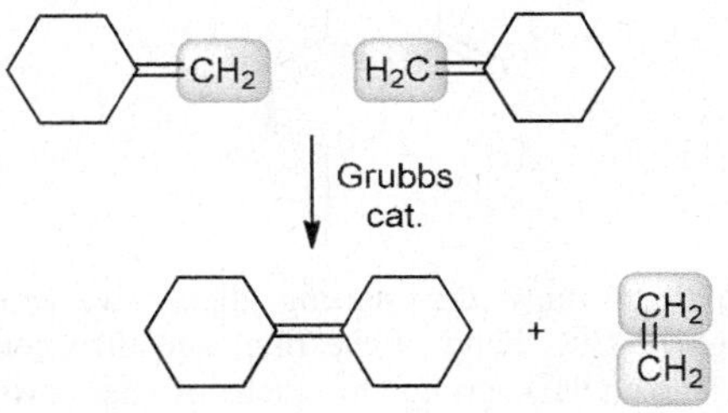

(d) The starting material is a monosubstituted alkene, which is an excellent starting material for alkene metathesis. To draw the metathesis product, we begin by drawing two molecules of the starting alkene, and we remove a methylene group from the C=C unit of each molecule (these methylene groups ultimately combine to give ethylene gas). We then draw a double bond between the remaining vinyl positions, as shown. Don't forget that a mixture of *E* and *Z* isomers is expected.

CH_2 H_2C Grubbs cat. (*E* + *Z* isomers) + CH_2 || CH_2

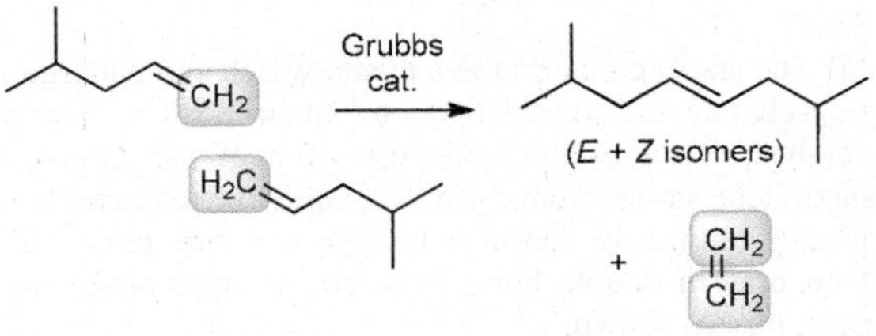

(e) The starting material is a monosubstituted alkene, which is an excellent starting material for alkene metathesis. To draw the metathesis product, we begin by drawing two molecules of the starting alkene, and we remove a methylene group from the C=C unit of each molecule (these methylene groups ultimately combine to give ethylene gas). We then draw a double bond between the remaining vinyl positions, as shown. Don't forget that a mixture of *E* and *Z* isomers is expected.

O CH_2 H_2C O Grubbs cat. O O (*E* + *Z* isomers) + CH_2 || CH_2

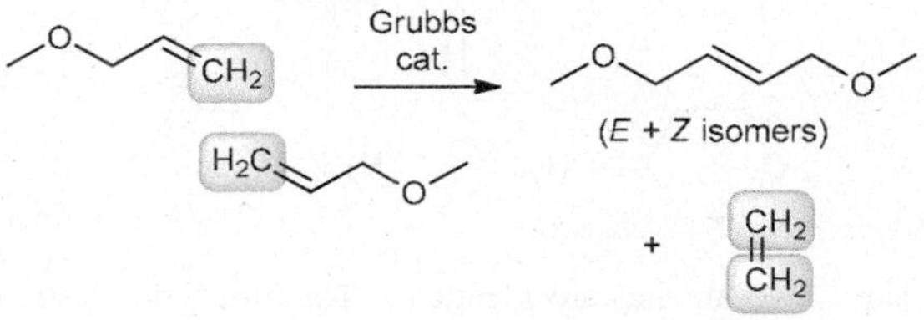

(f) The starting material is a 2,2-disubstituted alkene, which is an excellent starting material for alkene metathesis. To draw the metathesis product, we begin by drawing two molecules of the starting alkene, and we remove a methylene group from the C=C unit of each molecule (these methylene groups ultimately combine to give ethylene gas). We then draw a double bond between the remaining vinyl positions, as shown.

CH_2 H_2C

Grubbs cat.

+ CH_2 || CH_2

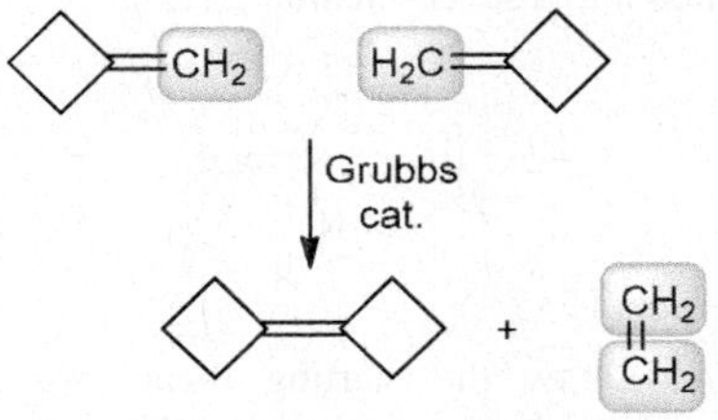

23.28. The starting alkene is *trans*-2-pentene, which is comprised of two different alkylidene fragments, highlighted here:

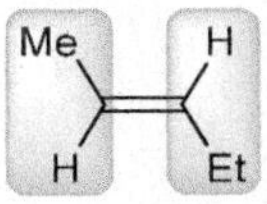

Me H H Et

During metathesis, these alkylidene fragments are separated from each other, and then recombined in all possible permutations. In this case, there are three possible alkenes produced, each of which is expected as a mixture of *E* and *Z* isomers (for a total of six different products).

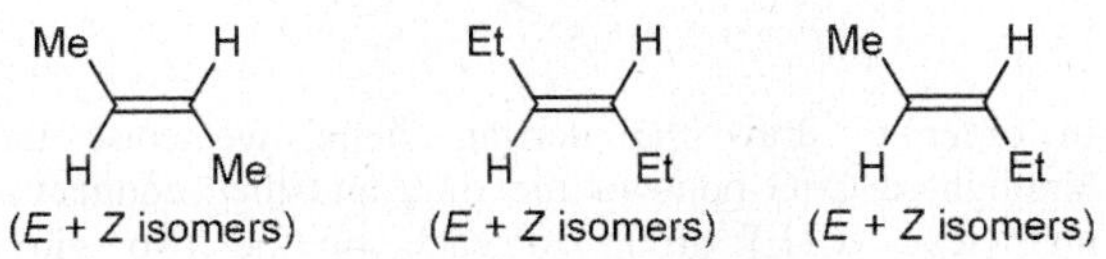

Me H / H Me	Et H / H Et	Me H / H Et
(*E* + *Z* isomers)	(*E* + *Z* isomers)	(*E* + *Z* isomers)

(Notice that the *E* isomer of the third product is the same as the starting alkene)

23.29. To draw the starting materials necessary to make the desired compound via alkene metathesis, we must separate the alkylidene fragments found in the product, and then attach a methylene group to each fragment:

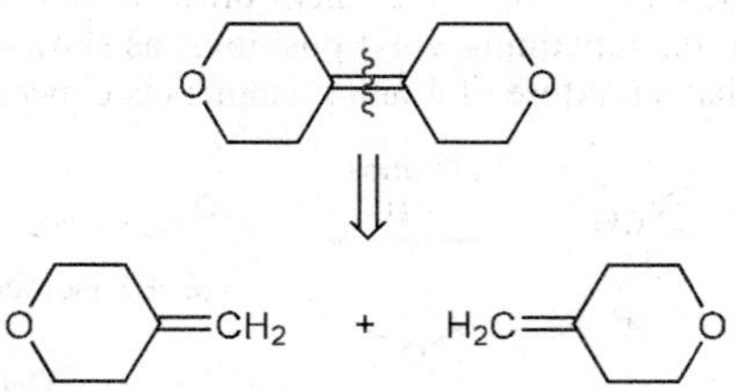

These two alkenes are identical. Therefore, the desired product can be made by treating this starting alkene with a Grubbs catalyst:

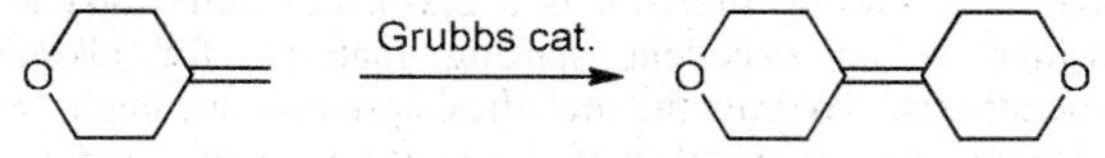

Grubbs cat.

23.30.

(a) First, identify any C=C unit that is incorporated into a ring. In this case, the highlighted π bond is incorporated into a seven-membered ring:

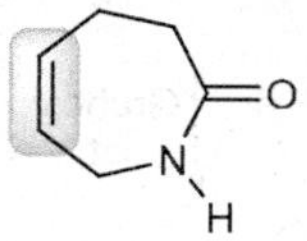

In order to draw the starting diene, we erase the highlighted C=C bond of the ring, and then connect a methylene (CH_2) group to each of the two vinyl positions, as shown:

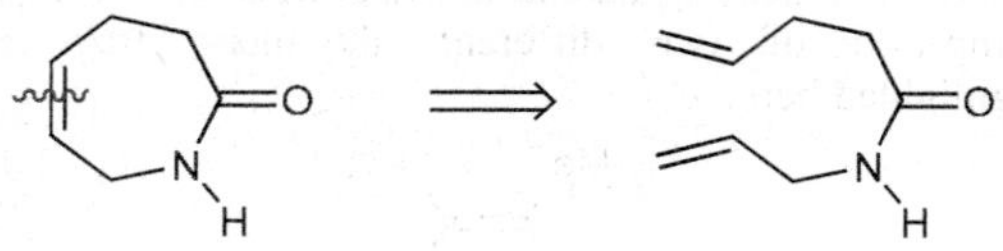

(b) First, identify any C=C unit that is incorporated into a ring. In this case, the highlighted π bond is incorporated into a six-membered ring:

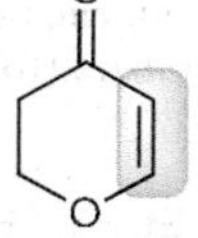

In order to draw the starting diene, we erase the highlighted C=C bond of the ring, and then connect a methylene (CH_2) group to each of the two vinyl positions, as shown:

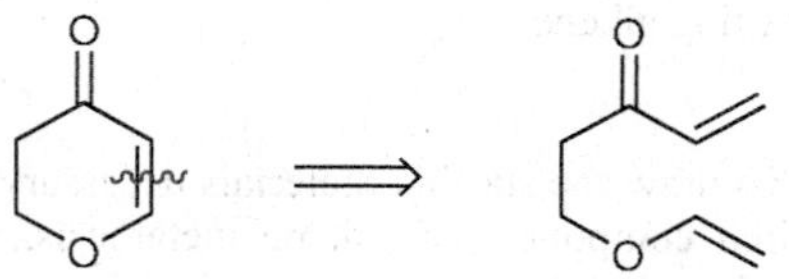

(c) First, identify any C=C unit that is incorporated into a ring. In this case, the highlighted π bond is incorporated into a five-membered ring:

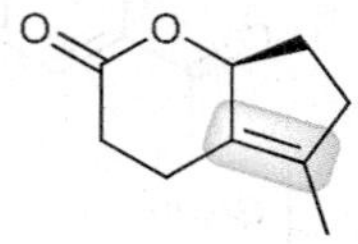

In order to draw the starting diene, we erase the highlighted C=C bond of the ring, and then connect a methylene (CH_2) group to each of the two vinyl positions, as shown:

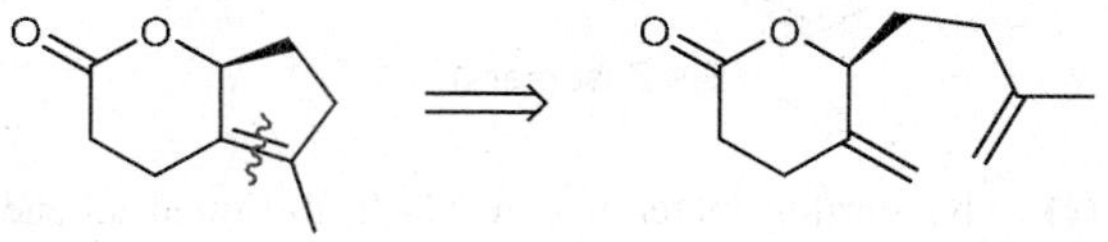

(d) First, identify any C=C unit that is incorporated into a ring. In this case, the highlighted π bond is incorporated into a six-membered ring:

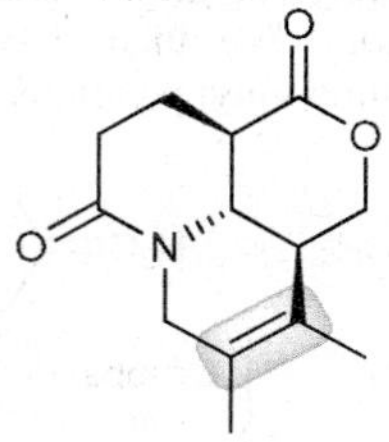

In order to draw the starting diene, we erase the highlighted C=C bond of the ring, and then connect a methylene (CH_2) group to each of the two vinyl positions, as shown:

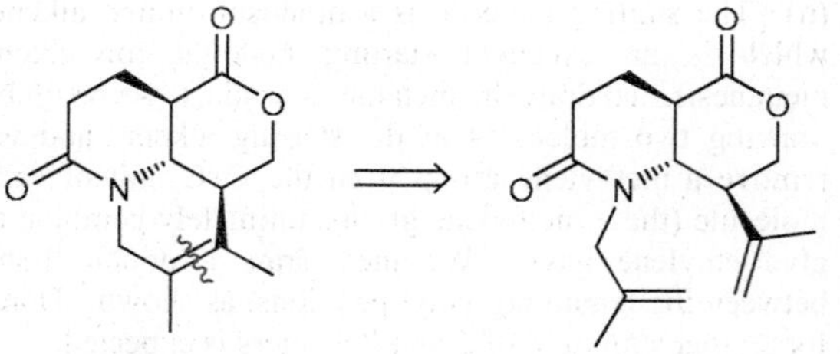

23.31.

(a) The starting material is a diene, which can undergo a ring-closing metathesis upon treatment with a Grubbs catalyst. To draw the metathesis product, we remove a methylene group from each C=C unit (these methylene groups ultimately combine to give ethylene gas). We then draw a double bond between the remaining vinyl positions, as shown:

Grubbs cat.

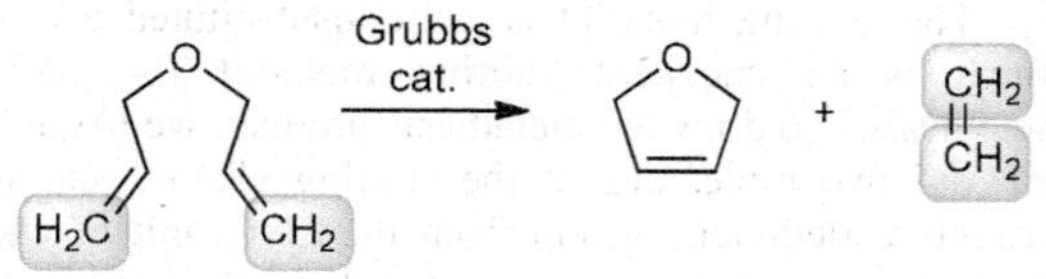

(b) The starting material is a diene, which can undergo a ring-closing metathesis upon treatment with a Grubbs catalyst. To draw the metathesis product, we remove a methylene group from each C=C unit (these methylene groups ultimately combine to give ethylene gas). We then draw a double bond between the remaining vinyl positions, as shown:

CH_2 CH_2 Grubbs cat. + $CH_2=CH_2$

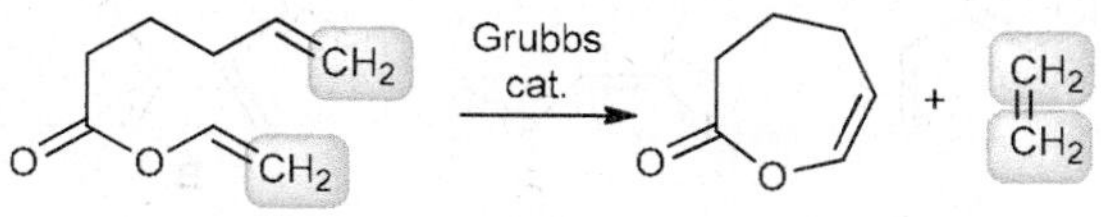

(c) The starting material is a diene, which can undergo a ring-closing metathesis upon treatment with a Grubbs catalyst. To draw the metathesis product, we remove a methylene group from each C=C unit (these methylene groups ultimately combine to give ethylene gas). We then draw a double bond between the remaining vinyl positions, as shown:

H_2C CH_2 Grubbs cat. + $CH_2=CH_2$

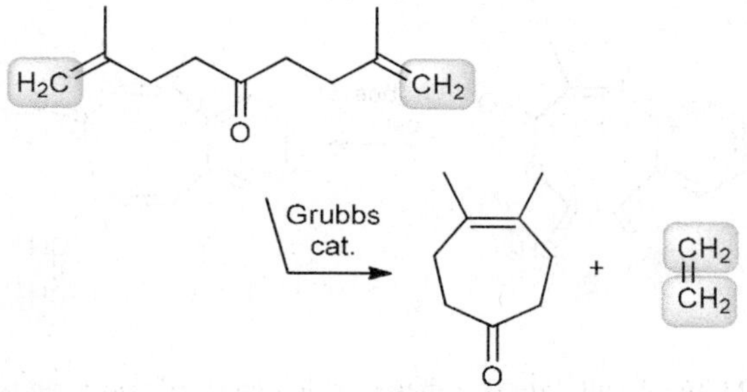

(d) The starting material is a diene, which can undergo a ring-closing metathesis upon treatment with a Grubbs catalyst. To draw the metathesis product, we remove a methylene group from each C=C unit (these methylene groups ultimately combine to give ethylene gas). We then draw a double bond between the remaining vinyl positions, as shown:

CH_2 CH_2 Grubbs cat. + $CH_2=CH_2$

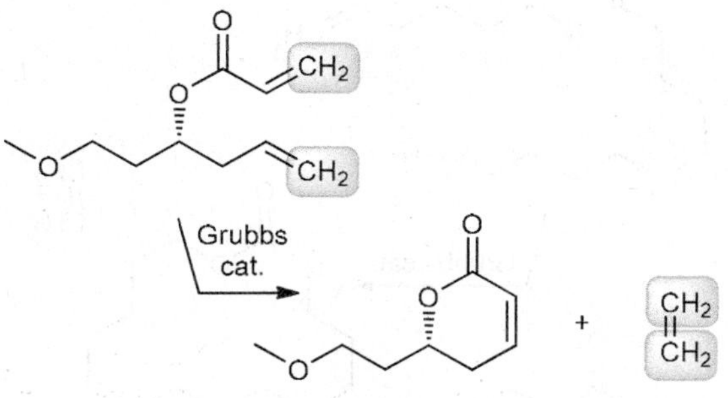

23.32.

(a) As described in the problem statement, olefin metathesis of this triene results in cleavage of the cyclic C=C unit and formation of two new C=C units, giving a product with two new five-membered rings:

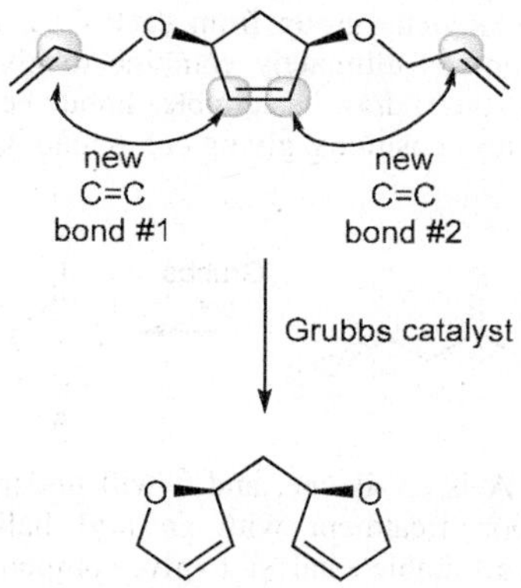

new C=C bond #1 new C=C bond #2 Grubbs catalyst

(b) As described in the problem statement, olefin metathesis of this triene results in cleavage of the cyclic C=C unit and formation of two new C=C units, giving a product with a five-membered ring and a six-membered ring:

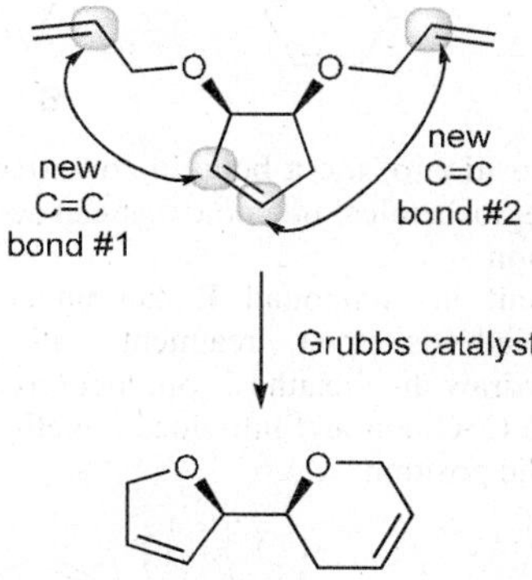

new C=C bond #1 new C=C bond #2 Grubbs catalyst

23.33. First identify the carbon atoms that will be joined. The carbon atom connected directly to the bromide (in the aryl bromide) will be joined with the carbon atom connected directly to tin (Sn) in the organostannane, highlighted here:

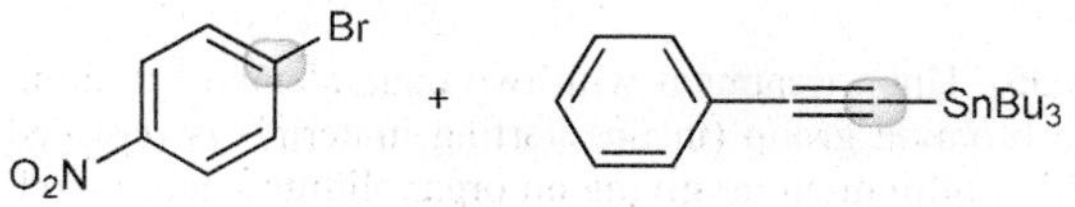

Br O_2N + $SnBu_3$

Then, rotate the coupling partners so that they are aligned to form a bond, and draw the product:

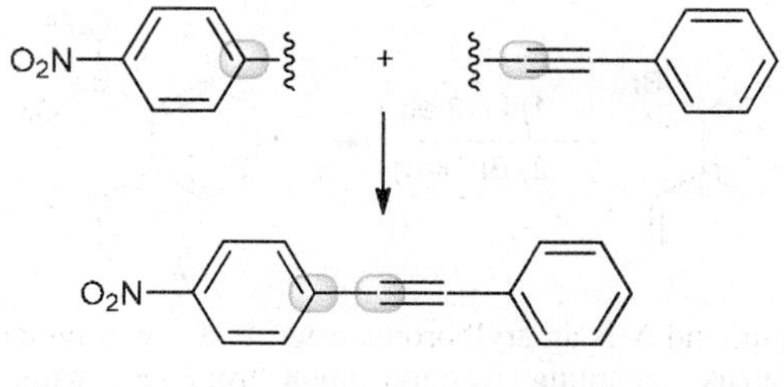

O_2N + O_2N

23.34. The starting material is a diene, which can undergo a ring-closing metathesis upon treatment with a

Grubbs catalyst. To draw the metathesis product, we remove a methylene group from each C=C unit (these methylene groups ultimately combine to give ethylene gas). We then draw a double bond between the remaining vinyl positions, giving compound **A**:

H_2C CH_2 Grubbs cat. **A** + CH_2=CH_2

Compound **A** is an alkene, and it will undergo a Heck reaction upon treatment with an aryl halide in the presence of a suitable catalyst, to give compound **B**:

A $Pd(PPh_3)_4$ I **B**

Notice the position of the π bond in compound **B**. This is the only regiochemical outcome that can accommodate *syn* elimination.

The C=C unit in compound **B** can undergo a ring-opening metathesis upon treatment with a Grubbs catalyst. To draw the metathesis product (compound **C**), we break the C=C unit and introduce a methylene group at each vinylic position:

B Grubbs cat. H H C=C H H **C**

23.35. Upon treatment with two equivalents of lithium, the bromide group (in the starting material) is replaced with a lithium atom, giving an organolithium compound. This organolithium compound is then converted into the corresponding arylboronic ester (compound **A**) upon treatment with trimethylborate, $B(OMe)_3$:

Br 1) Li (2 eq.) 2) $B(OMe)_3$ OMe B OMe **A**

Compound **A** is an arylboronic ester, and it will undergo a Suzuki coupling reaction upon treatment with an organohalide in the presence of $Pd(PPh_3)_4$. In order to draw the coupling product (compound **B**) more easily, the organohalide has been rotated in the following drawing. Notice that the organohalide has a C=C unit with the *Z* configuration. So, when drawing **B**, we must be careful that this C=C unit maintains its configuration.

OMe B OMe **A** $Pd(PPh_3)_4$ I O O **B**

Compound **B** has two monosubstituted C=C units that can undergo a ring-closing metathesis upon treatment with a Grubbs catalyst. To draw the metathesis product, we remove a methylene group from each C=C unit (these methylene groups ultimately combine to give ethylene gas). We then draw a double bond between the remaining vinyl positions, giving compound **C**:

O O H_2C CH_2 **B** Grubbs cat. O O **C** + CH_2=CH_2

23.36. Exaltolide is a diene, which can undergo a ring-closing metathesis upon treatment with a Grubbs catalyst. To draw the metathesis product, we remove a methylene group from each C=C unit (these methylene groups ultimately combine to give ethylene gas). We then draw a double bond between the remaining vinyl positions. This metathesis product has a C=C unit that undergoes hydrogenation to give exaltolide, as shown here:

O O CH_2 CH_2 Grubbs cat. O O + CH_2=CH_2 H_2, Pt O O

Exaltolide
($C_{15}H_{28}O_2$)

23.37.

(a) The starting materials are an aryl diiodide and two equivalents of an alkene. These compounds will serve as coupling partners in two successive Heck reactions, as shown below. The coupling partners have been rotated in order to draw the coupling product more easily. Notice that the aromatic ring is coupled to each of the C=C units regioselectively (at the less substituted position) and stereoselectively (to give an *E* alkene), as shown.

Pd(OAc)$_2$
PPh$_3$
Et$_3$N

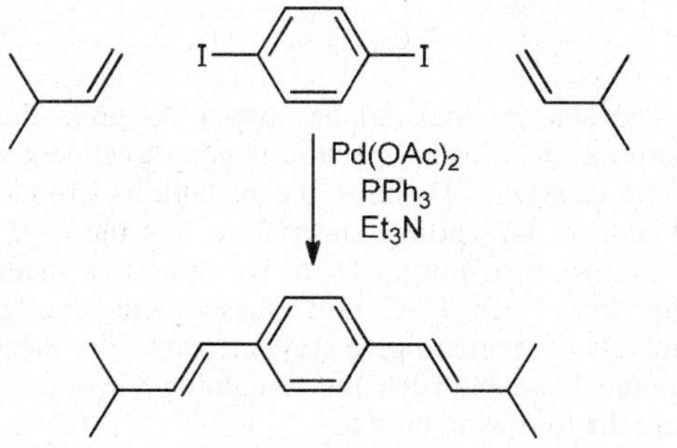

(b) The starting materials are an aryl iodide and an alkene. These compounds will serve as coupling partners in a Heck reaction, as shown below. The coupling partners have been rotated in order to draw the coupling product more easily. Notice that the aromatic ring is coupled to the C=C unit regioselectively (at the less substituted position) and stereoselectively (to give an *E* alkene), as shown.

Pd(OAc)$_2$
PPh$_3$
Et$_3$N

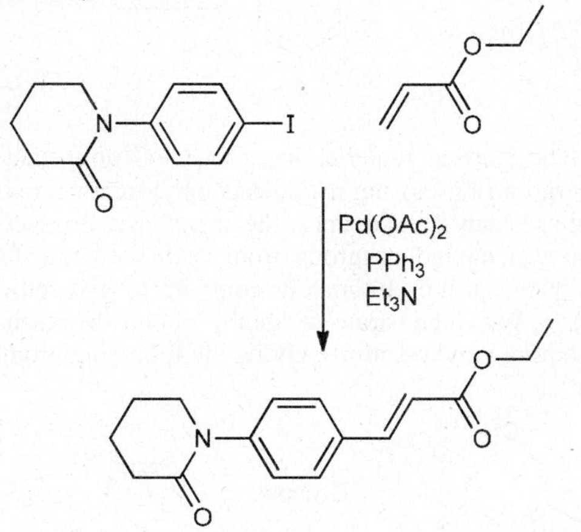

(c) The starting materials are an aryl triflate and an alkene. These compounds will serve as coupling partners in a Heck reaction, as shown. The coupling partners have been rotated in order to draw the coupling product more easily. Notice the position of the π bond in the product (this is the only regiochemical outcome that can accommodate *syn* elimination).

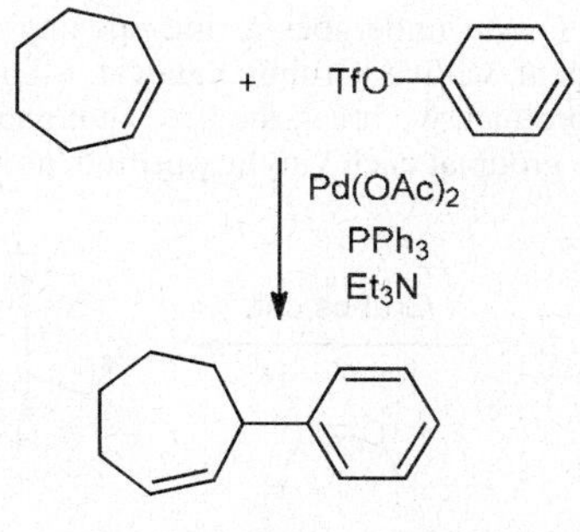

TfO
Pd(OAc)$_2$
PPh$_3$
Et$_3$N

(d) The starting materials are an aryl iodide and an aryl alkene. These compounds will serve as coupling partners in a Heck reaction, as shown below. The coupling partners have been rotated in order to draw the coupling product more easily. Notice that the aryl iodide is coupled to the C=C unit regioselectively (at the less substituted position) and stereoselectively (to give an *E* alkene), as shown.

Pd(OAc)$_2$
PPh$_3$
Et$_3$N

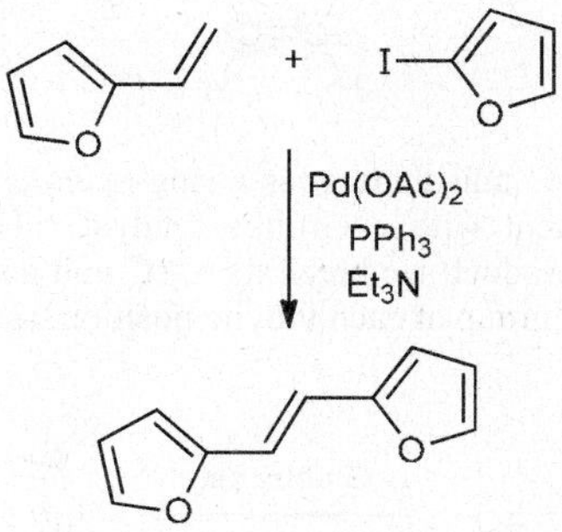

23.38.

(a) The C=C unit undergoes a ring-opening metathesis upon treatment with a Grubbs catalyst. To draw the metathesis product, we break the C=C unit and introduce a methylene group at each vinylic position, as shown:

Grubbs cat.
$H_2C=CH_2$

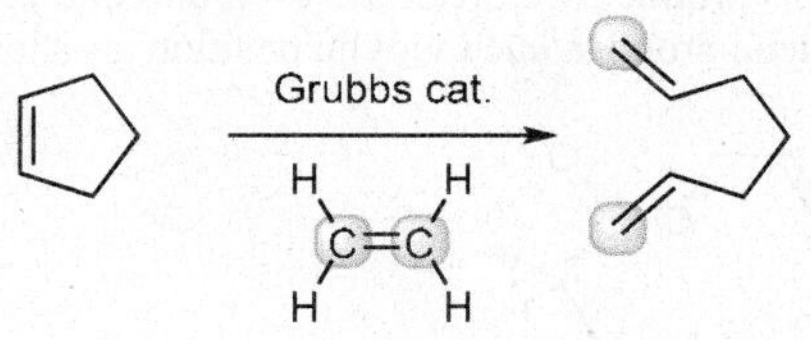

(b) The C=C unit undergoes a ring-opening metathesis upon treatment with a Grubbs catalyst. To draw the metathesis product, we break the C=C unit and introduce a methylene group at each vinylic position, as shown:

Grubbs cat.
$H_2C=CH_2$

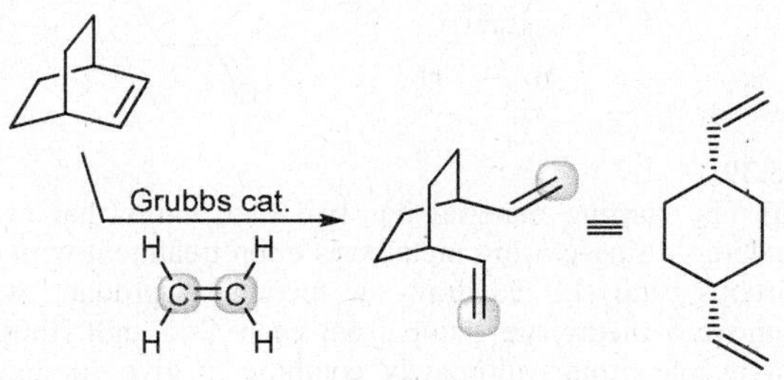

(c) The C=C unit undergoes a ring-opening metathesis upon treatment with a Grubbs catalyst. To draw the metathesis product, we break the C=C unit and introduce a methylene group at each vinylic position, as shown:

Grubbs cat.

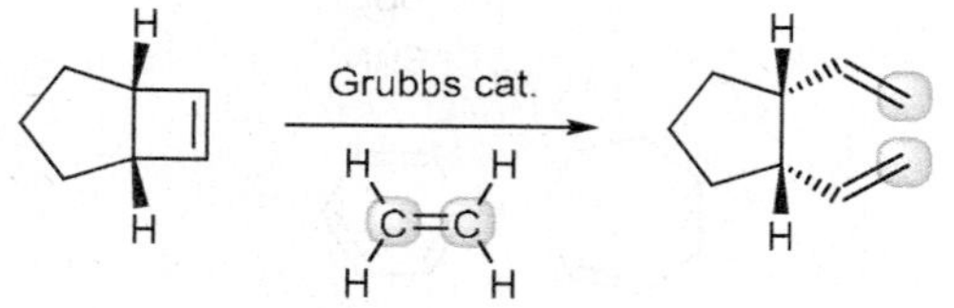

(d) The C=C unit undergoes a ring-opening metathesis upon treatment with a Grubbs catalyst. To draw the metathesis product, we break the C=C unit and introduce a methylene group at each vinylic position, as shown:

Grubbs cat.

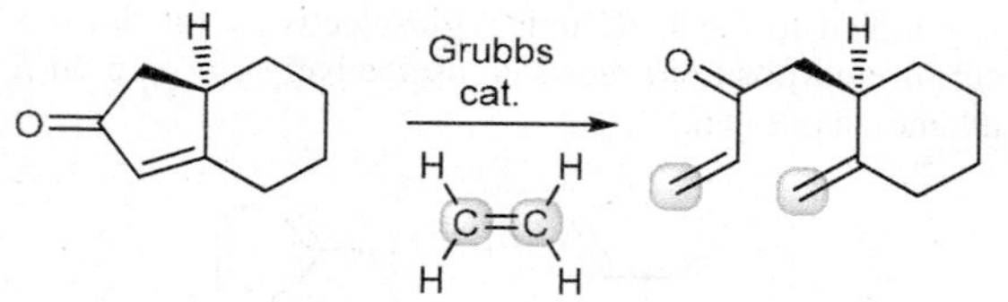

(e) The C=C unit undergoes a ring-opening metathesis upon treatment with a Grubbs catalyst. To draw the metathesis product, we break the C=C unit and introduce a methylene group at each vinylic position, as shown:

Grubbs cat.

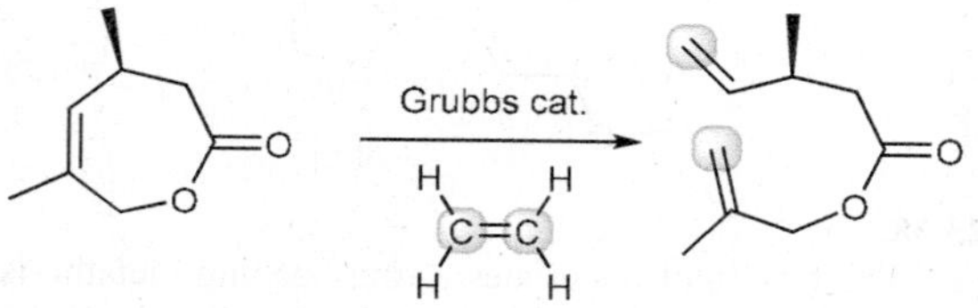

(f) The C=C unit undergoes a ring-opening metathesis upon treatment with a Grubbs catalyst. To draw the metathesis product, we break the C=C unit and introduce a methylene group at each vinylic position, as shown:

Grubbs cat.

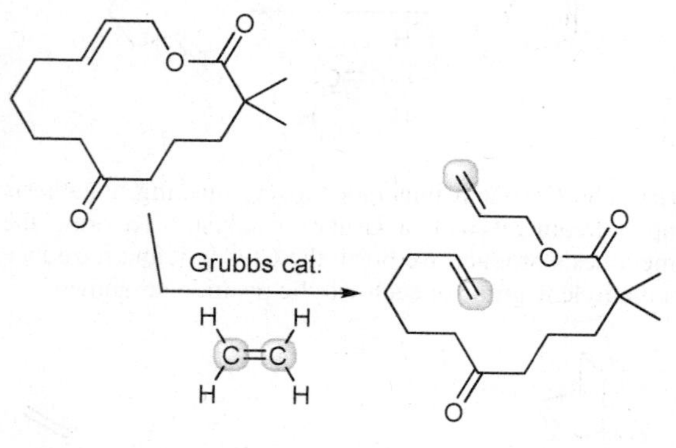

23.39.

(a) The starting material has two C=C units that can undergo a ring-closing metathesis upon treatment with a Grubbs catalyst. To draw the metathesis product, we remove a methylene group from each C=C unit (these methylene groups ultimately combine to give ethylene gas). We then draw a double bond between the remaining vinyl positions, giving the following product:

Grubbs cat. + $CH_2{=}CH_2$

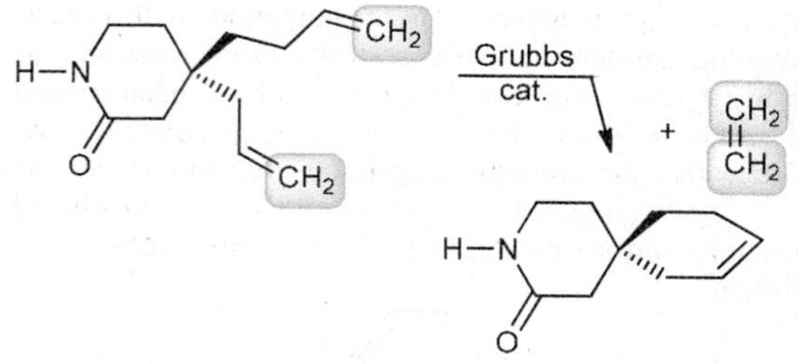

(b) The starting material has two C=C units that can undergo a ring-closing metathesis upon treatment with a Grubbs catalyst. To draw the metathesis product, we first redraw the starting material so that the C=C units are in close proximity. Then, we remove a methylene group from each C=C unit (these methylene groups ultimately combine to give ethylene gas). We then draw a double bond between the remaining vinyl positions, giving the following product:

Grubbs cat. $CH_2{=}CH_2$ +

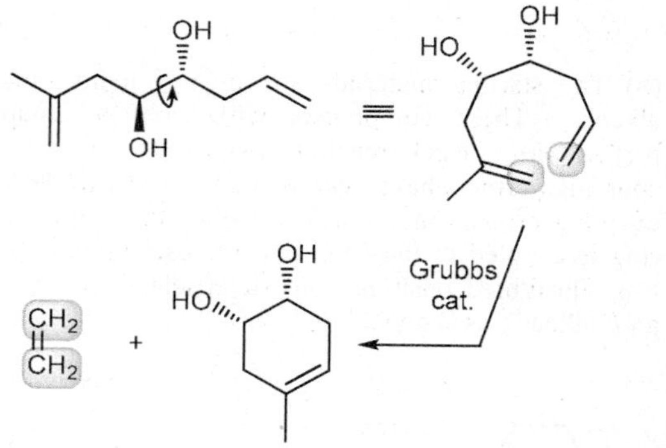

(c) The starting material has two C=C units that can undergo a ring-closing metathesis upon treatment with a Grubbs catalyst. To draw the metathesis product, we remove a methylene group from each C=C unit (these methylene groups ultimately combine to give ethylene gas). We then draw a double bond between the remaining vinyl positions, giving the following product:

Grubbs cat. + $CH_2{=}CH_2$

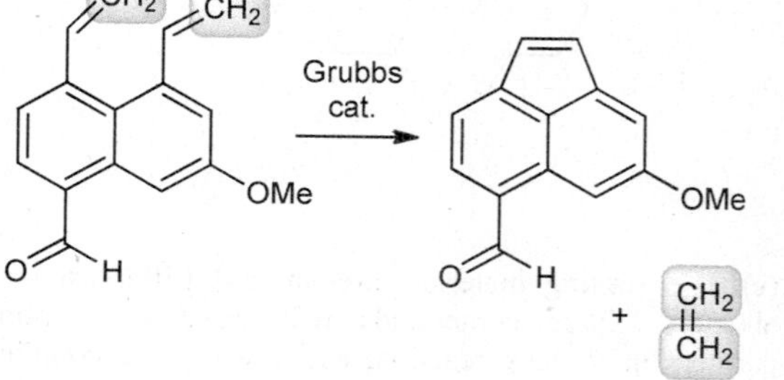

(d) The starting material has two C=C units that can undergo a ring-closing metathesis upon treatment with a Grubbs catalyst. To draw the metathesis product, we remove a methylene group from each C=C unit (these methylene groups ultimately combine to give ethylene gas). We then draw a double bond between the remaining vinyl positions, giving the following product:

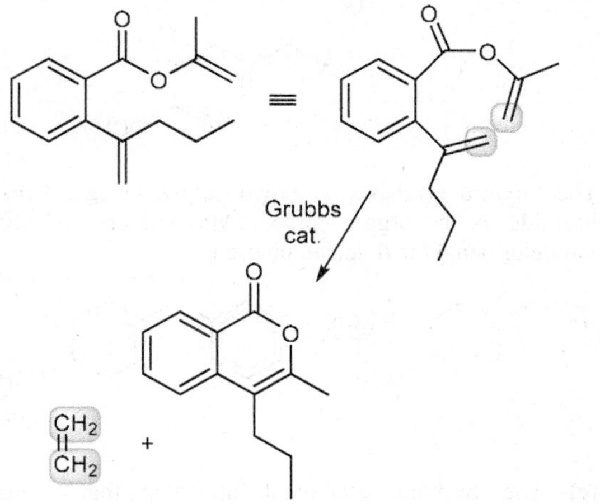

Grubbs cat.

CH_2 = CH_2 +

23.40.

(a) Lithium diphenyl cuprate is a Gilman reagent, and it will react with the following vinyl iodide to give a coupling product, as shown here:

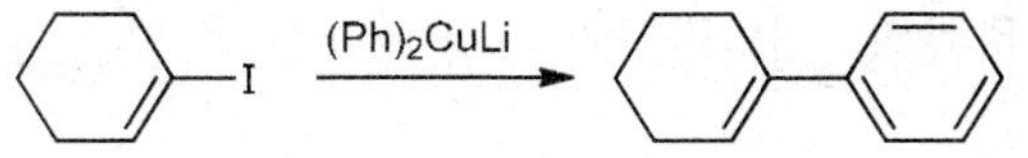

$(Ph)_2CuLi$

(b) Upon treatment with catecholborane, the alkyne is converted into a vinyl boronic ester with the *E* configuration. This compound then serves as a coupling partner in a Suzuki coupling reaction with bromobenzene. The carbon atom connected directly to the bromine atom will be joined with the carbon atom connected directly to boron in the organoboron compound. Notice that the *E* configuration is preserved during the process:

(catecholborane)

C_6H_5Br, $Pd(PPh_3)_4$, NaOH

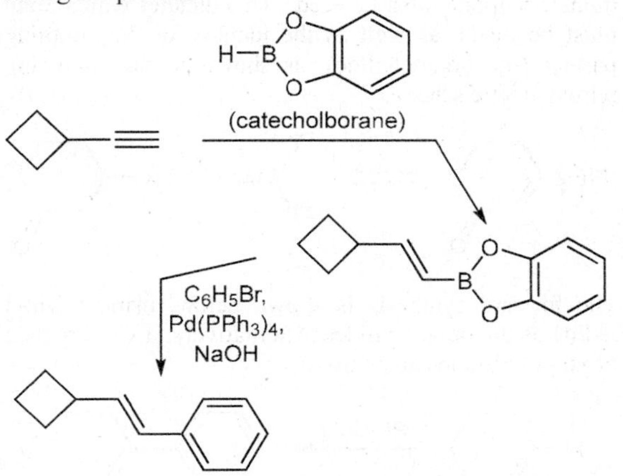

(c) The starting materials are a vinyl iodide and a vinyl boronic ester, indicating a Suzuki coupling reaction. The carbon atom connected directly to the iodide group will be joined with the carbon atom connected directly to boron in the organoboron compound, giving the product shown. The coupling partners have been rotated in order to draw the coupling product more easily. Notice that the *E* configuration is preserved during the process:

$Pd(PPh_3)_4$, NaOH

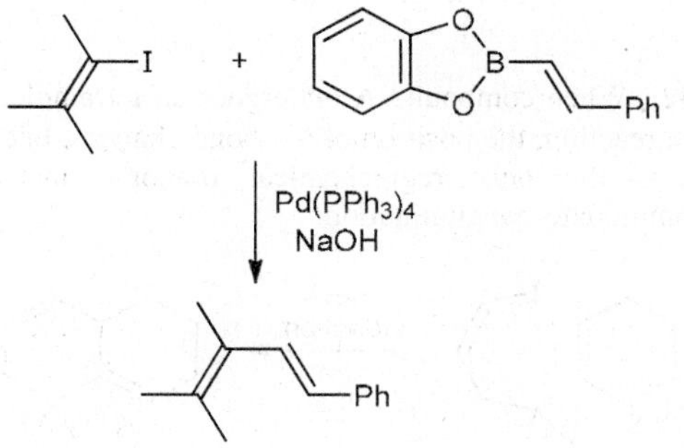

(d) The starting materials are an aryl triflate and an alkene, indicating a Heck reaction. The coupling partners have been rotated in order to draw the coupling product more easily. Notice that the aryl triflate is coupled to the C=C unit regioselectively (at the less substituted position) and stereoselectively (to give an *E* alkene), as shown:

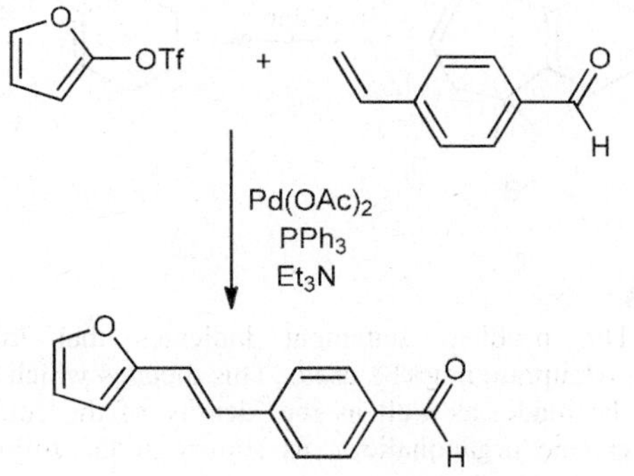

$Pd(OAc)_2$, PPh_3, Et_3N

23.41.

(a) In a Simmons-Smith reaction, the C=C unit is converted into a three-membered ring. This newly formed ring can be *cis* or *trans* to the *t*-butyl group, as shown:

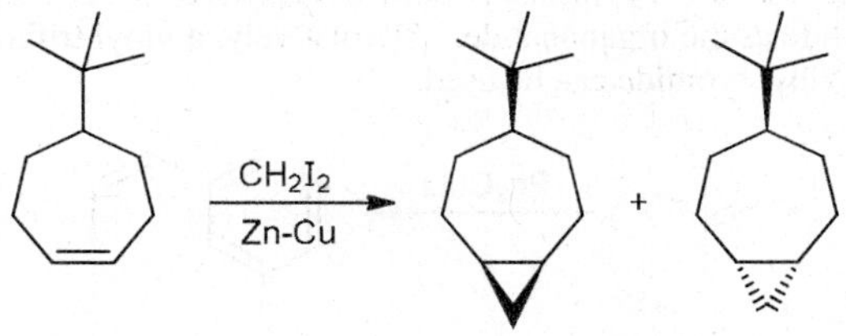

CH_2I_2, Zn-Cu

A B

(b) Compounds **A** and **B** have the same constitution (connectivity of atoms), but they differ from each other in their configuration, so they are stereoisomers. More specifically, compounds **A** and **B** are stereoisomers that are not mirror images, so they are diastereomers.

(c) If we take steric considerations into account, we would expect that cyclopropanation will occur more readily on the face that is less sterically hindered (the face that is *trans* to the *t*-butyl group). Therefore, compound **B** is expected to predominate.

23.42. When compound **A** undergoes an intramolecular Heck reaction, the position of a π bond changes, because this is the only regiochemical outcome that can accommodate *syn* elimination:

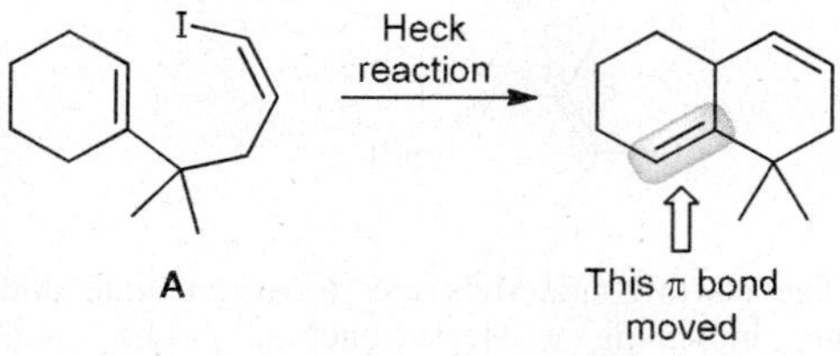

Heck reaction

A

This π bond moved

In contrast, when compound **B** undergoes an intramolecular Heck reaction, the π bonds remain in the same location:

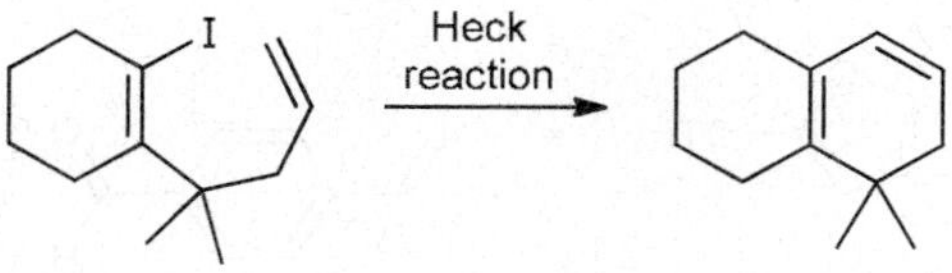

Heck reaction

B

23.43.

(a) The problem statement indicates that lithium diphenylcuprate must be used. This dictates which bond must be made, as well as the identity of the coupling partner (the organohalide), as shown in the following retrosynthetic scheme:

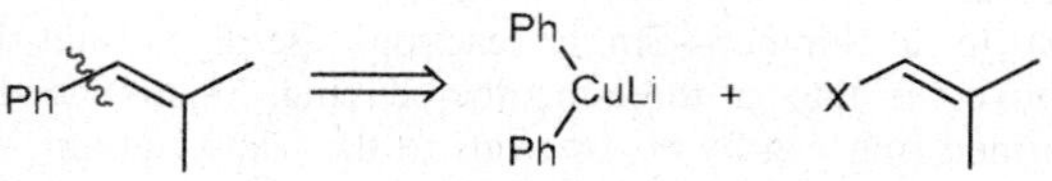

The forward synthesis is shown below, using a vinyl iodide as the organohalide. Alternatively, a vinyl triflate or vinyl bromide can be used.

Ph_2CuLi

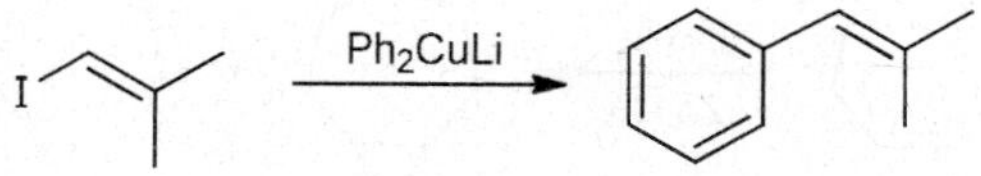

(b) The problem statement indicates that lithium diphenylcuprate must be used. This dictates which bond must be made, as well as the identity of the coupling partner (the organohalide), as shown in the following retrosynthetic scheme:

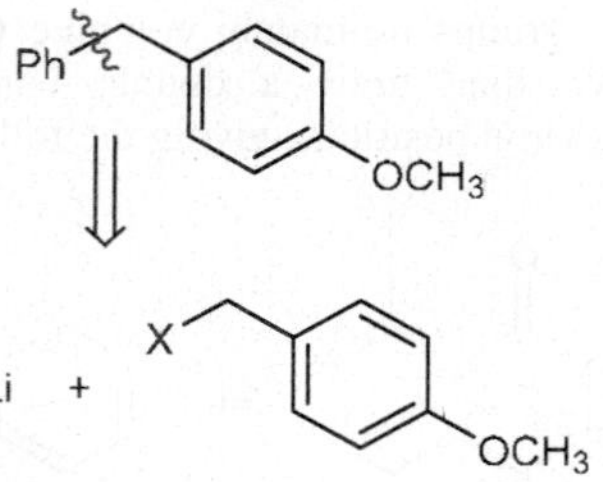

The forward synthesis is shown below, using a benzyl bromide as the organohalide. Alternatively, a benzyl iodide or benzyl triflate can be used.

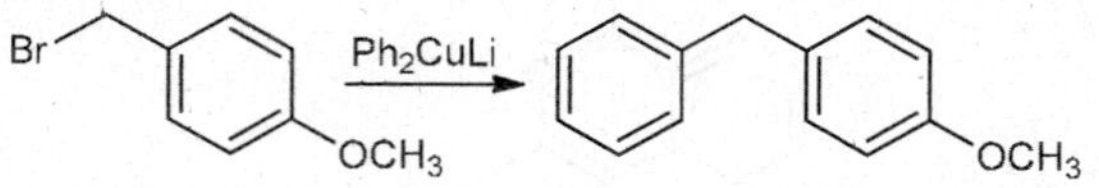

Ph_2CuLi

(c) The problem statement indicates that lithium diphenylcuprate must be used. This dictates which bond must be made, as well as the identity of the coupling partner (the organohalide), as shown in the following retrosynthetic scheme:

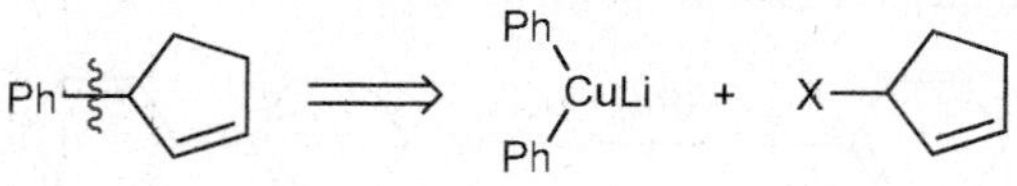

The forward synthesis is shown below, using an allyl bromide as the organohalide. Alternatively, an allyl iodide or allyl triflate can be used.

Ph_2CuLi

(d) The problem statement indicates that lithium diphenylcuprate must be used. This dictates which bond must be made, as well as the identity of the coupling partner (the organohalide), as shown in the following retrosynthetic scheme:

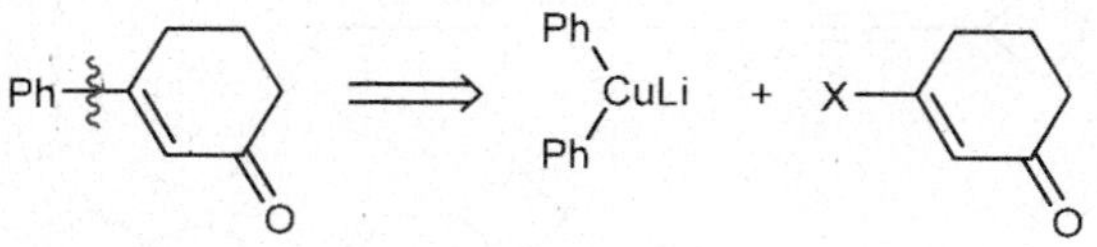

The forward synthesis is shown below, using a vinyl iodide as the organohalide. Alternatively, a vinyl triflate or vinyl bromide can be used.

Ph_2CuLi

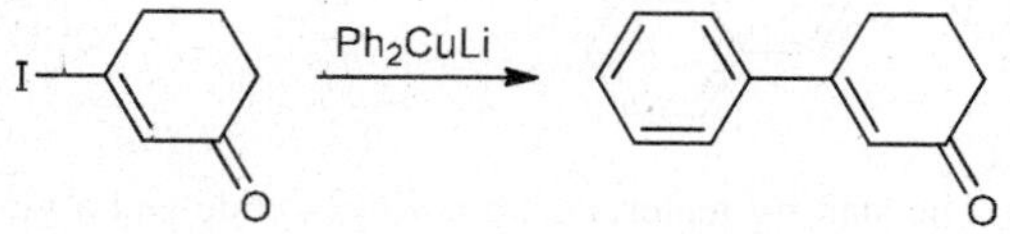

23.44.

(a) The following retrosynthetic analysis shows how the desired compound can be made from 1-butyne via a Suzuki coupling reaction:

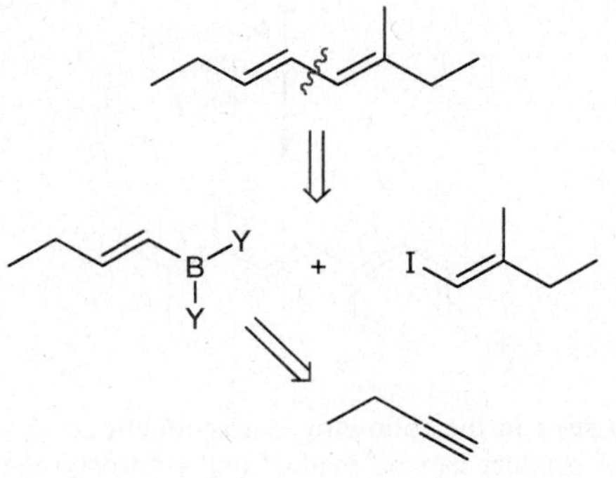

B–Y, Y + I

The forward scheme is shown here:

H–B (catecholborane)

B–O, O

I

$Pd(PPh_3)_4$
NaOEt

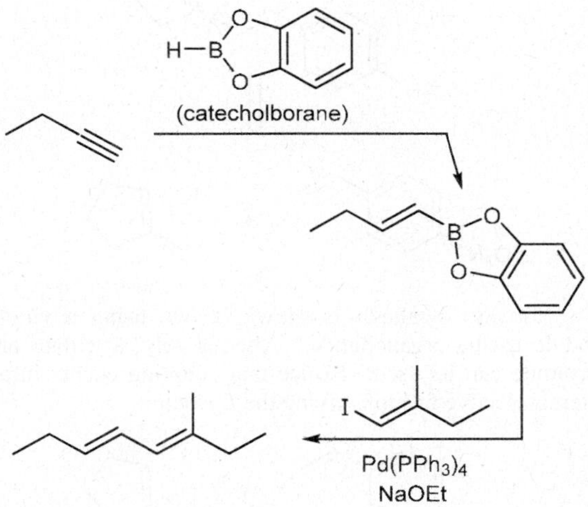

(b) The following retrosynthetic analysis shows how the desired compound can be made from 1-butyne via a Suzuki coupling reaction:

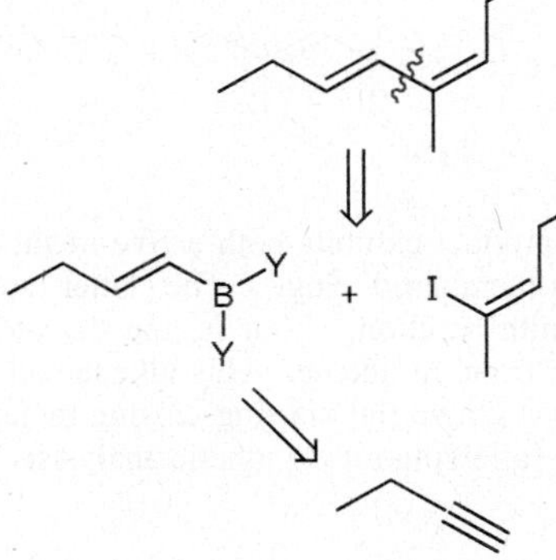

B–Y, Y + I

The forward scheme is shown here:

H–B (catecholborane)

B–O, O

I

$Pd(PPh_3)_4$
NaOEt

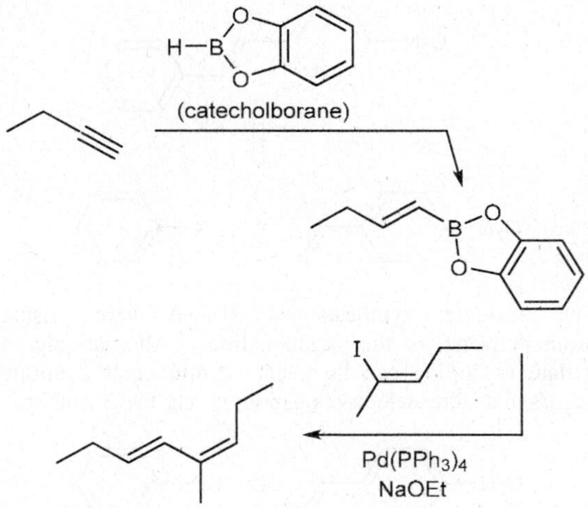

(c) The following retrosynthetic analysis shows how the desired compound can be made from 1-butyne via a Suzuki coupling reaction:

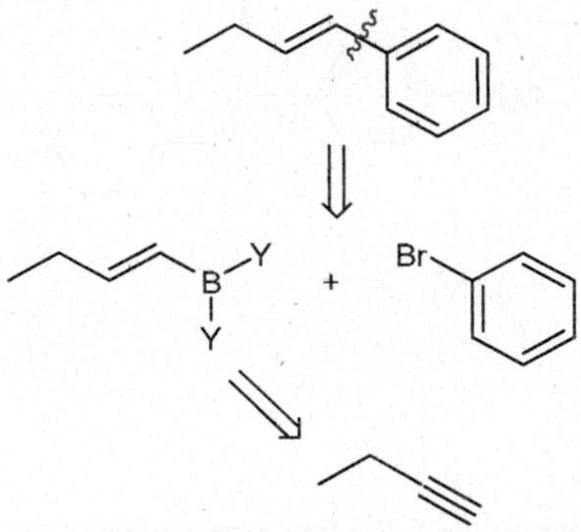

B–Y, Y + Br

The forward scheme is shown here:

H–B (catecholborane)

B–O, O

Br

$Pd(PPh_3)_4$
NaOEt

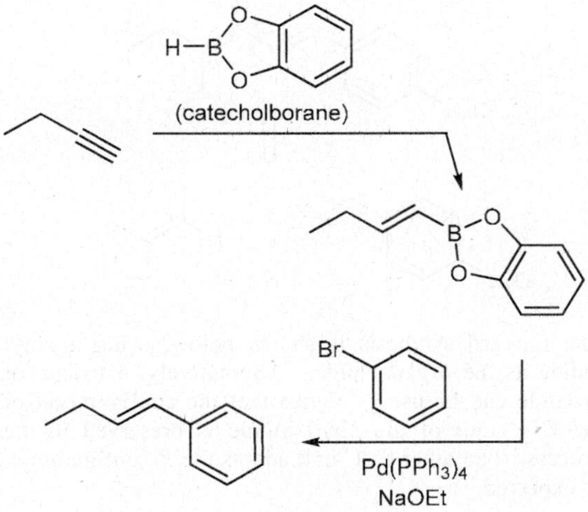

23.45.

(a) As seen in the following retrosynthetic analysis, the desired product can be made from 4-nitrostyrene via a Heck reaction:

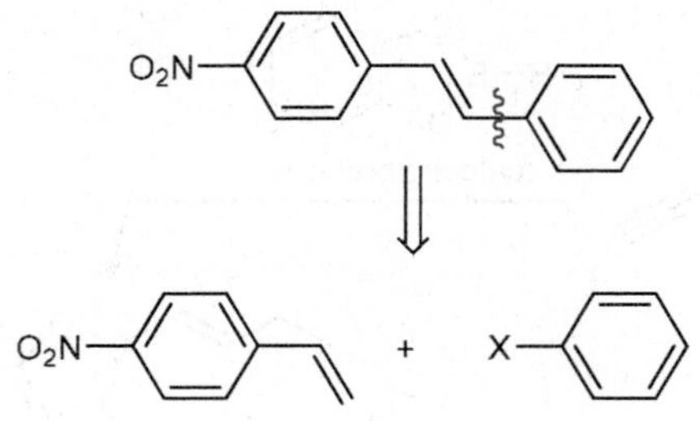

The forward synthesis is shown here, using bromobenzene as the organohalide. Alternatively, a triflate or iodide can be used. Notice that coupling occurs in a stereoselective fashion, giving the *E* isomer.

$Pd(OAc)_2$, PPh_3, Et_3N

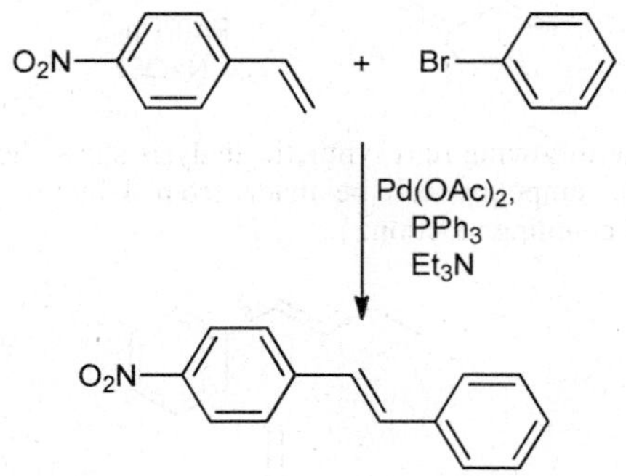

(b) As seen in the following retrosynthetic analysis, the desired product can be made from 4-nitrostyrene via a Heck reaction:

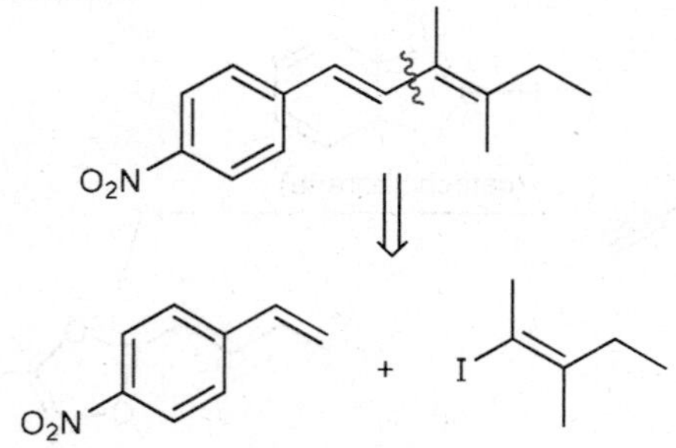

The forward synthesis is shown below, using a vinyl iodide as the organohalide. Alternatively, a triflate or bromide can be used. Notice that the configuration of the C=C unit of the vinyl iodide is preserved in the process. The other C=C unit adopts the *E* configuration, as expected.

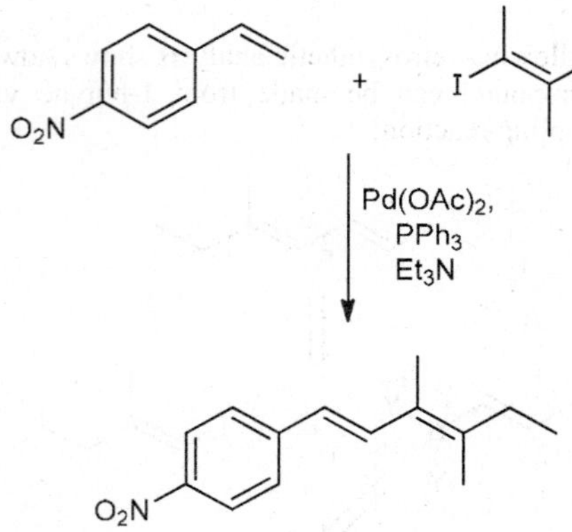

$Pd(OAc)_2$, PPh_3, Et_3N

(c) As seen in the following retrosynthetic analysis, the desired product can be made from 4-nitrostyrene via a Heck reaction:

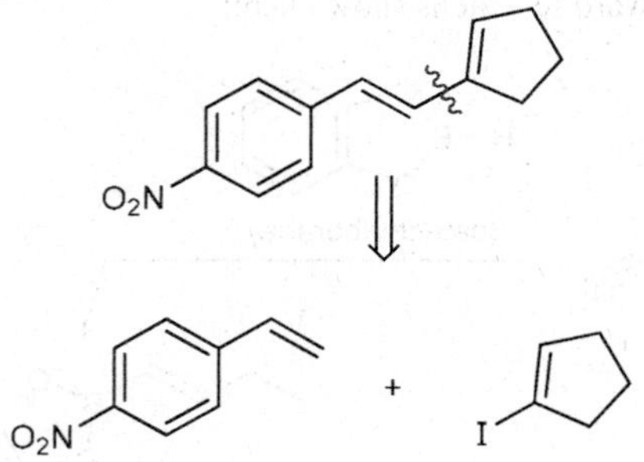

The forward synthesis is shown below, using a vinyl iodide as the organohalide. Alternatively, a triflate or bromide can be used. Notice that coupling occurs in a stereoselective fashion, giving the *E* isomer.

$Pd(OAc)_2$, PPh_3, Et_3N

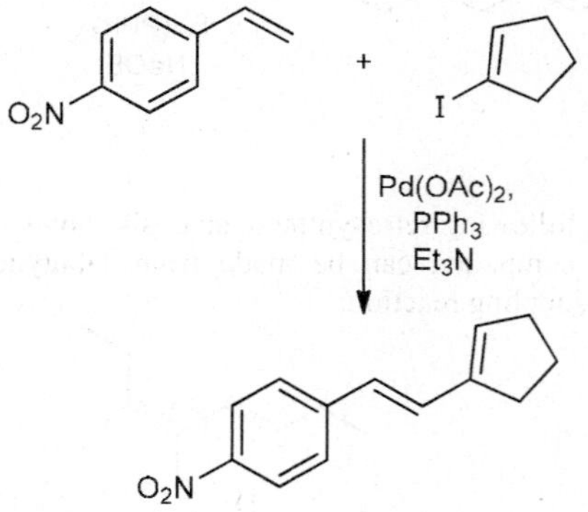

23.46. The product exhibits both a five-membered ring and a three-membered ring. The latter suggests a Simmons-Smith reaction. That is, the desired product can be made from an alkene. This alkene can be made from the starting material via ring-closing metathesis, as shown in the following retrosynthetic analysis:

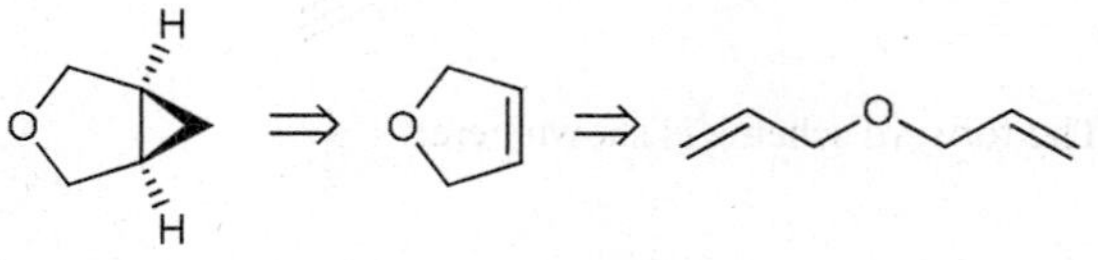

The forward synthesis is shown here:

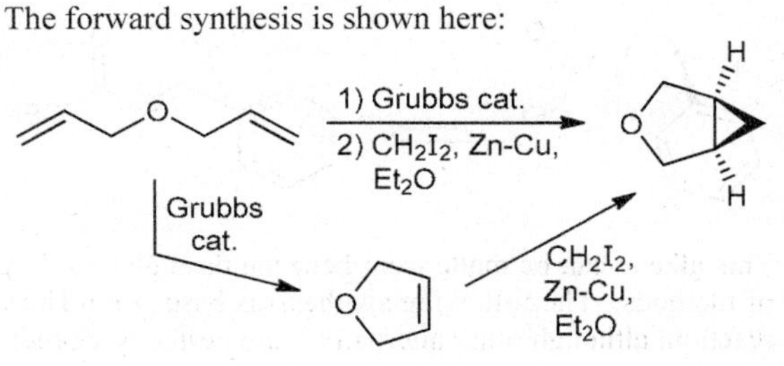
1) Grubbs cat.
2) CH2I2, Zn-Cu, Et2O
Grubbs cat.
CH2I2, Zn-Cu, Et2O
H
H

23.47. We must first determine which bond (in the product) will be made via a coupling process. The product has ten carbon atoms, and the starting material has five carbon atoms, which means that we must form the bond between C5 and C6 via a coupling reaction. This requires that we use organohalides, as shown in the following retrosynthetic analysis:

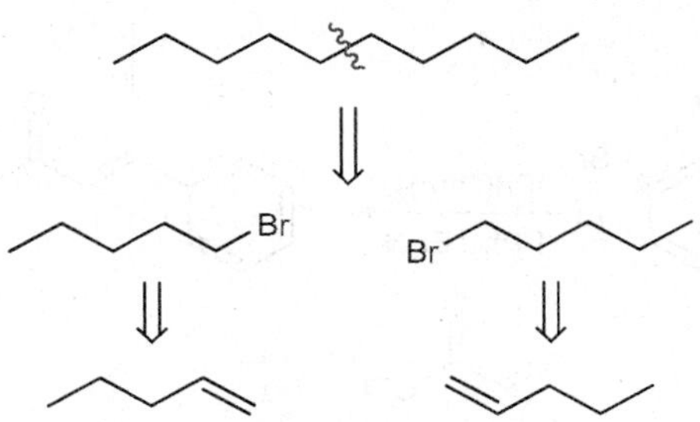
Br
Br

The organohalides can be made directly from the starting material via addition of HBr in the presence of peroxides. The forward synthesis is shown here:

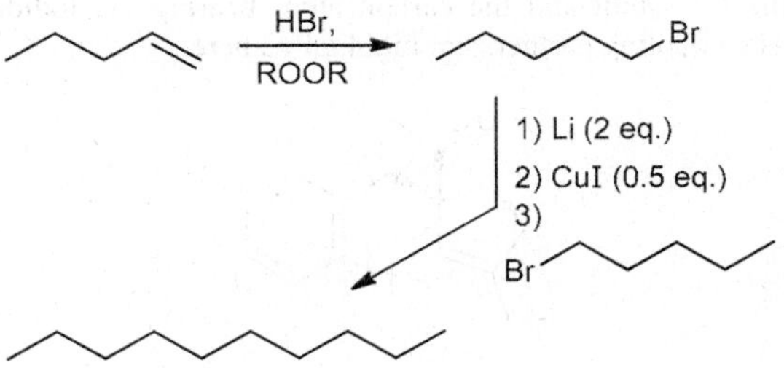
HBr,
ROOR
Br
1) Li (2 eq.)
2) CuI (0.5 eq.)
3)
Br

23.48.

(a) The problem statement indicates that we must start with benzene. This determines the bond (in the product) that must be made:

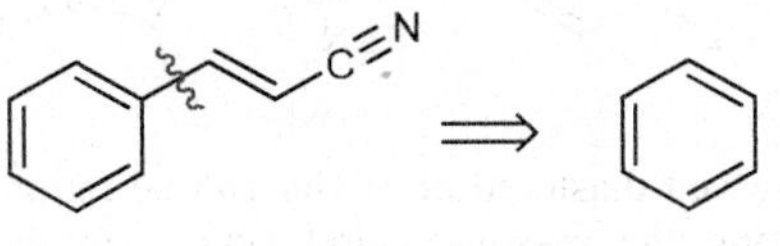
C≡N

This bond can be made via a Heck reaction, as shown below. This strategy requires that we first convert benzene into bromobenzene, which can be achieved via electrophilic aromatic substitution (See Section 18.2):

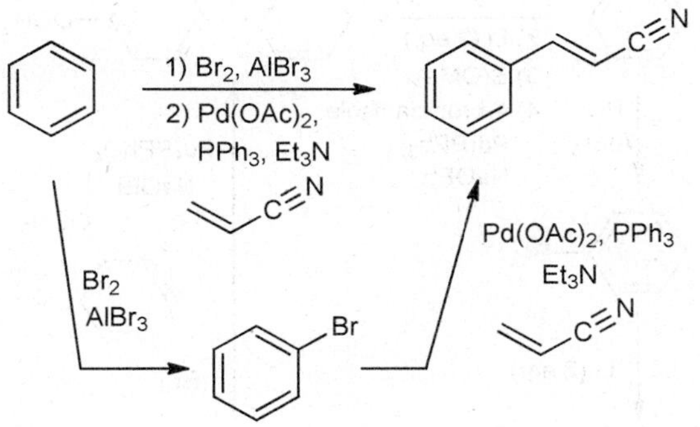
1) Br2, AlBr3
2) Pd(OAc)2, PPh3, Et3N
C≡N
Br2
AlBr3
Br
Pd(OAc)2, PPh3
Et3N
C≡N

(b) The problem statement indicates that we must start with benzene. This determines the bond (in the product) that must be made:

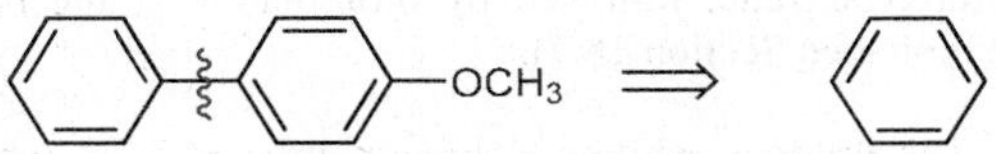
OCH3

We learned many coupling reactions in this chapter, so there are many ways to make this bond. The following retrosynthetic analysis employs a Suzuki coupling reaction:

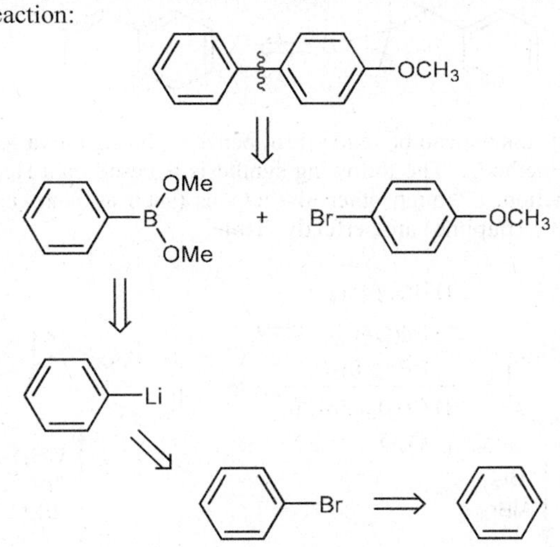
OCH3
OMe
B
OMe
+
Br
OCH3
Li
Br

The organohalide coupling partner (4-bromoanisole) can also be made from benzene, although the problem statement indicates that we can use benzene *and any other reagents of our choice*. So, we can choose to use 4-bromoanisole and benzene as our starting materials, and we have satisfied the requirement of the problem. The forward synthesis is shown here:

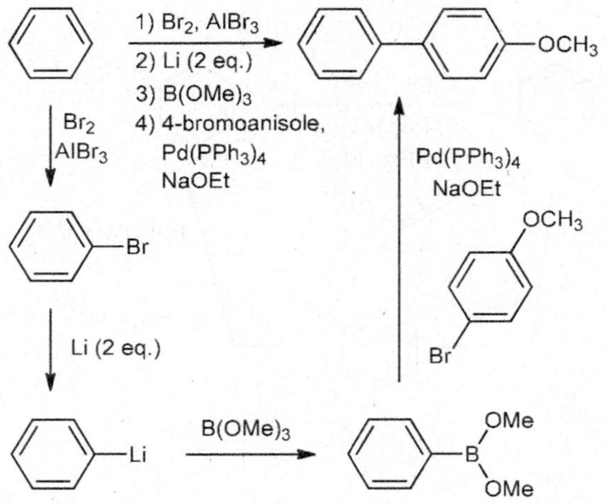

For those who are interested, 4-bromoanisole can be made using reactions that we have learned in previous chapters: chlorination of benzene to give chlorobenzene, followed by the Dow process to give phenol (see Section 18.14), followed by a Williamson ether synthesis to give methoxybenzene, followed by bromination at the *para* position (see Section 18.7).

(c) The product exhibits a three-membered ring, which suggests a Simmons-Smith reaction. That is, the desired product can be made from the following alkene:

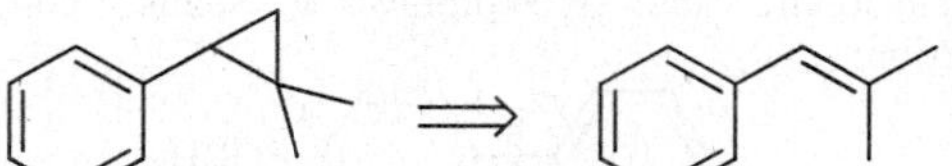

This alkene can be made from benzene through a variety of methods. The following synthesis is based on a Heck reaction, although other alternatives (such as Suzuki or Stille coupling) are perfectly viable:

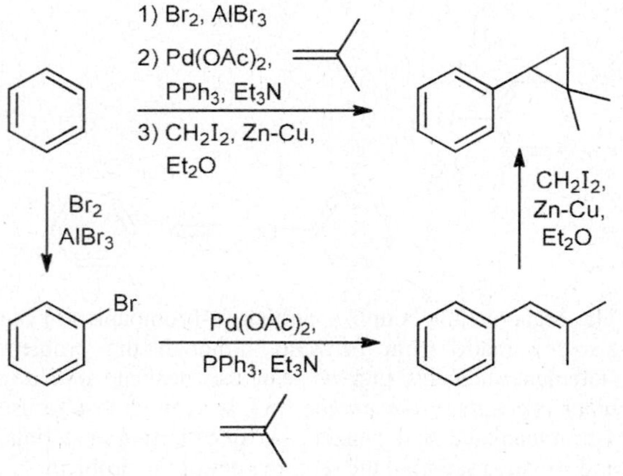

(d) The product exhibits a three-membered ring, which suggests a Simmons-Smith reaction. That is, the desired product can be made from the following *E* alkene:

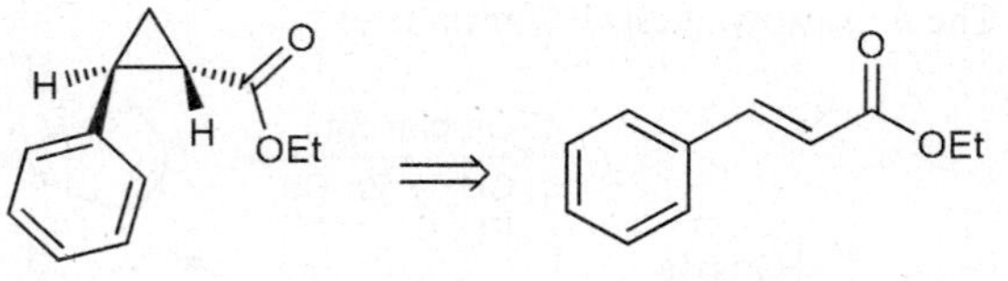

This alkene can be made from benzene through a variety of methods. The following synthesis is based on a Heck reaction, although other alternatives are perfectly viable:

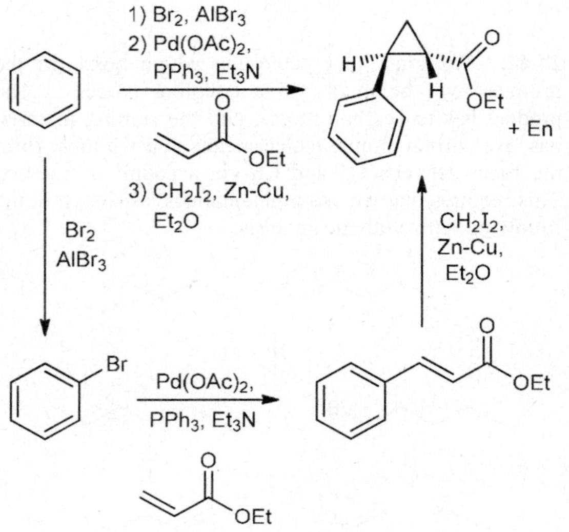

23.49. The starting material is an organohalide that also exhibits a C=C unit. Upon treatment with a base and a suitable palladium catalyst, an intramolecular Heck reaction can occur, in which a bond is formed between the C=C unit and the carbon atom bearing the iodide. The coupling partners are highlighted here:

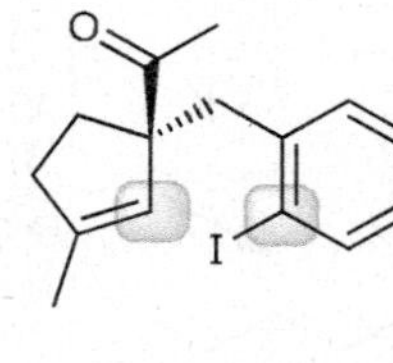

During the course of the reaction, the position of the π bond changes, because this is the only regiochemical outcome that can accommodate *syn* elimination:

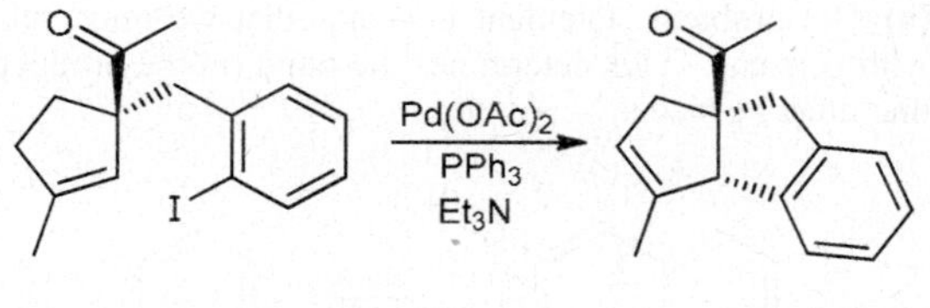

The observed diastereoselectivity can be rationalized by considering the existing chiral center (in the starting material). As a result of the configuration of this chiral center, the organohalide partner must approach the C=C unit from one particular face of the C=C bond (behind the plane of the five-membered ring). This can best be visualized through the construction of a molecular model. As a result, the newly formed chiral center has the *S* configuration (the *R* configuration is not formed).

23.50.

(a) The desired product can be made from cyclopentene via a Heck reaction, as shown below. Notice the position of the π bond in the product (this is the only regiochemical outcome that can accommodate *syn* elimination).

Pd(OAc)$_2$, PPh$_3$, Et$_3$N

(b) The product exhibits a three-membered ring, which suggests a Simmons-Smith reaction. That is, the desired product can be made from the alkene shown here:

This alkene can be made from cyclopentene via a Heck reaction, as shown in the following synthesis:

Pd(OAc)$_2$, PPh$_3$, Et$_3$N

Br

CH$_2$I$_2$, Zn-Cu, Et$_2$O

(c) The product is acyclic (no ring), but the starting material is cyclic, so our synthesis must involve breaking a C-C bond. We could perform an ozonolysis (of cyclopentene) to open the ring, but that would give us a compound with only five carbon atoms. We need a compound with seven carbon atoms, which suggests that we should open the ring with a ring-opening metathesis, thereby converting a five-membered cyclic compound into a seven-membered acyclic compound:

Grubbs cat. H$_2$C=CH$_2$

This compound can then be converted into the desired product via hydroboration/oxidation of each C=C unit.

1) BH$_3$•THF (xs)
2) H$_2$O$_2$, NaOH (xs)

HO OH

23.51. Compound **A** has two C=C units that can undergo a ring-closing metathesis upon treatment with a Grubbs catalyst. To draw the metathesis product, we remove a methylene group from each C=C unit (these methylene groups ultimately combine to give ethylene gas). We then draw a double bond between the remaining vinyl positions, giving compound **1**:

Grubbs cat.

A

1 ($C_6H_8O_2$)

Compound **1** has a C=C unit with the *Z* configuration. The *E* configuration cannot be formed in this case, because a seven-membered ring cannot accommodate a *trans* alkene (see Bredt's rule, Section 7.6).

In contrast, compound **B** undergoes a ring-closing metathesis to give a nine-membered ring. A nine-membered ring is large enough to accommodate a *trans* alkene, so two products are possible (*cis* or *trans*):

B

Grubbs cat.

2 ($C_8H_{12}O_2$) or **3** ($C_8H_{12}O_2$)

23.52.

(a) Using the procedure described in the problem statement, the desired product can be assembled from bromobenzene, carbon monoxide, and an organotin coupling partner, as shown:

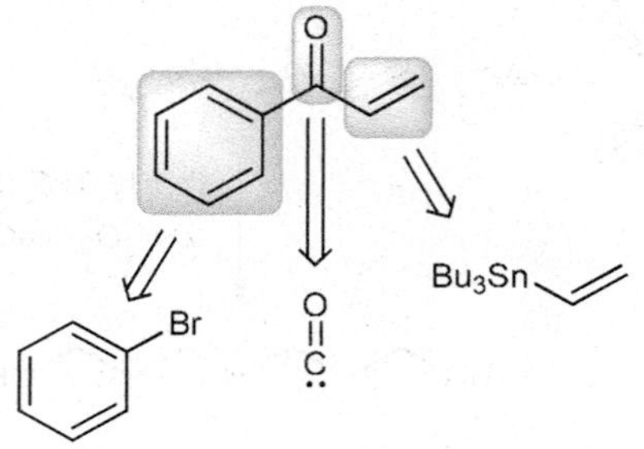

The forward synthesis is shown here:

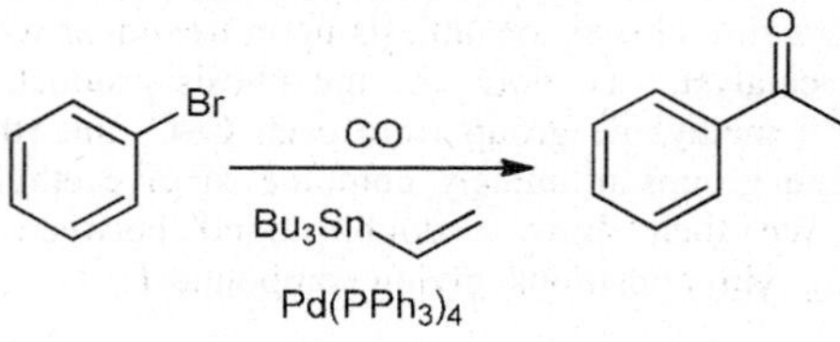

(b) Using the procedure described in the problem statement, the desired product can be assembled from bromobenzene, carbon monoxide, and an organotin coupling partner, as shown:

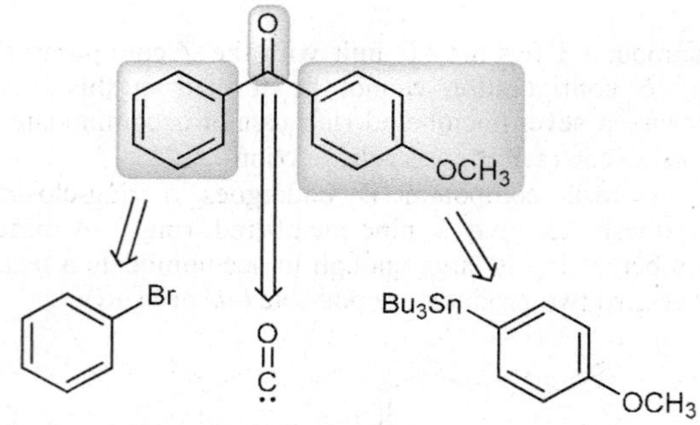

The forward synthesis is shown here:

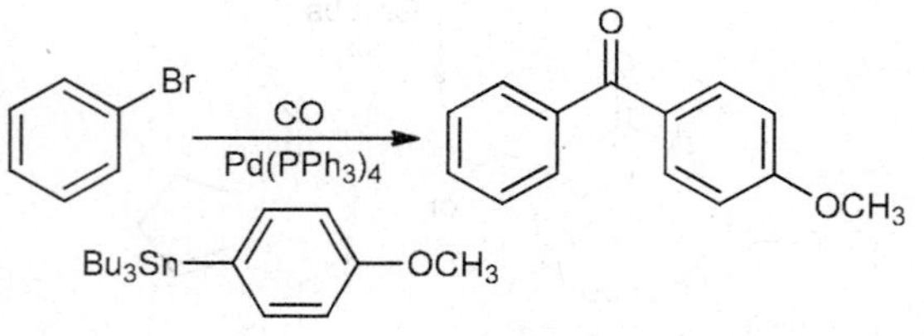

(c) Using the procedure described in the problem statement, the desired product can be assembled from bromobenzene, carbon monoxide, and an organotin coupling partner, as shown:

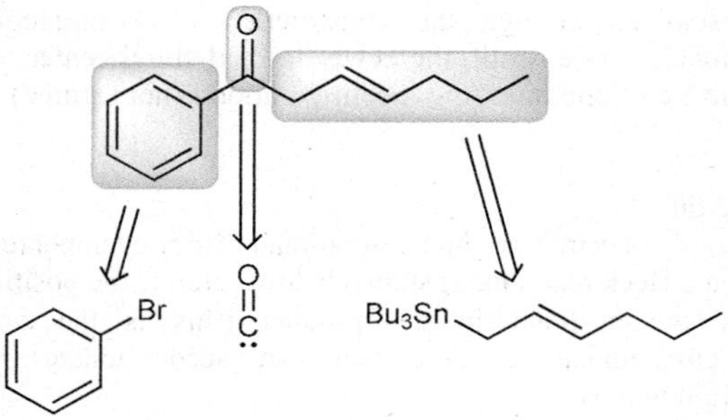

The forward synthesis is shown here:

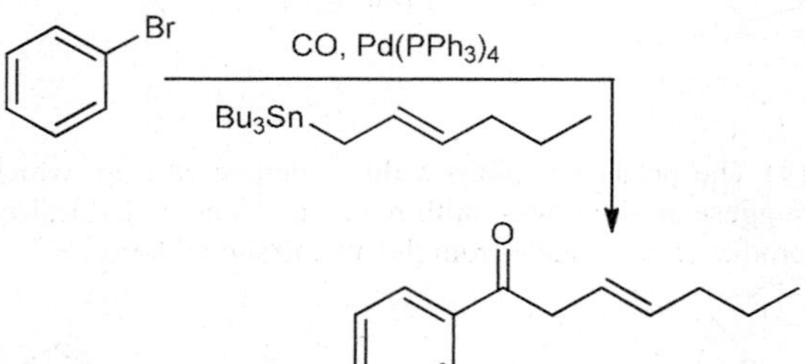

(d) Using the procedure described in the problem statement, the desired product can be assembled from bromobenzene, carbon monoxide, and an organotin coupling partner, as shown:

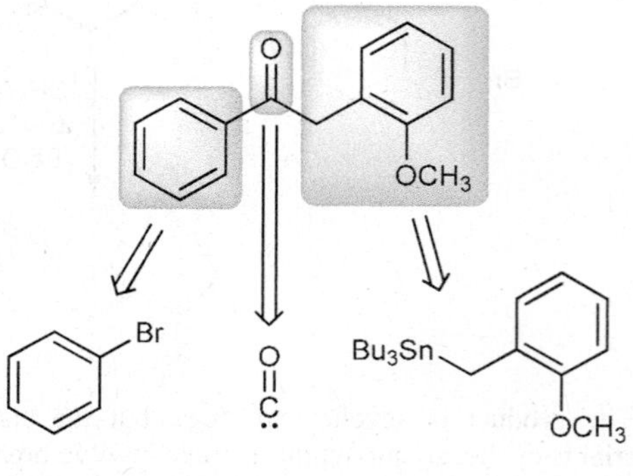

The forward synthesis is shown here:

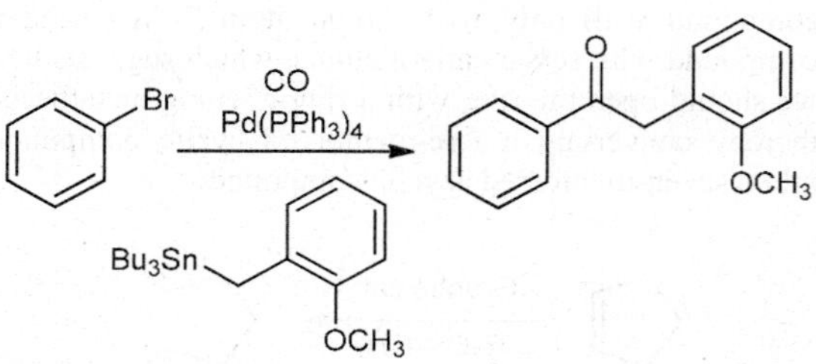

23.53.

(a) The starting material (1-pentene) is comprised of two alkylidene fragments:

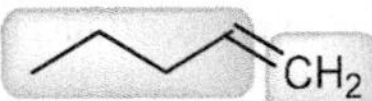

One of these fragments has four carbon atoms, and the other fragment has only one carbon atom (a methylene group). When 1-pentene is treated with a Grubbs catalyst, the methylene fragments are recombined to give ethylene gas (which evolves as a gas and is removed from the reaction vessel), while the other alkylidene fragments (with four carbon atoms) are recombined to give 4-octene. Two stereoisomers are possible (*cis* and *trans*), and both are obtained (**A** and **B**):

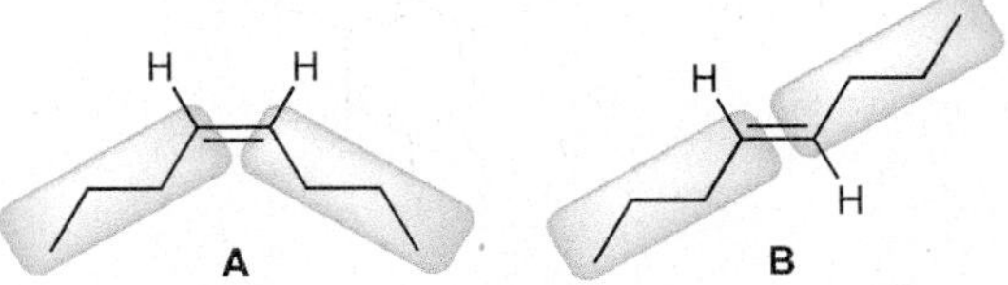

Compound **A** is a *cis*-disubstituted alkene, so it undergoes a Simmons-Smith reaction to give a *cis*-disubstituted cyclopropane ring (a *meso* compound):

In contrast, compound **B** is a *trans*-disubstituted alkene, so it undergoes a Simmons-Smith reaction to give a *trans*-disubstituted cyclopropane ring. Two stereoisomers (enantiomers) are possible (compounds **2** and **3**), and both are obtained:

(b) Compounds **2** and **3** are nonsuperimposable mirror images of each other, so they are enantiomers.

23.54. In the first step of the synthesis, Zn metal inserts into the C-I bond to produce organozinc **A**:

Next, identify the carbon atoms that will be joined in the Negishi coupling. The carbon atom connected directly to the triflate group (which is more reactive than the aryl chloride position) will be joined with the carbon atom connected directly to Zn in the organozinc, highlighted here:

Then, rotate the coupling partners so that they are aligned to form a bond, and draw the product, **B**:

23.55. In the first step of the synthesis, excess *t*-BuLi converts each of the four aryl bromide groups to aryllithium groups:

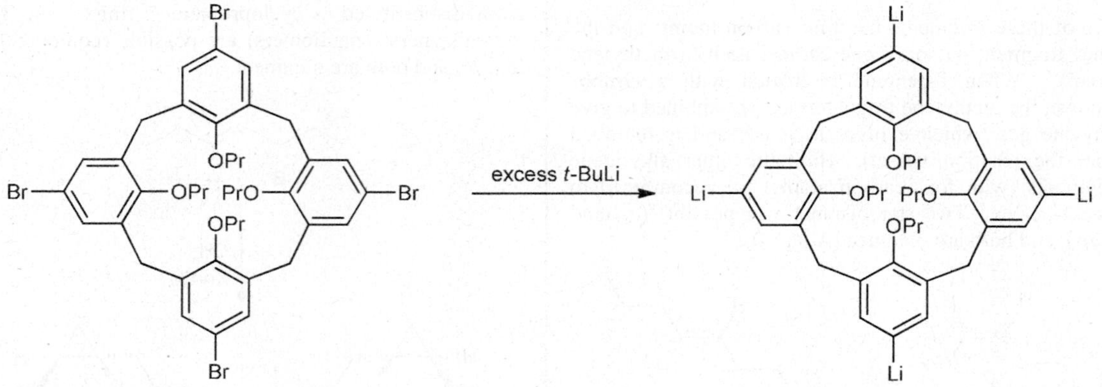

In step 2, $ZnCl_2$ converts each aryllithium group to an arylzinc, thus producing the following tetraarylzinc compound:

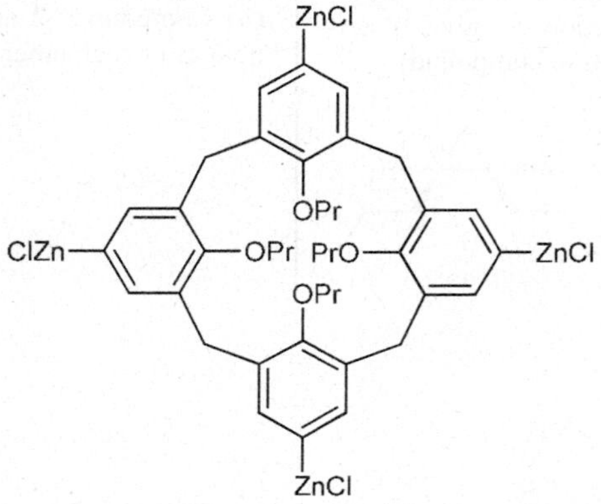

Next, identify the carbon atoms that will be joined in the Negishi coupling. The carbon atom connected directly to the iodide group will be joined with the carbon atom connected directly to Zn in the organozinc, highlighted below. In this case, four equivalents of *p*-iodotoluene will couple to the tetraiodozinc calixarene.

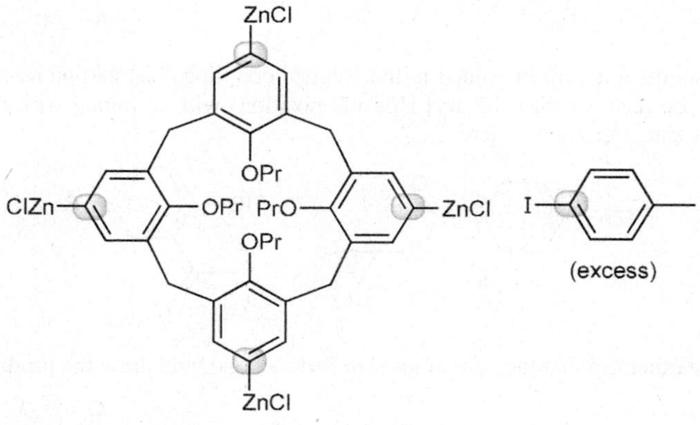

Then, rotate the coupling partners so that they are aligned to form a bond, and draw the product:

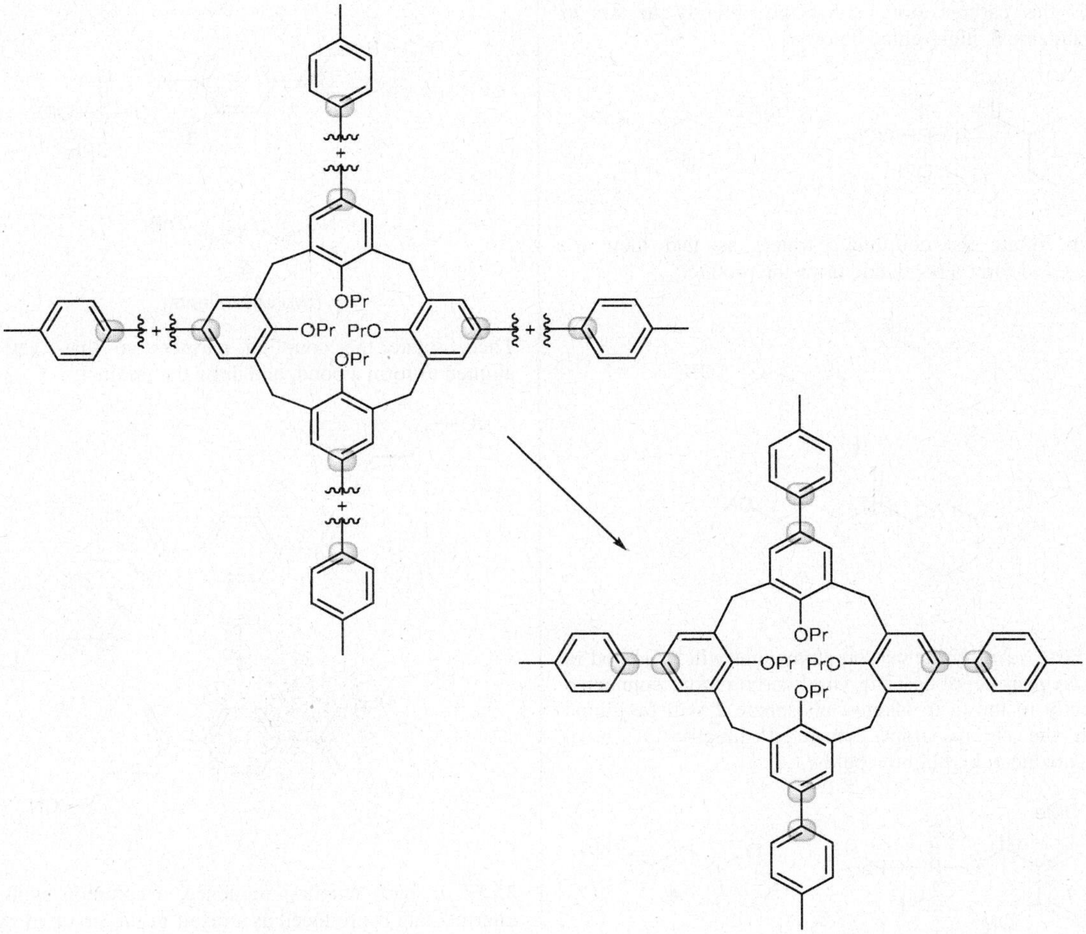

23.56.

(a) First, identify the carbon atoms that will be joined in the Negishi coupling. The vinyl carbon atom connected directly to the O in alkenyl phosphate **1** will be joined with the carbon atom connected directly to Zn in organozinc **4**, highlighted below.

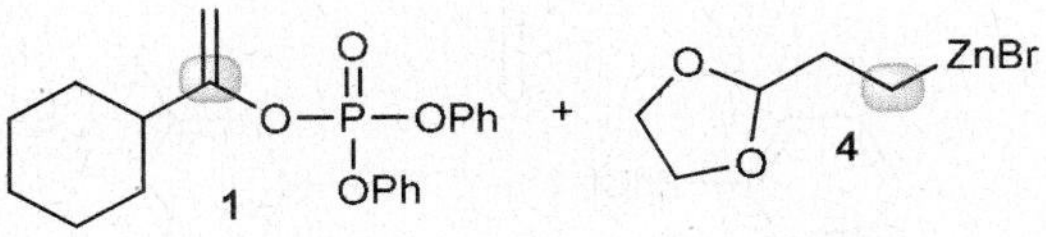

Then, rotate the coupling partners so that they are aligned to form a bond, and draw the product:

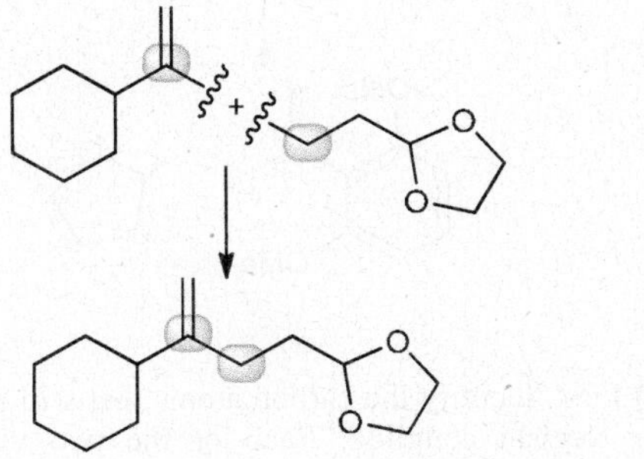

(b) First, identify the carbon atoms that will be joined in the Negishi coupling. The vinyl carbon atom connected directly to the O in alkenyl phosphate **1** will be joined with the carbon atom connected directly to Zn in organozinc **5**, highlighted below.

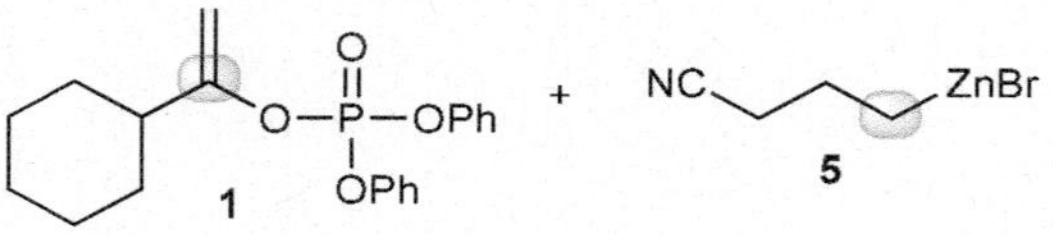

Then, rotate the coupling partners so that they are aligned to form a bond, and draw the product:

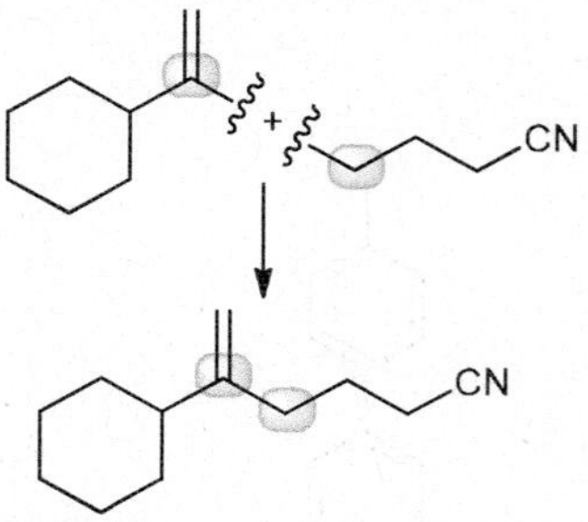

(c) First, identify the carbon atoms that will be joined in the Negishi coupling. The vinyl carbon atom connected directly to the O in alkenyl phosphate **2** will be joined with the carbon atom connected directly to Zn in organozinc **4**, highlighted below.

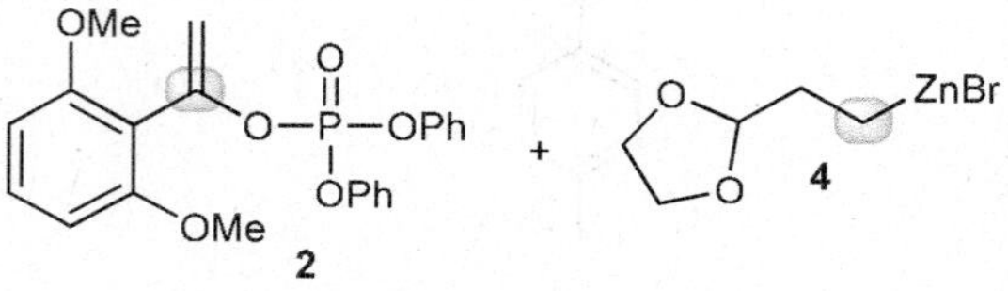

Then, rotate the coupling partners so that they are aligned to form a bond, and draw the product:

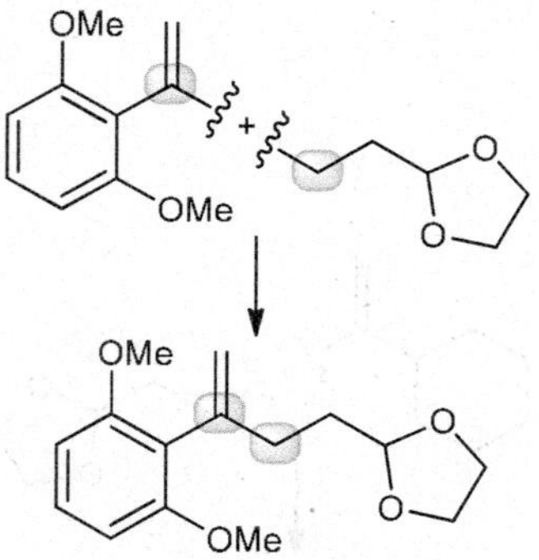

(d) First, identify the carbon atoms that will be joined in the Negishi coupling. Each of the two vinyl carbon atoms connected directly to an O in alkenyl phosphate **3** will be joined with the carbon atom connected directly to Zn in each equivalent of organozinc **5**, highlighted below.

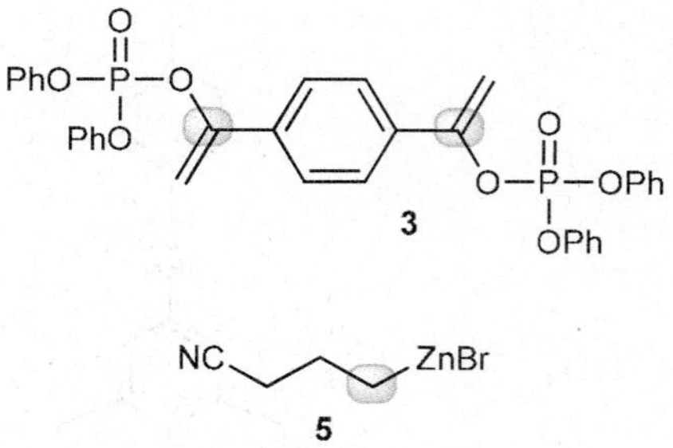

(two equivalents)

Then, rotate the coupling partners so that they are aligned to form a bond, and draw the product:

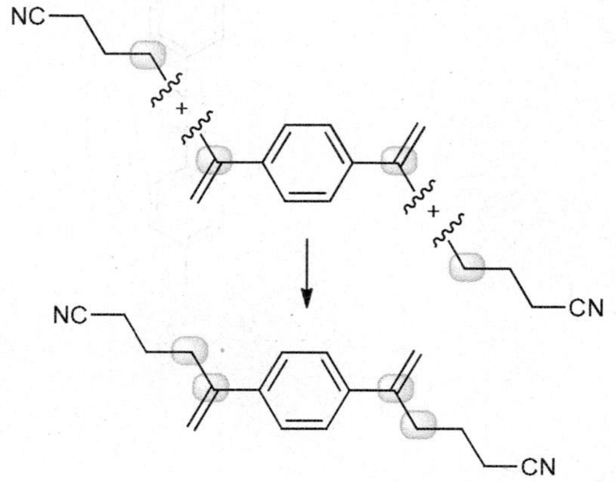

23.57. In each reaction sequence, a common synthetic intermediate is produced as a result of the first two steps: lithium-halogen exchange, followed by conversion of the resulting aryllithium to the arylzinc using $ZnCl_2$.

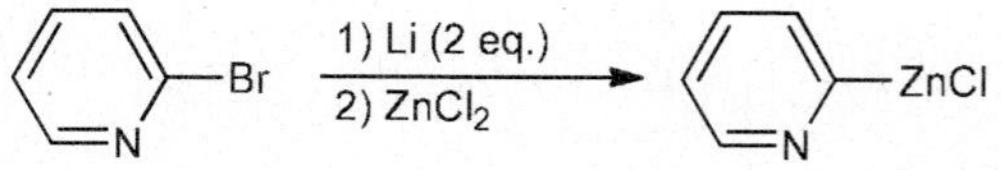

1) Li (2 eq.)
2) $ZnCl_2$

(a) Identify the C–C bond in the product that was produced in the Negishi coupling (indicated below). Disconnect the bond to determine the arylzinc and aryl triflate required for the reaction.

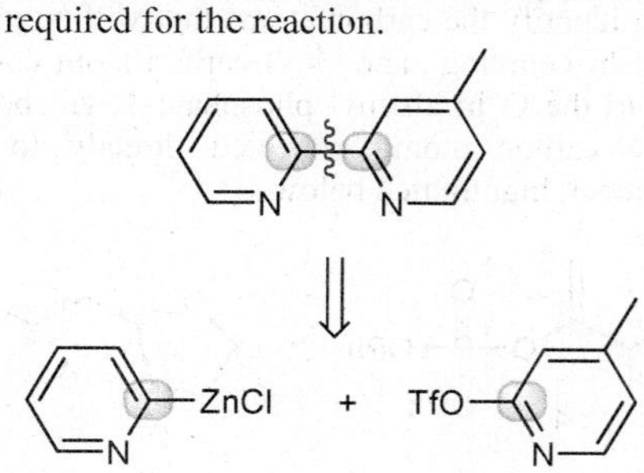

(b) Identify the C–C bond in the product that was produced in the Negishi coupling (indicated below). Disconnect the bond to determine the arylzinc and aryl triflate required for the reaction.

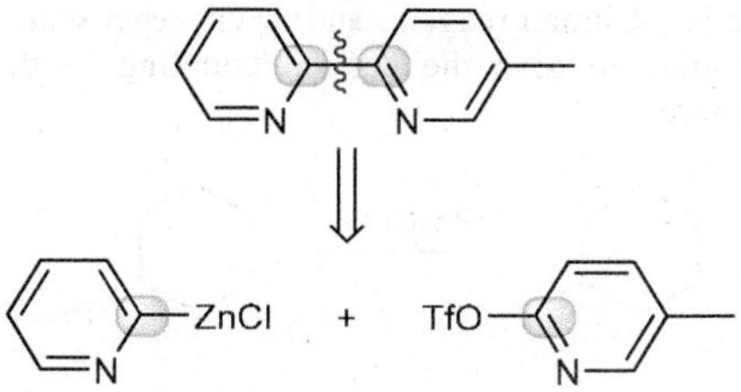

(c) Identify the C–C bond in the product that was produced in the Negishi coupling (indicated below). Disconnect the bond to determine the arylzinc and aryl triflate required for the reaction.

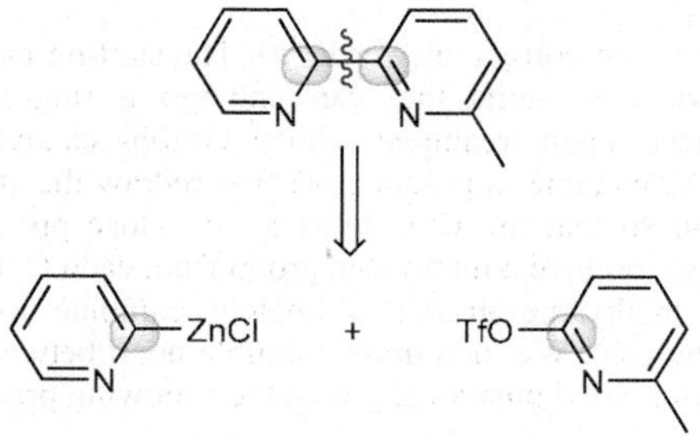

23.58. First identify the carbon atoms that will be joined. This reaction is an intramolecular Negishi coupling, in which both coupling partners are in the same molecule. The carbon atom connected directly to the iodide group will be joined with the carbon atom connected directly to Zn in the organozinc, highlighted here:

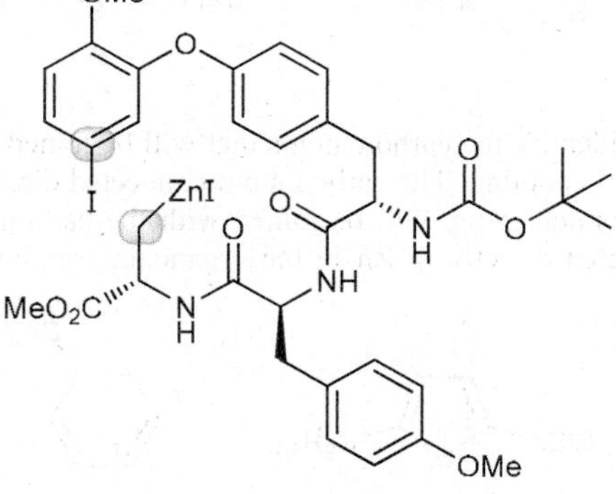

Next, realign the coupling partners so that they are aligned to form a bond, and draw the product, in which a new macrocycle (large ring) has formed:

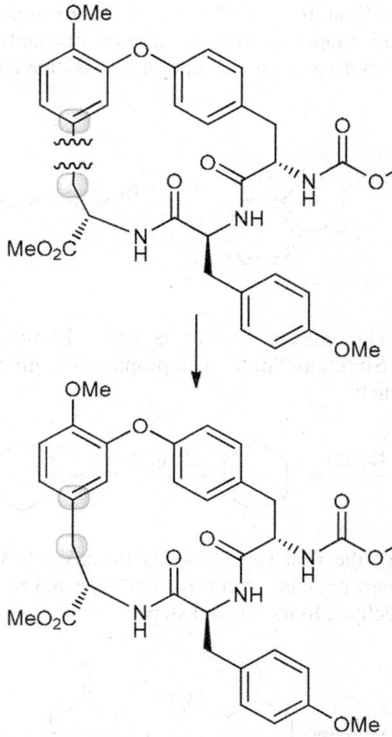

23.59. The correct answer is (c). In a Heck cross-coupling reaction, an alkene is coupled with an organohalide. The following highlighted carbon atoms will be connected to each other as a result of the coupling process:

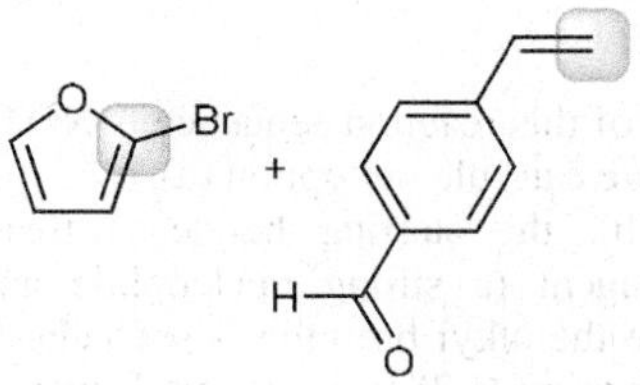

The coupling process affords a disubstituted alkene, and we expect the *trans* isomer:

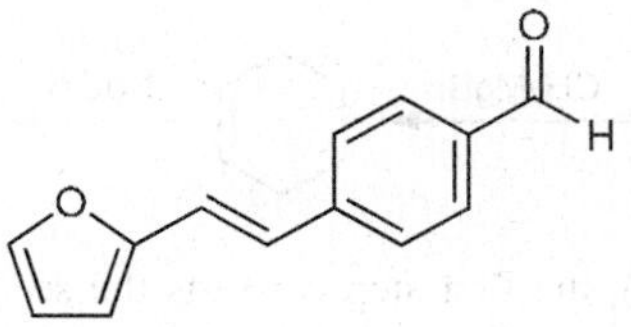

23.60. The correct answer is (d). All of the options involve a vinyl bromide and a vinyl boronic ester or acid, but we must choose the coupling partners that give the desired stereochemical outcome. In the product, both C=C units have the *E* configuration:

Since the configuration of all C=C units are preserved during Suzuki coupling, we must choose reactants that have the *E* configuration, so option (d) is the correct answer:

(E) (E) B O O + Br

23.61. The correct answer is (d). Elimination, followed by Simmons-Smith cyclopropanation, gives the desired product:

Br NaOEt CH_2I_2, Zn-Cu

In option (a), the first step converts the alkyl bromide into a Grignard reagent, which is then expected to react with formaldehyde in the second step:

Br Mg MgBr O H H O⊖ H H

The product of this reaction sequence is NOT the desired product, so we can rule out option (a).

In option (b), the starting halide is treated with a Grignard reagent (a strong nucleophile and a strong base). Since the alkyl bromide is secondary, we would expect elimination (E2) to be favored over substitution, giving an alkene. This alkene is not expected to react with NaOH, so option (b) does not give the desired product:

Br CH_3MgBr NaOH no reaction

In option (c), the first step converts the secondary alkyl bromide into an alkene (because elimination is favored over substitution for secondary alkyl halides), and the alkene is then converted into an epoxide upon treatment with a peroxyacid:

Br NaOEt RCO_3H O

This product is similar in structure to the desired product. But we are not trying to make an epoxide. Rather, we must form a cyclopropane ring, so option (c) is not correct.

23.62. The correct answer is (b). Lithium diphenyl cuprate is a Gilman reagent, and it will react with the an alkyl iodide to give the desired coupling product, as shown here:

I Ph_2CuLi Ph

23.63. The correct answer is (a). The electrophile (PhBr) has reacted with a palladium complex to give a new complex in which both Ph and Br are each bonded directly to palladium. During this addition process, the oxidation state of the palladium atom increases, from Pd(0) to Pd(II), and the process is therefore called an oxidative addition.

23.64. The correct answer is (c). The starting material has two C=C units that can undergo a ring-closing metathesis upon treatment with a Grubbs catalyst. To draw the metathesis product, we first redraw the starting material so that the C=C units are in close proximity. Then, we remove a methylene group from each C=C unit (these methylene groups ultimately combine to give ethylene gas). We then draw a double bond between the remaining vinyl positions, giving the following product:

CH_2 CH_2 Grubbs catalyst + CH_2 = CH_2

23.65. In the first step of the synthesis, $ZnCl_2$ converts the aryllithium to arylzinc **A**:

Li S C_9H_{19} $ZnCl_2$ ClZn S C_9H_{19} **A**

Next, identify the carbon atoms that will be joined in the Negishi coupling. The carbon atom connected directly to the bromide group will be joined with the carbon atom connected directly to Zn in the organozinc, highlighted here:

ClZn S C_9H_{19} + Br Bu_3Sn

Then, rotate the coupling partners so that they are aligned to form a bond, and draw the product:

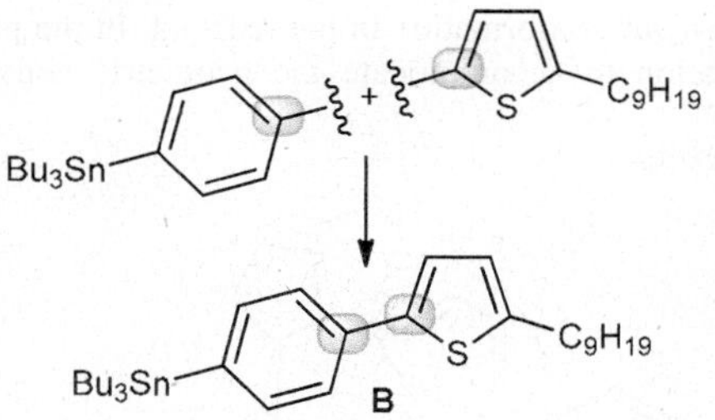

Next, identify the carbon atoms that will be joined in the Stille coupling. The carbon atom connected directly to the bromide group will be joined with the aryl carbon atom connected directly to Sn in the organostannane, highlighted here:

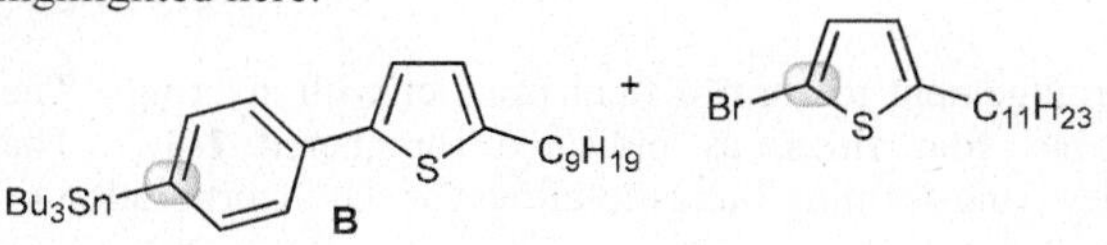

Then, rotate the coupling partners so that they are aligned to form a bond, and draw the product:

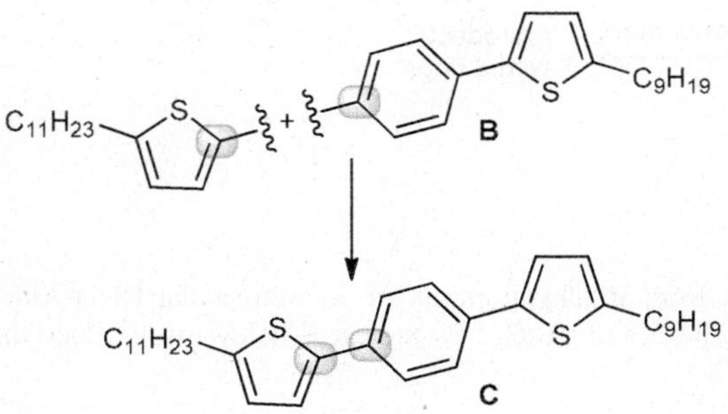

23.66. First identify the carbon atoms that will be joined. The carbon atom connected directly to the iodide group will be joined with the carbon atom connected directly to the tributylstannyl group, highlighted below:

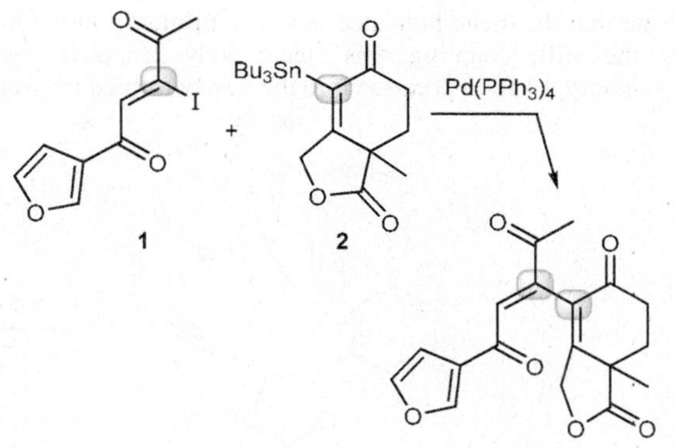

The product of the Stille reaction subsequently undergoes an electrocyclic reaction (Section 16.9) involving the flow of 6 electrons in a cycle, as shown below:

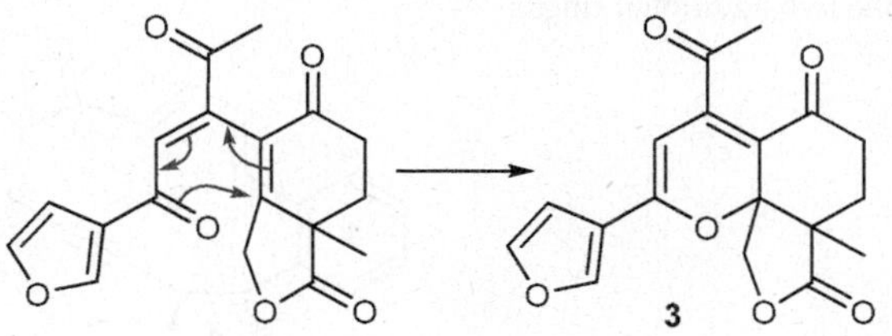

The transition state of the electrocyclic rearrangement is shown below. Three π bonds (two C=C plus one C=O) are breaking, while two π bonds (C=C) and one σ bond (C–O) are forming, resulting in the net conversion of one C=O π bond to one C–O σ bond.

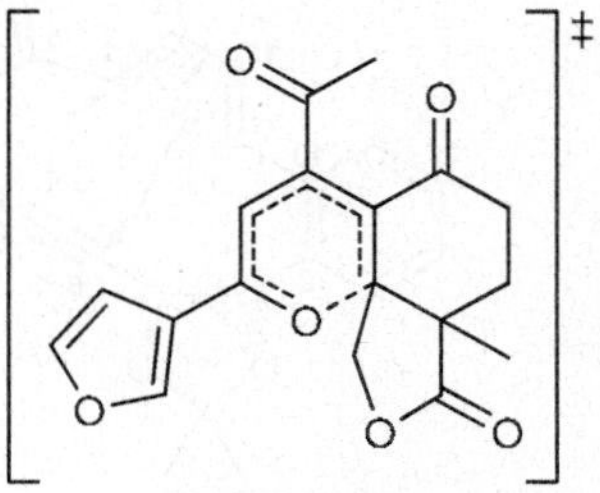

23.67. The reactant undergoes an intramolecular Stille coupling with a palladium catalyst to produce the macrocycle shown. In this reaction, the sp^2-hybridized carbon atom connected to the tin atom (in the reactant) forms a new bond to the sp^2-hybridized carbon atom connected to the triflate leaving group (in the reactant), thereby producing a new 13-membered ring.

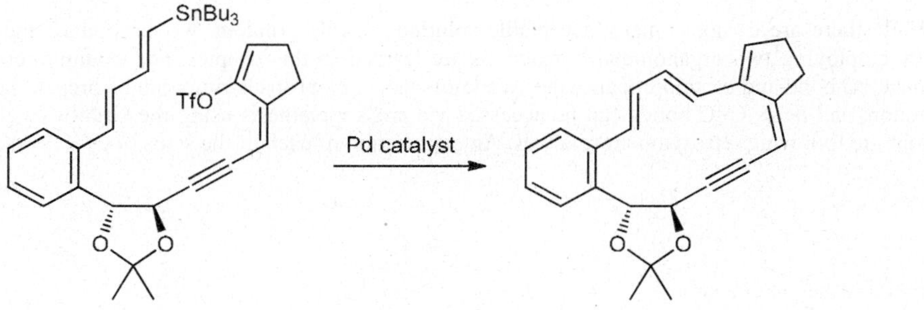

Note that the diene indicated below is primarily in the more stable s-*trans* conformation in the reactant. In the product of the Stille coupling, this diene likely adopts an s-*cis* conformation to accommodate the geometric constraints necessary for incorporation into the newly formed macrocycle.

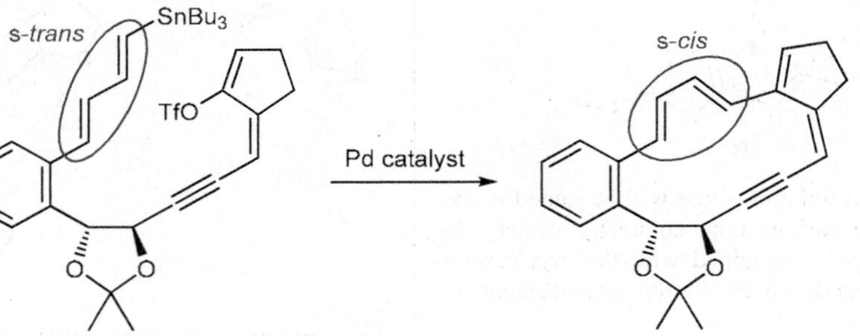

s-*trans* $SnBu_3$ TfO Pd catalyst s-*cis*

The question indicates that this macrocycle undergoes a rearrangement to form a final product with six rings. The macrocycle product of the intramolecular Stille coupling has four rings, as indicated (numbered **1-4**). The rearrangement must therefore result in the formation of two new ring-forming bonds to allow for the incorporation of the two additional rings.

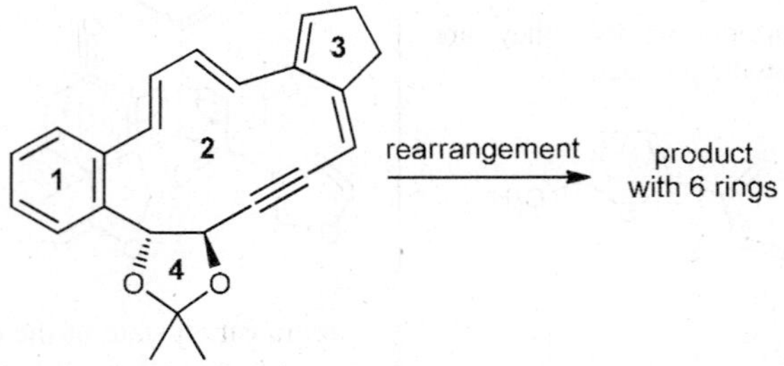

rearrangement → product with 6 rings

The macrocycle has an s-*cis* diene group across the macrocycle from an alkyne group. An intramolecular Diels-Alder reaction between these two entities forms two new σ bonds (C1-C5 and C6-C11 as indicated below) to produce the hexacyclic product shown.

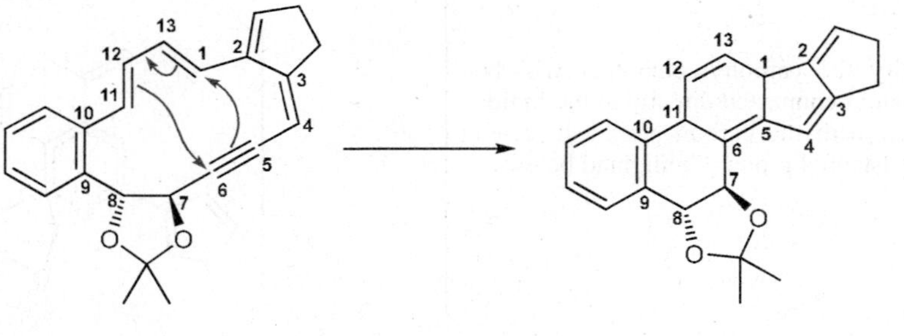

23.68. While there are certainly many acceptable solutions to this problem, we can reduce the complexity of this problem by employing two organometallic reactions we learned in this chapter. For example, compound **1** contains both a *trans* C=C bond and a cyclopropane ring. We know that cyclopropane rings can be prepared using the Simmons-Smith reaction, and *trans* C=C bonds can be accessed via cross metathesis using the Grubbs catalyst. One synthesis derives from the following retrosynthetic analysis. An explanation of each of the steps (*a-d*) follows.

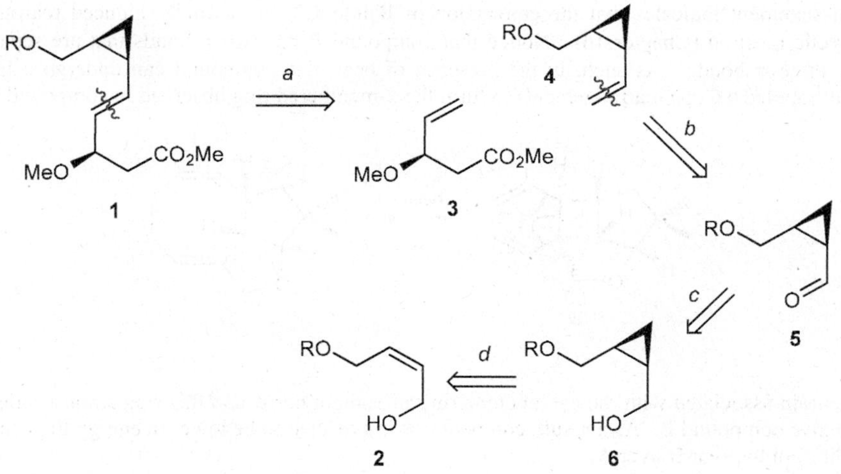

a. Compound **1** can be synthesized from the union of two terminal alkenes (**3** & **4**) using a Grubbs catalyst.
b. Alkene **4** can be made from aldehyde **5** employing the Wittig reaction.
c. Aldehyde **5** can be accessed from the corresponding alcohol (**6**), using PCC (or DMP or a Swern oxidation).
d. The cyclopropane ring can be installed via the Simmons-Smith reaction on substrate **2.**

Now let's draw the forward scheme. Alcohol **2** is treated with diiodomethane in the presence of a Zn-Cu couple to produce cyclopropane **6** (as a racemic mixture). The alcohol in **6** is converted to aldehyde **5** using PCC and is subsequently treated with a phosphorus ylide to produce alkene **4**. Finally, union of alkene **4** with compound **3** using a Grubbs catalyst will produce the desired product (as a mixture of diastereomers that can be separated).

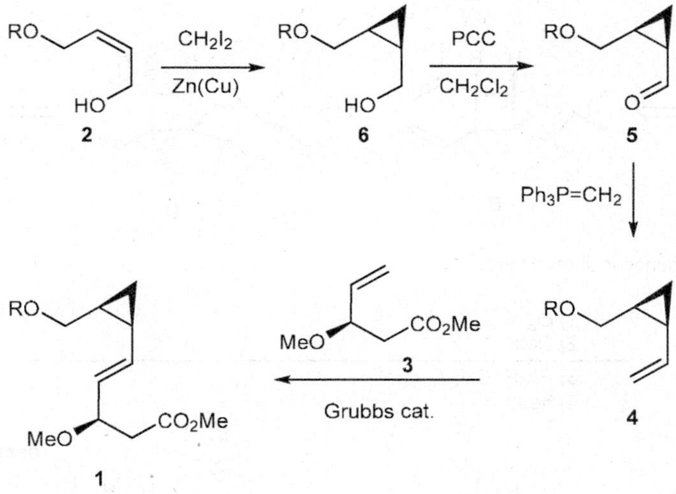

23.69.
(a) In the presence of a Grubbs catalyst and ethylene, a cycloalkene (compound **A**) is expected undergo a ring-opening metathesis. To draw the product of this process (compound **B**), we erase the C=C bond of the ring, and then connect a methylene (CH_2) group to each of the two vinyl positions, like this:

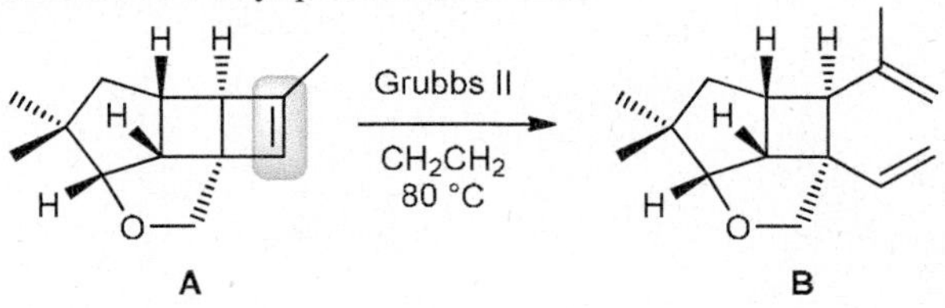

(b) The problem statement indicates that the conversion of **B** into **C** is a thermally induced rearrangement. This suggests a pericyclic reaction (Chapter 16). Notice that compound **B** has two π bonds that are separated from each other by exactly three σ bonds. As such, in the presence of heat, this compound can undergo a [3,3] sigmatropic rearrangement (also called a Cope rearrangement) to form the 8-membered ring observed in compound **C**.

B —heat→ C

(c) There is ring strain associated with the cyclobutene ring of compound **A**, and this ring strain is alleviated when the ring is opened to give compound **B**. As a result, compound **B** is expected to be lower in energy than compound **A**, thus causing the equilibrium to favor **B** over **A**.

(d) There is ring strain associated with the cyclobutane ring of compound **B**, and this ring strain is alleviated when the ring is opened to give the eight-membered ring in compound **C**. As a result, compound **C** is expected to be lower in energy than compound **B**, thus causing the equilibrium to favor **C** over **B**.

(e) We have seen that compound **C** can be made from compound **B** via a [3,3] sigmatropic rearrangement (Cope rearrangement). Compound **B** can be prepared from compound **A** via a number of synthetic routes. The following is one possible solution. Compound **B** can be made from compound **D** via Wittig reactions, and compound **D** can be generated from **A** via ozonolysis of the C=C bond in **A**:

C ⟹ B ⟹ D ⟹ A

The forward reaction sequence is shown here:

A → C: 1) O_3; 2) DMS; 3) $Ph_3P{=}CH_2$ (2 eq.); 4) heat

A → D: 1) O_3; 2) DMS

D → B: $Ph_3P{=}CH_2$ (2 eq.)

B → C: heat

23.70. Rapamycin has a large ring (it is a macrolide) which is assembled via two successive (tandem) Stille coupling reactions. The first reaction is intermolecular, and the second is intramolecular. The reactive centers are highlighted:

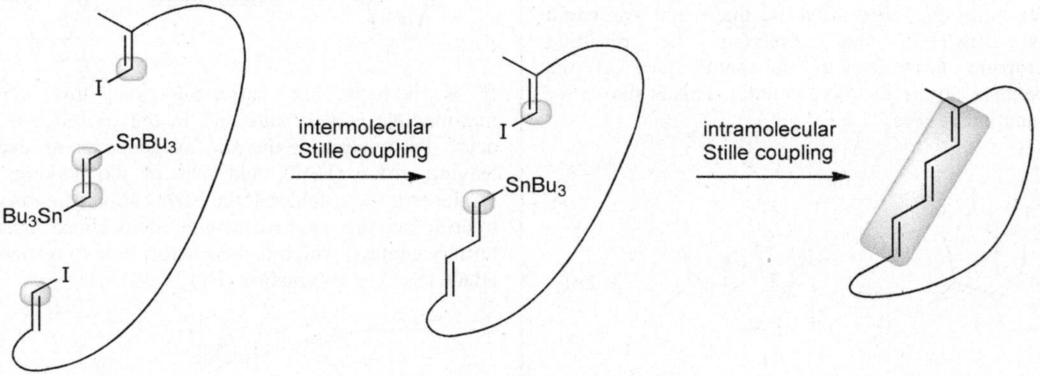

When stitched together, the large ring contains a triene group (highlighted). Notice that each of the three C=C units retains the configuration that it had in the starting materials. The rest of the structure (all of the functional groups and chiral centers) remains unaffected by these reaction conditions, so the structure of rapamycin is as follows:

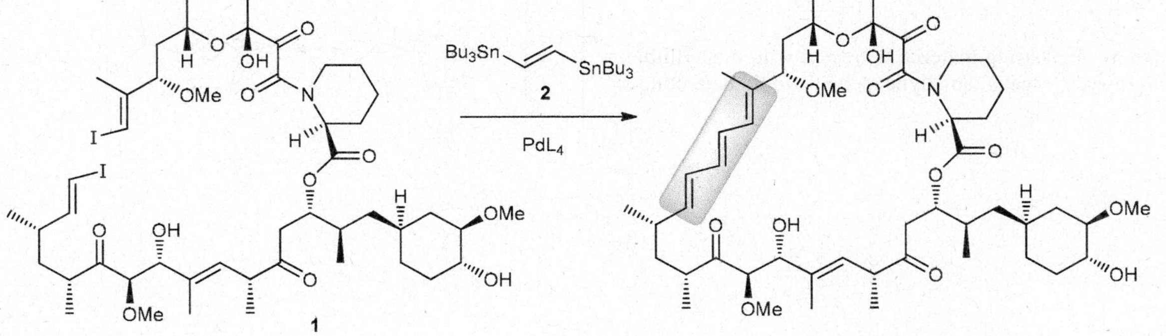

23.71.

(a) In step 1, methyllithium attacks the carbonyl carbon producing a racemic allylic alkoxide (note: the wavy line indicates both possible configurations at that chiral center).

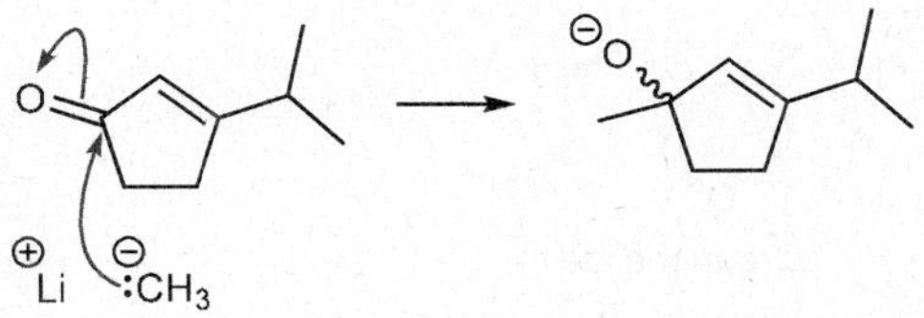

In step 2, the zinc carbenoid (formed by the reaction of the zinc-copper couple with diiodomethane) reacts in a Simmons-Smith cyclopropanation with the alkene to install the cyclopropane ring.

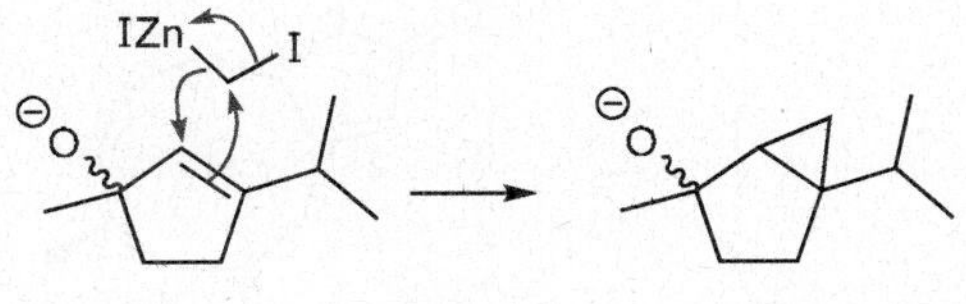

Finally, in step 3, the alkoxide ion is protonated (upon treatment with water), giving the product:

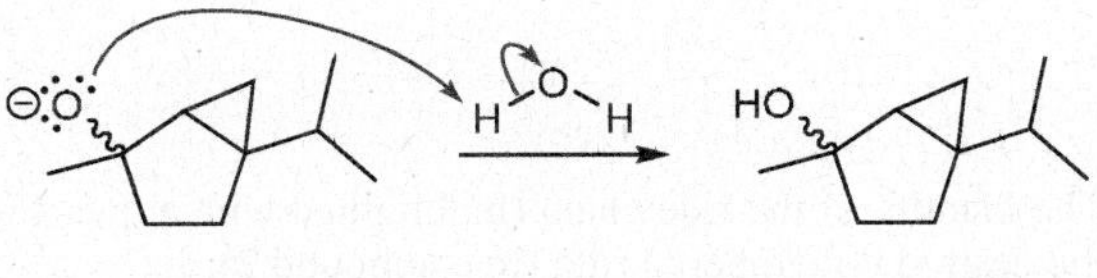

(b) In the product, the OH group and the fused cyclopropane ring are both on the same face of the cyclopentane ring (both "up" in the stereoisomer shown in the problem statement). This stereochemical outcome can be rationalized if we recognize that the anionic oxygen atom can form a Lewis acid / Lewis base complex with the zinc atom of the metal carbenoid species $IZnCH_2I$, thus directing the resulting cyclopropane formation to the same face of the cyclopentene ring as the oxygen atom. This is shown for each enantiomer here:

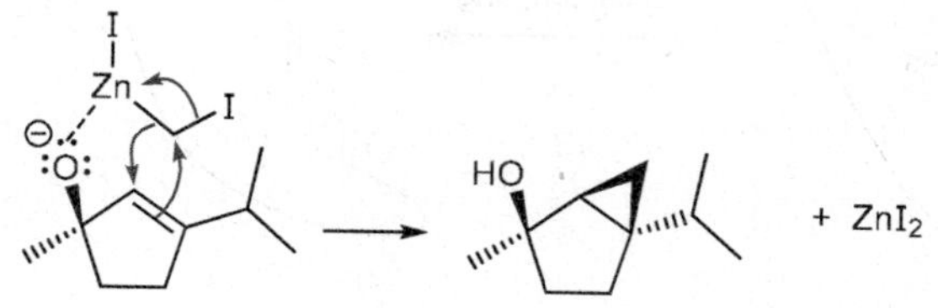

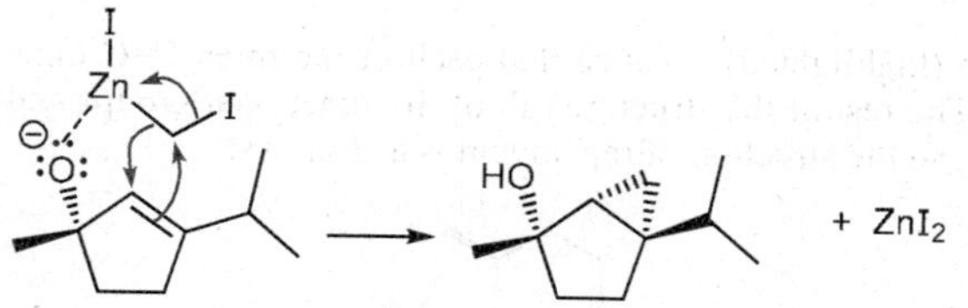

(c) If the starting material is treated with methyllithium followed by water, an allylic alcohol would be produced:

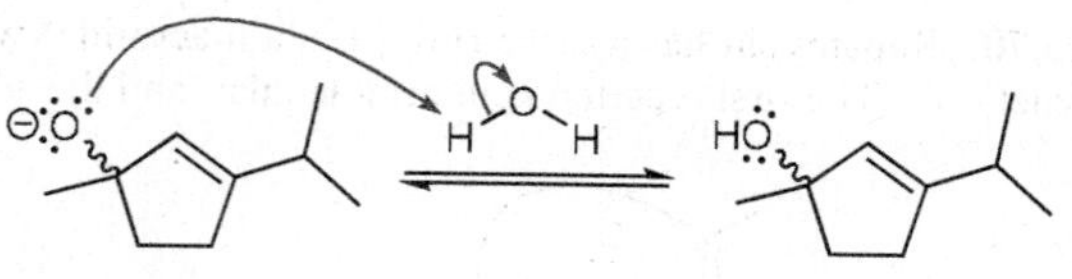

It is perhaps not surprising that this synthetic intermediate is very unstable in the presence of trace acid. Protonation of the alcohol generates an excellent leaving group (H_2O), and loss of the leaving group produces a very stable allylic carbocation (the resonance hybrid has the positive charge delocalized over two tertiary centers) which is then susceptible to nucleophilic attack (S_N1) or elimination (E1).

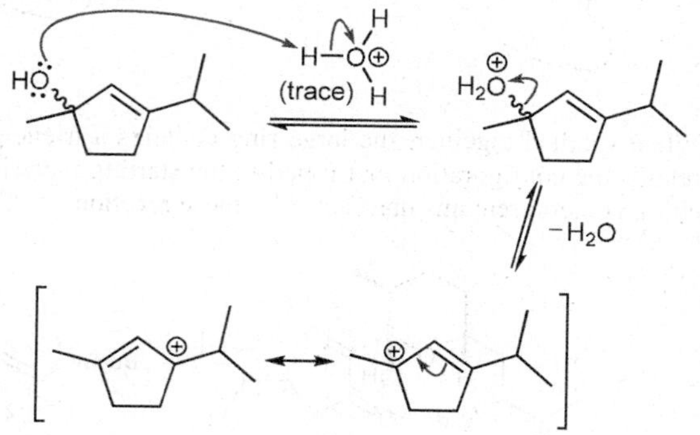

23.72. The starting material (compound **3**) is an organostannane, which indicates a Stille coupling reaction. To determine the structure of the starting organohalide, we analyze the structure of the product (compound **2**) and determine the part of the structure that corresponds with organostannane **3** (highlighted). The rest of compound **2** (not highlighted) must come from the organohalide, as shown:

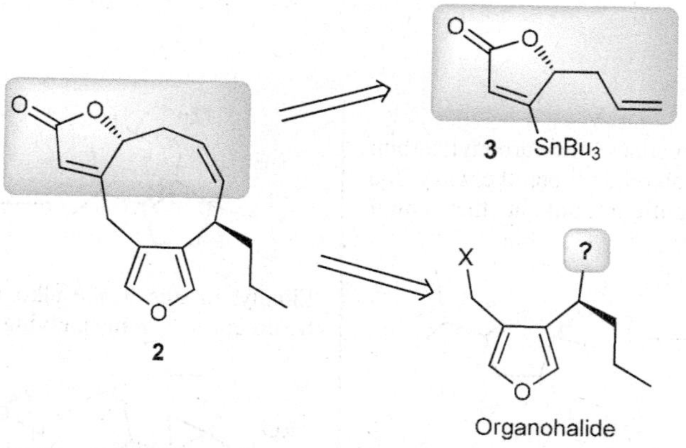

The identity of the side chain (highlighted with a question mark) must be chosen to enable the subsequent formation of the desired 9-membered ring (in compound **2**):

Must close ring

Stille coupling

3 SnBu$_3$

2

This can be achieved via a ring-closing metathesis, as shown in the following retrosynthetic analysis:

SnBu$_3$

I

The forward process is summarized here:

1) Pd(PPh$_3$)$_4$

2) Grubbs catalyst

SnBu$_3$

Stille Conditions

Grubbs cat.

23.73. The product of the intramolecular Simmons-Smith reaction is shown below. The carbenoid carbon of the intermediate (labeled as C1 below) forms two new C–C bonds: one to each of the sp^2-hybridized carbon atoms of the C=C unit (C6 and C7), thus producing the fused bicyclic structure shown (**3**).

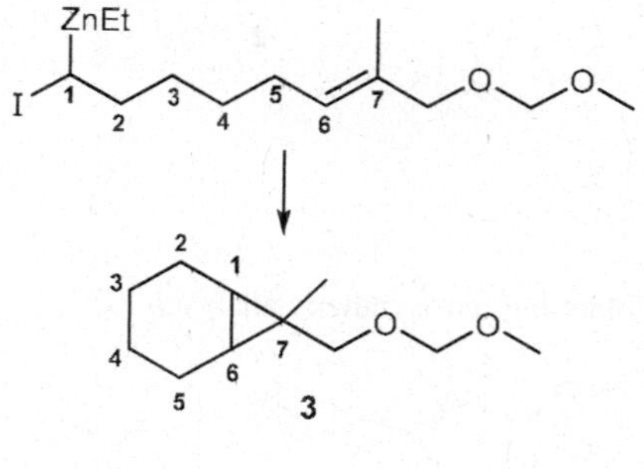

The observation that compound **1** successfully undergoes intramolecular Simmons-Smith reaction (while compound **2** does not) suggests that the oxygen atoms in **1** (which are absent in **2**) play a role. The zinc atom in the carbenoid intermediate is electron poor and can thus serve as a Lewis acid. The lone pairs of electrons on the oxygen atoms allow them to serve as Lewis bases. An intramolecular Lewis acid / Lewis base complex between one or both oxygen atoms and the zinc atom may align the reactive groups and facilitate the Simmons-Smith reaction. The transition state below is consistent with this explanation.

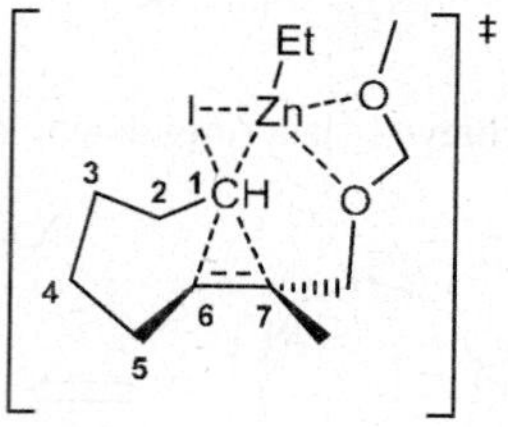

23.74. The desired product (dibenzo[*a,c*]cyclohepten-5-one) contains 15 carbon atoms, so in order to propose a synthesis from starting materials with 8 or fewer carbons, we will need to make several new C-C bonds in our synthesis. Recognition of the α,β-unsaturated ketone in the central ring suggests the possibility of utilizing an aldol condensation as part of our approach.

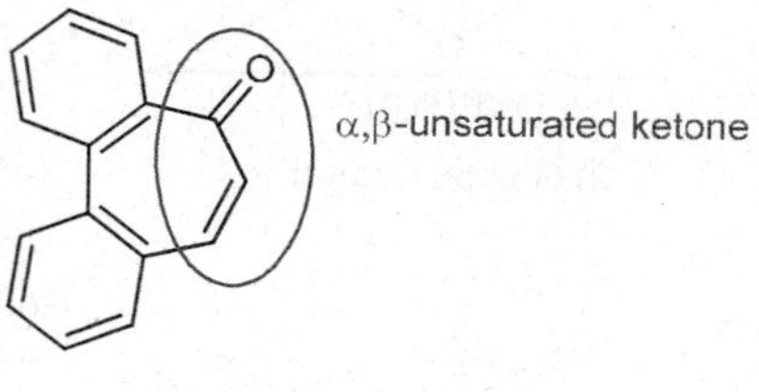

The following retrosynthesis outlines one approach to make the desired compound in two steps. The α,β-unsaturated ketone can be prepared via an intramolecular aldol condensation between the methyl ketone group and the aldehyde group on the biphenyl intermediate below. The bond between the two aromatic rings can be constructed using a Suzuki coupling reaction between the aryl bromide and the aryl boronic acid shown.

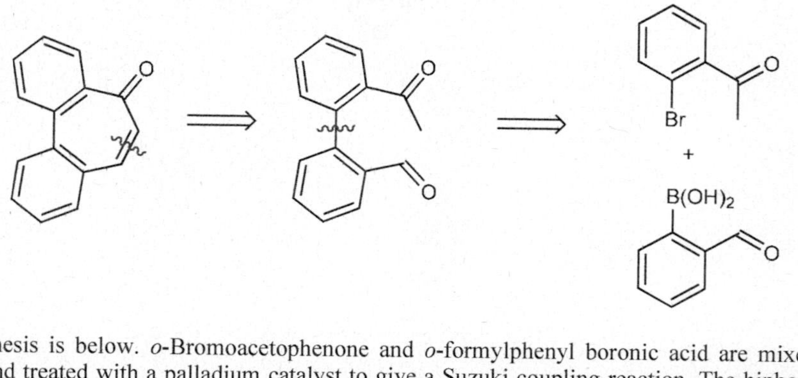

The forward synthesis is below. *o*-Bromoacetophenone and *o*-formylphenyl boronic acid are mixed together under basic conditions and treated with a palladium catalyst to give a Suzuki coupling reaction. The biphenyl keto-aldehyde intermediate is then heated under basic conditions to afford the desired aldol condensation product (via an intramolecular aldol condensation).

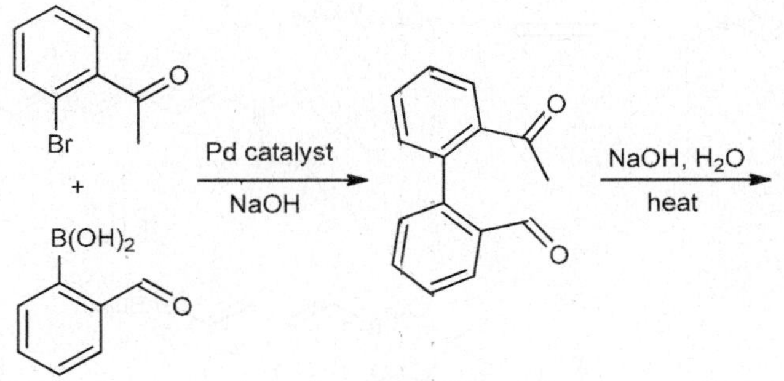

Br

+

$B(OH)_2$

Pd catalyst

NaOH

$NaOH, H_2O$

heat

23.75.

(a) Reaction of 1 with 4-iodoanisole in the presence of a palladium catalyst affords the Stille coupling product shown. Subsequent addition of 4-*tert*-butyliodobenzene and CsF to the reaction mixture produces the final product as the result of a palladium-catalyzed Suzuki coupling between the vinyl boronic ester and the aryl iodide. Note that each of these coupling reactions proceeds with retention of configuration: the *trans/trans* configuration of 1 is retained in the product.

Bu_3Sn

1

MeO

Pd catalyst

then

CsF

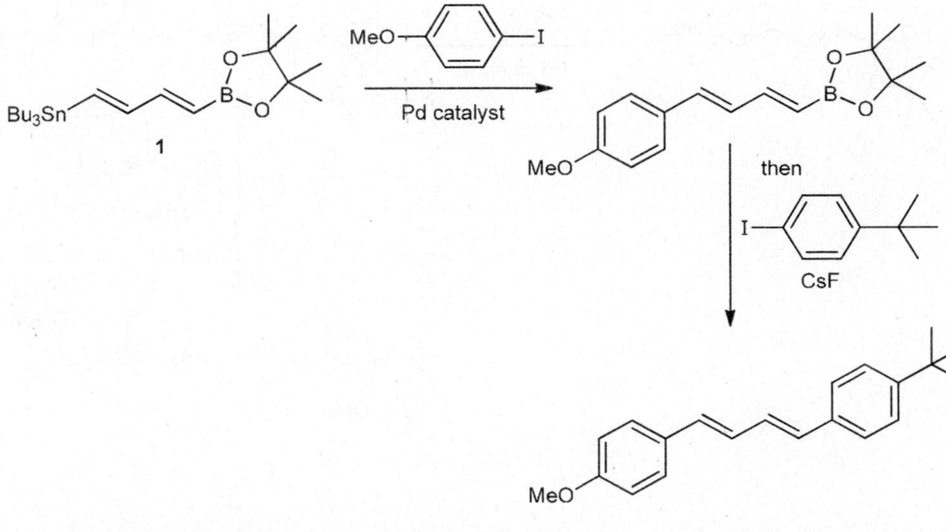

(b) An analysis of **2** reveals that it contains the *trans/trans* diene present in **1**, highlighted below.

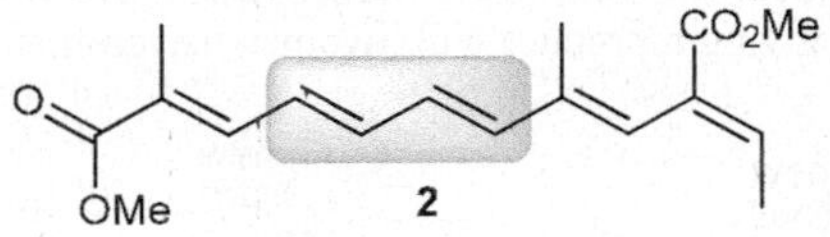

CO_2Me

OMe

2

One retrosynthetic analysis of **2** is presented below.

a. Disconnection of the indicated bond in **2** suggests that it can be produced from the vinyl iodide and the vinyl boronic ester shown. Note that the configuration of the C=C unit in each coupling partner is retained in the reaction.

b. This tetraene intermediate can be produced via Stille coupling between **1** and the vinyl iodide shown. Again, the configuration of each coupling partner is retained in the product.

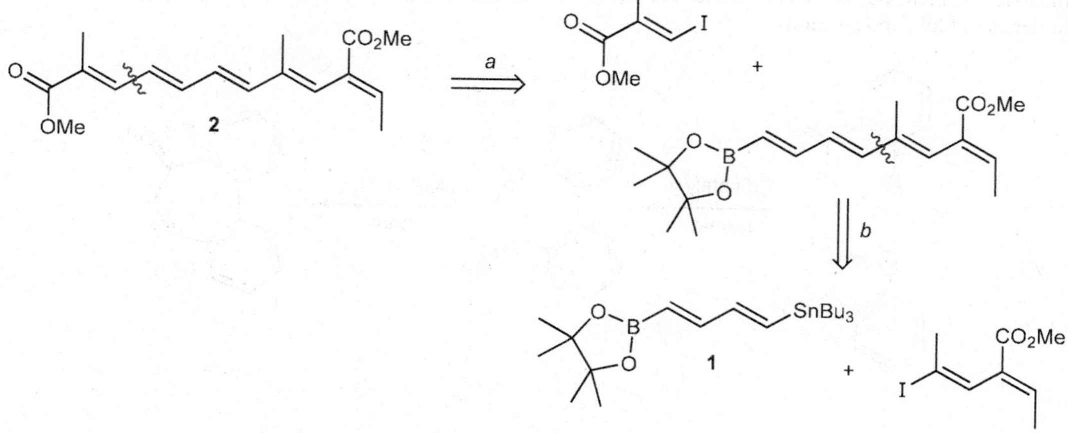

The forward scheme is presented below. Diene organostannane **1** reacts with the vinyl iodide in a palladium-catalyzed Stille reaction to afford the tetraene intermediate. Subsequent addition of the base CsF and a different vinyl iodide produces compound **2** via Suzuki coupling.

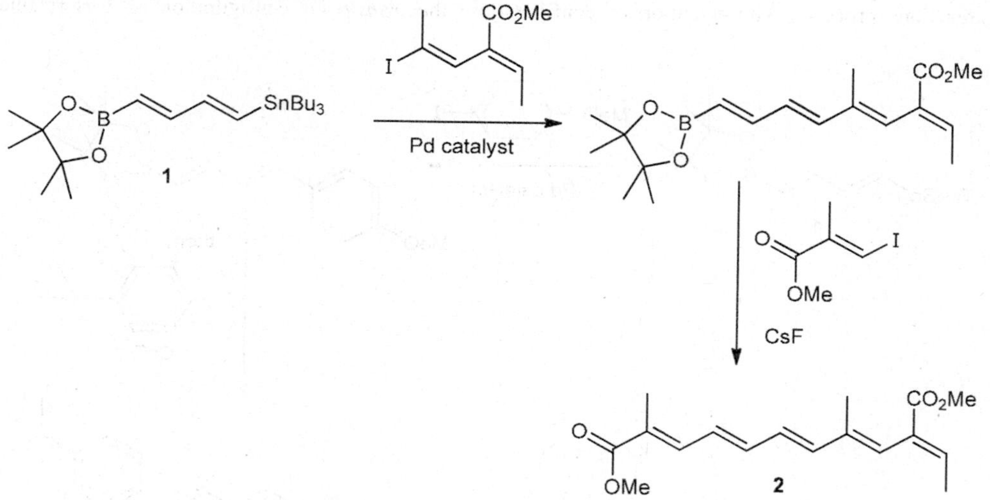

23.76. As described in the problem statement, olefin metathesis of this triene results in cleavage of the cyclic C=C unit and formation of two new C=C units, giving a product with two new seven-membered rings as part of a fused tricyclic system:

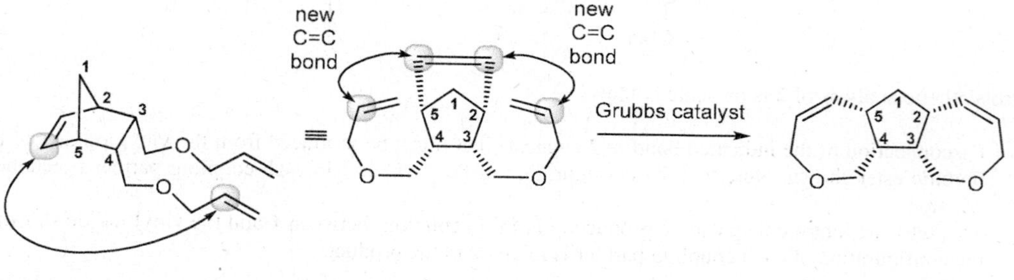

23.77. We first begin by taking inventory of the four building blocks shown. We are given a vinyl boronic acid, an aryl bromide, and two compounds which each contain both an aryl bromide and a MIDA boronate. The latter two compounds are designed to be used in iterative Suzuki coupling reactions. Next, consider the structure of ratanhine.

The following illustration shows how each of the starting materials corresponds with a portion of the natural product, and the squiggly lines represent the bonds that are formed via Suzuki coupling.

The forward reaction sequence is shown below. After the first Suzuki reaction, treatment with NaOH will reveal the boronic acid needed for the second Suzuki coupling. Likewise, after the second cross coupling, treatment with NaOH will reveal the final boronic acid. After the last Suzuki reaction, treatment with HCl cleaves the R protecting group, giving ratanhine.

23.78. We begin by identifying the carbon atoms in compound **1** that will be joined via an intramolecular Heck reaction, highlighted here:

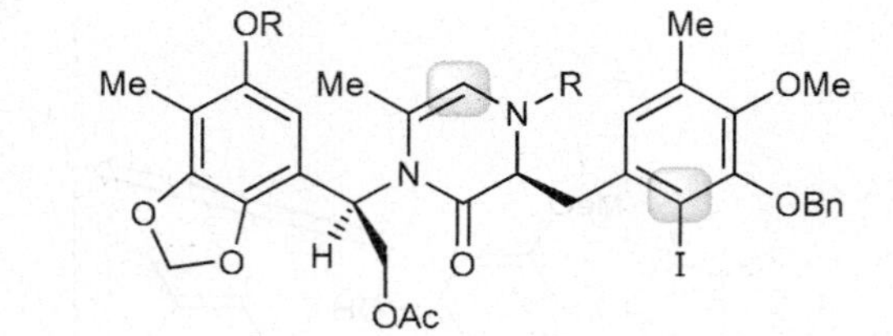

In order to draw the compound that is obtained when a new C–C bond is formed between these two locations, we must first redraw compound **1** so that these two centers are near each other. To do this, we first rotate the aryl group and redraw it in close proximity to the double bond. Due to the inherent stereochemistry already present in the molecule, the aryl group will be positioned above the plane of the alkene.

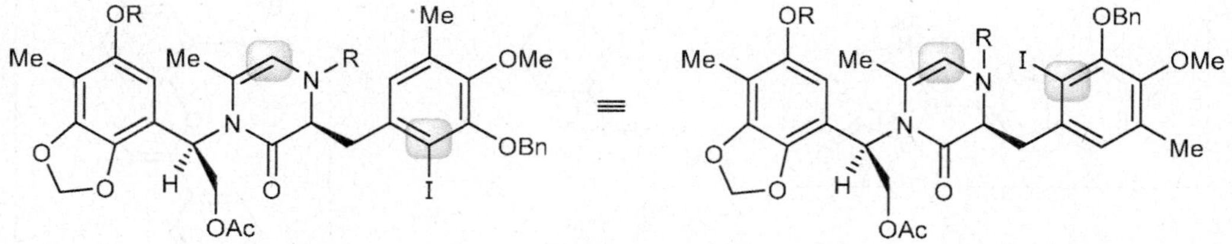

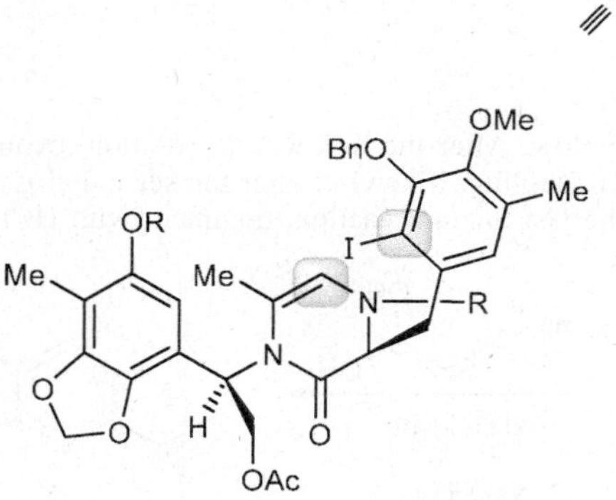

When the Pd catalyst inserts into the aryl iodide bond and then adds across the alkene in a *syn*-fashion, the new C-C bond is formed from the top face of the molecule, resulting in a *cis* junction for the newly formed 6-membered ring. The new C–Pd adduct (compound **B** below) cannot undergo a conformational change via C–C bond rotation. Therefore, the *syn*-reductive elimination can only occur at one possible site within the molecule with an accessible *syn* hydrogen atom *beta* to the Pd, resulting in compound **A**. The newly formed C=C bond is exocyclic to the 6-membered ring:

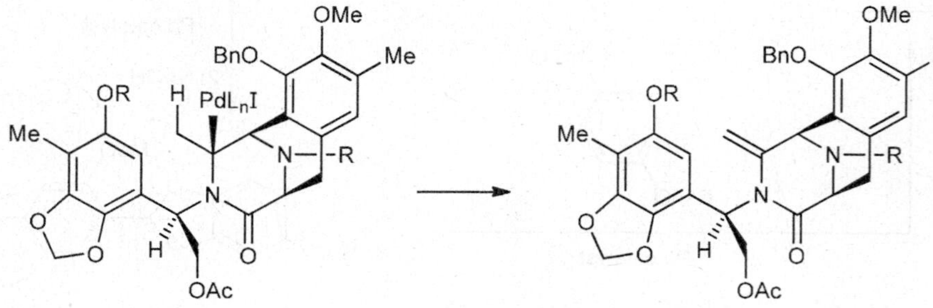

23.79.

(a) Compound **1** has three monosubstituted C=C units, giving rise to three possible ways in which ring-closing metathesis can occur:

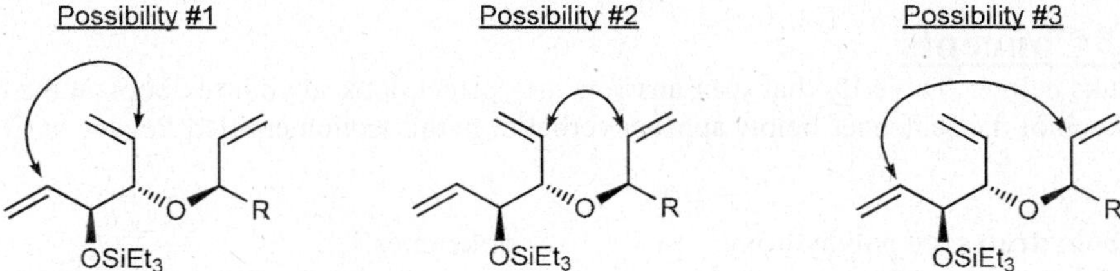

We can exclude the first possibility, because that would lead to a four-membered ring, and the problem statement indicates that a four-membered ring will not form when a five- or six-membered ring can form instead. Therefore, we consider only the latter two possibilities, which lead to a five-membered ring and a six-membered ring, respectively:

Possibility #2

ring-closing metathesis

Possibility #3

ring-closing metathesis

(b) When compound **2** is used instead, it reacts with the Grubbs catalyst to give intermediate **3**, which is then converted to intermediate **4**, as shown in the problem statement. Intermediate **4** will give rise to a five-membered cyclic ether. So this cyclic ether must be compound **A**.

ring-closing metathesis

4 **A**

The relay event is accompanied by the loss of compound **5** (via ring-closing metathesis) as shown:

3 **4** + **5**

Chapter 24
Carbohydrates

Review of Concepts

Fill in the blanks below. To verify that your answers are correct, look in your textbook at the end of Chapter 24. Each of the sentences below appears verbatim in the section entitled *Review of Concepts and Vocabulary*.

- **Carbohydrates** are polyhydroxy ___________ or ketones.
- Simple sugars are called _______________ and are generally classified as **aldoses** and ________.
- For all **D sugars**, the chiral center farthest from the carbonyl group has the ___ configuration.
- Aldohexoses can form cyclic hemi_______ that exhibit a **pyranose** ring.
- Cyclization produces two stereoisomeric hemiacetals, called ___________. The newly created chiral center is called the ___________ **carbon.**
- In the **α anomer**, the hydroxyl group at the anomeric position is ______ to the CH_2OH group, while in the **β anomer**, the hydroxyl group is ______ to the CH_2OH group.
- Anomers equilibrate by a process called ____________, which is catalyzed by either ______ or _______.
- Some carbohydrates, such as D-fructose, can also form five-membered rings, called ___________ rings.
- Monosaccharides are converted into their ester derivatives when treated with excess ____________________________.
- Monosaccharides are converted into their ether derivatives when treated with excess __________________ and silver oxide.
- When treated with an alcohol under acid-catalyzed conditions, monosaccharides are converted into acetals, called ___________. Both anomers are formed.
- Upon treatment with sodium borohydride an aldose or ketose can be reduced to yield an ________.
- When treated with a suitable oxidizing agent, an aldose can be oxidized to yield an _________.
- When treated with HNO_3, an aldose is oxidized to give a dicarboxylic acid called an _______________.
- D-Glucose and D-mannose are **epimers** and are interconverted under strongly __________ conditions.
- The **Kiliani-Fischer synthesis** can be used to lengthen the chain of an ________.
- The **Wohl degradation** can be used to shorten the chain of an ________.
- ________________ are comprised of two monosaccharide units, joined together via a glycosidic linkage.
- **Polysaccharides** are polymers consisting of repeating monosaccharide units linked by ___________ bonds.
- When treated with an ________ in the presence of an acid catalyst, monosaccharides are converted into their corresponding **N-glycosides**.

Review of Skills

Fill in the blanks and empty boxes below. To verify that your answers are correct, look in your textbook at the end of Chapter 24. The answers appear in the section entitled *SkillBuilder Review.*

24.1 Drawing the Cyclic Hemiacetal of a Hydroxyaldehyde

Draw the cyclic hemiacetal that is formed when the following hydroxyaldehyde undergoes cyclization under acidic conditions.

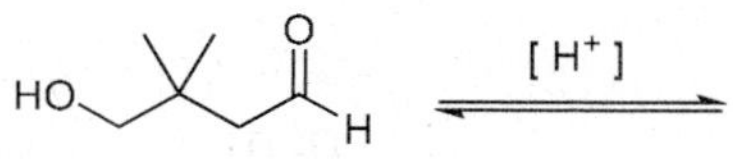

24.2: Drawing a Haworth Projection of an Aldohexose in the Pyranose Form

Draw a Haworth projection of α-D-galactopyranose.

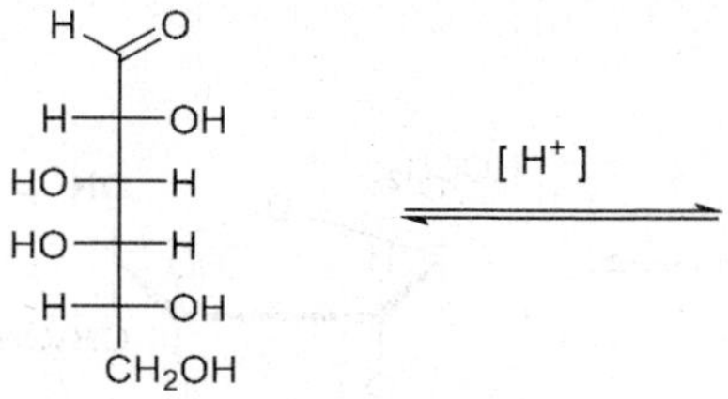

D-Galactose

24.3: Drawing the More Stable Chair Conformation of a Pyranose Ring

Draw the more stable chair conformation of α-D-galactopyranose.

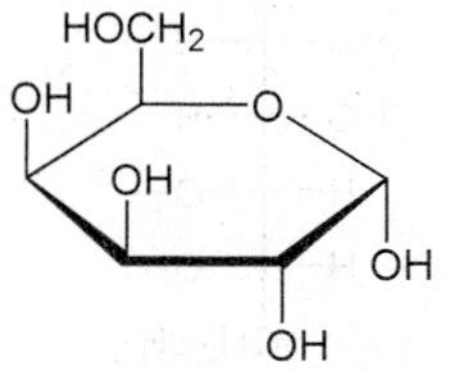

α-D-Galactopyranose

More stable chair conformation

24.4 Identifying a Reducing Sugar

Step 1 - Identify the ______________ position.

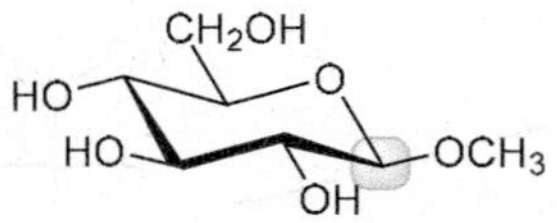

Step 2 - Determine if the group attached to the anomeric position is a __________ group or __________ group.

- If ____________ group, then the compound is a reducing sugar.
- If ____________ group, then the compound is not a reducing sugar.

24.5 Determining Whether a Disaccharide Is a Reducing Sugar

Step 1 - Identify the ______________ positions.

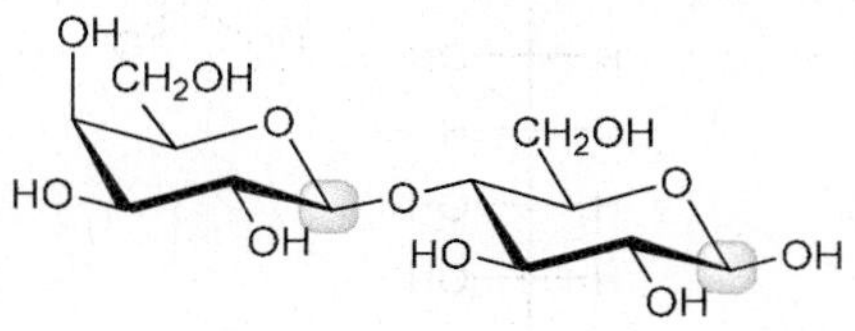

Step 2 - Determine if the groups attached to the anomeric positions are hydroxy groups or alkoxy groups.

- If one is a ____________ group, then the compound is a reducing sugar.
- If neither are ____________ groups, then the compound is not a reducing sugar.

Review of Reactions

Identify the reagents necessary to achieve each of the following transformations. To verify that your answers are correct, look in your textbook at the end of Chapter 24. The answers appear in the section entitled *Review of Reactions*.

Hemiacetal Formation

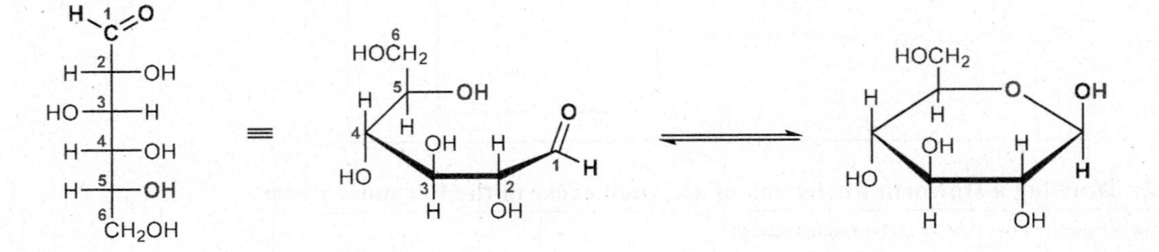

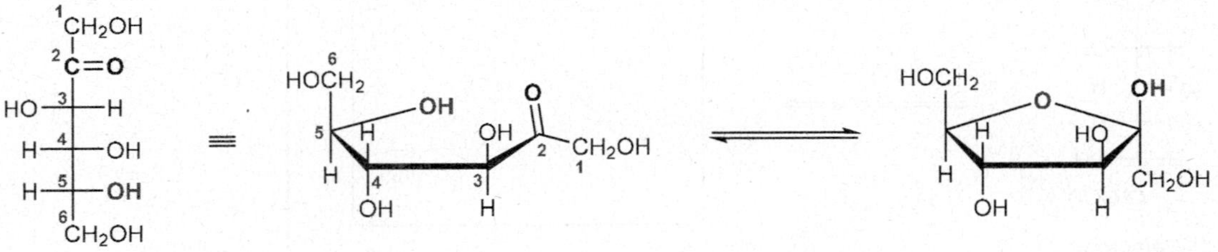

Chain Lengthening and Chain Shortening

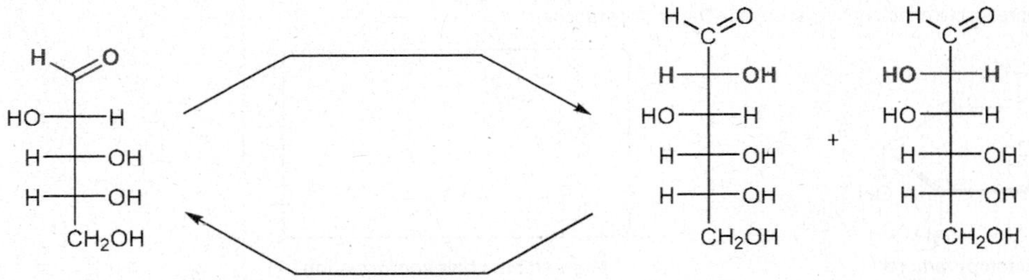

Reactions of Monosaccharides

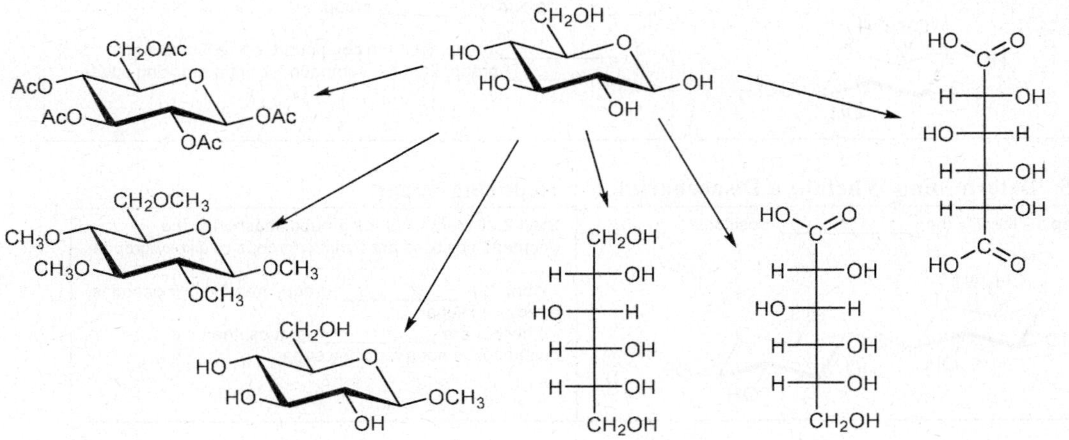

Solutions

24.1.
(a) This compound is an aldehyde and it is comprised of six carbon atoms, so it is an aldohexose.
(b) This compound is an aldehyde and it is comprised of five carbon atoms, so it is an aldopentose.
(c) This compound is a ketone and it is comprised of five carbon atoms, so it is a ketopentose.
(d) This compound is an aldehyde and it is comprised of four carbon atoms, so it is an aldotetrose.
(e) This compound is a ketone and it is comprised of six carbon atoms, so it is a ketohexose.

24.2. Both are hexoses so both have the molecular formula ($C_6H_{12}O_6$). Although they have the same molecular formula, they have different constitution – one is an aldehyde and the other is a ketone. Therefore, they are constitutional isomers.

24.3. All are D sugars except for (b), which is an L sugar. The configuration of each chiral center is shown here:
(a) 2*S*, 3*S*, 4*R*, 5*R*
(b) 2*R*, 3*S*, 4*S*
(c) 3*R*, 4*R*
(d) 2*S*, 3*R*
(e) 3*S*, 4*S*, 5*R*
Pay special attention to the following trend: The configuration of each chiral center is *R* when the OH group is on the right side of the Fischer projection (**R**ight = ***R***), and the configuration is *S* when the OH group is on the left side. This trend is observed because the OH group is always the #1 priority, the H atom is always the #4 priority, and the group including the top of the Fischer projection is always the #2 priority. However, these priorities may change for sugar derivatives, so configuration of chiral centers should be confirmed in those cases.

24.4.
(a) As seen in the solution to the previous problem, the *R* configuration is characterized by an OH group on the right side of the Fischer projection. D-Allose has all of the OH groups on the right side. As with all D sugars, the OH group on the chiral center farthest from the carbonyl group is on the right:

H–C=O
H–C–OH
H–C–OH
H–C–OH
H–C–OH
CH_2OH

(b) L-Allose is the enantiomer of D-allose, so all of the OH groups are on the *left* side of the Fischer projection. As with any L sugar, the highlighted OH group (the OH group connected to the chiral center that is farthest from the carbonyl group) is on the left:

H–C=O
HO–C–H
HO–C–H
HO–C–H
HO–C–H
CH_2OH

24.5. A ketotetrose has four carbon atoms and a ketone group, leaving only one chiral center, giving the following two enantiomers. The D and L identities are determined by the location of the OH group on the chiral center farthest from the carbonyl group (on the right for D sugars, and on the left for L sugars):

D: CH_2OH / C=O / H–C–OH / CH_2OH
L: CH_2OH / C=O / HO–C–H / CH_2OH
enantiomers

24.6. An aldotetraose has four carbon atoms and an aldehyde group. As such, there are two chiral centers, giving the following four stereoisomers. The D and L sugars are determined by the location of the OH group on the chiral center farthest from the carbonyl group (on the right for D sugars, and on the left for L sugars)

D: H–C=O / H–C–OH / H–C–OH / CH_2OH
L: H–C=O / HO–C–H / HO–C–H / CH_2OH
enantiomers

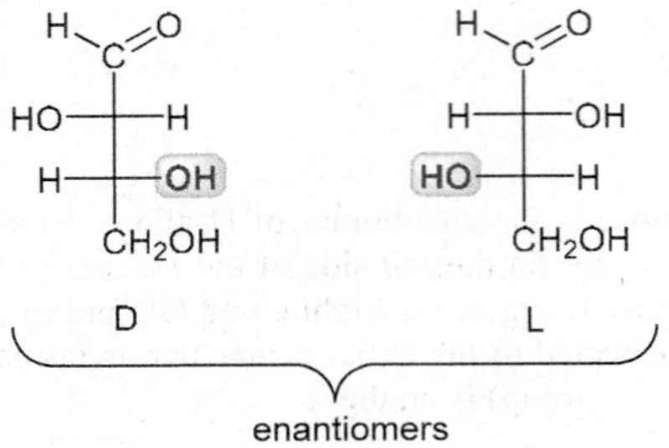

enantiomers

24.7. The enantiomer of D-fructose is L-fructose, in which all of the chiral centers have opposite configuration (as compared with D-fructose). As with all L sugars, the OH group on the chiral center farthest from the carbonyl group is on the left:

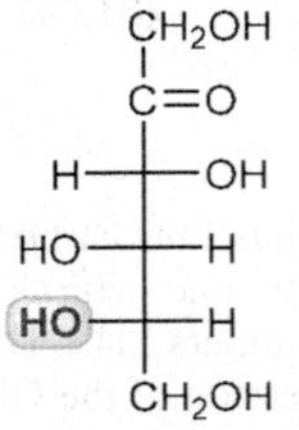

L-Fructose

24.8. Both D-fructose and D-glucose have the molecular formula ($C_6H_{12}O_6$). However, they have different constitution – one is a ketone, and the other is an aldehyde. Therefore, they are constitutional isomers.

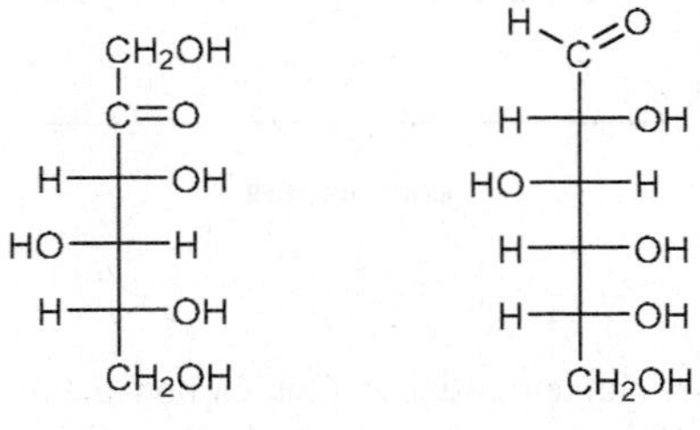

D-Fructose D-Glucose

24.9.

(a) We use a numbering system to determine the size of the ring that is formed. Four carbon atoms and one oxygen atom are incorporated into a five-membered ring. There are two methyl groups at C4, which must be drawn in the product:

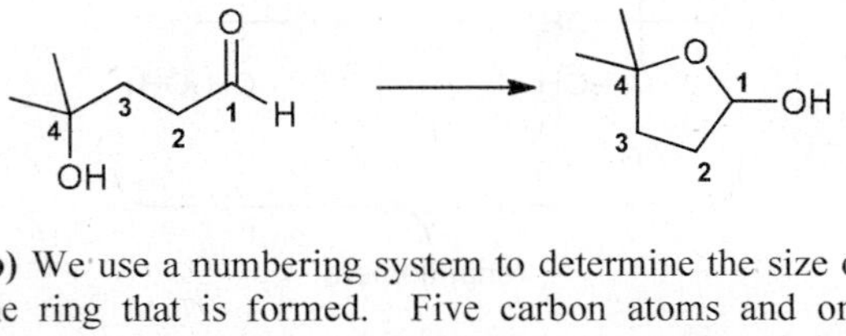

(b) We use a numbering system to determine the size of the ring that is formed. Five carbon atoms and one oxygen atom are incorporated into a six-membered ring, as shown. There are two methyl groups at C5, which must be drawn in the product:

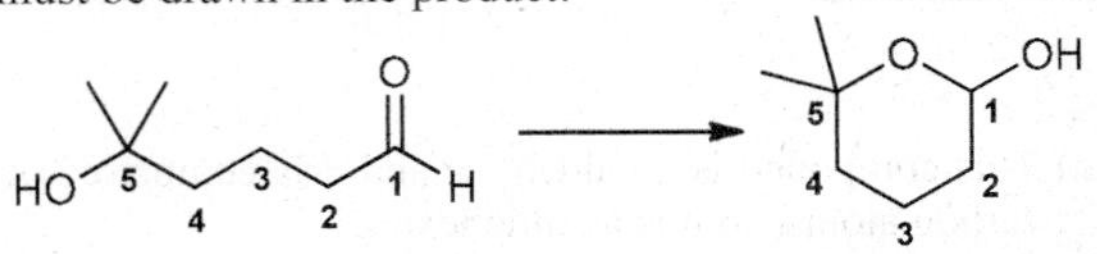

(c) We use a numbering system to determine the size of the ring that is formed. Four carbon atoms and one oxygen atom are incorporated into a five-membered ring, as shown. There are two methyl groups at C4 of the ring and one methyl group at C1 of the ring, which must be drawn in the product:

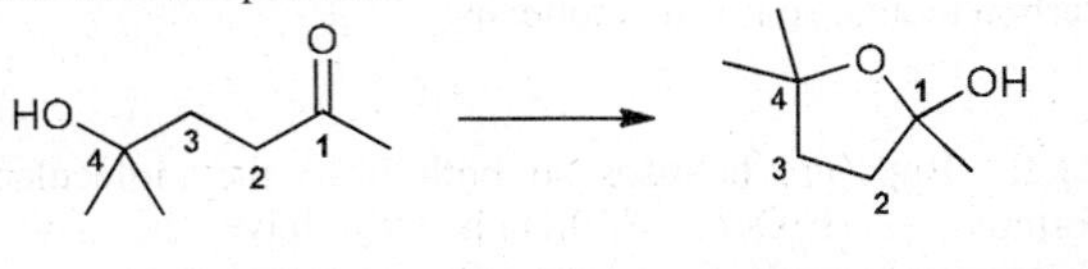

24.10. Use a numbering system, starting on the carbon atom that is bonded to the two oxygen atoms. Number in the direction to include the next carbon atom and continue around the ring. Open the ring by disconnecting the C–O bond within the ring. The carbon that was bound to the two oxygen atoms is a carbonyl group in the starting material. There are two methyl groups at C4, which must be drawn in the starting material:

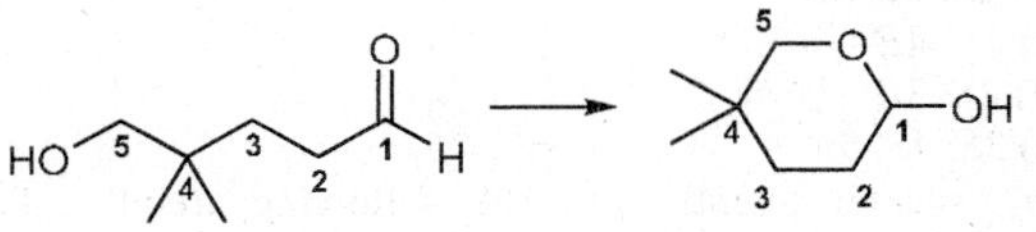

24.11.

(a) The carbonyl group can be attacked by the OH group that is connected to C4, giving a five-membered ring;

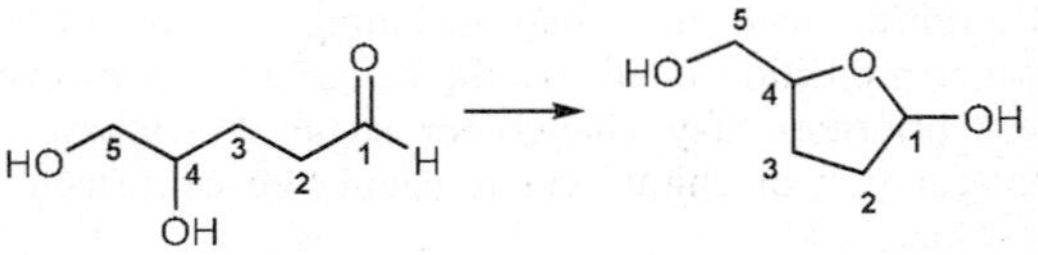

or the carbonyl group can be attacked by the OH group that is connected to C5, giving a six-membered ring:

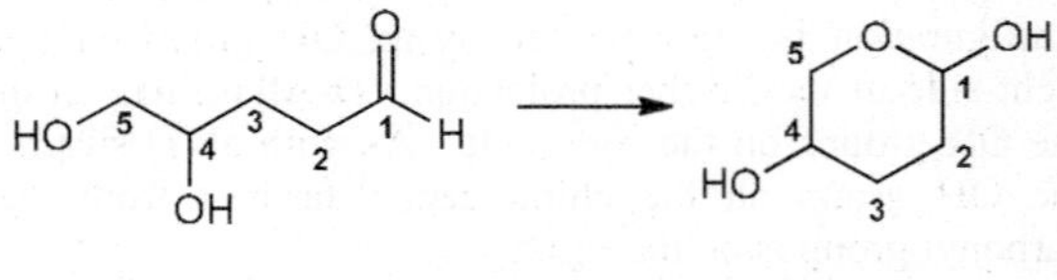

(b) The six-membered ring is expected to predominate because it has less ring strain than a five-membered ring.

24.12.

(a) We begin by drawing the skeleton of the Haworth projection. Then, we draw the CH_2OH group (at C6) pointing up. The anomeric OH group is then drawn pointing up (the β anomer has a *cis* relationship between the anomeric OH group and the CH_2OH group). Finally, the remaining groups are drawn. All OH groups on the right side of the Fischer projection will be pointing down in the Haworth projection (and all OH groups on the left side of the Fischer projection will be pointing up in the Haworth projection).

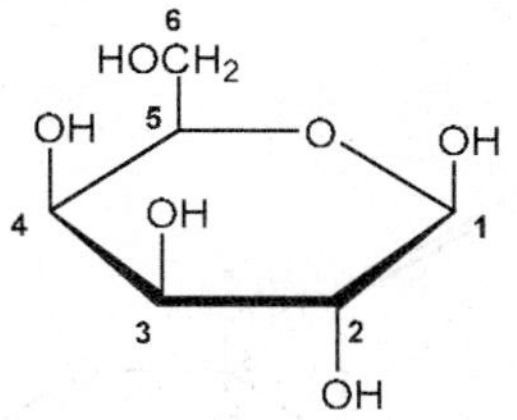

(b) We begin by drawing the skeleton of the Haworth projection. Then, we draw the CH_2OH group (at C6) pointing up. The anomeric OH group is then drawn pointing down (the α anomer has a *trans* relationship between the anomeric OH group and the CH_2OH group). Finally, the remaining groups are drawn. All OH groups on the right side of the Fischer projection will be pointing down in the Haworth projection (and all OH groups on the left side of the Fischer projection will be pointing up in the Haworth projection).

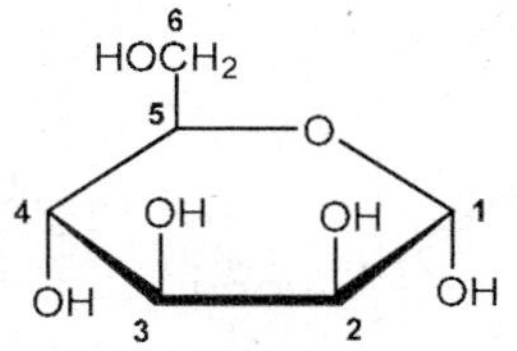

(c) We begin by drawing the skeleton of the Haworth projection. Then, we draw the CH_2OH group (at C6) pointing up. The anomeric OH group is then drawn pointing down (the α anomer has a *trans* relationship between the anomeric OH group and the CH_2OH group). Finally, the remaining groups are drawn. All OH groups on the right side of the Fischer projection will be pointing down in the Haworth projection (and all OH groups on the left side of the Fischer projection will be pointing up in the Haworth projection).

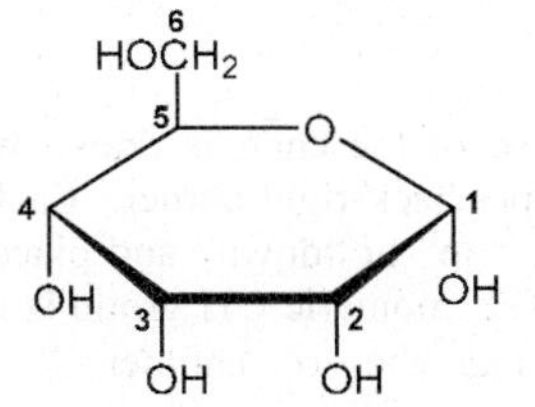

(d) We begin by drawing the skeleton of the Haworth projection. Then, we draw the CH_2OH group (at C6) pointing up. The anomeric OH group is then drawn pointing up (the β anomer has a *cis* relationship between the anomeric OH group and the CH_2OH group). Finally, the remaining groups are drawn. All OH groups on the right side of the Fischer projection will be pointing down in the Haworth projection (and all OH groups on the left side of the Fischer projection will be pointing up in the Haworth projection).

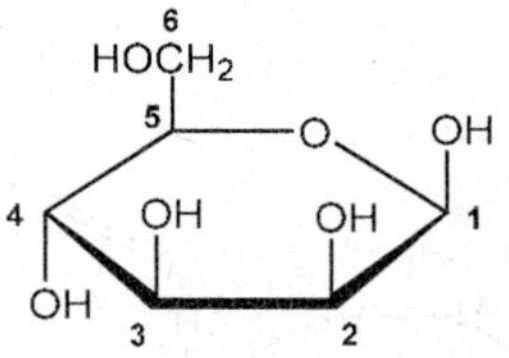

(e) We begin by drawing the skeleton of the Haworth projection. Then, we draw the CH_2OH group (at C6) pointing up. The anomeric OH group is then drawn pointing up (the β anomer has a *cis* relationship between the anomeric OH group and the CH_2OH group). Finally, the remaining groups are drawn. All OH groups on the right side of the Fischer projection will be pointing down in the Haworth projection (and all OH groups on the left side of the Fischer projection will be pointing up in the Haworth projection).

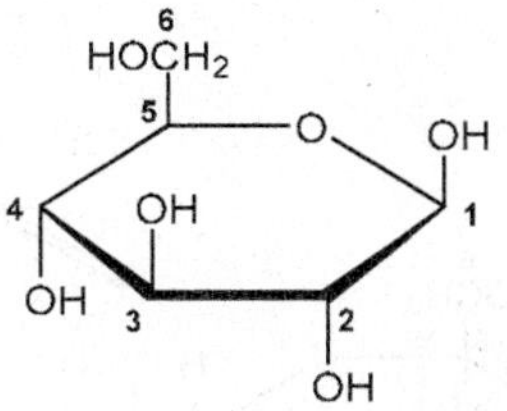

(f) We begin by drawing the skeleton of the Haworth projection. Then, we draw the CH_2OH group (at C6) pointing up. The anomeric OH group is then drawn pointing down (the α anomer has a *trans* relationship between the anomeric OH group and the CH_2OH group). Finally, the remaining groups are drawn. All OH groups on the right side of the Fischer projection will be pointing down in the Haworth projection (and all OH groups on the left side of the Fischer projection will be pointing up in the Haworth projection).

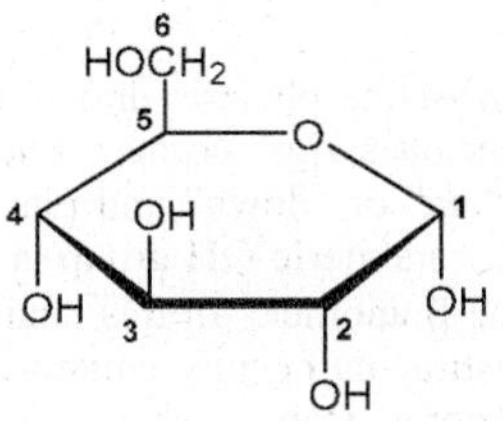

24.13. This structure represents the β anomer of the cyclic form of D-galactose (as a six-membered ring) and is therefore called β-D-galactopyranose.

24.14. The two pyranose forms are in equilibrium with each other, via the open-chain form, as shown:

α-D-Mannopyranose ⇌ D-Mannose ⇌ β-D-Mannopyranose

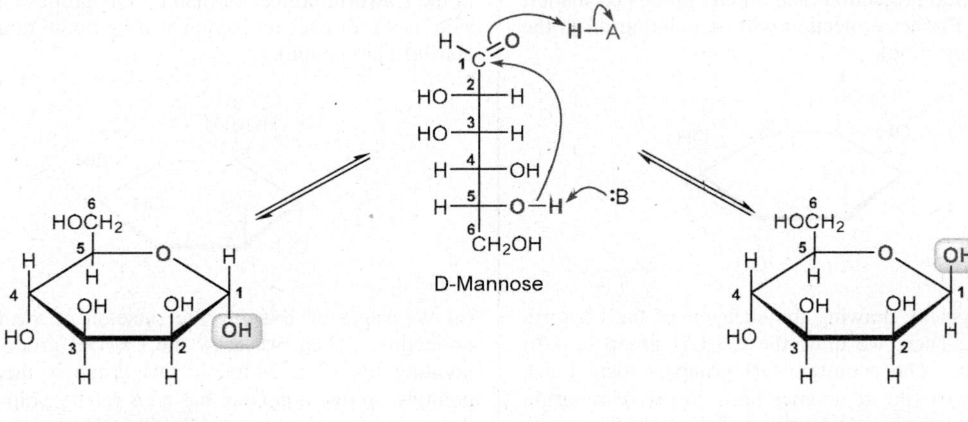

24.15. The two pyranose forms are in equilibrium with each other, via the open-chain form, as shown:

α-D-Talopyranose ⇌ D-Talose ⇌ β-D-Talopyranose

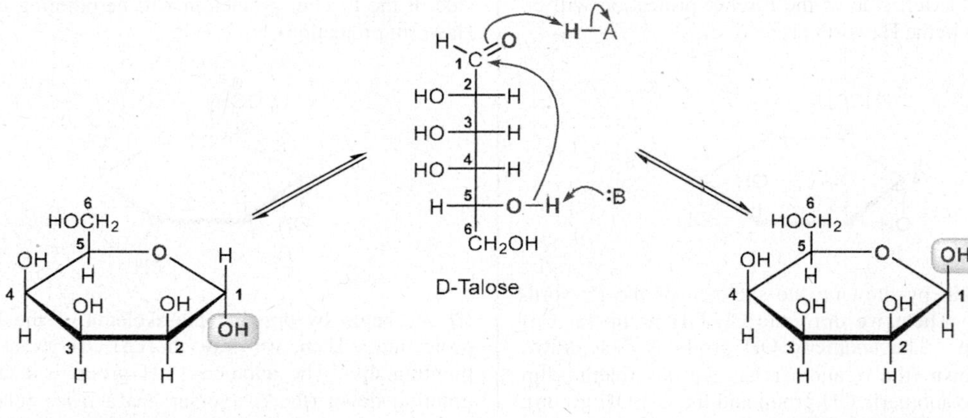

24.16.

(a) The skeleton of the chair is drawn with an oxygen atom in the upper back-right corner. Each substituent is then labeled as "up" or "down" and placed on the chair accordingly. The anomeric OH group is drawn pointing up, indicating the β anomer. In this chair conformation, most of the substituents occupy equatorial positions, so this is the more stable chair conformation of β-D-galactopyranose.

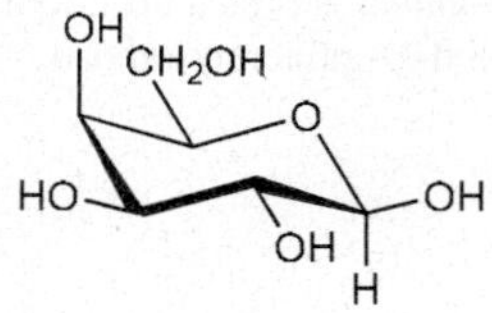

(b) The skeleton of the chair is drawn with an oxygen atom in the upper back-right corner. Each substituent is then labeled as "up" or "down" and placed on the chair accordingly. The anomeric OH group is drawn pointing down, indicating the α anomer. In this chair conformation, most of the substituents occupy equatorial positions, so this is the more stable chair conformation of α-D-glucopyranose.

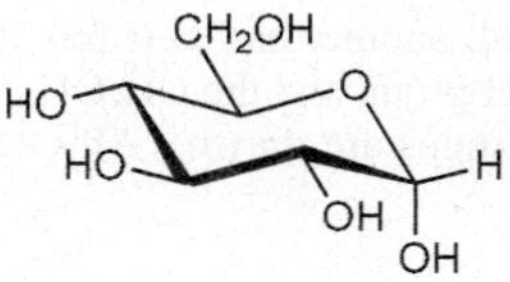

(c) The skeleton of the chair is drawn with an oxygen atom in the upper back-right corner. Each substituent is then labeled as "up" or "down" and placed on the chair accordingly. The anomeric OH group is drawn pointing up, indicating the β anomer. In this chair conformation, most of the substituents occupy equatorial positions, so this is the more stable chair conformation of β-D-glucopyranose.

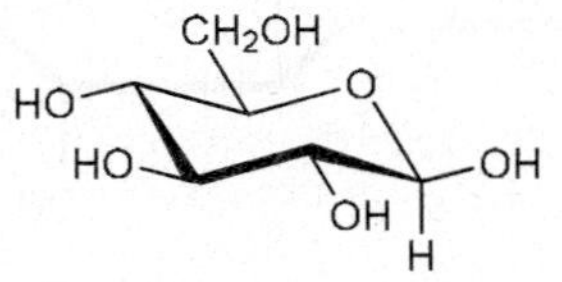

24.17. The anomeric position becomes an aldehyde group in the open-chain form. The OH group at C5 (of the open-chain form) must be pointing to the right, because this is a D sugar (CH_2OH is "up" in the chair conformation). The remaining three OH groups (at C2, C3 and C4) are all pointing "down," so they must be on the right side of the Fischer projection of the open-chain form. This structure represents D-allose.

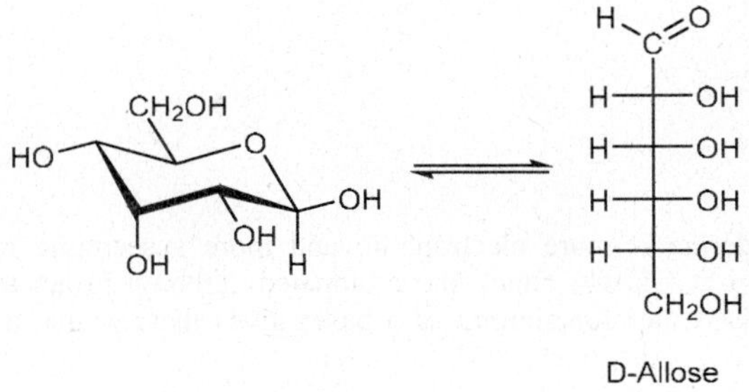

D-Allose

24.18. Both chair conformations of β-D-glucopyranose are shown below. The less stable conformation is the one in which all substituents occupy axial positions.

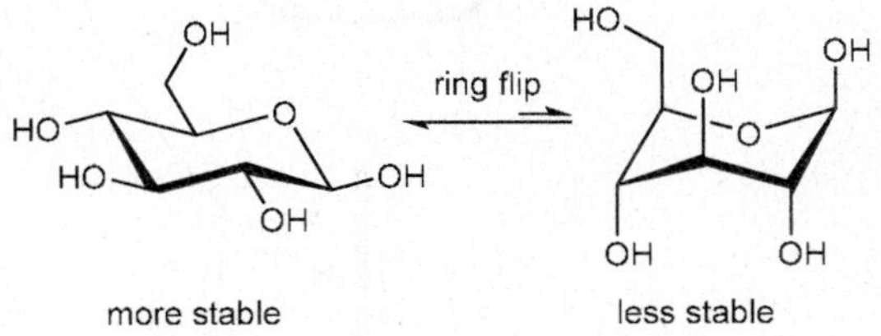

ring flip

more stable

less stable

24.19.

(a) The anomeric OH group is drawn pointing down (thus, the α anomer). The remaining two OH groups (at C2 and C3) are both on the right side of the Fischer projection of D-erythrose, so both OH groups are both drawn pointing down in the Haworth projection.

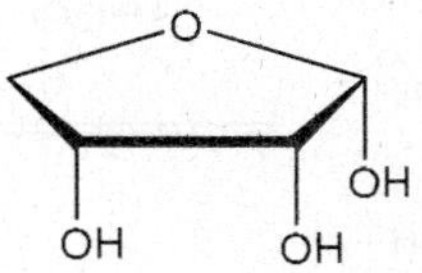

(b) The anomeric OH group is drawn pointing up (thus, the β anomer). The remaining two OH groups (at C2 and C3) are both on the right side of the Fischer projection of D-erythrose, so both OH groups are both drawn pointing down in the Haworth projection.

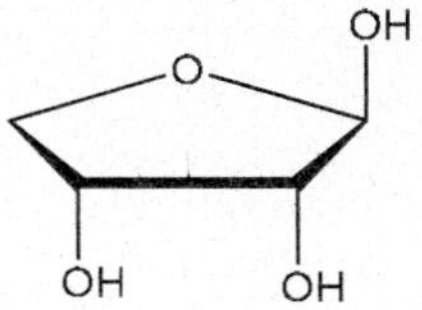

(c) The anomeric OH group is drawn pointing down (thus, the α anomer). The OH group on the left side of the Fischer projection (at C2 of D-threose) is drawn pointing up in the Haworth projection (at C2). Similarly, the OH group on the right side of the Fischer projection (at C3 of D-threose) is drawn pointing down in the Haworth projection (at C3).

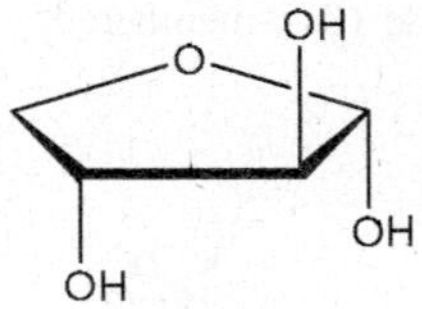

(d) The anomeric OH group is drawn pointing up (thus, the β anomer). The OH group on the left side of the Fischer projection (at C2 of D-threose) is drawn pointing up in the Haworth projection (at C2). Similarly, the OH group on the right side of the Fischer projection (at C3 of D-threose) is drawn pointing down in the Haworth projection (at C3).

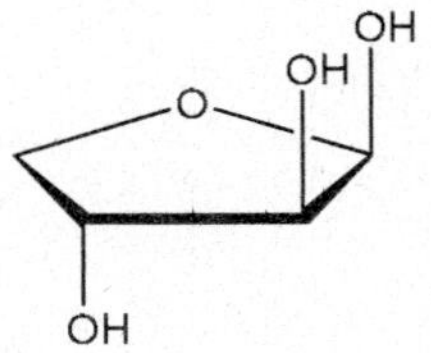

24.20. The carbonyl group is first protonated, thereby rendering it more electrophilic and more susceptible to nucleophilic attack by one of the OH groups. The OH group at C4 will attack the protonated carbonyl group to generate a furanose (five-membered) ring. Deprotonation (with water functioning as a base) gives the product, as shown.

L-Threose

β-L-Threofuranose

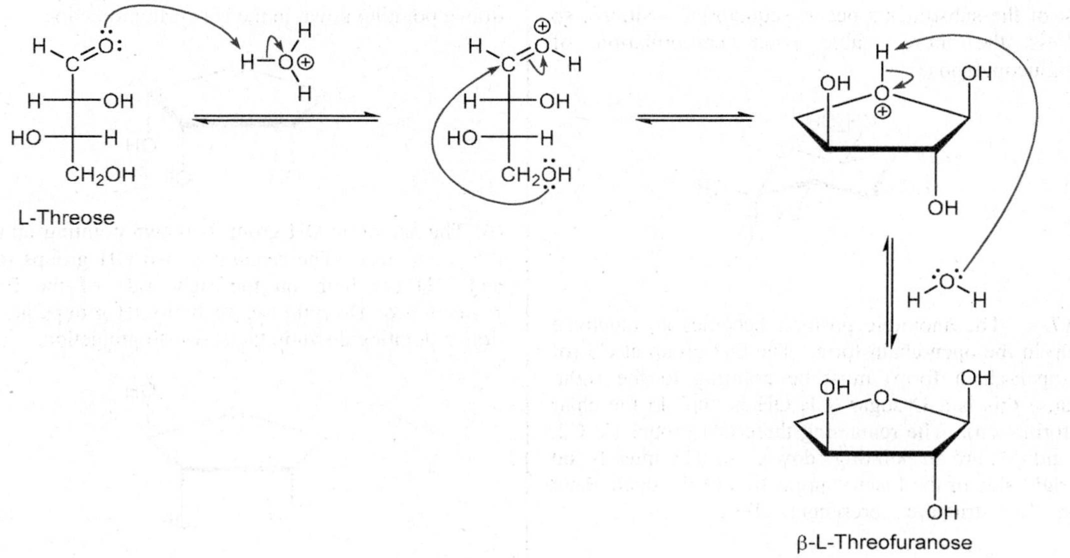

24.21. The carbonyl group is first protonated, thereby rendering it more electrophilic and more susceptible to nucleophilic attack by one of the OH groups. The OH group at C5 will attack the protonated carbonyl group to generate a furanose (five-membered) ring. Deprotonation (with water functioning as a base) gives the product, as shown.

D-Fructose

β-D-Fructofuranose

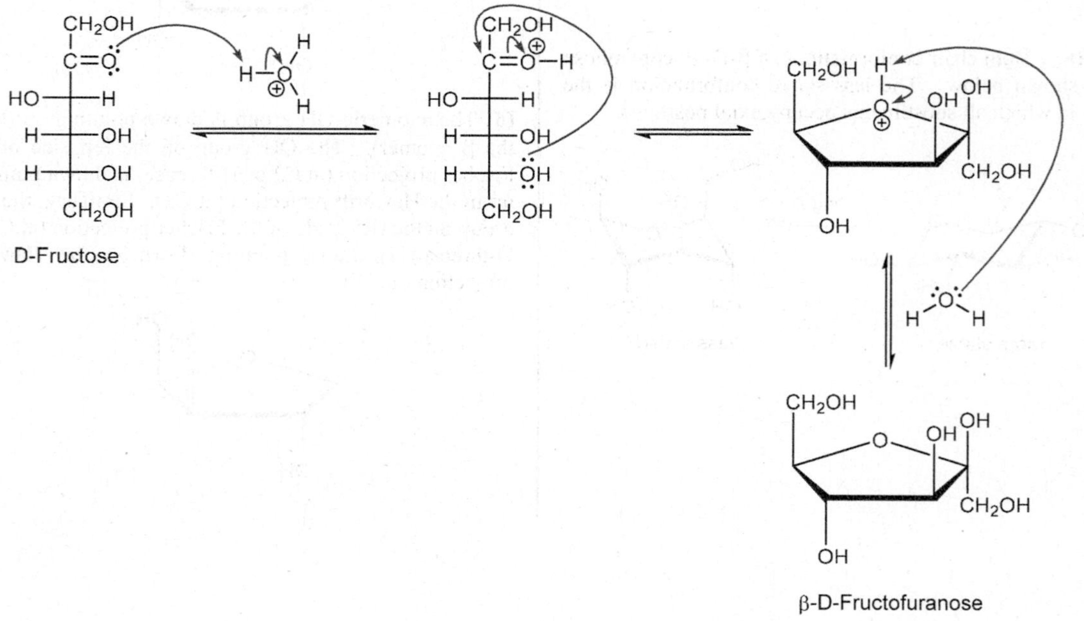

24.22. The name (α-D-fructopyranose) indicates that the open chain form is D-fructose, shown here:

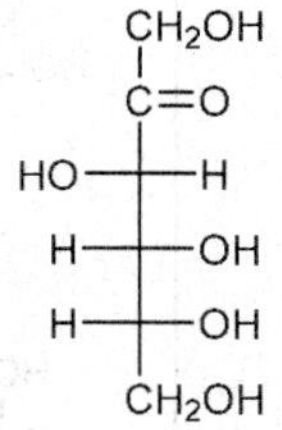

D-Fructose

24.23.

(a) Upon treatment with excess acetic anhydride and pyridine, all of the OH groups undergo acetylation, as shown:

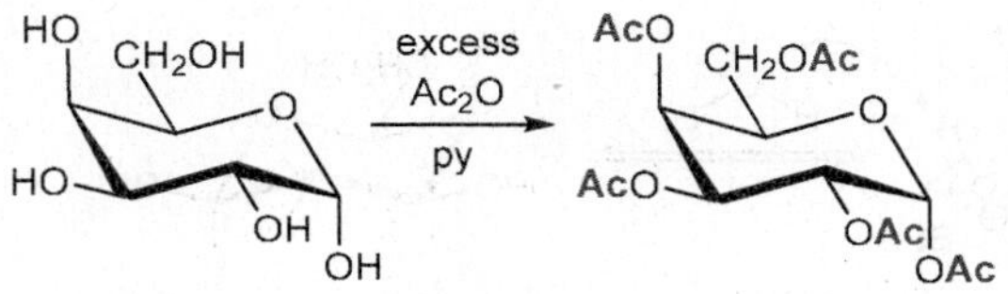

(b) Upon treatment with excess acetic anhydride and pyridine, all of the OH groups undergo acetylation, as shown:

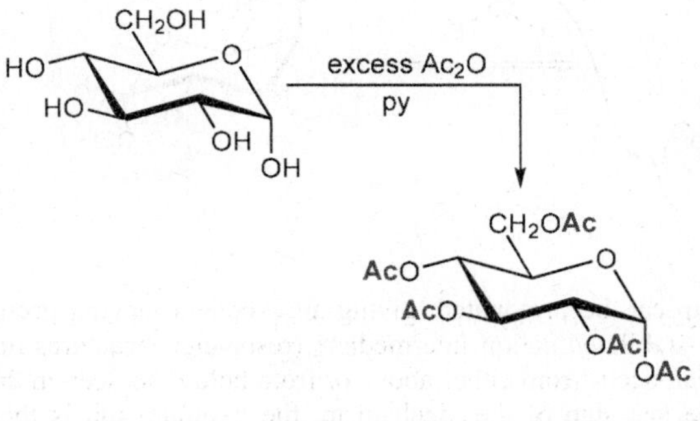

(c) Upon treatment with excess acetic anhydride and pyridine, all of the OH groups undergo acetylation, as shown:

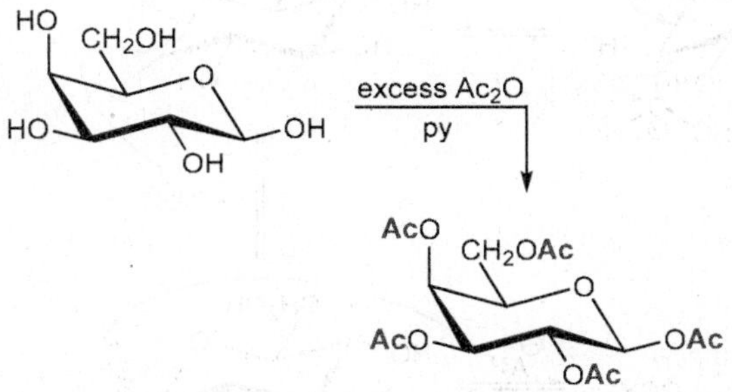

24.24.

(a) Upon treatment with excess methyl iodide in the presence of silver oxide, all of the OH groups are converted into methoxy groups, as shown:

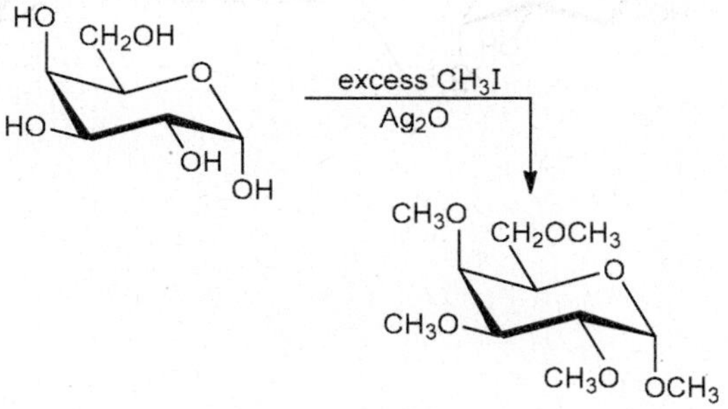

(b) Upon treatment with excess methyl iodide in the presence of silver oxide, all of the OH groups are converted into methoxy groups, as shown:

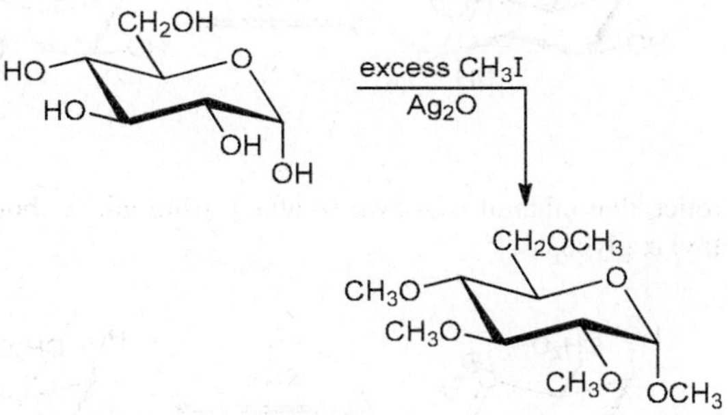

(c) Upon treatment with excess methyl iodide in the presence of silver oxide, all of the OH groups are converted into methoxy groups, as shown:

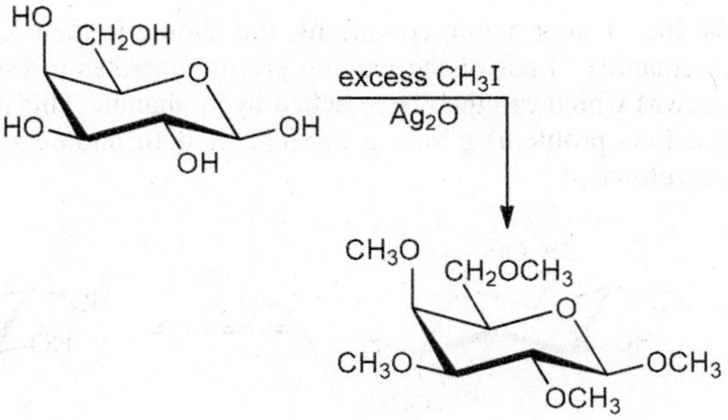

24.25. Under acidic conditions, the anomeric OH group can be protonated, giving an excellent leaving group (water). Loss of the leaving group generates a resonance-stabilized cation intermediate.

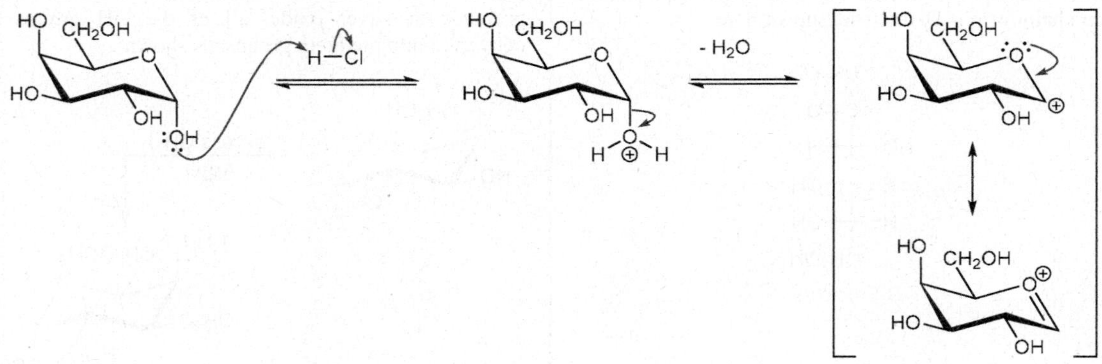

This intermediate can then be attacked by ethanol, giving an oxonium ion, which is then deprotonated to give an acetal (an ethyl β glycoside):

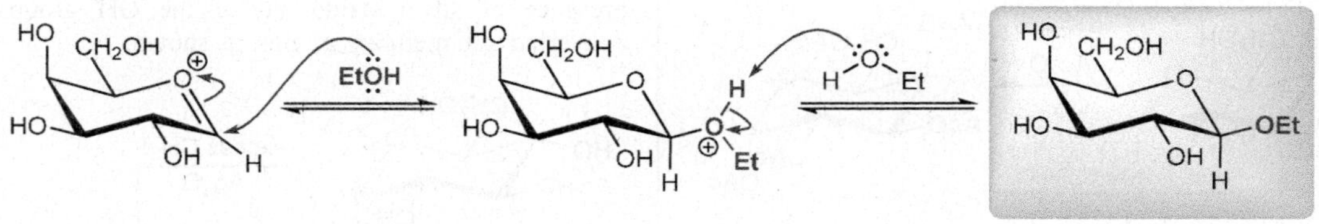

Notice that ethanol is shown to attack from above, but it can also attack from below, giving the following acetal (an ethyl α glycoside):

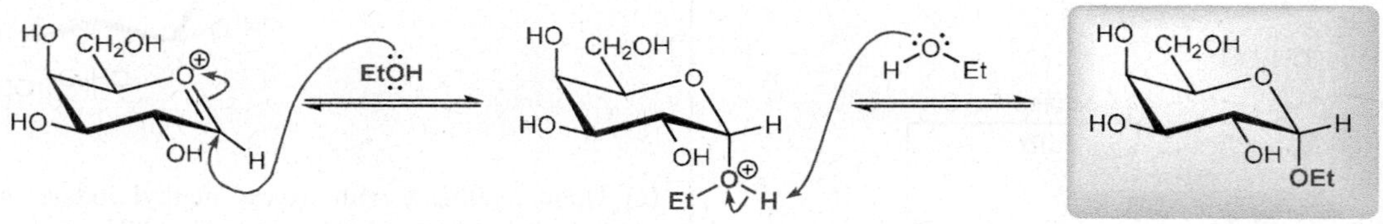

24.26. Under acidic conditions, the anomeric methoxy group can be protonated, giving an excellent leaving group (methanol). Loss of the leaving group generates a resonance-stabilized cation intermediate (resonance structures not shown) which can then be attacked by methanol. This attack can occur from either above or from below (as seen in the previous problem) giving a mixture of both anomers. In the last step of the mechanism, the oxonium ion is then deprotonated:

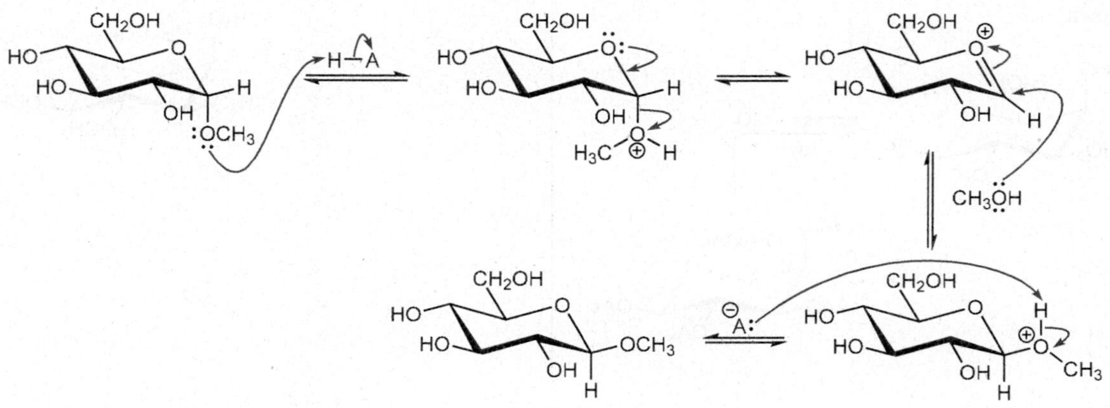

24.27.

(a) D-Mannose is epimeric with D-glucose at C2, as shown:

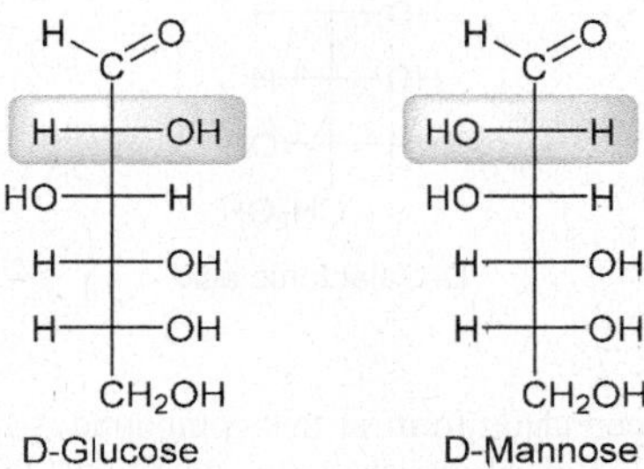

(b) D-Allose is epimeric with D-glucose at C3, as shown:

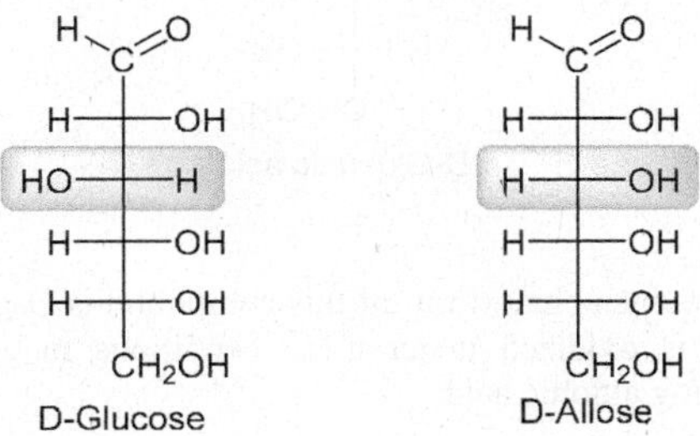

(c) D-Galactose is epimeric with D-glucose at C4, as shown:

D-Glucose D-Galactose

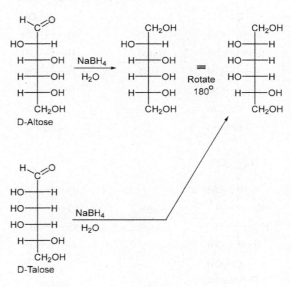

24.28. Reduction of the carbonyl group generates the same product in each case. This can be seen by rotating one of the products by 180°, as shown:

D-Altose $\xrightarrow[H_2O]{NaBH_4}$ ≡ Rotate 180°

D-Talose $\xrightarrow[H_2O]{NaBH_4}$

24.29. Reduction of the carbonyl group generates the same *meso* product in both cases. This can be seen by rotating one of the products by 180°, as shown:

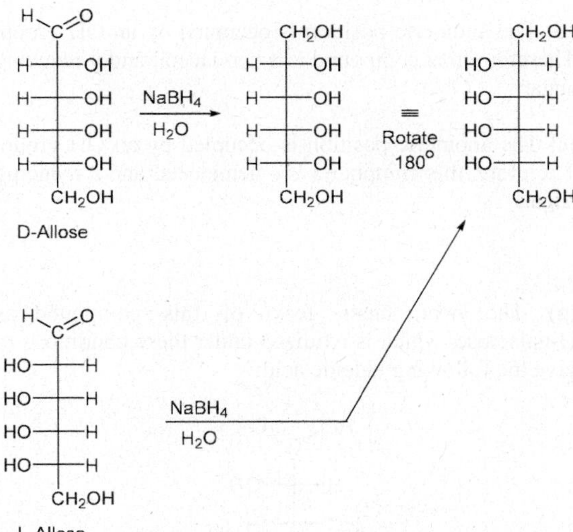

24.30. Reduction of either D-allose or D-galactose will produce a *meso* alditol. *Meso* compounds are optically inactive:

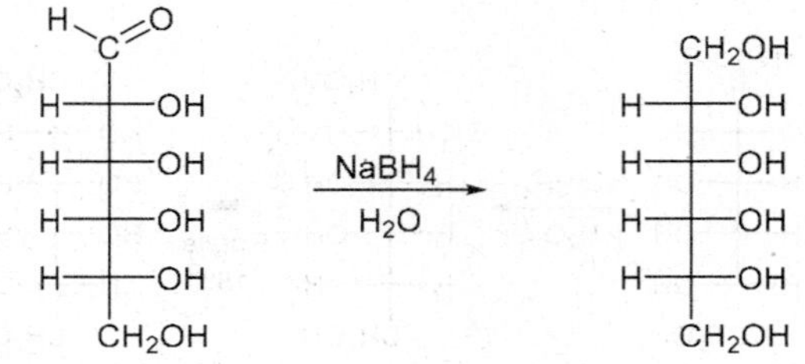

D-Allose

$NaBH_4$, H_2O

(*meso*)

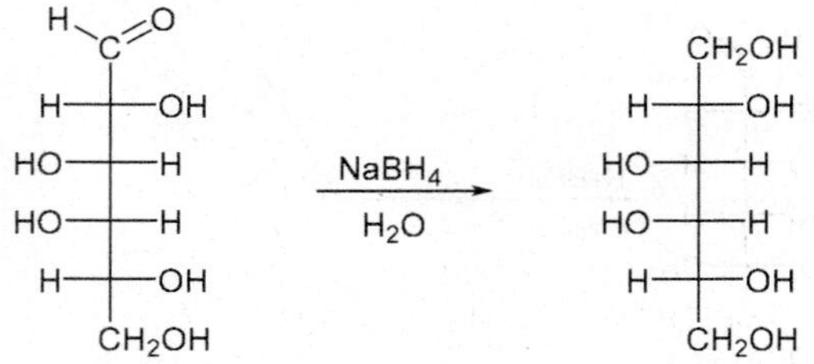

D-Galactose

$NaBH_4$, H_2O

(*meso*)

24.31.

(a) The anomeric position is occupied by a methoxy group. Therefore, this compound is an acetal and is not a reducing sugar.

(b) The anomeric position is occupied by an OH group. Therefore, this compound is a hemiacetal and a reducing sugar.

(c) The anomeric position is occupied by an OH group. Therefore, this compound is a hemiacetal and a reducing sugar.

24.32.

(a) The open chain form of this compound is D-galactose, which is oxidized under these conditions to give the following aldonic acid:

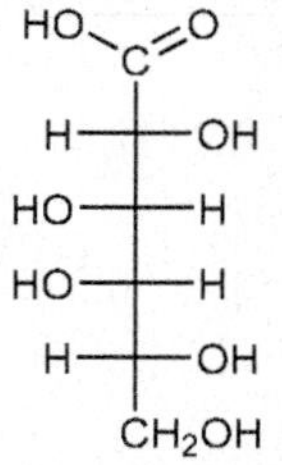

D-Galactonic acid

(b) The open chain form of this compound is D-galactose, which is oxidized under these conditions to give the following aldonic acid:

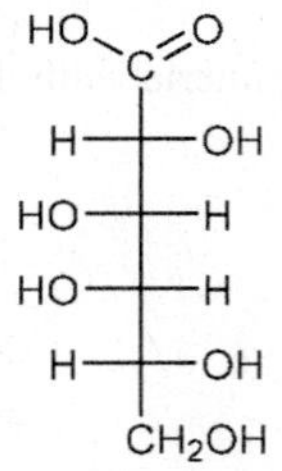

D-Galactonic acid

(c) The open chain form of this compound is D-glucose, which is oxidized under these conditions to give the following aldonic acid:

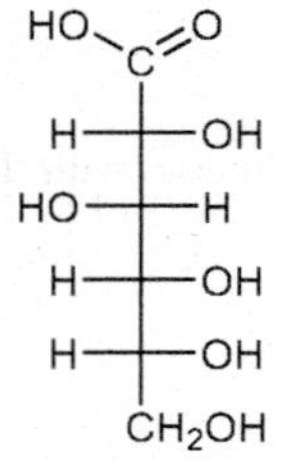

D-Gluconic acid

(d) The open chain form of this compound is D-glucose, which is oxidized under these conditions to give the following aldonic acid:

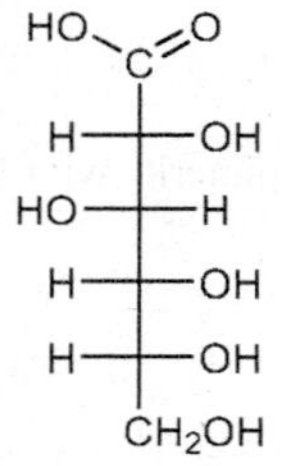

D-Gluconic acid

24.33. This compound will not be a reducing sugar because the anomeric position is an acetal group.

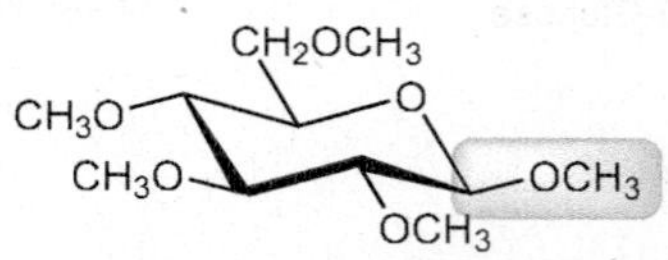

β-D-Glucopyranose pentamethyl ether

24.34.

(a) In a Kiliani-Fischer synthesis, the chain is lengthened, with C1 becoming C2 in the product. Both possible configurations of the C2 position are obtained, giving the following epimers:

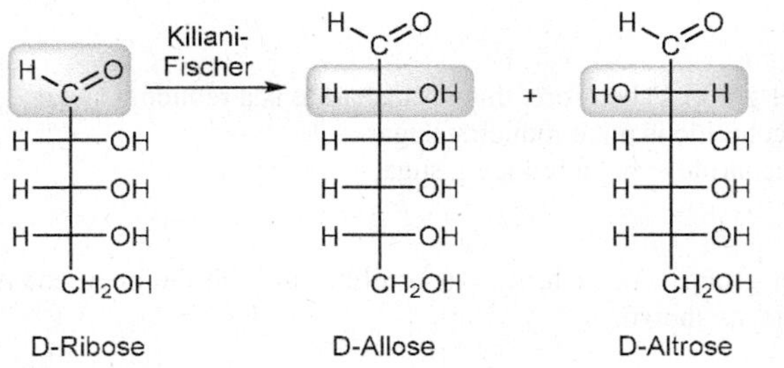

Kiliani-Fischer

D-Ribose D-Allose D-Altrose

(b) In a Kiliani-Fischer synthesis, the chain is lengthened, with C1 becoming C2 in the product. Both possible configurations of the C2 position are obtained, giving the following epimers:

Kiliani-Fischer

D-Xylose D-Gulose D-Idose

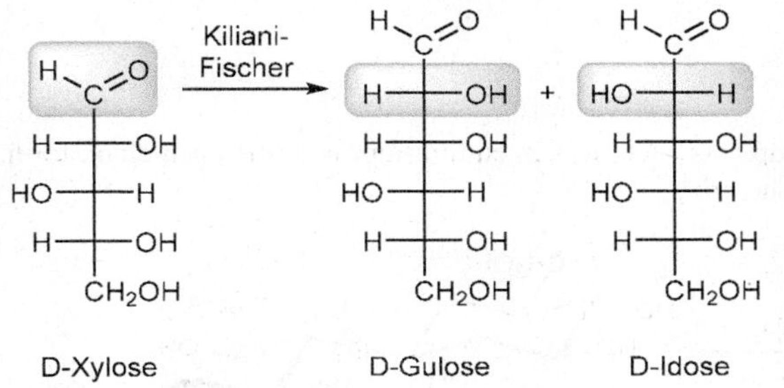

(c) In a Kiliani-Fischer synthesis, the chain is lengthened, with C1 becoming C2 in the product. Both possible configurations of the C2 position are obtained, giving the following epimers:

Kiliani-Fischer

D-Lyxose D-Galactose D-Talose

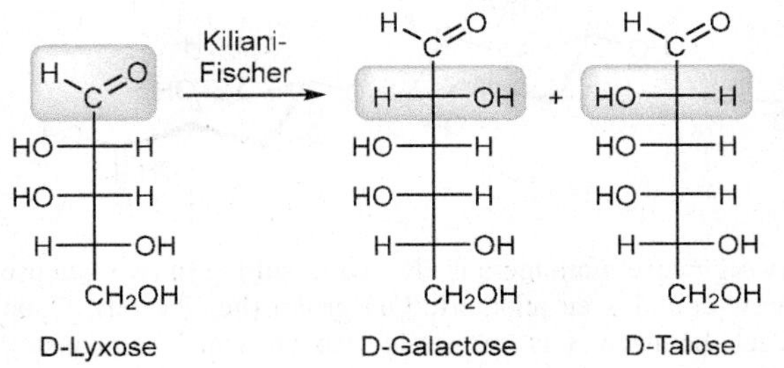

24.35. Conversion of D-erythrose (which has four carbon atoms) to D-ribose (which has five carbon atoms) requires a chain-lengthening process. This process will produce D-ribose together with its C2 epimer, D-arabinose, as shown:

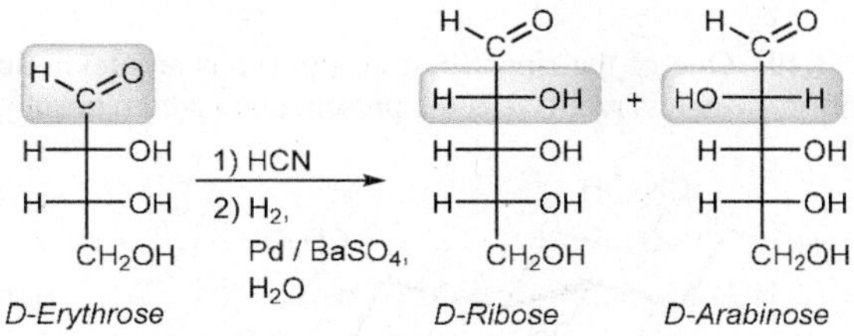

1) HCN
2) H_2, Pd / $BaSO_4$, H_2O

D-Erythrose *D-Ribose* *D-Arabinose*

24.36. A Wohl degradation involves the removal of a carbon atom from an aldose. As shown below, D-ribose can be made from either D-allose or D-altrose:

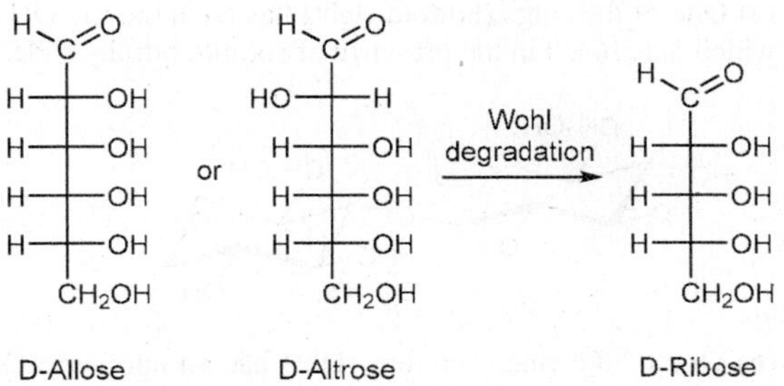

Wohl degradation

D-Allose or D-Altrose D-Ribose

24.37. A Wohl degradation will remove a carbon atom from an aldose. This method can be used to convert D-ribose into D-erythrose, as shown:

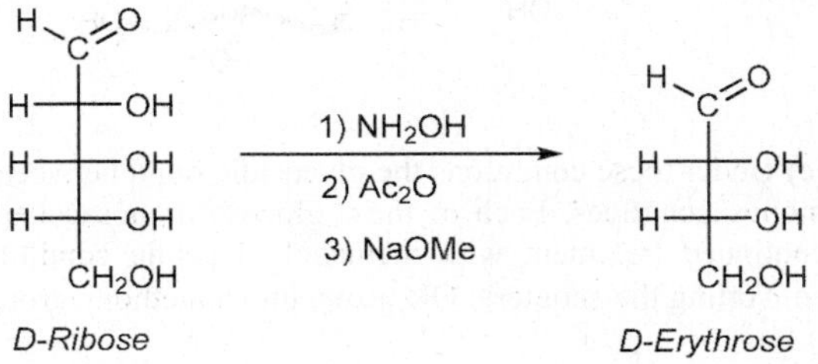

1) NH_2OH
2) Ac_2O
3) NaOMe

D-Ribose *D-Erythrose*

24.38. A Wohl degradation will remove a carbon atom from D-glucose. This carbon atom is then restored with a Kiliani-Fischer synthesis, giving D-glucose and its C2 epimer, D-mannose.

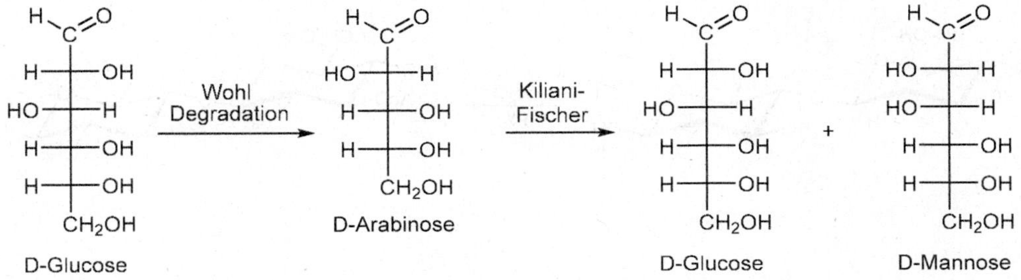

Wohl Degradation Kiliani-Fischer

D-Glucose D-Arabinose D-Glucose D-Mannose

24.39.
(a) One of the anomeric positions (bottom right) bears an OH group. Therefore, this disaccharide is a reducing sugar.
(b) Both anomeric positions bear alkoxy groups, so this disaccharide is not a reducing sugar.
(c) Both anomeric positions bear alkoxy groups, so this disaccharide is not a reducing sugar.

24.40. One of the rings (bottom right) has an anomeric OH group. As such, it is in equilibrium with the open chain form, which is reduced in the presence of sodium borohydride, as shown:

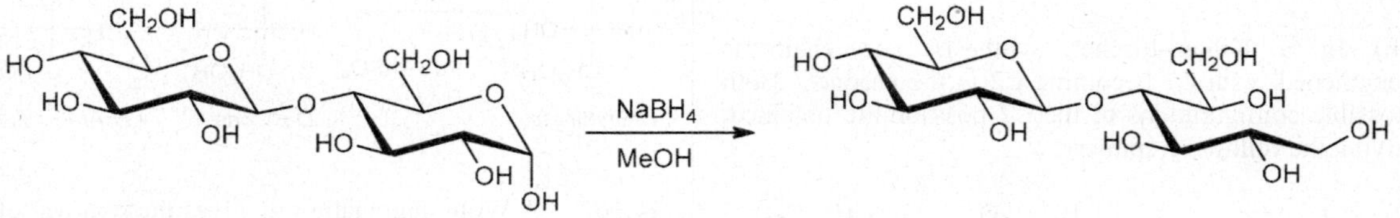

24.41.
(a) One of the rings (bottom right) has an anomeric OH group. As such, it is in equilibrium with the open chain form, which is reduced in the presence of sodium borohydride, as shown:

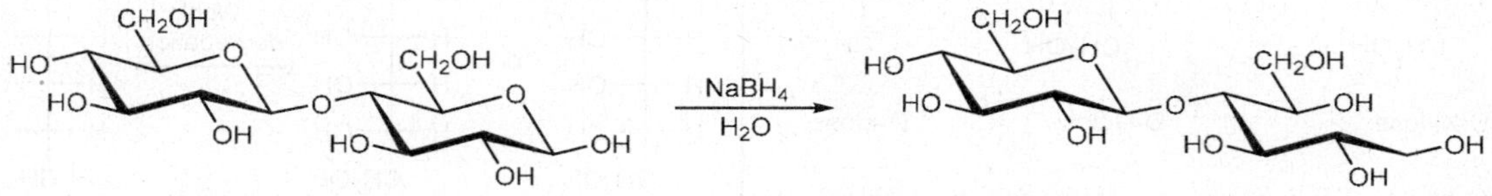

(b) One of the rings (bottom right) has an anomeric OH group, and is therefore in equilibrium with the open chain form, which is oxidized in the presence of Br_2 and H_2O (at pH = 6), as shown:

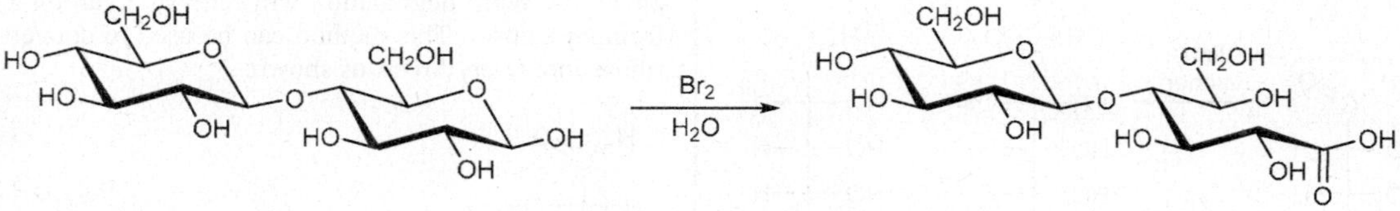

(c) Under these conditions the glycosidic bond between the two glucose monomers is cleaved, resulting in two glucose monosaccharides. Each of these glucose monosaccharides now contains an anomeric OH group (hemiacetal). Upon continued treatment with methanol in acidic conditions, each hemiacetal is converted into an acetal (glycoside), converting the anomeric OH group into a methoxy group. This results in a mixture of α and β anomers, as shown:

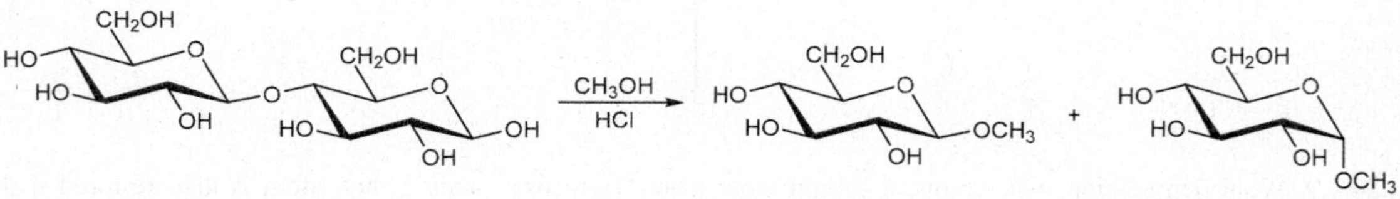

(d) Upon treatment with excess acetic anhydride and pyridine, each of the OH groups undergoes acetylation, as shown:

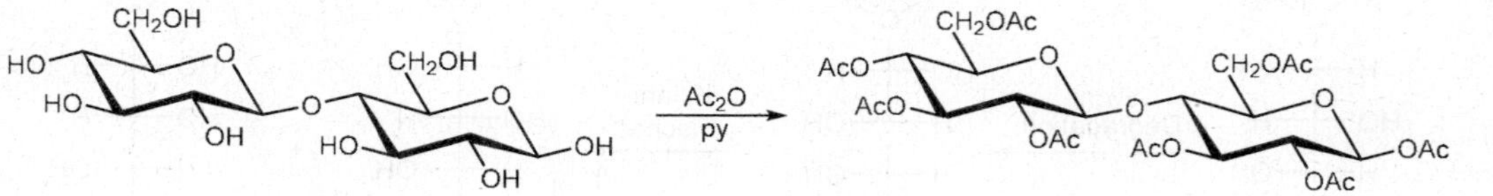

24.42.

(a) The OH group connected to C3 is pointing to the right, so this is a D-sugar. The functional group at C1 is an aldehyde group, so the compound is an aldose. And finally, the compound has four carbon atoms, so it is a tetrose. In summary, this compound is a D-aldotetrose.

(b) The OH group connected to C4 is pointing to the left, so this is an L-sugar. The functional group at C1 is an aldehyde group, so the compound is an aldose. And finally, the compound has five carbon atoms, so it is a pentose. In summary, this compound is an L-aldopentose.

(c) The OH group connected to C4 is pointing to the right, so this is a D-sugar. The functional group at C1 is an aldehyde group, so the compound is an aldose. And finally, the compound has five carbon atoms, so it is a pentose. In summary, this compound is a D-aldopentose.

(d) The OH group connected to C5 is pointing to the right, so this is a D-sugar. The functional group at C1 is an aldehyde group, so the compound is an aldose. And finally, the compound has six carbon atoms, so it is a hexose. In summary, this compound is a D-aldohexose.

(e) The OH group connected to C4 is pointing to the right, so this is a D-sugar. The functional group at C2 is a ketone group, so the compound is a ketose. And finally, the compound has five carbon atoms, so it is a pentose. In summary, this compound is a D-ketopentose

24.43. D-Glyceraldehyde has the *R* configuration, while L-glyceraldehyde has the *S* configuration. Therefore:

(a) The chiral center has the *R* configuration. This compound is D-glyceraldehyde.

(b) The chiral center has the *S* configuration. This compound is L-glyceraldehyde.

(c) The chiral center has the *R* configuration. This compound is D-glyceraldehyde.

(d) The chiral center has the *S* configuration. This compound is L-glyceraldehyde.

24.44.

(a) This compound is D-Glucose (see Figure 24.4).

(b) This compound is D-Mannose (see Figure 24.4).

(c) This compound is D- Galactose (see Figure 24.4).

(d) This compound is L-Glucose (see Figure 24.4).

24.45.

(a) D-Ribose is epimeric with D-arabinose at C2.

(b) D-Arabinose is epimeric with D-lyxose at C2.

(c) The enantiomer of D-ribose has the opposite configuration (*S*, rather than *R*) for all three chiral centers:

L-Ribose

(d) The enantiomer of D-arabinose is L-arabinose,

D-Arabinose L-Arabinose

and the C2 epimer of L-ribose is also L-arabinose, shown here:

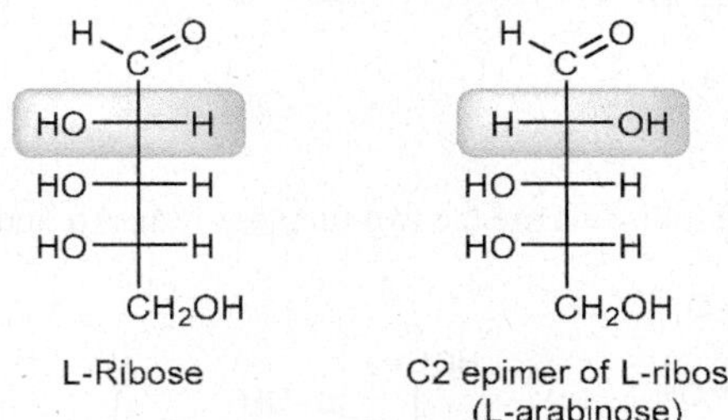

L-Ribose

C2 epimer of L-ribose (L-arabinose)

Therefore, the enantiomer of D-arabinose and the C2 epimer of L-ribose are the same compound (they are both L-arabinose).

(e) They are diastereomers because they are stereoisomers that are not mirror images.

24.46.

(a) We use a numbering system to determine the size of the ring that is formed. The carbonyl group can be attacked by the OH group that is connected to C4, giving a cyclic hemiacetal that is a five-membered ring:

(b) We use a numbering system to determine the size of the ring that is formed. The carbonyl group can be attacked by the OH group that is connected to C5, giving a cyclic hemiacetal that is a six-membered ring. Notice that there is a methyl group connected to C1 of the ring, as well as a methyl group connected to C5 of the ring:

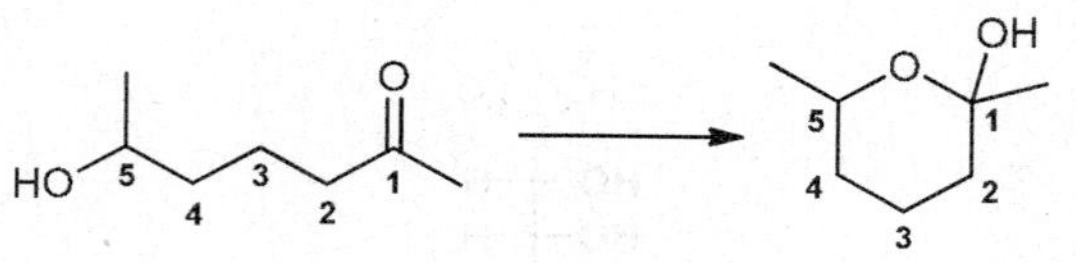

HO 5 3 1 4 2 → 5 1 4 2 3 OH

24.47. Use a numbering system starting on the carbon atom that is bonded to the two oxygen atoms. Number in the direction to include the next carbon atom and continue around the ring. Open the ring by disconnecting the C–O bond within the ring. The carbon atom that was bound to the two oxygen atoms is a carbonyl group in the starting material. There are two methyl groups at C3, which must be drawn in the starting material:

HO 4 3 2 1 H → 4 1 OH 3 2

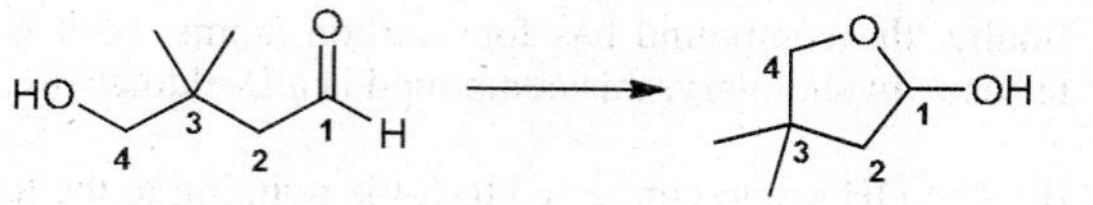

24.48.

(a) The following are the two pyranose forms (α and β anomers) of D-ribose:

D-ribose

α pyranose ring

β pyranose ring

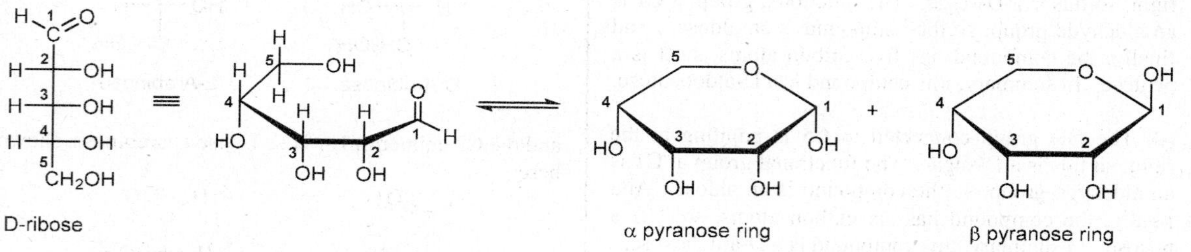

(b) The following are the two furanose forms (α and β anomers) of D-ribose:

D-ribose

α furanose ring

β furanose ring

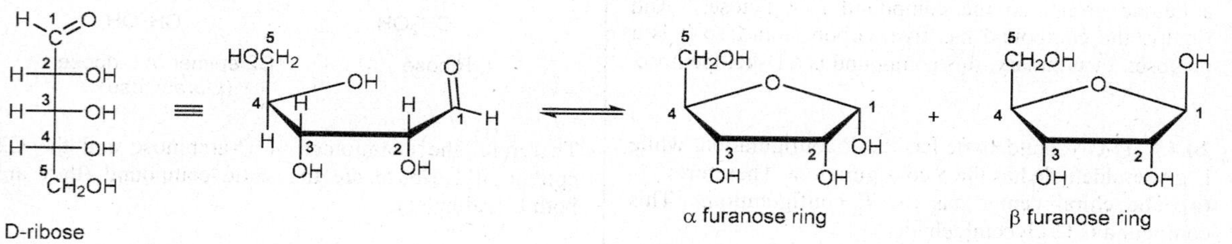

24.49.

(a) These compounds are diastereomers that differ from each other in the configuration of only one chiral center. Therefore, they are epimers.

(b) These compounds are stereoisomers that are not mirror images of one another. Therefore, they are diastereomers.

(c) These compounds are non-superimposable mirror images of one another. Therefore, they are enantiomers.

(d) These structures are two different representations of the same compound (β-D-glucopyranose), so they are identical compounds.

24.50. Upon treatment with aqueous acid, the acetal undergoes hydrolysis to give a hemiacetal (the anomeric methoxy group is replaced with an anomeric hydroxy group). The open chain form of the cyclic hemiacetal is D-glucose.

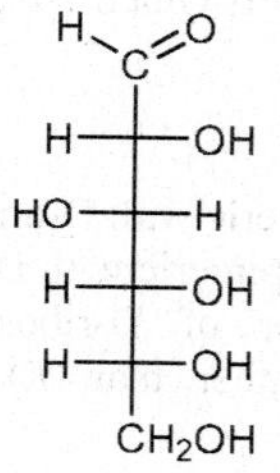

D-Glucose

24.51. To assign the configuration of each chiral center, we can use the rule of thumb that was pointed out in the solution to Problem **24.3**. Specifically, an OH on the right side of the Fischer projection indicates the *R* configuration, while an OH on the left side of the Fischer projection indicates the *S* configuration:

(a) **(b)** 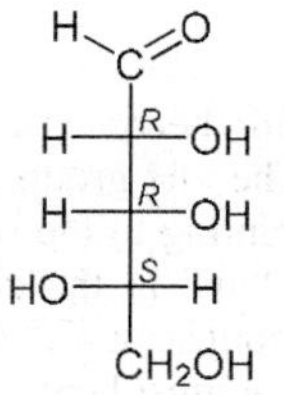 **(c)**

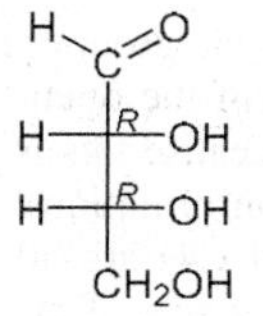

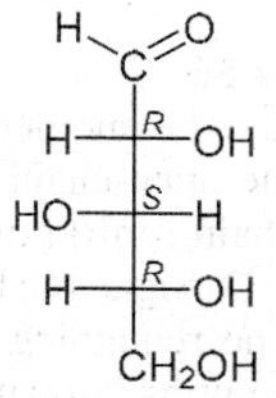

(d) 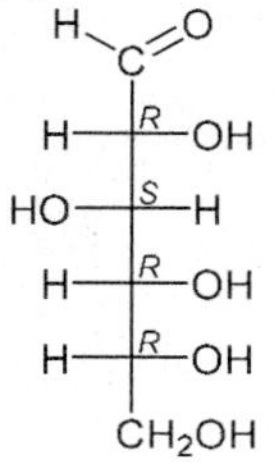 **(e)**

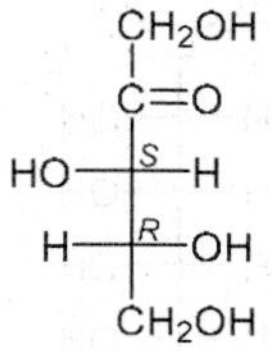

24.52. The structures below can be found in Figure 24.5:

(a) **(b)**

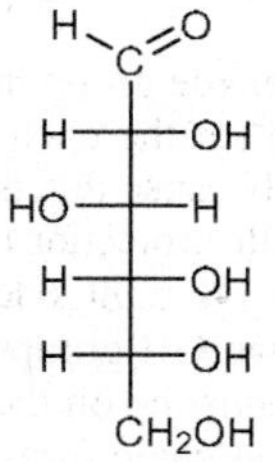

D-Glucose D-Galactose

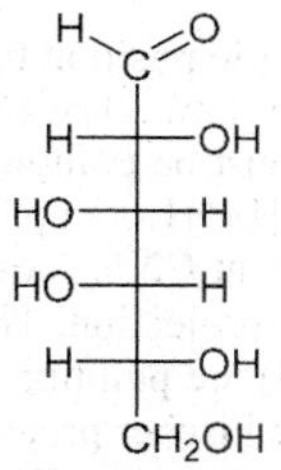

(c) **(d)**

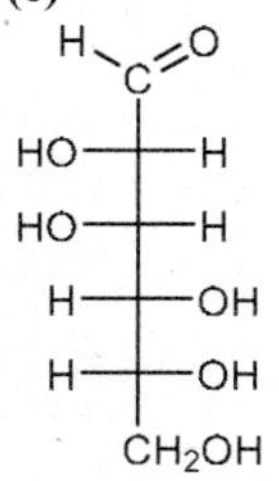

D-Mannose D-Allose

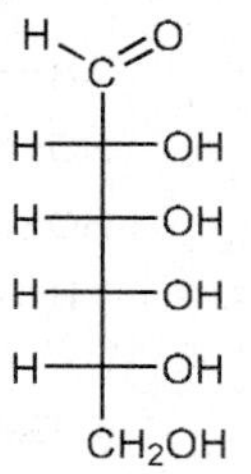

24.53.

(a) We begin by drawing the open chain form of D-fructose (see Figure 24.6). This compound is closed into a furanose form, so we draw a Haworth projection of a furanose skeleton (a five-membered ring with the oxygen atom in the back). We use a numbering system to assist us. Next, we draw the CH_2OH group (connected to C5) pointing up, because this is a D sugar. The anomeric OH group is then drawn pointing up (the β anomer has a *cis* relationship between the anomeric OH group and the CH_2OH group). Finally, the remaining OH groups are drawn. Any OH groups on the right side of the Fischer projection will be pointing down in the Haworth projection, while any OH groups on the left side of the Fischer projection will be pointing up in the Haworth projection):

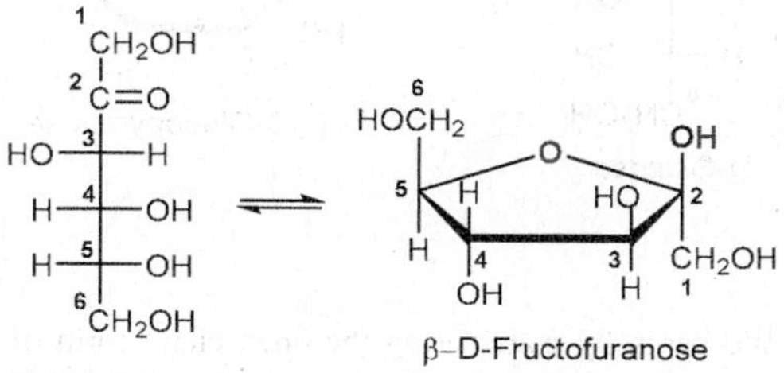

D-Fructose β-D-Fructofuranose

(b) We begin by considering the open chain form of D-galactose (see Figure 24.5). This compound is closed into a pyranose form, so we draw a Haworth projection of a pyranose skeleton (a six-membered ring with the oxygen atom in the back right corner). Next, we draw the CH_2OH group (connected to C5) pointing up, because this is a D sugar. The anomeric OH group is then drawn pointing up (the β anomer has a *cis* relationship between the anomeric OH group and the CH_2OH group). Finally, the remaining OH groups are drawn. Any OH groups on the right side of the Fischer projection will be pointing down in the Haworth projection, while any OH groups on the left side of the Fischer projection will be pointing up in the Haworth projection):

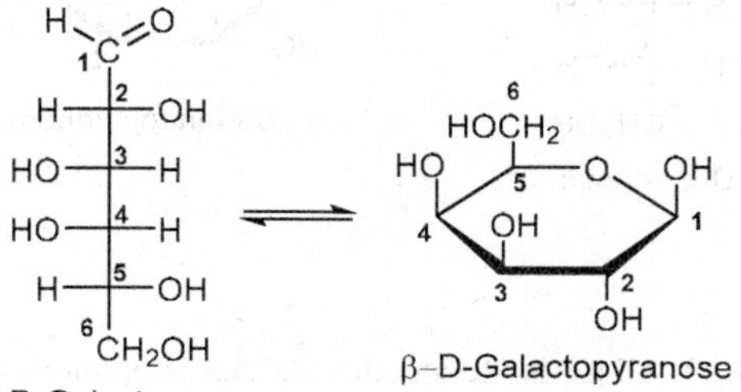

D-Galactose β-D-Galactopyranose

(c) We begin by considering the open chain form of D-glucose (see Figure 24.5). This compound is closed into a pyranose form, so we draw a Haworth projection of a pyranose skeleton (a six-membered ring with the oxygen atom in the back right corner). Next, we draw the CH_2OH group (connected to C5) pointing up, because this is a D sugar. The anomeric OH group is then drawn pointing up (the β anomer has a *cis* relationship between the anomeric OH group and the CH_2OH group). Finally, the remaining OH groups are drawn. Any OH groups on the right side of the Fischer projection will be pointing down in the Haworth projection, while any OH groups on the left side of the Fischer projection will be pointing up in the Haworth projection):

D-Glucose

β-D-Glucopyranose

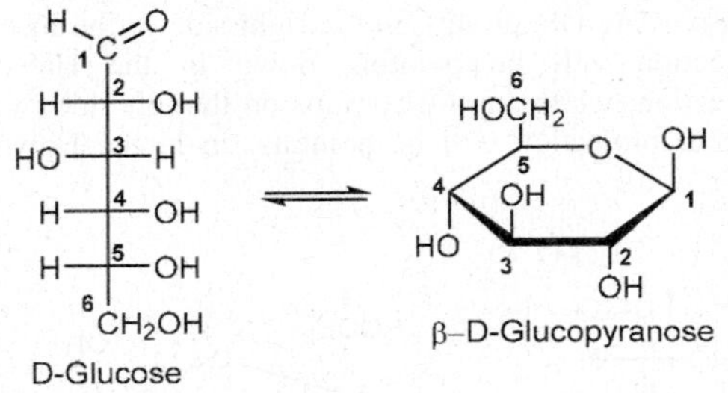

(d) We begin by considering the open chain form of D-mannose (see Figure 24.5). This compound is closed into a pyranose form, so we draw a Haworth projection of a pyranose skeleton (a six-membered ring with the oxygen atom in the back right corner). Next, we draw the CH_2OH group (connected to C5) pointing up, because this is a D sugar. The anomeric OH group is then drawn pointing up (the β anomer has a *cis* relationship between the anomeric OH group and the CH_2OH group). Finally, the remaining OH groups are drawn. Any OH groups on the right side of the Fischer projection will be pointing down in the Haworth projection, while any OH groups on the left side of the Fischer projection will be pointing up in the Haworth projection):

D-Mannose

β-D-Mannopyranose

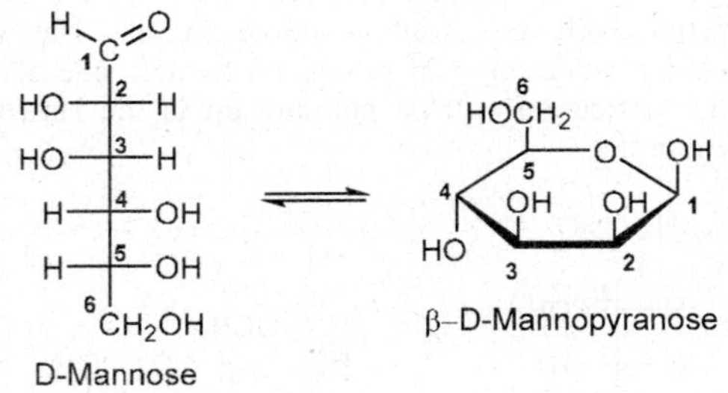

24.54. D-allose is the aldohexose that is epimeric with D-glucose at C3. The α-pyranose form of D-allose is shown here:

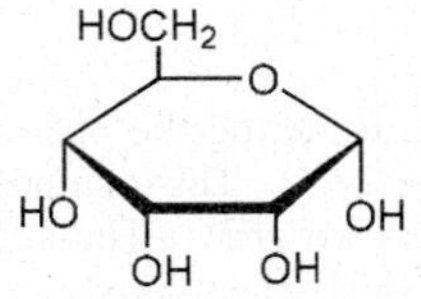

24.55.

(a) This structure represents the α anomer of the pyranose form of D-allose and is therefore called α-D-allopyranose.

(b) This structure represents the β anomer of the pyranose form of D-galactose and is therefore called β-D-galactopyranose.

(c) This structure represents a methyl glycoside. It is an acetal of the β anomer of the pyranose form of D-glucose and is called methyl β-D-glucopyranoside.

24.56.

(a) The anomeric position becomes an aldehyde group in the open-chain form. The OH group at C5 (of the open-chain form) must be pointing to the right, because this is a D sugar (CH_2OH is "up" in the chair conformation). The remaining three groups (at C2, C3 and C4) are all pointing "down," so they must be on the right side of the Fischer projection of the open-chain form. This structure represents D-allose:

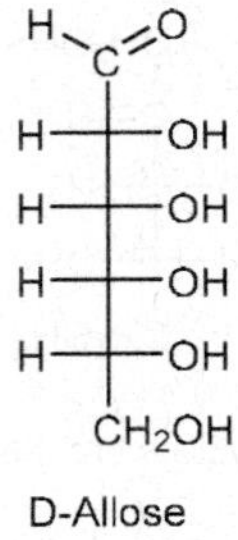

D-Allose

(b) The anomeric position becomes an aldehyde group in the open-chain form. The OH group at C5 (of the open-chain form) must be pointing to the right, because this is a D sugar (CH_2OH is "up" in the Haworth projection). The OH group at C2 is "down," so it is on the right side of the Fischer projection. The remaining two OH groups (at C3 and C4) are pointing "up," so they must be on the left side of the Fischer projection of the open-chain form. This structure represents D-galactose.

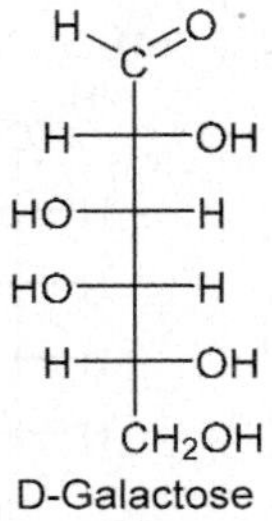

D-Galactose

(c) The anomeric position becomes an aldehyde group in the open-chain form. The OH group at C5 (of the open-chain form) must be pointing to the right, because this is a D sugar (CH_2OH is "up" in the chair conformation). The OH groups at C2 and C4 are "down," so they appear on the right side of the Fischer projection. The remaining OH group at C3 is pointing "up," so it appears on the left side of the Fischer projection of the open-chain form. This structure represents D-glucose:

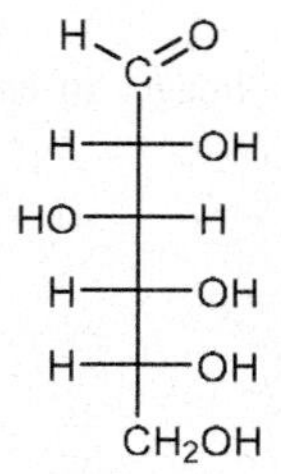

H–C=O; H–OH; HO–H; H–OH; H–OH; CH_2OH

D-Glucose

24.57.

(a) When treated with excess methyl iodide and silver oxide, all of the OH groups in the β-pyranose form of D-allose are converted into methoxy groups (via methylation), giving the following compound:

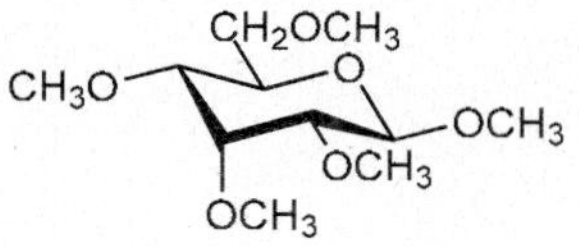

(b) When treated with excess acetic anhydride and pyridine, each of the OH groups in the β-pyranose form of D-allose will undergo acetylation to give the following compound:

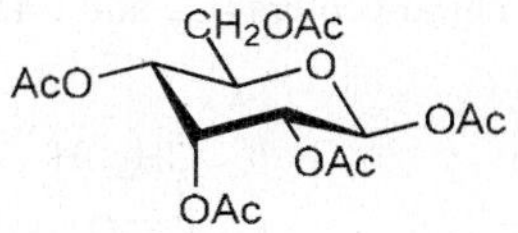

(c) When treated with methanol and HCl, the anomeric OH group in the β-pyranose form of D-allose is converted into a methoxy group. Both possible anomers of the methyl glycoside are formed:

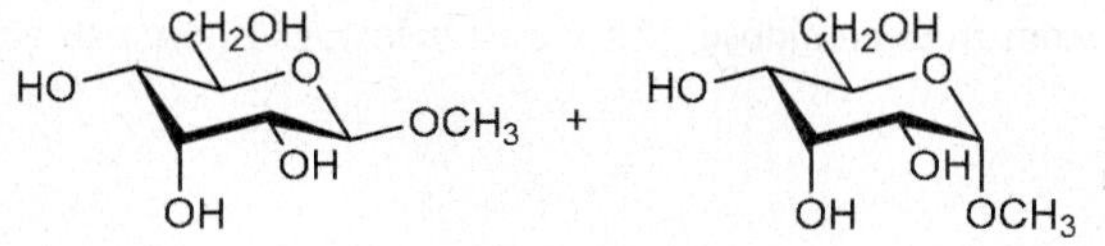

24.58. The product, shown below, is optically inactive because it is a *meso* compound:

meso compound

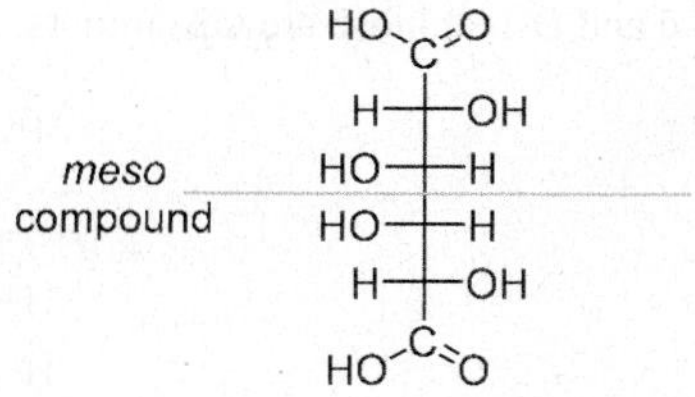

24.59. Upon treatment with nitric acid, D-allose undergoes oxidation to give an aldaric acid that is optically inactive because it is a *meso* compound:

HNO_3, H_2O, heat

D-Allose

optically inactive (*meso* compound)

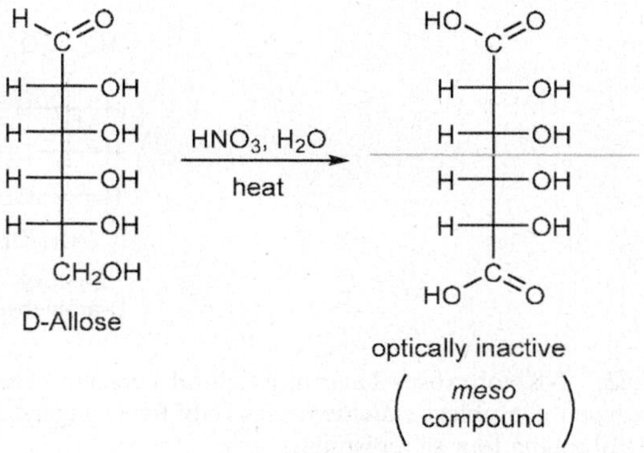

24.60. The skeleton of the chair is drawn with an oxygen atom in the upper back-right corner. Each substituent is then labeled as "up" or "down" and placed on the chair accordingly. The anomeric OH group is drawn pointing down, indicating the α anomer. In this chair conformation, the largest substituent (CH_2OH) occupies an equatorial position, so this is the more stable chair conformation of α-D-altropyranose.

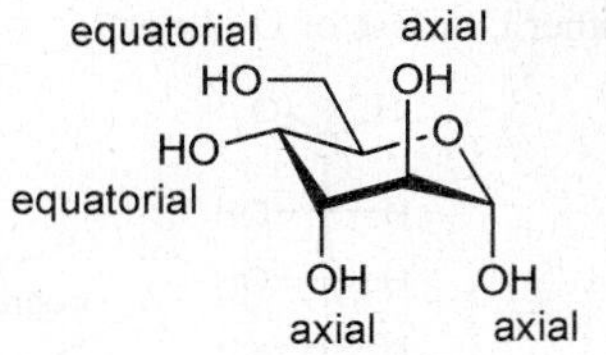

equatorial, axial, equatorial, axial, axial

24.61. Upon treatment with excess methyl iodide in the presence of silver oxide, all of the OH groups in D-galactopyranose are converted into methoxy groups, as shown. Upon treatment with aqueous acid, the acetal is hydrolyzed, giving both anomers of the resulting hemiacetal:

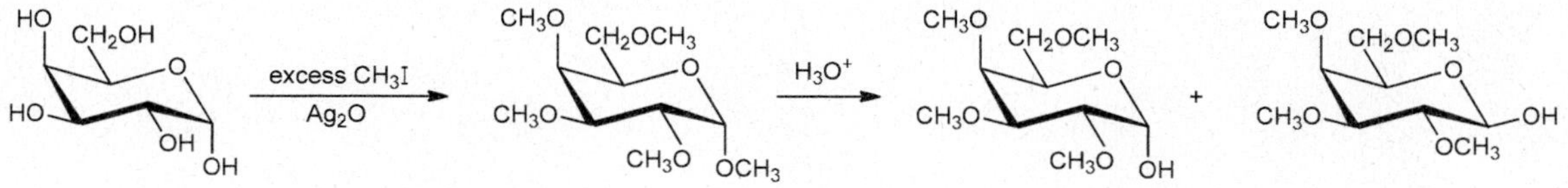

excess CH_3I, Ag_2O → H_3O^+ →

24.62.
(a) D-Glucose and D-gulose (see Figure 24.5) are stereoisomers that are not mirror images of each other, so they are diastereomers.
(b) D-Ribose and D-arabinose are C2 epimers:

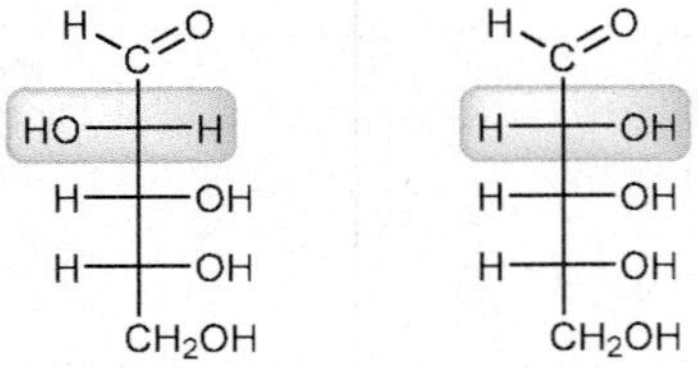

D-arabinose R-ribose

Therefore, removing the OH group from C2 of either compound will result in the same structure. That is, 2-deoxy-D-ribose and 2-deoxy-D-arabinose are the same compound.

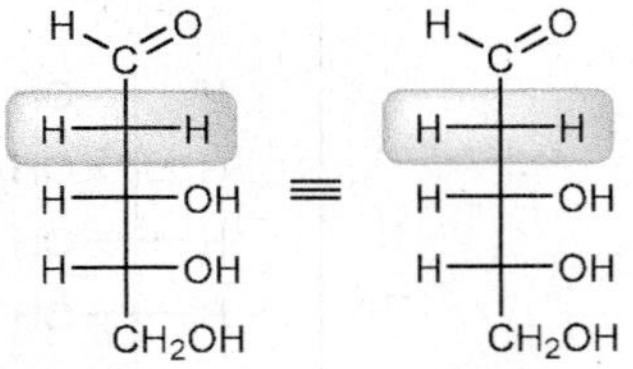

2-deoxy-D-arabinose ≡ 2-deoxy-R-ribose

24.63. 2-Ketohexoses have three chiral centers. The configuration of one of these chiral centers (at C5) is fixed because the problem statement asks only for D sugars. That leaves two other chiral centers (C3 and C4), giving rise to the following four stereoisomers:

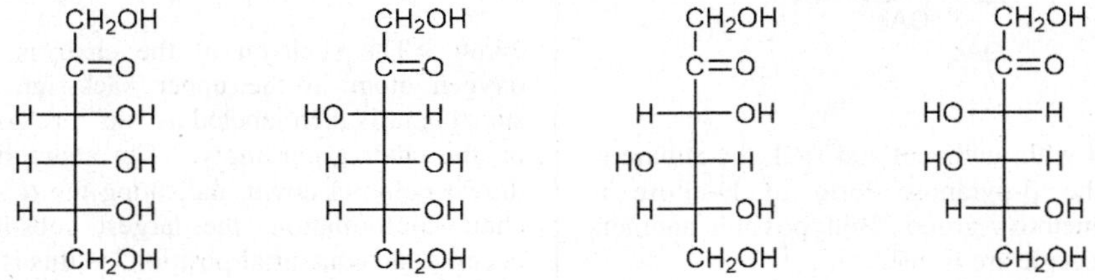

24.64. A Wohl degradation involves the removal of a carbon atom from an aldose. As shown below, D-ribose can be made from either D-allose or D-altrose:

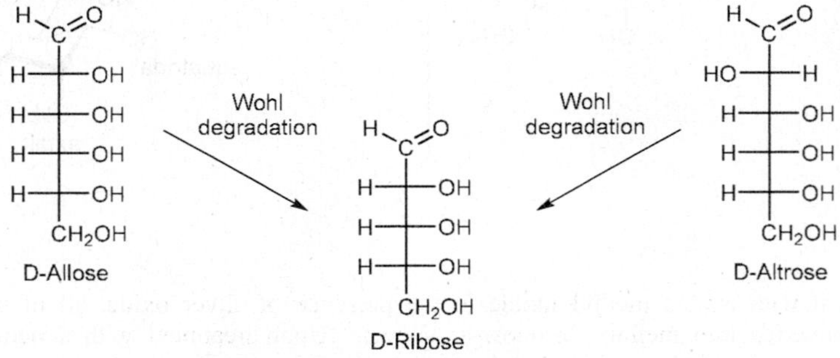

D-Allose —Wohl degradation→ D-Ribose ←Wohl degradation— D-Altrose

24.65. In a Kiliani-Fischer synthesis, the chain is lengthened, with C1 becoming C2 in the product. Both possible configurations of the C2 position are obtained, giving the following epimers:

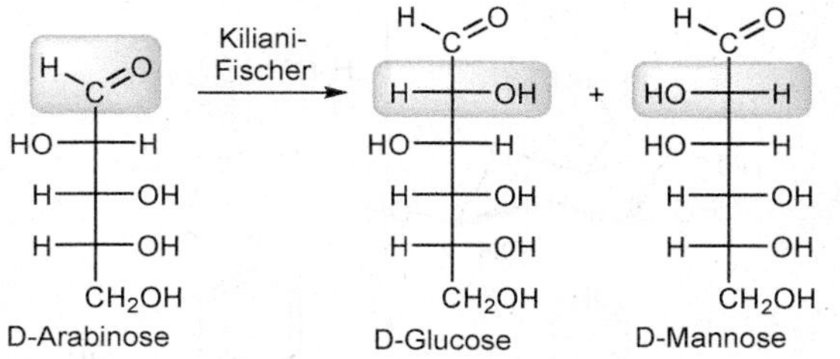

Kiliani-Fischer

D-Arabinose D-Glucose D-Mannose

24.66. The aldehyde group is converted into a cyanohydrin. The newly installed chiral center can have either *R* or *S* configuration, giving the following diastereomers:

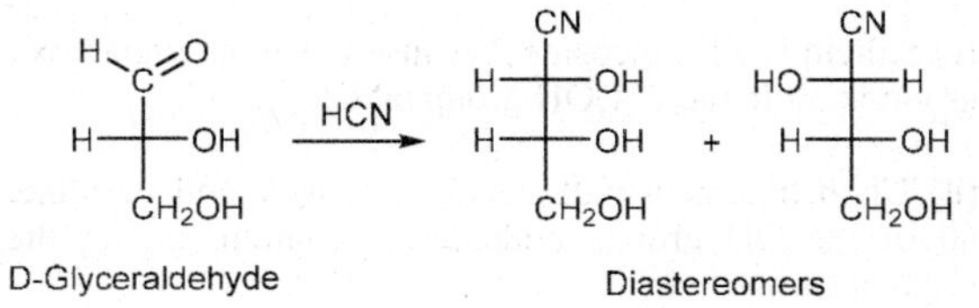

HCN

D-Glyceraldehyde Diastereomers

24.67.
(a) Upon treatment with sodium borohydride, the aldehyde group of D-glucose is reduced to an alditol, giving the following alditol:

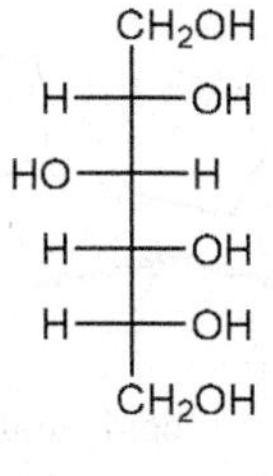

(b) As shown, treatment of L-gulose with sodium borohydride gives the same alditol as above (the two drawings are superimposable when rotated 180°).

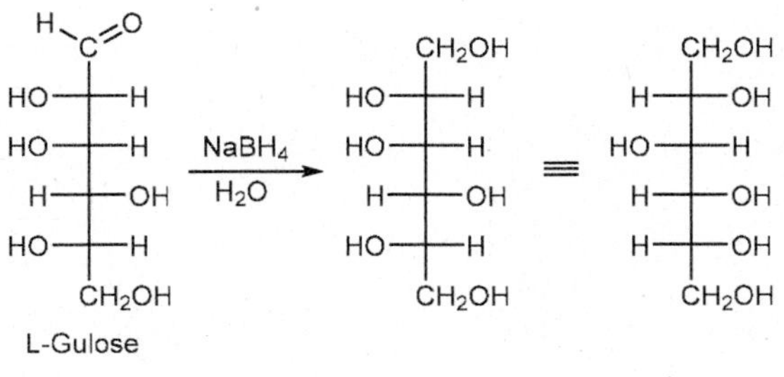

$NaBH_4$ / H_2O

L-Gulose

24.68. As seen in the solution to Problem **24.30**, D-allose and D-galactose are converted into optically inactive alditols (*meso* compounds) upon treatment with sodium borohydride.

24.69.
(a) This compound is not a reducing sugar because the anomeric position bears an alkoxy group (the compound is a cyclic acetal).
(b) This compound is a reducing sugar because the anomeric position bears an OH group (the compound is a cyclic hemiacetal).

24.70.
(a) Treatment with CH_3OH and HCl converts the hemiacetal to an acetal (a methyl glycoside).
(b) Treatment with CH_3OH and HCl converts the hemiacetal to an acetal (a methyl glycoside).
(c) Treatment with HNO_3, H_2O and heat oxidizes the primary alcohol and aldehyde to carboxylic acids.
(d) Treatment with excess CH_3I and Ag_2O methylates all of the OH groups, and subsequent hydrolysis with H_3O^+ converts the acetal to a hemiacetal, leaving the methyl ethers intact.

24.71.
(a) The cyclic acetal undergoes hydrolysis to give a cyclic hemiacetal (the methoxy group at the anomeric carbon atom is replaced with an OH group). Under these conditions, both anomers are formed, giving α-D-glucopyranose and β-D-glucopyranose.
(b) The cyclic acetal undergoes hydrolysis to give a cyclic hemiacetal (the ethoxy group at the anomeric carbon atom is replaced with an OH group). Under these conditions, both anomers are formed, giving α-D-galactopyranose and β-D-galactopyranose.

24.72.
(a) Oxidation of D-arabinose with nitric acid gives the same aldaric acid as oxidation of D-lyxose.
(b) D-Ribose and D-xylose yield optically inactive alditols (*meso* compounds) when treated with sodium borohydride.
(c) Reduction of D-xylose yields the same alditol as reduction of L-xylose.
(d) D-xylose has a β-pyranose form in which all substituents are equatorial.

24.73. Trehalose is a disaccharide assembled from two equivalents of the α-pyranose form of D-glucose. Trehalose is not a reducing sugar, which means that the two rings must be fused at the anomeric positions (so there is no anomeric OH group). The disaccharide is assembled from the α-pyranose form of each subunit of D-glucose, with the anomeric alkoxy group *trans* to the CH_2OH group on C5 of each glucose subunit:

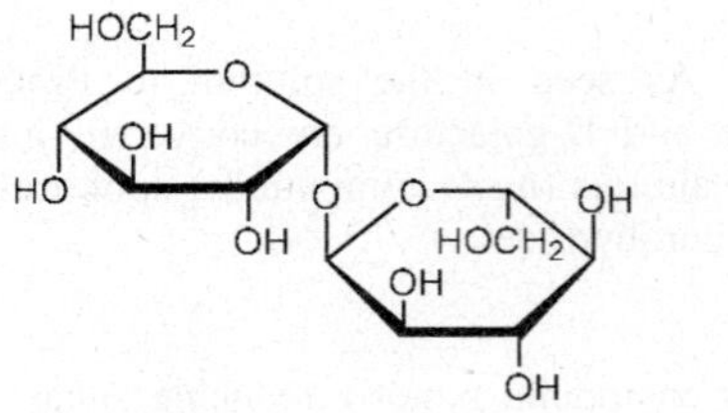

24.74. Reduction of D-xylose gives the following structure (D-xylitol):

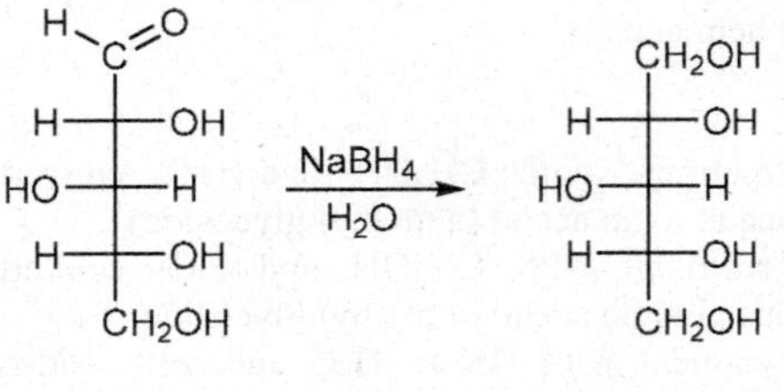

D-Xylose D-Xylitol

24.75. The alpha-1,6 glycoside linkage of isomaltose is illustrated below:

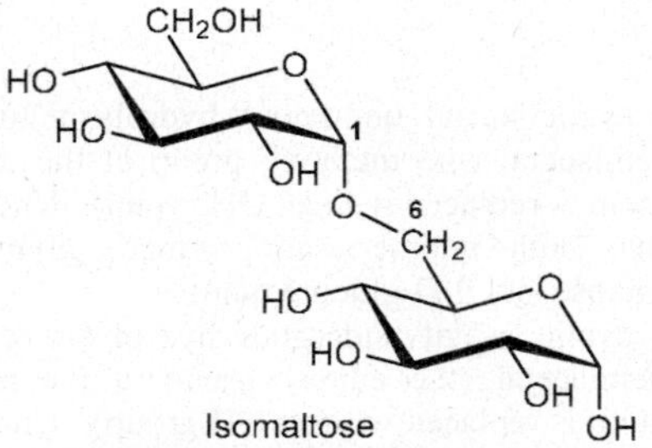

Isomaltose
(an α-1,6 glycoside)

24.76.

(a) No, it is not a reducing sugar because the anomeric position has a phenoxy group (it is a cyclic acetal).

(b) The acetal group is hydrolyzed, giving both anomers of the cyclic hemiacetal:

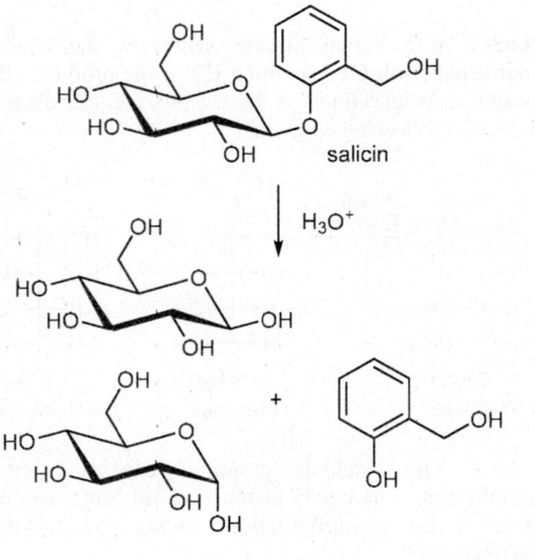

(c) Salicin is a β-glycoside, because the anomeric alkoxy group is *cis* to the CH_2OH group on C5.

(d) Upon treatment with acetic anhydride and pyridine, all of the OH groups undergo acetylation, giving the product shown:

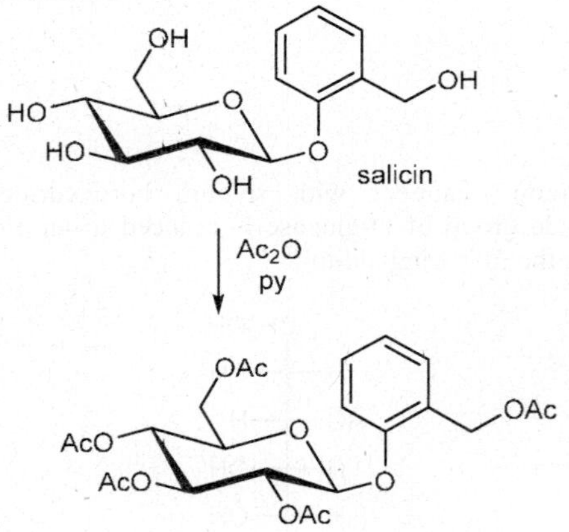

(e) No. In the absence of acid catalysis, the acetal group is not readily hydrolyzed.

24.77. This transformation involves the conversion of a cyclic hemiacetal to an acetal. The anomeric OH group is protonated under acidic conditions, giving an excellent leaving group (water). Loss of the leaving group gives a resonance-stabilized cation (the most significant resonance structure is shown). This cation is then attacked by phenol, giving an oxonium ion that is deprotonated to give the product:

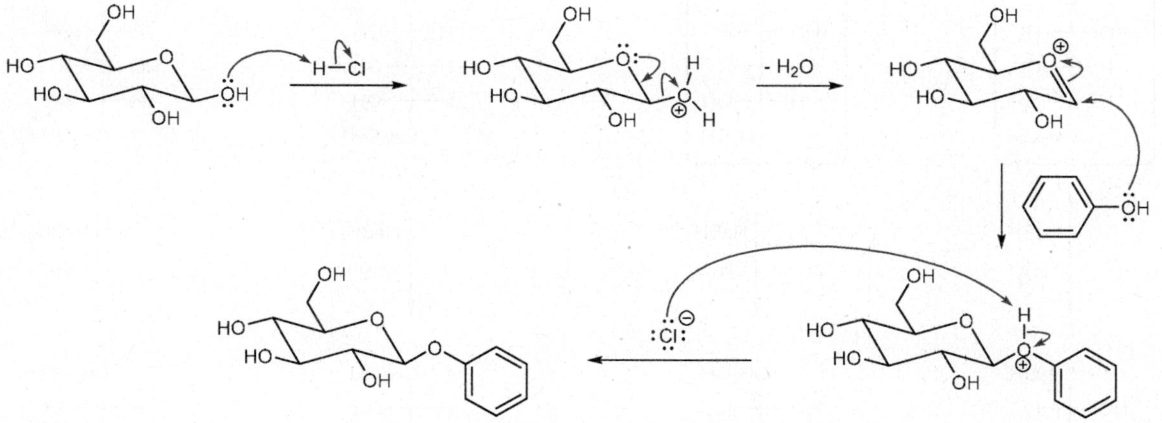

24.78. The following α and β anomers are obtained when D-glucose is treated with aniline:

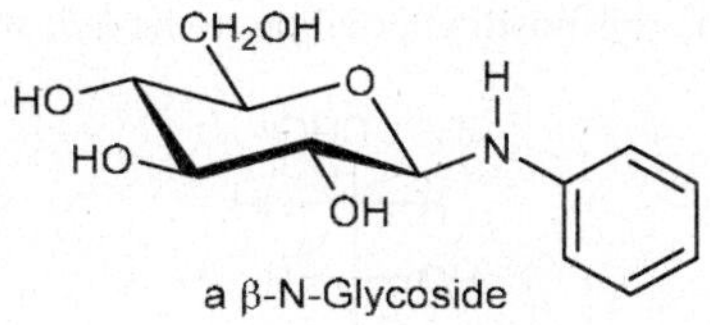

a β-N-Glycoside

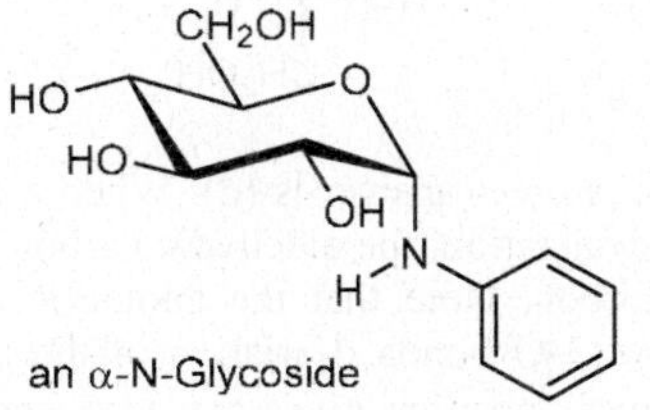

an α-N-Glycoside

24.79.

(a) The following nucleoside is formed from 2-deoxy-D-ribose and adenine. Notice that this structure differs from adenosine (see Figure 24.13) only at the C2 position. Specifically, the C2 position in this structure does not bear an OH group (as compared with adenosine):

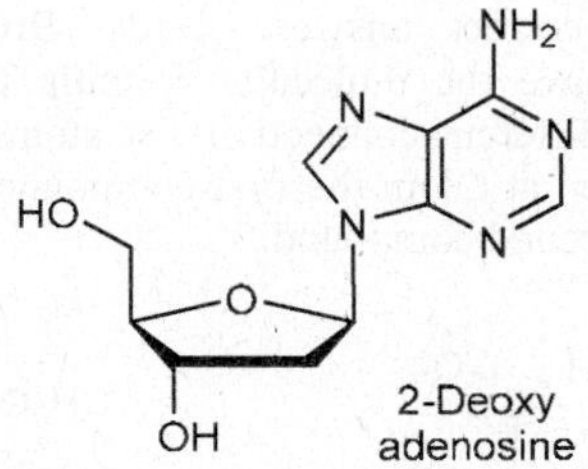

2-Deoxy adenosine

(b) The following nucleoside, called guanosine (see Figure 24.13), is formed from D-ribose and guanine:

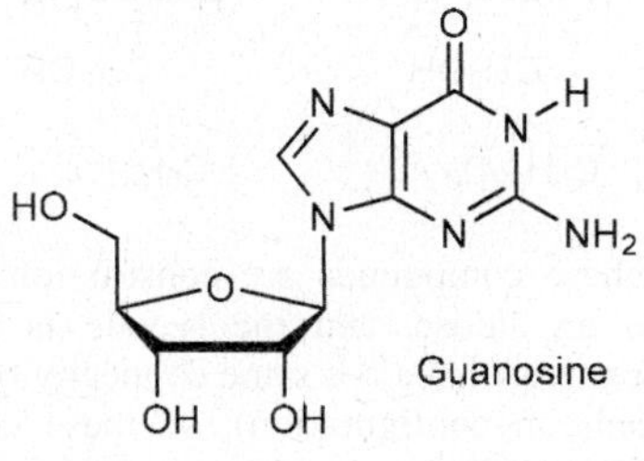

Guanosine

24.80. There are only four D-aldopentoses, all which are shown below. Only the first two are reduced to give an optically active alditol, as shown. The latter two are reduced to give *meso* compounds (not optically active):

$\xrightarrow[H_2O]{NaBH_4}$

(optically active) (optically active) (optically inactive) (optically inactive)

24.81. The correct answer is (d). Both of these compounds have the molecular formula $C_6H_{12}O_6$, but they have a different connectivity of atoms. The C=O bond is located at C1 in the first compound, but located at C2 in the second compound:

$C_6H_{12}O_6$ $C_6H_{12}O_6$

Therefore, these compounds are constitutional isomers (the first is an aldose, and the second is a ketose). Stereoisomers must have the same connectivity of atoms (differing only in configuration), so these compounds cannot be stereoisomers.

24.82. The pyranose form of glucose has a six-membered ring. The correct answer is (c).

24.83. The correct answer is (b). The classification of a sugar as D or L depends on the configuration of the chiral center farthest from the carbonyl group. All D sugars will have an OH group in this position pointing to the *right* in the Fischer projection. Structure (b) has an OH group in this position pointing to the *left*, so it is an L sugar:

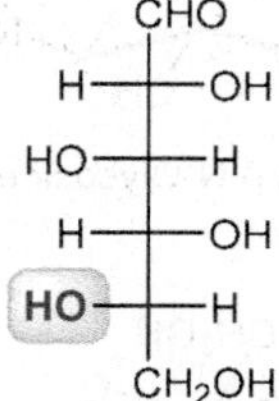

24.84. The correct answer is (d). When a carbohydrate undergoes cyclization, the aldehydic carbon becomes the anomeric carbon. Note that the anomeric carbon atom still has two C–O bonds. Variation of the configuration at the anomeric position gives rise to α and β forms of the cyclic sugar, called anomers. The structure below represents the β anomer:

anomeric carbon was C=O carbon in acyclic form

24.85. The correct answer is (a). Epimers are diastereomers that differ from each other in the configuration of only one chiral center, as shown below:

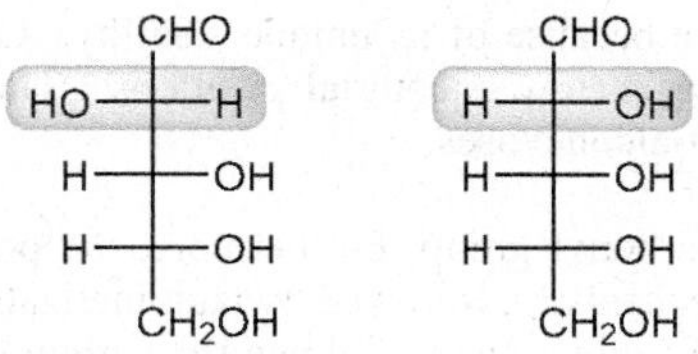

CHO | CHO
HO—H | H—OH
H—OH | H—OH
H—OH | H—OH
CH_2OH | CH_2OH

24.86.

(a) D-Gluconic acid is formed when the C1 position of D-glucose undergoes oxidation to give a carboxylic acid, shown here:

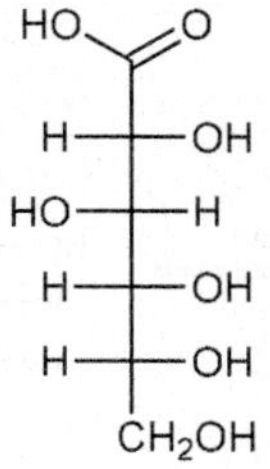

HO–C=O; H—OH; HO—H; H—OH; H—OH; CH_2OH

(b) A numbering system is used to help draw the product. The CH_2OH group (connected to C5) is drawn pointing up, because the starting material is a D sugar. Any OH groups on the right side of the Fischer projection will be pointing down in the Haworth projection, while any OH groups on the left side of the Fischer projection will be pointing up in the Haworth projection):

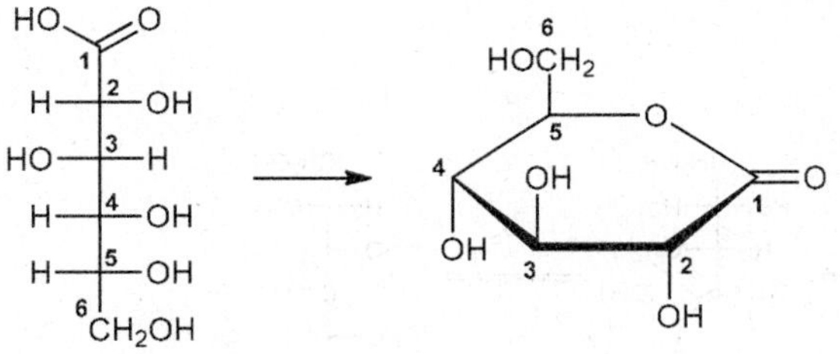

(c) Yes. The compound has chiral centers, and it is not a *meso* compound. Therefore, it will be optically active.

(d) The gluconic acid is a carboxylic acid and its IR spectrum is expected to have a broad signal between 2200 and 3600 cm^{-1}. The IR spectrum of the lactone will not have this broad signal.

24.87. In order for the CH_2OH group to occupy an equatorial position, all of the OH groups on the ring must occupy axial positions. The total energy cost associated with the steric interactions of the axial OH groups is more than the energy cost associated with one (albeit larger) CH_2OH group in an axial position. Therefore, the equilibrium will favor the form in which the CH_2OH group occupies an axial position. The structure of L-idose is shown here:

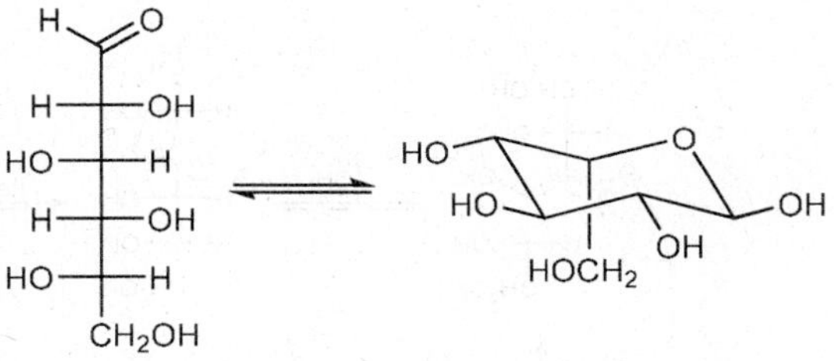

L-Idose

24.88. The molecular formula ($C_6H_{12}O_6$) indicates that compound **A** is a hexose (carbohydrate with six carbon atoms). Compound **A** is a reducing sugar, so it must be an aldohexose (rather than a ketohexose). Two successive Wohl degradations of compound **A** gives D-erythrose, which indicates the configurations of C4 and C5 of the aldohexose (both positions have the *R* configuration, just as in D-erythrose). Compound **A** is epimeric with glucose at C3, which indicates the *R* configuration at C3 (D-glucose has the *S* configuration at C3). Finally, C2 has the *R* configuration, giving the structure shown below (D-allose). The β-pyranose form of D-allose has also been drawn below. When treated with excess ethyl iodide in the presence of silver oxide, all of the OH groups in the β-pyranose form of compound **A** undergo alkylation, thereby converting them into ethoxy groups, as shown:

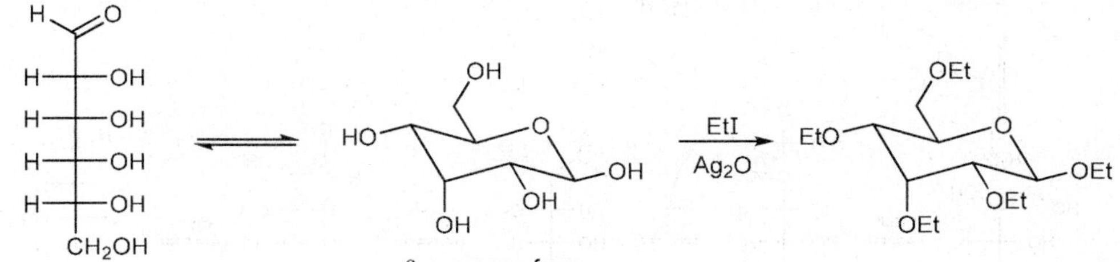

EtI / Ag_2O

β-pyranose form

D-Allose
(Compound **A**)

24.89. Glucose is the most common monosaccharide observed in nature because of its unique stability. Glucose can adopt a chair conformation in which all of the substituents on the ring occupy equatorial positions. Therefore, D-glucose can achieve a lower energy conformation than any of the other D-aldohexoses.

24.90. Epimerization of a chiral center can occur if it is alpha to a carbonyl group. Furthermore, the position of a carbonyl group can be changed by tautomerization to a dienol (2-step mechanism), followed by tautomerization back to a carbonyl group in a new position (2-step mechanism). As the carbonyl group "walks" down the carbon chain, each chiral center can be inverted. Once the carbonyl group "walks" back to its original position, the enantiomer of the original sugar is obtained. The following mechanism consists entirely of deprotonation and protonation steps. Notice that for each deprotonation step, hydroxide is used as the base, while water is used as the proton source for each protonation step (consistent with basic conditions):

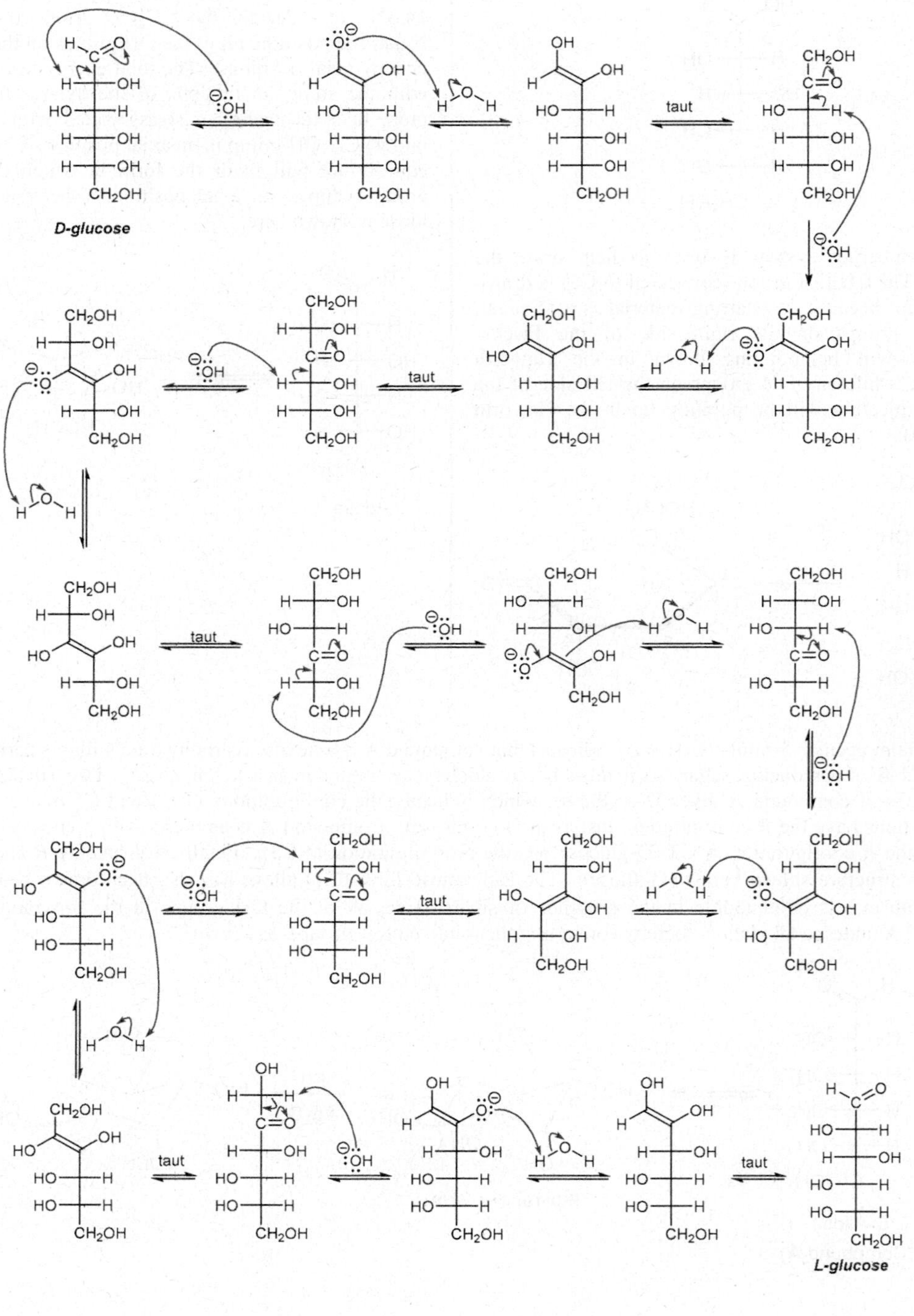

D-glucose

L-glucose

24.91. Compound **X** is a D-aldohexose that can adopt a β-pyranose form with only one axial substituent. Recall that D-glucose has all substituents in equatorial positions, so compound **X** must be epimeric with D-glucose either at C2 (D-mannose), C3 (D-allose), or C4 (D-galactose).

Compound **X** undergoes a Wohl degradation to produce an aldopentose, which is converted into an optically active alditol when treated with sodium borohydride. Therefore, compound **X** cannot be D-allose, because a Wohl degradation of D-allose followed by reduction produces an optically *inactive* alditol (a *meso* compound).

We conclude that compound **X** must be either D-mannose or D-galactose.

The identity of compound **X** can be determined by treating compound **X** with sodium borohydride. Reduction of D-mannose should give an optically active alditol, while reduction of D-galactose gives an optically inactive alditol (a *meso* compound).

24.92.

(a) Compound **A** is a D-aldopentose. Therefore, there are four possible structures to consider (Figure 24.3).

When treated with sodium borohydride, compound **A** is converted into an alditol that exhibits three signals in its ^{13}C NMR spectrum. Therefore, compound **A** must be D-ribose or D-xylose both of which are reduced to give symmetrical alditols (thus, three signals for five carbon atoms).

When compound **A** undergoes a Kiliani-Fischer synthesis, both products can be treated with nitric acid to give optically active aldaric acids. Therefore, compound **A** cannot be D-ribose, because when D-ribose undergoes a Kiliani-Fischer synthesis, one of the products is D-allose, which is oxidized to give an optically inactive aldaric acid (a *meso* compound). We conclude that the structure of compound **A** must be D-xylose.

H–C=O
H–|–OH
HO–|–H
H–|–OH
CH_2OH

D-Xylose

(b) Compound D is expected have six signals in its ^{13}C NMR spectrum, while compound E is expected to have only three signals in its ^{13}C NMR spectrum.

Compound D: HO–C=O; H–|–OH; H–|–OH; HO–|–H; H–|–OH; HO–C=O

Compound D

Compound E: HO–C=O; HO–|–H; H–|–OH; HO–|–H; H–|–OH; HO–C=O

Compound E

Chapter 25
Amino Acids, Peptides, and Proteins

Review of Concepts

Fill in the blanks below. To verify that your answers are correct, look in your textbook at the end of Chapter 25. Each of the sentences below appears verbatim in the section entitled *Review of Concepts and Vocabulary*.

- Amino acids in which the two functional groups are separated by exactly one carbon atom are called _______ **amino acids.**
- Amino acids are coupled together by amide linkages called ____________ **bonds.**
- Relatively short chains of amino acids are called ___________.
- Only 20 amino acids are abundantly found in proteins, all of which are ___ **amino acids**, except for ____________, which lacks a chiral center.
- Amino acids exist primarily as ______________ at physiological pH.
- The __________________ of an amino acid is the pH at which the concentration of the zwitterionic form reaches its maximum value.
- Peptides are comprised of **amino acid** _________ joined by peptide bonds.
- Peptide bonds experience restricted rotation, giving rise to two possible conformations, called _______ and _______. The _______ conformation is generally more stable.
- Cysteine residues are uniquely capable of being joined to one another via _____________ **bridges.**
- ______ is commonly used to form peptide bonds.
- In the **Merrifield synthesis**, a peptide chain is assembled while tethered to ________________________.
- The **primary structure** of a protein is the sequence of____________________.
- The **secondary structure** of a protein refers to the ______________________ ____________________ of localized regions of the protein. Two particularly stable arrangements are the ___ **helix** and ____ **pleated sheet**.
- The **tertiary structure** of a protein refers to its ________________________.
- Under conditions of mild heating, a protein can unfold, a process called ________________.
- **Quaternary structure** arises when a protein consists of two or more folded polypeptide chains, called _____________, that aggregate to form one protein complex.

Review of Skills

Fill in the blanks and empty boxes below. To verify that your answers are correct, look in your textbook at the end of Chapter 25. The answers appear in the section entitled *SkillBuilder Review*.

25.1 Determining the Predominant Form of an Amino Acid at a Specific pH

Consider the following amino acid, and draw the form that predominates at physiological pH.

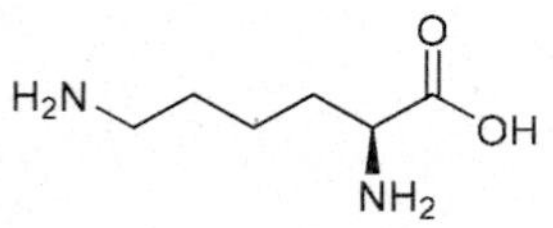

H_2N O OH NH_2

25.2 Using the Amidomalonate Synthesis

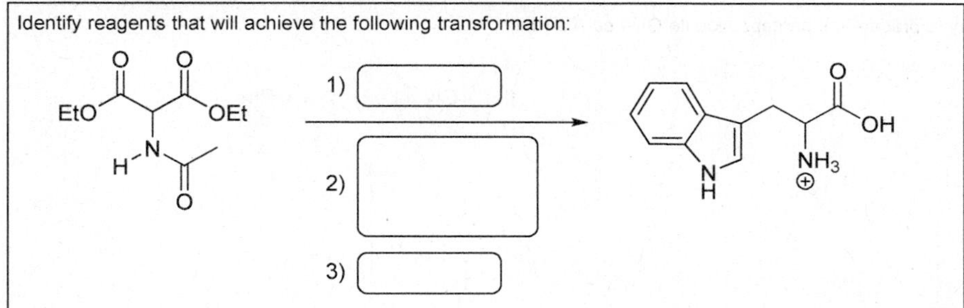

25.3 Drawing a Peptide

Draw a bond-line structure for the tripeptide Phe-Val-Trp

25.4 Sequencing a Peptide via Enzymatic Cleavage

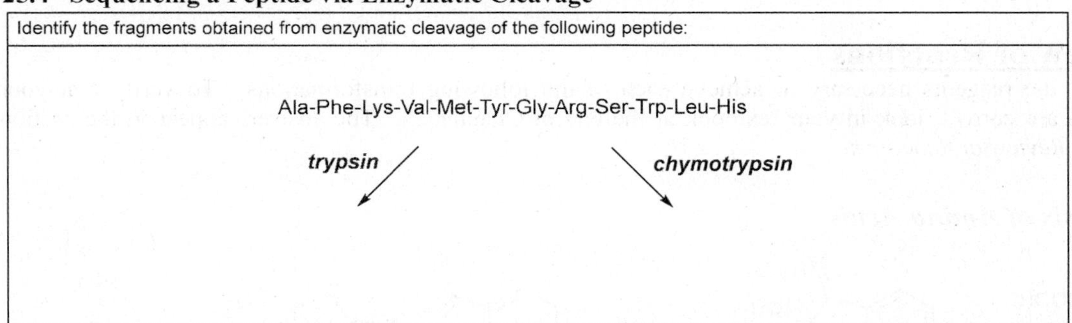

Identify the fragments obtained from enzymatic cleavage of the following peptide:
Ala-Phe-Lys-Val-Met-Tyr-Gly-Arg-Ser-Trp-Leu-His
trypsin
chymotrypsin

25.5 Planning the Synthesis of a Dipeptide

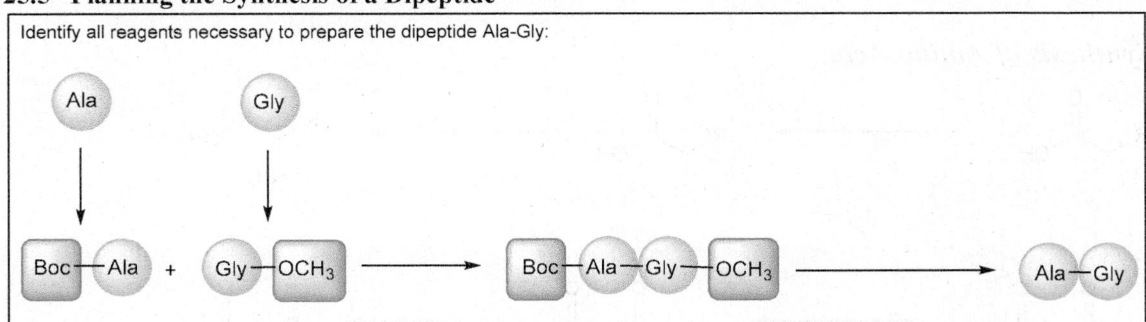

25.6 Preparing a Peptide using the Merrifield Synthesis

Identify all reagents necessary to prepare the pentapeptide Ile-Gly-Leu-Ala-Phe:

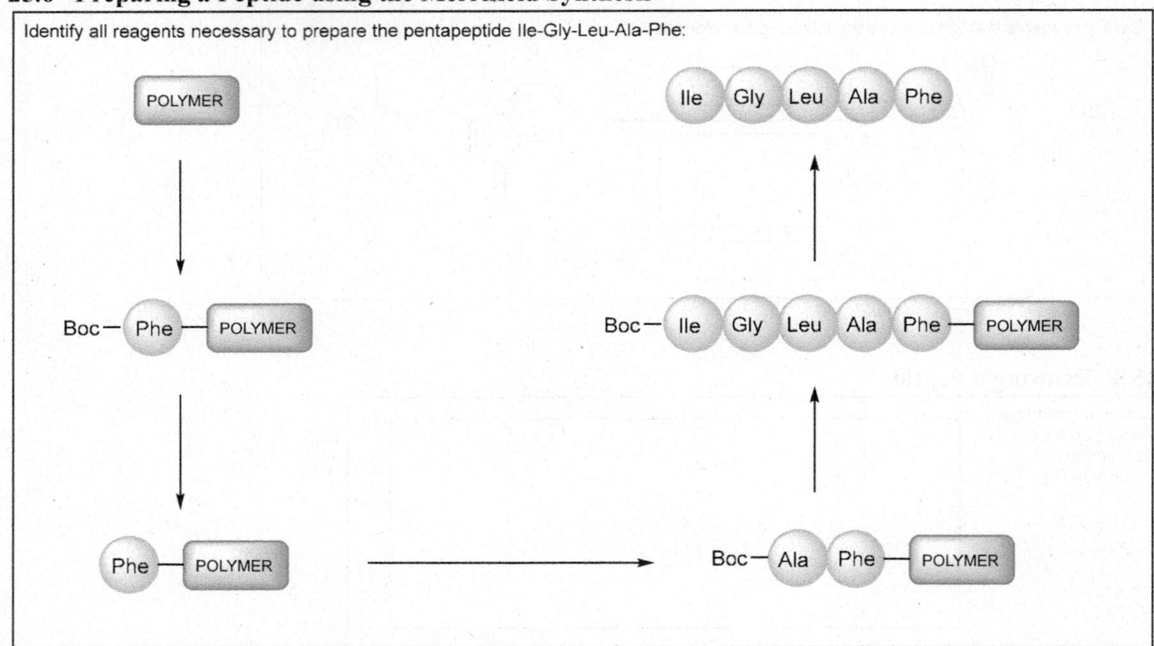

Review of Reactions

Identify the reagents necessary to achieve each of the following transformations. To verify that your answers are correct, look in your textbook at the end of Chapter 25. The answers appear in the section entitled *Review of Reactions*.

Analysis of Amino Acids

Synthesis of Amino Acids

O; R, H; H2N, C≡N, R, H; H3N⊕, O, C, OH, R, H

O, OH, NHAc; O, OH, NHAc; O, OH, NH2

Analysis of Amino Acids

PEPTIDE, O, N, H, NH2, R; PEPTIDE, N, H, H; +; Ph, S, N, NH, O, R

Synthesis of Peptides

O, OH; +; H2N; O, N, H

H2N, O, OH, R; O, H, N, O, O, OH, R

H2N, O, OH, R; H2N, O, OR, R

Review of Mechanisms

Complete each of the following mechanisms by drawing the missing curved arrows. To verify that your curved arrows are drawn correctly, compare them to the curved arrows in the mechanism boxes for Mechanisms 25.1 – 25.6, which can be found throughout Chapter 25 of your text.

Mechanism 25.1 Formation of an α-Amino Nitrile

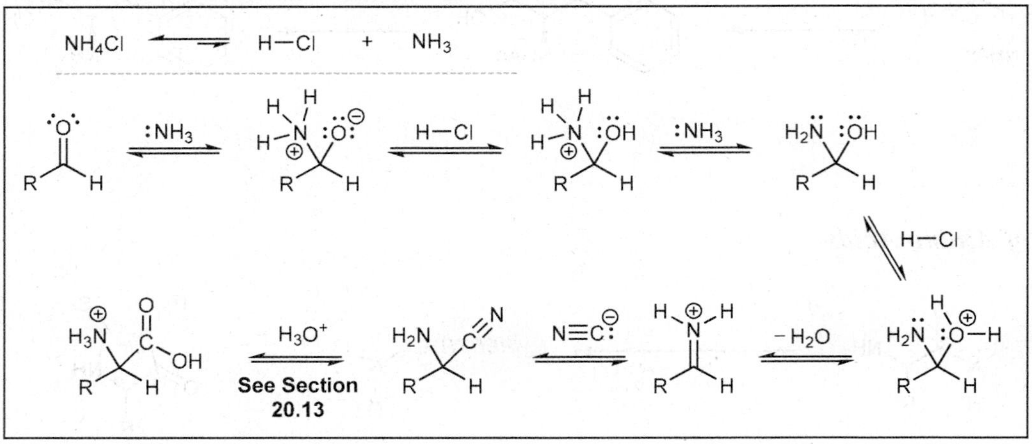

$NH_4Cl \rightleftharpoons H{-}Cl + NH_3$

:NH₃ H–Cl :NH₃ H–Cl NH₃ N≡C:⁻ – H₂O H₃O⁺

See Section 20.13

Mechanism 25.2 The Edman Degradation

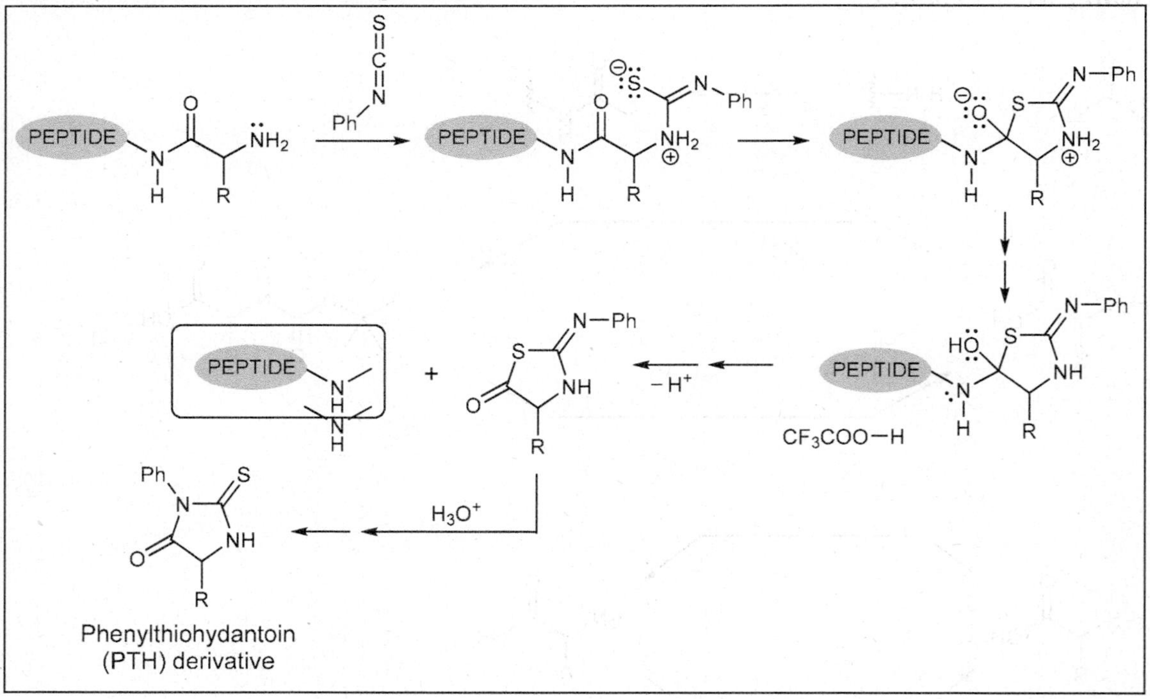

PEPTIDE PEPTIDE PEPTIDE PEPTIDE PEPTIDE

– H⁺ CF₃COO–H H₃O⁺

Phenylthiohydantoin (PTH) derivative

Mechanism 25.3 Peptide Bond Formation via DCC

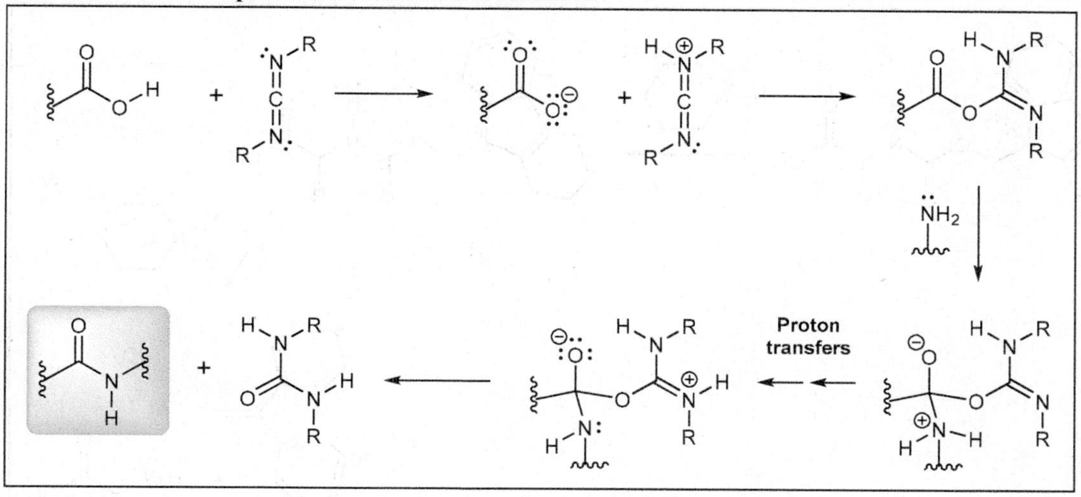

Proton transfers

Mechanism 25.4 Installation of a Boc Protecting Group

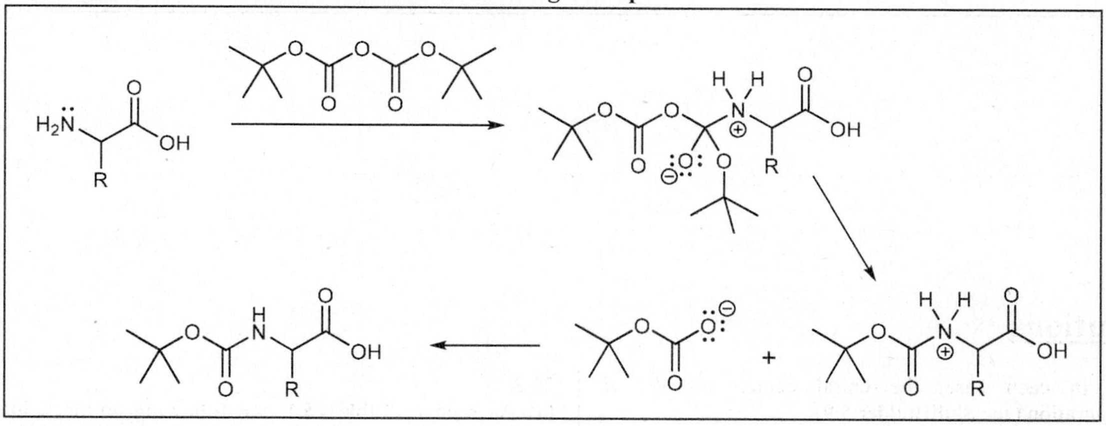

Mechanism 25.5 Removal of a Boc Protecting Group

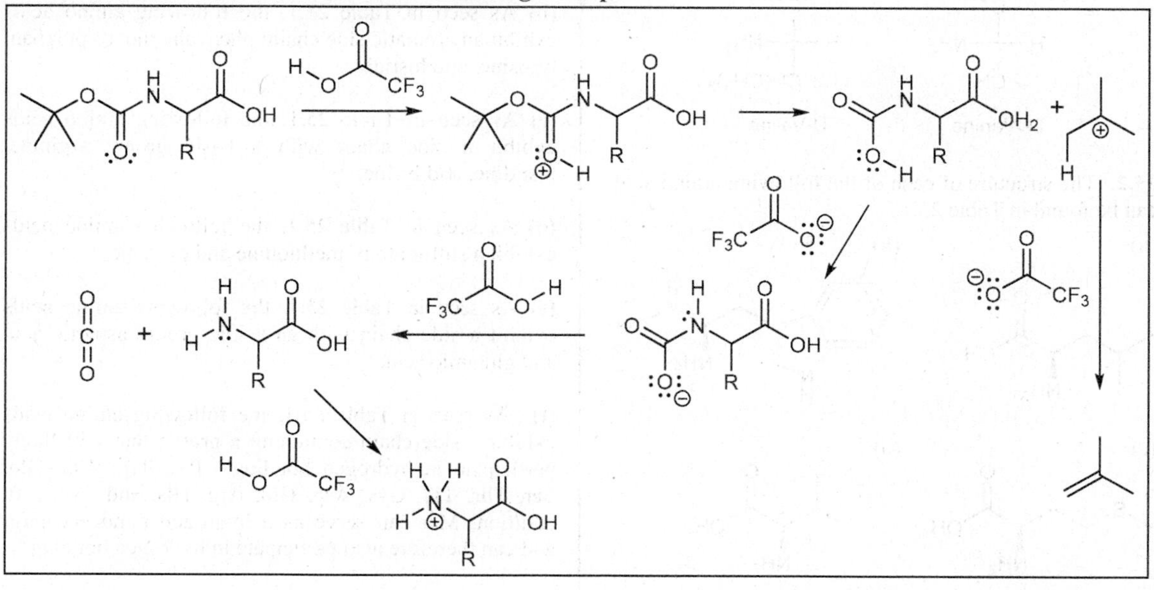

Mechanism 25.6 Removal of an Fmoc Protecting Group

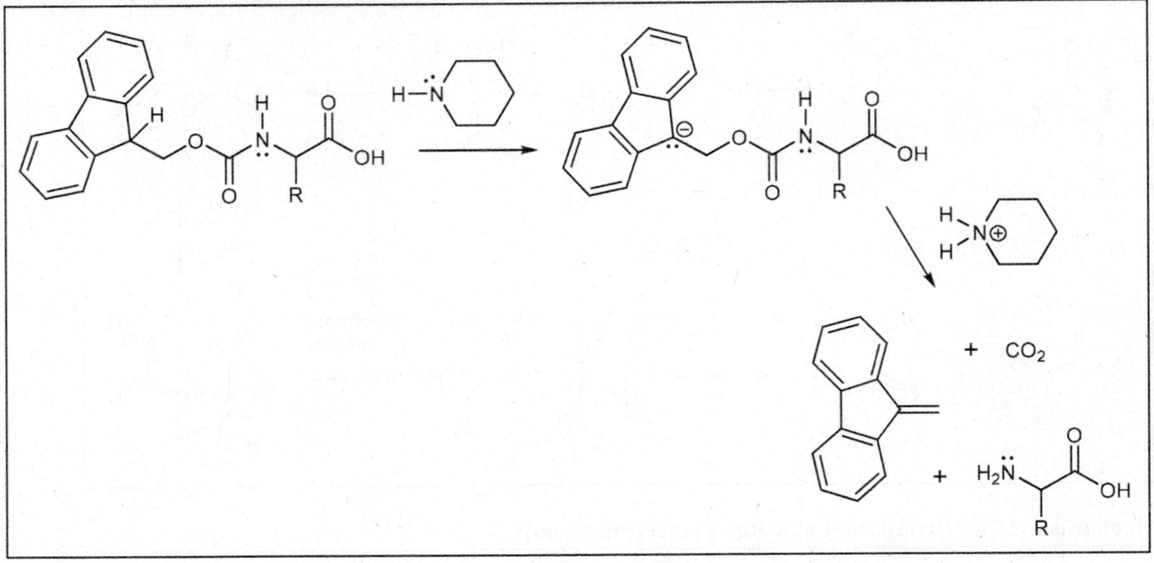

Solutions

25.1. In each case, the chiral center has the *R* configuration (see SkillBuilder 5.9).

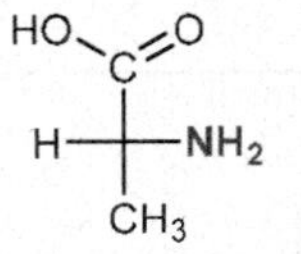

D-Alanine 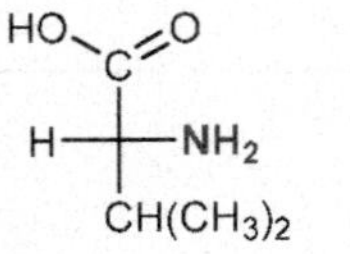 D-Valine

25.2. The structure of each of the following amino acids can be found in Table 25.1.

(a) **(b)**

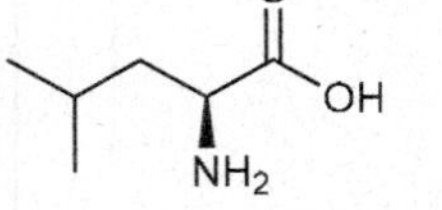

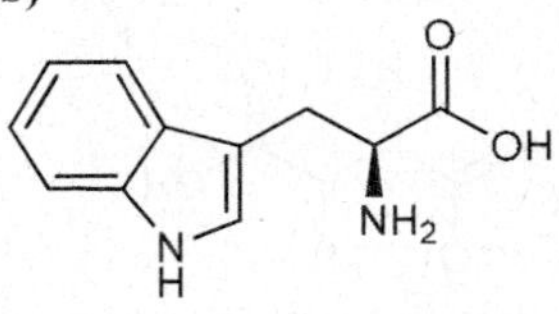

(c) 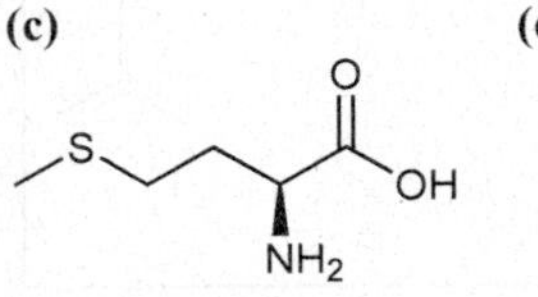 **(d)**

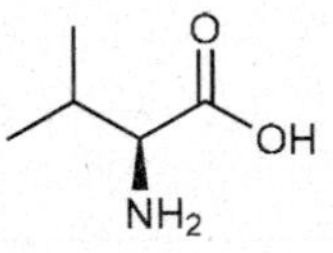

25.3.

(a) As seen in Table 25.1, the following amino acids exhibit a cyclic structure: proline, phenylalanine, tryptophan, tyrosine, and histidine.

(b) As seen in Table 25.1, the following amino acids exhibit an aromatic side chain: phenyalanine, tryptophan, tyrosine, and histidine.

(c) As seen in Table 25.1, the following amino acids exhibit a side chain with a basic group: arginine, histidine, and lysine.

(d) As seen in Table 25.1, the following amino acids exhibit a sulfur atom: methionine and cysteine.

(e) As seen in Table 25.1, the following amino acids exhibit a side chain with an acidic group: aspartic acid and glutamic acid.

(f) As seen in Table 25.1, the following amino acids exhibit a side chain containing a proton that will likely participate in hydrogen bonding: Pro, Trp, Asn, Gln, Ser, Thr, Tyr, Cys, Asp, Glu, Arg, His, and Lys. In addition, Met can serve as a hydrogen bond acceptor, and can therefore also participate in hydrogen bonding.

25.4. In each case, we first identify the pK_a of the carboxylic acid group and determine which form predominates. The protonated form (RCOOH) will predominate if pH < pK_a, while the carboxylate ion will predominate if pH > pK_a. Next, we identify the pK_a of the α-amino group and determine which form predominates. The protonated form (RNH_3^+) will predominate if pH < pK_a, while the uncharged form (RNH_2) will predominate if pH > pK_a. Finally, a similar analysis is performed for the side chain (if necessary). See Table 25.2 for pK_a values.

(a)

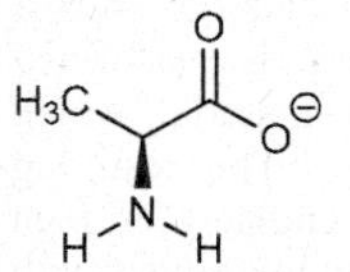

(b)

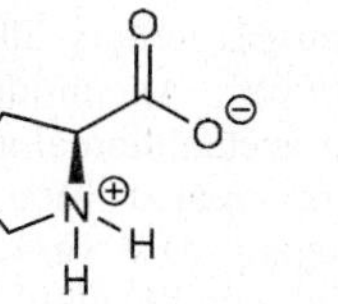

(c)

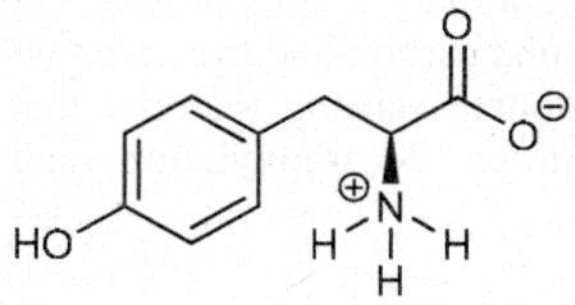

(d)

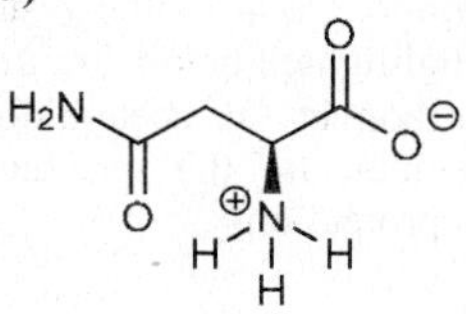

(e)

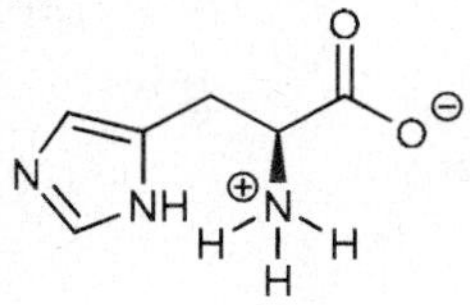

(f)

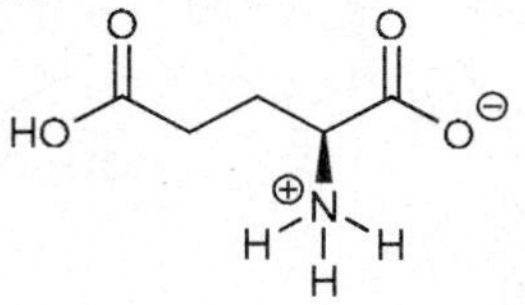

25.5. Arginine has a basic side chain, while asparagine does not. At a pH of 11, arginine exists predominantly in a form in which the side chain is protonated. Therefore, it can serve as a proton donor.

25.6. Tyrosine possesses a phenolic proton which is more readily deprotonated because deprotonation forms a resonance-stabilized phenolate ion. In contrast, deprotonation of the OH group of serine gives an alkoxide ion that is not resonance-stabilized. As a result, the OH group of tyrosine is more acidic than the OH group of serine.

25.7.
(a) Aspartic acid has two carboxylic acid groups, so the pI of aspartic acid is calculated using the pK_a values of the two carboxylic acid groups, as shown here (pK_a values can be found in Table 25.2):

$$pI = \frac{1.88 + 3.65}{2} = 2.77$$

(b) Leucine does not have an acidic side chain or a basic side chain, so the pI of leucine is calculated using the pK_a value of the carboxylic acid group and the pK_a value of the ammonium group, as shown here (pK_a values can be found in Table 25.2):

$$pI = \frac{2.36 + 9.60}{2} = 5.98$$

(c) Lysine has two ammonium groups, so the pI of lysine is calculated using the pK_a values of the two ammonium groups, as shown here (pK_a values can be found in Table 25.2):

$$pI = \frac{8.95 + 10.53}{2} = 9.74$$

(d) Proline does not have an acidic side chain or a basic side chain, so the pI of proline is calculated using the pK_a value of the carboxylic acid group and the pK_a value of the ammonium group, as shown here (pK_a values can be found in Table 25.2):

$$pI = \frac{1.99 + 10.60}{2} = 6.30$$

25.8.
(a) The pK_a values for carboxylic acid groups (~2-4) are significantly lower than the pK_a values for ammonium groups (~9-10). Aspartic acid has two carboxylic acid groups, so it is expected to have the lowest pI.

(b) The pK_a values for carboxylic acid groups (~2-4) are significantly lower than the pK_a values for ammonium groups (~9-10). Glutamic acid has two carboxylic acid groups, so it is expected to have the lowest pI.

25.9. Leucine and isoleucine both exhibit the same pK_a value for the carboxylic acid group. Similarly, both leucine and isoleucine exhibit the same pK_a value for the amino group. As such, the pI value of leucine is expected to be the same as the pI value of isoleucine.

25.10. The greater the difference between pI and pH, the faster an amino acid will migrate. The pI of Phe = 5.48, the pI of Trp = 6.11, and the pI of Leu = 6.00. Using these values, we make the following predictions:
(a) At pH = 6.0, Phe will travel the farthest distance.
(b) At pH = 5.0, Trp will travel the farthest distance.

25.11. The following aldehyde is expected when L-leucine is treated with ninhydrin:

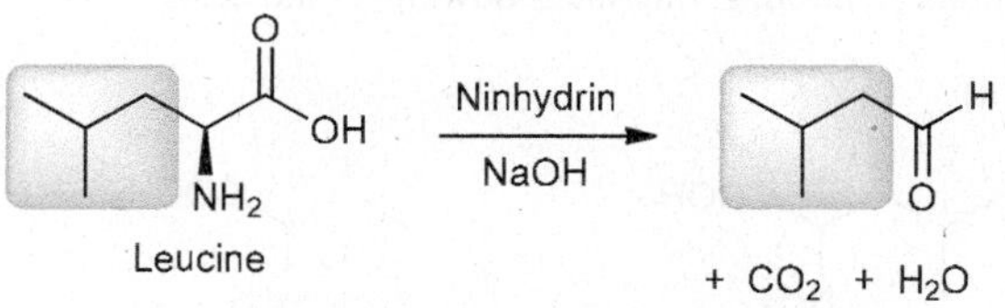

Leucine —Ninhydrin, NaOH→ aldehyde + CO_2 + H_2O

25.12.

(a) Racemic leucine can be made using a Hell-Volhard-Zelinsky reaction, as shown:

1) Br_2, PBr_3 2) H_2O 3) xs NH_3 → NH_2 (racemic)

1) Br_2, PBr_3 2) H_2O → Br; xs NH_3

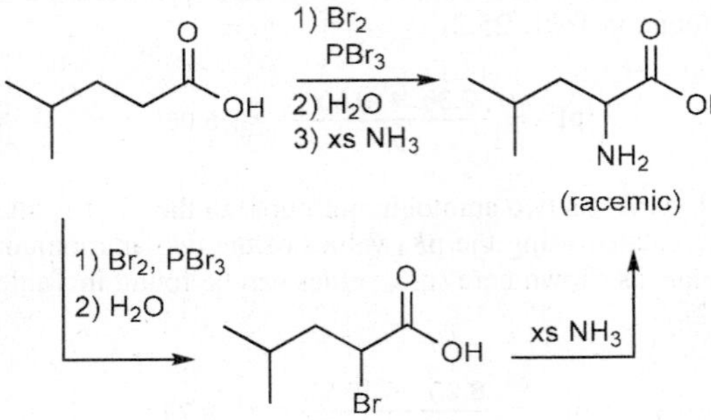

(b) Racemic alanine can be made using a Hell-Volhard-Zelinsky reaction, as shown:

1) Br_2, PBr_3 2) H_2O 3) xs NH_3 → NH_2 (racemic)

1) Br_2, PBr_3 2) H_2O → Br; xs NH_3

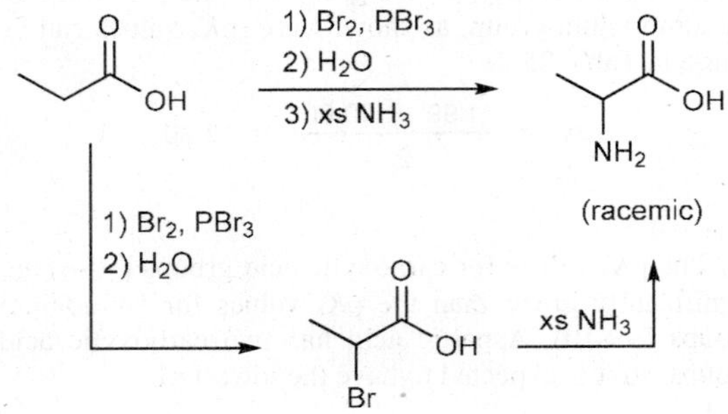

(c) Racemic valine can be made using a Hell-Volhard-Zelinsky reaction, as shown:

1) Br_2, PBr_3 2) H_2O 3) xs NH_3 → NH_2 (racemic)

1) Br_2, PBr_3 2) H_2O → Br; xs NH_3

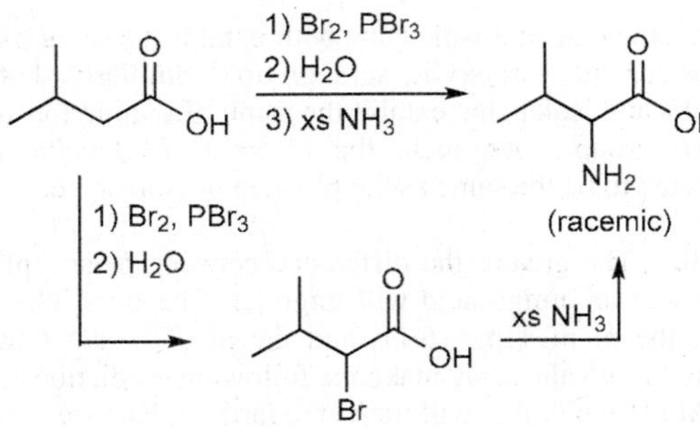

25.13. In each case, the process is a Hell-Volhard-Zelinsky reaction, which installs an amino group at the alpha position, giving the following amino acids:

(a) Leucine **(b)** Valine

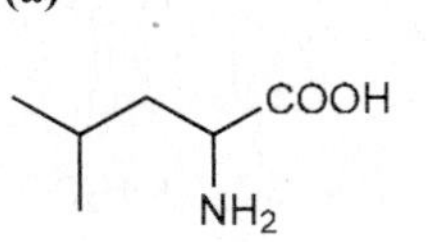

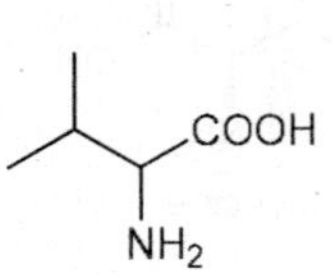

(c) Phenylalanine **(d)** Glycine (H_2N–COOH)

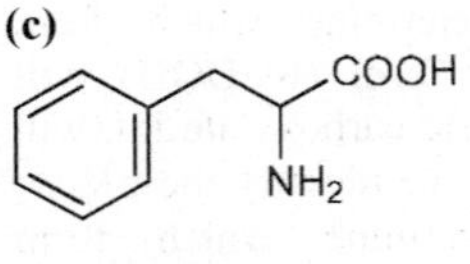

25.14. In each case, we begin by identifying the side chain connected to the α position. Then, we identify the necessary alkyl halide and ensure that it is not tertiary (because a tertiary alkyl halide will not undergo an S_N2 reaction). An amidomalonate synthesis is performed using acetamidomalonate as the starting material, which is first treated with sodium ethoxide. The resulting conjugate base (a doubly stabilized enolate) is then treated with the alkyl halide, followed by hydrolysis with aqueous acid and heat, to give the desired amino acid. Note that the final step employs acidic conditions, so the amino group of the resulting amino acid is protonated. Highlighted below are the alpha carbon and the carbonyl carbon in the acetamidomalonate starting material that remain in the product (after decarboxylation and deprotection):

(a) 1) NaOEt 2) (2-bromobutane) 3) H_3O^+, heat → NH_3^+ product

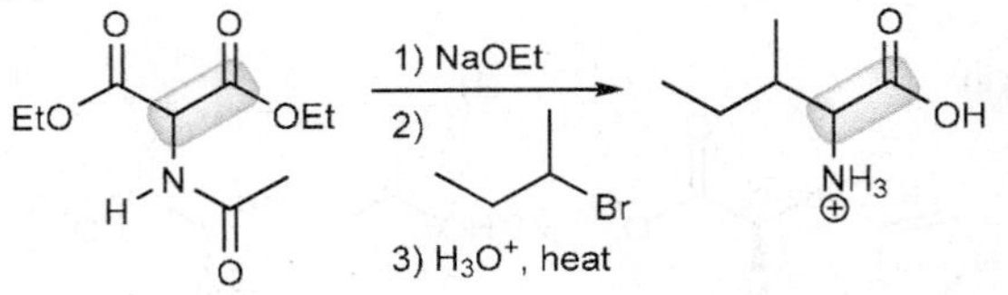

(b) 1) NaOEt 2) CH_3Br 3) H_3O^+, heat → NH_3^+ product

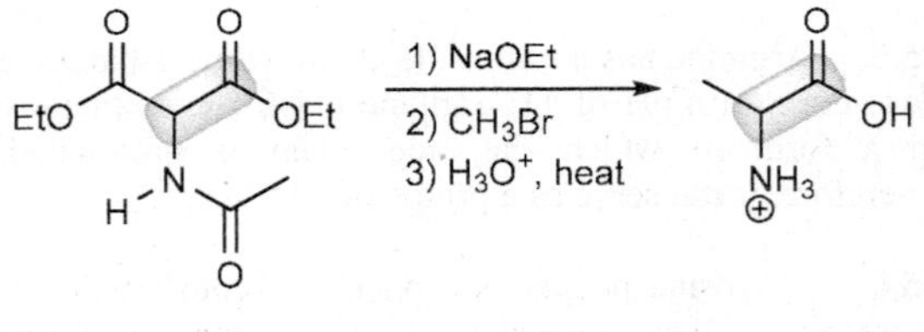

(c) 1) NaOEt 2) (2-bromopropane) 3) H_3O^+, heat → NH_3^+ product

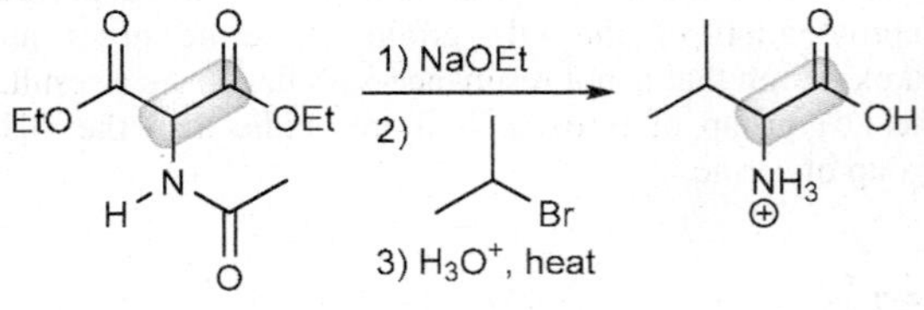

25.15.

(a) Alanine is obtained when methyl iodide is used as the alkyl halide in an amidomalonate synthesis. The methyl group (from methyl iodide) is highlighted in the product:

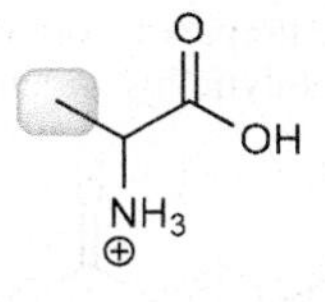

(b) Valine is obtained when isopropyl chloride is used as the alkyl halide in an amidomalonate synthesis. The isopropyl group (from isopropyl chloride) is highlighted in the product:

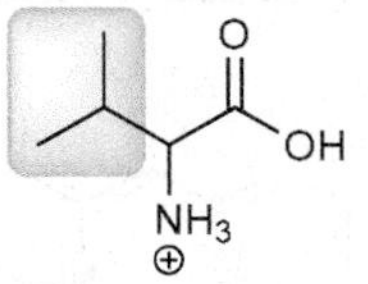

(c) Leucine is obtained when 2-methyl-1-chloropropane is used as the alkyl halide in an amidomalonate synthesis. The alkyl group (from the alkyl chloride) is highlighted in the product:

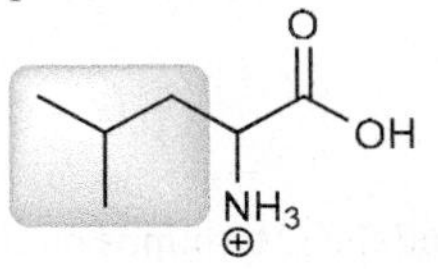

25.16. Leucine can be prepared via the amidomalonate synthesis with higher yields than isoleucine, because leucine requires an SN2 reaction with a primary alkyl halide, while isoleucine requires an SN2 reaction with a secondary (more hindered) alkyl halide.

25.17.

(a) Methionine can be prepared from the aldehyde below via a Strecker synthesis, as shown:

1) NH_4Cl, NaCN
2) H_3O^+

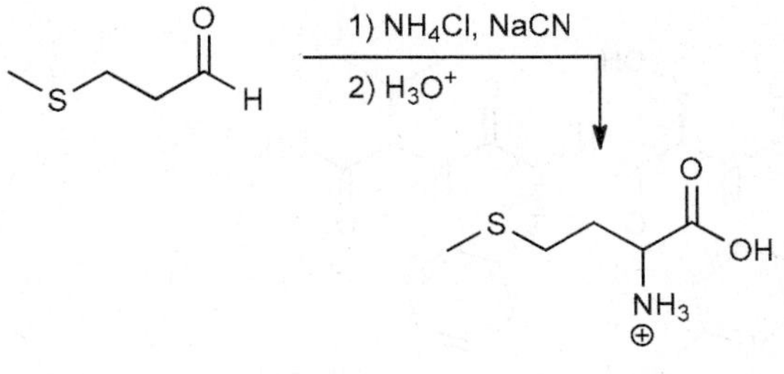

(b) Histidine can be prepared from the aldehyde below via a Strecker synthesis, as shown:

1) NH_4Cl, NaCN
2) H_3O^+

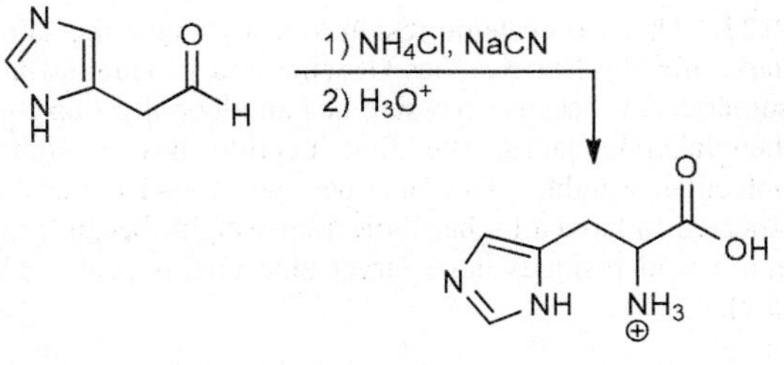

(c) Phenylalanine can be prepared from the aldehyde below via a Strecker synthesis, as shown:

1) NH_4Cl, NaCN
2) H_3O^+

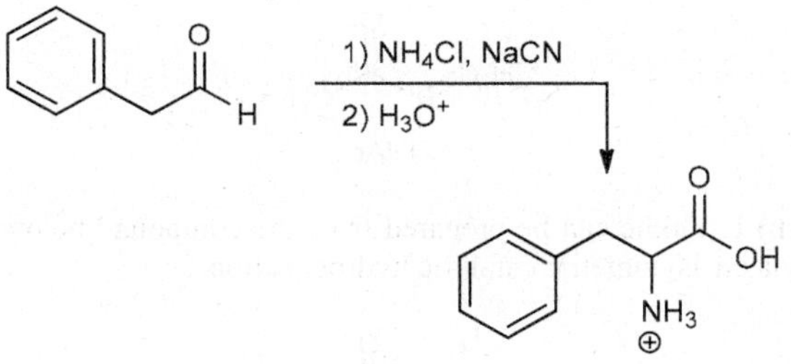

(d) Leucine can be prepared from the aldehyde below via a Strecker synthesis, as shown:

1) NH_4Cl, NaCN
2) H_3O^+

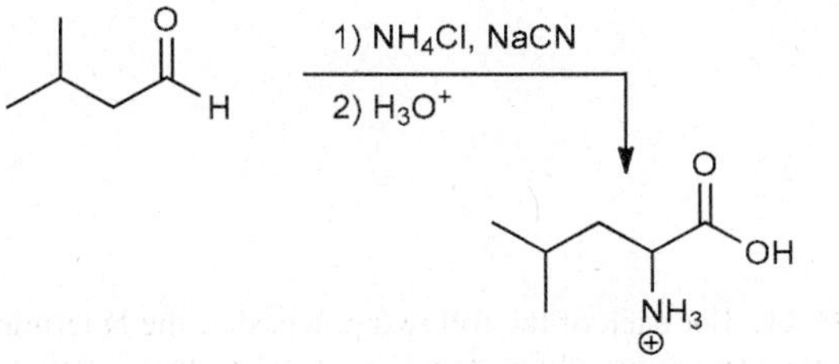

25.18.

(a) Acetaldehyde is converted into a racemic mixture of alanine (via a Strecker synthesis), as shown:

1) NH_4Cl, NaCN
2) H_3O^+

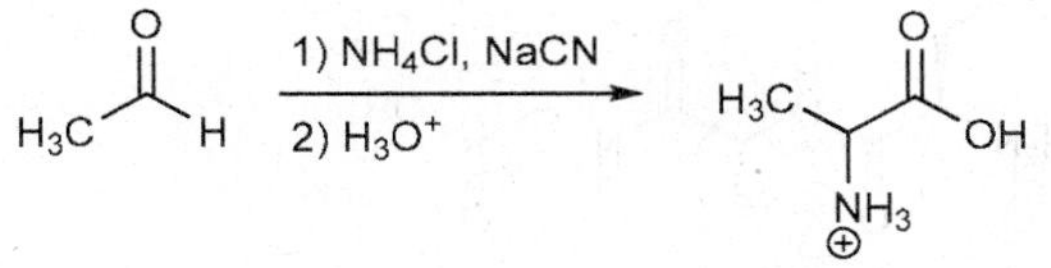

Alanine

(b) 3-Methylbutanal is converted into a racemic mixture of leucine (via a Strecker synthesis), as shown:

1) NH_4Cl
NaCN
2) H_3O^+

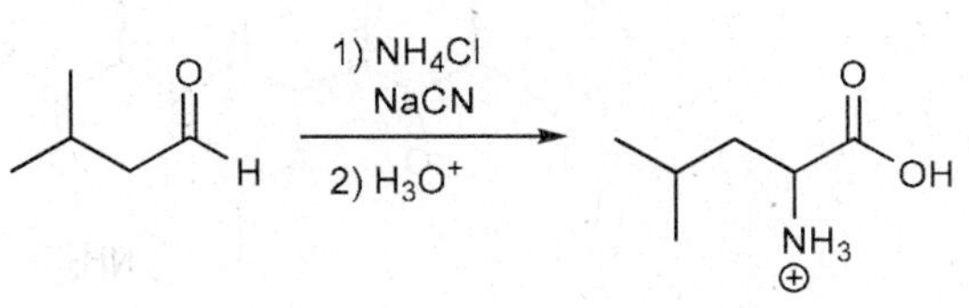

Leucine

(c) 2-Methylpropanal is converted into a racemic mixture of valine (via a Strecker synthesis), as shown:

1) NH_4Cl, NaCN
2) H_3O^+

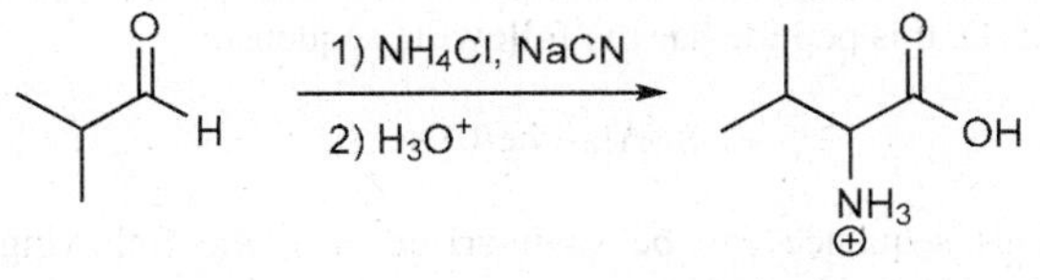

Valine

25.19.

(a) L-Alanine can be prepared from the following compound via an asymmetric catalytic hydrogenation:

H_2C OH NHAc

(b) L-Valine can be prepared from the compound below via an asymmetric catalytic hydrogenation:

OH NHAc

(c) L-Leucine can be prepared from the compound below via an asymmetric catalytic hydrogenation:

OH NHAc

(d) L-Tyrosine can be prepared from the compound below via an asymmetric catalytic hydrogenation:

HO OH NHAc

25.20. Glycine does not possess a chiral center, so the use of a chiral catalyst is unnecessary. Also, there is no alkene that would lead to glycine upon hydrogenation.

25.21. For each of the following peptides, the N terminus is drawn on the left and the C terminus on the right. Side chains at the top of the drawing are on wedges, while side chains on the bottom of the drawing are on dashes. The identity of each side chain can be found in Table 25.1.

(a) Leu-Ala-Gly

H_2N OH

(b) Cys-Asp-Ala-Gly

HS H_2N COOH OH

(c) Met-Lys-His-Tyr-Ser-Phe-Val

S HN N OH H_2N NH_2 OH OH

25.22. Based on the identities of side chains (see Table 25.1), this peptide has the following sequence:

Leu-Ala-Phe-Cys-Asp

This sequence can be summarized with the following one-letter abbreviations:

LAFCD

25.23. The two cysteine residues clearly have the same mass, and the last residues (leucine and isoleucine) are isomeric. So, because tyrosine has an alcohol group that phenylalanine lacks, the first peptide has a higher molecular weight. The first peptide (Cys-Tyr-Leu) is expected to have a higher molecular weight, because the amino acid residues have larger side chains (see Table 25.1).

25.24. These peptides have the same molecular formula but they differ from each other in their connectivity of atoms (or constitution). As such, they are constitutional isomers.

25.25. The following is the *s-trans* conformation of the dipeptide Phe-Leu. Notice that the N terminus is on the left, while the C terminus is on the right, as per accepted convention. Also notice that the side chain at the top of the drawing is on a wedge, while the side chain at the bottom of the drawing is on a dash (see SkillBuilder 25.3).

C terminus

N terminus

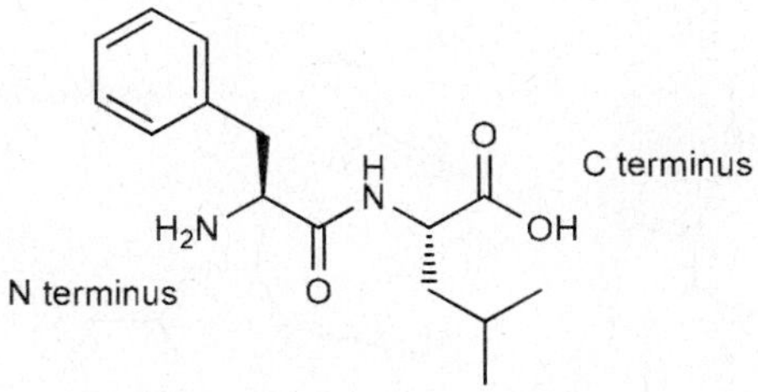

25.26. The following is the *s-cis* conformation of the dipeptide Phe-Phe. In this conformation, the phenyl groups will experience a severe steric interaction, thereby causing the *s-cis* conformation to be extremely high in energy:

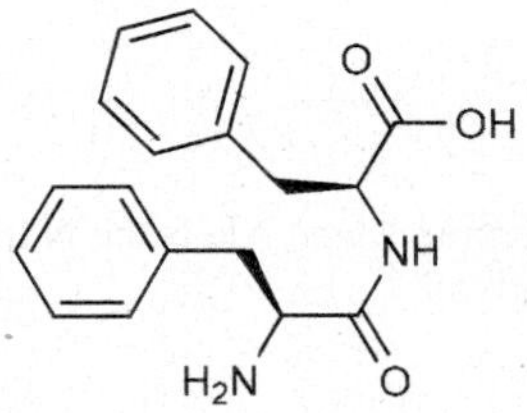

25.27. Two equivalents of the Cys-Phe dipeptide are drawn, and they are then connected by a disulfide bridge, as shown:

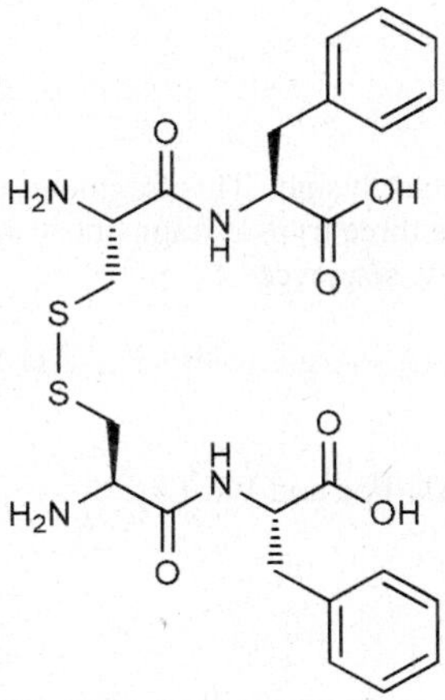

25.28.

(a) Below is the structure of aspartame (Asp-Phe-OCH_3). Notice that the N terminus is on the left, while the C terminus is on the right, as per accepted convention. In this case, the C terminus is a methyl ester (rather than a carboxylic acid). Also notice that the side chain at the top of the drawing is on a wedge, while the side chain at the bottom of the drawing is on a dash (see SkillBuilder 25.3).

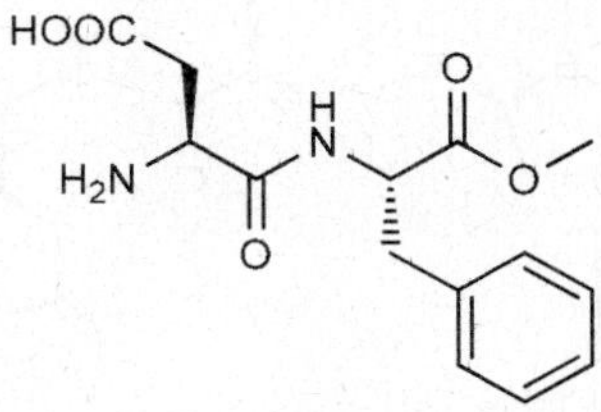

(b) The compound above has two chiral centers, giving rise to a total of four possible stereoisomers. The structure above represents one of these stereoisomers, with both amino acids having the naturally occurring L configuration. The other three stereoisomers, having a bitter taste, are shown here:

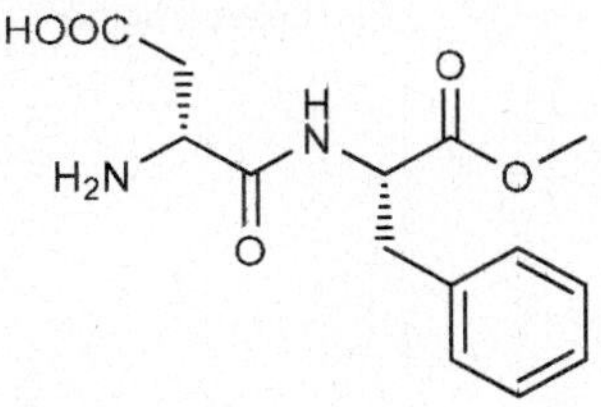

(D-Asp and L-Phe)

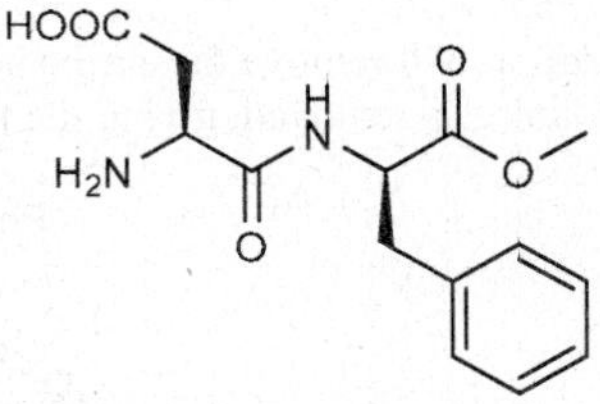

(L-Asp and D-Phe)

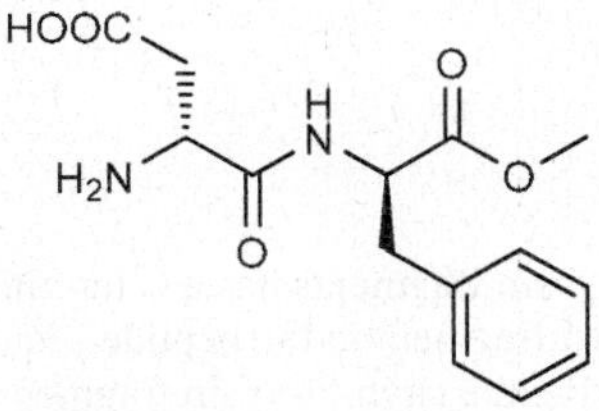

(D-Asp and D-Phe)

25.29. In the following structure, each of the amino acid residues has been highlighted and labeled:

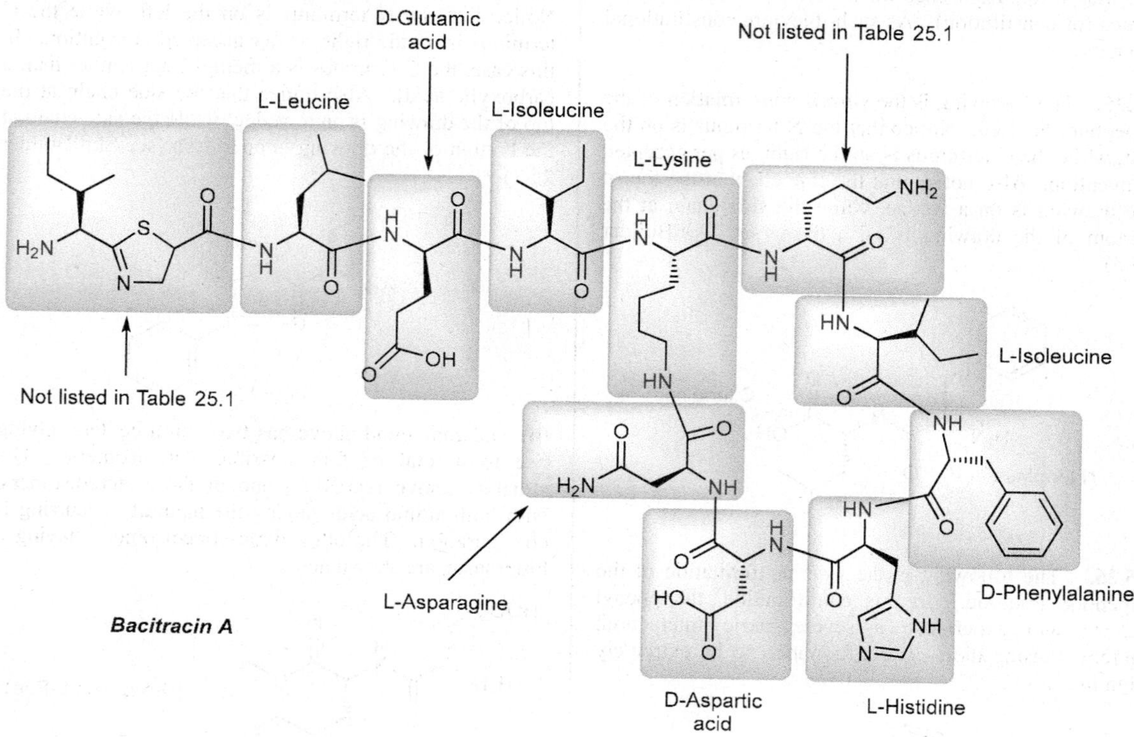

25.30. An Edman degradation will remove the amino acid residue at the N terminus, and Ala is the N terminus in Ala-Phe-Val. Therefore, alanine is removed, giving the following PTH derivative:

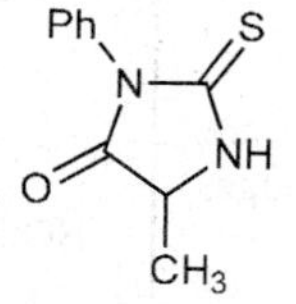

25.31. Only one of the trypsin fragments has a C terminus that is not arginine or lysine. This fragment, which ends with valine, must be the last fragment in the peptide sequence. The remaining three trypsin fragments can be placed in the proper order by analyzing the chymotrypsin fragments. The correct peptide sequence is:

Met-Phe-Val-Ala-Tyr-Lys-Ser-Val-Ile-Leu-Arg-Trp-His-Phe-Met-Cys-Arg-Gly-Pro-Phe-Ala-Val

25.32. The following tetrapeptide will be cleaved by chymotrypsin to give Ala-Phe and Val-Lys:

Ala-Phe-Val-Lys

25.33. Cleavage with trypsin will produce Phe-Arg, while cleavage with chymotrypsin will produce Arg-Phe. These dipeptides are not the same. They are constitutional isomers.

25.34.

(a) We begin by installing the appropriate protecting groups. Then, upon treatment with DCC, the protected amino acids are coupled. And finally, the protecting groups are removed, as shown:

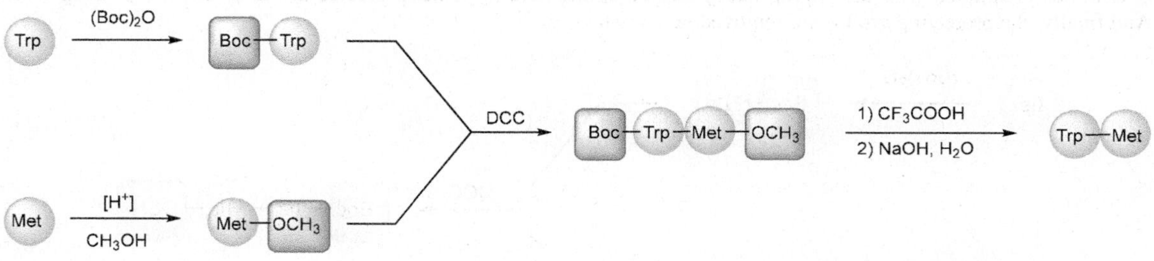

(b) We begin by installing the appropriate protecting groups. Then, upon treatment with DCC, the protected amino acids are coupled. And finally, the protecting groups are removed, as shown:

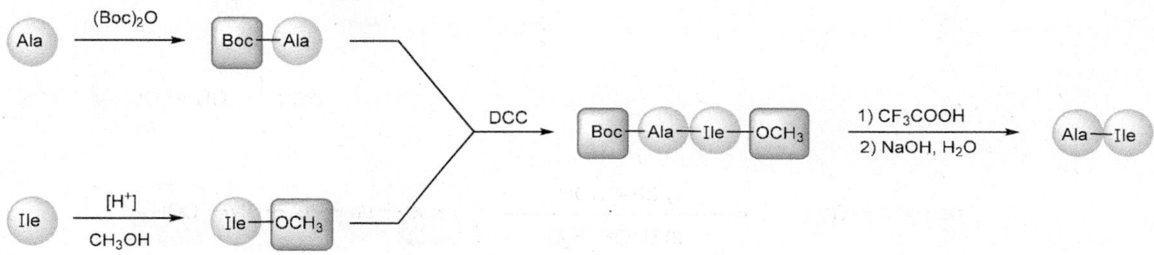

(c) We begin by installing the appropriate protecting groups. Then, upon treatment with DCC, the protected amino acids are coupled. And finally, the protecting groups are removed, as shown:

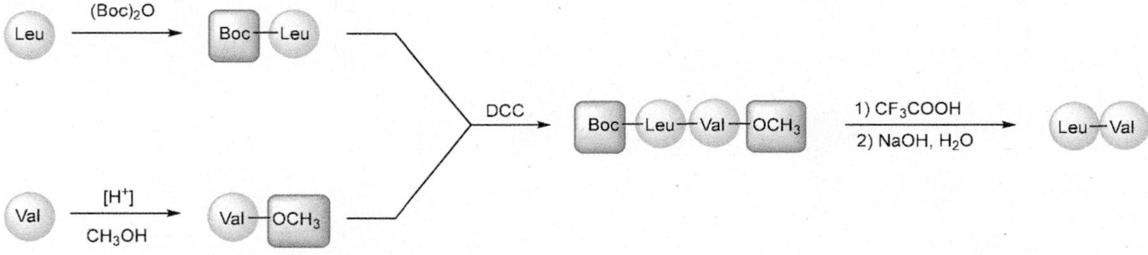

25.35. The first two amino acid residues (in the desired peptide sequence) are Ile and Phe. So we must begin with those amino acids. We first install the appropriate protecting groups. Then, upon treatment with DCC, the protected amino acids are coupled. The protecting group at the C terminus is then removed and the resulting unprotected C terminus is coupled with the appropriate protected amino acid (glycine, protected at the C terminus), using DCC. And finally, the protecting groups are removed, as shown:

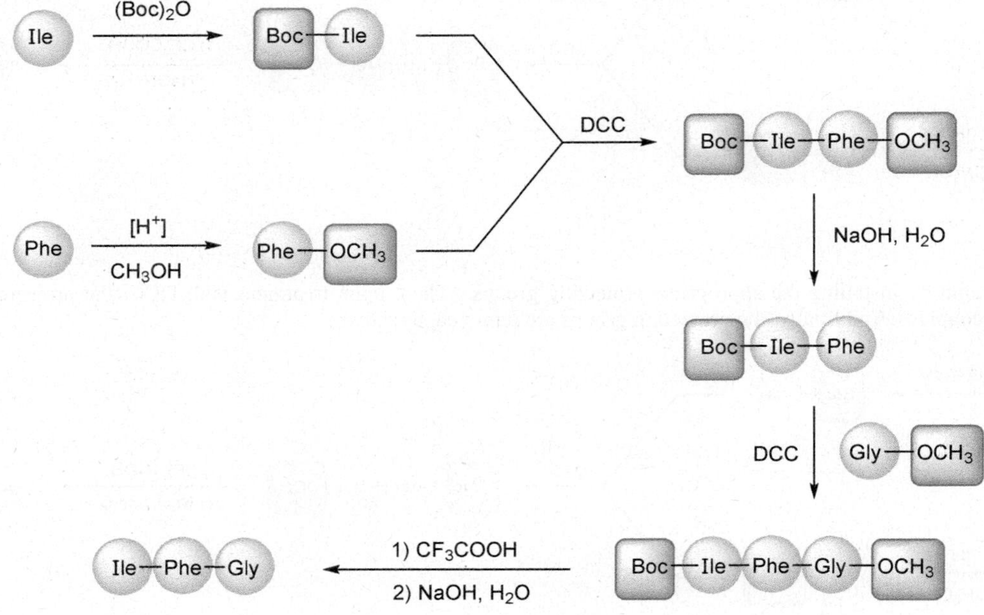

25.36. The first two amino acid residues (in the desired peptide sequence) are Leu and Val. So we must begin with those amino acids. We first install the appropriate protecting groups. Then, upon treatment with DCC, the protected amino acids are coupled. The protecting group at the C terminus is then removed and the resulting unprotected C terminus is coupled with the appropriate protected amino acid, using DCC. Each additional residue is installed via deprotection of the C terminus followed by coupling with the appropriate protected amino acid, as shown. Finally, the protecting groups are removed, giving the desired pentapeptide:

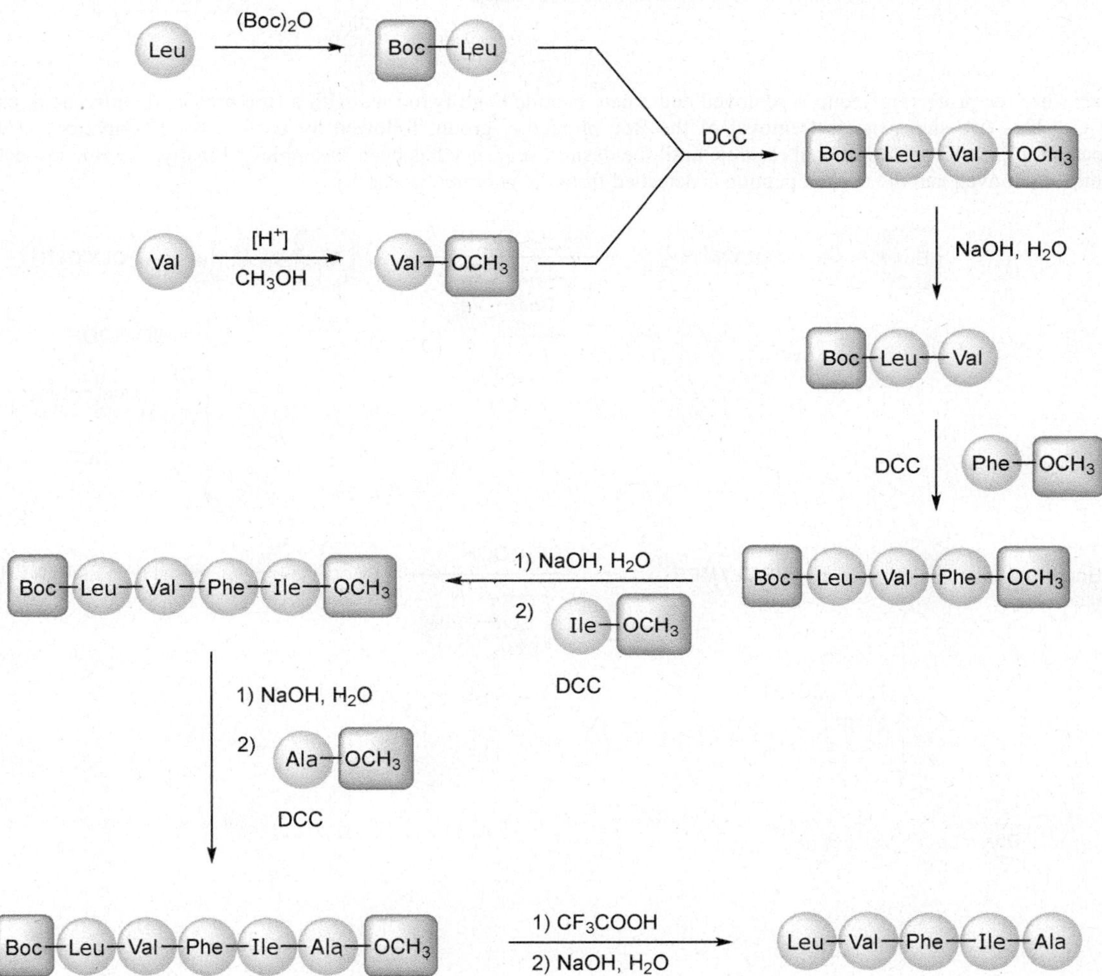

25.37.

(a) First, we attach the appropriate Boc-protected residue to the polymer, starting with the amino acid residue that corresponds to the C terminus of the target peptide:

Then, the Boc protecting group is removed and a new peptide bond is formed with a Boc-protected amino acid, using DCC. This two-step process (removal of the Boc protecting group, followed by peptide bond formation) is then repeated to install each additional residue, until the desired sequence has been assembled. Finally, the Boc protecting group is removed and the desired peptide is detached from the polymer, as shown:

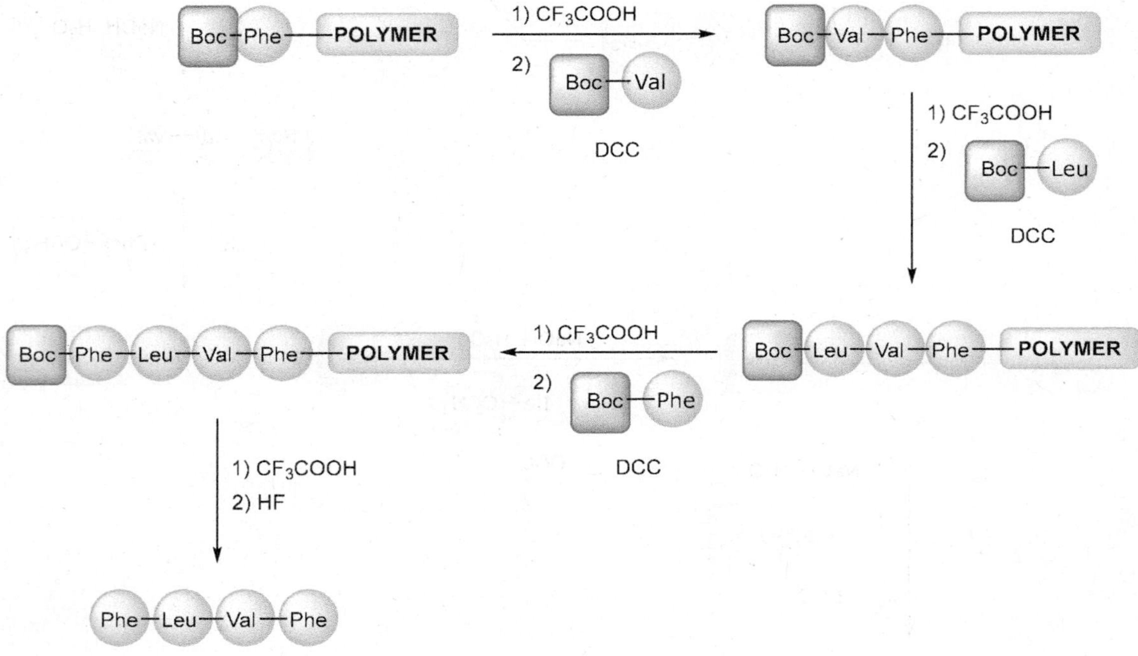

(b) First, we attach the appropriate Boc-protected residue to the polymer, starting with the amino acid residue that corresponds to the C terminus of the target peptide:

Then, the Boc protecting group is removed and a new peptide bond is formed with a Boc-protected amino acid, using DCC. This two-step process (removal of the Boc protecting group, followed by peptide bond formation) is then repeated to install each additional residue, until the desired sequence has been assembled. Finally, the Boc protecting group is removed and the desired peptide is detached from the polymer, as shown:

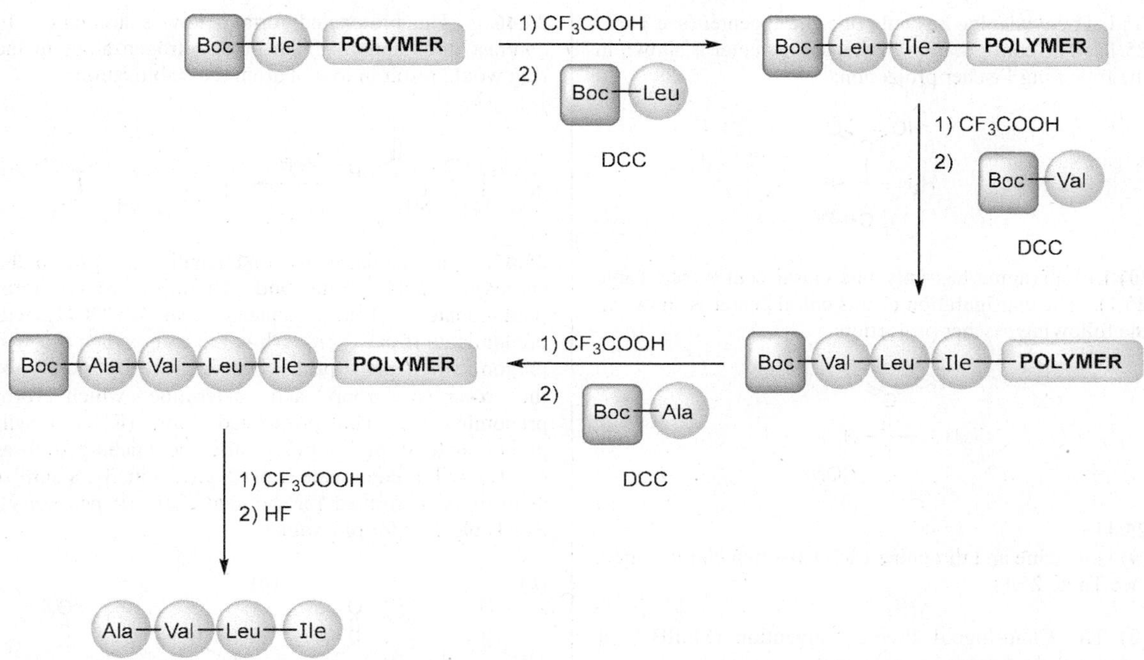

25.38. First, a protected phenylalanine residue is connected to the polymer. After deprotection, a protected alanine residue is installed. Then, after deprotection again, a protected valine residue is installed. Deprotection, followed by detachment from the polymer, gives the following tripeptide:

(N terminus) Val-Ala-Phe (C terminus)

25.39. The regions that contain repeating glycine and/or alanine units are the most likely regions to form β sheets:

Trp-His-Pro-Ala-Gly-Gly-Ala-Val-His-Cyst-Asp-Ser-Arg-Arg-Ala-Gly-Ala-Phe

25.40. In each case, the carboxylic acid group is drawn in its deprotonated form (as a carboxylate ion), and the amino group is drawn in its protonated form (as an ammonium ion):

(a)

(b)

(c)

(d)

25.41. When applying the Cahn-Ingold-Prelog convention for assigning the configuration of a chiral center, the amino group generally receives the highest priority (1), followed by the carboxylic acid group (2), followed by the side chain (3), and finally the H (4). Accordingly, the *S* configuration is assigned to L amino acids. Cysteine is the one exception because the sulfur atom makes the side chain a higher priority than the carboxylic acid group. As a result, the *R* configuration is assigned.

25.42.

(a) L-threonine has two chiral centers (see Table 25.1). The configuration of each of these chiral centers is shown in the following Fischer projection:

(b) L-Serine has only one chiral center (see Table 25.1). The configuration of this chiral center is shown in the following Fischer projection:

(c) L-Phenylalanine has only one chiral center (see Table 25.1). The configuration of this chiral center is shown in the following Fischer projection:

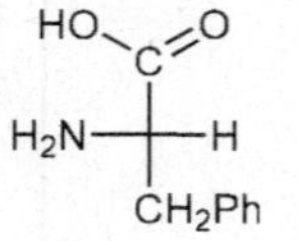

(d) L-Asparagine has only one chiral center (see Table 25.1). The configuration of this chiral center is shown in the following Fischer projection:

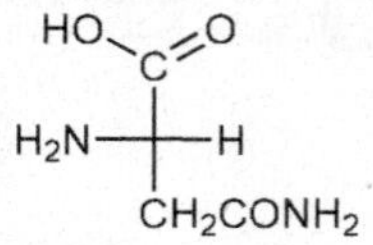

25.43.

(a) Isoleucine and threonine each have two chiral centers (see Table 25.1).

(b) The Cahn-Ingold Prelog convention (SkillBuilder 5.4) gives the following configurations:

Isoleucine = 2*S*, 3*S*

Threonine = 2*S*, 3*R*

25.44. Isoleucine has two chiral centers, so we expect four possible stereoisomers, shown here. The configuration of each chiral center is shown.

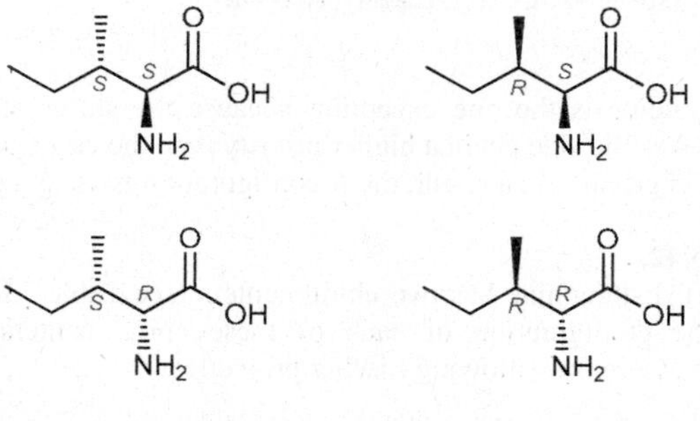

25.45. Protonation of the highlighted nitrogen atom gives a conjugate acid that is highly stabilized by resonance (the positive charge is highly delocalized).

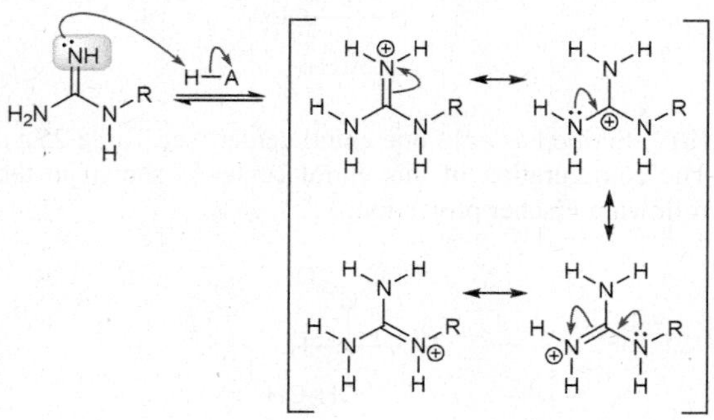

25.46. The protonated form below is aromatic. In contrast, protonation of the other nitrogen atom in the ring would result in loss of aromatic stabilization.

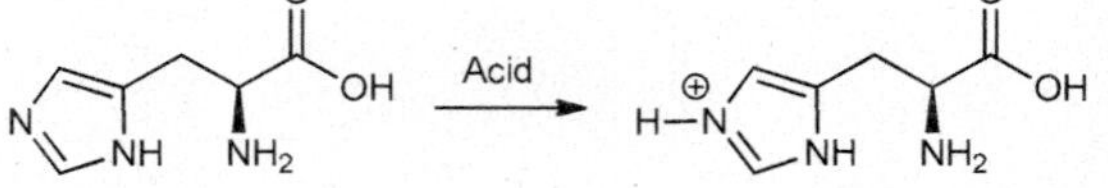

25.47. In each case, we first identify the pK_a of the carboxylic acid group and determine which form predominates. The protonated form (RCOOH) will predominate if pH < pK_a, while the carboxylate ion will predominate if pH > pK_a. Next, we identify the pK_a of the α-amino group and determine which form predominates. The protonated form (RNH_3^+) will predominate if pH < pK_a, while the uncharged form (RNH_2) will predominate if pH > pK_a. Finally, a similar analysis is performed for the side chain (if necessary). See Table 25.2 for pK_a values.

(a) **(b)**

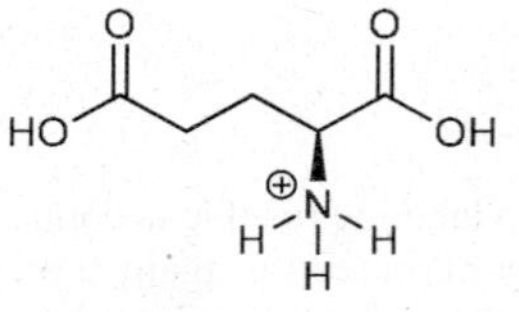

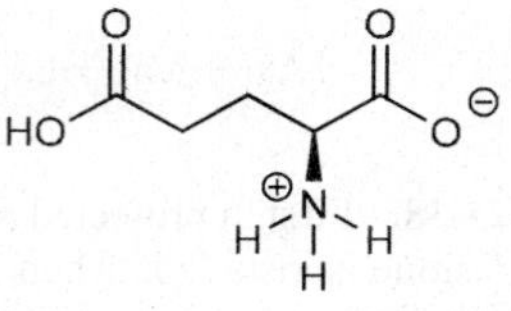

(c) 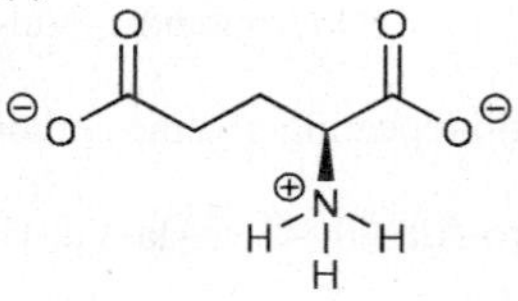 **(d)**

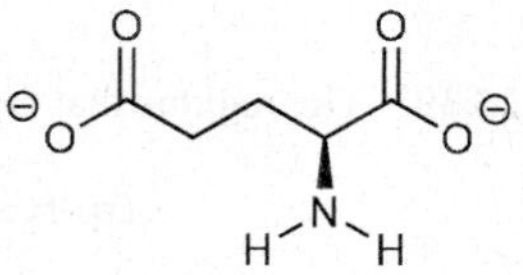

25.48. At physiological pH, each of the carboxylic acid groups is deprotonated (and will exist primarily as a carboxylate ion), while each of the amino groups is protonated (and will exist primarily as an ammonium ion):

(a) **(b)**

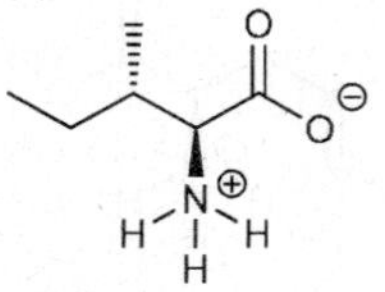

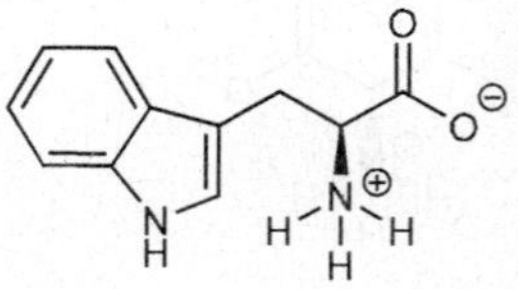

(c) 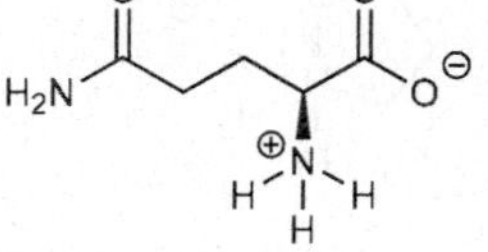 **(d)**

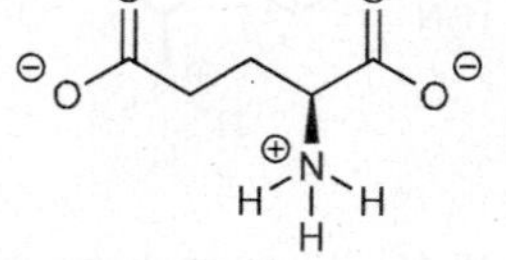

25.49.

(a) L-Alanine does not have an acidic side chain or a basic side chain, so the pI of L-alanine is calculated using the pK_a value of the carboxylic acid group and the pK_a value of the ammonium group, as shown here (pK_a values can be found in Table 25.2):

$$pI = \frac{2.34 + 9.69}{2} = 6.02$$

(b) L-Asparagine does not have an acidic side chain or a basic side chain, so the pI of L-asparagine is calculated using the pK_a value of the carboxylic acid group and the pK_a value of the ammonium group, as shown here (pK_a values can be found in Table 25.2):

$$pI = \frac{2.02 + 8.80}{2} = 5.41$$

(c) L-Histidine has a basic side chain. As such, the pI of L-histidine is calculated using the pK_a values of the two ammonium groups, as shown here (pK_a values can be found in Table 25.2):

$$pI = \frac{9.17 + 6.00}{2} = 7.58$$

(d) L-Glutamic acid has two carboxylic acid groups, so the pI of L-glutamic acid is calculated using the pK_a values of the two carboxylic acid groups, as shown here (pK_a values can be found in Table 25.2):

$$pI = \frac{2.19 + 4.25}{2} = 3.22$$

25.50. Lysozyme is likely to be comprised primarily of amino acid residues that contain basic side chains (arginine, histidine, and lysine), while pepsin is comprised primarily of amino acid residues that contain acidic side chains (aspartic acid and glutamic acid).

25.51. First, we must calculate the pI for each amino acid (using the procedure shown in the solution to Problem 25.49). Next, we identify the pK_a of the carboxylic acid group and determine which form predominates. The protonated form (RCOOH) will predominate if pI < pK_a, while the carboxylate ion will predominate if pI > pK_a.
Then, we identify the pK_a of the α-amino group and determine which form predominates. The protonated form (RNH_3^+) will predominate if pI < pK_a, while the uncharged form (RNH_2) will predominate if pI > pK_a. Finally, a similar analysis is performed for the side chain (if necessary). See Table 25.2 for pK_a values.

(a) **(b)**

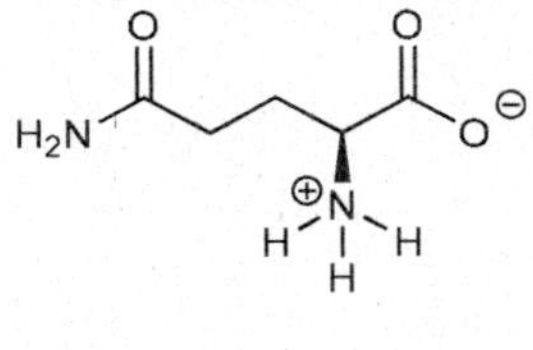

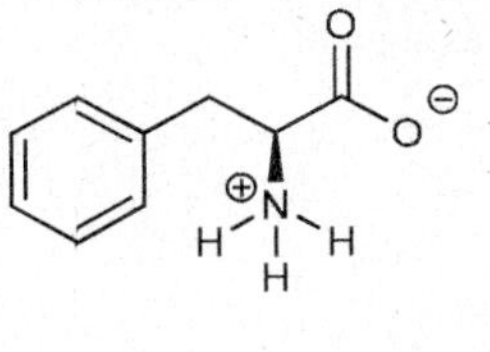

(c) 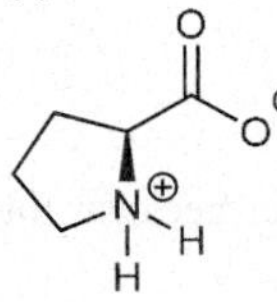 **(d)**

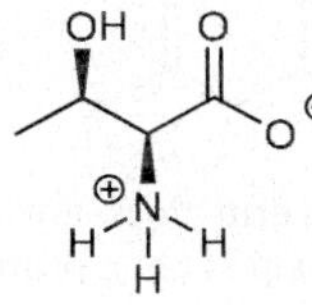

25.52. Under strongly basic conditions (NaOH), the carboxylic acid group exists as a carboxylate ion.
Under these conditions, the α position can be deprotonated, giving a resonance-stabilized enolate that is a dianion. This enolate can be protonated by water (at the α position), thereby regenerating the carboxylate ion. In the process, racemization occurs at the α position because the α position is sp^2 hybridized (trigonal planar) in the dianion intermediate. Protonation of the dianion can occur on either face of the plane (with equal likelihood), giving a racemic mixture.

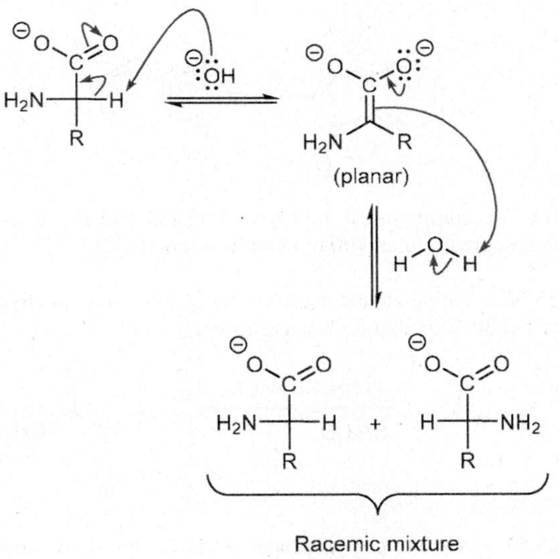

(planar)

Racemic mixture

25.53. The greater the difference between pI and pH, the faster an amino acid will migrate. The pI of Gly = 5.97, the pI of Gln = 5.65, and the pI of Asn = 5.41. Using these values, we make the following predictions:
(a) At pH = 6.0, Asn will travel the farthest distance.
(b) At pH = 5.0, Gly will travel the farthest distance.

25.54. When treated with ninhydrin, the carboxylic acid group (connected to the α position) and the amino group (connected to the α position) are both removed, and the α position becomes an aldehyde group, giving the following products:
(a) **(b)**

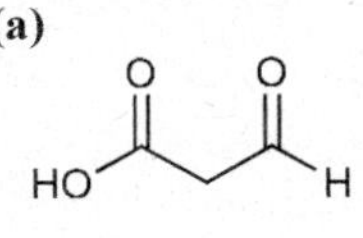

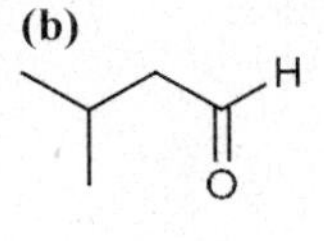

(c)

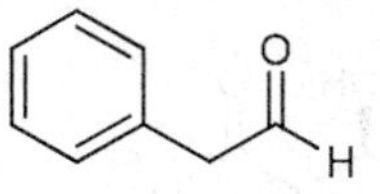
O
H

(d) Ninhydrin does not react with proline because the amino group is not a primary amine.

25.55.

(a) When treated with ninhydrin, each of the amino acids (except proline) is converted into an aldehyde with complete removal of the carboxylic acid group and the amino group (that are connected to the α position). Therefore, the identity of each aldehyde indicates the side chain of the corresponding amino acid from which it was made (see Table 25.1). This analysis reveals that the starting mixture must have contained methionine, valine, and glycine.

(b) The following purple product is obtained whenever ninhydrin reacts with an amino acid (except for proline).

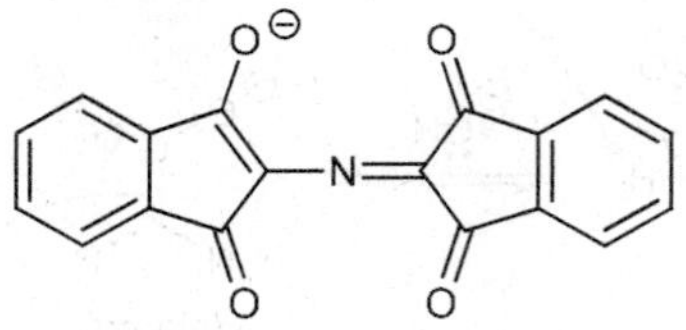
O
O
N
O
O

(c) The compound is highly conjugated and has a λ_{max} that is greater than 400 nm (see Section 16.12)

25.56. Valine can be made from the following aldehyde (via a Strecker synthesis), as shown:

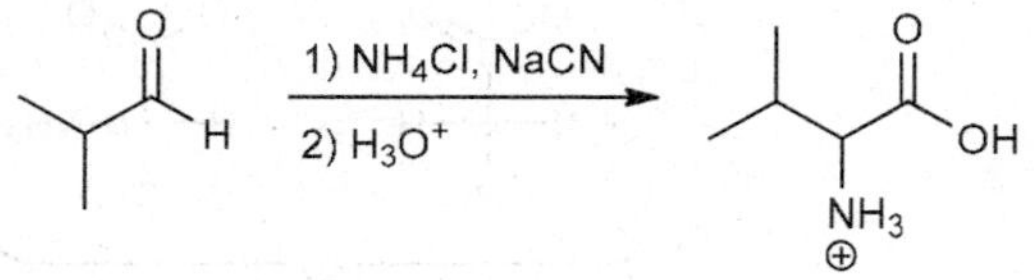
O
H
1) NH4Cl, NaCN
2) H3O+
O
OH
NH3

25.57. Alanine can be prepared via the amidomalonate synthesis with higher yields than valine, because the former requires an S_N2 reaction with a primary alkyl halide, while the latter requires an S_N2 reaction with a secondary (more hindered) alkyl halide.

25.58. Glycine has no side chain. Therefore, no alkyl group needs to be installed at the α position and its preparation via the amidomalonate synthesis would not be reasonable. The goal of a synthesis is to begin with a starting material that is simpler than the product, but in this case the starting material would be significantly more complex than the product, as shown below. The alpha carbon and the carbonyl carbon in the acetamidomalonate starting material that remain in the product (after decarboxylation and deprotection) have been highlighted:

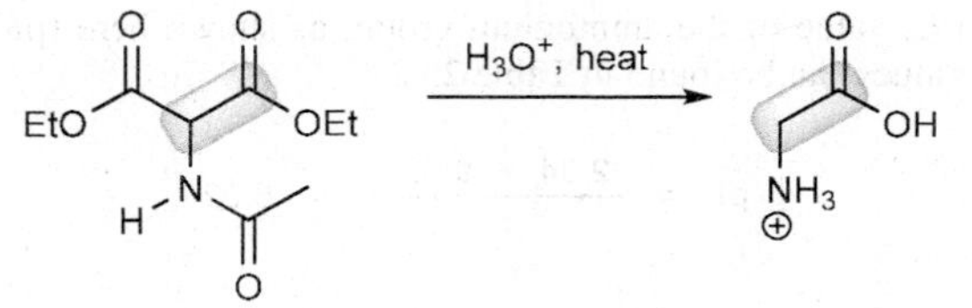
O
O
EtO
OEt
H
N
O
H3O+, heat
O
OH
NH3

25.59.

(a) The reagents indicate a Hell-Volhard-Zelinsky reaction (thereby installing a bromine atom at the α position), followed by an S_N2 reaction (thereby replacing the bromine atom with an amino group). The product is a racemic amino acid (phenylalanine), as shown:

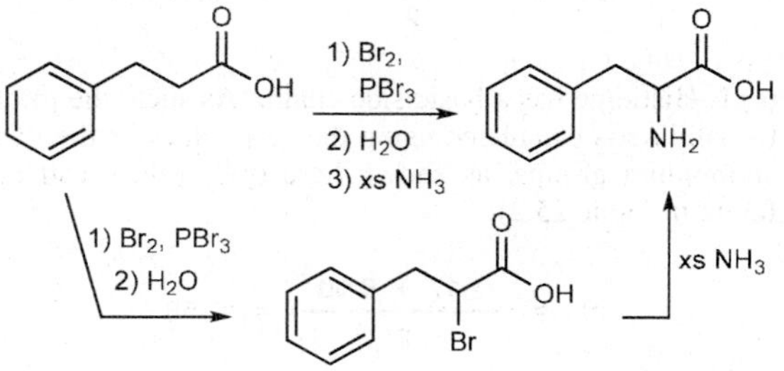
O
OH
1) Br2, PBr3
2) H2O
3) xs NH3
O
OH
NH2
1) Br2, PBr3
2) H2O
O
OH
Br
xs NH3

(b) The reagents indicate a Strecker synthesis, giving a racemic amino acid (phenylalanine), as shown:

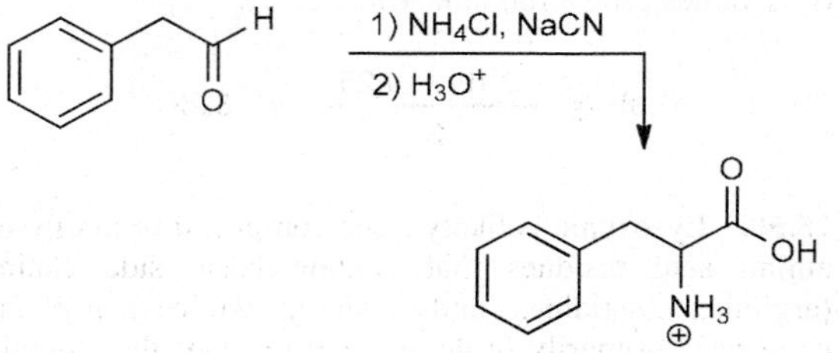
H
O
1) NH4Cl, NaCN
2) H3O+
O
OH
NH3

(c) The reagents indicate an amidomalonate synthesis, giving a racemic amino acid (alanine), as shown. The alpha carbon and the carbonyl carbon in the acetamidomalonate starting material that remain in the product (after decarboxylation and deprotection) have been highlighted:

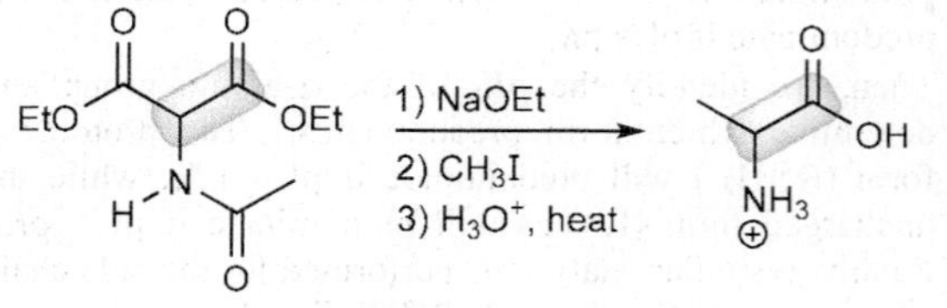
O
O
EtO
OEt
H
N
O
1) NaOEt
2) CH3I
3) H3O+, heat
O
OH
NH3

25.60.

(a) Racemic valine can be made from the corresponding carboxylic acid using a Hell-Volhard-Zelinsky reaction, as shown:

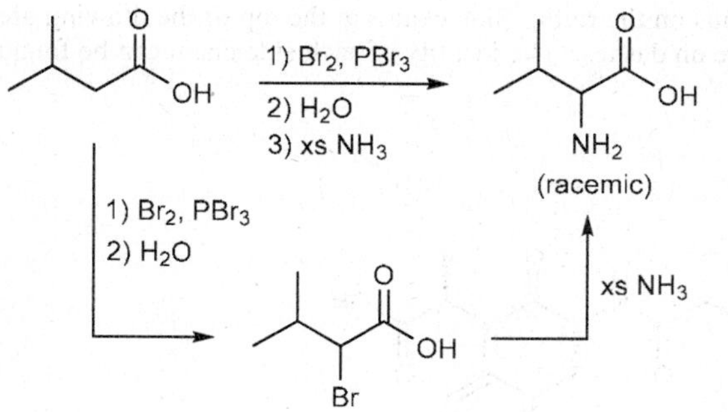

(racemic)

(b) Racemic valine can be made from acetamidomalonate using an amidomalonate synthesis, as shown. The alpha carbon and the carbonyl carbon in the starting material that remain in the product (after decarboxylation and deprotection) have been highlighted:

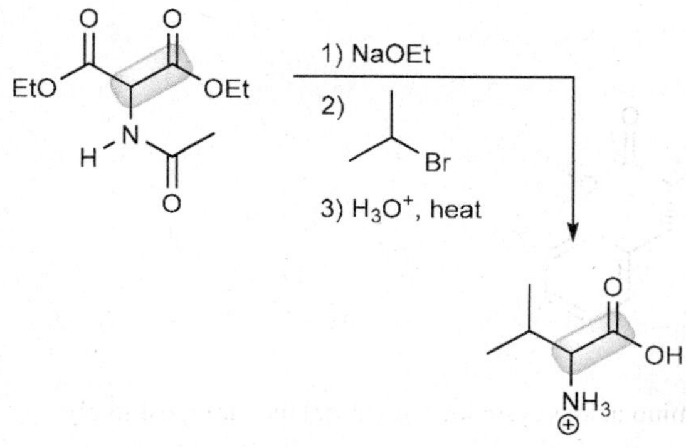

(c) Racemic valine can be made from 2-methylpropanal using a Strecker synthesis, as shown:

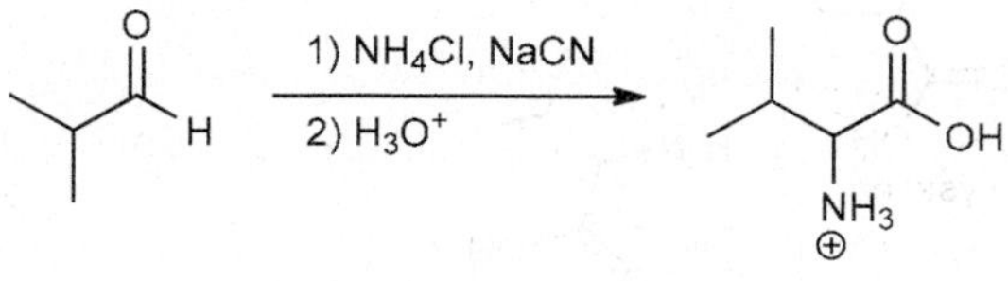

25.61. A pentapeptide has five amino acid residues, each of which can be any of the 20 naturally occurring amino acids. Therefore, there are $20 \times 20 \times 20 \times 20 \times 20 = 20^5 = 3{,}200{,}000$ possible pentapeptides that can be made from the naturally occurring amino acids!

25.62. L-Histidine can be prepared from the compound below via an asymmetric catalytic hydrogenation:

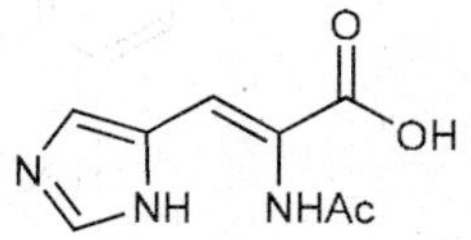

25.63. Below are the six possible sequences for a tripeptide containing L-leucine, L-methionine, and L-valine:

1) Leu-Met-Val
2) Leu-Val-Met
3) Met-Val-Leu
4) Met-Leu-Val
5) Val-Met-Leu
6) Val-Leu-Met

25.64. The N terminus of the tripeptide Asp-Lys-Phe is drawn on the left and the C terminus on the right. Side chains at the top of the drawing are on wedges, while side chains on the bottom of the drawing are on dashes. The identity of each side chain can be found in Table 25.1.

At physiological pH, the amino groups exist primarily in their protonated form (ammonium ions) while the carboxylic acid groups exist primarily in their deprotonated form (carboxylate ions):

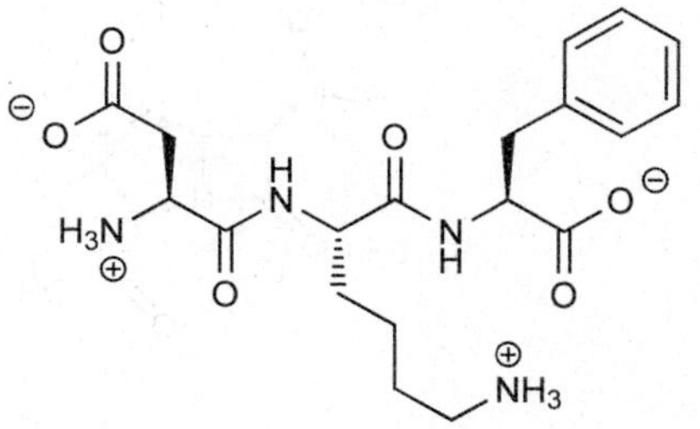

25.65. The N terminus is drawn on the left and the C terminus on the right. Side chains at the top of the drawing are on wedges, while side chains on the bottom of the drawing are on dashes. The identity of each side chain can be found in Table 25.1.

N Terminus

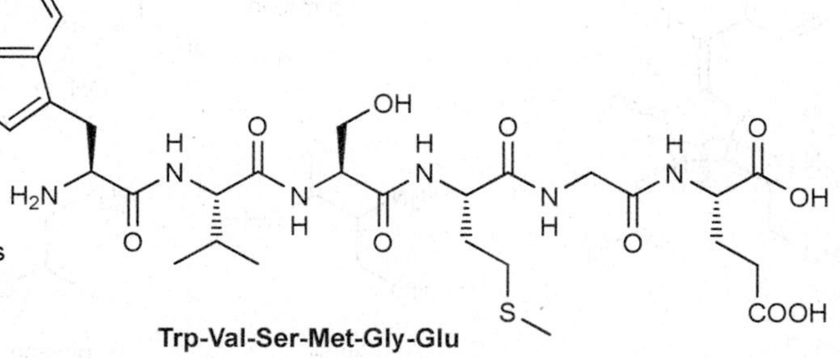

C Terminus

Trp-Val-Ser-Met-Gly-Glu

25.66. The N terminus is drawn on the left and the C terminus on the right. Side chains at the top of the drawing are on wedges, while side chains on the bottom of the drawing are on dashes. The identity of each side chain can be found in Table 25.1.

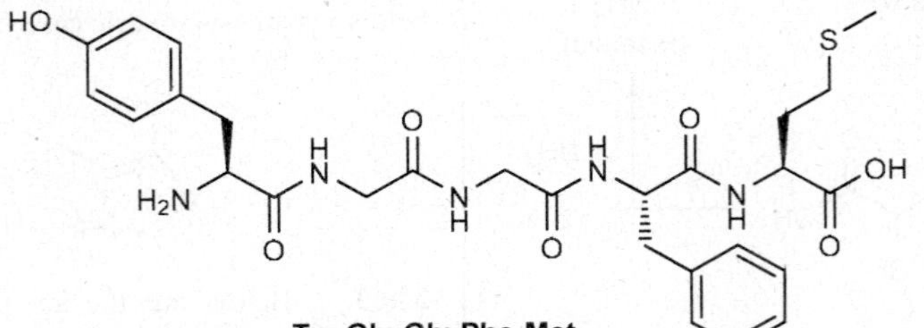

HO H O H O S OH H₂N N N N N H H O O O

Tyr-Gly-Gly-Phe-Met

25.67. The structure of aspartame (Asp-Phe-OCH_3) is shown in the solution to Problem 25.28. At physiological pH, the carboxylic acid group is deprotonated (and will exist primarily as a carboxylate ion), while the amino group is protonated (and will exist primarily as an ammonium ion):

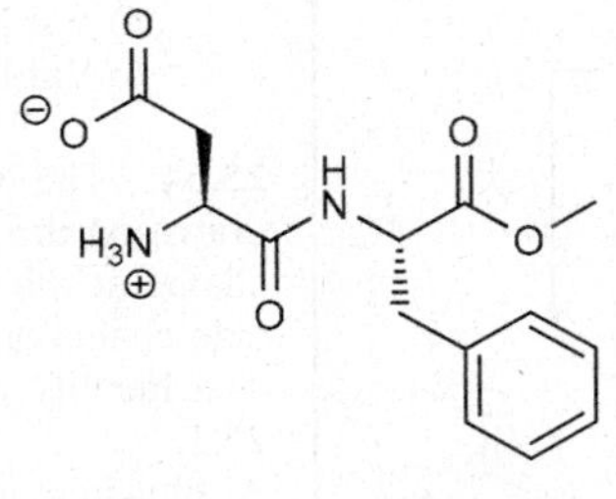

O ⊖O H O N H₃N ⊕ O O

25.68. The following retrosynthetic analysis reveals the two amino acids (cysteine and valine) that are most likely utilized during the biosynthesis of penicillin antibiotics:

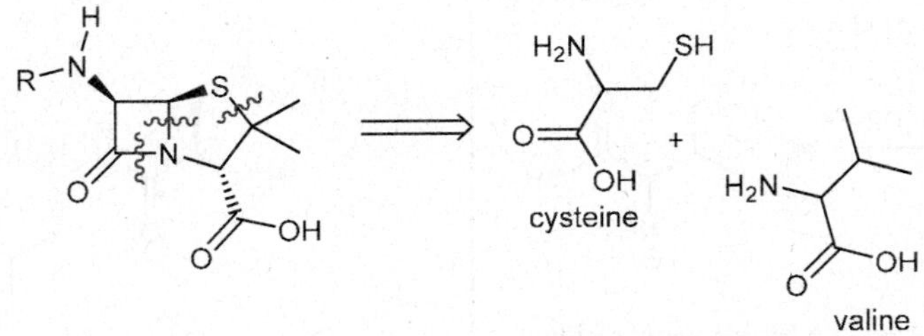

H R–N S N O O OH ⟹ H₂N SH O OH cysteine + H₂N O OH valine

25.69. The following retrosynthetic analysis reveals the three amino acids (tyrosine, serine, and glycine) that are necessary for biosynthesis of the fluorophore:

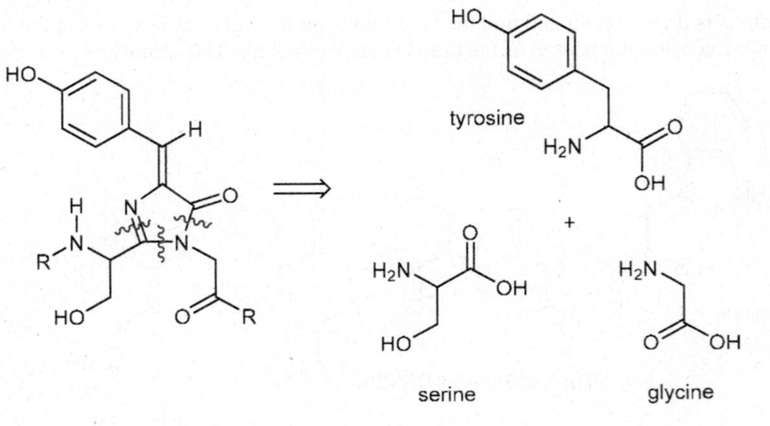

HO H H N O R–N N HO O R ⟹ HO tyrosine H₂N O OH + H₂N O OH HO serine H₂N O OH glycine

25.70. If a tripeptide does not react with phenyl isothiocyanate, then it must not have a free N terminus. It must be a cyclic tripeptide. Below are the two possible cyclic tripeptides:

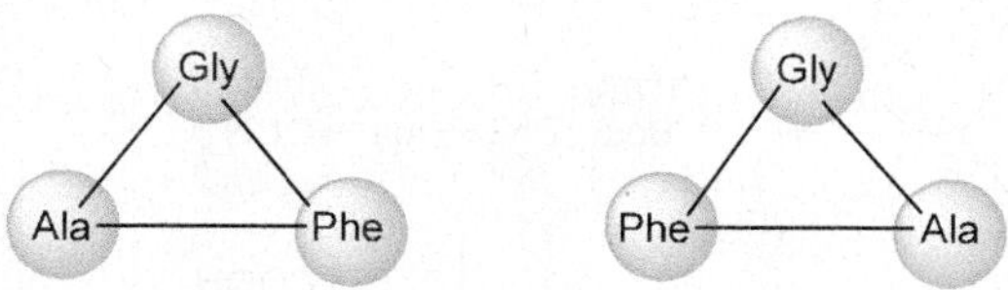

25.71.
(a) Trypsin catalyzes the hydrolysis of the peptide bond at the carboxyl side of arginine, giving the following two fragments:

Arg + Val-Pro-Gly-Phe-Ser-Pro-Phe-Arg

(b) Chymotrypsin catalyzes the hydrolysis of the peptide bonds at the carboxyl side of phenylalanine, giving the following three fragments:

Arg-Val-Pro-Gly-Phe + Ser-Pro-Phe + Arg

25.72. The R group (highlighted) in the PTH derivative indicates the identity of the N-terminal residue. Since this R group is a benzylic group ($-CH_2Ph$), the N-terminal residue must be phenylalanine.

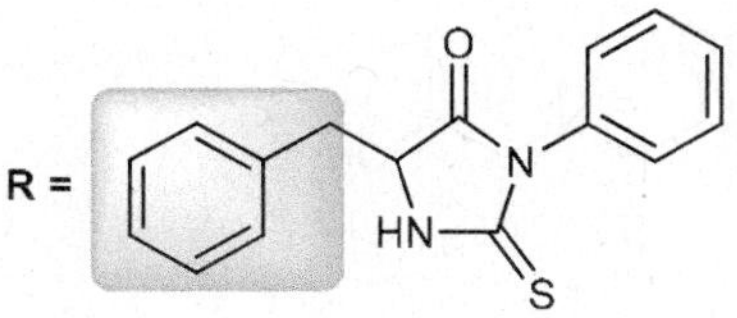

25.73. The first Edman degradation indicates that the N-terminal residue is valine (the R group is isopropyl). The second Edman degradation indicates that the N-terminal residue of the dipeptide is alanine. And finally, the remaining amino acid (glycine) must be at the C terminus of the tripeptide. In summary, the tripeptide is Val-Ala-Gly, drawn here:

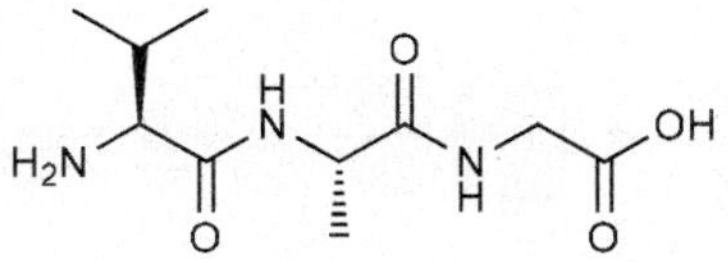

25.74. Only one of the trypsin fragments has a C terminus that is not arginine or lysine. This fragment (which ends with threonine), must be the last fragment in the peptide sequence. The remaining three trypsin fragments can be placed in the proper order by analyzing the chymotrypsin fragments. The correct peptide sequence is:

His-Ser-Gln-Gly-Thr-Phe-Thr-Ser-Asp-Tyr-Ser-Lys-Tyr-Leu-Asp-Ser-Arg-Arg-Ala-Gln-Asp-Phe-Val-Gln-Trp-Leu-Met-Asn-Thr

There cannot be any disulfide bridges in this peptide, because it has no cysteine residues, and only cysteine residues form disulfide bridges.

25.75. Prior to acetylation, the nitrogen atom of the amino group is sufficiently nucleophilic to attack phenyl isothiocyanate. Acetylation converts the amino group into an amide group, and the lone pair of the nitrogen atom is delocalized via resonance, rendering it much less nucleophilic.

25.76.
(a) When treated with acid and methanol, the carboxylic acid group is converted into a methyl ester (via a Fischer esterification process), and the amino group is protonated, giving the following compound:

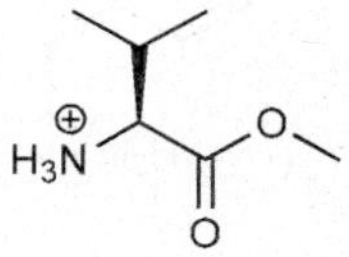

(b) When treated with di-*tert*-butyl dicarbonate, $(Boc)_2O$, the amino group is protected with a Boc protecting group, giving the following product:

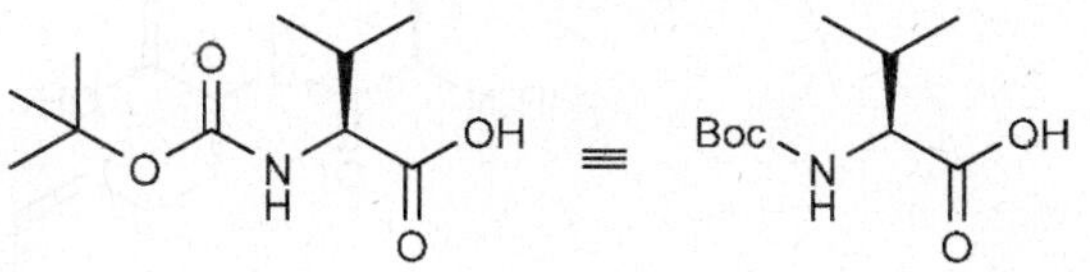

(c) Under basic conditions, the amino group is not protonated, and the carboxylic acid group is deprotonated, giving a carboxylate ion:

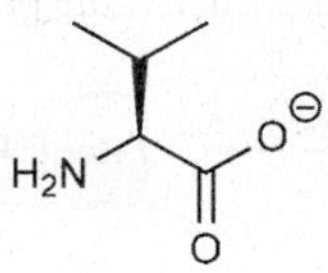

(d) Under acidic conditions, the amino group is protonated to give an ammonium ion, and the carboxylic acid group will be in its protonated form:

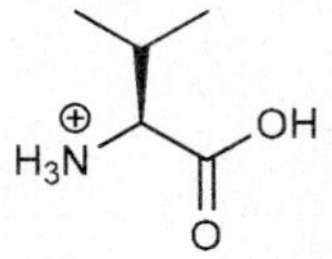

25.77. We begin by installing the appropriate protecting groups. Then, upon treatment with DCC, the protected amino acids are coupled. And finally, the protecting groups are removed, as shown:

Phe —$(Boc)_2O$→ Boc–Phe

Ala —$[H^+]$, CH_3OH→ Ala–OCH_3

DCC → Boc–Phe–Ala–OCH_3

1) CF_3COOH
2) NaOH, H_2O

Phe–Ala

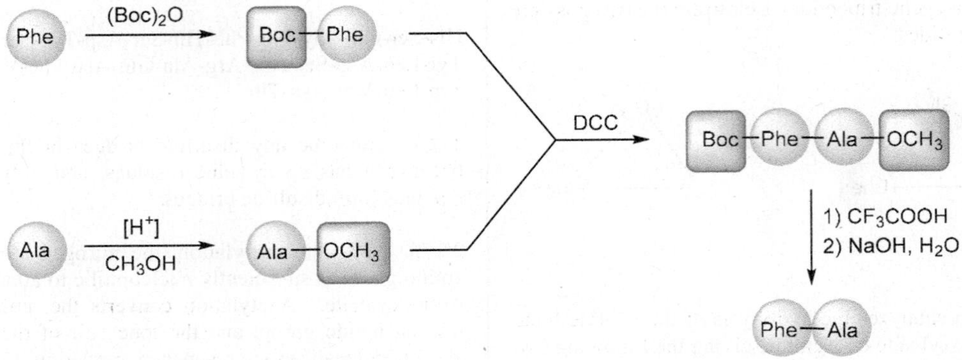

25.78. When a mixture of L-phenylalanine and L-alanine is treated with DCC, there are four possible dipeptides: 1) Phe-Ala, or 2) Ala-Phe, or 3) Phe-Phe, or 4) Ala-Ala. These four possibilities are drawn below:

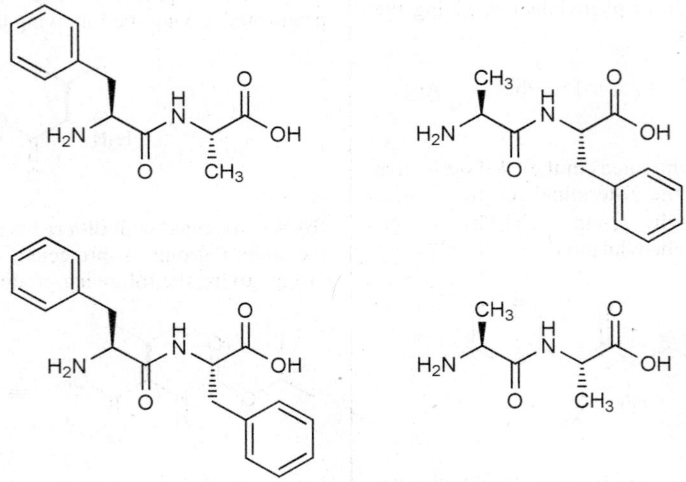

25.79. We begin by installing the appropriate protecting groups. Then, upon treatment with DCC, the protected amino acids are coupled. And finally, the protecting groups are removed, as shown:

Phe–Ile —$(Boc)_2O$→ Boc–Phe–Ile

Val–Leu —$[H^+]$, CH_3OH→ Val–Leu–OCH_3

DCC → Boc–Phe–Ile–Val–Leu–OCH_3

1) CF_3COOH
2) NaOH, H_2O

Phe–Ile–Val–Leu

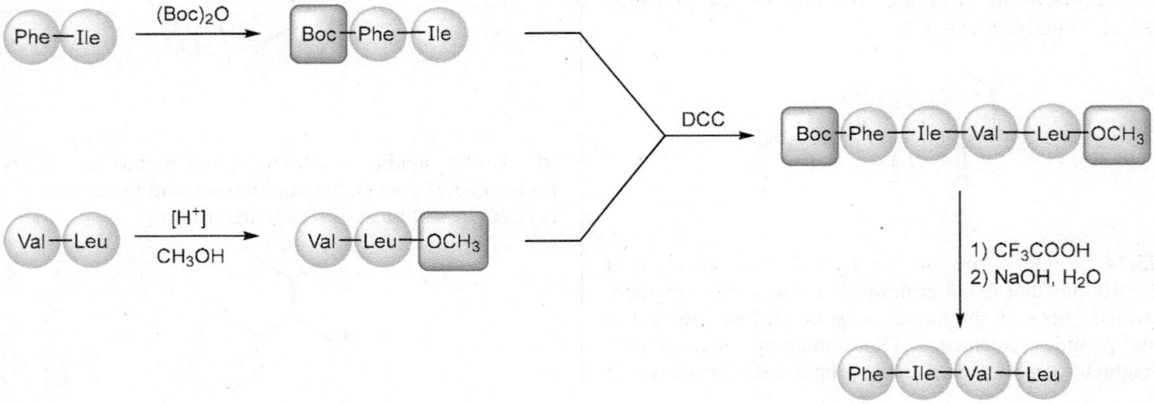

25.80. First, we attach the appropriate Boc-protected residue to the polymer:

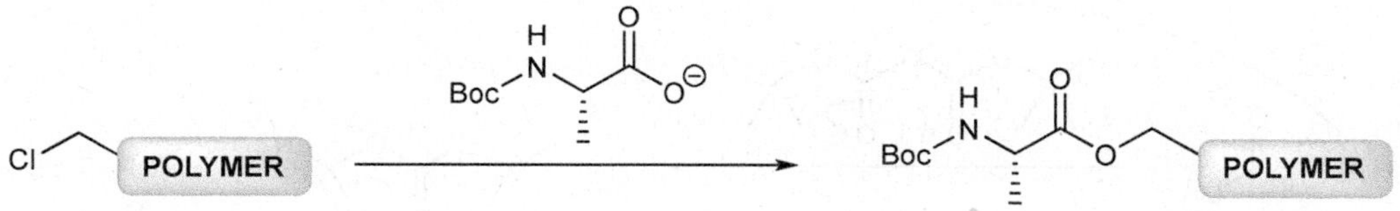

Then, the Boc protecting group is removed and a new peptide bond is formed with a Boc-protected amino acid (valine), using DCC. This two-step process (removal of the Boc protecting group, followed by peptide bond formation) is then repeated to install the leucine residue. Finally, the Boc protecting group is removed and the desired tripeptide is detached from the polymer, as shown:

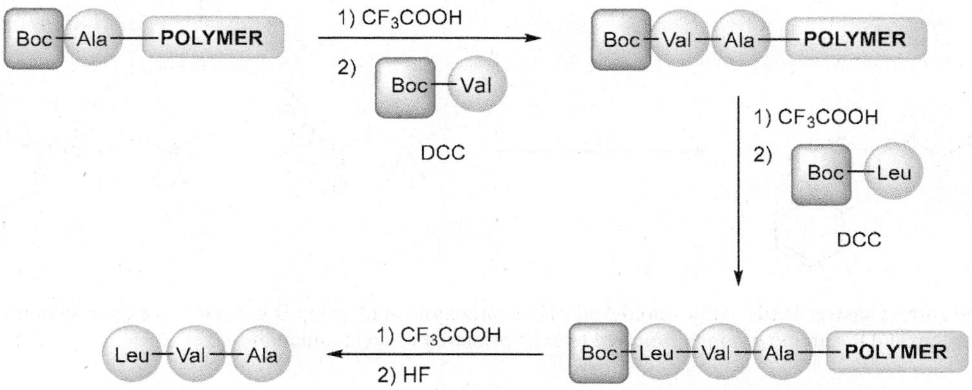

25.81. During a Merrifield synthesis, the C terminus of the growing peptide chain remains anchored to the polymer. The C terminus of the desired peptide (leucine enkephalin, Tyr-Gly-Gly-Phe-Leu) is occupied by a leucine residue. Therefore, the following Boc-protected amino acid (Boc-Leu) must be anchored to the polymer in order to prepare leucine enkephalin via a Merrifield synthesis:

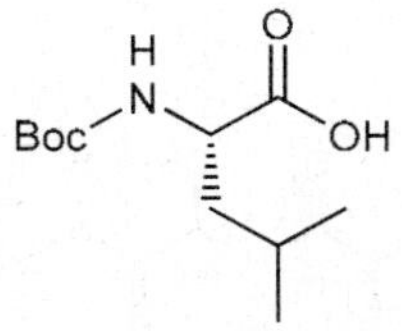

25.82. A proline residue cannot be part of an α helix, because it lacks an N-H proton and does not participate in hydrogen bonding. (The amino acid proline does indeed have an N-H group, but when incorporated into a peptide, the proline residue does not have an N-H group)

25.83. The amino group attacks one of the carbonyl groups of DCC, giving a tetrahedral intermediate. The carbonyl group is then reformed upon expulsion of a resonance-stabilized leaving group. The resulting cation is then deprotonated to give the product:

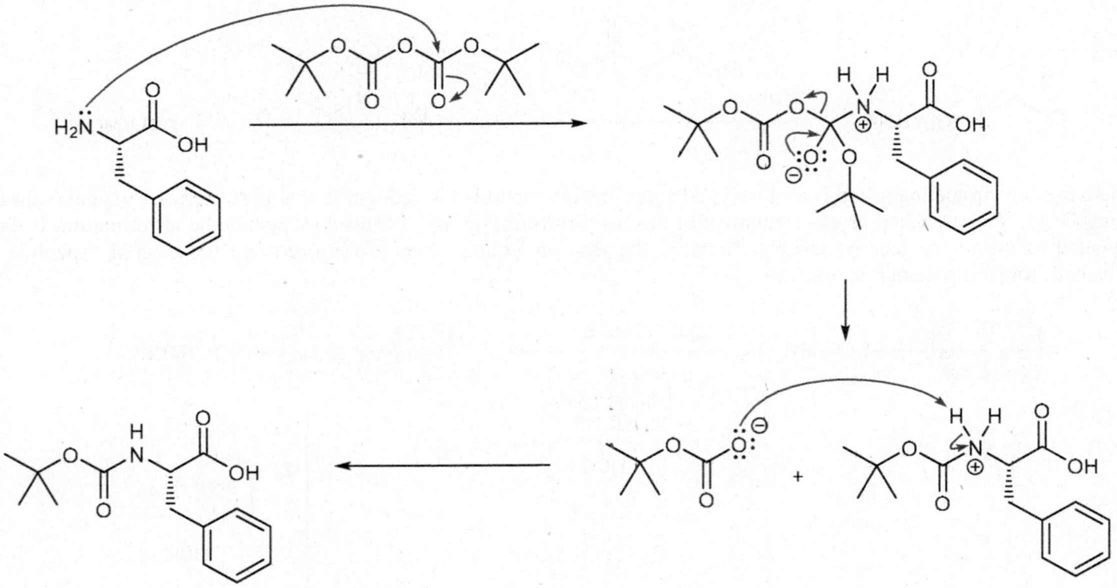

25.84. The correct answer is (d). At physiological pH, a carboxylic acid group is expected to exist predominantly as a carboxylate ion, and the amino group is expected to exist primarily as an ammonium ion:

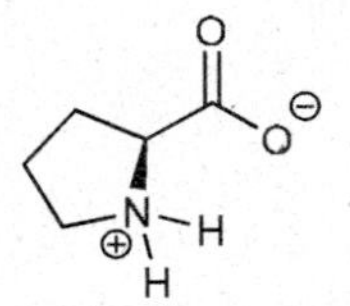

25.85. The correct answer is (c). The compound is constructed from five amino acid residues, as highlighted below, and is therefore a pentapeptide:

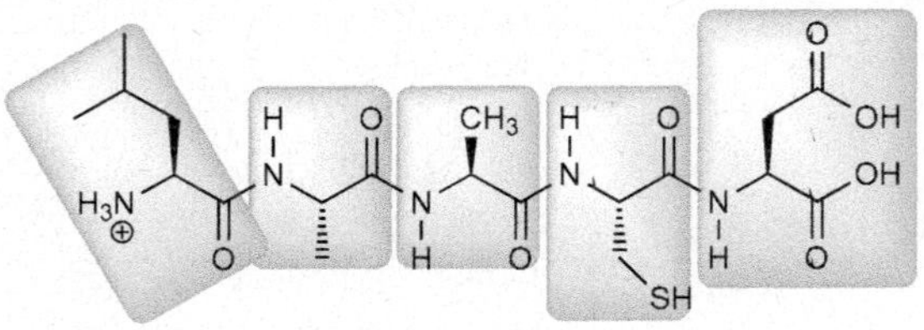

25.86. The correct answer is (c). Serine does not have an acidic side chain or a basic side chain, so the pI of serine is calculated by finding the average of the pK_a value of the carboxylic acid group and the pK_a value of the ammonium group, as shown here:

$$pI = \frac{2.21 + 9.15}{2} = 5.68$$

25.87. The correct answer is (b). Two amino acids are linked together by forming an amide. The lone pair on the nitrogen atom of an amide is adjacent to a π bond and, therefore, is delocalized by resonance. Because of this resonance, the peptide bond has substantial double-bond character. The carbonyl carbon and the nitrogen atom of the amide group are both sp^2 hybridized, making the six atoms highlighted below coplanar:

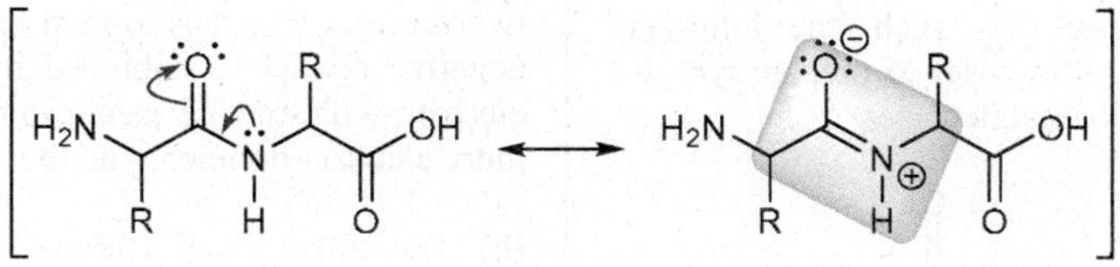

25.88. The correct answer is (a). The Boc group (*tert*-butoxycarbonyl) is a protecting group for amines, so Boc-glycine has a protected amino group and a reactive carboxylic acid group. The phenylalanine has the opposite reactivity: the carboxylic acid is protected as a methyl ester, and the amino group is reactive. The two protected amino acids could be coupled with the help of DCC to prepare a protected dipeptide:

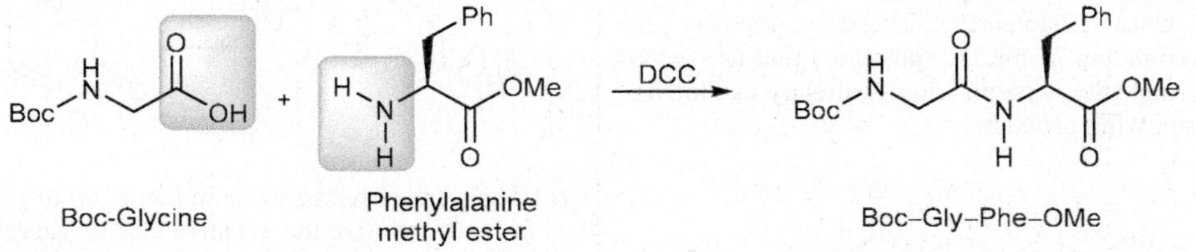

Boc-Glycine + Phenylalanine methyl ester → (DCC) Boc–Gly–Phe–OMe

25.89. The following alkyl halide would be necessary in order to prepare tyrosine via an amidomalonate synthesis. This starting material possesses both a nucleophilic center (the OH group) as well as an electrophilic center (it is a primary benzylic bromide). As such, the molecules can react with each other via an SN2 process, thereby forming the polymer shown:

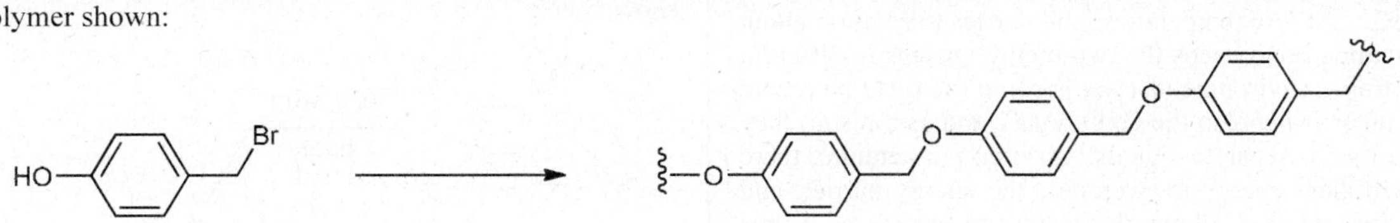

25.90. The stabilized enolate ion (formed in the first step) can function as a base, rather than a nucleophile, giving an E2 reaction:

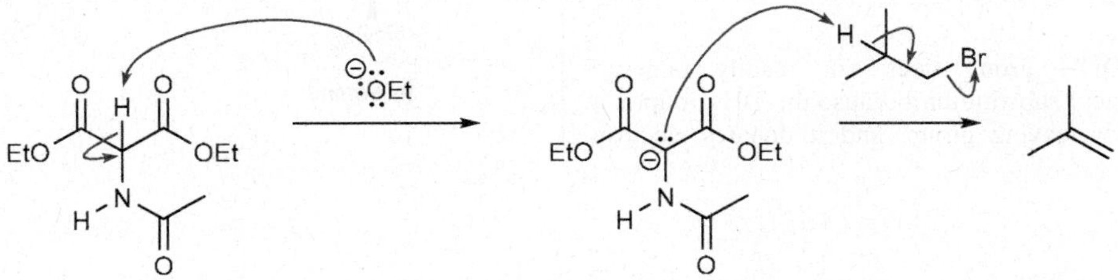

25.91. The lone pair on that nitrogen atom is highly delocalized via resonance and is participating in aromaticity. Accordingly, the lone pair is not available to function as a base.

25.92.

(a) The Hell-Volhard-Zelinsky reaction will install a bromine atom at the α position of a carboxylic acid. This bromine atom can then be replaced with an amino group via an S_N2 process. As such, the following carboxylic acid is necessary in order to prepare tyrosine via a Hell-Volhard-Zelinsky reaction:

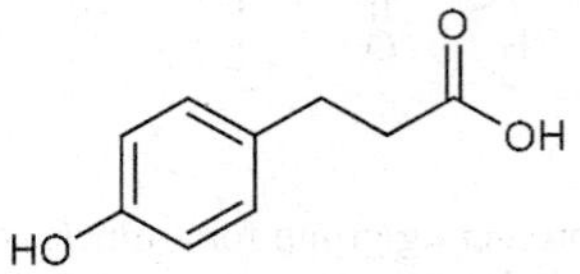

(b) The aromatic ring is highly activated toward electrophilic aromatic substitution, as a result of the presence of the OH group, which is an *ortho-para* director (as seen in Chapter 18). Therefore, the ring can undergo bromination in the two positions that are *ortho* to the OH group (the *para* position is already occupied), giving the following product:

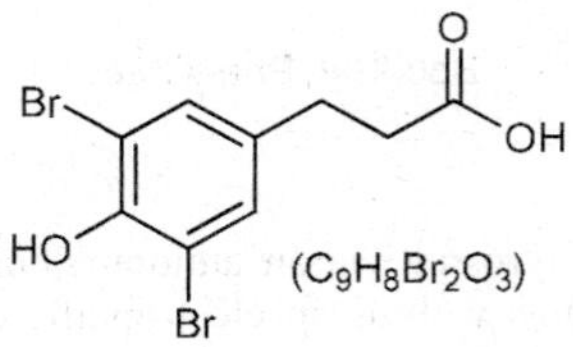

($C_9H_8Br_2O_3$)

25.93. At low temperature, the barrier to rotation about the amide bond keeps the two methyl groups in different electronic environments (one is *cis* to the C=O bond and the other is *trans* to the C=O bond), and as a result, they give rise to separate signals. At high temperature, there is sufficient energy to overcome the energy barrier, and the free rotation allows the methyl groups to exchange positions on a timescale that is faster than the timescale of the NMR spectrometer. The result is an averaging effect which gives rise to only one signal.

25.94.

(a) The COOH group does not readily undergo nucleophilic acyl substitution because the OH group is a relatively poor leaving group, and it donates electron density to the carbonyl group (making it more electron-rich). By converting the COOH group into an ester with an attached electron-withdrawing group, the compound is now activated toward nucleophilic acyl substitution for two reasons: 1) it has a good leaving group because the negative charge is stabilized by resonance, and 2) the electron-withdrawing group makes the carbonyl carbon more electron-deficient and therefore more electrophilic.

(b) The nitro group stabilizes the leaving group via resonance. As described in Chapter 18, the electron-withdrawing nitro group serves as a reservoir for electron density:

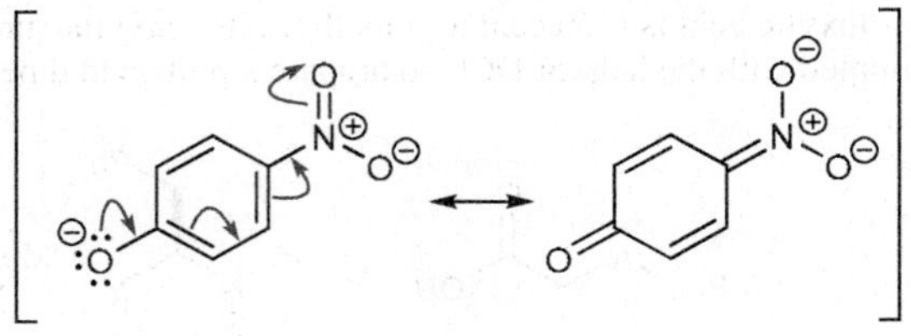

(c) The nitro group must be in the *ortho* or *para* position in order to stabilize the negative charge via resonance (as shown above). If the nitro group is in the *meta* position, the negative charge cannot be pushed onto the nitro group.

25.95. Hydrolysis of the ester group gives threonine, as shown here:

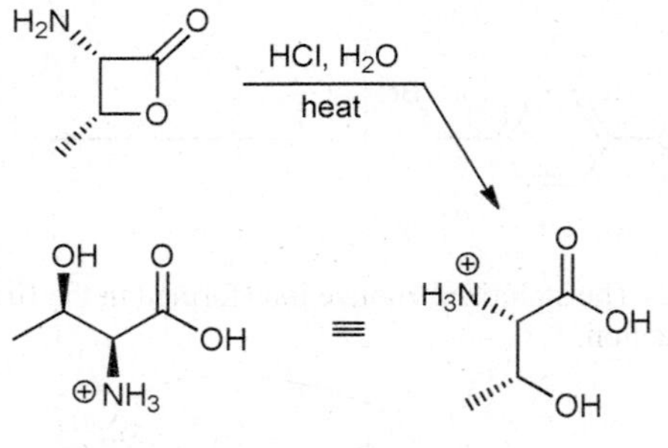

HCl, H_2O

heat

threonine

Chapter 26
Lipids

Review of Concepts

Fill in the blanks below. To verify that your answers are correct, look in your textbook at the end of Chapter 26. Each of the sentences below appears verbatim in the section entitled *Review of Concepts and Vocabulary*.

- **Lipids** are naturally occurring compounds that are extracted from cells using _______________ solvents.
- **Complex lipids** readily undergo ______________, while **simple lipids** do not.
- _________ are high-molecular-weight esters that are constructed from carboxylic acids and alcohols.
- ___________________ are the triesters formed from glycerol and three long-chain carboxylic acids, called **fatty acids.** The resulting triglyceride is said to contain three fatty acid _____________.
- For saturated fatty acids, the melting point increases with increasing __________ ____________. The presence of a _____ double bond causes a decrease in the melting point.
- Triglycerides that are solids at room temperature are called ______, while those that are liquids at room temperature are called _______.
- Triglycerides containing unsaturated fatty acid residues will undergo hydrogenation. During the hydrogenation process, some of the double bonds can isomerize to give _______ π bonds
- In the presence of molecular oxygen, triglycerides are particularly susceptible to oxidation at the ___________ position to produce hydroperoxides.
- Transesterification of triglycerides can be achieved via either _____ catalysis or _____ catalysis to produce biodiesel.
- ________________ are similar in structure to triglycerides except that one of the three fatty acid residues is replaced by a phosphoester group.
- The structures of **steroids** are based on a tetracyclic ring system, involving three six-membered rings and one ______-membered ring.
- The ring fusions are all _______ in most steroids, giving steroids their rigid geometry.
- All steroids, including cholesterol, are biosynthesized from ____________.
- Prostaglandins contain 20 carbon atoms and are characterized by a ______-membered ring with two side chains.
- **Terpenes** are a class of naturally occurring compounds that can be thought of as being assembled from _________ units.
- A terpene with 10 carbon atoms is called a ______________, while a terpene with 20 carbon atoms is called a _________________.

Review of Skills

Fill in the blanks and empty boxes below. To verify that your answers are correct, look in your textbook at the end of Chapter 26. The answers appear in the section entitled *SkillBuilder Review*.

26.1 Comparing Molecular Properties of Triglycerides

Circle the triglyceride below that is expected to have a higher melting point.

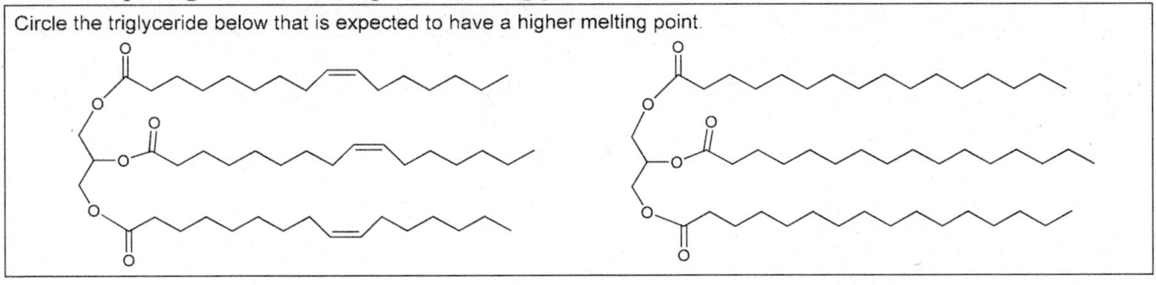

26.2 Identifying the Products of Triglyceride Hydrolysis

Draw the products obtained when the following triglyceride is treated with aqueous sodium hydroxide.

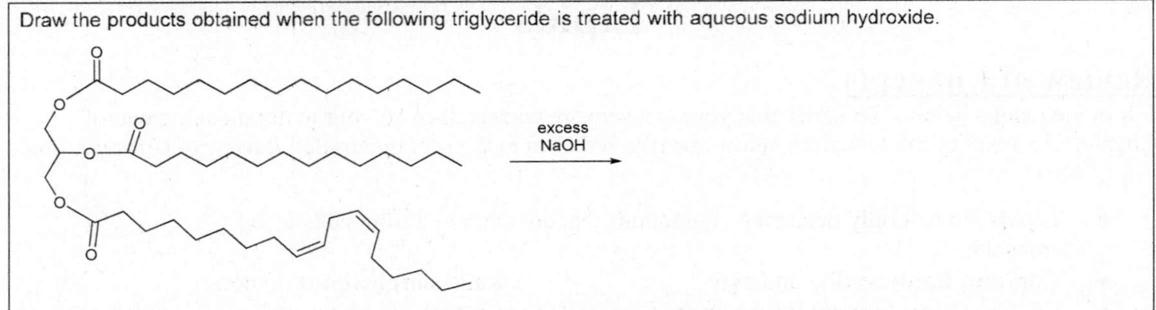

26.3 Drawing a Mechanism for Transesterification of a Triglyceride

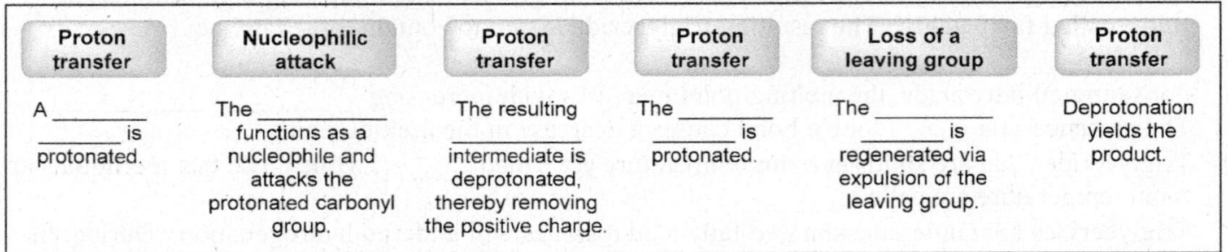

Proton transfer
A ________ is ________ protonated.
Nucleophilic attack
The ________ functions as a nucleophile and attacks the protonated carbonyl group.
Proton transfer
The resulting ________ intermediate is deprotonated, thereby removing the positive charge.
Proton transfer
The ________ is ________ protonated.
Loss of a leaving group
The ________ is ________ regenerated via expulsion of the leaving group.
Proton transfer
Deprotonation yields the product.

26.4 Identifying Isoprene Units in a Terpene

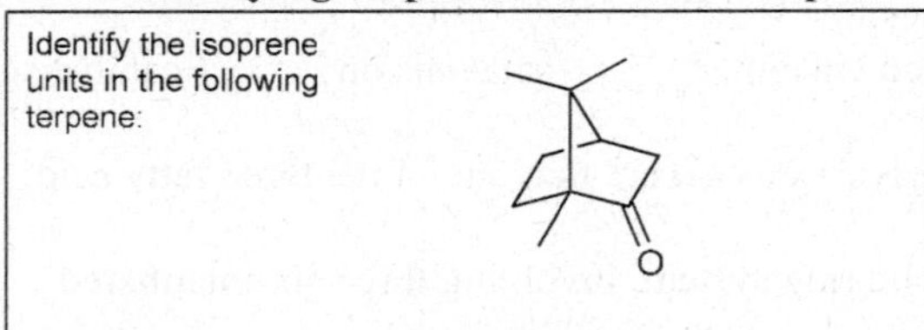

Identify the isoprene units in the following terpene:

Review of Reactions

Identify the reagents necessary to achieve each of the following transformations. To verify that your answers are correct, look in your textbook at the end of Chapter 26. The answers appear in the section entitled *Review of Reactions*.

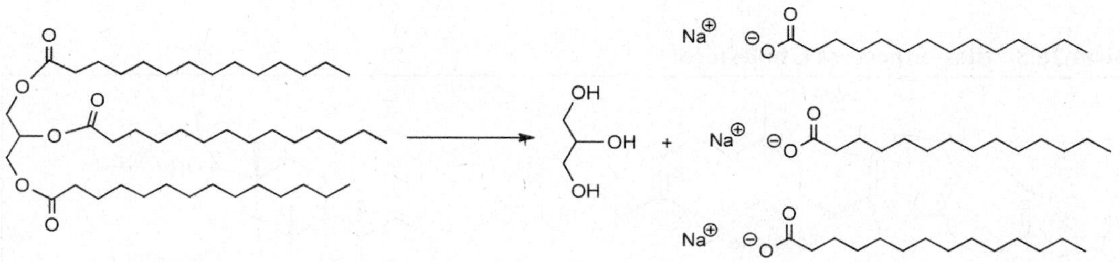

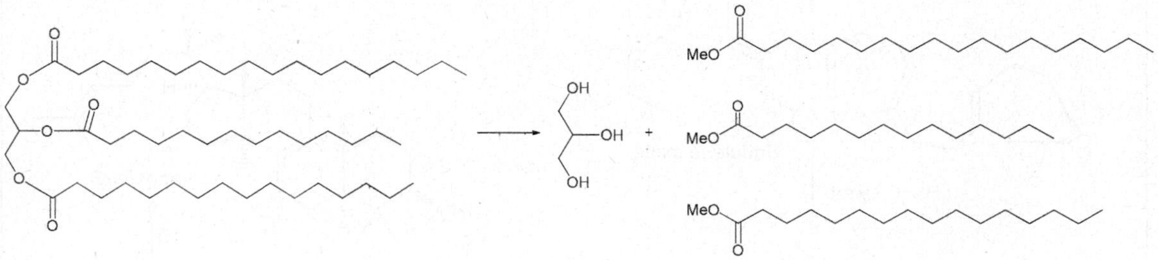

Review of Mechanisms

Complete each of the following mechanisms by drawing the missing curved arrows. To verify that your curved arrows are drawn correctly, compare them to the curved arrows in the mechanism boxes for Mechanisms 26.1 – 26.6, which can be found throughout Chapter 26 of your text.

Mechanism 26.1 Hydrolysis of an Ester Under Basic Conditions

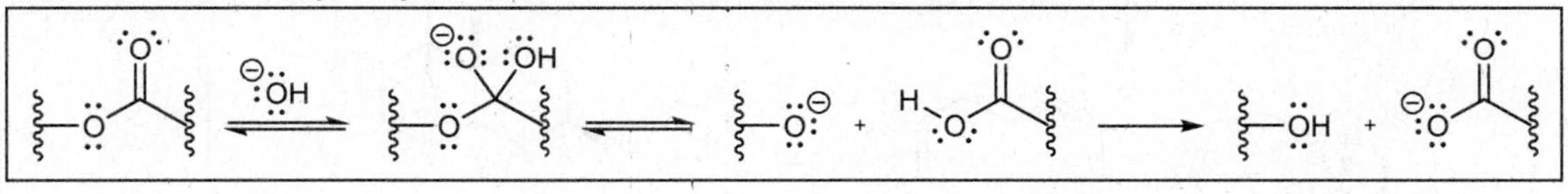

Mechanism 26.2 Acid-Catalyzed Transesterification

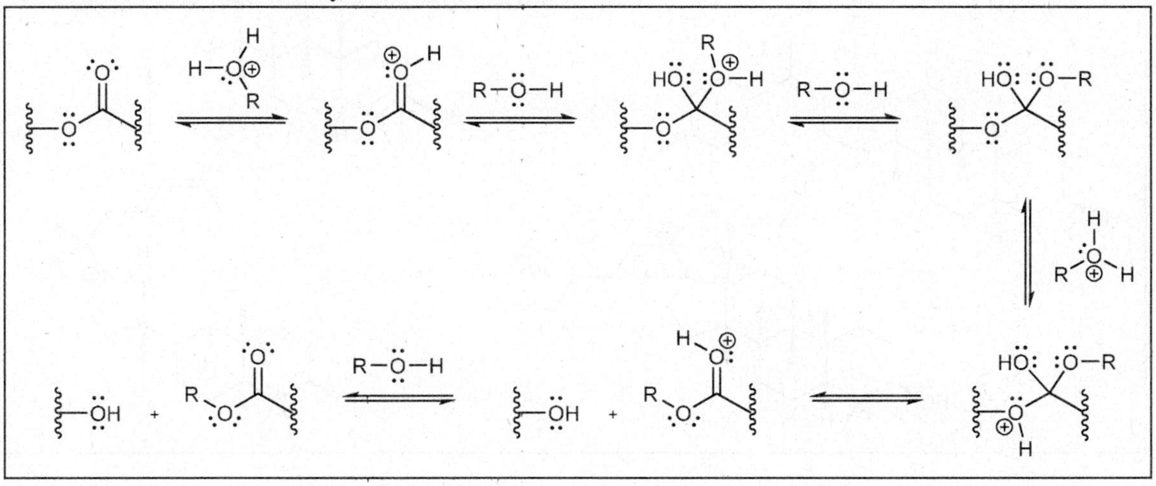

Mechanism 26.3 Biosynthesis of Cholesterol

Squalene

Squalene oxide

H—Enzyme

Cholesterol

Lanosterol

:B

Mechanism 26.4 Biosynthesis of Prostaglandins from Arachidonic Acid

Arachidonic acid

Cyclooxygenase

COOH

PGG_2

OH

Prostaglandins

Mechanism 26.5 Biosynthesis of Geranyl Pyrophosphate

OPP

:B

geranyl pyrophosphate

All Monoterpenes

Mechanism 26.6 Biosynthesis of Farnesyl Pyrophosphate

OPP

:B

farnesyl pyrophosphate

All sesquiterpenes and diterpenes

Solutions

26.1 Hydrolysis of triacontyl hexadecanoate gives a carboxylic acid with sixteen carbon atoms and an alcohol with thirty carbon atoms, as shown:

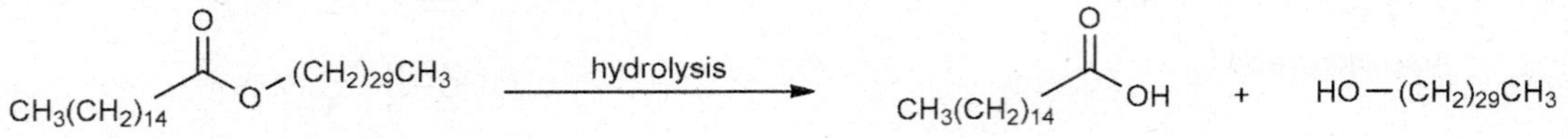

$$CH_3(CH_2)_{14}C(=O)O(CH_2)_{29}CH_3 \xrightarrow{\text{hydrolysis}} CH_3(CH_2)_{14}C(=O)OH + HO-(CH_2)_{29}CH_3$$

26.2. The products of hydrolysis suggest the following ester:

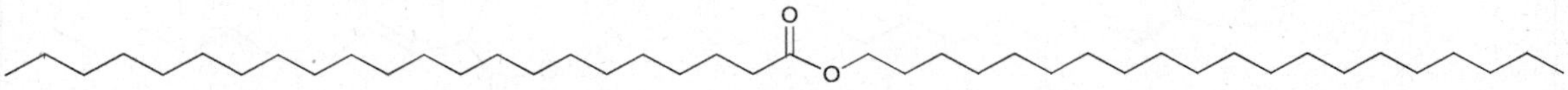

26.3.
(a) Trimyristin is expected to have a higher melting point because the fatty acid residues in trimyristin have more carbon atoms than the fatty acid residues of trilaurin.

(b) Triarachidin is expected to have a higher melting point because the fatty acid residues in triarachidin have more carbon atoms and less unsaturation than the fatty acid residues of trilinolein.

(c) Triolein is expected to have a higher melting point because the fatty acid residues in triolein have less unsaturation than the fatty acid residues of trilinolein.

(d) Tristearin is expected to have a higher melting point because the fatty acid residues in tristearin have more carbon atoms than the fatty acid residues of trimyristin.

26.4. Of the three triglycerides, tristearin is expected to have the highest melting point because the fatty acid residues in tristearin have more carbon atoms and less unsaturation than the fatty acid residues of tripalmitolein or tripalmitin. Tripalmitolein is expected to have the lowest melting point because the fatty acid residues in tripalmitolein have fewer carbon atoms than the fatty acid residues in tristearin, and it is the only triglyceride of the three that has unsaturated fatty acid residues..

26.5. The fatty acid residues in triarachidin have more carbon atoms than the fatty acid residues in tristearin. Therefore, triarachadin is expected to have a higher melting point. It should be a solid at room temperature, and should therefore be classified as a fat, rather than an oil.

26.6.
(a) All three fatty acid residues are saturated, with either 16 or 18 carbon atoms, so the triglyceride is expected to have a high melting point. It should be a solid at room temperature, so it is a fat.
(b) All three fatty acid residues are unsaturated, so the triglyceride is expected to have a low melting point. It should be a liquid at room temperature, so it is an oil.

26.7.
(a) The C=C bond in each oleic acid residue undergoes hydrogenation, giving the following saturated triglyceride:

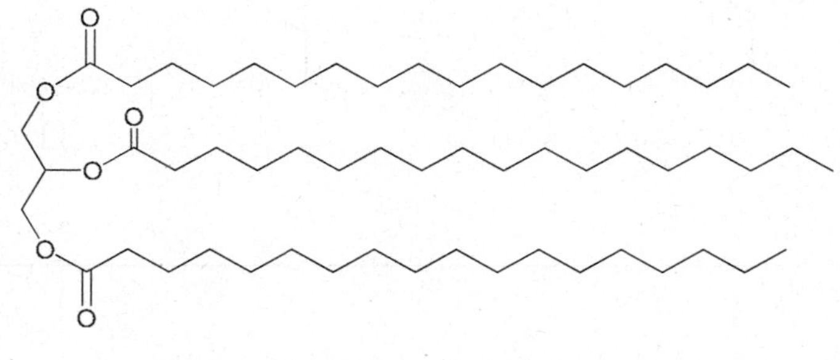

(b) The triglyceride obtained from hydrogenation (shown above) has three stearic acid residues, so it is called tristearin.
(c) Tristearin is expected to have a higher melting point because the fatty acid residues in tristearin are saturated, while the fatty acid residues of triolein are unsaturated.
(d) Upon treatment with aqueous base, each of the ester groups will undergo hydrolysis, giving three equivalents of stearic acid, as the carboxylate form (stearate) until protonated.

26.8. There are three fatty acid residues. Partial hydrogenation indicates that either one or two of these residues has a double bond (in the *trans* configuration). There are two isomers that exhibit one C=C bond, and there are two isomers that exhibit two C=C bonds, as shown here:

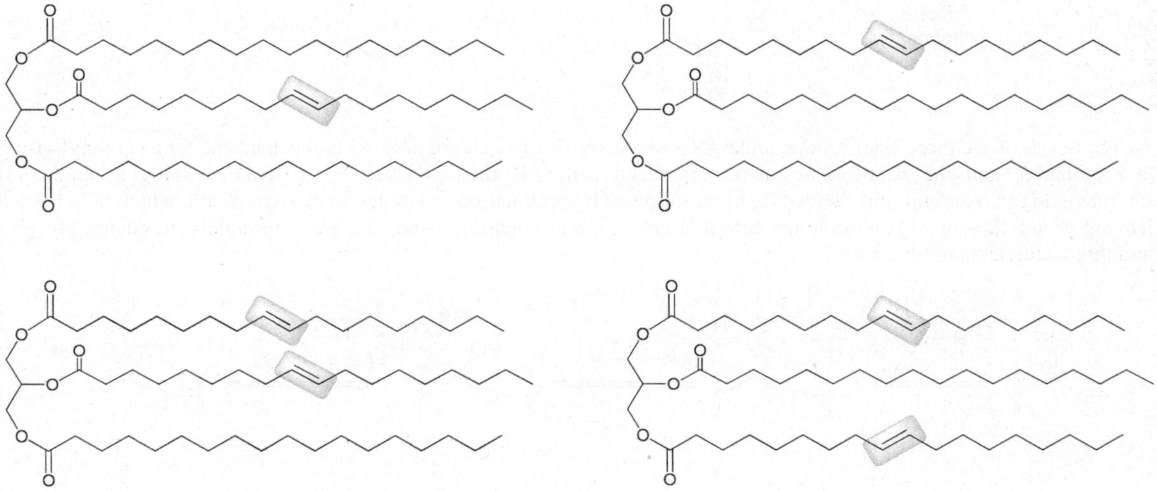

26.9. When a triglyceride is treated with aqueous base, each of the ester groups is hydrolyzed, giving glycerol and three carboxylate ions, as shown below. Notice that two of these carboxylate ions are identical, so the reaction gives two carboxylate ions, in a 2:1 ratio:

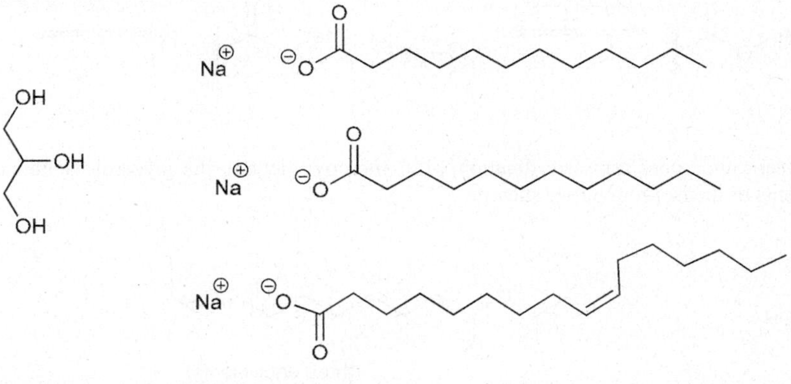

26.10. The products of hydrolysis indicate that the starting triglyceride has three lauric acid residues, as shown:

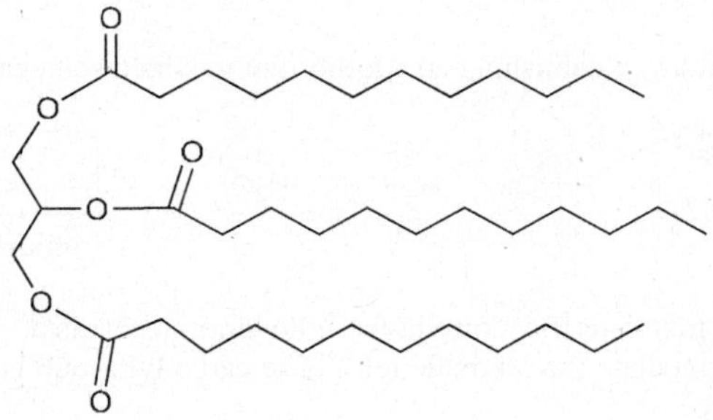

26.11. The products of hydrolysis indicate that the starting triglyceride has two lauric acid residues and one palmitic acid residue. In order to be achiral, the palmitic acid residue must be connected to C2 of the glycerol backbone, as shown below. Otherwise, there is no symmetry and the highlighted position (C2 of the glycerol backbone) would be a chiral center:

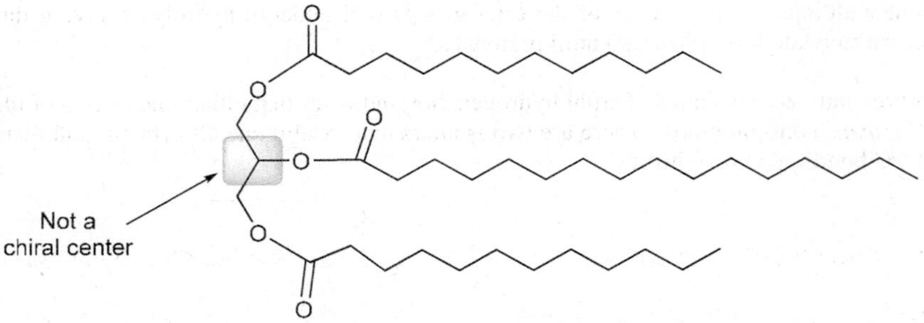

Not a chiral center

26.12. Each of the three ester groups undergoes transesterification via the following mechanism. The carbonyl group is protonated, and the resulting resonance-stabilized cation is then attacked by methanol, giving a tetrahedral intermediate (an oxonium ion). Deprotonation, followed by protonation, gives another oxonium ion, which then loses a leaving group, thereby regenerating the carbonyl group. Deprotonation is the final step, ultimately producing glycerol and three equivalents of the ester:

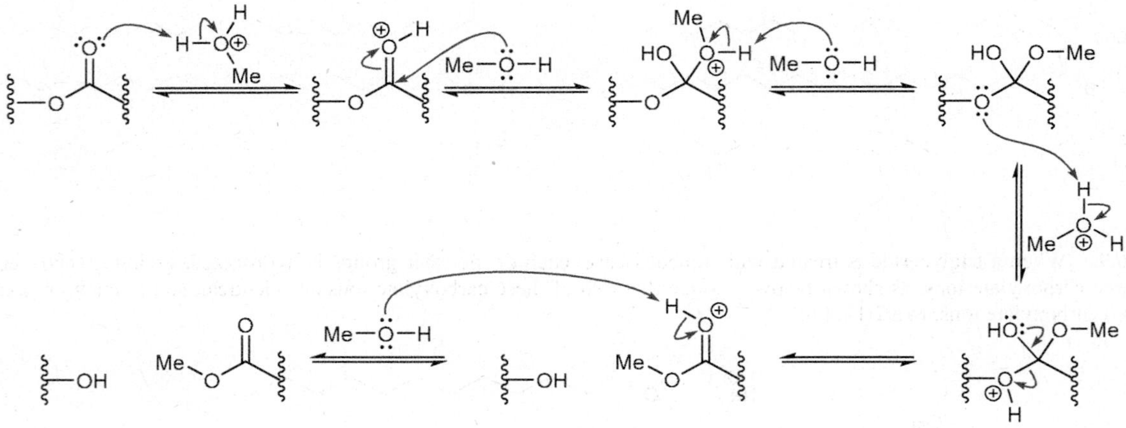

26.13. When triolein undergoes transesterification with isopropyl alcohol, the glycerol backbone is released, along with three equivalents of the isopropyl ester shown:

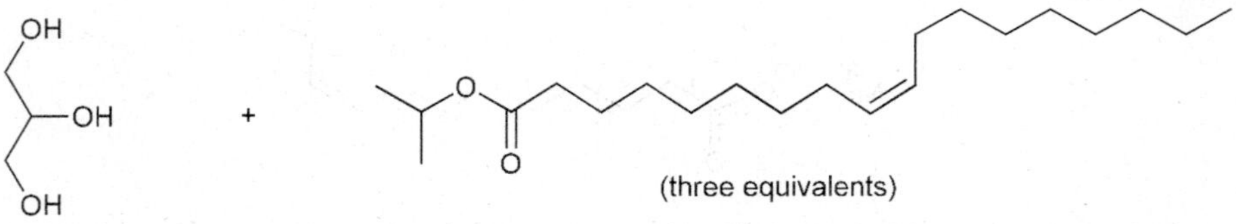

(three equivalents)

26.14.

(a) Hydroxide functions as a catalyst by establishing an equilibrium in which some ethoxide ions are present.

$HO^- + H{-}OEt \rightleftharpoons H_2O + {}^-OEt$

ethoxide

Then, each ester group undergoes transesterification via the following mechanism. The carbonyl group is attacked by ethoxide, giving a tetrahedral intermediate (an alkoxide ion). The carbonyl group is then reformed via expulsion of an alkoxide leaving group, which is then protonated by water, ultimately producing glycerol and three equivalents of an ethyl ester, as shown.

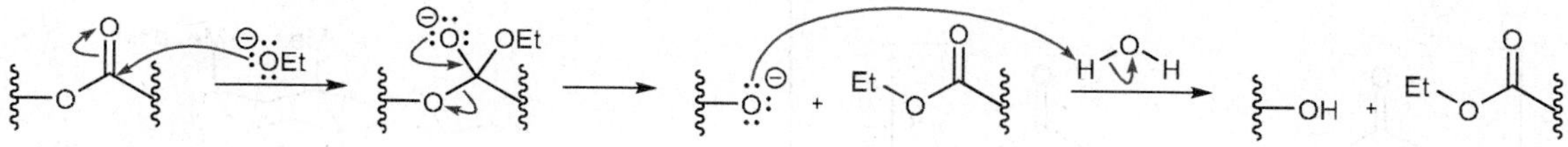

(b) If there is too much hydroxide present, it can function as a nucleophile and attack each ester group directly, giving hydrolysis (saponification) rather than transesterification. A mixture of products would be formed that included both esters and carboxylate salts.

26.15.

(a) The following lecithin has two myristic acid residues:

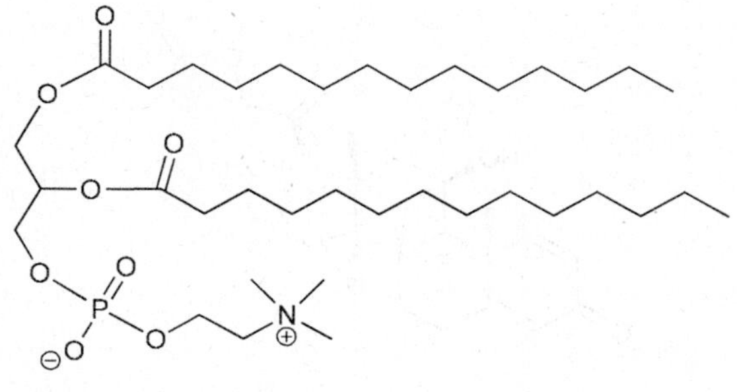

(b) The C2 position (of a lecithin) will be a chiral center and will generally have the *R* configuration, as shown:

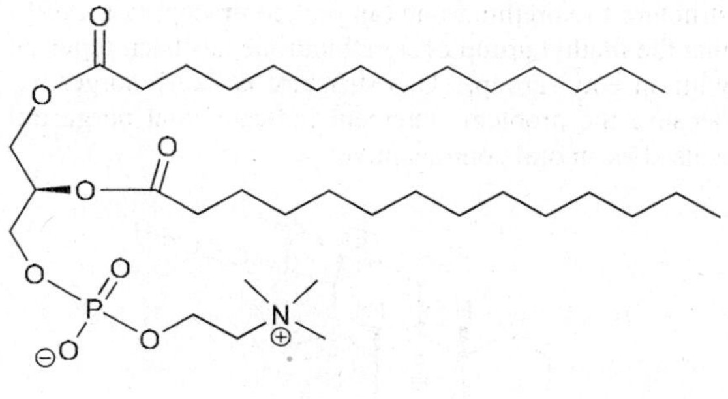

(c) If the phosphodiester was located at C2, that position would no longer be a chiral center, because it would be connected to two identical groups. Therefore, the molecule would no longer be chiral.

26.16.

(a) Each of the following two cephalins has one palmitic acid residue and one oleic acid residue. In each case, the C2 position is a chiral center (shown here with the *R* configuration).

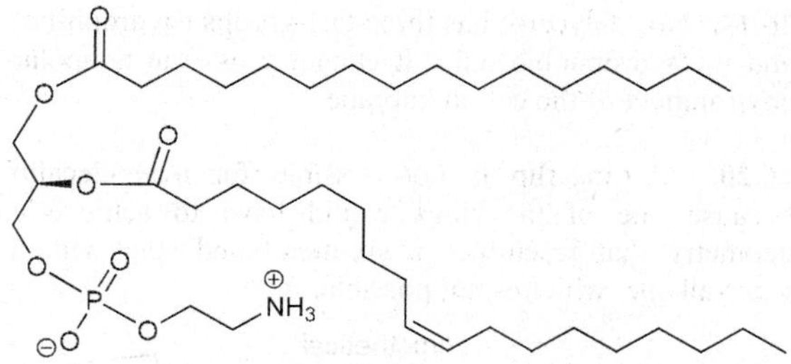

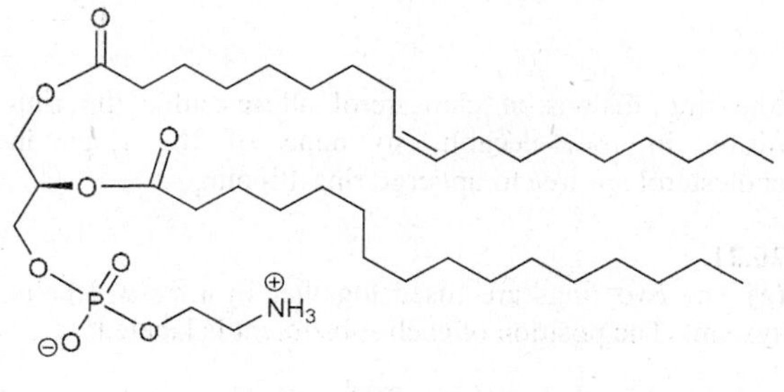

(b) If the phosphodiester was located at C2, then that carbon atom would still be connected to four different groups so it would still be a chiral center. Therefore, the compound would still be chiral.

26.17. The conjugate base of a phosphatidic acid contains a phosphate group that has three resonance structures, as shown:

26.18. Octanol has a longer hydrophobic tail than hexanol and is therefore more efficient at crossing the nonpolar environment of the cell membrane.

26.19. No. Glycerol has three OH groups (hydrophilic) and no hydrophobic tail. It cannot cross the nonpolar environment of the cell membrane.

26.20. A ring-flip is not possible for *trans*-decalin because one of the rings would have to achieve a geometry that resembles a six-membered ring with a *trans*-alkene, which is not possible.

Hypothetical ring flip

The ring fusions of cholesterol all resemble the ring fusion in *trans*-decalin, so none of the rings in cholesterol are free to undergo ring-flipping.

26.21.

(a) The two rings are fused together in a *trans*-decalin system. The position of each substituent is labeled:

(b) The two rings are fused together in a *trans*-decalin system. The position of each substituent is labeled:

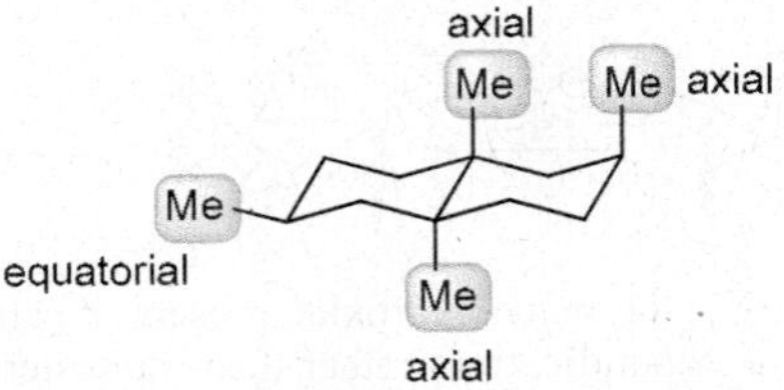

(c) Each set of neighboring rings are fused together, much like in a *trans*-decalin system. The position of each substituent is labeled:

26.22. The structure of prednisolone acetate is shown below. As described in the problem statement, this structure is different from the structure of cortisol in two ways, highlighted below:

26.23. One of these compounds is extremely similar in structure to norethindrone (an oral contraceptive), except that the methyl group of norethindrone has been replaced with an ethyl group. This structure is likely norgestrel, because the problem statement indicates that norgestrel is used as an oral contraceptive.

Oxymetholone is an anabolic steroid and it is similar in structure to nandrolone (a synthetic androgen analogue):

oxymetholone

26.24.

(a) This prostaglandin has the PGE substitution pattern, and there is only one π bond in the side chains, so this compound is PGE_1.

(b) This prostaglandin has the PGF substitution pattern, and there is only one π bond in the side chains. For a PGF substitution pattern, an additional descriptor is added to the name to indicate the configuration of the OH groups. A *cis* diol is designated as α, so this compound is $PGF_{1\alpha}$.

26.25.

(a) This terpene has ten carbon atoms and is therefore comprised of two isoprene units, shown here in one possible isoprene arrangement:

menthol

(b) This terpene has ten carbon atoms and is therefore comprised of two isoprene units, shown here:

grandisol

(c) This terpene has ten carbon atoms and is therefore comprised of two isoprene units, shown here in one possible isoprene arrangement:

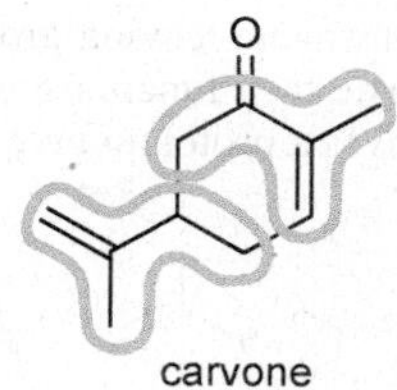

carvone

26.26.

(a) Yes, this compound has 10 carbon atoms and is comprised of two isoprene units.

(b) No, this compound is not a terpene, because it has 11 carbon atoms. In order to be a terpene, the number of carbon atoms must be divisible by 5.

(c) No, this compound is not a terpene, because it has 11 carbon atoms. In order to be a terpene, the number of carbon atoms must be divisible by 5.

(d) This compound has 10 carbon atoms, but the branching pattern cannot be achieved by joining two isoprene units, so this compound is not a terpene.

26.27. The pyrophosphate leaving group is expelled to give a resonance-stabilized (allylic) carbocation. The π bond of isopentyl phosphate then functions as a nucleophile and attacks the carbocation. Finally, a basic amino acid residue of the enzyme removes a proton to give the product:

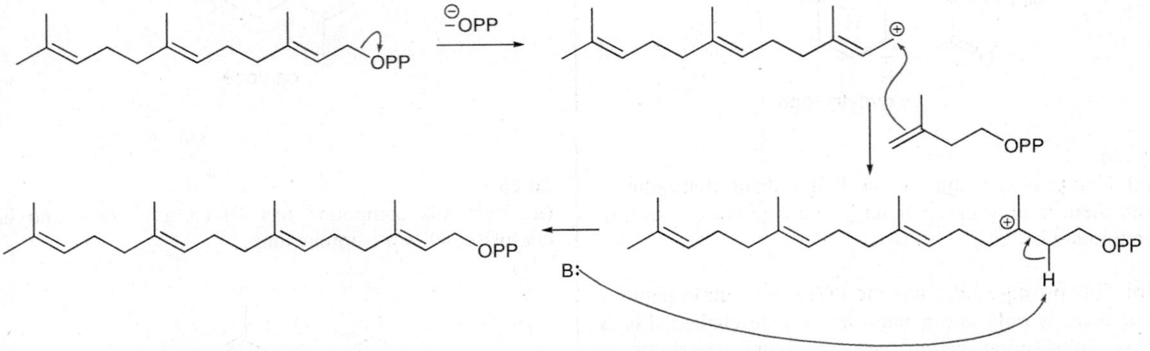

26.28. The pyrophosphate leaving group is expelled to give a resonance-stabilized (allylic) carbocation. The π bond of isopentyl phosphate then functions as a nucleophile and attacks the carbocation. A basic amino acid residue of the enzyme then removes a proton to give geranyl pyrophosphate. The previous three steps are then repeated to give farnesyl pyrophosphate, followed by an elimination process to give α-farnesene, as shown:

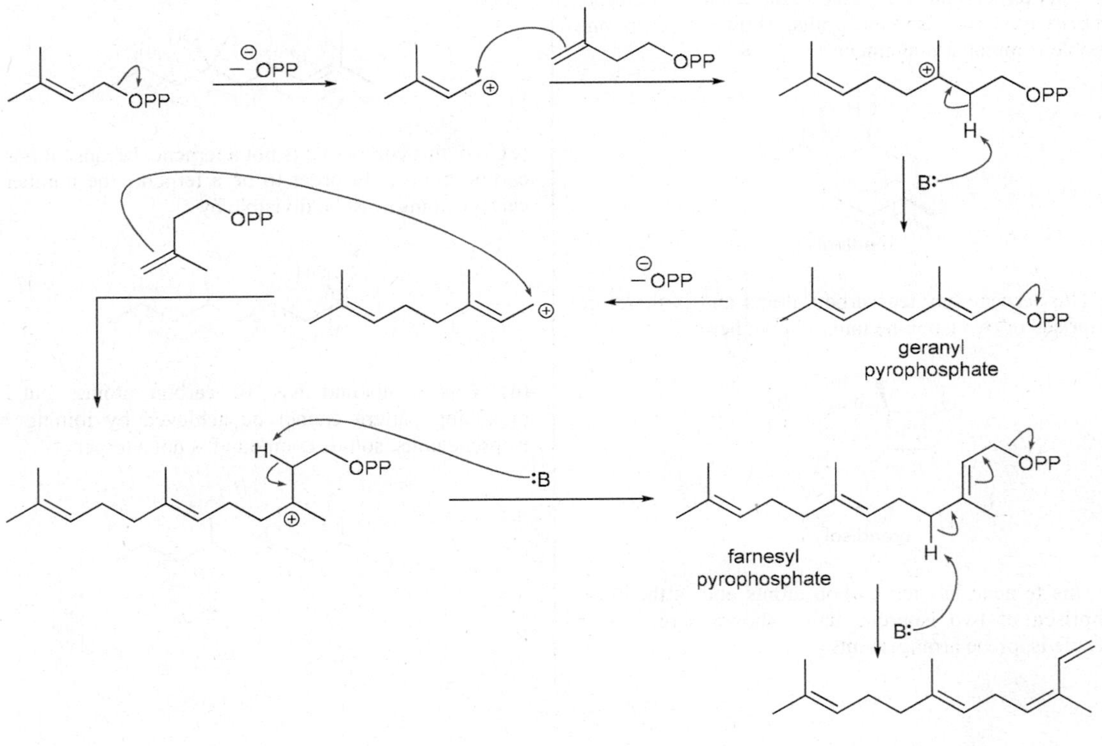

geranyl pyrophosphate

farnesyl pyrophosphate

α-farnesene

26.29.
(a) As seen in Section 26.6, stanozolol is a steroid.
(b) As seen in Section 26.8, lycopene is a terpene.
(c) As seen in Section 26.3, tristearin is a triglyceride.
(d) As seen in Section 26.5, lecithins are phospholipids.
(e) As seen in Section 26.7, PGF_2 is a prostaglandin.
(f) As seen in Section 26.2, pentadecyl octadecanoate is a wax.

26.30.
(a) When treated with excess molecular hydrogen and a metal catalyst (Ni), the C=C bond in each palmitoleic acid residue undergoes hydrogenation, giving the saturated triglyceride tripalmitin:

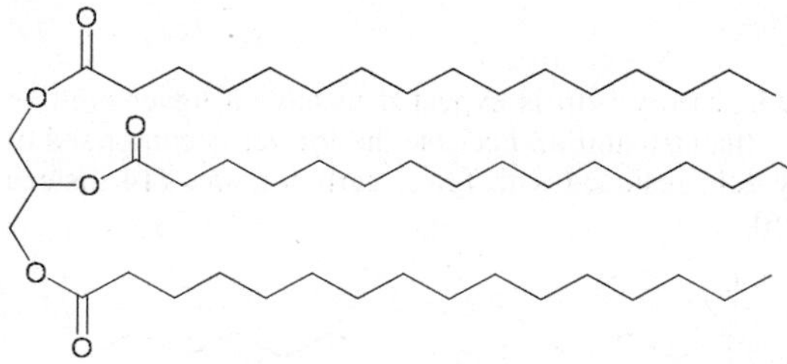

(b) When a triglyceride is treated with aqueous base, each of the ester groups is hydrolyzed (saponification), thereby releasing glycerol:

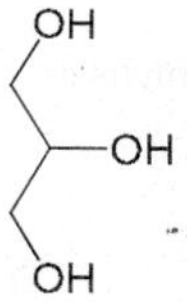

and three carboxylate ions (palmitate ions), as shown:

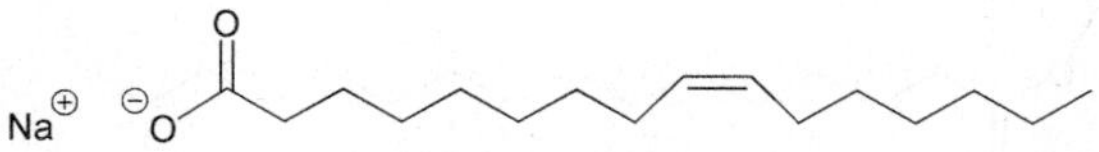

(three equivalents)

26.31. Each of the following two cephalins has one lauric acid residue and one myristic acid residue. In each case, the C2 position is a chiral center with the *R* configuration. Both compounds are chiral:

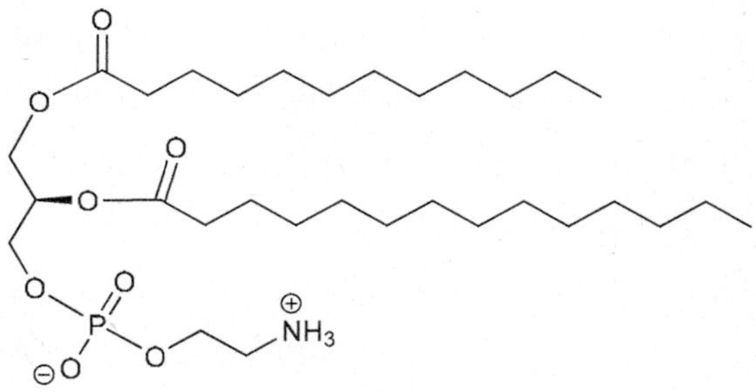

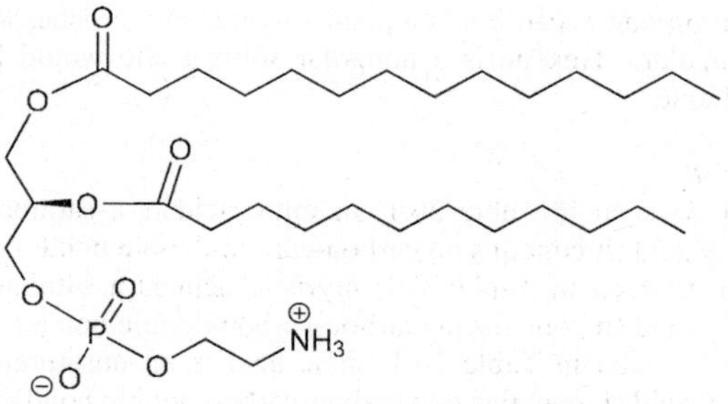

26.32. The fatty acid residues in this triglyceride are saturated. Due to the lack of C=C π bonds, it will not react with molecular hydrogen.

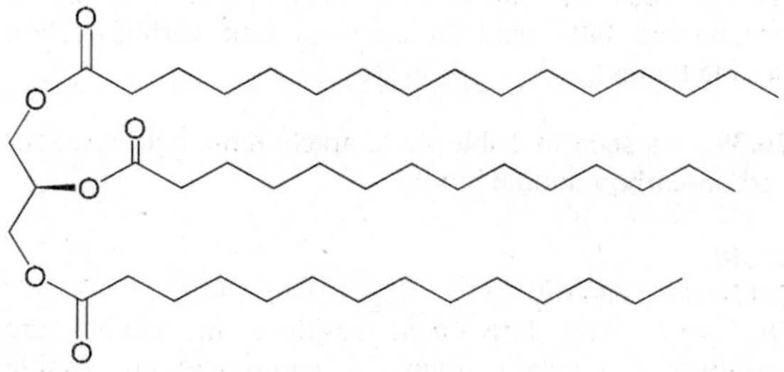

26.33.
(a) This compound is an amino acid. It is not a lipid.
(b) This compound has a large hydrophobic tail and is therefore a lipid.
(c) Lycopene is terpene, which is a type of lipid.
(d) Trimyristin is a triglyceride, which is a type of lipid.
(e) Palmitic acid has a large hydrophobic tail and is therefore a lipid.
(f) D-Glucose is a carbohydrate. It is not a lipid.
(g) Testosterone is a steroid, which is a type of lipid.
(h) D-Mannose is a carbohydrate. It is not a lipid.

26.34. *trans*-Oleic acid has 18 carbon atoms and a *trans* π bond between C9 and C10, as shown:

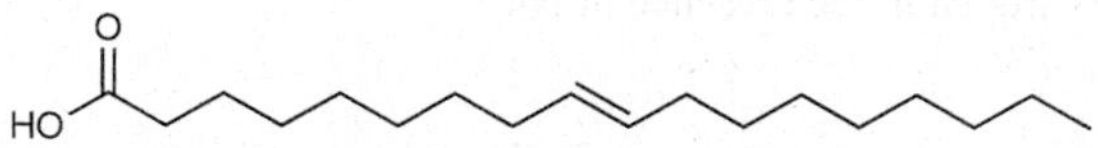

26.35. Auto-oxidation causes oils to become rancid, and this reaction occurs primarily at positions allylic to carbon-carbon double bonds. The fatty acid residues of tristearin are saturated and are therefore less susceptible to auto-oxidation than the unsaturated fatty acid residues in triolein.

26.36. A monoglyceride exhibits two OH groups (of the glycerol backbone) and is therefore expected to be the most water soluble of the three compounds. A triglyceride has no OH groups (all three positions of the glycerol backbone are occupied), so a triglyceride will be the least water soluble.

26.37. The phrase "like dissolves like" helps us to identify a suitable solvent. Water would not be appropriate because it is a polar solvent, and terpenes are nonpolar. Hexane is a nonpolar solvent and would be suitable.

26.38.
(a) As seen in Table 26.1, palmitic acid is a saturated fatty acid (it contains no carbon-carbon double bonds).
(b) As seen in Table 26.1, myristic acid is a saturated fatty acid (it contains no carbon-carbon double bonds).
(c) As seen in Table 26.1, oleic acid is an unsaturated fatty acid (it contains one carbon-carbon double bond).
(d) As seen in Table 26.1, lauric acid is a saturated fatty acid (it contains no carbon-carbon double bonds).
(e) As seen in Table 26.1, linoleic acid is an unsaturated fatty acid (it contains two carbon-carbon double bonds).
(f) As seen in Table 26.1, arachidonic acid is an unsaturated fatty acid (it contains four carbon-carbon double bonds).

26.39. As seen in Table 26.1, arachidonic acid has four carbon-carbon double bonds.

26.40.
(a) No. It is an oil.
(b) No. The fatty acid residues in triolein are unsaturated (each contains a carbon-carbon double bond), so triolein is reactive towards molecular hydrogen in the presence of Ni.
(c) Yes. It undergoes hydrolysis to produce unsaturated fatty acids (each contains a carbon-carbon double bond).
(d) Yes. It is a complex lipid because it undergoes hydrolysis.
(e) No. It is not an ester with a high molecular weight. It is not a wax.
(f) No. It does not have a phosphate group.

26.41.
(a) Yes. It is a fat.
(b) Yes. The fatty acid residues in tristearin are saturated (they do not contain any carbon-carbon double bonds), so tristearin is not reactive towards molecular hydrogen in the presence of Ni.
(c) No. It undergoes hydrolysis to produce fatty acids that are saturated (they do not contain any carbon-carbon double bonds).
(d) Yes. It is a complex lipid because it undergoes hydrolysis.
(e) No. It is not an ester with a high molecular weight. It is not a wax.
(f) No. It does not have a phosphate group.

26.42. The products of hydrolysis suggest the following ester:

$$H_3C{-}(CH_2)_{18}{-}\overset{O}{\overset{\|}{C}}{-}O{-}(CH_2)_{29}{-}CH_3$$

20 carbon atoms   30 carbon atoms

26.43. Trimyristin is expected to have a lower melting point than tripalmitin because the former is comprised of fatty acid residues with fewer carbon atoms (14 instead of 16).

trimyristin

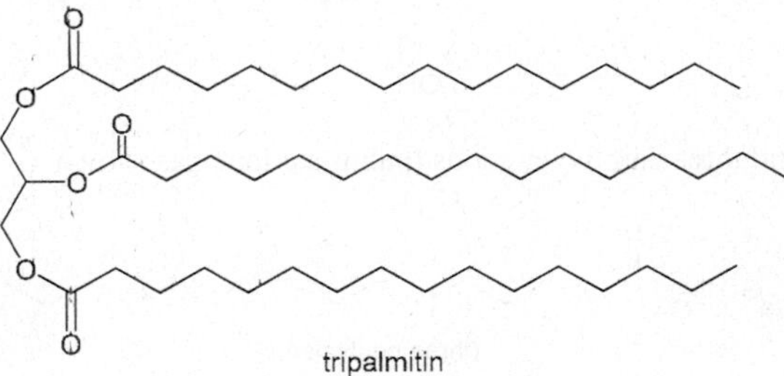

tripalmitin

26.44. Each of the three ester groups undergoes transesterification via the following mechanism. The carbonyl group is protonated, and the resulting resonance-stabilized cation is then attacked by isopropanol, giving a tetrahedral intermediate (an oxonium ion). Deprotonation, followed by protonation, gives another oxonium ion, which then loses a leaving group, thereby regenerating the carbonyl group. Deprotonation is the final step, ultimately producing glycerol and three equivalents of the isopropyl ester:

26.45. See the solution to Problem 26.14.

26.46. In order to be achiral (possessing a plane of symmetry), the palmitic acid residue must be connected to C2 of the glycerol backbone (shown below). Otherwise, C2 would be a chiral center. In the triglyceride shown below, C2 is connected to two identical groups, so it is not a chiral center.

26.47. In order for the triglyceride to be chiral (lacking a plane of symmetry), the palmitic acid residue cannot be connected to C2 of the glycerol backbone (as explained in the previous problem).

26.48. The carbon skeleton of cholesterol is redrawn, but all wedges are replaced with dashes, and all dashes are replaced with wedges, giving the following structure (the enantiomer of cholesterol):

26.49.

(a) This terpene has fifteen carbon atoms and is therefore comprised of three isoprene units, shown here:

bisabolene

(b) This terpene has twenty carbon atoms and is therefore comprised of four isoprene units, shown here:

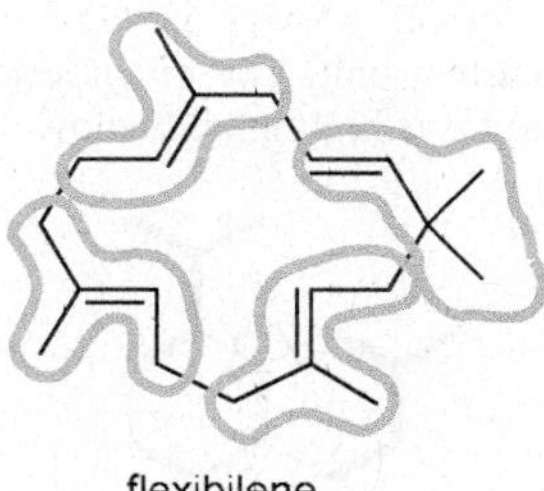

flexibilene

(c) This terpene has fifteen carbon atoms and is therefore comprised of three isoprene units, shown here:

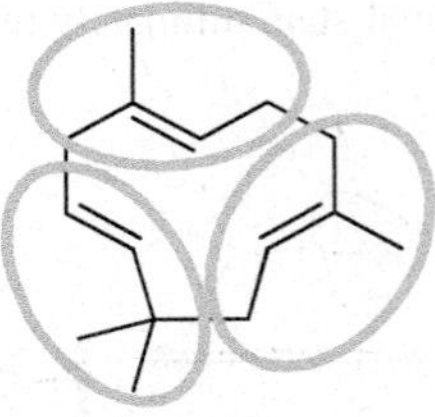

humulene

(d) This terpene has twenty carbon atoms and is therefore comprised of four isoprene units, shown here:

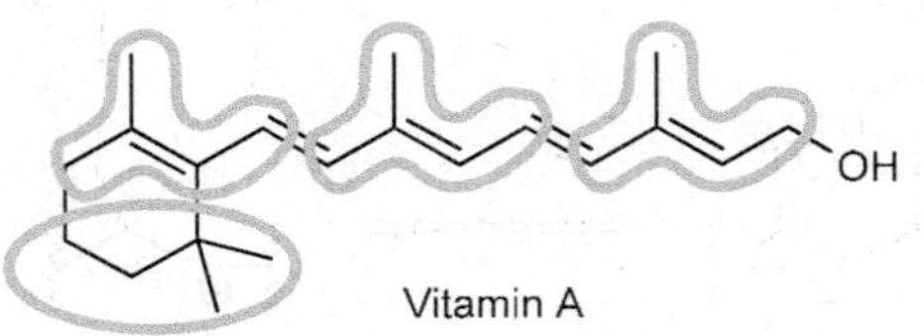

Vitamin A

(e) This terpene has ten carbon atoms and is therefore comprised of two isoprene units, shown here:

OH

geraniol

(f) This terpene has ten carbon atoms and is therefore comprised of two isoprene units, shown here:

sabinene

26.50.
(a) The polar head and the two hydrophobic tails are labeled in the following structure:

Hydrophobic tails

Polar head

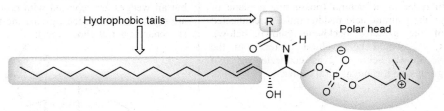

(b) Yes, they have one polar head and two hydrophobic tails. See Figure 26.6.

26.51. The correct answer is (c). The given compound is a triglyceride that would produce glycerin (and three fatty acids) upon hydrolysis. The presence of double bonds indicates that it will undergo catalytic hydrogenation. The *cis* double bonds also cause kinks in the carbon chains, thus lowering the melting point of the compound. A *polyunsaturated* triglyceride is described as an oil because it is a liquid at room temperature. It is not a solid.

26.52. The correct answer is (a). Isoprene is a branched, five-carbon unit. The given sesquiterpene has three isoprene units, as highlighted below:

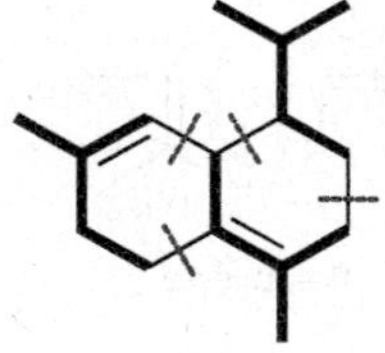

26.53. The correct answer is (b). See Figure 26.6, which illustrates that phospholipids have the appropriate geometry for bilayer formation, while fatty acids and triglycerides do not have the appropriate geometry for bilayer formation.

26.54. The correct answer is (b). Fatty acids are long-chain carboxylic acids, and structure (b) contains no carbon-carbon double bonds, so it is described as *saturated*.

26.55. The correct answer is (b). Lipids are naturally occurring compounds that can be extracted from cells using nonpolar organic solvents. Triglycerides, terpenes and steroids are all lipids, but a tripeptide is not. A tripeptide has many polar functional groups and is not classified as a lipid.

26.56.

(a) The epoxide can be formed on the top face of the π bond, as shown:

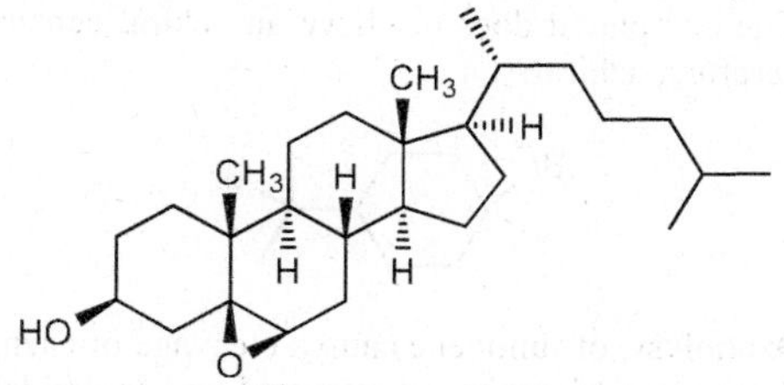

or the epoxide can be formed on the bottom face of the π bond, as shown:

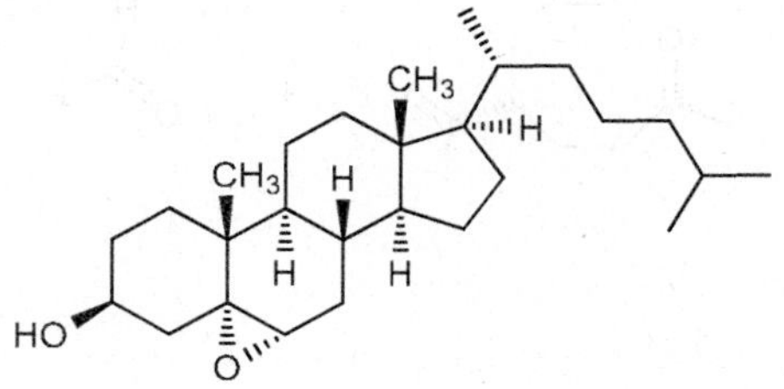

(b) The methyl group (C19) provides steric hindrance that blocks one side of the π bond, and only the following epoxide is obtained:

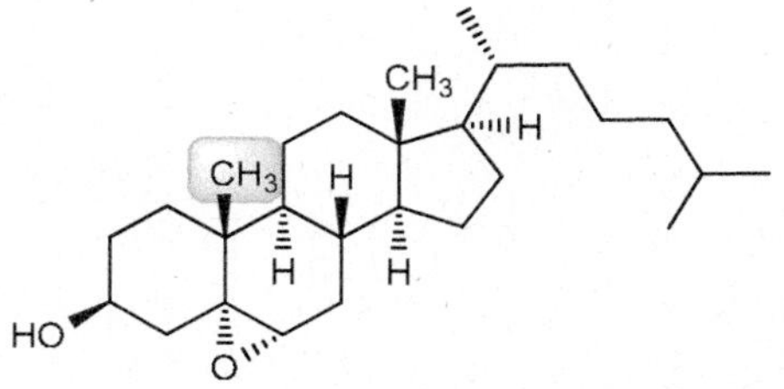

26.57.

(a) Estradiol has an aromatic ring that bears an OH group. As such, the ring is strongly activated toward electrophilic aromatic substitution. Upon treatment with excess Br_2, bromination occurs at the two positions that are *ortho* to the OH group, which is an *ortho-para* director.

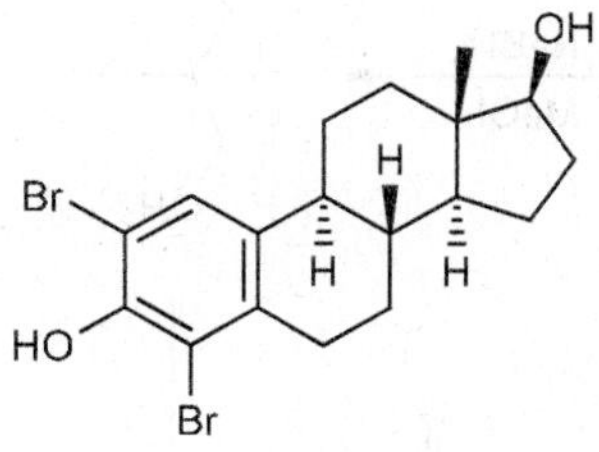

The *para* position is already occupied so bromination does not occur at that location.

(b) Upon treatment with PCC, the secondary alcohol is oxidized to give the following ketone:

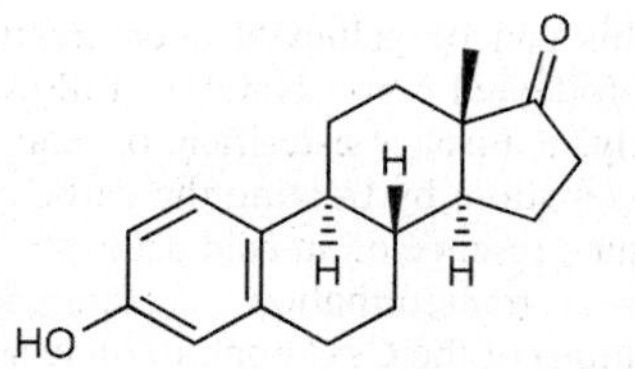

(c) Upon treatment with a strong base, followed by excess ethyl iodide, each of the OH group undergoes alkylation, thereby converting the OH groups into ethoxy groups, as shown:

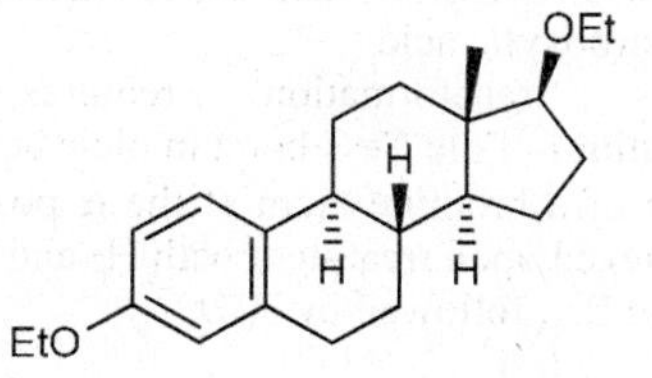

(d) Upon treatment with excess acetyl chloride in the presence of pyridine, each of the OH group undergoes acetylation, giving the following product:

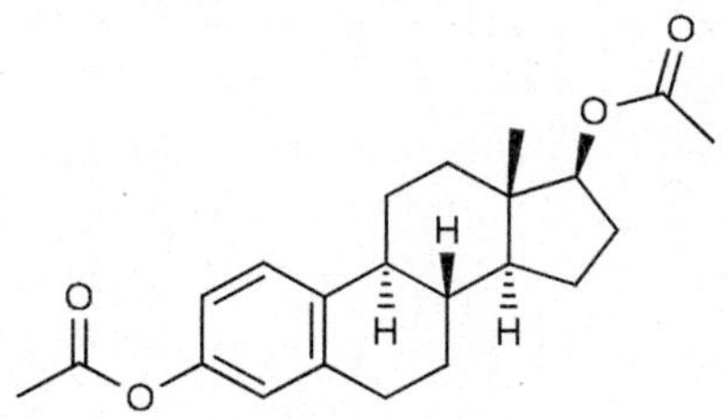

26.58. Every one of the OH groups in sucrose (see Section 24.7) is converted into an ester group, where R is used to represent the hydrophobic tail of each lauric acid residue. This compound is not superimposable on its mirror image, so it is chiral (much like sucrose).

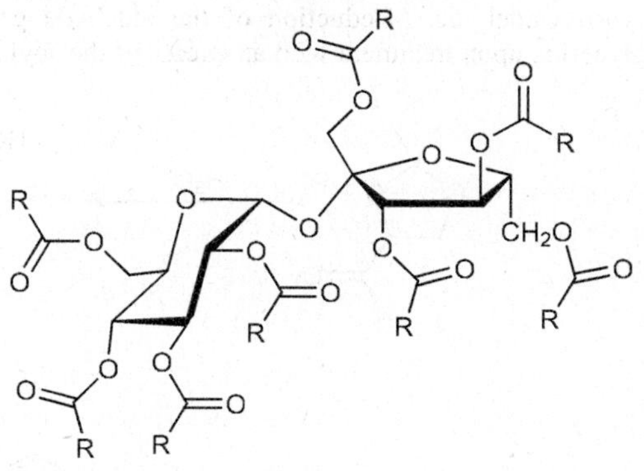

26.59.

(a) This transformation requires reduction (hydrogenation) of the C=C bond in oleic acid, which can be achieved upon treatment with H_2 in the presence of Ni.

(b) This transformation requires reduction (hydrogenation) of the C=C bond in oleic acid, as well as esterification (conversion of the OH group to an ethoxy group). This can be achieved upon treatment with H_2 and Ni, followed by NaOH, followed by EtI. Alternatively, a Fischer esterification can be employed after hydrogenation, by treating the carboxylic acid with ethanol in the presence of an acid catalyst.

(c) This transformation requires reduction (hydrogenation) of the C=C bond in oleic acid, as well as reduction of the carboxylic acid group to give a primary alcohol. This can be achieved upon treatment with H_2 and Ni, followed by $LiAlH_4$, followed by aqueous acidic work-up.

(d) Ozonolysis (O_3, followed by DMS) followed by oxidation with $Na_2Cr_2O_7$ and H_2SO_4 will generate the desired dicarboxylic acid.

(e) This transformation requires reduction (hydrogenation) of the C=C bond in oleic acid, as well as installation of a bromine atom at the α position. This can be achieved upon treatment with H_2 and Ni, followed by PBr_3 and Br_2, followed by H_2O.

26.60.

(a) Limonene is comprised of 10 carbon atoms and is therefore a monoterpene.

(b) The compound does not have any chiral centers and is, therefore, achiral:

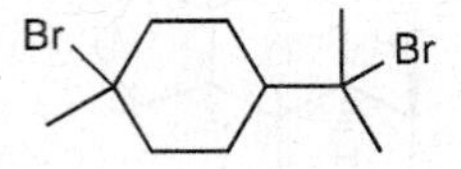

Br Br

(c) Ozonolysis of limonene causes cleavage of each C=C bond, giving a tricarbonyl compound and formaldehyde, as shown:

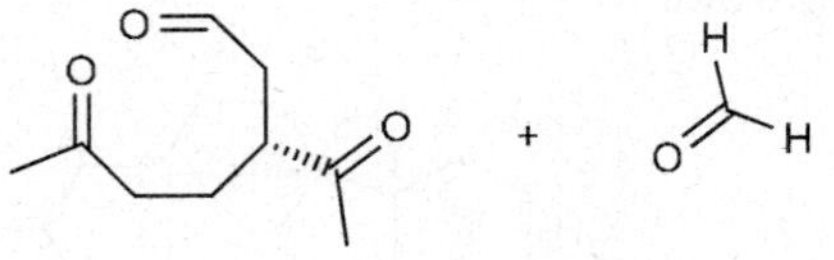

O O O + H O H

26.61. The starting material is a cyclic acetal. Upon treatment with aqueous acid, the acetal is opened to give a dihydroxyaldehyde. Reduction of the aldehyde group gives glycerol, which is then converted into the desired triglyceride upon treatment with an excess of the acyl halide in the presence of pyridine:

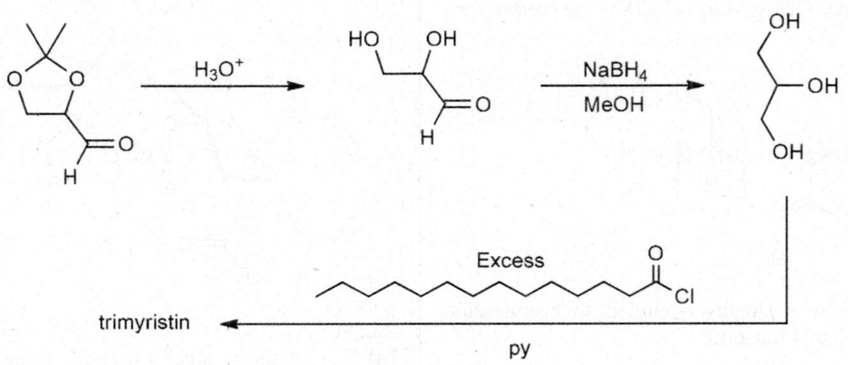

O O H O → H_3O^+ → HO OH O H → $NaBH_4$ / MeOH → OH OH OH → Excess (acyl chloride, O, Cl) / py → trimyristin

26.62.

(a) Fats and oils have a glycerol backbone connected to three fatty acid residues. This compound also has a glycerol backbone, but it is only connected to two fatty acid residues. The third group (left) is not a fatty acid residue.

(b) Both ester groups are hydrolyzed upon treatment with aqueous base, giving the following products:

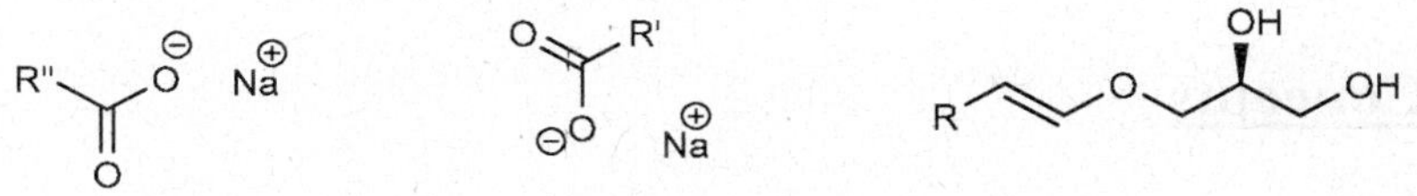

(c) In aqueous acid, the two ester groups undergo hydrolysis, just as we saw in basic conditions. Under these conditions, the enol ether also undergoes acidic cleavage, thereby freeing glycerol and an enol. Upon its formation, the enol rapidly tautomerizes to give an aldehyde. In summary, we expect the following products:

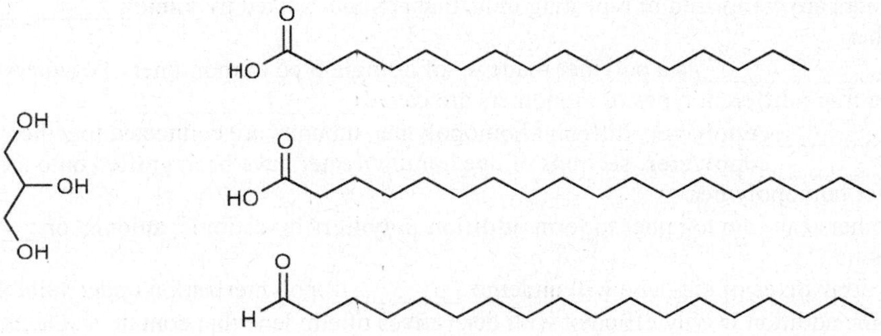

Chapter 27
Synthetic Polymers

Review of Concepts

Fill in the blanks below. To verify that your answers are correct, look in your textbook at the end of Chapter 27. Each of the sentences below appears verbatim in the section entitled *Review of Concepts and Vocabulary*.

- Polymers are comprised of repeating units that are constructed by joining _______________ together.
- A _______________ is a polymer made up of a single type of monomer. Polymers made from two or more different types of monomers are called _________________.
- In a ________ **copolymer**, different homopolymer subunits are connected together in one chain. In a ________ **copolymer**, sections of one homopolymer have been grafted onto a chain of another homopolymer.
- Monomers can join together to form **addition polymers** by cationic, anionic, or _____________ addition.
- Most derivatives of ethylene will undergo __________ polymerization under suitable conditions.
- Cationic addition is only efficient with derivatives of ethylene that contain an electron-____________ group.
- Anionic addition is only efficient with derivatives of ethylene that contain an electron-____________ group.
- Polymers generated via condensation reactions are called ________________ **polymers**.
- _______**-growth polymers** are formed under conditions in which each monomer is added to the growing chain one at a time. The monomers do not react directly with each other.
- _______**-growth polymers** are formed under conditions in which the individual monomers react with each other to form __________, which are then joined together to form polymers.
- **Crossed-linked polymers** contain ________ bridges or branches that connect neighboring chains.
- **Thermoplastics** are polymers that are _____ at room temperature but ______ when heated. They are often prepared in the presence of ___________ to prevent the polymer from being brittle.
- ___________ are polymers that return to their original shape after being stretched.
- ________________ **polymers** can be broken down by enzymes produced by microorganisms in the soil.

Review of Skills

Fill in the blanks and empty boxes below. To verify that your answers are correct, look in your textbook at the end of Chapter 27. The answers appear in the section entitled *SkillBuilder Review*.

27.1 Determining Which Polymerization Technique is More Efficient

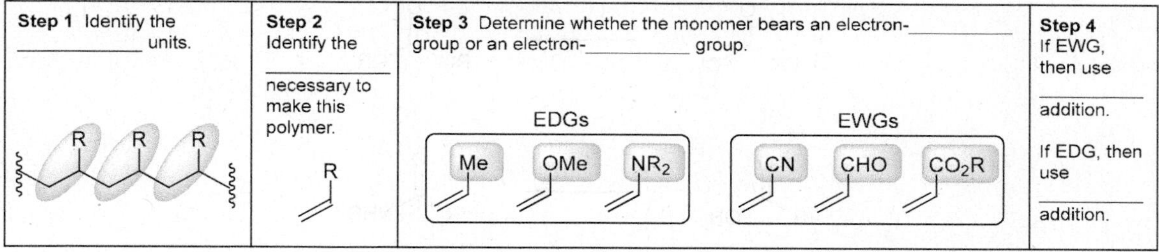

27.2 Identifying the Monomers Required to Produce a Desired Condensation Polymer

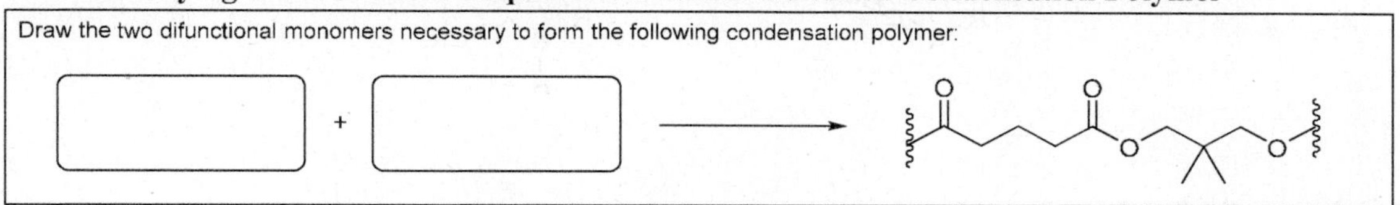

Review of Reactions

Identify the reagents necessary to achieve each of the following transformations. To verify that your answers are correct, look in your textbook at the end of Chapter 27. The answers appear in the section entitled *Review of Reactions*.

Reactions for Formation of Chain-Growth Polymers

R → R R R

EDG → EDG EDG EDG

EWG → EWG EWG EWG

Reactions for Formation of Step-Growth Polymers

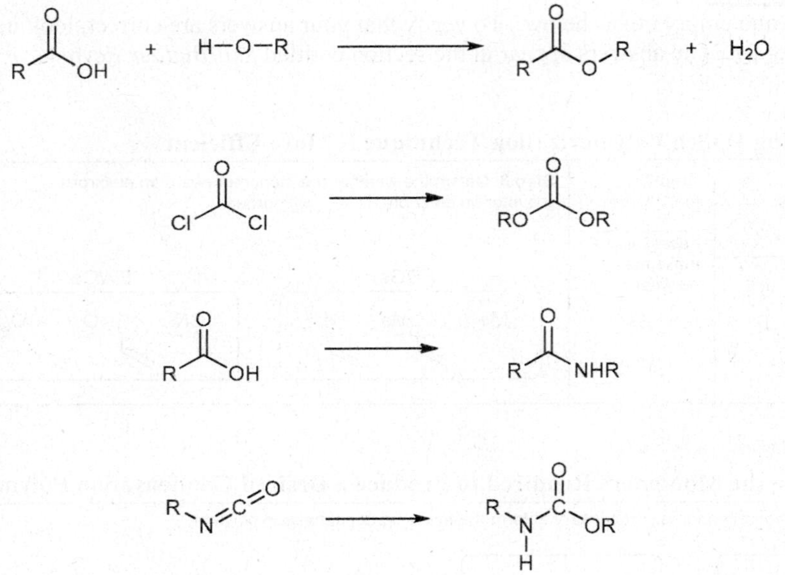

Review of Mechanisms

Complete each of the following mechanisms by drawing the missing curved arrows. To verify that your curved arrows are drawn correctly, compare them to the curved arrows in the mechanism boxes for Mechanisms 27.1 – 27.3, which can be found throughout Chapter 27 of your text.

Mechanism 27.1 Radical Polymerization

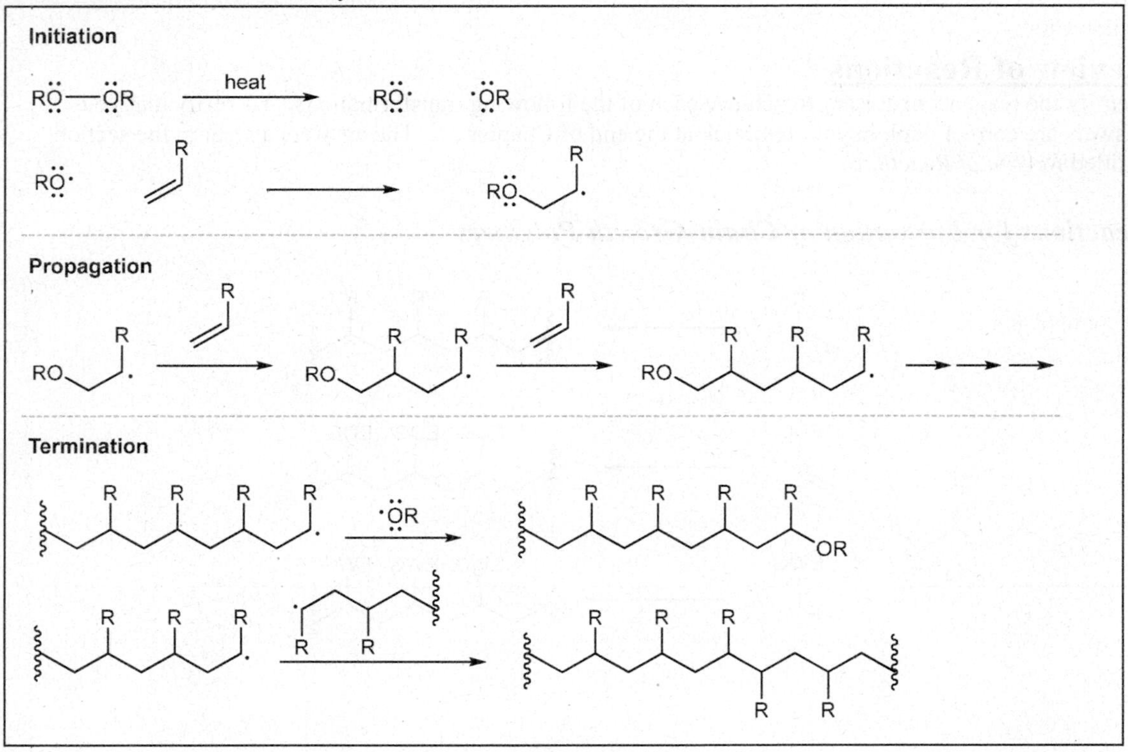

Initiation

heat

Propagation

Termination

Mechanism 27.2 Cationic Polymerization

Initiation

H—A

Propagation

Termination

:Base

:Nuc

Mechanism 27.3 Anionic Polymerization

Initiation

Bu: Li +

Propagation

Termination

H_2O

O=C=O

Solutions

27.1.

(a) Polymerization of vinyl acetate gives poly(vinyl acetate):

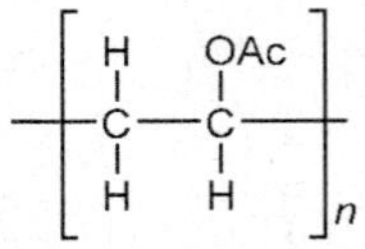

poly(vinyl acetate)

(b) Polymerization of vinyl bromide gives poly(vinyl bromide):

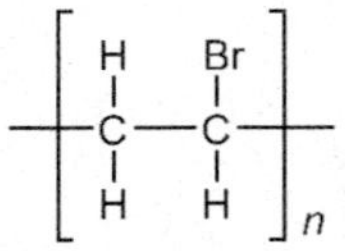

poly(vinyl bromide)

(c) Polymerization of α-butylene gives poly-α-butylene:

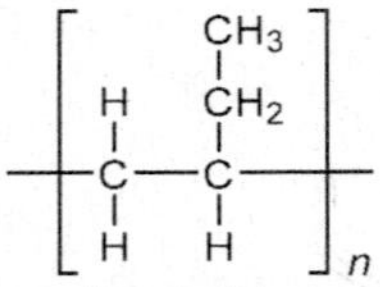

poly-α-butylene

27.2. Poly(methyl acrylate) can be made from methyl acrylate, shown here:

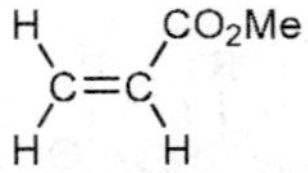

methyl acrylate

27.3. The following structure represents an alternating copolymer constructed from styrene and ethylene. The units produced from styrene and ethylene are highlighted:

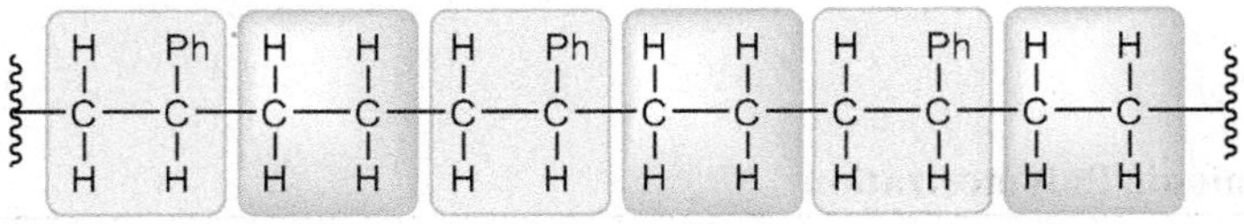

27.4. The following structure represents a block copolymer constructed from propylene and vinyl chloride. The units produced from propylene and vinyl chloride are highlighted:

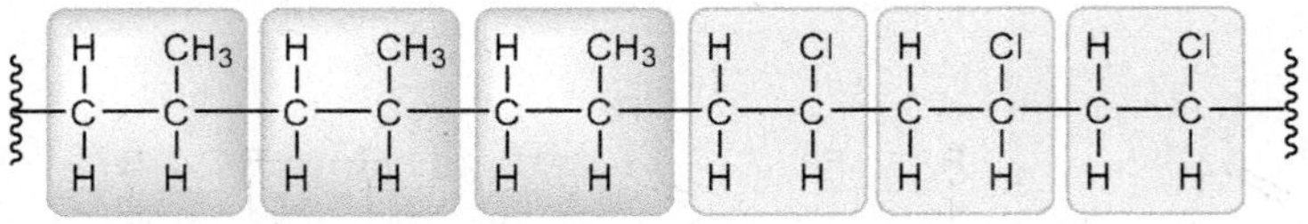

27.5. This copolymer can be made from isobutylene and styrene, as shown:

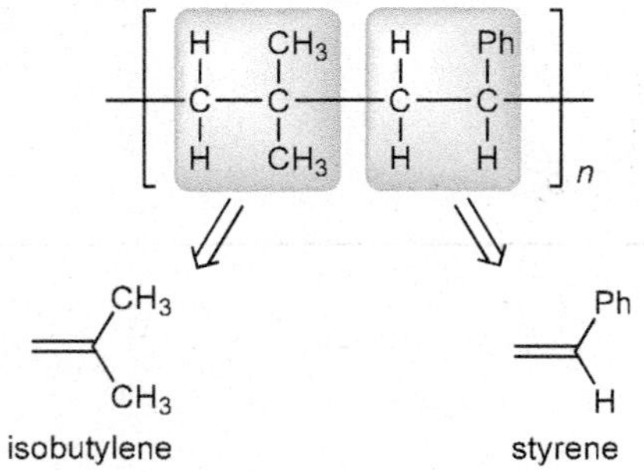

isobutylene styrene

27.6. In each case, we identify the nature of the vinylic group, which determines the conditions to use. Anionic conditions are used if the vinylic group is electron-withdrawing, while cationic conditions are used for an electron-donating group:

(a) A cyano group is an electron-withdrawing substituent (see Table 18.1 and associated discussion), so preparation of this compound would be best achieved via anionic addition.

(b) A methoxy group is an electron-donating substituent (see Table 18.1 and associated discussion), so preparation of this compound would be best achieved via cationic addition.

(c) Methyl groups are electron-donating substituents (see Table 18.1 and associated discussion), so preparation of this compound would be best achieved via cationic addition.

(d) An acetate group is an electron-donating substituent (see Table 18.1 and associated discussion), so preparation of this compound would be best achieved via cationic addition.

(e) A nitro group is an electron-withdrawing substituent (see Table 18.1 and associated discussion), so preparation of this compound would be best achieved via anionic addition.

(f) A trichloromethyl group is an electron-withdrawing substituent (see Table 18.1 and associated discussion), so preparation of this compound would be best achieved via anionic addition.

27.7. An acetate group is more powerfully electron-donating (via resonance) than a methyl group (via hyperconjugation), so vinyl acetate is expected be the most reactive toward cationic polymerization. A nitro group is electron-withdrawing, so nitroethylene is expected to be the least reactive toward cationic polymerization.

least reactive → most reactive

NO_2 CH_3 OAc

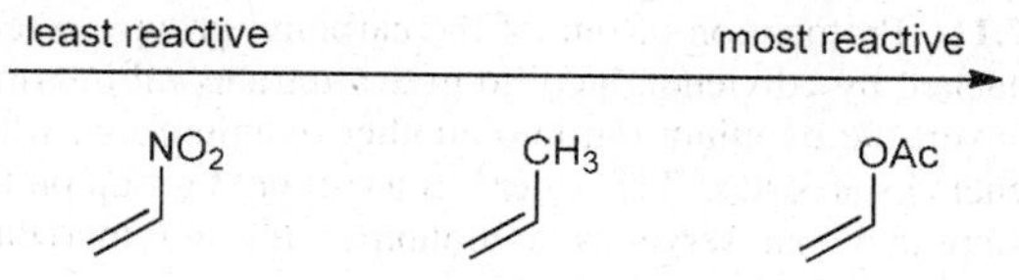

27.8. A nitro group is a very powerful electron-withdrawing group (see Table 18.1 and associated discussion), so nitroethylene is expected to be the most reactive toward anionic polymerization. A chlorine atom is only weakly electron-withdrawing, as compared with a nitro group or a carbonyl group (see Table 18.1 and associated discussion), so vinyl chloride is expected to be the least reactive toward anionic polymerization.

least reactive → most reactive

Cl O NO_2

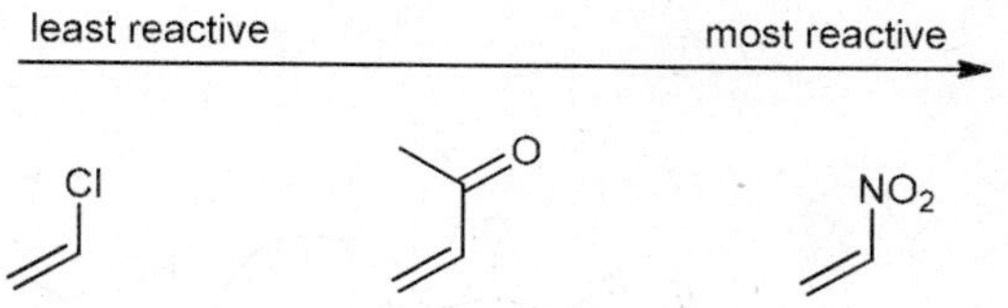

27.9. A benzylic anion, cation or radical will be stabilized by resonance, so styrene can be effectively polymerized under anionic, cationic or radical-initiated addition conditions.

27.10. In the initiation step, water attacks one molecule of the monomer, giving a carbanion. This carbanion then attacks another molecule of the monomer in a propagation step. This propagation step repeats itself, thereby growing the polymer chain. A termination step can occur if the carbanion is protonated by water, as shown:

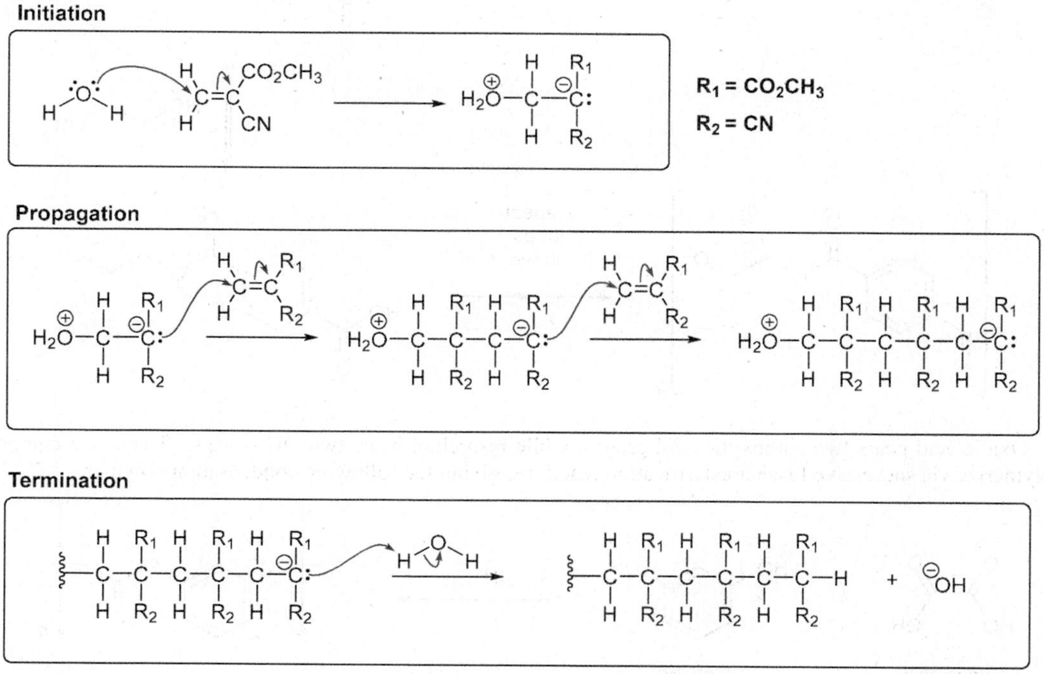

Initiation

$R_1 = CO_2CH_3$

$R_2 = CN$

Propagation

Termination

27.11. Protonation of one of the carbonyl groups in terephthalic acid renders it even more electrophilic, and it is then attacked by ethylene glycol to give a tetrahedral intermediate (an oxonium ion). Two successive proton transfer steps convert the oxonium ion into another oxonium ion, which can lose water to regenerate the C=O bond. Deprotonation generates an ester. This ester has a carbonyl group on the left side, and an OH group on the right side. As a result, this compound can serve as a monomer for polymerization (via a repetition of the Fischer esterification steps just described).

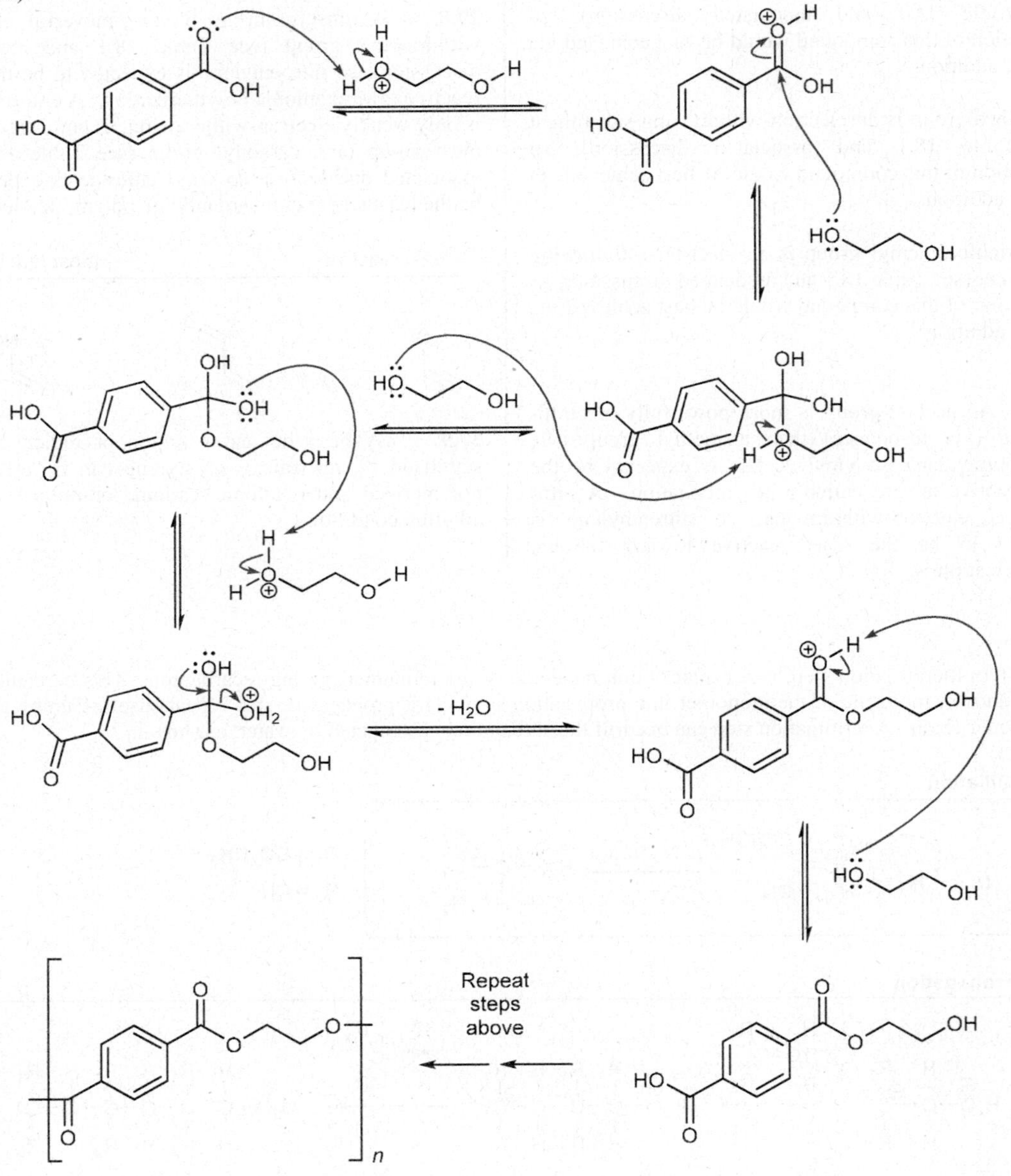

- H_2O

Repeat steps above

n

27.12. Oxalic acid bears two carboxylic acid groups, while resorcinol bears two OH groups. These two compounds can polymerize via successive Fischer esterification reactions, giving the following condensation polymer:

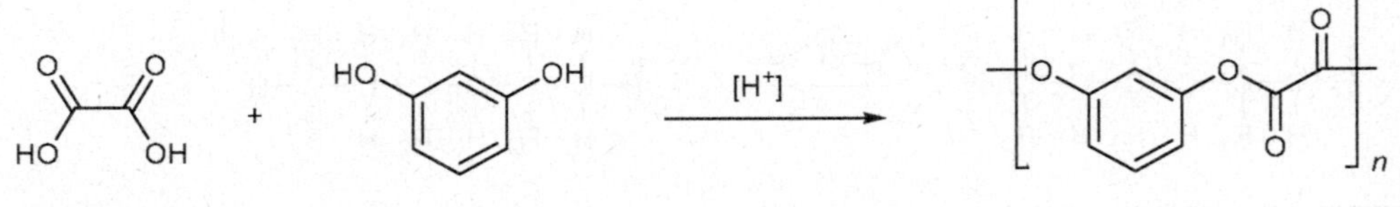

[H^+]

oxalic acid

resorcinol

n

27.13. Each of the amide groups can be made via the reaction between a carboxylic acid group and an amino group. Therefore, Kevlar can be made from the following dicarboxylic acid and diamine monomers:

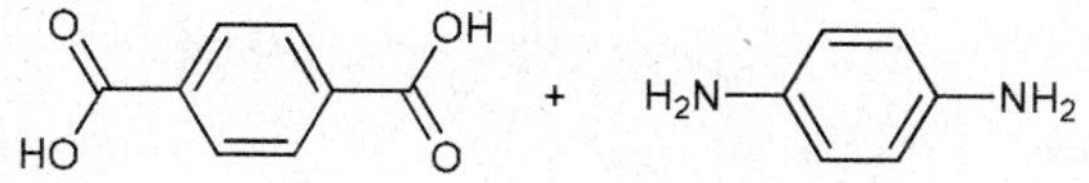

27.14.
(a) Each of the ester groups can be made from the reaction between a carboxylic acid group and an alcohol (via a Fischer esterification reaction). Therefore, the desired polymer can be made from the following diol and dicarboxylic acid monomers:

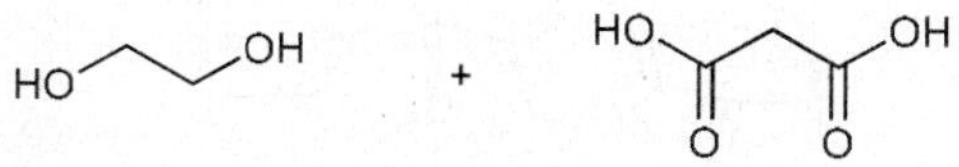

(b) Each carbonate group can be made from the reaction between phosgene and two alcohols. Therefore, the desired polymer can be made from the following diol and phosgene:

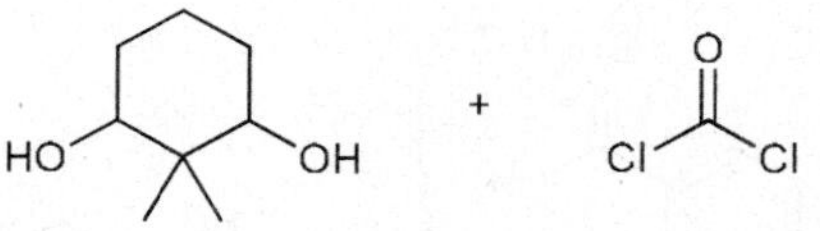

27.15. Each of the OH groups in 1,4-butanediol can attack phosgene, giving the following polymer:

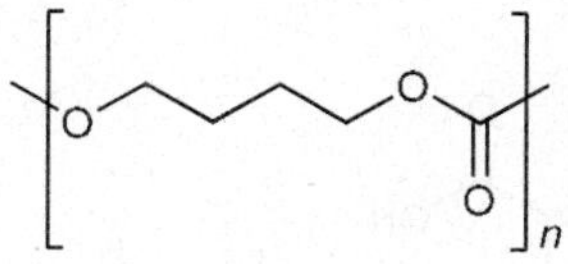

27.16.
(a) ε-Aminocaproic acid has both a carboxylic acid group and an amino group (which can react with each, in an intermolecular fashion, to give an amide linkage). As such, this compound will polymerize to form the following polymer:

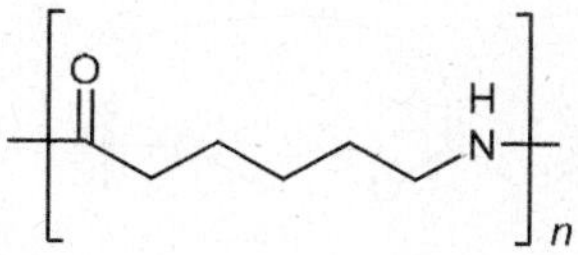

(b) Nylon 6 exhibits a smaller repeating unit than Nylon 6,6.

27.17.
(a) Each monomer has two growth points, so we expect that polymerization will generate a step-growth polymer.

(b) When these monomers are used to form a copolymer, the growing polymer chain has only one growth point, so we expect that polymerization will generate a chain-growth polymer.

27.18. This polymer exhibits repeating carbonate groups, so it can be made from phosgene and the appropriate diol. Since the diol has two growth points (and since the growing oligomers also have two growth points), this polymer can be classified as a step-growth polymer.

27.19. Polyisobutylene does not have any chiral centers, so it cannot be described as isotactic, syndiotactic, or atactic.

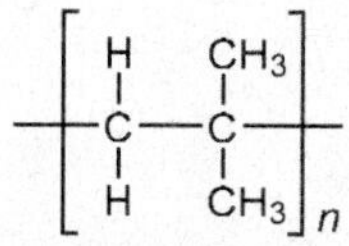

polyisobutylene

27.20. LDPE is used to make Ziploc bags (a flexible product, like trash bags) and HDPE is used to make folding tables (an inflexible product, like Tupperware).

27.21. Protonation of an ester group in PET renders it even more electrophilic, and it is then attacked by water to give a tetrahedral intermediate (an oxonium ion). Two successive proton transfer steps convert the oxonium ion into another oxonium ion, which can lose a leaving group to regenerate the C=O bond. Deprotonation generates a carboxylic acid (and an alcohol). These steps are then repeated to give terephthalic acid and ethylene glycol, as shown:

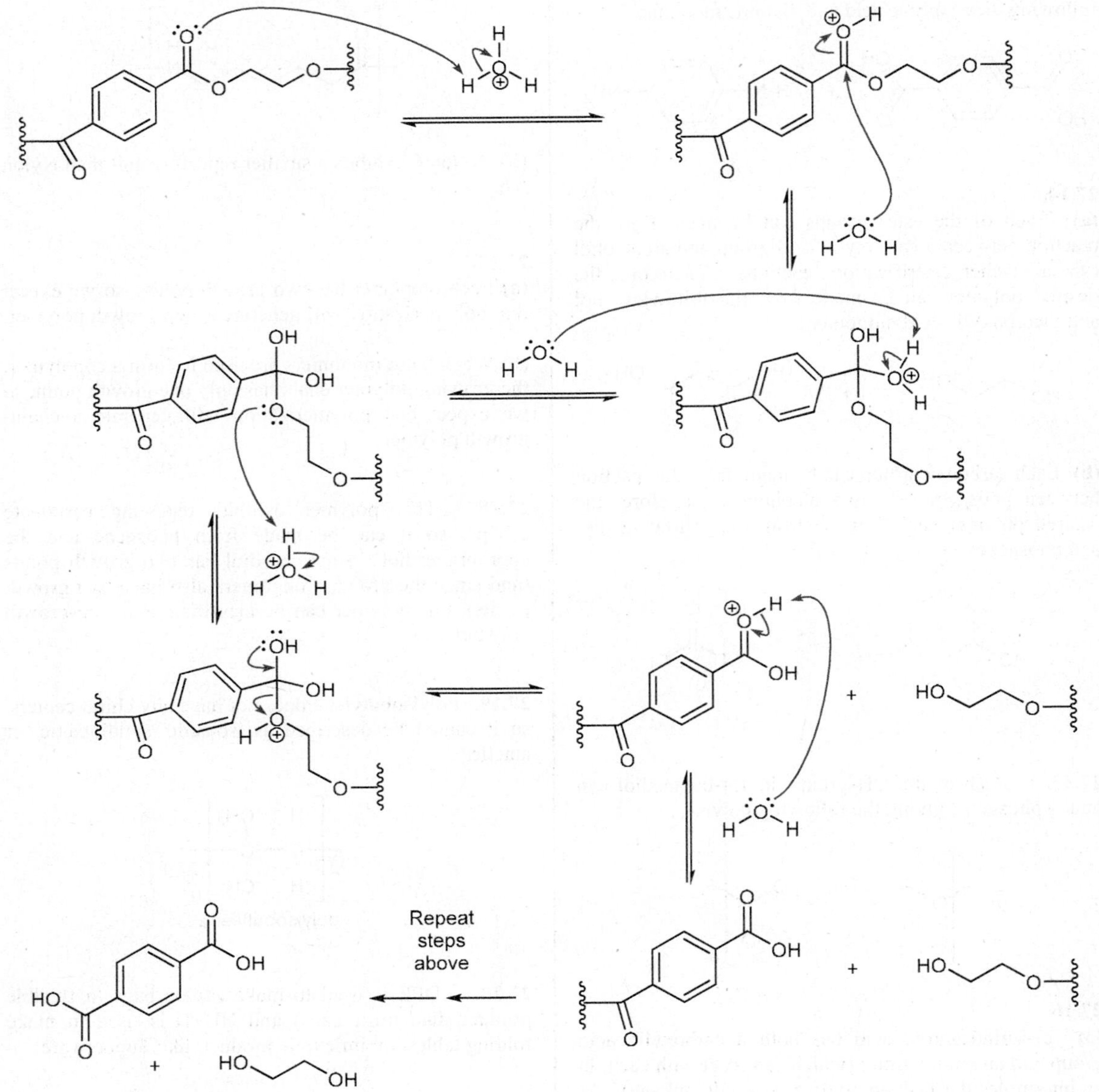

Repeat steps above

27.22.

(a) Polymerization of nitroethylene gives polynitroethylene:

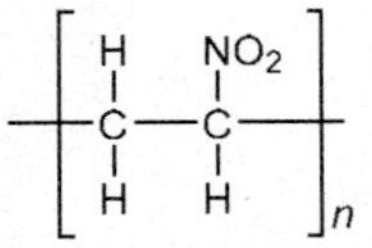

polynitroethylene

(b) Polymerization of acrylonitrile gives polyacrylonitrile:

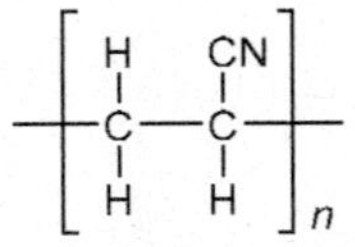

polyacrylonitrile

(c) Polymerization of vinylidene fluoride gives poly(vinylidene fluoride):

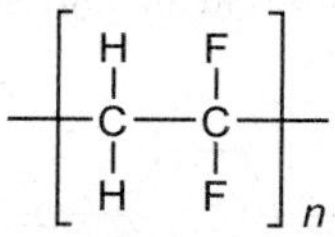

poly(vinylidene fluoride)

27.23.

(a) Each of the ester groups can be made from the reaction between a carboxylic acid group and an alcohol (via a Fischer esterification reaction). Therefore, the desired polymer can be made from the following dicarboxylic acid and diol monomers:

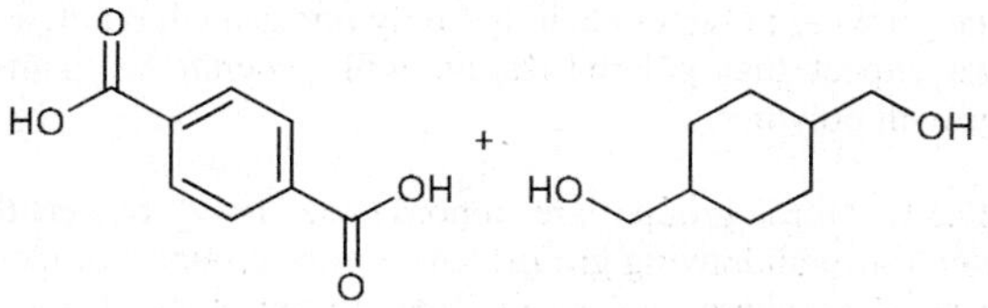

(b) The reaction of a carboxylic acid with an alcohol is a Fischer esterification, which requires acidic conditions. The acid protonates the carbonyl of the ester group to make it a strong electrophile that can be subsequently attacked by the weakly nucleophilic alcohol.

27.24. This copolymer can be made from the following monomers:

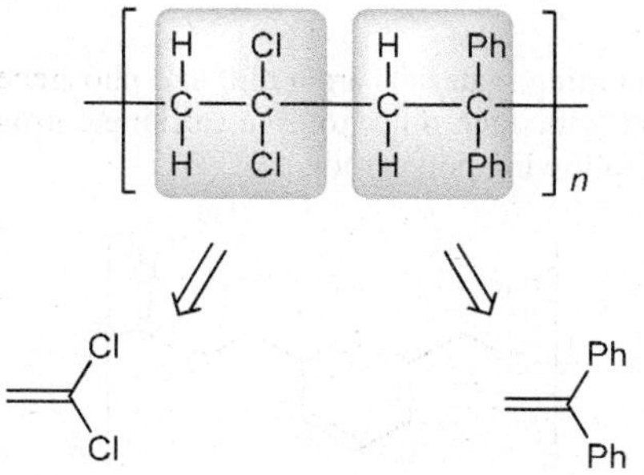

27.25. The following structure represents a block copolymer constructed from isobutylene and styrene. The isobutylene and styrene units are highlighted:

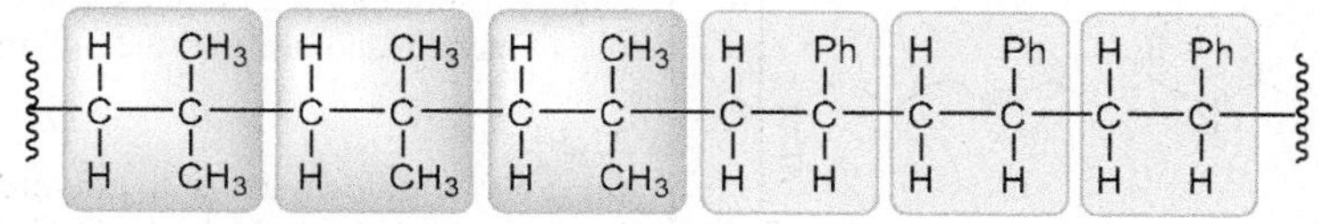

27.26. The following structure represents an alternating copolymer constructed from vinyl chloride and ethylene. The vinyl chloride and ethylene units are highlighted:

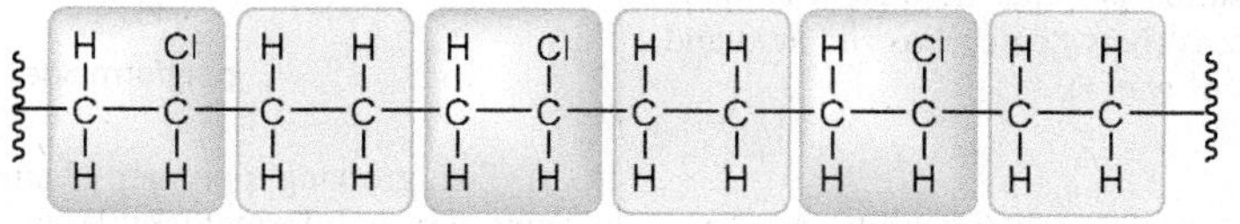

27.27. An acetate group is an electron-donating substituent (via resonance), while the other two groups (CN and Cl) are electron-withdrawing substituents. Therefore, vinyl acetate is expected be the most reactive toward cationic polymerization.

27.28. A cyano group is an electron-withdrawing substituent (via resonance), while the other two groups (acetate and methyl) are both electron-donating substituents. Therefore, the compound bearing the cyano group is expected be the most reactive toward anionic polymerization.

27.29. All three polymers are step-growth polymers, because in each case, the growing oligomers have two growth points.

(a) The starting materials are a diacid and a diamine, which can be linked together via amide groups, giving the following polymer:

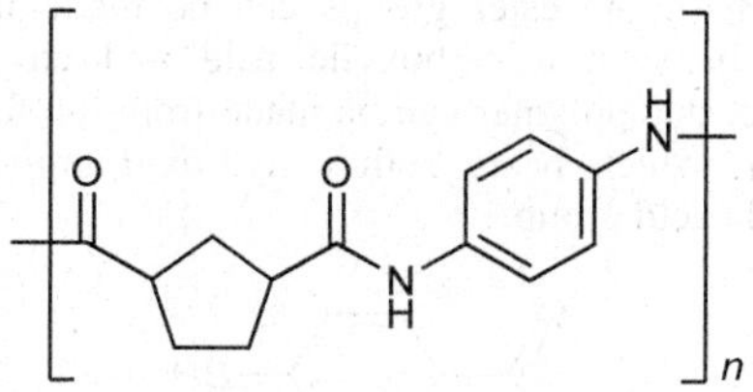

(b) The starting materials are a diol and a diisocyanate, which can be linked together as carbamate groups, giving the following polyurethane:

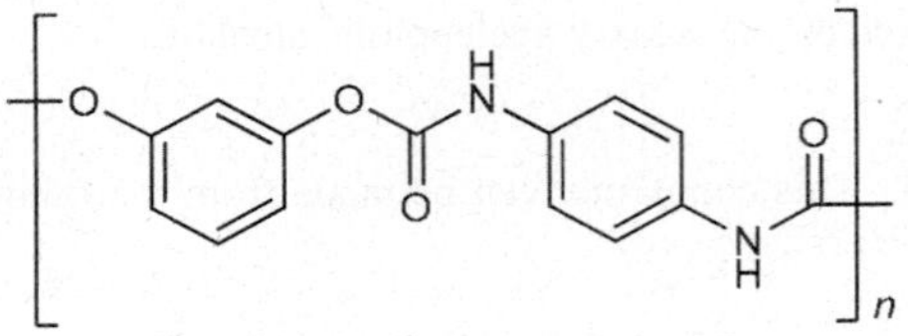

(c) The starting materials are a diol and phosgene, which will react with each other to give carbonate groups, and thus the following polycarbonate:

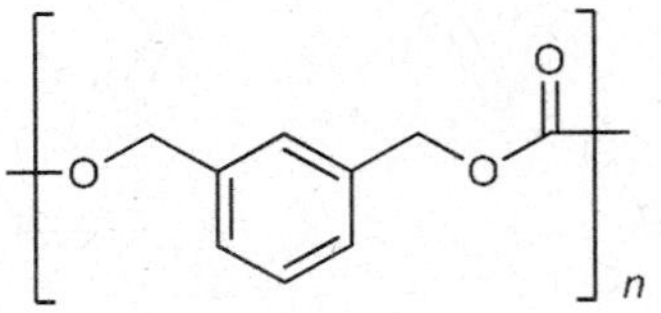

27.30.

(a) The starting materials are a diacid and a diamine, which can be linked together via amide groups, giving the following polymer:

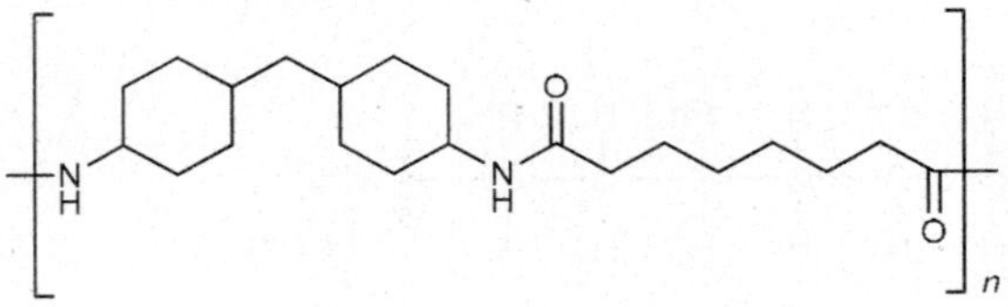

(b) Quiana is a polyamide.

(c) Quiana is a step-growth polymer, because each of the growing oligomers has two growth points.

(d) Quiana is a condensation polymer because it is made via a condensation process (between carboxylic acid and amino groups, with loss of water).

27.31.

(a) Each of the amide groups can be made from the reaction between a carboxylic acid and an amino group. Therefore, this polymer can be made from the following monomer, which bears both the amino group and the carboxylic acid group:

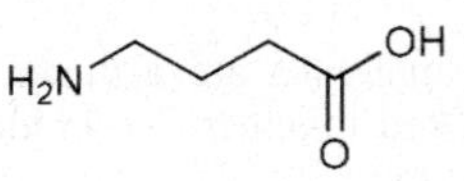

(b) Each of the ester groups can be made from the reaction between a carboxylic acid and an alcohol. Therefore, this polymer can be made from the following monomer, which bears both a hydroxyl group and a carboxylic acid group:

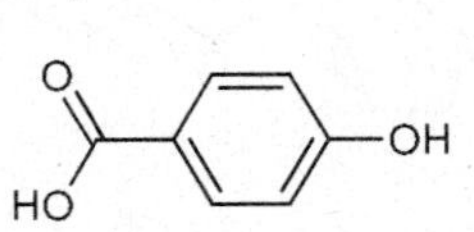

27.32. The starting materials are a diol and phosgene, which will react with each other to give carbonate groups, and thus the following polycarbonate:

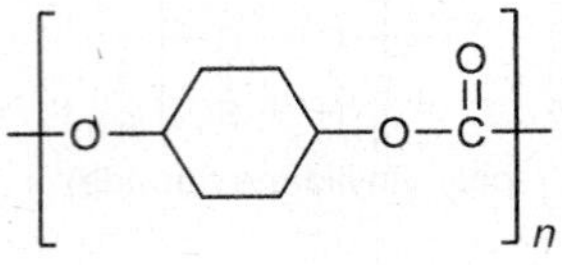

27.33.

(a) Each monomer has two growth points, so we expect that polymerization will generate a step-growth polymer.

(b) When these monomers are used to form a copolymer, the growing polymer chain has only one growth point, so we expect that polymerization will generate a chain-growth polymer.

27.34. Nitro groups are among the most powerful electron-withdrawing groups, and a nitro group stabilizes a negative charge on an adjacent carbon atom, thereby facilitating anionic polymerization.

27.35. Shower curtains are made from PVC, which is a thermoplastic polymer. To prevent the polymer from being brittle, the polymer is prepared in the presence of plasticizers which become trapped between the polymer chains where they function as lubricants. Over time, the plasticizers evaporate, and the polymer becomes brittle.

27.36.

(a) Polyformaldehyde is a polymer that is assembled from repeating formaldehyde ($H_2C{=}O$) units, as shown:

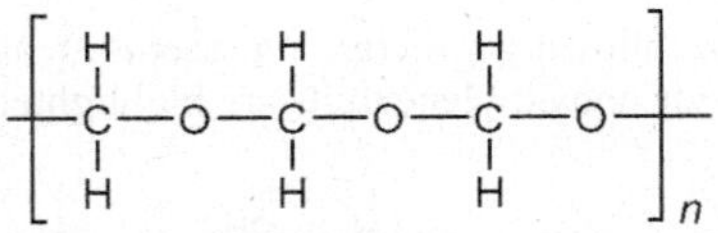

polyformaldehyde

(b) The growing polymer chain has only one growth point, so polyformaldehyde is classified as a chain-growth polymer.

(c) Polyformaldehyde is an addition polymer, because it is formed via successive addition reactions (involving the π bond in each molecule of formaldehyde). No small molecule (such as water) is produced in the polymerization process, so it is not a condensation polymer.

27.37. The monomer bears an electron-withdrawing group (CN) that can stabilize a negative charge via resonance, making it suitable for anionic polymerization. However, it also bears an electron-donating group (OMe) that can stabilize a positive charge via resonance, making it suitable for cationic polymerization as well.

27.38. The nitro group serves as a reservoir of electron density that stabilizes a negative charge via resonance (see Chapter 18). Resonance involving a nitro group delocalizes the negative charge onto an electronegative oxygen atom (highly effective stabilization), while resonance with a benzene ring delocalizes the negative charge onto carbon atoms (providing less effective stabilization).

27.39. The methoxy group is an electron-donating group that stabilizes an adjacent carbocation via resonance (see Chapter 18).

27.40. A methoxy group can only delocalize the positive charge via resonance if it is located in an *ortho* or *para* position. A methoxy group cannot stabilize the carbocation if it is located in the *meta* position (see Chapter 18).

27.41.

(a) In a syndiotactic polymer, the chiral centers exhibit alternating configuration, as shown:

Cl Cl Cl Cl Cl Cl Cl

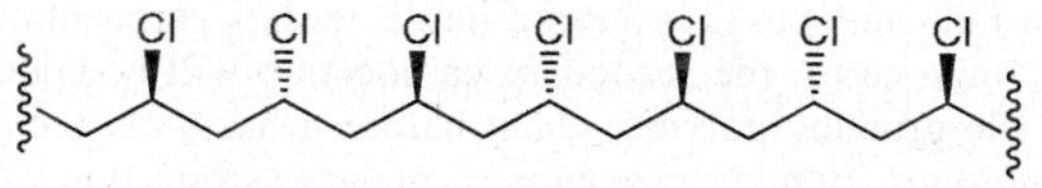

(b) In an isotactic polymer, the chiral centers all exhibit the same configuration, as shown:

Ph Ph Ph Ph Ph Ph Ph

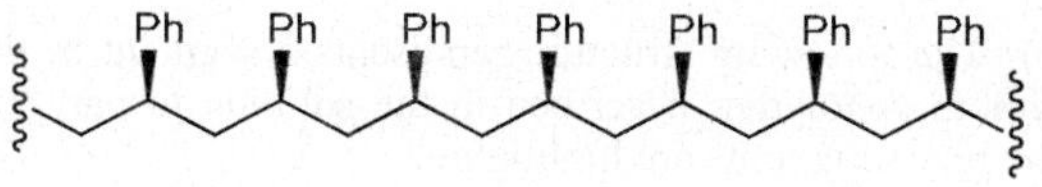

27.42.

(a) The desired polymer is a polyurethane, which can be prepared from the following diisocyanate and the following diol:

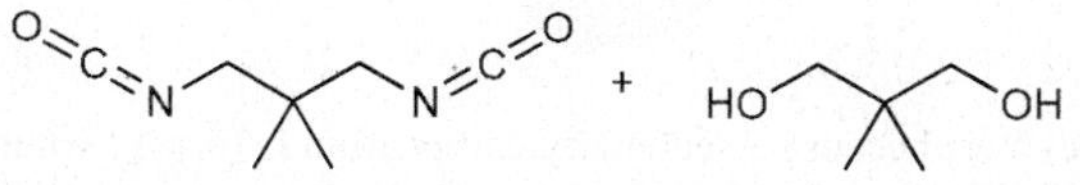

O=C=N N=C=O + HO OH

(b) Each monomer has two growth points, so we expect that polymerization will generate a step-growth polymer.

(c) Polyurethanes are classified as addition polymers (see end of Section 27.5), because they are formed via successive addition reactions. No small molecule (such as water) is produced in the polymerization process, so it is not a condensation polymer.

27.43. The correct answer is (b). The repeating units are comprised of four carbon atoms:

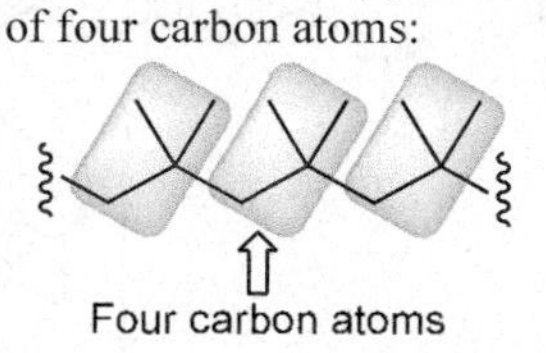

Four carbon atoms

Therefore, the required monomer must also have four carbon atoms. So we can rule out options (a) and (c). The remaining two options, (b) and (d), have the same carbon skeleton, but only option (b) has a functional group (a π bond). This π bond is necessary for the polymerization process to occur under acid-catalyzed conditions (see Mechanism 27.2).

27.44. The correct answer is (b). There are two ester groups that are hydrolyzed. One ester group can be seen in the center of the structure:

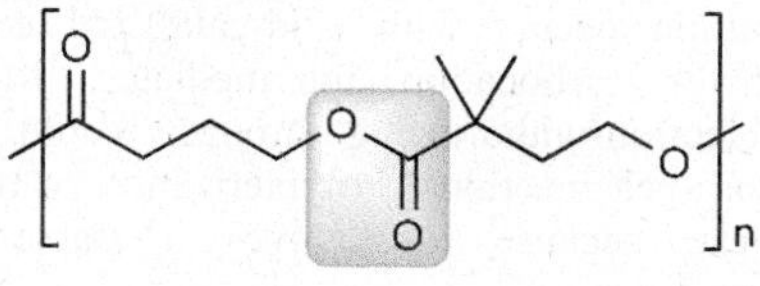

And the other ester group is located at the connection between the repeating units:

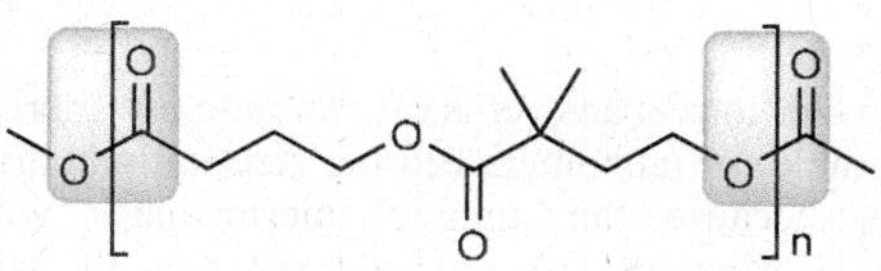

Each ester group is hydrolyzed to give a carboxylic acid and an alcohol:

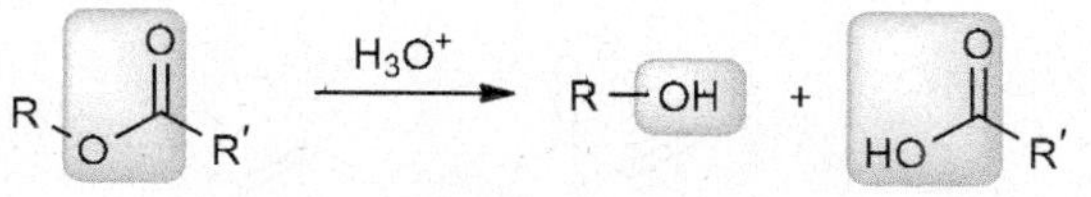

$$\text{R-O-C(=O)-R'} \xrightarrow{H_3O^+} \text{R-OH} + \text{HO-C(=O)-R'}$$

Therefore, option (b) is the correct answer, because the structures have the correct functional groups. In all of the other options, one or both of the carboxylic acid groups have been replaced by an aldehydic group.

27.45. The correct answer is (b). An alcohol will react with a carboxylic acid to give an ester:

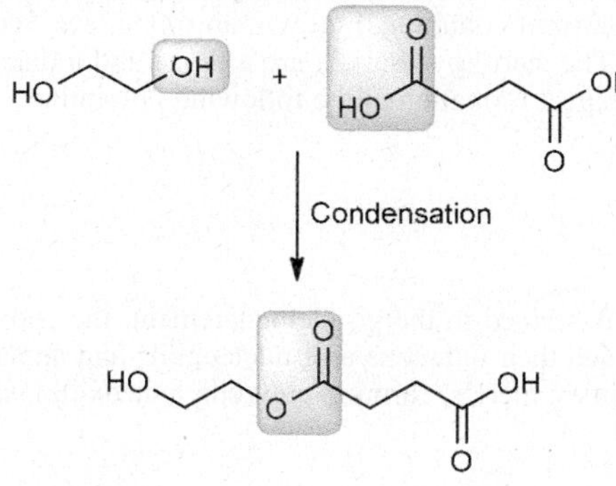

HO–CH₂CH₂–OH + HO–C(=O)–CH₂CH₂–C(=O)–OH

Condensation

This initial condensation product still has one OH group and one carboxylic acid group (highlighted),

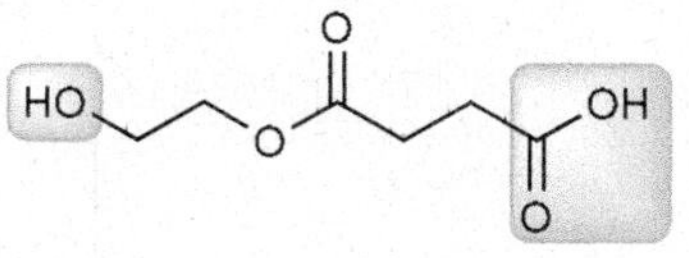

so it can undergo further condensation to give the following polymer:

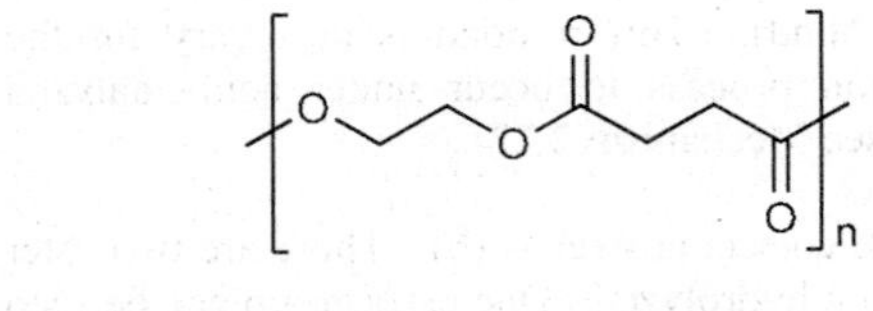

27.46. The correct answer is (d). Cationic polymerization occurs with acid-catalyzed conditions and involves carbocation intermediates. Monomers bearing electron-withdrawing groups would not be suitable for such reactions. Polymerization of monomer (d) is best, because it involves a stable tertiary carbocation:

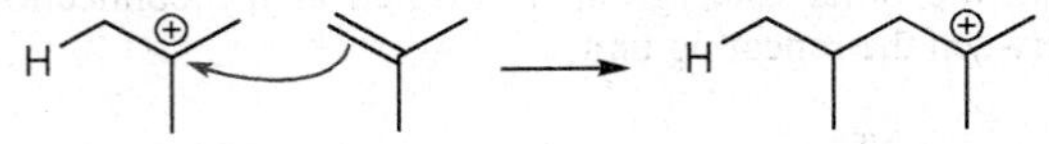

27.47. The correct answer is (a). Anionic polymerization is favorable for monomers bearing electron-withdrawing groups, because the anionic intermediates will be stabilized. Choices (b), (c) and (d) are all Michael acceptors and are, therefore, suitable for anionic polymerization. Choice (a) contains an electron-*donating* group that cannot stabilize an adjacent anion.

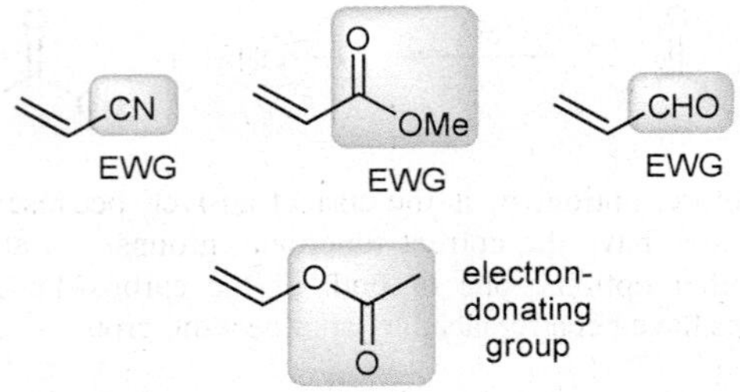

EWG EWG EWG

electron-donating group

27.48. A ketone will react with a primary amine (under acid-catalyzed conditions) to give an imine (see Section 19.6). The starting materials are a dione and a diamine, so we expect formation of the following polyimine:

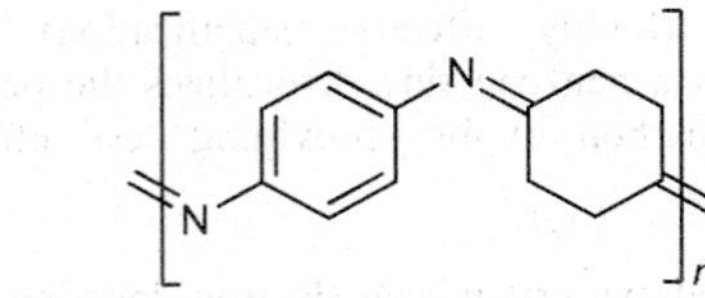

27.49. Vinyl alcohol is an enol, which is not stable. If it is prepared, it undergoes rapid tautomerization to give an aldehyde, which will not produce the desired product upon polymerization.

27.50. The ester groups undergo hydrolysis in basic conditions (saponification), which cleaves the polymer chain.

27.51.

(a) The carbocation that is initially formed is a secondary carbocation, and it can undergo a carbocation rearrangement to give a more stable, tertiary carbocation. In some cases, the secondary carbocation will be added to the growing polymer chain before it has a chance to rearrange. In other cases, the secondary carbocation will rearrange first and then be added to the growing polymer chain. The result is the incorporation of two different repeating units in the growing polymer chain.

(b) The following structure represents a segment of the random copolymer described in the solution to part (a). The repeating units are highlighted:

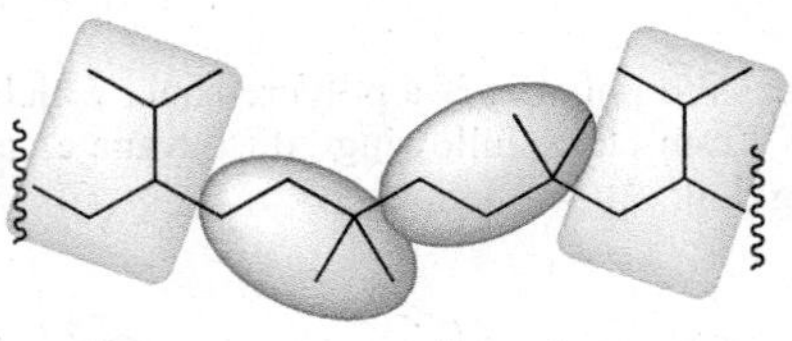

(c) Yes, because a secondary carbocation is formed when 3,3-dimethyl-1-butene is protonated, and a methyl shift can occur that converts the secondary carbocation into a tertiary carbocation.

27.52.

(a) As described in the problem statement, the epoxide ring is opened with a strong nucleophile to form an alkoxide ion, which then functions as a nucleophile and attacks another molecule of ethylene oxide. This process repeats itself many times, thereby forming poly(ethylene oxide), as shown:

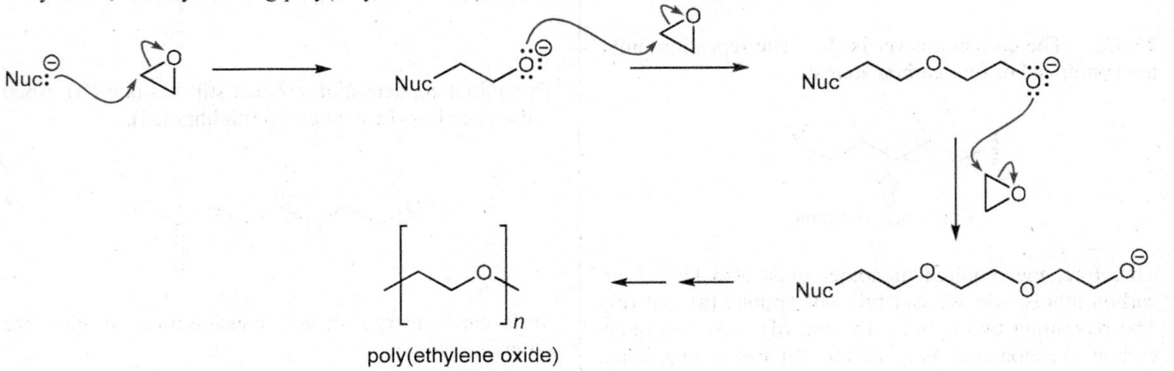

poly(ethylene oxide)

(b) Under acidic conditions, an epoxide can be protonated. A nucleophile can then attack the protonated epoxide, thereby opening the ring, and forming an alcohol. An alcohol is a weak nucleophile and it can then attack another protonated epoxide, once again opening the ring. The resulting oxonium ion is then deprotonated. If the base for this proton transfer step is a molecule of the epoxide, the resulting protonated epoxide can then serve as the electrophile for the next step. This process can repeat itself many times, thereby forming poly(ethylene oxide), as shown:

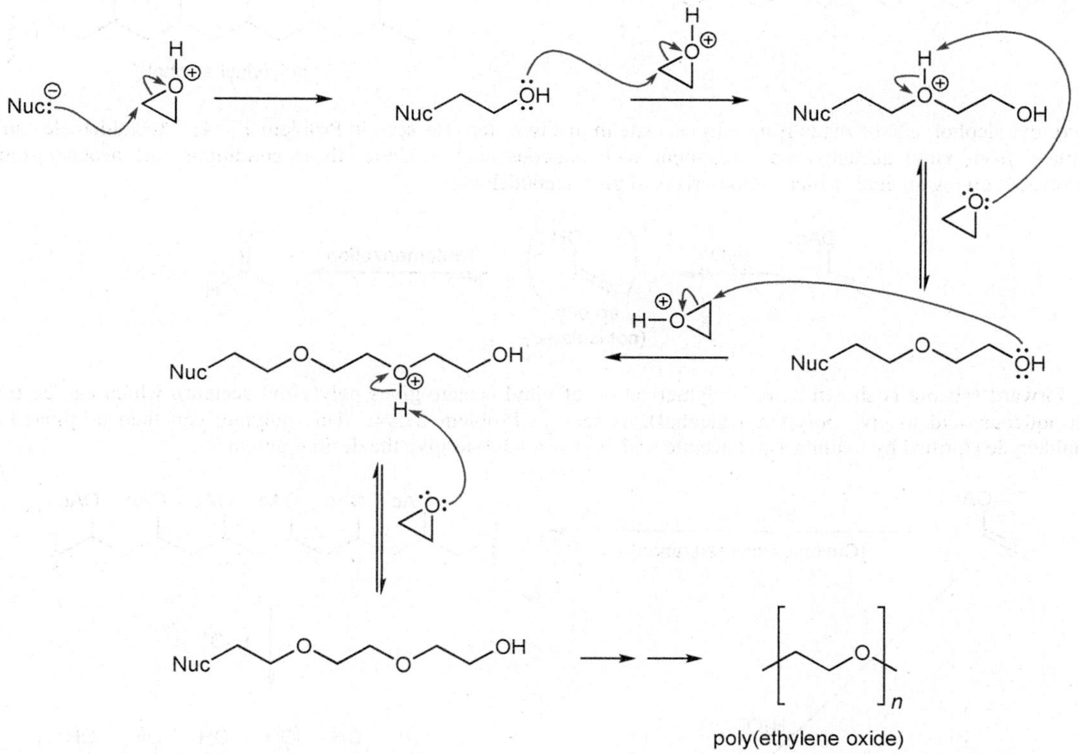

poly(ethylene oxide)

(c) The desired polymer is similar in structure to poly(ethylene oxide), but there is a *gem*-dimethyl group present in the repeating unit. This polymer can be made if the starting epoxide also bears a *gem*-dimethyl group:

(d) Preparation of this polymer would require the following epoxide:

Acidic conditions will be required, because the epoxide is too sterically hindered to be attacked under basic conditions (see Section 13.10).

27.53. Each of the highlighted positions represents an acetal group (see Section 20.5):

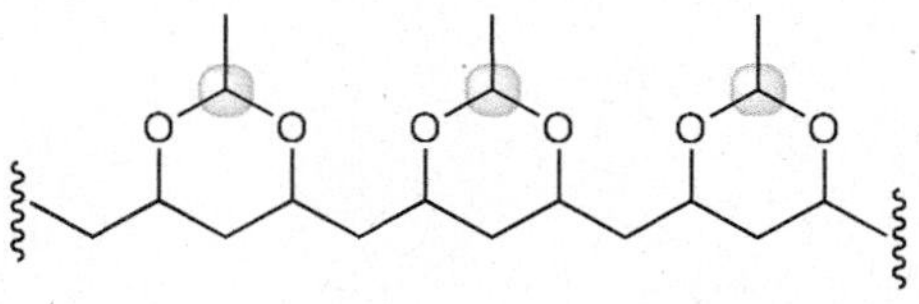

Therefore, this polymer can be made via acetal formation, from poly(vinyl alcohol) and acetaldehyde (in the presence of an acid catalyst, and with the removal of water):

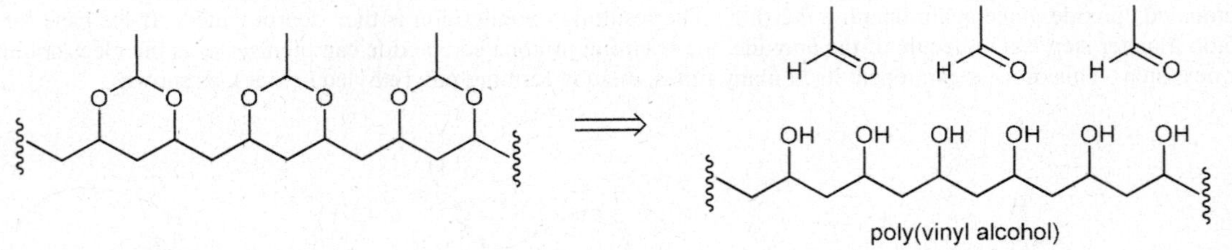

poly(vinyl alcohol)

Poly(vinyl alcohol) can be made from vinyl acetate in just two steps (as seen in Problem 27.44). Acetaldehyde can also be made from vinyl acetate (upon treatment with aqueous acid). Under these conditions, the acetate group is hydrolyzed, giving an enol, which tautomerizes to give acetaldehyde:

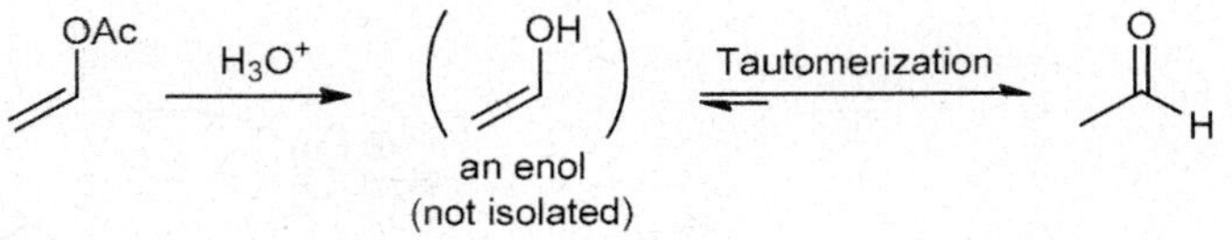

OAc $\xrightarrow{H_3O^+}$ (OH) an enol (not isolated) $\xrightleftharpoons{\text{Tautomerization}}$ acetaldehyde

The forward scheme is shown here. Polymerization of vinyl acetate gives poly(vinyl acetate), which can be treated with aqueous acid to give poly(vinyl alcohol), as seen in Problem 27.44. This polymer can then be treated with acetaldehyde (formed by treating vinyl acetate with aqueous acid) to give the desired polymer:

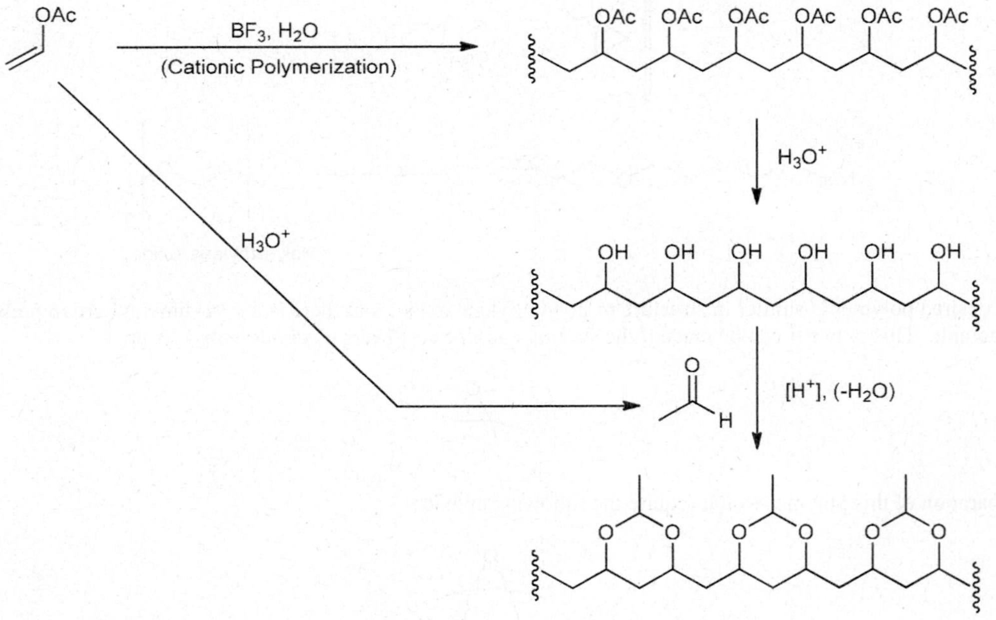

BF_3, H_2O (Cationic Polymerization)

H_3O^+

H_3O^+

$[H^+]$, $(-H_2O)$

APPENDIX
Nomenclature of Polyfunctional Compounds

A.1. This compound is a primary amine because there is only one alkyl group attached to the nitrogen, but it also has a hydroxyl group. Because alcohol functionalities have a higher suffix priority than amine groups, the compound will be named as an alcohol, and the amine will be named as a substituent (an amino group). Referencing Table 4.1, we can infer that an 18-carbon chain is called an octadecane, so this alcohol is an octadecanol. Numbering from left to right gives the alcohol and the amine the lowest possible numbers,

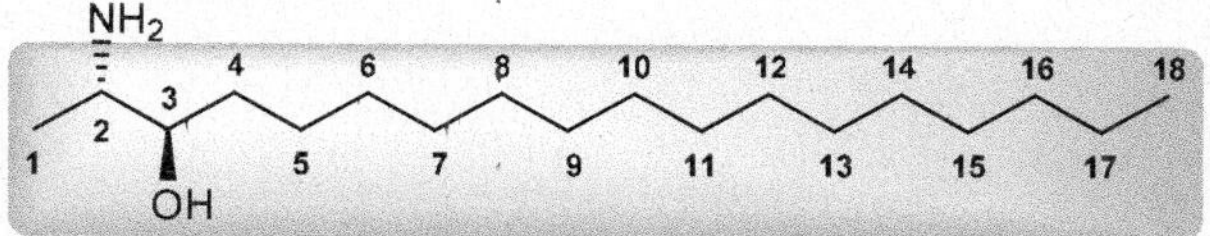

The name so far is 2-amino-3-octadecanol (or 2-aminoctadecan-3-ol using the new IUPAC rules). Lastly, the configuration of each chiral center must be specified.

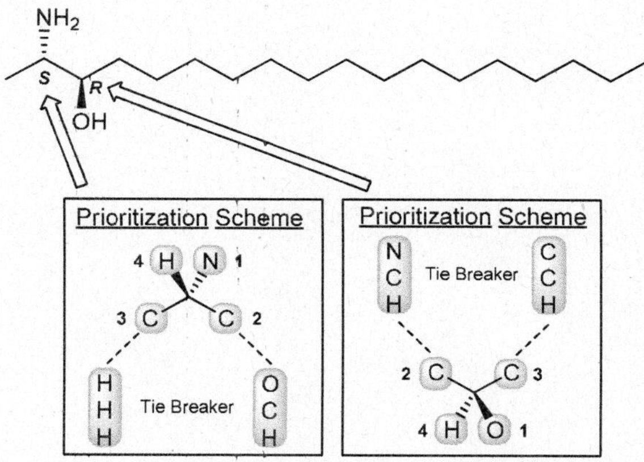

The full IUPAC name is thus (2*S*,3*R*)-2-amino-3-octadecanol.